SFPE Handbook of
Fire Protection Engineering

SFPE Handbook of Fire Protection Engineering

Second Edition

Editorial Staff

Philip J. DiNenno, P.E., Hughes Associates, Inc.
Craig L. Beyler, PhD., Hughes Associates, Inc.
Richard L. P. Custer, Custer Powell, Inc.
W. Douglas Walton, P.E., National Institute of Standards and Technology
John M. Watts, Jr., PhD., The Fire Safety Institute
Dougal Drysdale, PhD., University of Edinburgh
John R. Hall, Jr., PhD., National Fire Protection Association

National Fire Protection Association
Quincy, Massachusetts

Society of Fire Protection Engineers
Boston, Massachusetts

NFPA Editorial Staff

Managing Editor:	Jim L. Linville
Project Editor:	Susan Merrifield
Copy Editor:	Hilary Davis
Artwork:	George Nichols
	José R. Díaz
Composition:	Debra Rose
	Publication Services, Inc.
Cover Design: ·	Groppi Design
Manufacturing:	Donald C. McGonagle
	R. R. Donnelley & Sons

First impression June 1995

Library of Congress Catalog Card Number: 95-68247
NFPA Publication Number: HFPE—95
ISBN: 0-87765-354-2
Published by the National Fire Protection Association
One Batterymarch Park, Quincy, MA 02269-9101

Preface

This second edition of the *SFPE Handbook of Fire Protection Engineering* represents an effort by the editors to accomplish four objectives. The first objective was to update the information in existing chapters to include more recent developments and current thinking, particularly in the areas of fire dynamics and fire risk analysis. A second objective was to cover a broader range of topics, particularly those representing relatively new developments in the field. The Handbook now includes chapters on calorimetry, smoke control in atria, halon replacement agents, and risk assessment methods. A third objective was to promote widespread assimilation of computer-based hazard modeling into practical use. To accomplish this, three chapters were added: fundamentals of compartment modeling, zone modeling, and field modeling. The fourth objective was to make the Handbook more useful for practicing engineers by including additional reference material.

While the first edition concentrated on fundamentals and treatment of topics for which discrete closed form calculation procedures exist, this second edition has a broader scope. The trends toward more rigorous and complete engineering bases for fire safety design reinforces the requirement for a continuing breadth and depth of coverage. There is a clear need for documenting and integrating the technical bases and methods in all aspects of fire protection engineering. This edition represents incremental progress toward that goal.

There are many people to recognize and thank for their contributions to this edition; it would not have been possible without the sustained dedication required of a largely volunteer effort. The work of two new editors is gratefully recognized: Dr. John Hall of NFPA and Dr. Dougal Drysdale of the University of Edinburgh, whose expertise and energy greatly facilitated the publication of this edition. The continued support of the publisher, the National Fire Protection Association, is acknowledged. Particular recognition is due to Mr. Jim Linville, who, with energy and talent, gracefully applied the necessary discipline of a schedule and its deadlines; his patience and skill is gratefully acknowledged. But the largest share of gratitude is owed to the authors, particularly to those who contributed new chapters, who gave their knowledge, time, and energy so unselfishly. Without their generous contributions, this book would not have been possible. While they are owed a debt that cannot be paid outright, we trust that the application of their work in solving fire engineering problems worldwide will serve as some reward for their efforts.

The editors and the Society of Fire Protection Engineers welcome the comments and suggestions of readers as we seek to improve the coverage and utility of future editions.

Philip J. DiNenno, P.E.
April 13, 1995

Table of Contents

SFPE Handbook of
Fire Protection Engineering

Section One
Fundamentals

Section 1 Fundamentals

INTRODUCTION TO MECHANICS OF FLUIDS

B. S. Kandola

FLUID PROPERTIES

A fluid is defined as a substance that has the capacity to flow freely and as a consequence deform continuously when subjected to a shear stress. A fluid can either be a liquid, vapor, or gas.

For the purposes of fluid flow studies, a very important distinction is made between compressible fluids and incompressible fluids. In general, the compressibility effects of liquids are so small that they can be regarded as incompressible, whereas gases and vapors can be both compressible and incompressible depending on the forces involved.

To simplify analytical investigation of fluid motion, the intermolecular forces of the fluids are ignored, and such a fluid is known as *inviscid* (i.e., zero viscosity).

An incompressible, inviscid fluid is called a *perfect fluid*. In reality no real fluid is a perfect fluid, but the effects of viscosity are so small in a perfect fluid that they can be ignored.

Density. The density of a fluid is defined as the mass of the fluid per unit volume. The density, ρ, is therefore defined as

$$\rho = \text{mass/volume} = m/v$$

where m is the mass of fluid of volume, v. If the units of mass are kilograms (kg) and the volume m^3, then the units of density are kg/m^3.

Specific Volume. This is the reciprocal of density, i.e., specific volume (m^3/kg)

$$v = 1/\rho$$

Shear Force. The component of total force, F, in a direction tangential to the surface of a body is called the shear force. Similarly, the component perpendicular to the tangent is called the normal force. Force is measured in newtons (N, or 1 kg m/s^2).

Shear Stress. The shear stress, τ, at a point is defined as the limiting value of shear force per unit area as the area is reduced to a point, or

Dr. B. S. Kandola is a member of the Unit of Fire Safety Engineering, University of Edinburgh. He has worked on various fluid mechanics projects in industry and at Imperial College, London.

$$\tau = \text{shear force/area}$$

Pressure. The pressure, P, at a point in a fluid is defined as the limiting value of normal force to area as the area is reduced to the point, or

$$P = \text{normal force/area}$$

where the units are N/m^2, or pascals (Pa).

Physical Properties of Fluids

Viscosity. All real fluids offer some resistance, however small, to applied shear stresses. This resistance results from the property of the fluid called viscosity. According to Newton's law, the rate at which a fluid element deforms for a given shear stress is inversely proportional to the fluid viscosity.

For a two-dimensional flow between two parallel plates, the rate of deformation is the rate of change of x-component of velocity, u, with y-direction, i.e., $\partial u/\partial y$. If τ is the frictional shearing stress, then according to the above definition

$$\tau \propto \partial u/\partial y \tag{1}$$

or

$$\tau = \mu \, \partial u/\partial y \tag{1a}$$

where μ is the coefficient of viscosity. The units of viscosity are Ns/m^2. In general, the viscosity of a gas increases with temperature while the viscosity of a liquid decreases with temperature.

Kinematic Viscosity. In all real fluid motions the frictional and inertia forces interact. The ratio of μ to ρ is important and is known as the kinematic viscosity, ν.

$$\nu = \mu/\rho \tag{2}$$

The units of ν are m^2/s.

The Equation of State

For a fluid in which the properties are the same at all points (i.e., uniform fluid) and having definite chemical composition, experimental evidence indicates that the fluid density, ρ, is a function only of the pressure, p, and temperature, T.

$$\rho = f(p, T) \tag{3}$$

According to this relation, any one property, i.e., p, T or ρ, is determined by any of the other two. For a gas at temperatures and pressures well away from liquefaction or dissociation, the following relationship holds with good approximation.

$$\frac{p}{\rho} = \frac{\bar{R}}{M} \cdot T \tag{4}$$

where $\bar{R}$ is the universal gas constant, M is the molecular weight of the particular gas, and T is its absolute temperature (K). The value of the universal gas constant, $\bar{R}$, is 8.315×10^3 m^2/s^{2}K (or J/kg $\cdot$ K), and the molecular weights for various gases are shown in Table 1-1.1.

A gas that obeys Equation 4 is called a *perfect gas* (also sometimes referred to as *ideal gas*). With an acceptable degree of accuracy, this relationship is also assumed to apply for the calculation of air flows, although air is, in reality, a mixture of various gases. It is assumed that the equation of state holds when the gas is in motion as well as at rest.

Compressibility and Thermal Expansion

Equation 4 describes the compressibility of a gas. According to this equation, the volume is decreased with the increase in pressure, provided the process takes place under isothermal conditions.

If the volume changes from v to $v + \partial v$ due to the change in pressure from p to $p + \partial p$, the coefficient of compressibility is then defined as

$$\beta = -\frac{1}{v}\left(\frac{\partial v}{\partial p}\right)_T \tag{5}$$

Similarly, at constant pressure

$$\beta_1 = \frac{1}{v}\left(\frac{\partial v}{\partial T}\right)_p \tag{6}$$

where β_1 is called the coefficient of thermal expansion. This equation describes the resulting change of volume due to any change in temperature occurring under isobaric conditions.

The compressibility of a liquid is described by its bulk modulus, K, which is the inverse of coefficient of compressibility, β;

i.e.,
$$K = 1/\beta \tag{7}$$

For a perfect gas (i.e., a gas that obeys Equation 4)

TABLE 1-1.1 *Molecular Weight of Gases*

Gas	Molecular Weight (M)
Hydrogen (H$_2$)	2.0
Carbon Monoxide (CO)	28.0
Methane (CH$_4$)	16.0
Ethane (C$_2$H$_6$)	30.0
Propane (C$_3$H$_8$)	44.0
Air	28.9

$$\beta = \beta_1 = 1/T. \tag{8}$$

Specific Heat. The specific heat of a substance (or gas) is defined as the quantity of heat required to raise the temperature of a unit mass of the substance by one degree. The specific heat depends on whether the process takes place under conditions of constant pressure, C_p, or constant volume, C_v. If q is the quantity of heat supplied per unit mass of gas, then

$$C_p = \left(\frac{\partial q}{\partial T}\right)_p \quad \text{constant pressure} \tag{9}$$

$$C_v = \left(\frac{\partial q}{\partial T}\right)_v \quad \text{constant volume} \tag{10}$$

Vapor Pressure. When the temperature of a liquid increases, its molecular activity is also increased. With the sufficient rise in temperature, the molecules begin to escape from the surface of the liquid. The pressure exerted by these molecules just above the surface is called the vapor pressure of the liquid. The vapor pressure increases with the increase in temperature, and boiling occurs when the vapor pressure becomes equal to the pressure above the liquid surface.

Surface Tension. The molecular attraction of liquids causes the free surface of a liquid to act as a stretched membrane in such a way that work is required to change the shape of its free surface.

The surface tension coefficient of a liquid is then defined as the force per unit length of any line on the free surface necessary to hold that surface together at that line

$$F = \int \sigma \, dl \tag{11}$$

The quantity σ is the surface tension coefficient, and depends on the properties of the free surface of a liquid and its surroundings.

FLUID STATICS AND BUOYANCY

When a fluid system is in motion, shear stresses develop in the fluid if one layer of fluid moves at a different velocity from an adjacent layer. These stresses are proportional to the velocity gradient. In the case of a static fluid (zero velocity) the shear stresses are zero.

Forces Acting on Fluid Systems

Forces acting on fluid systems are classified according to the geometry on which they act.

1. Body force is the force resulting from the total mass of the fluid system. For example, the body force of a solid object at rest equals its weight acting at its center of gravity.
2. Surface force, e.g., shear force, acts on a surface and is proportional to the extent of the surface. Such a force can be resolved into its normal and tangential components.
3. Line force, e.g., surface tension, depends on the extent of the line perpendicular to which the force acts.

Pressure at a Point

The pressure at a point is defined as the normal component of surface force acting on a unit area of the surface; or, the limit of the ratio of normal force per unit area as the area

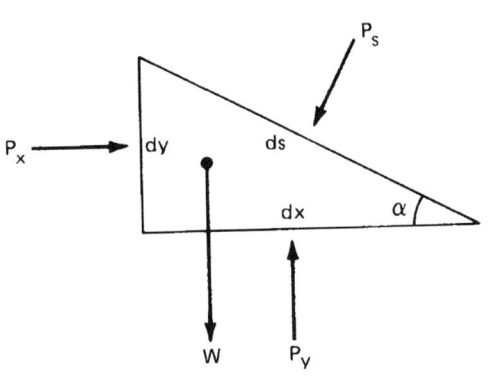

Fig. 1-1.1. Wedge-shaped imaginary fluid element.

approaches zero at the point. For a fluid system in equilibrium, the pressure is a scalar quantity, as it is independent of orientation.

For an imaginary wedge-shaped element in a static fluid, the forces acting on each side are due to body and surface forces, as shown in Figure 1-1.1.

Since the fluid is in equilibrium, the sum of the horizontal forces and the vertical forces is zero, i.e.,

$$\sum F_x = 0 \qquad (12)$$

$$\sum F_y = 0 \qquad (13)$$

Resolving the forces shown in Figure 1-1.1, horizontally and vertically, it can be shown that in the limit as $\partial x \to 0$, $\partial y \to 0$

$$p_x = p_y = p_s \qquad (14)$$

This result is sometimes referred to as Pascal's law, which states that for a fluid system in static equilibrium the pressure at a point is the same in all directions.

Hydrostatic Equation

Within a static fluid, the rate of change of static pressure with height, z, from a datum level is given by

$$dp/dz = -\rho g \qquad (15)$$

Integrating Equation 15 for constant ρ gives

$$p + \rho gz = C \qquad (16)$$

where g is the acceleration due to gravity, and C is the constant of integration.

This equation is generally known as the hydrostatic equation or Torricelli's principle, which states that, at every elevation within a static, homogeneous, and incompressible fluid, the static pressure plus the head of fluid above a given datum line, ρgz, is constant.

Forces on Submerged Surfaces

In numerous engineering design problems involving submerged surfaces, such as containers, off-shore oil rigs, walls of a dam, the walls of a liquid-filled tank, etc., it is necessary to know the magnitude of forces, and the point of action of these forces that act on the surfaces. The total force acting on a submerged surface is obtained by integrating the pressure over the entire surface area.

$$F = \int_A p \, dA \qquad (17)$$

If the pressure is constant, then the force on a submerged horizontal surface is given by

$$F = pA \qquad (18)$$

where A is the total surface area.

The point on the surface at which this resultant force acts is called the center of pressure. It can be shown that the center of pressure for a horizontal submerged surface is at the centroid of the area. This result is arrived at by considering the moment of the distributed force about any axis through the center of pressure to be zero.

Forces on Plane-Inclined and Submerged Curved Surfaces

Consider an inclined plane surface, as shown in Figure 1-1.2.

Force, dF, acting on elemental area, dA, at depth z from the free surface is given by

$$dF = \rho gz \, dA = \rho gy \sin \alpha \, dA \qquad (19)$$

and the total force is given by

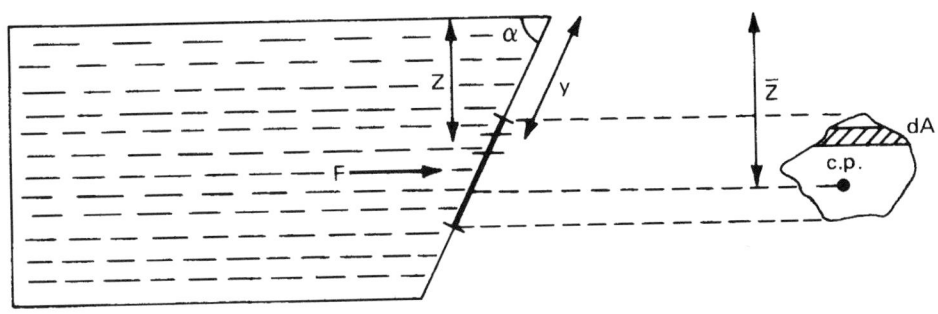

Fig. 1-1.2. Plane-inclined surface.

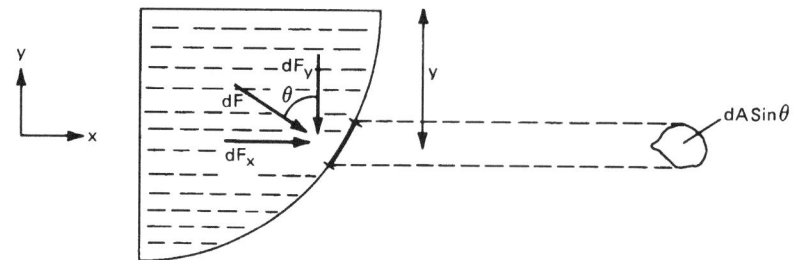

Fig. 1-1.3. Submerged curved surface.

$$F = \rho g \sin \alpha \int_A y \, dA \qquad (20)$$

If the y-coordinate of the centroid is defined as

$$\bar{y} = \frac{1}{A} \int_A y \, dA \qquad (21)$$

then the total force can be written as

$$F = \rho g \bar{y} A \sin \alpha = \rho g \bar{z} A$$

or

$$F = p_{\bar{z}} A \qquad (22)$$

This equation shows that for a plane-inclined surface the magnitude of the force on the surface is the product of the pressure at the centroid of the surface and its area.

For a curved surface, as shown in Figure 1-1.3, it can be shown that the horizontal and vertical components of force are

$$dF_x = \rho g y \sin \theta \, dA \qquad (23)$$

$$dF_y = \rho g y \cos \theta \, dA \qquad (24)$$

However, $\sin \theta \, dA$ is the projection of the elemental area, dA, onto a plane perpendicular to x-direction. Equation 23, therefore, shows that the horizontal component of pressure force on a curved surface is equal to the pressure force exerted on a projection of the curved surface.

The y-component of the force reduces to

$$F_y = \rho g \int_A dV \qquad (25)$$

where $dV = y \cos \theta \, dA$.

Integration of Equation 25 gives

$$F_y = \rho g V \qquad (26)$$

where V is the volume of fluid above the surface.

Equation 26 shows that the vertical component of pressure force on a curved surface is equal to the weight of liquid that is vertically above the curved surface extending up to free surface.

Buoyant Force on Submerged Bodies

As seen earlier in this chapter, a body submerged in a static fluid experiences a net force as a result of the pressure variation at its surface. The vertical component of this force is called the buoyant force, and always acts vertically upward.

The magnitude of the buoyant force is expressed as the difference between the vertical component of the pressure force on the upper and the lower surfaces of the body. This fact is expressed in the form of Archimedes' principle, which states: "Any body submerged in a fluid experiences a lift in a direction opposite to its weight and equal in magnitude to the weight of displaced volume of fluid."[1] If a submerged body is in equilibrium, it can be shown that

$$\rho = \rho_b \qquad (27)$$

where ρ is the density of fluid and ρ_b is the density of body.

If the buoyant force is greater than the weight of the body, the body will float; but if the force is less, the body will sink.

Center of buoyancy: The center of buoyancy is the point on the body at which the buoyant force acts. The precise location of this point can be found by taking the integrated moments of the elementary buoyant forces and equating them to the moment of the total buoyant force or the weight of the displaced fluid.

It can be shown that the center of buoyany lies at the centroid of the displaced volume of fluid.

Stability of Floating and Submerged Bodies

From the previous text it is clear that a floating body in a static fluid is in equilibrium, provided:

1. Buoyant force is equal to the weight of the body, and
2. Center of buoyancy lies on the same vertical line as the center of gravity.

If a body is in such a stable equilibrium, with respect to vertical displacement, then any upward or downward displacement sets up a force that tends to return the body to its original position.

In general, floating bodies are in a stable configuration if the center of gravity, *CG*, is lower than the center of buoyancy, *CB*. This does not imply that stability may not exist when the opposite situation occurs.

Conversely, a completely submerged body is only rotationally stable if its center of gravity is below the center of buoyancy, as illustrated in Figure 1-1.4.

KINEMATICS OF FLUID MOTION

In general, kinematics of particle or fluid motion is concerned with the effects of motion on quantities derivable

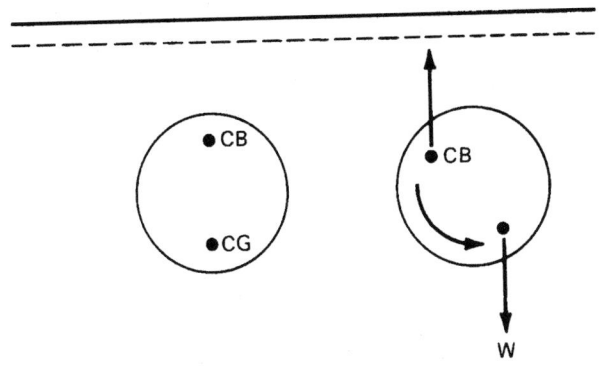

Fig. 1-1.4. Rotationally stable submerged body.

from displacement and time, such as velocity and acceleration. Although these motions are caused by external forces, they are studied in isolation from these forces.

In analyzing the motion of a rigid body or a fluid system, the space-time quantities are expressed in terms of a convenient coordinate system. The choice of this system is quite arbitrary and depends only on the nature of the problem investigated.

Two methods can be used to describe fluid motion: (1) the Langrangian method describes the motion of a particle in terms of a coordinate system that moves with the particle; and (2) the Eulerian method describes the motion of a particle in terms of a fixed coordinate system. The Eulerian method is the more commonly used to analyze fluid mechanics problems.

Classification of Fluid Motion

To simplify the analysis of fluid motion, various assumptions are made about the nature of fluid and its motion.

A *perfect fluid* is defined as that fluid which does not sustain shear stresses. A perfect fluid is inviscid ($\mu = 0$), and irrotational with constant density.

A *real fluid* is one in which the effects of viscosity are of paramount importance (i.e., $\mu \neq 0$). It is only in real fluids that a boundary layer is developed adjacent to a body over which the fluid flows.

Incompressible fluid motion is a motion in which the density remains unaltered, (i.e., the time derivative of density is zero, $\partial\rho/\partial t = 0$).

With a *compressible fluid motion*, if the density of a fluid in motion is a function of time, the flow is termed as compressible (time derivative of density is nonzero, i.e., $\partial\rho/\partial t \neq 0$).

When a real fluid is in motion, two basic forces determine the nature of flow: (1) inertial forces, which arise as a result of the velocity of the fluid, and (2) viscous forces, which arise as a result of the viscosity of the fluid.

The ratio of inertial to viscous forces is known as the Reynolds number and is defined as

$$\mathrm{Re} = Vd/v$$

where V is the velocity, d is some characteristic length, and v is the kinematic viscosity.

For a fluid motion in which the value of Re is so small that the layers of flow are smooth and laminar, the flow is then referred to as laminar. As the velocity is increased (i.e., the inertial forces are increased), the layers start to break up and the flow is then said to be turbulent.

At a critical Reynolds number the fluid flow becomes fully turbulent. For example, for a sharp-edged entry flow in a pipe the critical Reynolds number is 2700.

Steady and unsteady motion: A flow in which the properties of the fluid do not vary with time is called the steady flow (i.e., $\partial V/\partial t = 0$). On the other hand, a flow regime in which these properties vary with time as well as space is characterized as unsteady (e.g., $\partial V/\partial t \neq 0$). A turbulent flow can either be steady or unsteady, as illustrated in Figure 1-1.5.

A turbulent flow is composed of the mean (average) component and a fluctuating component.

$$u = \bar{u} + u' \tag{28a}$$

$$v = \bar{v} + v' \tag{28b}$$

where $\bar{u}$ and $\bar{v}$ are mean values of u and v, and u' and v' are the fluctuating components. The fluctuations u' and v' are such that their time average is zero.

Flow Concepts

Various flow concepts are used to make the mathematical analysis of fluid motion easier.

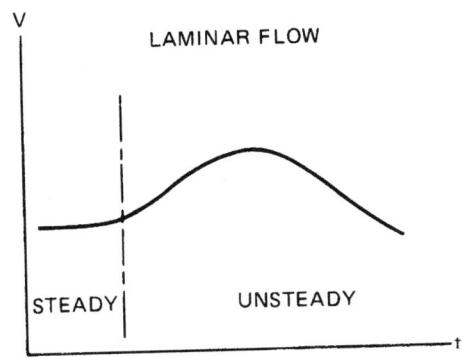

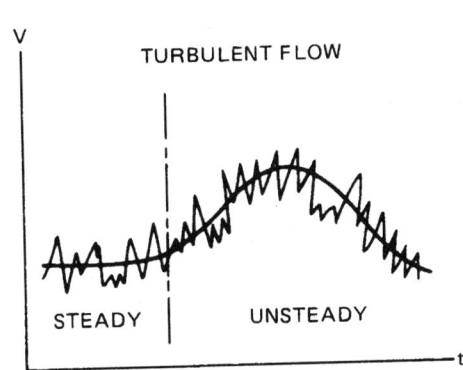

Fig. 1-1.5. Steady and unsteady flows.

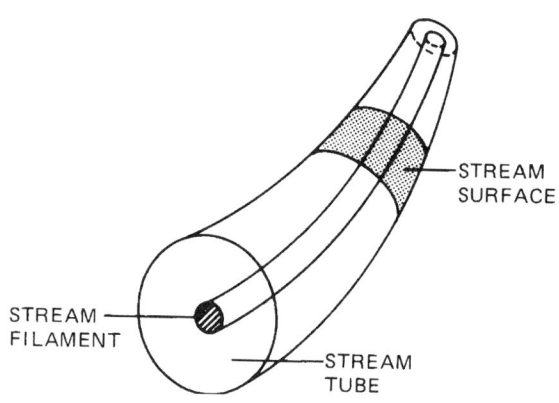

Fig. 1-1.6. *Basic concepts of stream filaments.*

A *streamline* in a flow is defined as an imaginary curve in which the tangent at every point on this curve gives the direction of the fluid velocity at that point. This means that there is no flow in the direction normal to the streamline. When the motion is steady the streamlines are the same at all times.

Pathline is defined as a line in the flow field describing the path or trajectory of a given fluid element.

Streakline is defined as the locus in a flow field, at a given time, of all the fluid particles that have passed through a given point within the fluid. A smoke trail in the atmosphere at a given time is an example of a streakline.

For a steady motion the streamlines, streaklines, and pathlines are identical.

A stream filament is a group of neighboring streamlines forming a cylindrical passage with an infinitesimal cross-section. A group of stream filaments is called a streamtube. The cross-section of a streamtube is finite, and its surface across which there is no flow is called a stream surface. These concepts are illustrated in Figure 1-1.6.

Equation of Continuity

The flow through a streamtube, i.e., no flow through the stream surface, is shown in Figure 1-1.7.

Since there is no flow through the surface of the tube, and if the flow is steady, the mass of fluid flowing through area A_1 must equal the mass flowing through area A_2; i.e., mass inflow equals mass outflow, or

$$\rho_1 A_1 u_1 = \rho_2 A_2 u_2 \tag{29}$$

For an incompressible flow, $\rho_1 = \rho_2$ and the previous equation then becomes

$$u_1 A_1 = u_2 A_2 \tag{30}$$

This relationship is known as the equation of continuity and, in general, can also be written as

$$\frac{\partial u}{\partial x} + \frac{\partial v}{\partial y} + \frac{\partial w}{\partial z} = 0 \tag{31}$$

where for an incompressible flow, u, v, and w are the velocity components in the x, y, and z directions.

The Stream Function

Consider a two-dimensional incompressible flow in which the equation of continuity can be written as

$$\frac{\partial u}{\partial x} + \frac{\partial v}{\partial y} = 0 \tag{32}$$

From this equation it follows that there exists a function, ψ, such that

$$u = \frac{\partial \psi}{\partial y}, \quad \text{and} \tag{33a}$$

$$v = -\frac{\partial \psi}{\partial x} \tag{33b}$$

The function ψ is called the stream function, which satisfies the continuity equation. In cylindrical coordinates this relationship can be written as

$$v_r = \frac{1}{r}\frac{\partial \psi}{\partial \theta}, \quad v_\theta = -\frac{\partial \psi}{\partial r} \tag{34}$$

where v_r and v_θ are the velocity components in the r and θ directions.

Sources, Sinks, and Doublets

A point within a fluid from which fluid emanates and spreads radially outward in a uniform fashion is called a

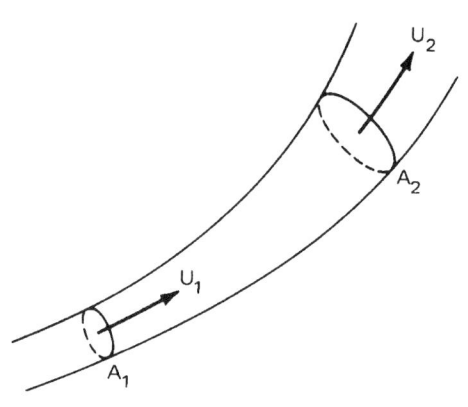

Fig. 1-1.7. *A streamtube.*

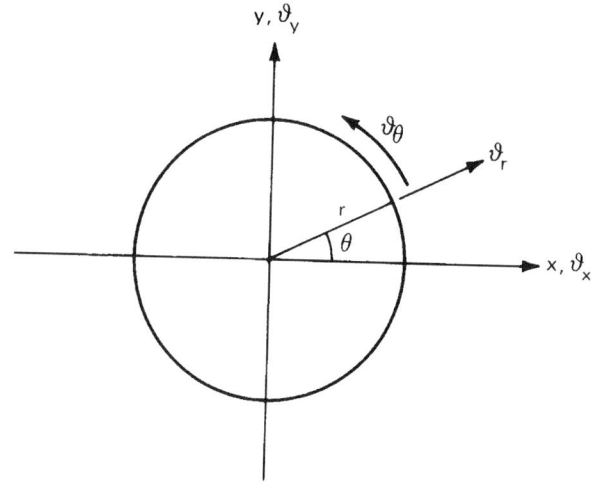

Fig. 1-1.8. *A two-dimensional source.*

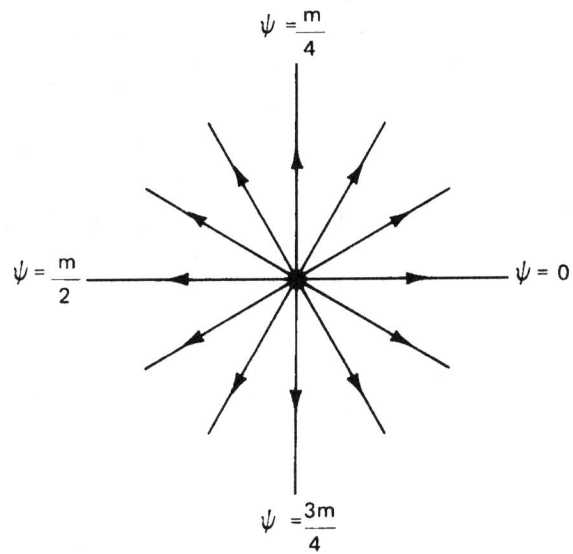

Fig. 1-1.9. Streamlines of a two-dimensional source.

source. On the other hand, a sink is the opposite of a source, and the flow is inward into the point.

The strength of a source or sink is equal to the volume of fluid that issues from a source or to a sink per unit time. When a source-sink pair is separated by a distance, δx, such that $\delta x \rightarrow 0$, then this arrangement is called a doublet or a dipole.

Stream Function of a Two-Dimensional Source

Since the flow from a source is radially outward, the flux per unit time is the strength of the source. (See Figure 1-1.8.)

If the flow is steady and incompressible, the strength, m, per unit length of cylinder is given by

$$m = 2\pi r v_r \tag{35}$$

where v_r is the radial velocity component.

Since the flow is completely radial, $v_\theta = 0$.

The radial component, v_r, is given by

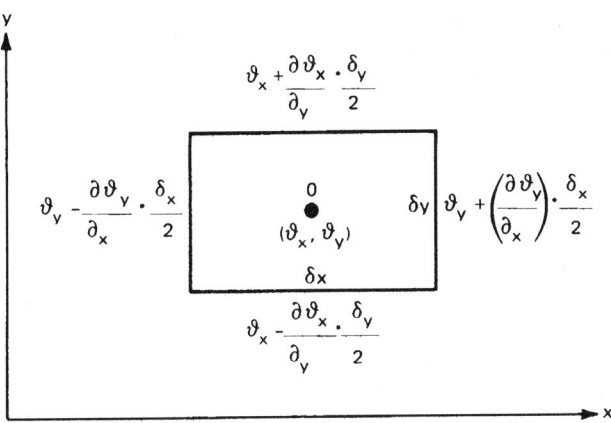

Fig. 1-1.11. A fluid element.

$$v_r = \frac{1}{r}\frac{\partial \psi}{\partial \theta} = \frac{1}{r}\frac{d\psi}{d\theta} \tag{36}$$

Combining the two equations and integrating for ψ gives

$$\psi = \frac{m}{2\pi}\theta \tag{37}$$

This equation shows that the stream lines for a source are radial lines as shown in Figure 1-1.9.

Similarly, the stream lines for a two-dimensional source-sink pair and a two-dimensional doublet are shown in Figure 1-1.10.

Fluid Rotation

For a rigid body, if each particle of the body describes a circle about its axis of rotation, the body is said to be rotating.

But since in a fluid each particle is free to move in any direction, the fluid may not describe a perfect circle about the axis of rotation. Consequently, the rotation of a fluid element at a point is defined as the average angular velocity of the element.

An element of fluid is shown in Figure 1-1.11.

The average angular velocity of the element can be shown to be

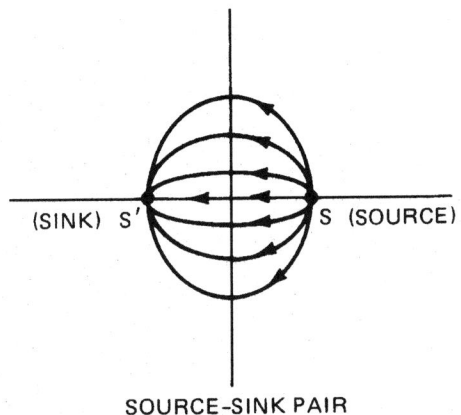

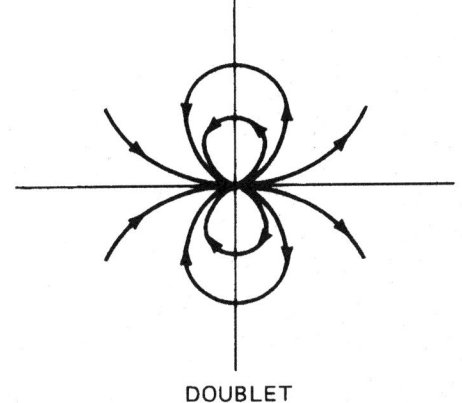

Fig. 1-1.10. Streamlines of a source-sink pair and a doublet.

$$2\omega_z = \frac{\partial v_y}{\partial x} - \frac{\partial v_x}{\partial y} \qquad (38)$$

where $2\omega_z$ is called the *vorticity*, ζ_z.

The direction of the vorticity, by convention, is given by the right-hand screw rule. In this case ζ_z is positive in an upward direction, perpendicular to the plane of paper.

Similarly, the other components can be written as

$$\zeta_y = \frac{\partial v_x}{\partial z} - \frac{\partial v_z}{\partial x} \qquad (39a)$$

$$\zeta_z = \frac{\partial v_z}{\partial y} - \frac{\partial v_y}{\partial z} \qquad (39b)$$

The components of vorticity in cylindrical coordinates are

$$\zeta_z = \frac{1}{r}\left[\frac{\partial}{\partial r}(rv_\theta) - \frac{\partial v_\theta}{\partial \theta}\right] \qquad (40a)$$

$$\zeta_\theta = \frac{\partial v_r}{\partial z} - \frac{\partial v_z}{\partial r} \qquad (40b)$$

$$\zeta_r = \frac{1}{r}\frac{\partial v_z}{\partial \theta} - \frac{\partial v_\theta}{\partial z} \qquad (40c)$$

The total vorticity of the fluid at a given point is obtained by vectorially adding the three components.

Irrotational fluid motion: If the vorticity at a given point within the fluid is zero, the fluid motion is said to be irrotational. For example, a uniform parallel flow with no velocity gradients is irrotational. The condition for irrotationality of flow in terms of stream function is

$$v_r = 0 \qquad (41)$$

$$\frac{\partial^2 \psi}{\partial x^2} + \frac{\partial^2 \psi}{\partial y^2} = 0 \qquad (42)$$

This is known as the Laplace equation.

Free Vortex

The type of motion in which the vorticity is zero and the peripheral velocity varies inversely with the radial distance is called the free vortex, i.e.,

$$v_r = 0 \qquad (43a)$$

$$\zeta_z = 0, \quad \text{i.e.,} \quad \frac{\partial v_\theta}{\partial r} + \frac{v_\theta}{r} = 0 \qquad (43b)$$

This gives

$$v_\theta r = \text{constant} \qquad (44)$$

From Equation 34,

$$v_\theta = -\frac{\partial \psi}{\partial r} \qquad (45)$$

Combining Equations 44 and 45 gives

$$\psi = k \ln r \qquad (46)$$

where k is a constant.

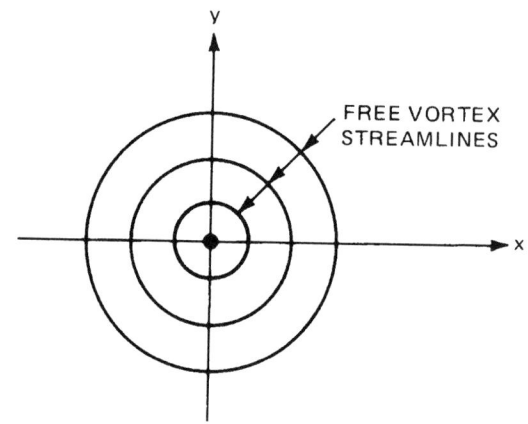

Fig. 1-1.12. Free vortex streamlines.

The stream lines for a vortex, then, are circles as shown in Figure 1-1.12.

Velocity Potential

For an irrotational flow, all the components of vorticity are zero, i.e.,

$$\zeta_z = \zeta_y = \zeta_x = 0 \qquad (47)$$

From Equations 38, 39a and 39b, and 47 it follows that there exists a function $\phi(x,y)$ such that

$$v_x = \frac{\partial \phi}{\partial x}, \qquad v_y = \frac{\partial \phi}{\partial y}, \qquad v_z = \frac{\partial \phi}{\partial z} \qquad (48)$$

in cylindrical coordinates

$$v_r = \frac{\partial \phi}{\partial r}, \qquad v_\theta = \frac{1}{r}\frac{\partial \phi}{\partial \theta} \qquad (49)$$

These relationships generate

$$\frac{\partial^2 \phi}{\partial x^2} + \frac{\partial^2 \phi}{\partial y^2} = 0 \qquad (50)$$

This shows that the velocity potential, ϕ, also satisfies Laplace's law.

It must be noted that the velocity components can only be expressed in terms of a velocity potential if the flow is irrotational. For ψ to exist continuity must be satisfied, and for ϕ to exist flow must be irrotational.

For a pure vortex

$$v_r = 0 \qquad (51)$$

$$\text{therefore} \quad v_\theta = \frac{1}{r}\frac{\partial \phi}{\partial \theta}$$

or

$$\phi = k\theta \qquad (52)$$

where k is a constant.

The equipotential (constant ϕ) lines for a vortex are shown in Figure 1-1.13.

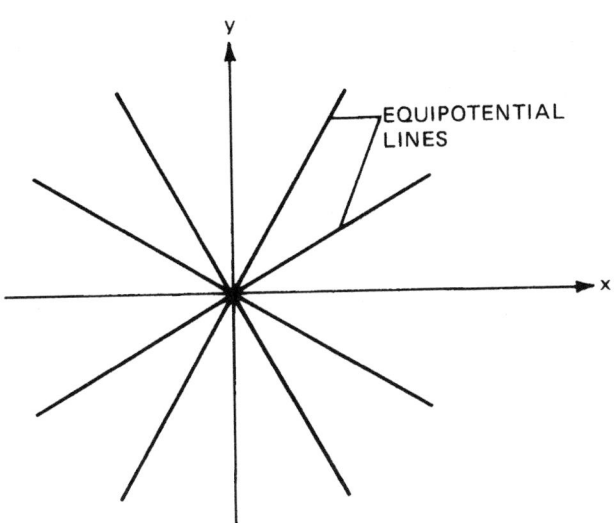

Fig. 1-1.13. *Equipotential lines for a vortex.*

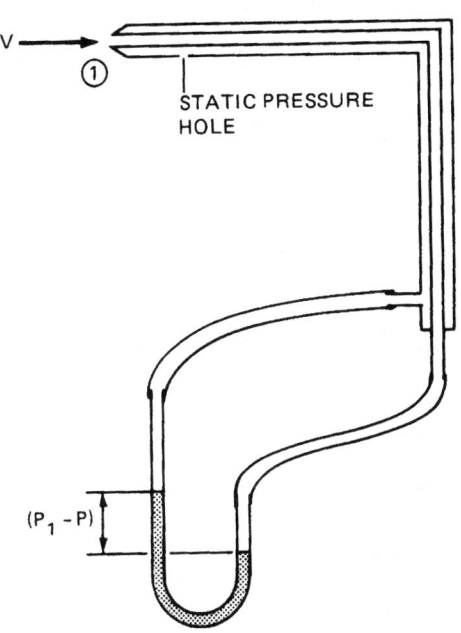

Fig. 1-1.15. *Pitot-static tube.*

Circulation

The circulation around a closed contour within a flow field is defined as the sum of the product of the tangential velocity and the elemental length at every point on the contour. (See Figure 1-1.14.)

For a closed contour, C, the circulation is given by

$$\Gamma = \oint_c U \cos \theta \, ds \qquad (53)$$

It can be shown that for an element of fluid (δx, δy) the total circulation is given by

$$\Gamma = \iint \zeta_z \, dx \, dy \qquad (54)$$

This result is known as Stokes' theorem.

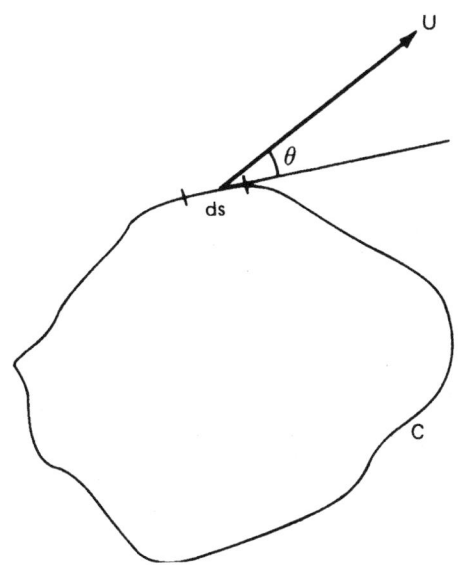

Fig. 1-1.14. *Circulation around a circuit, C.*

DYNAMICS OF INCOMPRESSIBLE FLUIDS

Kinematics of fluid motion only deals with the general characteristics of flow of fluids. It does not answer the question of how fluids will move under given conditions, e.g., flow of real fluids over bodies of varying shapes, etc. In other words, it is necessary to establish the relationship between the force and the resulting motion. This relationship is covered under the general heading of fluid dynamics.

The Bernoulli Equation

The Bernoulli equation expresses the relationship between the pressure and velocity within a fluid flow; the general form of the equation is

$$p + \frac{1}{2}\rho V^2 + \rho gz = \text{constant} \qquad (55)$$

where p is the static pressure at a point called the pressure head, $\frac{1}{2}(\rho V^2)$, is called the dynamic head, gz is the position head, and the constant is usually called the Bernoulli constant.

In some applications the position head term is considerably smaller, and therefore, the Bernoulli equation simplifies to

$$p + \frac{1}{2}\rho V^2 = \text{constant} \qquad (56)$$

From Equation 56 it is clear that the static pressure, p, decreases with the increase in velocity, V. It should be remembered that the Bernoulli equation only applies to a stream line. It follows that this relationship enables the flow velocity to be calculated from the measurement of pressure. This is the principle of the pitot-static tube.

Figure 1-1.15 shows a typical arrangement of a pitot-static tube. When pointing in the direction of the flow, this tube allows the magnitude of the velocity to be determined through the measurement of pressures.

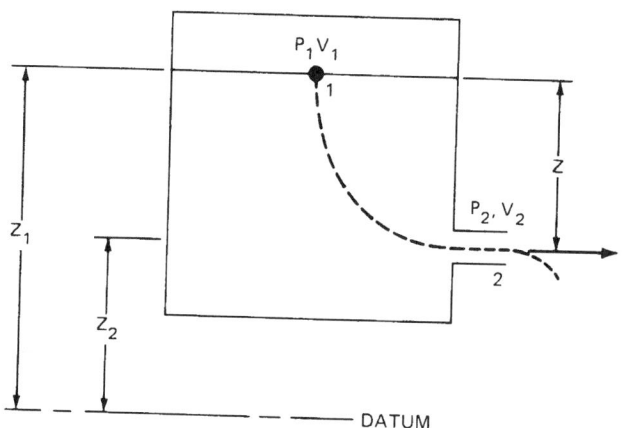

Fig. 1-1.16. Efflux from a large tank.

If V is the free stream velocity, then according to the Bernoulli equation

$$\frac{1}{2}\rho V^2 + p = \frac{1}{2}\rho V_1^2 + p_1 \qquad (57)$$

At Point 1 at the mouth of the tube the velocity $V_1 = 0$; i.e., the stagnation point and p_1 is the stagnation pressure at that point.

Equation 57 then becomes

$$V = \sqrt{\frac{2(p_1 - p)}{\rho}} \qquad (58)$$

where $(p_1 - p)$ is pressure measured by the manometer.

Another application of the Bernoulli relation is to the flow of water under pressure from a tank. (See Figure 1-1.16.)

In Figure 1-1.16, p_1 is the atmospheric pressure.

Applying the Bernoulli equation at Points 1 and 2 gives

$$p_1 + \frac{1}{2}\rho V_1^2 + \rho g z_1 = p_2 + \frac{1}{2}\rho V_2^2 + \rho g z_2 \qquad (59)$$

If $V_1 \ll V_2$ then

$$\frac{1}{2}\rho V_2 = (p_2 - p_1 + \rho g(z_2 - z_1)) \qquad (60a)$$

but $p_1 = p_2 =$ atmospheric pressure.
Therefore,

$$\frac{1}{2}\rho V_2^2 = \rho g(z_2 - z_1) = \rho g z \qquad (60b)$$

or

$$V_2 = \sqrt{2gz} \qquad (61)$$

This is known as Torricelli's law for efflux from a container.

The Venturi meter, which measures the volume flow rate in a pipe, basically consists of a converging cone that merges into a parallel throat with a minimal cross-sectional area. (See Figure 1-1.17.) From the measurement of the static pressures in the parallel region and the converging region, the volume flow rate can be calculated as shown below.

Applying the Bernoulli equation at Points 1 and 2 gives

$$\frac{1}{2}\rho V_1^2 + p_1 = \frac{1}{2}\rho V_2^2 + p_2 \qquad (62)$$

But from continuity

$$\rho V_1 A_1 = \rho V_2 A_2 \qquad (63)$$

Combining Equations 62 and 63 results in

$$V_2 = \sqrt{\frac{2}{\rho} \frac{A_1^2(p_2 - p_1)}{(A_2^2 - A_1^2)}} \qquad (64)$$

Therefore the volume flow rate, Q, is given by

$$Q = V_2 A_2$$

or

$$Q = A_2 \left[\frac{2A_1^2(p_2 - p_1)}{\rho(A_2^2 - A_1^2)} \right]^{1/2} \qquad (65)$$

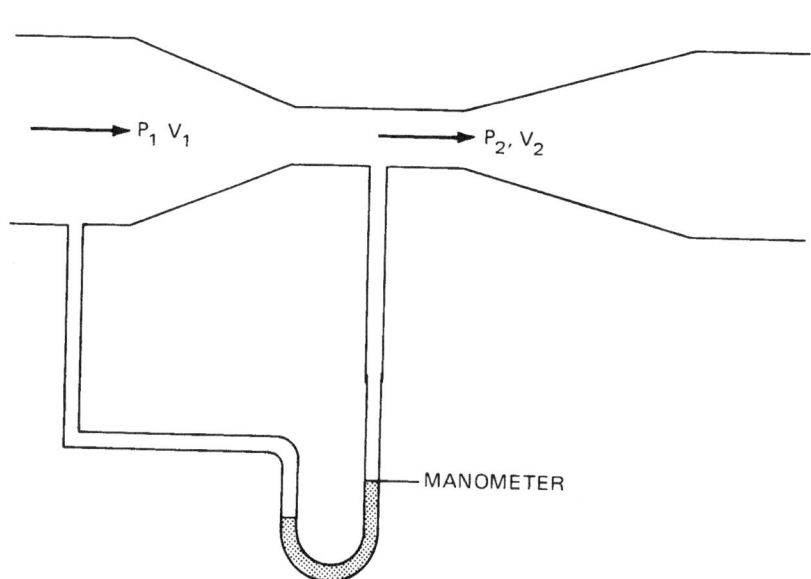

Fig. 1-1.17. Venturi meter.

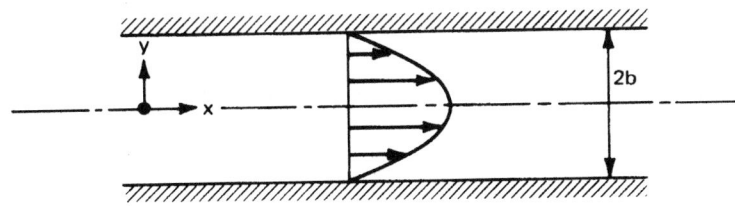

Fig. 1-1.18. Flow in a channel.

The Navier-Stokes Equations

The Navier-Stokes (N-S) equations are the exact equations describing the fluid motion. They are valid for both the laminar and turbulent flows. They are derived from Newton's second law of motion, which states that the sum of the external forces acting on a body is equal to the product of the mass and acceleration of the body. In the case of a fluid, this body is assumed to be a fixed control volume within which the fluid properties remain unchanged. To account for the fluid viscosity (i.e., stickiness of the fluid), it is further assumed that the instantaneous rate of strain (distortion) of the fluid element (body) is a simple linear function of the stresses (forces) in the fluid. Two types of forces are considered important: (1) the body forces (e.g., gravitation) and (2) the surface forces (e.g., pressure and friction).

The Navier-Stokes equations can be viewed as the transport equations that equate the net rate of transport of some quantity Q (momentum or enthalpy). For momentum transport the second law of motion is utilized; and for the enthalpy transport, the principle of first law of thermodynamics is used.

For momentum rate of transport, the general form of the N-S equations is

$$\frac{Du}{Dt} = X - \frac{1}{\rho}\frac{\partial p}{\partial x} + \nu\nabla^2 u \qquad (66a)$$

$$\frac{Dv}{Dt} = Y - \frac{1}{\rho}\frac{\partial p}{\partial y} + \nu\nabla^2 v \qquad (66b)$$

$$\frac{Du}{Dt} = Z - \frac{1}{\rho}\frac{\partial p}{\partial z} + \nu\nabla^2 w \qquad (66c)$$

where X, Y, Z are the body forces in the x, y, z directions; ρ is the fluid density; p is pressure; ν is the fluid kinematic viscosity; u, v, w are the velocity components; and D/Dt is the substantive derivative consisting of the non-steady and convective components.

For example,

$$\frac{Du}{Dt} = \frac{\partial u}{\partial t} + u\frac{\partial u}{\partial x} + v\frac{\partial u}{\partial y} + w\frac{\partial u}{\partial z}$$

The Laplace's operator ∇^2 (also known as del squared) is defined as

$$\nabla^2 = \frac{\partial^2}{\partial x^2} + \frac{\partial^2}{\partial y^2} + \frac{\partial^2}{\partial z^2}$$

The Navier-Stokes equations, together with the continuity Equation 31, form four simultaneous differential equations from which the four unknowns (u, v, w and p) could, in principle, be solved. However, the non-linear nature of these equations make the task prohibitively complex. But for very simple cases of laminar flow, analytical solutions are possible, as demonstrated in Example 1. For more practical applications involving complex turbulent flows, the computer-based numerical techniques are used. In computational fire modeling, these computer codes are generally referred to as *field models*.

EXAMPLE 1: Flow in a Channel.

Consider the steady, incompressible viscous flow in an infinitely long two-dimensional stationary channel with two parallel flat plates. (See Figure 1-1.18.)

The flow everywhere is parallel to the x-axis. Since the flow is considered to be steady, the components of velocity in the y and z directions are zero; i.e., $v = 0$, $w = 0$. The Navier-Stokes equations then become

$$-\frac{1}{\rho}\frac{\partial p}{\partial x} + \nu\frac{\partial^2 u}{\partial y^2} = 0 \qquad (67)$$

$$-\frac{1}{\rho}\frac{\partial p}{\partial y} = 0 \qquad (68)$$

From the continuity equation

$$\frac{\partial u}{\partial x} = 0 \qquad (69)$$

These two equations are solved, subject to the boundary conditions

$$u = 0, \qquad y = b \qquad (70)$$
$$u = 0, \qquad y = -b$$

Equation 68 shows that the pressure, p, is a function of x only, while Equation 69 shows that u only varies with y. Equation 67 therefore becomes

$$-\frac{1}{\rho}\frac{dp}{dx} + \nu\frac{d^2 u}{dy^2} = 0$$

When solved for the given boundary conditions (Equation 70), the velocity component u becomes

$$\frac{dp}{dx} = \mu\frac{d^2 u}{dy^2} \qquad (71)$$

$$u = -\frac{1}{2\mu}\frac{dp}{dx}(b^2 - y^2) \qquad (72)$$

Thus, the velocity profile is parabolic and the corresponding shear stress is given by

$$\tau = \mu\frac{du}{dy} = -\frac{1}{2\mu}\frac{dp}{dx}(b^2 - 2y) \qquad (73)$$

The Energy Equation

As stated, the N-S equations are basically the transport equations. As such, they apply to both the transport of momentum as well as the transport of heat (enthalpy). The transfer of heat between a solid body and a gaseous flow involves the conservation equation of motion and that of heat. In fire problems, the transfer of heat energy from a fire source and the resulting rise in temperature is of great importance; for example, the assessment of a fire barrier performance and fire detection. In the smoke transport problems, the temperature distribution throughout a building, contributing to the stack effect, is also of major importance. To calculate this distribution, it is necessary to solve the energy conservation equation along with the momentum transfer N-S equations.

For an incompressible fluid the energy balance is determined by the internal energy, the conduction of heat, the convection of heat and the generation of heat through friction, and by a heat source.

According to the First Law of Thermodynamics, the energy balance can be written as

$$\underset{\substack{\text{rate heat} \\ \text{input}}}{\frac{dQ}{dt}} = \underset{\substack{\text{rate of change} \\ \text{of internal} \\ \text{energy}}}{\frac{dE}{dt}} + \underset{\substack{\text{work done}}}{\frac{dW}{dt}} \quad (74)$$

In Equation 74, the radiative heat transfer is neglected.

For a constant property fluid, the energy equation is given by

$$\frac{\partial \theta}{\partial t} + u \frac{\partial \theta}{\partial x} + v \frac{\partial \theta}{\partial y} + w \frac{\partial \theta}{\partial z} = \frac{k}{\rho c_p} \nabla^2 \theta \quad (75)$$

where θ is the temperature rise above datum value (e.g., ambient), k is thermal conductivity, and c_p is the specific heat at constant pressure.

Turbulence

Randomness is the necessary and sufficient condition for turbulent motion. A fluid flow that is highly disordered, rotational, and three dimensional has all the essential characteristics of turbulence. These features, at first glance, may point to the unpredictability of turbulent flow. Indeed, in reality, it is not possible either to define turbulence in precise terms or to predict precisely its flow characteristics. However, in practice, what is of importance is the effect of turbulence; i.e., the way it manifests itself in the flow phenomenon. From this standpoint its definition in precise terms is neither essential nor desirable.

In nature, there is underlying order and regularity; i.e., perpetual striving for local equilibrium (harmony). Instability is only a transition (intermediate stage) from one stable state to another. The macroscopic worldview of chaos and randomness pertains to human perception, and is a consequence of human inability to relate the intrinsic stability at the infinitesimal scale to the finite world of the observed physical phenomena. It is, for this reason, that the empirical option of "lumped parameter" is the approach used for statistical descriptions.

On this basis and in order to make any meaningful progress in the treatment of turbulent flow, it is important to recognize that turbulent motion is effectively made up of: (1) a mean component, which is intelligible, and (2) a fluctuating component, which is random. In terms of the mean flow, there are very little qualitative differences between the laminar and turbulent flows; i.e., the mean motion is fully described. In contrast, fluctuations are based solely on statistical information.

One of the main observed features of turbulence is that is causes diffusion. In other words, it transports the fluid itself and any characteristic associated with it, such as the airborne pollution or smoke particles. In this respect, turbulence is a feature of the flow and not of the fluid. Experimental evidence shows that turbulent fluctuations result from the highly disordered array of eddies of widely different sizes that transport the fluid elements. These turbulent eddies, as they are swept along by the mean flow, undergo both the translational and rotational motion. During this process the larger eddies are distorted and stretched, and consequently break up into smaller ones.

As described, the turbulent motion can be broken down into mean and fluctuating components. Thus, the instantaneous value of a quantity, q, can be written as

$$q = \bar{q} + q'$$

where $\bar{q}$ is the mean with respect to time (time-average), defined as

$$\bar{q} = \frac{1}{\Delta t} \cdot \int_{t_0}^{t_0 + \Delta t} q(t) \, dt$$

According to this definition, the time-average of all fluctuating quantities is equal to zero, i.e.,

$$\bar{q}' = 0$$

Accordingly, the instantaneous velocity and pressure components can be written as

$$u = \bar{u} + u', \quad v = \bar{v} + v', \quad w = \bar{w} + w', \quad p = \bar{p} + p'$$

By substituting these quantities into the continuity equation, it can readily be shown that

$$\frac{\partial \bar{u}}{\partial x} + \frac{\partial \bar{v}}{\partial y} + \frac{\partial \bar{w}}{\partial z} = 0$$

and

$$\frac{\partial u'}{\partial x} + \frac{\partial v'}{\partial y} + \frac{\partial w'}{\partial z} = 0$$

This result shows that the time-average velocity field and the fluctuating velocity components satisfy the same continuity equation as the actual velocity field.

Now, when the above definitions of the instantaneous velocity are substituted in the Navier-Stokes equations and the continuity equation is utilized, the x-direction turbulent form of the Navier-Stokes equation reduces to

$$\bar{u} \frac{\partial \bar{u}}{\partial x} + \bar{v} \frac{\partial \bar{u}}{\partial y} + \bar{w} \frac{\partial \bar{u}}{\partial z} = X - \frac{1}{\rho} \frac{\partial \bar{p}}{\partial x} + \nu \nabla^2 \bar{u}$$
$$- \left(\frac{\partial \bar{u}'^2}{\partial x^2} + \frac{\partial \bar{u}'\bar{v}'}{\partial y^2} + \frac{\partial \bar{u}'\bar{w}'}{\partial z^2} \right)$$

The y and z direction equations are of similar form.

From the comparison of the laminar and turbulent forms of these equations, it is clear that, in addition to the usual non-linearities, extra terms involving fluctuating velocity products ($\bar{u}'^2$, $\bar{u}'\bar{v}'$, $\bar{u}'\bar{w}'$) appear on the right-hand

TABLE 1-1.2 *Important Dimensionless Groups*

Name	Group	Physical Significance
Reynolds Number	$Re = \dfrac{\rho u l}{\nu}$	Ratio of the inertia force to the friction force.
Froude Number	$Fr = \dfrac{u^2}{lg}$	Ratio of the inertia force to the gravity force. Relevant to buoyant flows associated with fires.
Grashof Number	$Gr = \dfrac{gl^3 \beta T}{\nu^2}$	Ratio of buoyancy force to viscous forces, as in fire plumes.
Prandtl Number	$Pr = \dfrac{\mu c_p}{k}$	Ratio of momentum diffusivity to thermal diffusivity.

side. These terms, which account for the effects of turbulence, are generally known as the *Reynolds stresses* (also sometimes referred to as *apparent* or *virtual stresses*). These additional stresses arise from the turbulent fluctuations, and have a similar influence as the viscous terms in the laminar flow case. It is for this reason they are often said to be caused by the *eddy viscosity*. In almost all turbulent flows of practical interest, Reynolds stresses are much larger than the viscous stresses. This is one reason why turbulence is of such great practical importance.

The Navier-Stokes equations, as described, are the exact equations describing the fluid motion. However, in the solution of these equations a formidable mathematical difficulty arises due to the non-linearity of the relation between the velocity and momentum flux, as reflected in the Reynolds stress terms. An approach is therefore used that solves these equations over a numerical grid or mesh within the specified region having prescribed boundary conditions. The exact equations are averaged over a time scale. The Reynolds stresses are expressed in terms of known quantities under the framework of a "turbulence model." Appropriate turbulence models are selected depending on the nature and complexity of the flow phenomena. The so-called k-ε turbulence model is by far the most common and is shown to give the satisfactory answers for engineering applications. For fire applications, this approach is referred to as "field modeling" in contrast to "zone modeling," which is purely an empirical approach. A detailed discussion of the numerical techniques and the turbulence models is beyond the scope of this text. However, a useful and comprehensive review of these models is given by Kumar.[2]

FLOW SIMILARITY AND DIMENSIONAL ANALYSIS

It is clear so far that the flow characteristics of any fluid flow system are determined not only by the properties of the fluid but also by the geometry of flow. For example, the flow through an open channel and tube will differ because the two flow regimes are not geometrically similar. This does not imply that any geometric similarity will produce flow similarity (i.e., dynamic similarity). For a process or physical system to be dynamically similar, the ratios of the forces involved, in addition to the geometrical similarity, must be equal.

In the experimental investigation of the underlying fluid flow phenomena, small-scale "physical modeling" approach is often used. The basis of this approach relies on the hypothesis that the full-scale physical phenomena can be simulated in a model scale (i.e., small scale) experiment, provided certain non-dimensional parameters (or ratios) are preserved. In other words, provided the physical similarity (e.g., geometric, kinematic, and dynamic similarity) is maintained, the results of the small-scale experiments are assumed to be equally valid for the full-scale case. Table 1-1.2 outlines the physical significance of important dimensionless groups.

Kinematic similarity exists if the particle paths are geometrically similar. Dynamic similarity exists between two geometrically and kinematically similar systems if the ratios of all the forces are equal.

If F_p, F_μ, and F_u are denoted as the pressure, viscous, and inertial forces, respectively, for dynamic similarity

$$\frac{(F_p)_1}{(F_p)_2} = \frac{(F_\mu)_1}{(F_\mu)_2} = \frac{(F_u)_1}{(F_u)_2} \quad \text{or} \quad \left(\frac{F_p}{F_u}\right)_1 = \left(\frac{F_p}{F_u}\right)_2 = \text{constant}$$

or

$$\left(\frac{F_\mu}{F_u}\right)_1 = \left(\frac{F_\mu}{F_u}\right)_2 = \text{constant}$$

This means that a flow over two spheres of different radii will be dynamically similar, provided

$$\left(\frac{F_u}{F_\mu}\right)_{\text{sphere 1}} = \left(\frac{F_u}{F_\mu}\right)_{\text{sphere 2}} = Re$$

where Re is a constant.

This constant is generally known as the Reynolds number and is defined as

$$Re = \frac{\rho V L}{\mu}$$

where ρ = density of fluid, V = velocity, μ = viscosity, and L = sphere diameter.

If Re for two kinematically similar flows is the same, the flows are then said to be dynamically similar. Other similarity parameters are given in Table 1-1.2.

Dimensions and Units

There are three fundamental units of measure in fluid mechanics. These fundamental units are mass, M; length, L; and time, T. All the other quantities, such as force, pressure, etc., can be expressed in terms of these fundamental units.

If a quantity is capable of being expressed in these fundamental units, the resulting function of M, L, and T is then termed the dimensions of the quantity.

From Newton's second law of motion, force is given by $F = ma$, where m = mass, and a = acceleration.

The dimensions of m are [M] and are written as

$$[m] \; \hat{=} \; [M]$$

the symbol $\hat{=}$ means "has the dimensions of."

Similarly

$$[a] = [L/T^2] \; \hat{=} \; [LT^{-2}]$$

and, therefore

TABLE 1-1.3. *Measure Formula or Physical Dimensions of Quantities Occurring in Mechanics. (Based on mass, length, and time as fundamental units.)*

Quantity	Measure formula	Quantity	Measure formula
Mass	M	Mass per unit area	ML^{-2}
Length	L	Mass moment	ML
Time	T	Moment of inertia and product of inertia	ML^2
Speed or velocity	LT^{-1}	Stress and pressure	$ML^{-1}T^{-2}$
Acceleration	LT^{-2}	Strain	$M^0L^0T^0$
Momentum and impulse	MLT^{-1}	Elastic modulus	$ML^{-1}T^{-2}$
Force	MLT^{-2}	Flexural rigidity of a beam $E1$	ML^3T^{-2}
Energy and work	ML^2T^{-3}	Torsional rigidity of a shaft GJ	ML^3T^{-2}
Power	ML^2T^{-2}	Linear stiffness (force per unit displacement)	MT^{-2}
Moment of force	ML^2T^{-2}	Angular stiffness (moment per radian)	ML^2T^{-2}
Angular momentum or moment of momentum	ML^2T^{-1}	Linear flexibility or receptance (displacement per unit force)	$M^{-1}T^2$
Angle	$M^0L^0T^0$	Vorticity	T^{-1}
Angular velocity	T^{-1}	Circulation (hydrodynamics)	L^2T^{-1}
Angular acceleration	T^{-2}	Viscosity	$ML^{-1}T^{-1}$
Area	L^2	Kinematic viscosity	L^2T^{-1}
Volume and first moment of area	L^3	Diffusivity of any quantity	L^2T^{-1}
Second moment of area	L^4	Coefficient of solid friction	$M^0L^0T^0$
Density	ML^{-3}	Coefficient of restitution	$M^0L^0T^0$

$$F \stackrel{\wedge}{=} MLT^{-2}$$

A table of dimensions for other physical quantities is given in Table 1-1.3.

Dimensional Analysis

When a physical phenomenon is represented by an equation, it is absolutely necessary that all the terms in the equation have the same units, i.e., that the equation be dimensionally homogeneous. A quick dimensional analysis of the parameters involved in an equation provides a powerful clue to the homogeneities of the equation.

EXAMPLE 2:

Find the expression for discharge, Q, through a horizontal capillary tube.

The discharge, Q, depends on the following parameters:

pressure drop per unit length	$\Delta p/\ell$
diameter of the tube	D
fluid viscosity	μ

These parameters have the following dimensions:

$$\frac{\Delta p}{\ell} \stackrel{\wedge}{=} ML^{-2}T^{-2}, \quad D \stackrel{\wedge}{=} L$$

$$\mu \stackrel{\wedge}{=} ML^{-1}T^{-1}$$

Therefore,

$$Q = k\left(\frac{\Delta p}{\ell}\right)^\alpha \cdot D^\beta \cdot \mu^\gamma$$

i.e.,

$$[L^3T^{-1}] = k[ML^{-2}T^{-2}]^\alpha \cdot [L]^\beta \cdot [ML^{-1}T^{-1}]^\gamma$$

From the principle of homogeneity and by comparing the exponents of each dimension

$$\gamma + \alpha = 0 \quad \text{from } [M]$$
$$\beta - 2\alpha - \gamma = 3 \quad \text{from } [L]$$
$$2\alpha + \gamma = 1 \quad \text{from } [T]$$

Solving these equations gives

$$\alpha = 1, \quad \gamma = 1, \quad \beta = 4$$

Therefore, the equation for discharge, Q, becomes

$$Q = k\left(\frac{\Delta p}{\ell}\right)^1 \cdot D^4 \cdot \mu^{-1} = k\frac{\Delta p}{\ell} \cdot \frac{D^4}{\mu}$$

No information about the numerical value of the dimensionless constant k can be obtained from the equation. This information can be obtained, however, from the experiment.

BOUNDARY LAYERS

When a fluid flows over a solid boundary (surface), the velocity at the surface is zero (no-slip condition). This velocity increases with the distance away from the surface and eventually becomes equal to the free stream velocity. Therefore, a region exists close to the surface of the body in which the velocity varies with distance. This region is called the boundary layer. Within this region each layer of fluid moves relative to the adjacent layer, and as a result large shear stresses are set up within the boundary layer. An important approximation can be made in solving the Navier-Stokes equations such that viscosity only plays a role within the boundary layer, while outside this layer the fluid can be treated as inviscid ($\mu = 0$).

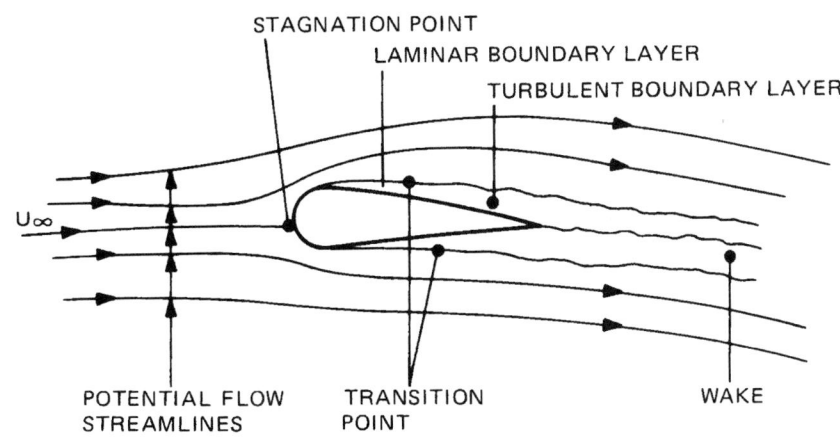

Fig. 1-1.19. Flow over a two-dimensional aerofoil.

Physics of Boundary Layers

The flow over a two-dimensional aerofoil is shown in Figure 1-1.19.

It can be observed that over the forward part the flow is smooth and the streamlines are parallel to the surface. This region is known as the laminar boundary layer.

But as the flow progresses over the surface of the aerofoil toward the trailing edge, the streamlines break up and there are random fluctuations in velocity direction and magnitude, even though the general mean motion remains roughly parallel to the surface. This region is called the turbulent boundary layer. As the flow leaves the surface behind the body, it merges to form a stream of relatively slow-moving fluid, which is known as the wake. Consequently, the velocity profiles in the turbulent and laminar boundary layers are quite different. The measurements of velocity profiles on a flat plate are shown in Figure 1-1.20.

This figure shows that the velocity gradient at the surface ($\partial u/\partial y$) is much greater in the turbulent boundary layer than in the laminar boundary layer. It follows that the frictional stresses, τ_w,

$$\tau_w = \mu\left(\frac{\partial u}{\partial y}\right)_{y=0} \qquad (76)$$

at the wall surface, i.e., the drag, are much greater for the turbulent than the laminar boundary layer.

As the pressure increases downstream along the surface, the velocity is decreased within the boundary layer. Thus, as $\partial p/\partial x$ increases, $\partial u/\partial y$ decreases. With any further increases in this pressure gradient a stage may be reached where $\partial u/\partial y$ at the wall is zero and the fluid adjacent to the surface is then on the point of reversing its direction of flow. The boundary layer is then said to be on the point of separation, as illustrated in Figure 1-1.21.

Subsequently, the velocity gradient (du/dy) at the wall becomes negative downstream of separation and an inner portion of the boundary layer then flows against the stream. This reversed flow forms a large eddy under the outer part of the boundary layer. These eddies carry a large amount of energy, which results in the increase of drag on the body.

A body with extensive boundary layer separation and a wake with large scale eddies is referred to as a bluff body. In contrast, a body on which the boundary layer remains attached over the whole surface to the rearmost point and then merges smoothly into the wake without the formation of any large scale eddies is referred to as a streamline body.

The nature and magnitude of drag on these two shapes of bodies is quite different. The drag of a bluff body is usually very large, and is almost entirely due to the energy wasted in the castoff eddies. Drag can be determined fairly accurately from the pressure distribution acting normal to its surface. The drag of a streamline body is relatively small, and is mainly due to the viscous or frictional stresses that act tangential to its surface.

Boundary Layer Thickness

The velocity within a boundary layer varies from zero at the surface to free-stream velocity just outside the boundary layer. The boundary layer thickness, δ, is defined as the distance from the wall at which the velocity reaches 99 percent of the free-stream velocity (at $y = \delta$, $u = 0.99U_\infty$).

The displacement thickness, δ^*, is defined as that distance by which the external potential field of flow is displaced outward as a consequence of the decrease in velocity in the boundary layer. The precise form of displacement thickness is

$$\delta^* = \int_{y=0}^{\infty} \left(1 - \frac{u}{U_\infty}\right) dy \qquad (77)$$

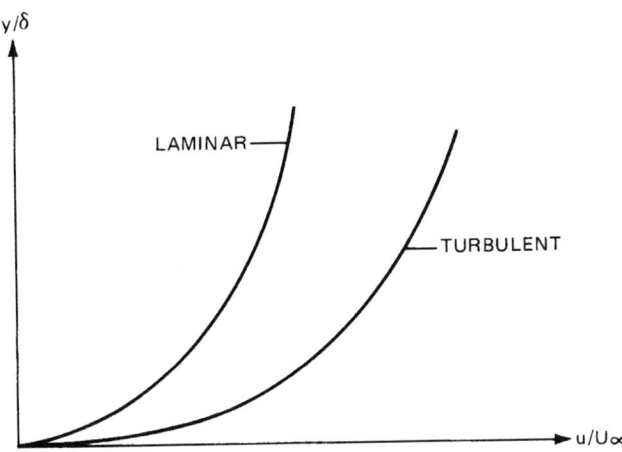

Fig. 1-1.20. Velocity profiles over a flat plate.

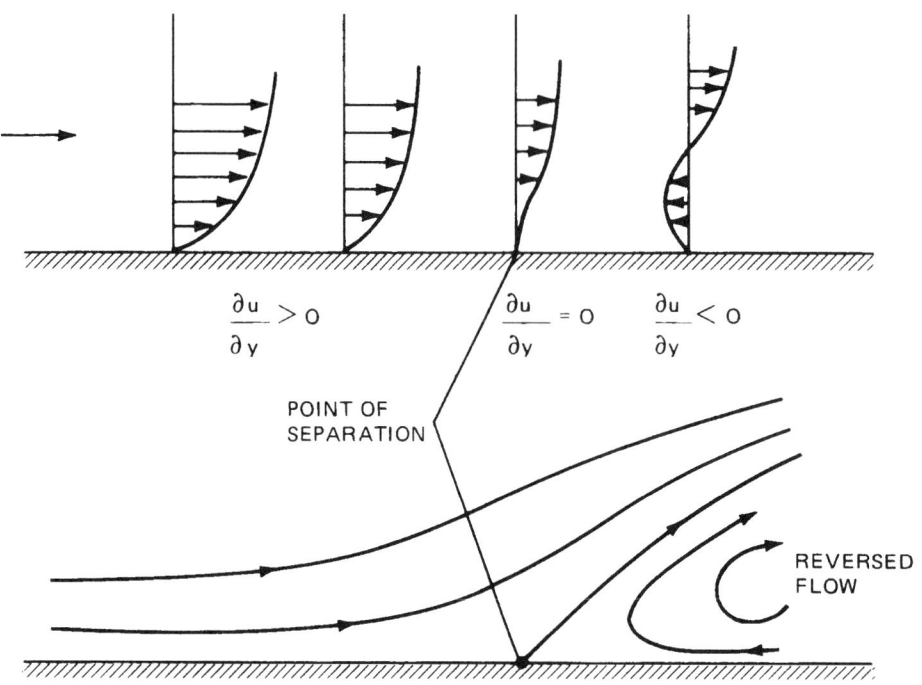

Fig. 1-1.21. *Point of separation and flow reversal.*

Momentum thickness, θ, is defined as

$$\theta = \int_{y=0}^{\infty} \frac{u}{U_\infty} \left(1 - \frac{u}{U_\infty}\right) dy \qquad (78)$$

From this equation it is clear that $\rho U_\infty^2 \theta$ represents the defect in the rate of transport of momentum in the boundary layer as compared with the rate of transport of momentum in the absence of the boundary layer.

Energy thickness, δ_E, is defined as

$$\delta_E = \int_{y=0}^{\infty} \frac{u}{U_\infty}\left[1 - \left(\frac{u}{U_\infty}\right)^2\right] dy \qquad (79)$$

The quantity $\rho U_\infty^3 \delta_E/2$ represents the defect in rate of transport of kinetic energy in the boundary layer when compared with the rate of transport in the absence of the boundary layer.

FLOWS IN PIPES AND DUCTS

The flow in the inlet length of a circular pipe of constant cross-section is shown in Figure 1-1.22. Initially, at distances very close to the inlet, the flow is uniform with a

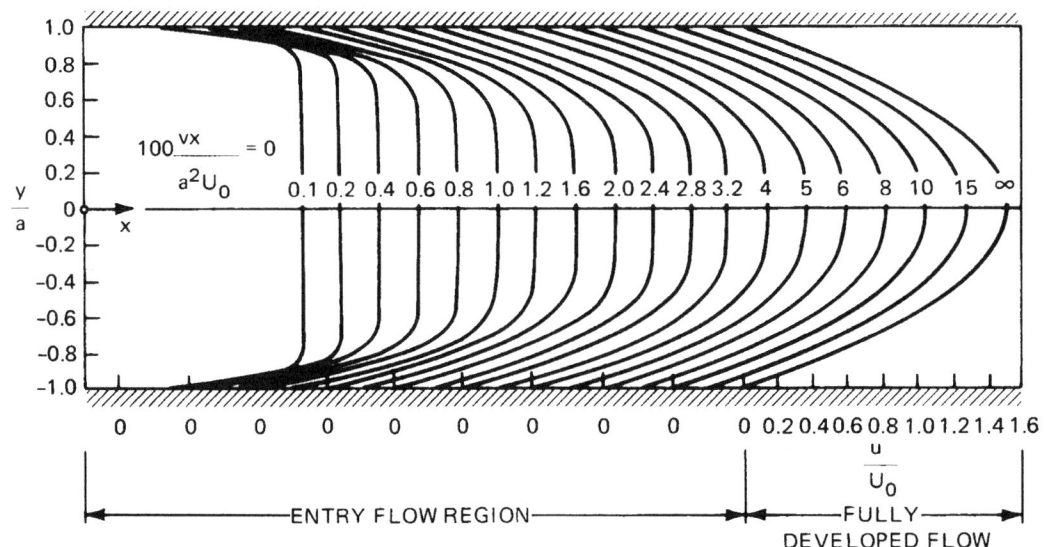

Fig. 1-1.22. *Development of laminar boundary layer in a pipe.*

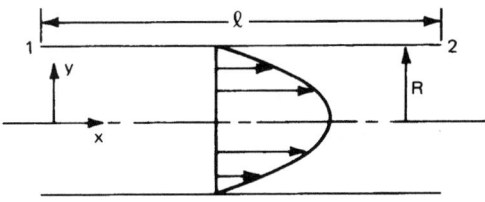

Fig. 1-1.23. **Fully developed laminar flow in a pipe.**

boundary layer of very small thickness forming very close to the wall surface.

The velocity profile across the pipe changes continuously downstream. As the boundary layer on the wall grows downstream, the velocity in the central region increases to compensate for this growth, since the equation of continuity must hold true. At some distance from the inlet the boundary layers merge and the flow becomes fully developed. Onward from this distance, the velocity profile remains the same. Since the velocity profile in the fully developed flow is constant, it follows that the force due to friction at the pipe wall exactly balances that force due to the pressure gradient down the pipe required to maintain the flow.

Laminar Flow in a Pipe

With fully developed laminar flow in a smooth-walled pipe, the fluid moves under the influence of the pressure gradient and is retarded by the frictional forces acting on the wall surface. (See Figure 1-1.23.)

For equilibrium in the x-direction, the pressure force must balance the shear force, i.e.

$$2\pi y \ell \cdot \tau = (p_2 - p_1)\pi y^2 \qquad (80)$$

since

$$\tau = -\mu \frac{du}{dy} \qquad (81)$$

Therefore

$$\frac{du}{dy} = -\frac{p_2 - p_1}{\mu \ell} \cdot \frac{y}{2} \qquad (82)$$

On integrating Equation 80 with 81 and 82 gives

$$u(y) = \frac{p_1 - p_2}{\mu \ell}\left(C - \frac{y^2}{4}\right) \qquad (83)$$

The constant of integration, C, is obtained from the no-slip condition at the wall.

Thus $u = 0$, $y = R$, so that $C = R^2/4$.

Hence

$$u(y) = \frac{p_1 - p_2}{4\mu \ell}(R^2 - y^2) \qquad (84)$$

This shows that the velocity is distributed parabolically over the radius. The same result is arrived at using the method of dimensional analysis (see the dimensional analysis example in this chapter) and also using the Bernoulli equation (Equation 55).

The maximum velocity on the pipe axis is

$$u_{\max} = \frac{p_1 - p_2}{4\mu \ell} R^2 \qquad (85)$$

The volume of fluid passing in unit time is

$$Q = \frac{\pi}{2} R^2 U_{\max} = \frac{\pi R^4}{8\mu \ell} (p_1 - p_2) \qquad (86)$$

The average velocity of flow is

$$\bar{u} = \frac{Q}{\pi R^2} \qquad (87)$$

The expression for Q is known as the Hagen-Poiseuille equation of laminar flow through a pipe. This expression is only valid if the flow Reynolds number is less than the critical Reynolds number for a pipe (i.e., $\bar{u}d/\nu \leq 2300$, with $d = 2R$).

Turbulent Flow in a Pipe

Measurement of velocity profiles for a fully developed turbulent flow in a pipe shows that it is very different from the parabolic laminar profile just discussed. This difference results because in turbulent flow extra shear stresses are developed due to the turbulent momentum transfer from one layer to the other.

The shear stresses in the turbulent flow can be expressed as

$$\frac{\tau}{\rho} = \nu \frac{d\bar{u}}{dy} + \varepsilon \frac{d\bar{u}}{dy} \qquad (88)$$

where ν is kinematic viscosity, ε is the eddy viscosity, and $\bar{u}$ is the mean velocity.

For a smooth pipe, ε is very nearly equal to zero.

From the experimental measurements of turbulent flow in pipes conducted by Nikuradse, three regions of flow can be identified. (See Figure 1-1.24.)

The profile is expressed in terms of the shearing or friction velocity, defined as

$$U_* = \left(\frac{\tau_w}{\rho}\right)^{1/2} \qquad (89)$$

In a region very close to the surface, the velocity is proportional to the radial distance measured from the wall, y, and

$$\frac{u}{U_*} = \frac{yU_*}{\nu} \qquad (90)$$

This region is called the laminar sublayer. In this region

$$\mu\left(\frac{d\bar{u}}{dy}\right) \gg \varepsilon\left(\frac{du}{dy}\right) \qquad (91)$$

In the transition region the turbulent friction and the laminar friction are of the same order of magnitude. This is true for the following range.

$$5 < \frac{yU_*}{\nu} < 70 \qquad (92)$$

In the turbulent region the laminar contribution is negligible compared with the turbulent friction. This occurs when

$$\frac{yU_*}{\nu} = > 70 \qquad (93)$$

For this region the following semi-empirical relation holds very well for the velocity profile. The contribution from eddy viscosity is also higher.

$$\frac{\bar{u}(y)}{U_\tau} = 5.75 \log_{10}\left(\frac{yU_\tau}{\nu}\right) + 5.5 \qquad (94)$$

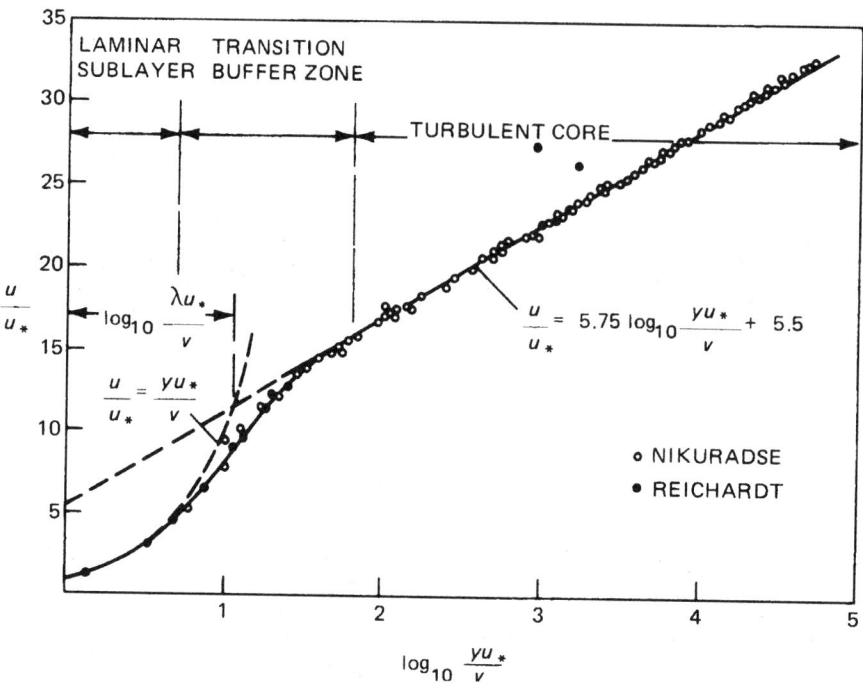

Fig. 1-1.24. *Measurement of turbulent velocity profile in a pipe.*[1]

Duct Flows

In dealing with flow through noncircular cross-section ducts, a parameter called the hydraulic radius, R_h, is defined as

$$R_h = \frac{2A}{C} \qquad (95)$$

where A denotes the cross-sectional area and C is the wetted perimeter of the duct.

In the case of a circular cross-section the hydraulic radius is equal to the radius of the circle.

Laminar flow: For a circular cross-section pipe of diameter, d, of length, L, with pressure difference $(p_1 - p_2)$, the friction factor, λ, is defined as

$$\lambda = \frac{p_1 - p_2}{L} \cdot d / \frac{1}{2} \rho \bar{u}^2 \qquad (96)$$

Similarly, for noncircular cross-section pipe, friction factor can be defined based on the hydraulic radius as

$$\lambda' = \frac{p_1 - p_2}{L} \cdot R_h / \frac{1}{2} \rho \bar{u}^2 \qquad (97)$$

TABLE 1-1.4. *Laminar Flow Through Rectangular Ducts*[4]

a/b or b/a	C_1
0.00	1.50
0.05	1.40
0.10	1.32
0.12	1.28
0.16	1.23
0.25	1.14
0.40	1.02
0.50	0.97
0.75	0.90
1.00	0.89

For any noncircular shape the relationship between the friction factor and Reynolds number can be expressed in the form

$$\lambda' = \lambda C_1 \qquad (98)$$

where C_1 is a function of geometry and λ is the friction factor for the flow in a circular pipe of the same Reynolds number, $Re = u\,R_h/\nu$.

The values of C_1 for rectangular cross-section ducts are given in Table 1-1.4.

Turbulent flow: Turbulent flow measurements show that for noncircular cross-section ducts, the velocities at the corners are comparatively very large. This results because of secondary flows, which arise because all three components of the fluctuating velocity are nonzero. These flows are not present in the corresponding laminar motion because the fluctuating velocity components are not present. The secondary flows continually transport momentum from the center to the corners and generate high velocities there as shown in Figure 1-1.25.

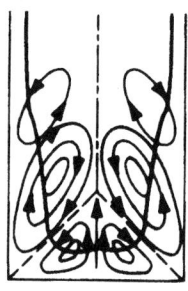

Fig. 1-1.25. *Secondary flows in the corners of a rectangular duct.*[3]

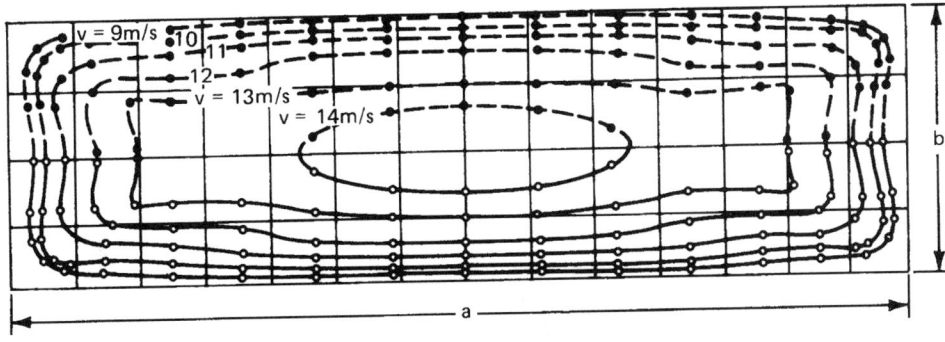

Fig. 1-1.26. Axial turbulent contours in a rectangular duct.[3]

Turbulent velocity measurements in a rectangular duct made by Nikuradse are shown in Figure 1-1.26.

BUILDING AERODYNAMICS AND APPLICATIONS TO FIRE ENGINEERING

In general, building aerodynamics is concerned with the study of atmospheric wind flows and wind loading on buildings and structures. Such a study is important in building design, as large wind-induced loads may result in the instability and eventual collapse of a structure. Collapse of chimney stacks and suspension bridges are examples of such failures.

In fire engineering, obvious applications of such studies are in the problems relating to smoke control and buoyancy-driven roof vent design. Other applications may include the ingress of smoke and other contaminants from external sources and the effects of wind on fire spread.

Over the past two decades a large amount of experimental data, both in the wind-tunnel studies and full-scale measurements, has been collected for a wide range of building shapes and structures.[5-7] Initially, because of their immediate applications to structural design, most of the information was presented in the form of the wind loads. However, in recent years, attention has also been focused on the effects of wind on internal environment systems, such as ventilation and air conditioning. This is important in controlling the spread of contaminants (such as smoke and other airborne particles generated by fire or explosion) to other parts of the building. Various smoke-control measures that are currently in use require the effects of wind to be taken into account in their design. It is essential that large pressure differentials do not hinder escape while serving to confine smoke to the fire areas. The design of a pressurization system, therefore, requires information on the flow field around, as well as inside, the building.

The wind-induced internal and external pressure distribution for a given building is determined not only by the building shape or size, but also by the wind characteristics. These characteristics include the wind velocity profile (which is terrain specific) and the level of turbulence (generated by the adjacent buildings) of the approaching wind.

Natural Wind Characteristics

On the global scale the movement of wind and weather systems is dependent on such factors as the location, vertical temperature gradient, the earth's rotation, etc.[8] In a region very close to the earth's surface the variation of wind velocity with height is considerable and is influenced by the surface roughness. This region is usually referred to as the terrestrial boundary layer. The height at which the velocity becomes independent of the surface roughness is called the gradient height (boundary layer thickness), Z_G, and the corresponding wind speed is called the gradient wind speed.

Figure 1-1.27 shows the terrestrial boundary layer velocity profile for a range of terrain (surface roughness) categories. Category 1 refers to plain areas such as deserts and the open sea, while Category 4 refers to the high-roughness city centers with numerous tall buildings. The other categories lie between these two limits.[9]

For the purposes of wind-tunnel investigations, the wind velocity profile is usually represented by a power law profile of the type,

$$\frac{\bar{V}_Z}{V_{Z_G}} = \left(\frac{Z}{Z_G}\right)^\beta \tag{99}$$

where $\bar{V}_Z$ is the mean wind speed at height Z and β is a constant, which varies for different terrain categories. For example, for Category 1, $\beta = 0.11$; and for Category 4, $\beta = 0.36$.

Wind Flow over Buildings

Wind-tunnel studies show that, when buildings of rectangular cross-section are subjected to a power law velocity profile as shown in Figure 1-1.27, the windward surface of the building experiences positive pressures (i.e., higher than the atmospheric pressure) while all the other surfaces, including the roof, experience negative pressures (i.e., lower than the atmospheric pressure). This is illustrated in Figure 1-1.28.

As the air flows toward the building it is slowed down and comes to rest at a point "S," called the stagnation point, on the windward surface. The total pressure increases with height up to the stagnation point and begins to decrease toward the roof. Below this point the pressure gradient down the face of the building causes air to flow down the building. At the ground level the flow turns and starts to flow upwind. At some distance from the building the flow is opposed by the oncoming wind and once again turns and starts to flow up toward the building. This way a vortex is set up in front of the building as illustrated in Figure 1-1.28.

The pressure measurements are usually expressed in terms of a pressure coefficient defined as

$$C_p = \frac{p - p_0}{\frac{1}{2}\rho V_H^2} \tag{100}$$

where p = total pressure at a point, p_0 = static pressure at that point, V_H = wind speed at building height, and ρ is air density.

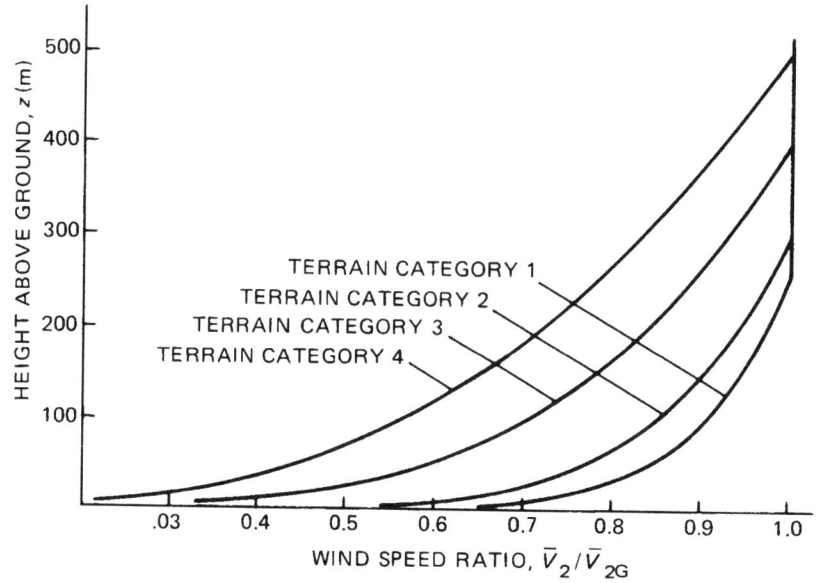

Fig. 1-1.27. Velocity profile in the terrestrial boundary layer.[8]

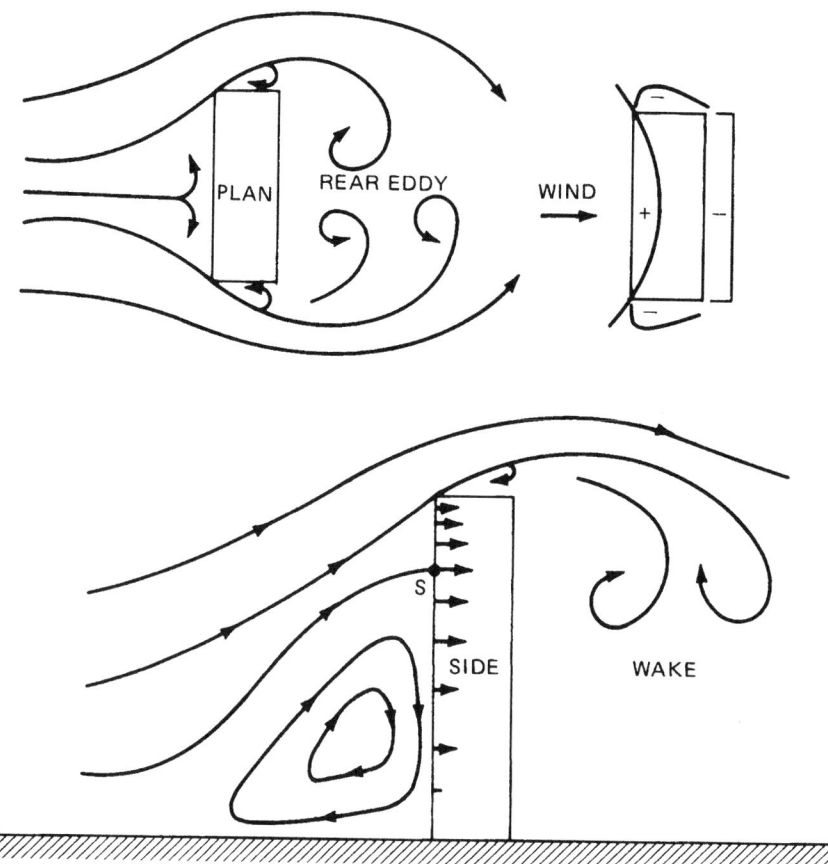

Fig. 1-1.28. Flow over a tall building.

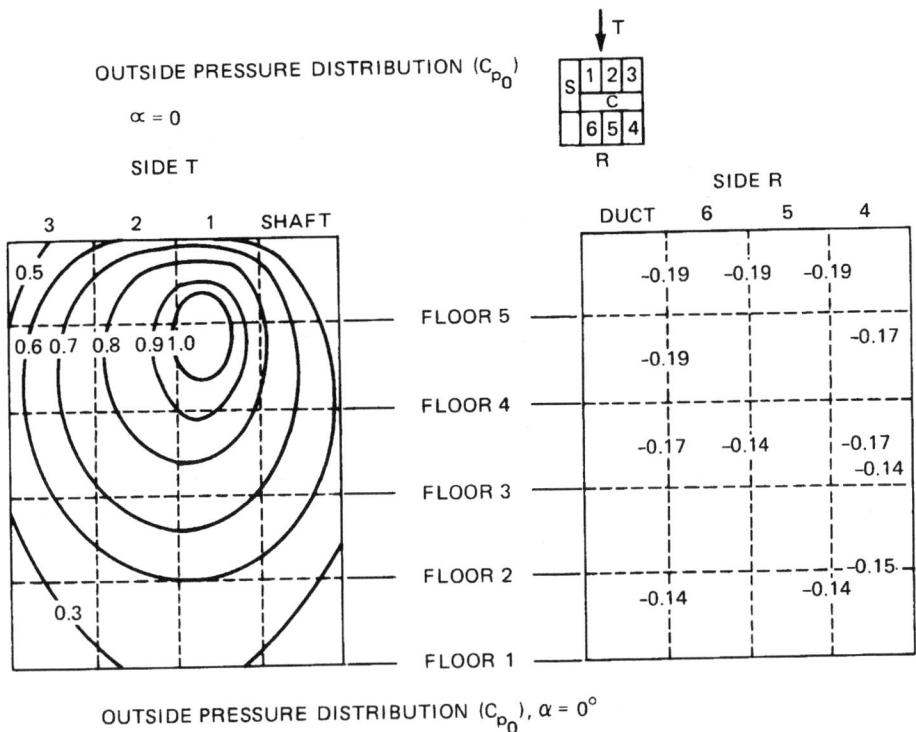

Fig. 1-1.29(a). *Outside pressure distribution (Cp),* $\alpha = 0$ *degrees.*

Figures 1-1.29(a) and 1-1.29(b) show wind-tunnel pressure measurements on the walls of a model building shown in Figure 1-1.29(c).[10] It is clear that, on the windward side [Figure 1-1.29(b)], the pressure is the highest at a point on the fourth-floor level (the stagnation point). This point usually occurs at about four-fifths the height of the building. On the rear wall (side R, for wind incidence $\alpha = 0$ degrees) the pressure distribution is negative but fairly constant. These

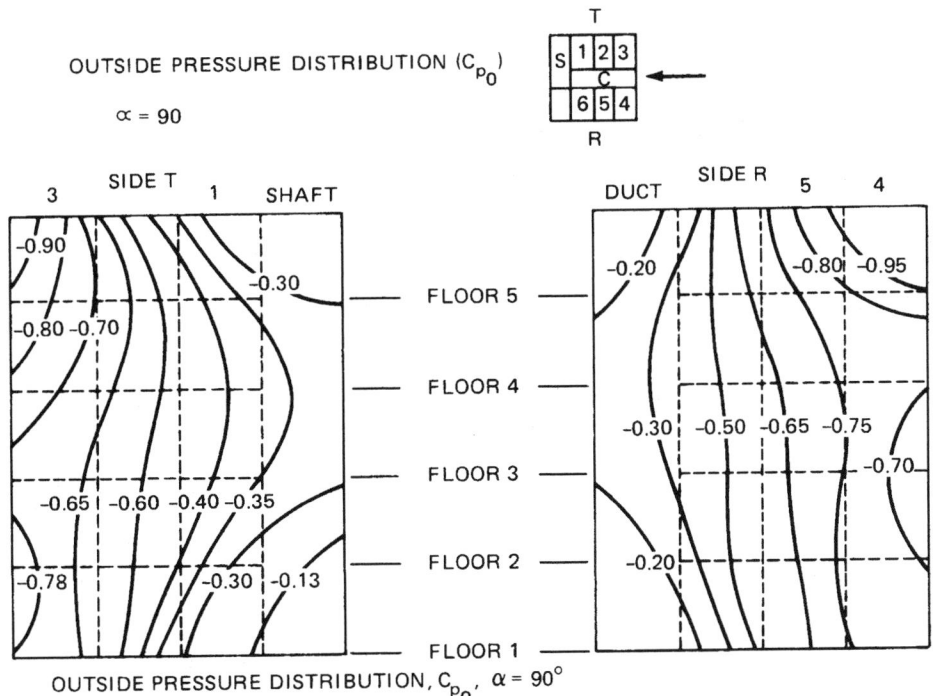

Fig. 1-1.29(b). *Outside pressure distribution (Cp),* $\alpha = 90$ *degrees.*

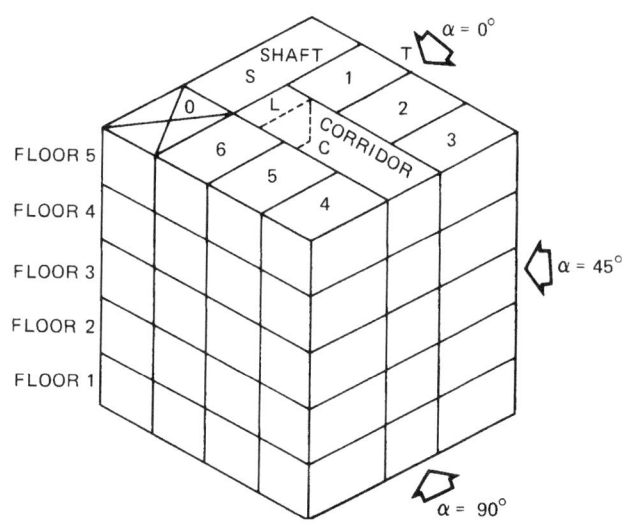

Fig. 1-1.29(c). *Five-story wind-tunnel model.*[10]

negative pressures on the walls are caused by the eddies generated as a result of flow separation at the sharp corners of the building. Because of these marked pressure variations on the walls, internal flow patterns are set according to these variations.

Wind-Induced Internal Flows

For completely sealed buildings, the outside wind has no effect on the internal flow patterns in a building. The internal flows are greatly influenced by the outside wind, for a building with outside doors and windows. The magnitude of internal pressure distribution is a function of wind speed, building geometry, and building leakage characteristics. Wind-tunnel experiments[11] show that the wind-induced internal pressures are determined by the outside pressures on the wall with highest leakage. The results are summarized in Figure 1-1.30, in which the leakage ratio, R, is defined as the ratio of the total leakage area of the windward wall to the total leakage area of the rear wall.

The curve for $\alpha = 0$ degrees clearly shows that as the windward wall leakage area is increased the internal pressure is also increased. For the case of $\alpha = 90$ degrees, both the leaky walls are under negative outside pressure and the internal pressure is negative as expected and does not change very much with the leakage ratio.

Such a marked change in internal pressure due to the outside wind can have serious implications for the functioning of smoke extract vents used in these types of buildings.

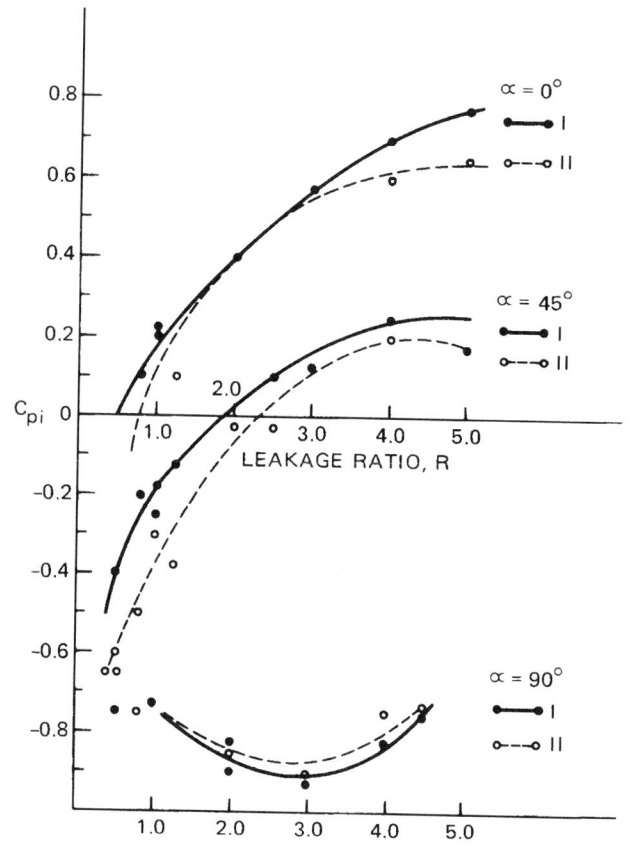

Fig. 1-1.30. *Variation of internal pressure,* C_{p_i}, *with leakage ratio,* R, *for wind direction* $\alpha = 0$ *degrees, 45 degrees, 90 degrees.*[11]

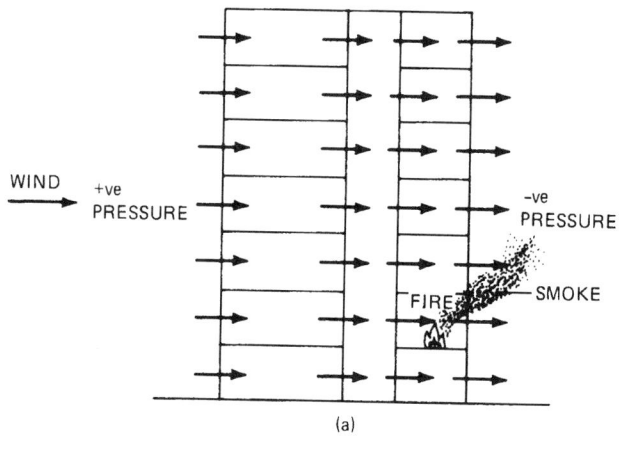

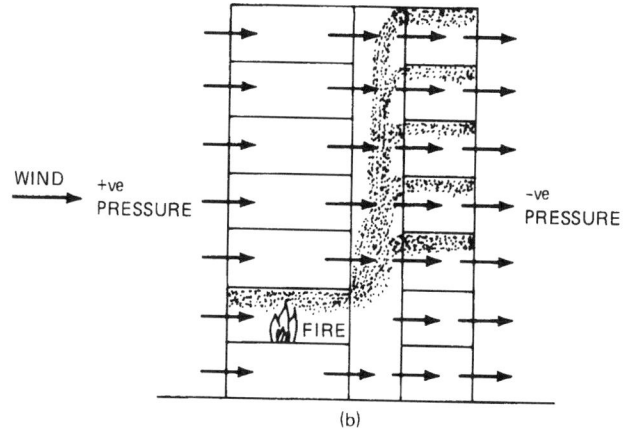

Fig. 1-1.31. *Smoke movement in a high-rise building.*

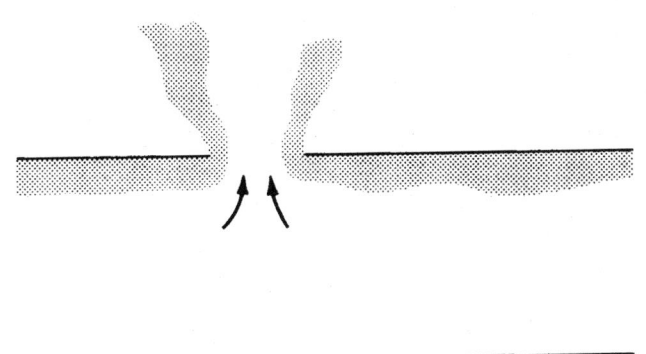

Fig. 1-1.32. *Operation of a roof vent.*[12]

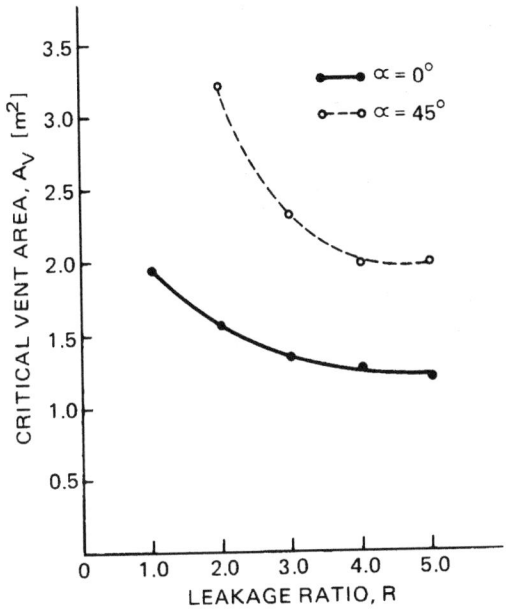

Fig. 1-1.33. *Variation of critical vent area (Av_{exit}) with leakage ratio R, for α = 0 degrees, 45 degrees.*[11]

Smoke Movement in Tall Buildings

When fire occurs in a building, smoke or toxic gases are carried to other parts of the building interior by the pre-existing airflows. These flows are a function of the outside wind as well as the building leakage characteristics. For the example of a high-rise building (Figure 1-1.31), if fire occurs in a room that has window openings on a wall subjected to negative wind-induced pressures, most of the smoke will be drawn out of the building by the prevailing negative pressures on that wall. Smoke will be prevented from flowing into the building by the positive pressure there (Figure 1-1.31b).

However, the situation is reversed if the fire room has its window openings on the windward wall. The prevailing positive pressure on this wall will push the smoke into the building, where it will spread to other parts of the building, carried there by pre-existing flows.

If under such conditions a smoke control pressurization system is used, the design of such a system must account for the wind-induced pressures. Otherwise the system may cause overpressure (preventing escape of occupants) or underpressure (rendering the system ineffective) within the building.

Roof Vent Design

For large single-space buildings (e.g., warehouses and factories, etc.) roof vents are used to extract smoke in fire situations. Usually, the design of these vents is based on the assumption that smoke buoyancy is the only force driving the smoke out. This may not be true for very tall buildings because the smoke cools very quickly as it moves up toward the ceiling. When this happens the buoyant force of the smoke can be negligibly small, reducing the rate of flow through the vent.

If a sufficiently strong wind is blowing, the roof is subjected to considerable, high, wind-induced, negative pressures. These pressures can add to the buoyancy of the smoke with a subsequent increase in the extract flow velocity and mass rate. A situation may arise when the vent starts extracting clear air from underneath the hot layer. Under such conditions (Figure 1-1.32) the efficiency of the vent is decreased.

Therefore, a critical mass flow rate or critical vent area exists which must not be exceeded or the vent will be ineffective.[11] Figure 1-1.33 shows the variation of calculated critical vent area with leakage ratio. These calculations were performed on the basis of roof pressure distribution shown in Figure 1-1.34.

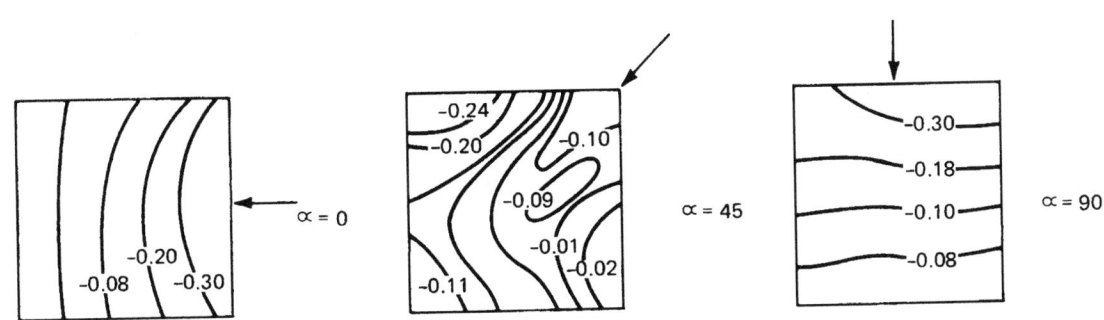

Fig. 1-1.34. *Pressure contours on the roof (C_{p_0}) for wind directions α = 0 degrees, α = 45 degrees, and α = 90 degrees.*[11]

REFERENCES CITED

1. E. Salamon, *Principles of Fluid Mechanics*, Allyn and Bacon, Boston (1963).
2. S. Kumar, "Mathematical Modeling of Natural Convection in Fire—A State-of-the-Art Review of the Field Modeling of Variable-Density Turbulent Flow," *Fire and Materials*, Vol. 7, No. 1 (1983).
3. H. Schlichting, *Boundary Layer Theory*, Pergamon, London (1955).
4. A.J. Ward-Smith, *Internal Fluid Flow*, Oxford University, Oxford (1980).
5. T.W. Lawson, *Wind Effects on Buildings*, Applied Science, London (1980).
6. E. Naudascher, ed., *Flow-induced Structural Vibrations*, Symposium, Berlin (1972).
7. *Proceedings of the 3rd International Conference of Wind Effects on Buildings and Structures*, Saikon, Tokyo (1971).
8. R.S. Scorer, *Natural Aerodynamics*, Pergamon, London (1958).
9. R.M. Aynsley, W. Melbourne and B.J. Vickery, *Architectural Aerodynamics*, Applied Science, London (1977).
10. B.S. Kandola, *F. Safety J.*, 10, 211 (1986).
11. B.S. Kandola, *J. of Indus. Aerodyn.*, 3, 267 (1978).
12. D.D. Drysdale, *Introduction to Fire Dynamics*, John Wiley & Sons, New York (1985).

CONDUCTION OF HEAT IN SOLIDS

John A. Rockett and James A. Milke

INTRODUCTION

There are three mechanisms by which heat (thermal energy) is transferred from one object to another: (1) radiation, (2) convection, and (3) conduction. This chapter addresses both heat conduction and the historic development of heat conduction theory.[1-5]

Before discussing heat conduction, it is useful to consider the heat content of an object; i.e., the property of the object subject to movement by conduction. The heat content of an object, Q, and its temperature are closely linked. When the heat content of one gram of water at 0°C is increased by one calorie, its temperature increases one degree Celsius. Heat content is associated with the kinetic energy of vibration of the atomic particles of which an object is composed. Since at absolute zero temperature all motion stops, heat content is zero at zero (absolute) temperature. In metals, where there are free electrons (those electrons not localized on a particular atom, but free to move through the crystal lattice), some of the heat content is associated with this electron "gas." In all materials the atoms vibrate about their average position in the crystal lattice. Atomic vibration within the crystal lattice (standing acoustic waves called "phonons") accounts for the remainder of the heat content. The hotter the object, the more violent is the vibration of the electrons and atoms. Thus the transfer of heat from one body to another, or from one part of a body to another part of the same body, is equivalent to transferring the kinetic energy of vibration from particles in one location to those in another, adjacent location.

The heat content of an object is expressed as the integral of the material's specific heat with respect to temperature from absolute zero to the temperature in question, i.e.,

$$q''' = \int_0^T \rho c \, dT$$

The inability to correctly calculate specific heat was one of the singular failures of classical physics. The roots of modern quantum physics lie in the work of Planck, Lorentz, and Einstein to resolve this problem. Building on their work, Debye, and (independently) Born and von Karman, published a theory of specific heat in 1912, which is still accepted today with only refinements in detail. According to this theory, the specific heat rises from zero at zero temperature and approaches a constant at high temperature. A temperature characteristic of the material, or the "Debye temperature," determines the region where transition from rising to constant specific heat occurs. At a point well above the Debye temperature, the specific heat is given by the classical, constant value; below it, quantum effects must be included. For heavy atoms the Debye temperature is well below room temperature, but it is higher for light atoms. Beryllium, for example, is light enough that its specific heat varies noticeably near room temperature. For aluminum at room temperature the specific heat is 93 percent of the classical value; for copper, 95 percent. For some minerals, however, the Debye temperature is quite high and the assumption of constant specific heat may lead to significant errors. Nevertheless, constant specific heat is generally assumed, usually without discussion, in heat conduction studies.

Heat conduction is observed when a hot object is brought in physical contact with a cold one: the hot object cools and the cold one is heated. This process was studied by Newton but, in the modern sense, first quantified by Fourier in 1812. Fourier's equation states that the quantity of heat transferred per unit time across an area, A, is proportional to the area and the temperature gradient, dT/dx. The heat flows from the hotter to the cooler material, so

$$\dot{q} = -kA \, dT/dx$$

where

A = surface area across which heat is transferred (m^2)
d = differential operator
k = thermal conductivity of the solid (W/m · K)
$\dot{q}$ = the rate of heat flow across the area A (W)
T = temperature (K)
x = distance normal to the surface A (m).

Dr. John A. Rockett, formerly Senior Scientist, U.S. National Bureau of Standards, is a consultant based in Washington, DC. His research has focused on the growth and spread of fire and the development of analytic models for describing fire in buildings.

James A. Milke is Assistant Professor of Fire Protection Engineering at the University of Maryland. His recent research activities have included the impact of fires on the structural response of steel and composite members, as well as the design of smoke management systems.

A more modern and more precise statement of Fourier's law is

$$dQ/dt = -\int_A (k_{ij}T_{,j})n_i \, dA$$

where in addition to the above definitions, Q is the heat content (Joules) of the material inside the closed surface A, n_i is the outward directed vector normal to the surface element dA, t is time, $T_{,j}$ is the vector gradient of the temperature of the object, k_{ij} is the thermal conductivity (written here in its most general form as a tensor), and the integral is taken over the entire surface, A.

The thermal conductivity of a material, k, will play a central role in what follows. Wiedemann and Franz observed in 1853 that electrical and thermal conductivity were proportional for metals. This suggests that the kinetic energy of free electrons is primarily responsible for heat transfer. This is a good first approximation for good electrical conductors, where free electrons are abundant. Sommerfeld, in 1928, calculated the Wiedemann-Franz ratio (ratio of thermal to electrical conductivity) based on a quantum mechanical model for the free electron gas. He found

$$k/T\sigma = 2.45 \times 10^{-8}$$

where

σ is electrical conductivity, and T is absolute temperature.

This relation is obeyed (error less than about 10 percent) by gold, silver, copper, copper-silver alloys, tungsten, and molybdenum for absolute temperatures from 50 to 360 K.

If this were all that was involved, electrical insulators would conduct virtually no heat. Actually, although heat is poorly conducted by electrical insulators, the amount of heat conducted is orders of magnitude too great to be accounted for solely by free electron conduction. Thus, the ratio of electrical conduction of glass to that of gold is 1.44×10^{-19}, but the ratio of their thermal conduction is 2.5×10^{-2}. Phonons are present in all solids heated above absolute zero. However, if the crystal lattice were perfect, and the material perfectly pure, the material would conduct no heat because these standing, acoustic waves would experience no attenuation. But imperfections within the solid (e.g., missing atoms-vacancies, extra atoms-interstitials, and lattice imperfections-dislocations) scatter the phonons and free electrons when they are present. It is this scattering that diffuses the phonon energy to produce heat conduction in electrical insulators.

Because cold-working a metal increases dislocations and heat treatment (e.g., precipitation hardening), lattice imperfections, thermal conductivity, as well as physical properties, can be significantly altered by these treatments. In general, treatments that increase hardness will decrease thermal (and electrical) conduction.

For some materials, e.g., some laminates and fibrous materials such as wood and reinforced plastics, the thermal conductivity differs depending on whether the heat flow is parallel or across the grain. For these materials, the thermal conductivity, in general, differs with the direction of the heat flow. This is the most general case, and will be addressed in this chapter. For these materials (orthotropic solids), the conductivity is a diagonal matrix (its only nonzero elements are the diagonals, k_{ii}). These are normally written simply k_1, k_2, k_3. In other materials, most notably some types of crystals, the off-diagonal elements of the k matrix are nonzero.

For these materials the conductivity must be expressed as a general matrix (tensor).

From this brief discussion it is clear that the theoretical basis for thermal analysis is comparatively recent. The mathematical analysis of heat conduction used today was, however, well-established by the end of the last century.

EQUATION OF HEAT CONDUCTION

Fourier's equation of heat conduction (as given in the introduction to this chapter) has been generalized to the basic, modern statement of conductive heat transfer. To reproduce this, consider a small rectangular block of material with density, ρ, specific heat, c, thermal conductivity, k, and average temperature, T. This small block is a segment of a much larger solid. Let the small block be oriented parallel to a Cartesian coordinate system, x, y, z, and let its dimensions be dx, dy, and dz measured along the respective axes. The center of the block is at location x, y, z.

If the temperature within the solid varies nonuniformly, the heat flux entering the small block across the face with area $dydz$ located at $x - dx/2$ may be different from the heat leaving across the parallel face at $x + dx/2$. According to Fourier's law and using a Taylor-series expansion for k and T, the difference in heat flux would be

$$\dot{q}_{x-net} = \left[\dot{q}''\left(x + \frac{dx}{2}\right) - \dot{q}''\left(x - \frac{dx}{2}\right)\right] dy \, dz$$

$$= \left\{\left[\left(k_1 + \frac{\partial k_1}{\partial x}\frac{dx}{2}\right)\left(\frac{\partial T}{\partial x} + \frac{\partial^2 T}{\partial x^2}\frac{dx}{2}\right)\right]\right.$$

$$\left. - \left[\left(k_1 - \frac{\partial k_1}{\partial x}\frac{dx}{2}\right)\left(\frac{\partial T}{\partial x} - \frac{\partial^2 T}{\partial x^2}\frac{dx}{2}\right)\right]\right\} dy \, dz \quad (1)$$

Expanding this and neglecting terms containing (dx^2), the result is

$$\dot{q}_x = \frac{\partial}{\partial x}\left(k_1 \frac{\partial T}{\partial x}\right)dx \, dy \, dz$$

Similarly,

$$\dot{q}_y = \frac{\partial}{\partial y}\left(k_2 \frac{\partial T}{\partial y}\right)dx \, dy \, dz$$

$$\dot{q}_z = \frac{\partial}{\partial z}\left(k_3 \frac{\partial T}{\partial z}\right)dx \, dy \, dz$$

The sum of these three terms is the net rate of heat gain by the small block. Dividing by the volume of the block, the net gain per unit volume is

$$\dot{q}''' = \frac{\partial}{\partial x}\left(k_1 \frac{\partial T}{\partial x}\right) + \frac{\partial}{\partial y}\left(k_2 \frac{\partial T}{\partial y}\right) + \frac{\partial}{\partial z}\left(k_3 \frac{\partial T}{\partial z}\right) \quad (2)$$

As stated, the energy stored in a material at temperature, T, is

$$q''' = \int_0^T \rho c \, dT \quad (3)$$

For temperatures well above the Debye temperature, this can be written as

$$q''' = q_0''' + \rho c(T - T_0) \quad (4)$$

where

$$q_0''' = \int_0^{T_0} \rho c \, dT \qquad (5)$$

is the heat stored in the material as a result of raising its temperature from absolute zero to T_0, a temperature high enough that c, in Equation 4, can be assumed constant. T_0 may be conveniently assumed to be zero Celsius, 273.16 K. Further, since changes in energy are the concern, the reference energy, q_0''', can be dropped from further consideration. Thus, the rate of change of thermal energy stored in the block per unit volume can be written as

$$\dot{q}''' = \rho c \, \frac{dT}{dt} \qquad (6)$$

Sometimes it is required to analyze materials that undergo an internal change of state. For example, in heating porous materials, such as wood or gypsum plaster, absorbed water may evaporate in one part of the solid, diffuse to a cooler part, and condense. The evaporation will remove heat from the locality where it occurs due to the latent heat of vaporization of the moisture, and the condensation will add heat locally where it occurs. Let the rate of such local heat addition be $\dot{q}_i'''$.

Combining Equations 2 and 6 and accounting for heat addition due to chemical or physical changes in the material results in

$$\rho c \, \frac{\partial T}{\partial t} = \frac{\partial}{\partial x}\left(k_1 \frac{\partial T}{\partial x}\right) + \frac{\partial}{\partial y}\left(k_2 \frac{\partial T}{\partial y}\right) + \frac{\partial}{\partial z}\left(k_3 \frac{\partial T}{\partial z}\right) + \dot{q}_i''' \quad (7)$$

Equation 7 is the basic equation describing heat conduction in a solid.

To solve the heat conduction equation, initial and boundary conditions must be provided. These conditions distinguish one problem of interest from another. The conditions that must be specified are:

1. The initial temperature throughout the solid, and
2. The temperature or heat flux at the surface of the solid for all times. Note that along any part of the surface, either the temperature or the heat flux (but not both) should be specified; temperature may be specified for some areas, and heat flux for others. The temperature and flux need not be constant in space or time.

In formulating Equations 2 and 7, it was assumed that the thermal conductivity might be different in each of the three directions and further, that each of these values might vary with position. There are cases where these variations are important.[6,7] However, for the remainder of this chapter (unless specifically mentioned) the solid will be assumed isotropic (i.e., k the same in each direction) and independent of temperature. Also, for simplicity, the heat addition term, $\dot{q}_i'''$, will be assumed zero, except for a few specific examples.

If a dimensionless time ($\tau = t/t_0$) and a dimensionless distance ($\xi = x/b$) are substituted into Equation 7, and t_0 is chosen as

$$t_0 = \rho c \, b^2/k$$

the equation contains no parameters (for the case of an isotropic solid without local heat addition). Since t_0 and b are related, the choice of b is significant. Normally b would be a dimension of the object in question, with the most useful dimension being that along the principal direction of heat transfer. The quantity $(k/\rho c)$ is called the "thermal diffusivity" of the solid; it has dimensions (m²/s) and is often represented by the Greek letter α.

Steady-State Solutions

If the thermal environment of a solid has been constant for a sufficient time (i.e., the boundary conditions have not changed in value), it will achieve a steady temperature distribution. In this case, the time derivative of T will vanish and the left side of Equation 7 will be zero. Five examples of steady-state heat conduction will be considered.

EXAMPLE 1:

Flat, rectangular plate, insulated along its edges, heat flowing from one face to the other.

In this case Equation 7 reduces to

$$\frac{\partial}{\partial x}\left(k \frac{\partial T}{\partial x}\right) = 0 \qquad (7a)$$

The first integral with respect to x yields

$$k\frac{\partial T}{\partial x} = -\dot{q}'' \qquad (7b)$$

where $\dot{q}''$ is the constant of integration; it is the heat flux per unit area through the plate. Equation 7b is a variation of Fourier's law. The constant heat flux is characteristic of the one-dimensional, steady-state problem. Consequently, dT/dx is also a constant for cases of uniform conductivity, yielding a linear temperature profile through the solid. Equation 7b, in turn, integrates to

$$T = T_0 - x\frac{\dot{q}''}{k} \qquad (8)$$

T_0 is the second integration constant. Two integration constants, $\dot{q}''$ and T_0, have been introduced, but no boundary conditions used. The boundary conditions selected will fix the values of these two constants. If the thickness of the plate is b, T_0 is the temperature of the left face, and the heat flux, $\dot{q}''$, is specified, the temperature on the right face must be

$$T_b = T_0 - b\frac{\dot{q}''}{k}$$

Or, if the two surface temperatures are known, the heat flux through the plate will be

$$\dot{q}'' = (T_0 - T_b)k/b \qquad (8a)$$

Figure 1-2.1 illustrates this example.

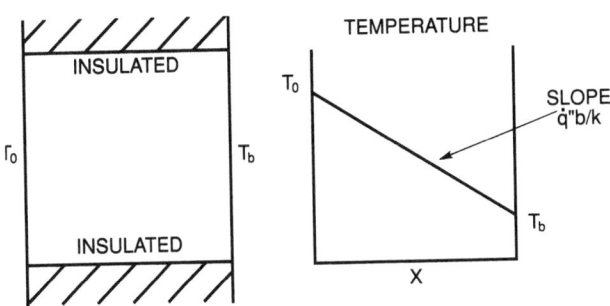

Fig. 1-2.1. *Flat plate of thickness b, insulated at its edges. Steady-state surface temperatures T_0 and T_b.*

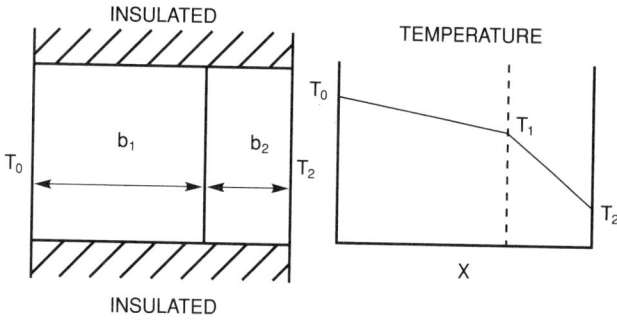

Fig. 1-2.2. Two-layer, flat laminated plate insulated at its edges. Layer thickness b_1 and b_2. Layer 2 has the smaller thermal conductivity. Steady-state surface temperatures T_0 and T_2. Temperature of the interface between the plates, T_1.

EXAMPLE 2: Flat, rectangular, laminated plate, insulated along its edges.

Here Equation 8 is appropriate to each layer separately, but there may be different constants for each layer. For a two-layer laminate with layers of thickness b_1 and b_2, and thermal conductivities k_1 and k_2, the temperature at the interface will be

$$T_1 = T_0 - b_1 \frac{\dot{q}''}{k_1} \qquad (8b)$$

A key principle in the solution of this problem is that the heat flux is a constant across any plane parallel to the surface of the plate. (See first integration of the steady-state heat conduction, i.e., Equation 7b.) Thus the temperature of the rear face will be

$$T_2 = T_0 - \dot{q}''(b_{1_1}/k_1 + b_2/k_2) \qquad (8c)$$

See Figure 1-2.2.

EXAMPLE 3: Radial heat flow between two concentric cylindrical surfaces (heat loss through the wall of a pipe with inside radius, r_0, and thickness, b).

Rewriting Equation 7 in cylindrical coordinates and assuming radial symmetry yields

$$\rho c \frac{\partial T}{\partial t} = \frac{1}{r} \frac{\partial}{\partial r} \left(kr \frac{\partial T}{\partial r} \right) + \frac{\partial}{\partial z} \left(k \frac{\partial T}{\partial z} \right) + \dot{q}''' \qquad (9)$$

For the present problem assume no heat addition, $\dot{q}''' = 0$, and no variation of T along the length of the pipe (z direction); hence the last two terms on the right vanish. For steady-state the left side also vanishes. Thus, for purely radial heat flow, Equation 9 can be integrated to

$$r \frac{\partial T}{\partial r} = -\frac{\dot{q}'}{2\pi k} \qquad (9a)$$

where $\dot{q}'$ is a constant of integration whose significance will be discussed. A second integration yields

$$T = T_0 - \frac{\dot{q}'}{2\pi k}[\ln(r/r_0)] \qquad (9b)$$

The integration constant, $\dot{q}'$, can be understood by considering a very thin-walled pipe, thickness b much less than r_0. Then $r_1 = r_0 + b$ and

$$\ln(r_1/r_0) = \ln(1 + b/r_0) \approx b/r_0 + \frac{1}{2}(b/r_0)^2 + \cdots$$
$$\approx b/r_0$$

or, with this approximation and rewriting Equation 9b,

$$b/r_0 = 2\pi k (T_0 - T_b)/\dot{q}'$$

$$\frac{\dot{q}'}{2\pi r_0} = k(T_0 - T_b)/b$$

Comparing this with Equation 8a it can be seen that

$$\dot{q}' = 2\pi r_0 \dot{q}''$$

or that $\dot{q}'$ equals the heat flux through the pipe per unit length. [See Figure 1-2.3(a).]

It can be verified that the heat flux per unit length of a pipe with conductivity k_1 and radii (r_0, r_1) surrounded by insulation, conductivity k_2, and radii (r_1, r_2) is

$$\dot{q}' = \frac{2\pi(T_0 - T_2)}{\left[\frac{1}{k_1}\ln(r_1/r_0) + \frac{1}{k_2}\ln(r_2/r_1) \right]} \qquad (9c)$$

where T_0 is the temperature at the inside of the pipe and T_2 the temperature of the outside of the insulation. [See Figure 1-2.3(b).] In this case integration of 9a gives the temperatures in the pipe and insulation. For the pipe, the integral 9b applies, but with $k = k_1$ and $\dot{q}'$ given by 9c, T_1, the temperature at the pipe-insulation interface, is given by this version of 9b with $r = r_1$. For the insulation, Equation 9b again applies with $\dot{q}'$ given by Equation 9c, and T_0 replaced by T_1, k by k_2 and r_0 by r_1.

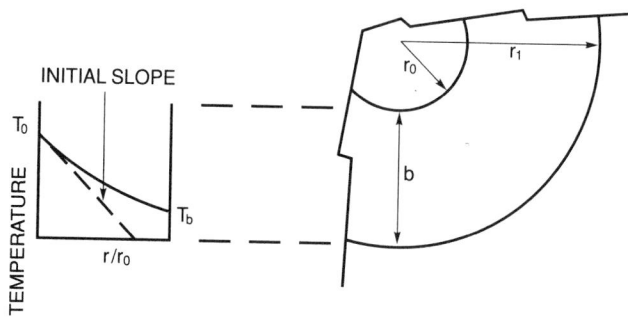

Fig. 1-2.3(a). Steady-state radial heat flow through a thick-walled pipe, internal radius r_0, thickness b. The logarithmic curvature of the temperature vs radius is shown by comparing the initial slope of the temperature curve with its final value.

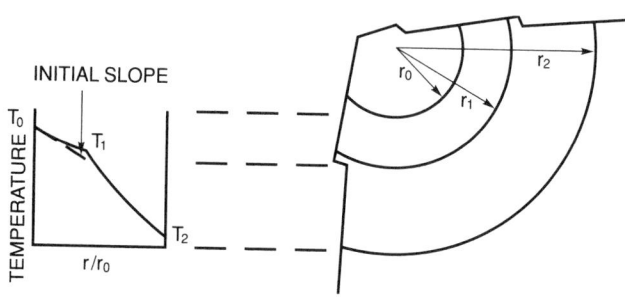

Fig. 1-2.3(b). Steady-state radial heat flow through two concentric pipes or one pipe covered with insulation. The insulation has a lower conductivity than the pipe.

EXAMPLE 4:   Cavity wall.

Many building walls consist of internal framing sheathed on both sides. For example, an interior wall of a residential building might be framed with wood "2 × 4" studs (41.3 × 92.1 mm), spaced 406 mm (16 in.) and sheathed with gypsum wallboard. An exterior wall might be similar except that the sheathing on the exterior side would be a weather-resistant material perhaps backed by insulation board. Between the two sheathing layers there will be a cavity unless the spaces between the studs are filled with insulation or other material.

For this example, assume that the two coverings are identical 15.9-mm (⅝ in.) gypsum board and the cavity is, for Case 1, filled with fiberglass insulation, and for Case 2, unfilled. The thermal conductivity of gypsum board is taken as 0.8 W/m-K, of wood 0.17 W/m-K, and of fiberglass 0.04 W/m-K.[8] Let the hot side wall surface temperature be 200°C and the cold side 20°C. [See Figure 1-2.4(a).]

Case 1: A correct calculation of this case requires solution of the two-dimensional steady-state heat flow equation (specialized from Equation 7). Either an analytical solution or numerical techniques (described later in this chapter) might be used. Here a small correction is anticipated for the studs. Equation 8 is used for the two layers of gypsum board and either the fiberglass insulation or the wood studs. An effective conductivity is calculated for the insulation or stud-filled cavity using an arithmetic average based on the space occupied by studs and insulation. The effective conductivity for the 92.1-mm deep by 406-mm wide cavity is

$$(0.04 \cdot 0.3647 + 0.17 \cdot 0.0413)/0.406 = 0.0532 \text{ W/m-K}$$

Referring back to Example 2, the reader can verify that the heat conduction through the wall would be

$$\dot{q}'' = (200 - 20)/(2 \times 0.0159/0.8 + 0.0921/0.0532)$$
$$= 180/(0.0398 + 1.7312)$$
$$= 0.1016 \text{ kW/m}^2$$

In spite of its low conductivity, the insulation in the cavity conducts the majority of the heat through the wall because of the large area occupied by it. The studs conduct about 30 percent of the heat for this particular case. Had the gypsum board not been present, the heat flow would have increased 2.3 percent for the same overall ΔT. Compared to the insulation, the gypsum board offers little resistance to the heat flow.

Had the steady-state, two-dimensional heat flow equation been solved for this case, the wood and insulation temperatures would reflect a small flow of heat from the insulation into the wood on the warm side of the cavity and an equal, reverse flow on the cool side. The temperature deviations (from one-dimensional heat flow) for the insulation

are the larger, about 1.5°C, while those in the wood are less than 1°C. A small transverse heat flow in the gypsum board, toward the studs on the warm side of the cavity and away from the studs on the cool side, would be revealed. [See Figure 1-2.4(b).]. In this figure, use has been made of symmetry about the center of the cavity. The total 180°C temperature difference is divided in half: 90°C on the hot side and −90°C on the cool side. Only the area around the insulation-stud-gypsum board junction, on the hot side, is shown.

In the upper diagram in Figure 1-2.4(b), the stud and gypsum board opposite it are considered as one isolated unit, and the gypsum board and insulation as another. Each are separately analyzed using a one-dimensional model. There is a physically unrealistic temperature discontinuity within the gypsum board and between the stud and the insulation. The lower diagram is a two-dimensional treatment of the assembly. The temperatures at the interface between the insulation and the stud are now equal. There is a discontinuity in the slope of the temperature profile across the insulation-to-stud joint. This is expected, as the thermal

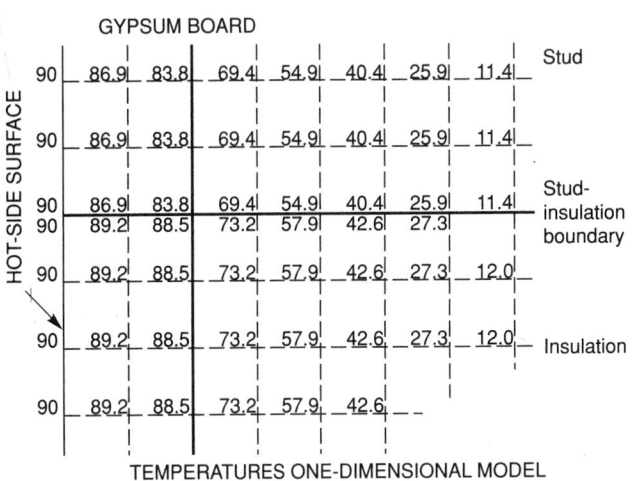

TEMPERATURES ONE-DIMENSIONAL MODEL

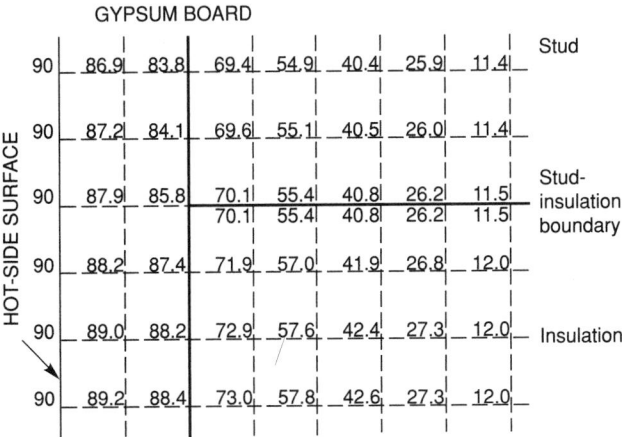

Fig. 1-2.4b. *Comparison of wall and cavity temperatures—one- and two-dimensional models.*

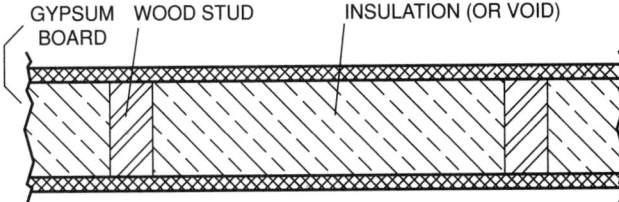

Fig. 1-2.4a. *Geometry of the cavity wall.*

conductivity of the stud and insulation are different. (See Example 2.)

Case 2: The two layers of gypsum board are treated as in Case 1. However, the cavity is empty and must be treated differently. With no insulation in the void, radiation across, and convection within it, will dominate the heat transfer. Again, conduction through the studs will yield a minor correction.

Radiation is addressed in Section 1, Chapter 4; the radiation exchange across the void is simple, but depends on the temperatures of the gypsum board on either side of the cavity, which must be approximated or found by an iterative calculation. Referring to Section 1, Chapter 4, the reader can verify that the effective conductivity of the cavity, if radiation alone were active and the emissivity of the gypsum board were 1.0, would be about 1.1723 W/m-K, more than 10 times that of the insulated cavity. (See Case 1.) If the wall were two sheets of gypsum board back to back (no cavity), about 4.5 kW/m² would be conducted. With the cavity present and only radiation active within it, this would be reduced to 1.52 kW/m².

Section 1, Chapter 3 addresses convection. In the absence of insulation to break up the buoyancy-induced flow in the cavity, air rises along the warmer side, crosses over, and descends along the cooler side. The calculation of the heat transferred by this circulation can be complex and is beyond the scope of this chapter. In a typical cavity, convective heat transfer is more than the heat transferred by radiation.

Is Case 2 consistent with what is known about vacuum bottles? The answer is yes. First, in the vacuum bottle, convective heat transfer is suppressed by removing the air. Second, the inner walls of the bottle are silvered and have a low emissivity. Had it been assumed that the cavity side of the gypsum board was aluminized and used an emissivity of 0.1 instead of 1.0, the radiative contribution to the heat transfer would have been comparable to that of the insulation.

EXAMPLE 5: Spherical solid with internal heat generation.

Rewriting Equation 7 in spherical polar coordinates and assuming spherical symmetry yields

$$\rho c \frac{\partial T}{\partial t} = \frac{1}{r^2} \frac{\partial}{\partial r}\left(kr^2 \frac{\partial T}{\partial r}\right) + \dot{q}''' \tag{10}$$

For steady-state, purely radial heat flow, Equation 10 can be integrated to

$$-4\pi r^2 k \frac{\partial T}{\partial r} = \int_0^r \dot{q}''' 4\pi r^2 \, dr \tag{10a}$$

Here the term on the left is the outward heat flux across the spherical surface at r, while the term on the right is the total heat released inside the sphere bounded by this surface. A second integration yields

$$T = T_0 - \int_0^r \frac{1}{kr^2}\left(\int_0^r \dot{q}''' r^2 \, dr\right) dr$$

If $\dot{q}'''$ and k are constant and R is the outer radius of the sphere, the sphere's surface temperature will be

$$T_R = T_0 - \dot{q}''' R^2/6k$$

Conversely, the sphere is hottest at its center with temperature

$$T_0 = T_R + \dot{q}''' R^2/6k \tag{10b}$$

This analysis could be generalized to illustrate self-heating

in a spherical object. However, it would involve a differential equation requiring more advanced mathematics than is desirable here. Instead, self-heating for a simpler, plane geometry is addressed later in this chapter.

ONE-DIMENSIONAL, TRANSIENT EQUATION

Thermally Thin Material

The simplest transient problem of interest is derived from the basic equation of heat conduction, Equation 7, in its one-dimensional form

$$\rho c \frac{\partial T}{\partial t} = \frac{\partial}{\partial x}\left(k \frac{\partial T}{\partial x}\right) \tag{11}$$

This is integrated once with respect to x. The result is

$$\rho c b \frac{\partial \bar{T}}{\partial t} = k \frac{\partial T}{\partial x}\bigg|_{x_1} - k \frac{\partial T}{\partial x}\bigg|_{x_0} \tag{12}$$

where $\bar{T}$ is the average plate temperature

$$\bar{T} = \frac{1}{b}\int_{x_0}^{x_1} T \, dx \tag{12a}$$

and $b = x_1 - x_0$.

The right side of Equation 12 is the net heat transferred into the plate; i.e., the difference between the heat entering one side and that leaving by the other side. The left side of the equation is the rate of change of the average thermal energy stored in the plate.

For many problems it is sufficient to know only the average temperature, e.g., where thermal conduction within the plate is fast compared to the rate at which heat can be supplied to it. This will occur either for thin materials (small d) or ones of high thermal conductivity (large k).

As an example consider a plate with initial temperature of T_0, insulated on one face and suddenly exposed, at time $t = 0$, to a hot, moving air stream with temperature T_f. The heat transfer from such a hot stream is known, experimentally, to depend on the difference between the temperature of the air stream far from the plate (T_f) and the surface temperature of the plate ($\bar{T}$ in the present case, since it is assumed that the average temperature is a good approximation for the temperature anywhere in the plate). The constant of proportionality is the "convective heat transfer coefficient" and is usually designated with the letter h. (For more detail on convection, see Section 1, Chapter 3.) Thus, the heat transferred from the air to the plate is

$$h(T_f - \bar{T})$$

where h has units (W/m²-K). Substituting this convective heating assumption for one term on the right of Equation 12 and zero (insulated rear face) for the other term yields the equation

$$\frac{\partial \bar{T}}{\partial t} = \frac{h}{\rho c b}(T_f - \bar{T})$$

This simple, first-order differential equation is easily integrated to give

$$\bar{T} = T_0 e^{-ht/\rho cb} + \frac{h}{\rho cb}\int_0^t T_f(\tau) e^{-(h/\rho cb)(t-\tau)} \, d\tau \tag{13}$$

The characteristic time for this solution is

$$t_c = \rho c \, b/h \quad \text{(seconds)}$$

It is noted that the transient heat flow equation, Equation 7, leads to the characteristic time

$$t_0 = \rho c \, b^2/k$$

Comparing these two characteristic times yields

$$t_0/t_c = hb/k = \text{(Biot Number)}$$

When Bi is small (e.g., if thermal conductivity, k, is large or the sample thin) the material is thermally thin. Further, transient heat conduction within the solid is fast compared to heat transfer to its surface. Conversely, when Bi is large (e.g., if the surface heat transfer coefficient is large or the sample thick), it is not thermally thin and the above analysis would be inappropriate.

If T_f is a constant, the integral in Equation 13 can be evaluated. The result is

$$\bar{T} = T_0 e^{-t/t_e} + T_f(1 - e^{-t/t_c}) \qquad \text{(13a)}$$

After a long time, $\bar{T} = T_f$.

The result, Equation 13a, can be recast in terms of the characteristic time mentioned above in the discussion of Equation 7. Recall that this time was

$$t_0 = \rho c \, b^2/k$$

Using this in Equation 13 gives

$$
\begin{aligned}
ht/\rho cb &= (ht_0/\rho cb)(t/t_0) \\
&= [h(\rho cb^2/k)/(\rho cb)](t/t_0) \\
&= (hb/k_{\text{solid}})(t/t_0) \\
&= (h\ell/k_{\text{air}})(k_{\text{air}}/k_{\text{solid}})(d/\ell)(t/t_0)
\end{aligned}
$$

where ℓ is the length of the solid in the direction of the air flow. This length is needed to determine the value of h from published heat transfer data, such as McAdams.[9] The quantity, $(h\ell/k_{\text{air}}) = \text{Nu}$, is the Nusselt number, the ratio of the actual heat transferred under the given flow conditions to what would be conducted through still, stably stratified air, provided the temperature difference $(T_f - \bar{T})$ were applied over a distance ℓ. For natural convection, such as is often found in fire situations, the Nusselt number would be on the order of 100. For wood $(k_{\text{air}}/k_{\text{solid}})$ is about 0.1, and it may be supposed for the present that d/ℓ is also 0.1. Then the assumption that the sample is "thermally thin" would be true so long as t/t_0 were large compared to 1. For a metal, k_{solid} is 10^4 times that of wood, so the thermally thin assumption would hold for t/t_0 large compared to 10^{-3}.

A more interesting, but also more difficult problem arises if the hot air temperature is high enough that the plate begins to radiate significantly as it warms in response to the air's heating. This would arise if the heating were due to flames passing over the plate. For simplicity it can be assumed that the hot gas absorbs negligible radiant energy and all the radiation escapes to a distant, surrounding region, which is at the initial plate temperature, T_0. In this case, one must subtract the heat lost by radiation from the plate's surface from the heat transferred to the plate from the gas.

(For more detail on radiation see Section 1, Chapter 4.) The re-radiation will be

$$\dot{q}_{RR} = \epsilon\sigma(T^4 - T_0^4)$$

where σ = Boltzmann's constant = 5.6696×10^{-11} (kW/m^{2}K); ϵ = plate emissivity; and the equation to be integrated is

$$\rho cd\frac{\partial \bar{T}}{\partial t} = h(T_f - \bar{T}) - \epsilon\sigma(\bar{T}^4 - T_0^4)$$

Because of the fourth power of $\bar{T}$ on the right side of the equation, this cannot be integrated in closed form. An approximate integral is obtained by linearizing the radiation term

$$(\bar{T}^4 - T_0^4) = [(T_f^4 - T_0^4)/(T_f - T_0)] \times (\bar{T} - T_0)$$

Linearized in this fashion, the equation can be integrated. After integration the result is

$$\bar{T} = T_0 + (h/H)(T_f - T_0)(1 - e^{-Ht/\rho cd})$$

where

$$H = h + \epsilon\sigma[(T_f^4 - T_0^4)/(T_f - T_0)]$$

With significant effort, higher order analytic approximations can be obtained, or numerical methods used if a more accurate answer is needed. Note that, without the radiation loss, the plate temperature approaches the gas temperature, T_f (Equation 13a), but with the radiation loss, since $h/H < 1$, the final temperature is lower. (See Equation 14.) It also retains a dependence on the ambient temperature.

$$\bar{T} \rightarrow T_0 + (h/H)(T_f - T_0) \qquad \text{(14)}$$

Thick Plates

The theory of thick plates leads to much more involved analytic techniques than were needed for thermally thin plates. Only a brief outline of the theory will be given here; more complete treatments are available in such standard heat transfer texts as Carslaw and Jaeger.[10]

The most common approach to transient heating of thick plates uses the linearity of Equation 7. Equation 7 is linear only for the particular case where (1) the thermal diffusivity is either constant or a function of position, but not temperature, and (2) heat addition, if present, is a function of position and/or linear in temperature. These linearity restrictions can often be met while still giving useful results for practical problems. A linear equation can, of course, be solved by the linear superposition (sum) of primitive solutions. To use this technique a suitable primitive solution is needed. Two ways to obtain and use primitive solutions will be discussed.

Separation of variables: A technique that may be used with Equation 7 for multidimensional problems is called "separation of variables."* Here the technique will be

---

*The method of separation of variables for solution of linear, partial differential equations is discussed in many graduate-level texts on mathematical methods of physics and chemistry. Recent texts tend to emphasize numerical methods to the exclusion of older, analytic ones. Texts predating the computer revolution may be more useful.

illustrated for only one dimension. The above linearity restrictions will be met by assuming constant density, specific heat, and thermal conductivity (more restrictive than just constant thermal diffusivity); and a heat addition term linear in temperature.

Consider a flat, rectangular plate of thickness $2b$, insulated along its edges, with heat generated within the plate at a rate that increases with temperature.

The following is a very simple example illustrating some aspects of self-heating. For an object to self-heat, it is necessary for an exothermic chemical reaction to occur within the solid. Further, the rate of this reaction should accelerate as the temperature rises. For a more complete discussion of self-heating see Thomas[11] or Section 2, Chapter 12 of this handbook.

The rate of chemical reactions typically increases exponentially with temperature, but to keep within the restrictions of the linear equation mathematics, it is assumed that the heat released per unit volume, $\dot{q}'''$, varies linearly with temperature

$$\dot{q}''' = C + DT$$

where C and D are presumably known and D is positive. For small temperature changes, this may be accepted as a valid linearization of an exponential increase. C may be eliminated by a suitable choice of reference temperature, $T_0 = -C/D$. It is assumed that the two surfaces of the plate are kept at the same temperature, T_s. The center of the plate is at $x = 0$. Because of the geometric symmetry, the boundary conditions are symmetric and T must be a symmetric function of x.

Using the separation of variables technique, assume that T can be written in the form

$$T = \theta(t)X(x)$$

Substituting this in the one-dimensional version of Equation 7, yields

$$X\partial\theta/\partial t = \alpha\theta\partial^2 X/\partial x^2 + (D\alpha/k)\theta X$$

Dividing by θX yields

$$\frac{[\partial\theta(t)/\partial t]}{\theta(t)} = \alpha\frac{[\partial^2 X(x)/\partial x^2]}{X(x)} + D\alpha/k \tag{15}$$

The left side of the equation depends only on time while the right depends only on location, x. This can only be true if both sides are independently constant. Let this constant be λ and write two equations expressing the constancy of the two sides

$$\theta' + \lambda\theta = 0$$

$$X'' + (D/k + \lambda/\alpha)X = 0$$

If $\lambda = 0$ the first of these equations says $\theta = $ constant and the second that

$$\theta X = T = A\sin(\sqrt{D/k}\,x) + B\cos(\sqrt{D/k}\,x) \tag{15a}$$

Invoking the required symmetry and using the boundary condition $T = T_s$ at the surface, $x = b$, gives (remembering the reference temperature)

$$T = T_0 + (T_s - T_0)\frac{\cos(\sqrt{D/k}\,x)}{\cos(\sqrt{D/k}\,b)} \tag{15b}$$

This would be the steady-state (time-independent) solution, provided one exists.

If λ is not 0 the result is

$$T(x, t, \lambda) = e^{-\lambda t}[A(\lambda)\sin(\sqrt{D/k + \lambda/\alpha}\,x) + B(\lambda)\cos(\sqrt{D/k + \lambda/\alpha}\,x)] \tag{15c}$$

Equations 15a and 15c are the primitive solutions that are superimposed (summed). The separation parameter, λ, and the coefficients A and B are chosen to satisfy the initial and boundary conditions. The two major terms of Equation 15c decay exponentially with time (for positive λ), with those with the most rapid oscillation in x decaying most quickly. Clearly, this technique requires inclusion of many terms if the behavior for small time is needed. For hand calculations this is a serious drawback, but with the advent of very efficient "fast Fourier transform" numerical packages to evaluate the coefficients A_n and B_n, the separation of variables method has acquired some new appeal.

Combining Equation 15b (which is Equation 15a adjusted to satisfy the boundary conditions) with Equation 15c allows the evaluation of λ. Because the boundary conditions are already satisfied by Equation 15b, Equation 15c must be symmetric in X, and zero at $x = b$, hence $A_n = 0$ and

$$\sqrt{D/k + \lambda/\alpha} = (2n - 1)\pi/2b$$

$$\lambda = \{[(2n - 1)\pi/2b]^2 - D/k\}\alpha$$

and

$$T = T_0 + (T_s - T_0)\frac{\cos(\sqrt{D/k}\,x)}{\cos(\sqrt{D/k}\,b)}$$
$$+ \sum_{n=1}^{\infty} B_n\cos\{[(2n - 1)\pi/2](x/b)\}$$
$$\times e - (\{[(2n - 1)\pi/2b]^2 - D/k\}\alpha t) \tag{15d}$$

To satisfy the initial conditions, B_n are evaluated using the standard technique of multiplying both sides of Equation 15d, with $t = 0$, by the appropriate cosine and integrating with respect to x from 0 to b. To illustrate self-heating, however, this step is not necessary as long as it is realized that in general the B_n are nonzero. It will be assumed that the initial condition was almost the steady-state solution, Equation 15b; i.e., almost all the $B_n = 0$. But suppose that instead of the initial condition being exactly the steady-state distribution, a small perturbation away from the steady-state temperature distribution existed at $t = 0$. In other words, suppose that one of the B_n were a small, nonzero value. In this case Equation 15d could show "thermal runaway"; i.e., the temperature would grow exponentially with time, if D exceeded

$$D = k[(2n - 1)\pi/2b]^2 \tag{15e}$$

In particular, if B_1 is nonzero, Equation 15e gives the critical value for D for which thermal runaway will occur. Where thermal runaway occurs, there is no steady-state solution. The use of Equation 15b is still valid, but its interpretation is altered—it describes a stable state if D is less than $k(\pi/2b)^2$; it is a metastable state if D is larger.

This example does not require evaluation of a large number of terms of the series. In fact, only one term is sufficient, since the lowest order term (smallest n) has the largest growth rate. Thus the separation of variables technique is particularly suited to this problem.

The analysis of self-heating involves many factors not considered in this simple calculation. One factor is determining the acceleration of the rate of chemical reaction as local temperature increases; i.e., determining the constant D for the material in question. Another factor is the effect of size. From the present simple example it is clear that for positive D, there is some thickness of material, $2b$, near which the potential for disastrous self-heating exists. The larger D, the smaller that b need be for self-heating runaway. Although the critical size is often large, materials, such as coal, wood chips, pelletized animal feeds, etc., are occasionally stored in piles large enough for self-heating to be important.

Integral (Duhammel's) method: Where a very large number of terms are needed to yield sufficient accuracy using the separation of variables technique, integral solutions may be more efficient. If an integral is considered as a special kind of sum, the technique is another type of superposition. Just as the separation of variables technique required a primitive solution, which could be summed, so the integral technique requires a suitable primitive solution. One such function is

$$f(x,t) = q_1'' 2\rho c \sqrt{\pi \alpha t}\, \exp(-x^2/4\alpha t) \tag{16}$$

where α is the thermal diffusivity referred to earlier. Also, $f(x, t)$ is the temperature response of the plate to an instantaneous heat addition per unit area of plate, q_1'' (units J/m^2) applied at time $t = 0$, and no heat is transferred to or from the plate at subsequent times. That f is a solution to the one-dimensional version of Equation 7 can be readily determined by differentiation. Note that at any time after the heat addition, the heat in the plate, assumed to extend to infinity in both directions from $x = 0$, will be

$$\rho c \int_{-\infty}^{\infty} f(x, t)\, dx = q_1'' \sqrt{\pi} \int_{-\infty}^{\infty} \exp(-x^2/\sqrt{4\alpha t})\, d(x\sqrt{4\alpha t})$$

$$= q_1'' 2/\sqrt{\pi} \int_{-\infty}^{\infty} \exp(-\xi^2)\, d\xi$$

$$= q_1''$$

This shows that the quantity of heat added per unit area to the plate is just q'' which, with $q'' = 1$, gives 1 for the integral; i.e., the total heat added to the plate is one Joule per unit area of plate.

Equation 16 yields a singular temperature at $x = 0$, $t = 0$. For all later times the temperature has a "Gaussian" profile with respect to x with a half width $\sqrt{4\alpha t}$. Thus the half width of the temperature distribution increases with time. The heat spreads out, so that, while the peak temperature drops, the heated part of the plate includes an ever thicker region adjacent to $x = 0$.

The principle of superposition, mentioned previously, is now used to construct a particular solution to the one-dimensional heat flow equation. Let

$$T(x, t) = T_0 + \int_{-\infty}^{\infty} [q'''(\xi)/q_1'']f_1(x - \xi, t)\, d\xi \tag{17}$$

where f_1 is used in place of the f of Equation 16 to emphasize the choice of $q_1'' = 1$ J per unit area. The quantity $q'''(\xi)d\,\xi/q_1''$

in Equation 17 is dimensionless. If q''' is chosen such that $q'''(-\xi) = q'''(\xi)$, the temperature at $x = 0$ will be T_0 and $[T(-x, t) - T_0] = -[T(x, t) - T_0]$, shifting from an infinite solid to a semi-infinite one. Further, writing $g = q'''/q_1''$ (g has dimension $1/m$)

$$T(x, t) = T_0 + \int_{-\infty}^{0} g(\xi)f_1(x - \xi, t)\, d\xi$$

$$+ \int_{0}^{\infty} g(\xi)f_1(x - \xi, t)\, d\xi$$

$$= T_0 - \int_{0}^{\infty} g(+\xi)f_1(x + \xi, t)\, d\xi$$

$$+ \int_{0}^{\infty} g(\xi)f_1(x - \xi, t)\, d\xi$$

Recalling the definition of f_1, substitute in the first integral

$$\xi = -x + 2\eta \sqrt{\alpha t}$$

and in the second

$$\xi = x + 2\eta \sqrt{\alpha t}$$

The result is

$$T(x, t) = T_0 + q_1''/(\rho c \sqrt{\pi}) \left[\int_{-x/2\sqrt{\alpha t}}^{\infty} g(x + 2\eta \sqrt{\alpha t})\exp(-\eta^2)\, d\eta \right.$$
$$\left. - \int_{x/2\sqrt{\alpha t}}^{\infty} g(-x + 2\eta \sqrt{\alpha t})\exp(-\eta^2)\, d\eta \right] \tag{17a}$$

Finally, if g is constant,

$$T(x, t) = T_0 + (q_1'' g/\rho c)\left[2\sqrt{\pi} \int_{0}^{x/2\sqrt{\alpha t}} \exp(-\eta^2) d\eta \right] \tag{17b}$$

$T(x, t)$ is the temperature in a semi-infinite plate with initial temperature $T_0 + T_{c(x)}$ whose surface is brought to T_0 at time zero and held there for all subsequent time.* If $x = 0$ the integral is zero and thus $T(0, t) = T_0$. The surface temperature of the plate is a constant for all time. Also, because of the t in the denominator of the limit of integration of the integral, it is clear that, for very small time, for any finite x, the integral tends to 1 and the result is $T_0 + q''\sqrt{\rho c} = T_0 + T_c$ (for simplicity). Thus Equation 17 is the temperature distribution in a solid, initially at temperature $(T_0 + T_c)$ which, at time 0, has its surface brought to temperature T_0 and held there for all future time. Figures 1-2.5(a) and (b) show the temperature in the plate according to Equation 17b.

What heat input to the material must be supplied in order to achieve this; i.e., to bring the surface instantly from $(T_0 + T_c)$ to T_0 and hold it there? The heat input to the surface is just

$$-k\frac{\partial T}{\partial x}\bigg|_{x=0} = kT_c(1/\sqrt{\pi \alpha t})\exp(-x^2/4\alpha t\big|_{x=0}$$
$$= kT_c(1/\sqrt{\pi \alpha t}) = T_c(k^2/\pi \alpha t)^{1/2} \tag{18}$$

Thus the necessary heat flux to be applied at the surface is initially infinite (at $t = 0$) and for finite times, decreases as $1\sqrt{t}$.

This is a useful approximation for fire studies as many materials may be rapidly heated to a high temperature and their surface temperature stabilized there.

---

*The quantity within the square brackets is the "error function." Its value may be found in mathematical tables just as for the sine or cosine functions.

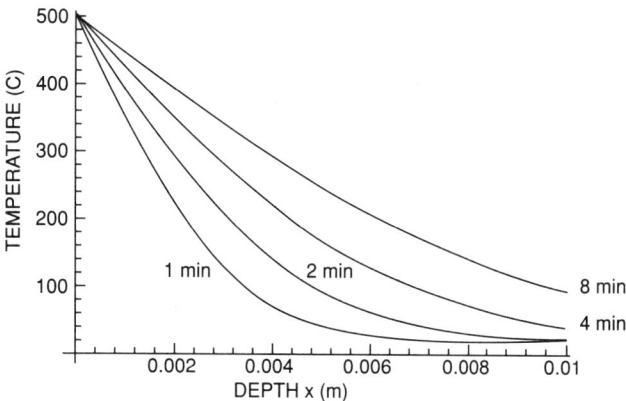

Fig. 1-2.5(a).   Equation 17b temperature vs depth into the plate for four times.

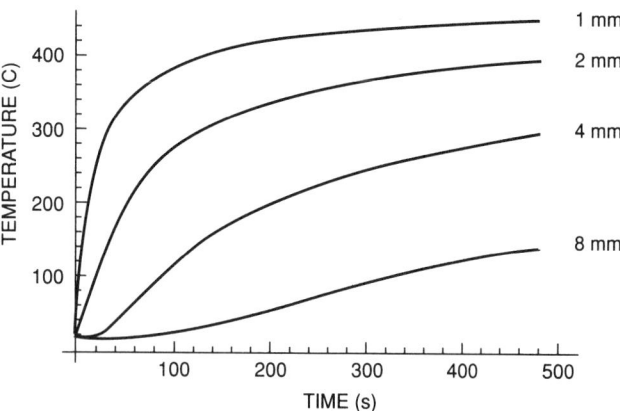

Fig. 1-2.5(b).   Equation 17b temperature vs time for four depths into the plate.

The quantity $k^2/\alpha = k\rho c$, which appears in Equation 18, is an important parameter of a material and is often referred to as the "thermal inertia." The ratio of the thermal inertia to the square of the convective heat transfer coefficient, h, has the units of time

$$k\rho c/h^2$$

$$(W/m\text{-}K)(gm/m^3)(J/gm\text{-}K)/(W/m^2\text{-}K)^2 = (sec)$$

It is the characteristic time for the surface of a solid to respond to heating on exposure to a hot gas with film resistance h. With other modes of heating, an analogous ratio would be appropriate, and would have the same significance—the characteristic time for the solid to respond to the heating.

The next example of heat conduction in a solid is from thermal flame spread theory. The problem is to determine the temperature of the surface of a thick fuel as a function of time, as the fuel is heated from ambient conditions to its ignition temperature. This example of the thermally thick solid is analogous to the thermally thin case discussed earlier.

If heat is applied uniformly to the surface of a flat material (fuel) only one-dimensional heat conduction needs to be considered in calculating the surface temperature of the

material. If the heat is applied nonuniformly, but in such a way that temperature gradients along the surface are small compared to those normal to the surface, lateral heat conduction in the fuel will be small compared to conduction normal to its surface, and a one-dimensional heat transfer model may still be adequate. For many cases this approach is indeed suitable. The principal exception is near extinction (very slow flame spread), where the flame spread rate is comparable to or less than the thermal penetration rate. Assuming that extinction is not a present concern, the one-dimensional approach may be adopted.

The analysis is complicated, but available in many standard heat transfer references, such as Carslaw and Jaeger.[10] It will not be repeated here. If the solid is exposed to a constant external heat flux, $\dot{q}''$, and loses heat by radiation and convection, its temperature will be given by

$$T(t) = T_0 + \frac{\dot{q}''}{H}\left[erfc\left(\frac{x}{2\sqrt{\alpha t}}\right)\right.$$
$$\left. - e^{-(Hx/k + H^2t/\beta)}erfc\left(\frac{x}{2\sqrt{\alpha t}} + \sqrt{H^2t/\beta}\right)\right] \quad (19)$$

where *erfc* is the complementary error function $(1 - erf)$ and has a value of 1 for an argument of 0.

Figure 1-2.6 shows the temperature of the solid as given by Equation 19 for $x = 0$ (surface) and a depth of 1 mm below the surface $(x > 0)$. In preparing this plot, properties for oak were used.[8] The applied external flux was 30 kW/m^2. Ambient temperature was 20°C. When the surface of the sample reached the approximate ignition temperature of wood (420°C), the temperature 1 mm below the surface was 60°C cooler. The heat was diffusing into the sample, but was far from equilibrium conditions. [See Figure 1-2.5(a).] If pyrolysis products from the sample were exposed to a nonimpinging pilot flame, Equation 19 suggests that ignition would occur after about 3 sec exposure to the applied flux. If the applied flux were abruptly switched off as soon as ignition occurred, the sample would probably self-extinguish because of the heat loss to its interior. This would rapidly cool the surface to below its piloted ignition temperature. Conversely, Figure 1-2.6 illustrates that, to achieve sustained ignition, the surface must be: (1) heated to ignition temperature, but heating must continue until the bulk of the sample has been sufficiently heated and (2) that quenching will not occur due to heat conduction into the bulk material.

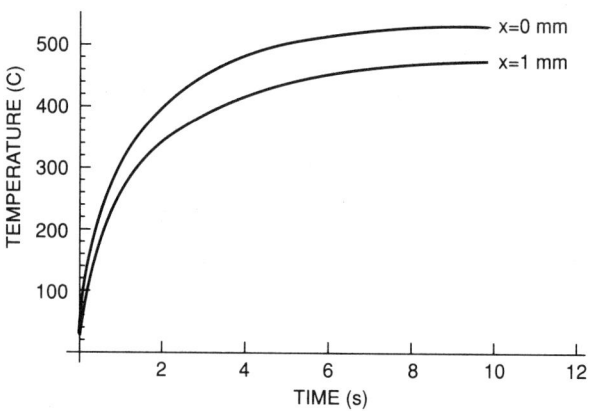

Fig. 1-2.6.   Transient heating of a solid exposed to a constant incident heat flux. The material loses heat to its surroundings by convection and a linearized approximation of radiation.

Many other integral solutions based on Equation 16, but satisfying other initial and boundary conditions, can be found in the heat conduction literature.[10] However, the solutions involve special mathematical functions and use more advanced analytic techniques than are appropriate to this text. The reader is urged to become acquainted with these powerful tools.

The above analysis is qualitatively correct, but, for fire problems, may not yield satisfactory quantitative results. The assumptions of constant material thermal properties and linearization of the convective/radiative heat transfer coefficient may not be valid for the temperature range encountered. In addition, for porous materials containing absorbed moisture (e.g., wood), thermal effects associated with migration of the moisture within the solid may lead to significant deviation from temperatures calculated on the assumption that heat conduction alone is occurring.

Considerable algebraic manipulation is required to get analytic solutions to heat conduction problems. The advantage of an analytic solution includes the insight that it gives about thermal behavior. However, even when this labor is completed, the solution may be limited by the necessity to linearize it. In many cases, numerical techniques, discussed next, may be the simplest route. A limitation of numerical methods is that, to understand the full thermal behavior, many solutions may be needed before a reasonable understanding of the behavior is attained.

NUMERICAL TECHNIQUES

The heat conduction literature contains many analytic solutions to transient heat transfer problems for various geometries. However, rather than use the complex analytic forms, it is often easier to obtain solutions numerically, capitalizing on the recent increase in the availability of powerful, inexpensive microcomputers. Further, for solutions in which material properties vary significantly with direction or temperature, or heat addition within the solid is important, numerical methods are likely to be the most efficient approach. Two numerical formulations are available for solving heat conduction problems: (1) finite difference method and (2) finite element method. A comprehensive overview of numerical methods for heat transfer is available in Minkowycz.[12] An extensive review of the finite difference method is available in most introductory heat transfer texts. The finite element method is reviewed by Zienkiewicz.[13]

Both numerical formulations involve the *discretization* of the object being analyzed. *Discretization* divides the section into small segments using a set of nodes connected by lines referred to as grid or mesh. (See Figure 1-2.7.) Consequently, both methods approximate the geometry of the section being modeled.

Smaller grids provide increased accuracy, though extra nodes result in an increase in computation time. Grid spacing may vary through the assembly, being smallest where the temperature gradient is expected to be the greatest, e.g., near the heated boundary or in insulating materials. A variable grid spacing requires an additional amount of effort to formulate the governing equations for each node.

Generally, the finite difference method uses a "Taylor series" expansion to reformulate the partial differential equation, resulting in a set of algebraic equations. The finite difference form of Equation 7 for the case of transient one-dimensional heat conduction with no internal heat addition is

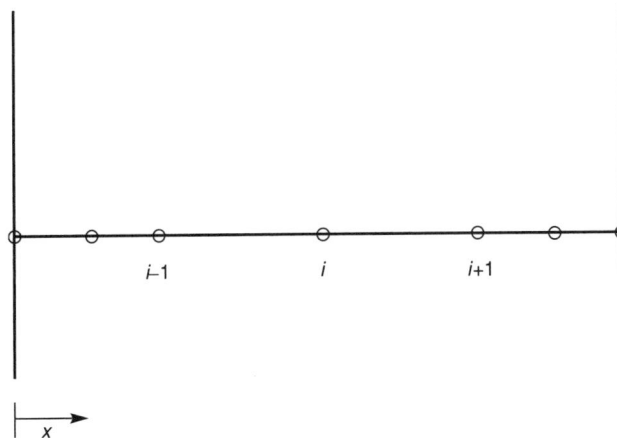

Fig. 1-2.7. Nomenclature of finite difference grid.

$$C_i \frac{(T_i' - T_i)}{\Delta t} = \frac{k_m (T_{i+1} - T_i)/\Delta x - k_n (T_i - T_{i-1})/\Delta x}{\Delta x} \quad (20)$$

Equation 20 is applicable to cases where the grid spacing is the same on each side of node i. Alternative expressions can be readily developed for nonuniform grids. A version of Equation 20 must be developed for each node. Conditions at each node are assumed to represent the conditions in a small region around each node. In particular:

T_i = temperature at node i at time t (at beginning of time step), and
T_i' = temperature at node i after time step, Δt.

Where points $i + 1$, i, and $i - 1$ are all in the interior of the section, and

$$k_m = \frac{1}{2}(k_{i+1} + k_i)$$
$$k_n = \frac{1}{2}(k_i + k_{i-1})$$
$$C_i = \rho_i c_p \Delta x$$

Alternatively, for points on a surface, a heat balance needs to be formulated to develop the finite difference equation. Introductory heat transfer texts include finite difference equations for surface nodes exposed to a variety of conditions.

Equation 20 is expressed as an explicit formulation, i.e., T_i', can be solved directly, as long as the nodal temperatures are known at the beginning of the time step. Consequently, the equations are uncoupled and can be readily solved. However, this formulation can provide numerically unstable results (computed temperatures may become greater than the exposure temperature or may decrease within an object even though it is being heated). Stability limits are based on the time step and grid size. For constant material properties, the stability limit for one-dimensional problems is expressed as

$$\frac{\alpha \Delta t}{(\Delta x)^2} < 0.5 \quad (21)$$

In order to satisfy the stability limit, a sufficiently small time step must be selected for a given grid spacing. As the grid spacing is decreased, the time step also needs to be decreased. The stability criterion is applied at each node, with the resulting time step for the analysis being less than the smallest time step required for stability.

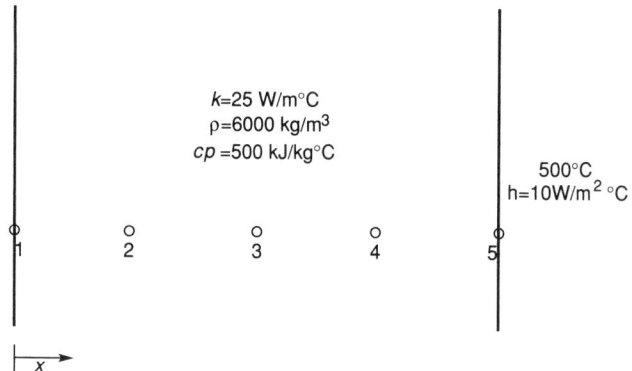

Fig. 1-2.8. *Finite difference grid, Example 6.*

Another formulation for the finite difference equation, referred to as the implicit formulation, is presented as

$$C_i \frac{(T_i' - T_i)}{\Delta t} = \frac{k_m(T_{i+1}' - T_i')/\Delta x - k_n(T_i' - T_{i-1}')/\Delta x}{\Delta x} \quad (22)$$

The implicit formulation has the characteristic of being numerically stable for any time step or grid size. In this formulation, the time step and grid size are selected solely based on concerns for accuracy. However, the disadvantage of the implicit formulation is that all of the equations are coupled as the right-hand side of the equation includes nodal temperatures at the end of the time step. Thus, all of the equations must be solved simultaneously.

EXAMPLE 6: A 10-cm thick wall is initially at a uniform temperature of 20°C. The properties of the wall are: conductivity = 25 W/m-K, density = 6000 kg/m³, and the specific heat = 500 J/kg-K. Assume that the material properties are independent of temperature. The right side is suddenly exposed to a convective environment, with a coefficient of 10 W/m²-K and a temperature of 500°C. The left side of the wall is maintained at 20°C. A grid size of 2.5 cm is selected. Determine the temperature within the wall after 5 min. (See Figure 1-2.8.)

Node 1 has a known temperature due to the maintained surface temperature. Thus, only the temperatures at the other nodes need to be determined.

The finite difference equation for the interior nodes (nodes 2 through 4), based on the formulation of Equation 20, is

$$6000 \cdot 500 \frac{(T_i' - T_i)}{\Delta t} = \frac{25(T_{i+1} - 2T_i + T_{i-1})}{(0.025)^2}$$

For node 5, the following equation is applicable

$$6000 \cdot 500 \frac{(T_i' - T_i)}{\Delta t} = \frac{10(500 - T_i) - 25(T_i - T_{i-1})/0.025}{0.025}$$

The stability criterion for nodes 2 through 4 is given as

$$\frac{\dfrac{25}{6000 \cdot 500} \Delta t}{(0.025)^2} < 0.5$$

According to this criterion, the time step must be less than 37 s. For node 5, the stability criterion is expressed as

$$\frac{\left(\dfrac{25}{0.025} + 10\right) \Delta t}{(6000 \cdot 500 \cdot 0.025)} < 1.0$$

The maximum time step according to this criterion is 2,970 s. Consequently, a time step of 30 s is selected, less than the smallest time step of 37 s determined for the wall. The resulting calculations are presented in Table 1-2.1.

The finite element method also starts with a discretized object. However, one of the principal differences from the finite difference method is the use of the exact governing equation. Most finite element formulations use a polynomial fit of the temperature profile within an element to solve the equation. The partial differential equation is reformulated as an ordinary differential equation often using principles of variational calculus. Because an exact equation is used, and only the geometry is being approximated, the finite element method typically provides a more accurate analysis for course grids than the finite difference method. The equation is applied for each element, with the resulting set of equations assembled into a matrix.

$$[C]\{\dot{T}\} + [K]\{T\} = \{Q\}$$

where

C = capacitance matrix, accounting for ρc product associated with each element;
K = conductivity matrix, accounting for conductivity of element;
T = vector (column matrix), which represents temperature at each node; and
Q = vector (column matrix), which represents heat generation at each node.

Determination of the temperature profile within the wall requires solving the matrix-based equations. Several typical matrix solutions can be used, such as Gaussian elimination, though such methods are not efficient due to the K matrix being sparsely populated. Consequently, advancements in finite element modeling are related to more efficient solution techniques.

As with finite difference models, explicit and implicit formulations can be developed. However, an elementary expression for the stability criterion for the explicit formulation is not available.

Several finite element codes are available for heat transfer analysis. FIRES-T3 and TASEF are of particular interest due to their capability to account for time-varying exposure conditions and temperature-dependent material properties.[14,15] Both codes were formulated to predict the temperature distribution in structural members resulting from exposure to a fire. Both models are based on the same principles and have virtually the same explicit formulation. They differ primarily in their input requirements. Some additional information on FIRES-T3 and

TABLE 1-2.1 *Temperatures within Wall Assembly (°C)*

Time	Node 1	Node 2	Node 3	Node 4	Node 5
0	20.0	20.0	20.0	20.0	20.0
30	20.0	20.0	20.0	20.0	21.9
60	20.0	20.0	20.0	20.8	23.1
90	20.0	20.0	20.3	21.4	24.1
120	20.0	20.1	20.6	22.0	24.9
150	20.0	20.3	21.0	22.6	25.6
180	20.0	20.5	21.3	23.2	26.3
210	20.0	20.6	21.7	23.7	27.0
240	20.0	20.8	22.1	24.2	27.6
270	20.0	21.0	22.4	24.7	28.1
300	20.0	21.2	22.8	25.2	28.6

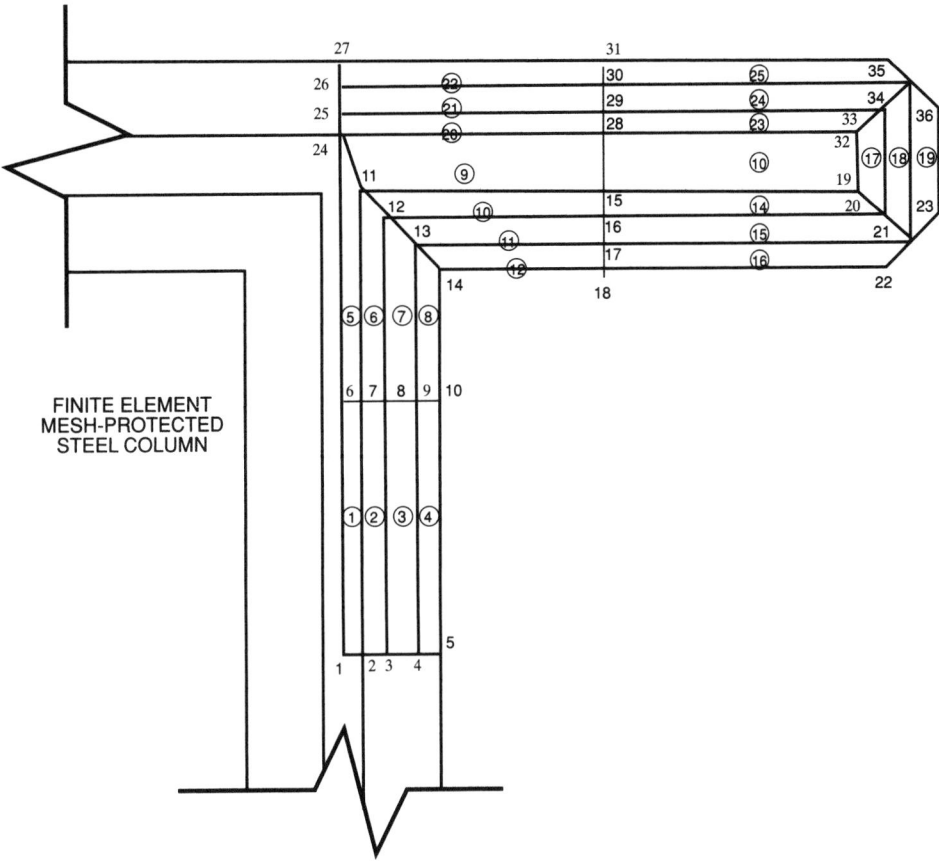

Fig. 1-2.9. *Finite element mesh—protected steel column.*

TASEF is available in Section 4, Chapter 9 of this handbook. A comparison of FIRES-T3 and TASEF was provided elsewhere by Milke.[16] TASEF is available from Sterner and Wickström at the Swedish National Testing Institute.[14] FIRES-T3 is available from Milke at the University of Maryland.

A typical diagram of a finite element mesh for a steel column with spray-applied, contour protection is presented in Figure 1-2.9. Element numbers are included within the circles, nodal numbers are encircled. Elements representing the steel are 1, 5, 9, and 13. As a result of symmetry, only one quadrant of the column needs to be modeled.

LIMITATIONS

This short summary of heat conduction has included only a few of the better known solutions of the heat conduction equation. It is very difficult to obtain analytic solutions for objects with complex geometric contours. Published analytic solutions all relate to simple geometric forms. Numerical methods can be used for more complex shapes, but coding the boundary conditions for complex shapes is awkward and tedious. It is seldom done unless there is strong economic motivation to do so.

For the temperature ranges found in fire studies, there are often significant changes in k, the thermal conductivity.[8] Where k is a function of T, the heat conduction equation becomes nonlinear and numerical methods are required. Using a variable k produces a quantitatively superior result, but usually not one that is qualitatively different than would be found with constant k.

Some materials, such as wood, are not isotropic and in these cases k is a tensor quantity. Although results calculated assuming a constant, scalar k may not differ qualitatively from more accurate calculations using a tensor k, in some cases (for example, flame spread), the difference may be important.[17]

Specific heat may also vary, though this is less often a problem than changes in k. When the specific heat appears to vary, it is well to suspect chemical changes, as they are more likely to occur than the Debye temperature being too high.

Transmission of heat in gases or liquids often plays a major role in fire analysis. Occasionally, conduction is important for these materials, but in making calculations, care must be taken that thermally induced convection or radiation does not dominate the heat transfer. If the hotter material is above the cooler it will be stable against convective motions, while in the reverse case, it will be unstable and convection will surely dominate. Radiation may be more important than conduction even in the thermally stable case because for many gases (e.g., air), conduction is low (since gases are electrical insulators), but they absorb radiation only weakly. The radiant energy passes directly through these gases. For example, near room temperature a layer of air 1 cm thick transmits about as much heat by radiation as by conduction.

Where liquids are involved, changes in surface tension with temperature have been shown to induce subsurface convective flows that dominate the heat transfer process. In this case, heat transfer calculations including only conduction (based on the assumption of a quiescent liquid) may be incorrect.[18]

An example where heat transmission in a "solid" may be influenced by convection is a large-pore, closed-cell cellular foam. When a temperature difference is imposed across the sides of a vertical slab, a small temperature difference is imposed across each cell. This results in small density differences in the gas in each cell. In the presence of gravity these differences will induce a convective circulation of the gas in each cell, which may transfer more heat across the cell than can be conducted through its thin walls.

Radiation may also be important in these materials. An example where radiation is an important mode of heat transfer within a "solid" is in a low-density porous, smoldering material or in porous char. The forward transmission of heat, which is necessary for the spread of smoldering combustion, against a thermally induced airflow through the porous material occurs almost entirely by radiation across the pores, not by conduction through the solid pore walls.[19]

One of the most serious problems in heat conduction calculations occurs when the material is not thermally stable. Changes in thermal conductivity may be important; but often more important, there may be the release or absorption of heat within the solid. This release or absorption is often accompanied by the release of material, which then migrates (diffuses) through the (porous) solid. The heat conduction equation allows for this release of heat, but the accompanying physical/chemical processes must be treated by separate, coupled, simultaneous equations. The total computation becomes complex, e.g., in the theoretical treatment of smoldering or the pyrolysis of wood. Theoretical studies of the fire performance of reinforced concrete require that the effect of adsorbed and chemically bound water be included, and today this is regularly done.[20,21] Similarly, the effect of moisture should be included in the analysis of the fire performance of gypsum wallboard. Unfortunately, this is not often done, although the same techniques that are used successfully with concrete apply.

NOMENCLATURE

A	surface area across which heat is transferred (m^2)
A	constant
B	constant
b	plate thickness (m)
C	capacitance matrix
C_i	volumetric thermal capacitance of node i(J/m^3 K)
c	specific heat (J/gm K)
D	rate of heat release parameter
H	convective heat transfer coefficient plus linearized radiation term (W/m^2 K)
	$h_c + \epsilon\sigma[(T_s + T_a)^4 - (T_0 + T_a)^4]/[(T_s + T_a) - (T_0 + T_a)]$
h	convective film coefficient (W/m^2 K)
h_c	convective film coefficient (W/m^2 K)
K	conductivity matrix
k	thermal conductivity of the solid (W/m K)
k_i	thermal conductivity evaluated at temperature of node i(W/m K)
$k_{m,n}$	average thermal conductivity (W/m K)
ℓ	length of plate in direction of airflow across its surface
Q	heat content (J)
Q	heat generation vector
$\dot{q}''$	radiative heat flux per unit area to the surface (W/m^2)
T_a	ambient temperature (K)
T_i	temperature at node i(°C)
T_i'	temperature at node i at end of next time step (°C)
T_s	temperature of the material surface (K)
T	temperature (K)
T	temperature of the solid in Eq 19 (K)

T	temperature vector
T_0	reference temperature, T_0 (K)
t	time (s)
Δt	time step
x	distance normal to the surface (m)
x	coordinate normal to the fuel surface
Δx	grid spacing

Greek Letters

α	the material thermal diffusivity, $k/\rho c$ (m^2/sec)
β	thermal inertia, $k\rho c$ (W^{2}s/m^4 K^2)
ϵ	material surface emissivity
ξ	dimensionless distance
τ	dimensionless time
θ	temperature (K)
ρ	density (kg/m^3)
ρ_i	density evaluated at temperature of node i(kg/m^3)
σ	electrical conductivity (ohms)
σ	Boltzmann's constant, $\sigma = 5.67 \times 10^{-11}$ (kW/m^2 K^4)
η	dummy integration variable
μ	dummy integration variable
∂	differential operator

REFERENCES CITED

1. M. Born, *Atomic Physics*, Hafner, New York (1969).
2. H.L. Callendar, *Ency. Britannica*, 11th ed., 6, 890.
3. E.R.G. Eckert, *Heat and Mass Transfer*, McGraw-Hill, New York (1963).
4. F. Seitz, *The Modern Theory of Solids*, McGraw-Hill, New York (1940).
5. J.C. Slater, *Quantum Theory of Matter*, McGraw-Hill, New York (1951).
6. T. Handa, M. Morita, O. Sugawa, T. Ishii, and K. Hayashi, *Fire Sci. and Tech.*, 2 (1982).
7. J. Milke and A.J. Vizzini, "Thermal Response of Fire-Exposed Composites," *Jour. of Composites Technology and Research*, Vol. 13, No. 3, Fall 1991, pp. 145–151.
8. D. Gross, *NBSIR 85–3223*, National Bureau of Standards, Washington (1985).
9. W.H. McAdams, *Heat Transmission*, McGraw-Hill, New York (1942).
10. H.S. Carslaw and J.C. Jaeger, *Conduction of Heat in Solids*, Oxford University, Oxford (1959).
11. P.H. Thomas, *Special Technical Pub. No 502*, American Society of Testing and Materials, Philadelphia (1972).
12. W.J. Minkowycz, E.M. Sparrow, G.E. Schneider, and R.H. Pletcher, *Handbook of Numerical Heat Transfer*, New York: John Wiley and Sons (1988).
13. Zienkiewicz, O.C., *The Finite Element Method*, 3rd ed., New York: McGraw-Hill (1977).
14. E.S. Sterner and U. Wickström, *TASEF—Temperature Analysis of Structures Exposed to Fire—Users Manual*, Report 1990:05, Borås, Sweden, Swedish National Testing Institute (1990).
15. R.H. Iding, Z. Nizamuddin, and B. Bresler, *Computer Program for the FIre Response of Structures—Thermal Three-Dimensional Version*, University of California, Berkeley, CA, October 1977.
16. J.A. Milke, "Temperature Analysis of Structures Exposed to Fire," *Fire Technology*, 28, 2, 1993, pp. 184–189.
17. A. Atreya, *NBS-GCR-83-449*, National Bureau of Standards, Gaithersburg, MD (1984).
18. W.A. Sirignano and I. Glassman, *Comb. Sci. and Tech.*, 1, 307 (1974).
19. T. Ohlmiller, J. Bellan, and F. Rogers, *Comb. and Flame*, 36, 197 (1979).
20. B. Bressler, *F. Safety J.*, (1984).
21. K. Harada, T. Terai, "Dependence of Thermal Responses of Composite Slabs Subject to Fire on Cross Sectional Shapes," *Fire Safety Science—Proceedings* of the 4th International Symposium, IAFSS, (1994) pp. 1159–1170.

CONVECTION HEAT TRANSFER

Arvind Atreya

INTRODUCTION

There are only two fundamental physical modes of energy transfer, conduction and radiation. In conduction, energy slowly diffuses through a *medium* from a point of higher temperature to a point of lower temperature, whereas in radiation, energy is transmitted with the speed of light by electromagnetic waves (or photons), and a transmitting medium is not required. Thus from a conceptual viewpoint, convection is not a basic mode of heat transfer. Instead, it occurs by a combined effect of conduction (and/or radiation) and the motion of the transmitting medium.

Nevertheless, convection plays a very important role in fires. It transports the enormous amount of chemical energy released during a fire to the surrounding environment by the motion of hot gases. This motion may be induced naturally by the fire itself (hot gases rise and cold air rushes to replace them) or by a source external to the fire, such as a prevailing wind. Based on this distinction, the subject of convective heat transfer is usually subdivided into *natural* (free) and *forced* convection. Obviously, both natural and forced convection may occur simultaneously, resulting in a mixed mode of convective heat transfer. A further subdivision based on whether the flow occurs inside (e.g., in a pipe) or outside the body under consideration is also often made. For application of convective heat transfer to fire science, natural convection around objects is clearly far more important than forced convection inside a pipe. Thus greater attention is devoted here to natural convective heat transfer and external flows.

The objective of this chapter is to provide a firm understanding of the physical mechanisms that underlie convective heat transfer, as well as to develop the means to perform convection heat transfer calculations. In the first section of the chapter, basic concepts and relations are developed, while calculation methods are illustrated with the help of examples in the second section. Tables of empirical and theoretical results (including their range of applicability) are also provided for quick reference.

Dr. Arvind Atreya is Associate Professor of Mechanical Engineering at University of Michigan. He has been actively involved in fire research since 1979. This chapter is respectfully dedicated to the author's father Dr. Dharam Dev Atreya.

CONCEPTS AND BASIC RELATIONS

A simple, everyday problem of drying a wet body in a stream of warm dry air is shown in Figure 1-3.1. From our knowledge of fluid mechanics, we expect the flow of air to slow down next to the surface of the wet body, thus transferring some of its momentum to the body. Conversely, the body will experience a drag force if it moves through stationary air. In addition to this exchange of momentum, the body also losses some of its moisture; i.e., transfer of mass takes place from the wet surface to the warm air. Furthermore, for moisture to evaporate at the wet surface the necessary heat must also be transferred from the warm air to the wet body. Hence, the body experiences a simultaneous transfer of momentum, mass, and energy.

To obtain a quantitative description of the above process, note first that all the quantities (mass, momentum, and energy) being transported in the example are conserved, so conservation laws govern their rate of transfer.

These conservation laws need to be supplemented by basic or constitutive relations that relate the rate of transfer to the driving forces and fluid properties. These basic laws are: (1) Newton's law of viscosity, which relates the rate of change of momentum to velocity gradients; (2) Fourier's law of heat conduction, which relates the rate of heat transfer to temperature gradients; and (3) Fick's law of mass diffusion, which relates the rate of mass transfer to concentration gradients. With this framework of conservation and basic laws a majority of laminar convective heat transfer problems can be analyzed, at least in principle. For the turbulent case this framework provides guidance for developing useful empirical correlations.

This section covers the basic laws in the context of laminar flows and their relationships to the more familiar heat transfer coefficients. Later, we see how these heat transfer coefficients can be determined by the application of conservation laws. The effect of turbulence is also discussed and empirical correlations presented.

Basic Laws of Molecular Transfer

Newton's law of viscosity: An isothermal system is shown in Figure 1-3.2. It consists of a fluid trapped between two impervious flat plates that are infinite in extent and separated by a distance, δ. Experiments show that if the

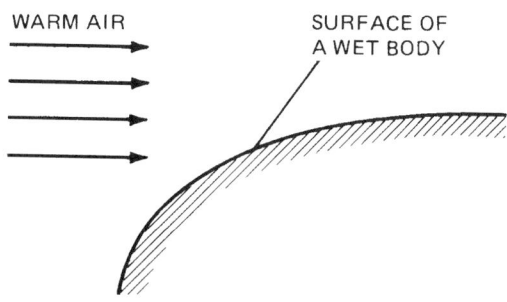

Fig. 1-3.1. Drying of a wet body in a stream of warm air.

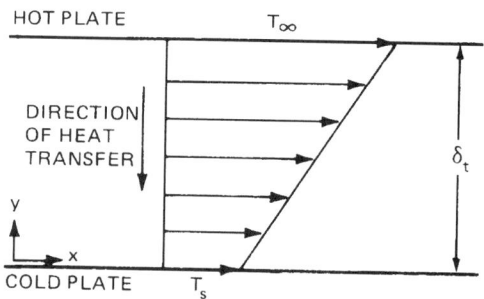

Fig. 1-3.3. Steady-state temperature distribution.

lower plate is fixed and the upper plate is moved at a constant velocity, u_∞, then the velocity of the fluid between the two plates varies from u_∞ near the top plate to zero at the fixed plate. Under steady laminar conditions, a linear velocity profile is established as shown in Figure 1-3.2. The fluid thus exerts a shear force on the stationary plate. Experiments also show that τ, the shear force exerted on the bottom plate per unit area, is directly proportional to u_∞ and inversely proportional to the separation distance, δ; i.e.,

$$\tau \, \alpha \, \frac{u_\infty}{\delta}. \quad \text{or} \quad \tau = \mu \, \frac{u_\infty}{\delta}$$

The constant of proportionality, μ, is called the dynamic viscosity of the fluid (units: N·s/m²) and the force per unit area, τ, is called the shear stress. (In some texts a negative sign is introduced to emphasize the direction of net momentum transfer, i.e., from the fluid at higher velocity to the fluid at lower velocity.) In differential form this relationship expresses the shear stress at any location, y, in the fluid as

$$\tau = \mu \, \frac{du}{dy} \tag{1}$$

This is Newton's law of viscosity. Equation 1 states that the shear stress experienced by a fluid layer is directly proportional to the velocity gradient inside the fluid at that location.

Fluids that behave according to Equation 1 (i.e., τ is linearly related to the velocity gradient, du/dy, and μ is not a function of the velocity gradient) are called Newtonian fluids. Fortunately, all gases and most simple liquids such as water obey this simple law. For gases, μ roughly increases as the square root of temperature as predicted by the kinetic theory of dilute gases. Liquids, on the other hand, become "thinner" (less viscous), i.e., μ decreases with increase in temperature. For non-Newtonian fluids (e.g., pastes, slur-

ries, blood, etc.) the dynamic viscosity μ also depends on the velocity gradient or the rate of shear.

Fourier's law of heat conduction: Two stationary parallel plates separated by a distance, δ_t, are shown in Figure 1-3.3. Let the temperature of the upper plate be T_∞ and that of the lower plate be T_s. Under steady conditions and for temperature independent properties of the trapped fluid, a linear temperature distribution as shown in Figure 1-3.3 is obtained. Thus, as expected, heat is transferred by the stationary fluid from the hot to the cold plate. The heat flow per unit area per unit time through the fluid ($\dot{q}''$ J/m²s) is found to be directly proportional to the temperature difference, $T_\infty - T_s$, and inversely proportional to the separation distance, δ_t; i.e.,

$$\dot{q}'' \, \alpha \, \frac{T_\infty - T_s}{\delta_t} = -k \frac{(T_\infty - T_s)}{\delta_t}$$

The constant of proportionality, k, is called the thermal conductivity (units: J/mKs). The minus sign is a consequence of the second law of thermodynamics, which requires the heat to flow in the direction of decreasing temperature. In differential form the heat across any fluid layer is given by

$$\dot{q}'' = -k \frac{dT}{dy}. \tag{2}$$

This is known as Fourier's law of heat conduction, which states that the heat flux is directly proportional to the temperature gradient and the heat flux vector is oriented in the direction of decreasing temperature. The thermal conductivity, k, like viscosity, μ, is a physical property of the fluid. The thermal conductivity of gases at low densities increases with increasing temperature (roughly as $\sqrt{T}$ according to the kinetic theory of gases) whereas the thermal conductivity of most liquids decreases with increasing temperature.

Fick's law of mass diffusion: Once again consider the parallel plate example, this time there are no temperature gradients and no directed motion of the plates. Instead, the top plate is maintained at a higher concentration of species A ($C_{A\infty}$, kg of species A/m³), assume that it is wet, and the bottom plate is maintained at a fixed but lower concentration of species A (C_{Aw}, kg of A/m³). Then, under steady conditions a concentration profile, as shown in Figure 1-3.4, is established. This nonuniform concentration field is the driving force for species A to diffuse from the top to the bottom plate. The mass flux of species A, $\dot{m}''$, (units: kg/m²s) leaving the top plate and arriving at the bottom plate through the fluid, B, is found to be directly proportional to the concentration difference and inversely proportional to the separation distance δ_d, i.e.,

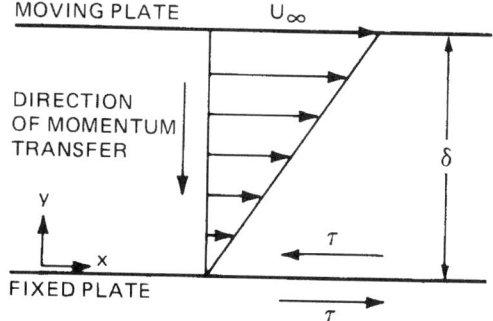

Fig. 1-3.2. Steady-state velocity distribution in a Newtonian fluid.

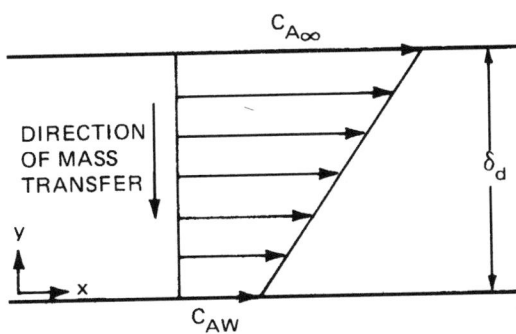

Fig. 1-3.4. *Steady-state concentration distribution.*

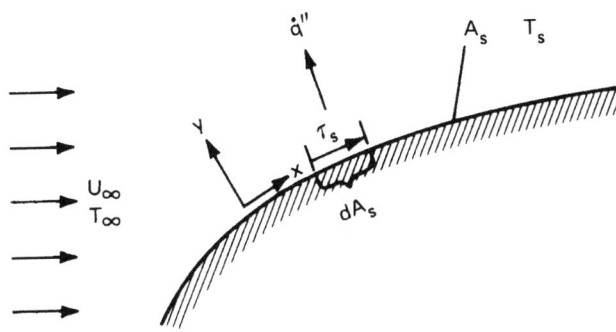

Fig. 1-3.5. *Example of the convection heat transfer problem.*

$$\dot{m}''_A \propto \frac{C_{A\infty} - C_{AW}}{\delta_d} = -D_{AB} \frac{(C_{A\infty} - C_{AW})}{\delta_d}$$

The constant of proportionality, D_{AB}, is called diffusivity of species A through species B and has units of m^2/s. The negative sign once again indicates that the net mass transfer of species A occurs in the direction of decreasing concentration. In the differential form, the mass flux of species A across any layer of fluid B is given by

$$\dot{m}'' = -D_{AB} \frac{dC_A}{dy} \qquad (3)$$

This is known as Fick's law of mass diffusion, which states that the mass flux across a fluid layer is directly proportional to the local concentration gradient. For binary gas mixtures, D_{AB} increases with increasing temperature roughly as three-halves power, as predicted by the kinetic theory of dilute gases.

Discussion: Equations 1, 2, and 3 look very similar—they all relate the flux of the transported quantity to their respective local gradients. Actually these equations may be considered definitions of the three macroscopic physical properties, μ, k, and D_{AB} of the fluid. In general, these properties are functions of temperature, pressure, and composition. As noted earlier, for low pressure binary gas mixtures (< 10 atm) the pressure, temperature, and composition dependence of these properties can be approximately predicted by the kinetic theory of gases. In fact, the physical mechanism of all three transport processes is easily understood by considering the random motion of molecules in an ideal gas.

Molecules of a gas, even in the absence of bulk fluid motion, move around randomly at high speeds and bump into each other. Thus, a given molecule may be found anywhere between the two parallel plates. The problem, however, is how to distinguish one molecule from another. When the upper plate is moving relative to the bottom plate, the molecules adjacent to the upper plate attain a directed velocity over and above their random motion. Consequently, as the molecules near the upper plate find themselves in lower fluid layers (and vice versa) they exchange directed motion (or momentum) by bumping into each other. Similarly, the gas molecules near the hot upper plate are distinguished from those adjacent to the cold lower plate because they possess a higher kinetic energy. Once again, by virtue of random motion, these higher kinetic energy molecules find themselves near the cold plate and collide with low kinetic energy molecules (or vice versa), thus transporting energy. In the case of mass diffusion, the molecules are chemically labeled and their random motion results in mass transfer. Since increasing the gas temperature increases the random

molecular motion, the transport processes become more efficient at higher gas temperatures. For gases, the macroscopic physical properties (viscosity, thermal conductivity, and diffusivity) that characterize momentum, heat, and mass transport must also increase with temperature.

Assuming constant properties, Equations 1 and 2 can be rewritten in the following forms

$$\tau = \left(\frac{\mu}{\rho}\right)\frac{d(\rho u)}{dy} \qquad (1a)$$

$$\dot{q}'' = -\left(\frac{k}{\rho c_p}\right)\frac{d(\rho c_p T)}{dy} \qquad (1b)$$

Here, μ/ρ is known as the kinematic viscosity, v, and has units of m^2/s. The product, ρu, has units of $(kg\ m/s)/m^3$; i.e., momentum per unit volume. The quantity, $k/\rho c_p$, is known as the thermal diffusivity, α, and it too has units of m^2/s. The product, $\rho c_p T$, then becomes enthalpy per unit volume and has units of J/m^3. Comparing these with Equation 3, where the mass diffusivity, D_{AB}, also has units of m^2/s and C_A is expressed in kg of species A/m^3, we find that the fluxes are related to their corresponding gradients of volumetric concentration. Furthermore, the ratios of various physical constants yield the familiar nondimensional numbers. These are:

Prandtl number, $\mathrm{Pr} = v/\alpha$
Schmidt number, $\mathrm{Sc} = v/D_{AB}$
and Lewis number, $\mathrm{Le} = \alpha/D_{AB}$

Here, the Prandtl number compares the relative magnitude of momentum transfer to heat transfer, the Schmidt number compares momentum transfer to mass transfer, and the Lewis number compares heat transfer to mass transfer. The significance of these nondimensional numbers will become obvious when we discuss boundary layer transfer processes. The balance of this section will discuss primarily heat transfer. The treatment of mass transfer is similar and will not be discussed. However, since fluid motion is central to convective heat transfer, it is necessary to understand momentum transport to solve convective heat transfer problems.

Relationship of Basic Laws to Transfer Coefficients

A flow condition is shown in Figure 1-3.5. A fluid of velocity, u_∞, and temperature, T_∞, flows over an arbitrarily shaped stationary surface of area, A_s. If the surface conditions are such that $T_s \neq T_\infty$, we know that convection heat transfer will occur. The convection heat transfer problem is to relate the local heat flux, $\dot{q}''$, to its driving force, $T_s - T_\infty$. By expressing the heat flux as

$$\dot{q}'' = h(T_s - T_\infty) \qquad (4)$$

the problem is reduced to determining h, which is called the heat transfer coefficient. From Equation 2 it is clear that the heat transfer coefficient may be expressed as

$$h = \frac{-k\frac{\partial T}{\partial y}\Big|_{y=0}}{(T_s - T_\infty)} \left(\frac{J}{m^2\,s\,K}\right) \qquad (5)$$

In Equation 5 the partial derivative is used because, in general, temperature is a function of x, y, z, and time. Thus, if the thermal conductivity of the fluid is known, the problem of determining the local heat transfer coefficient is reduced to that of determining the local temperature gradient in the fluid adjacent to the surface. This local temperature gradient can be experimentally measured or obtained theoretically from the solution of the conservation laws. Obviously, the temperature gradient will vary from point to point along the surface of the body. Often, such detail is not required and it may only be necessary to determine an average heat transfer coefficient, $\bar{h}$. This is obtained by integrating over the entire surface area, A_s. The total rate of heat transfer from the body to the fluid is given by

$$\dot{q}\,(J/s) = \int_{A_s} \dot{q}''(J/m^2\,s) \cdot dA_s(m^2) \qquad (6)$$

Defining an average heat transfer coefficient, $\bar{h}$, as:

$$\dot{q} = \bar{h}A_s(T_s - T_\infty) \qquad (7)$$

The following is obtained from the use of Equations 4, 6 and 7

$$\bar{h} = \frac{1}{A_s(T_s - T_\infty)} \int_{A_s} h(T_s - T_\infty)\,dA_s. \qquad (8)$$

If the surface temperature, T_s, is held constant, then

$$\bar{h} = \frac{1}{A_s} \int_{A_s} h\,dA_s\left(\frac{J}{m^2\,s\,K}\right) \qquad (9)$$

Similarly, the local shear stress at the surface, τ_s, can be related to its cause, the fluid velocity, u_∞. This is done by defining a local nondimensional friction coefficient, C_f, according to the equation

$$\tau_s \equiv C_f\left(\frac{1}{2}\rho u_\infty^2\right) \qquad (10)$$

Once again, the problem is reduced to the determination of C_f. Using Equation 1 we obtain

$$C_f = \frac{\mu\frac{\partial u}{\partial y}\Big|_{y=0}}{\left(\frac{1}{2}\rho u_\infty^2\right)} \qquad (11)$$

Thus, the local friction coefficient can be evaluated from the knowledge of the local velocity gradient in the fluid adjacent to the surface. The average friction coefficient can easily be obtained by integrating the local shear stress, τ_s, over the entire surface area, A_s. The total drag force, D, experienced by the body is given by the product of average shear stress, $\bar{\tau}_s$, and the surface area, A_s. In other words,

$$D \equiv \bar{\tau}_s A_s = \int_{A_s}^{\tau_s} dA_s \qquad (12)$$

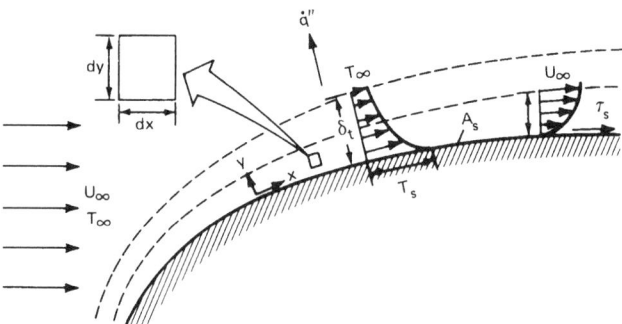

Fig. 1-3.6. *Velocity and temperature distributions inside the fluid for flow over a hot surface.*

Assuming u_∞ to be the same at all locations,

$$\bar{\tau}_s = \frac{\left(\frac{1}{2}\rho u_\infty^2\right)}{A_s} \int_{A_s} C_f A_s \qquad (13)$$

Defining the average friction coefficient, $\bar{C}_f$, as

$$\bar{C}_f \equiv \frac{\bar{\tau}_s}{\left(\frac{1}{2}\rho u_\infty^2\right)} \qquad (14)$$

we get

$$\bar{C}_f = \frac{1}{A_s} \int_{A_s} C_f\,dA_s \qquad (15)$$

To obtain h, $\bar{h}$, C_f, and $\bar{C}_f$ a knowledge of fluid properties and temperature and velocity gradients in the fluid adjacent to the surface is required. To obtain these gradients it is necessary to be more specific about the surface geometry and the flow conditions. The governing conservation equations for an arbitrary surface geometry will be first derived and then applied to a flat plate to illustrate the methodology.

Conservation Equations for Convection Heat Transfer

It has been shown that to determine the heat transfer and friction coefficients, the temperature and velocity distributions in the flow are needed. In principle, these may be obtained from the solution of the conservation equations with appropriate boundary conditions. Although in practice it is difficult, the very knowledge of conservation equations and their solutions for simple cases (such as a flat plate) provide considerable insight about the parameters influencing the heat transfer and friction coefficients. Thus, the necessary equations will be first developed and then applied to a flat plate.

Consider the flow over the surface shown in Figure 1-3.6. To simplify the development assume two-dimensional flow conditions, for which x is the direction along the surface and y is normal to the surface. Extension of this to three-dimensional flows is available in the literature.[1-4]

Conservation of mass (continuity equation): The first conservation law that is pertinent to the problem is that matter is neither created nor destroyed. When applied to the differential control volume shown in Figures 1-3.6 and 1-3.7 it states that the net rate of mass flow entering the elemental control volume in the x-direction, plus the net rate of mass flow entering the elemental control volume in the y-direction equals the net rate of increase of mass stored in the control

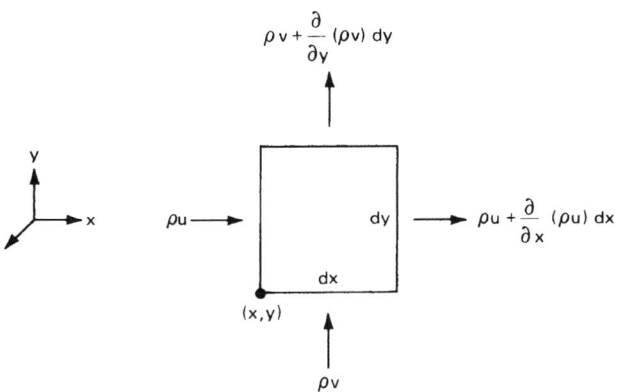

Fig. 1-3.7. *Differential control volume for mass conservation in a two-dimensional flow.*

volume. Mass enters and leaves the control volume exclusively through gross fluid motions. Such a transport is often referred to as convective transport.

For a control volume of unit depth in the z-direction, mass entering the left face per unit time, $\dot{M}_x$, is given by

$$\dot{M}_x = \rho \text{ (kg/m}^3) \, u \text{ (m/sec)} \, dy \text{ (m)} \cdot 1 \text{ (m)}$$

Similarly, mass entering the bottom face per unit time, $\dot{M}_y$, is given by

$$\dot{M}_y = \rho \cdot v \cdot dx \cdot 1$$

where v is the fluid velocity in the y-direction.

Since ρu and ρv are continuous functions of x, y, and time, in general they will be different at different locations. To determine ρu at $(x + dx)$, expand ρu about the point x in the Taylor series as

$$\rho u(x + dx, y) = \rho u(x, y) + \left(\frac{\partial(\rho u)}{\partial x}\right)_{x, y} dx$$

Similarly,   (16)

$$\rho v(x, y + dy) = \rho v(x, y) + \left(\frac{\partial(\rho v)}{\partial y}\right)_{x, y} dy$$

Thus, mass leaving the right face per unit time, $\dot{M}_{x+dx}$, is given by

$$\dot{M}_{x+dx} = \left[\rho u + \frac{\partial(\rho u)}{\partial x} dx\right] \cdot dy \cdot 1 \qquad (17)$$

and mass leaving the top face per unit time, $\dot{M}_{y+dy}$, is

$$\dot{M}_{y+dy} = \left[\rho v + \frac{\partial(\rho v)}{\partial y} dy\right] \cdot dx \cdot 1 \qquad (18)$$

Finally, the rate of increase (or decrease) of mass stored in the control volume, $\dot{M}_s$, is of the form

$$\dot{M}_s = \frac{\partial}{\partial t} (\rho \cdot dx \cdot dy \cdot 1) = \frac{\partial \rho}{\partial t} dx \, dy \qquad (19)$$

Thus, conservation of mass requirement may now be expressed as

$$(\rho u) \, dy + (\rho v) \, dx - \left[\rho u + \frac{\partial(\rho u)}{\partial x} dx\right] dy$$

$$- \left[\rho v + \frac{\partial(\rho v)}{\partial y} dy\right] dx = \frac{\partial \rho}{\partial t} dx \, dy \qquad (20)$$

After canceling terms and dividing by $dx \, dy$

$$\frac{\partial \rho}{\partial t} + \frac{\partial(\rho u)}{\partial x} + \frac{\partial(\rho v)}{\partial y} = 0 \qquad (21)$$

This is the *continuity equation*, which is an expression of the overall mass conservation requirement and must be satisfied at every point in the flow. This equation applies for a single species fluid, as well as for mixtures in which species diffusion and chemical reactions may be occurring.

Conservation of momentum: The second conservation law pertinent to the convection heat transfer problem is Newton's second law of motion. For a differential control volume in a flow field, this requirement states that the sum of all forces acting on the control volume must equal the rate of increase of the fluid momentum within the control volume, plus the net rate at which momentum leaves the control volume (outflow-inflow).

The forces acting on the fluid may be categorized into *body forces* that are proportional to the volume, and *surface forces*, which are proportional to the area. Gravitational, centrifugal, magnetic, and electric fields are familiar examples of body forces. Of these, gravitational body force is the most important from the fire science point of view. The x- and y-components of this body force per unit volume of the fluid will be designated as F_{Bx} and F_{By}, respectively.

The surface forces, F_s, acting on the fluid are called stresses (force/area). These are due to fluid static pressure, p, and viscous stresses. Since pressure is always normal to the surface, the viscous stresses are also resolved into normal stresses, σ_{ii}, which act normal to the surface, and shear stresses, τ_{ij}, which act along or parallel to the surface. Figure 1-3.8 shows the various viscous stresses acting on the surface of a differential control volume. A double subscript notation is used to specify the stress components. The first subscript indicates the direction of the outward normal to the surface, and the second subscript indicates the direction of the force component. Accordingly, the stress τ_{xy}, acting on the left face, corresponds to the viscous shear force per unit area in the negative y-direction on a face whose normal is in the negative x-direction—resulting in a positive shear stress. All the viscous stresses shown in Figure 1-3.8 are

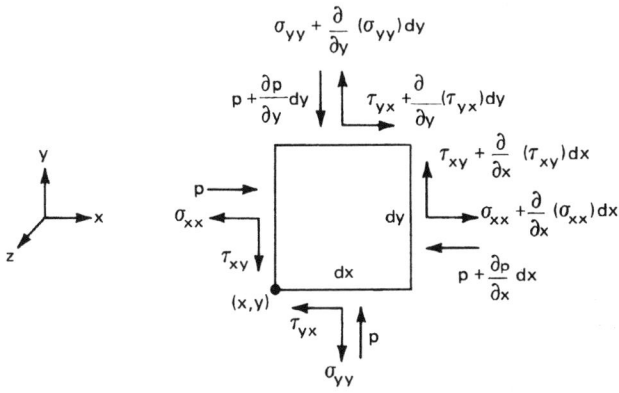

Fig. 1-3.8. *Static pressure, p, normal and shear viscous stresses acting on a differential control volume in a two-dimensional flow.*

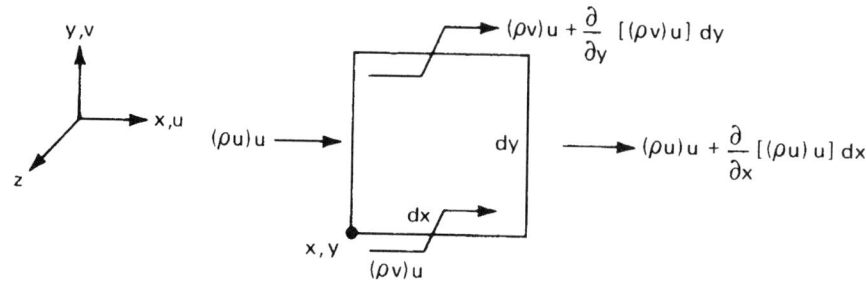

Fig. 1-3.9. *Influx and outflux of x-momentum in the control volume.*

positive according to the adopted convention. It should be noted that these forces act on the fluid inside the control volume and are caused by its interaction with the surrounding fluid. Thus, these viscous stresses will vanish if the fluid velocity, or more specifically the velocity gradient in the fluid, becomes zero. The normal viscous stresses shown in Figure 1-3.8 must not be confused with static pressure, p, which does not vanish for zero velocity. Since these stresses are continuous functions of x, y, and time, the customary Taylor's expansion is used to express the stresses on the top and right faces of the control volume shown in Figure 1-3.8. Thus, the net surface force in the x- and y-directions may be expressed as

$$F_{sx} = \left(\frac{\partial \sigma_{xx}}{\partial x} - \frac{\partial p}{\partial x} + \frac{\partial \tau_{yx}}{\partial y}\right) dx\, dy \qquad (22)$$

$$F_{sy} = \left(\frac{\partial \sigma_{yy}}{\partial y} - \frac{\partial p}{\partial y} + \frac{\partial \tau_{xy}}{\partial x}\right) dx\, dy \qquad (23)$$

To use Newton's second law, the time rate of change of momentum and the momentum influx and outflux must also be evaluated. To focus on the x-direction, the relevant momentum fluxes are shown in Figure 1-3.9. The mass flux through the left face is ρu and hence the corresponding momentum flux is $(\rho u)u$. Similarly, the x-momentum flux due to mass flow in the y-direction through the bottom face is $(\rho v)u$. Thus the net rate at which momentum leaves the control volume is given by (x-momentum outflow − inflow)

$$\frac{\partial}{\partial x}[(\rho u)u]\, dx\, dy + \frac{\partial[(\rho v)u]}{\partial y}\, dy\, dx$$

In addition, the time rate of change of x-momentum of the fluid within the control volume is given by

$$\frac{\partial}{\partial t}(\rho u)\, dx\, dy$$

Equating the total rate of change in the x-direction to the sum of forces in the x-direction, we obtain

$$\frac{\partial(\rho u)}{\partial t} + \frac{\partial[(\rho u)u]}{\partial x} + \frac{\partial[(\rho v)u]}{\partial y} = \frac{\partial \sigma_{xx}}{\partial x} - \frac{\partial p}{\partial x} + \frac{\partial \tau_{yx}}{\partial y} + F_{Bx} \qquad (24)$$

By using Equation 21, Equation 24 may be expressed in a more convenient form as

$$\rho\left(\frac{\partial u}{\partial t} + u\frac{\partial u}{\partial x} + v\frac{\partial u}{\partial y}\right) = \frac{\partial}{\partial x}(\sigma_{xx} - p) + \frac{\partial \tau_{yx}}{\partial y} + F_{Bx} \qquad (25)$$

A similar expression is obtained for the y-direction. This is

$$\rho\left(\frac{\partial v}{\partial t} + u\frac{\partial v}{\partial x} + v\frac{\partial v}{\partial y}\right) = \frac{\partial}{\partial y}(\sigma_{yy} - p) + \frac{\partial \tau_{xy}}{\partial x} + F_{By} \qquad (26)$$

In Equations 25 and 26, the first term on the left side represents the increase in momentum of the fluid inside the control volume, and the remaining terms represent the net rate of momentum efflux from the control volume. The terms on the right side of the equations account for the net viscous, pressure, and body forces acting on the control volume. These equations must be satisfied at every point in the fluid. A solution of Equations 21, 25, and 26 along with appropriate boundary conditions yields the velocity field needed to determine the friction coefficient.

Before a solution to the above equations can be obtained, it is necessary to relate the viscous stresses to the velocity gradients. For a one-dimensional flow of a Newtonian fluid, Equation 1 relates the shear stress to the velocity gradient in the fluid. For a two-dimensional flow of Newtonian fluid[2] the required stress-velocity gradient expressions are

$$\sigma_{xx} = 2\mu\frac{\partial u}{\partial x} - \frac{2}{3}\mu\left(\frac{\partial u}{\partial x} + \frac{\partial v}{\partial y}\right) \qquad (27)$$

$$\sigma_{yy} = 2\mu\frac{\partial v}{\partial y} - \frac{2}{3}\mu\left(\frac{\partial u}{\partial x} + \frac{\partial v}{\partial y}\right) \qquad (28)$$

$$\tau_{xy} = \tau_{yx} = \mu\left(\frac{\partial u}{\partial y} + \frac{\partial v}{\partial x}\right) \qquad (29)$$

On substituting Equations 27, 28, and 29 into Equations 25 and 26 the desired form of the x- and y-momentum equations is obtained. These are

x-momentum equation

$$\rho\left(\frac{\partial u}{\partial t} + u\frac{\partial u}{\partial x} + v\frac{\partial u}{\partial y}\right) = -\frac{\partial p}{\partial x} + \mu\left(\frac{\partial^2 u}{\partial x^2} + \frac{\partial^2 u}{\partial y^2}\right)$$
$$+ \frac{1}{3}\mu\frac{\partial}{\partial x}\left(\frac{\partial u}{\partial x} + \frac{\partial v}{\partial y}\right) + F_{Bx} \qquad (30)$$

y-momentum equation

$$\rho\left(\frac{\partial v}{\partial t} + u\frac{\partial v}{\partial x} + v\frac{\partial v}{\partial y}\right) = -\frac{\partial p}{\partial y} + \mu\left(\frac{\partial^2 v}{\partial x^2} + \frac{\partial^2 v}{\partial y^2}\right)$$
$$+ \frac{1}{3}\mu\frac{\partial}{\partial y}\left(\frac{\partial u}{\partial x} + \frac{\partial v}{\partial y}\right) + F_{By} \qquad (31)$$

For an isothermal system, Equations 21, 30, and 31 along with the equation of state ($p = \rho RT$ for an ideal gas) provide a complete set for determining the four dependent variables (u, v, p, and ρ) as a function of the three independent variables x, y, and t. However, for a nonisothermal system such as a fire, energy balance must also be considered.

Conservation of energy: The temperature field inside the fluid, $T(x, y, t)$, needed to determine the heat transfer coefficient is obtained by applying the first law of thermodynamics to the differential control volume shown in Figure 1-3.6. Before writing the energy balance for this control volume it is necessary to identify the items that must be included in the energy budget. These are:

1. The stored energy. This includes the specific internal or thermal energy, e; J/kg, and the kinetic energy of the fluid per unit mass, $V^2/2 = (u^2 + v^2)/2$. Potential energy is neglected because for most problems in convective heat transfer it is substantially smaller than thermal and kinetic energy. Hence, the total energy content per unit volume is given by: $\rho(e + V^2/2)$.

2. Conduction of thermal energy across the surfaces of the control volume. Here, the rate of energy transported per unit area per unit time across the control surface is given by Equation 2 as $\dot{q}_x'' = -k\,\partial T/\partial x$ for the x-direction, and $\dot{q}_y'' = -k\,\partial T/\partial y$ for the y-direction.

3. Energy generated per unit volume per unit time inside the control volume (Q watts/m³). This may be due to chemical reactions (endothermic or exothermic) or may be caused by the radiative loss of heat. Although the specific form of Q will depend upon the nature of the physical process, here we will treat it only as a rate of heat loss or gain per unit volume.

4. The rate of work done by surface or body forces. For surface forces, $\vec{F}_S$, it is given by: [$\vec{F}_S \cdot$ (velocity vector)] (surface area): and for the body forces, $\vec{F}_B$, it is given by: [$\vec{F}_B \cdot$ (velocity vector)](volume). Both these expressions have units of 'watts' or work done per unit time.

With these definitions, consider the control volume shown in Figures 1-3.6 and 1-3.10. The conservation of energy for this control volume can be simply stated as:

$$\begin{pmatrix} \text{rate of increase} \\ \text{of energy inside} \\ \text{the energy volume} \end{pmatrix} = \begin{pmatrix} \text{net rate of energy flow} \\ \text{into the control volume} \\ \text{by bulk fluid motion} \end{pmatrix}$$

$$+ \begin{pmatrix} \text{flow of heat through} \\ \text{the control surface} \\ \text{by conduction} \end{pmatrix}$$

$$+ \begin{pmatrix} \text{rate of work done by body} \\ \text{and surface forces} \end{pmatrix}$$

$$+ \begin{pmatrix} \text{energy generated inside} \\ \text{the control volume} \end{pmatrix} \quad (32)$$

In Equation 32 the rate o f increase of energy inside the control volume is given by

$$\frac{\partial}{\partial t}\left[\rho\left(e + \frac{V^2}{2}\right)\right] dx\, dy$$

The net rate at which the energy enters the control volume by convection or bulk fluid motion is obtained by subtracting the energy going out from that coming in, to yield

$$-\frac{\partial}{\partial x}\left[\rho u\left(e + \frac{V^2}{2}\right)\right] dx\, dy - \frac{\partial}{\partial y}\left[\rho v\left(e + \frac{V^2}{2}\right)\right] dx\, dy$$

Similarly, the heat flowing into the control volume by conduction is given by

$$-\left(\frac{\partial \dot{q}_x''}{\partial x} + \frac{\partial \dot{q}_y''}{\partial y}\right) dx\, dy = \left[\frac{\partial}{\partial x}\left(k\frac{\partial T}{\partial x}\right) + \frac{\partial}{\partial y}\left(k\frac{\partial T}{\partial y}\right)\right] dx\, dy$$

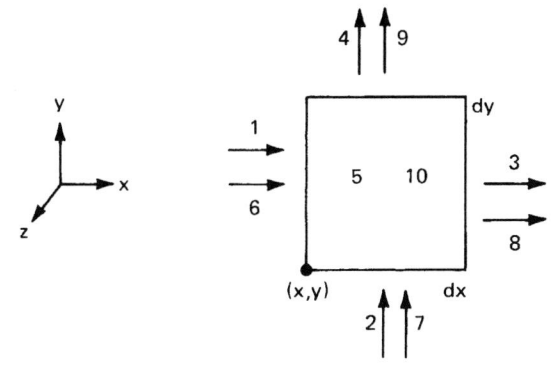

1. $\rho\left(e + \dfrac{V^2}{2}\right)u \cdot dy \cdot 1$

2. $\rho\left(e + \dfrac{V^2}{2}\right)v \cdot dx \cdot 1$

3. $\left\{\rho\left(e + \dfrac{V^2}{2}\right)u + \dfrac{\partial}{\partial x}\left[\rho u\left(e + \dfrac{V^2}{2}\right)\right]dx\right\} dy \cdot 1$

4. $\left\{\rho\left(e + \dfrac{V^2}{2}\right)v + \dfrac{\partial}{\partial x}\left[\rho v\left(e + \dfrac{V^2}{2}\right)\right]dy\right\} dx \cdot 1$

5. $\dfrac{\partial}{\partial t}\left[\rho\left(e + \dfrac{V^2}{2}\right)dx \cdot dy \cdot 1\right]$

6. $\dot{q}_x'' \cdot dy \cdot 1$

7. $\dot{q}_y'' \cdot dx \cdot 1$

8. $\left(\dot{q}_x'' + \dfrac{\partial \dot{q}_x''}{\partial x}dx\right)dy \cdot 1$

9. $\left(\dot{q}_y'' + \dfrac{\partial \dot{q}_y''}{\partial y}dy\right)dx \cdot 1$

10. Energy generated $\quad Q\,dx\,dy \cdot 1$

Fig. 1.3-10. A control volume showing the energy conducted and convected through its control surfaces.

Finally, the net rate at which work is done on the fluid inside the control volume (Figure 1-3.11) is given by the expression

$$(uF_{Bx} + vF_{By})\,dx\,dy - \left[\frac{\partial(pu)}{\partial x} + \frac{\partial(pv)}{\partial y}\right]dx\,dy$$

$$+ \left[\frac{\partial}{\partial x}(u\sigma_{xx}) + \frac{\partial}{\partial y}(v\sigma_{yy}) + \frac{\partial}{\partial y}(u\tau_{yx}) + \frac{\partial}{\partial x}(v\tau_{xy})\right]dx\,dy$$

On substituting these expressions into Equation 32, simplifying and using Equations 27 through 31 we obtain

$$\rho C_p \frac{\partial T}{\partial t} + \rho u C_p \frac{\partial T}{\partial x} + \rho v C_p \frac{\partial T}{\partial y} = \frac{\partial}{\partial x}\left(k\frac{\partial T}{\partial x}\right) + \frac{\partial}{\partial y}\left(k\frac{\partial T}{\partial y}\right)$$

$$+ \left(\frac{\partial p}{\partial t} + u\frac{\partial p}{\partial x} + v\frac{\partial p}{\partial y}\right) + \mu\Phi \quad (33)$$

In Equation 33, the thermodynamic definition of enthalpy ($i = e + p/\rho$ and $di = C_p dT$) has been used. Also, the term $\mu\Phi$ is called the viscous dissipation and is given by:

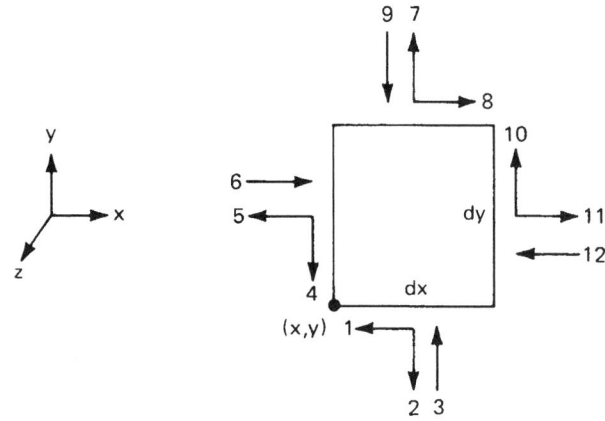

1. $-u\tau_{yx}\,dx$

2. $-v\sigma_{yy}\,dx$

3. $\rho v\,dx$

4. $-v\tau_{zy}\,dy$

5. $-u\sigma_{xx}\,dy$

6. $\rho u\,dy$

7. $\left[v\sigma_{yy} + \dfrac{\partial}{\partial y}\,(v\sigma_{yy})\,dy\right]dx$

8. $\left[u\tau_{yx} + \dfrac{\partial}{\partial y}\,(u\tau_{yx})\,dy\right]dx$

9. $\left[-\rho v - \dfrac{\partial}{\partial y}\,(\rho v)\,dy\right]dx$

10. $\left[v\tau_{xy} + \dfrac{\partial}{\partial x}\,(v\tau_{xy})\,dx\right]dy$

11. $\left[u\sigma_{xx} + \dfrac{\partial}{\partial x}\,(u\sigma_{xx})\,dx\right]dy$

12. $\left[-\rho u - \dfrac{\partial}{\partial x}\,(\rho u)\,dx\right]dy$

Fig. 1-3.11.   Control volume showing the work done by various surface forces. All units are watts.

$$u\Phi = 2\mu\left[\left(\frac{\partial u}{\partial x}\right)^2 + \left(\frac{\partial v}{\partial y}\right)^2\right] + \left(\frac{\partial u}{\partial y} + \frac{\partial v}{\partial x}\right)^2$$
$$-\frac{2}{3}\left(\frac{\partial u}{\partial x} + \frac{\partial v}{\partial y}\right)^2 \quad (34)$$

Equations 21, 30, 31, and 33, along with the equation of state, $(p = \rho RT)$, provide a complete set for determining the temperature and velocity field $[T(x, y, t), u(x, y, t), v(x, y, t)]$ inside the fluid. However, it is not possible to solve the above set of coupled nonlinear partial differential equations. Therefore, several simplifying approximations are made. These are discussed below.

Simplifications:

1. *Low velocity*: For most problems encountered in convective heat transfer, the flow velocity is low enough (Mach

$\# < \frac{1}{3}$) to ignore the contribution of viscous work in the energy equation. This allows the term $\mu\Phi$ in Equation 33 to be dropped.

2. *Incompressible flow*: Fluid density is assumed to be constant except in the buoyancy terms (F_{Bx}, F_{By}) of Equations 30 and 31. This is called the Boussinesq approximation and will be discussed later in greater detail.

3. *Steady flow*: This approximation allows all the time derivative terms in the above equations to be dropped.

4. *Constant properties*: Specific heat, thermal conductivity, and viscosity are all assumed to be constant; i.e., independent of temperature and pressure.

With these simplifications and assuming that the body force is only due to gravity (i.e., $F_{Bx} = -\rho g_x$ and $F_{By} = -\rho g_y$), Equations 21, 30, 31, and 33 become

Continuity

$$\frac{\partial u}{\partial x} + \frac{\partial v}{\partial y} = 0 \quad (35)$$

x-momentum

$$u\,\frac{\partial u}{\partial x} + v\,\frac{\partial u}{\partial y} = -\frac{1}{\rho}\,\frac{\partial p}{\partial x} + v\left(\frac{\partial^2 u}{\partial x^2} + \frac{\partial^2 u}{\partial y^2}\right) - g_x \quad (36)$$

y-momentum

$$u\,\frac{\partial v}{\partial x} + v\,\frac{\partial v}{\partial y} = -\frac{1}{\rho}\,\frac{\partial p}{\partial y} + v\left(\frac{\partial^2 v}{\partial x^2} + \frac{\partial^2 v}{\partial y^2}\right) - g_y \quad (37)$$

Energy equation

$$u\,\frac{\partial T}{\partial x} + v\,\frac{\partial T}{\partial y} = \alpha\left(\frac{\partial^2 T}{\partial x^2} + \frac{\partial^2 T}{\partial y^2}\right) + \frac{1}{\rho C_p}\left(u\,\frac{\partial p}{\partial x} + v\,\frac{\partial p}{\partial y}\right) \quad (38)$$

Often, the energy equation is further simplified by assuming that the terms $(u\,\partial p/\partial x)$ and $(v\,\partial p/\partial y)$ are negligible. This assumption is justified since most processes of interest are nearly isobaric. Thus the energy equation becomes

$$u\,\frac{\partial T}{\partial x} + v\,\frac{\partial T}{\partial y} = \alpha\left(\frac{\partial^2 T}{\partial x^2} + \frac{\partial^2 T}{\partial y^2}\right). \quad (39)$$

Equations 35, 36, 37, and 39, along with the equation of state, provide a complete set for determining $u(x, y)$, $v(x, y)$, $T(x, y)$, $\rho(x, y)$, and $p(x, y)$. Once these dependent variables are known, the desired heat transfer coefficient and friction factor are obtained from Equations 5 and 11, respectively. However, the above equations are still too difficult to solve and a further simplification, known as the boundary layer approximation, is often made.

The Boundary Layer Concept

In 1904, Prandtl proposed that all the viscous effects are concentrated in a thin layer near the boundary and that outside this layer the fluid behaves as though it is inviscid. Thus, the flow over a body, such as the one shown in Figure 1-3.6, can be divided into two zones: (1) a thin viscous layer near the surface, called the boundary layer; and (2) inviscid external flow, which can be closely approximated by the potential flow theory. As will be seen later, the fact that the boundary layer is thin compared to the characteristic dimensions of the object is exploited to simplify the governing equations and obtain a useful solution. This boundary layer approximation plays an important role in convective heat transfer, since the gradients of velocity and temperature at

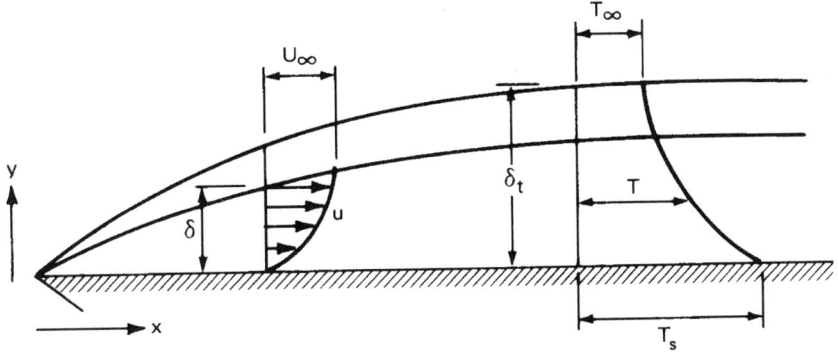

Fig. 1-3.12. Velocity and thermal boundary layers on a flat plate.

the surface of the body are required to determine the heat transfer coefficient and the friction factor.

To illustrate these ideas, consider fluid flow over a flat plate as shown in Figure 1-3.12. The fluid particles in contact with the plate surface must assume zero velocity because of no slip at the wall, whereas the fluid particles far away from the wall continue to move at the free stream velocity, u_∞. The transition of fluid velocity from zero to u_∞ takes place in a small distance, δ, which is known as the boundary layer thickness and is defined as the value of y for which $u = 0.99\,u_\infty$. As is intuitively obvious, the thickness of the boundary layer increases with fluid viscosity and decreases with increasing free-stream velocity. By defining the Reynolds number, Re, as Re $= u_\infty L/v$, where L is the characteristic length of the plate, the boundary layer thickness decreases with increasing Re. For most flows of practical interest, the Reynolds number is large enough such that δ is much less than the characteristic length, L $(\delta << L)$.

Just as a velocity boundary layer develops because of viscous effects near the surface, a thermal boundary layer develops due to heat transfer between the free stream and the surface if their temperatures are different. The fluid particles that come into contact with the plate surface achieve thermal equilibrium at the plate's surface temperature. In turn, these particles exchange energy with those in the adjoining fluid layer, and temperature gradients develop in the fluid. As shown in Figure 1-3.12, the region of the fluid in which these temperature gradients exist is the thermal boundary layer, and its thickness, δ_t, is defined as the value of y for which the ratio $[(T - T_S)/(T_\infty - T_S)] = 0.99$. The thermal boundary layer thickness increases with the thermal diffusivity, α, of the fluid and decreases with increasing free stream velocity. In other words, δ_t is inversely proportional to the product of Reynolds number and Prandtl number [Re Pr $= (u_\infty L/v)(v/\alpha) = u_\infty L/\alpha$]. For air, Pr ≈ 0.7 and the Reynolds number is sufficiently large for flows of practical interest, consequently $\delta_t << L$.

Boundary layer approximation: The governing Equations 35 through 37, and 39 can be further simplified for the case when the Reynolds number is reasonably large [Re $\sim (L/\delta)^2$; i.e., Re is of the order $(L/\delta)^2$] such that $\delta << L$. To compare the various terms in the governing equations, first normalize all the variables so that they are of the order of magnitude unity. By defining

$$x^* = x/L; \qquad y^* = y/\delta; \qquad u^* = u/u_\infty$$

and

$$T^* = \frac{T - T_S}{T_\infty - T_S} \qquad (40)$$

variables that change from 0 to 1 inside the boundary layer are obtained. Substituting these into Equation 35 we find that

$$-\frac{\partial u^*}{\partial x^*} = \left(\frac{L}{\delta u_\infty}\right)\frac{\partial v}{\partial y^*}$$

This suggests that

$$v^* = \frac{Lv}{\delta u_\infty}$$

so that

$$\frac{\partial u^*}{\partial x^*} + \frac{\partial v^*}{\partial y^*} = 0 \qquad (41)$$

Substituting x^*, y^*, u^*, and v^* into Equation 36 and simplifying

$$u^* \frac{\partial u^*}{\partial x^*} + v^* \frac{\partial u^*}{\partial y^*} = -\frac{\partial p^*}{\partial x^*} - g_x^*$$
$$+ \left(\frac{v}{Lu_\infty}\right)\left[\frac{\partial^2 u^*}{\partial x^{*2}} + \left(\frac{L}{\delta}\right)^2 \frac{\partial^2 u^*}{\partial y^{*2}}\right] \quad (42)$$

where $p^* \equiv p/\rho u_\infty^2$ and $g_x^* \equiv g_x L/u_\infty^2$.

In Equation 42, the quantity v/Lu_∞ is recognized as $1/\mathrm{Re}$ which is of the order (δ/L^2). Thus all terms in Equation 42 are of order of magnitude unity except the term $[(v/Lu_\infty) \partial^2 u^*/\partial x^{*2}]$, which is much less than 1 and can be ignored. Thus, Equation 36 is simplified to

$$u \frac{\partial u}{\partial x} + v \frac{\partial u}{\partial y} = -\frac{1}{\rho}\frac{\partial p}{\partial x} - g_x + v \frac{\partial^2 u}{\partial y^2} \qquad (43)$$

Similarly, Equations 37 and 39 reduce to

$$\frac{\partial p}{\partial y} \approx 0 \qquad (44)$$

and

$$u \frac{\partial T}{\partial x} + v \frac{\partial T}{\partial y} = \alpha \frac{\partial^2 T}{\partial y^2} \qquad (45)$$

Equation 44 simply implies that $p = p(x)$, i.e., the pressure at any plane where x = constant does not vary with y inside the boundary layer and hence is equal to the free stream pressure. To summarize, the boundary layer approximation yields a simpler set of governing equations that are valid inside the boundary layer. These equations for steady flow of an incompressible fluid with constant properties are:

Continuity

$$\frac{\partial u}{\partial x} + \frac{\partial v}{\partial y} = 0 \tag{35}$$

x-momentum

$$u\frac{\partial u}{\partial x} + v\,\frac{\partial u}{\partial y} = -\frac{1}{\rho}\,\frac{\partial p}{\partial x} - g_x + v\,\frac{\partial^2 u}{\partial y^2} \tag{43}$$

Energy

$$u\,\frac{\partial T}{\partial x} + v\,\frac{\partial T}{\partial y} = \alpha\,\frac{\partial^2 T}{\partial y^2} \tag{45}$$

To illustrate the use of these equations in determining the heat transfer coefficient, consider two classical examples: (1) laminar forced convection over a flat surface; and (2) laminar free convection on a vertical flat surface. Forced convection is chosen as a precursor to free convection because it is simpler and also allows us to illustrate the difference between them. A flat geometry is also chosen in both cases for simplicity.

Laminar forced convection over a flat surface: A schematic of this problem is presented in Figure 1-3.12, and the objective here is to obtain the gradients of temperature and velocity profile at $y = 0$. By applying the Bernoulli Equation in the potential flow region outside the boundary layer we obtain

$$\frac{u_\infty^2}{2} + \frac{p}{\rho} + gh = \text{constant} \tag{46}$$

Since the free stream velocity, u_∞, is constant, for a given height $y = h$ above the flat surface we obtain that $p =$ constant, i.e., $p \ne p(x)$ outside the boundary layer in the potential flow region. From Equation 44 note that $p \ne p(y)$ inside the boundary layer. Hence, $p =$ constant both inside and outside the boundary layer over a flat surface. This implies that the term $\partial p/\partial x$ equals zero in Equation 43. Also, since the flow is forced (i.e., generated by an external agent such as a fan, rather than by buoyancy) the gravitational force, g_x, in Equation 43 does not contribute to the increase in momentum represented by the left side of the equation, and $g_x = 0$. Thus Equation 43 becomes

$$u\,\frac{\partial u}{\partial x} + v\,\frac{\partial u}{\partial y} = v\,\frac{\partial^2 u}{\partial y^2} \tag{47}$$

Equations 35, 47, and 45 govern the temperature and velocity distributions inside the boundary layer shown in Figure 1-3.12. The associated boundary conditions are:

no-slip $u = v = 0 \quad \text{at} \quad y = 0$

and $T = T_S \quad \text{at} \quad y = 0 \tag{48}$

also $u = u_\infty \quad \text{and} \quad T = T_\infty \quad \text{as} \quad y \to \infty$

Nondimensionalizing Equations 35, 45, 47, and 48 according to Equation 40† we obtain

---

† A more convenient definition of $y^* = y/L$ and $v^* = v/u_\infty$ has been used since we are no longer interested in quantities of order of magnitude unity; instead we are simply interested in eliminating units.

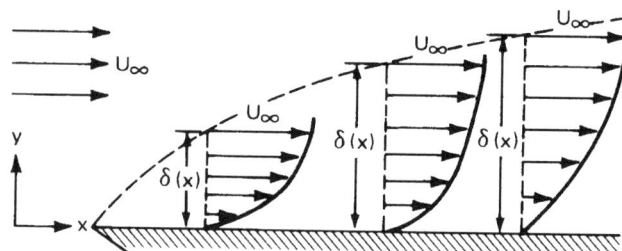

Fig. 1-3.13.   *Observed velocity profiles at different values of x.*

$$\frac{\partial u^*}{\partial x^*} + \frac{\partial v^*}{\partial y^*} = 0 \tag{49}$$

$$u^*\,\frac{\partial u^*}{\partial x^*} + v^*\,\frac{\partial u^*}{\partial y^*} = \frac{1}{\text{Re}_L}\,\frac{\partial^2 u^*}{\partial y^{*2}} \tag{50}$$

$$u^*\,\frac{\partial T^*}{\partial x^*} + v^*\,\frac{\partial T^*}{\partial y^*} = \frac{1}{\text{Re}_L\,\text{Pr}}\,\frac{\partial^2 T^*}{\partial y^{*2}} \tag{51}$$

along with the boundary conditions

$$u^* = v^* = T^* = 0 \quad \text{at} \quad y^* = 0$$

and $u^* = T^* = 1 \quad \text{as} \quad y^* \to \infty \tag{52}$

where

$$\text{Re}_L \equiv \frac{u_\infty L}{v}$$

is the Reynolds number based on length, L, and $\text{Pr} \equiv v/\alpha$ is the Prandtl number. Note that Equations 49 and 50 are sufficient for determining $u^*(x^*, y^*)$ and $v^*(x^*, y^*)$ and that once these are known, Equation 51 can be independently solved for $T^*(x^*, y^*)$. Also note that for $\text{Pr} = 1$, Equations 50 and 51 as well as their corresponding boundary conditions are identical. Thus for $\text{Pr} = 1$ only Equations 49 and 50 need to be solved.

A similarity solution of Equations 49 and 50 along with the boundary conditions (Equation 52) was obtained by Blasius.[2] Blasius observed that since the system under consideration has no preferred length, it is reasonable to suppose that the velocity profiles at different values of x have similar shapes; i.e., if u and y are suitably scaled then the velocity profile may be expressed by a single function for all values of x. (See Figure 1-3.13.) An obvious choice is

$$\frac{u}{u_\infty} = \phi[y/\delta(x)] \tag{53}$$

This choice, as it stands, is not very useful because $\delta(x)$ is not known. However, in accordance with the boundary layer approximation, $\text{Re}_x \equiv u_\infty x/v \sim (x/\delta)^2$. Therefore,

$$\delta \sim \sqrt{vx/u_\infty}$$

can be expected. Substituting into Equation 53 we obtain

$$u^* = u/u_\infty = \phi\!\left[\frac{y}{x}\sqrt{\text{Re}_x}\right] = \phi(\eta) \tag{54}$$

where

$$\eta \equiv \frac{y}{x}\sqrt{\text{Re}_x} = [(y^*/\sqrt{x^*})\sqrt{\text{Re}_L}]$$

is the similarity variable. By introducing a stream function, ψ, such that

$$u^* = \frac{\partial \psi}{\partial y^*} \quad \text{and} \quad v^* = -\frac{\partial \psi}{\partial x^*} \quad (55)$$

Equation 49 is identically satisfied. Substituting Equation 54 into Equation 55 and integrating, we get

$$\psi = \int \phi(\eta)\, dy^* + f_1(x^*)$$
$$= \frac{\sqrt{x^*}}{\sqrt{Re_L}} \int \phi(\eta)\, d\eta + f_1(x^*).$$

Since $v^* = 0$ at $y^* = 0$, $f_1(x^*)$ is at best an arbitrary constant which is taken as zero. Also, defining a new function $f(\eta) \equiv \int \phi(\eta)\, d\eta$, we obtain

$$\psi = \sqrt{x^*}\, f(\eta)/\sqrt{Re_L} \quad (56)$$

therefore

$$u^* = \frac{\partial \psi}{\partial y^*}\bigg)_{x^*} = \frac{\partial \psi}{\partial \eta}\bigg)_{x^*} \frac{\partial \eta}{\partial y^*}\bigg)_{x^*} = f'(\eta) = \frac{df}{d\eta} \quad (57)$$

and

$$-v^* = \left(\frac{\partial \psi}{\partial x^*}\right)_{y^*} = \left(\frac{\partial \psi}{\partial x^*}\right)_\eta + \left(\frac{\partial \psi}{\partial \eta}\right)_{x^*}\left(\frac{\partial \eta}{\partial x^*}\right)_{y^*}$$
$$= \frac{1}{2}(f - \eta f')\sqrt{x^* \, Re_L} \quad (58)$$

On substituting u^*, v^* into Equation 50 and simplifying we obtain

$$2f''' + ff'' = 0 \quad (59)$$

where primes represent differentiation with respect to η. Equation 59 is a third-order nonlinear ordinary differential equation. Recall that η was a combination of two independent variables, x^* and y^*, and it was assumed that $u^* = \phi(\eta)$. If this similarity assumption was incorrect, then the partial differential Equation 50 would not have reduced to an ordinary differential Equation 59—i.e., x^* would not have completely disappeared from the governing equation. Note also that even though Equation 59 is nonlinear and has to be solved numerically, there are no parameters and therefore it needs to be solved only once. Boundary conditions corresponding to Equation 59 become

$$f = f' = 0 \quad \text{at} \quad \eta = 0, \quad \text{and} \quad f' = 1 \quad \text{as} \quad \eta \to \infty \quad (60)$$

A numerical solution of Equation 59 along with the boundary conditions, Equation 60 is shown in Figure 1-3.14. Note that for Pr = 1, the solution for T^* is the same as that for u^*. Also, once $T^*(x^*, y^*)$ and $u^*(x^*, y^*)$ are known the heat transfer coefficient and friction factor can easily be obtained from Equations 5 and 11. Furthermore, from the definition of thermal and velocity boundary layer thickness ($T^* = u^* = 0.99$), we find that $\eta = 5$.
Therefore

$$\eta = \frac{y}{x}\sqrt{\frac{u_\infty x}{\nu}} = 5 \quad \text{for} \quad y = \delta = \delta_t$$

or for Pr = 1

$$\delta = \delta_t = \frac{5x}{\sqrt{Re_x}} \quad (61)$$

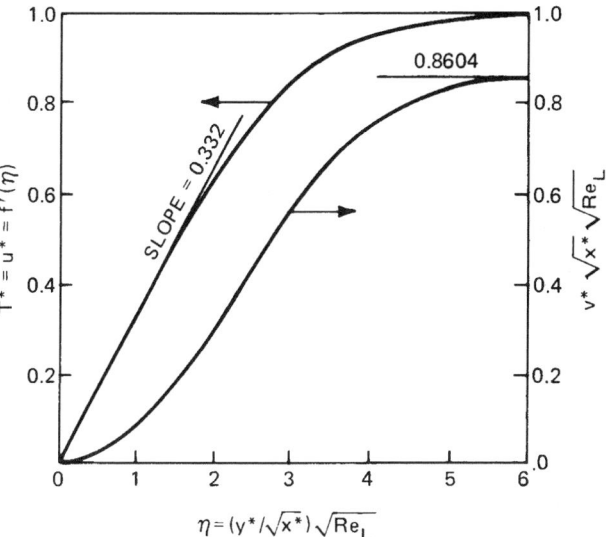

Fig. 1-3.14. *Nondimensional velocity profiles in laminar boundary layer over a flat plate.*

From Equation 61 it is clear that δ and δ_t increase with x but decrease with increasing u_∞ (the larger the free stream velocity, the thinner the boundary layer). Now, to determine the heat transfer coefficient and the friction factor we need $\partial T/\partial y)_{y=0}$ and $\partial u/\partial y)_{y=0}$. From Figure 1-3.14, we have

$$\frac{\partial T^*}{\partial \eta}\bigg|_{\eta=0} = \frac{\partial u^*}{\partial \eta}\bigg|_{\eta=0} = 0.332$$

Thus,

$$\tau_s = \mu\,\frac{\partial u}{\partial y}\bigg|_{y=0} = 0.332 u_\infty\,\sqrt{\rho\mu u_\infty/x} \quad (62)$$

and

$$\dot{q}_s'' = -k\,\frac{\partial T}{\partial y}\bigg|_{y=0} = 0.332(T_s - T_\infty)k\sqrt{\rho u_\infty/\mu x} \quad (63)$$

Hence the *local* friction and heat transfer coefficients are

$$C_f = \frac{0.664}{\sqrt{Re_x}} \quad (64)$$

and

$$h = 0.332\,\frac{k}{x}\sqrt{Re_x} \quad (65)$$

Equation 65 is often rewritten in terms of a nondimensional heat transfer coefficient called the Nusselt number, Nu, as

$$Nu = \frac{hx}{k} = 0.332\sqrt{Re_x} \quad (66)$$

All the above results are for the case when Pr = 1. When Pr $\neq$ 1, Equation 51 must also be solved with the help of the solution just obtained for Equations 49 and 50. This solution does not change the expressions for δ and C_f given by Equations 61 and 64. However δ_t and Nu become[2]

$$\delta_t = \frac{5x}{\sqrt{Re_x}}\,Pr^{-1/3} = \delta\,Pr^{-1/3} \quad (67)$$

and

$$Nu = 0.332\sqrt{Re_x}\,Pr^{1/3}(Pr \gtrsim 0.6) \quad (68)$$

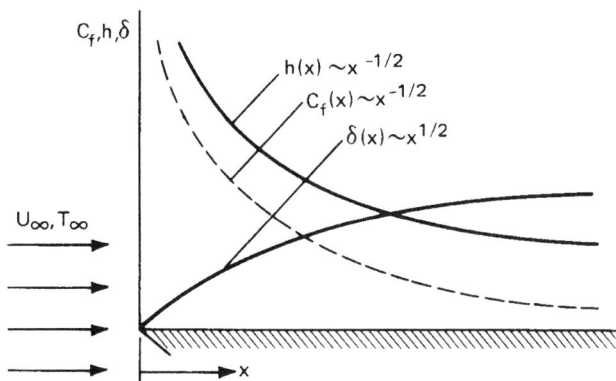

Fig. 1-3.15. *Variation of C_f, h, and δ with x for flow over a flat plate.*

Note that for Pr < 1 (usually true for gases) $\delta_t > \delta$, i.e., the thermal boundary layer is thicker than the momentum boundary layer. This is to be expected since Pr < 1 implies that $v < \alpha$.

The results for the friction factor, C_f, and the Nusselt number, Nu, given by Equations 64 and 68 are for local values; i.e., C_f and Nu change with x. This variation is shown in Figure 1-3.15. At $x = 0$, both C_f and h tend to infinity. This is physically incorrect and happens because near $x = 0$ the boundary layer approximation breaks down since δ is no longer much less than x.

For many applications, only average values of the heat transfer coefficient, $\bar{h}$, and friction factor, $\bar{C}_f$, are required. These are obtained by using Equations 9 and 15. In these equations $dA_s = dx \cdot$ (the unit width of the flat plate), and the average can be obtained from $x = 0$ to any length, L (which may be the total length of the plate). Simple integration leads to the following results

$$\bar{C}_{fL} = 1.328 \, \text{Re}_L^{-1/2} = 2C_f \left(\begin{array}{c}\text{evaluated}\\\text{at } x=L\end{array}\right) \qquad (69)$$

and

$$\overline{\text{Nu}}_L \equiv \frac{\bar{h}_L L}{k} = 0.664 \, \text{Re}_L^{1/2} \text{Pr}^{1/3} = 2\text{Nu}\left(\begin{array}{c}\text{evaluated}\\\text{at } x=L\end{array}\right) \qquad (70)$$

It is interesting to note that C_f and Nu are closely related. For example, from Equations 69 and 70 one can easily obtain

$$\overline{\text{Nu}}_L = \frac{C_{fL} \, \text{Re}_L \, \text{Pr}^{1/3}}{2}$$

or

$$\text{St} \equiv \frac{\overline{\text{Nu}}_L}{\text{Re}_L \, \text{Pr}} = \frac{\bar{C}_{fL}}{2} \, \text{Pr}^{-2/3} \qquad (71)$$

where St is known as the Stanton number.

This analogy between heat and momentum transfer is called the *Reynolds analogy* which is significant because the heat transfer coefficient can be determined from the knowledge of the friction factor. This analogy is especially useful for cases where mathematical solutions are not available.

Laminar free convection: In contrast with forced convection, where the fluid motion is externally imposed, for free convection the fluid motion is caused by the buoyancy forces. Buoyancy is due to the combined effect of *density gradients* within the fluid and a *body force* that is proportional to the fluid density. In practice the relevant body force

is usually gravitational, although it may be centrifugal, magnetic, or electric. Of the several ways in which a density gradient may arise in a fluid, the two most common situations are due to: (1) the presence of temperature gradients; and (2) the presence of concentration gradients in a multicomponent system such as a fire. Here, the focus will be on free convection problems in which the density gradient is due to temperature and the body force is gravitational. Note, however, that the presence of density gradients in a gravitational field does not ensure the existence of free convection currents. For example, the high temperature lighter fluid may be on top of a low temperature, denser fluid, resulting in a stable situation. It is only when the condition is unstable that convection currents are generated. An example of an unstable situation would be a denser fluid on top of a lighter fluid. In a stable situation there is no fluid motion and, therefore, heat transfer occurs purely by conduction. Here we will only consider the unstable situation that results in convection currents.

Free convection flow may be further classified according to whether or not the flow is bounded by a surface. In the absence of an adjoining surface, free boundary flows may occur in the form of a plume or a buoyant jet. A buoyant plume above a fire is a familiar example. However, here we will focus on free convection flows that are bounded by a surface. A classical example of boundary layer development on a heated vertical flat plate, is discussed below.

Heated, vertical flat plate: Consider the flat plate shown in Figure 1-3.16. The plate is immersed in an extensive, quiescent fluid, with $T_s > T_\infty$. The density of the fluid close to the plate is less than that of the fluid that is farther from the plate. Buoyancy forces therefore induce a free convection

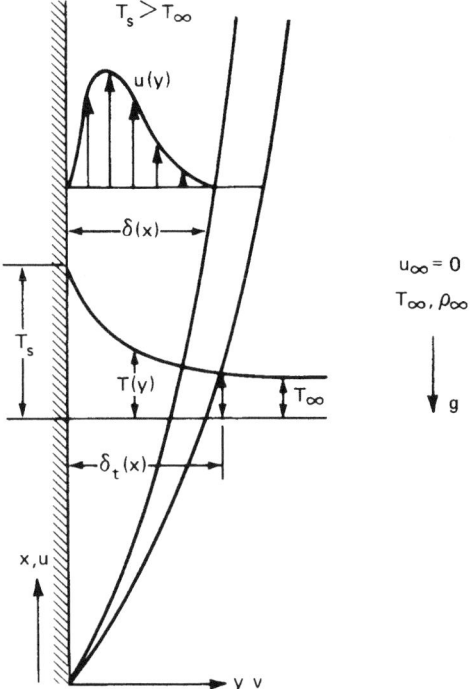

Fig. 1-3.16. *Boundary layer development on a heated vertical plate.*

boundary layer in which the heated fluid rises vertically, entraining fluid from the quiescent region.

Under steady-state laminar flow conditions, Equations 35, 43, and 45 describe the mass, momentum, and energy balances for the two-dimensional boundary layer shown in Figure 1-3.16. Assume that the temperature differences are moderate, such that the fluid may be treated as having constant properties. Also, with the exception of the buoyancy force term (g_x in Equation 43), the fluid can be assumed to be incompressible. Outside the boundary layer, Equation 36 is valid, and since $u = v = 0$ outside the boundary layer we obtain: $\partial p/\partial x = -\rho_\infty g_x$. Since $\partial p/\partial y = 0$ because of the boundary layer approximation [i.e., $p \neq p(y)$ inside the boundary layer; Equation 44], $\partial p/\partial x$ inside the boundary layer must be equal to its corresponding value outside, i.e., $\partial p/\partial x = -\rho_\infty g_x$.

Substituting this into Equation 43

$$u \frac{\partial u}{\partial x} + v \frac{\partial u}{\partial y} = g_x \frac{(\rho_\infty - \rho)}{\rho} + v \frac{\partial^2 u}{\partial y^2} \quad (72)$$

The first term on the right side of Equation 72 is the buoyancy force, and the flow originates because the density ρ is variable. By introducing the coefficient of volumetric thermal expansion, β,

$$\beta = -\frac{1}{\rho} \left(\frac{\partial \rho}{\partial T} \right)_\rho \approx -\frac{1}{\rho} \frac{(\rho_\infty - \rho)}{(T_\infty - T)} \quad (73)$$

it follows that

$$\frac{(\rho_\infty - \rho)}{\rho} = \beta(T - T_\infty) \quad (74)$$

Substituting Equation 74 into Equation 72 a useful form of the x-momentum is obtained as

$$u \frac{\partial u}{\partial x} + v \frac{\partial u}{\partial y} = g_x \beta(T - T_\infty) + v \frac{\partial^2 u}{\partial y^2} \quad (75)$$

From Equation 75 it is now apparent how buoyancy force is related to temperature difference. Note that the appearance of the buoyancy term in the momentum equation mathematically complicates the situation. The decoupling between the hydrodynamic and the thermal problems achieved in forced convection is no longer possible, since T appears in both Equations 45 and 75. The boundary conditions associated with the governing equations, Equations 35, 45, and 75, are

$$u = v = 0. \quad T = T_\infty, \quad as \quad y \to \infty \quad (76)$$

Nondimensionalizing Equations 35, 45, 75, and 76 with $x^* = x/L$, $y^* \equiv y/L$, $u = u/u_0$, $v^* \equiv v/u_0$ and $T^* \equiv (T - T_\infty)/(T_s - T_\infty)$, we obtain

$$\frac{\partial u^*}{\partial x^*} + \frac{\partial v^*}{\partial y^*} = 0 \quad (77)$$

$$u^* \frac{\partial u^*}{\partial x^*} + v^* \frac{\partial u^*}{\partial y^*} = \frac{g\beta T^*(T_s - T_\infty)L}{u_0^2} + \frac{1}{Re_L} \frac{\partial^2 u^*}{\partial y^{*2}} \quad (78)$$

and

$$u^* \frac{\partial T^*}{\partial x^*} + v^* \frac{\partial T^*}{\partial y^*} = \frac{1}{Re_L Pr} \frac{\partial^2 T^*}{\partial y^{*2}} \quad (79)$$

Note that u_0 in Equation 78 is an unknown reference velocity and not the free stream velocity as in the case of forced convection. Also, the dimensionless parameter

$$\frac{g\beta(T_s - T_\infty)L}{u_0^2}$$

is a direct result of buoyancy forces. To eliminate the unknown reference velocity, u_0 from the dimensionless parameter, we define

$$\text{Grashof Number, } Gr_L \equiv \frac{g\beta(T_s - T_\infty)L}{u_0^2} \left(\frac{Lu_0}{v} \right)^2 = \frac{g\beta(T_s - T_\infty)L^3}{v^2}$$

Thus, the first term on the right side of Equation 78 becomes $Gr_L/(Re_L)^2$. The Grashof number plays the same role in free convection as the Reynolds number does in forced convection. Gr is the ratio of buoyancy and viscous forces. The governing equations now contain three parameters—the Grashof number, Reynolds number, and Prandtl number. For the forced convection case it is seen (Equation 68) that Nu = Nu (Re, Pr); thus for the free convection case, we expect Nu = Nu (Re, Gr, Pr). If the buoyancy term in Equation 79 is $Gr/(Re)^2 >> 1$, then we primarily have free convection; i.e., Nu = Nu(Gr, Pr). For $Gr/(Re)^2 << 1$, the forced convective case exists, where as has already been seen, Nu = Nu (Re, Pr). However, when $Gr/(Re)^2 \sim 1$ a mixed (free and forced) convection case is obtained. For the present problem we will assume that $Gr >> (Re)^2$, thus, Nu must be a function of only Gr and Pr.

Since $Gr >> Re^2$, it follows that buoyancy forces are much larger than inertia forces; in other words, the primary balance is between the buoyancy and viscous forces. Since the left side of Equation 78 represents the inertia forces, the primary balance is between the two terms on the right side; i.e.,

$$-\frac{g\beta T^*(T_s - T_\infty)L}{u_0^2} \approx \left(\frac{v}{u_0 L} \right) \frac{\partial^2 u^*}{\partial y^{*2}}$$

Crudely approximating the various terms, we have in dimensional variables

$$g\beta(T_\infty - T) \approx v \frac{u}{\delta^2} \quad (a)$$

Similarly approximating Equations 77 and 79 and expressing the result in dimensional form, (it is more convenient to use Equations 35 and 45) we get from Equation 35 or 77

$$\frac{u}{x} \approx \frac{v}{\delta} \quad or \quad v \approx \frac{\delta u}{x} \quad (b)$$

and from Equation 79 or 45 along with relation (b)

$$u \frac{(T_\infty - T)}{x} \approx \alpha \frac{(T_\infty - T)}{\delta^2} \quad or \quad u \approx \frac{\alpha x}{\delta^2} \quad (c)$$

Combining (a) and (c) we obtain an expression for the boundary layer thickness, δ,

$$\delta \approx \left(\frac{v\alpha x}{g\beta(T_\infty - T)} \right)^{1/4}$$

Thus, we expect δ to scale with $x^{1/4}$ and u to scale with $x^{1/2}$. (Note that in the forced convective case we found that $\delta \sim x^{1/2}$; Figure 1-3.15). Following a reasoning similar to the forced convective case, a similarity variable $\xi \approx y/\delta(x)$ or $\xi = Ay/x^{1/4}$ may be found, where A is an arbitrary constant. Also, motivated by Equation 57 for forced convection, it is hoped that $u = B x^{1/2} f'(\xi)$, where B is an arbitrary constant. Expressing these in nondimensional variables, we get

$$\xi = Ay^*/x^{*1/4} \tag{80}$$

and

$$u^* = Bx^{*1/2}f'(\xi)$$

where $f'(\xi) = df/d\xi$. Note that the definitions of the arbitrary constants A and B have been changed during nondimensionalization. By introducing a stream function, ψ, as in Equation 55, Equation 77 is identically satisfied.

Thus

$$\begin{aligned}
\psi &= \int Bx^{*1/2}f'(\xi)\,dy^* + f_1(x^*) \\
&= \int \frac{B}{A}x^{*3/4}f'(\xi)\,d\xi + f_1(x^*) \\
&= \frac{B}{A}x^{*3/4}f(\xi) + f_1(x^*)
\end{aligned} \tag{81}$$

Since $v^* = 0$ at $y^* = 0$ (or $\xi = 0$), $f_1(x^*)$ is at best an arbitrary constant which is taken to be zero without any loss of generality. From Equations 55 and 81 we get

$$v^* = -\frac{B}{4Ax^{*1/4}}\left[3f(\xi) - \xi f'(\xi)\right] \tag{82}$$

By using Equations 80 and 82, Equations 78 and 79 can be reduced to

$$f''' + 3ff'' - 2(f')^2 + T^* = 0 \tag{83}$$

and

$$T^{*''} + 3\,\mathrm{Pr}\,fT^{*'} = 0 \tag{84}$$

where the following definitions of the arbitrary constants A and B have been used:

$$B = \left(\frac{4g\beta(T_s - T_\infty)L}{u_\infty^2}\right)^{1/2}$$
$$A = \left(\frac{g\beta(T_s - T_\infty)L^3}{4v^2}\right)^{1/4} \tag{85}$$

Note that in Equation 84 it has been assumed that T^* is a function of ξ only. From Equation 85 it follows that

$$\xi = \frac{y^*}{x^{*1/4}}\left(\frac{g\beta(T_s - T_\infty)L^3}{4v^2}\right)^{1/4} = \frac{y^*}{x^{*1/4}}\left(\frac{\mathrm{Gr}_L}{4}\right)^{1/4} \tag{86}$$

The associated boundary conditions given by Equation 76 become

$$\begin{aligned}
&f = f' = 0 \quad\text{and}\quad T^* = 1 \quad\text{when}\quad \xi = 0 \\
&f' = 0 \qquad\qquad T^* = 0 \quad\text{at}\quad \xi = \infty
\end{aligned} \tag{87}$$

A numerical solution of Equations 83 and 84 along with the boundary conditions given by Equation 87 is shown in Figure 1-3.17. Note that the nondimensional x-velocity component, u^*, may be readily obtained from Figure 1-3.17 part (a) through the use of Equations 80 and 85. Note also that, through the definition of the similarity variable, ξ, Figure 1-3.17 may be used to obtain values of u^* and T^* for any value of x^* and y^*. Once $u^*(x^*, y^*)$ and $T^*(x^*, y^*)$ are known, the heat transfer coefficient can easily be obtained from Equation 5. Thus, the temperature gradient at $y = 0$ after using Equation 86, becomes

$$\left.\frac{\partial T}{\partial y}\right|_{y=0} = \frac{(T_s - T_\infty)}{L}\left.\frac{\partial T^*}{\partial y^*}\right|_{y^*=0} = \frac{(T_s - T_\infty)}{Lx^{*1/4}}\left(\frac{\mathrm{Gr}_L}{4}\right)^{1/4}\left.\frac{dT^*}{d\xi}\right|_{\xi=0}$$

TABLE 1-3.1 *Dimensionless Temperature Gradient for Free Convection on a Vertical Flat Plate*

Pr	0.01	0.72	1	2	10	100	1000
g(Pr)	0.081	0.505	0.567	0.716	1.169	2.191	3.966

The *local* heat transfer coefficient is

$$h = \frac{-k}{Lx^{*1/4}}\left(\frac{\mathrm{Gr}_L}{4}\right)^{1/4}\left.\frac{dT^*}{d\xi}\right|_{\xi=0} \tag{88}$$

or

$$\mathrm{Nu} = \frac{hx}{k} = -x^{*3/4}\left(\frac{\mathrm{Gr}_L}{4}\right)^{1/4}\left.\frac{dT^*}{d\xi}\right|_{\xi=0} = \left(\frac{\mathrm{Gr}_x}{4}\right)^{1/4}g(\mathrm{Pr}) \tag{89}$$

As is evident from Figure 1-3.17, the dimensionless temperature gradient at $\xi = 0$ is a function of the Prandtl number. In Equation 89 this function is expressed as $-g(\mathrm{Pr})$. Values of $g(\mathrm{Pr})$ obtained from the numerical solution are listed in Table 1-3.1.

From Equation 88 for the local heat transfer coefficient the average heat transfer coefficient for a surface of length L is obtained by using Equation 9 as follows

$$\begin{aligned}
\bar{h}_L &= \frac{1}{L}\int_0^L h(dx \cdot 1) = \frac{k}{L^{7/4}}\left(\frac{\mathrm{Gr}_L}{4}\right)^{1/4}g(\mathrm{Pr}) \times \int_0^L\frac{dx}{x^{1/4}} \\
&= \frac{4}{3}\frac{k}{L}\left(\frac{\mathrm{Gr}_L}{4}\right)^{1/4}g(\mathrm{Pr})
\end{aligned} \tag{90}$$

Thus,

$$\overline{\mathrm{Nu}}_L = \frac{\bar{h}_L L}{k} = \frac{4}{3}\left(\frac{\mathrm{Gr}_L}{4}\right)^{1/4}g(\mathrm{Pr}) \tag{91}$$

or from Equation 89, with $x = L$ we get

$$\overline{\mathrm{Nu}}_L = \frac{4}{3}\,\mathrm{Nu}\left(\begin{array}{c}\text{evaluated}\\\text{at }x = L\end{array}\right) \tag{92}$$

It should be noted that the foregoing results apply irrespective of whether $T_s > T_\infty$ or $T_s < T_\infty$. If $T_s < T_\infty$, the conditions are inverted from those shown in Figure 1-3.16. The loading edge is on the top of the plate, and positive x is defined in the direction of the gravity force.

Complications in Practical Problems

In the previous section two relatively simple problems of laminar forced and free convection on a flat surface were solved. These solutions illustrate the methodology for determining the heat transfer coefficient and provide the necessary insight regarding the relationship between the various dimensionless parameters. Most practical situations are often more complex, and mathematical solutions, such as those presented in the previous section, are not always possible. Complexities arise due to more complex geometry, onset of turbulence, changes in fluid properties with temperature, and because of simultaneous mass transfer from the surface as illustrated in Figure 1-3.1. For such cases, empirical correlations are obtained. These correlations are discussed in the next section and the various complications are individually discussed below.

Effect of turbulence: In both forced and free convective flows, small disturbances may be amplified downstream,

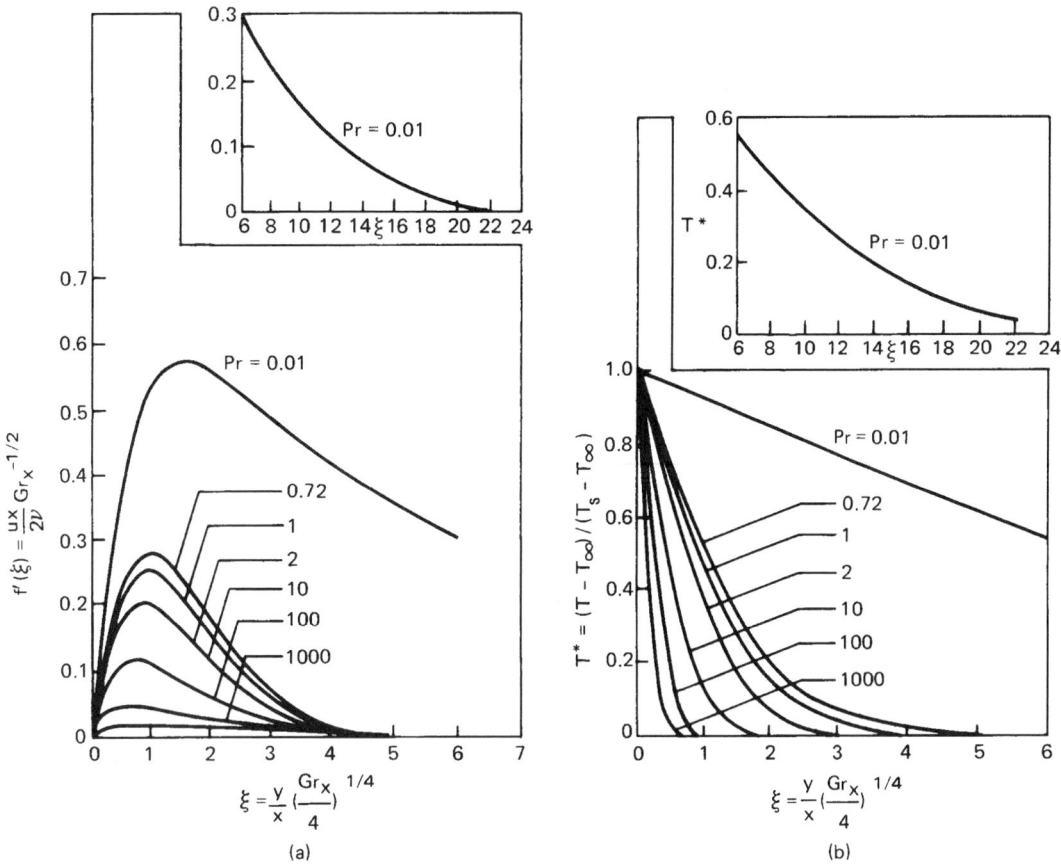

Fig. 1-3.17. *Laminar free convection boundary layer on an isothermal, vertical surface.*

leading to transition from laminar to turbulent flow conditions. These disturbances may originate from the free stream or be induced by surface roughness. Whether these disturbances are amplified or attenuated depends upon the ratio of inertia to viscous forces for forced flows (the Reynolds number), and the ratio of buoyancy to viscous forces for free convective flows (the Grashof number). Note that in both Reynolds and Grashof numbers, viscosity appears in the denominator. Thus for relatively large viscous forces or small Reynolds and Grashof numbers, the naturally occurring disturbances are dissipated, and the flow remains laminar. However, for sufficiently large Reynolds and Grashof numbers (Re $> 5 \times 10^5$ and Gr $> 4 \times 10^8$, for flow over a flat plate) disturbances are amplified, and a transition to turbulence occurs.

The onset of turbulence is associated with the existence of *random fluctuations* in the fluid, and on a small scale the flow is *unsteady*. As shown in Figure 1-3.18, there are sharp differences between laminar and turbulent flows. In the laminar boundary layer, fluid motion is highly ordered and it is possible to identify streamlines along which fluid particles move. In contrast, fluid motion in the turbulent boundary layer is highly irregular and is characterized by velocity fluctuations. These fluctuations enhance the momentum and energy transfers and hence increase the surface friction and convection heat transfer rate. Also, due to the mixing of fluid resulting from the turbulent fluctuations, the turbulent

boundary layer is thicker and the boundary layer profiles (of velocity, temperature, and concentration) are flatter than in laminar flow.

In a fully turbulent flow, the primary mechanism of momentum and heat transfer involves macroscopic lumps of fluid randomly moving about in the flow. This contrasts with the random molecular motion resulting in molecular properties discussed at the beginning of this chapter. In the turbulent region *eddy viscosity* and *eddy thermal conductivity* are important. These eddy properties may be ten times as large as their molecular counterparts.

If one measures the variation of an arbitrary flow variable, P, as a function of time at some location in a turbulent boundary layer, then the typical behavior observed is shown in Figure 1-3.19. The variable P, which may be a velocity component, fluid temperature, pressure, or species concentration, can be represented as the sum of a time-mean value, $\bar{P}$, and a fluctuating component, P'. The average is taken over a time interval that is large compared with the period of a typical fluctuation, and if $\bar{P}$ is time independent then the mean flow is steady. Thus, the instantaneous values of each of the velocity components, pressure, and temperature are given by

$$u = \bar{u} + u', \qquad v = \bar{v} + v', \qquad p = \bar{p} + p'$$

$$T = \bar{T} + T' \quad \text{and} \quad \rho = \bar{\rho} + \rho' \tag{93}$$

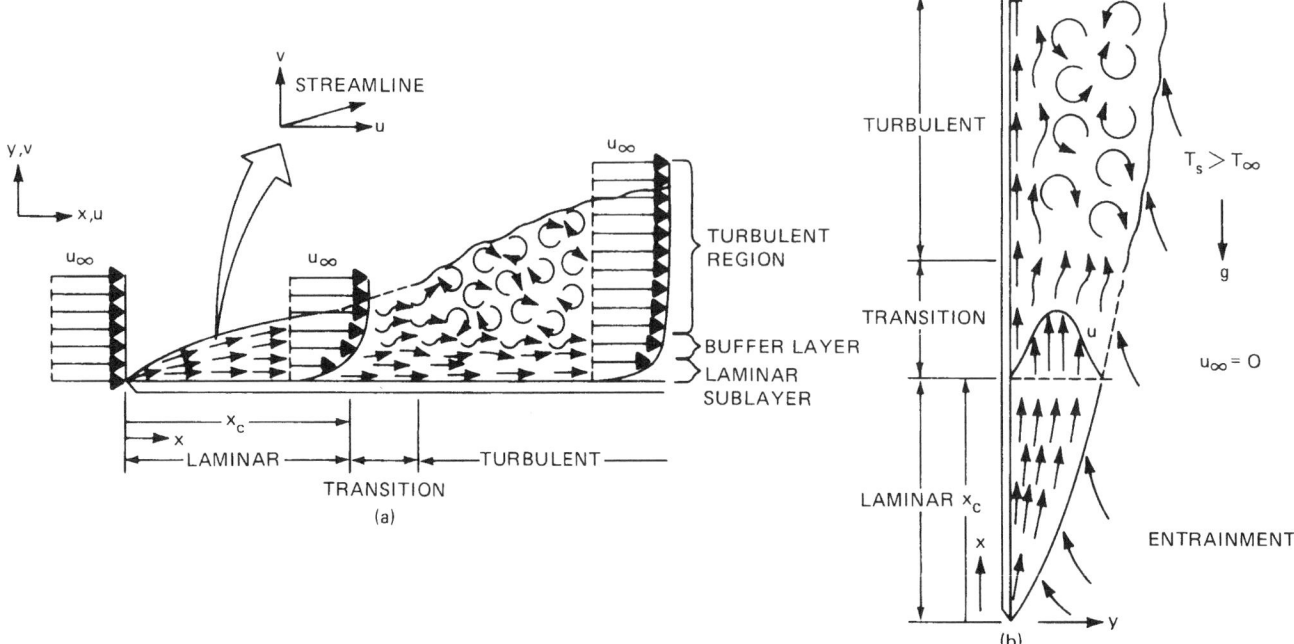

Fig. 1-3.18. *(a) Velocity boundary layer development on a flat plate for forced flow; (b) Velocity boundary layer development on a vertical flat plate for free convective flow.*

Substituting these expressions for each of the flow variables into the boundary layer equations (Equations 35, 43, and 45) and assuming the mean flow to be steady, incompressible (ρ = constant) with constant properties, and using the well established time averaging procedures,[1-4] the following governing equations are obtained

Continuity

$$\frac{\partial \bar{u}}{\partial x} + \frac{\partial \bar{v}}{\partial y} = 0 \tag{94}$$

x-momentum

$$\rho\left(\bar{u}\frac{\partial \bar{u}}{\partial z} + \bar{v}\frac{\partial \bar{u}}{\partial y}\right) = \frac{\partial}{\partial y}\left(\mu\frac{\partial \bar{u}}{\partial y} - \rho\,\bar{u}'\bar{v}'\right) - \frac{\partial \bar{p}}{\partial x} - \rho g_x \tag{95}$$

Energy

$$\rho C_p\left(\bar{u}\frac{\partial \bar{T}}{\partial x} + \bar{v}\frac{\partial \bar{T}}{\partial y}\right) = \frac{\partial \bar{T}}{\partial y}\left(k\frac{\partial \bar{T}}{\partial y} - \rho C_p\,\bar{v}'\bar{T}'\right) \tag{96}$$

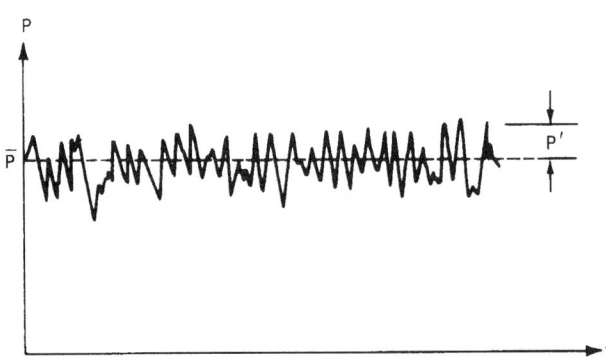

Fig. 1-3.19. *Variation in the variable P with time at some point in a turbulent boundary layer.*

Equations 94 through 96 are similar to the laminar boundary layer equations expressed in mean flow variables, except for the presence of additional terms $\rho\,\bar{u}'\bar{v}'$ and $\rho C_p\,\bar{v}'\bar{T}'$. Physical arguments[2] show that these terms result from the motion of macroscopic fluid lumps and account for the effect of the turbulent fluctuations on momentum and energy transport.

On the basis of the foregoing result it is customary to speak of total shear stress and total heat flux, which are defined as

$$\tau_{tot} \equiv \left(\mu\frac{\partial \bar{u}}{\partial y} - \rho\,\bar{u}'\bar{v}'\right)$$

and

$$\dot{q}_{tot} \equiv -\left(k\frac{\partial \bar{T}}{\partial y} - \rho C_p\,\bar{v}'\bar{T}'\right) \tag{97}$$

The terms $\rho\bar{u}'\bar{v}'$ and $\rho\,C_p\,\bar{v}'\bar{T}'$ are always negative and so result in a positive contribution to total shear stress and heat flux. The term $\rho\,\bar{u}'\bar{v}'$ represents the transport of momentum flux due to turbulent fluctuations (or eddies), and it is known as the Reynolds stress. The notion of transport of heat and momentum by turbulent eddies has prompted the introduction of transport coefficients, which are defined as the eddy diffusivity for momentum transfer, ε_M, and eddy diffusivity for heat transfer, ε_H, and have the form

$$\varepsilon_M\frac{\partial \bar{u}}{\partial y} \equiv -\bar{u}'\bar{v}'$$

$$\varepsilon_H\frac{\partial \bar{T}}{\partial y} \equiv -\bar{v}'\bar{T}' \tag{98}$$

Thus Equation 97 becomes

$$\tau_{tot} \equiv \rho(v + \varepsilon_M)\frac{\partial \bar{u}}{\partial y}$$

and

$$\dot{q}''_{tot} \equiv -\rho C_p(\alpha + \varepsilon_H)\frac{\partial \bar{T}}{\partial y} \tag{99}$$

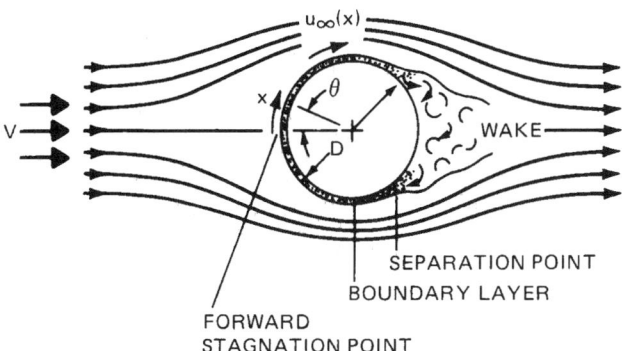

Fig. 1-3.20. Boundary layer formation and separation on a circular cylinder in cross flow.

As noted earlier, eddy diffusivities are much larger than molecular diffusivities, therefore the heat and momentum transfer rates are much larger for turbulent flow than for laminar flow. A fundamental problem in performing turbulent boundary layer analysis involves determining the eddy diffusivities as a function of the mean properties of the flow. Unlike the molecular diffusivities, which are strictly fluid properties, the eddy diffusivities depend strongly on the nature of the flow. They vary across the boundary layer and the variation can only be determined from experimental data. This is an important point, because all analyses of turbulent flow must eventually rely on experimental data. To date, there is no adequate theory for predicting turbulent flow behavior.

Complex geometry: In a previous section on the boundary layer concept, analysis was limited to the simplest possible geometry, i.e., a flat plate. This provided considerable simplification because $dp/dx = 0$ in Equation 43 for the forced flow case. However, the situation is not as simple for fluid flow over bodies with a finite radius of curvature.

Consider a common example of flow across a circular cylinder shown in Figure 1-3.20. Boundary layer formation is initiated at the forward stagnation point, where the fluid is brought to rest with an accompanying rise in pressure. The pressure is a maximum at this point and decreases with increasing x, the streamline coordinate, and θ, the angular coordinate. (Note: In the boundary layer approximation, the pressure is the same inside and outside the boundary layer. This can be seen from Equation 44.) The boundary layer then develops under the influence of a favorable pressure gradient ($dp/dx < 0$). At the top of the cylinder, i.e., at $\theta = 90°$, the pressure eventually reaches a minimum and then begins to increase toward the rear of the cylinder. Thus, for $90° < \theta < 180°$, the boundary layer development occurs in the presence of an adverse pressure gradient ($dp/dx > 0$).

Unlike parallel flow over a flat plate, for curved surfaces the free stream velocity, u_∞, varies with x. [Note that in Figure 1-3.20 a distinction has been made between the fluid velocity upstream of the cylinder, V, and the velocity outside the boundary layer, $u_\infty(x)$.] At the stagnation point, $\theta = 0°$, $u_\infty = 0$. As the pressure decreases for $\theta > 0°$, u_∞ increases according to the Bernoulli equation, Equation 46, and becomes maximum at $\theta = 90°$. For $\theta > 90°$, the adverse pressure gradient decelerates the fluid, and conversion of kinetic energy to pressure occurs in accordance with Equation 46, which applies only to the inviscid flow outside the boundary layer. The fluid inside the boundary layer has considerably slowed down because of viscous friction and does not have enough momentum to overcome the adverse pressure gradient. This eventually leads to boundary layer separation which is illustrated more clearly in Figure 1-3.21. At some location in the fluid, the velocity gradient at the surface becomes zero and the boundary layer detaches or separates from the surface. Farther downstream of the separation point, flow reversal occurs and a wake is formed behind the solid. Flow in this region is characterized by vortex formation and is highly irregular. The separation point is defined as the location at which $(\partial u/\partial y)_{y=0} = 0$. If the boundary layer transition to turbulence occurs prior to separation, the separation is delayed and the separation point moves farther downstream. This happens because the turbulent boundary layer has more momentum than the laminar boundary layer to overcome the adverse pressure gradient.

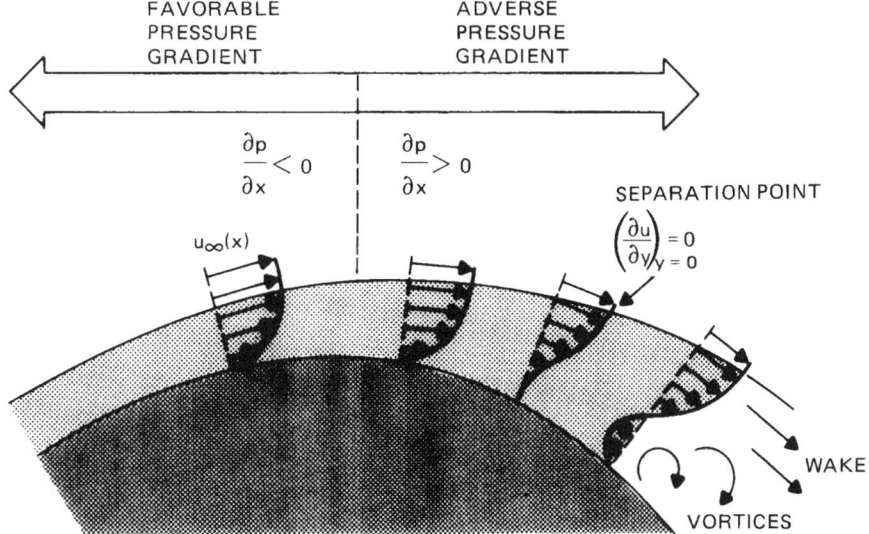

Fig. 1-3.21. Velocity profiles associated with separation on a circular cylinder in cross flow.

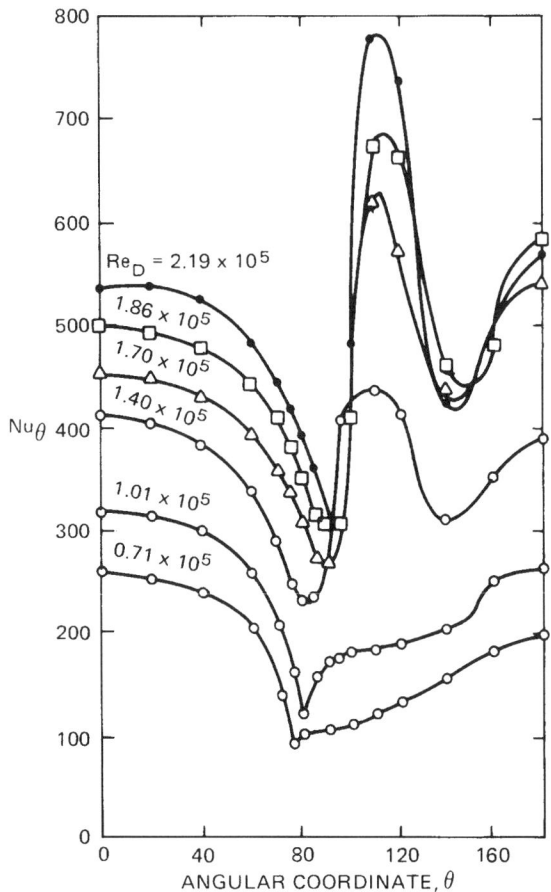

Fig. 1-3.22. *Local Nusselt number for airflow normal to a circular cylinder.*

The foregoing processes strongly influence both the rate of heat transfer from the cylinder surface and the drag force acting on the cylinder. Because of the complexities associated with flow over a cylinder, experimental methods are used to determine the heat transfer coefficient. Such experimental results for the variation of the local Nusselt number with θ are shown in Figure 1-3.22 for a cylinder in a cross flow of air. Consider the results for $Re_D \leq 10^5$ (note: Re_D is defined as VD/ν). Starting at the stagnation point, Nu_θ decreases with increasing θ due to the development of the laminar boundary layer. However, a minimum is reached at $\theta \approx 80°$. At this point separation occurs, and Nu_θ increases with θ due to the mixing associated with vortex formation in the wake. For $Re_D \geq 10^5$, the variation of Nu_θ with θ is characterized by two minima. The decline in Nu_θ from the value at the stagnation point is again due to laminar boundary layer development, but the sharp increase that occurs between 80° and 100° is now due to boundary layer transition to turbulence. With further development of the turbulent boundary layer, Nu_θ must again begin to decline. However, separation eventually occurs ($\theta \approx 140°$), and Nu_θ increases due to considerable mixing associated with the wake region.

The foregoing example clearly illustrates the complications introduced by nonplanar geometry. Heat transfer correlations for these cases are often based on experimental data. Fortunately, for most engineering calculations the local variation in the heat transfer coefficient such as that presented in Figure 1-3.22 is not required; only the overall average conditions are needed. Empirical correlations for average heat transfer coefficients will be presented in the next section.

Changes in fluid properties: In the analysis and discussion presented thus far, fluid properties were assumed to be constant. However, fluid properties vary with temperature across the boundary layer and this variation will have a significant impact on the heat transfer rate. In the empirical heat transfer correlations this influence is accounted for in one of two ways: (1) in correlating the experimental data all properties are evaluated at the mean boundary layer temperature, $T_f = (T_s + T_\infty)/2$, called the film temperature; and (2) alternatively, all properties are evaluated at T_∞ and an additional parameter is used to account for the property variation. This parameter is commonly of the form $(Pr_\infty/Pr_s)^r$ or $(\mu_\infty/\mu_s)^r$, where the subscripts ∞ and s designate evaluation of properties at the free stream and surface temperatures, respectively, and r is an empirically determined constant.

It is important to note that in the empirical correlations to be presented in the next section, the same method that is employed in deriving the correlation should be used when applying the correlation.

Effect of mass transfer: Special attention needs to be given to the effect that species mass transfer from the surface of the solid has on the velocity and thermal boundary layers. Recall that the velocity boundary layer development is generally characterized by the existence of zero fluid velocity at the surface. This condition applies to the velocity component v normal to the surface, as well as to the velocity component u parallel to the surface. However, if there is simultaneous mass transfer to or from the surface, it is evident that v can no longer be zero at the surface. Nevertheless, for the problems discussed in this chapter, mass transfer is assumed to have a negligible effect, i.e., $v \approx 0$. This assumption is reasonable for problems involving some evaporation from gas-liquid or sublimation from gas-solid interfaces. For larger surface mass transfer rates a correction factor (often called the blowing correction) is utilized. This correction factor is simply stated here, and discussed in greater detail by Bird et al.[1] The correction factor is defined as $E(\phi) \equiv h^*/h$, where h^* is the corrected heat transfer coefficient and h is the heat transfer coefficient in the absence of mass transfer. According to film theory, $E(\phi)$ is given by

$$E(\phi) = \frac{\phi}{(e^\phi - 1)} \qquad (100)$$

where

$$\phi = \frac{\dot{m}'' C_{pg}}{h}$$

$\dot{m}'' = \rho_s v_s$ is the mass flux coming out of the surface and C_{pg} is the specific heat of the gas.

Empirical Relations of Convection Heat Transfer

The analysis and discussion presented in the section on the boundary layer concept have shown that for simple cases the convection heat transfer coefficient may be determined directly from the conservation equations. In the previous section it was noted that the complications inherent to most practical problems do not always permit analytical solutions, and that it is necessary to resort to experimental methods. Experimental results are usually expressed in the form of either empirical formulas or graphical charts so that they may be utilized with maximum generality. Difficulties are encountered in the process of trying to generalize the experimental results in the form of empirical correlations. The availability of an analytical solution for a simpler but similar problem greatly assists in guessing the functional form of the results. Experimental data is then used to obtain values of constants or exponents for certain significant parameters, such as the Reynolds or Prandtl numbers. If an analytical solution for a similar problem is not available, it is necessary to rely on the physical understanding of the problem and on dimensional or order-of-magnitude analysis. In this section the experimental methods, the dimensionless groups, and the functional form of the relationships expected between them will be discussed; in addition the empirical formulas that will be used in the "Applications" section of this chapter will be summarized.

Functional form of solutions: The nondimensional Equations 49, 50, 51, and 78 are extremely useful from the standpoint of suggesting how important boundary layer results can be generalized. For example, the momentum equation, Equation 50, suggests that although conditions in the velocity boundary layer depend on the fluid properties, ρ and μ, the velocity, u_∞, and the length scale, L, this dependence may be simplified by grouping these variables in a nondimensional form called the Reynolds number. We therefore anticipate that the solution of Equation 50 will be of the form

$$u^* = f_1\left(x^*, y^*, \mathrm{Re}_L, \frac{dp^*}{dx^*}\right) \tag{101}$$

Note that the pressure distribution, $p^*(x^*)$, depends on the surface geometry and may be obtained independently by considering flow conditions outside the boundary layer in the free stream. Hence, as discussed in the section on complex geometry, the appearance of dp^*/dx^* in Equation 101 represents the influence of geometry on the velocity distribution. Note also that in Equation 50 the term dp^*/dx^* did not appear because it was equal to zero for a flat plate.

Similarly we anticipate that the solution of Equation 78 will be the form

$$u^* = f_2(x^*, y^*, \mathrm{Gr}_L, \mathrm{Pr}) \tag{102}$$

Here, the Prandtl number is included because of the coupling between Equations 78 and 79. If the flow is mixed, i.e., buoyant as well as forced, then the Reynolds number must also be included in the functional relationship expressed by Equation 102.

From Equation 1, the shear stress at the surface, $y^* = 0$, may be expressed as

$$\tau_s = \mu \left.\frac{\partial u}{\partial y}\right|_{y=0} = \left(\frac{\mu u_\infty}{L}\right) \left.\frac{\partial u^*}{\partial y^*}\right|_{y^*=0}$$

and from Equation 10 it follows that the friction coefficient is

$$C_f = \frac{\tau_s}{\frac{1}{2}\rho u_\infty^2} = \frac{2}{\mathrm{Re}_L} \left.\frac{\partial u^*}{\partial y^*}\right|_{y^*=0} \tag{103}$$

From Equation 101 it is clear that

$$\left.\frac{\partial u^*}{\partial y^*}\right|_{y^*=0} = f_3\left(x^*, \mathrm{Re}_L, \frac{dp^*}{dx^*}\right) \tag{104}$$

Hence, for a prescribed geometry (i.e., dp^*/dx^* is known from the free stream conditions) we have

$$C_f = \frac{2}{\mathrm{Re}_L} f_3(x^*, \mathrm{Re}_L) \tag{105}$$

Equation 105 is very significant because it states that the friction coefficient may be expressed exclusively in terms of a dimensionless space coordinate and the Reynolds number. For a prescribed geometry, the function that relates C_f to x^* and Re_L can be expected to be *universally* applicable. That is, it can be expected to apply to different fluids and over a wide range of values for u_∞ and L.

Similar results may be obtained for the heat transfer coefficient. Equation 51 suggests that the solution may be expressed in the form

$$T^* = f_4\left(x^*, y^*, \mathrm{Re}_L, \frac{dp^*}{dx^*}\right) \tag{106}$$

for forced flow, and

$$T^* = f_5(x^*, y^*, \mathrm{Gr}_L, \mathrm{Pr}) \tag{107}$$

for free convective flow. Here Re_L, Gr_L, and dp^*/dx^* originate from the influence of fluid motion (u^* and v^*) on Equation 51.

From the definition of the convection heat transfer coefficient, Equation 5, and Equation 40 with $y^* = y/L$ we obtain

$$h = -\frac{k \left.\frac{\partial T}{\partial y}\right|_{y=0}}{(T_s - T_\infty)} = +\frac{k}{L} \left.\frac{\partial T^*}{\partial y^*}\right|_{y^*=0} \tag{108}$$

Thus $\qquad \mathrm{Nu} \equiv \dfrac{hL}{k} = \left.\dfrac{\partial T^*}{\partial y^*}\right|_{y^*=0}$

Note that the Nusselt number, Nu, is equal to the dimensionless temperature gradient at the surface. From Equation 106 or 107 it follows that for a prescribed geometry, i.e., known dp^*/dx^*

$$\mathrm{Nu} = f_6(x^*, \mathrm{Re}_L, \mathrm{Pr}) \tag{109}$$

for forced flow, and

$$\mathrm{Nu} = f_7(x^*, \mathrm{Gr}_L, \mathrm{Pr}) \tag{110}$$

for free convective flow. The Nusselt number is to the thermal boundary layer what the friction factor is to the velocity boundary layer. Equations 109 and 110 imply that for a given geometry, the Nusselt number must be some *universal* function of x^*, Re_L, and Pr. If this function were known, it could be used to compute the value of Nu for different fluids and different values of u_∞, T_∞ and L. Furthermore, since the average heat transfer coefficient is obtained by integrating over the surface of the body, it must be independent of the spatial variable, x^*. Hence, the functional dependence of the average Nusselt number is

$$\overline{\mathrm{Nu}} = \frac{\bar{h}L}{k} = f_8(\mathrm{Re}_L, \mathrm{Pr}) \tag{111}$$

for forced flow, and

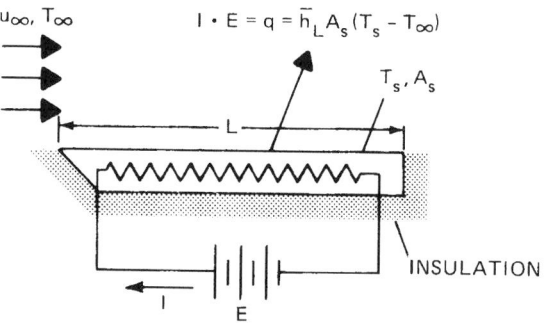

Fig. 1-3.23. Experiment for measuring the average convection heat transfer coefficient, h_L.

$$\overline{Nu} = f_9(Gr_L, Pr) \tag{112}$$

for free convective flows.

Although it is very helpful to know the functional dependence of Nu, the task is far from complete, because the function may be any of millions of possibilities. It may be a sine, exponential, or a logarithmic function. The exact form of this function can only be determined by an analytical solution of the governing equations, such as Equations 70 and 91.

Experimental determination of heat transfer coefficient: The manner in which a convection heat transfer correlation may be obtained experimentally is illustrated in Figure 1-3.23. If a prescribed geometry, such as the flat plate in parallel flow, is heated electrically to maintain $T_s > T_\infty$, convection heat transfer occurs from the surface to the fluid. It would be a simple matter to measure T_s and T_∞ as well as the electrical power, $E \cdot I$, which is equal to the total heat transfer rate, $\dot{q}$. The average convection coefficient, $\bar{h}_L$, can now easily be computed from Equation 7. Also, from the knowledge of the characteristic length, L, and the fluid properties, the values of the various nondimensional numbers—such as the Nusselt, Reynolds, Grashof, and Prandtl numbers—can be easily computed from their definitions.

The foregoing procedure is repeated for a variety of test conditions. We could vary the velocity, u_∞, the plate length, L, and the temperature difference $(T_s - T_\infty)$, as well as the fluid properties, using, for example, fluids such as air, water, and engine oil, which have substantially different Prandtl

numbers. Many different values of the Nusselt number would result, corresponding to a wide range of Reynolds and Prandtl numbers. At this stage, an analytical solution to a similar but simpler problem proves very useful in guiding how the various nondimensional numbers should be correlated. For laminar flow over a flat plate it has been seen that in Equation 70 the relationship is of the form

$$\overline{Nu} = C\,Re_L^m\,Pr^n$$

Thus, we plot the results on a log-log graph as shown in Figure 1-3.24 and determine the values of C, m, and n. Because such a relationship is inferred from experimental measurements, it is called an empirical correlation. Along with this empirical correlation it is specified how the temperature-dependent properties were determined for calculating the various nondimensional numbers. When such a correlation is used, it is important that the properties must be calculated in exactly the manner specified. If they are not specified, then the mean boundary layer temperature, T_f, called the film temperature, must be used.

$$T_f \equiv \frac{T_s + T_\infty}{2} \tag{113}$$

A summary of empirical and practical formulas: In this section a variety of convection correlations (Tables 1-3.3 and 1-3.4) and the associated dimensionless groups (Table 1-3.2) for external flow conditions are tabulated. Correlations for both forced and free convection are presented along with their range of applicability. The contents of this section are more or less a collection of "recipes." Proper use of these recipes is essential to solving practical problems. The reader should not view these correlations as sacrosanct; each correlation is reasonable over the range of conditions specified, but for most engineering calculations one should not expect the accuracy to be much better than 20 percent.

For proper use of the foregoing correlations it is important to note that the flow may not be laminar or turbulent over the entire length of the plate under consideration. Instead, transition to turbulence may occur at a distance, x_c, $(x_c < L$, where L is the plate length) from the leading edge of the plate. In this mixed boundary layer situation, the average convection heat transfer coefficient for the entire plate is obtained by integrating first over the laminar region $(0 \le x \le x_c)$ and then over the turbulent region $(x_c < x \le L)$ as follows

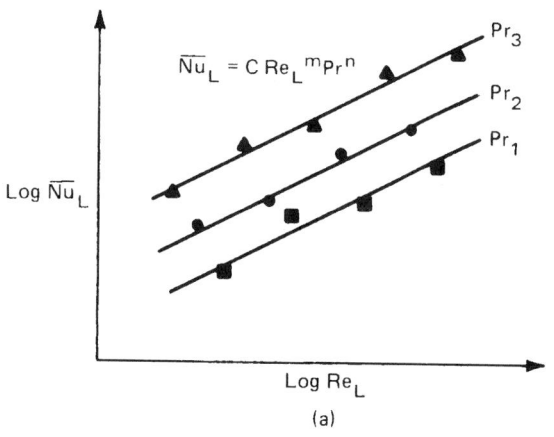

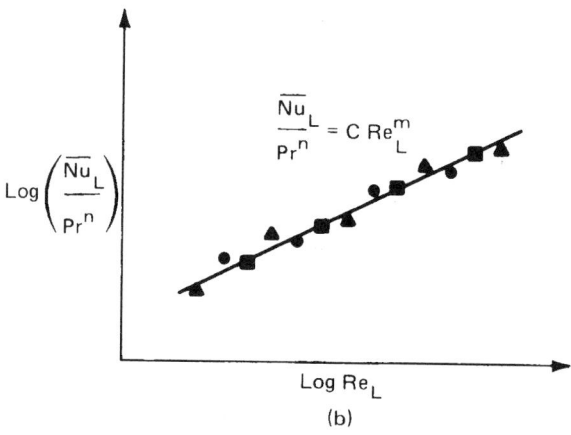

Fig. 1-3.24. Dimensionless representation of convection heat transfer measurements.

TABLE 1-3.2 *Selected Dimensionless Groups*

Group		Definition	Interpretation
Friction coefficient	local	$C_f = \dfrac{\tau_2}{\rho u_\infty^2 / 2}$	dimensionless surface shear stress
	average	$\bar{C}_f = \dfrac{\bar{\tau}_s}{\rho u_\infty^2 / 2}$	
Reynolds number	location x	$Re_x = \dfrac{u_\infty x}{\nu}$	ratio of inertia and viscous forces
	length L	$Re_L = \dfrac{u_\infty L}{\nu}$	
	diameter D	$Re_D = \dfrac{u_\infty D}{\nu}$	
Prandtl number		$Pr = \dfrac{c_p \mu}{k} = \dfrac{\nu}{\alpha}$	ratio of molecular momentum and thermal diffusivities
Grashof number	location x	$Gr_x = \dfrac{g\beta(T_s - T_\infty)x^3}{\nu^2}$	ratio of buoyancy to viscous forces
	length L	$Gr_L = \dfrac{g\beta(T_s - T_\infty)L^3}{\nu^2}$	
	diameter D	$Gr_D = \dfrac{g\beta(T_s - T_\infty)D^3}{\nu^2}$	
Rayleigh number	location x	$Ra_x = Gr_x Pr = \dfrac{g\beta(T_s - T_\infty)x^3}{\nu\alpha}$	product of Grashof and Prandtl numbers
		replace x by L and D to get Ra_L and Ra_D	
Nusselt number	location x	$Nu_x = \dfrac{hx}{k}$	ratio of convection heat transfer to conduction in a fluid slab of thickness x
		replace x by L and D to get Nu_L and Nu_D	
Modified Grashof number	location x	$Gr_x^\star = Gr_x Nu_x = \dfrac{g\beta > \dot{q}_s'' \, x^4}{k\nu^2}$	product of Grashof and Nusselt numbers
Stanton number		$St = \dfrac{h}{\rho u_\infty c_p} = \dfrac{Nu}{RePr}$	dimensionless heat transfer coefficient

$$\bar{h}_L = \frac{1}{L}\left(\int_0^{x_c} h_{lam}\,dx + \int_{x_c}^{L} h_{turb}\,dx \right) \qquad (114)$$

where x_c may be obtained from the critical Reynolds or Grashof numbers.

Also, several correlations given in Tables 1-3.3 and 1-3.4 are for the constant heat flux ($\dot{q}_s'' = $ constant) boundary condition. Thus, the surface temperature of the object is unknown and yet the fluid properties are to be determined at $T_f = (T_s + T_\infty)/2$. For such cases an iterative procedure is employed and the average surface temperature can be determined as follows

$$\dot{q}_s''(\text{known}) = \bar{h}(T_s - T_\infty) = \frac{\overline{Nu}_L}{(L/k)}(T_s - T_\infty)$$

thus

$$T_s(\text{average}) = T_\infty + \frac{\dot{q}_s''(L/k)}{\overline{Nu}_L} \qquad (115)$$

The use of correlations given in Tables 1-3.3 and 1-3.4 is illustrated via examples in the next section.

Applications

This section briefly summarizes the methodology for convection calculations and then presents examples to illustrate the use of various correlations.

Methodology for convection calculations: The application of a convection correlation for any flow situation is facilitated by following a few simple rules:

TABLE 1-3.3　*Summary of Forced Convection Correlations for External Flow Geometries*

Geometry/Flow	Type	Equation	Restrictions	Comments
Flat plate/laminar (T_s = constant)	local:	$Nu_x = 0.332\ Re_x^{1/2} Pr^{1/3}$	$Re_x < 5 \times 10^5$	Properties evaluated at $T_t = (T_s + T_\infty)/2$
	average:	$\overline{Nu}_L = 0.664\ Re_L^{1/2}\ Pr^{1/3}$	$0.6 \leq Pr \leq 50$	
	boundary layer thickness:	$\dfrac{\delta}{x} = 5\ Re_x^{-1/2}$		
Flat plate/laminar ($\dot{q}_s'' $ = constant)	local:	$Nu_x = 0.453\ Re_x^{1/2}\ Pr^{1/3}$	$Re_x < 5 \times 10^5$ $0.6 \leq Pr \leq 50$	Properties evaluated at T_f. However, T_s is not known. Instead, q_s'' is known. Thus, $T_f = T_\infty + (\bar{T}_s - \bar{T}_\infty)/2$. where, $(\bar{T}_s - \bar{T}_\infty)$ $= \dfrac{q_s L/k}{0.6795\ Re_L^{1/2}\ Pr^{1/3}}$
Flat plate/turbulent (T_s = constant)	local:	$Nu_x = 0.0296\ Re_x^{4/5}\ Pr^{1/3}$	$Re_x < 10^8$	Properties evaluated at T_f
	boundary layer thickness:	$\dfrac{\delta}{x} = 0.37\ Re_x^{-1/5}$	$0.6 \leq Pr \leq 60$	
	mixed average (laminar-turbulent):	$\overline{Nu}_L = (0.037\ Re_L^{4/5} - 871)Pr^{1/3}$	transition to turbulence at $Re_{crit} = 5 \times 10^5$	

| Flow across cylinders Circular cylinder | average: | $\overline{Nu}_D = C\ Re_D^m\ Pr^{1/3}$ | $0.4 < Re_D < 4 \times 10^5$ | Properties evaluated at T_f. |

Re_D	C	m
$0.4 - 4$	0.989	0.330
$4 - 40$	0.911	0.385
$40 - 4000$	0.683	0.466
$4 \times 10^3 - 4 \times 10^4$	0.193	0.618
$4 \times 10^4 - 4 \times 10^5$	0.027	0.805

OTHER GEOMETRIES

SQUARE

| $5 \times 10^3 - 10^5$ | 0.246 | 0.588 |
| $5 \times 10^3 - 10^5$ | 0.102 | 0.675 |

HEXAGON

| $5 \times 10^3 - 1.95 \times 10^4$ | 0.160 | 0.638 |
| $1.95 \times 10^4 - 10^5$ | 0.0385 | 0.782 |

VERTICAL PLATE

| $5 \times 10^3 - 10^5$ | 0.153 | 0.638 |
| $4 \times 10^3 - 1.5 \times 10^4$ | 0.228 | 0.731 |

Geometry/Flow	Type	Equation	Restrictions	Comments
Flow across spheres	average:	$\overline{Nu}_D = 2 + (0.4\ Re_D^{1/2}$ $+ 0.06\ Re_D^{2/3})\ Pr^{0.4} \left[\dfrac{\mu_\infty}{\mu_s}\right]^{1/4}$	$3.5 < Re_D < 7.6 \times 10^4$ $0.71 < Pr < 380$ $1.0 < \left[\dfrac{\mu_\infty}{\mu_s}\right] < 3.2$	Properties evaluated at T_∞.
Falling drop	average:	$\overline{Nu}_D = 2 + 0.6\ Re_D^{1/2}\ Pr^{1/3}.$ $\cdot [25(x/D)^{-0.7}]$	where x is the falling distance measured from rest.	Properties evaluated at T_∞.

TABLE 1-3.4 *Summary of Free Convection Correlations for External Flow Geometries*

Geometry/Flow	Type	Equation	Restrictions	Comments
Vertical Plates OR	local: (T_s = const)	$Nu_x g(Pr)$ from Table 1-3.1 $= \left[\dfrac{Gr_x}{4}\right]^{1/4} g(Pr)$	$Gr_x \le 4 \times 10^8$ (Laminar)	Properties evaluated at $(T_f = T_s + T_\infty)/2$
	average: (T_s = const)	$\overline{Nu}_L = \dfrac{4}{3}\left[\dfrac{Gr_L}{4}\right]^{1/4} g(Pr)$	$Gr_x \le 4 \times 10^8$ (Laminar)	Properties at T_f
	average: (T_s = const)	$\overline{Nu}_L = \left\{0.825 + \dfrac{0.387\,Ra_L^{1/6}}{[1 + (0.492/Pr)^{9/16}]^{8/27}}\right\}^2$	none	Properties at T_f This correlation may be applied to vertical cylinders if $\left(\dfrac{D}{L}\right) \ge (35/Gr_L^{1/4})$
	local: ($\dot{q}_s''$ = const)	$Nu_x = 0.6\,(Gr_x^\star\,Pr)^{1/5}$	$10^5 < Gr_x^\star\,10^{11}$ (Laminar)	Properties at T_f
	local: ($\dot{q}_s''$ = const)	$Nu_x = 0.17\,(Gr_x^\star\,Pr)^{1/4}$	$2 \times 10^{13} < Gr_x^\star\,Pr < 10^{16}$	Properties at T_f
	average: ($\dot{q}_s''$ = const)	$\overline{Nu}_L = 0.75\,(Gr_L^\star\,Pr)^{1/5}$	$10^5 < Gr_x^\star < 10^{11}$ (Laminar)	Properties at T_f
Horizontal plates (hot surface up or cold surface down)	average: (T_s = const)	$\overline{Nu}_L = 0.54\,Ra_L^{1/4}$ $\overline{Nu}_L = 0.15\,Ra_L^{1/3}$	$10^5 \le Ra_L \le 10^7$ $10^7 \le Ra_L \le 10^{10}$	Properties at T_f characteristic length L is defined as $L = A_s/P$ where, A_s = plate surface area F = perimeter of the plate (1) All properties except β are evaluated at $T_e = T_s - \dfrac{1}{4}(T_s - T_\infty)$. β is evaluated at T_f
	average: ($\dot{q}_s''$ = const)	$\overline{Nu}_L^{(1)} = 0.16\,Ra_L^{1/3}$	$Ra_L \le 2 \times 10^8$	
Horizontal plates (cold surface up or hot surface down)	average: (T_s = const)	$\overline{Nu}_L = 0.27\,Ra_L^{1/4}$	$10^5 \le Ra_L \le 10^{10}$	
	average: ($\dot{q}_s''$ = const)	$\overline{Nu}_L^{(1)} = 0.16\,Ra_L^{1/3}$	$2 \times 10^8 \le Ra_L \le 10^{11}$	

Inclined plates Hot Surface	average: ($\dot{q}_s''$ = const)	$\overline{Nu}_L = 0.56\,(Ra_L \cos\theta)^{1/4}$ (hot surface facing down)
		 For hot surface facing up
		 $\overline{Nu}_L = 0.14\,[(Gr_L\,Pr)^{1/3} - (Gr_c\,Pr)^{1/3}]$ $+ 0.56\,(Ra_L \cos\theta)^{1/4}$
		 $\theta = -15^\star$; $Gr_c = 5 \times 10^9$
		   $-30^\star$; 2×10^9
		   $-60^\star$; 10^8
		   $-75^\star$; 10^5
			$\theta < 88''$
			 $10^5 < Ra_L \cos\theta < 10^{11}$
			 $-15^\star > \theta > -75^\star$
			 $10^5 < Ra_L \cos\theta < 10^{11}$
				Properties evaluated at $T_e = T_s - \dfrac{1}{4}(T_s - T_\infty)$. Grashof number

Horizontal cylinders	average: (T_s = const)	$\overline{Nu}_D = \left\{0.6 + \dfrac{0.387\,Ra_D^{1/6}}{[1 + (0.559/Pr)^{9/16}]^{8/27}}\right\}^2$	$10^{-5} < Ra_D < 10^{12}$	Properties evaluated at T_f.
Spheres	average: (T_s = const)	$\overline{Nu}_D = 2 + 0.43\,Ra_D^{1/4}$ $\overline{Nu}_D = 2 + 0.5\,Ra_D^{1/4}$	$1 < Ra_D < 10^5$ $Pr \approx 1$ $3 \times 10^5 < Ra < 8 \times 10^8$	Properties evaluated at T_f.

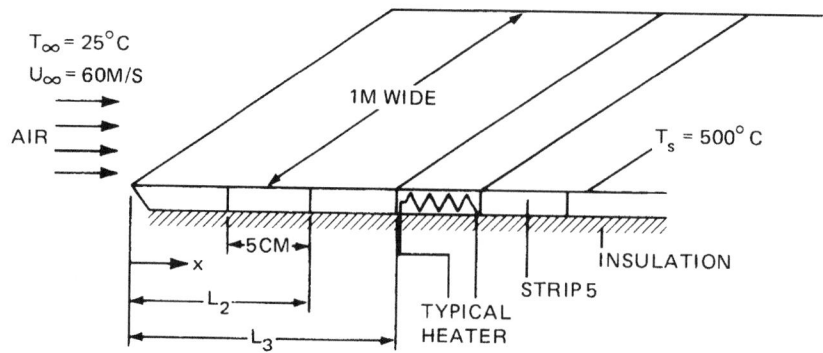

Fig. 1-3.25. Construction details for wind tunnel experiments.

1. Become immediately cognizant of the flow geometry. Does the problem involve flow over a flat plate, a sphere, a cylinder, etc.? The specific form of the convection correlation depends, of course, on the geometry.
2. Specify the appropriate reference temperature and then evaluate the pertinent fluid properties at that temperature. For moderate boundary layer temperature differences, it has been found that the film temperature may beused for this purpose. However, there are correlations that require property evaluation at the free stream temperature and include a property ratio to account for the nonconstant property effect.
3. Determine whether the flow is laminar or turbulent. This determination is made by calculating the Reynolds number and comparing the value with the appropriate transition criterion. For example, if a problem involves parallel flow over a flat plate for which the Reynolds number is $Re_L = 10^6$ and the transition criterion is $Re_{crit} = 5 \times 10^5$, it is obvious that a mixed boundary layer condition exists.
4. Decide whether a local or surface average coefficient is required. Recall that the local coefficient is used to determine the flux at a particular point on the surface, whereas the average coefficient determines the transfer rate for the entire surface.

Having complied with the foregoing rules, sufficient information will be available to select the appropriate correlation for the problem.

EXAMPLE 1:

Electrical strip heaters are assembled to construct a flat radiant heater 1 m wide for conducting fire experiments in a wind tunnel. The heater strips are 5 cm wide and are independently controlled to maintain the surface temperature at 500°C. Construction details are shown in Figure 1-3.25. If air at 25°C and 60 m/s flows over the plate, at which strip is the electrical input maximum? What is the value of this input? The radiative heat loss is ignored.

SOLUTION:

Assumptions

Steady-state conditions, neglect radiation losses, and no heat loss through the bottom surface.

Properties

$T_f = 535$ K: $\rho = 1$ atm. From air property Table 1-3.5, $k = 42.9 \times 10^{-3}$ W/m K; $\nu = 43.5 \times 10^{-6}$ m²/s; $Pr = 0.683$.

Analysis

The strip heater requiring the maximum electrical power is that for which the average convection coefficient is the largest. From the knowledge of variation of the local convection coefficient with distance from the leading edge, the local maximum can be found. Figure 1-3.15 shows that a possible location is the leading edge on the first plate. A second likely location is where the flow becomes turbulent. To determine the point of boundary layer transition to turbulence assume that the critical Reynolds number is 5×10^5. It follows that transition will occur at x_c, where

$$x_c = \frac{\nu \, Re_{crit}}{u_\infty} = \frac{43.5 \times 10^{-6} \times 5 \times 10^5}{60} \text{ m}$$
$$= 0.36 \text{ m} \quad \text{or on the eight strip.}$$

Thus there are three possibilities:

1. Heater strip 1, since it corresponds to the largest local, laminar convection coefficient.
2. Heater strip 8, since it corresponds to the largest local turbulent convection coefficient.
3. Heater strip 9, since turbulent conditions exist over the entire heater.

For the first heater strip

$$q_{conv, \, 1} = \bar{h}_1 L_1 W(T_s - T_\infty)$$

where $\bar{h}_1$ is determined from the equation below (see also Table 1-3.3).

$$\overline{Nu}_1 = 0.664 \, Re_1^{1/2} \, Pr^{1/3}$$
$$= 0.664 \left(\frac{60 \times 0.05}{43.5 \times 10^{-6}} \right)^{1/2} (0.683)^{1/3}$$
$$= 153.6$$

hence

$$\bar{h}_1 = \frac{\overline{Nu}_1 \, k}{L_1} = \frac{153.6 \times 42.9 \times 10^{-3}}{0.05}$$
$$= 131.8 \text{ W/m}^2 \, K$$

hence

$$q_{conv, \, 1} = (131.8)(0.05)(1 \text{ m})(500 - 25)$$
$$= 3129 \text{ W}$$

The power requirement for the eighth strip may be obtained by subtracting the total heat loss associated with the first seven heaters from that associated with the first eight heaters. Thus

$$q_{conv,\,8} = \bar{h}_{1-8}\, L_8\, W(T_s - T_\infty) - \bar{h}_{1-7}\, L_7\, W(T_s - T_\infty)$$

The value of $\bar{h}_{1-7}$ is obtained from the equation applicable to laminar conditions (Table 1-3.3). Thus

$$\overline{\mathrm{Nu}}_{1-7} = 0.664\, \mathrm{Re}_7^{1/2}\, \mathrm{Pr}^{1/3}$$
$$= 0.664\left(\frac{60 \times 7 \times 0.05}{43.5 \times 10^{-6}}\right)^{1/2} (0.683)^{1/3}$$
$$= 406.3$$
$$\bar{h}_{1-7} = \frac{\overline{\mathrm{Nu}}_{1-7}\, k}{L_7} = \frac{406.3 \times 42.9 \times 10^{-3}}{7 \times 0.05}$$
$$= 49.8\ \mathrm{W/m^2\ K}$$

By contrast, the eighth heater is characterized by mixed boundary layer conditions. Thus use the formula (Table 1-3.3).

$$\overline{\mathrm{Nu}}_{1-8} = (0.037\, \mathrm{Re}_8^{4/5} - 871)\, \mathrm{Pr}^{1/3}$$
$$\mathrm{Re}_8 = 8 \times \mathrm{Re}_1 = 5.52 \times 10^5$$
$$\overline{\mathrm{Nu}}_{1-8} = 510.5$$
$$\bar{h}_{1-8} = \frac{\overline{\mathrm{Nu}}_{1-8}\, k}{L_8} = 54.7\ \mathrm{W/m^2\ K}$$

The rate of heat transfer from the eighth strip is then

$$q_{conv,\,8} = (54.7 \times 8 \times 0.05 - 49.8 \times 7 \times 0.05)(500 - 25)$$
$$= 2113.8\ \mathrm{W}$$

The power requirement for the ninth heater strip may be obtained by either subtracting the total heat loss associated with the first eight from that associated with the first nine, or by integrating over the local turbulent expression, since the flow is completely turbulent over the entire width of the strip. The latter approach produces

$$\bar{h}_9 = \left(\frac{k}{L_9 - L_8}\right) 0.0296 \left(\frac{u_\infty}{v}\right)^{4/5} \mathrm{Pr}^{1/3} \int_{L_8}^{L_9} \frac{dx}{x^{1/5}}$$
$$\bar{h}_9 = \left(\frac{42.9 \times 10^{-3}}{0.05}\right) 0.0296 \left(\frac{60}{43.5 \times 10^{-6}}\right)^{4/5}$$
$$\times (0.683)^{1/3} \int_{L_8}^{L_9} \frac{dx}{x^{1/5}}$$
$$= 1825.22[(0.45)^{0.8} - (0.4)^{0.8}] = 86.7$$
$$q_{conv,\,9} = 86.7 \times 0.05 \times 1 \times (500 - 25) = 2059\ \mathrm{W}$$

hence $\qquad q_{conv,\,1} > q_{conv,\,8} > q_{conv,\,9}$

and the first heater strip has the largest power requirement.

EXAMPLE 2:

A glass-door fire screen, shown in Figure 1-3.26, is used to reduce exfiltration of room air through a chimney. It has a height of 0.71 m, a width of 1.02 m, and reaches a temperature of 232°C. If the room temperature is 23°C, estimate the convection heat transfer rate from the fireplace to the room.

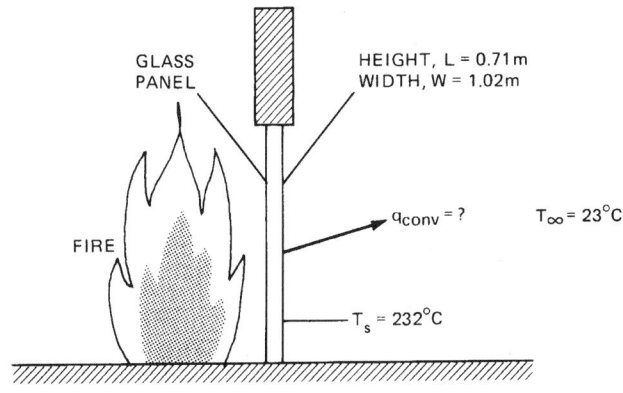

Fig. 1-3.26. Glass panel fire screen.

SOLUTION:

Assumptions

The screen is at a uniform temperature, T_s, and room air is quiescent.

Properties

$T_f = 400$ K, $P = 1$ atm. From air property table (Table 1-3.5):

$$k = 33.8 \times 10^{-3}\ \mathrm{W/m\ K}; \qquad v = 26.41 \times 10^6\ \mathrm{m^2/s};$$
$$\alpha = 38.3 \times 10^{-6}\ \mathrm{m^2/s};\ \ \mathrm{Pr} = 0.69;\ \ \beta = 1/T_f = 0.0025\ \mathrm{K^{-1}}$$

Analysis

The rate of heat transfer by free convection from the panel to the room is given by:

$$q = \bar{h} A_s (T_s - T_\infty)$$

where $\bar{h}$ is obtained from the following equation from Table 1-3.4.

$$\overline{\mathrm{Nu}}_L = \left\{ 0.825 + \frac{0.387\, \mathrm{Re}_L^{1/6}}{[1 + (0.492/\mathrm{Pr})^{9/16}]^{8/27}} \right\}^2$$

here $\qquad \mathrm{Ra}_L = \dfrac{g\beta(T_s - T_\infty)L^3}{\alpha v}$

$$= \frac{9.8 \times 0.0025 \times (232 - 23) \times (0.71)^3}{38.3 \times 10^{-6} \times 26.4 \times 10^{-6}}$$
$$= 1.813 \times 10^9$$

Since $\mathrm{Ra}_L > 10^9$, transition to turbulence will occur on the glass panel and the appropriate correlation from Table 1-3.4 has been chosen

$$\overline{\mathrm{Nu}}_L = \left\{ 0.825 + \frac{0.387(1.813 \times 10^9)^{1/6}}{[1 + (0.492/0.69)^{9/16}]^{8/27}} \right\}^2$$
$$= 147$$

Hence $\qquad \bar{h} = \dfrac{\overline{\mathrm{Nu}}_L \times k}{L} = \dfrac{147 \times 33.8 \times 10^{-3}}{0.71} = 7\ \mathrm{W/m^2\ K}$

and $\qquad q = 7\ \dfrac{\mathrm{W}}{\mathrm{m^2\ K}}\ (1.02\ \mathrm{m} \times 0.71\ \mathrm{m}) \times (232 - 23)°\mathrm{C}$

$$= 1060\ \mathrm{W}$$

TABLE 1-3.5 *Thermophysical Properties of Air at Atmospheric Pressure*

T K	ρ kg/m³	c_p kJ/kg · K	$\mu \cdot 10^7$ N · s/m²	$v \cdot 10^6$ m²/s	$k \cdot 10^3$ W/m · K	$\alpha \cdot 10^6$ m²/s	Pr
100	3.5562	1.032	71.1	2.00	9.34	2.54	0.786
150	2.3364	1.012	103.4	4.426	13.8	5.84	0.758
200	1.7458	1.007	132.5	7.590	18.1	10.3	0.737
250	1.3947	1.006	159.6	11.44	22.3	15.9	0.720
300	1.1614	1.007	184.6	15.89	26.3	22.5	0.707
350	0.9950	1.009	208.2	20.92	30.0	29.9	0.700
400	0.8711	1.014	230.1	26.41	33.8	38.3	0.690
450	0.7740	1.021	250.7	32.39	37.3	47.2	0.686
500	0.6964	1.030	270.1	38.79	40.7	56.7	0.684
550	0.6329	1.040	288.4	45.57	43.9	66.7	0.683
600	0.5804	1.051	305.8	52.69	46.9	76.9	0.685
650	0.5356	1.063	322.5	60.21	49.7	87.3	0.690
700	0.4975	1.075	338.8	68.10	52.4	98.0	0.695
750	0.4643	1.087	354.6	76.37	54.9	109	0.702
800	0.4354	1.099	369.8	84.93	57.3	120	0.709
850	0.4097	1.110	384.3	93.80	59.6	131	0.716
900	0.3868	1.121	398.1	102.9	62.0	143	0.720
950	0.3666	1.131	411.3	112.2	64.3	155	0.723
1000	0.3482	1.141	424.4	121.9	66.7	168	0.726
1100	0.3166	1.159	449.0	141.8	71.5	195	0.728
1200	0.2902	1.175	473.0	162.9	76.3	224	0.728
1300	0.2679	1.189	496.0	185.1	82	238	0.719
1400	0.2488	1.207	530	213	91	303	0.703
1500	0.2322	1.230	557	240	100	350	0.685
1600	0.2177	1.248	584	268	106	390	0.688
1700	0.2049	1.267	611	298	113	435	0.685
1800	0.1935	1.286	637	329	120	482	0.683
1900	0.1833	1.307	663	362	128	534	0.677
2000	0.1741	1.337	689	396	137	589	0.672
2100	0.1658	1.372	715	431	147	646	0.667
2200	0.1582	1.417	740	468	160	714	0.655
2300	0.1513	1.478	766	506	175	783	0.647
2400	0.1448	1.558	792	547	196	869	0.630
2500	0.1389	1.665	818	589	222	960	0.613
3000	0.1135	2.726	955	841	486	1570	0.536

Note: in this case radiation heat transfer calculations would show that radiant heat transfer is greater than free convection heat transfer.

REFERENCES CITED

1. R.B. Bird, W.E. Stewart, and E.N. Lightfoot, *Transport Phenomena*, Wiley, New York (1966).
2. H. Schlichting, *Boundary Layer Theory*, McGraw-Hill, New York (1979).
3. V.S. Arpaci and P.S. Larsen, *Convection Heat Transfer*, Prentice-Hall, Englewood Cliffs, NJ (1984).
4. E.R.G. Eckert and R.M. Drake, *Analysis of Heat and Mass Transfer*, McGraw-Hill, New York (1973).
5. F.P. Imcropera and D.P. Dewitt, *Fundamentals of Heat Transfer*, Wiley, New York (1981).

RADIATION HEAT TRANSFER

C. L. Tien, K. Y. Lee, and A. J. Stretton

INTRODUCTION

Researchers have become increasingly aware of the role of thermal radiation in fires, and a significant amount of recent research work has been published on the subject. An overview of the current understanding in this area can be found in the literature.[1-3] Human safety has made the assessment of fire hazard one of the most important concerns of fire protection engineers. Many fire research experiments in the past were performed in laboratories with very few attempts to simulate actual fire situations, due to the inherent expense and difficulty of controlling large fires. However, these experiments lacked information on thermal radiation, since reduced fire scales often reduce the proportion of radiation as compared to the other modes of heat transfer. It is now recognized that radiation is the dominant mode of heat transfer in flames with characteristic lengths exceeding 0.2 m, while convection is more significant in smaller flames.

Thermal radiation in fires involves energy exchange between surfaces (i.e., walls, ceilings, floors, furniture, etc.) as well as emission and absorption by various gases and soot particles. Among those gases of great practical importance to fire engineers are water vapor and carbon dioxide, which are strongly absorbing-emitting in the major thermal radiation spectrum of 1 to 100 μm (1 μm = 10^{-6} m). Many petroleum-based materials, such as plastics, evolve hydrocarbon gases upon heating, which are also strongly absorbing. In addition, the contribution of the soot particles is very important in evaluating the properties of the participating media and in most situations, soot radiation contributes more than gaseous radiation. Exact calculations of radiative exchanges in fire systems are often prohibitively expensive, even under idealized conditions, due to the dependence of the radiation properties of each material on geometry and wavelength. Many of the simplifying assumptions used in current analytical methods will be covered in this chapter.

This chapter will introduce the fundamentals of thermal radiation and offer simple methods of calculating radiant heat transfer in fires. The first section of the chapter, on basic concepts, deals with the theoretical framework for radiative heat transfer and is followed by the engineering assumptions and simple equations used for practical heat transfer calculations. The third section, on thermal radiation properties of combustion products, covers the properties of various gases and soot present in fires. The last section applies the preceding methods to several fire systems and shows some of the directions of current research.

BASIC CONCEPTS

Radiation Intensity and Energy Flux

Thermal radiation transport can be described by electromagnetic wave theory or by quantum mechanics. In the general quantum mechanical consideration, electromagnetic radiation is interpreted in terms of photons. Each photon possesses energy, $h\nu$, and momentum, $h\nu/c$, with h as the Planck constant (6.6256×10^{-34} J·s), ν the frequency of the radiation, and c the speed of light in the medium. A radiation field is fully described when the flux of photons (or energy) is known for all points in the field for all directions and for all frequencies. The net flow of thermal radiative energy for a single frequency, across a surface of an arbitrary orientation, is represented by the spectral radiative energy flux[4-6]

$$q_\nu = \int_0^{4\pi} I_\nu \vec{n} \cdot \vec{R} \, d\Omega = \int_0^{4\pi} I_\nu \cos\theta \, d\Omega \tag{1}$$

In Equation 1, Ω denotes solid angle ($d\Omega = \sin\theta d\theta d\phi$) and I_ν is the intensity of radiation expressed as energy per unit area per unit solid angle (Figure 1-4.1) within a unit frequency interval. Intensity is a useful measure for thermal radiation because the intensity of a radiant beam remains constant if it is traveling through a nonparticipating medium.

Planck's Law

The energy spectrum of the radiation given off by a surface that is a perfect emitter and absorber can be calculated by Planck's quantum theory. This theoretical surface is called a blackbody radiator, and is best simulated by a small opening into an enclosed cavity. The isotropic equilibrium radiation field within a uniform temperature enclosure is

Dr. C.L. Tien is Professor of Mechanical Engineering at the University of California at Berkeley. Dr. K.Y. Lee is Manager, New Engine Development, Daewoo Technical Center, Inchon, Korea. A.J. Stretton is Assistant Professor of Mechanical Engineering at the University of Toronto.

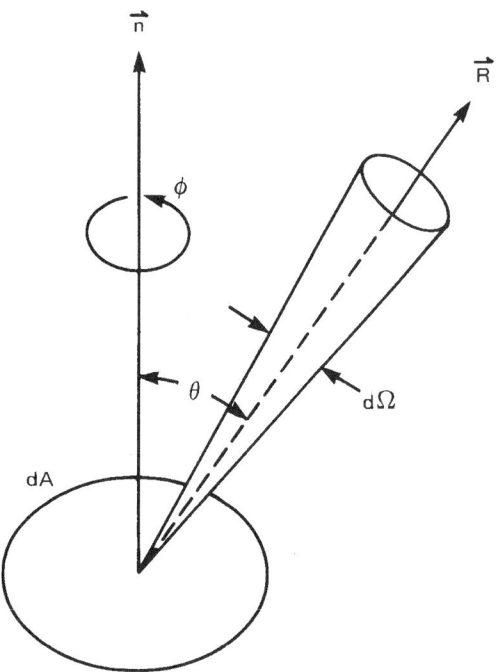

Fig. 1-4.1. Coordinate system for radiation intensity.

called blackbody radiation. The spectral (or monochromatic) intensity of blackbody radiation, $I_{b\nu}$, is often called the Planck function, illustrated in Figure 1-4.2, and is given by Equation 2.

$$I_{b\nu}(T) = \frac{2h\nu^3 n^2}{c_0^2 [\exp(h\nu/kT) - 1]} \quad (2)$$

where k is the Boltzmann constant ($k = 1.3806 \times 10^{-23}$ J/K), c_0 is the speed of light in vacuum ($c_0 = 2.998 \times 10^8$ m/s), and n is the index of refraction for the medium ($n = c_0/c$ is very close to one for most gases of interest in fires). In many engineering applications and experimental measurements of thermal radiation properties, it is advantageous to use wavelength λ instead of ν. Equation 2 can then be expressed in the form

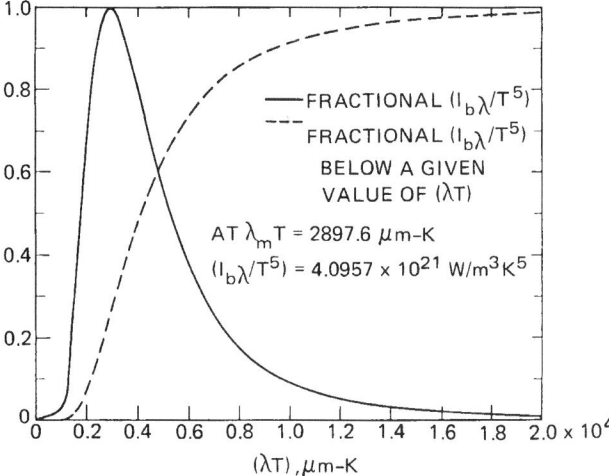

Fig. 1-4.2. Planck's function.

$$\frac{I_{b\lambda}}{T^5} = \frac{2hc_0^2}{n^2(\lambda T)^5 [\exp(hc_0/n\lambda kT) - 1]} \quad (3)$$

where the relations $\nu = c_0/n\lambda$ and $I_{b\lambda}d\lambda = -I_{b\nu}d\nu$ have been used. The wavelength at which radiation intensity becomes the maximum is readily obtainable by simple differentiation as

$$(n\lambda T)_{max} = 2897.8 \ \mu m - K \quad (4)$$

This relationship is known as Wien's displacement law, which shows that the maximum monochromatic emmisive power of a blackbody shifts to shorter wavelengths as its temperature increases. The total radiant intensity from a blackbody radiator can be obtained by integrating Equation 3 over the entire range of wavelengths, giving

$$I_b = \int_0^\infty I_{b\lambda} \ d\lambda = \frac{n^2\sigma T^4}{\pi} \quad (5)$$

where σ is the Stefan-Boltzmann constant (5.6696×10^{-8} W/m^{2}K^4). The intensity of radiation from a blackbody is independent of direction, which allows integration of Equation 1 in a simple manner to give the total hemispherical emissive power per unit area of a blackbody

$$E_b = \int_0^{4\pi} I_b \cos \theta \ d\Omega \quad (6)$$

Kirchhoff's Law

If a fire in an isolated, uniform temperature enclosure that contains different media inside has reached its equilibrium state, the relation

$$\alpha_\nu + \rho_\nu + \tau_\nu = 1 \quad (7)$$

will hold at the interface between each medium, where α, ρ, and τ denote the fraction of energy absorbed by, reflected at, and transmitted through the interface, respectively. The assumption of the local thermodynamic equilibrium can be used to derive more extensive results and is used extensively in radiation heat transfer calculations. Kirchhoff's law states that in order to maintain equilibrium, the spectral absorptivity and spectral emissivity must be related by

$$\alpha_\nu = \frac{I_\nu}{I_{b\nu}} = \epsilon_\nu \quad (8)$$

More importantly, when Equation 8 is applied to total properties,

$$\alpha_t = \epsilon_t \quad (9)$$

is valid for the special case when the incident radiation is independent of the incident angle and has the same spectral proportions as a blackbody radiator (i.e., a "gray body"). Fortunately, this is the case in many radiation heat transfer engineering models for participating media in fire applications. Although gas emissivity is only dependent on the state of the gas, the absorptivity is also a function of the source temperature of the incident beam of radiation, which may originate outside the gas body (e.g., wall temperature).

The Equation of Transfer

The equation of transfer describes the variation in intensity of a radiant beam at any position along its path in an absorbing-emitting-scattering medium. This equation is the foundation upon which detailed radiation analyses are based, and the source of approximate solutions when simplifying assumptions are made. For a given direction line in the medium, the equation of transfer is

$$\frac{1}{\kappa_\lambda(T, S)} \frac{dI_\lambda(S)}{dS} + I_\lambda(S) = I_{b\lambda}(T) \qquad (10)$$

where S represents the physical pathlength and κ_λ represents the spectral extinction coefficient, which includes the effects of both absorption and scattering within the medium. The intensity, $I_\lambda(S)$, is coupled with the spatial distribution of the extinction coefficient and with temperature through conservation of energy in the medium. The contributions of intensity passing through an area must be integrated over all directions to calculate a net radiative energy flux. The integral nature of radiation makes analysis difficult and simplifications necessary for engineering practice.

BASIC CALCULATION METHODS

Energy Exchange in a Nonparticipating Medium

In this section, cases are examined where the surfaces are separated by a medium that does not emit, absorb, or scatter radiation. A vacuum meets this requirement exactly, and common diatomic gases of symmetric molecular structure such as N_2, O_2, and H_2 are very nearly nonparticipating media within the thermal radiation spectrum. The radiative energy transfer between the surfaces depends on the geometry, orientation, and temperature of the surfaces, while the material surface radiative properties are a function of temperature, bounding medium, direction, and polarization of radiation. In practice, most surfaces (either entire or subdivided) are assumed to be isothermal, surface radiation properties are approximated by those of ideal diffuse surfaces, and polarization effects are neglected. The geometry and orientation of each surface is commonly accounted for in calculations by one or more configuration factors, which are also known as view factors, shape factors, angle factors, and geometric factors.[5-8] A configuration factor is a purely geometrical relation between two surfaces, and is defined as the fraction of radiation leaving one surface which is intercepted by the other surface.

Configuration factors and their algebra: Consider the two arbitrarily oriented surfaces A_1 and A_2 in Figure 1-4.3. Assuming that the radiosity from dA_1 on A_1 is diffusely distributed, the configuration factor for the differential area dA_1 to the finite area A_2, F_{d1-2}, is given by

$$F_{d\,1-2} = \int_{A_2} \frac{\cos \beta_1 \cos \beta_2}{\pi |\vec{R}|^2} \, dA_2 \qquad (11)$$

where the separation distance between the two surfaces is denoted by $|\vec{R}|$, β is the angle between the line of sight $\vec{R}$ and the surface normal $\vec{n}$, and A_2 is the area of surface 2. If the assumption of the radiosity distribution on surface A_1 is extended to include a uniform radiosity distribution over A_1 (physically, a uniform radiant heat flux from an isothermal surface), then the configuration factor for the finite area A_1 to A_2, F_{1-2}, is simply

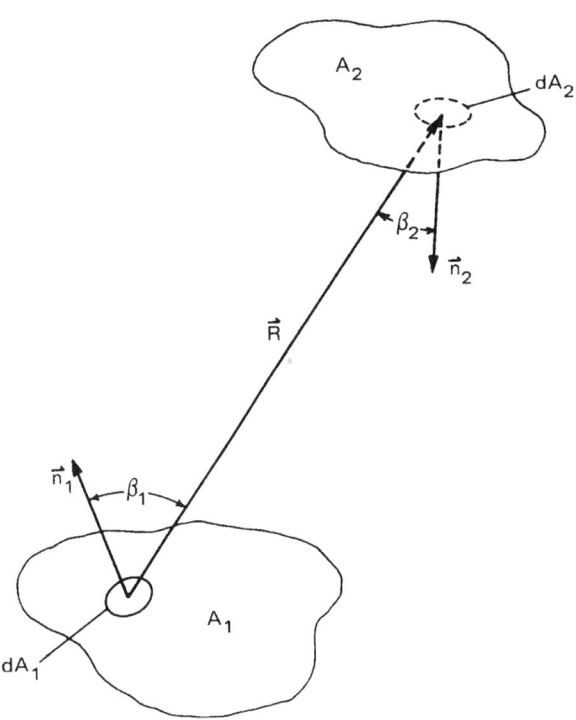

Fig. 1-4.3. *Coordinate system for shape factors.*

$$F_{1-2} = \frac{1}{A_1} \int_{A_1} \int_{A_2} \frac{\cos \beta_1 \cos \beta_2}{\pi |\vec{R}|^2} \, dA_1 \, dA_2 \qquad (12)$$

When the radiant fluxes from both surfaces are uniformly and diffusely distributed (a common engineering assumption), a reciprocity relation for the configuration factors for any given pair in a group of exchanging surfaces can readily be obtained to be

$$A_i F_{i-j} = A_j F_{j-i} \qquad (13)$$

The summation rule is another useful relation for calculating unknown configuration factors

$$\sum_j F_{i-j} = 1 \qquad (14)$$

where F_{i-j} relate to surfaces that subtend a closed system. Note that it is possible for a concave surface to "see" itself, which can make F_{i-i} important.

All configuration factors can be derived using the multiple integration of Equations 11 and 12, but this is generally very tedious except for simple geometries. A large number of cases have already been tabulated with the numerical results or algebraic formulas available in various references.[5-7] A catalog of common configuration factors is provided in Table 1-4.1. This data base can be extended to cover many other situations by the use of configuration algebra and the method of surface decomposition. In surface decomposition, unknown factors can be determined from known factors for convenient areas or for imaginary surfaces which can extend real surfaces or form an enclosure.[5,6]

Gray diffuse surfaces: For engineering purposes, the emittance from most surfaces is treated as having diffuse directional characteristics independent of wavelength and temperature. Real surfaces exhibit radiation properties that are

TABLE 1-4.1 *Common Configuration Factors*

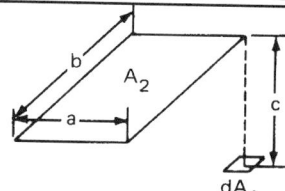

$X = a/c, \quad Y = b/c$

$$F_{d1-2} = \frac{1}{2\pi} \left\{ \frac{X}{\sqrt{1+X^2}} \tan^{-1}\left[\frac{Y}{\sqrt{1+X^2}}\right] + \frac{Y}{\sqrt{1+Y^2}} \tan^{-1}\left[\frac{X}{\sqrt{1+Y^2}}\right] \right\}$$

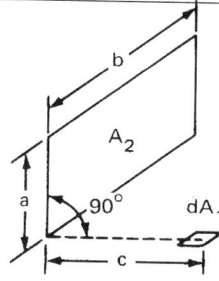

$X = a/b, \quad Y = c/b, \quad A = 1/\sqrt{X^2 + Y^2}$

$$F_{d1-2} = \frac{1}{2\pi} \left\{ \tan^{-1}(1/Y) - AY \tan^{-1}A \right\}$$

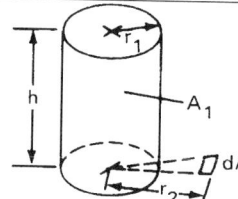

$R = r_1/r_2, \quad L = h/r_2, \quad X = \sqrt{(1 + L^2 + R^2)^2 - 4R^2}$

$$F_{d1-2} = \frac{1}{2\pi} \cos^{-1}R + \frac{1}{\pi}\left\{ \tan^{-1}\left[\frac{R}{\sqrt{1-R^2}}\right] - \frac{1+L^2-R^2}{X} \tan^{-1}\left[\frac{X\tan(0.5\cos^{-1}R)}{1+L^2+R^2-2R}\right] \right\}$$

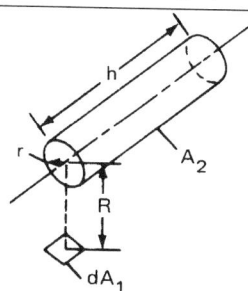

$L = h/r, \quad H = R/r, \quad X = (1 + H)^2 + L^2, \quad Y = (1 - H)^2 + L^2$

$$F_{d1-2} = \frac{1}{\pi H} \tan^{-1}\left[\frac{L}{\sqrt{H^2-1}}\right] + \frac{L}{\pi}\left\{ \frac{X-2H}{H\sqrt{XY}} \tan^{-1}\sqrt{\frac{X(H-1)}{Y(H+1)}} - \frac{1}{H}\tan^{-1}\sqrt{\frac{H-1}{H+1}} \right\}$$

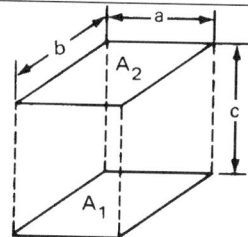

$X = a/c, \quad Y = b/c$

$$F_{1-2} = \frac{2}{\pi XY} \left\{ \ln\sqrt{\frac{(1+X^2)(1+Y^2)}{1+X^2+Y^2}} + X\sqrt{1+Y^2}\tan^{-1}\left[\frac{X}{\sqrt{1+Y^2}}\right] \right.$$
$$\left. + Y\sqrt{1+X^2}\tan^{-1}\left[\frac{Y}{\sqrt{1+X^2}}\right] - X\tan^{-1}X - Y\tan^{-1}Y \right\}$$

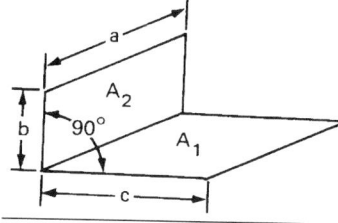

$X = b/a, \quad Y = c/a$

$$F_{1-2} = \frac{1}{\pi Y} \left\{ Y\tan^{-1}\left(\frac{1}{Y}\right) + X\tan^{-1}\left(\frac{1}{X}\right) - \sqrt{X^2+Y^2}\tan^{-1}\left[\frac{1}{\sqrt{X^2+Y^2}}\right] \right.$$
$$\left. + \frac{1}{4}\ln\left[\frac{(1+X^2)(1+Y^2)}{(1+X^2+Y^2)}\left(\frac{X^2(1+X^2+Y^2)}{(1+X^2)(X^2+Y^2)}\right)^{X^2}\left(\frac{Y^2(1+X^2+Y^2)}{(1+Y^2)(X^2+Y^2)}\right)^{Y^2}\right] \right\}$$

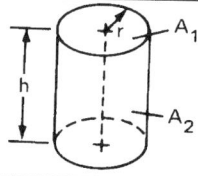

$H = h2r$
$F_{1-2} = 2\{\sqrt{1+H^2} - H\}$

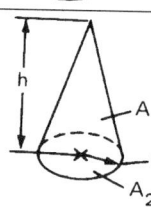

$H = h/r$
$F_{1-2} = 1/\sqrt{1+H^2}$

so complex that information about these property measurements for many common materials is not available. The uncertainties associated with the property measurements, combined with the simplifying assumptions used in the calculations, usually reduce the knowledge of the radiative energy transfer to a simple overall flux. The gray diffuse surface is a useful model that alleviates many of the complexities associated with a detailed radiation analysis, while providing reasonably accurate results in many practical situations. The advantage of diffuse surface analysis is that radiation leaving the surface is independent of the direction of the incoming radiation, which greatly reduces the amount of computation required to solve the governing equations. Discussions for specularly reflecting surfaces and nongray surfaces can be found in the literature.[5,6]

A convenient method to analyze radiative energy exchange in an enclosure of diffuse gray surfaces is based on the concept of radiosity and irradiation. The irradiation, G_i, represents the radiative flux reaching the ith surface regardless of its origin

$$G_i = \sum_j F_{i-j} J_j \qquad (15)$$

where J_j is the surface radiosity, defined as the total radiative flux leaving the jth surface (including both emission and reflection)

$$J_i = \epsilon_i E_{bi} + \rho_i G_i \qquad (16)$$

The net loss of radiative energy is then given by

$$Q_i = (J_i - G_i) A_i \qquad (17)$$

It should be reemphasized that the radiosity-irradiation formulation is based on the assumption that each surface has uniform radiosity and irradiation (or equivalently, uniform temperature and uniform heat flux). Physically unrealistic calculations can result if each surface does not approximately satisfy this condition. Larger surfaces should be subdivided into smaller surfaces if necessary.

Resistance network method: The radiosity-irradiation formulation allows a more physical and graphic interpretation using the resistance network analogy. Eliminating the irradiation G_i from Equations 15 through 17, and substituting $\rho_i = 1 - \epsilon_i$ gives

$$Q_i = \frac{E_{bi} - J_i}{(1 - \epsilon_i)/(\epsilon_i A_i)} = \sum_j \frac{J_i - J_j}{1/(A_i F_{i-j})} \qquad (18)$$

The denominator in the last term of Equation 18 corresponds to resistance in electric circuits. As illustrated in Figure 1-4.4, the diffuse-gray surface has a radiation potential difference $(E_{bi} - J_i)$ and a resistance $(1 - \epsilon_i)/\epsilon_i A_i$. This simple example also illustrates that an adiabatic surface, such as a reradiating or refractory wall, exhibits a surface temperature that is independent of the surface emissivity or reflectivity.

Thermal Radiation in Participating Media

Spectral emissivity and absorptivity: From a microscopic viewpoint, emission and absorption of radiation are caused by the change in energy levels of atoms and molecules due to interactions with photons. A summary and discussion of these effects in gases from an engineering perspective has been written by Tien.[9] Consider a monochromatic beam of

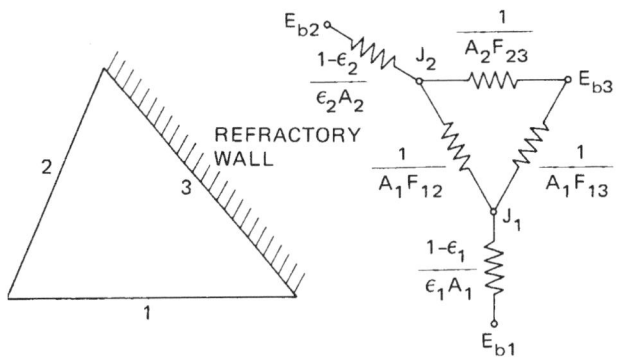

Fig. 1-4.4. Network analogy for radiative exchange.

radiation passing through a radiating layer of thickness L; provided that the temperature and properties of the medium are uniform along the path, the intensity of radiant beam at point x is given by integration of Equation 10 as

$$I_\lambda(x) = I_\lambda(0)e^{-\kappa_\lambda x} + I_{b\lambda}(1 - e^{-\kappa_\lambda x}) \qquad (19)$$

which accounts for the loss of intensity by absorption and the gain by emission, and where κ_λ denotes the extinction coefficient. The extinction coefficient is generally the sum of two parts: the absorption coefficient and the scattering coefficient. In many engineering applications, the effects of scattering are negligible and the extinction coefficient represents only absorption. The spectral emissivity for pathlength S in a uniform gas volume can be readily expressed by considering the case of no incident radiation (or $I_\lambda(0) = 0$)

$$\epsilon_\lambda = \frac{I_\lambda}{I_{b\lambda}} = 1 - e^{-\kappa_\lambda S} \qquad (20)$$

which compares the fraction of energy emitted to the maximum (blackbody) emission at the same temperature for the pathlength S through the material.

The term $\kappa_\lambda S$ in Equation 20, called the optical pathlength or opacity, can be defined more generally for nonhomogeneous media as

$$\tau_\lambda = \int_0^S \kappa_\lambda(x) \, dx \qquad (21)$$

If $\tau_\lambda << 1$, the medium is optically thin at wavelength λ and the properties of the participating medium can generally be expressed with the very simple approximation $\epsilon_\lambda \approx \tau_\lambda$. The medium is considered optically thick when $\tau_\lambda >> 1$, which implies that the mean penetration distance is much less than the characteristic length of the medium. In optically thick media, the local radiant intensity results only from local emission and the equation of transfer can be approximated by a diffusion equation.

Total emissivity: Total emissivity is an average property over all wavelengths, defined by

$$\epsilon_t = \frac{\pi}{\sigma T^4} \int_0^\infty \epsilon_\lambda I_{b\lambda} \, d\lambda \qquad (22)$$

At moderate temperatures up to about 2000 K (which is the range of interest for fire protection engineers), the total emissivity of combustion gases consists of contributions from discrete bands, with negligible contributions from wavelengths between the bands. It is thus convenient to use

$$\epsilon(T, S, P_e) \approx \pi \sum_i \left(\frac{I_{\lambda bi}}{\sigma T^4} \right) \epsilon_i(T, X, P_e) \qquad (23)$$

where $I_{\lambda bi}$ is the blackbody intensity evaluated at the center of the ith band, and ϵ_i is the total band absorption defined by

$$\epsilon_i = \int_{(\Delta\lambda)_i} (1 - e^{-\kappa_\lambda S}) \, d\lambda \qquad (24)$$

Recent progress on band structure and absorption has made it possible to determine total emissivity information by both theoretical and experimental means. Engineers traditionally determine the total emissivity of a homogeneous gas by graphical interpolation from charts with temperature and pressure-pathlength as parameters. The total emissivity of a mixture of gases cannot be determined simply by adding the total emissivities of the various components, because the active spectral bands for a combination of gases will often overlap. The correction for band overlapping should be calculated from spectral information for each gas, which can be estimated from the wide-band model of Edwards.[10,11]

Mean absorption coefficient: The mean absorption coefficient is often useful when radiative energy transport theory must be used to describe the local state of a gas at various locations. The mathematical complexity involved in the calculations often dictates a solution based on the gray-gas assumption, where all radiation parameters are considered to be wavelength independent. Thus solutions are given in terms of mean (gray-gas) absorption coefficients representing average properties over the whole spectrum of wavelengths. It has been well established that the appropriate mean absorption coefficients are the Planck mean, κ_P, for optically thin mediums, and the Rosseland mean, κ_R, for optically thick mediums.[4-6,9]

The Planck mean absorption coefficient is defined as

$$\kappa_P \equiv \frac{\int_0^\infty I_{b\lambda} \kappa_\lambda \, d\lambda}{\int_0^\infty I_{b\lambda} \, d\lambda} = \frac{\pi}{\sigma T^4} \int_0^\infty I_{b\lambda} \kappa_\lambda \, d\lambda \qquad (25)$$

It is important to note that this form of the absorption coefficient is a function of temperature alone and is independent of pressure. The effect of the beam source temperature (e.g., a hot or cold wall) in the gas absorptivity is approximated by a simple ratio correction[9,11]

$$\kappa_m = \kappa_P(T_s) \frac{T_s}{T_g} \qquad (26)$$

where T_s is the source temperature and T_g is the gas temperature. When the Planck mean absorption coefficient is used to estimate the emissivity of a gas, the source temperature is set equal to the gas temperature.

The formulation of radiative transfer becomes relatively simple when the medium is optically thick. In this case, the radiative transfer can be regarded as a diffusion process (the Rosseland or diffusion approximation), and the governing equation is approximated by

$$q_\lambda \approx -\frac{4}{3\kappa_\lambda} \frac{\partial e_{b\lambda}}{\partial S} \qquad (27)$$

Evaluation of the total heat flux in an optically thick medium is simplified by defining an average absorption coefficient which is independent of wavelength

$$\frac{1}{\kappa_R} \equiv \int_0^\infty \frac{1}{\kappa_\lambda} \frac{d e_{b\lambda}}{d e_b} \, d\lambda \qquad (28)$$

The Rosseland mean absorption coefficient is not well de-

fined for gases under ordinary conditions, because astronomically long pathlengths are required to make the windows between the bands optically thick. The Rosseland limit is, however, useful when dealing with gases in the presence of soot particles, which are characterized by a continuous spectrum. The source temperature effect is accounted for by using Equation 26 in the same manner as for the Planck mean absorption coefficient.

The radiating gas in many actual fire systems is neither optically thin nor optically thick, so it is necessary to use band theory to calculate a mean absorption coefficient, κ_m. However, with a reasonable estimate of the mean absorption coefficient radiative transport calculations are much more convenient.

Mean beam length for homogeneous gas bodies: The concept of mean beam length is a powerful and convenient tool to calculate the energy flux from a radiating homogeneous gas volume to its boundary surface. It may also be used to approximate radiative energy flux for a nonhomogeneous gas, especially when more elaborate calculations are not feasible. Consider the coordinate system given in Figure 1-4.1, where dA is a differential area on the boundary surface of the gas body. The radiative heat flux from the gas body to dA is

$$q = \int_0^\infty \int_\Omega \epsilon_\lambda(X) I_{b\lambda} \cos\theta \, d\lambda \, d\Omega \qquad (29)$$

where the spectral emissivity, ϵ_λ, is a function of pressure pathlength

$$X \equiv \int_0^S P_a \, x(\xi) \, d\xi$$

which in turn varies with solid angle Ω according to the gas body geometry.

In practical situations, the calculation of q is more convenient in terms of total emissivity, which is often available in chart form. From the definition of total emissivity, Equation 29 can be expressed as

$$q = \frac{\sigma T^4}{\pi} \int_\Omega \epsilon_t(X) \cos\theta \, d\Omega \equiv \sigma T^4 \epsilon_t(L) \qquad (30)$$

which gives the definition of mean beam length, L, for a gas body, where $\epsilon_t(L)$ has the same functional form as $\epsilon_t(X)$. Physically, the mean beam length represents the equivalent radius of a hemispherical gas body such that it radiates a flux to the center of its base equal to the average flux radiated to the boundary surface by the actual volume of gas. The determination of the mean beam length is considerably simplified when the gas is optically thin and only the geometry of the gas body enters the calculation. In the optically thin limit, it is convenient to define

$$L = L_0 \equiv \frac{1}{\pi} \int_\Omega X \cos\theta \, d\Omega \qquad (31)$$

where L_0 is called the geometric mean beam length. In the optically thick limit, it has been found that the use of a simple correction factor provides reasonable radiative fluxes

$$L \approx C L_0 \qquad (32)$$

In Table 1-4.2, L_0 and C have been provided for a variety of gas body shapes. For an arbitrarily shaped gas volume, the geometric beam length from the gas volume to the entire boundary surface can be estimated by

TABLE 1-4.2 *Mean Beam Lengths for Various Gas Body Shapes*

Geometry of Gas Body	Radiating to	Geometric Mean Beam Length L_0	Correction Factor C
SPHERE	Entire surface	0.66 D	0.97
CYLINDER H = 0.5D	Plane end surface Concave surface Entire surface	0.48 D 0.52 D 0.50 D	0.90 0.88 0.90
CYLINDER H = D	Center of base Entire surface	0.77 D 0.66 D	0.92 0.90
CYLINDER H = 2D	Plane end surface Concave surface Entire surface	0.73 D 0.82 D 0.80 D	0.82 0.93 0.91
SEMI-INFINITE CYLINDER H → ∞	Center of base Entire base	1.00 D 0.81 D	0.90 0.80
INFINITE SLAB	Surface element Both bounding planes	2.00 D 2.00 D	0.90 0.90
CUBE D x D x D	Single face	0.66 D	0.90
BLOCK D x D x 4D	1 × 4 face 1 × 1 face Entire surface	0.90 D 0.86 D 0.89 D	0.91 0.83 0.91

$$L_0 = \frac{4V}{A} \qquad (33)$$

where V and A are the volume and the area of the boundary surface of the gas body, respectively. The correction factor C should be estimated as 0.9, which is close to the known values for a wide range of geometries.

THERMAL RADIATION PROPERTIES OF COMBUSTION PRODUCTS

Radiation Properties of Gases

The emissivity of any gas is a strong function of wavelength, varying by as much as several orders of magnitude over minute changes in wavenumber. However, the level of accuracy required in engineering calculations, where many of the parameters are difficult to measure or estimate, seldom requires high resolution spectra of emissivity. Where wavelength dependence of the radiative heat flux is a concern, the properties of the gas may be calculated by means of the exponential wide-band model.[10] The uncertainties involved in estimating parameters to calculate radiative heat flux make average properties such as total emissivity a useful tool. The first comprehensive total emissivity charts were formulated by H. C. Hottel and coworkers to summarize work performed up to about 1945. Modern formulations for the emissivity of gases have been summarized by Edwards.[11]

New total emissivity charts for water vapor and carbon dioxide[11] have been provided in Figures 1-4.5 and 1-4.6, respectively. The gas emittance can be found from the charts by knowing the partial pressure and temperature of each gas and the mean beam length for the gas volume geometry. Correction factors for the chart emissivities are available in the

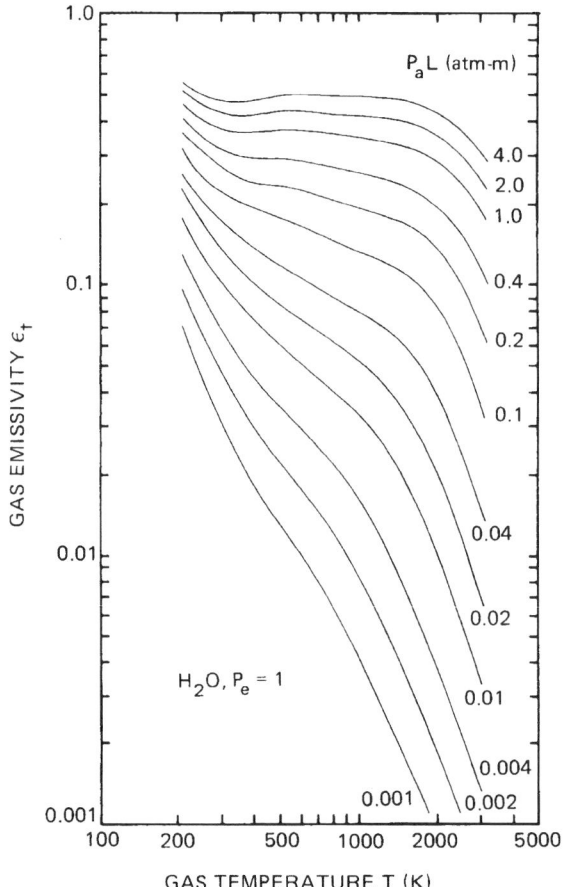

Fig. 1-4.5. Total emittance of water vapor.

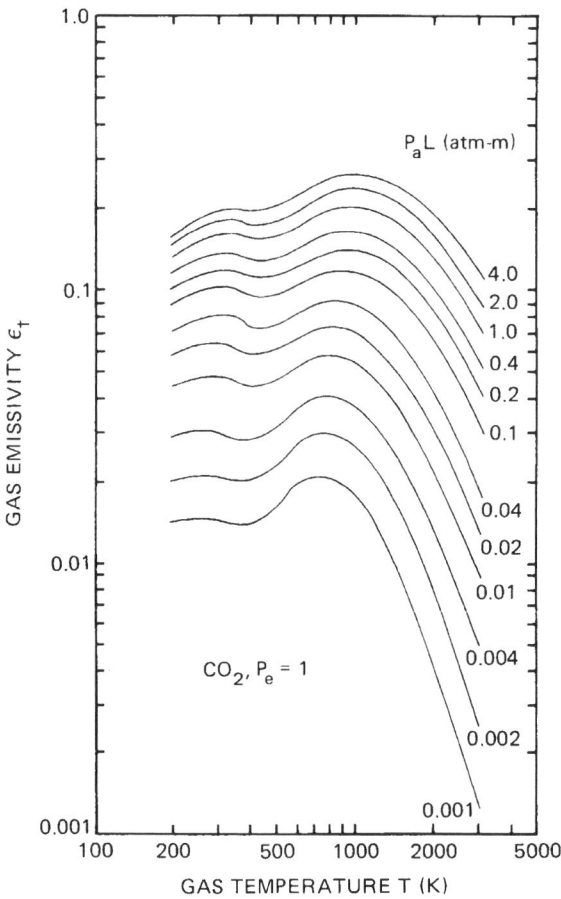

Fig. 1-4.6. Total emittance of carbon dioxide.

literature for the pressure effect on water vapor emissivity,[12] the pressure effect on carbon dioxide emissivity,[4,6] and the band overlap for mixtures of the two gases.[13] For most fire protection engineering applications, the pressure correction factors are 1.0 and the band overlap correction is approximately $\Delta\epsilon \approx \frac{1}{2}\epsilon_{CO_2}$ for medium to large fires. Assuming the carrier gas is transparent (e.g., air), the emittance is given by

$$\epsilon_g = C_{H_2O}\,\epsilon_{H_2O} + C_{CO_2}\,\epsilon_{CO_2} - \Delta\epsilon \approx \epsilon_{H_2O} + \frac{1}{2}\epsilon_{CO_2} \quad (34)$$

At temperatures below 400 K, the older charts by Hottel[4,6] may be more reliable than the new charts used in Figure 1-4.5 and Figure 1-4.6, and the use of wide-band models is advised to estimate the band overlap correction instead of using the correction charts at these lower temperatures.[14] For crucial engineering decisions, wide-band model block calculations as detailed by Edwards[11] are recommended over the graphical chart method to determine total emissivity.

Other gases such as sulphur dioxide, ammonia, hydrogen chloride, nitric oxide, and methane have been summarized in chart form.[4] The carbon monoxide chart by Hottel is not recommended for use according to recent measurements[15] and other theoretical investigations, probably due to traces of carbon dioxide in the original experiments. Recent results, including both spectral and total properties, have recently been published for some of the important hydrocarbon gases, e.g., methane, acetylene, and propylene.[16–18] Mixtures of several hydrocarbon gases are subject to band overlapping, and appro-

priate corrections must be made to avoid overestimating total emissivity of a mixture of fuels.

The total emissivity for a gas in the optically thin limit can be calculated from the Planck mean absorption coefficient. Graphs of the Planck mean absorption coefficient for various gases that are important in fires are shown in Figure 1-4.7, which can be used with Equation 20 to estimate the total emissivity (by assuming that total properties represent a spectral average value).

Radiation Properties of Soot

In a nonhomogeneous (e.g., with soot) medium, scattering becomes an important radiative mechanism in addition to absorption and emission. The absorption and scattering behavior of a single particle can be described by solving the electromagnetic field equations; however, many physical idealizations and mathematical approximations are necessary. The most common assumptions include perfectly spherical particles, uniformly or randomly distributed particles, and interparticle spacing so large that the radiation for each particle can be treated independently.

Soot particles are produced as a result of incomplete soot combustion and are usually observed to be in the form of spheres, agglomerated chunks, and long chains. They are generally very small (50–1000 Å where 1 Å = 10^{-10} m = 10^{-4} μm) compared to infrared wavelengths, so that the Rayleigh limit is applicable to the calculation of radiation

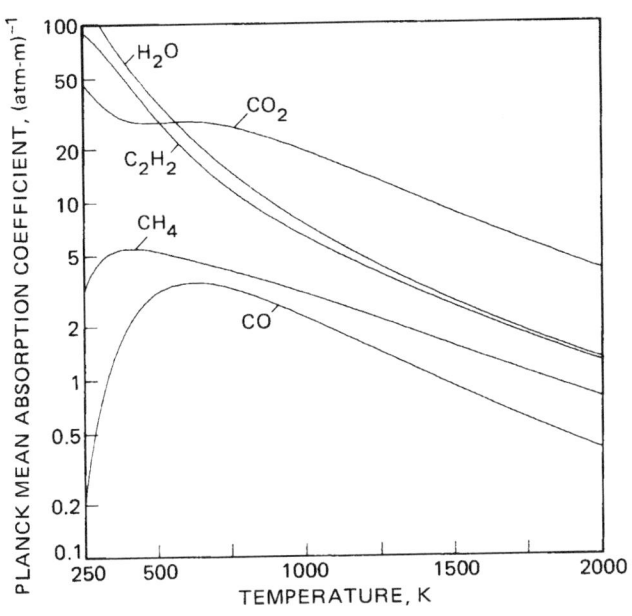

Fig. 1-4.7. Planck mean absorption coefficient for various gases.

properties.[19,20] Soot particles are normally characterized by their optical properties, size, shape, and chemical composition (hydrogen-carbon ratio). From a heat transfer viewpoint, radiation from a soot cloud is predominantly affected by the particle size distribution and can be considered independent of the chemical composition.[19] Soot optical properties are relatively insensitive to temperature changes at elevated temperatures, but as shown in Figure 1-4.8, room temperature values representative of soot in smoke do show appreciable deviations. By choosing appropriate values of optical constants for soot, the solution for the electromagnetic field equations gives[21]

$$k_\lambda = \frac{C_0}{\lambda} f_v \tag{35}$$

where f_v is the soot volume fraction (generally about 10^{-6}),

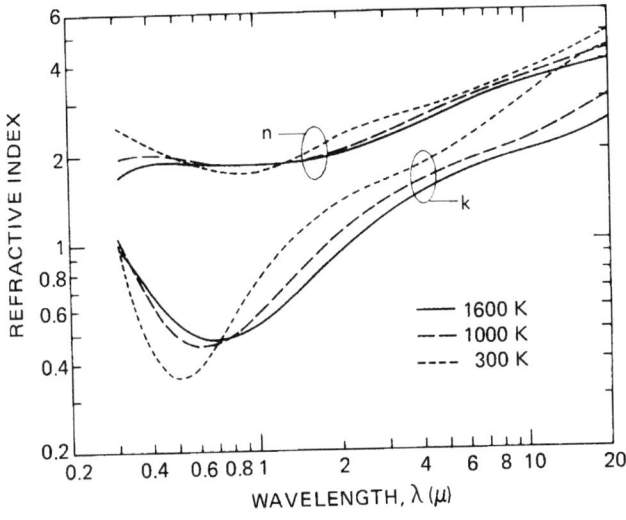

Fig. 1-4.8. Optical constants for soot.

and C_0, a constant between 2 and 6 dependent on the complex index of refraction $m = n - ik$, is given by

$$C_0 = \frac{36\pi nk}{(n^2 - k^2 + 2)^2 + 4n^2k^2} \tag{36}$$

Equations 35 and 36 can be used to evaluate the Planck mean absorption coefficient in the optically thin limit,[22] giving

$$\kappa_P = 3.83 \frac{C_0}{C_2} f_v T \tag{37}$$

where C_2 is Planck's second constant (1.4388×10^{-2} m·K). The Rosseland mean absorption coefficient in the optically thick limit is

$$\kappa_R = 3.6 \frac{C_0}{C_2} f_v T \tag{38}$$

A mean coefficient that may be used for the entire range of optical thickness is suggested as

$$\kappa_m = 3.72 \frac{C_0}{C_2} f_v T \tag{39}$$

to be used in Equation 40 for the soot radiation calculations. Typical temperatures, volume fractions, and mean absorption coefficients for soot particles in the luminous flames of various fuels are tabulated in Table 1-4.3.[21,24]

Radiation Properties of Gas-Soot Mixtures

The calculation of the total emissivity of a gas-soot mixture requires information on basic flame parameters such as soot volume fraction, the absorption coefficient of the soot, the temperature and geometric length of the flame, and the partial pressure of the participating gas components.[23] These parameters can be estimated for various types of fuel when actual measurements are unavailable for a particular situation. Recent research to develop simple accurate formulas to predict total emissivities for homogeneous gas-soot mixtures has found the following equation to be an excellent approximation[24]

$$\epsilon_t = (1 - e^{-\kappa S}) + \epsilon_g e^{-\kappa_s S} \tag{40}$$

where S is the physical pathlength, ϵ_g is the total emissivity of the gas alone, and κ_s is the effective absorption coefficient of the soot. The Planck mean absorption coefficients for gas-soot mixtures in luminous flames and smoke are shown in Figure 1-4.9. In situ measurements are currently the only way other than estimation to obtain the soot volume fraction in smoke, since the soot particle concentration can be either diluted or concentrated by the gas movements within the smoke region.

APPLICATION TO FLAME AND FIRE

Heat Flux Calculation from a Flame

Prediction of the radiative heat flux from a flame is important in determining ignition and fire spread hazard, and in the development of fire detection devices. The shape of flames under actual conditions is arbitrary and time dependent, which makes detailed radiation analysis very cumbersome and uneconomical. In most calculations, flames are idealized as simple geometric shapes such as plane layers or axisymmetric cylinders and cones. A cylindrical geometry will be analyzed here and used in a sample calculation.

TABLE 1-4.3. Radiative Properties for Soot Particles

	Fuel, composition	κ_s (m^{-1})	$f_v \times 10^6$	T_s(K)
Gas Fuels	methane, CH$_4$	6.45	4.49	1289
	ethane, C$_2$H$_6$	6.39	3.30	1590
	propane, C$_2$H$_8$	13.32	7.09	1561
	isobutane, (CH$_3$)$_3$CH	16.81	9.17	1554
	ethylene, C$_2$H$_4$	11.92	5.55	1722
	propylene, C$_3$H$_6$	24.07	13.6	1490
	n-butane, (CH$_3$)(CH$_2$)$_2$(CH$_3$)	12.59	6.41	1612
	isobutylene, (CH$_3$)$_2$CCH$_2$	30.72	18.7	1409
	1,3-butadiene, CH$_2$CHCHCH$_2$	45.42	29.5	1348
Solid Fuels	wood, $\approx$ (CH$_2$O)$_n$	0.8	0.362	1732
	plexiglas, (C$_5$H$_8$O$_2$)$_n$	0.5	0.272	1538
	polystyrene, (C$_8$H$_8$)$_n$	1.2	0.674	1486

Assuming κ_λ is independent of pathlength, integration of the radiative transport of Equation 10 yields[25]

$$I_\lambda = I_{b\lambda}\left\{1 - \exp\left[\frac{-2\kappa_\lambda}{\sin\theta}\sqrt{r^2 - L^2\cos^2\phi}\right]\right\} \quad (41)$$

where θ, ϕ, r and L are geometric variables defined in Figure 1-4.10. The monochromatic radiative heat flux on the target element is given by

$$\frac{dq}{d\lambda} = \int_\Omega \frac{I_\lambda}{|\vec{R}|}(\vec{n}\cdot\vec{R})\,d\Omega \quad (42)$$

where $\vec{n}$ is a unit vector normal to the target element dA and $\vec{R}$ is the line-of-sight vector extending between dA and the far side of the flame cylinder. The evaluation of Equation 42 is quite lengthy, but under the condition of $L/r \gtrsim 3$, it can be simplified to[25]

$$\frac{dq}{d\lambda} = \pi I_{b\lambda}\,\epsilon_\lambda(F_1 + F_2 + F_3) \quad (43)$$

where the shape factor constants and emittance are defined as

$$F_1 = \frac{u}{4\pi}\left(\frac{r}{L}\right)^2[\pi - 2\theta_0 + \sin(2\theta_0)] \quad (44a)$$

$$F_2 = \frac{v}{2\pi}\left(\frac{r}{L}\right)[\pi - 2\theta_0 + \sin(2\theta_0)] \quad (44b)$$

$$F_3 = \frac{w}{\pi}\left(\frac{r}{L}\right)\cos^2\theta_0 \quad (44c)$$

$$\epsilon_\lambda = 1 - \exp(-0.7\mu_\lambda) \quad (45)$$

The parameters in the definitions are given by

$$\theta_0 = \tan^{-1}(L/H) \quad (46a)$$

$$\mu_\lambda = 2r\frac{\kappa_\lambda}{\sin\left(\frac{1}{2}\theta_0 + \frac{1}{4}\pi\right)} \quad (46b)$$

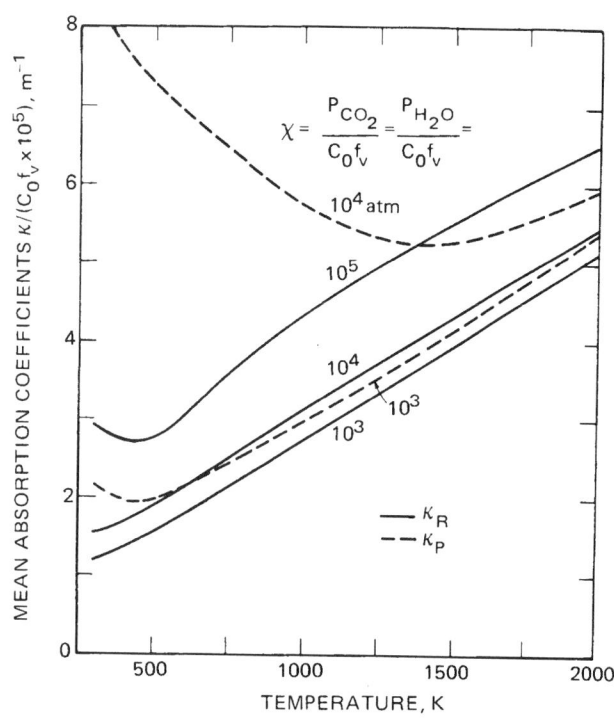

Fig. 1-4.9. Mean absorption coefficients for luminous flames and smoke.

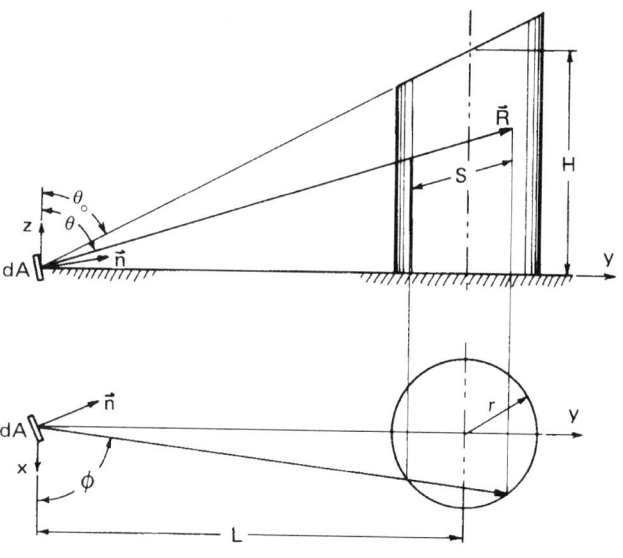

Fig. 1-4.10. Schematic of a cylindrical flame.

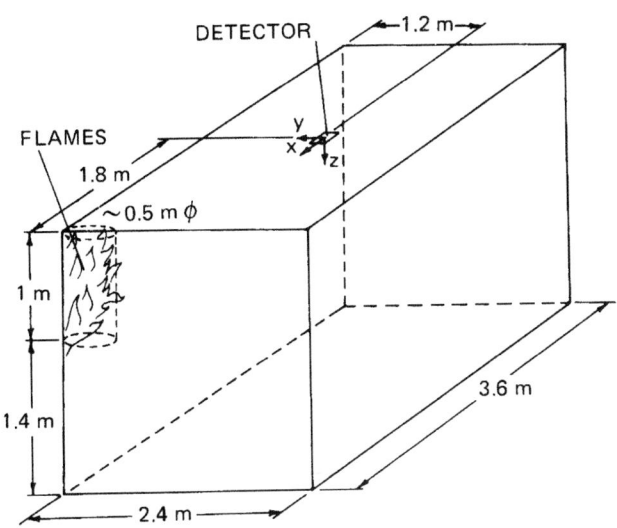

Fig. 1-4.11. *Example calculation for flux to target element from flame.*

$$\vec{n} = u\vec{i} + v\vec{j} + w\vec{k} \qquad (46c)$$

If the flame is considered to be homogeneous and Equation 43 is integrated over all wavelengths, the total heat flux is simply

$$q = \epsilon_m E_b \sum_{j=1}^{3} F_j \qquad (47)$$

EXAMPLE 1:

A fire detector is located at the center of the ceiling in a room ($2.4 \times 3.6 \times 2.4$ m) constructed of wood. (See Figure 1-4.11.) The sprinkler system is capable of extinguishing fires smaller than 0.5 m in diameter $\times$ 1.0 m high. For this example, determine the appropriate heat flux setting for the detector, using a worst case scenario of ignition in one of the upper ceiling corners.

SOLUTION:

First, the condition of $L/r \geq 3$ should be checked to verify that the previous analysis is applicable.

$$L/r = \frac{\sqrt{(1.2^2) + (1.8^2)}}{0.25} = 8.65 > 3$$

The unit normal vector to the detector is given by $\vec{n} = \vec{k}$, the polar angle $\theta_0 = \tan^{-1}(1.818) = 1.068$ is determined from Equation 46a, and the shape factors are evaluated from Equations 44a, b, and c

$$F_1 = F_2 = 0.0$$

$$F_3 = \frac{1}{\pi}\left(\frac{0.25}{1.818}\right)\cos^2(1.068) = 0.0102$$

From Equation 47, the required heat flux can be obtained as

$$q = (1 - e^{-\kappa_m S})(\sigma T_f^4)F_3$$
$$= (1 - e^{0.8 \times 0.5})[5.67 \times 10^{-11} \times (1730)^4](0.0102)$$
$$= 1.7 \text{ kW/m}^2$$

If the geometry of the example had been $L/r < 3$, it would have been necessary to interpolate between the $L/r = 3$ case

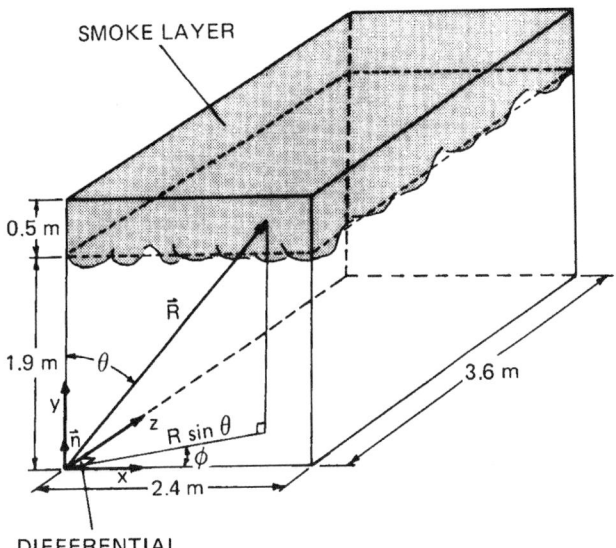

Fig. 1-4.12. *Example calculation for flux to target element from smoke layer.*

and the $L/r = 0$ case, which has been obtained accurately.[6,25] If the detector is pointed directly at the burning corner in this example (i.e., $\vec{n} = 0.55\,\vec{i} + 0.83\,\vec{j}$), the calculated heat flux jumps to 9.0 kW/m^2, showing the strong influence of direction in calculations of radiation heat transfer.

Heat Flux Calculation from a Smoke Layer

Consider radiative heat transfer in a room fire situation where a smoke layer is built up below the ceiling. Typical smoke layers are generally at temperatures ranging up to 1100–1500 K, and are composed of strongly participating media such as carbon dioxide, water vapor, and soot particles. Heat flux from the smoke layer has been directly related to ignition of remote surface locations such as furniture or floor carpets. The schematic in Figure 1-4.12 will be considered in a radiative transport analysis and example calculation. The calculation is based on a considerably simplified formulation which provides reasonable results with only a small penalty in accuracy.

Integration of Equation 10 over the pathlength S through the smoke layer yields

$$I(S) = \frac{\sigma T^4}{\pi}\{1 - (T_w/T)^4\}e^{-\kappa S} \qquad (48)$$

The monochromatic radiative heat flux on a differential target element is again given by Equation 42. However, for the present geometry of the ceiling layer and enclosure surface, integration of Equation 42 is quite time-consuming since the upper and lower bounds of the integral vary with the angle of the pathlength. The calculation can be simplified by assuming as a first order approximation that the lower face of the smoke layer is an isothermal surface. Using this assumption, the problem can be handled using the simple relations of radiative exchange in a nonparticipating medium between gray surfaces (the absorption of the clear air below the smoke layer is negligible). From basic calculation methods we have the radiosity and the irradiation of each surface in the enclosure:

$$J_i = \varepsilon_i \sigma T_i^4 + (1 - \varepsilon_i)G_i \qquad (49a)$$

$$G_i = \sum_j F_{i-j} J_j \qquad (49b)$$

After solving the simultaneous equations for all J_i and G_i, the net heat flux on any of the surfaces can be calculated from

$$q_i = J_i - G_i \qquad (50)$$

EXAMPLE 2:

A smoke layer 0.5 m thick is floating near the ceiling of a room with dimensions of $3.6 \times 2.4 \times 2.4$ m. (See Figure 1-4.12.) The floor is made from wood, and the four side walls are concrete covered with zinc white oil paint. The calculation will determine the heat flux in a bottom corner of the room, assuming that each surface in the enclosure is kept at constant temperature: the smoke layer at 1400 K, the side walls at 800 K, and the floor at 300 K. Assume there is a differential target area 0.01 m^2 in one of the corners of the floor, and also at the floor temperature of 300 K.

SOLUTION:

The bottom of the smoke layer will be designated surface 1, the floor will be surface 2, and the differential target area in the bottom corner will be surface 3. Only four surfaces are required since the four side walls can be treated as a single surface 4. Shape factors F_{12} and F_{31} can be found in Table 1-4.1, and from these two factors, the remaining shape factors are determined by shape factor algebra.

$$F_{12} = 0.3242,$$

$$F_{31} = 0.1831,$$

$$F_{13} = \frac{A_3}{A_1} F_{31} = 0.0002,$$

$$F_{14} = 1 - F_{12} - F_{13} = 0.6756$$

Continuing in a similar fashion, the other shape factors are obtained as

$F_{21} = 0.3242$	$F_{31} = 0.1831$	$F_{41} = 0.2560$
$F_{22} = 0$	$F_{32} = 0$	$F_{42} = 0.2561$
$F_{23} = 0$	$F_{33} = 0$	$F_{43} = 0.0003$
$F_{24} = 0.6758$	$F_{34} = 0.8169$	$F_{44} = 0.4876$

The emissivity for wood and white zinc paint are 0.9 and 0.94, respectively,[6] and the emissivity for the smoke layer can be estimated from the mean absorption coefficient for a wood flame (Table 1-4.3) as

$$\varepsilon_1 = 1 - e^{-\kappa_m S} = 1 - e^{-0.8 \times 0.5} = 0.33$$

The blackbody emission flux from each surface is calculated by the simple relation of Equation 6, for example

$$(\sigma T^4)_1 = 5.6696 \times 10^{-8}(1400)^4 = 217.8 \text{ kW/m}^2$$

From Equations 49a and 49b, the radiative fluxes to and from each surface are determined by solving the eight simultaneous equations

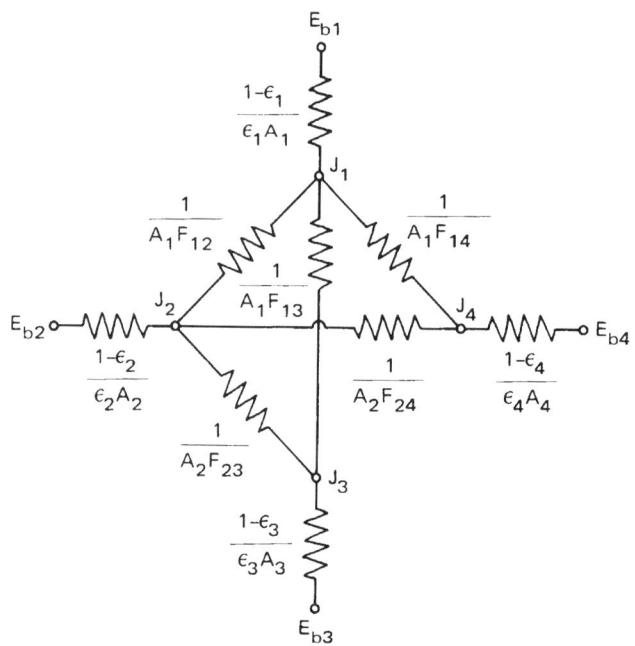

Fig. 1-4.13. Equivalent resistance network for an enclosure problem.

$J_1 = 88.7 \text{ kW/m}^2$	$G_1 = 17.7 \text{ kW/m}^2$
$J_2 = 4.7 \text{ kW/m}^2$	$G_2 = 43.3 \text{ kW/m}^2$
$J_3 = 3.9 \text{ kW/m}^2$	$G_3 = 34.8 \text{ kW/m}^2$
$J_4 = 23.9 \text{ kW/m}^2$	$G_4 = 34.3 \text{ kW/m}^2$

The net radiative heat flux on the target element from Equation 50 is

$$q_3 = J_3 - G_3 = -30.9 \text{ kW/m}^2$$

where the negative sign indicates that heat must be removed from the target element so it remains in equilibrium. This example also could have been solved by the resistance network method. (See Figure 1-4.13.)

Fuel Pyrolysis Rate

Fuel pyrolysis is an important concern in the combustion of condensed fuels, which upon heating undergo gasification (sometimes preceded by liquefication for solid fuels) before combustion in the gaseous phase.[26] This process is often strongly influenced by radiative heat flux. Unlike an internal combustion engine or burner where the fuel is supplied externally, the fuel must be supplied by gasification of the material itself. The rate of gasification is sometimes called the pyrolysis rate or burning rate, and can serve as a measure of fire hazard since it is directly proportional to the growth rate of the fire. Because determination of the pyrolysis rate is based on conservation of energy and mass at the surface of the material, it is essential to know the total heat flux reaching the fuel surface. Assuming steady-state conditions, the energy balance can be expressed as

$$q_e + q_c + q_r + q_{rr} = \dot{m}'' \Delta H \qquad (51)$$

where q represents the heat flux (the subscripts are external, convective, radiative, and reradiative, respectively), $\dot{m}''$ is the pyrolysis rate, and ΔH is the latent heat of gasification. The configuration of the fire and the thermophysical properties of the fuel are required to calculate the terms in the energy balance, with the exception of the external flux term, which represents heat exchanged with the environment away from the fire.

Analysis of turbulent combustion with radiation in three-dimensional systems has been an ultimate research goal in the field of combustion for many years. Despite recent progress, the current state of the art is capable of handling only very limited problems in laminar combustion and turbulent combustion in simple geometries. Thermal radiation has been included in only a few special cases such as analysis of stagnation point combustion in boundary layer type flows,[27,28] and empirical studies of pool fire configurations.[29] In this section two cases will outline incorporating radiation into modeling the fuel pyrolysis rate, which can be applied to fire growth rate estimates.

Some basic concepts of the combustion phenomenon should be reviewed. Flames are often categorized as either diffusion or premixed, depending on the dominant physical processes controlling the burning. In a diffusion flame, the characteristic time for transport of the species is much longer than that required for the chemical reaction. Flames in which the oxygen initially separated from the fuel are generally considered to be diffusion flames. In a premixed flame, the fuel and oxygen are mixed together before reaching the combustion zone, so the characteristic times for transport and reaction are comparable in magnitude. The details of the chemical reaction, which even for simple reactions often involve many intermediate reactions and species conservation of intermediate products, are usually simplified in radiation analysis to a one-step irreversible global relationship such as

$$\text{fuel} + \text{oxygen} \rightarrow \text{products} + \text{heat} \qquad (52)$$

Another major simplification that is frequently used is the flame sheet approximation, where it is assumed that the fuel and oxygen react nearly instantaneously upon contact, thus forming an infinitely thin reaction zone. This approximation is quite useful in the study of flames where the chemical reaction kinetics are dominated by the physical process of diffusion, such as a typical room fire. The counterpart to the flame sheet approximation is the flame layer approximation, where the chemical reaction is assumed to take place at a finite rate and creates a reaction zone of finite thickness. The flame layer approximation is applicable to the study of ignition, extinction, flame stability, and other transient flame phenomena.

Pyrolysis rate in boundary layer combustion: Due to the complicated nature of radiative calculations, only one-dimensional radiation in the limit of an optically thin medium has been attempted in boundary layer analysis. Kinoshita and Pagni[27] analyzed stagnation point flow under the approximations of the flame sheet model, unity Lewis number, and film optical depth of less than 0.1. The net effect of radiation heat transfer on the pyrolysis process was small, which is to be expected in an optically thin convective environment; however, the relative importance of the di-

mensionless parameters governing pyrolysis was dramatically altered by the inclusion of radiation in the analysis. The pyrolysis rate and excess (unburned) pyrolyzed gases were strongly dependent on both the wall temperature and the heat of combustion, which had been of secondary importance in the nonradiative analysis. In general, the effect of radiation is to reduce the pyrolysis rate by compensatory surface emission and by radiative loss from the flame to the cold environment, which lowers the flame temperature and decreases the conduction heat flux.

Pyrolysis rate in a pool fire: Applying energy conservation to find the fuel pyrolysis rate in a pool fire has been difficult due to a lack of appropriate information on both convective heat transfer and radiative feedback. Prediction of the pyrolysis rate is still largely dependent on correlations of limited experimental data, but effort has been made to theoretically formulate the convective and radiative contributions.[29] The pyrolysis rate can be calculated from Equation 51, assuming that the external heat supply can be neglected and the radiation terms are given by

$$q_{rr} = \epsilon_s \sigma (T_s^4 - T_\infty^4) \qquad (53)$$

$$q_r = \sigma T_f^4 [1 - e^{\kappa_f L_m}] \qquad (54)$$

where ϵ_s is the fuel surface emissivity (typically 1.0 for liquids and char) and T_f is the flame temperature as represented by a homogeneous isothermal gas volume. The accuracy of Equation 54 is dependent on the values chosen for the flame absorption coefficient, κ_f, and the mean beam length, L. Orloff and De Ris[29] have proposed the use of

$$\kappa_f = \frac{-1}{L_m} \ln \left[1 - \frac{\chi r \dot{q}_a''' L_m}{36 \chi_a \sigma T_f^4} \right] \qquad (55)$$

$$L_m = 3.6 \frac{V_f}{A_b} \qquad (56)$$

where χ_a is a fractional measure of the completeness of combustion lying in the range 0.6 to 0.95 depending on the type of fuel,[30] χ_r is the fraction of heat lost by radiation in the flame,[1,29,30] $\dot{q}_a'''$ is the volumetric heat output of the flame (typically on the order of 1200 kW/m³ in many flames), V_f is the flame volume, T_f is the flame temperature, and A_b is the area of the pool of fuel. Orloff and De Ris also proposed an expression for the convective heat flux

$$q_c = \frac{h_c}{C_p} \left[\frac{\Delta H (\chi_a - \chi_r) r}{\chi_a} - C_p (T_s - T_\infty) \right] \left(\frac{y}{e^y - 1} \right) \qquad (57)$$

where y is defined to be

$$\left(\frac{m'' C_p}{h_c} \right)$$

and r is the stoichiometric mass ratio of fuel to air.

Large-scale fires are distinctly nonhomogeneous in both temperature and gas species concentrations, which makes single-zone flame models difficult to correlate to the available experimental data. A two-zone model has recently been proposed;[31] it successfully predicts the pyrolysis rate of large PMMA (Plexiglas) fires. The flame is modeled as two conical homogeneous layers: a lower cool layer of pyrolyzed fuel gases, and an upper hot layer of product gases and soot.

More experimental data on large-scale pool fires is required to verify the model for fuels other than PMMA.

Ignition Applications

Ignition is a branch of flammability-limiting behavior concerned with the initiation of burning. Ignition is a rate-controlled mechanism in which chemical reaction kinetics play an important role. Prediction of ignition phenomena is largely dependent on the ignition criteria chosen in the analysis.[26] These criteria are currently the center of a vigorous controversy and far from being uniquely defined. Many practical applications of ignition theory are based on knowledge of the ignition temperature, which in turn makes the heat flux directed at the fuel surface the most important physical quantity. Fire prediction often requires the determination of the ignition delay time after the fuel surface is exposed to a given heat flux. The transient nature of ignition makes it necessary to consider full transient energy equations unless the quasi-steady assumption can be invoked, which makes radiation analysis extremely difficult for many ignition applications.

Pilot ignition and spontaneous ignition are two of the main classes in the broad category of ignition. Pilot ignition is generally achieved through localized heating such as a spark or pilot flame, and the flame then propagates into the rest of the fuel material. In contrast, spontaneous or self-ignition occurs as a result of raising the bulk temperature of a combustible gas mixture, and does not require any further external heat supply once combustion has started. Spontaneous ignition requires a higher temperature for the same material than pilot ignition. Radiation heat transfer has generally been neglected in analyses of these mechanisms due to a lack of physical understanding and practical calculation methods,[32] and more work is required in this area to make the radiation calculations worthwhile.

A somewhat different phenomenon occurs in enclosure fires, where excessive radiant heat supply from the fire ignites material away from the flames. This is called secondary or remote ignition and is of special interest to fire protection engineers as a significant source of flame propagation. Quasi-steady analysis, where the gas is treated with a steady analysis and the solid fuel is handled with a transient analysis, has been shown to yield reasonably accurate results.[33] The chemical reaction terms can be neglected for a first order analysis, although they often play an important role in higher order models.

The relatively simple geometry of a semi-infinite solid bounded by a gas can illustrate a one-dimensional radiative analysis.[34] Attention will be focused on the solid region near the interface, so that the transient energy conservation equation is expressed as

$$\frac{\partial T}{\partial t} = \alpha \frac{\partial^2 T}{\partial x^2} \qquad (58)$$

where α is the thermal diffusivity of the solid. The boundary conditions for Equation 58 are given by

$$T = T_i \qquad \text{at } t = 0, \quad x \rightarrow \infty \qquad (59a)$$

$$k \frac{\partial T}{\partial x} + \epsilon_s q_r = h_c (T - T_i) \quad \text{at } x = 0 \qquad (59b)$$

Equation 59b states that conduction, convection, and radiation will be balanced at the fuel surface, and Equation 59a dictates the temperature level. Solution of Equation 58 is straightforward with the Laplace transform technique, giving the result[35]

$$T(x, t) = \left(T_i + \frac{\epsilon_s q_r}{h_c} \right) \left[\text{erfc} \left(\frac{x}{2\sqrt{\alpha t}} \right) \right.$$
$$\left. - \exp \left(h_c x + \alpha h_c^2 t \right) \text{erfc} \left(\frac{x}{2\sqrt{\alpha t}} + \frac{h_c}{k} \sqrt{\alpha t} \right) \right] \quad (60)$$

The ignition delay time (from initial application of the heat flux) can be accurately calculated from Equation 60 if the radiant heat flux, q_r, is known. The ignition delay time calculation is significantly affected by changes in q_r, which is dependent on the radiation properties of the smoke layer beneath the ceiling flames and the relatively cool pyrolyzed gases near the fuel surface. This effect is called radiation blockage or radiation blanketing, and is a current area of attention in the field of flame radiation research.[36] The blockage effect can be accurately calculated if the composition and properties of the smoke layer are known.[34] Another form of thermal energy blockage to the fuel surface is the surface emissivity, ϵ_s, which can have strong wavelength dependence. For example, a fuel such as PMMA is a poor absorber of radiation in wavelengths below 2.5 μm, where the radiant intensity is strongest from typical flame and smoke temperatures, and is an excellent absorber at wavelengths above 2.5 μm. In addition, the total emissivity of a surface can change as the fuel surface liquefies or begins to char due to pyrolysis. Care should be taken when considering the radiative properties of the fuel surface, which can be strongly dependent on the surface conditions.

NOMENCLATURE

A	area (m^2)
C	correction factor for mean beam length
C_0	soot concentration parameter
C_p	specific heat (J/kg·K)
C_2	Planck's second constant (1.4388 × 10^{-2} m · K)
c	speed of light in the medium (m/s)
c_0	speed of light in a vacuum (2.998 × 10^8 m/s)
E	radiative emmisive power (W/m^2)
F_{i-j}	configuration factor from surface i to surface j
f_v	soot volume fraction
G	irradiation or radiative heat flux received by surface (W/m^2)
H	height (m)
h	Planck's constant (6.6256 × 10^{-34} J · s)
h_c	convective heat transfer coefficient (W/m^2 · K)
I	radiation intensity (W/m^2)
$\vec{i}, \vec{j}, \vec{k}$	Cartesian coordinate direction vectors
J	radiosity or radiative heat flux leaving surface (W/m^2)
k	Boltzmann constant (1.3806 × 10^{-23} J/K), or infrared optical constant of soot (imaginary component), or thermal conductivity (W/m · K)
L	mean beam length or distance (m)
L_0	geometrical mean beam length (m)
M_i	molecular weight of species i
m''	mass loss rate or pyrolysis rate (kg/m^2 · s)
n	index of refraction (c_0/c) or infrared optical constant of soot (real component)
$\bar{n}$	unit normal vector
P_a	partial pressure of absorbing gas (Pa)

P_e	effective pressure (Pa)
Q	energy rate (W)
q	heat flux (W/m^2)
q_a'''	volumetric heat output (W/m^3)
$\vec{R}$	line of sight vector
r	radius of cylinder (m) or fuel/air stoichiometric mass ratio
S	pathlength (m)
T	temperature (K)
t	time (s)
u, v, w	Cartesian components of unit vector $\vec{n}$ volume (m^3)
X	Pressure pathlength, $\int_0^s P_a\, x(\xi)\, d\xi$ (atm-m)
x	spatial coordinate (m)
y	defined parameter, Equation 57

Greek Symbols

α	absorptivity or thermal diffusivity $k/\rho C_p$ (m^2/s)
β	angle from normal (radians)
ΔH	latent heat of gasification (J/kg)
ϵ	emissivity
θ	polar angle (radians)
κ	extinction coefficient or absorption coefficient (m^{-1})
λ	wavelength (m)
μ	micron (10^{-6} m)
μ_λ	defined parameter, Equation 46b
ν	frequency (s^{-1})
ξ	integration dummy variable
ρ	reflectivity or density (kg/m^3)
Ω	solid angle (steradians)
σ	Stefan-Boltzmann constant (5.6696 $\times$ 10^{-8} W/m^{2}K^4)
τ	transmissivity or optical pathlength
ϕ	azimuthal angle (radians)
χ	fractional measure

Subscripts

a	actual
b	blackbody or base
c	convective
e	external
f	flame or fuel
g	gas
i	initial or ith surface
j	summation variable or jth surface
m	mean value
0	original
P	Planck mean
R	Rosseland mean
r	radiative
rr	reradiative
s	surface or soot
t	total
w	wall
λ	spectral wavelength
ν	spectral frequency
∞	ambient

REFERENCES CITED

1. J. deRis, *17th Symp. (Int) Comb.*, 1003, Comb. Inst. (1979).
2. S.C. Lee and C.L. Tien, *Prog. Energy Comb.Sci.*, 8, 41 (1982).
3. G.M. Faeth, S.M. Jeng, and J. Gore, in *Heat Transfer in Fire and Combustion Systems*, American Society of Mechanical Engineers, New York (1985).
4. H.C. Hottel and A.F. Sarofim, *Radiative Heat Transfer*, McGraw-Hill, New York (1967).
5. E.M. Sparrow and R.D. Cess, *Radiation Heat Transfer*, McGraw-Hill, New York (1978).
6. R. Siegel and H.R. Howell, *Thermal Radiation Heat Transfer*, McGraw-Hill, New York (1981).
7. J.R. Howell, *A Catalog of Radiation Configuration Factors*, McGraw-Hill, New York (1982).
8. C.L. Tien, in *Handbook of Heat Transfer Fundamentals*, McGraw-Hill, pp 14.36, New York (1985).
9. C.L. Tien, *Advances in Heat Trans.*, 5, 253 (1968).
10. D.K. Edwards, *Advances in Heat Trans.*, 12, 115 (1976).
11. D.K. Edwards, in *Handbook of Heat Transfer Fundamentals*, McGraw-Hill, pp 14.53, New York (1985).
12. C.B. Ludwig, W. Malkmus, J.E. Reardon, and J.A.L. Thompson, *Handbook of Radiation from Combustion Gases*, NASA SP-3080, Washington (1973).
13. T.F. Smith, Z.F. Shen, and J.N. Friedman, *J. Heat Trans.*, 104, 602 (1982).
14. J.D. Felske and C.L. Tien, *Comb. Sci. Tech.*, 11, 111 (1975).
15. M.M. Abu-Romia and C.L. Tien, *J. Quant. Spec. Radiat. Trans.*, 107, 143 (1966).
16. M.A. Brosmer and C.L. Tien, *J. Quant. Spec. Radia. Trans.*, 33, 521 (1985).
17. M.A. Brosmer and C.L. Tien, *J. Heat Trans.*, 107, 943 (1985).
18. M.A. Brosmer and C.L. Tien, *Comb. Sci. Tech.*, 48, 163 (1986).
19. S.C. Lee and C.L. Tien, *18th Symp. (Int) Comb.*, Comb. Inst., 1159, Pittsburgh (1981).
20. C.L. Tien, in *Handbook of Heat Transfer Fundamentals*, McGraw-Hill, pp 14.83, New York (1985).
21. G.L. Hubbard and C.L. Tien, *J. Heat Trans.*, 100, 235 (1978).
22. J.D. Felske and C.L. Tien, *J. Heat Trans.*, 99, 458 (1977).
23. J.D. Felske and C.L. Tien, *Comb. Sci. Tech.*, 7, 25 (1977).
24. W.W. Yuen and C.L. Tien, *16th Symp. (Int) Comb.*, Comb. Inst., 1481, Pittsburgh (1977).
25. A. Dayan and C.L. Tien, *Comb. Sci. Tech.*, 9, 41 (1974).
26. A.M. Kanury, *Introduction to Combustion Phenomenon*, Gordon and Breach, New York (1975).
27. C.M. Kinoshita and P.J. Pagni, *18th Symp. (Int) Comb.*, Comb. Inst., 1415, Pittsburgh (1981).
28. D.E. Negrelli, J.R. Lloyd, and J.L. Novotny, *J. Heat Trans.*, 99, 212 (1977).
29. L. Orloff and J. deRis, *19th Symp. (Int) Comb.*, Comb. Inst., 885, Pittsburgh (1982).
30. A. Tewarson, J.L. Lee, and R.F. Pion, *18th Symp. (Int) Comb.*, Comb. Inst., 563, Pittsburgh (1981).
31. M.A. Brosmer and C.L. Tien, *Comb. Sci. Tech.*, **51**, 21 (1987).
32. I. Glassman, *Combustion*, Academic, New York (1971).
33. T. Kashiwagi, B.W. MacDonald, H. Isoda, and M. Summerfield, *13th Symp. (Int) Comb.*, 1073, Pittsburgh (1971).
34. K.Y. Lee and C.L. Tien, *Int. J. Heat and Mass Trans.*, 29, 1237 (1986).
35. H.S. Carslaw and J.C. Jaeger, *Conduction of Heat in Solids*, Oxford University, Oxford (1959).
36. T. Kashiwagi, *Comb. Sci. Tech.*, 20, 225 (1979).

THERMOCHEMISTRY

D. D. Drysdale

THE RELEVANCE OF THERMOCHEMISTRY IN FIRE PROTECTION ENGINEERING

Thermochemistry is the branch of physical chemistry that is concerned with the amounts of energy released or absorbed when a chemical change (reaction) takes place.[1,2] Inasmuch as fire is fundamentally a manifestation of a particular type of chemical reaction, viz., combustion, thermochemistry provides methods by which the energy released during fire processes can be calculated from data available in the scientific and technical literature.

To place it in context, thermochemistry is a major derivative of the first law of thermodynamics, which is a statement of the principle of conservation of energy. However, while concerned with chemical change, thermodynamics does not indicate anything about the rate at which such a change takes place or about the mechanism of conversion. Consequently, the information it provides is normally used in association with other data, e.g., to enable the rate of heat release to be calculated from the rate of burning.

THE FIRST LAW OF THERMODYNAMICS

It is convenient to limit the present discussion to chemical and physical changes involving gases; this is not unreasonable, as flaming combustion takes place in the gas phase. It may also be assumed that the ideal gas law applies, i.e.,

$$PV = n \cdot RT \qquad (1)$$

where P and V are the pressure and volume of n moles of gas at a temperature, T (in degrees Kelvin); values of the universal gas constant (R) in various sets of units are summarized in Table 1-5.1. At ambient temperatures, deviations from "ideal behavior" can be detected with most gases and vapors, while at elevated temperatures such deviations become less significant.

Dr. D. D. Drysdale is Reader in the Unit of Fire Safety Engineering at the University of Edinburgh, Scotland. His research interests lie in fire science, fire dynamics, and the fire behavior of combustible materials.

Internal Energy

As a statement of the principle of conservation of energy, the first law of thermodynamics deals with the relationship between work and heat. Confining our attention to a "closed system"—for which there is no exchange of matter with the surroundings—it is known that there will be a change if heat is added or taken away, or if work is done on or by "the system" (e.g., by compression). This change is usually accompanied by an increase or decrease in temperature and can be quantified if we first define a function of state known as the internal energy of the system, E. Any change in the internal energy of the system (ΔE) is then given by

$$\Delta E = q - w \qquad (2)$$

where q is the heat transferred to the system, and w is the work done by the system. This can be expressed in differential form

$$dE = dq - dw \qquad (3)$$

Being a function of state, E varies with temperature and pressure, i.e., $E = E(T, P)$.

According to the standard definition, work, w, is done when a force, F, moves its point of application through a distance, x, thus, in the limit

$$dw = F \cdot dx \qquad (4)$$

The work done during the expansion of a gas can be derived by considering a cylinder/piston assembly (see Figure 1-5.1); thus

$$dw = P \cdot A \cdot dx = P \, dV \qquad (5)$$

where P is the pressure of the gas, A is the area of the piston, and dx is the distance through which the piston is moved; the increment in volume is therefore $dV = A \cdot dx$. The total work done is obtained by integrating Equation 5 from the initial to the final state; i.e.,

$$w = \int_{\text{initial}}^{\text{final}} P \cdot dV \qquad (6)$$

Combining Equations 3 and 5, the differential change in internal energy can be written

TABLE 1-5.1 *Values of the Ideal Gas Constant, R*

Units of Pressure	Units of Volume	Units of R	Value of R
Pa (N/m^2)	m^3	J/K · mol	8.31431
atm	cm^3	cm^3 · atm/K · mol	82.0575
atm	ℓ	ℓ · atm/K · mol	0.0820575
atm	m^3	m^3 · atm/K · mol	8.20575×10^{-5}

$$dE = dq - P \cdot dV \tag{7}$$

This shows that if the volume remains constant, as $P \cdot dV = 0$, then $dE = dq$; if this is integrated, we obtain

$$\Delta E = q_v \tag{8}$$

where q_v is the heat transferred to the constant volume system; i.e., the change in internal energy is equal to the heat absorbed (or lost) at constant volume.

Enthalpy

With the exception of explosions in closed vessels, fires occur under conditions of constant pressure. Consequently, the work done as a result of expansion of the fire gases must be taken into account. At constant pressure, Equation 5 may be integrated to give

$$w = P \cdot (V_2 - V_1) \tag{9}$$

where V_1 and V_2 are the initial and final volumes, respectively. Equation 2 then becomes

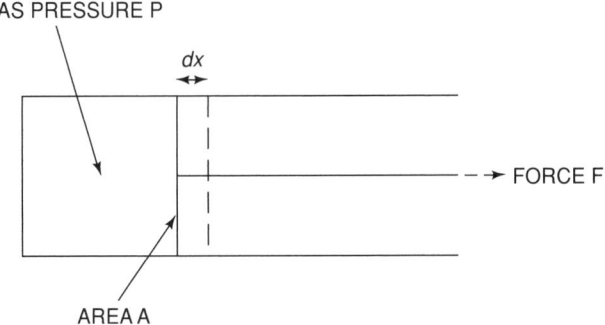

GAS PRESSURE P

FORCE F

AREA A

Fig. 1-5.1. Cylinder/piston assembly.

$$\Delta E = E_2 - E_1 = q_p + PV_1 - PV_2 \tag{10}$$

or, rearranging,

$$q_p = (E_2 + PV_2) - (E_1 + PV_1)$$
$$= H_2 - H_1 \tag{11}$$

where q_p is the heat transferred at constant pressure, and H is known as the enthalpy ($H \equiv E + PV$). The change in enthalpy is therefore the heat absorbed (or lost) at constant pressure (provided that only $P - V$ work is done), and consequently it is the change in enthalpy that must be considered in fire-related problems.

Specific Heat

Specific heat, or heat capacity, of a body or "system" is defined as the amount of heat required to raise the temperature of unit mass by one degree Celsius; the units are J/kg·K, although for most thermochemical problems the units J/mol·K are more convenient. The formal definition of the "mole" is the amount of a substance (solid, liquid, or gas) which contains as many elementary units (atoms or molecules) as there are carbon atoms in exactly 0.012 kg of carbon-12 (C^{12}). This number—known as Avogadro's number—is actually 6.023×10^{23}; in its original form, Avogadro's Hypothesis was applied to gases and stated that equal numbers of molecules of different gases at the same temperature and pressure occupy the same volume. Thus, the quantity of a substance which corresponds to a mole is simply the gram-molecular weight, but expressed in kilograms to conform with SI units. For example, the following quantities of the gases N$_2$, O$_2$, CO$_2$, and CO represent one mole of the respective gas and, according to Avogadro's Hypothesis, will each occupy 0.022414 m^3 at 273 K and 760 mm Hg (101.1 kPa):

0.028 kg nitrogen (N$_2$)
0.032 kg oxygen (O$_2$)
0.044 kg carbon dioxide (CO$_2$)
0.028 kg carbon monoxide (CO)
0.016 kg methane (CH$_4$)
0.044 kg propane (C$_3$H$_8$).

The concept of specific heat is normally associated with solids and liquids, but it is equally applicable to gases. Such specific heats are required for calculating flame temperatures, as described below. Values for a number of important gases at constant pressure and a range of temperatures are given in Table 1-5.2.

TABLE 1-5.2 *Heat Capacities of Selected Gases at Constant Pressure (101.1 kN/m^2).* *

Temperature (K)	C_p (J/mol · K)				
	298	500	1000	1500	2000
Species					
CO	29.14	29.79	33.18	35.22	36.25
CO$_2$	37.129	44.626	54.308	58.379	60.350
H$_2$O(g)	33.577	35.208	41.217	46.999	51.103
N$_2$	29.125	29.577	32.698	34.852	35.987
O$_2$	29.372	31.091	34.878	36.560	37.777
He	20.786	20.786	20.786	20.786	20.786
CH$_4$	35.639	46.342	71.797	86.559	94.399

* From JANAF *Thermochemical Tables.*[5]

(It is important to note that there are two distinct heat capacities; at constant pressure, C_p, and at constant volume, C_v. Thus, at constant pressure)

$$dq_p = dH = C_p \cdot dT \qquad (12)$$

while at constant volume

$$dq_v = dE = C_v \cdot dT \qquad (13)$$

For an ideal gas, $C_p = C_v + R$.

While the concept of specific heat is normally associated with solids and liquids, it is equally applicable to gases. Indeed, such specific heats are required for calculating flame temperatures. (See the section on calculation of adiabatic flame temperatures.)

HEATS OF COMBUSTION

Chemical Reactions and Stoichiometry

When chemical reactions occur, they are normally accompanied by the release or absorption of heat. Thermochemistry deals with the quantification of the associated energy changes. This requires a definition of the initial and final states, normally expressed in terms of an appropriate chemical equation, e.g.,

$$C_3H_8 + 5O_2 \rightarrow 3CO_2 + 4H_2O \qquad (R1)$$

in which the reactants (propane and oxygen) and products (carbon dioxide and water) are specified. This balanced chemical equation defines the *stoichiometry* of the reaction, i.e., the exact proportions of the two reactants (propane and oxygen) for complete conversion to products (no reactants remaining). Note that the physical states of the reactants and products should also be specified. In most cases, the initial conditions correspond to ambient (i.e., 25°C and atmospheric pressure) so that there should be no doubt about the state of the reactants. In this case both are gaseous, but it is more common in fires for the "fuel" to be in a condensed state, either liquid or solid. As an example, the oxidation of n-hexane can be written

$$C_6H_{14} + 9.5O_2 \rightarrow 6CO_2 + 7H_2O \qquad (R2)$$

but the fuel may be in either the liquid or the vapor state. The consequences of this will be discussed below.

Reaction 1 may be used to calculate the mass of oxygen or air required for the complete oxidation of a given mass of propane. Thus, we deduce that one mole of propane (44 g) reacts completely with five moles of oxygen ($5 \times 32 = 160$ g); i.e., 1 g propane requires 3.64 g oxygen. If the propane is burning in air, then the presence of nitrogen needs to be taken into account, although it does not participate to any significant extent in the chemical change. As the ratio of oxygen to nitrogen in air is approximately 21:79 (or 1:3.76), Reaction 1 can be rewritten

$$C_3H_8 + 5O_2 + 18.8N_2 \rightarrow 3CO_2 + 4H_2O + 18.8N_2 \qquad (R3)$$

(where $18.8 = 5 \times 3.76$), showing that 44 g propane requires ($160 + 18.8 \times 28$), or 686.4 g of "air" for complete combustion, i.e., 15.6 g air/g propane. Calculations of this type are valuable in assessing the air requirements of fires.

Thus, on the assumption that wood has the empirical formula[3] $CH_{1.5}O_{0.75}$, it can be shown that its stoichiometric air requirement is 5.38 g air for each gram of fuel, assuming complete combustion of wood to CO_2 and H_2O. In this calculation no distinction is made of the fact that flaming combustion of wood involves oxidation of the volatile gases and vapors produced by the pyrolysis of wood, while the residual char burns much more slowly by surface oxidation.

Measurement of Heats of Combustion

The heat of combustion of a fuel is defined as the amount of heat released when unit quantity is oxidized completely to yield stable end products. In the present context, the relevant combustion processes occur at constant pressure so that we are concerned with an enthalpy change, ΔH_c. It should be remembered that as oxidation reactions are exothermic, ΔH_c is always negative, by convention.

Heats of combustion are measured by combustion bomb calorimetry in which a precise amount of fuel is burned in pure oxygen inside a pressure vessel whose temperature is strictly monitored. The apparatus is designed to reduce heat losses to a minimum so that the amount of heat released can be calculated from the rise in temperature and the total thermal capacity of the system; corrections can be made for any residual heat loss. Combustion bomb calorimetry has received a great deal of attention within physical chemistry[1] as the technique has provided a wealth of information relevant to thermochemistry. However, the experiment gives the heat released at constant volume; i.e., the change in internal energy, ΔE (Equation 8). The change in enthalpy is given by

$$\Delta H = \Delta E + \Delta(PV) \qquad (14)$$

where $\Delta(PV)$ is calculated using the ideal gas law

$$\Delta(PV) = \Delta(nRT) \qquad (15)$$

The method gives the gross heat of combustion; i.e., in which the reactants and products are in their standard states. The net heat of combustion, on the other hand, refers specifically to the situation in which water as a product is in the vapor state. Net heat of combustion is less than the gross heat of combustion by an amount equal to the latent heat of evaporation (2.26 kJ/g of water) and is the value that should be used in fire calculations. It should be remembered that there is a heat of gasification associated with any condensed fuel (liquid or vapor); a correction must be made for this if the heat of combustion of the fuel vapor is required.

Table 1-5.3 contains the heats of combustion (ΔH_c) of a number of combustible gases, liquids, and solids, expressed in various ways, viz., kJ/mole (fuel), kJ/g (fuel), kJ/g (oxygen), and kJ/g (air). The first of these is the form normally encountered in chemistry texts and reference books, while the second is more commonly found in sources relating to chemical engineering and fuel technology and is more useful to the fire protection engineer. However, the third and, particularly, the fourth have very specific uses in relation to fire problems. It is immediately apparent from Table 1-5.3 that $\Delta H_c(O_2)$ and ΔH_c (air) are approximately constant for most of the fuels listed, having average values of 13.1 kJ/g and 3 kJ/g, respectively. (See the section on rate of heat release in fires.)

The data quoted in Table 1-5.3 refer to heats of combustion measured at ambient temperature, normally 25°C. These data will be satisfactory for virtually all relevant fire

TABLE 1-5.3 *Heats of Combustion of Selected Fuels at 25°C (298 K)* *

	ΔH_c (kJ/mol)	ΔH_c (kJ/g)	ΔH_c‡ [kJ/g(O_2)]	ΔH_c [kJ/g(air)]
Carbon monoxide (CO)	283	10.10	17.69	4.10
Methane (CH_4)	800	50.00	12.54	2.91
Ethane (C_2H_6)	1423	47.45	11.21	2.96
Ethene (C_2H_4)	1411	50.53	14.74	3.42
Ethyne (C_2H_2)	1253	48.20	15.73	3.65
Propane (C_3H_8)	2044	46.45	12.80	2.97
n-Butane (n-C_4H_{10})	2650	45.69	12.80	2.97
n-Pentane (n-C_5H_{12})	3259	45.27	12.80	2.97
n-Octane (n-C_8H_{18})	5104	44.77	12.80	2.97
c-Hexane (c-C_6H_{12})	3680	43.81	12.80	2.97
Benzene (C_6H_6)	3120	40.00	13.06	3.03
Methanol (CH_3OH)	635	19.83	13.22	3.07
Ethanol (C_2H_5OH)	1232	26.78	12.88	2.99
Acetone (CH_3COCH_3)	1786	30.79	14.00	3.25
D-glucose ($C_6H_{12}O_6$)	2772	15.40	13.27	3.08
Cellulose†	-	16.09	13.59	3.15
Polyethylene	-	43.28	12.65	2.93
Polypropylene	-	43.31	12.66	2.94
Polystyrene	-	39.85	12.97	3.01
Polyvinylchloride	-	16.43	12.84	2.98
Polymethylmethacrylate	-	24.89	12.98	3.01
Polyacrylonitrile	-	30.80	13.61	3.16
Polyoxymethylene	-	15.46	14.50	3.36
Polyethyleneterephthalate	-	22.00	13.21	3.06
Polycarbonate	-	29.72	13.12	3.04
Nylon 6,6	-	29.58	12.67	2.94
Polyester	-	23.8	-	-
Wool	-	20.5	-	-
Wood (European Beech)	-	19.5	-	-
Wood volatiles (European Beech)	-	16.6	-	-
Wood char (European Beech)	-	34.3	-	-
Wood (Ponderosa Pine)	-	19.4	-	-

* Apart from the solids (d-glucose, etc.), the initial state of the fuel and of all the products is taken to be gaseous.

† Cotton and rayon are virtually pure cellulose and can be assumed to have the same heat of combustion.

‡ ΔH_c (O_2) = 13.1 kJ/g is used in the oxygen consumption method for calculating rate of heat release.

problems, but occasionally it may be necessary to consider the heat released when combustion takes place at higher temperatures. This requires a simple application of the first law of thermodynamics. If the reaction involves reactants at temperature T_0 reacting to give products at the final temperature T_F, the process can be regarded in two ways:

1. The products are formed at T_0, absorb the heat of combustion, and are heated to the final temperature T_F.
2. The heat of combustion is imagined first to heat the reactants to T_F, then the reaction proceeds to completion, with no further temperature rise.

By the first law, we can write

$$(\Delta H_c)^{T_0} + C_p^{Pr} \cdot (T_F - T_0) = (\Delta H_c)^{T_F} + C_p^{R} \cdot (T_F - T_0) \quad (16)$$

where C_p^{Pr} and C_p^{R} are the total heat capacities of the products and reactants, respectively. This may be rearranged to give

$$\frac{(\Delta H_c)^{T_F} - (\Delta H_c)^{T_0}}{T_F - T_0} = \Delta C_p \quad (17)$$

or, in differential form, we have Kirchoff's equation

$$d(\Delta H_c)/dT = \Delta C_p \quad (18)$$

where $\Delta C_p = C_p^{Pr} - C_p^{R}$. This may be used in integrated form to calculate the heat of combustion at temperature T_2 if ΔH_c is known at temperature T_1 and information is available on the heat capacities of the reactants and products, thus

$$(\Delta H_c)^{T_2} = (\Delta H_c)^{T_1} + \int_{T_1}^{T_2} \Delta C_p \cdot dT \quad (19)$$

where

$$\Delta C_p = \sum C_p(\text{products}) - \sum C_p(\text{reactants}) \quad (20)$$

and C_p is a function of temperature, which can normally be expressed as a power series in T, e.g.,

$$C_p = a + bT + cT^2 + \cdots \quad (21)$$

Information on heat capacities of a number of species and their variation with temperature may be found in references 4 and 5. Some data are summarized in Table 1-5.2.

HEATS OF FORMATION

The first law of thermodynamics implies that the change in internal energy (or enthalpy) of a system depends only on the initial and final states of the system and is thus independent of the intermediate stages. This is embodied in thermochemistry as Hess' Law, which applies directly to chemical reactions. From this, we can develop the concept of heat of formation, which provides a means of comparing the relative stabilities of different chemical compounds and may be used to calculate heats of chemical reactions which cannot be measured directly.

The heat of formation of a compound is defined as the enthalpy change when 1 mole of that compound is formed from its constituent elements in their standard state (at 1 atm pressure and 298 K). Thus, the heat of formation of liquid water is the enthalpy change of the reaction (at 298 K)

$$H_2(g) + 0.5O_2(g) \rightarrow H_2O(l) \; \Delta H_f = -285.8 \text{ kJ/mol} \quad \text{(R4)}$$

so that $\Delta H_f(H_2O)(l) = -285.8$ kJ/mole at 25°C. This differs from the heat of formation of water vapor [$\Delta H_f(H_2O)g = -241.84$ kJ/mol] by the latent heat of evaporation of water at 25°C (43.96 kJ/mol).

By definition, the heats of formation of all the elements are set arbitrarily to zero at all temperatures. This then allows the heats of reaction to be calculated from the heats of formation of the reactants and products, thus

$$\Delta H = \Delta H_f(\text{products}) - \Delta H_f(\text{reactants}) \quad (22)$$

where ΔH is the heat (enthalpy) of the relevant reaction. However, most heats of formation cannot be obtained as easily as heats of combustion. The example given is unusual in that the heat of formation of water also happens to be the heat of combustion of hydrogen. Similarly, the heat of combustion of carbon in its most stable form under ambient conditions (graphite) is the heat of formation of carbon dioxide. Fortunately, combustion calorimetry can be used indirectly to calculate heats of formation.

The heat of formation of ethyne (acetylene), which is the enthalpy change of the reaction

$$2C(\text{graphite}) + H_2 \rightarrow C_2H_2 \quad \text{(R5)}$$

can be deduced in the following way: the heat of combustion of ethyne has been determined by bomb calorimetry as -1255.5 kJ/mol at 25°C (298 K). This is the heat of the reaction

$$C_2H_2 + 2.5O_2 \rightarrow 2CO_2 + H_2O \quad \text{(R6)}$$

which, by Hess' Law (see Equation 22), can be equated to

$$(\Delta H_c)^{298}(C_2H_2) = 2(\Delta H_f)^{298}(CO_2) + (\Delta H_f)^{298}(H_2O)$$
$$- (\Delta H_f)^{298}(C_2H_2) - 2.5(\Delta H_f)^{298}(O_2) \quad (23)$$

We know that

$$(\Delta H_c)^{298}(C_2H_2) = -1255.5 \text{ kJ/mol}$$

$$(\Delta H_f)^{298}(CO_2) = -393.5 \text{ kJ/mol}$$

$$(\Delta H_f)^{298}(H_2O) = -241.8 \text{ kJ/mol}$$

and $(\Delta H_f)^{298}(O_2) = 0.0$ kJ/mol (by definition),

TABLE 1-5.4 *Heats of Formation at 25°C (298 K)*

Compound	$(\Delta H_f)^{298}$ (kJ/mol)
Hydrogen (atomic)	+218.00
Oxygen (atomic)	+249.17
Hydroxyl (OH)	+38.99
Chlorine (atomic)	+121.29
Carbon Monoxide	−110.53
Carbon Dioxide	−393.52
Water (liquid)	−285.8
Water (vapor)	−241.83
Hydrogen Chloride	−92.31
Hydrogen Cyanide (gas)	+135.14
Nitric Oxide	+90.29
Nitrogen Dioxide	+33.85
Ammonia	−45.90
Methane	−74.87
Ethane	−84.5
Ethene	+52.6
Ethyne (Acetylene)	+226.9
Propane	−103.6
n-Butane	−124.3
iso-Butane*	−131.2
Methanol	−242.1

* Heats of formation of other hydrocarbons are tabulated in reference 6.

so that by rearrangement, Equation 23 yields

$$(\Delta H_f)^{298}(C_2H_2) = +226.7 \text{ kJ/mol}$$

This compound has a positive heat of formation, unlike CO_2 and H_2O. This indicates that it is an endothermic compound and is therefore less stable than the parent elements. Under appropriate conditions, ethyne can decompose violently to give more stable species.

The heats of formation of a number of compounds are given in Table 1-5.4. The most stable compounds (CO_2 and H_2O) have the largest negative values, while positive values tend to indicate an instability with respect to the parent elements. While this can indicate a high chemical reactivity, it gives no information about the rates at which chemical changes might take place (i.e., kinetics are ignored). However, heats of formation have been used in preliminary hazard assessment to provide an indication of the risks associated with new processes in the chemical industry. It should be noted that the heats of combustion of endothermic compounds do not give any indication of any associated reactivity (compare Tables 1-5.3 and 1-5.4).

RATE OF HEAT RELEASE IN FIRES

While thermochemistry can give information relating to the total amount of energy that can be released when a fuel is burned to completion, it is rarely possible to use heats of combustion directly to calculate the heat released in "real" fires. However, it can be argued that the *rate* of heat release is more important than the total available. When a single item is burning in isolation, the rate of burning and the rate of heat release in the flame are coupled. Nevertheless, it is convenient to express the rate of heat release in terms of the burning rate, which is expressed as a rate of mass loss, $\dot{m}$ (kg/s)

$$\dot{Q}_c = \dot{m} \cdot \Delta H_c \quad (24)$$

where ΔH_c is the net heat of combustion of the fuel (kJ/kg). However, this assumes that combustion is complete, although it is known that this is rarely so. Even under conditions of unrestricted ventilation, the products of combustion will contain some species which are only partially oxidized, such as carbon monoxide, aldehydes, ketones, and particulate matter in the form of soot or smoke. Their presence indicates that not all the available combustion energy has been released. The "combustion efficiency" is likely to vary from around 0.3 to 0.4 for heavily fire-retarded materials to 0.9 or higher in the case of oxygen-containing products (e.g., polyoxymethylene).[7]

Fires burning in compartments present a completely different problem. In the first place, there is likely to be a range of different fuels present, each with a different stoichiometric air requirement. These will all burn at different rates, dictated not just by the nature of the fuel but also by the levels of radiant heat existing within the compartment during the fire. The rate of heat release during the fully developed stage of a compartment fire is required for calculating post-flashover temperature-time histories for estimating fire exposure of elements of structures, as in the method developed by Pettersson et al.[8] This is apparently complicated by the fact that not all of the fuel may burn within the compartment; some of the fuel volatiles can escape to burn outside as they mix with fresh air. The proportion of the heat of combustion that is effectively lost in this way cannot easily be estimated.

However, if it is assumed that the fire is ventilation controlled and that all of the air that enters the compartment is "burned" therein, then the rate of heat release within the compartment can be calculated from the expression

$$\dot{Q}_c = \dot{m}_{air} \cdot \Delta H_c(air) \quad (25)$$

where $\dot{m}_{air}$ is the mass flow rate of air into the compartment, and $\Delta H_c(air)$ is the heat of combustion per unit mass of air consumed (3000 kJ/g, see Table 1-5.3). The mass flow rate of air can be approximated by the expression

$$\dot{m}_{air} = 0.52 A_w h^{1/2} \quad (kg/s) \quad (26)$$

where A_w is the effective area of ventilation (m^2) and h is the height of the ventilation opening (m).

In this, it is tacitly assumed that the combustion process is stoichiometric, although in fact the rate of supply of air may not be sufficient to burn all the fuel vapors within the compartment. Indeed, if the equivalence ratio $\dot{m}_{air}/\dot{m}$ is less than the stoichiometric ratio, excess fuel will escape from the compartment and mix with air to give flames whose length will depend inter alia on the equivalence ratio.[9] Furthermore, in using Equation 26 to calculate the temperature-time course of a fire, it is implied that the fire remains at its maximum rate of burning for its duration, the latter being controlled by the quantity of fuel present (the fire load). This method will overestimate the severity of fuel-controlled fires in which the ventilation openings are large.[10]

Much useful data on the fire behavior of combustible materials can be obtained by using the technique of "oxygen consumption calorimetry." This is the basis of the "cone calorimeter," in which the rate of heat release from a small sample of material burning under an imposed radiant heat flux is determined by measuring the rate of oxygen con-

sumption. The latter can be converted into a rate of heat release using the conversion factor 13.1 kJ/g of oxygen consumed. (A small correction is required for incomplete combustion, based on the yield of CO.) This technique can be used on a larger scale to measure the rate of heat release from items of furniture, wall lining materials, etc.[11]

CALCULATION OF ADIABATIC FLAME TEMPERATURES

In the previous sections, no consideration has been given to the fate of the energy released by the combustion reactions. Initially it will be absorbed within the reaction system itself by (1) unreacted reactants, (2) combustion products, and (3) diluents, although it will ultimately be lost from the system by various heat transfer processes. If we consider a premixed reaction system, such as a flammable vapor/air mixture, and assume it to be adiabatic, i.e., there is no transfer of heat to or from the system, then it is possible to calculate an adiabatic flame temperature.

Consider a flame propagating through a stoichiometric propane/air mixture of infinite extent (i.e., there are no surfaces to which heat may be transferred) and which is initially at 25°C. The appropriate equation is given by Reaction 7:

$$C_3H_8 + 5O_2 + 18.8N_2 \rightarrow 3CO_2 + 4H_2O + 18.8N_2 \quad (R7)$$

This reaction releases 2044 kJ for every mole of propane consumed. This quantity of energy goes toward heating the reaction products, i.e., 3 moles of carbon dioxide, 4 moles of water (vapor), and 18.8 moles of nitrogen for every mole of propane burned. The thermal capacity of this mixture can be calculated from the thermal capacities of the individual gases, which are available in the literature (e.g., JANAF).[5] The procedure is straightforward, provided that an average value of C_p is taken for each gas in the temperature range involved. (See Table 1-5.5.)

As 2044 kJ are released at the same time as these species are formed, the maximum temperature rise will be

$$\Delta T = 2044000/942.5 = 2169 \text{ K}$$

giving the final (adiabatic) temperature as 2169 + 298 = 2467 K. In fact, this figure is approximate for the following reasons:

1. Thermal capacities change with temperature, and average values over the range of temperatures appropriate to the problem have been used.
2. The system cannot be adiabatic as there will be heat loss by radiation from the hot gases (CO_2 and H_2O).

TABLE 1-5.5 Thermal Capacity of the Products of Combustion of a Stoichiometric Propane/Air Mixture

	No. of moles	Thermal Capacity at 1000 K (J/mole · K)	(J/K)
CO_2	3	54.3	162.9
H_2O	4	41.2	164.8
N_2	18.8	32.7	614.8
Total thermal capacity (per mole of propane) =			942.5 J/K

Gas	Adiabatic Flame Temperature at Lower Flammability Limit (K)
Methane	1446
Ethane	1502
Propane	1554
n-Butane	1612
n-Pentane	1564
n-Heptane	1692
n-Octane	1632

3. At high temperatures, dissociation of the products will occur; as these are endothermic processes, there will be a reduction in the final temperature.

Of these, (2) and (3) determine that the actual flame temperature will be much lower than predicted. Thus, with propane burning in air, the final temperature may not exceed 2000 K.

If the propane were burning as a stoichiometric mixture in pure oxygen, then in the absence of nitrogen as a "heat sink," much higher temperatures would be achieved. The total thermal capacity would be $(942.5 - 614.8) = 327.7$ J/K. However, the amount of heat released remains unchanged (2044 kJ) so that the maximum temperature rise would be

$$\Delta T = 2044000/327.7 = 6238 \text{ K}$$

predicting a final temperature of 6263°C. Because dissociation will be a dominant factor, this cannot be achieved and the temperature of the flame will not exceed ca. 3500 K.

The occurrence of dissociation at temperatures in the region of 2000 K and above makes it necessary to take dissociation into account. This is discussed in Unit I, Chapter 2 of this handbook. However, the simple calculation outlined above can be used to estimate the temperatures of near-limit flames, when the temperature is significantly lower and dissociation can be neglected.

It is known that the lower flammability limit of propane is 2.2 percent. The oxidation reaction taking place in this mixture can be described by the following equation

$$0.022C_3H_8 + 0.978(0.21O_2 + 0.79N_2) \rightarrow \text{Products}$$

Dividing through by 0.022 allows this to be written

$$C_3H_8 + 9.34O_2 + 35.12N_2 \rightarrow 3CO_2$$
$$+ 4H_2O + 4.34O_2 + 35.12N_2 \quad \text{(R8)}$$

showing that the heat released by the oxidation of 1 mole of propane is now absorbed by excess oxygen (4.34 moles) as well as by an increased amount of nitrogen. Carrying out the same calculation as before, it can be shown that the adiabatic flame temperature for this limiting mixture is 1281°C (1554 K). If the same calculation is carried out for the other hydrocarbon gases, it is found that the adiabatic limiting flame temperature lies in a fairly narrow band, 1600 ± 100 K. (See Table 1-5.6.) This can be interpreted by assuming that the limit exists because heat losses (by radiation from the flame) exceed the rate of heat production (within the flame) below a certain threshold temperature. As a consequence, the

flame cannot sustain itself. This concept can be applied to certain practical problems relating to the lower flammability limit.

EXAMPLE:

A mechanical engineering research laboratory contains a six-cylinder internal combustion engine which is being used for research into the performance of spark plugs. The fuel being used is methane, CH_4, and the fuel/air mixture can be adjusted at will. The combustion products are extracted from the exhaust manifold through a 30 cm square duct, 20 m long. It is found that the engine will continue to operate with a stoichiometric mixture when only three of the cylinders are firing. If under these conditions the average temperature of the gases entering the duct from the manifold is 700 K, is there a risk of an explosion in the duct?

SOLUTION:

The stoichiometric reaction for methane in air is

$$CH_4 + 2O_2 + 7.52N_2 \rightarrow CO_2 + 2H_2O + 7.52N_2 \quad \text{(R9)}$$

If we consider that one mole of fuel passes through each of the six cylinders, but of the six moles only three are burned, we have overall

$$6CH_4 + 12O_2 + 45.12N_2 \rightarrow 3CH_4 + 3CO_2 + 6H_2O$$
$$+ 6O_2 + 45.12N_2 \quad \text{(R10)}$$

Dividing through by 3 gives

$$2CH_4 + 4O_2 + 15.04N_2 \rightarrow CH_4 + 2O_2 + CO_2$$
$$+ 2H_2O + 15.04N_2 \quad \text{(R11)}$$

The mixture discharged into the exhaust manifold has the composition given by the right-hand side of Reaction 10. If this "burns" at 700 K, the final abiabatic flame temperature may be calculated on the basis of the reaction

$$CH_4 + 2O_2 + CO_2 + 2H_2O + 15.04N_2$$
$$\rightarrow 2CO_2 + 4H_2O + 15.04N_2 \quad \text{(R12)}$$

The total thermal capacity of the product gases ($2CO_2 + 4H_2O + 15.04N_2$) (at 1000 K) can be shown to be 765.3 J per mole of methane burned. Using Kirchoff's Equation (Equation 19), $\Delta H_c(CH_4)$ at 700 K is calculated as 802.8 kJ/mol, giving $T = 802800/765.3 = 1049$ K. This gives a final temperature of 1749 K, which is significantly higher than the limiting flame temperature (1600 K) discussed above. This indicates that there is a risk of explosion, and measures should be applied to prevent this mixture being discharged into the duct.

It should be noted that at 700 K there will be a "slow" reaction between methane and the oxygen present, which could invalidate the tacit assumption that the duct becomes completely filled with the mixture described by the right-hand side of Reaction 10. However, slow oxidation of the methane will tend to make the mixture less flammable, and so the calculation gives a conservative answer.

NOMENCLATURE

A	area (Equation 5)
A_w	area of ventilation opening
C_p	specific heat
E	internal energy
F	force (Equation 4)
h	height of ventilation opening
H	enthalpy
ΔH_c	heat of combustion
ΔH_f	heat of formation
$\dot{m}$	mass rate of burning
$\dot{m}_{air}$	mass flow rate of air
n	number of moles
P	pressure
q	energy
Q_c	rate of heat release
R	universal gas constant
T	temperature
V	volume
w	work

Subscripts

c	combustion
F	final
f	formation
o	initial
p	constant pressure
v	constant volume

Superscripts

Pr	products
R	reactants

REFERENCES CITED

1. W.J. Moore, *Physical Chemistry*, 5th ed, Longman, London (1974).
2. D.D. Drysdale, *Introduction to Fire Dynamics*, John Wiley & Sons, Chichester (1985).
3. A.F. Roberts, *Comb. and Flame*, 8, 245 (1964).
4. R.A. Strehlow, *Combustion Fundamentals*, McGraw Hill, New York (1984).
5. *JANAF Thermochemical Tables*, National Bureau of Standards, Washington (1970).
6. R.C. Weast, *Handbook of Chemistry and Physics*, Chemical Rubber Co., Cleveland (1973).
7. A. Tewarson, in *Flame Retardant Polymeric Materials*, M. Lewin, ed., Plenum, New York (1982).
8. O. Pettersson, S.E. Magnusson, and J. Thor, *Fire Engineering Design of Structures*, Swedish Institute of Steel Construction, Publication 50 (1976).
9. M.L. Bullen and P.H. Thomas, 17th Symposium (International) on Combustion, Pittsburgh (1979).
10. P.H. Thomas and A.J.M. Heselden, "Fully Developed Fires in Compartments," *CIB Report No. 20; Fire Research Note No. 923*, Conseil International du Batiment, France (1972).
11. V. Babrauskas and S.J. Grayson (eds.) *Heat Release in Fires* Elsevier Applied Science, London (1992).

CHEMICAL EQUILIBRIUM

Raymond Friedman

RELEVANCE OF CHEMICAL EQUILIBRIUM TO FIRE PROTECTION

The temperature of a flame must be known in order to calculate convective and radiative heat transfer rates, which control pool-fire burning rates, flame spread rates, remote ignitions, damage to exposed items (e.g., structural steel, wiring), and response of thermal fire detectors or automatic sprinklers.

The chapter on *Thermochemistry* provides a simple technique for calculating flame temperature, based on ignoring the dissociations that occur at high temperature. This technique gives answers that are too high. For example, if propane (C_3H_8) burns in stoichiometric proportions with air at 300 K, and it is assumed that the only products are CO_2, H_2O, and N_2, then the simple thermochemical calculation yields a flame temperature of 2394 K. On the other hand, if chemical equilibrium is considered, so that the species CO, O_2, H_2, OH, H, O, and NO are assumed present in the products, then the flame temperature, calculated by methods described in this section, comes out to be 2268 K. Flame temperature measurements in laminar premixed propane-air flames agree with the latter value. (The discrepancy in flame temperature caused by neglecting dissociation would be even greater for fires in oxygen-enriched atmospheres.)

The chemical equilibrium calculation yields not only the temperature but the equilibrium composition of the products. Thus, the generation rate of certain toxic or corrosive products such as carbon monoxide, nitric oxide, or hydrogen chloride may be calculated, insofar as the assumption of equilibrium is valid.

For a fire in a closed volume, the final pressure as well as the temperature will depend on the dissociations and therefore require a calculation taking chemical equilibrium into account.

From a fire research viewpoint, there is interest in correlating flammability limits, extinguishment, soot formation, toxicity, flame radiation, or other phenomena; and chemical equilibrium calculations in some cases will be a useful tool in such correlations.

In a later part of this chapter, departure of actual fires from chemical equilibrium will be discussed.

INTRODUCTION TO THE CHEMICAL EQUILIBRIUM CONSTANT

Consider a chemical transformation, such as

$$2CO + O_2 \rightarrow 2CO_2 \tag{1}$$

If this process can occur, presumably the reverse process can also occur (principle of microscopic reversibility, or principle of detailed balancing)

$$2CO_2 \rightarrow 2CO + O_2 \tag{2}$$

If both processes occur at finite rates in a closed system, then, after a sufficient time, a condition of *chemical equilibrium* will be reached, after which no further change occurs as long as the temperature and pressure remain constant and no additional reactants are introduced. This condition of equilibrium can be expressed as a *mathematical constraint* on the system, which, for the gaseous reaction $2CO + O_2 \leftrightarrows 2\,CO_2$, can be written

$$K_3 = p_{CO_2}^2 / p_{CO}^2\, p_{O_2} \tag{3}$$

where the p_i are partial pressures (atm), and K_3 is the equilibrium constant. This expression can be rationalized by the following argument.

According to the chemical "law of mass action," first stated a century ago, the rate of the forward reaction (Equation 1) at a given temperature is given by $k_f\, p_{CO}^2\, p_{O_2}$, while the rate of the reverse reaction (Equation 2) is given by $k_r\, p_{CO_2}^2$. At equilibrium, the forward rate must be equal to the reverse rate

$$k_f\, p_{CO}^2 p_{O_2} = k_r\, p_{CO_2}^2 \tag{4}$$

which may be rearranged to

Dr. Raymond Friedman was with Factory Mutual Research from 1969 through 1993. During most of this time he was vice president and manager of their Research Division. Currently he is an independent consultant. He has past experience at Westinghouse Research Laboratories and Atlantic Research Corporation. He is a past president of The Combustion Institute, current vice chairman of the International Association for Fire Safety Science, and an expert in fire research and combustion.

$$\frac{p_{CO_2}^2}{p_{CO}^2 \, p_{O_2}} = \frac{k_f}{k_r} = K_3 \tag{5}$$

While this appears to be a satisfactory explanation, research over the past hundred years has shown that chemical reactions in fact rarely proceed as suggested by the stoichiometric equation. For example, the three-body collision of two CO molecules and an O_2 molecule, resulting in the formation of two CO_2 molecules, simply does not happen. Rather, the reaction would occur as follows

$$O_2 + M \rightarrow 2O + M \tag{6}$$

(where M is any molecule) followed by

$$O + CO + M \rightarrow CO_2 + M \tag{7}$$

Now, observe how Equation 3 can be obtained from this reaction sequence.

The reverse of $O_2 + M \rightarrow 2O + M$, namely $2O + M \rightarrow O_2 + M$, can also occur, and the equilibrium constant for this pair of reactions which actually do occur is $K_6 = p_O^2 p_M / p_{O_2} p_M = p_O^2 / p_{O_2}$. (The p_M term is seen to cancel.)

Similarly the reverse reaction $CO_2 + M \rightarrow O + CO + M$ can occur, and the equilibrium constant is $K_7 = p_{CO_2} / p_{CO} p_O$. If we now multiply K_7^2 by K_6, we obtain

$$K_7^2 K_6 = \left(\frac{p_{CO_2}}{p_{CO} \, p_O}\right)^2 \frac{p_O^2}{p_{O_2}} = \frac{p_{CO_2}^2}{p_{CO}^2 \, p_{O_2}} = K_3 \tag{8}$$

Thus, Equation 3 is perfectly valid, even if the "law of mass action" does not correctly describe the reaction process involving CO and O_2.

To get a further understanding of the validity of the equilibrium constant concept, consider the following facts: CO will not react with O_2 even by the above mechanism involving O atoms unless first heated to quite high temperatures. However, at least a trace of moisture is usually present, and in such cases the reaction occurs by the following process, which can occur at lower temperatures. First, H and OH are formed by dissociation of H_2O. Then, the CO is converted by

$$CO + OH \rightleftarrows CO_2 + H \quad K_9 = p_{CO_2} \, p_H / p_{CO} \, p_{OH} \tag{9}$$

while the O_2 reacts with H

$$O_2 + H \rightleftarrows OH + O \quad K_{10} = p_{OH} \, p_O / p_{O_2} \, p_H \tag{10}$$

If the quantity $K_9^2 K_{10}$ is now calculated

$$K_9^2 K_{10} = \frac{p_{CO_2}^2}{p_{CO}^2 \, p_{O_2}} \frac{p_H \, p_O}{p_{OH}} \tag{11}$$

But, it is known that the reaction $H + O + M \rightarrow OH + M$ can occur, as well as its reverse, $OH + M \rightarrow H + O + M$. It does not matter if these reactions are actually important in the rate of oxidation of CO in the presence of H_2O; as long as these reactions can occur, then at equilibrium

$$k_f \, p_H \, p_O \, p_M = k_r \, p_{OH} \, p_M,$$

and

$$k_f / k_r = K_{12} = p_{OH} / p_H \, p_O \tag{12}$$

Substituting this into Equation 11

$$\frac{p_{CO_2}^2}{p_{CO}^2 \, p_{O_2}} = K_9^2 \, K_{10} \, K^{12} = K_3 \tag{13}$$

Thus, the ratio $p_{CO_2}^2 / p_{CO}^2 p_{O_2}$ is a constant at equilibrium (at a given temperature) regardless of the reaction mechanism, even if other (hydrogen-containing) species are involved, because by the principle of microscopic reversibility, these other species (catalysts) affect the reverse reaction as well as the forward reaction.

Let us now consider the mathematical specification of the $CO–CO_2–O_2$ system at equilibrium. The system, at a given temperature and pressure, may be described by three variables, namely the partial pressures of the three species: p_{CO}, p_{O_2}, and p_{CO_2}. There are already two well-known constraints on the system:

The sum of the partial pressures must equal the total pressure, p

$$p_{CO} + p_{O_2} + p_{CO_2} = p \tag{14}$$

The ratio of carbon atoms to oxygen atoms in the system must remain at the original, presumably known, value of C/O:

$$C/O = \frac{p_{CO} + p_{CO_2}}{p_{CO} + 2p_{O_2} + 2p_{CO_2}}. \tag{15}$$

A third constraint, that of chemical equilibrium, provides a third equation involving p_{CO}, p_{O_2}, and p_{CO_2}

$$\frac{p_{CO_2}^2}{p_{CO}^2 \, p_{O_2}} = K_3 \tag{3}$$

Now, the system is completely defined by the simultaneous solution of these three equations. The equilibrium constant varies with temperature but is independent of pressure (except at rather high pressures). It is also independent of the presence of other reactive chemical species.

GENERALIZED DEFINITION OF EQUILIBRIUM CONSTANT

For a generalized reaction

$$aX_1 + bX_2 \rightleftarrows cY_1 + dY_2$$

K would be given by

$$K = \frac{(p_{Y_1})^c (p_{Y_2})^d}{(p_{X_1})^a (p_{X_2})^b}$$

Notice that, instead of writing $2CO + O_2 \rightleftarrows 2CO_2$, one could equally well have written $CO + \frac{1}{2}O_2 \rightleftarrows CO_2$. The equilibrium constant for the latter formulation is

$$K_{16} = \frac{p_{CO_2}}{p_{CO} \, p_{O_2}^{1/2}} \tag{16}$$

By comparison of Equation 16 with Equation 3, it is clear that $K_{16} = \sqrt{K_3}$. Again, the equilibrium constant for the reaction, if written $2CO_2 \rightleftarrows 2CO + O_2$, would be equal to $1/K_3$.

SIMULTANEOUS EQUILIBRIA

In most real chemical systems, one must deal with a number of simultaneous chemical equilibria. For example, air at 2500 K will contain the species N_2, O_2, NO, and O. The following simultaneous equilibria may be considered

$$O_2 = 2O; \qquad K_{17} = p_O^2/p_{O_2} \qquad (17)$$

$$N_2 + O_2 = 2NO; \qquad K_{18} = p_{NO}^2/p_{N_2} p_{O_2} \qquad (18)$$

$$N_2 + 2O = 2NO; \qquad K_{19} = p_{NO}^2/p_{N_2} p_O^2 \qquad (19)$$

It is easily seen from the above relations that $K_{19} = K_{18}/K_{17}$. Hence, Equations 17, 18, and 19 are not three independent equations, and *any two* may be used to describe the equilibrium condition; the third would be redundant. To determine the four unknowns, p_{N_2}, p_{O_2}, p_{NO}, and p_O, one would simultaneously solve the selected two equilibrium relations plus the following two relations

$$p_{NO} + p_{N_2} + p_{O_2} + p_O = p \qquad (20)$$

and

$$\frac{p_{NO} + 2p_{N_2}}{2p_{O_2} + p_O} = 3.76 \qquad (21)$$

where 3.76 is the ratio of nitrogen atoms to oxygen atoms in air.

If one knows the temperature, the equilibrium constants may be calculated from the thermodynamic properties of the reactants and products, as discussed in the next section. However, since the various equilibrium reactions release or absorb energy, and accordingly raise or lower the temperature of an adiabatic system, the determination of equilibrium composition of an adiabatic system must proceed simultaneously with the calculation of its temperature; i.e., an energy balance must be satisfied as well as the equilibrium equations, the atom-ratio equations, and the $p = \Sigma p_i$ equation.

As a general rule, a gaseous chemical system *at a given temperature*, containing *s* kinds of chemical species involving *e* chemical elements, will require *s-e* equilibrium relations, as well as *e*-1 atom-ratio relations, and a $p = \Sigma p_i$ equation, in order to specify it. If the temperature is also unknown, an energy balance equation is also needed. (If the pressure is unknown, but the volume is known, then the equation of state must be used in the pressure equation.)

In order to solve an actual problem, one must select the species to be considered. The more species one includes, the more difficult is the calculation. There is no need to include any species that will be present in very small quantity at equilibrium. Some guidelines can be provided.

For combustion of a C–H–O compound in air, it is usually sufficient to include the species CO_2, H_2O, N_2, O_2, CO, H_2, OH, H, O, and NO. This is adequate if the air-fuel ratio is sufficiently large so that the O/C atomic ratio is greater than one. If the O/C atomic ratio is less than one, then solid carbon must be considered, as well as many additional gaseous species. If chlorine is present, then HCl, Cl_2, and Cl must be added. If sulfur is present, then SO_2 and SO_3 are the primary species, unless there is a deficiency of oxygen.

THE QUANTIFICATION OF EQUILIBRIUM CONSTANTS

While a chemist might establish the numerical value of an equilibrium constant for A $\leftrightarrows$ 2B by direct measurement of the partial pressures of A and B in a system at equilibrium, this is rarely done, because it is difficult to make such measurements in a high-temperature system, and it takes a long time to establish equilibrium in a low-temperature system. Instead, the equilibrium constant is generally determined from the thermodynamic relation first deduced by van't Hoff in 1886[1]

$$\Delta F° = -RT \ln K \qquad (22)$$

If this is applied to A $\leftrightarrows$ 2B at absolute temperature T, then $K = p_B^2/p_A$, and $\Delta F°$ is the free energy of two moles of B at one atmosphere and temperature T, minus the free energy of one mol of A at one atmosphere and temperature T. (The superscript o designates that each substance is in its "standard state," i.e., an ideal gas at one atmosphere.) By definition

$$\Delta F° = \Delta H° - T\Delta S° = \Delta E° + \Delta(pV°) - T\Delta S° \qquad (23)$$

Accordingly, if $\Delta S°$, the entropy difference, and either $\Delta H°$, the enthalpy difference, or $\Delta E°$, the energy difference, is known for the substances involved in an equilibrium at temperature T, then the equilibrium constant, K, may be calculated. It happens that $\Delta S°$, $\Delta H°$, and $\Delta E°$ are well known for almost all substances expected to be present at equilibrium in combustion gases at any temperature up to 4000 K, so the calculation of equilibrium constants is straightforward.

The variation of the equilibrium constant with temperature was shown by van't Hoff[1] to be given by

$$\frac{d \ln K}{dT} = \frac{\Delta H°}{RT^2} \left[= \frac{\Delta H}{RT^2} \text{ for ideal gases} \right] \qquad (24)$$

Thus, for an exothermic reaction occurring at temperature T, ΔH is negative, and K decreases as T increases. The converse is true for endothermic reactions.

It is appropriate to inquire about the underlying physical reason for the value of K to be governed by $\Delta F°$ (actually by $\Delta H°$ and $\Delta S°$). An explanation is as follows:

Any chemical system being held at constant temperature will seek to reduce its energy, E, and to increase its entropy, S. The reduction of energy is analogous to a ball rolling downhill. The increase of entropy is analogous to shuffling a sequentially arranged deck of cards, yielding a random arrangement. These two tendencies will often affect the equilibrium constant in opposite directions. Consider the equation

$$\ln K = \frac{\Delta S°}{R} - \frac{\Delta E°}{RT} - \Delta n \qquad (25)$$

where Δn is the increase in the number of moles of product relative to reactant. Equation 25 is obtained by combining Equations 22 and 23 with the ideal gas law at constant temperature $\Delta(pV°) = \Delta nRT$. Inspection of Equation 25 shows that, if $\Delta S°$ is a large positive quantity and $\Delta S°/R$ dominates the other terms, K will be large, i.e., the reaction is driven by the "urge" to increase entropy. Again, if the reaction is highly endothermic, then $(-\Delta E°/RT)$ will be a large negative number and can dominate the other terms to cause K to be small; i.e., the reaction prefers to go in the reverse, or

TABLE 1-6.1 *Values of Log$_{10}$ K for Selected Reactions*

TEMP (K)	K_A $\frac{1}{2}O_2 = O$	K_B $\frac{1}{2}H_2 = H$	K_C $H_2 + \frac{1}{2}O_2 = H_2O$	K_D $\frac{1}{2}H_2 + \frac{1}{2}O_2 = OH$	K_E $C_{(S)} + O_2 = CO_2$	K_F $C_{(S)} + \frac{1}{2}O_2 = CO$	K_G $\frac{1}{2}N_2 + \frac{1}{2}O_2 = NO$	K_H $\frac{1}{2}F_2 = F$	K_I $\frac{1}{2}Cl_2 = Cl$	K_J $\frac{1}{2}Br_2 = Br$	K_K $\frac{1}{2}H_2 + \frac{1}{2}F_2 = HF$	K_L $\frac{1}{2}H_2 + \frac{1}{2}Cl_2 = HCl$	K_M $\frac{1}{2}H_2 + \frac{1}{2}Br_2 = HBr$
600	−18.574	−16.336	18.633	−2.568	34.405	14.318	−7.210	−3.814	−7.710	−5.641	24.077	8.530	5.036
700	−15.449	−13.599	15.583	−2.085	29.506	12.946	−6.086	−2.810	−6.182	−4.431	20.677	7.368	4.374
800	−13.101	−11.539	13.289	−1.724	25.830	11.914	−5.243	−2.053	−5.031	−3.522	18.125	6.494	3.876
900	−11.272	−9.934	11.498	−1.444	22.970	11.108	−4.587	−1.462	−4.133	−2.814	16.137	5.812	3.486
1000	−9.807	−8.646	10.062	−1.222	20.680	10.459	−4.062	−.988	−3.413	−2.245	14.544	5.265	3.173
1100	−8.606	−7.589	8.883	−1.041	18.806	9.926	−3.633	−.599	−2.822	−1.799	13.240	4.816	2.917
1200	−7.604	−6.707	7.899	−.890	17.243	9.479	−3.275	−.273	−2.328	−1.389	12.152	4.442	2.702
1300	−6.755	−5.958	7.064	−.764	15.920	9.099	−2.972	.003	−1.909	−1.059	11.230	4.124	2.520
1400	−6.027	−5.315	6.347	−.656	14.785	8.771	−2.712	.240	−1.549	−.775	10.438	3.852	2.364
1500	−5.395	−4.756	5.725	−.563	13.801	8.485	−2.487	.447	−1.236	−.527	9.752	3.615	2.229
1600	−4.842	−4.266	5.180	−.482	12.940	8.234	−2.290	.627	−.962	−.311	9.191	3.408	2.110
1700	−4.353	−3.833	4.699	−.410	12.180	8.011	−2.116	.788	−.720	−.119	8.420	3.225	2.006
1800	−3.918	−3.448	4.270	−.347	11.504	7.811	−1.962	.930	−.504	.053	8.147	3.062	1.913
1900	−3.529	−3.102	3.886	−.291	10.898	7.631	−1.823	1.058	−.310	.207	7.724	2.916	1.829
2000	−3.178	−2.790	3.540	−.240	10.353	7.469	−1.699	1.173	−.136	.346	6.998	2.785	1.754
2100	−2.860	−2.508	3.227	−.195	9.860	7.321	−1.586	1.277	.022	.472	6.684	2.666	1.686
2200	−2.571	−2.251	2.942	−.153	9.411	7.185	−1.484	1.372	.166	.587	6.396	2.558	1.625
2300	−2.307	−2.016	2.682	−.116	9.001	7.061	−1.391	1.459	.298	.692	6.134	2.459	1.568
2400	−2.065	−1.800	2.443	−.082	8.625	6.946	−1.305	1.539	.419	.789	5.892	2.368	1.517
2500	−1.842	−1.601	2.224	−.050	8.280	6.840	−1.227	1.613	.530	.879	5.668	2.285	1.469
2600	−1.636	−1.417	2.021	−.021	7.960	6.741	−1.154	1.681	.633	.962	5.460	2.208	1.425
2700	−1.446	−1.247	1.833	.005	7.664	6.649	−1.087	1.744	.729	1.039	5.268	2.070	1.384
2800	−1.268	−1.089	1.658	.030	7.388	6.563	−1.025	1.802	.818	1.110	5.088	2.008	1.311
2900	−1.103	−.941	1.495	.053	7.132	6.483	−.967	1.857	.900	1.178	4.920	1.950	1.278
3000	−.949	−.803	1.343	.074	6.892	6.407	−.913	1.908	.978	1.240	4.763	1.896	1.248
3100	−.805	−.674	1.201	.094	6.668	6.336	−.863	1.956	1.050	1.299	4.616	1.845	1.219
3200	−.670	−.553	1.067	.112	6.458	6.269	−.815	2.001	1.118	1.355	4.478	1.798	1.192
3300	−.543	−.439	.942	.129	6.260	6.206	−.771	2.043	1.182	1.407	4.347	1.753	1.166
3400	−.423	−.332	.824	.145	6.074	6.145	−.729	2.082	1.242	1.459	4.224	1.710	1.142
3500	−.310	−.231	.712	.160	5.898	6.088	−.690	2.120	1.299	1.503	4.108	1.670	1.119
3600	−.204	−.135	.607	.174	5.732	6.034	−.653	2.155	1.353	1.547	3.998	1.632	1.098
3700	−.103	−.044	.507	.188	5.574	5.982	−.618	2.189	1.404	1.589	3.894	1.596	1.077
3800	−.007	.042	.413	.200	5.425	5.933	−.585	2.220	1.452	1.629	3.795	1.562	1.058
3900	.084	.123	.323	.212	5.283	5.886	−.554	2.251	1.498	1.666	3.700	1.529	1.039
4000	.170	.201	.238	.223	5.149	5.841	−.524	2.280	1.541	1.703	—	—	—

Note: Partial pressures of all gases are expressed in atmospheres (Pascals/101,325). Graphite, $C_{(S)}$, is assigned a value of unity in the equilibrium expressions for K_E and K_F.

exothermic direction, so as to reduce the energy of the system. (Most spontaneous reactions are exothermic.) The Δn term is generally small compared with the other terms and represents the work done by the expanding system on the surroundings, or the work done on the contracting system by the surroundings. In summary, Equation 25 represents the balance of these various tendencies and determines the relative proportions of reactants and products at equilibrium. Notice that the term $\Delta E°/RT$ will become small at sufficiently high temperature, and the entropy term will then dominate. In other words, all molecules will break down into atoms at sufficiently high temperature, to maximize entropy.

The important conclusion from this discussion is that there is no need to consider rates of forward and reverse processes to determine equilibrium.

Table 1-6.1 provides values of equilibrium constants for 13 reactions involving most species found in fire products at equilibrium, over a temperature range from 600 K to 4000 K. Equilibrium constants for other reactions involving the same species may be obtained by combining these constants, as in Equation 13, or as will be illustrated in the examples below.

Table 1-6.1 does not include the $\frac{1}{2}N_2 = N$ equilibrium, because fire temperatures are generally not high enough for significant N to form. Tables 1-6.2 and 1-6.3 present information on the degree to which various gases are dissociated at various temperatures.

In performing calculations, it should be remembered that even if a relatively small fraction of dissociation occurs, a rather large amount of energy may be absorbed in the dissociation, with a corresponding large increase in the energy of the system. For example, if water vapor initially at 2800 K is allowed to dissociate adiabatically at one atmosphere, only 5.7 percent of the H_2O molecules will

TABLE 1-6.2 *Temperature (K) at Which a Given Fraction of a Pure Gas at One ATM is Dissociated*

Fraction	CO_2	H_2O	H_2	O_2	N_2
0.001	1600	1700	2050	2200	4000
0.004	1800	1900	2300	2400	—
0.01	1950	2100	2450	2600	—
0.04	2200	2400	2700	2900	—
0.1	2450	2700	2900	3200	—
0.4	2950	3200	3350	3700	—

dissociate, but the temperature will drop from 2800 K to 2491 K; i.e., the temperature relative to a 300 K baseline is lower by 12.4 percent.

CARBON FORMATION IN OXYGEN-DEFICIENT SYSTEMS

Solid carbon (soot) may be expected to form in oxygen-deficient combustion products, under some conditions. Since solid carbon does not melt or boil until extremely high temperatures ($\sim$ 4000 K), we only need concern ourselves with solid carbon $C_{(s)}$, not liquid $C_{(l)}$ or gaseous carbon $C_{(g)}$.

Consider pure carbon monoxide at 2000 K. There are three conceivable ways in which it might form solid carbon

α. $\qquad CO \rightleftarrows \frac{1}{2}C_{(S)} + \frac{1}{2}CO_2;$ $\qquad K_\alpha = (p_{CO_2})^{1/2}/p_{CO}$

β. $\qquad CO \rightleftarrows C_{(S)} + \frac{1}{2}O_2;$ $\qquad K_\beta = (p_{O_2})^{1/2}/p_{CO}$

γ. $\qquad CO \rightleftarrows C_{(S)} + O;$ $\qquad K_\gamma = p_O/p_{CO}$

Note that solid carbon does not appear in any of the equilibrium expressions. (By convention, a solid in equilibrium with gases is assigned a value of unity.)

From Table 1-6.1, we see that, at 2000 K,

$K_\alpha = K_E^{1/2}/K_F = $ antilog$_{10}[(10.353/2) - 7.469] = 5.1 \times 10^{-3}$

$K_\beta = 1/K_F = $ antilog$_{10}[0 - 7.469] = 3.4 \times 10^{-8}$

$K_\gamma = K_A/K_F = $ antilog$_{10}[-3.178 - 7.469] = 2.2 \times 10^{-11}$

We see that K_α, K_β, and K_γ are all small compared with unity, so very little of the CO would decompose by any of these modes. However, K_α is much larger than either K_β or K_γ, so it is the dominant mode for whatever decomposition may occur.

Thus, from the expression $p_{CO_2} = (K_\alpha p_{CO})^2$, and taking p_{CO} as 1 atm, we calculate $p_{CO_2} = (5.1 \times 10^{-3})^2 = 2.6 \times 10^{-5}$ atm. Since, by process α, two moles of CO must decompose for each mole of CO_2 formed, we conclude that $2 \times 2.6 \times 10^{-5}$ or 5.2×10^{-5} moles of CO will decompose to

$C_{(S)}$ plus CO_2, per mole of CO originally present, after which we will have reached an equilibrium state. In other words, about 1/20,000 of the CO will decompose.

If the original mixture had consisted of CO at 1 atm plus CO_2 at any pressure greater than 2.6×10^{-5} atm, at 2000 K, then we could conclude that no carbon whatsoever would form.

It can also be shown that addition of a trace of O_2 or a trace of H_2O to CO at 2000 K would completely suppress carbon formation. As a general statement, for a chemical system containing fewer carbon atoms than oxygen atoms, the equilibrium condition will favor CO formation rather than that of solid carbon.

For a carbon-containing system with little or no oxygen, carbon may or may not form, depending on the hydrogen partial pressure. For example, carbon may form according to $C_2H_2 \leftrightarrows C_{(s)} + H_2$. The equilibrium expression for this reaction is written

$$p_{H_2}/p_{C_2H_2} = K(= 13.9 \text{ at } 3000 \text{ K})$$

Again, note that solid carbon does not appear in the expression. If we rewrite the expression in the form $p_{H_2} > 13.9$ $p_{C_2H_2}$, this becomes the criterion for suppression of carbon formation at 3000 K. In other words, as long as p_{H_2} is more than 13.9 times as large as $p_{C_2H_2}$, no carbon will form at 3000 K and any carbon present will be converted to C_2H_2. On the other hand, pure C_2H_2 will decompose to $C_{(S)}$ plus H_2 until the H_2/C_2H_2 ratio reaches 13.9, after which no further decomposition will occur at 300 K.

Another way to view this is to say that H_2, C_2H_2, and solid carbon at 3000 K will be in a state of equilibrium if and only if the ratio $p_{H_2}/p_{C_2H_2} = 13.9$, and this is true regardless of the quantity of solid carbon present, and also regardless of the presence of other gases.

For a C–H–O–N system, the threshold conditions for equilibrium carbon formation are somewhat more complicated, but the trends are illustrated by the calculated values shown in Table 1-6.4 for carbon formation thresholds in carbon-hydrogen-air systems at 1 atm.

TABLE 1-6.3 *Temperature at Which Air at Equilibrium Contains a Given Fraction of Nitric Oxide, at One ATM*

Fraction	Temperature (K)
0.001	1450
0.004	1750
0.01	2100
0.04	2800

TABLE 1-6.4 *Threshold Atomic C/O Ratios for Carbon Formation (Equilibrium at one ATM, N/O = 3.76)*

Temperature (K)	1600	2000	2400	2800
Atomic H/C Ratio				
0	1.00	1.00	1.00	1.00
2	1.00	1.02	1.09	1.30
4	1.00	1.05	1.16	1.56

It must be noted that carbon forms more readily in actual flames than Table 1-6.4 indicates, because of non-equilibrium effects. In premixed laminar flames, incipient carbon formation occurs at a C/O ratio roughly 60 percent of the values shown in Table 1-6.4. See the next section for further comments on nonequilibrium.

DEPARTURE FROM EQUILIBRIUM

This procedure of specifying chemical systems by equilibrium equations will only yield correct results if the system is truly in equilibrium. If one prepares a mixture of H_2 and O_2 at room temperature, and then ages the mixture for a year, it will be found that essentially nothing has happened and the system will still be very far from equilibrium. On the other hand, such a system at a high temperature characteristic of combustion will reach equilibrium in a small fraction of a second. For example, a hydrogen atom, H, in the presence of O_2 at partial pressure 0.1 atm will react so fast at 1400 K that its half-life is only about 2 microseconds. (At room temperature, the half-life of this reaction is about 300 days.)

Since peak flame temperatures are almost always above 1400 K, and sometimes as high as 2400 K, it would appear that equilibrium should always be reached in flames. However, luminous (yellow) flames rapidly lose heat by radiation, turbulent flames may be partially quenched by the action of steep velocity gradients, and flames burning very close to a cold wall may be partially quenched by heat conductivity to the wall. Thus, the equilibrium condition is only a limiting case that real flames may approach. The products of a nonluminous laminar flame more than a few millimeters from any cold surface will always be very nearly in equilibrium.

SAMPLE PROBLEMS

EXAMPLE 1:
Given a mixture of an equal number of moles of steam and carbon monoxide, what will be the equilibrium composition, at 1700 K and 1 atm?

SOLUTION:
We would expect the species CO, H_2O, CO_2, and H_2 to be present. From Table 1-6.2, we see that the equilibria $H_2 \leftrightharpoons 2H$, $O_2 \leftrightharpoons 2O$, and $H_2O \leftrightharpoons \frac{1}{2}H_2 + OH$ can all be neglected at 1700 K, so the species H, O, and OH will not be present in significant quantities.

Since we have four species involving three chemical elements, we will require four minus three or one equilibrium relationship, for the equilibrium $H_2O + CO \leftrightharpoons H_2 + CO_2$.

The relationship is

$$\frac{p_{H_2} \cdot p_{CO_2}}{p_{H_2O} \cdot p_{CO}} = K \qquad (26)$$

In addition, we need three minus one or two atom-ratio relations, which are

H/C:
$$\frac{2p_{H_2} + 2p_{H_2O}}{p_{CO} + p_{CO_2}} = 2 \qquad (27)$$

(because the original mixture of H_2O + CO contains two H atoms per C atom) and

O/C:
$$\frac{p_{H_2O} + p_{CO} + 2p_{CO_2}}{p_{CO} + p_{CO_2}} = 2 \qquad (28)$$

(because the original mixture of H_2O + CO contains two O atoms per C atom). Finally, the sum of the partial pressures equals 1 atm

$$p_{H_2O} + p_{CO} + p_{H_2} + p_{CO_2} = 1 \qquad (29)$$

We now have a well-set problem, four equations and four unknowns, which may be solved as soon as K is quantified.

We do not find the equilibrium $H_2O + CO \leftrightharpoons H_2 + CO_2$ in Table 1-6.1. However, if we calculate $K_E/(K_F K_C)$ from Table 1-6.1, we see that

$$\frac{K_E}{K_F K_C} = \frac{p_{CO_2}}{(1) \cdot p_{O_2}} \cdot \frac{(1) \cdot (p_{O_2})^{1/2}}{p_{CO}} \cdot \frac{p_{H_2} \cdot (p_{O_2})^{1/2}}{p_{H_2O}}$$

$$= \frac{p_{CO_2} \cdot p_{H_2}}{p_{CO} \cdot p_{H_2O}} = K$$

From Table 1-6.1, $\log_{10} (K_E/K_F K_C)$ at 1700 K = 12.180 − 8.011 − 4.699 = −0.51, and K = antilog$_{10}$ (−0.51) = 0.309.

Upon substituting K = 0.309 into Equation 26, and then simultaneously solving Equations 26 through 29, we obtain

$$p_{CO_2} = p_{H_2} = 0.179 \text{ atm}$$

and
$$p_{H_2O} = p_{CO} = 0.321 \text{ atm}$$

EXAMPLE 2:
One mole of hydrogen is introduced into a 50 L vessel which is maintained at 2500 K. How much dissociation will occur, and what will the pressure be?

SOLUTION:
Let α be the degree of dissociation of the hydrogen defined by $\alpha = (p_H/2)/[p_{H_2} + (p_H/2)]$. Thus, α ranges from zero to one. One mole of H_2 partially dissociates to produce 2α moles of H, leaving $1 - \alpha$ moles of H_2. The total number of moles is then $2\alpha + 1 - \alpha$, or $\alpha + 1$. In view of the definition of α, the total number of moles present is $(p_H + p_{H_2})/[p_{H_2} + (p_H/2)]$.

By the ideal gas law, $PV = n R T$.

$$(p_H + p_{H_2})(50) = \frac{p_H + p_{H_2}}{p_{H_2} + (p_H/2)} (0.08206)(2500) \qquad (30)$$

which reduces to

$$p_{H_2} + (p_H/2) = 4.103 \qquad (31)$$

The equilibrium equation is

$$p_H/(p_{H_2})^{1/2} = K_B \qquad (32)$$

From Table 1-6.1, $\log_{10} K_B$ = −1.601 at 2500 K, and therefore K_B = 0.0251. Upon substitution into Equation 32 and elimination of p_{H_2} between Equations 31 and 32, one obtains

$$p_H^2 + 0.000315 p_H - 0.00258 = 0 \qquad (33)$$

This yields a positive and a negative root. The negative root has no physical significance. The positive root is $p_H = 0.0506$ atm. Then, Equation 32 yields $p_{H_2} = 4.08$ atm, and the total final pressure is $4.08 + 0.0506 = 4.13$ atm. The degree of dissociation, α, comes out to be 0.0062. (This is less dissociation than indicated by Table 1-6.2 because the pressure is well above 1 atm.)

EXAMPLE 3:

Propane is burned adiabatically at 1 atm with a stoichiometric proportion of air. Calculate the final temperature and composition. The initial temperature is 300 K.

SOLUTION:

The problem must be solved by a series of iterations. The first step is to assume a final temperature, either based on experience or by selecting a temperature substantially below the value calculated by assuming that CO_2 and H_2O are the only products of combustion. The second step is to solve the set of equations that specify the equilibrium composition at the assumed final temperature. The third step is to consult an overall enthalpy balance equation, which will show that the assumed final temperature was either too high or too low. The fourth step is to assume an appropriate new final temperature. The fifth and sixth steps are repeats of the second and third steps. If the correct final temperature is now found to be bracketed between these two assumed temperatures, then an interpolation should give a fairly accurate value of the true final temperature. Additional iterations may be made to improve the accuracy of the results to the degree desired.

As a guess, the final temperature is assumed to be 2300 K.

Now, the equilibrium equations at 2300 K are set up. The species to be considered are three principal species: CO_2, H_2O, and N_2, and seven minor species: H_2, O_2, OH, H, O, CO, and NO. (Based on chemical experience, the following possible species may be neglected at 2300 K when stoichiometric oxygen is present: N, $C_{(g)}$, NH, CN, CH, C_2, HO_2, HCN, O_3, C_3, NO_2, HNO, C_2H, CH_2, C_2O, CHO, and NH_2.) Thus, we consider ten species involving four elements, so ten minus four or six equilibrium equations are needed. Any six *independent* equilibria may be selected. We can assure independence by requiring that each successive equilibrium expression we write will introduce at least one new chemical species. Observe that this is so in the following list

$$CO + \frac{1}{2}O_2 = CO_2; \quad \frac{p_{CO_2}}{p_{CO} \cdot (p_{O_2})^{1/2}} = K = K_E/K_F \quad (34)$$

$$\frac{1}{2}O_2 = O; \quad p_O/(p_{O_2})^{1/2} = K_A \quad (35)$$

$$\frac{1}{2}O_2 + \frac{1}{2}N_2 = NO; \quad \frac{p_{NO}}{(p_{O_2} \cdot p_{N_2})^{1/2}} = K_G \quad (36)$$

$$H_2 + \frac{1}{2}O_2 = H_2O; \quad \frac{p_{H_2O}}{p_{H_2}(p_{O_2})^{1/2}} = K_C \quad (37)$$

$$\frac{1}{2}H_2 + \frac{1}{2}O_2 = OH; \quad \frac{p_{OH}}{(p_{H_2} \cdot p_{O_2})^{1/2}} = K_D \quad (38)$$

$$\frac{1}{2}H_2 = H; \quad p_H/(p_{H_2})^{1/2} = K_B. \quad (39)$$

Four additional equations are needed to determine the ten unknown partial pressures. These are three atom-ratio equations and a summation of the partial pressures to equal the total pressure. To obtain the atom ratios, we take air to consist of 3.76 parts of N_2 (by volume) per part of O_2, neglecting argon, etc. Then, from stoichiometry, $C_3H_8 + 5O_2 + (5 \cdot 3.76)N_2 \rightarrow 3CO_2 + 4H_2O + 18.8N_2$.

$$\frac{H}{C}: \quad \frac{8}{3} = \frac{p_H + p_{OH} + 2p_{H_2O} + 2p_{H_2}}{p_{CO_2} + p_{CO}} \quad (40)$$

$$\frac{H}{N}: \quad \frac{8}{37.6} = \frac{p_H + p_{OH} + 2p_{H_2O} + 2p_{H_2}}{2p_{N_2} + p_{NO}} \quad (41)$$

$$\frac{O}{C}: \quad \frac{10}{3} = \frac{p_O + p_{OH} + p_{NO} + p_{CO} + p_{H_2O} + 2p_{O_2} + 2p_{CO_2}}{p_{CO_2} + p_{CO}} \quad (42)$$

Finally,

$$p_{CO_2} + p_{H_2O} + p_{N_2} + p_{H_2} + p_{O_2} + p_{OH}$$
$$+ p_H + p_O + p_{CO} + p_{NO} = 1 \quad (43)$$

From Table 1-6.1 at 2300 K:

x	$\log_{10} x$	x
K_E	9.001	—
K_F	7.061	—
K_E/K_F	9.001–7.061	87.1
K_A	−2.307	0.00493
K_G	−1.391	0.0406
K_C	2.682	481
K_D	−0.116	0.766
K_B	−2.016	0.00964

We insert these K values into Equations 34 through 39, and then solve the set of 10 equations, Equations 34 through 43, for the equilibrium values of the 10 partial pressures at 2300 K. This solution may be obtained by a tedious set of successive approximations. The first approximation is obtained by solving for the three principal species N_2, CO_2, and H_2O, assuming the partial pressures of the remaining species are zero. Then, using this trial value of p_{CO_2}, solve for p_{CO} and p_{O_2}, using Equation 34 and assuming that $p_{CO} = 2 p_{O_2}$. Next, using p_{H_2O} and p_{O_2} as determined, use Equation 37 to determine a trial value of p_{H_2}. Then, using all the foregoing partial pressures, determine p_O from Equation 35, p_{NO} from Equation 36, p_{OH} from Equation 38, and p_H from Equation 39. Thus, ten trial values of the partial pressures are found. However, upon substitution into Equations 40, 41, 42, and 43, none of these equations will be quite satisfied. The partial pressures of the principal species must then be adjusted so as to satisfy Equations 40 through 43, and then a second iteration with the equilibrium equations must be carried out to establish new values for the minor species. After four or five such iterations, the results should converge to a set of partial pressures satisfying all equations.

A faster method is to use a computer program to solve the equations. (See the following section.)

The equilibrium partial pressures at 2300 K will come out to be:

P_{N_2}	0.7195 atm
P_{H_2O}	0.1474 atm
P_{CO_2}	0.1006 atm
P_{CO}	0.0143 atm
P_{O_2}	0.0066 atm
P_{H_2}	0.0038 atm
P_{OH}	0.0037 atm
P_{NO}	0.0028 atm
P_H	0.0006 atm
P_O	0.0004 atm

Now, we must determine if 2300 K was too high or too low a guess, by writing the enthalpy balance equation (cf. the chapter of this handbook on thermochemistry).

As a basis for the enthalpy balance, we assume that we have exactly one mole of products, at one atmosphere. Then, if $p_{CO_2} = 0.1006$ atm (see above), we must have 0.1006 moles of CO_2. Similarly, we have 0.0143 moles of CO. Since these are the only two carbon compounds in the products, and since three moles of CO_2 plus CO must form from each mole of C_3H_8 burned, it follows that $(0.1006 + 0.0143)/3 =$

0.0383 moles of C_3H_8 must have burned. Since the original C_3H_8-air mixture was stoichiometric, it follows that the reactants also consisted of $5 \times 0.0383 = 0.1915$ moles of O_2 and $3.76 \times 0.1915 = 0.7200$ moles of N_2. (Thus a total of 0.9498 moles of reactant form one mole of product, if the product is indeed at equilibrium at 2300 K.)

The enthalpy balance equation is

$$\sum n_i H_{i,\, T_r} = \sum n_j H_{j,\, T_p} \qquad (44)$$

where n_i and H_i are the number of moles and the enthalpy per mole of each reactant species at reactant temperature T_r, and n_j and H_j are the number of moles and the enthalpy per mole of each product species at product temperature T_p.

The enthalpy of each reactant or product species x at temperature T is given by

$$H_{x,\, T} = (\Delta H_f^\circ)_{298.15} + H^\circ - H_{298}^\circ \qquad (45)$$

where $(\Delta H_{f,\, 298.15}^\circ)_x$ is the enthalpy of formation of a mole of species x from its constituent elements in their standard states at 298 K. These constituent elements are H_2, O_2, N_2, and $C_{(s)}$, so $\Delta H_{f,\, 298.15}^\circ$ for each of these four species is zero, by definition.

Values of $(\Delta H_f^\circ)_{298.15}$ and $H^\circ - H_{298}^\circ$* for various species are contained in Table 1-6.5. Substitution of numerical values into Equation 44 yields:

Reactant species	$(\Delta H_f^\circ)_{298.15}$ (kJ/mole)	$H_{300}^\circ - H_{298}^\circ$ (kJ/mole)	H_{300}° (kJ/mole)	n_i (moles)	$n_i H_{i,\, 300}^\circ$ (kJ)
C_3H_8	−103.85	0.16	−103.69	0.0383	−3.971
O_2	0	0.05	0.05	0.1915	+0.010
N_2	0	0.05	0.05	0.7200	+0.036
					−3.925

and

Product species	$(\Delta H_f^\circ)_{298.15}$ (kJ/mole)	$H_{2300}^\circ - H_{298}^\circ$ (kJ/mole)	H_{2300}° (kJ/mole)	n_i (moles)	$n_i H_{i,2300}^\circ$ (kJ)
N_2	0	66.99	66.99	0.7195	+48.199
H_2O	−241.83	88.29	−153.54	0.1474	−22.632
CO_2	−393.52	109.67	−283.85	0.1006	−28.555
CO	−110.53	67.68	−42.85	0.0143	−0.613
O_2	0	70.60	70.60	0.0066	+0.466
H_2	0	63.39	63.39	0.0038	+0.241
OH	38.99	64.28	103.27	0.0037	+0.382
NO	90.29	68.91	159.20	0.0028	+0.446
H	218.00	41.61	259.61	0.0006	+0.156
O	249.17	41.96	291.13	0.0004	0.116
					−1.794

* If $H^\circ - H_{298}^\circ$ is not available from a table, it may be evaluated from the equation $H^\circ - H_{298}^\circ = \int_{298}^T C_p dT$. For C_3H_8, $C_p = 0.09$ kJ/mol·K at 298 K.

TABLE 1-6.5 *Enthalpies of Selected Combustion Products*

Species	N_2	O_2	O	NO	H_2	H	H_2O (g)	OH	CO_2	CO
$(\Delta H°_f)_{298.15}$	0.00 kJ/mole	0.00 kJ/mole	249.17 kJ/mole	90.29 kJ/mole	0.00 kJ/mole	218.00 kJ/mole	−241.83 kJ/mole	38.99 kJ/mole	−393.52 kJ/mole	−110.53 kJ/mole
Temp (K)	$H°-H°_{298}$, kJ/mole	$H°-H°_{298}$, kJ/mole	$H°-H°_{298}$, kJ/mole	$H°-H°_{298}$, kJ/mole	$H°-H°_{298}$, kJ/mole	$H°-H°_{298}$, kJ/mole	$H°-H°_{298}$, kJ/mole	$H°-H°_{298}$, kJ/mole	$H°-H°_{298}$, kJ/mole	$H°-H°_{298}$, kJ/mole
100	−5.77	−5.78	−4.52	−6.07	−5.47	−4.12	−6.61	−6.14	−6.46	−5.77
200	−2.86	−2.87	−2.19	−2.95	−2.77	−2.04	−3.28	−2.97	−3.41	−2.87
298	0.00	0.00	0.00	0.00	0.00	0.00	0.00	0.00	0.00	0.00
300	.05	.05	.04	.05	.05	.04	.06	.05	.07	.05
400	2.97	3.03	2.21	3.04	2.96	2.12	3.45	3.03	4.01	2.97
500	5.91	6.08	4.34	6.06	5.88	4.20	6.92	5.99	8.31	5.93
600	8.90	9.24	6.46	9.15	8.81	6.28	10.50	8.94	12.92	8.94
700	11.94	12.50	8.57	12.31	11.75	8.35	14.18	11.90	17.76	12.02
800	15.05	15.84	10.67	15.55	14.70	10.43	17.99	14.88	22.82	15.18
900	18.22	19.24	12.77	18.86	17.68	12.51	21.92	17.89	28.04	18.40
1000	21.46	22.70	14.86	22.23	20.68	14.59	25.98	20.94	33.41	21.69
1100	24.76	26.21	16.95	25.65	23.72	16.67	30.17	24.02	38.89	25.03
1200	28.11	29.76	19.04	29.12	26.80	18.74	34.48	27.16	44.48	28.43
1300	31.50	33.34	21.13	32.63	29.92	20.82	38.90	30.34	50.16	31.87
1400	34.94	36.96	23.21	36.17	33.08	22.90	43.45	33.57	55.91	35.34
1500	38.40	40.60	25.30	39.73	36.29	24.98	48.10	36.84	61.71	38.85
1600	41.90	44.27	27.38	43.32	39.54	27.06	52.84	40.15	67.58	42.38
1700	45.43	47.96	29.46	46.93	42.84	29.14	57.68	43.50	73.49	45.94
1800	48.98	51.67	31.55	50.56	46.17	31.22	62.61	46.89	79.44	49.52
1900	52.55	55.41	33.63	54.20	49.54	33.30	67.61	50.31	85.43	53.12
2000	56.14	59.17	35.71	57.86	52.95	35.38	72.69	53.76	91.45	56.74
2100	59.74	62.96	37.79	61.53	56.40	37.46	77.83	57.25	97.50	60.38
2200	63.36	66.77	39.88	65.22	59.88	39.53	83.04	60.75	103.57	64.02
2300	66.99	70.60	41.96	68.91	63.39	41.61	88.29	64.28	109.67	67.68
2400	70.64	74.45	44.04	72.61	66.93	43.69	93.60	67.84	115.79	71.35
2500	74.30	78.33	46.13	76.32	70.50	45.77	98.96	71.42	121.93	75.02
2600	77.96	82.22	48.22	80.04	74.09	47.85	104.37	75.01	128.08	78.71
2700	81.64	86.14	50.30	83.76	77.72	49.92	109.81	78.63	134.26	82.41
2800	85.32	90.08	52.39	87.49	81.37	52.00	115.29	82.27	140.44	86.12
2900	89.01	94.04	54.48	91.23	85.04	54.08	120.81	85.92	146.65	89.83
3000	92.71	98.01	56.58	94.98	88.74	56.16	126.36	89.58	152.86	93.54
3100	96.42	102.01	58.67	98.73	92.46	58.24	131.94	93.27	159.09	97.27
3200	100.14	106.02	60.77	102.48	96.20	60.32	137.55	96.96	165.33	101.00
3300	103.85	110.05	62.87	106.24	99.96	62.40	143.19	100.67	171.59	104.73
3400	107.57	114.10	64.97	110.00	103.75	64.48	148.85	104.39	177.85	108.48
3500	111.31	118.16	67.08	113.77	107.55	66.55	154.54	108.12	184.12	112.22

The enthalpy of the products (−1.794 kJ) is seen to be 2.131 kJ larger than the enthalpy of the reactants (−3.925 kJ). To put this 2.131 kJ difference in perspective, note that the heat of combustion of 0.0383 moles of propane at 298 K, to form 3 moles of CO_2 and 4 moles of H_2O per mole of propane, is 0.0383 (3 × 393.52 + 4 × 241.83 − 103.85) = 78.29 kJ. Thus, the 2.131 kJ discrepancy when compared with 78.29 kJ is rather small, showing that the 2300 K "first guess" was very close. Since the products, at 2300 K, are seen to have a slightly higher enthalpy than the reactants, the correct temperature must be slightly less than 2300 K.

To continue the calculation, the next step is to assume that the final temperature is 2200 K instead of 2300 K. The details will not be presented, but this will yield a new and slightly different set of values of the ten partial pressures of the products. Thus, a new enthalpy balance may be attempted, in the same manner as before. When this is done, the result will be that this time the enthalpy of the reactants will come out to be slightly higher than the enthalpy of the products, showing that the correct temperature is above 2200 K.

An interpolation may be made between the 2200 K enthalpy discrepancy and the 2300 K enthalpy discrepancy, which will show that the correct final temperature is *2268 K*. Furthermore, the partial pressures of each product species may be obtained by interpolating between the 2200 K partial pressures and the 2300 K partial pressures, with results as follows:

T = 2268 K		
	P_{N_2}	0.7207 atm
	P_{H_2O}	0.1484 atm
	P_{CO_2}	0.1026 atm
	P_{CO}	0.0125 atm
	P_{O_2}	0.0059 atm
	P_{H_2}	0.0034 atm
	P_{OH}	0.0032 atm
	P_{NO}	0.0025 atm
	p_H	0.0005 atm
	P_O	0.0003 atm

TABLE 1-6.5 *Enthalpies of Selected Combustion Products (Continued)*

$C_{(s)}$	F_2	F	HF	Cl_2	Cl	HCl	Br_2	Br	HBr
0.00 kJ/mole	0.00 kJ/mole	78.91 kJ/mole	−272.55 kJ/mole	0.00 kJ/mole	121.29 kJ/mole	−92.31 kJ/mole	0.00 kJ/mole	111.86 kJ/mole	−36.44 kJ/mole
$H°-H°_{298}$, kJ/mole	$H°-H°_{298}$, kJ/mole	$H°-H°_{298}$, kJ/mole	$H°-H°_{298}$, kJ/mole	$H°-H°_{298}$, kJ/mole	$H°-H°_{298}$, kJ/mole	$H°-H°_{298}$, kJ/mole	$H°-H°_{298}$, kJ/mole	$H°-H°_{298}$, kJ/mole	$H°-H°_{298}$, kJ/mole
−.99	−5.92	−4.43	−5.77	−6.27	−4.19	−5.77	−21.72	−4.12	−5.77
−.67	−2.99	−2.23	−2.86	−3.23	−2.10	−2.86	−16.82	−2.04	−2.86
0.00	0.00	0.00	0.00	0.00	0.00	0.00	0.00	0.00	0.00
.02	.06	.04	.05	.06	.04	.05	.14	.04	.05
1.04	3.28	2.30	2.97	3.54	2.26	2.97	34.61	2.12	2.97
2.36	6.64	4.53	5.88	7.10	4.52	5.89	38.31	4.20	5.90
3.94	10.11	6.72	8.80	10.74	6.80	8.84	42.02	6.28	8.87
5.72	13.66	8.90	11.73	14.41	9.08	11.81	45.76	8.36	11.88
7.64	17.27	11.05	14.68	18.12	11.34	14.84	49.51	10.46	14.96
9.67	20.91	13.19	17.64	21.84	13.59	17.91	53.27	12.57	18.10
11.79	24.59	15.33	20.64	25.59	15.82	21.05	57.03	14.70	21.30
13.99	28.30	17.45	23.68	29.34	18.03	24.24	60.81	16.84	24.56
16.24	32.03	19.56	26.76	33.10	20.23	27.48	64.58	19.01	27.87
18.54	35.77	21.67	29.87	36.88	22.41	30.78	68.37	21.20	31.24
20.88	39.54	23.78	33.04	40.66	24.60	34.12	72.16	23.40	34.65
23.25	43.32	25.89	36.24	44.45	26.77	37.51	75.96	25.61	38.10
25.66	47.11	27.99	39.48	48.25	28.93	40.93	79.76	27.85	41.59
28.09	50.91	30.09	42.76	52.05	31.09	44.39	83.57	30.09	45.11
30.55	54.72	32.18	46.09	55.86	33.23	47.89	87.38	32.35	48.66
33.02	58.54	34.28	49.44	59.68	35.38	51.41	91.20	34.61	52.24
35.53	62.38	36.37	52.83	63.51	37.51	54.96	95.02	36.88	55.84
38.05	66.22	38.46	56.25	67.34	39.64	58.53	98.85	39.15	59.46
40.58	70.07	40.55	59.69	71.18	41.77	62.12	102.68	41.43	63.10
43.13	73.93	42.64	63.17	75.02	43.89	65.73	106.52	43.70	66.76
45.71	77.80	44.73	66.66	78.88	46.02	69.37	110.36	45.98	70.44
48.29	81.67	46.82	70.18	82.74	48.13	73.01	114.20	48.26	74.13
50.89	85.55	48.91	73.73	86.61	50.25	76.68	118.05	50.54	77.83
53.50	89.45	50.99	77.29	90.50	52.36	80.36	121.91	52.81	81.55
56.13	93.35	53.08	80.87	94.39	54.48	84.06	125.77	55.09	85.28
58.77	97.25	55.17	84.47	98.29	56.58	87.76	129.63	57.36	89.02
61.43	101.16	57.25	88.09	102.21	58.69	91.48	133.49	59.63	92.77
64.09	105.08	59.34	91.72	106.14	60.79	95.21	137.37	61.89	96.53
66.78	109.01	61.42	95.37	110.08	62.90	98.95	141.24	64.15	100.31
69.47	112.94	63.50	99.03	114.03	65.00	102.70	145.13	66.41	104.09
72.17	116.88	65.59	102.71	118.00	67.10	106.46	149.01	68.67	107.88
74.89	120.83	67.67	106.39	121.98	69.20	110.23	152.90	70.92	111.68

COMPUTER PROGRAMS FOR CHEMICAL EQUILIBRIUM CALCULATIONS

In view of the extremely tedious calculations needed for determination of the equilibrium temperature and composition in a combustion process, a computer program for executing these calculations would be desirable. Fortunately, such programs have been developed.

However, the user of a computer program should be warned that thorough understanding of the material in this chapter is needed to avoid misinterpreting the computer output. Further, given such understanding, simple manual calculations can be performed to obtain independent checks of the computer output.

In the United States, the most widely used chemical equilibrium program for combustion is a National Aeronautics and Space Administration (NASA) program, on magnetic tape, which is described in Report SP-273,[2] issued in 1976, by S. Gordon and B. J. McBride, of NASA's Lewis Research Center, Cleveland, Ohio. The FORTRAN code, known as the CET 89 program, updated in 1989, can be obtained from COSMIC, The University of Georgia, 382 East Broad St., Athens, GA, 30602-4272, USA; telephone (706) 542-3265. The order number for the code is LEW-15113. The program is machine-independent and requires about 423 Kb of memory.

The program will calculate the final equilibrium conditions for adiabatic combustion at either constant pressure or constant volume, given the initial conditions. For the constant-pressure calculations, one specifies the initial temperature, the pressure, and the identities and relative proportions of the reactants. The computer program contains the properties of selected reactants including: air, oxygen, nitrogen, hydrogen, graphite, methane, acetylene, ethylene, ethane, propane, butane, 1-butene, heptane, octane, benzene, toluene, JP-4, JP-5, methanol, ethanol, and polyethylene, so if the fire only involves reactants from this list, no further input is necessary. If the fire involves a reactant not on this list, the input data must include the elemental composition and the enthalpy of formation of the reactant at 298 K, as well as enthalpy versus temperature data for the reactant over the temperature range from 298 K to the initial temperature. (If the initial temperature is 298 K, the last item is not needed.)

The computer program can handle reactants containing any of the following elements: A, Al, B, Br, C, Cl, F, Fe, H, He, K, Li, Mg, N, Na, Ne, O, P, S, Si, and Xe. Data are included in the program on all known compounds, including liquids and solids, that can form at elevated temperatures from combinations of these elements. It is not necessary for the user to specify which product species to consider. The program can consider them all, and will print out all equilibrium species present with mole fractions greater than 5×10^{-6}, unless instructed to print out trace values down to some lower specified level.

The program can calculate Chapman-Jouguet detonation products as well as constant-pressure or constant-volume combustion products, if desired.

An addition to the program permits calculation of viscosity and thermal conductivity of gaseous mixtures, selected from 154 gaseous species, at temperatures from 300 K to 5000 K.[3,4]

NOMENCLATURE

C_p heat capacity at constant pressure (kJ/mol·K)
$\Delta E°$ energy of products relative to energy of reactants, all at temperature T and 1 atm (kJ/mol)

$\Delta F°$ free energy of products relative to free energy of reactants, all at temperature T and 1 atm (kJ/mol)
$\Delta H°$ enthalpy of products relative to enthalpy of reactants, all at temperature T and 1 atm (kJ/mol)
K equilibrium constant (based on partial pressures expressed in atmospheres)
K   degrees Kelvin
n number of moles (e.g., a mole of oxygen is 32 g)
p_i partial pressure of ith species (atm)
p total pressure (atm)
R   gas constant (kJ/mol-K)
$\Delta S°$ entropy of products relative to entropy of reactants, all at temperature T and 1 atm (kJ/mol)
T absolute temperature (K)

REFERENCES CITED

1. J. van't Hoff, cf. G. Lewis, M. Randall, K. Pitzer, and L. Brewer, *Thermodynamics*, McGraw-Hill, NY, (1961).
2. A. Gordon and B. McBride, "Computer Program for Calculation of Complex Chemical Equilibrium and Application," *NASA Report SP-273* (1971).
3. A. Svehla and B. McBride, "Fortran Computer Program for Calculation of Thermodynamic and Transport Properties of Complex Chemical Systems," *NASA TN-D-7056* (1971).
4. A. Gordon and B. McBride, "Computer Program for Calculation of Complex Chemical Equilibrium and Application, Supplement 1," *NASA TM-86885* (1984).

THERMAL DECOMPOSITION OF POLYMERS

Craig L. Beyler and Marcelo M. Hirschler

INTRODUCTION

Solid polymeric materials undergo both physical and chemical changes when heat is applied; this will usually result in undesirable changes to the properties of the material. A clear distinction needs to be made between thermal decomposition and thermal degradation. The American Society for Testing and Materials' (ASTM) definitions should provide helpful guidelines. Thermal decomposition is "a process of extensive chemical species change caused by heat." Thermal degradation is "a process whereby the action of heat or elevated temperature on a material, product, or assembly causes a loss of physical, mechanical, or electrical properties."[1] In terms of fire, the important change is thermal decomposition, whereby the chemical decomposition of a solid material generates gaseous fuel vapors, which can burn above the solid material. In order for the process to be self-sustaining, it is necessary for the burning gases to feed back sufficient heat to the material to continue the production of gaseous fuel vapors or volatiles. As such, the process can be a continuous feedback loop if the material continues burning. In that case, heat transferred to the polymer causes the generation of flammable volatiles; these volatiles react with the oxygen in the air above the polymer to generate heat, and a part of this heat is transferred back to the polymer to continue the process. (See Figure 1-7.1.) This chapter is concerned with chemical and physical aspects of thermal decomposition of polymers. The chemical processes are responsible for the generation of flammable volatiles while physical changes, such as melting and charring, can markedly alter the decomposition and burning characteristics of a material.

Dr. Craig L. Beyler is the Technical Director of Hughes Associates, Fire Science and Engineering. He was the founding editor of the *Journal of Fire Protection Engineering* and serves on a wide range of committees in the fire safety community.

Dr. Marcelo M. Hirschler has two decades of experience researching fire and polymers and has published extensively. He managed a US industrial fire testing and research laboratory for seven years, while chairing plastics industry committees on fire issues. He is also active in developing fire standards, and is an associate editor of the journal *Fire and Materials*. At present he is an independent consultant on fire safety.

The gasification of polymers is generally much more complicated than that of flammable liquids. For most flammable liquids, the gasification process is simply evaporation. The liquid evaporates at a rate required to maintain the equilibrium vapor pressure above the liquid. In the case of polymeric materials, the original material itself is essentially involatile, and the quite large molecules must be broken down into smaller molecules that can vaporize. In most cases, a solid polymer breaks down into a variety of smaller molecular fragments made up of a number of different chemical species. Hence, each of the fragments has a different equilibrium vapor pressure. The lighter of the molecular fragments will vaporize immediately upon their creation while other heavier molecules will remain in the condensed phase (solid or liquid) for some time. While remaining in the condensed phase, these heavier molecules may undergo further decomposition to lighter fragments which are more easily vaporized. Some polymers break down completely so that virtually no solid residue remains. More often, however, not all the original fuel becomes fuel vapors since solid residues are left behind. These residues can be carbonaceous (char), inorganic (originating from heteroatoms contained in the original polymer, either within the structure or as a result of additive incorporations), or a combination of both. Charring materials, such as wood, leave large fractions of the original carbon content as carbonaceous residue, often as a porous char. When thermal decomposition of deeper layers of such a material continues, the volatiles produced must pass through the char above them to reach the surface. During this travel, the hot char may cause secondary reactions to occur in the volatiles. Carbonaceous chars can be intumescent layers, when appropriately formed, which slow down further thermal decomposition considerably. Inorganic residues, on the other hand, can form glassy layers that may then become impenetrable to volatiles and protect the underlying layers from any further thermal breakdown. Unless such inorganic barriers form, purely carbonaceous chars can always be burned by surface oxidation at higher temperatures.

As this brief description of the thermal decomposition process indicates, the chemical processes are varied and complex. The rate, mechanism, and product composition of these thermal decomposition processes depend both on the physical properties of the original material and on its chemical composition.

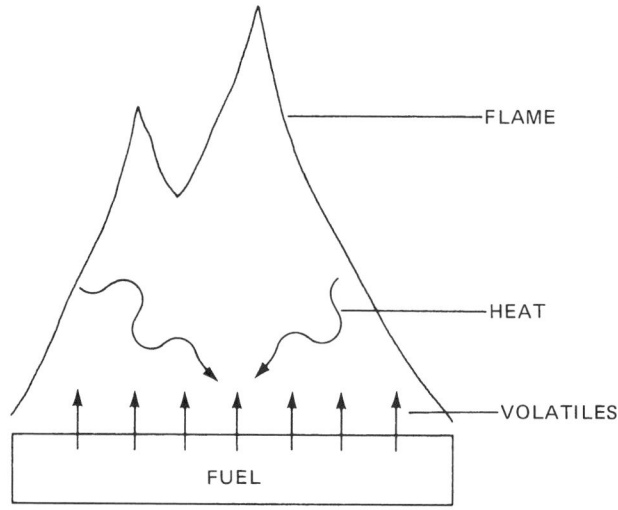

Fig. 1-7.1. *Energy feedback loop required for sustained burning.*

POLYMERIC MATERIALS

Polymeric materials can be classified in a variety of ways.[2] First, polymers are often classified, based on their origin, into natural and synthetic (and sometimes including a third category of semi-natural or synthetic modifications of natural polymers). However, more useful is a classification based on physical properties, in particular the elastic modulus and the degree of elongation. Following this criterion, polymers can be classified into elastomers, plastics, and fibers. Elastomers (or rubbers) are characterized by a long-range extensibility that is almost completely reversible at room temperature. Plastics have only partially reversible deformability, while fibers have very high tensile strength but low extensibility. Plastics can be further subdivided into thermoplastics (whose deformation at elevated temperature is reversible) and thermosets (which undergo irreversible changes when heated). Elastomers have elastic moduli between 10^5 and 10^6 N/m^2, while plastics have moduli between 10^7 and 10^8 N/m^2, and fibers have moduli between 10^9 and 10^{10} N/m^2. In terms of the elongation, elastomers can be stretched roughly up to 500 to 1000 percent, plastics between 100 to 200 percent, and fibers only 10 to 30 percent before fracture of the material is complete.

Polymers can also be classified in terms of their chemical composition; this gives a very important indication as to their reactivity, including their mechanism of thermal decomposition and their fire performance.

The main carbonaceous polymers with no heteroatoms are polyolefins, polydienes, and aromatic hydrocarbon polymers (typically styrenics). The main polyolefins are thermoplastics: polyethylene [repeating unit: $-(CH_2-CH_2)-$] and polypropylene {repeating unit: $-[CH(CH_3)-CH_2]-$}, which are two of the three most widely used synthetic polymers. Polydienes are generally elastomeric and contain one double bond per repeating unit. Other than polyisoprene (which can be synthetic or natural, e.g., natural rubber) and polybutadiene (used mostly as substitutes for rubber), most other polydienes are used as copolymers or blends with other materials [e.g., in ABS (acrylonitrile butadiene styrene terpolymers), SBR (styrene butadiene rubbers), MBS (methyl

methacrylate butadiene styrene terpolymers), and EPDM (ethylene propylene diene rubbers)]. They are primarily used for their high abrasion resistance and high impact strength. The most important aromatic hydrocarbon polymers are based on polystyrene {repeating unit: $-[CH(phenyl)-CH_2]-$}. It is extensively used as a foam and as a plastic for injection-molded articles. A number of styrenic copolymers also have tremendous usage, e.g., principally, (ABS), styrene acrylonitrile polymers (SAN), and (MBS).

The most important oxygen-containing polymers are cellulosics, polyacrylics, and polyesters. Polyacrylics are the only major oxygen-containing polymers with carbon-carbon chains. The most important oxygen-containing natural materials are cellulosics, mostly wood and paper products. Different grades of wood contain 20 to 50 percent cellulose. The most widely used polyacrylic is poly(methyl methacrylate)(PMMA) {repeating unit: $-[CH_2-C(CH_3)-CO-OCH_3]-$}. PMMA is valued for its high light transmittance, dyeability, and transparency. The most important polyesters are manufactured from glycols, for example, polyethylene terephthalate (PET) or polybutylene terephthalate (PBT), or from biphenol A (polycarbonate). They are used as engineering thermoplastics, as fibers, for injection-molded articles, and unbreakable replacements for glass. Other oxygenated polymers include phenolic resins (produced by the condensation of phenols and aldehydes, which are often used as polymeric additives), polyethers [such as polyphenylene oxide (PPO), a very thermally stable engineering polymer], and polyacetals (such as polyformaldehyde, used for its intense hardness and resistance to solvents).

Nitrogen-containing materials include nylons, polyurethanes, polyamides, and polyacrylonitrile. Nylons, having repeating units containing the characteristic group $-CO-NH-$, are made into fibers and also into a number of injection-molded articles. Nylons are synthetic aliphatic polyamides. There are also natural polyamides (e.g., wool, silk, and leather) and synthetic aromatic polyamides (of exceptionally high thermal stability and used for protective clothing). Polyurethanes (PU), with repeating units containing the characteristic group $-NH-COO-$, are normally manufactured from the condensation of polyisocyanates and polyols. Their principal area of application is as foams (flexible and rigid, or as thermal insulation). Other polyurethanes are made into thermoplastic elastomers, which are chemically very inert. Both these types of polymers have carbon-nitrogen chains, but nitrogen can also be contained in materials with carbon-carbon chains, the main example being polyacrylonitrile [repeating unit: $-(CH_2-CH-CN-)$]. It is used mostly to make into fibers and as a constituent of engineering copolymers (e.g., SAN, ABS).

Chlorine-containing polymers are exemplified by poly(vinyl chloride) [PVC, repeating unit: $-(CH_2-CHCl)-$]. It is the most widely used synthetic polymer, together with polyethylene and polypropylene. It is unique in that it is used both as a rigid material (unplasticized) and as a flexible material (plasticized). Flexibility is achieved by adding plasticizers or flexibilizers. Through the additional chlorination of PVC, another member of the family of vinyl materials is made: chlorinated poly(vinyl chloride) (CPVC) with very different physical and fire properties from PVC. Two other chlorinated materials are of commercial interest: (1) polychloroprene (a polydiene, used for oil-resistant wire and cable materials and resilient foams) and (2) poly(vinylidene chloride) [PVDC, with a repeating unit: $-(CH_2-CCl_2)-$ used for making films and fibers]. All these polymers have carbon-carbon chains.

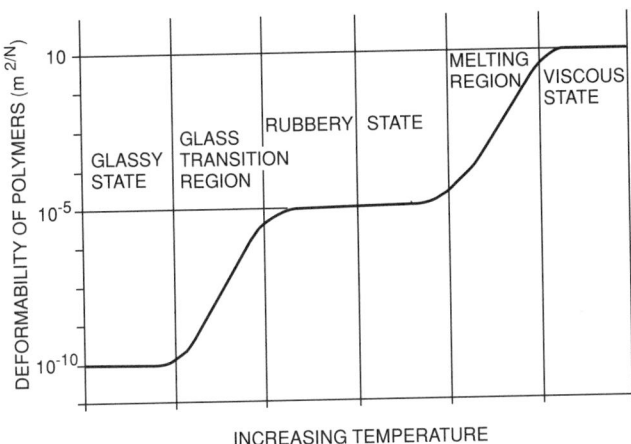

Fig. 1-7.2. Idealized view of effect on deformability of thermoplastics with increasing temperature.

Fluorine-containing polymers are characterized by high thermal and chemical inertness and low coefficient of friction. The most important material in the family is polytetrafluoroethylene (PTFE); others are poly(vinylidene fluoride) (PVDF), poly(vinyl fluoride) (PVF), and fluorinated ethylene polymers (FEP).

PHYSICAL PROCESSES

The various physical processes that occur during thermal decomposition can depend on the nature of the material. For example, as thermosetting polymeric materials are infusible and insoluble once they have been formed, simple phase changes upon heating are not possible. Thermoplastics, on the other hand, can be softened by heating without irreversible changes to the material, provided heating does not exceed the minimum thermal decomposition temperature. This provides a major advantage for thermoplastic materials in the ease of molding or thermoforming of products.

The physical behavior of thermoplastics in heating is dependent on the degree of order in molecular packing, i.e., the degree of crystallinity. For crystalline materials, there exists a well-defined melting temperature. Materials that do not possess this ordered internal packing are amorphous. An example of an amorphous material is window glass. While it appears to be a solid, it is in fact a fluid that over long periods of time (centuries) will flow noticeably. Despite this, at low temperatures amorphous materials do have structural properties of normal solids. At a temperature known as the glass transition temperature in polymers, the material starts a transition toward a soft and rubbery state. For example, when using a rubber band, one would hope to use the material above its glass transition temperature. However, for materials requiring rigidity and compressive strength, the glass transition temperature is an upper limit for practical use. In theoretical terms, this "deformability" of a polymer can be expressed as the ratio of the deformation (strain) resulting from a constant stress applied. Figure 1-7.2 shows an idealized view of the effect on the deformability of thermoplastics of increasing the temperature: a two-step increase. In practice, it can be stated that the glass transition temperature is the upper

limit for use of a plastic material (as defined above, based on its elastic modulus and elongation) and the lower limit for use of an elastomeric material. Furthermore, many materials may not achieve a viscous state since they begin undergoing thermal decomposition before the polymer melts. Some typical glass transition temperatures are given in Table 1-7.1. As this type of physical transformation is less well defined than a phase transformation, it is known as a second order transition. Typically, materials are only partially crystalline, and, hence, the melting temperature is less well defined, usually extending over a range of 10°C or more.

Neither thermosetting nor cellulosic materials have a fluid state. Due to their structure, it is not possible for the original material to change state at temperatures below that at which thermal decomposition occurs. Hence, there are no notable physical transformations in the material before decomposition. In cellulosic materials, there is an important semi-physical change that always occurs on heating: desorption of the adsorbed water. As the water is both physically and chemically adsorbed, the temperature and rate of desorption will vary with the material. The activation energy for physical desorption of water is 30 to 40 kJ/mol, and it starts occurring at temperatures somewhat lower than the boiling point of water (100°C).

Many materials (whether cellulosic, thermosetting, or thermoplastic) produce carbonaceous chars on thermal decomposition. The physical structure of these chars will strongly affect the continued thermal decomposition process. Very often the physical characteristics of the char will dictate the rate of thermal decomposition of the remainder of the polymer. Among the most important characteristics of char are density, continuity, coherence, adherence, oxidation-resistance, thermal insulation properties, and permeability.[3] Low-density-high-porosity chars tend to be good thermal insulators; they can significantly inhibit the flow of heat from the gaseous combustion zone back to the condensed phase behind it, and thus slow down the thermal decomposition process. This is one of the better means of decreasing the flammability of a polymer (through additive or reactive flame retardants).[1,3,4] As the char layer thickens, the heat flux to the virgin material decreases, and the decomposition rate is reduced. The char itself can undergo glowing combustion when it is exposed to air. However, it is unlikely that both glowing combustion of the char and significant gas-phase combustion can occur simultaneously in the same zone above the surface, since the flow of volatiles through the char will tend to exclude air from direct contact with the char. Therefore, in general, solid-phase char combustion tends to occur after volatilization has largely ended.

CHEMICAL PROCESSES

The thermal decomposition of polymers may proceed by oxidative processes or simply by the action of heat. In many polymers, the thermal decomposition processes are accelerated by oxidants (such as air or oxygen). In that case, the minimum decomposition temperatures are lower in the presence of an oxidant. This significantly complicates the problem of predicting thermal decomposition rates, as the prediction of the concentration of oxygen at the polymer surface during thermal decomposition or combustion is quite difficult. Despite its importance to fire, there have been many fewer studies of thermal decomposition processes in oxygen or air than in inert atmospheres.

TABLE 1-7.1   *Glass Transition and Crystalline Melting Temperatures*

Polymer	% Crystalline	Glass Transition Temperature (°C)	Crystalline Melting Temperature (°C)
Acetal	high		175–181
Acrylonitrile-butadiene-styrene	low	91–110	110–125
Cellulose	high		decomposes
Ethylene-vinyl acetate	high		65–110
Fluorinated ethylene propylene	high		275
High-density polyethylene	95	−125	130–135
Low-density polyethylene	60	−25	109–125
Natural rubber	low		30
Nylon 11	high		185–195
Nylon 6		75	215–220
Nylon 6-10		50	215
Nylon 6-6		57	250–260
Polyacrylonitrile	low	140	317
Poly(butene 1)		−25	124–142
Polybutylene		−25	126
Poly(butylene terephthalate)	high	40	232–267
Polycarbonate	low	145–150	215–230
Polychlorotrifluoroethylene	high	45	220
Poly(ether ether ketone)	high	143	334
Poly(ether imide)		217	
Poly(ethylene terephthalate)	high	70	265
Poly(hexene 1)			55
Poly(methylbutene 1)			300
Polymethylene	100		136
Poly(methyl methacrylate)	low	50	90–105
Polyoxymethylene	75–80	−85	175–180
Poly(pentene 1)			130
Poly(3-phenylbutene 1)			360
Poly(phenylene oxide)/polystyrene	low	100–135	110–135
Poly(phenylene sulphide)	high	88–93	277–282
Polypropylene	65	−20	170
Polystyrene	low	>80	230
Polysulphone	low	190	190
Polytetrafluoroethylene	100	125	327
Poly(vinyl chloride)	5–15	80–85	75–105 (212)
Poly(vinylidene chloride)	high	−18	210
Poly(vinylidene fluoride)	high	−30– −20	160–170
Poly(p-xylene)			>400
Styrene-acrylonitrile	low	100–120	120

It is worthwhile highlighting, however, that some very detailed measurements of oxygen concentrations and of the effects of oxidants have been made by Stuetz et al. in the 1970s[5] and more recently by Kashiwagi et al.,[6–10] Brauman,[11] and Gijsman et al.[12] Stuetz found that oxygen can penetrate down to at least 10 mm below the surface of polypropylene. Moreover, for both polyethylene and polypropylene, this access to oxygen is very important in determining thermal decomposition rates and mechanisms. Another study of oxygen concentration inside polymers during thermal decomposition, by Brauman,[11] suggests that the thermal decomposition of polypropylene is affected by the presence of oxygen (a fact confirmed more recently by Gijsman et al.[12]) while poly(methyl methacrylate) thermal decomposition is not. Kashiwagi found that a number of properties affect the thermal and oxidative decomposition of thermoplastics, particularly molecular weight, prior thermal damage, weak linkages, and primary radicals. Of particular interest is the fact that the effect of oxygen (or air) on thermal decomposition depends on the mechanism of polymerization: free-radical polymerization leads to a neutralization of the effect of oxygen. A study on poly(vinylidene fluo-

ride) indicated that the effect of oxygen can lead to changes in both reaction rate and kinetic order of reaction.[13]

Kashiwagi's work in particular has resulted in the development of models for the kinetics of general random-chain scission thermal decomposition,[14] as well as for the thermal decomposition of cellulosics[15] and thermoplastics.[16]

There are a number of general classes of chemical mechanisms important in the thermal decomposition of polymers: (1) random-chain scission, in which chain scissions occur at apparently random locations in the polymer chain; (2) end-chain scission, in which individual monomer units are successively removed at the chain end; (3) chain-stripping, in which atoms or groups not part of the polymer chain (or backbone) are cleaved; and (4) cross-linking, in which bonds are created between polymer chains. These are discussed in some detail under "General Chemical Mechanisms," later in this text. It is sufficient here to note that thermal decomposition of a polymer generally involves more than one of these classes of reactions. Nonetheless, these general classes provide a conceptual framework useful for understanding and classifying polymer decomposition behavior.

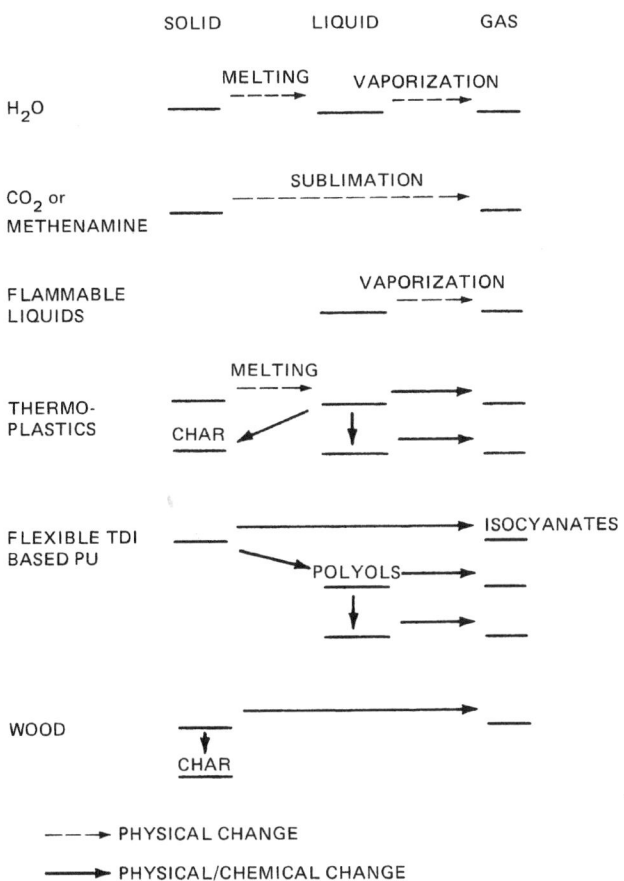

---→ PHYSICAL CHANGE

——→ PHYSICAL/CHEMICAL CHANGE

Fig. 1-7.3. Physical and chemical changes during thermal decomposition.

INTERACTION OF CHEMICAL AND PHYSICAL PROCESSES

The nature of the volatile products of thermal decomposition is dictated by the chemical and physical properties of both the polymer and the products of decomposition. The size of the molecular fragments must be small enough to be volatile at the decomposition temperature. This effectively sets an upper limit on the molecular weight of the volatiles. If larger chain fragments are created, they will remain in the condensed phase and will be further decomposed to smaller fragments, which can vaporize.

Figure 1-7.3 shows examples of the range of chemical or physical changes that can occur when a solid polymer is volatilized. The changes range from simple phase transformations (solid going to liquid and then to gas, at the top of the figure), to complex combinations of chemical and physical changes (in the lower part of the figure). Water and many other liquids forming crystalline solids on freezing (e.g., most flammable liquids) undergo straightforward physical phase changes. Sublimation, i.e., the direct phase change from a solid to a gas, without going through the liquid phase, will happen with materials such as carbon dioxide (e.g., CO_2, dry gas) or methenamine at normal temperatures and pressures. Methenamine is of interest in fires because methenamine pills are the ignition source in a standard test for carpets, ASTM D2859,[17] used in mandatory national regula-

tions.[18,19] Thermoplastics can melt without chemical reaction to form a viscous state (polymer melt), but they often decompose thermally before melting. This polymer melt can then decompose into smaller liquid or gaseous fragments. The liquid fragments will then decompose further until they, too, are sufficiently volatile to vaporize. Some polymers, especially thermosets or cellulosics, have even more complex decomposition mechanisms. Polyurethanes (particularly flexible foams) can decompose by three different mechanisms. One of them involves the formation of gaseous isocyanates, which can then repolymerize in the gas phase and condense as a "yellow smoke." These isocyanates are usually accompanied by liquid polyols, which can then continue to decompose. Cellulosics, such as wood, decompose into three types of products: (1) laevoglucosan, which quickly breaks down to yield small volatile compounds; (2) a new solid, char; and (3) a series of high molecular weight semi-liquid materials generally known as tars. Figure 1-7.3 illustrates the complex and varied physico-chemical decomposition pathways available, depending on the properties of the material in question. These varied thermal degradation/decomposition mechanisms have clear effects on fire behavior.

EXPERIMENTAL METHODS

By far, the most commonly used thermal decomposition test is thermogravimetric analysis (TGA). In TGA experiments, the sample (mg size) is brought quickly up to the desired temperature (isothermal procedure) and the weight of the sample is monitored during the course of thermal decomposition. Because it is impossible in practice to bring the sample up to the desired temperature before significant thermal decomposition occurs, it is common to subject the sample to a linearly increasing temperature at a predetermined rate of temperature rise. One might hope to obtain the same results from one nonisothermal test that were possible only in a series of isothermal tests. In practice, this is not possible since the thermogram (plot of weight *vs* temperature) obtained in a nonisothermal test is dependent on the heating rate chosen. Traditional equipment rarely exceed heating rates of 0.5 K/s, but modifications can be made to obtain rates of up to 10 K/s.[20,21] This dependence of thermal decomposition on heating rate is due to the fact that the rate of thermal decomposition is not only a function of the temperature, but also of the amount and nature of the decomposition process that has preceded it.

There are several reasons why the relevance of thermogravimetric studies to fire performance can be questioned: heating rate, amount of material, and lack of heat feedback are the major ones. For example, it is well known that heating rates of 10 to 100 K/s are common under fire conditions but are rare in thermal analysis. However, low heating rates *can* occur in real fires. More seriously, thermogravimetric studies are incapable of simulating the thermal effects due to large amounts of material burning and resupplying energy to the decomposing materials at different rates. However, analytical thermogravimetric studies do give important information about the decomposition process even though extreme caution must be exercised in their direct application to fire behavior.

Differential thermogravimetry (DTG) is exactly the same as TGA, except the mass loss *vs* time output is differentiated automatically to give the mass loss rate *vs* time. Often, both the mass loss and the mass loss rate *vs* time are produced automatically. This is, of course, quite convenient as the rate of thermal

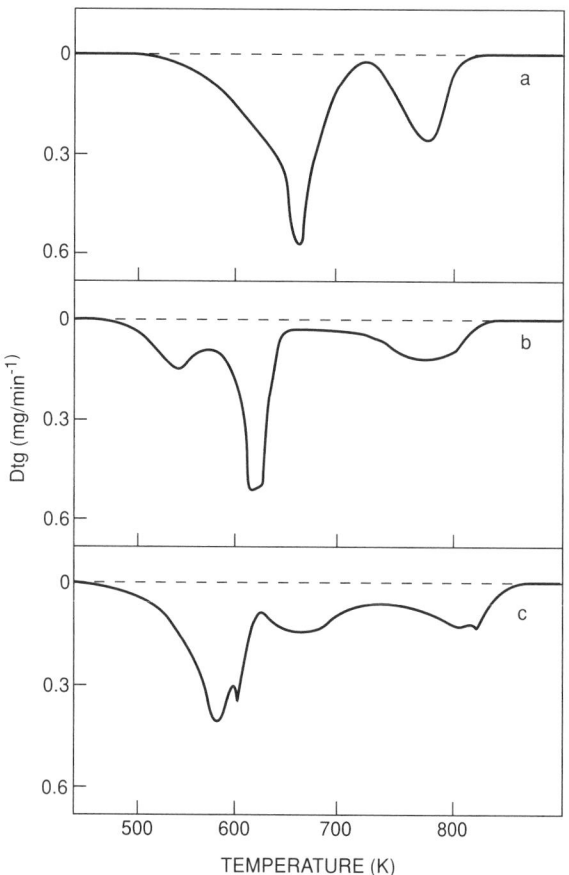

Fig. 1-7.4. *Effect of hydrated alumina and of antimony oxide-decabromobiphenyl (DBB) on DTG of ABS copolymer: (a) ABS; (b) ABS (60%) Aℓ_2O$_3$, 3 H$_2$O (40%); (c) ABS (70%) + DBB (22.5%) + Sb$_2$O$_3$ (7.5%).*

tilization rate, and a pressure transducer, rather than a sample microbalance, is used to measure the decomposition rate.

In addition to the rate of decomposition, it is also of interest to determine the heat of reaction of the decomposition process. In almost all cases, heat must be supplied to the sample to get it to a temperature where significant thermal decomposition will occur. However, once at such a temperature, the thermal decomposition process may either generate or utilize additional heat. The magnitude of this energy generation (exothermicity) or energy requirement (endothermicity) can be determined in the following ways.

In differential thermal analysis (DTA), a sample and a reference inert material with approximately the same heat capacity are both subjected to the same linear temperature program. The sample and reference material temperatures are measured and compared. If the thermal decomposition of the sample is endothermic, the temperature of the sample will lag behind the reference material; if the decomposition is exothermic, the temperature of the sample will exceed the reference material temperature. Very often, the sample is held in a crucible, and an empty crucible is used as a reference. Such a test can be quite difficult to calibrate to get quantitative heats of reaction.

In view of the considerable importance of the exact process of thermal decomposition, it is advantageous to carry out simultaneously the measurements of TGA, DTG, and DTA. This can be achieved by using a simultaneous thermal analyzer (STA), which uses a dual sample/reference material system. In the majority of cases, polymeric materials are best represented by a reference material which is simply air, i.e., an empty crucible. STA instruments can then determine, at the same time, the amounts of polymer decomposed, the rates at which these stages/processes occur, and the amount of heat evolved or absorbed in each stage. Examples of the application of this technique are contained in references 20 and 21. Recently, STA equipment is often being connected to Fourier transform infrared spectrometers (FTIR) for a complete chemical identification and analysis of the gases evolved at each stage, making the technique even more powerful.

Another method, which yields quantitative results more easily than DTA, is differential scanning calorimetry (DSC). In this test procedure, both the sample and a reference material are kept at the same temperature during the linear temperature program, and the heat of reaction is measured as the difference in heat input required by the sample and the reference material. The system is calibrated using standard materials, such as melting salts, with well-defined melting temperatures and heats of fusion. In view of the fact that DSC experiments are normally carried out by placing the sample inside sealed sample holders, this technique is seldom suitable for thermal decomposition processes. Thus, it is ideally suited for physical changes, but not for chemical processes. Interestingly, some of the commercial STA apparatuses are, in fact, based on DSC rather than DTA techniques for obtaining the heat input.

So far the experimental methods discussed have been concerned with the kinetics and thermodynamics of the thermal decomposition process. There is also concern with the nature of the decomposition process from the viewpoints of combustibility and toxicity. Chemical analysis of the volatiles exiting from any of the above instruments is possible. However, it is often convenient to design a special decomposition apparatus to attach directly to an existing analytical instrument. This is particularly important when the heating rate to be studied is much higher than that which traditional instruments can achieve. Thermal breakdown of cellulosic

decomposition is proportional to the volatilization or mass loss rate. One of the main roles where DTG is useful is in mechanistic studies. For example, it is the best indicator of the temperatures at which the various stages of thermal decomposition take place and the order in which they occur as illustrated in Figure 1-7.4. Part "a" of this figure shows the DTG of a thermoplastic polymer, acrylonitrile-butadiene-styrene (ABS), and part "b" shows the same polymer containing 40 percent alumina trihydrate.[22] The polymer decomposes in two main stages. The addition of alumina trihydrate has a dual effect: (1) it makes the material less thermally stable, and (2) it introduces a third thermal decomposition stage. Moreover, the first stage is now the elimination of alumina trihydrate. A more complex example is shown in Figure 1-7.5, where the effects of a variety of additives are shown;[23] some of these additives are effective flame retardants and others are not: the amount of overlap between the thermal decomposition stages of polymer and additives is an indication of the effectiveness of the additive.

Another method for determining the rate of mass loss is thermal volatilization analysis (TVA).[24] In this method, a sample is heated in a vacuum system (0.001 Pa) equipped with a liquid nitrogen trap (77 K) between the sample and the vacuum pump. Any volatiles produced will increase the pressure in the system until they reach the liquid nitrogen and condense out. The pressure is proportional to the mass vola-

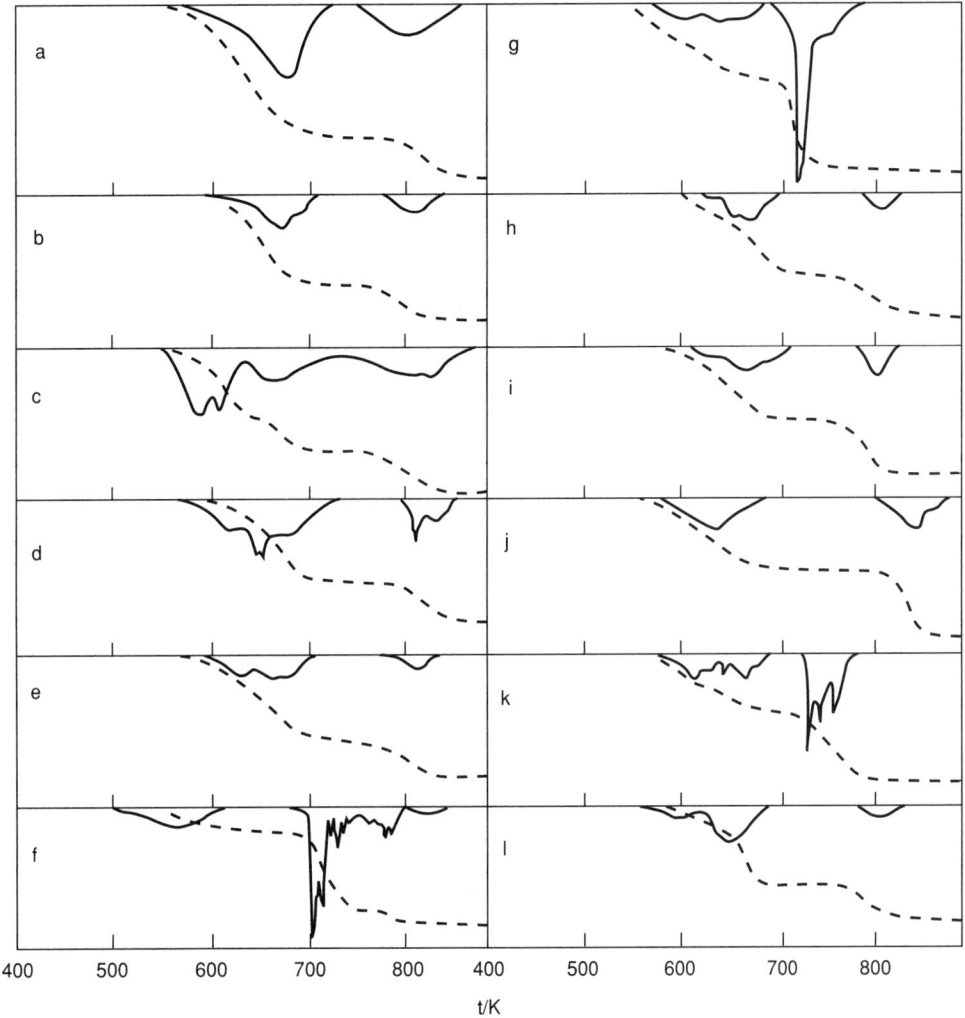

Fig. 1-7.5. **Thermal analyses of systems containing ABS, decabromobiphenyl (DBB) and one metal oxide: (a) ABS; (b) ABS + DBB; (c) ABS + DBB + Sb$_2$O$_3$; (d) ABS + DBB + SnO; (e) ABS + DBB + SnO$_2$. H$_2$O; (f) ABS + DBB + ZnO; (g) ABS + DBB + Fe$_2$O$_3$; (h) ABS + DBB + AℓOOH; (i) ABS + DBB + Aℓ_2O$_3$; (j) ABS + DBB + Aℓ_2O$_3$. 3 H$_2$O; (k) ABS + DBB + ammonium molybdate; (l) ABS + DBB + talc. DTG (——); TGA (----).**

materials, for example, has been investigated at heating rates as high as 10 K/s[25,26] or even up to over 1000 K/s[27-29] in specialized equipment. The major reason this was done was in order to simulate the processes involved in "smoking," but the results are readily applicable to fire safety.

Given the vast numbers of different products that can result from the decomposition in a single experiment, separation of the products is often required. Hence, the pyrolysis is often carried out in the injector of a gas chromatograph (PGC). In its simplest but rarely used form, a gas chromatograph consists of a long tube with a well-controlled flow of a carrier gas through it. The tube or column is packed with a solid/liquid that will absorb and desorb constituents in the sample. A small sample of the decomposition products is injected into the carrier gas flow. If a particular decomposition product spends a lot of time adsorbed on the column packing, it will take a long time for it to reach the end of the column. Products with different adsorption properties relative to the column packing will reach the end of the column at different times. A detector placed at the exit of the gas chromatograph will respond to the flow rate of gases other than the carrier gas, and if separation is successful, the detector output will be a series of peaks. For a single peak, the time from injection is characteristic of the chemical species, and the area under the peak is proportional to the amount of the chemical species. Column packing, column temperature programming, carrier gas flow rate, sample size, and detector type must all be chosen and adjusted to achieve optimal discrimination of the decomposition products.

Once the gases have been separated, any number of analytical techniques can be used for identification. Perhaps the most powerful has been mass spectrometry (MS). Again speaking in very simple terms, in MS the chemical species is ionized, and the atomic mass of the ion can be determined by the deflection of the ion in a magnetic field. Generally, the ionization process will also result in the fragmentation of the molecule, so the "fingerprint" of the range of fragments and their masses must be interpreted to determine the identity of the original molecule. Gas chromatography and mass spectrometry are the subject of a vast literature, and many textbooks and specialized journals exist.

TABLE 1-7.2 *Analytical Methods*

Method	Isothermal	Nonisothermal	In Vacuo	Inert	Air
Thermogravimetric Analysis (TGA)	X	X	X	X	X
Differential Thermogravimetry (DTG)	X	X	X	X	X
Thermal Volatilization Analysis TVA)	X	X	X		
Differential Thermal Analysis (DTA)		X		X	X
Differential Scanning Calorimetry (DSC)		X		X	X
Pyrolysis Gas Chromatography (PGC)	X			X	
Thermomechanical Analysis (TMA)	X	X		X	X

Useful physical data can be obtained by thermomechanical analysis (TMA). This is really a general name for the determination of a physical/mechanical property of a material subjected to high temperatures. Compressive and tensile strength, softening, shrinking, thermal expansion, glass transition, and melting can be studied by using TMA.

As displayed in Table 1-7.2, many of these tests can be performed *in vacuo*, in inert atmospheres, and in oxidizing atmospheres. Each has its place in the determination of the decomposition mechanism. Experiments performed *in vacuo* are of little practical value, but under vacuum the products of decomposition are efficiently carried away from the sample and its hot environment. Thus, secondary reactions are minimized so that the original decomposition product may reach a trap or analytical instrument intact. The practical significance of studies of thermal decomposition carried out in inert atmospheres may be argued. However, when a material burns, the flow of combustible volatiles from the surface and the flame above the surface effectively exclude oxygen at the material's surface. Under these conditions, oxidative processes may be unimportant. In other situations, such as ignition where no flame yet exists, oxidative processes may be critical. Whether or not oxygen plays a role in decomposition can be determined by the effect of using air rather than nitrogen in thermal decomposition experiments.

The decomposition reactions in the tests of Table 1-7.2 are generally monitored by the mass loss of the sample. With the exception of charring materials (e.g., wood or thermosets), analysis of the partially decomposed solid sample is rarely carried out. When it is done, it usually involves the search for heteroatom components due to additives. Analysis of the composition of the volatiles can be carried out by a wide range of analytical procedures. Perhaps the simplest characterization of the products is the determination of the fraction of the volatiles that will condense at various trap temperatures. Typically, convenient temperatures are room temperature (298 K), dry-ice temperature (193 K), and liquid nitrogen temperature (77 K). The products are classified as to the fraction of the sample remaining as residue; the fraction volatile at the pyrolysis temperature, but not at room temperature, V_{pyr}; the fraction volatile at room temperature, but not at dry-ice temperature, V_{298}; the fraction volatile at dry-ice temperature, but not at liquid nitrogen temperature, V_{-193}; and the fraction volatile at liquid nitrogen temperature, V_{-77}. This characterization gives a general picture of the range of molecular weights of the decomposition products. The contents of each trap can also be analyzed further, perhaps by mass spectroscopy.

The residual polymer can be analyzed to determine the distribution of molecular weights of the remaining polymer chains. This information can be of great value in determining the mechanism of decomposition. The presence of free radicals in the residual polymer can be determined by electron spin resonance spectroscopy (ESR, EPR), which simplistically can be considered the determination of the concentration of unpaired electrons in the sample. Other techniques, like infrared spectroscopy (IR), can be usefully employed to detect the formation of bonds not present in the original polymer. Such changes in bonding may be due to double-bond formation due to chain-stripping or the incorporation of oxygen into the polymer, for example.

GENERAL CHEMICAL MECHANISMS

Four general mechanisms common in polymer decomposition are illustrated in Figure 1-7.6. These reactions can be divided into those involving atoms in the main polymer chain and those involving principally side chains or groups. While the decomposition of some polymers can be explained by one of these general mechanisms, others involve combinations of these four general mechanisms. Nonetheless, these categorizations are useful in the identification and understanding of particular decomposition mechanisms.

Among simple thermoplastics, the most common reaction mechanism involves the breaking of bonds in the main polymer chain. These chain scissions may occur at the chain end or at random locations in the chain. End-chain scissions result in the production of monomer, and the process is often known as "unzipping." Random-chain scissions generally result in the generation of both monomers and oligomers (polymer units with ten or fewer monomer units) as well as a variety of other chemical species. The type and distribution of volatile products depend on the relative volatility of the resulting molecules.

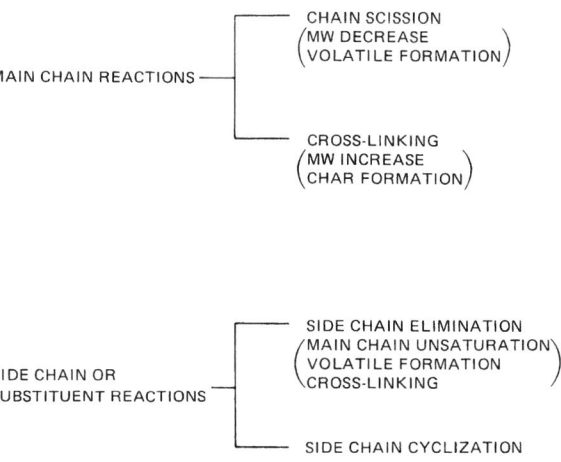

Fig. 1-7.6. General decomposition mechanisms.

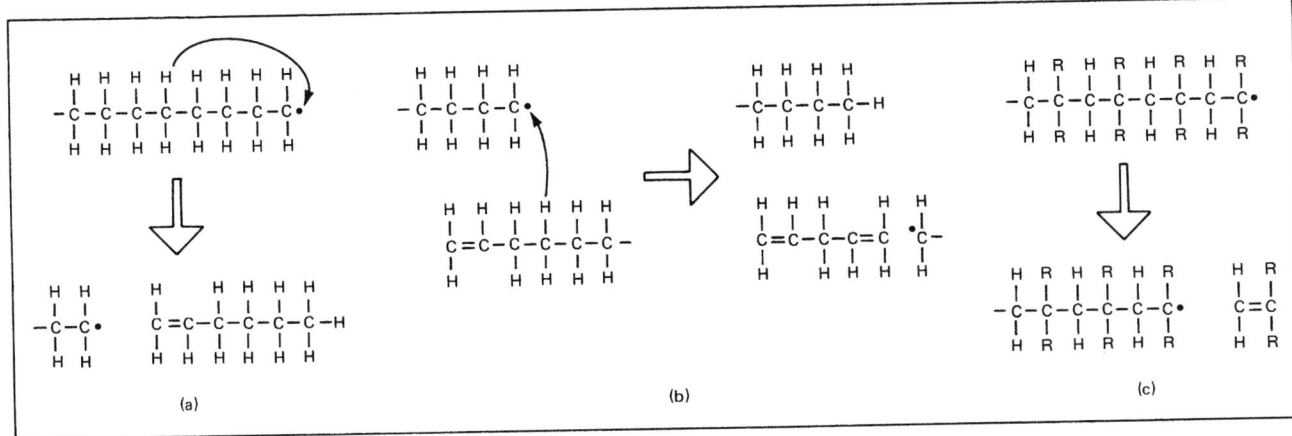

Fig. 1-7.7. (a) Intramolecular H transfer, (b) intermolecular H transfer, (c) unzipping.

Cross-linking is another reaction involving the main chain. It generally occurs after some stripping of substituents and involves the creation of bonds between two adjacent polymer chains. This process is very important in the formation of chars, since it generates a structure with a higher molecular weight which is less easily volatilized.

The main reaction types involving side chains or groups are elimination reactions and cyclization reactions. In elimination reactions, the bonds connecting side groups of the polymer chain to the chain itself are broken, with the side groups often reacting with other eliminated side groups. The products of these reactions are generally small enough to be volatile. In cyclization reactions, two adjacent side groups react to form a bond between them, resulting in the production of a cyclic structure. This process is also important in char formation because, as the reaction scheme shows, the residue is much richer in carbon than the original polymer as seen, for example, for poly(vinyl chloride):

$$—CH_2—CHCl— \Rightarrow —CH{=}CH— + HCl$$

which leads to a hydrogenated char or for poly(vinylidene chloride):

$$—CHCl—CHCl— \Rightarrow —C{\equiv}C— + 2HCl$$

which yields a purely carbonaceous char with an almost graphitic structure. These chars will tend to continue breaking down by chain scission, but only at very high temperatures.

Chain-Scission Mechanisms

Decomposition by chain scission is a very typical mechanism for polymer decomposition. The process is a multi-step radical chain reaction with all the general features of such reaction mechanisms: initiation, propagation, branching, and termination steps.

Initiation reactions are of two basic types: (1) random-chain scission and (2) end-chain scission. Both, of course, result in the production of free radicals. The random scission, as the name suggests, involves the breaking of a main chain bond at a seemingly random location, all such main chain bonds being equal in strength. End-chain initiation involves the breaking off of a small unit or group at the end of the chain. This may be a monomer unit or some smaller

substituent. These two types of initiation reactions may be represented by the following generalized reactions:

$$P_n \Rightarrow R_r + R_{n-r} \quad \text{(random-chain scission)}$$

$$P_n \Rightarrow R_n + R_E \quad \text{(end-chain initiation)}$$

where P_n is a polymer containing n monomer units, and R_r is a radical containing r monomer units. R_E refers to an end group radical.

Propagation reactions in polymer decomposition are often called depropagation reactions, no doubt due to the polymer chemist's normal orientation toward polymer formation (polymerization) rather than decomposition. Regardless, there are several types of reactions in this class [see Figure 1-7.7, parts (a),(b), and (c)]:

$$R_n \Rightarrow R_{n-m} + P_m \quad \begin{array}{l}\text{(intramolecular H transfer,} \\ \text{random-chain scission),}\end{array}$$

$$P_m + R_n \Rightarrow P_{m-j} + P_n + R_j \quad \begin{array}{l}\text{(intermolecular} \\ \text{H transfer),}\end{array}$$

$$R_n \Rightarrow R_{n-1} + P_1 \quad \begin{array}{l}\text{(unzipping, depropagation,} \\ \text{depolymerization).}\end{array}$$

The first of these reactions involves the transfer of a hydrogen atom within a single polymer chain, i.e., intramolecular hydrogen atom transfer. The value of m is usually between one and four as polymer molecules are often oriented such that the location of the nearest available H within the chain is one to four monomer units away from the radical site. The value of m need not be a constant for a specific polymer as the closest available hydrogen atom in the chain may vary due to conformational variations. Decomposition mechanisms based on this reaction are sometimes known as random-chain scission mechanisms. The second reaction involves the transfer of a hydrogen atom between polymer chains, i.e., intermolecular hydrogen atom transfer. The original radical, R_n, abstracts a hydrogen atom from the polymer, P_m. As this makes P_m a radical with the radical site more often than not within the chain itself (i.e., not a terminal radical site), the newly formed radical breaks up into an unsaturated polymer, P_{m-j}, and a radical, R_j. In the final reaction, no hydrogen transfer occurs. It is essentially the reverse of the polymerization step and, hence, is called unzipping, depropagation, or depolymerization. Whether the decomposition involves principally hydrogen transfer reactions or unzipping can be determined by examining the structure of the polymer, at least for polymers with only carbon in the main chain. If hydrogen transfer is impeded, then it is likely that the unzipping reaction will occur.

Vinyl polymers, strictly speaking, are those derived from a vinyl repeating unit, namely

$$-[CH_2-CH_2]_n-$$

where n is the number of repeating monomers. Here, the hydrogen atoms can be substituted, leading to a repeating unit of the following form:

$$-[\underset{X}{\overset{W}{C}}-\underset{Z}{\overset{Y}{C}}]_n-$$

where W, X, Y, and Z are substituent groups, perhaps hydrogen, methyl groups, or larger groups. Consider that the C-C bond connecting monomer units is broken and that a radical site results from the scission shown as

$$-[\underset{X}{\overset{W}{C}}-\underset{Z}{\overset{Y}{C}}]_j-\underset{X}{\overset{W}{C}}-\underset{Z}{\overset{Y}{C}} \bullet$$

where the symbol "•" indicates an unpaired electron and, hence, a radical site. In order for a hydrogen atom to be transferred from the chain to the radical site, it must pass around either Y or Z. If Y and Z are hydrogens, this is not at all difficult due to their small size. However, if the alpha carbon has larger substituents bound to it (i.e., Y and Z are larger groups), the transfer of hydrogen to the radical site is more difficult. This type of interference with hydrogen transfer is known as "steric hindrance." Table 1-7.3 shows this effect.[2] Polymers near the top of Table 1-7.3 have Y and Z substituents that are generally large, with a resulting high monomer yield, characteristic of unzipping reactions. Near the bottom of Table 1-7.3, where Y and Z are small, the polymers form negligible amounts of monomer as other mechanisms dominate.

While chain-branching reactions seem to be of little importance in polymer decomposition, termination reactions are required in all chain mechanisms. Several types of termination reactions are common.

$$R_m \Rightarrow P_m \qquad \text{(unimolecular termination);}$$

$$R_m + R_n \Rightarrow P_{m+n} \qquad \text{(recombination); and}$$

$$R_n + R_m \Rightarrow P_m + P_n \qquad \text{(disproportionation).}$$

The first of these reactions is, strictly speaking, not generally possible. Nonetheless, there are instances where the observed termination reaction appears to be first order (at least empirically). It is impossible to remove the radical site from a polymer radical without adding or subtracting at least one hydrogen atom while still satisfying the valence require-

ments of the atoms. What probably occurs is that the termination reaction is, in fact, second order, but the other species involved is so little depleted by the termination reaction that the termination reaction appears not to be affected by the concentration of that species. This is known as a pseudo-first order reaction. The recombination reaction is a classical termination step that is actually just the reverse of the random-chain scission initiation reaction. Finally, the disproportionation reaction involves the transfer of a hydrogen atom from one radical to the other. The hydrogen donor forms a double bond as a result of the hydrogen loss, and the acceptor is fully saturated. If this sort of reaction occurs immediately after an initiation reaction, no unzipping or other propagation reaction occurs, and the polymer decomposition is fully characterized by a random process of bond scissions.

There is a natural tendency to regard all materials with the same generic name, such as poly(methyl methacrylate), as being the same material with the same properties. As these are commercial products, the preparation methods (including the polymerization process) are dictated by the required physical and chemical properties of the material for normal use. Additives, both intentional and inadvertent, may be present, and the method of polymerization and the molecular weight of the polymer chains may vary. This is particularly important in the case of polymeric "compounds" (the actual polymeric material that is used commercially to fabricate a product of any kind) that contain a large fraction of additives. In some polyolefins, the fraction of polymer (known as "resin") may be much less than half of the total mass of the "compound," because of the presence of large amounts of fillers. In some "compounds" derived from poly(vinyl chloride), flexibility is introduced by means of "plasticizers."

In this regard, it is interesting and important to note that polymers tend to be less stable than their oligomer counterparts. This results from several effects involved in the production and aging of polymers as well as simply the chain length itself. Initiation reactions in a polymer can lead to far more monomer units being involved in decomposition reactions, relative to the polymer's short-chain oligomeric analog. In the production and aging of polymers, there are opportunities for the production of abnormalities in the polymer chains due to the mode of synthesis and thermal, mechanical, and radiation effects during aging.

In the synthesis of the polymer, abnormalities may result from several sources. Unsaturated bonds result from chain termination by free-radical termination reactions. End-chain unsaturation results from second-order disproportionation reactions, and midchain unsaturation often occurs due to chain-transfer reactions with subsequent intramolecular hydrogen transfer. Chain branching may result from the formation of midchain radicals. During synthesis, chain transfer reactions may cause midchain radicals that then go on to react with monomers or polymers to create a branched polymer structure. Termination of the polymerization reaction may also result in "head-to-head linkages"; i.e., monomer units are attached such that some of the monomers are oriented opposite to the remainder of the chain. Lastly, foreign atoms or groups may be incorporated into the polymer chain. This may occur due to impurities, polymerization initiators, or catalysts. Oxygen is often a problem in this regard.

The purity and the molecular weight of the polymer can markedly affect not only the decomposition rates, but also

TABLE 1-7.3 *Monomer Yield from Thermal Decomposition of Polymers of the General Form [CWX-CYZ]$_n$[2]*

Polymer	W	X	Y	Z	Monomer Yield (wt. %)	Decomposition Mechanism*
PMMA	H	H	CH$_3$	CO$_2$CH$_3$	91–98	E
Polymethacrylonitrile	H	H	CH$_3$	CN	90	E
Poly (a-methylstyrene)	H	H	CH$_3$	C$_6$H$_5$	95	E
Polyoxymethylene**	-	-	-	-	100	E
Polytetrafluoroethylene	F	F	F	F	95	E
Poly (methyl atropate)	H	H	C$_6$H$_5$	CO$_2$CH$_3$	>99	E
Poly (p-bromostyrene)†	H	H	H	C$_6$H$_4$Br	91–93	E
Poly (p-chlorostyrene)†	H	H	H	C$_6$H$_4$Cl	82–94	E
Poly (p-methyoxystyrene)†	H	H	H	C$_7$H$_7$O	84–97	E
Poly (p-methylstyrene)	H	H	H	C$_7$H$_7$	82–94	E
Poly (a-deuterostyrene)	H	H	D	C$_6$H$_5$	70	E
Poly (a,β,β-trifluorostyrene)	F	F	F	C$_6$H$_5$	44	E/R
Polystyrene	H	H	H	C$_6$H$_5$	42–45	E/R
Poly (m-methylstyrene)	H	H	H	C$_7$H$_8$	44	E/R
Poly (β-deuerostyrene)	H	D	H	C$_6$H$_5$	42	E/R
Poly (β-methylstyrene)	H	CH$_3$	H	C$_6$H$_5$		E/R
Poly (p-methoxystyrene)‡	H	H	H	C$_7$H$_7$O	36–40	E/R
Polyisobutene	H	H	CH$_3$	CH$_3$	18–25	E/R
Polychlorotrifluoroethylene	F	F	Cl	F	28	E/S
Poly (ethylene oxide)**	-	-	-	-	4	R/E
Poly (propylene oxide)**	-	-	-	-	4	R/E
Poly (4-methyl pent-1-ene)	H	H	H	C$_4$H$_9$	2	R/E
Polyethylene	H	H	H	H	0.03	R
Polypropylene	H	H	H	CH$_3$	0.17	R
Poly (methyl acrylate)	H	H	H	CO$_2$CH$_3$	0.7	R
Polytrifluoroethylene	F	F	H	F	-	R
Polybutadiene**	-	-	-	-	1	R
Polyisoprene**	-	-	-	-	5	R
Poly (vinyl chloride)	H	H	H	Cl	0–0.07	S
Poly (vinylidene chloride)	H	H	Cl	Cl	-	S
Poly (vinylidene fluoride)	H	H	F	F	-	S
Poly (vinyl fluoride)	H	H	H	F	-	S
Poly (vinyl alcohol)	H	H	H	OH	-	S
Polyacrylonitrile	H	H	H	CN	5	C

* R, random-chain scission; E, end-chain scission (unzipping); S, chain-stripping; C, cross-linking.

** not of general form [CWX-CYZ]$_n$.

† cationic polymerization.

‡ free-radical polymerization.

the mechanism of decomposition. An example of such a change might involve chain initiations occurring at the location of impurities in the chain of a polymer which, if pure, would principally be subject to end-chain initiation. Both the mechanism and the decomposition rate would be affected. Not all polymer "defects" degrade polymer thermal performance. In a polymer that decomposes by unzipping, a head-to-head linkage can stop the unzipping process. Thus, for an initiation that would have led to the full polymer being decomposed, only the part between the initiation site and the head-to-head link is affected. At least one additional initiation step is required to fully decompose the chain. This has been studied in detail by Kashiwagi *et al.*[6–10]

Kinetics

Eight generic types of reaction involved in simple decomposition processes have been addressed in the previous sections. Even if only a subset of these reaction types are required and the reaction rates are not a strong function of the size of the polymer chains and radicals, the kinetics describing the process can be quite complex. In engineering applications, such complex reaction mechanisms are not used. Rather, simple overall kinetic expressions are generally utilized if, in fact, decomposition kinetics are considered at all. The most common assumption is that the reactions can be described by an Arrhenius expression of first-order in the remaining polymer mass. Often one goes even further and ignores any dependence of the reaction rate on the remaining polymer or the thickness of the decomposition zone and simply expresses the volatilization rate per unit surface area as a zero-order Arrhenius expression. This effectively assumes that the decomposition zone is of constant thickness and fresh polymer replaces the decomposed polymer by surface regression. Such an approach would clearly not be satisfactory for charring materials where decomposition is clearly not a surface phenomenon. As some of the work quoted earlier has indicated (e.g., reference 13), it is also not suitable for many thermoplastic polymers.

Despite the fact that detailed kinetic models are not used in engineering calculations, it is instructive to consider some very simple cases, by the use of overall kinetic expressions, to indicate what is being lost. The effect of the initiation mechanism on decomposition kinetics can be easily demonstrated by considering either random- or end-chain initiation with propagation by unzipping and no termination reactions other than exhaustion of the polymer chain by unzipping. The rate of weight loss for random-chain initiation can be expressed as

$$dW/dt = D_p \cdot k_{ir} \cdot W$$

where D_p, the degree of polymerization, is the number of monomer units per polymer chain; and k_{ir} is the rate constant for the random-chain initiation reaction. Notice that the rate constant of the propagation reaction is not included in the expression. A further assumption that the propagation rate is much faster than the initiation rate has also been made. The initiation reaction is said to be the "rate-limiting step." The degree of polymerization arises in the equation since, for each initiation, D_p monomer units will be released; and the remaining weight, W, arises because the number of bonds available for scission is proportional to W. Since the polymer unzips completely, the molecular weight of all remaining polymer chains is the same as the initial molecular weight.

Considering end-chain initiation, the rate of mass loss is given by

$$dW/dt = D_p \cdot (2n) \cdot k_{ie}$$

where n is the number of polymer chains, and, hence, $2n$ is the number of chain ends, and k_{ie} is the rate constant for end-chain initiation. The number of polymer chains is simply the mass of the sample divided by the molecular weight of each chain, or

$$n = \frac{W}{D_p \cdot MW_m}$$

where MW_m is the molecular weight of the monomer. Using this expression yields

$$dW/dt = \frac{2 \cdot k_{ie} \cdot W}{MW_m}$$

Comparing this with the random initiation expression, one can see that, for random initiation, the rate is dependent on the original degree of polymerization; whereas for end-chain initiation the rate is independent of the degree of polymerization or, equivalently, the original molecular weight of the polymer. In both cases, however, the rate is first order in the mass of the sample. This derivation has been for a monodisperse polymer; i.e., all chains have been considered to be the same length initially.

Returning to the random-chain initiation expression, it is clear that longer chains are decomposed preferentially. If the initial sample had a range of molecular weights, the longer chains would disappear more quickly than shorter chains, and the molecular weight distribution would change with time, unlike in the monodisperse case. It can be shown that in this case the reaction order is no longer unity, but is between one and two, depending on the breadth of the distribution.[30] Thus,

$$dW/dt \sim W^n$$

with

$$1 < n < 2$$

for random-chain initiation and complete unzipping of a polydisperse system.

This simple comparison illustrates some of the ways in which the details of the polymerization process, which control variables like the molecular weight distribution, can alter the decomposition process. For a particular polymer sample, no single initiation reaction need be dominant, in general. The activation energies for the different initiation steps may be quite different, leading to large variations in the relative rates with temperature. For instance, in PMMA, the dominant initiation step at low temperatures (around 570 K) is end-chain initiation. At higher temperatures (around 770 K), the random-chain initiation step dominates. In a single nonisothermal TGA experiment, this temperature range can easily be traversed, and overall interpretation of the results in terms of a single mechanism would be unsatisfactory and misleading.

Nonetheless, simple overall kinetic expressions are likely to be dominant in engineering for some time. The pitfalls with this approach simply serve to reinforce the need to determine the kinetic parameters in an experiment that is as similar to the end use as is practical. This is one of the major reasons why the use of TGA results has been brought into question. As stated before, the heating rates often are far less than those generally found in fire situations. The low heating rates in TGA experiments tend to emphasize lower-temperature kinetics which may be much less important at the heating rates characteristic of most fire situations.

One interesting study worth presenting here is a theoretical analysis of thermal decomposition that presents a technique for calculating the temperature at the beginning and end of thermal decomposition, based on structural data from the polymer and on scission at the weakest bond, with considerable degree of success, particularly for successive members of a polymeric family.[31] A subsequent analysis has also been published that is much simpler, but it has not been validated against experimental data.[32]

GENERAL PHYSICAL CHANGES DURING DECOMPOSITION

The physical changes that occur on heating a material are both important in their own right and also impact the course of chemical decomposition significantly. The nature of the physical changes and their impact on decomposition vary widely with material type. This section addresses the general physical changes that occur for thermoplastic (glass transition, melting) and thermosetting (charring, water desorption) materials.

Melting and Glass Transition

On heating a thermoplastic material, the principal physical change is the transformation from a glass or solid to the fluid state. (See Figure 1-7.2.) If this transformation occurs at temperatures well below the decomposition temperature, it becomes more likely that the material will drip and/or flow. While such behavior is a complication, in terms of fire safety it can either improve or degrade the performance of the material. In some configurations, flowing of the material can remove it from the source of heat and thus avoid ignition or further fire growth. In other situations, the flow of material may be toward the heat source, leading to a worsened fire situation. Many standard fire tests that allow materials to flow away from the heat source have been shown to be unsuitable for assessing the hazards of flowing or dripping materials. Care must be taken in the evaluation of standard test results in this regard. However, many thermoplastics do not show marked tendencies to flow during heating and combustion. Whereas polyethylene melts and flows readily, high-quality cast poly(methyl methacrylate) shows only slight tendencies to flow under fire conditions.

When designing a material, there are several techniques that can be utilized to increase the temperature at which physical transformations occur. These strategies are generally aimed at increasing the stiffness of the polymer or increasing the interactions between polymer chains. It is clear that increasing the crystallinity of the polymer increases the interaction between polymer chains. In the highly ordered state associated with crystalline materials, it is less possible for polymer chains to move relative to one another, as additional forces must be overcome in the transformation to the unordered fluid state. Crystallinity is enhanced by symmetric regular polymer structure and highly polar side groups. Regular structure allows adjacent polymer chains to pack in a regular and tight fashion. As such, isotactic polymers are more likely to crystallize than atactic polymers, and random copolymers do not tend to crystallize. Polar side groups enhance the intermolecular forces. Regular polar polymers, such as polyesters and polyamides, crystallize readily. Even atactic polymers with OH and CN side groups will crystallize due to polarity. The melting temperature of a polymer is also increased with increasing molecular weight up to a molecular weight of about 10,000 to 20,000 g/mol.

Melting temperatures can also be increased by increasing the stiffness of the polymer chain. Aromatic polyamides melt at much higher temperatures than their aliphatic analogs due to stiffness effects. Aromatics are particularly useful for chain stiffening, as they provide stiffness without bulk which would hinder crystallinity. At the opposite extreme, the increased flexibility of the oxygen atom links in polyethers is responsible for a lowering of the melting temperature of polyethers relative to polymethylene. Chain stiffening must be accompanied by suitable thermal stability and oxidation resistance in order to achieve increased service temperatures. Many aromatic polymers have melting temperatures in excess of their decomposition temperatures, making these materials thermosetting.

Cross-linking also increases the melting temperature and, like chain stiffening, can render a material infusible. Cross-links created in fabrication or during heating are also important in thermoplastics. The glass transition temperature can be increased in amorphous polymers by the inclusion of cross-links during fabrication. Random-chain scissions can quickly render a material unusable by affecting its physical properties unless cross-linking occurs. Such cross-linking in thermoplastics on heating may be regarded as a form of repolymerization. The temperature above which depolymerization reactions are faster than polymerization reactions is known as the ceiling temperature. Clearly, above this temperature catastrophic decomposition will occur.

Charring

While char formation is a chemical process, the significance of char formation is largely due to its physical properties. Clearly, if material is left in the solid phase as char, less flammable gas is given off during decomposition. More importantly, the remaining char can be a low-density material and is a barrier between the source of heat and the virgin polymer material. As such, the flow of heat to the virgin material is reduced as the char layer thickens, and the rate of decomposition is reduced, depending on the properties of the char.[3] If the heat source is the combustion energy of the burning volatiles, not only will the fraction of the incident heat flux flowing into the material be reduced, but the incident heat flux as a whole will be reduced as well. Unfortunately, char formation is not always an advantageous process. The solid-phase combustion of char can cause sustained smoldering combustion. Thus, by enhancing the charring tendency of a material, flaming combustion rates may be reduced, but perhaps at the expense of creating a source of smoldering combustion that would not otherwise have existed.

Charring is enhanced by many of the same methods used to increase the melting temperature. Thermosetting materials are typically highly cross-linked and/or chain-stiffened. However, charring is not restricted to thermosetting materials. Cross-linking may occur as a part of the decomposition process, as is the case in poly(vinyl chloride) and polyacrylonitrile.

IMPLICATIONS FOR FIRE PERFORMANCE

As explained earlier, one of the major reasons why thermal decomposition of polymers is studied is because of its importance in terms of fire performance. This issue has been studied extensively.

Early on, Van Krevelen[33,34] showed that, for many polymers, the limiting oxygen index (LOI, an early measure of flammability)[35] could be linearly related to char yield as measured by TGA under specified conditions. Then, since Van Krevelen showed how to compute char yield to a good approximation from structural parameters, LOI should be computable; and for pure polymers having substantial char yields, it is fairly computable. Somewhat later, comparisons were made between the minimum decomposition temperature (or, even better, the temperature for 1 percent thermal decomposition) and the LOI.[2,22] The conclusion was that, although in general low flammability resulted from high minimum thermal decomposition temperatures, no easy comparison could be found between the two. There were some notable cases of polymers with both low thermal stability and low flammability. This type of approach has since fallen into disrepute, particularly in view of the lack of confidence remaining today in the LOI technique.[36] Table 1-7.4 shows some thermal decomposition temperatures and limiting oxygen indices[22] as well as heat release rate values, the latter as measured in the cone calorimeter.[37,38] It is clear from the data in Table 1-7.4 that thermal decomposition is *not* a stand-alone means of predicting fire performance. Promising work in this regard is being made by Lyon,[39] who appears to be able to preliminarily predict some heat release information from thermoanalytical data.

However, mechanisms of action of fire retardants and potential effectiveness of fire retardants can be well predicted from thermal decomposition activity (for example, see Figures 1-7.4 and 1-7.5).[22,23] It is often necessary to have some additional understanding of the chemical reactions

TABLE 1-7.4 *Thermal Stability and Flammability of Polymers*

Polymer	T_d* (K)	$T_{1\%}$† (K)	LOI‡ (-)	Pk RHR§ (kW/m^2)
Polyacetal	503	548	15.7	360
Poly(methyl methacrylate)	528	555	17.3	670
Polypropylene	531	588	17.4	1500
Polyethylene (LDPE)	490	591	17.4	800**
Polyethylene (HDPE)	506	548	17.4	1400
Polystyrene	436	603	17.8	1100
ABS copolymer	440	557	18.0	950**
Polybutadiene	482	507	18.3	
Polyisoprene	460	513	18.5	
Cotton	379	488	19.9	450**
Poly(vinyl acohol)	337	379	22.5	
Wool	413	463	25.2	310**
Nylon-6	583		25.6	1300
Silicone oil	418	450	32	140**
Poly(vinylidene fluoride)	628	683	43.7	30**
Poly(vinyl chloride)	356	457	47	180
Polytetrafluoroethylene	746	775	95	13

Notes:
*T_d: Minimum thermal decomposition temperature from TGA (10 mg sample, 10 K/min heating rate, nitrogen atmosphere[22]
†$T_{1\%}$: Temperature for 1% thermal decomposition, conditions as above[22]
‡LOI: Limiting oxygen index[22]
§Pk RHR: Peak rate of heat release in the cone calorimeter, at 40 kW/m^2 incident flux, at a thickness of 6 mm,[35] all under the same conditions, except for those marked **, which have various sources.

involved. In Figure 1-7.5, for example, the systems containing ABS, *decabromobiphenyl*, and either antimony oxide (c) or ferric oxide (g) have very similar TGA/DTG curves, with continuous weight loss. This indicates that the Sb system is effective but the iron one is not, because antimony bromide can volatilize while iron bromide does not. On the other hand, the system containing zinc oxide (f) is inefficient because the zinc bromide volatilizes "too early," i.e., before the polymer starts breaking down. Some authors have used thermal decomposition techniques *via* the study of the resulting products to understand the mechanism of fire retardance (e.g., Grassie[40]), or together with a variety of other techniques (e.g., Camino *et al.*[41,42]).

Whatever the detailed degree of predictability of fire performance data from thermal decomposition data, its importance should not be underestimated: *polymers cannot burn if they do not break down.*

BEHAVIOR OF INDIVIDUAL POLYMERS

The discussion, thus far, has been general, focusing on the essential aspects of thermal decomposition without the complications that inevitably arise in the treatment of a particular polymer. This approach may also tend to make the concepts abstract. Through the treatment of individual polymers by polymer class, this section provides an opportunity to apply the general concepts to real materials. In general, the section is restricted to polymers of commercial importance. More complete and detailed surveys of polymers and their thermal decomposition can be found in the literature.[2,30,43–52]

Polyolefins

Of the polyolefins, low-density polyethylene (LDPE), high-density polyethylene (HDPE), and polypropylene (PP) are of the greatest commercial importance because of their production volume. Upon thermal decomposition, very little monomer formation is observed for any of these polymers; they form a large number of different small molecules (up to 70), mostly hydrocarbons. Thermal stability of polyolefins is strongly affected by branching, with linear polyethylene most stable and polymers with branching less stable. The order of stability is illustrated as follows:

$$
\begin{array}{ccccccc}
\text{H} & \text{H} & & \text{H} & \text{CH}_3 & & \text{H} & \text{R} & & \text{H} & \text{R} \\
| & | & & | & | & & | & | & & | & | \\
-\text{C}-\text{C} & > & \text{C}-\text{C} & > & \text{C}-\text{C} & > & \text{C}-\text{C}- \\
| & | & & | & | & & | & | & & | & | \\
\text{H} & \text{H} & & \text{H} & \text{H} & & \text{H} & \text{R} & & X & Z
\end{array}
$$

where R is any hydrocarbon group larger than a methyl group.

Polyethylene (PE): In an inert atmosphere, polyethylene begins to cross-link at 475 K and to decompose (reductions in molecular weight) at 565 K though extensive weight loss is not observed below 645 K. Piloted ignition of polyethylene due to radiative heating has been observed at a surface temperature of 640 K. The products of decomposition include a wide range of alkanes and alkenes. Branching of polyethylene causes enhanced intramolecular hydrogen transfer and results in lower thermal stability. The low-temperature molecular weight changes without volatilization are principally due to the scission of weak links, such as oxygen, incorporated into the main chain as impurities. Initiation reactions at higher temperatures involve scission of tertiary carbon bonds or ordinary carbon-carbon bonds in the beta position to tertiary carbons. The major products of decomposition are propane, propene, ethane, ethene, butene, hexene-1, and butene-1. Propene is generated by intramolecular transfer to the second carbon and by scission of the bond beta to terminal =CH$_2$ groups.

The intramolecular transfer route is most important, with molecular coiling effects contributing to its significance. A broad range of activation energies has been reported, depending on the percent conversion, the initial molecular weight, and whether the remaining mass or its molecular weight were monitored. Decomposition is strongly enhanced by the presence of oxygen, with significant effects detectable at 423 K in air.

Polypropylene (PP): In polypropylene, every other carbon atom in the main chain is a tertiary carbon which is thus prone to attack. This lowers the stability of polypropylene as compared to polyethylene. As with polyethylene, chain scission and chain transfer reactions are important during decomposition. By far, secondary radicals (i.e., radical sites on the secondary carbon) are more important than primary radicals. This is shown by the major products formed, i.e., pentane (24 percent), 2 methyl-1-pentene (15 percent), and 2–4 dimethyl-1-heptene (19 percent). These are more easily formed from intramolecular hydrogen transfer involving secondary radicals. Reductions in molecular weight are first observed at 500 to 520 K and volatilization becomes significant above 575 K. Piloted ignition of polypropylene due to radiative heating has been observed at a surface temperature of 610 K. Oxygen drastically affects both the mechanism and rate of decomposition. The decomposition temperature is reduced by about 200 K, and the products of oxidative decomposition include mainly ketones. Unless the polymer samples are very thin (less than 0.25–0.30 mm or 0.010–0.012 in. thick), oxidative pyrolysis can be limited by diffusion of oxygen into the material. At temperatures below the melting point, polypropylene is more resistant to oxidative pyrolysis as oxygen diffusion into the material is inhibited by the higher density and crystallinity of polypropylene. Most authors have assumed that the oxidation mechanism is based on hydrocarbon oxidation, but recent work suggests that it may actually be due to the decomposition of peracids resulting from the oxidation of primary decomposition products.[12]

Polyacrylics

Poly(methyl methacrylate) (PMMA): PMMA is a favorite material for use in fire research since it decomposes almost solely to monomer, and burns at a very steady rate. Methyl groups effectively block intramolecular H transfer as discussed in "General Chemical Mechanisms," leading to a high monomer yield. The method of polymerization can markedly affect the temperatures at which decomposition begins. Free radical polymerized PMMA decomposes around 545 K, with initiation occurring at double bonds at chain ends. A second peak between 625 and 675 K in dynamic TGA thermograms is the result of a second initiation reaction. At these temperatures, initiation is by both end-chain and random-chain initiation processes. Anionically produced PMMA decomposes at about 625 K because the end-chain initiation step does not occur due to the lack of double bonds at the chain end when PMMA is polymerized by this method. This may explain the range of observed piloted ignition temperatures (550 to 600 K). Decomposition of PMMA is first order with an activation energy of 120 to 200 kJ/mol, depending on the end group. The rate of decomposition is also dependent on the tacticity of the polymer and on its molecular weight. These effects can also have a profound effect on the flame spread rate.

It is interesting to note that a chemically cross-linked copolymer of PMMA was found to decompose by forming an extensive char, rather than undergoing end-chain scission which resulted in a polymer with greater thermal stability.[53]

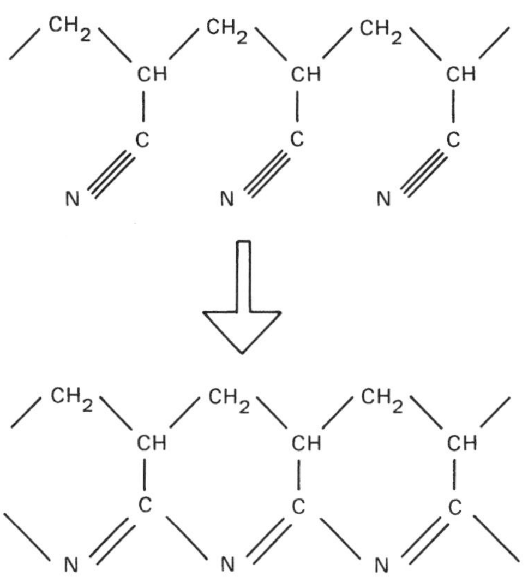

Fig. 1-7.8. PAN cyclization.

Poly(methyl acrylate) (PMA): Poly(methyl acrylate) decomposes by random-chain scission rather than end-chain scission, with almost no monomer formation. This results because of the lack of a methyl group blocking intramolecular hydrogen transfer as occurs in PMMA. Initiation is followed by intra- and intermolecular hydrogen transfer.

Polyacrylonitrile (PAN): PAN begins to decompose exothermically between 525 K and 625 K with the evolution of small amounts of ammonia and hydrogen cyanide. These products accompany cyclization reactions involving the creation of linkages between nitrogen and carbon on adjacent side groups. (See Figure 1-7.8.) The gaseous products are not the result of the cyclization itself, but arise from the splitting off of side or end groups not involved in the cyclization. The ammonia is derived principally from terminal imine groups (NH) while HCN results from side groups that do not participate in the polymerization-like cyclization reactions. When the polymer is not isotactic, the cyclization process is terminated when hydrogen is abstracted by the nitrogen atom. The cyclization process is reinitiated as shown in Figure 1-7.9. This leaves CN groups not involved in the cyclization which are ultimately removed and appear among the products as HCN. Typically, there are between 0 and 5 chain polymerization steps between each hydrogen abstraction. At temperatures of 625 to 975 K, hydrogen is evolved as the cyclic structures carbonize. At higher temperatures, nitrogen is evolved as the char becomes nearly pure carbon. In fact, with adequate control of the process, this method can be used to produce carbon fibers. Oxygen stabilizes PAN, probably by reacting with initiation sites for the nitrile polymerization. The products of oxidative decomposition are highly conjugated and contain ketonic groups.

Halogenated Polymers

Poly(vinyl chloride) (PVC): The most common halogenated polymer is PVC; it is one of the three most widely used polymers in the world, with polyethylene and polypropylene. Between 500 and 550 K, hydrogen chloride gas is

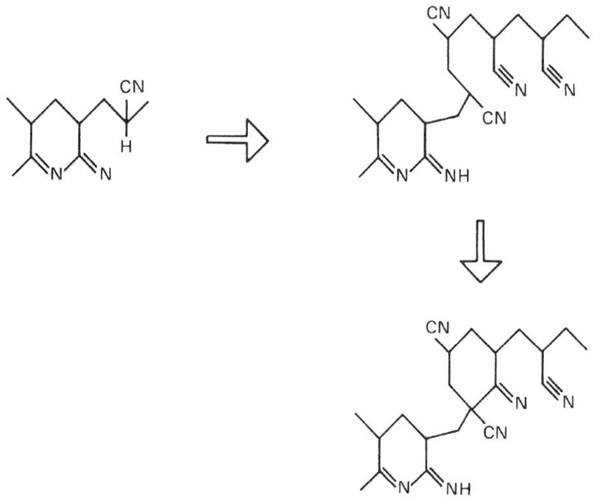

Fig. 1-7.9. Reinitiation of PAN side-chain cyclization.

evolved nearly quantitatively, by a chain-stripping mechanism. It is very important to point out, however, that the temperature at which hydrogen chloride starts being evolved in any measurable way is heavily dependent on the stabilization package used. Thus, commercial PVC "compounds" have been shown, in recent work, not to evolve hydrogen chloride until temperatures are in excess of 520 K and to have a dehydrochlorination stage starting at 600 K.[54] Between 700 and 750 K, hydrogen is evolved during carbonization, following cyclization of the species evolved. At higher temperatures, cross-linking between chains results in a fully carbonized residue. The rate of dehydrochlorination depends on the molecular weight, crystallinity, presence of oxygen, hydrogen chloride gas, and stabilizers. The presence of oxygen accelerates the dehydrochlorination process, produces main-chain scissions, and reduces cross-linking. At temperatures above 700 K, the char (resulting from dehydrochlorination and further dehydrogenation) is oxidized, leaving no residue. Lower molecular weight increases the rate of dehydrochlorination. Dehydrochlorination stabilizers include zinc, cadmium, lead, calcium, and barium soaps and organotin derivatives. The stability of model compounds indicates that weak links are important in decomposition. The thermal decomposition of this polymer has been one of the most widely studied ones. It has been the matter of considerable controversy, particularly in terms of explaining the evolution of aromatics in the second decomposition stage. The most recent evidence seems to point to a simultaneous cross-linking and intramolecular decomposition of the polyene segments resulting from dehydrochlorination, *via* polyene free radicals.[54] Earlier evidence suggested a Diels-Alder cyclization process (which can only be intramolecular if the double bond ends up in a "cis" orientation).[55] Evidence for this was given by the fact that smoke formation (inevitable consequence of the emission of aromatic hydrocarbons) was decreased by introducing cross-linking additives into the polymer.[56] Thus, it has now become clear that formation of any aromatic hydrocarbon occurs intramolecularly. The chemical mechanism for the initiation of dehydrochlorination was also reviewed a few years ago.[57] More recently, a series of papers was published investigating the kinetics of chain stripping, based on PVC.[58]

Chlorinated poly(vinyl chloride) (CPVC): One interesting derivative of PVC is chlorinated PVC (CPVC), resulting from post-polymerization chlorination of PVC. It decomposes at a much higher temperature than PVC, but by the same chain-stripping mechanism. The resulting solid is a polyacetylene, which gives off much less smoke than PVC and is also more difficult to burn.[59]

Poly(tetrafluoroethylene) (PTFE): PTFE is a very stable polymer due to the strength of C-F bonds and shielding by the very electronegative fluorine atoms. Decomposition starts occurring between 750 and 800 K. The principal product of decomposition is the monomer, CF_4, with small amounts of hydrogen fluoride and hexafluoropropene. Decomposition is initiated by random-chain scission, followed by depolymerization. Termination is by disproportionation. It is possible that the actual product of decomposition is CF_2, which immediately forms in the gas phase. The stability of the polymer can be further enhanced by promoting chain transfer reactions that can effectively limit the zip length. Under conditions of oxidative pyrolysis, no monomer is formed. Oxygen reacts with the polymeric radical, releasing carbon monoxide, carbon dioxide, and other products.

Other fluorinated polymers are less stable than PTFE and are generally no more stable than their unfluorinated analogs. However, the fluorinated polymers are more stable in an oxidizing atmosphere. Hydrofluorinated polymers produce hydrogen fluoride directly by chain stripping reactions, but the source of hydrogen fluoride by perfluorinated polymers, such as PTFE, is less clear. It is related to the reaction of the decomposition products (including tetrafluoroethylene) with atmospheric humidity.

Other Vinyl Polymers

Several other vinyl polymers decompose by mechanisms similar to that of PVC: all those that have a single substituent other than a hydrogen atom on the basic repeating unit. These include poly(vinyl acetate), poly(vinyl alcohol), and poly(vinyl bromide), and result in gas evolution of acetic acid, water, and hydrogen bromide, respectively. While the chain stripping reactions of each of these polymers occur at different temperatures, all of them aromatize by hydrogen evolution at roughly 720 K.

Styrenics: *Polystyrene (PS).* Polystyrene shows no appreciable weight loss below 575 K, though there is a decrease in molecular weight due to scission of "weak" links. Above this temperature, the products are primarily monomer with decreasing amounts of dimer, trimer, and tetramer. There is an initial sharp decrease in molecular weight followed by slower rates of molecular weight decrease. The mechanism is thought to be dominated by end-chain initiation, depolymerization, intramolecular hydrogen transfer, and bimolecular termination. The changes in molecular weight are principally due to intermolecular transfer reactions while volatilization is dominated by intramolecular transfer reactions. Depropagation is prevalent despite the lack of steric hindrance due to the stabilizing effect of the electron delocalization associated with the aromatic side group. The addition of an alpha methyl group to form poly(α-methylstyrene) provides additional steric hindrance such that only monomer is produced during decomposition while the thermal stability of the polymer is lessened. Free radical polymerized polystyrene is less stable than anionic polystyrene with the rate of decomposition dependent on the end group.

Other styrenics tend to be copolymers of polstyrene with acrylonitrile (SAN), acrylonitrile and butadiene (ABS), or methyl methacrylate and butadiene (MBS), and their decomposition mechanisms are hybrids between those of the individual polymers.

Synthetic Carbon-Oxygen Chain Polymers

Poly(ethylene terephthalate) (PET): PET decomposition is initiated by scission of an alkyl-oxygen bond. The decomposition kinetics suggest a random-chain scission. Principal gaseous products observed are acetaldehyde, water, carbon monoxide, carbon dioxide, and compounds with acid and anhydride end groups. The decomposition is accelerated by the presence of oxygen. Recent evidence indicates that both PET and PBT [poly(butylene terephthalate)] decompose *via* the formation of cyclic or open-chain oligomers, with olefinic or carboxylic end groups.[60]

Polycarbonates (PC): Polycarbonates yield substantial amounts of char if products of decomposition can be removed (the normal situation). If volatile products are not removed, no cross-linking is observed due to competition between condensation and hydrolysis reactions. The decomposition is initiated by scission of the weak $O-CO_2$ bond, and the volatile products include 35 percent carbon dioxide. Other major products include bisphenol A and phenol. The decomposition mechanism seems to be a mixture of random-chain scission and cross-linking, initiated intramolecularly.[61] Decomposition begins at 650 to 735 K, depending on the exact structure of the polycarbonate in question.

Blends of polycarbonate and styrenics (such as ABS) make up a set of "engineered thermoplastics." Their properties are intermediate between those of the forming individual polymers, both in terms of physical properties (and processability) and in terms of their modes of thermal breakdown.

Phenolic resins: Phenolic resin decomposition begins at 575 K and is initiated by the scission of the methylene-benzene ring bond. At 633 K, the major products are C_3 compounds. In continued heating (725 K and higher), char (carbonization), carbon oxides, and water are formed. Above 770 K, a range of aromatic, condensable products are evolved. Above 1075 K, ring breaking yields methane and carbon oxides. In TGA experiments at 3.3°C/min, the char yield is 50 to 60 percent. The weight loss at 700 K is 10 percent. All decomposition is oxidative in nature (oxygen provided by the polymer itself).

Polyoxymethylene (POM): Polyoxymethylene decomposition yields formaldehyde almost quantitatively. The decomposition results from end-chain initiation followed by depolymerization. The presence of oxygen in the chain prevents intramolecular hydrogen transfer quite effectively. With hydroxyl end groups, decomposition may begin at temperatures as low as 360 K while with ester end groups decomposition may be delayed to 525 K. Piloted ignition due to radiative heating has been observed at a surface temperature of 550 K. Acetylation of the chain end group also improves stability. Upon blocking the chain ends, decomposition is by random-chain initiation, followed by depolymerization with the zip length less than the degree of polymerization. Some chain transfer occurs. Amorphous polyoxymethylene decomposes faster than crystalline polyoxymethylene, presumably due to the lack of stabilizing intermolecular forces associated with the crystalline state (below the melting temperature). Incorporating oxyethylene in polyoxymethylene improves stability, presumably due to H transfer reactions that stop unzipping. Oxidative pyrolysis begins at 430 K and leads to formaldehyde, carbon monoxide, carbon dioxide, hydrogen, and water vapor.

Epoxy resins: Epoxy resins are less stable than phenolic resins, polycarbonate, polyphenylene sulphide, and polytetrafluoroethelyne. The decomposition mechanism is complex and varied and usually yields mainly phenolic compounds. A review of epoxy resin decomposition can be found in Lee.[43]

Polyamide Polymers

Nylons: The principal gaseous products of decomposition of nylons are carbon dioxide and water. Nylon 6 produces small amounts of various simple hydrocarbons while Nylon 6–10 produces notable amounts of hexadienes and hexene. As a class, nylons do not notably decompose below 615 K. Nylon 6–6 melts between 529 and 532 K, and decomposition begins at 615 K in air and 695 K in nitrogen. At temperatures in the range 625 to 650 K, random-chain scissions lead to oligomers. The C–N bonds are the weakest in the chain, but the $CO-CH_2$ bond is also quite weak, and both are involved in decomposition. At low temperatures, most of the decomposition products are nonvolatile, though above 660 K main chain scissions lead to monomer and some dimer and trimer production. Nylon 6–6 is less stable than nylon 6–10, due to the ring closure tendency of the adipic acid component. At 675 K, if products are removed, gelation and discoloration begin.

Aromatic polyamides have good thermal stability, as exemplified by Nomex®, which is generally stable in air to 725 K. The major gaseous products of decomposition at low temperatures are water and carbon oxides. At higher temperatures, carbon monoxide, benzene, hydrogen cyanide (HCN), toluene, and benzonitrile are produced. Above 825 K, hydrogen and ammonia are formed. The remaining residue is highly cross-linked.

Wool: On decomposition of wool, a natural polyamide, approximately 30 percent is left as a residue. The first step in decomposition is the loss of water. Around 435 K, some cross-linking of amino acids occurs. Between 485 and 565 K, the disulphide bond in the amino acid cystine is cleaved with carbon disulphide and carbon dioxide being evolved. Pyrolysis at higher temperatures (873 to 1198 K) yields large amounts of hydrogen cyanide, benzene, toluene, and carbon oxides.

Polyurethanes

As a class, polyurethanes do not break down below 475 K, and air tends to slow decomposition. The production of hydrogen cyanide and carbon monoxide increases with the pyrolysis temperature. Other toxic products formed include nitrogen oxides, nitriles, and tolylene diisocyanate (TDI) (and other isocyanates). A major breakdown mechanism in urethanes is the scission of the polyol-isocyanate bond formed during polymerization. The isocyanate vaporizes and recondenses as a smoke, and liquid polyol remains to further decompose.

Polydienes and Rubbers

Polyisoprene: Synthetic rubber or polyisoprene decomposes by random-chain scission with intramolecular hydrogen transfer. This, of course, gives small yields of monomer. Other polydienes appear to decompose similarly though the thermal stability can be considerably different. The average

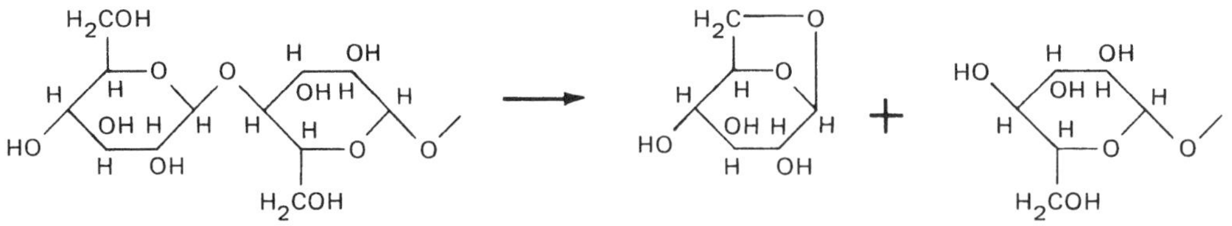

Fig. 1-7.10.   Formation of laevoglucosan from cellulose.

size of fragments collected from isoprene decomposition are 8 to 10 monomer units long. This supports the theory that random-chain scission and intermolecular transfer reactions are dominant in the decomposition mechanism. In nitrogen, decomposition begins at 475 K. At temperatures above 675 K, increases in monomer yield are attributable to secondary reaction of volatile products to form monomer. Between 475 and 575 K, low molecular weight material is formed, and the residual material is progressively more insoluble and intractable. Preheating at between 475 and 575 K lowers the monomer yield at higher temperatures. Decomposition at less than 575 K results in a viscous liquid and, ultimately, a dry solid. The monomer is prone to dimerize to dipentene as it cools. There seems to be little significant difference in the decomposition of natural rubber and synthetic polyisoprene.

Polybutadiene: Polybutadiene is more thermally stable than polyisoprene due to the lack of branching. Decomposition at 600 K can lead to monomer yields of up to 60 percent, with lower conversions at higher temperatures. Some cyclization occurs in the products. Decomposition in air at 525 K leads to a dark impermeable crust, which excludes further air. Continued heating hardens the elastomer.

Polychloroprene: Polychloroprene decomposes in a manner similar to PVC, with initial evolution of hydrogen chloride at around 615 K and subsequent breakdown of the residual polyene. The sequences of the polyene are typically around three (trienes), much shorter than PVC. Polychloroprene melts at around 50 °C.

Cellulosics

The decomposition of cellulose involves at least four processes in addition to simple desorption of physically bound water. The first is the cross-linking of cellulose chains, with the evolution of water (dehydration). The second concurrent reaction is the unzipping of the cellulose chain. Laevoglucosan is formed from the monomer unit. (See Figure 1-7.10.) The third reaction is the decomposition of the dehydrated product (dehydrocellulose) to yield char and volatile products. Finally, the laevoglucosan can further decompose to yield smaller volatile products, including tars and, eventually, carbon monoxide. Some laevoglucosan may also repolymerize.

Below 550 K, the dehydration reaction and the unzipping reaction proceed at comparable rates, and the basic skeletal structure of the cellulose is retained. At higher temperatures, unzipping is faster, and the original structure of the cellulose begins to disappear. The cross-linked dehydrated cellulose and the repolymerized laevoglucosan begin to yield polynuclear aromatic structures, and graphite carbon structures form at around 770 K. It is well known that the char yield is quite dependent on the rate of heating of the sample. At very high rates of heating, no char is formed. On

the other hand, preheating the sample at 520 K will lead to 30 percent char yields. This is due both to the importance of the low-temperature dehydration reactions for ultimate char formation and the increased opportunity for repolymerization of laevoglucosan that accompanies slower heating rates.

Wood is made up of 50 percent cellulose, 25 percent hemicellulose, and 25 percent lignin. The yields of gaseous products and kinetic data indicate that the decomposition may be regarded as the superposition of the individual constituent's decomposition mechanisms. On heating, the hemicellulose decomposes first (475 to 535 K), followed by cellulose (525 to 625 K), and lignin (555 to 775 K). The decomposition of lignin contributes significantly to the overall char yield. Piloted ignition of woods due to radiative heating has been observed at a surface temperature of 620 to 650 K.

Polysulfides and Polysulphones

Polysulfides are generally stable to 675 K. Poly(1, 4 phenylene sulfide) decomposes at 775 K. Below this temperature, the principal volatile product is hydrogen sulfide. Above 775 K, hydrogen, evolved in the course of cross-linking, is the major volatile product. In air, the gaseous products include carbon oxides and sulfur dioxide.

The decomposition of polysulphones is analogous to polycarbonates. Below 575 K, decomposition is by heteroatom bridge cleavage, and above 575 K, sulfur dioxide is evolved from the polymer backbone.

Thermally Stable Polymers

The development of thermally stable polymers is an area of extensive ongoing interest. Relative to many other materials, polymers have fairly low use temperatures, which can reduce the utility of the product. This probable improvement in fire properties is, often, counterbalanced by a decrease in processability and in favorable physical properties. Of course, materials that are stable at high temperatures are likely to be better performers as far as fire properties are concerned. The high-temperature physical properties of polymers can be improved by increasing interactions between polymer chains or by chain-stiffening.

Chain interactions can be enhanced by several means. As noted previously, crystalline materials are more stable than their amorphous counterparts as a result of chain interactions. Of course, if a material melts before volatilization occurs, this difference will not affect chemical decomposition. Isotactic polymers are more likely to be crystalline due to increased regularity of structure. Polar side groups can also increase the interaction of polymer chains. The melting point of some crystalline polymers is shown in Table 1-7.1.

The softening temperature can also be increased by chain-stiffening. This is accomplished by the use of aromatic or heterocyclic structures in the polymer backbone. Some

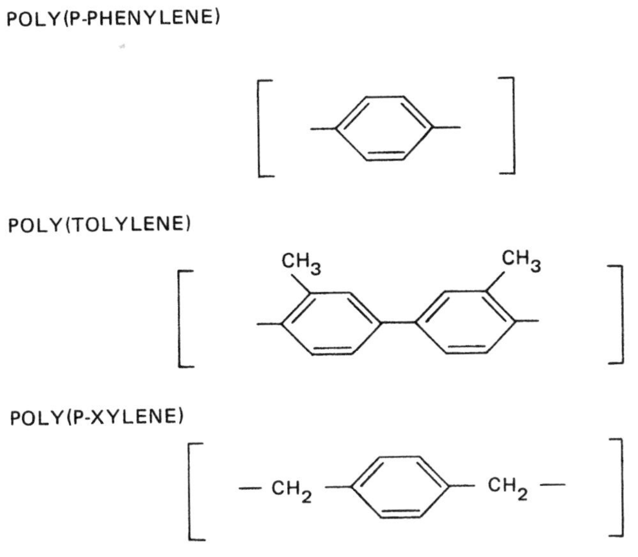

POLY(P-PHENYLENE)

POLY(TOLYLENE)

POLY(P-XYLENE)

Fig. 1-7.11. Thermostable aromatic polymers.

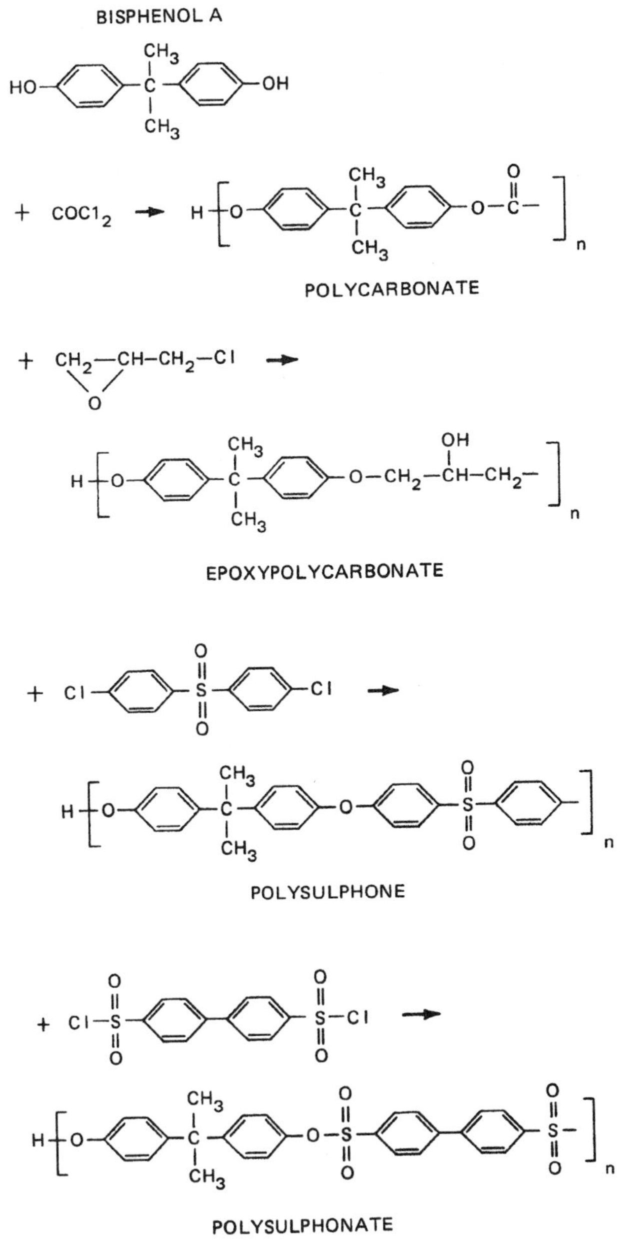

BISPHENOL A

POLYCARBONATE

EPOXYPOLYCARBONATE

POLYSULPHONE

POLYSULPHONATE

Fig. 1-7.12. Bisphenol A polymers.

aromatic polymers are shown in Figure 1-7.11. Poly(p-phenylene) is quite thermally stable but is brittle, insoluble, and infusible. Thermal decomposition begins at 870 to 920 K; and up to 1170 K, only 20 to 30 percent of the original weight is lost. Introduction of the following groups:

$$-O-, \; -CO-, \; -NH-, \; -CH_2-, \; -O-CO-, \; -O-CO-O-$$

into the chain can improve workability though at the cost of some loss of oxidative resistance. Poly(p-xylene) melts at 675 K and has good mechanical properties though it is insoluble and cannot be thermoprocessed. Substitution of halogen, acetyl, alkyl, or ester groups on aromatic rings can help the solubility of these polymers at the expense of some stability. Several relatively thermostable polymers can be formed by condensation of bisphenol A with a second reagent. Some of these are shown in Figure 1-7.12. The stability of such polymers can be improved if aliphatic groups are not included in the backbone, as the $-C(CH_3)_2-$ groups are weak links.

Other thermostable polymers include ladder polymers and extensively cross-linked polymers. Cyclized PAN is an example of a ladder polymer where two chains are periodically interlinked. Other polymers, such as rigid polyurethanes, are sufficiently cross-linked that it becomes impossible to speak of a molecular weight or definitive molecular repeating structure. As in polymers that gel or cross-link during decomposition, cross-linking of the original polymer yields a carbonized char residue upon decomposition, which can be oxidized at temperatures over 775 K.

REFERENCES CITED

1. ASTM E176-92, "Standard Terminology of Fire Standards," E176, Annual Book of ASTM Standards, Vol. 4.07, American Society for Testing and Materials, Philadelphia, PA.
2. C.F. Cullis and M.M. Hirschler, The Combustion of Organic Polymers, Oxford University Press, Oxford, UK (1981).
3. E.D. Weil, R.N. Hansen, and N. Patel, "Prospective Approaches to More Efficient Flame-Retardant Systems," in Fire and Polymers: Hazards Identification and Prevention, (ed. G.L. Nelson), ACS Symposium Series 425, Developed from Symp. at 197th. ACS Mtg. Dallas, TX. April 9–14, 1989, Amer. Chem. Soc., Washington, DC, Chapter 8, p. 97–108 (1990).
4. M.M. Hirschler, "Recent Developments in Flame-Retardant Mechanisms," in Developments in Polymer Stabilisation, Vol. 5, (ed. G. Scott), pp. 107–52, Applied Science Publ., London (1982).
5. D.E. Stuetz, A.H. DiEdwardo, F. Zitomer, and B.F. Barnes, "Polymer Flammability II," J. Polym. Sci., Polym. Chem. Ed., 18, p. 987–1009 (1980).
6. T. Kashiwagi and T.J. Ohlemiller, "A Study of Oxygen Effects on Flaming Transient Gasification of PMMA and PE during Thermal Irradiation," Nineteenth Symp. (Int.) on Combustion, The Combustion Institute, Pittsburgh, PA, pp. 1647–54 (1982).
7. T. Kashiwagi, T. Hirata, and J.E. Brown, "Thermal and Oxidative Degradation of Poly(methyl methacrylate), Molecular Weight," Macromolecules, 18, Feb., pp. 131–138 (1985).

8. T. Hirata, T. Kashiwagi, and J.E. Brown, "Thermal and Oxidative Degradation of Poly(methyl methacrylate), Weight Loss," *Macromolecules*, 18, July, pp. 1410–1418 (1985).

9. T. Kashiwagi, A. Inabi, J.E. Brown, K. Hatada, T. Kitayama, and E. Masuda, "Effects of Weak Linkages on the Thermal and Oxidative Degradation of Poly(methyl methacrylates)," *Macromolecules*, 19, 2160–68 (1986).

10. T. Kashiwagi and A. Inabi, "Behavior of Primary Radicals during Thermal Degradation of Poly(methyl methacrylate)," *Polymer Degradation and Stability*, 26, 161–184 (1989).

11. S.K. Brauman, "Polymer Degradation during Combustion," *J. Polymer Sci.*, B, 26, 1159–71 (1988).

12. P. Gijsman, J. Hennekens, and J. Vincent, "The Mechanism of the Low-Temperature Oxidation of Polypropylene," *Polymer Degradation and Stability*, 42, 95–105 (1993).

13. M.M. Hirschler, "Effect of Oxygen on the Thermal Decomposition of Poly(vinylidene fluoride)," *Europ. Polymer J.*, 18, 463–67 (1982).

14. A. Inabi and T. Kashiwagi, "A Calculation of Thermal Degradation Initiated by Random Scission, Unsteady Radical Concentration," *Eur. Polym. J.*, 23 (11), 871–881 (1987).

15. T. Kashiwagi and H. Nambu, "Global Kinetic Constants for Thermal Oxidative Degradation of a Cellulosic Paper," *Combust. Flame*, 88, 345–68 (1992).

16. K.D. Steckler, T. Kashiwagi, H.R. Baum, and K. Kanemaru, "Analytical model for transient Gasification of Noncharring Thermoplastic Materials," *Fire Safety Sci., Proc. Third Int. Symp.*, eds. G. Cox and B. Langford, Elsevier, London (1991).

17. ASTM D2859, "Standard Test Method for Flammability of Finished Textile Floor Covering Materials," *Annual Book of ASTM Standards*, Vol. 7.01, American Society for Testing and Materials, Philadelphia, PA.

18. CFR 1630, 16 CFR Part 1630, "Standard for the Surface Flammability of Carpets and Rugs (FF 1–70)," *Code of Federal Regulations, Commercial Practices, Subchapter D: Flammable Fabrics Act Regulations*, Vol. 16, Parts 1602–1632, pp. 573–582, Jan. 1 (1990).

19. CFR 1631, 16 CFR Part 1631, "Standard for the Surface Flammability of Small Carpets and Rugs (FF 2–70)," *Code of Federal Regulations, Commercial Practices, Subchapter D: Flammable Fabrics Act Regulations*, Vol. 16, Parts 1602–1632, pp. 582–591, Jan. 1 (1990).

20. L.A. Chandler, M.M. Hirschler, and G.F. Smith, "A Heated Tube Furnace Test for the Emission of Acid Gas from PVC Wire Coating Materials: Effects of Experimental Procedures and Mechanistic Considerations," *Europ. Polymer J.*, 23, 51–61 (1987).

21. M.M. Hirschler, "Thermal Decomposition (STA and DSC) of Poly(vinyl chloride) Compounds under a Variety of Atmospheres and Heating Rates," *Europ. Polymer J.*, 22, 153–60 (1986).

22. C.F. Cullis and M.M. Hirschler, "The Significance of Thermoanalytical Measurements in the Assessment of Polymer Flammability," *Polymer*, 24, 834–40 (1983).

23. M.M. Hirschler, "Thermal Analysis and Flammability of Polymers: Effect of Halogen-Metal Additive Systems," *Europ. Polymer J.*, 19, 121–9 (1983).

24. I.C. McNeill, "The Application of Thermal Volatilization Analysis to Studies of Polymer Degradation," in *Developments in Polymer Degradation*, Vol. 1, (ed. N. Grassie), p. 43, Applied Science, London, UK (1977).

25. C.F. Cullis, M.M. Hirschler, R.P. Townsend, and V. Visanuvimol, "The Pyrolysis of Cellulose under Conditions of Rapid Heating," *Combust. Flame*, 49, 235–48 (1983).

26. C.F. Cullis, M.M. Hirschler, R.P. Townsend, and V. Visanuvimol, "The Combustion of Cellulose under Conditions of Rapid Heating," *Combust. Flame*, 49, 249–54 (1983).

27. C.F. Cullis, D. Goring, and M.M. Hirschler, "Combustion of Cigarette Paper under Conditions Similar to Those during Smoking," *Cellucon '84* (Macro Group U.K.), Wrexham (Wales), Chapter 35, pp. 401–10, July 16–20, Ellis Horwood, Chichester, (1984).

28. P.J. Baldry, C.F. Cullis, D. Goring, and M.M. Hirschler, "The Pyrolysis and Combustion of Cigarette Constituents," *Proc. Int. Conf. on "Physical and Chemical Processes Occurring in a Burning Cigarette,"* R.J. Reynolds Tobacco Co., Winston-Salem, Apr. 26–29, pp. 280–301 (1987).

29. P.J. Baldry, C.F. Cullis, D. Goring, and M.M. Hirschler, "The Combustion of Cigarette Paper," *Fire and Materials*, 12, 25–33 (1988).

30. L. Reich and S.S. Stivala, *Elements of Polymer Degradation*, McGraw-Hill, New York, (1971).

31. A.A. Miroshnichenko, M.S. Platitsa, and T.P. Nikolayeva, "Technique for Calculating the Temperature at the Beginning and End of Polymer Thermal Degradation from Structural Data," *Polymer Science*, USSR, 30 (12), 2707–16 (1988).

32. O.F. Shlenskii and N.N. Lyasnikova, "Predicting the Temperature of Thermal Decomposition of Linear Polymers," *Intern. Polymer Sci. Technol.*, 16 (3), T55-T56 (1989).

33. D.W. Van Krevelen, "Thermal Decomposition," Chapter 21 and "Product Properties (II), Environmental Behavior and Failure," in *Properties of Polymers*, 3rd ed., Elsevier, Amsterdam, pp. 641–525–535 (1990).

34. D.W. Van Krevelen, "Some Basic Aspects of Flame Resistance of Polymeric Materials," *Polymer*, 16, 615–620 (1975).

35. ASTM D2863-87, "Standard Method for Measuring the Minimum Oxygen Concentration to Support Candle-like Combustion of Plastics (Oxygen Index)," *Fire Test Standards*, 3rd ed., American Society for Testing and Materials, Philadelphia, PA (1990).

36. E.D. Weil, M.M. Hirschler, N.G. Patel, M.M. Said, and S. Shakir, "Oxygen Index: Correlation to Other Tests," *Fire Materials*, 16, 159–67 (1992).

37. ASTM E1354, "Standard Test Method for Heat and Visible Smoke Release Rates for Materials and Products Using an Oxygen Consumption Calorimeter," *Annual Book of ASTM Standards*, Volume 4.07, American Society for Testing and Materials, Philadelphia, PA.

38. M.M. Hirschler, "Heat Release from Plastic Materials," Chapter 12a, in *Heat Release in Fires*, Elsevier, London, UK, eds. V. Babrauskas and S.J. Grayson, pp. 375–422 (1992).

39. R.E. Lyon, "Fire-Safe Aircraft Materials" Fire and Materials, Proceeding 3rd Int. Conference and Exhibition, Crystal City, VA, Oct. 27–28, Interscience Communications, pp. 167–177 (1994).

40. N. Grassie, "Polymer Degradation and Fire Hazard," *Polymer Degradation and Stability*, 30, 3–12 (1990).

41. G. Camino and L. Costa, "Performance and Mechanisms of Fire Retardants in Polymers—A Review," *Polymer Degradation and Stability*, 20, 271–294 (1988).

42. G. Bertelli, L. Costa, S. Fenza, F.E. Marchetti, G. Camino, and R. Locatelli, "Thermal Behaviour of Bromine-Metal Fire Retardant Systems," *Polymer Degradation and Stability*, 20, 295–314 (1988).

43. L.H. Lee, *J. Polymer Sci.*, 3, 859 (1965).

44. R.T. Conley, ed., *Thermal Stability of Polymers*, Marcel Dekker, New York, (1970).

45. W.J. Roff and J.R. Scott, *Fibres, Films, Plastics, and Rubbers*, Butterworths, London, (1971).

46. F.A. Williams, in *Heat Transfer in Fires*, Scripta, Washington (1974).

47. C. David, in *Comprehensive Chemical Kinetics*, Elsevier, Amsterdam, The Netherlands (1975).

48. S.L. Madorsky, *Thermal Degradation of Polymers*, reprinted by Robert E. Kreiger, New York (1975).

49. D.W. Van Krevelen, *Properties of Polymers*, Elsevier, Amsterdam, The Netherlands (1976).

50. T. Kelen, *Polymer Degradation*, Van Nostrand Reinhold, New York, (1983).

51. W.L. Hawkins, *Polymer Degradation and Stabilization*, Springer Verlag, Berlin, Germany (1984).

52. N. Grassie and G. Scott, *Polymer Degradation and Stabilisation*, Cambridge University Press, Cambridge, UK (1985).

53. S.M. Lomakin, J.E. Brown, R.S. Breese, and M.R. Nyden, "An Investigation of the Thermal Stability and Char-Forming Tendency of Cross-Linked Poly(methyl methacrylate)," *Polymer Degradation Stability*, 41, 229–43 (1993).

54. G. Montaudo and C. Puglisi, "Evolution of Aromatics in the Thermal Degradation of Poly(vinyl chloride): A Mechanistic Study," *Polymer Degradation Stability*, 33, 229–62 (1991).

55. W.H. Starnes, Jr., and D. Edelson, *Macromolecules*, 12, 797 (1979).

56. D. Edelson, R.M. Lum, W.D. Reents, Jr., W.H. Starnes, Jr., and L.D. Westcott, Jr., "New Insights into the Flame Retardance Chemistry of Poly(vinyl chloride)," in *Proc. Nineteenth (Int.) Symp. on Combustion*, The Combustion Institute, Pittsburgh, PA, pp. 807–14 (1982).

57. K.S. Minsker, S.V. Klesov, V.M. Yanborisov, A.A. Berlin, and G.E. Zaikov, "The Reason for the Low Stability of Poly(vinyl chloride)—A Review," *Polymer Degradation Stability*, 16, 99–133 (1986).

58. P. Simon *et al*, "Kinetics of Polymer Degradation Involving the Splitting off of Small Molecules. Parts 1–7," *Polymer Degradation Stability*, 29, 155; 253; 263 (1990); 35, 45; 157; 249 (1992); 36, 85 (1992).

59. L.A. Chandler and M.M. Hirschler, "Further Chlorination of Poly(vinyl chloride): Effects on Flammability and Smoke Production Tendency," *Europ. Polymer J.*, 23, 677–83 (1987).

60. G. Montaudo, C. Puglisi, and F. Samperi, "Primary Thermal Degradation Mechanisms of PET and PBT," *Polymer Degradation Stability*, 42, 13–28 (1993).

61. G. Montaudo, C. Puglisi, R. Rapisardi, and F. Samperi, "Further Studies on the Thermal Decomposition Processes in Polycarbonates," *Polymer Degradation Stability*, 31, 229–46 (1991).

STRUCTURAL
MECHANICS

Robert W. Fitzgerald

INTRODUCTION

Structural mechanics is the analysis of the external and internal force systems of structural members, as well as the behavior of those members under loading conditions. Before describing the different types of members and their structural characteristics, it is helpful to describe briefly the structural design process.

Structural design follows roughly the same stages of design as the architectural process. During the schematic stage when the building layout is being created, the structural engineer and the architect identify column locations. Then, a number of different framing schemes utilizing the different structural materials are considered. A design is made for each potential framing alternative for a part of the building that is representative of a major segment of the structure. Economic and functional analyses are made with the different alternatives. The architect and structural engineer select the framing system that is best for the specific building being designed.

After the schematic design has been completed and accepted, the detail design and contract documents stages are undertaken. During these stages, all of the structural members and the important details are designed. Critical design connections, significant construction details, and specifications are developed to ensure a complete and adequate structural system.

The structural design must conform to accepted professional practice at the time. Regardless of materials, this involves three major interrelated considerations. They are:

1. The appropriate loading conditions and combinations,
2. Structural mechanics, and
3. Control parameter limits.

The objective of structural design is to select materials and dimensions so that economy is achieved and the building will perform satisfactorily. Performance here means that the structure is compatible with architectural needs and is free from excessive deflection and vibration. Prevention of collapse under expected or reasonably forseen conditions is included in performance, and safety is a major part of the professional responsibility.

A major aspect of structural engineering is the recognition of conditions that can lead to failure. When these conditions are present, the designer must proportion the members or take other measures to ensure that failure under design conditions will not occur. The identification of loads, selection of engineering calculation models, and the establishment of control parameter limits are all interwoven. Figure 1-8.1 shows a schematic relationship of these components. Although each component may be addressed separately, their interrelationships comprise the unification of the design methodology. Together they allow performance to be monitored.

The loading conditions of Figure 1-8.1 are generally specified in the building code. They include live load values for floor systems, snow and ice, wind, and earthquake. The engineer also will include the dead load for the framing system and any special loading that may be expected for the structure being considered.

The engineering models involve two considerations. One is the mechanics of computing the internal forces that result from the loading, dimensions, and support conditions. The other is the relationship between these internal forces and the performance function. This performance function relates the internal forces to control parameters. Stress is the most convenient control parameter, although others, such as deflection, are used also.

Structural mechanics is the engineering science that enables the engineer to calculate the internal shear, moment, and axial force and the related stresses at any location in the structural member for any combination of loads. In addition, it describes the behavior of the member as loads are increased up to failure. This is dependent on the materials involved, the type of loading, and the geometric and support conditions of the member. The behavior includes deformations, vibrations, and failure modes. Structural mechanics may be considered as the "exact" analytical part of the design process.

Another consideration in Figure 1-8.1 involves the specific design requirements for the materials and assembly. These are developed by the different products industries. For example, the American Institute of Steel Construction

Dr. Robert W. Fitzgerald is Professor of Civil and Fire Protection Engineering at Worcester Polytechnic Institute. His major activities are in structural engineering and in building design and technology for fire-safety.

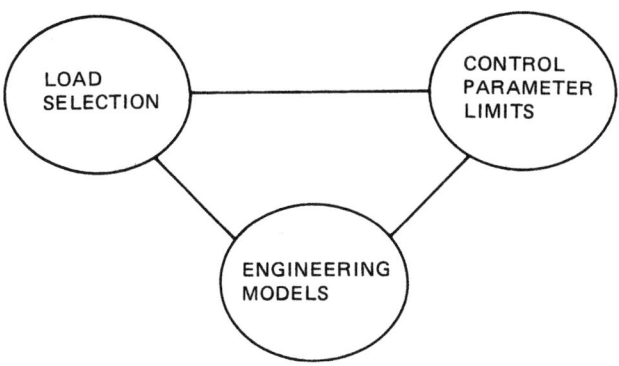

Fig. 1-8.1. *Components of the design methodology.*

(AISC) publishes its code of practice,[1] and the American Concrete Institute (ACI) publishes its building code.[2] These publications, often called "codes" by the engineers, give requirements on design and construction that will avoid failure for normal usage. The values for allowable stress are the most common limits. These values are empirically selected considering theoretical mechanical behavior for the material and practical applications. When a designer uses building code loadings with allowable stresses and other control criteria through the mechanism of the engineering models, one can have confidence that the member probably will not fail. The control performance consideration is normally deflection, even though the calculations usually involve stress. Stress and deformation are, of course, related.

The reliability of structural design has evolved through consideration of the entire process. Although individual parts can be examined by in-depth research, the process as a whole is considered in design. Values for loading and codi-

fied limits of the parameters are established by the end performance to be achieved.

Professional structural practice integrates the loadings, usually obtained from the local building code or from the conditions that may reasonably be expected by the engineer, with the structural (mechanical) analysis and the design procedures of the structural code. The structural codes are updated periodically, usually about every five to ten years. The literature of the profession can keep the engineer aware of new developments in the field.

With this brief discussion of the structural design process, we will now describe briefly the elements of structural mechanics. In general, this may be grouped as the calculation of external reactions and internal forces and the prediction of failure modes for different materials, geometry, support conditions, and loads.

STATICAL ANALYSIS FOR REACTIONS

The calculation of external reactions of a defined structural element for a given loading condition is the first part of the statical analysis. For planar structures, the available equations of statics are $\Sigma F_x = 0$, $\Sigma F_y = 0$, and $\Sigma M = 0$. For three-dimensional structures, $\Sigma F_z = 0$, $\Sigma M = 0$ about the other axes are added. Therefore, for planar structures, one can calculate as many as three unknown reactions on each free body diagram by statical analysis. For three-dimensional structures, one can calculate as many as six unknowns. For this discussion, we will consider only planar structures.

EXAMPLE 1:

To illustrate this process, consider the beam *ABC* of Figure 1-8.2. The supports include a pin at *B* and a roller at *C*. Figure 1-8.2, part (b) shows the free body of this beam. The reactions are computed as follows

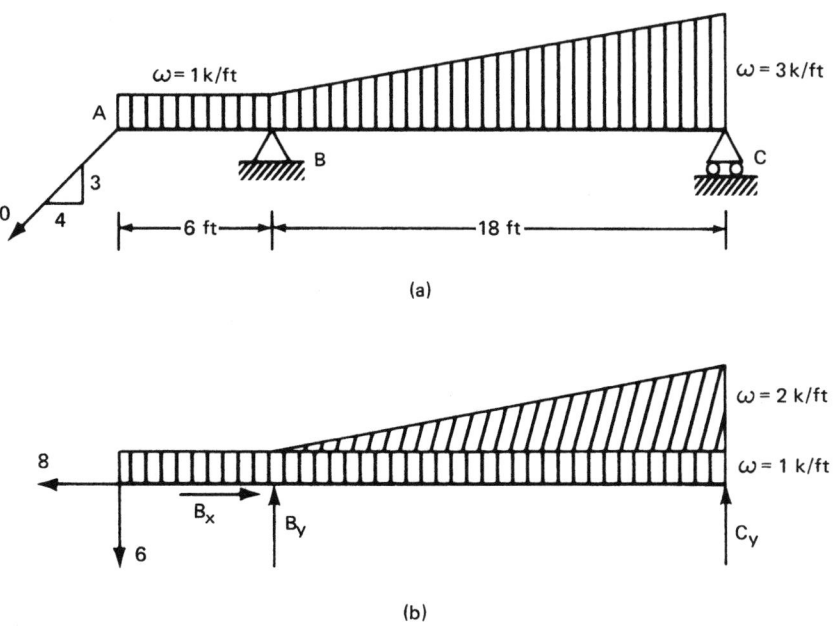

Fig. 1-8.2. *Statically determinate beam.*

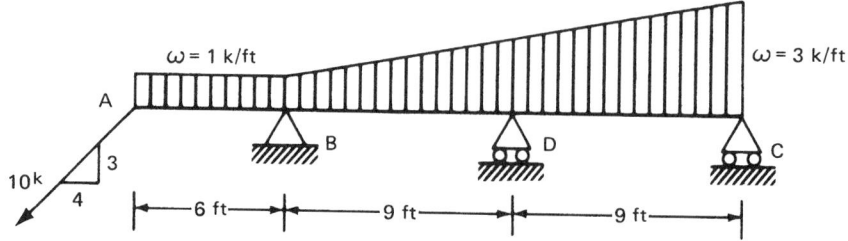

Fig. 1-8.3. Statically indeterminate beam.

$$\sum M_B = 0: C_y(18) - (1)(24)(6) - (1/2)(2)(18)(12) + 6(6) = 0$$
$$C_y = 18 \text{ k}$$

$$\sum M_C = 0: B_y(18) - (1)(24)(12) - (1/2)(2)(18)(6) - 6(24) = 0$$
$$B_y = 30 \text{ k}$$

$$\sum F_x = 0: B_x = 8 \text{ k}$$

Since there were only three reaction components, one could calculate all three by means of statics alone. The structure would be described as statically determinate. However, if an additional support were introduced, as shown in Figure 1-8.3, four reaction components would exist. Since only three equations of statics are available, all of these reactions cannot be calculated by means of statics alone. This structure of Figure 1-8.3 would be described as statically indeterminate. Means other than statics alone are needed to calculate the reactions. Generally, these techniques involve either superposition or relaxation methods of analysis.

STATICAL ANALYSIS FOR INTERNAL FORCES

After the external reactions have been calculated, the characteristics of the internal shear, moment, and axial force are determined. This may be computed by cutting the member at the desired location and drawing a free body diagram of one segment.

EXAMPLE 2: Figure 1-8.4 shows a free body diagram for a section a distance, x, between B and C of the beam of Figure 1-8.2. The internal forces are the shear, V; the bending moment, M; and the normal force, N. These forces are calculated from the free body diagram of Figure 1-8.4 as follows:

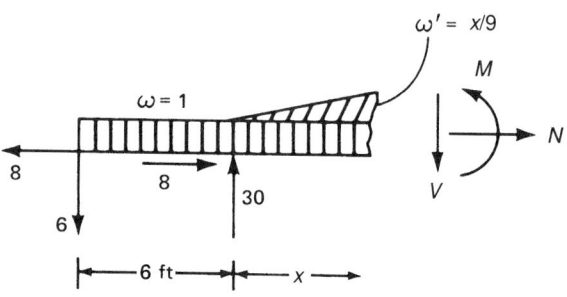

Fig. 1-8.4. Free body diagram.

$$\sum M_{cut} = 0: \quad M + 1(6 + x)(6 + x)/2 + 1/2(x/9)(x)(x/3)$$
$$+ 6(x + 6) - 30(x) = 0$$
$$M = -x^3/54 - x^2/2 + 18x - 54$$

$$\sum F_y = 0: \quad V + 1(x + 6) + 1/2(x/9)(x) + 6 - 30 = 0$$
$$V = -x^2/18 - x + 18$$
$$\Sigma F_x = 0: \quad N = 8$$

The distribution of the internal forces may be plotted on diagrams that show the change in values throughout the length of the beam. Figure 1-8.5 shows the N, V, and M diagrams for the beam of Figure 1-8.2.

FAILURE MODES

Structural design consists of identifying all of the potential failure modes and providing resistance to avoid failure. Both safety and economy are considerations. A major part of the professional engineering services is the skill in

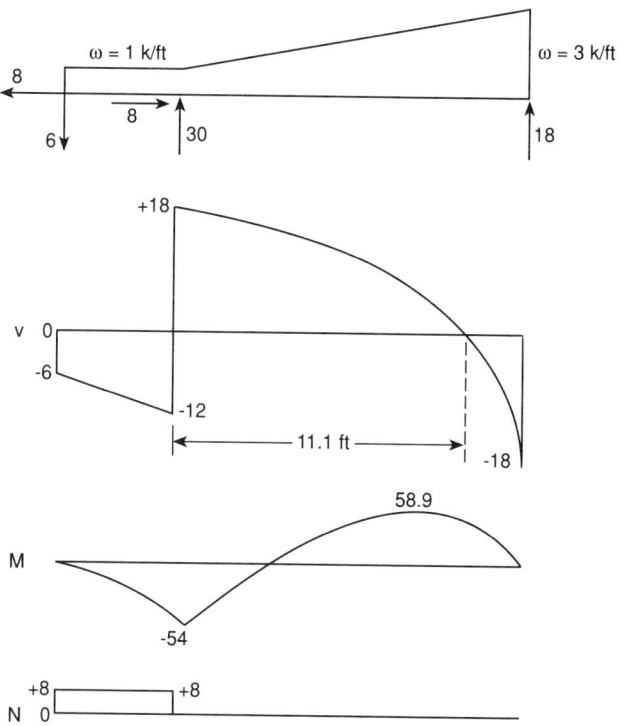

Fig. 1-8.5. Shear, moment and normal force diagram.

Fig. 1-8.6. *Tensile loading of a straight member.*

identifying appropriate loading conditions and the associated failure modes for the construction conditions.

The ways in which members fail depend upon the materials, geometry, loading conditions, and support conditions. This section will describe the common structural forms and the failure modes generally associated with those forms.

Tension Members

Figure 1-8.6 illustrates tensile loading on a straight member. The stress in the member is defined as $\sigma = P/A$. The load must be applied through the centroid of the cross-section for this equation to be valid. When loads are applied eccentrically to the cross-section, a combined bending and axial condition exists. This will be described later.

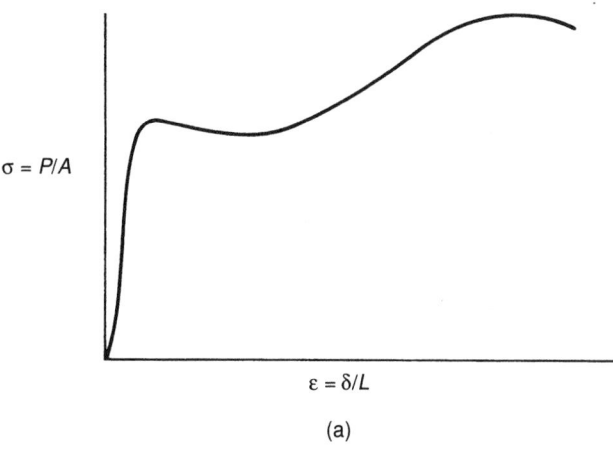

(a)

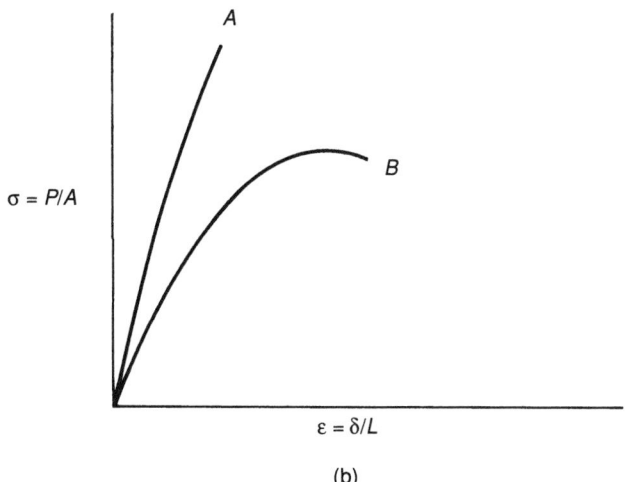

(b)

Fig. 1-8.7. *Stress-strain diagram for (a) ductile and (b) brittle materials.*

Figure 1-8.7, part (a) shows relationship between unit stress, $\sigma = P/A$, and unit strain, $\varepsilon = \delta/L$ for a coupon of mild structural steel. This stress-strain diagram is obtained experimentally and depicts only mild structural steel loaded in tension. Stress-strain diagrams for other materials are also obtained experimentally and show distinctly different load-deformation characteristics.

The stress-strain diagram provides an indication of the expected failure mode of the member. A ductile material, such as the mild steel in Figure 1-8.7, part (a), will elongate significantly under tensile loading. Frequently, the deformations are so great that the structure becomes unusable long before actual rupture. Rupture eventually will occur if loads are increased to the ultimate stress. Brittle materials, such as those shown in Figure 1-8.7, part (b), will fail by sudden rupture. Little or no warning of impending failure may be present with materials of this type.

There are situations in which a normally ductile material will exhibit a brittle type of failure. This occurs under conditions of low temperature or repeated, fatigue loading conditions.

High temperatures, such as those present in building fires, will cause an increase in the elongation of tension members because of creep. Creep is the phenomenon in which a member will continue to deform after the applied load becomes steady. The magnitude of creep depends upon the material being loaded, the level of stress, the temperature, and the time duration.

Other potential failure modes for tension members include connection failures, excessive stress concentrations due to changes in cross-sections, and twisting when unsymmetrical members are excessively long.

Compression Members

Figure 1-8.8 illustrates compressive loading on a member. When the loading is applied along the centroidal axis of the member, the stress may be calculated as $\sigma = P/A$. The importance of centroidal loading is even more critical for compressive forces than for tensile forces because of the magnification effect of eccentricity. This will be discussed more completely later.

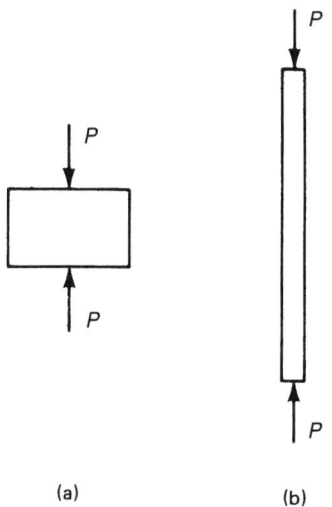

(a)   (b)

Fig. 1-8.8. *Compressive loading of (a) short and (b) long columns.*

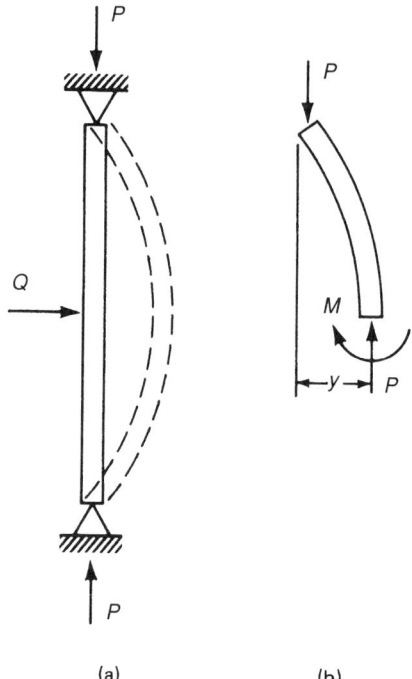

(a)   **(b)**

Fig. 1-8.9. Column buckling.

Compression members, unlike tension members, have no single general failure mode, regardless of their length. Short columns, as illustrated by Figure 1-8.8, part (a), fail by general yielding. Long columns, as illustrated by Figure 1-8.8, part (b), fail by buckling. Buckling is the rapid collapse of a compression member due to instability. To describe the nature of column behavior, the following discussion may be helpful.

Consider the column shown in Figure 1-8.8, part (b). Assume that the axial compressive load P starts at a low value and gradually increases in magnitude. Assume a small lateral force is applied, as shown in Figure 1-8.9, part (a). The bar will deflect laterally by a small amount. When Q is removed, the bar returns to its original position. When a particular value of P is reached, the bar will remain in the deflected position after Q is removed. That load for which the bar is indifferent to its position is defined as the critical buckling load, P_{cr}. If P were increased above P_{cr}, the bar would collapse. If P were decreased below P_{cr}, the bar would return to its straight P position. The critical buckling load is, therefore, the particular load at which neutral equilibrium occurs.

Considering the equilibrium condition when the bar is deformed, we may determine the bending moment from Figure 1-8.9, part (b), as

$$\sum M_{cut} = 0: \quad M = -Py$$

The equation of the elastic curve of a beam[3] is $d^2y/dx^2 = M/EI$. Substituting this for M above, we obtain

$$\frac{d^2y}{dx^2} + \frac{Py}{EI} = 0$$

Letting $\kappa^2 = P/EI$ and solving this differential equation yields $y = A \cos(\kappa x) + B \sin(\kappa x)$.

Using the boundary conditions of $y = 0$ at $x = 0$ and $y = 0$ at $x = L$, we obtain $A = 0$ and $B \sin(\kappa L) = 0$. Since B cannot be zero, $\sin(\kappa L) = 0$. This eigenvalue equation has solutions of

$$\kappa L = 0, \pi, 2\pi, 3\pi, \ldots, n\pi$$

Taking the general solution we obtain

$$\kappa L = n\pi, \qquad P/EI = n\pi/L$$

$$P = \frac{n^2\pi^2 EI}{L^2}$$

The n-term describes the number of modes of buckling. Since the first mode of buckling will cause failure unless special construction features exist, buckling will occur at $P_{cr} = \pi^2 EI/L^2$.

This column equation was originally described in 1757 by Leonhard Euler, a Swiss mathematician. Controversy about its validity for predicting column loads raged for sixty years. In 1820 it was recognized that the derivation incorporated the bending equation, $\sigma = Mc/I$. Consequently, all assumptions of elastic behavior are intrinsic to the use of the Euler column equation. Therefore, the limit of validity is the proportional limit of the material.

Two clearly identifiable compression failure conditions can exist. One is the yielding condition for short columns where $P = \sigma_y A$, as illustrated in Figure 1-8.8, part (a). The second is the buckling of long, slender columns, where $P_{cr} = \pi^2 EI/L^2$. This equation may be converted into one involving axial stress by recognizing that $I = Ar^2$, where r is the radius of gyration of the cross-section. Dividing both sides by A gives

$$\sigma_{cr} = \pi^2 E/(L/r)^2$$

The term L/r is defined as the slenderness ratio. Therefore, column slenderness is a function both of the length and the cross-sectional geometry, as described by the radius of gyration, r.

If the critical stress versus the slenderness ratio were plotted for columns, the graph of Figure 1-8.10 would result. The segment from C to D describes long columns in which the critical buckling load may be calculated from Euler's equation. The failure mode is pure buckling, and the limit of

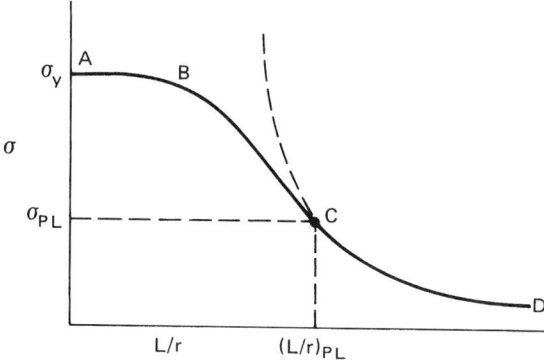

Fig. 1-8.10. Critical stress of columns as a function of the slenderness ratio, L/r.

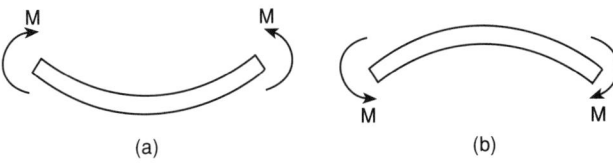

Fig. 1-8.11. Deflection of flexural members under load.

validity of the equation is the proportional limit of the material, σ_{PL}. The segment from A to B identifies short columns, which fail by yielding. The maximum load is $P = \sigma_y A$.

The segment from B to C is described as the intermediate column range where failure may be considered a combination of buckling and yielding.

Considerable controversy and research has been associated with attempts to relate theory and experimental validation in this intermediate range. While the history of these studies is fascinating, the major interest here relates to design equations. The importance of the intermediate column range is that, from a practical viewpoint, most columns have a slenderness ratios within this range. Therefore, the readily derived and theoretically accurate Euler's equation, $P_{cr} = \pi^2 EI/L^2$, is inappropriate for slenderness ratios less than $(L/r)_{PL}$.

From a historical and practical viewpoint, intermediate column formulas have been obtained by curve fitting experimental results. Therefore, one obtains equations that are material dependent, rather than an equation analogous to $\sigma = Mc/I$ that may be valid for a variety of materials. Most column equations have been parabolic or straight line expressions for ease of design calculations. These expressions may be used because a factor of safety is incorporated for design purposes. The material product industries publish equations appropriate for their materials. Therefore, one must be careful to select column equations that are appropriate to the materials and conditions for the construction.

The most prevalent failure mode for columns is due to general buckling, as described above. It may be seen from Figure 1-8.10 that the load carrying capacity is reduced significantly as the slenderness ratio increases. Consequently, a long column will buckle at axial loads considerably lower than those for a shorter column of the same cross-section. In addition to the slenderness ratio, the strength of columns is dependent upon the modulus of elasticity. In fire conditions, the modulus of elasticity is reduced. This reduction causes a loss in strength of columns.

Although general buckling is the most common type of failure, local buckling can occur on platelike elements in compression. This occurs when the plates are too thin for the applied load and premature localized buckling takes place. Because this type of behavior is also related to the modulus of elasticity, fire conditions can cause an earlier localized buckling to members, such as wide-flange steel shapes or angles that are made of thin-plate elements.

Flexural Members

The third type of structural loading is flexural. This occurs when loads are applied perpendicular to the longitudinal axis of the member. These members are described as being in flexure or bending. In structural use they are described as beams, girders, slabs, plates, and rigid frames. Although each of these types of members act in flexure, their behaviors will differ.

Figure 1-8.11 shows flexural members with couples as the applied load. The top fibers of Figure 1-8.11, part (a) are in compression, and the bottom fibers are in tension. This is defined as positive bending. The opposite occurs in Figure 1-8.11, part (b), and this condition is described as negative bending.

Figure 1-8.3 showed a beam supporting transverse loads. The reactions of the beam were calculated in Example 1. The internal shear, moment, and axial force were computed for a general distance x in Example 2. Diagrams that describe the change in vertical shear, V, and the change in internal moment, M, are constructed to show the distribution of these changes throughout the beam. These are called "shear and moment diagrams." Every textbook on mechanics of materials and most texts on statics cover procedures for constructing V and M diagrams for beams. From these shear and moment diagrams the design values for those parameters are selected.

The relationship between the fiber stresses in the member and the internal resisting moment can be obtained in the following manner. Consider a homogeneous beam in pure bending as shown in Figure 1-8.12, part (a). Two lines, parallel before bending, would assume the position shown after the couples are applied. It is assumed that plane sections before bending remain plane after bending. Figure 1-8.12, part (b) shows the strain distribution of the fibers throughout the cross-section. The top fibers have shortened, and the bottom fibers have elongated. One layer of fibers has not changed in length. This plane is called the neutral plane of the member.

Hooke's law states that stress is proportional to strain. When the proportionality is linear, the stress distribution is as shown in Figure 1-8.12, part (c). The maximum stress, σ, will occur at the fibers located farthest from the neutral axis. The stress at the neutral axis is zero.

If we consider the stress, σ_1, in a single fiber of area, dA, at a distance, y, above the neutral axis, the force exerted by that fiber due to the stress is $dP = \sigma' dA$. This stress may be related to the stress, σ, in the extreme fibers, by the similar triangles of Figure 1-8.12, part (c).

$$\frac{\sigma'}{y} = \frac{\sigma'}{c}, \quad \text{or} \quad \sigma' = \sigma(y/c)$$

The force, dP, exerted by this fiber may be expressed as

$$dP = \sigma' dA = \sigma(y/c)dA$$

The moment of this force about the neutral axis is

$$dM = dPy = \sigma(y/c)dA(y)$$

$$dM = \frac{\sigma}{c}y^2 dA$$

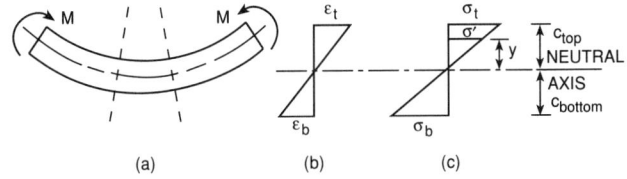

Fig. 1-8.12. Homogeneous beam in (a) pure bending, and the resulting (b) strain distribution, and (c) stress distribution.

Summing the moments of each of the fibers of the member yields

$$\int_0^M dM = \int_{-c}^{+c} \frac{\sigma}{c} y^2 \, dA = \frac{\sigma}{c} \int_{-c}^{+c} y^2 \, dA$$

The moment of inertia, I, of the cross-section is defined as

$$I = \int_{-c}^{+c} y^2 \, dA$$

The flexure formula, therefore, may be expressed as

$$\sigma = \frac{Mc}{I} \qquad (1)$$

where

σ = flexural stress at the extreme fibers;
M = bending moment at the section of the beam being considered;
c = distance from the neutral axis to the extreme fibers; and
I = moment of inertia of the cross-section.

Equation 1 has several limitations that have been incorporated into the assumptions of its derivation. These include: (1) the beam is initially straight and of constant cross-section; (2) all stresses are below the proportional limit, and Hooke's law applies; (3) the modulus of elasticity in compression is equal to that in tension; (4) loads are applied through the shear center so that torsion will not occur; and (5) the compression fibers are laterally restrained.

The design of flexural members for bending loads involves: (1) determining the dead and live loading for the member; (2) calculating the maximum moment in the beam; (3) selecting the materials and obtaining the allowable stresses; (4) calculating a required section modulus; and (5) selecting a beam to provide for that section modulus efficiently and economically; and (6) ensuring that all other failure modes will not occur.

Because many beams have common loading and support conditions, it is possible to develop standard conditions to obtain the maximum shear and moment by formula. Figures 1-8.13 and 1-8.14 illustrate two common conditions. Most handbooks and mechanics of materials textbooks provide several additional cases. The maximum moment from Figure 1-8.13 is $M = \frac{1}{8}\omega L^2$, and that from Figure 1-8.14 is $M = Pab/L$. Loading conditions may be combined by superpo-

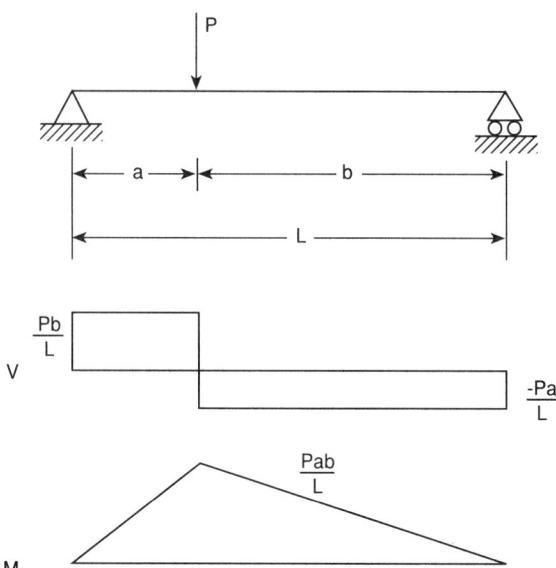

Fig. 1-8.14. *Shear and moment diagram for a simply supported beam under concentrated loading.*

sition. However, it is important to conform with the conditions where superposition is valid. For example, Figure 1-8.15 shows a beam with a uniformly distributed load, ω, and a concentrated load, P, at the center. Because the maximum moment of each load occurs at the same location, it is possible to compute the maximum moment as $M = \omega L^2/8 + PL/4$. However, if the concentrated load were at another location, as in Figure 1-8.16, superposition would not be

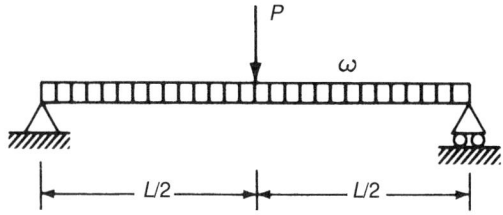

Fig. 1-8.15. *Uniformly distributed load with a concentrated load at the center.*

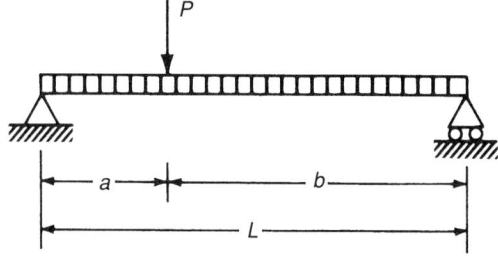

Fig. 1-8.16. *Nonsymmetric loading where superposition is not valid.*

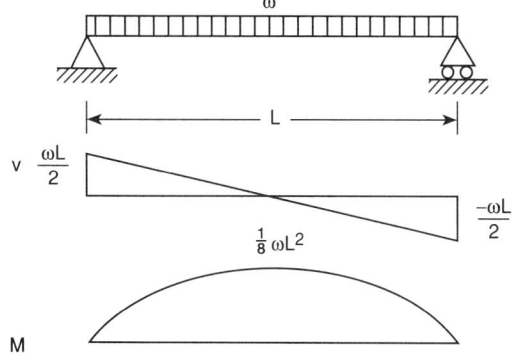

Fig. 1-8.13. *Shear and moment diagram for a simply supported beam under uniform loading.*

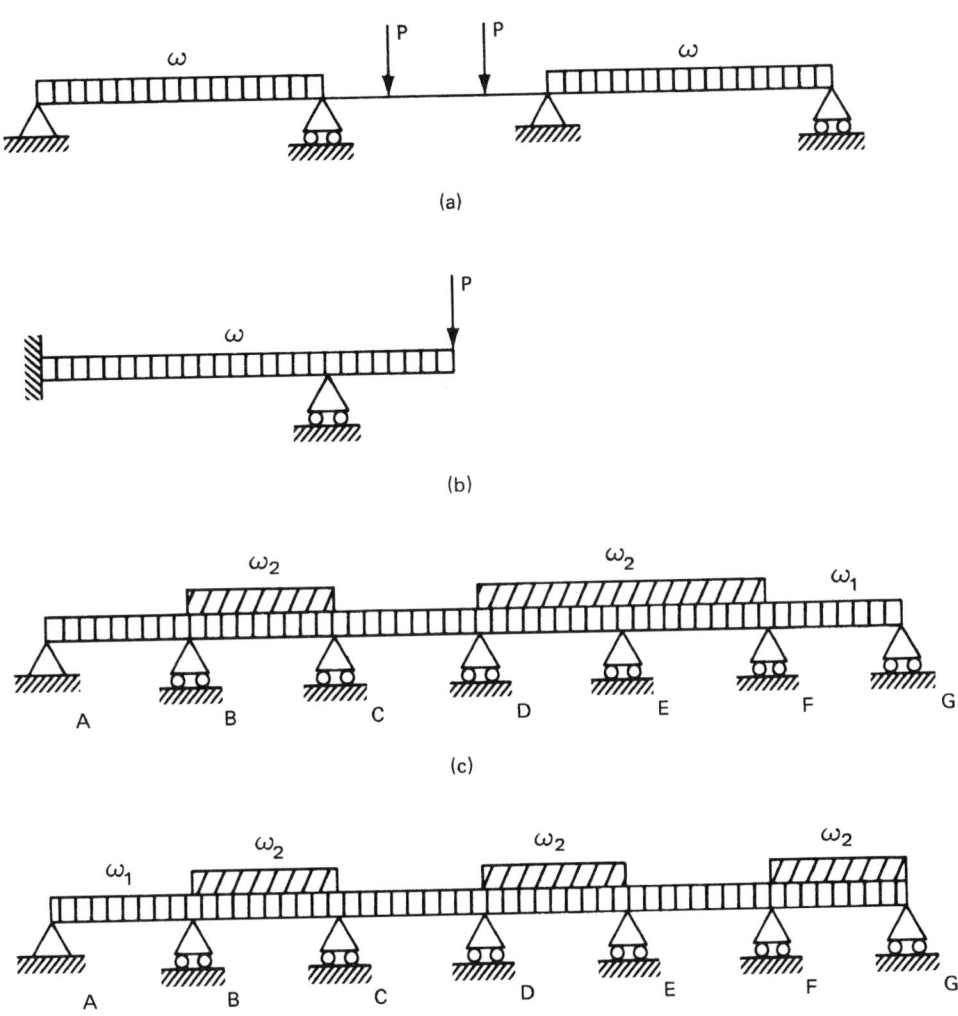

Fig. 1-8.17. Statically indeterminate beams.

valid. In those cases, the engineer must compute the maximum V and M by using the basic principles of statics. Example 2 illustrated that technique.

Statically Indeterminate Beams

It is common to construct beams with more reactions than are necessary for statical stability alone. These members are statically indeterminate because the calculation of reactions requires means in addition to statics alone. Figure 1-8.17 illustrates some statically indeterminate beams, and Figure 1-8.18 shows some statically indeterminate frames.

The procedure for designing statically indeterminate beams is similar to that described for statically determinate beams. An increased complication arises, however, when determining the maximum shear and moment for a beam such as that shown in Figure 1-8.17, part (c). The dead load is applied over the entire span and is fixed. The live load is movable, and may be applied to any or all spans. An integral part of the computation of the design shear and moment is to place the movable live load at positions that produce the most severe values. This may be done by constructing influence lines for the design functions. An influence line is a graph of the function as a unit load moves across the struc-

ture. The influence line shows where loads must be placed to produce the most severe conditions. To illustrate this concept, the loading condition shown in Figure 1-8.17, part (c) would be used to determine the maximum negative moment over support E, while the loading condition shown in Figure 1-8.17, part (d) would be used to determine the maximum positive moment at the midpoint of span DE.

Statically indeterminate structures are inherently stronger than statically determinate structures. This occurs because of the additional load-carrying capacity due to the redistribution of moments. The amount of this increased load capacity depends upon the type and location of load, the support conditions, the material properties, and the geometry and dimensions of the cross-section.

To illustrate this concept, consider the simply supported beam of Figure 1-8.19. The maximum moment is $M = \frac{1}{8}\omega L^2$. The stress may be computed as $\sigma = Mc/I$ as long as the fibers are stressed below the proportional limit. Figure 1-8.20, part (a) shows a wide flange cross-section. Figure 1-8.20, part (b) shows a stress variation that is valid up to the value where the extreme fibers reach σ_y. The moment that causes that stress is M_y, the bending moment that will just cause yielding to be imminent at the extreme fibers. That value is the limit of validity for the flexure formula, $\sigma = Mc/I$.

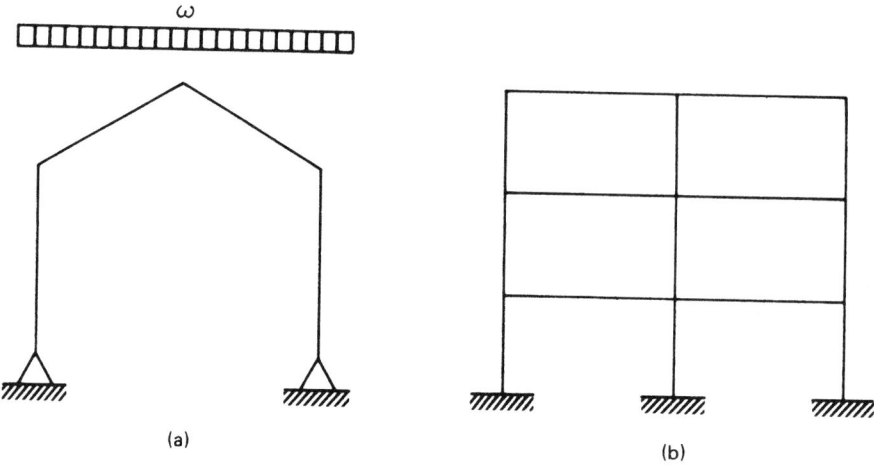

Fig. 1-8.18. Statically indeterminate frames.

The beam, however, has an increased load-bearing capacity beyond that value. Excessive deformation (i.e., collapse) will not occur until the entire cross-section has yielded. The stress distribution for this condition is shown in Figure 1-8.20, part (c). The moment capacity at that point is called the fully plastic moment, M_p. The increase in moment is dependent upon the geometry of the cross-section. The ratio of M_p/M_y is the shape factor. For steel wide flange beams, the shape factor averages 1.14. The shape factor will be different for other geometrical shapes and dimensions.

If the simply supported beam of Figure 1-8.19 were a steel wide flange shape, we would expect the collapse load to be 14 percent higher than the yield load. The design load usually has a factor of safety of 1.5 over the yield load. Therefore, the factor of safety for collapse above the design value is $1.50 \times 1.14 = 1.71$ for normal design conditions.

If the support conditions are fixed, as shown in Figure 1-8.21, the maximum moment occurs at the support. For elastic conditions, the moment at the support is $M = (1/12)\omega L^2$, and the moment at the center is $M = (1/24)\omega L^2$. As the load is increased to the point where σ_y is first reached (at the ends), the value of $M_y = (1/12)\omega_y L^2$.

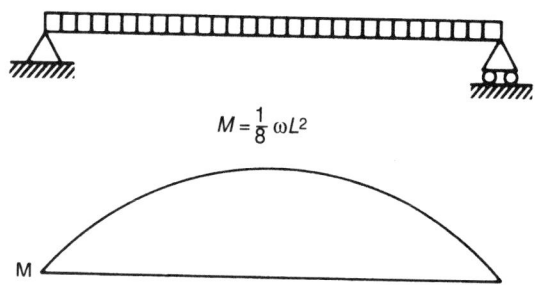

$$M = \frac{1}{8}\omega L^2$$

Fig. 1-8.19. Uniformly loaded simply supported beam.

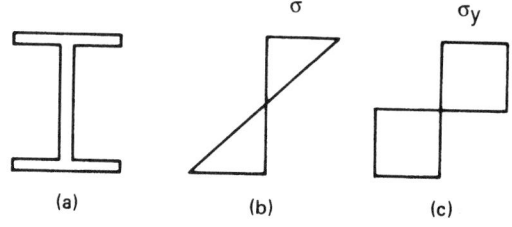

Fig. 1-8.20. (a) Wide flange cross-section and stress distributions, (b) prior to yielding of extreme fibers, and (c) after complete yielding.

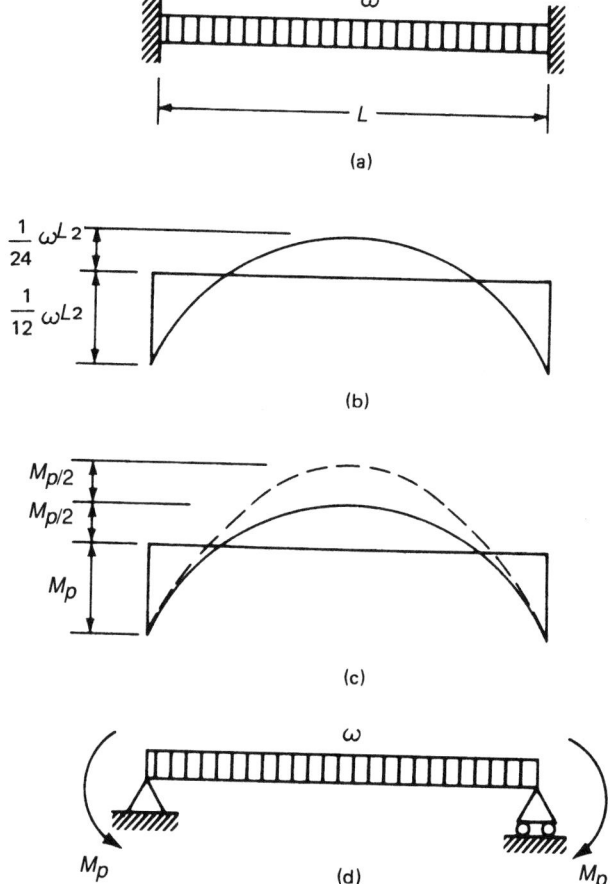

Fig. 1-8.21. Moment redistribution in a fixed-ended beam.

As the load continues to increase, the location of greatest stress (the ends) will reach their fully plastic value, M_p. However, the beam still has additional carrying capacity because a collapse mechanism will not occur until three hinges form. At the time M_p occurs at the ends, the other location of maximum moment, the center, is still in the elastic range. The value of M_p at the ends cannot increase. Therefore, any increase in load must be carried by the elastic portion. The moments redistribute, as illustrated by the dashed line of Figure 1-8.21, part (c). They will increase until M_p occurs at the center. At that time, collapse is imminent.

The collapse moment for the beam of Figure 1-8.21 is $2M_p = \frac{1}{8}\omega_u L^2$; therefore $\omega_u = 16\ M_p/L^2$. The collapse moment for the simply supported beam of Figure 1-8.19 is $M_p = \frac{1}{8}\omega_u L^2$; therefore $\omega_u = 8\ M_p/L^2$. Therefore, the ultimate load-carrying capacity of the beam with fixed ends is twice that of a beam with simply supported ends.

This concept is sometimes described as limit state design, ultimate design, inelastic design, or plastic design. Limit state design seems more appropriate to a variety of materials.

The concept of ductility and its behavior is intrinsic to safe structural design because a ductile structure will deform considerably before collapse. This deformation warns occupants of impending danger before failure. Brittle design and elastic instability are not as desirable because failure can occur with relatively little warning. Therefore, structural engineers attempt to incorporate ductility into their designs as much as possible. This ductility is evident in most structural building materials.

Flexural Failure Modes

Depending upon the magnitude, type of loading, and support conditions, flexural members may exhibit different types of failure modes. The most evident type of failure is the overstress that contributes to the development of a plastic hinge. This was described in the previous section. A statically determinate structure will collapse when the first plastic hinge forms. A statically indeterminate structure requires two or three hinges to form before collapse. The support conditions determine the number of hinges needed for collapse.

Another common mode of failure is lateral instability. The compression flange of the beam must be supported laterally at sufficient intervals to prevent lateral buckling. Lateral buckling is similar to column buckling, and it can occur when supports are spaced too far apart. When lateral supports are spaced farther apart than the distance needed to avoid lateral buckling, the allowable stress is reduced to compensate for the reduction in local carrying capacity.

A third mode of beam failure is through torsional loading. An open cross-section is particularly weak when subjected to torsional loads. A torsional load exists whenever the line of action of the applied loads does not intersect the shear center of a beam. The shear center is a particular location on the cross-section. For symmetrical members, such as that of Figure 1-8.22, part (a), the shear center coincides with the centroid. For unsymmetrical members, the shear center may be calculated. Figure 1-8.22, parts (b) and (c) illustrate the location of the shear center for this type of cross-section. Whenever loads do not pass through the shear center, construction features must be introduced to counterbalance the rotational effect of the loads.

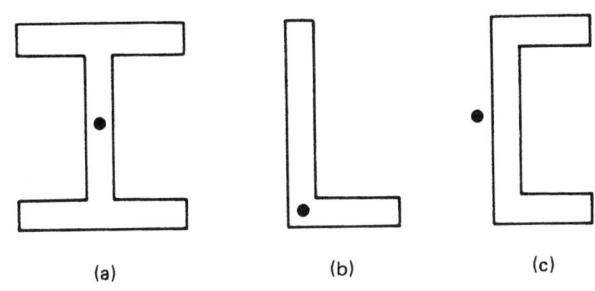

Fig. 1-8.22. *Location of shear center for (a) symmetrical and (b,c) unsymmetrical sections.*

Another mode of failure is excessive deflection. This can occur when the stiffness is insufficient, and can occur at relatively low stresses. In addition to excessive deflection, unwanted vibrations or sway, such as wind loads on tall buildings, may occur.

It is not uncommon for members to be loaded with both axial and flexural forces. Rigid frames and chord members of trusses, where the load is applied directly to a chord, are common examples of this condition. Depending upon the construction conditions, failure may occur due to premature formation of plastic hinges or buckling of the compression flange.

STRUCTURAL DESIGN FOR FIRE CONDITIONS

The theory and procedures for structural analysis and design at normal temperatures have been well studied. Understanding of theoretical and empirical relationships, and the ability to predict performance for practical applications and conditions, is relatively clear to most practitioners. However, the understanding of the structural behavior for fire conditions becomes theoretically and practically a far more complex problem.

Structural analysis and design for fire conditions can take one of several forms. The simplest application can be described by the procedures used in most conventional building codes. In this case, a representative sample is tested in a standard fire endurance test, such as ISO 834 or ASTM E119, *Standard Test Methods for Fire Tests of Building Construction and Materials.* The length of time in the laboratory test before failure occurs produces a fire endurance time. Building codes specify the fire endurance required for structural members and barriers in identified occupancies and classifications of construction. The engineer or architect need only incorporate the standard construction features identified by the test results and published documentation to satisfy the code requirements.

This procedure requires no knowledge of fire or structural engineering by the practitioner. A catalog of construction assemblies and their fire endurance ratings is the only data needed to satisfy code compliance. Unfortunately, the actual structural performance in fires is not known or investigated. The sophisticated knowledge of the interrelationships that lead to an understanding of structural performance and economical design is submerged in the process. The increased structural strength achieved by continuity in construction is not a consideration in this procedure. The rated fire endurance of a beam and its value from the building code viewpoint is the same whether it is constructed as a statically determinate structure or as continuous construction, even though the performance may vary quite significantly.

Another procedure involves the calculation of structural fire endurance based on the standard ASTM E119 or ISO 834 fire test. Empirical and theoretical relationships are used to predict the fire endurance, based on the standard fire time-temperature relationship. Two advantages of this procedure are: (1) it allows the building codes to retain their present form and (2) leaves undisturbed the interrelationship between construction classifications and other fire defense measures. Also, it provides more flexibility: fire endurance of different types of assemblies can be obtained by calculation rather than by test. However, the same limitations present in the traditional test procedure and code format remain.

A third procedure can be described as a "rational" approach to structural design for fire conditions. This approach is exemplified by the procedures for (1) structural steel design and (2) reinforced concrete design developed in Sweden. In these procedures, the design incorporates the structural performance at elevated temperatures in a manner analogous to design at normal temperatures. The mechanical properties of the structural materials at elevated temperatures are incorporated into the traditional structural theory to develop a rational analytical procedure for predicting structural behavior. Further, the natural room fire temperature-time relationship is used instead of the standard test time-temperature relationship, and the thermal properties and heat transfer through the insulating materials are incorporated into the analysis. The procedures follow more closely the traditional structural engineering methods for predicting structural behavior.

SUMMARY

The ability of structural members to withstand failure of excessive deflection, insufficient strength, and instability is a major requirement of any structure. While the analysis and design process for normal loads and conditions is not particularly difficult, it does require care in application. The care relates to the type and validity of the assumptions made and to the form of construction used.

The anatomy of the entire structural system is an important aspect of the analysis and design process. Unless care is taken to specify clearly the construction details, inappropriate design calculations can result. To the lay person, one form of construction often appears to be the same as another. To the student who often has insufficient opportunity and training to recognize the construction details, analysis and design may appear to be an academic exercise. Normally, much of the ability to recognize the essential details is obtained through engineering practice with a professional engineer.

Fires in buildings create an added dimension of complexity to the analysis of the behavior of structural members at elevated temperatures. The fire design of structural members must include the same attention to details as the design of members at normal temperatures.

NOMENCLATURE

A = area
c = distance from neutral axis to extreme fibers
E = modulus of elasticity
F = force
I = moment of inertia
L = length of beam or member
M = moment
P = concentrated or point load
r = radius of gyration
V = shear
x = space coordinate along the beam
y = space coordinate normal to the beam
ϵ = strain
δ = deformation
σ = stress
ω = uniform load density

REFERENCES CITED

1. *Manual of Steel Construction*, American Institute of Steel Construction, Chicago (1980).
2. *ACI 318-83*, American Concrete Institute, Detroit (1983).
3. R.W. Fitzgerald, *Mechanics of Materials*, Addison-Wesley, Reading (1982).

ADDITIONAL READING

ASTM E119, *Standard Test Methods for Fire Tests of Building Construction and Materials*, American Society for Testing and Materials, Philadelphia, PA

ISO 834, International Organization for Standardization, Geneva, Switzerland

PREMIXED BURNING

Robert F. Simmons

INTRODUCTION[1]

When a mixture of fuel vapor and oxidant burns on a cylindrical tube, as with a Bunsen burner, the resulting premixed flame has the characteristic structure of a luminous inner cone and an outer sheath of hot combustion gases.[1] This inner cone depicts the end of the primary reaction zone of the flame, in which a fast oxidation reaction occurs, so that in this part of the flame the temperature rises very rapidly. When the initial mixture of fuel and oxidant is fuel rich, i.e., there is a deficiency of oxidant in terms of the stoichiometric conversion of the fuel to its oxidation products, the outer sheath is essentially a diffusion flame in which the hot combustion products from the primary reaction zone burn in the surrounding atmosphere. In contrast, with lean flames this outer sheath is indistinct as there is sufficient oxidant in the initial mixture for the complete combustion of the fuel, and the surrounding atmosphere is entrained into the burnt gases.

A stable premixed flame can only be obtained over a limited range of mixture compositions and flow rates, and the flow conditions for a given initial mixture can be seen from a consideration of an idealized flat flame, as shown in Figure 1-9.1. For such a flame, the flow of the initial combustible mixture is normal to the flame front. If the flow is too fast the flame is "blown off," while if it is too slow the flame "flashes back." A stable flame is only obtained when the flow velocity of the incident mixture is just equal to the *burning velocity* of the mixture. This fundamental parameter is defined as the velocity with which a premixed flame moves normal to its surface through the adjacent unburnt mixture. While a flat flame can only be stabilized over a narrow range of flow velocities, a conical flame can be established over a much wider range of flows, as the area of the inner cone can change to maintain the balance between the burning velocity and the flow velocity normal to the flame front. A typical flow pattern and temperature distribution in such a flame is shown in Figure 1-9.2.[2]

Dr. Robert Simmons has a Ph.D. in physical chemistry and his major research interests have centered around flame propagation, the chemistry of combustion reactions, and industrial safety. Until his retirement, he was a senior lecturer in chemistry at the University of Manchester Institute of Science and Technology, and he served as Deputy Editor of *Combustion and Flame*.

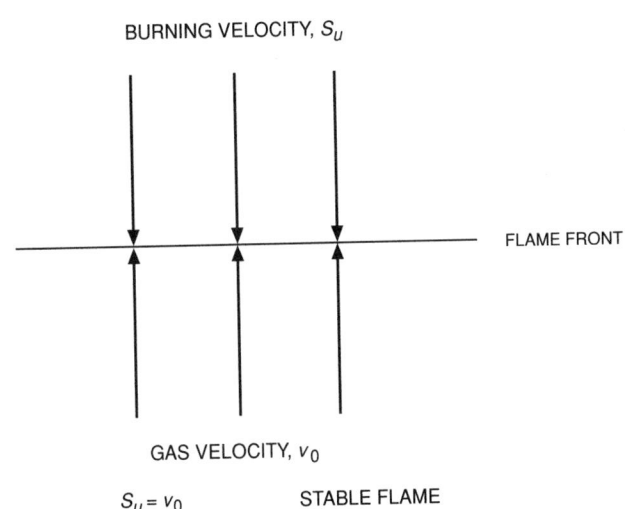

BURNING VELOCITY, S_u

FLAME FRONT

GAS VELOCITY, v_0

$S_u = v_0$	STABLE FLAME
$S_u < v_0$	FLAME "BLOWN OFF"
$S_u > v_0$	FLAME "FLASHES BACK"

Fig. 1-9.1. Diagrammatic representation of a flat premixed flame.

The maximum temperature in the flame is usually reached a little downstream of the inner cone. If the flame is sufficiently large, the adiabatic flame temperature is reached in the middle part of the gas stream; thus, with the temperature distribution given in Figure 1-9.2, the calculated adiabatic flame temperature is nearly reached just above the inner cone of the flame. Toward the outside of the flame, however, heat losses lead to a steep temperature gradient between the combustion products and surroundings. This *adiabatic flame temperature* is given by the balance between the heat released in the combustion reaction (ΔH_r^θ) at the initial temperature of the reactants (T_i) and the heat required to raise the temperature of the products to the final flame temperature (T_f), i.e.,

$$\Delta H_r^\theta = \int_{T_i}^{T_f} C_p \, dT$$

where C_p is the heat capacity of the combustion products. This calculation assumes that heat losses from the flame by

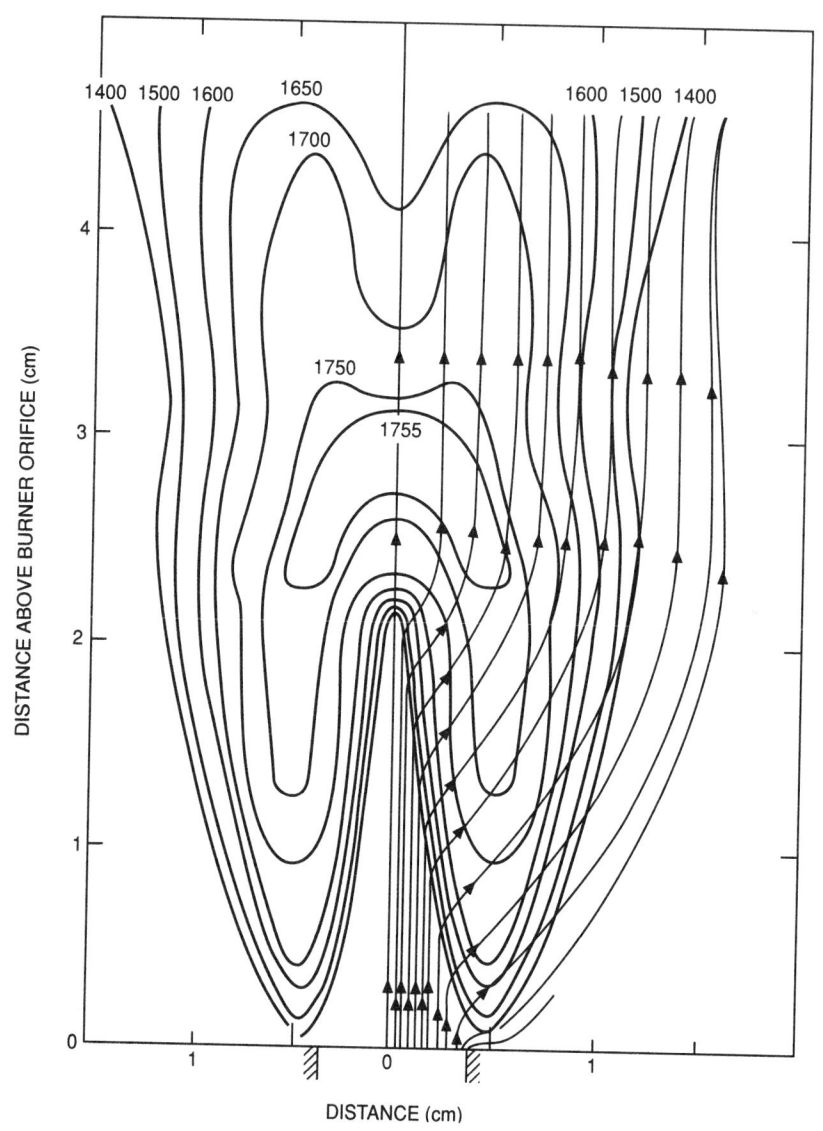

Fig. 1-9.2. Temperature distribution and flow pattern in a premixed flame (7.5 percent natural gas and air burning on a rectangular burner 0.755 × 2.19 cm). The quoted temperatures are °C.[2]

radiation, thermal conduction, or diffusion to a wall can be neglected. Thus, small premixed flames and turbulent flames of all kinds normally fail to reach their adiabatic flame temperature as the heat loss from such flames is appreciable.

Calculation of the adiabatic flame temperature always assumes chemical equilibrium has been reached in the burnt gas. For lean flames with a relatively low adiabatic flame temperature, the calculation is relatively straightforward, in that the combustion products are given by the simple stoichiometry for the overall combustion process, but for temperatures above 1800 K allowance must be made for the heat used up in the dissociation of carbon dioxide, steam, oxygen, etc. However, the composition of the products can only be calculated if the temperature is known, and the temperature depends on this composition. As a result, a method of successive approximations must be employed in the calculation, and this can most conveniently be done using a computer program, such as NASA SP-273.[3] Although the adiabatic flame temperature is only

reached in a restricted region of a large premixed flame, it is a useful combustion parameter and especially useful in the calculation of limits of flammability. (See Section 1, Chapter 5.)

All experimental determinations of burning velocity involve measuring the area of the flame front for a particular flow of unburnt mixture. Much of the discrepancy between different determinations can be ascribed to the method used to specify the position of the flame front; further, when burner flames have been used there is also the complication of quenching near the burner rim and the increase in burning velocity near the tip of the cone (because the heat flow in this region is strongly convergent). For example, the maximum values reported for the laminar burning velocity of propane-air mixtures mainly lie between 37 and 45 cm s^{-1}, but the majority of the values lie in the range of 41 ± 2 cm s^{-1}. All the saturated hydrocarbons have about the same maximum burning velocity, and Table 1-9.1 lists the maximum values for some other fuel-oxidant combinations.

TABLE 1-9.1 *Maximum Burning Velocities for Laminar Fuel-Oxidant Mixtures*

Mixture	Maximum Burning Velocity S_u/cm s^{-1}	Reference
Propane-air	41	1
Ethene-air	68	1
Acetylene-air	175	1
Hydrogen-air	320	4
Propane-oxygen	360	5
Acetylene-oxygen	1120	4
Hydrogen-oxygen	1180	4

The values in Table 1-9.1 refer to initial conditions of room temperature and atmospheric pressure. In general, hotter flames have higher burning velocities, and thus increasing the initial temperature of the mixture increases the burning velocity. For example, when the initial temperature of propane-air mixtures is increased from 300 to 480 K, the maximum burning velocity doubles.[6] The effect of pressure on burning velocity is simple and is frequently expressed as a simple power law

$$S_{u,a}/S_{u,b} = (p_a/p_b)^n \qquad (1)$$

where $S_{u,a}$ and $S_{u,b}$ are the burning velocities with respect to the unburnt gas at pressures p_a and p_b, respectively, and n is a constant for the flame. Values of n have been reported for a number of flames ranging from 0.25 (for hot flames with oxygen as the oxidant) to -0.33 (for cooler flames supported by air).[7]

Another factor that affects the burning velocity is the degree of turbulence in the flame. In laminar flames the flow lines in any given volume are parallel, but for turbulent flow the velocities have components normal to the average flow direction. The state of flow is usually characterized in terms of the Reynolds number (R_e) which is the dimensionless quantity

$$R_e = vd\rho/\eta \qquad (2)$$

where v is the average gas velocity, d the diameter of the tube, ρ the density of the gas stream, and η its viscosity. For $R_e < 2300$ the flow is always laminar, and for $R_e > 3200$ it is usually turbulent.[8] In the intermediate region the flow alternates between laminar and turbulent flow, the periods of each depending on whether R_e is nearer the lower or higher value.

When the flow is laminar the flame front is sharply defined, but as the Reynolds number of the flow increases the flame front becomes progressively more and more blurred, so that the whole volume in which the primary reaction occurs has the appearance of a "brush." This arises because of the fluctuations in the local gas velocity. At points where the velocity is high the flame front moves away from the burner, while in regions of low velocity it moves toward the burner. Thus the net effect of turbulence is to increase the effective area of the flame front, with a resulting increase in burning velocity. For example, with propane-air, ethene-air, and acetylene-air flames the burning velocity approximately doubles as the Reynolds number is increased to 40,000.[9] A theoretical treatment of turbulent combustion suggests that the burning velocity can increase by a factor of five when the degree of turbulence in the flow is very high.[10]

The above discussion has been concerned with stationary flames burning on a burner, but if the local flow velocity in a tube or duct is too low to sustain a stationary flame, the flame propagates through the incident mixture, provided it is flammable, i.e., its composition lies within the *limits of flammability*. These are the limits of composition over which a self-sustaining flame can propagate and, as such, they are important parameters in any consideration of the fire and explosion risk associated with a particular fuel-oxidant system. They are normally measured for upward propagation of the flame (since this gives the widest limits) in a tube sufficiently wide to minimize quenching effects and sufficiently long to ensure that it is the self-propagation of the flame that is being studied, i.e., the measured limits are independent of the energy input from the ignition source. The dimensions of the tube are typically 100 cm long × 5 cm inside diameter (ID); this is quite satisfactory for hydrocarbon fuels, but a larger diameter is necessary for fuels, such as ethyl chloride, that have a large quenching distance.[11]

It should be noted that if burning occurs in a closed vessel the resulting temperature rise produces a corresponding rise in pressure. For example, a stoichiometric hydrocarbon-air mixture at an initial temperature of 300 K has an adiabatic flame temperature of about 2200 K, so that the pressure can reach a little over 7 bar under adiabatic conditions. In practice, the maximum pressure is likely to be somewhat lower, because of heat loss to the walls of the vessel, but unless the vessel has been designed to withstand such pressures it will rupture explosively. The situation is further complicated if connected vessels are involved, as burning in one vessel leads to an increased pressure in the connected vessel; if the flame then propagates into this second vessel, a correspondingly higher pressure is produced.

These limits of flammability widen as the temperature increases, but at sufficiently low temperatures the flammable range is limited by the vapor pressure of the liquid fuel, as shown in Figure 1-9.3. It follows that the *flash point* is the temperature at which the vapor pressure is just sufficient to

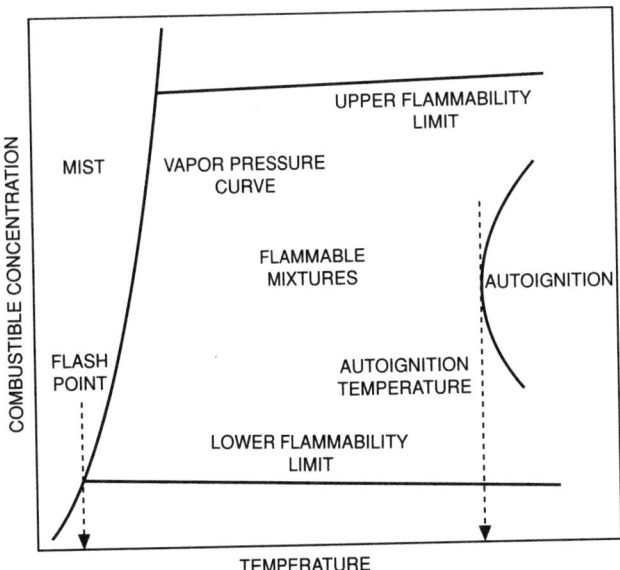

Fig. 1-9.3. Effect of temperature on the limits of flammability of a combustible vapor in air at a constant initial pressure.[12]

give a lean limit mixture of the fuel vapor in air. However, a somewhat higher temperature (the *fire point*) is needed before the fuel is ignited by the burning gas above the surface of the material. At sufficiently high temperatures autoignition occurs and the minimum temperature at which this can happen is termed the *autoignition temperature*. (See Section 2, Chapter 10.) Although the flame in a real fire is essentially a diffusion flame (addressed later in this chapter), the initial ignition of the combustible material, whether liquid or solid, involves the ignition of a mixture of fuel vapor and air in the boundary layer above its surface.

If the flame tries to propagate through a gap that is too small it is quenched. These limiting distances are usually measured using spark ignition at the center of a pair of parallel plates[13] or a rectangular burner,[14] and the *quenching distance* is the maximum distance that will just prevent the propagation of a flame through any mixture of the fuel-oxidant mixture. The quenching is probably due to a combination of heat loss to the walls and the removal of free radicals which are important for the propagation of the flame. These quenching effects are utilized in flame traps, but here it is also necessary that any hot gas forced through the trap must be sufficiently cooled so that it does not ignite any flammable mixture present on the other side.

A flame propagating along a duct away from an opening usually proceeds at first at a fairly uniform speed which is controlled by the burning velocity of the mixture and the area of the flame front. This linear velocity, V, is related to the burning velocity through the relation

$$V = S_u A_f/A_d \tag{3}$$

where A_f and A_d are the areas of the flame front and the cross section of the duct, respectively. Since A_f is always greater than A_d (typically by a factor of two or three), it follows that the linear velocity of the flame is correspondingly larger than the burning velocity. This distinction is important when the design of automatic protection is considered.

If the hot combustion products cannot vent to maintain an approximately constant pressure in the system, they force the flame and the unburnt gases forward with increasing velocity. In turn, this induces increased turbulence ahead of the flame front, with a consequential further increase in burning velocity. If the duct is sufficiently long and the resulting acceleration is sufficiently rapid, the flame front acts as an accelerating piston and a shock wave is formed ahead of the flame front. Under such conditions the flame becomes a detonation propagating at supersonic velocity, typically between 1500 and 3000 m s^{-1}.[15] For gases initially at atmospheric pressure, the pressure immediately behind the detonation can be up to 20 bar, and up to 100 bar, if it is reflected from the end of the duct. As a result, detonations are much more destructive than a propagating premixed flame.

MECHANISM OF FLAME PROPAGATION

The above discussion has been concerned with phenomena associated with the propagation of the flame and, while these are important from the practical viewpoint, they give no insight into how the flame propagates from one layer of gaseous mixture to the next. This comes from using the continuity equations for flame propagation in a laminar flow[16] and, for convenience and simplicity, the following discussion is based on the equations for a flat flame.

First, there must be *conservation of mass* through the flame, so that

$$\rho v A = \rho_0 v_0 A_0 = M \tag{4}$$

where ρ is the density, v the gas velocity, A the cross-sectional area of the gas flow, and M the mass burning rate (mass per unit time).

Conservation of energy requires that the heat conducted into a gaseous element of the flame plus the heat liberated by chemical reaction within the element is used up in raising its temperature, i.e.,

$$d(k \cdot dT/dz)/dz + Q \cdot R - d(c_p \cdot T \cdot \rho \cdot v)/dz = 0 \tag{5}$$

where k is the thermal conductivity of the mixture, c_p its heat capacity, T the temperature at distance z, Q the heat of reaction and R its rate.

There must also be *conservation of the individual atomic species* through the flame; i.e., for a given chemical species i there must be a balance between its rate of production (or removal) in a given element of the flame and its transport by diffusion and convection. Thus,

$$R_i + d(D_i \cdot dn_i/dz)/dz - d(n_i \cdot v)/dz = 0 \tag{6}$$

where D_i is its diffusion coefficient, and n_i its concentration. Equation 6 leads to the following expression for the rate of reaction of species i

$$R_i = (\rho_0 v_0/M_i)(dG_i/dz) \tag{7}$$

where M_i is the molecular weight of species i, V_i its diffusion velocity, and G_i its mass flux fraction. The latter is given by

$$G_i = M_i(v + V_i)/v$$

where the diffusion velocity, V_i is given by

$$V_i = -D_i/X_i(dX_i/dz) \tag{8}$$

In principle, Equations 4 through 6 can be solved to give the burning velocity (v_0), plus the composition and temperature profiles through the flame, but it will be obvious that a detailed reaction mechanism is needed before this can be done. Dixon-Lewis has used the established mechanism for the hydrogen-oxygen reaction to do this for hydrogen-air flames,[17] and similar calculations have been carried out for other hydrocarbon-air flames, such as those presented by Warnatz[18]; such numerical computations for the structure of one-dimensional flames have now become quite commonplace.[19]

The present detailed understanding of the important chemical processes occurring in a premixed flame has come from an analysis of the experimental temperature and concentration profiles through a flat flame; some typical results for a lean propane-oxygen-argon flame are given in Figure 1-9.4.[20] Such analyses show that the flame can be divided up into a number of distinct regions, as shown at the top of Figure 1-9.4. In the initial *pre-heating zone* the temperature rise is that expected from conduction of heat back from the hotter parts of the flame, and chemical reaction does not start until the temperature has reached about 700 K. There is some depletion of fuel at lower temperatures than this, but it is the result of its forward diffusion to a higher temperature

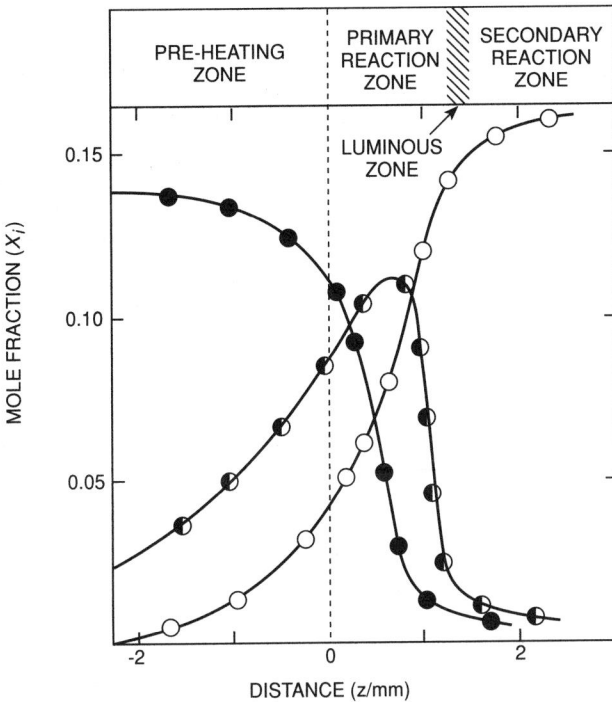

Fig. 1-9.4. *Typical composition profiles through a flat propane-oxygen-argon flame (1.38 percent propane; O_2:Ar = 15:85). Note: $\odot$ 4 × CO_2; ● 10 × C_3H_8; ◐ 10 × CO.*[20]

of the flame and not chemical reaction. Similarly, there is back-diffusion of carbon monoxide, carbon dioxide, and water vapor into this region of the flame.

The reaction in the *primary reaction zone* is induced by the diffusion of free radical species, X, back from the hotter parts of the flame. These react with the fuel to give alkyl radicals in reaction (1). At the start of the primary reaction zone in lean flames, X is probably mainly the hydroxyl radical,[20] but in rich

$$X + C_3H_8 = HX + C_3H_7 \qquad (R1)$$

$$n - C_3H_7 = C_3H_6 + H \qquad (R2)$$

$$sec - C_3H_7 = CH_3 + C_2H_4 \qquad (R3)$$

flames the hydrogen atom is likely to be the predominant species. It should be noted that the reaction of these alkyl radicals with oxygen is not important in flames,[21] and there is direct experimental evidence that octyl radicals (for example) break down into smaller fragments (mainly C_1 and C_2) in this region of the flame.[22] In the case of propane, both n-propyl and sec-propyl radicals are formed in reaction (1), and these react by reactions (2) and (3), respectively.

Typical reaction rate profiles are given in Figure 1-9.5. This shows that the maximum rate of removal of propane and the maximum rate of formation of carbon monoxide both occur at about the same temperature (1160 K), and, at this temperature, about 90 percent of the original propane has been consumed. In this region of the flame, the ratio R_{CO} + $R_{CO_2}/ - R_{C_3H_8}$ has a value of 3.1, which is reasonably close to the expected value of 3.0; such simple checks give confi-

dence that the analysis of the experimental data is essentially correct. A detailed discussion of the chemistry involved in the formation of carbon monoxide from the hydrocarbon fragments is inappropriate for the present purposes, but it is generally agreed that one very important route is *via* the reaction of methyl radicals with oxygen atoms. This produces formaldehyde, which subsequently gives carbon monoxide by reactions such as (5) and (6).

$$CH_3 + O = HCHO + H \qquad (R4)$$

$$X + HCHO = HX + CHO \qquad (R5)$$

$$CHO = H + CO \qquad (R6)$$

$$OH + CO = CO_2 + H \qquad (R7)$$

Figure 1-9.5 shows that there is significant conversion of carbon monoxide to carbon dioxide in the primary reaction zone of lean flames, and this arises through reaction (7). Many researchers have used the experimentally measured rate of this formation, in conjunction with the experimental local concentration and temperature (and, hence, known rate constant),[23] to derive a mole fraction profile for hydroxyl radicals. Such a profile is shown in Figure 1-9.6, together with the profile obtained for lower temperatures, assuming the removal of propane is solely due to its reaction with hydroxyl radicals. While the two profiles cover different parts of the flame, they are in excellent agreement where they overlap. A further indication that the removal of propane is *via* reaction with hydroxyl radicals comes from the temperature dependence of the rate of heat release in the early part of the flame; this is about 10 kJ mol^{-1}, which is

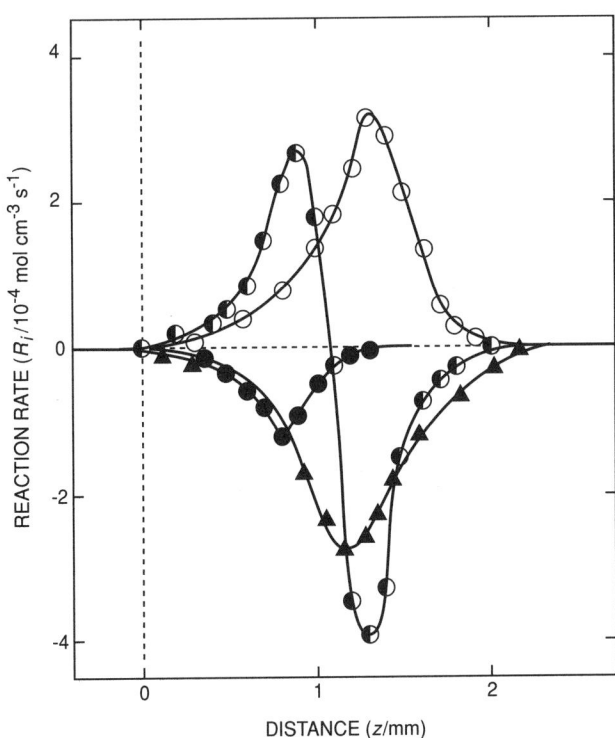

Fig. 1-9.5. *Reaction rate profiles through the flame of Figure 1-9.4. Note: ● C_3H_8; ▲ O_2; ◐ CO; $\odot$ CO_2.*[20]

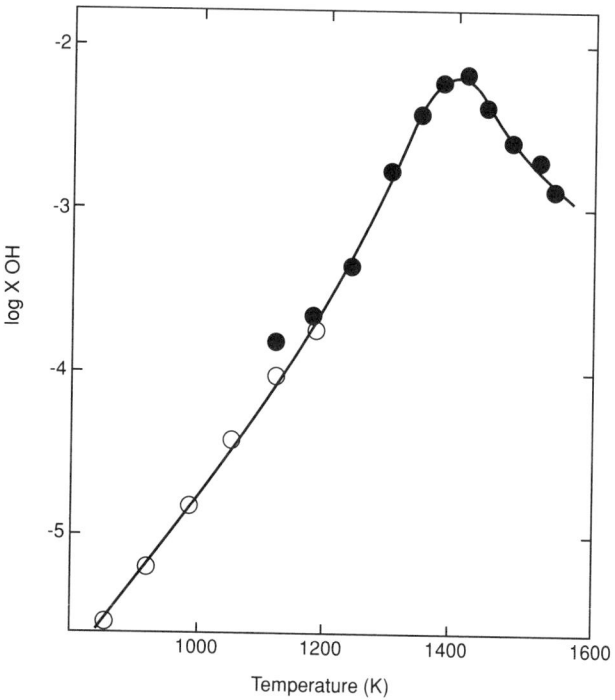

Fig. 1-9.6. Variation in X_{OH} with temperature through the flame of Figure 1-9.4. Note: ● from R_{CO_2}; ⊙ from $-R_{C_3H_8}$.[20]

close to that expected for the reaction of propane and hydroxyl radicals and much less than that expected for the reaction of propane with either hydrogen atoms or HO_2 radicals (37 and 78 kJ mol^{-1}, respectively). It must be stressed, however, that this conclusion comes from the analysis of data for very lean flames and that reaction with hydrogen atoms will become increasingly important as the mole fraction of propane in the initial mixture increases.

$$H + O_2 = OH + O \qquad (R8)$$

$$H + O_2 + M = HO_2 + M \qquad (R9)$$

Figure 1-9.5 shows that the maximum rate of removal of oxygen occurs somewhat later in the flame (at 1280 K) than the maximum rate of removal of propane (1160 K). It is instructive to examine the relative rates of reaction of hydrogen atoms with oxygen through the pre-heating and primary reaction zones of the flame. (See Table 1-9.2.) It can be seen that, at the lower temperatures, the thermolecular reaction (9) predominates over the branching reaction (8); however, as the temperature rises above 1000 K, reaction (8) becomes increasingly important. Similarly, Table 1-9.2 also shows that, between 1000 and 1200 K, reaction (8) becomes faster than the removal of hydrogen atoms by reaction with propane. As a result, there is a rapid increase in the concentration of free atoms and radicals, i.e., chain centers, toward the end of the primary reaction zone, so that the expected thermal equilibrium level can be exceeded by more than an order of magnitude.[26]

This rapid increase is in accordance with chain reaction theory.[27] This shows that, with a reaction involving linearly branched chains and linear chain termination, there is an exponential increase in the concentration of chain centers,

even under isothermal conditions, as soon as the rate of chain branching exceeds the rate of chain termination. The situation is more complicated in the case of a hydrocarbon-air flame, as there has already been a major consumption of reactants by the time the branching reaction (8) becomes important. This limits the concentration of chain centers, as shown in Figure A-1-9.1. In principle, another limiting factor to the growth in the concentration of chain centers in the flame is the occurrence of quadratic termination processes, such as reaction (10), which are known to be important in the secondary reaction zone of the flame. A combination of linearly branched chains and quadratic termination must lead to a stationary-state concentration.

$$H + OH + M = H_2O + M \qquad (R10)$$

This high concentration of chain centers produces the luminous flame front, i.e., the characteristic inner cone of the Bunsen burner flame. The radiation from this region of the flame includes that from electronically excited hydroxyl radicals, which are believed to be formed partly by radical-radical reactions, such as reactions (11) and (12), and partly by reaction (13).[1]

$$O + H = OH^* \qquad (R11)$$

$$H + OH + OH = OH^* + H_2O \qquad (R12)$$

$$CH + O_2 = CO + OH^* \qquad (R13)$$

At this point in the flame all the fuel has effectively been consumed, some of the resulting carbon monoxide has been converted to carbon dioxide, and the radical concentration exceeds the corresponding thermal equilibrium level. Thus, in the *secondary reaction zone* the important processes are the conversion of the major part of the carbon monoxide to carbon dioxide, plus the decay in the concentration of radical species by recombination reactions. This leads to a further, but slower, rise in temperature until the final thermodynamic equilibrium has been reached with the burnt gas at the final flame temperature.

The detailed computations for a one-dimensional laminar flame structure typically involve more than 120 elementary reaction steps even for simple hydrocarbon fuels; with more realistic fuels, the potential number can become very large. While this is practicable for such systems, it is out of

TABLE 1-9.2 Relative Rates of Reaction of Hydrogen Atoms with Oxygen and Propane Through a Premixed Propane-Oxygen-Argon Flame

Initial mixture composition: 1.38 percent C_3H_8, 14.8 percent O_2, 83.82 percent Ar

Temp (K)	$R_{H+O_2+M_2}/R_{H+O_2}$	$R_{H+C_3H_8}/R_{H+O_2}$
600	450	38
800	12	7.5
1000	1.2	1.5
1200	0.25	0.4

k_{H+O_2} and $k_{H+O_2} + M$ have been taken from Baulch *et al.*[24]
$k_{H+C_3H_8}$ has been taken from Walker.[25]

$$CH_4 + 2O_2 = CO_2 + H_2 + H_2O \qquad (R14)$$

$$CO + H_2O = CO_2 + H_2 \qquad (R15)$$

$$O_2 + 2H_2 = 2H_2O \qquad (R16)$$

the question for many engineering applications where three-dimensional and time-dependent effects arise, e.g., turbulent flames. Here, a substantial reduction in the number of reaction steps is needed before inclusion of the chemistry in the computations becomes practicable. The early attempts at producing a simplified global reaction scheme involved adjusting rate coefficients and reaction orders to fit the experimental observations, but this has too many unsatisfactory aspects. A much more satisfactory approach involves the systematic use of steady-state and partial-equilibrium approximations to reduce the number of independent reaction steps.[28] Under these circumstances the rate constants for these global reaction steps can be expressed as a combination of the known rate constants of elementary reactions. For example, Peters and Williams have shown that the three-step mechanism comprising reactions (14) through (16) gives a good representation of a stoichiometric methane-air flame burning at atmospheric pressure and above.[28] Similar mechanisms have also been derived for methanol and propane flames.[29]

EFFECT OF ADDITIVES ON FLAME PROPAGATION

When inert diluents such as nitrogen, argon, and carbon dioxide are introduced into a premixed flame, they reduce the final flame temperature and, if the corresponding reduction in burning velocity is sufficiently large, the flame is extinguished. The limits of flammability data for propane[30] in Figure 1-9.7 show that the "peak" concentration of nitrogen is quite high [about 43 percent (by vol)], so that the oxygen content of an air-nitrogen mixture must be reduced to below 12 percent to ensure that no mixture of propane and air will burn. For such systems, the adiabatic flame temperature at the limit of flammability is not only remarkably constant, but this temperature is effectively the same all around the limit curve, so that the additive must act only as a diluent.

Hydrogen chloride must also act predominantly as an inert diluent as its effect on the limits of flammability of hydrogen-air mixtures is almost identical to that of nitrogen.[31] This arises because, although the formation of chlorine atoms by reaction (17) is fast, the subsequent abstraction of a hydrogen atom by reaction (18) is also fast. However, the effective equilibrium position lies over on the side of hydrogen chloride, so that even though the additive is involved chemically, it has no overall chemical effect on the combustion process.

$$X + HCl = HX + Cl \qquad (R17)$$

$$Cl + RH = HCl + R \qquad (R18)$$

In contrast, bromine compounds are much more effective than the inert diluents in preventing flame propagation as they act as chemical inhibitors.[32,33] (See Figure 1-9.7.) Even a trace amount of such compounds in a premixed flame

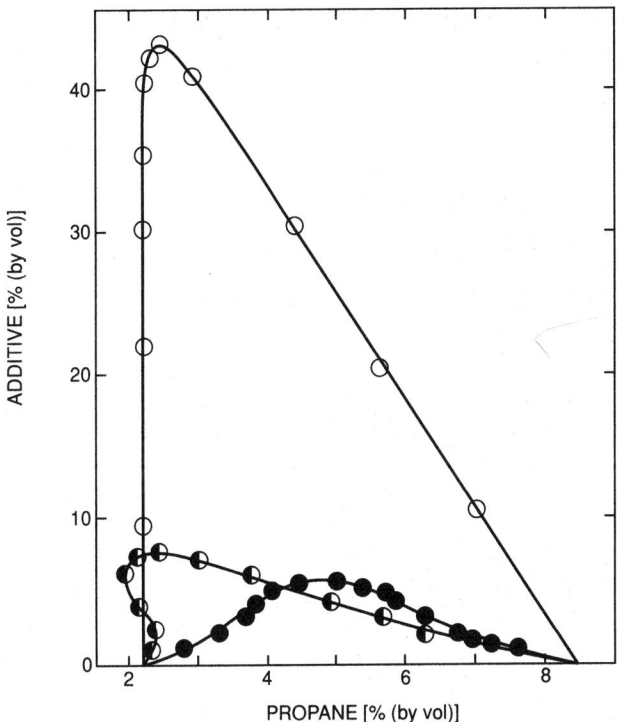

Fig. 1-9.7. *Effect of nitrogen, hydrogen bromide, and methyl bromide on the limits of flammability of propane in air (pressure 380 mmHg). Note: $\odot N_2$; $\bullet$ HBr; $\mathbb{O}$ CH$_3$Br.*

markedly reduces its burning velocity, and it is particularly striking that, when sufficient compound has been added to extinguish a stoichiometric hydrocarbon-air flame, the adiabatic flame temperature of the resulting limit mixture is only slightly lower than that in its absence.[32] The action of such inhibitors can be illustrated by considering the action of hydrogen bromide in a hydrogen flame.[34] In this case, hydrogen atoms are removed by reaction (19) in preference to reacting in the chain branching reaction (8), so that the reactive hydrogen atoms are converted into relatively unreactive bromine atoms. Computations show that apart from in

$$H + O_2 = OH + O \qquad (R8)$$

$$H + HBr = H_2 + Br \qquad (R19)$$

$$H + Br_2 = HBr + Br \qquad (R20)$$

$$Br + Br + M = Br_2 + M \qquad (R21)$$

the very early stages of the flame, reactions (19) and (20) are effectively in equilibrium locally, with the equilibrium lying over on the side of bromine atoms. This produces some reduction in burning velocity, but it does not explain quantitatively the observed results. This requires the inclusion of an additional chain termination step, such as reaction (21), so that the rate of chain termination is increased relative to the rate of chain branching, in accord with the theory of chain reactions. (See Appendix A to this chapter.)

APPLICATION TO "REAL" FIRES

A basic understanding of premixed burning is an important prerequisite in a number of applications concerning "real fires," even though the latter are essentially diffusion flames by nature. In such systems the rate-controlling process is normally the diffusion of fuel and oxygen from their respective sides of the flame and not the rate of chemical reaction (1) and, in the case of a jet of fuel gas burning in air, the stability of the flame depends on the burning of a pocket of premixed gas at the base of the flame. Immediately above the burner rim there is a region where the gas velocity is low and where the fuel and air mix; it is a combination of the burning velocity of this mixture and the local gas velocity that determines whether the diffusion flame is stable or is "blown off." Similarly, diffusion flames can be stabilized behind an obstruction in a gas stream, since the recirculation zone behind the obstruction produces a region of low gas velocity. This has important practical implications concerning the extinction of fires where the source of the fuel lies behind an obstruction, since such flames can be highly stable.[35]

In a diffusion flame, the fuel and oxidant react overall in stoichiometric proportions, but the local stoichiometric ratio ranges from very fuel rich in the yellow carbon zone to excess oxygen in the hot blue zone on the air side of the flame. The basic chemistry of the combustion process in a hydrocarbon-air diffusion flame is essentially the same as in a premixed flame, but the detailed mechanism reflects the change in local conditions across the flame. Thus, on the fuel side of the flame, thermal decomposition reactions are the most important processes and, owing to the lack of oxygen in this region, this leads to carbon formation and the characteristic yellow color associated with such flames. The maximum temperature is reached in the main reaction zone on the air side of the flame. The oxygen consumption occurs mainly on the air side of this zone by reaction (8),[36] and diffusion of the resulting hydroxyl radicals toward the rich side of the flame leads to the conversion of carbon monoxide to carbon dioxide by reaction (7). Since this latter reaction produces hydrogen atoms, reactions (7) and (8) constitute a self-sustaining sequence for this part of the combustion process. In addition, reaction (8) is also the source of the oxygen atoms required for the formation of carbon monoxide by reactions (4) through (6).

$$OH + CO = CO_2 + H \qquad (R7)$$

$$H + O_2 = OH + O \qquad (R8)$$

$$H + H_2O = H_2 + OH \qquad (R9)$$

Many researchers have assumed that a state of quasi-equilibrium exists in the main reaction zone of a diffusion flame, but this is only strictly true for a limited region of the flame. For example, Mitchell et al have shown that the water-gas reaction ($CO + H_2 = CO_2 + H_2$) approaches equilibrium on the fuel side of the main reaction zone of a methane-air flame, which implies that reactions (7) and (9) are effectively balanced in this part of the flame.[37] Similarly, comparison of experimental data for a propane flame with that expected from thermodynamic equilibrium shows that the situation is close to equilibrium around the stoichiometric plane of the flame.[36] On the rich side of the flame, how-ever, the carbon dioxide and water vapor levels exceed their thermodynamic equilibrium values, while the carbon monoxide level is much lower. On the air side, the conversion of carbon monoxide to carbon dioxide has not reached complete equilibrium. The final "burnout" of carbon monoxide to carbon dioxide on the air side of the flame is effectively stopped when the temperature falls below 1250 K, provided a critical temperature gradient is also exceeded.[38] Since this conversion occurs by reaction (7), this quenching must reflect a correspondingly sharp drop in the concentration of hydroxyl radicals, as it only has a low-temperature dependence with an activation energy of 3 kJ mol^{-1}.[23]

As far as the elementary reactions are concerned, Bilger et al has shown that reactions (22) and (23) are effectively balanced on the fuel side of a methane-air diffusion flame, and that reactions (8), (24), and (25) only approach equilibrium in a very narrow region of the flame.[39]

$$CH_4 + H = CH_3 + H_2 \qquad (R22)$$

$$H + H_2O = H_2 + OH \qquad (R23)$$

$$O + H_2O = 2OH \qquad (R24)$$

$$O + H_2 = OH + H \qquad (R25)$$

This lack of equilibrium is much more pronounced in turbulent diffusion flames, where it has long been recognized as a problem in the modeling of such systems from first principles. With "real fires" there is the added complication of the feed-back mechanism responsible for the growth of the fire. Such modeling involves a highly complex interaction of chemistry, heat transfer, and fluid dynamics and, to date, such simulations have effectively ignored the chemistry by either concentrating on the steady-state situation or assuming an exponential rate of growth for the fire as observed experimentally. Since the flow is usually dominated by buoyancy, the rate of mixing (and, hence, chemical reaction) is controlled by the resulting turbulent motion. If it is also assumed that chemical equilibrium exists through the flame, the problem reduces to the solution of the classical equations for conservation of mass, momentum, heat, and species to obtain gas velocities and temperatures for discrete points in space and moments in time.

One promising way of avoiding the equilibrium assumption is the use of laminar flamelets, in which a given microscopic element in the turbulent flow is assumed to have the same composition as an element of the same overall stoichiometry in a laminar flame.[40,41] The advantage of this approach can be seen from the predictions for carbon monoxide for a turbulent methane-air diffusion flame; when thermodynamic equilibrium was assumed the peak mole fraction of carbon monoxide was about 2.5 times that observed experimentally, whereas when the laminar flamelet approach was used the agreement was within 10 percent.[42] Lack of appropriate experimental data may restrict the use of this approach, however, unless it can be shown experimentally that essentially the same variation in composition with stoichiometry is obtained for a range of fuels.

$$CH_4 + 2H + H_2O = CO + 4H_2 \qquad (R26)$$

$$CO + H_2O = CO_2 + H_2 \qquad (R15)$$

$$O_2 + 2H_2 = 2H_2O \qquad \text{(R16)}$$

$$O_2 + 3H_2 = 2H + 2H_2O \qquad \text{(R27)}$$

An alternative approach is to simplify the chemical component of the computation by using a reduced reaction mechanism.[38,43] Full calculations for a laminar methane-air diffusion flame shows that the steady-state approximation can be applied to the species:

$$OH, O, HO_2, CH_3, CH_2O, \text{ and } CHO.$$

Using this approximation enables the full mechanism to be reduced to four global reaction steps, for example[37] reactions (26), (15), (16), and (27) whose rate can be represented by a combination of the rates of individual elementary reactions. The computational advantage of this approach comes from reducing the number of chemical species, since it is the number of species rather than the number of reaction steps that determines the complexity of the calculations.

APPENDIX A
MATHEMATICAL TREATMENT OF BRANCHING CHAIN REACTIONS

Linearly Branched Chains with Linear Gas-Phase Termination under Isothermal Conditions

With such a system, the rate of change of radical concentration can be represented by an equation of the form

$$dn/dt = \Theta + fn - gn \qquad \text{(A1)}$$

where n is the radical concentration and Θ the rate of chain initiation, while f and g are the coefficients of linear branching and linear termination, respectively. The classic example of such a system is the thermal reaction between hydrogen and oxygen, where the second explosion limit is controlled by a competition between the reactions

$$H + O_2 = OH + O \qquad \text{(R8)}$$

$$H + O_2 + M = HO_2 + M \qquad \text{(R9)}$$

In this case: $n = (H)$, $f = k_2(O_2)$, and $g = k_3(O_2)(M)$, where i represents the concentration of species i.

Three cases can be distinguished:

1. $g > f$:

In this case the rate of chain termination is greater than the rate of chain branching, and integration of Equation A1 gives

$$n = \frac{\Theta}{(f-g)} \cdot \{1 - \exp^{-(g-f) \cdot t}\} \qquad \text{(A2)}$$

When t is large $\exp^{-(g-f) \cdot t}$ approaches zero, so that n tends to the stationary state value

$$n = \Theta/(g - f) \qquad \text{(A3)}$$

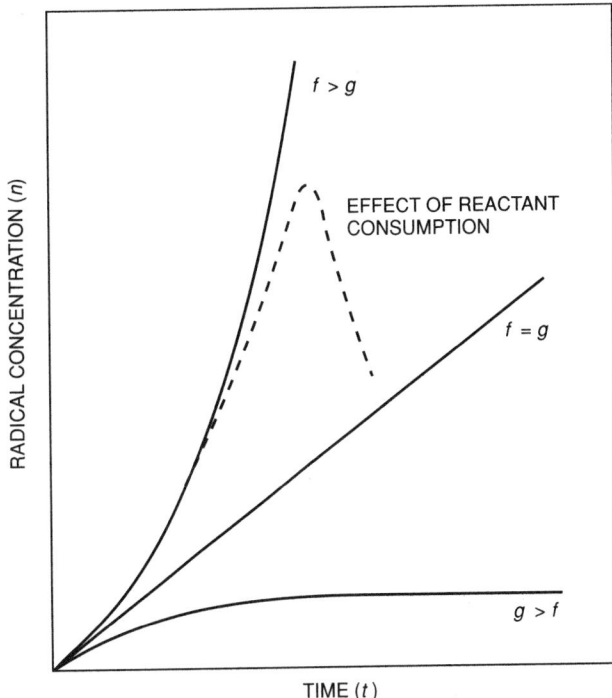

Fig. A-1-9.1. *Growth of free radical concentration, n, with time, t, in a linearly branched and terminated reaction. Boundary condition between a steady-state and exponential growth is f = g.*

2. $f = g$:

In this case n increases linearly with time, i.e.,

$$n = \Theta \cdot t \qquad \text{(A4)}$$

3. $f > g$:

In this case the rate of chain branching is always greater than the rate of chain termination, and so n grows exponentially with time. Integration of Equation A1 gives

$$n = \frac{\Theta}{(f-g)} \cdot \{\exp^{(f-g) \cdot t} - 1\}$$

These three cases are represented graphically in Figure A-1-9.1 and, in the latter case, the exponential increase in the radical concentration must lead to explosion unless this is prevented by consumption of reactants.

In the case of the hydrogen-oxygen reaction, where the hydrogen atom is the slowest reacting species, it has been shown that the ratios [H]/[OH] and [H]/[O] maintain their stationary-state values even during the exponential growth in radical concentration which leads to explosion. As a result, with such systems the usual practice is to consider the change in concentration of the slowest reacting species and assume that the concentration of all other species is the corresponding stationary-state value.

REFERENCES CITED

1. A.G. Gaydon and H.G. Wolfhard, *Flames: Their Structure, Radiation, and Temperature*, 4th ed., Chapman and Hall, London (1979).
2. B. Lewis and G. von Elbe, *Combustion, Flames, and Explosions of Gases*, 2nd ed., p. 280, Academic Press, New York (1961).

3. S. Gordon and B.J. McBride, NASA SP-273 (1971).

4. E. Bartholome, *Z. Elektrochem.* 54, 165 (1950).

5. J.M. Singer, J. Grumer, and E.B. Cook, *Proc. Gas Dynamics Symposium on Aerothermochemistry* (1956). Data reported in ref 2, p. 390.

6. B. Lewis, *Selected Combustion Problems*, p. 177, Butterworth, London (1954).

7. G.L. Dugger and S. Heimel. *NACA Tech. Note 2624* (1952). Data reported in ref 1, p. 82.

8. H. Mache, *Forsch. Ing. Wes.*, 14, 77 (1943).

9. D.T. Williams and L.M. Bollinger, *Third Symposium* on Combustion, Flames and Explosion Phenomena, p. 176, Williams and Wilkins, Baltimore (1949).

10. B. Karlovitz, D.W. Denniston, and F. E. Wells, *J. Chem. Phys.*, 19, 541 (1951).

11. H.F. Coward and G.W. Jones, US Bureau of Mines, *Bulletin 503* (1952).

12. M.G. Zabetakis. US Bureau of Mines, *Bulletin 627* (1964).

13. M.V. Blanc, P.G. Guest, B. Lewis, and G. von Elbe, *J. Chem. Phys.*, 15, 798 (1947).

14. R. Freidman, *3rd Symposium* on Combustion, Flames, and Explosion Phenomena, p. 110, Williams and Wilkins, Baltimore (1949).

15. B. Lewis and G. von Elbe, *Combustion, Flames, and Explosions of Gases*, 2nd ed., p. 511 ff, Academic Press, New York (1961).

16. R.M. Fristrom and A.A. Westenberg. *Flame Structure*, McGraw-Hill, New York (1965).

17. G. Dixon-Lewis, *Phil. Trans. Roy. Soc.*, A292, 45 (1979).

18. J. Warnatz, *Eighteenth Symposium on Combustion*, p. 369, The Combustion Institute, Pittsburgh (1981).

19. N. Peters and J. Warantz (eds.), *Numerical Methods in Laminar Flame Propagation*, Vieweg, Braunschweig (1982).

20. S.J. Cook and R.F. Simmons, *Combust. and Flame*, 46, 177 (1982).

21. J. Warnatz, *Combust. Sci. & Tech.*, 34, 177 (1983).

22. E. Axelssohn and L. G. Rosengren, *Combust. and Flame*, 64, 229 (1986).

23. D.L. Baulch, D.D. Drysdale, J. Duxbury, and S.J. Grant, *Evaluated Kinetic Data for High Temperatures*, Vol. 3, Butterworths, London (1976).

24. D.L. Baulch, C.J. Cobos, R.A. Cox, P. Frank, G. Hayman, T.H. Jost, J.A. Kerr, T. Murrells, M.J. Pilling, J. Troe, R.W. Walker, and J. Warnatz. *Combust. and Flame*, In press.

25. R.W. Walker, *Reaction Kinetics I*, p. 161, The Chemical Society, London (1975).

26. E.M. Bulewicz and T.M. Sugden, *Proc. Roy. Soc.*, A277, 143 (1964).

27. F.S. Dainton, *Chain Reactions*, Methuen, London (1956).

28. N. Peters and F.A. Williams, *Combust. and Flame*, 68, 185 (1987).

29. G. Paczko, P.M. Lefdal, and N. Peters, *Twenty-first Symposium on Combustion*, p. 739, The Combustion Institute, Pittsburgh (1986).

30. R.F. Simmons and N. Wright, Unpublished Results.

31. R.N. Butlin and R.F. Simmons, *Combust. and Flame*, 12, 447 (1968).

32. R.F. Simmons and H.G. Wolfhard, *Trans.* Faraday Soc., 51, 1211 (1955).

33. W.A. Rosser, H. Wise, and J. Miller, *Seventh Symposium on Combustion*, p. 175, Butterworths, London (1959).

34. M.J. Day, D.V. Stamp, K. Thompson, and G. Dixon-Lewis, *Thirteenth Symposium on Combustion*, p. 705, The Combustion Institute, Pittsburgh (1971).

35. R. Hirst and D. Sutton, *Combust. and Flame*, 5, 317 (1961).

36. S. Evans and R.F. Simmons, *Twenty-second Symposium on Combustion*, p. 1433, The Combustion Institute, Pittsburgh (1988).

37. R.E. Mitchell, A.F. Sarofim, and L.A. Clomburg, *Combust. and Flame*, 37, 201 (1980).

38. C.P. Fenimore and J. Moore, *Combust. and Flame*, 22, 343 (1974).

39. R.W. Bilger, S.H. Starmer, and R.J. Kee, *Combust. and Flame*, 80, 135 (1990).

40. F.A. Williams, In *Turbulent Mixing in Non-Reactive and Reactive Flows* (ed. Muthy, S.N.), p. 189, Plenum Press (1975).

41. R.W. Bilger, "Turbulent Reacting Flows" (ed. Libby, P.A. and Williams, F.A.), p. 65, Springer (1980).

42. S.K. Liew, K.N.C. Bray and J.B. Moss, *Combust. Sci. & Tech.*, 27, 69 (1981).

43. N. Peters and R.J. Kee, *Combust. and Flame*, 68, 17 (1987).

PROPERTIES OF BUILDING MATERIALS

Tibor Z. Harmathy

INTRODUCTION

Homogeneous materials, i.e., materials that have the same composition and properties throughout their volume, are rarely found in nature. Most construction materials are *heterogeneous*, yet their heterogeneity is often glossed over when dealing with practical problems.

The heterogeneity of concrete is easily noticeable. Other heterogeneities related to the microstructure of materials, i.e., their *grain* and *pore structures*, are rarely detectable by the naked eye. The microstructure depends greatly on the way the materials are formed. In general, materials formed by solidification from a melt show the highest degree of homogeneity. The result of the solidification is normally a *polycrystalline* material, comprising polyhedral *grains* of crystals which, in general, are equiaxed and randomly oriented. Severe *cold-working* in metals may produce an elongated grain structure and crystals with preferred orientations.

Non-crystalline solids are called *amorphous* materials. *Gels* and *glasses* are amorphous materials. Gels are formed by the coagulation of a colloidal solution. Glasses (vitreous materials) are solids with a liquid-like, grainless submicroscopic structure with low crystalline order. On heating they will go through a series of phases of decreasing viscosity.

Synthetic polymers (plastics) are made up of long macromolecules created by polymerization from smaller repeating units (monomers). In the case of *thermoplastic* materials, the mobility of the molecular chains increases on heating. Such materials soften, much like glasses. In some other types of plastics, called *thermosetting* materials, polymerization also produces cross-bonds between the molecular chains. These cross-bonds prevent the loosening of the molecular structure and the transition of the material into a liquid-like state.

Some building materials are formed from a wet, plastic mass or from compacted powders by firing. The resulting product is a polycrystalline solid with a well-developed pore structure.

Two important building materials, i.e., concrete and gypsum, are formed by mixing finely ground powders (and aggregates) with water. The mixture solidifies by hydration. The cement paste in a concrete has a highly complex microstructure, interspersed with very fine, elaborate pores.

Most building materials can be treated as *isotropic* materials, i.e., as though they possessed the same properties in all directions. Among the material properties, those that are unambiguously defined by the current composition and phase are referred to as *structure-insensitive*. Some others depend on the microstructure of the solid or on its previous history. These properties are *structure-sensitive*.

POROSITY AND MOISTURE SORPTION

What is commonly referred to as a solid object is actually all the material within its visible boundaries. Clearly, if the solid is porous—and most building materials are—the so-called solid consists of at least two phases: (1) a solid-phase matrix, and (2) a gaseous phase (namely, air) in the pores within the matrix. Usually, however, there is also a liquid or liquid-like phase present: moisture either adsorbed from the atmosphere to the pore surfaces, or held in the pores by capillary condensation. This third phase is always present if the pore structure is *continuous*; *discontinuous* pores (like the pores of some foamed plastics) are not readily accessible to atmospheric moisture.

The pore structure of materials is characterized by two properties: *porosity*, P (m^3 m^{-3}), the volume fraction of pores within the visible boundaries of the solid; and *specific surface*, S (m^2 m^{-3}), the surface area of the pores per unit volume of the material. For a solid with continuous pore structure, the porosity is a measure of the maximum amount of water the solid can hold when saturated. The specific surface and (to a lesser degree) porosity together determine the moisture content the solid holds in equilibrium with given atmospheric conditions.

The *sorption isotherm* shows the relationship at constant temperature between the equilibrium moisture content of a porous material and the relative humidity of the atmosphere. A sorption isotherm usually has two branches: (1) an *adsorption branch*, obtained by monotonically increasing the relative humidity of the atmosphere from 0 to 100 percent through very small equilibrium steps; and (2) *desorption branch*, obtained by monotonically lowering the relative humidity from 100 to 0 percent. Derived experimentally, the

Dr. Tibor Z. Harmathy was Head of the Fire Research Section, Institute of Research in Construction, National Research Council of Canada, until his retirement in 1988. His research centered on materials science and the spread potential of compartment fires.

sorption isotherms offer some insight into the nature of the material's pore structure.[1,2]

For heterogeneous materials consisting of solids of different sorption characteristics (e.g., concrete, consisting of cement paste and aggregates), the sorption isotherms can be estimated using the simple mixture rule (with $m = 1$; see Equation 1).

Among the common building materials, only concrete (or more accurately, the cement paste in the concrete) and wood, because of their large specific surfaces, can hold water in amounts substantial enough to be taken into consideration in fire performance assessments.

MIXTURE RULES

Some properties of materials of mixed composition or mixed phase can be calculated by simple rules if the material properties for the constituents are known. The simplest mixture rule is[3]

$$\pi^m = \sum_i v_i \pi_i^m \tag{1}$$

where π is a material property for the composite, π_i is that for the composite's ith constituent, v_i ($m^3\ m^{-3}$) is the volume fraction of the ith constituent, and m (dimensionless) is a constant that has a value between -1 and $+1$.

Hamilton and Crosser recommended the following rather versatile formula for two-phase solids[4]

$$\pi = \frac{v_1 \pi_1 + \beta v_2 \pi_2}{v_1 + \beta v_2} \tag{2}$$

where

$$\beta = \frac{n \pi_1}{(n-1)\pi_1 + \pi_2} \tag{3}$$

Here phase 1 must always be the principal continuous phase. n (dimensionless) is a function of the geometry of phase distribution. With $n \to \infty$ and $n = 1$, Equations 2 and 3 convert into Equation 1 with $m = 1$ and $m = -1$, respectively. With $n = 3$, a relation is obtained for a two-phase system where the discontinuous phase consists of spherical inclusions.[5]

By repeated application, Equations 2 and 3 can be extended to a three-phase system,[6] e.g., to a moist, porous solid that consists of three essentially continuous phases (the solid matrix, with moisture and air in its pores).

SURVEY OF BUILDING MATERIALS

There are *burnable* and *nonburnable* building materials.* To a designer concerned with the structural performance of a building during a fire, the mechanical and thermal properties of these materials are of principal interest. Yet, burnable building materials may become ignited, and thereby the positive role assigned to them by design (i.e., functioning as structural elements of the building) may change into a negative role: i.e., becoming fuel and adding to the severity of fire. Those properties of burnable building materials that are related to the latter role are discussed in other chapters of this handbook and in Reference 2.

Combustible and *noncombustible* are the commonly used words. The reason for preferring the use of the words *burnable* and *nonburnable* has been discussed in Reference 2.

From the point of view of their performance in fire, building materials can be divided into the following groups:

1. Group L (load-bearing) materials: materials capable of carrying high stresses, usually in tension. With these materials the mechanical properties related to behavior in tension are of principal interest.
2. Group L/I (load-bearing/insulating) materials: materials capable of carrying moderate stresses and, in fire, providing thermal protection to Group L materials. With Group L/I materials, the mechanical properties (related mainly to behavior in compression) and the thermal properties are of equal interest.
3. Group I (insulating) materials: materials not designed to carry load. Their role in fire is to resist the transmission of heat through building elements and/or to provide insulation to Group L or Group L/I materials. With Group I materials only the thermal properties are of interest.
4. Group L/I/F (load-bearing/insulating/fuel) materials: Group L/I materials that may become fuel in fire.
5. Group I/F (insulating/fuel) materials: Group I materials that may become fuel in fire.

The number of building materials has been increasing dramatically during the past few decades. By necessity, only a few of those commonly used will be discussed in this chapter in some detail. These materials are: in Group L—structural steel and prestressing steel, in Group L/I—concrete and brick, in Group L/I/F (or Group I/F)—wood, and in Group I—gypsum.

BUILDING MATERIALS AT ELEVATED TEMPERATURES

While calculation techniques for predicting the process of deterioration of building elements in fire have developed rapidly in recent years, research related to supplying input information into these calculations has not kept pace. The designer of the firesafety features of buildings will find that information on the properties of building materials for the temperature range of interest, 20 to 700°C, is not easy to come by. Most building materials are not stable throughout this temperature range. On heating they undergo physicochemical changes ("reactions" in a generalized sense), accompanied by transformations in their microstructure and changes in their properties. A concrete at 500°C is completely different from the material at room temperature.

Clearly, the generic information available on the properties of building materials at room temperature is seldom applicable in firesafety design. It is imperative, therefore, that the firesafety practitioner know how to extend, based on *a priori* considerations, the utility of the scanty data that can be gathered from the technical literature.

REFERENCE CONDITION

Most building materials are porous and therefore capable of holding moisture, the amount of which depends on the atmospheric conditions. Since the presence of moisture may have a significant and often unpredictable effect on the properties of materials at any temperature below 100°C, it is imperative to conduct all property tests on specimens brought into a moistureless "reference condition" by some drying technique prior to the test. The reference condition is normally interpreted as that attained by heating the test

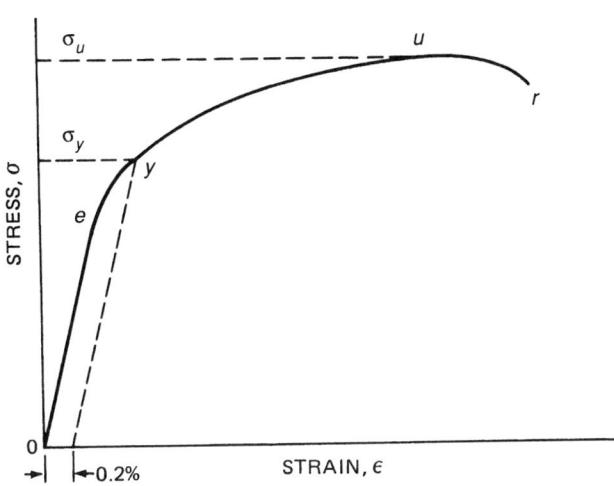

Fig. 1-10.1. *Stress-strain curve (strain rate is roughly constant).*

specimen in an oven at 105°C until its weight shows no change. A few building materials, however, among them all gypsum products, may undergo irreversible physicochemical changes when held at that temperature for an extended period. To bring them to a reference condition, specimens of these materials should be heated in a vacuum oven at some lower temperature level (e.g., at 40°C in the case of gypsum products).

MECHANICAL PROPERTIES

Stress-Strain Curve

The mechanical properties of solids are usually derived from conventional tensile or compressive tests. Figure 1-10.1 shows for a metallic material the variation of stress, σ (Pa), with increasing strain (deformation), ε (m m^{-1}), while the material is strained (deformed) in a tensile test at more or less constant rate (i.e., constant crosshead speed), usually of the order of 1 mm min^{-1}.

Modulus of Elasticity, Yield Strength, Ultimate Strength

Section 0-e of the curve in Figure 1-10.1 represents the elastic deformation of the material, which is instantaneous and reversible. The modulus of elasticity, E (Pa), is the slope of that section. Between points e and u the deformation is plastic, non-recoverable, and quasi-instantaneous. The plastic behavior of the material is characterized by the yield strength at 0.2 percent offset, σ_y (Pa), and the ultimate strength, σ_u (Pa). After some localized necking (i.e., reduction of cross-sectional area), the test specimen ruptures at point r. The modulus of elasticity is more or less a structure-insensitive property.

For metals of similar metallurgical characteristics, the stress-strain curve can be reproduced at room temperature at a reasonable tolerance, and the shape of the curve does not depend significantly on the crosshead speed. At sufficiently high temperatures, however, the material undergoes plastic deformation even at constant stress, and the e-r section of

the stress-strain curve will depend markedly on the cross-head speed.

Creep

The time-dependent plastic deformation of the material is referred to as creep strain, and is denoted by ε_t (m m^{-1}). In a creep test the variation of ε_t is recorded against time, t (h), at constant stress (more accurately, at constant load) and at constant temperature T (K). A typical strain-time curve is shown in Figure 1-10.2, part (a). The total strain, ε (m m^{-1}), is

$$\varepsilon = \frac{\sigma}{E} + \varepsilon_t \qquad (4)$$

The 0-e section of the strain-time curve represents the instantaneous elastic (and reversible) part of the curve; the rest is creep, which is essentially nonrecoverable. The creep is fast at first [primary creep, section e-s_1 in Figure 1.10.2, part (a)], then proceeds for a long time at an approximately constant rate (secondary creep, section s_1-s_2), and finally accelerates until rupture occurs (tertiary creep, section s_2-r). The curve becomes steeper if the test is conducted either at a higher load (stress) or at a higher temperature.

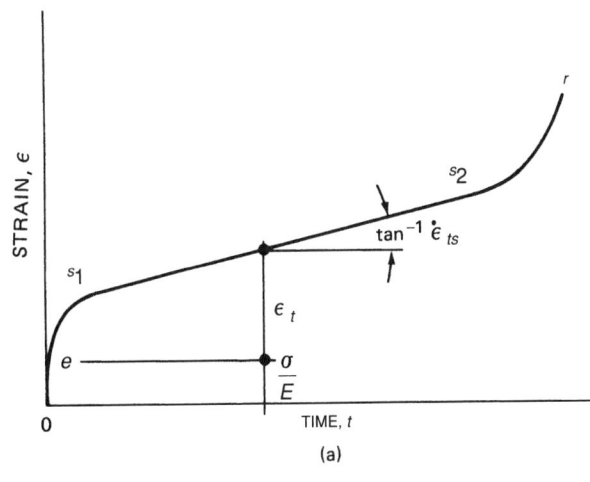

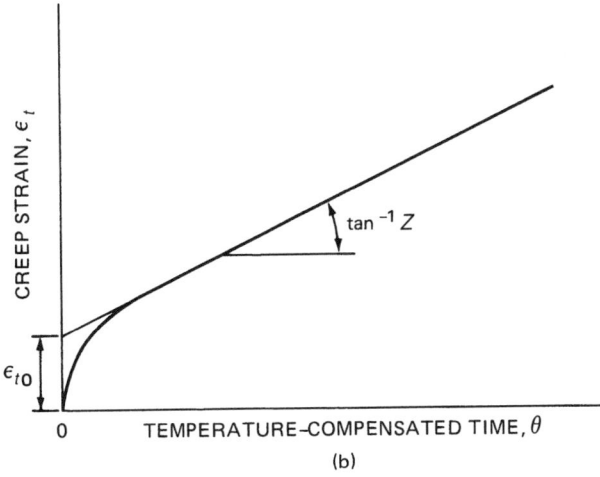

Fig. 1-10-2. *(a) Creep strain vs time curve (T = constant; $\sigma \simeq$ constant); (b) creep strain vs temperature-compensated time curve ($\sigma \simeq$ constant).*

Dorn's concept is particularly suitable for dealing with deformation processes developing at varying temperatures.[7] Dorn eliminated the temperature as a separate variable by the introduction of a new variable: the "temperature-compensated time," θ (h), defined as

$$\theta = \int_0^t e^{-\Delta H_c/RT} \, dt \qquad (5)$$

where ΔH_c (J kmol^{-1}) is the activation energy of creep, and R (J kmol^1 K^{-1}) is the gas constant.

From a practical point of view, only the primary and the secondary creeps are of importance. It has been shown that the creep strain in these two regimes can be satisfactorily described by the following equation[8]

$$\varepsilon_t = \frac{\varepsilon_{t0}}{\ln 2} \cosh^{-1}(2^{Z\theta/\varepsilon_{t0}}) \qquad (\sigma \simeq \text{constant}) \qquad (6)$$

or approximated by the simple formula[9]

$$\varepsilon_t \simeq \varepsilon_{t0} + Z\theta \qquad (\sigma \simeq \text{constant}) \qquad (7)$$

where Z (h^{-1}) is the Zener-Hollomon parameter, and ε_{t0} (m m^{-1}) is another creep parameter, the meaning of which is explained in Figure 1-10.2, part (b). The Zener-Hollomon parameter is defined as[10]

$$Z = \dot{\varepsilon}_{ts} \, e^{\Delta H_c/RT} \qquad (8)$$

where $\dot{\varepsilon}_{ts}$ (m m^{-1} h^{-1}) is the rate of secondary creep at a temperature, T. The two creep parameters, Z and ε_{t0}, are functions of the applied stress only (i.e., independent of the temperature).

For most materials, creep becomes noticeable only if the temperature is higher than about one-third of the melting temperature (on the absolute scale).

The creep of concrete is due to the presence of water in its microstructure.[11] There is no satisfactory explanation for the creep of concrete at elevated temperatures. Anderberg and Thelandersson[12] and Schneider[13] suggested techniques for the calculation of the deformation of concrete under conditions characteristic of fire exposure.

THERMAL PROPERTIES

Dilatometric Curve

The dilatometric curve is a record of the fractional change of a linear dimension of a solid at steadily increasing or decreasing temperature. With mathematical symbolism, the dilatometric curve is a plot of

$$\frac{\Delta \ell}{\ell_0} \quad \text{against} \quad T$$

where $\Delta \ell = \ell - \ell_0$, and ℓ (m) and ℓ_0 (m) are the changed and original dimensions of the solid, respectively, the latter usually taken at room temperature. $\Delta \ell$ reflects not only the linear expansion or shrinkage of the material, but also the dimensional effects brought on by possible physicochemical changes (i.e., "reactions").

The heating of the solid usually takes place at an agreed-upon rate, 5°C min^{-1} as a rule. Because the physicochemical changes proceed at a finite rate and some of them are irreversible, a dilatometric curve obtained by heating rarely coincides with that obtained during the cooling cycle. Sluggish reactions may bring about a steady rise or decline in the slope of the dilatometric curve. Discontinuities in the slope indicate very fast reactions. Heating the material at a rate higher than 5°C min^{-1} usually causes the reactions to shift to higher temperatures and to develop faster.

The coefficient of linear thermal expansion, α (m m^{-1} K^{-1}), is defined as

$$\alpha = \frac{1}{\ell} \frac{d\ell}{dT} \qquad (9)$$

Since $\ell \simeq \ell_0$, the coefficient of linear thermal expansion is, for all intents, the tangent to the dilatometric curve. For solids that are isotropic in a macroscopic sense, the coefficient of volume expansion is approximately equal to 3α.

Most of the dilatometric curves to be shown in this chapter were recorded by the author.[14] A horizontal dilatometric apparatus was used, made to the design of the British Ceramic Research Association by a British manufacturer. It was modified to make it suitable for automatic operation.[15] The sample was 76.2 mm long and about 13 by 13 mm in cross section. It was subjected to a small spring load which varied during the test. Unfortunately, even this small load caused creep shrinkage with those materials that tended to soften at higher temperatures. Furthermore, since the apparatus did not provide a means for placing the sample in a nitrogen atmosphere, in certain cases oxidation may also have had some effect on the shape of the curves.

Thermogravimetric Curve

The thermogravimetric curve is a record of the fractional variation of the mass of a solid at steadily increasing or decreasing temperature. Again, with mathematical symbolism, a thermogravimetric curve is a plot of

$$\frac{M}{M_0} \quad \text{against} \quad T$$

where M and M_0 (kg) are the changed and original masses of the solid, respectively, the latter usually taken at room temperature. If the curve is obtained by heating, the agreed-upon rate of heating is, again, 5°C min^{-1}.

A thermogravimetric curve reflects reactions accompanied by loss or gain of mass but, naturally, it does not reflect changes in the materials' microstructure or crystalline order. $M/M_0 = 1$ is the thermogravimetric curve for a chemically inert material. Again, an increase in the rate of heating usually causes those features of the curve that are related to chemical reactions to shift to higher temperatures and to develop faster.

The thermogravimetric curves to be shown were obtained by a DuPont 951 thermogravimetric analyzer,[16] using specimens of 10 to 30 mg in mass, placed in a nitrogen atmosphere.[14] The rate of temperature rise was 5°C min^{-1}.

Density, Porosity

The density, ρ (kg m^{-3}), in oven-dry condition, is the mass of a unit volume of the material, comprising the solid itself and the air-filled pores. Assuming that the material is isotropic with respect to its dilatometric behavior, its density at any temperature can be calculated from the thermogravimetric and dilatometric curves.

$$\rho = \rho_0 \frac{\left(\dfrac{M}{M_0}\right)_T}{\left[1 + \left(\dfrac{\Delta \ell}{\ell_0}\right)_T\right]^3} \qquad (10)$$

where ρ_0 (kg m^{-3}) is the density of the solid at the reference temperature (usually room temperature), and the T subscript indicates values pertaining to temperature T in the thermogravimetric and dilatometric records.

The density of composite solids at room temperature can be calculated by means of the mixture rule in its simplest form (Equation 1 with $m = 1$).

$$\rho = \sum_i v_i \rho_i \qquad (11)$$

where the i subscript relates to information on the ith component. At elevated temperatures the expansion of the components is subject to constraints, and therefore the mixture rule can only yield a crude approximation.

If, as usual, the composition is given in mass fractions rather than in volume fractions, the volume fractions can be obtained as

$$v_i = \frac{\dfrac{w_i}{\rho_i}}{\sum_i \dfrac{w_i}{\rho_i}} \qquad (12)$$

where w_i is the mass fraction of the ith component (kg kg^{-1}).

True density, ρ_t (kg m^{-3}), is the density of the solid in a poreless condition. Such a condition is nonexistent for many building materials, and, therefore, ρ_t may be a theoretical value derived on crystallographic considerations, or determined by some standard technique, e.g., ASTM C135.[17] The relationship between the porosity and density is

$$P = \frac{\rho_t - \rho}{\rho_t} \qquad (13)$$

The overall porosity of a composite material consisting of porous components is

$$P = \sum_i v_i P_i \qquad (14)$$

where, again, the i subscript relates to the ith component of the material.

Calorimetric Curve

A calorimetric curve describes the variation with temperature of the apparent specific heat of a material at constant pressure, c_p (J kg^{-1} K^{-1}). The apparent specific heat is defined as

$$c_p = \left(\frac{\partial h}{\partial T}\right)_p \qquad (15)$$

where h is enthalpy (J kg^{-1}), and the p subscripts indicate the constancy of pressure. If the heating of the solid is accompanied by physicochemical changes (i.e., "reactions"), the enthalpy becomes a function of the reaction progress variable, ξ (dimensionless), i.e., the degree of conversion at a particular temperature from reactant(s) into product(s). For any temperature interval where physicochemical change takes place,[2,6,18] $0 \leq \xi \leq 1$, and

$$c_p = \bar{c}_p + \Delta h_p \frac{d\xi}{dT} \qquad (16)$$

where $\bar{c}_p$ (J kg^{-1} K^{-1}) is the specific heat for that mixture of reactants and (solid) products that the material consists of at a given stage of the conversion (as characterized by ξ), and Δh_p (J kg^{-1}) is the latent heat associated with the physicochemical change.

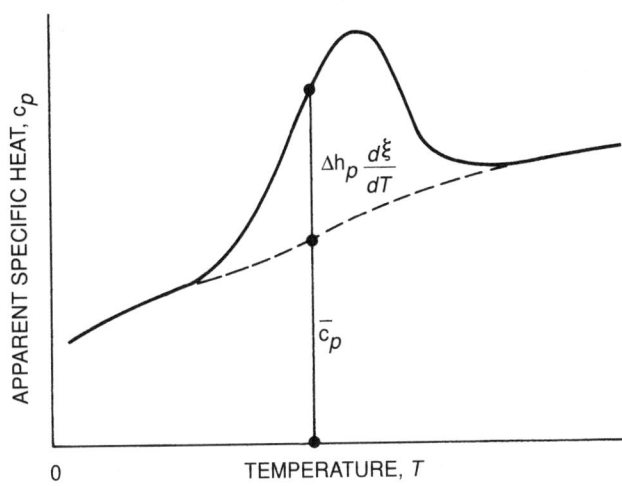

Fig. 1-10.3.   *The apparent specific heat.*

As Equation 16 and Figure 1-10.3 show, in temperature intervals of physicochemical instability the apparent specific heat consists of sensible heat and latent heat contributions. The latter contribution will result in an extremal in the calorimetric curve: a maximum if the reaction is endothermic, a minimum if it is exothermic.

In heat flow studies it is usually the ρc_p product (J m^{-3} K^{-1}) rather than c_p that is needed as input information. This product is referred to as "volume specific heat."

For a long time adiabatic calorimetry was the principal method to study the shape of the c_p vs T relationship. Today, differential scanning calorimetry (DSC) is the most commonly used technique for mapping the curve in a single temperature sweep at a desired rate of heating. Unfortunately, the accuracy of the DSC technique in determining the sensible heat contribution to the apparent specific heat may not be particularly good (sometimes it may be as low as ±20 percent).

The rate of heating is, again, usually 5°C min^{-1}. At higher heating rates the peaks in the DSC curves tend to shift to higher temperatures and become sharper.

Materials that undergo exothermic reactions may yield negative values in the calorimetric curve. A negative value for c_p indicates that, at the applied (and enforced) rate of heating, the rate of evolution of reaction heat exceeds the rate of absorption of sensible heat by the material. In natural processes the apparent specific heat can never be negative, because the heat evolving from the reaction is either scattered to the surroundings or, if absorbed by the material, causes a very fast temperature rise. If the heat of reaction is not very high, obtaining non-negative values for c_p can be achieved by suitably raising the scanning rate. For this reason, some materials undergoing exothermic reactions must be tested at rates of heating higher than 5°C min^{-1}, often as high as 50°C min^{-1}.

Some of the information in this chapter was developed with the aid of a DuPont 910 differential scanning calorimeter.[14,19] The samples, 10 to 30 mg in mass, were placed in a nitrogen atmosphere. The rate of temperature rise was usually 5°C min^{-1}.

If experimental information is not available, the c_p vs T relationship can be calculated from data on heat capacity and heat of formation for all the components of the material (including reactants and products), tabulated in a number of handbooks.[20,21] Examples of calculations are presented in References 2 and 6, where information is developed for the apparent specific heat *vs* temperature relation for a cement paste and four kinds of concrete.

Thermal Conductivity

Heat transmission solely by conduction can occur only in poreless, non-transparent solids. In porous solids (most building materials) the mechanism of heat transmission is a combination of conduction, radiation, and convection.* The thermal conductivity of porous materials is, in a strict sense, merely a convenient empirical factor that makes it possible to describe the heat transmission process with the aid of the Fourier law. That empirical factor will depend not only on the conductivity of the solid matrix, but also on the porosity of the solid and the size and shape of the pores. At elevated temperatures, because of the increasing importance of radiant heat transmission through the pores, conductivity becomes sensitive to the temperature gradient.

Since measured values of the thermal conductivity depend to some extent on the temperature gradient employed in the test, great discrepancies may be found in thermal conductivity data reported by various laboratories. A thermal conductivity value yielded by a particular technique is, in a strict sense, applicable only to heat flow patterns similar to that characteristic of the technique employed.

Experimental data indicate that porosity is not a greatly complicating factor as long as it is not larger than about 0.1. With insulating materials, however, the porosity may be 0.8 or higher. Conduction through the solid matrix may be an insignificant part of the overall heat transmission process; therefore, using the Fourier law of heat conduction in analyzing heat transmission may lead to deceptive conclusions.

If the solid is not oven-dry, a temperature gradient will induce migration of moisture, mainly by an evaporation-condensation mechanism.[22] The migration of moisture is usually, but not necessarily, in the direction of heat flow, and manifests itself as an increase in the apparent thermal conductivity of the solid. Furthermore, even oven-dry solids may undergo decomposition (mainly dehydration) reactions at elevated temperatures. The sensible heat carried by the gaseous decomposition products as they move in the pores adds to the complexity of the heat flow process. At present there is no way of satisfactorily accounting for the effect of simultaneous mass transfer on heat flow processes occurring under fire conditions.

The thermal conductivity of layered, multiphase solid mixtures depends on whether the phases lie in the direction of, or normal to, the direction of heat flow. The simple mixture rule is applicable, with $m = 1$ in the former case and with $m = -1$ in the latter. Thus, in these cases

$$k = \sum_i \nu_i \, k_i \qquad (17)$$

and

$$k = \sum_i \frac{\nu_i}{k_i} \qquad (18)$$

where k and k_i (J m^{-1} K^{-1}) are the thermal conductivities of the mixture and its constituents, respectively.

---

*If the pore size is less than about 5 mm, the contribution of the pores to convective heat transmission is negligible.

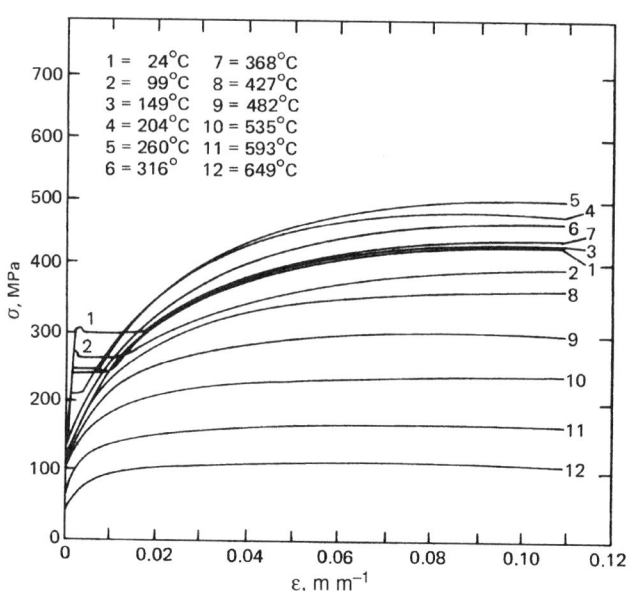

Fig. 1-10.4. *Stress-strain curves for a structural steel (ASTM A36) at room temperature and elevated temperatures.*[31]

There is substantial evidence[4,23] that Equations 2 and 3, with $n = 3$, are applicable to any two-phase material consisting of a continuous and a discontinuous solid phase, irrespective of the geometry of the discontinuous phase.

$$k = \frac{\nu_1 k_1 + \beta \nu_2 k_2}{\nu_1 + \beta \nu_2} \qquad (19)$$

where

$$\beta = \frac{3k_1}{2k_1 + k_2} \qquad (20)$$

and the subscripts 1 and 2 relate to the continuous and discontinuous phases, respectively. When both phases are essentially continuous, as with most porous materials (air as the second phase), a lower value of n in Equation 3 seems to be applicable: about $n = 1.5$. If the conductivity of air is negligibly small in comparison with that of the solid, the following is a fair approximation[2]

$$k = k_s \frac{1-P}{1+aP} \qquad (21)$$

where k_s (W m^{-1} K^{-1}) is the conductivity of the (possibly hypothetical) non-porous solid, and a (dimensionless) is a material constant. If the pores are discontinuous $a \cong 0.5$, and $a \cong 2$ if the pores are fully interconnected.

At higher temperatures, because of radiative heat transfer through the pores, the contribution of the pores to the thermal conductivity of the solid must not be disregarded. The thermal conductivity is customarily expressed as a sum of two terms, i.e.[24]

$$k = k_s \frac{1-P}{1+aP} + P(k_a + 4\sigma\epsilon b \delta T^3) \qquad (22)$$

where k_a (J m^{-1} K^{-1}) is the thermal conductivity of air at temperature T (K), σ (W m^{-2} K^{-4}) is the Stefan-Boltzmann constant, ϵ (dimensionless) is the emissivity of the pores (probably between 0.7 and 1), b (dimensionless) is a constant characteristic of the pore geometry (probably between 0.6 and 1), and δ (m) is the characteristic pore size.

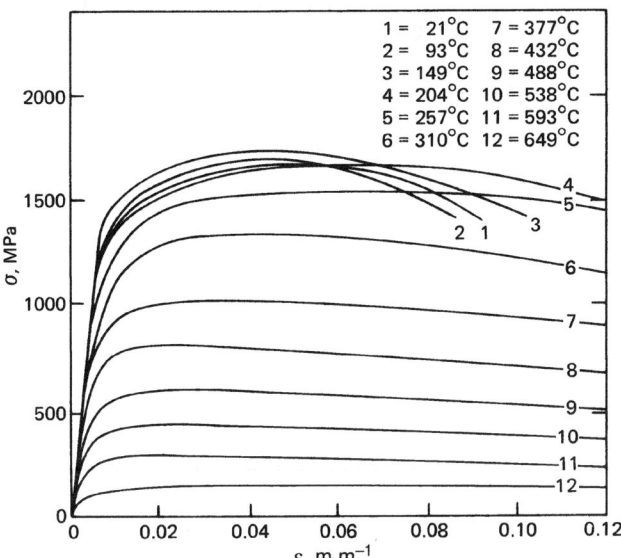

Fig. 1-10.5. Stress-strain curves for prestressing steel (ASTM A421) at room temperature and elevated temperatures.[31]

The thermal conductivity of solids is a structure-sensitive property. For crystalline solids the thermal conductivity is relatively high at room temperature, and gradually decreases as the temperature rises. For predominantly amorphous solids, on the other hand, the conductivity is low at room temperature and increases slightly with the rise of temperature. The conductivity of porous crystalline materials may also increase at very high temperatures because of the radiant conductivity of the pores.

Unfortunately, to the author's knowledge, no scanning technique exists for acquiring a continuous thermal conductivity *vs* temperature curve from a single temperature sweep. Such a curve must be estimated from the results of tests conducted at discrete temperature levels.

Special problems arise with the estimation of the thermal conductivity for temperature intervals of physicochemical instability. Both the steady-state and variable-state techniques of measuring thermal conductivity require the stabilization of a pattern of temperature distribution (and thereby a certain microstructural pattern) in the test sample prior to the test. The test results can be viewed as points on a continuous thermal conductivity *vs* temperature curve obtained by an imaginary scanning technique performed at an extremely slow scanning rate. Since each point pertains to a more or less stabilized microstructural pattern, there is no way of knowing how the thermal conductivity would vary in the course of a physicochemical process developing at a finite rate and varying microstructure.

The points plotted as open or solid circles in the figures to be shown represent information developed using a variable-state technique[25] characterized by relatively small temperature gradients during the test. The accuracy of the method is estimated to be about ±7 percent for physicochemically stable, low-porosity materials. Because of the difficulties described earlier, the accuracy of this method in temperature intervals of physicochemical instability cannot be firmly stated.

On account of the non-reversible microstructural changes brought about by heating, the thermal conductivity of building materials (and perhaps most other materials) is usually different in the heating and cooling cycles. Open and solid circles are used in the figures to identify thermal conductivity values obtained by stepwise increasing and stepwise decreasing the temperature of the sample, respectively.

SOURCE OF INFORMATION

Information on the properties of building materials at elevated temperatures is scattered throughout the literature. There are a few publications, however, that may be particularly valuable for fire safety practitioners. A book by Harmathy presents a wealth of information on concrete, steel, wood, brick, and gypsum, and on various plastics.[2] The thermal properties of 31 building materials are surveyed in an NRCC report.[14] The mechanical and thermal properties of concrete are discussed in an ACI guide,[26] and in reports by Bennetts[27] and Schneider.[28] Those of steel are surveyed in the ACI guide, in Bennetts's report, and in a report by Anderberg.[29] Information on the thermal conductivity of more than 50 rocks (potential concrete aggregates) is presented in a paper by Birch and Clark.[30]

STEEL

Steel is a Group L material. The steels most often used in the building industry are either hot-rolled or cold-drawn. The structural steels and concrete-reinforcing bars are hot-rolled, low-carbon, ferrite-pearlite steels. They have a randomly oriented grain structure, and their strength depends mainly on their carbon content. The prestressing steel wires and strands for concrete are usually made from cold-drawn, high-carbon, pearlitic steels with an elongated grain structure, oriented in the direction of the cold work.

Information on the mechanical properties of two typical steels [a structural steel (ASTM A36) and a prestressing wire (ASTM A421)] is presented in Figures 1-10.4 through 1-10.6, and in Table 1-10.1.[31] Figures 1-10.4 and 1-10.5 are stress-strain curves at room temperature (24°C and 21°C, respectively) and at a number of elevated temperature levels. Figure 1-10.6 shows the effect of temperature on the yield and ultimate strengths of the two steels.

Table 1-10.1 presents information on the effect of stress on the two creep parameters, Z and ε_{t0}. Since creep is a very structure-sensitive property, the creep parameters may show a substantial spread even for steels with similar characteristics at room temperature. The application of the creep parameters to the calculation of the time of structural failure in fire is discussed in Reference 2.

The modulus of elasticity (E) is about 210×10^3 MPa for a variety of common steels at room temperature. Figure 1-10.7 shows its variation with temperature for structural steels[32] and steel reinforcing bars.[33] (E_0 in Figure 1-10.7 is the modulus of elasticity at room temperature.)

The density (ρ) of steel is about 7850 kg/m^{-3}. Its coefficient of thermal expansion (α) is a structure-insensitive property. For an average carbon steel it is 11.4×10^{-6} m m^{-1} K^{-1} at room temperature. The dilatometric curve shown in Figure 1-10.8 is applicable to most of the common steels. The curve reveals substantial contraction of the material at about 700°C, which is associated with the transformation of the ferrite-pearlite structure into austenite.

Being a structure-sensitive property, the thermal conductivity of steel is not easy to define. For carbon steels it usually varies within the range of 46 to 65 W m^{-1} K^{-1}.

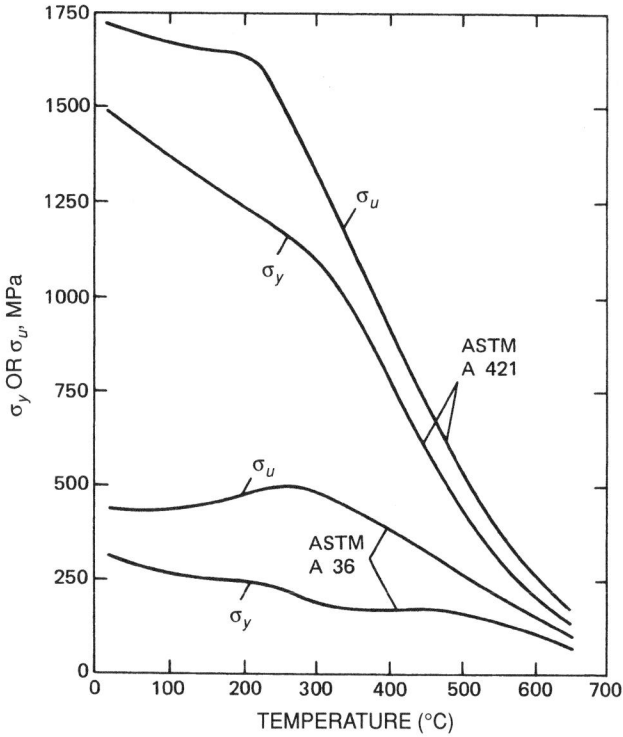

Fig. 1-10.6. The ultimate and yield strengths for a structural steel (ASTM A36) and a prestressing steel (ASTM A421) at elevated temperatures.[31]

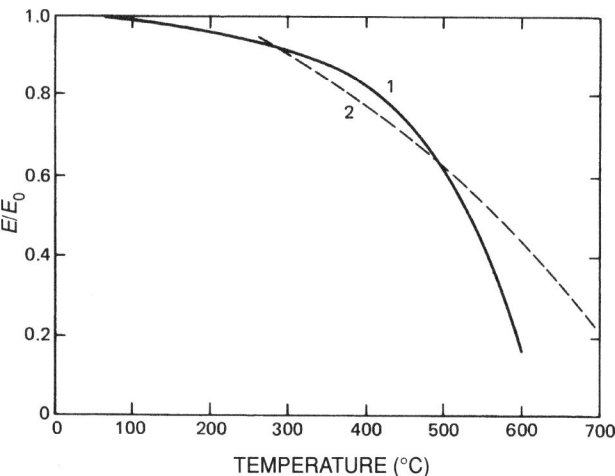

Fig. 1-10.7. The effect of temperature on the modulus of elasticity of (1) structural steels[32] and (2) steel reinforcing bars.[33]

CONCRETE

Concrete is a Group L/I material. The word concrete covers a large number of different materials, with the single common feature that they are formed by the hydration of portland cement. Since the hydrated cement paste amounts to only 24 to 43 volume percent of the materials present, the properties of concrete may vary widely with the aggregates used.

Concretes are usually subdivided into two major groups: (1) normal-weight concretes with densities in the 2150 to 2450 kg m^{-3} range, and (2) lightweight concretes with densities between 1350 and 1850 kg m^{-3}. Fire safety practitioners again subdivide the normal-weight concretes into silicate (siliceous) and carbonate aggregate concretes, according to the composition of the principal aggregate.

A great deal of information is available in the literature on the mechanical properties of various concretes. It is summarized in reports by Bennetts[27] and Schneider,[28] in the ACI guide,[26] and in Harmathy's book.[2] Figure 1-10.9 shows the stress-strain curves for a lightweight concrete with expanded shale aggregate at room temperature (24°C) and a few elevated temperature levels.[34] The shape of the curves

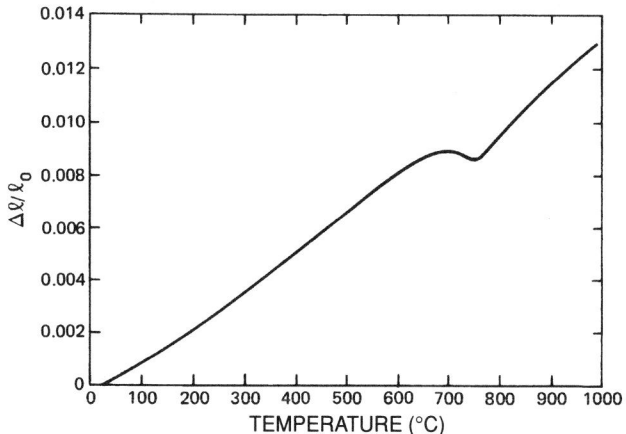

Fig. 1-10.8. Dilatometric curve for steel.

TABLE 1-10.1 Creep Parameters for a Structural Steel and a Prestressing Steel[31]*

Steel	$\Delta H_c/R$ (K)	$\varepsilon_{to}(\sigma)$ (m m^{-1})	$Z(\sigma)$ (h^{-1})
ASTM A36	38890	$3.258 \times 10^{-17}\ \sigma^{1.75}$	$2.365 \times 10^{-20}\ \sigma^{4.7}$ if $\sigma \leq 103.4 \times 10^6$ $1.23 \times 10^{16}\ \exp(4.35 \times 10^{-8}\ \sigma)$ if $103.4 \times 10^6 \leq \sigma \leq 310 \times 10^6$
ASTM A421	30560	$8.845 \times 10^{-9}\ \sigma^{0.67}$	$1.952 \times 10^{-10}\ \sigma^3$ if $\sigma \leq 172.4 \times 10^6$ $8.21 \times 10^{13}\ \exp(1.45 \times 10^{-8}\ \sigma)$ if $172.4 \times 10^6 \leq \sigma \leq 690 \times 10^6$

*σ in Pa.

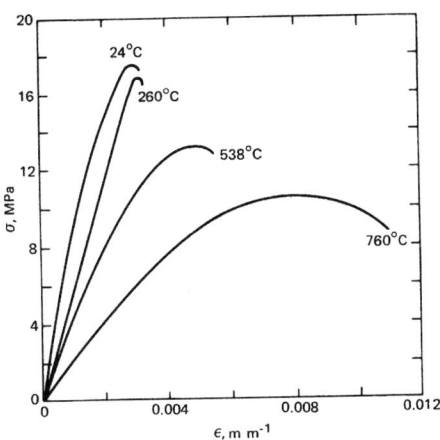

Fig. 1-10.9. Stress-strain curves for a lightweight masonry concrete at room and elevated temperatures.[34]

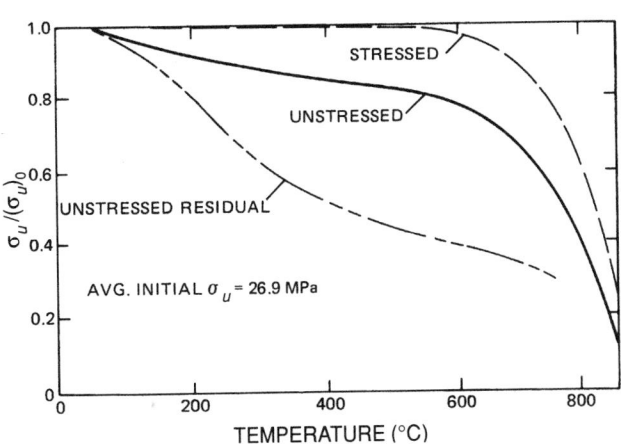

Fig. 1-10.11. The effect of temperature on the compressive strength of a normal-weight concrete with silicate aggregate.[36]

may depend on the time of holding the test specimen at the target temperature level before the compression test.

The modulus of elasticity (E) of various concretes at room temperature may fall within a very wide range, 5.0×10^3 to 35.0×10^3 MPa, dependent mainly on the water-cement ratio in the mixture, the age of concrete, the method of conditioning, and the amount and nature of the aggregates. Cruz found that the modulus of elasticity decreases rapidly with the rise of temperature, and the fractional decline does not depend significantly on the type of aggregate.[35] (See Figure 1-10.10; E_0 in Figure 1-10.10 is the modulus of elasticity at room temperature.) From other surveys[2,27] it appears, however, that the modulus of elasticity of normal-weight concretes decreases faster with the rise of temperature than that of lightweight concretes.

The compressive strength (σ_u) of concrete may also vary within a wide range. It is influenced by the same factors as the modulus of elasticity. For conventionally produced normal-weight concretes, the strength at room temperature is usually between 20 and 50 MPa. For lightweight concretes it is usually between 10 and 30 MPa.

Information on the variation of the compressive strength with temperature is presented in Figures 1-10.11 (for a silicate aggregate concrete), 1-10.12 (for a carbonate aggregate concrete), and 1-10.13 (for two lightweight aggregate concretes; one made with the addition of natural sand).[36] [$(\sigma_u)_0$ in the figures stands for the compressive strengths of concrete at room temperature.] In some experiments the specimens were heated to the test temperature without load (see curves "unstressed"). In others they were heated under a load amounting to 40 percent of the ultimate strength (curves "stressed"). Again, in others they were heated to the target temperature without load, then cooled to room temperature and stored at 75 percent relative humidity for six days, and finally tested at room temperature (curves "unstressed residual").

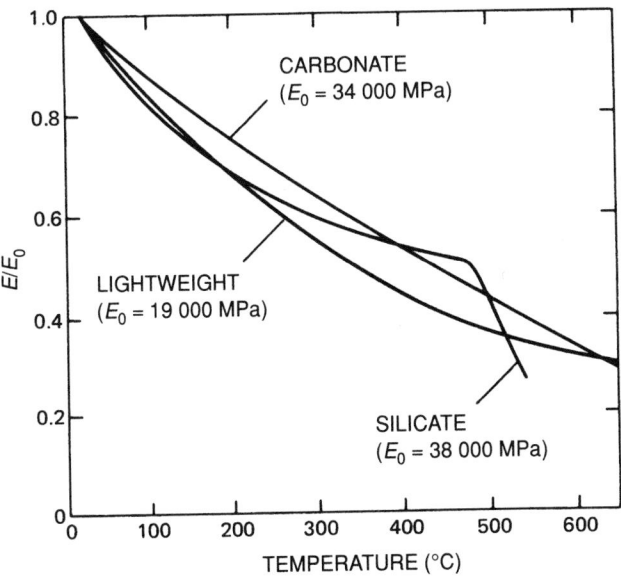

Fig. 1-10.10. The effect of temperature on the modulus of elasticity of concretes with various aggregates.[35]

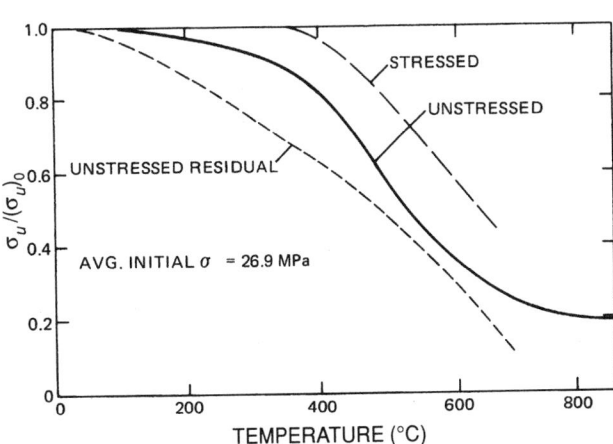

Fig. 1-10.12. The effect of temperature on the compressive strength of a normal-weight concrete with carbonate aggregate.[36]

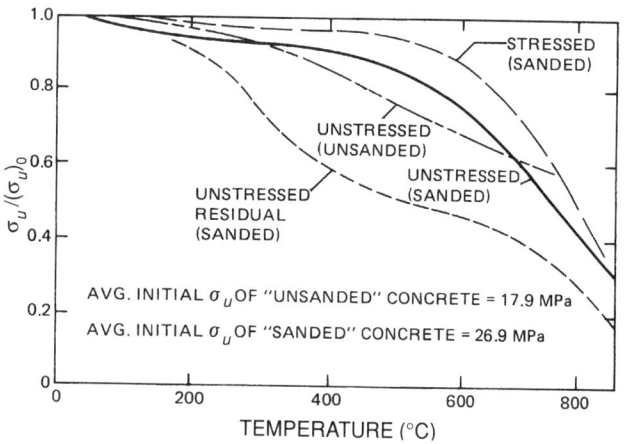

Fig. 1-10.13. The effect of temperature on the compressive strength of two lightweight concretes (one with added natural sand).[36]

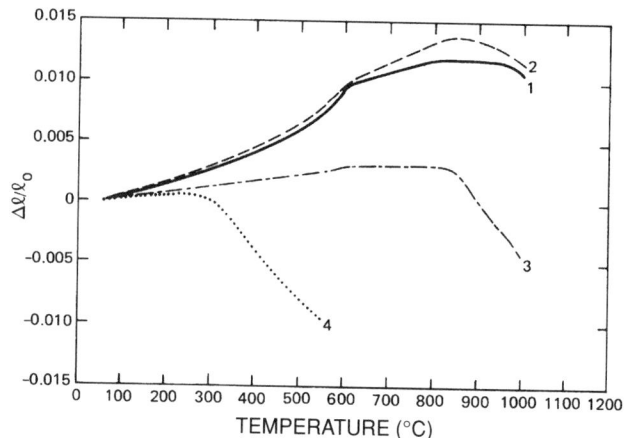

Fig. 1-10.15. Dilatometric curves for two normal-weight and two lightweight concretes.[18] *(1) normal-weight concrete with silicate aggregate, (2) normal-weight concrete with carbonate aggregate, (3) lightweight concrete with expanded shale aggregate, (4) lightweight concrete with pumice aggregate.*

Some information on the creep of concrete at elevated temperatures is available from the work of Cruz,[37] Maréchal,[38] Gross,[39] and Schneider *et al.*[40] The creep curves shown in Figure 1-10.14 are those recorded by Cruz for a normal-weight concrete with carbonate aggregates.

Since the aggregates amount to 60 to 75 percent of the volume of concrete, the dilatometric curve usually resembles that of the principal aggregate. However, some lightweight aggregates, e.g., pearlite and vermiculite, are unable to resist the almost continuous shrinkage of the cement

paste on heating, and therefore their dilatometric curves bear the characteristic features of the curve for the paste.

The dilatometric curves of two normal-weight concretes (with silicate and carbonate aggregates) and two lightweight concretes (with expanded shale and pumice aggregates) are shown in Figure 1-10.15.[18] These curves were obtained in the course of a comprehensive study performed on 16 concretes.

The results of dilatometric and thermogravimetric tests were combined to calculate the density (ρ) *vs* temperature relation for these four concretes. They are shown in Figure 1-10.16. The partial decomposition of the aggregate is responsible for a substantial drop (above 700°C) in the density of concretes made with carbonate aggregate.

The usual ranges of variation of the volume-specific heat (i.e., the product ρc_p) for normal-weight and lightweight concretes is shown in Figure 1-10.17. This information, derived by

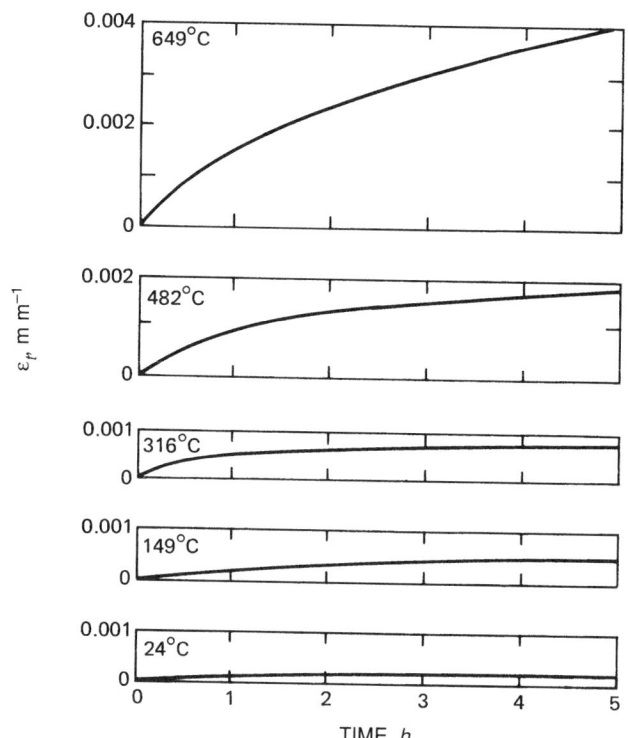

Fig. 1-10.14. Creep of a carbonate aggregate concrete at various temperature levels (applied stress: 12.4 MPa; compressive strength of the material at room temperature: 27.6 MPa).[37]

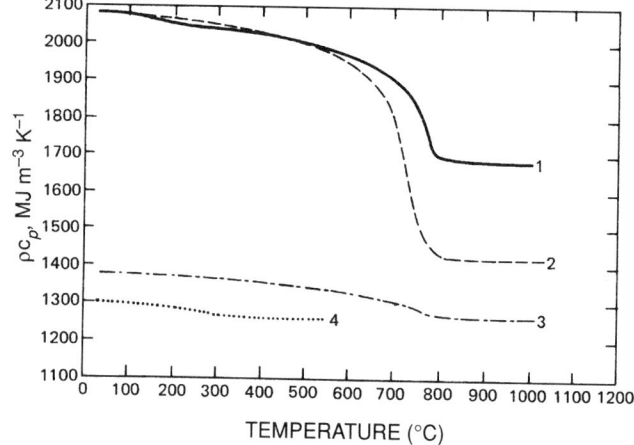

Fig. 1-10.16. Density of two normal-weight and two lightweight concretes.[18] *(1) normal-weight concrete with silicate aggregate, (2) normal-weight concrete with carbonate aggregate, (3) lightweight concrete with expanded shale aggregate, (4) lightweight concrete with pumice aggregate.*

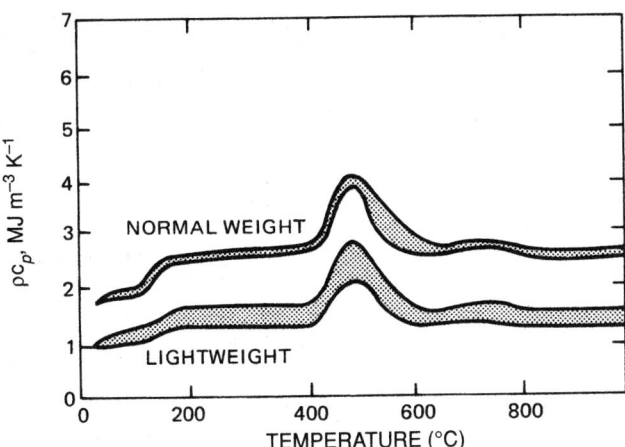

Fig. 1-10.17. Usual ranges of variation for the volume-specific heat of normal-weight and lightweight concretes.[6]

combining thermodynamic data with thermogravimetric observations,[2,6] has since been confirmed by differential scanning calorimetry.[14] Experimental data are also available on a few concretes and some of their constituents.[2,14]

The thermal conductivity (k) of concrete depends mainly on the nature of its aggregates. In general, concretes made with dense, crystalline aggregates show higher conductivities than those made with amorphous or porous aggregates. Among common aggregates, quartz has the highest conductivity; therefore, concretes made with siliceous aggregates are on the whole more conductive than those made with other silicate and carbonate aggregates.

Derived from theoretical considerations,[6] the solid curves in Figure 1-10.18 describe the variation with temperature of the thermal conductivity of four concretes. In deriving these curves, two concretes (Nos. 1 and 2) were visualized to represent limiting cases among normal-weight concretes, and the other two (Nos. 3 and 4), limiting cases among lightweight concretes. The points in Figure 1-10.18 stand for experimental data. They reveal that the upper limiting case is probably never reached with aggregates in com-

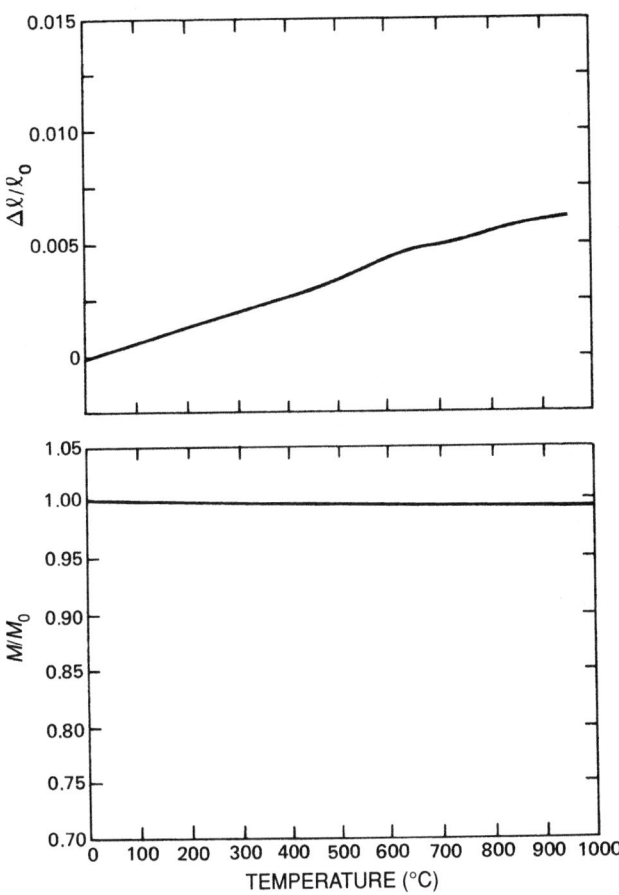

Fig. 1-10.19. Dilatometric and thermogravimetric curves for a clay brick.[14]

mon use, and that the thermal conductivity of lightweight concretes may be somewhat higher than predicted on theoretical considerations.

Further experimental information on the thermal conductivity of some normal-weight and many lightweight concretes is available from the literature.[6,14,18]

BRICK

Building brick belongs in the L/I group of materials. The density (ρ) of bricks ranges from 1660 to 2270 kg m^{-3}, depending on the raw materials used in the manufacture, and on the molding and firing technique. The true density of the material (ρ_t) is somewhere between 2600 and 2800 kg m^{-3}.

The modulus of elasticity of brick (E) is usually between 10×10^3 and 20×10^3 MPa. Its compressive strength (σ_u) varies in a very wide range, from 9 to 110 MPa. 50 MPa may be regarded as average.[41] This value is an order of magnitude greater than the stresses allowed in the design of grouted brickwork. Since brick is rarely considered for important load-bearing roles in buildings, there has been little interest in the mechanical properties of bricks at elevated temperatures.

At room temperature the coefficient of thermal expansion (α) for clay bricks is about 5.5×10^{-6} m m^{-1} K^{-1}. The dilatometric and thermogravimetric curves for a clay brick of 2180 kg m^{-3} density are shown in Figure 1-10.19.[14] The

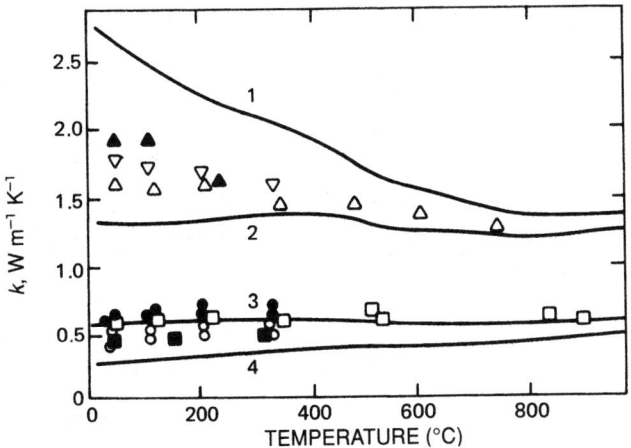

Fig. 1-10.18. Thermal conductivity of four "limiting" concretes and some experimental thermal conductivity data.[6,18] Symbols: ▲, ▽—various gravel concretes; ●—expanded slag concretes; ■, □—expanded shale concretes; ○—pumice concrete.

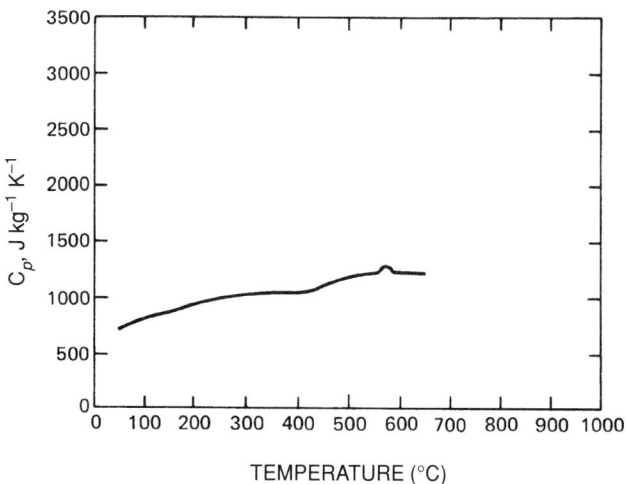

Fig. 1-10.20.   Apparent specific heat of a clay brick.[14]

variation with temperature of the specific heat and the thermal conductivity of this brick is shown in Figures 1-10.20 and 1-10.21, respectively.[14]

WOOD

Wood is a Group L/I/F or I/F material. Although still favored in residential construction, wood structures have lost ground to the two contemporary materials: concrete and steel.

The use of wood as a load-bearing material must not be taken for granted. Although about 180 wood species are commercially grown in the United States, only about 25 species have been assigned working stresses. The two groups most extensively used as structural lumber are the Douglas firs and the southern pines.

The oven-dry density (ρ) of commercially important woods ranges from 300 kg m^{-3} (white cedar) to 700 kg m^{-3} (hickory, black locust). The density of Douglas firs varies from 430 to 480 kg m^{-3}, and that of southern pines, from 510 to 580 kg m^{-3}. The true density of the solid material

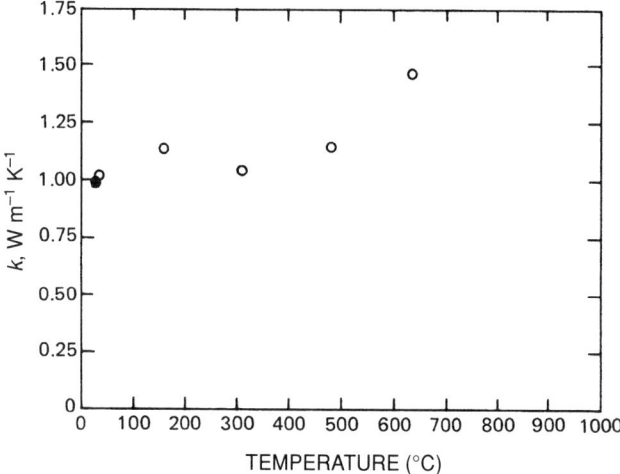

Fig. 1-10.21.   Thermal conductivity of a clay brick. Symbols: ○—heating cycle, ●—after cooling.[14]

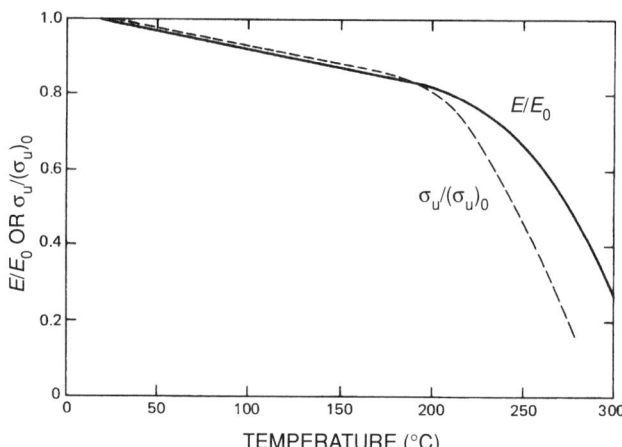

Fig. 1-10.22.   The effect of temperature on the modulus of elasticity and compressive strength of wood.[43–45]

that forms the walls of wood cells (ρ_t) is about 1500 kg m^{-3} for all kinds of wood.

The modulus of elasticity (E) of air-dry, clear wood along the grain varies from 5.5×10^3 to 15.0×10^3 MPa, and its crushing strength (σ_u) from 13 to 70 MPa. These properties are related and roughly proportional to the density, regardless of the species.[42]

Figure 1-10.22 shows the variation of the modulus of elasticity and compressive strength of oven-dry, clear wood with temperature.[43–45] [E_0 and $(\sigma_u)_0$ in the figure are modulus of elasticity and compressive strength at room temperature, respectively.]

The coefficient of linear thermal expansion (α) ranges from 3.2×10^{-6} to $4.6 \times 10^{-6} \text{ m m}^{-1} \text{ K}^{-1}$ along the grain, and from 21.6×10^{-6} to $39.4 \times 10^{-6} \text{ m m}^{-1} \text{ K}^{-1}$ across the grain.[46]

The dilatometric and thermogravimetric curves of a pine with a 400 kg m^{-3} oven-dry density are shown in Figure 1-10.23.[14]

The thermal conductivity (k) across the grain of this pine was measured as 0.86 to $0.107 \text{ W m}^{-1} \text{ K}^{-1}$ between room temperature and 140°C.[14] Figure 1-10.24 shows the apparent specific heat for the same pine, as a function of temperature.[14] The accuracy of the curve [developed by differential scanning calorimeter (DSC)] is somewhat questionable. However, it provides useful information on the nature of decomposition reactions that take place between 150 and 370°C.

GYPSUM

Gypsum (calcium sulfate dihydrate: $CaSO_4 \cdot 2H_2O$) is a Group I material. Gypsum board is produced by mixing water with plaster of paris (calcium sulfate hemihydrate: $CaSO_4 \cdot \frac{1}{2} H_2O$) or with Keene's cement (calcium sulfate anhydrate: $CaSO_4$). The interlocking crystals of $CaSO_4 \cdot 2H_2O$ are responsible for the hardening of the material.

Gypsum products are extensively used in the building industry in the form of boards, including wallboard, formboard, and sheathing. The core of the boards is fabricated with plaster of paris, to which weight- and set-controlling additives are mixed. Furthermore, plaster of paris, with the addition of aggregates (such as sand, pearlite, vermiculite, wood fiber) is used in wall plaster as base coat, and Keene's cement (neat or mixed with lime putty) as finishing coat.

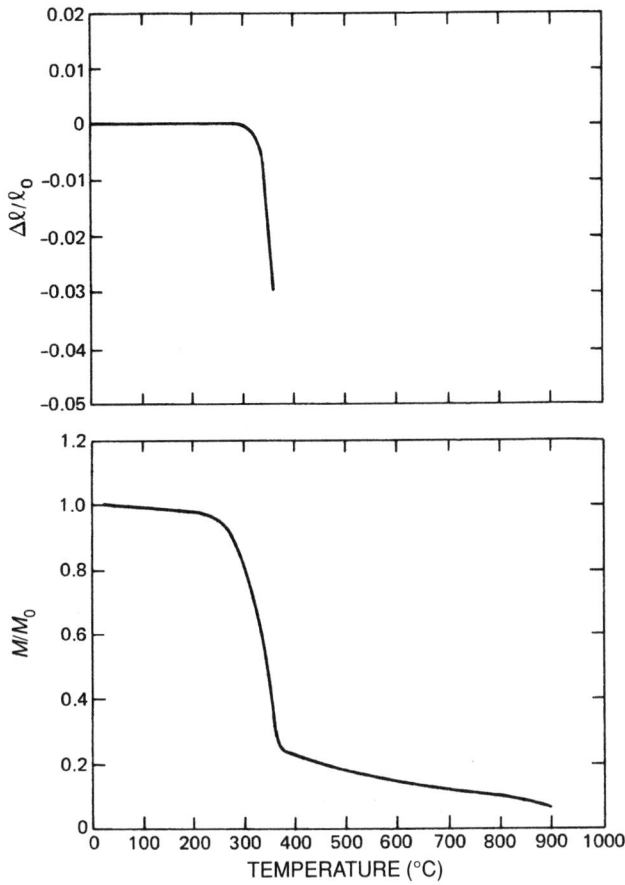

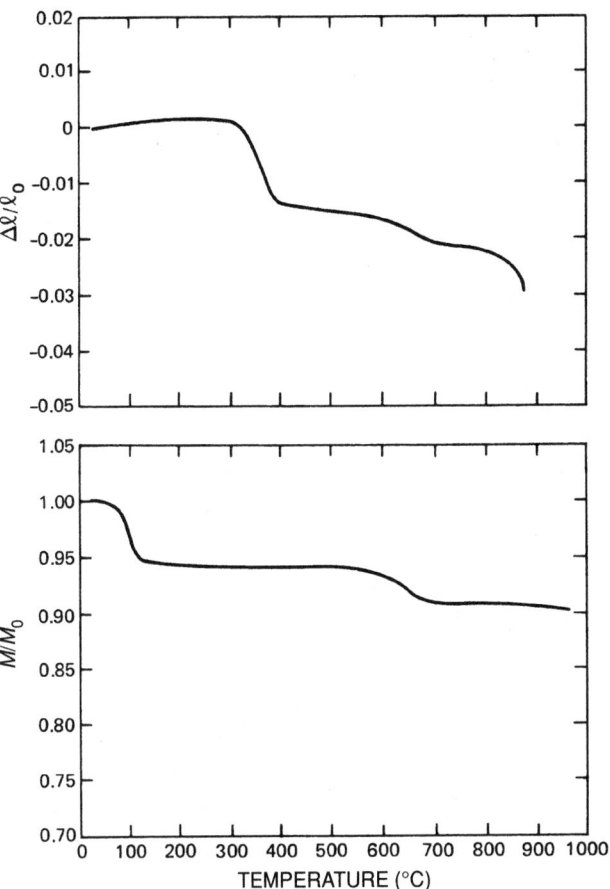

Fig. 1-10.23. Dilatometric and thermogravimetric curves for a pine of 400 kg m^{-3} density.[14]

Fig. 1-10.25. Dilatometric and thermogravimetric curves for a gypsum board of 678 kg m^{-3} density.[14]

Gypsum is an ideal fire protection material. On heating it will lose the two H_2O molecules at temperatures between 125 and 200°C. The heat of complete dehydration is 0.61×10^6 J per kg gypsum. Due to the substantial absorption of energy in the dehydration process, a gypsum layer applied to

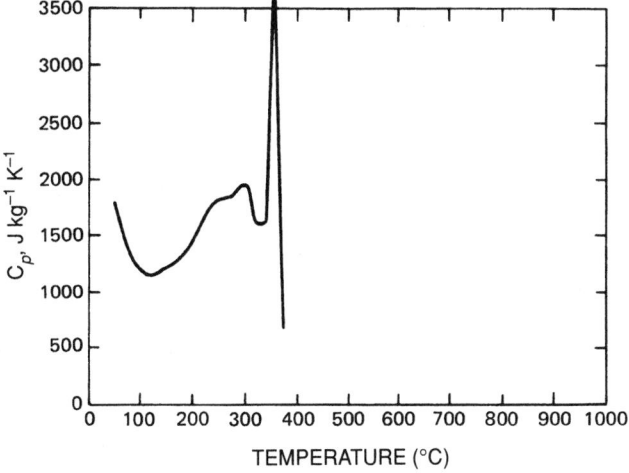

Fig. 1-10.24. Apparent specific heat for a pine of 400 kg m^{-3} density.[14]

the surface of a building element is capable of markedly delaying the penetration of heat into the underlying load-bearing construction.

The coefficient of thermal expansion (α) of gypsum products may vary between 11.0×10^{-6} and 17×10^{-6} m m^{-1} K^{-1} at room temperature, depending on the nature and amount of aggregates used.

The variation with temperature of the volume specific heat (ρc_p) of pure gypsum has been illustrated in reference 47, based on information reported in the literature.[48,49] The dilatometric and thermogravimetric curves of a so-called fire-resistant gypsum board of 678 kg m^{-3} density are shown in Figure 1-10.25. The apparent specific heat (c_p) vs temperature curve for the same gypsum board is shown in Figure 1-10.26. Figure 1-10.27 gives some information on its thermal conductivity (k) on heating and cooling.[14]

The thermal conductivity of gypsum products is difficult to assess, owing to large variations in their porosities and the nature of the aggregates. A typical value for plaster boards of about 700 kg m^{-3} density is 0.25 W m^{-1} K^{-1}.

MISCELLANEOUS OTHER MATERIALS

Further information is available from the literature on the dilatometric and thermogravimetric behavior, apparent specific heat, and thermal conductivity of a number of materials in Group I, including asbestos cement board, expanded

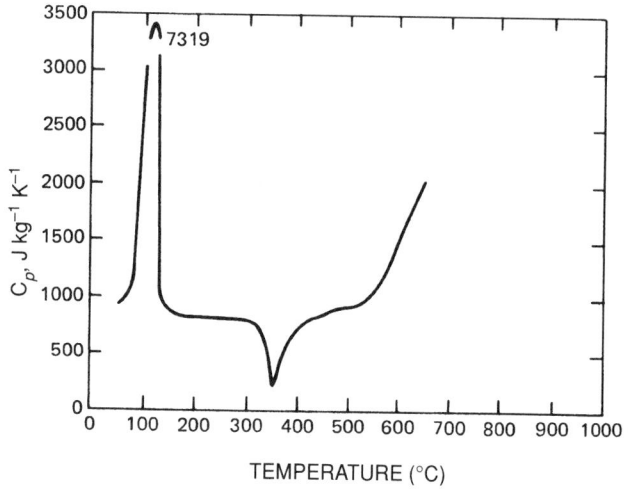

Fig. 1-10.26. *Apparent specific heat of a gypsum board of 678 kg* m^{-3} *density.*[14]

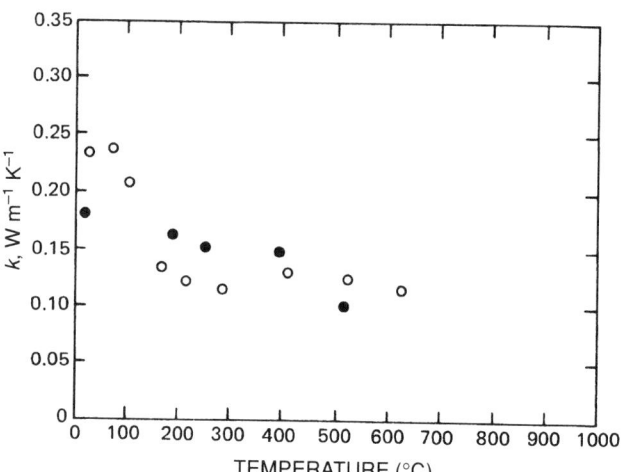

Fig. 1-10.27. *Thermal conductivity of a gypsum board of 678 kg* m^{-3} *density.*[14] *Symbols:* ○—*values pertaining to the heating cycle,* ●—*values pertaining to the cooling cycle.*

plastic insulating boards, mineral fiber fireproofing, arborite, and glass-reinforced cement board.[14] The properties of plastics and their behavior in fire are discussed in other chapters of this handbook and in reference 2.

NOMENCLATURE

a	material constant, dimensionless
b	constant, characteristic of pore geometry, dimensionless
c	specific heat, $J\ kg^{-1}\ K^{-1}$
$\bar{c}$	specific heat for a mixture of reactants and solid products, $J\ kg^{-1}\ K^{-1}$
E	modulus of elasticity, Pa
h	enthalpy, $J\ kg^{-1}$
Δh	latent heat associated with a "reaction," $J\ kg^{-1}$
ΔH_c	activation energy for creep, $J\ kmol^{-1}$
k	thermal conductivity, $W\ m^{-1}\ K^{-1}$
ℓ	dimension, m
$\Delta\ell$	$= \ell - \ell_0$
m	exponent, dimensionless
M	mass, kg
n	material constant, dimensionless
P	porosity, $m^3\ m^{-3}$
R	gas constant, $8315\ J\ kmol^{-1}\ K^{-1}$
S	specific surface area, $m^2\ m^{-3}$
t	time, h
T	temperature, K (or °C)
v	volume fraction, $m^{-3}\ m^3$
w	mass fraction, $kg\ kg^{-1}$
Z	Zener-Hollomon parameter h^{-1}

Greek letters

α	coefficient of linear thermal expansion, $m\ m^{-1}$
β	expression defined by Equation 3, dimensionless
δ	characteristic pore size, m
ε	emissivity of pores, dimensionless
ε	strain (deformation), $m\ m^{-1}$
ε_{t0}	creep parameter, $m\ m^{-1}$
$\dot{\varepsilon}_{ts}$	rate of secondary creep, $m\ m^{-1}\ h^{-1}$
θ	temperature-compensated time, h
ξ	reaction progress variable, dimensionless

π	material property (any)
ρ	density, $kg\ m^{-3}$
σ	stress; strength, Pa
σ	Stefan-Boltzmann constant, $5.67 \times 10^{-8}\ W\ m^{-2}\ K^{-4}$

Subscripts

a	of air
i	of the ith constituent
p	at constant pressure
s	of the solid matrix
t	true
t	time-dependent (creep)
T	at temperature T
u	ultimate
y	yield
0	original value, at reference temperature

REFERENCES CITED

1. T.Z. Harmathy, *Technical Paper No. 242*, National Research Council of Canada, Ottawa (1967).
2. T.Z. Harmathy, *Fire Safety Design and Concrete*, Longman Scientific and Technical, Harlow, U.K.(1993).
3. D.A.G. Bruggeman, *Physik. Zeitschr.*, 37, 906 (1936).
4. R.L. Hamilton and O.K. Crosser, *I & EC Fundamen.*, 7, 187 (1962).
5. J.C. Maxwell, *A Treatise on Electricity and Magnetism*, Vol 1, 3rd ed. Clarendon Press, Oxford (1904).
6. T.Z. Harmathy, *J. Matls.*, 5, 47 (1970).
7. J.E. Dorn, *J. Mech., Phys. Solids*, 3, 85 (1954).
8. T.Z. Harmathy, in *ASTM STP 422*, American Society for Testing and Materials, Philadelphia (1967).
9. T.Z. Harmathy, *Trans. Am. Soc. Mech. Eng., J. Basic Eng.*, 89, 496 (1967).
10. C. Zener and J.H. Hollomon, *J. Appl. Phys.*, 15, 22 (1944).
11. F.H. Wittmann (ed.), *Fundamental Research on Creep and Shrinkage of Concrete*, Martinus Nijhoff, The Hague, The Netherlands (1982).
12. Y. Anderberg and S. Thelandersson, *Bulletin 54*, Lund Institute of Technology, Lund, Sweden (1976).
13. U. Schneider, *Fire & Matls.*, 1, 103 (1976).
14. T.Z. Harmathy, *DBR Paper No. 1080, NRCC 20956*, National Research Council of Canada, Ottawa (1983).

15. T.Z. Harmathy, *J. Am. Concr. Inst.*, 65, 959 (1968).
16. *951 Thermogravimetric Analyzer (TGA)*, DuPont Instruments, Wilmington (1977).
17. Test Method C 135–86, *1990 Annual Book of ASTM Standards*, Vol. 15.01, American Society for Testing and Materials, Philadelphia (1990).
18. T.Z. Harmathy and L.W. Allen, *J. Am. Concr. Inst.*, 70, 132 (1973).
19. *910 Differential Scanning Calorimeter (DSC)*, DuPont Instruments, Wilmington (1977).
20. J.H. Perry (ed.), *Chemical Engineers' Handbook*, 3rd ed., McGraw-Hill, New York (1950).
21. W. Eitel, *Thermochemical Methods in Silicate Investigation*, Rutgers Univ., New Brunswick (1952).
22. T.Z. Harmathy, *I & EC Fundamen.*, 8, 92 (1969).
23. D.A. DeVries, in *Problems Relating to Thermal Conductivity*, Bulletin de l'Institute International du Froid, Annexe 1952–1, Louvain, Belgique, p. 115 (1952).
24. W.D. Kingery, *Introduction to Ceramics*, John Wiley & Sons, New York (1960).
25. T.Z. Harmathy, *J. Appl. Phys.*, 35, 1190 (1964).
26. *Guide for Determining the Fire Endurance of Concrete Elements*, ACI-216-89, American Concrete Institute, Detroit (1989).
27. I.D. Bennetts, *Report No. MRL/PS23/81/001*, BHP Melbourne Research Laboratories, Clayton, Australia (1981).
28. U. Schneider (ed.), *Properties of Materials at High Temperatures—Concrete*, Kassel Univ., Kassel, Germany (1985).
29. Y. Anderberg (ed.), *Properties of Materials at High Temperatures—Steel*, Lund University, Lund, Sweden (1983).
30. F. Birch and H. Clark, *Am. J. Sci.*, 238, 542 (1940).
31. T.Z. Harmathy and W.W. Stanzak, in *ASTM STP 464*, American Society for Testing and Materials, Philadelphia (1970).
32. *European Recommendations for the Fire Safety of Steel Structures*, European Convention for Construction Steelwork, Tech. Comm. 3, Elsevier, New York (1983).
33. Y. Anderberg, "Mechanical Properties of Reinforcing Steel at Elevated Temperatures," *Tekniska Meddelande*, nr. 36, Sweden (1978).
34. T.Z. Harmathy and J.E. Berndt, *J. Am. Concr. Inst.*, 63, 93 (1966).
35. C.R. Cruz, *J. PCA Res. Devel. Labs.*, 8, 37 (1966).
36. M.S. Abrams, in *ACI SP 25*, American Concrete Institute, Detroit (1971).
37. C.R. Cruz, *J. PCA Res. Devel. Labs.*, 10, 36 (1968).
38. J.C. Maréchal, in *ACI SP 34*, American Concrete Institute, Detroit (1972).
39. H. Gross, *Nucl. Eng. Design*, 32, 129 (1975).
40. U. Schneider, U. Diedrichs, W. Rosenberger, and R. Weiss, *Sonderforschungsbereich 148*, Arbeitsbericht 1978–1980, Teil II, B 3, Technical University of Braunschweig, Braunschweig, Germany (1980).
41. J.W. McBurney and C.E. Lovewell, ASTM *Proceedings of the Thirty-sixth Annual Meeting*, 33 (II), 636 (1933), American Society for Testing and Materials, Detroit.
42. *Wood Handbook: Wood as an Engineering Material*, Agriculture Handbook No. 72, Forest Products Laboratory, U.S. Gov. Printing Office, Washington (1974).
43. C.C. Gerhards, *Wood & Fiber*, 14, 4 (1981).
44. E.L. Schaffer, *Wood & Fiber*, 9, 145 (1977).
45. E.L. Schaffer, *Res. Paper FPL 450*, U.S. Dept. Agric., Forest Products Lab., Madison (1984).
46. F.F. Wangaard, in Section 29 of *Engineering Materials Handbook*, C.L. Mantell (ed.), McGraw Hill, New York (1958).
47. T.Z. Harmathy, in *ASTM STP 301*, American Society for Testing and Materials, Philadelphia (1961).
48. R.R. West and W.J. Sutton, *J. Am. Ceram. Soc.*, 37, 221 (1954).
49. P. Ljunggren, *J. Am. Ceram. Soc.*, 43, 227 (1960).

PROBABILITY CONCEPTS

John R. Hall, Jr.

INTRODUCTION

This chapter introduces the basic definitions and methods of probability theory, which are the foundation for all work on statistics, fire risk evaluation, reliability analysis, and the other topics of this section. With increased availability of sizeable quantities of reliable data on a whole range of topics related to fire protection engineering, it is essential that the analysis of this data be based on sound mathematical principles from probability theory.

BASIC CONCEPTS OF PROBABILITY THEORY

Probability Theory

Probability theory is a branch of mathematics dealing with the modeling of uncertainty through measures of the relative likelihood of alternative occurrences, whether specifically or generally defined.

Set

A set is a collection of elements; to be well-defined it must be possible, for any object that can be defined or described, to say with certainty whether that object is or is not an element or part of the set.

Set Theory

The theory of sets is the most fundamental branch of mathematics and is relevant to probability theory, because all probabilities are built up from sets.

Subsets

A set, *A*, that consists entirely of elements that all are also contained in set *B* is called a subset of *B*. Each element in a set may also be considered a subset of that set.

Dr. John R. Hall, Jr. is Assistant Vice President for Fire Analysis and Research at the National Fire Protection Association. He has been involved in studies of fire experience patterns and trends, models of fire risk, and studies of fire department management experiences since 1974 at NFPA, the National Bureau of Standards, the U.S. Fire Administration, and the Urban Institute.

Set Operators

There are three basic operators essential to the algebraic manipulation of sets:

Complement ($\sim$): This operator applies to a single set *A* and produces the set of all elements that are *not* in *A*. Such an operator is always applied relative to some specification of the set of all elements, which is called the *universal set*, (U). The complement of the universal set is the *null set* (ϕ) or *empty set*, the set with no elements.

Union ($\cup$): This operator is applied to two sets, as in $A \cup B$. It produces the set consisting of all elements that are members of *either A* or *B* or both.

Intersection ($\cap$): This operator is applied to two sets, as in $A \cap B$. It produces the set consisting of all elements that are members of *both A* and *B*.

Relationships Among the Operators

$$\sim (A \cup B) = \sim A \cap \sim B$$

$$\sim (A \cap B) = \sim A \cup \sim B$$

$$(A \cup B) \cap C = (A \cap C) \cup (B \cap C)$$

$$(A \cap B) \cup C = (A \cup C) \cap (B \cup C)$$

$$(A \cup B) \cup C = A \cup (B \cup C)$$

$$(A \cap B) \cap C = A \cap (B \cap C)$$

Venn diagrams: These are a graphical technique for displaying relationships among sets (represented by circles) and operators, within a rectangle that represents the universal set, $\cup$. (See Figures 1-11.1 through 1-11.3.)

Sample Space

This is a set of mutually exclusive elements, each representing a possible outcome or occurrence and collectively representing all possible outcomes or occurrences for the experiment or problem under consideration. A sample space must also have the property that the set operators defined previously, if applied to the subsets of the sample space in

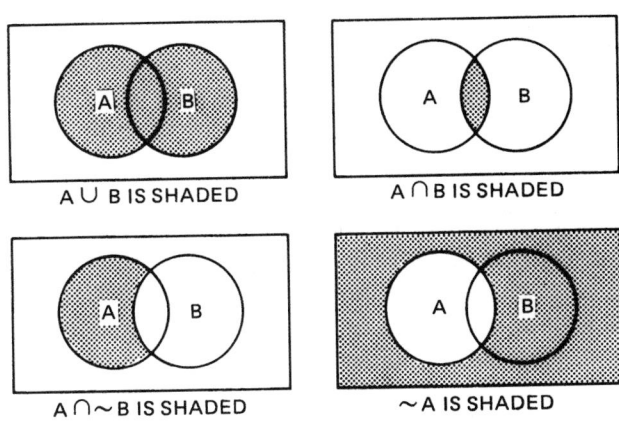

Fig. 1-11.1

any combination, will always produce subsets of the sample space. Subsets of a sample space are called *events*.

Probability Measure

This is a mathematical function, *P*, defined on the subsets (events) of a sample space, *U*, and satisfying the following rules:

1. $P(A) \geq 0$ for any *A*, where *A* is an event subset of *U*.
2. $P(\phi) = 0$.
3. $P(U) = 1$.
4. If $A \cap B = \phi$, then $P(A \cup B) = P(A) + P(B)$.

In the classical theory of probability, it was assumed that all probability measures must be based on experiments (actual or at least imaginable), which could be run repeatedly so that for each outcome *e* (an element of the sample space of possible outcomes), $P(e)$ would be given asymptotically as the ratio between the number of times outcome *e*

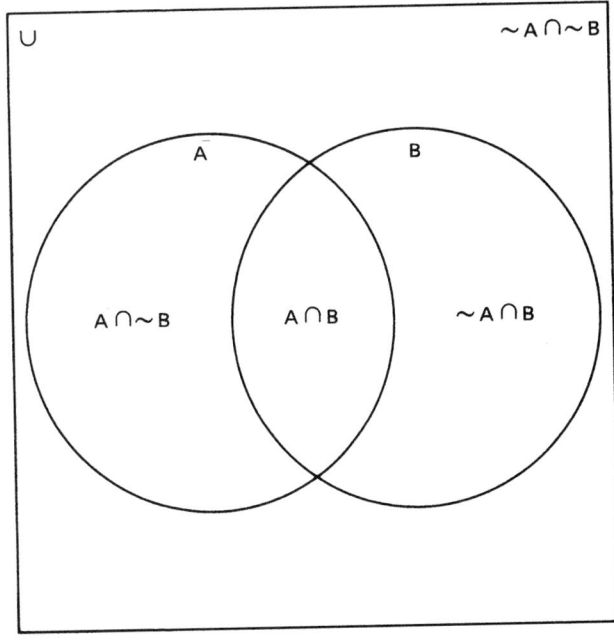

Fig. 1-11.2

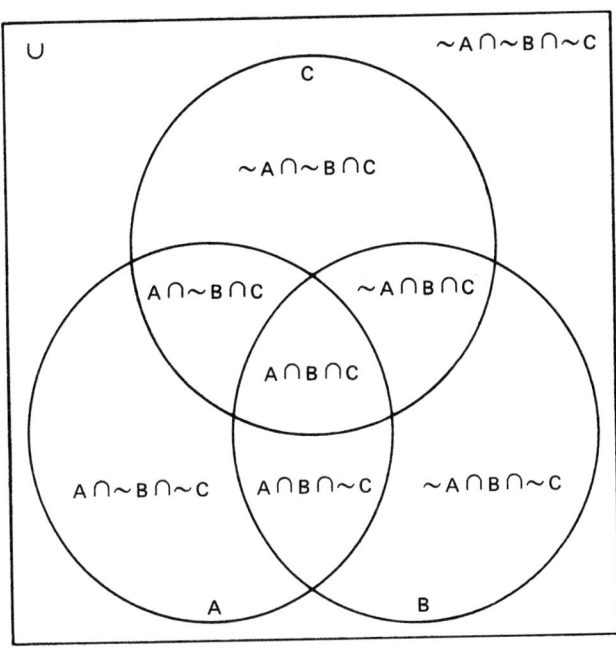

Fig. 1-11.3

occurs and the number of times the experiment is performed. This is called the *frequency interpretation of probability*. More recently, theorists associated with the Bayesian school of statistical inference have argued for the interpretation of probability only as a measure of the individual's strength of belief in the likelihood of an outcome. This is called *subjective probability*. Each of these two schools represents both an underlying conceptual model and an approach that makes practical sense in some but not all situations. In assigning probabilities to the outcomes of heads and tails for a single coin, for example, a relatively brief frequency experiment is easy to conduct. In assigning probabilities to the possible values of the annual inflation rate for next year, the requisite experiment cannot be performed repeatedly. The mathematics of probability theory applies regardless of the source of the probability measure.

Probability Formulas Related to Set Operators

1. $P(A \cup B) = P(A) + P(B) - P(A \cap B)$
2. $P(\sim A) = 1 - P(A)$

These two formulas state, respectively, that (1) the probability that either (inclusive version) of two events will occur is equal to the sum of the probabilities that each event will occur minus the probability that both will occur; and (2) the probability that an event will not occur is equal to one minus the probability that it will occur.

INDEPENDENCE AND CONDITIONALITY

The two events, *A* and *B*, are called *independent* if $P(A \cap B) = P(A) \times P(B)$. Two events that are not independent are called *dependent*.

The *conditional probability of A given B*, $P(A \mid B)$, is defined as $P(A \cap B)/P(B)$. It is normally interpreted to mean the probability that A will occur, given that B has occurred or will occur. If A and B are independent, then $P(A \mid B) = P(A)$ and $P(B \mid A) = P(B)$; in other words, the occurrence of A does not affect the likelihood of B, and vice versa.

It is important to note that two events may be dependent without either being the cause of the other and without any apparent logical connection. A common phenomenon involves two apparently unrelated variables (e.g., annual fire department expenditures on gasoline, annual sales revenue from plastics and petrochemicals) that are dependent because each is related in an understandable way to a third variable (e.g., price per barrel of oil).

Bayes' Law (also called *Bayes' Theorem* and *Bayes' Formula*) states that

1. If B_i, $i = 1, \ldots, N$, are sets (events), and
2. If $B_1 \cup B_2 \cup \ldots \cup B_N = U$, and
3. If $B_i \cap B_j = \phi$ for all $i \neq j$ between 1 and N, and
4. If $P(B_i) \neq 0$, $i = 1, 2, \ldots, N$,

then

$P(B_i \mid A) =$

$$\frac{P(B_i) \times P(A \mid B_i)}{[P(B_1) \times P(A \mid B_1) + P(B_2) \times P(A \mid B_2) + \cdots + P(B_N) \times P(A \mid B_n)]}$$

This is a particularly powerful consequence of the laws of conditional probability and is the foundation for the modern theory of statistical decision theory. What makes it so powerful is this application. Suppose $P(B_1), \ldots, P(B_N)$ represent the current best estimates of the probabilities of various events of interest prior to the performance of an experiment (or the collection of some data on experience). These are called *prior probabilities*. Suppose A is a possible outcome of that experiment whose probability of occurrence, given each of the events $B_1, \ldots, B_N$, can be derived. Then Bayes' law can be used to develop a new set of probabilities, $P(B_1 \mid A), \ldots, P(B_N \mid A)$, that incorporate the information provided by the experiment. These are called *posterior probabilities* because they are probabilities calculated *after* the gathering of information (e.g., through an experiment).

EXAMPLE:

Suppose you have ten coins, nine of which are fair (0.5 probability of heads) and one of which is fixed (1.0 probability of heads). Choose one coin. With no other information, the probability that you have a fair coin (B_1) is 0.9 and the probability that you have the fixed coin (B_2) is 0.1. Suppose you flip the coin once.

If it comes up tails, you know it is a fair coin and Bayes' law confirms this. Let A be the event of getting tails on the one coin flip. Then

$$P(A \mid B_1) = 0.5 \quad \text{and} \quad P(A \mid B_2) = 0$$

Therefore

$$P(B_1 \mid A) = \frac{(0.5)(0.9)}{(0.5)(0.9) + (0)(0.1)} = 1$$

If the coin comes up heads, you still do not know whether it is the fixed or a fair coin. Since heads is more likely with the fixed coin, the evidence points slightly in that

direction. Let A' be the event of getting heads on the single coin flip. Then

$$P(A' \mid B_1) = 0.5 \quad \text{and} \quad P(A' \mid B_2) = 1$$

Therefore

$$P(B_1 \mid A') = \frac{(0.5)(0.9)}{(0.5)(0.9) + (1.0)(0.1)} = 0.82$$

Thus the result of flipping the coin once and obtaining heads has lowered the estimate of the probability that you hold a fair coin from 0.9 to 0.82; correspondingly, your estimate that you hold the fixed coin has risen from 0.1 to 0.18.

RANDOM VARIABLES AND PROBABILITY DISTRIBUTIONS

A *random variable* is a real-number-valued function defined on the elements of a sample space. In some cases, the elements of a sample space may lend themselves to association with a particular random variable (e.g., the sample space consists of outcomes of tossing a die; the random variable is the number of spots on the exposed face). In other cases, the random variable may be only one of many that could easily have been associated with the sample space (e.g., the sample space consists of all citizens of the U.S.; the random variable is the weight to the nearest pound).

Each value of a random variable corresponds to an event subset of the sample space consisting of all elements for which the random variable takes on that value. The *probability of a value of the random variable*, then, is the probability of that event subset.

A *discrete probability distribution* is one for which the random variable has a finite or countably infinite number of possible values (e.g., values can be any integer from 0 to 10; values can be any integer).

A *continuous probability distribution* is one for which the random variable can take on an uncountably infinite number of possible values (e.g., values can be any real number from 0 to 10; values can be any real number).

A *probability distribution function* (also called *probability density*, *probability density function*, and *probability distribution*) is a mathematical function, f, that gives the probability associated with each value of a random variable, $f(y) = P(x = y)$. The term *density* is usually reserved for random variables that can take on an uncountably infinite range of values, so that the probability of a range of values of the variable must be computed through integral calculus.

Because each value, y, of a random variable, x, is associated with a subset of the sample space, $f(y) \geq 0$ for all y.

Because no element of a sample space can take on two or more values of a random variable and each element must take on some value, the values of the random variable collectively correspond to a set of mutually exclusive subsets that exhaust all elements of the sample space, and so

$$\sum_{\text{all } x} f(x) = 1$$

for discrete probability distributions, and

$$\int_{\text{all } x} f(x) dx = 1$$

for continuous probability distributions.

A *cumulative distribution* is a mathematical function that, for each value of a random variable, gives the probability that the random variable will take on that value or any lesser value,

$$F(y) = P(x \leq y) = \sum_{x \leq y} f(x)$$

for discrete probability distributions, and

$$F(y) = \int_{x \leq y} f(x)\, dx$$

for continuous probability distributions.

Note that some references use the term "probability distribution" to refer to the cumulative distribution function, F, of a continuous probability distribution, while referring to the probability density function, f, only as a probability density function.

A *survival function* is a mathematical function that, for each value of a random variable, gives the probability that the random variable will exceed that value

$$S(y) = P(x > y) = \sum_{x > y} f(x)$$

for discrete probability distributions, and

$$S(y) = \int_{x > y} f(x)\, dx$$

for continuous probability distributions.

Therefore for any probability distribution function $P(x)$ and any value y, the cumulative distribution and the survival function based on $P(x)$ sum to one for all values of y,

$$F(y) + S(y) = 1.$$

A *multivariate probability distribution* gives the probability for all combinations of values of two or more random values, e.g., $f(u,v) = P(x = u \text{ and } y = v)$.

KEY PARAMETERS
OF PROBABILITY DISTRIBUTIONS

Certain key parameters of probability distributions are of use because (1) they help to provide essential summary information about the random variable and its probability distributions, and (2) they are included in the functional forms of certain probability distributions that are of use in many practical situations.

The mean, μ, of a random variable (also called its *expected value* or *average*) is defined as

$$\mu = \sum_{\text{all } x} x f(x)$$

for discrete probability distributions, and

$$\mu = \int_{\text{all } x} x f(x)\, dx$$

for continuous probability distributions.

It is also written as $E(x)$, which stands for *expected value of x*. This is the most commonly used of several parameters that relate to some concept of the most typical or average value of a random variable.

The expected value can also be calculated for a function of the random variable, as follows

$$E[g(x)] = \sum_{\text{all } x} g(x) f(x)$$

for discrete probability distributions, and

$$E[g(x)] = \int_{\text{all } x} g(x) f(x)\, dx$$

for continuous probability distributions.

The *variance*, σ^2, of a random variable is a measure of the likelihood that a random variable will take on values far from its mean value. It is a parameter used in the functional form of some commonly occurring probability distributions.

$$\sigma^2 = \sum_{\text{all } x} (x - \mu)^2 f(x)$$

for discrete probability distributions, and

$$\sigma^2 = \int_{\text{all } x} (x - \mu)^2 f(x)\, dx$$

for continuous probability distributions.

The variance can also be expressed as the expected value of a function of the random variable, as follows

$$\sigma^2 = E[(x - \mu)^2] = E(x^2) - \mu^2 = E(x^2) - [E(x)]^2$$

The variance is expressed as σ^2 because most calculations use the square root of the variance, which is called the *standard deviation*, σ.

The *moments* of a probability distribution are defined as the expected values of powers of the random variable. The nth moment is $E(x^n)$. Thus, the mean is the first moment, and the variance is the second moment minus the square of the first moment. The value given by $E[(x - \mu)^n]$ is defined as the nth moment about the mean.

The function defined by $E[e^{\theta x}]$ is called the moment generating function because it is equivalent to an infinite series whose terms consist of, for all k, the kth moment of x times $(\theta^k/k!)$.

For continuous probability distributions, the *median* is that value, y, for which the cumulative distribution, $F(y)$, is equal to 0.5. For discrete probability distributions, the median is that value, y, for which $f(x < y) = f(x > y)$. If the random variable can take on only a finite number of values, the median may not be uniquely defined. The median is less sensitive than the mean to extreme values of the random variable and is the "average" of choice for certain kinds of analyses.

Skewness refers to the symmetry of a probability distribution function around its mean. The median is not equal to the mean in a skewed distribution. An age distribution of fire department uniformed personnel will be skewed, for example, because the small number of personnel in their 50s and 60s will raise the average (mean) age well above the typical age (middle to late 20s). The term is used more frequently than is any specific measure of it. A symmetric distribution has a skewness of zero, no matter how skewness is measured.

Kurtosis is a rarely used term for the relative flatness of a distribution.

The *failure rate* or *hazard rate*, $r(x)$, is defined as: $r(x) = f(x) / S(x)$.

When $f(x)$ is a probability density function for the time to failure, then $r(x)$ will give the conditional probability of time to failure, given survival to time x.

Degrees of freedom is the term given to certain parameters in many commonly used distributions (e.g., Student's t, chi square, F). The distributions that use these parameters are used in tests of the variance of samples. In those tests the parameters always correspond to positive integer values based on the size of the sample (e.g., n, $n - 1$, $n - 2$). Since increasing sample size gives the sample more freedom to vary, it is natural to call those parameters measures of the "degrees of freedom" to vary in the sample.

COMMONLY USED PROBABILITY DISTRIBUTIONS

Uniform and Rectangular Distributions

These distributions give equal probability to all values. The term *rectangular distribution* is reserved for the continuous probability distribution case.

1. $f(x) = 1/N$, for $x_1, \ldots, x_N$, if $f(x)$ is a discrete probability distribution over N values of a random variable.
2. $f(x) = 1/(b - a)$, for $a \leq x \leq b$, if $f(x)$ is a continuous probability distribution over a finite range.

Multivariate versions of the uniform distribution can be readily constructed for both the discrete and the continuous cases.

The uniform and rectangular distributions are used when every outcome is equally likely. As such, they tend to be useful, for example, as a first estimate of the probability distribution if nothing is known; i.e., if nothing is known, treat every possibility the same.

EXAMPLE 1:

One of the 30 fire protection engineers in a firm is to be selected at random to accompany the local fire department on a fire code inspection. Each engineer is assigned a playing card, the reduced deck of 30 cards is shuffled and cut several times, and the top card is selected. Here, N is 30, so $f(x) = \frac{1}{30}$, for each engineer.

EXAMPLE 2:

When the winning engineer arrives at the fire department, a random procedure is used to select one point on the city map. Whatever point is selected, they will inspect the buildings on the property of which that point is part.

Suppose A is the total area of the city. Then $f(x) = 1/A$, for every point in the city. For a given occupancy B, whose lot has area a, the probability of the event of choosing B (which corresponds to choosing any point on B's property) is equal to

$$\int_{\text{all points in } B} (1/A) \, dx = a/A$$

Note that while this is a uniform (rectangular) distribution over all *area* in the city, it is not a uniform distribution over all occupancies of the city, because an occupancy's probability of being chosen will be proportional to the size of its lot. In any analysis, there may be several different, incompatible ways of treating all possibilities "equally."

Normal Distribution (also called Gaussian Distribution)

This distribution, the familiar "bell shaped" curve, is the most commonly used continuous probability density function in statistics; its density is a function of its mean, μ, and standard deviation, σ, as follows

$$f(x) = \frac{1}{\sigma\sqrt{2\pi}} \exp\left[-\frac{1}{2}\left(\frac{x - \mu}{\sigma}\right)^2\right], \quad \text{for } -\infty < x + \infty.$$

The *Central Limit Theorem* establishes that for *any* probability density function, the distribution of the sample mean, $\bar{x}$, of a sample from that density asymptotically approaches a normal distribution as the size of the sample increases. This means that the normal distribution can be used validly to test hypotheses about the means of any population, even if nothing is known or can be assumed about the population's underlying distribution. Also, the *Law of Large Numbers* establishes that the standard deviation of the distribution of the sample mean is inversely proportional to the square root of the sample size, which means that larger samples always produce more precise estimates of the sample mean. These two results are the cornerstones of sample-based statistical inference.

In addition to proving a valid distribution for sample *means* in all situations, the normal distribution also directly characterizes many *populations* of interest, including experimental measurement errors and quality control variations in materials properties.

A sample size of at least 30 should be used to obtain an acceptable fit of the sample mean distribution to the normal distribution.

The standard tables of the normal distribution are for a random variable with mean 0 and variance 1. They can be used for values from any normal distribution by subtracting the mean, then dividing the result by the standard deviation.

The multivariate form of the normal distribution is also commonly used. Its parameters are given by a vector of the means of all the variables and a matrix with both the variances of all the variables and the covariances of pairs of variables (which are functions of the variances and the correlation coefficients).

EXAMPLE 1:

The promotional examination for lieutenant is taken by 100 fire fighters, whose test scores, shown in Table 1-11.1 below, fit a normal distribution with mean score of 50 and standard deviation of 15. The fit is not exact because strictly speaking, the 100 scores comprise a discrete distribution, not a continuous distribution, and the possible scores are bounded by 0 and 100. Also, with only 100 scores, the fit to a normal distribution can be seen in this grouped data but might not be apparent if every score had its own frequency entered separately. (See Figure 1-11.4.)

TABLE 1-11.1 *Normal Distribution Sample Test Scores*

Score	Number of Fire Fighters Receiving That Score
0–9	1
10–19	2
20–29	7
30–39	15
40–49	25
50–59	25
60–69	15
70–79	7
80–89	2
90–100	1

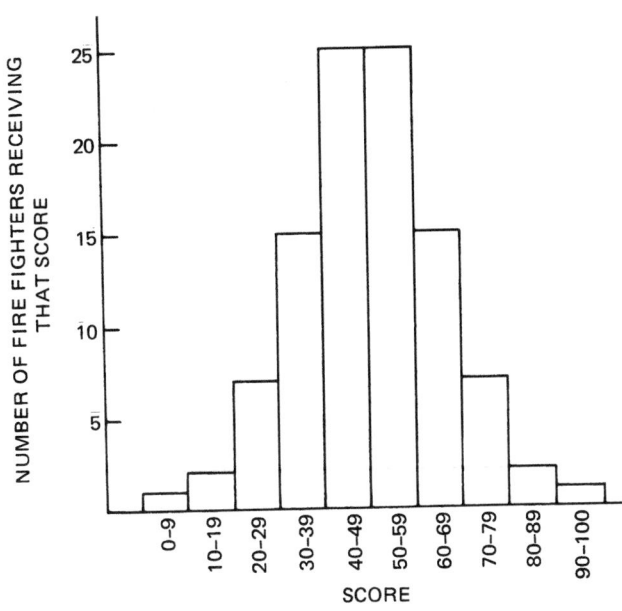

Fig. 1-11.4

EXAMPLE 2:

Suppose the widths of U.S. adults, fully clothed (including overcoats), at their widest points are normally distributed with mean 0.5 m and standard deviation of 0.053 m. Then, a door width equal to the mean (0.5 m) would accommodate 50.0 percent of the population [$F(x \leq \mu) = 0.50$]. A door width equal to the mean plus one standard deviation (0.553 m) would accommodate 84.1 percent of the population [$F(x \leq \mu + \sigma) = 0.841$]. A door width equal to the mean plus two standard deviations (0.606 m) would accommodate 97.7 percent of the population [$F(x \leq \mu + 2\sigma) = 0.977$].

But some buildings hold 10,000 persons, so suppose it is desired to construct a door width that will be too narrow for only one of every 20,000 persons. Then the value of a is desired, such that $F(x \leq \mu + a\sigma) = 0.99995$. That value of a is 3.87, which translates to a door width of 0.705 m, or more than 40 percent wider than the door width that sufficed for one-half the population.

All basic statistics texts contain tables of the cumulative distribution function for the normal distribution.

Log-Normal Distribution

It is not unusual to deal with random variables whose logarithms (to any base) are normally distributed. In such cases, the original variables are said to be log-normally distributed. For example, fire load density (i.e., mass of combustibles per unit floor area) typically has a log-normal distribution.

Student's *t* Distribution

For small samples, the distribution of the sample mean is not well approximated by the normal distribution. Even for somewhat larger samples, the population variance is typically not known, and the sample variance must be used instead. The Student's *t* distribution may be used instead of the normal distribution, but it does assume that the population is normally distributed. Its distribution is a function of its degrees of freedom, *m*.

$$f(t) = \left[\Gamma\left(\frac{m+1}{2}\right)\right]\left[\left(1 + \frac{t^2}{m}\right)^{-(m+1)/2}\right] / \left(\sqrt{\pi m}\left[\Gamma\left(\frac{m}{2}\right)\right]\right),$$
$$\text{for } -\infty < t < +\infty$$

where

$$\Gamma(u) = \int_0^\infty y^{u-1}e^{-y}\,dy$$

Expressed in this standard form, the *t* distribution has a mean of zero and a variance of $m/(m - 2)$. Since the Student's *t* distribution is used primarily in statistical testing, an example of its use is included in Section 4, Chapter 3.

Chi Square Distribution

Whereas the normal and *t* distributions may be used to test hypotheses about means, the chi square distribution may be used to test hypotheses about variances or entire distributions. Its density is a function of its degrees of freedom, *m*.

$$f(x) = x^{(m-2)/2}e^{-x/2}/(2^{m/2}\Gamma(m/2)), \qquad \text{for } x \geq 0$$

where

$$\Gamma(u) = \int_0^\infty y^{u-1}e^{-y}\,dy$$

Expressed in this standard form, the chi square distribution has its mean equal to *m*, the number of degrees of freedom, and its variance equal to $2m$.

F Distribution

Whereas the normal distribution may be used to test hypotheses about the means of samples of a single random variable, the *F* distribution permits simultaneous testing of hypotheses about the means of samples reflecting several random variables, each with its own variance, and each pair of variables correlated to some unknown degree. Its density is a function of two noninterchangeable degrees-of-freedom parameters, m_1 and m_2.

$$f(x) = \frac{\left(\frac{m_1}{m_2}\right)^{m_1/2}x^{(m_1-2)/2}\left[\Gamma\left(\frac{m_1+m_2}{2}\right)\right]}{\left[\Gamma\left(\frac{m_1}{2}\right)\right]\left[\Gamma\left(\frac{m_2}{2}\right)\right]\left[\left(1 + \frac{m_1 x}{m_2}\right)^{(m_1+m_2)/2}\right]}$$

where

$$\Gamma(u) = \int_0^\infty y^{u-1}e^{-y}\,dy$$

The mean of the *F* distribution is $m_2/(m_2 - 2)$, and the variance is given by

$$\sigma^2 = \frac{2m_2^2(m_1+m_2-2)}{m_1(m_2-2)^2(m_2-4)} \qquad \text{if } m_2 > 4$$

Exponential Distribution

This is the simplest distribution for use in reliability analysis, where it can be used to model the time to failure. Its density is a function of a parameter, θ, that is equal to its mean and its standard deviation.

$$f(x) = \left(\frac{1}{\theta}\right)e^{-x/\theta}, \qquad \text{for } x \geq 0$$

Its hazard rate is a constant, $1/\theta$, so the exponential distribution is the one to use if the expected time to failure is the same, regardless of how much time has already elapsed. This distribution also is commonly used to represent the time required to serve customers waiting in a queue.

EXAMPLE:

A smoke detector is installed in a private home and is powered by a battery from a lot with average life of six months. Suppose the time until the battery dies can be represented by an exponential distribution. (In practice, retailed batteries have a more complex failure rate function.) Then the time until failure might look like that shown in Table 1-11.2.

Note that there is a high probability of failure in the first month and a high probability of survival past one year.

Poisson Distribution

If a system has exponentially distributed time to failure with mean time θ, then the distribution of the total number of failures, n, in time, t, has a Poisson distribution. Its distribution is given by a parameter, λ, that is equal to both its mean and its variance.

$$f(n) = \frac{\lambda^n e^{-\lambda}}{n!}$$

for

$$n = 0, 1, 2, \ldots, +\infty$$

where

$$\lambda = t/\theta \quad \text{and} \quad n! = n(n-1)(n-2) \cdots (3)(2)(1)$$

This distribution also is commonly used to represent the number of customers entering a queue for service in a unit of time. It assumes that the expected number of arriving customers in any short interval of time is proportional to the length of time.

EXAMPLE:

Using the smoke detector scenario in the previous example, suppose each time the battery fails, it is detected immediately and immediately replaced with a new battery of

TABLE 1-11.2 *Example of Exponential Distribution (detector batteries)*

Months Old	Probability of Failure at This Age
0–1	0.154
1–2	0.129
2–3	0.110
3–4	0.094
4–5	0.078
5–6	0.067
6–7	0.057
7–8	0.047
8–9	0.041
9–10	0.034
10–11	0.029
11–12	0.025
Over 12	0.135

TABLE 1-11.3 *Poisson Distribution*

Number of Times Detector Will Have Dead Batteries In One Year	Probability
0	0.135
1	0.271
2	0.271
3	0.181
4	0.090
5	0.036
6 or more	0.016

similar expected life. Then the number of times the batteries will fail in the first year is given by a Poisson distribution. (See Table 1-11.3.) Here t is 12 months and θ is 6 months, so λ is 2.

Gamma Distribution (also called Erlang Distribution)

This distribution is also commonly used to represent time to failure for a system, particularly in a situation where m independent faults, all with identical exponential distributions of time to occur, are required before the system fails. Its density is a function of two parameters, m and θ, which must both be greater than zero; m need not be an integer.

$$f(x) = x^{m-1} e^{-x/\theta} / (\theta^m \Gamma(m)), \qquad \text{for } x \geq 0$$

where

$$\Gamma(m) = \int_0^\infty y^{m-1} e^{-y} \, dy$$

The mean is $m\theta$ and the variance is $m\theta^2$.

Weibull Distribution

Another distribution commonly used in reliability studies to represent time to failure, the Weibull distribution is flexible enough to permit failure rates that increase or decrease with system age. Its density is a function of two parameters, a and b, which must both be greater than zero.

$$f(x) = abx^{b-1} e^{-ax^b}, \qquad \text{for } x \geq 0$$

$$\mu = a^{-(1/b)} \Gamma\left(\frac{b+1}{b}\right)$$

$$\sigma^2 = a^{-(2/b)} \left\{ \left[\Gamma\left(\frac{b+2}{b}\right) \right] - \left[\Gamma\left(\frac{b+1}{b}\right) \right]^2 \right\}$$

where

$$\Gamma(u) = \int_0^\infty y^{u-1} e^{-y} dy$$

The cumulative distribution can be expressed in closed form, as follows

$$F(x) = 1 - e^{-ax^b}$$

Therefore, the failure rate has a simple form

$$h(x) = abx^{b-1}$$

The failure rate increases with system age if $b > 1$ and decreases if $b < 1$. If $b = 1$, the Weibull distribution becomes an exponential distribution, with $\theta = 1/a$.

Pareto Distribution

This distribution is not as commonly used but does provide a simple form for a distribution whose failure rate decreases with system age. Its density is a function of two parameters, a and b, which must both be greater than zero.

$$f(x) = ab^a x^{-(a+1)}, \quad \text{for } x > b$$

$$\mu = ab/(a - 1)$$

$$\sigma^2 = ab^2/[(a - 1)^2(a - 2)]$$

$$F(x) = 1 - b^a x^{-a}$$

$$h(x) = a/x$$

The parameter a must be greater than 2 for the mean and variance to converge to the values shown above. In general, a must be greater than k for the kth moment to converge.

Bernoulli Distribution

This distribution is the most basic of the discrete probability distributions and it represents a single trial or experiment in which there are only two possible outcomes—success (with probability p) and failure. The random variable is the number of successes.

$$f(x) = p^x(1 - p)^{(1 - x)}, \quad \text{for } x = 0, 1$$

Therefore $f(x) = p$ if $x = 1$ and $f(x) = (1 - p)$ if $x = 0$. The mean is p and the variance is $p(1 - p)$.

EXAMPLE:

Suppose there are 100 fire fighters in a department, 15 of whom are minorities. If all fire fighters are equally qualified, the probability that a minority fire fighter will be chosen as the next lieutenant is give by a Bernoulli distribution, with $p = 15/100 = 0.15$.

Binomial Distribution

This is the probability distribution for the number of successes in n independent Bernoulli trials, all having the same probability of success.

$$f(x) = \binom{n}{x} p^x(1 - p)^{(n - x)}, \quad \text{for } x = 0, 1, \ldots, n$$

where

$$\binom{n}{x} = \frac{n!}{x!(n - x)!}$$

and

$$x! = x(x - 1)(x - 2) \cdots (3)(2)(1)$$

The mean is np and the variance is $np(1 - p)$.

The use of *factorials* (e.g., $x!$) can lead to time consuming calculations. It is possible for large values of n to approximate the binomial distribution by a normal distribution [with $\mu = np$ and $\sigma^2 = np(1 - p)$]. This approximation will work acceptably if $np \geq 5$ and $n(1 - p) \geq 5$. For small values of p, μ and σ^2 become very close, and one can approximate the binomial distribution by a Poisson distribution (with $\lambda = np$). This works acceptably if $n > 100$ and $p < 0.05$.

TABLE 1-11.4 *Example of Binomial Distribution*

Number of Minority Fire Fighters Promoted	Probability
0	0.444
1	0.392
2	0.138
3	0.024
4	0.002
5	0.000

EXAMPLE:

Suppose in the fire fighter promotion example just used, five lieutenants have been selected sequentially. Also suppose that each time a fire fighter is promoted to lieutenant, that slot is filled with another fire fighter of the same race before the next lieutenant is selected. Under these conditions, the five promotions represent five Bernoulli trials, all having the same probability that a minority fire fighter will be promoted. The number of minority fire fighters promoted will then be governed by a binomial distribution, as shown in Table 1-11.4.

Geometric Distribution

In the case of a potentially unlimited number of independent Bernoulli trials with identical probabilities of success, the geometric distribution gives the distribution of the trial on which the first success will occur.

$$f(x) = p(1 - p)^{(x - 1)}, \quad \text{for } x = 1, 2, 3, \ldots, +\infty$$

The mean is $(1/p)$ and the variance is $(1 - p)/p^2$.

EXAMPLE

Continuing the example of serial promotions in which each open slot is filled by a new fire fighter of the same race, the geometric distribution would give the probability of which of the promotions will be the first to involve a minority fire fighter. (See Table 1-11.5.)

Note the high probability that chance alone will delay the first minority promotion past the tenth promotion.

TABLE 1-11.5 *Geometric Distribution with Serial Promotion Example*

First Promotion to Involve a Minority Fire Fighter	Probability
First	0.150
Second	0.128
Third	0.108
Fourth	0.092
Fifth	0.078
Sixth	0.067
Seventh	0.057
Eighth	0.048
Ninth	0.041
Tenth	0.035
Later than Tenth	0.196

Negative Binomial Distribution (also called Pascal Distribution)

This generalization of the geometric distribution gives the probability distribution for the trial on which the kth success will occur.

$$f(x) = \binom{x-1}{k-1} p^k (1-p)^{-(x-1)},$$
$$\text{for } x = k, k+1, k+2, \ldots, +\infty$$

where

$$\binom{x-1}{k-1} = \frac{(x-1)!}{(k-1)!(x-k)!}$$

and

$$x! = x(x-1)(x-2) \cdots (3)(2)(1)$$

Hypergeometric Distribution

This variation on the binominal distribution applies to cases where the initial probability of success, p, reflects a fixed number of total successes and failures, N, available for selection so that each trial reduces either the number of successes remaining or the number of failures remaining. (For example, imagine an urn filled with balls of two different colors. If each trial consists of removing a ball, then replacing it in the urn, the binomial distribution applies. If each trial consists of removing a ball and keeping it out, the hypergeometric distribution applies.)

$$f(x) = \frac{\binom{Np}{x}\binom{N(1-p)}{n-x}}{\binom{N}{n}} \quad \text{for } x = 0, 1, 2, \ldots, n$$

where

N is the total number of successes and failures possible, $n \leq N$, Np and $N(1-p)$ are integers,

$$\binom{m}{y} = \frac{m!}{y!(m-y)!}$$

and

$$y! = y(y-1)(y-2) \cdots (3)(2)(1)$$

The mean is np and the variance is $np(1-p)[(N-n)/(N-1)]$. For very large values of N (relative to n), the hypergeometric distribution asymptotically approaches the binomial distribution.

EXAMPLE:

Continuing the fire fighter promotion example, suppose five promotions are carried out all at once. (See Table 1-11.6.) The hypergeometric distribution then gives the

TABLE 1-11.6 *Example of Hypergeometric Distribution*

Number of Minority Fire Fighters Promoted	Probability
0	0.436
1	0.403
2	0.138
3	0.022
4	0.001
5	0.000

TABLE 1-11.7 *Example of Multinomial Distribution*

Number of Fire Fighters Promoted			Probability
Minority Males	Female	White Males	
0	0	2	0.640
0	1	1	0.080
0	2	0	0.002
1	0	1	0.240
1	1	0	0.015
2	0	0	0.023

probability distribution for the number of minorities promoted; note how its probabilities differ from those generated by the binomial distribution.

For example,

$$0.436 = \frac{\binom{15}{0}\binom{85}{5}}{\binom{100}{5}} = \frac{\left(\frac{15!}{15!0!}\right)\left(\frac{85!}{80!5!}\right)}{\left(\frac{100!}{95!5!}\right)}$$

$$= \frac{(85)(84)(83)(82)(81)}{(100)(99)(98)(97)(96)}$$

Multinomial Distribution

This generalization of the binomial distribution addresses the case where there are more than two possible outcomes. Given k possible outcomes, such that the probability of the ith outcome is always p_i and the p_i collectively sum to unity, then for a series of n independent trials

$$f(x_1, \ldots, x_k) = \frac{n!}{x_1! x_2! \ldots x_k!} p_1^{x_1} p_2^{x_2} \cdots p_k^{x_k}$$

for all cases of $x_1 = 0, 1, 2, \ldots, n$, for $i = 1, 2, \ldots, k$, subject to

$$\sum_{i=1}^{k} x_i = n$$

$$\mu_i = np_i$$

$$\sigma_i^2 = np_i(1 - p_i)$$

EXAMPLE:

Continuing the fire department example, suppose that the department's 100 fire fighters include 15 black fire fighters and 5 female fire fighters, none of whom is black. Suppose two promotions are made, and the slot vacated for the first promotion is filled by a fire fighter of the same race and sex before the second promotion is made. Then the multinomial distribution (Table 1-11.7) describes the possible outcomes of interest.

For example, this is the probability that the promotions will go to one white male and one white female:

$$0.080 = \frac{2!}{0!1!1!}(0.15)^0(0.05)^1(0.80)^1 = 2 \times 0.05 \times 0.80$$

Beta Distribution

In Bayesian statistical inference, if the phenomenon of interest is governed by a Bernoulli distribution, then one needs a probability distribution for the parameter, p, of that Bernoulli distribution, and a Beta distribution is typically used.

$$f(p) = \frac{\Gamma(a+b)}{[\Gamma(a)][\Gamma(b)]} p^{a-1} \, 1 - p^{(b-1)}$$

where

$$\Gamma(u) = \int_0^\infty y^{u-1} e^{-y} \, dy$$

The mean is $a/(a + b)$ and the variance is given by:

$$\Gamma^2 = \frac{ab}{(a+b)^2(a+b+1)}$$

If $a = b = 1$, this becomes a *uniform distribution*. Larger values of b correspond to smaller variances, hence tighter confidence bands around the mean estimate of the parameter.

ADDITIONAL READING

1. J.R. Benjamin and C.A. Cornell, *Probability, Statistics and Decision for Civil Engineers*, McGraw-Hill, New York (1970).
2. W. Feller, *An Introduction to Probability Theory and Its Applications*, John Wiley and Sons, New York (1957).
3. J.E. Freund and F.J. Williams, *Dictionary/Outline of Basic Statistics*, McGraw-Hill, New York (1966).
4. N.A.J. Hastings and J.B. Peacock, *Statistical Distributions: A Handbook for Students*, Butterworths, London (1975).
5. M.R. Spiegel, *Probability and Statistics*, McGraw-Hill, New York (1975).
6. R.E. Walpole and R.H. Myers, *Probability and Statistics for Engineers and Scientists*, Macmillan, New York (1972).

STATISTICS

John R. Hall, Jr.

INTRODUCTION

Statistical analysis is basic to all aspects of fire protection engineering that involve abstracting results from experiments or real experience. Statistical analysis is the applied side of the mathematics of probability theory.

BASIC CONCEPTS OF STATISTICAL ANALYSIS

Statistic

A statistic is (a) any item of numerical data, or (b) a quantity (e.g., mean) computed as a function on a body of numerical data, or the function itself.

Statistical Analysis

This is the use of mathematical methods to condense sizeable bodies of numerical data into a small number of summary statistics from which useful conclusions may be drawn.

Statistical Inference

Statistical inference is statistical analysis that consists of using methods based on the mathematics of probability theory to reason from properties of a body of numerical data, regarded as a sample from a larger population, to properties of that larger population.

In *classical statistical inference*, a single best estimate of each statistic of interest is developed from available data, the uncertainty of that statistic is estimated, and hypotheses are tested and conclusions drawn from those bases.

In *Bayesian statistical inference*, a probability distribution for each statistic of interest is developed, using a form that permits new information, when it is acquired, to be used to adjust that distribution. Bayes' law, which was described

in Section 1, Chapter 11, is used to adjust the distribution in light of the new information.

EXAMPLE:

Suppose there are 100 fire fighters in a department, 15 of them black, and in a group of 5 recent promotions, all the promotions were given to whites. How likely is it that the department never selects blacks for promotions?

A Bayesian analysis uses (a) a prior estimate of the probability that the department discriminated, made before considering the evidence of the recent promotions; (b) a computed probability that the promotions would have had this pattern if the department never selects blacks; and (c) a computed probability that the promotions would have had the result if promotions are random with respect to race. For (b), the probability is 1.0, because an all white promotion list is the only possible outcome under the hypothesis that blacks are never selected. For (c), the probability is given by the hypergeometric distribution

$$\frac{\binom{15}{0}\binom{85}{5}}{\binom{100}{5}} = 0.44.$$

By Bayes' law, then, given a prior probability, q, that blacks are never selected, the posterior probability is $q/(0.56q + 0.44)$. The new evidence produces some shift in the estimated likelihood of prejudice. If prejudice was considered an even proposition before ($q = 0.5$), then the new estimate is 0.69. If prejudice was considered certain ($q = 1.0$) or impossible ($q = 0.0$) before, no new evidence will alter those estimates. If prejudice was considered very unlikely before (say, $q = 0.01$), then it will still be considered very unlikely (new value of 0.022).

Suppose in the same example there had been 25 promotions with no blacks selected. This more extensive evidence would have produced a more dramatic shift in the estimated probability. Instead of 0.44, the probability of this outcome (given no prejudice) would be 0.009. If the prior probability was 0.5, the posterior probability would be 0.991. Even if the prior probability is 0.01, the posterior probability would be 0.529. In other words, the new evidence changes the estimate of the likelihood that the department never selects blacks from one chance in a hundred to a better than even chance.

Dr. John R. Hall, Jr. is Assistant Vice President for Fire Analysis and Research at the National Fire Protection Association. He has been involved in studies of fire experience patterns and trends, models of fire risk, and studies of fire department management experiences since 1974 at NFPA, the National Bureau of Standards, the U.S. Fire Administration, and the Urban Institute.

In more sophisticated Bayesian statistical analysis, the prior probability is given not as a single probability but as a probability distribution, which permits the analyst to reflect the strength of the evidence that went into choosing the prior probability distribution.

Exploratory Data Analysis

This analysis is the development of *descriptive statistics*, i.e., statistical analysis that does not make inferences to a population.

KEY PARAMETERS OF DESCRIPTIVE STATISTICS

The *mean*, *median*, *variance*, and *standard deviation*, as described in the previous chapter, can all be applied here, using the relative frequency of occurrence of each value in the body of data to define a discrete probability distribution.

The *mode* is the value that occurs most frequently, i.e., the value of x for which $f(x) > f(y)$ for all $y \neq x$.

A body of data is called *unimodal* if $f(z) < f(y)$ in all cases where $|z - x| > |y - x|$, that is, if the probability distribution function steadily decreases as one moves away from the mode.

A body of data is called *multimodal* if it is not unimodal. In such cases there will be two or more values of x for which $f(x) > f(y)$ for all $y \neq x$ and $|y - x| < \varepsilon$, where ε is some small value. Although there may be only one mode in the sense of a most frequently occurring value, the existence of local maximums in the probability distribution function is sufficient to make the distribution multimodal. Multimodal data usually occurs when data are combined from two or more populations, each having an underlying unimodal distribution. For example, if data were collected on the lengths of fire department vehicles, it probably would be multimodal, having one peak each for automobiles, ambulances/vans, engines, and ladders.

A *geometric mean* is another type of average:

$$\text{G.M.} = (x_1 x_2 x_3 \ldots x_n)^{1/n}$$

The geometric mean is useful in averaging index numbers reflecting rates of change. For example, suppose a, b, and c are annual rates of increase in the fire department budget for three successive years. Then $A = 1 + a$, $B = 1 + b$, and $C = 1 + c$ would be index numbers reflecting those three rates. The index number, D, reflecting the cumulative increase over all three years, would be given by $D = ABC$, and so an index number yielding an "average" rate of inflation for the three-year period would be given by $(ABC)^{1/3}$, or the geometric mean of the index numbers. This geometric mean is the index number that could be compounded over the three years to obtain the actual cumulative increase. Note that the geometric mean is equivalent to computing the arithmetic mean of the logarithms of the data values, then exponentiating the result, i.e., using the result as an exponential power to be applied to the base used in computing the logarithms.

The *harmonic mean* is a less commonly used average that consists of the reciprocal of the arithmetic mean of the reciprocals of the data values. For example, suppose $V_1, \ldots, V_n$ are a set of n values of the speed achieved by an engine company on a set of test runs from the firehouse to a single location. Then these speeds can also be represented as d/t_1, $d/t_2, \ldots, d/t_n$, where d is the constant distance and $t_1, \ldots, t_n$ are the times of the n runs. The average speed would be given by $nd/(t_1 + t_2 + \ldots + t_n)$, or total distance divided by total time. That value will also be given by the harmonic mean of the speed values. This example also helps illustrate why the harmonic mean is rarely used. It is likely that anyone who had access to the speed values would also have access to time values, $t_1, \ldots, t_n$, and could compute the average more quickly by using them directly.

The *range* is the difference between the highest and lowest values, or the term may be used to refer to those two values and the interval between them.

Quartiles, *deciles*, and *percentiles* are useful measures of the dispersion of the data. If the data are arranged in ascending or descending order, the three quartiles, Q_1, Q_2, and Q_3, are the values that mark off 25 percent, 50 percent, and 75 percent, respectively, of the data set. In other words:

Q_1 is chosen so that $F(Q_1) = 0.25$

Q_2 is chosen so that $F(Q_2) = 0.50$; Q_2 is also the median

Q_3 is chosen so that $F(Q_3) = 0.75$.

Deciles and percentiles are defined analogously so as to divide the data set into tenths or hundredths, respectively, rather than fourths. Like the second quartile, the fifth decile equals the median. The *interquartile range*, or $Q_3 - Q_1$, is an alternative to the full range that is less sensitive to extreme values.

A *histogram* is a technique of exploratory data analysis for displaying the frequency of occurrence of a finite set of data. The data values are arrayed along the x-axis of a graph, and the y-axis is used to plot the frequency, usually as number of occurrences or percentage of total occurrences.

A *scatter plot* or *scatter diagram* is a technique of exploratory data analysis for displaying the patterns of a finite set of bivariate data. Each pair of data values is plotted on an (x,y) graph. This technique works best if both dimensions of the data are continuous so that the same pair of values does not occur more than once.

The *coefficient of variation* is given by the standard deviation divided by the mean. When the result is multiplied by 100, it gives the scatter about the mean in percentage terms relative to the mean.

CORRELATION, REGRESSION, AND ANALYSIS OF VARIANCE

Correlation

In qualitative terms, correlation refers to the degree of association between two or more random variables. (Random variables with discrete and continuous probabilty distributions were defined in the previous chapter.)

The most common quantitative measure of correlation specifically addresses the extent to which two random variables are linearly related.

Correlation Coefficient (also called the Pearson product-moment correlation coefficient)

Let two discrete random variables, X and Y, have a joint probability distribution given by $f(x_i, y_j) = $ probability ($X = x_i$ and $Y = y_j$).

Then the correlation coefficient of X and Y is given by

$$\rho_{XY} = \frac{\left[\sum_{i=1}^{\infty}\sum_{j=1}^{\infty}(x_i-\mu_X)(y_j-\mu_Y)f(x_i,y_j)\right]}{\sqrt{\sum_{i=1}^{\infty}\sum_{j=1}^{\infty}[(x_i-\mu_X)^2 f(x_i,y_j)]}\sqrt{\sum_{i=1}^{\infty}\sum_{j=1}^{\infty}[(y_i-\mu_Y)^2 f(x_i,y_j)]}}$$

where

$$\mu_X = \sum_{i=1}^{\infty}\sum_{j=1}^{\infty} x_i f(x_i, y_j)$$

and

$$\mu_Y = \sum_{i=1}^{\infty}\sum_{j=1}^{\infty} y_j f(x_i, y_j)$$

Let two continuous random variables, X and Y, have a joint probability density function given by $f(x, y)$ such that

$$\int_{-\infty}^{y}\int_{-\infty}^{x} f(u, v)du\, dv = \text{probability } (X \leq x \text{ and } Y \leq y)$$

Then the correlation coefficient of X and Y is given by

$$\rho_{XY} = \frac{\left[\int_{-\infty}^{\infty}\int_{-\infty}^{\infty}(x-\mu_X)(y-\mu_Y)f(x,y)dxdy\right]}{\sqrt{\int_{-\infty}^{\infty}\int_{-\infty}^{\infty}(x-\mu_X)^2 f(x,y)dxdy}\sqrt{\int_{-\infty}^{\infty}\int_{-\infty}^{\infty}(y-\mu_Y)^2 f(x,y)dxdy}}$$

where

$$\mu_X = \int_{-\infty}^{\infty}\int_{-\infty}^{\infty} xf(x, y)dx\, dy$$

and

$$\mu_X = \int_{-\infty}^{\infty}\int_{-\infty}^{\infty} yf(x, y)dx\, dy$$

If $y = ax + b$, then $\rho = 1$ if $a > 0$ and $\rho = -1$ if $a < 0$.

It is possible for one variable to be a function of another, yet have zero correlation with it (e.g., $y = x$ for $x \geq 0$ and $y = -x$ for $x < 0$).

If two random variables are independent, they will have zero correlation. However, zero correlation can occur without independence.

Even if two variables are highly correlated, it is not necessary for either to be the cause of the other. Many so-called spurious correlations occur. An example is a case of two variables (e.g., sales of fire extinguishers, sales of chewing gum) that are both strongly influenced by a third variable (e.g., disposable income) and so will be highly correlated with each other because each is correlated with the third variable.

In the case of a multimodal joint probability distribution, the correlation may be quite different at a macro- and a micro-level. Consider the variables of fire rate per household and average income per household with regard to census tracts in a city. A small number of tracts typically will have high fire rates and low incomes; the rest will have low fire rates and high incomes. The two variables will be highly correlated if all census tracts are considered together, but if the two relatively homogeneous areas are analyzed separately, there may be little correlation.

If a sample of size n consists of pairs of values (x_i, y_i), then the sample correlation coefficient is

$$r_{XY} = \frac{\sum_{i=1}^{\infty}(x_i-\bar{x})(y_i-\bar{y})}{\sqrt{\sum_{i=1}^{n}(x_i-\bar{x})^2}\sqrt{\sum_{i=1}^{n}(y_i-\bar{y})^2}}$$

EXAMPLE:

Suppose the scores of ten fire fighters on a promotional exam are compared to their numbers of years with the fire service, with results shown in Table 1-12.1 and in Figure 1-12.1.

Then the mean age is 23.7 and the mean score is 72. The correlation coefficient is 0.67, indicating moderate correlation. If the second individual's score, which is the farthest from the group pattern, were changed from 85 to 60, the correlation coefficient would rise to 0.89, indicating high correlation.

The *coefficient of determination* (also called the *percentage of variation explained*) is given by the square of the correlation coefficient.

Regression

Regression analysis consists of fitting a relationship, usually a linear relationship ($Y = aX + b$), to two random variables, X and Y. The term "regression" is left over from one of the findings in one of the earliest applications of the theory, where it was discovered that heights of parents are good predictors of heights of children but that heights of children tend to "regress" toward the mean. (In other words, for this problem, the best fit was $Y = a(X - \mu_x) + \mu_y$, where $a < 1$.)

Method of Least Squares

This method assumes that the best fit is obtained by minimizing the weighted sum of the squared differences between predicted and observed values of Y. In other words:

For two discrete random variables, X and Y, with joint probability distribution $f(x_i, y_j)$, choose a and b to minimize

$$\sum_{i=1}^{\infty}\sum_{j=1}^{\infty}(y_i - ax_i - b)^2 f(x_i, y_j)$$

For two continuous random variables, X and Y, with joint density function $f(x,y)$, choose a and b to minimize

$$\int_{-\infty}^{\infty}\int_{-\infty}^{\infty} (y - ax - b)^2 f(x, y)dx\, dy$$

For a sample of size n of pairs of values (x_i, y_i), choose a and b to minimize

$$\sum_{i=1}^{n} (y_i - ax_i - b)^2$$

TABLE 1-12.1 *Distribution of Test Scores*

Age	Score
18	54
20	85
20	62
20	60
22	66
25	70
25	75
28	88
29	70
30	90

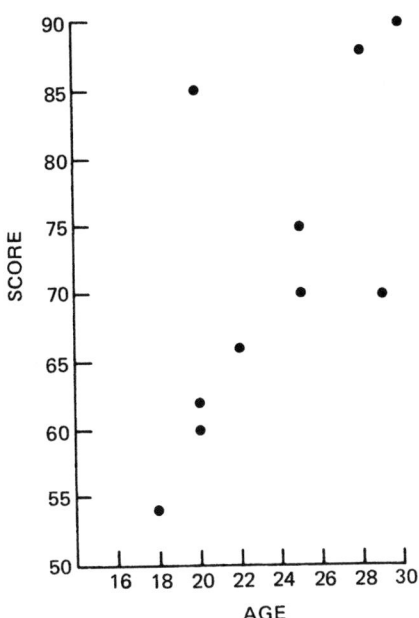

Fig. 1-12.1. *Distribution of test scores.*

The method of least squares is the best method if the deviations between observed and expected values of Y are themselves normally distributed, independent random variables. This condition would be satisfied, for example, in most experiments if the only source of deviation was error in reading a measuring device. The deviations are also called *residuals*.

Analysis of patterns in residuals can be done to confirm the normality assumptions cited above. Also, data points may be selected with extremely large residuals and studied for common characteristics as a means of trying to identify other factors that may be correlated to the outcomes, y. These results, in turn, may lead to a more sophisticated, multivariate regression analysis.

Regression Coefficients

The least-squares fit of a relationship of the form $Y = aX + B$ will be given by

$$a = \rho_{xy}\sigma_y/\sigma_x$$

$$b = \mu_y - a\mu_x$$

For a sample of size n of pairs (x_i, y_i), the formulas are

$$a = \frac{\left[n\sum_{i=1}^n x_i y_i\right] - \left[\left(\sum_{i=1}^n x_i\right)\left(\sum_{i=1}^n y_i\right)\right]}{\left[n\sum_{i=1}^n x_i^2\right] - \left[\left(\sum_{i=1}^n x_i\right)^2\right]}$$

$$b = \frac{\left[\left(\sum_{i=1}^n x_i^2\right)\left(\sum_{i=1}^n y_i\right)\right] - \left[\left(\sum_{i=1}^n x_i\right)\left(\sum_{i=1}^n x_i y_i\right)\right]}{\left[n\sum_{i=1}^n x_i^2\right] - \left[\left(\sum_{i=1}^n x_i\right)^2\right]}$$

EXAMPLE:

Re-examine the case of age versus test score examined earlier under the discussion of correlation. In that case, as noted, the correlation coefficient was 0.67, the mean age was 23.7, and the mean score was 72. The ratio of standard deviations can be calculated as 2.87. Therefore, $a = 1.92$ and $b = 26.5$. This means that the predicted score for age 20 would be 64.9, compared to the 60, 62, and 85 scored by persons of that age, while the predicted score for age 30 would be 84.1, compared to the 90 scored by the person of that age.

This regression line tends to overpredict scores for younger persons because the line is tipped as it tries to accommodate the 85 score achieved by one 20 year old. If that score had been a 60, then as noted the correlation coefficient would be 0.89; also, the mean score would be 69.5 and the ratio of standard deviations would be 2.78. Therefore, a would be 2.47 and b would be 11.0. The predicted score for age 20 would change from 64.9 to 60.4, and the predicted score for age 18 would change from 61.1 to 55.5, much closer to the score actually achieved by the 18 year old.

While it is theoretically possible to fit any relationship, not just a linear one, between X and Y, it is rarely possible to develop least-square formulas for a and b if the relationship is not linear. Accordingly, the analyst will usually want to try to transform problems into linear regression problems. For example, if the true relationship is believed to be of the form $y = c^{(x+d)}$, one would set up a linear regression of $\log y$ versus x. Then $d \log c = b$ and $\log c = a$.

HYPOTHESIS TESTING IN CLASSICAL STATISTICAL INFERENCE

Hypothesis and Test

A statistical *hypothesis* is a well-defined statement about a probability distribution or, more frequently, one of its parameters. A classical *test* of a statistical hypothesis is based on the use of several concepts to organize the uncertainty inherent in any probabilistic situation.

The hypothesis being considered is called the *null hypothesis* and implies a probability distribution. Classical statistical inference asks whether the probability of having obtained the statistics actually collected, given the null hypothesis, is so low that the null hypothesis must be rejected.

The test works on the basis of a statistic computed from a sample. That statistic is compared to a reference value. If the statistic falls to one side of the reference value, then the null hypothesis is rejected; if the statistic falls to the other side, then the null hypothesis is not rejected.

For the reasons given above, a statistical test resolves doubts in favor of the null hypothesis. Therefore, an analyst may choose to say that the null hypothesis was "not rejected" rather than say it was "accepted." The analogy is to a criminal trial, which may find a defendant "not guilty" but does not make findings of "innocent."

A *Type I error* occurs when the null hypothesis is really true, but the test says that it should be rejected. A *Type II error* occurs when the null hypothesis is really false, but the test says it should not be rejected. (Informally, many analysts use the term *Type III error* to refer to analyses that set up the initial problem incorrectly, thereby producing results that, however precise, are irrelevant to the real issue.)

A *confidence coefficient*, or measure of the *degree of confidence*, is used to indicate the maximum acceptable probability of Type I error. In most cases, the null hypothesis corresponds to a single, well-defined probability distribution. Therefore, the probability of the sample statistic falling on the reject side of the reference value can be calculated precisely, and the reference value can be selected so as to set that probability equal to the confidence coefficient.

One way of using the confidence coefficient is to set *confidence limits* or define a confidence interval. These limits or internal boundaries are set so that if the null hypothesis is true, then the probability of obtaining a sample whose test statistic is outside the limits (or interval) is equal to the confidence coefficient. These confidence limits indicate to the user how precisely the probability distribution or its parameter can be defined, given the size of the sample and its variability.

The value of the confidence coefficient can be set at any of certain standard levels (90 percent, 95 percent, and 99 percent are often used), or it can be derived from an analysis that seeks to balance Type I and Type II errors. The latter approach is more comprehensive, but it is much more difficult because the alternative(s) to the null hypothesis rarely correspond(s) to a single probability distribution. In a typical case, the null hypothesis states a single value for a population parameter ($\mu = a$) and the alternative corresponds to all other values ($\mu \neq a$). Each specific alternative defines a specific probability distribution with a specific probability of Type II error. The *power function of the test* is that function which gives the probability of *not* committing a Type II error for each parameter value covered by the alternative(s) to the null hypothesis.

As the parameter value approaches the value in the null hypothesis (e.g., $\mu \rightarrow a$), the power of the test drops toward the confidence coefficient.

Test of Mean—z Test

If a sample has been collected from a population with known standard deviation σ, the central limit theorem indicates that the sample mean has an approximately normal distribution about the true population mean μ_0.
Let

$$z = (\sqrt{n})(\bar{x} - \mu_0)\sigma$$

where n is the sample size, μ_0 is the hypothesized true value of μ, and $\bar{x}$ is the sample mean. Let z_α be the value for which $F(z_\alpha) = 1 - \alpha$, where F is the cumulative distribution function of a normal distribution with mean zero and variance one. (Note that $z_{1-\alpha} = -z_\alpha$.)

A *two-sided test* presumes that, if the true population mean is not μ_0, then it is equally likely to be greater than or less than μ_0. In that case positive and negative values of z are treated the same, and the confidence coefficient must be divided between the two sides of the confidence interval. Then if α is the confidence coefficient, the two-sided test says accept the null hypothesis if

$$-z_{\alpha/2} \leq z \leq z_{\alpha/2}$$

A *one-sided test* presumes that if the true population mean is not μ_0, then it must be greater than (or less than) μ_0. If α is the confidence coefficient, then the one-sided test says accept the null hypothesis if

$$z \geq -z_\alpha \qquad \text{if the alternative to } \mu = \mu_0 \text{ is } \mu < \mu_0.$$

or

$$z \leq z_\alpha \qquad \text{if the alternative to } \mu = \mu_0 \text{ is } \mu > \mu_0$$

The value

$$\sigma/\sqrt{n}$$

is called the *standard error of the mean.*

Test of Difference Between Two Means—z Test

If two samples from populations with known standard deviations, σ_1 and σ_2, have been collected, a null hypothesis might be that they are from the same population, which means their means would be the same ($\mu_1 = \mu_2$). Then a two-sided test is applied, using the following statistic

$$z = (\bar{x}_1 - \bar{x}_2)/\sqrt{\frac{\sigma_1^2}{n_1} + \frac{\sigma_2^2}{n_2}}$$

where $\bar{x}_i$, σ_i^2, and n_i are the sample mean, population variance, and sample size, respectively, of the ith sample.

Test of Proportion—z Test

If a sample has been drawn from a population governed by a binominal distribution, then the normal approximation gives the following statistic, to be used in one- or two-sided tests

$$z = n(\bar{p} - p_0)/\sqrt{p_0(1-p_0)}$$

Where p_0 is the hypothesized true proportion and p is the sample proportion.

Test of Difference Between Two Proportions—z Test

Again the normal approximation to the binomial distribution gives the test statistic

$$z = (\bar{p}_1 - \bar{p}_2)/\sqrt{\frac{\bar{p}_1(1-\bar{p}_1)}{n_1} + \frac{\bar{p}_2(1-\bar{p}_2)}{n_2}}$$

Test of Mean—t Test

The z tests assume known variance(s), so if variances are not known, a test based on Student's t distribution must be used. Let $t_{\alpha,m}$ be defined such that $F(t_{\alpha,m}) = 1 - \alpha$, where F is the cumulative distribution function for a Student's t distribution with m degrees of freedom. Note that $t_{(1-\alpha),m} = -t_{\alpha,m}$.
Then

$$t = \sqrt{n}(\bar{x} - \mu_0)/s$$

where n is the sample size, μ_0 is the hypothesized population mean, and s is the sample standard deviation.

A two-sided test says accept the null hypothesis if

$$-t_{(\alpha/2),(n-1)} \leq t \leq t_{(\alpha/2),(n-1)}$$

TABLE 1-12.2 *Distribution of Test Scores*

Score	Number of Fire Fighters With That Score in	
	Department A	Department B
10	0	5
20	6	5
30	5	10
40	20	20
50	29	25
60	20	20
70	9	10
80	5	3
90	3	2
100	3	0
Total	100	100
Sample mean	53.2	48.2
Sample standard deviation	17.54	17.34
Proportion of sample passing	0.20	0.15

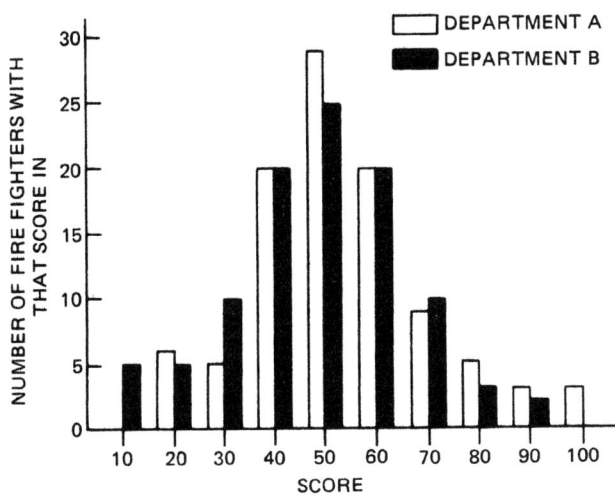

Fig. 1-12.2. *Distribution of test scores in two fire departments.*

Note that the number of degrees of freedom is one less than the sample size. An informal method of remembering this is that one degree of freedom is used to estimate the standard deviation.

A one-sided test says accept the null hypothesis if

$$t \geq -t_{\alpha,(n-1)} \quad \text{if the alternative to } \mu = \mu_0 \text{ is } \mu < \mu_0$$

or

$$t \leq t_{\alpha,(n-1)} \quad \text{if the alternative to } \mu = \mu_0 \text{ is } \mu > \mu_0$$

A test of differences between two means is constructed analogously.

EXAMPLE:

This example will illustrate all the tests described thus far. Suppose there are two fire departments, each of which has given promotional tests to 100 fire fighters. Scores can range from 0 to 100, and the passing score is 70. The actual distributions of scores are shown in Table 1-12.2 and Figure 1-12.2.

Suppose that nationwide the standard deviation for this test is 17.45, and the mean score is 50, and the proportion who pass is 0.17. Is Department A an average department? This suggests a two-sided test of the mean score, using the z test because the standard deviation is known. Let α be 0.05, so $z_{\alpha/2} = 1.96$. The z statistic for the test is

$$(\sqrt{100})(53.2 - 50)/17.60 = 1.83$$

which is between -1.96 and $+1.96$, so the null hypothesis is accepted. Department A is average.

Suppose it is asked instead whether Department A is a better-than-average department. This suggests a one-sided test of the mean score, again using the z test because the standard deviation is known. Again, let $\alpha = 0.05$, so $Z_{\alpha/2} = 1.64$. The z statistic for the test is again 1.83, which is greater than 1.64, so we reject the null hypothesis and conclude that Department A is above average.

The results for these two tests seem contradictory because one concludes that Department A is average and the other concludes that Department A is above average. This is the nature of statistical tests. They can be sensitive to the choice of α. (If α were 0.10, both null hypotheses would be rejected, while if α were 0.01, both null hypotheses would be accepted. In either case, the two tests would give consistent results.) They can be sensitive to how the alternatives were posed, as was true here.

Suppose it is asked whether Departments A and B have significantly different mean scores. This is a z test of the difference between two means. In this case the standard deviations are the same and the sample sizes are the same, so the z statistic reduces to

$$(53.2 - 48.2)/\sqrt{2 \times \frac{17.45}{100}} = 8.46$$

The two-sided test reference value of 1.96, calculated earlier, is easily exceeded, and we conclude that the two departments do have statistically significant differences in their mean scores.

Suppose it is asked whether Department A's proportion of students passing (0.20) is statistically significantly greater than the overall average of 0.17. This is a z test of a proportion, and the z statistic is

$$(\sqrt{100})(0.20) - 0.17/\sqrt{(0.17)(0.83)} = 0.80$$

This is not statistically significant under either a one-sided or a two-sided test.

Are the proportions passing in Departments A and B different? This is a z test of the difference between two proportions, and the z statistic is

$$(0.20 - 0.15)/\sqrt{\frac{(0.20)(0.80)}{100} + \frac{(0.15)(0.85)}{100}} = 0.93$$

Even though the average scores for Departments A and B were found to be different by a statistically significant margin, their percentage of test takers passing were not found to be significantly different.

Suppose the value of the overall standard deviation for the test was not known, or it was not known whether it applied to these departments, but it was known that the overall average score was 50. Is Department A's score significantly better? This is a one-sided t test. The t statistic is

$$\sqrt{100}(53.2 - 50.0)/17.54 = 1.82$$

For a one-sided t test with a 0.05 confidence level and 99 degrees of freedom, the reference value is the same as for a one-sided z test with a 0.05 confidence level, namely 1.64. Because the sample standard deviation is also nearly equal to the overall standard deviation used earlier, the test results are virtually the same, and the null hypothesis is rejected. This would not have been the case if the sample size had been considerably smaller, leading to a larger reference value. The smaller the difference you are examining, the larger the sample size required to be sure that difference is real and not just the result of random variation.

Test of Variance—Chi Square Test

Assuming a normal population, one can test the hypothesis $\sigma = \sigma_0$ with the following

$$\psi^2 = (n - 1)s^2/\sigma_0^2$$

where n is the sample size and s^2 is the sample variance.

A two-sided test accepts the null hypothesis if

$$\psi^2_{(1-\alpha/2),(n-1)} \leq \psi^2 \leq \psi^2_{(\alpha/2),(n-1)}$$

where $\psi^2_{\alpha,m}$ is the value such that $F(\psi^2_{\alpha,m}) = 1 - \alpha$, where F is the cumulative distribution function of a chi square distribution with m degrees of freedom. Note that the degrees of freedom used in the test are one less than the sample size. One-sided tests can be constructed analogously.

Test of Goodness of Fit to a Distribution—Chi Square Test

A special use of the test of variance is to test how well a set of experimental data fit a presumed theoretical probability distribution. Suppose the distribution in question is represented as a set of k values or ranges of values for the random variable. Let $p_i,\ldots,p_k$ be the hypothesized probabilities for those k values or ranges; let $p_i,\ldots,p_k$ be the sample estimates of those probabilities; and let n be the sample size. Then the statistic is

$$\psi^2 = \sum_{i=1}^{k} \left\{ \frac{[u(p_i,\bar{p}_i,n)]^2}{np_i} \right\}$$

where

$$u(p_i, \bar{p}_i, n) = \begin{cases} 0 & \text{if } -\frac{1}{2} \leq np_i - n\bar{p}_i \leq \frac{1}{2} \\ np_i - n\bar{p}_i - \frac{1}{2} & \text{if } np_i - n\bar{p}_i > \frac{1}{2} \\ n\bar{p}_i - np_i - \frac{1}{2} & \text{if } n\bar{p}_i - np_i > \frac{1}{2} \end{cases}$$

This process of reducing the gap between np_i and $n\bar{p}_i$ by $\frac{1}{2}$ is called the *Yates continuity correction*, and it compensates for the fact that the chi square distribution, a continuous function, is being used to approximate a discrete probability distribution. Also, to apply this test validly, one must make sure that the k classes are grouped sufficiently that $np_i \geq 5$ for all $i = 1, \ldots, k$.

The null hypothesis says this sample came from the distribution represented by $p_i, \ldots, p_k$. That hypothesis is accepted if

$$\psi^2 \leq \psi^2_{\alpha,j}$$

where j is at most $k - 1$ and may be less if the p_i are based in part on the sample.

For example, suppose an analyst wishes to test goodness of fit to a binomial distribution but has no prior estimate of which binomial distribution should be used. The analyst would select the particular binomial distribution that has p equal to the sample proportion. In that case one parameter has been estimated from the sample, and j would be reduced by one to $(k - 2)$. If the analyst were testing goodness of fit to a normal distribution and estimated both mean and variance from the sample, then j would drop by two to $(k - 3)$.

Contingency Test of Independence—Chi Square Test

A special case of the goodness of fit test is a test of the hypothesis that two random variables are independent, in which case the goodness of fit test is displayed in a *contingency table*, as follows:

Let $X_1, \ldots, X_m$ be the m values or subranges of a random variable, X; and let $Y_1, \ldots, Y_k$ be the k values or subranges of a random variable, Y.

Let p_i be the estimated probability of X_i, for $i = 1, \ldots, m$, and let q_j be the estimated probability of Y_j, for $j = 1, \ldots, k$.

Let n be the size of a sample such that, for each sample entry, a value of X and a value of Y are provided. (Be sure $np_iq_j \geq 5$ for all i and j.)

Let r_{ij} be the number of sample entries for which $X = X_i$ and $Y = Y_j$.

Therefore

$$\sum_{i=1}^{m} \sum_{j=1}^{k} r_{ij} = n$$

Then the sample will provide estimated values of p_i and q_j as follows

$$p_i = \left(\sum_{j=1}^{k} r_{ij} \right) / n$$

and

$$q_j = \left(\sum_{i=1}^{m} r_{ij} \right) / n$$

If the two random variables are independent, then the expected values for r_{ij} are given by np_iq_j.

The test statistic, therefore, is given by

$$\psi^2 = \sum_{i=1}^{m} \sum_{j=1}^{k} \{[u(r_{ij}, p_i, q_j, n)]^2/np_iq_j\}$$

where

$$u(r_{ij}, p_i, q_j, n) = \begin{cases} 0 & \text{if } -\frac{1}{2} \leq r_{ij} - np_iq_j \leq \frac{1}{2} \\ r_{ij} - np_iq_j - \frac{1}{2} & \text{if } r_{ij} - np_iq_j > \frac{1}{2} \\ np_iq_j - r_{ij} - \frac{1}{2} & \text{if } np_iq_j - r_{ij} > \frac{1}{2} \end{cases}$$

The null hypothesis of independence is accepted if

$$\psi^2 \leq \psi^2_{\alpha,[(m-1)(k-1)]}$$

The number of degrees of freedom comes from the formula given for goodness-of-fit tests. One begins with $mk - 1$ degrees of freedom. There are m values of p_i, but they sum to one, so only $m - 1$ need be estimated from the sample, and similarly $(k - 1)$ values of q_j must be estimated. Therefore the degrees of freedom for the test equal $(mk - 1) - (m - 1) - (k - 1) = (m - 1)(k - 1)$.

Nonparametric Tests

There are a large number of nonparametric tests, so called because they use no sample or population parameters and make no assumptions about the type of probability distribution that produced the sample.

SAMPLING THEORY

A *random sample* is a sample chosen in accordance with a well-defined procedure that assures (a) each equal item (e.g., each person) has an equal chance of being selected; or (b) each value of a random variable (e.g., height) has a likelihood of being selected that is the same as its probability of occurrence in the full population. A sample that is selected with no conscious biases still may not be truly random; the burden of proof is on the procedure that claims to produce a random sample. A random sample may not be as representative as a sample that is chosen to be representative, but a sample chosen to be representative on a few characteristics may not be random and may not be representative with respect to other important characteristics.

In addition to requiring that each item have an equal chance of being selected, a random sample must assure that every combination of items also has an equal chance of being selected. For example, a random sample of currently married couples would not be a random sample of currently married persons, because spouses would be either selected or not selected together in the former but not necessarily in the latter.

A *sampling frame* is a basis for reaching any member of a population for sampling in a way that preserves the randomness of selection. An example would be a mailing list, although if it had missing names or duplicate names it would be deficient as a sampling frame, because each equally likely name would not have an equal chance at selection. A *sample design* is a procedure for drawing a sample from a sampling frame so that the desired randomness properties are achieved.

A *simple random sample* is a sample that is drawn by a procedure assuming complete randomness from a population all of whose elements are equally likely. If they are not all equally likely, a procedure that assures complete randomness is called a *probability sample*.

A *stratified random sample* is a sample that achieves greater precision than a simple random sample by taking advantage of existing knowledge about the variance structure of subpopulations. By concentrating a disproportionate share of the sample in subpopulations that account for disproportionate shares of the total variance, a stratified random sample produces lower total variance for a given sample size. The annual National Fire Protection Association survey of fire departments that produces the annual estimates of total U.S. fire loss is a stratified random sample.

A *cluster sample* is a sample that randomly selects certain subpopulations, then samples only them. This approach often involves subpopulations that consist of geographical areas, in which case it is also called an *area sample*. The purpose of cluster sampling is to hold down the cost of sampling. It is not as statistically acceptable as a simple or stratified random sample.

A *systematic sample* begins with a listing of the population, then random selection of the first sample member, and finally selection of the remaining members at fixed intervals (e.g., every kth name on the list). This approach is simpler than true random sampling but not as acceptable.

A *representative sample* is one chosen to guarantee representation from each of several groups. If properly designed, it is a special case of stratified random sampling, but often the term is used for samples where the need for representation is the only part of the procedure specified. If the size of the representation is also specified, it is called a *quota sample*. The statistical properties of a sample constructed in this way cannot be determined, and nothing useful can be said about its accuracy or precision. That is also true of a *judgment sample*, in which the only rules governing sample selection are the statistician's judgments.

ADDITIONAL READING

1. R. Baldwin, *Some Notes on the Mathematical Analysis of Safety* Fire Research Note 909, Joint Fire Research Organization, Borehamwood (1972).
2. J.R. Benjamin and C.A. Cornell, *Probability, Statistics and Decision for Civil Engineers*, McGraw-Hill, New York (1970).
3. M.H. DeGroot, *Optimal Statistical Decisions*, McGraw-Hill, New York (1970).
4. J.E. Freund, *Mathematical Statistics*, Prentice Hall, Englewood Cliffs (1962).
5. J.E. Freund and F.J. Williams, *Dictionary/Outline of Basic Statistics*, McGraw-Hill, New York (1966).
6. G.J. Hahn and S.S. Shapiro, *Statistical Models in Engineering*, John Wiley and Sons, New York (1966).
7. N.A.J. Hastings and J.B. Peacock, *Statistical Distributions: A Handbook for Students*, Butterworths, London (1975).
8. P.G. Hoel, *Introduction to Mathematical Statistics*, John Wiley and Sons, New York (1962).
9. D. Huff, *How to Lie with Statistics*, Penguin, Middlesex (1973).
10. K.C. Kapur and L.R. Lamberson, *Reliability in Engineering Design*, John Wiley and Sons, New York (1977).
11. A.M. Mood and F.A. Graybill, *Introduction to the Theory of Statistics*, McGraw-Hill, New York (1963).
12. D.T. Phillips, *Applied Goodness of Fit Testing*, American Institute of Industrial Engineers, Atlanta (1972).
13. S. Siegel, *Nonparametric Statistics for the Behavioral Sciences*, McGraw-Hill, New York (1956).
14. M.R. Spiegel, *Probability and Statistics*, McGraw-Hill, New York (1975).
15. R.E. Walpole and R.H. Myers, *Probability and Statistics for Engineers and Scientists*, Macmillan, New York (1972).

Section Two
Fire Dynamics

Section 2 Fire Dynamics

FLAME HEIGHT

Bernard McCaffrey

Man has known for a long time that the more wood that is added to a fire, the bigger the fire or flames get. The understanding, quantification, and implications of this well-known fact are the subjects of this chapter.

DIFFUSION VS PREMIXED FLAMES

For fire applications, we are mainly concerned with what are called diffusion flames, as contrasted to premixed flames where fuel and oxidant are mixed or brought together prior to the combustion region. The word "diffusion" is used because the oxidant (most of the time ordinary air) and the fuel (the combustible vapors coming from the surface of the solid or liquid that is burning, or from a burner or other orifice, in the case of a gaseous fuel) come together or diffuse into a region in space where the rapid chemical reactions associated with combustion are occurring. The resulting luminosity is called a flame.

A welder's torch using oxygen and acetylene when adjusted properly yields a short, very hot premixed flame (it is not as blue as other premixed flames, such as a domestic range using natural gas, because of the very large relative amount of carbon in acetylene). One can turn off the oxygen valve and leave the same flow of acetylene exiting from the nozzle. The short, hot, premixed flame is seen to be replaced by a long, very luminously yellow, meandering, geometrically irregular, sooty diffusion flame. The latter configuration is analogous to a fire; the former, to a well-designed, useful combustion tool.

If a given amount of fuel is burned in a premixed manner, the result will be a small mass of very hot gas. If the same amount of fuel is allowed to burn in a diffusion mode, the result will be a larger amount of cooler gas. The product of mass and temperature rise are comparable in the two cases and equal, when suitably expressed, to the original energy content of the fuel, assuming relatively equivalent combustion efficiencies, which, in general, will not be the case.

The late Dr. Bernard McCaffrey was a professor at the University of Maryland at the time of his death in 1990. Prior to that he was a member of the research staff of the National Bureau of Standards Center for Fire Research. The chapter has been updated from the first edition by the editors.

Additionally, the oxygen depletion in the two modes can be identical. Combustion efficiency refers to the amount of energy released compared to the starting chemical potential of the fuel. Any part of the fuel not fully burned (such as soot or carbon monoxide, for example, which still contain relative chemical energy) decreases combustion efficiency.

The energy released from the burning of the fuel can be thought of as divided into two parts: (1) convective energy and (2) radiant energy. The convective portion is reflective of the temperature of the products and is manifested by the hot gases rising above the fire and heating the ceiling or other surface or obstacle that is in their path. The radiant energy can be felt without actually touching the flames or hot gases. In general, diffusion flames radiate a lot more than premixed flames, and, consequently, the convective portion is less, meaning their temperatures are lower and their ability to do work is less, unless a particular apparatus is specifically designed for a high radiative component. The amount of luminosity is generally proportional to the radiation output of the fire. For reasonably sized diffusion flames, most of the radiation is due to incandescent soot particles, mainly glowing fuel pyrolysis agglomerated fragments. The black smoke or soot seen above the flame regions is a portion of the particulate that escaped the flame without being oxidized. It is a net loss as far as combustion efficiency is concerned, but the real significance of this smoke is from a safety point of view in that it is responsible for obstructing the view of escape routes, and causes incapacitation from inhalation, etc. Luminosity, soot, and related problems are due to the fact that fires are poorly mixed diffusion flames.

FROUDE NUMBER:
MOMENTUM VS BUOYANCY

Consider a surface that can be a solid, liquid, or the exit plane of a pipe, burner, or other orifice issuing fuel gas. The size of the surface can be characterized by a single parameter, D, the diameter in the case of a circular surface, the length of a side in the case of a square source, or some other dimension that determines the size or area of the surface. Situations exist in which this single parameter completely determines the size of the fire. For example, a pool of liquid fuel burning under normal conditions is completely characterized by the type of fuel and the size of the pan or the extent of the spill. Many solids and synthetic polymers, burning in a horizontal configuration, behave similarly. Blinov and Khudiakov ran a series of liquid pool fire experiments encompassing very small, 35 mm D, to very large, 25 m, diameter pools for a variety of industrial liquids.[1] They were able to reduce all their flame height data when scaled with D and burning rate data scaled with D^2 (or burning rate per unit area) to single curves that were functions only of the pool size, D, with only a slight fuel structure dependence at small D.

Other situations, especially in the case of a gaseous fuel, require an additional specification, that is, the rate at which the gases are evolving or flowing out of the surface. We let U stand for that velocity. For most cases, D and U and the type of fuel completely determine the fire. Where the ambient is exposed to high winds, less-than-normal oxygen concentrations, or very high radiation levels, additional considerations will be necessary; i.e., U can be increased or decreased from its non-disturbed value, even for solid or liquid fuels, by external forces.

The amount of mass or fuel coming off the surface per unit time, often referred to as the burning rate or mass loss rate, $\dot{m}$, is simply U multiplied by the area, proportional to D^2 times the fuel vapor density, ρ_v, numerically similar for most fuels.

$$\dot{m} = \rho_v A U$$

From the burning rate, one can calculate a nominal or theoretical heat release rate, $\dot{Q}$, by simply multiplying the burning rate by a heat of combustion, ΔH, available in tables for most fuels.

$$\dot{Q} = \dot{m} \cdot \Delta H \qquad (1)$$

Inexactness enters these calculations in two ways. First, the amount of fuel leaving the surface may not all be taking part in the fire. That is to say, the burning rate may be somewhat less than the mass loss rate. More fuel is leaving the surface than is participating in the burning; probing above the flame region would detect a nonzero fuel concentration. The second involves the completeness of combustion. The heat of combustion found in Tables 2-1.1 and 2-1.2 assumes that the reaction goes to completion, reflected chemically as the products CO_2 and H_2O. Significant amounts of products of incomplete combustion, such as CO and soot, leaving the flame contain chemical energy that could have added to the heat release rate had they been oxidized in the flame. Incidentally, just because a flame is very luminous does not necessarily mean it is chemically inefficient; it only means that there is a nontrivial radiative component of the energy released. Those radiating carbon particles can be consumed in the upper, intermittent portions of the flame region, with little soot escape. A large methane flame on a refractory burner can be quite luminous and still exhibit a very high completeness of combustion. If, however, lots of soot and other visible aerosols are leaving the flame region, then that is usually a good indication that something less than complete combustion is taking place, and the calculated heat release rate, $\dot{Q}$, will be on the high side.

From hydraulics, when both momentum and gravity or buoyancy are of similar importance in a problem, one defines the nondimensional Froude number with which to quantify the relative magnitude of the two effects

TABLE 2-1.1 *Estimates of the Range of $\dot{Q}^*$*

Material	m'' (g/s m^2)	ΔH (MJ/kg)	$\dot{Q}''$ (kW/m^2)	$\dot{Q}^*$
Polyurethane foam				
(Flexible $D = 1$ m)			420–510	0.36
(Rigid $D = 1$ m)			280–440	0.20
Polypropylene ($D = 1$ m)			890	0.61
Polyethylene ($D = 1$ m)			960	0.67
Polystyrene	8.3	42	330	0.3
Cotton bales (190 kg/m^3)	2.4	17	37	
Paper in loose form	6.7	13	84	
Wood (moisture content 14%)	14	14	180	0.16
Lumber and sawn timber, stacked in open	110	14	1450	1.3
Ethyl Alcohol	27–33	27	650–800	0.7
Kerosene	49	44	1900	1.7
Gasoline	44–53	42	1700–2000	1.7
Heptane ($D = 0.3 - 0.5$ m)	28	33[†] 42	1000–1200	1.2
Crude oil ($D = 0.6$ m)	20.5	29[†] 32	620	0.6
Wood cribs	70–80	13[†]	900–1000	1.1
Urethane cribs	100	20[†]	2000	2.4

From references 2, 3, 4, 5.

[†]$\dot{Q}^*$ = Effective heat of combustion—heat release rate divided by rate of weight loss.

$$\mathrm{Fr} = U^2/gD \qquad (2)$$

where g is the acceleration of gravity. The numerator is in proportion to momentum, the denominator to the gravity or buoyancy forces. From the previous definitions, we can express Froude number in terms of heat release rate, $\dot{Q}$,

$$\mathrm{Fr} \propto \dot{Q}^2/D^5 = C(\dot{Q}/D^{5/2})^2 \qquad (3)$$

where C is a constant composed of fuel properties, source geometry, and gravity. For diffusion flames resulting from gas flow out of a pipe, D and U can be varied independently. Let us fix the heat release rate, $\dot{Q}$. The amount of gas flowing per unit time is then a constant; i.e., so many grams per second of fuel or a fire of so many kW. Now we can choose our pipe size, D, from very small to very large and keep $\dot{Q}$, or the heat release rate, the same. First, consider the small diameter. To maintain $\dot{Q}$ for gas flowing through a small hole, the velocity of the gas must be quite large. We end up with a high-velocity jet diffusion flame characterized by a large Froude number. These are similar to flares seen in the chemical process industry or the flame from a leak in a high-pressure storage vessel. The flame height can be several hundred times the size of the pipe diameter and the luminous portion is long and thin.

Now, if the same flow rate of gas is directed through a large-diameter aperture, the velocity of the gas will be very small, and a different kind of diffusion flame results. Here the Froude number is very small and buoyancy becomes important, and the all-important mixing process is totally different from the previous small-hole jet configuration where turbulent shear forces did most of the entraining. Flame heights for this buoyancy-dominated case are the same order of magnitude as the diameter. These are the

TABLE 2-1.2 *Flame Height Correlations*

Symbol (See Figure 2-1.1)	Reference (Fuel, Geometry)	$\dot{Q}*$ Range	H/D	Comments
(Z)	Zukoski[6] (Natural gas, 10–50 cmD Burner)	$\dot{Q}* < 0.15$ $0.15 < \dot{Q}* < 1$ $1 < \dot{Q}* < 40$	$40\,\dot{Q}*^2$ $3.3\,\dot{Q}*^{2/3}$ $3.3\,\dot{Q}*^{2/5}$	
(C)	Cox & Chitty[2] (Natural gas, 45, 60 cm square)	$0.13 < \dot{Q}* < 0.28$ $0.28 < \dot{Q}* < 0.55$	$15.1\,\dot{Q}*^2$ $3.2\,\dot{Q}*$	
(T)	Thomas[7] (Wood cribs, 10–200 cm side)	$0.75 < \dot{Q}* < 8.8$	$3.4\,\dot{Q}*^{0.61}$	$\Delta H = 18{,}600\,\dfrac{kJ}{kg}$ (wood)
(H)	Heskestad[8] (Gas, liquid, solid, literature)	$0.12 < \dot{Q}* < 1.2 \times 10^4$	$3.7\,\dot{Q}*^{2/5} - 1.02$	$\dfrac{\Delta H}{r} = 3185\,\dfrac{kJ}{kg\ air}$ (C_2H_4)
(S)	Steward[9] (Literature & gas jets)	$1 < \dot{Q}* < 10^4$	$4.16\,\dot{Q}*^{2/5}$	$\dfrac{\Delta H}{r} = 3185\,\dfrac{kJ}{kg\ air}$ $\omega = .0833$
(B)	Becker & Liang[10] (Literature & 0.7–4.6 mm tubes, various gases)	$\dot{Q}* < 1.7$ $1.7 < \dot{Q}* < 21$ $33 < \dot{Q}* < 10^3$ $(20 < \xi_L < 40)$ $10^3 < \dot{Q}* < 10^6$ $(1 < \xi_L < 20)$ $\dot{Q}* > 10^6$ $(\xi_L < 1)$	$1.52\,\dot{Q}*^2$ $3.6\,\dot{Q}*^{2/5}$ $\psi = 0.064\,\xi_L - 0.58$ $\psi = 0.18 + 0.022\,\xi_L$ $H/D < 11(\beta r)\sqrt{\rho_0/\rho_\infty}$	$\dfrac{\Delta H}{r}\dfrac{1}{\beta} = 1121\,\dfrac{kJ}{kg\ air}$ (C_2H_4)
(K)	Kalghatgi[11] (Various gases, 1–10 mm tubes)	$(2 < \xi_L < 11)$	$\psi = 0.2 + 0.024\,\xi_L$	
(W)	Hawthorne et al[12] (Various gases from small nozzles)		$\dfrac{H - \ell}{D} = 5.3(\beta r)\sqrt{\rho_0/\rho_\infty}$	

flames with which researchers interested in domestic fires are generally concerned. Most flames from upholstered furnishings, sofas, beds, wastepaper baskets, chip pan fires, spilled gasoline, etc., are of this type. The combustible vapors come off a large surface with negligible velocity, Froude numbers are small, and buoyancy dominates these pool fire-like configurations.

FLAME HEIGHT CORRELATIONS

The entire world of diffusion flames can be laid out on a plot using the Froude number or some of its variations as a scale. That world is essentially divided into three parts: (1) small Fr, (2) intermediate Fr, and (3) large Fr. In each of the regimes, we can note the behavior of two easily observed and measured quantities: (1) flame height and (2) the radiative fraction.[13] Although arbitrary in exact determination, most researchers can agree on what constitutes a measure of flame height; i.e., most of the time the top of the luminous portion is at a certain height above the surface. The second quantity, the radiative fraction, is the amount of energy radiated away from the flame zone, expressed as a fraction of the nominal heat release rate of the fire. This is an easily measured quan-

tity, and, hence, there is less arbitrariness about quantification, although agreement among various observers is not that great, perhaps due to the lack of a traceable thermal radiation standard at fire output levels.

Changes in behavior of flame height and the radiative fraction with fuel flow rate or fire size demarcate the various regimes. A more or less one-to-one correspondence between plots of flame height and radiative fraction has been observed for the three Froude number regimes.[13] In the intermediate Froude number regime, consisting of small diameter sources and intermediate velocity jets, the radiative fraction is constant, independent of both D and U. Flame height normalized by diameter rises with Froude number to the ⅕ power, which means that flame height becomes independent of diameter and varies with heat release rate $\dot{Q}$ to the ⅖ power. There is some theoretical justification for a ⅖ power variation.[9]

In the small and large Fr regimes, things are not so straightforward. Radiative fraction depends on both U and D, and the flame height/diameter variation with Fr deviates from the ⅖ power, exhibiting a larger variation in the low Fr pool configuration regime, and a smaller variation, approaching zero with Fr in the high-momentum jet end of the

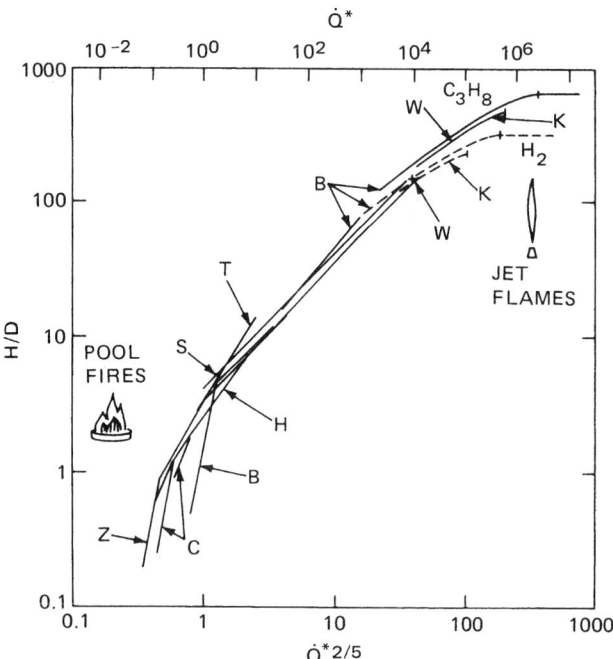

Fig. 2-1.1.   Flame height for the entire Froude number spectrum. Capital letters correspond to studies listed in Table 2-1.2.

Froude spectrum. These are the more scientifically interesting and, in terms of practical problems, the more important regimes.

Figure 2-1.1 shows the flame height as a function of the Froude number for nonlaminar diffusion flames. The form of the Froude number chosen for plotting is $\dot{Q}^*$, or rather $\dot{Q}^{*\,2/5}$, used to compress the horizontal scale. Plotting the data in Figure 2-1.1 in terms of actual Froude number (Equation 3) would require 17 cycles, and, hence, the vertical sensitivity encompassing only 4 cycles would appear diminished. $\dot{Q}^*$ is a nondimensional square root of Froude number given in terms of heat release rate, e.g., Equation 3

$$\dot{Q}^* = \frac{\dot{Q}}{\rho_\infty C_p T_\infty \sqrt{gD} D^2} \qquad (4)$$

where ρ_∞, C_p, and T_∞ are the density, specific heat, and absolute temperature of some reference gas, generally, the ambient. If $\dot{Q}$ is broken into its components through Equation 1, individual pieces can be scaled, and D^2 can be associated with area , $\sqrt{gD}$ with velocity, and $\rho_\infty C_p T_\infty$ with a heat release rate per unit volume. (Normally, $\dot{Q}^* = \dot{Q}/D^{5/2}/1110$, where $\dot{Q}$ is in kW and D is in m.) Candlelike flames and those in bench-scale experiments using sheltered apparatus and employing low Reynolds number flows are excluded. What are included might be termed "free," since they are subject to subtle laboratory disturbances. On the left are pool-configured fires with flame heights of the same order of magnitude as the base dimension D. In the middle is the intermediate regime where all flames are similar and the $\dot{Q}^{2/5}$ is seen as a 45-degree line in the figure. Finally, in the upper right is the high Froude number, high-momentum jet flame regime where flame height ceases to vary with fuel flow rate and is several hundred times the size of the source diameter. H is the flame height normalized by D, the actual physical dimension.

An idea of what $\dot{Q}^*$ is numerically for various fuels and configurations can be obtained from Table 2-1.1 and Figure 2-1.2, taken directly from Hasemi and Tokunaga.[3] Of course gaseous fuels can take on any value of $\dot{Q}^*$.

Shown in Figure 2-1.1 are lines representing the data correlations available in the literature. Table 2-1.2 shows the values or expressions for H/D as a function of $\dot{Q}^*$, along with the researcher; the letters correspond to the circled letters on the lines in Figure 2-1.1. Different researchers, guided by different theoretical insights, have used different means to plot their data. The necessary conversion as well as a discussion of the various correlations follows.

Pool Fires

Zukoski, whose concern was primarily for pool-configured flames, took the lead in correlating his data in terms of $\dot{Q}^*$.[6] He has further subdivided the low Froude number end into two regimes: (1) a transition from the intermediate, $2/5$ regime, using the $2/3$ power down to about $\dot{Q}^* \sim 0.15$; and (2) into the very steep $\dot{Q}^{*\,2}$ region. Caution should be exercised in interpreting data where H is much smaller than D. Most of these laboratory flamlets are of a laminar character and do not necessarily represent mass fire.[2,14] Mass fires are composed of many individual turbulent fire plumes which may merge at some elevation above the ground.

Cox and Chitty, also plotting their data in terms of $\dot{Q}^*$, were also able to subdivide their data into two increasingly larger slopes as $\dot{Q}^*$ was reduced.[2] Thomas, early on, recognized a larger than $2/5$ power variation and correlated wood crib fire data with a 0.61 power variation.[7] Wood crib fires, due to their geometrical structure, tend to exhibit somewhat larger flame heights than pool fires but are still similar enough to be classified together in this review. The Thomas correlation is given in terms of an actual mass loss rate per unit area and, therefore, like most of the rest of the correlations, must be converted to $\dot{Q}^*$ by a somewhat arbitrary but (hopefully) reasonable choice of thermochemical properties. These are presented clearly in Table 2-1.2, in order that the reader may make contact with the original reference. For

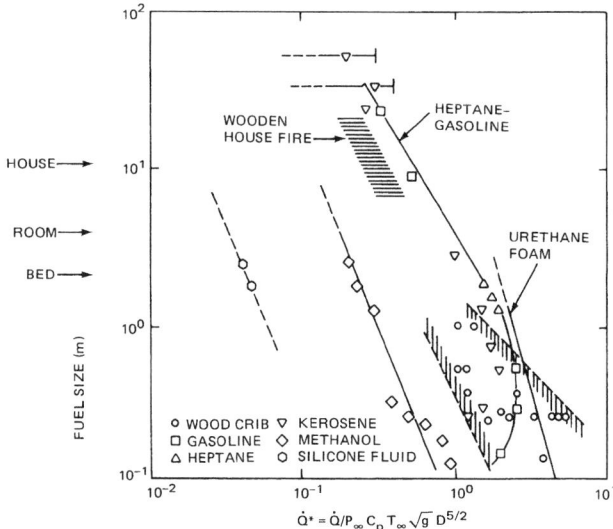

Fig. 2-1.2.   Fire size vs. $\dot{Q}^*$[17]

Thomas, the heat of combustion of wood was chosen as 18,600 kJ/kg.[7] The position of his correlation shown in Figure 2-1.1 will therefore move up or down according to $\Delta H^{-0.61}$ for alternate choice of heats of combustion.

The other two correlations that extend into the buoyant end were developed by Heskestad[8] and Becker and Liang[10] and are part of more grand schemes attempting universal correlations of all available literature data and extend into the next or intermediate regime at higher $\dot{Q}^*$.

Heskestad[8] uses a combustion number to correlate the data. The number contains an interesting thermochemical property, the heat of combustion per unit amount of oxygen consumed, $\Delta H/r$, which is very nearly the same for most fuels and is the basis for oxygen consumption calorimetry. Instead of the heat of combustion being expressed as a quantity of energy released per unit fuel burned, which varies widely, Heskestad divides this number by r, the stoichiometric air-to-fuel ratio (mass basis), and arrives at the heat released per unit air consumed, which is equal to about 3000 kJ/kg$_{air}$ for most common fuels.

The Heskestad correlation will, therefore, vary weakly with fuel type, H/D being proportional to $\Delta H/r$ to the $-\frac{3}{5}$ power. Note the $\frac{2}{5}$ form of the correlation except for $\dot{Q}^*$ near order 1 where the steeper slope begins to show.

Becker and Liang, who have attempted the most ambitious task of trying to correlate the entire Froude regime, suggest a squared variation in $\dot{Q}^*$ for the region of $H/D < 4.5$.[10] Besides the usual $\Delta H/r$, their correlating variable contains an additional fuel parameter, β, defined as the ambient molecular weight times the adiabatic flame temperature divided by the product of the molecular weight of the stoichiometric combustion products and the ambient temperature, all raised to the $\frac{1}{2}$ power. Like $\Delta H/r$, it is practically independent of fuel type. Here the correlated results depend on $\Delta H/r \cdot 1/\beta$ to the -2 power.

Wall/Corner Geometry

Investigations of flame heights of pool-type fires located next to a wall or in a corner have been made by Hasemi and Tokunaga.[15] They found that the flame height of pool fires against a wall have the same flame height as the same fire away from the wall. Back et al[16] confirmed this result, and Mizuno and Kawagoe[17] confirmed this result for chairs burning against a wall. As such, the flame height correlations shown in Table 2-1.2 can be applied to fires against walls.

Hasemi and Tokunaga[15] and Kokkala[18] have investigated pool-type fires in a corner configuration. These investigators found that their data could be correlated as $\dot{Q}^{*2/3}$, but Kokkala[18] found that his flame heights were somewhat larger than those of Hasemi and Tokunaga.[15] The Hasemi and Tokunaga correlation is given by

$$\frac{H}{D} = 3.0 \, \dot{Q}^{*2/3} \text{ (continuous flame height)}$$

$$\frac{H}{D} = 4.3 \, \dot{Q}^{*2/3} \text{ (flame tip height)}$$

The continuous flame height is the height to which flame exists 100 percent of the time; the flame tip height is the greatest height that flame is observed. The average flame height is obviously between these two limits. Many (e.g., Kokkala[18]) have interpreted the above equations to indicate an average flame height is halfway between these limits, yielding 3.65 for the average flame height.

Wall Flames

The flame height of line fires against walls and fires on walls have been studied by several investigators. Hasemi studied line fires against walls.[19] He correlated his observed flame heights on the basis of a modified $\dot{Q}^*$, $\dot{Q}_l^*$,

$$\dot{Q}_l^* = \frac{\dot{Q}_l}{\rho_\infty C_p T_o g^{1/2} D^{3/2}}$$

where $\dot{Q}_l$ is the heat release per unit length of the fire source, D is the length of the fire source, and other variables are as previously defined. Hasemi's[19] data yielded the following correlations:

$$H = 2.8 \, \dot{Q}_l^{*2/3} D \text{ (continuous flame height)}$$

$$H = 6.0 \, \dot{Q}_l^{*2/3} D \text{ (flame tip height)}$$

Delichatsios[20] interpreted the experimental results of Ahmad and Faeth[21] in this form and found that the visible flame height was correlated by

$$H = 4.2 \, \dot{Q}_l^{*2/3} D$$

which is quite close to the average flame height obtained by averaging Hasemi's[19] continuous flame and flame tip heights, i.e., $4.4 \, \dot{Q}_l^{*2/3} D$. Quintiere, Harkelroad, and Hasemi[22] measured flame heights from small-scale wall fires for a range of materials and found that the results generally followed the Delichatsios[20] correlation.

All the available work for line fires against walls and wall fires indicate that the flame height scales with the heat release rate per unit length is to the $\frac{2}{3}$ power. Further, the flame height is independent of D.

Intermediate $\dot{Q}^*$: Constant Radiative Fraction

Steward constructed an analytical model, based on conservation equations used in prior plume studies, which attempted to predict the flame height of tall diffusion flames.[9] With some adjustment for poor mixing of the air and fuel, the results were quite successful for the intermediate, $\frac{2}{5}$ power Froude regime. With the adjustment, the prediction was able to capture for various gaseous fuels the gross features over many orders of magnitude in flow rate. Steward uses a combustion number that is slightly more complicated than the previous expressions but does reduce to virtually $\Delta H/r$ to the $-\frac{2}{5}$ power.

Becker and Liang, in the lower end of intermediate regime, exhibit the requisite $\frac{2}{5}$ power with heat release rate for flame height-to-diameter ratio.[10] As for property variation, $\Delta H/r/\beta$, which is the same as the pool fire data property, varies here with the $-\frac{2}{5}$ power vs the -2 power for the pool.

Note in Table 2-1.2, under "Comments," the fuel chosen for all the correlations discussed so far in converting to $\dot{Q}^*$ was ethylene, C_2H_4. Although having a $\Delta H/r$ slightly on the high side ($\Delta H/r = 2900$, 2950 kJ/kg$_{air}$ for CH_4, and C_3H_8 as well as C_4H_{10} respectively), and, it is not atypical for hydrocarbons and offers the advantage of being neutrally buoyant; i.e., before burning, it has practically the same density as the surrounding ambient. At larger $\dot{Q}^*$ fuel density is a determining factor for estimating flame height. Up until this point, only the thermochemical fuel properties ΔH, r, β, and

ω, which includes C_p and T, were required for determining the flame heights of different fuels.

At the upper end of the intermediate regime, the variation of flame height with $\dot{Q}^*$ begins to decrease, resulting in powers less than $2/5$. To bridge the gap between the $2/5$ power regime on the left and the high-momentum regime on the right (see Figure 2-1.1), Becker and Liang[10] have chosen to plot the inverse of flame height and Froude number, using ξ_L and ψ to represent flame height correlations in this regime; i.e.,

$$\psi = A + B\xi_L \tag{5}$$

where A and B are numeric constants determined from the flame height *vs* flow rate data. ξ_L is the cube root of a Richardson number having the flame height as a characteristic length scale

$$\xi_L = \left(\frac{gD}{U^2}\right)^{1/3}\left(\frac{\rho_0}{\rho_\infty}\right)^{-1/3}\frac{H}{D} \tag{6}$$

and

$$\psi = (\beta r)^{2/3}\left(\frac{\rho_0}{\rho_\infty}\right)^{1/3}\left(\frac{H}{D}\right)^{-2/3} \tag{7}$$

Note the source-to-ambient density ratio is now appearing prominently. Unfortunately, H/D cannot be solved explicitly in terms of $\dot{Q}^*$ in the simple manner presented previously for small H/D and $\dot{Q}^*$. Based on the Becker and Liang correlation,[10] a power-law expression no longer appears adequate. Solving the inverse problem,

$$\dot{Q}^* = \frac{B^{3/2}\frac{\pi}{4}\frac{\Delta H}{C_p T_\infty}(H/D)^{5/2}}{\left[(\beta r)^{2/3} - A\left(\frac{H/D}{\sqrt{\rho_0/\rho_\infty}}\right)^{2/3}\right]^{3/2}} \tag{8}$$

where A and B are the data-correlating constants from Equation 5. (See Table 2-1.2.)

Although explicitly contained in Equation 8, the effect of the square root of the density ratio does not begin to dominate the behavior until $\dot{Q}^*$ equals about 10^3; i.e., the second of the Becker and Liang[10] $\psi(\xi_L)$ expressions in Table 2-1.2. (This is actually the third $\psi(\xi_L)$ expression for what we are calling here the intermediate or $2/5$ regime. The first of the three is, in fact, a $2/5$ variation and is given in terms of $\dot{Q}^*$.)

High-Velocity Jet Flames

Where the variation of flame height with fuel flow rate begins growing weaker and disappearing in the limit, we have the high-momentum limit. Here, like the pool fire end of the Froude spectrum, the radiative fraction begins to vary with flow rate. The luminosity decreases as flow rate is increased and liftoff height increases until blowoff occurs.[23] Becker and Liang[10] and Kalghatgi[11] have developed correlations beyond the classical Hawthorne *et al*[12] expression for nonvarying flame height. That work[12] utilized small diameters, and blowoff occurred at rather small $\dot{Q}^*$.

At the high-momentum limit, exit jet momentum effects are all-important, and, for illustrative purposes, two sets of curves are presented in Figure 2-1.1: one for a very light gas, H_2, and one for a heavier gas, C_3H_8, in order to appreciate the expected variation. H/D is a function of the square root of the jet-to-ambient density ratio, a direct consequence of momentum conservation between the jet and surroundings. There is also a fuel property effect, βr, according to Becker

and Liang,[10] but, as seen previously for most fuels, this tends not to vary markedly, CO being a notable exception. (H_2 is also an exception, but the large βr tends to mitigate somewhat the very small density ratio.)

Kalghatgi's[11] data would dictate a considerably smaller H/D than Becker and Liang[10] and would indicate a transition into the H/D = constant regime at considerably lower $\dot{Q}^*$, more in line with the classical results indicated in Figure 2-1.1 (one each for C_3H_8 and H_2) by a W. Distortion can result in this regime, depending on whether or not a stabilizing gas is used with the fuel in order to prevent liftoff and subsequent blowoff.

The subjects of liftoff, blowoff and flame stability, some of the characteristics of the high-momentum regime, have been recently studied by Kalghatgi.[24] Additional work, including large CH_4 jet diffusion flames to 500 MW including underexpanded supersonic jet flames, has been covered by McCaffrey and Evans.[25]

USE OF FLAME HEIGHT CORRELATIONS

Figure 2-1.1 and the corresponding expressions in Table 2-1.2 can stand on their own, meaning that one correlation will not be deemed better than another. It must be presumed that all researchers are equally competent and that the variation reflects the nature of the phenomenon itself. Of course, each researcher's definition of flame height constitutes a considerable variable accounting for some of the imprecision. There are plenty of recipes suggested in the literature for determining an unambiguous definition, which will not be debated; but it is felt that there is some advantage to presenting the available data as it stands in the forms of Figure 2-1.1 and Table 2-1.2.

The user can determine or estimate a $\dot{Q}^*$ based on fuel flow rate or mass loss rate data available for the fire. (See Section 3, Chapter 1, "Burning Rates.") At that point, the various expressions can be solved for flame height (unfortunately for high $\dot{Q}^*$ using ψ and ξ_L, the solution must be determined iteratively) and the range of expected results studied. The user will ultimately be the best judge in determining how to apply or utilize these "bracketed" results. For routine use, an easy-to-apply expression,[8] valid everywhere except at the jet flame end

$$H/D = 3.7\,\dot{Q}^{*\,2/5} - 1.02 \tag{9}$$

can be recommended. It maintains a $2/5$ variation in line with the other equations over the large intermediate regime while exhibiting an increasing slope at small $\dot{Q}^*$.

A note of caution when trying to solve the inverse problem, that is, attempting to determine the heat release rate based on the flame height. The exponents like the $2/5$ power become the inverse, i.e., $5/2$, and inaccuracies associated with flame height become exaggerated. They are certainly useful as guides to what might be expected, but any quantitative use ought to be heavily qualified.

EXAMPLE:

In the study of certain scale-model enclosure fires, it is desired to evaluate ignition sources that will yield flames just reaching or touching the ceiling, a height of approximately 1 m above the floor. It is desired to consider various fuels and configurations.

SOLUTION:

If a well-controlled and constant fire source is desired, a gas burner is ideal. The diameter of the burner must be of sufficient size to ensure nonlaminar burning—we desire the irregular (but coherent) erratic patterns of luminosity associated with real fires as opposed to the kind of flames seen on candles. (This requirement is functional not aesthetic. The entrainment and, hence, mixing and burning will be different.) Anything greater than about 0.3 m will suffice. Solving Equation 9,

$$\dot{Q}^* = \left[\left(\frac{H}{D} + 1.02\right)\Big/3.7\right]^{5/2} = \left[\left(\frac{1}{3} + 1.02\right)\Big/3.7\right]^{5/2} = 1.5$$

and

$$\dot{Q} = 1110 \times 1.5 \times 0.3^{5/2} = 82 \text{ kW}$$

This number is now related back to a steady gas flow rate of fuel, e.g., about 1.7 g/s of methane or propane, through the burner. Note from Table 2-1.2, at $\dot{Q}^*$ near 1 for $H/D = 1/.3$, values as low as $\dot{Q} = 30$ kW are obtainable although the consensus seems to be between 50 and 80 kW.

A liquid pool fire could be used to study the effect the environment might have on the source or to exhibit a growing and decaying source; heptane, for example, could be chosen. From Table 2-1.1, the heat release rate per unit area for these size pools for heptane is about 1100 kW/m². Dividing the number into 82 kW yields a diameter again of about 0.3 m. Using gasoline or kerosene, on the other hand, would dictate the use of a smaller pan, since the burning rate or heat release rate per unit area for these approaches 2000 kW/m². For the gas burner, D was not important (provided $D > \sim 0.3$ m). We controlled $\dot{Q}$ with a valve. For a liquid, D is all-important, since D controls $\dot{Q}$, and $\dot{Q}$ for both liquid and gas determines flame height.

For a solid source, configuration becomes important. Compare the entries "wood" and "lumber stacked" in Table 2-1.1 to see the variation in burning rates, a fact well known to campfire builders. Chairs and other irregularly shaped items are difficult to characterize accurately. Knowing or estimating the heat release rate will, however, go a long way in allowing one to be able to predict flame height, since, for values of $\dot{Q}^*$ near unity, flame height doesn't vary greatly with diameter or fuel composition.

APPENDIX

Disturbances to Free-Burn Fire Behavior

The literature[27,28,29] contains various estimates of the amount of air entrained by a free-burning pool-configured or buoyant diffusion flame, from the base of the flame or surface of the pool up to the flame tip. The numbers range from about $m_{ent}/\dot{Q} = 0.0045$ to 0.0060 kg/s/kW. The smaller value, based on point source plume theory, represents about 13 times the stoichiometric air requirements of the flame, assuming a hydrocarbonlike fuel of $\Delta H = 45$ MJ/kg and $r = 15$ kg/kg. Zukoski's[6] hood collection technique (0.005) yields about 15 times stoichiometric and, finally, the upper value, based on point measurement techniques coupled to an integral model of the fluid mechanical behavior which comes out to be about 20 times the stoichiometric requirement.[29] One might suspect, then, that something between 15 and 20 would be a "reasonable" value, since these were based on actual measurements in flames. From this number, one can estimate what the ambient velocities generated by the fire are likely to be.

Imagine a cylinder of radius, r, and height, H, equal to the flame height, surrounding a pool fire of diameter, D. If air comes in uniformly through the surface of this cylinder to "feed" the fire, we can easily calculate the average value of the velocity of that air in SI units

$$\bar{V} = \frac{\dot{m}}{\rho A} = \frac{(0.005 - 0.006) \cdot \dot{Q}}{1.2 \times \pi D \cdot H}$$

We can now relate flame height, H, to heat release rate, $\dot{Q}$, and heat release rate with pool diameter, D, or *vice versa*.

The burning rate for realistic materials in reasonably sized samples ($D \sim 1$ m) ranges from about 300 to 1500 kW/m². (See Table 2-1.1.) Beyond this size, burning rates generally do not increase with pan diameter. Similarly, H/D becomes constant, independent of diameter and equal to about $1 - 3$. Substituting yields

$$\bar{V} = \frac{(0.005 - 0.006)}{1.2} \frac{(300 - 1500)}{4(1 - 3)}[\text{m/s}]$$

$$\bar{V} = 0.1 \text{ to } 2 \text{ m/s}$$

In McCaffrey's experiments[29] a more-detailed analysis and comparison of experiments yields inflow velocities at about 0.2 m/s, which is at the lower end of the above simple calculation. Note that V turns out to be independent of fire size for the present conditions; i.e., for those pools of D of order 1 m or larger. Incidentally, $\dot{Q} \sim D^2$, and $H/D \sim$ constant implies that $H \sim \dot{Q}^{1/2}$, not $\dot{Q}^{2/5}$, which means that these flames fall into the pool-fire regime, i.e., slopes larger than ⅖, which is precisely where they would be expected to fall.

Any external velocity (from winds, air-conditioning systems in fire experiment laboratories, or other sources) equal to or greater than the above value will cause some disturbance compared to a completely quiescent ambient. The decision on how much disturbance can be tolerated in an experiment or what values of wind velocity will warrant different fire-fighting techniques will (obviously) be left to the user of the information. Suffice it to say that the lower end of the 0.1 to 2 m/s range represents a very small velocity. For example, if an overhead sprinkler suddenly comes on in a room containing a fire, the flow field developed by the spray entrainment will totally overwhelm that previously developed solely by the buoyant forces of the fire. The gases in the room will become totally mixed in the spray case rather than two layers—hot upper and cold lower—in the developing fire case.

NOMENCLATURE

A, B constants in correlation of flame height data near high-momentum end of spectrum, Equation 5

C constant defined in Equation 3 =

for CH$_4$ $\left(\frac{4/\pi}{\sqrt{g}} \frac{1}{\rho \Delta H}\right)^2 \sim 1.4 \times 10^{-10}$ m⁵/kW²

C_p specific heat—for air ~ 1 kJ/kg · K

D diameter or characteristic dimension of burner or pyrolyzing surface (m)

Fr   Froude number $U^2/(g \cdot D)$

g   gravitational acceleration (9.8 m/s²)

ΔH theoretical lower heat of combustion (MJ/kg)

H visible flame height—vertical distance from burner to extremity of luminosity

$\dot{m}$ mass loss rate (approximately equal to burning rate) (g/s)

M molecular weight

r stoichiometric air-to-fuel mass ratio

$\dot{Q}$ energy or heat release rate (kW)

$\dot{Q}^*$ nondimensional, Froude-like flame source characterization—Equation 4

T temperature (K)

U velocity of gaseous effluent from burner or pyrolyzing surface (m/s)

β $(M_\infty T_{ab}/M_{comb} T_\infty)^{1/2}$

ψ correlating variable, Equation 7

ξ_L correlating variable, Equation 6

ρ gas, fuel, or air density

Subscripts

$COMB$ stoichiometric combustion products

0 flame or burner source conditions

∞ ambient conditions, usually air

ab adiabatic

ent entrained

ℓ per unit length of fire source

REFERENCES CITED

1. V.I. Blinov and G. N. Khudyakov, *T-1490 a-c*, U.S. Army R & D Laboratories, Fort Belvoir (1961).
2. G. Cox and R. Chitty, *Comb. and Flame*, 60, 219 (1985).
3. Y. Hasemi and T. Tokunaga, *Fire Sci. and Tech.*, 4, 15 (1984).
4. M.Y. Roytman, *Principles of Fire Safety Standards for Building Construction*, Translation, Amerind Publishing, New Delhi, p. 23, (1975).
5. J.G. Quintiere and B.J. McCaffrey, "The Burning of Wood and Plastic Cribs in an Enclosure," *NBSIR 80-2054*, National Bureau of Standards, Washington (1980).
6. E. Zukoski, in *Fire Safety Science, Proceedings of the First International Symposium on Fire Safety Science*, Hemisphere, New York (1984).
7. P.H. Thomas, Ninth Symposium, The Combustion Institute (1963).
8. G. Heskestad, *F. Safety J.*, 5, 103 (1983).
9. F.R. Steward, *Comb. Sci. and Tech.*, 2, 203 (1970).
10. H.A. Becker and D. Liang, *Comb. and Flame*, 32, 115 (1978).
11. G.T. Kalghatgi, *Comb. Sci. and Tech.*, 41, 17 (1984).
12. W.R. Hawthorne, D.S. Weddell, and H.C. Hottel, *Third Symposium on Combustion, Flame, and Explosion Phenomena*, Williams and Wilkins, Baltimore (1949).
13. B. McCaffrey, *Some Measurements of the Radiative Power Output of Diffusion Flames*, WSS Combustion Meeting, Pullman (1981).
14. B.D. Wood, P.L. Blackshear, and E.R.G. Eckert, *Comb. Sci. and Tech.*, 4, 113 (1971).
15. Y. Hasemi and T. Tokunaga, "Some Experimental Aspects of Turbulent Diffusion Flames and Buoyant Plumes from Fire Sources Against a Wall and in a Corner of Walls," *Combustion Science and Technology*, 40, pp. 1–17 (1984).
16. J. Back, C. Beyler, and P. DiNenno, "Wall Incident Heat Flux Distributions Resulting from Adjacent Fire," *Fourth International Symposium on Fire Safety Science*, International Association of Fire Safety Science (1994).
17. T. Mizuno and K. Kawagoe, "Burning Rate of Upholstered Chairs in the Center, Alongside a Wall and in a Corner of a Compartment," *Fire Safety Science—Proceedings of the First International Symposium*, pp. 849–857 (1985).
18. M.A. Kokkala, "Characteristics of a Flame in an Open Corner of Walls," *Interflam 1993*, Interscience Communications Limited, pp. 13–29 (1993).
19. Y. Hasemi, "Experimental Wall Flame Heat Transfer Correlations for the Analysis of Upward Flame Spread," *Fire Science and Technology*, 4(2) (1984).
20. M.A. Delichatsios, "Flame Heights in Turbulent Wall Fires with Significant Flame Radiation," *Combustion Science and Technology*, 39, pp. 195–214 (1984).
21. T. Ahmad and G.M. Faeth, "Turbulent Wall Fires," *Seventeenth Symposium (International) on Combustion*, The Combustion Institute (1978).
22. J.G. Quintiere, M. Harkelroad, and Y. Hasemi, "Wall Flames and Implications for Upward Flame Spread," *Combust. Sci. and Tech.*, Vol. 48, pp. 191–222 (1986).
23. B.J. McCaffrey, *Momentum Diffusion Flame Characteristics and the Effect of Water Sprays*, NBSIR 86-3442, National Bureau of Standards, Washington (1986).
24. G.T. Kalghatgi, *Comb. Sci. and Tech.*, 26, 233 (1981).
25. B.J. McCaffrey and D.D. Evans, *Very Large Methane Jet Diffusion Flames*, 21st Combustion Symposium, Munich (1986).
26. B.M. Cetegen, E.E. Zukoski, and T. Kubota, *Comb. Sci. and Tech.*, 39, 305 (1984).
27. C.L. Beyler, *Entrainment in Pool-Type Diffusion Flames*, ESS Combustion Meeting, Providence (1983).
28. B.J. McCaffrey, *Comb. and Flame*, 52, 149 (1983).
29. V. Babrauskas, *F. Safety J.*, 11, 53 (1986).

FIRE PLUMES

Gunnar Heskestad

INTRODUCTION

Practically all fires go through an important, initial stage in which a coherent, buoyant gas stream rises above a localized area undergoing combustion into surrounding space of essentially uncontaminated air. This stage begins at ignition, continues through a possible smoldering interval, into a flaming interval, and may be said to end prior to flashover. The buoyant gas stream is generally turbulent, except when the fire source is very small. The buoyant flow, including any flames, is referred to as a fire plume.

The properties of fire plumes are important in dealing with problems related to fire detection, fire heating of building structures, smoke filling rates, fire venting, etc. They can also be important in fire suppression system design.

This chapter deals with axisymmetric, turbulent fire plumes and reviews some relations for predicting the properties of such plumes. It is assumed throughout the chapter that the surrounding air is uncontaminated by fire products and uniform in temperature. The relations cease to be valid beyond the elevation where the plume enters a smoke layer.

FIRE PLUME FEATURES

Figure 2-2.1 shows a schematic representation of a fire plume originating at a flaming source. Volatiles driven off from the combustible by heat fed back from the fire mix with the surrounding air and form a diffusion flame. The mean height of the flame is L. Surrounding the flame and extending upward is a boundary (broken lines) that confines the entire buoyant flow of combustion products and entrained air. The air is entrained across this boundary, which instantaneously is very sharp, highly convoluted, and easily discernible in smoky fires. The flow profile could be the time-averaged temperature rise above the ambient temperature, or

the concentration of a gas (such as CO_2) generated by the fire, or the axial velocity in the fire plume.

Figure 2-2.1 suggests qualitatively, based on experimental observations,[1-5] how the temperature rise on the centerline, ΔT_0, and the velocity on the centerline, u_0, might behave along the plume axis. In this example, using a relatively tall flame, the temperatures are nearly constant in the lower portion of the flame. Temperatures begin to decay in the intermittent, upper portion of the flame as the combustion reactions trail off and air entrained from the surroundings cools the flow. The centerline velocities, u_0, tend to have their maxima slightly below the mean flame height and always decay toward higher elevations. If the combustible is porous and supports internal combustion, there may not be as pronounced a falloff in the gas velocity toward the top of the combustible, as suggested in Figure 2-2.1.[3]

The total heat release rate of a fire source, $\dot{Q}$, is either convected, $\dot{Q}_c$, or radiated, $\dot{Q}_r$, away from the combustion region. In a fire deep in a porous combustible pile (e.g., a stack of wood pallets), some of the total heat generated is trapped by and stored in the not yet burning material; the rest escapes from the combustible array as either convective or radiative energy flux. If most of the volatiles released undergo combustion above the fuel array, as in pool fires of

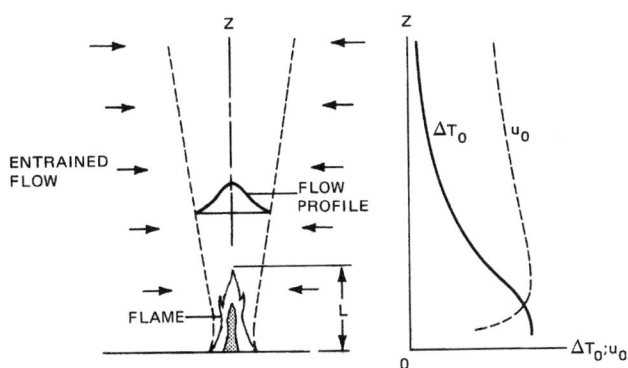

Fig. 2-2.1. *Features of a fire plume, including axial variations on the centerline of mean excess temperature, ΔT_0, and mean velocity, u_0.*[15]

Dr. Gunnar Heskestad is Principal Research Scientist at Factory Mutual Research Corporation, specializing in fluid mechanics and heat transfer of fire, with applications to fire protection.

liquids and other horizontal-surface fires, and even in well-developed porous pile fires, then the convective fraction of the total heat release rate is rarely measured at less than 60 to 70 percent of the total heat release rate.[6,7] The convective flux, $\dot{Q}_c$, is carried away by the plume above the flames, while the remainder of the total heat liberated, $\dot{Q}_r$, is radiated away in all directions.

The total heat release rate, $\dot{Q}$, is often assumed to be equal to the *theoretical* heat release rate, which is based on complete combustion of the burning material. The theoretical heat release rate in kW is evaluated as the mass burning rate in kg/s multiplied by the lower heat of complete combustion in kJ/kg. The ratio of the total heat release rate to the theoretical heat release rate, which is the combustion efficiency, is indeed close to unity for some fire sources (e.g., methanol and heptane pools),[6] but may deviate significantly from unity for others (e.g., a polystyrene fire, for which the combustion efficiency is about 45 percent,[7] and a fully involved stack of wood pallets, for which the combustion efficiency is 63 percent[6]).

CALCULATION METHODS

Flame Heights

The visible flames above a fire source includes the combustion reaction zone. Typically, the luminosity of the lower part of the flaming region appears fairly steady, while the upper part appears to be intermittent. The intermittent flaming region is associated with shedding of vortex structures.[8-10]

Figure 2-2.2 helps to define the mean flame height, L.[10] It shows schematically the variation of intermittency, I, versus distance above the fire source, z, where $I(z)$ is defined as the fraction of time that at least part of the flame lies above the elevation, z. The intermittency decreases from unity deep in the flame to smaller values in the intermittent flame region, eventually reaching zero. The mean flame height, L, is the distance above the fire source where the intermittency has declined to 0.5. Objective determinations of mean flame height according to intermittency measurements are fairly consistent with (although tending to be slightly lower than) flame heights that are averaged by the human eye.[10]

The mean flame height is an important quantity that marks the level where the combustion reactions are essentially complete and the inert plume can be considered to begin. Several expressions for mean flame height have been proposed. (See Section 1, Chapter 1.) The following correlation of experimental data[11] is convenient and quite general (excluding high-momentum jet discharge corresponding to the nondimensional parameter, N, well beyond 10^5)

$$L/D = -1.02 + 15.6N^{1/5} \qquad (1)$$

where D is the diameter of the fire source (or effective diameter for noncircular fire sources such that $\pi D^2/4 = $ area of fire source) and N is the nondimensional parameter defined by

$$N = \left[\frac{c_p T_\infty}{g\rho_\infty^2 (H_c/r)^3}\right]\frac{\dot{Q}^2}{D^5} \qquad (2)$$

where c_p is the specific heat of air; T_∞ and ρ_∞ are the ambient temperature and density, respectively; g is the acceleration of gravity; H_c is the actual lower heat of combustion; r is the actual mass stoichiometric ratio of air to volatiles; and $\dot{Q}$ is the total heat release rate given by

$$\dot{Q} = \dot{m}_f H_c \qquad (3)$$

where $\dot{m}_f$ is the mass burning rate. This flame-height relationship does not include fire sources with substantial in-depth combustion, but does include liquid pool fires and other horizontal-surface fires. A fire source may not have substantial in-depth combustion if a major fraction (perhaps ⅔ or greater) of the volatiles released undergoes combustion above the fuel array, i.e., if only a minor fraction (perhaps ⅓ or smaller) of the volatiles is oxidized within the fuel array by air entering the array. Most types of wood cribs used in laboratory fire investigations may be considered to lack substantial in-depth combustion when fully involved.[3] Similarly, commonly designed wood pallets can also be considered to lack substantial in-depth combustion when fully involved. However, fire sources such as very openly constructed wood cribs do have substantial in-depth combustion.[3]

A convenient form of Equation 1 can be developed. Let

$$A = 15.6\left[\frac{c_p T_\infty}{g\rho_\infty^2 (H_c/r)^3}\right]^{1/5} \qquad (4)$$

Then Equation 1 can be written in the dimensional form

$$L = -1.02D + A\dot{Q}^{2/5} \qquad (5)$$

The coefficient, A, varies over a rather narrow range, associated with the fact that H_c/r, the heat liberated per unit mass of air entering the combustion reactions, does not vary appreciably among various combustibles. For a large number of gaseous and liquid fuels, H_c/r remains within the range of 2900 to 3200 kJ/kg, for which the associated range of A under normal atmospheric conditions (293 K, 760 mm Hg) is 0.240 to 0.226 (m kW$^{-2/5}$), with a typical value of $A = 0.235$. Hence, under normal atmospheric conditions

$$L = -1.02D + 0.235\dot{Q}^{2/5} \qquad (6)$$

(L and D in m; $\dot{Q}$ in kW).

Fairly common fuels that deviate significantly from the cited range 0.240 to 0.226 for A include acetylene, hydrogen (0.211), and gasoline (0.200). In general, the coefficient

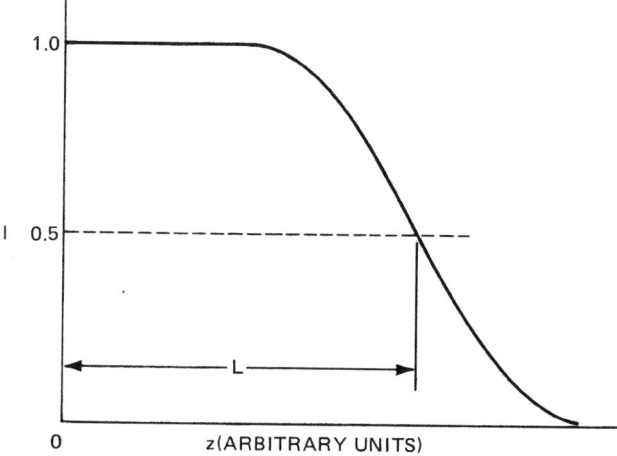

***Fig. 2-2.2.** Definition by Zukoski et al[10] of mean flame height, L, from measurements of intermittency, I.*

$A = 0.235$ in Equation 6 may be considered adequate unless actual values of H_C and r are known that indicate otherwise, and/or atmospheric conditions deviate significantly from normal.

Referring to any of the flame-height relations (Equations 1, 5, and 6), it can be seen that negative flame heights are calculated for sufficiently small values of the heat release rate. Of course, this situation is unphysical and the correlation is not valid here. There are indications that a single flaming area breaks down into several zones when heat release rates decrease to the point where negative flame height (L) is calculated.

EXAMPLE 1:

Consider a 1.5-m diameter pan fire of methyl alcohol with a heat release intensity of 500 kW/m^2 of surface area. Normal atmospheric conditions prevail (760 mm Hg, 293 K). Calculate the mean flame height.

SOLUTION:

Available values of the lower heat of combustion ($H_c = 21\,100$ kJ/kg) and stoichiometric ratio ($r = 6.48$) give $H_c/r = 3260$ kJ/kg. With this value for H_c/r substituted in Equation 4, together with $c_p = 1.00$ kJ/kg K, $T_\infty = 293$ K, $g = 9.81$ m/s^2 and $\rho_\infty = 1.20$ kg/m^3, the coefficient A is calculated as 0.223 (m kW$^{-2/5}$). The total heat release rate is $\dot{Q} = 500\pi 1.5^2/4 = 884$ kW. Equation 5 gives a mean flame height of $L = -1.02 \cdot 1.5 + 0.223 \cdot 884^{2/5} = 1.83$ m.

EXAMPLE 2:

Similar to Example 1, except for new atmospheric conditions representative of Denver, Colorado on a hot day: 630 mm Hg pressure and 310 K temperature.

SOLUTION:

Using Equation 4, the new coefficient, A, increases from 0.223 to 0.249 [most readily calculated from $(310/293)^{3/5}$ $(760/630)^{2/5}$ 0.223 = 0.249, where the equation of state for a perfect gas has been used]. Using Equation 5, the new flame height is $L = 2.23$ m, increased from 1.83 m for normal atmospheric conditions.

EXAMPLE 3:

One 1.2-m high stack of wood pallets (1.07 × 1.07 m) burns at a total heat release rate of 2600 kW under normal atmospheric conditions. Calculate the mean flame height above the top of the pallet stack.

SOLUTION:

The square flaming area can be converted to an equivalent diameter: $\pi D^2/4 = 1.07^2$, which gives the equivalent diameter, D, of 1.21 m. Since the combustion efficiency of wood is considerably less than 100 percent, it is difficult to select reliable and consistent values for H_c and r to form the ratio H_c/r. Instead, it can be assumed that $A = 0.235$, the typical value. Using Equation 5, the mean flame height is calculated as $L = -1.02 \cdot 1.21 + 0.235 \cdot 2600^{2/5} = 4.22$ m.

Plume Temperatures and Velocities

The first plume theories assumed

1. A point source of buoyancy,

2. That variations of density in the field of motion are small compared to the ambient density,
3. That the air entrainment velocity at the edge of the plume is proportional to the local vertical plume velocity, and
4. That the profiles of vertical velocity and buoyancy force in horizontal sections are of similar form at all heights.

On the further assumption that the profiles are uniform, the mean motion is then governed by the following three conservation equations for continuity, momentum, and buoyancy[12]

Continuity:
$$\frac{d}{dz}(b^2 u) = 2\alpha b u \tag{7}$$

Momentum:
$$\frac{d}{dz}(b^2 u^2) = b^2 g \frac{\rho_\infty - \rho}{\rho_\infty} \tag{8}$$

Buoyancy:
$$\frac{d}{dz}\left(b^2 u g \frac{\rho_\infty - \rho}{\rho_\infty}\right) = 0 \tag{9}$$

In these equations, z is the elevation above the point source of buoyancy; b is the radius to the edge of the plume; u is the vertical velocity in the plume; α is the entrainment coefficient (the proportionality constant relating the inflow velocity due to entrainment at the edge of the plume to u); ρ is the density in the plume; and ρ_∞ is the ambient density. Equation 9 can be integrated immediately to

$$b^2 u g \frac{\rho_\infty - \rho}{\rho_\infty} = B = \text{constant} \tag{10}$$

Here, B is the buoyancy flux in the plume which remains constant at all heights. The flux can be related to the convective heat in the plume, $\dot{Q}_c$, by noting

$$\dot{Q}_c = \rho u \pi b^2 c_p (T - T_\infty) = \pi u b^2 c_p (\rho_\infty - \rho) T_\infty \tag{11}$$

where the ideal gas law has been used. In this equation, T is the plume temperature and T_∞ is the ambient temperature. Combining Equations 10 and 11 gives

$$B = g(\pi c_p T_\infty \rho_\infty)^{-1} \dot{Q}_c \tag{12}$$

Solutions to Equations 7, 8, and 10 can be determined[12] in the form (expressing B in terms of $\dot{Q}_c$ using Equation 12).

$$b = \frac{6\alpha}{5} z \tag{13}$$

$$u = \frac{5}{6}\left(\frac{9}{10\pi\alpha^2}\right)^{1/3} g^{1/3} (c_p \rho_\infty T_\infty)^{-1/3} \dot{Q}_c^{1/3} z^{-1/3} \tag{14}$$

$$\frac{\Delta\rho}{\rho_\infty} = \frac{5}{6}\left(\frac{9\pi^2\alpha^4}{10}\right)^{-1/3} g^{-1/3} (c_p \rho_\infty T_\infty)^{-2/3} \dot{Q}_c^{2/3} z^{-5/3} \tag{15}$$

Equations 13 through 15 are the weak plume (small density deficiency) relations for point sources. To account for area sources, a virtual source location or virtual origin, z_0, is introduced[12,13] and z in Equations 13 through 15 is replaced by $z - z_0$. In addition, to accommodate large density deficiencies as are present in fire plumes, Morton's extension of the weak-plume theory[14] leads to the result that $\Delta\rho/\rho_\infty$ in Equation 15 should be replaced by $\Delta\rho/\rho$ [$= \Delta T/T_\infty$ using the ideal gas law]. Also, Equation 13, for growth in plume radius should incorporate the additional factor $(\rho_\infty/\rho)^{1/2}$ [$= (T/T_\infty)^{1/2}$ using the ideal gas law] on the right side of the equation. Relaxing the assumption that the flow profiles are uniform, renders the numerical coefficients in the resulting equations in doubt.

Measurements in fire plumes above the flames have to a large extent supported the theory. The plume radius and centerline values of mean excess temperature and mean velocity have been found[15] to obey the following relations

$$b_{\Delta T} = 0.12(T_0/T_\infty)^{1/2}(z - z_0) \qquad (16)$$

$$\Delta T_0 = 9.1[T_\infty/(gc_p^2\rho_\infty^2)]^{1/3}\dot{Q}_c^{2/3}(z - z_0)^{-5/3} \qquad (17)$$

$$u_0 = 3.4[g/(c_p\rho_\infty T_\infty)]^{1/3}\dot{Q}_c^{1/3}(z - z_0)^{-1/3} \qquad (18)$$

Here, $b_{\Delta T}$ is the plume radius to the point where the temperature rise has declined to 0.5 ΔT_0; T_0 is the centerline temperature; $\dot{Q}_c$ is the convective heat release rate; z is the elevation above the fire source; and z_0 is the elevation of the virtual origin above the fire source.[†] (If z_0 is negative, the virtual origin lies below the top of the fire source.)

The virtual origin is the equivalent point source height of a finite area fire. This origin is usually located near the fuel surface and may be assumed coincident with the fuel surface when the plume flow is predicted at high elevations. Near the fire source, however, it is important to know the location of the virtual origin for accurate predictions. Calculation of the virtual origin is discussed in the following section.

Equations 16 through 18 are known as the *strong plume relations*. The numerical coefficients for the relations have been determined from data sets for which the locations of the virtual origin, z_0, have been established and the convective heat release rates, $\dot{Q}_c$, are known.[4,16]

In addition to the temperature radius of a plume, $b_{\Delta T}$, a velocity radius, b_u, can also be defined. The velocity radius is the plume radius to the point where the gas velocity has declined to 0.5 u_0. The most reliable measurements[16] indicate that b_u is perhaps ten percent larger than $b_{\Delta T}$. Other measurements indicate ratios $b_u/b_{\Delta T}$ of 0.86,[17] 1.00,[18] 1.08 and 1.24,[19] 1.31,[5] 1.05,[1] and 1.5.[2] The widely differing results can probably be attributed to the difficulty of positioning the measuring probes accurately with respect to the plume centerline, and to different, intrinsic errors associated with the diverse types of anemometers used (pitot tube, bidirectional flow probe, hot wire, vane anemometer, cross-correlation techniques, laser Doppler anemometer).

Often, profiles of temperature rise and velocity are represented as Gaussian in shape, although there is no theoretical foundation for this distribution

$$\Delta T = \Delta T_0 \exp[-(R/\sigma_{\Delta T})^2] \qquad (19)$$

$$u = u_0 \exp[-(R/\sigma_u)^2] \qquad (20)$$

Here, ΔT and u are the local values, at the radius, R, in the plume of temperature rise and gas velocity. The quantities $\sigma_{\Delta T}$ and σ_u are measures of the plume width, corresponding to the radii where local values of temperature rise and velocity are $e^{-1} = 0.368$ multiplied by the centerline values. For

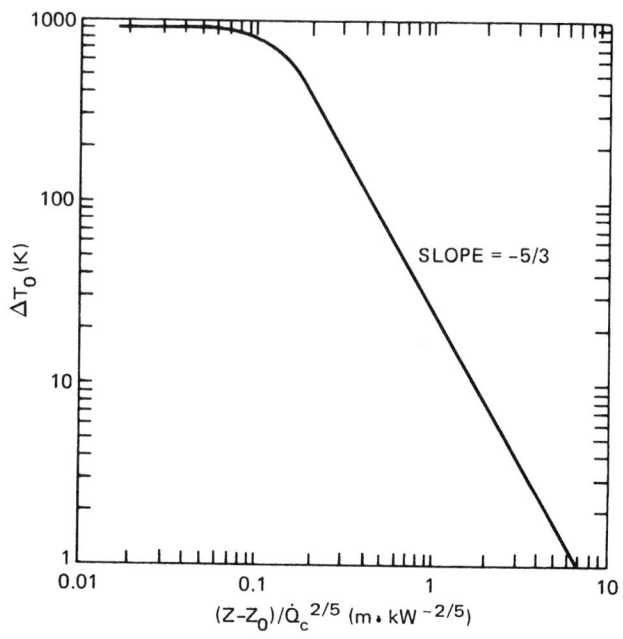

Fig. 2-2.3. Temperature rise on the plume centerline for normal atmospheric conditions[15] in a form attributable to McCaffrey,[1] and Kung and Stavrianidis.[4]

Gaussian profiles, the plume radii $\sigma_{\Delta T}$ and σ_u are 1.201 multiplied by the plume radii, $b_{\Delta T}$ and b_u, discussed previously.

Equations 16 through 18 cease to be valid at and below the mean flame height as defined by Equation 1 for fire sources with no substantial in-depth combustion. However, it is possible to represent ΔT_0 such that a general plot of experimental temperature variations is produced throughout the length of the plume, including the flames. The method is based on the observation that $\dot{Q}_c^{2/3}(z - z_0)^{-5/3}$ in Equation 17 can be written as $[(z - z_0)/\dot{Q}_c^{2/5}]^{-5/3}$. This result suggests plotting ΔT_0 versus $(z - z_0)/\dot{Q}_c^{2/5}$. Figure 2-2.3 shows the result in logarithmic coordinates for normal atmospheric conditions. For values of the abscissa greater than 0.15 to 0.20 $(m/kW^{2/5})$, the centerline temperature rise falls off with the $-\frac{5}{3}$ power of the abscissa, in accordance with the plume law for temperature (Equation 17). Abscissa values in the 0.15 to 0.20 range correspond to the mean flame height; an associated temperature rise of about 500 K is indicated in Figure 2-2.3. At smaller abscissa values, the experimentally observed temperature rise increases more slowly, approaching a value deep in the flame of approximately $\Delta T_0 = 900$ K. When closer to the fuel surface than represented in Figure 2-2.3, the temperatures on the plume axis tend to decrease again.[1,2,5]

The plume law for velocity, Equation 18, may be combined with the plume law for temperature, Equation 17, to produce the following useful nondimensional parameter[3]

$$\xi = \left[\frac{T_\infty^{2/5}(c_p\rho_\infty)^{1/5}}{g^{2/5}}\right]\frac{u_0}{(\Delta T_0 \dot{Q}_c)^{1/5}} \qquad (21)$$

In the plume region where Equations 17 and 18 are valid, their numerical coefficients correspond to a constant value $\xi = 2.2$. This value has been confirmed for a number of test fires,[3] at heights as low as the mean flame height and even somewhat lower. Equation 21 with $\xi = 2.2$ is a useful

---

[†]For normal atmospheric conditions ($T_\infty = 293$ K, $g = 9.81$ m/s², $c_p = 1.00$ kJ/kg K, $\rho_\infty = 1.2$ kg/m³), the factor $9.1[T_\infty/(gc_p^2\rho_\infty^2)]^{1/3}$ has the numerical value 25.0 K m$^{5/3}$ kW$^{-2/3}$ and the factor $3.4[g/(c_p\rho_\infty T_\infty)]^{1/3}$ has the numerical value 1.03 m$^{4/3}$s^{-1}kW$^{-1/3}$.

relation for determining the maximum velocity in the plume, which occurs slightly below the mean flame height where the temperature rise may be taken at approximately $\Delta T_0 = 650$ K. For normal atmospheric conditions and the value $\xi = 2.2$, Equation 21 becomes

$$u_0/(\Delta T_0 \dot{Q}_c)^{1/5} = 0.54 \qquad (22)$$

The maximum velocity just under the mean flame height, u_{0m}, is obtained by setting $\Delta T_0 = 650$ K

$$u_{0m} = 1.97\dot{Q}_c^{1/5} \qquad (23)$$

Fires with low flame height-to-diameter ratios (Equation 1) have not been investigated extensively and may require special consideration. For one particular fire with very low flame height,[4] in which a proprietary silicone transformer fluid was burned in a 2.44-m diameter pool, a flame height ratio of $L/D = 0.14$ was measured[†] at a convective heat release rate of $\dot{Q}_c = 327$ kW. Using the results in the next section, the virtual origin is calculated at $z_0 = -1.5$ m, assuming $\dot{Q}_c/\dot{Q} = 0.7$. With respect to the abscissa in Figure 2-2.3, the lowest possible value is $-z_0/\dot{Q}^{2/5}$ corresponding to the fuel surface, $z = 0$. For the present case, $-z_0/\dot{Q}_c^{2/5} = 1.5/327^{2/5} = 0.15$ (m kW$^{-2/5}$). At this abscissa value, a centerline temperature rise of 580 K is indicated in Figure 2-2.3. From the experiment,[4] a near surface ΔT_0 of 440 K can be determined by slight extrapolation, fairly close to the prediction from Figure 2-2.3. Fires with very low flame height-to-diameter ratios may generally be expected to produce lower maximum mean temperatures than other fires. However, it is not yet clear whether the type of prediction attempted here for a particular low L/D fire is generally valid.

There is also uncertainty associated with assuming that ξ in Equation 21 remains completely constant down to the flame level in low L/D fires. It may be found that ξ still remains approximately constant down to the height where the maximum gas velocity occurs, although this maximum will probably occur above the flames. The associated temperatures at this height cannot as yet be predicted. Consequently, the relation in Equation 23 becomes somewhat uncertain as L/D decreases (ΔT_0 is overestimated, resulting in u_{0m} being overestimated, although the effect is probably not very large because of the slow, ⅕th power dependence on ΔT_0).

The turbulence intensities in a fire plume are quite high. On the axis, George et al[16] report an intensity of temperature fluctuations of approximately $T'/\Delta T_0 = 0.38$, where T' is the root mean square (rms) temperature fluctuation. Centerline values of the intensity of axial velocity fluctuations were measured near $u'/u_0 = 0.27$ by George et al[16] and near $u'/u_0 = 0.33$ by Gengembre et al,[5] where u' is the rms velocity fluctuation in the axial direction.

EXAMPLE 4:

Example 1 concerned a 1.5-m diameter methyl alcohol fire burning under normal atmospheric conditions, generating $\dot{Q} = 884$ kW with a calculated mean flame height of 1.83 m. For an elevation of 5 m and given a virtual origin $z_0 = -0.3$ m (from Example 6), calculate the temperature radius, $b_{\Delta T}$, as well as the centerline value of temperature rise, ΔT_0, and gas velocity, u_0. Also calculate the maximum gas velocity in the flame.

SOLUTION:

Assume[‡] $\dot{Q}_c = 0.8 \dot{Q}$ and first calculate the temperature rise, using Equation 17 and properties for normal atmospheric conditions ($T_\infty = 293$ K, $g = 9.81$ m/s^2, $c_p = 1.00$ kJ/kg K, $\rho_\infty = 1.20$ kg/m^3)

$$\Delta T_0 = 9.1[293/(9.81 \; 1.00^2 \; 1.20^2)]^{1/3}(0.8 \; 884)^{2/3}(5 + 0.3)^{-5/3}$$
$$= 123 \text{ K}$$

The temperature radius can now be calculated from Equation 16

$$b_{\Delta T} = 0.12[(123 + 293)/293]^{1/2}(5 + 0.3)$$
$$= 0.76 \text{ m}$$

The velocity is calculated from Equation 18

$$u_0 = 3.4[9.81/(1.00 \; 1.20 \; 293)]^{1/3}(0.8 \; 884)^{1/3}(5 + 0.3)^{-1/3}$$
$$= 5.3 \text{ m/s}$$

Instead of Equation 18, the velocity can also be calculated from Equation 21 in this case, since ΔT_0 is already known. Actually, because normal ambient conditions prevail, Equation 22 can be used

$$u_0 = 0.54(123 \; 0.8 \; 884)^{1/5} = 5.3 \text{ m/s}$$

Finally, the maximum velocity in the flame is given by Equation 23

$$u_{0m} = 1.97(0.8 \; 884)^{1/5} = 7.3 \text{ m/s}$$

EXAMPLE 5:

Recalculate the quantities called for in Example 4 using ambient conditions representative of Denver, Colorado on a hot day: 630 mm Hg pressure and 310 K temperature.

SOLUTION:

Changed ambient variables entering the equations include $T_\infty = 310$ K and $\rho_\infty = 0.78$ kg/m^3. From Equation 17, the new temperature rise is

$$\Delta T_0 = 167 \text{ K} \quad \text{(versus 123 K in Example 4)}$$

The new velocity from Equation 18 is

$$u_0 = 6.0 \text{ m/s} \quad \text{(versus 5.3 m/s)}$$

[†]A ratio $L/D = 0.02$ can be calculated from Equation 5 assuming $H_c/r = 3470$ kJ/kg, an average for silicone oils from values reported by Tewarson[20] and assuming a convective heat fraction $\dot{Q}_c/\dot{Q} = 0.7$. If a value of H_c/r near the bottom of the reported range[20] is selected, 3230 kJ/kg, the observed value $L/D = 0.14$ is reproduced; slight changes in the assumed convective fraction will also reproduce the measured value.

[‡]Without specific knowledge, $\dot{Q}_c/\dot{Q}$ may usually be assumed at 0.7. However, methyl alcohol produces a fire of low luminosity and radiation, for which $\dot{Q}_c/\dot{Q} = 0.8$ is a good estimate.

For the new ambient conditions, the relation analogous to Equation 23 is calculated as

$$u_{0m} = 2.10 \dot{Q}_c^{1/5}$$

from which the new maximum velocity in the flame is

$$u_{0m} = 7.8 \text{ m/s} \quad (\text{versus } 7.3 \text{ m/s})$$

Virtual Origin

As pointed out earlier in this chapter, knowledge of the virtual origin of fire plumes is important for predicting the near source plume behavior.

The virtual origin of a test fire is most conveniently determined from temperature data above the flames along the plume axis. According to Equation 17, a plot of $\Delta T_0^{-3/5}$ versus z should produce a straight line which intercepts the z-axis at z_0. Despite this apparent simplicity of obtaining z_0, the task is very difficult in practice. Slight inaccuracies in the determinations of centerline temperatures have large effects on the intercept, z_0; such inaccuracies may be associated with off axis placement of sensors, radiation-induced errors in the temperature signal, or inadequate averaging of the signal.

Data obtained in this manner on the virtual origin for pool fires varying in diameter from 0.16 to 2.4 m,[1,3,4] were examined for consistency with a theoretical model by Heskestad.[21] The model relied heavily on the flame-height correlation represented by Equation 1 and led to the prediction

$$\frac{z_0}{D} = -1.02 + F\frac{\dot{Q}^{2/5}}{D} \tag{24}$$

where F is a rather complex dimensional function of environmental variables c_p, T_∞, ρ_∞, g; H_c/r for the combustible, and the fraction of the total heat release carried away by convection.[21] It appeared that F could be considered a constant for rather wide variations in ambient temperature and pressure, but might be affected by wide swings in the fuel variables, H_c/r and convective fraction. The available data did not reflect any sensitivity to fuel identity within their scatter, and led to the determination $F = 0.083$ m kW$^{-2/5}$, with Equation 24 becoming

$$\frac{z_0}{D} = -1.02 + 0.083\frac{\dot{Q}^{2/5}}{D} \quad (\dot{Q} \text{ in kW}, D \text{ in m}) \tag{25}$$

Later, Hasemi and Tokunaga[22] analyzed their temperature measurements in plumes from gas burners of diameters in the 0.2 to 0.5 m range and established alternative correlations for the virtual origin. In terms of the nondimensional parameter

$$\dot{Q}^\star = \dot{Q}/(\rho_\infty c_p T_\infty g^{1/2} D^{5/2}) \tag{26}$$

their correlations are

$$z_0/D = 2.4(\dot{Q}^{\star 2/5} - 1), \quad \dot{Q}^\star \geq 1.0$$
$$z_0/D = 2.4(\dot{Q}^{\star 2/3} - \dot{Q}^{\star 2/5}), \quad \dot{Q}^\star < 1.0 \tag{27}$$

For normal ambient conditions, these correlations can be written in terms of the variable $\dot{Q}^{2/5}/D$ (cf. Equation 25):

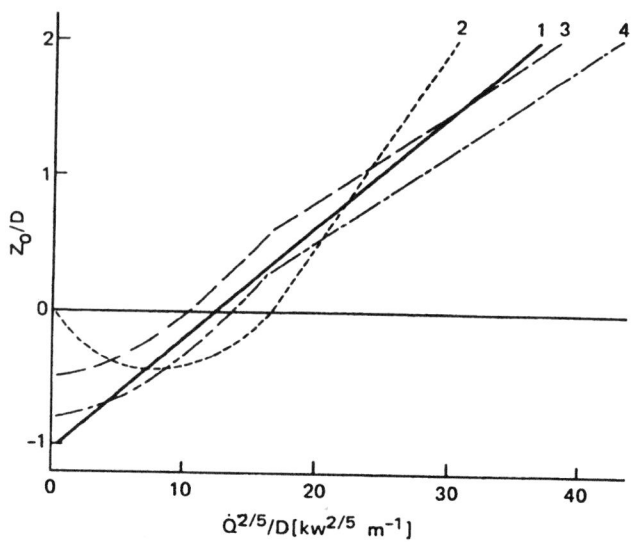

Fig. 2-2.4. Correlations for the virtual origin. Curve 1—Equation 25; Curve 2—Equation 28; Curve 3—Equation 30 with floor; Curve 4—Equation 30 without floor.

$$\frac{z_0}{D} = -2.4 + 0.145\frac{\dot{Q}^{2/5}}{D}, \quad \frac{\dot{Q}^{2/5}}{D} \geq 16.5$$
$$\frac{z_0}{D} = 0.0224\left(\frac{\dot{Q}^{2/5}}{D}\right)^{5/3} - 0.145\frac{\dot{Q}^{2/5}}{D}, \quad \frac{\dot{Q}^{2/5}}{D} < 16.5 \tag{28}$$

Cetegen et al[23] have proposed correlations for the virtual origin on the basis of their air entrainment measurements in fire plumes and attempts to apply entrainment theory for a point source to the laboratory fires. Their experiments involved gas burners (natural gas) with diameters of 0.10, 0.19, 0.30, and 0.50 m. The experiments were performed with and without a floor mounted flush with the upper surface of the burners located some distance above the floor of the laboratory. Their correlations for the virtual origin are

$$z_0/D = c + 1.09\dot{Q}^{\star 2/5}, \quad \dot{Q}^\star > 1$$
$$z_0/D = c + 1.09\dot{Q}^{\star 2/3}, \quad \dot{Q}^\star \leq 1 \tag{29}$$

where $\dot{Q}^\star$ has been defined by Equation 26, and where $c = -0.50$ with a flush floor around the burners and $c = -0.80$ without a flush floor. Using Equation 26, Equation 29 can be written in terms of $\dot{Q}^{2/5}/D$ yielding

$$\frac{z_0}{D} = c + 0.0659\frac{\dot{Q}^{2/5}}{D}, \quad \frac{\dot{Q}^{2/5}}{D} > 16.5$$
$$\frac{z_0}{D} = c + 0.01015\left(\frac{\dot{Q}^{2/5}}{D}\right)^{5/3}, \quad \frac{\dot{Q}^{2/5}}{D} \leq 16.5 \tag{30}$$

where $c = -0.50$ and $c = -0.80$ with and without a flush floor, respectively.

Figure 2-2.4 is a composite plot of the various correlations for the virtual origin, plotted as z_0/D versus $\dot{Q}^{2/5}/D$. Despite the diverse approaches, the overall correlations are surprisingly similar. Precise measurements are not yet available to clearly identify an optimal correlation. In the meantime, curve 1 in Figure 2-2.4 (i.e., Equation 25) is recommended for its simplicity, clear foundation in theory,[21] and central position among the other correlations.

In some cases, a significant fraction of volatiles produced by the fire source may undergo combustion below the top of the combustible array, for instance in the case of well-ventilated wood cribs in the laboratory and rack storage of combustibles in the field. The virtual source relation, Equation 25, cannot be justified under these circumstances.[21] You and Kung[24] have reported on virtual origins measured in experimental rack storage fires of standard Class II commodity and standard plastic commodity, involving cubic pallet loads measuring 1.1 m on the side. The pallet loads were arranged four per tier (horizontal level) in a two wide by two deep configuration incorporating 0.15 m flue spaces separating the loads. Storages of two, three, and four tiers were investigated, with a clear space of 0.35 m between the top of one tier of storage and the bottom of the next. Ignition was near the central flue at the bottom of the lowest tier. Using measurements of the instantaneous convective heat flux above the flames and of temperature rise on the plume axis, the following correlations were established:

For two-tier storage, Class II and plastic commodity

$$z_0 = -1.6 + 0.094\dot{Q}_c^{2/5}; \tag{31}$$

For three- and four-tier storage, Class II and plastic commodity

$$z_0 = -2.4 + 0.095\dot{Q}_c^{2/5} \quad (z_0 \text{ in m and } \dot{Q}_c \text{ in kW}) \tag{32}$$

These correlations applied from early in the fire, when combustion was confined to the central flue and flames had not yet reached to the top of the storage ($\dot{Q}_c < 20$ kW), until a later stage when typically the entire 2.3×2.3 m top surface of the array had become involved ($\dot{Q}_c > 5000$ kW). Using Equation 25, assuming a ratio of convective heat flux to total heat release rate of $\dot{Q}_c/\dot{Q} = 0.7$, and considering the entire top surface of the array to be involved throughout [effective diameter of $D = (4/\pi)^{1/2} 2.3$], one would predict

$$z_0 = -2.7 + 0.096\dot{Q}_c^{2/5} \quad (z_0 \text{ in m and } \dot{Q}_c \text{ in kW}) \tag{33}$$

The coefficient for $\dot{Q}_c^{2/5}$ is similar to the experimental coefficient in Equations 31 and 32, but the constant, -2.7 (m), is lower than the experimental values in these equations. It is encouraging that there is close similarity, but further research is needed to establish credible general predictions for such fires. Meantime, virtual source predictions are considered feasible only for pool-type fires and other fires with no substantial in-depth combustion. (See discussion after Equation 3.) For other types of fires, $z_0 = 0$ may be assumed in order to perform an approximate calculation.

EXAMPLE 6:

Example 1 concerned a 1.5-m diameter methyl alcohol fire generating $\dot{Q} = 884$ kW. Calculate the virtual origin.

SOLUTION:

In this example, $D = 1.5$ m. Direct substitution into Equation 25 gives

$$z_0/D = -1.02 + 0.083(884)^{2/5}/1.5$$
$$= -1.02 + 0.83 = -0.19$$

from which

$$z_0 = -0.19 \cdot 1.5 = -0.29 \text{ m}$$

This is the value for z_0 (rounded off) used in Example 1.

EXAMPLE 7:

Negative values for z_0 are often calculated for low-heat-release fires and sufficiently large fire diameters, as in Example 6. Positive virtual origins are often found for high-heat-release fires. Substituting heptane for methyl alcohol in Example 2 (2500 kW/m² rather than 500 kW/m² measured for methyl alcohol),[6] calculate the new virtual origin.

SOLUTION:

The new heat release rate is

$$\dot{Q} = (\pi 1.5^2/4)2500 = 4420 \text{ kW}$$

From Equation 25

$$z_0/D = -1.02 + 0.083 \cdot 4420^{2/5}/1.5 = 0.57$$

from which

$$z_0 = 0.57 \cdot 1.5 = 0.85 \text{ m}$$

Entrainment

After ignition, the fire plume carries fire products diluted in entrained air to the ceiling. A layer of diluted fire products, or "smoke," forms under the ceiling, which thickens and generally becomes hotter with time. The fire environment is intimately tied to the behavior of this layer which, in turn, depends to a major extent on the mass flow rate of plume fluid into the layer. Consequently, it is important to be able to predict the mass flow rate that may occur in a fire plume.

The mass flow at a particular elevation in a fire plume is almost completely attributable to air entrained by the plume at lower elevations. The mass flow contributed by the fire source itself is insignificant in comparison.

For a weak plume, the mass flow rate at a cross section can be written

$$\dot{m}_{ent} = E\rho_\infty u_0 b_u^2 \tag{34}$$

where E is a nondimensional constant of proportionality. With the aid of Equation 18 and the equivalent of Equation 16 written for b_u (setting $T_0/T_\infty = 1$ because of the weak plume assumption), Equation 34 becomes

$$\dot{m}_{ent} = E\left(\frac{g\rho_\infty^2}{c_p T_\infty}\right)^{1/3} \dot{Q}_c^{1/3}(z - z_0)^{5/3} \tag{35}$$

Early measurements by Yih[25] indicated a value $E = 0.153$.

Cetegen et al[23,26] concluded from theoretical analysis that Equation 35 also applies to strongly buoyant plumes. From extensive entrained-flow measurements for natural gas burners of several diameters, these authors proposed a coefficient $E = 0.21$. However, the plume flow rates at large heights were somewhat overpredicted and those at low heights, approaching the flames, were somewhat underpredicted.

Heskestad[27] reconsidered the entrainment problem for strong plumes, assuming self-preserving density deficiency profiles instead of self-preserving excess temperature profiles as traditionally assumed. This approach led to the following extension of Equation 35

$$\dot{m}_{ent} = E\left(\frac{g\rho_\infty^2}{c_p T_\infty}\right)^{1/3} \dot{Q}_c^{1/3}(z - z_0)^{5/3}$$

$$\cdot \left[1 + \frac{G\dot{Q}_c^{2/3}}{(g^{1/2}c_p\rho_\infty T_\infty)^{2/3}(z-z_0)^{5/3}}\right] \qquad (36)$$

Equation 36, with $E = 0.196$ and $G = 2.9$, was found to represent the data of Cetegen et al[23,26] very well over the entire nonreacting plume.[27] At large heights, the bracketed term involving G approaches unity, and at levels approaching the flame tip (Equation 1), this term approaches 1.5, approximately. Equation 36, with $E = 0.196$ and $G = 2.9$, is the recommended relation for calculating mass flow rates in plumes, at and above the mean flame height.

The entrained flow at the mean flame height, $\dot{m}_{ent,L}$, follows from setting $z - z_0 = L - z_0$ in Equation 36 and substituting the expression for L in Equation 1, which leads to

$$\dot{m}_{ent,L}(\text{kg/s}) = J\dot{Q}_c(\text{kW}) \qquad (37)$$

Here, J is a coefficient that varies slightly with ρ_∞ and H_c/r; effects of T_∞ are much smaller. For ρ_∞ in the range 0.78 to 1.20 kg/m^3 and H_c/r in the range 2900 to 3200 kJ/kg, it can be verified that J varies in the range 0.0055 to 0.0066 kg/kJ (at $T_\infty = 293$ K).

Mass flow rates in fire plumes at levels below the flame tip have been found to increase linearly with height for fire diameters of 0.3 m and greater,[27] where the flames are substantially turbulent, from zero (essentially) at the fire base to the flame-tip value in Equation 37, i.e.,

$$\dot{m}_{ent} = \dot{m}_{ent,L}z/L \qquad (38)$$

Delichatsios has analyzed mass flow rates in this region, with data primarily from smaller fire diameters than 0.3 m. (See Section 2, Chapter 3.)

For normal atmospheric conditions and typical fuels ($H_c/r = 3100$ kJ/kg), Equations 36 through 38 can be written as follows for the plume mass flow rate at various heights:

Above the mean flame height, L ($\dot{Q}_c$ in kW, z and z_0 in m):

$$\dot{m}_{ent}(\text{kg/s}) = 0.071\dot{Q}_c^{1/3}(z - z_0)^{5/3}$$

$$\cdot [1 + 0.027 \dot{Q}_c^{2/3}(z - z_0)^{-5/3}] \qquad (39)$$

At the mean flame height, L:

$$\dot{m}_{ent,L}(\text{kg/s}) = 0.0056\dot{Q}_c \text{ (kW)} \qquad (40)$$

At and below the mean flame height, L:

$$\dot{m}_{ent}(\text{kg/s}) = 0.0056\dot{Q}_c \text{ (kW)} \cdot z/L \qquad (41)$$

Under the prevailing assumptions, and the further assumption $\dot{Q}_c/\dot{Q} = 0.7$, Equation 40 implies that the mass flow at the flame tip is 12 times the mass stoichiometric requirement of the fuel.[27]

Fires with very low flame height-to-diameter (L/D) ratios have not been investigated extensively. It is not clear to what L/D limit the entrained-flow relations presented here apply, but this limit is smaller than 0.9, the lowest L/D ratio associated with the data of Cetegen et al.[23,26] For plume mass flows above the flames, there is no L/D limit for predictions at the higher elevations, but predictions of mass flows at levels just above the flames may begin to deteriorate before $L/D = 0.14$ is reached, as seems to be implied in the observations following Equation 23.

Further, mention should be made of a plume mass flow formula often used because of its simplicity, originally developed for the flaming region of large fires by Thomas et al[28]

$$\dot{m}_{ent} = 0.096(g\rho_\infty \rho_{fl})^{1/2} W_f z^{3/2} \qquad (42)$$

Here ρ_{fl} is the gas density in the flames and W_f is the fire perimeter. This formula has also been tested against mass flow data above the flames by Hinkley,[29] who claims it is very satisfactory for heights up to 10 times the linear dimension (or diameter) of a fire, although there is little theoretical justification for its use above the flames. The following version of Equation 42 is often used[29] (based on normal atmospheric conditions and an assumed flame temperature)

$$\dot{m}_{ent} \text{ (kg/s)} = 0.188W_f \text{ (m) } z \text{ (m)}^{3/2} \qquad (43)$$

It is instructive to compare the predictions of Equations 39 and 43 for plume regions above the flames. In a number of comparisons for heat release rates in the range 1000 to 8000 kW, heat release rates per unit area in the range 250 to 1000 kW/m^2, and heights varying from the flame level to 128 m, the predictions of Equation 43 range from 0.64 to 1.38 times the predictions of Equation 39.

Cetegen et al,[23,26] whose data have contributed most to the mass flow recommendations in this text, have carefully pointed out that their fire plumes were produced in as quiet an atmosphere as could be maintained in their laboratory. They report that small ambient disturbances could provide 20 to 50 percent increases in the measured plume mass flows. Clearly, there is need for further research.

EXAMPLE 8:

Calculate plume mass flow rates for the methyl alcohol fire of Examples 1, 4, and 6.

SOLUTION:

Equations 39 through 41 can be used to sufficient accuracy. From Example 1, $\dot{Q} = 884$ kW and $L = 1.83$ m; from Example 4, $\dot{Q}_c = 0.8 \dot{Q} = 707$ kW; from Example 6, $z_0 = -0.29$ m. At the mean flame height, 1.83 m, the mass flow rate follows from Equation 40

$$\dot{m}_{ent,L} = 0.0056 \cdot 707 = 4.0 \text{ kg/s}$$

Mass flow rates in the flaming region are calculated using Equation 41

$$\dot{m}_{ent} \text{ (kg/s)} = 4.0 \cdot z/1.83 = 2.2z \text{ (m)}$$

Mass flow rates above the flames are obtained from Equation 39; e.g., at a height of 3.66 m (twice the flame height)

$$\dot{m}_{ent} = 0.071 \cdot 707^{1/3}(3.66 + 0.29)^{5/3}$$
$$\cdot [1 + 0.027 \cdot 707^{2/3}(3.66 + 0.29)^{-5/3}]$$
$$= 6.24(1 + 0.22)$$
$$= 7.6 \text{ kg/s}$$

Note that Equations 40 and 41 incorporate $H_c/r = 3100$ kJ/kg, whereas the value for methyl alcohol is $H_c/r = 3260$ kJ/kg. (See Example 1.) Had the latter, more correct value been used, the coefficient 0.0056 in the expression for $\dot{m}_{ent,L}$ in this example would have changed to 0.0054.

ILLUSTRATION

In addition to the previous examples, it is instructive to work through a somewhat larger problem to illustrate handling of the equations and their limitations. Units used throughout this section are: kW for heat release rate; m for length; s for time; K for temperature; and m/s for velocity.

The example can be used of a large building that will allow clear, uncontaminated air to exist around a particular growing fire for at least ten minutes before smoke begins to recirculate into the region. Normal atmospheric conditions prevail. Wood pallets are stored in a large, continuous array on the floor to a height of 1.2 m. This array is ignited locally at an interior point, and the fire spreads in a circular pattern at constant radial speed (as predicted and observed for wood cribs),[30] such that the heat release rate grows with the second power of time

$$\dot{Q} \text{ (kW)} = 1000(t/t_g)^2 \tag{44}$$

Here, t is time and t_g is the so-called growth time. When t_g is 60 s, the fire grows through a magnitude of 1000 kW in 60 s. When t_g is 600 s, the fire grows through a magnitude of 1000 kW in 600 s, a much slower growth rate. In this illustration, it is assumed that the growth time is $t_g = 140$ s.[31] It is also assumed that the fully involved pallet storage generates a total heat release rate of 2270 kW/m² of floor area.[6] The objective is to determine flame height as a function of time, as well as the variation of plume centerline temperature, plume centerline velocity, and plume width at a given elevation of 5 m above the fuel array where a structural member may cross and be heated by the plume.[†]

For the assumed growth time, $t_g = 140$ s, the variation of total heat release rate with time comes from Equation 44

$$\dot{Q} = 5.10 \times 10^{-2}t^2 \tag{45}$$

The convective heat release rate is assumed at 70 percent of the total heat release rate

$$\dot{Q}_c = 3.57 \times 10^{-2}t^2 \tag{46}$$

The instantaneous fire diameter, D, is determined as follows. Since the heat release rate per unit floor area is 2270 kW/m²

$$\dot{Q} = 2270(\pi D^2/4) \tag{47}$$

Upon eliminating $\dot{Q}$ between Equations 45 and 47, the following can be obtained

$$D = 5.35 \times 10^{-3}t \tag{48}$$

[†]In addition to convective heating, which depends on gas temperature and velocity, radiative heating would also be important in such cases and might even dominate over convective heating if the structure is immersed in flames.

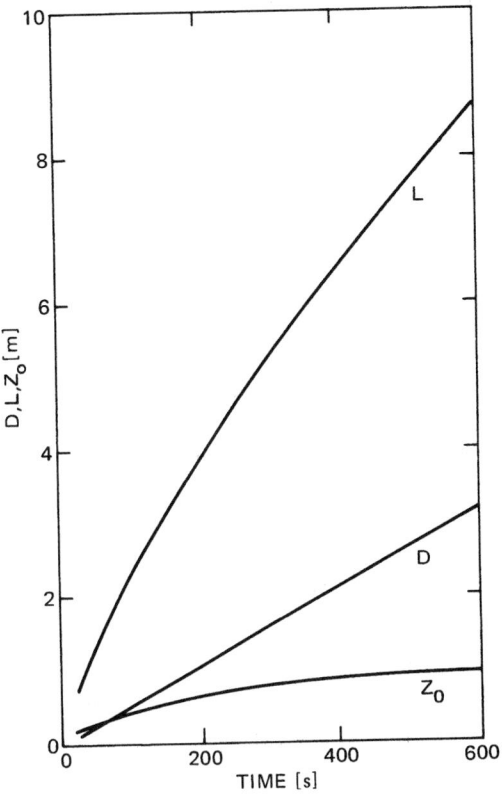

Fig. 2-2.5. *Growing fire illustration: fire diameter, D, flame height, L, and virtual origin, z_0.*

First, the behavior of flame height may be calculated using Equation 6. Substitution of Equations 45 and 48 into Equation 6 gives the following relation of flame height as a function of time

$$L = -5.46 \times 10^{-3}t + 7.15 \times 10^{-2}t^{4/5} \tag{49}$$

This relation is plotted in Figure 2-2.5 for the 10-min (600-s) fire interval and is labeled L. The fire diameter, D, is also plotted in Figure 2-2.5, based on Equation 48.

The virtual origin, z_0, is determined from Equation 25, with substitutions for $\dot{Q}$ from Equation 45 and for D from Equation 48

$$z_0 = -5.46 \times 10^{-3}t + 2.52 \times 10^{-2}t^{4/5} \tag{50}$$

The curve labeled z_0 in Figure 2-2.5 represents the virtual origin according to Equation 50. It is seen that z_0 nearly levels off in the time interval plotted in the Figure; actually, z_0 begins to decrease again at somewhat larger times.

With this foundation, there is sufficient information to calculate gas temperatures, velocities, and plume widths at the 5-m height above the fuel array.

The temperature rise on the plume centerline at the selected height is determined from Equation 17 by substituting $z = 5$ (m), z_0 from Equation 50, $\dot{Q}_c$ from Equation 46, and values of T_∞, g, c_p, and ρ_∞ for the normal atmosphere, yielding

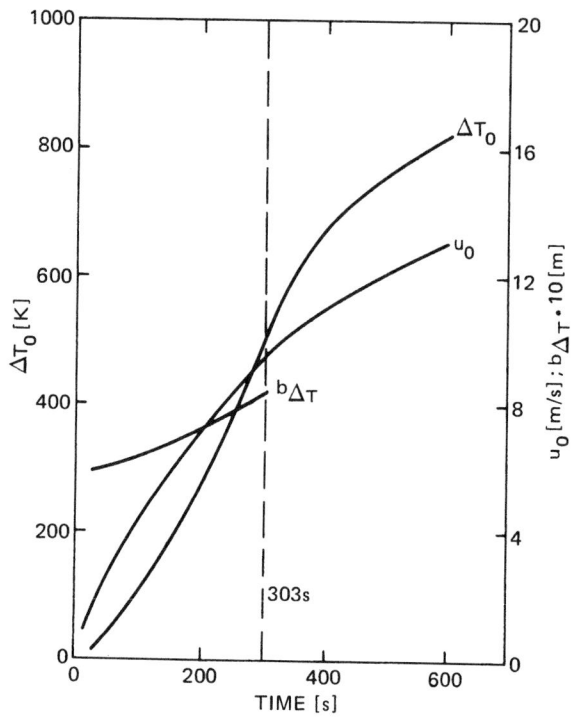

Fig. 2-2.6. *Growing fire illustration: plume width, $b_{\Delta T}$, and centerline values of temperature rise, ΔT_0, and velocity, u_0, at 5 m above fuel.*

$$\Delta T_0 = 2.71t^{4/3}/(5 + 5.46 \times 10^{-3}t - 2.52 \times 10^{-2}t^{4/5})^{5/3} \quad (51)$$

This relation is valid up to the time that a temperature rise associated with the flame tip, $\Delta T_0 = 500$ K, is felt at the selected height, which occurs at $t = 303$ s. The plot of ΔT_0 in Figure 2-2.6 is according to Equation 51 up to the time $t = 303$ s. At larger times, ΔT_0 is determined from Figure 2-2.3 in the following manner: at each selected time, $z - z_0$ is calculated using Equation 50; $\dot{Q}_c$ is calculated from Equation 46; the quantity $(z - z_0)/\dot{Q}_c^{2/5}$ is determined and ΔT_0 is read from Figure 2-2.3. The resulting extension of the ΔT_0 curve is seen in Figure 2-2.6.

The centerline gas velocity at the 5 m height above the fuel array can then be considered. Equation 18 can be used up to the moment that the flame tip reaches the 5-m height; i.e., at $t = 303$ s. After substitution of $z = 5$ (m), z_0 from Equation 50, $\dot{Q}_c$ from Equation 46, and normal ambient conditions, Equation 18 becomes

$$u_0 = 0.339t^{2/3}/(5 + 5.46 \times 10^{-3}t - 2.52 \times 10^{-2}t^{4/5})^{1/3} \quad (52)$$

The u_0 curve in Figure 2-2.6 follows Equation 52 to the limit, $t = 303$ s. As stated in conjunction with Equation 22, the maximum velocity (for a given size fire) occurs just below the mean flame height where $\Delta T_0 = 650$ K, which corresponds to $(z - z_0)/\dot{Q}_c^{2/5} = 0.135$ according to Figure 2-2.3. Using $z = 5$ (m), z_0 and $\dot{Q}_c$ from Equations 50 and 46, the 0.135 limit is found to correspond to a time of $t = 385$ s, where Equation 23 gives the centerline velocity in terms of $\dot{Q}_c$. In fact, it appears that Equation 23 can be used with good accuracy to even larger times, at least to times associated with a lower limit of $(z - z_0)/\dot{Q}_c^{2/5} = 0.08$, according to available measurements.[1,5] Since the largest time in Figure

2-2.6 corresponds to $(z - z_0)/\dot{Q}_c^{2/5} = 0.092$, Equation 23 has been used to calculate the entire extension of the u_0 curve in Figure 2-2.6.

The temperature radius of the plume at the 5-m height above the fuel array is calculated from Equation 16, which can be written

$$b_{\Delta T} = 0.12(1 + \Delta T_0/T_\infty)^{1/2}(z - z_0) \quad (53)$$

With substitution of $z = 5$ (m), ΔT_0 from Equation 51 and z_0 from Equation 50, Equation 53 becomes

$$b_{\Delta T} = 0.12\left[\frac{1 + 9.25 \cdot 10^{-3}t^{4/3}}{(5 + 5.46 \cdot 10^{-3}t - 2.52 \cdot 10^{-2}t^{4/5})^{5/3}}\right]^{1/2}$$
$$\cdot (5 + 5.46 \cdot 10^{-3}t - 2.52 \cdot 10^{-2}t^{4/5}) \quad (54)$$

This equation is plotted in Figure 2-2.6 up to the time the flames reach the 5-m height at $t = 303$ s. The temperature radius at the 5-m height is seen to vary from 0.59 m early in the fire to 0.83 m at 303 s. Plume fluid will reach a minimum of twice the temperature radius, $b_{\Delta T}$; hence, the total width of the plume in this example will be at least four times $b_{\Delta T}$, growing from a minimum of 2.4 m early in the fire to a minimum of 3.3 m as the flames reach the 5-m height.

DATA SOURCES

NFPA 204M, *Guide for Smoke and Heat Venting*,[31] was referenced in this chapter for tables of heat release rate per unit floor area, kW/m², and growth times, t_g, of a number of fuel arrays. The same information has been incorporated by Alpert and Ward,[32] together with additional data.

In Section 3, Chapter 4 by Tewarson, tables are included to estimate combustion efficiencies as well as total and convective heat release rates per unit exposed area of materials under full-scale burning conditions.

NOMENCLATURE

A	defined in Equation 4 (m·kW$^{-2/5}$)
B	buoyancy flux defined in Equation 10 (m^4·s^{-3})
b	plume radius (m)
$b_{\Delta T}$	plume radius to point where $\Delta T/\Delta T_0 = 0.5$ (m)
b_u	plume radius to point where $u/u_0 = 0.5$ (m)
c	adjustable constant, Equation 29 ($-$)
c_p	specific heat of air (kJ/kg·K)
D	diameter (m)
F	function (c_p, T_∞, ρ_∞, g); see Equation 24 (m·kW$^{-2/5}$)
g	acceleration due to gravity (m/s^2)
H_c	actual lower heat of combustion (kJ/kg)
I	intermittency ($-$)
L	mean flame height (m)
$\dot{m}_{ent}$	entrained mass flow rate in plume (kg/s)
$\dot{m}_{ent,L}$	$\dot{m}_{ent}$ at the mean flame height, L (kg/s)
$\dot{m}_f$	mass burning rate (kg/s)
N	nondimensional parameter defined in Equation 2 ($-$)
$\dot{Q}$	$\dot{m}_f H_c$, total heat release rate (kW)
$\dot{Q}_c$	convective heat release rate (kW)
$\dot{Q}_r$	radiative heat release rate (kW)
$\dot{Q}^*$	nondimensional parameter defined in Equation 26 ($-$)
R	radius (m)

r actual mass stoichiometric ratio, air to fuel volatiles

T mean temperature (K)

T_0 mean centerline temperature in plume (K)

T_∞ ambient temperature (K)

T' rms temperature fluctuation (K)

ΔT $T - T_\infty$, mean temperature rise above ambient (K)

ΔT_0 value of ΔT on plume centerline (K)

t time (s)

t_g growth time; see Equation 34 (s)

u mean axial velocity (m/s)

u_0 mean axial velocity on centerline (m/s)

u_{0m} maximum value of u_0, near flame tip (m/s)

u' rms velocity fluctutation in axial direction (m/s)

W_f fire perimeter (m)

z height above top of combustible (m)

z_0 height of virtual origin above top of combustible (m)

α entrainment coefficient ($-$)

ξ nondimensional parameter defined in Equation 21 ($-$)

ρ mean density (kg/m^3)

$\rho_{f\ell}$ mean density in flames (kg/m^3)

ρ_∞ ambient density (kg/m^3)

$\Delta\rho$ $\rho_\infty - \rho$, mean density deficiency (kg/m^3)

$\sigma_{\Delta T}$ plume radius to point where $\Delta T/\Delta T_0 = e^{-1}$ (m)

σ_u plume radius to point where $u/u_0 = e^{-1}$ (m)

REFERENCES CITED

1. B.J. McCaffrey, *NBSIR 79-1910*, National Bureau of Standards, Washington (1979).
2. G. Cox and R. Chitty, *Comb. and Flame*, 39, 191 (1980).
3. G. Heskestad, *18th Symposium on Combustion*, Comb. Inst., Pittsburgh (1981).
4. H.C. Kung and P. Stavrianidis, *19th Symposium on Combustion*, Comb. Inst., Pittsburgh (1983).
5. E. Gengembre, P. Cambray, D. Karmed and J.C. Bellet, *Comb. Sci. and Tech.*, 41, 55 (1984).
6. G. Heskestad, *Report OC2E1.RA*, Factory Mutual Research Corp., Norwood (1981).
7. A. Tewarson, *NBS-GGR-80-295*, National Bureau of Standards, Washington (1982).
8. R. Portscht, *Comb. Sci. and Tech.*, 10, 73 (1975).
9. E.E. Zukoski, T. Kubota, and B. Cetegen, *F. Safety J.*, 3, 107 (1980-81).
10. E.E. Zukoski, B.M. Cetegen, and T. Kubota, *20th Symposium on Combustion*, Comb. Inst., Pittsburgh (1985).
11. G. Heskestad, *F. Safety J.*, 5, 103 (1983).
12. B.R. Morton, G.I. Taylor, and J.S. Turner, *Proc. Roy. Soc.*, A234, 1 (1956).
13. B.R. Morton, *J. Fluid Mech.*, 5, 151 (1959).
14. B.R. Morton, *10th Symposium on Combustion*, Comb. Inst., Pittsburgh (1965).
15. G. Heskestad, *F. Safety J.*, 7, 25 (1984).
16. W.K. George, R.L. Alpert, and F. Tamanini, *Int. J. Heat Mass Trans.*, 20, 1145 (1977).
17. H. Rouse, C.S. Yih, and H.W. Humphreys, *Tellus*, 4, 201 (1952).
18. S. Yokoi, *Report No. 34*, Building Research Institute, Japan (1960).
19. G. Heskestad, *Report 18792*, Factory Mutual Research Corp., Norwood (1974).
20. A. Tewarson, in *Flame-Retardant Polymeric Materials*, Plenum, New York (1982).
21. G. Heskestad, *F. Safety J.*, 5, 109 (1983).
22. Y. Hasemi and T. Tokunaga, *Fire Sci. and Tech.*, 4, 15 (1984).
23. B.M. Cetegen, E.E. Zukoski, and T. Kubota, *Comb. Sci. and Tech.*, 39, 305 (1984).
24. H.Z. You and H.C. Kung, *20th Symposium on Combustion*, Comb. Inst., Pittsburgh (1985).
25. C-S Yih, *Proc. U.S. National Cong. App. Mech.* (1952).
26. B.M. Cetegen, E.E. Zukoski, and T. Kubota, *Report G8-9014*, California Institute of Technology, Daniel and Florence Guggenheim Jet Propulsion Center (1982).
27. G. Heskestad, *21st Symposium on Combustion*, Comb. Inst., Pittsburgh (1986).
28. P.H. Thomas, P.L. Hinkley, C.R. Theobald, and D.L. Sims, *Fire Technical Paper No. 7*, Joint Fire Research Organization, London: H.M. Stationery Office (1963).
29. P.L. Hinkley, *F. Safety J.*, 10, 57 (1986).
30. M.A. Delichatsios, *Comb. and Flame*, 27, 267 (1976).
31. NFPA 204M, *Guide for Smoke and Heat Venting*, National Fire Protection Association, Quincy (1991).
32. R.L. Alpert and E.J. Ward, *F. Safety J.*, 7, 127 (1984).

AIR ENTRAINMENT INTO BUOYANT JET FLAMES AND POOL FIRES

M. A. Delichatsios

Abstract

A simple model for the fire dynamics of turbulent fires is proposed and then used to derive correlations for air entrainment into turbulent jet flames and pool fires. Five parameters are identified to control the fire dynamics: (1) the pool diameter, D; (2) the total buoyancy flow, B_∞; (3) a buoyancy force per unit mass at the adiabatic stoichiometric temperature, $(\Delta T_{ad}/T_\infty)g$; (4) the ambient density, ρ_∞; and (5) the stoichiometric mass air to fuel ratio, S. These parameters lead, through dimensional arguments, to correlations for air entrainment rates, which are further refined by using a simplified physical model. Three distinct regimes for air entrainment are identified in turbulent pool fires: (1) close to pool surface where entrainment rates vary as $Z^{1/2}$ (where Z is the vertical distance from the source), (2) near the neck-in area where entrainment rates vary as $Z^{3/2}$, and (3) farther downstream, but below the flame tip, where entrainment rates vary as $Z^{5/2}$. The proposed correlations are used to plot experimental data from which proportionality coefficients are determined. It is shown, also, that there is a relationship between entrainment rates up to the flame tip and visible flame heights; such a relationship allows one to estimate entrainment rates by measuring visible flame heights. A procedure for calculating air entrainment into turbulent pool and jet fires is included in the Appendix to this chapter.

INTRODUCTION

Fires entrain the surrounding air, which is partly used to burn the flammable gases issuing from a fuel source. Most of the air (about 90 percent) entrained between the base of a fire and its flame tip dilutes the products of combustion. Global computer codes need the magnitude of the air entrainment into a fire to predict fire growth and smoke distribution in a multiroom building. Furthermore, air entrainment also controls the burning rate of the flammable gases.

Entrainment of air into fires is induced by the buoyant upward flow of the hot gases, which are replaced by the cold ambient air drawn into the fire column. Mixing of flammable gases and air occurs as large eddies emanate from the base of the fire then roll over and engulf fresh ambient air into the fire. Concurrently, the fire column oscillates vertically at a natural frequency which decreases as the fire source diameter increases. Pool fires are very sensitive to slight ambient disturbances which cause them to wobble or wander relative to their symmetry axis. Figure 2-3.1 shows a few large eddies entraining ambient air into a pool fire.

Understanding the air entrainment process in fires might explain the sensitivity of the entrainment rates to the geometry of the fire as well as to interactions with environmental disturbances that might induce extraneous oscillations in the fire plume. Air entrainment into a fire varies with the fire's geometry (e.g., pool fires, wall fires, ceiling fires) and the degree of its confinement (e.g., pool fire against a wall or in a room). This chapter addresses air entrainment rates in turbulent buoyant jet flames and pool fires that burn in unrestricted and quiescent environments.

During the last five years, several experiments have provided air entrainment rate measurements in the burning region of turbulent buoyant diffusion flames.[1-5] These experiments span a range from turbulent flames originating from small nozzles ($\geq$ 10 mm in diameter) to pool fires (up to 50 cm in diameter). Entrainment rates were obtained by radially integrating point mean temperature and mean velocity data[1,4] or by using an integral type procedure that directly measured the entrainment rates.[2,3,5] The latter measurement technique is inherently more accurate, since it involves a single measurement of the entrainment rate at a given distance from the fire source. The former measurement technique involves many data integrated radially while typically neglecting unavailable (but necessary) cross correlations between temperature (density) and velocity.

There is no complete agreement among the reported entrainment measurements, especially for pool fires having large diameters. The discrepancies stem not only from differences in the experimental procedures but also from a lack of understanding of the fluid dynamics and combustion in pool fires. This is probably the reason that no generally accepted correlation for entrainment rates has appeared in the literature for varying heat-release rates and pool diameters.

Scientists have made theoretical predictions and correlations for turbulent fires by using three different approaches.

Dr. Michael A. Delichatsios is Senior Research Specialist at Factory Mutual Research Corporation. His research includes two-phase flows, turbulent flows, buoyant flows, radiation in turbulent flows, combustion, heat transfer, pyrolysis and fires.

Fig. 2-3.1. A (gaseous) turbulent pool fire with large flame eddies (pool diameter 19 cm).

First, the k-ε-g method[6,7] is a useful engineering tool for many applications not involving combustion, but it is deficient in describing large-scale buoyant eddies, which control fire dynamics. It has been extensively used for turbulent forced and buoyant jets but not for large pool fires. Second, Steward[8] was the first to propose a simple integral model for turbulent buoyant jet flames. This model has led to a widely accepted correlation for flame heights, but it fails to adequately describe combustion rates, mixing rates, entrainment, and effects of pool diameter. And third, simple physical (Froude) modeling and dimensional analysis have been used by several investigators[2,3,4,9] to model turbulent jet fires. However, no one has discussed in depth: (1) the effects of large density variations, (2) the dominant mechanisms for mixing and entrainment, (3) an appropriate model for turbulent combustion rates, and (4) the effects on entrainment of increasing pool diameter [e.g., in "mass" (very large-scale) fires].

A method for characterizing the flow dynamics of fires over a wide range, from small diameter turbulent jet flames to intermediate and large-scale pool fires of varying heat release rates, is presented in this chapter. For this purpose, the selection of the controlling independent parameters to develop nondimensional groups that characterize the fire dynamics is discussed. The effects of chemistry and gaseous reactions on the independent parameters are then analyzed. The results provide correlations for entrainment rates in both turbulent buoyant jet and pool fires, where proportionality coefficients are determined by comparison with experimental results.

THE DYNAMICS OF TURBULENT BUOYANT DIFFUSION FLAMES

One may distinguish from observation three types of turbulent buoyant diffusion flames: (1) buoyant jet flames, (2) pool fires of intermediate scale, and (3) very large pool fires which are called "mass" fires. (See Figure 2-3.2.) In

buoyant jets, the fuel is supplied from a nozzle having a diameter much smaller than the flame height; the flame height is independent of the nozzle diameter, and burning occurs almost uniformly along the jet flames after fuel mixes with entrained air. This mixing is caused by flame-generated buoyancy. In both intermediate and "mass" pool fires, the source diameter also affects the upward flow dynamics; typically, 90 percent of the fuel burns within half a diameter from the pool surface, well below the visible flame tip.[10,11] As the source diameter increases or at low fuel mass transfer rates, a transition to a "percolating" type of burning occurs, wherein fuel burns uniformly over the fire base without significant flame merging. [See Figure 2-3.2, part (c).]

The objective now is to provide fundamental physical parameters for (1) delineating the different types of turbulent buoyant diffusion flames (see Figure 2-3.2) and (2) predicting their properties, such as combustion rates, air entrainment rates, species concentrations, and species yields.

For any given fixed point in the flow field, only one explicit length-scale affects the flow properties: the source diameter, D. Turbulent buoyant jets and pool fires are by definition dominated by mixing of large-scale eddies so that

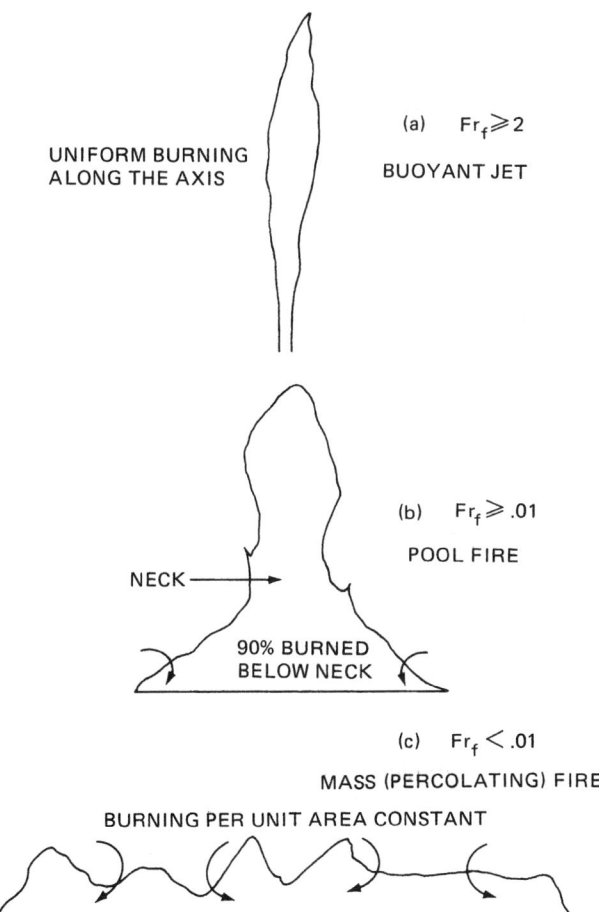

Fig. 2-3.2. Three characteristic types of turbulent buoyant diffusion flames: (a) buoyant jet flames: $Fr_f \geq 2$; (b) pool fires of intermediate size: $Fr_f \geq 0.01$; and (c) "mass" (percolating) fires: $Fr_f < 0.01$, where the fire Froude number, Fr_f, is defined by Equation 8b.

one may neglect molecular diffusion processes, even though the flow can be laminar around the rim over a small region relative to the total extent of the flames. Buoyancy provides the energy for the formation of large eddies, which engulf and mix fuel and ambient air; subsequently, these eddies rapidly break up into smaller eddies by a turbulent cascade process. There are two buoyant "driving forces" generating turbulence in turbulent diffusion flames: (1) the overall buoyancy flow, which is similar to noncombusting thermal plumes; in the present case this buoyancy flow reaches a maximum at the flame tip; and (2) the local variation of buoyancy attributable to the steep local temperature variations created by the local heat release rates.

The buoyancy flow at height Z from the source is related to the convective flow in the following way

$$B(Z) = g \int 2\pi r \overline{u(\rho_\infty - \rho)} \, dr$$
$$= \frac{g}{C_p T_\infty} \int 2\pi r C_p \overline{\rho u \Delta T} \, dr \qquad (1)$$
$$= \frac{g \dot{Q}_c(Z)}{C_p T_\infty}$$

It attains its maximum value at the end of combustion near the flame tip

$$B_\infty = \frac{g \dot{Q}_{c,\infty}}{C_p T_\infty} = \frac{g \dot{Q}(\chi_A - \chi_R)}{C_p T_\infty} \qquad (2)$$

where $\dot{Q}$ is the theoretical heat release rate, $\dot{Q}_c$ is the convective heat release rate, χ_A is the efficiency of combustion, and υ_R is the fraction of the theoretical heat release rate that is lost by radiation. The specific heat of the gases, C_p, is assumed constant, while T_∞ is the ambient temperature, and g is the gravitational acceleration. In manipulating Equation 1, these common assumptions were used: (1) the same pressure in the fire plume as in the ambient, (2) perfect gas laws, and (3) equal molecular weight gases. The assumption of equal molecular weight breaks down near the fuel source, if the fuel has different molecular weight than the air. This effect will be important only in the initial laminar region before turbulence starts. Just at the transition to turbulence mixing, entrainment by large-scale eddies increases so fast that the molecular weight becomes equal everywhere to the molecular weight of the air. Molecular weight effects as well as laminar effects have been neglected in this work because, as we have emphasized before, for turbulent fires and jets dominated by buoyancy, the laminar flow region is small relative to the flame height.[3,4,21]

The other buoyancy source, which is distinctive for combusting flows, arises from the buoyancy of individual hot eddies burning at flame temperatures. These eddies generate a maximum instantaneous force per unit mass or an effective acceleration which is equal to

$$\frac{\Delta \rho_f}{\rho_f} g = \frac{T_f - T_{0\infty}}{T_\infty} g = \frac{\Delta T_f}{T_\infty} g \qquad (3)$$

where T_f and ρ_f are the peak instantaneous temperature and the corresponding density of individual laminar flamelets comprising the turbulent flame. First, note that *the peak laminar flame temperature*[12,13,14] is uniquely characterized when radiation is neglected. Furthermore, we suggest that the parameter in Equation 3 incorporates by construction variable density effects associated with a turbulent combusting flow. This hypothesis is also supported by analysis and

experiments in strong (variable density) buoyant plumes sustaining no combustion.[15,16] We may conclude that density variations in the fire plume are accounted for by using the force per unit mass (see Equation 3) at peak temperature conditions. It is well known, however, that the ambient density, ρ_∞, also influences the fire dynamics.

The collective effects of the laminar flamelets in the fire dynamics should also depend on the stoichiometric mass air to fuel ratio (S) and the turbulence scales—Kolmogorov (molecular) and macroscale. Observe that the peak laminar temperature (see Equation 3) together with the stoichiometric ratio characterize the chemical structure of a laminar flamelet.[17,18] (The chemical structure determines species concentrations and temperatures across a laminar flame from the air to the fuel side.) We also note that existing evidence[12,13,14] with gaseous flames as well as analysis of turbulent combustion flows[13,14] suggests that for fast chemical reactions, the Kolmogorov scales do not affect the mean volumetric reaction rates, which depend only on macroscale mixing. We propose, therefore, that only the turbulent macroscales, which are determined by the fire geometry, affect the fire dynamics.

Finally, we may also conclude, based on the previous discussion, that the vertical distribution of the buoyancy flow normalized by the maximum buoyancy flow satisfies the following relationship

$$\frac{B(Z)}{B_\infty} = fcn \quad \begin{array}{l} \text{flow field geometry (e.g., } Z, D), \\ \text{(stoichiometric ratio } \equiv S) \end{array} \qquad (4)$$

which implies that the buoyancy flow variation with height may be substituted by the maximum buoyancy flow, B_∞, and the stoichiometric ratio, S.

We summarize now the independent parameters that should control the flow dynamics *at any given point* of a turbulent pool fire:

1. the pool diameter, D;
2. the maximum buoyancy flow, B_∞;
3. the force per unit mass at the peak laminar flame temperature, $(\Delta T_f / T_\infty)$g;
4. the stoichiometric ratio, S; and
5. the ambient density, ρ_∞.

Note that the stoichiometric mass air to fuel ratio, S, used throughout this chapter is considered to be an effective stoichiometric ratio equal to its theoretical value multiplied by the combustion efficiency ($S \equiv \chi_A S_{th}$).

Radiation losses can be incorporated in the buoyancy parameters, B_∞, $(\Delta T_f / T_\infty)$g, by using the overall radiant fraction, χ_R, characteristic for turbulent buoyant fires of a given fuel. Similarly, incompleteness of combustion is introduced by using an empirical efficiency coefficient, χ_A. Both these modifications have been already included in the definition of the total buoyancy flow, B_∞ (see Equation 2). In a similar fashion, the parameter that represents the buoyancy per unit mass of hot gases at the flame temperature becomes

$$\frac{\Delta T_f (\chi_A - \chi_R)}{T_\infty} g \qquad (5)$$

We point out that radiation losses and incompleteness of combustion have been introduced in an *ad hoc* manner by using empirical coefficients for the completed combustion process. Such a procedure is partially justified, since these coefficients depend essentially on the type of the fuel for turbulent pool fires.[19,20]

From the controlling parameters, one can form two dimensionless groups: (1) the stoichiometric ratio, S, and (2) a characteristic Froude number of the fire

$$\mathrm{Fr}_f = \frac{B_\infty/\rho_\infty}{D^{5/2}\left[\dfrac{\Delta T_f}{T_\infty}(\chi_A - \chi_R)g\right]^{3/2}} \qquad (6)$$

The present analysis proposes that any mean property normalized by a combination of the independent parameters can be expressed in terms of the two fundamental dimensionless groups at a given location in the flow field normalized by the source diameter.

Three additional remarks concerning the proposed group of controlling parameters in turbulent pool fires can be made:

1. For common fuels burning in air, the stoichiometric ratio (mass ratio of air to fuel at stoichiometric conditions) has large values, e.g., $S > 10$, which means that reactions are controlled principally by the supply of air to the diffusion flames; thus, it is intuitively expected that the stoichiometric ratio, S, will have a negligible influence on the fire dynamics. This hypothesis is supported for large S by applying a flamelet model for turbulent combustion with appropriate probability distribution functions to represent turbulent mixing.[12,13,14] More importantly, however, this hypothesis is justified from the empirical correlations of flame heights in turbulent jet diffusion flames[2,3] of common fuels; such correlations do not show a dependence on the stoichiometric ratio.[2,3] It is important to clarify here a source of possible confusion for the reader. The stoichiometric ratio, S, will appear later in fire dynamic correlations even when it takes large values; this presence, however, originates from the fact that the buoyancy parameter, which includes the peak laminar flame temperature rise (see Equation 3), depends on the stoichiometric ratio (see, e.g., Equation 7).

2. For practical reasons, we adopt a simple method for characterizing the laminar flame temperature

$$\frac{\Delta T_f}{T_\infty} \simeq \frac{\Delta T_{ad}}{T_\infty} \cong \frac{\Delta H_c}{C_p T_\infty (S+1)} \qquad (7)$$

where ΔH_c is the heat of combustion per unit mass of the fuel, and C_p is the specific heat at ambient temperature. Here, we have equated the peak laminar flame temperature involving no radiation loss to the stoichiometric adiabatic temperature. In addition, we have approximated, naively, the stoichiometric adiabatic temperature by using a simplified energy balance. The correct stoichiometric flame temperature can be calculated by using the NASA equilibrium code[18] for fuels of known chemical compositions (see Section 1, Chapter 6); Equation 7 overestimates the true adiabatic stoichiometric temperature, since it ignores dissociation effects and variation of specific heat with temperature.

3. For hydrocarbon fuels and other fuels having large stoichiometric ratios, the fire Froude number is sufficient to characterize the overall flow field (Equations 2, 6, and 7):

$$\mathrm{Fr}_f = \frac{\dot{Q}(\chi_A - \chi_R)g}{\rho_\infty C_p T_\infty D^{5/2}\left[\dfrac{\Delta T_{ad}}{T_\infty}(\chi_A - \chi_R)g\right]^{3/2}} \qquad (8a)$$

or, by a simple approximation of the adiabatic temperature (Equation 7),

$$\mathrm{Fr}_f = \frac{\dot{Q}}{\rho_\infty \left(\dfrac{\Delta H_c}{S+1}\right)D^2 \sqrt{\dfrac{\Delta H_c}{(S+1)C_p T_\infty}gD(\chi_A - \chi_R)}} \qquad (8b)$$

Figure 2-3.2 indicates the range of fire Froude numbers characterizing the three types of turbulent diffusion flames, based on experience with gaseous pool fires and jet flames.[2,4,5]

Finally, it should be noted that this theory has ignored as having negligible effects on the flow dynamics: (1) the laminar flow near the source, and (2) any Kolmogorov scale effects on the reaction rates. In the next section, we apply our analysis to predicting air entrainment rates in turbulent buoyant jet flames and pool fires. Similar analysis has been used to predict mean species concentrations and the conserved scalar distribution in turbulent fires.[21]

AIR ENTRAINMENT INTO BUOYANT FIRES

Air entrainment correlations for the burning region of turbulent pool fires and for the buoyant plume beyond the flame tip are developed by employing dimensional analysis and the modeling hypothesis described in the previous section. The entrainment correlations will be used subsequently for plotting various experimental data.

Air Entrainment into the Burning Region

The air entrainment rate, $\dot{m}_{ent}$, up to a vertical distance Z above the fuel source is determined in terms of the controlling parameters listed prior to Equation 5. It follows by dimensionless analysis that

$$\frac{\dfrac{\dot{m}_{ent}}{B_\infty}}{\dfrac{\Delta T_{ad}}{T_\infty}g(\chi_A - \chi_R)} = fcn\left(\frac{Z}{D}, \mathrm{Fr}_f, S\right) \qquad (9)$$

wherein Equation 7 has been used and the fire Froude number is defined by Equation 6.

By using the simple expression for the flame temperature (see Equation 7), the definition of buoyancy flow, B_∞ (Equation 2), and the relationship for the theoretical heat release rate, $\dot{Q} = \dot{m}_f \Delta H_c$, where $\dot{m}_f$ is the fuel mass supply rate, one finds that

$$\frac{\dot{m}_{ent}}{(S+1)\dot{m}_f} = fcn\left(\frac{Z}{D}, \mathrm{Fr}_f, S\right) \qquad (10)$$

This general entrainment expression provides the ratio of entrained air to the total combustion product flow under stoichiometric conditions. It was argued at the end of the previous section (see discussion following Equation 6), that the stoichiometric ratio, S, is not a controlling parameter when it takes large values, as for hydrocarbon fuels or vapor gases from most combusting materials. In these circumstances, one can drop the dependence on stoichiometric ratio at the right-hand sides of Equations 9 and 10 to obtain

$$\frac{\dot{m}_{ent}}{(S+1)\dot{m}_f} = fcn\left(\frac{Z}{D}, \mathrm{Fr}_f\right) \qquad \text{for large } S \qquad (11a)$$

Equation 11a can be further simplified by using a generally reported experimental observation[2-5] that entrainment rates over the entire burning region of pool fires of common fuels are independent of the heat release rate $\dot{Q}$. This experimental fact leads to the following relationship for entrainment rates

$$\frac{\dot{m}_{ent}}{(S+1)\dot{m}_f}\mathrm{Fr}_f = fcn\left(\frac{Z}{D}\right), \tag{11b}$$

as can be shown by using the definition of fire Froude number, Fr_f (see Equation 8a) and the theoretical heat release rate $\dot{Q} = \dot{m}_f \Delta H_c$.

Equation 11b and Equation 8a are the basic results of the present analysis. They have been derived through physical reasoning and dimensional analysis by systematically stating all assumptions and limitations. One may further propose specific forms for the function at the right-hand side of Equation 11b by devising a heuristic model for entrainment (see Figure 2-3.3). First observe that a characteristic entrainment velocity, U_e, generated by eddies at stoichiometric flame temperature, is

$$U_e \sim \sqrt{\frac{\Delta T_{ad}}{T_\infty}(\chi_A - \chi_R)gZ} \tag{12}$$

while the area around the flames available for entrainment up to a height Z is

1. $A_e \sim Z^2$ for turbulent jet fires or upper part of pool fires $\qquad\qquad$ (13a)

2. $A_e \sim DZ$ for pool fires close to the neck-in area $\quad$ (13b)

3. $A_e \sim D^2$ or near the base of pool fires for large mass or "percolating" fires $\qquad\qquad$ (13c)

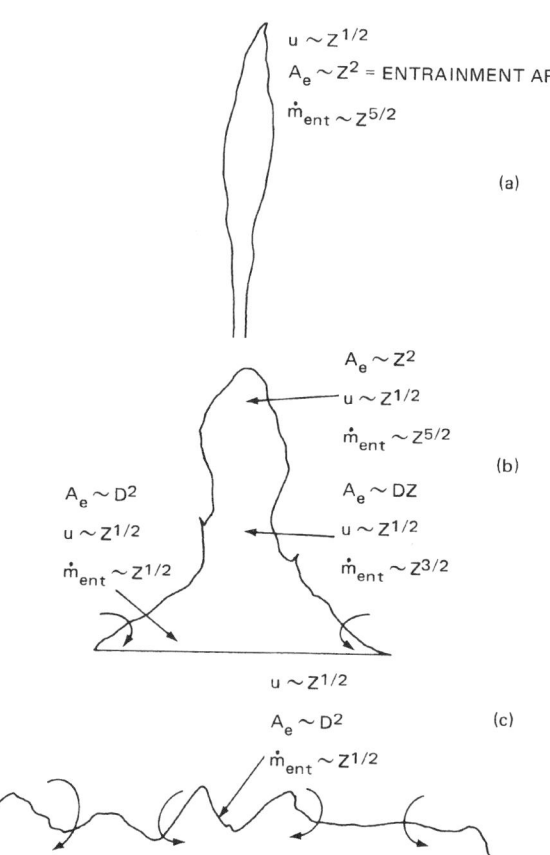

Fig. 2-3.3. *Entrainment rate relationships for turbulent jet flames and pool fires. (See the section, "Air Entrainment into the Turbulent Buoyant Plume above the Flames.")*

Since the entrainment rate is $\sim \rho_\infty A_e U_e$, one finds by comparison with Equation 11a and by using Equation 7 for the adiabatic temperature and Equation 8b for the fire Froude number

see also Figure 2-3.3 and note that

$$\left(\frac{\mathrm{Fr}_f}{(S+1)\dot{m}_f} = \frac{1}{\rho_\infty D^2 \sqrt{\frac{\Delta T_{ad}}{T_\infty}gD(\chi_A - \chi_R)}}\right):$$

$$\frac{\dot{m}_{ent}}{(S+1)\dot{m}_f}\mathrm{Fr}_f \sim \left(\frac{Z}{D}\right)^{5/2}$$

for buoyant jets or upper part of pool fires $\qquad$ (14a)

$$\frac{\dot{m}_{ent}}{(S+1)\dot{m}_f}\mathrm{Fr}_f \sim \left(\frac{Z}{D}\right)^{3/2}$$

close to the neck-in area of pool fires $\qquad$ (14b)

or

$$\frac{\dot{m}_{ent}}{(S+1)\dot{m}_f}\mathrm{Fr}_f \sim \left(\frac{Z}{D}\right)^{1/2}$$

near the base of pool fires or floor mass fires. (14c)

An entrainment equation similar to Equation 14a for turbulent buoyant jets could be derived from Steward's integral model[8] and has also been used by Cetegen *et al*[2] and Delichatsios and Orloff.[4] These expressions,[2,4] however, did not include the specific fuel effects through the fuel adiabatic temperature, radiative losses, or incompleteness of combustion, all of which are included in Equation 14a. The other correlations, Equations 14b and 14c, or the general form given by Equation 11b, differ from entrainment relationships proposed by Cetegen *et al*[2] in the lower part of the flames. These authors assumed that the flow in the lower half of the flames could be described by a cylindrical laminar shear layer model.[2] Finally, McCaffrey has also proposed some completely *empirical* entrainment relationships[1] which were presented in dimensional form. These relationships do not include pool diameter effects and are different from Equations 14a, b, and c and other results.[2,4]

Air Entrainment into the Turbulent Buoyant Plume above the Flames

After burning ends, a turbulent buoyant plume develops beyond the flame tip. The buoyant plume (see Figure 2-3.4) is entirely characterized by its constant buoyancy flow, B_∞, (see Equation 2), the source diameter, D, the ambient density, ρ_∞, and the initial conditions near the flame tip at the end of combustion; these initial conditions depend in general on the fire Froude number and the stoichiometry. Note that in contrast to the burning region, the temperature excess parameter (see Equation 3) is not an independent parameter. If flames *are very short* (see Figure 2-3.4) or at large distances from the source, *the buoyancy plume properties are independent of the initial conditions at the flame tip.* In these circumstances, the entrainment rate up to a height Z is given by dimensional analysis as

$$\frac{\dot{m}_{ent}}{\rho_\infty \left(\frac{B_\infty}{\rho_\infty}\right)^{1/3}D^{5/3}} = fcn\left(\frac{Z}{D}\right) \tag{15}$$

Far away from the source of the buoyancy flow, the plume properties are independent of the diameter, so that one finds (see Figure 2-3.4)

$$\frac{\dot{m}_{ent}}{\rho_\infty \left(\frac{B_\infty}{\rho_\infty}\right)^{1/3} D^{5/3}} \sim \left(\frac{Z}{D}\right)^{5/3} \text{ far away from the base,} \quad (16a)$$

which agrees with buoyant plume theories and experiments.

Furthermore, a physical heuristic model, similar to the one employed in the burning region (see Equations 12 and 13), provides expressions for the entrainment rate near the neck-in area and the base of a purely buoyant plume. Buoyancy flow conservation determines a characteristic centerline temperature excess ($\dot{m}_{ent} \cdot \Delta T_c \sim \dot{Q} \sim B_\infty$) which, together with velocity and entrainment area equations similar to Equations 12

$$\left(u \sim \sqrt{\frac{\Delta T_c}{T_\infty} g Z}\right)$$

and 13, produces the entrainment rates (see Figure 2-3.4)

$$\frac{\dot{m}_{ent}}{\rho_\infty \left(\frac{B_\infty}{\rho_\infty}\right)^{1/3} D^{5/3}} \sim \left(\frac{Z}{D}\right)^{1/3} \quad \text{near the base, and} \quad (16b)$$

$$\frac{\dot{m}_{ent}}{\rho_\infty \left(\frac{B_\infty}{\rho_\infty}\right)^{1/3} D^{5/3}} \sim \frac{Z}{D} \quad \text{near the neck-in area.} \quad (16c)$$

It is important to mention that Equation 16c is consistent with experimental data in the near region of a plume originating from square heat sources,[25] wherein it is shown experimentally that the centerline mean temperature rise decreases inversely proportional to height from the source and is proportional to the equivalent source diameter; i.e., $\Delta T_c \sim D/Z$.

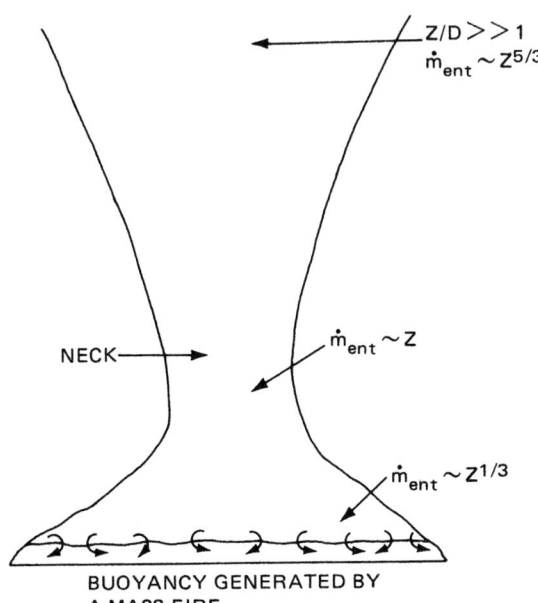

BUOYANCY GENERATED BY
A MASS FIRE

Fig. 2-3.4. Entrainment regimes for turbulent buoyant flows (without flames) from large-area sources (see the section on air entrainment into the burning region).

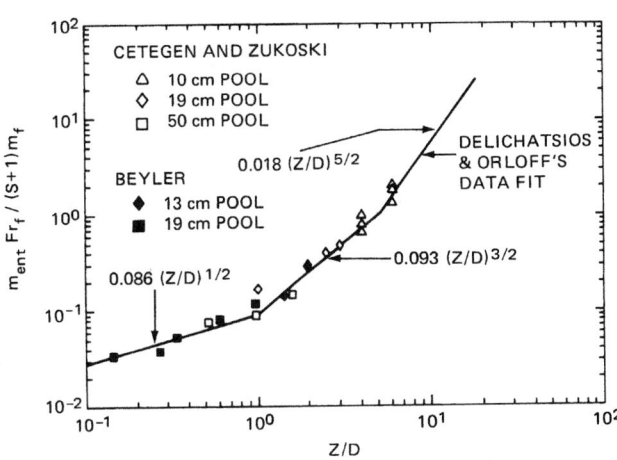

Fig. 2-3.5. Entrainment measurements in turbulent jet flames and pool fires, plotted in terms of a normalized entrainment rate up to a vertical distance, Z, above the source normalized by its diameter, D, (the fire Froude number, Fr_f, in the ordinate is defined by Equation 8b, S is the stoichiometric mass air to fuel ratio, and $\dot{m}_f$ is the fuel mass supply rate).

COMPARISON WITH EXPERIMENTAL DATA

Entrainment correlations of experimental data are presented only for a baseline situation which comprises circular pool fires, without any circumferential lip, burning in a nominally quiescent open environment. There are not enough data to implement correlations of entrainment rates for other practical situations, such as square pool fires, pool fires with lips, or pool fires in enclosures. In the present analysis, the experimental data reported by Cetegen et al,[2] Beyler,[5] and Delichatsios and Orloff,[4] have been used because these data were obtained by direct integral measurements, in contrast to a generally less accurate evaluation of entrainment rates by integrating point measurements of temperature and velocities.[1,3]

Entrainment measurements in the burning region of pool fires and buoyant jets for hydrocarbon fuels are plotted in Figure 2-3.5 on a log-log plot by using the coordinates suggested by Equation 11b, which is strictly applicable for fuels having high stoichiometric ratios. The fire Froude number defined by Equation 8b applies to those fuels.

The solid lines on Figure 2-3.5 are fair curves through the data of Beyler[5] and Delichatsios and Orloff,[4] while the symbols are measurements of Beyler[5] and Cetegen et al.[2]* The upper branch represents (within ±5 percent) data[4] for entrainment rates in turbulent jet fires from nozzles 25 mm in diameter, wherein the entrainment rates were measured by using a Ricou-Spaulding technique.[4] Both Cetegen et al[2] and Beyler[5] used a collection hood to obtain the entrainment measurements for pool diameters 10 cm, 19 cm, and 50 cm; and 13 cm and 19 cm, respectively; the reference distance,

---

*Cetegen et al[2] data were reduced by using in the definition of the fire Froude number (Equation 8b) the values $\chi_A = 1.0$ and $\chi_R = 0.1$ for the methane fires employed by these authors. For the other data,[4,5] where propane was the fuel, these values were taken to be equal to $\chi_A = 1.0$ and $\chi_R = 0.28$. For the rest of the properties, standard reference values were used.

Z, from the fuel source to the stable hot layer interface in the collection hood was determined either visually[2] or by measuring a vertical distribution of species or temperatures in the hot layer.[5] Beyler[5] demonstrates that there is large uncertainty and scatter in the determination of the interface (approximately ± 4 cm). Moreover, Beyler's[5] hood was not large enough to allow entrainment measurements at distances greater than 26 cm from the sources, while many of his data were for distances less than 10 cm. For those reasons, Figure 2-3.5 uses: (1) the entrainment data for the 19 cm diameter pool fire[5] and (2) only those data for the 13 cm diameter pool fire[5] at relative distances, Z/D, from the source greater than unity.

There is good agreement among the various data plotted as in Figure 2-3.5, even though the Cetegen et al[2] entrainment data are, in general, higher than Beyler's,[5] and, in some cases, almost twice as high. However, for values of dimensionless heights greater than unity $(Z/D > 1)$, the differences become smaller. Following our previous discussion, one may attribute the discrepancies to the uncertainties concerning the location of the interface (i.e., Z); such uncertainties may cause a large error in the abscissa (i.e., Z/D), especially for small pool diameters.

The experimental results in Figure 2-3.5 support the theoretical predictions for entrainment rates as shown by Equations 14a, 14b, and 14c. Moreover, they allow the determination of the proportionality coefficients for these equations. Three regimes are distinguished

(a) Close to the pool surface $\left(\dfrac{Z}{D} < 1.0\right)$

$$\frac{\dot{m}_{ent}}{(S+1)\dot{m}_f}\mathrm{Fr}_f = 0.086\left(\frac{Z}{D}\right)^{1/2} \qquad (17a)$$

(b) Around and beyond the neck-in area $\left(1.0 < \dfrac{Z}{D} < 4.0\right)$

$$\frac{\dot{m}_{ent}}{(S+1)\dot{m}_f}\mathrm{Fr}_f = 0.093\left(\frac{Z}{D}\right)^{3/2} \qquad (17b)$$

(c) Farther downstream but before the flame tip

$$\frac{\dot{m}_{ent}}{(S+1)\dot{m}_f}\mathrm{Fr}_f = 0.018\left(\frac{Z}{D}\right)^{5/2} \qquad (17c)$$

Here, the fire Froude number is given by (see Equation 8b):

$$\mathrm{Fr}_f = \frac{\dot{Q}}{\rho_\infty \dfrac{\Delta H_c}{S+1}D^2\sqrt{\dfrac{\Delta H_c}{(S+1)C_pT_\infty}gD(\chi_A - \chi_R)}}$$

$$= Q_D^*\left[\frac{1}{\dfrac{\Delta H_c}{(S+1)C_pT_\infty}\sqrt{\dfrac{\Delta H_c}{(S+1)C_pT_\infty}(\chi_A - \chi_R)}}\right] \qquad (18a)$$

where

$$Q_D^* = \frac{\dot{Q}}{\rho_\infty C_pT_\infty D^2\sqrt{gD}} \qquad (18b)$$

is the dimensionless number used by Cetegen et al[2] and others in correlations for the visible flame heights in turbulent pool fires.

We can make the following comments concerning the applicability of entrainment correlations (see Equations 17a, 17b, and 17c, and Figure 2-3.5):

1. The present results apply to fuels having high mass air-to-fuel ratios at stoichiometric burning conditions. $(S > \sim 10)$;

2. The three distinct regimes shown in Figure 2-3.5 exist together only for pool fires [see Figure 2-3.2(b)]; for turbulent buoyant jets [see Figure 2-3.2(a)] only the upper part of the conversion in Figure 2-3.5 is applicable (in this case, $Z/D < \sim 10$ corresponds to the transition from laminar flow to turbulent flow); for mass fires [see Figure 2-3.2(c)], only the lower part of the curves in Figure 2-3.5 is applicable.

3. (a) The initial laminar region of the flow should occupy a small fraction of the flame height; i.e., the flow should be predominantly turbulent; and (b) the turbulent flow is primarily dominated by buoyancy; i.e., source momentum is negligible. These effects are discussed in detail in a recent report.[26]

Also, all experimental data used in the present correlations satisfy all the previous requirements.

Note, also, that there is a direct relationship between entrainment rates and flame heights in turbulent diffusion flames. This relationship becomes clear if one uses the experimental observation reported by various investigators[2,4] that, at the visible flame tip, the flow rate in the plume is about ten times the flow rate corresponding to the stoichiometric mass requirement for combustion:[†]

$$\dot{m}_{ent} \simeq (\dot{m}_{ent} + \dot{m}_f) = 10(S+1)\dot{m}_f \qquad (19)$$

It is not difficult to see how Figure 2-3.5 together with Equation 19 can be used to determine flame heights. The flame height of a given fire can be determined by the abscissa of Figure 2-3.5 corresponding to an ordinate equal to

$$\frac{\dot{m}_{ent}}{(S+1)\dot{m}_f}\cdot\mathrm{Fr}_f = 10\cdot\mathrm{Fr}_f$$

It is interesting to compare flame heights determined by the present method and visible flame heights reported by Cetegen et al[2] and others.[2,22]

Following the prescribed procedure, we can invert Equations 17a, 17b, and 17c to obtain the flame heights

$$\frac{Z_f}{D} = 1.35 \times 10^4\,\mathrm{Fr}_f^2 \quad \text{for } \mathrm{Fr}_f \leq 8.6 \times 10^{-3} \qquad (20a)$$

$$\frac{Z_f}{D} = 22.54\,\mathrm{Fr}_f^{2\mathrm{K}3} \quad \text{for } \mathrm{Fr}_f < 7.44 \times 10^{-2} \qquad (20b)$$

[†]It is important to note that one can show by using the physical turbulent model with flamelet combustion, presented in the subsection "The Dynamics of Turbulent Buoyant Diffusion Flames," that the entrainment rate up to the flame tip is proportional to the stoichiometric mass requirements for combustion (see Equation 19). One should use:[15] (1) the integral conservation equation for the conserved scalar, (2) an assumed probability distribution for flamelet mixing, and (3) a definition of the flame tip as the vertical location from the source where the mean centerline fuel concentration is zero. Moreover, one can argue (with the assistance of experiments[21]) that for large stoichiometric ratios, the ratio of entrainment rate up to the flame tip to the stoichiometric mass flow requirement is independent of the pool size or fire intensity.

$$\frac{Z_f}{D} = 12.52\, \mathrm{Fr}_f^{2/5} \quad \text{for } \mathrm{Fr}_f > 7.44 \times 10^{-2} \quad (20c)$$

or, in terms of the more commonly used dimensionless parameter Q_D^* (see Equation 18b) one obtains for *methane* (see Equation 18a):

$$\frac{Z_f}{D} = 18.8\, Q_D^{*2} \quad \text{for } Q_D^* < 0.23 \quad (21a)$$

[very large (mass) pool fires]

$$\frac{Z_f}{D} = 2.6 Q_D^{*2/3} \quad \text{for } 0.23 < Q_D^* < 1.9 \quad (21b)$$

(intermediate-scale pool fires)

$$\frac{Z_f}{D} = 3.4 Q_D^{*2/5} \quad \text{for } Q_D^* > 1.9 \quad (21c)$$

(turbulent jet fires)

To judge the proposed factor of 10 at the flame height, we make the following comparisons with experiments:

1. Cetegen et al[2] have reported flame height correlations for relatively small pool fires [$D < 50$ cm, Q_D^* (methane) ≥ 0.2]; they have cast their equations in a form like that of Equations 21b and 21c, while their reported coefficients are 3.3 and 3.3, respectively.
2. Recently, Cox and Chitty[22] reported flame heights for pool fires that correspond to the conditions of Equation 21a; i.e., $Q_D^* \geq 0.2$; (Note that the form of Equation 21a for very large pool fires was first suggested by Thomas et al.[23]) Cox and Chitty[22] found a coefficient 9.3 instead of 18.8, a discrepancy that can probably be attributed to the fact that they tested square pool fires, while the data presented here apply to circular pool fires. In square pool fires, larger eddies are generated near the corners so that entrainment increases, burning speeds up, and, therefore, flame heights become smaller; note that differences between square and circular pool fires are larger for small flame heights.

Beyond the flame tip (given by Equation 21) one has to apply entrainment relationships valid for purely buoyant plumes. For relatively tall flames, where the flame tip is beyond the neck-in area, Equation 16a has been extensively used to calculate and correlate experimental data, with the help of a virtual origin.[2,24] If the plumes are short (see Figure 2-3.4), however, new data are required to test and apply the newly proposed Equations 16b and 16c.

Additional indirect evidence for the validity of the present entrainment relationships stems from the correlations of the normalized mixture fraction, ξ, with heat release rate and size inside turbulent pool fires.[21] The mixture fraction, ξ, is the fraction of atomic species originating in the supplied fuel and, therefore, it is a conserved quantity. The centerline mean value of the mixture fraction is inversely proportional to the mass flow rate

$$\dot{m}_{ent} + \dot{m}_f \simeq \dot{m}_{ent} \quad (22)$$

$$\xi \sim \frac{1}{\dot{m}_{ent}}$$

where we have neglected the cross-correlations of fluctuations between mixture fraction and velocity.

Figures 2-3.6(a) and 2-3.6(b) include the mixture fraction data taken from Orloff et al.[21] Figures 2-3.5 and 2-3.6(a)

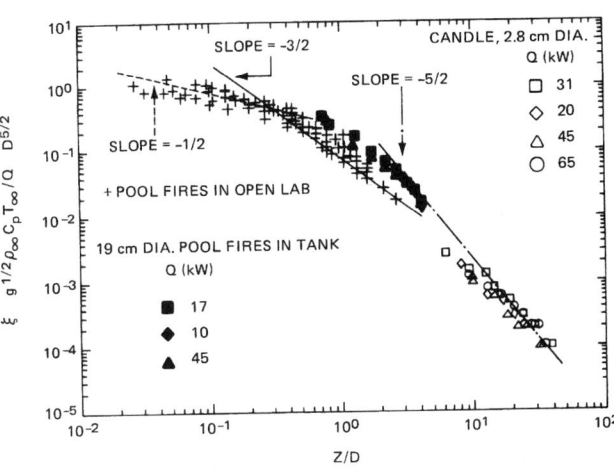

Fig. 2-3.6(a). Pool and buoyant turbulent jet propane mixture fraction data illustrating entrainment regimes below the flame tip.

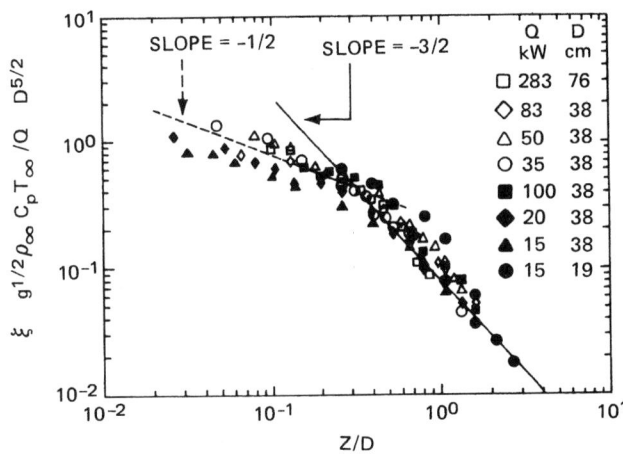

Fig. 2-3.6(b). Results from pool fires (without lips) as correlated in two of the entrainment regimes shown in Figure 2-3.6(a).

and (b) demonstrate the correspondence between the mixture fraction regimes in Figures 2-3.6(a) and (b) and the entrainment regimes in Figure 2-3.5 as they are related by means of Equation 22.

CONCLUSIONS

The major accomplishments presented in this chapter are the following:

1. A model for the fire dynamics of turbulent diffusion flames has been developed by identifying the controlling parameters (see list prior to Equation 5). Variable density effects have been incorporated by using the mean buoyancy force per unit mass at mean peak temperatures (see Equation 3). Turbulent combustion is modeled by a statistical ensemble of laminar flamelets whose structure depends essentially on the *stoichiometric ratio* and *values* of species concentrations at the air side, fuel side, and near stoichiometric conditions.

2. The model led to correlations of entrainment rates (see Equation 14) in turbulent pool fires that agree with experimental data (see Figure 2-3.5 and Equation 17). Three regimes for air entrainment have been identified in turbulent pool fires: (1) close to pool surface, where entrainment rates vary as $Z^{1/2}$ (where Z is the vertical distance from the source); (2) near the neck-in area, where entrainment rates vary as $Z^{3/2}$; and (3) farther downstream, but below the flame tip, where entrainment rates vary as $Z^{5/2}$.

3. There is a simple relationship between entrainment rates up to the flame tip and visible flame heights (cf. Equations 17 and 20), which had been previously measured by other investigators. One may pursue the possibility of estimating entrainment rates by measuring visible flame heights in pool fires.

Although the turbulent model is applicable for arbitrary mass stoichiometric ratios (air to fuel), the entrainment rate correlations are restricted to fuels having large mass stoichiometric values ($S \geq 10.0$), which, fortunately, is true for most practical fuels. Additional theoretical work is in progress for refining the effects of the mass stoichiometric ratio on the fire dynamics. It would be desirable, of course, to perform experiments with fuels having low mass stoichiometric ratios (air to fuel); this condition can be obtained, for example, by inert gas dilution of a fuel or by oxygen enrichment of ambient air.

APPENDIX

PROCEDURE FOR CALCULATING AIR ENTRAINMENT INTO TURBULENT POOL AND JET FIRES*

INTRODUCTION

The air entrainment into a fire is a necessary ingredient in room fire models for calculating, for example, the growth of the ceiling smoke layer and vent sizes.

Correlations for air entrainment into (unrestricted) turbulent pool fires of varying source diameter have been presented in the main text and in Figure 2-3.5. These correlations were developed based on (1) numerous experimental data (Beyler,[5] Cetegen et al,[2] Becker and Yamataki,[3] Toner,[27] Delichatsios and Orloff[4]), and (2) similarity arguments. An important experimental input for the similarity analysis was the fact that the air entrainment below the vertical flame extent is independent of the heat release rate, $\dot{Q}$, (References 5, 2, 3, 4, 27, see also Figure A-2-3.1) and depends only on the pool diameter, D, and the distance from the pool surface, Z. These entrainment results and correlations are reproduced in Figure 2-3.5, which is applicable for $Z < Z_f$ = flame height.

Far above the vertical flame extent ($Z \gg Z_f$) the air entrainment into the buoyant plume is given by the following relation[2,3,27]

$$\dot{m} = C\rho_\infty \left(\frac{Q}{\rho_\infty C_p T_\infty g^{1/2}} \right)^{1/3} g^{1/2} Z^{5/3} \quad \text{for } Z \gg Z_f \quad \text{(A1)}$$

*Extracted from: M.A. Delichatsios, *J. of Fire Prot. Engr.*, 2(3), 1990, pp. 93–98.

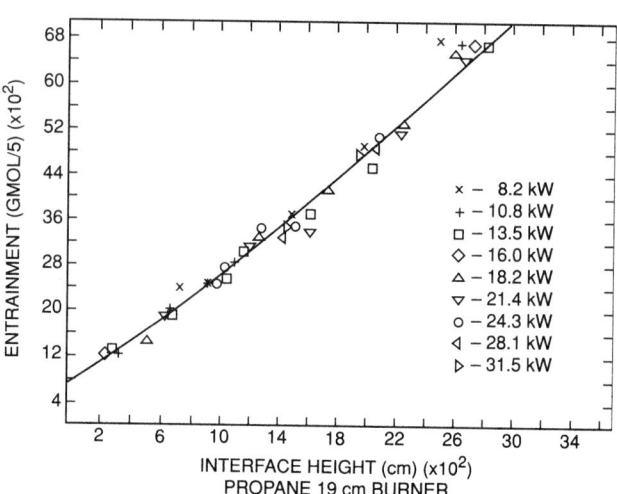

Fig. A-2-3.1. *Air entrainment into a 19 cm propane burner pool fire below the flame tip. These results illustrate that air entrainment is independent of heat release rate, Q (taken from Reference 5).*

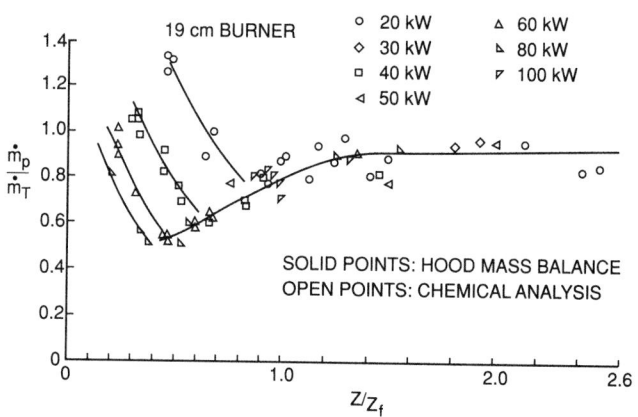

Fig. A-2-3.2. *Air entrainment into the buoyant plume above the flame tip (taken from Reference 2). The reference value, $\dot{m}_T$, in the ordinate is given by Equation A1.*

wherein $\dot{Q}_{conv}$ is the convective heat release from the fire, and the denominator contains ambient air properties and the gravitational acceleration.

Experimental results[2,3,27,28] show that the dimensionless coefficient C in Equation A1 is independent of density even for strong density variations, an observation that is also consistent with air entrainment measurements in momentum jets having strong density variations (Ricou-Spalding[29])

$$C = 0.21 \quad \text{(A2)}$$

The experimental results of Cetegen et al[2] that provide Equation A1, are shown in Figure A-2-3.2.

We present in this work an example to illustrate how to use the correlations in Figure 2-3.5 to predict air entrainment into pool fires below the vertical flame extent. In addition, we also demonstrate how to use the results in Figure 2-3.5 in conjunction with the air entrainment Equation A1 in the buoyant plume above the flames in order to determine a virtual fire origin so that one can also use Equation A1 for intermediate heights $Z \gtrsim Z_f$.

COMMENTS ON CORRELATIONS AND LIMITATIONS

The examples we present here are applicable only for turbulent buoyant jet flames and pool fires that do not represent a mass fire situation. More precisely the present work is limited to pool fires wherein the visible flame height extends beyond the neck-in area of the flame (see Figure 2-3.3) in contrast to a large-area pool fire that may produce relatively short flames (see Figure 2-3.4).

For the present case of Figure 2-3.3, one can assume that there are two regimes: (1) below the vertical extent of the flame for which the entrainment relationships of Figure 2-3.5 are applicable, and (2) above and far above the flame extent wherein entrainment relationships are applicable for turbulent point source buoyant releases, such as Equation A1. Finally, we propose that these entrainment relationships have an overlap at $Z \cong Z_f$.

Next, we illustrate an algorithm based on the results of the main text (see also Figure 2-3.5) for calculating entrainment rates in pool fires represented by Figure 2-3.3.

An Algorithm for Calculating Entrainment Rates into Pool Fires

Computation steps: Define the characteristics of a pool fire.

Diameter: D (m)

Fuel Type
ΔH_c = heat of combustion per unit fuel mass (kJ/kg)
S = stoichiometric ratio air to fuel
χ_A = efficiency of combustion
T_∞ = ambient temperature in K (e.g., 300 K)
χ_R = radiant fraction (based on the theoretical heat release rate)
C_p = specific heat at ambient conditions (1 kJ/kg)
ρ_∞ = ambient density (1.2 kg/m^3)
$\dot{Q}_{th}$ = theoretical heat release rate
$$\dot{Q}_{th} = \dot{m}_f \Delta H_c,$$
Chemical heat release rate
$$\dot{Q}_{chem} = \chi_A \dot{Q}_{th},$$
Convective heat release rate,
$$\dot{Q}_{conv} = \dot{Q}_{th}(\chi_A - \chi_R)$$
Mass fuel flow rate (kg/s)

1. Calculate the fire Froude number

$$\text{Fr} = Q_D^* \frac{1}{\left[\dfrac{\Delta H_c \chi_A}{(S\chi_A + 1)C_p T_\infty}\right]^{3/2} \sqrt{(1 - \chi_R/\chi_A)}}$$

where:

$$Q_D^* = \frac{\dot{Q}_{th}\chi_A}{\rho_\infty C_p T_\infty D^2 \sqrt{gD}}$$

2. Find the flame height based on the following equations.

$$\frac{Z_f}{D} = 1.35 \times 10^4 \, \text{Fr}^2 \quad \text{for Fr} \le 8.6 \times 10^{-3}$$

$$\frac{Z_f}{D} = 22.54 \, \text{Fr}^{2/3} \quad 8.6 \times 10^{-3} \le \text{Fr} \le 10 \times 10^{-2}$$

$$\frac{Z_f}{D} = 12.52 \, \text{Fr}^{2/5} \quad \text{for Fr} > 10 \times 10^{-2}$$

3. Find entrainment rate from the top of the pool fire to a height Z if $Z \le Z_f$.

a) $\quad \dfrac{\dot{m}_{ent} \, \text{Fr}}{(\chi_A S + 1)\dot{m}_f} = 0.086(Z/D)^{1/2} \quad Z/D < 1.0$

b) $\quad \dfrac{\dot{m}_{ent} \, \text{Fr}}{(\chi_A S + 1)\dot{m}_f} = 0.093(Z/D)^{3/2} \quad 1.0 < Z/D < 5.0$

c) $\quad \dfrac{\dot{m}_{ent} \, \text{Fr}}{(\chi_A S + 1)\dot{m}_f} = 0.018(Z/D)^{5/2} \quad Z/D > 5.0$

4. Find the entrainment rate for $Z \ge Z_f$.

$$\frac{\dot{m}_{ent} \, \text{Fr}}{(\chi_A S + 1)\dot{m}_f} = 0.21[(Z + Z_v)/D]^{5/3}$$

where the virtual source is defined by

$$\frac{Z_v}{D} = -\frac{Z_f}{D} + 10.21 \, \text{Fr}^{2/5}$$

and the flame height has been determined earlier in Step 2, above.

A note on the derivation of Step 4 in the algorithm calculation: The derivation of the entrainment relationship in Step 4 has been obtained in the following way. First notice that for large distances from the source, the entrainment for the buoyant plume becomes[2]

$$\begin{array}{c}(Z \gg D)\\[4pt] \dfrac{\dot{m}_{ent} \, \text{Fr}}{(\chi_A S + 1)\dot{m}_f} = 0.21 \, \text{Fr}^{1/3}(Z/D)^{5/3}\end{array} \qquad \text{(A3)}$$

One can show that Equation A3 is exactly the same as Equation A1 by noticing that

$$\dot{Q}_{conv} = \dot{Q}_{th}(\chi_A - \chi_R).$$

We assume that this equation can be extended down to the flame tip by using a virtual origin correction, Z_v, for the distance Z from the source

$$\frac{\dot{m}_{ent} \, \text{Fr}}{(\chi_A S + 1)\dot{m}_f} = 0.21 \, \text{Fr}^{1/3}\left(\frac{Z + Z_v}{D}\right)^{5/3} \qquad \text{(A4)}$$

By requiring that at the flame tip the entrainment rates are the same as calculated in Step 3, $(Z \le Z_{fl})$ and by using Equation A3 $(Z \ge Z_{fl})$, one can obtain an equation for the location of the virtual origin. Furthermore, one would note that at the flame tip

$$\frac{\dot{m}_{ent}}{\dot{m}_f(\chi_A S + 1)} = 10$$

so that Equation A4 gives when applied at $Z = Z_{fl}$

$$\frac{Z_v}{D} = -\frac{Z_f}{D} + 10.2 \, \text{Fr}^{2/5} \qquad \text{(A5)}$$

A note on the calculation of fire Froude number in the algorithm calculation: In case the properties of the fuel (i.e., ΔH_c, S, χ_R, χ_A) are not known one has to use default values, as for example, assuming that the fuel is propane.

SAMPLE CALCULATIONS FOR POOL FIRES

We follow the procedure outlined under Comments on Correlations and Limitations.

First Case

Kerosene pool fire
D = 1.33 m
$\dot{Q}_{th}$ = 4 MW
S = 15
ΔH_c = 44 kJ/g (see main text)
χ_R = 0.4 (assumed)
χ_A = 1 (assumed)
$\dot{m}_f$ = $\dot{Q}_{th}/\Delta H_c$ = 100 g/s

1. Calculate fire Froude number.

$$\text{Fr} = Q_D^* \frac{1}{\left(\dfrac{\Delta H_c}{(S+1)C_p T_\infty}\right)^{3/2}(1-\chi_R)^{1/2}}$$

$$Q_D^* = \frac{4000}{1.2 \times 1 \times 300(1.33)^2\sqrt{9.8 \times 1.33}} = 1.74$$

$$\left[\frac{\Delta H_c}{(S+1)C_p T_\infty}\right]^{3/2}(1-\chi_R)^{1/2} = 20.11$$

$$\text{Fr} = \frac{Q_D^*}{20.11} = 0.0864$$

2. Find flame height.

Then $$\frac{Z_f}{D} = (22.54)[0.0864]^{2/3} = 4.4$$

3. The required location for entrainment is at $Z = 15$ m $> Z_f$ = 5.8 m. One must go to Step 4.
4. First find virtual origin.

$$\frac{Z_v}{D} = -\frac{Z_f}{D} + 10.21(0.0864)^{2/5}$$

$$= -4.4 + 3.83 = -0.57$$

Hence

$$\frac{\dot{m}_{ent}}{(S+1)\dot{m}_f}\,\text{Fr} = 0.21\,\text{Fr}^{1/3}\left(\frac{15}{1.33} - 0.570\right)^{5/3}$$

$$\frac{\dot{m}_{ent}}{(S+1)\dot{m}_f} = \frac{10.82}{\text{Fr}^{2/3}} = \frac{10.82}{(0.0864)^{2/3}} = 55.27$$

Hence

$$\dot{m}_{ent} = 16 \cdot (100)(55.27) = 88.5 \text{ kg/s}$$

Second Case

Fuel and its properties as before
D = 4 m
$\dot{Q}_{th}$ = 36 MW
$\dot{m}_f$ = $\dot{Q}_{th}/\Delta H_c$ = 900 g/s

1. Calculate fire Froude number.

$$\text{Fr} = Q_D^* \frac{1}{\left(\dfrac{\Delta H_c}{(S+1)C_p T_\infty}\right)^{3/2}(1-\chi_R)^{1/2}}$$

$$\text{Fr} = \frac{0.99}{20.11} = .0496$$

2. Find flame height.

$$\frac{Z_f}{D} = (22.54)(0.0496)^{0.666} = 3.04$$

3. The required location for entrainment is at $Z = 15$ m $> Z_f$ = 12 m. One must go to Step 4.
4. First find virtual origin

$$\frac{Z_v}{D} = -\frac{Z_f}{D} + 10.21(0.0496)^{0.4} = 0.3056$$

Then

$$\frac{\dot{m}_{ent}}{(S+1)\dot{m}_f}\,\text{Fr} = 0.208\,\text{Fr}^{1/3}\left(\frac{15}{4} + 0.03\right)^{5/3}$$

$$\frac{\dot{m}_{ent}}{(S+1)\dot{m}_f} = \frac{1.872}{\text{Fr}^{2/3}} = 13.83$$

Hence

$$\dot{m}_{ent} = 199.1 \text{ kg/s.}$$

CONCLUSIONS

In this appendix, an algorithm and sample examples have been presented for calculating entrainment rates into (round) pool fires based on correlations that appear in the main text. Well-established experimental evidence for supporting those correlations has been assembled here by using original sources. A virtual source origin for the plume above the flame tip has been determined by matching mass entrainment measurements below and above the flame tip (only for cases represented by Figure 2-3.3). No attempt was made here to compare this virtual source expression with other virtual source expressions in the literature. (See, for example, Cox and Chitty[22] wherein various virtual source expressions are summarized.) Finally, there is not enough quantitative evidence to establish how to extend the present correlations below the flame tip (see Figure 2-3.5) to noncircular fires. (See Cox and Chitty[22] for some limited data.)

NOMENCLATURE

C	dimensionless constant for Equation A1, $C = 0.21$
C_p	specific heat of air
D	pool diameter
Fr	fire Froude number
g	gravitational acceleration
$\dot{m}_{ent}, \dot{m}_p$	entrainment rate up to a height Z
$\dot{m}_f$	mass flow rate
$\dot{Q}_{chem}$	chemical heat release rate = $\chi_A \dot{Q}_{th}$
$\dot{Q}_{conv}$	convective heat release rate $\equiv (\chi_A - \chi_R)\dot{Q}_{th}$
$\dot{Q}_{th}$	theoretical heat release rate
S	stoichiometric oxidant to fuel mass ratio
T_∞	ambient air absolute temperature
Z	distance from pool surface
Z_f	"visual" flame height
Z_v	virtual source origin based on entrainment rates
ΔH_c	theoretical heat of combustion per unit fuel mass
ρ_∞	ambient air density
χ_A	efficiency of combustion
χ_R	radiant fraction

REFERENCES CITED

1. B.J. McCaffrey, *Comb. and Flame*, 52, 149 (1983); see also B. McCaffrey and G. Cox, *NBSIR 82-2473*, National Bureau of Standards, Washington (1983).
2. B.M. Cetegen, E.E. Zukoski, and T. Kubota, *Comb. Sci. and Tech.*, 39, 305 (1984); see also E. Zukoski, T. Kubota, and B. Cetegen, *NBS CFRR Grant GB-9014 Report*, California Institute of Technology, Pasadena (1984).
3. H.A. Becker and S. Yamotaki, *Comb. and Flame*, 33, 123 (1978).
4. M.A. Delichatsios and L. Orloff, *Entrainment Measurements in Turbulent Buoyant Jet Flames and Implications for Modeling*, *Twentieth Symp. (Int.) on Comb.*, Pittsburgh (1984).
5. C.L. Beyler, *Development and Burning of a Layer of Products of Incomplete Combustion Generated by a Buoyant Diffusion Flame*, Thesis, Cambridge (1983).
6. S.M. Jeng, L.D. Chen, and G.M. Faeth, *The Structure of Buoyant Methane and Propane Diffusion Flames, Nineteenth Symp. (Int.) on Comb.*, Pittsburgh (1982).
7. F. Tamanini, *Comb. and Flame*, 30, 85 (1977).
8. F. Steward, *Comb. Sci. and Tech.*, 2, 203 (1970).
9. G. Heskestad, *Eighteenth Symp. (Int.) on Comb.*, Pittsburgh (1981).
10. F. Tamanini, *Comb. and Flame*, 51, 231 (1983).
11. G. Santo and M.A. Delichatsios, *F. Safety J.*, 7, 159 (1984).
12. M.A. Delichatsios, *Comb. Sci. and Tech.*, 24, 191 (1981).
13. M.A. Delichatsios, *On the Modeling of Strong (Variable Density) Turbulent Buoyant Plumes and Ceiling Jets*, Meeting, The Combustion Institute (1985).
14. P. Demotakis, *J. Fluid Mech.*, 148, 349 (1984).
15. M.A. Delichatsios, *A PDF Method for Predicting Major Species Concentrations in Turbulent Fires, Eastern Section Meeting of Comb. Inst.*, The Combustion Institute (1985).
16. N. Peters, *Prog. in Energy and Comb. Sci.*, 1985.
17. R.W. Bilger, *Comb. and Flame*, 30, 277 (1977).
18. J.A. Miller, R.J. Kee, M.D. Smooke, and J.F. Grcar, *The Computation of the Structure and Extinction Limit of a Methane-Air Stagnation Point Diffusion Flame*, Meeting, The Combustion Institute (1984).
19. A.T. Tewarson, *Technical Report RC80-7-79*, Factory Mutual Research Corp., Norwood (1980).
20. G.H. Markstein, *Radiative Properties of Plastic Fires, Seventeenth Symp. (Int.) on Comb.*, The Combustion Institute (1979).
21. L. Orloff, J. de Ris, and M.A. Delichatsios, *Chemical Modeling of Gaseous Species in Turbulent Flames, Eastern Section Meeting of Comb. Inst.*, The Combustion Institute (1979).
22. G. Cox and R. Chitty, *Comb. and Flame*, 60, p. 219 (1985).
23. P.H. Thomas, C.T. Webster, and M.M. Raftery, *Comb. and Flame*, 5, 359 (1961).
24. G. Heskestad, *F. Safety J.*, 5, 109 (1983).
25. M. Law, *F. Safety J.*, 10, 197 (1986).
26. J. de Ris, *GRI-86/0111*, Gas Research Institute, (1986); see also M.A. Delichatsios, *Transition from Momentum to Buoyancy Controlled Turbulent Jet Diffusion Flames*, Meeting, The Combustion Institute (1985).
27. J.S. Toner, Ph.D. Thesis, California Institute of Technology (1987).
28. M.A. Delichatsios, *Comb. Sci. and Tech.*, 60, p. 253 (1988).
29. F.P. Ricou and D.B. Spaulding, *J. Fluid Mech.*, 11, p. 21 (1961).

CEILING JET FLOWS

David D. Evans

INTRODUCTION

Much of the hardware associated with detection and suppression of fires in commercial, manufacturing, storage, and recently constructed residential buildings is located near the ceiling surfaces. In the event of a fire, hot gases in the fire plume rise directly above the burning fuel and impinge on the ceiling. The ceiling surface causes the flow to turn and move horizontally under the ceiling to other areas of the building remote from the fire position. The response of smoke detectors, heat detectors, and sprinklers installed below the ceiling so as to be submerged in this hot flow of combustion products from a fire provides the basis for the building fire protection.

Studies quantifying the flow of hot gases under a ceiling resulting from the impingement of a fire plume have been conducted since the 1950s. Early studies at the Fire Research Station in Great Britain,[1,2] and more recently at Factory Mutual Research Corporation,[3-6] the National Institute of Standards and Technology (NIST),[7,8] and at other research laboratories,[9,10] have sought to quantify the gas temperatures and velocities in the hottest portion of the flow produced by steady fires beneath smooth, unconfined horizontal ceilings.

"Ceiling jet" refers to the relatively rapid gas flow in a shallow layer beneath the ceiling surface which is driven by the buoyancy of the hot combustion products. Figure 2-4.1 shows an idealization of the ceiling jet flow beneath an unconfined ceiling. In actual fires within buildings, the simple conditions pictured—a hot rapidly moving gas layer between the ceiling surface and the tranquil ambient air at room temperature—exist only at the beginning of a fire when the quantity of hot gases produced is not sufficient to accumulate into a stagnant warm gas layer in the upper portion of the compartment. The accumulation of this warm gas layer can be retarded by venting the ceiling jet flow through openings in the ceiling surface or edges. As shown in Figure 2-4.1, the ceiling jet flow emerges from the region of plume impingement on the ceiling, flowing away from the fire. As it does, the layer grows thicker by entraining room air at the lower boundary. This entrained air cools the gases in the jet and reduces its velocity. As the hot gases move out across the ceiling, the portion adjacent to the ceiling surface is cooled by heat transfer.

Quantification of the heat transfer from fire plumes impinging on ceiling surfaces is an area of recent research activity.[7-11] As a rule of thumb,[3] the thickness of the ceiling jet flow is 5 to 12 percent of the ceiling-to-fire-source height. Within this ceiling jet flow, the maximum temperature and velocity occurs within 1 percent of the distance from the ceiling to the fire source.[3] Detailed measurement and analysis of the temperature and velocity distributions in the ceiling jet flow for the region $r/H < 2$ has been performed by Motevalli and Marks.[12]

Much of the work that is collected below deals with means to predict the temperature and velocities in the ceiling jet flow both above and remote from the fire source. In most cases, the reported information deals only with predictions of the maximum temperature and velocity in the flow at positions normally one percent of the fire-source distance below the ceiling. Often fire detectors or sprinklers are placed at ceiling standoff distances which are outside of this region and therefore will experience cooler temperatures and lower velocities than predicted. In facilities with very high ceilings, the detectors could be closer to the ceiling than one percent of the ceiling-to-fire-source separation and will fall in the ceiling jet thermal and viscous boundary layers. In low-ceiling facilities, it is possible for sprinklers or detectors to be placed outside of the ceiling jet flow entirely if the standoff is greater than 12 percent of the ceiling-to-fire-source height. In this case response time could be drastically increased.

STEADY FIRES

A generalized theory to predict gas velocities, gas temperatures, and the depth of steady fire-driven ceiling jet flows has been developed by Alpert.[4] This work involved the use of several idealizations in the construction of the theoretical model, but results are likely to provide reasonable estimates over radial distances of one or two ceiling heights from the point of fire plume impingement on the ceiling.

Dr. David D. Evans is the acting-chief of the Fire Safety Engineering Division at the National Institute of Standards and Technology, Building and Fire Research Laboratory. He is engaged in research to support the development of performance-based fire standards and means to predict and mitigate the impact of large fires.

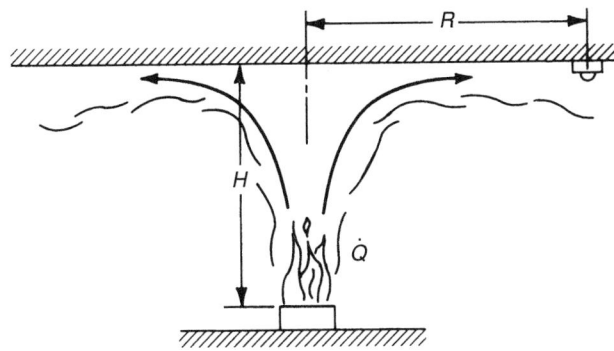

Fig. 2-4.1. Ceiling jet flow beneath an unconfined ceiling.

Alpert[3] has also developed easy-to-use correlations to quantify the maximum gas temperature and velocity at a given position in a ceiling jet flow produced by a steady fire. These correlations are widely used in hazard analysis calculations. They have been employed by Evans and Stroup[13] in the development of a generalized program for prediction of heat detector response for the case of the detector totally submerged in the ceiling jet flow. The correlations are based on measurements collected during test burns of fuel arrays of wood and plastic pallets, cardboard boxes, plastic materials in cardboard boxes, and liquid fuels with energy release rates ranging from 668 kW to 98 MW under ceiling heights from 4.6 to 15.5 m. The correlations developed by Alpert[3] for determining maximum ceiling jet temperatures and velocities in S.I. units are

$$T - T_\infty = \frac{16.9 \dot{Q}^{2/3}}{H^{5/3}} \qquad \text{for } r/H \leq 0.18 \qquad (1)$$

$$T - T_\infty = \frac{5.38(\dot{Q}/r)^{2/3}}{H} \qquad \text{for } r/H > 0.18 \qquad (2)$$

$$U = 0.96\left(\frac{\dot{Q}}{H}\right)^{1/3} \qquad \text{for } r/H \leq 0.15 \qquad (3)$$

$$U = \frac{0.195\dot{Q}^{1/3}H^{1/2}}{r^{5/6}} \qquad \text{for } r/H > 0.15 \qquad (4)$$

where temperature, T, is in °C; velocity, U, is in m/s; and total energy release rate, $\dot{Q}$, is in kW; and ceiling height and radial position (r and H) are in m.

Data from these tests were correlated using the total energy release rate of the fire. Even though it is the convective fraction of the total energy release rate that is directly related to the buoyancy of the fire, most available data is correlated using the total energy release rate. For common materials, such as those used by Alpert, the convective energy release rate, $\dot{Q}_c$, is considered to be proportional to the total energy release rate, $\dot{Q}$.

The correlations for both temperatures and velocities (Equations 1 through 4) are broken into two parts. One part applies for the ceiling jet in the area of the impingement point where the upward flow of gas in the plume turns to flow out beneath the ceiling horizontally. These correlations (Equations 1 and 3) are independent of radius and are actually axial plume flow temperatures and velocities calculated at the ceiling height above the fire source. The other corre-

lations apply outside of this turning region as the flow moves away from the impingement area. Certain constraints should be understood when applying these correlations in the analysis of fire flows. The correlations apply only during times after fire ignition when the ceiling flow may be considered unconfined; i.e., no accumulated warm upper layer is present. Walls close to the fire affect the temperatures and velocity in the ceiling jet. The correlations were developed from test data to apply in cases where the fire source is at least a distance 1.8 times the ceiling height from the enclosure walls. Ideally, for the special cases where burning fuel is located against wall surfaces or two wall surfaces forming a 90-degree corner, the correlations may be adjusted based on method of reflection making use of symmetry to account for the effects of the walls blocking entrainment of air into the fire plume. For the case of a fire adjacent to a flat wall, $2\dot{Q}$ is substituted for $\dot{Q}$ in the correlations. For a fire in a 90-degree corner, $4\dot{Q}$ is substituted for $\dot{Q}$ in the correlations.[3]

Experiments have shown that, unless great care is taken to ensure that the fuel perimeter is in contact with the wall surfaces, the method of reflection used to estimate the effects of the walls on ceiling jet temperature will be inaccurate. For example, Zukoski *et al*[14] found that a circular burner placed against a wall so that only one point on the perimeter contacted the wall, behaved almost identically to a fire far from the wall with plume entrainment only decreasing by 3 percent. When using Equations 1 through 4, this fire would be represented by replacing $\dot{Q}$ with $1.05\dot{Q}$ and not $2\dot{Q}$ as would be predicted by the method of reflections. The value of $2\dot{Q}$ would be appropriate for a semicircular burner with the entire flat side pushed against the wall surface.

Consider the following calculations, which demonstrate typical uses of the correlations, using Equations 1 through 4.

(a) The maximum temperature rise under a ceiling 10 m directly above a 1.0 MW energy release rate fire is calculated using Equation 1 as

$$T - T_\infty = \frac{16.9(1000)^{2/3}}{10^{5/3}}$$

$$= \frac{16.9(100)}{46.42}$$

$$\Delta T = 36.4°C$$

(b) The minimum energy release rate of a fire against noncombustible walls in the corner of a building 12 m below the ceiling needed to raise the temperature of the gas below the ceiling 50°C at a distance 5 m from the corner is calculated using Equation 2 and the symmetry substitution of $4\dot{Q}$ for $\dot{Q}$ to account for the effects of the corner as

$$T - T_\infty = \frac{5.38(4\dot{Q}/r)^{2/3}}{H}$$

$$50 = \frac{5.38(4\dot{Q}/5)^{2/3}}{12}$$

$$\dot{Q} = \frac{5}{4}\left[\frac{50(12)}{5.38}\right]^{3/2}$$

$$\dot{Q} = 1472 \text{ kW}$$

$$\dot{Q} = 1.472 \text{ MW}$$

(c) The maximum velocity at this position is calculated from Equation 4, modified to account for the effects of the corner as

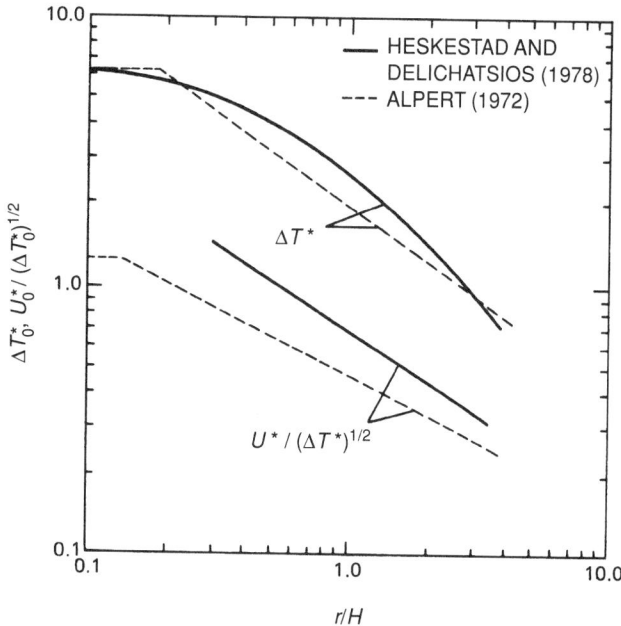

Fig. 2-4.2. *Dimensionless correlations for maximum ceiling jet temperatures and velocities produced by steady fires. Solid line: Heskestad and Delichatsios[15]; dotted line: Alpert.[3]*

$$U = \frac{0.197(4\dot{Q})^{1/3}H^{1/2}}{r^{5/6}}$$

$$= \frac{0.197(5888)^{1/3}(12)^{1/2}}{5^{5/6}}$$

$$U = 3.2 \text{ m/s}$$

Heskestad and Delichatsios[15] have developed correlations for maximum ceiling jet temperature rise and velocities that are based on testing completed subsequent to Alpert's analysis.[3] Their correlations are cast in generalized variables (indicated by the superscript asterisk) for energy release rate, temperature rise, and velocity as

$$\dot{Q}_0^* = \dot{Q}/(\rho_\infty C_p T_\infty g^{1/2} H^{5/2}) \quad (5)$$

$$\Delta T_0^* = \Delta T/T_\infty/(\dot{Q}_0^*)^{2/3} = [0.188 + 0.313r/H]^{-4/3} \quad (6)$$

$$U_0^* = 0.68(\Delta T_0^*)^{1/2}(r/H)^{-0.63} \quad \text{for } r/H \geq 0.3 \quad (7)$$

For the case of steady fires under unconfined ceilings, Figure 2-4.2 shows the plot of the Heskestad and Delichatsios correlation for temperature rise and velocity as solid line curves. The correlations developed by Alpert are plotted as broken curves, using the same dimensionless parameters with assumed ambient temperature of 293 K (20°C), normal atmospheric pressure, and convective energy release rate equal to the total energy release rate, $\dot{Q}_c = \dot{Q}$. Generally, the results of Heskestad and Delichatsios predict larger temperature rises and gas velocities than Alpert's results.

Other methods used to calculate estimates of ceiling jet velocity distributions and maximum possible (adiabatic) ceiling jet temperatures are reported by Cooper and Woodhouse.[8] A recent review of ceiling jet correlations for temperature rise and velocity has been assembled by Beyler.[16]

TIME DEPENDENT FIRES

For time dependent fires, all estimates from the previous section may still be used, but with the constant energy release rate, $\dot{Q}$, replaced by an appropriate time dependent $\dot{Q}(t)$. In making this replacement, a "quasisteady" flow has been assumed. This assumption implies that when a change in energy release rate occurs at the fire source, its full effects are felt everywhere in the flow field immediately. In a relatively small room-size enclosure, under conditions where the fire is growing slowly, this assumption is reasonable. In large industrial facilities, where travel times of the fire gases from the burning fuel to a detector or sprinkler submerged in the ceiling jet flow may be 10 s or longer, this may not be an appropriate assumption, depending largely on the rate of fire growth and desired accuracy of the gas temperature and velocity predictions.

Testing has shown that the energy release rate during the growth phase of many fires can often be characterized by simple time dependent polynominal or exponential functions. The most extensive research and analysis have been performed with energy release rates that vary with the second power of time.

t^2 Fire Growth

The growth phase of many fires can be characterized as increasing proportionally with the square of time, measured from an ignition reference time, t_i, as

$$\dot{Q} = \propto (t - t_i)^2 \quad (8)$$

Figure 2-4.3 shows one case where the energy release rate for a burning foam sofa during the growth phase of the fire more than 80 seconds (t_i) after ignition[17] can be represented by the equation

$$\dot{Q} = 0.1736(t - 80)^2 \quad (9)$$

In an extensive series of tests conducted by Factory Mutual Research Corporation,[15,18] measurements were made of maximum ceiling jet temperatures and velocities during the growth of fires in which various wood cribs were burned. The energy release rate, $\dot{Q}$, from these fires was calculated as the product of measured mass loss rate and oxygen bomb calorimeter values for the heat of combustion

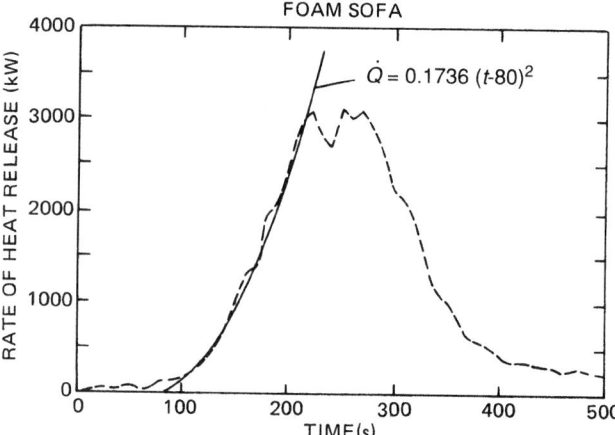

Fig. 2-4.3. *Energy release rate history for a burning foam sofa.[17]*

of the wood, which was found to be 20.9 MJ/kg. The resulting dimensionless correlations for maximum ceiling jet temperatures and velocities are

$$\Delta T_2^* = \begin{cases} 0, & t^* \leq t_f^* \quad (10a) \\ \left(\dfrac{t_2^* - t_f^*}{0.188 + 0.313 \, r/H}\right)^{4/3} & t^* > t_f^* \quad (10b) \end{cases}$$

$$U_2^* / \sqrt{\Delta T_2^*} = 0.59(r/H)^{-0.63} \qquad (11)$$

where

$$t_2^* = (t - t_i)/(A^{-1/5}\alpha^{-1/5}H^{4/5}) \qquad (12)$$

$$U_2^* = U/(A^{1/5}\alpha^{1/5}H^{1/5}) \qquad (13)$$

$$\Delta T_2^* = (T - T_\infty)/[A^{2/5}(T_\infty/g)\alpha^{2/5}H^{-3/5}] \qquad (14)$$

$$A = g/(c_p T_\infty \rho_\infty) \qquad (15)$$

$$\alpha = \dot{Q}/(t - t_i)^2 \qquad (16)$$

$$t_f^* = 0.954(1 + r/H) \qquad (17)$$

where dimensionless variables are indicated with the superscript asterisk. Notice that in Equation 10b the dimensionless time, t_2^*, has been reduced by the time t_f^*. This reduction accounts for the gas travel time between the fire source and the location of interest along the ceiling at the specified r/H.

The dimensional temperature rise ΔT_2 (Equations 10b and 14) for the t^2 fire growth is related to the temperature rise from the steady fire analysis, ΔT_0, (Equation 6) by the simple relationship

$$\frac{\Delta T_2}{\Delta T_0} = \left(\frac{t - t_i - t_f}{t - t_i}\right)^{4/3} \qquad (18)$$

This relationship may be used to evaluate the extent to which a quasisteady analysis of a growing t^2 fire is appropriate.

These correlations of ceiling jet temperatures and velocities are the basis for the calculated values of fire detector spacing found in NFPA 72, *National Fire Alarm Code*, Appendix B, Engineering Guide for Automatic Fire Detector Spacing.[19] In NFPA 72, three or four selected fire energy release rates assumed to increase proportionally with the square of time were used as the basis for the evaluation. These fire energy release rate histories were chosen to be representative of actual fire situations involving different commodities and geometric storage arrangements. These idealized fire energy release rates are

Slow,	$\dot{Q} = 0.00293t^2$	(19)
Medium,	$\dot{Q} = 0.01172t^2$	(20)
Fast,	$\dot{Q} = 0.0469t^2$	(21)
Ultrafast,	$\dot{Q} = 0.1876t^2$	(22)

where $\dot{Q}$ is in kW and t is in s. Consider the following calculation which demonstrates a use of the correlation (Equations 10b and 11) for calculation of ceiling jet maximum temperature and velocity produced by a t^2 fire growth:

A foam sofa, of the type analyzed in Figure 2-4.3, is burning in a showroom 5 m below a suspended ceiling. The showroom temperature remote from the fire remains at 20°C at floor level as the fire begins to grow. Determine the gas temperature and velocity at the position of a ceiling-mounted fire detector submerged in the ceiling jet flow 4 m away from the fire axis when the fire energy release rate first reaches 2.5 MW.

Figure 2-4.3 shows that the energy release rate from the sofa first reaches 2.5 MW (2500 kW) at about 200 s after ignition. Using the analytic formula for the time dependent energy release rate, Equation 9, the time from the virtual ignition of the sofa at 80 s to reach 2500 kW is

$$2500 = 0.1736 \, (t - 80)^2$$
$$(t - 80) = 120 \text{ s}$$

In this problem, the low-level energy release rate up to 80 s after actual ignition of the sofa is ignored. Thus, the sofa fire can be treated as having started at $t = 80$ seconds and grown to 2.5 MW in the following 120 seconds. Equations 12 through 17 are used to evaluate parameters of the problem, using the dimensionless correlations for ceiling jet temperature and velocity.

For the sofa fire in the showroom example, $T = 293$ K, $\rho = 1.204$ kg/m^3, $c_p = 1$ kJ/kg K, $g = 9.8$ m/s^2, $\alpha = 0.1736$ kW/s^2, $A = 0.0278$ m^4/kJ s^2, $r = 4$ m, $H = 5$ m, $t_f^* = 1.72$, $t - t_i = 120$ s, and $t_2^* = 11.40$. For the conditions of interest $t_2^* > t_f^*$, so the correlation (Equation 10b) is used to evaluate the dimensionless ceiling jet temperature

$$\Delta T_2^* = \left[\frac{11.40 - 1.72}{0.188 + 0.313(4/5)}\right]^{4/3}$$
$$\Delta T_2^* = 61.9$$

Equation 11 is used to calculate the dimensionless ceiling jet velocity

$$U_2^* = 0.59(4/5)^{-0.63}(61.9)^{1/2}$$
$$= 5.34$$

The dimensional temperature rise and velocity are calculated using Equations 14 and 13, respectively, to yield

$$\Delta T = 83.5 \text{ K}$$
$$T = 83.5 \text{ K} + 293 \text{ K} = 376.5 \text{ K} = 103.5°C$$
$$U = 2.54 \text{ m/s}$$

CONFINED CEILINGS

The corresponding gas temperature calculated with the quasi-steady analysis Equations 6 or 18 instead of the t^2-fire analysis is 124°C.

Previous discussions of ceiling jets in this chapter have all dealt with unconfined radial spread of the gas flow away from a ceiling impingement point. In practice this flow may be interrupted by ceiling beams or walls in a corridor situation creating a long channel that partially confines the flow. In this case, the flow near the impingement point will remain radial, but after spreading to the walls or beams that bound the ceiling, the flow will be altered into a channel flow. Delichatsios[20] has developed correlations for ceiling jet temperatures and velocity which apply to the channel flow between beams and down corridors. In the case of corridors, the correlations apply when the corridor half-width, ℓ_b, is greater than 0.2 times the distance from the fire source to the ceiling, H, or ($\ell_b/H > 0.2$). In the case of beams, the flows must also be contained fully, so that a channel flow results without "spillage" over the beams. In order for this condition to be satisfied, the beam depth, h_b, must be greater than the quantity $(H/10)(\ell_b/H)^{-1/3}$ or

$$h_b/H > 0.1(\ell_b/H)^{-1/3}$$

then

$$\frac{\Delta T}{\Delta T_{imp}} = 0.29\left(\frac{H}{\ell_b}\right)^{1/3}\exp[-0.20(Y/H)(\ell_b/H)^{1/3}]$$

$$\text{for } Y > \ell_b$$

where ΔT_{imp} is the temperature in the gas near the ceiling directly over the fire, and Y is the distance along the channel measured from the plume impingement point.

Generally, for large industrial or commercial storage facilities, the analysis for unconfined ceiling jet flows will be sufficient for most purposes. In smaller rooms, or for very long times after fire ignition in larger industrial facilities, a quiescent warm layer of gas will accumulate in the upper portion of the enclosure. This warm layer can be deep enough to totally submerge the ceiling jet flow. In that case, temperatures in the ceiling jet can be expected to be greater than if the ceiling jet was entraining gas from the cooler room ambient temperature layer. Quantitative methods for the prediction of temperature and velocity in a two-layer room environment in which the ceiling jet is contained totally in a warm upper layer and the fire is burning totally in the lower cool layer have been formulated. Contributions to this area have been made by Evans,[21,22] Cooper,[23] and Zukoski and Kubota.[11]

In these methods, the flow of the ceiling jet within the warm upper layer of the enclosure is imagined to result from a fire totally contained in a uniform ambient environment with temperature equal to that of the warm upper layer. This substitute fire has an energy release rate, $\dot{Q}_2$, and location below the ceiling, H_2, differing from the original fire. Calculation of the substitute quantities $\dot{Q}_2$ and H_2 depends on the energy release rate and location of the original fire as well as the depths and temperatures of the upper and lower layers within the enclosure.

Following the development by Evans,[22] the substitute source energy release rate and distance below the ceiling are calculated from Equations 23 through 26. Originally developed for the purpose of sprinkler and heat detector response time calculations, these equations are applicable during the growth phase of enclosure fires.

$$\dot{Q}_{I,2}^* = [(1 + C_T\dot{Q}_{I,1}^{*2/3})/\xi C_T - 1/C_T]^{3/2} \tag{23}$$

$$Z_{I,2} = \left\{\frac{\xi\dot{Q}_{I,1}^*C_T}{\dot{Q}_{I,2}^{*1/3}[(\xi-1)(\beta^2+1)+\xi C_T\dot{Q}_{I,2}^{*2/3}]}\right\}^{2/5} Z_{I,1} \tag{24}$$

$$\dot{Q}_{c,2} = \dot{Q}_{I,2}^*\rho_{\infty,2}c_{p\infty}T_{\infty,2}g^{1/2}Z_{I,2}^{5/2} \tag{25}$$

$$H_2 = H_1 - Z_{I,1} + Z_{I,2} \tag{26}$$

Further explanation of variables is contained in the nomenclature section.

Cooper[23] has formulated an alternative calculation of substitute source energy release rate and position below the ceiling that provides for generalization to situations in which portions of the time averaged plume flow in the lower layer are at temperatures below the upper layer temperature. In these cases, only part of the plume flow may penetrate the upper layer sufficiently to impact on the ceiling. The remaining portion at low temperature may not penetrate into the hotter upper layer. In the extreme, when the maximum temperature in the lower layer plume flow is less than the upper layer temperature, none of the plume flow will penetrate significantly into the upper layer. This could be the case during the decay phases of an enclosure fire, when the energy release rate is small compared to earlier times in the fire growth and spread. In this calculation of substitute fire source quantities, the first step is to calculate the fraction of the plume mass flow penetrating the upper layer, m_2^*, from Equations 27 and 28.

$$m_2^* = \frac{1.04599\sigma + 0.360391\sigma^2}{1 + 1.37748\sigma + 0.360391\sigma^2} \tag{27}$$

where

$$\sigma = [\xi/(\xi - 1)][(1 + C_T(\dot{Q}_{I,1}^*)^{2/3})/\xi) - 1] \tag{28}$$

Then, analogous to Equations 24, 25, and 26 of the previous method

$$Z_{I,2} = Z_{I,1}\xi^{3/5}(m_2^*)^{2/5}[(1 + \sigma)/\sigma]^{1/3} \tag{29}$$

$$\dot{Q}_{c,2} = \dot{Q}_{c,1}[\sigma m_2^*/(1 + \sigma)] \tag{30}$$

$$H_2 = H_1 - Z_{I,1} + Z_{I,2} \tag{31}$$

After the substitute values of energy release rate and distance to the ceiling are calculated, the warm upper layer gas temperature and density are used in the previous correlations developed for ceiling jet flows in uniform ambient environments to predict ceiling jet temperature and velocity values.

Using a substitute fire source technique and a series of steady fires to represent growing fires in an enclosure, Evans has calculated the effects of warm upper layer depth on temperatures in the ceiling jet in an analysis of detector response in compartments.[22]

To demonstrate the use of the techniques, the previous example in which a sofa was imagined to be burning in a showroom may be expanded. Let all the parameters of the problem remain the same except that at 200 s after ignition $(t - t_i = 120\,\text{s})$, when the fire energy release rate has reached 2.5 MW, a quiescent warm layer of gas at a temperature of 50°C is assumed to have accumulated under the ceiling to a depth of 2 m. For this case, the two-layer analysis is needed to determine the ceiling jet maximum temperature at the same position as calculated previously (4 m radially distant from the plume impingement point on the ceiling).

All of the two-layer calculations presented assume quasisteady conditions. Using Equation 18 and the values of parameters in the single-layer calculation, it can be shown that $\Delta T_2 = 0.85\,\Delta T_0$. So in the uniform ambient case, the quasisteady analysis should be adequate. It will be assumed that this finding will carry over to the two-layer case.

Using Equations 23 through 26 from the work of Evans,[22] values of the energy release rate and position of the substitute fire source which compensates for the two-layer effects on the plume flow can be calculated. The dimensionless energy release rate of the fire source evaluated at the position of the upper and lower layer interface is

$$\dot{Q}_{I,1}^* = \dot{Q}/(\rho_\infty C p_\infty T_\infty g^{1/2} Z_{I,1}^{5/2})$$

For an actual energy release rate of 2500 KW, ambient temperature of 293 K, and distance between the fire source and an interface between the lower and upper layers of 3 m this becomes

$$\dot{Q}_{I,1}^* = 2500/(1.204 * 1 * 293 * 9.8^{1/2} * 3^{5/2})$$

$$= 0.1452$$

Using the ratio of upper-layer temperature to lower-layer temperature $\xi = 323/293 = 1.1024$ and the constant $C_T = 9.115$, the dimensionless energy release for the substitute fire source is

$$\dot{Q}_{I,2}^* = 0.1179.$$

Using the value for the constant $\beta^2 = 0.913$, the position of the substitute fire source relative to the interface is

$$Z_{I,2} = 3.161$$

Using Equations 25 and 26, the dimensional energy release rate and position relative to the ceiling are found to be

$$\dot{Q}_2 = 2313 \text{ kW}$$

$$H_2 = 5.161 \text{ m}$$

The analogous calculations for substitute fire-source energy release rate and position following the analysis of Cooper[23] Equations 27 through 31 are

$$T = 23.60 \qquad m_2^* = 0.962 \qquad Z_{I,2} = 3.176$$

so

$$\dot{Q}_2 = 2308 \text{ kW}$$

$$H_2 = 5.176 \text{ m}$$

These two results are essentially identical for the purpose of ceiling jet flow analysis.

The dimensionless maximum temperature in the ceiling jet flow, 4 m from the impingement point, is calculated from Equation 6, using the ceiling height above the substitute source as

$$\Delta T_0^* = [0.188 + 0.313(4/5.161)]^{-4/3}$$

$$\Delta T_0^* = 3.076$$

Using the corresponding energy release rate for the substitute source and the upper layer ambient temperature, the dimensional temperature elevation at the position in the ceiling jet is

$$\Delta T = \Delta T_0^* T_\infty (\dot{Q}_0^*)^{2/3}$$

$$= 3.076 * 323 * [2313/(1.092 * 1 * 323 * 9.8^{1/2} * 5.161^{5/2})]^{2/3}$$

$$\Delta T = 106 \text{ K}$$

$$T = 106 \text{ K} + 323 \text{ K} = 429 \text{ K} = 156°C$$

This is 52°C above the temperature calculated previously using the quasi-steady analysis and a uniform 20°C ambient.

TRANSIENT CEILING JETS

At the beginning of a fire, the initial buoyant flow from the fire must spread across the ceiling, driven by buoyancy, to penetrate the cooler ambient air ahead of the flow. Research studies designed to quantify the temperatures and velocities of this initial spreading flow have only recently been started.[24] At a minimum, it is useful to become aware of the many fluid mechanical phenomena embodied in a description of the ceiling jet flow in a corridor up to the time when the ceiling jet is totally submerged in a quiescent warm upper layer. Borrowing heavily from a description of this flow provided by Zukoski *et al*,[24] the process is as follows.

A fire starts in a small room with an open door to a long corridor having a small vent near the floor at the opposite end. As the fire starts, smoke and hot gases rise to form a layer near the fire room ceiling. The layer is contained in the small room by the door soffit. [See Figure 2-4.4, part (a).] As the fire continues, hot gas from the room begins to spill out under the soffit into the hallway. The fire grows to a relatively constant energy release rate.

The outflowing gas forms a short, buoyant plume [see Figure 2-4.4, part (b)] that impinges on the hallway ceiling, producing a thin jet that flows away from the fire room in the same manner as the plume within the room flows over the interior ceiling. The gas flow in this jet is supercritical, analogous to the shooting flow of liquids over a weir. The velocity of the gas in this flow is greater than the speed of gravity waves on the interface between the hot gas and the cooler ambient air. The interaction of the leading edge of this flow with the ambient air ahead of it produces a hydraulic jump-like condition, as shown in Figure 2-4.4, part (c). A substantial amount of ambient air is entrained at this jump. Downstream of the jump, the velocity of the gas flow is reduced and mass flow is increased due to the entrainment at the jump. A head is formed at the leading edge of the flow. Mixing between this ceiling-layer flow and the ambient cooler air occurs behind this head.

The flow that is formed travels along the hallway ceiling [see Figures 2-4.4, part (c) and 2-4.4, part (d)] with constant velocity and depth until it impinges on the end wall [see Figure 2-4.4, part (e)]. A group of waves are reflected back toward the jump near the fire room, traveling on the interface. Mixing occurs during the wall impingement process [see Figure 2-4.4, part (f)], but no significant entrainment occurs during the travel of the nonbreaking reflected wave. When these waves reach the jump near the fire room door, the jump is submerged in the warm gas layer, eliminating the entrainment of ambient lower layer air at this position. [See Figure 2-4.4, part (g).] After several wave reflections up and down the corridor along the interface, the wave motion dies out, and a ceiling layer more uniform in depth is produced. This layer slowly grows deeper as the hot gas continues to flow into the hallway from the fire room.

It is clear from this description that quantification of effects in the transient ceiling jet flow is quite complex.

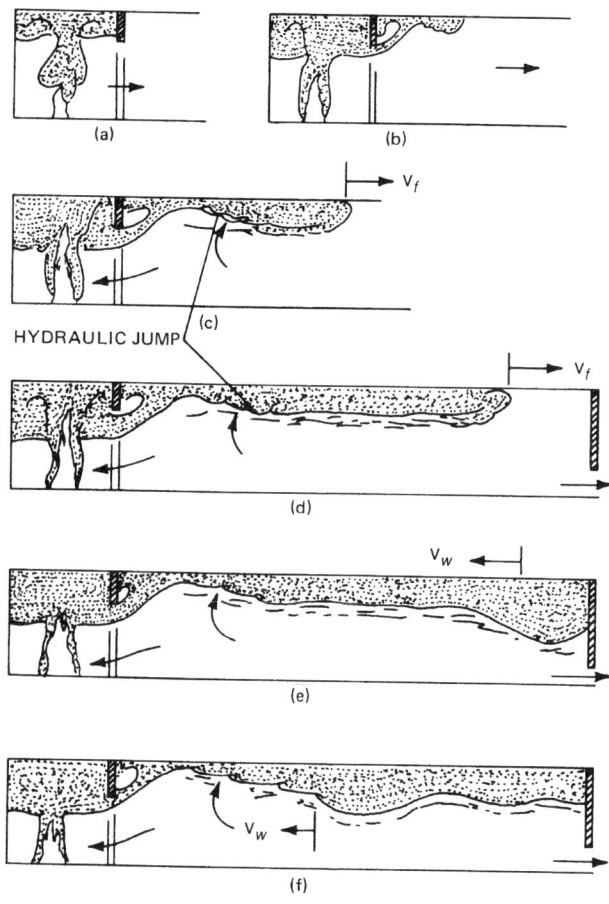

Fig. 2-4.4. Transient ceiling jet flow in a room and corridor. [24]

Analysis and experiments are under way to understand better the major features of a ceiling jet flow in a corridor.[25,26]

SUMMARY

Reliable means are available to predict the temperatures and velocities of gases in fire-driven ceiling jet flows beneath unobstructed ceilings for both steady and t^2 fire growth. These predictive methods apply to quantifying the maximum temperature and maximum velocity at a given position in the ceiling jet flow and apply to situations where the flow can be considered unconfined. These methods are the basis for acceptable design methods exemplified by Appendix B of NFPA 72, *National Fire Alarm Code*.[19]

NOMENCLATURE

A	$g/(c_p T_\infty \rho_\infty)$
c_p	heat capacity at constant pressure
C_T	constant related to plume-flow value 9.115 [14]
g	gravitational acceleration
H	ceiling height above fire source

m_2^* fraction of the fire-plume mass flux penetrating upper layer
$\dot{Q}$ total energy release rate
$\dot{Q}^*$ $\dot{Q}/\rho_\infty c_p T_\infty g^{1/2} H^{5/2}$
r radial distance
t time
T gas temperature
ΔT $T - T_\infty$
U gas velocity
z distance above fire source
α fire growth parameter for t^2 fires
β^2 constant related to plume flow value 0.913 [14]
ρ gas density
ξ ratio of temperatures $T_{\infty,2}/T_{\infty,1}$
σ parameter defined in Equation 16

Subscripts

0 based on steady fire
1 lower layer
2 upper layer
∞ ambient, outside ceiling jet or plume flows
c convective fraction
f associate with gas travel delay
I value at the interface position between the warm upper layer and cool lower layer
i reference value at ignition

Superscripts

* dimensionless quantity

REFERENCES CITED

1. R.W. Pickard, D. Hird, and P. Nash, *F.R. Note 247*, Building Research Establishment, Borehamwood (1957).
2. P.H. Thomas, *F.R. Note 141*, Building Research Establishment, Borehamwood (1955).
3. R.L. Alpert, *Fire Tech.*, 8, 181 (1972).
4. R.L. Alpert, *Comb. Sci. and Tech.*, 11, 197 (1975).
5. H.Z. You, *Fire and Matls.*, 9, 46 (1985).
6. G. Heskestad and T. Hamada, *FMRC J.I. OKOEI.RU 070(A)*, Factory Mutual Research Corp., Norwood (1984).
7. L.Y. Cooper, NBSIR 87-3535, *J. of Heat Trans.*, 104, 446 (1982).
8. L.Y. Cooper and A. Woodhouse, *J. of Heat Trans.*, 108, 822 (1986).
9. H.Z. You and G.M. Faeth, *Fire and Matls.*, 3, 140 (1979).
10. C.C. Veldman, T. Kubota, and E.E. Zukoski, *NBS-GCR-77-97*, U.S. National Bureau of Standards, Gaithersburg (1977).
11. E.E. Zukoski and T. Kubota, *NBS-GCR-77-98*, National Bureau of Standards, Gaithersburg (1977).
12. V. Motevalli and C.H. Marks, "Characterizing the Unconfined Ceiling Jet under Steady-State Conditions: A Reassessment," *Fire Safety Science Proceedings* of the 3rd International Symposium, G. Cox and B. Langford, eds., Elsevier Applied Science, New York, 301 (1991).
13. D.D. Evans and D.W. Stroup, *Fire Tech.*, 22, 54 (1986).
14. E.E. Zukoski, T. Kubota, and B. Cetegen, *F. Safety J.*, 3, 107 (1981).
15. G. Heskestad and M.A. Delichatsios, *The Initial Convective Flow in Fire*, 17th International Symposium on Combustion, Combustion Institute, Pittsburgh (1978).
16. C.L. Beyler, *F. Safety J.*, 11, 53 (1986).
17. R.P. Schifilliti, *Use of Fire Plume Theory in the Design and Analysis of Fire Detector and Sprinkler Response*, Thesis, Worcester Polytechnic Institute (1986).
18. G. Heskestad and M.A. Delichatsios, *NBS-GCR-77-86 and NBS-GCR-77-95*, National Bureau of Standards, Gaithersburg (1977).

19. NFPA 72, *National Fire Alarm Code*, National Fire Protection Association, Quincy, MA (1993).

20. M.A. Delichatsios, *Comb. and Flame*, 43, 1 (1981).

21. D.D. Evans, *Comb. Sci. and Tech.*, 40, 79(1984).

22. D.D. Evans, *F. Safety J.*, 9, 147 (1985).

23. L.Y. Cooper, *A Buoyant Source in the Lower of Two, Homogeneous, Stably Stratified Layers, 20th International Symposium on Combustion*, Combustion Institute, Pittsburgh (1984).

24. E.E. Zukoski, T. Kubota, and C.S. Lim, *NBS-GCR-85-493*, National Bureau of Standards, Gaithersburg (1985).

25. H.W. Emmons, "The Ceiling Jet in Fires," *Fire Safety Science, Proceedings* of the 3rd International Symposium, G. Cox and B. Langford, eds., Elsevier Applied Science, New York, 249 (1991).

26. W.R. Chan, E.E. Zukowski, and T. Kubota, "Experimental and Numerical Studies on Two-Dimensional Gravity Currents in a Horizontal Channel," NIST-GCR-93-630, National Institute of Standards and Technology, Gaithersburg (1993).

VENT FLOWS

Howard W. Emmons

INTRODUCTION

Fire releases a great amount of heat which causes the heated gas to expand. The expansion produced by a fire in a room drives some of the gas out of the room. Any opening through which gas can flow out of the fire room is called a *vent*.

The most obvious vents in a fire room are open doors and open or broken windows. Ventilation ducts also provide important routes for gas release. A room in an average building may have all of its doors and windows closed and if ventilation ducts are also closed, the gas will leak around normal closed doors and windows and through any holes made for pipes or wires. These holes will act as vents. (If a room were hermetically sealed, a relatively small fire would raise the pressure in the room and burst the window, door, or walls.)

Gas will move only if it is pushed. The only forces acting on the gas are the gas pressure and gravity. Since gravity acts vertically, it might seem that gas could only flow through a hole in the floor or ceiling. Gravity, however, can produce horizontal pressure changes which will be explained in detail below. A gas flow which is caused directly or indirectly by gravity is called a buoyant flow.

When a pressure difference exists across a vent, fluid (liquid or gas) will be pushed through. Precise calculation of such flows from the basic laws of nature can only be performed today by the largest computers. For fire purposes, and all engineering purposes, calculations are carried out with sufficient precision using the methods of hydraulics. Since these formulas are only approximate, they are made sufficiently accurate (often to within a few percent) by a flow coefficient. These coefficients are determined by experimental measurements.

CALCULATION METHODS FOR NONBUOYANT FLOWS

If a pressure drop, $\Delta p = p_1 - p_2$, exists across a vent of area, A, with a fluid density, ρ, the flow through the vent has (see Figure 2-5.1)[1]

Velocity
$$V = \sqrt{\frac{2\Delta p}{\rho}} \tag{1}$$

Volume flow
$$Q = CA\sqrt{\frac{2\Delta p}{\rho}} \tag{2}$$

and

Mass flow
$$\dot{m} = CA\sqrt{2\rho\Delta p} \tag{3}$$

In these formulas the SI units are: $\Delta p = $ (Pa) = (N/m^2), $A = $ (m^2), $\rho = $ (kg/m^3), $V = $ (m/s), $Q = $ (m^3/s), $\dot{m} = $ (kg/s).

If the flow of water from a fire hose or sprinkler (Figure 2-5.2) is to be calculated and the pressure, p_g, is read on a gauge (in pounds/in^2) at the entrance to the nozzle where the area is A_1, the previous formulas provide the velocity, volume flow, and mass flow by using

$$\Delta p = 6895 p_g / [1 - (A/A_1)^2] \tag{4}$$

where

$A = $ area of vent and $A_1 = $ area of supply pipe.

The factor 6895 converts pressure in lb/in^2 to Pascals while the factor $[1 - (A/A_1)^2]$ corrects Δp for the dynamic effect of the inlet velocity in the supply hose or pipe.

In the atmosphere, the pressure at the ground is p_a, which is just sufficient to support the weight of the air above. If the air density is ρ_a, the pressure, p, at height, h, is less than p_a by the weight of the air in height, h. Thus the pressure difference is

$$\Delta p = p_a - p = \rho_a g h \tag{5}$$

It is sometimes convenient when considering fire gases to use $h = \Delta p / \rho_a g$, the pressure head, in meters of ambient air, in the velocity and flow rate formulas given above.

Dr. Howard W. Emmons is Professor Emeritus of Mechanical Engineering at Harvard University. His research has focused on heat transfer, supersonic aerodynamics, numerical computation, gas turbine compressors, combustion, and fire.

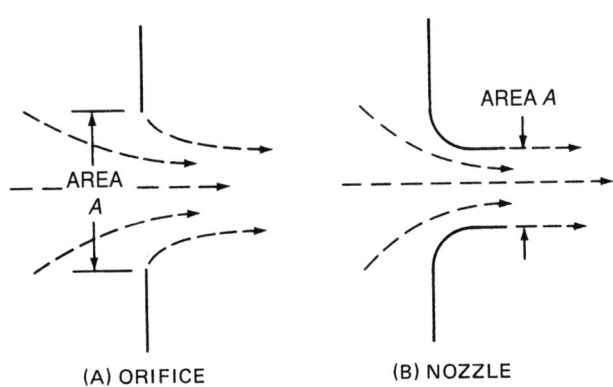

(A) ORIFICE   (B) NOZZLE

Fig. 2-5.1. *Most fire vents are orifices.*

The previous discussion supposes that the flowing fluid is of constant density. For liquids this is true for all practical situations. The density of air or fire gases will not change significantly during the flow through the vent so long as the pressure change is small, so they can also be treated as constant density fluids.

If the pressure drop is large, the equations become more complicated.[2] If the pressure and density upstream of the vent are p_1, ρ_1 while the pressure after the vent is p_2, the equations for velocity and mass flow become

$$V = \sqrt{\frac{2p_1}{\rho_1}} \left\{ \frac{\gamma}{\gamma-1} \left(\frac{p_2}{p_1}\right)^{2/\gamma} \left[1 - \left(\frac{p_2}{p_1}\right)^{(\gamma-1)/\gamma} \right] \right\}^{1/2} \quad (6)$$

$$\dot{m} = CA\sqrt{2\rho_1 p_1} \left\{ \frac{\gamma}{\gamma-1} \left(\frac{p_2}{p_1}\right)^{2/\gamma} \left[1 - \left(\frac{p_2}{p_1}\right)^{(\gamma-1)/\gamma} \right] \right\}^{1/2} \quad (7)$$

where $\gamma = c_p/c_v$.

The value of γ depends upon the complexity of the molecules of the flowing gas. For fire gases (which always contain a large amount of air) the value of γ will fall between 1.33 and 1.40. For most fire purposes the diatomic gas value (air) of 1.40 is sufficiently accurate.

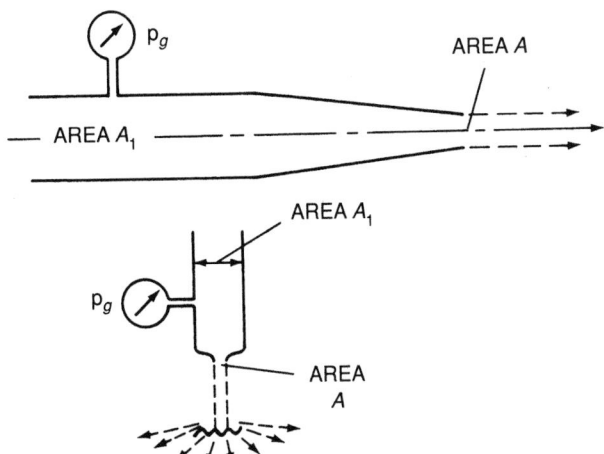

Fig. 2-5.2. *A hose nozzle and a sprinkler nozzle.*

The mass flow given by the previous equation has a maximum at

$$\frac{p_2}{p_1} = \left(\frac{2}{\gamma+1}\right)^{\gamma/(\gamma-1)} \quad (8)$$

For $\gamma = 1.40$, the maximum flow is reached for a downstream pressure $p_2 = 0.528p_1$. For all lower back pressures the flow remains constant at its maximum

$$\dot{m} = CA\sqrt{\rho_1 p_1} \left[\gamma \left(\frac{2}{\gamma+1}\right)^{(\gamma+1)/(\gamma-1)} \right]^{1/2} \quad (9)$$

With these equations, the mathematical description of the rate of flow of liquids and gases through holes is complete as soon as the appropriate flow coefficients are known. The coefficients, found by experiment, correct the formulas for the effect of the fluid viscosity, the nonuniformity of the velocity over the vent, turbulence and heat transfer effects, the details of nozzle shape, the location of the pressure measurement points, etc. The corrections also depend upon the properties and velocity of the fluid. The most important coefficient corrections for any given vent geometry is the dimensionless combination of variables which is called the Reynolds number, Re, and

$$\text{Re} = \frac{VD\rho}{\mu} \quad (10)$$

where

V = velocity of the fluid given by the previous equations
D = diameter of the nozzle or orifice
ρ = density of the fluid approaching the vent
μ = viscosity of the fluid approaching the vent

A door or window vent is almost always rectangular, not circular. The D to be used in the Reynolds number should be the hydraulic diameter

$$D = \frac{4A}{P} \quad (11)$$

where

A = area of the vent
P = perimeter of vent

For a rectangular vent, a wide and b high, $A = ab$, $P = 2(a + b)$.

$$D = \frac{2ab}{(a+b)} \quad (12)$$

The experimental values of the flow coefficients for nozzles and orifices are given in Figure 2-5.3.[2] Flow coefficients for nozzles are near unity while for orifices are approximately 0.6; the reason for this can be seen from Figure 2-5.1, wherein the flow from an orifice separates from the edge of the orifice and decreases to a much smaller area, in fact about 0.6 of the orifice area.

For most fire applications the Reynolds number will be about 10^6. Sprinklers and fire nozzles are small but the velocity is quite high. Conversely, ventilation systems of buildings are larger but have a lower velocity. Finally, doors and windows in the areas of a building not too near the fire are still larger but the velocity is still smaller. For most

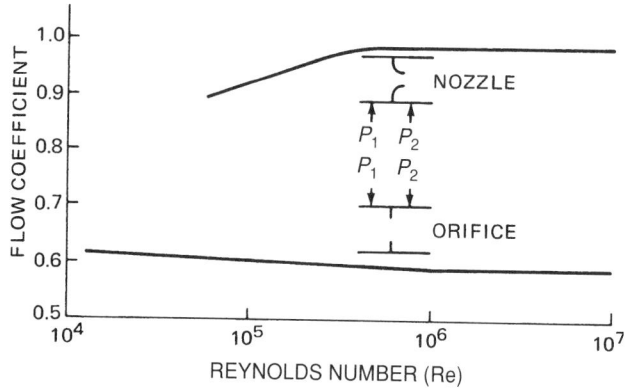

Fig. 2-5.3. Orifice and nozzle flow coefficients.

purposes the flow coefficient can be set as $C = 0.98$ for a nozzle and $C = 0.60$ for an orifice.

Buoyant Flows Through Vertical Vents

A fire in a room causes gases to flow out through a vent by two processes. The heating of the air in a room causes the air to expand, pushing other air out through all available vents and hence throughout the entire building. At the same time, the heated air, with products of combustion and smoke, rises in a plume to the ceiling. When the hot layer of gas at the ceiling becomes deep enough to fall below the top of a vent, some hot gas will flow out through the vent. As the fire grows, the buoyant flow out will exceed the gas expansion by the fire. Thus the pressure in the fire room at the floor will fall below atmospheric, and outside air will flow in at the bottom. A familiar sight develops, where smoke and perhaps flames issue out the top of a window while fresh air flows in near the bottom. This buoyant flow mechanism allows a fire to draw in new oxygen so essential for its continuation.

For these buoyantly driven flows to occur, there must be a pressure difference across the vent. Figure 2-5.4 illustrates how these pressure differences are produced. The pressure difference at the floor is

$$\Delta p_f = p_f - p_a \tag{13}$$

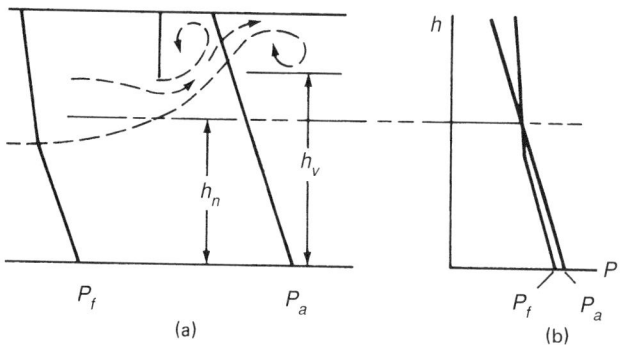

Fig. 2-5.4. Pressure gradients: (a) each side of a door; (b) super-imposed on a pressure vs height graph.

where

p_f = pressure at the floor inside the room in front of the vent

p_a = pressure at the floor level outside of the room just beyond the vent

The pressure at height y is less than the pressure at the floor and can be found by the following hydrostatic equations

Inside $$p_1 = p_f - \int_0^y \rho_1 g\, dy \tag{14}$$

Outside $$p_2 = p_a - \int_0^y \rho_2 g\, dy \tag{15}$$

The pressure difference at height, h, is

$$\Delta p = p_1 - p_2 = \Delta p_f + \int_0^h (\rho_2 - \rho_1) g\, dy \tag{16}$$

Since the outside density, ρ_2, is greater than the inside density, ρ_1, the integral is positive so that Δp is often positive (outflow) at the top of the vent and negative (inflow) at the bottom. The flow properties at the elevation, h, are the same as previously given.

$$V = \sqrt{\frac{2\Delta p}{\rho}} \tag{17}$$

$$\frac{Q}{A} = C\sqrt{\frac{2\Delta p}{\rho}} \tag{18}$$

$$\frac{\dot{m}}{A} = C\sqrt{2\rho\Delta p} \tag{19}$$

Since they are not the same at different heights in the vent, the volume and mass flow are given as flow per unit area.

Measuring Vent Flows in a Fire Experiment

Sufficient measurements must be made to evaluate ρ and Δp to allow use of Equation 19. There are four different available methods which differ in simplicity, accuracy, and cost.

Method 1.   The dynamic pressure distribution can be measured in the plane of the vent. This requires a sensitive pressure meter. The pressure difference is almost always less than the atmospheric pressure difference between the floor, p_f, and the ceiling, p_c. For a room 2.5 m in height the atmospheric pressure difference is

$$p_f - p_c = \rho_a g H = 1.176 \times 9.81 \times 2.5$$
$$= 28.84 \text{ Pascals} \quad (3.0 \text{ mm H}_2\text{O})$$

This is only

$$\frac{p_f - p_c}{p_a} = \frac{28.84}{101,325}$$

$$= 0.00028 \quad \text{fraction of atmospheric pressure.}$$

SUPPORT TUBES
(CONNECT PRESS. TAPS TO INDICATING INSTRUMENT)

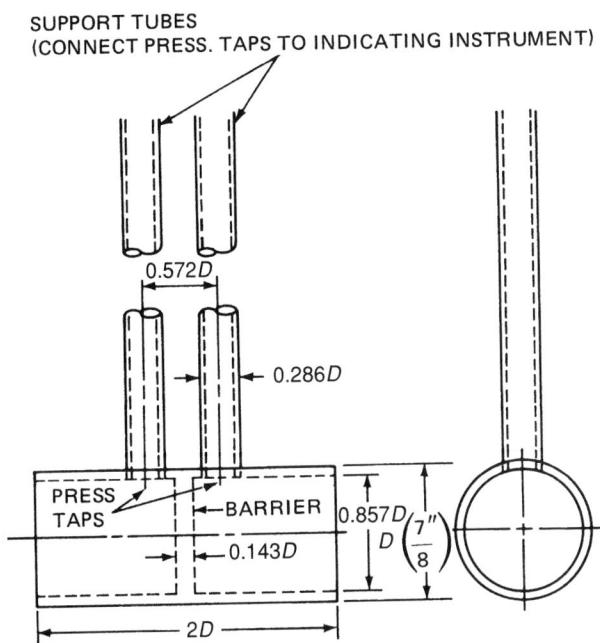

Fig. 2-5.5. A bidirectional flow probe.

Thus the buoyantly driven flow velocities induced by a room fire could be as high as

$$V = \sqrt{\frac{2\Delta p}{\rho}} = \sqrt{\frac{2 \times 28.84}{1.176}} = 7.00 \text{ m/s} \quad (23 \text{ ft/s})$$

Since the pressure varies with height and time, a series of pressure probes are required and each should have its own meter or a rapid activation switch. Although standard pitot tubes are the most accurate dynamic pressure probes, they are sensitive to flow direction and would have to be adjusted at each location for the direction of the local flow, especially for outflow and inflow. The probe orientation would need to be continually changed as the fire progressed.

A single string of fixed orientation pressure probes arranged vertically down the center of the door increases convenience of the measurement but forces a decrease in accuracy. The out-in flow problem is avoided by use of bidirectional probes in place of pitot tubes.[3] (See Figure 2-5.5.) These probes give velocities within 10 percent over an angular range of ±50 degrees of the probe axis in any direction.

Determination of the local velocity also requires the measurement of the local gas density. The density of fire gases can be determined from measured gas temperatures with sufficient accuracy by the ideal gas law

where

$$\rho = \frac{Mp}{RT} \quad (20)$$

M = Avg. molecular weight of flowing gas

$R = 8314 \dfrac{\text{J}}{\text{kg mol K}}$ = universal gas constant.

As noted previously, the pressure changes by only a very small percentage throughout a building so its effect on gas density is negligible.

Fire gases contain large quantities of nitrogen from the air and a variety of other compounds. The average molecular weight of the mixture will be close to but somewhat larger than that of air. Incomplete knowledge of the actual composition of fire gas prevents high accuracy calculations. For most fire calculations, it is accurate enough to neglect the effect of the change of molecular weight from that of air (M_a = 28.95). Density of gas is determined primarily by its temperature (which may vary by a factor of 4 in a fire). Thus

$$\rho = \frac{352.8}{T} \frac{\text{kg}}{\text{m}^3} \quad (21)$$

where

T = temperature in Kelvin (= °C + 273).

A string of thermocouples must be included along with the bidirectional probes to measure vent flows. For higher accuracy, aspirated thermocouples must be used or a correction made for the effect of fire radiation.[3] The temperature, and hence the gas density, will vary over the entire hot vent outflow. To determine the temperature distribution so completely would require an impractically large number of thermocouples. Fortunately the temperature in the vent is a reflection of the temperature distribution in the hot layer inside the room, which normally is stratified, and hence varies most strongly with the distance from the ceiling. Thus, a string of thermocouples hanging vertically on the centerline of the vent is usually considered to be the best that can be done in a practical fire test. Special care must be exercised to keep the test fire some distance away from the entrance to the vent. Since a fire near a vent has effects at present unknown, fire model calculations of real fire vent flows under such conditions will be of unknown accuracy. The velocity distribution vertically in the vent is given by

$$V = 0.93\sqrt{\frac{2\Delta p}{\rho}} \quad (22)$$

where ρ follows from Equation 21 using the temperature distribution *in the vent* with a calibration factor of 0.93 for the bidirectional probes.[4] Using ρ from Equation 21 gives the directly useful forms

$$V = 0.070\sqrt{T\Delta p} \qquad \Delta p\left[\frac{\text{N}}{\text{m}^2}\right]$$

$$V = 5.81\sqrt{T\Delta p} \qquad \Delta p\left[\frac{\text{lb}}{\text{in}^2}\right]$$

pressure measured with bidirectional probe (23)

where

V is in [m/s] $\left(V\left[\dfrac{\text{ft}}{\text{s}}\right] = 3.281 \ V\left[\dfrac{\text{m}}{\text{s}}\right]\right)$

T is in [K]

Except for very early stages of a room fire, there will be flow out at the top (V, $\Delta p > 0$) and flow in at the bottom (V, $\Delta p < 0$).* Thus there is a position in the vent at which

*Equation 23 should be written $V = (\text{sign } \Delta p)K\sqrt{T|\Delta p|}$ since when $\Delta p < 0$ the absolute value must be used to avoid the square root of a negative number and the sign of the velocity changes since the flow is in and not out.

$V = 0$; this is the vertical location where the pressure inside is equal to that outside. This elevation, h_n, is called the neutral axis. Defining the elevation of the vent sill as h_b ($h_b = 0$ for a door) and the elevation of the soffit as h_t, the flows are given by

Flow out

$$\dot{m}_u = C\int_{h_n}^{h_t} \rho V b \, dy \qquad (24)$$

Flow in

$$\dot{m}_d = C\int_{h_b}^{h_n} \rho V b \, dy \qquad (25)$$

where

b = width of the vent
C = experimentally determined flow coefficient ($= 0.68$)[7]

These equations in the most convenient form are

Flow out

$$\dot{m}_u = 16.79\int_{h_n}^{h_t} b\sqrt{\frac{\Delta p}{T_V}} \, dy \quad [\text{kg/s}] \qquad (26)$$

Flow in

$$\dot{m}_d = 16.79\int_{h_b}^{h_n} b\sqrt{\frac{\Delta p}{T_V}} \, dy \quad [\text{kg/s}] \qquad (27)$$

where

Δp = pressure drop in Pascals measured with bidirectional probe as a function of y
b = width of the vent in m
T_V = vertical distribution of temperature (K) in the vent

If the bidirectional probe pressures are measured in psi the coefficient 16.79 must be replaced by 1394.

Method 2. A somewhat simpler but less accurate procedure to measure vent flows requires the measurement of the pressure difference at the floor (or some other height). One pressure difference measurement together with the vertical temperature distribution measurement, T_1, *inside the room* (about one vent width in from the vent) and T_2, *outside the vent* (well away from the vent flow) provides the density information required to find the pressure drop at all elevations (Equation 16).

$$\Delta p = \Delta p_f + 3461\int_0^y \left(\frac{1}{T_2} - \frac{1}{T_1}\right) dy \qquad (28)$$

For most fires, Δp_f will be negative; i.e., the pressure at the floor inside the fire room will be less than the pressure outside. This is only true for a fire room with a normal size vent (door, window). For a completely closed room the inside pressure is well above the outside pressure. Since the temperature inside the fire room is higher than that outside, Equation 28 gives a Δp which becomes less negative, passes through zero at the neutral axis, h_n, and becomes positive at higher levels in the fire room. The vertical location of the neutral axis is therefore readily found from Equation 28.

The calculation of the pressure distribution requires measurement of the temperature distribution both inside, T_1, and outside, T_2, of the vent. However, calculation of the flow requires a knowledge of the density distribution in the vent itself. Thus a third thermocouple string is required to measure the temperature distribution, T_V, in the vent. The desired flow properties[6] are

Velocity

$$V = \sqrt{\frac{2\Delta p}{\rho}} = 4.43\sqrt{T_V\int_{h_n}^y \left(\frac{1}{T_2} - \frac{1}{T_1}\right)dy} \quad [\text{m/s}] \quad (29)$$

Flow out

$$\dot{m}_u = C\int_{h_n}^{h_t} \rho b V \, dy$$
$$= 1063\int_{h_n}^{h_t} b\left[\frac{1}{T_V}\int_{h_n}^y \left(\frac{1}{T_2} - \frac{1}{T_1}\right)dy\right]^{1/2} dy \qquad (30)$$

Flow in

$$\dot{m}_d = C\int_{h_b}^{h_n} \rho b V \, dy$$
$$= 1063\int_{h_b}^{h_n} b\left[\frac{1}{T_V}\int_{h_b}^y \left(\frac{1}{T_2} - \frac{1}{T_1}\right)dy\right]^{1/2} dy \qquad (31)$$

where

b = width of the vent at height y
Δp = calculated from Equation 16 using the temperatures (and thus densities) *inside and outside of the room*
ρ = density computed from the temperature *in the vent*.

(Note that for inflow Δp is negative. Therefore the equation takes the square root of the magnitude $|\Delta p|$ while its sign gives the flow direction.)

Method 3. The use of a sensitive pressure meter can be avoided entirely by visually (or better, photographically) locating the bottom of the outflow *in the vent* during the test. This is at the position of the neutral axis, h_n, where $\Delta p = 0$. Method 3 is the same as Method 2 except that the neutral axis location is found directly by experiment, rather than being deduced from the pressures. The distribution of pressure drop across the vent is found by integrating Equation 16 above ($\Delta p > 0$) and below ($\Delta p < 0$) h_n using the density distribution *inside*, ρ_1, and *outside*, ρ_2, *the room*. The flow properties are computed as before from Equations 29 through 31.

Method 4. A simpler but less accurate method uses the fair assumption that the gas in the fire room soon separates into a nearly uniform hot layer of density, ρ, with a nearly uniform cold layer below density, ρ_d. This separation with appropriate notation is shown in Figure 2-5.6. In this approximation the appropriate flow formulas[7] are

Outflow

$$V_u = \left(2g\frac{\rho_a - \rho}{\rho}y\right)^{1/2} \qquad (32)$$

where y is distance above the neutral plane

$$\dot{m}_u = \frac{\sqrt{8}}{3}Cb\sqrt{g\rho(\rho_a - \rho)}(h_v - h_n)^{3/2} \qquad (33)$$

The inflow by this two-layer method depends upon δ, which is small and cannot be determined with sufficient accuracy because of the effect of gas motions in the fire room.

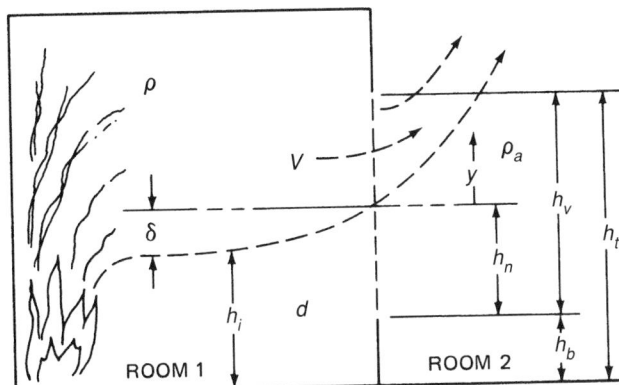

Fig. 2-5.6. Buoyant flow out of the window of a fire room.

The neutral axis may be found in several ways:

1. It may be located visually or photographically during the test; or
2. It may be found from the vent temperature distribution by locating [visually on a plot of $T_V(y)$] the position just below the most rapid temperature rise from bottom to top of the vent.

The low temperature, T_d, of the two-layer model is taken as the gas temperature just above the vent sill. The high temperature, T_u, is chosen so that the two-layer model has the same total mass, i.e., the same mean density, in the vent as the real flow.*

$$\left(\frac{1}{T}\right) = \frac{1}{h_v}\int_0^{h_v}\frac{dy}{T} = \frac{h_n}{h_v T_d} + \frac{h_v - h_n}{h_v T_u} \qquad (34)$$

The densities ρ_a and ρ are found using Equation 21 from the temperatures T_a and T_u, respectively.

The outflow velocity and mass flow are found from Equations 32 and 33.

An estimate of the air inflow rate can be found if the test has included the measurement of the oxygen concentration in the gases leaving the fire room. The gas outflow rate is equal to the inflow rate plus the fuel vaporized, except for the effect of transient variations in the hot layer depth. Thus

$$\dot{m}_d = \dot{m}_u\left(\frac{1+y_{O_2}\lambda}{1+0.23\lambda}\right) \qquad (35)$$

where $\qquad \lambda$ = effective fuel-air ratio

The flow coefficient to be used for buoyant flows is 0.68 as determined by specific experiments designed for the purpose. For nonbuoyant flows (nozzles and orifices), the flow coefficients are determined to better than 1 percent and presented as a function of the Reynolds number as in Figure 2-5.3. This accuracy is possible because the fluid can be collected and measured (by weight or volume).

For buoyant flows the experiments are much more difficult because the hot outflow and cold inflow cannot be collected and weighed. The best fire gas vent flow coefficient

measurements to date[5,6] have ± 10 percent accuracy with occasional values as bad as ± 100 percent (for inflow). The most accurate buoyant flow coefficients were measured not for fire gases but for two nonmiscible liquids (kerosene and water).[7] In this case the two fluids could be separated and measured, and the value 0.68 was found except for the very low flow rates (near the beginning of a fire). When buoyant flow coefficients can be measured within a few percent accuracy, they will be a function of the Reynolds number, Re $= Vh_v\rho/\mu$; the Froude number, $F_r = V^2\rho_a/gh_v(\rho - \rho_a)$; and the depth parameter, h_n/h_v.

The best option now available is to use $C = 0.68$ and expect ± 10 percent errors in flow calculations.

Note that all of the above four methods require a knowledge of h_n, the dividing line between outflow above and inflow below. It would be useful to have a simple formula by which h_n could be calculated without any special measurements. What determines h_n?

The fire at the start sends a plume of heated gas toward the ceiling and, by gas expansion, pushes some gas out of the vent. The hot plume gases accumulate at the ceiling with little, if any, flowing out the vent. After a time, dependent on the size of the room, the hot layer depth becomes so large that its lower surface falls below the top of the vent. Hot gas begins to flow out.

When a fire has progressed to a second room, there is a hot layer on each side of a connecting vent. Thus, (with two layers on each side) there are as many as four different gas densities: $\rho_{d_1} > \rho_1$, densities below and above in room 1, and $\rho_{d_2} > \rho_2$, densities below and above in room 2. There are also four pertinent levels: h_b, sill height (0 if the vent is a door); h_t, soffit height; h_{i_1}, interface height in room 1; h_{i_2}, interface height in room 2. There are many different flow situations possible depending upon these eight values.

The pressure variation from floor to ceiling in each room depends upon the densities and layer heights in that room. In addition, the pressure difference between the two rooms (at the floor, for example) may have any value depending upon the fire in each room, all the room vents, and especially the vent (or vents) connecting the two rooms. Figure 2-5.7 shows a few of the possible pressure distributions. The pressure distribution in room 1 is shown with a dotted line while that in room 2 is shown as a solid line.

In Figure 2-5.7 (part a), there are no hot layers, the pressure in room 1 at every level is higher than that in room 2, and the flow is everywhere out (positive) (room 1 to room 2).

In Figure 2-5.7 (part b), a common situation exists. The density in room 2 is uniform (perhaps the outside atmosphere). Room 1 has a hot layer and a floor pressure difference such that there is outflow at the top, inflow at the

---

*Sometimes the mean temperatures, $\overline{T}$, of the two-layer model and the real flow are also used and both h_n and T_u are determined (using T_d as above). The requirement of identical $\overline{T}$ is arbitrary, sometimes leads to impractical results, and is not recommended.

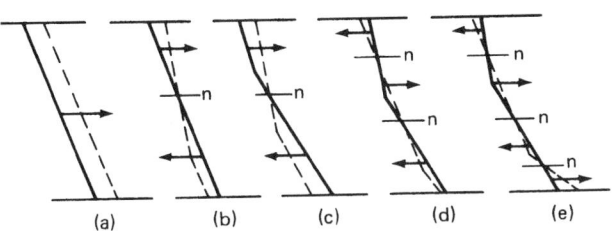

Fig. 2-5.7. Some selected two-layer vent pressure drop distributions. Dotted line is pressure distribution in room 1; solid line is pressure distribution in room 2.

bottom, and a single neutral axis somewhat above the hot-cold interface in the room.

In Figure 2-5.7 (part c), the flow situation is similar to that in Figure 2-5.7 (part b), although there are hot layers in both rooms (but with a neutral axis above the interface in room 1 and below the interface in room 2).

In Figure 2-5.7 (part d), the densities (slopes of pressure distribution lines) are somewhat different than those in Figure 2-5.7 (part c) (the hot layer in room 2 is less deep but hotter than that in room 1). Consequently there are two neutral axes with a new small inflow layer at the top; three flow layers in all—two in and one out.

In Figure 2-5.7 (part e), the densities and floor level pressure difference are such that there are four flow layers, two out and two in, with three neutral axes.

These five cases do not exhaust the possible vent flow situations.

Figure 2-5.7 (parts a and b) account for all cases early in a fire and all cases of vents from inside to outside of a building. They are also the only cases for which experimental data is available. The case illustrated in Figure 2-5.7 (part c) is common inside a building after a fire has progressed to the point that hot layers exit in the two rooms on each side of a vent. The cases illustrated in Figures 2-5.7 (parts d and e) have not been directly observed but probably account for an occasional confused flow pattern. (In fact, the above discussion assumes two distinct layers in each room.) The layers are seldom sharply defined and in this case there may be many neutral axes, or regions, with a confusing array of in-out flow layers. These confused flow situations are probably not of much importance in a fire since they seldom occur and when they do they don't last very long.

The previous discussion of the possible two-layer flow situation is very important for the zone modeling of a fire. Fire models to date are all two-layer (a three or more layer model will present far more complex vent flows than those pictured in Figure 2-5.7). In fire computation by a zone model, e.g., cases (d) and (e) in Figure 2-5.7 will be unimportant to fire development. However, since these situations can arise, they should be handled via fire computation; i.e., by computing the flow layer by layer. Each layer has a linear pressure variation from sill, interface, or neutral axis up to the next interface, neutral axis, or soffit.

By use of the pressure drop at the floor and the room densities on each side of the vent in Equation 16, the position, h_j, of all layers and the sill, interfaces, neutral axes, and soffit will be known. Thus, for each layer (defined as j) the pressure drop at the bottom, Δp_j, and at the top, Δp_{j+1}, will be known. Since the room densities are constant in each room for each layer, the vent pressure drop will vary linearly from Δp_j to Δp_{j+1}. The flow in each layer from room 1 to room 2, found by integration,[8] is given by

$$\dot{m}_i = (\text{sign } \alpha) C \frac{\sqrt{8}}{3} b(h_{j+1} - h_j)\sqrt{\rho}$$

$$\times \left(\frac{|\Delta p_j| + \sqrt{|\Delta p_j \Delta p_{j+1}|} + |\Delta p_{j+1}|}{\sqrt{|\Delta p_j|} + \sqrt{|\Delta p_{j+1}|}} \right) \quad (36)$$

where

$\alpha = \left(\dfrac{\Delta p_j + \Delta p_{j+1}}{2} \right)$ whose sign determines the in-out direction of the flow

ρ = density of the gas flowing in the flow layer i

Thus

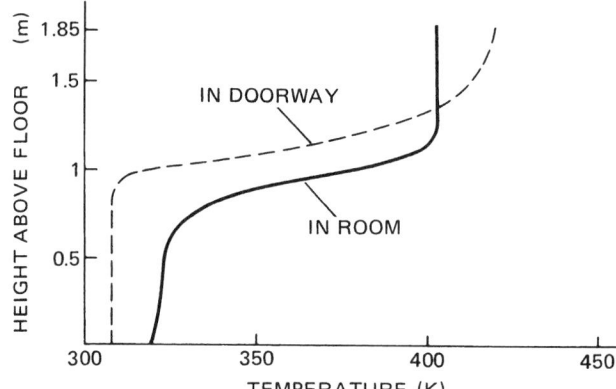

Fig. 2-5.8.   Sample fire room and doorway temperature distributions.

$$\rho = \begin{cases} \text{density in room 1 at height } h_j^+ & \text{if } \alpha > 0 \\ \text{density in room 2 at height } h_j^+ & \text{if } \alpha < 0 \end{cases}$$

This flow calculation appears complex but can be coded quite easily for computer use and then used to calculate all the possible cases.

Although all vent flows can now be calculated, the path of each layer of gas flow when it enters a room is still needed for fire modeling. If the two-layer model is to be preserved, each inflow must mix with the hot layer or the cold layer, or be divided between them. No information is yet available as to the best solution of this problem.

To illustrate these various methods of flow calculation, some test data from a steady burner fire in a room at the U.S. Bureau of Standards[6] is used. Some typical data are shown in Figure 2-5.8. Accurate results, even in a steady-state fire, are difficult to obtain and questions about the data in this figure will be noted as appropriate. The vent temperatures were measured by small diameter bare thermocouples for which there is some unknown radiation correction. This may account for the top vent temperature being higher than that in the fire room.

The vent was 1.83 m high, 0.737 m wide and the outflow measured with bidirectional probes (not corrected for flow angle) was 0.588 kg/sec for a fire output of 0.63 kW. The ambient temperature was 21.3°C (= 294.3 K). This flow was determined by using Method 1.

Method 2 uses the known location of the neutral axis and requires the integration of Equations 30 and 31. In this way the data of Figure 2-5.8 gives outflow of 0.599 kg/s, 1.8 percent higher compared to Method 1 and inflow of 0.652 kg/s. A measured (by bidirectional probes) inflow is not given, but it seems odd that the inflow is greater than the outflow since inflow must be smaller than the outflow by the mass rate of fuel burned at steady state.

Data for use of Method 3 is not available.

Method 4 requires the selection from Figure 2-5.8, of a neutral axis location and inlet temperature. In the figure the rapid temperature rise *in the vent* begins at about 1 m. Hence this height is chosen as the neutral axis. The lowest inlet temperature is $T_d = 308$ K. By computing $(1/T_V)$ the average value was found to be $(1/T_V) = 2.875 \times 10^{-3}$. Now by Equation 34

$$2.875 \times 10^{-3} = \frac{1.00 - 0}{1.83 \times 308} + \frac{1.83 - 1.00}{1.83 T_u}$$

Thus $T_u = 411.9$ K. The corresponding density is $\rho = 352.8/411.9 = 0.8565$ kg/m^3. From the ambient temperature, T_a, we find $\rho_a = 352.8/294.3 = 1.199$ kg/m^3. Thus the outflow by Equation 33 is

$$\dot{m}_u = \frac{\sqrt{8}}{3}0.68 \times 0.737[9.81 \times 0.8565(1.199 - 0.8565)]^{1/2}$$
$$\times (1.83 - 1)^{3/2} = 0.607 \text{ kg/s}$$

This value is 3.2 percent higher compared to Method 1.

Buoyant Flows Through Horizontal Vents

Unlike nonbuoyant flows through orifices or flow through vents in a vertical wall, very little quantitative work has been done on flow through vents in horizontal (floors or flat roofs) or slightly sloped (inclined roofs) surfaces. The following discussion is included to clarify the present status of our knowledge and to provide flow calculation formulas of unknown accuracy in lieu of nothing.

Consider the flow through a hole in a horizontal surface. The velocity and flow rate are determined by the pressure drop from the upstream side of the vent to the *vena contracta*. Therefore, the buoyancy of the fluid from the vent to the *vena contracta* influences the flow. Thus, for upward flow of the lower fluid the velocity is given by

$$v_H = \left[\frac{2}{\rho_H}(gh\Delta\rho + \Delta p)\right]^{1/2} \quad (37)$$

where

$\Delta\rho = \rho_c - \rho_H$
$\Delta p = p_H - p_c$ measured at the vent's lower and upper surfaces
h = the vertical distance from the vent lower surface to the *vena contracta* (about equal to the orifice diameter D)

If the unidirectional flow were down, the velocity would be

$$v_c = -\left[\frac{2}{\rho_c}(gh\Delta\rho - \Delta p)\right]^{1/2} \quad (38)$$

The magnitude of the buoyancy effect is 8.6 pascals (for a fire density ratio of 4 to 1 and a 1-m diameter horizontal vent), and a buoyant velocity of 4m/s (about ⅛ of the velocity) is produced by the fire room buoyancy. The plume above the *vena contracta* stirs the fluid on the upper surface but does not influence the flow.

As the flow nears zero, the interface between the lower (hot) and upper (cold) gases becomes flat and is unstable. The unidirectional flow is replaced by simultaneous up and down flows usually oscillating in time and location.

At present there are no measurements of effective values of h. There are only a couple of quantitative studies of horizontal vent flows in which the pressure drop-flow information has been adequately measured.[9,10] These are for very small holes (a diameter of 2 in. or less), and in many cases the holes were fitted with a short pipe. Ceiling or roof holes in fires are usually irregular in shape and have a length to "diameter" ratio of 0.13 or less. There are a number of studies[11,12,13] of rooms with a ceiling hole with a fire either under the hole or on a wall. These supply interesting fire data but are not useful as horizontal vent studies, since the results do not include adequate orifice pressure and flow measurements. Epstein and Kenton[9] have measured the transfer of fluid from the lower to the upper chambers using water below and a brine above (density ratio 1.1 or less).

They found that, at zero net volume flow (the lower chamber was closed except for a ceiling hole 2 in. in diameter or less), the fluid transfer to the upper chamber was

$$\frac{Q_H}{\left(D^5 g \frac{\Delta\rho}{\rho}\right)^{1/2}} = 0.055 \quad (39)$$

while the unidirectional volume flow, q, that just prevented reverse flow was

$$\frac{q}{\left(D^5 g \frac{\Delta\rho}{\rho}\right)^{1/2}} = 0.20 \quad (40)$$

The unidirectional volume flows that follow from the velocity Equations 37 and 38 are

$$\frac{Q_H}{\left(D^5 g \frac{\Delta\rho}{\rho_H}\right)^{1/2}} = \frac{\pi}{8}C_D\left(\frac{h}{D} + \frac{\Delta p}{\Delta\rho g D}\right)^{1/2} \quad (41)$$

$$\frac{Q_C}{\left(D^5 g \frac{\Delta\rho}{\rho_c}\right)^{1/2}} = \frac{\pi}{8}C_D\left(\frac{h}{D} - \frac{\Delta p}{\Delta\rho g D}\right)^{1/2} \quad (42)$$

If we assume that when $Q_H = q$, the flooding value, then $\Delta p/\Delta\rho g D$ has the value for which $Q_C = 0$. Then $\Delta p/\Delta\rho g D = h/D$, and by Equations 40 and 41

$$\frac{\Delta p}{\Delta\rho g D} = 0.045 \quad (43)$$

With this value as the limit of unidirectional flow, $-.045 < \Delta p/\Delta\rho g D < .045$ is the pressure range in which flows occur simultaneously in both directions. There is no current theory nor measurements to compute these low flows so each flow squared up and down are assumed to vary linearly in this range. The resultant flows are

$$\dot{m} = cD[2\rho_H(\Delta\rho g D + \Delta p)]^{1/2} \quad \text{for } \Delta p > .045\Delta\rho g D \quad (44)$$

$$\left.\begin{array}{l} \dot{m}_u = cD[23.22\rho_H(.045\Delta\rho g D + \Delta p)]^{1/2} \\ \dot{m}_D = -cD[23.22\rho_c(.045\Delta\rho g D - \Delta p)]^{1/2} \end{array}\right\} \begin{array}{l} \text{simultaneous} \\ \text{up down} \\ \text{flow for} \\ -.045\Delta\rho g D \\ < \Delta p < \\ .045\Delta\rho g D \end{array}$$
$$(45)$$

$$\dot{m} = -cD[2\rho_c(\Delta\rho g D - \Delta p)]^{1/2} \quad \text{for } p < -.045\Delta\rho g D \quad (46)$$

Note: The use of the dimensionless form $\Delta p/\Delta\rho g D$ has been changed in Equations 44 thru 46 so that numerical computations when there is no density change ($\Delta\rho = 0$) does not encounter division by zero.

Equations 44 through 46 describe the positive upward flow through a horizontal vent over the entire pressure range from $-\infty$ inflow to $+\infty$ outflow. This theory is shown in Figure 2-5.9 for a density ratio of 2 with coordinates using the average density.

The theory of Cooper[14] omitted the buoyancy effect on the vertical flow and was developed before the Epstein flood data were available. However, in view of present horizontal vent data uncertainty, it is a useful alternative.

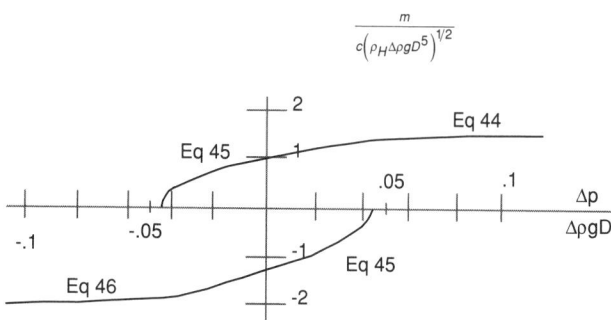

Fig. 2-5.9. *Theory of flow based on Equations 44 through 46.*

Accuracy of Vent Flow Calculations

For nonbuoyant flows (using nozzles or orifices in a straight run of pipe made and calibrated with a specific geometry over a known Reynolds number range) one easily obtains 2 percent accuracy. Thus, Equations 1 through 9 are capable of high accuracy.

For vents in vertical walls with limited internal room fire circulations, the best methods of measurement may get 5 percent accuracy. However, in real fires, induced circulations are often severe and unknown. Thus, errors of 10 percent or higher must be expected. Even if flow instrumentation is located in the vent itself, there is never enough to really account for variations over the vent surface and time fluctuations originating in the fire phenomena inside the fire room.

For vents in a horizontal surface, the accuracy is completely unknown. Equations 45 through 48 reproduce the water-brine experiments in small holes. The experimental accuracy is 10 percent. However, for a real fire, the errors are probably much higher. A typical case is a hole in the ceiling burned through by the flames from below. The hole geometry is very irregular and is completely unknown. Furthermore, a fire directly below the hole supplies hot gas with a considerable vertical velocity. Also, the ceiling jet flow often provides considerable cross flow.

Full-scale experimental results determining the effects of fire circulation, large density ratios, and large Reynolds numbers are needed. The present formulas are given as "better than nothing."

VENTS AS PART OF THE BUILDING FLOW NETWORK

A building is an enclosed space generally with floors, walls that divide the space both vertically and horizontally into rooms, corridors, and stairwells. A fire that starts at any place in the building causes gas expansion, which raises the local pressure and pushes air throughout the building through all pathways leading to the outside. If a window is open in the room of fire origin, and there is little or no wind, little flow moves through the remainder of the building. If there is no open window, the flow will move toward cracks and leaks wherever they may be in the building. All these flows are initially nonbuoyant. The flow through the building is simply flow through a complex system of "pipes" and "orifices." As the fire grows larger, hot gas flows buoyantly out of the place of origin, while cold gas flows in below.

Thus, while the net flow (out-in) is just sufficient to accomodate the fire gas expansion, the actual volumetric hot gas outflow may be 2.5 times larger than the inflow. A layer of hot gas moves along the ceiling of connected spaces and at the first opportunity proceeds up a stairwell or other ceiling (roof) opening into regions above.[15] The accumulating hot gas will help spread the fire while the newly created hot fire gases build a new hot layer in the adjacent spaces. The flow and pressure drop across each vent will then progress through a succession of situations as previously discussed. The flow throughout the building is therefore determined by the vent and flow friction drops along *all* of the available flow paths from the fire to the outside of the building.

The vent flow calculation procedures described in this section are sufficiently accurate and general to compute the required flow-pressure drop relations for building flow networks (except slow buoyant flows through horizontal vents).

NOMENCLATURE

A	area (m^2)
a	length (m)
b	width (m)
C	flow coefficient ($-$)
D	orifice diameter (m)
Fr	Froude number ($-$)
g	gravity constant (m/s^2)
h	height (m)
M	molecular weight (kg/kg mol)
$\dot{m}$	mass flow rate (kg/s)
P	perimeter (m)
p	pressure (Pa)
Q	volume flow rate (m^3/s)
R	gas constant (J/kg mol K)
Re	Reynolds number ($-$)
T	temperature (K)
V	velocity (m/s)
y	vertical coordinate (m)
Δ	increment of
δ	depth (see Figure 2-5.6) (m)
$\gamma = c_p/c_v$	isentropic exponent ($-$)
ρ	density (kg/m^3)
μ	viscosity (N s/m^2)

Subscripts

a	atmosphere
b	sill of vent
c	ceiling of room
d	lower
f	floor
g	gauge
i	hot-cold interface
j	index of layer
n	neutral axis
O_2	oxygen
t	soffit of vent
u	upper
v	in the vent
1	upstream of orifice
2	downstream of orifice

REFERENCES CITED

1. H. Rouse, *Fluid Mechanics for Hydraulic Engineers*, McGraw-Hill, New York (1938).

2. *Mark's Mechanical Engineers Handbook*, McGraw-Hill, New York (1958).

3. J.S. Newman and P.A. Croce, *Serial No. 21011.4*, Factory Mutual Research Corp., Norwood (1985).

4. D.J. McCaffrey and G. Heskestad, *Comb. and Flame*, 26, 125 (1976).

5. J. Quintiere and K. DenBraven, *NBSIR 78-1512*, National Bureau of Standards, Washington (1978).

6. K.D. Steckler, H.R. Baum, and J. Quintiere, *20th Symposium on Combustion*, Pittsburgh (1984).

7. J. Prahl and H.W. Emmons, *Comb. and Flame*, 25, 369 (1975).

8. H.E. Mitler and H.W. Emmons, *NBS-GCR-81-344*, National Bureau of Standards, Gaithersburg (1981).

9. M. Epstein and M.A. Kenton, *Jour. of Heat Trans.*, 111, 980 (1989).

10. Q. Tan and Y. Jaluria, *NIST-G&R-92-607*, Nat. Inst. of Stds. and Tech. (1992).

11. C.F. Than and B.J. Savilonis, *Fire Safety Jour.* 20, 151 (1993).

12. J.L. Bailey, F.W. Williams, and P.A. Tatum, *NRL Report 6811*, Naval Research Lab. (1991).

13. R. Jansson, B. Onnermark, and K. Halvarsson, *FAO Report C 20606-D6*, Nat. Defense Research Inst., Stockholm (1986).

14. L.Y. Cooper, *NISTIR 89-4052*, Nat. Inst. of Stds. and Tech. (1989).

15. T. Tanaka, *Fire Sci. and Tech.*, 3, 105 (1983).

NATURAL CONVECTION WALL FLOWS

Yogesh Jaluria

INTRODUCTION

In recent years there has been considerable interest and research in the buoyancy-driven flow of air and gases in room fires, with a view to determine the changing environment within the enclosure and to provide inputs for the eventual spread of the combustion products into other connected spaces.[1-3] Much of the present information on the development of the fire concerning the temperatures in the room and the movement of gases has been obtained from zone modeling, which divides the room into distinct homogeneous regions. The regions are generally two layers consisting of the hot upper gas zone and the cooler lower zone.[4-6] The zone models have received corroboration from detailed experimental measurements of temperature and velocity distributions in enclosures ventilated from single openings.[7-9] The zone models predict the average gas and wall temperatures in the various regions but do not provide information on local conditions, which may be obtained from recent field model studies.[10,11]

The basic features of the zone model for fire in a room with an opening, such as a door or a window, are shown in Figure 2-6.1. As a fire starts in a room, buoyant, heated fluid rises from the fire, entrains fresh air from the room, and moves toward the ceiling in the form of a buoyant plume. This flow eventually generates a distinct layer of hot, and thus lighter, gas below the ceiling. This layer increases in thickness as the plume brings more hot fluid into the layer. The hot gases eventually start flowing out of the room while the cold air from outside flows into the room due to the resulting pressure difference. As estimated by zone modeling studies and also observed in several experiments, fires of any thermal significance (> 10 kW) in typical residential rooms display a rapid increase in the upper layer thickness. The upper layer interface with the lower layer reaches the opening in a matter of seconds. However, conventional walls have a much slower time response than the gas layers,[2] so despite the rapid increase in the temperatures in the two layers, the walls continue to increase in temperature at a

Dr. Yogesh Jaluria is Professor of Mechanical Engineering at Rutgers—The State University of New Jersey. His research has focused on various facets of heat transfer, including natural convection, enclosure fires, manufacturing processes, environmental transport processes, and computational heat transfer.

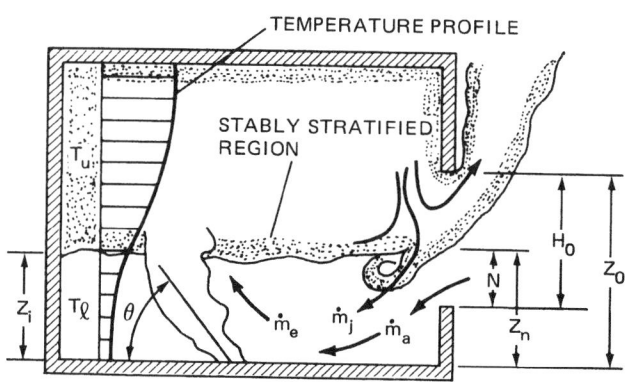

Fig. 2-6.1. Buoyancy-induced flow due to fire in an enclosure with a single opening.

very gradual rate. An initially rapid increase in the gas temperatures is therefore followed by slowing varying wall and gas temperatures. This had led to the modeling of many practical circumstances as quasi-steady, with the flow and temperature distributions progressing from one essentially steady situation to another, until rapid pyrolysis of combustible material in the room occurs, followed by room flashover.

Due to the difference between the temperatures of the gas and the adjacent walls that are gradually being heated due to convective and radiative heat transfer, a significant buoyancy-induced flow may arise adjacent to the walls and may lead to additional mixing across the interface. At the initial stages of the fire in a room, the upper layer generates a downward flow adjacent to the cooler upper region of the wall. The flow emerges from the upper layer, becomes upwardly buoyant in the lower layer, and rises toward the interface to cause greater mixing in the lower zone, near the interface.[12] At later stages of the fire, the lower region of the wall, in contact with the lower layer, heats and an upward flow may also be generated in the cooler lower layer. These opposing streams of wall flows may be important in the mass and energy balances for the zones.

This chapter discusses the nature and importance of wall flows thus generated in enclosure fires. The basic characteristics of these flows are outlined, followed by a detailed consideration of the analytical methods for computing the

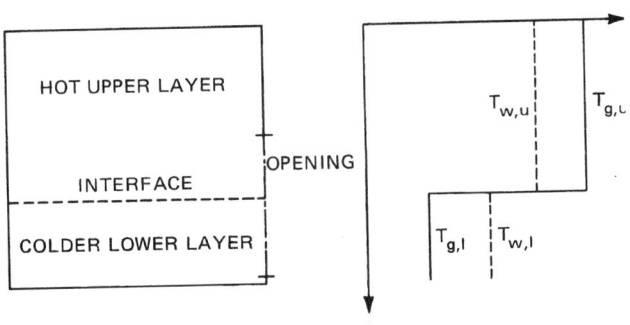

Fig. 2-6.2. *Sketch of the wall and gas temperature distributions T_w and T_g, respectively, in a two-layer zone model for an enclosure fire.*

mass, momentum, and energy transfer resulting from these flows. The procedure for incorporating these effects into existing zone models for enclosure fires is also discussed.

NATURE OF WALL FLOWS GENERATED IN ENCLOSURE FIRES

There are important physical aspects relevant to a study of the temperatures and flows arising due to an enclosure fire. For a two-layer zone model, the upper layer is characterized by a uniform average temperature, $T_{g,u}$, and the lower layer by $T_{g,\ell}$, as shown in Figure 2-6.2. Similarly, the wall temperatures may be considered in terms of an upper region value, $T_{w,u}$, and a lower region value, $T_{w,\ell}$. Thus, a step change in the temperatures is assumed to occur at the interface height, Z_i. In actual practice, of course, a step change is an idealization of a more gradual variation (as shown in Figure 2-6.1) and is discussed later in this chapter in terms of temperature measurements in such an enclosure.[13] As shown in Figure 2-6.2, the wall temperature in the lower layer is higher than the adjacent gas temperature, and lower than the gas temperature in the upper layer. Therefore, a buoyancy-induced wall flow rises in the lower layer and descends in the upper layer.[13,14] Such boundary layer flows that arise adjacent to heated or cooled vertical surfaces have been extensively studied.[15,16] A qualitative sketch of these flows is shown in Figure 2-6.3, with the corresponding velocity and temperature distributions. Sim-

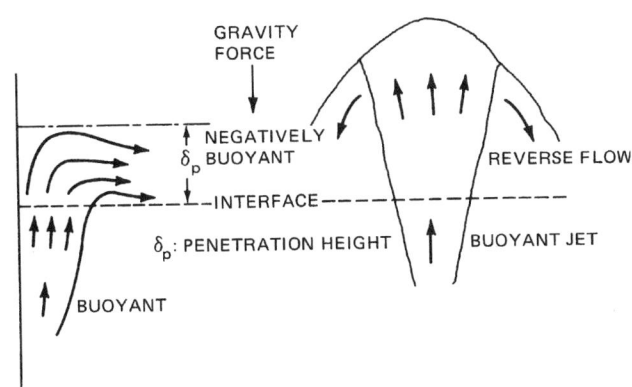

Fig. 2-6.4. *Flow in negatively buoyant wall and free jets.*

ilarly, at various stages in the fire growth, downward or upward flows may arise individually, without the corresponding opposing flow.

In the absence of the lower opposing flow, a downward flow arises in the upper layer, penetrates the interface, becomes upwardly buoyant as it encounters the lower temperature in the lower layer, stagnates at a certain depth, and rises toward the interface due to this buoyancy effect. This is similar to the case of a negatively buoyant jet as shown in Figure 2-6.4.[17-20] Such a circumstance arises at the initial stages of fire growth when the wall is essentially unheated and the adjacent gases are at a relatively high temperature.[12] Figure 2-6.5 shows the nature of such a wall effect. At later stages, the walls also become heated and an upward stream may also arise in the lower layer.[21] In the presence of the opposing stream, the flow stagnates and separates from the wall, either above or below the interface, as shown in Figure 2-6.6.

There are several important considerations with such wall flows; the most significant one relates to the magnitude of the flow generated in an actual room fire situation. To determine the relative importance of this effect, the wall flow rate may be compared with the flow in the fire plume and also with the flow entering through the opening. The second consideration concerns the modification of the present two-layer zone models. The wall flows could also affect the interpretation of the experimental results on enclosure fires. Two flow configurations will be considered. The first configuration relates to the buoyancy-induced downward flow

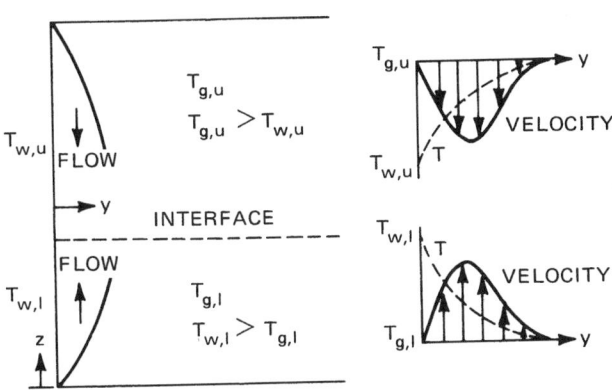

Fig. 2-6.3. *Sketch of the velocity and temperature profiles in the natural convection wall flows.*

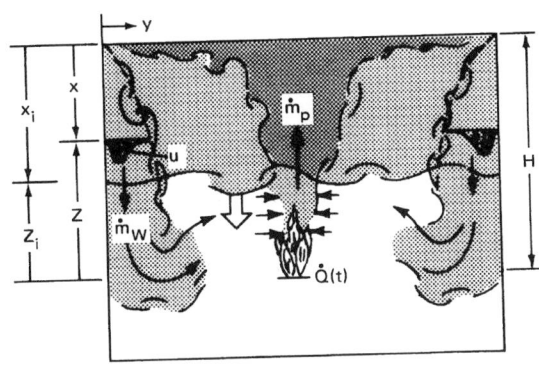

Fig. 2-6.5. *Sketch of the significant features of the wall effect at the early stages of a room fire.[12]*

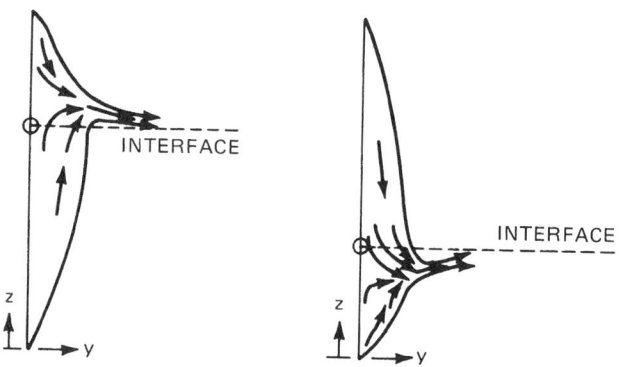

Fig. 2-6.6.   Interaction of opposing wall flows.

in the upper region, with no opposing flow in the lower layer. This circumstance arises at the early stages of the fire, when the walls have not heated.[22] The second flow configuration involves the two opposing streams, as shown in Figure 2-6.3.

These two configurations characterize the natural convection wall flows that arise at various stages in the development of an enclosure fire. These flows, however, are not restricted to rooms where the fire originates and may arise in spaces which communicate with the rooms containing the fire. The walls in such adjacent enclosures would generally remain close to their original ambient temperature until relatively late into the fire, and would, therefore, generate downward natural convection wall flows which could affect the redistribution of combustion products in these spaces.

CALCULATION OF TRANSPORT DUE TO WALL FLOWS

The wall flows generated in the two isothermal regions may be analyzed as boundary layer flows adjacent to isothermal surfaces immersed in uniform temperature media. In the upper region, the wall is colder than the ambient fluid and a downward flow arises, with no opposing flow at the initial stages. Eventually, the wall in the lower region heats and an

upward flow may also be generated. Both these flows encounter a stable stratified region as they go from one layer to the other, whether through the step change of Figure 2-6.7, part (a) or the more gradual variation of Figure 2-6.7, part (b). The actual profiles are typical of experimental measurements, and are shown in Figure 2-6.7, part (c).[13] The flow should be considered in the two isothermal regions, followed by a consideration of the penetration of a negative buoyant flow or of the flow in stable stratified regions.

A considerable amount of research has been done on steady, two-dimensional, laminar, and turbulent natural convection flows adjacent to isothermal vertical surfaces in isothermal ambient media.[15,16] For the coordinate system shown in Figure 2-6.8, the vertical velocity, u, and the temperature distribution may be determined for laminar flow from similarity analysis.[23,24] The mass flow rate, $\dot{m}$, rate of convected energy, $\dot{q}$, and momentum flow rate, $\dot{M}$, in the boundary layer could then be calculated. These are obtained, per unit flow perimeter, from the following equations

$$\dot{m} = \int_0^\delta \rho u \, dy \tag{1}$$

$$\dot{q} = \int_0^\delta \rho C_p u (T - T_\infty) \, dy \tag{2}$$

$$\dot{M} = \int_0^\delta \rho u^2 \, dy \tag{3}$$

where C_p is the fluid specific heat at constant pressure, T_∞ is the local gas temperature, ρ is the fluid density (which varies as $1/T$ for a perfect gas), and δ is the boundary layer thickness.

For the exact boundary layer similarity analysis, δ is replaced by infinity and the velocity and temperature distributions from the numerical solution of the governing equations are employed.[15] An approximate integral analysis[15,25] yields results quite close to the exact analysis for laminar flow. The integral analysis is probably more appropriate here, since there is no satisfactory exact analysis for turbulent flow, and since the numerical values of the integrals required to determine the previous physical quantities are not readily available from the exact similarity analysis.

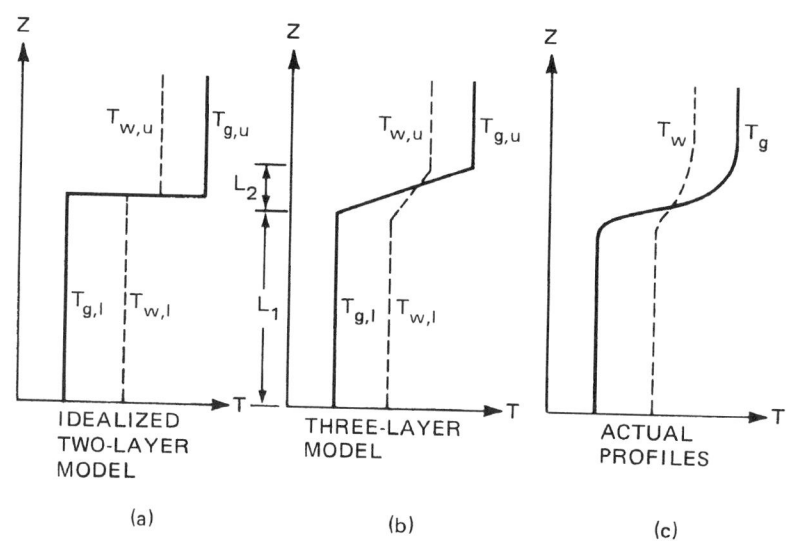

Fig. 2-6.7.   Actual and idealized temperature profiles.

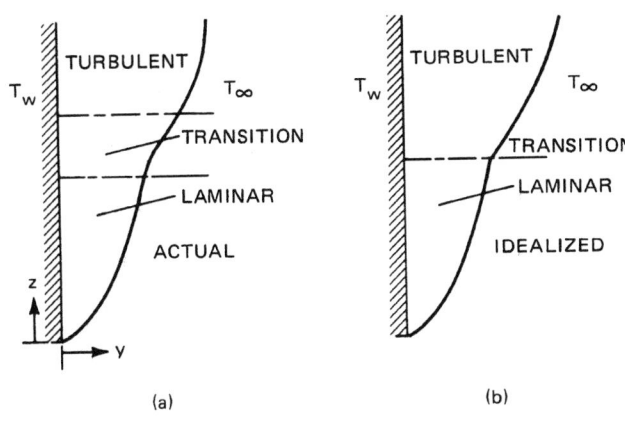

*Fig. 2-6.8. **Actual and idealized natural convection boundary layer wall flow.***

The integral analysis assumes the velocity and temperature distributions, substitutes these in the governing integral equations, and solves for the unknown quantities such as δ, maximum velocity, etc. The distributions employed for laminar flow are

$$u = U\frac{y}{\delta}\left(1 - \frac{y}{\delta}\right)^2, \quad \frac{T-T_\infty}{T_w-T_\infty} = \left(1 - \frac{y}{\delta}\right)^2 \quad (4)$$

where U and δ are functions of the height, Z. The governing equations are obtained as ordinary differential equations for U and δ. These equations are

$$\frac{d}{dZ}(\delta U^2/105) = \frac{g\beta(T_w - T_\infty)}{3} - \frac{\nu U}{\delta} \quad (5)$$

$$\frac{d}{dZ}(\delta U/30) = \frac{2\alpha}{\delta} \quad (6)$$

In the previous equations, the Boussinesq approximation, $\rho_\infty - \rho = \beta\rho(T - T_\infty)$, has been employed. Here, g is the acceleration due to gravity, ν the fluid kinematic viscosity, α its thermal diffusivity, and β the coefficient of thermal expansion. For the temperature differences, $\Delta T = T_w - T_\infty$, generally observed in experimental studies, the Boussinesq approximation is quite satisfactory. These equations yield U and δ as

$$U = 5.17\nu\left(\text{Pr} + \frac{20}{21}\right)^{-1/2}\left(\frac{g\beta\Delta T}{\nu^2}\right)^{1/2}Z^{1/2} \quad (7)$$

$$\delta = 3.93 \, \text{Pr}^{-1/2}\left(\text{Pr} + \frac{20}{21}\right)^{1/4}\left(\frac{g\beta\Delta T}{\nu^2}\right)^{-1/4}Z^{1/4} \quad (8)$$

where Pr is the Prandtl number and is given by $\text{Pr} = \nu/\alpha$. Similarly, $\dot{m}$, $\dot{M}$, and $\dot{q}$ may be determined from these integral expressions for the constant density approximation as

$$\dot{m} = \rho\delta U/12 \quad (9)$$

$$\dot{M} = \rho\delta U^2/105 \quad (10)$$

$$\dot{q} = \rho C_p\delta U\Delta T/30 \quad (11)$$

The maximum velocity, u_{max}, occurs at $y = \delta/3$ and $(4/27)U$. Here, ρ may be taken as constant in each of the two regions,

with different values evaluated at the average of the wall and gas temperatures being employed for each layer.

Similarly, various analyses have been carried out[26-28] for turbulent flow, supported by various experimental studies.[16,29] The profiles employed in the analyses are

$$u = U\left(\frac{y}{\delta}\right)^{1/7}\left(1 - \frac{y}{\delta}\right)^4, \quad \frac{T-T_\infty}{T_w-T_\infty} = 1 - \left(\frac{y}{\delta}\right)^{1/7} \quad (12)$$

The governing ordinary differential equations are obtained as[26]

$$0.0523\frac{d}{dZ}(\delta U^2) = 0.125g\beta\Delta T\delta - 0.0225U^2\left(\frac{\nu}{U\delta}\right)^{1/4} \quad (13)$$

$$0.0366\frac{d}{dZ}(\delta U) = 0.0225U\left(\frac{\nu}{U\delta}\right)^{1/4}\text{Pr}^{-2/3} \quad (14)$$

If the boundary layer is assumed to be fully turbulent from the leading edge, $Z = 0$, the expressions for U and δ are obtained as

$$U = 1.185\frac{\nu}{Z}(\text{Gr})^{1/2}[1 + 0.494(\text{Pr})^{2/3}]^{-1/2} \quad (15)$$

$$\delta = 0.565Z(\text{Gr})^{-1/10}\text{Pr}^{-8/15}[1 + 0.494(\text{Pr})^{2/3}]^{1/10} \quad (16)$$

where Gr, the Grashof number, is given by

$$\text{Gr} = g\beta\Delta T \, Z^3/\nu^2 \quad (17)$$

Again, the mass flow rate, momentum flow rate, and rate of convected energy in the boundary layer may be obtained per unit perimeter of the flow region as

$$\dot{m} = 0.1463\rho\delta U \quad (18)$$

$$\dot{M} = 0.0523\rho\delta U^2 \quad (19)$$

$$\dot{q} = 0.0366\rho C_p\delta U\Delta T \quad (20)$$

The maximum vertical velocity, u_{max}, is obtained as $0.537\,U$.

As an example, consider these equations if all the gas properties are taken as those of air at standard ambient conditions. The above physical quantities may then be calculated from the following equations, obtained by taking $\text{Pr} = 0.72$

$$\dot{m} = \begin{cases} 1.755\mu\text{Gr}^{1/4}; & \text{Laminar, Gr} \lesssim 10^9 \\ 0.101\mu\text{Gr}^{2/5}; & \text{Turbulent, Gr} \gtrsim 10^9 \end{cases} \quad (21)$$

$$\dot{M} = \begin{cases} 0.802\rho\nu^2\text{Gr}^{3/4}/Z; & \text{Laminar} \\ 0.036\rho\nu^2\text{Gr}^{9/10}/Z; & \text{Turbulent} \end{cases} \quad (22)$$

$$\dot{q} = \begin{cases} 0.702\mu C_p\text{Gr}^{1/4}\Delta T; & \text{Laminar} \\ 0.025\mu C_p\text{Gr}^{2/5}\Delta T; & \text{Turbulent} \end{cases} \quad (23)$$

Also, the maximum velocity, u_{max}, in the boundary layer is given by

$$u_{max} = \begin{cases} 0.592\nu\text{Gr}^{1/2}/Z; & \text{Laminar} \\ 0.538\nu\text{Gr}^{1/2}/Z; & \text{Turbulent} \end{cases} \quad (24)$$

Therefore, the physical quantities of interest may be obtained from these expressions for mass, momentum, and energy transport. However, the results are obtained with the assumption of a steady boundary layer flow. Also, the expressions for turbulent flow are based on the assumption of fully turbulent flow over the entire wall, starting with the leading edge.

A question of accuracy may be raised regarding the application of the above boundary layer analysis since considerable turbulent flow and mixing occur in the region away from the walls. To some extent, this effect can be taken into consideration by determining the location where the boundary layer flow undergoes transition to turbulence.[15,16] Transition occurs earlier if the flow away from the walls is highly turbulent. In addition, the turbulent flow outside the boundary layer gives rise to a nonzero velocity in the ambient region and consequently a pure natural convection flow is not obtained and the mixed convection effects may have to be considered. As reviewed by Jaluria,[15] this effect may be considered and the resulting flow rate, momentum flow rate, and the convected thermal energy may be determined. For the range of velocities generally encountered in these flows, as estimated from the flow in the fire plume, it can be shown that the effect is not very large.

In the previous equations, fluid properties are taken as constant and the values may be determined at the average of the local wall and gas temperatures. For the range of ΔT observed experimentally, this approach is quite accurate.[15] If variations in fluid properties are to be incorporated into the analysis, the governing equations and the results would need to be suitably modified. Numerically, however, the variations have only a small effect for the range of temperature difference generally encountered in these flows.

In actual practice, the boundary layer is laminar near the leading edge, which is the top or bottom edge where the flow starts. (See Figure 2-6.3.) This flow undergoes transition downstream and eventually attains fully developed turbulence as shown in Figure 2-6.8. This is obviously a complicated problem and a detailed study becomes quite involved. Laminar flow may be assumed up to a certain point, followed by turbulent flow downstream. This implies a step change from laminar to turbulent flow, as shown in Figure 2-6.8, part (b), is an approximation to the actual circumstance. However, this flow circumstance requires a solution of the governing equations (Equations 13 and 14) for turbulent flow, with inputs from the laminar flow (governed by Equations 7 and 8), which applies up to transition. A numerical solution, employing the Runge-Kutta method, may be obtained.[13] The values for laminar flow up to transition are obtained from the expressions for δ and U previously given, and the numerical solution of the equations for turbulent flow yields the downstream values. For various values chosen for the Grashof number, Gr_c, at which transition occurs, the numerical solution may be obtained by matching the flow rate and the momentum flow across the transition location. Figure 2-6.9 shows the flow rate in kg/m·s over the flow perimeter for various conditions. The results are shown for $\Delta T = 60°C$ and $30°C$.

It is obvious from Figure 2-6.9 that the boundary layer mass flow rate depends greatly on the transitional Grashof number. Many earlier studies have indicated the value of Gr_c to be around 5×10^9; however, the value can be as large as 10^{10} and as low as 10^9, depending on the background

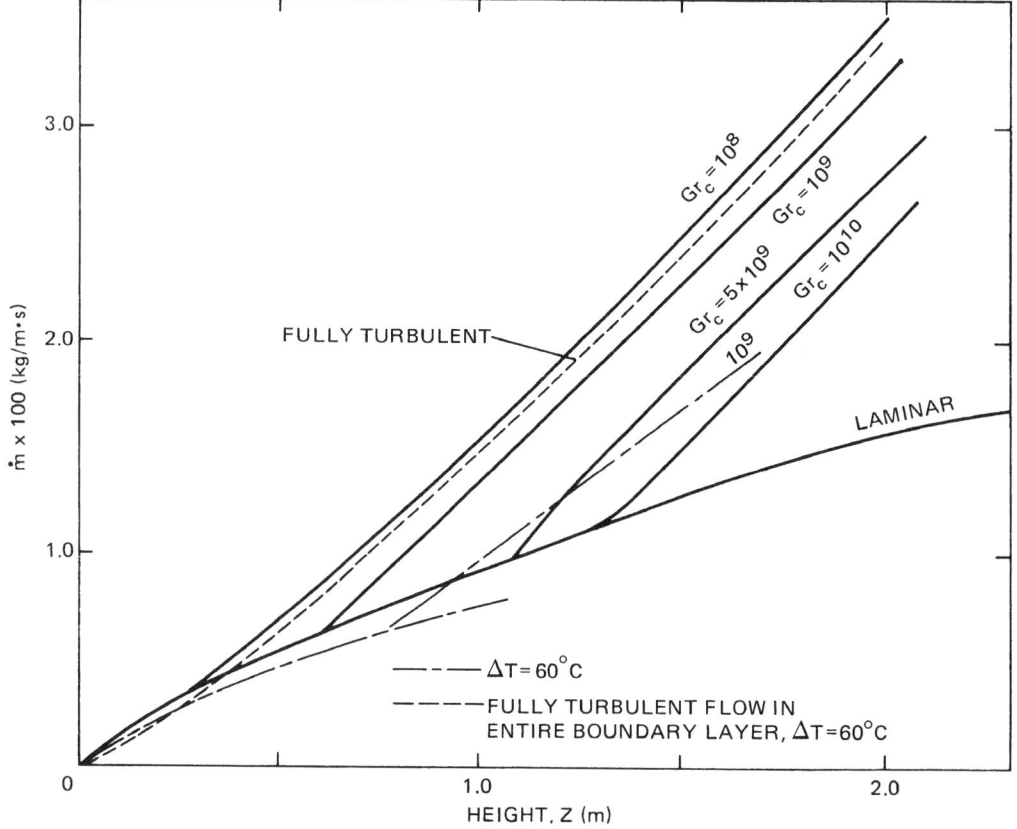

Fig. 2-6.9.   Mass flow rate in the wall flow, $\dot{m}$, as a function of the height Z.

disturbance. For the problem of a relatively quiescent compartment fire, 5×10^9 is expected to apply and for a disturbed environment a fully turbulent flow may be expected over much of the region. It is clear, therefore, that there is uncertainty regarding the onset of turbulence. The two extreme cases of laminar and completely turbulent flow may be considered and an average taken to represent the resulting values. This approach appears reasonable in terms of the results shown in Figure 2-6.9. One needs to know the position of the interface so that the expressions for $\dot{m}$, $\dot{M}$, and $\dot{q}$, corresponding to laminar and turbulent flow, may be employed to obtain the flow rate, momentum flow rate, and convected energy due to the wall flow. The location of the interface may be obtained experimentally or analytically, from zone model analysis, as discussed later in this chapter.

It is important to remember that the foregoing analysis assumes steady flow. Though steady-state conditions may be obtained in some cases, particularly in the laboratory,[9,13,30] the general problem of the development of fire in an enclosure is a transient circumstance and, thus, the interface position generally varies with time.[12] However, the problem may be treated as quasi-steady in most cases, after the establishment of the two zones, and the above expressions may then be employed to determine the transport quantities. This analysis would not be expected to apply at very early stages of the fire, when the two layers are being established.[22]

RELATIVE MAGNITUDE OF WALL FLOWS

To determine the relative importance of the wall flows generated in enclosure fires, the flow rates generated by these flows may be compared with the corresponding flow rate in either a fire plume or through the openings of the enclosure. Experimental results obtained in studies of steady enclosure fire situations[9,13,30] determined the wall and gas temperature profiles generated by fire in a room. The experiments were conducted in a $2.8 \times 2.8 \times 2.18$ m room with an opening with variable dimensions. Essentially steady-state conditions were attained within about 30 minutes following ignition of the burner. The experiments employed circular and line burners, the latter consisting of a long slot with the gas supplied at an array of circular holes. The gas flow rate and the length of the line burner could be adjusted. Velocity and temperature measurements were taken by means of bidirectional velocity probes and thermocouples. Bare thermocouples were embedded in the walls and the ceiling to obtain wall temperature data.

Several experiments were carried out with various burner locations and configurations, door openings, and fire heat inputs. Some of the typical results obtained from these experiments are shown in Figures 2-6.10 and 2-6.11. As expected, it was found that the results depended on the heat input, the size of the door opening, and the burner location. However, the basic form of the profiles was similar in all the cases considered. The results showed that wall temperature increases gradually from the floor and then more sharply as the upper layer is approached, attaining essentially uniform temperature in the upper layer in most cases. The gas temperature shows greater uniformity in the two layers and a much sharper temperature rise from the lower to the upper layer. It starts increasing gradually as the upper layer is approached, then rises very rapidly over a relatively very short distance, and gradually attains the upper layer temper-

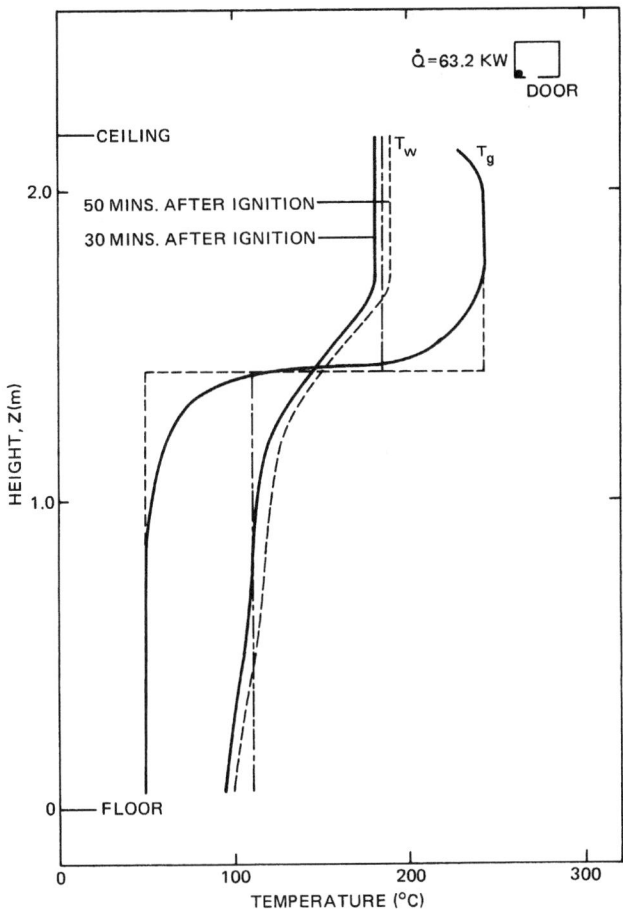

Fig. 2-6.10. *Measured gas and wall temperatures in the room with a 1.83 m × 0.74 m door opening, for a circular burner and Q = 63.2 kW.*

ature. The region of rapid temperature variation spreads over a distance of about 20 cm; the region is smaller in some cases and much larger in others, particularly at small door openings which give rise to much greater mixing in the room.[9] Two cases are shown in this text. (Similar trends were observed at other operating conditions.) It is evident from the results presented here that a two-layer zone model can be employed to obtain average temperatures in the two zones. Gas temperatures are fairly uniform in the two layers and a sharp change in the temperature occurs between one layer and the other.

The natural convection flow generated due to the temperature difference between the walls and the adjacent fluid can be computed. If the data of Figure 2-6.10 is taken as typical, the upward flow generated in the lower region may be calculated; and similarly, the downward flow that arises in the upper zone may be determined. From Figure 2-6.10, the interface for a two-layer model may be taken at a height of 1.42 m, for which the laminar solution gives a flow rate for the upward flow in the lower region at the interface height of 0.0121 kg/m·s and the turbulent solution yields 0.0225 kg/m·s. The opening flow measured in this case was 0.424 kg/s.[30] For a flow perimeter of about 10 m, which applies for the experimental arrangement, the upward wall flow in the

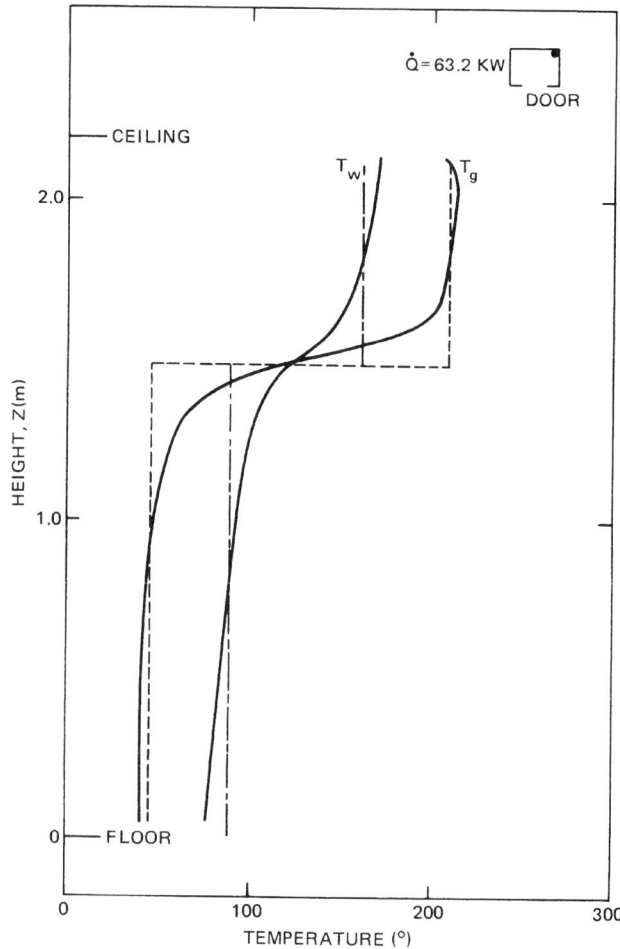

Fig. 2-6.11. Measured gas and wall temperatures in the room, with a 1.83 m × 0.74 m door opening, for a circular burner and Q = 63.2 kW, at a different location from that in Fig. 2-6.10.

lower region ranges from 28.5 percent for laminar flow to 53 percent for turbulent flow, of the opening flow rate. An average of the two gives a value of 41 percent, which indicates the level of importance of the wall flow in this case. This average value of the upward flow at the interface was obtained as 42.5 percent, 39.4 percent, and 21.6 percent in three other experiments. Thus, the wall flow was found to be significant for these experimental conditions.

A similar comparison of the estimated wall flow rate may be made with the corresponding fire plume flow rate. Following the model of Morton, Taylor, and Turner[31] for a point heat source, the plume width, d, centerline velocity, U_c, and the centerline density difference, $\Delta\rho_c$, may be obtained in terms of the heat input, $\dot{Q}_c$, into the convective flow. The relevant equations are

$$d = \frac{6\tilde{\alpha}}{5}Z \qquad (25)$$

$$U_c = \frac{5}{6\tilde{\alpha}}\left(\frac{9}{5}F\tilde{\alpha}\right)^{1/3}Z^{-1/3} \qquad (26)$$

$$\rho_\infty - \rho_c = \rho_0\frac{5F}{6\tilde{\alpha}\lambda^2 g}\left(\frac{9}{5}F\tilde{\alpha}\right)^{-1/3}Z^{-5/3} \qquad (27)$$

where

$$F = \frac{g(\lambda^2+1)\dot{Q}_c}{\pi\rho_0 T_0 C_p} \qquad (28)$$

Here, the subscript $_0$ refers to the ambient conditions at the level of the heat source, $\tilde{\alpha}$ is a constant, termed the entrainment coefficient,[31] and λ a constant that compares the spread of the thermal field with that of the velocity field in the plume. The radial velocity and density distributions at any height, Z, have been assumed to be Gaussian, i.e.,

$$u = U_c\exp(-r^2/d^2), \quad \rho_\infty - \rho = (\rho_\infty - \rho_c)\exp(-r^2/\lambda^2 d^2) \qquad (29)$$

The constants $\tilde{\alpha}$ and λ may be taken as 0.1042 and 1.15, based on the work of Yokoi[32] and Zukoski *et al.*[33] The numerical results from this model were also compared with the measurements of McCaffrey,[34] indicating a fairly good agreement far from the source.

As measured by McCaffrey,[35] about 20 percent of the input energy is lost as radiation for a methane flame. Employing this result to determine the convected energy, $\dot{Q}_c$, the flow rate in the fire plume was determined. Figure 2-6.12 shows a comparison between the wall flow rate and the plume flow rate as a function of the height, Z. For the experimental interface location at 1.42 m, corresponding to the data of Figure 2-6.10, the fractional strength of the wall flow varies from 27.5 percent in laminar flow to 51 percent in fully turbulent flow, resulting in an average value of 39.3 percent. At two values of Gr_c, the corresponding results are also shown in Figure 2-6.12. The relative significance of the wall flow increases as the interface height decreases, indicating the dominant effect of a decrease in the height on the entrainment into the plume and on the consequent plume flow rate. These results are shown for $\dot{Q} = 63.2$ kW, with the conditions in Figure 2-6.10, and with the interface height as 1.42 m. Therefore, it can be seen that a significant flow rate,

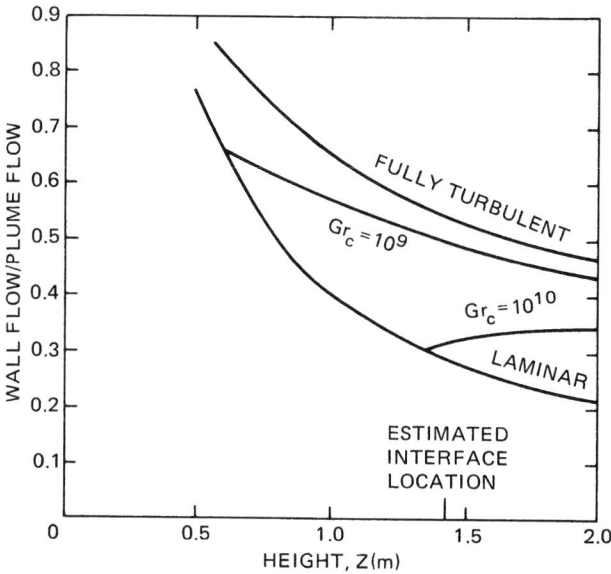

Fig. 2-6.12. A comparison between the wall flow rate and the plume flow rate, for a circular burner at Q = 63.2 kW and a 1.83 m × 0.74 m door opening.

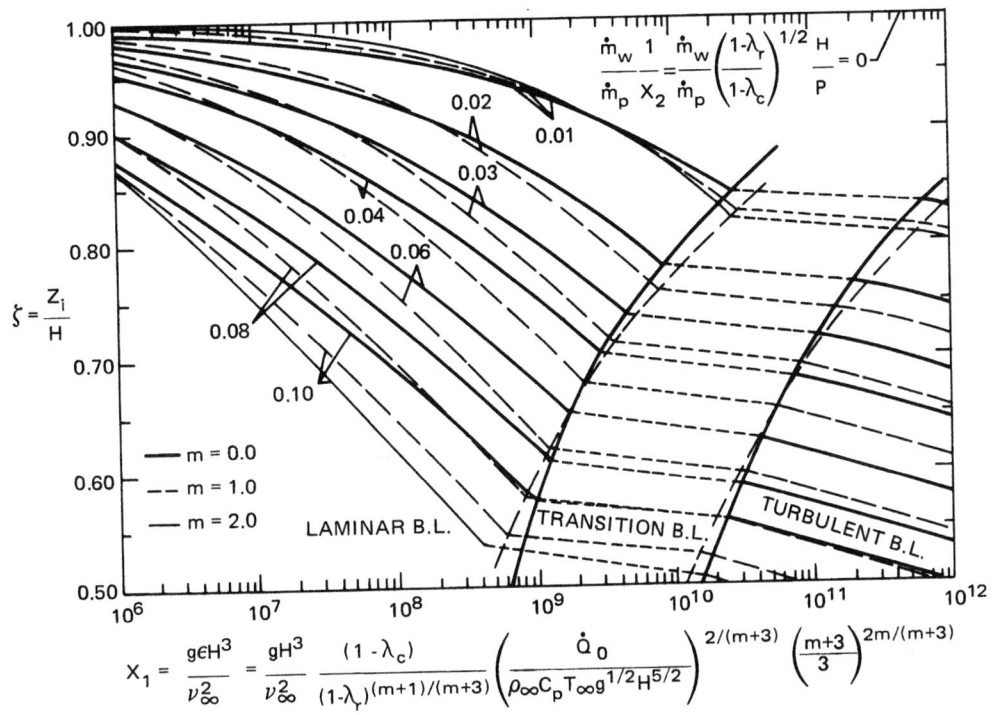

$$X_1 = \frac{g\epsilon H^3}{\nu_\infty^2} = \frac{gH^3}{\nu_\infty^2} \frac{(1-\lambda_c)}{(1-\lambda_r)^{(m+1)/(m+3)}} \left(\frac{\dot{Q}_0}{\rho_\infty C_p T_\infty g^{1/2} H^{5/2}}\right)^{2/(m+3)} \left(\frac{m+3}{3}\right)^{2m/(m+3)}$$

Fig. 2-6.13. Dependence of the relative importance of the wall flow on the physical parameters of an enclosure fire.[12]

as compared to the flow in the fire plume and also to the flow through the opening, is generated adjacent to the walls. If this flow penetrates into the upper layer from below, or into the lower layer from the upper region, it conveys air and gases across the interface. This condition implies a transport of mass, momentum, and energy from one layer to the other and should be included with other transport mechanisms across the interface.

A detailed study of the wall effect during the early stages of fire growth in an enclosure has been carried out by Cooper,[12] considering the circumstance shown in Figure 2-6.5—the downward injection of the hot upper layer gas into the relatively cool, uncontaminated lower layer. No opposing upward flow arises in the lower region, since the walls are taken as being close to the initial temperature during these early stages of fire growth. The energy release, $\dot{Q}(t)$, by the fire (where t is time), is taken as proportional to t^m, with $m \geq 0$. The mass flux $\dot{m}_w$ in the wall flow is estimated as given in the preceding section, and compared with that in the fire plume, $\dot{m}_p$, at the upper layer-lower layer interface.[33] Figure 2-6.13 shows the dependence of $\dot{m}_w/\dot{m}_p$ on the physical parameters of the problem; the variables X_1 and X_2 are defined in the figure. Here, Z_i is the height of the interface above the fire, H the height of the ceiling, $\dot{Q}_0$ is a characteristic energy release rate, λ_r is the fraction of $\dot{Q}$ lost by radiation, λ_c is the fraction of $\dot{Q}$ which is instantaneously transferred to the internal bounding surfaces of the enclosure, and P is the perimeter of the enclosure. For further details on the problem and the analysis carried out, see Ref. 12. Similar trends have been observed in the experimental study reported in Ref. 22, wherein the early stages of a fire, without an upward wall flow, were investigated.

Assuming that the wall effect may be taken as significant when $\dot{m}_w/\dot{m}_p > 0.3$, Cooper[12] considered several experimental studies and determined the height of the inter-

face when the wall effect must be taken into account. For instance, for a 25 kW fire located 2.12 m below the ceiling in a room with a total wall perimeter of 81 m and floor area of 89.6 m², the wall effect was found to be significant once the upper layer thickness exceeded 0.32 m. Similarly, for a 225 kW fire in the smallest test space area, 40.6 m², the wall effect was found to be significant when the upper layer thickness exceeded about 0.51 m. In these calculations, the wall temperature was taken as close to the ambient temperature and the enclosures were fully enclosed, with leakage at the bottom.

PENETRATION ACROSS THE INTERFACE

The next step in the consideration of natural convection wall flows is a study of the penetration of these flows across the interface. If the flows do penetrate, conditions under which this occurs, plus the magnitude of the penetrating flow, should be determined. For the idealized two-layer system in Figure 2-6.7, part (a), there are two isothermal regions at different temperatures and a step change in temperature is assumed. This implies an opposing buoyancy effect as the flow moves from one layer to the other. Both the three- and two-layer models have been considered.[13] The three-layer profile is of interest when experimental data are available and the two-layer system is applicable to the zone-modeling studies. The two are interrelated and the ultimate focus for modeling purposes is on the two-layer approximation.

In flow penetration in a two-zone system such as the one shown in Figure 2-6.7, part (a), as the wall flow from below crosses the interface it is subjected to a step change in the ambient temperature. Under the action of the strong buoyancy forces and the large temperature gradients that arise

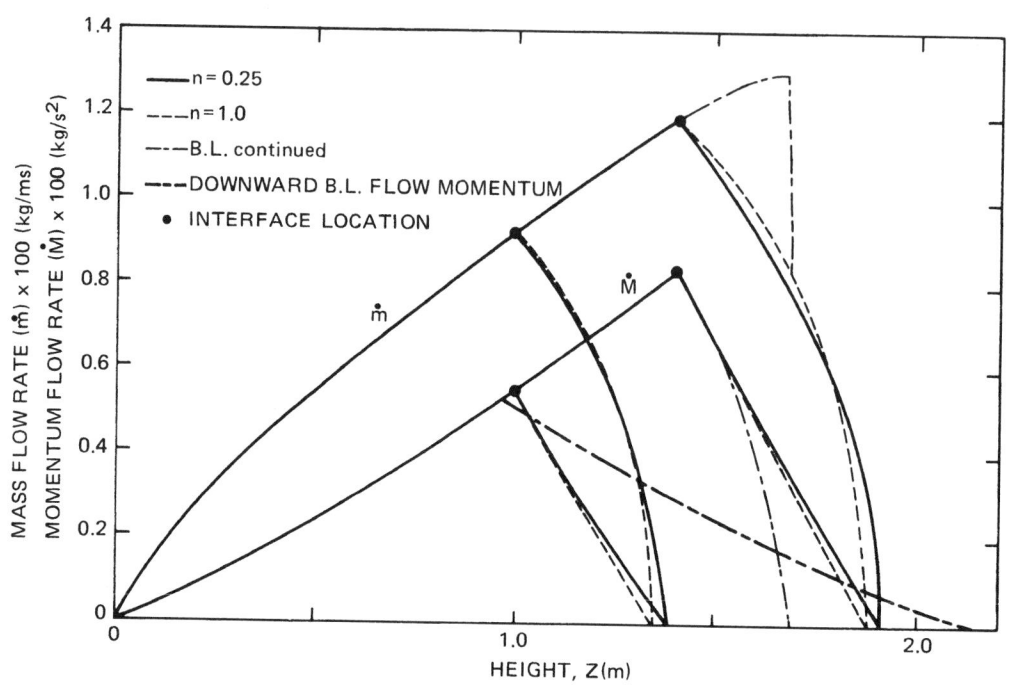

Fig. 2-6.14. Mass and momentum flow for a two-layer system in laminar flow, considering various models for the negatively buoyant circumstance.

across the flow, the temperature profile will adjust in a very short distance to attain the wall temperature, $T_{w,u}$, at $y = 0$ and the ambient temperature, $T_{g,u}$, at $y = \delta$. In actual practice, of course, there is a finite region over which this adjustment occurs. A three-layer model as seen in Figure 2-6.7, part (b), may be employed to consider the adjustment. Here, because of the step change, $\dot{m}$ and $\dot{M}$ may be taken as unchanged while T_∞ changes from $T_{g,\ell}$ to $T_{g,u}$. This temperature change implies a considerable energy flux into the flow due to convective flow and diffusion. This flux can be achieved by a substantial amount of mixing at the interface, as observed in the experiments. The resulting thermal transport raises the temperature in the outer region of the flow to the upper layer gas temperature. Such a mixing process will lead to a finite amount of entrainment into the flow, and since the flow is negatively buoyant it is also expected to shed flow further downstream. In view of the disturbed nature of this penetration, an analytical or experimental study of the flow in the vicinity of the interface is desirable; unfortunately, very little information is presently available on such penetrative flows. However, some recent work on the penetration of wall jets and buoyancy-induced flows across the interface of a two-layer stratified region has been reported.[36,37] The basic mechanisms outlined here are corroborated by these experiments, lending support to the simple model presented here.

This flow may be analyzed by integral methods, as discussed earlier in this chapter. However, boundary layer analysis is not strictly valid, since reverse flow arises downstream. A solution of the full governing equations would be more appropriate. If both the upward and downward fluid streams are analyzed by the integral approach, the variation of the mass flux and the momentum flux with the height may be obtained. A simple model developed for these configurations[13] assumes that the location where the flows separate from the wall occurs where the momentum fluxes of the two

streams are equal. It could then be determined whether the upward flow penetrates into the upper layer or the downward flow penetrates into the lower layer.

Figure 2-6.14 shows the mass flow rate, $\dot{m}$, and the momentum flow rate, $\dot{M}$, per unit perimeter for laminar flow, considering the data from Figure 2-6.10. Two values of n, where δ is assumed to vary as Z^n beyond the interface, along with the results from the corresponding boundary layer analysis are shown. The separation point may be determined by employing the criterion of equal downward and upward momentum. Figure 2-6.15 shows the results for a two-layer system in turbulent flow. From such calculations, the penetration distance, δ_p, may be obtained as a function of the interface height, Z_i. Figure 2-6.16 shows the results for turbulent flow in a two-zone system. A considerable difference between the values with and without opposing flow is seen. Without the opposing stream, the steady-state upward flow always penetrates, with δ_p becoming smaller as the interface height, Z_i, becomes smaller. But with opposing flow, the penetration is restricted. For $Z_i \leq 0.9$ m, the flow from the upper layer penetrates into the lower layer. Also note that with a large Z_i (> 1.6 m), the flow from below does not stagnate at any height, but ultimately hits the ceiling (2.18 m height).

The penetration of the wall flow from one isothermal layer into another at a different temperature is a very involved process. The penetration depends on the temperature profile and the opposing flow that may arise in the other layer. The flow separates from the wall when the two opposing streams have equal momentum, as observed in similar flow configurations.[38,39] The flow turns away from the wall and descends or rises toward the interface, since the flow is negatively buoyant. A considerable amount of mixing occurs in this region and flow stability considerations are important.[36,37] The actual profiles may be considered if appropriate experimental data are available. A two-zone idealized

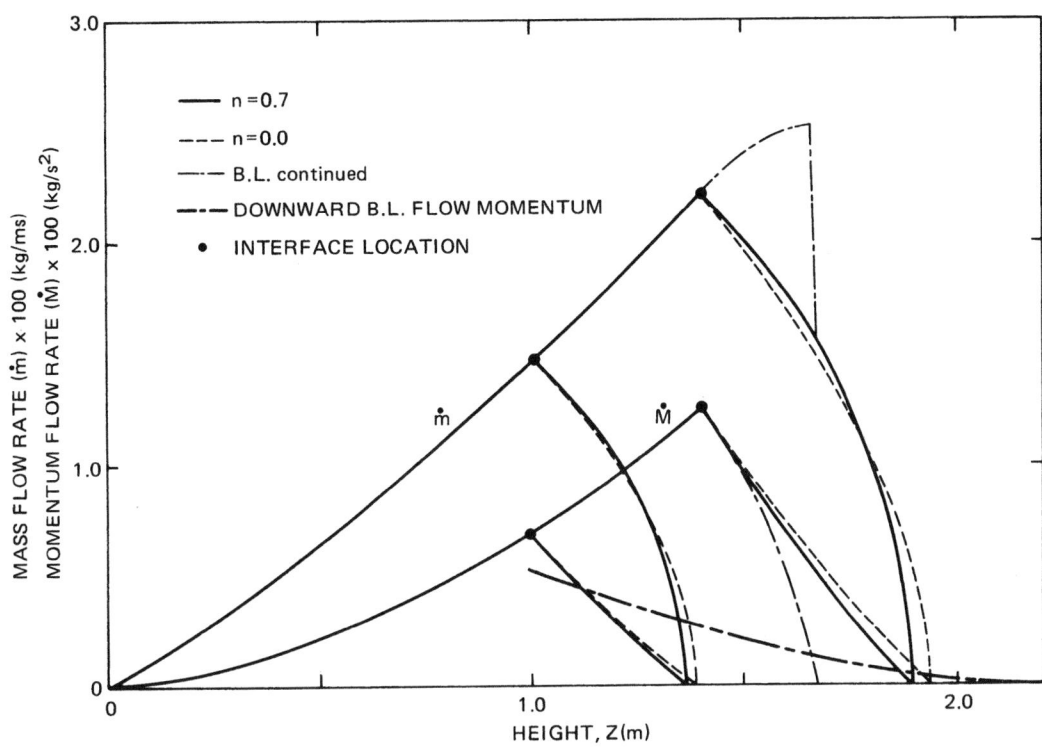

Fig. 2-6.15. *Mass and momentum flow for a two-layer system in turbulent flow.*

profile is appropriate for zone modeling studies. For these studies, computation of the momentum for the two streams indicates which zone the flow enters and is mixed with the fluid there, conveying mass and energy. If measured profiles are available, a three-layer system may be considered and the location of the separation point determined.[13] The flow rates and the energy transported across the interface may be determined in order to provide inputs into the zone model.

For the flow circumstance in Figure 2-6.5, with no opposing flow, Cooper[12] took the criterion for penetration as

$$(u_{max})_{Z_i} \geq 5\frac{dZ_i}{dt} \qquad (30)$$

where the maximum velocity is computed at the interface. This penetration criterion is based on a comparison between the velocity in the wall flow and the movement of the interface. It was shown that the flow did penetrate according to this criterion in the various experimental conditions considered for estimation of the wall effects.

Detailed studies on the nature of such negatively buoyant and penetrative flows have been conducted.[40-42] The penetration depth and the entrainment in negatively buoyant two-dimensional jets have been determined experimentally and it was found that the flow characteristics largely depend on the parameter Gr/Re^2, where Gr and Re are the Grashof and Reynolds numbers, defined as

$$Gr = g\beta(\hat{T}_0 - T_\infty)D^3/\nu^2, \quad Re = \hat{U}_0 D/\nu \qquad (31)$$

Here, $\hat{T}_0$ and $\hat{U}_0$ are inlet temperature and velocity of the negatively buoyant jet, and D is the width of the discharge slot. Figures 2-6.17(a) and 2-6.17(b) show the penetration depth, δ_p, of such a wall jet as a function of Gr/Re^2 and

Figure 2-6.18 shows the total flow rate, $\dot{m}_{out}$, which is buoyed back out of the enclosure.

These results are considerably important in determining the nature of penetration of wall flows. At the interface, the flow rate, average temperature, and boundary layer thickness due to the wall flow may be computed to yield representative values for Gr and Re. The penetration distance and the total entrainment of the lower layer fluid into the downward wall flow may then be determined. This flow

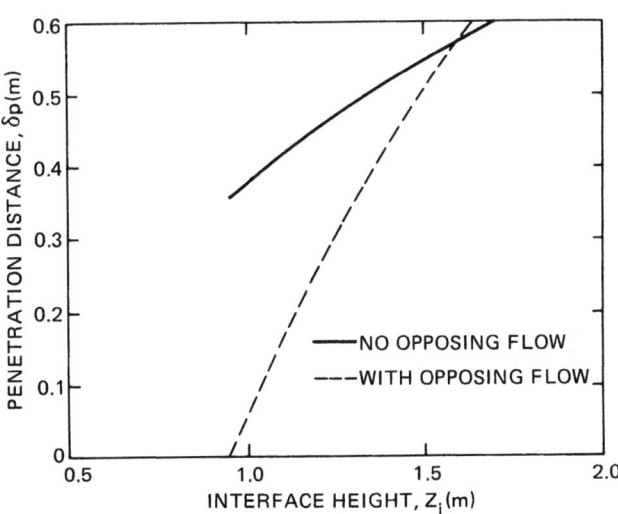

Fig. 2-6.16. *Penetration depth, δ_p, as a function of the interface height, Z_i, for turbulent flow in a two-layer system.*

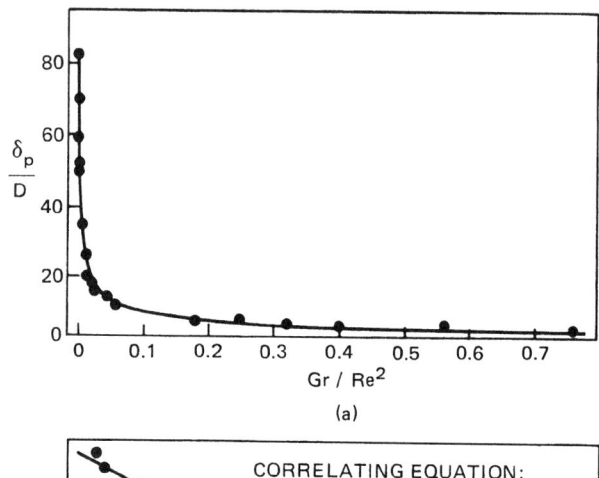

(a)

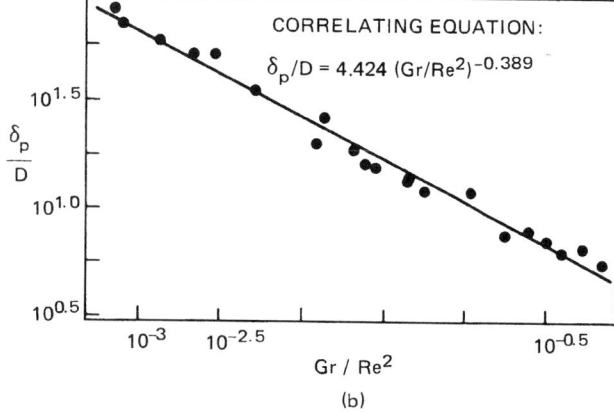

CORRELATING EQUATION:

$\delta_p/D = 4.424 \, (Gr/Re^2)^{-0.389}$

(b)

Fig. 2-6.17. *Downward penetration distance, δ_p, of a negatively buoyant wall jet as a function of Gr/Re^2.[40-42]*

spreads out horizontally and rises toward the interface because of the buoyancy effect and again penetrates into the upper layer (to a much shorter distance due to the much lower velocity level),[40,43] and then drops back toward the interface. The wall flow thus results in a mixing of the fluid in the two zones. Using such information on the resulting transport due to the wall flow, the effect of the flow on the redistribution of the combustion products may be estimated by determining the appropriate penetration distance and entrainment.

As pointed out by Cooper,[12] for the flow circumstance of Figure 2-6.5, once it is in the lower layer, the downward-directed wall jet would be buoyed back upward and away from the wall to either mix with and contaminate the lower-layer air or to convey additional lower-layer air into the upper zone. The basic wall flow could significantly alter the rate of development of life-threatening conditions in the enclosure. If, for example, the flow is buoyed upward as a plume and re-enters the upper layer, then, having entrained additional lower-layer air, the depth of the upper layer will grow more rapidly, though at a reduced temperature and product concentration. On the other hand, if the wall flow is mixed with the lower-layer gas, the net effect would be a contamination of the lower layer and less rapid growth in upper layer depth at increased temperature and concentration.

When the mass flux leaving the upper layer in the wall flow is large enough to significantly retard the velocity of descent, dZ_i/dt, of the upper layer-lower layer interface or to lead to significant increases in temperature and/or product of com-

bustion contamination of the lower layer, this flow must be taken into account in a mathematical simulation of the overall enclosure environment. As mentioned earlier, $\dot{m}_w/\dot{m}_p > 0.3$ is taken as the criterion for significant wall effect.[12]

RESULTING EFFECT ON ZONE MODELING

In addition to the flow rates shown in Figure 2-6.1, the total wall flow rate can be denoted by $\dot{m}_w$, the plume flow rate that crosses the interface by $\dot{m}_p$, and the plume flow that is peeled off and is unable to penetrate by $\dot{m}_p'$. For steady-state conditions in a lower layer control volume that excludes the fire plume, a mass balance gives

$$\dot{m}_f + \dot{m}_a + \dot{m}_j = \dot{m}_p + \dot{m}_w \qquad (32)$$

where $\dot{m}_f$ is the mass flow rate of the fuel. Therefore

$$\dot{m}_p = \dot{m}_a + \dot{m}_j + \dot{m}_f - \dot{m}_w \qquad (33)$$

and

$$\dot{m}_e = \dot{m}_p + \dot{m}_p' \qquad (34)$$

This gives the plume flow entering the upper layer, normalized by the flow through the opening, $\dot{m}_a$, as

$$\frac{\dot{m}_p}{\dot{m}_a} = 1 + \frac{\dot{m}_j}{\dot{m}_a} + \frac{\dot{m}_f}{\dot{m}_a} - \frac{\dot{m}_w}{\dot{m}_a} \qquad (35)$$

and the rate of entrainment over height Z_i, $\dot{m}_e$, as

$$\frac{\dot{m}_e}{\dot{m}_a} = 1 + \frac{\dot{m}_j}{\dot{m}_a} + \frac{\dot{m}_f}{\dot{m}_a} - \frac{\dot{m}_w}{\dot{m}_a} + \frac{\dot{m}_p'}{\dot{m}_a} \qquad (36)$$

All these flow rates may be defined in terms of the circumstance shown in Figure 2-6.1.

In a three-layer system, some of the plume flow is shed in the intermediate stratified region. However, the effect is not large for the range of physical variables considered here, and for the two-layer approximation, the flow rate may be

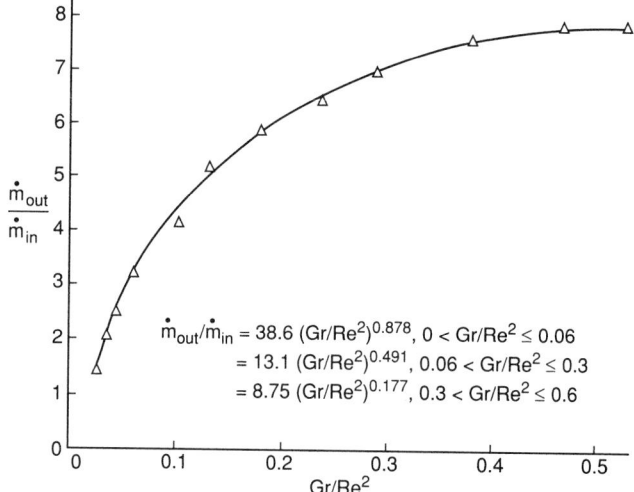

$\dot{m}_{out}/\dot{m}_{in} = 38.6 \, (Gr/Re^2)^{0.878}, \, 0 < Gr/Re^2 \le 0.06$
$= 13.1 \, (Gr/Re^2)^{0.491}, \, 0.06 < Gr/Re^2 \le 0.3$
$= 8.75 \, (Gr/Re^2)^{0.177}, \, 0.3 < Gr/Re^2 \le 0.6$

Fig. 2-6.18. *Variation of the net flow out of the enclosure, $\dot{m}_{out}$, normalized by jet discharge flow rate, $\dot{m}_{in}$, as a function of Gr/Re^2.[40-42]*

assumed to remain essentially unaltered. The fuel flow rate is generally much smaller than the opening flow rate. Neglecting these two quantities, the rate of entrainment becomes

$$\frac{\dot{m}_e}{\dot{m}_a} = 1 + \frac{\dot{m}_j}{\dot{m}_a} - \frac{\dot{m}_w}{\dot{m}_a} \qquad (37)$$

The airflow rate may be measured experimentally with good accuracy.[9,30] If an energy balance is considered for the lower layer, the energy input into the plume flow at the source and into the wall flow is conveyed to the upper layer unless a significant amount of these flows peel off and get mixed with the lower layer. This is a negligible effect according to the results obtained herein.

Thus, the energy balance gives

$$\dot{m}_a(T_{g,\ell} - T_0) = \dot{m}_j(T_{g,u} - T_{g,\ell}) + \frac{\dot{Q}_{floor}}{C_p} \qquad (38)$$

where T_0 is the outside air temperature and $\dot{Q}_{floor}$ is the heat transfer between the floor and the gas. Also, $\dot{Q}_{floor}$ may be estimated as

$$\dot{Q}_{floor} = hA(T_{floor} - T_{g,\ell}) \qquad (39)$$

where h is the convective heat transfer coefficient, A the floor area, and T_{floor} the floor temperature, which is greater than the gas temperature in the lower layer. If $\dot{Q}_{floor}$ is neglected, an upper bound for $\dot{m}_j$ is obtained as

$$(\dot{m}_j)_{max} = \frac{T_{g,\ell} - T_0}{T_{g,u} - T_{g,\ell}} \dot{m}_a \qquad (40)$$

If T_{floor} is measured, available heat transfer correlations[15,16] may be employed to determine $\dot{m}_j$ more accurately. As discussed,[9,30] $\dot{m}_j/\dot{m}_a$ increases as the door opening is decreased. It must be noted that if the wall flow from the upper layer penetrates into the lower layer, the energy added to the lower layer due to this flow and the energy due to heat transfer from the lower wall should be included in Equation 38 for greater accuracy.

Therefore, the wall flow rate enters into the calculation for $\dot{m}_e$ and this flow is often quite significant as compared to the flow through the opening $\dot{m}_a$. The two can be compared by considering the results shown in Figure 2-6.9 for $T_{w,\ell} - T_{g,\ell} = 60°C$ and $30°C$, along with those from other experiments. The estimated value of $\dot{m}_e$ obtained from the measured value of $\dot{m}_a$ is reduced by 20 to 40 percent by considering the wall flow. This additional consideration could lead to a better agreement between the theoretical and experimental predictions of the plume entrainment; however, this assumes that the entire wall flow penetrates into the other zone. As shown earlier in this chapter, the flow buoys back toward the interface and some flow redistribution between the two zones occurs. In the absence of more detailed information, the above treatment would overpredict the wall flow effects.[30] An approximate distribution of flow between the two zones may be based on the momenta of the two fluid streams. However, more work is clearly needed on the quantification of flow distribution in the penetration of wall flows.

The wall flows not only transport mass across the interface, but also lead to an energy transfer. The energy transfer for a two-layer system is given by

$$q_w = p\dot{q} = P\int_0^\delta \rho C_p u(T - T_{g,\ell})\,dy$$

$$= \frac{P\rho\delta U C_p}{30}(T_{w,\ell} - T_{g,\ell}), \qquad \text{for laminar flow,}$$

$$= 0.0366P\rho\delta U C_p(T_{w,\ell} - T_{g,\ell}), \quad \text{for turbulent flow} \quad (41)$$

where the physical variables δ and U are determined in the lower layer just below the interface, and P is the perimeter of the wall flow. The perimeter, P, is the horizontal perimeter of the room minus the door width. Similarly, the mass flow rate due to the wall flow is

$$\dot{m}_w = P\dot{m} = P\int_0^\delta \rho u\,dy$$

$$= P\rho\delta U/12, \qquad \text{for laminar flow}$$
$$= 0.1463P\rho\delta U, \qquad \text{for turbulent flow} \qquad (42)$$

The variables δ and U are determined from Equations 7 and 8 for laminar flow and Equations 15 and 16 for fully turbulent flow. As explained, the governing equations may be solved with either the transition Grashof number $Gr_c \approx 5 \times 10^9$ to obtain these variables, or an average from the laminar and fully turbulent analyses may be employed as an approximation. The above mass and energy transported across the interface must therefore be included in the mass and energy balance equations for the two layers. The energy transport is relatively small compared to the energy input into the upper layer due to the fire plume because of the much larger temperature levels in the plume. The energy transfer across the interface due to the fire plume is $\dot{Q}_c$, which is the total energy input minus the radiative loss from the fire.

The wall flow, $\dot{m}_w$, may be positive or negative depending on the wall and gas temperature levels and the location of the interface. If the downward flow momentum is larger than the upward flow momentum, the flow may be assumed to penetrate from the upper layer into the lower layer and to mix with the fluid there. If the upward flow is stronger, it is assumed to penetrate into the upper layer. The flow momentum is given by

$$\dot{M}_w = P\dot{M} = P\int_0^\infty \rho u^2\,dy$$

$$= P\rho\delta U^2/105, \qquad \text{for laminar flow,}$$
$$= 0.0523P\rho\delta U^2, \qquad \text{for turbulent flow} \qquad (43)$$

If the downward flow momentum in the upper zone, $(\dot{M}_w)_u$, is much larger than the upward flow momentum in the lower zone, $(\dot{M}_w)_\ell$, both being calculated at the interface height Z_i, i.e., $(\dot{M}_w)_u >> (\dot{M}_w)_\ell$, the flow penetrates into the lower zone from the upper. Similarly, the flow penetrates into the upper zone if $(\dot{M}_w)_u << (\dot{M}_w)_\ell$. Thus, $\dot{m}_w$ is negative in the first case and positive in the second. It is difficult to predict penetration when $(\dot{M}_w)_u \approx (\dot{M}_w)_\ell$ because the penetrative flow presents a stability problem and a considerable disturbance is expected to arise in the region where the flow turns away from the wall. When there is a large difference between the two flow momenta, this disturbed region is well within one of the zones. But when the two are close in magnitude, it is very difficult to predict the penetrative flows for the two zones (a detailed analysis is also not available at present).

The interaction region in which the flow turns is expected to be of order δ. If the separation point, as determined from the analysis of the penetration interface, is farther than distance δ from the interface, a complete penetration into

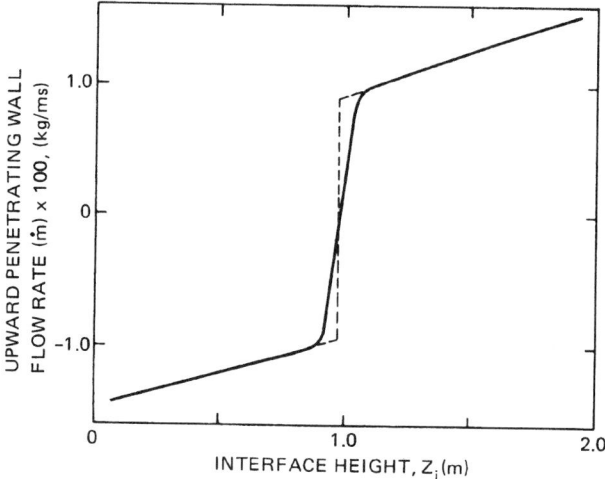

Fig. 2-6.19.   Variation of the upward penetrating wall flow rate with the interface height, Z_i, for a two-zone model.

that zone may be assumed since the turning region is beyond the interface. If the separation point is less than δ, a proportional distribution of the flow in the two zones may be considered, assuming a uniform flow away from the wall over a flow height, δ. If the two momenta are exactly equal (which is rather unlikely), the wall flows are assumed not to penetrate, i.e., the net flow across the interface is taken as zero. Similar considerations have been outlined in related studies,[38–42] and the experimental observations support these conjectures. In many cases information on the separation point, as a function of the physical variables of the problem, may not be available. In zone modeling, for instance, only the temperature levels and the interface height at a given time is known. For such cases, $\dot{m}_w$ may be determined for various interface heights, ranging from a large positive value to a large negative value as Z_i decreases. The value of Z_i at which the two momenta are equal may be taken as zero $\dot{m}_w$ and an interpolation employed to determine $\dot{m}_w$ at other values of Z_i. Depending on the accuracy desired in considering the wall flow effects, the scheme may be constructed to determine the penetrative flow. The simplest scheme assumes complete penetration when the two momenta are different, as shown in Figure 2-6.19 for the experimental conditions of Figure 2-6.10.

Similar considerations prevail for circumstances in which wall flows, without an opposing flow, arise in either of the two regions. The mass and energy transfer across the interface due to the wall flow must be taken into account if the effect is significant. Building this effect into the mathematical model improves the accuracy of the prediction of the interface location and temperatures in the two zones. The heat transfer to the walls, due to radiation and convection, may be considered in computing the wall temperature as a function of time.[12] This allows incorporation of the wall effect into the model for the prediction of changing environment in the enclosure. The wall flows may also be coupled with the ceiling jet and the flows that arise at the corners of the compartment.[44] These considerations are particularly important at the very early stages following the onset of the fire.

SUMMARY

This chapter discusses the nature of natural convection wall flows that arise in enclosure fires. Methods that may be adopted to calculate the transport quantities, such as mass, momentum, and energy fluxes, due to the wall flow are presented. The flow rates in the fire plume and those at the opening of the enclosure are compared to determine the conditions under which the wall effect would be significant. Two flow configurations are considered. The first, involving a downward injection of hot gases across the interface, is expected to arise at the early stages of the fire growth. The second configuration involves opposing fluid streams in the two regions of an enclosure fire and arises at later stages when the walls heat up.

The transport across the interface due to such flows and conditions under which the flows penetrate from one layer into the other are considered. Following such a penetration, opposing buoyancy effects arise which direct the flow toward the interface. Considerable ambiguity exists regarding the final distribution of the combustion products and gases conveyed by the wall flow between the two zones. Although in some cases most of the flow may remain in either of the two zones, it is expected that the fluid would generally be distributed between the two zones. A considerable amount of mixing arises near the interface and a third, intermediate layer is created between the two zones. (In this chapter the wall flow is assumed to distribute between the two zones.) More work needs to be done on this aspect but the distribution of fluid between the two regions can be approximated, based on the momenta of the two opposing streams or on the penetration distance of the wall flow.

The effect of the incorporation of the wall flow into zone-model analysis is often significant and should be taken into account for an accurate prediction of the temperatures in the enclosure and of the position of the interface.

NOMENCLATURE

A	horizontal cross-sectional area of the room, m^2
C_p	specific heat at constant pressure, J/kg·K
D	width of the discharge slot of a negatively buoyant jet, m
d	width of the fire plume, m
F	heat input parameter, defined in Equation 28
g	magnitude of gravitational acceleration, m/s^2
Gr	Grashof number, defined in Equations 17 and 31
Gr_c	Grashof number at transition from laminar to turbulent flow
h	convective heat transfer coefficient at the floor, W/m^2·K
L	height of the room, m
L_1	height of lower isothermal layer, m
L_2	height of intermediate stratified layer, m
$\dot{m}$	mass flow rate in the wall flow per unit flow perimeter, kg/ms
$\dot{m}_w$	total mass flow rate in the wall flow, kg/s
$\dot{m}_a$	mass flow rates defined in Figure 2-6.1, kg/s
$\dot{m}_e$	mass flow rates defined in Figure 2-6.1, kg/s
$\dot{m}_j$	mass flow rates defined in Figure 2-6.1, kg/s
$\dot{m}_f$	mass flow rate of the fuel, kg/s
$\dot{m}_{in}$	net discharge flow rate due to a negatively buoyant jet
$\dot{m}_0$	outflow mass flow rate at the opening, kg/s

$\dot{m}_{out}$ net flow rate out of the enclosure due to a negatively buoyant jet

$\dot{m}_p'$ mass flow rate in the plume entering the upper layer, kg/s

$\dot{m}_p$ mass flow peeled off from the plume at the interface, kg/s

$\dot{M}$ momentum flow rate in the wall flow per unit perimeter, kg/s^2

$\dot{M}_w$ total momentum flow rate in the wall flow, $kg\ m/s^2$

n coefficient which gives the variation of flow region thickness with height, δ proportional to Z^n

P flow perimeter for the wall flow, m

Pr Prandtl number, $Pr = \nu/\alpha$

$\dot{Q}$ heat input into the fire, kW

$\dot{Q}_c$ convective heat input into the plume, kW

$\dot{q}$ convective energy flux in wall flow per unit perimeter, W/m

$\dot{q}_w$ total convective energy flux in wall flow, W

Re Reynolds number defined in Equation 31

r radial distance from fire plume axis, m

T temperature, K

ΔT difference between wall and gas temperatures, K

T_f floor temperature, K

T_0 outside ambient temperature, K

$\hat{T}_0$ discharge temperature of a negatively buoyant jet, K

t time, s

u vertical velocity, m/s

U vertical velocity parameter in Equations 4 and 12, m/s

U_c centerline velocity in the plume

$\hat{U}_0$ inlet velocity of a negatively buoyant jet, m/s

x vertical coordinate distance, m

y horizontal coordinate distance from the walls, m

Z vertical coordinate distance from room floor or from the fire, m

Z_i interface height, m

α entrainment coefficient, Equation 25

α thermal diffusivity, m^2/s

β coefficient of thermal expansion, $\beta = 1/T$ for perfect gas, K^{-1}

δ boundary layer thickness, m

δ_p penetration distance of a negatively buoyant flow

λ constant, defined in Equation 29

ν kinematic viscosity, m^2/s

ρ density, kg/m^3

Subscripts

u upper layer

ℓ lower layer

g gas

w wall

∞ local conditions outside the flow region

0 conditions at the level of the heat source

c centerline conditions in the plume

REFERENCES CITED

1. J.G. Quintiere, *Comb. Sci. Tech.*, 39, 11 (1984).
2. J.G. Quintiere, *F. Safety J.*, 3, 201 (1981).
3. L.Y. Cooper, M. Harkleroad, J.G. Quintiere, and W. Rinkinen, *J. Heat Trans.*, 104, 741 (1982).
4. E.E. Zukoski, *Fire and Matl.*, 2, 54 (1978).
5. H. Emmons, *17th Symp. (Int.) on Combustion*, Combustion Institute, Pittsburgh, 1101 (1978).
6. E.E. Zukoski and T. Kubota, *Fire and Matl.*, 4 (1980).
7. J.G. Quintiere, B.J. McCaffrey, and K. DenBraven, *17th Symp. (Int.) on Combustion*, Combustion Institute, Pittsburgh, 1125 (1978).
8. J.G. Quintiere, K.D. Steckler, and B.J. McCaffrey, *A Model to Predict the Conditions in a Room Subject to Crib Fires*, 1st Special Meeting (Int.), Combustion Institute, France (1981).
9. K.D. Steckler, J.G. Quintiere, and W.J. Rinkinen, *19th Symp. (Int.) on Combustion*, Combustion Institute, Pittsburgh, 913 (1982).
10. R.G. Rehm and H.R. Baum, *J. Res. Nat. Bur. Stds.*, 83, 297 (1978).
11. H.R. Baum and R.G. Rehm, *Comb. Sci. Tech.*, 45, 55 (1984).
12. L.Y. Cooper, *Comb. Sci. Tech.*, 40, 19 (1984).
13. Y. Jaluria, *NBSIR 84-2841*, National Bureau of Standards, Washington (1984).
14. Y. Jaluria and K.D. Steckler, *Wall Flow Due to Fire in a Room*, Proceedings East Coast Sect. Meeting, Combustion Institute, Pittsburgh, Paper No. 42 (1982).
15. Y. Jaluria, *Natural Convection Heat and Mass Transfer*, Pergamon Press, Oxford (1980).
16. B. Gebhart, Y. Jaluria, R.L. Mahajan, and B. Sammakia, *Buoyancy-Induced Flows and Transport*, Taylor & Francis, Washington (1988).
17. J.S. Turner, *J. Fluid Mech.*, 26, 779 (1966).
18. J.S. Turner, *Buoyancy Effects in Fluids*, Cambridge University, Cambridge (1973).
19. B.R. Morton, *J. Fluid Mech.*, 5, 151 (1966).
20. R.A. Seban, M.M. Behnia, and J.E. Abreu, *Int. J. Heat Mass Trans.*, 21, 1453 (1978).
21. Y. Jaluria and L.Y. Cooper, *Prog. Energy Combust. Sci.*, 15, 159 (1989).
22. Y. Jaluria and K. Kapoor, *Combust. Sci. Tech.*, 59, 355 (1988).
23. S. Ostrach, *NACA TR-1111*, National Advisory Committee for National Aeronautics, Washington (1953).
24. E.M. Sparrow and J.L. Gregg, *J. Heat Trans.*, 30, 379 (1958).
25. H.B. Squire, in *Modern Developments in Fluid Dynamics*, Oxford University, New York (1938).
26. E.R.G. Eckert and T.W. Jackson, *NACA TN-2207*, National Advisory Committee for National Aeronautics (1950).
27. H. Kato, N. Nichiwaki, and M. Hirata, *Int. J. Heat Mass Tran.*, 11, 1117 (1968).
28. K. Noto and R. Matsumoto, *J. Heat Trans.*, 97, 621 (1975).
29. R. Cheesewright, *J. Heat Trans.*, 90, 1 (1968).
30. J.G. Quintiere, K. Steckler, and D. Corley, *Fire Sci. Tech.*, 4, A234, 1, (1984).
31. B.R. Morton, G.I. Taylor, and J.S. Turner, *Proc. Roy. Soc.*, (1956).
32. S. Yokoi, *Bul. J. Assoc. Fire Sci. Eng.*, 5, 53 (1956).
33. E.E. Zukoski, T. Kubota, and B. Cetegen, *F. Safety J.*, 3, 107 (1981).
34. B.J. McCaffrey, *NBSIR 79-1910*, National Bureau of Standards, Washington (1979).
35. B.J. McCaffrey, *Some Measurements of the Radiative Power Output of Diffusion Flames*, Western States Meeting, Combustion Institute, Pullman (1981).
36. K. Kapoor and Y. Jaluria, *Int. J. Heat Mass Transfer*, 36, 155 (1993).
37. K. Kapoor and Y. Jaluria, "Penetrative Natural Convection Flow Due to an Isothermal Vertical Surface Immersed in a Thermally Stable Two-Layer Environment," in *Funda. Nat. Conv.*, ASME Heat. Tr. Div., 178, 15 (1991).
38. J.A. Schetz and R. Eichhorn, *J. Fluid Mech.*, 18, 167 (1964).
39. J.A.C. Humphrey and C. Bleinc, *Int. Comm. Heat Mass Trans.*, 12 (1985).
40. Y. Jaluria and D. Goldman, *NBS-GCR-85-487*, National Bureau of Standards, Washington (1985).
41. D. Goldman and Y. Jaluria, *J. Fluid Mech.*, 166, 41 (1986).
42. K. Kapoor and Y. Jaluria, *Int. J. Heat Mass Transfer*, 32, 697 (1989).
43. C.C. Chen and R. Eichhorn, *J. Heat Trans.*, 98, 446 (1976).
44. Y. Jaluria and K. Kapoor, *Combust. Sci. Tech.*, 86, 311 (1992).

EFFECT OF COMBUSTION CONDITIONS ON SPECIES PRODUCTION

D.T. Gottuk and R.J. Roby

INTRODUCTION

A complete compartment fire hazard assessment requires a knowledge of toxic chemical species production. Although combustion products include a vast number of chemical species, in practical circumstances the bulk of the product gas mixture can be characterized by less than ten species. Of these, carbon monoxide (CO) represents the most common fire toxicant (see Section 2, Chapter 8), as over half of all fire fatalities have been attributed to CO inhalation.[1,2] Concentrations as low as 4000 ppm (0.4 percent by volume) can be fatal in less than an hour; and carbon monoxide levels of several percent have been observed in full-scale compartment fires. The ability to accurately predict CO production in a compartment fire dramatically extends the scope of a fire hazard analysis.

The transport of CO from the room of fire origin is ultimately of most importance, because many people who die in fires are in rooms remote from the room of fire origin. Compared to the compartment of fire origin, where heat and oxygen depletion are as hazardous as CO, toxic gases are the primary hazard in remote areas.

The compartment geometry, ventilation, fluid dynamics, thermal environment, chemistry, and mode of burning are examples of combustion conditions that effect species production. The mode of burning and ventilation are two of the key conditions that dictate product formation. These conditions can be used to classify fires into three general categories: (1) smoldering, (2) free- (or open-) burning fires, and (3) ventilation-limited fires. Smoldering is a slow combustion process characterized by low temperatures and no flaming. Under these conditions high yields of CO can be generated. Section 2, Chapters 8 and 11, and Section 3, Chapter 4, discuss this mode of burning in detail; thus, it

Dr. Daniel T. Gottuk is a Senior Engineer at the Fire Science and Engineering Firm of Hughes Associates, Inc. He received his Ph.D. from Virginia Tech in the area of Carbon Monoxide generation in compartment fires. He continues to work in the area of species generation and transport in compartment fires.

Dr. Richard J. Roby is Director of Combustion Research at Hughes Associates, Inc. Previously, he was an Associate Professor in the Mechanical Engineering Department of Virginia Tech and he currently retains Adjunct Professor status. He has performed research in a wide range of combustion areas including toxic species generation in fires and pollutant formation and control.

will not be discussed further here. Free-burning fires are flaming fires that have an excess supply of air. These well-ventilated fires (discussed in Section 3, Chapter 4) are generally of little concern in terms of generating toxic species. This chapter focuses on the third category, ventilation-limited flaming fires. These fires consist of burning materials inside an enclosure, such as a room, in which airflow to the fire is restricted due to limited ventilation openings in the space. As a fire grows, conditions in the space will transition from overventilated to underventilated (fuel rich). It is normally during underventilated conditions that production of high levels of CO creates a major fire hazard.

This chapter discusses the production of species within a compartment fire and the transport of these gases out of the fire compartment to adjacent areas. Engineering correlations are presented along with brief reviews of pertinent work in this field of study.

BASIC CONCEPTS

In a typical compartment fire, a two-layer system is created in which the upper layer consists of hot products of combustion that collect below the ceiling, and the lower layer consists of primarily ambient air that is entrained into the base of the fire. [See Figure 2-7.1(a).] Initially, the fire plume is totally in the lower layer, and the fire burns in an overventilated mode similar to open burning. Due to excess air, near-complete combustion and little CO formation is expected in this mode (See Section 3, Chapter 4 for yields.) As the fire grows, ventilation paths in the room restrict airflow, creating underventilated (fuel-rich) burning conditions. It is generally under these conditions that products of incomplete combustion are created. Typically the fire plume extends into the upper layer such that layer gases recirculate through the upper part of the plume.

As the upper layer descends it will spill below the top of a doorway or other opening into adjacent areas. As hot, vitiated fuel-rich gases flow into adjacent spaces, air with relatively high O_2 concentrations can mix with these gases to create a secondary burning zone outside of the compartment. [See Figure 2-7.1(b).] This is referred to as external burning. In some circumstances, external burning can decrease human fire hazard through the oxidation of CO and smoke leaving the compartment.

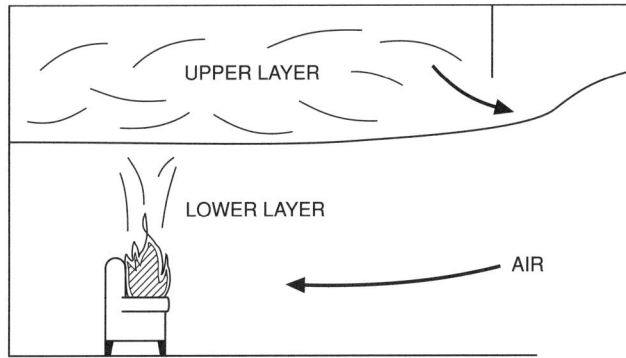

Fig. 2-7.1(a). *An overventilated compartment fire with the fire plume below the layer interface.*

The generation of fire products in compartment fires can be quantified in terms of species yields, Y_i, defined as the mass of species i produced per mass of fuel burned (g/g). Similarly, oxygen is expressed as the depletion of O_2, (i.e., D_{O_2}), which is the grams of O_2 consumed per gram of fuel burned. The normalized yield, f_i, is the yield divided by the theoretical maximum yield of species i for the given fuel, k_i. For the case of oxygen, f_{O_2} is the normalized depletion rate, where k_i is the theoretical maximum depletion of oxygen for the given amount of fuel. As a matter of convenience, the use of the term "yield" throughout this chapter will also include the concept of oxygen depletion. The normalized yield is also aptly referred to as the "generation efficiency" of compound i. By definition, the normalized yields range from 0 to 1 and are, thus, good indicators of the completeness of combustion. For example, under complete combustion conditions the normalized yields of CO_2, H_2O, and O_2 are 1. As a fire burns more inefficiently, these yields decrease. The use of normalized yields is also useful for establishing mass balances. The conservation of carbon requires that

$$f_{CO} + f_{CO_2} + f_{THC} + f_{resid.C} = 1 \qquad (1)$$

where f_{THC} is the normalized yield of gas-phase total hydrocarbons and $f_{resid.C}$ is the normalized yield of residual carbon, such as soot in smoke or high molecular weight hydrocarbons that condense out of the gas sample.

For two-layer systems the yield of all species except oxygen can be calculated as follows:

$$Y_i = \frac{X_{i_{wet}} \, (\dot{m}_f + \dot{m}_a) \, M_i}{\dot{m}_f M_{mix}} \qquad (2)$$

where $X_{i_{wet}}$ is the wet mole fraction of species i, $\dot{m}_a$ is the mass air entrainment rate into the upper layer, $\dot{m}_f$ is the mass volatilization rate of fuel, M_i is the molecular weight of species i, and M_{mix} is the molecular weight of the mixture (typically assumed to be that of air).

The depletion rate of oxygen is calculated as

$$D_{O_2} = \frac{0.21 \dot{m}_a \, \dfrac{M_{O_2}}{M_a} - X_{O_2 \, wet} \, (\dot{m}_f + \dot{m}_a) \, \dfrac{M_{O_2}}{M_{mix}}}{\dot{m}_f} \qquad (3)$$

The normalized yield, f_i, is simply calculated by dividing the yield by the maximum theoretical yield

$$f_i = \frac{Y_i}{k_i} \qquad (4)$$

Typical operation of common gas analyzers requires that water be removed from the gas sample before entering the instrument. Consequently, the measured gas concentration is considered dry and will be higher than the actual "wet" concentration. Equation 5 can be used to calculate the wet mole fraction of species i, $X_{i_{wet}}$, from the measured dry mole fraction, $X_{i_{dry}}$. As can be seen from Equation 5, the percent difference between $X_{i_{dry}}$ and $X_{i_{wet}}$ is on the order of the actual H_2O concentration which, depending on conditions, is typically between 10 to 20 percent by volume.

$$X_{i_{wet}} = (1 - X_{H_2O_{wet}}) X_{i_{dry}} \qquad (5)$$

Reliable water concentration measurements are difficult to obtain. Therefore, investigators have calculated wet species concentrations using the above relationship with the assumption that the molar ratio, C, of H_2O to CO_2 at any equivalence ratio is equal to the calculated molar ratio at stoichiometric conditions.[3,4] Based on this assumption, Equation 6 can be used to calculate wet species concentrations from dry concentration measurements.

$$X_{i_{wet}} = \frac{X_{i_{dry}}}{1 + C X_{CO_{2_{dry}}}} \qquad (6)$$

The concept of a global equivalence ratio can be used to express the overall ventilation of the fire compartment. However, due to the complex interaction between the plume and the upper and lower layers, a unique definition for the global equivalence ratio does not exist. The two main definitions are presented below.

The global upper layer equivalence ratio, ϕ_{ul}, is the ratio of the mass of the upper layer that originates from fuel sources to the mass of the upper layer that originates from the air streams divided by the stoichiometric fuel-to-air ratio. The global plume equivalence ratio, ϕ_p, is the ratio of the mass of fuel burning, m_f, to the mass of oxygen entrained, m_a, into the fire plume normalized by the stoichiometric fuel-to-oxygen ratio, r_{O_2}.

$$\phi_p = \frac{m_f/m_{O_2}}{r_{O_2}} \qquad (7a)$$

Since oxygen is typically entrained into a fire plume in the form of air, ϕ_p is commonly defined as

$$\phi_p = \frac{m_f/m_a}{r} \qquad (7b)$$

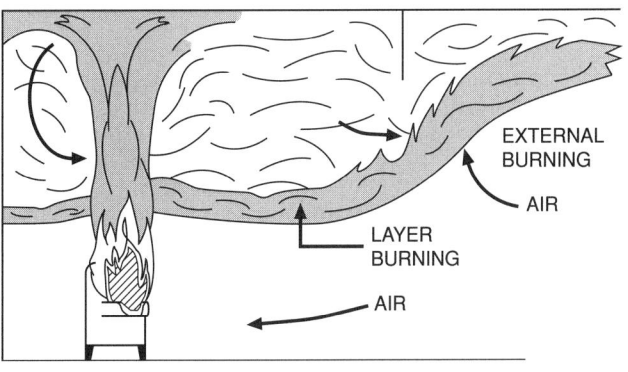

Fig. 2-7.1(b). *An underventilated compartment fire with external burning of fuel-rich upper layer gases.*

where m_a is the mass of air entrained into the plume and r is the stoichiometric fuel-to-air ratio.

The two equivalence ratios (ϕ_p and ϕ_{ul}) are not necessarily the same. As a fire grows, the upper layer composition represents a collective time history of products. In an ideal two-layer fire, where all air enters the upper layer through the plume, ϕ_{ul} is the same as ϕ_p only during steady burning conditions. If the burning rate of the fire changes quickly compared to the residence time of the gases in the upper layer, the upper layer equivalence ratio lags behind the plume equivalence ratio. The residence time, t_R, can be defined as the time required for a unit volume of air to move through the upper layer volume and can be characterized according to Equation 8.

$$t_R = \frac{V_{ul}\rho_{ul}}{\dot{m}_{\text{exhaust}}} \qquad (8)$$

where $\dot{m}_{exhaust}$ is the mass flow rate of gases out of the layer, ρ is the density of the hot layer gases, and V_{ul} is the volume of the upper layer.

For example, consider a compartment fire burning with a plume equivalence ratio of 0.5 with upper layer gases that have a residence time of 20 s. If the fire grows quickly such that ϕ_p increases to a value of 1.5 in about 5 s, ϕ_{ul} would now lag behind (i.e., still be less than 1.5). The fuel-rich (ϕ = 1.5) gas mixture from the plume is effectively diluted by the upper layer gases since there has not been sufficient time (greater than 20 s) for the layer gases to change over. The result is that ϕ_{ul} will have a value between 0.5 and 1.5. Another instance when ϕ_{ul} can differ from ϕ_p is when additional fuel or air enters the upper layer directly.

The calculation of ϕ_{ul} can be a complex task. Either a fairly complete knowledge of the gas composition is needed[5] or time histories of ventilation flows and layer residence times are needed to be able to calculate ϕ_{ul}. Toner et al[5] and Morehart et al[6] present detailed methodologies for calculating ϕ_{ul} from gas composition measurements. Equation 9 can be used to calculate ϕ_{ul} if the mass flow rates can be expressed as a function of time.

$$\phi_{ul}(t) = \frac{1}{r} \frac{\int_{t-t_R}^{t} \dot{m}_f(t')\, dt'}{\int_{t-t_R}^{t} \dot{m}_a(t')\, dt'} \qquad (9)$$

Experimentally the calculation of ϕ_p is based on measurements of the instantaneous fuel volatilization rate, $\dot{m}_f$, and air entrainment rate, $\dot{m}_a$, into the plume (Equation 10).

$$\phi_p = \frac{\dot{m}_f/\dot{m}_a}{r} \qquad (10)$$

The global equivalence ratio is an indicator of two distinct burning regimes: (1) overventilated (fuel lean) and (2) underventilated (fuel rich). Overventilated conditions are represented by global equivalence ratios less than 1 while underventilated conditions are represented by global equivalence ratios greater than 1. An equivalence ratio of unity signifies stoichiometric burning which, in an ideal process, represents complete combustion of the fuel to CO_2 and H_2O with no excess oxygen. During underventilated conditions there is insufficient oxygen to completely burn the fuel; therefore, products of combustion will also include excess fuel (hydrocarbons), carbon monoxide, and hydrogen. It follows that the highest levels of CO production in flaming fires is expected when underventilated conditions occur in the compartment on fire. This basic chemistry also suggests that

species production can be correlated with respect to a global equivalence ratio. Although the non-ideal behavior of actual fires prevents accurate theoretical prediction of products of combustion, experimental correlations have been established.

A simple model for the most complete combustion of a fuel can be represented by the following expressions[7]

$$f_{CO_2} = f_{O_2} = f_{H_2O} = 1 \qquad \text{for } \phi < 1 \qquad (11a)$$
$$f_{CO_2} = f_{O_2} = f_{H_2O} = 1/\phi \qquad \text{for } \phi > 1 \qquad (11b)$$
$$f_{CO} = f_{H_2} = 0 \qquad \text{for all } \phi \qquad (11c)$$
$$f_{THC} = 0 \qquad \text{for } \phi < 1 \qquad (11d)$$
$$f_{THC} = 1 - 1/\phi \qquad \text{for } \phi > 1 \qquad (11e)$$

This assumes that for $\phi > 1$, all excess fuel can be characterized as unburned hydrocarbons. Since compartment fire experiments have shown that significant levels of both CO and H_2 are produced at higher equivalence ratios, Equation 11c is not always representative and reveals a shortcoming of assuming this simple ideal behavior. However, for the products of complete combustion (CO_2, O_2, and H_2O), this model serves as a good benchmark for comparison of experimental results.

EXAMPLE 1:

Consider a piece of cushioned furniture to be primarily polyurethane foam. The nominal chemical formula of the foam is $CH_{1.74}O_{0.323}N_{0.07}$. Calculate the stoichiometric fuel-to-air ratio; the maximum yields of CO, CO_2, and H_2O; and the maximum depletion of O_2.

SOLUTION:

For complete combustion of the fuel to CO_2 and H_2O, the following chemical equation can be written

$$CH_{1.74}O_{0.323}N_{0.07} + 1.2735(O_2 + 3.76N_2)$$
$$\rightarrow 1.0CO_2 + 0.87H_2O + 4.823N_2$$

The molecular weight of the fuel, M_f, = 1(12) + 1.74(1) + 0.323(16) + 0.07(14) = 19.888.

The stoichiometric fuel-to-air ratio is

$$r = \frac{(1 \text{ mole fuel}) (M_f)}{(\text{moles of air}) (m_a)} = \frac{19.888}{1.2735 (4.76) (28.8)}$$

$$r = 0.1139$$

The stoichiometric air-to-fuel ratio is

$$\frac{1}{r} = 8.78$$

The maximum yield of CO (i.e., k_{CO}), is calculated by assuming that all carbon in the fuel is converted to CO. Therefore, the number of moles of CO formed, n_{CO}, equals the number of moles of carbon in one mole of fuel. For the polyurethane foam, n_{CO} = 1.

$$k_{CO} = \frac{n_{CO} (M_{CO})}{n_f (M_{\text{fuel}})} = \frac{(1) (28)}{(1) (19.888)} = 1.41$$

Similarly, k_{CO_2} and k_{H_2O} are calculated as

$$k_{CO_2} = \frac{(1) (44)}{19.888} = 2.21$$

$$k_{H_2O} = \frac{(0.87)(18)}{19.888} = 0.787$$

The maximum depletion of oxygen, k_{O_2}, refers to the mass of oxygen needed to completely combust one mole of fuel to CO_2 and H_2O. This is the same as the stoichiometric requirement of oxygen.

$$k_{O_2} = \frac{n_{O_2}(M_{O_2})}{n_f M_f} = \frac{(1.2735)(32)}{(1)\,19.888} = 2.05$$

EXAMPLE 2:

The fuel specified in Example 1 is burning at a rate of 9 g/s and entraining air at a rate of 56 g/s. Measurements of the upper layer gas composition reveal dry concentrations of 3.7 percent by volume CO, 14.3 percent CO_2, and 0.49 percent O_2. Correct the concentrations for the water removed during the gas analysis process (i.e., calculate the wet concentrations).

SOLUTION:

In order to use Equation 6 to calculate the wet mole fractions, the stoichiometric molar ratio of H_2O to CO_2 for C, needs to be calculated. This ratio is simply obtained from the stoichiometric chemical equation in Example 1.

$$C = \frac{n_{H_2O}}{n_{CO_2}} = \frac{0.87}{1} = 0.87$$

Once C is obtained, the wet mole fractions can be calculated as

$$X_{CO_{wet}} = \frac{0.037}{1 + 0.87\,(0.143)} = 0.033$$

$$X_{CO_{2wet}} = \frac{0.143}{1 + 0.87\,(0.143)} = 0.127$$

$$X_{O_{2wet}} = \frac{0.0049}{1 + 0.87\,(0.143)} = 0.0044$$

The estimated mole fraction of water is

$$X_{H_2O} = C X_{CO_{2wet}} = 0.87\,(0.127) = 0.11$$

Therefore, the corrected gas concentrations on a percent volume basis are 3.3 percent CO, 12.7 percent CO_2, and 0.44 percent O_2.

EXAMPLE 3:

Continuing from Example 2, calculate the yields and normalized yields for each species measured. The wet mole fractions are $X_{CO_{wet}} = 0.033$, $X_{CO_{2wet}} = 0.127$, and $X_{O_{2wet}} = 0.0044$.

SOLUTION:

Using Equations 2 and 4, the yield and normalized yield of CO, CO_2, and H_2O can be calculated. The maximum yields calculated in Example 1 are $k_{CO} = 1.41$, $k_{CO_2} = 2.21$, $k_{H_2O} = 0.787$, and $k_{O_2} = 2.05$.

$$Y_{CO} = \frac{X_{CO_{wet}}(\dot{m}_f + \dot{m}_a)M_{CO}}{\dot{m}_f M_a} = \frac{(0.033)(9+56)(28)}{9\,(28.8)}$$
$$= 0.23$$

$$f_{CO} = \frac{Y_{CO}}{k_{CO}} = \frac{0.23}{1.41} = 0.16$$

$$Y_{CO_2} = \frac{(0.127)(9+56)(44)}{9\,(28.8)} = 1.40$$

$$f_{CO_2} = \frac{1.40}{2.21} = 0.63$$

$$Y_{H_2O} = \frac{(0.11)(9+56)(18)}{9\,(28.8)} = 0.50$$

$$f_{H_2O} = \frac{0.50}{0.787} = 0.63$$

The depletion of oxygen is calculated using Equation 3, assuming the molecular weight of the gas mixture, M_{mix}, to be approximately that of air.

$$D_{O_2} = \frac{0.21\,\dot{m}_a\,\dfrac{M_{O_2}}{M_a} - X_{O_{2wet}}(\dot{m}_f + \dot{m}_a)\dfrac{M_{O_2}}{M_{mix}}}{\dot{m}_f}$$

$$D_{O_2} = \frac{0.21\,(56)\,\dfrac{32}{28.8} - 0.0044\,(9+56)\dfrac{32}{28.8}}{9}$$

$$D_{O_2} = 1.42$$

The normalized yield is calculated as

$$f_{O_2} = \frac{D_{O_2}}{k_{O_2}} = \frac{1.42}{2.05} = 0.69$$

HOOD EXPERIMENTS

Beyler[7,8] was the first to publish major species production rates in a small-scale two-layer environment. The experiments performed consisted of situating a burner under a 1-m diameter, insulated collection hood. The result was the formation of a layer of combustion products in the hood similar to that found in a two-layer compartment fire. (See Figure 2-7.2.) By varying the fuel supply rates and the distance between the burner and layer interface, and, consequently, the air entrainment rate, a range of equivalence ratios was obtained. Layer gases were exhausted at a constant, metered flow rate from the periphery of the hood at a depth of 15 cm below the ceiling. The general procedure was to allow steady-state burning conditions to develop, so the layer maintained a constant depth below the exhaust flow location. Tests revealed a reasonably well-mixed uniform layer both in temperature and chemical composition during the steady-state conditions. Gas analysis was performed on

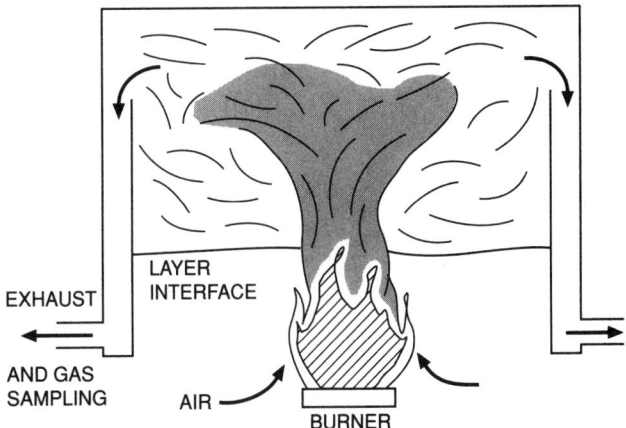

Fig. 2-7.2. Schematic of the two-layer system created in the hood experiments of Beyler.[7,8]

TABLE 2-7.1 *Physicochemical Data for Selected Fuels*

Fuel	Empirical Chemical Formula of Volatiles	Empirical Molecular Weight	Maximum Theoretical Yields				
			k_{CO}	k_{CO_2}	k_{O_2}	k_{H_2O}	$1/r^c$
Acetone	C_3H_6O	58	1.45	2.28	2.21	0.93	9.45
Ethanol	C_2H_5OH	46	1.22	1.91	2.09	1.17	8.94
Hexane	C_6H_{14}	86	1.95	3.07	3.53	1.47	15.1
Isopropanol	C_3H_7OH	60	1.40	2.20	2.40	1.20	10.3
Methane	CH_4	16	1.75	2.75	4.00	2.25	17.2
Methanol	CH_3OH	32	0.88	1.38	1.50	1.13	6.43
Propane	C_3H_8	44	1.91	3.00	3.64	1.64	15.6
Propene	C_3H_6	42	2.00	3.14	3.43	1.29	14.7
Polyurethane foam	$CH_{1.74}O_{0.323}N_{0.0698}$	20	1.41	2.21	2.05	0.79	8.78
Polymethylmethacrylate	$C_5H_8O_2$	100	1.40	2.20	1.92	0.72	8.23
Toluene	C_7H_8	92	2.13	3.35	3.13	0.78	13.4
Wood (ponderosa pine[a])	$C_{0.95}H_{2.4}O$	30	0.89	1.40	1.13	0.73	4.83
Wood (spruce[b])	$CH_{3.584}O_{1.55}$	40	0.69	1.09	0.89	0.80	3.87

[a] Ref. 8, chemical formula estimated from $\phi < 1$ yield data
[b] Ref. 3
[c] r is the stoichiometric fuel to air ratio

samples taken from the exhaust stream. Table 2-7.1 shows the physicochemical properties of the fuels tested.

Beyler's results show that species yields correlate very well with the plume equivalence ratio. Figure 2-7.3 shows normalized yields of major species for propane fires plotted against the plume equivalence ratio. The trends seen in these plots for propane are fairly representative of the other fuels tested. For overventilated conditions, the yield of CO_2 and H_2O and depletion of O_2 are at a maximum, and there is virtually no production of CO, H_2, or unburned hydrocarbons (*THC*). As underventilated burning conditions ($\phi \geq 1$) are approached, products of incomplete combustion (CO, H_2, and *THC*) are generated.

For comparison, the corresponding expressions of Equations 11a through e are shown on each plot in Figure 2-7.3. The CO_2 yield departs from Equation 11b, as CO production increases at higher equivalence ratios. This departure, which is fairly independent of ϕ for $\phi > 1$, has been described by the yield coefficient.[3] The ratios of the normalized yield of CO_2, H_2O, or normalized depletion of O_2 to the theoretical maximums expressed by Equations 11a through e are defined as the yield coefficients, B_{CO_2}, B_{H_2O} and B_{O_2}, respectively.[3] These terms are useful in discussing characteristics of the combustion efficiency; for example, an O_2 yield coefficient of 1 indicates complete utilization of available O_2. In the case of CO_2 and H_2O, deviation from the model (as indicated by B_{CO_2} or $B_{H_2O} < 1$) is a measure of the degree of incomplete combustion. It can be seen from Figure 2-7.3 that the production of CO is primarily at the expense of CO_2 (i.e., B_{O_2} and B_{H_2O} remain nearly 1, while B_{CO_2} is about 0.8). Table 2-7.2 shows average yield coefficients for underventilated fires.

Figure 2-7.4 shows CO yields plotted against the plume equivalence ratio for fuels tested by Beyler.[7,8] The correlations agree quite well for all fuels. Below an equivalence ratio of 0.6 minimal CO production is observed. Above ϕ_p equal to 0.6, carbon monoxide yield increases with ϕ_p and, for most fuels, tends to level out at equivalence ratios greater than 1.2. Toluene, which creates large amounts of soot, is

anomalous compared to the other fuels studied. As can be seen in Figure 2-7.4, the CO yields from toluene fires remain fairly constant at about 0.09 for both overventilated and underventilated burning conditions.

It should be noted that Beyler originally presented all correlations with normalized yields, f_{CO}. However, better agreement is found between CO yield-equivalence ratio correlations for different fuels when the data is presented as unnormalized yields, Y_{CO} (shown in Figure 2-7.4), rather than normalized yields. One point of interest, though, is that, when CO production is correlated as normalized yields, a more distinct separation of the data occurs for ϕ_p greater than one. The degree of carbon monoxide production (represented as f_{CO}) during underventilated conditions can be ranked by chemical structure according to oxygenated hydrocarbons > hydrocarbons > aromatics. This ranking is not observed for unnormalized yield correlations.

Toner *et al*[5] and Zukoski *et al*[9,10] performed similar hood experiments with a different experimental setup. The hood used was a 1.2 m cube, insulated on the inside with ceramic fiber insulation board. The layer in the hood formed to the lower edges where layer gases were allowed to spill out. Gas sampling was done using an uncooled stainless-steel probe inserted into the layer. Detailed gas species measurements were made using a gas chromatograph system. The upper layer equivalence ratio was determined from conservation of atoms using the chemical species measurements, the measured composition of the fuel, and the metered fuel flow rate. Natural gas flames with heat release rates of 20 to 200 kW on a 19-cm-diameter burner were studied. The layer in the hood was allowed to form and reach a steady-state condition before gas sampling was performed.

It was concluded that species concentrations were well correlated to the upper layer equivalence ratio, ϕ_{ul}, and insensitive to temperatures for the range studied (490 to 870 K). Since these experiments were conducted during steady-state conditions with mean upper layer residence times of about 25 to 180 s, it can be concluded that ϕ_p and ϕ_{ul} were equal.

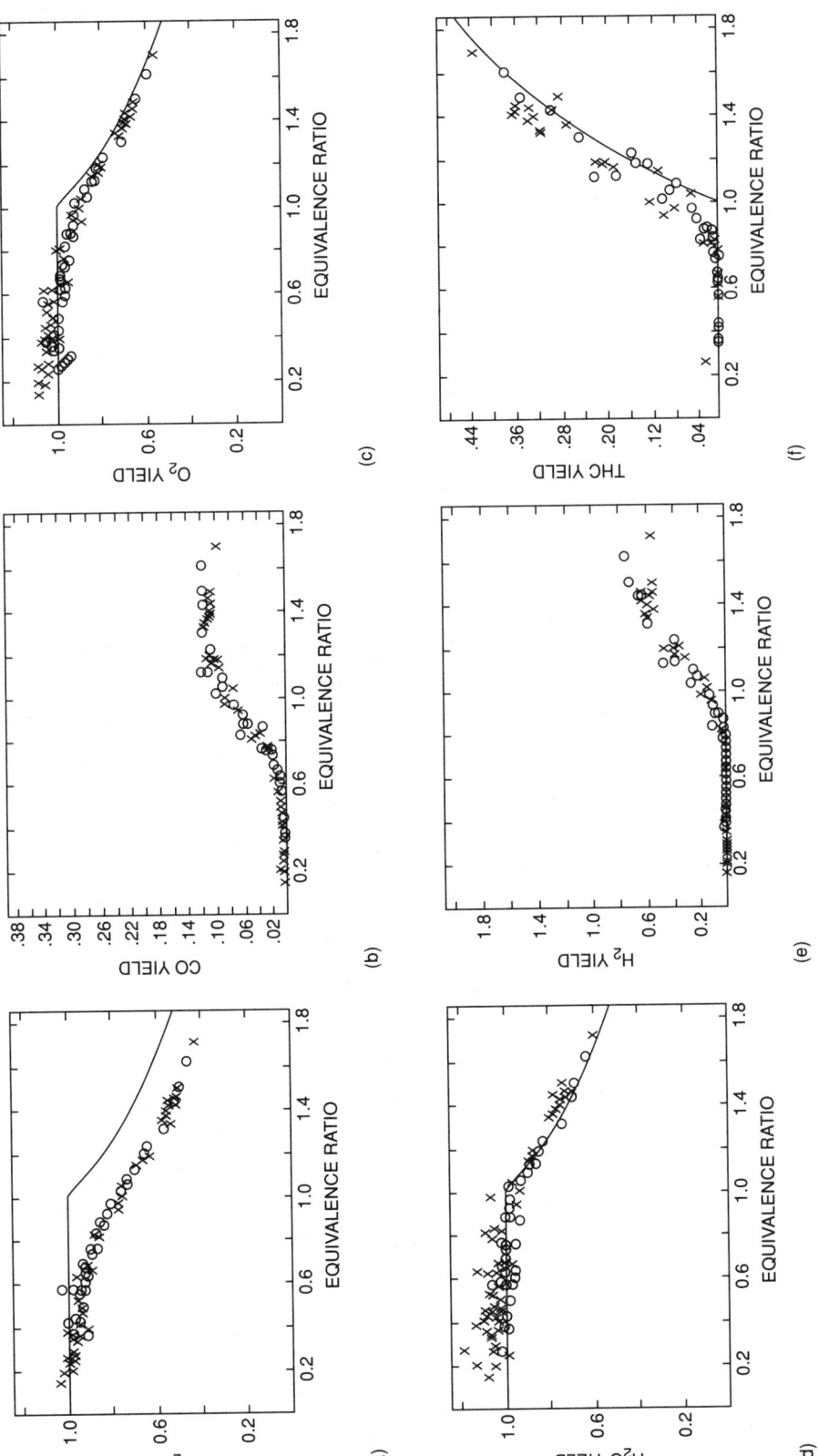

Fig. 2-7.3. *Normalized yields of measured chemical species as a function of the equivalence ratio for propane experiments using a 13-cm (o) or 19-cm (x) burner with supply rates corresponding to 8 to 32 kW theoretical heat release rate.[7]*

TABLE 2-7.2 *Average Yield Coefficients and Upper Layer Temperatures for Underventilated Fires* (Values in parenthesis are standard deviations.)*

Fuel	B_{CO_2}	B_{O_2}	B_{H_2O}	Temperature (K)	Ref.
Acetone	0.78 (0.03)	0.92 (0.04)	0.99 (0.04)	529 (76)	7
Ethanol	0.79 (0.01)	0.97 (0.01)	1.00 (0.04)	523 (72)	7
Hexane	0.61 (0.10)	0.82 (0.02)	0.87 (0.03)	529 (25)	7
Hexane	0.83 (0.05)	0.96 (0.06)	NA	1038 (62)	3
Isopropanol	0.75 (0.01)	0.89 (0.01)	0.96 (0.01)	513 (33)	7
Methane	0.80 (0.05)	1.00 (0.04)	1.01 (0.03)	713 (101)	5
Methane	0.69 (0.03)	0.87 (0.07)	0.86 (0.06)	547 (12)	6
Methanol	0.79 (0.03)	0.99 (0.00)	0.94 (0.02)	566 (53)	7
Propane	0.78 (0.05)	0.97 (0.03)	1.05 (0.04)	557 (62)	7
Propene	0.77 (0.08)	0.92 (0.08)	1.02 (0.10)	629 (51)	7
Polyurethane foam	0.87 (0.04)	0.97 (0.02)	NA	910 (122)	3
Polymethylmethacrylate	0.77 (0.06)	0.92 (0.19)	0.72 (0.04)	525 (37)	8
Polymethylmethacrylate	0.93 (0.04)	0.98 (0.04)	NA	1165 (126)	3
Toluene	0.57 (0.04)	0.62 (0.05)	0.78 (0.03)	509 (23)	7
Wood (ponderosa pine)	0.85 (0.05)	0.89 (0.03)	0.79 (0.10)	537 (37)	8
Wood (spruce)	0.90 (0.00)	0.95 (0.00)	NA	890 (0)	3

*Values have been calculated from data found in the cited references. Values for References 5 to 8 are from hood experiments, and values for References 3 are for a reduced-scale enclosure.

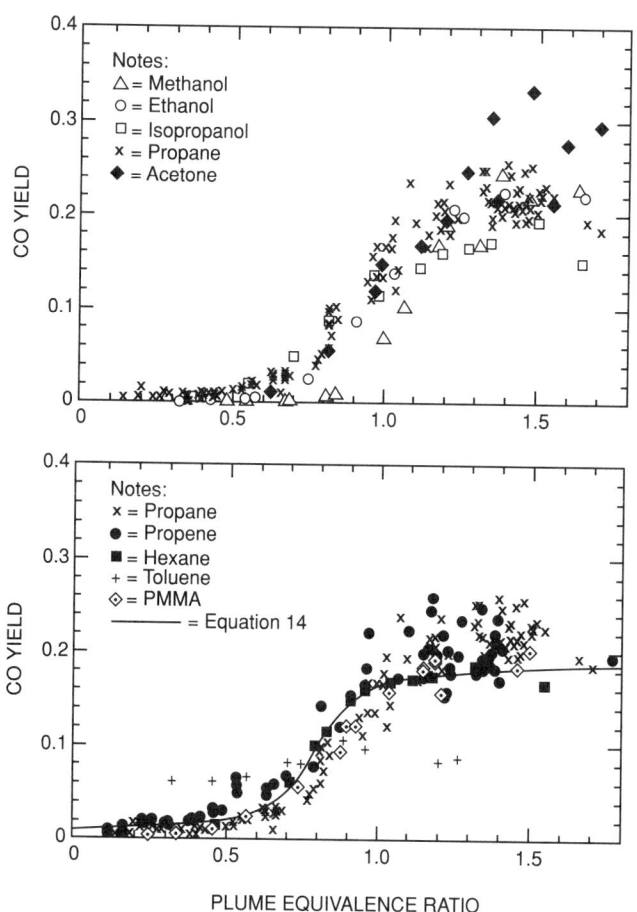

Fig. 2-7.4. *Carbon monoxide yields as a function of the plume equivalence ratio for various fuels studied by Beyler in a hood apparatus.*[7,8]

The data of Toner *et al*[5] has been used to plot CO and CH_4 yields *versus* upper layer equivalence ratio in Figures 2-7.5 and 2-7.6, respectively. The correlations are qualitatively similar to the correlations obtained by Beyler for different fuels. An analysis of these test results also showed that normalized CO_2 and O_2 yield *versus* equivalence ratio data is represented reasonably well by Equations 11a through e. Similar to Beyler's propane results, the average B_{O_2} value is about 1 and B_{CO_2} is 0.8 for underventilated burning conditions (the use of yield coefficients is discussed further in the section on engineering correlations).

Toner compared the measured species concentrations to the calculated equilibrium composition of the reactants at constant temperature and pressure. The layer composition was modeled quite well by the chemical equilibrium composition for very overventilated conditions but not for underventilated conditions. His observance of CO production

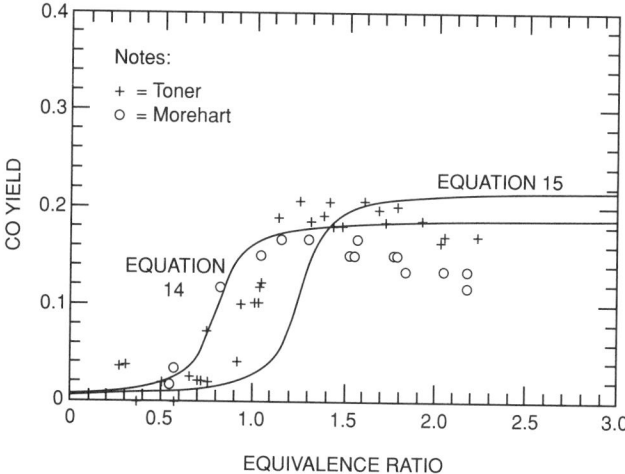

Fig. 2-7.5. *Carbon monoxide yields as a function of equivalence ratio for methane fires studied by Toner* et al[5] *and Morehart* et al[6] *in hood experiments. Yields were calculated from data in Refs. 5 and 6.*

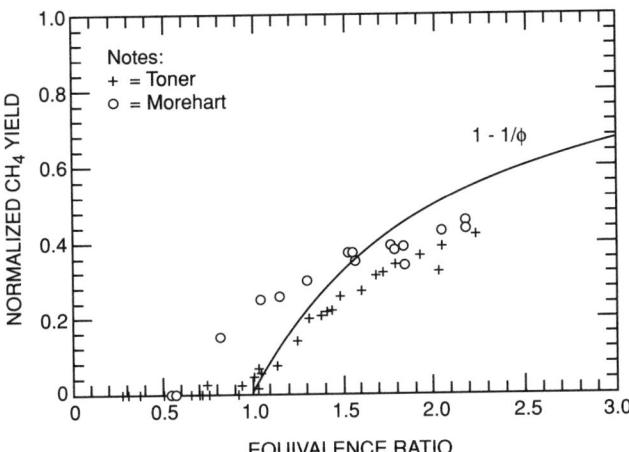

Fig. 2-7.6. Normalized yields as a function of equivalence ratio for methane fires studied by Toner et al[5] and Morehart et al[6] in hood experiments. Yields were calculated from data in Refs. 5 and 6.

for near-stoichiometric and underventilated fires, at the expense of CO_2 production, led them to suggest that the oxidation of CO was "frozen out" before completion (i.e., at low temperatures, there is insufficient energy for CO to chemically react to CO_2).[5] Since the results showed that species production was independent of temperature for the range studied (490 to 870 K), Toner et al concluded that, if a freeze-out temperature existed, it must be higher than 900 K. Work by Pitts[11] and by Gottuk et al,[12] discussed later, shows that a freeze-out temperature does exist in the range of 800 to 900 K, depending on other factors.

Zukoski, Morehart et al[10] performed a second series of tests similar to that described above for Zukoski et al and Toner et al.[9] Much of the same apparatus was used except for a different collection hood. The hood, 1.8 m square by 1.2 m high, was larger than that used by Toner et al and was uninsulated.

Morehart et al experiments consisted of establishing steady-state burning conditions such that the burner-to-layer interface height was constant. A constant ϕ_p was maintained based on this constant interface height in conjunction with the fact that the mass burning rate of fuel was metered at a constant rate. Additional air was then injected into the upper layer at a known flow rate until a new steady-state condition was achieved. This procedure established a ϕ_{ul} that was lower than the ϕ_p, since ϕ_p was based on the ratio of the mass burning rate to the mass of air entrained into the plume from room air below the layer interface. By increasing the air supply rate to the upper layer, a range of ϕ_{ul} was established while maintaining a constant ϕ_p.

Although similar, the correlations obtained by Morehart et al deviated from those obtained by Toner et al. Figures 2-7.5 and 2-7.6 compare the CO and CH_4 yields calculated from the data of Morehart et al with the yields calculated from the data of Toner et al. For overventilated conditions, Morehart et al observed higher CO and CH_4 yields, signifying that the fires conducted by Morehart et al burned less completely. For underventilated methane fires, Morehart et al observed lower CO, CO_2, and H_2O and higher CH_4 and O_2 concentrations than Toner et al. The only apparent differences between experiments was that Morehart

et al found layer temperatures were 120 to 200 K lower for fires with the same equivalence ratio as those observed by Toner et al, i.e., they ranged from 488 to 675 K. Due to the similarity in experimental apparatus except for the hood, Morehart et al concluded that the temperature difference resulted from having a larger uninsulated hood.

Morehart et al studied the effect of increasing temperature on layer composition by adding different levels of insulation to the hood. Except for the insulation, the test conditions (e.g., ϕ of 1.45 and layer interface height) were held constant. Residence times of layer gases in the hood were in the range of 200 to 300 s. For the range of temperatures studied (500 to 675 K), substantial increases in products of complete combustion (i.e., CO_2 and H_2O) and decreases in fuel and oxygen occurred with increasing layer temperature. Upper layer oxygen mass fraction was reduced by approximately 70 percent and methane was reduced by 25 percent.[6,10] Excluding one outlier data point, CO concentrations increased by 25 percent. This is an important result. Although the gas temperatures were well below 800 K, an increase in the layer temperature resulted in more fuel being combusted to products of complete combustion and additional CO. (See section on chemical kinetics.)

COMPARTMENT FIRE EXPERIMENTS

The hood experiments performed by Beyler and Zukoski et al differ from actual compartment fires in several ways. The hood setup allowed considerable radiation to the lab space below. Conversely, a real compartment would contain most of the radiation, thus, resulting in higher wall and upper layer temperatures. Consequently, higher fuel volatilization rates for pool fires would be expected for an actual compartment fire. Also, the hood setup results in a lower layer that has an infinite supply of air which is neither vitiated nor heated. In a real compartment fire, the air supply is limited by the ventilation openings (doors, windows, etc.) and the depth of the upper layer. The air that is entrained into the lower layer of an actual compartment fire can be convectively heated by hot compartment surfaces prior to fire plume entrainment. Third, the hood experiments did not include any significant ceiling and wall flame jets. These dynamic flame structures enhance mixing of the upper layer in actual compartment fires and extend the flame zone beyond the plume. Lastly, the hood experiment correlations were developed from sustained steady-state burning conditions. Actual fires of interest are usually in a continual growth stage, and, thus, more transient in nature.

Tewarson reported that CO and CO_2 yields and O_2 depletion were correlated well by the air-to-fuel stoichiometric fraction (i.e., the reciprocal of the equivalence ratio) for wood crib enclosure fires.[13] Enclosure fire data was taken from previous work in the literature for cellulosic-base fiberboard and pine wood cribs burned in various compartment geometries, 0.21 to 21.8 m³ in volume, with single and dual horizontal and vertical openings centered on the end walls. Additional data were obtained for pine wood cribs burned in a small-scale flammability apparatus that exposed the samples to variable external radiant heat fluxes with either natural or forced airflow from below.

The characteristics of the correlations presented by Tewarson are similar to the correlations developed by Beyler. The CO_2 yield and O_2 depletion are relatively constant for low equivalence ratios and decrease sharply as the equivalence ratio increases for underventilated conditions. The

CO yield correlates with the equivalence ratio but with a fair amount of scatter in the data.

Due to the lack of measurements, the air entrainment rate used to calculate the mass air-to-fuel ratio was estimated from the ventilation parameter, $Ah^{1/2}$, where A is the cross-sectional area and h is the height of the vent. Although the general shape of the correlations are valid, the use of the ventilation parameter assumption causes the equivalence ratio data to be suspect. In addition, the elemental composition of the fuel volatiles for the wood was not corrected for char yield. A correction of this sort would tend to decrease the calculated equivalence ratio and increase the CO and CO_2 yields.

Gottuk et al[3] and Gottuk[14] conducted reduced-scale compartment fire tests specifically designed to determine the yield-equivalence ratio correlations that exist for various fuels burning in a compartment fire. A 2.2 m^3 (1.2 m × 1.5 m × 1.2 m high) test compartment was used to investigate the burning of hexane, PMMA, spruce, and flexible polyurethane foam. The test compartment was specially designed with a two-ventilation path system that allowed the direct measurement of the air entrainment rate and the fuel volatilization rate. The setup created a two-layer system by establishing a buoyancy-driven flow of air from inlet vents along the floor, up through the plume, and exhausting through a window-style exhaust vent in the upper layer. There was no inflow of air through the exhaust vent. The upper layer gas mixture was sampled using an uncooled stainless steel probe placed into the compartment through the center of the exhaust vent. This location for the probe was chosen after species concentration and temperature measurements, taken at several locations in the upper layer, showed a well-mixed, uniform layer.

Table 2-7.1 shows the physicochemical properties used for the four fuels. It should be noted that in determining properties of a fuel, such as maximum yields or the stoichiometric fuel-to-air ratio, the chemical formula must characterize the fuel not necessarily the base fuel. For liquid fuels or "simple" polymers, such as PMMA, the composition of the volatiles is the same as the base fuel. However, more complex fuels can char or contain nonvolatile fillers, as found in polyurethane foams. As a result, the composition of the volatiles differs from that of the base material. As an example, the composition of the wood volatiles used in this study was obtained by adjusting the analyzed wood composition for an observed average of 25 percent char.[3]

The results of these compartment tests showed quite similar correlations as were observed for Beyler's hood correlations. However, some significant quantitative differences exist. Figure 2-7.7 compares the CO yield correlations from Beyler's hood study and that of these compartment tests for hexane fires. This plot illustrates the primary difference observed between the hood and compartment hexane and PMMA fire test results. An offset exists between the rise in CO yield for the two studies. For the hood experiments, higher CO production was observed for overventilated ($\phi_p < 1$) and slightly underventilated burning conditions. Whereas, for the compartment fire experiments, negligible CO was produced until underventilated conditions were reached. Consistent with the increased CO production and the conservation of carbon, CO_2 yields were lower for the hood experiments compared to the compartment fires. The spruce and polyurethane compartment fires produced similar CO yield-equivalence ratio correlations to those observed by Beyler in hood experiments (i.e., high CO yields were observed for overventilated fires).

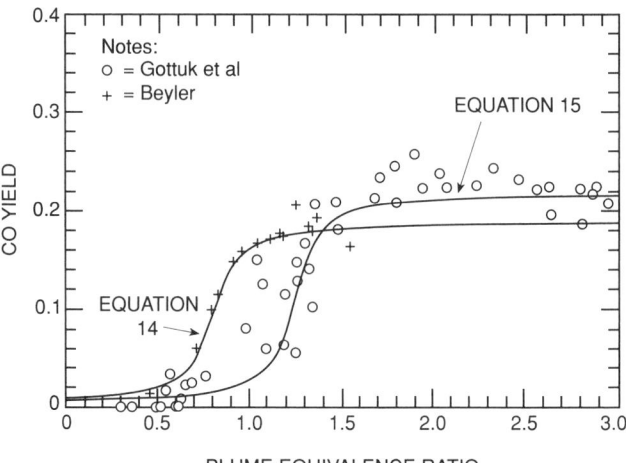

Fig. 2-7.7. *Comparison of CO yield correlations for hexane fires in a compartment and under a hood apparatus. (Figure taken from Ref. 3.)*

The above differences in CO formation can be explained in terms of temperature effects. For the region of discrepancy between equivalence ratios of 0.5 and 1.5, upper layer temperatures in Beyler's hood experiments and the spruce and polyurethane compartment fire experiments were typically below 850 K, whereas temperatures for the hexane and PMMA fires were above 920 K (temperatures typically associated with postflashover fires).[15]

As is detailed in the section Chemical Kinetics, the temperature range between 800 and 900 K is a transition range over which the oxidation of CO to CO_2 changes from a very slow to a fast reaction. That is, for upper layer temperatures below 800 K, the conversion of CO to CO_2 does not occur at an appreciable rate to affect CO yields. Since the oxidation of a fuel first results in the production of CO which then further reacts to form CO_2, the low temperatures (< 800 K) prevent CO from oxidizing. This results in high CO yields. For temperatures greater than 900 K, the reactions that convert CO to CO_2 occur faster as temperature increases. Therefore, for the overventilated conditions discussed above, the high temperatures associated with the hexane and PMMA compartment fires resulted in virtually all CO being oxidized to CO_2 for $\phi_p < 1$.

Overall, the compartment fire test results revealed that the production of CO is primarily dependent on the compartment flow dynamics (i.e., the equivalence ratio) and the upper layer temperature.

The National Institute of Standards and Technology, Building and Fire Research Laboratory, has also performed reduced-scale compartment fire experiments using a natural gas burner for the heat source.[4] The compartment (0.98 m × 1.46 m × 0.98 m high) had a single ventilation opening consisting of a 0.48-m wide by 0.81-m high doorway. A large number of tests were conducted covering a range of heat release rates from 7 to 650 kW. Fires greater than 150 kW resulted in upper layer temperatures greater than 870 K and flames 0.5 to 1.5 m out of the door. This single ventilation opening and the large fires (up to 650 kW) produced non-uniform upper layer conditions. For fires with heat release rates greater than about 250 kW ($\phi > 1.5$), carbon monoxide concentrations in the front of the compartment were approximately 30 to 60 percent higher than in the rear. Temperature gradients of 200 to 300°C were observed from the back to the front of the compartment. Due to the non-uniform air entrainment at the base of the fire and possible mixing of additional air near the front, it is difficult to determine the local equivalence ratio for each region. The concentration gradient from front to rear of the compartment may have been due to differences in the local equivalence ratios. Nonetheless, plots of concentration measurements in the rear of the compartment *versus* equivalence ratio are quite similar to the data of Zukoski *et al* and Toner *et al*. Yield data for these results have not yet been reported.

A second set of experiments was performed to investigate the generation of CO in wood-lined compartments.[16] Douglas fir plywood (6.4 mm thick) was lined on the ceiling and on the top 36 cm of the walls of the compartment described above. Natural gas fires ranging from 40 to 600 kW were burned in the compartment. The results showed that, for tests in which wood pyrolysis occurred, increased levels of CO were observed compared to burning the natural gas alone. Carbon monoxide concentrations (dry) reached levels of 7 percent in the front and 14 percent in the rear of the compartment. These are extremely high concentrations compared to the peak levels of 2 to 4 percent observed in the unlined compartment fire tests with the methane burner only. Typical peak CO concentrations observed for a range of fuels (including wood) in hood experiments[7-10] and the compartment fire experiments of Gottuk *et al*[3] also ranged from 2 to 4 percent. However, concentrations greater than 5 percent have also been reported for cellulosic fuels burning in enclosures.[13,17]

Since wood is an oxygenated fuel, it does not require additional oxygen from entrained air to pyrolyze. This enhances the ability of the wood to pyrolyze in a vitiated atmosphere. Therefore, there are two reasons that high CO concentrations can result in fires with oxygenated fuels in the upper layer. First, the fuel-bound oxygen allows the fuel to generate CO by pyrolysis. Second, due to preferential oxidation of hydrocarbons over CO, the limited oxygen in the upper layer reacts with the pyrolyzing wood to form additional CO. Aspects of this chemistry are discussed in the next section.

These test results emphasize the importance of adding additional fuel to the upper layer. The practical implications are significant, as many structures have cellulosic-based wall coverings and other combustible interior finishes. Further work is needed to determine the applicability of CO yield-equivalence ratio correlations under conditions of multiple fuel sources, particularly fuel within the vitiated upper layer.

CHEMICAL KINETICS

The field of chemical kinetics can be used to describe the changes in gas composition with time that result from chemical reactions. The kinetics of actual combusting flows is dependent on the initial species present, temperature, pressure, and the fluid dynamics of the gases. Due to the inability to adequately characterize the complex mixing processes and the significant temperature gradients in turbulent flames, the use of kinetic models is restricted to simplified combusting flow processes. Consequently, the fire plume in a compartment fire is beyond current chemical kinetics models. However, the reactivity of the upper layer gas composition can be reasonably modeled if one assumes that the layer can be characterized as a perfectly stirred reactor or that the layer gases flow away from the fire plume in a plug-flow-type process.[11,12] Pitts has shown that no significant differences between results exist for either modeling approach when applied to these upper layers.[11]

Several kinetics studies have been performed to examine aspects of the reactivity of upper layer gases.[6,11,12] Comparisons between different hood experiments and between hood and compartment fire experiments have indicated that upper layer temperatures have an effect on CO production. The results of these chemical kinetics studies provide insights into CO generation in compartment fires, which also serve to explain the differences in CO yields between experiments with respect to temperature effects. These studies primarily focused on the question: "What would the resulting composition be if the upper layer gases in the hood experiments existed at different isothermal conditions (constant temperature)?" A particular focus was to examine the resulting compositions for cases modeled under the high temperatures characteristic of compartment fires. Chemical kinetics models calculate the change in species concentrations with respect to time. Calculations are dependent on the reaction mechanism (i.e., the set of elementary reactions and associated kinetic data) and the thermodynamic data base used.

Thermodynamic data is fairly well known and introduces little uncertainty into the modeling. However, reaction mechanisms do vary, and this is an area of active research. Pitts presents a comparison of the use of various mechanisms in the literature.[11] The comparison indicates that reaction kinetics for high temperatures (greater than 1100 K) are fairly well understood. However, the elementary reactions for the range of 800 to 1000 K are not as certain; therefore, quantitative modeling results in this range may be suspect. Nevertheless, the general trends presented below are valid despite any uncertainty associated with the mechanisms used.

Chemical kinetics modeling shows that significantly different trends occur for overventilated and underventilated burning conditions. This can be seen in Figures 2-7.8 and 2-7.9 which present major species concentrations with respect to time for an overventilated and underventilated condition, respectively. Figure 2-7.8 shows a modeled case for ϕ equal to 0.91 and a temperature of 900 K. The initial composition is taken from Beyler's data for a fire with a layer temperature of 587 K. The temperature of 900 K corresponds to the temperature observed by Gottuk *et al* for a hexane compartment fire at the same global equivalence ratio. For overventilated conditions, increased temperatures cause CO concentrations to initially increase. As can be seen in Figure 2-7.8, this is due to the incomplete oxidation of hydrocarbons (modeled as C_2H_4). Once the hydrocarbons are consumed, available O_2 is used in the oxidation of CO to CO_2. Since overventilated conditions indicate excess oxygen, CO concentrations are reduced to zero given sufficient time. This is representative of the case of the overventilated hexane and PMMA compartment fires studied by Gottuk *et al*, in which the higher compartment temperatures compared to the hood tests of Beyler resulted in near-zero CO yields for $\phi < 1$.

Figure 2-7.9 shows an underventilated case for ϕ equal to 2.17 and a temperature of 1300 K. The initial composition is taken from Morehart *et al* for a methane hood experiment.[6] Similar to the overventilated conditions, CO increases due to the oxidation of hydrocarbons (CH_4). However, the available oxygen is depleted before the hydrocarbons are fully

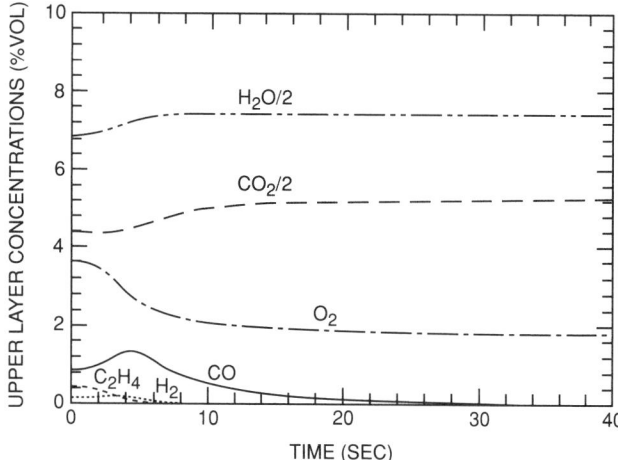

Fig. 2-7.8. *Chemical kinetics model calculated species concentrations versus time for an overventilated (ϕ = 0.91) burning condition with an upper layer temperature of 900 K. (Figure taken from Ref. 12.)*

oxidized. The resulting composition consists of higher levels of CO and H_2 and decreased levels of unburned fuel. Carbon dioxide levels remain virtually unchanged. The much higher temperature studied in this case results in much quicker reaction rates, as is reflected in the 2-s time scale for Figure 2-7.9 compared to 30 s for Figure 2-7.8.

It is clear from Figures 2-7.8 and 2-7.9 that hydrocarbon oxidation to CO and H_2 is much faster than CO and H_2 oxidation to CO_2 and H_2O, respectively. This is a result of the preferential combination of free radicals, such as OH, with hydrocarbons over CO. Carbon monoxide is oxidized almost exclusively by OH to CO_2.[18] Therefore, it is not until the hydrocarbons are consumed that free radicals are able to oxidize CO to CO_2.

The formation and consumption of CO in a reactive gas mixture is dependent on both the temperature of the mixture and the duration of time over which the mixture reacts. This point is illustrated in Figure 2-7.10 which shows the resulting CO concentrations at different isothermal conditions from an initial gas mixture taken from an underventilated fire (ϕ = 2.17). Pitts noted that there are three distinct temperature regimes. At temperatures under 800 K, the gas mixture is unreactive and the CO to CO_2 reactions are said to be "frozen out." As the temperature increases in the range of 800 to 1000 K, the mixture becomes more reactive and CO is formed at faster rates due to the oxidation of unburned hydrocarbons. For the time period shown, it is interesting to note that the ultimate concentration is approximately constant* for each case in this temperature range. The third regime of high temperatures above 1100 K is characterized by fast reaction rates and much higher CO production for the 20-s reaction time shown. With sufficient time, the ultimate CO concentration for the 800 to 1000 K conditions would approach the same value as that seen for the higher temperatures.

Results of Zukoski *et al*[9] and Gottuk *et al*[12] indicated that layer temperatures of 850 to 900 K or higher are needed

---

*Note that, although the ultimate CO concentration is roughly constant, the value of 2.1 percent for this illustration is not to be taken as a universal limit for this temperature range. In general, the resulting CO concentration will depend on the initial gas composition and the time to which the mixture is allowed to react.

for the layer gases to be reactive. Considering that the minimum (freeze out) temperature above which a gas mixture is reactive is dependent on the time scale evaluated, these values are consistent with the results shown in Figure 2-7.10. In terms of compartment fires, the time over which the gases react can be taken as the residence time of the gases in the upper layer, which is calculated according to Equation 8. In many practical cases of high-temperature compartment fires,

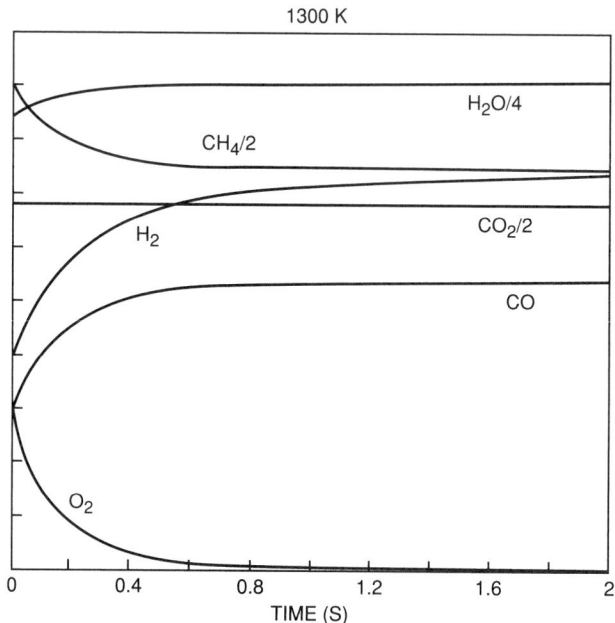

Fig. 2-7.9. *Chemical kinetics model calculated species concentrations versus time for an underventilated (ϕ = 2.17) burning condition with an upper layer temperature of 1300 K. (Figure taken from Ref. 11.)*

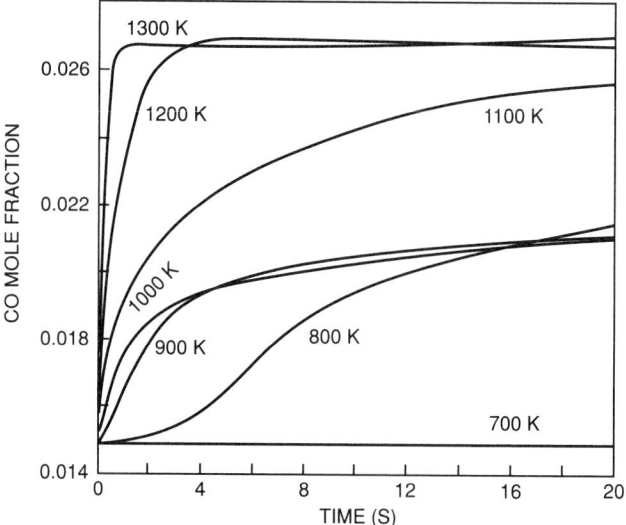

Fig. 2-7.10. *Carbon monoxide concentrations as a function of time for a range of isothermal conditions. Initial concentrations from a methane hood fire at ϕ = 2.17.[11]*

it would be reasonable to assume that the residence time of layer gases would be longer than the time needed for the gas mixture to react fully.

FIRE PLUME EFFECTS

Although a fire plume is too complex to adequately model the chemistry, the hood experiments discussed earlier provide significant insights with respect to the fire plume and species production in compartment fires. Results of Beyler's hood experiments suggest that the production of upper layer gases is independent of the structure and fluid dynamics of the flame.

Beyler modified a 19-cm propane burner by including a 2.8-cm lip to enhance turbulence and the large-scale structure of the flame.[7] Compared to the no-lip burner, he reported that the flame was markedly changed and that air entrainment was increased by 30 percent. Yet, the upper layer species-equivalence ratio correlations were the same for both burners. Additionally, as shown in Figure 2-7.3, correlations for different size burners are also identical.

The insensitivity of species yields to the details of the flame structure is also suggested by the compartment fire hexane results of Gottuk et al.[3] The correlations include data from fires utilizing various size burn pans and with a wide range of air entrainment rates. In several cases, nearly equal steady-state equivalence ratio fires were obtained with quite different burning rates and air entrainment rates. Although the conditions varied significantly, the good correlation between yields and equivalence ratio suggests that the yields are not sensitive to the details of the flame structure.

The temperature of the fire plume has a significant effect on species production from the fire plume. It is reasonable to assume that differences in upper layer temperature are also reflective of a similar trend in the average temperature of the fire plume gases. An increase in the upper layer temperature can increase the fire plume temperature in two ways. For plumes that extend into the upper layer, entrainment of hotter upper layer gases will result in increased plume temperatures compared to plumes in layers with lower temperature gases. Secondly, an increase in the surrounding temperature (both gases and compartment surfaces) reduces the radiant heat loss from the plume, thus resulting in a higher plume temperature.

The effect of temperature on species generation in a fire plume can be found in the methane hood experiments of Morehard et al[6] and Zukoski et al.[10]. Morehart studied the effect of increasing temperature on layer composition by adding different levels of insulation to his hood. Except for the insulation, the test conditions (e.g., ϕ of 1.45 and layer interface height) were held constant. For the range of temperatures studied (500 to 675 K), substantial increases in products of complete combustion and decreases in fuel and oxygen occurred with increasing layer temperature. Upper layer oxygen mass fraction was reduced by approximately 70 percent and methane was reduced by 25 percent. Excluding one outlier data point, CO concentrations increased by 25 percent. The temperatures of the Morehart et al upper layer were well below 700 K. Therefore, based on kinetics modeling, these layers were unreactive at these low temperatures. It follows that the change in layer composition must have been due to changes in the plume chemistry. The more complete combustion can be attributed to an extension of the flammability limits (or reaction zone) in the plume due to raising the flame temperature. The above discussion demonstrates that increasing the plume temperature substantially increases the consumption of O_2 and fuel and, primarily, increases the levels of products of complete combustion.

The effect of changing temperature on a compartment fire upper layer composition is twofold: (1) the generation of species in the fire plume is changed, and (2) oxidation of post-flame gases in the layer is affected. Elevated compartment temperatures correlate with increased fire plume temperatures and more complete oxidation of the fuel to CO_2 and H_2O within the fire plume. The layer temperature dictates post-flame oxidation in the upper layer.

Upper layer temperatures below about 800 K indicate chemically unreactive layers. As such, combustion within the fire plume controls the final CO levels that would be measured in the upper layer. At these low temperatures significant levels of CO can be generated even for some overventilated conditions ($0.5 < \phi < 1$). The yield of CO is inversely proportional to temperature for overventilated conditions and directly proportional to temperature for underventilated conditions.

Upper layer temperatures of about 900 K and higher indicate chemically reactive layer gases. As such, reactions in the layer dictate final CO production. Temperatures above 900 K allow nearly complete oxidation of CO to CO_2 for overventilated conditions. For underventilated fires, chemical kinetics modeling indicates that higher temperature environments may result in slightly higher CO yields due to preferentially accelerated hydrocarbon oxidation compared to CO oxidation.

During underventilated conditions, two mechanisms affecting net CO formation compete (i.e., CO and hydrocarbon oxidation). Increasing gas temperatures above 900 K depletes CO by accelerating the CO to CO_2 conversion. However, incomplete oxidation of unburned hydrocarbons increases the CO production. Since hydrocarbon oxidation is much faster than CO oxidation, net CO levels increase until all available oxygen is consumed.

ENGINEERING CORRELATIONS

In light of the experimental work and chemical kinetics considerations discussed above, several correlations can be used as guidelines for fire protection engineering. The production of chemical species in compartment fires has been shown to be best correlated with the global equivalence ratio. For most purposes the global equivalence ratio can be taken as the plume equivalence ratio, ϕ_p, defined in Equation 10. The upper layer equivalence ratio, ϕ_{ul}, should be used for cases in which fuel or air enters a reactive upper layer other than through the fire plume. However, even for these cases, the use of ϕ_p may be appropriate if most of the fuel or air added to the upper layer is circulated into the fire plume.

Due to its toxicity, CO production is of primary importance. Four correlations (see Equations 12 through 15) are presented below representing varying degrees of complexity. In each case, the correlations basically represent a lower bound for the yield of CO. Equation 12 (parts a and b) represents a "zeroth order" correlation between CO yield and equivalence ratio. For overventilated burning conditions, there is no CO production and for underventilated conditions CO is produced at a yield of 0.2 grams per gram of fuel burned. This correlation applies best to fires with average upper layer temperatures greater than 900 K.

$$f_{CO} = 0 \qquad \text{for } \phi < 1 \qquad (12a)$$

$$f_{CO} = 0.2 \qquad \text{for } \phi > 1 \qquad (12b)$$

Equation 13 (parts a, b, and c) accounts for some of the temperature effect by including a linear rise in CO yield over the transition region from ϕ of 0.5 to 1.5.

$$f_{CO} = 0 \qquad \text{for } \phi < 0.5 \qquad (13a)$$

$$f_{CO} = 0.2\phi - 0.1 \qquad \text{for } 0.5 < \phi < 1.5 \qquad (13b)$$

$$f_{CO} = 0.2 \qquad \text{for } \phi > 1.5 \qquad (13c)$$

The temperature effect on CO production is best represented in the following two correlations. Equation 14, which represents a fit to the hexane data of Beyler's hood experiments, is suggested for compartment fires with average upper layer temperatures below 800 K. Equation 15 is used for fires with upper layer temperatures above 900 K. Equation 15 is an approximate fit to the compartment fire hexane data of Gottuk *et al.* For the most part, CO yields from hexane fires represent lower limits observed for the fuels studied to date.[3,7] Therefore, these equations provide a *minimum* CO production that can be used for hazard analysis.

$$Y_{CO} = (0.19/180) * \tan^{-1}(X) + 0.095$$
$$\text{for } T < 800 \text{ K} \qquad (14)$$

where $X = 10(\phi - 0.8)$ in degrees

$$Y_{CO} = (0.22/180) * \tan^{-1}(X) + 0.11$$
$$\text{for } T > 900 \text{ K} \qquad (15)$$

where $X = 10 (\phi - 1.25)$ in degrees

The figures presented earlier of CO yield *versus* equivalence ratio also show plots of Equations 14 and 15. Figure 2-7.5 shows the CO yield data for methane hood experiment fires in which upper layer temperatures ranged from 490 to 870 K. The CO yield data of Zukoski *et al* and Toner *et al* lies between the curves of Equation 14 and 15, particularly for slightly overventilated and stoichiometric conditions. This is consistent with the fact that these fires had temperatures that were higher than those represented by Equation 14 and some were within the transition range of 800 to 900 K.

The simple model presented as Equation 11, parts a through e, with the inclusion of the empirically determined yield coefficients, is fairly adequate for predicting CO_2, O_2, and H_2O normalized yields. (See Equations 16 through 18.) Suggested average yield coefficients for compartment fires of elevated temperatures ($T > 900$ K) are 0.88 for B_{CO_2} and 0.97 for B_{O_2}.[3] Suggested values for low upper layer temperatures ($T < 800$ K) are 0.77 for B_{CO_2}, 0.92 for B_{O_2}, and 0.95 for B_{H_2O}. Average yield coefficients for underventilated fires are shown in Table 2-7.2.

$$f_{CO_2} = 1 \qquad \text{for } \phi < 1 \qquad (16a)$$

$$f_{CO_2} = B_{CO_2}/\phi \qquad \text{for } \phi > 1 \qquad (16b)$$

$$f_{O_2} = 1 \qquad \text{for } \phi < 1 \qquad (17a)$$

$$f_{O_2} = B_{O_2}/\phi \qquad \text{for } \phi > 1 \qquad (17b)$$

$$f_{H_2O} = 1 \qquad \text{for } \phi < 1 \qquad (18a)$$

$$f_{H_2O} = B_{H_2O}/\phi \qquad \text{for } \phi > 1 \qquad (18b)$$

As presented in Equations 16 through 18, normalized chemical species yields, f, can be correlated quite well by the global equivalence ratio. This is true for a wide range of fuel types. However, it is worthwhile to point out that for different fuels, the CO_2, O_2, and H_2O yields to equivalence ratio correlations only collapse down to a single curve when the yields are normalized by the maximum possible yield for a given fuel (i.e., presented as f rather than Y). Although complete combustion does not occur, combustion efficiencies with respect to equivalence ratio are similar enough between fuels that the stoichiometry of a particular fuel will dictate the generation of CO_2 and the depletion of O_2. Therefore, the species associated with complete combustion (CO_2, O_2, and H_2O) are not expected to have equal yields for different fuels, since varying fuel compositions will dictate different limits of CO_2 and H_2O that can be generated and O_2 that can be consumed for a gram of fuel burned. By normalizing the yields, the variability of fuel composition is removed.

On the other hand, carbon monoxide production is best correlated by the equivalence ratio when represented as a simple yield, Y_{CO}, rather than a normalized yield, f_{CO}. This is one indicator that CO production is not strongly dependent on fuel type, as is production of CO_2 and O_2. The reason for this is believed to be due to the fact that CO is effectively an intermediate product that depends more on the elementary chemistry than on fuel composition which determines products of complete combustion.

Once yields are determined using the above correlations, species gas concentrations can be calculated. Equation 19 can be used to calculate the concentration of species i for all species except oxygen. Oxygen concentrations can be calculated from the depletion of oxygen using Equation 20.

$$X_{i_{wet}} = \frac{Y_i \dot{m}_f M_{mix}}{(\dot{m}_f + \dot{m}_a) M_i} \qquad (19)$$

$$X_{O_{2wet}} = \frac{0.21 \dot{m}_a \dfrac{M_{O_2}}{M_a} - D_{O_2} \dot{m}_f}{(\dot{m}_f + \dot{m}_a) \dfrac{M_{O_2}}{M_a}} \qquad (20)$$

The yield-equivalence ratio correlations shown in Figure 2-7.7, which are also represented by Equations 14 and 15, have been replotted as CO concentration *versus* equivalence ratio in Figure 2-7.11. As indicated previously, the yield correlations in Figure 2-7.7 (and thus, the concentrations in Fig. 2-7.11) represent a reasonable lower bound for a range of typical fuels. Higher concentrations of CO can be created, particularly when additional fuel is added to a vitiated upper layer. High CO concentrations from 5 to 15 percent by volume have been reported for wood and cellulosic materials burning in reduced-scale enclosures.[13,17] However, problems in the positioning of sampling probes within reaction zones[8] and in maintaining carbon balances raises questions about the validity of these high concentrations. Nonetheless, others have reported CO concentrations over 5 percent for full-scale compartment fires.[19–23]

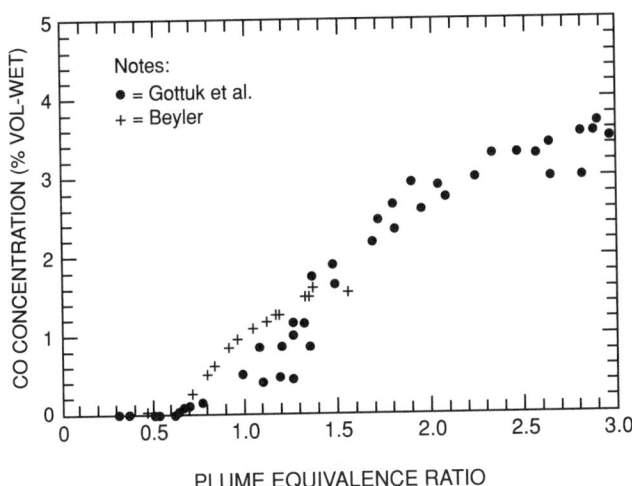

Fig. 2-7.11. *Carbon monoxide concentrations as a function of equivalence ratio for hexane fires in a compartment (•) and under a hood (+). Data represent the same tests shown in Figure 2-7.7 as yields.*

The ratio of CO to CO_2 concentrations can be used as an indicator of the combustion mode. Higher combustion efficiency is obtained as more fuel is burned completely to CO_2 and H_2O and is indicated by a ratio of CO to CO_2 near zero. Since CO is a product of incomplete combustion, the ratio of CO to CO_2 concentrations will increase as fires burn less efficiently. The ratio increases with equivalence ratio even for well-underventilated conditions, as evidenced by experimental data [e.g., reference 3] and the engineering correlations presented above.

EXAMPLE 4:

Consider that the piece of furniture described in Example 1 is burning in a room such that a two-layer system develops. The only ventilation to the room is an open doorway through which 217 g/s of air is being entrained. The material is burning at a rate of 37 g/s, and the average temperature of the upper layer is 700°C. Calculate the plume equivalence ratio and determine the yield of CO and depletion of O_2.

SOLUTION:

The plume equivalence ratio is calculated using Equation 10. The stoichiometric fuel-to-air ratio, r, has already been calculated in Example 1.

$$\phi_p = \frac{\dot{m}_f/\dot{m}_a}{r} = \frac{\frac{37}{217}}{0.1139} = 1.5$$

Since the average upper layer temperature (700°C + 273 = 973 K) is above 900 K, Equation 15 is used to calculate the yield of CO. The argument, X, of the inverse tangent is

$$X = 10\,\phi_p - 1.25 = 10\,(1.5) - 1.25 = 13.75$$

$$Y_{CO} = \left(\frac{0.22}{180}\right)\tan^{-1}(X) + 0.11$$

$$Y_{CO} = \left(\frac{0.22}{180}\right)\tan^{-1}(13.75) + 0.11$$

$$Y_{CO} = 0.21$$

Therefore, 0.21 grams of CO are produced for every gram of polyurethane foam that burns. The production rate of CO is equal to that yield, Y_{CO}, multiplied by the fuel burning rate (0.21 * 37 g/s = 7.8 g/s).

The normalized yield of oxygen is determined using Equation 17b, and the recommended yield coefficient, B_{O_2}, of 0.97.

$$f_{O_2} = \frac{B_{O_2}}{\phi} = \frac{0.97}{1.5} = 0.65$$

From Example 1, we obtain the maximum theoretical depletion of oxygen, k_{O_2}, and calculate the depletion of oxygen as

$$D_{O_2} = f_{O_2}\,k_{O_2} = 0.65\,(2.05)$$
$$= 1.33g \; of \; O_2 \; per \; gram \; of \; fuel \; burned.$$

The depletion rate of oxygen is (1.33 * 37 g/s) 49.2 g/s.

EXAMPLE 5:

For the piece of furniture burning in Example 4, calculate the CO and O_2 concentrations in the upper layer.

Gas concentrations can be calculated from the yields determined in Example 4 using Equation 19 for CO and Equation 20 for O_2.

$$X_{CO_{wet}} = \frac{Y_{CO}\dot{m}_f M_{mix}}{(\dot{m}_f + \dot{m}_a)M_{CO}} = \frac{0.21\,(37)\,(28.8)}{(37 + 217)\,(28)} = 0.031$$

$$X_{O_2} = \frac{0.21\dot{m}_a\dfrac{M_{O_2}}{M_a} - D_{O_2}\dot{m}_f}{(\dot{m}_a + \dot{m}_f)\dfrac{M_{O_2}}{M_a}}$$

$$= \frac{0.21\,(217)\dfrac{32}{28.8} - 1.33(37)}{(37 + 217)\dfrac{32}{28.8}}$$

$$X_{O_2} = 0.005$$

The resulting concentrations of CO and O_2 are 3.1 and 0.5 percent by volume, respectively.

TRANSIENT CONDITIONS

Transient conditions cause the upper layer equivalence ratio to differ from the plume equivalence ratio. A fast-growing fire will tend to have a ϕ_{ul} that is less than ϕ_p. Conversely, a fire that is dying down quickly, such that ϕ_p is decreasing rapidly, will have a ϕ_{ul} that is higher than ϕ_p. These trends result due to the upper layer being a temporary collection reservoir for the gases from the fire plume.

In an effort to characterize transient conditions, Gottuk *et al* defined a steady-state time ratio, τ_{SS}, as the ratio of the residence time, t_R, to a characteristic growth time of the fire. Since fire growth is directly related to the fuel volatilization rate, a representative growth time of the fire was defined as the ratio of the fuel volatilization rate, $\dot{m}_f$, to the derivative of the fuel volatilization rate, $\ddot{m}_f$. An increase in τ_{SS} is indicative of more transient conditions.

$$\tau_{SS} = \frac{t_R}{\dot{m}_f/\ddot{m}_f} \tag{21}$$

An analysis of the transient nature of the compartment fires

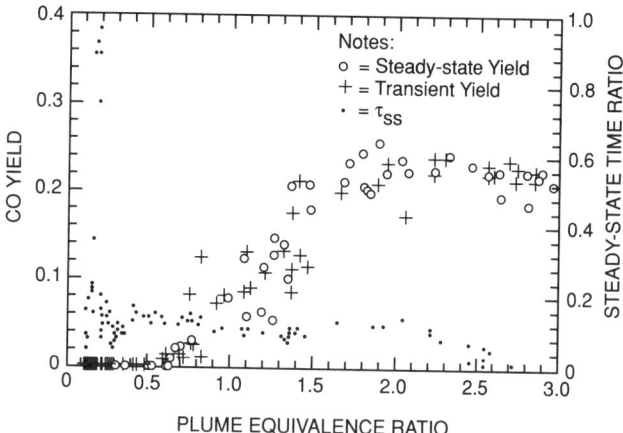

Fig. 2-7.12. Comparison between a transient, CO yield corre-laion for a hexane fire with an average steady-state ϕ_p of 3 and the steady-state correlation for all hexane fires studied by Got-tuk et al. The steady-state time ratio, τ_{SS}, data are shown as solid dots. (Figure taken from Ref. 14.)

studied by Gottuk *et al* showed that values well below 1 indicated near steady-state conditions, such that the plume and upper layer equivalence ratios could be considered equal. Investigation of individual fires showed that the steady-state time ratio decreased below 1.0 at very early times in the fire. Typically, the ratio was 0.1 or less for the quasi-steady-state periods over which data was averaged. For some fires, during the highly transient transition from overventilated to underventilated conditions, the τ_{SS} increased quickly approaching a value of 1.

The correlations presented above represent data that has been averaged over steady-state (hood experiments) or quasi-steady-state (compartment fires) periods. For the purpose of modeling fires with respect to time it is of interest to know how the species yields correlate with the equivalence ratio during transient conditions (i.e., as the fire is growing). This was accomplished by plotting the yield to equivalence ratio data for individual fires from the time of ignition to the steady-state period. These transient correlations were compared to the steady-state correlations obtained from steady-state averaged data from all tests (e.g., the CO yield correlation shown in Figure 2-7.7). An example of one comparison is shown in Figure 2-7.12. Figure 2-7.12 shows the steady-state hexane CO yield correlation along with the transient yield *versus* equivalence ratio data for a hexane compartment fire that obtained a steady-state average ϕ_p of 3. The solid dots in Figure 2-7.12 represent the steady-state time ratio data, τ_{SS}. For this example, τ_{SS} remained fairly constant at about 0.1 for the entire fire. And as can be seen, the agreement between the transient and steady-state correlations is quite good, even for the transition to underventilated conditions. Good agreement between transient and steady-state data was also observed for CO_2 and O_2 yield correlations.

Although more transient in nature than the hood experiments, the compartment fires are characterized as primarily quasi-steady and, therefore, do not differ significantly from Beyler's hood experiments in this respect. This analysis also shows that the species yield correlations developed for steady-state conditions are representative of the transient growth periods of these fires.

In terms of full-scale application, these results suggest that ϕ_p and ϕ_{ul} are approximately equal for compartment fires characterized by relatively slow growth compared to the upper layer residence time (i.e., $\tau_{SS} \ll 1$). However, the low τ_{SS} values observed in the reduced-scale compartment fires may not always be representative of full-scale fires. The reduced-scale compartment fires had residence times typically between 4 and 20 s. These short residence times were a result of having relatively large fires compared to the compartment volume. Until flashover conditions are approached, a full-scale compartment fire will most likely have smaller fires compared to the volume of the space. As a result, the residence time of gases in the upper layer of a full-scale fire may be much longer. Times on the order of 5 to 10 min. may not be unrealistic in some cases. Therefore, in the case of a fast-growing full-scale fire, values of ϕ_p could increase relative to ϕ_{ul}.

In applying the yield-equivalence ratio correlations presented above, the following general guidelines can be used. The plume equivalence ratio, ϕ_p, should be used when all oxygen and fuel in the upper layer pass through the fire plume, regardless of the transient nature of the fire. In the case when τ_{SS} approaches or exceeds unity, species yields should still be calculated based on ϕ_p. But this is done with the consideration that the instantaneous upper layer concentrations will be less (O_2 higher) than if at steady-state, because the long residence time of the upper layer gases causes the gases from the fire plume to be diluted. The upper layer equivalence ratio, ϕ_{ul}, should be used in the yield correlations when additional oxygen or fuel enters the upper layer and is not accounted for by the plume equivalence ratio.

TRANSPORT OF TOXIC GASES

The previous sections of this chapter have dealt with determining the production of carbon monoxide in compartments under various fire conditions. However, since CO is an odorless and colorless gas, the transport of this toxic gas to regions remote from the burning compartment also poses a serious threat. Due to the inability of the human senses to detect CO, most carbon monoxide deaths in compartment fires occur in rooms remote from the fire.

Some studies have been conducted to investigate the transport of exhaust gases to rooms other than the burning room, but there is still much more work to be done. These studies are summarized and discussed below. The transport of exhaust gases to other locations is a complex phenomenon due to the coupling between the chemical, thermodynamic, heat transfer, and fluid dynamic transport processes involved. Adding to the complexity of the problem, a large number of varied building geometries and orientations exist that may significantly affect all of these processes in different ways.

During the postflashover stage of a fire, a layer of hot, buoyant, exhaust gases collects in the top of the compartment. The depth of this layer grows during the fire, and eventually drops below the height of the soffits of the room exits, i.e., doorways and windows. The exhaust gases then spill through these exits into the neighboring space to either the exterior of the building or another room or hallway. This causes the transport of toxic exhaust gases, posing a threat to anyone in an adjacent space. As the fuel-rich exhaust gases escape from the compartment, ambient air in the adjacent space is entrained into the exhausting gases. If the exhaust gas mixture becomes flammable, and an ignition source is present, burning of the exhausting gases can occur. This phenomenon is termed "external burning."

External burning can result in the nearly complete oxidation of the toxic exhaust gases to less-threatening carbon dioxide (CO_2) and water, provided the correct conditions exist. However, incomplete oxidation may occur if air entrainment is insufficient or unfavorable conditions exist, possibly resulting in a more toxic environment.

As discussed above, studies have been performed to investigate the environment generated inside a burning compartment. However, only a few experimental studies have touched on the subject of exhaust gas transport from a burning compartment to other locations within the confines of a building. These few studies have only begun to investigate the exhaust gas transport phenomenon.

The work that has been completed on exhaust gas transport to date has focused on two main areas. First, studies have been performed to investigate the characteristics of a flame as it impinges on a ceiling above the fire.[24,25] The second area of focus has been on the species environment produced in a remote location from the burning compartment within the same confined structure.[26,27] In keeping with the theme of this chapter, only those studies that examined CO transport from the compartment of fire origin will be discussed below.

Fardell et al[26] conducted large-scale experiments with a burning compartment exhausting perpendicular to the axial direction of a 11.4-m long, 1.2-m wide corridor with full-height walls. This study investigated the environments produced just outside of the compartment, and at the end of the corridor, by burning four fuels: wood (pine), polymethylmethacrylate (PMMA), polypropylene homopolymer (PP), and expanded polystyrene (EPS). Only two ventilation conditions were used with a 2-m-high doorway between the burning compartment and the corridor, i.e., 0.76- and 0.2-m-wide openings. The doorway provided the single ventilation path between the burning compartment and the corridor for both airflow into the compartment and exhaust gases flowing out of the compartment. Gas sampling was performed at two locations in the corridor, both 15 cm from the ceiling; one location was just outside of the burning compartment, and the other was at the end of the corridor. The gas samples were continuously analyzed for carbon monoxide (CO), carbon dioxide (CO_2), and oxygen (O_2). Gas chromatography was used to analyze samples taken at three different stages of the fire: development, postflashover, and decay. Between 20 and 40 hydrocarbon compounds were found in analysis of the spot samples. Other measurements made included gas temperatures, smoke measurements at the exit of the corridor, and crude air inlet velocity measurements at the open end of the corridor.

The main focus of the study by Fardell et al[26] was to investigate the toxicity of the environment produced at the sampled locations, and the dependence on the fuel burned and the stage of the fire. Many types of hydrocarbons were found with spot sampling, including oxygenated organics, saturated and unsaturated hydrocarbons, and aromatic hydrocarbons. The types of compounds found for each fuel were similar, although their concentrations varied between fuels, ventilation conditions, and stages of the fire. Overall, CO was found to be by far the most significant gas, in concentration and toxicity, at both locations sampled. The next significant toxic gas found was acrolein, an aldehyde; however, a high concentration of CO was indicated whenever

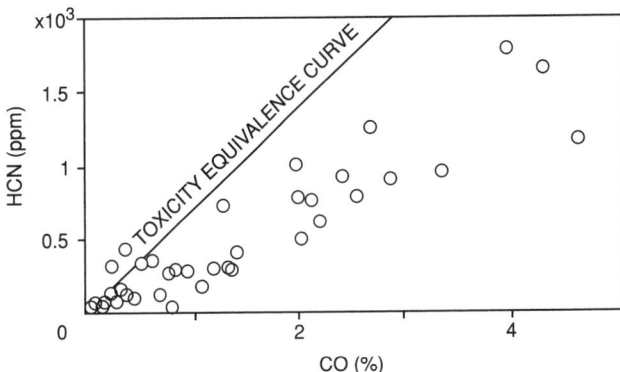

Fig. 2-7.13. HCN concentrations versus CO concentrations measured at the exit from the burn room to the hallway leading to the second floor and in the second floor room for all tests. (Figure taken from Ref. 28.)

significantly high concentrations of acrolein were present. It was noted that most hydrocarbons analyzed did not pose an immediate lethal hazard at any time during the fire, although they would have contributed to the irritancy of the gases.

Morikawa et al[27] performed experiments with a fire-resistant two-story house with a first floor burn room vented to the open atmosphere and to a hallway attached to stairs leading to the second floor. A room on the second floor had a door opening whose size was varied for different experiments. The burn room was equipped with typical room contents of furniture and draperies, representing a wide range of materials.

Continuous monitoring of CO, CO_2, and O_2 was performed in the first floor hallway just outside of the burn room, and in the second floor room, at both high and low locations. Spot sampling of gases allowed chromatographic analysis of hydrogen cyanide (HCN) and acrolein concentrations at the same sampling locations. Other toxic gas concentrations were not analyzed, since a previous study by the authors indicated these two to be the most significant toxic species other than CO.[28] Other measurements made included gas temperatures, smoke measurements, and gas velocities with bi-directional pitot tubes.

The results indicated, as with Fardell et al, that CO was the toxic gas with the most significant concentration in the fire exhaust. Figures 2-7.13 and 2-7.14 show reproductions from Ref. 28 of a comparison of the toxicity of CO versus HCN and acrolein, respectively, for all of the sampled locations and for all experiments presented. The toxicity equivalence lines are based on lethal concentrations over a time period of 5 to 10 min., such that lethal levels are 5000 ppm for CO, 350 ppm for HCN, and 30 ppm for acrolein. Note that the majority of the data points presented in these figures are located below the toxicity equivalence line, on the CO side. Thus, the data in these figures show that CO was the most toxic gas for these experiments by a significant margin.

However, investigation indicated that the CO levels were not completely responsible for the lethal toxicity of the atmosphere in the second floor room; HCN also had a significant contribution. HCN was found to be generated only during the burning of nitrogen-containing fuels, which was highly dependent on how the fire spread within the burning compartment. The levels of acrolein in the second floor room were always found to be less than toxic on a 10-min. exposure time scale. One interesting point to note from these

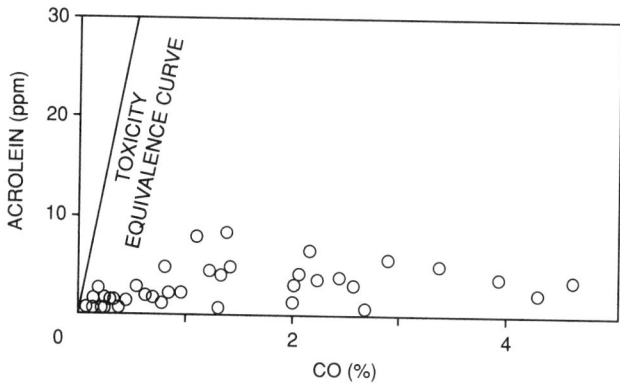

Fig. 2-7.14. *Acrolein concentrations versus CO concentrations measured in the second floor room for all tests. (Figure taken from Ref. 28.)*

experiments is that a toxic environment in the second floor room was generated even when the door was completely closed, leaving only small leaks around the door.

Both of the studies, by Fardell et al[26] and by Morikawa et al,[28] focused on the toxicological effects of the atmosphere generated. Neither study investigated the details of how the composition of the gases exhausting from the burning compartment changed during transport to the other remote locations. Even though sampling was made in two locations, comparisons between these locations were not extensively discussed in either paper.

These studies only investigated a very limited number of ventilation cases, with no clear, well-defined classification scheme for the differences in ventilation. Neither study indicated whether external burning of the exhaust gases occurred during transport. Both studies also picked a single building geometry for all experiments, with no systematic investigation of the effect of room and corridor geometry on the environments produced.

Studies by Gottuk et al[14,29] focused on the effect of open-jet external burning, on the burning of exhaust gases from a compartment fire as they vented to the open atmosphere, and on destroying carbon monoxide and soot escaping from the burning compartment. These tests represented the case of exhaust gases venting to the atmosphere through a window-style exhaust vent. These experiments were compared to the previous study by Gottuk et al[3] on the gas composition inside the compartment during similar fires to determine the efficiency of external burning in oxidizing CO and soot. A summary of these studies is given below.

Two distinct types of external flames were observed during compartment fires. First, external flame jets appeared as ceiling jets extended from the main fire within the compartment, and out through the exhaust vent. During significantly underventilated fires in the compartment, the second type of external flames occurred when the exhausting flammable gases from the compartment mixed with a sufficient amount of ambient air, and were ignited causing external burning. Three different types of external burning were observed: (1) quick flashes, (2) short bursts lasting greater than 1 s, and (3) sustained external burning.

Overventilated fires never produced external burning, because the excess oxygen in the compartment completely oxidized all of the fuel inside the compartment. In this case, no flammable gases existed in the upper layer of the com-

partment to be exhausted and burned. For underventilated fires, Gottuk et al defined characteristic plume equivalence ratios that marked the onset of external flashes (ϕ_{flash}) and then sustained external burning (ϕ_{ss}).

Flashes were reported to occur at a ϕ_{flash} of 1.4 ± 0.4. The characteristic plume equivalence ratio for sustained external burning showed a slight dependence on the exhaust vent area, reported as 2.1 ± 0.3 for exhaust vents of 400 cm² area, and 1.8 ± 0.2 for exhaust vents of area in the range of 800 to 1600 cm². This exhaust vent dependence was explained in terms of the smaller flame jets observed with the smaller area exhaust vents, reducing the ability to ignite the exhausting gases. Although the flammability of the exhaust gases mixing with ambient air was determined by the global plume equivalence ratio, the occurrence of sustained external burning was found to be controlled by the presence of the ignition source.

An instantaneous plume equivalence ratio of 1.8 (2.1 for small vents) was required for sustained external burning to begin; however, compartment fires that produced a "quasi-steady-state" average plume equivalence ratio equal to or greater than 1.7 produced sustained external burning. The occurrence of sustained external burning was the only form of external burning observed to reduce CO and soot levels significantly. Figure 2-7.15 shows the results of external burning on the CO yields as compared to the in-compartment values as a function of the plume equivalence ratio. As can be seen in the figure, the downstream (exhaust duct sampled) CO yields during sustained external burning were reduced to 10–25 percent of the yields measured in the compartment (an average value of 0.22) for underventilated fires. Carbon dioxide yields downstream also approached their theoretical maximum during sustained external burning, indicating near-complete oxidation of all carbon to CO_2.

The effect of sustained external burning on soot yields followed the same trends as CO oxidation, but with a larger amount of scatter in the data. Soot was oxidized to 0–50 percent of levels observed just prior to sustained external

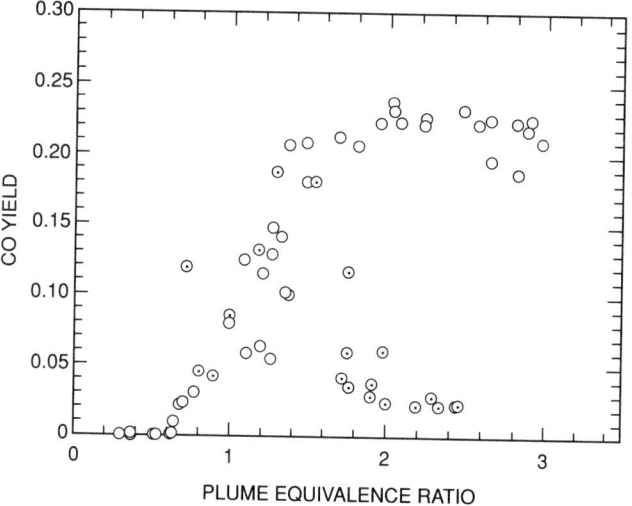

Fig. 2-7.15. *CO yield versus plume equivalence ratio for compartment fires with an exhaust jet to the open atmosphere through a window-style exhaust vent; (○) compartment sampled, (◉) exhaust duct sampled. (Data taken from Refs. 14 and 29.)*

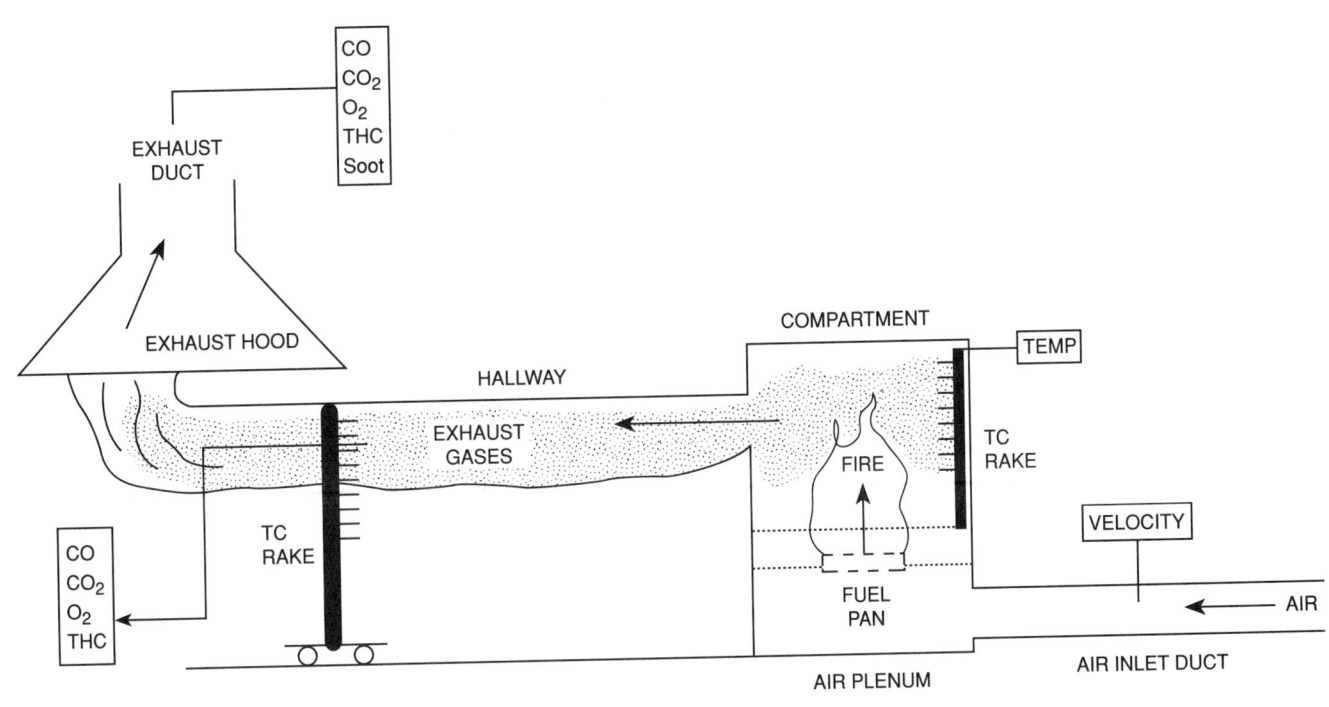

Fig. 2-7.16. *Schematic of experimental apparatus.*

burning. On average, soot yields prior to sustained external burning reached about 0.015.[14]

A study by Ewens[30] and Ewens *et al*[31] investigated the evolution of compartment fire exhaust gases during transport through a hallway. The fuel-rich plume exhausting from the compartment fire, into an adjacent enclosed space, mixed with ambient air allowing oxidation of incomplete products of combustion. Emphasis in this study was placed on investigation of the physical phenomena responsible for the overall species oxidation efficiencies during transport of the exhaust gases.

Experiments were performed with a compartment fire exhausting into a 3.7-m-long hallway, with exhaust gases flowing directly along the hallway axis and exiting the hallway to the open atmosphere. (See Figure 2-7.16.) Four soffit combinations, consisting of 0- and 20-cm soffit heights at both ends of the hallway, were investigated. For each soffit case, experiments were conducted for several different underventilated compartment fire cases. The compartment used was the same as used by Gottuk *et al*. The design allowed direct measurement of the equivalence ratio by separating the air inlet and exhaust gas exit ventilation paths. Measuring both the air inlet mass flow rate and the fuel vaporization rate allowed ϕ_p to be calculated.

Exhaust gases were sampled both well downstream of the hallway exit and inside the hallway. Downstream sampling allowed investigation of overall species levels exiting the hallway. In-hallway sampling allowed detailed investigation of species consumption and production occurring in the hallway. Gas species concentrations measured included carbon monoxide, carbon dioxide, oxygen, and total hydrocarbons measured as ethylene. Soot measurements were taken downstream of the hallway only. Vertical gas temperature profile measurements were taken both in the compartment, and in the hallway at various locations.

Results of CO yields as a function of global equivalence ratio for the 0-cm inlet/0-cm outlet soffit case are shown in Figure 2-7.17 together with the data for Gottuk *et al* for the case of exhaust directly to the outside. The post-hallway CO yields for this soffit case varied between 0.100 and 0.070, with an average of 0.089. These levels indicate a 54 to 68 percent reduction from in-compartment levels, averaging

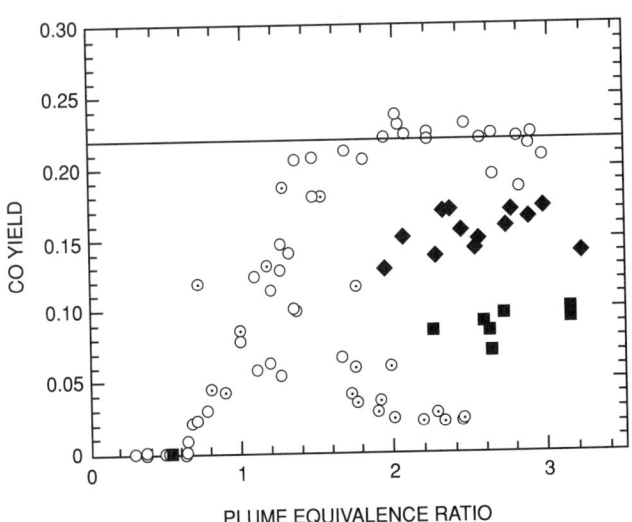

Fig. 2-7.17. *CO yield versus quasi-steady-state plume equivalence ratio for post-hallway sampled tests. (■) 0-cm inlet/0-cm exit soffits, (◆) 0-cm inlet/20-cm exit soffit and open jet tests, (○) compartment sampled, (⊙) exhaust duct sampled. (Data taken from Refs. 14, 29, 30, 31.)*

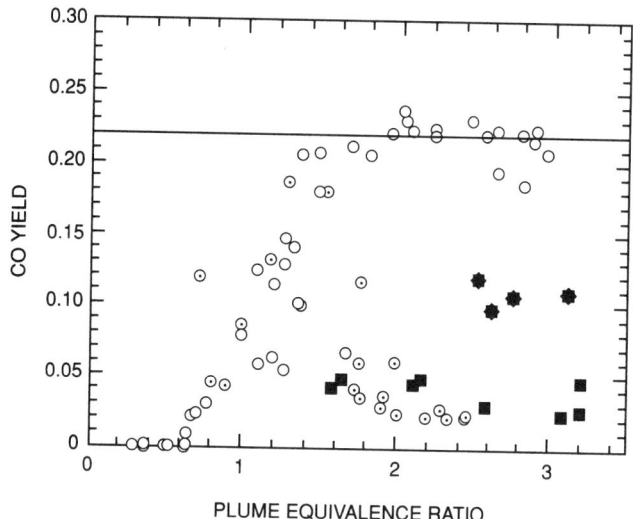

Fig. 2-7.18. *CO yield* versus *quasi-steady-state plume equivalence ratio for post-hallway sampled tests with 20-cm inlet/0-cm exit soffits.* (■) $\dot{m}_f < 10\ g/s$, (✱) $\dot{m}_f > 10\ g/s$ for open-jet tests, (○) *compartment sampled,* (⊙) *exhaust duct sampled. (Data taken from Refs. 14, 29, 30, and 31.)*

60 percent. These results demonstrate that oxidation of CO did occur in the hallway, but less efficiently than for the open-jet experiments.

Figure 2-7.17 shows that a similar result was obtained with a 0-cm inlet soffit and a 20-cm outlet soffit. However, in this case even less CO was consumed due to the deeper layer in the hallway caused by the presence of the exit soffit.

When an entrance soffit of 20-cm was used, Ewens *et al* reported significantly different results as shown in Figure 2-7.18. For fuel flow rates of less than 10 g/s, CO oxidation efficiencies approaching those of Gottuk *et al* for open jets were obtained. However, at higher fuel flow rates, CO yields similar to those for the 0-cm inlet soffit were observed.

Ewens *et al* concluded that the presence of the inlet soffit significantly enhanced entrainment of air from the corridor into the exhaust plume. For low fuel flow rates, this air entrainment resulted in nearly complete burnout of the CO similar to the open-jet experiments. At higher fuel flow rates, the air entrainment was insufficient compared to the open-jet cases and, thus, incomplete burnout of the CO was observed.

Another important aspect of the Ewens *et al* study was the importance of the relationship between hydrocarbon and carbon monoxide (CO) oxidation. Figure 2-7.19 shows profiles of normalized species concentrations and temperature taken along the centerline of the hallway for 0 inlet/0 outlet soffit case. The normalized concentrations represent the species concentration measured in the hallway divided by the upper layer concentration measured inside the compartment. The figure shows that the hydrocarbons oxidized much faster than CO, and in the environment with limited air (0 to 100 cm^3) caused poor CO oxidation. This effect was ascribed to the fact that hydrocarbons react more readily with available hydroxyl radical than does CO. This result, which was also reported by Pitts[11] and Gottuk *et al*[12] for chemical kinetics modeling, leads to a delay in CO oxidation in the hallway until most of the hydrocarbons are oxidized. Thus, for low air entrainment, lethal concentrations of CO can survive the external flame and be transported to other locations.

In summary, the fate of carbon monoxide leaving an underventilated compartment fire to an adjacent space has

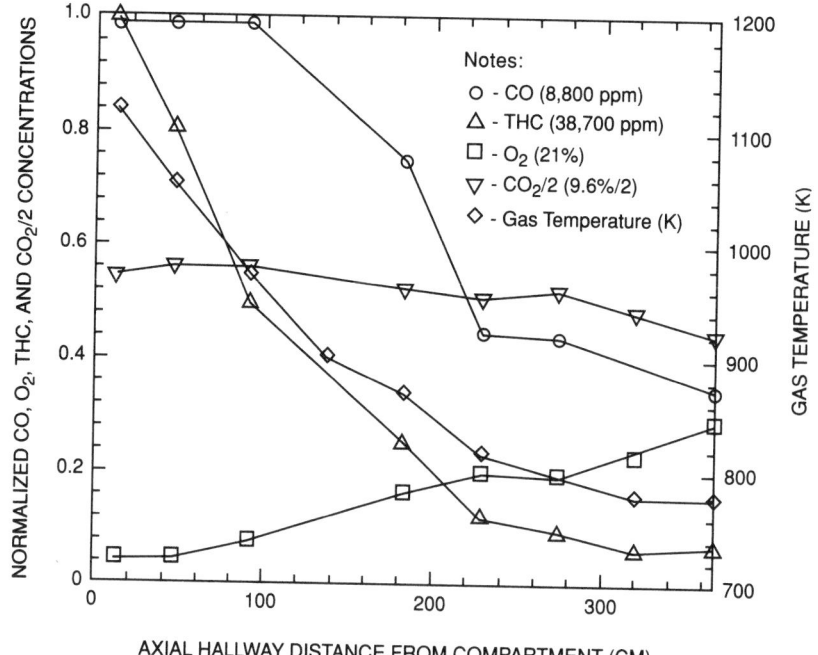

Fig. 2-7.19. *Normalized species concentrations and gas temperature versus hallway axial distance. Experimental conditions: 0/0 soffit case, 20-cm-diameter fuel pan, 1200 cm^2 exhaust vent. Sample location: 5.1 cm from ceiling, center width. Normalizing concentrations given in legend. (Figure taken from Refs. 30 and 31.)*

TABLE 2-7.3 *Post-compartment Oxidation Efficiency Ranges for Underventilated Compartment Fires (Data for hallway tests with various soffit heights[30,31] and open-jet tests.[14,29])*

Species	Open-Jet Experiments [14,29]	Hallway Experiments			
		Soffits: 0 cm inlet 0 cm exit	Soffits: 0 cm inlet 20 cm exit	Soffits[a]: 20 cm inlet 0 cm exit	Soffits[b]: 20 cm inlet 20 cm exit
CO	75–90 %	54–68 %	22–42 %	78–89 % (45–55 %)	71–91 % (39–51 %)
THCs	N/A	83–93 %	75–89 %	92–97 % (88–91 %)	91–98 % (86–93 %)
Soot	50–100 %	48–86 %	22–57 %	61–92 % (42–56 %)	56–88 % (28–46 %)

[a] values in parenthesis are for fuel volatilization rates > 10 g/s

[b] values in parenthesis are for fuel volatilization rates > 9 g/s

been shown to be very sensitive to the hydrodynamic mixing between the rich exhaust plume and the cooler ambient air in the hallway. Variations in the hallway inlet and exit soffits produced significant effects on the hydrodynamic structure of the exhaust plume and oxidation efficiencies. A summary of the oxidation efficiencies obtained for experiments of different soffit arrangements, including the open-jet case, is presented in Table 2-7.3.

For all soffit configurations, the overall oxidation of THCs was much more complete than for CO. This result was explained by two effects. First, at the local temperatures measured in the hallway, the oxidation rate of THCs is much faster than that for CO, causing the entrained oxygen to be depleted by THCs. Second, oxidation of THCs produced CO, which decreased the net CO oxidation rate. The oxidation of THCs essentially inhibited the oxidation of CO with the oxygen-limited environment in the hallway. Due to this favorable chemical kinetic situation, THC oxidation was very insensitive to changes in soffit heights, whereas the oxidation of CO was dramatically affected.

The fate of carbon monoxide as it is transported to areas outside the compartment of fire origin is an area of continuing research; however, some conclusions can be drawn from the research to date. In none of the studies reported has the CO concentration in a location outside the compartment of origin been observed to be higher than that in the compartment. Also, the presence of external burning can lead to a reduction in the amount of CO transported from the room of fire origin. However, since external burning cannot be well predicted currently, a conservative estimate for evaluating life safety hazards would be to assume that no external burning occurs and, thus, that the concentration of CO in adjacent spaces is governed by the rate of production of CO in the compartment of fire origin.

NOMENCLATURE

B_i yield coefficients of species i
C stoichiometric molar ratio of water to carbon dioxide
f normalized yield or generation efficiency
D_{O_2} mass depletion of oxygen per gram of fuel burned (g/g)
k maximum theoretical yield
$\dot{m}_a$ mass flow rate of air
$\dot{m}_f$ mass volatilization rate of fuel
$\ddot{m}_f$ the derivative of the fuel volatilization rate
M molecular weight

n molar quantity
r stoichiometric fuel-to-air ratio
t_r residence time of gases in the upper layer
τ_{SS} steady-state time ratio
ϕ_p global plume equivalence ratio
ϕ_{ul} global upper layer equivalence ratio
$X_{i_{dry}}$ dry mole fraction of species i (H_2O removed from sample)
$X_{i_{wet}}$ wet mole fraction of species i
X mole fraction
Y yield (g/g) also refers to D_{O_2}

Subscripts

a air
f fuel

Acknowledgements
The authors thank Mr. David Ewens and Dr. Uri Vandsburger for their contribution to the section, Transport of Toxic Gases.

REFERENCES CITED

1. R.A. Anderson, A.A. Watson, and W.A. Harland, "Fire Deaths in Glasgow Area: II The Role of Carbon Monoxide," *Med. Sci. Law*, 21, 289–294 (1981).
2. B. Harwood and J.R. Hall, "What Kills in Fires: Smoke Inhalation or Burns?" *Fire Journal*, 83, 29–34, May/June (1989).
3. D.T. Gottuk, R.J. Roby, M.J. Peatross, and C.L. Beyler, *J. of Fire Prot. Engr.*, 4:133–150 (1992).
4. N.P. Bryner, E.L. Johnsson, and W. M. Pitts, "Carbon Monoxide Production in Compartment Fires—Reduced-Scale Enclosure Test Facility," NISTIR 5568, National Institute of Standards and Technology, Gaithersburg, MD (1995).
5. S.J. Toner, E.E. Zukoski, and T. Kubota, "Entrainment, Chemistry, and Structure of Fire Plumes," National Institute of Standards and Technology, Report NBS-GCR-87–528, Gaithersburg, MD (1987).
6. J.H. Morehart, E.E. Zukoski, and T. Kubota, "Species Produced in Fires Burning in Two-Layered and Homogeneous Vitiated Environments," National Institute of Standards and Technology, Report NBS-GCR-90–585, Gaithersburg, MD (1990).
7. C.L. Beyler, *Fire Safety Journal* 10:47–56 (1986).
8. C.L. Beyler, *Fire Safety Science—Proceedings of First International Symposium*, Hemisphere, Washington, DC, 431–430 (1986).
9. E.E. Zukoski, S.J. Toner, J.H. Morehart, and T. Kubota, *Fire Safety Science—Proceedings of the Second International Symposium*, Hemisphere, Washington, D.C., 295–304 (1989).

10. E.E. Zukoski, J.H. Morehart, T. Kubota, and S.J. Toner, *Combustion and Flame* 83:324–332 (1991).

11. W.M. Pitts, *Twenty-Fourth Symposium (International) on Combustion*, The Combustion Institute, Pittsburgh, PA (1992).

12. D.T. Gottuk, R.J. Roby, and C.L. Beyler, "The Role of Temperature on Carbon Monoxide Production in Compartment Fires," submitted to *Fire Safety Journal* (1994).

13. A. Tewarson, "Fully Enveloped Enclosure Fires of Wood Cribs," *Twentieth Symposium (International) on Combustion*, The Combustion Institute, Pittsburgh, PA, 1555 (1984).

14. D.T. Gottuk, "The Generation of Carbon Monoxide in Compartment Fires," National Institute of Standards and Technology, Report NBS-GCR-92-619, Gaithersburg, MD (1992).

15. W.D. Walton, and P.H. Thomas, "Estimating Temperatures in Compartment Fires," *The SFPE Handbook of Fire Protection Engineering*, Chapter 2–2, National Fire Protection Association, Quincy, MA (1988).

16. W.M. Pitts, E.L. Johnsson, and N.P. Bryner, "Carbon Monoxide Formation in Fires by High-Temperature Anaerobic Wood Pyrolysis," presented at the *Twenty-Fifth Symposium (International) on Combustion* (1994).

17. D. Gross, and A.F. Robertson, *Tenth Symposium (International) on Combustion*, The Combustion Institute, Pittsburgh, PA, 931–942 (1965).

18. J. Warnatz, in *Combustion Chemistry* (W.C. Gardiner, ed.), Springer-Verlag, New York, 224–232 (1984).

19. E.K. Budnick, "Mobile Home Living Room Fire Studies: The Role of Interior Finish," NBSIR 78–1530, National Bureau of Standards, Gaithersburg, MD (1978).

20. E.K. Budnick, D.P. Klein, and R.H. O'Laughlin, "Mobile Home Bedroom Fire Studies: The Role of Interior Finish," NBSIR 78–1531, National Bureau of Standards, Gaithersburg, MD (1978).

21. R.S. Levine, and H.E. Nelson, "Full-Scale Simulation of a Fatal Fire and Comparison of Results with Two Multi-Room Models," National Institute of Standards and Technology, Internal Report 90–4268, Gaithersburg, MD (1990).

22. T. Morikawa, E. Yanai, T. Okada, T. Watanabe, and Y. Sato, "Toxic Gases from House Fires Involving Natural and Synthetic Polymers under Various Conditions," *Fire Safety Journal*, 20 257–274 (1993).

23. F.W. Williams, C.L. Beyler, D.T. Gottuk, T.A. Toomey, and J.L. Scheffey, "1993 Fleet Doctrine Evaluation Workshop: Phase II, Class B Fire Dynamics Test Series," NRL Ltr. Rpt. Ser. 6180/148.1, March 23, 1994.

24. P.L. Hinkley, H.G.H. Wraight, and C.R. Theobald, "The Contribution of Flames under Ceilings to Fire Spread in Compartments," *Fire Safety Journal*, 7, 227–242 (1984).

25. V. Babrauskas, "Flame Lengths under Ceilings," *Fire and Materials*, 4, 119–126 (1980).

26. P.J. Fardell, J.M. Murell, and J.V. Murell, "Chemical 'Fingerprint' Studies of Fire Atmospheres," *Fire and Materials*, 10, 21–28 (1986).

27. T. Morikawa, E. Yanai, T. Okada, and K. Sato, "Toxicity of the Atmosphere in an Upstairs Room Caused by Inflow of Fire Effluent Gases Rising from a Burn Room," *Journal of Fire Sciences*, 11, 195–209 (1993).

28. T. Morikawa, E. Yanai, T. Watanabe, T. Okada, and Y. Sato, "Toxic Gases from House Fires of Natural Polymers or Both Synthetic and Natural Polymers under Different Conditions," *Proceedings of Interflam '90*, 249–255 (1990).

29. D.T. Gottuk, R.J. Roby, and C.L. Beyler, "A Study of Carbon Monoxide and Smoke Yields from Compartment Fires with External Burning," *Twenty-fourth Symposium (International) on Combustion*, The Combustion Institute, Pittsburgh, PA, 1729–1735 (1992).

30. D.S. Ewens, "The Transport and Remote Oxidation of Compartment Fire Exhaust Gases," M.S. Thesis, Virginia Polytechnic Institute and State University, Department of Mechanical Engineering, Blacksburg, VA (1994).

31. D.S. Ewens, U. Vandsburger, and R.J. Roby, "Transport and Oxidation of Compartment Fire Exhaust Gases," submitted to the *Journal of Fire Protection Engineering* (1994).

TOXICITY ASSESSMENT OF COMBUSTION PRODUCTS

David A. Purser

INTRODUCTION

It has long been recognized that exposure to toxic smoke products is one of the hazards confronting people in fires, but over the last ten years there has been a considerable increase in concern in this area. In the United States, attention has been focused on the toxicity problem by a number of large fire disasters where victims have died from exposure to toxic smoke products.[1] Although this is also true to some extent for the United Kingdom, a major impetus for work in this area followed statistical surveys of fire casualties carried out in the mid-1970s. These surveys of casualties from all fires, and particularly from fires in domestic dwellings, revealed that not only were a large proportion of fatal and nonfatal fire casualties being reported in the category "overcome by smoke and toxic gases" rather than heat and burns, but also that there was a fourfold increase in the former category between 1955 and 1971.[2] This increasing trend has continued into the 1980s, so that now approximately half of all fatal casualties and a third of all nonfatal casualties of dwelling fires (the majority caused by fires in furniture and bedding) are reported as being "overcome by smoke and toxic gases."[3] This has occurred despite the fact that the total numbers of fires have remained approximately constant over this same period of time.

A number of possible reasons have been suggested for this increase in smoke-related casualties. They have been linked with the increased use of modern synthetic materials in furnishings. Another view is that the increase may not be directly related to modern materials but to changes in living styles over the period which have led to more furnishing and upholstery material being used in the average British home, and, therefore, a greater fire load. It has also been suggested that the increase is not real, but a statistical anomaly resulting from an increased awareness by the emergency services

of toxic effects on victims. Epidemiological data are often difficult to interpret, but many of those working in this area are convinced that the increase in smoke-related casualties is real. The situation in the United States is difficult to interpret since it contains a larger and more diverse population, and statistics may not have been collected so as to reveal such a trend. The U.S. fire death rate is more than twice that of the U.K. and other European countries, but over the period from 1965 to 1978 the annual numbers of U.S. fire deaths have been decreasing.[4] However, in the U.S. and the U.K., toxic smoke products are recognized as being the major cause of incapacitation and death in fires.[5]

Materials-Based and Combustion Product-Based Approaches to Toxicity Assessment

There are two main views as to why smoke toxicity appears to be an increasing problem, which in turn have led to two rather different approaches to the evaluation of toxicity:

1. One view held that smoke from modern synthetic materials contained new toxic products that were not present previously, and that in some cases these products might be very potent, exerting novel toxic effects at very low doses (the so-called 'supertoxicants'). Such effects could therefore be detected by means of simple, small-scale toxicity tests which could be used for regulatory purposes.[6-8] To some extent this approach followed the discovery that two materials, a flexible polyurethane foam containing a phosphorus-based fire retardant, and polytetrafluoroethylene (PTFE), could under certain laboratory conditions evolve products with a very high toxic potency.[6,7] This led to the use of rather simplistic materials-based toxicity tests, where the toxicity of materials is ranked in terms of the rodent LC_{50} (the concentration of combustion products expressed in terms of mg of material per liter of air causing the deaths of 50 percent of animals exposed).[8,9] This approach implies that the engineer should design by using those materials with the better performance in toxicity tests and that are consistent with good performance in other types of small-scale fire tests.

2. The other main view was that the basic toxic products of fires were much the same as always, but that in many modern fires the rate of fire growth and the rate of evolution of

Dr. David Purser has a Ph.D in neurophysiology, and a diploma in toxicology from the Royal College of Pathologists. As former director of combustion toxicology research at Huntingdon Research Centre, England, and currently, as head of the People and Fire Risk Section at the Building Research Establishment, Fire Research Station, Borehamwood, England, he has worked on fire toxicology and life assessment for 20 years, and serves as an expert for the International Organization for Standardization (ISO) fire hazard working groups.

the common toxic products were much greater than they had been previously. Therefore, the best way to mitigate toxic hazard in fires was to control such factors as ignition, flame spread, and rate of smoke evolution rather than the qualitative nature of toxic products. For this approach, which is favored in the U.K., there is more interest in estimating toxicity by making a chemical determination of the main toxic products given off by materials, and then following the time/concentration profiles of these few basic toxic products in large-scale fire tests. These data are then used to estimate time to incapacitation and death. In this context the main function of small-scale tests would be to confirm that the toxicity associated with particular burning materials was indeed due to the common toxic fire products via chemical atmosphere analysis in conjunction with animal exposures, and to identify those cases where unusual toxic effects occurred. This approach enables a firesafety engineer to design to a set of fire scenarios in a system (for example, a hotel bedroom or an aircraft cabin) and, by a simple chemical analysis of atmospheres produced during small- and large-scale fire tests, predict likely toxic hazard. The difficulty with these models is that they are often based on simplistic and erroneous assumptions concerning the effects and interaction of toxic products.

In practice there is a need for both small-scale materials-based toxicity tests and for profile modeling based on a few major toxic fire products. Existing information is often inadequate and misleading, and a better standard of research and testing is needed if data are to be produced for practical use. This chapter will give the reader an understanding of what is known about combustion product toxicity, the extent to which effects can be predicted from a knowledge of common fire products, and how small-scale tests should be performed and the results interpreted and used.

The Significance of Toxicity as Part of Total Fire Hazard

It is possible to consider the effects of fires on victims in three phases:

1. The first phase consists of the period when the fire is growing, but before the victim is affected by heat or smoke. During this phase the important factors influencing escape and ultimate survival are largely psychobehavioral or logistical factors, such as how the victim is alerted to the fire and reacts to that knowledge, whether he/she responds to alarms, attempts to leave or stays to fight the fire, interacts with other individuals, and how the person responds to the geography of the fire environment in effecting an escape.
2. The middle phase is the period when the victim is exposed to smoke, heat and toxic products, and where physiological factors such as irritancy and narcosis affect the victim's escape capability. During this phase such factors as the toxic nature of fire products and the dynamics of their production become critically important to escape.
3. The third phase is death in the fire, which may be caused by the major factors of toxicity and burns or a number of other factors.

The toxic effects of fire products are therefore most important during the second and third phases of fire growth. Most studies of fire toxicity have been confined to aspects of

lethality. The ultimate causes of death in fires have been studied through pathological investigations of fire fatalities such as the Strathclyde study in the U.K.[10] Also, the majority of combustion toxicity studies on laboratory animals have been used to measure lethality, principally in terms of the LC_{50} for individual fire products such as carbon monoxide (CO) or hydrogen chloride (HCl), or mixtures of thermal decomposition products from individual materials.[11]

The middle phase, of incapacitation in fires, can be studied either by animal experimentation or by investigations of the circumstances surrounding real fire casualties, particularly survivors of serious smoke exposure. However, this crucial area of toxicity has been largely neglected.

One particular series of studies has been carried out on the sublethal effects of combustion atmospheres on animals, mainly primates, to examine the mechanisms whereby people become incapacitated in fires.[12] Incapacitation rather than death has been studied because most fires are potentially lethal due to heat or CO if the victim is exposed to these for sufficient time. The two major determinants of whether a potential victim escapes are (1) the point at which incapacitation by toxic products is reached, and (2) how these products affect escape capability during the window of time available for escape between ignition and the development of lethal conditions.

The physiological effects of exposure to toxic smoke and heat in fires result in varying degrees of incapacitation which may also lead to death or permanent injury. Incapacitating effects include:

1. Impaired vision resulting from the optical opacity of smoke and from the painful effects of irritant smoke products and heat on the eyes.
2. Respiratory tract pain and breathing difficulties or even respiratory tract injury resulting from the inhalation of irritant smoke which may be very hot. In extreme cases this can lead to collapse within a few minutes from asphyxia due to laryngeal spasm and/or bronchoconstriction. Lung inflammation may also occur, usually after some hours, which can also lead to varying degrees of respiratory distress.
3. Narcosis from the inhalation of toxic gases, resulting in confusion and loss of consciousness.
4. Pain to exposed skin and the upper respiratory tract followed by burns, or hyperthermia, due to the effects of heat preventing escape; this can lead to collapse.

All of these effects can lead to permanent injury, and all except item 1 can be fatal if the degree of exposure is sufficient.

With regard to hazard assessment the major considerations are:

1. The time when partially incapacitating effects are likely to occur which might delay escape.
2. The time when incapacitating effects are likely to occur which might prevent escape, compared with the time required for escape.
3. Whether exposure is likely to result in permanent injury or death.

Up to a certain level of severity, the hazards listed in items 1 through 4 under physiological effects cause a partial incapacitation, by reducing the efficiency and speed of escape. These effects lie on a continuum from little or no effect at low levels to relatively severe incapacitation at high levels, with a variable response from different individuals. It is important to make some estimate of effects that are likely to delay escape, which may result in fewer occupants being able to escape during the short time before conditions become so bad that escape is no longer possible. Most important in this context is exposure to optically dense and irritant smoke, which tends to be the first hazard confronting fire victims. For more severe exposures a moment may be reached when incapacitation is predicted to be sufficiently bad as to prevent escape. For some forms of incapacitation, such as when narcosis leads to a rapid change from near normality to loss of consciousness, this moment is relatively easy to define. For other effects a defining moment is less easily characterized; e.g., when smoke becomes so irritant that pain and breathing difficulties lead to the cessation of effective escape attempts, or when pain and burns prevent movement. Nevertheless it is considered important to attempt some estimate of the moment when conditions become so severe in terms of these hazards that effective escape attempts are likely to cease, and when occupants are likely to suffer severe incapacitation or injuries. In the following text the mechanisms whereby these various factors cause incapacitation and death, and how they are likely to produce partial incapacitation and affect escape capability, are examined in detail. Toxicological and physiological data and models are presented to enable calculations to be made of time (or exposure dose) to incapacitation and death.

Basic Toxicity Patterns of Fire Products

As a result of chemical studies of large- and small-scale experimental fires[13] and animal exposures to the thermal decomposition products from a wide range of materials,[12] two important basic points have emerged concerning the nature of fire product toxicity:

1. Atmospheres of thermal decomposition products, even from single materials, contained large numbers of potentially toxic products. The chemical composition of the products could vary considerably depending upon the different conditions of temperature and oxygen supply under which they were decomposed.[12] When animals were exposed to these atmospheres, similar variations in toxicity were observed. In many cases, however, similar basic ranges of products were evolved from quite different materials.[13]

2. Despite the great complexity in chemical composition of smoke atmospheres, the basic toxic effects were relatively simple. For each individual smoke atmosphere the toxicity was dominated either by a narcotic (asphyxiant) gas (CO or HCN) or by irritants. Also, interactions between individual narcotic gases or, between narcotics and irritants, were found to be approximately additive, so that a reasonably good predictive model for incapacitation could be developed by summing the effects of each individual toxic component as acting separately.[12]

This work and that done by others using rodents[11] seems to indicate that the theory that smoke casualties have increased because new, highly toxic products are formed from modern materials is unlikely to be correct. Also, the finding that a small number of basic products are particularly important leads to the possibility of predicting toxic effects from a relatively simple chemical analysis of fire products. (This is not to say that unusual highly toxic products cannot occur, as at least two examples have been discovered in the laboratory[6,7]; and this is one important rea-

son that the toxicity of thermal decomposition products from materials should be submitted to an animal screen.)

DOSE/RESPONSE RELATIONSHIPS AND DOSE ESTIMATION IN THE EVALUATION OF TOXICITY

Before considering the particular effects of individual toxic fire products it is necessary to determine the basic parameters required to quantify exposure. Ultimately, the degree of toxicity is determined by such factors as the concentration of toxic product in the target organ of the body, and the time period for which a toxic concentration is maintained. For a narcotic product, the most important criterion is the concentration in the cerebral blood supply or inside the brain cells, while for an irritant product the most important factor is the concentration in the lining of the nose, throat, or lung. In some cases it is important and feasible to measure such parameters directly. For CO it is not the concentration in the smoke that directly determines how someone will be affected, but the concentration that has accumulated in the blood in the form of carboxyhemoglobin, and this can be determined relatively simply from a drop of blood.[14,15]

In practice, however, it is often not feasible to measure the amount of toxic product directly accumulated in the subject. Also, relating observed toxic effects to measurements of toxic products in the smoke itself is preferable, since it enables predictions of toxicity to be made based on chemical measurements of fire atmospheres without necessarily exposing animals. A series of useful secondary measurements can, therefore, be made that can be related to toxic effects in animals, but it must be remembered that these indirect measurements of exposure always involve some degree of error or uncertainty.

The Relationships Between Concentration Inhaled, Duration of Exposure, and Toxicity

In inhalation toxicology two parameters that are always measured and reported are the actual analyzed concentration of the test material per unit volume of air in the animals' exposure chamber, and the duration of exposure. For droplet aerosols or dusts the particle size range in the atmosphere is also measured so the respirable fraction (the part capable of entering the body) can be calculated. Where toxic effects can occur rapidly, as with the narcotic gases, it is also important to measure the rate of uptake of the toxicant. This can be estimated by measuring the volume of air breathed by the animal per minute (the respiratory minute volume, or RMV), although for accurate calculation of uptake and dose, further measurements, such as of blood levels, must be made. Variations in RMV can have dramatic effects on toxicity as will be described in the following section.

Although such parameters as respiration and particle size of aerosols are important, the most basic parameters reported are the concentration of the toxicant and the duration of exposure, which enables a rudimentary estimation of the dose. Thus the product of concentration and time (Ct product) gives an estimate of the dose available to the animal. In general safety evaluation of novel chemicals for an acute exposure, a standard single 4-hour exposure time is used and toxicity is expressed in terms of the concentration of test material causing the death of 50 percent of the animals during exposure or within 14 days after exposure. This is known as the 4-hour LC_{50} concentration.

In practice, however, it may be necessary to predict what will happen to a subject exposed to a higher concentration for a shorter period of time, or a lower concentration for longer time. Although this can be done by carrying out more LC_{50} experiments using different exposure durations, as an approximation toxicologists often resort to Haber's rule, which states that the toxicity depends upon the dose accumulated, and that the product of time and concentration is a constant,[16] so that

$$W = C \times t \qquad (1)$$

where W is a constant dose, specific for any given effect. In practice, dose in inhalation toxicology is often expressed in terms of Ct product. In the case of the LC_{50} the effect is death of 50 percent of the animals and

$$W = LC \cdot t_{50} \qquad (2)$$

expressed in mg·min/liter (i.e., the product of the concentration and the duration of exposure causing lethality). This relationship implies a linear uptake of the toxic substance with time. (See Figure 2-8.1.) It holds true for many substances where the primary target organ is the lung; in the context of combustion toxicology this relationship can be applied to estimates of the dose of a lung irritant likely to cause postexposure fire deaths from lung inflammatory responses. An example of such an irritant is carbonyl fluoride, a highly toxic lung irritant produced during the thermal decomposition of PTFE, which has a 1-hour LC_{50} of 0.990 mg/L, which is exactly four times the 4-hour LC_{50} of 0.248 mg/L.[17]

Unfortunately, this simple principle does not always hold true. In particular, some volatile substances (such as CO) are both taken up and excreted via the lungs. In this case

LETHAL TARGET ORGAN CONCENTRATION FOR 50% OF ANIMALS

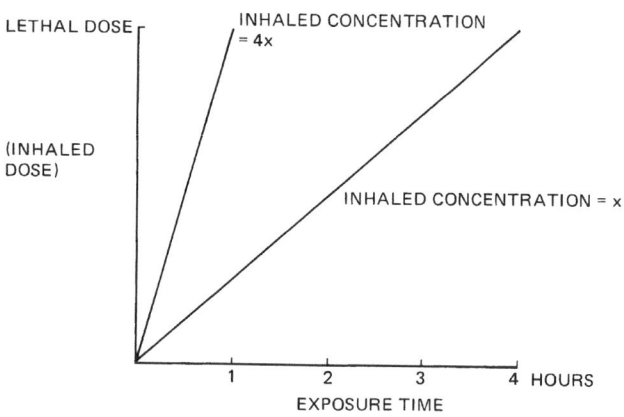

UPTAKE OF SUBSTANCE OBEYING HARBER'S RULE

Fig. 2-8.1. Uptake of a substance obeying Haber's rule (i.e., with a long halflife of detoxification or excretion), where the rate of uptake is directly proportional to the inhaled concentration, so that 1-hour LC_{50} is four times the 4-hour LC_{50}. The lethal dose represents a lethal target organ concentration for 50% of the animals.

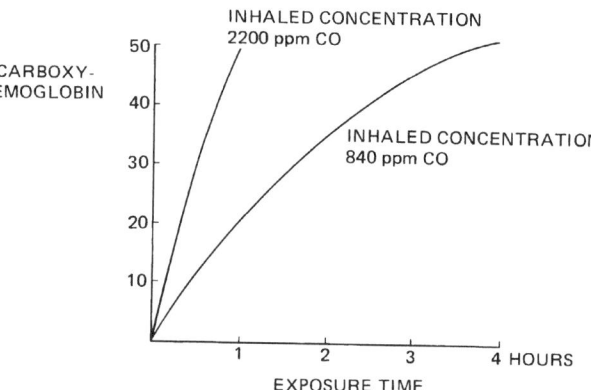

Fig. 2-8.2. Uptake of a substance (carbon monoxide) which is both absorbed and excreted via the lungs, and where the rate of uptake depends upon the difference between the concentration inhaled and that in the body. For short exposures at high concentrations, uptake to a lethal dose is almost linear, obeying Haber's rule. For longer exposure times, uptake follows a curve so that the inhaled concentration necessary to achieve a lethal dose (50 percent carboxyhemoglobin) at 4 hours (840 ppm), is 0.38 times that required for deaths at 1 hour, as opposed to 0.25 times (550 ppm) as predicted by Haber's rule. Uptake was calculated for a 70 kg human at rest (RMV 8.5 L/min) using the CFK equation.

the rate of uptake depends upon the difference between the concentration inhaled and that in the body, giving an exponential uptake so that

$$W = C(1 - e^{-tk}) \qquad (3)$$

which is the basis for the Coburn-Forster-Kane (CFK) equation[18,19] describing the uptake of CO in humans. This relationship approaches the linear Haber's rule (Equation 1) when the concentration, C, in the atmosphere is high with respect to the concentration in the body required to cause incapacitation or death (Figure 2-8.2), and for short exposures to high CO concentrations, uptake is approximately linear. This is illustrated by the results from CO exposure experiments in primates. At a constant level of activity, and thus respiration, the animals became unconscious when exposed to approximately 27,000 ppm·min of CO at concentrations between 1000 and 8000 ppm (Figure 2-8.3). For such situations it is therefore possible to use linear models for CO uptake without serious error.

Some toxic effects, however, are not dependent upon a dose acquired over a period of time, but are concentration related. Thus the irritant effects of smoke products on the eyes and upper respiratory tract (sensory irritation) occur immediately upon exposure, with the severity depending upon the exposure concentration. In fact, far from increasing as exposure continues, the effects usually lessen, as the subject adapts to the painful stimulus even though the dose is increasing.[20]

Other cases where concentration is an important determinant of toxicity as well as duration of exposure are the narcotic effects of hypoxic hypoxia (oxygen lack) and hypercapnia (high CO_2 concentrations). If a subject is exposed suddenly to a low oxygen concentration, a finite time is required for the air in the lungs and gases in the blood to equilibrate to the new conditions, so to some extent a "dose"

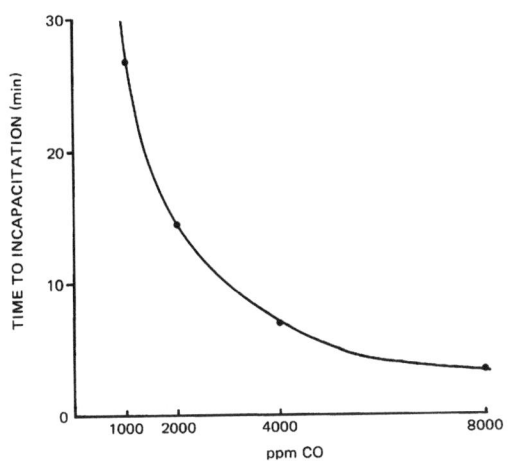

RELATIONSHIP BETWEEN TIME TO INCAPACITATION AND CARBON MONOXIDE CONCENTRATION IN ACTIVE MONKEYS

Fig. 2-8.3. Relationship between time to incapacitation and carbon monoxide concentration in active monkeys. 1000 ppm Ct = 26,600 ppm·min; 2000 ppm Ct = 28,097 ppm·min; 4000 ppm Ct = 26,868 ppm·min; 8000 ppm Ct = 26,086 ppm·min.

of hypoxia is acquired over a period of time. Once equilibrium is established, usually within a few minutes, the severity of the effects depend upon the oxygen concentration and do not then change appreciably with time.[21,22] This also applies to high CO_2 concentrations. Equilibrium is established within a few minutes and concentration related effects then determine the pattern of toxicity.[21]

For the other main narcotic gas in smoke, hydrogen cyanide (HCN), although accumulation of a dose is one factor, the most important determinant of toxicity appears to be the rate of uptake, which in turn depends upon the concentration. Thus as shown in Figure 2-8.4 incapacitation occurs rapidly (after 2 minutes) at the high concentration of 180 ppm (*Ct* product 400 ppm·min), but at the lower concentration of 100 ppm incapacitation occurs only after approximately 20 minutes, requiring a much higher *Ct* product dose (2000 ppm·min). This effect leads to the unusual kinked HCN time/concentration curve shown in Figure 2-8.4 compared to the smooth curve for CO.

In attempting to predict what will happen to a subject exposed to a smoke atmosphere containing all these products it is therefore important to allow for these different concentration/time/effect relationships.

Ct Product and Fractional Effective Dose

The basic concept established in the previous section is that for the majority of toxic products in a fire atmosphere a given toxic endpoint such as incapacitation or death occurs when the victim has inhaled a particular *Ct* product dose of toxicant. In order to make some estimate of the likely toxic hazard in a particular fire it is therefore necessary to determine at what point in time during the course of the fire exposure the victim will have inhaled a toxic dose. This can be achieved by integrating the area under the fire profile curve for the toxicant under consideration. When the integral is equal to the toxic dose the victim can be assumed to have received a dose capable of producing that toxic effect.

A practical method for making this calculation is the concept of Fractional Effective Dose (FED).[43] The *Ct* product

doses for small periods of time during the fire are divided by the *Ct* product dose causing the toxic effect. These Fractional Effective Doses are then summed during the exposure until the fraction reaches unity, when the toxic effect is predicted to occur. Thus:

$$FED = \frac{\text{dose received at time } t(Ct)}{\text{effective } Ct \text{ dose to cause incapacitation or death}} \quad (4)$$

For substances obeying Haber's rule the denominator of the equation is a constant for any particular toxic effect. For substances deviating from Haber's rule the denominator for each time segment during the fire is the *Ct* product dose at which incapacitation or death would occur at the concentration during that time segment. For the hazard model presented in this chapter the denominator is presented in the form of equations giving the required *Ct* product doses predicted for man, which have been derived for each toxic gas and are presented in the following sections. Special cases of the Fractional Effective Dose are referred to as the Fractional Incapacitating Dose (FID) and the Fractional Lethal Dose (FLD).

The Nominal Atmosphere Concentration

There are occasions in combustion toxicology when it may be desirable to relate the toxic effects of an exposure to the material being decomposed rather than to its individual toxic products. This applies particularly to small-scale test results, where, for example, the LC_{50} of wood when decomposed in a particular way might be considered. This is a somewhat unsatisfactory approach, since it cannot be predicted that any material will evolve the same products with the same yields in a large-scale fire as in a small-scale test, and therefore exhibit the same degree of toxicity. However, this parameter does have some value when calculated in

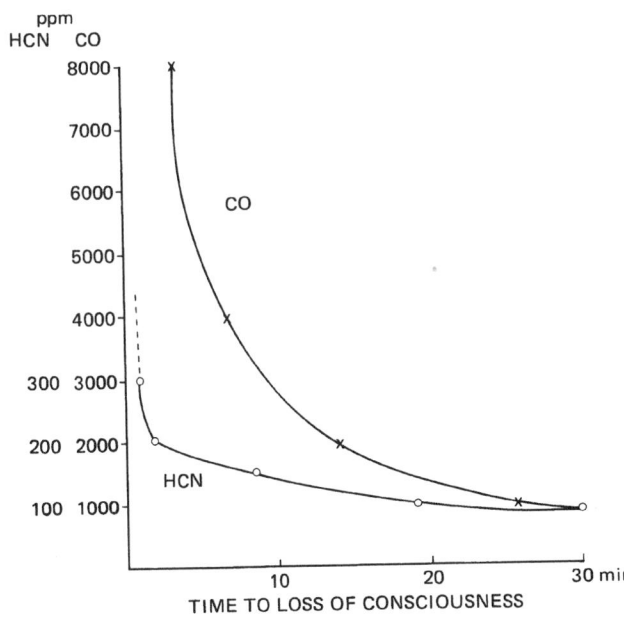

Fig. 2-8.4. Comparison of the relationship between time to incapacitation and concentration for HCN and CO exposures in primates. Time and concentration are equivalent for CO; for HCN, a small increase in concentration causes a large decrease in time to incapacitation.

conjunction with measurements of the actual toxic products. This approach is related to another concept used in inhalation toxicology, that of the nominal atmosphere concentration, NAC. This theoretical concentration of test material in the test atmosphere is calculated from the amount of material produced from the atmosphere generation system each minute, divided by the diluent airflow rate. This concept is not strictly applicable in combustion toxicology since the test material is decomposed in the fire or furnace system, but two analogous concepts are very useful with regard to small-scale test methods since they enable some relationship to be established between the test material and the degree of toxicity. These are the nominal atmosphere concentation in terms of mass charged into the furnace, and the nominal atmosphere concentration in terms of mass decomposed, as follows:

1. Nominal atmosphere concentration (mass charge) equals mass of material placed in the furnace system divided by volume of air into which it is dispersed.
2. Nominal atmosphere concentration (mass loss) equals mass lost by material during decomposition divided by volume of air into which it is dispersed.

In practice the calculation of these parameters depends upon the particular decomposition system used (see the section of this chapter on small-scale test methods) and a shortcoming of some systems is that these parameters cannot easily be estimated.

If predictions of toxicity are to be made from large-scale fire test atmospheres, or if toxicity data from animals are to be interpreted, the following data should be available:

1. The nominal atmosphere concentration(s) of the test material(s) mass charge $(g \cdot m^{-3})$ (NAC mass charge).
2. The nominal atmosphere concentration of the test material mass loss $(g \cdot m^{-3})$ (NAC mass loss).
3. The concentration of each major toxic product in the atmosphere and the anticipated duration of exposure, i.e., the concentration/time profile.
4. The rate of uptake of the atmosphere, or RMV.
5. Measurement of the blood concentration of certain toxicants.
6. Particle size range. This is also important in determining the respirability and site of deposition of atmospheric products. However, smoke from nonflaming decomposition in small-scale tests is usually highly respirable.
7. The nature of the effects of toxic products and the time/concentration relationships of these effects.

SIMPLE MASS LOSS-BASED FRACTIONAL EFFECTIVE DOSE HAZARD ASSESSMENTS

One way of performing a preliminary simple hazard analysis for a fire is to consider what exposure in terms of a single criterion, the mass loss concentration profile of products in a fire, is likely to be lethal to a victim. For such a calculation use can be made of mass loss lethality data from small-scale rodent toxicity test data for the material or materials involved in the fire. This makes the assumption that the lethal concentration to a human would be similar to that in a rat, but this is standard practice in toxicology for making approximate classifications of hazard for the acute effects of industrial chemicals. It is also necessary to make the assumption that mass loss lethality data follow Haber's rule, but any inaccuracies introduced by deviations from ideal behavior should not be important in such an approximate analysis. The three items of data needed for such an assessment are:

1. The basic fire condition (smoldering, early flaming, or post-flashover).
2. The mass loss/dispersal volume-time curve for the fire.
3. The rodent LCt_{50} concentrations for the materials involved in the fire in terms of the mass loss concentration for a quoted exposure time, determined under the same conditions as those in the fire.

LCt_{50} data for common materials have been derived using a number of small-scale test methods under a variety of decomposition conditions. Most published data relate to nonflaming oxidative decomposition conditions, well-ventilated flaming, or mixed flaming and nonflaming conditions. Very few data are available for post-flashover fire decomposition conditions. Table 2-8.1 shows examples of data sets obtained using three well-known test methods. The range of LC_{50} for the references quoted (shown in Table 2-8.1) was approximately 5 to 60 $g \cdot m^{-3}$ mass loss for a 30-minute exposure, which is equivalent to an LCt_{50} dose range of approximately 150 to 1800 $g \cdot m^{-3} \cdot min$, with an average value of 23 $g \cdot m^{-3}$ (690 $g \cdot m^{-3} \cdot min$). Allowing for a small margin of safety, it has been suggested within British standards that, for a simple hazard assessment, a single figure of 500 $g \cdot m^{-3} \cdot min$ might be considered as a single average figure for the approximate toxic potency of the thermal decomposition products from common materials. For the purpose of carrying out hazard calculations, the toxic

TABLE 2-8.1 *Lethal Toxicity Data for Combustion Products from a Range of Materials*

Author	n	Test method	30-minute LC$_{50}$ (mg/l mass loss)			L(Ct$_{50}$) (mg · min/l mass loss)	
			mean	range	wood	mean	range
Levin[7]	11	NBS (NF)	24	5–40	25	720	150–1200
		(F)	27	4–57	49	810	120–1710
Prager[104]	18	DIN (NF)	23	6–60	20–50	690	180–1500
Alexeef[118]	46	UPIT (mixed)	19	4–88	68	580	117–2648
Average			23	5–61	42	700	142–1765

NF = nonflaming
F = flaming

potency of any individual material can then be expressed in terms of a potency factor relating the actual LCt_{50} to 500, as follows:

Toxic potency factor for a material under a defined fire condition = $500/LCt_{50}$ g·m^{-3}·min

General Pattern of Toxic Potency for Common Materials under Three Fire Conditions

Using a wider data base than that for Table 2-8.1, a survey of the toxic potency data for common materials in nonflaming, early flaming, and post-flashover fire conditions revealed an inadequate data base, but it was possible to derive approximate LCt_{50} for common materials.[119] The results for individual materials range over approximately 2 orders of magnitude from 20 to 3750 g·m^{-3}·min; but when the data are reduced to basic types of materials under each decomposition condition, a relatively simple pattern can be described, as presented in Table 2-8.2. The table shows the approximate average lethal exposure doses (LCt_{50}) for classes of materials, the LC_{50} for 30-minute exposures, and a potency factor (based upon a figure of 500 g·m^{-3}·min for the "normal" lethal potency for combustion products), respectively. The findings are as follows:

Under nonflaming oxidative decomposition conditions at >400°C, most materials have a similar potency close to 500 g·m^{-3}·min, i.e., potency factor of approximately 1, due mainly to the effects of carbon monoxide and irritants. The main exceptions are nitrogen-containing materials releasing significant HCN at low temperatures (e.g., polyacrylonitrile, modacrylic, and rigid polyurethane foam), which have toxic potency factors of 3 to 8.

Under early flaming conditions most nonfire-retarded materials are substantially less toxic than under nonflaming conditions. Cellulosics (wood and cotton) are the least toxic with LCt_{50} of >3000 g·m^{-3}·min (potency factor 0.2). Plastics containing carbon, hydrogen, and/or oxygen are somewhat more toxic with a potency factor of 0.4 (LCt_{50} ~1200), and those containing low percentages of nitrogen (e.g., flexible polyurethanes, wool, and nylon) also fall into this area. PVC and fire-retarded materials have a similar toxic po-

tency factor to that under nonflaming conditions of approximately 1. Rigid polyurethanes and nitrogen-containing acrylics have high potencies similar to those under nonflaming conditions.

Under post-flashover conditions, the potency of all materials increases due to the increased yields of HCN and/or CO. More smoke and irritants are also present than under early flaming conditions, which may add somewhat to the potency, particularly of the non-nitrogen-containing materials. For cellulosic materials and hydrocarbon plastics, the potency is similar to that under nonflaming conditions (potency factor close to 1). For all nitrogen-containing materials, the toxic potency is high, ranging from approximately 2.5 for flexible polyurethane foam to approximately 11 for modacrylic and polyacrylonitrile. It is suggested that PVC would have a potency factor of approximately 2.5 under these conditions.

It is suggested that the data in Table 2-8.2 provide a mechanism whereby small-scale toxicity test data, obtained under appropriate decomposition conditions, can be applied in fire engineering calculations. A simple, first estimate could be based upon a single toxic potency figure of 500 g·m^{-3}·min for all materials, using total fire load or heat release as the source of the mass term. When more detailed information on the nature of the materials likely to be involved in a fire is known, the calculations for particular fire scenarios can be based upon the predicted mass loss rate for each material, adjusted by the appropriate toxic potencies. The range of toxic potencies of common materials decomposed under conditions occurring in flaming fires is approximately two orders of magnitude. Data obtained so far that relate to nonflaming fires show a relatively narrow range of potencies around 500 g·m^{-3}·min. Early flaming, well-ventilated fires show toxic potency to be generally low where combustion is efficient, ranging from approximately 75 to 3750 g·m^{-3}·min, while data obtained so far that relate to vitiated, post-flashover fires show that potencies are generally higher, due to increased yields of CO and HCN, ranging from approximately 21 to 3000 g·m^{-3}·min.

Difficulties in making estimates of the specific toxicities and toxic potencies of common materials arise from the very poor data base of both small- and large-scale tests conducted under appropriate conditions. This is particularly true of the vitiated post-flashover condition, which is not well simulated by existing test methods, except possibly the DIN method, for which very few data are available.

Incapacitation

This rather crude method could be used to give an approximate indication of when conditions in a fire are likely to be lethal, but in practice the effects of fires on exposed victims are not so simple. In many cases death is not due to the immediate toxic effects of exposure, but results from the victim being trapped in the fire, either because irritant and optically obscure smoke prevents escape, or because narcotic gases cause incapacitation, so that the victim remains in the fire to die either from a fatal dose of toxic products acquired during the prolonged exposure, or from burns. One way of taking these factors into account would be to determine the Ct product dose at which effects such as incapacitation due to narcosis occur in small-scale toxicity tests. These could then be applied to the fire hazard analysis to estimate the Fractional Incapacitating Dose, rather than the Fractional Lethal Dose. However, because of differences in

TABLE 2-8.2 *Approximate Lethal Exposure Doses (LCt_{50} g·m^{-3}·min), LC_{50} (g·m^{-3}), and Toxic Potency Factors for Common Materials under Different Fire Conditions**

Material	Nonflaming			Early Flaming			Post-Flashover		
Cellulosics	730	24	(0.7)	3120	104	(0.2)	750	25	(0.7)
C, H, O plastics	500	17	(1.0)	1200	40	(0.4)	530	18	(0.9)
PVC	500	17	(1.0)	300	10	(1.6)	200	7	(2.5)
Wool/nylon (low N₂)	500	17	(1.0)	920	31	(0.5)	70	2	(7)
Flexible Polyurethane	680	23	(0.7)	1390	46	(0.4)	200	7	(2.5)
Rigid Polyurethane	63	2	(8)	100	3	(5)	54	2	(8)
Modacrylic/PAN	160	5	(3)	140	5	(3.6)	45	1.5	(11)

*Toxic potency factors are calculated from the LCt_{50} based upon a "normal" potency of 500 g·m^{-3}·min. LC_{50} is for a 30-min exposure time with 14 days observation period.

generating small-scale fire test atmospheres similar to those occurring in large-scale fires, a potentially much more effective way of predicting toxic hazard would be to measure the concentration/time profiles of the important toxic products in the fire and to determine their effects from toxicity data derived from experiments in man and primates (and to a lesser extent also from rodents). In the following sections the characteristics of the major narcotic and irritant fire products are described, together with methods for calculating their uptake and predicting their toxic effects.

NARCOSIS BY FIRE GASES AND PREDICTION OF TIME TO INCAPACITATION

Narcotic (asphyxiant) gases cause incapacitation mainly by effects on the central nervous system and to some extent, the cardiovascular system.[21] In general, time to incapacitation and its severity are predictable in that usually a short period of intoxication is followed by a relatively sharp decline into severe incapacitation (i.e., loss of consciousness).[12,25] Most narcotic gases produce their effects by causing brain tissue hypoxia.[21,22] Since the body possesses powerful adaptive mechanisms designed to maximize oxygen delivery to the brain, it is usually possible to maintain normal body function up to a certain dose of narcotic, and the victim is often unaware of the impending intoxication. Once a point is reached where normal function can no longer be maintained, however, deterioration is rapid and severe—beginning with signs similar to the effects of severe alcohol intoxication, consisting of lethargy or euphoria with poor physical coordination, and followed rapidly by unconsciousness and death if exposure continues.[22,25]

Narcotic Fire Products

The two major narcotic gases in fires are: (1) carbon monoxide (CO) and (2) hydrogen cyanide (HCN). Carbon monoxide is always present to some extent in all fires, irrespective of the materials involved or the stage (or type) of fire, so that there is almost always some degree of risk of narcosis from CO exposure.[13] Hydrogen cyanide is always present to some extent when nitrogen-containing materials are involved in fires. These include materials such as acrylics, polyurethane foams, melamine, nylon and wool, which are likely to be involved to some extent in most fires in buildings. Hydrogen cyanide is likely to be present at high concentrations in large, post-flashover fires. Unlike carboxyhemoglobin, which is routinely measured in the blood of fire victims, blood cyanide is often not measured. However, it has been detected at high concentrations in the blood of some fire victims, particularly when blood samples have been taken immediately after exposure.[10] In addition, low concentrations of oxygen (less than 15 percent)[26] and very high concentrations of carbon dioxide, CO_2, (greater than 5 percent) can have narcotic effects.[27]

Carbon Monoxide

Carbon monoxide combines with hemoglobin in the blood to form carboxyhemoglobin (COHb), which results in a toxic narcosis because it reduces the amount of oxygen supplied to the tissues of the body, particularly brain tissue. Tissue oxygen supply is reduced because the amount of hemoglobin available for the carriage of oxygen (in the form

of oxyhemoglobin) is reduced, and also because the ability of the remaining oxyhemoglobin to release oxygen to the tissues is impaired (due to a leftward shift of the oxygen dissociation curve).

The affinity of hemoglobin for CO is extremely high, so that the proportion of hemoglobin in the form of carboxyhemoglobin increases steadily as CO is inhaled. The toxicity of CO therefore depends upon the accumulated dose of carboxyhemoglobin, which is expressed in terms of the percentage of total hemoglobin in the form of carboxyhemoglobin, (% COHb).[15]

There is little doubt that CO is the most important narcotic agent formed in fires. In the Strathclyde pathology study[10] lethal levels (> 50% COHb) were found in 54 percent of all fatalities, while some 69 percent of fatalities had carboxyhemoglobin levels capable of causing incapacitation (> 30% COHb). Incapacitating levels of carboxyhemoglobin are also common in victims surviving immediate fire exposure. Carbon monoxide is therefore particularly important because:

1. It is always present in fires, often at high concentrations.
2. It causes confusion and loss of consciousness, thereby impairing or preventing escape.
3. It is the major ultimate cause of death in fires.

To understand the effects of CO exposure on fire victims and to predict the likely consequences of a particular exposure, it is essential to know a number of features of CO intoxication. (To some extent these apply to an evaluation of the toxicity of any fire product.) It is necessary to:

1. Determine which types of toxic effects occur at different dose levels.
2. Determine the concentration/time relationships of these toxic effects, whether they occur immediately or some time after exposure, and whether the effects of a short high concentration exposure are the same as those of a longer, low concentration, exposure.
3. Quantify the parameters that determine the rate of uptake and removal of CO from the body.

Some information on these points is available from accidental exposures and low-level experimental exposures in humans,[15] and data are available on the symptoms experienced in humans at various carboxyhemoglobin concentrations at rest.[15,26,28] Loss of consciousness is predicted at approximately 40% COHb, but can occur at lower levels (~ 30% COHb), and lower levels can be dangerous for subjects with compromised cardiac function.[15] Death is predicted at COHb concentrations of 50 to 70% COHb,[26,29,30] but it has been suggested that death can occur at lower concentrations in susceptible subjects.[15,31] The severe incapacitation caused by high-level exposures, such as those encountered in fires, has been studied experimentally in animals.[21,25] The effects of experimental exposures on cynomolgus monkeys are shown in Figure 2-8.5 in terms of physiological parameters (respiration, cardiovascular parameters, and brain electroencephalogram or EEG) monitored in sedentary animals sitting in chairs, and in Figure 2-8.3, where free-moving animals were trained to perform a behavioral task designed to simulate some aspects of the escape maneuvers of human fire victims (i.e., tasks involving bodily movements with a certain amount of exercise, requiring the use of coordinated movements and the application of psychomotor skills).

The first important characteristic of CO poisoning illustrated by these experiments is that CO uptake and intoxication are extremely insidious. During the early stages as the carboxyhemoglobin concentration builds up gradually in the blood the effects are minimal. Thus Figure 2-8.5 shows no detectible changes in physiological parameters until the end of the exposure when the COHb concentration approached 40 percent. In active animals the first minor signs of behavioral performance deficits did not occur until concentrations of 15 to 20% COHb were achieved. Similar results have been obtained in humans, where O'Donnel et al[32] could find no effects upon psychomotor performance at levels of up to 12% COHb, and Stewart et al[28] reported the first symptoms (consisting of a headache) to occur at 15 to 20% COHb, while objective tests at these levels showed only minor deficits in behavioral performance.

Another major characteristic of narcosis shown clearly by the animal experiments is that when significant effects do occur their onset is sudden and the degree of incapacitation rapidly becomes severe, so that by the time a victim is aware that he/she is affected effective action is probably not possible. Thus Figure 2-8.5 shows the physiology of a monkey passing from a normal state to unconsciousness within a few minutes after 25 minutes of exposure, with decreased respiration, a severe decrease in heart rate, and greatly increased slow-wave EEG activity (indicative of cerebral depression).

For active animals there was a sudden rapid decline in behavioral task performance accompanied briefly by signs similar to severe alcohol intoxication, which led rapidly to a state of deep coma.

These findings may explain why deaths from CO derived from defective heating appliances are so common. Survivors of such situations often report that they, or other victims that died, experienced headaches or nausea, but had no idea of the cause, so they did not attempt to leave the area until overcome by fumes.[33]

During the early stages of incapacitation the main effects appear to be on motivation and psychomotor ability, with a tendency for the victim to sleep if left undisturbed.[25] Under these conditions one might expect a subject, if alerted by a sudden noise such as of breaking glass (often reported by fire survivors), to "sober up" and awake sufficiently to make an escape attempt. However, such a victim is likely to fail for three reasons:

1. This stage is rapidly followed by unconsciousness and coma.
2. Active subjects are seriously affected by carboxyhemoglobin concentrations that have only minor effects on sedentary subjects. Thus, while sedentary primates were often unaffected at carboxyhemoglobin levels of up to 40 percent, those engaged in light activity were seriously affected at carboxyhemoglobin levels in the 25 to 35 percent range.[25] Similarly, in one study of humans, although a sedentary subject could perform such tasks as writing, even at the exceptionally high level of 55 percent carboxyhemoglobin, the subject collapsed and became unconscious immediately when attempting to rise and walk.[34] Therefore, a victim in a bed or chair attempting to escape not only would be in danger of a rapid collapse due to continued CO uptake, but even if no further uptake occurred the ability to perform even light work or exercise would be severely impaired. Even the simple act of rising from a horizontal to an upright position could precipitate loss of consciousness.
3. The rate of uptake of CO depends on the respiration (respiratory minute volume) and hence the activity of the subject. When the subject becomes active the blood carboxyhemoglobin is therefore likely to increase rapidly to an incapacitating level.

A Model for the Prediction of Time to Incapacitation by CO in Fires

Incapacitation by CO depends upon a dose accumulated over a period of time until a carboxyhemoglobin concentration is reached where compensatory mechanisms fail and collapse occurs. To predict time to incapacition of fire victims due to CO it is necessary to know the carboxyhemoglobin concentrations at which incapacitation is likely to occur, and the rate of uptake of CO so that the time to achieve this concentration can be calculated. The carboxyhemoglobin concentrations likely to cause incapacitation depend upon the activity of the victim and should be similar to the concentrations causing incapacitation in primates at similar levels of activity.[25]

Since CO is both inhaled and excreted via the lungs, the rate of uptake depends upon the difference between the CO concentration in the blood, W, and that in the inhaled air, C, and is an exponential function described by the general equation (Equation 3):

PHYSIOLOGICAL EFFECTS OF AN ATMOSPHERE CONTAINING CO (1850 ppm) — WOOD PYROLYSED AT 900°C

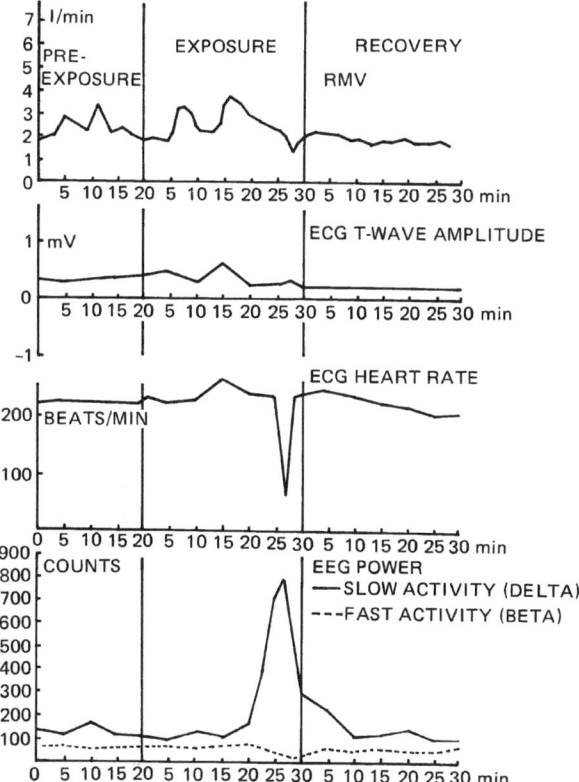

Fig. 2-8.5. Physiological effects of an atmosphere containing CO (1850 ppm)—wood pyrolyzed at 900°C. From Purser and Woolley.[12]

$$W = C(1 - e^{-tk}) \qquad (3)$$

where t is the time exposed and k is a constant determined by a number of factors, so that uptake is rapid initially, but gradually levels off as uptake and removal from the blood reach equilibrium. This relationship is described fully by the Coburn-Forster-Kane (CFK) equation,[18,19] which takes into account a whole range of variables, including RMV, body size, exposure duration, and parameters related to lung and blood physiology (see Appendix 2-8A). When all these various factors are known this equation enables accurate predictions of CO uptake to be made that agree well with experimental data.[19,35] The uptake pattern for CO is illustrated in Figure 2-8.2, which predicts time to achieve a potentially lethal blood CO concentration (50% COHb) for a 70 kg human at rest (RMV 8.5 L/min) at two CO concentrations, 2200 and 840 ppm.

When the inhaled concentration is high compared to that in the blood as during short duration, high concentration exposures such as those that occur in flaming fires, the departure from linear uptake is not great as shown in Figure 2-8.2, and the deviation from Haber's rule ($W = C \times t$) is small. However, over long periods at lower concentrations, as equilibrium is approached, uptake deviates considerably from linearity. For short exposures at high concentrations when the blood concentration is well below saturation level, an approximate prediction of COHb concentration can therefore be made assuming a linear relationship. Such an equation is derived from experimental human exposures by Stewart et al[29]

$$\%COHb = (3.317 \times 10^{-5})(ppm\ CO)^{1.036}(RMV)(t) \qquad (5)$$

where

$$ppm\ CO = CO\ concentration\ (ppm)$$
$$RMV = volume\ of\ air\ breathed\ (L/min);\ and$$
$$t = exposure\ time\ (min)$$

Thus, for the examples shown in Figure 2-8.2, which were calculated using the CFK equation, a concentration of 50% COHb is predicted following a 1-hour exposure to 2200 ppm CO, while the Stewart equation predicts 49% COHb. However, for a 4-hour exposure the CFK equation predicts 50% COHb from 840 ppm, while the Stewart equation would predict 50% COHb from 550 ppm over 4 hours, or a concentration of 72% COHb from 840 ppm.

Justification of the linear uptake relationship under high concentration/short exposure duration circumstances is illustrated by a series of primate exposures carried out over a 1000 to 8000 ppm concentration range. These experiments were performed using an active behavioral model,[25] with the endpoint being loss of consciousness, which occurred at approximately 34% COHb. The results are illustrated in Figure 2-8.3 which shows time to incapacitation for different inhaled CO concentrations. At each concentration the inhaled dose (Ct product) required to produce incapacitation in ppm CO · minutes is constant, as predicted by the linear uptake model of Haber's rule.

As stated, it is possible to make accurate predictions of CO uptake for a range of situations by using the CFK equation, provided that a number of variables are taken into account. For a particular individual the most important variable is the RMV, which varies considerably depending upon the level of activity of the subject. Figures for this and other

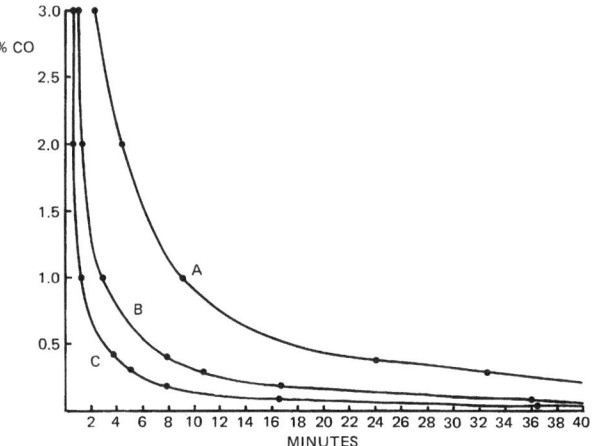

TIME TO INCAPACITATION BY CARBON MONOXIDE FOR A 70 kg MAN AT DIFFERENT LEVELS OF ACTIVITY

Fig. 2-8.6. *Time to incapacitation by carbon monoxide for a 70 kg human at different levels of activity. Curve A—40 percent carboxyhemoglobin RMV 8.5 L/min at rest sitting; Curve B—30 percent carboxyhemoglobin RMV 25 L/min, light work (e.g., walking 6.4 km/h); Curve C—20 percent carboxyhemoglobin RMV 50 L/min, heavy work (e.g., slow running 8.5 km/h) (or for walking 5.6 km/h up a 17 percent gradient).*

variables can be obtained from standard reference data.[36] Figure 2-8.6, generated from these data, shows the probable time to incapacitation (loss of consciousness) for a 70 kg human exposed to different CO concentrations at three levels of activity. The figure shows that the degree of activity can have a major effect on time to incapacitation. It must also be remembered that RMV per kilogram of body weight is greater for small subjects, which means that children will take up CO much more rapidly than adults and succumb much earlier, while uptake in small laboratory animals is even more rapid. An assumption made in these calculations is that the level of activity and hence the RMV remain constant during the exposure. In practice there is a tendency for the level of activity and ventilation to decrease slightly as the point of incapacitation is approached. It is considered that with a model for predicting time to incapacitation (unconsciousness), errors due to reduced ventilation will be minor, since the primate experiments demonstrate that there is little change in RMV until the point of incapacitation. Once the subject becomes unconscious, the RMV and hence the rate of CO uptake will be considerably reduced, particularly if the subject was previously engaged in heavy work. It is therefore possible that for calculating time to death allowance could be made for a low RMV (~6 L/min) once incapacitation has occurred. Not making this allowance does err slightly on the side of safety.

The Stewart[29] and CFK[18] equations enable reasonably good predictions of time to incapacitation or death for short (less than one hour) or long (greater than one hour) exposures, respectively, to constant concentrations of CO in air. In fires, however, victims are exposed to concentrations of CO which change during the course of the fire. For smoldering fires the CO concentration may grow slowly and remain fairly constant over long periods, but for early flaming fires and many fully developed fires where the victim is in a remote location, the CO concentration may increase rapidly over a short period of time as in Figure 2-8.7. It is therefore

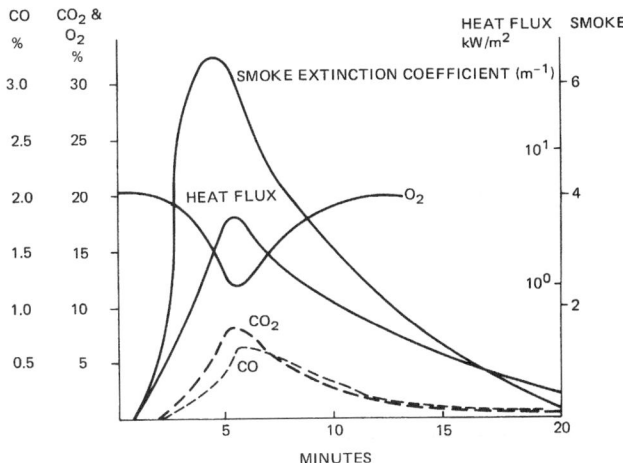

Fig. 2-8.7. Smoke, heat, and gases during single armchair room burn. Armchair is polystyrene with polyurethane cushions and covers. Room is 39 m³ with open doorway. Gases measured in doorway at 2.1 m height. From Babrauskas.[77]

necessary to be able to apply the uptake models to situations where the CO concentration is not constant. For most situations, fluctuations in CO concentration do not present a problem since incapacitation depends upon the total dose of CO inhaled, in the form of COHb, and is not affected by the immediate CO concentration. The COHb concentration is thus dependent upon the average CO concentration over the period of exposure, and significant errors will not occur even if the CO concentration falls somewhat at certain stages of the exposure. Errors in predicting the COHb concentration are possible if the CO concentration drops dramatically toward the end of an exposure, or if it decreases moderately during a prolonged exposure of several hours duration, when COHb concentrations can approach equilibrium with the inhaled CO concentration. In this case a fall in CO may result in a decrease in COHb concentration. The basic rule to apply is that fluctuations are unlikely to cause the COHb concentration to deviate from that predicted by assuming the constant average concentration throughout, providing the CO concentration is on a rising trend, is stable, or is well above the equilibrium concentration with the blood COHb.

Ct Product and Fractional Incapacitating Dose

Although the average CO concentration during a fire exposure can be used to predict COHb concentration and time to incapacitation, another useful concept for predicting incapacitation or death is the relationship with concentration-time (*Ct*) product, which is a representation of CO "dose." In changing fire conditions the *Ct* product may be obtained by integrating the CO concentration/time curve. The dose inhaled may then be related to the dose required to cause incapacitation, and the fraction of an incapacitating dose at any time, *t*, may be calculated, incapacitation occurring when the fractional dose reaches 1.0. Since the *Ct* "dose" actually represents the COHb concentration, the fractional dose would be better represented by the ratio of the COHb concentration at time, *t*, with the COHb concentration known to cause incapacitation or death, rather than by simple *Ct* product ratios. The Stewart equation[29] can be rewrit-

ten in the form of COHb ratios, requiring only a knowledge of the CO concentration and the exposure time, as follows

$$F_{Ico} = \frac{K(CO^{1.036})(t)}{D} \tag{6}$$

where

F_{Ico} = fraction of incapacitating dose
t = exposure time
K = 8.2925×10^{-4} for 25 L/min RMV (light activity); and
D = COHb concentration at incapacitation (30 percent for light activity)

This concept of *Ct* product fractional dose is also useful for predicting incapacitation and death from other fire products, and combinations of products, as will be discussed later in this chapter.

Hydrogen Cyanide

Hydrogen cyanide (HCN) has been measured in the blood of both fatal[37] and nonfatal[38] fire victims. However, in the Strathclyde fire fatality study high concentrations of hydrogen cyanide in the blood of victims were usually associated with lethal levels of carboxyhemoglobin, so that the role of hydrogen cyanide as a cause of incapacitation was difficult to determine.[37] It is also difficult to relate blood cyanide levels from samples collected after a fire to likely HCN exposure, since the dynamics of HCN uptake and removal from the blood are poorly understood.[23,24]

Although the ultimate effects of HCN exposure (consisting of unconsciousness with cerebral depression) are similar to those produced by CO, the pattern of toxicity during the early stages is very different. While the onset of CO intoxication is slow and insidious, HCN intoxication tends to be rapid and dramatic. The physiological signs of incapacitation produced in monkeys by an atmosphere containing HCN are shown in Figure 2-8.8.[23] As with CO the immediate effects were relatively minor, consisting of a slightly raised ventilation, but at some time during a 30 minute exposure period there was a marked increase in respiration (hyperventilation), the RMV increasing up to four times. Within one to five minutes of the start of this episode of hyperventilation the animals lost consciousness. This was accompanied by EEG signs of severe cerebral depression; loss of muscle tone; and marked effects upon the heart and circulation, including a significant decrease in heart rate, arrhythmias and changes in the EKG waveform indicative of cardiac hypoxia. This hyperventilatory episode was caused by the stimulatory effects of cyanide upon respiration. Since the cyanide was taken in via inhalation, a positive feedback situation resulted, and inhaled cyanide caused hyperventilation which increased the rate of HCN uptake and in turn provided a stronger hyperventilatory stimulus. Once the animals became unconscious the hyperventilation subsided and they went into a slow decline for the remainder of the exposure. This led eventually to a cessation of breathing in some cases, which would have proved fatal if exposure had not been terminated. It was therefore possible for an animal to survive a continuous HCN exposure for some time after the point of incapacitation. Once exposure was terminated the recovery was rapid and almost complete within five to ten minutes.

The pattern of incapacitation for HCN is somewhat different from that produced by CO in that the effects occur

PHYSIOLOGICAL EFFECTS OF AN ATMOSPHERE
OF HCN GAS (147 ppm)

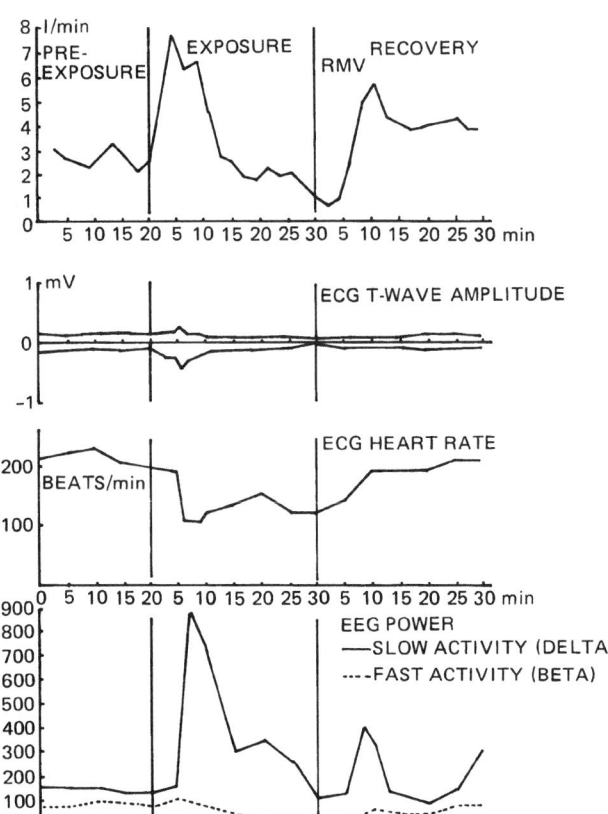

Fig. 2-8.8. *Physiological effects of an atmosphere of HCN gas (147 ppm) on monkeys. From Purser* et al.[23]

more rapidly, as unlike CO, HCN is not held almost exclusively in the blood but carried rapidly to the brain.[39] Although the accumulation of a dose is one factor, the most important determinant of incapacitation with HCN appears to be the rate of uptake, which in turn depends upon the HCN concentration in the smoke and the subjects' respiration. Thus in the animal experiments,[21,23] it was found that at HCN concentrations below approximately 80 ppm the effects were minor over periods of up to 1 hour, with mild background hyperventilation. At concentrations above 80 ppm up to approximately 180 ppm, an episode of hyperventilation with subsequent unconsciousness occurred at some time during a 30 minute period; there was a loose linear relationship between HCN concentration and time to incapacitation. Above 180 ppm the hyperventilatory episode began immediately with unconsciousness occurring within a few minutes. Data on human exposures to HCN are limited but Kimmerle[26] does quote some approximate data showing a similar effect in humans, with incapacitation occurring after 20 to 30 minutes at 100 ppm HCN and after 2 minutes at 200 ppm, death occurring rapidly at concentrations exceeding approximately 300 ppm.

Other data suggest that human victims might be able to survive higher concentrations of HCN for shorter periods. McNamara[40] suggests 539 ppm as the 10-minute LC_{50} for humans, and there is a report of a survival from an accidental exposure to 444 ppm.[41] An experimental human exposure to 530 ppm HCN was survived without immediate symptoms for 1.5 minutes, although a dog exposed at the same time suffered respiratory arrest.[42] Dogs are known to be particularly susceptible to cyanide poisoning,[40] but it does seem likely that to some extent with HCN (as with CO) body size influences time to incapacitation, and that a human would be able to tolerate exposure to a given concentration longer than a cynomolgus monkey. With HCN and CO, physical activity would be likely to cause more rapid uptake in adults, and uptake would be more rapid in children because of their smaller body size. The primate data therefore seem to provide a reasonable model for humans, possibly erring slightly on the side of safety.

The differences between CO and HCN in terms of the relationship between inhaled concentration and time to incapacitation are illustrated in Figure 2-8.4. While CO gives a smooth curve with incapacitation occurring at a constant Ct product of approximately 27,000 ppm · min for all CO concentrations, the almost linear portion of the HCN curve results in a Ct product of approximately 2000 ppm · min at 100 ppm HCN and 400 ppm·min at 200 ppm, with very rapid incapacitation at higher HCN concentrations. This deviation from Haber's rule (which predicts a constant Ct product) was recognized by Haber himself in 1924, when he stated that the Ct product for HCN depended upon the exposure concentration.[16] The exact reason for this is not known, but appears to be related to relationship between the rate of uptake of HCN and the dynamics of its distribution between different body fluid compartments.[24] The effect is to render concentrations in the range greater than 150 ppm more toxic than would be predicted from the effects of longer exposures to lower concentrations.

A Model for the Prediction of Time to Incapacitation by HCN in Fires

From these results it is possible to predict that HCN concentrations below a threshold concentration of approximately 80 ppm will have only minor effects over periods of up to 1 hour.

From 80 to 180 ppm the time to incapacitation (unconsciousness) t_{Icn} will be between 2 and 30 minutes approximately, according to the relationship

$$(t_{Icn})(min) = (185 - ppm \; HCN)/4.4 \qquad (7)$$

For concentrations above approximately 180 ppm incapacitation will occur very rapidly (within two minutes). This linear expression gives a reasonably good fit with the primate data over the range 80–180 ppm, but it would be preferable to derive a more general expression to include the effects of higher and lower concentrations for making fractional dose estimations. An exponential expression has therefore been derived which also gives a reasonably good fit with the data (regression coefficient 0.984) as follows

$$(t_{Icn})(min) = \exp(5.396 - 0.023 \times ppm \; HCN) \qquad (8)$$

HCN could be particularly dangerous in fires due to its rapid "knockdown" effect, and low HCN levels in the 100 to 200 ppm range could cause fire victims to lose consciousness rapidly and consequently to die later as a result of accumulation of CO or some other factor. A small change in HCN concentration could also cause a large decrease in time to incapacitation; for example, doubling the concentration from 100 to 200 ppm could bring the incapacitation time down from approximately 20 min to approximately 2 min.[23,26]

Although this relationship should enable reasonable predictions to be made of time to unconsciousness for subjects exposed continuously to HCN, especially over the critical range between 180 ppm (above which incapacitation will be rapid) and 80 ppm (below which incapacitation is unlikely over periods of up to 1 hour), real fires will involve exposures to changing concentrations. One method of predicting time to incapacitation or death would be to take the average concentration over the period of exposure. Although approximate times to incapacitation can be estimated with this method it is prone to error because of the departure from Haber's rule. In practice, since short exposures to high concentrations are more likely to cause incapacitation than longer exposures to lower concentrations, and averaging the concentration tends to give longer estimates of time to incapacitation than would be expected. This method also does not include the concept of fractional dose. A better model would include some degree of weighting to allow for the enhanced effect of high concentration exposure, and also enable incapacitation to be estimated in terms of Ct product fractional dose.

A method for estimating fractional dose to incapacitation has been developed for the rat by Hartzell et al,[43] based on this concept. In that model the Ct product over short periods of time is expressed as a fraction of the Ct product required to cause incapacitation or death at that concentration. The fractions for each short time interval are summed until the fraction reaches unity, which indicates incapacitation.[43] This approach should enable reasonable predictions of time and dose to incapacitation and death to be made, provided that the HCN concentration is stable or increasing (Hartzell et al have found a good correspondence between calculated predictions and experimental data in rats). This approach can be used to derive a fractional dose model for humans based upon the time to incapacitation equation, (Equation 8), derived from primate and human data, as follows

For a constant HCN concentration

$$t_{Icn} = \exp(5.396 - 0.023 \times \text{ppm HCN})$$

Dose to incapacitation = (ppm HCN)(t_{Icn})
Therefore, for a short exposure time, t, to a given HCN concentration

$$F_{Icn} = \frac{(\text{ppm HCN})(t)}{(\text{ppm HCN})(t_{Icn})}$$

where F'_{Icn} = fraction of an incapacitating dose.
Taking t = 1 minute, this simplifies to

$$F_{Icn} = 1/t_{Icn} \qquad (9)$$

If the fractional doses per minute, F'_{Icn}, are summed throughout the exposure, the dose and time to incapacitation can be predicted.

EXAMPLE:

A subject is exposed to 90 ppm HCN for 15 minutes, then to 180 ppm HCN for 2 minutes

t_{Icn} for 90 ppm = $\exp(5.396 - 0.023 \times 90)$ = 27.83
t_{Icn} for 180 ppm = 3.51 min

$$F_{Icn} = (1/27.83) \times 15 + (1/3.51) \times 2$$
$$= 1.111$$

Incapacitation is therefore predicted at between 17 and 18 minutes.

Cyanide in Blood

The importance of HCN as a cause of incapacitation and death in fires can be underestimated, due to poor understanding of the dynamics of cyanide uptake, dispersal and metabolism in the body, and the inadequate data base of blood cyanide measurements from both injured and dead fire victims. Carboxyhemoglobin is the only blood toxin routinely measured in fire victims, and when blood cyanide is measured (usually post-mortem), the sample and measurement are often taken a day or more after exposure.

Evidence from the primate experiments reported above, and from further experiments where measurements were made of arterial blood cyanide during and after exposure,[24] shows that when HCN is inhaled for short periods at air concentrations above approximately 150 to 200 ppm, loss of consciousness results from a transient high plasma cyanide concentration. HCN uptake rate is then greatly reduced when the subject loses consciousness (or dies) and the cyanide in the plasma disperses throughout the body fluids, leaving a low immediate post-exposure plasma concentration. Also, cyanide decomposes rapidly in cadavers,[120] by approximately 50 percent in 1 to 2 days, and may subsequently decrease further, or even increase slightly in stored blood. For these reasons, blood cyanide concentrations measured in fire victims are often relatively low, but when blood samples are obtained immediately after exposure,[121] higher, toxicologically significant or life-threatening levels are detected. It is suggested that, in freshly obtained whole blood samples, levels of 2.0 to 2.5 μg CN/ml should be considered capable of causing incapacitation and 3.0 μg CN/ml should be considered lethal, while for samples not taken and analyzed immediately after exposure, these concentrations/effect ranges should be at least halved, depending upon the time of storage.

Hypoxia

Apart from the tissue hypoxia caused by CO and HCN, hypoxia in fires can also be caused by exposure to low oxygen concentrations. To some extent, a lowered oxygen concentration in the inspired air or a lowered oxygen concentration in the lungs (during exercise for example) is a normal physiological occurrence, and there are compensatory mechanisms that tend to maximize the supply of oxygen to the brain. When a subject is placed in a hypoxic situation there is a reflex increase in cerebral blood flow and also, up to a point, the unloading of oxygen from the blood is more efficient at lower arterial and venous blood oxygen concentrations.[44] These factors compensate to a large degree for any decrease in the oxygen concentration of the inspired air. When cynomolgus monkeys were exposed to atmospheres containing 15 percent oxygen no deleterious effects occurred beyond a slight increase in heart rate.[21]

However, a time is reached where these compensatory mechanisms fail; a 10 percent oxygen atmosphere produced a marked cerebral depression in monkeys.[21] In humans hypoxia due to lack of oxygen (hypoxic hypoxia) has been studied extensively, particularly hypoxia that occurs at high altitudes.[22,44] As in monkeys there is little effect down to

15 percent O_2 beyond a slightly reduced exercise tolerance, but at approximately 10 percent O_2 effects suddenly become severe. It is however possible to identify a number of degrees of physiological and behavioral decrement, and for low-oxygen hypoxia certain signs can be related to particular exposure concentrations. From experiments in humans the effects have been classified into four phases as follows,[22,26] the appropriate altitude ranges and equivalent sea-level oxygen concentrations being given for each phase:

1. Indifferent phase (sea level–3,000 m or 20.9–14.4% O_2): Minor effects on visual dark adaptation and beginnings of effects on exercise tolerance toward 15% O_2.
2. Compensated phase (3,000–4,500 m or 11.8–14.4% O_2): Slightly increased ventilation and heart rate, slight loss of efficiency in performance of complex psychomotor tasks and short term memory, some effects on judgment. Maximal exercise work capacity is reduced.
3. Manifest hypoxia (4,500–6000 m or 9.6–11.8% O_2): Degradation of higher mental processes and neuromuscular control, loss of critical judgment and volition, with dulling of the senses. Emotional behavior may vary from lethargy and indifference to excitation with euphoria and hallucinations. Marked increase in cardiovascular and respiratory activity. This is the region likely to be particularly dangerous during fire exposures, representing the catastrophe point as a victim passes from this stage into the fourth stage at approximately 10% O_2 (or COHb or blood cyanide concentrations producing an equivalent degree of brain hypoxia).
4. Critical hypoxia (6000–7,600 m or 7.8–9.6% O_2): Rapid deterioration of judgment and comprehension leading to unconsciousness followed by cessation of respiration and finally of circulation at death.

If a subject is suddenly placed into a low-oxygen environment a finite time elapses before the blood gas concentration equilibrates with the new conditions, and a certain degree of physiological effect then occurs, depending upon the equilibrium blood concentration attained. The time to effect functions described in the next section have been based on the concept to a certain "dose" of hypoxia being taken up over a period of time to reach equilibrium at the chosen endpoint of severe incapacitation (the catastrophe point). The concept of the catastrophe point relates to observations, mainly during exposures of primates to the narcotic gases CO, HCN and low-oxygen hypoxia[21,25] that due to physiological compensatory mechanisms there is very little decrement in physiological status or behavioral task performance as the severity of an exposure increases, until a certain point is reached when tissue hypoxia becomes critical and deterioration becomes very marked and very rapid, usually leading to unconsciousness. This endpoint therefore marks the sudden change in a potential fire victim from a condition of near normality to a condition in which escape would not be possible.

A Model for the Prediction of Time to Incapacitation by Hypoxia in Fires

Incapacitation due to oxygen lack, consisting of loss of consciousness, occurs when the oxygen supply to cerebral tissue falls below a certain critical value, which in turn occurs when the partial pressure of oxygen in the cerebral venous blood falls below 20 mmHg.[22] Due to the effects of the compensatory mechanisms, to residual oxygen in the

lungs, and to oxygen stores available from the blood, a certain period of time elapses before the oxygen tension of venous blood declines to this critical level when a subject is suddenly faced with a reduced oxygen atmosphere after breathing normal air.[22] The time taken for this depletion depends upon the level to which the oxygen concentration falls, but also on the activity of the subject (which affects oxygen demand) and the RMV. It is therefore possible to plot time to loss of consciousness against oxygen concentration. Studies of this kind have been performed on human subjects, principally for hypoxia caused by exposure to reduced atmospheric pressures simulating the effects of high altitudes, which has similar effects to those of exposure to reduced oxygen concentrations at sea level. Figure 2-8.9 shows such a plot of time of useful consciousness for humans at rest following sudden decompression (less than one second transition time) to a range of simulated altitudes. The data are adapted from Luft,[22] and are expressed in terms of altitude. The equivalent sea-level oxygen concentrations have been added to the figure, and also the percentage oxygen vitiation (i.e., the equivalent decrease in percentage oxygen concentration at sea level below the normal concentration of 20.9 percent oxygen).

From this curve it is possible to derive an equation that should give a reasonable prediction of time to loss of consciousness (t_{Io}) for a victim exposed to a hypoxic fire environment, as follows

$$(t_{Io})\text{min} = \exp[8.13 - 0.54(20.9 - \%O_2)] \qquad (10)$$

where $(20.9 - \text{percent } O_2) = \text{percent } O_2\text{Vit}$ (percent oxygen vitiation).

As with exposure to HCN, time to incapacitation for exposure to low oxygen concentrations does not follow Haber's rule, since short exposures to severe hypoxia cause incapacitation very rapidly, and long exposures to modest hypoxia have little effect (e.g., at 17 percent O_2Vit, $Ct = 17 \times 0.33 = 5.61$ percent · min, while at 11.3 percent O_2Vit, $Ct = 11.3 \times 7.73 = 87.3$ percent · min). In attempting to predict time or dose to incapacitation or death for a subject exposed to changing oxygen concentrations, it is therefore necessary to apply a weighting factor to allow for these deviations from ideal behavior. As with HCN this may be achieved by using the fractional effective dose concept as follows:

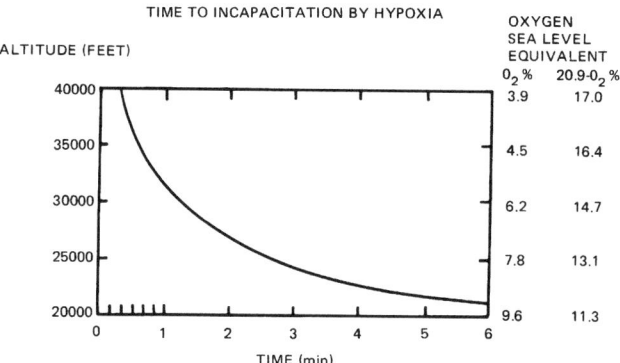

Fig. 2-8.9. Time of useful consciousness on sudden exposure to high altitudes (less than 1 s transition time). Scales also show equivalent sea-level percent oxygen concentration and percent oxygen vitiation (decrease below 20.9 percent O_2). Adapted from Luft.[22]

For a constant level of hypoxia the time to incapacitation due to oxygen depletion is given by

$$t_{Io} = \exp[8.13 - 0.54(20.9 - \text{percent } O_2)]$$

Dose to incapacitation = $(20.9 - \text{percent } O_2)(t_{Io})$
Therefore, for a short exposure time, t, to a given level of oxygen vitiation

$$F_{Io} = \frac{(20.9 - \%O_2)(t)}{(20.9 - \%O_2)(t_{Io})}$$

Where F_{Io} = fraction of an incapacitating dose of hypoxia, and where $t = 1$ min this simplifies to

$$F'_{Io} = 1/t_{Io} \qquad (11)$$

If the fractional doses per each minute are summed throughout the exposure, the dose and time to incapacitation can be predicted.

EXAMPLE:

A subject is exposed to a concentration of 10 percent oxygen for 5 min followed by 7.8 percent oxygen for 1.5 min. For 10% O_2 (10.9% O_2 Vit)

$$t_{Io} = \exp[8.13 - 0.54(20.9 - 10)]$$
$$1/t_{Io} = 0.106$$

For 7.8% O_2 (13.1% O_2 Vit)

$$t_{Io} = 2.8748$$
$$1/t_{Io} = 0.3478$$
$$F_{Io} = 0.106 \times 5 + 0.3478 \times 1.5$$
$$= 1.05$$

Therefore loss of consciousness is predicted at 6.5 minutes.

A Model for the Prediction of Hyperventilation and Time to Incapacitation by Carbon Dioxide

Carbon dioxide (CO_2), like carbon monoxide, is universally present in fires. Although carbon dioxide is not toxic at concentrations of up to 5 percent it stimulates breathing, so that at 3 percent the RMV is approximately doubled, and at 5 percent tripled.[21] This hyperventilation, apart from being stressful, can increase the rate at which other toxic fire products (such as CO) are taken up.

For narcotic gases such as CO or HCN it is likely that the increased uptake resulting from carbon dioxide induced hyperventilation will significantly reduce time to incapacitation and death. The ventilatory response to carbon dioxide varies among individuals and reported data also vary. An average curve has been constructed from data given in three sources[45-47] and is presented in Figure 2-8.10, giving the following regression equation

$$\text{RMV(L/min)} = \exp(0.2496 \times \%CO_2 + 1.9086) \qquad (12)$$

From this expression a multiplication factor (VCO_2) can be calculated for the enhanced uptake of other narcotic gases as follows

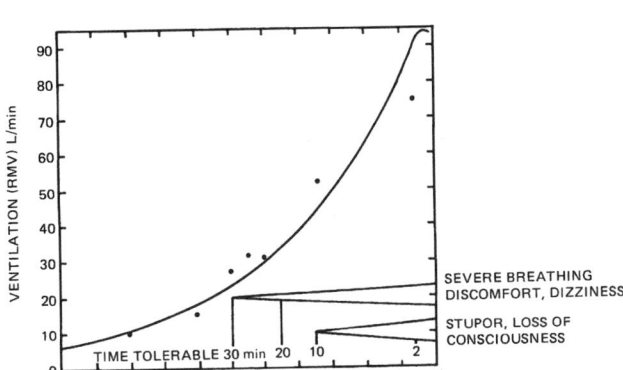

THE VENTILATORY RESPONSE TO CARBON DIOXIDE

Fig. 2-8.10. *Ventilating response to carbon dioxide. Compiled from data in: Lambertson,[45] Comroe et al,[46] Altman and Ditter[47] and King.[48]*

$$VCO_2 = \frac{\exp(0.2496 \times \%CO_2 + 1.9086)}{6.8} \qquad (13)$$

Where 6.8 L/min is a suggested figure for the resting RMV at the background CO_2 concentration.

Upon further examination of this relationship it is considered that, although it provides a reasonable estimate of the change in VCO_2, it gives an exaggerated value for the increase in uptake rate for other gases. This is primarily because the efficiency of uptake decreases as ventilation increases. A modified expression has therefore been derived, based upon that used in the CFK equation,[18] which gives a somewhat lower prediction for the increase in uptake rate of other gases. A slightly higher figure has also been used for the resting RMV. The modified equation is as follows

$$VCO_2 = \frac{\exp(0.1903 \times \%CO_2 + 2.0004)}{7.1}$$

At concentrations of approximately 5 percent and above carbon dioxide is itself a narcotic, but for elevated CO_2 concentrations (hypercapnia) the change in degree of incapacitation with exposure concentration is more gradual than with hypoxia. From approximately 3 percent up to 6 percent there is a gradually increasing degree of respiratory distress. This becomes severe at approximately 5 to 6 percent, with clinical comments from subjects such as "breathing fails to satisfy intense longing for air" or "much discomfort, severe symptoms impending," with headache and vomiting also occurring.[48] Although due to the gradual equilibration process these signs tend to worsen during exposure, it seems unlikely that they would proceed as far as loss of consciousness over the course of a 30 or even a 60 minute exposure period. However, once the concentration of carbon dioxide is in the 7 to 10 percent plus range, a new set of signs consisting of dizziness, drowsiness and unconsciousness is superimposed on the severe respiratory effects. A time factor does enter here due to gradual uptake, with loss of consciousness being more certain, and occurring earlier (over a period of a few minutes) as the exposure concentration approaches and exceeds 10 percent.[26,27,48,49] Approximate tolerance times for the distressing effects on breathing and the onset of narcosis for humans are shown in Figure 2-8.10. The effects are perceptible to subjects from 3 percent as increasingly rapid breathing, and at approximately 6 percent become intolerable within 20 minutes. Symptoms of dizziness, headache, and fatigue start to occur at concentrations above 7 percent, with danger of unconsciousness occurring within a few minutes increasing from 7 to 10 percent. Loss of consciousness is likely within 2 minutes at 10 percent CO_2 in humans.[48]

As with HCN and low-oxygen hypoxia, intoxication by carbon dioxide does not follow Haber's rule (Ct for 10 percent CO_2 = 20 percent $\cdot$ min, Ct for 5 percent CO_2 = 175 percent $\cdot$ min). From the approximate data in Figure 2-8.10 an expression predicting approximate time to incapacitation (t_{Ico_2}) has been derived as follows

$$t_{Ico_2} = \exp(6.1623 - 0.5189 \times \%CO_2) \qquad (14)$$

Using the fractional dose concept previously described for HCN and hypoxia, it is possible to predict approximate dose to incapacitation, provided that the CO_2 concentration is stable or increasing, as follows

For a constant CO_2 concentration

$$t_{Ico_2} = \exp(6.1623 - 0.5189 \times \%CO_2)$$

Dose to incapacitation = $(\% CO_2)(t_{Ico_2})$

Therefore, for a short exposure time, t, to a given CO_2 concentration

$$F_{Ico_2} = \frac{(\%CO_2)(t)}{(\%CO_2)(t_{Ico_2})}$$

Where F_{Ico_2} = fraction of an incapacitating dose, and where t = 1 min this simplifies to

$$F'_{Ico_2} = 1/t_{Ico_2} \qquad (15)$$

If the fractional doses per minute are summed throughout the exposure, the dose and time to incapacitation can be predicted.

EXAMPLE:

A subject is exposed to a concentration of 5 percent CO_2 for 20 minutes, followed by 9 percent CO_2 for 2 min.

For 5% CO_2, t_{Ico_2} = 35.44; $1/t_{Ico_2}$ = 0.0282
For 9% CO_2, t_{Ico_2} = 4.45; $1/t_{Ico_2}$ = 0.2247

$$F_{Ico_2} = 0.0282 \times 20 + 0.2247 \times 2 = 1.01$$

Severe incapacitation with probable loss of consciousness is therefore predicted at approximately 22 minutes.

Interactions Between Toxic Fire Gases

Although data on the concentration/time/dose relationships of the dangerous and lethal narcotic effects in humans of individual fire gases are necessarily limited, they are adequate for the construction of a usable incapacitation model. However, the effect of interactions between combinations of these gases on time to incapacitation in fires is an area that requires further investigation, as very little information is currently available. The best that can be done currently is to suggest likely degrees of interaction based on physiological

data from individual gases and on such experimental data for gas combinations as do exist.

Effect of Carbon Dioxide on Effects of CO, HCN, and Low-Oxygen Hypoxia

The interaction likely to be most important is that hyperventilation due to carbon dioxide exposure will increase the rate of uptake of other toxic gases and thus decrease the time to incapacitation (or the time taken to inhale a lethal dose), in proportion to the increase in ventilation. This is likely to be most important with respect to CO intoxication, particularly for a subject at rest, and also to some extent for active subjects. An expression for calculating the increase in RMV resulting from exposure to different carbon dioxide concentrations is given in the section on CO_2, but as an approximation it should be assumed that there would be little effect below 3 percent CO_2, while at 3 percent CO_2, RMV would be doubled, so time to incapacitation by CO should be halved. At 5 percent CO_2, RMV would be approximately tripled and time to incapacitation would be approximately one-third of that in the absence of carbon dioxide. There is a possibility that the effects on time to incapacitation would not be as dramatic as this, since there is evidence that the presence of carbon dioxide may counteract the leftward shift in the oxygen dissociation curve caused by carbon monoxide, somewhat counteracting its deleterious effects.[50] However, in the absence of experimental data on combination exposures it is best to ignore this possible beneficial effect, since the effect on uptake rate is likely to be dominant. A similar effect on uptake may also occur with HCN. With regard to low oxygen, carbon dioxide has been shown to have a marked beneficial effect on resistance to incapacitation. This is partly due to the hyperventilatory effect that increases the rate of oxygen uptake, and partly due to the rightward shift in the oxygen dissociation curve caused by carbon dioxide. This improves the delivery of oxygen to the tissues, counteracting the respiratory alkalosis that otherwise occurs.[51,52] New evidence is currently being obtained from experiments on the effects of combinations of narcotic gases with CO_2 in rodents, that with severe exposures, postexposure lethality is increased by the presence of CO_2. When animals are severely affected and suffering from a hypoxia-induced metabolic acidosis, this appears to be enhanced by the further acidotic effect of CO_2 inhalation, and the animals then fail to recover after exposure under conditions when they would otherwise be expected to do so. It is also to be expected that hyperventilation induced by CO_2 would increase the uptake of substances that irritate the lung, which also tend to cause toxic effects some time after exposure, and recent experiments in rodents are providing evidence that this is so, with increased deaths possibly caused by postexposure acidosis and increased lung damage. Exercise also causes a CO_2-driven hyperventilation, and there is new evidence that this may also cause deaths when rodents are exposed to, irritants at normally sub-lethal concentrations.[116]

Interactions Between CO and HCN

Some studies have been made of interactions between CO and HCN, with varying results.[53] On theoretical grounds little interaction is to be expected, since CO diminishes the carriage of oxygen in the blood and its delivery to the tissues, while HCN diminishes the ability to use oxygen once delivered to the tissues. It is therefore to be expected that either one or the other gas would constitute the rate-limiting step in oxygen supply and utilization. However the consensus view is that there is at least some additive effect between these two gases. Experiments in primates have shown that time to incapacitation by HCN is slightly reduced by the presence of near-toxic concentrations of CO;[53] also, the rate of uptake of CO may be increased by the hyperventilatory effect of HCN. In these circumstances it is probably safest to assume that these gases are additive in terms of time to incapacitation and dose to death, and that incapacitation or death will occur when the fraction of the toxic dose of each one adds up to unity.

Interactions Between CO and Low-Oxygen Hypoxia

The most likely interaction between CO and low-oxygen hypoxia would be some degree of addition, since both reduce the percentage oxygen saturation of arterial blood, and CO also impairs the delivery of oxygen to the tissue by causing a leftward shift of the oxygen dissociation curve.[50] It is possible that during the early stages of CO exposure in hypoxic subjects the CO occupies the upper, oxygen-free part of the oxygen dissociation curve, and therefore has little effect. Von Leggenhager[34] reports that subjects at rest at altitude remain symptom free at low levels of CO saturation. However, it is likely that the effect of more severe exposure to CO in a hypoxic subject would be additive to some extent, as reported by Heim[54] and McFarland et al.[55]

An important point is the possible interaction between irritant smoke products and narcotic gases. This effect is particularly strong when rats and mice are exposed to smoke, since the rodent response to irritation of the upper respiratory tract is a marked decrease in respiratory rate and RMV. Thus if CO is present in the smoke, the rate of uptake will be considerbly reduced if the smoke is irritant. This sometimes leads to misleading results in combustion product toxicity tests, where a material producing irritant smoke will have an apparently low LC$_{50}$, although high CO concentrations are present in the atmosphere.

This marked, prolonged decrease in respiratory rate does not occur in humans or nonhuman primates; indeed, in primate smoke experiments, irritant products tend to increase rather than decrease ventilation (although not sufficiently to increase CO toxicity).[12,21]

In summary, data on interactions between the narcotic gases CO, HCN, low oxygen, and CO$_2$ are limited, but where deleterious interactions are likely it would be prudent to include them in the incapacitation model, if only to err on the side of safety. For this reason it is proposed that the interactions should be quantified in the incapacitation model as follows:

1. Assume that CO and HCN are directly additive (1:1) on a fractional dose basis (the evidence suggests that they are additive, but that the additive interaction may actually be less than unity).
2. Assume that the rates of uptake of CO and HCN and their fractional doses are increased in proportion to any increase in ventilation (RMV) caused by carbon dioxide.
3. Assume that the fractional doses of CO and HCN, adapted for carbon dioxide, are additive with the fractional dose of low-oxygen hypoxia.
4. Assume that narcosis by carbon dioxide is independent of that induced by CO, HCN, and hypoxia.

5. Assume that irritancy is independent of narcosis, but that uptake of irritants is increased by carbon dioxide (see next section of this chapter.)

Implications of Interactions for Predicting Time to Incapacitation in Smoke Atmospheres

In general, there is evidence for some possible interactions between toxic fire gases, but whether these are likely to be important in practice depends upon the composition of actual fire atmospheres, which is discussed in a later section of this chapter. For most practical situations, the composition of fire atmospheres will be such that for narcotic effects CO will be the most important toxic product, and that the most important interaction will be an increased rate of CO uptake due to hyperventilation caused by CO$_2$. The additional effects of HCN and low-oxygen hypoxia will contribute to the effects of CO-induced narcosis, and may significantly reduce time to incapacitation in some situations.

On this basis the fractional dose equation for narcosis would be

$$\text{Total } F_{IN} = [(F_{Ico} + F_{Icn}) \times VCO_2 + F_{Io}] \text{ or } F_{Ico_2} \quad (16)$$

where:

F_{IN} = fraction of an incapacitating dose of all narcotic gases
F_{Ico} = fraction of an incapacitating dose of CO
F_{Icn} = fraction of an incapacitating dose of HCN
VCO_2 = multiplication factor for CO$_2$-induced hyperventilation
F_{Io} = fraction of an incapacitating dose of low-oxygen hypoxia
F_{Ico_2} = fraction of an incapacitating dose of CO$_2$

For a 1-minute exposure to each gas at a concentration C

$$F'_{Ico} = \frac{8.2925 \times 10^{-4} \times \text{ppm CO}^{1.036}}{30}$$

$$F'_{Icn} = \frac{1}{\exp(5.396 - 0.023 \times \text{ppm HCN})}$$

$$VCO_2 = \frac{\exp(0.1903 \times \%CO_2 + 2.0004)}{7.1}$$

$$F'_{Io} = \frac{1}{\exp[8.13 - 0.54(20.9 - \%O_2)]}$$

$$F'_{Ico_2} = \frac{1}{\exp[6.1623 - 0.5189 \times \%CO_2]}$$

Figure 2-8.11 shows an expanded detail of narcotic gas profiles during the first 10 minutes of the single armchair room burn that is presented in more detail in Figure 2-8.7. The histograms show the average concentrations of each gas at minute intervals during the first 6 min of the fire, the figures for which are given in Table 2-8.3. The HCN concentration was not measured in this fire, but it is likely to have been present as a major toxic product. Possible HCN concentrations have therefore been suggested for inclusion in the model and are shown in a histogram.

Applying the expressions for the fractional incapacitating dose of each gas to the data in Table 2-8.3, the total fractional dose of all narcotic gases for each minute during the fire (F_{IN}) has been calculated according to Equation 16,

NARCOTIC GASES DURING EARLY STAGES OF SINGLE ARMCHAIR ROOM BURN

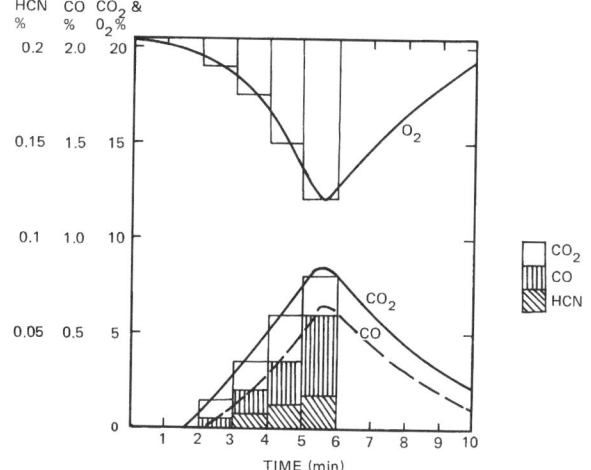

Fig. 2-8.11. Expanded detail from Figure 2-8.7—average concentrations of narcotic gases each minute (histograms) from gas profiles (curves) measured during the first 10 minutes of a single armchair (polystyrene with polyurethane cushions and covers) room burn.[77] Since HCN was not measured, but was likely to have been present as an important toxic product, possible concentrations have been suggested for inclusion in the model.

and summed for each successive minute during the fire, as shown in Table 2-8.4. Incapacitation (loss of consciousness) is predicted at 5 minutes when the fractional incapacitating dose exceeds unity (F_{IN} = 1.2).

IRRITANT FIRE PRODUCTS

Unlike the incapacitating effects of narcotics, which are clear-cut and well understood, the incapacitative effects of irritants are much more difficult to determine. Irritant fire products produce incapacitation during and after exposure in two distinct ways. During exposure the most important form of incapacitation is sensory irritation, which causes painful effects to the eyes and upper respiratory tract, and to some extent also the lungs. Although exposure may be painful and thus incapacitating, it is unlikely to be directly lethal during exposure unless exceptionally high concentrations of irritants are present. However, the second effect of irritants penetrating into the lungs is an acute pulmonary irritant response, consisting of edema and inflammation which can cause respiratory difficulties and may lead to death 6 to 24 hours after exposure.[56,57] The effects do not show the sharp cut off of narcosis, but lie on a continuum from mild eye irritation to severe pain, depending upon the concentration of the irritant and its potency.[20,58,59] For sensory irritation the effects do not depend upon an accumulated dose but occur immediately upon exposure, and usually lessen somewhat if exposure continues.[12,20] For the later inflammatory reaction the effect does depend upon an accumulated dose, approximately following Haber's rule, and there seems to be a threshold below which the consequences are minor, but when this dose is exceeded severe respiratory difficulties and often death occur, usually 6 to 24 hours after exposure. However, for most sensory irritants the ratio between the concentration producing severe irritation and the

TABLE 2-8.3 *Average Concentrations of Narcotic Gases Each Minute During the First Six Minutes of the Single Armchair Room Burn Shown in Figure 2-8.11*

Time (min)	1	2	3	4	5	6
CO ppm	0	0	500	2000	3500	6000
HCN ppm	0	0	0	75	125	174
CO_2%	0	0	1.5	3.5	6	8
O_2%	20.9	20.9	19	17.5	15	12

dose causing death is usually large (15 to 500 times)[60,61] for 30-minute exposure times. (See Table 2-8.5.)

The effects of low concentrations of irritants can best be considered as adding to the obscurational effects of smoke by producing mild eye and upper respiratory tract irritation. In this situation irritants may have some effect by impairing the speed of an individual's movement through a building (as would simple visual obscuration), but the combined effects of eye irritation and direct visual obscuration may be more serious, and it has been shown that human volunteers moved more slowly through irritant smoke than through nonirritant smoke.[62] The limitation of escape capability may not be simply restricted to direct physiological effects, but also to psychological and behavioral effects such as the willingness of an individual to enter a smoke-filled corridor.[62]

At the other end of the scale, when irritants are present at high concentrations, there is some disagreement about the likely degree of incapacitation. Some investigators believe that the painful effects on the eyes and upper respiratory tract would be severely incapacitating, so that, for example, escape from a building would be rendered extremely difficult.[63] Others believe that the effects peak at moderate concentrations, and that although the effects may be very unpleasant they would not significantly impair the ability to escape from a building, and would provide a strong stimulus to escape that might almost be beneficial.[61]

TABLE 2-8.4 *Fractions of an Incapacitating Dose of Narcotic Gases Calculated from the Data in Table 2-8.3 According to Equation 16 for One-Minute Intervals During the Single Armchair Room Burn Shown in Figure 2-8.11.*

Time (min)	1	2	3	4	5
F'_{Ico}	0	0	0.017	0.074	0.130
+ F'_{Icn}	0	0	0.000	0.025	0.080
× VCO_2	0	0	1.442	2.376	4.434
=	0	0	0.025	0.233	0.931
+ F_{Io}	0	0	0.001	0.002	0.007
= Total	0	0	0.026	0.235	0.938
Running total (F_{IN})	0	0	0.026	0.261	1.199
or:					
F'_{Ico_2}	0	0	0.005	0.013	0.047
Running total (F_{IN})	0	0	0.005	0.018	0.065

F_{IN} = 1.2 at 5 minutes due to the combined effects of CO, HCN, and low-oxygen hypoxia, the uptake of which was increased by CO_2, and incapacitation is predicted at between 4 and 5 minutes. Although carbon dioxide was present at concentrations sufficient to have caused significant hyperventilation, the fractional incapacitating dose for narcosis by carbon dioxide was only 0.065 at 5 minutes, and this is therefore unlikely to have had any effect.

TABLE 2-8.5 *Sensory and Pulmonary Irritancy of Combustion Products*

Irritant	RD$_{50}$ (ppm) mouse[†]	Severe sensory irritancy in humans (ppm)	30-minute LC$_{50}$ (ppm)[‡] mammal	LC$_{50}$/RD$_{50}$
	0.1–1.0			
Toluene diisocyanate	0.20	1.0[17]	100[17]	500
O-chlorobenzylidene -malonitrile (CS)*	0.52	0.5[20]	150–400[111,112]	529
α-chloroacetophenone (CN)*	0.96	6–50[17]	300–400[58,110]	365
	1.0–10			
Acrolein	1.7	1–5.5[17,58]	140–170[58,110,61]	91
Formaldehyde	3.1	5–10[17,58]	700–800[60,110]	242
Chlorine[1]	9.3	9–20[58]	100[60]	11
	10–100			
Crotonaldehyde		4–45[17]	200–1500[110,17,113]	
Acrylonitrile		>20[17]	4,000–4,600[110]	
Penteneone			1,000[110]	
Phenol		>50[17]	400–700[110]	
	100–1,000			
SO$_2$	117	50–100[17,58]	300–500[60,110]	3
NH$_3$	303	700–1700[68]	1,400–8,000[60,110]	16
HF		120[17]	900–3,600[110]	
HCl	309	100[17,58]	1,600–6,000[110,61,60]	12
HBr		100[17]	1,600–6,000[110]	
NO$_2$	349	80[17,58]	60–250[60,110]	0.4
Styrene	980	>700[17]	10,000–80,000[110]	46
	1,000–10,000			
Acetaldehyde	4946	>1,500[17]	20,000–128,000[110]	15
	10,000–100,000			
Ethanol	27,314	>5,000[17]	400,000[110]	15
Acetone	77,516	>12,000[17]	128,000–250,000[17,110]	2

The potential for causing sensory irritation spans six orders of magnitude, while that for causing death spans approximately three orders of magnitude. For substances down to NO$_2$ death is likely to be due to lung irritation, while for the remainder from styrene to acetone death is likely to be due to narcosis.

*Substances not detected in combustion atmospheres.

[†]RD$_{50}$ from Alarie,[70] where no data exist substances have been ranked according to their reported irritancy in man.

[‡]LC$_{50}$ concentrations have been normalized to a 30-minute exposure time according to Haber's rule.

One of the main difficulties in attempting to predict the consequences of exposure to irritants is the poor quality of data available on humans. Obviously very few controlled studies have been made of the effects of severe irritancy in humans, so that most data are anecdotal, derived from accidental industrial exposures, with only a vague knowledge of exposure concentrations. Reports of the severity of the effects also tend to be very subjective, so that the term "severe irritation" could cover a wide range of sensations with varying degrees of actual incapacitation. Another problem is that since sensory irritation covers a continuous range from mild eye and upper respiratory tract irritation to severe pain, there are no simple objective end points or thresholds. The most extensive studies of the effects of severe irritancy in humans have been performed on volunteers exposed to riot-control agents such as CS (o-chlorobenzylidine malonitrile) or CN (α-chloroacetophenone). Even these studies do not really show how the ability to escape from a building might be affected, but they do to some extent convey the severity of the effects.

The effects of CS, which are probably similar to those of any severe sensory irritant, have been described by Beswick et al.[20] They consist of an almost instantaneous severe inflammation of the eyes accompanied by pain, excessive lacrimation (tearing), and blepharospasm (spasm of the eyelids). There is irritation and running of the nose with a burning sensation in the nose, mouth, and throat, and a feeling of intense discomfort during which the subjects cough, often violently. If the exposure continues, the discomfort spreads to the chest and there is difficulty in breathing. Many subjects describe a tightness of the chest or chest pain as the worst symptom of CS exposure. The respiration pattern is irregular and the breath is held for short periods. Attempts to avoid the irritation by breath-holding, followed eventually by fairly deep breaths, are reported as being extremely unpleasant. At this stage most individuals are acutely apprehensive and highly motivated to escape from the smoke. However, if exposure continues there is some remission of signs and symptoms. When subjects were exposed to 0.08 ppm CS, they found the immediate effects very unpleasant but after 4 to 5 min were able to play cards. Another finding with CS, perhaps related to the development of tolerance, was that subjects could endure a relatively high concentration (~0.8 ppm CS) if it was achieved gradually even over as short a period as ten minutes (as is likely to occur in fires) while they were totally unable to bear an immediate exposure to the same concentration.[20,59]

Reports are conflicting among fire victims. Some persons say they went through dense smoke without experiencing any great discomfort, while others say that respiratory difficulties prevented them from entering smoke-filled areas.[64] This seems to depend upon the type of fire. For example, the smoke from some well-ventilated fires involving primarily cellulosic materials has been reported as irritant but not seriously incapacitating, while that from some plastic materials (e.g., the interior of a burning car) was found to cause severe effects when only a small amount of smoke was inhaled. Anyone who has had bonfire smoke in their eyes will know the pain of the experience. However, the effects can be mitigated by blinking or shutting the eyes, and the effects on the nose can be mitigated by mouth breathing and breath-holding. Also, it is known that people in emergency situations are often unaware of painful stimuli.[65] It is therefore likely that irritant smoke products do have some severe effects on the escape capability of fire victims, but it is difficult at present to predict accurately the likely degree of incapacitation.

Animal Models for the Assessment of Irritancy and Their Extrapolation to Humans

Having established that irritancy is likely to be a major cause of incapacitation in fires, it is important to find some way of assessing the potential of fire atmospheres for causing irritancy. The basic effects, consisting of an acute inflammatory reaction of the tissues accompanied by stimulation of pain receptors, are common to all mammals, so that animals can realistically be used to assess the potential for irritancy in humans.

The characteristic response to eye irritation is stimulation of trigeminal nerve endings in the cornea leading to pain, blepharospasm (reflex, or more or less involuntary closure of the eyelids) and lacrimation. Severe damage may also lead to corneal opacity.

The effects of irritants on the upper respiratory tract have been studied in a number of species, including humans, for a wide range of airborne irritants including atmospheres of combustion products.[66,67] The characteristic physiological response (also due to trigeminal nerve stimulation), is a reflex decrease in respiratory rate accompanied by a prickling or burning sensation in the nose, mouth, and throat, often accompanied by mucus secretion. At very high concentrations rapidly developing inflammatory reactions in the upper respiratory tract and laryngeal spasm may cause death during or soon after exposure, but in humans the lung is likely to be the more seriously affected target organ.

In contrast to these effects, the characteristic response to irritants penetrating into the lung is an increase in respiratory rate, generally accompanied by coughing and a slight decrease in tidal volume. There may also be bronchial constriction and increased pulmonary flow resistance, particularly if the victim is hypersensitive to the irritant.[60,66] This is accompanied by tissue inflammation and edema, which at very high concentrations can cause death during, or more often after, exposure.[10,56]

The effect (upper respiratory tract or lung irritancy) that predominates depends upon a number of factors such as the physical characteristics of the aerosol, the aqueous solubility of the irritant, the animal species, and the duration of the insult. An important difference between mice and rats

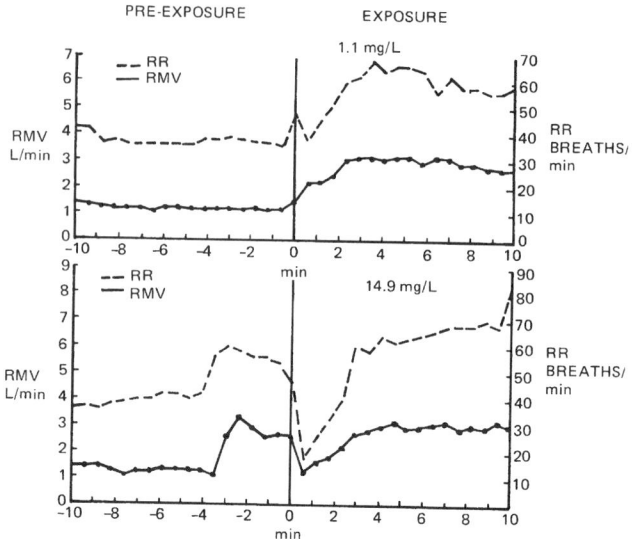

Fig. 2-8.12. Respiratory effects of polypropylene decomposed under nonflaming oxidative conditions—cynomolgus monkey.

(which are often used to measure irritancy) and primates, including humans, is that in rodents the nasal passages are complex in structure and have a large surface area, so that soluble gases are readily taken up and particulates are readily deposited. A decrease in respiratory rate in the upper respiratory tract is the major response in rodents which tends to protect them from exposure. Their great tolerance to hypoxia (and possible circulatory adaptations similar to the diving reflex) also enables them to maintain greatly reduced breathing for long periods. In humans and other primates the nasal passages are simply structured with a relatively small surface area, and in humans particularly, mouth breathing is common. Thus, although upper respiratory tract irritation occurs initially and is accompanied by some respiratory rate decrease, a greater proportion of the inhaled irritant is carried into the lungs, so that lung irritation is generally a more pronounced effect. In primates, including humans, a transient respiratory rate depression is followed rapidly by the increased respiratory rate characteristic of lung irritation.[12,55] (See Figures 2-8.12 and 2-8.13.)

Rodent Respiratory Rate Depression Test

The method most commonly used to quantify upper respiratory tract irritancy both for pedigree chemicals and combustion products is the mouse respiratory rate depression test.[67,69] The basic method involves measurement of the percentage decrease in respiratory rate during exposure over a range of different atmosphere concentrations. From this measurement the RD_{50} (the concentration required to produce a 50 percent decrease in respiratory rate) is calculated. Since the basic irritant mechanisms are the same in all mammals, it is certainly possible to identify an individual substance, or mixture (such as a combustion atmosphere), that is likely to be irritant to humans; however, it may also be possible to predict the degree of irritation in humans. As a result of comparisons of data from humans with results from mice, Alarie[70] has demonstrated a relationship between the potency of known sensory irritants in humans and a derivative of the mouse RD_{50}. When the log of the TLV (often based on symptoms of irritancy in man) is plotted against the

THE EFFECTS UPON RESPIRATION OF EXPOSURE TO AN IRRITANT
SMOKE ATMOSPHERE-WOOD PYROLYSED AT 300°C.

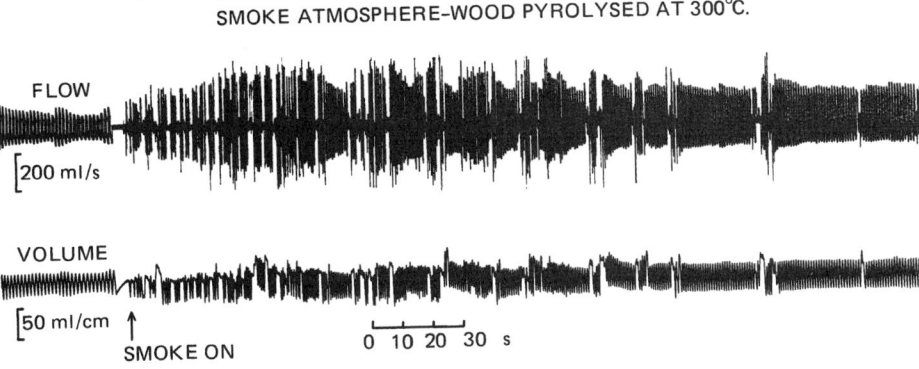

Fig. 2-8.13. Effects upon respiration of exposure to an irritant smoke atmosphere; wood pyrolyzed at 300°C. When the smoke reaches the primate there is an initial sensory irritant response consisting of a decrease in respiratory rate with pauses between breaths. This rapidly gives way to a pulmonary irritant response consisting of an increase in respiratory rate and volume, which is maintained for the duration of the exposure. From Purser and Woolley.[12]

log of 0.03 multiplied by the RD_{50} in ppm, most chemicals known to be sensory irritants in humans fall into a linear relationship with the respiratory effects in mice. Chemicals that are highly irritant in humans such as acrolein or chlorine have low RD_{50} measures in mice while mild irritants such as ethanol have a high mouse RD_{50}.[70] (See Table 2-8.5.)

Such methods appear to give good predictions of irritancy at low levels that are suitable for fixing hygiene standards, but for the high concentrations occurring in fires it is neccessary to predict which concentrations will produce sufficient incapacitation to cause serious impairment of escape capability, and also which concentrations will cause serious lung damage after exposure. As stated previously, such predictions of incapacitation in humans are difficult because of the variable and subjective nature of irritancy. However, Alarie states that a human exposed at the mouse RD_{50} concentration of any substance would find the atmosphere severely irritating and would be seriously incapacitated within 3 min[63] Certainly this seems to be justified in terms of some individual chemicals. For example, the work with CS[20,59] shows that, although it is very difficult to measure intolerable irritancy in humans, there does seem to be a reasonable agreement between the human and mouse data. Thus the mouse RD_{50} (4 mg/m^3, 0.52 ppm) is very close to the concentration found to be immediately intolerable when humans were exposed for up to 12 min (5 mg/m^3, 0.6 ppm), and the list of RD_{50} levels and concentrations reported as highly irritant in humans[17] given in Table 2-8.5 shows a reasonable correspondence, particularly for the irritants commonly found in fires.

However these apparent irritancy correlations between rodents and humans still do not enable the exact prediction of the degree of incapacitation likely in humans, and in some cases there is evidence that the mouse model does not even give a good prediction of the degree of sensory irritation, although it can be used to demonstrate when sensory irritation is likely to occur in humans. Three experiments, two involving nonhuman primates and the other involving humans may illustrate the difficulties of extrapolating from the RD_{50} and physiological effects in mice to the degree of irritancy, incapacitation, and physiological effects in primates. In a series of experiments the irritant effects of smoke produced by the nonflaming oxidative decomposition of polypropylane were evaluated in cynomolgus monkeys.[12,71] The effects of breathing smoke through a face mask were mild at

concentrations (NAC mass charge) of up to 4 mg polypropylene/L, consisting of a transient decrease in respiratory rate lasting approximately 30 seconds (a sensory irritant response), followed by an increased respiratory rate (lung irritant response) with a slight increase in RMV. (See Figure 2-8.12.) The respiratory response pattern is illustrated in Figure 2-8.13, where similar effects occurred during exposure to pyrolysis products from wood. For polypropylene the lung irritant response was the most sensitive effect with a threshold of 1 mg polypropylene/L. At concentrations above 6 mg/L the irritant effects were more marked, and although recovery appeared complete immediately after exposure, signs of nasal and pulmonary inflammation occured some hours later. One animal died following an exposure of 30 minutes at 8 mg polypropylene/L. When free-moving monkeys trained to perform a behavioral task were exposed, there was evidence of some eye irritation and mild disruption of behavioral performance at a concentration of 1.85 mg polypropylene/L, the effects of exposure at 0.92 mg/L being very slight. However, the mouse RD_{50} for the same polypropylene atmosphere was found to be 0.1 mg/L. According to the model this concentration should have been highly irritant to the monkey, and yet in practice only the mildest of signs occurred at concentrations more than an order of magnitude higher. Similarly, in another study Potts and Lederer[72] exposed mice and humans simultaneously to smoke from the pyrolysis of red oak (mouse RD_{50} 0.37 mg/L). At this concentration the smoke was barely visible and all subjects stated that although the smoke was unpleasant and irritating, in no sense were they physically incapacitated, and they were quite capable of performing tasks such as threading nuts and bolts of various sizes.

The third experiment was perfomed on two pedigree substances, hydrogen chloride and acrolein, often regarded as important irritants in smoke. In these experiments baboons were trained to press a lever in order to escape from a chamber after a 5-minute exposure.[61] It was found that the animals could perform this task efficiently even at the incredibly high concentrations of 2780 ppm acrolein or 16,570 ppm HCl, although in both cases the animals died from lung inflammation after exposure. These concentrations compare with mouse RD_{50}s of 1.68 ppm for acrolein and 309 ppm for HCl, and at these concentrations both substances are highly irritant in humans. (See Table 2-8.6.)

TABLE 2-8.6　*Irritancy Data. Compiled from References 17, 58, 61, 63, and 110*

Acrolein*
Mouse RD_{50} = 1.68 ppm[63]

Marked irritation of eyes and nose in humans—1 ppm[17]

Severe irritation of eyes and nose in humans—5.5 ppm[17]

Henderson and Haggard[17] state that 10+ ppm is lethal in humans within a short time due to pulmonary irritation. However 10 ppm for 3.5 hours in cats was nonlethal[17]

Kaplan[61] has reported that baboons can escape from a chamber after 5 minutes exposure at up to 2780 ppm. One animal died due to pulmonary effects following exposure at 1025 ppm, and another following 2780 ppm. No signs of pulmonary effects were observed following exposure at 505 ppm and below

A case has been reported of a man dying following exposure to 153 ppm for 10 minutes[110]

The 6 hour mouse LC_{50} is 66 ppm[110]

The 30 minute rat LC_{50} is 135 ppm[110]

Hydrogen chloride**
Mouse RD_{50} = 309 ppm[63]

Strongly irritant to humans at 50 to 100 ppm for 1 hour[17]

Brief exposure at 1,000 to 2000 ppm is regarded as dangerous to lethal in humans[58];

Humans LCLO 1300 ppm for 30 minutes.[110]

Kaplan exposed baboons for 5 minutes to concentrations of up to 16,570 ppm and found that they were able to perform escape maneuvers. One animal suffered permanent lung damage at 11,400 ppm and two died at approximately 17,000 ppm (≃ 2830 ppm for 30 minutes).

---

*From these rather variable data the concentration lethal to humans following a 20 minute exposure would be 80 to 260 ppm. A severe irritant effect on the upper respiratory tract would be expected at around 5 ppm, but from Kaplan's work[61] this may not be unbearable, even up to several hundred ppm.

**Therefore anything over 100 ppm is likely to be highly irritant and over a 20-minute period approximately ten times this concentration may cause permanent lung damage or endanger life.

As a result of this work on irritants in rodents, nonhuman primates, and humans it would seem that the rodent models are good methods for identifying smoke atmospheres or individual substances likely to be irritant to humans, and even for ranking irritants in order of potency and setting hygiene standards. However, when predicting concentrations of smoke atmospheres that would seriously impair the ability to escape from a fire, there is a need for more work to establish the effects of known irritants in humans, and to establish the relationship between the rodent response and human incapacitation.

Lung Inflammatory Reactions

Just as it is difficult to predict the degree of incapacitation from sensory irritation likely to occur in victims during fire exposure, it is also difficult to predict concentrations likely to cause death in humans from lung inflammatory reactions, although experiments in rodents should enable some estimates to be made from postexposure LC_{50} data. When rodents are exposed to smoke atmospheres in small-scale combustion toxicity experiments, death occurs principally either during exposure due to narcotic gases (CO, HCN) or some time after exposure, due to lung irritation. In cases where the majority of deaths occur after exposure and are accompanied by signs of lung irritation, measurements of the concentration causing postexposure deaths give some indication of concentrations likely to be hazardous to humans.

Since the effects of narcotic gases can be predicted without animal exposure, while the potential for causing sensory irritation and lung inflammation cannot, measurements of sensory irritancy by the respiratory depression test (RD_{50}) and measurements of the concentrations causing postexposure deaths (LC_{50}) are important uses for small-scale toxicity tests. It is also important to stress that whereas sensory irritation occurs immediately upon exposure, and is concentration related, the inflammatory reactions resulting from

lung irritancy are dose related and depend approximately upon the product of exposure concentration and duration (*Ct* product). When an LC_{50} concentration is quoted, it is also important to quote the exposure duration. For combustion toxicity experiments exposure times are usually 10 or 30 min, and the 30 min LC_{50} should be approximately one-third of the 10 min LC_{50} when postexposure irritancy is the cause of death. It is also important to know the time over which the deaths were scored. Thus some studies quote LC_{50} levels only in terms of animals dying during exposure (which will be due to narcosis or very high concentrations of irritants). Other studies include deaths occuring both during or up to 24 hours after exposure, while other studies use the standard method for inhalation toxicology studies which includes deaths during exposure and for up to 14 days after. For the assessment of toxic hazard in possible fire scenarios, it is important to take all these factors into account when considering different building designs or applications of materials.

Irritant Components of Thermal Decomposition Product Atmospheres

If mathematical models are to be constructed to predict the potential for sensory irritancy and later lung inflammation of exposures in fires, it is important to attempt to identify the main irritant chemical species occurring in fires and to measure their potency individually and in combination. This is an area where knowledge is still inadequate, but large numbers of known irritant chemicals have been found to occur in fire atmospheres.[12,13] (See Table 2-8.5.) The irritant chemicals released in fires are formed during the pyrolysis and partial oxidation of materials, and the combinations of products from different materials are often remarkably similar.[13] Some materials release irritant components simply upon pyrolysis, such as HC1 from PVC, isocyanates from

TABLE 2-8.7 *Pyrolytic and Oxidative Decomposition Products of Polypropylene at 500°C Showing Percentage Yields of Major Irritants as Indicated, and Threshold Limit Values (TLVs) where Available from American Conference of Governmental Industrial Hygienists[17]*

MS interpretation	Pyrolysis yield** (%) ($\times 10^{-1}$)	Oxidation yield** (%) ($\times 10^{-1}$)	TLV (ppm)
Ethylene	10.4	8.1	
Ethane	3.7	2.1	
Propene	18.6	18.4	
Cyclopropane	0.5	0.3	
Formaldehyde*	—	33.2	2
Propyne	0.2	—	
Acetaldehyde*	—	35.0	100
Butene	9.6	20.1	
Cyclobutene	0.3	0.8	
Methyl vinyl ether*	—	10.4	
Acetone*	—	38.4	750
Butane	1.2	—	
Methyl propane	0.4	—	
Methyl butane	4.0	—	
Butenone*	—	1.3	
Methyl butene	29.7	12.9	
Pentanol*	—	12.5	
Cyclopentane	0.5	1.4	
Pentadiene	1.3	—	
Crotonaldehyde*	—	7.7	2
Ethylcyclopropane	0.1	—	
Methyl vinyl ketone*	—	2.8	
Methyl ethyl ketone*	—	4.7	200
Hexane	0.9	1.2	
Cyclohexane	32.2	19.3	
Hexadiene	3.7	2.2	
Hexyne	—	1.3	
Benzene	6.7	5.1	
Methyl propyl ketone*	—	1.9	
Pent-2-ene-4-one*	—	7.5	
Phenol*	—	11.6	5
Toluene	2.4	16.1	
Methyl cyclohexadiene	2.1	0.1	
Xylene	6.1	0.2	
Styrene	5.6	4.0	

*Oxygen-containing products
**Weight percentage conversion of polymer

flexible polyurethanes, and various substances from natural materials such as wood. However, for all organic materials and particularly for simple hydrocarbon polymers such as polypropylene or polyethylene, the main pyrolysis products, which consist of various hydrocarbon fragments, are innocuous. Thus when polypropylene is pyrolyzed in nitrogen the products listed in Table 2-8.7 are produced, and such an atmosphere was found to have no effect upon primates.[12,56] However when these products are oxidized during nonflaming decomposition in air, some are converted to highly irritant products as shown in the table, and such atmospheres were indeed found to be highly irritant to both mice and primates.[71]

These atmospheres produced by the nonflaming oxidative decomposition of materials are always the "worst case" for any material in terms of irritant potency. Both the chemical profile and the irritant potency as determined by the mouse RD_{50} test are often similar for different materials, the majority lying within a range of approximately one order of magnitude.[73] However, when materials flame these organic irritants are destroyed in the flame to produce CO_2 and water, so that the irritancy of the atmosphere depends upon how much irritant escapes the actual flame zone and the efficiency of combustion. Thus "clean," smoke-free flames, involving efficient combustion such as those that occur in a gas burner or well-ventilated fire, are relatively nonirritant, whereas "dirty," smoky flames resulting from inefficient combustion may contain high concentrations of irritants, and produce these irritants at a greater rate than under smoldering or nonflaming conditions. In primates the atmosphere produced by flaming polypropylene was found to retain some irritancy, although considerably less than under nonflaming conditions.[12] In mouse experiments, some fire retardant materials, which could be induced to flame only intermittently, with considerable smoke production, were found to produce atmospheres up to 300 times more irritant than the same polymer in its non-fire retardant state, which burned cleanly.[73]

The picture is again confused regarding the role of specific chemical products in smoke. Table 2-8.5 shows some irritants identified in smoke atmospheres in order of their sensory irritancy, and includes data on the LC_{50}, due principally to lung inflammation, where these were obtainable from the literature. In some cases there was a considerable range of estimates for both sensory irritancy and lethality, and these are indicated. The lethality data are mostly for rodents, and have been normalized to a 30 min exposure period assuming that lethality is dose-related according to Haber's rule. Table 2-8.6 gives more detailed data for the effects in humans and animals of the best known fire irritants, hydrogen chloride and acrolein. One point to note from Table 2-8.5 is that irritants vary enormously in their potency, over five orders of magnitude. The most important irritants are probably the ones near the top of the list, including isocyanates (from polyurethanes), the unsaturated aledehydes acrolein and crotonaldehyde, and the first of the saturated aldehyde series—formaldehyde (which is produced by nearly all materials when decomposed). Irritants with a moderate potency, such as phenol and the halogen acid gases HCl, HF, HBr may be important in some fires if they are present in high concentrations, while it is difficult to conceive of any product with an RD_{50} of more than 1000 ppm, such as acetaldehyde, methanol, or any hydrocarbon, being of significance as a smoke irritant.

A difficulty in predicting the irritancy potential of fire atmospheres is that it is not known exactly how the various irritant components of an atmosphere interact, although there are indications that some degree of additive effect occurs. However, a more serious problem is that where comparisons have been made between the mouse RD_{50} of combustion atmospheres and their chemical composition as revealed by GC-MS analysis, the atmospheres in most cases turn out to be much more irritant than can be accounted for by a knowledge of their components. It is possible that small amounts of short-lived reactive chemical species with a very high irritant potency (RD_{50} <1 ppm) are responsible, and more work is needed in this area.[73,74]

Prediction of Incapacitation
Due to Sensory and Lung Irritation

In previous sections a model for the predicition of narcosis in fires has been presented. This model is realistic because it is based on the following facts:

1. Narcosis in fires gives two well-defined endpoints, loss of consciousness and death, both of which are likely to occur during the fire exposure.
2. Time and dose to these endpoints can be predicted from established data.
3. Narcosis in fires can be shown to depend upon a few known products.
4. The effects of each product are well known, and so to some extent are their interactions, so that predictions of narcosis based on gas profile measurements show a good agreement with observations of animal exposures.

However, it is very difficult to develop a predictive model for irritancy because:

1. Sensory irritation does not have a clear endpoint, but lies on a continuum of increasing eye and respiratory tract pain. Although this pain may be considered incapacitating, sensory irritation does not cause obvious incapacitative effects such as loss of consciousness, and it is not lethal, except under extreme conditions.

   Lung inflammation appears to cause relatively minor effects until near lethal levels of exposure are reached, so the main predictable endpoint is death, although this does not usually occur until several hours after exposure.
2. The identity and number of the irritant products important in fires are unknown. There is also a poor correlation between the composition of experimental fire atmospheres in terms of known irritants and their actual irritant effects on animals.
3. The concentration/time/dose effects of irritants and the degree of interaction between different irritants are also unknown.

For these reasons it is currently possible only to develop an approximate mathematical model to predict irritant effects from a knowledge of fire profiles in terms of known irritant products. An alternative is to base a model empirically on the effects on animals of smoke atmospheres produced in small-scale tests (which are described in a later section of this chapter).

To use small-scale test data, it is first necessary to ensure that the test fire model reasonably represents the decomposition conditions in the type of fire of interest (smoldering, early flaming, or postflashover). The concentration of irritants may then be represented in terms of mass charge or mass loss of the test material per liter of diluent air.

Sensory irritation may then be measured in terms of the mouse RD_{50} test. Although this test relates primarily to upper respiratory tract irritation, the results also show some correlation with eye irritation[70]; sensory irritation may also be assessed directly if required in terms of the severity of eye lesions occurring in the mice (ranging from lacrimation through chromodacryorrhoea to ocular opacity).

Lung edema and inflammation may be measured in terms of the mouse or rat LC_{50} where deaths occur principally after exposure and are accompanied by signs of inflammation such as increased lung weights or histopathological lung lesions.

Attempts may then be made to relate the RD_{50} and LC_{50} to possible effects in humans, but as described there are difficulties with these extrapolations. Fortunately, for most different materials the rodent RD_{50} and LC_{50} levels cover relatively narrow ranges (approximately one order of magnitude) under some given thermal decomposition conditions. For these reasons it is difficult with current knowledge to give guidance on irritancy for hazard modeling. Rough, general tenability limits can be presented, based upon observations of the sensory and lethal lung irritant effects of exposures of rodents and primates (and humans) to combustion product atmospheres from a range of materials decomposed under a range of conditions. It is therefore suggested that irritancy should be treated as follows:

1. Sensory irritation occurs immediately on exposure and is primarily concentration-related, not increasing with exposure duration.
2. All fire atmospheres are likely to be highly irritant, and at low concentrations the irritant effects should be regarded as adding to the obscurational effects of smoke through eye irritation, and possibly causing mild behavioral disruption by effects on the respiratory tract. At higher concentrations the disruptive effects of severe sensory irritation on vision and breathing may seriously limit escape capability, and as a rough approximation nonflaming atmospheres are likely to cause severe sensory irritation at a nominal atmosphere concentration (mass loss) of around 1 mg material/L air. This may occur when the optical density of the smoke is very low (<0.01/m) or even where there is no detectable smoke. It is therefore proposed that an NAC (mass loss) of 1 mg/L should be used as a tenability limit, although this may be modified in the light of RD_{50} or other data on the particular material and fire condition of interest.
3. Lung edema and inflammation are most likely to occur some hours after fire exposure, and severity of these is basically dose-related, approximately following Haber's rule.
4. Dangerous lung edema and inflammation are likely to occur following a 30 min exposure to an NAC (mass loss) of approximately 10 mg material/L, representing a Ct product of approximately 300 mg/L min. This figure may be replaced by LCt_{50} data on the materials involved in individual fires.
5. These exposure limits for sensory and lung irritation are based largely on exposures to nonflaming atmospheres. Flaming atmospheres may be somewhat less of an irritant, depending upon the efficiency of combustion and the severity of the fire.
6. Data for some common fire irritants are shown in Tables 2-8.5 and 2-8.6.

A Model for the Prediction of the Effects of Optically Dense, Irritant Smoke on the Eyes and Respiratory Tract

Although it is difficult to make precise predictions of the effects exposure to irritant smoke has on escape behavior, the degree of incapacitation, and the post-exposure effects on lungs, it is important to consider these effects in engineering design, and to be able to make some estimates of the likely effects of exposure. Herein is presented a current "best estimate" model for likely effects. Optically dense smoke affects wayfinding ability and the speed of movement of occupants, and a smoke barrier may be perceived as being impenetrable: effects that depend upon the concentration (optical density) of the smoke and its irritancy to the eyes and respiratory tract. Jin[62] found that, at an optical density of 0.5/m for non-irritant smoke, people behaved as if in darkness in terms of movement speed and behavior, and that irritant smoke was worse than non-irritant smoke.

Smoke irritants consist of inorganic acid gases (e.g., hydrogen chloride) and organic compounds, particularly low-molecular-weight aldehydes (e.g., formaldehyde and acrolein). More than 20 irritant substances have been detected in smoke and it is considered that others remain to be identified.

The first effect of exposure to smoke irritants is sensory irritation, which consists of painful stimulation of the eyes, nose, throat, and lungs. Sensory irritation depends upon the immediate concentration of irritants to which the subject is exposed rather than a dose acquired over a period of time, the effects lying on a continuum from mild eye irritation to severe eye and respiratory tract pain. In evaluating this aspect of irritancy, the aim is to predict what concentration of mixed irritant products is likely to cause such pain and difficulty in breathing that escape attempts would be slowed or rendered less efficient, and what concentration is likely to seriously disrupt or prevent escape (a degree of incapacitation equivalent to that at the point of collapse resulting from exposure to narcotic gases). For example, with regard to hydrogen chloride, it is considered that concentrations from approximately 100 to 500 ppm would be painfully irritant, and that the effects may slow escape but probably would not prevent it. However at approximately 1000 ppm and above, it is suggested that the effects might be so severe as to prevent escape.[122] In the absence of detailed information on irritant mixtures, it is assumed that all irritants would be additive in their effects, since they are all capable of causing damage to lung tissue.

In large-scale fire tests it is possible to measure inorganic irritants directly, but it is difficult to assess the degree of irritancy from organic products, which form a very important component. In general, the effects of organic irritants depend on the concentration of partially oxidized organic species in the smoke. For example, smokes from smoldering wood or polyolefines have a high organic content and are highly irritant; they are characterized by low CO_2/CO ratios and high smoke yields. Whereas under well-ventilated flaming conditions, the organic content of the effluents is low, and irritancy is low. In general, it is predicted that smoke from a mixed fuel source with an optical density/m of 0.5 would be strongly irritant to the eyes and respiratory tract.[5] However, for a given smoke density, there are differences between different types of fires; e.g., some people report that smoke from some fires, while dense optically, is relatively low in irritancy, while that, from other fires, is extremely irritant.

In order to assess the combined effects of irritants, a concept of fractional irritant concentration (FIC) has been developed, whereby the concentration of each irritant present is expressed as a fraction of the concentration considered to be severely irritant.[123] The FICs for each irritant are then summed to give a total FIC. If the total FIC reaches unity, then it is predicted that the smoke atmosphere would be highly irritant, sufficient to slow escape attempts. If the total greatly exceeds unity, then it is likely that escape would be prevented, and possible that collapse might occur due to static hypoxia from bronchoconstriction or laryngeal spasm. It is difficult to quantify these effects, because the data base on the effects of individual irritants or irritant mixtures on escape behavior in humans is poor, and there are no precise endpoints. On the basis of available data, current estimates of the concentrations of each gas likely to be highly irritant are as follows:

Toxic gas	Concentration
HCl	200 ppm
HBr	200 ppm
HF	120 ppm
SO_2	30 ppm
NO_2	80 ppm for 5 minutes; 25 ppm for 30 minutes
Total organics	0.5 OD/m^3

On the basis of the assumption that all irritants capable of damaging lung tissue are additive in their effects, the overall irritant concentration, FIC_{irr}, is then given by

$$FIC_{irr} = FIC_{HCl} + FIC_{HBr} + FIC_{HF} + FIC_{SO_2} + FIC_{NO_2} + FIC_{org}$$

The other important effect of irritants is that a proportion of those inhaled penetrate into the deep lung. If a sufficient dose is inhaled over a period of time a lung inflammatory response can occur, usually some hours after exposure. This may cause respiratory failure and death, or permanent lung damage in survivors. The 30-minute exposure concentrations likely to be lethal used for each irritant gas are as follows:

Toxic gas	Concentration
HCl	3800 ppm
HBr	3800 ppm
HF	2900 ppm
SO_2	400 ppm
NO_2	170 ppm for 30 minutes; 375 ppm for 5 minutes
Total organics	OD/m^3

The effects depend upon the exposure dose, which can be quantified approximately in terms of the product of concentration (c) and exposure time (t) to give the Ct product exposure dose (ppm · min). During a fire, when the concentrations of the toxic products vary with time, it is possible to predict when an incapacitating or lethal dose has been received by using the fractional effective dose (FED) method. For this method the Ct product doses for small periods of time during the fire are expressed as a fraction of the dose causing a toxic effect, and these FEDs are summed until the fraction reaches unity, i.e., when the toxic effect is predicted.

The fraction of lethal dose (FLD) for each irritant is calculated as the Ct product exposure dose during a period in the fire (e.g., ppm · min) expressed as a fraction of the lethal exposure dose. The lethal effects of the different irritants are assumed to be additive, on the same basis as the irritant effects, so that the total FLD_{irr} for each time period is given by

$$FLD_{irr} = FLD_{HCl} + FLD_{HBr} + FLD_{HF} + FLD_{SO_2} + FLD_{NO_2} + FLD_{org}$$

TABLE 2-8.8 *Classification of Toxic Hazards in Fires as Revealed by Large-Scale Fire Simulation Tests*

Fire	Rate of Growth	CO_2/CO	Toxic Hazard	Time to Incapacitation	Escape Time Available
1. Smoldering/non-flaming: victim in room of origin or remote	Slow	$\simeq 1$	CO 0 to 1500-ppm low O_2 15 to 21%, irritants, smoke	Hours	Ample if alerted
2. Flaming: victim in room of origin	Rapid	1000 decreasing toward 50	CO 0 to 1%, CO_2 0 to 10%, low O_2 10 to 21%, irritants, heat, smoke	A few minutes	A few minutes
3. Fully developed: (Postflashover) victim remote	Rapid	<10	*CO 0 to 3%, HCN 0 to 500 ppm some irritants, smoke and possibly heat	<1 min near fire, elsewhere depends on degree of smoke dilution	Escape may be impossible, or time very restricted. More time at remote locations.

*Concentrations depend on position relative to fire compartment.

CHEMICAL COMPOSITION AND TOXICITY OF COMBUSTION PRODUCT ATMOSPHERES

In the preceding sections the effects of individual narcotic gases and irritant chemical products known to occur in fires have been described together with their interactions. In this section the occurrence of these products in large-scale fires and laboratory-scale thermal decomposition experiments and the extent to which the toxicity can be interpreted in terms of these common products will be examined. In the next section the application of these data to fire scenarios and the interpretation of fire victim statistics will be examined.

From the point of view both of product composition and toxic hazard, it is possible to distinguish three basic types of fire situations:

1. Nonflaming thermal decomposition/smoldering fires.
2. Early/developing flaming fires.
3. Fully developed or postflashover fires.

The work of Woolley and his colleagues at the U.K. Fire Research Station[13] has shown that at midrange temperatures (400 to 700°C) such as those found in smoldering fires or in the vicinity of early flaming fires, materials are decomposed into pyrolysis products and oxidation fragments containing a mixture of narcotic and irritant gases and particulates. Under these conditions the highest yields of a variety of potentially toxic products are formed (Tables 2-8.7 and 2-8.8), and since incomplete oxidation is favored CO yields are high with CO_2:CO ratios approaching unity. For many materials the product mix remains fairly constant over this temperature range although the yields may increase somewhat with temperature.

Once flaming occurs, the high-temperature well-oxygenated flames of early flaming fires consume most of these combustion products to form simple, mostly innocuous products such as carbon dioxide and water. The CO_2 to CO ratios are very high initially, even up to the 500 to 1000 range. Since CO is only approximately 10 to 50 times as toxic as CO_2, it is conceivable that in this type of fire CO_2 could present more of a toxic hazard than CO. However, as the CO_2 concentration in the fire compartment approaches 5 percent and the O_2 concentration decreases towards 15 percent, the com-

bustion becomes less efficient and the CO_2 to CO ratios decrease to the region of 50 to 100, CO becoming a more important toxic factor. Nevertheless, as Figure 2-8.7 shows, the atmosphere obtained in a rapidly growing fire can contain narcotic concentrations of CO_2 (>5 percent), CO (>1000 ppm) and low oxygen (<15 percent O_2). With such flaming fires the yields of irritant, oxidized fragments and of CO are generally lower than under nonflaming conditions. However, due to the rapid growth and development of such fires the rate of production of these toxic products is often very high. Another factor that has been found to influence the yield and rate of production of toxic products (at least in small-scale experiments)[74] is the efficiency of combustion. When materials burn with efficient, nonsmoky flames, most CO, HCN, and organic irritants are consumed so the resultant atmosphere is relatively nontoxic. However immediately after ignition, and during intermittent flaming of some materials when combustion is inefficient and smoke production tends to be high, it has been found that high yields of CO, HCN, and irritants are produced, giving highly toxic atmospheres.

Under the severe conditions found in high temperature postflashover fires (exceeding 800°C) where oxygen concentrations are low, there is a major qualitative change in decomposition in that the main pyrolysis products break down into low molecular weight fragments and can contain high concentrations of narcotic substances such as CO and HCN, with low CO_2 to CO ratios. (See Table 2-8.8.)

Chemical product formation and the resultant toxicity can vary in different types and stages of fires,[12] as shown in a series of experiments. In the experiments, a number of materials were decomposed under a range of temperatures and oxygen supplies designed to simulate some of the conditions described previously. In one series of experiments, a study was made of the effects of temperature on the pyrolysis of three materials—polyacrylonitrile, flexible polyurethane foam, and wood. These were decomposed at 300, 600, and 900°C to cover as far as possible the full temperature range known to occur in fires, simulating conditions where oxygen supply might be limited. Another set of experiments were then carried out on a further range of polymers to examine the effects of nonflaming and flaming oxidative decomposition on toxicity over midrange temperatures (440 to 700°C). These conditions were chosen to embrace the conditions known to give the greatest complexity of products, particularly with polymers that are known to be sensitive to oxidation in product formation. In this series the test materials were rigid polyurethane foam, polypropylene, polystyrene, and nylon. In all cases, the sublethal toxic effects of exposure were evaluated using primates.

With the series of pyrolysis experiments, it was found that up to midrange temperatures a rich product mix occurred, while at 900°C clear atmospheres containing high yields of simple, low molecular weight products occurred. Thus for wood, which contains oxygen in its structure, the principal products at lower temperatures were oxidized organics, which caused upper respiratory tract and lung irritancy. There was a relatively low CO yield with wood, causing signs of narcosis at sufficiently high wood nominal atmosphere concentrations. Polyacrylonitrile produced relatively low yields of HCN and other nitriles, causing signs of cyanide intoxication.[12] With the flexible polyurethane foam, the lower-temperature atmospheres consisted principally of a dense yellow isocyanate smoke which was a powerful lung irritant capable of causing severe pulmonary inflammation after exposure.[12,56,75,76] Low yields of CO and HCN resulted with this foam.

At 900°C, the isocyanate smoke from flexible polyurethane foam was destroyed, giving a clear atmosphere consisting mainly of HCN and CO, the main signs being of cyanide poisoning. Polyacrylonitrile also produced high yields of HCN, causing cyanide intoxication, and wood produced clear atmospheres consisting principally of CO.

An example of the effects of oxidation on product formation and toxicity is given by the effects of polypropylene decomposition products. Under pyrolytic conditions, polypropylene decomposed to give a nontoxic atmosphere consisting of hydrocarbon fragments. (See Table 2-8.7.) Under nonflaming oxidative conditions, some of these fragments are oxidized to form a highly irritant atmosphere that caused upper respiratory tract and lung irritation during exposure and severe pulmonary inflammation after exposure.[12,56] There were also minor signs of CO-induced narcosis. Under flaming conditions the pyrolysis and oxidation products were partially destroyed, with the most important products being carbon monoxide, carbon dioxide, water, and some irritants. The principal toxic effect of this atmosphere was CO intoxication with some signs of irritancy. With flaming atmospheres, however, the nature and degree of toxicity depends upon the efficiency of combustion. By varying the decomposition conditions it is possible to substantially alter the toxicity.

From the animal exposures to these thermal decomposition and combustion product atmospheres, the major findings were that despite the great complexity in chemical composition of the test atmospheres, the basic toxic effects were relatively simple. For each individual atmosphere, the toxicity was always dominated either by a narcotic gas (CO or HCN) or by irritants. The toxicity of each atmosphere was therefore basically that of the major component that was present at its most toxic concentration. For narcosis, the effects of individual atmospheres were virtually identical to those of either CO or HCN as individual gases at the same concentrations as they were present in the smoke atmosphere.

There were some minor exceptions. For those atmospheres containing HCN, the narcosis appeared to be marginally less than that for an equivalent HCN/air atmosphere. However, the narcotic effects of atmospheres containing CO were identical to those of equivalent CO/air mixtures. The presence of low concentrations of hydrocarbons (which are potentially narcotic at high concentrations), and of CO_2 in the 1000 to 7000 ppm range did not contribute to the narcosis. The effects of irritant products on respiratory pattern for primates did not affect the pattern of CO or HCN narcosis. There may have been a marginal additive effect of CO on HCN toxicity, but in these cases the pattern of narcosis and times to incapacitation were very similar to those produced by equivalent HCN/air mixtures. Irritant effects did, of course, occur in these exposures in conjunction with narcosis, but there was no interaction between them, in that irritancy did not affect the progress of narcosis, and narcotic gases did not affect the response to irritants.

Implications for Human Narcosis Models for Fire Exposures

As a result of this work, it can be stated with some confidence that the narcotic effects of fire atmospheres should be predictable on the basis of the common narcotic gases CO, HCN, low O_2, and CO_2, and that the models derived from the work on individual gases should be valid. It is not possible with current knowledge to predict exactly the potential of a fire atmosphere to cause sensory irritation or lung inflammation from a chemical analysis of the product composition, but the model presented is considered a reasonable best estimate.

SMOKE

Smoke comprises the total effluents from a fire and consists of two parts: the invisible vapor phase and the visible particulate phase. From a toxicological standpoint, all of the narcotic fire products occur in the vapor phase, while irritant products may occur in both phases. The particulate phase consists of solid and liquid particles covering a wide range of particle sizes, depending upon the nature and age of the smoke. These particles may contain condensed liquid or solid irritant products; or irritant products including gaseous ones may be dissolved in liquid particles (as in acid mists), or be absorbed onto the surface of solid, carbonaceous particles. Particle size is of great toxicological importance since it determines how "deeply" particles penetrate into the respiratory tract and the patterns of subsequent deposition. Particles with a mean aerodynamic diameter of less than 5 μm are capable of penetrating deep into the lung, while larger particles tend to deposit in the nasal passages and upper airways. Generally speaking, the smoke from smoldering or nonflaming decomposition tends to consist mainly of small particles (less than 1 μm mean aerodynamic diameter, as in cigarette smoke) which are highly respirable. Smoke from flaming fires contains larger particles, particularly as it ages and the particles agglomerate; however, reports on fire victims usually record smoke penetration well into the lung.[30] At very high concentrations, smoke deposits may physically clog the airways. This could occur even with biologically inert particles at concentrations in excess of 5 mg/L, and is more probable with irritant smoke particles that are likely to acutely inflame tissues. Apart from the toxic effects of these particles on the lung, they may also be important in increasing the thermal capacity of the smoke and increasing the likelihood of lung burns (see next section of this chapter).

The other important physiological effect of the particulate phase of smoke is visual obscuration, which in conjunction with irritant effects on the eyes, may impair the ability of victims to escape from fires. As with sensory irritation this effect is concentration-related, and does not increase with prolonged exposure. It is best represented in a hazard model in terms of a tenability limit concentration. Smoke obscuration is usually expressed in terms of OD/m or extinction coefficient, K, (K = OD/m × 2.3). Jin[62] suggests tenability limits of extinction coefficient equal to 0.15/m (OD/m = 0.06) for subjects unfamiliar with an escape route, or 0.5/m (OD/m = 0.2) for subjects familiar with the escape route. Rasbash[117] suggests a 10 m visibility limit (equivalent to OD/m = 0.08), while Babrauskas[77] suggests a tenability limit of extinction coefficient 1.2/m (OD/m = 0.5).

THE EXPOSURE OF FIRE VICTIMS TO HEAT

There are three basic ways in which exposure of fire victims to heat may lead to incapacitation and death: by (1) heat stroke, (2) body surface burns, and (3) respiratory tract burns.

Heat Stroke (Hyperthermia)

If a subject is exposed to a hot environment, especially if the humidity is high and the subject is active, there is a danger of incapacitation and death due to hyperthermia. The time to effect and the type of hyperthermia depend principally upon the heat flux to which the subject is exposed, and are greatly affected by factors such as the amount and type of clothing and degree of work performed. A detailed analysis of the parameters that determine heat transfer to subjects over a range of environmental conditions and levels of activity, and the protective effects of different types of clothing, is given by Berenson and Robertson,[78] and Simms and Hinkley.[79]

Simple hyperthermia involves prolonged exposure (approximately 15 minutes or more) to heated environments at ambient temperatures too low to cause burns. Under such conditions, where the air temperature is less than approximately 120°C for dry air or 80°C for saturated air, the main effect is a gradual increase in the body core temperature.[80] Increases above the normal core temperature of 37°C up to approximately 39°C are within the physiological range and can occur at normal ambient temperatures during hard exercise, but once 40°C is reached consciousness becomes blurred and the subject becomes seriously ill. Further increase causes irreversible damage, with temperatures above 42.5°C being fatal unless treated within minutes.[78,81] The time taken to reach such a state depends upon a number of variables including those mentioned. Figure 2-8.14, adapted from Blockley,[80] shows approximate tolerance times for unclothed subjects at rest, under conditions of low air movement (30 m/min). At temperatures below 120°C tolerance is

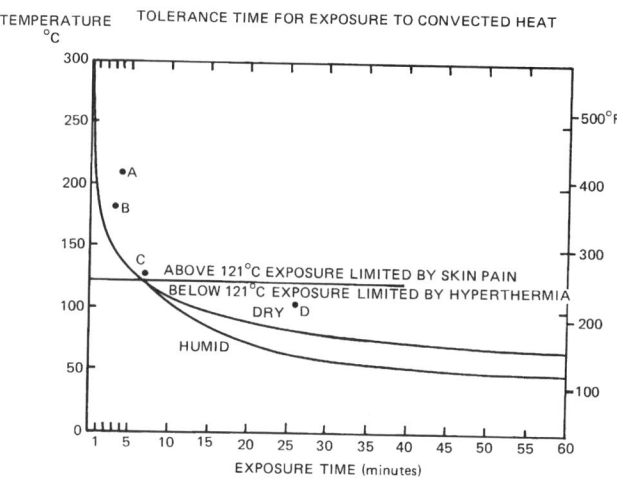

Fig. 2-8.14. *Thermal tolerance for humans at rest, naked, with low air movement (less than 30 m/min). Adapted from Blockley,[80] with added data from references 79, 82, and 83. See Table 2-8.9 for discussion of data points A to D.*

limited by hyperthermia, whereas above this temperature pain followed by burns become important. The data points a to d (for clothed subjects) were taken from various authors, and are added for comparison. At temperatures below 120°C evaporative cooling from sweat is important, so that humidity has a considerable influence on tolerance time. Clothing, therefore, offers some immediate protection at temperatures above 120°C but at lower temperatures may reduce tolerance time by impeding heat loss due to evaporative cooling. Details of the data points and authorship are given in Table 2-8.9.

Experiments conducted with pigs by Moritz et al[85] confirm the basic signs of hyperthermia, with death occurring principally due to circulatory collapse associated with severe cardiac irregularities (ventricular tachycardia).

A second situation described by Moritz et al[85] involves exposure to high temperatures for short periods (less than 15 minutes), and here hyperthermia is accompanied by cutaneous burns (in pigs at temperatures above 120°C). When deaths occurred soon after exposure to severe heat (within 30 minutes) the cause was considered to be due not to burns but to a rise in blood temperature. In this situation the exposure duration was insufficient to raise the body core temperature greatly, but if the temperature of the blood in the heart reached 42.5°C, the animal died within a few minutes from circulatory collapse.

It therefore seems that a victim exposed for more than a few minutes to high temperatures and heat fluxes (exceeding 120°C) in a fire is likely to suffer burns and die either during

TABLE 2-8.9 *Reported Tolerance Times for Exposures to Hot Air*

	Temperature (°C)	Time (min)		Reference	Letter in Figure 2-8.14
Dry air:					
	110	25		79	D
	180	3		79	B
	205	4	bare headed, protected	82	A
	126	7		83	C
Humid air:	32 at 100% RH	32	men working	84	

TIME TO SKIN BURN FOR EXPOSURE TO CONDUCTED HEAT

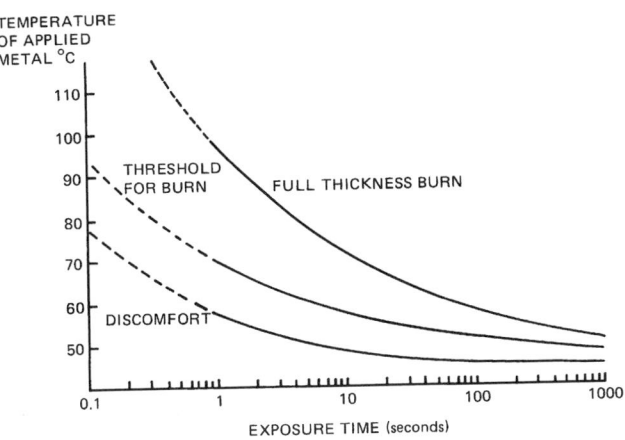

Fig. 2-8.15. *The relation between time and the temperature of metal to cause thermal injury to skin (values below 1 s are extrapolated). From Bull and Lawrence.*[88]

or immediately after exposure, due principally to hyperthermia. Victims surviving the hyperthermia phase may die later due to burns of the upper respiratory tract, particularly the larynx, or due to the secondary effects of skin burns. A victim or fire fighter exposed to temperatures unlikely to cause burns (less than 120°C) may also suffer heat stroke after a prolonged exposure (exceeding 15 min), especially if the humidity is high and the person is working hard.

Skin Burns

According to Buettner,[86] pain from the application of heat to the skin occurs when the skin temperature at a depth of 0.1 mm reaches 44.8°C, which agrees with the finding of Lawrence and Bull[87] that discomfort was experienced when the interface between a hot handle and the skin of the hand reached 43°C. The sensation of pain is followed soon afterward by burns, causing incapacitation, severe injury, or death depending upon their severity. The time from the application of heat to the sensation of pain, and from pain to the occurrence of burns of various degrees of severity, depends upon the temperature, or more properly, the heat flux to which the skin is exposed. The effects of heating the skin are essentially the same whether the heat is supplied by conduction from a hot body, convection from air contact, or by direct radiation.[86,88] Curves for the relationship between time and effect have been published for conducted heat from a "hot handle"[88] and radiant heat. The relationship between time and effect is exponential. (See Figure 2-8.15.) Thus for conduction from heated metal at 60°C, pain occurs after 1 s and a burn after 10 s, while at 80°C pain occurs at 100 ms and a burn after 1 s contact.

Pain therefore occurs when the difference between the rate of supply of heat to the skin surface exceeds the rate at which heat is conducted away by an amount sufficient to raise the skin temperature to 44.8°C. The thermal inertia of human skin is similar to that of water[86] or wood[88] with a value of $k\rho c$ for the surface (depth 0.1 mm) of 1.05 W/m·k. For the skin surface the rate of heat removal is not considered to be affected by blood supply[86] except for the fingertips, where blood flow may be sufficient to remove a significant amount of heat.[86] However, blood supply may have

some effect on the occurrence of burns, especially to the deeper layers of the skin.[88] Obviously rates of heating and the occurrence of pain and burns are greatly affected by the extent and type of clothing,[79,85,89] but only effects on naked skin are considered here. The temperature increase of the skin for the situation in which constant radiant heat is absorbed by the upper surface of the skin, or heat from a hot air current is applied to the skin, may be calculated as follows[86]

$$T - T_0 = 2Q\sqrt{t}/\sqrt{\pi k\rho c} \qquad (17)$$

where

T = final temperature of skin at 0.1 mm depth
T_0 = starting temperature of skin at 0.1 mm depth
Q = heat supply (W/m^2)
$k\rho c$ = 1.05 W/m·k; and
t = time (s)

Conducted heat: The effect of conducted heat is related to the temperature of the hot object and its thermal inertia, depending upon the interface temperature between the object and the body tissue at the skin surface,[88,89] as illustrated by the examples in Table 2-8.10.

A skin temperature of 43°C causes pain and some cellular damage, while a temperature of 60°C coagulates tissue protein. A brass block heated to 60°C will produce a partial thickness skin burn within 10 s, pain within 1 s and a full thickness burn after approximately 100 s.[88] The time/temperature relationships for these effects of conducted heat are shown in Figure 2-8.15.

Convected heat: For a victim attempting to escape from a fire, the most important sources of heat exposure are radiation from hot areas and convection from hot gases. Pain and the likelihood of skin burns occur at air temperatures above approximately 120°C. The rate of heat transfer from hot air to the skin depends upon the rate of ventilation, humidity, and the protective value of clothing as well as air temperature. The effects of these parameters are described by Berenson and Robertson,[78] and Simms and Hinkley.[79] However, for unprotected areas such as the head, data on naked skin are relevant, and the data shown in Figure 2-8.14 for temperatures above 120°C provide limits for tolerance to the painful effects of contact with hot air.

Apart from the problem of hyperthermia, dry air has been tolerated by humans as shown in Table 2-8.9. Moritz et al[85] state that dry air at 300°C injured unprotected skin within 30 s in pigs and dogs. Pigs also suffered burns at 150°C after 100 s and after 400 s at 100°C. However, it was considered that humans would be more resistant to burns, especially at temperatures below 120°C, due to the protective effect of sweating. Air with a high level of humidity not

TABLE 2-8.10 *Theoretical Contact Temperatures Between Skin at 35°C and a Selection of Hot Bodies at 100°C. From Bull and Lawrence*[88]

Material of hot body	Contact temperature (°C)
Mild steel	98
Glass	82
Wood	65
Cork	46

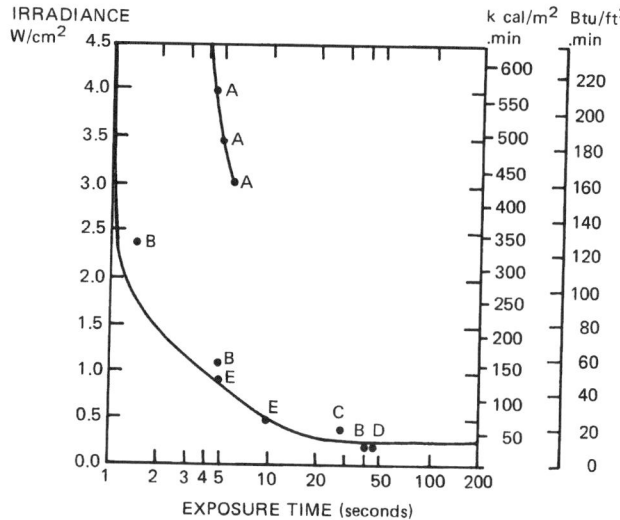

Fig. 2-8.16. *Time to severe skin pain from radiant heat. Adapted from Berenson and Robertson,[78] with added data points from references 77 through 79, 82, 86, 91, and 92. See Table 2-8.11 for discussion of data points A to E.*

only reduces or prevents heat loss through sweat, but also delivers more heat to the skin. Thus Moritz et al[90] found that steam at 100°C destroyed the epidermis of dogs within a few seconds.

Figure 2-8.14 shows curves for tolerance time of convected heat for both dry and humid air. A search for the original data used to produce these curves has not been successful, but upon careful consideration it seems likely that the humid curve must represent air that was humid (perhaps saturated) at *room* temperature, which was then heated subsequently, and was therefore nowhere near saturated with water vapor at higher temperatures. This must be the case because the capacity of air for water vapor increases dramatically at temperatures above 60°C, so that the amount of deliverable latent heat also increases. In practice 60°C has been found to be the highest temperature at which 100 percent water-vapor saturated air can be breathed. Since all fires produce a considerable amount of water from combustion, it is possible that the presence of water vapor may be an important neglected hazard in fires.

Radiant heat: For radiant heat, clothing also greatly influences tolerance times, but again data on naked skin are relevant to exposure of unprotected areas such as the head. Figure 2-8.16,[78] shows the relationship between time to skin pain and radiant heat flux. Data points A to E, taken from a number of authors (detailed in Table 2-8.11) have been added for comparison. Points B through E agree with the curve presented by Berenson, but data from one source (Perkins et al)[91] (points labelled A) deviate somewhat from the rest. From Perkins's data (which were produced by experiments where thermal injury was caused by exposing subjects to radiant heat from a searchlight), the heat fluxes for erythema (reddening of the skin said to coincide with pain[88]) appear rather higher than the heat flux limits for pain supplied by the other authors. This is possibly due to differences in the wavelength, and thus degree of penetration, of the radiation.[86] The searchlight data do, however, show the relationship between time to erythema, time to partial skin burn, and time to full thickness skin burn. The shape of the radiant heat tolerance curve suggests a fairly obvious tolerance limit for exposure to radiant heat of 0.25 W/cm² (2.5 kW/m²) which is that suggested by Babrauskas.[77]

Consequences of body surface burns: Apart from the immediate pain caused by exposure to heat and by skin burns, as well as the accompanying psychological shock and fear, incapacitation may result from body surface burns during or

TABLE 2-8.11   *Data on the Effects of Exposure to Radiant Heat*

Reference source	Heat flux W/cm²	Time to effect(s)			Letter in Figure 2-8.16
		erythema (or pain)	burn	full burn	
Perkins et al[91]	15	1	2.5	4	
	10	2	4	6	
	5	4	7	>15	
	4	4.5	9	>15	A
	3.5	5	9.5	>15	A
	3	6	10	>15	A
Buettner[86]	2.35	1.6			B
	1.05	5			B
	0.25	40			B
Veghte[82]			blisters		
	0.42		30		C
Simms and Hinkley[79]		unbearable pain			
	0.126	600			
	0.252	30 to 60			D
Dinman[92]	0.24	lower limit for pain after a long period			
	0.82	5			E
	0.48	10			E
Berenson and Robertson[78]	0.34	limit for blood to carry away heat			
Babrauskas[77]	0.25	tenability limit			

after a fire due to physiological shock. In this situation loss of body fluids into the burn results in circulatory failure and a fall in blood pressure, which may lead to collapse and even loss of consciousness.[10,85] The immediate effect of burns and the later chances of recovery depend upon a number of factors such as the site and extent of the burn, the depth of the burn, the age of the victim, and the treatment received.[88] While victims may continue to function for some time with severe burns, and survivals have occurred with up to 80 percent body surface area burns,[88] in general, if 35 percent or more of the body surface area is burned the chances of survival are low. Young adults generally have the best chance of survival, with a 50 percent chance of surviving a 50 percent body surface area burn, while children and old people are the most vulnerable, with a 50 percent chance of surviving a 20 percent body surface area burn in the elderly.[10,88] The depth of burn is classified on a scale of six degrees. First-, second-, and third-degree burns involve damage to the skin from which it can recover, while fourth-degree burns require skin grafts. Fifth- and sixth-degree burns involve destruction of muscle and/or bones, respectively.[10] Another scheme classifies burns as partial thickness skin burns, which will heal, or full thickness burns, which require grafts.[88]

If the victim survives the initial period of shock, death may occur over a period of up to a few weeks due to secondary effects on the brain, heart, lungs, liver, and kidneys.[10] The most common secondary effect and cause of death involves the lungs,[10,93-95] consisting of pulmonary edema resulting from effects on the circulatory system secondary to shock and metabolic acidosis. Postexposure treatment to replace body fluids and control acidosis are important in improving the prognosis for survival. If the victim survives the respiratory distress resulting from edema during the first week after exposure, pneumonia may then develop as a further, possibly fatal complication.[10,96-98]

This fatal damage to the lungs may occur following body surface burns when there has been no inhalation of heat or toxic gases. In many fire victims, however, damage to the respiratory tract and lungs results from a combination of all three causes.[10,96-98]

Thermal Damage to the Respiratory Tract

Thermal (as opposed to chemical) burns to the respiratory tract never occur in the absence of burns to the skin of the face.[57,98] Heat damage to the respiratory tract is even more dependent upon the humidity of inhaled hot gases than are skin burns. As a result of the low thermal capacity of dry air and the large surface area of the airways, which are lined with a wet surface and good blood supply, thermal burns are not induced by dry air below the top of the trachea. However, steam at around 100°C is capable of causing severe burns to the entire respiratory tract down to the deep lung, due to its higher thermal capacity and the latent heat released during condensation. These effects of inhaled hot gases are demonstrated by the work of Moritz et al[85] in which anesthetized dogs and pigs breathed hot air, flame from a burner, or steam, supplied through a cannula to the larynx. Dry air at 350°C and flame from a blast burner at 500°C caused damage to the larynx and trachea, but had no effect on the lung, while steam at 100°C caused burns at all levels. In these experiments the most important site of damage was the larynx, and death resulted from obstructive edema of the laryngopharynx within a few hours of expo-

sure. This work was taken further by Zikria[98] using steam burns in dogs, induced by a 15 s application of steam at 100°C via an endotracheal tube. The animals survived the initial affects and a number of phases of reaction were observed. The first phase consisted of necrosis and edema in the tracheobronchial airway, and early lung parenchymal edema within one hour. This was followed by increasing parenchymal edema, sloughing of the mucosa, and collapse of lung segments. The next phase after 24 hours consisted of bronchopneumonia behind respiratory tract obstructions.

All these features occur in fire victims, but it is difficult to separate the effects of thermal inhalation burns from edema and inflammation due to burns caused by irritant chemical smoke products, or edema secondary to body surface burns, all of which may be involved.[10,96-98] Thus fire victims with facial burns subjected to endoscopy have been found to have burns well into the respiratory tract.[57] If these lesions are caused by heat, it would imply that fire atmospheres resemble steam rather than dry air in terms of their thermal capacity. However, it is possible that such lesions are caused by chemical smoke irritants, which have been shown to produce fatal pulmonary edema and inflammation in the absence of heat.[56] Unfortunately, data on the thermal capacity and latent heat of actual fire atmospheres are not readily available, although it may be possible to calculate probable values from a knowledge of fire atmosphere temperature and composition.

The situation is therefore complicated, but from a fire engineering standpoint a number of basic points may be useful.

1. Thermal burns to the respiratory tract will not occur unless the air temperature and/or humidity are sufficient to cause facial skin burns.
2. Dry air at around 300°C may cause burns at the larynx after a few minutes. This may result in life-threatening obstructive edema of the larynx within an hour if not treated, although damage to the deeper structures of the lung is unlikely. It is possible that such laryngial burns may occur at lower temperatures down to approximately 120°C depending upon the duration of exposure, and breathing dry air at these temperatures would be painful. Laryngial burns followed by obstructive edema are common findings in fire victims, and are important causes of incapacitation and death during and immediately after fires.[10,96-98]
3. Humid air, steam, or smoke with a high thermal capacity or latent heat (due to vapor content or suspended liquid or solid particles) may be dangerous at temperatures of around 100°C, causing burns throughout the respiratory tract. It may be possible to predict the likely effects of hot-smoke atmospheres if thermal capacity or latent heat were measured.
4. In practice, fire victims may be affected by the inhalation of chemically irritant smoke, by hot humid gases, and by the secondary effects on the lung of body surface burns, all of which may combine to cause fatal respiratory tract lesions during the hours, days, or weeks following the fire exposure. However, these effects are probably less likely to be fatal during exposure to the fire atmosphere over periods of less than 30 minutes.
5. Heat flux and temperature tenability limits designed to protect victims from incapacitation by skin burns should be adequate to protect them from burns to the respiratory tract.

Model for the Prediction of Time to Incapacitation by Exposure to Heat in Fires

For use in the model, incapacitation by heat is regarded as (1) the point where exposure of the skin becomes painful, just before the burn threshold; or (2) the point where hyperthemia is sufficient to cause mental deterioration and threaten survival. The most important sources of heat in this context are radiant heat from the fire or hot surfaces, and convected heat from fire gases. From the curve of time to pain for exposure to radiant heat (Figure 2-8.16), there is an obvious cutoff point at 0.25 W/cm^2 below which exposure can be tolerated for at least several minutes, but above which the tolerance time is less than 20 s. It is, therefore, proposed that in the absence of data for beyond 3.3 min, 0.25 W/cm^2 (2.5 kW/m^2) should be set as a tenability limit.

For exposure to convected heat, Figure 2-8.14 shows a more obvious time-tolerance curve. Taking average values from the curve between humid and dry air the equation for time to incapacitation, t_{Ih}, at a temperature T is as follows

$$t_{Ih}(min) = \exp[5.1849 - 0.0273\,T(°C)] \qquad (18)$$

As with the toxic gas exposures, it would seem reasonable to regard the body of a fire victim as acquiring a "dose" of heat over a period of time during exposure, with a short exposure to a high temperature being more incapacitating than a longer exposure to a lower temperature (as in Figure 2-8.14). It is therefore possible to apply the same fractional incapacitating dose model as with the toxic gases, and providing that the temperature in the fire is stable or increasing, the fractional dose of heat acquired each minute during the exposure can be calculated as follows

$$F'_{Ih} = 1/t_{Ih} \qquad (19)$$

Although the radiant heat tenability limit and convected heat curve provide reasonable models, it would be better if the time curves for both heat sources were expressed in terms of heat flux, and treated as additive in terms of dose to incapacitation. For a more detailed treatment of these aspects, the reader should consult the literature.[78,79,86]

Example of a Calculation of Time to Incapacitation for Physical Fire Parameters and Irritancy

In a previous section the single armchair room burn shown in Figure 2-8.7 was used to illustrate how the model for prediction of narcosis could be applied to a practical fire scenario. To complete the incapacitation model it is necessary to include calculations for the effects of physical parameters (heat, smoke optical density), and mass loss concentration as an indication of irritancy. The curves for radiant heat, air temperature, smoke extinction coefficient, and mass loss during the first ten minutes of the armchair burn are shown in Figure 2-8.17.

Of the physical factors likely to affect a victim during the fire exposure, the majority are basically concentration or intensity-related rather than dose-related, and for these factors tenability limits have been set (radiant heat, smoke optical density, and sensory irritancy). The other two factors, convected heat and lung irritancy, are primarily dose-related, but lung irritant effects are likely to be relatively minor until after exposure. This leaves the fractional incapacitating dose of convected heat to be calculated. The average temperatures per minute during the first minutes of the fire are shown as histograms in Figure 2-8.17, with Table 2-8.12 showing the fractional incapacitating dose calculation.

Convected heat: The effects of exposure to convected heat increase dramatically in this type of fire as shown in Figure 2-8.17. Incapacitation, mainly due to skin pain and burns, is predicted sometime during the fourth minute, when the air temperature is 220°C. The situation then rapidly worsens, and it would seem likely that severe and probably fatal burns and/or fatal hyperthermia would be sustained by any victim remaining in the fire during the fifth minute. Even if the victims were protected to some extent by clothing, they would sustain burns to the face and probably fatal burns to the larynx. The occurrence of lung burns would depend upon the thermal capacity (principally the latent heat) of the smoke.

Radiant heat: From Figure 2-8.17, it would seem that the effects of radiant heat would be relatively minor in this fire compared to the effects of convected heat. The radiant heat peaks at just above 3kW/m^2 during the sixth minute, and therefore just exceeds the tenability limit. Nevertheless, the radiation alone would probably be sufficient to cause some burns and seriously inhibit escape during the sixth minute, and there would almost certainly be some degree of additive effect with convected heat.

Smoke: From the point of view of its obscurational effects, incapacitation by smoke is concentration-related rather than dose-related. For this series of chair burns, Babrauskas sets a tenability limit of extinction coefficient 1.2/m (OD/m 0.52).[77] This would give approximately 2 m visibility, which should be adequate for escaping from a room, and could be used as a tenability limit for input into the model. Incapacitation due to visual obscuration would occur at the end of the second minute. The smoke curve is rising very steeply at this point, with an OD/m of 1 at the beginning of the third minute. Escape would therefore become extremely difficult, and certainly slow during the third minute, unless the victim was familiar with the surroundings and able to find the exit in the dark.

Irritancy: As stated in the section of this chapter on irritancy, there are two factors to consider: the immediate incapacitation due to the painful effects of sensory irritation of the eyes and respiratory tract, adding to the obscurational effects of smoke and disrupting escape behavior; and the later inflammatory effects on the lung which may cause death after exposure.

The first consideration is whether the victim would be able to escape from the fire. In this context sensory irritation is the most important. This is concentration-related; to predict the irritancy of the smoke, it is necessary to know the RD$_{50}$ concentration of the atmosphere produced by the materials involved under the particular decomposition conditions existing in the fire. Most importantly, it is necessary to know the concentration/time profile of the fire products in terms of mass loss per liter of air (NAC mass loss). Although the mass loss curve for the armchair is shown in Figure 2-8.17, there are no data on the volume of air into which this mass was dispersed during the fire; so for the purposes of this example it will be necessary to make an estimate of possible mass loss concentration. Also, since the RD$_{50}$ of the polyurethane and polystyrene components of the chair under

TABLE 2-8.12 *Calculation of Fractional Incapacitating Accumulation of Convected Heat for the Single Armchair Room Burn Data shown in Figure 2-8.17. (Calculated according to Equation 19.)*

Time (min)	1	2	3	4	5	6
Average temp. (°C)	20	65	125	220	405	510
F_{lh}	0	0.033	0.170	2.273	355	6236
Cumulative F_{lh}			0.203	2.476		

Incapacitation occurs when F_{lh} = 1, and is therefore predicted during the fourth minute of exposure.

flaming conditions are unknown, it will be necessary to use estimated values.

In the discussion of irritancy, a general tenability limit for severe sensory irritation was set at a concentration of 1 mg/L NAC mass loss, and an incapacitating dose for serious postexposure lung inflammation was set at 10 mg/L NAC mass loss for 30 minutes (a Ct product of 300 mg·min/L). From the general conditions, the smoke curve, and the CO concentration curve, it is estimated that the tenability limit for sensory irritancy would be exceeded during the third minute, greatly adding to the deleterious effects of smoke on vision and escape behavior.

With regard to lung irritation, it is estimated that the average mass loss concentration over the first five minutes of the fire would be approximately 10 mg/L. If so, this would represent a fractional incapacitating dose of 50 mg·min/L, which would probably be insufficient to cause significant lung damage after exposure, compared to the more serious effects of heat exposure. However, if the average mass loss concentration over the first five minutes should reach

60 mg/L, serious effects on the lung would likely occur after and probably during exposure.

Interactions: In terms of physiological effects, it is likely that there would be some degree of interaction between narcosis and several of these physical factors, but it is likely that most would be relatively minor during the fire, except for some possible enhancement of pulmonary irritation due to the hyperventilatory effect of CO_2 during the fourth to sixth minute of the fire. A reasonable model can be used in which narcosis, sensory irritancy, and the effects of heat and visual obscuration can be treated separately. Interactions may be more important at the behavioral level. The interaction between sensory irritation and visual obscuration has been mentioned and there is some experimental evidence for such an interaction in humans.[62] After exposure, as mentioned in the section on heat, the effects of skin burns, respiratory tract burns, and chemical irritation (and even possibly CO narcosis) all combine to increase the probability of fatal pulmonary edema and inflammation.

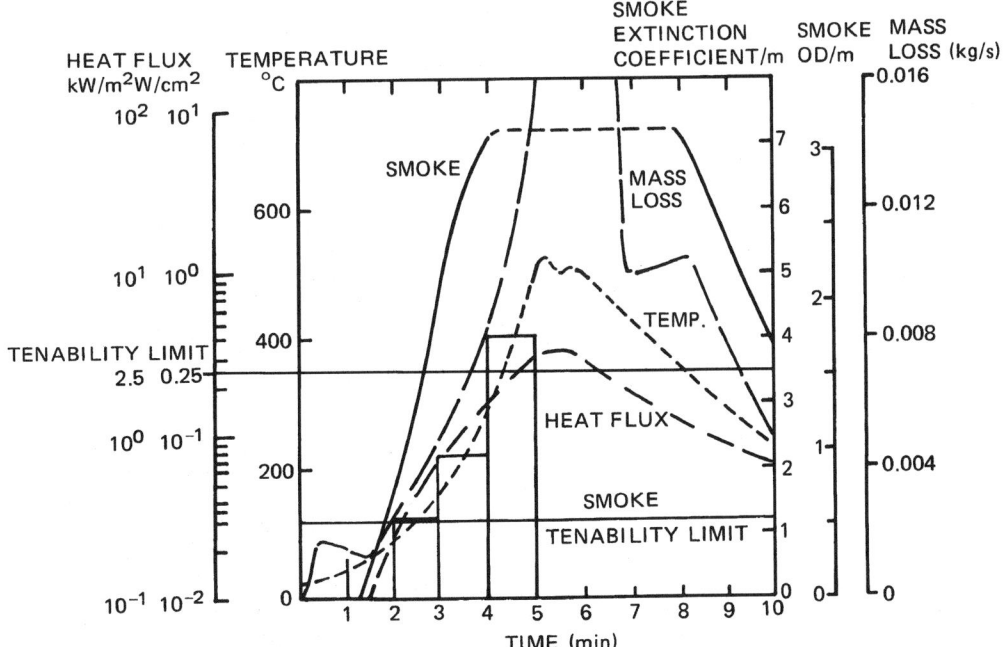

PHYSICAL PARAMETERS DURING EARLY STAGES OF SINGLE ARMCHAIR ROOM BURN

Fig. 2-8.17. Profiles for heat (radiant flux and temperature), smoke and mass loss rate during the first ten minutes of a single armchair (polystyrene, with polyurethane cushions and covers) room burn.[77] (Expanded detail from Figure 2-8.7.) Histogram shows average temperature each minute during the first five minutes.

Summary: From the analyses performed, the effects on a victim exposed to the conditions in the armchair room burn (Figure 2-8.7) are predicted as follows:

1. Toward the end of the second minute and beginning of the third minute, the smoke optical density and mass loss/liter would sufficiently exceed the tenability limits for visual obscuration and sensory irritancy to severely inhibit escape from the room.
2. During the fourth minute, the average temperature was 220°C, and sufficient heat would be accumulated in the skin surface to cause skin burns resulting in incapacitation.
3. During the fifth minute, a victim is likely to lose consciousness due to the combined effects of the accumulated doses of narcotic gases.
4. It is predicted that a victim escaping or rescued after the fourth minute would suffer severe postexposure effects due to skin burns, plus pulmonary edema and inflammation which might well be fatal (due to the combined effects of inhaled hot gases, chemical irritants, and the pulmonary secondary effects of skin burns). After the sixth minute, it is likely that a rescued victim would die at some time between a few minutes and one hour due to the effects of narcosis and circulatory shock.

It is unlikely that an otherwise healthy adult would be able to escape from a fire such as this if he/she remained longer that three minutes after ignition. However, three minutes is a long time in which to leave a room, so that providing the victim is awake and aware of the fire, is not otherwise incapacitated, and does not stay after two minutes in an attempt to fight the fire or rescue belongings, it is likely that he/she would be able to escape without serious injury. In the next section, data on real fire victims is examined in an attempt to relate fire conditions to actual injury and death statistics.

FIRE SCENARIOS AND VICTIM INCAPACITATION

From the point of view of both product composition and toxic hazard, it is possible to distinguish three basic types of fire situations (see Table 2-8.8):

1. Smoldering fires where the victim may be in the room of origin of the fire or a remote location.
2. Early flaming fires where the victim is in the room of origin.
3. Fully developed or postflashover fires where the victim is remote from the fire.

In the U.K., 80 percent of fire deaths and injuries occur in domestic dwellings, and in most cases the casualties occur in the compartment or origin of the fire. This class of fire is responsible for the highest incidence of deaths (60 percent) and a high incidence of injuries (39 percent). These fires occur mostly in living rooms or bedrooms, and in upholstery or bedding.[3] In these cases, fire is often confined to the material first ignited. The toxic hazard in such fires depends upon whether there is a long period of smoldering or a rapidly growing flaming fire.

Smoldering Fires

With smoldering fires, the decomposition temperatures are relatively low (~400°C) and materials are decomposed into a mixture of pyrolysis and oxidation fragments containing mixtures of narcotic and irritant gases and particulates.

Under these conditions, the highest yields of a great variety of products are formed, many of which are irritant.[12,13] Incomplete oxidation is favored and CO_2 to CO ratios approach unity, so CO is likely to be an important toxic factor. The formation of high yields of HCN is, however, not normally favored.[13]

Although toxic products are formed under these conditions, the rate of evolution is slow, smoke is seldom dense, and room temperatures are relatively low. A potential victim therefore has ample time to escape if alerted sufficiently early, but may be overcome by fumes after a long period of time if unaware of the danger, particularly if asleep. The main danger here is almost certainly narcosis by CO, with possibly a small contribution from low oxygen if the victim is in a room with a poor air supply.[12,99,100] It is not possible from fire statistics to determine how often this type of fire occurs, since in many cases smoldering fires become flaming fires before they are detected. However, it is likely that fires estimated to have burned for 30 min or more before discovery have involved long-term smoldering, and it may be relevant that for victims in the 19-49 age group deaths are 20 times more likely in this situation than for fires discovered within 5 min of ignition, which are often rapidly growing flaming fires.[3]

The ability of smoldering fires to build up concentrations of CO capable of causing incapacitation and death in potential victims has been shown in large-scale fire tests.[101] A good example of such a situation is presented by a series of tests carried out at NIST,[123,124] where two armchair types made from a standard and a fire-retarded polyurethane foam with cotton covers (combustible mass 5.7 kg) were burned in a simulated small apartment (volume 101 m^3) consisting of a burn room (11.8 m^3), connected via a corridor 12 m long to a target room (volume 12.08 m^3). (See Tables 2-8.13 and 2-8.14.) The armchairs were tested by flaming ignition of the seat back, and also by smoldering caused by one or two cigarettes placed in the seat angle for approximately 1 hour, followed either by spontaneous flaming or ignition from a flaming source. Under smoldering conditions, approximately 1 kg of foam was decomposed in just over 1 hour. The smoke layer had reached the floor after 1 hour, but there was a concentration gradient for smoke and toxic gases between the burn room and the target room. The major narcotic gas present was CO, which gradually increased in concentration in the burn room from 180 ppm during the first 13 min to 1000 ppm between 67 to 75 min. This was sufficient to have caused incapacitation (i.e., loss of consciousness) in just over 1 hour in the burn room, but probably not in the target room where the concentration was lower. When flaming ignition occurred the armchair burned very rapidly, and produced high concentrations of narcotic gases that would have been almost immediately fatal in the burn room. Within the target room an occupant would have become unconscious within less than 1 min and received a fatal dose within 2 min. The smoke in the system was also very irritant, and it is likely that anyone spending more than 1 hr in the burn room would have suffered serious and possibly fatal lung damage, even if that person had been rescued. This example illustrates the dangers of smoldering conditions, which can continue for several hours and spread lethal products throughout a building, creating danger for the sleeping, trapped, or otherwise incapacitated occupant. Since such fires often change to flaming before they are discovered, it is difficult to know the true incidence of incapacitation and death occurring during the non-flaming

TABLE 2-8.13 *Concentrations of Toxic Gases and FEDs in Burn Room for Smoldering Followed by Flaming Ignition of Standard Foam Armchairs*

Fractional effective doses of narcotic gases

Time (min)	0–13	13–27	27–40	40–53	53–67	67–75	75–76	76–77
Gas concentrations—burn room								
CO ppm	180	300	360	700	700	1000	10000	
HCN ppm	0	0	0	0	0	0	1320	
CO_2%	0.11	0.16	0.18	0.30	0.30	0.40	15.0	
O_2%	21	21	21	21	21	21	3	
Fractional effective doses for incapacitation								
FED_{CO}	0.006	0.010	0.012	0.024	0.024	0.035		
FED_{HCN}	0	0	0	0	0	0	Immed.	
VCO_2	1.019	1.032	1.037	1.069	1.069	1.096	fatal	
FED_O	0	0	0	0	0	0		
FED/min	0.006	0.010	0.013	0.026	0.026	0.039		
Σ FED	0.078	0.218	0.387	0.725	**1.089**	1.401		

Note: By 71 min the mass loss exposure dose of irritants was 600 g · m³ · min, which is likely to cause fatal lung damage.

phase of fires. For this example, both the standard and FR chairs would have caused incapacitation after 1 hr in the burn room, but due to its higher yield of CO and irritants, the FR chair would also cause incapacitation in the target room soon after, and death in the burn room after 1.5 hrs of smoldering. The standard foam death at both locations would occur within 1 min of the spontaneous transition to flaming after 75 min. These dangers can be overcome by the provision of detection and warnings, but the siting of detectors can be important for effective operation.

Flaming Fires

For flaming fires where the victim is in the room of origin, the hazard relates to the early stages of fire growth. Such fires often grow quickly, but even the most rapidly growing flaming fires take approximately 3 min to reach levels of heat and gases hazardous to life,[77] which should allow ample time to escape from a room, and of course most people do escape. As Figure 2-8.7 shows, the hazard in this situation relates to a number of factors, all of which may reach life-threatening levels simultaneously as the fire reaches the rapid phase of exponential growth. As stated previously, in the high temperature, well-oxygenated flames of early flaming fires, much of the thermal decomposition products are consumed to form simple. comparatively innocuous products such as CO_2 and water, the CO_2 to CO ratios being high initially and then decreasing to the region of 50 to 100 as the oxygen concentration decreases toward 15 percent. The yields and relative rates of production of CO_2, CO, and irritant smoke depend upon the rate and efficiency of combustion. As Figure 2-8.7 and Table 2-8.8

TABLE 2-8.14 *Concentrations of Toxic Gases and FEDs in Target Room for Smoldering Followed by Flaming Ignition of Standard Foam Armchairs*

Fractional effective doses of narcotic gases

Time (min)	0–13	13–27	27–40	40–53	53–67	67–75	75–76	76–77
Gas concentrations—target room								
CO ppm	0	0	100	270	550	800	2700	2000
HCN ppm	0	0	0	0	0	0	125	120
CO_2%	0.04	0.04	0.08	0.15	0.20	0.30	9.00	8.50
O_2%	21	21	21	21	21	21	13	14
Fractional effective doses for incapacitation								
FED_{CO}			0.003	0.009	0.019	0.028	0.099	0.073
FED_{HCN}			0	0	0	0	0.080	0.072
VCO_2			1.012	1.030	1.042	1.069	5.772	5.248
FED_O			0	0	0	0	0.021	0.012
FED/min	0	0	0.003	0.009	0.020	0.030	1.054	0.773
ΣFED	0	0	0.039	0.156	0.436	0.676	**1.730**	2.503

Note: By 71 min the mass loss exposure dose of irritants was approximately 300 g · m³ · min, which may cause some lung damage.

show, the atmospheres obtained in a rapidly growing fire can contain narcotic concentrations of CO_2 (greater than 5 percent), CO (greater than 1000 ppm) and low oxygen (less than 15 percent O_2), as well as dense irritant smoke from products escaping the flame zone. A victim in this situation is therefore likely to be confronted simultaneously by high temperatures and heat radiation, smoke, and high concentrations of CO and CO_2 accompanied by low O_2, any one of which could incapacitate and prevent escape.

The inability of victims to escape from such fires seems to depend upon a number of factors. Casualties include a higher proportion of young children and the elderly than does the general population. (In 1978, fatalities in bedding fires for those over 65 were seven times those expected from the distribution in the 1978 population.)[3,10] People who are incapacitated by a previous period of smoldering or by some other infirmity (such as a physical disability or alcohol or drug intoxication) are obviously more at risk.[10] However, there seem to be two other factors of importance: (1) the behavior of the victim and (2) the exponential rate of fire development.

In many cases, the victim has a short period in which to carry out the correct actions enabling escape, after which he/she is rapidly trapped. Some victims may be asleep during this critical escape "window," but there are also reports of situations where the victim was aware of the fire from time of ignition, but remained in an attempt to extinguish the fire, or for some other reason failed to leave before the phase of very rapid fire growth when heat and narcotic gases rapidly reach life-threatening levels. Another, perhaps surprising, finding is that victims often appear to be unaware of the fire, and remain to be discovered in a burned out chair or bed. The insidious nature of CO intoxication has been described; it also seems that irritant smoke products often fail to wake sleeping victims, although a sudden noise such as of breaking glass may do so. It may seem odd that acrid fumes may fail to alert sleeping victims, but a possible explanation may lie in the adaptation to sensory irritation during continuous exposure reported for the experiments with CS gas.[20] In smoldering fires, the concentration buildup of irritants is slow, allowing time for adaptation to occur. There may be no subsequent response to a high concentration which, if presented suddenly (as for example with smelling salts), would rouse the subject. Other victims appear to have roused themselves at some stage of the fire but have been overcome, again probably by CO or HCN before they are able to escape, and are found behind a door. There are also cases reported by survivors where a victim has attempted to extinguish a rapidly growing flaming fire, but failed to leave in time and is discovered near the fire having been overcome by fumes.[74,99] Unfortunately, reports of such effects on victims are largely anecdotal, and systematic studies of fire victim experience are lacking. The apparent anomaly of why so many casualties occur in the room of fire origin when theoretically there should be ample time to escape would seem to be a particular area that needs further investigation. The following pilot study has therefore been made of this problem.

Pilot Study of "Room of Origin" Deaths

As stated, the results of large-scale fire tests suggest that even in a worst-case situation there should be a period of at least three minutes before conditions in a rapidly growing flaming fire become untenable, and in many cases fire growth will take considerably longer. From a knowledge of the toxicity of fire products, it would therefore seem that a normal, healthy, waking individual should be able to escape from such situations without much difficulty, while a sleeping or otherwise incapacitated victim may be overcome by a smoldering or slowly growing fire, as well as by a rapidly growing fire. To test this hypothesis, an examination was made of the 1981 U.K. statistics, specifically for textile fires in dwellings for casualties in the 19 to 49 year age group, a group most likely to be active and able bodied. The data are summarized in Table 2-8.15 for fires estimated to have been discovered within 5 min of ignition (most likely to have been rapidly growing flaming fires) and for fires where the time to discovery is estimated to have been 30 min or more (most likely to have involved a period of prolonged smoldering before severe flaming).

The preliminary data show that there were 23,082 fires in the first category, but only 4 fatalities, while for the second category there were fewer fires (5,870) but 20 fatalities, a ratio per fire of 1:20. Obviously, a number of interpretations could be put on these data, but it does seem that people in this active age group are able to escape from rapidly growing fires in domestic-sized compartments. Fatalities are much more likely in fires that have undergone a period of prolonged smoldering, when victims may have been overcome by prolonged, low level exposure to narcotic fumes. If this is the case, perhaps there should be more concern about the ability of materials to continue smoldering, with toxic gas buildup over a long period of time, at least in the context of this class of fire.

Small Restricted Ventilation Fires in Closed Compartments

Closed room fires are hazardous situations, wherein a smoldering, or especially a flaming, fire quickly uses up the available oxygen, and as the oxygen concentration falls after a minute or so of burning the combustion becomes inefficient, producing a dense smoke rich in carbon monoxide and other toxic products. These, together with the lowered oxygen concentration in a room, can produce a rapidly lethal atmosphere. An example is a recent fire involving an adult and a four-year-old child. Both were in a small bedroom for a short time during which the adult went to sleep and the child is thought to have ignited a small piece of foam using a cigarette lighter. The fire was discovered after a few minutes, when the door was opened by a family member, who extinguished the very small fire with a bucket of water. Both the adult and child were dead, with blood carboxyhemoglobin concentrations of about half a lethal level. Based upon the dimensions of the room, it is calculated that the decomposition of approximately 0.5 kg of material would be sufficient to lower the oxygen concentration to 10 percent and give carbon monoxide concentrations of approximately 1 percent or more, which together with other toxic products would cause incapacitation and death within a few minutes.

TABLE 2-8.15 *Fatal Casualties in Room of Fire Origin (Textile Fires in Dwellings for 1981[3] Age Group 19–49 years)*

Time to Discovery (min)	Number of Fires	Number of Fatalities
< 5	23,082	4
> 30	5,870	20
		Ratio 1:20

With such small fires it would seem that early detection and warning, coupled with materials giving a slow fire growth, would greatly increase the probability of escape and survival.

Fully Developed Fires

The third scenario involves large, fully developed fires where casualties occur remote from the source of the fire. This type of fire has progressed beyond the stage of local growth and has spread from the material first ignited to others. The fire may still be largely confined to the compartment or area of origin, but large amounts of toxic smoke are formed, which spread throughout buildings, giving rise to lethal atmospheres remote from the actual fire. Apart from being a common occurrence in domestic dwellings, such situations often occur in public buildings where there may be major loss of life in a single incident. Materials in such fires are subjected to substantial external heat flux and in some cases to oxygen deficient environments. Under the severe conditions found in such high temperature, post-flashover fires with low oxygen concentrations, the basic pyrolysis products break down into low molecular weight fragments and can contain high concentrations of narcotic substances such as CO and HCN, with CO_2 to CO ratios of less than 10.[13]

Under such conditions, a building can fill rapidly with a lethal smoke capable of causing incapacitation and death within minutes. Fires where the victim is remote from the compartment of origin are responsible for the highest incidence of nonfatal casualties (48 percent) and a large proportion of deaths (37 percent) in the United Kingdom.[3] The victim is five times more likely to be killed by smoke than by burns, and is often unaware of the fire during the crucial early phase, so that the gases may not penetrate to the victim until the fire has reached its rapid growth phase and the victim is already trapped. The major causes of incapacitation and death in this type of fire are almost certainly narcotic gases, particularly CO, which can build rapidly to high concentrations (although the role of irritants in causing incapacitation and impeding escape attempts may be crucial to victim survival).

An example of such a fire is provided by some studies of the effects of the penetration of a large external fuel fire into the cabin of an airplane, as happened in the Manchester Airtours fire.[125] Table 2-8.16 shows the results obtained inside the cabin of a Boeing 707 containing a few rows of seats opposite an open doorway, outside which was 50 gal of burning aviation fuel.[123] The rapid involvement of the cabin contents gave rise to a dense smoke containing large amounts of carbon monoxide and hydrogen cyanide at a measurement point half way down the fuselage. Incapacitation is predicted just after 2 min followed rapidly by death, mainly from the effects of hydrogen cyanide (high concentrations of which were found in the blood of the Manchester victims).

Although in many large fires the original fuel and the major source of heat and toxic products may be the contents, a significant contribution may be made by construction products. Of great importance in some cases are surface coverings or components with a large surface area, such as doors or partitions. Surface coverings may contribute to flashover spread (as in the Dublin Stardust disco fire) and may release a bolus of toxic products very quickly, which may have a serious incapacitating effect on victims. An example would be vinyl wall coverings or the vinyl laminates used in aircraft cabins. PVC releases all its hydrogen chloride at a low temperature (approximately 250 to 300°C), so that, as a fire develops and the hot layer reaches this temperature, HCl may be suddenly released. In another aircraft fire test conducted by the FAA, high concentrations of HCl and HF occurred in the cabin atmosphere before other gases reached toxic levels.[122]

In general, although in some cases fire and heat may eventually kill victims, this is usually preceded by dense, highly toxic, smoke that can spread rapidly throughout a space or a building, and it is this that is usually responsible for the initial incapacitation of occupants, as well as being the cause of many deaths.

A common feature in many major fire disasters is a failure in early detection and effective warnings. In many cases the fire was detected at an early stage, and some attempts were made to deal with the fire, but often there was a failure to instruct people to leave while the fire was small and a failure to realize that a small fire grows with exponential speed into a life-threatening one. For example, in Summerland[126] the initial fire was considered non-threatening, and occupants were encouraged to remain seated rather than to leave. At the Manchester Woolworth's fire[127] people continued to eat in the restaurant area while the fire was growing on the other side of the sales floor. At the Bradford stadium fire people watched the early fire development at the end of the stand and did not begin to move until a late stage. At the Dupont Plaza hotel, the fire began in an unoccupied furniture storage area, but was discovered and fought at an early stage; however, mass evacuation of the hotel did not occur until after the fire went to flashover and started to spread. At the Beverly Hills Supper Club fire,[126] 20 min elapsed between discovery and evacuation instructions. At the Kings Crossing fire,[126] the escalator fire was burning for some time before attempts were made to close the station. At the Dublin Stardust disco, the fire grew unnoticed behind a partition. It could be said that, when a person is able to perceive a fire as a possible threat to their life, it may already be too late to escape. In the majority of these cases, evacuation occurred at a late stage and there were failures in the provision of accurate, authoritative, and informative evacuation instructions.

General Comment

The severe narcotic incapacitation and subsequent death of many fire victims are almost certainly due to the common narcotic gases. However, the importance of irritants in impeding escape is an important consideration, and it is not obvious from narcotic gas profiles why so many fatalities occur in the room of fire origin. Useful information may be obtainable from survivors who have experienced exposure to dense, irritant smokes, and from case studies of room of origin fires.

Possible Routes to Mitigation of Toxic Hazard

For smoldering fires it would be advantageous if materials were designed to self-extinguish, and if the formation of products other than CO during decomposition (such as oxidized hydrocarbon fragments or CO_2) could be encouraged. Early audible warning by smoke alarms may be particularly advantageous, as sound often appears to alert victims where the presence of irritant smoke or heat fails.

TABLE 2-8.16 *Average Concentrations of Toxic and Physical Hazards and Fractional Incapacitating Doses over 30-second Periods During Aircraft Cabin Fire*

Fractional effective doses of narcotic gases							
Time (min)	0.5	1.0	1.5	2.0	2.5	3.0	3.5
Gas concentrations							
CO ppm	8	34	282	1157	3326	8410	19490
HCN ppm	0	10	38	143	340	740	1380
CO_2%	0.0	0.0	0.4	1.2	2.8	4.1	6.0
O_2%	21	21	21	20	18	16	13
Fractional incapacitating doses							
FED_{CO}	0.00	0.00	0.00	0.02	0.06	0.16	0.38
FED_{HCN}	0.00	0.00	0.01	0.06	5.65	>10	>10
VCO_2	1.00	1.00	1.12	1.31	1.77	2.27	3.26
FED_O	0.00	0.00	0.00	0.00	0.00	0.00	0.01
FED/30s	0.00	0.00	0.00	0.10	10.10	>10	>10
ΣFED	0.00	0.00	0.00	0.11	11.10	>10	>10
Fractional effective doses of convected heat							
Temp °C	12	14	28	81	156	274	408
FED/30s	0.00	0.00	0.01	0.03	0.20	4.96	>10
ΣFED	0.00	0.00	0.01	0.04	0.24	5.20	>10
Radiant heat flux							
W/cm^2	0.10	0.12	0.14	0.18	0.23	0.28	0.57

Time to exceed smoke tenability limit: 1 min 40 sec
Time to incapacitation by narcotic gases: 2 min 15 sec
Time to incapacitation by convected heat: 2 min 45 sec
Time to tenability limit for radiant heat: 2 min 45 sec
Effects of irritants:
Over period between 1 and 4 min:　　average respirable particulates 6.7 mg/l
　　　　　　　　　　　　　　　　　average total particulates　　11.6 mg/l
　　　　　　　　　　　　　　　　　average HCl concentration 1027 ppm
　　　　　　　　　　　　　　　　　average HBr concentration 1228 ppm
It is considered that the oily, organically rich, particulate collected, with its very high acid gas content, would be highly irritant and extremely painful to eyes and breathing, causing incapacitation and impairing escape attempts. It is considered likely that these irritants reached high concentrations (approaching 1000 ppm total acid gases) early in the fire at approximately 1 to 1.5 min, from which time escape capability would be significantly impaired. It is likely that sufficient irritants would be inhaled up to 4 min to cause life-threatening post-exposure lung damage.

For early flaming fires where the victim is in the room of origin, any measure that limits the rate of growth once ignition has occurred will give a victim more time to extinguish a small fire or escape from a growing one.

For fully developed fires where the victim is remote from the point of origin, the most important mitigating factors are probably early warning and containment of the fire and gases within the original fire compartment.

The development of hazardous situations in a fire involves a whole range of factors, including fire development from ignition to the post-flashover spread of fire and smoke, toxicity, and the interaction of the fire with the structure and with passive and active fire protection, as well as escape-related factors, including detection, warnings, the provision of escape routes, wayfinding, physiological and behavioral impairment, and escape movements or rescue. In designing a system to be safe in fire all these factors should be considered, and the ultimate evaluation of safety depends upon whether it is possible to ensure, by performing a life-threat hazard and risk assessment, that the occupants can reasonably be expected to have escaped before they are exposed to levels of heat and smoke that may endanger health and threaten life.

THE USE OF SMALL-SCALE COMBUSTION PRODUCT TOXICITY TESTS FOR ESTIMATING TOXIC POTENCY AND TOXIC HAZARD IN FIRES

Essential Criteria for Test Methods

The evolution of toxic products in large-scale fires is determined essentially by two sets of parameters:

One set of parameters relates to the large-scale structure and development of fire scenarios that determine the growth and spread of the fire, the decomposition conditions under which products are formed, and their rate of evolution. It is not possible to model these features in a single small-scale test, although they are the most important factors determining the toxic hazard in a fire. This is the main reason why it has not been possible for experts to agree on standard test methods suitable for regulatory control of materials.

The other set of parameters that contributes to the determination of toxic hazard, and that can be modeled using small-scale fire tests (albeit only in a general way) is the set of thermal decomposition conditions encountered by materials at different stages or in different types of fires. Basically, the thermal decomposition products given off by any material depend upon the temperature and oxygen supply to which it is subjected, and whether it is flaming or nonflaming. Although it is possible to identify a number of subsets, there are three basic thermal decomposition conditions, namely:

1. Nonflaming oxidative/smoldering decomposition at mid-range temperatures.
2. Flaming decomposition in a local environment at mid-range temperatures.
3. High temperature/low oxygen conditions encountered in fully developed or postflashover fires.

If it can be demonstrated that a particular small-scale test method can model one or more of these conditions, then it has the potential, if used with a suitable bioassay procedure, to produce toxicity information that can be used as one item in a toxic hazard evaluation. To produce such information, a number of essential test criteria must be fulfilled:

1. The most important criterion for any small-scale fire model is that it should be capable of producing atmospheres with broadly similar compositions to those formed in one or more of the basic stages of a fire.
2. Following from this, it is essential that measurements should be made of the temperature and/or radiant flux to which the sample is subjected, the air/fuel ratio, the smoke optical density, and whether the sample is flaming or nonflaming. The chemical composition of the atmosphere must also be characterized as fully as possible, the minimum being in terms of CO, CO_2 (with calculation of the CO_2 to CO ratio), O_2, and HCN (if the material contains nitrogen), or hydrogen halides (if F, Cl, or Br are present). These measurements are important for two reasons: first, because they define whether one of the basic fire conditions is being adequately simulated; and second, because they should enable the key products responsible for the animal toxicity to be identified, or at least make it possible to decide if unusual toxic products are present.

    The time/concentration profiles of the key toxic products can then be measured in large-scale fire tests, or predicted by mathematical fire modeling. Estimates of likely toxic hazard in realistic fire scenarios can then be attempted.

    If possible, a fuller product analysis including GC-MS measurements of the profiles of organic products should be made for more accurate correlations with the observed toxicity. In addition, comprehensive measurements of product composition in small-scale tests enable comparisons to be made with product profiles in large-scale fire tests, and a judgment of how well the small-scale decompositions are able to reproduce the chemical "cocktails" present in different large-scale fire conditions.[13]

3. The bioassay method must be capable of detecting and measuring the particular types of toxic effects experienced by human fire victims. It must also be possible to make a reasonabe extrapolation from the toxic effects in the animal model to those likely to occur in humans. It is therefore essential to determine both the qualitative and quantitative aspects of toxicity, and the time during or after exposure when they occur. People in fires experience incapacitating effects that may crucially impair their ability to escape, causing them to remain in the fire to be killed later by heat or CO, so a simple body count (LC_{50}) animal test without any description of preterminal toxicity is of very limited usefulness for input to toxic hazard models.

4. Some estimate of dose must be made. This should be reported in terms of concentration and duration of exposure as mass charged and mass consumed of material per liter of diluent air, and concentrations of basic toxic products as described previously. By carrying out tests at different concentrations, estimates should then be made of dose/time/response relationships.

Practical Methods for Toxic Hazard Assessment

There are essentially two ways in which toxic hazard in fire can be assessed:

1. From large-scale fire tests that include measurements of the concentration/time profiles of the major toxic gases, and existing knowledge of the toxic effects of these gases.
2. From a battery of small-scale tests and mathematical models, or simple large-scale tests. The essential components are:
    (a) the toxic potency data for materials [lethal mass loss exposure dose ($g \cdot m^{-3} \cdot min$)] obtained from small-scale tests using animal exposures or analytical methods, and
    (b) the mass loss/concentration curve for the fire.

Of these, full-scale simulations and large-scale tests are the most valuable, since they enable the first two major parameters (fire growth and product yield) to be measured directly. For the third parameter (toxicity), an algorithm for calculating times to incapacitation and death for humans from toxic gases, heat, and smoke obscuration from large-scale fire data has been developed, and a simplified version is presented in ISO/IEC TR 9122-5.[128] Full-scale tests are already in use for a number of purposes (such as the furniture calorimeter and room calorimeter used for heat release measurements). With the addition of a few simple gas measurements, they can be extended to provide the data for full-scale hazard assessment. Although the use of full-scale tests for regulating fire performance has been criticized on grounds of cost, this approach has been used successfully in California.

The other notable method is to make use mainly of small-scale test data. Small-scale tests suffer from the difficulty that they are several steps removed from a full-scale fire, so that it is important to determine the validity of the parameters measured with respect to real fire conditions.

It has been claimed that some small-scale tests can be used to give a direct indication of the toxic hazard presented by a material in a full-scale fire, including elements of ignitability, fire growth and toxic potency. The UPITT test and the NIBS test are claimed to be usable in this way to produce hazard rankings for materials. However, small-scale tests cannot successfully model the complex growth and development of a full-scale fire, so that such methods are considered invalid. It may however be possible to extract a number of characteristics of the fire behavior of a material from a single test protocol, and use these as inputs to mathematical models for making hazard assessments. An example is the use of the cone calorimeter for measurement of both heat release and toxic gas yield from a material.

Nevertheless, the main job of small-scale toxicity tests is to provide data on the toxic *potency* of the combustion products from materials so that the data can be used as input to toxic hazard analysis in conjunction with fire growth data obtained from other sources.

The lethal toxic potency of combustion products from materials can be measured in small-scale tests in terms of the LCt_{50}, the lethal exposure dose to rodents. This is expressed in terms of the mass loss concentration of products (the mass of material decomposed in the test) divided by the volume of air into which they are dispersed (g/m^3) and multiplied by the time for which the animals were exposed, to give the exposure dose ($g \cdot m^{-3} \cdot min$). It is possible to quote the toxic potencies of materials in these terms for most small-scale test protocols, so that the results from different tests can be compared directly and so that the data can be used in conjunction with mass loss data from full-scale fire tests or model calculations to make toxic hazard assessments. The toxic hazard presented by the materials in full-scale fires can then be estimated in terms of the time during a fire when a victim will have received a lethal exposure dose, provided that the decomposition conditions in the small-scale test are the same as those in the full-scale fire. This is done by the fractional effective dose (FED) method, in which the mass loss exposure dose generated each minute during the full-scale fire (either measured or calculated) is expressed as a fraction of the lethal mass loss exposure dose for the material, obtained from the small-scale toxicity tests. When the integrated fractions during the fire reach unity, death is predicted. This method has been developed for use in toxic hazard assessments in British Standards[129] and in ISO/IEC.[128]

Examples of Small-Scale Test Methods

A variety of different methods have been used to generate combustion product atmospheres and evaluate their toxicity, and it is not possible here to describe these methods or the results obtained from them in detail. There are, however, three test methods that are in relatively wide use in the United States and in Europe—the NBS test method,[7,102] the DIN test method,[103–105] and the University of Pittsburgh test method.[69] An excellent description of these and other methods, and of the results obtained from them, is given in Kaplan *et al.*[11] A brief description of the principles of these three methods and their validity in predicting effects of exposure to large-scale fire environments follows. The methods are illustrated diagramatically in Figure 2-8.18, parts (a) and (b).

National Bureau of Standards (NBS) Test Method

The NBS test method[7,102] is the simplest of the three. The test apparatus consists of a sealed rectangular polymethylmethacrylate chamber (200 L volume) containing a cup furnace set into the floor at one end. The animals (rats) are exposed nose only and placed in restraining tubes that are plugged into the sides of the chamber. Exposures are carried out under nonflaming conditions 25°C below the ignition temperature of the sample and under flaming conditions 25°C above the ignition temperature. The sample is placed into the preheated cup with the animals already in position,

and they are exposed to the atmosphere in the chamber for 30 minutes. For this test there is thus an initial period when the concentration of decomposition products in the box increases as the sample decomposes, followed by a period when the atmospheric composition remains relatively constant for the remainder of the exposure. The concentration of products in the chamber is varied by placing different amounts of sample in the cup. This type of method is referred to as a "static" method, since there is no forced airflow over the sample and no ventilation of the chamber. It is possible to characterize the atmosphere approximately in terms of mass charge per liter of air by assuming a constant composition for the duration of the test. The atmosphere also can be characterized in terms of mass loss by weighing the residue after the test, or by placing the cup on a load cell. Measurements are also made of the chemical composition of the atmosphere in terms of the principle products. The animal toxicity is usually quoted in terms of the LC_{50} concentration of mass charge/L but it is also common to monitor the narcotic effects on the animals by means of the leg flexion test. Effects upon respiration (such as irritant effects) could be monitored if desired.

The main advantage of this method is its simplicity, but it has the disadvantage that there is little control over the decomposition conditions. Another disadvantage is that the composition of the atmosphere does not remain constant, since there is a variable period during which the sample is decomposing, followed by a period when the smoke is aging and some products may condense onto the chamber walls. In tests where the decomposition product atmosphere has been compared to that produced in large-scale tests,[13] it has been found that the method gives a reasonably good simulation of the products from a smoldering or nonflaming fire, and also of the conditions during the early stages of a flaming fire. However, it was not possible to model the decomposition conditions of a high-temperature, fully developed fire, since when the cup is heated to high temperatures the pyrolysis products tend to escape from the cup before they are fully decomposed. In general, it should therefore be possible to use data from this method as one item in a toxic hazard assessment for nonflaming or early flaming fires.

University of Pittsburgh Test Method

In this method[69] the sample is also heated statically in a cylindrical furnace (early versions of the test used a tube furnace). However, in this case a flow of air is maintained over the sample and then mixed with diluent air before being passed through the animal (mouse) exposure chamber. In this way fresh products are continually passed to the animals under dynamic airflow conditions as they are generated over a 30-min exposure period. The unique feature of this method is that the sample is not maintained at a constant temperature, but is heated by means of a ramped temperature profile, the temperature increasing at 20°C/min. The animal exposure starts when the sample begins to decompose and lose weight as indicated by a load cell. The mice are exposed in the head-only configuration, as with the NBS method. In the University of Pittsburgh method the composition of the atmosphere changes continuously throughout the 30-minute animal exposure, and the atmosphere usually changes from nonflaming to flaming at some stage. The concentration of the decomposition products can be varied by changing the mass of material placed in the furnace (although presumably this could also be achieved by altering

(a) University of Pittsburg Apparatus.

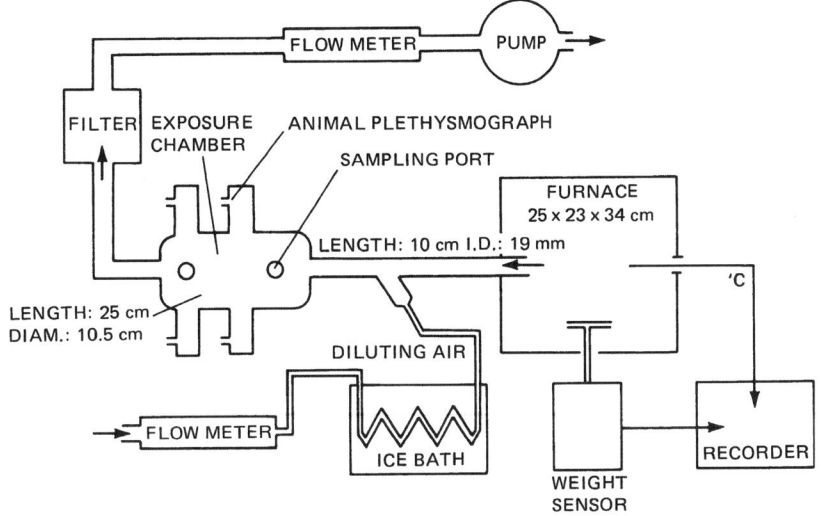

(b) National Bureau of Standards (NBS) Apparatus.[7]

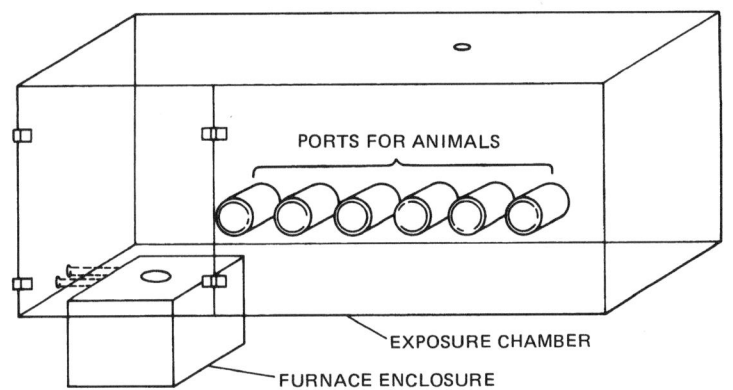

(c) DIN 53436 Apparatus.[115]

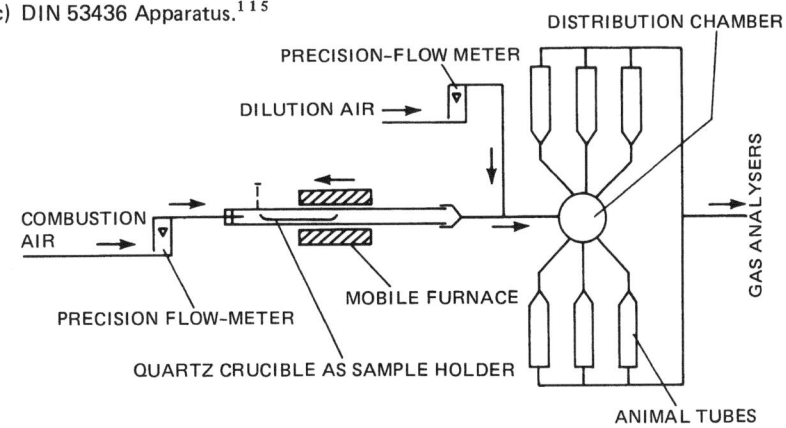

Fig. 2-8.18.   Laboratory-scale combustion toxicity test methods.

the flow of diluent air). It is possible to measure the changing profile of products to which the animals are exposed throughout the test, and also to monitor toxic effects such as narcosis or irritancy, mainly from recordings of respiratory pattern. However, in its more recent application, the results of the test are usually expressed in terms of the LC_{50} in grams of material charged, which is used to rank different materials.

The main advantage of this method is that it theoretically covers a number of different decomposition conditions within a single test run, ranging from low-temperature nonflaming to high-temperature flaming. It is also said that this situation mimics the conditions in a real fire where materials begin by being cold, and are then heated up until they pyrolize and eventually flame. A feature of the test is that the time/temperature increase taken from the start of the decomposition run to the occurrence of the evolution of smoke, toxic effects, and flaming may be used as criteria for judging materials. Thus a material that flames early and/or produces smoke early may be judged more hazardous than one that does not start to decompose until a high temperature is attained.

Although ramped heating of a sample may provide a useful model for the specific situation where a material is subjected to a slowly rising temperature, it cannot be said to mimic the changing conditions in a fire. In a fire, a material or its immediate pyrolysis products may be subjected to any of a variety of conditions of temperature or oxygen supply under nonflaming or flaming conditions, and the way in which these conditions change is governed by the nature of the large-scale fire scenario. In order to model these various conditions, it is necessary to subject separate samples of test material to a range of different temperatures (or radiant fluxes) and oxygen supplies. The main problem with ramped heating of a sample is that it does not submit the whole material to the necessary range of conditions, since it causes fractional decomposition of the material. Products evolved at relatively low temperatures will not therefore be present at a later stage to be involved in flame; neither will they be subjected to high temperatures, since they will have left the furnace before higher temperatures are achieved. This may have a profound effect upon the kind of products evolved and hence the toxicity.

Another disadvantage of this test method is the difficulty of characterizing dose, since the composition of the atmosphere changes throughout the test. For comparison with other methods, it would be possible to calculate a very approximate nominal atmospheric concentration in terms of mass charged/L if it is assumed that the decomposition averages out over the duration of the test. A better estimate may be made of the atmosphere concentration in terms of mass loss/L if the mass loss as measured by a load cell is integrated over the exposure period. This might give a reasonable measure of dose to enable an LC_{50} to be calculated, but other estimates of toxicity are complicated by the changing nature of the atmosphere. Thus, if death or narcosis were to occur at 30 min for example it is not possible to determine whether this would be due to the delayed effect of a product evolved at 5 min or an immediate effect of a product from 29 min.

There are thus some difficulties with this method both as a fire model and as a toxicity assay. But, if the method is backed by a full profile of material mass loss and product concentrations, plus qualitative and quantitative estimates of toxic effects throughout and after the exposure, it may be possible to apply the data in certain special situations.

German DIN 53 436 Test Method

This method[103-105] is widely used both in the United States and in Europe. It employs a fully dynamic system in that fresh material is decomposed at a constant rate throughout the test and fresh atmosphere is supplied continuously to the animals. The method has been used with rats, mice, and primates; rats being the principal test animals. The principle of the method is that a strip of test material is placed in a silica tube under a current of air (or nitrogen for studies of pyrolysis) and a traveling annular furnace is moved along the outside of the tube at a constant rate, thereby continuously decomposing the sample. The products are expelled from the silica tube, whereupon they are mixed with diluent air and passed through the animal exposure chamber for 30 min. The rats are exposed in the head-only manner as with the other methods. The sample weight and volume, the primary air supply to the sample, the diluent air, the decomposition temperature, and the concentrations of products in the exposure chamber are measured. The test may be repeated over a range of temperatures under nonflaming or flaming conditions, and the concentration of the products is varied principally by changing the diluent airflow. This has the advantage that the decomposition conditions remain constant as the product concentration is changed, which is not necessarily the case with the other two methods. However, it is also possible to change the product concentration by altering the furnace load, if necessary. The animal toxicity is usually reported in terms of the LC_{50}, which is expressed in terms of mass charged (mass entering the furnace/min)/L diluent air. It is also possible to calculate the concentration in terms of mass loss from the mass of residue remaining after the test run, since the rate of decomposition is constant. Sometimes the results are expressed in terms of the volume of material consumed. In addition, it is possible to monitor the toxic effects on the animals during and after the test as with the other methods.

The great advantage of this method is its versatility, and in theory it can model any of the three basic fire conditions: nonflaming, flaming, and high temperature, as well as variations on them. It is also possible to carry out static ramped temperature decompositions similar to the Pittsburgh method, if required. In practice the method has been found to provide a good correlation with large-scale nonflaming decomposition conditions, and at high temperatures it provides a reasonable model for the conditions of a fully developed flaming or nonflaming fire (high temperature-low oxygen).[13] However, under standard conditions it does not provide a good model for an early flaming fire (which is reasonably modeled by the NBS method) as the oxygen concentration in the furnace tube tends to be low. Another problem is that the flame tends to travel along the sample more rapidly than the furnace, giving an uneven decomposition. These problems can be overcome by varying the fuel-to-air ratio in the furnace tube and by altering the rate of travel of the furnace.

On balance, the DIN method is reasonably good, and the data obtained could be applied to a number of stages of a fire as one item in a hazard assessment. However, like all small-scale tests it cannot mimic the changing and developing conditions of a large-scale fire; neither can it test materials or objects in their end-use configuration, all of which can have an enormous impact on a developing toxic hazard.

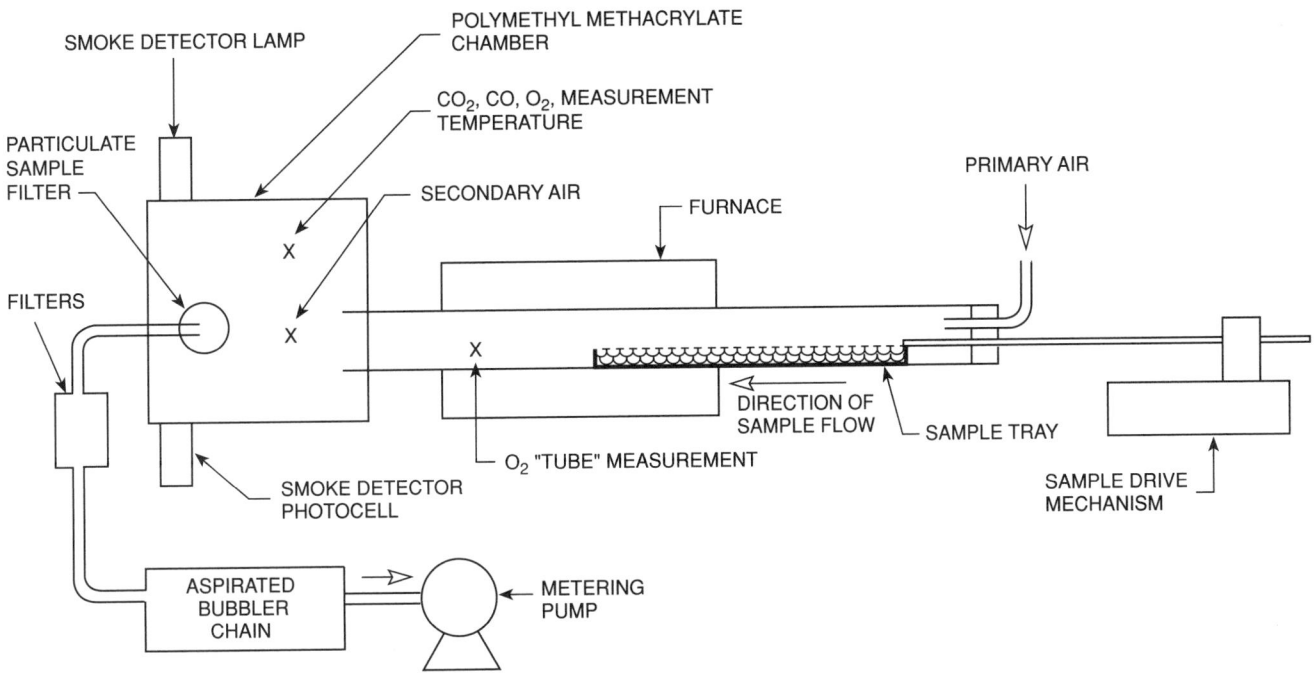

Fig. 2-8.19. Improved tube furnace method.

Second Generation Test Methods

Since these small-scale test methods were developed, there has been considerable progress in understanding of how toxic hazard develops in fires, and in particular that hazard depends to a large extent on the general fire properties of materials (in terms of ignitibility, flame spread, rate of fire growth, and smoke evolution), as well as the specific toxicity of the combustion products. Also, critical examination of the fire models used for toxicity test methods (and for small-scale tests for other fire properties such as smoke) has led to the recognition that the models are somewhat inadequate, particularly for the main fire condition of flaming. Another difficulty is that toxicity data are of little value unless they can be related to a range of physical and chemical parameters necessary to characterize the thermal decomposition process, as described in the previous section of this chapter.

There is, therefore, a case for arguing that for many material-based fire properties a good small-scale test method depends upon a decomposition model that can be convincingly related to the essential features of large-scale decomposition conditions. If such a model could be developed, it could be used to measure simultaneously a number of material-based fire performance parameters ranging from ease of ignition, through growth and heat release characteristics, to smoke and toxicity.

A second generation of small-scale fire test methods is being developed, incorporating, hopefully, some of the best features of existing methods, and designed to measure a range of parameters.

Cone calorimeter: A second generation small-scale test method currently under development is the National Bureau of Standards (NBS—now NIST) Cone Calorimeter,[106] which has been developed primarily as a heat release apparatus, but which may also offer the possibility of measuring ig-

nitibility, smoke evolution, and toxicity. This would not only enable a range of parameters to be measured simultaneously, but would enable the separate parameters to be related, hopefully providing a more comprehensive data set for comparison with, and inputs for modeling of, large-scale fire conditions.

In practice the Cone Calorimeter has not proved to be very suitable for measuring toxic potency. It is capable only of reproducing the decomposition conditions in very well-ventilated fires, and the products are subjected to a very large dilution, so that measurement of toxic species is difficult. Attempts are being made to enlarge the range of fire types and stages addressed by modification of the apparatus.

NIST radiant method: A second generation version of the NABS cup furnace method has been developed in which the cup furnace was replaced by a radiant panel heater unit. This method has been developed to address the conditions in post-flashover fires, but as with the cup furnace version, and the Cone Calorimeter, the combustion process tends to be too well ventilated to reproduce the conditions typical of these fires, which are usually rather vitiated.

Tube furnace method developed at the UK Fire Research Station:[130] The most recently developed method intended to address some of the deficiencies of older methods is a tube furnace method based upon the same concept as the DIN tube furnace method. The new method employs a strip of sample being advanced through a standard tube furnace under a stream of air. (See Figure 2-8-19.) The products are expelled into a chamber where they are diluted with secondary air, and analytical measurements are made or animals can be exposed nose-only. By using a range of different temperatures and airflow rates it is possible to reproduce all the different fire stages and types defined by ISO (and others), including low-temperature non-flaming oxidative decomposition, well-ventilated flaming, and high-temperature

TABLE 2-8.17 *ISO/IEC TR 9122-4 General Classification of Fires*[131]

Fire	Oxygen* %	Ratio† CO_2/CO	Temperature*	Irradiance‡ kW/m^2
1. Decomposition				
(a) Smoldering (self-sustained)	21	N/A	<100	N/A
(b) Non-flaming (oxidative)	5 to 21	N/A	<500	<25
(c) Non-flaming (pyrolytic)	<5	N/A	<1000	N/A
2. Developing fire (flaming)	10 to 15	100 to 200	400 to 600	20 to 40
3. Fully developed (flaming)				
(a) Relatively low ventilation	1 to 5	<10	600 to 900	40 to 70
(b) Relatively high ventilation	5 to 10	<100	600 to 1200	50 to 150

*General environmental condition (average) within compartment.

†Mean value in fire plume near to fire.

‡Incident irradiance on to a sample (average).

vitiated (post-flashover) flaming decomposition conditions. The method also provides stable flaming conditions, solving the problem of mixed, intermittent periods of flaming and nonflaming common with the original DIN method.

Relationship Between Toxic Potencies of Materials in Small-Scale Tests and Full-Scale Fires

When the toxic potency of the combustion products from a material are expressed empirically in mass loss terms, the data relate to the toxic effects of the total mixed combustion product evolved. This depends upon the type of toxic products evolved and their yields. The most difficult problem in estimating the toxic potency of a material in a fire is that the yields of toxic products depend very much upon the decomposition conditions, which vary considerably at different stages and between different types of fires. If small-scale test data are to be used as estimates of the likely toxic potency of products evolved in full-scale fires, it is essential that the decomposition conditions in the test be shown to be the same as those in the type or stage of full-scale fire being modeled, otherwise the small-scale test data are not valid.

The decomposition conditions existing in full-scale fires depend mainly upon temperature and oxygen concentration in the fire environment and whether or not the material is flaming. In ISO/IEC TR 9122-4[131] an attempt has been made to define the major categories of fire in these terms, the type of decomposition for flaming fires being expressed in terms of the CO_2/CO ratio. The 6 fire types (shown in Table 2-8.17) contain three major categories: (1) non-flaming, (2) the early or developing flaming fire, and (3) the fully developed fire. In ISO/IEC TR 9122-4 small-scale toxicity test protocols are judged by the extent to which the test conditions are relatable to one of these categories in terms of temperature or radiant heat flux, oxygen concentration, and CO_2/CO ratio. If they are to be considered useful to measure the toxic potency of the combustion products from materials, the decomposition conditions must relate to one of these fire stages or types, and the results of any small-scale test are then only valid for the particular category being modeled. Based upon the results of full-scale fire tests, and fire death statistics, it is suggested here that the most important toxic hazard situations that should be assessed for all materials are:

1. Non-flaming oxidative/smoldering decomposition at low/mid-range temperatures where the potential hazard relates mainly to victims in the compartment of fire origin.
2. Early/well-ventilated flaming conditions at mid-range temperatures, and later small vitiated fires in closed compartments, where the potential hazard relates mainly to victims in the compartment of fire origin.
3. Fully developed/post-flashover, vitiated decomposition at high temperatures, where the potential hazard relates mainly to victims remote from the fire.

In the UK just over half of all fire deaths in buildings occur in the room of fire origin and most result from exposure to toxic smoke evolved from small fires (which may involve periods of non-flaming and both early and later flaming decomposition). The other major category, particularly related to deaths from smoke exposure, consists of victims in remote locations following fully developed fires. It is this second category that has been identified as the major problem in the United States, particularly in relation to fires in multi-occupation buildings.

Unfortunately, existing small-scale test protocols do not cover the necessary range of decomposition conditions found in full-scale fires, especially the third category of the fully developed, high-temperature, oxygen vitiated fire. However, a new technique is presented here whereby certain predictions about the toxicity of combustion product atmosphere can be made, based entirely upon analytical data. This enables use of data both from small-scale experiments and full-scale fire tests.

It is considered that the major toxic effects of fire effluents can be explained in terms of a small number of well-known fire gases, so that the effects of fire gases on human fire victims can be predicted to a large extent if the concentrations of these gases during a fire are known. In a similar way it is now possible to a large extent to predict the exposure dose of combustion products generated in small-scale tests that would be lethal to rodents, if the concentrations of the major toxic gases are measured. If necessary it is then possible to verify the prediction by carrying out the animal exposures. Experiments of this kind, carried out by Hartzell et al[132] and Levin et al[133] have shown that toxic gases are basically additive in their effects, so that, for example, an exposure to an atmosphere containing half a lethal dose of carbon monoxide mixed with half a lethal dose of hydrogen cyanide constitutes a lethal mixed atmosphere.

The toxic effects of combustion products result mainly from narcosis and irritancy. Narcosis is caused by carbon monoxide, hydrogen cyanide, low-oxygen hypoxia, and carbon dioxide, and so can be quite well predicted if the concentrations of these gases are known. Irritancy is somewhat harder to predict because many irritant organic products and inorganic acid gases occur in fire atmospheres. Where acid gases are present the concentrations can be measured and their effects added to those of the narcotic gases. In small-scale tests where both chemical analysis and animal exposures are used it is possible to calculate the contribution to the overall toxicity made by the measured narcotic gases and acid gas irritants. Any residual lethal toxicity can then be reasonably considered to be due to the effects of organic irritants, except in very rare cases where unusual toxic effects occur. For small-scale tests, or even large-scale fires, where analytical data only are available it is possible to calculate a theoretical LCt_{50} in terms of the main narcotic fire gases and acid gases. This then represents the *highest* estimate of what constitutes a lethal dose for that atmosphere (i.e., the smoke atmosphere must be at least as toxic as this, and could be somewhat more toxic if substantial amounts of organic irritants are present, or if unusual toxic effects are present. In small-scale tests there is little oxygen vitiation, so this effect can be ignored. On this basis an equation has been developed to predict the lethal FED to rats of a combustion product atmosphere as follows:

Fractional Effective Dose (FED) =
[(CO dose/171150 + HCN dose/4920 + IRR dose/LCt_{50}
IRR) × VCO_2] + A

where × VCO_2] + A = × 1.5] + 0.25 for 5% CO_2
× 1.4] + 0.2 for 4%
× 1.25] + 0.1 for 3%
× 1 for less than 2.5%

where

LCt_{50} = the lethal exposure dose for the material in mass loss terms
dose = the exposure dose in ppm · min
IRR dose = exposure dose of irritants (inorganic and organic)
LCt_{50}IRR = lethal exposure dose of irritants
VCO_2] + A = multiplication factor for CO_2 driven hyperventilation plus acidosis factor

When the FED for this equation equals unity, then death is predicted, and the mass loss exposure dose for the material producing these gas concentrations is then equal to the LCt_{50} for that material decomposed under the conditions of the test. If the concentrations of the irritants present and their lethal exposure doses are known, then the equation can be solved fully (e.g., the LCt_{50} for HCl is 112,980 ppm · min).[134,135] Where unknown irritants are present the equation enables the maximum LCt_{50} to be predicted based upon the narcotic gases and any known irritants.

This is a powerful technique because it enables a number of things to be done, as follows:

1. Where a material is tested in a small-scale test using only chemical atmosphere analysis, it enables an estimate to be made of the likely approximate toxic potency of the combustion products from the material, without the use of animal exposures.

2. Where a material is tested in a small-scale test using both chemical atmosphere analysis and animal exposures, then it is possible to determine the extent to which the toxicity of the combustion products can be accounted for in terms of the common toxic gases, or if additional toxicants are present.

3. If the toxic effects are almost entirely accountable in terms of the common toxic gases (as is often the case), then it enables the toxic effects of full-scale test atmospheres to be predicted with confidence, without animal exposures, if these gases are measured.

4. Where the LCt_{50} in a small-scale test is estimated from analytical data, or where it is measured using animals, it enables estimates of toxic potency of full-scale fires to be made simply from the mass loss rate and dispersal of products in the fire, provided the full-scale fire is of the same type as the small-scale test decomposition.

5. Where analytical data are available from full-scale tests, they enable some estimates to be made of the toxic potencies of the materials involved.

The following examples show how this technique can be applied, using data from experiments with wood. When samples of Douglas fir were decomposed under flaming conditions in the NABS cup furnace, the LCt_{50} for a 30-min exposure of rats was 1194 g · m^{-3} · min. In the test, the CO concentration was 3400 ppm and the CO_2 concentration 3.71 percent, a CO_2/CO ratio of 11/1. According to the FED equation given above, this represents a FED of 1.0. It can therefore be concluded that in this test the observed toxic potency can be fully accounted for in terms of CO and CO_2, and that there was little or no contribution from irritants or other toxic products on lethality. This result is to be expected, since the NBS cup furnace method generally simulates reasonably well-ventilated early flaming conditions where combustion is usually efficient, so that the yield of organic irritants would be expected to be low. However, in this test it is surprising that the CO_2/CO ratio is so low, and more representative of somewhat vitiated burning conditions.

It is now possible to examine some full-scale test data on wood fires of this type, in the knowledge that CO and CO_2 are the main toxic products to consider. Such a test was performed at the U.K. Fire Research Station, where a 5-kg wood crib (Scotch pine) was burned in a closed 26 m^3 room. At 6 min into this fire the CO_2/CO ratio in the smoke was 60/1. Based upon a 44 percent carbon content for wood, and assuming that all carbon in the mass lost was converted to COx, it is now possible to calculate what mass loss of wood in a small-scale test would be required for a 50 percent rodent lethality at a CO_2/CO ratio of 60/1. It is to be expected that under these conditions the wood smoke would be less toxic than in the reported NBS test, since the major toxicant is the CO, and the CO yield is low at this moment in the fire. This is indeed likely to be the case, since the FED equation predicts an LCt_{50} of 3750 g · m^{-3} · min under these decomposition conditions, approximately 3 times less toxic than in the small-scale test.

Major Determinants of Toxicity in Fires and Small-Scale Tests

The toxicity of the combustion products from individual materials in fire, in terms of the type and yields of the major narcotic, irritant, and other toxic products depends principally upon three factors:

TABLE 2-8.18 *Toxic Potency Analysis of Materials Decomposed under Non-Flaming Oxidative Conditions in the NBS Cup Furnace*

	Douglas fir 440°C		Flexible polyurethane foam 400°C	
	Conc.	FED	Conc.	FED
Carbon monoxide	2700 ppm	0.47	1261 ppm	0.22
Hydrogen cyanide	0 ppm	0.00	11 ppm	0.07
Carbon dioxide	0.69%	$\times$ VCO$_2$ 1.0	0.4%	$\times$ VCO$_2$ 1.0
Total FED narcotics		0.47		0.29
FED presumed due to irritants		0.53		0.71
LCt$_{50}$ calculated	1455 gm $\cdot$ m^{-3} $\cdot$ min		3621 g $\cdot$ m^{-3} $\cdot$ min	
LCt$_{50}$ observed	684 g $\cdot$ m^{-3} $\cdot$ min		1050 g $\cdot$ m^{-3} $\cdot$ min	

1. The elemental composition of the material,
2. The organic composition of the material, and
3. The decomposition conditions.

The most important toxic products from fires are usually carbon monoxide and hydrogen cyanide, so that the most important elemental determinants of toxic potency are normally the carbon and nitrogen content of the fuel, with the halogen content being important to a lesser extent in some cases, and organic irritants in others.

Non-Flaming Oxidative/Smoldering Fires

Non-flaming decomposition is slow, so that a long time is required for the development of hazardous conditions. However, of the small masses of materials decomposed during non-flaming oxidative decomposition, the yield of CO can be quite high, and these conditions generally provide the highest yields or organic products, including irritants, the identity of which is often unknown. In small-scale tests conducted under these conditions, only a small proportion of the observed toxic potency can be accounted for in terms of the common toxic gases. Table 2-8.18 shows two examples of this type, from experiments using the NBS cup furnace method to decompose Douglas fir and a flexible polyurethane foam under non-flaming oxidative conditions. The results show that at the LCt$_{50}$ of wood to rats of 684 g $\cdot$ m^{-3} $\cdot$ min only 0.47 of the observed toxicity could be accounted for in terms of common toxic gases, and for the flexible polyurethane foam only 0.29 of the observed toxicity could be accounted for. This means that for these (and many other) materials decomposed under non-flaming oxidative conditions, a large part of the toxic potency is due to products other than those normally measured, almost certainly organic irritants, so that the FED method tends to underestimate the toxicity, unless allowance is made for irritancy (such as assuming that approximately half of the toxic potency is likely to be due to irritants for most common materials under these decomposition conditions).

However, the non-flaming condition is adequately replicated by a number of small-scale test methods, and there is a large toxicity data base available for many materials, since by far the greatest amount of published test results are obtained under these conditions.

From a recent review of published toxic potency data from common materials[14] including data from a number of small- and large-scale test methods it is possible to make some general observations regarding the toxic potencies of the decomposition products from materials under a range of fire conditions. For non-flaming oxidative decomposition conditions the range of LCt$_{50}$ for individual materials covers approximately a factor of 12 from 63 to 767 g $\cdot$ m^{-3} $\cdot$ min.

Early or Well-Ventilated Flaming Fires

In flaming fires, the yields of carbon oxides and nitrogen compounds depend mostly upon the decomposition conditions, particularly the air/fuel ratio. With regard to carbon the main consideration is the CO$_2$/CO ratio, which not only determines the toxic potency of the smoke, since CO is approximately 20 times more toxic than CO$_2$, but to a large extent defines the fire type. In early well-ventilated fires combustion is usually efficient, and the CO$_2$/CO ratio may be as high as 200/1, although in practice somewhat lower ratios around 60/1 are more typical. Under these conditions the yield of organic irritants is usually low, since combustion is efficient, and the yield of CO is so low that the overall toxic potency of materials containing principally hydrogen and carbon can be expected to be low. The exceptions tend to be fire-retardant materials, "naturally" fire-retardant materials such as PVC, and some largely aromatic materials such as polystyrene, all of which tend to burn inefficiently and give low CO$_2$/CO ratios even under well-ventilated conditions. This results in high yields of CO and usually of irritants, and somewhat higher toxic potencies than for more easily and cleanly burning materials.

With nitrogen-containing materials, the situation is somewhat analogous to that with carbon, since in well-ventilated, early flaming fires, most nitrogen in materials is oxidized to nitrogen oxides and N$_2$. The yield of HCN is generally low (with the exception of acrylic materials, and to some extent rigid polyurethanes). Although NO$_2$, which is a potent lung irritant, can be expected to be present at high yields under these conditions, the general effects seems to be that, since the HCN yield is low, toxic potency tends to be low. With materials like PVC, almost all the chloride is released as HCl under almost all decomposition conditions, including flaming conditions.

The general picture is then that the toxic potency of combustion products from most materials is lowest under early, well-ventilated flaming conditions. Materials that tend to perform comparatively less well under these conditions are FR-materials, and materials like PVC, where the halogen acid gases cannot be destroyed by the flames.

Small-Scale Tests Replicating Well-Ventilated Flaming Conditions

To study the toxic potencies of materials under the decomposition conditions similar to those during the early stages of flaming fires it is necessary to use a test method that provides flaming combustion throughout the test until the material is fully decomposed. The test method must also provide good ventilation of non-vitiated air to the specimen, and have a general temperature environment of around 400 to 700°C (or equivalent radiant flux). Most important, however, it must produce high CO_2/CO ratios (in the range approximately 200/1 to 50/1) from normally combustible (i.e., non-FR) materials.

NBS cup furnace test: Of the small-scale test methods commonly in use, a number perform well in this area. The one for which the most toxicity data are available is the NBS cup furnace test. In this test the material specimen is decomposed in a crucible furnace, and the products are evolved into a 200-L box. The key feature of this test is that it is normally used with quite small specimens, so that the oxygen concentration in the box is not significantly lowered during the test, and studies of the combustion process have shown that air circulates rapidly down into the cup furnace during decomposition, so that combustion tends to be reasonably efficient. The CO_2/CO ratios typically produced in tests are in the 40/1 to 60/1 region, so that, although perhaps not representing the most efficient combustion, they are generally a reasonable model for the results obtained in small, well-ventilated full-scale fire tests.

NIST U.S. radiant (NIBS) test: A more recent development of the NBS cup furnace test is the NIBS test or NIST radiant test. The two versions of the test use the same apparatus, but somewhat different test protocols. For this test a radiant heating unit is placed in a cavity under the NBS chamber, and connected to it by a slit-shaped chimney. Investigations of the combustion mode of this test[17] have also shown that under flaming combustion conditions the circulation of air is such that the specimen is very well ventilated, so that CO_2/CO ratios are generally reasonably high. Data from this test method suggest that it may best represent the decomposition conditions in a well-ventilated fully developed (possibly post-flashover) fire. However in its present form it does not appear to generate the very low CO_2/CO ratios and high CO and HCN yields found in typical post-flashover fires in compartments.

Cone calorimeter: The cone calorimeter has not been used very successfully with animal exposures, but using the FED model presented in this chapter, it is possible to make some useful estimates of likely toxic potency based on the toxic gas yields and the mass loss of the specimen. The cone calorimeter gives the most efficient combustion conditions of any test method, typically producing CO_2/CO ratios in the 200/1 to 100/1 range for non-FR materials. It can also therefore be used as representative of the decomposition conditions during very early and very well-ventilated fires. So far, attempts to modify the combustion process and decrease the combustion efficiency to model other stages of fire have not proved very successful.

DIN method: For the DIN 53 436 method, decomposition occurs in a tube furnace, the furnace passing over a strip of sample, with decomposition achieved by passing a stream of air through the furnace and over the sample. The products from the tube furnace may then be diluted with secondary air for animal exposures, if required. The atmosphere produced is in a "dynamic steady state," in that the concentrations of decomposition products remain constant because the test material is decomposed at a constant rate throughout the test. The epithet "DIN" has come to represent a number of tube furnace methods based upon the same principle, generally accepted as DIN test results, the important point being to demonstrate that the decomposition atmosphere generated is relatable to real fire conditions, rather than that the apparatus design is standard.

The important feature of this design is its versatility, since the decomposition conditions can be varied over a wide range by varying the sample load, air supply, and furnace temperature. This contrasts with the other methods described, which are very restricted in the range of conditions that can be modeled. The improved tube furnace method, recently developed at the UK Fire Research Station (FRS),[130] is based upon the DIN method concept, and is the only one so far developed that can simulate the decomposition conditions for all fire types, including non-flaming, early flaming, and fully developed fires with restricted ventilation, particularly post-flashover fires. (See Figure 2-8.19.) A limitation with the official method, in regard to flaming combustion, is that the decomposition conditions tend to be rather vitiated, giving low CO_2/CO ratios, and flaming is unstable. However this is remedied in the FRS method by increasing the air/fuel ratio in the furnace tube, and the rate of sample advance. In a recent series of experiments it has been possible to increase the ratios to those occurring in early, well-ventilated, full-scale fires, while achieving stable flaming conditions.

UPITT method: The University of Pittsburgh method is considered to give a poor representation of actual fire conditions, and it is considered that it is not possible to relate it to any of the fire conditions shown in Table 2-8.17. For this reason (and others), UPITT data are considered unsuitable for the assessment of toxic potency of materials involved in fires, except in very special cases where the conditions can be shown to be similar to the sequence of events occurring in the UPITT apparatus.

Toxic Potency Data Obtained from Tests under Early, Well-Ventilated Flaming Conditions

Because well-ventilated flaming conditions tend to destroy compounds such as organic irritants, it is to be expected that the toxic potency will be more completely due to the common toxic gases than for the non-flaming fires shown in Table 2-8.18. Table 2-8.19 illustrates this with some examples taken from NBS cup furnace test data. The data for Douglas fir shown that, unlike the non-flaming situation illustrated in Table 2-8.18, the toxic potency can be fully accounted for in terms of carbon oxides, and an LCt_{50} for wood calculated on this basis would be very close to the observed value. For flexible polyurethane (FPU) foam it was not possible to obtain a lethal concentration in the cup furnace under flaming conditions due to limits on the capacity of the apparatus for the size of sample required, but in other experiments a mixture of polyester and FPU were tested. With this mixture of materials it was possible to obtain lethal exposure conditions, and the data are also shown in Table 2-8.19. As with the wood, it was possible to account fully for

TABLE 2-8.19   *Toxic Potency Analysis of Materials Decomposed under Early, Well-Ventilated Flaming Conditions in the NBS Cup Furnace*

| | Douglas fir 485°C | | Flexible polyurethane foam and polyester 525°C | |
	Conc.	FED	Conc.	FED
Carbon monoxide	3400 ppm	0.60	2270 ppm	0.40
Hydrogen cyanide	0 ppm	0.00	63 ppm	0.38
Carbon dioxide	3.71%	$\times$ VCO$_2$ 1.4 + 0.2	3.36%	$\times$ VCO$_2$ 1.25 + 0.1
Total FED narcotics		1.04		1.08
FED presumed due to irritants		0		0
LCt$_{50}$ calculated	1148 g · m^{-3} · min		1038 g · m^{-3} · min	
LCt$_{50}$ observed	1194 g · m^{-3} · min		1170 g · m^{-3} · min	

the observed toxic potency on the basis of carbon oxides and hydrogen cyanide.

For materials that burn less efficiently under these conditions, or which produce inorganic acid gases, the data analyses indicate contributions to lethality from irritants. For example, Table 2-8.20 shows data on PVC and a FR polyurethane foam obtained using the NBS cup furnace method. For the PVC test the contribution to the total FED from carbon oxides was only 0.19, so that the major cause of death had to be some other factor. Unfortunately the HCl concentration was not measured, but from the mass of PVC decomposed, it can be estimated at approximately 5000 ppm. As the analysis shows this would have been more than enough to have accounted for the observed lethality. With regard to the FR FPU, the yield of common toxic gases was significantly greater than that from the untreated foam, so that it was possible to obtain an LCt$_{50}$ using the cup furnace. The concentrations of carbon oxides and hydrogen cyanide were sufficient to account for approximately 0.7 of the observed toxic potency, but it is possible that the remaining 0.3 represents the effects of unidentified irritants evolved due to the less efficient combustion occurring from this foam compared to an untreated foam.

Based upon available published small- and large-scale test data, it is possible to make some general observations regarding the early, well-ventilated, flaming condition. The basic finding is that the published data base is very poor, there being only a few tests or none on quite common materials. The only materials for which a reasonable number of tests have been performed under flaming conditions are wood, flexible polyurethanes, and PVC. Needless to say

these involve a variety of wood species and polymer formulations. Based upon this inadequate data base, the pattern that emerges is that the range of toxic potencies of common materials covers approximately a factor of 50, with LCt$_{50}$ exposure doses of from approximately 75 to 3750 g · m^{-3} · min. As could be predicted, the least toxic materials are the cellulosics and simple hydrocarbon polymers, such as polypropylene. Flexible polyurethanes are of low to intermediate toxic potency within this range. The most toxic materials are the acrylonitriles, which release quite large amounts of HCN even under well-ventilated flaming conditions. PVCs are generally somewhat more toxic than the cellulosic materials under these conditions, due to their relatively low combustion efficiency and high HCl yield.

Small-Scale Tests Replicating Fully Developed Fire Conditions— Especially Post-Flashover Fires

The decomposition conditions in fully developed fires depend very heavily upon the conditions in fire compartments and in particular the air supply. A general principle would be that in the common situation of a fire in a building, which would typically contain large amounts of combustible fuel, the fire growth will depend upon the rate of involvement of the fuel in the early stages, with efficient combustion and high CO$_2$/CO ratios (100/1 to 200/1). Then as the fire grows, combustion becomes increasingly ventilation controlled, so that fully developed fires tend to be oxygen vitiated,

TABLE 2-8.20   *Toxic Potency Analysis of Materials Decomposed Less Efficiently under Early, Well-Ventilated Flaming Conditions in the NBS Cup Furnace*

| | PVC 625°C | | FR flexible polyurethane foam 425°C | |
	Conc.	FED	Conc.	FED
Carbon monoxide	1100 ppm	0.19	1040 ppm	0.18
Hydrogen chloride	5000 ppm	1.33	86 ppm	0.52
Carbon dioxide	0.55%	$\times$ VCO$_2$ 1	2.1%	$\times$ VCO$_2$ 1
Total FED narcotics		0.19		0.70
FED presumed due to irritants		1		0.30
LCt$_{50}$ calculated	341 g · m^{-3} · min		1157 g · m^{-3} · min	
LCt$_{50}$ observed	519 g · m^{-3} · min		810 g · m^{-3} · min	

with low CO_2/CO ratios ($<10/1$). However, it is possible to have fully developed, well-ventilated fires with high CO_2/CO ratios (up to 100/1), as indicated in Table 2-8.17. These conditions can commonly occur during some stages of test fires in large-scale test rigs. Such large-scale tests usually have a relatively small amount of fuel (such as a single chair), and have a rig with an open side or doorway and/or window openings, which are, in turn, supplied freely with air from outside or from a large test facility building. Another factor that may provide high CO_2/CO ratios in the effluents from a primarily vitiated post-flashover fire is secondary combustion outside the fire compartment, where the products mix with air and are sufficiently hot to support further combustion. This effect has been observed at the DIN furnace outlet when attempting to simulate post-flashover decomposition conditions.

Accidental fires in real occupied buildings often have access to a much larger amount of fuel than test fires, and often have access to a more restricted air supply (such as air from inside the building). Thus in the Boston Fire Department study of accidental fires,[136] 50 percent had CO_2/CO ratios of less than 10/1 and a further 22 percent had ratios of approximately 10 to 20/1. Only 17 percent fell into the well-ventilated category, with ratios above 40/1. When full-scale tests are more closely related to real buildings or contents, then low ratios occur. For example, a simulation of a fire in a fully furnished hotel bedroom, opening on an open corridor with a side room attached, gave CO_2/CO ratios of 2/1 in the burn room and 3/1 in the side room at the fire peak.[124]

Once a fire has passed beyond the very early, well-ventilated stage, there are basically two paths for continued development, depending largely on the type of fire compartment. First, where the fire occurs in a room-sized compartment, and the room doors and windows are shut, the combustion becomes vitiated from a very early stage, since a typical domestic room will not support complete combustion of more than approximately 1 kg of fuel before the oxygen concentration in the room is reduced to approximately 10 percent and the fire extinguishes or dies down. Fires of this type, involved in many deaths, tend not to develop beyond a small size as long as the compartment is closed, but the CO_2/CO ratio decreases from a very early stage. An example is a burning 5-kg wood crib in a closed room. Table 2-8.21 shows the gas concentrations in the room during this fire, the atmosphere becoming progressively more vitiated and the CO_2/CO ratio decreasing as the fire progresses. The last column shows the influence this process has on the toxic potency of wood, assuming that carbon oxides are the only important toxic products (and also ignoring any toxic effects of low-oxygen hypoxia). The data show that, if a sample of wood was decomposed in a number of runs of a small-scale test under conditions giving the range of CO_2/CO ratios recorded at different stages of the full-scale fire, then the toxic potency of the wood would increase from very low levels as shown.

The other common situation is where a window or door is open, or where the compartment is large, so that there is sufficient ventilation to support a much bigger fire before the air supply becomes the controlling factor. Such fires, typically the cause of smoke deaths in locations remote from the fire, become both hot and vitiated, and constitute the post-flashover situation that chiefly needs to be simulated in small-scale tests, with a temperature of 800°C or more, and CO_2/CO ratios of less than 10/1, and as low as 2/1. The lower part of Table 2-8.21 illustrates this with data from the devel-

TABLE 2-8.21 *Toxic Gas Concentrations and Calculated Toxic Potencies during Full-Scale Wood Fires in a Room-Corridor Test Rig*

1. Fire of 5-kg wood in a closed room—total mass loss 3.5 kg; room temperature approximately 200°C

Time (min)	CO ppm	CO_2 ppm	CO_2/CO ratio	$O_2\%$	LCt_{50}* for COx $g \cdot m^{-3} \cdot min$
6	750	45000	60	18	3750
8	1500	75000	50	13.5	3461
10	2500	88500	35	11	2857
12	4000	95000	24	10	2222
20	9000	75000	8.3	11.5	1034

2. Fires of 44-kg wood in open room with high and low ventilation; room temperature approximately 800°C

High vent.	10000	150000	15	9	1800
Low vent.	50000	150000	3 to 1.5†	4	750

*LCt_{50} calculated in terms of mass loss concentration of wood, assuming carbon oxides to be only toxic products of importance.

†Second figure shows ratio in corridor, all other figures in room.

oped stage of larger (44 kg) wood fires run in the same rig as the 5-kg test. These fires were run with high and low ventilation from the corridor, to simulate well-ventilated and oxygen-vitiated fully developed (post-flashover) fires. The results show that both fires become vitiated when fully developed, the poorly ventilated fire giving very low CO_2/CO ratios of 3/1 in the room and 1.5/1 in the corridor.

If the typical post-flashover fire is hot, and oxygen vitiated, with low CO_2/CO ratios, the next consideration is what effects these decomposition conditions have on the toxic potency of the products. The most obvious effect is that the toxic potency is increased compared to well-ventilated fires due to the higher concentrations of CO produced as the CO_2/CO ratio falls. The series of room-corridor fires performed at the Fire Research Station using 44-kg wood cribs provides a good example. When these were burned with restricted ventilation, the CO_2/CO ratio fell to the low values mentioned, of 3/1 in the room and 1.5/1 in the corridor after 3.5 min, with very high concentrations of CO (5 percent) in the room. Assuming as before that toxicity would be due solely to carbon oxides, then the theoretical LCt_{50} would be $750 \ g \cdot m^{-3} \cdot min$, which is approximately 5 times more toxic than that in the well-ventilated, early flaming fire, and is similar to results obtained for pine and sipo wood in the DIN apparatus at 850°C.[16] Table 2-8.22 shows the toxic potency analysis for this fire.

A very important aspect of post-flashover fire conditions is the fate of nitrogen in materials. Under hot, vitiated conditions the yield of HCN from all nitrogen-containing materials increases dramatically. Hydrogen cyanide can, therefore, be an important toxic product in post-flashover fires where the fuel has a high nitrogen content. Another problem with such fires is the yield of organic irritants. Vitiated post-flashover fires produce large quantities of smoke, and recent experiments with some common materials decomposed under these conditions in a DIN-style tube furnace have shown that the dense smoke is rich in organic products, which are irritant to mice. With regard to inorganic irritants, such as HCl, these are produced at the same

TABLE 2-8.22 *Toxic Potency Analysis of Materials Decomposed under High-Temperature, Vitiated Conditions in Large-Scale Fires and in the DIN Apparatus*

| | Scotch pine 850°C room corridor low ventilation | | Wool 700°C DIN | |
	Conc.	FED	Conc.	FED
Carbon monoxide	5515 ppm	0.97	379 ppm	0.07
Hydrogen cyanide	0 ppm	0	153 ppm	0.93
Carbon dioxide	1.7%	$\times VCO_2$	0.17%	$\times VCO_2$
		1		1
Total FED narcotics		0.97		1
FED presumed due to irritants		0		0
LCt_{50} calculated	750 g · m^{-3} · min		81 g · m^{-3} · min	
LCt_{50} observed (DIN test under similar conditions)	876 g · m^{-3} · min		no data	

high yield as with early flaming fires, except that their effects are less prominent in the fully developed fire in comparison with the high yields of other toxic products.

Results from DIN and Other Tube Furnace Methods and Full-Scale Tests

If the data base of small-scale toxicity test results on materials tested under early flaming conditions is poor, that on materials tested under post-flashover conditions is almost non-existent. The only small-scale apparatus that can be used to replicate these conditions is the DIN tube furnace, when it is run at high temperatures. A small amount of rodent lethality data is available from tests run using this method at temperatures above 800°C. Apart from this, other data are from a number of small-scale and large-scale tests where analytical measurements were made, from which it is possible to make toxic potency assessments, assuming that toxic effects were due only to carbon oxides and HCN. In such vitiated post-flashover conditions, it is to be expected that carbon monoxide and hydrogen cyanide would be the dominant toxic species, since carbon monoxide will be present at high yields in all fires and hydrogen cyanide can also be expected at high yields if the materials being burned contain nitrogen. Added to this will be an uncertain contribution from organic irritants, and a contribution from inorganic irritants if these are present. While recent work suggests that organic irritants may be more important under these conditions than was thought previously, it is considered likely that they are less important than under non-flaming conditions, and are unlikely to be the dominant factor in the toxic potency.

Table 2-8.22 shows some examples of toxic potency analyses for this fire condition. The table shows analytical data from a large-scale wood fire, compared with animal data from a DIN test on wood for which no analytical data are available, and data from a small-scale (DIN) furnace test on a nitrogen-containing material, i.e., wool. Rats were exposed in the test at an exposure dose of 18 g · m^{-3} · min, and all died. The data shown are projected gas concentrations at the calculated LCt_{50}. For the wood data, the point illustrated is that, based upon the measured carbon oxide concentrations, the toxic potency is likely to be dominated by CO, and on this basis the potency of wood is greater than under well-ventilated flaming conditions. That this projection is reasonable is supported by the results of DIN work on Scotch pine, carried out under nominally similar conditions. This work gave a rat LCt_{50} of

875 g · m^{-3} · min, which is similar to the predicted figure for wood decomposed under these general conditions. Unfortunately, no analytical data were published for the DIN test results. For the wool data, the main point illustrated is that even at 700°C, which is somewhat below what would be considered a post-flashover temperature, the toxicity is likely to be dominated by the high HCN yield, and this is considered also to be true for most other nitrogen-containing materials. The other point made strongly by these examples is the paucity of available data for this type of fire.

Based upon the available data it is estimated that the toxic potency range for common materials decomposed under vitiated post-flashover conditions covers an LCt_{50} range from approximately 21 g · m^{-3} · min for materials with a high nitrogen content decomposed at temperatures around 1000°C, up to 3000 g · m^{-3} · min for certain cellulosic or hydrocarbon-based polymers, a range of more than two orders of magnitude for the small sample of published data.

Adaptation of Data from Other Small-Scale Tests

It has been recognized that tests other than the DIN, such as the cone calorimeter and the U.S. radiant method, are incapable of simulating post-flashover decomposition conditions, producing the wrong yields of CO and the wrong CO_2/CO ratios. However, since it is also considered in the U.S. that this fire condition is the most important to study, suggestions have been made that a calculation method can be applied to cone and U.S. radiant data toxicity data to allow for the low CO yield in the tests, relative to those in post-flashover fires. This is obviously not a realistic suggestion, since even if a calculation factor could be used to correct the CO data, the result would still be wrong if no factor were used to correct for the differences in the yields of HCN and other nitrogen-containing products, and for the yields of the many other organic irritants. Rather, if it is wished to study the behavior of materials under vitiated, high-temperature post-flashover conditions, small-scale tests should be used that create such decomposition conditions, so that the chemistry and toxicity of the decomposition products evolved under these conditions can be studied. Since tube-furnace methods similar to the DIN are cheap and very effective means of simulating this fire condition, it is recommended that this method be used, under appropriate conditions of temperature and airflow, for this purpose.

Where calculation methods are to be used, it is better to base them on the elemental composition of the material and knowledge of full-scale fire conditions rather than on small-scale tests conducted under inappropriate conditions.

General Pattern of Toxic Potency for Common Materials under Three Fire Conditions

The survey of the toxic potency data for common materials under three fire conditions: (1) non-flaming, (2) early flaming, and (3) post-flashover has revealed an inadequate data base, but it has been possible to derive approximate LCt_{50} for common materials. The results for individual materials range over approximately 2 orders of magnitude from 20 to 3750 $g \cdot m^{-3} \cdot min$, but when the data are reduced to basic types of materials under each decomposition condition a relatively simple pattern can be described. (See Table 2-8.2.) The table shows the approximate average lethal exposure doses (LCt_{50}) for classes of materials, the LC_{50} for 30-min exposures, and a potency factor (based upon a figure of 500 $g \cdot m^{-3} \cdot min$ for the "normal" lethal potency for combustion products). The findings are as follows.

Under non-flaming oxidative decomposition conditions at >400°C most materials have a similar potency close to 500 $g \cdot m^{-3} \cdot min$, i.e., a potency factor of approximately 1, due mainly to the effects of carbon monoxide and irritants. The main exceptions are nitrogen-containing materials releasing significant HCN at low temperatures (e.g., polyacrylonitrile, modacrylic, and rigid polyurethane foam), which have toxic potency factors of 3 to 8.

Under early flaming conditions most non-fire-retardant materials are substantially less toxic than under non-flaming conditions. Cellulosics (e.g., wood and cotton) are the least toxic with LCt_{50} of >3000 $g \cdot m^{-3} \cdot min$ (potency factor 0.2). Plastics containing carbon, hydrogen, and/or oxygen are somewhat more toxic with a potency factor of 0.4 (LCt_{50} ~1200), and those containing low percentages of nitrogen (e.g., flexible polyurethanes, wool, and nylon) also fall into this area. Both PVC and fire-retardant materials have toxic potencies similar to those under non-flaming conditions of approximately 1. Rigid polyurethanes and nitrogen-containing acrylics have high potencies similar to those under non-flaming conditions.

Under post-flashover conditions the potency of all materials increases due to the increased yields of HCN and/or CO. More smoke and irritants are also present than under early flaming conditions, which may add somewhat to the potency, particularly of the non-nitrogen-containing materials. For cellulosic materials and hydrocarbon plastics, the potency is similar to that under non-flaming conditions (potency factor close to 1). For all nitrogen-containing materials the toxic potency is high, ranging from approximately 2.5 for flexible polyurethane foam to approximately 11 for polyacrylonitrile and modacrylic. It is suggested that PVC would have a potency of approximately 2.5 under these conditions.

THE CONDUCT AND APPLICATION OF SMALL-SCALE TESTS IN THE ASSESSMENT OF TOXICITY AND TOXIC HAZARD

Small, laboratory-scale toxicity tests are of necessity capable only of investigating materials. Investigation of toxic fire hazard associated with actual items such as furnishings can be investigated only in large-scale tests, although it may be possible to a limited extent to study some composite materials using small-scale tests. The potential usefulness of these tests is then to examine the toxicity of the decomposition products from materials. This information can be used in conjunction with other small-scale test data on such characteristics of materials as ease of ignition, rate of flame spread, heat release, and smoke production to judge the suitability of one material *vs* another for a particular application, and ideally as a prelude to large-scale fire tests. From these it should be possible to draw some conclusions as to likely fire scenarios as well as the toxic and general fire hazards involved. A sensible approach to the use and application of such toxicity tests should involve the following steps:

1. Decide what kinds of fire scenarios are of interest and likely to involve the material under investigation, and what types of fire conditions it may be subjected to—smoldering/overheat, small flaming fire, or fully developed/postflashover.

2. Choose a small-scale test method or methods capable of simulating these conditions.

3. Run the test without animals and measure as many as possible of the common fire products important with respect to toxicity. A minimum that should be measured include CO, CO_2, O_2, HCN (if nitrogen present in material), HCl (or other appropriate acid gases if likely to be present), smoke optical density, and particulate concentration. All tests should be characterized in terms of NAC mass charge, NAC mass loss, decomposition temperature, and whether the decomposition is flaming or non-flaming. Calculate an approximate LCt_{50} at this point for use in hazard modeling. If more information is required, proceed to carry out animal experiments.

4. Set up a test atmosphere at a concentration that should be just sublethal for the known toxic atmosphere constituents (in most cases the determining factor will be the CO concentration). Then, expose a group of animals and measure the toxic effects in terms of type (narcotic or irritant), time of onset, severity, and duration, noting in particular the degree of incapacitation and the occurrence of any deaths. In the first instance this should involve a 30-min exposure followed by a 14-day observation period.

5. Decide from this whether the observed effects are consistent with the toxicity due to common fire products, or whether there were any unusual or severe toxic effects. If the toxicity can be interpreted in terms of the common narcotic products CO and HCN, then it should be possible to attempt modeling of toxic hazard on the large-scale. However, if the products are irritant, as in most cases they will be, or if some unexpected toxic effect should occur, then further investigations are indicated.

6. If some unexpected toxic effect should occur, attempt to identify the toxic product or products responsible, and the conditions under which they are likely to be formed. The minimum necessary is to establish the 30-min exposure LC_{50} concentration to give some indication of the possible toxic potency of the material when decomposed in a fire. However, if the identity of the toxic product and the conditions of its formation are not understood, it is unwise to assume that small-scale tests will adequately predict of what might happen in a large-scale fire. A good example is PTFE (teflon). In one small-scale test method

(the NBS method), PTFE decomposes to form a highly toxic lung irritant which causes death at concentrations of two to three orders of magnitude less than that of other polymeric materials.[7] In the Pittsburgh method the material is approximately 20 percent less toxic,[107] and in a tube furnace method similar to the DIN method a further three times less toxic,[108] although still somewhat more toxic than most other materials. However, when decomposed in a way different from any of these tests, the high toxic potency is lost,[109] and it is possible that under real fire conditions the products may not be significantly more toxic than those of other materials, although this is yet to be established.

7. Assess the irritancy potential of a material by measuring the effects of its thermal decomposition and combustion products in animals. With regard to the assessment of irritancy, although many known irritants have been identified in combustion product atmospheres it is still not possible to predict the irritancy of an atmosphere from an analysis of its composition. The potential for causing upper respiratory tract and eye irritation (sensory irritancy) should be assessed by measuring the mouse RD_{50} concentration of the material. The potential for causing lung irritation with serious or lethal lung inflammation should be assessed by examining postexposure lethality in rats or mice. Thus if carbon monoxide concentrations are relatively low in relation to irritant products, a concentration of decomposition products may occur when the animals die either during, or in most cases after, exposure due to lung inflammation. A LC_{50} concentration for these non-narcotic deaths should then be determined to indicate the potency of the material in terms of causing lung inflammation under specific decomposition conditions. If it is not possible to identify the product or products responsible for these irritant effects, it will be necessary to use the material RD_{50} and LC_{50} data in an attempt to predict likely large-scale toxic hazard. Although this measurement is only approximate, there are indications that both the RD_{50} and LC_{50} levels of most materials fall into relatively narrow bands, each effect spanning approximately one order of magnitude, with one to two orders of magnitude between the effects, at least under nonflaming conditions. Under nonflaming conditions it is likely that most materials may cause potentially serious lung inflammation following a 30 min exposure at an NAC mass loss of approximately 10 mg/L, and severe sensory irritation at somewhat lower concentrations, possibly around 1 mg/L, although tenability limits for humans are difficult to estimate. Under flaming conditions the degree of irritancy is likely to be less, sometimes considerably so, depending upon the efficiency of combustion.

8. Having evaluated the toxicity of the combustion products from the material in this way, it should be possible to use the data in conjunction with other information from small- and large-scale fire tests, and/or mathematical models, to assess potential toxic hazard for the material or materials in question in their end use configuration.

Misuse of Toxicity Test Data

Another way of using toxicity test results is to rank materials in order of toxicity and choose the least "toxic" material, as if toxic hazard were an inherent property of the material. This is not a realistic approach, since the toxic hazard associated with any material is not an inherent property but depends upon how the material is used and how it may be decomposed in a real fire scenario. Indeed, it is very easy to alter the toxicity of a material in a small-scale fire test simply by altering the test conditions, particularly with respect to CO and HCN yields. Thus, when wood or most other hydrocarbon polymers are decomposed under flaming conditions with restricted oxygen, the CO_2 to CO ratio in the products can be lowered (to around a value of 4)[12,13] and the toxic potency is high. However, under well-oxygenated efficient combustion conditions, the CO_2 to CO ratios in a fire may be as high as 1000, and under such conditions the toxic potency of the products is very low. Although such anomalies can be overcome to some extent by careful control of the small-scale decomposition conditions and by relating them to conditions known to occur in large-scale fires, it is still difficult to predict the CO concentration profile, and hence the toxic hazard, for a large-scale fire from the small-scale fire model in a toxicity test. For this reason, ranking materials in order of their performance in small-scale toxicity tests does not have much meaning or usefulness. The best use of small-scale toxicity tests is to identify the products responsible for the major toxic effects so that the concentration/time profiles of these products in large-scale fires can be measured or modeled, and the likely toxic hazard can be assessed.

SUMMARY OF TOXIC AND PHYSICAL HAZARD ASSESSMENT MODEL

Having identified the main toxic products evolved in particular fire scenarios in terms of the main types of fire conditions—smoldering/nonflaming, early flaming, and fully developed/postflashover—estimates of potential toxic hazard can be made by a consideration of two sets of information:

1. The concentration/time profiles of the major toxic products in the full-scale fire.
2. The time/concentration/toxicity relationships for these toxic products when they occur individually and in combination.

From these two data sets, it is possible to construct a model to predict probable time to incapacitation and/or death due to toxicity for a victim exposed in such a fire.

The first data set, the large-scale fire profile, may be determined by a combination of small- and large-scale fire tests and mathematical modeling. A guide to the general characteristics of fires is shown in Table 2-8.8, and an example of a fire profile in Figure 2-8.7.

The fire profile should be characterized in terms of the following minimum range of parameters, measured or estimated at the breathing zone of a potential victim:

1. Mass loss of material divided by the volume of air into which the material is dispersed.
2. Carbon monoxide concentration.
3. Hydrogen cyanide concentration (if materials containing nitrogen are present).
4. Carbon dioxide concentration.
5. Oxygen concentration.
6. Radiant heat flux.
7. Air temperature.
8. Smoke optical density.

Ideally, some measure should also be made of:

9. Mass charge concentration of material divided by the volume of air into which the products are dispersed.
10. Acid gas concentrations (HF, HCl, HBr, SO_2, NO_2).
11. Organic product profile, particularly oxidized organic species (especially acrolein, formaldehyde, and crotonaldehyde).

Any properly conducted large-scale fire test or practical fire model should be able to provide concentration (or intensity)/time profiles for all of the first eight parameters. These can then be used as input data for the toxic and physical hazard assessment model.

The aim of the hazard assessment model is to determine the point in time during exposure when potential victims are predicted to become incapacitated, such that they would be unable to escape or their ability to escape would be severely compromised. The model also determines the point where the exposure would be sufficient to cause death either during the fire or later, as a result of the injuries sustained. The hazard assessment is based on a "step through" approach whereby the degree of hazard is calculated for each successive minute (or other appropriate time interval) during the fire, until a point is reached when incapacitation or death is predicted.

For some fire parameters the occurrence of incapitation (or death) is related primarily to the concentration or intensity of the agent to which the victim is exposed (radiant heat, smoke optical density, sensory irritation). Tenability limits have been set for these parameters, and incapacitation is predicted to occur rapidly at the point where the limits are met.

For other parameters (carbon monoxide, hydrogen cyanide, carbon dioxide, low-oxygen hypoxia, convected heat, lung irritants) a dose accumulated over a period of time is the primary concern. For these parameters, the fraction of an incapacitating dose acquired each minute is calculated, and the fractions added for each successive minute until the fractional dose reaches unity, when incapacitation is predicted.

Preliminary Rough Estimate of Toxic Hazard

Although it is strongly recommended that a hazard assessment for any particular fire scenario be based on all of the eight major parameters previously listed, it is possible to make a very crude assessment of the existence of a serious toxic hazard by reference to a single parameter, the mass loss/volume dispersed profile. The three items of data needed for such an assessment are:

1. The basic fire condition (smoldering, early flaming, or fully developed).
2. The mass loss/dispersal volume time curve for the fire.
3. The rodent LC_{50} concentrations for the materials involved in the fire in terms of NAC mass loss, determined from small-scale tests performed under the same decomposition conditions as those in the fire.

The dose of products available in the fire per minute in terms of mg min/L of mass loss is expressed as a fraction of the rodent LC_{50} dose in mg · min/L. The fractions of a lethal dose available each minute are added until the fraction reaches unity, when it is predicted that a human fire victim would probably have inhaled a lethal dose of toxic products.

Since the toxic potencies of combustion products from most materials fall into a relatively narrow range of approximately one order of magnitude,[7,104,107] (Table 2-8.1) it is possible to use a single toxic potency figure in simple hazard

models. The range of LC_{50} for the preceding references quoted was 5 to 61 mg/L NAC mass loss for a 30-min exposure, which is equivalent to a dose range of approximately 150 to 1800 mg min/L, with an average value of approximately 30 mg · min/L. Allowing for a margin of safety and for the possibility (derived from primate data)[56] that humans may be more sensitive than rodents to lung damage from fire products, it is recommended that a figure of 300 mg min/L should be used as a probable lethal dose of combustion products in man for preliminary modeling purposes.

Main Toxic and Physical Hazard Assessment Model

Narcosis: The first task is to assess the point at which a victim is likely to become incapacitated by loss of consciousness due to the effects of narcotic (asphyxiant) gases. For this assessment, it is neccessary to calculate the fractional incapacitating doses of each narcotic gas (CO, HCN, O_2, CO_2) individually, and interactions between them, for each successive minute of the fire.

Carbon monoxide: The most important toxic fire product is CO. Toxic effects occur when a certain dose in the form of carboxyhemoglobin (COHb) has been inhaled. Time to achieve a COHb concentration causing incapacitation (unconsciousness) for a 70-kg human engaged in three levels of activity is shown in Figure 2-8.6 and calculated from primate incapacitation data and the Coburn-Forster-Kane equation.

For short exposures to high concentrations of CO, COHb concentration can be calculated approximately using the Stewart equation (Equation 5)

$$\%COHb = (3.317 \times 10^{-5})(ppm\ CO)^{1.036}(RMV)(t) \quad (5)$$

where

$$\begin{aligned} CO &= \text{CO concentration (ppm)} \\ RMV &= \text{volume of air breathed (L/min)} \\ t &= \text{exposure time (min)} \end{aligned}$$

From this equation, the expression for the fractional incapacitating dose each minute for a 70-kg human engaged in light activity over periods of up to one hour is derived

$$F_{Ico} = \frac{K(ppmCO^{1.036})(t)}{D} \quad (6)$$

where

$$\begin{aligned} F_{Ico} &= \text{fraction of incapacitating dose} \\ t &= \text{exposure time (min)} \\ K &= 8.2925 \times 10^{-4} \text{ for 25 L/min RMV (light activity)} \\ D &= \text{COHb concentration at incapacitation (30 percent for light activity)} \end{aligned}$$

For a 70-kg human engaged in light work an RMV of approximately 25 L/min can be expected with loss of consciousness at around 30 percent COHb. Therefore, this equation predicts incapacitation after approximately 5.3 minutes at a concentration of 5000 ppm CO. For a subject at rest, the RMV will be approximately 8.5 L/min giving a value for K of 2.8195×10^{-4}, and incapacitation is likely at approximately 40 percent COHb. Thus, at 5000 ppm, time to incapacitation will be approximately 21 min.

For smaller adults and especially for children, or for CO concentrations below approximately 2000 ppm and exposure durations above one hour, or also for estimating time to death (at approximately 50 percent COHb), the rate of uptake departs significantly from the Stewart equation and predictions should be based on the CFK equation. (See Appendix 2-8A.) Time to incapacitation and death for small children may be one-half that for adults.

Hydrogen cyanide: Hydrogen cyanide is the next most important toxic gas causing incapacitation by narcosis in fires. Time to incapacitation depends partly on rate of uptake and partly on dose. (See Figure 2-8.4.) Below a threshold of approximately 80 ppm HCN only minor effects should occur over periods of up to one hour. From 80 to 180 ppm, time to incapacitation t_{Ico} (loss of consciousness) will be between 2 and 30 min approximately according to the relationship (Equation 7)

$$(t_{Icn})(\text{min}) = (185 - \text{ppm HCN})/4.4 \quad (7)$$

For concentrations above approximately 180 ppm, incapacitation will occur rapidly (0 to 2 min).

Deriving an exponential expression for the data gives

$$t_{Icn} = \exp(5.396 - 0.023 \times \text{ppm HCN}) \quad (8)$$

From Equation 8, the fractional incapacitating dose per minute equation for HCN is derived

$$F'_{Icn} = \frac{1}{\exp(5.396 - 0.023 \times \text{ppm HCN})} \quad (9)$$

Low-oxygen hypoxia: The effects of low-oxygen hypoxia are partly concentration-related and partly dose-related. When a subject reaches equilibrium with respect to different oxygen concentrations, the effects are approximately as follows:

20.9–14.4%—no significant effects, slight loss of exercise tolerance.
14.4–11.8%—slight effects on memory and mental task performance, reduced exercise tolerance.
11.8–9.6%—severe incapacitation, lethargy, euphoria, loss of consciousness.
9.6–7.8%—loss of consciousness, death.

The time taken to achieve a blood oxygen concentration causing incapacitation depends on the dose of hypoxia acquired over that period. For input into the model, time to loss of consciousness is given by

$$(t_{Io})\text{min} = \exp[8.13 - 0.54(20.9 - \%O_2)] \quad (10)$$

where $20.9 - \%O_2 = \%O_2\text{Vit}$ (percent oxygen vitiation).

From this the expression for fractional incapacitating dose per minute is derived

$$F'_{Io} = \frac{1}{\exp[8.13 - 0.54(20.9 - \%O_2)]} \quad (11)$$

Carbon dioxide: As with hypoxia, the effects of carbon dioxide are partly concentration-related, but it is also possible to calculate the time taken to acquire a dose capable of causing loss of consciousness. The concentration-related effects of carbon dioxide are approximately:

3–6%—respiratory distress, increasing with concentration.
6–7%—severe respiratory distress, dizziness, bordering on loss of consciousness.
7–10%—loss of consciousness.

There are, therefore, two important considerations with respect to carbon dioxide: (1) it greatly increases the RMV, which will increase the rate of uptake of other toxic gases; and (2) it is itself a narcotic.

It is therefore necessary to calculate a multiplication factor (VCO_2) to allow for the effect of the increased RMV caused by carbon dioxide on the rate of uptake of other toxic gases. The expression for this is

$$VCO_2 = \frac{\exp(0.2496 \times \%CO_2 + 1.9086)}{6.8} \quad (13)$$

It is also necessary to calculate the fractional incapacitating dose of carbon dioxide. Time to unconsciousness by carbon dioxide is given by

$$t_{Ico_2} = \exp(6.1623 - 0.5189 \times \%CO_2) \quad (14)$$

From this the fractional incapacitating dose per minute expression is derived:

$$F'_{Ico_2} = \frac{1}{\exp(6.1623 - 0.5189 \times \%CO_2)} \quad (15)$$

Interactions between narcotic gases: For the purposes of the hazard model, the following interaction factors are used:

1. CO and HCN are considered to be directly additive.
2. CO_2 increases the rate of uptake of CO and HCN in proportion to its effect on the RMV.
3. The narcotic effect of low oxygen hypoxia is considered to be directly additive to the combined effects of CO and HCN.
4. The narcotic effect of CO_2 is considered to act independently of the effect of the other gases.

On this basis, it is possible to derive a fractional incapacitating dose equation for narcosis (Equation 16)

$$\text{Total } F_{IN} = [(F_{Ico} + F_{Icn}) \times VCO_2 + F_{Io}] \quad \text{or} \quad F_{Ico_2} \quad (16)$$

where

F_{IN} = fraction of an incapacitating dose of all narcotic gases
F_{Ico} = fraction of an incapacitating dose of CO
F_{Icn} = fraction of an incapacitating dose of HCN
VCO_2 = multiplication factor for CO_2-induced hyperventilation
F_{Io} = fraction of an incapacitating dose of low-oxygen hypoxia; and
F_{Ico_2} = fraction of an incapacitating dose of CO_2

For a one-minute exposure to each gas at a concentration, C,

$$F'_{Ico} = \frac{8.2925 \times 10^{-4} \times \text{ppm CO}^{1.036}}{30}$$

$$F'_{Icn} = \frac{1}{\exp(5.396 - 0.023 \times \text{ppm HCN})}$$

$$VCO_2 = \frac{\exp(1.903 \times \%CO_2 + 2.0004)}{7.1}$$

$$F'_{Io} = \frac{1}{\exp[8.13 - 0.54(20.9 - \%O_2)]}$$

$$F'_{Ico_2} = \frac{1}{\exp(6.1623 - 0.5189 \times \%CO_2)}$$

An example using data obtained from a large-scale fire test (the single armchair room burn shown in Figures 2-8.7 and 2-8.11) is given in Table 2-8.3. The histograms in Figure 2-8.11 show the average concentrations of narcotic gases each minute for the first six minutes of the fire, which are shown numerically in Table 2-8.3. Using the equations given above, the fractional incapacitating dose for each narcotic gas is calculated for each minute and these data are shown in Table 2-8.4. The first two rows give the fractional doses for CO and HCN, which are added together and multiplied by the carbon dioxide hyperventilation factor VCO_2. The fractional dose of low-oxygen hypoxia is added to give a total fractional dose for narcosis for each minute. The running total summed each minute exceeds unity during the fifth minute, giving a figure of 1.199, and indicating the onset of incapacitation (loss of consciousness). Alternatively, narcosis may occur due to the effects of carbon dioxide, but the cumulative dose of this gas is only 0.065 during the fifth minute, which is insufficient to have any narcotic effect.

Physical factors and irritancy: Having calculated the effects of narcotic gases the next steps are to assess the effects of radiant and convected heat, smoke obscuration, sensory irritation, and lung irritation.

Radiant heat: As shown in Figure 2-8.16, there is a fairly obvious intensity limit for tolerance of radiant heat at 0.25 W/cm² (2.5 kW/m²). Below this intensity, radiant heat can be tolerated for at least several minutes, but above this intensity for a few seconds only. The curves of radiant heat flux and of the other physical parameters for the first ten minutes of the armchair fire are shown in Figure 2-8.17. The tenability limit is exceeded for approximately one minute during the sixth minute of the fire, and it is predicted that some degree of pain and skin burns might be sustained during that minute due solely to the effects of radiant heat.

Convected heat: The curve for tolerance time of convected heat is shown in Figure 2-8.13, and from this is derived the expression for time to incapacitation

$$t_{Ih}(\text{min}) = \exp[5.1849 - 0.0273\ T(°C)] \qquad (18)$$

As with the narcotic gases, it is possible to use the concept of a fractional incapacitating dose of heat acquired each minute as follows

$$F'_{Ih} = \frac{1}{\exp[5.1849 - 0.0273\ T(°C)]} \qquad (19)$$

The average temperature each minute during the fire and the fractional incapacitating dose of heat are shown in Table 2-8.12. The cumulative fractional dose exceeds unity during the fourth minute (2.273) as the temperature exceeds 220°C, and then continues to increase dramatically during the fifth and sixth minutes. There will also be some degree of added effect from the radiant heat which would further increase the fractional dose. Incapacitation due to skin pain and burns is therefore predicted during the fourth minute, with severe and probably fatal burns of the skin and upper respiratory tract being a strong possibility, particularly after the fourth minute.

Smoke: Visual obscuration by smoke is obviously concentration-related, and a tenability limit of extinction coefficient 1.2/m (OD/m = 0.5) has been set. As Figure 2-8.17 shows, this limit is exceeded during the second minute of the armchair fire, and an extinction coefficient 2.4/m (OD/m = 1.0), or approximately 1 m visibility, is exceeded at the beginning of the third minute, at which point the smoke curve rises very steeply.

Sensory and lung irritancy: As stated in the section on irritancy, there are two factors to consider: (1) the immediate incapacitation due to the painful effects of sensory irritation of the eyes and respiratory tract, adding to the obscurational effects of smoke and disrupting escape behavior; and (2) the later inflammatory effects on the lung which may cause death after exposure.

The first consideration is whether the victim would be able to escape from the fire. In this context sensory irritation is the most important. This is concentration-related, and in order to predict the irritancy of the smoke, it is necessary to know the RD_{50} concentration of the atmosphere produced by the materials involved under the particular decomposition conditions existing in the fire. Most importantly, it is necessary to know the concentration/time profile of the fire products in terms of mass loss per liter of air (NAC mass loss). Although the mass loss curve for the armchair is shown in Figure 2-8.17, there are no data on the volume of air into which this mass was dispersed during the fire. For the purposes of this example it will be necessary to make an estimate of possible mass loss concentration. Also, since the RD_{50} of the polyurethane and polystyrene components of the chair under flaming conditions are unknown, it will be necessary to use estimated values.

In the section on irritancy, a general tenability limit for severe sensory irritation was set at a concentration of 1 mg/L NAC mass loss, and an incapacitating dose for serious postexposure lung inflammation was set at 10 mg/L NAC mass loss for 30 min (a Ct product of 300 mg · min/L). From the general conditions, the smoke curve, and the CO concentration curve, it is estimated that the tenability limit for sensory irritancy would be exceeded during the third minute, greatly adding to the deleterious effects of smoke on vision and escape behavior.

With regard to lung irritation, it is estimated that the average mass loss concentration over the first five minutes of the fire would be approximately 10 mg/L. If so, this would represent a fractional incapacitating dose of 50 mg · min/L, which would probably be insufficient compared with the more serious effects of heat exposure, to cause significant lung damage after exposure. However, if the average mass loss concentration over the first five minutes should reach 60 mg/L, then serious effects on the lung would likely occur after, and probably during exposure.

Interactions: In terms of physiological effects, it is likely that there would be some degree of interaction between narcosis and several of these physical factors, but it is likely that most would be relatively minor during the fire, and a reasonable model can be used in which narcosis, sensory irritancy, and the effects of heat and visual obscuration can be treated separately (with the possible exception of the effects of CO_2-induced hyperventilation on the uptake of irritants). At the behavioral level interactions may be more important. The interaction between sensory irritation and visual obscuration has been mentioned and there is some experimental evidence for such an interaction in humans.[62] After exposure, as mentioned in the section of this chapter on heat, the effects of skin burns, respiratory tract burns, and chemical irritation (and even possibly CO narcosis) all combine to increase the probability of fatal pulmonary edema and inflammation.

Summary of Model Predictions for Armchair Fire

From the analysis, the effects on a victim exposed to the conditions in the armchair room burn (Figure 2-8.7) are predicted as follows:

1. Toward the end of the second minute and the beginning of the third minute, the smoke optical density and mass loss/liter would exceed the tenability limits for visual obscuration and sensory irritancy sufficiently to severely inhibit escape from the room.
2. During the fourth minute, the average temperature is 220°C, and sufficient heat would be accumulated in the skin surface to cause skin burns resulting in incapacitation.
3. During the fifth minute, a victim is likely to lose consciousness due to the combined effects of the accumulated doses of narcotic gases.
4. It is predicted that a victim escaping or rescued after the fourth minute would suffer severe postexposure effects due to skin burns, possible laryngeal burns with accompanying edema and danger of obstructive asphyxia, and also pulmonary edema and inflammation which might well be fatal (due to the combined effects of inhaled hot gases, chemical irritants, and the pulmonary secondary effects of skin burns). After the sixth minute, it is likely that a victim would die at some time between a few minutes and one hour due to the effects of narcosis, circulatory shock, and possibly hyperthermia.

It is unlikely that an otherwise healthy adult would be able to escape from a fire such as this if he/she remained longer than three minutes after ignition. Three minutes is a long time in which to leave a room, however, so that providing the victim is awake and aware of the fire, is not otherwise incapacitated, and does not stay after two minutes in an attempt to fight the fire or rescue belongings, it is likely that he or she would be able to escape without serious injury. Tenability limits for some 5- and 30-minute exposures to common toxic fire products in terms of time to incapacitation and death are shown in Appendix 2-8B.

APPENDIX 2-8A

Coburn-Forster-Kane Equation for the Uptake of Carbon Monoxide in Man

The Coburn-Forster-Kane (CFK) equation[18] provides an accurate method for predicting the blood carboxyhemoglobin concentration in humans (or various animal species) resulting from exposure to a given concentration of carbon monoxide. The theoretical predictions of the equation have been validated for humans by experimental human exposures to carbon monoxide.[19,35] The strength of the equation is that it is based upon numerical values for all the main constants and variables that determine the uptake of CO into the blood; it is therefore not based on a simple empirical fit to observed uptake data in individuals, as are other CO uptake equations.[15,43] The result is a powerful equation that can be used to predict CO uptake over wide ranges of con-

centrations and time scales, and can accommodate variables such as the degree of activity of the subject, body size (for men, women, or children), blood volume, hemoglobin concentration, and lung function status, all of which can affect CO uptake and therefore time to incapacitation or death for a given subject. The equation should be equally applicable to animals, providing of course that data for the various constants and variables are available. The disadvantage of the CFK equation is its complexity. In particular, several of the variables need to be calculated from other equations, which in turn contain variables which must be calculated from further equations. The data in the text and figures in this chapter are all based on CO uptake for a 70-kg human, either at rest [RMV approximately 8.5 L/min, engaged in light work (e.g., walking 6.4 km/h—RMV approximately 25 L/min)] or engaged in heavy work (e.g., slow running 8.5 km/h or walking 5.6 km/h up a 17 percent gradient—RMV approximately 50 L/min).[36] In this Appendix, data for necessary constants and equations for the derivation of all variables, with their sources, have been provided to enable uptake calculations to be made for any particular situation.

The basic form of the CFK equation[18] is as follows:

$$\frac{A[HbCO]_t - BV_{CO} - PI_{CO}}{A[HbCO]_0 - BV_{CO} - PI_{CO}} = e^{-tAV_bB}$$

where

$$\begin{aligned}
A &= \bar{P}_{C,O_2}/M[HbO_2] \\
\bar{P}_{C,O_2} &= PI_{O_2} - 49 \\
PI_{O_2} &= 148.304 - 0.0208 \times PI_{CO} \\
M &= 218 \\
[HBO_2] &= 0.22 - [HbCO]_t \\
[HbCO]_t &= [COHb\%t] \times 0.0022 \\
B &= 1/DL_{CO} + PL/VA \\
DL_{CO} &= 35VO_2 \times e^{0.33\ \dagger} \\
VO_2 &= RMV/22.274 - 0.0309^{\ddagger} \\
PL &= 713 \\
V_A &= 0.933\ V_E - 132f \\
f &= \exp[0.0165 \times RMV + 2.3293]^\dagger
\end{aligned}$$

where

$[HbO_2]$ = mL of O_2 per mL of blood

$[HbCO]_t$ = mL of CO per mL of blood at time of t

$[HbCO]_0$ = mL of CO/mL blood at the beginning of the exposure. Can be taken as 0.8 percent COHb = 0.0176 mL CO/mL blood for a nonsmoker[17]

M = ratio of the affinity of blood for CO to that for O_2

P_{C,O_2} = average partial pressure of oxygen in lung capillaries, mmHg

$\dot{V}_{CO}$ = rate of endogenous CO production, mL/min very small set at 0.007 mL/min[19]

DL_{CO} = diffusivity of the lung for CO, mL/min · mmHg

PL = barometric pressure minus the vapor pressure of water at body temperature, mmHg

V_b = blood volume, mL; 74 mL/kg body weight (approximately 5,500 mL for a 70 kg human)[19]

PI_{CO} = partial pressure of CO in inhaled air, mmHg

V_A = alveolar ventilation rate, mL/min

t = exposure duration, min

e = 2.7182

---

†Expression derived from Reference 114.

‡Expression derived from Reference 36.

TABLE 2-8B(a) *Tenability Limits for Incapacitation or Death from Exposures to Common Narcotic Products of Combustion*

	5 minutes		30 minutes	
	Incapacitation	Death	Incapacitation	Death
CO	6000–8000 ppm	12000–16000 ppm	1400–1700 ppm	2500–4000 ppm[25,26]
HCN	150–200 ppm	250–400 ppm	90–120 ppm	170–230 ppm[23,26]
Low O_2	10–13%	<5%	<12%	6–7%[17,21,26]
CO_2	7–8%	>10%	6–7%	>9%[17,21,26]

PIO_2 = partial pressure of oxygen in inspired air
$COHb\%t$ = percent carboxyhemoglobin at time of t
V_{O_2} = oxygen consumption, L/min
RMV = respiratory minute volume (volume of air breathed/min, liters)
V_E = RMV, mL
f = respiratory frequency, breaths/min

APPENDIX 2-8B

Tenability Limits

Narcotics: Concentrations at which there would be danger of incapacitation (loss of consciousness) and death after approximately 5 and 30 minutes exposure in a person engaged in light activity are shown in Table 2-8B (a).

Irritants: The initial painful effects of irritants (sensory irritation) are mainly upon the eyes and upper respiratory tract. These affects do not worsen with prolonged exposure and may even lessen. The toxic effects on the lungs increase with prolonged exposure, are often most serious some hours after exposure, and may cause death.

For sensory irritation two levels are presented: level a represents unpleasant and quite severely disturbing eye and upper respiratory tract irritation; and level b represents severe eye and upper respiratory tract irritation with severe pain, blepharospasm, copious lacrymation, and mucus secretion accompanied by chest pain. For deaths, the levels represent concentrations at which there is danger of death occurring during or immediately after exposure.

In general, smokes are irritating when they contain oxidized organic products.[12] The most irritating of these substances known to occur commonly in smokes from a number of different materials is acrolein. Another well-known irritant is the acid gas hydrogen chloride, which is evolved during the thermal decomposition of polyvinyl chloride (PVC). Data on these two products are presented in Table 2-8B (b) as examples of irritant effects.

APPENDIX 2-8C

Glossary of Terms

Acidosis: A condition in which the pH of the blood is lowered (i.e., becomes more acidic). Respiratory acidosis in fire exposures results from excess carbon dioxide uptake. Metabolic acidosis results from impaired tissue respiration (due to tissue hypoxia) caused by burns or narcosis. (See alkalosis.)

Addition: Two or more toxic substances are considered to exert an additive effect when they act in concert, such that the effect in combination is greater than the effect of either substance acting alone, but not greater than the sum of the effects of either substance acting alone (when they may be said to be directly additive). (See also synergism.)

Aerodynamic diameter: The aerodynamic diameter of a particle is an expression of particle size, and represents the diameter of a spherical particle of unit density with the same aerodynamic properties as the particle under consideration.

Aerosol: Solid or liquid particles dispersed in air.

Alkalosis: Respiratory alkalosis occurs when the pH of the blood is increased (i.e., becomes more alkaline). It is caused by excess removal of carbon dioxide from the blood via the lungs during hyperventilation, and may cause a loss of consciousness.

Asphyxia: Suffocation, decrease in the oxygen content, and increase in the carbon dioxide content of the blood. This may occur due to laryngeal spasm caused by burns or irritant gases, or to impairment of breathing or gas exchange in the lung. The term has been extended to include all causes of tissue hypoxia, including exposure to asphyxiant gases (low oxygen concentration due to the excess of any other gas, or exposure to the narcotic gases carbon monoxide and hydrogen cyanide, which produce asphyxia chemically).

Atmosphere: (fire atmosphere or test atmosphere). The total airborne medium to which a victim or experimental animal is exposed, consisting of solid and liquid particles and vapors dispersed in air.

Behavioral effects/incapacitation: The extent to which exposure to fire products affects the ability or willingness of a subject or experimental animal to perform coordinated movements or tasks, particularly movements or tasks similar to those required to escape from a fire. (See incapacitation.)

Bioassay: Originally a term reserved for the use of a biological system to detect and/or measure the amount of a

TABLE 2-8B(b) *Tenability Limits for Sensory Irritation or Death from Irritant Substances*

	Sensory irritation		Death (minutes)		
	a	b	5	10	30
Acrolein (ppm)	1–5[17,60]	5–95[60,61]	500–1000[61]	150–690[60,110]	50–135[60,110]
HCl (ppm)	75–300[17]	300–1100[60,61]	12000–16000[60,61]	10000[60]	2000–4000[17]

biologically active material. In the fire context, it refers to the use of animal exposures rather than chemical analysis to determine the toxicity of a combustion product atmosphere.

Blepharospasm: Involuntary and sustained closure (spasm) of the eyelids. In fires this is due to the painful stimulation of the cornea by combustion products that are sensory irritants.

Bronchoconstriction: Constriction of the conducting airways in the lung due to the contraction of smooth muscle in the airway walls in response to an agonist or to stimulation of irritant receptors acting through the vagus nerve.

Burn: Tissue lesion caused by heat or chemicals. For description of burn types and degrees, see text.

Carboxyhemoglobin (COHb): Combination of carbon monoxide with hemoglobin in the blood, which limits the combination of hemoglobin with oxygen (oxyhemoglobin), and therefore the carriage of oxygen in the blood.

Cerebral depression: Condition in which the electrical activity of the cerebral cortex as revealed in the electroencephalogram consists mainly of slow wave (or delta wave) activity. This is typical of a semiconscious or unconscious state.

Combustion products: Strictly speaking, this means the products of flaming decomposition, and is used in this sense when contrasted with thermal decomposition products. However, in general usage, the term "combustion products" may be taken to include all fire products, whether produced by flaming or nonflaming thermal decomposition.

Concentration: The amount of a contaminant in the atmosphere per unit volume of the atmosphere, usually quoted as mass/volume (mg/L or mg/m^3) or volume/volume (ppm or percent). (See nominal atmosphere concentration.)

Dose: The amount of a toxicant to which a fire victim or test animal is exposed. The simplest estimation of dose for inhalation toxicology is to multiply the atmosphere concentration by the duration of exposure (Ct product). A lethal dose may be expressed in terms of the LC t$_{50}$. However, other factors may affect the amount of toxicant actually entering the body, and for fires it may be necessary to express dose in terms of the material in the fire. (See nominal atmosphere concentration.)

Edema: Accumulation of an excessive amount of fluid in cells, tissues, or body cavities. Pulmonary edema occurs when a fluid exudate leaks out of blood vessels as a result of inflammation or circulatory insufficiency, and the lung tissues become swollen and waterlogged. Further development results in a fluid exuded within the alveolar spaces. This fluid accumulation seriously affects gas exchange in the lung and may be fatal.

Electroencephalogram: Waves of electrical activity in the cerebral cortex recorded from the surface of the head, which give an indication of the physiological state of the brain and the degree of alertness of the subject. A preponderance of fast (beta and alpha) activity indicates a conscious and normal state, whereas a preponderance of slow (theta and delta) activity signifies a physiologically depressed or unconscious state.

Erythema: Reddening of the skin in response to heat. This change coincides with pain and just precedes a skin burn.

Fire profile: Record of the changes with time of the concentrations of important fire products and intensities of physical parameters during the course of a fire.

Flaming fire: In the context of this chapter, this refers to the early stages of fire growth (preflashover), when the fire is still confined to burning items within a well-defined area.

Flashover: Point in growth of a flaming fire where the flames are no longer confined to burning items, but also occur within the fire effluent, remote from the seat of the fire.

Fractional incapacitating dose: The dose of a toxic product acquired during a short period of time, expressed as a fraction of the dose required to cause incapacitation at the average exposure concentration during that time interval. The fractional incapacitating doses acquired during each short time period are summed throughout the exposure, incapacitation occurring when the fraction reaches unity.

Fully developed fire: A fire that has reached its maximum extent of growth, usually extending throughout the fire compartment.

Haber's rule: Principle that toxicity in inhalation toxicology depends on the dose available and that the product of concentration and exposure time is a constant.

Hazard: A toxic fire hazard exists when a toxic product is present at a sufficient concentration and over a sufficient period of time to cause a toxic effect. A physical fire hazard exists when a physical fire parameter (heat or smoke), is present at an intensity and over a period sufficient to cause injury or seriously inhibit the ability to escape from a fire.

Hypercapnia: Increased blood carbon dioxide concentration.

Hyperthermia (heat stroke): An increase in body temperature above 37°C. Hyperthermia is life-threatening if the body core temperature, or temperature of the blood entering the heart, exceeds 42.5°C.

Hyperventilation: Increased rate and depth of breathing (increased respiratory minute volume, or RMV), in response to increased carbon dioxide, hypoxic hypoxia, hydrogen cyanide, exercise, heat, or stimulation of pulmonary irritant receptors.

Hypoxia: A reduction in the amount of oxygen available for tissue respiration. This can occur in four ways:

Anemic hypoxia: in which the arterial PO$_2$ is normal, but the amount of hemoglobin available to carry oxygen is reduced and/or the ability to release oxygen to the tissues is impaired. For fire exposures this results mainly from the formation of carboxyhemoglobin following exposure to CO, but an anemic subject would be at increased risk.

Histotoxic hypoxia: in which the amount of oxygen delivered to the tissues is adequate, but due to the action of a toxic agent such as HCN, the tissue cells cannot make use of the oxygen supplied to them.

Hypoxia hypoxia (low-oxygen hypoxia): in which the PO$_2$ of the arterial blood is reduced as a result of a low atmospheric oxygen concentration or impairment of gas exchange in the lung, due to bronchoconstriction or respiratory tract damage or disease.

Ischaemic hypoxia: in which blood flow to a tissue is so low that adequate oxygen is not delivered to it despite a normal PO_2 and hemoglobin concentration. This occurs during shock following burns and in cerebral tissue due to alkalosis, or briefly during postural hypotension.

Incapacitation: An inability to perform a task (related to escape from a fire) caused by exposure to a toxic substance or physical agent in a fire. A distinction is sometimes made between severe physiological incapacitation, in which the subject is unable to move normally, such as might occur in an unconscious or badly burned victim, and the more behavioral incapacitation, such as that caused by visual obscuration or eye irritation from smoke, in which the victim is more or less intact, but still unable to escape from the fire.

Inflammation: A complex of reactions occurring in blood vessels and adjacent tissues around the site of an injury. The initial reaction is congestion (engorgement of local blood vessels), exudation of fluid into the tissues (edema), and pain. This is followed by a phase of destruction and removal of injured tissue by inflammatory cells, and then a phase of repair.

Intensity: Level of a harmful physical fire parameter (such as radiant heat flux, air temperature, or smoke optical density).

Intoxication: A state in which a subject is adversely affected by a toxic substance. Specifically, the time at which a subject has taken up sufficient of a narcotic (asphyxiant) gas that he or she behaves like someone severely affected by alcohol.

Irritation and irritancy: Irritation is the action of an irritant substance, irritancy is the response. This response takes two forms:

Pulmonary (lung) irritant: response occurs when an irritant penetrates into the lower respiratory tract. This may result in breathing discomfort (dyspnea), bronchoconstriction, and an increase in respiratory rate during the fire exposure. In severe cases it is followed after a period (usually of a few hours) by pulmonary inflammation and edema, which may be fatal.

Sensory irritant: response occurs when an irritant substance comes in contact with the eyes and upper respiratory tract (and sometimes the skin), causing a painful sensation accompanied by inflammation with lacrimation or mucus secretion. At low concentrations, this adds to the visual obscuration caused by smoke, but at high concentrations the severe effects may cause behavioral, and to some extent physiological, incapacitation. Sensory irritation causes a decrease in respiratory rate which is transient in humans, but continuous in rodents.

Lacrimation: The production of tears in response to sensory irritation of the eyes.

LC$_{50}$: Lethal concentration—50 percent. The concentration statistically calculated to cause the deaths of one half of the animals exposed to a toxicant for a specified time. It may be expressed as volume/volume (ppm, percent) or mass/volume (mg/L). Care must be taken in comparing LC$_{50}$s of both the exposure duration and the postexposure period over which deaths were scored. In combustion toxicology, the LC$_{50}$ may be related to the test material rather than its products, and expressed in terms of the nominal atmosphere concentration of material either of mass charge or mass loss. (See nominal atmosphere concentration.)

LC t$_{50}$: The product of exposure concentration and duration causing the deaths of 50 percent of animals.

Narcosis: Literally "sleep induction," but used in combustion toxicology to describe central nervous system depression causing reduced awareness, intoxication, and reduced escape capability, leading to loss of consciousness and death in extreme cases. The narcotic gases CO, HCN, and CO_2 cause narcosis, as does lack of oxygen due to the inhalation of an atmosphere low in oxygen, an impairment of breathing, or an impairment of gas exchange in the lung. The terms "narcosis" and "narcotic gases" are used synonomously with the terms "asphyxia" and "asphyxiant gases."

Nominal Atmosphere Concentration, NAC: The theoretical concentration of test substance in a test atmosphere, calculated from the mass of test substance produced from the atmosphere generation system each minute divided by the air volume into which it is generated. This concept is not directly applicable to combustion toxicology since the test material is decomposed in the fire or furnace system, but two derivative concepts are used to relate the test material to the degree of toxicity as follows:

Nominal Atmosphere Concentration mass charge, NAC mass charge: The mass of material placed in the furnace system per volume of air into which it is dispersed (mg material/liter).

Nominal Atmosphere Concentration mass loss, NAC mass loss: The mass loss of material during decomposition per volume of air into which it is dispersed (mg material/liter).

Physiological effects: Effects of chemical fire products or physical fire parameters on the functioning of the body, as opposed to parameters affecting the mind. Thus, a physiological effect of smoke is that it obscures vision, which might have a psychological effect on the willingness of a victim to enter a smoke-filled corridor.

Pneumonia (Pneumonitis): Inflammation of the lungs, in fire victims due to the direct effects of inhaled chemicals or hot gases, or secondarily to skin burns. The initial inflammatory phase may be followed by infection. As it passes through different phases, pneumonia may be life-threatening at any time from one hour after exposure in a fire to several weeks after exposure.

Potency: The toxic potency is a measure of the amount of a toxic substance required to elicit a specific toxic effect—the smaller the amount required, the greater the potency.

Psychological effects: Psychological effects of exposure to fire scenarios are on the mind of the victim, and may result in a variety of behavioral effects. These are distinct from physiological effects on body function (see above). A fire victim is likely to suffer both types of effect at various stages of a fire, and interactions between psychological and physiological effects are likely.

Psychomotor: Psychomotor skills are required to perform behavioral tasks involving a series of coordinated movements of the type required to escape from a fire in a compartment (such as a building).

Pyrolysis: In this chapter, the term "pyrolysis" is restricted to the thermal decomposition of materials without oxidation. In small-scale tests pyrolysis may be achieved by heating the material in a stream of nitrogen.

RD$_{50}$: Respiratory depression 50 percent—statistically calculated concentration of a sensory irritant required to reduce the breathing rate of laboratory rodents (usually mice) by 50 percent.

Respiratory Minute Volume, RMV: Volume of air breathed each minute (liters/minute). RMV = TV × RR.

Respiratory Rate, RR: Respiratory frequency, i.e., number of breaths per minute.

Respiratory tract: The nose, pharynx, larynx, trachea, and large bronchi are termed the upper respiratory tract, and the bronchioli, alveolar ducts, and alveoli are termed the lower respiratory tract.

Shock: A reduction in the circulating blood volume with a fall in blood pressure.

Smoke: Total fire effluents, consisting of solid and liquid particles and vapors.

Smoldering/Nonflaming Oxidative Decomposition: Thermal decomposition in which there is partial oxidation of the pyrolysis products, but no flame. This may result from overheating of materials by means of an external heat source, or from self-sustained smoldering.

Specific toxicity: A particular adverse effect caused by a toxicant (e.g., narcosis, irritancy).

Supertoxicant: A term used to describe a toxicant with an unusual specific toxicity not usually associated with fire effluents, often with a high potency.

Synergism: Situation where the toxic potency of two or more substances acting in concert is greater than the sum of the potencies of each substance acting alone.

Tenability limit: Maximum concentration of a toxic fire product or intensity of a physical fire parameter that can be tolerated without causing incapacitation.

Thermal decomposition: Chemical breakdown of a material induced by the application of heat.

Tidal Volume, TV: Volume of air exhaled in each breath.

Toxicity: The nature and extent of adverse effects of a substance upon a living organism.

Ventilation, lung: The volume of air breathed each minute (synonymous with respiratory minute volume).

REFERENCES CITED

1. NFPA, *The Reconstruction of a Tragedy, the Beverly Hills Supper Club Fire.* National Fire Protection Association, Quincy MA (1978).
2. P.C. Bowes, *Ann. Occup. Hyg.,* 17, 143 (1974).
3. Home Office, *United Kingdom Fire Statistics 1983,* London (1983) (Published annually).
4. B.F. Clarke, in *Fire Deaths, Causes and Strategies for Control,* Technomic, Lancaster (1984).
5. V. Berl and B. Halpin, *NBS Grant Contract Report NBS-GCR 168,* National Bureau of Standards, Gaithersburg (1979).
6. J.H. Petajan, K.L. Voorhees, S.C. Packham, R.C. Baldwin, I.N. Einhorn, M.L. Grunnet, B.G. Dinger, and M.M. Birky, *Sci.,* 187, 742 (1975).
7. B.C. Levin, A.J. Fowell, M.M. Birky, M. Paabo, S. Stolte, and D. Malek, *NBSIR 82-2532,* National Bureau of Standards, Washington (1982).

8. R.C. Anderson, P.A. Croce, F.G. Feeley, and J.D. Sakura, *Reference 88712,* Arthur D. Little, Cambridge (1983).
9. C.J. Hilado, *Mod. Plastics,* July (1977).
10. R.A. Anderson, A.A. Watson, and W.A. Harland, *Med. Sci. Law,* 21, 175 (1981).
11. H.L. Kaplan, A.F. Grand, and G.E. Hartzell, *Combustion Toxicology: Principles and Test Methods,* Technomic, Lancaster (1983).
12. D.A. Purser and W.D. Woolley, *J. Fire Sci.,* 1, 118 (1983).
13. W.D. Woolley and P.J. Fardell, *F. Safety J.,* 5, 29 (1982).
14. B.T. Commins and P.J. Lawther, *Br. J. Ind. Med.,* 22, 139 (1965).
15. R.D. Stewart, *J. Comb. Toxicol.,* 1, 167 (1974).
16. F. Haber, *Funf Vortrange aus den jaren 1920-1923,* Verlag von Julius Springer, Germany (1924).
17. *Documentation of the Threshold Limit Values for Substances in Workroom Air,* American Conference of Governmental Industrial Hygienists, Cincinnati (1980).
18. R.F. Coburn, R.E. Forster, and P.B. Kane, *J. Clin. Invest.,* 44, 1899 (1965).
19. J. E. Peterson and R.D. Stewart, *J. Appl. Physiol.,* 39, 633 (1975).
20. F.W. Beswick, P. Holland, and K.H. Kemp, *Br. J. Ind. Med.,* 29, 298 (1972).
21. D.A. Purser, *J. Fire. Sci.,* 2, 20 (1984).
22. U.C. Luft, in *Handbook of Physiology,* American Physiology Society, Washington (1965).
23. D.A. Purser, P. Grimshaw, and K.R. Berrill, *Arch. Environ. Hlth.,* 39, 394 (1984).
24. D.A. Purser, *Determination of blood cyanide and its role in producing incapacitation in fire victims,* Royal Society of Chemistry Meeting, Huntingdon (1984).
25. D.A. Purser and K.R. Berrill, *Arch. Environ. Hlth.,* 38, 308 (1983).
26. G. Kimmerle, *J. Comb. Toxicol.,* 1, 4 (1974).
27. J.S. Haldane, in *Respiration,* Yale University, New Haven (1922).
28. R.D. Stewart, J.E. Peterson, E.D. Baretta, H.C. Dodd, and A.A. Herrmann, *Arch. Environ. Hlth.,* 21, 154 (1970).
29. R.D. Stewart, J.E. Peterson, T.N. Fisher, M.J. Hosko, E.D. Baretta, H.C. Dodd, and A.A. Herrmann, *Arch. Environ. Hlth.,* 26, 1 (1973).
30. R.A. Anderson, A.A. Watson, and W.A. Harland, *Med. Sci. Law,* 21, 175 (1981).
31. Consumer Product Safety Commission, *Federal Register,* 45-182, 61925 (1980).
32. R.D. O'Donnel, P. Miluka, P. Heinig, and J. Theodorem, *Toxicol. App. Pharmacol.,* 18, 593 (1971).
33. *Carbon monoxide poisoning due to faulty gas water heaters in British holidaymakers on the Algarve,* Sunday Times, London (1983).
34. K. Von Leggenhager, *Acta. Med. Scand.,* 196/suppl. 563, 1 (1974).
35. H. Hauck and M. Neuberger, *Eur. J. Appl. Physiol.,* 53, 186 (1984).
36. *Handbook of Respiration,* Saunders, Philadelphia (1958).
37. R.A. Anderson, I. Thompson, and W.A. Harland, *Fire and Matls.,* 3, 91 (1979).
38. C.J. Clark, D. Campbell, and W.H. Reid, *Lancet,* June 20th, 1332 (1981).
39. B. Ballantyne, *Fund. Appl. Toxicol.,* 3, 400 (1983).
40. B.P. McNamara, *EA-TR-76023,* Edgewood Arsenal, Aberdeen Proving Ground, Maryland (1976).
41. J.L. Bonsall, *Human Toxicol.,* 3, 57 (1984).
42. J. Barcroft, *J. Hyg.,* 31, 1 (1931).
43. G.E. Hartzell, D.N. Priest, and W.G. Switzer, *J. Fire Sci.,* 3, 115 (1985).
44. L.C. Luft, in *Clinical Cardiopulmonary Physiology,* Grune and Stratton, Orlando (1969).
45. C.J. Lambertson, *Anaesthesiology,* 21, 642 (1960).
46. J. H. Comroe, R.E. Forster, A.B. Dubois, W.A. Briscoe, and E. Carlsen, in *The Lung,* Year Book Medical Publishers, Chicago (1962).

47. P. Altman and D.S. Ditter, ed., *Environmental Biology*, Federation of American Societies for Experimental Biology, Bethesda (1966).
48. B.G. King, *J. Ind. Hyg. Toxicol.*, 31, 365 (1949).
49. J.E. Schulte, *Arch. Environ. Hlth.*, 8, 427 (1964).
50. W.S. Root, in *Handbook of Physiology*, American Physiology Society, Washington, (1985).
51. F.A. Gibbs, E.L. Gibbs, W.G. Lennox, and L.F. Nims, *J. of Aviation Med.*, 14, 250 (1943).
52. A.A. Karl, G.R. McMillan, S.L. Ward, A.T. Kissen, and M.E. Souder, *Aviation, Space, and Environ. Med.*, 49, 984 (1978).
53. D.A. Purser and P. Grimshaw, *Fire and Matls.*, 8, 10 (1984).
54. J.W. Heim, *J. Aviation Med.*, 10, 211 (1939).
55. R.A. McFarland, F.J. Roughton, M.H. Halperin, and J.I. Niven, *J. Aviation Med.*, 15, 381 (1944).
56. D.A. Purser and P. Buckley, *Med. Sci. Law*, 23, 142 (1983).
57. D. Campbell, *Respiratory tract trauma in burned patients*, Colloquium, Borehamwood (1985).
58. L. Kane, C.S. Barrow, and Y. Alarie, *Am. Ind. Hyg. Assoc. J.*, 40, 207 (1979).
59. C.L. Punte, E.J. Owens, and P.J. Gutentag, *Arch. Environ. Hlth.*, 6, 366 (1963).
60. Y. Alarie, *Proceedings of the Inhalation Toxicology Symposium, Upjohn Company, October 23-24 1980*, Ann Arbor Science (The Butterworth Group) Ann Arbor (1980).
61. H.L. Kaplan, A.F. Grand, W.R. Rogers, W.G. Switzer, and G.C. Hartzell, *DOT/FAA/CT-84/16*, Federal Aviation Administration, Washington (1984).
62. T. Jin, *J. Fire Flamm.*, 12, 130 (1981).
63. Y. Alarie, *CRC Crit. Rev. Toxicol.*, 2, 299 (1973).
64. D. Canter, *Studies of human behavior in fire: Empirical results and their implications for education and design*, University of Surrey, Surrey, (1983).
65. R. Melzack and P.D. Wall, *Sci.*, 150, 971 (1965).
66. D.G. Clark, S. Buch, J.E. Doe, H. Frith, and D.H. Pullinger, *Pharmacol. Ther.*, 5, 149 (1979).
67. C.S. Barrow, H. Lucia, M.F. Stock, and M.F. and Y. Alarie, *Am. Ind. Hyg. Assoc. J.*, 40, 408 (1979).
68. T.J. Cole, J.E. Cotes, G.R. Johnson, H. deV. Martin, J.W. Reed and M.J. Saunders, *Q. J. Exp. Physiol.*, 62, 130 (1977).
69. Y. Alarie, and R.C. Anderson, *Toxicol. Appl. Pharmacol.*, 51, 341 (1979).
70. Y. Alarie, *Fd. Cosmet. Toxicol.*, 19, 623 (1981).
71. D.A. Purser, *Combustion toxicity research and animal models*, Society of Plastics Industry Meeting, Hilton Head (1983).
72. W.J. Potts and T.S. Lederer, *J. Comb. Toxicol.*, 5, 182 (1978).
73. D.A. Purser, unpublished data (1984).
74. D.A. Purser, unpublished data (1982-85).
75. W.D. Woolley, *Br. Polym. J.*, 4, 27 (1972).
76. W.D. Woolley, S.A. Ames, and P.J. Fardell, *Fire Matls.*, 3, 110 (1979).
77. V. Babrauskas, *Technical Note 1103*, National Bureau of Standards, Washington (1979).
78. P.J. Berenson and W.G. Robertson, in *Bioastronautics Data Book*, Biotechnology, Virginia (1972).
79. D.L. Simms and P.L. Hinkley, *Fire Research Special Report No. 3*, Her Majesty's Stationery Office, London (1963).
80. W.V. Blockley, in *Biology Data Book*, Federation of American Societies for Experimental Biology, Bethesda (1973).
81. D.C. Edholm, in *A Companion To Medical Studies*, Blackwell, Oxford (1968).
82. J.H. Veghte, *F. Service Today*, 49,16 (1982).
83. H. Elneil, in *A Companion To Medical Studies*, Blackwell, Oxford (1968).
84. C.S. Leithead and A.R. Lind, in *Heat Stress and Heat Disorders*, Cassel and Co., London (1963).
85. A.R. Moritz, F.C. Henriques, F.R. Dutra, and J.R. Weisiger, *Arch. Pathol.*, 43, 466 (1947).
86. K. Buettner, *J. Appl. Physiol.*, 3, 703 (1951).
87. J.P. Lawrence and J.C. Bull, *Eng. Med.*, 5, 61 (1976).
88. J.P. Bull and J.C. Lawrence, *Fire Matls.*, 3, 100 (1979).
89. *PD 6504*, British Standards Institution, London (1983).
90. A.R. Moritz, F.C. Henriques, and R. McLean, *Am. J. Path.*, 21, 311 (1945).
91. J.B. Perkins, H.E. Pearse, and H.D. Kingsley, *UR217 Atomic Energy Project*, University of Rochester (1952).
92. B.D. Dinman, *J. Am. Med. Assoc.*, 235, 2874 (1976).
93. Y. Enson, R.M. Harvey, M.L. Lewis, W.B. Greenough, K.M. Ally, and R.A. Panno, *J. Ann. NY Acad. Sci.*, 150, 577 (1968).
94. W.E. Zimmermann, *J. Ann. NY Acad. Sci.*, 150, 584 (1968).
95. S.M. Rosenthal and R.C. Millican, *J. Ann. NY Acad. Sci.*, 150, 604 (1968).
96. B.C. Zawacki, R.C. Jung, J. Joyce, and E. Rincon, *Ann. Surg.*, 185, 100 (1977).
97. H.N. Harrison, *J. Ann. NY Acad. Sci.*, 150, 627 (1968).
98. B.A. Zikria, W.O. Sturner, N.K. Astarjian, C.L. Fox, Jr., and J.M. Ferrer, *J. Ann. NY Acad. Sci.*, 150, 618 (1968).
99. A. Silcock, D. Robinson, and N.P. Savage, *Current Paper CP80/78*, Building Research Establishment, Borehamwood (1978).
100. D.A. Purser, unpublished experimental data (1985).
101. J.G. Quintiere, M.M. Birky, F. Macdonald, and B. Smith, *Fire Matls.*, 6, 99 (1982).
102. W.J. Potts and T.S. Lederer, *J. Comb. Toxicol.*, 4, 113 (1977).
103. *DIN 53 436.*
104. G. Kimmerle and F.C. Prager, *J. Comb. Toxicol.*, 7, 54 (1980).
105. H. Klimisch, H.W. Hollander, and J. Thyssen, *J. Comb. Toxicol.*, 7, 209 (1980).
106. V. Babrauskas, *NBSIR 82-2611*, National Bureau of Standards, Washington (1982).
107. R. C. Anderson and Y.C. Alarie, *J. Comb. Toxicol.*, 5, 54 (1978).
108. W.E. Coleman, L.D. Scheel, R.E. Kupel, and R.L. Larkin, *Am. Ind. Hyg. Assoc. J.*, 29, 33 (1968).
109. S.J. Williams and F.B. Clarke, *Fire Matls.*, 6, 161 (1982).
110. *Registry of toxic effects of chemical substances*, National Institute for Occupational Safety and Health, Washington (1982).
111. C.L. Punte, J.T. Weimer, T.A. Ballard, and J.L. Wilding, *Toxicol. Appl. Pharmacol.*, 4, 656 (1962).
112. B. Ballantyne and S. Calloway, *Med. Sci. Law*, 12, 43 (1972).
113. H. Salem, and H. Cullumbine, *Toxicol. Appl. Pharmacol.*, 2, 183 (1960).
114. T.E. Bernard and J. Duker, *Am. Ind. Hyg. Assoc. J.*, 42, 271 (1981).
115. C. Herpol and P. Vandevelde, *F. Safety J.*, 4, 271 (1981).
116. G.E. Hartzell, B.C. Levin, and Y. Alarie, *Personal Communications* (1988).
117. D.J. Rasbash, *Fire Intl.*, 5 (40) 30 (1975).
118. G.V. Alexeev and S.C.Packham, *J. Fire Sci.*, 2, 362 (1984).
119. D.A. Purser, *Proceedings of First International Fire and Materials Conference*, Interscience Communications, London (1992).
120. B. Ballantyne, in *Forensic Toxicology*, Wright, Bristol (1974).
121. F.J. Baud, P. Barriot, and V. Toffis, *et al*, *New Eng. J. Med.*, 325, 1761 (1991).
122. D.A. Purser, "Modelling Time to Incapacitation and Death from Toxic and Physical Hazards in Aircraft Fires." In: Conference *Proceedings* No. 467, Aircraft Fire Safety, NATO-AGARD, Sintra, Portugal, 22–26 May, pp 41–1 to 41–13 (1989).
123. D.A. Purser, in *Proceedings* of Interflam '93, Interscience Communications, London (1993).
124. D.A. Purser, "The Development of Toxic Hazard in Fires from Polyurethane Foams and the Effects of Fire Retardants," in *Proceedings of Flame Retardants 90*, British Plastics Federation, ed, Elsevier, London pp. 206–221 (1990).
125. D.F. King, "*Aircraft Accident Report 8/88*," Air Accidents Investigation Branch, Her Majesty's Stationery Office, London (1989).
126. J. Sime, "Crowd Safety Design, Communications and Management: The Psychology of Escape Behaviour," Easinwold Papers No. 4, pp. 16–29, Home Office Emergency Planning College, York (1992).
127. *The Fire at Woolworth's, Piccadilly, Manchester, on 8 May 1979*, Home Office, Fire Department (1980).

128. "Toxicity Testing of Fire Effluents—Part 5. Prediction of Toxic Effects of Fire Effluents," ISO/IEC TR 9122-5 (in press).

129. "Guide for the Assessment of Toxic Hazard of Materials in Fire in Buildings and Transport," British Standards Institution, BSI DD 180 (1989).

130. D.A. Purser, in *Proceedings* of Flame Retardants '94 Conference, Interscience Communications, London (1994).

131. "Toxicity Testing of Fire Effluents—Part 4, The Fire Model." ISO/IEC TR 9122-4, (in press).

132. G.E. Hartzell, W.G. Switzer, and D.N. Priest, "Modeling of Toxicological Effects of Fire Gases," *J. Fire Sci.* 3:330–342 (1985).

133. B.C. Levin, M. Paabo, J.L. Gurman, and S.C. Harris, "Effects of Exposure to Single or Multiple Combinations of the Predominant Toxic Gases and Low-Oxygen Atmospheres Produced in Fires," *Fund. Appl. Toxicol.* 9:236–250 (1987).

134. G.E. Hartzell, S.C. Packham, A.F. Grand, and W.G. Switzer, "Modeling of Toxicological Effects of Fire Gases: III. Quantification of Post-Exposure Lethality of Rats from Exposure to HCl Atmospheres," *J. Fire Sci.* 3:196–207 (1985).

135. G.E. Hartzell, A.F. Grand, and W.G. Switzer, "Modeling of Toxicological Effects of Fire Gases: VI. Further Studies on the Toxicity of Smoke Containing Hydrogen Chloride. *Advances in Combustion Toxicology, Vol. II,* G.E. Hartzell, ed., Technomic, Lancaster, PA, pp. 285–308 (1989).

136. W.A. Burgess, R.D. Trietman, and A. Gold, "IR Contaminants in Structural Firefighting," *Final Report to the National Fire Prevention and Control Administration and the Society of Plastics Industry Inc.,* Harvard School of Public Health, Cambridge, MA (1979).

ADDITIONAL READING

V. Babrauskas, B.C. Levin, R.G. Gann, *et al*, Toxic Potency Measurement for Fire Hazard Analysis, NIST Special Publication 827 (1991).

FLAMMABILITY LIMITS OF PREMIXED AND DIFFUSION FLAMES

Craig Beyler

INTRODUCTION

It is well known that not all fuel/oxidant/diluent mixtures can propagate flame. There exist limits definable in terms of fuel/oxidant/diluent composition outside which normal flame-type combustion cannot be sustained. Definition of these limits has received a great deal of attention in premixed combustion conditions, that is in systems where the fuel and oxidant are mixed prior to combustion. Despite scientific interest in the subject dating back to the 19th century, the mechanism responsible for flammable limits is not yet understood. Nonetheless, a great deal has been learned that has practical application.

Much less investigation into the nature and cause of limits in diffusion flames has been undertaken. Empirically, clear parallels exist between diffusion and premixed limits, and these will be explored in the latter portion of this chapter.

PREMIXED COMBUSTION

Premixed flame fronts can only propagate within a range of compositions of fuel and oxidant. The composition limits within which a flame can propagate are known as the upper and lower flammable limits, and are expressed as concentrations of the fuel in a specified oxidant/diluent mixture at a specified temperature and pressure. For instance, the lower flammable limit (LFL) of methane in air at normal temperature and pressure is 5 percent by volume, and the upper flammable limit (UFL) is 15 percent by volume. As such, only methane/air mixtures with methane concentrations between 5 and 15 percent methane will support propagation of flame. For most simple hydrocarbons, the lower and upper flammable limits in air correspond to an equivalence ratio of approximately 0.5 and 3, respectively. The lower flammable limit concentrations for these fuels is approximately 48 g/m^3. (See Figure 2-9.1.)

The most widely used method of measuring flammable limits was developed by the U.S. Bureau of Mines.[1] The apparatus consists of a 1.5 m long, 0.05 m diameter vertical

tube which is filled with the fuel/oxidant/diluent mixture to be tested. The top of the tube is closed, and the base of the tube can be closed until the start of the test to prevent diffusion of the mixture from the tube. With the base of the tube open, the mixture is ignited by a spark or small pilot flame at the base of the tube, and the travel of the flame front up the tube is observed. The mixture is deemed to be within the flammable limits if the flame can propagate halfway up the 1.5 m tube. The test is designed to identify the range of mixture compositions capable of flame propagation remote from the ignition source.

The apparatus can be used with ignition at the top of the tube, but the flammable limits determined for downward propagation are narrower than for upward propagation. The 0.05 m diameter of the tube was chosen as the smallest diameter at which the heat losses from the flame to the tube wall had minimal effect on the flammable limits determined. (See Figure 2-9.2.)

Mixtures are capable of combustion outside the flammable limits, but external energy must be provided throughout the mixture volume in order to allow propagation of a flame.[2] An example of this behavior is shown in Figure 2-9.3. A small hydrogen diffusion flame is used as a pilot source in a lean methane/air mixture. At methane concentrations less than 5 percent, combustion occurs only in the

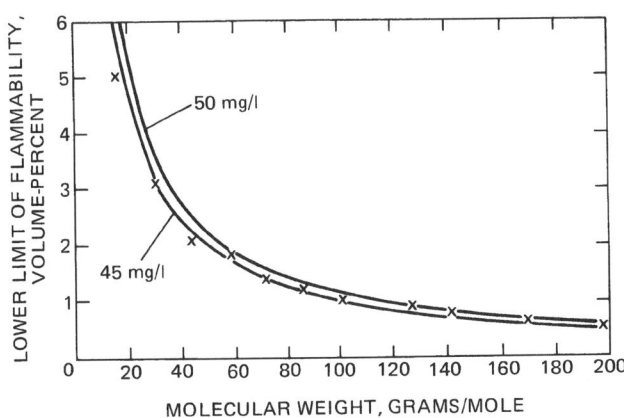

Fig. 2-9.1. Effect of molecular weight on lower limits of flammability of alkanes at 25°C.[3]

Dr. Beyler earned a Ph.D. in Engineering Science at Harvard University under the direction of Professor Howard Emmons, and served on the faculty of Worcester Polytechnic Institute's Center for Firesafety Studies. Dr. Beyler is currently Technical Director of Hughes Associates, Inc., Fire Science and Engineering.

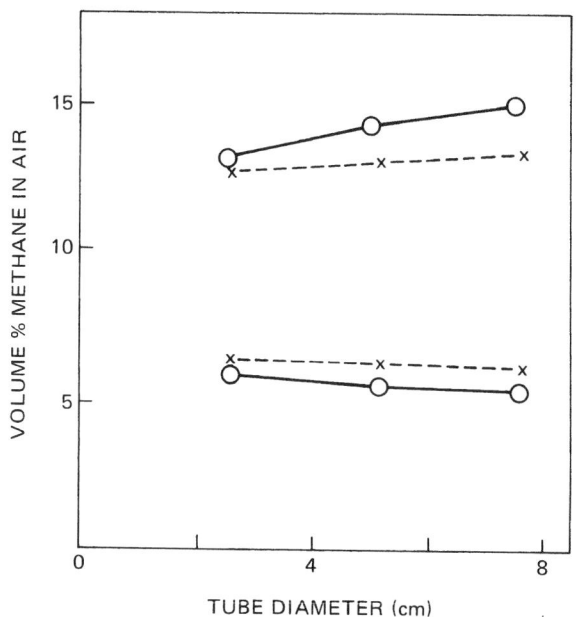

Fig. 2-9.2. *Upper and lower flammable limits of methane as determined in a vertical tube apparatus for upward propagation (circles), and for downward propagation (crosses).*[20]

wake of the pilot flame. Above 5 percent, the flame can propagate away from the pilot flame, regardless of the orientation of the pilot flame.

Flammable limits are a function of the oxygen and inert concentrations, as well as the mixture temperature and pressure. As the concentration of inerts is reduced and the oxygen concentration is increased, the upper flammable limit is increased, while the lower limit is relatively unchanged. This result can be understood by observing that at the lower flammable limit there is always more than enough oxygen present for complete combustion, while at the upper limit less than the stoichiometrically required oxygen is present. Hence, at the upper limit the additional oxygen participates in the combustion process, while at the lower limit the additional oxygen simply replaces inert gas.

The lower flammable limit is also insensitive to the pressure, except at pressures well below atmospheric. The upper limit shares this insensitivity at subatmospheric pressures, but the upper limit increases with increasing pressure above atmospheric. (See Figure 2-9.4.)

The flammable limits widen with increases in mixture temperature as illustrated in Figure 2-9.5; this will be discussed further later in this chapter. Figure 2-9.5 also relates flammable limits with the saturation vapor curve and the autoignition temperature (AIT). The flashpoint of a liquid is given in the figure as T_L. At that temperature, the vapor pressure at the liquid surface is at the lower flammable limit. The corresponding upper limit temperature is given as T_U. If a liquid is contained within a closed vessel and the vapors are allowed to come into equilibrium at temperatures above the upper limit temperature, the vapors in the vessel will be above the upper flammable limit. This typically occurs in an automobile gas tank. If the liquid is not enclosed fully, there will be a location above from the surface of the liquid where

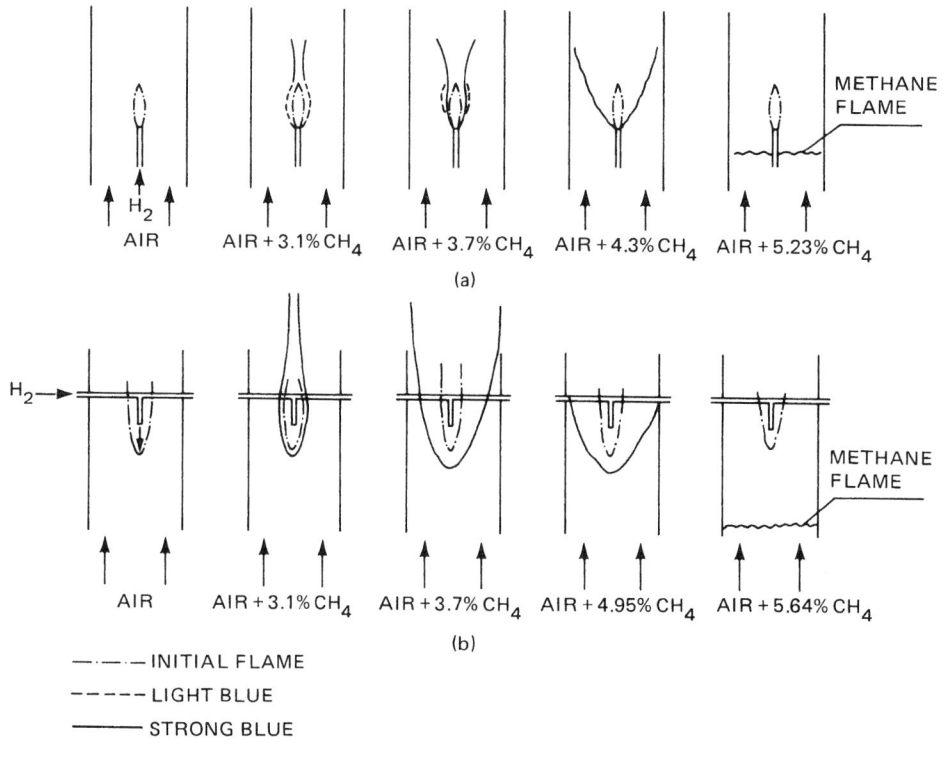

—·—·— INITIAL FLAME

- - - - - LIGHT BLUE

———— STRONG BLUE

Fig. 2-9.3. *A small jet diffusion flame in a coflowing (a) and contraflowing (b) stream as the concentration of the fuel in the stream is gradually increased up to ignition. The stream velocity is 0.222 m/s, and the hydrogen jet diameter is 1.52 mm.*[2]

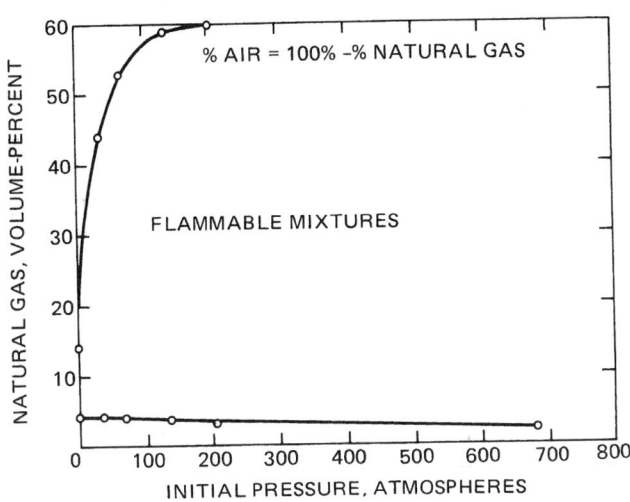

Fig. 2-9.4. Effect of pressure on the limits of flammability of natural gas in air at 28°C.[3]

the fuel/air mixture will be diluted below the upper flammable limit and will ignite if an ignition source is present.

Predicting Lower Flammable Limits of Mixtures of Flammable Gases (Le Chatelier's Rule)

Based on an empirical rule developed by Le Chatelier in the late 19th century, the lower flammable limit of mixtures of multiple flammable gases in air can be determined. A generalization of Le Chatelier's rule was given by Coward *et al.*[4]

$$\sum_{i=1}^{n} (C_i/\text{LFL}_i) \geq 1 \qquad (1)$$

where C_i is the volume percent of fuel gas, i, in the fuel/air mixture and LFL_i is the volume percent of fuel gas, i, at its lower flammable limit in air alone. If the indicated sum is

greater than unity, the mixture is above the lower flammable limit. This can be restated in terms of the lower flammable limit concentration of the fuel mixture, LFL_m, as follows:

$$\text{LFL}_m = \frac{100}{\sum_{i=1}^{n} (C_{fi}/\text{LFL}_i)} \qquad (2)$$

where C_{fi} is the volume percent of fuel gas i in the fuel gas mixture.

EXAMPLE 1:

A mixture of 50 percent methane, 25 percent carbon monoxide, and 25 percent hydrogen is mixed with air. Calculate the lower flammmable limit of this fuel gas mixture.

SOLUTION:

Referring to Table 2-9.1, LFLs of methane, carbon monoxide, and hydrogen are 5.0 percent, 12.5 percent, and 4.0 percent by volume, respectively. Using Equation 2 we find

$$\text{LFL}_m = \frac{100}{50/5 + 25/12.5 + 25/4} = 5.48\%$$

The composition of the lower flammable limit fuel/air mixture is 2.74 percent methane, 1.37 percent carbon monoxide, 1.37 percent hydrogen, and 94.5 percent air.

Critical Adiabatic Flame Temperature at the Lower Flammable Limit

As early as 1911, Burgess and Wheeler[5] noted the constancy of the potential heat release rate per unit volume of normal alkane/air lower flammable mixtures at room temperature. Since the heat capacity of the products of complete combustion are nearly the same for all hydrocarbons, their observation also implies that the adiabatic flame temperature at the lower flammable limit is a constant. Examination of a wide range of C,H,O-containing fuels indicates that the adiabatic flame temperature at the LFL is approximately 1600 K (± 150 K) for most C,H,O-containing fuels, with the following notable exceptions: hydrogen, 980 K; carbon monoxide, 1300 K; and acetylene, 1280 K. This indicates that the adiabatic flame temperature at the lower flammable limit is an indication of the reactivity of the fuel; the lower the adiabatic flame temperature, the more reactive the fuel.

The utility of the concept of a critical adiabatic flame temperature at the lower flammable limit goes beyond that outlined above. It has been demonstrated that the adiabatic flame temperature at the lower flammable limit is relatively insensitive (± 100 K) to the diluent used and to the initial temperature of the mixture.[6–8]

The adiabatic flame temperature at the limit is insensitive to initial temperature only so long as significant preflame combustion reactions do not occur. As such, for a mixture near or above its autoignition temperature (AIT) for a significant length of time, the adiabatic flame temperature at the limit is not expected to be constant. Weinberg[8] has shown that a mixture of 1 percent methane (LFL = 5 percent at 293 K) in air can burn if it is preheated to 1270 K, even though the flame only increases that temperature by about 250 K, in accordance with the expected adiabatic flame temperature. This was achieved by mixing the methane and air just before the flame so that preflame reactions were not allowed to proceed significantly.

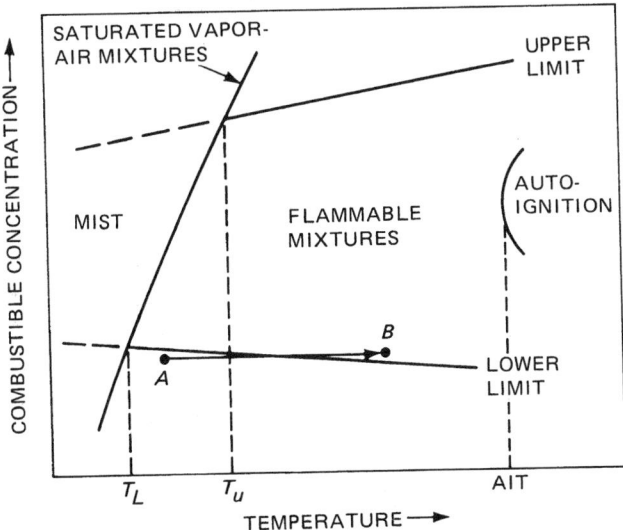

Fig. 2-9.5. Effect of temperature on limits of flammability of a combustible vapor in air at constant initial pressure.[3]

TABLE 2-9.1   *Summary of Limits of Flammability, Lower Temperature Limits (T_L), and Minimum Autoignition Temperatures (AIT) of Individual Gases and Vapors in Air at Atmospheric Pressure[3]*

Combustible	Limits of flammability (volume-percent)		T_L(°C)	AIT (°C)	Combustible	Limits of flammability (volume-percent)		T_L(°C)	AIT (°C)
	L_{25}	U_{25}				L_{25}	U_{25}		
Acetal	1.6	10	37	230	m-Cresol	[8]1.1	--------	------	-----
Acetaldehyde	4.0	60	------	175	Crotonaldehyde	2.1	[11]16	------	-----
Acetic acid	[1]5.4	------	40	465	Cumene	[1]0.88	[1]6.5	------	425
Acetic anhydride	[2]2.7	[3]10	47	390	Cyanogen	6.6	--------	------	-----
Acetanilide	[4]1.0	------	------	545	Cycloheptane	1.1	6.7	------	-----
Acetone	2.6	13	------	465	Cyclohexane	1.3	7.8	------	245
Acetophenone	[4]1.1	--------	------	570	Cyclohexanol	[4]1.2	--------	------	300
Acetylacetone	[4]1.7	--------	------	340	Cyclohexene	[1]1.2	--------	------	-----
Acetyl chloride	[4]5.0	--------	------	390	Cyclohexyl acetate	[4]1.0	--------	------	335
Acetylene	2.5	100	------	305	Cyclopropane	2.4	10.4	------	500
Acrolein	2.8	31	------	235	Cymene	[1]0.85	[1]6.5	------	435
Acrylonitrile	3.0	--------	−6	-----	Decaborane	0.2	--------	------	-----
Acetone-cyanohydrin	2.2	12	------	-----	Decalin	[1]0.74	[1]4.9	57	250
Adipic acid	[4]1.6	--------	------	420	n-Decane	[12]0.75	[13]5.6	46	210
Aldol	[4]2.0	--------	------	250	Deuterium	4.9	75	------	-----
Allyl alcohol	2.5	18	22	-----	Diborane	0.8	88	------	-----
Allyl amine	2.2	22	------	375	Diesel fuel (60 cetane)	--------	--------	------	225
Allyl bromide	[4]2.7	--------	------	295	Diethyl amine	1.8	10	------	-----
Allyl chloride	2.9	--------	−32	485	Diethyl analine	[4]0.8	--------	80	630
o-Aminodiphenyl	0.66	4.1	------	450	1,4-Diethyl benzene	[1]0.8	--------	------	430
Ammonia	15	28	------	-----	Diethyl cyclohexene	0.75	--------	------	240
n-Amyl acetate	[1]1.0	[1]7.1	25	360	Diethyl ether	1.9	36	------	160
n-Amyl alcohol	[1]1.4	[1]10	38	300	3,3-Diethyl pentane	[1]0.7	--------	------	290
tert-Amyl alcohol	[4]1.4	--------	------	435	Diethyl ketone	1.6	--------	------	450
n-Amyl chloride	[5]1.6	[1]8.6	------	260	Diisobutyl carbinol	[1]0.82	[10]6.1	------	-----
tert-Amyl chloride	[6]1.5	--------	−12	345	Diisobutyl ketone	[1]0.79	[1]6.2	------	-----
n-Amyl ether	[4]0.7	--------	------	170	2-4,Diisocyanate	--------	--------	120	-----
Amyl nitrite	[4]1.0	--------	------	210	Diisopropyl ether	1.4	7.9	------	-----
n-Amyl propionate	[4]1.0	--------	------	380	Dimethyl amine	2.8	--------	------	400
Amylene	1.4	8.7	------	275	2,2-Dimethyl butane	1.2	7.0	------	-----
Aniline	[7]1.2	[7]8.3	------	615	2,3-Dimethyl butane	1.2	7.0	------	-----
Anthracene	[4]0.65	--------	------	540	Dimethyl decalin	[1]0.69	[9]5.3	------	235
n-Amyl nitrate	1.1	--------	------	195	Dimethyl dichlorosilane	3.4	--------	------	-----
Benzene	[1]1.3	[1]7.9	------	560	Dimethyl ether	3.4	27	------	350
Benzyl benzoate	[4]0.7	--------	------	480	n,n-Dimethyl formamide	[1]1.8	[1]14	57	435
Benzyl chloride	[4]1.2	--------	------	585	2,3-Dimethyl pentane	1.1	6.8	------	335
Bicyclohexyl	[1]0.65	[8]5.1	74	245	2,2-Dimethyl propane	1.4	7.5	------	450
Biphenyl	[9]0.70	--------	110	540	Dimethyl sulfide	2.2	20	------	205
2-Biphenylamine	[4]0.8	--------	------	450	Dimethyl sulfoxide	--------	--------	84	-----
Bromobenzene	[4]1.6	--------	------	565	Dioxane	2.0	22	------	265
Butadiene (1,3)	2.0	12	------	420	Dipentene	[8]0.75	[8]6.1	45	237
n-Butane	1.8	8.4	−72	405	Diphenylamine	[4]0.7	--------	------	635
1,3-Butandiol	[4]1.9	--------	------	395	Diphenyl ether	[4]0.8	--------	------	620
Butene-1	1.6	10	------	385	Diphenyl methane	[4]0.7	--------	------	485
Butene-2	1.7	9.7	------	325	Divinyl ether	1.7	27	------	-----
n-Butyl acetate	[5]1.4	[1]8.0	------	425	n-Dodecane	[4]0.60	--------	74	205
n-Butyl alcohol	[1]1.7	[1]12	------	----	Ethane	3.0	12.4	−130	515
sec-Butyl alcohol	[1]1.7	[1]9.8	21	405	Ethyl acetate	2.2	11	------	-----
tert-Butyl alcohol	[1]1.9	[1]9.0	11	480	Ethyl alcohol	3.3	[11]19	------	365
tert-Butyl amine	[1]1.7	[1]8.9	------	380	Ethyl amine	3.5	--------	------	385
n-Butyl benzene	[1]0.82	[1]5.8	------	410	Ethyl benzene	[1]1.0	[1]6.7	------	430
sec-Butyl benzene	[1]0.77	[1]5.8	------	420	Ethyl chloride	3.8	--------	------	-----
tert-Butyl benzene	[1]0.77	[1]5.8	------	450	Ethyl cyclobutane	1.2	7.7	------	210
n-Butyl bromide	[1]2.5	--------	------	265	Ethyl cyclohexane	[14]2.0	[14]6.6	------	260
Butyl cellosolve	[8]1.1	[10]11	------	245	Ethyl cyclopentane	1.1	6.7	------	260
n-Butyl chloride	1.8	[1]10	------	-----	Ethyl formate	2.8	16	------	455
n-Butyl formate	1.7	8.2	------	-----	Ethyl lactate	1.5	--------	------	400
n-Butyl stearate	[4]0.3	--------	------	355	Ethyl mercaptan	2.8	18	------	300
Butyric acid	[4]2.1	--------	------	450	Ethyl nitrate	4.0	--------	------	-----
α-Butyrolactone	[8]2.0	--------	------	-----	Ethyl nitrite	3.0	50	------	-----
Carbon disulfide	1.3	50	------	90	Ethyl propionate	1.8	11	------	440
Carbon monoxide	12.5	74	------	-----	Ethyl propyl ether	1.7	9	------	-----
Chlorobenzene	1.4	--------	21	640	Ethylene	2.7	36	------	490

[1] $T = 100°$ C.   [5] $T = 50°$ C.   [9] $T = 110°$ C.   [13] $T = 86°$ C.   [17] $T = 125°$ C.   [21] $T = 43°$ C.   [24] $T = 96°$ C.   [27] $T = 247°$ C.
[2] $T = 75°$ C.   [6] $T = 85°$ C.   [10] $T = 175°$ C.   [14] $T = 130°$ C.   [18] $T = 200°$ C.   [22] $T = 195°$ C   [25] $T = 70°$ C.   [28] $T = 30°$ C.
[3] $T = 75°$ C.   [7] $T = 140°$ C.   [11] $T = 60°$ C.   [15] $T = 72°$ C.   [19] $T = 78°$ C.   [23] $T = 160°$ C.   [26] $T = 29°$ C.   [29] $T = 203°$ C.
[4] Calculated.   [8] $T = 150°$ C.   [12] $T = 53°$ C.   [16] $T = 117°$ C.   [20] $T = 122°$ C.

TABLE 2-9.1 *Summary of Limits of Flammability, Lower Temperature Limits (T_L), and Minimum Autoignition Temperatures (AIT) of Individual Gases and Vapors in Air at Atmospheric Pressure[3] (Continued)*

Combustible	Limits of flammability (volume-percent)		T_L(°C)	AIT (°C)	Combustible	Limits of flammability (volume-percent)		T_L(°C)	AIT (°C)
	L_{25}	U_{25}				L_{25}	U_{25}		
Ethyleneimine	3.6	46	------	320	Monoisopropyl bicyclohexyl	0.52	[18]4.1	124	230
Ethylene glycol	[4]3.5	--------	------	400	2-Monoisopropyl biphenyl	[10]0.53	[18]3.2	141	435
Ethylene oxide	3.6	100	------	-----	Monomethylhydrazine	4	--------	------	-----
Furfural alcohol	[15]1.8	[16]16	72	390	Naphthalene	[19]0.88	[20]5.9	------	526
Gasoline:					Nicotine	[1]0.75	--------	------	-----
100/130	1.3	7.1	------	440	Nitroethane	3.4	--------	30	-----
115/145	1.2	7.1	------	470	Nitromethane	7.3	--------	33	-----
Glycerine	--------	--------	------	370	1-Nitropropane	2.2	--------	34	-----
n-Heptane	1.05	6.7	−4	215	2-Nitropropane	2.5	--------	27	-----
n-Hexadecane	[4]0.43	--------	126	205	n-Nonane	[21]0.85	--------	31	205
n-Hexane	1.2	7.4	−26	225	n-Octane	0.95	--------	13	220
n-Hexyl alcohol	[1]1.2		------	-----	Paraldehyde	1.3	--------	------	-----
n-Hexyl ether	[4]0.6	--------	------	185	Pentaborane	0.42	--------	------	-----
Hydrazine	4.7	100	------	-----	n-Pentane	1.4	7.8	−48	260
Hydrogen	4.0	75	------	400	Pentamethylene glycol	--------	--------	------	335
Hydrogen cyanide	5.6	40	------	-----	Phthalic anhydride	[7]1.2	[22]9.2	140	570
Hydrogen sulfide	4.0	44	------	-----	3-Picoline	[4]1.4	--------	------	500
Isoamyl acetate[1]	1.1	[17]7.0	25	360	Pinane	[23]0.74	[23]7.2	------	-----
Isoamyl alcohol[1]	1.4	[19]9.0	------	350	Propadiene	2.16	--------	------	-----
Isobutane	1.8	8.4	−81	460	Propane	2.1	9.5	−102	450
Isobutyl alcohol	[1]1.7	[1]11	------	-----	1,2-Propandiol	[4]2.5	--------	------	410
Isobutyl benzene	[10]0.82	[10]6.0	------	430	β-Propiolactone	[3]2.9	--------	------	-----
Isobutyl formate	2.0	8.9	------	-----	Propionaldehyde	2.9	17	------	-----
Isobutylene	1.8	9.6	------	465	n-Propyl acetate	1.8	8	------	-----
Isopentane	1.4	--------	------	-----	n-Propyl alcohol	[12]2.2	[1]14	------	440
Isophorone	0.84	--------	------	460	Propyl amine	2.0	--------	------	-----
Isopropylacetate	[4]1.7	--------	------	-----	Propyl chloride	[4]2.4	--------	------	-----
Isopropyl alcohol	2.2	--------	------	-----	n-Propyl nitrate	[17]1.8	[17]100	21	175
Isopropyl biphenyl	[4]0.6	--------	------	440	Propylene	2.4	11	------	460
Jet fuel:					Propylene dichloride	[4]3.1	--------	------	-----
JP-4	1.3	8	------	240	Propylene glycol	[24]2.6	--------	------	-----
JP-6	--------	--------	------	230	Propylene oxide	2.8	37	------	-----
Kerosine	--------	--------	------	210	Pyridine	[11]1.8	[25]12	------	-----
Methane	5.0	15.0	−187	540	Propargyl alcohol	[5]2.4	--------	------	-----
Methyl acetate	3.2	16	------	-----	Quinoline	[4]1.0	--------	------	-----
Methyl acetylene	1.7	--------	------	-----	Styrene	[26]1.1	--------	------	-----
Methyl alcohol	6.7	[11]36	------	385	Sulfur	[27]2.0	--------	247	-----
Methyl amine	[4]4.2	--------	------	430	p-Terphenyl	[4]0.96	--------	------	535
Methyl bromide	10	15	------	-----	n-Tetradecane	[4]0.5	--------	------	200
3-Methyl butene-1	1.5	9.1	------	-----	Tetrahydrofurane	2.0	--------	------	-----
Methyl butyl ketone	S51.2	[1]8.0	------	-----	Tetralin	[1]0.84	[8]5.0	71	385
Methyl cellosolve	[17]2.5	[7]20	------	380	2,2,3,3-Tetramethyl pentane	0.8	--------	------	430
Methyl cellosolve acetate	[8]1.7	--------	46	-----	Tetramethylene glycol	--------	--------	------	390
Methyl ethyl ether	[4]2.2	--------	------	-----	Toluene	[1]1.2	[1]7.1	------	480
Methyl chloride	[4]7	--------	------	-----	Trichloroethane	--------	--------	------	500
Methyl cyclohexane	1.1	6.7	------	250	Trichloroethylene	[28]12	[25]40	30	420
Methyl cyclopentadiene	[1]1.3	[1]7.6	49	445	Triethyl amine	1.2	8.0	------	-----
Methyl ethyl ketone	1.9	10	------	-----	Triethylene glycol	[8]0.9	[28]9.2	------	-----
Methyl ethyl ketone peroxide	--------	--------	40	390	2,2,3-Trimethyl butane	1.0	--------	------	420
Methyl formate	5.0	23	------	465	Trimethyl amine	2.0	12	------	-----
Methyl cyclohexanol	[4]1.0	--------	------	295	2,2,4-Trimethyl pentane	0.95	--------	------	415
Methyl isobutyl carbinol	[4]1.3	--------	40	-----	Trimethylene glycol	[4]1.7	--------	------	400
Methyl isopropenyl ketone	[5]1.8	[5]9.0	------	-----	Trioxane	[4]3.2	--------	------	-----
Methyl lactate	[12]2.2	--------	------	-----	Turpentine	[1]0.7	--------	------	-----
α-Methyl naphthalene	[4]0.8	--------	------	530	Unsymmetrical dimethylhydrazine	2.0	95	------	-----
2, Methyl pentane	[4]1.2	--------	------	-----	Vinyl acetate	2.6	--------	------	-----
Methyl propionate	2.4	13	------	-----	Vinyl chloride	3.6	33	------	-----
Methyl propyl ketone	1.6	8.2	------	-----	m-Xylene	[1]1.1	[16]6.4	------	530
Methyl styrene	[4]1.0	--------	49	495	o-Xylene	[1]1.1	[16]6.4	------	465
Methyl vinyl ether	2.6	39	------	-----	p-Xylene	[1]1.1	[16]6.6	------	530
Methylene chloride	--------	--------	------	615					

[1] $T = 100°$ C.   [5] $T = 50°$ C.   [9] $T = 110°$ C.   [13] $T = 86°$ C.   [17] $T = 125°$ C.   [21] $T = 43°$ C.   [24] $T = 96°$ C.   [27] $T = 247°$ C.

[2] $T = 75°$ C.   [6] $T = 85°$ C.   [10] $T = 175°$ C.   [14] $T = 130°$ C.   [18] $T = 200°$ C.   [22] $T = 195°$ C   [25] $T = 70°$ C.   [28] $T = 30°$ C.

[3] $T = 75°$ C.   [7] $T = 140°$ C.   [11] $T = 60°$ C.   [15] $T = 72°$ C.   [19] $T = 78°$ C.   [23] $T = 160°$ C.   [26] $T = 29°$ C.   [29] $T = 203°$ C.

[4] Calculated.   [8] $T = 150°$ C.   [12] $T = 53°$ C.   [16] $T = 117°$ C.   [20] $T = 122°$ C.

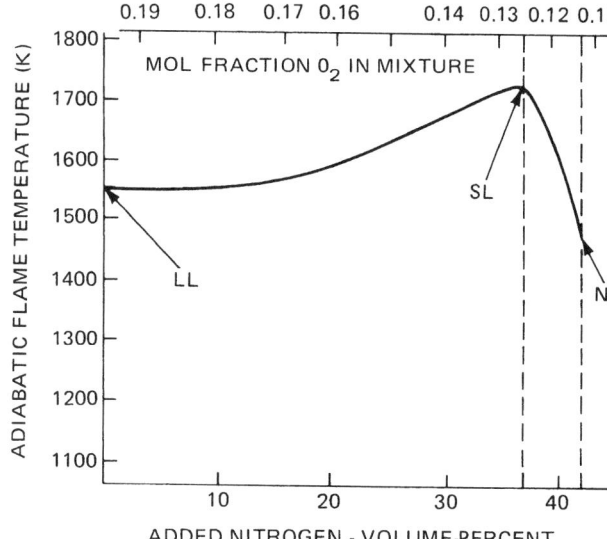

Fig. 2-9.6. *Computed adiabatic flame temperature along the lower branch of the flammability limits of propane (adapted from reference 11). SL and NP are defined in Figure 2-9.9.*

Due to the constancy of the adiabatic flame temperature at the lower limit, the concept can be utilized to predict the effect of variable mixture temperature and diluents on the flammable limits of a mixture. Coward and Jones[1] have examined variable oxygen/diluent ratios, using nitrogen, carbon dioxide, water, argon, and helium as diluents. Their work shows that the limit temperature is insensitive to the oxygen/diluent ratio. Figure 2-9.6 illustrates the change in adiabatic flame temperature at the lower flammable limit as additional nitrogen is added to decrease the oxygen/nitrogen ratio. The figure shows an increase in the adiabatic flame temperature at the lower flammable limit from 1550 K to over 1700 K as we move from normal air to the stoichiometric limit. Beyond the stoichiometric limit, no fuel-lean mixture can burn. The region beyond the stoichiometric limit can be best understood in the context of flammability diagrams and upper flammable limits. We will examine these later in the chapter.

The insensitivity of the limit temperature to the chemical structure of C,H,O-containing fuels contributes significantly to the utility of the concept of a critical adiabatic flame temperature at the lower flammable limit. No systematic evaluation of the limit temperature concept for fuels containing sulfer, nitrogen, or halogens has been undertaken. Existing data indicates that halogen-containing fuels have limit temperatures several hundred degrees higher than C,H,O fuels. Since halogens are combustion inhibitors, this is consistent with the idea that the adiabatic flame temperature at the lower flammable limit is indicative of the reactivity of the fuel. Thus, possible exceptions to the generalization that the adiabatic flame temperature at the lower flammable limit is approximately 1600 K may be identifiable by considering the reactivity of the fuel gas.

Egerton and Powling[9] have shown that the limit temperatures at the upper flammable limit for hydrogen and carbon monoxide are equal to their limit temperatures at the lower flammable limit. Stull[10] has reported the same result for methane. However, it is not generally possible to calculate the adiabatic flame temperature for other fuels, since the products of combustion under fuel-rich conditions include a mixture of products of combustion and pyrolysis, which cannot be predicted by assuming chemical equilibrium is achieved or by detailed chemical kinetics calculations. Equilibrium calculations indicate that the only carbon-containing species that should be produced are CO, CO_2, CH_4, and solid carbon. This is not generally a good approximation under fuel-rich conditions.

EXAMPLE 2:

The lower flammable limit of propane at 20°C is 2.1 percent by volume. Find the lower flammable limit at 200°C.

SOLUTION:

For adiabatic combustion, all the heat released is absorbed by the products of combustion:

$$(LFL/100)\Delta H_c = \int_{T_0}^{T_{f,LFL}} nC_p dT \tag{3}$$

where ΔH_c is the heat of combustion of the fuel, LFL/100 is the mole fraction of fuel, n is the number of moles of products of combustion per mole of fuel/air mixture, C_p is the heat capacity of the products of combustion, T_0 is the initial temperature of the fuel/air mixture, and $T_{f,LFL}$ is the adiabatic flame temperature of a lower flammable limit mixture. This is a utilization of concepts developed in the chapter on thermochemistry. For the present purposes, it is suitable to use an average value of the heat capacity. This reduces Equation 3 to

$$(LFL/100)\Delta H_c = nC_p(T_{f,LFL} - T_0) \tag{4}$$

We know that $T_{f,LFL} = 1600$ K and for $T_0 = 20°C$ we also know that LFL = 2.1 percent. Rearranging Equation 4 yields

$$\Delta H_c/(nC_p) = (T_{F,LFL} - T_0)/(LFL/100)$$
$$= (1600\ K - 293\ K)/(2.1/100)$$
$$= 6.22 \times 10^4\ K$$

Both the heat of combustion and the heat capacity are weak functions of temperature, and these effects will be ignored. As such we can use the above expression to predict the lower flammable limit for an initial temperature of 200°C.

$$(T_{F,LFL} - T_0)/(LFL/100) = (1600\ K - 473\ K)/(LFL/100)$$
$$= 6.22 \times 10^4\ K$$

This gives LFL = 1.8 percent.

Flammability Diagrams

While the flammable limits of a fuel in air can be characterized by the lower and upper flammable limits, it is necessary to represent flammable limits of more general fuel/oxidant/inert mixtures, using flammability diagrams. Examples of flammability diagrams for methane/oxygen/nitrogen mixtures are shown in Figures 2-9.7(a) and (b). Based on an extensive series of tests with a range of mixture compositions, a flammability diagram can be constructed indicating the regions of mixture compositions within the flammable limits.

Two types of flammability diagrams are often used: one utilizing three axes in which each of the three constituent

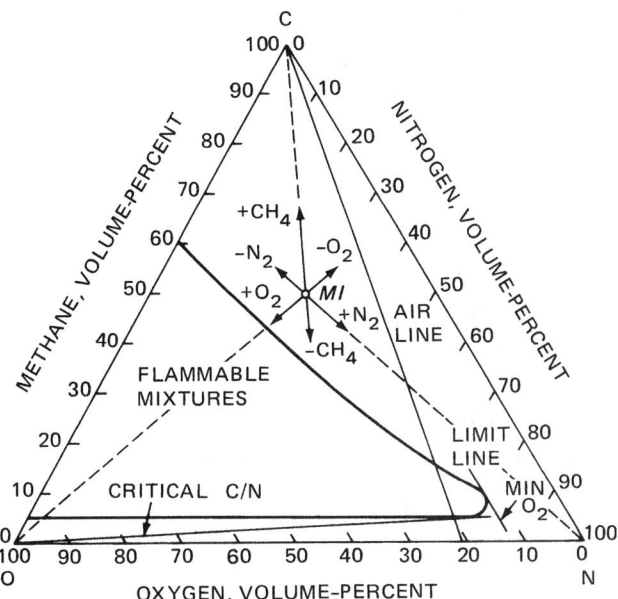

Fig. 2-9.7(a). *Three-axis flammability diagram for the system methane/oxygen/nitrogen at atmospheric pressure and 26°C.*[3]

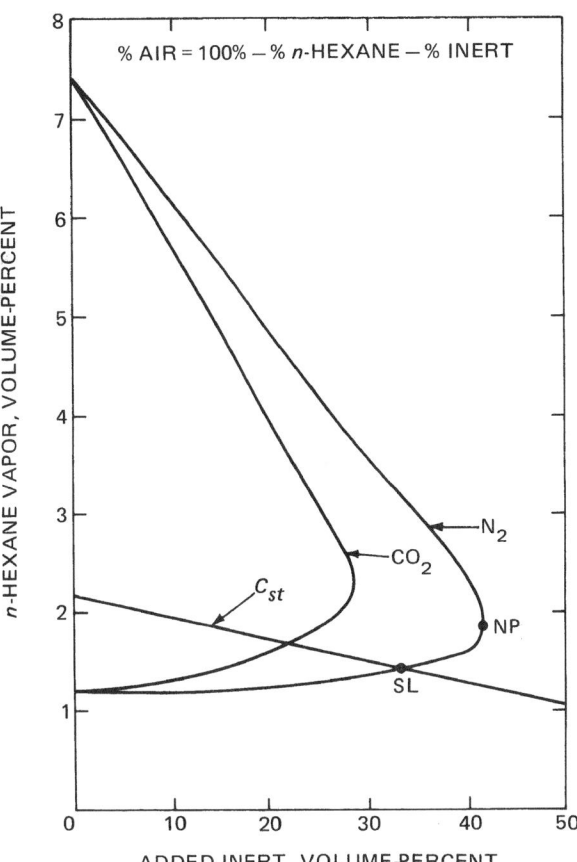

Fig. 2-9.8. *Limits of flammability of various n-hexane/inert gas/air mixtures at 25°C and atmospheric pressure.*[3]

gases is explicitly represented, and another utilizing only two axes in which the third gas concentration is determined by the difference between the sum of the other two gases and 100 percent. Both types give the same information.

Shown in Figures 2-9.7(a) and (b) are the air and limit lines. Anywhere along the air line the ratio of oxygen to nitrogen is the same as in air. The limit line represents a range of mixtures with a fixed oxygen/nitrogen ratio which is tangent to the flammable region. Any oxygen/nitrogen mixture, with an oxygen-to-nitrogen ratio less than that of

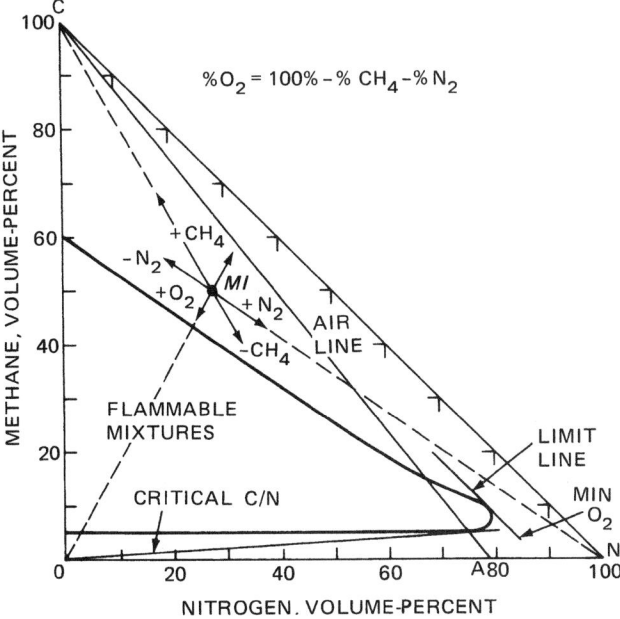

Fig. 2-9.7(b). *Two-axis flammability diagram for the system methane/oxygen/nitrogen at atmospheric pressure and 26°C.*[3]

the limit line, will not support flame propagation when mixed with any amount of methane.

Figure 2-9.8 is yet another representation of the flammable limits of fuel/oxidant/inert mixtures. The dilution of a fuel/air mixture is given by the percent of inert gas in excess of the nitrogen present in air. Figure 2-9.8 includes only mixtures that lie to the right of the air line, and as such is a magnification of a portion of the region included in Figures 2-9.7(a) and (b). Also shown in Figure 2-9.8 are several lines and points of specific interest. The highest concentration of nitrogen that will allow propagation of a flame is known as the nitrogen point (NP). Of course, this concept can be generalized to any inert (IP). If the concentration of the inert is greater than that at the inert point, no mixture of fuel and oxidant will propagate a flame remote from the ignition source.

As shown in Figure 2-9.8, the stoichiometric line passes through the flammable region. The point at which the stoichiometric line intersects the boundary of the flammable region is known as the stoichiometric limit (SL). The SL limit is the most dilute stoichiometric mixture that will propagate a flame remote from the ignition source. In the case of methane, the peak of the flammable region occurs near the stoichiometric limit. (See Figure 2-9.9.) For longer chain alkanes, the peak shifts to the rich side of the stoichiometric line. (See Figure 2-9.8.) For C_5 and higher hydrocarbons, the peak of the flammable region is bisected by the stoichiometric line defined by combustion to CO rather than to products of complete combustion. This shift has been attributed to incomplete combustion[11] and to

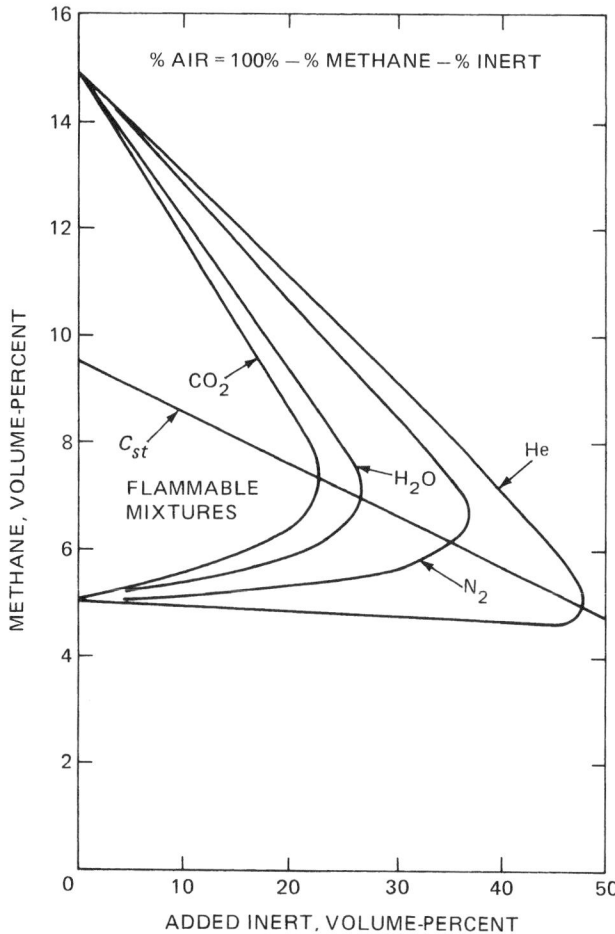

Fig. 2-9.9. *Limits of flammability of various methane/inert gas/ air mixtures at 25°C and atmospheric pressure.*[3]

Figure 2-9.9 shows the effect of various inert diluents on the flammable region. As indicated by the critical adiabatic flame temperature concept, the lower flammable limit is increased in proportion to the heat capacity of the diluent. (See Section 1, Chapter 5.)

EXAMPLE 3:

A methane leak fills a 200 m^3 room until the methane concentration is 30 percent by volume. Calculate how much nitrogen must be added to the room before air can be allowed in the space.

SOLUTION:

The initial mixture in the room is given by the point B in Figure 2-9.10. Adding nitrogen moves along the line toward pure nitrogen (the N point). Drawing the line from the air point, A, tangent to the flammable region defines the mixture C: the mixture with the least nitrogen added that on mixing with air will not form a flammable mixture. Referring to Figure 2-9.10 we see that point C corresponds to a methane concentration of 13 percent. In order to reduce the methane concentration from 30 percent to 13 percent, an as yet unknown amount of nitrogen must be added. If we could remove only the initial mixture and replace it with nitrogen, the amount of nitrogen would simply be

$$[(30 - 13)/30] \times 200 \text{ m}^3 = 113 \text{ m}^3$$

However, there is generally no way to prevent mixing of the initial mixture to be exhausted and the nitrogen being introduced to replace it. As such, inerting nitrogen is also lost. We can model this by assuming that the room is well mixed during nitrogen injection so that the concentrations are uniform everywhere. Under these conditions the methane concentration, C, is given by

$$C = C_0 \exp(-V_N/V)$$

preferential diffusion of reactants.[12] A similar shift of the maximum burning velocity to the rich side of stoichiometry is also observed. In this case, preferential diffusion of reactants has been shown to be the responsible factor.

Flammability diagrams are useful not only in determining the flammability of a given mixture, but also in developing strategies for avoiding flammable mixtures while diluting fuel-rich mixtures. In order to make use of the diagrams in this fashion, we must examine the change in position on the diagram when fuel, oxygen, or inert gas is added to the mixture. Consider a mixture given by point *MI* in the three-axis diagram, Figure 2-9.7(a). The arrows indicate the change in the mixture composition with the addition or removal of each gas species. In the three-axis diagram, moving toward the vertex corresponding to 100 percent of any one the gases corresponds to the addition of that gas, since adding an infinite amount of a single gas will reduce the concentrations of the other gases to zero. Adding air corresponds to moving toward the point on the air line at which there is no fuel. Clearly, following these examples, the effect of adding any gas or gas mixture can be plotted in the three-axis diagram. In the two-axis diagram, moving toward the vertex with 0 percent inert, 0 percent fuel corresponds to the addition of oxygen. In Figure 2-9.8 moving toward the 0 percent inert, 0 percent fuel vertex corresponds to adding air.

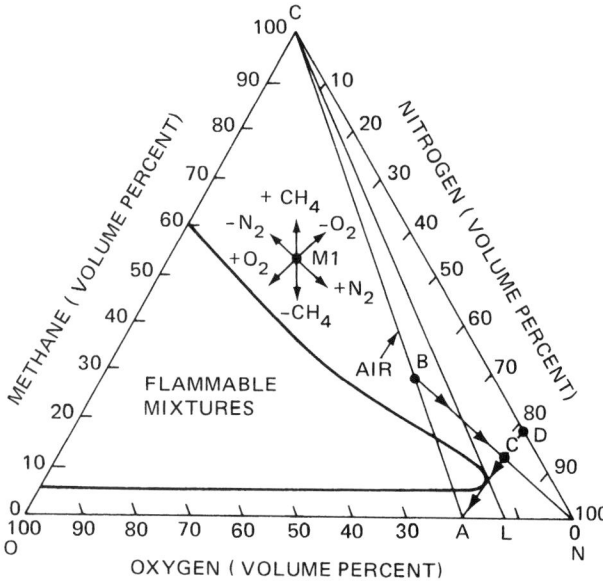

Fig. 2-9.10. *Graphic solution of Example 3 (adapted from reference 3).*

where C_0 is the initial methane concentration, V_N is the volume of nitrogen added, and V is the volume of the room. Rearranging this equation we find

$$V_N = -V \ln(C/C_0) = -200 \text{ m}^3 \ln(13/30) = 167 \text{ m}^3$$

Of course, the flow of gases out of the room contains methane and may burn on mixing with air. Mixing air and the initial gases in the room results in mixtures along the line AB (see Figure 2-9.10), some of which are clearly flammable. As such, ignition sources must be excluded near the room exhaust, or the exhaust also needs to be inerted.

EXAMPLE 4:

A 1 kg/s flow of methane is being dumped into the atmosphere. How much nitrogen must be mixed with methane to avoid a flammable mixture in the open?

SOLUTION:

In order to make the methane nonflammable, it needs to be diluted with enough nitrogen so that on further addition of air the flammable region is missed. Such a mixture of methane and nitrogen is given by extrapolating the line AC back to zero oxygen; i.e., point D on Figure 2-9.10, where the mixture is 82 percent nitrogen, 18 percent methane. The ratio of the flow rates of nitrogen to methane must equal the ratio of the concentrations of nitrogen and methane. Since concentrations expressed as volume percent are directly related to mole fractions, the flow rates of nitrogen and methane must be expressed as molar flow rates, $\dot{n}$,

$$\dot{n}_{N_2}/\dot{n}_{CH_4} = C_{N_2}/C_{CH_4}$$

The molar flow rate of methane is given by

$$\dot{n}_{CH_4} = \dot{m}_{CH_4}/MW_{CH_4}$$

where MW is the molecular weight and $\dot{m}$ is the mass flow rate.

$$\dot{n}_{N_2} = (\dot{m}_{CH_4}/MW_{CH_4})(C_{N_2}/C_{CH_4})$$

$$= (1000 \text{ g/s}/16 \text{ g/mol})(82\%/18\%)$$

$$= 285 \text{ mol/s}$$

$$\dot{m}_{N_2} = \dot{n}_{N_2}MW_{N_2}$$

$$= (285 \text{ mol/s})(28 \text{ g/mol})$$

$$= 7970 \text{ g/s or } 7.97 \text{ kg/s}$$

Ignition Energies and Quenching Diameters

The energy required to ignite flammable mixtures is generally quite low, on the order of a few tenths of a millijoule for near-stoichiometric mixtures in air and as low as a few thousandths of a millijoule in oxygen. Here again, preferential diffusion causes the minimum to occur for rich mixtures for fuels with molecular weights greater than that of air.[12] As the flammable limits are approached, the ignition energy increases sharply.

Several methods exist for preventing the initiation of an explosion. These include: avoiding flammable mixtures, excluding ignition sources whose energy is greater than the minimum ignition energy, and enclosing any ignition sources in an enclosure that will not allow the propagation

of the flame to the outside. We have already discussed the first of these. Some low-power electrical equipment can be designed such that the worst fault condition cannot produce the minimum ignition energy for a specified gas. Such equipment is termed "intrinsically safe" and may be used where there is a risk of a flammable atmosphere being formed.

Where this is not feasible, the electrical equipment may be housed in an "explosion-proof" enclosure, which will not allow propagation of the flame out of the enclosure. This is accomplished by making the size of the openings small enough that sufficient heat is lost by the flame as it passes through the opening that it is quenched. The quenching distance is most often determined by placing a pair of flanged electrodes in a gas mixture and attempting to ignite the gases. The flanges are parallel plates, and if the mixture can be ignited in the presence of the plates, the separation of the plates is greater than the quenching distance. The quenching distance with parallel plates, $d_\parallel$, is 65 percent of the quenching diameter in circular tubes. Figure 2-9.11

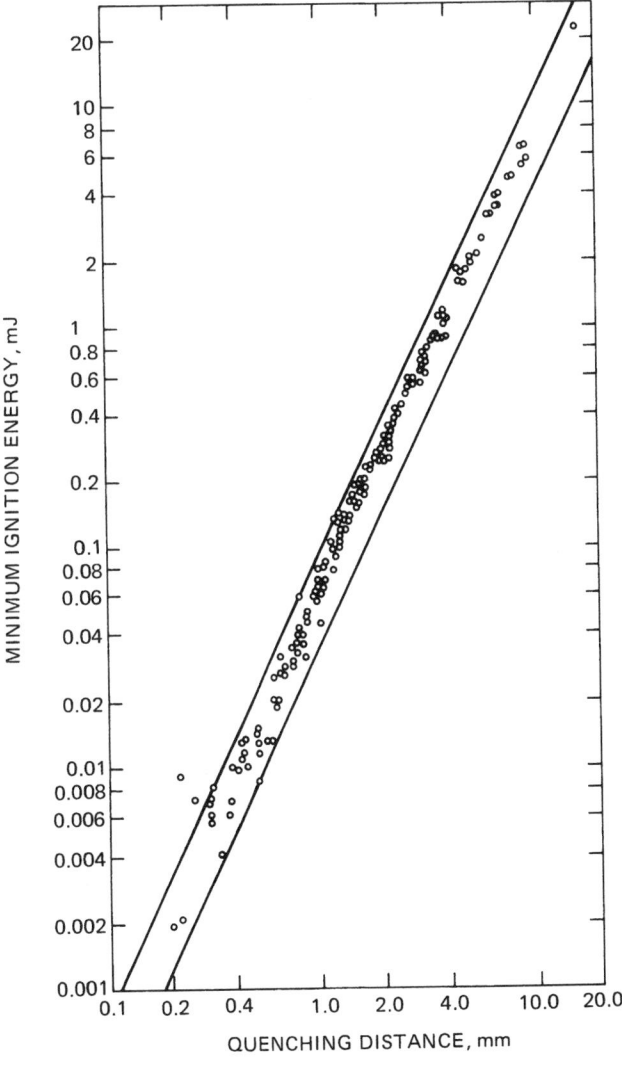

Fig. 2-9.11. The relation between flat-plate quenching and spark minimum ignition energies for a number of hydrocarbon-air mixtures.[13]

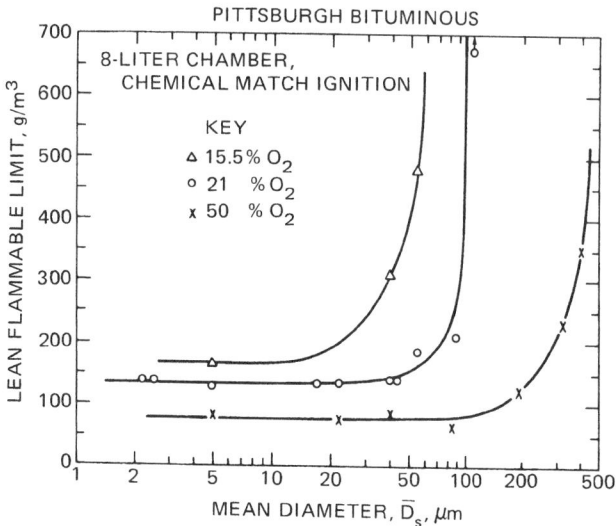

Fig. 2-9.12.   Lean flammability limit data for Pittsburgh bituminous coal as a function of particle size for three oxygen concentrations.[14]

shows the relation of the quenching distance to the minimum ignition energy for a number of hydrocarbon/air mixtures. The relation can be expressed as $E_{min} = 0.06 \, d_{\parallel}^2$, where E_{min} is the minimum ignition energy in air given in mJ and $d_{\parallel}$ is the quenching distance in air given in mm.

Because the hot quenched flame gases in an enclosure will expand through the opening, they may autoignite outside the enclosure. It has been found that the minimum experimental safe gap (MESG) for most hydrocarbons is approximately half the quenching distance.[13]

Dusts and Mists

The lower flammable limit of dusts and mists would be expected to be higher than their gaseous counterparts due to the need to volatilize the dust or mist. For very small particles with high surface-area-to-volume ratios, the lower flammable limit is independent of particle diameter, and the limit concentrations are approximately the same as the analogous gaseous fuel for fuels that volatilize completely. Hertzberg *et al*[14] have shown that bituminous coal dusts with particle diameters of 50 μm or less and polyethylene dusts with particle diameters of 100 μm or less have lower flammable limits in air which are independent of particle diameter. Figure 2-9.12 shows the measured lower flammable limit concentration for Pittsburgh bituminous coal as a function of average particle diameter and oxygen concentration. Notice that the lower flammable limit in the small-particle limit is a function of the oxygen concentration, unlike gaseous fuels. Also note that the lower flammable limit concentration is much higher than 48 g/m³, typical of gaseous hydrocarbons. These effects are due to the fact that not all the coal dust is volatilized. The fraction of dust which is volatilized is a function of the particle diameter and the oxygen concentration. As the oxygen concentration affects the maximum flame temperature and, hence, the heat flux to the particle, both the ability of heat to penetrate the particle and the rate of heating are affected. It is well known that the fraction of the material volatilized increases with the rate of

heating. It is not expected that the lower flammable limit can be reduced below 50 g/m³, even at 100 percent oxygen.

As the particle size increases, it would be expected that the lower flammable limit would also increase due to the difficulty of getting the fuel into the gas phase where combustion will take place. This does in fact occur, but depending on the geometry of the test, the apparent lower flammable limit of mists can actually decrease with increasing particle diameter due to the effects of gravity.[15] If the ignition source is at the bottom of the container and the aerosol is not kept well mixed, the particles can begin to settle out, causing the local concentration in the lower portions of the apparatus to be higher. This laboratory effect can also be expected to operate under actual conditions, depending on the degree of mixing of the aerosol.

While it is in principle possible for flame propagation to occur as a result of heterogeneous combustion of particles, this appears not to be an important mechanism for organic materials. Lower flammable limits of anthracite coal dusts with only a 20 percent volatile yield can be explained solely on the basis of gas-phase combustion.[16] Flame propagation by heterogeneous combustion is important for metal and graphite dusts.

DIFFUSION FLAME LIMITS

The limits of flammability for diffusion flames were first examined by Simmons and Wolfhard.[17] In their experiments, they determined the minimum level of dilution of the oxidant stream necessary to prevent the stabilization of a diffusion flame for a variety of gas and liquid fuels. The oxygen mole fraction, X_{O_2}, of the oxidant stream at the flammability limit is known as the limiting oxygen index (LOI), or simply the oxygen index (OI). Simmons and Wolfhard's results are included in Table 2-9.2. They observed that the oxygen index of their diffusion flames equaled the ratio, $X_{O_2}/(X_{O_2} + X_{diluent})$, found in a premixed stoichiometric-limit mixture involving the same fuel. This implies that the adiabatic flame temperature for the limit diffusion flame, calculated on the basis of stoichiometric combustion of the fuel and oxidant streams, is equal to the adiabatic flame temperature at the stoichiometric limit of a premixed system involving the same fuel, oxidant, and diluent.

Figure 2-9.13 graphically illustrates the relationship of the adiabatic flame temperatures at the lean, premixed limit in air, at the stoichiometric limit (premixed), and at the oxygen index (premixed). As the figure shows, the adiabatic flame temperature at the stoichiometric limit and the oxygen index are essentially equal, and the adiabatic flame temperature at the lower flammable limit in air is approximately 150 K less. Ishizuka and Tsuji[18] verified Simmons and Wolfhard's results for methane and hydrogen, and showed that the adiabatic flame temperature at the limit is the same whether dilution is of the fuel or oxidizer stream.

The information in Figure 2-9.13 forms the basis of a method for the evaluation of diffusion flame limits for fuel mixtures. In essence, the ability of a fuel and oxidant pair to react in a diffusion flame is evaluated by examining the flammability of a premixed stoichiometric mixture of the fuel and oxidant.

To do this, we assume that Le Chatelier's rule holds at the stoichiometric limit; i.e.,

$$\sum_{i=1}^{n} (C_i/\text{SL}_i) \geq 1 \qquad (5)$$

TABLE 2-9.2 *Thermodynamic Equilibrium Properties at Extinction (Adapted from References 11 and 13)*

Fuel	LL (vol %)	T(LL) (K)	X(SL)*	T(SL) (K)	X(NP)*	T(NP) (K)	OI*	X(OI)*	T(OI) (K)
CH_4	5.0	1480	0.123	1720	0.117	1610	0.139	0.130	1780
C_2H_2	2.7						0.085		1540
C_2H_4	2.7						0.105		1610
C_2H_6	3.0	1530	0.114	1620	0.111	1540	0.118	0.114	1630
C_3H_8	2.1	1540	0.125	1730	0.114	1470	0.127	0.124	1720
n-C_4H_{10}	1.8	1640	0.134	1830	0.121	1490	—	—	—
n-C_5H_{12}	1.4	1590	0.135	1810	0.115	1410	0.1325	0.130	1760
n-C_6H_{14}	1.2	1610	0.135	1800	0.117	1420	0.1335	0.132	1770
n-C_7H_{16}	1.05	1620	0.134	1770	0.118	1430	—	—	—
n-C_8H_{18}	0.90	1650	0.134	1770	0.118	1440	0.134	0.133	1780
n-$C_{10}H_{22}$							0.1345	0.133	1780
CH_3COCH_3	2.6						0.1285		1730
CH_3OH	6.7	1550	0.112	1690	0.085	1430	0.111	0.103	1530
C_2H_5OH	3.3	1490	0.118	1700	0.106	1430	0.126	0.121	1670
n-C_3H_7OH	2.2	1490					0.128	0.124	1700
n-C_4H_9OH	1.7	1510					0.129	0.126	1710
n-$C_5H_{11}OH$	1.4	1550					0.130	0.128	1730
n-$C_6H_{13}OH$	1.2	1490					0.1315	0.130	1740
n-$C_8H_{17}OH$							0.1315	0.130	1750
C_6H_6	1.3						0.133		1810
C_6H_{12}	1.2						0.134		1770
H_2	4.0						0.054		1080
CO	12.5						0.076		1450

*Expressed as mole fraction of oxygen.

and that the adiabatic flame temperature at the stoichiometric limit for each fuel is a constant. These lead to the expression

$$\sum_{i=1}^{n} \frac{(C_i/100)\Delta H_{c,i}}{\int_{T_0}^{T_{f,\text{SL},i}} n_p C_p \, dT} \geq 1 \qquad (6)$$

where

C_i = the volume percent of fuel species, i, when the fuel stream is mixed stoichiometrically with the oxidant stream

$T_{f,\text{SL},i}$ = adiabatic flame temperature of the stoichiometric limit mixture for fuel species i
= 1700 K for most hydrocarbons
= 1450 K for carbon monoxide
= 1080 K for hydrogen

T_0 = temperature of the stoichiometric mixture prior to reaction

$\Delta H_{c,i}$ = heat of combustion of fuel species
= 620 kJ/mol for hydrocarbons (per carbon, assuming $H/C = 2$)
= 283 kJ/mol for carbon monoxide
= 242 kJ/mol for hydrogen

n_p = number of moles of products of combustion per mole of reactants (stoichiometric mixture of the fuel and oxidant streams)

C_p = heat capacity of the products of combustion.

This approach has been successfully used to predict the flammability of the hot gas layer formed in enclosure fires.[19]

EXAMPLE 5:

As part of a hazard analysis of a particular room fire, the composition of the hot layer during fire development has

been estimated. The results of the analysis indicate that the following composition represents the highest concentration of fuel gases expected.

Hot Layer—700 K, 10 percent total hydrocarbons (THC), in the form of CH_2, 2 percent CO, 1 percent H_2, 15 percent CO_2, 2 percent O_2, 70 percent N_2;

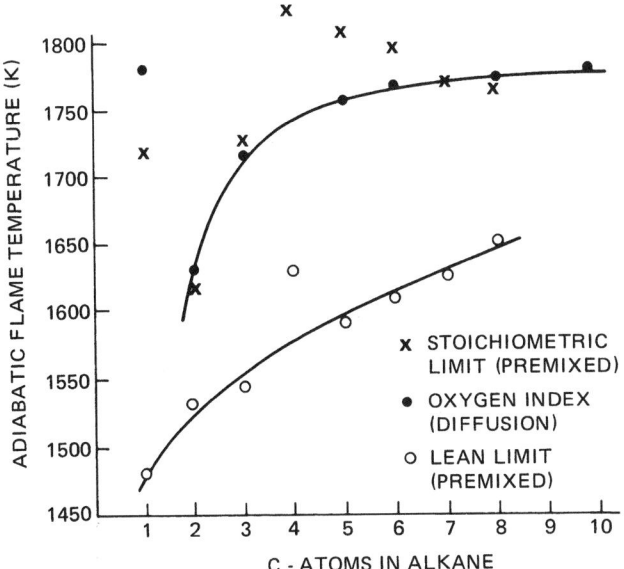

Fig. 2-9.13. *Computed adiabatic flame temperatures at flammability limits for n-alkanes (adapted from reference 11).*

Cold Layer—300 K, 21 percent O_2, 79 percent N_2. Will the hot layer burn?

SOLUTION:

The working equation is Equation 6. The first step is to write a balanced chemical equation for stoichiometric burning.

$$0.1CH_2 + 0.02CO + 0.01H_2 + 0.02O_2 + 0.7N_2 + 0.15CO_2$$
$$+ x(O_2 + 3.78N_2) \rightarrow 0.27CO_2 + 0.11H_2O + (0.7 + 3.78X)N_2$$

We can find x by requiring that both sides of this equation have the same amount of oxygen

$$\underset{0.02/2}{\overset{CO}{}} + \underset{0.02}{\overset{O_2}{}} + \underset{0.15}{\overset{CO_2}{}} + \underset{x}{\overset{air}{}} = \underset{0.27}{\overset{CO_2}{}} + \underset{0.11/2}{\overset{H_2O}{}} \rightarrow x = 0.145$$

The concentrations in the stoichiometric mixture can be determined from the balanced chemical equation

$$C_i = (n_i/n_T) \times 100\%$$

$$n_T = 0.1 + 0.02 + 0.01 + 0.02 + 0.7 + 0.15$$
$$+ 0.145 + 0.145(3.78) = 1.693$$

$$C_{THC} = (0.1/1.693) \times 100\% = 5.9\%$$

$$C_{CO} = (0.02/1.693) \times 100\% = 1.2\%$$

$$C_{H_2} = (0.01/1.693) \times 100\% = 0.6\%$$

Similarly, the number of moles of products per mole of reactants can be determined from the chemical equation

$$n_p = [0.27 + 0.11 + 0.7 + 0.145(3.78)]/1.693 = 0.962$$

This is lower than typical values of 1 to 1.1, because the unknown hydrocarbon mixture is taken as CH_2. This is not an error, since CH_2 has been consistently used for the heat release and heat capacity as well. For convenience, we will use constant average specific heats taken from Drysdale:[20]

	C_p (J/molK)	C (%)*
CO_2	54.3	16.2
H_2O	41.2	6.6
N_2	32.7	77.2

*Calculated by the same method as the fuel gas concentrations.

$$n_p C_p = n_p \sum (C_i/100)C_{p,i}$$
$$= 0.96[(0.162)(54.3) + (0.066)(41.2)$$
$$+ (0.772)(32.7)]$$
$$= 35.3 \text{ J/mol K}$$

Notice that the average specific heat is near that of nitrogen, since it is the major constituent of the mixture. In calculating T_0, the initial temperature of the mixture, we will ignore variations in C_p between the hot and cold layers.

$$T_0 = (n_h Th + n_c Tc)/(n_h + n_c) \quad n_h + n_c = n_T$$
$$= [(1)(700 \text{ K}) + (0.69)(300 \text{ K})]/1.69 = 537 \text{ K}$$

where n_h and n_c are the number of moles originating in the hot and cold layers, respectively. Substituting into Equation 5,

$$\sum_{i=1}^{n} \frac{(C_i/100)\Delta H_{c,i}}{C_p(T_{f,SL,i} - T_0)} = \frac{(0.059)(620)10^3}{35.3(1700 - 537)}$$
$$+ \frac{(0.012)(283)10^3}{35.3(1450 - 537)}$$
$$+ \frac{(0.006)(242)10^3}{35.3(1080 - 537)} = 1.07 \quad (7)$$

Since the result is greater than one, the hot layer *will* ignite and burn.

While the approach to the onset of layer burning used in Example 5 has a great deal of generality, it requires a very detailed characterization of the upper and lower layers. It has been shown by Beyler[19] that a much simpler method can be used to evaluate the conditions required for layer burning.

The method[19] is based on the very simple chemical model

$$\text{Fuel + Oxidizer} \rightarrow \begin{cases} \text{Products + Excess Oxidizer for } \phi < 1 \\ \text{Products + Excess Fuel for } \phi > 1 \end{cases} \quad (8)$$

where the equivalence ratio, ϕ, is given by

$$\phi = \frac{\dot{m}_f}{\dot{m}_{air} \cdot r}$$
$$r = \left(\frac{m_f}{m_{air}}\right)_{\text{Stoichiometric}} \quad (9)$$

According to this model, the fuel mass fraction in the upper layer is

$$Y_f = 0 \qquad \text{for } \phi < 1$$
$$Y_f = \frac{1 - 1/\phi}{1 + 1/\phi r} \qquad \text{for } \phi > 1 \quad (10)$$

Equation 6 can be expressed on a mass basis for this application as

$$\frac{Y_f \Delta H_c}{m_p C_p (T_{SL} - T_0)} \geq 1 \quad (11)$$

where ΔH_c is the heat of combustion of the fuel, and m_p is the mass of products resulting from burning a unit mass of upper layer gases.

Substituting the $\phi > 1$ relationship for Y_f into Equation 6, expressing the heat release in terms of oxygen consumed using

$$\Delta H_c = \frac{\Delta H_{O_2} Y_{O_2}}{r} \quad (12)$$

and recognizing that

$$m_p = 1 + \frac{Y_f}{r} \quad (13)$$

yields

$$\left(\frac{1 - 1/\phi}{1 + r}\right)\left[\frac{\Delta H_{O_2} Y_{O_2}}{C_p(T_{SL} - T_0)}\right] \geq 1 \quad (14)$$

Equation 14 can be solved for the equality condition to give the equivalence ratio at which layer burning begins, ϕ_{ig},

$$\phi_{ig} = \frac{k}{k - r - 1} \quad (15)$$

where

$$k = \frac{\Delta H_{O_2} Y_{O_2}}{C_p(T_{SL} - T_0)}$$

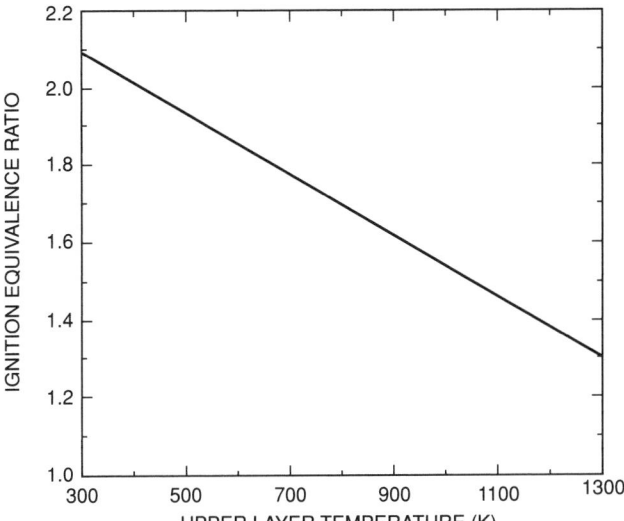

Fig. 2-9.14.   *Equivalence ratio required for upper layer ignition as a function of the upper layer temperature determined using Equations 15 and 16 with typical properties: Using normal values for the semi-universal constants: $\Delta H_{O_2} = 13.4$ MJ/kg, $C_p = 1.1$ kJ/kg K, $T_{SL} = 1700$ K; using air properties for the lower layer: $Y_{O_2} = 0.233$, $T_1 = 300$ K; and using a typical $r = 0.07$ yields the relationship between ϕ_{ig} and T_u.*

T_0 is the precombustion temperature resulting from stoichiometric mixing of the air and fuel streams. Here, the upper layer contains the fuel and the lower layer contains the air. T_0 can be expressed as

$$T_0 = \frac{T_u + \left(\frac{Y_f}{r}\right)T_l}{1 + \frac{Y_f}{r}} \qquad (16)$$

Using Equations 15 and 16, a relationship between the critical ignition equivalence ratio and the layer temperatures can be developed. Using normal values for the semi-universal constants: $\Delta H_{O_2} = 13.4$ MJ/kg, $C_p = 1.1$ kJ/kg K, $T_{SL} = 1700$ K; using air properties for the lower layer: $Y_{O_2} = 0.233$, $T_1 = 300$ K; and using a typical $r = 0.07$ yields the relationship between ϕ_{ig} and T_u shown in Figure 2-9.14. The results shown in Figure 2-9.14 are consistent with the measurements of Beyler,[19] where ϕ_{ig} was found to be 1.7 for T_u of 500 to 600 K. Gottuk[21] found that external burning was first observed in flashes at $\phi = 1.4 \pm 0.4$, and sustained external burning was first observed at $\phi = 1.9 \pm 0.3$ when T_u was in the range 900 to 1100 K. While in Gottuk's[21] experiments it was difficult to observe burning at the layer interface due to soot deposits on the viewing ports, layer interface burning was generally observed shortly after the initiation of flashes in the exhaust. Because the exhaust flow was isolated from the inflow in the experiment, there is some issue of the availability of a pilot flame which does not arise in normal two-directional vents found in most fires. Thus, Gottuk's work is generally consistent with Figure 2-9.14.

Oxygen Index Test Method

The original oxygen index test method, used to determine the oxygen index of liquid and gas fuels, utilizes a counterflow diffusion flame formed at the stagnation region of a porous cylinder or sphere through which fuel vapors are fed. A low-velocity oxidant stream passes over the porous body. This arrangement yields the most favorable aerodynamic conditions for flame stabilization. As such, fuel and oxidant streams that can burn in the low-velocity counterflow system may not burn under less favorable aerodynamic conditions characterized by higher velocities and shear.

It is also important to point out the difference between the oxygen index as measured for gas and liquid fuels and the oxygen index of solids as measured using a candle-type test.[22,23] The oxygen indexes of the gas and liquid fuels as tested by Simmons and Wolfhard[17] were governed by gas-phase effects. In the American Society for Testing and Materials (ASTM) test[23] for solids, the extinction can be caused by gas- and solid-phase effects. As such, the oxygen index of a solid fuel is not directly relevant to gas-phase diffusion flame limits and should not be used to calculate adiabatic flame temperature at the limit for use in the expressions presented here.

REFERENCES CITED

1. H.F. Coward and G.W. Jones, *Bulletin, No. 503*, Bureau of Mines, Washington (1952).
2. G.A. Karim, I. Wierzba, M. Metwally, and K. Mohon, *Eighteenth Symposium (International) on Combustion*, The Combustion Institute, PA
3. M.G. Zabetakis, *Bulletin No. 627*, Bureau of Mines, Washington (1965).
4. H.F. Coward, C.W. Carpenter, and W. Payman, *J. Chem. Soc.*, 115, 27 (1919).
5. M.J. Burgess and R.V. Wheeler, *J. Chem. Soc.*, 99, 2013 (1911).
6. A.G. White, *J. Chem. Soc.*, 127, 672 (1925).
7. M.G. Zabetakis, S. Lambiris, and G.S. Scott, *Seventh Symposium (International) on Combustion*, The Combustion Institute, Pittsburgh, PA
8. F.J. Weinberg, *Nature*, 283, 239 (1971).
9. A. Egerton and J. Powling, *Proc. Roy. Soc.*, 193, London (1948).
10. D.R. Stull, *F. Res. Abst. and Rev.*, 13, 161 (1971).
11. A. Macek, *Comb. Sci. and Tech.*, 21, 43 (1979).
12. B. Lewis and G. Von Elbe, *Combustion, Flame, and Explosions of Gases*, Academic, New York (1961).
13. R.A. Strehlow, *Combustion Fundamentals*, McGraw-Hill, New York (1984).
14. M. Hertzberg, K. Cashdollar, and R. Conti, *Nineteenth Symposium (International) on Combustion*, The Combustion Institute, Pittsburgh, PA
15. J.H. Burgoyne and L. Cohen, *Proc. Roy. Soc.*, A225, 375 (1954).
16. M. Hertzberg, K. Cashdollar, and C. Lazzara, *Eighteenth Symposium (International) on Combustion*, The Combustion Institute, Pittsburgh, PA
17. R.F. Simmons and H.G. Wolfhard, *Comb. and Flame*, 1, 155 (1957).
18. S. Ishizuka and H. Tsuji, *Eighteenth Symposium (International) on Combustion*, The Combustion Institute, Pittsburgh, PA (1981).
19. C.L. Beyler, *Comb. Sci. and Tech.*, 39, 287 (1984).
20. D.D. Drysdale, *An Introduction to Fire Dynamics*, John Wiley and Sons, New York (1985).
21. D.T. Gottuk, "The Generation of Carbon Monoxide in Compartment Fires," Ph.D. Dissertation, Virginia Polytechnic and State University, Blacksburg, VA (Sept. 1992). [Also in NIST-GCR-92-619, National Institute of Standards and Technology, Gaithersburg, MD (1992).]
22. C.P. Fenimore and F.J. Martin, *Comb. and Flame*, 10, 135 (1966).
23. American Society for Testing and Materials, *ASTM D2863-77*, Philadelphia (1977).

IGNITION OF LIQUID FUELS

A. Murty Kanury

INTRODUCTION

This chapter introduces the topic of liquid fuel ignition in air. A qualitative description of the physical and chemical steps leading to ignition makes possible a systematic identification of the factors of concern. Different measures of liquid fuel ignitibility are defined, and a comprehensive collection of data is presented. Simple mechanistic models are formulated for the piloted ignition (i.e., flash point) of liquids as well as for the autoignition of vapor and air mixtures. The results of these models are employed to highlight: (1) the manner in which physics and chemistry play a complicated role to culminate in ignition; (2) the resultant dependency of the measurements on the specific apparatus test method; and (3) a number of dependencies of the measured ignition temperature on experimental conditions. These predicted dependencies are found to be consistent with the trends in reported data.

Liquid combustibles are ever-present in the context of mobile and stationary power plants as well as in an innumerable array of industrial processes. A variety of physical forms in which they arise can be identified. Dip-tanks of paints or solvents, sprays, spills from a storage tank, flowing or creeping films, and pools in open pits are but a few examples. In almost all situations, air is in contact, or readily comes into contact, with the fuel to make the associated fire hazard obvious. A number of questions arise in relation to firesafety in these situations:

1. Under what circumstances is ignition possible?
2. Under the conditions of possible ignition, what is the external impulse required to produce ignition within a given time? Alternately stated, given a set of conditions under which ignition is possible, how long an exposure is required to produce ignition?

The first of these questions is addressed here along with some cursory observations on the second question.

Ignition is defined as the onset or initiation of combustion, usually flaming. As such, ignition is indicated by the oxidation reaction attaining a rapidly increasing rate. The rapidity often makes the ignition phenomenon an abrupt event. In practice, ignition is noted by the appearance of a flame, by a significant increase in oxidative energy release, or by a corresponding large rise in temperature.

Three sets of conditions have to be fulfilled to ignite a condensed-phase material. First, sufficient quantities of combustible vapors and gases have to be emanated as a result of preheating the solid or liquid; second, these vapors and gases have to be mixed with the oxidant in the gas phase; and third, the mixture has to be either at a high enough temperature to induce self-accelerative oxidation (i.e., *spontaneous* or *autoignition*) or to be provided with a pilot source (e.g., a small ignitor flame, a heated wire, or an electric spark) to locally heat a minimum quantity of the mixture to a temperature approaching the adiabatic flame temperature (i.e., *piloted* or *forced ignition*). These three sets of conditions enable a systematic enumeration of the factors influencing the ignition process.

VAPORIZATION: A CONTRAST BETWEEN LIQUID AND SOLID COMBUSTIBLES

All condensed-phase materials release vapors and gases in response to heating. As related to the first of the three steps mentioned above, this response is dramatically different between solids and liquids.

Liquids generally* vaporize in the sense of a thermodynamic phase change in which the chemical structure of vapor remains the same as that of liquid. Such a vaporization is usually a surface mass transfer phenomenon, although intense (and internal) heating may produce bubbles within the liquid. In contrast, heating of solids generally** results in complicated destructive distillation (also known as thermochemical degradation or *pyrolysis*) to yield a complex combustible mixture of gases and vapors. Known also as pyrolyzate, this mixture is produced *within* the solid at a rate dependent upon the local instantaneous density and

---

*Some high molecular weight liquids are exceptional in that they exhibit a chemical breakdown in response to heating.

**Some solids, such as waxes, are exceptional in that they melt and vaporize in the manner of a simple liquid. Some other exceptional solids, such as camphor and sulfur, sublime to a gas phase.

Dr. A. Murty Kanury is Professor of Mechanical Engineering at Oregon State University. His professional activities are centered around teaching thermal science and research in fire and combustion.

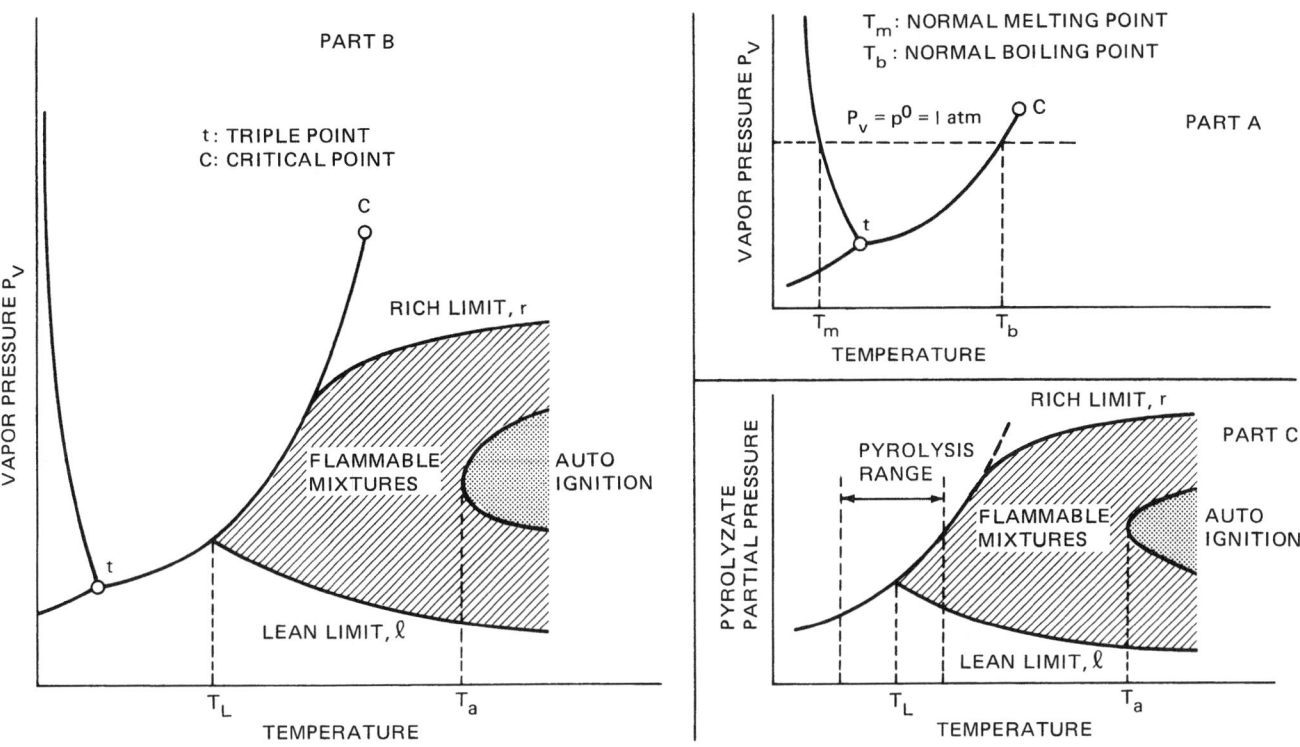

Fig. 2-10.1. *Phase-change diagram for: (a) an inert liquid; (b) a combustible liquid; and (c) a combustible solid.*

temperature. The pyrolysis process, thus, is volumetric and chemical rather than superficial and physical.

Pyrolysis of most solids leaves a carbon-rich porous residue, known as char, behind. Vaporization of most common liquids, in contrast, leaves no residue behind.

Upon heating a liquid body, internal convective currents develop with an intensity dependent on the heating rate, viscosity, surface-tension, gravity, and geometry of the body. The heating of a solid, by definition, does not invoke the same sort of bulk flows; there does arise, however, an internal convection process associated with the outflow of the pyrolyzates through the partially or totally degraded porous solid. The temperature distribution within a liquid body can be made artificially uniform by mixing with a stirrer; the same is not possible within a solid body.

The ease with which vapors can be produced by heating a liquid is known as its volatility. A liquid is said to be highly volatile if its vapor pressure at a given temperature is high (i.e., its boiling point at a given pressure is low), and its "latent heat" (i.e., enthalpy of vaporization) is low.

MIXING OF VAPORS WITH AIR

The vapors emanating from the surface of a heated liquid mix with the ambient gas, which is usually air at normal pressure. This mixing is dependent upon such factors as whether the air is quiescent or in motion, whether the liquid reservoir is closed or open, the geometry relative to the gravity vector, the temperature of the liquid surface relative to the ambient, the molecular weight of the vapors relative to that of air, and the height and nature of the lip of the vessel containing the liquid.

IGNITION OF THE MIXTURE

In order to exhibit a flame, the mixture thus formed has to be rich enough in combustible content to be over the so-called *lean limit of flammability* to make a propagable flame possible from a localized pilot source. The location and nature of the pilot is of obvious importance. Most generally, when ignition is marginally possible, the flame would simply flash through the mixture, consume the combustibles in its sweep and then go out. Only if the vaporization is copious enough, due to a higher bulk temperature of the liquid, does the flame become self-sustained. These flashing and self-sustainment concepts are central to the liquid fuel ignition process.

If the gaseous mixture temperature is sufficiently high, ignition may occur even without a pilot source. This is known as auto- or spontaneous ignition.

Figure 2-10.1, the rudiments of which were first formulated by Zabetakis,[1] indicates the essence of the foregoing descriptions. Part (a) of Figure 2-10.1, familiar from elementary thermodynamics, exhibits the exponential dependency of equilibrium vapor pressure on the liquid temperature. The normal melting and boiling point temperatures are also shown in this figure. Part (b) shows that as the temperature is gradually raised, the vapor pressure of the combustible liquid gradually rises to yield, at a limit temperature, T_L, a lean limit mixture capable of supporting a propagable flame. At a higher temperature, the partial pressure may become too high and make the mixture too rich to support a flame. Furthermore, at relatively high temperatures exceeding T_a, ignition can occur even in the absence of a pilot source. As indicated, the lean and rich limits, as well as the autoignition limits, are generally temperature dependent. Part (c)

TABLE 2-10.1　*Flash Points of Several Combustible Liquids Measured by Several Different Tests in Air**

	Flash Point (°C) as Measured in						
	Abel	Tagliabue	Elliot	Pensky-Martens	Cleveland	Luchaire	Open Cup
Naptha	34.4	39.4	36.4	40.6	46.1	50.0	
Kerosene	52.8	54.4	53.3	57.2	60.0	58.9	
Petrolite	61.1	59.4	60.0	65.6	68.3	62.3	
Gas oil				90.6	93.3	92.2	
300 oil				123.9	129.4	130.0	
Straw oil				157.2	162.8	161.1	
Engine oil				221.1	226.7	221.1	
Heavy oil				265.6	293.3	265.5	
Mexican crude				19.0			35.0
Petrol				26.5			33.5
Tar oil				87.0			92.5

* Excerpted from the *International Critical Tables*.[5]

shows that the features of ignition of a pyrolyzing combustible solid are similar to those of a combustible liquid, except for the fact that the pyrolyzates are evolved due to degradation over a range of temperatures.

SOME EXPERIMENTAL TECHNIQUES AND DEFINITIONS

Piloted ignition data are mostly measured in a number of American Society for Testing and Materials (ASTM) tests.[2] These tests involve either an open-cup (e.g., ASTM D 1310 open-cup test or D 92 Cleveland Test) or a closed-cup test (e.g., ASTM D 56 Tag Closed Tester and D 93 Pensky-Martens test) to hold the fuel to a prescribed level. This cup is then heated either directly by a prescribed, Bunsen-type laboratory flame or indirectly by a water (or water and glycol mixture) bath. There are similar international tests under a variety of names, including Engler and Haass, Danish open, Tagliabue, Elliot, Luchaire, Treumann, Albrecht, Abel, Saybolt, and Parish, among others.

Mullins[3] presents a comprehensive review of a variety of methods by which the spontaneous ignition temperature, T_a, of combustible liquids is experimentally determined. A number of variations, in which the reaction vessel is a crucible, flask, furnace, bomb, or a compression device, are employed to precisely control the experimental conditions, to observe whether or not ignition will occur, and to measure the time to ignition (i.e., the ignition delay). Flow methods, as well as heating of the liquid by hot surfaces, are also frequently employed. The ASTM test standard D 2155 holds the liquid in a flask, which is heated in a furnace to determine the autoignition temperature and ignition delay.

The finer details of these tests are not materially relevant to the scope of this chapter. Zabetakis[1] and the ASTM standards[2] do an excellent job of giving these details. It is important, however, to define certain measures of the volatility and ignitibility of liquid fuels as obtained from these experimental techniques.

The *bubble point (R:1)* of a liquid fuel[4] is the temperature at which equilibrium exists between the wholly condensed fuel and an infinitesimal quantity of its vapor mixed with air in the ratio of 1:R. When R is zero, the bubble point is the same as the normal boiling point.

The *flash point* of a liquid fuel is its temperature (presumed to be uniform) at which the vapor and air mixture lying just above its vaporizing surface is capable of margin-

ally supporting a momentarily flashing propagation of a flame when prompted by a quick sweep of a small gas flame pilot near the surface. It is expected that the flash point is somehow related to the minimum temperature, T_L, for piloted ignition indicated in Figure 2-10.1, part (b).

The *fire point* of a liquid fuel is very similar in definition to the flash point, except that the flame does not merely flash and cease but must also be marginally self-sustained so as to continue burning the liquid.

The *autoignition temperature*, T_a, of a vapor (or gas) and air mixture is the minimum temperature at which the mixture is marginally self-igniting.

Another property relevant in interpreting the significance of flash point is the *lean limit of flammability*, defined as the lowest *volume* percentage of fuel vapor (or gas) in the mixture with air (or oxygen) that will barely support propagation of a flame away from a pilot ignition source.

EXAMPLE DATA

The most extensive flash- and fire-point data sources are the *International Critical Tables*[5] and the *Loss Prevention Handbook*.[6] Tables 2-10.1, 2-10.2, and 2-10.3 are excerpted from these two sources to note the typical values and to make several important observations.

Table 2-10.1 shows the flash points of a number of liquids in air measured by various test methods to indicate that the measurements depend quite strongly upon the apparatus employed. Although, as expected, the closed-cup tests (e.g., Pensky-Martens) produce flashing at a lower temperature than the open-cup tests (e.g., Cleveland), other differences are neither small nor systematic.

Table 2-10.2 gives open- and closed-cup flash and fire points for a number of real-world combustibles of varying specific gravity. Two points are noteworthy: first, the fire point consistently exceeds the flash point by about 20 to 40°C; and second, heavier fuels tend to have higher flash points. Recalling that a higher flash point implies a lower saturation pressure and, hence, a lower volatility, this second observation is reasonable.

Table 2-10.3, adapted from *ICT*,[5] shows the closed-cup flash points as well as the minimum autoignition temperatures for a number of real liquids. Noteworthy is the decrease in T_a due to high pressure.

Table 2-10.4, is a comprehensive collection of the flash point, lean limit, and minimum-most measured autoignition

TABLE 2-10.2 *Flash and Fire Points of Some Hydrocarbon Liquids in Air**

Liquid	Open Cup		
	Flash Point (°C)	Fire Point (°C)	Specific Gravity
Fuel oil**	133	164	0.921
Crude**	125	155	0.923
Light fuel**	187	220	0.900
Black oil**	144	172	0.928
Refined oil**	122	135	0.904
Texas solar oil**	92	97	0.862
Shale oil**	130	150	0.862
Gas oil**	90	109	1.067
Neutral oils**	135	163	0.843
	216	252	0.878
Paraffin oils**	163	193	0.870
	216	254	0.912
Paraffin oil†	98	112	0.916
Napthene base oil**	196	227	0.937
Diesel fuel†			
Russia	53–138	78–180	0.876–0.950
N. America	82–166	103–200	0.865–0.950
India	92–150	120–174	0.890–0.950
Tar oil for diesels†			
Coke oven	90–135	108–166	1.14–1.18
Water gas	34–91	50–155	0.97–1.13
Oil gas	18–69	20–89	1.05–1.07
Coal tar oil	66–121	84–160	1.00–1.11

* Extracted from the *International Critical Tables*.[5]

** = Open cup

† = Pensky closed cup

temperature for a host of liquid fuels in air at a nominal pressure of 1 atm. A number of other properties of use in data intepretations of the following section are also included. Besides serving as a practical reference dictionary, Table 2-10.4 also indicates a number of trends that are often vague or weak. Most of these trends are conceptually unsubstantiated to date. For example, in a given family of fuels, an increase in fuel molecular weight generally (but not always) indicates: (1) a marked increase in the normal boiling point T_b^o; (2) a slight decrease in the normal enthalpy of vaporization h_{fg}^o; (3) an apparent increase in the flash point T_F^o; (4) a

TABLE 2-10.3 *Autoignition Temperatures at Two Pressures and Closed-Cup Flash Points for a Number of Liquid Fuels in Air**

Fuel	Flash point (K)	Autoignition Temp. T_a (K)	
		$P° = 1$ atm	$P° = 33$ atm
Sperm oil	509	581	413
Lard oil	513	546	417
Castor oil	536	598	426
Glycerine	—	685	478
Kerosene	328	528	448
Spindle oil	467	521	451
Turbine oils	—	—	526–564
Compressor oil, A	—	582	461
Compressor oil, B	469	546	460
Compressor oil, C	489	559	430

* Adapted from the *International Critical Tables*.[5]

modest and erratic decrease in the lean limit ℓ; and (5) a noticeable decrease in the minimum temperature for autoignition, T_a.

THEORY AND DISCUSSION

Theoretical models are required in order to substantiate and understand these and other trends that may lie buried in the elaborate data. The mechanisms culminating in the global effects of flashing and autoignition also elicit the way in which the measured properties depend on the apparatus used and the testing procedure followed. However, the physics and chemistry involved in open- or closed-cup flash point and autoignition testers is much too complicated with numerous interacting thermodynamic, fluid mechanical, heat and mass transport, and chemical kinetic processes. Complete mathematical models capable of predicting the outcome of a specific test do not exist and are nearly impossible. Simple syntheses given below of the essential physical and chemical processes of the tests are not only possible but also enlightening.

Flash Point and Lean Flammability Limit

In principle, the problem involves transient analysis of heat and mass transfer with oxidative reactions in the gas phase lying above the liquid surface. By focusing attention on the cup tests, the boundaries of the three-dimensional space of interest are the vaporizing liquid surface at the bottom, the open or closed top, and the side walls which are probably cylindrical, impervious to mass but conductive to heat. (The flash points and autoignition temperatures measured in cups or flasks whose walls are highly conductive are known to be higher.) If the cup were open, the nature of airflow and drafts in the room would be expected to alter the heat and mass transport processes of concern, and, hence, the measurement. A closed cup minimizes these alterations by allowing air to stealthily leak into the region of interest.

If the vapor were to be released at a constant rate at the hot liquid surface, beginning with time, t, equal to zero, it would transiently diffuse in the vertical direction, leading to a concentration profile* with the highest vapor concentration at the fuel surface and the least at the mouth of the cup. In the initial stages, the concentration everywhere is too low to support a flame. As time progresses, however, the temperature of the gas mixture near the surface rises beyond the minimum temperature limit T_L. (See Figure 2-10.1.) Equally important, the vapor concentration also increases to, and beyond, the lean limit. Later still, the thickness of the mixture layer that is beyond the lean limit increases due to diffusion. (If a pilot source were presented to this layer, a propagable flame could be generated. Thus, it is evident that the location and time at which pilot is introduced are of consequence in a test seeking the conditions of a propagable flame.) With continued time, the mixture at the surface becomes much too fuel-rich (oxygen-poor) to permit generation of a propagable flame with a pilot source. At larger times, the spatial belt containing mixtures between the lean and rich limits broadens and propagates away from the surface.

---

*The steady-state, one-dimensional diffusion version of this problem is known as the *Stefan problem* and can be found in Bird, Stewart, and Lightfoot.[8]

TABLE 2-10.4   Selected Ignition, Flammability, and Autoignition Properties of Some Fuels in Air*

Fuel	Formula	Weight (kg/kmol)	T_b° (K)	h_{fg}° (kJ/kg)	h_c (MJ/kg)	T_F° (K) closed	T_F° (K) open	Fl. limits by vol. lean	Fl. limits by vol. rich	T_a° (K)	h_{fg}°/RT_b°
Alkanes:											
Methane	CH_4	16	111	509	50.2	—	—	0.053	0.150	910	8.81
Ethane	C_2H_6	30	184	489	47.6	—	138	0.030	0.125	745	9.57
Propane	C_3H_8	44	231	426	46.4	—	169	0.022	0.095	723	9.76
n-Butane	C_4H_{10}	58	273	386	45.9	—	213	0.019	0.084	561	9.88
i-Butane			263	366	—	156	—	0.018	0.084	735	9.71
n-Pentane	C_5H_{12}	72	309	365	45.5	—	224	0.014	0.078	516	10.22
i-Pentane			286	371	—	—	222	0.014	0.076	693	11.23
n-Hexane	C_6H_{14}	86	342	365	45.2	251	—	0.012	0.074	498	11.04
i-Hexane			—	—	—	244	—	0.010	0.070	—	—
n-Heptane	C_7H_{16}	100	371	365	45.0	269	—	0.012	0.067	477	11.83
i-Heptane			—	—	—	255	—	0.010	0.060	—	—
n-Octane	C_8H_{18}	114	398	298	44.9	286	—	0.008	0.032	479	10.26
i-Octane			—	—	—	261	—	0.010	0.060	—	—
n-Nonane	C_8H_{20}	128	424	288	44.8	304	—	0.007	0.029	478	10.46
n-Decane	$C_{10}H_{22}$	142	447	360	44.7	317	—	0.006	0.054	474	13.76
n-Undecane	$C_{11}H_{24}$	156	469	308	44.6	—	338	0.007*	0.123	—	12.32
n-Dodecane	$C_{12}H_{26}$	170	489	293	44.6	345	—	0.006	0.123	476	12.25
Kerosene ~	$C_{14}H_{30}$	198	505	291	44.0*	322	—	0.006	0.056	533	13.72
Alkenes:											
Ethylene	C_2H_4	29	169	516	47.3	152	—	0.027	0.286	763	10.28
Propene	C_3H_6	42	225	437	45.9	165	—	0.021	0.111	728	9.81
1-Butene	C_4H_8	56	267	398*	45.4	193	—	0.016	0.099	658	10.04
1-Pentene	C_5H_{10}	70	303	314	46.9	—	255	0.014	0.097	548	8.72
Hexelene	C_6H_{12}	84	340*	388	47.5	—	—	—	—	518	11.53
Cycloparaffins:											
cycloPropane	C_3H_6	42	239	588	46.3	178	—	0.024	0.104	771	12.43
cycloButane	C_4H_8	56	286	483	44.8	208	—	0.011*	—	483	11.38
cycloPentane	C_5H_{10}	70	322	443	44.3	236	—	0.020*	—	634	11.58
cycloHexane	C_6H_{12}	84	354	358	43.9	253	—	0.013	0.078	518	10.22
cycloHeptane	C_7H_{14}	99	392	376	43.7	282*	—	0.012*	—	—	11.31
dimethyl-cycloHexane	C_8H_{16}	112	392	300	46.3*	284	—	—	—	505	10.31
Aromatics:											
Benzene	C_6H_6	78	353	432	40.7	262	—	0.012	0.071	771	11.48
Toluene	C_7H_8	92	383	362	41.0	277	280	0.013	0.068	753	10.46
m-Xylene	C_8H_{10}	106	412	343	41.3	298	—	0.011	0.070	801	10.61
o-Xylene			414	347	41.3	290	297	0.010	0.060	737	10.69
p-Xylene			410	339	41.3	298	—	0.011	0.070	802	10.54
bi-Phenyl	$C_{12}H_{10}$	154	527	—	40.6	386	397	—	—	813	—
Napthalene	$C_{10}H_8$	128	491s	316s	40.3	352	361	—	—	813	—
Anthracene	$C_{13}H_{10}$	166	613	310*	40.0*	394	469	0.009	0.059	799	9.91
ethylBenzene	C_8H_{10}	106	409	320*	43.1	288	297	0.006	—	813	10.10
butylBenzene	$C_{10}H_{14}$	134	446	277*	43.7	322	336	0.010	—	705	9.98
								0.008	0.059	683	10.01
Alcohols:											
Methanol	CH_3OH	32	337	1101	20.8	285	289	0.067	0.365	658	12.57
Ethanol	C_2H_5OH	46	351	837	27.8	286	295	0.033	0.190	636	13.19
n-Propanol	C_3H_7OH	60	370	686	31.3	288	302	0.022	0.135	705	13.38
i-Propanol			355	667	33.1	285	—	0.020	0.118	672	13.56
n-Butanol	C_4H_9OH	74	390	621	36.1	302	316	0.014	0.113	616	14.17
i-Butanol			380	578	36.1	301	—	0.017	—	678	13.54
2-Pentanol	$C_5H_{11}OH$	88	392	575	—	—	314	0.012	—	616	15.52
i-Amyl alcohol	$C_5H_{11}OH$	88	403	501	35.3	316	319	0.012	0.100	573	13.16
3-Pentanol			391	575*	—	307	312	0.012*	—	708	15.56
n-Hexanol	$C_6H_{13}OH$	102	430	458	36.4	318	347	0.012*	—	—	13.07
cyclohexanol			434	460*	36.6	341	—	0.012*	—	573	13.00
n-Heptanol	$O_7H_{15}OH$	116	449	439	39.8	—	344	—	—	—	13.64
1n-Octanol	$C_8H_{17}OH$	130	469	408	40.6	354	—	—	—	—	13.60
2n-Octanol			453	419	—	347	355	—	—	—	14.46
Nonanol	$C_9H_{19}OH$	144	487	403	40.3	—	—	—	—	—	14.33
i-Decanol	$C_{10}H_{21}OH$	158	508	373	—	—	—	—	—	561	14.12
Carbonyls:											
Formaldehyde	CH_2O	30	370	826	18.7	366	—	0.070	0.73	703	8.05
" 37% in H_2O		—	370*	826*	—	327	366	0.070*	—	697	8.05
Acetaldehyde	C_2H_4O	44	294	570	25.1	235	—	0.040	0.570	477	10.26
Allyl alcohol	C_3H_6O	58	368	684	31.9	294	297	0.025	0.180	651	12.93

TABLE 2-10.4   *(Continued)*

Fuel	Formula	Weight (kg/kmol)	T_b^o (K)	h_{fg}^o (kJ/kg)	h_c (MJ/kg)	T_F^o (K) closed	T_F^o (K) open	Fl. limits by vol. lean	Fl. limits by vol. rich	T_a^o (K)	h_{fg}^o/RT_b^o
i-Butyraldehyde	C_4H_8O	72	334	444*	33.8	233	249	0.025	—	503	11.51
Crotonaldehyde diethyl	C_4H_6O	70	375	490*	34.8	286	—	0.021	0.155	505	11.00
Acetaldehyde ethyl	$C_4H_{12}O$	76	391	500*	—	294	—	—	—	—	11.70
Hexaldehyde	$C_8H_{16}O$	128	436	325*	39.4	—	325	—	—	—	11.48
Paraldehyde	$C_6H_{12}O_3$	132	397	328	—	290	309	0.013	—	511	13.11
Salicyl aldehyde	$C_7H_6O_2$	122	469s	396	—	351	—	—	—	—	12.39
Benzaldehyde	C_7H_6O	106	452	362	—	337	347	—	—	465	10.21
Ketones:											
Acetone	C_3H_6O	58	329	521	29.1	255	264	0.026	0.128	738	11.05
2-Butanone	C_4H_8O	72	353	443	33.8	271	274	0.018	0.095	677	10.87
diethyl ketone	$C_5H_{10}O$	86	374	380	33.7	—	286	—	—	723	10.51
methyl i-Butyl ketone	$C_6H_{12}O$	100	389	345*	35.2	296	297	0.014	0.076	806	10.66
dipropyl ketone	$C_7H_{14}O$	114	417	317	38.6	—	—	—	—	806	10.42
methyl n-Propyl ketone	$C_5H_{10}O$	86	375	376*	33.7	280	289	0.015	0.082	613	10.37
methyl vinyl ketone	C_4H_6O	70	354	440*	—	266*	—	—	—	—	10.46
Acids:											
Formic acid	CH_2O_2	46	374	502	5.7	342	—	—	—	813	7.42
Acetic acid	$C_2H_4O_2$	60	391	405	14.6	313	330	0.054	—	737	7.48
Benzoic acid	$C_7H_6O_2$	122	523s	270*	24.4	394	—	—	—	843	7.58
Miscellany:											
Camphor	$C_{10}H_{16}O$	152	477s	265*	38.8	339	366	0.006	0.035	739	10.16
Carbon disulfide	CS_2	76	320s	—	13.6	303	—	0.013	0.500	398	—
m-Creosol	C_7H_8O	108	476	—	34.6	359	—	0.011	—	832	—
o-Creosol			464	—	34.1	354	—	0.013	—	872	—
p-Creosol			475	—	34.1	359	—	0.010	—	832	—
Furan	C_4H_4O	68	304	399	—	238	—	0.023	0.143	—	10.73
Pyridine	C_5H_5N	79	387	449	35.0	293	—	0.018	0.124	755	11.02
Aniline	C_6H_7N	93	456	434	36.5	349	364	0.013	—	890	10.64
Acetal	$C_6H_{14}O_2$	118	376	277	31.8	252	—	0.016	0.104	503	10.46
p-Cymene	$C_{10}H_{14}$	134	449	283	43.9	320	336	0.007	0.056	709	10.16
o-Dichloro Benzene	$C_6H_4Cl_2$	146	453	—	19.3	339	347	0.022	0.092	921	—
1.1-Dichloro Ethylene	$C_2H_2Cl_2$	96	310	—	—	—	263	0.056	0.114	733	—
1.2-Dichloro Ethylene			334	—	—	279	—	0.097	0.128	—	—
monochloro Benzene	C_6H_5Cl	112	405	—	—	305	311	0.018	—	947	—
Resorcinol	$C_6H_6O_2$	110	549	—	26.0	400	—	0.014	—	881	—
ethylFormate	$C_3H_6O_2$	74	327	—	22.5	253	261	0.027	0.164	728	—
ethylAcetate	$C_4H_8O_2$	88	350	—	25.9	269	272	0.022	0.114	700	—
methyl Propionate	$C_4H_8O_3$	104	353	—	22.2	271	—	0.024	0.130	742	—
Acrolein	C_3H_4O	56	326	—	29.1	—	247	0.028	0.310	508	—
Acrylonitrile	C_3H_3N	53	350	—	24.5	—	273	0.030	0.170	754	—
n-Amyl Acetate	$C_7H_{14}O_2$	130	422	—	33.5	297	300	0.011	0.075	633	—
1-Amyl Acetate			416	—	—	298	311	0.010	0.075	652	—
1,3-Butadiene	C_4H_6	54	269	—	—	197	—	0.020	0.115	693	—
n-Butyl Acetate	$C_6H_{12}O_2$	116	400	—	30.0	295	305	0.014	0.076	694	—
n-Butyl Ether	$C_8H_{18}O$	130	414	—	39.7	298	311	0.015	—	467	—
dimethyl Ether	C_2H_6O	46	249	—	31.6	232	—	0.034	0.267	623	—
divinyl Ether	C_4H_4O	70	312	—	—	243*	—	0.017	0.270	633	—
diethyl Ether	$C_4H_{10}O$	74	308	—	37.4	228	—	0.019	0.365	433	—
Gasoline ~	—	—	306	—	44.1	228	—	0.014	0.068	644*	—
Naptha ~	—	—	450	—	—	314	—	0.008	0.050	519	—
Petroleum Ether ~	—	—	351	—	—	255	—	0.014	0.059	561	—

* Estimate; ~ approximate line; — unavailable; R = $\overline{R}$/M; $\overline{R}$ = 8.314 kJ/kmol K.

** Compiled from references 3–7 and 15.

This description of transient diffusion of vapor issued at a constant rate at the surface also holds, qualitatively, to a situation of a temporally variable vapor release rate, as in a typical flash point test.

As pointed out by Burgoyne and Williams-Leir[9] and by Kanury,[10] the vapor release boundary condition is quite crucial in the flash point phenomenon. For a flame to flashover the surface of liquid fuel when presented with an ignition source, the vapor and air mixture (at the location and instant of the ignition source) has to contain sufficient fuel to be above the lean limit of flammability, ℓ. The partial pressure of the vapor in the mixture at the liquid surface is approximately equal to the saturation pressure, P_v, of the fuel at the surface temperature, T. The error associated with the assumption of thermodynamic equilibrium (at the surface across which mass transfer occurs at a nonnegligible rate) is known to be quite small. The $P_v(T)$ relation is given by the integrated form of the well-known[11] Clausius-Clapeyron equation of state*

$$\frac{P_{vs}}{P^\circ} = \exp\left[\left(\frac{h^\circ_{fg}}{RT^\circ}\right)\left(1 - \frac{T^\circ}{T_s}\right)\right] \tag{1}$$

where T° and P° are reference temperature and pressure. The subscript s denotes the surface. When T° is equal to the normal boiling point temperature, T°_b, of the liquid, the reference vapor pressure, P°, will become equal to the atmospheric total pressure. Equation 1 is generally written in the form $\ln P_{vs} = -\alpha/T_s + \beta$, where $\alpha \equiv h^\circ_{fg}/R$ and $\beta \equiv \ln P^\circ + \alpha/T^\circ$ are constants of the liquid. The *CRC Handbook of Physics and Chemistry*[7] gives extensive $P_v(T)$ tabulations for several thousand liquids.

For many nonpolar liquids, the molar specific entropy of vaporization is empirically found to be a universal constant, a consequence of the empirical invariance of the ratio of normal boiling point to the critical temperature, so that

$$\frac{h^\circ_{fg}}{RT^\circ_b} \approx \text{a constant} = 10.18$$

Known as *Trouton's rule*,[12] this powerful relation enables one to estimate the normal enthalpy of vaporization h°_{fg}/R from a knowledge of the normal boiling point T°_b and the liquid molecular** weight, M. The last column of Table 2-10.4 indicates the validity of Trouton's rule; in general, Trouton's constant is larger than 10.18 for alcohols and smaller for acids.

Due to noninfinite diffusion effects, the molar (or volume) fraction, $X_v \equiv P_v/P^\circ$, of fuel vapor at stations away from the surface will be smaller than that at the surface, given by Equation 1. The solution of the diffusion equation, with Equation 1 serving as one of the boundary conditions,

gives the following distribution. With y denoting the distance in the gas phase from, and normal to, the surface and t denoting time, let $g(y,t)$ be a distribution solution such that

$$\frac{P_v(y, t)}{P^\circ} = g(y, t)\,\frac{P_{vs}}{P^\circ} \tag{2}$$

The function $g(y, t)$ is dimensionless and smaller than unity in magnitude.

Suppose that a large enough pilot source is introduced at a height, y_p, and time, t_p. Recalling the definition of the lean limit, ℓ, the flame would then flash if the local instantaneous mole fraction, $P_v(y_p, t_p)/P^\circ$, is marginally in excess of the lean limit, ℓ. Thus, the flash point and the lean limit are related by

$$g(y_p, t_p)\, \exp\left[\left(\frac{h^\circ_{fg}}{RT^\circ}\right)\left(1 - \frac{T^\circ}{T_F}\right)\right] \geq \ell \tag{3}$$

Taking the total pressure to be 1 atm, $T^\circ = T^\circ_b$, the normal boiling point. We denote the flash point in 1 atm air by T°_F. The inequality sign in this equation is quite close to an equality. Noting that g and ℓ are smaller than unity, Equation 2 can be rewritten as

$$\ln\left(\frac{1}{\ell}\right) \approx \left(\frac{h^\circ_{fg}}{RT^\circ_b}\right)\left(\frac{T^\circ_b}{T^\circ_F} - 1\right) + \ln\frac{1}{g(y_p, t_p)} \tag{4}$$

The following comments can be made of Equation 4:

1. The lean limit of flammability, ℓ, as listed in the literature and in Table 2-10.4, is itself an apparatus-dependent "property." The most notable factor influencing it is gravity (causing buoyancy in the limit mixtures and inhomogeneity of the fuel-air mixture). The approximate equality sign in Equation 4 is meant to highlight this point.

2. Most of the flash point data seldom point out explicitly that they have been determined in air and nominally at 1 atm total pressure. The superscript "°" is a reminder of these conditions.

3. If the ambient total pressure, P°, is higher, the flash point is expected to be higher also. A plot of $\ln P^\circ$ versus $1/T^\circ_F$ is known to be a straight line, apparently parallel to the boiling point line for the same liquid.[13] Thus, correction for the flash point dependency on pressure can be made the same way as for the boiling point dependency. Leslie and Geniesse refer in *ICT*[5] (p. 161) to a method by which the flash point is calculated as the temperature at which the liquid's vapor pressure is equal to P°/kN, where P° is the total pressure, N is the number of moles of oxygen stoichiometrically required for complete combustion of 1 mole of the liquid, and k is an apparatus constant of about 8. The stoichiometry ratio, N, apparently is introduced through the lean limit, ℓ.

4. The question of the dependency of lean limit on whether the ambient gas is normal air or pure oxygen is a complicated one. If the oxygen/fuel ratio at the lean limit is assumed to be the same for both cases, then ℓ is expected to be numerically larger in pure oxygen than in air. Lewis and von Elbe[15] report a few values to confirm this expectation. A consensus exists in the combustion literature that the percent fuel in the limit mixtures is, more or less, independent of the presence of nitrogen; this, too, leads to a larger value of ℓ in pure oxygen than in air. Both these arguments, coupled with Equation 4, lead to the conclusion that the flash point of a liquid will be higher (and closer to the boiling point) in pure oxygen than in air.

---

*With its genesis in the Maxwell relation $(\partial P/\partial T)_v = (\partial s/\partial v)_T$, Equation 1 arises from the facts: (1) equilibrium evaporation is a reversible process of constant pressure and temperature; (2) the heat addition in a constant temperature reversible process is equal to $T\Delta s$; and (3) the constant pressure heat addition is equal to Δh; and the assumptions: (1) the enthalpy change associated with the change of phase from saturated liquid to saturated vapor is independent of temperature so that $\Delta h = h^\circ_{fg}$, a constant; (2) the vapor volume far exceeds the liquid volume; and (3) the vapor behaves nearly as an ideal gas. The basic constraint of equilibrium, of course, is the equality of Gibbs' free energy of the saturated liquid and that of the saturated vapor.

**Recall that the gas constant R is related to the universal gas constant $\bar{R} = 8.314$ (kJ/kmol K) through the molecular weight M (kg/kmol) by $R \equiv \bar{R}/M$ (kJ/kg K).

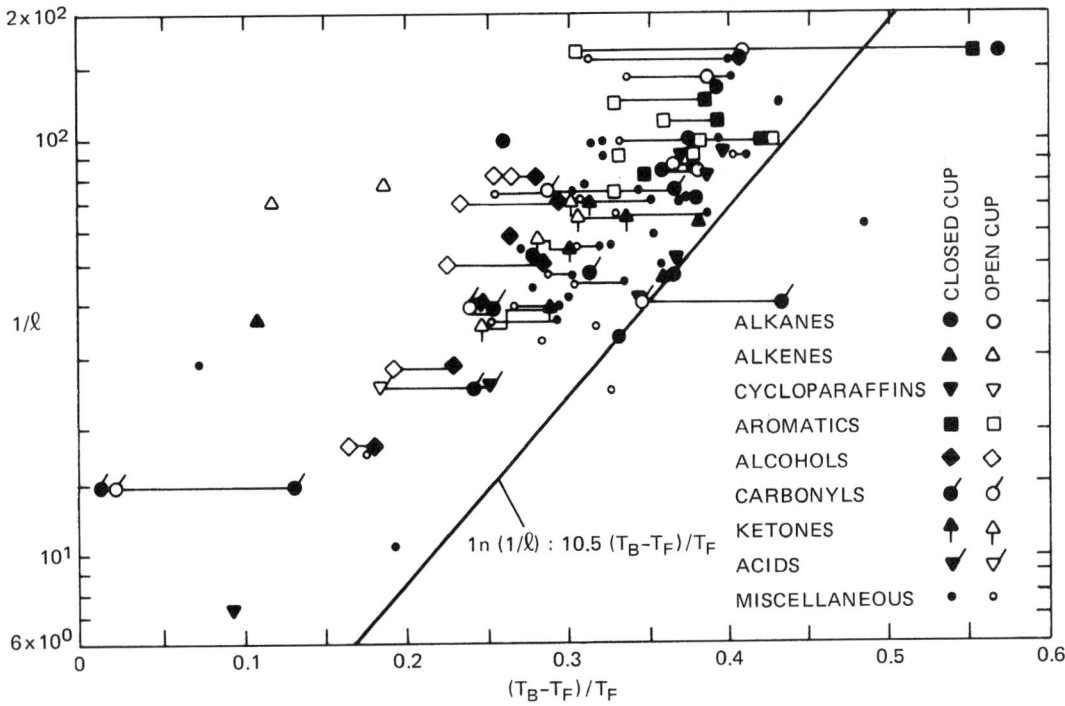

Fig. 2-10.2. *A relation between the lean limit, flash point, and boiling point.*[10]

This is, however, contrary to common sense; ignition is expected to occur more easily in the absence of the impeding inert nitrogen. Experience indicates that in pure oxygen atmospheres, the minimum ignition energies of combustible mixtures are lower, quenching distances are smaller, and flame temperatures and speeds are larger. Based on this experience, it is logical to surmise that the chemical kinetics of oxidation and thermochemistry underlying the concept of the lean limit are far more complex than mere mass conservation. The paper of Mullins[14] on combustion in vitiated air and the tome of Lewis and von Elbe[15] serve well as starting points for a much-needed research investigation into this important issue of limit mixtures.

5. When the cup is closed, the escape of the fuel vapor to the environment is prevented or minimized. Appearing as the second boundary condition in the diffusion equation, this results in a higher value of $g(y, t)$ at any location and time. For a given lean limit, Trouton's constant, and normal boiling point, Equation 4 indicates that the observed flash point will be lower than if the cup were open.

6. The effects of the cup being open, the air being in a state of motion, and the room being gusty are all in the direction of increasing the observed flash point, all due to reduced $g(y, t)$.

7. Burgoyne and Williams-Leir[9] assume the dispersion function $g(y_p, t_p)$ to be unity to use a variant form of Equation 4 to predict the flash point from given ℓ, h_{fg}°, and T_b°.

8. Such a use of Equation 4 as a predictive tool is probably premature for several reasons. First, the transient convective-diffusion process underlying the g-function has a crucial role to play in determining T_F°. Second, even if one works out the solution to obtain $g(y, t)$, the flash point experiments do not clearly stipulate the $y = y_p$ and $t = t_p$ values; i.e., they do not state the time and location at which the pilot is introduced. Thus, the estimates of $g(y_p, t_p)$ require subjective judgments. Last, as discussed in item (4) above, the lean limit, ℓ, is still an empirical extrinsic parameter whose dependence on fundamental properties of the system is not quantitatively known.

Notwithstanding these reasons, by merely noting that the last term in Equation 4 is always positive, one concludes that

$$\ln\left(\frac{1}{\ell}\right) \geq \left(\frac{h_{fg}^\circ}{RT_b^\circ}\right)\left(\frac{T_b^\circ}{T_F^\circ} - 1\right)$$

A plot of $\ln(1/\ell)$ on the y-axis and $(T_b^\circ/T_F^\circ - 1)$ on the x-axis should then yield all experimental points lying above a straight line of slope $(h_{fg}^\circ/RT_b^\circ) \approx 10.18$ and intercept zero. Figure 2-10.2 shows most of the data of Table 2-10.4 thus plotted. It is obvious that the hypothesis underlying Equation 4 bears merit. The general trend of data (especially for each family of fuels) indicates that Trouton's constant (or a like one) holds to describe the slope. The intercept, expected to be zero if diffusional effects are absent, indicates that $g(y_p, t_p)$ lies in the approximate range of 0.07 to 0.25, substantially smaller than unity, indicating the need for consideration of diffusion.

9. All through this discussion energy conservation has not been considered although it may be important.

Autoignition Temperature

Whereas the piloted ignition in flash point tests involves a localized initiation of the flame and observation to see if it would propagate away from the ignitor, the autoignition is a self-induced initiation of the flame over a relatively larger volume of the reacting gases. As such, the autoignition is perhaps amenable to an easier description than the forced ignition. At the outset, it should be noted that the autoignition process pertains to the vapor (or gas) and air mixture, irrespective of the source of the vapor or gas. Figure 2-10.1, part (b), indicates that by the time the autoignition process becomes evident, the liquid fuel will be vaporizing vigorously and even boiling.

Consider the oxidation reaction in a vessel of known geometry. The initial temperature and mixture composition are usually known. The mechanism of heat exchange between the reactive medium and the vessel walls constitutes the most important of the boundary conditions. Although the oxidation mechanism is probably composed of a complex set of elementary reaction steps, a global rate law for the fuel consumption rate generally suffices.

$$-\dot{W}_A''' = k_0\, Y_A^n\, Y_B^m\, \exp(-E/\bar{R}T) \qquad (5)$$

The quantity on the left-hand side of Equation 5 is fuel consumption rate in $kg/m^3 \cdot s$. The empirical kinetic constants k_0, n, m, and E are, respectively, the collision factor, orders with respect to the fuel (A) and oxygen (B), and activation energy (kJ/kmol). In principle, then, the equations of conservation of mass, momentum, energy, and species can be solved to obtain the transient fields of velocity, temperature, and composition. According to thermal theory,[16-18] ignition is inferred to have occurred at the instant when the temperature (usually, in the middle of the reactor) increases rapidly. Known as *thermal runaway*, this event also marks the onset of a rapid decrease in the reactant mass fractions, Y_A and Y_B. If the reactant consumption in the preignition period is ignored, then the autoignition is called a *thermal explosion*.

Solutions are available in the literature for this problem, with and without reactant depletion, for a number of geometries of the reactant mass and of boundary conditions. When the reactant depletion is ignored, the species equations become unnecessary. The convective mixing in gaseous mixtures is usually taken into account by defining an effective, augmented, thermal conductivity so as to obviate the momentum conservation and reduce the problem to one involving merely transient, reactive conduction. By considering the effective conductivity to be large (due to good mixing) and the heat loss from the reacting mixture to the vessel boundaries to be by convection, Semenov[16] solves the resultant transient, uniform reaction problem to predict thermal runaway conditions as

$$E(T_a - T_w)/\bar{R}T_w^2 = 1$$
$$\mathrm{Da}_0 \equiv [h_c V(-\dot{W}_A''')_0]/[hS(T_a - T_w)] \geq 1/e \qquad (6)$$

where T_a is the (critical) temperature of the reactive mass at the instant of explosion (see Figure 2-10.1), T_w is the vessel wall temperature, h_c is the enthalpy of combustion (kJ/kg fuel), h is the heat transfer coefficient between the mixture and the walls (kW/m^2 K), V/S is the volume-to-surface ratio of the reactive mass, and $(-\dot{W}_A''')_0$ is the reaction rate given by Equation 5 at the initial conditions $T = T_0$, $Y_A = Y_{A0}$, and $Y_B = T_{B0}$. e is the Naperian constant.

Two points are immediately notable from Equation 6. First, the critical temperature T_a is quite close to T_w, since $\bar{R}T_w^2/E$ is, generally, quite small. Thus, $T_a \approx T_w$. Second, due to the mixture being well mixed, the precise geometry of the reacting body is irrelevant; its V to S ratio suffices; $V/S =$

$a/(j + 1)$, where $j = 0$, 1, and 2, respectively, for an infinite plane slab geometry of thickness, $2a$, for an infinitely long cylinder of diameter, $2a$, and a sphere of diameter, $2a$. Other implications of Equation 6 will be discussed later.

Frank-Kamenetskii[17] solves the problem by assuming that the effective conductivity, K, is small and seeks the conditions under which steady-state solutions are impossible. Thermal runaway is found to occur persuant to the following critical conditions

$$\begin{array}{cccc} & j=0 & j=1 & j=2 \\ E(T_a - T_w)/\bar{R}T_w^2 & 1.20 & 1.37 & 1.60 \end{array} \qquad (7)$$

$$\mathrm{Da}_\infty \equiv \dfrac{a^3 h_c k_0 Y_{A0}^n Y_{B0}^m \exp(-E/\bar{R}T_w)}{a^2 K(T_a - T_w)/a} \quad\;\; 0.88 \;\; 2.00 \;\; 3.32$$

This solution, as does Equation 6, indicates that $(T_a - T_w) \approx \bar{R}T_w^2/E$ is small. Vessels with a geometry resulting in a larger surface-to-volume ratio (e.g., a sphere) exhibit a larger critical temperature, T_a.

The nondimensional parameters Da_0 and Da_∞ of Equations 6 and 7, respectively, are known as Damköhler numbers. A Damköhler number[18] is the ratio of the characteristic rate of energy release due to chemical reaction to that of physical dissipation (by convection in the Semenov problem and by conduction in the Frank-Kamenetskii problem).

The Semenov problem corresponds to the zero Biot number limit, while the Frank-Kamenetskii problem deals with the infinite Biot number situation. The finite Biot number problem has been solved by Thomas[19] and Kanury and Gandhi.[20] (An excellent review of this subject is presented by Gray and Lee.[21]) Equations 6 and 7 can be seen to be in mutual agreement if one notes that the heat transfer coefficient, h, is linearly proportional to the ratio of gas conductivity, K, to the vessel half-size, a.

Even more importantly, the pressure dependency of the collision factor is given[18] by

$$k_0 \equiv k_0' M_A^{(1-n)} M_B^{-m} \left(\frac{PM}{\bar{R}T}\right)^{(n+m)} \qquad (8)$$

where k_0' is a function only of the molecular collisional nature of species A and B; M_A and M_B are molecular weights, respectively, of the fuel (A) and oxygen (B); M is the mean molecular weight of the mixture; and P is the total pressure of the reacting mixture. The Damköhler number, Da_∞, from the criterion for explosion given by Equation 7 can then be recast to the form

$$\mathrm{Da}_\infty \equiv \left[\frac{k_0' M_B^{-m} M^{(n+m)} Y_{A0}^n Y_{B0}^m}{\bar{R}^3 E^{(1+n+m)}}\right]\left[\frac{a^2 h_c P^{(n+m)} M_A^{1-n}}{K} f(T_a)\right] \qquad (9)$$

where $f(T_a) \equiv [\exp(-E/\bar{R}T_a)](\bar{R}T_a/E)^4$. If the initial mixture composition (Y_{A0}, Y_{B0}) is kept fixed, the first bracketed quantity of Equation 9 is, more or less, a constant. For typical hydrocarbons reacting with air, $n \approx 0.5$, $m \approx 1.5$, and $E/\bar{R} \approx 15{,}000$ K. The function $f(T_a)$ monotonically increases with ($\bar{R}T_a/E$), as indicated in Figure 2-10.3. The second bracketed term of Equation 9 and the explosion criterion constant of Equation 7 thus give the influence of a, h_c, P, M_A, and K on the minimum temperature for autoignition, T_a. Specifically, it is clear that the autoignition temperature is lower if the vessel size is larger, heat of combustion is larger, pressure is

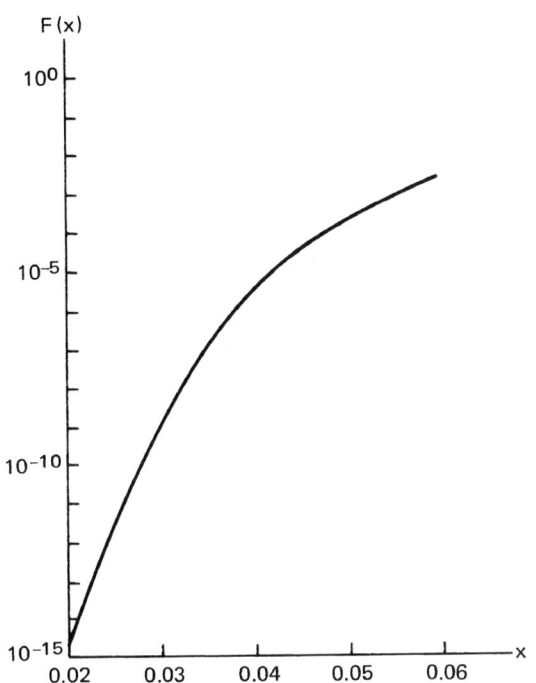

Fig. 2-10.3. *The monotonic nature of* $f(x) = [exp(-1/x)]/x^4$.

higher, fuel molecular weight is larger, or the mixture conductivity is smaller. This predicted pressure dependency is in excellent quantitative agreement with the data in Table 2-10.3. The influence of molecular weight appears corroborated by the data in Table 2-10.4. Not so evident from Table 2-10.4 is the effect of the heat of combustion, mainly due to the relatively minor variance of h_c within a family of fuels.

Ignition Delay

The discussion so far has centered on the threshold conditions below which ignition is not possible (i.e., question 1 of the Introduction). So far as the flash point and fire point are concerned, a more detailed analysis is required to go further. Concerning autoignition, however, a few further steps can be taken related to question 2 of the Introduction. First, note that the critical temperature T_a from Equations 6, 7, and 9 and Table 2-10.4 is the minimum temperature of the mixture that can produce autoignition, but marginally. This marginality implies that time to ignition at T_a is infinite. Any temperature greater than T_a will result in ignition within a finite time. The larger the temperature, the sooner the ignition.

Furthermore, as indicated in Figure 2-10.1, there exist upper and lower composition limits to autoignition at any $T > T_a$. The most quickly ignitible mixture at any T will perhaps be near stoichiometric in composition. Thus, energy balance on the reacting mixture leads to the approximate ignition delay, t_i, of

$$t_i \approx \frac{(\rho C_p)(T - T_0)}{h_c(-\dot{W}_A''')_{ref}} \qquad (10)$$

where (ρC_p) is mixture volumetric heat capacity (kJ/m³ K), T_0 is initial temperature of the mixture, and the reaction rate is estimated at the reference conditions of 0 for the Semenov (Biot number = 0) problem and w for the Frank-Kamenetskii

(Biot number = ∞) problem. This rate also accounts for the mixture composition Y_{A0} and Y_{B0}. The Arrhenius dependency of the reaction rate on temperature makes the ignition delay, t_i, proportional to $\exp(-E/\overline{R}T_{ref})$. Thus, measurement of t_i at various T_{ref} values leads to determination of the activation energy, E. In the light of Equation 8, t_i will be approximately proportional to the inverse of the mixture pressure and to the square root of the molecular weight of the fuel. Since Da_0 and Da_∞ are about unity, the denominator of Equation 10 is proportional to $K(T - T_w)/a^2$ so that ignition delay is longer for higher conductivity mixtures, if all else is kept fixed.

If the consumption of reactants in the preignition reactions is significant, the problem becomes more difficult. The outcome, however, is that reactant consumption makes the limit temperature, T_a, somewhat higher than the estimates from Equations 6, 7, and 9, and the ignition delay is longer than that given by Equation 10.

Concluding Remarks

The concepts of piloted ignition of liquid fuels and the spontaneous ignition of their vapors mixed with air are examined in this chapter. Starting with a physical description of the ignition process and definitions, a number of experimental techniques were alluded to and typical data were presented. The empirical trends were shown to be consistent with the predictions based on simple mechanistic models.

Although the flash point concept is useful in rating the volatility and ease of ignition of a variety of liquid combustibles, it must be noted that many practical situations involve heating of the liquid by a heat flux imposed on the evaporating surface. The heating process itself might then become a major topic of study to estimate the liquid surface temperature before examining whether this temperature is above or below the flash point at a given instant.

NOMENCLATURE

a	vessel half-size (m)
C_p	specific heat (kJ/kg K)
Da_0	Semenov's Damköhler number $(-)$
Da_∞	Frank-Kamenetskii's Damköhler number $(-)$
E	activation energy (kJ/kmol)
e	Naperian constant $(-)$
g	dispersion function $(-)$
h	heat transfer coefficient (W/m² K)
h_c	enthalpy of combustion (kJ/kg)
h_{fg}°	enthalpy of vaporization (kJ/kg)
j	geometry index
K	mixture thermal conductivity (W/m K)
k	apparatus constant $(-)$
k_0	collision factor (kg/m³ s)
k_0'	collision parameter
ℓ	lean limit of flammability $(-)$
M	molecular weight (unsubscripted for the mixture) (kg/kmol)
m	a reaction order $(-)$
N	Stoichiometric molar O_2/fuel ratio $(-)$
n	reaction order $(-)$
P	pressure (atm)
P°	total pressure (atm)
P_v	vapor pressure (atm)
R	gas constant (kJ/kg K)

$\bar{R}$	universal gas constant (kJ/kmol K)
S	surface area (m^2)
s	specific entropy (kJ/kg K)
T	temperature (K)
T_a	the least autoignition temperature (K)
T_b°	normal boiling point (K)
T_F°	flash point in air at 1 atm (K)
t	time (s)
V	volume (m^3)
v	specific volume (m^3/kg)
$\dot{W}'''$	reaction rate (kg/m^3 s)
Y	mass fraction ($-$)
y	distance into gas phase from, and normal to, the vaporizing surface (m)

Greek

α	property constant (K)
β	property constant ($-$)
ρ	density (kg/m^3)

Subscripts

A	fuel
a	autoignition
B	oxygen
b	boiling point
c	combustion
F	flash point
i	ignition
0	initial
p	pilot
s	liquid surface
v	vapor
w	vessel wall surface

REFERENCES CITED

1. M.G. Zabetakis, *Bulletin No. 627*, Bureau of Mines, Pittsburgh (1965).
2. ASTM, *Standard, Vol. 17, 20, and 22*, American Society for Testing and Materials, Philadelphia (1972).
3. B.P. Mullins, *Spontaneous Ignition of Liquid Fuels*, Butterworths, London (1955).
4. B.P. Mullins, in *Combustion Researches and Reviews*, Butterworths, London (1957).
5. *International Critical Tables of Numerical Data, Physics, Chemistry, and Technology*, McGraw-Hill, New York (1927).
6. Factory Mutual Research Corp., *Factory Mutual Handbook of Loss Prevention*, McGraw-Hill, New York (1967).
7. *CRC Handbook of Physics and Chemistry*, The Chemical Rubber Co., Cleveland (1973).
8. R.B. Bird, W.E. Stewart, and E.N. Lightfoot, *Transport Phenomena*, Wiley, New York (1960).
9. J.H. Burgoyne and G. Williams-Leir, *Fuel*, 28-7, 145 (1949).
10. A.M. Kanury, *Comb. Sci. and Tech.*, 31, 297 (1983).
11. W.C. Reynolds and H.C. Perkins, *Engineering Thermodynamics*, McGraw-Hill, New York (1977).
12. F. Daniels and R.A. Alberty, *Physical Chemistry*, Wiley, New York (1967).
13. Anonymous, *Physikalisch-Technische Reich-Sanstalt* (1910).
14. B.P. Mullins, in *Selected Combustion Problems*, Butterworths, London (1954).
15. B. Lewis and G. von Elbe, *Combustion, Flames and Explosions of Gases*, 3rd ed., Academic, New York (1987).
16. N. Semenov, *Chemical Kinetics and Chain Reactions*, Oxford, London (1935).
17. D.A. Frank-Kamenetskii, *Diffusion and Heat Exchange in Chemical Kinetics*, Princeton University, Princeton (1955).
18. A.M. Kanury, *Combustion*, unpublished manuscript (1995).
19. P.H. Thomas, *Trans. Faraday Soc.*, 54(421), 60 (1958).
20. A.M. Kanury and P.D. Gandhi, *Theory of Spontaneous Heating and Thermal Ignition: A Review*, unpublished manuscript (1989).
21. P. Gray and P.R. Lee, in *Oxidation and Combustion Reviews*, Elsevier, Amsterdam (1967).

SMOLDERING COMBUSTION

T. J. Ohlemiller

INTRODUCTION

Smoldering is a slow, low-temperature, flameless form of combustion, sustained by the heat evolved when oxygen directly attacks the surface of a condensed-phase fuel. Smoldering constitutes a serious fire hazard for two reasons. First, it typically yields a substantially higher conversion of a fuel to toxic compounds than does flaming (though this occurs more slowly). Second, smoldering provides a pathway to flaming that can be initiated by heat sources much too weak to directly produce a flame.

The term smoldering is sometimes inappropriately used to describe a non-flaming response of condensed-phase organic materials to an external heat flux. Any organic material, when subjected to a sufficient heat flux, will degrade, gasify, and give off smoke. There usually is little or no oxidation involved in this gasification process, and thus it is endothermic. This is more appropriately referred to as forced pyrolysis, not smoldering.

A burning cigarette is a familiar example of true smoldering combustion; it is also one of the most common initiators of smoldering in other materials, especially upholstery and bedding.[1] A cigarette also has several characteristics common to most materials that smolder. The finely divided fuel particles provide a large surface area per unit mass of fuel, which facilitates the surface attack by oxygen. The permeable nature of the aggregate of fuel particles permits oxygen transport to the reaction site by diffusion and convection. At the same time, such particle aggregates typically form fairly effective thermal insulators that help slow heat losses, permitting sustained combustion despite low heat release rates.

The physical factors that favor smoldering must be complemented by chemical factors as well. Like virtually all other cellulosic materials, tobacco in a cigarette, when degraded thermally, forms a char. A char is not a well-defined material, but typically it is considerably richer in carbon content than the original fuel; its surface area per unit mass is also enhanced. This char has a rather high heat of oxidation and is susceptible to rapid oxygen attack at moderate temperatures (≥ 670 K). The attack of oxygen (to form mainly carbon monoxide and carbon dioxide) is facilitated not only by the enhanced surface area but also by alkali metal impurities (present in virtually all cellulosic materials derived from plants) which catalyze the oxidation process.[2] Char oxidation is the principal heat source in most self-sustained smolder propagation processes; the potential for smoldering combustion thus exists with any material that forms a significant amount of char during thermal decomposition. (Char oxidation is not always the only heat source and it may not be involved at all in some cases of smolder initiation.)[3]

Various quantitative combinations of these physical and chemical factors can produce a material that will undergo sustained smoldering in some conditions. The enormous range of factors results in materials that will only smolder when formed into fuel aggregates many meters across, at one extreme, to materials that smolder when formed into aggregates only a few tens of microns across. Unfortunately, a theory that allows for the calculation of materials and conditions that are conducive to smoldering has been developed only for certain types of smolder initiation. (See Section 2, Chapter 12.) Conditions sufficient to yield smolder initiation, especially near an external heat source, are not necessarily sufficient to assure self-sustained smolder spread away from the initiation region. The potential transition of the smolder process into flaming combustion is even less correlated with factors determining smolder initiation.

This chapter is restricted to consideration of post-initiation behavior of smoldering. There are a few models of smolder propagation in the literature but none sheds much light on any practical smolder problem. The state of modeling is reviewed elsewhere.[4] Lacking any definitive theoretical description, this chapter is largely restricted to examining typical experimentally determined behavior. In this overview of smoldering, an attempt is made to convey some of the qualitative interplay of processes that determines overall behavior together with specific experimental results.

SELF-SUSTAINED SMOLDER PROPAGATION

The smolder initiation process is dominated by the kinetics of the oxidation of the solid. Subsequent propagation of smolder is controlled to a large degree, however, by the

Dr. T. J. Ohlemiller is a member of the research staff of the National Institute of Standards and Technology, Building and Fire Research Laboratory. He has investigated a variety of solid fuel combustion problems, specializing in smoldering and the flammability of solid fuels.

rate of oxygen transport to the reaction zone. The control via transport rate occurs because the heat evolved during smolder initiation raises the local temperature and thus the local reaction rate, until all of the neighboring oxygen is consumed. Subsequently, the reaction continues to consume oxygen as fast as it reaches the reaction zone, yielding a very low oxygen level locally, which limits the reaction rate.

The subsequent evolution of the smoldering zone away from the initiation region is heavily influenced by oxygen supply conditions. If initiation occurs deep within a layer of fine particles (sawdust, coal dust), for example, it will slowly work its way to the nearest free surface at a rate dictated by oxygen diffusion through the particle layer. (The more coarse and loosely packed the particles, the greater the influence of buoyant flow through the fuel leading to predominant upward spread.) When the smolder zone reaches the free surface region, it will spread more rapidly over this region in response to local convective and diffusive oxygen supply conditions. As will be seen, when smolder spread over the surface region of a fuel layer is forced by airflow, its response also depends on heat transfer considerations.

In examining self-sustained smolder propagation and its response to oxygen supply conditions, dimensionality is important. It is necessary to distinguish one-dimensional from multi-dimensional configurations. It is further necessary to discern whether the smolder zone is spreading in the same or opposite direction as the net movement of oxygen.

One-Dimensional Smolder Spread

One-dimensional smolder spread is an idealized situation that is sometimes approximated in real fires. For example, the spread outward or upward from deep in a layer of fuel particles approaches this one-dimensional limit when oxygen diffusion dominates convection and any curvature of the reaction front is small compared to the reaction zone thickness. In practice, this curvature requirement would likely be met by spread about 0.10 to 0.20 m away from the ignition source. One-dimensional smolder can be characterized by the direction of smolder propagation relative to the direction of oxygen flow: forward and reverse propagation.

Reverse propagation: When oxygen diffuses to the reaction zone from the outer surface of the fuel layer, through the unburned fuel and toward the reaction front, it is moving opposite to the direction of smolder propagation; such a case of relative movement is called reverse smolder.

Palmer[5] examined this diffusive reverse smolder case using layers of wood sawdust of various depths; the configuration was only roughly one-dimensional. Some of his results are shown in Figure 2-11.1. Note that the time scale is in hours; the time to smolder up through a layer 1 m deep is about two weeks, a surprisingly long time. Palmer noted that in this configuration the smoldering process gave little hint of its presence until it was close to the surface of the fuel layer.

The slope in Figure 2-11.1 indicates that the time for smolder to penetrate a fuel layer in this mode is nearly proportional to the square of the layer depth.[5] Palmer showed that a second power dependence on layer depth would be expected if it is assumed that the smolder reaction zone propagation velocity is proportional to the one-dimensional diffusion rate of oxygen from the surroundings, through the unburned fuel, to the reaction zone. This results in

$$t_L = AL^2 \qquad (1)$$

where t_L is the time for the smolder zone to penetrate the layer of thickness, L; and A is a constant that can, at present, only be determined by experimental measurement of at least one layer thickness.

This relation and Figure 2-11.1 imply that a 10-m fuel layer, such as might be encountered in a landfill or coal mine tailing pile, would require more than four years for smolder penetration. Such a deep layer is unlikely to be uniform in practice and the smolder front movement would be dominated by buoyant convective flow in regions of lesser flow resistance. However, this does illustrate how very slow some smolder processes can be.

A well-insulated reaction zone is a key factor in the existence of stable, self-sustaining smolder at such extremely low rates. The heat loss rate cannot exceed the heat

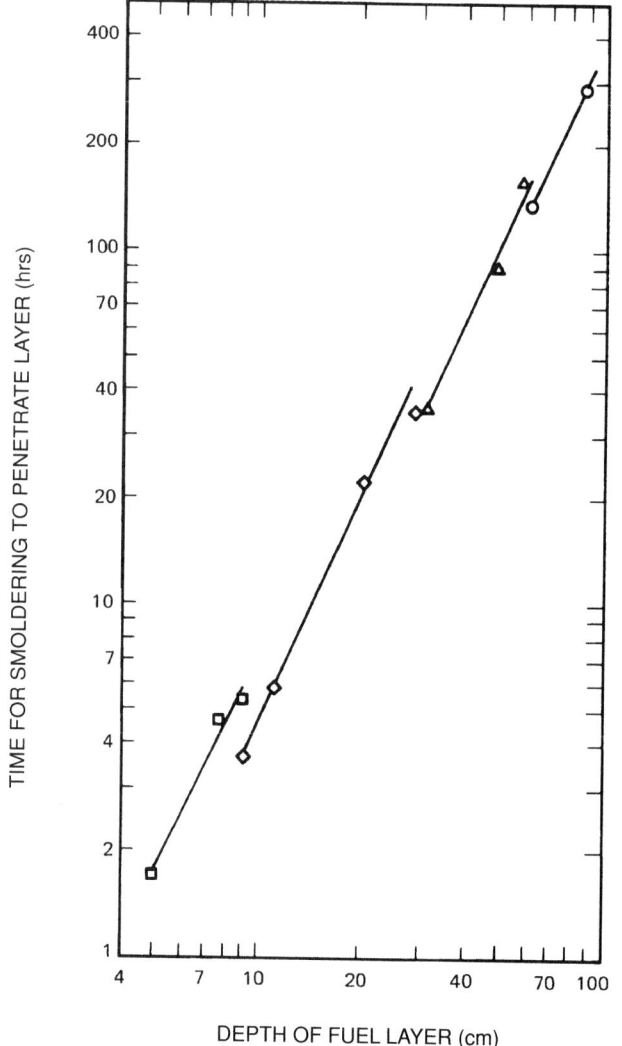

Fig. 2-11.1. Smoldering upward from bottom within thick layers of mixed wood sawdust.[5] Squares: initiating layer 0.025 m deep, 0.3 m square box; diamonds: initiating layer 0.052 m deep, 0.3 m square box; triangles: initiation layer 0.052 m deep, 0.6 m square box; circles: initiating layer 0.052 m deep, 0.9 m square box.

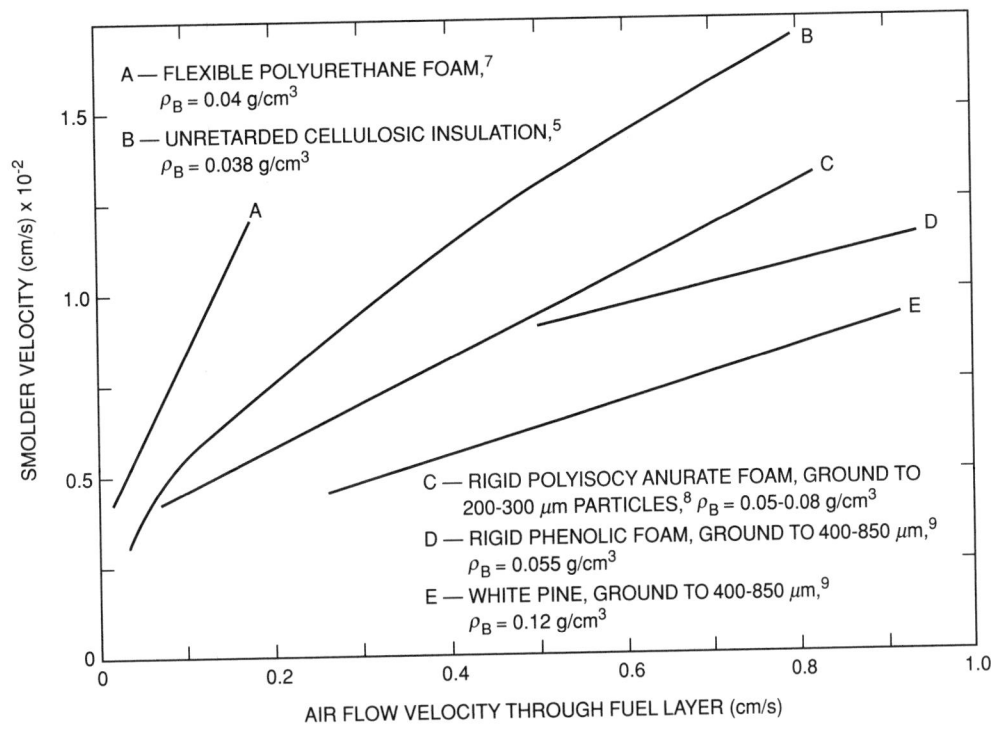

Fig. 2-11.2. **Smolder velocity vs airflow velocity into reaction zone for nearly one-dimensional reverse smolder.** ρ_B **is bulk density.**

generation rate. In this case, the same factor that is slowing the oxygen supply rate, and therefore the heat generation rate (i.e., the thick layer of wood particles over the reaction zone), is also slowing the heat loss rate.

In the previous example, the smolder propagation process is inherently unsteady because of the time-dependent oxygen supply process. If oxygen is instead continually supplied by a forced convective flow *through* the fuel layer, nearly steady propagation occurs. Such a configuration is encountered in some incinerators and coal burners but rarely in a smoldering fire. This configuration has been examined experimentally[6] and modeled,[7,8] and is a relatively well-understood smolder mode with underlying mechanisms qualitatively similar to the transient case just discussed.

In this mode of reverse smolder propagation, oxygen surrounds the fuel particles as they are heated by the advancing smolder reaction zone. Thermal degradation of some fuels in the presence of oxygen is exothermic. This is particularly true of cellulosic materials and this heat can be sufficient to drive the smolder wave without any char oxidation.[6] In flexible polyurethane foams, the presence of oxygen during degradation plays another key role. Without oxygen many foams do not form any char,[7] although char oxidation is a necessary source of heat for these materials. In the reverse smolder mode, the net oxidation rate and net heat release rate are again directly proportional to the oxygen supply rate; the smolder zone spreads to adjacent material as fast as this generated heat can be conducted and radiated to it. An increased oxygen supply rate causes a greater rate of heat release and increased peak temperature in the reaction zone which, in turn, increases the heat transfer rate to adjacent fuel, thus accelerating the smolder spread rate. This sequence implies that the smolder reaction zone may well move through a layer of fuel without fully consuming the solid at any point. This unconsumed material,

in fact, acts like an insulator for the reaction zone, increasing its stability. On the other hand, Dosanjh et al,[8] point out that this mode of smolder propagation can achieve a steady-state only if, as a minimum, the energy released is sufficient to heat the incoming air supply; otherwise it will extinguish.

Figure 2-11.2 shows measured reverse smolder velocities for several types of fuel as a function of airflow velocity through the fuel bed. The bulk densities of the fuel bed are all low but typical for these types of materials. Note that the airflow velocity range is also quite low, although higher flows are sufficient to move the fuel particles in the bed (i.e., an upward flow higher than approximately 0.01 to 0.02 m/s would fluidize the fuel bed).

Despite the considerable variation in the chemical nature of these fuels, the smolder velocity is always of order 10^{-4} m/s. For the same air supply rate, the smolder velocities do not vary much more than a factor of 2. This is consistent with the idea that the oxygen supply rate, not reaction kinetics, dominates the propagation process. The differences with fuel nature that do exist mainly appear to reflect variations in available heat and effective thermal conductivity.

Only limited information is available on toxic gas production from this mode of smoldering. The molar percentage of carbon monoxide in the evolved gases has been examined for two of the fuels in Figure 2-11.2. For the flexible polyurethane foam, the carbon monoxide is 6 to 7 percent for an air velocity of 1.5×10^{-3} m/s; the flow rate dependency was not examined.[9] For the cellulosic insulation material,[10] the carbon monoxide mole fraction varies from about 10 to 22 percent from the lowest to the highest air flow velocity in Figure 2-11.2. The mass flux of carbon monoxide from such a smoldering process (grams of CO/m^2 of smolder front/second) then is estimated as follows

$$Y_{co}(\dot{m}_{air} + \dot{m}_{GS}) \qquad (2)$$

or

$$Y_{co}[\rho_{air}v_{air} + (1 - \phi)\Delta\rho_s v_s] \qquad (3)$$

Here Y_{CO} is the mass fraction of carbon monoxide in the evolved product gases (approximately equal to the mole fraction); m_{air} is the mass flux of air entering the smolder zone; m_{GS} is the mass flux of gaseous material evolved from the solid fuel; ρ_{air} is the density of the air at the point where its velocity, v_{air}, is measured; ϕ is the initial void fraction of the fuel bed; $\Delta\rho_s$ is the change in density of the fuel bed (for reverse smolder, typically 65 to 95 percent of the original mass is gasified); and v_s is the smolder front velocity.

Limited information is also available on the aerosol emitted by a reverse smolder source;[11] this is pertinent to detection of a smoldering fire. The source studied was essentially identical to that used to obtain the data for curve B in Figure 2-11.2; the fuel again was an unretarded cellulosic insulation. The mass mean particle size of the aerosol was 2 to 3 μm; this is about 5 times larger than cigarette smoke and 50 to 200 times larger than the sooty particulate produced by flaming combustion. This large size explains the relatively poor sensitivity of ionization smoke detectors to realistic smolder sources. The residual solid left in the smolder wave and the original fuel both were found to be effective filters for this aerosol; this helps explain the observation by Palmer[5] that smoldering in a thick layer of fuel was not detectable until it neared the surface exposed to the ambient atmosphere.

The rate of heat release for this mode of smolder can be estimated from the total mass flux of products and their heat content (gas temperature typically 670 to 970 K). The result is a few kW/m^2 of smolder front. This translates to a few hundredths of a kW for a reverse smolder source 0.1 to 0.15 m in diameter. The strength of the heat source has a bearing on the behavior of the buoyant plume. (See Section 2, Chapter 2.) Sources as weak as those considered here generate plumes that may not reach the ceiling of a room.[12]

Forward propagation: The second limiting case of one-dimensional smolder propagation is called forward smolder; in this case the oxygen flow is in the same direction as the movement of the smolder front. The most familiar example (though not one-dimensional) of forward propagation is a cigarette during a draw. This limiting case is encountered in some industrial combustion processes but is unlikely to be found in its pure, one-dimensional form in a fire context (some elements of this mode are encountered in realistic cases, however). An approximate model of this process (in one dimension) has been presented by Dosanjh and Pagni.[13] They point out that this smolder mode will die out if the heat generated by char oxidation is insufficient to drive the drying and fuel pyrolysis reactions that precede char formation in the reaction zone.

Some characteristics of forward propagation are briefly mentioned here to describe the major effects that reversing the direction of oxygen flow can have on smolder propagation characteristics.

Forward and reverse smolder propagation have been compared experimentally;[6,10] the fuel was an unretarded cellulosic insulation. Forward smolder through this same fuel at the same air supply rate is about ten times slower than reverse smolder. The carbon monoxide mole fraction is independent of air supply rate and is about 9 percent. Forward smolder also allows for more complete combustion of the fuel. These and other differences between the two smolder modes can be explained in terms of the differing wave structures.[6]

Frandsen[14] investigated the downward propagation of smoldering in horizontal layers of peat as a model fuel for the complex duff layer found on the floor of a forest. No external flow was imposed. This is essentially a diffusion-driven forward smolder process forced to be one-dimensional in this study; it normally is multidimensional in character. The influence of both moisture and inorganic diluents on the limits of smolder propagation was measured. At the extremes, it was found that this cellulosic fuel will just smolder when it contains 50 percent water by weight and no inorganic diluents; when dry it will just smolder when the mix contains 80 percent inorganic diluents. These results should be roughly indicative of the limits for other cellulosic fuels in the absence of a crossflow over the fuel layer.

Multi-Dimensional Smolder Spread

Factors such as ignition source geometry, fuel geometry, and the strong influence of buoyant flow on oxygen supply usually interact to assure that a smolder reaction zone has significant gradients of temperature and species in two or three dimensions. The number of possible configurations becomes virtually limitless. The practical configurations that have been studied are few and they are usually two-dimensional; they do shed some light on most cases likely to be of interest.

Horizontal fuel layer: The configuration that has been studied most extensively is two-dimensional smolder propagation in a uniform horizontal layer of particles or fibers. Ohlemiller[15] examined the structure of the smolder zone in a thick (0.18-m) horizontal layer of cellulosic insulation in the absence of any forced airflow over the fuel layer. In these conditions, the flow induced by the buoyant plume rising above the smolder zone assures a constant supply of oxygen to the space above the layer; oxygen penetrates the layer largely by diffusion.

If such a layer is ignited uniformly on one end, the smolder reaction zone soon evolves into a new shape dictated by oxygen supply rates.[15] The uppermost elements of the reaction zone, being closest to the free surface and hence, ambient air, spread away from the ignition source the fastest; successively deeper elements spread in the same direction but more slowly. The result is a smolder reaction zone that (viewed in vertical cross-section) slopes upward from the bottom of the layer to the top, in the direction of movement. The steady-state length of this inclined smolder front is roughly twice the depth of the original fuel layer. This inclined reaction zone is several centimeters thick, and across this thickness there is a smooth transition from unburned fuel to ash. On the ash side (the free surface adjacent to air) oxygen diffuses down and inward in the same direction as the smolder front is moving and attacks the charred fuel; this is analogous to forward smolder discussed earlier. On the unburned fuel side of the inclined smolder front, oxygen diffuses in from the region ahead of the front to react with the fuel as it is thermally degraded by heat conducted from the char oxidation region. Oxygen here is moving opposite to the direction of smolder propagation, so this aspect of the overall reaction zone is analogous to reverse smolder. Remember that in cellulosic materials, this oxidative/thermal degradation is exothermic. Thus the two-dimensional

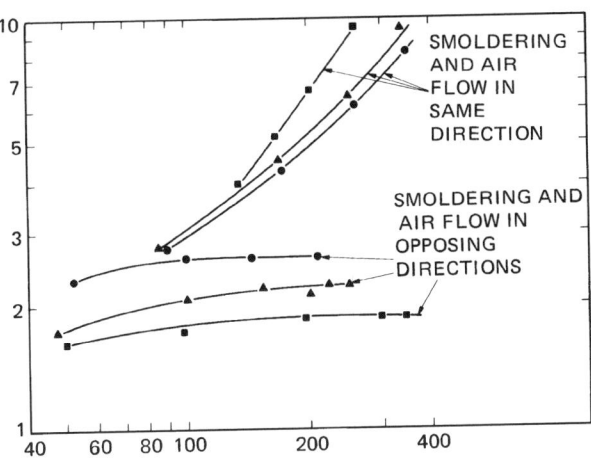

Fig. 2-11.3. Dependence of smolder propagation rate through horizontal layers of beech sawdust on air velocity over top of layer.[4] Circles: 120 μm mean particle size; triangles: 190 μm mean particle size; squares: 480 μm mean particle size.

horizontal smolder zone incorporates features of both forward and reverse smolder and is driven forward by the combined heat release from char oxidation and oxidative/thermal degradation.

The participation of oxidative/thermal degradation in driving the smolder process requires that oxygen have free access to the thermal degradation region. For a low-permeability fuel such as solid wood, this is not the case. Even though solid wood has basically the same reaction chemistry as cellulosic insulation (which consists mostly of wood fibers) and smolders with a qualitatively similar inclined reaction zone, it must be driven solely by char oxidation.

The low permeability and corresponding high density of solid wood has another consequence with regard to smolder. The self-insulating quality of the reaction zone is much less than with a low-density layer of fuel particles or fibers. A single layer of wood will not sustain smolder unless it is subjected to an additional heat input of about 10 kW/m²;[16] this heat could come from some external radiant source or from another piece of smoldering wood that has an adequate radiative view factor with respect to the first piece.

In view of the strong role of oxygen supply rate in shaping the smolder process in a horizontal fuel layer, it is not surprising that smolder also accelerates in response to an increased oxygen supply rate produced by an airflow over the top of the smoldering layer. As with the one-dimensional propagation situation, two possibilities again exist: the airflow can travel in the same direction as the smolder front (again called forward smolder) or in the opposite direction (reverse smolder). Note, however, that now the actual fluxes of oxygen *within* the smoldering fuel bed may go in various directions; they are no longer constrained to being parallel to the smolder wave movement, as in the one-dimensional cases.

Palmer[5] examined both of the flow direction possibilities for relatively thin horizontal layers (3×10^{-3} to 5.7×10^{-2} m) of various cellulosic particles (cork, pine, beech, grass). Figure 2-11.3 shows some typical results. Note that the smolder velocities are less than or equal to those in Figure 2-11.2, despite the much higher air velocities. This is probably due to differing rates of actual oxygen delivery to the reaction zone, and to the fact that the near-surface re-

gion, which receives the best oxygen supply, is also subjected to the highest heat losses.

The influence of two factors, fuel particle size and relative direction of airflow and smolder propagation is shown in Figure 2-11.3. Particle size has a relatively weak effect on smolder velocity but its effect depends on whether the smolder configuration is forward or reverse. The configuration itself has a much greater effect.

Ohlemiller[17] obtained comparable smolder velocities and dependence on configuration for 0.10- to 0.11-m thick layers of cellulosic insulation. It was found that the configuration dependence cannot be explained solely on the basis of oxygen supply rates. The mass transfer rate to the surface of the fuel bed was measured for forward and reverse configurations; it differs by only 20 to 30 percent (these differences are caused by changes in the bed shape due to shrinkage during smolder). It was pointed out that the observed dependence on relative direction of the airflow is consistent with there being a prominent role for convective heat transfer along the top surface of the fuel layer. This dependence occurs only if part of the smolder wave, i.e., the region near the leading edge, is kinetically limited (and therefore highly temperature sensitive) rather than oxygen supply rate limited. This explains the qualitative impact of both relative airflow direction and combustion retardants on smolder velocity; it also explains why forward smolder is faster than reverse smolder in the horizontal layer configuration, whereas the opposite was true for one-dimensional propagation. The role played by fuel particle size may be implicit in this view, but a quantitative model is not yet available.

In contrast to the monotonic enhancement of forward smolder velocity with increased airflow rate found by Palmer and by Ohlemiller, Sato and Sega[18] observed more complex behavior with thin (0.004 to 0.01-m) layers of a cellulosic mixture. Smolder velocity increased up to freestream air velocities of about 3 m/s and then remained constant to the highest air velocity examined (6 m/s). This plateau correlated with erratic behavior at the leading edge of the smolder reaction zone involving both periodic extinctions and mechanical disruptions. These authors also examined the thermal structure of their forced smolder waves. The results were qualitatively similar to those of Ohlemiller for buoyant smolder,[15] but the peak temperatures were appreciably higher due to the enhanced oxygen supply rates.

There is a minimum thickness below which a horizontal fuel layer will not undergo self-sustained smolder propagation. As the thickness of a fuel layer decreases, its surface-to-volume ratio increases (inversely with thickness to the first power). The ratio of the rate of heat loss to the rate of heat generation also varies in this manner so that ultimately the losses are overwhelming and extinction occurs. The exact thickness will depend on factors such as bulk density, fuel type and particle size, rate of oxygen supply, etc., influencing the heat generation per unit volume at a given thickness. The same considerations apply to other thin layers of fuel such as fabrics on upholstery and sheets of paper, wood, or particle board. Palmer[5] found that the minimum depth for sustained smolder in still air increased linearly with particle size for beech, pine and cork; for cork this dependence ceased above 2 mm, apparently because more complete oxidation of the char stablized the process in the layers of larger particles. For very small particles, (<100 μm), the minimum depth dropped as low as 1 mm for cork dust; 0.01 m was typical of small particles of beech or pine sawdust. Ohlemiller and Rogers[19] found the minimum depth in still

TABLE 2-11.1 *Data on Multi-Dimensional Smolder in Various Fuels*

Fuel	Fuel/Smolder Configuration	Air Supply Condition/Rate	Smolder Velocity (cm/sec)	Maximum Temp. (°C)	Ref. No.	Comment
Pressed fiber insulation board, 0.23–0.29 g/cc	1.3 cm thick, horizontal strips, width large compared to thickness	Natural convection/diffusion	1.3–2.2 · 10^{-3}	NA	4	Smolder velocity increased ≈ 50% for strips with width ≈ thickness
Pressed fiber insulation board, 0.23–0.29 g/cc	1.3 cm × 1.3 cm strips varied angle to vertical	Natural convection/diffusion	2.7–4.7 · 10^{-3}	NA	4	Smolder velocity highest for upward spread; lowest for horizontal spread
Pressed fiber insulation board, 0.23–0.29 g/cc	1.3 cm × 5 cm strips forward smolder	Forced flow, 20 to 1500 cm/s	3.5 · 10^{-3} cm/s (20 cm/s air) 13.0 · 10^{-3} cm/s (1400 cm/s air)	770°C (200 cm/s) 790°C (900 cm/s)	4	Some samples extinguished due to air cooling at air velocity > 1450 cm/s
Pressed fiber insulation board, 0.23–0.29 g/cc	1.3 cm × 5 cm strips reverse smolder	Forced flow, 80–700 cm/s	2.8–3.5 · 10^{-3} cm/s	NA	4	Extinguishment indicated above 900 cm/s
Pressed fiberboard (pine or aspen) 0.24 g/cc	1.3 cm × 30 cm sheets, horizontal, forward smolder	Forced flow, 10–18 cm/s	0.7 · 10^{-3} cm/s	NA	28	
Cardboard	Vertical rolled cardboard cylinder, downward propagation, varied dia. 0.19–0.38 cm	Natural convection, diffusion	5.0–8.4 · 10^{-3} cm/s	NA	29	Small dia. ≈ 2× faster than large dia.; ambient temp. effect measured
Shredded tobacco	0.8 cm dia. cigarette, horizontal, in open air	Natural convection, diffusion	3.0–5.0 · 10^{-3} cm/s	820°C	30	
Cellulose fabric + 3% NaCl	Double fabric layer, 0.2 cm thick, horizontal, forward smolder	Forced flow, ≈ 10 cm/s	≈ 1.0 · 10^{-2} cm/s	770°C	31	Smolder behavior dependent on alkali metal content
Cellulosic fabric on substrates	Various weight fabrics horizontal on fiberglass, PU foam, cotton batting	Natural convection, diffusion	≈ 3.0–75 · 10^{-3} cm/s dependent on substrate and fabric	Reported values suspiciously low	32, 33	Smolder fastest on inert fiberglass substrate

air for an unretarded cellulosic insulation to be 0.035 m; a heavy loading of the smolder retardant boric acid roughly doubled this value. Since the insulation has a very small effective particle size and essentially the same chemistry as Palmer's sawdusts, most of the difference in minimum depth (for the unretarded material) probably lies in the bulk density, which is about four to five times less for the insulation compared to the sawdusts (40 kg/m^3 *vs* 180 kg/m^3). Palmer found that the minimum depth dropped rapidly with increased airflow over the sawdust layers, in keeping with the idea that a greater rate of heat release per unit volume stabilizes the smolder process.

Beever[20,21] has addressed a problem at the opposite extreme of layer thickness, that of underground fires in land fills, peat deposits, and mine tailings. These tend to be smoldering fires in roughly horizontal layers where the principal mode of oxygen access is from the top surface. Beever[21] studied this process on a laboratory scale using mixtures of fine sawdust or charcoal with an inert diluent, i.e., diatomaceous earth, in a trough that was insulated on the sides and bottom but open to quiescent air on the top. The trough was 0.13 m by 0.38 m in cross section and 0.14 m deep; a deeper trough was used in separate experiments in which pure layers of the inert diluent were placed atop the combustible layer. Local ignition near the top of a layer yielded steady propagation over a limited depth at rates that varied only weakly with inert content. However, while 25 percent fuel

content yielded smolder spread, 10 percent fuel content did not. The depth to which this spreading smolder zone reached increased with the cross-sectional dimensions of the fuel bed. Material below this depth, having been heated and partially decomposed by the smoldering zone above, could itself subsequently propagate a second wave moving in the opposite direction. It was pointed out that such behavior can make it possible for a landfill or similar fire to spread under a barrier intended to stop it. Similarly, inert covering layers may simply slow but not stop such fires. The true key to stopping a smoldering fire is getting the heat out of the fuel, but this can prove to be extraordinarily difficult.

Smolder propagation data on a few other fuels (including some that are inorganic) in horizontal layers can be found.[19] Unfortunately, no data are currently available on the evolved products of horizontal layer smolder. For crude estimates on cellulosic materials the previous results for reverse smolder are adequate, but they should be applied here with caution.

Other fuel configurations: Data on a few other multidimensional smolder configurations are summarized in Table 2-11.1. Again there is little more information available than the rate of smolder propagation. An exception to this is the smoldering cigarette, which has been extensively studied,[22,23,24] albeit usually in a manner most pertinent to its peculiar mode of cyclicly forced air supply.

All the materials in Table 2-11.1 are fairly porous. As noted previously, solid wood, a low-porosity fuel, also smolders, given a configuration that limits heat losses.[25]

Ohlemiller and Shaub[26] and Ohlemiller[27] examined smolder spread along the interior surface of a three-sided channel constructed of either white pine or red oak. A controlled flow of air was introduced at one end of the channel; the products evolved from the other end were monitored as was the rate of smolder spread. For both types of wood, stable smolder was observed for only a narrow range of inlet air velocities, 0.05 to 0.20 m/s. (From limited data this appeared true for both forward and reverse smolder.) Below this range the smolder process extinguished and above it flaming eventually erupted. Both of these limits, but particularly the lower limit, are probably dependent on the specific conditions of the tests. Carbon monoxide typically comprised 2 to 3 percent of the gases leaving the channel or about 10 to 15 percent of the gases leaving the surface of the wood. The rate of heat release during smoldering was estimated from the oxygen consumption rate, correcting for carbon monoxide. This ranged from about 0.5 to 2 kW or roughly 10 to 30 kW/m^2, based on the approximate area visibly glowing.

The last type of smolder configuration referenced in Table 2-11.1 is quite pertinent to the scenario that makes smoldering a major contributor to residential fire deaths, i.e., upholstery and bedding fires initiated by cigarettes. This is frequently a composite problem, with the smoldering tendency of both the fabric and the substrate (polyurethane foam, cotton batting) pertinent to the overall smolder behavior of the combined assembly.[28] Ortiz-Molina *et al* have shown that the combination of a cellulosic fabric plus a polyurethane foam can smolder over a substantially wider range of conditions than can the foam alone.[29] The fabric smolder process supplies added heat to the foam smolder zone while simultaneously competing for oxygen. The full complexity of this interaction is yet to be explored. A considerable amount of empirical data on the tendency of cigarettes to initiate this type of smolder is available.[30–36]

The life hazard posed by smoldering bedding or upholstery within a closed room has been studied to some extent.[37–39] Data have been presented[38] on the buildup of carbon monoxide (near the ceiling) in a 2.4 m room on a side due to cigarette-initiated smolder in a cotton mattress. The smolder front was reported to spread radially at a rate of 6.3 10^{-5} m/s independent of the size of the smoldering area. In two out of five tests the smolder process underwent a transition to flaming combustion after 65 to 80 minutes, which is close to the time at which total carbon monoxide exposure was estimated to be lethal. Similar data are reported[38] for a greater variety of bedding and upholstery materials; these were ignited by cigarettes (and by flaming sources) in a room 4.3 × 3.6 × 2.4 m. Carbon monoxide and several other gases were sampled at three locations. Flaming developed from smoldering in several of the tests; this usually required 2 to 3 hours of smoldering first. Again, the total exposure to carbon monoxide from the smolder smoke approached or exceeded lethal levels. Lethal conditions due to carbon monoxide were reached in much shorter times in some cases.

All available data on the hazards of smoldering in a closed room were evaluated;[39] it was concluded that the probability of a lethal carbon monoxide dose and of transition to flaming are comparable for a period from 1 to 2½ hours after cigarette initiation of smoldering. A model is presented for buildup of carbon monoxide due to a smoldering fire;[39] the results generally show reasonable agreement with experiment though some of the input parameters must be forced slightly.

In contrast to the above result, a more recent study of the fire risks associated with upholstered furniture implied that the toxic exposure from a smoldering chair in an "average" house was rarely fatal; transition to flaming brought with it death due to thermal causes.[40] The methodology was indirect; it involved using the Hazard I smoke movement and tenability models in a reasonably successful effort to reproduce national fire statistics for upholstery fires. There are not as yet sufficient data on the toxicity hazards of smoldering upholstery materials to definitively resolve this issue.

Transition to Flaming

The transition process from smolder to flaming in the above bedding and upholstery fires is essentially spontaneous. At room conditions both smoldering and flaming are possible in many such systems. Sato and Sega[41] explored the domain of overlapping smolder and flaming potential for cellulosic materials and noted a hysteresis in the spontaneous transition between these two combustion modes. The mechanism of such a spontaneous transition has not been investigated in detail. It has been suggested on the basis of small mock-up studies that a chimneylike effect develops in the crevice between the horizontal and vertical cushions of a smoldering chair;[42] the enhanced air supply presumably accelerates local char oxidation, heating the char to the point where it can ignite pyrolysis gases. Such a mechanism is plausible but it has not been demonstrated to be operable in real upholstery or bedding, where the chimney effect may not develop so readily.

Transition to flaming (fast exothermic gas-phase reactions) requires both a mixture of gases and air that are within their flammability limits and a sufficient heat source to ignite this mixture. Furthermore, these two requirements must be realized at the same locus in space and at the same time. Any factor that either enhances the net rate of heat generation or decreases the net rate of heat loss will move the smoldering material toward flaming ignition by increasing both local temperature and rate of pyrolysis gas generation. Such factors include an enhanced oxygen supply, an increase in scale (which usually implies lesser surface heat losses per unit volume of smoldering material), or an increasingly "concave" smolder front geometry, which reduces radiative losses to the surroundings and enhances gaseous fuel concentration buildup. All of these factors may be operating simultaneously in the case of upholstery and bedding smolder; sequential photos of smolder initiation, growth, and transition to flaming in an upholstered chair appear consistent with this idea.[42]

A further factor in this and in other systems involving cellulosic materials is secondary char oxidation. This process is quite similar to the afterglow seen in cellulosic chars left by flaming combustion. Intense, high-temperature (probably greater than 1070 K) reaction fronts propagate intermittently in seemingly random directions through the fibrous low-density char left by the main lower temperature smolder front. In charred fabrics, these glowing fronts can sometimes progress in a stable manner along the charred residue of a single fiber, despite very high heat losses per unit volume of fuel. Such a process requires the catalytic action of alkali metals that are frequently found naturally in cellulosics or left there during manufacture.[43,44] While in a very hot smolder front the size of a single fiber is unlikely to be sufficiently energetic to ignite flammable gases, the larger fronts (10^{-3} to 10^{-2} m in scale) may well be. An analogous process has been found to cause occasional flaming ignition of smoldering, unretarded cellulosic insulation.[15]

The transition from smolder to flaming can also be induced, for example, by a forced increase in oxygen supply rate to the smolder reaction zone.[5,9,41,45,46] This was first studied quantitatively by Palmer[5] for airflow over horizontal layers of wood sawdust; this process, of course, is familiar to anyone who has started a camp fire from tinder and sparks. Transition to flaming was noted by Palmer only for airflow in the same direction as smolder propagation (forward smolder); depending on the material, the transition occurred at airflow velocities from about 0.9 to 1.7 m/s. For these materials, flaming did not develop when the mean particle size was less than 1 mm. Ohlemiller[45] did obtain transition to flaming in layers of fibrous insulation materials of very small diameter (~ 25 μm) but again only with forward smolder; this occurred at air velocities of about 2 m/s for unretarded insulation. Leisch[46] utilized ignition sources placed midway along the length of grain and wood particle fuel layers so that forward and reverse smolder zones were simultaneously obtained; flaming was noted at 4 m/s air velocity only after the smoldering process produced a substantial depression or cavity in the surface of the fuel layer.

Ohlemiller[45] explained the weak response and lack of flaming transition in reverse smolder on the basis of heat transfer effects influencing the leading edge of the smolder reaction zone. These heat transfer effects intensify the smolder in the leading edge region for forward smolder. In the case of cellulosic insulation, the intensification leads to random development of small (a few cm) cavities near the leading edge which act as flame initiation regions and flame holders.

Ohlemiller[45] also found that both boric acid (a smolder retardant) and borax (a flame retardant) could each eliminate the transition to flaming when the retarded cellulosic insulation was the only fuel. However, the effectiveness of the acid and borax was substantially reduced if the smoldering fuel abutted unretarded wood; heat transferred from the smolder zone readily ignited the wood. Palmer[47] noted similarly that layers of fine dust that would not themselves undergo transition to flaming readily ignited adjacent flammable materials.

Smoldering solid wood undergoes a transition to flaming readily in a configuration that minimizes heat losses.[26,27] It was inferred that the limiting variable in the transition is the surface temperature of the smoldering wood, with the transition occurring when that temperature reached about 950 to 1000 K.

CONCLUSION

Smoldering is a branch of solid fuel combustion quite distinct in many aspects from flaming, but equally diverse and complex. Unfortunately it has not been studied nearly to the same extent as flaming. This is quite apparent in the lack of quantitative guidelines that can be provided here for estimating the behavior of realistic smolder propagation processes, smolder detection, toxic gas production, and the transition to flaming. The experimental data provided can be readily used for closely analogous situations; they must be used cautiously for dissimilar conditions. The reader should always bear in mind the strong role that the oxygen supply rate has on the smolder process. The other very important factor is the relative direction of movement of oxygen supply and smolder propagation; this can be somewhat obscure in many realistic configurations. The actual chemical nature of the fuel is relatively secondary, at least with regard to smolder rate. It may be important for toxic gas production rates, but the data here are quite limited.

REFERENCES CITED

1. F. Clarke and J. Ottoson, *Fire J.*, 20, May (1976).
2. R. McCarter, *J. Cons. Prod. Flamm.*, 4, 346 (1977).
3. P. Bowes, *Self-Heating: Evaluating and Controlling the Hazards*, Chap. 7, Elsevier, New York/Amsterdam (1984).
4. T. Ohlemiller, *Prog. in Energy and Comb. Sci.*, 11, 277 (1985).
5. K. Palmer, *Comb. and Flame*, 1, 129 (1957).
6. T. Ohlemiller and D. Lucca, *Comb. and Flame*, 54, 131 (1983).
7. T. Ohlemiller, J. Bellan, and F. Rogers, *Comb. and Flame*, 36, 197 (1979).
8. S. Dosanjh, P. Pagni, and C. Fernandez-Pello, *Comb. and Flame*, 68, 131 (1987).
9. F. Rogers and T. Ohlemiller, *J. Fire Flamm.*, 11, 32 (1980).
10. D. Lucca, *An Investigation of Co-Current and Counter-Current Smoldering Combustion in Particulated Fuel Beds*, MSE Thesis, Princeton (1979).
11. G. Mulholland and T. Ohlemiller, *Aero. Sci. and Tech.*, 1, 59 (1982).
12. H. Hotta, Y. Oka, and O. Sugawa, *Fire Sci. and Technol.*, 7, 17 (1987).
13. S. Dosanjh, and P. Pagni, *Proc. of the 1987 ASME/JSME Thermal Engineering Joint Conference—Volume 1*, (P. Marto and I. Tanasawa, eds.) Book No. 10219A, American Society of Mechanical Engineers, New York, (1987).
14. W. Frandsen, *Can. J. For. Res.*, 17, 1540, (1987).
15. T. Ohlemiller, *Comb. and Flame*, 81, 341 (1990).
16. T. Ohlemiller, unpublished test results.
17. T. Ohlemiller, *Comb. and Flame*, 81, 354 (1990).
18. K. Sato and S. Sega, *Fire Safety Science—Proceedings of the Second International Symposium*, Hemisphere Publishing Corp., New York, p. 87 (1989).
19. T. Ohlemiller and F. Rogers, *Comb. Sci. and Tech.*, 24 139 (1980).
20. P. Beever, "Subterranean Fires in the UK—the Problem," Building Research Establishment Paper IP 3/89, March, 1989.
21. P. Beever, "Initiation and Propagation of Smouldering Reactions," Ph. D. Thesis, Dept. of Physical Chemistry, University of Leeds, June, 1986.
22. R. Baker, *Nature*, 247, 405 (1974).
23. R. Baker and K. Kilburn, *Beitrage zur Tabakforschung*, 7, 79 (1973).
24. R. Baker, *ibid*, 11, 1 (1981).
25. T. Ohlemiller and F. Rogers, *Engineering Report No. 1432*, Princeton University, Princeton (1979).
26. T. Ohlemiller and W. Shaub, "Products of Wood Smolder and Their Relation to Wood-Burning Stoves," National Bureau of Standards NBSIR 88-3767, May, 1988.
27. T. Ohlemiller, *Fire Safety Science—Proceedings of the Third International Symposium*, Elsevier Science Publishing Co. Inc., New York, p. 565 (1991).
28. G. Tesoro, and T-Y. Toong, *Smoldering in Cotton Upholstery Fabrics and Fabric/Cushioning Assemblies*, Mass. Institute of Technology, Cambridge (1981).
29. M. Ortiz-Molina, T-Y. Toong, N. Moussa, and G. Tesoro, *17th Symp. (Int.)*, Combustion Institute, Pittsburgh (1979).
30. G. Damant, *J. Cons. Prod. Flamm.*, 2, March (1975).
31. G. Damant, *J. Cons. Prod. Flamm.*, 2, 140 (1975).
32. G. Damant, *J. Cons. Prod. Flamm.*, 6, 95 June (1979).
33. K. Palmer and W. Taylor, *J. Cons. Prod. Flamm.*, 1, 186 (1974).
34. J. Loftus, *NBSIR 78-1438*, National Bureau of Standards, Washington (1978).
35. J. Krasny, "Cigarette Ignition of Soft Furnishings—A Literature Review with Commentary," National Bureau of Standards NBSIR 86-3509, Oct. 1986.

36. R. Gann, R. Harris, J. Krasny, R. Levine, H. Mitler, and T. Ohlemiller, "The Effect of Cigarette Characteristics on the Ignition of Soft Furnishings," National Bureau of Standards Technical Note 1241, Jan. 1988.

37. K. Sumi and G. Williams-Leir, *Research Paper No. 402*, National Research Council, Ottawa (1969).

38. C. Hafer and C. Yuill, *Characterization of Bedding and Upholstery Fires*, Southwest Research Institute, San Antonio (1970).

39. J. Quintiere, M. Birky, F. McDonald, and G. Smith, *Fire and Matls.*, 6, 99 (1982).

40. W. Stiefel, R. Bukowski, J. Hall, and F. Clarke, "Fire Risk Assessment Method: Case Study 1, Upholstered Furniture in Residences," National Institute of Standards and Technology NISTIR 90-4243, Mar. 1990.

41. K. Sato and S. Sega, *J. Fire Sci.*, 3, 26 (1985).

42. R. Salig, *Smoldering Behavior of Upholstered Polyurethane Cushioning and Its Relevance to Home Furnishings Fires*, Master's Thesis, Dept. of Mech. Eng., MIT, Cambridge (1981).

43. A. Ihrig, A. Rhyne, V. Norman, and A. Spears, *J. Fire Sci.*, 4, 237 (1986).

44. R. McCarter, *J. Cons. Prod. Flamm.*, 4, 346 (1977).

45. T. Ohlemiller, *NBSIR 85-3212*, National Bureau of Standards, Washington (1985).

46. S. Leisch, *Smoldering Combustion in Horizontal Dust Layers*, Ph.D. Thesis, Dept. of Aero Eng., U of Mich., Ann Arbor, Michigan (1983).

47. K. Palmer, *Dust Explosions and Fires*, Chapman and Hall, London (1973).

SELF-HEATING AND SPONTANEOUS COMBUSTION

Paula F. Beever

INTRODUCTION

The term spontaneous combustion (or self-ignition) describes the culmination of a runaway temperature rise in a body of combustible material, which arises as a result of heat generated by some process taking place within the body. The theoretical background to the treatment of spontaneous combustion in this chapter was developed to describe the ignition of explosives, propellants, and unstable materials, and of liquid and gaseous systems. In such systems the reactants are essentially premixed. Emphasis will be given here, however, to the application of the theory to porous accumulations of material, which can react exothermically with oxygen in air. It is in such cases that spontaneous fires arise most unexpectedly. The approach described has been found to be useful in the study of self-ignition in bulk storage of materials at relatively low temperatures and in smaller accumulations of materials at higher temperatures.

Spontaneous combustion may occur in piles of moist organic material where heat is generated in the early stages by the respiration of bacteria, molds, and fungi. A high moisture content is required for vigorous activity, and heating is generally controlled by maintaining the moisture content below a predetermined level. This type of heating can only raise the material to the temperature range of 50 to 75°C, where the living organisms die. Beyond this point, some form of chemical reaction must take over if ignition is to occur. Ignition due primarily to biological causes has not, to date, proved amenable to theoretical treatment; this problem will not be pursued further in this chapter. A review of the subject is given in Bowes.[1]

THE CRITICAL PARAMETER δ_c

The self-ignition potential of a pile of material depends on the balance between the rate of heat generation within the pile and the rate at which heat is lost to the surroundings. The theoretical model (due originally to Frank-Kamenetskii[2]) highlighted the importance of a dimensionless group of terms, δ, known as the Frank-Kamenetskii parameter. This parameter

is fixed by the relevant physical and chemical properties of the material together with the size of the pile and a reference temperature. All these factors are important; more heat will be generated at elevated temperatures and by highly exothermic reactions, less heat will be lost from large piles with poor thermal conductivity. Material that is safe in one set of circumstances is not necessarily safe in another.

For a given system, it is generally possible to determine a critical value of the Frank-Kamenetskii parameter, δ_c. This value may be taken from the literature, calculated by known methods, or derived from first principles by solving the equations for heat balance. If the value of δ as evaluated for a given system is greater than δ_c, then the system will self-ignite; the heat generated at all times exceeds that which is lost. The temperature will rise, slowly at first, and then rapidly until ignition occurs.

If the calculated value of δ is less than the critical value, then only moderate self-heating can occur. The theory predicts that the maximum temperature rise that can safely be sustained in a body is low—of the order of a few tens of degrees Celsius in practice. Above this temperature rise, runaway self-heating to ignition will occur. The distinction between ignition and nonignition is therefore, in principle, sharp. This arises as a consequence of the assumption in the theory that the heat-generating reaction is highly sensitive to temperature. In general this is true and the distinction between subcritical and supercritical states is also sharp in practice.

ASSUMPTIONS OF THE THEORY

It is not necessary to present a theoretical derivation of critical values of δ. However, it is important to understand the assumptions of the theory to assess the extent to which the assumptions are borne out in practice and under what circumstances corrections must be made.

The basic assumptions made by Frank-Kamenetskii[2] are:

1. Heat is generated by a single reaction whose rate at a given temperature is not a function of time. The rate of reaction is assumed to be a function of temperature, T, according to the Arrhenius equation, namely

$$\text{rate} \propto \exp\left(-E/RT\right) \quad (1)$$

Dr. Paula F. Beever is a member of the staff of the Fire Research Station, Borehamwood, England. Her research has mainly focused on the fire problems associated with industrial handling of combustible powders and smoldering combustion.

Where E is a parameter of the reaction, known as the activation energy.

2. The activation energy is sufficiently high for the condition

$$\varepsilon = \frac{RT_R}{E} \ll 1 \qquad (2)$$

to hold. T_R is a reference temperature, taken problems to be ambient.

3. Heat transfer through the body is by conduction.
4. Heat transfer at the surface to the surroundings, by convection and radiation, is high, such that the surface temperature of the body is ambient.
5. The material is isotropic and homogeneous with physical properties that do not depend on temperature.

Though appearing rather restrictive, these assumptions hold sufficiently well in many cases for predictions to be based on them. They are rarely so unrealistic as to preclude useful estimates. Further discussion of the failure of these assumptions, and appropriate corrections that can be made to the parameter δ_c in that event, are given below.

The foregoing assumptions allow the heat balance in a body that is simultaneously generating and losing heat to be written, subject to a suitable boundary condition, as

$$\nabla^2\theta + \delta \exp(\theta) = 0 \qquad (3)$$

where $\exp \theta$ is an approximate form of the Arrhenius expression (Equation 1). θ is a dimensionless temperature given by

$$\theta = \frac{E}{RT_R^2}(T - T_R) \qquad (4)$$

where T_R is a reference temperature. Suitable forms for the differential operator, ∇, depend on the geometry of the body. The appropriate space variable is made dimensionless using a characteristic dimension, r, of the body.

Equation 3 has no time dependence and its solutions are steady-state temperature profiles. If the value of δ is increased, higher temperatures exist in the body. If δ is made sufficiently large, no solutions can be found to Equation 3. The value of δ at the transition, δ_c, is identified as the critical value for ignition, in practice. Solutions of the equation also yield a value, θ_0, which is the maximum central temperature attainable at the critical value of δ. At $\delta > \delta_c$, the central temperature theoretically becomes infinite.

CRITICAL VALUES OF THE FRANK-KAMENETSKII PARAMETER

Table 2-12.1 gives values of the critical Frank-Kamenetskii parameter for a range of shapes, calculated on the basis of the assumptions of the previous section, for bodies exposed to a steady, uniform ambient temperature. The reference temperature is ambient in all cases, and the characteristic dimension used in deriving δ_c is given. The final column of the table gives the maximum calculated value for the central temperature, in dimensionless form given by Equation 4.

Equation 3 is only amenable to exact solution for the simplest of geometries, and many of the values given in Table 2-12.1 are approximate. Where the values given are exact, this is indicated. It will be noted, for example, that the general expression for the cylinder does not give exactly the

TABLE 2-12.1 *Values of δ_c and θ_0 for Various Geometries*

Geometry	Dimensions	δ_c	θ_0
Infinite plane layer	Thickness $2r$	0.878*	1.12
Rectangular box	Sides $2r$, $2l$, $2m$ $r < l, m$	$0.873\left(1 + \frac{r^2}{l^2} + \frac{r^2}{m^2}\right)$	
Cube	Side $2r$	2.52*	1.89
Infinitely long cylinder	Radius r	2.00*	1.39
Short cylinder	Radius r Height $2l$	$2.00 + 0.841\frac{r^2}{l^2}$	
Equicylinder	Radius r Height $2r$	2.76*	1.78
Sphere	Radius r	3.32*	1.61
Tetrahedron	Radius of insphere r Side $2l = 2\sqrt{6r}$	2.23	2.06

* Exact value

correct value for the equicylinder. There is also some disagreement in the literature over the correct coefficient in the expression for the rectangular box. The expressions and values given in Table 2-12.1, however, all yield δ_c to within a few percent. As will be illustrated later, errors of this magnitude do not give rise to unacceptably large errors when applied to practical problems. If more accurate values are required in particular applications, then reference should be made to Bowes,[1] who gives more exact expressions for some geometries and also gives values for cases such as cones that are not discussed here.

Ultimately, if the required value cannot be found from the table, then δ_c must be calculated by studying solutions to Equation 3. This equation may be solved numerically using standard procedures, the value of δ being increased until solutions fail to converge. In principle this procedure can give δ_c to any required degree of accuracy. Numerical approaches can dispense with many of the assumptions outlined previously to yield values highly specific to the problem.

For less precise work on problems for which δ_c has not been calculated, a particularly flexible approximate method outlined by Hardee et al[3] may be used. This method assumes that solutions to Equation 3 are in the form of polynomials. The coefficients of the polynomial are determined by satisfying the boundary conditions and the differential equation at selected points. Integration of Equation 3 becomes unnecessary. Once an expression is obtained for an internal temperature as a function of the parameter δ, the critical condition can be obtained by applying the criterion that, at critical, a small change in δ produces a very large change in internal temperature. The method is described in detail by Hardee et al[3] and is covered by Bowes.[1] Quite complex shapes and boundary conditions may be handled in principle, but, although the method is straightforward, it can become very cumbersome. The accuracy of the results is not easy to assess except by comparison in the limits where exact results exist.

CORRECTIONS TO δ_c

Since the assumptions of simple theory generally will not be met in practice, corrections will have to be made to the value of δ_c where the departures are significant or where particularly accurate values are required. The corrections given below are judged to be of greatest use for practical problems. The correction expressions are all approximations and in each case the range of values over which they hold and the expected accuracy are indicated. Attention has been drawn, where appropriate, to experimental validation or to relevant numerical work for comparison. The expressions given here are very often not the only ones that could be used; other expressions in the literature may be better or provide more valid results in certain limits, but perhaps are less convenient to apply. For typical values of the relevant parameters, errors of up to 10 percent in the determination of δ_c will give errors of about 1°C in prediction of critical temperature, or 5 percent in critical size.

Finite Heat Transfer Coefficient

The fourth assumption of the Frank-Kamenetskii theory set the surface temperature of the body to ambient, which implies a high heat transfer coefficient, H, at the surface. If H is not sufficiently large, the surface temperature will be above ambient. It is important to correct δ_c for this failure, as a low heat transfer coefficient will make ignition more likely, heat losses being reduced. Predictions made using uncorrected values of δ_c will therefore fail to err on the safe side. The manner in which the rate of heat transfer from the surface affects the value of δ_c is embodied in a dimensionless group known as the Biot number, α, given by

$$\alpha = \frac{Hr}{\lambda} \tag{5}$$

where λ is the thermal conductivity and r the characteristic dimension from Table 2-12.1. The Biot number represents the ratio of external to internal heat transfer. If the thermal conductivity is low, the surface heat transfer coefficient high, or the pile large, then $\alpha \to \infty$ and no correction is necessary. The error in δ_c is about 2 percent when $\alpha = 100$ and still less than 10 percent when $\alpha = 25$.

At the other extreme, if thermal conductivity is high compared to the heat transfer coefficient at the surface, then $\alpha \to 0$. The body is effectively at uniform internal temperature with a step to ambient temperature occurring at the surface. This is called the Semenov condition, and as this limit is approached, δ_c is given for all geometries by

$$\delta_c = \frac{\alpha}{e} \frac{Sr}{V} \tag{6}$$

where S is the surface area and V the volume of the body. This expression is appropriate for low values of α. It overestimates δ_c for the layer geometry by about 3 percent when $\alpha = 0.1$ and 10 percent when $\alpha = 0.3$. Errors for other geometries would be less.[5]

For intermediate values of α, an expression derived by Barzykin et al[4] can be used for all geometries.

This gives

$$\delta_c(\alpha) = \delta_c(\alpha \to \infty) \frac{\alpha}{2} (\sqrt{\alpha^2 + 4} - \alpha)$$

$$\cdot \exp\left[\frac{(\sqrt{\alpha^2 + 4} - \alpha - 2)}{\alpha}\right] \tag{7}$$

where $\delta_c (\alpha \to \infty)$ is the appropriate value taken from Table 2-12.1. This expression gives results within 2 percent of the exact results of Thomas[5] for $\alpha > 2$. For $\alpha = 1$ the error is about 8 percent for the sphere and about 5 percent for the infinite cylinder.

There exists a region $0.3 < \alpha < 1$ where correction for α will be difficult. Thomas[5] gives analytical results for the layer and cylinder and tabulated results for the sphere for all α, but the expressions are cumbersome.

Low Activation Energy

In setting up Equation 3 it has to be assumed that the quantity $\varepsilon\theta$ is small. Since maximum values of θ are of the order unity (see Table 2-12.1) this is equivalent to the assumption that ε is small. Typical values of E/R are of order 10^4 so in most cases this assumption is justified. If correction for finite ε is necessary, the following expression derived by Boddington et al[6] may be used

$$\delta_c(\varepsilon) = \delta_c(\varepsilon = 0)(1 + 1.07\varepsilon) \tag{8}$$

This equation was derived by fitting the numerical results of Parks[7] and is valid for $\varepsilon < 0.05$. Boddington et al[8] calculate δ_c for the sphere as a function of ε for a range of Biot numbers. A proportional correction applied to other geometries from their results for the sphere would be reasonable.

Reactions that are not sufficiently sensitive to temperature, i.e., low E/R, cannot exhibit self-ignition. In fact, there is a sharp cutoff at definite values of ε, above which ignition will not be observed. These values vary slightly with geometry and Biot number but are in the region of $\varepsilon = 0.25$. Such large values of ε are very far from those normally encountered in practice.

The value of θ_0 also changes as ε is increased from zero. At $\varepsilon = 0.05$, for the sphere, $\theta_0 = 1.82$ as opposed to 1.61 at $\varepsilon = 0$ from Table 2-12.1. Further details, including dependence of θ_0 on α, are given by Boddington et al.[8]

Reactant Consumption

The first assumption of the theory is that the rate of the heat-generating reaction is a function of temperature only. In practice, of course, the reactants are inevitably depleted over a period of time and the reaction consequently slows down. The assumption is a good one if the reaction is sufficiently exothermic for negligible reactant consumption to have occurred at the point of ignition. The parameter that governs the effects of reactant consumption is the dimensionless adiabatic temperature rise, B, given by

$$B = \frac{E}{RT_R^2} \frac{Q}{C} \tag{9}$$

Highly exothermic reactions have high B values, while a low heat of reaction gives a low B value. The correction to be applied also depends on the order of reaction, n, where the rate of reaction depends on reactant concentration, c, such that

$$\text{rate} \propto c^n$$

B and n for the reaction must be determined before correction can be made. For materials undergoing a simple reaction, it may be possible to obtain n and Q, the heat of the

TABLE 2-12.2 *Values for a and b for use in Equation 10 as a Function of ε. Tyler and Wesley [9]*

ε	a	b
0.000	1.000	2.28
0.025	0.973	2.35
0.050	0.944	2.41
0.075	0.916	2.49
0.100	0.895	2.56

reaction, from the literature and to calculate B. In most cases, however, especially where natural materials are involved, the reaction responsible for self-heating will not be simple. The heat of reaction of a material is in general not the same as the heat of combustion. Techniques for establishing Q and n will be covered in the next section of this chapter. These techniques are not straightforward and the evaluation of B is not always easy.

Assuming that values for B and n are available, the correction for large B takes the form

$$\frac{\delta_c(B)}{\delta_c(\infty)} = \frac{1}{a + b(n/B)^{2/3}} \tag{10}$$

Tyler and Wesley[9] have obtained values for a and b for a range of ε. These values are given in Table 2-12.2. If the values given are used, Equation 10 reproduces numerically obtained values for δ_c to within 2 percent for $B > 25, n = 1$ and for $B > 100, n = 2$. For $B = 25, n = 2$ the error is about 10 percent. Tyler and Wesley[9] do not study the problem for values of B below 25 but very low values of B have been studied for Semenov conditions (i.e., uniform body temperature with a temperature step at the boundary). For large B the correction under these conditions is of the same form as Equation 10. For smaller B numerical results are available[10] as well as some approximate analytical results. Carter et al[11] suggest the following form for a first-order ($n = 1$) reaction

$$\frac{\delta_c(B, \varepsilon)}{\delta_c(\infty, 0)} = \frac{(3 - 4\varepsilon)(1 - 4\varepsilon)}{e(1 - 2\varepsilon)(1 - 4\varepsilon - 4/B)} \tag{11}$$

Though a little cumbersome, this expression does have the advantage of including a dependence of δ_c on ε. Under Semenov conditions, this expression predicts the numerical results of Adler and Enig[10] for $\varepsilon = 0$ to better than 7 percent for $100 > B > 25$. The error increases to 14 percent when $B = 10$. If the same correction is applied to values of δ_c under Frank-Kamenetskii conditions (surface temperature equal to ambient) and compared with the numerical work of Tyler and Wesley, the error is less than 5 percent for $B = 25$ and $0 < \varepsilon < 0.05$. For lower values of B, the error is probably no worse than under Semenov conditions and the above expression may be used. For more exact work at low values of B, a proportional correction to δ_c based on the numerical work of Adler and Enig[10] for Semenov conditions is probably quite accurate for Frank-Kamenetskii conditions.

The maximum central temperature attainable by a system without ignition increases as B is decreased. For $B > 100$ Tyler and Wesley[9] found the following to hold for $\varepsilon = 0$

$$\frac{\theta_0(B)}{\theta_0(\infty)} = 0.88 + 5.58(n/B)^{2/3} \tag{12}$$

The coefficients vary as ε is increased. As B decreases, θ_0 increases and the maximum temperature reached at ignition decreases. Ultimately, for a sufficiently small B, the subcritical temperature rise approaches the supercritical temperature rise and ignition cannot be recognized. Under such circumstances δ_c cannot be defined. In theory, ignition is not possible if $B < 4$ for $n = 1$. Tyler and Wesley[9] found that ignition was not recognizable in numerical work for $B < 14$. In real systems, however, it appears that ignition may be observed at lower values of B.

Oxygen Diffusion

If a reaction proceeding in a porous body requires oxygen, this must then diffuse into the body from the surrounding atmosphere. Takeno and Sato[12] have studied the effects of oxygen diffusion on ignition and subsequent extinction of a self-heating body of material. The effects of oxygen diffusion were found to be governed by a parameter, Φ, given by

$$\Phi = \frac{n\lambda T_R}{c_0 Q_0 D} \tag{13}$$

where c_0 is the concentration of oxygen in the voids by volume, Q_0 is the heat of reaction by volume of oxygen, and n is the order of reaction with respect to oxygen concentration. Classical Frank-Kamenetskii theory assumes that $\Phi \ll 1$. Takeno and Sato show that as Φ is increased, δ_c increases also and therefore the system ignites less easily. Thus, as would be expected, low concentrations of oxygen, low heat of reaction, and low diffusivity increase δ_c. Takeno and Sato[12] present an analytical method for the determination of $\delta_c(\Phi, \varepsilon)$ which is too complex to reproduce here. The method is compared with numerical work and agreement is excellent. Predictions are also compared with the results of experiments carried out under conditions of reduced oxygen concentration and again agreement is good. The evaluation of Φ in practice may present problems. Suitable values for Q_0 may be found in the literature, including Bowes.[1] Otherwise a method for measuring Q_0 directly must be sought. The diffusion coefficient, D, depends on the porosity of the powder and the temperature. Bowes[1] suggests the following form

$$D = D_0(T/273)^{1.75} p^{1.5} \tag{14}$$

where D_0 is the diffusion coefficient for oxygen in free air at 0°C and p is the porosity (i.e., fraction voids) in the material. For typical values of the parameters in Equation 13, Φ is frequently much less than unity and correction of δ_c will not be necessary. If a correction is necessary, Takeno and Sato[12] give the method, though the notation used differs from that given here.

Asymmetric Heating

Under many practical circumstances, the self-heating body is not exposed to uniform temperatures on all surfaces. Problems in asymmetric heating have been studied widely and the problem of the layer of material exposed to a high temperature on one face has received particular attention. As this is the most important problem from a practical standpoint, it is the only one which will be discussed here; other geometries are covered elsewhere.[1,3]

In dealing with problems where one face of a body is at a much higher temperature than the other, the reference temperature, T_R, which appears in the definitions of δ, θ, etc., is set to the hot surface temperature rather than to the ambient temperature. This device ensures that the product $\varepsilon\theta$ is kept small where heat generation is high. For the case of an infinite layer of material exposed to a high temperature,

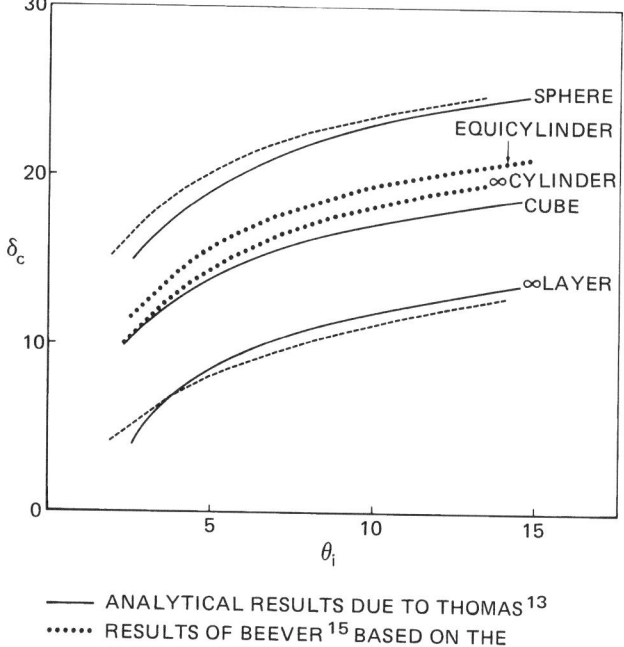

— ANALYTICAL RESULTS DUE TO THOMAS[13]

••••• RESULTS OF BEEVER[15] BASED ON THE METHOD OF THOMAS

----- NUMERICAL RESULTS OF MERZHANOV ET AL[14]

Fig. 2-12.1. δ_c as a function of θ_i for the problem of hot material in cool surroundings.

T_P, on one surface and losing heat from the other surface at T_S to surroundings at T_A, the value of δ_c is given by

$$\sqrt{2\delta_c}\ \tanh\ \sqrt{2\delta_c}\ +\ 2\alpha\ln\cosh\ \sqrt{2\delta_c} = -\alpha\theta_A \quad (15)$$

where δ_c and θ_A have T_P as reference temperature and θ_A will therefore be negative (see Equation 4). For $\delta_c > 5$, $\sqrt{2\delta_c}$ is close to unity and Equation 15 becomes

$$\delta_c = \frac{1}{2}\left(\frac{\alpha}{1+2\alpha}\right)^2 (1.4 - \theta_A)^2 \quad (16)$$

Equation 15 is based on approximate solution methods and is accurate to within about 4 percent for $\alpha < 4$ and $\theta_A = 10$. For low values of θ_A combined with low values of α, numerical solutions of the basic Equation 3 will be required for accurate results.

Hot Material

Thomas[13] has developed an approximate method for dealing with hot materials in cooler surroundings. It is included here because it is especially useful in many industrial situations where material hot from processing may be stored in bags or bins. It is possible that instead of cooling to ambient the material may self-heat to ignition.

This analysis involves the use of a parameter, θ_i, so that

$$\theta_i = \frac{E}{RT_i^2}(T_i - T_A) \quad (17)$$

where T_i is the initial temperature of the hot material. This parameter is then related to values of δ_c, which is defined with T_i as the reference temperature. Curves of δ_c as a function of θ_i for this problem are given in Figure 2-12.1. Numerical calculations[14] for layer and spherical geometries are

shown together with the work of Thomas for the layer, the sphere, and the cube. Some recent results obtained by Beever[15] for cylindrical geometries using Thomas' method are also included. Values for δ_c including a dependence on α are given by Bowes,[1] corrected for an error in the original paper by Thomas.[13] The method suggested by Thomas[13] is flexible and the derivation of $\delta_c(\theta_i)$ for other geometries is not too difficult in principle.

The analysis of Thomas assumes that $\varepsilon = 0$. Numerical results for the same type of problem and covering a range of values of ε are given by Gray and Scott.[16] Their formulation of the problem is rather different from that of Thomas, however, with δ and θ being defined with ambient temperature as reference. As a result, the curves of $\theta_i(\delta)$ become very steep for small δ giving inaccurate predictions of θ_i when the critical initial temperature is well above ambient. In many cases this will be the region of practical interest. Otherwise the results of Gray and Scott[16] are useful, particularly for large ε.

Summary

The corrections given in the present section can be applied successively to a value of δ_c to obtain a suitable value for a given set of circumstances. On occasion this procedure will increase the errors slightly. In most practical cases this will not matter greatly and rarely will all the corrections be necessary in a particular case.

PRACTICAL EVALUATION OF THE FRANK-KAMENETSKII PARAMETER
Direct Measurement of Material Properties

To determine whether self-heating in a body is destined to culminate in ignition, the value of parameter δ must be determined and compared to the value predicted in the previous two sections of the chapter. The definition of the Frank-Kamenetskii parameter is

$$\delta = \frac{E}{R}\ \frac{\rho Q}{\lambda}\ \frac{r^2}{T_R^2}\ A\ \exp\left(\frac{-E}{RT_R}\right) \quad (18)$$

where T_R is a suitable reference temperature and r a characteristic dimension, defined in the same way as when δ_c is calculated.

In principle the parameters in Equation 18 may be measured or taken from the literature and δ evaluated directly. Care is required in the choice of technique used to evaluate E, A, and Q (together with n if a correction for B is being applied). Such values are usually established using differential thermal analysis (DTA) or differential scanning calorimetry (DSC), possibly combined with thermo-gravimetric analysis (TGA). These techniques work best on materials that exhibit a single reaction that behaves according to a simple rate law. Complications arise when the material is a naturally occurring compound with various possible decomposition routes. The best course in practice is to obtain "effective" parameters that describe the reaction. It is particularly important that an accurate and precise value for the effective activation energy, E, be obtained. Because of the exponential dependence of δ on E, it is crucial to the success of the method that this parameter be determined with a high degree of reliability.

Therefore, the direct evaluation of δ may not be straightforward. However, if only small amounts of material are available the methods mentioned above may be the most suitable, and appropriate references should be consulted.[17,18] Ohlemiller and Rogers[19] have numerically calculated critical conditions using reaction parameters determined as outlined, and agreement with experiments is good. The analytical techniques used on the DSC and DTA curves to deduce E, A, and Q were, however, quite sophisticated. These are described in detail by Rogers and Ohlemiller.[20]

Indirect Evaluation

It is possible to obtain values for groups of parameters in Equation 18 by searching for critical behavior under controlled ambient conditions. The theory predicts that material will either undergo moderate self-heating for $\delta < \delta_c$ or will ignite for $\delta > \delta_c$. The critical condition is in principle and in practice very sharply defined, except for materials with very low B values.

In searching for critical behavior in a material, the parameters an experimenter can control are the size, r, and the ambient temperature, T_R. Thus the critical size may be obtained for a given temperature, or more simply in practice, the critical temperature may be determined for a known size. Equation 18 can be rewritten as

$$\ln \frac{\delta T_R^2}{r^2} = P - \frac{E}{RT_R} \qquad (19)$$

where

$$P = \ln \frac{E}{R} \rho \frac{QA}{\lambda} \qquad (20)$$

If T_R and r are known for the critical case then the appropriate value of δ_c may be substituted in Equation 19. A plot of $\ln(\delta_c T_R^2/r^2)$ against $1/T_R$ should yield a straight line of slope $-E/R$ and intercept P. Once these two parameters are known, δ can be calculated for any T_R and r from Equation 19.

Any system for which δ_c is known can be used in the experimental arrangement. In practice two setups have been found useful. The first setup involves exposing a sample of material held in a wire mesh basket to a uniform temperature in an oven. The baskets might be cubes or short cylinders and appropriate values of δ_c would be taken from Table 2-12.1. The central and surface temperatures are generally monitored with thermocouples. If at a given oven temperature the sample exhibits moderate self-heating, the test is repeated with a fresh sample at a higher temperature. Conversely, if ignition occurs the test is repeated at a lower temperature. In this way the critial value of T_R may be bracketed as closely as desired. It is usually found that ignition is very sharply defined and a difference in oven temperature of only 0.5°C will produce a rise in the recorded central temperature of several hundred degrees Celsius.

The closeness with which the critical temperature is determined is balanced between the amount of time and material available and the desired accuracy of the results. Clearly, if critical temperatures are only determined to ±5°C, the errors in E/R and P so calculated will be greater than if critical temperatures are bracketed to ±0.5°C. The former precision is sufficient for rough or preliminary work while the latter is essential if extrapolation over a wide range of sizes is envisaged.

The sample sizes that have been found to be useful in practice cover a range of r from 25 to 300 mm. Ovens of suitable sizes covering temperatures in the range of 60 to 300°C are usually adequate. Any of the corrections to δ_c outlined in the corrections section of this chapter may have to be applied to the value of δ_c used in Equation 19. Experimental conditions are chosen to minimize the necessary corrections. In particular, the ovens should have vigorous air recirculation to ensure a high heat transfer coefficient at the basket surface. If this cannot be ensured then correction will have to be made for finite α. An example of how this can be done is given in the section of this chapter on milk powder (Example 2 below).

As was noted in "Corrections to δ_c," the maximum possible subcritical temperature rise increases as B is increased (Equation 12). If the values measured are significantly greater than those which would be calculated from values for θ_0 (Table 2-12.1 and Equation 4), then a correction for B is required. (The maximum possible value of θ_0 may not be observed in practice; it depends on how closely the critical condition is bracketed.) To correct for B, estimates have to be made for C, Q, and n. The correction applied is not particularly sensitive to the value for Q, so this does not have to be obtained with a high degree of accuracy. Nevertheless, as outlined above, care has to be exercised in the choice of method adopted.

An alternative test method involves the determination of the critical temperature of a hotplate for a layer of known thickness of material upon it. The appropriate value of δ_c for this case is given by Equations 10 and 11. The reference temperature in the definition of δ_c and θ_A is the hotplate temperature, T_p. In practice, layer thicknesses of 5 to 25 mm are used. To calculate δ_c, which will be different for each layer thickness, α has to be estimated, and standard expressions for heat transfer from hot surfaces to cooler surroundings are used.[1] The diameter of the layer must be at least six times its thickness for it to be assumed that it is of infinite extent. Tests of this type have been used for many years to assess the ignition properties of dusts processed in hot environments, for example, dryers. Interpretation and extrapolation of the results in terms of the theory presented here have rarely been attempted. Because of the high temperatures, these tests are more susceptible to distortion as a result of reactant consumption and it is not easy to correct for this effect with this geometry. This type of test is useful if the problem being modeled is one of accumulations of powder on hot surfaces. Otherwise, an approach based on symmetrical heating is likely to yield better predictions of critical conditions.

Once the group, P, and a value for E/R in Equation 19 are known, prediction of δ under a wide range of circumstances is possible in principle. It is possible to decide whether a given arrangement is sub- or supercritical or to predict a critical temperature for a given size or shape of pile. Care must always be taken to ensure that the correct value for r from Table 2-12.1 is included in the evaluation of δ, together with the relevant value of T_R.

Quite apart from being unable to predict self-heating due to biological causes, tests of the type outlined may also fail to reflect the self-heating reaction where this depends on moisture content. Tests are almost always carried out above 100°C and moisture is driven out as the sample heats to oven temperature. This possibility must always be kept in mind, together with the apparent increase in thermal conductivity that can result from the movement of moisture in large piles of material.

TABLE 2-12.3 *Critical Temperatures for Self-Ignition of Animal Feedstuff*

Cube size (mm)	Critical temperature (°C ± 3°C)
50	158
75	140
150	114
300	85

TIME TO IGNITION

For many practical purposes it is not sufficient to know that a system is supercritical. It is usually important to have some idea of the time that will elapse before ignition occurs. Very large piles of material that are not much above critical can have very long times to ignition and storage times are often appreciably less such that no hazard arises in practice. Boddington et al[21,22] have derived an expression for time to ignition, t_i, for systems not too far above critical

$$t_i = M t_{ad} \left(\frac{\delta}{\delta_c} - 1 \right)^{-1/2} \tag{21}$$

where t_{ad} is the adiabatic time to ignition given by

$$t_{ad} = \frac{RT_R^2}{E} \frac{C}{QA} \exp\left(\frac{E}{RT_R} \right) \tag{22}$$

The adiabatic time is the ignition time in the absence of any heat losses and is the shortest possible time in which a system could ignite. M, a constant, depends on the geometry and also on α. Some values for M are given below. Boddington et al[21] indicate how M could be calculated for other conditions.

	$\alpha \to \infty$			$\alpha = 0$
	∞ layer	∞ cylinder	sphere	all geometries
M	1.534	1.429	1.316	1.634

The variation in M, however, is not large and for most work an estimate based on an average value would be perfectly adequate. M also depends on reactant consumption but Boddington et al[22] have shown that this variation is not large either.

It can be seen from Equation 21 that times tend to infinity under conditions that are close to critical and reduce sharply as δ is increased. The expression given predicts times to ignition well for $\delta/\delta_c < 2$ and to better than 20 percent for $\delta/\delta_c < 3$.

In the evaluation of t_{ad}, the product QA may be estimated from P, (Equation 20), if values can be obtained for λ and ρ. T_R is the critical value of the reference temperature.

By comparing Equation 22 with the definition of δ, Equation 18, it can be seen that

$$\delta_c = \frac{t_F}{t_{ad}} \tag{23}$$

where $t_F = \rho C r^2 / \lambda$ is known as the Fourier time and is a characteristic cooling time for the system. Bowes[1] suggests the correlation $t_i \propto r^2$ for estimating times to ignition from experimental results. If Equation 23 is substituted into Equation 21 it can be seen that this is true for a given value of δ/δ_c. In experimental work, however, the closeness of δ to the real value δ_c is never accurately known, and since times to igni-

tion vary most steeply when δ is close to δ_c, Equation 21 is preferable for all but order-of-magnitude estimates.

EXAMPLES

The examples that follow illustrate some of the methods discussed in previous sections. The examples are each based on genuine case histories although some details have been altered to simplify the analysis.

EXAMPLE 1:

Animal feedstuff: An organization wished to produce a new animal feedstuff—a processed byproduct from a distillery. Having had self-ignition problems with related materials, they were wary of stacking the material into large piles. The original plan involved the use of storage silos 3 m by 12 m high, and the problem was to determine whether this system presented a hazard.

Critical ignition temperatures were determined to within ±3°C for cubes of material at uniform ambient temperatures in ovens. The results are given in Table 2-12.3.

SOLUTION:

According to the theory outlined previously, plotting $\ln(\delta_c T_R^2 / r^2)$ as a function of $1/T_R$ should yield a straight line of slope $-E/R$ and intercept P (Equations 19 and 20).

Carrying out this procedure using linear regression with T_R in Kelvin and r (one-half side of a cube) in meters gives

$$\ln \frac{\delta_c T_R^2}{r^2} = 39.88 - \frac{8404}{T_R} \tag{24}$$

where the uncorrected value of $\delta_c = 2.52$ for a cube has been taken from Table 2-12.1.

For any system δ can therefore be calculated from Equation 24 and is given by

$$\delta = 2.088 \times 10^{17} \frac{r^2}{T_R^2} \exp\left(-\frac{8404}{T_R} \right) \tag{25}$$

For the silos in question, δ_c is taken from Table 2-12.1, the rectangular box shape. This gives

$$\delta_c = 0.873 \left(1 + \frac{r^2}{l^2} + \frac{r^2}{m^2} \right)$$

where r is the shortest half-side. Therefore, $r = 1.5$ m, $l = 1.5$ m, $m = 6$ m and hence, $\delta_c = 1.80$. For an ambient temperature of 20°C (293 K) in summer and $r = 1.5$, Equation 25 gives $\delta = 1.9$. Therefore, the system would be marginally supercritical in warm weather. The critical temperature can be determined from Equation 25 by trial to yield 19.5°C. Evaluating t_{ad} at 19.5°C using $\rho = 300$ kg/m[3], $\lambda = 0.05$ W/m K and $C = 1200$ J/kg K as suitable estimates for the material in question, Equations 20 and 22 give $t_{ad} = 103$ days. This is the estimate of the time to ignition with no cooling. At 20°C and $\delta = 1.9$, the time to ignition is given by Equation 21. Taking M for a cylinder as most nearly appropriate, $t_i = 624$ days. This shrinks to 181 days if the temperature rises to 25°C and $\delta = 2.99$. While these times are only estimates, it is clear that under the storage conditions of relatively cool summers, the time to ignition will be long enough to preclude any hazard. Should the temperature exceed 25°C for a period of a few months a problem might arise. Bearing in mind the chances of this occurring and the projected storage times for the material the company decided to proceed with

TABLE 2-12.4 *Critical Temperatures for Self-Ignition of Skimmed Milk Powder*

Cube size (mm)	Critical temperature (°C ± 1°C)
50	171
75	156
100	141.5

the original production plan while monitoring the silos with thermocouples.

EXAMPLE 2:

Milk powder: Following a series of fires in a milk spray-drying plant, an organization decided to find out whether self-heating of milk powder deposits could be a cause.

SOLUTION:

Self-ignition tests in ovens were carried out for the range of powders dried in the plant. The results for skimmed milk powder are presented in Table 2-12.4.

Cubic baskets were used. The correlation of this data using the uncorrected value $\delta_c = 2.52$ for a cube according to Equation 19 gives

$$\ln \frac{\delta_c T_R^2}{r^2} = 41.84 - \frac{9497}{T_R} \qquad (26)$$

Using the value obtained for E/R, the value for δ_c may be corrected using Equation 8. This gives a revised value of $\delta_c = 2.64$ and the new correlation gives

$$\ln \frac{\delta_c T_R^2}{r^2} = 41.90 - \frac{9504}{T_R} \qquad (27)$$

Air entering the top of the spray dryer in question was at approximately 200°C (473 K). In the worst case, areas of the dryer surface near to the air inlet might be close to this temperature. Near the base of the dryer where fairly large accumulations of powder might occur, the temperature was about 80°C (353 K).

For a layer of material uniformly heated, $\delta_c = 0.88$. Correcting this value for ε gives $\delta_c = 0.92$. Substituting into Equation 27 with $T_R = 473$ K gives a critical thickness of layer $2r = 17$ mm. This estimate assumes that there is good heat transfer at the surfaces of the layer. For material accumulating on a dryer wall, the surface in contact with the wall would have a lower heat transfer coefficient than the free surface. In the limit, if the wall acted as a perfect insulator, the critical thickness would be half that just calculated. This extreme condition is unlikely to be met, so the true critical thickness probably lies somewhere between the two. In this case, as in many others of practical interest, the exact prediction of critical conditions is not as important as establishing the principle that self-ignition is a possibility. It was known that deposits of about 20 mm thick could build up occasionally on the dryer roof. It was therefore probable that fires could arise. In the cooler parts of the dryer, at 80°C, the critical layer thickness from Equation 27 is about 40 cm. It was thought that accumulations of this order would be very unlikely to occur unnoticed and the much longer ignition time of the larger pile would allow time for it to be cleared safely.

After altering startup and cleaning routines such that deposits of critical thickness were not allowed to accumulate on the dryer walls, no further fires occurred.

EXAMPLE 3:

Bagasse: Correction for finite α: This problem is taken from Griffiths *et al*[23] and involves the study of self-ignition in piles of bagasse—a residue from sugar cane processing. Large piles of bagasse about 4 m thick were stored in the open for several months. Fires were fairly frequent and spontaneous combustion was a suggested cause.

SOLUTION:

Griffiths *et al*[23] undertook the investigation of the self-ignition properties of the material by using cubes of material exposed to constant temperatures in ovens. Unusually, however, they fitted each cube with an internal array of thermocouples such that temperature profiles across the cube could be obtained as a function of time. From these profiles, a measure of the temperature gradient within the cube close to its surface could be obtained. This allowed the Biot number, α, to be estimated directly since H is defined by

$$H = \frac{dT}{dn} \frac{\lambda}{T_s - T_A} \qquad (28)$$

where dT/dn is the temperature gradient normal to and at the surface. The Biot number is therefore given by

$$\alpha = \frac{dT}{dn} \frac{r}{(T_s - T_A)} \qquad (29)$$

Griffiths *et al* did not compare their measurements with methods by which the Biot number could have been calculated; this will be done here for completeness.

Bowes[1] suggests two methods for estimating the Biot number. The first method derives a coefficient from first principles. The convective and radiative heat transfer coefficients, h_c and h_r, respectively, from the sample are estimated utilizing well-established engineering relationships assuming natural convection. The Biot number is then given by

$$\alpha = \frac{(h_r + h_c)}{\lambda} r \qquad (30)$$

Examples are given by Bowes[1] and sufficient additional data is provided to enable further calculations to be made. The value obtained for α is sensitive to the choice of value for the thermal conductivity. This is often not known with any great accuracy.

Alternatively, Bowes suggests that if the central and surface temperatures are measured during self-heating tests in ovens, α then may be estimated from

$$\alpha = \beta \left(\frac{T_0 - T_s}{T_s - T_A} \right) \qquad (31)$$

where β is a geometrical constant equal to 2.36 for a cube and 2.70 for a long cylinder.

Using the data given by Griffiths, α has been estimated using Bowes[1] methods. The results, together with the measured values given by Griffiths, are given in Table 2-12.5.

Both calculation methods overestimate α as compared with the measured values. This is particularly true for Equation 31 and is more marked for the smaller cubes. The accuracy of the value derived from first principles very much depends on the value used for the thermal conductivity. Here the value used was $\lambda = 0.09$ W/m K as measured for

TABLE 2-12.5 *Critical Temperatures and Calculated Biot Numbers for Self-Ignition in Bagasse*

Cube size (mm)	Critical temperature (°C)	Biot number measured by Griffiths *et al*	Biot number Equation 30	Biot number Equation 31
50	192.5	5.75	8.8	8
75	177.5	9.75	12	14
100	171	14.0	15	22

TABLE 2-12.7 *Critical Temperatures for Self-Ignition in Yeast-Based Powder*

Cube size (mm)	Critical temperature (°C ± 3°C)
50	167
100	150
300	127

similar materials. This could conceivably be in error by a factor of two. On the other hand, the measurements and the approximate method suggested by Bowes depend on the accurate measurement of the fairly small quantity $(T_s - T_A)$. For ovens with fairly good air recirculation this may be less than 5 K.

The values of δ_c corrected from $\delta_c = 2.52$ for a cube, using Equation 7 for each set of Biot numbers, are given in Table 2-12.6, as well as the values derived for P and E/R. Griffiths uses a slightly different method of correction for α, so the results presented here differ a little from those in his paper.

It is most notable that the slope changes as the different estimates for δ_c are used. The slope for the uncorrected values of δ_c is steepest, meaning that overestimates of critical size would occur, on extrapolation to low temperatures. For example, estimating the critical cube size at 100°C (373 K) on the basis of uncorrected results gives 2.53 m and on the basis of Griffiths' results gives 1.76 m (the others give 2.40 m and 1.87 m, respectively). Griffiths *et al* deduced that it is important to correct for α because it changes systematically with r, and as a result, will significantly alter the value obtained for E/R and therefore the validity of any attempts at large-scale extrapolation. In this respect corrections for B and ε are not nearly as important because, in general, these parameters do not much alter the value of δ_c over a range of experiments and the error introduced into the estimate of E/R will be less.

On extrapolation of their corrected results, Griffiths *et al* noted that for stockpiles of bagasse stored at about 37°C (310 K) the critical sizes estimated for layers of material were still far in excess of those found in practice. The possibility of other factors, including biological heating at low temperatures, is mentioned.

EXAMPLE 4:

Hot material: A company was engaged in the design of a plant for the manufacture and storage of a yeast-derived product. The material was to be spray-dried and the resulting powder was to be stored for some days in hoppers. The proposed hoppers were to be cylindrical, 6 m diameter by 10 m high to hold 135 tons of powder at temperatures between 70 and 90°C at an average ambient temperature of

25°C. The company wished to know if there was a risk of self-heating to ignition during the normal period of storage, or if a longer-term risk would arise if material were left as a result of a production stoppage.

SOLUTION:

Ignition tests on cubes of material were carried out in the manner previously described. The results are given in Table 2-12.7.

High heat transfer at the cube surface was assumed and no correction for α or B was made. A preliminary plot allowed an estimate for E/R to be obtained and a correction for ε applied to δ_c. This gave $\delta_c = 2.60$. Correlating the results in the form of Equation 19 yielded

$$\ln \frac{\delta_c T_R^2}{r} = 58.33 - \frac{1.664 \times 10^4}{T_R} \quad (32)$$

This gives a critical size of around 275 m when the ambient temperature is 25°C. Under normal storage conditions, if the material were not hot it would be well below critical.

In order to predict the behavior of hot material the parameter θ_i in Equation 17 must be evaluated. Setting $T = 90°C$ (363 K) gives $\theta_i = 8.21$. Using Figure 2-12.1, the corresponding value for δ_c may be found. The bin in question is higher than its diameter so the curve for the equicylinder will overestimate δ_c, and that for the infinite cylinder will underestimate it. Taking the latter value, to err on the side of caution, $\delta_c = 17.1$. From Equation 32 δ for the system is given by

$$\delta = 2.150 \times 10^{25} \frac{r^2}{T_R^2} \exp\left(-\frac{1.664 \times 10^4}{T_R}\right) \quad (33)$$

where T_R is the initial (hot) temperature. For the bins in question the radius, r, is 3 m. Therefore, for material at 90°C, $\delta = 18.1$. The system is supercritical but only just. At an initial temperature of 89°C the material is subcritical. Bearing in mind that the value for δ_c is an underestimate, it is unlikely that ignitions would be observed in practice. However, it is worth estimating a time to ignition. The adiabatic time at 90°C can be calculated from Equation 22 using the value of P in Equation 32 to estimate QA, taking suitable values for ρ, C, and λ, say 250 kg/m^3, 1200 J/kg K, and

TABLE 2-12.6 *Corrected Values for δ_c Using the Values for Biot Number From Table 2-12.5*

Dimensions	δ_c uncorrected	δ_c measured α	δ_c, α from Equation 30	δ_c, α from Equation 31
50	2.52	1.83	2.03	1.99
75	2.52	2.07	2.15	2.20
100	2.52	2.20	2.22	2.31
P	50.10	46.11	48.89	46.89
E/R	1.410×10^4	1.234×10^4	1.361×10^4	1.268×10^4

0.05 W/m K. This gives a time of about 35 days. Times will of course be very much longer close to critical, so it is possible to say that in normal storage, ignition will not be expected; even in the event of a protracted delay, self-ignition would be extremely unlikely.

CONCLUSION

Other case histories involving the application of the model of thermal ignition theory to the self-heating of porous solids are given by Bowes.[1] Despite the many simplifying assumptions, the model appears to apply well, especially over restricted temperature extrapolation. Both Bowes[1] and Gray et al[24] from small-scale tests, have highlighted the risks of extrapolating over orders of magnitude in size. The theoretical developments that have proved most useful in tackling industrial problems have been highlighted here although there are many other developments that have not been mentioned. These include extensions to cover more than one reaction, to allow for time-dependent boundary conditions, to incorporate reactions that do not have an Arrhenius dependence on temperature, and many others. To solve a problem not covered in this text, reference should be made to Bowes[1] in the first instance. In addition, reference should be made to a recent series of papers by Boddington, Gray, and co-workers published largely in the *Proceedings of the Royal Society*. These papers[6,8,21,22] are not only a review of much of the early theoretical work (such that it can be compared) but have contributed to solving practical problems.

NOMENCLATURE

A pre-exponential factor in Arrhenius expression (s^{-1})

a constant, Table 2-12.2 (-)

B dimensionless adiabatic temperature rise (-)

b constant, Table 2-12.2 (-)

C specific heat $(J\ kg^{-1}K^{-1})$

c concentration of reactant (-)

c_0 oxygen concentration by volume (-)

D diffusion coefficient $(m^2 s^{-1})$

D_0 diffusion coefficient at 0°C

E activation energy $(J\ mol^{-1})$

H heat transfer coefficient $(W\ m^{-2}K^{-1})$

h_r radiative heat transfer coefficient $(W\ m^{-2}K^{-1})$

h_c convective heat transfer coefficient $(W\ m^{-2}K^{-1})$

l length, Table 2-12.1 (m)

m length, Table 2-12.1 (m)

n order of reaction (-)

P constant, Equation 20 (-)

p porosity (-)

Q heat of reaction $(J\ kg^{-1})$

Q_0 heat of reaction by volume of oxygen $(J\ m^{-3})$

R universal gas constant $(J\ mol^{-1}K^{-1})$

r characteristic length (m)

S surface area (m^2)

T temperature (K)

T_A ambient temperature (K)

T_i initial temperature (K)

T_p hot surface temperature (K)

T_R reference temperature (K)

t_{ad} adiabatic time to ignition, Equation 22 (s)

t_F Fourier time, Equation 23 (s)

t_i time to ignition (s)

V volume (m^3)

α Biot number, Equation 6 (-)

β geometric factor, Equation 31 (-)

∇ Laplacian operator (-)

δ Frank-Kamenetskii parameter (-)

δ_c critical value of δ (-)

ε small parameter, Equation 2 (-)

θ dimensionless temperature, Equation 4 (-)

θ_0 maximum subcritical value of θ (-)

ρ bulk density $(kg\ m^{-3})$

λ thermal conductivity $(W\ m^{-1}K^{-1})$

Φ oxygen diffusion parameter, defined by Equation 13 (-)

REFERENCES CITED

1. P.C. Bowes, *Self-Heating: Evaluating and Controlling the Hazards*, Her Majesty's Stationery Office, London (1984).
2. D.A. Frank-Kamenetskii, *Diffusion and Heat Transfer in Chemical Kinetics*, Plenum, New York (1969).
3. H.C. Hardee, P.R. Lee, and A.B. Donaldson, *Comb. and Flame*, 18, 403 (1971).
4. V.V. Barzykin, V.T. Gontlovskaya, A.G. Merzhanov, and S.I. Khudyaev, *Zh. Prik. Mekh. i Tekh. Fiz.*, 3, 118 (1964).
5. P. H. Thomas, *Trans. Faraday Soc.*, 54, 60 (1968).
6. T. Boddington, P. Gray, and D.I. Harvey, *Phil. Trans. Roy. Soc.*, 270, 467 (1971).
7. J.R. Parks, *J. Chem. Phys.*, 34, 46 (1961).
8. T. Boddington, C-G. Feng, and P. Gray, *Proc. Roy. Soc.*, A390, 247 (1983).
9. B.J. Tyler and T.A. Wesley, 11th Symp (Int.), The Combustion Institute, Pittsburgh (1967).
10. J. Alder and J.W. Enig, *Comb. and Flame*, 8, 97 (1964).
11. M.R. Carter, O.J. Druce, and G.C. Wake, *Proc. Roy. Soc.*, A367, 411 (1979).
12. T. Takeno and K. Sato, *Comb. and Flame*, 38, 75 (1980).
13. P.H. Thomas, *Comb. and Flame*, 21, 99 (1973).
14. A.G. Merzhanov, V.V. Barzykin, and V.T. Gontkovskaya, *Dokl. Acad. Nauk. USSR*, 148, 380 (1963).
15. P.F. Beever, forthcoming.
16. B.F. Gray and S.K. Scott, *Comb. and Flame*, 61, 227 (1985).
17. T. Daniels, *Thermal Analysis*, Kogan Page, London (1973).
18. W.W. Wendlandt, *Thermal Methods of Analysis*, Wiley, New York (1974).
19. T.J. Ohlemiller and F.E. Rogers, *Comb. Sci and Tech.*, 24, 139 (1980).
20. F.E. Rogers and T.J. Ohlemiller, *J. Macromol. Sci-chem.*, A24, 169 (1981).
21. T. Boddington, C-G. Feng, and P. Gray, *Proc. Roy. Soc.*, A385, 289 (1983).
22. T. Boddington, C-G. Feng, and P. Gray, *Proc. Roy. Soc.*, A391, 269 (1984).
23. J.F. Griffiths, S.M. Hasko, and A.W. Tong, *Comb. and Flame*, 59, 1 (1985).
24. B.F. Gray, J.F. Griffiths, and S.M. Hasko, *J. Chem. Tech. Biotech.*, 34A, 453 (1984).

FLAMING IGNITION OF SOLID FUELS

A. Murty Kanury

INTRODUCTION

This chapter concerns flaming ignition of solid combustibles that are heated by either thermal radiation or convection. Different kinds of ignitions encountered in practice are defined. The existing empirical knowledge is highlighted by describing Martin's map of spontaneous ignition of radiantly heated cellulosic solids. The various physical and chemical processes that culminate in ignition of a heated solid are identified through a qualitative description and a simplified mathematical model. The typical assumptions underlying a theoretical model are systematically enumerated to point out the complexities involved in the ignition problem.

A number of existing criteria for ignition are examined. Limiting case models are demonstrated to increase understanding of the ignition process. Analysis of the solid-conduction-controlled case, for example, leads to a prediction of the main features of Martin's map. A thumbnail sketch of the gas phase problem, in another limit, confirms the observed influences of the gas phase properties on ignition. Finally, a brief summary is given of a recent comprehensive analysis of Gandhi. Many existing fragments of knowledge on the ignition problem can now be synthesized into a coherent quantitative description.

THE PROCESS OF IGNITION

A number of aspects of unwanted combustion, in which an understanding of the ignition process is required, can be readily enumerated:

1. An obvious first step in all fire prevention strategies is the ignition-hardening of materials of construction, finishings, and furnishings;
2. Fire spread over combustibles is often viewed as a process of continuous ignition of the successively upstream material;
3. Room fire flashover is believed by many to be a process in which the contents of the room experience a nearly simultaneous ignition;

4. The jump of a forest fire across a firebreak is generally viewed to be a radiant ignition process; and
5. Fire growth over noncontiguous surfaces in arrays of combustibles, such as buildings in a city, involves radiant heating to ignition.

Most of the natural and synthetic organic (and some inorganic) solids in air will become "ignited" in response to an externally imposed heating source. The subtleties of the ignition process—the definition and delineation of different sorts of ignitions, and the qualitative and quantitative understanding of the influence of various physical and chemical factors on ignition—are not always familiar. Kanury[1] and Steward[2] have presented comprehensive reviews of this subject. The global objective of this chapter is to develop a concise description of the ignition process to achieve enough familiarity with the concepts involved, existing literature and implications of this knowledge in such real-world problems as enumerated above.

Ignition of a heated combustible body marks a stage beyond which the associated fuel/oxidant system is capable of supporting a sustained exothermic reaction. It is necessary to clarify this definition to develop a qualitative picture of the technical problem addressed here and to define the scope of this chapter.

The heated body (i.e., the target) is taken here to be a piece of a cellulosic solid (e.g., wood, paper, cloth, etc.). The concepts are generally adaptable[3] to synthetic solids (e.g., the numerous modern polymers popularly known as "plastics") as well. The physical and chemical characteristics of the body are assumed to be known. Thermophysical (conductivity, density, specific heat, etc.) and geometrical (dimensions, shape, and configuration) properties are physical. The intrinsic thermochemical properties (e.g., pyrolysis kinetics and energetics) are chemical.

Heating is presumed to be externally imposed. It can be radiative (as from a heat lamp, a bank of electrically heated incandescent tungsten filaments, a gas-fired radiant panel, or a nearby flame) and/or convective (as from a hot gas flowing about the body). These two modes of heating will be considered individually to note certain commonalities and differences, although many practical situations involve simultaneous convective and radiative heating.

When a porous reactive solid body is kept immersed in an ambient medium, a runaway self-heating may become

Dr. A. Murty Kanury is Professor of Mechanical Engineering at Oregon State University. His professional activities are centered around teaching thermal science and research in fire and combustion.

possible if the rate of internal reactive energy release exceeds the rate of energy loss from the body to the ambient medium. This sort of self-heating is known to arise in connection with contaminated sawdust, oily shop rags, grain silos, and certain unstable solid propellant storages. The energy loss mechanism is generally conductive and convective. While an enhanced convective transport tends to increase the energy loss rate, it also increases the oxygen mass transfer to the surface and the net effect is quite complex. This type of self-heating problem is addressed in Section 2, Chapter 12.

Smoldering, a familiar fire phenomenon, is defined as a relatively low temperature combustion process within a porous fuel bed. Its inception usually involves a localized ignition source (such as an overheated electrical conductor or a steam pipe) within the fuel bed. The fuel bed porosity plays an important role in smoldering. A highly porous solid (1) implies finely divided, thermally thin fuel elements that are more easily heated conductively, and (2) offers an effective diffusion of the oxidant gas into the interior of the bed. Furthermore, high porosity makes the overall thermal conductivity of a solid quite low and, hence, ignition is easier. Once smoldering ignition begins, the smolder combustion wave propagates into the fresh fuel at a rate governed by thermal conduction in the porous solid and oxidant gas diffusion. It is possible for convection to intrude into this propagation process. The intensity with which this occurs is dependent upon such factors as the porosity itself; if the bounding surfaces are sealed or unsealed; and the direction of propagation relative to the sealed surfaces and to the gravity vector. Under certain suitable conditions, a smolder may suddenly flare into flame. The ignition and propagation of the smoldering combustion and its transition to flaming are addressed in Section 2, Chapter 11.

The combustible may be supplied to the reaction system in a gaseous, liquid or solid state. It is with solids we are concerned with here. All cellulosic and some synthetic solids undergo thermal degradation to yield char and fuel gases. Some other synthetic solids, known as thermoplastics, melt and depolymerize to monomers or multimers, which serve as the fuel gases. Some depolymerization reactions proceed in a manner remarkably similar to the simple physical vaporization of pure liquids.

Because the focus here is on gas phase ignition of charring type solids, there arises a possibility of surface ignition of the char. This type of an ignition marks the inception of glowing combustion of the carbonaceous solid (such as that encountered in carbon and charcoal combustion); the glowing ignition problem is not covered in this chapter.

As an event, ignition occurs at a certain pronounced instant in the history of the exposure. The time to ignition in a given situation depends upon three broadly grouped factors: (1) the degradative thermal response of the solid to yield the combustible gases; (2) the mixing of these gases with the oxidant gas (generally the oxygen of the normal air); and (3) the induction of the temperature- and composition-dependent rate of the combustion reaction to a sufficiently high level to be measurable and self-supporting. Sustainment of the reaction implies that the reaction can self-perpetuate even if the external heating is removed, and can even self-accelerate to grow in space or in intensity at a fixed size. An "exothermic reaction" refers to the combustive oxidation reaction itself. This reaction is called smoldering if it is situated within the subsurface layers of the solid, glowing if at the solid-gas interface, and flaming if in the gas phase.

Ignition in the absence of a pilot source is known as spontaneous, or autoignition. Ignition in the presence of a pilot source (such as a small flame, a heated wire, or an electric spark) in the reactive fuel/air mixture flow is called piloted, or forced, ignition. A pilot source is not meant to heat the solid to generate the fuel gases nor to enhance the mixing of the fuel gases with air, but to locally induce the combustion reaction which would propagate into the mixture.

Finally, a distinction should be made between transient and persistent ignitions. Ignition can be produced with a minimum required heating, but the flame would not sustain itself if the external heating is removed. This is transient ignition, akin to the flash point phenomenon in liquid fuel ignition. Substantially longer heating is required to make a flame that would self-sustain, self-perpetuate, or persist even after the external heating is removed. This distinction becomes clearer in the next section of this chapter.

CONDUCTION-CONTROLLED SPONTANEOUS IGNITION OF CELLULOSE DUE TO RADIANT HEATING—MARTIN'S MAP

The large number of experiments on spontaneous ignition of radiantly heated cellulosic solids (insulated backface) has been synthesized by Martin.[4] (See Figure 2-13.1.) The x-axis is the quantity $i_0\ell/K_s$ where i_0 is the exposure irradiance, ℓ is the target thickness and K_s is the target solid conductivity. Martin calls this quantity normalized irradiance, and its units are those of temperature (K). The y-axis, termed normalized exposure, is $i_0 t/\rho_s C_s \ell$ (where t is exposure time and $\rho_s C_s$ is the volumetric heat capacity of the solid); its units are K.

This ignition map shown in Figure 2-13.1 can be delineated into four distinct regions. The lowermost boundary indicates the minimum exposure intensity and time required to produce ignition. When intensity is low and exposure is long, glowing ignition of a thermally thin body is accomplished. In the upper left region of the map, the thickness of the target specimen is noted to be of consequence; the thinner the specimen, the quicker the ignition. Furthermore, these thin bodies ignite not to yield a flame but to yield glowing combustion of the solid residue. The pyrolyzates not only are limited in quantity but are also released at a rate and time unsuitable for flame evolution.

The middle right region of the map deals with heating of thick targets with moderately high exposure intensities. Flaming ignition will occur within a moderate exposure time but will not persist to yield a self-supporting flame, presumably due to excessive conductive drain of energy into the thick solid and inadequate production of the pyrolyzates. At even higher exposure intensities, the surface is found to experience ablation without ignition.

For the ignition of thick bodies to persist into a flame, significantly longer exposures are required which are nearly independent of the exposure intensity. Martin considered many tests, which lead him to draw the boundary line between the transient and persistent ignition regions. The persistence of flames is expected to depend on a purely physical cause such as the sustained evolution of pyrolyzates due to attainment of a certain minimum relaxed temperature by the entire body. Further discussion of Martin's work is presented later in this chapter.

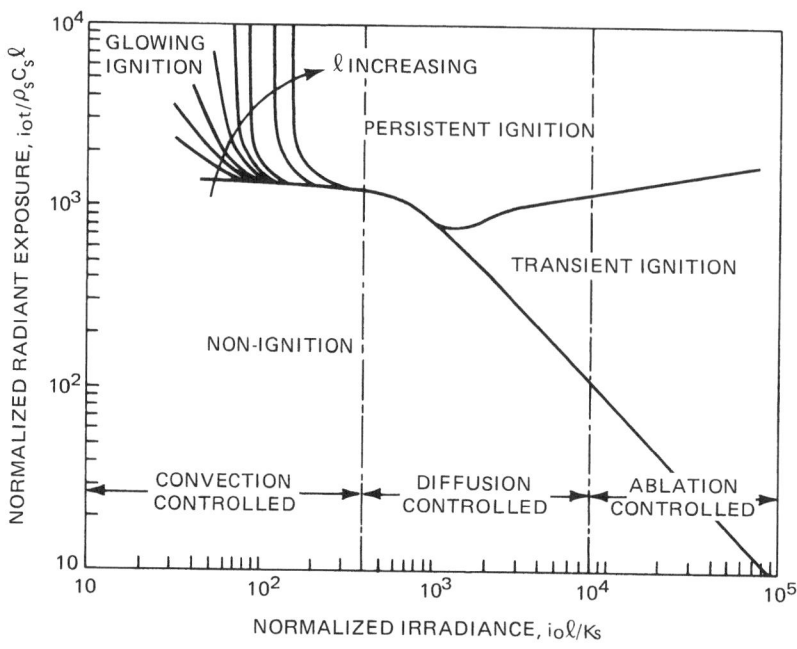

Fig. 2-13.1.   Ignition behavior of cellulose, showing areas controlled by convective cooling, diffusion of heat into the solid, and ablation of the exposed surface. Martin's map.[4]

A QUALITATIVE DESCRIPTION

The initial uniform temperature of a vertical cellulosic slab (of known dimensions standing in a quiescent atmosphere of normal air) is perhaps the same as the temperature of the ambient air. (See Figure 2-13.2.) Orientation, geometry, the kind of solid, and the composition and motion of the atmosphere are chosen to consolidate a picture of the physical and chemical processes leading to ignition. The concept elicited in this qualitative description can be adapted, in principle, to other situations.

The front face of the solid is exposed to a radiant flux at time, t, equal to and beyond zero. In general, this flux is unsteady in the period of exposure, nonuniform over the target surface, and composed of a spectrum of wavelengths.

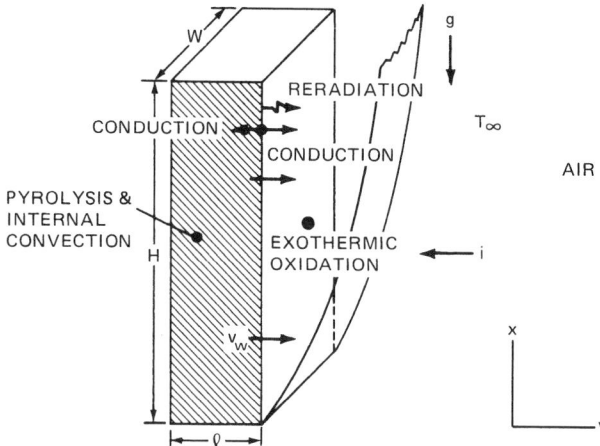

Fig. 2-13.2.   Schematic representation of the component processes.

It is assumed that the incident flux is steady and uniform, and that the target surface is a diffuse, grey absorber/emitter. Furthermore, the incident flux is radiative alone; the backface of the solid and its edges are taken to be impervious to both heat and mass flows.

Once the exposure is triggered, the solid is heated by transient conduction. The now hotter surface commences to reradiate energy to the surroundings. Energy is also imparted by the hot surface by transient conduction to the infinite, initially quiescent, adjacent, ambient gas. This heated gas soon responds to buoyancy and forms a transient natural convective boundary layer.

The incident radiant energy is thus partitioned into three components: conduction into the interior of the solid, reradiation to the surroundings, and transient natural convection to the adjacent gas. The energy conducted into the solid initially raises the solid temperature everywhere within a progressively thickening "conductive penetration layer." Any free moisture of the solid will be driven away from a layer when it attains a temperature of approximately 100°C. While most of the water vapor released in this manner flows out of the solid, some vapor may migrate towards the cooler interior to augment the heat flux by both convection and condensation. Subsequent evaporation will not only introduce a distinct heat sink but also physically and chemically modify the original solid, which is yet to experience pyrolysis.

There would soon ensue an instant at which layers of the solid near its surface would become sufficiently hot to undergo pyrolysis, leaving a carbonaceous residue ("char") behind. With continued heating, progressively deeper layers of the solid become pyrolyzed so that the release of the gas mixture is distributed in space and time. The gas mixture, thus originating at different rates at different depths and time, flows predominantly out through the porous char. This flow introduces an outwardly convective heat flux in the char. The outward flux opposes the inward conductive flux and tends to retard the heating.

Integration of the pyrolysis rate over the thickness of the entire solid yields the total pyrolyzate mass flux issuing at the solid surface into the gas phase. This transpiration tends to thicken the instantaneous boundary layer. The boundary layer flow is generally expected to be laminar since the target size is usually small. This flow will cause mixing of the pyrolyzates with the ambient air.

If the pyrolyzates are either absent or inert, the boundary layer continues to receive heat from the bounding hot solid surface by conduction. The gas temperature profile will then be a monotonically decreasing function of the distance normal to the surface into the gas. If the pyrolyzates are combustible, however, oxidative energy release alters the boundary layer temperature profile to such an extent as to exhibit a temperature maximum in the gas phase at some finite distance from the surface. This nonmonotonic gas temperature profile, at later times, indicates a gas phase conductive heat flux to, rather than from, the solid surface. As the oxidation reaction develops in the boundary layer, the nature of the temperature profile is drastically changed. Spontaneous ignition can be presumed to have occurred in the reactive boundary layer at the instant when the sign of the gas temperature gradient at the surface is reversed. This time to ignition obviously depends upon factors governing the surface temperature, pyrolyzate transpiration rate and composition, air induction rate into the boundary layer, mixing of air and pyrolysis gases and gas-phase reactions.

This qualitative description applies to spontaneous ignition in the boundary layer, which is defined as a situation where the boundary layer mixture is not only within the flammability limits of composition but also at such a thermal condition that it can react on its own in an accelerating manner to lead to a flame. This description can also be extended to piloted ignition, i.e., ignition in the presence of an ignition source such as a small flame, an electrical filament or spark, or an incandescent particle. A boundary layer mixture within the flammability limits of composition is a sufficient condition for piloted ignition. The required energetic strength of the ignition source is a function of several thermochemical and physical properties of the mixture in addition to its composition and temperature. Generally, the pilot source is so small that its contribution to heating and pyrolyzing the solid can be ignored; changes brought by the pilot source in the boundary layer of the chemical environment can also be ignored. It is customary to consider the ignition source as a localized initiator of the combustion reaction whereupon the reaction wave propagates into the mixture. The ignition theory of premixed gases, however, dictates that the pilot source cannot be infinitely small; the source must exceed the approximate characteristic quenching distance of the ignited mixture.

Mixing of the transpired pyrolyzates with the induced air is a necessary condition for ignition. Turbulence in the boundary layer is obviously desirable to produce this mixing. Spontaneous ignition sometimes occurs in the wake of the target with the (so initiated) flame propagating down into the boundary layer. While mixing the fuel gases and air is a prerequisite for ignition, mixing may also result in thermal dilution (i.e., lowering of the mean boundary layer temperature) which is contrary to the requirements of spontaneous ignition.

The possibility also exists that the pyrolyzate plus air mixture in the boundary layer can selectively attenuate the incoming radiant beam so that the irradiance actually experienced by the exposed surface is substantially less than the level calibrated without the absorbing gas. This phenomenon has implications yet to be understood. First, attenuation is expected to retard the rate of heating and pyrolysis of the solid, and consequently to delay the attainment of an ignitable boundary layer mixture. Second, the attenuation in the boundary layer is expected to become significant only after vigorous pyrolysis of the solid is established to furnish the boundary layer with the attenuating species. Third, the attenuation and absorption is expected to raise the local temperature in the reactive boundary layer to accelerate the oxidation reaction leading to ignition. Finally, this attenuation is expected to occur selectively in certain distinct wavelength regions and hence to depend upon the emission characteristics of the radiation source.

It is evident from the present qualitative description of flaming ignition that the total problem is complex and involves such difficult features as: (1) unsteady development of physical and chemical processes; (2) conjugate coupling of the solid and gas phases; (3) chemical kinetics and the associated strong nonlinearities of pyrolysis in the solid phase and oxidation in the gas phase; (4) moving boundaries; (5) coupling of conservation equations; (6) strongly variable properties; and others. When the lack of well-established input databases for the transport, thermodynamic (and thermochemical), and reaction kinetic properties is superimposed on this complexity, it is imminently clear that all attempts to solve the problem theoretically and apply the predictions to experimental data are prone to extensive approximation. Some of these attempts will be described in a subsequent section of this chapter.

CONSERVATION EQUATIONS

In this section, the preceding qualitative description is cast into a mathematical form to identify the assumptions involved. For example, consider a vertical plate of a solid fuel of thickness, ℓ, height, H, and width, W, standing in a quiescent air atmosphere. The solid will occupy $0 \leq y \leq \ell$ and the gas phase, $\ell \leq y < \infty$. (See Figure 2-13.2.) For time $t \geq 0$, the frontface $y = \ell$ of the solid is exposed to an irradiance, i; while a part, i_0, of this flux is absorbed by the solid, the rest of the irradiance is reflected. The absorbed flux is partially conducted into the solid, partially reradiated by the surface to the surroundings, and the rest of the flux is imparted by gaseous conduction to the gaseous boundary layer. These three processes are strongly transient. The heated solid undergoes volumetric, transient pyrolysis, and the pyrolyzates flow out of the solid to mix with air in the boundary layer. At the instant when the mixture composition and temperature are suitable, flaming ignition will occur in the boundary layer. This section of the chapter will describe the estimation of this time to ignition.

A number of simplifying assumptions make the problem formulation easier. The initial temperature of the solid is usually uniform and equal to the temperature of the ambient air. The backface, $y = 0$, as well as all edges of the solid slab, are assumed to be impervious to both heat and mass. The solid surface is taken to be grey and diffuse with uniform radiosity. The imposed irradiance is considered to be constant with respect to time, and uniform over the entire exposed surface. The transient gas boundary layer problem is simpler to deal with in two dimensions, i.e., $W \gg \ell$ and H. The transient solid conduction problem is taken to be one-dimensional in the direction of y. The solid is taken here to be initially dry. Absorption of radiation is taken to occur only at the solid surface and not in the layers beneath the surface; i.e., diathermancy of the solid is taken to be zero. The char and virgin solid properties are considered to be different but independent of temperature.

A single-step Arrhenius rate law with reaction order equal to unity is generally sufficient to describe the pyrolysis kinetics. The pyrolyzates are viewed to flow through the char with no resistance. Thermal equilibrium between the pyrolyzates and the porous matrix is assumed. Secondary chemical transformation of the pyrolyzates flowing through the char is negligible, as is migration of the pyrolyzates into the cooler interior of the solid.

The air, pyrolyzates, and products of oxidation are taken to behave as radiatively nonparticipating ideal gases. The gas phase density is considered to be constant in all respects except in producing the buoyancy force (in other words, the Boussinesq approximation is made). Boundary layer approximations are made for the transient free convective gas flow in the vicinity of the heated surface. All gas properties are taken in this discussion to be constants, independent of both the temperature and composition. (This appears to be an overly crude approximation but leads to reasonable results.) Viscous dissipation in the boundary layer is ignored. Cross-diffusion effects in the gas phase are assumed to be absent.

The gas phase oxidation reaction is assumed to follow a simple, single step, second order, Arrhenius rate law. (Since the pyrolyzate composition is known to vary not only with time but also with the heating conditions and precise chemical constituency of the target solid, any further sophistication of these kinetics appears to be unwarranted at present.) Simple stoichiometry relates the sources and sinks of energy, fuel pyrolyzate, oxygen, and the products.

The conservation equations describing this problem follow. For the solid phase, the continuity and energy equations suffice while the momentum equation is obviated by the assumption of no resistance to flow. Thus,

$$\frac{\partial \rho_s}{\partial t} + \frac{\partial \dot{m}_p''}{\partial y} = 0 \qquad (1)$$

$$\frac{\partial}{\partial t}\left(\rho_s\, C_s\, T_s\right) + \frac{\partial}{\partial y}\left(\dot{m}_p''\, C_p\, T_s\right) = \frac{\partial}{\partial y}\left(K_s\, \frac{\partial T_s}{\partial y}\right) + \dot{q}_p''' \qquad (2)$$

where, s is the solid, p is the pyrolyzate gas mixture, ρ is density, $\dot{m}''$ is mass flux, C is specific heat, T is temperature, K is thermal conductivity, t is time, and y is depth normal to the exposed surface. Equation 1 implies that the local spatial gradient of the pyrolyzate mass flux $\dot{m}_p''$ is equal in magnitude to the local and instantaneous rate of pyrolysis. The terms in the energy equation represent the unsteady energy accumulation rate, internal convective excess flux, conductive excess flux, and the pyrolysis sink, respectively. The constitutive equations for this part of the problem are given by the pyrolysis rate equation and the definition of the energy sink.

$$-\frac{\partial \rho_s}{\partial t} = Z_s(\rho_s - \rho_c)\exp\left(-E_s/\bar{R}T_s\right) \qquad (3)$$

$$\dot{q}_p''' = h_p\left(-\frac{\partial \rho_s}{\partial t}\right) \qquad (4)$$

Here ρ_s is the instantaneous, local solid density while ρ_c is the ultimate char density, Z_s is the pre-exponential factor, E_s is the activation energy of the pyrolysis reaction, and h_p is the pyrolysis enthalpy of reaction. The boundary and initial conditions will be discussed later in the chapter.

Let v and u denote the gas velocity components in the normal (y) and longitudinal (x) directions, respectively. With the enumerated assumptions the gas phase mass, momentum, energy, and species conservation equations then take the following forms.

$$\frac{\partial u}{\partial x} + \frac{\partial v}{\partial y} = 0 \qquad (5)$$

$$\frac{\partial u}{\partial t} + u\frac{\partial u}{\partial x} + v\frac{\partial u}{\partial y} = \nu_g\frac{\partial^2 u}{\partial y^2} + \frac{(\rho_\infty - \rho_g)}{\rho_g}g \qquad (6)$$

$$\frac{\partial P}{\partial y} = 0 \qquad (6a)$$

$$\frac{\partial T_g}{\partial t} + u\frac{\partial T_g}{\partial x} + v\frac{\partial T_g}{\partial y} = \alpha_g\frac{\partial^2 T_g}{\partial y^2} + \frac{\dot{q}_g'''}{\rho_g C_g} \qquad (7)$$

$$\frac{\partial Y_i}{\partial t} + u\frac{\partial Y_i}{\partial x} + v\frac{\partial Y_i}{\partial y} = D_{ig}\frac{\partial^2 Y_i}{\partial y^2} + \frac{\dot{m}_i'''}{\rho_g} \qquad (8)$$

Here, subscript g is the gas mixture, ∞ the ambient conditions, and ν, α, D_{ig} are momentum, heat and species-i-mass diffusivities, respectively; P is pressure, Y_i is the ith species mass fraction with $i = F$ for the fuel pyrolyzate gas (O for oxygen, P for product and I for the inert). The respective volumetric source-sink strength due to the oxidation reaction is $\dot{m}_i'''$. The corresponding energy source strength is $\dot{q}_g'''$. While the energy and species equations indicate a balance between the unsteady accumulation, convection, diffusion and reaction rates, Equation 6 shows that momentum is conserved under the balance of accelerative, inertial, viscous, and buoyant forces. The constitutive relations to complement Equations 5 through 8 are

$$\sum Y_i = 1 \qquad (9)$$

$$P = \rho_g(\bar{R}/M_g)T_g \qquad (10)$$

$$\dot{m}_F''' = -Z_{gF}\rho_g^2 Y_F Y_O \exp(-E_g/\bar{R}T_g) \qquad (11)$$

$$-\dot{m}_F'''/f = -\dot{m}_O'''/1 = +\dot{m}_P'''/(1 + f) = +\dot{q}_g'''/fh_c \qquad (12)$$

$$\dot{m}_I''' = 0 \qquad (12a)$$

where Z_{gF} is the pre-exponential factor, E_g is activation energy, f is fuel pyrolyzate to oxygen stoichiometric mass ratio, and h_c is enthalpy of oxidation. Besides the fact that all the edges of the solid are sealed for both mass and heat, the initial and boundary conditions are quite straightforward. In the solid $0 \le y \le \ell$, $0 \le x \le H$; for $t < 0$, $T_s = T_{s0}$ ($= T_\infty$) and $\rho_s = \rho_{s0}$, both uniform in space. At the backface of the solid $y = 0$, at all times $t \ge 0$, $\partial T_s/\partial y = 0$ as well as $\dot{m}_p'' = 0$. In the gas, at $t < 0$, $\ell \le y < \infty$, all x, and at $t \ge 0$, as $y \to \infty$, for all x, $u = v = 0$, $T_g = T_\infty$, $Y_F = Y_P = 0$ and $Y_O = Y_{O\infty}$.

The boundary conditions at the gas-solid interface ($y = \ell$, subscript w) play an important role in the physics of the problem. To avoid slip for the momentum equation, $u = 0$. The time-dependent normal transpiration velocity $v = v_w(t)$ is given by the transpiration mass flux of pyrolyzates $\dot{m}_{pw}''(t)$ [given by integration of the pyrolysis rate over the entire solid (Equation 1)] divided by the local gas density. The species balance at the interface indicates that the flux of species i

arriving from within the solid is carried away into the gas phase by combined convection and diffusion. This balance, known as the Dankwert boundary condition, is given by

$$\dot{m}''_{pw} Y_{iR} = \rho_g v_w Y_{iw} + (-\rho_g D_{ig} \partial Y_i / \partial y)_w \qquad (13)$$

where Y_{iR} is the mass fraction of species i in the pyrolyzate mixture prior to any dilution with air. The energy interface condition states that the absorbed irradiance, i_0, is conducted into the solid and the gas or reradiated.

The problem is now fully formulated. Relaxation of any of the simplifying assumptions can be done by appropriately reworking the model equations. The intention here is to present the skeletal model which can reasonably easily be adapted to suit special needs. The solution is straightforward in principle, although implementation is not. The transience, conjugate nature, the radiation nonlinearity, and the presence of the Arrhenius exponential prevent these coupled equations from possessing similarity solutions. It is possible to obtain some approximate solutions which are useful in understanding the ignition process.

IGNITION CRITERIA

Based upon experimental observations and intuition, attempts to identify the instant of ignition have led to the development of a number of criteria for ignition. The experiments of Bamford[5] with wood suggest that ignition can be expected when the pyrolyzate outflow rate reaches 2.5×10^{-4} g/cm²s. Martin,[4] Akita,[6] and others[7,8] advocate that the event of ignition can be described simply by the attainment of a critical temperature of the exposed surface. Martin[4] postulates that persistent ignition is possible only when the entire solid attains an average temperature that exceeds a critical value. All these criteria presumably have something to do with the ease of heating the solid to cause sufficiently intense combustible gas generation by pyrolysis, and with the thermal conditions of the boundary layer conducive to flame inception. Experimental evidence developed by Alvares[9] suggests that the surface temperature and the pyrolyzate efflux rate at ignition are not constants but depend upon the exposure flux and other factors.

Deverall and Lai[10] showed theoretically that for solids that undergo ignition through gas-phase exothermic oxidation reactions, a reversal in the sign of the boundary layer gas temperature gradient at the solid-gas interface was a definite indication of ignition. Consistent with some experimental data for premixed gases ignited by a heated wire,[11] the transition of the gas-phase boundary layer to flaming justifies the gradient reversal criterion. Employing high-speed motion picture photography, Simms[7] observed that the flame is indeed initiated in the relatively well-mixed wake of the finite height vertical target, and that this flame quickly propagates down into the boundary layer, altering the nature of the gas temperature distribution so as to reverse the gradient at the interface. Sauer's correlations[12] on thick slabs of wood showed that a certain minimum char depth was required before ignition could occur. The method of estimating the instantaneous char thickness, however, is experimentally subjective and analytically difficult.

Intuition suggests that a rapid rise of gas temperature is a sure sign of ignition. Since this abrupt rise is due to the oxidation reaction, the phenomenon of ignition should be associated somehow with the reaction rate in the gas bound-

ary layer. In pursuing this line of thought, Kashiwagi[13] suggested that ignition occurs when the reaction rate in the boundary layer exceeds an arbitrary value of approximately 10^{-5} g/cm³s. This type of criterion is consistent with ideas employed in solid propellant ignition studies,[14] in which a surface reaction rate greater than a prescribed critical value is taken to ensure ignition of the exothermic surface reactions.

The ignition criteria utilized by various investigators can thus be summarized:

1. $T_{sw} \geq T_1^*$—critical surface temperature (Simms, Martin)

2. $\bar{T}_s \geq T_2^*$—critical average solid temperature (Simms, Martin)

3. $\dot{m}''_w \geq R_1^*$—critical pyrolyzate mass flux rate (Bamford)

4. $\delta_c \geq \delta_c^*$—critical char depth (Sauer)

5. $\partial T_g / \partial t \geq R_3^*$—critical local gas temperature increase rate

6. $\int_i^{\infty}$ (Reaction rate) $dy \geq R_4^*$—critical total reaction rate in the boundary layer (Kashiwagi)

7. $(\partial T_g / \partial y)_{\hat{y}=0} = 0$—gas temperature gradient reversal at the solid-gas interface (Deverall and Lai)

Kashiwagi demonstrated that if arbitrary but "reasonable" values were selected for the critical condition, the ignition delay was not sensitive to any particular ignition criterion employed. This is fully expected, for each of Numbers 1 through 6 above indicates a feature of the incipient flame and none excludes or contradicts another. Number 7 is a trivial, although powerful, indicator of the arrival of the flame.

If it is accepted that flaming ignition is merely the onset of significant gas oxidation reactions in the boundary layer, then it is evident that Numbers 5 through 7 would not only necessarily but also sufficiently address the ignition process mechanistically by accounting for the gas oxidation reaction as well as its prerequisite, namely, the solid decomposition rate. The question of which of these three criteria is more desirable may be answered if the sequence of events that lead to ignition is followed qualitatively.

The previous section on conduction-controlled spontaneous ignition indicates that the transiently heated solid conducts heat to the gas phase in the early stages of ignition. As the solid pyrolyzes, oxidation reactions begin in the gas phase. The reaction rate at any time is a function of distance normal to the slab surface. The composition distribution dictates that the reaction rate be zero at the edge of the boundary layer, and maximum somewhere within it. Therefore, a shift in the temperature profile occurs such that the deviation from the inert temperature profile at any point in the boundary layer reflects the oxidation reaction rate at that point. If the exothermic gas reactions persist, the gradient at the wall will eventually become zero. Beyond that point the solid begins to receive energy from the hotter gas, and the pyrolysis and gas oxidation rates increase until flaming occurs. Upon flaming, the gas temperature increases from the solid surface to the flame and then decreases from the flame to the ambient. Hence, attainment of a zero gradient of gas temperature at the wall is not only an important event leading to ignition but also a necessary indicator of the evolution of the flame. It must be recognized, however, that the gas temperature gradient reversal criterion is conservative in that it estimates the time to ignition equal to or less than the actual measured time.

TABLE 2-13.1 *The First Six Roots[17] of the Transcendental Equation*

			$a \tan a$	$=$ Bi			
Bi	a_1	a_2	a_3	a_4	a_5	a_6	a_n
0	0	3.1416	6.2832	9.4248	12.5664	15.7080	$n\pi$
10^{-3}	0.0316	3.1419	6.2833	9.4249	12.5665	15.7080	
10^{-2}	0.0998	3.1448	6.2848	9.4258	12.5672	15.7086	
10^{-1}	0.3111	3.1731	6.2991	9.4354	12.5743	15.7143	
10^0	0.8603	3.4256	6.4373	9.5293	12.6453	15.7713	
10^1	1.4289	4.3058	7.2281	10.2003	13.2142	16.2594	
10^2	1.5552	4.6658	7.7764	10.8871	13.9981	17.1093	
∞	1.5708	4.7124	7.8540	10.9956	14.1372	17.2788	$(n - \frac{1}{2})\pi$

Employing an integral solution approach, Gandhi[15] solves the ignition model of the preceding section of this chapter, paying full attention to simultaneous developments in both the solid and gas phases. A comparison[15,16] is made of the ignition criteria 1, 3, and 6 in the framework of criterion 7 to note that the surface temperature, pyrolyzate efflux rate, and the total extent of gas phase reaction at ignition will all strongly depend upon such factors as the exposure intensity, target height and thickness, reradiant loss, and the chemical kinetics of both the pyrolysis and oxidation reactions. Alvares' experiments[9] are thus explained with the conclusion that the critical temperature, pyrolyzate efflux, and gas reaction criteria are not generally acceptable, although they are valid in narrowly defined circumstances.

SOLID CONDUCTION-CONTROLLED IGNITION

The complex problem of heterogeneous ignition has been tackled by various investigators to identify two limiting cases. When the intensity of exposure is relatively low and the pyrolyzate production is relatively slow, the time to ignition is found to be 10^{-1} second or longer and mainly governed by characteristics of the solid phase thermal response. Higher heating rates, higher fuel volatility, and low gas pressures are found to result in a small time to ignition (of the order of 10^{-2} second or smaller) and are mainly governed by the gas phase phenomena.

In the solid phase-controlled limit, the solid is approximated to be inert. Its ignition is presumed to occur when the exposed surface attains a prescribed critical temperature, a property of the substance. As discussed in the preceding section of this chapter, this criterion is simple although limited in the range of validity.

Let the inert slab of thickness, ℓ, conductivity, K_s, density, ρ_s, specific heat, C_s, and initial uniform temperature, T_0, experience heating at the front face $y = \ell$ by a calibrated constant uniform absorbed irradiance, i_0, while the backface $y = 0$ is insulated. (The solution will later be extended to ignition by convective heating.) As heating occurs, the front face is permitted to lose heat to the surroundings at T_∞ by convection. The convection loss coefficient, h, is assumed to be constant and uniform; the reradiative loss is ignored.

The solid phase continuity equation (Equation 1), the pyrolysis terms in the solid energy conservation (Equation 2), the gas-phase species conservation (Equation 3), and the energy source term in Equation 7 all become unnecessary. The gas phase continuity and the momentum and energy

equations also become unnecessary because their combined outcome is embodied in the prescribed loss heat transfer coefficient, h. The interface boundary condition is simplified by ignoring the reradiative loss term and setting the gas conduction flux equal to the convective loss flux, $h(T_{sw} - T_\infty)$. The resulting simple one-dimensional transient conduction problem has a well-known closed-form solution. Adapted from Carslaw and Jaeger[17] the solution is

$$\frac{T_s - T_{s0}}{(i_0/h) + T_\infty - T_{s0}}$$
$$= 1 - \sum_{n=1}^{\infty} \frac{2\text{Bi} \sec(a_n) \cos(a_n y/\ell)}{\text{Bi}(\text{Bi}+1) + a_n^2} \exp(-a_n^2 \alpha_s t/\ell^2) \quad (14)$$

where $\text{Bi} \equiv h\ell/K_s$ is the Biot number, indicating the characteristic ratio of convective and conductive fluxes. The constants a_n, $n = 1, 2, \cdots \infty$, are the positive roots of $a\tan(a) = \text{Bi}$; the first six of these roots are presented in Table 2-13.1. It is important to note that Equation 14 is applicable to: (1) heating by irradiation in conjunction with surface heat loss by convection to the surrounding gas at T_∞ (which may be equal to the initial temperature of the solid, T_{s0}); (2) heating (or cooling) by convection alone (i_0 is set equal to zero and T_∞ equal to the hot ambient gas temperature); and (3) heating by combined radiation and convection. The limiting form of Equation 14 when $h \rightarrow 0$ (i.e., radiant heating without the convective loss) is available explicitly as given in the final section of this chapter along with the versions of Equation 14 corresponding to thermally thin and semi-infinitely thick bodies.

Integration of Equation 14 over the space $0 \leq y \leq \ell$ leads to the determination of the average temperature, $\bar{T}_s$, of the solid at any point in time. The result[17] is

$$\frac{\bar{T}_s - T_{s0}}{(i_0/h) + T_\infty - T_{s0}}$$
$$= 1 - \sum_{n=1}^{\infty} \frac{2(\text{Bi}/a_n)^2}{\text{Bi}(\text{Bi}+1) + a_n^2} \exp(-a_n^2 \alpha_s t/\ell^2) \quad (15)$$

Other inert transient conduction solutions involving two- and three-dimensional effects, a variety of geometries, and boundary conditions can be found in a number of treatises. Equations 14 and 15 are sufficient for the present purposes.

Equations 14 and 15 are shown plotted in Figure 2-13.3, taking $T_\infty = T_0$. The solid curves show the surface temperature while the dashed curves indicate the mean temperature. The coordinates used in this figure are related to those obvious from

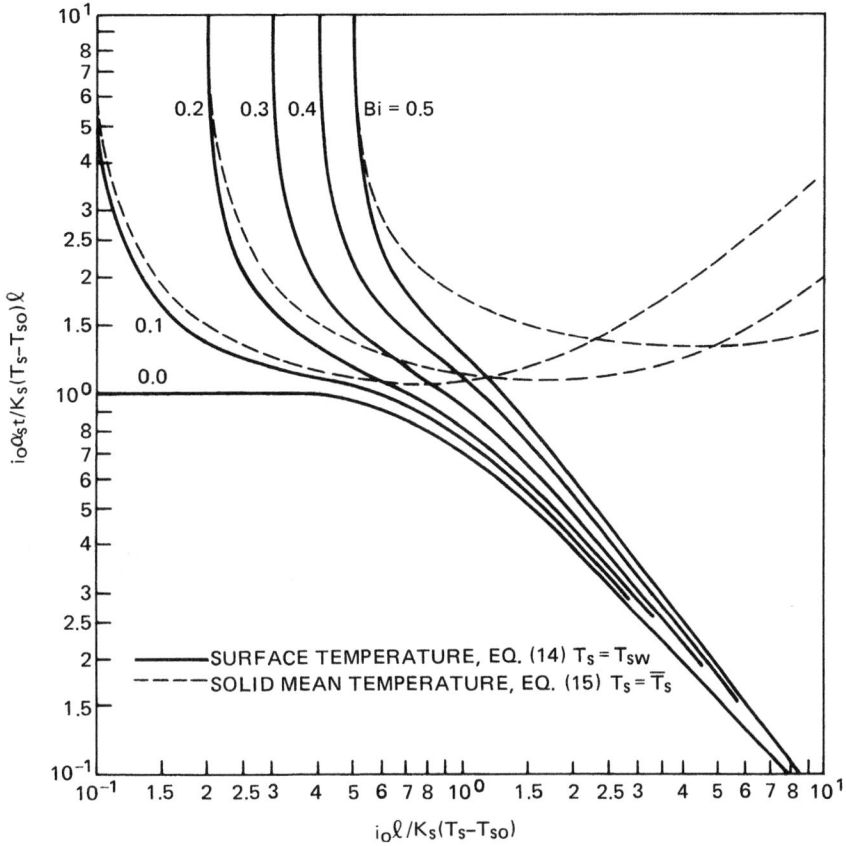

Fig. 2-13.3. *Surface and solid mean temperature histories for a slab heated by radiation and cooled by convective loss. Solid lines: surface temperature, Equation 14, $T_s = T_{sw}$. Dashed lines: solid mean temperature, Equation 15, $T_s = \overline{T}_s$. Numbers on curves are values of Bi.*

x-coordinate:
$$\frac{i_0\ell}{K_s(T_s - T_{so})} \equiv \text{Bi} \left| \frac{T_s - T_{s0}}{\dfrac{i_0}{h} + T_\infty - T_{s0}} \right.$$

y-coordinate:
$$\frac{i_0\alpha_s t}{K_s(T_s - T_{s0})\ell} \equiv \frac{\alpha_s t}{\ell^2} \cdot x\text{-coordinate}$$

The x-coordinate is the intensity of radiant exposure normalized with the instantaneous characteristic conduction flux. The y-coordinate has the meaning of the radiant fluence, $i_0 t$, nondimensionalized with the appropriate instantaneous conduction quantities. A quick comparison of Figures 2-13.1 and 2-13.3 points to the reasons underlying the choice of presenting Equations 14 and 15 in these coordinates. The following observations can be made from Figure 2-13.3.

The surface temperature is shown in Figure 2-13.3 for six different convective loss Biot numbers. The effect of convective heat loss is most pronounced when the specimen thickness is small, intensity of irradiance is low, and duration of exposure is long, i.e., in the top left region of the graph. The thin body limit solution given by Equation 22 shows the reason, in that convective loss plays an equally prominent role with the radiant heating of the solid.

As the slab thickens (i.e., in the lower right region of the graph), the convective loss exerts a less significant effect, compared to the effect of conduction, on the surface temperature history. In this thick regime, the semi-infinite solid

solution without surface heat loss given by Equation 25 results in a surface temperature history of

$$\frac{i_0\alpha_s t}{K_s(T_s - T_{s0})\ell} = \frac{\pi}{4} \left| \frac{i_0\ell}{K_s(T_s - T_{s0})} \right. \tag{16}$$

where the length, ℓ, is used to depict this limiting solution on the same plane as the thin and finitely thick slab solutions. The Bi = 0 line of Figure 2-13.3 in the lower right region is, in fact, Equation 16 indicating a slope of -1.

Expectedly, the mean temperature of a heated solid is always lower than its surface temperature. When the solid is thin, internal gradients become negligible, and the mean and surface temperatures are equal and relatively low. Convective loss obviously makes the surface temperature always lower than it is without losses. This is not always true, however, for the mean temperature. A complex compromise between the exposure irradiance and the histories of conductive drain and convective loss results in a minimum in the exhibited mean temperature curves.

Consider now the premise that spontaneous ignition of a radiantly heated organic solid is primarily a consequence of heat conduction in the solid and convective loss to the gas phase, and that all other physical and chemical processes occur promptly without exerting any resistance. Additionally, the symptom of the incipient ignition is the attainment of a prescribed critical temperature at the exposed surface.

Then, setting $T_{sw} = T_{sw}^*$, the solid curves of Figure 2-13.3 should give the time to ignition $t = t^*$. It may be postulated further that persistent flaming of thick bodies will occur only when the mean temperature of the body becomes sufficiently high to promise a continuous pyrolysis at an adequately high rate; then the mean temperature curves of Figure 2-13.3 are expected to indicate the minimum required exposure for persistent flaming. With this framework of an ignition criterion, a remarkable similarity between Figures 2-13.1 and 2-13.3 can be recognized. This similarity perhaps endorses the prescribed critical temperature criterion for ignition and the cumulative outcome of the numerous assumptions underlying the development of Figure 2-13.3. The limitations of this success will be discussed later in this chapter.

There is a particular physical meaning of the coordinates chosen for Figure 2-13.3. As mentioned earlier, the x-coordinate, $i_0\ell/K_s(T_s - T_{s0})$, is a ratio of the imposed flux, i_0, to the instantaneous characteristic conductive flux, $K_s(T_s - T_{s0})/\ell$. The x-coordinate can also be viewed as a ratio of the physical thickness, ℓ, to the thermal length characteristic, $K_s(T_s - T_{s0})/i_0$, which is approximately the depth to which the thermal wave would penetrate into the solid in the time taken by the surface to raise its temperature from T_{s0} to T_s. If the ratio $i_0\ell/K_s(T_s - T_{s0}) \ll 1$, the solid may be considered thermally thin. On the other hand, if this ratio is much greater than unity, either because the solid is in fact physically thick or because the thermal length is small, then the body behaves as thermally thick. Large temperature gradients then exist near the exposed surface, and the relatively well-insulated interior of the solid causes the surface temperature to be high.

The y-axis, $i_0\alpha_s t/K_s(T_s - T_{s0})\ell = i_0 t/\rho_s C_s \ell(T_s - T_{s0})$, can also be viewed in different ways. It can be seen as a ratio of total energy incident on a unit area within the time, t, (known as the "fluence") to the characteristic enthalpy rise of the mass lying within this area. It is equally interesting to view the y-axis as a ratio of the clock time, t, to the heating time $\rho_s C_s \ell(T_s - T_{s0})/i_0$ taken by the flux, i_0, to raise the temperature of the involved solid mass from T_{s0} to T_s. The y-axis is also recognizable as the product of the x-axis and the traditional Fourier modulus, $\alpha_s t/\ell^2$. It is thus clear that when the intensity and duration of exposure are low, ignition of the solid is impossible.

Where possible, ignition can be one of three kinds:

1. Thin bodies exposed to low intensity ignite, but only after a long exposure. The convective losses and the specimen thickness have a strong effect on this thin body ignition. In fact, experiments indicate that these thin bodies yield their small pyrolyzate content to the gas phase rather abruptly at a high rate and so quickly that the gas phase thermal and mixture conditions are mutually out of phase to exclude the development of a flame. By the time the surface reaches a sufficiently high temperature, the solid is already converted to char. This char experiences the observed thin body glowing ignition. Martin's experiments confirm the strong influence of the specimen thickness on this glowing ignition. While this description is valid for isolated single thin fuel elements, it should not imply that beds of shredded paper and clouds of minute fuel particles will fail to flame. Such beds or clouds will have to be considered to be the global fuel element with low conductivity and large thickness, belonging at moderate to high values of the x-axis of Figure

2-13.3. Consideration of the volumetric absorption of radiation may become crucial in such a scenario.
2. Ignition occurs at moderate values of the exposure and fluence parameters, yielding only a transient flame. If the externally controlled exposure were then interrupted, the flame would cease, presumably due to large conductive and convective losses and inadequately established pyrolyzate production.
3. At moderate values of exposure intensity and long exposure time, the flaming will persist even if the external irradiance is cut off. The longer exposure will raise the entire solid to a sufficient temperature, at which sustained production of the pyrolyzate is possible. Martin's experiments confirm that the effect of specimen thickness on this persistent flaming ignition is relatively weak.

Figures 2-13.1 and 2-13.3 also show that as the irradiant intensity is gradually reduced, a threshold (a lower limit of i_0) is approached below which ignition is impossible even with infinitely long exposure. While both the theory based on prescribed critical temperature criterion and the related experiments appear to adequately predict the nature of and time to ignition, they do not address the ignition threshold. A scrutiny of the gas phase reaction dynamics is required to gain an understanding of this issue. In fact, sufficient evidence exists[9] to indicate that the ignition temperature increases with decreasing irradiance.

This is not the first time the essence of Figure 2-13.1 has been developed by inert conduction theory. Kanury[1] and Steward[2] independently obtained plots similar to Figure 2-13.3 over a decade ago. The present Figure 2-13.3 is special because it is based on a single equation (Equation 14) rather than a patching of several solutions.

Martin's collection of experimental data, when examined in the perspective of Figure 2-13.3, indicate that for spontaneous ignition of radiantly heated thick cellulose, the critical surface temperature is in the vicinity of 900 K. This is in keeping with the measurements of Alvares[9] and Akita.[6] A comparison of the presently predicted and Martin's experimental results for persistent ignition leads to an estimation of the required mean solid temperature to be between 800 and 1200 K. Using physical reason, this temperature should be near, and slightly above, the temperature at which cellulose would pyrolyze profusely. Pyrolysis literature indicates this to be about 600 K. The present overestimation may be a consequence of the ignored reradiative heat loss from the surface.

Koohyar[18] demonstrates that the present concepts are valid for a variety of woods exposed to radiation from flames as well, provided corrections are made for the surface absorptivity differences. Koohyar[18] and Wesson[19] demonstrate that the ignition of radiantly heated wood in the presence of a small pilot flame in the reactive boundary layer also obeys the essence of the inert conduction theory. The critical surface temperature for this situation is about 600 K, near the pyrolysis temperature. This is consistent with the conclusions reached earlier by Akita.[6] It thus appears that two conditions have to be met for spontaneous ignition: first, heating has to be sufficiently intense and long to produce sufficient pyrolyzates, which would result in a boundary layer mixture of fuel content exceeding the lean limit of flammability; and second, the boundary layer has to become sufficiently hot enough to support significant oxidation reaction. For piloted ignition, the first condition is adequate.

Hallman[3] demonstrates that the inert conduction theory can be used to predict piloted ignition time for a variety of synthetic materials (i.e., plastics) as well. Smith[20] experimentally found that edges and corners of pieces of pine blocks exposed to irradiation from a quartz lamp ignite sooner but require a higher surface temperature than that required for surface ignition. Multidimensional conduction and the effect of orientation on the reactive boundary layer characteristics are two most relevant factors here.

Equations 14 and 15 indicate that if i_0 is replaced by $h(T_\infty - T_{s0})$, Figure 2-13.3 can describe ignition by convective heating[21] as well. The x- and y-coordinates then are

$$\text{Bi} \cdot \frac{T_\infty - T_{s0}}{T_s - T_{s0}} \quad \text{and} \quad \text{Bi} \cdot \frac{T_\infty - T_{s0}}{T_s - T_{s0}} \cdot \frac{\alpha_s t}{\ell^2}$$

respectively, with the same physical meaning as before. Several points should be noted:

1. The Biot number stands to represent the convective losses in radiant ignition. In convective ignition, however, it stands to represent the heating itself.
2. In practice, convective heating generally occurs as a result of flowing hot flame gases adjacent to the target surface. These gases are either devoid of, or diminished in, the oxygen content. Such a heating may produce pyrolytic damage to the target without ever producing ignition and flame. If the oxygen content of these heating gases is sufficiently high, potential ignition may be piloted by the hot gas flame itself. To avoid these real-world complexities, application of Figure 2-13.3 to convective heating should be made only when the convective heat source is hot air.
3. It may seem desirable to sort out the framework of the ignition map such that the heating Biot number is removed from the coordinates and made to appear only as a parameter in the curves. If this occurs, however, all the physical interpretations must be revised as well. (No substantial improvement in understanding seems to be gained.)
4. Akita's[6] experiments indicate that spontaneous and piloted ignition temperatures[1] of convectively heated wood are about 765 and 725 K, respectively.

ROLE OF THE GAS PHASE PROCESSES

Alvares and Martin[22] reported experiments in which cellulose was radiantly and spontaneously ignited in an atmosphere of oxygen-enriched air at high pressures. Eliciting a number of interesting gas phase processes that influence heterogeneous ignition, this work indicates that the time to ignition is dramatically reduced by an increase in the ambient oxygen mass fraction and/or ambient gas pressure, and by a reduction in its thermal conductivity. The surface temperature at ignition is also found to substantially decrease with an increase in the oxygen content and in the pressure of the ambient gas. A simple analysis can be made to predict these observations and attempt to develop an understanding of the activity in the gas phase.

Consider a situation in which attention is focused only on the gas phase. Assume, accordingly, that the solid phase problem furnishes to the boundary layer a pyrolyzate content corresponding to a mass fraction of Y_F and the ambient gas furnishes oxygen such that its mass fraction is Y_O. Let the solid surface at $T_g = T_{sw}$ and the edge of the boundary layer at $T_g = T_\infty < T_{sw}$ constitute two large parallel plates separated by a distance, δ. Assume further that the reactive fluid mixture in this layer is stagnant so that energy transfer

from the hot to cold boundary occurs only by conduction. Additionally, let the problem be steady state. Then, all equations become irrelevant, except for Equation 7, simplified. This simplification[23] is

$$K_g \frac{d^2 T_g}{d\hat{y}^2} + h_c[Z_{gF} \rho_g^2 Y_F Y_O \exp(-E_g/\bar{R}T_g)] = 0 \quad (17)$$

with the boundary conditions: $\hat{y} = 0$, $T_g = T_{sw}$ and $\hat{y} = \delta$, $T_g = T_\infty$. Recall that the square-bracketed term in this equation is the volumetric fuel consumption rate due to oxidation, whose kinetics are taken to exhibit an order of unity with respect to both the pyrolyzate fuel and the oxygen.

For an inert gas mixture with constant conductivity, steady-state conduction would occur across the gas layer with a linear temperature profile. On the other hand, if the mixture is reactive in a strongly temperature dependent way, the energy source prevalent in the proximity of the hot wall will locally reduce the temperature gradient. The stronger this energy source becomes, the greater is this gradient reduction. In fact, if the source is strong and T_{sw} is sufficiently high, the gradient near the hot wall will even be reversed in sign, signifying that heat would flow from the reacting gas to the hot wall as well as to the cold wall. Ignition is thus conceivably accomplished when T_{sw} is just high enough to result in a zero temperature gradient at the hot wall. This critical T_{sw} will depend upon the kinetics and energetics of the exothermic reaction and the gas thermal conductivity. To determine the critical T_{sw}, the energy conservation problem of Equation 17 must be solved, in which the preignition reactant consumption is ignored and the mixture conductivity is assumed constant. Upon defining

$$\theta \equiv \frac{E_g(T_{sw} - T_g)}{\bar{R}T_{sw}^2}; \qquad \eta \equiv \frac{\hat{y}}{\delta};$$

and

$$D \equiv \frac{E_g \delta^2 h_c \rho_g^2 Z_{gF} Y_F Y_O \exp(-E_g/\bar{R}T_{sw})}{\bar{R}T_{sw}^2 K_g}$$

Equation 17 reduces to $d^2\theta/d\eta^2 - D\exp(-\theta) = 0$ with the boundary conditions $\theta(0) = 0$ and $\theta(1) = \theta_\infty$. Integration of this equation is quite straightforward, although somewhat tedious. One integration gives

$$\frac{d\theta}{d\eta} = \sqrt{[4C_1^2 - 2D\exp(-\theta)]} \quad (18)$$

relating local temperature to its gradient. With one additional integration and application of the boundary conditions, $\theta = \theta(\eta; D, \theta_\infty)$ can be found and the conditions of $D = D^*(\theta_\infty)$ surrounding the ignition indicated by the zero temperature gradient at the hot wall can be deduced.

There is a simple approximation by which this ignition result can be obtained relatively easily. The temperature profile of a marginally igniting system is given by Equation 18 in which the constant, C_1, is evaluated by setting θ and $d\theta/d\eta$ equal to zero at $\eta = 0$. Thus, for the temperature profile in the system at ignition, $4C_1^2 = 2D$, so that $d\theta/d\eta = \{2D[1 - \exp(-\theta)]\}^{1/2}$. Change in this gradient at ignition from zero to a nearly constant value of $\{2D[1 - \exp(-\theta_\infty)]\}^{1/2} \approx (2D)^{1/2}$ occurs within a remarkably small distance from the hot wall. If one assumes that the linearity of the temperature profile holds in the entire range $0 \le \xi \le 1$, the gradient is noted to be equal to θ_∞. Thus, ignition by the hot surface is expected if $(2D)^{1/2} \ge \theta_\infty$; that is, if

$$D \geq \theta_\infty^2/2$$

Written in physical terms, ignition would occur if

$$\frac{2\delta^2 \bar{R} T_{sw}^2 \rho_g^2 h_c Z_{gF} Y_F Y_O \exp(-E_g/\bar{R} T_{sw})}{K_g E_g (T_{sw} - T_\infty)^2} \geq 1 \qquad (19)$$

If T_∞ is small, the hot wall temperature appears only in the Arrhenius term. Therefore, it is easy to see that a small gas layer thickness requires a higher T_{sw} to produce ignition.

If the reactable gas were to be flowing over a hot plate, Equation 19 can be adapted by viewing δ, the boundary layer thickness, as the average thermal boundary layer thickness and the ratio (K_g/δ) as the average heat transfer coefficient, h. Inasmuch as both the boundary layer thickness and the heat transfer coefficient depend upon the flow dynamics and the nature of the fluid, the ordinary heat transfer results may be incorporated into Equation 19. With $\delta = H/\text{Nu}$ where H is the flow characteristic length and Nu is the average Nusselt number (which depends on the Grashof or Reynolds number of the flow and the Prandtl number of the fluid), ignition occurs if

$$\frac{2H^2 \bar{R} T_{sw}^2 h_c \rho_g^2 Z_{gF} Y_F Y_O \exp(-E_g/\bar{R} T_{sw})}{\text{Nu}^2 K_g E_g (T_{sw} - T_\infty)^2} \geq 1 \qquad (20)$$

If all else is kept fixed, longer hot plates are capable of producing ignition with lower wall temperatures. Faster moving gas streams are not as likely to be ignited by the same T_{sw} hot plate as are the slower moving ones. This ignition-retarding effect of flow speed is more pronounced for laminar flows than for turbulent flows. (This deduction is, of course, based on the fact that Nu $\propto$ Re^b, where $b \approx \frac{1}{2}$ for laminar flow and $b \approx \frac{1}{5}$ for turbulent flow.) Note also that Equation 20 can be adapted to hold for combustible gas flows over a variety of geometries.

Equations 19 and 20 are entirely consistent with the observations of Alvares and Martin.[22] As the surface temperature, T_{sw}, develops slowly due to the solid phase conductive response to heating, the gas phase responds quickly. It is eminently clear from Equations 19 and 20 that taller targets in a slow-moving, poorly conducting, low viscosity, high pressure, oxygen-enriched ambient gas will ignite sooner at a lower surface temperature. Although the present analysis is only approximate, it powerfully depicts how gas phase oxidation reactions are induced by the composition and thermal conditions.

Gandhi[15] solved the comprehensive set of the conservation equations, accounting for the full unsteady, reactive, conjugate heat and mass transfer problem with reradiant heat loss at the surface of the target, which is vertical and subjected to external irradiance. An integral technique is employed to obtain the characteristics of the transiently developing, reactive, natural convective boundary layer averaged over the target height. The gas temperature gradient reversal at the gas-solid interface is used as the criterion for spontaneous ignition. This solution yielded valuable information on the ignition criterion as discussed previously. Additionally, it is possible to predict such radiant spontaneous ignition phenomena as the time to ignition and ignition thresholds.

Figure 2-13.4 indicates Gandhi's predicted influence of exposure irradiance and the target thickness on the time to ignition for fixed values of the plate height, H, and kinetics of oxidation and pyrolysis. The general characteristics of Figures 2-13.1 and 2-13.3 obviously appear to be captured.

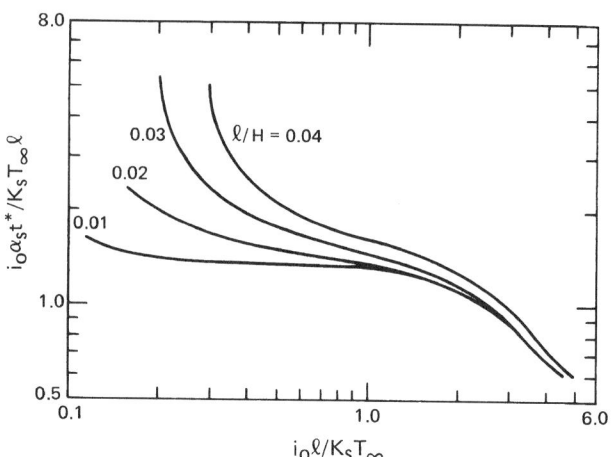

Fig. 2-13.4. *Influence of plate thickness on time of ignition.* $gH^3/\nu_g^2 = 10^6$; $E_g/\bar{R}T_\infty = 30$; $Z_{gF}\rho_g\ell^2/\alpha_s = 10^8$; $E_s/\bar{R}T_\infty = 50$; $Z_s\ell^2/\alpha_s = 10^7$.

(Recall that the surface reradiant energy loss has been ignored in deducing Figure 2-13.3.) Typical agreement of the predictions with measurements is shown in Figure 2-13.5. Further details of this predictive model are available.[15,16,24]

Most importantly, Gandhi predicts the lower limit exposure flux below which ignition of a given solid in a given situation is impossible. These limiting conditions, known as ignition thresholds, are difficult to measure experimentally. The threshold minimum flux is found to depend on the plate height and thickness, gas and solid phase reaction kinetics, and reradiant loss as well as the heat of combustion of the pyrolyzates. One typical prediction is indicated in Figure 2-13.6, where the threshold flux is plotted as dependent on the heat of combustion and the plate height for given reradiant loss and kinetics. Ignition will not occur in the area lying below the curve, while it will occur elsewhere. Taller targets will experience ignition even if the pyrolyzate is less combustible; for a

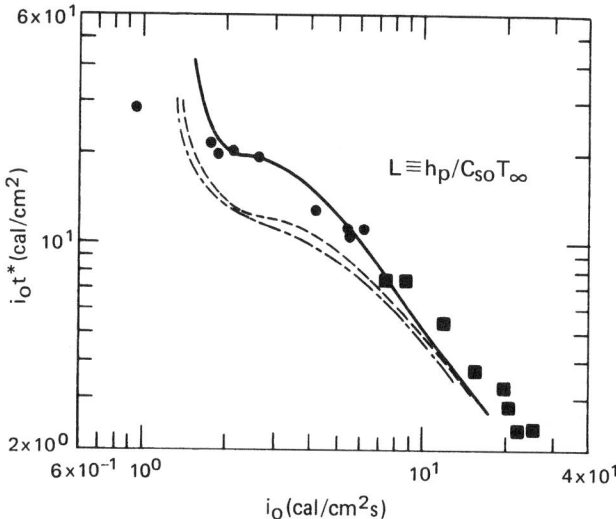

Fig. 2-13.5. *Comparison with Martin's data for α-cellulose.* ■ *and* ●: *data*[25]; *solid line: theory*[15] *($L = -1$); dashed line: theory*[15] *($L = 1$); dash-dotted line: theory*[15] *($Cp = 0$).*

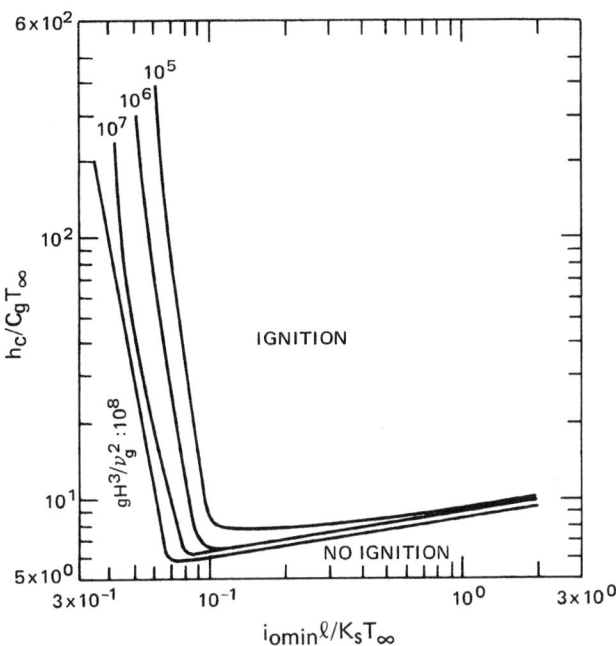

Fig. 2-13.6. Influence of plate height on threshold flux. $E_s/\bar{R}T_\infty$ = 50. $E_g/\bar{R}T_\infty$ = 30. $\sigma\varepsilon_{sw}T_\infty^3\ell/K_s$ = 10^{-3}.

fixed heat of combustion of the pyrolyzates, taller targets exhibit a lower threshold flux. The minima in these curves indicate that when the heat of combustion is very low, ignition is impossible irrespective of the exposure flux level. At enthalpies of combustion that are close to but slightly larger than the limiting value (which depends on the target height), two roots of the threshold flux exist between which ignition is possible. These are important findings and must be further studied for use in firesafety. While Figure 2-13.6 is presented here only as an example of Gandhi's predictions, much similar information is available.[15]

SOME PRACTICAL ISSUES

The understanding of the ignition process developed earlier in this chapter is useful in a number of practical firesafety problems, including those cited in the Introduction. These problems fall broadly into two groups: those pertaining to the very initiation of the fire, and those pertaining to the spatial (or temporal) growth of fire in combustible ensembles. A number of questions arise in attempts to assess the ignition hazard on the basis of the existing simple viewpoints. These questions point out the need for care and for a comprehensive model of ignition through which the assessment can be made with some confidence.

The possibility of ignition of an object under a given set of conditions is presently judged by evaluating whether or not the exposed surface would attain a critical ignition temperature, T^*, construed as a property (albeit, extrinsic) of the material. T^* for transient ignition of a broad range of natural and synthetic organic solids is taken to be approximately as follows: for spontaneous ignition, 600°C for radiant exposure and 500°C for convective exposure; for piloted ignition, 300 to 410°C for radiant exposure and 450°C for convective exposure. Note that these are approximate values, mostly deduced from experiments on small vertical specimens.

Persistent ignition would require heating of the solid, over a minimum thickness near its exposed surface, to a mean temperature in excess of the material's characteristic pyrolysis temperature to ensure continued pyrolysis. This would presumably mean higher surface temperatures than those just cited. It is noteworthy in this context that in spite of significant physical and chemical differences in structure and composition, most organic solids undergo pyrolysis in a rather narrow temperature range of 325 ± 50°C.

It is also important to note here that the size of the specimen and its orientation are expected to exert an influence on T^* by altering the convective patterns and rates of heat loss and gas mixing. The mechanisms involved in this influence are not straightforward. Reduction of heat loss from the surface of a radiantly heated solid will obviously tend to increase its heating and pyrolysis. It is, therefore, expected to result in a hastened ignition. This expectation, however, is not always valid since the gas phase ignition requires mixing of the pyrolyzates with air and heating of the mixture so as to induce the preignition oxidation reaction. These mixing and heating processes of the gas phase require a certain optimal boundary layer flow. Thus, while mixing and heating of the boundary layer gases is a condition required for ignition, excessive mixing will dilute the reactant mixture and deprive the solid from its energy demand for sustained production of pyrolyzate.

These ideas are also important from another practical viewpoint. Traditional strategies to delay or prevent ignition involve a physical and/or chemical alteration of the materials. One goal of such an alteration is usually to make the pyrolysis difficult; for example, by increasing the associated activation energy or endothermicity. If the heating were convective, however, an easy pyrolysis with an increased rate may be more desirable, for then the copious production of the pyrolyzate is expected to retard the convective heating rate by thickening the boundary layer.

Another material alteration that can delay or prevent ignition (irrespective of the radiative or convective heating mode) is to chemically tamper with the composition of the pyrolyzate mixture, rendering it abundant in such inert species as water vapor and carbon dioxide. Means of accomplishing this can be found in organic chemistry, wherein even trace quantities of certain additives are known to profoundly alter both the rate of production and composition of the pyrolyzate. However, complications arise due to the dependency of the action of these additives on the rate of heating of the sample.

The differences between spontaneous and piloted ignitions, and between radiant and convective heating to ignition, should be noted here. Spontaneous ignition requires: (1) sufficient heating of the solid to produce enough pyrolyzates continuously; (2) mixing of these pyrolyzates with air to produce a mixture above the lean limit; and (3) heating of this boundary layer mixture to a temperature sufficiently high to result in a thermochemical runaway to the state of flame. Requirement 3 is, perhaps, not necessary for piloted ignition, wherein an external ignition energy source is available to initiate the flame. If the heating were by irradiation of the specimen, the exposed surface of the solid would be at the highest temperature, invoking a transient natural convective boundary layer in which the induced air mixes with the effluent pyrolyzates. If forced blowing of air over a radiantly heated solid surface were involved, the surface then would be cooled to possibly influence the pyrolysis of the solid, mixing and heating would be enhanced in the boundary layer, and the resultant ignition behavior would be complicated.

If the heating were by convection, drastic differences would arise, depending upon the nature of the convecting hot gas. Heating by hot air is conducive to quick ignition. Heating by hot (nonflaming) combustion gases can result in copious pyrolysis, but ignition in the mixing layer is thwarted by the poor oxygen content. If the convective heating were by hot flames, the flames themselves would become extended in space due to the added pyrolyzate fuel; the very idea of ignition becomes vague.

The multitude of complexities and ambiguities involved in assessing the ignition hazard of a practical situation, even with the simple critical surface temperature criterion, is obvious. Care is therefore required in drawing quick conclusions from current simplified practices. The surface temperature at ignition is not a cause but an effect and, as such, depends on the conditions of an experiment (so, also, are most of the other known ignition criteria). A comprehensive quantitative model of physical chemistry of ignition, such as that presented by Gandhi for spontaneous ignition of radiantly heated vertical cellulosic solids, seems to be needed for piloted radiant ignition, for (spontaneous and piloted) ignition due to convective heating, and for a number of specimen orientations.

CONCLUSION

Great advances have been made in understanding the fundamental mechanisms involved in flaming ignition of radiantly and/or convectively heated solids—natural as well as synthetic—with and without the presence of a pilot source. The current practice of ignition hazard assessment employs such simplified concepts as a prescribed critical ignition temperature of the solid. This practice is prone to ambiguities and difficulties. Physico-chemical models of ignition are available and can be used to examine the ignition process in a broader and more complete perspective. Quantitative assessment of the conditions that differentiate between possible and impossible ignition is now feasible from these models. The time to ignition and energy required for ignition can be estimated with reasonable accuracy. The present understanding can be refined by relaxing most of the assumptions enumerated in the section of this chapter on conservation equations; but the degree of and the need for the necessary refinement must first be addressed.

To illustrate the practical utility of Figure 2-13.3, consider a firwood target 1 cm thick and 2 cm in height initially at 300 K in normal air, also at 300 K. Will spontaneous ignition occur if this target is exposed to a radiant flux of 2 W/cm^2 continuously for 1000 seconds? Will the ignition, if possible, be transient or sustained?

Approximate answers to such questions can be given using Figure 2-13.3. First, the conductivity, specific heat, density, and diffusivity of fir, respectively, are found to be 0.17 W/m · K, 2500 J/kg · K, 600 kg/m^3, and 11.3 × 10^{-8}m^2/s. From convection literature, the heat transfer coefficient, h, is estimated to be about 15 W/m^2 · K. Then the Biot number and x, y-coordinates for Figure 2-13.3 are calculated with the given exposure flux and duration.

$$\text{Bi} \equiv \frac{h\ell}{K_s} \approx \frac{15(\text{W/m}^2\text{K})\cdot 1(\text{cm})}{0.17(\text{W/mK})} \approx 0.88$$

$$x \equiv \frac{i_o\ell}{K_s(T^*-T_0)} \approx \frac{2(\text{W/cm}^2)1(\text{cm})}{0.17(\text{W/mK})(873-300)(\text{K})} \approx 2.05$$

$$y \equiv \frac{i_0\alpha_s t}{K_s(T^*-T_0)\ell} \approx \frac{2(\text{W/cm}^2)1.3\times 10^{-8}(\text{m}^2/\text{s})1000(\text{s})}{0.17(\text{W/mK})573(\text{K})1(\text{cm})} \approx 2.32$$

The critical temperature $T^* = 873$ K comes from the summary in the previous section and corresponds to spontaneous ignition due to radiant heating. Entering these values of Bi, x, and y on Figure 2-13.3 sustained spontaneous ignition is concluded to be possible with the given exposure and duration. If the exposure duration were shorter, say 400 seconds, the y-coordinate would be 0.93, indicating that the ignition would be transient. With even shorter exposure (less than 260 seconds), y will be less than approximately 0.6 and ignition would not occur.

Repeating the example but with a thinner target of thickness $\ell = 0.1$ cm, and keeping all else unchanged, the Bi, x, and y coordinates are estimated to be 0.088, 0.205, and 23.2, respectively. Sustained spontaneous ignition is obviously ensured on this thin body. With 400 seconds of exposure, $y \approx 9.28$, sustained ignition continues to be possible. The minimum exposure required of this thin body to produce ignition is 40 seconds, which is substantially shorter than the exposure of 260 seconds needed for transient ignition of the thicker body considered above.

If a pilot source were available, the appropriately lower critical temperature from the previous section of this chapter would enable similar estimates to be made of whether ignition would occur with a given exposure duration, whether the possible ignition would be transient or sustained, and of the minimum required exposures.

If convective heating from a hot gas flow at a temperature T_∞ were involved rather than radiant heating, the x- and y-coordinates of Figure 2-13.3 would be estimated by the definitions given in the end of the section of this chapter on solid conduction-controlled ignition. Throughout this example it should be remembered that Figure 2-13.3 has been developed on the basis of a greatly simplified view of the ignition process. The very concept of the prescribed critical temperature involves an oversimplification and limited validity. Detailed models, such as those underlying Figures 2-13.4 through 2-13.6, are more reliable, although considerably more difficult to apply.

Mathematical analyses lead to a predictive capability which is limited only by the accuracy and detail of such input data as the pyrolysis and combustion kinetic and thermodynamic properties, values of transport properties and their variability with temperature and composition, etc. Many real-world complexities exist: materials are often encountered in practice as composites with glues, stitches, and bonds; practical geometries of targets are seldom simple; aging and durability of a solid alter its physical and chemical characteristics; surfaces are almost always multicolored; etcetera, ad infinitum. An adherence to fixed sets of physical and chemical properties or to voluminous dictionaries of property values under different circumstances may not be wise for use in a predictive model. It may be more prudent to seek only a fundamental conceptual or mechanistic understanding of a phenomenon such as flaming ignition from a mathematical model.

Even more importantly, mathematical models suggest useful ways of correlating the experimental measurements. Meaningful correlations lead one to maximize the value of limited experiments, generalize the observations, and to postulate, interpret, and exploit the basic mechanisms of chemistry and physics.

A qualitative, comprehensive map of flaming ignition has been developed in the work presented here. The nature of the problem and the component processes are well identified, and the global physico-chemical behavior appears to be clear. Employing this understanding, it must be possible to draw some conclusions on the effects of such non-idealities as the edges and corners that reflect in three-dimensional conduction and in complicated boundary layer flows; surface roughness and color in conjunction with the nature of the radiation source; convective heating by vitiated hot gases; leakage of energy and mass through the backface and edges; and others.

EXPLICIT FORMS OF EQUATION 14 FOR SOME LIMITING CASES

1. *Explicit form of Equation 14 as convective losses tend to zero*: The transcendental equation becomes $\sin a = 0$ so that $a_n = n\pi$, $n = 1, 2, \ldots \infty$. This solution[17] is given as

$$\frac{K_s(T_s - T_{s0})}{i_0 \ell} = \frac{\alpha_s t}{\ell^2} + \frac{y^2}{2\ell^2}$$
$$- \frac{1}{6} - \sum_{n=1}^{\infty} \frac{2(-1)^n}{\pi^2 n^2} \cos\left(\frac{n\pi y}{\ell}\right) \exp\left(-\frac{n^2 \pi^2 \alpha_s t}{\ell^2}\right) \quad (21)$$

2. *Thermally thin slab*: If the solid were either physically so thin and/or so highly conductive that the temperature gradients within the solid are promptly relaxed by the vigorous conduction, then the solid temperature is a function of time alone. The energy equation for this case is obtainable by integrating Equation 2 as $\rho_s C_s (V/S)$ $dT_s/dt = i_0 - h(T_s - T_\infty)$ along with the initial condition $T_s = T_{s0}$ at time $t = 0$. The ratio (V/S) is the solid volume over its surface area, equal to the thickness, ℓ, for a slab. The solution is easily obtained:

$$\frac{T_s - T_{s0}}{(i_0/h) + T_\infty - T_{s0}} = 1 - \exp\left(-\frac{hSt}{\rho_s C_s V}\right) \quad (22)$$

which in fact is a limiting form of Equation 14. In the further limit $h \to 0$, as in Item (1) above, Equation 22 reduces to

$$\frac{\rho_s C_s V(T_s - T_{s0})}{i_0 St} = 1 \quad (23)$$

for pure radiant heating of a thin body without any convective loss.

3. *Semi-infinite slab*: If the slab were physically so thick and/or so poor a conductor that the backface does not realize the effect of thermal exposure at the frontface within the time period of interest, the slab then can be considered as a semi-infinite solid. A high frontface exposure flux and a low conductivity will tend to make even a physically thin sheet of a slab behave as though it is thermally infinitely thick. With radiant heating and convective loss at the only face, we know that $y' = 0$ (with y' measured into the interior from the surface), and the temperature-time-space distribution corresponds to the small time limit of Equation 14. This solution is given explicitly as:

$$\frac{T_s - T_{s0}}{(i_0/h) + T_\infty - T_{s0}} = erfc\left[\frac{y'}{2\sqrt{\alpha_s t}}\right]$$
$$- \left\{ erfc\left[\frac{y'}{2\sqrt{\alpha_s t}} + \frac{h\sqrt{\alpha_s t}}{K_s}\right] \exp\left(\frac{hy'}{K_s} + \frac{h^2 \alpha_s t}{K_s^2}\right) \right\} \quad (24)$$

(The manner in which the unavailability of a physical reference length, ℓ, is handled can be observed from Equation 24.) Heating by radiation alone without any

TABLE 2-13.2 *The Error Function and its Derivative*

ϕ	$erf\phi$	$erfc\phi = 1 - erf\phi$	$\frac{d}{d\phi}erf\phi = (2/\sqrt{\pi})\exp(-\phi^2)$
0	0	1.0	1.1284
0.05	0.05637	0.9436	1.1256
0.1	0.1125	0.8875	1.1172
0.2	0.2227	0.7773	1.0841
0.3	0.3286	0.6714	1.0313
0.4	0.4282	0.5716	0.9615
0.5	0.5205	0.4795	0.8788
0.6	0.6039	0.3961	0.7872
0.7	0.6778	0.3222	0.6913
0.8	0.7421	0.2579	0.5950
0.9	0.7969	0.2031	0.5020
1.0	0.8427	0.1573	0.4151
1.5	0.9661	0.0339	0.3568
2.0	0.9953	0.00468	0.0827
2.5	0.9996	0.00041	0.0109
3.0	0.99998	0.00002	0.0008

convective loss calls for the limit of Equation 24 as $h \to 0$. This limit solution[17] is given to be

$$\frac{K_s(T_s - T_{s0})}{2i_0\sqrt{\alpha_s t}}$$
$$= \frac{1}{\sqrt{\pi}} \exp\left(-\frac{y'^2}{4\alpha_s t}\right) - \left(\frac{y'}{2\sqrt{\alpha_s t}}\right) erfc\left(\frac{y'}{2\sqrt{\alpha_s t}}\right) \quad (25)$$

where *erfc* is the complementary error function. This and other related functions are synopsized in Table 2-13.2.

NOMENCLATURE

a_n constants, $n = 1, 2, \ldots \infty$ (-)
b index (-)
Bi   Biot number (-)
C specific heat $(kJ/Kg \cdot K)$
D Damköhler number (-)
D_{ig} species i mass diffusivity through the gas mixture (m^2/s)
E activation energy $(kJ/kmol)$
f fuel pyrolyzate/oxygen stoichiometric mass ratio (kg F/kg O)
g standard acceleration of gravity (m/s^2)
H height of the solid target plate (m)
h heat transfer coefficient $(W/m^2 \cdot K)$
h_c enthalpy of combustion of fuel pyrolyzate (kJ/kg)
h_p enthalpy of pyrolysis (kJ/kg)
i incoming exposure irradiance (W/m^2)
i_0 absorbed exposure irradiance (W/m^2)
K thermal conductivity $(W/m \cdot K)$
ℓ thickness of the solid target plate (m)
M molar mass (kg/kmol)
$\dot{m}''$ mass flux $(kg/m^2 \cdot s)$
$\dot{m}'''$ species i source-sink strength $(kg/m^3 \cdot s)$
Nu   Nusselt number (-)
P pressure (Pa)
$\dot{q}'''$ energy source-sink strength (W/m^3)
Re   Reynolds number (-)
$\bar{R}$ universal gas constant $(kJ/kmol \cdot K)$
S surface area (m^2)
T temperature (K)

$\bar{T}_s$ mean temperature of the solid (K)
t time (s)
u x-directional velocity component (m/s)
V volume (m^3)
v y-directional velocity component (m/s)
W width of the solid target plate (m)
x coordinate along the surface (m)
Y_i species i mass fraction (-)
y coordinate normal to the surface (m)
$\hat{y}$ $(y - \ell)$ (m)
y' $(\ell - y)$ (m)
Z pre-exponential (or collision) factor (1/s for solid pyrolysis, and m^3/kg · s for gas oxidation)

Greek Symbols

α thermal diffusivity (m^2/s)
δ reactive boundary layer thickness (m)
ε_s solid surface emissivity (-)
η nondimensional y (-)
θ nondimensional temperature (-)
ν kinematic viscosity (m^2/s)
ρ density (kg/m^3)

Subscripts

c char
F fuel pyrolyzate
g gas
I inert
O oxygen
0 initial
P products
p pyrolysis or fuel pyrolyzate
R reservoir
s solid
w interface
∞ ambient

Superscript

$*$ ignition

REFERENCES CITED

1. A.M. Kanury, *Fire Res. Abst. and Rev.*, 14, 24 (1971).
2. F.R. Steward, in *Heat Transfer in Fires*, Scripta, Washington, p. 379 (1974).
3. J.R. Hallman, J.R. Welker, and C.M. Sliepcevich, *Ignition of Polymers*, Proceedings of 30th Annual Conference, Soc. Plastics Engineers, p. 283 (1972).
4. S.B. Martin, *Diffusion-Controlled Ignition of Cellulosic Materials by Intense Radiant Energy*, 10th Symp. (Int.) on Comb., Combustion Institute, Pittsburgh, PA, p. 877 (1965).
5. C.H. Bamford, J. Crank and D.H. Malan, *Proc. Cambridge Phil. Soc.*, 42, 166 (1946).
6. K. Akita, *Report Fire Res. Inst.*, Japan, 9 (1959).
7. D.L. Simms, *Comb. and Flame*, 4, 293 (1960).
8. C.C. Ndibizu and P. Durbetaki, *Fire Res.*, 1, 281 (1977-78).
9. N.J. Alvares, *Tech. Report 735*, U.S.N.R.D.L. (1964).
10. L.I. Deverall and W. Lai, *Comb. and Flame*, 13, 8 (1969).
11. G. Adomeit, *Ignition of Gases at Hot Surfaces Under Nonsteady State Conditions*, 10th Symp. (Int.) on Comb., Combustion Institute, Pittsburgh, PA, p. 237 (1965).
12. F.M. Sauer, *Interim Technical Report AFSWP-868*, U.S. Department of Agriculture/Forest Service, (1956).
13. T. Kashiwagi, *Comb. Sci. and Tech.*, 8, 225 (1974).
14. E.W. Price, H.H. Bradely, Jr. and G.L. Dehority, *AIAA J.*, 4, 1153 (1966).
15. P.D. Gandhi, *Spontaneous Ignition of Organic Solids by Radiant Heating in Air*, Ph.D. dissertation, South Bend (1984).
16. P.D. Gandhi and A.M. Kanury, *Comb. Sci. and Tech.*, 50, 233 (1986).
17. H.S. Carslaw and J.C. Jaeger, *Conduction of Heat in Solids*, Oxford University, London (1959).
18. A.N. Koohyar, J.R. Welker, and C.M. Sliepcevich, *Fire Tech.*, 4, 284 (1968).
19. H.R. Wesson, J.R. Welker, and C.M. Sliepcevich, *Comb. and Flame*, 16, 303 (1971).
20. W.K. Smith, *Ignition of Edges and Corners of Wood*, WAM Paper No. 72-WA/HT-20, ASME (1972).
21. W.D. Weatherford and D.M. Sheppard, *Basic Studies of the Mechanics of Ignition of Cellulosic Materials*, 10th Symp. (Int.) on Comb., Combustion Institute, Pittsburgh, PA, p. 897 (1965).
22. N.J. Alvares and S.B. Martin, *Mechanism of Ignition of Thermally Irradiated Cellulose*, 13th Symp. (Int.) on Comb., Combustion Institute, Pittsburgh, PA, p. 905 (1971).
23. A.M. Kanury, *Combustion*, unpublished manuscript (1995).
24. P.D. Gandhi and A.M. Kanury, in *Dynamics of Reactive Systems, Part I: Flames and Configurations*, AIAA, New York, p. 208 (1986).
25. S.B. Martin and N.J. Alvares, *Tech. Report 1007*, U.S.N.R.D.L. (1968).

SURFACE FLAME SPREAD

James G. Quintiere

INTRODUCTION

This chapter covers both opposed flow and wind-aided flame spread over solids. Approximate formulas are developed using simple assumptions in order to illustrate the role of the relevant physical and chemical variables. The relationship between these approximate formulas and more exact results and data is discussed. Although an extensive review is not presented, an attempt is made to illustrate the extent of knowledge and opportunities for application. Flame spread over liquid fuels, in the forest, and in microgravity is also briefly discussed.

This chapter has been entitled "Surface Flame Spread" to ensure its distinction from flame propagation in premixed fuel and air systems. In the context here, flame spread applies to the phenomenon of a moving flame in close proximity to the source of its fuel originating from a condensed phase, i.e., solid or liquid. As in the premixed case, the flame propagates in the gas phase, with its front associated with the lower flammability limit mixture. The fuel in the mixture is the vaporized condensed-phase fuel. The vaporization process is caused by heat transfer from the advancing flame itself and is necessary to sustain the spreading flame. Thus, two advancing fronts are present: (1) the flame in the gas phase, and (2) the evaporation or pyrolysis region in the condensed phase. The former is easily perceived by the casual observer, while the evaporation front is actually a measure of the surface flame spread for the condensed phase. This front is not easily measured, but its rate of movement is defined as the "flame spread velocity." These processes are illustrated in Figures 2-14.1 and 2-14.5.

In natural fires, it is the flame-spread process that is critical to the fire's destiny. This applies whether the fire is an urban conflagration or the first growth after ignition of a room's draperies. The process of growth is unstable, with the time for flame spread competing with the time for burning. If the spread time is small compared to the burn time, fire growth is likely to accelerate; and, conversely, it could decelerate and stop if the spread time is large compared to the burn time.

An example might illustrate the significance of flame spread in fire growth. Consider the ignition of an uphol-stered chair over a region 6 cm in diameter. Typical unenhanced spread rates are order-of-magnitude 1 mm/s, so in one minute the pyrolyzing region would be 18 cm in diameter, since the front would have grown 6 cm. This is nearly a tenfold increase in the pyrolyzing region, and, for sufficient fuel (long burn time), the increased rate of energy release will provide enhanced heat transfer to increase the flame spread rate. The fire growth is manifested by the size of the evaporation region over the condensed phase, and the rate of movement of its boundary is the flame spread velocity. Thus, surface flame spread plays a significant role in natural fire growth.

Premixed flame propagation can also occur in natural fires, provided that, in the gas phase, fuel and air are within their flammable limits over an extended region. This might result from a combustible gas leak, or could also arise under sufficient heating or restrictive ventilation conditions during the development of a fire in a confined space. In such cases, a sufficient energy source would initiate the propagation of flames in the gas phase at velocities of the order of magnitude of 1 m/s compared to 1 mm/s associated with representative surface spread velocities. These values suggest the plausible limits of flame propagation in natural fires, but it is the lower end of the scale that governs the early development of fires on solid and liquid fuels. Subsequently, the nature and controlling variables can be described for such flame spread over solids and liquids. Table 2-14.1 gives the relative rates of flame spread for several fire phenomena, including horizontal surface spread and gas phase phenomena.

BACKGROUND

In recent years, research has focused on methods applying fundamental aspects of flame spread to real problems. Fundamental theory has provided a foundation for application, and formulas have been developed for practical use. However, these formulas are still limited by the lack of material data and phenomenological information about the fire conditions. For solid materials, the development of the cone calorimeter[1] and the lateral ignition and flame spread (LIFT)[2] tests has provided a basis for the needed material data. Attempts have been made to include configuraton effects on flame spread, such as fire spread in vertical corners and ceilings. These attempts have not led to fully

Dr. James G. Quintiere is head of the Fire Science and Engineering Division of the National Bureau of Standards Center for Fire Research. His research has focused on fire growth and flame spread.

TABLE 2-14.1 *Relative Flame Spread Rates (Order of Magnitude)*

Phenomenon	Rate (cm/s)
Smoldering	10^{-3} to 10^{-2}
Lateral or downward spread on thick solids	$\sim 10^{-1}$
Upward spread on thick solids	1 to 10^2
Horizontal spread on liquids	1 to 10^2
Forest and urban fire spread	1 to 10^2
Pre-mixed flame speeds	
Laminar deflagration	10 to 10^2
Detonation	$\sim 3 \times 10^5$

established methods, but have shown promise in replacing old flammability indices with more meaningful measures of flammability.

It is not possible to describe completely in a brief chapter all the research on flame spread. Hence, comprehensive reviews should be sought elsewhere. Recent reviews include Quintiere[3] on wall fire spread, and Hirano[4] on fundamental studies. A good source of material property data for wood species and products can be found from Janssens study.[5] Other noteworthy recent reviews on flame spread can also be found in the literature. For example, Fernandez-Pello and Hirano[6] give a detailed review of progress in flame spread over solids. Current knowledge on flame spread over liquid fuels is brought into focus by a critical review conducted by Glassman and Dryer.[7] For a more tutorial discussion, one might read the presentation by Drysdale[8] in the book, *An Introduction to Fire Dynamics*.

An excellent review and tutorial on flame spread was presented by Williams[9] in 1976. It organized the subject in terms of the nature of the fuel's configuration and the mechanisms controlling flame spread. Its clarity in describing these mechanisms in simple terms in valuable reading for any serious student of flame spread. It describes each type of flame spread by physical arguments and expressions based upon the "fundamental equation of flame spread." That equation considers the net energy (heat) transferred (per unit area per unit time), $\dot{q}''$, ahead of the advancing flame to heat the medium from its initial temperature, T_s, to its ignition temperature, T_{ig}. This energy is equated to the change in enthalpy (per unit area per unit time) that the medium experiences for an observer on the moving flame front. For steady conditions

$$\rho V \Delta h = \dot{q}'' \qquad (1)$$

where

ρ = density of the medium,

V = spread rate, and

Δh = change in enthalpy per unit mass of medium in going from T_s to T_{ig}.

The concept of an ignition temperature here represents the solid or liquid surface temperature that would cause a sufficient production of fuel in the gas phase to sustain piloted ignition. For liquid fuels, this temperature is called the firepoint, and for solids it would depend on the kinetics of thermal degradation. Usually, it is assumed that the condensed phase is heated to the ignition temperature as a constant-property homogeneous medium; therefore, we can write

$$\Delta h = c(T_{ig} - T_s) \qquad (2)$$

where c is the specific heat of the solid or liquid. Thus, a conceptual formulation has been established for the flame spread speed, V. It is interesting to observe that, for a given flame heat flux, V will increase as the fuel density decreases, and will increase without bound as T_s approaches T_{iq}. Thus, low-density solids could pose a potential flame-spread hazard; and, as T_s ahead of the advancing flame is increased by far-field convective and radiative effects of the fire, the spread will accelerate.

FLAME SPREAD OVER SOLIDS

In this section, some theoretical aspects will be sketched to illustrate the nature of flame spread on solids. In general, flame spread can occur in the presence of an ambient wind, such that the spread is upwind (opposed flow flame spread) or downwind (wind-aided flame spread). The wind might be due to external causes, such as might be meteorological in nature, or due to a fire-induced (natural convection) flow created by the spreading flame or an associated fire. These two categories will be treated subsequently. Also, flame spread on solids will depend on geometrical orientation, i.e., vertical or horizontal, facing upward or downward. All of these factors will affect the heat transfer ($\dot{q}''$) indicated in Equation 1. Indeed, the complex problem of computing $\dot{q}''$ from first principles has limited our ability to predict flame spread. Thus, issues of heat transfer pertaining to laminar and turbulent flow and flame radiation all play a role. Ultimately, chemical kinetic factors that control flame temperature, and indeed the survival of the flame itself, come into play. Hence, only the form of simple theoretical expressions will be derived. For more details and for the inclusion of other variables, specific references should be examined.

Opposed Flow Spread

Following along the concept of using an energy conservation principle to describe flame spread, i.e., Equation 1, some simple analyses will be described. Although these analyses will be heuristic, their final results will be consistent with more formal analyses found in the literature. First, opposed flow spread will be considered for a solid whose thickness is sufficiently thin so that the temperature is uniform across its thickness. In all cases, the spread rate is assumed steady and the reference frame is fixed to the pyrolysis front, i.e., the position where the solid (surface) temperature is T_{ig}. This is illustrated in Figure 2-14.1.

Thermally thin case:

For this *thermally thin case* of physical thickness, δ, and no heat loss from the bottom face, the energy equation for the control volume is developed. The control volume has been selected to extend from the region at the onset of pyrolysis, T_{ig}, to the region unaffected by the energy transported into the solid from the flame. This region is a distance Δ from the pyrolysis region to the position of initial solid temperature, T_s. The heat flux from the flame, $\dot{q}''$, will be assumed constant over Δ. It follows that for the solid moving steadily through the flame-fixed control volume at the flame spread speed, V

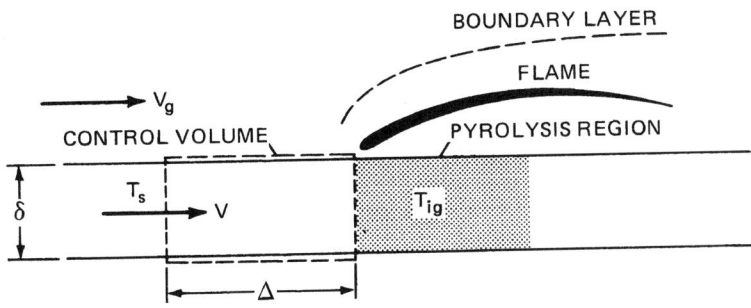

Fig. 2-14.1. Energy conservation analysis in opposed flow spread.

$$\rho c \delta (T_{ig} - T_s) V = \dot{q}'' \Delta \qquad (3)$$

This is a more complete version of Equation 1, with the length scales δ and Δ explicitly included. The net forward flame heat flux characterized as conduction in the gas phase suggests

$$\dot{q}'' \approx k_g \left(\frac{T_f - T_r}{\Delta} \right) \qquad (4)$$

k_g = gas phase conductivity,
T_f = flame temperature, and
T_r = reference temperature for the solid—either T_{ig} or T_s would suffice.

Combining Equations 3 and 4 yields an expression nearly identical to that more formally derived by deRis[10] for thermally thin solids

$$V = \frac{\sqrt{2}\, k_g (T_f - T_s)}{\rho c \delta (T_{ig} - T_s)} \qquad (5)$$

The flame temperature, T_f, here should ideally be taken as that due to adiabatic stoichiometric combustion but, in general, could be thought of as less due to heat losses and chemical kinetic effects. Under these ideal theoretical considerations, it can be shown[11] that

$$T_f - T_{ig} = \frac{(T_\infty - T_{ig}) + \dfrac{Y_{ox,\,\infty}}{r c_g}(\Delta H - L)}{1 - \dfrac{Y_{ox,\,\infty}}{r}} \qquad (6)$$

where

ΔH = heat of combustion of the solid fuel,
L = heat of gasification,
T_∞ = gas phase ambient temperature,
$Y_{ox,\,\infty}$ = gas phase ambient oxygen concentration,
r = stoichiometric mass ratio of oxygen to fuel, and
c_g = specific heat of the gas phase.

Because ΔH and $Y_{ox,\,\infty}/r$ are relatively large, and $\Delta H_{ox} = \Delta H / r$ is nearly a constant for most hydrocarbons (13 kJ/g), Equation 6 suggests that

$$T_f - T_{ig} \sim Y_{ox,\,\infty} \Delta H_{ox} / c_g \qquad (7)$$

Thus, the flame temperature for many solids is primarily only sensitive to the ambient oxygen concentration. This suggests that flame spread over a ceiling in a room would be reduced as the oxygen near the ceiling becomes reduced. Substitution of Equation 7 into 5, essentially yields the results of Magee and McAlevy;[12] however, they considered

pressure effects as well. Incidentally, their early work in flame spread[12] contains extensive data on factors affecting flame spread.

Up to now, no mention of the opposed flow speed, V_g, has been made; and the flame spread velocity appears to be independent of it. This is only the case as long as chemical effects are unimportant. Chemical kinetic effects become important when the time for chemical reactions to be completed in the flame (t_{chem}) becomes long compared to the fluid flow transit time (t_{flow}) through the flame. If the flow is too fast, chemical reaction would be incomplete. The Damköhler number (D) is a parameter used to express these effects and may be represented as Equation 8

$$D \sim \frac{t_{flow}}{t_{chem}} \sim \frac{1}{V_g^2} \qquad (8)$$

As a consequence, the flame heat transfer or theoretical flame temperature in Equation 5 is then modified by some function of D. The dependence of D on measured spread rates over thin paper sheets has been correlated by Fernandez-Pello *et al*[13] in forced flow conditions and by Altenkirch *et al*[11] under buoyant flow downward spread conditions. Recent theoretical analysis by Wichman[14] offers some explanation for these correlations. Each investigator defined a slightly different Damköhler number and selected different chemical kinetic property data. But in qualitative terms, Figure 2-14.2 represents the results given, with the

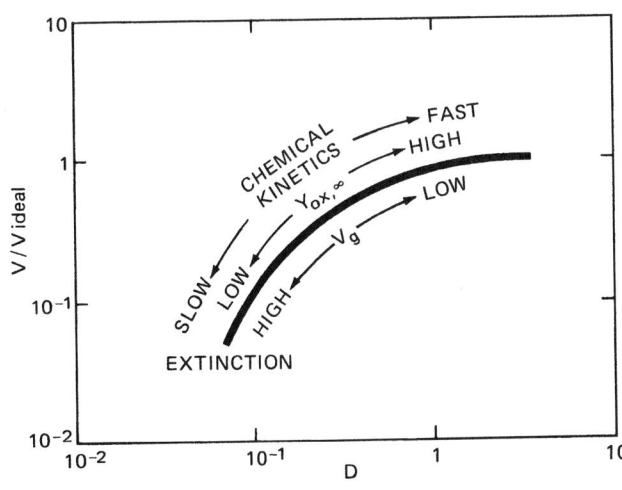

Fig. 2-14.2. Qualitative dependence of opposed flow flame speed with Damköhler number, D.

measured spread velocity normalized with the ideal theoretical velocity Equation 5 plotted against D. The results show that the flame spread decreases with D; and, therefore, decreases as the opposed flow velocity is increased for the thermally thin solid.

For flame-induced flows in pure natural convection, the ambient flow velocity is characteristically given as[11]

$$(V_g) \underset{\substack{natural \\ convection}}{} \sim \left(\frac{\nu_g g \Delta H Y_{ox, \infty}}{c_g r T_\infty} \right)^{1/3} \qquad (9)$$

where ν_g is the gas phase kinematic viscosity. But since $\Delta H/r = \Delta H_{ox}$, approximately constant at 13 kJ/g, V_g induced by buoyancy is primarily a function of $Y_{ox, \infty}$ to the $\frac{1}{3}$ power.

In summary, for thermally thin solids, the flame spread speed under an opposed flow velocity, V_g, with an ambient oxygen concentration, $Y_{ox, \infty}$, is from Equation 5

$$V = V_{ideal} f(D) \qquad (10)$$

where V_{ideal} is given by Equation 5 and the general behavior of $f(D)$ is shown in Figure 2-14.2. At some critical value of D, no further flame spread is possible and extinction occurs. For a given material, D depends on V_g and $Y_{ox, \infty}$, and, from Equation 7 the theoretical flame temperature depends only on $Y_{ox, \infty}$. Therefore, from Equations 5 and 10, the general form of the relationship in opposed flow flame spread on thin materials is

$$V = \frac{\phi(V_g, Y_{ox, \infty}, \text{material properties})}{\rho c \delta (T_{ig} - T_s)} \qquad (11)$$

where the function ϕ depends on such factors as the gas velocity, the local oxygen concentration, and the properties of the material. For common combustible solids, it has been reasoned[10] that the above analysis holds for $\delta \leq 1$ mm, approximately. In particular, Fernandez-Pello and Hirano[6] concluded that for polymethyl methacrylate (PMMA) downward flame spread could be considered thermally thin for $\delta < 2$ mm and thermally thick for $\delta > 2$ cm. In between, the solid would have a nonuniform but finite temperature rise throughout during flame spread.

Except for items like paper, garments, or draperies, most solids in practice will behave as thermally thick under flame spread conditions. It might appear, for engineering purposes, to regard solids with $\delta > 1$ mm as thermally thick; although up to thickness of 1 to 2 cm, flame spread could depend on thickness and on the substrate material adjacent to the solid. Based on these factors, it is apparent that the thermally thick case is more significant, and it will be derived below.

Thermally thick case:

The analysis based on Figure 2-14.1 still applies, except that for the *thermally thick case*, δ must be considered as the thermal penetration depth, which depends on time. From heat conduction theory, it can be reasoned that

$$\delta \approx \sqrt{\frac{k}{\rho c} t} \qquad (12)$$

where t is the time for the flame's pyrolysis front to traverse the heating length, Δ, of the control volume in Figure 2-14.1

$$t = \frac{\Delta}{V} \qquad (13)$$

Substitution of Equations 12 and 13 into Equation 3 yields

$$V = \frac{(\dot{q}'')^2 \Delta}{k \rho c (T_{ig} - T_s)^2} \qquad (14)$$

The length scale, Δ, depends on the nature of the forward heat transfer. In opposed flow spread, when forward conduction in the gas phase is the dominant mode of heat transfer, it can be reasoned that forward conduction must be balanced *with* convection, i.e.,

$$\rho c V_g \frac{\partial T}{\partial x} \sim k \frac{\partial^2 T}{\partial x^2}$$

in the gas phase.

As a consequence,

$$\Delta \sim \left(\frac{k}{\rho c} \right)_g / V_g \qquad (15)$$

and, on combining with Equations 4 and 14 yields

$$V = \frac{V_g (k \rho c)_g (T_f - T_{ig})^2}{k \rho c (T_{ig} - T_s)^2} \qquad (16)$$

as the flame speed for the thermally thick case—identical to the formally derived result by deRis.[10] This expression should be termed the "ideal value," with T_f given by its adiabatic stoichiometric value. With this viewpoint, the chemical kinetic effects enter through the Damköhler number, as in the thin case, with the actual velocity given as

$$\frac{V}{V_{ideal}} = F(D) \qquad (17)$$

and Figure 2-14.2 applies, as well. But, for the thick case, because the "ideal" flame speed (Equation 16) depends on the opposed velocity, V_g, directly as well as through the Damköhler number, D, the actual flame spread speed may either increase or decrease with V_g, depending on $Y_{ox, \infty}$. Fernandez-Pello et al[13] show for PMMA that V is insensitive to V_g for $V_g \leq 30$ cm/s at normal ambient oxygen levels ($Y_{ox, \infty} = 0.233$) and decreases as V_g is increased. These results are sketched in Figure 2-14.3. At higher-than-normal oxygen levels for air, the dependence of V on V_g is more complicated. Thus, the Damköhler number dependence is required to fully understand the effect of V_g on V for thick fuels. One consolation, at least for PMMA, is that under normal fire conditions (e.g., downward spread on a wall), the buoyancy-induced flow speed is roughly 30 cm/s. For this air speed or less, V is primarily dependent on $Y_{ox, \infty}$ only.

In previous studies, the opposed flow has been laminar. The effects of turbulence have been investigated by Zhou and Fernandez-Pello[15] who found that the spread rate decreased in air flows as the opposed turbulent intensity increased for thin paper, but had a maximum for thick PMMA at approximately 0.5 m/s.

A complete numerical simulation, including the effects of solid- and gas-phase kinetics, has been attempted by Di Blasi st al[16] with qualitative success. Such models will provide useful insight into the fundamental mechanisms, provided these effects can be computationally resolved.

A practical test procedure has been developed by Quintiere and Harkleroad[17] that allows one to determine the essential parameters needed to describe opposed flow flame spread on thick materials burning in air under natural convection conditions. As done for the thin case in Equation 11, Equation 17 is rewritten as follows

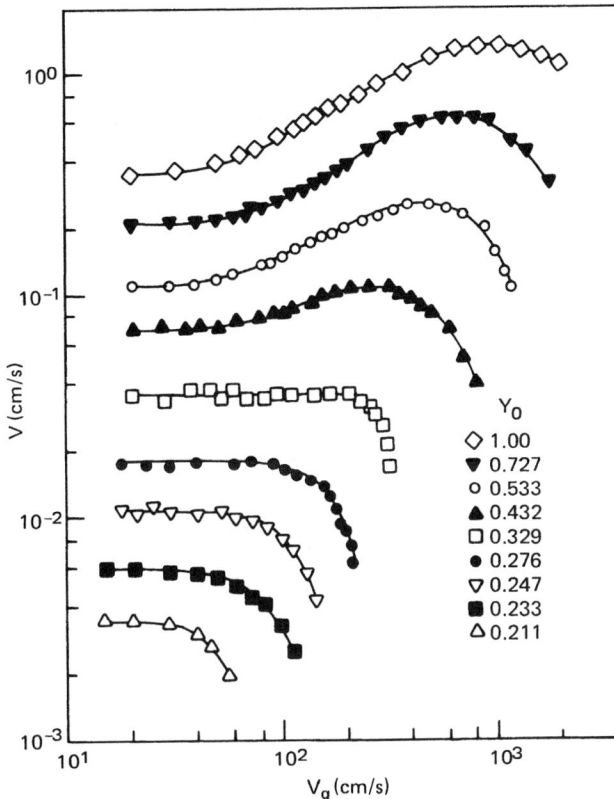

Fig. 2-14.3. *Effect of opposed velocity and oxygen concentration of flame spread speed for thick PMMA. (Taken from Fernandez-Pello et al.)[13]*

$$V = \frac{\Phi}{k\rho c(T_{ig} - T_s)^2} \qquad (18)$$

where Φ depends on V_g and $Y_{ox, \infty}$, in general, for a given material. By measuring the flame spread laterally on a vertical sample exposed to an external radiant heat flux, the parameters Φ, $k\rho c$, and T_{ig} are deduced by analysis. A wide range of tests indicated that T_{ig} ranged from 280°C for PMMA to 570°C for the paper on gypsum wallboard. The parameter, Φ, for these conditions in air ranged from about 1 to 15 $(kW)^2/m^3$. Also, a critical surface temperature was determined below which continued spread was not possible $(T_{s, min})$. Table 2-14.2 illustrates these deduced parameters for a range of materials. Results obtained in this lateral mode were shown to be similar to downward spread, except for materials with excessive melting and dripping,[17] and were shown to be similar to axisymmetric spread from a small (pool) fire on a horizontal surface.[18] Other data for wood products are given by Janssens.[5]

In these analyses, the ignition temperature has been assumed constant for each material, and its name implies a relationship between ignition and flame spread. It is interesting to illustrate this relationship by showing the behavior of spread velocity and time to ignite as a function of radiant heating. Consider downward or lateral spread on a vertical surface at various levels of irradiation, $\dot{q}_e''$. For each $\dot{q}_e''$ the surface is allowed to heat to a steady temperature, T_s, before flame spread is initiated. Thus, there is a unique T_s for a given $\dot{q}_e''$. Consider, also, piloted ignition of this same material under sufficiently high values of $\dot{q}_e''$. Such experiments

have been done for a wide range of materials[17] and the results all have the same characteristics. These trends are shown for a 1.27-cm-thick Douglas fir particleboard in Figure 2-14.4. It might be termed a "flammability diagram" for the material, since it shows at a glance its ease of ignition and its propensity for spread. Moreover, the minimum irradiance for ignition, $\dot{q}_e''(T_{ig})$, marks a critical condition for spread; and T_{ig} can be inferred from a knowledge of $\dot{q}_e''(T_s)$. This procedure has been adopted in the standard test known as LIFT (Lateral Ignition and Flamespread Test).[2]

Wind-Aided Spread

Flame spread in the same direction as the ambient flow is much different from opposed flow spread. This classification of spread could result from an external wind or the buoyancy-induced flow as a flame spreads up a wall or under a ceiling. Wind-aided spread can be acceleratory and, therefore, appears more rapid than opposed flow spread. Despite their distinct differences, the previous simplified analyses can still be used to explain wind-aided spread.

The illustration of Figure 2-14.1 can be used here with the control volume of interest selected downstream of the pyrolysis zone. This is shown in Figure 2-14.5, where the length of the control volume, Δ, is selected to span the heat transfer from the adjacent flame and hot combustion products in the boundary layer. Here, Δ is of the order of the flame length and could be in the range of 0.1 to 10 m as compared

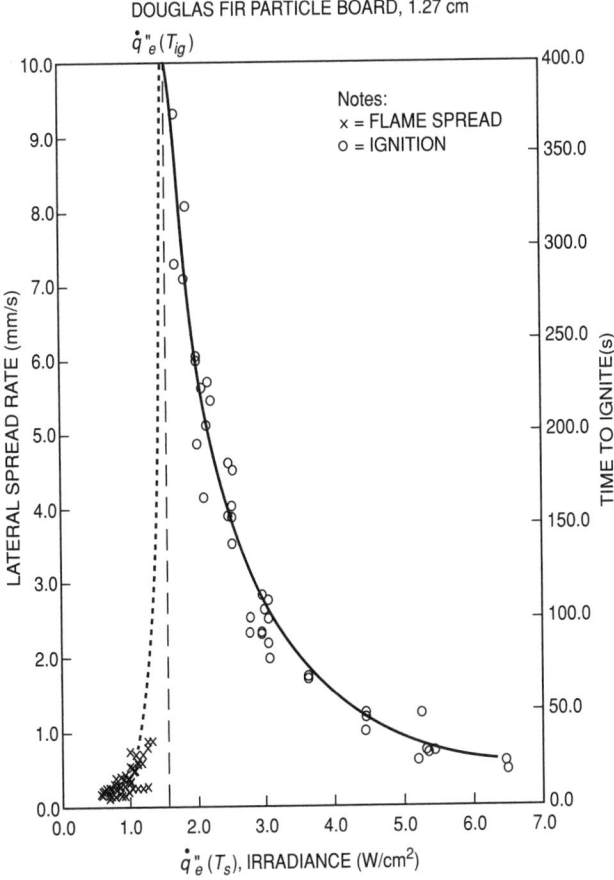

Fig. 2-14.4. *Relationship between piloted ignition and flame spread velocity.[17]*

TABLE 2-14.2   Effective Flame Spread Properties (Based on experimental correlations[17])

Material	T_{ig} (°C)	$k\rho c$ (kW²s/m⁴K²)	Φ (kW²/m³)	$T_{s,min}$ (°C)	$\Phi/k\rho c$ (mK²/s)
PMMA Polycast (1.59 mm)	278	0.73	5.4	120	8
Polyurethane (S353M)	280	—	—	105	82
Hardboard (6.35 mm)	298	1.87	4.5	170	2
Carpet (acrylic)	300	0.42	9.9	165	24
Fiberboard, low density (S119M)	330	—	—	90	42
Fiber Insulation Board	355	0.46	2.2	210	5
Hardboard (3.175 mm)	365	0.88	10.9	40	12
Hardboard (S159M)	372	—	—	80	18
PMMA Type G (1.27 cm)	378	1.02	14.4	90	14
Asphalt Shingle	378	0.70	5.3	140	8
Douglas Fir Particle Board (1.27 cm)	382	0.94	12.7	210	14
Wood Panel (S178M)	385	—	—	155	43
Plywood, Plain (1.27 cm)	390	0.54	12.9	120	24
Chipboard (S118M)	390	—	—	180	11
Plywood, Plain (0.635 cm)	390	0.46	7.4	170	16
Foam, Flexible (2.54 cm)	390	0.32	11.7	120	37
GRP (2.24 mm)	390	0.32	9.9	80	31
Mineral Wool, Textile Paper (S160M)	400	—	—	105	34
Hardboard (gloss paint) (3.4 mm)	400	1.22	3.5	320	3
Hardboard (nitrocellulose paint)	400	0.79	9.8	180	12
GRP (1.14 mm)	400	0.72	4.2	365	6
Particle Board (1.27 cm stock)	412	0.93	4.2	275	5
Gypsum Board, Wallpaper (S142M)	412	0.57	0.79	240	1
Carpet (Nylon/Wool Blend)	412	0.68	11.1	265	16
Carpet #2 (Wool, untreated)	435	0.25	7.3	335	30
Foam, Rigid (2.54 cm)	435	0.03	4.0	215	141
Polyisocyanurate (5.08 cm)	445	0.02	4.9	275	201
Fiberglass Shingle	445	0.50	9.0	415	18
Carpet #2 (Wool, treated)	455	0.24	0.8	365	4
Carpet #1 (Wool, stock)	465	0.11	1.8	450	17
Aircraft Panel Epoxy Fiberite	505	0.24	*	505	*
Gypsum Board, FR (1.27 cm)	510	0.40	9.2	300	23
Polycarbonate (1.52 mm)	528	1.16	14.7	455	13
Gypsum Board, (common) (1.27 mm)	565	0.45	14.4	425	32
Plywood, FR (1.27 cm)	620	0.76	*	620	*
Polystyrene (5.08 cm)	630	0.38	*	630	*

* Flame spread was not measureable.

— Data were not taken

Note: Values are only significant to two places

to Δ in opposed flow flame spread in the range of 1 to 3 mm. The control volume is fixed to the leading edge of the pyrolysis zone where the temperature has achieved the ignition temperature, T_{ig}. A distance Δ away, the solid is still at its initial temperature, T_s, and moves through the control volume at the flame spread speed V. An energy balance applied to this control volume will lead to the same equations as developed in the opposed flow analyses, but with a different interpretation for the terms $\dot{q}''$ and Δ. From Equations 3 and 14, those results are given as follows for the *thermally thin case*

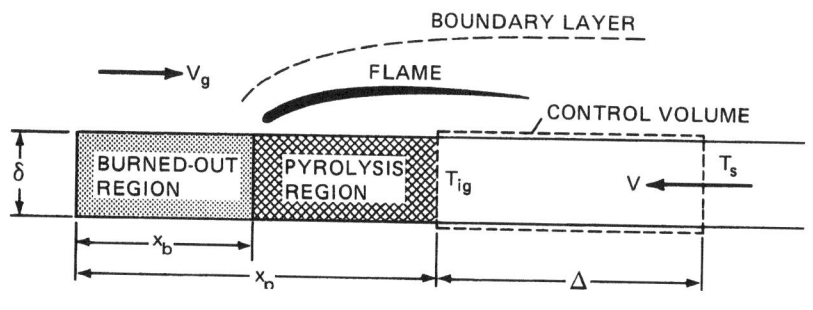

Fig. 2-14.5.   Energy conservation in wind-aided flame spread.

$$V = \frac{dx_p}{dt} = \frac{\dot{q}''\Delta}{\rho c \delta (T_{ig} - T_s)} \tag{19}$$

and for the *thermally thick case*

$$V = \frac{dx_p}{dt} = \frac{(\dot{q}'')^2 \Delta}{k \rho c (T_{ig} - T_s)^2} \tag{20}$$

These results should be taken as qualitatively illustrative, especially since $\dot{q}''$ has been assumed constant over Δ. Actually the heat flux distribution will vary, in general, with the downstream coordinate (x) in accordance with the convective boundary layer and flame radiative characteristics. Also, Δ would be intimately related to the nature of the heat flux distribution.

Alternatively, Equations 19 and 20 can be written as

$$V = \frac{dx_p}{dt} = \frac{x_f - x_p}{\tau} \tag{21}$$

where τ is an ignition time associated with the flame heat flux. Wichman and Agrawal[19] more carefully show the basis of this formulation for wind-aided spread. Here, x_f is the flame length measured from $x = 0$, and the flame heat flux is assumed constant over this region Δ. Saito et al[20] use this formulation to examine upward (or, in general, wind-aided) spread. From the definition of x_p,

$$x_p = x_p(0) + \int_0^t V(s)ds \tag{22}$$

It is postulated that x_f is directly related to energy released per unit width, $\dot{Q}'$, so that

$$x_f - x_b = K[\dot{Q}']^n \tag{23a}$$

where

$$\dot{Q}' = \int_{x_b}^{x_p} \dot{Q}''(\xi)d\xi \tag{23b}$$

with $\dot{Q}''$ the energy release per unit area. It should be noted that the cone calorimeter[1] potentially can provide material data to derive $\dot{Q}''$. However, this depends on the heat flux appropriate to the flame spread configuraton, not necessarily on the cone test irradiance. For example, for upward spread on a flat wall and under turbulent conditions, $K = 0.067 \text{ m}^{5/3} \cdot \text{kW}^{-2/3}$ and $n = \frac{2}{3}$.[21] It can be shown[20] that

$$\dot{Q}' = x_p(0)\dot{Q}''(t) + \int_0^t \dot{Q}''(t - s)V(s)ds \tag{24}$$

for the case of $x_b = 0$. Substituting Equations 22 through 24 yields an integral equation for V. This formulation gives a framework for solving particular wind-aided spread problems, provided that the flame heat flux and flame length can be expressed for that configuraton. Also, material data are needed to express τ and $\dot{Q}''$. Variations on this formulation have been applied successfully to predict upward turbulent flame spread.[22–25]

Fundamental studies have given us some limited results. A simpler result for V can be derived. The heat flux, $\dot{q}''$, could be reasoned to depend on the flame temperature, T_f, (Equation 7) and on the boundary layer thickness. The extent of heating, Δ, is related to flame length, and this has been shown to depend on the energy release rate to some power n (e.g., n is believed to be $\frac{1}{2}$ to 1 in upward turbulent spread). Since the energy release rate depends on the extent of the pyrolyzing region, $(x_p - x_b)$, then it suggests that

$$V = \frac{dx_p}{dt} \propto (x_p - x_b)^n \tag{25}$$

provided $\dot{q}''$ is not sensitive to position. Thus, depending on n and the relationships governing burnout, i.e., $x_b(t)$, the flame speed can accelerate to a limit or without bound if $n > 0$. Markstein and deRis[26] found for thin textiles that n varied from 0.5 to 0.7, with $x_p - x_b$ approaching an asymptotic steady-state limit after some time. The nature of their experiments was upward turbulent burning conditions. Under laminar upward spread, Fernandez-Pello[27] derived $n = \frac{3}{4}$ for the thermally thin case, $n = \frac{1}{2}$ for the thermally thick case, and $x_b \equiv 0$. Orloff et al[28] found that upward turbulent spread on thick PMMA followed Equation 25, with $n = 0.964$ and $x_b \equiv 0$. After spread extended over approximately 1 m of the PMMA, the speed was measured at roughly 0.5 cm/s. Note, for $n = 1$ and $x_b = 0$, the speed grows exponentially in time. This marks the potential hazard of wind-aided spread. It should be clear that the outcome depends on flame length which is controlled by configuration and flow conditions, and on burnout which is controlled by the material.

There are several notable wind-aided studies. Atreya and Mekki[29] have studied laminar flame spread on a ceiling-mounted sample. Di Blasi et al[30] mathematically modelled this spread problem, including unsteady and kinetic effects. They found that the spread speed reaches steady state at a given flow velocity, and also kinetic effects were only important at extinction. A turbulent flow study by Zhou and Fernandez-Pello[31] for ceiling- and floor-mounted PMMA (0.3 m in length) found flow speeds of 0.25 to 4.5 m/s with turbulent intensities 1 percent to 15 percent. They also found that the spread rate is steady at a given flow, but increases with flow speed. For their range of flow conditions, they found that Equation 20 agrees with a numerical coefficient of $1.4(4/\pi)$. At larger scales, radiation and buoyancy effects become important, especially for floor-mounted materials. Studies by Apte et al[32] for 2.4-m-long PMMA and by Perzak and Lazzara[33] for 9.8-m-long PMMA at flow speeds of 0.8 to 3.8 m/s show higher spread rates due to flame radiation effects. They measure rates of 1 to 6 cm/s, compared to approximately 0.01 to 0.3 cm/s for smaller samples.[31]

The ASTM E-84 Steiner Tunnel Test is the primary flammability test used in the United States. It represents a complex form of wind-aided spread on a ceiling-mounted material in a duct under forced flow conditions. Although that test is not capable of measuring x_p (nor x_b), a recent correlation showed its sensitivity to energy release rate.[34] It was found that tunnel ratings were low and insensitive to energy release rate per unit area, $\dot{Q}''$, for values less than approximately 70 kW/m^2 and highly sensitive and increasing above that value. In other words, some critical energy release rate appears to be required to permit sustained spread. It is likely that the ASTM E-84 Steiner Tunnel Test can be successfully modelled using Equation 20, provided tunnel relationships are developed for $\dot{q}''$ and Δ.

For upward turbulent flame spread on a vertical surface, it has been found that the flame heat flux ahead of the pyrolysis zone is approximately 25 kW/m^2 for moderate fires ($x_f < 1.4$ m), and approximately independent of material. This is shown in Figure 2-14.6 where the sketched region represents the data of Quintiere[3] and Quintiere et al.[35] It also displays the laminar and turbulent results of Ahmed and Faeth[36] without radiation effects, the large-scale PMMA results of Orloff et al,[37] and power-law correlations of the data by Hasemi.[38] It is seen that both laminar and large fires

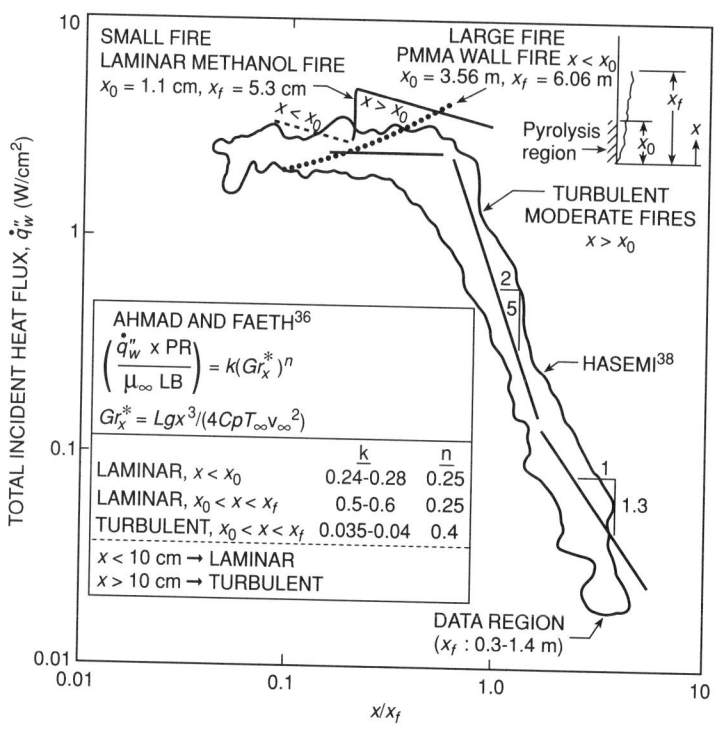

Fig. 2-14.6. *Wall flame incident heat flux for materials,[35] for laminar flames,[36] and for a large PMMA wall fire.[37]*

can exceed 25 kW/m². More recent data by Kulkarni *et al*[24] generally confirm this heat flux distribution, with some exceptions, possibly due to measurement difficulties for melting materials. Figure 2-14.7 shows results of a study by Ito and Kashiwagi[39] using interferometry to examine the temperature distributions at the leading edge of spreading flames on PMMA at orientations ranging from +90 degrees (upward vertical) to −90 degrees (downward vertical). Note the constant heat flux of approximately 25 kW/m² for upward spread, and a maximum of 70 kW/m² for downward spread. A notable orientation effect was found in the London King's Crossing fire involving spread up a wooden escalator. The sidewalls of the escalator caused the flames to "hug" the steps.[40]

The room-corner test has been developed as a realistic testing protocol for the flammability of interior finish materials. Several investigators have had good success at predicting the results of this configuration using data from the cone calorimeter test and the flame spread equations discussed above.[42–45] It appears that accurate simulation models for fire growth can be developed using the flame spread theories outlined here along with relevant data.

FLAME SPREAD OVER LIQUIDS

Flame spread over horizontal pools of liquid fuels differs from flame spread over solids, due to convective flows within the liquid that enhance spread. Also, whereas solid decomposition is more complex, involving kinetically driven pyrolysis, the liquid fuel evaporates under thermodynamic principles. The liquid temperature (T_s) controls the rate of spread relative to a critical ("ignition") temperature that corresponds to that necessary to produce a lean flammability limit over the fuel surface. This temperature is

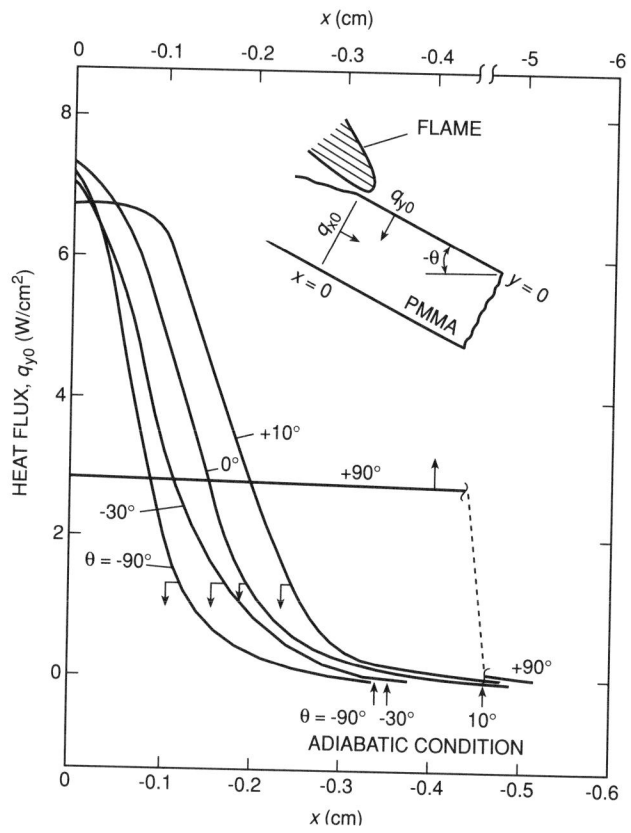

Fig. 2-14.7. *Distributions of normal heat flux at the surface along the distance ahead of the vaporization front, from Ito and Kashiwagi.[39]*

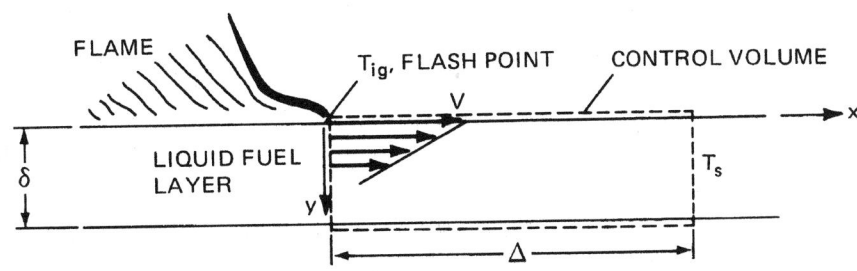

Fig. 2-14.8. **Enhanced flame spread speed in liquids due to surface-tension induced flow.**

usually referred to as the flash point for liquid fuels, and several methods exist for its measurement. Glassman and Dryer[7] discuss the implications of its measurement and its relationship to mechanisms associated with flame spread on liquids.

The convective phenomenon generated in the liquid is due to surface tension effects. The surface tension increases inversely with temperature and, thus, can "pull" the flame toward the unheated liquid. This is illustrated in Figure 2-14.8 for a thin liquid layer, δ. Under steady conditions, the viscous forces on the control volume are balanced by the surface tension forces. Thus, the shear stress, τ, at the bottom surface equals the surface tension gradient ($d\sigma/dx$) along the free surface

$$\tau = \frac{d\sigma}{dx} = \left(\frac{d\sigma}{dT}\right)\left(\frac{\partial T}{\partial x}\right) \qquad (26)$$

For a thin liquid layer, the surface tension effect results in nearly a Couette flow (constant shear) over the layer thickness, δ. Hence, it can be approximated that

$$\tau = \left(\mu\frac{\partial u}{\partial y}\right)_{y=0} \approx \mu\frac{V}{\delta} \qquad (27)$$

where μ is the liquid viscosity. By further approximating the surface tension gradient as a difference over length Δ, the flame speed can be estimated as

$$V = \frac{[\sigma(T_s) - \sigma(T_{ig})]\delta}{\mu\Delta} \qquad (28)$$

provided $\sigma(T)$, the surface tension, is known as a function of temperature for the liquid, and Δ can be estimated for the conditions of spread. Also, δ, as in the thermally thick case for solids, is only the physical liquid depth for pools less than about 1 mm, and, therefore, must be reinterpreted for pools of larger depth.[9] For example, one might estimate δ as

$$\sqrt{\left(\frac{\mu}{\rho}\right)\left(\frac{\Delta}{V}\right)}$$

for the deep-pool case.

Typical flame spread characteristics over a liquid fuel are sketched in Figure 2-14.9 for liquid methanol from the data of Akita.[46] Below the flash point, $T_s < T_{ig} \approx 11°C$, the spread is governed by transport phenomena within the liquid fuel. For initial liquid bulk temperatures above the flash point, a flammable mixture always exists everywhere above the surface so that propagation is governed by gas phase effects. Above a liquid temperature, which corresponds to stoichiometric conditions above the surface, the flame speed remains constant and usually above the normal premixed laminar flame speed. A recent study by Ito *et al*[47] used holographic interferometry to examine the liquid phase for sub-flashpoint liquid bulk temperatures. They examined the pulsating region depicted in Figure 2-14.9 and the adjacent "uniform" region of spread just below the flashpoint. They found that both a surface tension flow and a gravity-induced circulation flow in the liquid below the flame are present; both appear to contribute to flame spread rate in the uniform region.

FLAME SPREAD IN FORESTS

Flame spread in a forest has many characteristics related to fuel type and configuration, terrain, wind conditions, and humidity. Much research has been done, and no effort here will be made to describe that work. In recent years, based on its research, the Forest Service of the U.S. Department of Agriculture has implemented practical methods for estimating flame spread in forests under a variety of field conditions.[48]

Rothermel[49] presents a thorough review of the theory and models available for forest fire danger rating and behavior prediction. He also discusses a computer model (BE-HAVE) that is based on the instructional format previously presented.[48]

Flame spread in a forest depends on radiant heat transfer and convective heating due to wind or the slope of the terrain. In most cases, the porous brush along the floor of the

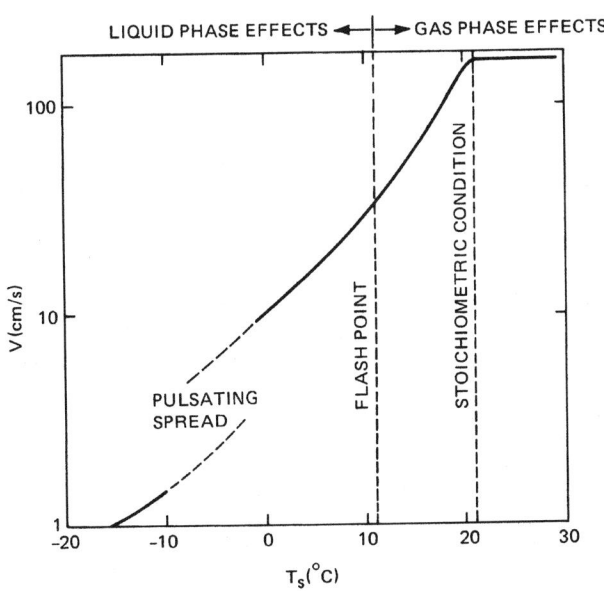

Fig. 2-14.9. **Results sketched for flame spread over liquid methanol, based on Akita.[3]**

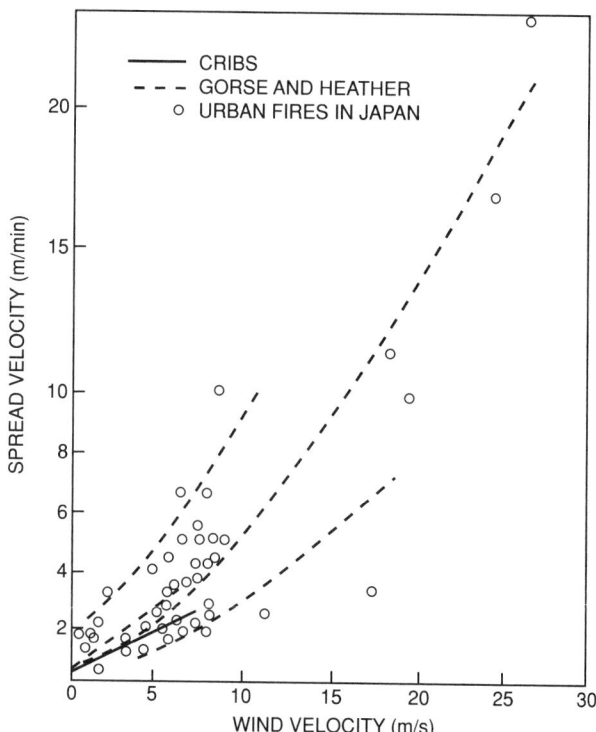

Fig. 2-14.10. *Comparison of rates of fire spread for urban and wildland fires as a function of wind speed.*[51]

forest is involved, but for severe fires the crowns of the trees also, or exclusively, become involved. In addition, wind currents can transport large embers (fire brands) great distances from the fire to start spot fires at other locations. Breaks in the fuel array can be used to control the spread, but the size of a break sufficient to interrupt the fire will depend on the size of the fire and its heat transfer characteristics. Albini[50] gives an excellent discussion of forest fire phenomena in conceptual terms. Many of these cases have been addressed, and the Forest Service and others have developed analytical models and data for their treatment.

An approximate formula applicable to wildland fires, wood cribs, and urban conflagration is given by Thomas[51] as

$$V\rho_b = k(1 + V_\infty)$$

where ρ_b is the bulk density of the fuel (kg/m³),
  V is the wind speed (concurrent) (m/s), and
  k = 0.07 for wildland fires (kg/m³)
    0.05 for wood cribs,
    0.046 for the Great Fire of London (V = 4.5 m/s)

Figure 2-14.10 shows the results for urban fires in Japan in relationship to this approximate formula.

FLAME SPREAD IN MICROGRAVITY

Because of an increased activity in space, flame spread research has been conducted under microgravity conditions. This has been done in drop towers and in the space shuttle Discovery, as reported by Bhattacharjee and Altenkirch.[52] They examined flame spread over this filter paper at 50 percent oxygen in nitrogen in a quiescent atmo-

sphere. The spread rate was measured at 0.44 cm/s compared to a predicted value of 0.42 cm/s. Opposed flow theory applies, but the inclusion of gas-phase radiation was an essential mechanism. A study of flame spread over liquid fuels was conducted by Miller and Ross.[53] They found similar results between microgravity and 1-g conditions, except that opposed flow appears to lower the limiting oxygen concentration at extinction for microgravity. Forced-flow wind-aided flame spread would not be expected to be significantly affected by gravity.

CONCLUDING REMARKS

This chapter has provided some insight into the nature of fire spread over materials for the practicing engineer. In general, flame spread depends on the heat transfer processes at the flame front. These transport processes depend not only on the fuel, but also the fuel's configuration and orientation, and on ambient environment conditions. Thus, estimates of flame spread require complex analysis and specific material data. The current state of knowledge does provide limited formulas and material data to make some estimates. In this chapter, the full scope of flame spread phenomena has not been addressed. For example, flame spread in enclosures, mines, ducts, and buildings presents an entirely new complex array of conditions. Thus, flame spread on materials must be evaluated in the context of their use, and appropriate data must be made available for proper assessments of materials.

NOMENCLATURE

c	specific heat
δ	fuel thickness
D	Damköhler number, Equation 8
Δh	enthalpy change
ΔH	heat of combustion
ΔH_{ox}	$\Delta H/r$
k	conductivity
μ	viscosity
ρ	density
Φ	numerator in Equation 18
ϕ	numerator in Equation 11
$\dot{q}''$	heat flux
Q	energy release
r	reference
r	stoichiometric mass ratio oxygen/fuel
t	time
T	temperature
V	spread velocity
V_g	gas velocity
x,y	coordinates
Y	mass fraction
Δ	length of flame heating
ν	kinematic viscosity, μ/ρ
σ	surface tension

Superscripts

˙	per unit time
′	per unit length
″	per unit area

Subscripts

b	burnout
f	flame
g	gas phase
i	ignition
ox	oxygen
p	pyrolysis
r	reference
s	surface
∞	ambient

REFERENCES CITED

1. ASTM E-1354-90, *Standard Test Method for Heat and Visible Smoke Release Rates for Materials and Products Using an Oxygen Consumption Calorimeter*, American Society for Testing and Materials, Philadelphia, (Aug. 1990).
2. ASTM E-1321-90, *Standard Method for Determining Material Ignition and Flame Spread Properties*, American Society for Testing and Materials, Philadelphia (May 1990).
3. J.G. Quintiere, "Application of Flame Spread Theory to Predict Material Performance," *Proceedings* Journal R8700433 QD381.9.C6 1987, *Journal of Research of the National Bureau of Standards*, Vol. 93, No. 1, pp. 61–70 (Jan./Feb. 1988) National Bureau of Standards, Gaithersburg, MD.
4. T. Hirano, "Physical Aspects of Combustion in Fires," Tokyo Univ., Japan, *Proceedings* R9200255 TH9112.F5626 1991, International Association for Fire Safety Science, *Proceedings* 3rd International Symposium, July 8–12, 1991, Edinburgh, Scotland, Elsevier Applied Science, New York, G. Cox, and B. Langford, ed., pp. 27–44 (1991).
5. M.L. Janssens, "Fundamental Thermophysical Characteristics of Wood and Their Role in Enclosure Fire Growth, National Forest Product Assoc., Washington, DC Thesis, p. 492 (Sept. 1991).
6. A.C. Fernandez-Pello and T. Hirano, *Comb. Sci. and Tech.*, 32, 1 (1983).
7. I. Glassman and F.L. Dryer, *F. Safety J.*, 3, 123 (1980).
8. D.D. Drysdale, *An Introduction to Fire Dynamics*, John Wiley and Sons, New York (1985).
9. F.A. Williams, 16th Symp. (Int) Comb., Pittsburgh (1976).
10. J.N. deRis, 12th Symp. (Int) Comb., Pittsburgh (1969).
11. R.A. Altenkirch, R. Eichhorn, and P.C. Shang, *Comb. and Flame*, 37, 71 (1980).
12. R.S. Magee and R.F. McAlevy III, *J. Fire and Flamm.*, 2, 271 (1971).
13. A.C. Fernandez-Pello, S.R. Ray, and I. Glassman, 18th Symp. (Int) Comb., Pittsburgh, (1981).
14. I.S. Wichman, *Comb. Sci. and Tech.*, 40, 223 (1984).
15. L. Zhou, A.C. Fernandez-Pello, and R. Cheng, "Flame Spread in an Opposed Turbulent Flow," Univ. California, Berkeley, Lawrence Livermore Lab., CA, *Journal Combustion and Flame*, Vol. 81, No. 1, 40-495, (July 1990).
16. C. DiBlasi, G. Continillo, S. Crescitelli, and G. Russo, "Numerical Simulation of Opposed Flow Flame Spread Over a Thermally Thick Solid Fuel," Universita di Napoli, Italy Istituto di Ricerche sulla Combustione-CNR, Napoli, Italy, *Journal Combustion Science and Technology*, Vol. 54, No. 1-6, 25–36 (1987).
17. J.G. Quintiere and M. Harkleroad, in *Fire Safety Science and Engineering*, American Society for Testing and Materials, Philadelphia (1985).
18. A. Atreya, C. Carpentier, and M. Harkleroad, in *Fire Safety Science, Proceedings of the First International Symposium on Fire Safety Science*, Hemisphere, New York (1984).
19. I.S. Wichman, and S. Agrawal, "Wind-Aided Flame Spread Over Thick Solids," Michigan State Univ., East Lansing, MI, *Journal Combustion and Flame*, Vol. 83, No. 1-2, 127–145, (Jan. 1991).
20. K. Saito, J.G. Quintiere, and F.A. Williams, "Upward Turbulent Flame Spread," Princeton Univ., N.J., National Bureau of Standards, Gaithersburg, MD, *Proceedings* R8601192 TH9112.F5626 1986, International Association for Fire Safety Science, *Fire*

21. K.-M. Tu and J.G. Quintiere, "Wall Flame Heights with External Radiation," *Fire Technol.*, 27 (3) 195–203 (1991).
22. Y. Hasemi, M. Yoshida, A. Nohara, and T. Nakabayashi, "Unsteady-State Upward Flame Spreading Velocity along Vertical Combustible Solid and Influence of External Radiation on the Flame Spread," Building Research Inst., Tokyo, Japan Science University of Tokyo, Japan, *Proceedings* R9200255 TH9112.F5626 1991, International Association for Fire Safety Science, *Fire Safety Science, Proceedings*, 3rd International Symposium, July 8–12, 1991, Edinburgh, Scotland, Elsevier Applied Science, New York, G. Cox and B. Langford, ed. pp. 197–206 (1991).
23. M.M. Delichatsios, M.K. Mathews, and M.A. Delichatsios, "Upward Fire Spread and Growth Simulation," Factory Mutual Research Corp., Norwood, MA, *Proceedings* R9200255 TH9112.F5626 1991, International Association for Fire Safety Science, *Fire Safety Science, Proceedings*, 3rd International Symposium, Jul. 8–12, 1991 Edinburgh, Scotland, Elsevier Appl. Science, New York, G. Cox and B. Langford, ed., pp. 207–216 (1991).
24. A.K. Kulkarni, C.I. Kim, and C.H. Kuo, "Turbulent Upward Flame Spread for Burning Vertical Walls Made of Finite Thickness," NIST-GCR-91-597, National Institute of Standards and Technology, Gaithersburg, MD, (May 1991).
25. D. Baroudi, and M. Kokkala, "Analysis of Upward Flame Spread," Project 5 of the EUREFIC Fire Research Program, VTT-Technical Research Center of Finland, Espoo Report, VTT Publications 89, p. 50 (Feb. 1992).
26. G.H. Markstein and J. deRis, 14th Symp. (Int) Comb., Pittsburgh, (1973).
27. A.C. Fernandez-Pello, *Comb. and Flame*, 31, 135 (1978).
28. L. Orloff, J. deRis, and G.H. Markstein, 15th Symp. (Int) Comb., Pittsburgh (1975).
29. A. Atreya, and K. Mekki, "Heat Transfer During Wind-Aided Flame Spread on a Ceiling-Mounted Sample," Michigan State Univ., East Lansing, Department of Agriculture, Washington, DC National Science Foundation, Washington, DC, *Proceedings* R9301113 QD516.S92 1992 Combustion Institute, *Symposium (International) on Combustion*, 24th, July 5–10, 1992, Sydney, Australia, Combustion Institute, Pittsburgh, PA, pp. 1677–1684 (1992).
30. C. DiBlasi, S. Crescitelli, and G. Russo, "Near-Limit Flame Spread Over Thick Fuels in a Concurrent Forced Flow," Universita di Napoli, Naples, Italy *Journal Combustion and Flame*, Vol. 72, No. 2, pp. 205–212 (May 1988).
31. L. Zhou, and A.C. Fernandez-Pello, "Turbulent, Concurrent, Ceiling Flame Spread: The Effect of Buoyancy," UCLA, Berkeley, National Institute of Standards and Technology, Gaithersburg, MD, *Journal Combustion and Flame*, Vol. 92, No. 1–2, pp. 45–59 (1993).
32. V.B. Apte, R.W. Bilger, A.R. Green, and J.G. Quintiere, "Wind-Aided Turbulent Flame Spread and Burning Over Large-Scale Horizontal PMMA Surfaces," Sydney Univ., Australia, Londonderry Occupational Safety Center, Australia, National Institute of Standards and Technology, Gaithersburg, MD *Journal Combustion and Flame*, Vol. 85, No. 1–2, pp. 169–184 (1991).
33. F.J. Perzak, and C.P. Lazzara, "Flame Spread Over Horizontal Surfaces of Polymethylmethacrylate," Bureau of Mines, Pittsburgh, PA, *Proceedings* R9301113 QD516.S92 1992, Combustion Institute, *Symposium (International) on Combustion*, 24th, July 5–10, 1992, Sydney, Australia, Combustion Institute, Pittsburgh, PA, pp. 1661–1668 (1992).
34. J. Quintiere, *Fire and Matls.*, 9, 65 (1985).
35. J.G. Quintiere, M.F. Harkleroad, and Y. Hasemi, "Wall Flames and Implicatons for Upward Flame Spread, Final Report," National Bureau of Standards, Gaithersburg, MD, Building Research Institute, Tsukuba, Japan, *Proceedings*, *Journal Combustion, Science and Technology*, Vol. 48, No. 3 & 4, pp. 191–222 (1986).

36. T. Ahmad and G.M. Faeth, *17th Symp.* (Int.) on Combustion, The Combuston Institute, Pittsburgh, PA, pp. 1149–1160 (1979).

37. L. Orloff, A.T. Modak, and R.L. Alpert, *16th Symp.* (Int.) on Combustion, The Combustion Institute, Pittsburgh, PA, 1345 (1976).

38. Y. Hasemi, "Thermal Modeling of Upward Wall Flame Spread," in *Proc. of the 1st Int. Symp.* on Fire Safety Science, ed. C.E. Grant and P.J. Pagni, Hemisphere Publishing Corp. NY, 87 (1985).

39. A. Ito, and T. Kashiwagi, "Characterization of Flame Spread Over PMMA Using Holographic Interferometry Sample Orientation Effects," National Bureau of Standards, Gaithersburg, MD, *Proceedings Journal* R8700670, Available from National Technical Information Services Combustion and Flame, Vol. 71, pp. 189–204 (1988).

40. D.D. Drysdale, A.J.R. Macmillan, and D. Shilitto, "King's Cross Fire: Experimental Verification of the 'Trench Effect'," Edinburgh Univ., UK, Cremer and Warner Ltd., London, England, *Fire Safety Journal*, Vol. 18, No. 1, pp. 75–82 (1992).

41. ASTM, "Proposed Method for Room Fire Test of Wall and Ceiling Materials and Assemblies," in *ASTM Annual Book of Standards Pt. 18*, American Society for Testing and Materials, Philadelphia, pp. 1618–38 (1982).

42. B. Karlsson, and S.E. Magnusson, "Combustible Wall Lining Materials: Numerical Simulation of Room Fire Growth and the Outline of a Reliability Based Classification Procedure," in *Fire Safety Science, Proc. of the 3rd Int. Symp.*, ed. G. Cox & B. Langford. Elsevier Applied Science, London (1991).

43. T.G. Cleary and J.G. Quintiere, "A Framework for Utilizing Fire Property Tests," in *Fire Safety Science, Proc. of the 3rd Int. Symp.*, ed. G. Cox & B. Langford, Elsevier Applied Science, London (1991).

44. U. Wickström, and U. Göransson, "Prediction of Heat Release Rates of Surface Materials in Large-Scale Fire Tests Based on Cone Calorimeter Results," *ASTM J. Testing and Evaluation*, 15(6), pp. 364–70 (1987).

45. J.G. Quintiere, "A Simulation Model for Fire Growth on Materials Subject to a Room-Corner Test," *Fire Safety Journal* 20, pp. 313–339 (1993).

46. K. Akita, *14th Symp.* (Int.) on Combustion, The Combustion Institute, Pittsburgh PA, p. 1075 (1975).

47. A. Ito, D. Masuda, and K. Saito, "Study of Flame Spread Over Alcohols Using Holographic Interferometry," Kentucky Univ., Lexington, *Journal Combustion and Flame*, Vol. 83, No. 3–4, pp. 375–389 (1991).

48. R.C. Rothermel, *Gen. Tech. Report INT-143*, Department of Agriculture/Forest Service, Madison (1983).

49. R.C. Rothermel, "Modeling Fire Behavior," in *Int. Conf. on Forest Fire Research*, Coimbra, Portugal, Nov. 1990.

50. F.A. Albini, "Dynamics and Modeling of Vegetation Fires: Observations," in *Fire in the Environment*, eds. P.J. Crutzen and J.G. Goldhammer, March 15–20, 1992 Berlin, John Wiley and Sons, NY, 39–52 (1992).

51. P.H. Thomas, "Rates of Spread for Some Wind-Driven Fires," *Forestry*, XLIV, 2 (1971).

52. S. Bhattacharjee, and R.A. Altenkirch, "Comparison of Theoretical and Experimental Results in Flame Spread Over Thin Condensed Fuels in a Quiescent, Microgravity Environment," San Diego State Univ., CA, Mississippi State Univ., Mississippi State, MI, Combustion Institute, *Symposium* (International) on Combustion, 24th, July 5–10, 1992, Sydney, Australia, Combustion Institute, Pittsburgh, PA, pp. 1669–1676 (1992).

53. F.S. Miller and H.D. Ross, "Further Observations of Flame Spread over Laboratory-Scale Alcohol Pools," *24th Symp.* (Int.) on Combustion, The Combustion Institute, Pittsburgh PA, pp. 1703–1711 (1992).

SMOKE PRODUCTION AND PROPERTIES

George W. Mulholland

INTRODUCTION

The term "smoke" is defined in this chapter as the smoke aerosol or condensed phase component of the products of combustion. This differs from the American Society for Testing and Materials (ASTM) definition of smoke, which includes the evolved gases as well. Smoke aerosols vary widely in appearance and structure, from light colored, for droplets produced during smoldering combustion and fuel pyrolysis, to black, for solid, carbonaceous particulate or soot produced during flaming combustion. A large fraction of the radiant energy emitted from a fire results from the blackbody emission from the soot in the flame. The subject of radiant heat transfer is of such importance that it is treated in a separate chapter. This chapter focuses on smoke aerosols outside the combustion zone.

The effects of the smoke produced by a fire depend on the amount of smoke produced and on the properties of the smoke. The following section presents experimental results on smoke emission for a variety of materials. The smoke emission, together with the flow pattern, determines the smoke concentration as smoke moves throughout a building.

The most basic physical property of smoke is the size distribution of its particles. Results on size distribution for various types of smoke and techniques used for measuring particle size are presented in the section "Size Distribution." The section "Smoke Properties" focuses on those properties of greatest concern to the fire protection community: light extinction coefficient of smoke, visibility through smoke, and detectability of smoke. These properties are primarily determined by the smoke concentration and the particle size distribution. References for other smoke aerosol properties, such as diffusion coefficient and sedimentation velocity, are also provided.

SMOKE PRODUCTION

Smoke emission is one of the basic elements for characterizing a fire environment. The combustion conditions under which smoke is produced—flaming, pyrolysis, and smoldering—affect the amount and character of the smoke. The smoke emission from a flame represents a balance between growth processes in the fuel-rich portion of the flame and burnout with oxygen. While it is not possible at the present time to predict the smoke emission as a function of fuel chemistry and combustion conditions, it is known that an aromatic polymer, such as polystyrene, produces more smoke than hydrocarbons with single carbon-carbon bonds, such as polypropylene. The smoke produced in flaming combustion tends to have a large content of elemental (graphitic) carbon.

Pyrolysis occurs at a fuel surface as a result of an elevated temperature; this may be due to a radiant flux heating the surface. The temperature of a pyrolyzing sample, 600 to 900 K, is much less than the gas phase flame temperature, 1200 to 1700 K. The vapor evolving from the surface may include fuel monomer, partially oxidized products, and polymer chains. As the vapor rises, the low vapor pressure constituents can condense, forming smoke droplets appearing as light-colored smoke.

Smoldering combustion also produces smoke droplets, but in this case the combustion is self-sustaining, whereas pyrolysis requires an external heat source. While most materials can be pyrolyzed, only a few materials, including cellulosic materials (wood, paper, cardboard, etc.) and flexible polyurethane foam, are able to smolder. The temperature during smoldering is typically 600 to 1100 K.

In Table 2-15.1 the smoke conversion factor, ε, is given for a variety of materials commonly found in buildings. The quantity ε is defined as the mass of smoke produced/mass of fuel burned.

The references cited in Table 2-15.1 should be consulted regarding the detailed description of the combustion conditions. In many instances,[1,3] ε was measured for a range of radiant fluxes, oxygen concentrations, sample orientations, and ambient temperatures. It is seen in Table 2-15.1 that ε has a greater range for flaming combustion, with values in the range 0.001 to 0.17, compared to pyrolysis and smoldering, with values in the range 0.01 to 0.17. The following factors should be taken into account when using this table for smoke emission estimates:

1. Most of the measurements reported in Table 2-15.1 were made on small-scale samples.

Dr. George W. Mulholland is Head of the Smoke Dynamics Research Group in the Center for Fire Research at the National Bureau of Standards. His research has focused on smoke aerosol phenomena and the development of accurate particle size standards.

TABLE 2-15.1 *Smoke Production for Wood and Plastics*

Type	Smoke Conversion Factor, ε	Combustion Conditions	Fuel Area, m²	Ref. No.
Douglas fir	0.03–0.17	pyrolysis	0.005	1
Douglas fir	<0.01–0.025	flaming	0.005	1
hardboard	0.0004–0.001	flaming*	0.0005	2
fiberboard	0.005–0.01	flaming*	0.0005	2
polyvinylchloride	0.03–0.12	pyrolysis	0.005	3
polyvinylchloride	0.12	flaming	0.005	1
polyurethane (flexible)	0.07–0.15	pyrolysis	0.005	3
polyurethane (flexible)	<0.01–0.035	flaming	0.005	1
polyurethane (rigid)	0.06–0.19	pyrolysis	0.005	1
polyurethane (rigid)	0.09	flaming	0.005	1
polystyrene	0.17 (m_{O_2} = 0.30)**	flaming	0.0005	4
polystyrene	0.15 (m_{O_2} = 0.23)	flaming	0.07	5
polypropylene	0.12	pyrolysis	0.005	1
polypropylene	0.016	flaming	0.005	1
polypropylene	0.08 (m_{O_2} = 0.23)	flaming	0.007	5
polypropylene	0.10 (m_{O_2} = 0.23)	flaming	0.07	5
polymethylmethacrylate	0.02 (m_{O_2} = 0.23)	flaming	0.07	5
polyoxymethylene	~0	flaming	0.007	5
cellulosic insulation	0.01–0.12	smoldering	0.02	6

*Sample smoldered for a period of time after the pilot flame was extinguished.
**m_{O_2} refers to mol fraction of O_2.

2. Most experiments were for free burning at ambient conditions; reduced ventilation can strongly affect the smoke production.
3. In transport, the smoke may coagulate, partially evaporate, and deposit on surfaces through diffusion and sedimentation. Also, additional smoke may be formed through condensation.

SIZE DISTRIBUTION

Smoke particle size distribution, together with the amount of smoke produced, primarily determines the properties of the smoke. A widely used representation of the size distribution is the geometric number distribution, $\Delta N/\Delta \log d$, versus $\log d$, where d represents the particle diameter. The quantity ΔN represents the number of particles per cm³, with diameter between $\log d$ and $\log d + \Delta \log d$. As an example, the particle size distribution of smoke produced by a smoldering incense stick is plotted in Figure 2-15.1, where $\Delta \log d$ for each discrete size range equals 0.25. In this case, the total number concentration for a given size range equals $0.25(\Delta N/\Delta \log d)$. It is seen that the logarithmic scale is necessitated by the wide range in particle size and concentration.

For many applications, the most important characteristics of a size distribution are the average particle size and the width of the distribution. A widely used measure of the average size is the geometric mean number diameter, d_{gn}, defined by

$$\log d_{gn} = \sum_{i=1}^{n} \frac{N_i \log d_i}{N} \qquad (1)$$

where N is the total number concentration, N_i is the number concentration in the ith interval, and log is to the base 10. For the size distribution plotted in Figure 2-15.1, $d_{gn} = 0.072$ μm.

The corresponding measure of the width of the size distribution is the geometric standard deviation, σ_g,

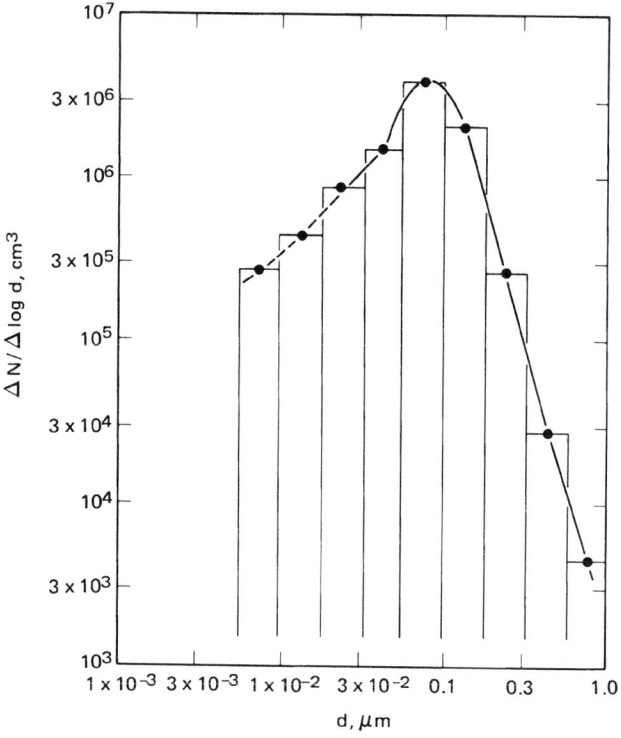

Fig. 2-15.1. Size distribution of incense smoke as measured by an electrical aerosol analyzer. There is a large uncertainty in the dashed portion of the curve.

$$\log \sigma_g = \left[\sum_{i=1}^{n} \frac{(\log d_i - \log d_{gn})^2 N_i}{N} \right]^{1/2} \quad (2)$$

For the size distribution plotted in Figure 2-15.1, $\sigma_g = 1.75$. A perfectly monodisperse distribution would correspond to $\sigma_g = 1$. The parameters d_{gn} and σ_g are useful because actual size distributions are observed to be approximately log-normal, which is the same as a normal or Gaussian distribution, except that $\log d$ is normally distributed instead of d. An important characteristic of the log-normal distribution is that 68.3 percent of the total particles are in the size range $\log d_{gn} \pm \log \sigma_g$; for $d_{gn} = 0.072$ μm and $\sigma_g = 1.75$, this corresponds to the size range of 0.041 to 0.126 μm.

EXAMPLE 1:

Compute d_{gn} and σ_g for the data given below:

Interval, μm	d_i	N_i, cm^{-3}	$\log d_i$	$N_i \times \log d_i$, cm^{-3}
0.0056–0.01	0.0078	6×10^4	−2.11	-1.27×10^5
0.010–0.018	0.014	2×10^5	−1.85	-3.7×10^5
0.018–0.032	0.025	4×10^5	−1.60	-6.40×10^5
0.032–0.056	0.044	9×10^4	−1.36	-1.22×10^5
0.056–0.10	0.078	3×10^4	−1.11	-3.33×10^4
0.10–0.18	0.14	1×10^3	−0.85	-0.85×10^3
		7.81×10^5		-1.30×10^6

SOLUTION:

$$d_{gn} = 10^{(-1.30 \times 10^6/7.81 \times 10^5)} = 0.022 \text{ μm}$$

Compute the geometric standard deviation:

N_i	$\log d_i$	$\log d_i - \log d_{gn}$	$N_i (\log d_i - \log d_{gn})^2$
6×10^4	−2.11	−0.45	1.22×10^4
2×10^5	−1.85	−0.19	7.2×10^3
4×10^5	−1.60	0.06	1.4×10^3
9×10^4	−1.36	0.30	8.1×10^3
3×10^4	−1.11	0.55	9.1×10^3
1×10^3	−0.85	0.81	6.5×10^2
7.81×10^5			3.87×10^4

SOLUTION:

$$\sigma_g = 10^{(3.87 \times 10^4/7.81 \times 10^5)^{0.5}} = 1.67$$

The size distribution plotted in Figure 2-15.1 is based on electrical mobility analysis of the smoke aerosol. Figures 2-15.2 and 2-15.3 show size distributions of droplet smoke produced by smoldering cellulosic insulation, as measured by an optical particle counter and by two cascade impactors.[6] The smoke volume distribution plotted in Figure 2-15.3 for the optical particle counter is obtained from the number distribution, using the following relation

$$V_i = N_i \frac{1}{6} \pi d_i^3 \quad (3)$$

For particles sized above 1 μm, impactors provide more reliable information on the smoke volume distribution than optical particle counters. An optical particle counter is the preferred instrument for the number distribution measurement.

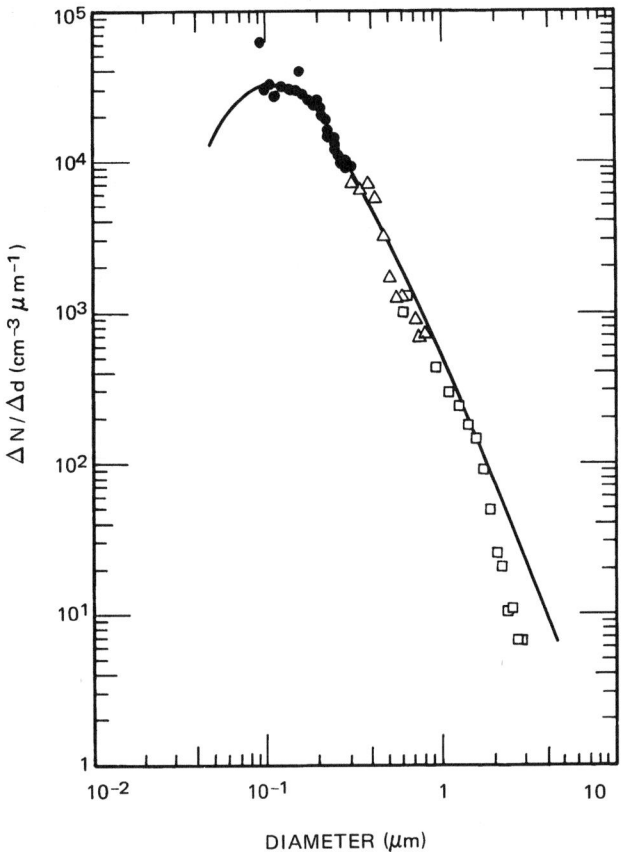

Fig. 2-15.2. *The number size distribution of smoke generated by smoldering cellulosic insulation as measured by an optical particle counter. The symbols correspond to the particle size range settings of the instrument, and the smooth curve is an exponentially truncated power law distribution fit to the data.*

To correlate the smoke volume/particle size distribution, the geometric mean volume diameter, d_{gv}, is a convenient measure of average particle size:

$$\log d_{gv} = \frac{\sum_{i=1}^{n} V_i \log d_i}{V_T} \quad (4)$$

where V_T is the total volume concentration of the smoke aerosol. For a log-normal distribution, there is the following relationship between the geometric mean volume diameter, d_{gv}, and the geometric mean number diameter, d_{gn}

$$\log d_{gv} = \log d_{gn} + 6.9(\log \sigma_g)^2 \quad (5)$$

In the case of smolder smoke, σ_g is above 2.4. This large value of σ_g results in a large difference between d_{gn} and d_{gv}, 0.2 μm versus 2 μm, respectively. Some devices, such as an ionization-type smoke detector, have an output depending primarily on d_{gn}, while others, such as light-scattering-type detectors, have an output depending more on d_{gv}. More than one instrument is necessary for a complete characterization of the smoke size distribution, because it is typically quite wide.

A list of commercially available instruments for measuring smoke aerosol concentration and particle size distribution is given in Table 2-15.2. Smoke measurements pose

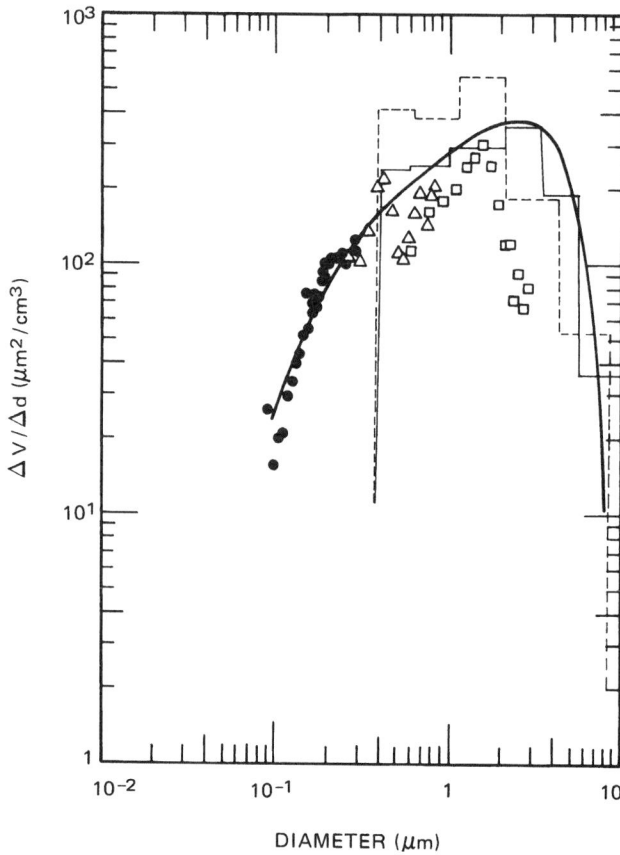

$\Delta V / \Delta d \; (\mu m^2 / cm^3)$

DIAMETER (μm)

Fig. 2-15.3. The volume size distribution of smoke obtained from the optical particle counter, quartz crystal microbalance cascade impactor (dashed histogram), and Andersen impactor (solid histogram). The smooth curve represents the exponentially truncated power law distribution.

special problems because of the high concentration, wide particle size range, and sometimes high temperature. In selecting an instrument it is important to make the following considerations:

1. Will the instrument respond to the smoke of interest? For example, the piezoelectric mass monitor does not respond well to soot.
2. Will dilution of the smoke be required?
3. Is the measurement size range of the instrument adequate?
4. Is a mass or number distribution measurement appropriate?
5. What is the particle size resolution needed?
6. Is real-time measurement capability needed?
7. Will the instrument perform at the temperature of the smoke environment?

In Table 2-15.3, average particle size and the width of the size distribution are presented for smoke generated by a variety of materials. The results are most meaningful for smoke droplets produced during pyrolyzing and smoldering combustion. In the case of flaming combustion, complex soot agglomerates are formed as shown in Figure 2-15.4. For soot agglomerates, the apparent particle size depends on the measurement technique, unlike the case for spherical smoke droplets.

Smoke aerosols are dynamic with respect to their particle size distribution function. Smoke particles or droplets undergoing Brownian motion collide and stick together. The result of this behavior is that, in a fixed volume of smoke-laden gas, the number of particles decreases while the total mass of the aerosol remains unchanged. This process is known as coagulation. The fundamental parameter for describing coagulation is the coagulation coefficient, Γ, the rate constant for the coagulation equation

$$\frac{dN}{dt} = -\Gamma N^2 \qquad (6)$$

For smoke produced from incense sticks, Γ was found to be about 4×10^{-10} cm^3/s and about 1×10^{-9} cm^3/s for smoke

TABLE 2-15.2 *Operational Characteristics of Commercially Available Instruments for Smoke Characterization*

Instrument Type	Function/Range	Advantage/Limitation for Smoke Measurements
filter-collection	mass conc.	accurate, slow
piezoelectric mass monitor	mass conc. $0.01 < d < 5$ μm	real-time output, but dilution required if >20 mg/m^3; does not respond well to soot
tapered element oscillating microbalance	mass conc. <5 μm	real time, 0.1–1000 mg/m^3; replace filter after 3–100 mg deposit
condensation nuclei counter	number conc. $0.005 < d < 2$ μm	$<3 \times 10^5$ particles/cm^3
photometer	scattered light 0.1–10 μm	1.1–1000 mg/m^3
nephelometer	total light scattered	<5 mg/m^3
electrical aerosol analyzer	size distribution $0.01 < d < 0.3$ μm	$<5 \times 10^5$ particles/cm^3; 2 min/scan
cascade impactor	mass size distribution* $0.5 < d < 10$ μm	no dilution needed, can be used at high temp., large sample required
optical particle counter	number distribution** $0.5 < d < 10$ μm	highest resolution, $<10^3$ particles/cm^3, large dilution

*Low-pressure impactor extends size range down to 0.05 μm.

**Laser model extends size range down to 0.1 μm and concentration up to 10^4 particles/cm^3.

TABLE 2-15.3 *Particle Size of Smoke from Burning Wood and Plastics*

Type	d_{gm}, µm*	d_{32}, µm**	σ_g	Combustion Conditions	Ref. No.
Douglas fir	0.5–0.9	0.75–0.8	2.0	pyrolysis	1, 3
Douglas fir	0.43	0.47–0.52	2.4	flaming	1, 3
polyvinylchloride	0.9–1.4	0.8–1.1	1.8	pyrolysis	3
polyvinylchloride	0.4	0.3–0.6	2.2	flaming	3
polyurethane (flexible)	0.8–1.8	0.8–1.0	1.8	pyrolysis	3
polyurethane (flexible)		0.5–0.7		flaming	3
polyurethane (rigid)	0.3–1.2	1.0	2.3	pyrolysis	3
polyurethane (rigid)	0.5	0.6	1.9	flaming	3
polystyrene		1.4		pyrolysis	1
polystyrene		1.3		flaming	1
polypropylene		1.6	1.9	pyrolysis	1
polypropylene		1.2	1.9	flaming	1
polymethylmethacrylate		0.6		pyrolysis	1
polymethylmethacrylate		1.2		flaming	1
cellulosic insulation	2–3		2.4	smoldering	6

*d_{gm} is analogous to d_{gv} but with mass replacing volume in Equation 4. Values of d_{gm} less than about 0.5 µm are probably overestimates arising from the minimum size resolution of the impactor at about 0.4 µm.

**The quantity d_{32} is obtained by optical measurements:

$$d_{32} = \frac{\sum_{i=1}^{n} N_i d_i^3}{\sum_{i=1}^{n} N_i d_i^2}$$

produced from flaming α-cellulose.[7] The coagulation process has a more pronounced effect on the number distribution than the mass distribution as small particles collide to form larger particles.

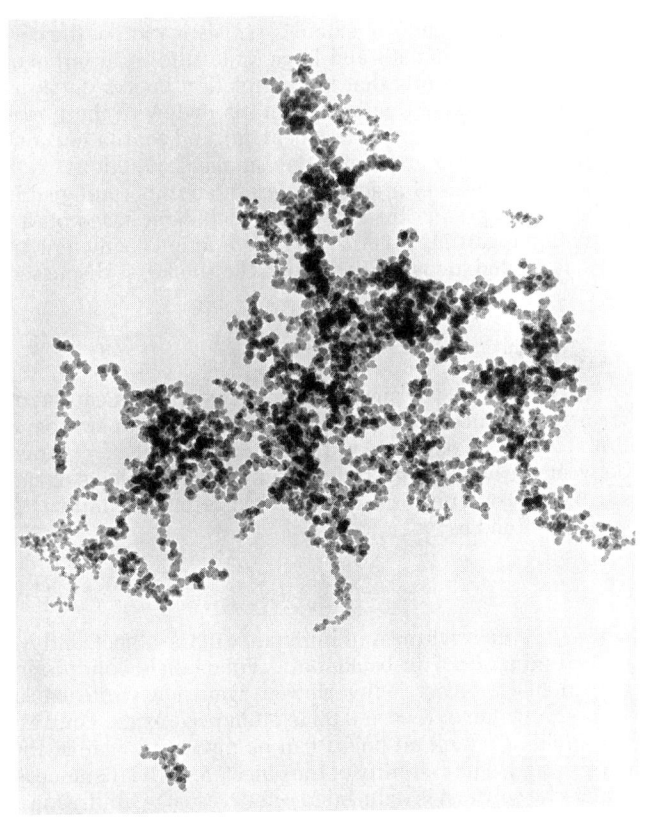

Fig. 2-15.4. Transmission electron micrograph of a soot particle. The overall size of the agglomerate is about 6 µm, and the diameter of the individual spherules is about 0.03 µm.

EXAMPLE 2:

Calculate the change in the number concentration over a 5 min time interval for a uniformly distributed smoke, generated from flaming α-cellulose given an initial concentration of 1×10^7 particles/cm³.

Integrating Equation 6, yields

$$N = \frac{N_0}{1 + \Gamma N_0 t} = \frac{1 \times 10^7}{1 + (10^{-9})(10^7)(300)} = \frac{10^7}{1 + 3}$$

$$N = 2.5 \times 10^6 \text{ particles/cm}^3$$

So in this example, there is a fourfold reduction in number concentration due to coagulation.

The effect of the decrease in number concentration on the size distribution is treated by Mulholland *et al.*[25] A general discussion of coagulation phenomena in aerosols is given by Friedlander.[8] In addition to coagulation, other smoke-aging processes, including condensation of vapor onto existing particles and evaporation of the volatile component of the smoke, can also take place. There is relatively little information on these processes. Also, smoke particles can be lost to the walls, ceiling, and floor of an enclosure through a variety of processes, including diffusion, sedimentation, and thermophoresis.

SMOKE PROPERTIES

The smoke properties of primary interest to the fire community are light extinction, visibility, and detection. For completeness, a list of other smoke aerosol properties and references is given in Table 2-15.4.

The most widely measured smoke property is the light extinction coefficient. The physical basis for light extinction measurements is Bouguer's law, which relates the intensity, I_0^λ, of the incident monochromatic light of wavelength λ and the intensity of the light, I_λ, transmitted through pathlength, L, of the smoke.

$$I_\lambda/I_\lambda^0 = e^{-KL} \qquad (7)$$

where K is the light extinction coefficient. When Equation 7 is expressed in terms of base 10

$$I_\lambda/I_\lambda^0 = 10^{-DL} \qquad (8)$$

The quantity D is defined as the optical density per meter, and $D = K/2.3$.

The extinction coefficient, K, is an extensive property and can be expressed as the product of an extinction coefficient per unit mass, K_m, and mass concentration of the smoke aerosol, m.

$$K = K_m m \qquad (9)$$

The specific extinction coefficient, K_m, depends on the size distribution and optical properties of the smoke through the relation

$$K_m = \frac{3}{2\rho m} \int_{d_{min}}^{d_{max}} \frac{1}{d} \frac{\delta m}{\delta d} Q_{ext}(d/\lambda, n_r) \delta d \qquad (10)$$

In Equation 10 the symbol $\delta m/\delta d$ represents the mass size distribution. The single particle extinction efficiency, Q_{ext}, is a function of the ratio of particle diameter to wavelength of light, d/λ, and of the complex refractive index of the particle, n_r.[8] The quantity ρ represents the particle density.

Seader and Einhorn[11] obtained K_m values of 7.6 m^2/g for smoke produced during flaming combustion of wood and plastics and a value of 4.4 m^2/g for smoke produced during pyrolysis of these materials. The experiments were small scale, utilizing samples of about 50 cm^2, and the value of K_m represents an integrated result for the entirety of the test. The light source used in the measurements was polychromatic, while Bouguer's law is strictly valid only for monochromatic light. Foster[12] predicted a 22 percent deviation from Bouguer's law over the mass concentration range from 0.06 to 2.8 g/m^3 as a result of using a polychromatic light source with wood smoke. Still, it is useful to use the Seader and Einhorn[11] result as a rough guide if more detailed optical data on the smoke of interest is not available.

Mulholland[13] has described the general design of a light extinction instrument that satisfies Bouguer's law. Two key features are the use of monochromatic light and the elimination of forward scattered light at the detector.

The specific optical density, D_s, is measured in a standard laboratory smoke test[14] for assessing the amount of visible smoke produced in a fire. The dimensionless quantity D_s is defined by

TABLE 2-15.4 *Smoke Aerosol Properties*

Property	Ref. No.
diffusion coefficient	8
sedimentation velocity	9
thermophoretic velocity	10
aerodynamic diameter	9
electrical mobility	9
thermal charging	9
scattering coefficient	8
extinction coefficient	8
condensation/evaporation	8

$$D_s = \frac{DV_c}{A} \qquad (11)$$

where V_c is the volume of the chamber, and A is the area of the sample. This is a convenient quantity to measure if the decomposed area is well defined. Since D_s depends on the sample thickness, the same thickness should be used for relative rating of materials tested. Table 2-15.5 includes results for D_s based on small-scale experiments with wood and plastics by Gross et al,[14] Seader and Chien,[15] and Breden and Meisters.[16] Lopez[17] demonstrated a correlation for D_s between small- and large-scale fires for aircraft interior construction materials.

If the mass loss of the sample is measured, then the mass optical density, D_m, is the appropriate measure of visible smoke.

$$D_m = \frac{DV_c}{\Delta M} \qquad (12)$$

This technique requires an accurate measurement of the mass loss of the sample, ΔM, in addition to a light extinction measurement. Table 2-15.5 includes results for D_m for a variety of materials studied by Seader and Chien,[15] Breden and Meisters,[16] Babrauskas,[18] and Evans.[19] The results of Babrauskas' study were expressed in terms of D_m by Quintiere.[20]

In two of the studies,[18,19] a comparison was made between D_m measured in small-scale tests and D_m measured in large-scale tests. The large-scale tests involved mattresses[18] in one case and plastic utility tables[19] in the other. In these two cases, there appeared to be a qualitative correlation between D_m measured for small- and large-scale tests. Quintiere[20] has made an extensive investigation of the correlation between small- and large-scale studies in terms of D_m and D_s and finds that the correlation breaks down as fires become more complex. From his review of the literature, Quintiere[20] suggests that heat flux and ventilation conditions can have a major effect on smoke production.

In most cases of practical interest, an important goal is to be able to predict the extinction coefficient based on information regarding D_s or D_m. The extinction coefficient, in turn, is related to visibility through the smoke, as discussed below.

Visibility

Visibility of exit signs, doors, and windows can be of great importance to an individual attempting to survive a fire. To see an object requires a certain level of contrast between the object and its background. For an isolated object surrounded by a uniform, extended background, contrast, C, can be defined as[21]

$$C = \frac{B}{B_0} - 1 \qquad (13)$$

where B is the brightness or luminance of the object, and B_0 is the luminance of the background. For daylight conditions, with a black object being viewed against a white background, a value of $C = -0.02$ is often used as the contrast threshold at which an object can be discerned against the background. The visibility of the object, S, is the distance at which the contrast is reduced to -0.02. Most visibility measurements through smoke have relied on test subjects to determine the distance at which the object was no longer visible rather than the actual measurement of C with a photometer.

TABLE 2-15.5 *Specific Optical Density and Mass Optical Density for Wood and Plastics*

Type (Sample #)	Maximum D_s	D_m (m^2/g)	Combustion Conditions	Sample* Thickness (cm)	Ref. No.
hardboard	6.7×10^1		flaming	0.6	14
hardboard	6.0×10^2		pyrolysis	0.6	14
plywood	1.1×10^2		flaming	0.6	14
plywood	2.9×10^2		pyrolysis	0.6	14
polystyrene	>660		flaming	0.6	14
polystyrene	3.7×10^2		pyrolysis	0.6	14
polyvinylchloride	>660		flaming	0.6	14
polyvinylchloride	3.0×10^2		pyrolysis	0.6	14
polyurethane foam	2.0×10^1		flaming	1.3	14
polyurethane foam	1.6×10^1		pyrolysis	1.3	14
nylon carpet	2.7×10^2		flaming	0.8	14
nylon carpet	3.2×10^2		pyrolysis	0.8	14
acrylic	1.1×10^2		flaming	0.6	14
acrylic	1.6×10^2		pyrolysis	0.6	14
plywood	5.3×10^2	0.29	pyrolysis	0.6	15
polymethylmethacrylate	7.2×10^2	0.15	pyrolysis	0.6	15
polyvinylchloride	1.8×10^2	0.12	pyrolysis	0.6	15
polyvinylchloride (with plasticizer)	3.5×10^2	0.64	pyrolysis	0.6	15
neoprene	8.8×10^2	0.55	pyrolysis	0.6	15
Douglas fir	6.2×10^2	0.28	pyrolysis	0.6	15
polypropylene	4.0×10^2	0.53	flaming**	0.4	16
polyethylene	2.9×10^2	0.29	flaming**	0.4	16
paraffin wax	2.3×10^2	0.23	flaming**	0.4	16
polystyrene		1.4	flaming**	0.4	16
sytrene		0.96	flaming**	0.4	16
polyvinylchloride		0.34	flaming**	0.4	16
polyoxymethylene		~0	flaming**	0.4	16
polyurethane (7A)	2.1×10^2		flaming	1.3	17
polyurethane (7A)	1.5×10^2		flaming***	1.3	17
wool (8A)	>5.5×10^2		flaming	0.9	17
wool (8A)	2.2×10^2		flaming***	0.9	17
acrylic (9B)	5.8×10^1		flaming	0.14	17
acrylic (9B)	1.2×10^2		flaming***	0.14	17
polyurethane (MO1)		0.33	flaming**		18†
polyurethane (MO1)		0.22	flaming‡		18
cotton (MO3)		0.17	flaming**		18
cotton (MO3)		0.12	flaming‡		18
latex (MO4)		0.65	flaming**		18
latex (MO4)		0.44	flaming‡		18
neoprene (MO8)		0.40	flaming**		18
neoprene (MO8)		0.20	flaming‡		18
polystyrene (7)		0.79	flaming**		19
polystyrene (7)		1.0	flaming§		19
polystyrene foam (16)		0.79	flaming**		19
polystyrene foam (16)		0.82	flaming§		19
ABS (18)		0.52	flaming**		19
ABS (18)		0.54	flaming§		19

*Sample area is 0.005 m² in vertical configuration, unless stated otherwise.

**Sample is in horizontal configuration (0.005 m²).

***0.09 m² sample size.

†The value of D_m is computed by Quintiere,[20] based on data in Babrauskas.[18]

‡The sample is a mattress.

§The sample is a plastic utility table.

Visibility depends on many factors, including the scattering and the absorption coefficient of the smoke, the illumination in the room, whether the sign is light-emitting or light-reflecting, and the wavelength of the light. Visibility also depends on the individual's visual acuity and on whether the eyes are "dark-," or "light-adapted." Nevertheless, a fair correlation between visibility of test subjects and the extinction coefficient of the smoke has been obtained in an extensive study by Jin[22] as illustrated in Figure 2-15.5. The visibility of light-emitting signs was found to be two to

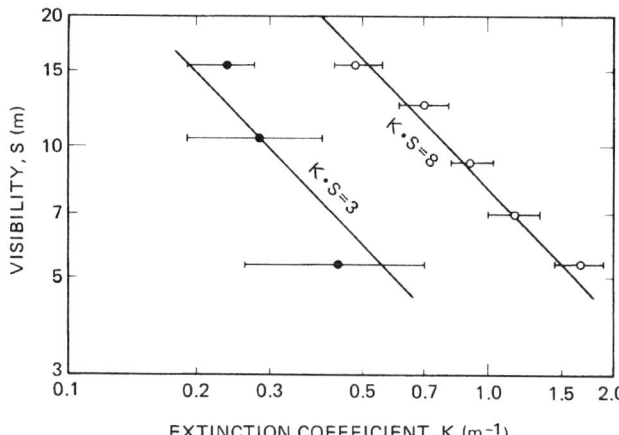

Fig. 2-15.5. *Visibility versus extinction coefficient for a light-emitting sign (○) and light-reflecting sign (●). The range bars include data for both flame- and smolder-generated smoke and sign illumination levels varying by about a factor of 4.*

four times greater than light-reflecting signs. The following expressions were found to correlate the data

$$KS = 8 \quad \text{light-emitting sign} \quad (14)$$

$$KS = 3 \quad \text{light-reflecting sign} \quad (15)$$

The data is based on the subjects viewing the smoke through glass so that the irritant effect of the smoke was eliminated. Jin and Yamada[23] have studied the visual acuity and eye-blink rate for highly irritant white smoke produced by burning wood cribs. They found that the ratio of visual acuity without goggles to acuity with goggles decreases markedly for smoke extinction coefficient, K, greater than 0.25 m^{-1}.

EXAMPLE 3:

Estimate the visibility of a light-reflecting exit sign in a 6 m square room with a 2.5-m height, as a result of flaming combustion of a 200-g polyurethane foam pillow.

The smoke yield for flexible polyurethane, according to Table 2-15.1, is about 0.03 for flaming combustion. This implies a smoke emission, M_s, given by

$$M_s = (0.03)(200) = 6 \text{ g}$$

The corresponding mass concentration in the room, m, is

$$m = \frac{6}{(6)^2(2.5)} = 0.067 \text{ g/m}^3$$

Taking K_m to be 7.6 m^2/g for flaming combustion, one obtains K using Equation 9

$$K = (7.6)(0.067) = 0.51 \text{ m}^{-1}$$

The visibility is next estimated using Equation 15

$$S = \frac{3}{K} = \frac{3}{0.51} = 5.9 \text{ m}$$

It is important to point out the approximations made in this analysis:

1. The smoke is confined to the room and is well-mixed. Actually the concentration will be higher near the ceiling and decrease abruptly below the flame.
2. The value of 0.03 for the smoke conversion factor, ε, is an estimated value in the upper part of the range (0.01 to 0.035) for generic flexible polyurethane foams measured in small-scale experiments and may not be appropriate for a pillow. In a realistic case, the pillow would probably smolder before flaming, and ε is much larger in the smolder mode.
3. The value of K_m is based on a limited number of small-scale experiments with a polychromatic light source.
4. The range of validity of Equation 15 has not been widely studied.

An alternative method for estimating the visibility is based on using the mass optical density data in Table 2-15.5. The quantity D_m for the pillow is estimated to be 0.22 m^2/g based on Babrauskas' results[18] given in Table 2-15.5 for polyurethane (m 01). On rearranging Equation 12, the following result is obtained

$$D = \frac{D_m \Delta M}{V_c} = \frac{(0.22)(200)}{(6)^2(2.5)} = 0.49 \text{ m}^{-1}$$

The smoke extinction coefficient, K, is 1.12 m^{-1} or 2.3 times D. Using Equation 15, we obtain $S = 2.7$ m compared to 5.9 m obtained by the first method. In principle, the second method is more reliable, because it is more direct.

Detection

In addition to their utility for estimating visibility, light extinction measurements are also widely used in characterizing smoke detector performance. Underwriters Laboratories' (U.L.) acceptance testing of smoke detectors[24] is based in part on a minimum sensitivity based on optical density per meter, D, of 0.06 (4 percent obscuration per ft for a 5 ft beam length) for grey-color (cellulosic) smoke and 0.14 (10 percent per ft) for black smoke (kerosene).

The electrical output of a detector, P, from a light-scattering or ionization-type smoke detector can be represented as an integrated product of the size distribution function and the basic response of the detector, $R(d)$.

$$P = \int_{d_{min}}^{d_{max}} R(d) \frac{\delta N}{\delta d} \delta d \quad (16)$$

The response functions for two smoke detectors are plotted in Figure 2-15.6. It is seen that the ionization-type smoke detector is more sensitive to smoke particles smaller than about 0.3 μm, and the light-scattering type more sensitive to particles larger than 0.3 μm.

The basic principle of ionization detectors is the interception of gaseous ions by smoke particles, reducing the ion current in the detector until a preset alarm point is reached. The detector response function is approximately proportional to the product of the number concentration and particle diameter.[25,26] For one detector[25] the response function is given by

$$R(d) = cd \quad (17)$$

where c has a value of 7 in units of μV per particle concentration per μm (μV cm^3/μm). Such detectors tend to be most sensitive to high concentrations of small particles, such as those produced by flaming paper and wood fires, and least sensitive to the low concentration of large smoke droplets produced in smoldering fires.

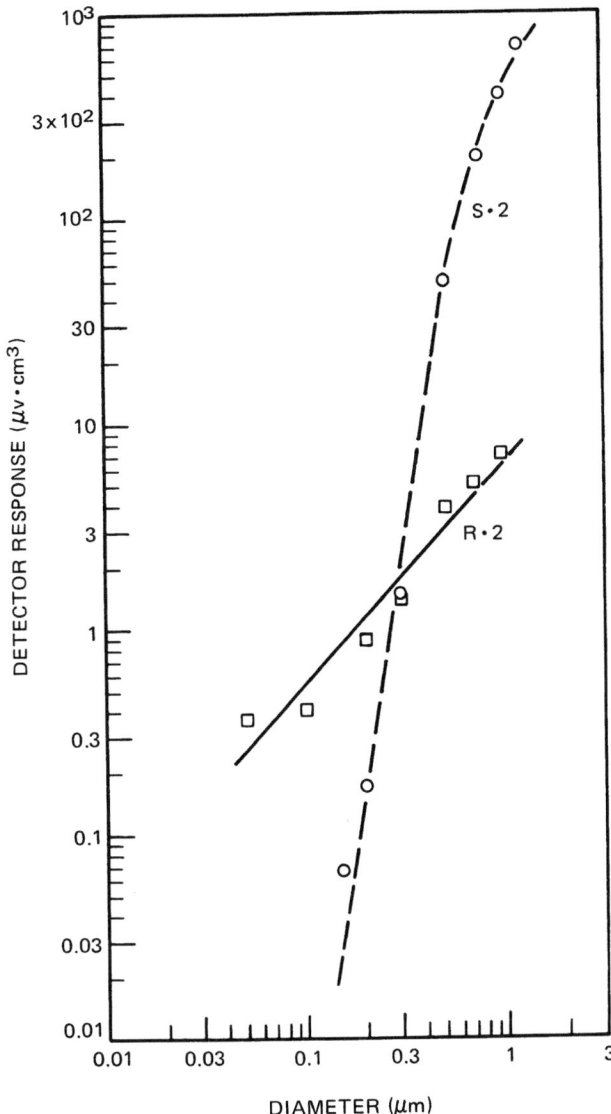

Fig. 2-15.6. The detector response function, R(d), is plotted versus particle size for detectors S-2 (light-scattering) and R-2 (ionization).

Light-scattering smoke detectors have a high sensitivity to smoke particles with diameters approximately equal to λ, the wavelength of light, and low sensitivity to particles much smaller than λ. The response function, $R(d)$, depends on the wavelength of the light source in the smoke detector, the scattering angle, and the scattering volume. For smoke particles with diameter greater than about 0.3 μm, the output of several light-scattering smoke detectors was found to be approximately proportional to the mass concentration of the smoke.[25] Light-scattering detectors complement ionization detectors in that they have high sensitivity to smoldering fires and low sensitivity to low-smoking flaming fires, such as paper and wood fires.

The purpose of smoke detectors is to give the occupants of a room adequate warning to escape a developing fire. The final examples of this chapter illustrate how to utilize all the concepts discussed above to estimate escape time.

EXAMPLE 4:

Suppose the pillow in the preceding example is burning at a steady rate of 50 g/min. How long would it take for an ionization detector with response function given by Equation 17 to alarm? Assume an alarm voltage of 2.5 V above background. How much time would an individual have before the visibility decreased to an unsafe level?

SOLUTION:

First consider a first principle analysis based on the size distribution of the smoke. From Equations 16 and 17

$$P = c \int_{d_{min}}^{d_{max}} d\frac{\delta N}{\delta d} \delta d$$

The following three identities[9] for the log-normal distribution are needed:

$$\int_0^\infty d\frac{\delta N}{\delta d} \delta d = N_0 d_{gn} \exp\left(\frac{1}{2} \ln^2 \sigma_g\right) \qquad (I-1)$$

$$d_{32} = \frac{\int_0^\infty d^3 \frac{\delta N}{\delta d}\delta d}{\int_0^\infty d^2 \frac{\delta N}{\delta d}\delta d} = d_{gn} \exp\left(\frac{5}{2} \ln^2 \sigma_g\right) \qquad (I-2)$$

$$\int_0^\infty d^3\frac{\delta N}{\delta d}\delta d = N_0 d_{gn}^3 \exp\left(\frac{3}{2} \ln^2 \sigma_g\right) \qquad (I-3)$$

Here N_0 refers to the number concentration. Taking $(\delta N/\delta d)$ to be log-normal and using Equation 1

$$P = c N_0 d_{gn} \exp\left(\frac{1}{2} \ln^2 \sigma_g\right)$$

Estimating σ_g to be 2.0 and d_{32} to be 0.6 μm from Table 2-15.3 for flexible polyurethane, d_{gn} is determined using Equation I-2

$$d_{gn} = d_{32} \exp\left(-\frac{5}{2} \ln^2 \sigma_g\right) = 0.6 \exp\left[-\frac{5}{2}(0.69)^2\right]$$

$$d_{gn} = 0.18 \ \mu\text{m}$$

Substituting for d_{gn} and for c in the expression for P yields

$$P = c N_0 (d_{gn}) \exp\left(\frac{1}{2} \ln \sigma_g^2\right) = 7 N_0 (0.18) \exp\left[\frac{1}{2}(0.69)^2\right]$$

$$P = 1.6 \ N_0$$

The final task is to estimate N_0 based on the mass generation rate of smoke. In one minute, 50 g of the pillow is consumed and 1.5 g of smoke are produced. This corresponds to a mass concentration, m, given by

$$m = \frac{1.5}{(6)^2(2.5)} = 0.0167 \ \text{g/m}^3 = 1.67 \times 10^{-8} \ \text{g/cm}^3$$

The quantity m is the third moment of the size distribution

$$m = \int_0^\infty \frac{1}{6}\pi\rho d^3 \frac{\delta N}{\delta d}\delta d$$

Using Equation 3,

$$m = \frac{1}{6}\pi\rho N_0 d_{gn}^3 \exp\left(\frac{3}{2} \ln^2 \sigma_g\right)$$

Finally, solving for N_0,

$$N_0 = \frac{6m}{\pi \rho d_{gn}^3} \exp\left(-\frac{3}{2} \ln^2 \sigma_g\right)$$

$$= \frac{(6)(1.67 \times 10^{-8})}{(3.14)(2)(1.8 \times 10^{-5})^3} \exp\left[-\frac{3}{2} \ln^2(2.0)\right]$$

$$N_0 = 1.3 \times 10^6 \text{ particles/cm}^3 \text{ (assuming } \rho = 2 \text{ g/cm}^3)$$

Substituting in the expression for P,

$$P = (1.6)(1.3 \times 10^6) = 2.1 \times 10^6 \quad \mu V = 2.1 \text{ volts}$$

This represents the voltage after 1 min. The estimated time to reach the alarm point, 2.5 V, will be 1.2 min. By the time the entire pillow is consumed in 4 min, the visibility has deteriorated to the point where escape is becoming less likely (visibility 5.9 m, according to Example 3, for a room 6 m across). So the individual's escape time is:

escape time = time to unsafe condition minus time
to detector alarm
= 4 − 1.2 = 2.8 min

Example 4 is intended to illustrate the complete method for estimating the alarm time of smoke detectors. However, there is not adequate information at this time to implement the method in a realistic manner. Information is lacking on the size distribution of smokes and on the detector response functions. The time for the smoke to reach the detector and the time lag for the smoke to enter the sensing zone of the detector are not included in this example, but should be included in a full analysis of the problem.

A simpler method for estimating the alarm time is to calculate the time at which the optical density per meter of the smoke exceeds the value of 0.06 (grey smoke) or 0.14 (black smoke), which correspond to the U.L. minimum sensitivity values. The limitation of this procedure is that a detector set to alarm at a particular optical density for one type of smoke may not respond in the same manner to another with a different size distribution and refractive index.

EXAMPLE 5:

Estimate the time to alarm for the conditions given in Example 4, using the simpler method described above.

SOLUTION:

In Example 3, the optical density was estimated to be 0.49 m^{-1}, based on D_m measured for polyurethane. This value corresponds to the burning of the entire pillow. Assuming a steady smoke generation rate, the alarm time [the time at which the minimum detector sensitivity value is exceeded (0.14 for black smoke)] is estimated to be given by

$$t = \frac{0.14}{0.49}(4) = 1.1 \text{ minutes}$$

This is comparable to the estimated 1.2 minutes in Example 4.

NOMENCLATURE

ε smoke conversion factor
d particle diameter, μm
d_i midpoint of the ith particle size channel, μm
d_{gn} geometric mean number diameter, μm
d_{gv} geometric mean volume diameter, μm
d_{32} volume surface mean diameter, μm
σ_g geometric standard deviation ($-$)
N number concentration, particles/cm^3
m mass concentration of smoke, mg/m^3 or g/m^3
V_T volume concentration of smoke, cm^3/m^3 or μm^3/cm^3

$\dfrac{\Delta N}{\Delta d}$ or $\dfrac{dN}{dd}$ number size distribution function, cm^{-3} μm^{-1}

$\dfrac{\Delta N}{\Delta \log d}$ or $\dfrac{dN}{d \log d}$ geometric number size distribution function, cm^{-3}

$\dfrac{\Delta m}{\Delta d}$ or $\dfrac{dm}{dd}$ mass size distribution function, mg μm^{-1} m^{-3}

Q_{ext} extinction efficiency ($-$)
λ wavelength of light, μm
n_r complex refractive index of smoke particles
K extinction coefficient, m^{-1}
D optical density per meter, m^{-1}
K_m specific extinction coefficient, m^2/g
D_s specific optical density ($-$)
D_m mass optical density, m^2/g
I_λ intensity of light at wavelength λ
B luminance
C contrast
s visibility range, m
L pathlength
Γ coagulation coefficient, cm^3/s
t time
ΔM mass loss of sample, g
P detector output, volts
$R(d)$ detector size response function, μv cm^3
V_c volume of chamber, m^3
A area of sample, m^2
M_s mass of smoke, g

REFERENCES CITED

1. C.P. Bankston, B.T. Zinn, R.F. Browner, and E.A. Powell, *Comb. and Flame*, 41, 273 (1981).
2. C.J. Hilado and A.M. Machado, *J. Fire and Flamm.*, 9, 240 (1978).
3. C.P. Bankston, R.A. Cassanova, E.A. Powell, and B.T. Zinn, *NBS-GCR G8-9000*, National Bureau of Standards, Gaithersburg (1978).
4. S.K. Brauman, N. Fishman, A.S. Brolly, and D.L. Chamberlain, *J. Fire and Flamm.*, 6, 41 (1976).
5. A. Tewarson, J.L. Lee, and R.F. Pion, 18th Symp. (Int) on Combustion, Pittsburgh (1981).
6. G.W. Mulholland and T.J. Ohlemiller, *Aerosol Sci. and Tech.*, 1, 59 (1982).
7. G.W. Mulholland, T.G. Lee, and H.R. Baum, *J. Colloid Interface Sci.*, 62, 406 (1977).
8. S.K. Friedlander, *Smoke, Dust and Haze*, Wiley and Sons, New York (1977).
9. P.C. Reist, *Introduction to Aerosol Science*, Macmillan, New York (1984).
10. T. Waldman and K.H. Schmitt, in *Aerosol Science*, Academic Press, New York (1966).
11. J.D. Seader and I.N. Einhorn, 16th Symp. (Int) on Combustion, Pittsburgh (1976).
12. W.W. Foster, *Br. J. App. Phys.*, 10, 416 (1959).
13. G.W. Mulholland, *Fire and Matls.*, 6, 65 (1982).
14. D. Gross, J.J. Loftus, and A.F. Robertson, *ASTM STP 422*, American Society for Testing and Materials, Philadelphia (1967).
15. J.D. Seader and W.P. Chien, *J. Fire and Flamm.*, 5, 151 (1974).

16. L.H. Breden and M. Meisters, *J. Fire and Flamm.*, 7, 234 (1976).
17. E.L. Lopez, *J. Fire and Flamm.*, 6, 405 (1975).
18. V. Babrauskas, *J. Fire and Flamm.*, 12, 51 (1981).
19. D.D. Evans, *NBSIR 81-2400*, National Bureau of Standards, Washington (1981).
20. J.G. Quintiere, *Fire and Matls.*, 6, 145 (1982).
21. E.J. McCartney, *Optics of the Atmosphere*, Wiley and Sons, New York (1976).
22. T. Jin, *J. Fire and Flamm.*, 9, 135 (1978).
23. T. Jin and T. Yamada, *Fire Sci. and Tech.*, 5, 79 (1985).
24. UL 217, *Standard for Single and Multiple Station Smoke Detectors*, Underwriters Laboratories, Northbrook (1993).
25. G.W. Mulholland and B.Y.H. Liu, *J. Res. NBS*, 85, 223 (1979).
26. C. Helsper, H. Fissan, J. Muggli, and A. Scheidweiler, *Fire Tech.*, 14 (1983).

Section Three
Hazard Calculations

Section 3 Hazard Calculations

BURNING RATES

Vytenis Babrauskas

INTRODUCTION

Calculations of fire behavior in buildings are generally not possible if it is not known what combustibles are burning, and at what rate. In some cases, it is possible to make a useful simplification, e.g., for postflashover fires (flashover is here simply defined as a time in the course of a room fire when the average upper gas temperature reaches 600°C), one can either take (1) a worst-case approach, where an exact burning rate is not required,[1] or (2) a schematized approach, where all of the burning rate information is expressed solely as a fuel loading.[2] In most cases, however, a realistic calculation of the effects of a fire requires that some more specific information about the burning rate be known. This may require only an average burning rate, or the peak burning rate, or the complete curve may be required. Furthermore, if the design concerns are detector or sprinkler actuation times, then emphasis is placed on the initial fire growth characteristics. If, on the other hand, the concerns are for overall fire hazard, then typically the peak burning rate is of essence. In this chapter, the main focus will be on the peak burning rate, although some characterizations of total fire growth will be presented.

MEASUREMENT TECHNIQUES

The "burning rate" is an ambiguous, though useful expression. Quantitatively, it is expressed either as a mass loss rate, $\dot{m}$, (kg/s) or as a heat release rate, $\dot{q}$, (kW), with the latter being more commonly desired. Mass loss rates have normally been measured with the use of load cells, or by volumetric techniques in the case of liquid pool fires. Useful techniques for measuring heat release rates in the open were not available until a few years ago, when the principle of oxygen consumption calorimetry was developed.[3] Earlier attempts required the direct measurement of sensible enthalpy, something which is very difficult to do correctly. The oxygen consumption technique, however, has enabled these measurements to be made easily and to good accuracy. The oxygen consumption principle states that, to within a small uncertainty band, the heat released from the combustion of any common combustible is uniquely related to the mass of oxygen removed from the combustion flow stream. The measurement technique then requires that only the flow rate and the oxygen concentration be determined, along with the knowledge of the oxygen consumption constant, 13.1×10^3 kJ heat released per kg of oxygen consumed. Large-scale calorimeters using this principle have been built for the measurement of warehouse commodities[4] and of free-standing combustibles.[5] (See Figure 3-1.1.)

For repeated testing, it is usually desirable to be able to obtain full-scale burning rate information from bench-scale test results. Such bench-scale measurements cannot be simply extrapolated in order to compute the full-scale fire. Instead, what is required is either a theory or an empirical relationship that connects the two. Bench-scale tests for measuring heat release rate and mass loss rate have followed a development similar to that for full-scale testing—early heat release rate tests were an attempt to simply capture the sensible enthalpy of the combustion stream;[6] more current

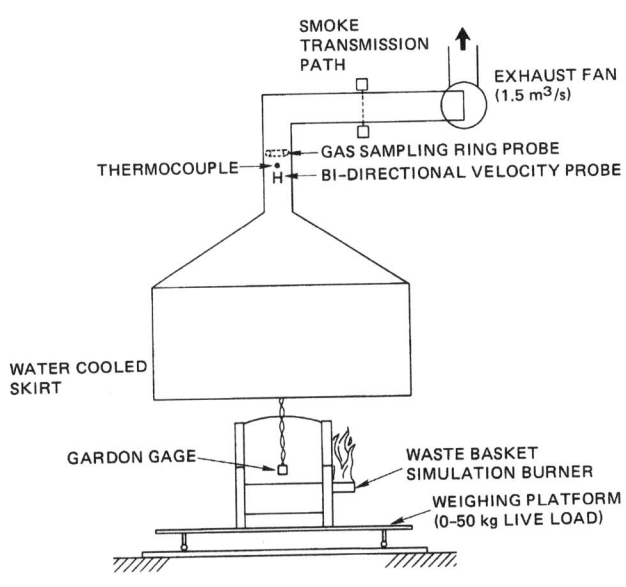

Fig. 3-1.1. The furniture calorimeter.

Dr. V. Babrauskas is President of Fire Science and Technology Inc., specializing in research and consulting on fire tests, hazard analysis, fire modeling, and related fire safety issues.

TABLE 3-1.1 *Dominant Heat Transfer Mechanism vs. Pool Diameter*

Diameter, D (m)	Burning mode
< 0.05	convective, laminar
0.05 to 0.2	convective, turbulent
0.2 to 1.0	radiative, optically thin
> 1.0	radiative, optically thick

methods[7,8] use oxygen depletion instead of sensible enthalpy measurement.

Attempts have often been made to determine burning rates during the course of room fires. Instrumentation issues aside, the interpretation of this kind of data can be difficult, since it may not be clear how far into the course of the fire the burning rate still corresponds to the open burning rate. This issue is discussed in more detail later.

AVAILABLE DATA

For a general burning item that may be of concern, it will typically be found that there have not been any experimental data taken on its burning rate. Certain few categories of combustibles, however, have been studied at length and systematic deductions are possible. An attempt is made here to give data on all product categories for which it is possible to find in the literature burning rate data that are nonproprietary and that pertain to adequately characterized specimens.

The following combustibles will be considered:

1. Pools, liquid or thermoplastic
2. Cribs (regular arrays of sticks)
3. Wood pallets
4. Upholstered furniture
5. Mattresses
6. Pillows
7. Wardrobes
8. Television sets
9. Christmas trees
10. Curtains
11. Electric cable trays
12. Trash bags and containers
13. Stored commodities.

Pools

Liquid pool fires have probably been studied in greater detail than any other combustible goods. The term "pool" shall be taken to include fires of spilled liquids, fires in diked or dammed areas, and fires in open-top tanks. These fires will be typically considered to be circular. Square or similar configurations can be expressed as equivalent circles, on an area basis. Highly elongated shapes, e.g., trench fires, require a different treatment.[9] In recent years, interest has

TABLE 3-1.2 *Data for Large-Pool Burning Rate Estimates*

Material	Density (kg/m³)	Δh_c (MJ/kg)	$\dot{m}''_\infty$ (kg/m²s)	$k\beta$ (m⁻¹)
Cryogenics*				
Liquid H_2	70	120.0	0.017	6.1
LNG (mostly CH_4)	415	50.0	0.078	1.1
LPG (mostly C_3H_8)	585	46.0	0.099	1.4
Alcohols				
methanol (CH_3OH)	796	20.0	0.017	†
ethanol (C_2H_5OH)	794	26.8	0.015	†
Simple organic fuels				
butane (C_4H_{10})	573	45.7	0.078	2.7
benzene (C_6H_6)	874	40.1	0.085	2.7
hexane (C_6H_{14})	650	44.7	0.074	1.9
heptane (C_7H_{16})	675	44.6	0.101	1.1
xylene (C_8H_{10})	870	40.8	0.090	1.4
acetone (C_3H_6O)	791	25.8	0.041	1.9
dioxane ($C_4H_8O_2$)	1035	26.2	0.018**	5.4**
diethyl ether ($C_4H_{10}O$)	714	34.2	0.085	0.7
Petroleum products				
benzine	740	44.7	0.048	3.6
gasoline	740	43.7	0.055	2.1
kerosine	820	43.2	0.039	3.5
JP-4	760	43.5	0.051	3.6
JP-5	810	43.0	0.054	1.6
transformer oil, hydrocarbon	760	46.4	0.039**	0.7**
fuel oil, heavy	940–1000	39.7	0.035	1.7
crude oil	830–880	42.5–42.7	0.022–0.045	2.8
Solids				
polymethylmethacrylate ($C_5H_8O_2$)n	1184	24.9	0.020	3.3
polypropylene (C_3H_6)n	905	43.2	0.018	
polystyrene (C_8H_8)n	1050	39.7	0.034	

*For pools on dry land, not over water.

**Estimate uncertain since only two data points available.

†Value independent of diameter in turbulent regime.

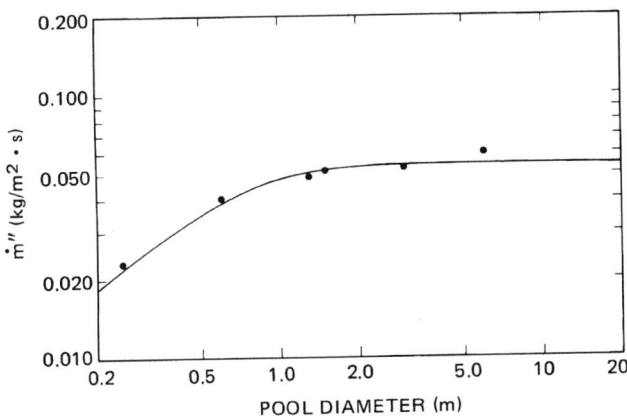

Fig. 3-1.2(a). Data correlations for typical gasoline pools.

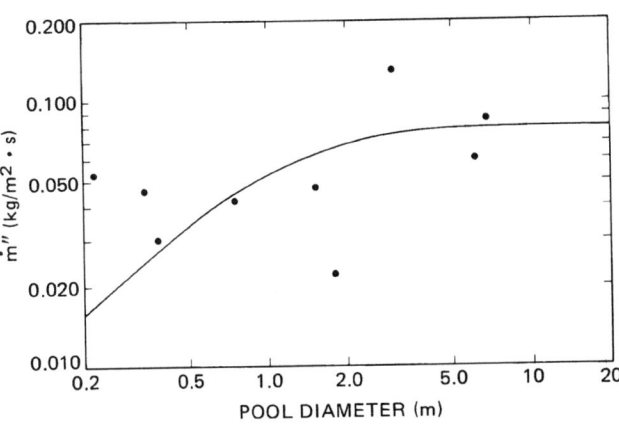

Fig. 3-1.2(b). Data correlations for typical LNG pools (illustrating the large data scatter for liquified gases).

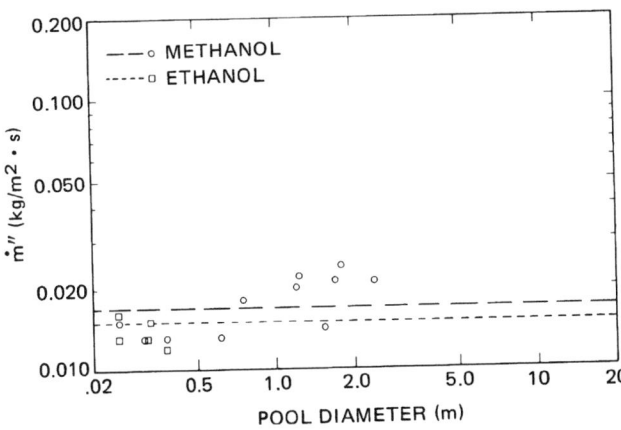

Fig. 3-1.2(c). Data correlations for typical alcohol pools (illustrating the use of a diameter-independent model for the burning rate).

arisen in pool fires of thermoplastic materials. These can be treated by exactly the same techniques as liquid pool fires, although relevent data are still scarce.

An engineering approach to pool fire burning characterization requires that a classification be made according to the dominant heat transfer mechanism, which can be expressed as dependent on pool diameter. (See Table 3-1.1.)

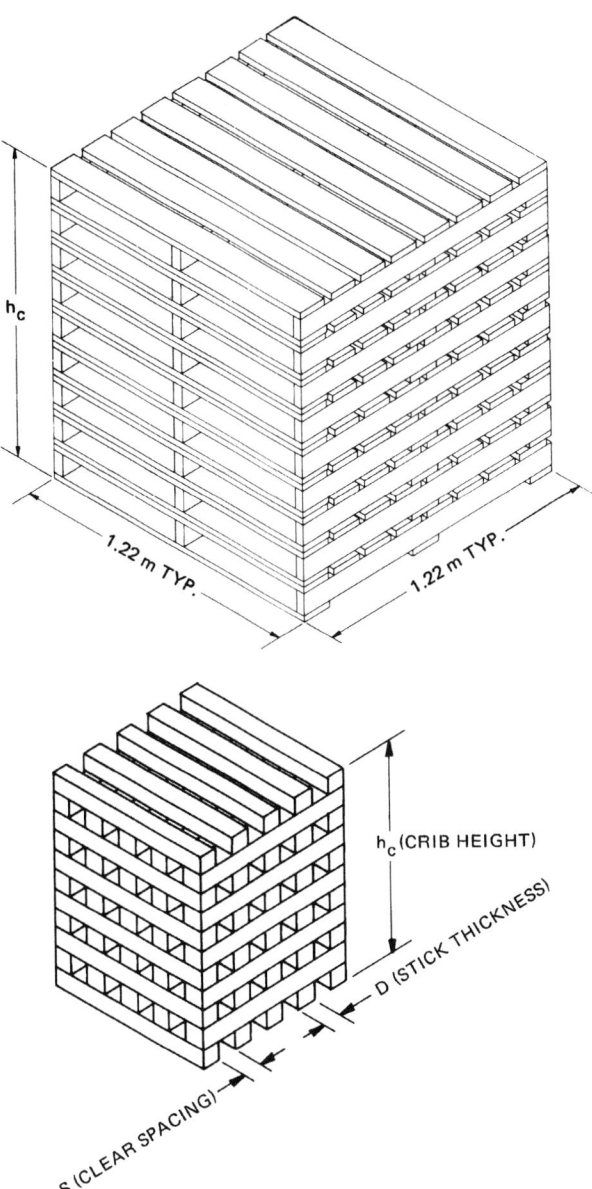

Fig. 3-1.3. Geometry of cribs and pallets.

Fires below 0.2 m diameter have often been studied in combustion experiments but do not show burning rates large enough to be of normal interest in building fires. Thus, attention here will be focused on fires with $D > 0.2$ m. A semitheoretical analysis, together with a study of the available experimental data, has shown that the following form can be used to represent the mass loss rate of pool fires burning in the open.[10]

$$\dot{m}'' = \dot{m}''_\infty (1 - e^{-k\beta D}) \qquad (1)$$

where in addition to the pool diameter, D, two empirical constants are necessary which characterize the particular fuel used: $\dot{m}''_\infty$, the mass loss rate for an infinite-diameter pool, and $(k\beta)$, the product of the extinction-absorption coefficient of the flame (k) and the mean-beam-length corrector

TABLE 3-1.3 *Fuel Type vs. v_p for Crib Fires*

Material	v_p
Wood	$2.2 \times 10^{-6}D^{-0.6}$
Polymethylmethacrylate	$1.4 \times 10^{-6}D^{-0.6}$
Thermosetting polyester	$3.1 \times 10^{-6}D^{-0.6}$
Rigid polyurethane foam	$3.8 \times 10^{-6}D^{-0.6}$

(β). The appropriate units for all variables are given in the "Nomenclature" section. Note that for pool calculation purposes it is not necessary to determine k and β separately, only their product. Table 3-1.2 lists these constants for those products where sufficient data are available.[10] It should be noted that there is a class of common combustibles, namely alcohols, for which radiatively dominant burning is not reached even at somewhat large diameters (although alcohol fire experiments have not been conducted for very large diameters). Thus, for the case of alcohols, a diameter-independent form is normally adopted

$$\dot{m}'' = \dot{m}''_\infty \qquad (2)$$

Figures 3-1.2(a), (b), and (c) show how well the mass loss rate versus pool diameter relationship corresponds to experimental data. Three cases are illustrated: gasoline, a typical fuel for which excellent correspondence is obtained; LNG, a difficult to measure fuel, where a great deal of experimental scatter can be seen; and the alcohols.

Once the mass loss rate is computed, the heat release rate can be obtained as

$$\dot{q} = \Delta h_c \cdot \dot{m}'' \cdot A \qquad (3)$$

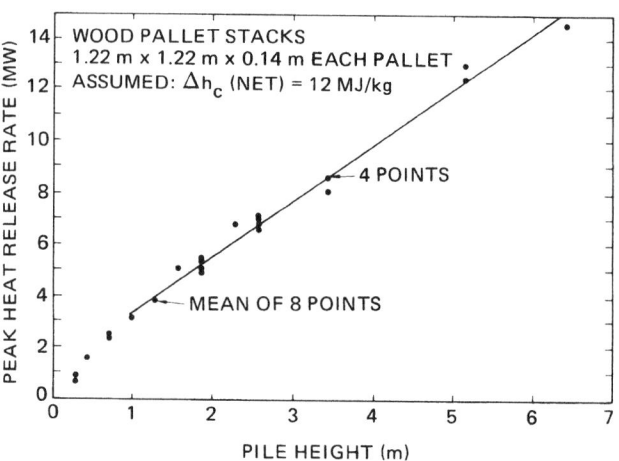

Fig. 3-1.5. *Dependence of pallet burning rate on stack height.*

Miscellaneous complicating effects: The Equations 1 through 3 represent a prediction of idealized pool burning. In many cases, complications can make such a simple prediction subject to an error of as much as a factor of 2. The best known of these complicating effects are the following:[10]

1. Boil-over. This is an effect where the pool does not establish steady-state burning but, rather, after some time of heating, erupts in a boiling manner, often overflowing the vessel, if located in a tank. This effect is rare but can occur for certain grades of petroleum crude and for petroleum products with moisture.
2. Transient Effects. The typical pool fire does not burn in a completely steady manner, increasing, instead, in burning rate as time goes on during the first stage of the fire. This is due to the heating of tank walls, vessel surfaces, the ground underneath, or other physical boundaries to the pool. For a conservative estimate, these effects can be ignored.
3. Lip Height. This is commonly referred to as ullage, when speaking of tanks. It is known that lip height can have a

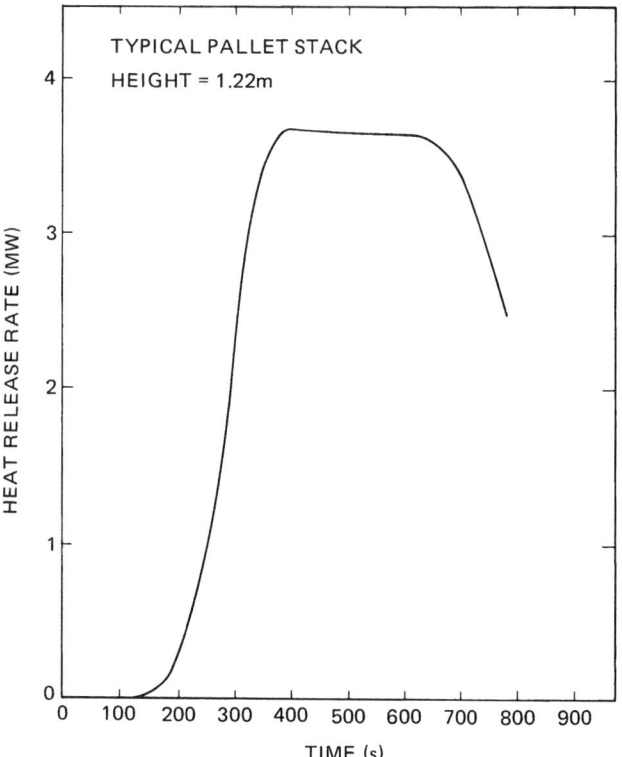

Fig. 3-1.4. *Typical heat release rate for a wood pallet stack.*

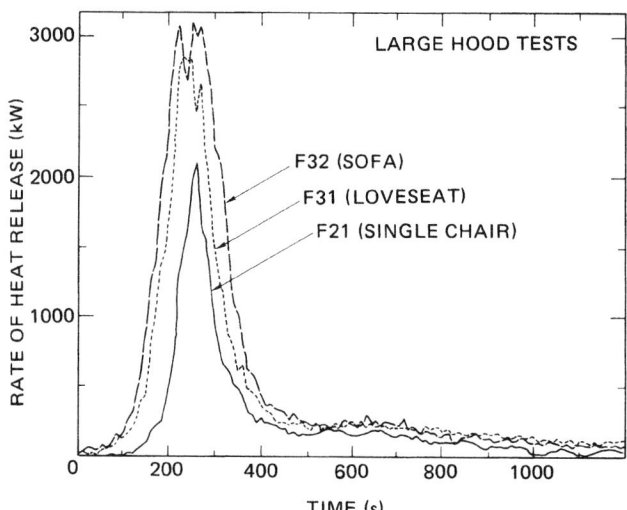

Fig. 3-1.6. *Typical upholstered chair heat release rates.*

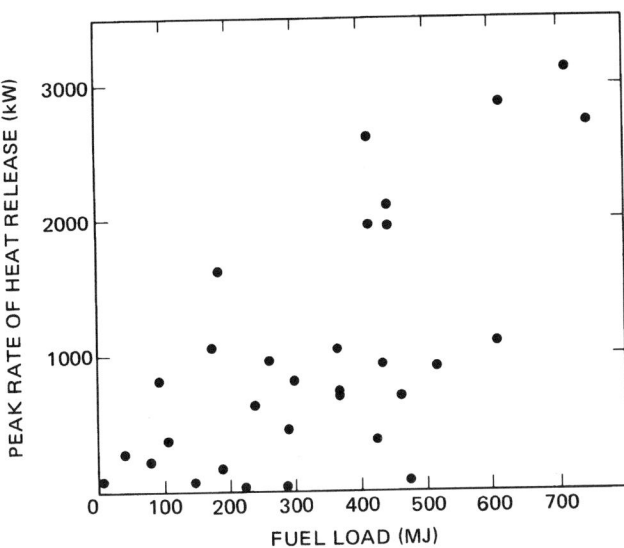

Fig. 3-1.7. *Data illustrating that total specimen heat content (fuel content) alone is not an adequate predictor of heat release rate.*

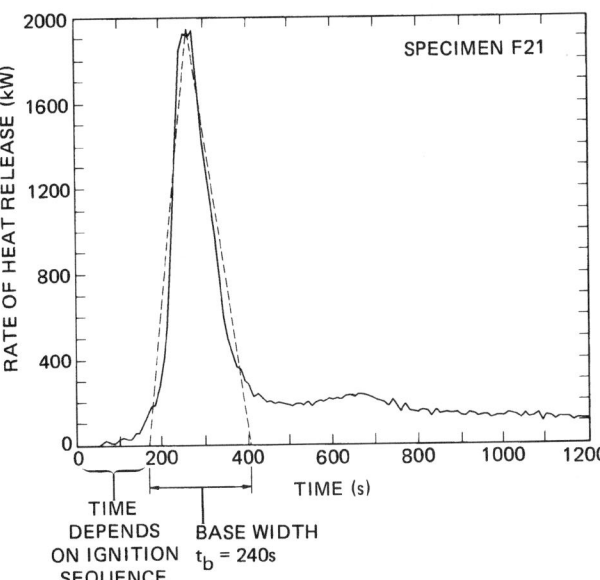

Fig. 3-1.9. *Fitting the heat release curve for upholstered furniture with a triangular curve shape.*

significant effect on the burning rate. However, the experimental data are scarce, and both increased and decreased burning rates have been reported with increased lip height.

4. Wind Effects. Wind effects can change a pool burning rate by up to a factor of 2; very large velocities, greater than about 5 m/s, can blow some pool fires out. Wind effects have been reported to both increase and decrease the burning rate.

5. Very Large Pool Diameters. For very large pool diameters, greater than 5 or 10 m, a decrease in the burning rate on the order of 20 percent is usually noted. This is attributed

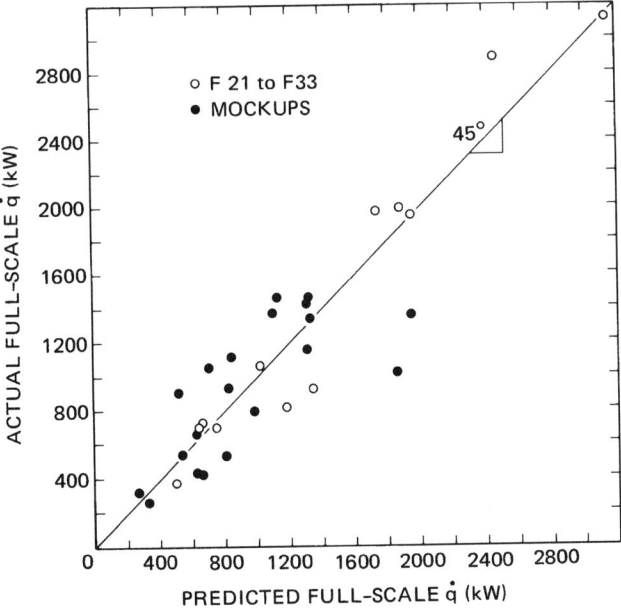

Fig. 3-1.8. *Correlation between bench-scale predictions and full-scale measurements for upholstered furniture.*

to poorer mixing, poorer combustion, and a layer of cold smoke above the surface.

Pools in room fires: The above analysis holds only for the case of pool fires in the open or else in a room where the temperature has not risen substantially. Pools are one of the few cases where a separate treatment is also available for the case of burning in postflashover fire. Thus, in a room fire, it is appropriate to use the relationships given above up to flashover (realizing that there may be a slight burning rate augmentation effect just prior to flashover), and a separate treatment after flashover. The postflashover analysis[11] requires solving calculations for the complete room fire, including the effects of ventilation, wall losses, and other heat balance terms. The steps required for a detailed analysis can be found in Babrauskas.[11]

Cribs

Cribs here are taken to mean regular, three-dimensional arrays of sticks. Each stick is of a square cross-section and of a length much greater than its thickness. The sticks are placed in alternatingly oriented rows, with an air space separating horizontally adjacent sticks. (See Figure 3-1.3). Wood crib burning rates have been studied longer than any other product, with early data available from the 1930s.[12] Different analysis formulas have been presented over the years by numerous authors. Here we present a method of analysis[11] based largely on the voluminous experimental data of Nilsson[13] on wood cribs and the functional form suggestions of Yamashika and Kurimoto.[14] The scant available data on plastic cribs are from Harmathy[15] and Quintiere and McCaffrey.[16] The conditions of most interest are when cribs are ignited instantaneously, as with the use of a small amount of combustible liquid underneath. The first group of equations below represents this case. There is occasionally an interest in a crib fire where only one end of a crib is ignited, and a slow fire propagation is seen. An analysis for this situation has also been made.[17] A similar analysis is also available for the center-ignited, fire-spreading crib scenario.[18]

TABLE 3-1.4 *Effective Heats of Combustion for Upholstered Furniture and Mattress Specimens*

Padding	Fabric	Frame	Average effective heat of combustion (MJ/kg)
	Construction		
PU† foam	polyolefin	wood	18.0
PU foam	cotton	wood	14.6
cotton	polyolefin	wood	16.1
cotton	cotton	wood	14.9
PU foam	polyolefin	polyurethane	20.9
PU foam	polyolefin	polypropylene	35.1
PU foam	polyolefin	*	30.4
PU foam	acrylic	*	18.4–22.9
PU foam	cotton	*	25.2
PU foam	wool	*	21.7
PU foam	cotton & rayon	*	14.4–23.0**
PU foam	PVC	*	12.7–24.9**
PU foam (hydrophilic)	PVC	*	8.8
cotton	cotton	*	5.7
cotton	PVC	*	7.5
latex	PVC	*	28.0
neoprene	cotton	*	6.7

*No frame, noncombustible frame, or mattress.

**Lower values for fire-retardant grades of foam, higher for nonretardant.

†Polyurethane.

For cribs ignited uniformly overall, it is observed that the burning rate can be governed by one of three conditions: (1) the natural limit of stick surfaces burning freely; this limit applies to cribs with wide interstick spacings; (2) the maximum flow rate of air and combustion products through the air holes in the crib; this governs for tightly packed cribs; and (3) the maximum oxygen that can be supplied to the room; this effect is discussed separately, below. The numerical expressions are as follows:

Fuel surface control:

$$\dot{m} = \frac{4}{D} m_0 v_p \left(1 - \frac{2v_p t}{D}\right) \quad (4)$$

or

$$\dot{m} = \frac{4}{D} m_0 v_p \left(\frac{m}{m_0}\right)^{1/2} \quad (4a)$$

with

$$m = m_0 - \sum_i^t \dot{m}_i(t_i) \, \Delta t$$

Crib porosity control:

$$\dot{m} = 4.4 \times 10^{-4} \left(\frac{S}{h_c}\right)\left(\frac{m}{D}\right) \quad (5)$$

Room ventilation control:

$$\dot{m} = 0.12 A_v \sqrt{h_v} \quad (6)$$

The *least* of Equations 4, 5, or 6 is to be taken as the governing rate (Equation 6 is discussed below): Equation 4a is necessary instead of the simpler Equation 4 when a switch of burning regime occurs during the course of the fire, e.g., the burning changes from porosity control to fuel surface control at some point. This can happen since Equation 4 [or (4a)] is a time-dependent expression. Thus, a crib may start burning under porosity or room ventilation-controlled conditions, then later switch to fuel surface control.

In the above equations, D is the stick thickness, m_0 is the crib initial mass, t is the time since ignition, h_c is crib height, S is the clear spacing between sticks, and room ventilation variables are A_v, the ventilation opening area, and h_v, the ventilation opening height. The fuel surface regression velocity, v_p, is taken as dependent on the stick thickness and on the fuel type, as shown in Table 3-1.3.

For the case of the center-ignited crib, the burning regimes are divided according to whether at a particular time the flame spread has reached the edge of the crib. This time is defined as t_0.

$$t_0 = 15.7n \quad (7)$$

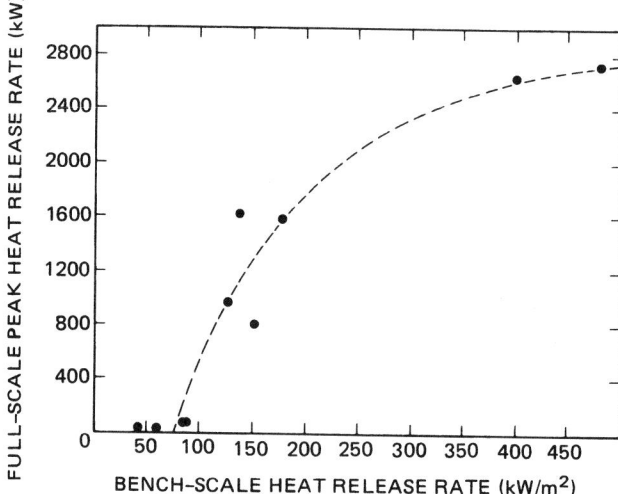

Fig. 3-1.10. *Correlation between bench-scale and full-scale heat release rate measurements on twin-size mattresses. The full-scale test mattresses included a standard set of bed linens.*

TABLE 3-1.5 *Heat Release Rate Data for Pillows*

Filling material	Fabric material	Pillow mass (kg)	Total mass* (kg)	Peak heat release rate (kW)	Total heat released (MJ)	Avg. heat of combustion (MJ/kg)
latex foam, one piece	50% cotton/ 50% polyester	1.003	1.238	117	27.5	27.6
polyurethane foam, shredded, #1	nonwoven	0.650	0.885	43	18.4	22.0
polyurethane foam, shredded, #2	nonwoven	0.628	0.863	35	18.9	23.7
polyester fiberfill	80% polyester/ 20% cotton	0.602	0.837	33	10.2	20.0
feathers	cotton	0.966	1.201	16**	8.9	18.3
polyester fiberfill	fiberglass	0.687	0.922	22	3.1	17.4

* Includes pillowcase and balled-up newspaper sheets used for ignition.

** Reading low due to slow ignition; otherwise expect $\cong$ 20 kW peak.

where n = the number of sticks per row. For time $t < t_0$, the following relation holds:[18]

$$\dot{m} = 0.0254 m_0 \frac{Dt^2}{n^2 v_p} \qquad (8)$$

For $t > t_0$, Equations 4 through 6 are used.

The heat release rate is determined as

$$\dot{q} = \Delta h_c \cdot \dot{m} \qquad (9)$$

where the heat of combustion, Δh_c, can be taken for wood as 12×10^3 kJ/kg, and values for other materials can be obtained from compilations.[19]

Room fire effects: Experimentally, it has long been observed[13] that, unlike a pool fire, which can burn in a room in a highly fuel-rich manner, a wood crib does not burn more than approximately 30 to 40 percent fuel rich. Conditions more fuel rich than that are not sustained, presumably, because of the highly vitiated air being supplied to the crib under those conditions. The stoichiometric fuel pyrolysis rate can be estimated as[11]

$$\dot{m}_p(\text{st}) = \frac{1}{r} \cdot 0.5 A \sqrt{h} \qquad (10)$$

where the stoichiometric air/fuel mass ratio, r, for wood can be taken as $r = 5.7$. Comparing, then, the maximum pyrolysis rate given by Equation 6 with the stoichiometric rate given by Equation 10, it can be seen that a limit of approximately 37 percent fuel rich is reached when Equation 6 becomes the governing limit to the burning rate. Similar limits may possibly exist for other classes of combustibles, but experimental data are only available for wood cribs.

Wood Pallets

Conceptually, a wood pallet is a similar arrangement to a wood crib. The dimensions, however, are substantially different. Instead of being composed of identical rows of square-section sticks, pallets are made up of rectangular elements in a traditionally dimensioned configuration. (See Figure 3-1.3.) When fire hazard of pallets alone is an issue, they are generally stacked many layers high. Krasner[20] has reported on a number of tests where the burning rate of pallets was measured. A typical experimental heat release rate curve is shown in Figure 3-1.4. This curve shows that, much like for a wood crib, a substantially constant plateau burning can be seen if the stack is reasonably high. The results for a standard pallet size of 1.22 m by 1.22 m can be given as a general heat release rate expression

$$\dot{q} = 1450(1 + 2.14 h_c)(1 - 0.027M) \qquad (11)$$

where h_c is stack height, M is moisture content, and a net heat of combustion of 12×10^3 kJ/kg has been assumed. For convenience in applying to nonstandard pallet sizes, this can be expressed on a per-unit-pallet-floor-area basis as

$$\dot{q}'' = 0.97(1 + 2.14 h_c)(1 - 0.027M) \qquad (12)$$

The agreement between Equations 11 and 12 and experimental data is seen to be good over a wide range of pallet heights (Figure 3-1.5), but the expressions do somewhat overpredict the burning rates if applied to short stacks, with stack height $h_c \lesssim 0.5$ m.

Upholstered Furniture

Within the category of upholstered furniture are included: upholstered chairs, sofas, two-seaters (love seats), and also upholstered transportation seating. Mattresses and beds share certain features of similarity, but are geometrically different and are treated separately below. Hassocks would be included as upholstered furniture, but there are no systematic burning rate data for them. A significant amount of both full-scale and bench-scale data is available on upholstered furniture. A NIST monograph reviews in depth the sources of experimental data;[21] thus only the summary predictive rules will be given here.

Figure 3-1.6 illustrates typical results from full-scale measurement of a burning upholstered chair in the furniture calorimeter. For modeling of fires and for hazard estimation, it is considered most important to predict the peak heat release rate. In addition, it is often useful to obtain an approximate representation of the curve shape, for which an extensive study has shown that a simple triangular representation is usually adequate. To fully define this triangular shape requires, in addition to the peak heat release height $\dot{q}$, the base width, t_b, and also the time from ignition (which is defined as $t = 0$) to the start of the triangular burning regime. This start-up time, however, is not a property of the upholstered item, but rather is mainly associated with the ignition

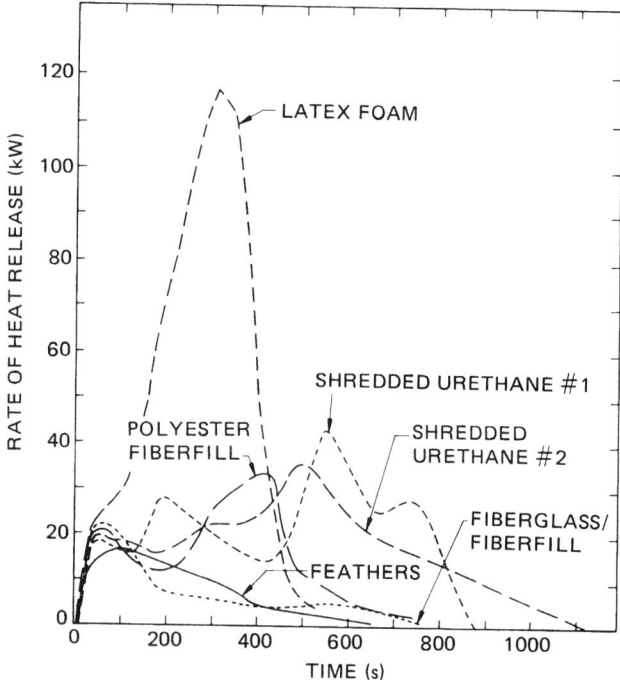

Fig. 3-1.11. *Pillow heat release rates.*

source and the ignition scenario. In other words, tests have shown that identical specimens ignited with different procedures (e.g., a cigarette versus a burning wastebasket) show the same $\dot{q}$ and the same t_b, but different start-up times.

To predict the peak heat release rate, a carefully developed analysis method is needed. It is interesting to examine what happens if a correlation is simply attempted with the total specimen fuel load (heat content), long considered in building codes as a good predictor of fire hazard.[22] Figure 3-1.7 shows that absolutely no correlation is achieved. To obtain a useful correlation, two different methods, one based on bench-scale testing and another, intended for field use, based solely on materials identification, have been developed.

Method based on bench-scale testing:

$$\dot{q} = 0.63[\dot{q}''_{bs}][\text{mass factor}][\text{frame factor}][\text{style factor}]$$

where

$$\left.\begin{array}{rl}
\dot{q} = & \text{estimated peak full-scale heat} \\
& \text{release rate} \\[4pt]
[\dot{q}''_{bs}] = & \text{rate of heat release in bench-scale} \\
& \text{test} \\[4pt]
[\text{mass factor}] = & \text{combustible mass, in kg} \\[4pt]
[\text{frame factor}] = & 1.66 \text{ for noncombustible frames} \\
& 0.58 \text{ for melting plastic} \\
& 0.30 \text{ for wood} \\
& 0.18 \text{ for charring plastic} \\[4pt]
[\text{style factor}] = & 1.0 \text{ for plain, primarily rectilinear} \\
& \quad \text{construction} \\
& 1.5 \text{ for ornate convoluted shapes} \\
& 1.25, \text{ or as appropriate, for inter-} \\
& \quad \text{mediate shapes}
\end{array}\right\} \quad (13)$$

Field method based only on generic materials identification:

$$\dot{q} = 210[\text{fabric factor}][\text{padding factor}][\text{mass factor}] \\
\cdot [\text{frame factor}][\text{style factor}]$$

where

$$\left.\begin{array}{rl}
[\text{fabric factor}] = & 1.0 \text{ for thermoplastic fabrics} \\
& \quad \text{(fabrics such as polyolefin,} \\
& \quad \text{which melt prior to burning)} \\
& 0.4 \text{ for cellulosic fabrics} \\
& \quad \text{(such as cotton)} \\
& 0.25 \text{ for PVC or polyurethane} \\
& \quad \text{film-type coverings} \\
& \quad \text{("imitation leather")} \\[4pt]
[\text{padding factor}] = & 1.0 \text{ for polyurethane foam or} \\
& \quad \text{latex foam} \\
& 0.4 \text{ for cotton batting} \\
& 1.0 \text{ for mixed materials} \\
& 0.4 \text{ for neoprene foam}
\end{array}\right\} \quad (14)$$

The method based on bench-scale testing is preferred whenever possible, since only by actually conducting laboratory tests is it possible to assess the effectiveness of any retardant formulations or improved treatments. The agreement between the predicted rate and the measured rate is shown in Figure 3-1.8 for data from two National Bureau of Standards (NBS) test series. For special applications, the alternate field method is presented for use in those instances where destructive testing is not feasible.

The procedures for the bench-scale testing, using the cone calorimeter, have been described in detail.[23] Briefly, the conditions entail an irradiance of 25 kW/m^2 for specimens oriented horizontally and prepared with certain edge protection precautions. The results reported are the average heat release rate for the first 180 s after ignition.

Neither method is applicable if $\dot{q}''_{bs}$ is less than approximately 75 kW/m^2 (or if the [fabric factor] × [padding factor] ≤ 0.23 for the field method). Specimens showing heat release rate values lower than this limit normally do not sustain flame spread over the complete surface. Since their total mass is not burned up, except possibly by smoldering, the predictive methods (Equations 13 and 14) would overestimate their actual burning rates.

Figure 3-1.9 illustrates for a typical upholstered furniture item how the heat release rate curve can be approximated by a triangular shape. The estimation for the triangle base width is taken as follows:[24]

$$t_b = \frac{Cm\Delta h_c}{\dot{q}} \quad (15)$$

where

$$\begin{array}{rl}
C = & 1.8 \text{ for noncombustible frames} \\
= & 1.3 \text{ for wood frames} \\
= & 1.8 \text{ for plastic frames} \\
m = & \text{combustible mass of item, and} \\
\Delta h_c = & \text{effective heat of combustion.}
\end{array}$$

A short tabulation of effective heats of combustion is given in Table 3-1.4. Additional data are available in the NFPA *Fire Protection Handbook.*[19]

Room fire effects: Only one test series has been reported where identical upholstered furniture items were tested in

TABLE 3-1.6 *Heat Release Rate Data for Wardrobes*

Construction	Wardrobe combustible mass (kg)	Clothing and paper (kg)	Peak heat release rate (kW)	Total heat content (MJ)	Avg. heat of combustion (MJ/kg)
metal	0	3.18*	770	70	14.8
metal	0	1.93**	270	52	18.8
plywood, 12.7 mm thick, #1	68.5	1.93**	3500	1067	NA
plywood, 12.7 mm thick, #2	68.3	1.93**	3100	1068	14.9
plywood, 3.2 mm thick	36.0	1.93**	6400	590	16.9
plywood, 3.2 mm thick, 1 coat FR paint	37.3	1.93**	5300	486	15.9
plywood, 3.2 mm thick, 2 coats FR paint	37.3	1.93**	2900	408	14.2
particleboard, 19 mm thick	120.3	0.81**	1900	1349	17.5

* Miscellaneous rags.

** Simulated hanging clothes.

TABLE 3-1.7 *Data for Television Set Burning Rates[30]*

Test No.	Total mass (kg)	Combustible mass burned (kg)	Peak q (kW)	Total heat content released (MJ)	Δh_c (MJ/kg)
1	32.7	10.2	230	146	14
2	27.2	5.8	120	NA	NA
3	39.8	10.2	290	150	15

open burning and in a room fire.[25] The room fires were all sized so that flashover would be reached and burning in a postflashover regime would occur, but that ventilation-limiting would not be reached. Under those conditions, to within experimental scatter, no effects of the room fire environment could be seen on the upholstered chair and love seat burning rates. This is interpreted to mean that the upholstered chair fire is relatively self-contained, with surfaces primarily viewing radiatively either other chair surfaces or an optically thick fire plume.

Mattresses

A large number of full-scale and bench-scale tests for mattresses have been conducted over the years at NBS and at other laboratories. Unfortunately, most of this testing predates the availability of current-day instrumentation. Nonetheless, by judicious assessment of existing data, a full-scale/bench-scale correlation could be deduced.[21] This correlation is shown in Figure 3-1.10 and is subject to the following limitations. The bench-scale specimens are exposed at a heating flux of 25 kW/m², with an average value being reported for the first 180 s after ignition. Specimen preparation is carried out in a similar manner as upholstered furniture specimens discussed above. The full-scale heat release rate values predicted hold only for twin-size mattresses, there being insufficient data for other mattress sizes. It is noted that, unlike for upholstered furniture, the correlation does not contain a mass factor. The reasons that a poorer prediction is obtained if a mass factor is introduced are not known at the present time.

Room fire effects: Of the room fire environment on burning mattresses only one test series[26] has explored the effects. The data were too limited to permit various fire effects to be separated out as well as may be desired; nonetheless, they

are suggestive of possibly a factor of 2 augmentation of the burning rate when a mattress is involved in a postflashover room fire. If supported by further tests, this would not be too surprising, since a mattress has a much more open geometry than an upholstered chair and may, thus, be more readily influenced by external radiant fluxes.

Pillows

Test data are available for pillows constructed of several common materials.[27] These data show that over 100 kW can be achieved with a latex foam pillow, and that common

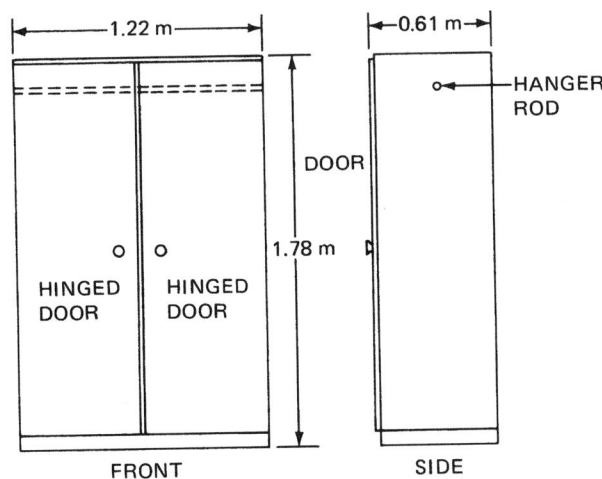

Fig. 3-1.12. Construction of a typical wardrobe tested.

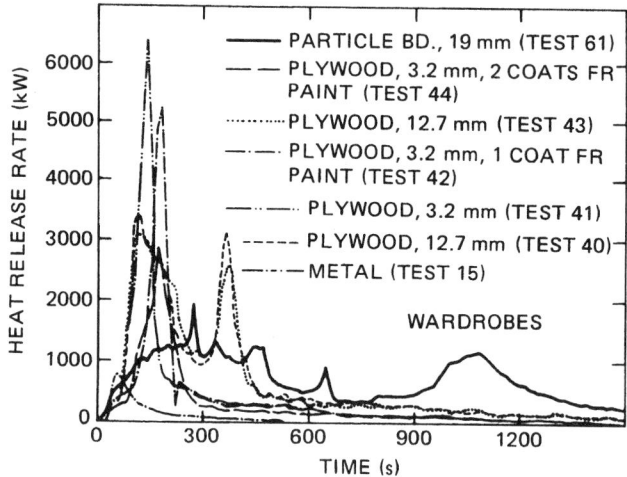

Fig. 3-1.13. Heat release rate results for wardrobes.

polyurethane pillows show heat release rate values of around 40 kW. (See Table 3-1.5 and Figure 3-1.11.) Thus, they can readily serve, for example, as a continued source of heating for a mattress.

Wardrobes

An exhaustive study of wardrobe fires is not available, nor is a specific prediction method. Nonetheless, data are available from a series of tests that can be useful in a design application if the design wardrobes are a reasonable match to the tested ones.[28] The test wardrobes are illustrated in

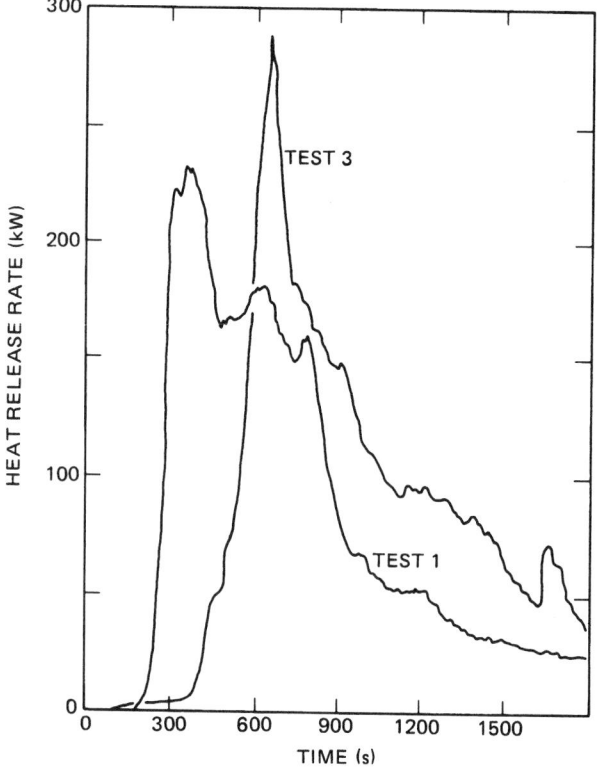

Fig. 3-1.14. Heat release rate results for television sets.

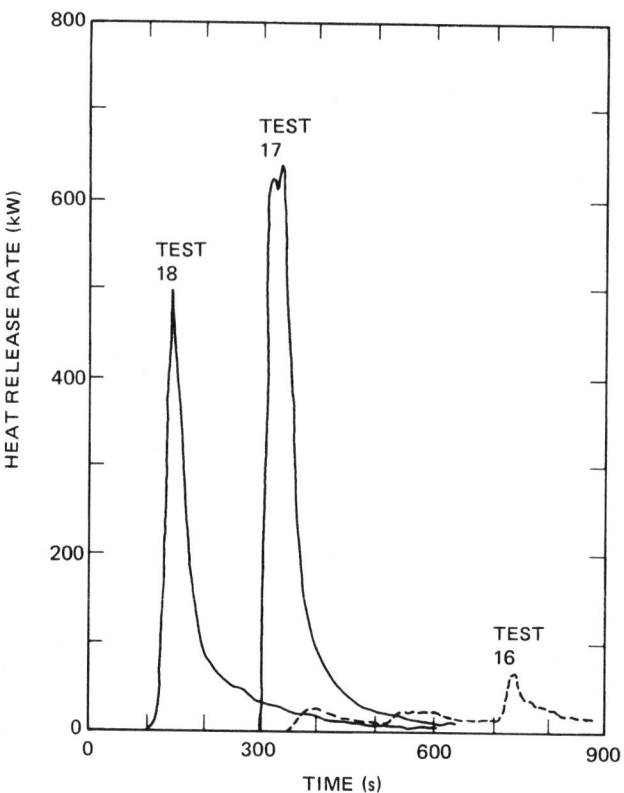

Fig. 3-1.15. Heat release rate results for Christmas trees.

Figure 3-1.12; data are given in Table 3-1.6. The wardrobes were outfitted with a small amount of clothing, or simulated clothing, and some paper, as listed in Table 3-1.6. Tests were not run on the clothes items by themselves. However, since in the case of the two metal wardrobe tests the only other combustible present was the paint on the metal, it is reasonable to assign a value of about 270 kW peak for the 1.93 kg clothes load. Ignition in all cases was by a match held to some of the contents' paper. The measured heat release curves are shown in Figure 3-1.13. From the results, it can be concluded that painting with flame-retardant (FR) paint is only effective under certain circumstances, i.e., when the paint is appropriately chosen and a sufficiently thick layer is developed. The most important conclusion, however, is that the peak heat release rate is inversely dependent on wardrobe panel thickness (and, by contrast, no simple connection to combustible specimen mass is seen). Thus, while the total heat content of the 19 mm particleboard specimen is high (see Table 3-1.6), its peak heat release rate is quite low, since flame spread and fire involvement proceed more slowly over a thick material. A similar effect was also noted for wall-lining insulation materials, when tested in varying thicknesses.[29]

Television Sets

Data on three tests of console-type television sets using wooden cabinets and 24 to 26 in. picture tube size were reported by the Valtion Teknillinen Tutkimuskeskus (VTT).[30] These sets were ignited with 100 mL of isopropanol; this ignition source alone constituted a small heat release rate of 4 kW. The data are summarized in Table 3-1.7,

TABLE 3-1.8 *Data for Christmas Tree Burning Rates[30]*

Test No.	Conditioning	Total mass (kg)	Peak $\dot{q}$ (kW)	Total heat content released (MJ)	Δh_c (MJ/kg)
16	green	6.5	69	11	NA
17	dry	7.0	650	41	NA
18	dry	7.4	500	30	NA

and typical curves are shown in Figure 3-1.14. It can be seen that, with peaks of only 120 to 290 kW, these fires are not severe by themselves but could serve to further involve other combustibles in a room.

Christmas Trees

Christmas trees are often associated with serious fires. In most scenarios, a number of additional combustibles are involved; no data are available on a living room, total fire. The VTT, however, has conducted tests on Christmas trees by themselves.[30] Three trees were tested (see Table 3-1.8). The first one was cut, left to stand outdoors, then brought in and kept in a 15°C room for two days prior to test. The two other trees were dried out with heat lamps to simulate a week's exposure in a warm, dry room. The trees were ignited

with a small amount of isopropanol. The first tree burned only sporadically and released a small amount of heat. The latter two, however, burned vigorously and produced heat release rate peaks in the vicinity of 600 kW. (See Figure 3-1.15.)

Curtains

The heat content and burning rate of curtains is generally moderate, but they are combustible and can contribute to the severity of fires by quickly propagating fire over large surfaces. Moore has done the most extensive study of curtains and draperies.[31] His test specimens were ignited with a match along the bottom. The results are summarized in Table 3-1.9 and Figure 3-1.16. His results show primarily the effect of fabric weight. Lightweight fabrics, of weight around 125 kg/m², can show heat release rate peaks almost as high

TABLE 3-1.9 *Data for Curtain Burning Rates.[31] Nominal Curtain Size: Two Curtains Each, 2.13 m High by 1.25 m Wide. Wall Area Covered: 2.13 m High by 1.0 m Wide (in Closed Position).*

Type of fiber	Weight (g/m²)	Configuration	Peak $\dot{q}$ (kW)	Number of wall and ceiling panels ignited*
cotton	124	closed	188	1
cotton	260	closed	130	7
cotton	124	open	157	0
cotton	260	open	152	7
cotton	313	closed	600	3
rayon/cotton	126	closed	214	0
rayon/cotton	288	closed	133	6
rayon/cotton	126	open	176	0
rayon/cotton	288	open	191	2
rayon/cotton	310	closed	177	8
rayon/acetate	296	closed	105	4
acetate	116	closed	155	0
cotton/polyester	117	closed	267	1
cotton/polyester	328	closed	338	5
cotton/polyester	117	open	303	0
rayon/polyester	367	closed	658	2
rayon/polyester	268	closed	329	7
rayon/polyester	53	closed	219	0
cotton/polyester	328	open	236	7
polyester	108	closed	202	0
acrylic	99	closed	231	0
acrylic	354	closed	1177	8
acrylic	99	open	360	0
acrylic	354	open	NA	7
cotton/polyester/foam	305	closed	385	1
rayon/polyester/foam	284	closed	326	0
rayon/fiberglass	371	closed	129	5
rayon/fiberglass	371	closed	106	5

* Maximum possible number of panels to ignite = 10.

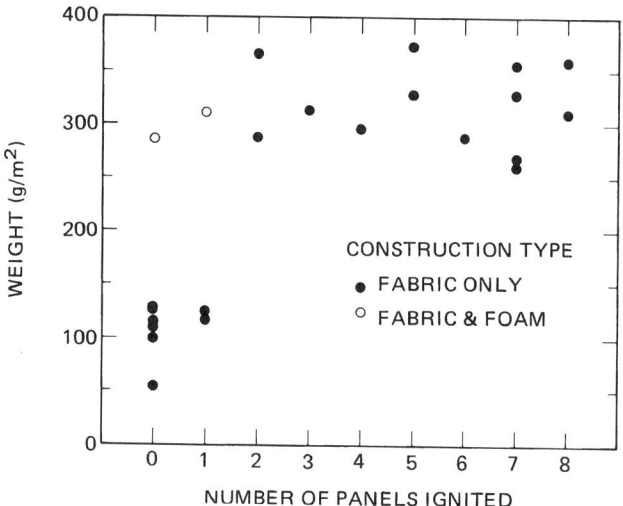

Fig. 3-1.16. *Heat release rate results for curtains.*

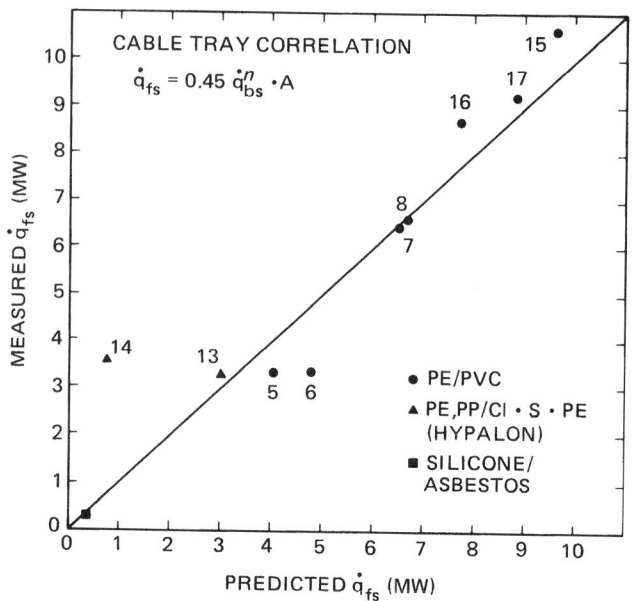

Fig. 3-1.17. *Heat release rate correlation for cable trays (numbers at data points identify full-scale tests).*

as heavy ones (around 300 kg/m^2); however, their potential to ignite surrounding objects is much smaller, as demonstrated in Figure 3-1.16. These conclusions hold for both thermoplastic and cellulosic materials, but not for constructions using foam backings, for which insufficient data were available. Whether the curtain was in the closed or in the open position seemed to make little difference. The reason for the more severe fire performance of the heavyweight curtains was largely due to their increased burning time, which was typically about twice that for the lightweight curtains.

Cable Trays

Cable tray fires present almost an endless plethora of combinations of cable materials, tray construction, stacking, ignition sources, etc. Only a very few of these have been explored. The most systematic studies available are those from Tewarson et al[32] and Sumitra.[33] A useful engineering analysis of their data has been prepared by Lee.[34] Lee provided a basic correlation of Tewarson's and Sumitra's data

TABLE 3-1.10 *Heat Release Rates of Typical Cables in Bench-Scale Tests*

Specimen number	Cable sample	IEEE 383 test	FM heat release rate test* (kW/m^2)
20	Teflon	pass	98
21	silicone, glass braid	pass	128
10	PE, PP/Cl · S · PE	pass	177
14	XPE/XPE	pass	178
22	silicone, glass braid, asbestos	pass	182
16	XPE/Cl · S · PE	pass	204
18	PE, nylon/PVC, nylon	**	218
19	PE, nylon/PVC, nylon	**	231
15	FRXPE/Cl · S · PE	pass	258
11	PE, PP/Cl · S · PE	pass	271
8	PE, PP/Cl · S · PE	pass	299
17	XPE/Neoprene	pass	302
3	PE/PVC	**	312
12	PE, PP/Cl · S · PE	pass	345
2	XPE/Neoprene	**	354
6	PE/PVC	**	359
4	PE/PVC	fail	395
13	XPE/FRXPE	pass	475
5	PE/PVC	fail	589
1	ldPE	**	1071

* Tested at an irradiance of 60 kW/m^2.

** Test not conducted.

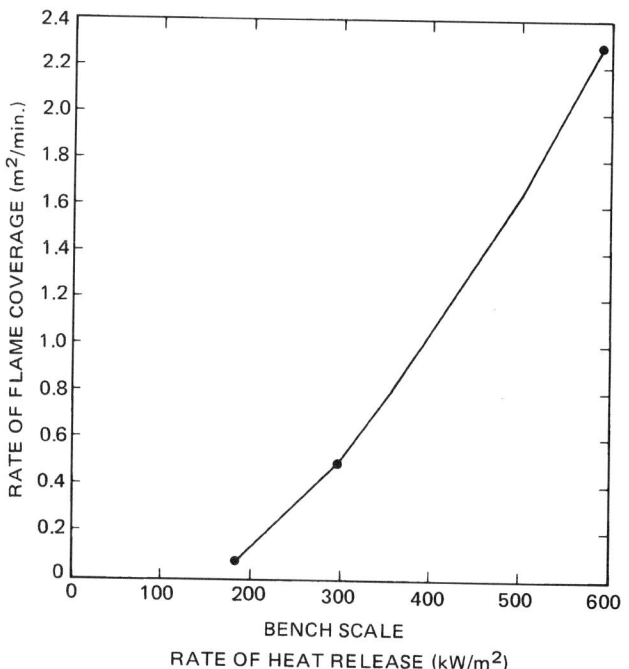

Fig. 3-1.18. *Effect of bench-scale cable heat release rate on full-scale rate of flame coverage.*

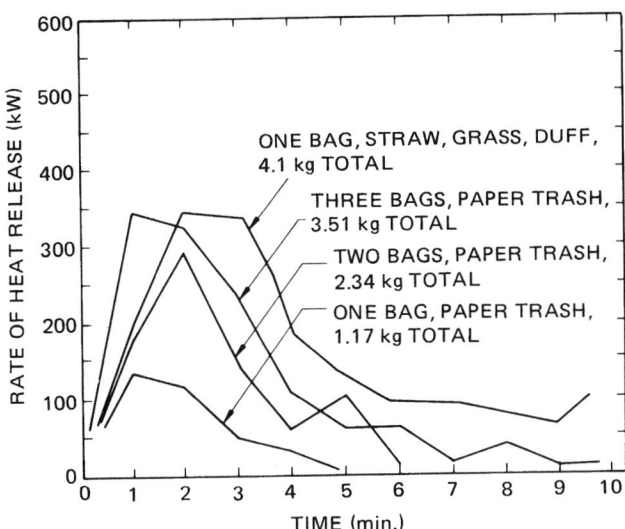

Fig. 3-1.19. *Heat release rates for trash bags.*[34]

Trash Bags and Containers

Bench-scale measurements of trash are not readily feasible, due to the naturally irregular arrangement of these combustibles. There are full-scale test results available, however, that can suggest appropriate values to be used in different circumstances. Some typical trash-bag fires are shown in Figure 3-1.19;[34] a typical small plastic wastebasket fire is shown in Figure 3-1.20.[5] Lee has correlated the peak heat release values according to the effective base diameter and packing density.[34] Figure 3-1.21(a) shows that the total burning rate (kW) increases with effective base diameter, but decreases with the tighter packing densities. Figure 3-1.21(b), conversely, illustrates that the increase with diameter is not to as great a degree as to be proportional to base area; thus, when per-unit-base-area results (kW/m^2) are expressed, a downward trend is seen. The correlations according to packing density should only be considered rough observations, and not firm guidelines. Table 3-1.11 shows some additional data,[30] where, over a certain range,

(see Figure 3-1.17), which shows that the peak full-scale heat release rate can be predicted according to bench-scale heat release rate measurements

$$\dot{q}_{fs} = 0.45 \dot{q}''_{bs} \cdot A \tag{16}$$

where the bench-scale heat release value is the peak, measured under irradiance conditions of 60 kW/m^2, and A is the exposed tray area actively pyrolyzing. The active pyrolysis area, in turn, is estimated from Figure 3-1.18, which gives dA/dt as a function of $\dot{q}''_{bs}$. Thus, at any given time, t,

$$A(t) = A_0 + \frac{dA}{dt} \cdot t \tag{17}$$

Finally, Table 3-1.10 gives a selection of measured values of $\dot{q}''_{bs}$ for various cable types.

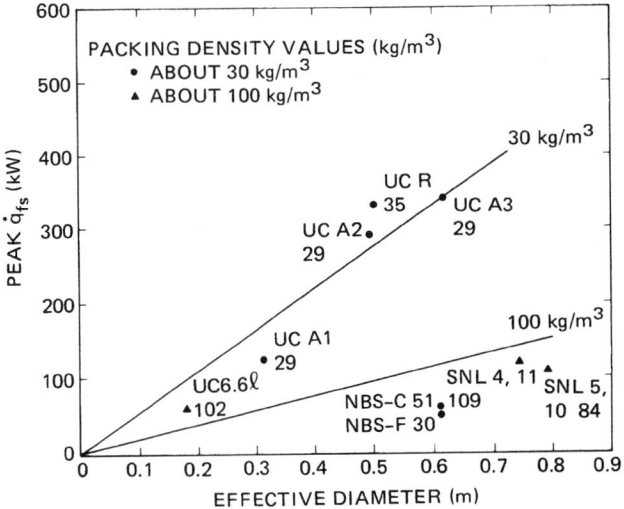

Fig. 3-1.21(a). *Summary of trash heat release rates—results for a total heat release rate.*

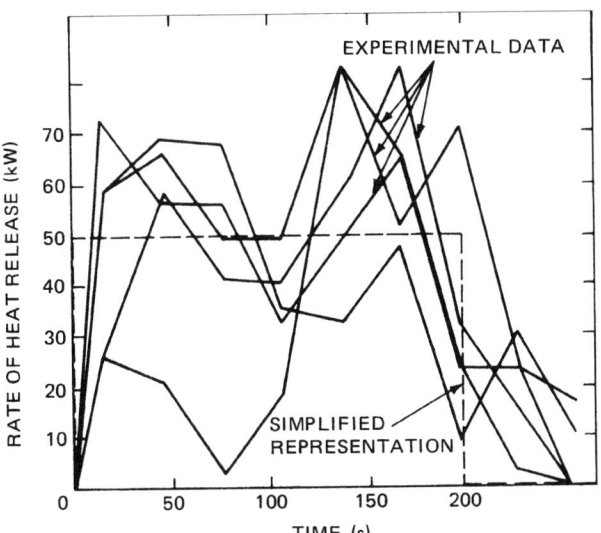

Fig. 3-1.20. *Heat release rates for a polyethylene 6.6 L wastebasket stuffed with 12 milk cartons.*

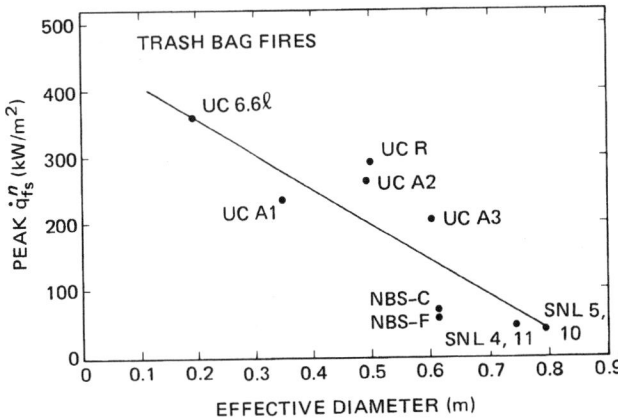

Fig. 3-1.21(b). *Summary of trash heat release rates—results for heat release rate per unit base area.*

TABLE 3-1.11   *Some 14 L Polyethylene Wastebasket Data Obtained at VTT Showing Effect of Packing Density and Basket Construction*

Basket sides	Basket mass (kg)	Filling type	Filling mass (kg)	Filling density (kg/m³)	Peak $\dot{q}$ (kW)	Total heat released (MJ)
solid	0.63	shredded paper	0.20	14	4	0.7
netted	0.63	milk cartons	0.41	29	13	3.0
solid	0.53	shredded paper	0.20	14	18	7.3
netted	0.53	milk cartons	0.41	29	15	5.8

increasing packing density increases the heat release rate. For design purposes, the range of 50 to 300 kW appears to cover the bulk of the expected fires from normal residential, office, airplane, or similar occupancy trash bags and trash baskets.

Stored Commodities

A large number of industrial commodities have been tested at the Factory Mutual Research Corporation and at other laboratories. Most of these data are not available to the designer for one or both of two reasons: (1) much of the data is proprietary; and (2) the tested configurations tend to be highly specific to one application. A certain amount of general utility data has been published[35-37] and is summarized in Table 3-1.12. The spread of values shown refers to the same test in each case, but a different method of estimating the peak heat release rate on the basis of recorded load cell values. In an actual application, the additional variations due to differences in ignition sources and locations, stacking arrangements, and so forth, would have to be considered.

LIMITATIONS

The ability to predict the burning rates of full-scale combustibles, at the moment, varies greatly with the com-bustible. For some, detailed studies and estimating methods are available, permitting rather close estimates. For others, only factor-of-4-type estimates may be possible. The predictive techniques are all based on an empirical analysis. In some cases the form of the relations is guided by a theoretical description, but in no case is it yet possible to predict the burning rates of practical combustibles simply on the basis of thermophysical and thermochemical data. Even in the "ideal" case of a liquid pool, the relationships require actual test data.

NOMENCLATURE

A	area (m²)
A_0	ignition area (m²)
A_v	area of ventilation opening (m²)
C	empirical constant
D	pool diameter; stick thickness (m)
h_c	crib or pallet height (m)
Δh_c	heat of combustion (kJ/kg)
h_v	height of ventilation opening (m)
k	extinction-absorption coefficient (m⁻¹)
m	mass (kg)
$\dot{m}$	mass loss rate (kg/s)
$\dot{m}''$	mass loss rate per unit area (kg/m² s)

TABLE 3-1.12   *Data for Stored Commodities Tested at Factory Mutual Research Corp.*

Commodity	Peak heat release rate per unit floor area* (kW/m²)	Growth time to reach 1 MW (s)
mail bags, filled, stored 1.5 m high	400 –	190
cartons, compartmented, stacked 4.6 m high	1700 – 4200	60
cartons, tri-wall cardboard, metal lined, stacked 4.9 m high	– 2800	
polyethylene bottles, packed in cartons as above	6200 – 7600	85
polystyrene jars, packed in cartons as above	14000 – 20900	55
PVC bottles, packed in cartons as above	3400 – 7000	95
polypropylene food tubs, packed in cartons as above	4400 – 9600	100
polyethylene letter trays, stacked 1.5 m high on cart	8500 –	180
polyethylene trash barrels in cartons, stacked 4.6 m high	2000 –	55
polyethylene bottles, random size, in cartons, stacked 4.6 m high	2000 – 4200	75
polyethylene bottles, large, in cartons, stacked 4.67 m high	– 7300	
polyethylene/fiberglass shower stalls in cartons, stacked 4.6 m high	1400 –	85
polyurethane rigid foam insulation board, stacked 4.6 m high	1900 – 3200	8
polystyrene rigid foam insulation board, stacked 4.3 m high	3300 –	6
polystyrene food tubs, with covers, nested in cartons, stacked 4.3 m high	5400 – 8200	120
polystyrene meat trays, wrapped in plastic sheet	– 12900	
polystyrene meat trays, wrapped in paper	– 13300	
polystyrene toy parts in cartons, stacked 4.6 m high	2000 – 6500	125
polyethylene and polypropylene film in rolls, stacked 4.3 m high	6200 –	40

* Estimates from Alpert and Ward[37] and Delichatsios.[36]

$\dot{m}''_\infty$ mass loss rate per unit area for very large pool diameters $(kg/m^2\,s)$
M moisture content (percent)
n number of sticks per row
$\dot{q}$ heat release rate (kW)
$\dot{q}''$ heat release rate per unit area (kW/m^2)
$\dot{q}''_{bs}$ bench-scale heat release rate (kW/m^2)
$\dot{q}_{fs}$ full-scale heat release rate (kW)
r stoichiometric air/fuel mass ratio $(-)$
S clear spacing between sticks (m)
t time (s)
t_b triangular burning time (s)
v_p fuel surface regression rate (m/s)
β mean-beam-length corrector $(-)$

REFERENCES CITED

1. V. Babrauskas and R.B. Williamson, "Post-flashover Compartment Fires—Application of a Theoretical Model," *Fire and Matls.*, 3, 1 (1979).
2. O. Pettersson, S.E. Magnusson, and J. Thor, *Fire Engineering Design of Steel Structures*, Stalbyggnadsinstitutet, Stockholm (1976).
3. M. Janssens and W.J. Parker, "Oxygen Consumption Calorimetry," *Heat Release in Fires*, V. Babrauskas and S.J. Grayson, eds., Elsevier Applied Science, London, pp. 31–59 (1992).
4. G. Heskestad, "A Fire Products Collector for Calorimetry into the MW Range," *FMRC J.I. OC2E1.RA*, Factory Mutual Research Corp., Norwood (1981).
5. V. Babrauskas, J.R. Lawson, W.D. Walton, and W.H. Twilley, "Upholstered Furniture Heat Release Rates Measured with a Furniture Calorimeter," *NBSIR 82-2604*, National Bureau of Standards, Washington (1982).
6. ASTM E906 83, *Standard Test Method for Heat and Visible Smoke Release Rates for Materials and Products*, American Society for Testing and Materials, Philadelphia (1993).
7. A. Tewarson, "Scale Effects on Fire Properties of Materials," *FMRC J.I. 0J4N2.RC*, Factory Mutual Research Corp., Norwood (1984).
8. V. Babrauskas, "Development of the Cone Calorimeter—A Bench-scale Heat Release Rate Apparatus Based on Oxygen Consumption," *NBSIR 82-2611*, U.S. National Bureau of Standards, Washington (1982).
9. P.A. Croce, *Calculating Impacts for Large Open Hydrocarbon Fires*, SFPE Symposium, College Park (1985).
10. V. Babrauskas, "Estimating Large Pool Fire Burning Rates," *Fire Tech.*, 19, 251 (1983).
11. V. Babrauskas, "A Closed-form Approximation for Post-flashover Compartment Fire Temperatures," *F. Safety J.*, 4, 63 (1981).
12. F. Folke, "Experiments in Fire Extinguishment, *NFPA Quarterly*, 31, 115 (1937).
13. L. Nilsson, "The Effect of Porosity and Air Flow on the Rate of Combustion of Fire in an Enclosed Space," *Bulletin 18*, Lund Institute of Technology, Lund (1971).
14. S. Yamashika and H. Kurimoto, "Burning Rate of Wood Crib," *Rept. of Fire Res. Inst. Japan*, 41, 8 (1976).
15. T.Z. Harmathy, "Experimental Study on the Effect of Ventilation on the Burning of Piles of Solid Fuels," *Comb. and Flame*, 31, 259 (1978).
16. J.G. Quintiere and B.J. McCaffrey, "The Burning of Wood and Plastic Cribs in an Enclosure, Vol. 1," *NBSIR 80-2054*, National Bureau of Standards, Washington (1980).
17. W.L. Fons, H.B. Clements, and P.M. George, "Scale Effects on Propagation Rate of Laboratory Crib Fires." in *Ninth International Symposium on Combustion*, The Combustion Institute, Pittsburgh (1962).
18. M.A. Delichatsios, "Fire Growth Rates in Wood Cribs," *Comb. and Flame*, 27, 267 (1976).
19. A.E. Cote, ed., "Tables and Charts," *Fire Protection Handbook*, 17th ed., National Fire Protection Association, Quincy (1991).
20. L.M. Krasner, "Burning Characteristics of Wooden Pallets as a Test Fuel," *Serial 16437*, Factory Mutual Research Corp., Norwood (1968).
21. V. Babrauskas and J.F. Krasny, *Fire Behavior of Upholstered Furniture* (NBS Monograph 173), National Bureau of Standards, Washington (1985).
22. S.H. Ingberg, "Tests of the Severity of Building Fires," *NFPA Quarterly*, 22, 43 (1928).
23. V. Babrauskas and J.F. Krasny, *Prediction of Upholstered Chair Heat Release Rates from Bench-Scale Measurements*, Fire Safety Science and Engineering (ASTM STP 882), American Society for Testing and Materials, Philadelphia, pp. 268–284 (1985).
24. V. Babrauskas and W.D. Walton, "A Simplified Characterization for Upholstered Furniture Heat Release Rates," *F. Safety J.*, 11, pp. 181–192 (1986).
25. V. Babrauskas, "Upholstered Furniture Room Fires—Measurements, Comparison with Furniture Calorimeter Data, and Flashover Predictions," *J. Fire Sci.*, 2, 5 (1984).
26. B.T. Lee, "Effect of Wall and Room Surfaces on the Rate of Heat, Smoke, and Carbon Monoxide Production in a Park Lodging Room Fire," *NBSIR 85-2998*, National Bureau of Standards, Washington (1985).
27. V. Babrauskas, "Pillow Burning Rates," *F. Safety J.*, 8, 199 (1984/85).
28. J.R. Lawson, W.D. Walton, and W.H. Twilley, "Fire Performance of Furnishings in the NBS Furniture Calorimeter. Part I," *NBSIR 83-2787*, National Bureau of Standards, Washington (1983).
29. B.T. Lee, "Fire Hazard Evaluation of Shipboard Hull Insulation and Documentation of a Quarter-Scale Room Fire Test Protocol," *NBSIR 83-2642*, National Bureau of Standards, Washington (1985).
30. A. Ahonen, M. Kokkala, and H. Weckman, "Burning Characteristics of Potential Ignition Sources of Room Fires," *Research Report 285*, Valtion Teknillinen Tutkimuskeskus, Espoo (1984).
31. L.D. Moore, "Full-scale Burning Behavior of Curtains and Drapes," *NBSIR 78-1448*, National Bureau of Standards, Washington (1978).
32. A. Tewarson, J.L. Lee, and R.F. Pion, "Categorization of Cable Flammability. Part I. Experimental Evaluation of Flammability Parameters of Cables Using Laboratory-scale Apparatus," *EPRI Project RP 1165-1*, Factory Mutual Research Corp., Norwood (1979).
33. P.S. Sumitra, "Categorization of Cable Flammability. Intermediate-Scale Fire Tests of Cable Tray Installations," *Interim Report NP-1881, Research Project 1165-1*, Factory Mutual Research Corp., Norwood (1982).
34. B.T. Lee, "Heat Release Rate Characteristics of Some Combustible Fuel Sources in Nuclear Power Plants," NBSIR 85-3195, National Bureau of Standards, Washington (1985).
35. R.K. Dean, "Stored Plastics Test Program," *Serial 20269*, Factory Mutual Research Corp., Norwood (1975).
36. M.A. Delichatsios, "A Scientific Analysis of Stored Plastic Fire Tests," *Fire Sci. and Tech.*, 3, 73 (1983).
37. R.L. Alpert and E.J. Ward, "Evaluation of Unsprinklered Fire Hazards," *F. Safety J.*, 7, 127 (1984).

CALORIMETRY

Marc Janssens

INTRODUCTION

Rate of heat release is the primary variable that determines the contribution to compartment fire hazard from materials. This was clearly demonstrated by Babrauskas and Peacock in a recent sensitivity study using the NIST fire hazard assessment software Hazard I.[1] However, the importance of heat release rate in fire hazard assessment was first recognized two decades ago by Smith at Ohio State University.[2] Smith and co-workers developed one of the first bench-scale heat release rate test methods.[3] They also proposed various procedures to assess compartment fire hazard on the basis of the bench-scale data, ranging from simple calculation methods[4] to a complex computer model.[5] This work was initiated at a time when the most accurate measuring techniques for heat release rate were not available, and when computer fire modeling was still in its infancy. Moreover, Smith advocated a practical approach based on engineering judgment and intuition rather than detailed science. Hence, his test and fire hazard assessment methods were far from perfect and received major criticism.[6,7] Nevertheless, Smith deserves recognition as one of the pioneers of heat release rate calorimetry.

There are two reasons that explain why heat release rate is so important. First, it is directly related to mass loss rate. The toxic fire hazard of a material is a function of the release rate of toxic gases, which is the product of total mass loss rate and yield of these gases. Thus, the fire hazard of material A, with a high yield of toxic gases, might not exceed that of material B with a lower yield, if the mass loss rate of A under identical exposure conditions is significantly lower. For example, many fire-retardant treatments increase the yield of toxic gases, but dramatically reduce the mass loss rate, resulting in a lower fire hazard. This was illustrated in the study by Babrauskas and Peacock.[1] Second, the heat released by a material burning in a compartment results in a temperature rise of the hot layer gases and compartment walls and ceiling. Part of the radiation from hot surfaces and gases strikes the fuel surface, resulting in an increase in the mass loss rate over that if the material were to burn outside a compartment. Increased heat release enhances this thermal feedback effect.

With compartment fire hazard assessment as the primary application, there is a need for high-quality heat release rate data, and, consequently, for devices and methods to measure it accurately. There are two basic approaches to evaluate the fire hazard of a material. The first option consists of an experimental evaluation in full scale. Typically, this approach requires multiple large-scale fire tests covering all relevant fire scenarios and end-use conditions. The second option is the use of bench-scale data, primarily heat release rate, in conjunction with a calculation procedure to estimate full-scale fire performance. This second approach is significantly more versatile, and time and cost efficient. With the continuous improvement of the predictive capability and accuracy of fire models and calculation methods, it has become the preferred approach. With this in mind, the emphasis of this chapter is on bench-scale calorimetry.

Many reaction-to-fire test methods include a (often crude) measure of heat release rate.[8,9,10,11,12] However, with one exception (the model box test in Ref. 12), these test methods were developed several decades ago and their measuring techniques are inaccurate and obsolete. Moreover, exposure conditions were not well defined and controlled, so that it is difficult to relate the results of these early tests to real fire performance. The discussion in this chapter does not include these reaction-to-fire tests, and is limited to calorimetric methods that were developed primarily for measuring heat release rate. These heat release test methods vary widely in concept and features. Four measuring techniques have been used: (1) sensible enthalpy rise method, (2) substitution method, (3) compensation method, and (4) oxygen consumption method. These are described in detail in the section "Techniques for Measuring Heat Release Rate." This is followed by a discussion of the effect of various calorimeter features and construction details. Then, a brief description is given of the many bench-scale calorimeters that were developed since the early 1970s. A comparison between several calorimeters is also included. The chapter is concluded with a discussion of some large-scale calorimeters. However, it is appropriate to start with a review of different methods for using heat release rate data in fire hazard assessment, illustrating what kind of data a calorimeter needs to provide.

Dr. Marc Janssens is Manager of Fire Technology at the American Forest & Paper Association.

USE OF BENCH-SCALE HEAT RELEASE RATE DATA

Because of the volume of available data, a general discussion of this subject would be either incomplete or lengthy. Therefore, the focus provided here is on one particular fire scenario; namely, wall linings in a room/corner test. Such a test consists of a small room with a single ventilation opening in the front narrow wall. Specimens of the material to be evaluated are attached to walls and/or ceiling. A gas burner ignition source is placed in contact with the walls in one of the rear corners. Products of combustion emerging through the ventilation opening are collected in a hood, and extracted through an exhaust duct. Instrumentation is located in the duct to measure the mole fraction of various gas species, total flow rate, smoke obscuration, etc. Room/corner test methods have become a popular tool to evaluate the fire hazard of linings, and have been standardized throughout the world.[13,14,15,16] These tests are discussed in more detail in the section on large-scale calorimeters.

Techniques for using bench-scale data to predict performance in the room/corner test differ widely in degree of complexity and sophistication. Requirements for the calorimeter that provide bench-scale data vary accordingly.

Interpretation of Bench-Scale Data in Terms of Material Properties

The most sophisticated technique to predict room/corner test performance uses heat release rate and other bench-scale measurements to obtain material properties. These properties (ideally) are apparatus-independent, and are used as input for a computer model that predicts full-scale fire performance. The two properties that are related to heat release rate are: (1) the effective heat of combustion, $\Delta h_{c,\,eff}$ (kJ $\cdot$ g^{-1}); and (2) the heat of gasification, Δh_g (kJ $\cdot$ g^{-1}). These properties are described below.

The effective heat of combustion is the ratio of heat release rate to mass loss rate

$$\Delta h_{c,\,eff} \equiv \frac{\dot{q}''}{\dot{m}''} \qquad (1)$$

where

$\dot{q}''$ = heat release rate per unit exposed area (kW $\cdot$ m^{-2}), and
$\dot{m}''$ = mass loss rate per unit exposed area (g $\cdot$ m^{-2} $\cdot$ s^{-1}).

The symbol $\Delta h_{c,\,eff}$ is used to make a distinction between this property and the lower calorific value measured in an oxygen bomb calorimeter, $\Delta h_{c,\,net}$. The latter is measured in a small container under high pressure and in pure oxygen, conditions that are not representative of real fires. The conditions in the new types of bench-scale calorimeters resemble those in real fires much more closely. For some fuels, in particular gases, $\Delta h_{c,\,eff}$ is nearly identical to $\Delta h_{c,\,net}$. However, for most solid materials $\Delta h_{c,\,eff}$ is significantly lower, and is the value to be used for fire hazard assessment. The issue is revisited in the next section.

The second material property is heat of gasification, Δh_g, defined as the net heat flow into the material required to convert one mass unit of solid material to volatiles. The net heat flux into the material can be obtained from an energy balance at the surface of the specimen. (See Figure 3-2.1.) Typically, a sample exposed in a calorimeter is heated by external heaters and by its own flame. Heat is lost from the

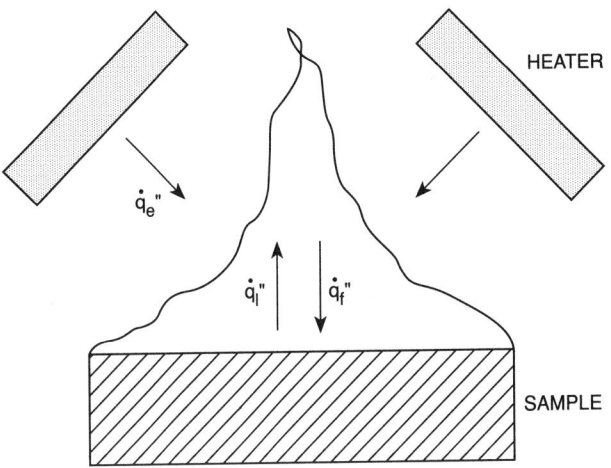

Fig. 3-2.1. Surface heat balance.

surface in the form of radiation. Due to the small sample size, the flame flux is primarily convective, and flame absorption of external heater and specimen surface radiation can be neglected. Hence, $\Delta h_{c,\,eff}$ can be defined as

$$\Delta h_g \equiv \frac{\dot{q}''_{net}}{\dot{m}''} = \frac{\dot{q}''_e + \dot{q}''_f - \dot{q}''_l}{\dot{m}''} \qquad (2)$$

If the flame is approximated as a homogeneous grey gas volume, the fluxes in the numerator of Equation 2 can be written as

$$\dot{q}''_f = \dot{q}''_{f,c} + \dot{q}''_{f,r} = h^*(T_f - T_s) + \sigma \epsilon_f T_f^4 \approx h^*(T_f - T_s) \qquad (3)$$

and

$$\dot{q}''_l = \sigma \epsilon_s (T_s^4 - T_\infty^4) \qquad (4)$$

where

h^* = convection coefficient, corrected for blowing (kW $\cdot$ m^{-2} $\cdot$ K^{-1}),
$\dot{q}''_e$ = incident heat flux from external heaters (kW $\cdot$ m^{-2}),
$\dot{q}''_f$ = flame flux (kW $\cdot$ m^{-2}),
$\dot{q}''_{f,c}$ = convective fraction of the flame flux (kW $\cdot$ m^{-2}),
$\dot{q}''_{f,r}$ = radiative fraction of the flame flux (kW $\cdot$ m^{-2}),
$\dot{q}''_l$ = radiative losses from the surface (kW $\cdot$ m^{-2}),
$\dot{q}''_{net}$ = net heat flux into the sample (kW $\cdot$ m^{-2}),
T_f = flame temperature (K),
T_s = surface temperature (K),
T_∞ = ambient temperature (K),
ϵ_f = flame emissivity,
ϵ_s = surface emissivity, and
σ = Boltzmann's constant (5.67 $\cdot$ 10^{-11} kW $\cdot$ m^{-2} $\cdot$ K^{-4}).

Some materials exhibit nearly steady mass loss rates when exposed to a fixed irradiance, $\dot{q}''_e$. For these materials, T_s reaches a steady value after a short initial transient period, and all terms in Equations 1 through 4 are approximately constant. Δh_g can then be obtained by measuring steady mass loss rates over a range of irradiance levels, and by plotting $\dot{m}''$ as a function of $\dot{q}''_e$. Δh_g is calculated as the reciprocal of the slope of a straight line fitted through the data points. $\dot{q}''_f - \dot{q}''_l$ is obtained as the intercept of this line with the abscissa. Tewarson and Pion[17] and Petrella[18] have used this technique to obtain average Δh_g values for a large number of materials. Tewarson and Pion also conducted tests in vitiated O_2/N_2 mixtures, and found $\dot{q}''_f$ to decrease linearly with oxygen concentration. Analysis of these additional experiments made it possible to separate $\dot{q}''_f$ and $\dot{q}''_l$.

Many materials, in particular those that form an insulating char layer as they burn, take a long time to reach steady burning conditions or do not ever reach steady conditions. Equations 1 through 4 are still valid for such materials, but the heat and mass fluxes and resulting $\Delta h_{c, eff}$ and Δh_g values vary with time. Tewarson and Petrella have used the same techniques to determine average Δh_g values for non-steady burning materials using average mass loss rates. They found that average $\dot{m}''$ is still an approximately linear function of $\dot{q}''_e$. However, the average heat of gasification values obtained in this manner may not have any physical meaning. Physically meaningful non-steady values for Δh_g can be obtained from Equation 2, with $\dot{q}''_f$ and $\dot{q}''_l$ calculated according to Equations 3 and 4, respectively. In order to do this, ϵ_s, T_s, ϵ_f, and T_f need to be measured, calculated, or estimated in some way. Surface temperature is the most important and most difficult to determine. Flame temperature and emissivities are nearly constant over the duration of a test and can be estimated more easily. Parker and Urbas used thermocouples to measure T_s for wood specimens tested in the cone calorimeter.[19] The problem with this technique is maintaining good contact between the thermocouple and the exposed surface. Parker and Urbas designed a special tension system to hold the thermocouple hot junction against the char surface, as this surface recedes during a test. The technique worked quite well, but the tension system made testing rather tedious. Urbas used an infrared pyrometer to measure T_s for wood specimens in an intermediate-scale calorimeter.[20] This technique is much more practical, but not without problems. First, the pyrometer must operate in a specific wavelength range so that the flame becomes transparent. The sensor used by Urbas responds to infrared radiation between 8 and 12 μm. This is well above the radiation bands of CO_2 and H_2O, and in a region where, given the high flame temperature (compared to the surface temperature), soot radiation is negligible. Second, a pyrometer measures emitted and reflected radiation from a surface. Hence, assuming the surface acts as a greybody radiator, surface emissivity must be known to calculate T_s from this measurement. For wood char, Urbas found $\epsilon_s \approx 1$ and confirmed this by comparing pyrometer and thermocouple measurements. For many other materials, $\epsilon_s < 1$, so that reflected radiation from the heater might be a problem. An alternative approach to determine T_s is by calculation. Janssens obtained T_s as a function of time for wood specimens exposed in the cone calorimeter by solving the equation for heat conduction through the char layer using an integral technique.[21] The resulting values for Δh_g compared well with those by Urbas. A drawback of this approach is that thermal properties of the material or its char are needed.

Another problem with non-steady values of $\Delta h_{c, eff}$ and Δh_g is the fact that time from ignition in a bench-scale test does not correspond directly to that in a full-scale room/corner fire. Considering a small element of the wall lining that is exposed in a room/corner test, incident heat flux

varies with time as the compartment fire grows. In a calorimeter, incident heat flux is nearly constant. Hence, the time axis of the non-steady $\Delta h_{c, eff}$ and $\Delta h_{c, net}$ curves must be transformed to a variable that also can be used for non-steady exposure conditions. Smith proposed using cumulative heat release rate to scale the time axis.[22] Mitler suggested cumulative mass loss.[23] Janssens plotted Δh_g curves for wood as a function of char depth.[21] Since char depth was defined as being proportional to cumulative mass loss, this approach is analogous to Mitler's.

With $\Delta h_{c, eff}$ and Δh_g for a material, performance in a room/corner test can be predicted. First, the exposed surface of the material is subdivided into smaller elements so that the incident heat flux to each element is approximately uniform. Then, a model is needed to determine the net heat flux into every element. This requires calculation of radiative and convective heat transfer between the different surfaces, gas volumes, and flames in the compartment coupled with the solution of compartment-wide mass, energy, and species conservation equations. $\dot{m}''$ can be calculated for every segment as the ratio of $\dot{q}''_{net}$ to Δh_g. The heat release rate from every segment is then obtained by multiplying $\dot{m}''$ with $\Delta h_{c, eff}$, with adjustment for oxygen vitiation or starvation. Note that, although $\Delta h_{c, eff}$ and Δh_g are the most important properties, other properties are needed to estimate the time to ignition of every element and the resulting flame spread over the surface. As the fire grows and conditions in the compartment change, the sequence of calculations must be repeated for subsequent time steps. Due to the required spatial subdivision of the exposed surfaces, which can be adapted to the gas-phase mesh, field models are ideally suited for this kind of application. Nevertheless, only one attempt to use such models for simulating room/corner tests has been reported.[24] This is probably because field models are very complex and costly, requiring a fast mini-computer or mainframe computer. Consequently, the number of field model users is small and most of them are not concerned with modeling room/corner tests. Fortunately, a simple model developed by Quintiere indicates such a level of detail is not needed to obtain reasonable predictions.[25] Quintiere assumed the total burning area is exposed to a uniform heat flux, and estimated this flux on the basis of experimental data for flame fluxes and hot layer temperatures as a function of heat release rate. Constant values were used for $\Delta h_{c, eff}$ and Δh_g, where the latter was determined from a reciprocal of the slope of a straight line fitted through data points of peak $\dot{m}''$ plotted against $\dot{q}''_e$ measured in the cone calorimeter.

If heat release measurements are first converted to material properties to use them for fire hazard assessment, the calorimeter that is used must have certain features. If unsteady values for $\Delta h_{c, eff}$ and Δh_g are used, the calorimeter must be capable of measuring $\dot{m}''$, and the addition of instrumentation to measure T_s must be practical (unless T_s is calculated or estimated in some other way). If average or steady values for $\Delta h_{c, eff}$ and Δh_g are sufficient, only the total mass loss over the duration of a test is needed, and measuring $\dot{m}''$ is not required. If the model addresses the effect of oxygen vitiation or starvation on $\Delta h_{c, eff}$, it must be possible to operate the calorimeter in atmospheres different from ambient. If the model keeps track of the concentration of gas species and soot, the calorimeter must be capable of measuring soot and gas species yields. The calorimeter must also be capable of running tests over a wide range of (steady) irradiance levels.

The discussion in this subsection is useful in clarifying a common misconception. Often it is believed that materials that are used in a particular orientation in practice should be tested in that orientation. This is not necessarily correct. Heat release rate as such is independent of specimen orientation. However, the heat release rate in a bench-scale calorimeter under otherwise identical exposure conditions is higher in the horizontal than in the vertical orientation. This is because the heat feedback from the flame is much greater in the horizontal orientation. In that orientation, the flame is a relatively large volume of hot gas located above the specimen. The flame is only a thin sheet in front of a vertical sample, leading to a much lower heat feedback. Neither of these situations are comparable to that in a real fire, where burning areas are much larger and heat flux from the flame is much greater regardless of orientation of the fuel surface. Hence, the best approach is to interpret bench-scale measurements in terms of material properties that are independent of the test apparatus. These material properties can then be used to predict full-scale performance using a method that accounts for the effect of the enhanced heat flux from large flames. The above reasoning indicates that bench-scale testing in the vertical orientation is preferable, because the heat feedback to the flame is smaller and errors of flame flux estimates are relatively less important. However, for practical reasons, it is often preferable to run bench-scale tests in the horizontal orientation to avoid problems with, for example, melting and dripping of the specimen.

Direct Use of Heat Release Rate Curves at Multiple Irradiance Levels

If the differences between surface heat losses and flame fluxes in full-scale *versus* bench-scale are ignored or compensated for by an adjustment to the calculated external full-scale heat flux, heat release curves obtained over a range of irradiance levels can be used without conversion to material properties. This approach was suggested by Smith and Satija.[5] Heat release rate is measured as a function of time at three or four irradiance levels in the range of 20 to 65 $kW \cdot m^{-2}$. Then, the time axis is transformed by using cumulative heat release. Smith and Satija also developed a computer model to predict the incident heat flux to the burning area, which is subdivided into smaller elements. Heat release from an element at a certain time is based on interpolation between the measured heat release curves, corresponding to the cumulative heat release from that element at that time and to the calculated incident heat flux. Smith and Green demonstrated that this interpolation method is reasonably accurate by running tests in their calorimeter while varying the incident heat flux.[26] The heat release rate curves measured under dynamic exposure conditions could be predicted quite accurately from interpolation between measurements of heat release rate under constant irradiance levels. A calorimeter that is intended to produce data suitable for this use must be capable of exposing specimens over a fairly wide range of constant irradiance levels.

Direct Use of Heat Release Rate Curves at a Single Irradiance Level

The approach outlined in the previous subsection can be simplified even further, if a heat release curve at a single level is used. This irradiance level is chosen so that it is a representative average (over space and time) of heat flux

levels occurring in room/corner tests, typically between 25 and 50 $kW \cdot m^{-2}$. Thus, the dynamic effects of the room fire on the exposure level are ignored while the dynamics of the heat release curve are largely maintained.

The single heat release curve is used in combination with a flame-spread algorithm to predict heat release rate as a function of time in the room/corner test. The flame-spread algorithm can be very simple, but needs at least some ignition data for the material. Transformation of the time axis for the heat release curve is not necessary. As time proceeds, the burning area expands as new sections of the material are ignited. The flame-spread algorithm calculates at every time step how large this newly ignited section is. Once ignited, the heat release rate from this section is obtained from the heat release curve for the same time relative to ignition. Thus, the total burning area and heat release at any time is obtained as a convolution integral. The most widely known room/corner test simulation of this nature was developed by Wickström and Göransson.[27] A calorimeter that is to provide data suitable for this use must be capable of providing heat release data at a single irradiance level, and some ignition data for the flame-spread algorithm.

Statistical Correlations

Statistical correlations are the least sophisticated method. A room/corner test result that is indicative of the fire hazard of a material, e.g., time to flashover, is correlated against a heat release parameter in combination with one or several other parameters measured in the calorimeter. This heat release parameter is typically the average heat release rate over a fixed period of time measured at one irradiance level. A (usually linear) statistical model is developed, and a regression analysis is performed to obtain model constants that give the best predictions for a given set of bench-scale and full-scale data. The resulting equation can then be used to predict full-scale performance on the basis of bench-scale data for materials that have not been tested in the room/corner test. Reasonable correlations of this type for the International Organization for Standardization (ISO) room/corner test were reported by Östman and Nussbaum[28] and Karlsson.[29] Clearly, with this approach the dynamics of the heat release curve are lost entirely. There is no difference in the prediction of full-scale performance for two materials with heat release curves of very different shape, provided the average heat release rate and other bench-scale measurements used in the correlation are identical. The predictions are valid for one scenario and geometry, only. A positive aspect of this method, however, is that the most rudimentary calorimeter is sufficient to provide the necessary bench-scale data.

Summation of Bench-Scale Test Data Use

It is thus obvious that there is a wide variety of methods that use bench-scale heat release rate data for predicting the real fire hazard of a material. In general, the versatility of the approach varies inversely with its complexity. A method using material properties can usually be applied over a range of conditions and configurations. Simple statistical correlations are locked into one specific full-scale fire scenario. As an example, Quintiere's model can be used for different ignition sources (e.g., ASTM *vs* ISO burner), specimen configurations (materials on walls only *vs* walls and ceiling), and geometries. The Östman and Nussbaum correlation only predicts time to flashover for one specific set of room/corner test conditions.

It must be stressed that, although heat release rate is the most important parameter, it is not the only parameter that needs to be included in a hazard analysis. Other factors related to piloted ignition, smoke release, opposed-flow flame spread, etc., must be considered also. The Babrauskas and Peacock[1] study illustrated that fire hazard is most sensitive to changes in the heat release rate of the burning fuel, but varies with other material characteristics, as well.

TECHNIQUES FOR MEASURING HEAT RELEASE RATE

Sensible Enthalpy Rise Method

Consider the energy balance of a gas-phase control volume enclosing the flame of a burning specimen. (See Figure 3-2.2.) Air enters the control volume at a flow rate $\dot{m}_a$ and temperature T_a. The enthalpy of this air can be written as

$$h_a = h_a^0 + c_p (T_a - T_0) \tag{5}$$

where

h_a = enthalpy of air at temperature T_a $(kJ \cdot kg^{-1})$,
h_a^0 = enthalpy of air at reference temperature T_0 $(kJ \cdot kg^{-1})$,
c_p = average specific heat of air between T_0 and T_a $(kJ \cdot kg^{-1} \cdot K^{-1})$,
T_a = temperature of the air entering the combustion zone (K), and
T_0 = reference temperature (K).

Part of the heat flux that strikes the exposed surface is conducted into the specimen. This heat flow raises the temperature of the solid, and decomposes some fraction into combustible fuel vapors. These vapors are generated at a rate $\dot{m}_v$, and enter the control volume at temperature T_v. Under the assumption that specific heat of all gases is approximately constant and temperature-independent (a reasonable approximation), the enthalpy of the fuel vapors can be written as

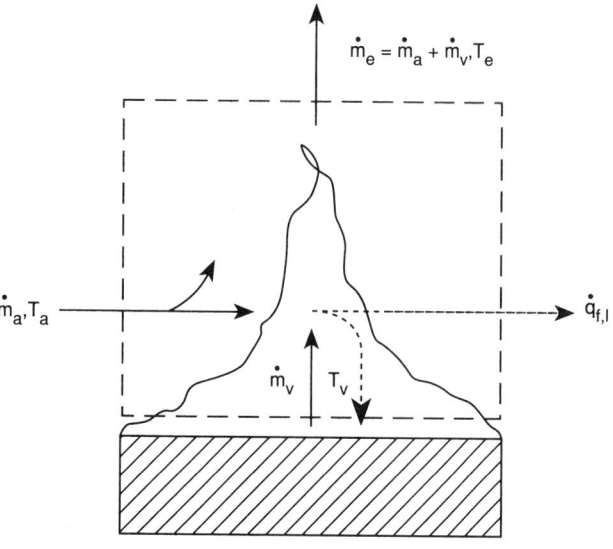

Fig. 3-2.2.   Gas-phase energy balance.

$$h_v = h_v^0 + c_p (T_v - T_0) \tag{6}$$

where

h_v = enthalpy of volatiles at temperature T_v $(kJ \cdot kg^{-1})$,
h_v^0 = enthalpy of volatiles at reference temperature T_0 $(kJ \cdot kg^{-1})$, and
T_v = temperature of volatiles entering the combustion zone (K).

The fuel vapors mix with air, and are converted in the flame to products of combustion. The total flow rate, $\dot{m}_e$, of combustion products, which includes some excess air, has a temperature T_e and enthalpy given by

$$h_e = h_e^0 + c_p (T_e - T_0) \tag{7}$$

where

h_e = enthalpy of combustion products at temperature T_e $(kJ \cdot kg^{-1})$,
h_e^0 = enthalpy of combustion products at reference temperature T_0 $(kJ \cdot kg^{-1})$, and
T_e = temperature of combustion products leaving the control volume (K).

T_e is higher than the mass-weighted average of T_a and T_v, because of the heat released by combustion in the flame, $\dot{q}$. However, only a fraction of this heat contributes to the temperature rise of the gases. This fraction is referred to as the convective fraction of the heat release rate. The remaining fraction of $\dot{q}$ is lost and is denoted as $\dot{q}_{f,l}$. For the most part, $\dot{q}_{f,l}$ is lost in the form of thermal radiation to the walls of the apparatus (closed configuration) or to the environment (open configuration). A small part of $\dot{q}_{f,l}$ consists of convective and radiative feedback to the fuel surface. Assuming gas-phase transients can be neglected, application of the first law of thermodynamics for the control volume in Figure 3-2.2 results in

$$\dot{q}_{f,l} = \dot{m}_a h_a + \dot{m}_v h_v - \dot{m}_e h_e \tag{8}$$

As an example, suppose now that the same flow rates of air and volatiles, both at temperature T_0, are mixed in a hypothetical combustion chamber. Furthermore, assume the combustion reactions are identical to those in the calorimeter in Figure 3-2.2, and the products of combustion are cooled down to the reference temperature T_0 without condensing water. This hypothetical situation is shown in Figure 3-2.3. Application of the first law of thermodynamics for the combustion chamber control volume in Figure 3-2.3 leads to

$$\dot{q} = \dot{m}_a h_a^0 + \dot{m}_v h_v^0 - \dot{m}_e h_e^0 \tag{9}$$

Here, $\dot{q}$ is equal to the total rate of heat released by combustion in the flame. This heat release rate is identical in Figure 3-2.2 and Figure 3-2.3, but it is distributed in different ways. By expressing the heat released per unit mass of volatiles, an effective heat combustion can be defined as

$$\dot{m}_v \Delta h_{c,\,eff} \equiv \dot{q} \tag{10}$$

or per unit exposed area

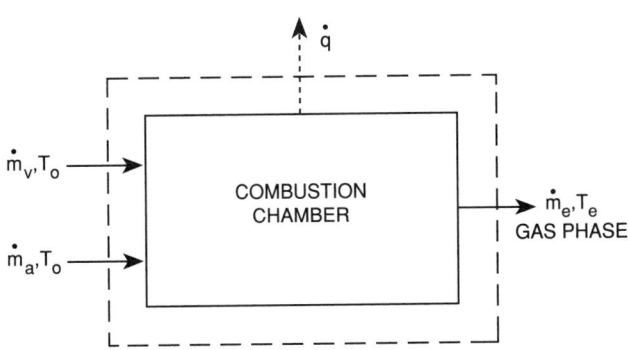

Fig. 3-2.3. Hypothetical combustion chamber.

$$\dot{m}_v'' \Delta h_{c,\,eff} \equiv \dot{q}'' \qquad (11)$$

Here, $\Delta h_{c,\,eff}$ is for the combustion reactions as they take place in the calorimeter. As explained in the previous section, $\Delta h_{c,\,eff}$ must be distinguished from the net heat of combustion, $\Delta h_{c,\,net}$, measured in an oxygen bomb calorimeter. The difference between $\Delta h_{c,\,eff}$ and $\Delta h_{c,\,net}$ is very significant for charring materials, such as wood. In an oxygen bomb calorimeter, nearly all the mass of wood is consumed, leaving a small fraction of noncombustible ash (usually less than 1 percent by mass). The net heat of combustion, $\Delta h_{c,\,net}$, of dry wood is in the range of 16 to 18 $kJ \cdot g^{-1}$. When exposed under real fire conditions, only 70 to 80 percent of the mass is converted to volatiles that burn almost completely. The heat of combustion of the volatiles, $\Delta h_{c,\,eff}$, measured in a calorimeter, is only 12 to 13 $kJ \cdot g^{-1}$. A solid char residue remains, primarily consisting of carbon, with a net heat of combustion of approximately 30 $kJ \cdot g^{-1}$. In an oxygen bomb calorimeter, most of this char is also burnt, explaining why $\Delta h_{c,\,net}$ exceeds $\Delta h_{c,\,eff}$ by 25 to 50 percent. Even for materials that do not form a char, $\Delta h_{c,\,eff}$ can be significantly lower than $\Delta h_{c,\,net}$ if combustion of the volatiles in the calorimeter is incomplete. In this case, the products of combustion contain measurable amounts of combustible components, such as CO, soot, unburnt hydrocarbons, etc. The ratio of $\Delta h_{c,\,eff}$ to $\Delta h_{c,\,net}$ is defined as combustion efficiency, χ. For clean-burning gaseous fuels, such as methane, χ is close to unity. For fuels that produce sooty flames, including gases, χ can be significantly lower. For example, χ for acetylene is approximately 0.75. χ values for a number of gases, liquids, and solids are listed in Section 3, Chapter 4.

Substitution of Equations 5, 6, 7, and 9 into 8 leads to

$$\dot{q} - \dot{q}_{f,l} = c_p \dot{m}_e(T_e - T_0) - c_p \dot{m}_a(T_a - T_0)$$
$$- c_p \dot{m}_v(T_v - T_0) \qquad (12)$$

For most combustible materials, the stoichiometric air-to-fuel ratio ranges between 3 and 16. Moreover, bench-scale calorimeters are usually operated with excess air. For example, the standard initial flow rate in the cone calorimeter is 30 $g \cdot s^{-1}$. Based on the oxygen consumption principle (see below), the stoichiometric flow rate of air for a 10-kW fire (practical upper limit in the cone calorimeter) can be calculated as $(10\ kW)/(3\ kJ\ per\ g\ of\ air) = 3.3\ g \cdot s^{-1}$. Thus, the air supply in the cone calorimeter is at least 9 times stoichiometric, or at least $9 \times 3 = 27$ times the generation rate of volatiles. Usually, the ratio is much greater. Hence, $\dot{m}_v$ is negligible compared to $\dot{m}_a$ and Equation 12 can be approximated as

$$\dot{q} - \dot{q}_{f,l} \approx \dot{m}_a c_p(T_e - T_a) \qquad (13)$$

This equation is the basis for the sensible enthalpy method. Heat release rate is calculated from the temperature rise $T_e - T_a$ of the gases flowing through a calorimeter. A schematic of a calorimeter based on this principle is shown in Figure 3-2.4.

There are a few problems with the practical implementation of this technique. The main concern is that only a fraction of the heat released in the flame is used to raise the sensible enthalpy or temperature of the gases. Therefore, another method is needed to recover or measure the loss term, $\dot{q}_{f,l}$. Some calorimeters have water-cooled walls that trap most of the losses. These losses can be estimated by measuring the enthalpy rise of the cooling water. However, due to the additional hardware and instrumentation, such calorimeters are rather complex and difficult to operate. A more popular method relies on a gas burner calibration to determine $\dot{q}_{f,l}$, under the assumption that the losses are fuel-independent. The loss fraction, χ_R, is defined by

$$\dot{q} - \dot{q}_{f,l} \equiv (1 - \chi_R)\dot{q} \qquad (14)$$

The symbol χ_R is chosen for this fraction, since $\dot{q}_{f,l}$ consists primarily of radiation. If the calorimeter is operated with a constant airflow rate $\dot{m}_a$, Equation 13 can then be written as

$$\dot{q} \approx \frac{\dot{m}_e c_p}{1 - \chi_R}(T_e - T_a) \equiv k(T_e - T_a) \qquad (15)$$

The calibration factor, k, is determined from a gas burner calibration with known $\dot{q}$. By repeating the calibration over a range of heat release rate levels, k can be determined as a function of $\dot{q}$ or T_e. If the specimen is enclosed with the heater, Equation 15 is still valid, provided a reference temperature T_r is used instead of T_a. The temperature difference $T_r - T_a$ results from the heat transfer between the heater and

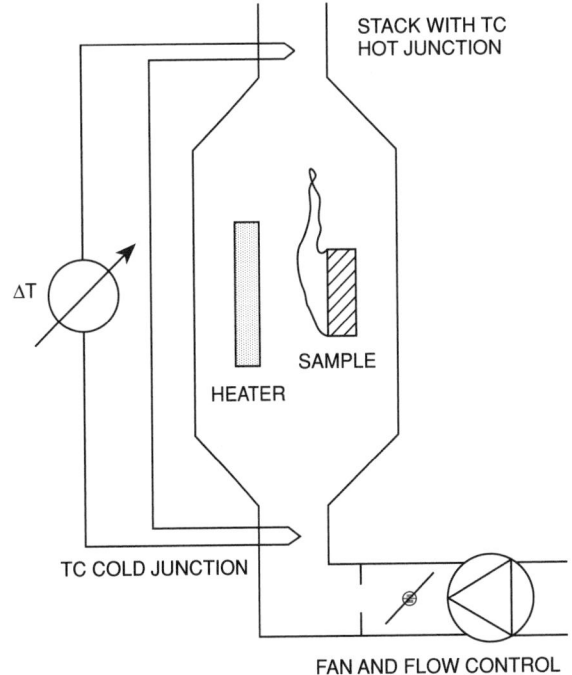

Fig. 3-2.4. Sensible enthalpy rise calorimeter.

the airflow through the enclosure. T_r is therefore a function of heater setting, to be determined *via* calibration.

Smith's rate of heat release test developed at Ohio State University is the most well-known and most widely used calorimeter based on the sensible enthalpy rise method.[3] The test is described in detail in a following section.

Substitution Method

For practical reasons, calorimeters based on the sensible enthalpy rise method use a closed configuration. The specimen and heater(s) are located inside a metal box, which may be (partly) insulated. The dynamic response of the enclosure to changes in the thermal environment creates major problems in the practical implementation of the sensible enthalpy rise method. After ignition, part of the heat released by a burning sample is transferred by radiation to the enclosure walls. A fraction of this heat is stored in the walls, causing an increase of its temperature, in turn resulting in an enhanced heat transfer with the air flowing through the box. The result is, for a material that quickly reaches steady burning conditions, there is a delay for T_e to reach the corresponding steady temperature. A similar phenomenon occurs when heat release rate from the specimen decreases, or after the specimen burns out and heat release rate goes back to zero. Under unsteady burning conditions, T_e constantly lags behind the temperature corresponding to the instantaneous heat release rate. There are various ways to address this problem, but none are completely satisfactory.

The substitution method was developed to eliminate problems associated with thermal lag. The method requires two runs to determine heat release rate of a material under a given set of conditions. The first run uses a similar arrangement as shown in Figure 3-2.4. The temperature difference $T_e - T_a$ is measured as a function of time. The second run uses the same apparatus, airflow rate, and irradiance. However, the specimen is replaced by a noncombustible dummy specimen and a substitution gas burner. The flow of gas to the burner is controlled in such a way that the temperature difference $T_e - T_a$ closely follows the curve measured during the first run. Figure 3-2.5 shows a schematic of the substitution run.

Presumably, the dynamics are identical in both runs. Hence, problems with thermal lag have been eliminated, and the heat release rate of the specimen can be determined from the fuel flow rate to the burner in the second run. Unfortunately, implementation of this method is not simple, since a sophisticated control system is needed for the second run. Moreover, due to the addition of substitution runs, the number of tests needed to evaluate a material are doubled.

The substitution method was first implemented at Factory Mutual Research Corporation.[30] The apparatus was designed to measure the heat release rate from roof assemblies. The test method is briefly discussed in the section on large-scale calorimeters. A bench-scale substitution calorimeter developed at the Forest Products Laboratory[31] is described in more detail in the section on bench-scale calorimeters.

Compensation Method

A compensation calorimeter is similar to a substitution calorimeter, except that the burner is operated while a specimen is exposed. A schematic is shown in Figure 3-2.6. Initially, the burner flow rate is chosen so that the corresponding heat release rate exceeds that of any material to be

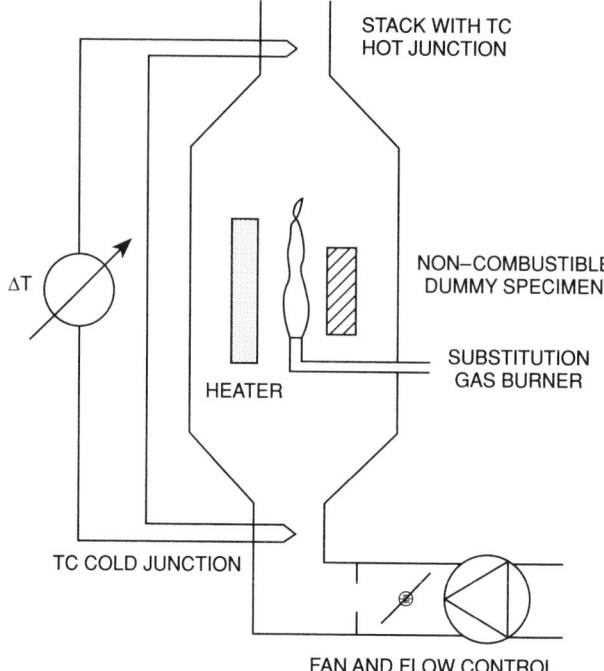

Fig. 3-2.5. *Second run with substitution calorimeter.*

tested. During a test, the gas flow rate to the burner is controlled so that $T_e - T_a$ remains constant. The heat release rate corresponding to the reduction in flow rate to the burner is equal to the heat release rate from the specimen.

The compensation method also eliminates problems with the dynamic response of the calorimeter enclosure. In theory, a compensation calorimeter is operated at a constant temperature. This would resolve another problem associated with the assumption that $\dot{q}_{f,l}$ is fuel-independent, while in reality it is not ($\dot{q}_{f,l}$ is a strong function of the sootiness of the flame).

In practice, however, the specimen and burner have to be separated to avoid radiation from the burner flame that enhances irradiance to the specimen. Hence, the calorimeter enclosure is not truly isothermal, and the problem remains unresolved. As with substitution calorimeters, the burner flow control system makes compensation calorimeters rather complex and difficult to operate. As a result, they are suitable only for research and not for routine testing.

Compensation calorimeters developed at the National Bureau of Standards[32,33] and Stanford Research Institute[34] are described in the section "Survey of Bench-Scale Calorimeters."

Oxygen Consumption Method

In 1917 Thornton showed that, for a large number of organic liquids and gases, a more or less constant net amount of heat is released per unit mass of oxygen consumed for complete combustion.[35] Huggett found this also to be true for organic solids, and obtained an average value for this constant of $13.1 \text{ kJ} \cdot \text{g}^{-1}$ for oxygen.[36] This value can be used for practical applications and is accurate with very few exceptions to within ± 5 percent. Thornton's rule implies that it is sufficient to measure the oxygen consumed in a combustion system in order to determine the net heat

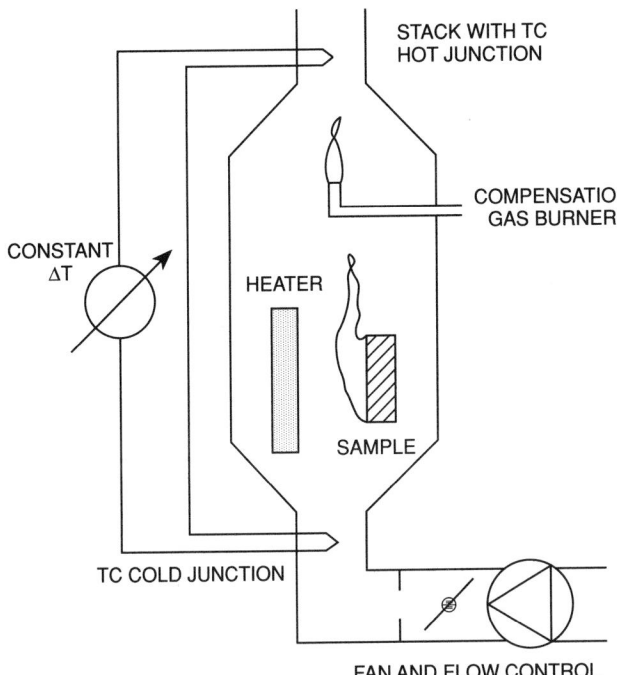

Fig. 3-2.6. *Compensation calorimeter.*

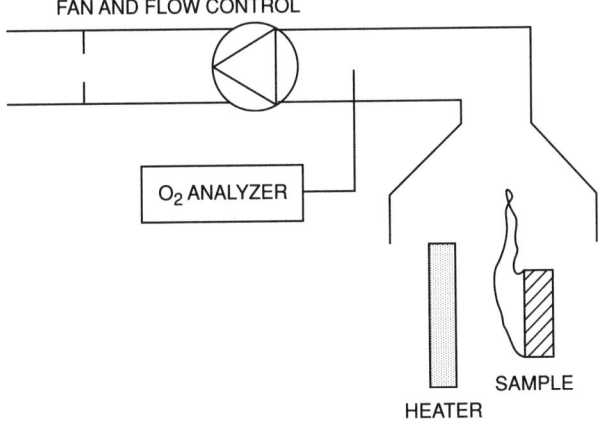

Fig. 3-2.7. *Oxygen consumption calorimeter.*

and composition of the gases are measured. A schematic of an oxygen consumption calorimeter is shown in Figure 3-2.7. It is not necessary to measure the inflow of air, provided the flow rate is measured in the exhaust duct. Therefore, oxygen consumption calorimeters are typically open, to avoid the part of $\dot{q}_{f,l}$ which is reflected by the calorimeter walls and reaches the specimen surface. This would result in an uncontrolled irradiance, in addition to that from the heater.

The practical implementation of the oxygen consumption method is not straightforward. Application of Thornton's rule to the combustion system shown in Figure 3-2.8 leads to the following equation for the heat release rate:

$$\dot{q} = E(\dot{m}_a Y_{O_2}^a - \dot{m}_e Y_{O_2}^e) \qquad (16)$$

where

E = heat release per mass unit of oxygen consumed ($\approx 13.1 \text{ kJ} \cdot \text{g}^{-1}$),
$Y_{O_2}^a$ = mass fraction of oxygen in the combustion air (0.232 g · g^{-1} in dry air), and
$Y_{O_2}^e$ = mass fraction of oxygen in the combustion products (g · g^{-1}).

The problems with the use of this equation are threefold. First, oxygen analyzers measure the mole fraction and not the mass fraction of oxygen in a gas sample. Mole fractions can be converted to mass fractions by multiplying the mole fraction with the ratio between molecular mass of oxygen and molecular mass of the gas sample. The latter is usually close to the molecular mass of air ($\approx 29 \text{ g} \cdot \text{mol}^{-1}$). Second, water vapor is removed from the sample before it passes through a paramagnetic analyzer, so that the resulting mole fraction is on a dry basis. This problem can be avoided by using a zirconium oxide analyzer, which measures oxygen mole fraction in a hot and wet sample. However, the performance of such analyzers is significantly inferior to that of paramagnetic instruments, making them unsuitable for accurate oxygen consumption calorimetry.[38] Third, flow meters measure volumetric rather than mass flow rates. The volumetric flow rate in the exhaust duct, normalized to the

released. This is the basis for the oxygen consumption method for measuring heat release rate in fire tests.

Perhaps the first application of the oxygen consumption principle in fire research was performed by Parker on the ASTM E-84 tunnel test.[37] During the late 1970s and early 1980s, the oxygen consumption technique was refined at the National Bureau of Standards (NBS) [currently the National Institute of Standards and Technology (NIST)]. The oxygen consumption method is now recognized as the most accurate and practical technique for measuring heat release rates from experimental fires. It is widely used throughout the world, both for bench-scale and large-scale applications.

The basic requirement to use the oxygen consumption technique is that all of the combustion products are collected and removed through an exhaust duct. At a distance downstream sufficient for adequate mixing, both flow rate

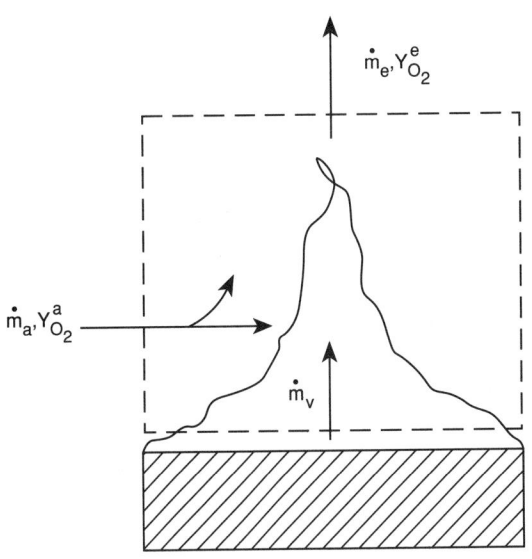

Fig. 3-2.8. *Gas-phase oxygen balance.*

same pressure and temperature, is usually slightly different from the inflow rate of air because of expansion due to the combustion reactions.

Equations for calculating rate of heat release by oxygen consumption for various applications were developed by Parker[39] and Janssens.[40] The differences in treatment and equations to be used are mainly due to the extent to which gas analysis is made. At a minimum, the oxygen concentration must be measured. However, accuracy can be improved by adding instrumentation for measuring the concentration of CO_2, CO, and H_2O. Equations for the most common configurations of the gas analysis system are given below. Detailed derivations are not repeated here, and can be found in the aforementioned references.

Only O_2 is measured: In this case all water vapor (by a cooling unit and a moisture sorbent) and CO_2 (by a chemical sorbent) must be removed from the exhaust gas sample stream before O_2 is measured. This leads to the assumption that the sample gas only consists of O_2 and N_2. The resulting equation for calculating heat release rate is

$$\dot{q} = E \frac{\phi}{1 + \phi (\alpha - 1)} \dot{m}_e \frac{M_{O_2}}{M_a} (1 - X_{H_2O}^a - X_{CO_2}^a) X_{O_2}^{Aa} \quad (17)$$

with

$$\phi = \frac{X_{O_2}^{Aa} - X_{O_2}^{Ae}}{(1 - X_{O_2}^{Ae}) X_{O_2}^{Aa}} \quad (18)$$

where

ϕ = oxygen depletion factor,
α = volumetric expansion factor,
M_{O_2} = molecular weight of oxygen ($28 \text{ g} \cdot \text{mol}^{-1}$),
M_a = molecular weight of the combustion air ($29 \text{ g} \cdot \text{mol}^{-1}$ for dry air),
$X_{H_2O}^a$ = actual mole fraction of water vapor in the combustion air,
$X_{CO_2}^a$ = actual mole fraction of carbon dioxide in the combustion air,
$X_{O_2}^{Aa}$ = measured mole fraction of oxygen in the combustion air, and
$X_{O_2}^{Ae}$ = measured mole fraction of oxygen in the exhaust flow.

As the composition of the fuel is usually not known, some average value has to be used for α. Complete combustion of carbon in dry air results in $\alpha = 1$. If the fuel is pure hydrogen, α is equal to 1.21. A recommended average value for α is 1.105. $X_{H_2O}^a$ can be calculated from the relative humidity and temperature in the laboratory. Typically it is less than a few percent in a temperature-controlled laboratory. $X_{CO_2}^a$ in dry air is 330 ppm. Note that the symbols for oxygen mole fraction measured in the combustion air (prior to a test) and the exhaust flow include a superscript A. This is to make a distinction between the actual and measured mole fractions of oxygen, with the latter on a dry gas sample basis.

Equation 17 is expected to be accurate to within ± 10 percent, provided combustion is complete; i.e., all of the carbon is converted to CO_2. The error might be larger if CO or soot production is considerable, or if a significant amount of combustion products consists of species other than CO_2 or H_2O (e.g., HCl). The error is partly due to the uncertainty of E and α. If more exact values are available, accuracy can be improved by using those instead of the generic values of $13.1 \text{ kJ} \cdot \text{g}^{-1}$ and 1.105.

O_2 and CO_2 are measured: In this case, only water vapor is trapped before the exhaust gas sample reaches the analyzers. The rate of heat release is given by Equation 17, with the minor modification that $X_{CO_2}^a$ is not included in the expression inside parentheses. In addition, ϕ is slightly different and follows from

$$\phi = \frac{X_{O_2}^{Aa} (1 - X_{CO_2}^{Ae}) - X_{O_2}^{Ae} (1 - X_{CO_2}^{Aa})}{(1 - X_{O_2}^{Ae} - X_{CO_2}^{Ae}) X_{O_2}^{Aa}} \quad (19)$$

where

$X_{CO_2}^{Aa}$ = measured mole fraction of carbon dioxide in the air (≈ 330 ppm), and

$X_{CO_2}^{Ae}$ = measured mole fraction of carbon dioxide in the exhaust flow.

Generally, adding the CO_2 measurement does not greatly improve accuracy of $\dot{q}$.

O_2, CO_2, and CO are measured: If a significant fraction of carbon in the fuel is converted to CO instead of CO_2, the equations may have to be corrected to take incomplete combustion into account. Heat release rate is now calculated from

$$\dot{q} = \left[E \phi - (E_{CO} - E) \frac{1 - \phi}{2} \frac{X_{CO}^{Ae}}{X_{O_2}^{Ae}} \right]$$
$$\cdot \frac{\dot{m}_e}{1 + \phi (\alpha - 1)} \frac{M_{O_2}}{M_a} (1 - X_{H_2O}^a) X_{O_2}^{Aa} \quad (20)$$

with

$$\phi = \frac{X_{O_2}^{Aa} (1 - X_{CO_2}^{Ae} - X_{CO}^{Ae}) - X_{O_2}^{Ae} (1 - X_{CO_2}^{Aa})}{(1 - X_{O_2}^{Ae} - X_{CO_2}^{Ae} - X_{CO}^{Ae}) X_{O_2}^{Aa}} \quad (21)$$

where

E_{CO} = heat release per mass unit of oxygen consumed for combustion of CO to CO_2 ($\approx 17.6 \text{ kJ} \cdot \text{g}^{-1}$)

X_{CO}^{Ae} = measured mole fraction of carbon monoxide in the exhaust flow.

One might wonder under what conditions the CO correction becomes significant. Figure 3-2.9 shows the ratio of heat release rate obtained by ignoring CO to the actual heat release rate, as a function of the ratio of measured CO to CO_2 mole fractions in the exhaust flow for methane (CH_4) and for a gaseous fuel of composition $(CH_2O)_n$. According to Roberts,[41] the molecular formula of the latter represents the thermal degradation products of beechwood. For the CO effect examined here, this fuel represents a "worst case," since it contains enough oxygen for combustion of all hydrogen. Methane gives a practical lower limit for the error, because it is the hydrocarbon with the highest hydrogen to carbon ratio. There is some experimental evidence that the yield of CO in underventilated fires reaches an upper limit approximately equal to 0.2 kg of CO per kg of fuel, when the equivalence ratio exceeds unity.[42] For the fuels considered here, the limit corresponds to a ratio of X_{CO}^{Ae} to $X_{CO_2}^{Ae}$ of 0.27. Figure 3-2.9 indicates that, even under the worst conditions, the error by ignoring CO generation is less than 5 percent.

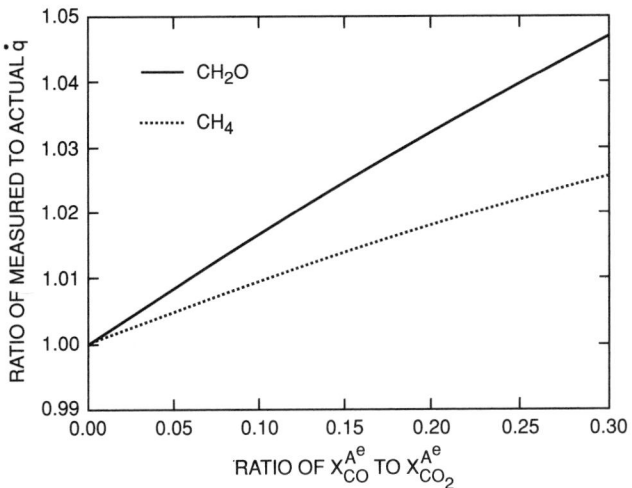

Fig. 3-2.9. *Effect of ignoring CO production on $\dot{q}$ error.*

O_2, CO_2, CO, and H_2O are measured: Often the combustion products comprise only O_2, CO_2, CO, and H_2O, with N_2 in insignificant amounts. In this case the expansion factor no longer has to be estimated, but can be calculated. Heat release rate is calculated from

$$\dot{q} = \left[E\phi - (E_{CO} - E) \frac{1-\phi}{2} \frac{X_{CO}^{Ae}}{X_{O_2}^{Ae}} \right] \dot{m}_a \frac{M_{O_2}}{M_a} (1 - X_{H_2O}^a) X_{O_2}^{Aa} \quad (22)$$

with

$$\frac{\dot{m}_a}{M_a} = \frac{(1 - X_{H_2O}^e)(1 - X_{O_2}^{Ae} - X_{CO_2}^{Ae} - X_{CO}^{Ae})}{(1 - X_{H_2O}^a)(1 - X_{O_2}^{Aa} - X_{CO_2}^{Aa})}$$
$$\cdot \frac{\dot{m}_e}{18 + 4(1 - X_{H_2O}^e)(X_{O_2}^{Ae} + 4X_{CO_2}^{Ae} + 2.5)} \quad (23)$$

where

$X_{H_2O}^e$ = actual mole fraction of water vapor in the exhaust flow.
ϕ is still determined according to Equation 21.

The oxygen consumption method is the most accurate and convenient way to measure net release rate. Problems due to thermal lag are eliminated. Two drawbacks of the method are: (1) the high cost of instrumentation (only the best available oxygen analyzers are adequate), and (2) the need for a rigorous calibration and maintenance schedule. Nevertheless, the benefits outweigh the disadvantages.

Solid or volatile heat release rate? It should be determined whether the oxygen consumption method measures heat release rate for the volatiles or the solid fuel. Thermal methods approximately measure heat release rate from the volatiles. However, Huggett's constant for oxygen of 13.1 kJ $\cdot$ g^{-1} is based on the average net heat of combustion for a large set of materials. Hence, one would expect that oxygen consumption calorimetry gives the heat released by the fuel in its natural state at ambient temperature, since that is how the fuel is supplied in an oxygen bomb calorimeter. The

question can be examined in more detail for some synthetic polymers, by comparing the net heat of combustion of the polymer to that of the corresponding monomer. If one were to burn a monomer in an oxygen consumption calorimeter, the products of complete combustion would be the same as for the corresponding polymer, provided test conditions are identical. Therefore, the measured heat release rate would be the same in the two cases. However, the net heat of combustion is higher for the monomer. The difference with the net heat of combustion of the polymer is the net heat released in the polymerization process. Table 3-2.1 gives values for the net heat of combustion of nine polymers and their monomers. The former are taken from Huggett;[36] the latter are obtained by adding the heat of polymerization as reported in the literature.[43] Table 3-2.1 confirms that the oxygen consumption method measures the net heat release rate of a solid fuel. The heat release rate from the volatiles is always higher, but not by as much as indicated in Table 3-2.1 because only a fraction of polymeric fuels decomposes back into the monomer. (See Section 1, Chapter 7).

EFFECTS OF CALORIMETER CONSTRUCTION DETAILS

In this section, the effects of some calorimeter construction details on quality and accuracy of the measurements are examined. The discussion results in some guidelines for building the "ideal" calorimeter for a certain application.

Open or Closed Configuration

Calorimeters that utilize a measuring technique other than oxygen consumption consist of a closed "box" configuration. Combustion air is supplied to one side of the box, and combustion products are removed from the opposite side. Sample, heater, and ignition device typically are located inside the box. Advantages of a closed configuration are: airflow rate can be measured at the inlet under clean and soot-free conditions, the combustion air can be heated, and the oxygen concentration in the air can be increased (by adding O_2) or decreased (by adding N_2) over ambient. Disadvantages are: thermal lag due to heating or cooling of the enclosure walls, and uncontrolled radiation feedback from the enclosure walls to the specimen.

To resolve the first problem, i.e., thermal lag, various numerical procedures have been proposed for correcting the temperature signal measured with calorimeters based on the sensible enthalpy rise method.[44,45,46,47] These procedures are based on a mathematical model of the calorimeter consisting of two first-order systems in series. The first system has a rather small time constant (between 8 and 30 s for various calorimeters), which is related to the heat capacity of the gases flowing through the calorimeter. The second system has a large time constant (200 to 930 s for various calorimeters), which is associated with the heat capacity of the calorimeter walls. The correction procedures adjust the output signal for thermal lag, using discrete forward and inverse Laplace transform techniques. In spite of the complex calculations, the resulting correction may not always be accurate due to the crude mathematical model for the calorimeter. A more convenient, and perhaps as accurate, correction method relies on an electronic compensator.[47,48,49] The compensator electronically corrects the output signal of the exhaust thermocouples, based on the negative feedback

TABLE 3-2.1 *Net Heat of Combustion of Some Polymers and Their Monomers*

Polymer	$\Delta h_{c,\,net}$ (kJ · g^{-1} fuel)	$\Delta h_{c,\,net}$ (kJ · g^{-1} O$_2$)	Monomer (state)*	$\Delta h_{c,\,net}$ (kJ · g^{-1} fuel)	$\Delta h_{c,\,net}$ (kJ · g^{-1} O$_2$)	Diff. (%)
Polyethylene	−43.3	−12.65	C$_2$H$_4$ (g)	−47.2	−13.78	8.9
Polypropylene	−43.3	−12.66	C$_3$H$_6$ (g)	−45.8	−13.39	5.8
Polybutadiene	−42.8	−13.14	C$_4$H$_6$ (l)	−44.1	−13.56	3.2
Polystyrene	−39.9	−12.97	C$_8$H$_8$ (l)	−40.5	−13.19	1.7
Polyvinylchloride	−16.4	−12.84	C$_2$H$_3$Cl (g)	−18.0	−14.10	9.8
Polyvinylidene chloride	−8.99	−13.61	C$_2$H$_2$Cl$_2$ (l)	−9.77	−14.79	8.7
Polyvinylidene fluoride	−13.3	−13.32	C$_2$H$_2$F$_2$ (g)	−15.6	−15.61	17.2
Polymethylmethacrylate	−24.9	−12.98	C$_5$H$_8$O$_2$ (l)	−25.4	−13.26	2.2
Polyacrylonitrile	−30.8	−13.61	C$_3$H$_3$N (l)	−32.2	−14.25	4.7
Average		−13.09			−13.99	6.9

*g = gaseous
 l = liquid

of a wall temperature signal. The oxygen consumption method also has a time delay, but with properly adjusted sampling flows and oxygen analyzer, this delay consists almost entirely of the transport time for a gas sample from the combustion zone to the analyzer.[50] Since flow rates in the exhaust duct and sampling lines do not change significantly during a test, this delay time is approximately constant. It can be determined with gas burner calibrations, and can be easily accounted for by shifting the gas analysis data over the appropriate time interval. When an inferior oxygen analyzer is used, its internal response time might become significant. In such a case, the analyzer output signal can be corrected numerically for time delays in a similar way as temperature measurements. Oxygen analyzers behave as a first- or second-order system, and the appropriate time constants can be obtained from step-response measurements. However, it is highly recommended that only the best and most accurate analyzers for oxygen consumption calorimetry be used, in which case internal delays are not a problem.

The second problem, i.e., uncontrolled radiation feedback, can only be eliminated by using blackened water-cooled calorimeter walls. If the walls are allowed to heat or cool freely, they emit radiation that varies with time. Part of this radiation reaches the specimen surface and enhances the irradiance from the heater in an uncontrolled fashion. Obviously, the need for water-cooled walls makes the apparatus much more complex and costly.

In conclusion, problems with thermal lag and radiation feedback to the specimen can be eliminated by using an open configuration. Solid objects must be water-cooled, or sufficiently remote from heater and specimen, so that they do not interfere with the controlled irradiance to the specimen. A closed configuration can only be recommended for specialized applications, e.g., to study the effect of oxygen concentration or temperature of the combustion air on heat release rate and burning behavior.

Type of Heater

Heat release rates must be measured at constant heat flux levels over a range that is relevant for the fire scenario of interest. The heat flux can be provided with a gas burner flame in contact with the specimen, or with a radiant panel remote from the specimen.

Incident heat flux from impinging gas burner flames can only be adjusted over a narrow range. To increase the heat flux from a gas burner, either flame size has to be increased, or a fuel with higher soot yield has to be used. Usually, these parameters can be adjusted only slightly, or not at all. It is very difficult to set and maintain a specific heat flux level since a major fraction of the heat transfer is convective. Moreover, the burner gas and combustion products mix with fuel volatiles, which affects burning behavior. In short, impinging flames are not desirable as the external heat source in heat release rate calorimeters.

It is much easier to create constant and uniform exposure conditions if the incident heat flux is primarily radiative. Porous gas panels as well as electrical heating elements are used for this purpose. The irradiance can be adjusted by changing the power of the heater, or by changing the distance between heater and specimen. If the second method is used, there are practical upper and lower limits to the range of irradiance levels that can be created. If the heater is too close to the specimen, convective heat transfer becomes significant. Therefore, the upper limit corresponds to the minimum distance that has to be maintained in order to ensure predominantly radiative heat transfer. The lower limit is determined by the uniformity of the incident irradiance, which drops with increasing distance between heater and specimen. The exact limits depend on the geometrical configuration, power of the heater, and the degree of non-uniformity of the incident heat flux profile that is deemed acceptable.

Another important aspect is the ability of the heater to maintain the irradiance at a constant level during a test. If the heater is operated at a constant power level, incident irradiance changes during testing. At the start of a test, a cold sample is inserted. The sample acts as a heat sink, resulting in a decrease of the heater temperature, and consequently a decrease of the incident irradiance. After ignition, the heat released by the specimen results in an increase to the heater temperature and incident irradiance. Therefore, in order to maintain incident irradiance during a test, it is necessary to keep the temperature of the heater constant. This is difficult with a gas panel, but relatively straightforward for electrical heating elements. With the oxygen consumption method, another drawback of using a gas panel is that its products of combustion result in an oxygen depletion that is usually much larger than the oxygen consumed for combustion of the specimen. Thus, small fluctuations in panel flow can result in significant error of the measured heat release rate.

This "baseline" problem can be avoided by using a separate exhaust system for the heater.

It is clear from the discussion above that an electrical heater is preferable over a gas panel. Two types of electrical heaters are used: (1) high temperature and (2) low temperature. The former commonly uses tungsten filament lamps that operate at temperatures close to 2300 K. According to Wien's displacement law, peak irradiance from such lamps is at a much shorter wavelength than for real fires, with temperatures in the range of 600 to 1400 K. Piloted ignition studies on plastics and wood have shown that these materials absorb much less radiation in the visible and near-infrared range, than at higher wavelengths.[51,52] On the basis of these findings, it can be concluded that high-temperature electrical heating elements are not suitable for use in fire testing. It is therefore recommended that commercially available low-temperature elements be used. Such elements typically operate between 800 and 1200 K, a range that is representative of real fire exposure conditions.

Type of Ignition Pilot

Heat release rate tests are most often conducted with an ignition pilot. The use of a pilot reduces the variation in time to sustained flaming between multiple tests conducted under identical test conditions. Because the duration of the preheat period prior to ignition affects burning rate after ignition, use of a pilot also improves repeatability of heat release rate measurements. Furthermore, piloted ignition is used because it is representative of most real fires, and conservative in other cases. The ignition pilot in bench-scale fire tests consists of a small gas burner flame, a glowing wire, or an electric spark. An impinging flame should not be used because it locally enhances irradiance to the specimen. Another problem with a pilot flame is that it is sometimes extinguished by fire retardants or halogens in the fuel volatiles. A glowing wire is not an efficient method for igniting fuel volatiles, sometimes leading to poor repeatability. An electric spark remains stable when fire retardants or halogens are present. However, it occupies a small volume, so that the positioning of the spark plug is more critical than with other types of ignition pilots. In conclusion, each type of pilot has its drawbacks. Nonetheless, a spark ignitor is probably the best type of ignition pilot for heat release rate tests.

Sample Size

The ideal situation would be if small-scale heat release rate data could be used directly to predict burning rate in real-scale fires. Unfortunately, the minimum bench-scale sample size that is required to allow for such a straightforward prediction is not practical. As described in previous sections, the burning rate of a specimen is a direct function of the net heat flux transferred to the fuel. The net flux is equal to the total of external heat flux, flame convection and flame radiation, minus radiative heat losses from the fuel surface and heat losses (or gains) at the specimen edges. The Russian work on the effect of diameter on pool fire burning rate by Blinov and Khudiakov gives some insight into this problem. A detailed discussion of this work and its implications is given by Drysdale.[53] If the pool diameter is less than 0.03 m, flame convection is laminar and burning rate increases with decreasing diameter. If the pool diameter exceeds 1 m, flame convection is turbulent and burning rate is independent of diameter. There is a transition region between these two limits, with a minimum burning rate for a

pool diameter of approximately 0.1 m. This work indicates that sample size in a heat release rate calorimeter must be at least 1 m for the results to be independent of scale. This is indeed not feasible in practice. The Russian pool fire data also indicate that heat transfer at the edges becomes excessive at diameters below 0.1 m. Therefore, sample size in bench-scale calorimeters should be at least 0.1 m. To predict real-scale burning rates, differences in flame heat transfer, and up to a lesser extent heat transfer at the edges, have to be accounted for. Nussbaum and Östman reported ignition and heat release data for 13 materials and 2 sample sizes.[54] Increase in sample size from 0.1 × 0.1 m to 0.2 × 0.2 m resulted in a slight reduction of piloted ignition time. Average heat release rate over the first minute after ignition on a per-unit-area basis increased by approximately 12 percent at exposure levels exceeding $25 \text{ kW} \cdot \text{m}^{-2}$. Larger increases were observed at the $25 \text{ kW} \cdot \text{m}^{-2}$ exposure level, and for peak heat release rate.

Depending on the specimen size in a bench-scale test, there is a limit on the degree of non-uniformity and irregularity of the product being tested, if the test conditions are to be representative of end-use conditions. Therefore, there might be some merit in choosing a specimen size that exceeds the minimum of 0.1 m. However, the main tradeoff is that a larger specimen requires a larger and more powerful heater to ensure uniform incident irradiance to the specimen. It should be recognized that, no matter what the specimen size is, there are assemblies and composites for which it is not possible to prepare representative bench-scale specimens. Intermediate-scale or full-scale tests are needed to evaluate the fire performance of such assemblies and composites.

An issue that is closely related to specimen size is that of edge effects. These effects have been studied extensively in the cone calorimeter. ASTM[55] and ISO[56] standards for the cone calorimeter prescribe that, except for calibrations with PMMA, the specimen is to be wrapped with aluminum foil on the sides and bottom. The main purpose of the foil is to eliminate mass transfer along all boundaries except the exposed face of the specimen. Furthermore, the ISO standard requires all tests be conducted in the horizontal orientation with the stainless-steel edge frame. Further details on the cone calorimeter specimen preparation and test conditions can be found in Section 3, Chapter 3. Toal *et al* tested several materials with and without foil wrapping, and with and without the edge frame.[57,58] They found that the edge frame reduces peak heat release rate, and lengthens the burning time. This is to be expected, because the edge frame is a relatively large mass of steel that acts as a heat sink, reducing the energy transferred to the specimen. Urbas and Sand were also concerned with the heat sink effect of the edge frame.[59] They designed an alternative edge frame, comprising an insulating collar made of medium-density or high-density refractory material. Their conclusion was that the best edge conditions are obtained using the insulating frame with insulation material that most closely resembles the specimen in thermal properties. Babrauskas *et al* conducted a very extensive study of the effects of specimen edge conditions on heat release rate.[60] The objective of this study was to further examine the issues raised by Toal *et al* and by Urbas and Sand, and to develop definitive recommendations. Specimens of 10 materials were tested in the horizontal orientation at $50 \text{ kW} \cdot \text{m}^{-2}$ using three configurations: (1) without edge frame, (2) with edge frame, and (3) with an insulated edge frame akin to that developed by Urbas and

Sand. All specimens were wrapped in aluminum foil. The study concluded that the use of an insulated frame gives heat release rate values that are slightly closer to the expected true values. However, the insulated frame makes the test procedure significantly more complicated, so that it is not recommended for routine testing. If the standard edge frame is used, Babrauskas *et al* recommend heat release rate data be expressed on the basis of an effective exposure area of 0.0081 m². The standard edge frame reduces the actual exposed area from 0.1 × 0.1 m to 0.094 × 0.094 m, or from 0.01 m² to 0.0088 m². The recommendation by Babrauskas *et al* to further reduce the exposed area to an effective value of 0.0081 m² indicates that the heat sink effect of the edge frame reduces heat release rate values by approximately 8 percent. Tsantaridis and Östman tested 11 products in the cone calorimeter in the horizontal orientation at 50 kW · m^{-2}, with and without the edge frame.[61] They also found that the use of the edge frame results in a reduction of heat release rate greater than what can be explained by the reduction of the exposed area. For the average heat release over the first three minutes following ignition, they found an average reduction of 8 percent, identical to Babrauskas *et al*. However, for maximum heat release rate they found reductions as high as 25 percent. It can be concluded from these studies that the sample holder configuration in a bench-scale heat release rate test may have a significant effect on the measurements, and that this should be accounted for if the test data are used to predict performance in real fires.

Specimen Orientation

As previously explained, products do not necessarily have to be tested in the same orientation as they are used. For practical reasons, the preferred orientation for bench-scale testing is horizontal facing upward. The vertical orientation might be preferable for collecting specialized data for research purposes.

Airflow

Standard rate of heat release test methods are operated under overventilated conditions. Plenty of excess air is supplied, so that the measurements are not affected by lack of oxygen. However, specialized studies have been conducted to evaluate the effect of ventilation and vitiation.[62] Such studies require a closed configuration. The original ISO heat release rate apparatus had the capability of heating the combustion air, so that the effect of air temperature on heat release rate could be studied.[63]

Other Measurements

Heat release rate calorimeters often include additional instrumentation to measure parameters that are important in characterizing the fire performance of materials. Perhaps the most important additional measurement is that of mass loss rate. Most calorimeters can be provided with a load cell to measure specimen mass loss, but this can be very difficult in a closed configuration. Mass loss rate is obtained from numerical differentiation of the mass loss measurements. Smoke meters are added to measure smoke obscuration in the exhaust duct. Both white light and laser light systems are being used. Toxic gas species can be measured in the exhaust duct with additional gas analysis equipment. Such equipment ranges from standard infrared CO and CO_2 ana-

lyzers to complex on-line Fourier Transform Infrared (FTIR) instrumentation. If and how additional instrumentation can be added depends largely on the design and construction details of the calorimeter.

SURVEY OF BENCH-SCALE CALORIMETERS

Numerous bench-scale heat release rate calorimeters have been developed since the early 1970s. Many of these calorimeters are described in detail below.

The Ohio State University (OSU) Apparatus

This apparatus, originally designed by Smith at Ohio State University,[3] is one of the most widely used and best-known bench-scale calorimeters. The test method was first published as a proposed ASTM standard in 1980. In 1983 it was adopted as ASTM Standard Test Method E 906, and the method has not changed since then.[48] A schematic view of the apparatus is shown in Figure 3-2.10. The apparatus consists of an insulated metal box. The conical wall section between the combustion chamber and the stack is hollow. Air flows through this cavity and mixes with the combustion products downstream of the thermocouple hot junctions. However, recovery of a significant fraction of wall heat losses is not fully accomplished. Some features of the OSU apparatus are described below.

Measuring technique: Heat release rate is determined by the sensible enthalpy rise method. The temperatures of inflowing air and outflowing gases are measured with a thermopile of three type K thermocouples. The hot junctions are located symmetrically along a diagonal of the stack cross section, above the baffle plate. The cold junctions are located below the air distributor plate. An electrical compensator is used to correct the temperature signals for thermal lag. The factor k in Equation 15 is obtained from line burner calibration runs.

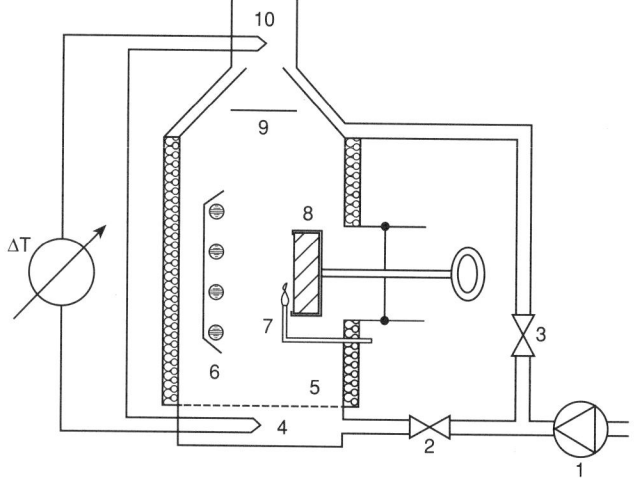

1 AIR SUPPLY FAN	6 HEATING ELEMENTS
2 MAIN FLOW CONTROL	7 GAS PILOT
3 BY-PASS FLOW CONTROL	8 SAMPLE AND HOLDER
4 TC COLD JUNCTIONS	9 BAFFLE PLATE
5 AIR DISTRIBUTOR PLATE	10 TC HOT JUNCTIONS

Fig. 3-2.10. Schematic of the OSU apparatus.

Configuration: Heater and specimen are located inside a box with approximate dimensions of $0.2 \times 0.41 \times 0.64$ m. The side walls of the box are insulated, and the hollow top wall section is cooled with air.

Heater: The vertical radiant heat source measures approximately 0.3×0.3 m, and consists of four silicon carbide heating elements. A steel masking plate is located in front of the elements to improve uniformity of the incident heat flux distribution over the specimen. The maximum incident heat flux to a vertical specimen is approximately 65 kW $\cdot$ m^{-2}.

Ignitor: The optional ignition source is a pilot flame of 2 ml $\cdot$ s^{-1} methane, premixed with 14 ml $\cdot$ s^{-1} air. The pilot flame is either impinging on the specimen at the bottom (point ignition), or is located in the gas phase at the top of the specimen (pilot ignition), or is not used.

Specimen size and orientation: For testing in the vertical orientation, specimens with an exposed area of 0.15×0.15 m are positioned parallel to the heating elements. Specimens can be tested in the horizontal orientation with the aid of an aluminum reflector foil, which reflects the radiation from the heating elements to the specimen. In this case, maximum irradiance is reduced to 50 kW $\cdot$ m^{-2}, and specimen size is 0.11×0.15 m. The use of the reflector plate is awkward and cumbersome, so that testing in the horizontal orientation with the OSU apparatus is not recommended.

Airflow: Total airflow rate is 40 l $\cdot$ s^{-1}, of which only 10 l $\cdot$ s^{-1} passes through the combustion chamber and the remaining 30 l $\cdot$ s^{-1} flows through the upper hollow wall section. Nevertheless, the airflow rate through the combustion chamber contains enough oxygen to feed a 36-kW fire. Because the heat release rate from test specimens rarely exceeds 20 kW, burning conditions in the OSU apparatus are always overventilated. The airflow rates are measured accurately with standard orifices.

Additional measurements: The ASTM E 906 standard does not include a mass loss measurement, but has a smoke measuring system with a white light source in the stack.

Federal Aviation Administration (FAA) version of the OSU apparatus: In 1978 the FAA established a committee to examine the factors affecting the ability of aircraft cabin occupants to survive in a post-crash environment. The committee recommended research to evaluate the fire performance of cabin materials, and development of a method using radiant heat for testing cabin materials. As a result, the FAA conducted an extensive series of full-scale fire tests and evaluated numerous bench-scale tests for their capability to provide results that correlate well with full-scale performance. The OSU apparatus, standardized as ASTM E 906, was found to be the most suitable for material qualification. Improved flammability standards and requirements for airplane cabin interior materials based on ASTM E 906 first went into effect in 1986.[64] The limits for acceptance were based on heat release rate measured at an irradiance level of 35 kW $\cdot$ m^{-2}. Peak heat release rate could not exceed 100 kW $\cdot$ m^{-2}, and average heat release rate over the first two minutes following ignition had to be 50 kW $\cdot$ m^{-2} or less. Originally, the test method used by the FAA was identical to ASTM E 906. Since then, modifications have been made.[65] The FAA method now uses a thermopile of five thermocouples, a lighter sample holder, and a modified test procedure to minimize problems associated with thermal lag.[47] The FAA criteria for acceptance were revised in 1990 to 65 kW $\cdot$ m^{-2} for peak heat release rate during the 5-minute test, and to 32.5 kW $\cdot$ m^{-2} for average heat release rate over the first 2 minutes following ignition.[65]

OSU Apparatus Modified for Oxygen Consumption

When oxygen consumption calorimetry became the preferred method for measuring rate of heat release, laboratories in the U.S., Canada, and Sweden modified their OSU apparatus. These modifications typically consisted of the elimination of the original thermopile, the addition of a gas sampling probe and gas analysis equipment, and some adjustments to the airflow rates. A brief summary of these investigations follows.

Modified OSU apparatus at NBS: As part of an investigation of combustion of mattresses exposed to flaming ignition sources, a number of mattress materials were tested at NBS in a modified OSU apparatus.[6,66] The following modifications were made to the standard OSU apparatus:

1. The oxygen consumption technique was used instead of the sensible enthalpy rise method,
2. The airflow through the hollow upper wall section was cut off to increase sensitivity of the oxygen measurement,
3. The airflow rate through the combustion chamber was increased to 12 l $\cdot$ s^{-1},
4. Specimen size was reduced to 0.1×0.1 m, and
5. The aluminum reflector foil was replaced by a more durable steel plate.

As a result of this study, good agreement was found between heat release rates measured in the modified OSU apparatus, and a classification of mattresses based on performance in full-scale experiments.

Modified OSU apparatus at the National Research Council of Canada (NRCC): The oxygen consumption method was compared with the three thermal methods, and was found to be preferable.[67] As a result, the OSU apparatus at NRCC was modified for oxygen consumption calorimetry and for measuring heat release rates at reduced airflow rates. Air regulators and flow meters were replaced so that the flow rates through the combustion chamber and through the upper hollow wall section could be set between 0.8 and 10 l $\cdot$ s^{-1}, and between 0.8 and 30 l $\cdot$ s^{-1}, respectively. At lower flow rates, longer flames were observed. The stack of the apparatus was extended to ensure all combustion took place inside the apparatus. For a wood specimen exposed to 25 kW $\cdot$ m^{-2}, it was found that both airflow rates could be reduced to 2.5 l $\cdot$ s^{-1} without significant incomplete combustion or flaming outside the apparatus.

Modified OSU apparatus at Lund University: The OSU apparatus at Lund University in Sweden was also modified for oxygen consumption calorimetry.[68] The airflow through the upper hollow wall section was cut off, and the stack of the apparatus was extended.

Modified OSU apparatus at the Forest Products Laboratory: Two significant modifications were made to the Forest Products Laboratory (FPL) apparatus, in addition to the inclusion of instrumentation for oxygen consumption calorimetry.[69] An auxiliary heat flux meter was added beneath the specimen, to monitor incident irradiance during a test. Measurements obtained with this auxiliary meter indicated that the incident irradiance to a burning wood specimen increases significantly during a test. For example, the incident irradiance to a Douglas fir plywood specimen at the end of a 10-minute burning period increased by 20 percent over the 35 kW · m^{-2} baseline. This is due to the fact that the heater elements in the OSU calorimeter are supplied with constant power and are not temperature controlled, and that the calorimeter walls are allowed to heat up (or cool down) during testing. The fact that exposure conditions in the OSU calorimeter are not constant is a major weakness of the apparatus. It is nearly impossible to remedy this problem. The addition of an auxiliary heat flux meter is highly recommended. Thus, at least the time-varying exposure conditions are known. Another modification at FPL was the addition of a load cell to measure specimen mass loss during a test. This was a rather difficult task due to the geometry of the apparatus and the mechanism for inserting specimens. The FPL load cell design seemed to be satisfactory, demonstrating the feasibility of measuring mass loss in the OSU apparatus.

The NBS I Calorimeter

This apparatus was developed by Parker and Long at NBS.[32] It consists of a combustion chamber at the bottom, a control chamber in the middle, and a mixing chamber at the top. Gases flow through the apparatus from bottom to top. The apparatus has the following features.

Measuring technique: Heat release rate is determined by the compensation method. The flow rate of propane to an auxiliary burner in the control chamber is adjusted to maintain the temperature in the mixing chamber at a constant level between 370 and 470°C. Additional air supplied to the control room greatly eliminates heat losses to the walls of this part of the apparatus.

Configuration: Three heaters and the specimen are located inside the combustion chamber with approximate dimensions of 0.33 × 0.33 × 0.36 m.

Heaters: Three walls of the combustion chamber consists of gas panels, capable of producing a heat flux to the specimen of up to 100 kW · m^{-2}. Incident irradiance to the specimen is checked with a copper disk heat flux meter.

Ignitor: Ignition of the specimen is non-piloted.

Specimen size and orientation: The vertical specimen measures 0.114 × 0.15 m with a maximum thickness of 25 mm. The specimen holder fits into an opening in the remaining combustion chamber wall that faces the three radiant panel walls.

Airflow: Air is supplied to the combustion chamber at a rate of 63 l · s^{-1}. The products of combustion are further diluted in the control chamber with air at a rate of 69 l · s^{-1}. The dilution air is induced through the porous walls of the control chamber, and serves to reduce the temperature of the exhaust gases. This reduces measurement errors and thermal stresses in the mixing chamber wall material. The errors result primarily from the assumption that constant temperature of the exhaust gases is equivalent to constant heat content. This is only approximately correct because of differences in specific heat between various gas species in the exhaust stream and air.

Additional measurements: The apparatus does not include any additional instrumentation, and specimen mass loss and smoke obscuration are not measured.

Stanford Research Institute calorimeter:　A scaled-up version of the NBS I calorimeter was constructed at Stanford Research Institute.[34] The redesigned combustion chamber measured approximately 0.86 × 0.68 × 0.98 m. Maximum

specimen size was increased to 0.46 × 0.61 m, and the radiant heat flux range was reduced to 15–70 kW · m^{-2}.

The FPL Calorimeter

This apparatus was developed by Brenden at the Forest Products Laboratory (FPL) in Madison, WI.[31] The apparatus has the following features.

Measuring technique: Heat release rate is determined by the substitution method. The flow rate of propane to the substitution burner in the combustion chamber is adjusted during a second run so that the temperature-time curve of the exhaust gases traces that measured during the experiment with the test specimen. Combustion chamber walls and the unexposed side of the specimen holder are water-cooled. Heat losses through walls and specimen are accounted for by measuring the enthalpy rise of the cooling water.

Configuration: The apparatus consists of a water-cooled combustion chamber with approximate dimensions of 0.76 × 0.43 × 1.09 m.

Heater: The specimen is exposed to the heat flux from a premixed gas burner flame. Experiments were conducted[70] using approximately 0.5 l · s^{-1} natural gas mixed with 5 l · s^{-1} air, generating a total heat flux to the specimen between 30 and 40 kW · m^{-2}.

Ignitor: The burner flame impinges on the specimen and also acts as ignition pilot.

Specimen size and orientation: The vertical specimen measures 0.45 × 0.45 m with a maximum thickness of 0.1 m.

Airflow: In addition to that supplied to the premixed burner, secondary air is supplied at the bottom of the combustion chamber. Various wood products were tested[70] with a secondary airflow rate of 14 l · s^{-1}. The maximum total flow rate of air that can be supplied is 130 l · s^{-1}.

Additional measurements: The apparatus does not include any additional instrumentation.

The FMRC Combustibility Apparatus

The Factory Mutual Research Corporation (FMRC) combustibility apparatus was initially developed to measure convective heat release rate and generation rates of smoke and combustion products.[71] Originally, only convective heat release rate was measured on the basis of enthalpy rise of the exhaust gases. Test results reported since the late 1970s also include total heat release rates calculated from oxygen consumption or carbon dioxide generation. In the 1980s, the FMRC apparatus was constructed in industrial laboratories in France, Germany, and the U.S. Tewarson used the apparatus to determine fire hazard indices[72] and material properties for fire modeling.[17] He also investigated the effect of environmental conditions (such as oxygen concentration in the combustion air) on heat release rate and burning behavior. The results of his extensive research are summarized in Section 3, Chapter 4. The FMRC combustibility apparatus has the following features.

Measuring technique: Total heat release rate is determined by the oxygen consumption method. Tewarson also used carbon dioxide generation to calculate heat release rate. However, the amount of energy generated per mass unit of carbon monoxide generated is much more fuel-dependent than the amount of energy produced per mass unit of oxygen consumed. Therefore, this technique is not as universally accepted as the oxygen consumption method.

Configuration: Tests are conducted in a semi-open environment. The specimen is located inside a quartz tube, 0.61 m in length and 0.17 m in diameter. A mixture of oxygen and nitrogen is supplied at the bottom of the tube. A stainless-steel funnel and vertical exhaust duct are located at some distance above the tube. Dilution air is entrained in the area between the tube and the exhaust system. The total flow of gases through the exhaust duct is determined by measuring pressure drop across a precalibrated orifice.

Heater: Four heaters, which are located coaxially outside the quartz tube, are used to generate an incident heat flux to the specimen with a maximum of 70 kW $\cdot$ m^{-2}. The electrical heaters operate at high temperatures, so that the spectral distribution of the emitted radiation is not representative of that present in most fires. This problem is discussed in a previous section.

Ignitor: A small pilot flame is located approximately 10 mm above the center of the specimen.

Specimen size and orientation: The specimen is circular and has a diameter of 0.1 m. Maximum thickness is 50 mm. The specimen is tested in the horizontal orientation.

Airflow: Total gas flow rate supplied to the bottom of the quartz tube can be set between 0 and 8.3 l $\cdot$ s^{-1}. Oxygen content of the combustion air can be varied between 0 and 60 percent. Oxygen concentrations below ambient are used for simulating ventilation-controlled fires. Oxygen concentrations above ambient are used to increase flame radiation, simulating larger fires.[73] Pure nitrogen is used to determine the heat of gasification.

Additional measurements: The apparatus includes instrumentation to measure specimen mass loss, smoke obscuration, and soot and gas species concentrations in the exhaust flow.

The First ISO Rate of Heat Release Apparatus

The apparatus discussed here never made it beyond a prototype. In 1984, ISO/TC92/SC1 initiated development of an international standard test method for measuring heat release rate of building products on the basis of the cone calorimeter. The cone calorimeter is discussed in detail in Section 3, Chapter 3. The development of an ISO heat release rate test method started more than ten years earlier, in the early 1970s. The original design was based on the British fire propagation test, BS 476, part 6. Much of the development work was conducted by three laboratories: two in England (Fire Research Station, and Timber Research and Development Association) and one in Denmark (Danish National Testing Institute). The prototype apparatus based on the sensible enthalpy rise method was finalized in 1980. The apparatus was modified for oxygen consumption calorimetry a few years later, and some of the technical problems with the prototype were resolved. The resulting apparatus is described in detail in a report issued by the laboratory that conducted the work between 1980 and 1982.[63] The features of this apparatus are briefly discussed below.

Measuring technique: Heat release rate is determined by the oxygen consumption method.

Configuration: Specimen and heater are located inside an insulated box with approximate dimensions 0.1 × 0.38 × 0.38 m.

Heater: The radiant panel measures 0.3 × 0.38 m, and consists of nine electrical elements. Maximum irradiance to the specimen is between 60 and 70 kW $\cdot$ m^{-2}. The power to the heating elements is controlled to keep panel temperature constant during testing.

Ignitor: A small hydrogen pilot flame is located in front of the specimen.

Specimen size and orientation: The exposed area of the specimen is 0.205 × 0.205 m. Maximum thickness is 125 mm. The specimen can be tested in any orientation by rotating the entire apparatus.

Airflow: Experiments have been conducted at an airflow rate of 8.3 l $\cdot$ s^{-1}. Air can be preheated to 100°C with heating elements located in the inlet duct. When testing specimens in the vertical orientation, direction of the airflow is from top to bottom. This results in a more uniform heat feedback from the flame to the specimen surface. When testing specimens in the horizontal orientation, air flows from one side to the other.

Additional measurements: The apparatus does not include instrumentation to measure specimen mass loss or other parameters.

The NBS II Calorimeter

This apparatus was derived from the NBS I calorimeter to accommodate larger specimens.[33] It has the following features.

Measuring technique: Heat release rate is determined by the compensation method. Gas flow to the burner is adjusted to maintain the temperature in the stack at a constant value of approximately 400°C. The same apparatus without a compensation burner was used to explore its suitability for oxygen consumption calorimetry.

Configuration: Radiant panels, specimen, and compensation burner are located in the same chamber with approximate dimensions of 1.07 × 1.37 × 1.07 m. When testing in the vertical orientation, two specimens are positioned back-to-back to approximate adiabatic rear boundary conditions.

Heaters: Two gas panels are located in front of two of the combustion chamber walls, forming an array of 0.3 × 0.6 m. Four panels are located in front of the remaining two walls, forming an array of 0.6 × 0.6 m. The specimen is positioned in the center between all panel arrays. Irradiance to the specimen can be set between 25 and 80 kW $\cdot$ m^{-2}.

Ignitor: An electric spark is used as the ignition pilot at the top of vertical specimens, and over the edge of horizontal specimens. Various other pilot flame configurations can be used as an option.

Specimen size and orientation: Initially, specimens of 0.3 × 0.3 m were tested in the vertical orientation. Babrauskas used the apparatus to test mattress samples of 0.15 × 0.3 m in the horizontal orientation.[66] A heat flux meter adjacent to the specimen was used to monitor incident irradiance during tests.

Airflow: Air is supplied at the bottom of the combustion chamber and to the premixed gas panel burners. The products are extracted through the stack at the top of the combustion chamber.

Additional measurements: The specimen is mounted on a load cell. Instrumentation is provided in the stack for measuring smoke obscuration and concentration of oxygen, carbon dioxide, and carbon monoxide.

The Cone Calorimeter

The cone calorimeter was developed at NBS by Dr. Vytenis Babrauskas in the early 1980s.[50] It is presently the most commonly used bench-scale rate of heat release apparatus.

The apparatus and test procedure are standardized in the U.S.[55] and internationally.[56] The cone calorimeter is described in detail in Section 3, Chapter 3, and its main features are summarized below.

Measuring technique: Heat release rate is determined by the oxygen consumption method. The gas flow rate in the exhaust duct is calculated from the pressure drop across, and temperature at, an orifice plate in the duct.

Configuration: Cone heater, spark ignitor, sample holder, and load cell are located underneath a hood. The standard configuration is open, with free access of air to the combustion zone.

Heater: The heater consists of a 5-kW electrical heating element, wound inside an insulated stainless-steel conical shell. Hence, the apparatus is named cone calorimeter. The heater can be oriented horizontally or vertically, to perform tests in either orientation. When tests are performed in the horizontal orientation, the specimen is positioned approximately 25 mm beneath the bottom plate of the cone heater. Flames and products of combustion rise and emerge through a circular opening at the top of the heater. Maximum irradiance to the specimen exceeds 100 kW $\cdot$ m^{-2}.

Ignitor: An electric spark is used as the ignition pilot at the top of vertical specimens, and over the center of horizontal specimens.

Specimen size and orientation: Specimen size in both orientations is 0.1 × 0.1 m.

Airflow: Combustion products and dilution air are extracted through the hood and exhaust duct by a high-temperature fan. The initial flow rate can be adjusted between 10 and 32 l $\cdot$ s^{-1}. Volumetric flow rate remains relatively constant during testing. Some cone calorimeters include additional instrumentation to optionally control and maintain the mass flow rate through the exhaust duct.

Additional measurements: The specimen is mounted on a load cell. Most cone calorimeters include instrumentation for measuring smoke obscuration (using a laser light source, described in ASTM E 1354 and ISO 5660-2) and concentration of soot, carbon dioxide, carbon monoxide, and other gases.

Larger cone heater at SWRI: A conical heater was designed at the Swedish Wood Research Institute (SWRI) double the size of that in the cone calorimeter.[54] The larger heater was used to study the effect of specimen size on heat release rate.

The SWRI Calorimeter

The Swedish Wood Research Institute (SWRI) developed an open oxygen consumption calorimeter using the apparatus built by Sensenig at NBS as a model.[74,75] The apparatus, which now has been replaced by a cone calorimeter, has the following features.

Measuring technique: Heat release rate is determined by the oxygen consumption method. Oxygen concentration is measured with a zirconium oxide cell. This analyzer has the advantage that oxygen concentration is measured in a hot and wet gas sample, which greatly simplifies the calculations. However, the accuracy of zirconium oxide cells is inferior to that of paramagnetic oxygen analyzers. For this reason, the laboratory later changed to a paramagnetic analyzer after it acquired a cone calorimeter.[38] The gas flow rate in the exhaust duct is calculated from centerline velocity measured with a pitot tube and gas temperature.

Configuration: Heater, pilot flame, sample holder, and balance are located underneath a hood. The standard configuration is open, with free access of air to the combustion zone.

Heater: The heater measures approximately 0.3 × 0.3 m, and consists of tubular medium-wave infrared lamps. Maximum irradiance to the specimen is 50 kW $\cdot$ m^{-2}. The irradiance level is adjusted by changing the distance between heater and specimen. Various heater operating conditions were examined, and conditions were chosen that result in the closest match with the spectral distribution of radiation from wood and oil flames.

Ignitor: A small pilot flame at the top of the specimen was found to result in the quickest involvement of the entire exposed specimen area.

Specimen size and orientation: Specimen size is 0.15 × 0.15 m. Specimens are positioned in the vertical orientation, parallel to the heating panel.

Airflow: Combustion products and dilution air are extracted through the hood and exhaust duct. The initial flow rate through the duct can be varied, but tests were conducted with a flow rate of approximately 20 l $\cdot$ s^{-1}.

Additional measurements: The specimen is mounted on an electronic balance to measure mass loss. A white light source system is used for measuring smoke obscuration in the exhaust duct.

Comparison Between Bench-Scale Calorimeters

How do results obtained with different calorimeters for the same material compare? A number of comparisons are reported in the literature. Sensenig tested particleboard, Type X gypsum board, medium-density fiberboard, and polyurethane foam in his oxygen consumption calorimeter, and according to ASTM E 9.6.[75] Specimen holder, orientation, specimen size (0.15 × 0.15 m), mounting method, ignition pilot, and heat flux were identical to eliminate some of the possible sources of variation. The heat release curves from the two methods agreed quite well. In general, results from the OSU apparatus were slightly higher. Chamberlain compared heat release rates for 10 wood products measured in a slightly modified NBS I calorimeter to results from the NBS II calorimeter.[76] Poor correlations were found for peak heat release rate and average heat release rate over the first minute. Agreement was much better for 5-minute and 10-minute averages. This indicates that the dynamic response of the two methods is very different. Östman *et al* reported on a comparison of heat release data for 13 building materials, obtained with the modified OSU apparatus at Lund, the cone calorimeter, and the SWRI apparatus.[77] Agreement was remarkably good with a correlation coefficient exceeding 90 percent for average heat release rate over the first minute following ignition. Babrauskas compared peak heat release rate from various calorimeters for 5 aircraft wall-paneling materials.[78] He found good agreement between the FMRC apparatus and the cone calorimeter. However, he also found that the peak heat release rate from the OSU apparatus was approximately 50 percent of the peak from the cone calorimeter. Whether thermopile or oxygen consumption was employed seemed to have only a minor effect on the results from the OSU apparatus. Unfortunately, correlation of average heat release rate was not reported, so that a comparison with the Swedish work is not possible.

Tran compared heat release rate curves for Douglas fir plywood from the cone calorimeter, and the OSU apparatus at FPL modified for oxygen consumption.[69] First and second peaks agreed well, but the OSU data exceeded the cone calorimeter data by up to 20 percent between the peaks. The increased burning rate can be explained by the enhanced irradiance to the specimen due to temperature rise of the calorimeter walls and heater during a test. Tran tested the same material in the OSU apparatus with the vertical specimen holder from the cone calorimeter and found no effect.

Heat release data from bench-scale calorimeters are always apparatus-dependent. Differences in geometry, test conditions, and mounting methods explain discrepancies between the results from different calorimeters. Apparatus-specific factors must be considered and accounted for in a comparison between different calorimeters, or when the data are used to predict performance in real scale.

LARGE-SCALE HEAT RELEASE RATE CALORIMETERS

There are two primary reasons for conducting a full-scale fire test: (1) to validate computer fire models that are used for hazard assessment and (2) to evaluate the fire performance of products and assemblies for which bench-scale tests are not representative. Many types of full-scale fire test rigs have been equipped with instrumentation to measure heat release rate. This became even more desirable with the development of the oxygen consumption method. These experimental arrangements can be categorized in four different groups: (1) scaled-up heat release rate calorimeters, (2) furniture calorimeters, (3) room tests, and (4) other large-scale fire tests equipped with instrumentation for measuring heat release rate.

Scaled-Up Heat Release Rate Calorimeters

These are larger scale versions of the bench-scale calorimeters discussed previously. A sample of one to several square meters is exposed to a constant or preprogrammed thermal exposure.

Perhaps the oldest calorimeter of this type is the Factory Mutual Research Corporation Construction Materials Calorimeter.[30] A specimen of 1.22×1.22 m is exposed for 10 to 30 minutes to the standard ASTM E 119 temperature-time curve, compressed into a much shorter time. Maximum heat flux is $150 \, kW \cdot m^{-2}$. The specimen is horizontal facing downward, and is exposed to a furnace operated with premixed heptane burners. Heat release rate is determined with the substitution method.

In recent years, a number of large-scale panel tests have been developed. These tests consist of a large radiant gas panel and a flat specimen parallel to the panel. Panel and specimen are located beneath a hood and exhaust system. Heat release rate is measured by oxygen consumption. A calorimeter of this type was developed in Canada for facade systems.[79] The test is capable of exposing specimens of 1.22×2.08 m to a maximum irradiance of $30 \, kW \cdot m^{-2}$. An intermediate-scale calorimeter of the same type was developed in the U.S., primarily to evaluate the heat release rate from construction assemblies.[80] Sample size is 1×1 m, with a maximum irradiance of $60 \, kW \cdot m^{-2}$.

Furniture Calorimeters

Furniture calorimeters have been developed in the U.S.,[81] the Nordic countries,[82] and England.[83] These calorimeters measure heat release rate from objects, such as chairs, sofas, mattresses, etc. The primary use of this kind of information is for input into compartment fire models and smoke transport models, which form part of fire hazard assessment. The specimen is placed on a load cell platform, beneath a hood and exhaust system. Instrumentation is provided in the exhaust duct to measure flow rate, oxygen and other gas species concentrations, and smoke obscuration.

Room Tests

Every year, numerous room fire experiments are conducted throughout the world for various reasons. Since the development of oxygen consumption calorimetry, heat release has become one of the routine measurements. Until a few years ago there was little or no standardization in the way room tests were carried out. In the 1970s it was recognized that there was a need for a standard room fire test procedure, in particular for evaluating wall and ceiling lining materials. On the basis of observations made in earlier tests, it was concluded that a room/corner scenario was the best choice for such a standardized procedure. Two major developments greatly contributed to the standardization of a room/corner fire test. The first development was the use of gas burner ignition sources over wood cribs. Gas burners are repeatable and reproducible. They have the advantage that a wide range of steady or time-varying exposure levels can easily be obtained. The second development was that of oxygen consumption calorimetry.

Most of the work toward the development of a standard room/corner test was done in the late 1970s and early 1980s in the U.S. The need for a standard room fire test and some aspects of its design were discussed by Benjamin in 1977.[84] Subsequent research in North America to arrive at a standard full-scale test was conducted primarily by Fisher and Williamson at the University of California (UCB)[85] and by Lee at NBS.[86]

Considerable research has also been conducted in the Nordic countries. A detailed project to construct a full-scale room calorimeter was undertaken in Sweden.[87,88] No oxygen consumption measurements were made at the time. A heat balance was obtained by comparing the theoretical heat release from combustion of gaseous fuel to the sum of the heat losses. The heat losses consisted of convection through the doorway, conduction through the walls and ceiling, and radiation through the doorway. Heat convection through the doorway was estimated by measuring gas velocity and temperature at many points in the doorway. Heat conduction through the surrounding surfaces was calculated using total heat flux, radiation, and surface temperature data. Heat loss by radiation through the door was calculated from radiometer measurements. Initially, a series of quasi-steady calibration tests were conducted in an inert room. Three different circular propane gas burners were used with diameters of 0.2, 0.3, and 0.4 m, respectively. Heat balance calculations showed reasonable agreement, with convection losses being dominant. In subsequent tests with surface finishes, a heptane pool fire with a heat release rate of about 50 kW was used as the ignition source.

Ahonen et al at the Technical Research Center of Finland studied the effects of different gas burner ignition sources on room/corner fire growth.[89] Three different burner sizes (0.17 × 0.17 m, 0.305 × 0.305 m, and 0.5 × 0.5 m) and three square wave output levels (40 kW, 160 kW, and 300 kW) were used, leading to a total of 9 combinations. Oxygen consumption calorimetry was implemented for measuring heat release rate. The burner was placed in a rear corner of the room. In all tests, the walls and ceiling, except for the front wall, were lined with a 10-mm particleboard of fairly high density (720 kg · m^{-3}). The following six criteria were used to determine the time to flashover:

1. Flames emerging through the door (flameover),
2. A total heat release rate of 1 MW,
3. Total heat flux to the floor of 20 kW · m^{-2},
4. A minimum rate of smoke production,
5. 600°C at the geometric center of the room, and
6. Total heat flux to the floor of 50 kW · m^{-2}.

With the time to flashover defined as the average of the six criteria, the following remarkable results were obtained:

1. At the 40-kW level, the medium-size burner resulted in flashover first, followed by the smaller burner and then the larger burner.
2. At 160 kW, the largest burner resulted in flashover first, quickly followed by the other two configurations.
3. At 300 kW, the trend was the same as at 160 kW with an even smaller spread between the three results.

The effect of burner size was most significant at the lowest power level, where the medium-size burner was the most severe. At higher exposure levels, the size of the burner was not important. Radiative and convective heat transfer from the burner flame was shown to depend on burner size and power level and had a significant effect on the performance of the material tested. On the basis of the results, the medium burner size and power level were recommended.

Based on the work at NBS and UCB, ASTM drafted a proposal for a standard room/corner test. The proposal was printed in the "grey" pages of the *ASTM Annual Book of Standards* in 1982 and 1983.[13] The draft specifies dimensions and geometry of the test compartment, burner size, exposure level, and basic instrumentation and measurements required. The compartment has a floor area of 2.44 m by 3.66 m, interior room height is 2.44 m. A single door in the middle of a square wall is 0.76 m wide and 2.03 m high. A sand burner ignition source is located in a corner opposite from the door. All combustion products emerging through the door are collected in a hood. Measurements are made in the exhaust duct to calculate heat release rate *via* oxygen consumption. The ASTM proposed standard is designed to look at fire growth of wall and ceiling lining materials only, without any other combustible objects in the compartment.

Parallel to the development of the ASTM proposal for a room/corner test, ISO has developed an international room test standard. The ISO standard was published in 1992[16] and is based primarily on the Nordic research described above. The concept is identical to that of the draft ASTM method. The dimensions of the test room and ventilation opening are basically the same, but the ISO standard allows ignition sources and specimen configurations other than those described in the ASTM document.

Other Large-Scale Fire Tests

The oxygen consumption method makes it possible to measure heat release rate on almost any fire test. Measurements are

reported in the literature for very small fires (kW range) up to 100 MW. Standard fire endurance tests have been instrumented for oxygen consumption calorimetry.[90,91] Exhaust systems of entire laboratories have been equipped for the technique. There are virtually no limits.

REFERENCES CITED

1. V. Babrauskas and R. Peacock, "Heat Release Rate: The Single Most Important Variable in Fire Hazard," *Fire Safety J.*, Vol. 18, pp. 255–272 (1992).
2. E. Smith, "An Experimental Determination of Combustibility," *Fire Technology*, Vol. 7, pp. 109–119 (1971).
3. E. Smith, "Heat Release Rate of Building Materials," in *Ignition, Heat Release and Noncombustibility of Materials*, A. Robertson, ed., American Society for Testing and Materials, STP 502, pp. 119–134 (1972).
4. E. Smith, "Application of Release Rate Data to Hazard Load Calculations," *Fire Technology*, Vol. 10, pp. 181–186 (1974).
5. E. Smith and S. Satija, "Release Rate Model for Developing Fires," *ASME Journal of Heat Transfer*, Vol. 105, pp. 282–287 (1983).
6. V. Babrauskas, "Performance of the OSU Rate of Heat Release Apparatus Using PMMA and Gaseous Fuels," *Fire Safety J.*, Vol. 5, pp. 9–20 (1982).
7. M. Janssens, "Critical Analysis of the OSU Room Fire Model for Simulating Corner Fires," in *Fire and Flammability of Furnishings and Contents*, A. Fowell, ed., ASTM, STP 1233, pp. 169–185 (1994).
8. "BS476: Fire Tests on Building Materials and Structures. Part 6: Method for Test for Fire Propagation for Products," British Standards Institution, London, England (1981).
9. "DIN 4102, Part 1: Fire Behavior of Building Materials and Building Components; Building Materials; Terminology, Calibration, and Test Procedures," DIN, Berlin, Germany (1984).
10. "NF P 92-501: Safety Against Fire—Building Materials—Reaction to Fire Tests—Radiation Test Used for Rigid Materials on Rigid Substrate (Flooring and Finishes) of All Thicknesses, and for Flexible Materials Thicker than 5 mm," AFNOR, Paris, France (1985).
11. "NT Fire 0004, Building Products: Heat Release and Smoke Generation," NORDTEST, Helsinki, Finland (1985).
12. "Designation of Semi-Non-Combustible Materials and Fire Retardant Materials," Notification No. 1231, Ministry of Construction, Tokyo, Japan (1986).
13. "Proposed Method for Room Fire Test of Wall and Ceiling Materials and Assemblies," *ASTM Annual Book of Standards*, Vol. 04.07, American Society for Testing and Materials, Philadelphia, pp. 958–978 (1983).
14. "UBC 8-2: Standard Test Method for Evaluating Room Fire Growth Contribution of Textile Wall Covering," International Conference of Building Code Officials, Uniform Building Code, Vol. 3, pp. 147–164, (1994).
15. "NT Fire 030, Building Products: Fire Spread and Smoke Production—Full-Scale Test," NORDTEST, Helsinki, Finland (1987).
16. "ISO 9705: Full-Scale Room Test," International Standards Organization, Geneva, Switzerland (1992).
17. A. Tewarson and R. Pion, "Flammability of Plastics, I. Burning Intensity," *Combustion and Flame*, Vol. 26, pp. 85–103 (1976).
18. V. Petrella, "The Mass Burning Rate of Polymers, Wood and Liquids," *J. of Fire and Flammability*, Vol. 11, pp. 3–21 (1980).
19. W. Parker and J. Urbas, "Surface Temperature Measurements on Burning Wood Specimens in the Cone Calorimeter and the Effect of Grain Orientation," *Wood Burning '92 Conference*, Štrebské Pleso, Czechoslovakia (1991).
20. J. Urbas, "Non-Dimensional Heat of Gasification Measurements in the Intermediate Scale Heat Release Apparatus," *J. of Fire and Materials*, Vol. 17, pp. 119–123 (1993).
21. M. Janssens, "Cone Calorimeter Measurements of the Heat of

Gasification of Wood," *Proceedings of the Interflam '93 Conference*, Interscience Communications, London, England, pp. 549–558 (1993).

22. E. Smith, "Math Model of a Fire in a Compartment Having Combustible Walls and Ceiling," *AIChE Symposium Series 246*, Vol. 81, pp. 64–74 (1985).

23. H. Mitler, "Predicting the Spread Rates of Fires on Vertical Surfaces," *23rd Symposium (International) on Combustion*, The Combustion Institute, pp. 1715–1721 (1990).

24. K. Opstad, "Fire Modeling Using Cone Calorimeter Results," *Proceedings of the Eurefic Seminar*, Interscience Communications, London, England, pp. 65–71 (1991).

25. J. Quintiere, "A Simulation Model for Fire Growth on Materials Subject to a Room/Corner Test," *Fire Safety J.*, Vol. 20, pp. 313–339 (1993).

26. E. Smith and T. Green, "Release Rate Tests for a Mathematical Model," *Mathematical Modeling of Fires*, J. Mehaffey, ed., American Society for Testing and Materials, STP 983, pp. 7–20 (1987).

27. U. Wickström and U. Göransson, "Full-Scale/Bench-Scale Correlations of Wall and Ceiling Linings," *J. of Fire and Materials*, Vol. 16, pp. 15–22 (1992).

28. B. Östman and R. Nussbaum, "Correlation Between Small-Scale Rate of Heat Release and Full-Scale Room Flashover for Surface Linings," in *Fire Safety Science, Proceedings of the 2nd International Symposium*, Hemisphere Publ. Co., NY, pp. 823–832 (1989).

29. B. Karlsson, "Modeling Fire Growth on Combustible Lining Materials in Enclosures," Ph.D. Thesis, University of Lund, Sweden (1992).

30. N. Thompson and E. Cousins, "The FM Construction Materials Calorimeter," *NFPA Quarterly*, Vol. 52, pp. 186–192 (1959).

31. J. Brenden, "Apparatus for Measuring Rate of Heat Release from Building Materials," *J. of Fire & Flammability*, Vol. 6, pp. 50–64 (1975).

32. W. Parker and M. Long, "Development of a Heat Release Rate Calorimeter at NBS," in *Ignition, Heat Release, and Noncombustibility of Materials*, A. Robertson, ed., American Society for Testing and Materials, STP 502, pp. 135–151 (1972).

33. J. Tordella and W. Twilley, "Development of a Calorimeter for Simultaneously Measuring Heat Release and Mass Loss Rate," NBSIR 83-2708, National Bureau of Standards, Gaithersburg, MD (1983).

34. S. Martin, "Characterization of the Stanford Research Institute Large-Scale Heat-Release Rate Calorimeter," NBS-GCR 76-54, National Bureau of Standards, Gaithersburg, MD (1975).

35. W. Thornton, "The Relation of Oxygen to the Heat of Combustion of Organic Compounds," *Philosophical Magazine and J. of Science*, Vol. 33 (1917).

36. C. Huggett, "Estimation of the Rate of Heat Release by Means of Oxygen Consumption," *J. of Fire and Materials*, Vol. 12, pp. 61–65 (1980).

37. W. Parker, "An Investigation of the Fire Environment in the ASTM E 84 Tunnel Test," NBS Technical Note 945, National Bureau of Standards, Gaithersburg, MD (1977).

38. R. Nussbaum, "Oxygen Consumption Measurements in the Cone Calorimeter—A Direct Comparison Between a Paramagnetic Cell and a High-Temperature Cell," *J. of Fire and Materials*, Vol. 11, pp. 201–203 (1987).

39. W. Parker, "Calculations of the Heat Release Rate by Oxygen Consumption for Various Applications," NBSIR 81-2427, National Bureau of Standards, Gaithersburg, MD (1982).

40. M. Janssens, "Measuring Rate of Heat Release by Oxygen Consumption," *Fire Technology*, Vol. 27, pp. 234–249 (1991).

41. A. Roberts, "Ultimate Analysis of Partially Decomposed Wood Samples," *Combustion and Flame*, Vol. 8, pp. 345–346 (1964).

42. V. Babrauskas, "The Generation of CO in Bench-Scale Fire Tests and the Prediction for Real-Scale Fires," *Proceedings of the 1st Fire and Materials Conference*, Interscience Communications, London, England, pp. 155–177 (1992).

43. H. Sawada, "Thermodynamics of Polymerization," Marcel Dekker, New York (1976).

44. D. Evans and L. Breden, "Time Delay Correction for Heat Release Rate Data," *Fire Technology*, Vol. 14, pp. 85–96 (1978).

45. D. Bluhme and R. Getka, "Rate of Heat Release Test—Calibration, Sensitivity, and Time Constants of ISO RHR Apparatus," NORDTEST Project 115-77, National Institute for Testing of Materials, Copenhagen, Denmark (1979).

46. P. Vandevelde, "An Evaluation of Heat Release Criteria in Reaction-to-Fire Tests," *J. of Fire and Materials*, Vol. 4, pp. 157–162 (1980).

47. A. Abramowitz and R. Lyon, "Comparison of Heat Release Rate from Deconvoluted OSU and Oxygen Consumption Principle Calorimeter Signals," *Proceedings of the 2nd Fire and Materials Conference*, Interscience Communications, London, England, pp. 161–170 (1993).

48. "ASTM E 906: Test Method for Heat and Visible Smoke Release Rates for Materials and Products," *ASTM Fire Test Standards*, 4th ed., ASTM, Philadelphia, PA, pp. 877–892 (1993).

49. "ASTM E 1317: Standard Test Method for Flammability of Marine Surface Finishes," *ASTM Fire Test Standards*, 4th ed., ASTM, Philadelphia, PA, pp. 921–938 (1993).

50. V. Babrauskas, "Development of the Cone Calorimeter—A Bench-Scale Heat Release Rate Apparatus Based on Oxygen Consumption," *J. of Fire and Materials*, Vol. 8, pp. 81–95 (1984).

51. A. Koohyar, "Ignition of Wood by Flame Radiation," Ph.D. Thesis, University of Oklahoma, Norman, OK (1967).

52. J. Hallman, "Ignition Characteristics of Plastics and Rubber," Ph.D. Thesis, University of Oklahoma, Norman, OK (1971).

53. D. Drysdale, "An Introduction to Fire Dynamics," John Wiley and Sons, Chichester, England, pp. 153–160 (1985).

54. R. Nussbaum and B. Östman, "Larger Specimens for Determining Rate of Heat Release in the Cone Calorimeter," *J. of Fire and Materials*, Vol. 10, pp. 151–160 (1986).

55. "ASTM E 1354: Test Method for Heat and Visible Smoke Release Rates for Materials and Products Using an Oxygen Consumption Calorimeter," *ASTM Fire Test Standards*, 4th ed., ASTM, Philadelphia, PA, pp. 968–984 (1993).

56. "ISO 5660-1: Rate of Heat Release of Building Products (Cone Calorimeter)," International Organization for Standardization, Geneva, Switzerland (1992).

57. B. Toal, T. Shields, and W. Silcock, "Observations on the Cone Calorimeter," *J. of Fire and Materials*, Vol. 14, pp. 73–76 (1989).

58. B. Toal, T. Shields, and W. Silcock, "Suitability and Preparation of Samples for the Cone Calorimeter," *Fire Safety J.*, Vol. 16, pp. 85–88 (1990).

59. J. Urbas and H. Sand, "Some Investigations on Ignition and Heat Release of Building Materials Using the Cone Calorimeter," *Proceedings of the Interflam '90 Conference*, Interscience Communications, London, England, pp. 183–192 (1990).

60. V. Babrauskas, W. Twilley, and W. Parker, "The Effect of Specimen Edge Conditions on Heat Release Rate," *J. of Fire and Materials*, Vol. 17, pp. 51–63 (1993).

61. L. Tsantaridis and B. Östman, "Communication: Retainer Frame Effects on Cone Calorimeter Results for Building Products," *J. of Fire and Materials*, Vol. 17, pp. 43–46 (1993).

62. V. Babrauskas, W. Twilley, M. Janssens, and S. Yusa, "A Cone Calorimeter for Controlled-Atmosphere Studies," *J. of Fire and Materials*, Vol. 16, pp. 37–43 (1992).

63. D. Bluhme, "ISO RHR Test Apparatus with Oxygen Consumption Technique," NORDTEST Project 115-77, Dantest, Copenhagen, Denmark (1982).

64. *Federal Register*, Vol. 51, pp. 26206–26221 (1986).

65. "Aircraft Material Fire Test Handbook," DOT/FAA/CT-89/15, US Department of Transportation, Federal Aviation Administration, Atlantic City, NJ (1990).

66. V. Babrauskas, "Combustion of Mattresses Exposed to Flaming Ignition Sources, Part II: Bench-Scale Tests and Recommended Standard Test," NBSIR 80-2186, National Bureau of Standards, Gaithersburg, MD (1981).

67. Y. Tsuchiya, "Methods of Determining Heat Release Rate: State-of-the Art," *Fire Safety J.*, Vol. 5, pp. 49–57 (1982).

68. J. Blomqvist, "Rate of Heat Release of Building Materials, Experiments with an OSU Apparatus Using Oxygen Consumption," Report LUTVDG/(TVBB-3017), Lund University, Sweden (1983).

69. H. Tran, "Modifications to an Ohio State University Apparatus and Comparison with Cone Calorimeter Results," in *Heat and Mass Transfer in Fires, Proceedings* of the AIAA/ASME Thermophysics and Heat Transfer Conference, HTD Vol. 141, ASME, pp. 131–139 (1990).

70. J. Brenden, "Wood-Base Building Materials: Rate of Heat Release," *J. of Fire & Flammability*, Vol. 6, pp. 274–293 (1975).

71. A. Tewarson, "Heat Release Rates from Burning Plastic," *J. of Fire & Flammability*, Vol. 8, pp. 115–130 (1977).

72. A. Tewarson, "Reliable Small-Scale Fire Testing Apparatus," *Modern Plastics*, pp. 58–62 (1980).

73. A. Tewarson, J. Lee, and R. Pion, "The Influence of Oxygen Concentration on Fuel Parameters for Fire Modeling," *18th Symposium (International) on Combustion*, The Combustion Institute (1981).

74. G. Svensson and B. Östman, "Rate of Heat Release by Oxygen Consumption in an Open Test Arrangement," *J. of Fire & Materials*, Vol. 8, pp. 206–216 (1984)

75. D. Sensenig, "An Oxygen Consumption Technique for Determining the Contribution of Interior Wall Finishes to Room Fires," NBS Technical Note 1128, National Bureau of Standards, Gaithersburg, MD (1980).

76. D. Chamberlain, "Heat Release Rate Properties of Wood-Based Materials," NBSIR 82-2597, National Bureau of Standards, Gaithersburg, MD (1983).

77. B. Östman, G. Svensson, and J. Blomqvist, "Comparison of Three Test Methods for Measuring Rate of Heat Release," *J. of Fire & Materials*, Vol. 9, pp. 176–184 (1985).

78. V. Babrauskas, "Comparative Rates of Heat Release from Five Different Types of Test Apparatuses," *J. of Fire Sciences*, Vol. 4, pp. 148–159 (1986).

79. I. Oleszkiewicz and G. Crampton, "Heat Release from Combustion of Large-Scale Wall Assemblies," *Abstracts of the 1st Heat Release and Fire Hazard Symposium*, Interscience Communications, London, England, pp. 18–22 (1991).

80. "ASTM E1623: Standard Test Method for Determination of Fire and Thermal Parameters of Materials, Products, and Systems Using an Intermediate Scale Calorimeter (ICAL)," *ASTM Annual Book of Standards*, Vol. 0407, ASTM, Philadelphia, pp. 1373–1392, 1994.

81. V. Babrauskas, J. Lawson, W. Walton, and W. Twilley, "Upholstered Furniture Heat Release Rates Measured with a Furniture Calorimeter," NBSIR 82-2604, National Bureau of Standards, Gaithersburg, MD (1982).

82. "NT Fire 032, Upholstered Furniture: Burning Behavior—Full-Scale Test," NORDTEST, Helsinki, Finland (1987).

83. S. Ames and S. Rogers, "Large- and Small-Scale Fire Calorimetry Assessment of Upholstered Furniture," *Proceedings of the Interflam '90 Conference*, Interscience Communications, London, England, pp. 221–232 (1990).

84. I. Benjamin, "Development of a Room Fire Test," *Fire Standards and Safety*, STP 614, ASTM, Philadelphia, PA, pp. 300–311 (1977).

85. F. Fisher and R. Williamson, "Intralaboratory Evaluation of a Room Fire Test Method," NBS GCR 83-421, National Bureau of Standards, Gaithersburg, MD (1983).

86. B. Lee, "Standard Room Fire Test Development at the National Bureau of Standards," *Fire Safety Science and Engineering*, STP 882, ASTM, Philadelphia, PA, pp. 29–42 (1985).

87. B. Sundström and U. Wickström, "Fire: Full-Scale Tests," Technical Report SP-RAPP 1980:14, National Testing Institute, Borås, Sweden (1980).

88. B. Sundström and U. Wickström, "Fire: Full-Scale Tests, Calibration of Test Room—Part 1," Technical Report SP-RAPP 1980:48, National Testing Institute, Borås, Sweden (1981).

89. A. Ahonen, C. Holmlund, and M. Kokkala, "Effects of Ignition Source in Room Fire Tests," *Fire Science & Technology*, Vol. 7, pp. 1–13 (1987).

90. J. Brenden and D. Chamberlain, "Heat Release Rates from Wall Assemblies," Research Paper FPL-RP-476, Forest Products Laboratory, Madison, WI (1986).

91. H. Tran and R. White, "Heat Release from Wood Wall Assemblies Using Oxygen Consumption Method," *Fire and Polymers: Hazards, Identification, and Prevention*, ACS Symposium Series 425, American Chemical Society, Washington, DC, pp. 411–428 (1990).

THE CONE CALORIMETER

Vytenis Babrauskas

INTRODUCTION

Section 3, Chapter 2 describes the history and development of techniques for measuring heat release rate (HRR). This chapter outlines features and details of today's preferred instrument for measuring bench-scale HRR—the cone calorimeter. Other cone calorimeter measuring functions are:

1. Effective heat of combustion,
2. Mass loss rate,
3. Ignitability,
4. Smoke and soot, and
5. Toxic gases.

The cone calorimeter is based on the concept of oxygen consumption calorimetry which is also presented in Section 3, Chapter 2.

This chapter provides both an introduction to and description of cone calorimeter measurement technology. The cone calorimeter has recently assumed a dominant role in bench-scale fire testing of various products; therefore, an emphasis will be placed on the *why* of various design features. When conducting tests, the cone calorimeter operator needs to consult several other documents; testing will, presumably, be in conformance with either ISO 5660[1] or ASTM E 1354.[2] In addition, the "User's Guide for the Cone Calorimeter"[3] should be consulted. This chapter, thus, does not emphasize the operational aspects, which are documented in the above references, but provides the reader with an overall feel for the equipment. Space is not available in the Handbook to fully discuss the applications of cone calorimeter data, apart from the review of data given in Section 3, Chapter 1. Extensive guidance on using cone calorimeter data is given in a new textbook on this subject.[4] It also provides example data compilations and information on using cone calorimeter data for predictions of fires.

SUMMARY OF FEATURES

A schematic view of the cone calorimeter is shown in Figure 3-3.1. Figure 3-3.2 shows a commercial instrument,

Dr. V. Babrauskas is President of Fire Science and Technology Inc., specializing in research and consulting on fire tests, hazard analysis, fire modeling, and related fire safety issues.

while Figure 3-3.3 identifies some of the major components. The more salient operational features and limits of the apparatus are:

- Specimen size: 100 mm × 100 mm, thickness of 6 to 50 mm
- Specimen orientation: Horizontal, face up (standard testing) or vertical (reserved for exploratory studies)
- Specimen back-face conditions: Very low loss insulating ceramic fibrous material
- Load cell live load capacity: 500 g
- Load cell tare capacity: 3.5 kg
- Load cell resolution: 0.005 g
- Ignition: Electric spark
- Heating flux range: 0 to 110 $kW \cdot m^{-2}$
- Flux uniformity, horizontal: Typically 2 percent
- Flux uniformity, vertical: Typically 7 percent
- Sensing principle: Oxygen consumption, only
- Maximum instantaneous output: In excess of 20 kW
- Normally calibrated range: 0 to 12 kW
- Linearity over 0 to 12 kW range: 5 percent
- Noise intrinsic to oxygen meter: 20 ppm O_2
- Noise in HRR measurement, over 0 to 12 kW range: 2.5 percent
- Smoke meter operating range: 0 to 20 m^{-1} (linear)
- Smoke meter resolution: 0.01 m^{-1}
- Soot sampler mass fraction range: 0 to 1 part in 200 (of exhaust gas flow)

Operating Principle

It is emphasized at this point that the cone calorimeter has been designed to use *only* oxygen consumption calorimetry as its measurement principle. Other calorimeters that on occasion use oxygen consumption principles, e.g., the Factory Mutual Research Corporation (FMRC) flammability apparatus (see Section 3, Chapter 2), also sometimes incorporate a sensible enthalpy flow measurement technique to arrive at the "convective component" of the heat release rate.

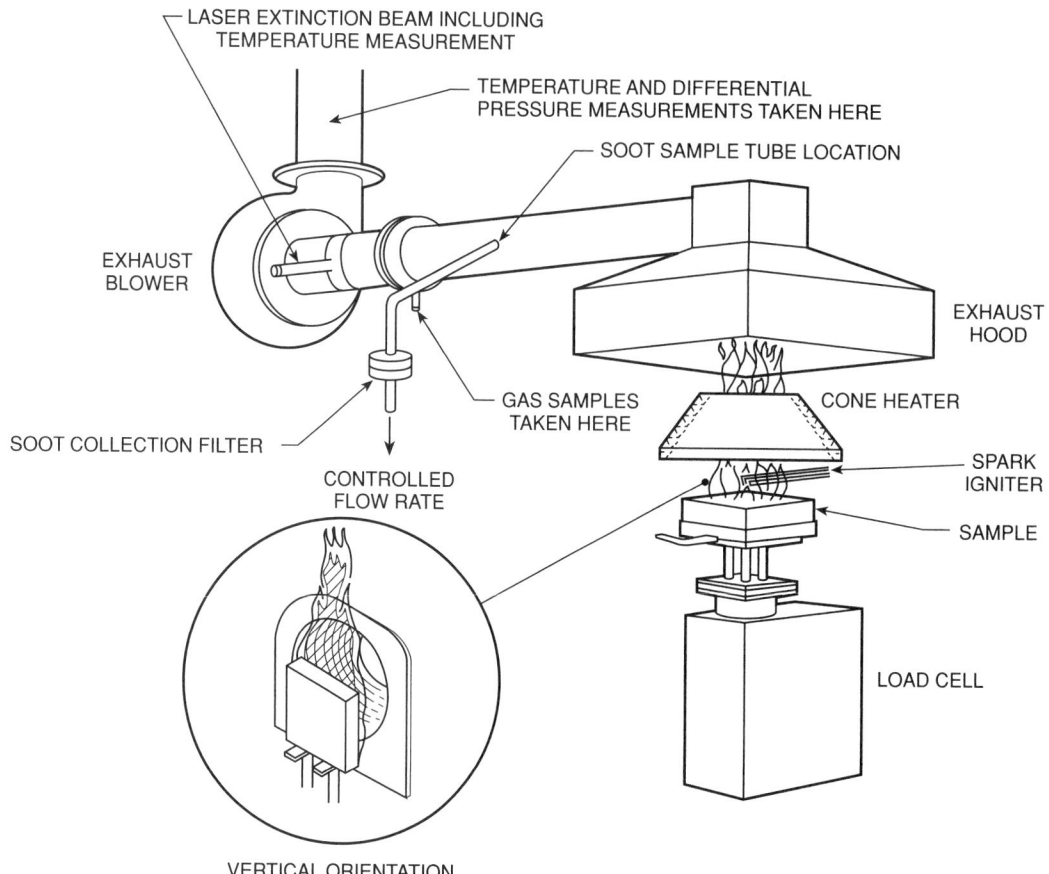

LASER EXTINCTION BEAM INCLUDING
TEMPERATURE MEASUREMENT

TEMPERATURE AND DIFFERENTIAL
PRESSURE MEASUREMENTS TAKEN HERE

SOOT SAMPLE TUBE LOCATION

EXHAUST
BLOWER

EXHAUST
HOOD

CONE HEATER

SPARK
IGNITER

SAMPLE

SOOT COLLECTION FILTER

GAS SAMPLES
TAKEN HERE

CONTROLLED
FLOW RATE

LOAD CELL

VERTICAL ORIENTATION

Fig. 3-3.1.   Schematic view of the cone calorimeter.

In the design of the cone calorimeter, such an approach was deemed to be misleading. The implicit assumption behind this type of measurement is that the fraction of the total heat release being manifest as the sensible flow enthalpy is a property of the material being tested. Such is not, in fact, the case. The convective fraction is dependent on details of the apparatus design, and also on the scale of the specimen.[5]

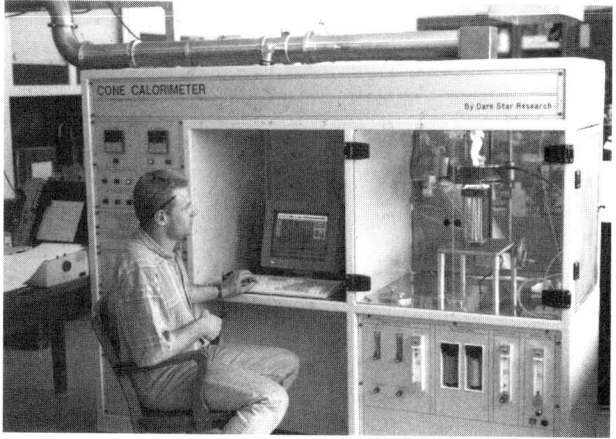

Fig. 3-3.2.   A commercial cone calorimeter. (Photo courtesy Dark Star Research.)

Where high-quality results are required, such as in the cone calorimeter, current-day practice demands that a paramagnetic oxygen analyzer be used. The various manufacturers use measuring schemes that differ in detail, but all rely on the same paramagnetic principle whereby the sensing element is sensitive to the partial pressure of oxygen in the cell. The most significant interferents to this detection principle are NO and NO_2, both of which show a strong, but not as strong as oxygen, paramagnetic response. This is never a problem in fire testing, however, since O_2 levels measured are 10 to 21 percent, while concentrations of NO_x are rarely above 100 ppm.

Unlike in applications where oxygen levels are monitored as simply one of many indications of fire hazard, in HRR work it is essential that the instrumentation be designed for the highest possible resolution. Thus, both the ASTM and the ISO standards specify that the short-term noise + drift of the oxygen analyzer must be ≤ 50 ppm O_2; the best-grade commercial instruments are able to meet a 20 ppm O_2 limit. In addition, the standards provide a significant amount of details on the layout of the gas sampling system, including desiccation, mass flow control, and bypass flows. All of these aspects have to be in conformance with the specifications for good repeatability and reproducibility performance (see below) to be achieved.

Because the detection principle responds to oxygen partial pressure, there needs to be a compensation for changes in atmospheric pressure. This can be done by a mechanical

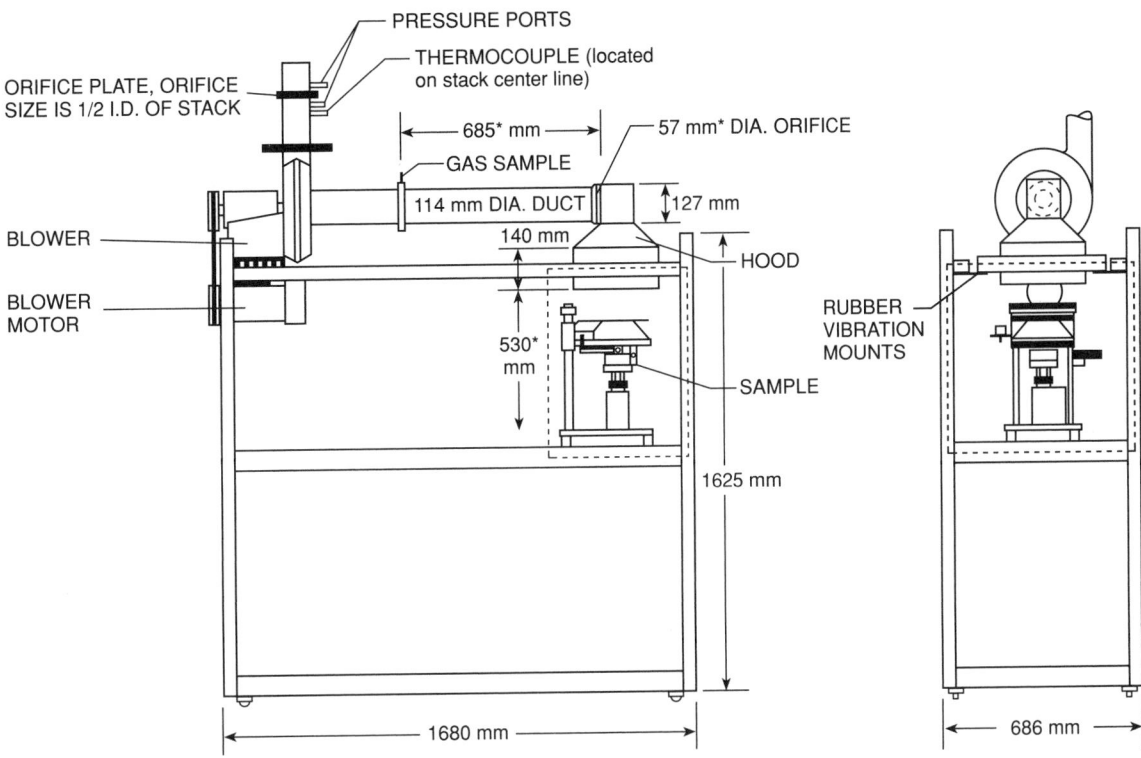

*Indicates a critical dimension

Fig. 3-3.3. View of major components of the cone calorimeter.

back-pressure regulator or by measuring the pressure and correcting electrically. If this compensation is not made, there can be significant error in the calculated heat release rate. Carbon dioxide, the other major component expected to be in the oxygen analyzer, causes less than 0.3 percent error in the oxygen reading. Extensive practice advice on selecting, setting up, and calibrating oxygen analysis systems is given in references 3 and 4.

The Radiant Heater

After establishing the operating principle the next most important feature is the type of heater. In general, such a heater should be able to achieve adequately high irradiances, have a relatively small convective heating component, present a highly uniform irradiance over the entire exposed face of the specimen, and be designed so as not to change its irradiance when the main voltage varies, when heater element aging occurs, or when the apparatus retains some residual heat from the exposure given to a prior specimen.

Range of heat fluxes needed for testing: A room fire burning near its maximum rate can show gas temperatures over 1000°C, producing corresponding irradiances to walls and contents of 150 kW · m^{-2}. Testing under such extreme conditions may not be required; nonetheless, if post-flashover fires are to be stimulated, irradiances of over 75 kW · m^{-2} should be available, and preferably closer to 100 kW · m^{-2}. A significant convective component would negate the purpose of having a radiant ignition test. Rather low convective fluxes can be achieved for specimens oriented horizontally, face up, and with the prevailing airflow being upwards. When a vertical specimen; orientation is consid-

ered, it becomes evident that a boundary layer will normally be expected to develop that will add some convective component. The convective boundary layer component is not uniform over the height of a specimen; thus it is seen that better uniformity can also be expected under conditions where the convective component is minimized.

Choice of type of heater: In a real fire scenario, the ignition source is, in most cases, in the vicinity of a combustible. The radiation spectrum depends on the size of the fire. A very small fire can show a substantial fraction of its radiation at wavelengths characteristic of H$_2$O, CO$_2$, and other combustion products.[6] For larger fires—certainly for room fires reaching a hazardous condition—the radiation from the soot tends to dominate; the result is an approximation to a grey-body radiation.[7] For such a grey-body radiation the temperature is typically in the vicinity of 1000°C.[8] Experimentally, heater choices for test apparatuses have included gas-fired panels, electric resistance heaters, flames, and high-temperature lamps. Electrical heaters tend to have a near-grey-body characteristic and, assuming a dull or oxidized surface condition, a high emissivity. Gas-fired panels derive a substantial portion of their radiation from the ceramic face; thus, while there are discrete molecular wavelength peaks, overall the radiation shows a grey-body continuum, typically in the range of 700 to 1000°C.[9] High-temperature lamps, which have been used by several investigators,[6,10] typically have radiating temperatures of 2200 to 3000°C. The spectral distribution of such a source—further limited by a translucent enclosure—is much different from one operating at 1000°C. Whether this change in spectral characteristics is important depends on the surface of the material to be

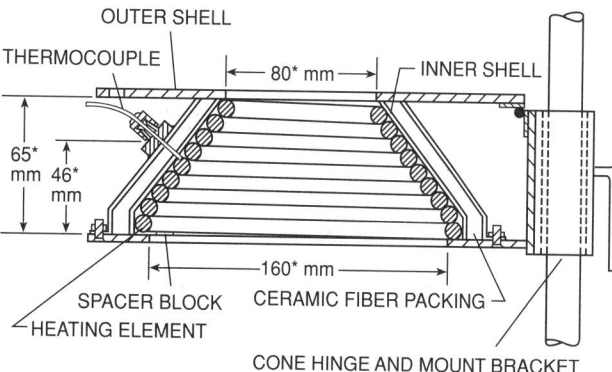

Fig. 3-3.4. Cross-sectional view through the cone heater.

ignited. For a material with a radiant absorbance independent of wavelength, this source variation would not matter. Hallman, however, has reported data for a large number of plastics and shows that, while there are some specimens with negligible wavelength dependence to their absorbance, the majority show strong variations.[6] Hallman also made measurements of ignition times of plastics with both a flame source and high-temperature lamps. The effect on ignition times ranges from negligible to more than an order of magnitude, depending on the specimen. For a general-purpose test, flames would probably be the least desirable source of heating. For a bench-scale test, flame size has to be kept small. This means that such flames are optically thin, their emissivity is low, and higher heat fluxes cannot be achieved unless a strong convective component were added.

Design details: Once an electrical radiant heater had been decided upon, design details were also influenced by work at NIST with earlier types of calorimeters. One of the primary requirements of the heater is that it not change the irradiance impressed on the specimen when the specimen ignites. This undesired event is, of course, exactly what happens with several of the older types of calorimeters. The specimen's flames directly heat up nearby ironwork, which, in turn, radiates to the specimen. The heater, also, which had been "viewing" a cold specimen prior to ignition, starts to view a hot flame afterwards. The result is that its efficiency increases drastically, giving a rise to its radiating temperature. Based on these observations, guidelines were formulated so that the specimen must, as much as possible, view only:

1. A temperature-controlled heater,
2. A water-cooled plate, or
3. The open-air, ambient-temperature environment.

Reliance on item 2 increased costs significantly; thus, it was desired to make use of only items 1 and 3. Prior to the development of the cone calorimeter, fire test apparatuses typically controlled the power (or fuel rate) into the heater, but did not maintain it at a fixed temperature.

The conical shape: The cone calorimeter derives its name from the conical shape of the heater. (See Figure 3-3.4.) Once the decision had been made to use an electric resistance heater, running at a realistic maximum temperature of about 950°C, its material and shape still had to be determined. The material was simply decided, based on poor experiences with exposed-wire resistance heaters and with silicon car-

bide rod-type heaters. That left the tube heater, which consists of a resistive wire element inside a protective tube, swaged over a packing of inorganic insulation. The tube is made of Incoloy™ and can be bent to a desired shape.

To determine the best shape, the conical heater used in the ISO 5657 ignitability apparatus[11] was examined. This was seen to be promising from the point of view of the shape. The proper shape had to have a hole in the middle, since otherwise a hot spot would occur at the sample center, where the radiation view factor is the highest. The same heater had to serve in both horizontal and vertical orientations. In the horizontal orientation, it was essential that all the products of combustion flow out the hole in the middle, and (1) not "splash" on the heater coil itself, nor (2) escape from the underside. The original ISO 5657 design proved to be unsuitable in the former respect; it also had problems with durability and assembly. Thus a totally new design was created, one that looks, however, superficially similar to the ISO 5657 cone. With the actual cone calorimeter design, the flames from the specimen do not splash on the heater coil; instead, a sheath of cold air is pulled up, surrounding the flame plume. Thus, there is not a concern that any surface reactions occur on the heater coil.

The space between the inner and outer cones is packed with refractory fiber. This arrangement helps keep the outside of the unit cool and also helps bring up the heater to operating temperature rapidly.

Emissivity of the heater: The emissivity was characterized by Janssens.[12] The heater coil, once installed and fired up a few times, becomes essentially radiatively black. The emissivity itself cannot be readily measured directly; however, it is possible to compute an approximate view factor, F, for the cone heater. The possibility of measurements is based on a simultaneous determination of the heater surface temperature and the heat flux falling on the heat flux meter, with the meter held in place at the same location where a specimen is situated. Over the range of fluxes of 10 to 90 kW $\cdot$ m^{-2}, Janssens determined the $\varepsilon \times F$ product to be 0.73, with F being computed as 0.78. Then, solving for ε gives $\varepsilon = 0.91$. Since the temperatures of the heater closely resemble those in room fires, and the emissivity approaches 1.0, this means that the spectral distribution is likely to be very close to that expected from room fires (neglecting the molecular radiation contribution from CO_2 and H_2O).

Convective fraction of the heating flux: During the development of the cone calorimeter at NIST, a study was conducted to determine the fraction of the heating flux, which is accounted for by the convective contribution.[13] When measured with respect to a water-cooled heat flux meter, the results showed that, in the horizontal specimen orientation, the convective contribution was immeasurably small. In the vertical orientation, the fraction was typically 8 to 12 percent. Janssens later remeasured the vertical configuration[12] using a more accurately calibrated heat flux meter and found that, even for the vertical orientation, the convective transfer is immeasurably small. Thus, it can be stated that the objective of having a test method where the heating is primarily radiant was successfully met. For modeling of test results, however, one may be more interested in the possibility of convective heat transfer to a specimen that is heated, or even burning, not to a calibration meter constrained by its water-cooling jacket at near-room temperature. Janssens also made some determinations of such actual specimen heating. The direction of the heat transfer was such as to represent a heat loss from the specimen in all cases. A single convective heat transfer coefficient could not be derived, however, since the value was dependent on the irradiance level from the heater. Janssens' results could be represented by:

Irradiance from Heater (kW · m^{-2})	Convective Heat Transfer Coefficient h_c (W m^{-2} K^{-1})
20	9.0
40	18.0
60	27.0

For practical work, Janssens recommended that an average value of $h_c = 13.5$ W m^{-2} K^{-1} should be appropriate for work over the common irradiance range of 20 to 40 kW · m^{-2}. The actual details of this small amount of convective heat transfer are pertinent only to certain specialized studies. For most work, it is entirely adequate to assume that the specimen heating is entirely radiative.

Uniformity of the heating flux: The uniformity of the heating flux over the face of the specimen in the cone calorimeter has been described.[13] Over the range of irradiances from 25 to 100 kW · m^{-2}, the ratio of the flux at the specimen center to average flux varied only from 1.00 to 1.06. The peak deviations from average were typically 2 percent in the horizontal orientation and 7 percent in the vertical. Deviations are, perforce, higher in the vertical orientation, since the effect of convective fluxes, due to the boundary layer flow, is more pronounced there. Additional measurements have been made in the specimen-depth plane. This has been a special concern to the designers of the ISO apparatus, where a special compressive loading mechanism is provided that attempts to re-level the exposed surface, in case the specimen recedes due to melting. In the cone calorimeter, measurements have been made in the horizontal orientation using a small, 6-mm diameter Gardon-type heat flux gage. A flux mapping was obtained starting at the initial surface, and progressing down to the maximum depth of a specimen, which is 50 mm. A normal aluminum foil rectangular specimen wrap was used for these tests, but without any specimen. The results show that, at heating fluxes of both 25 and 50 kW · m^{-2}, the deviations over the entire specimen depth are less than 10 percent, and can, therefore, be neglected. (See Figure 3-3.5.) At the lower depths, reflection from the aluminum foil probably assists in maintaining this uniformity.

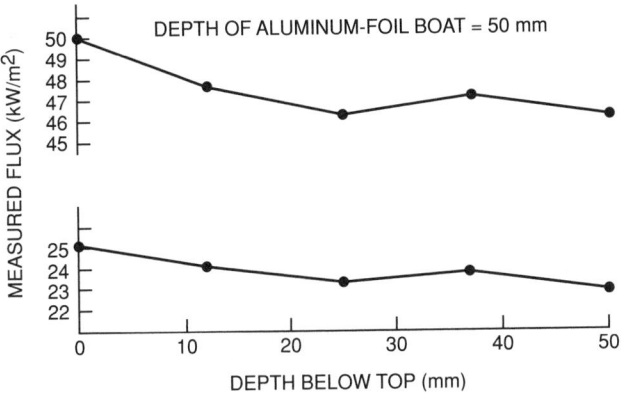

Fig. 3-3.5. Measured flux at various positions below the top surface of a specimen.

TABLE 3-3.1 *Effect of Exhaust Hood Airflow on Ignition Times in the Cone Calorimeter*[*]

Material	Thickness (mm)	Orientation	Fan Setting	Ignition Time[*] (s)
PMMA	13	Horizontal	no fan	71
PMMA	13	Horizontal	24 ℓ · s^{-1}	76
PMMA	13	Horizontal	41 ℓ · s^{-1}	67
PMMA	13	Vertical	no fan	86
PMMA	13	Vertical	24 ℓ · s^{-1}	84
PMMA	13	Vertical	41 ℓ · s^{-1}	77
Redwood	13	Horizontal	no fan	23
Redwood	13	Horizontal	24 ℓ · s^{-1}	24
Redwood	13	Horizontal	41 ℓ · s^{-1}	31
Redwood	13	Vertical	no fan	22
Redwood	13	Vertical	24 ℓ · s^{-1}	27
Redwood	13	Vertical	41 ℓ · s^{-1}	29

[*]At an irradiance of 35 kW · m^{-2}
[†]Typical ignition time scatter was on the order of ± 10% (1 σ, N = 3)

Orientation of the heater and specimen: It is seen that the normal orientation of the specimen should be horizontal, face up, with the heater being parallel, face down. This allows thermoplastics, liquids, and other melting or dripping samples to be successfully tested. For certain application exploratory studies it was considered desirable to allow testing in a vertical orientation. Thus, provision was made to swing the heater 90 degrees into a vertical orientation. Vertical orientation testing may be preferable when it is desired to probe the flame regions, or measure specimen surface temperatures. Figures 3-3.6 and 3-3.7 show the comparative horizontal and vertical heater orientations, respectively. It is especially emphasized that no standard testing should be specified for the vertical orientation, *even for products that are normally used in a vertical orientation.* The ASTM standard[2] was amended in 1992 to clarify that the vertical orientation is only for special research studies and not for product testing.

Airflow

The feasible airflow rate through the system is bounded by certain limits. It must not be so fast that ignition results are improperly affected. It must also not be so slow that products of combustion spill out of the hood. If this were a closed system, one would also be concerned about airflow being so slow that the air/fuel ratio drops into the fuel-rich regime; the standard cone calorimeter, however, has been designed for ambient air testing, and this consideration does not apply.

Systematic guidance in this area was not available. However, as an example of the effect of airflow, measurements were made at NIST using the OSU apparatus. Specimens of black polymethylmethacrylate (PMMA) were exposed in the horizontal orientation to a heating flux of 35 kW · m^{-2}. With an airflow rate of 12 ℓ · s^{-1} through the combustion chamber, the ignition time was 209 s. When the airflow rate was doubled to 24 ℓ · s^{-1}, the specimen ignition time increased to 403 s. By contrast, Table 3-3.1 shows comparative results with the cone calorimeter; it can be seen a flow rate of 24 ℓ · s^{-1} was found to be satisfactory. That flow rate was also about a factor of 2 greater than the minimum at which no spill out of the hood occurs.

The exhaust system uses a high-temperature cast-iron blower to exhaust the gases, and an orifice plate flowmeter.

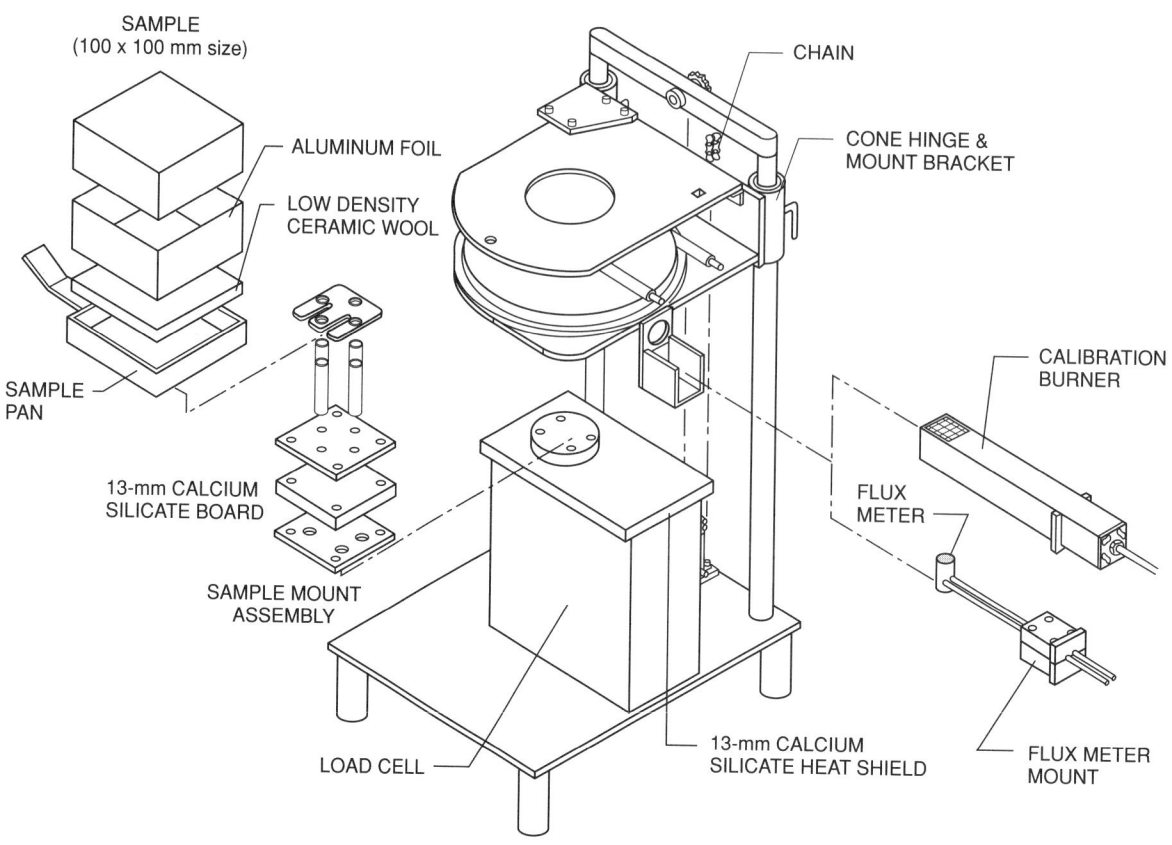

SAMPLE
(100 x 100 mm size)

ALUMINUM FOIL

LOW DENSITY
CERAMIC WOOL

SAMPLE
PAN

13-mm CALCIUM
SILICATE BOARD

SAMPLE MOUNT
ASSEMBLY

LOAD CELL

CHAIN

CONE HINGE &
MOUNT BRACKET

CALIBRATION
BURNER

FLUX
METER

13-mm CALCIUM
SILICATE HEAT SHIELD

FLUX METER
MOUNT

Fig. 3-3.6.   Heater in the horizontal (standard) orientation.

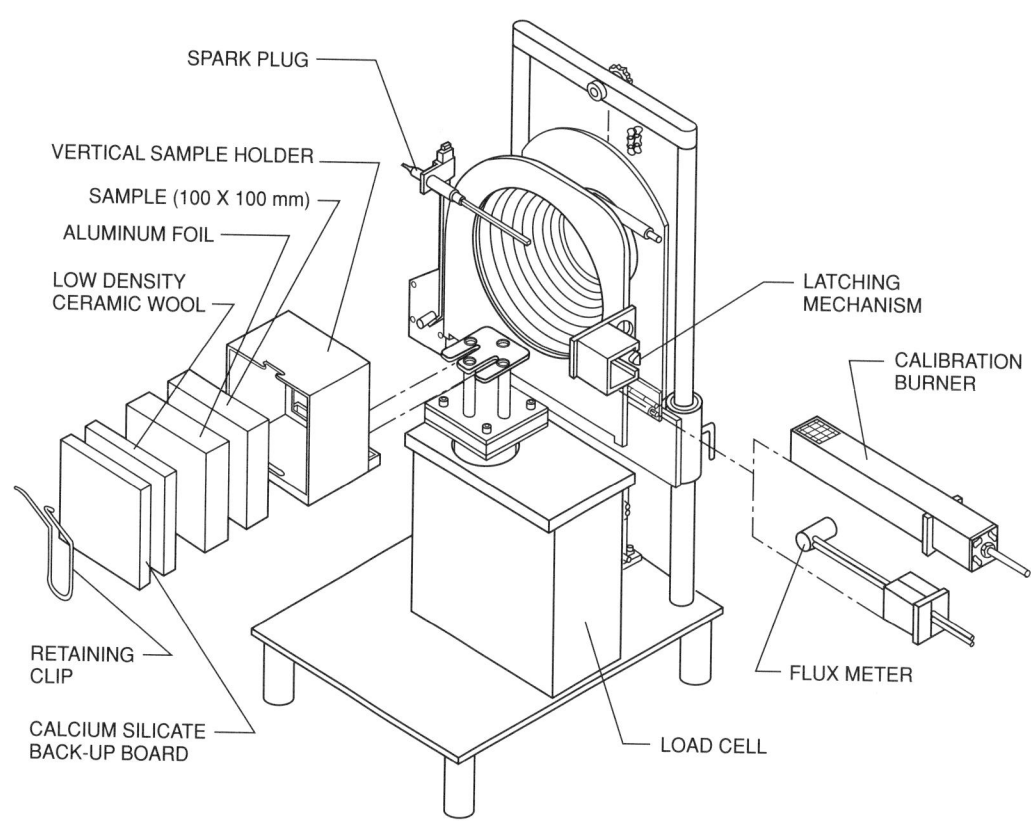

SPARK PLUG

VERTICAL SAMPLE HOLDER

SAMPLE (100 X 100 mm)

ALUMINUM FOIL

LOW DENSITY
CERAMIC WOOL

RETAINING
CLIP

CALCIUM SILICATE
BACK-UP BOARD

LATCHING
MECHANISM

CALIBRATION
BURNER

FLUX METER

LOAD CELL

Fig. 3-3.7.   Heater in the vertical orientation.

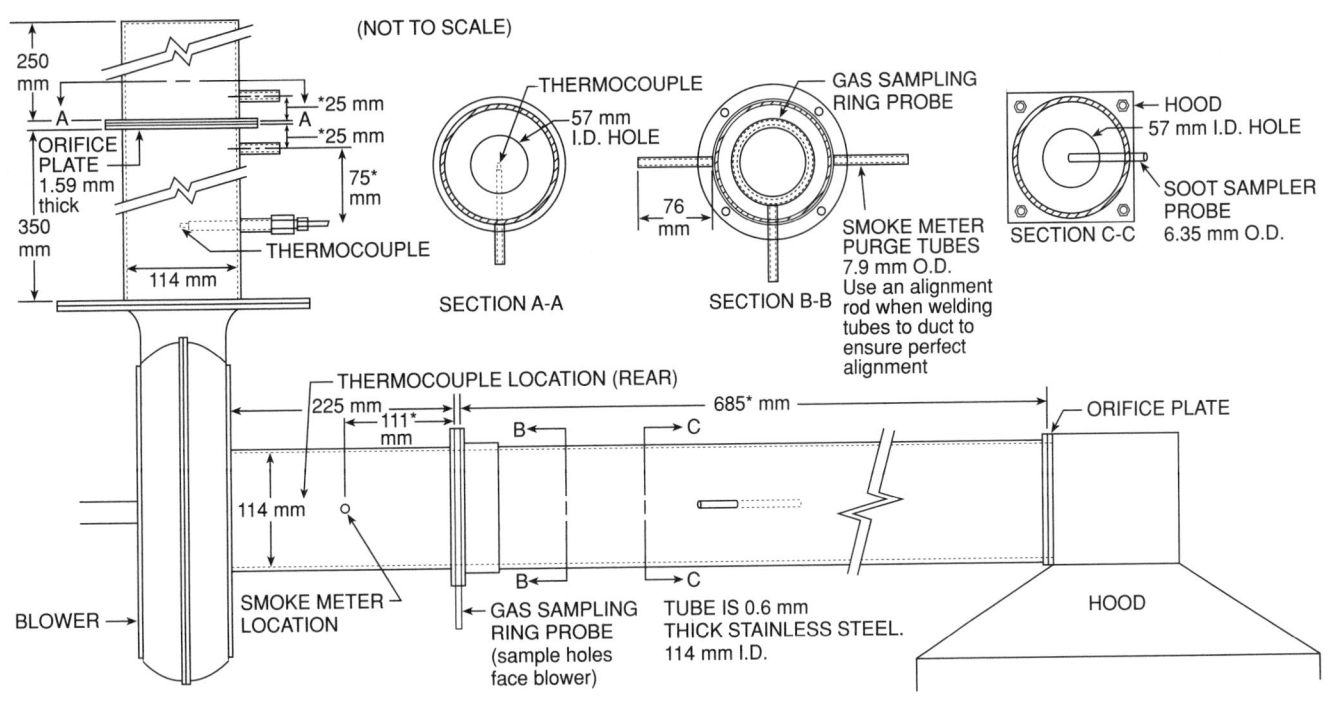

*Indicates a critical dimension

Fig. 3-3.8.   Exhaust duct.

(See Figure 3-3.8.) The orifice plate flowmeter is instrumented with a differential pressure transducer and a thermocouple. For specialized studies, where the entire combustion system is glass-enclosed,[14] it is possible to go to flow rates below 12 $\ell \cdot s^{-1}$. With such enclosed systems, accurate measurements can be made down to about 9 $\ell \cdot s^{-1}$ using the standard orifice plate; for lower flow rates, down to about 5 $\ell \cdot s^{-1}$, the standard orifice plate is replaced by one having a smaller opening.

Means of Ignition

In some cases no external ignition source is desired, and specimen testing is to be done solely on the basis of autoignition. In most cases, however, an external ignition source is desirable. This ignition source should, in general, not impose any additional localized heating flux on the specimen. Apparatus designs have been developed, with impinging pilots that can, in some cases, produce such high localized heat fluxes as to burn a hole through the specimen at the point of impingement, yet not ignite it outside of that region. (See reference 15.) Applications for such devices tend to be specialized, since the general objective of radiant ignition testing is to produce data that can be analyzed in the context of an assumed one-dimensional heat flow. A design using an impinging pilot has an additional difficulty. Since most of the specimen face is not yet heated to the ignition temperature when ignition first begins in the vicinity of the pilot, no unique ignition time can be determined. Instead, there is a significant time spread between when ignition first occurs at the initial location to when the final portions of the face have been ignited.

The ignitor should reliably ignite a combustible gas mixture in its vicinity. Thus, the location of the ignitor must be chosen so that it is near the place where maximum

evolution of pyrolysate gases is expected. Some materials are highly fire-retardant treated, and, when heated, emit vapors that tend to extinguish a pilot flame. The ignitor has to be designed so as not to be extinguished by fire-retardant compounds coming from the specimen, nor by airflows within the test apparatus.

The ISO 5657 apparatus was designed with a "dipping" gas pilot, which is periodically thrust for a short while down close to the specimen face, then retracted. This solution, however, introduces an uncertainty into ignition times and provides further complexity. A gas pilot, based on experience, also requires oxygen premix to be used if a flame that is both small and resistant to blowout is to be achieved.[16] With highly fire-retardant-bearing products, even such precautions are not likely to lead to a reliable pilot; thus, for instance, the ISO 5657 apparatus uses a second pilot to reignite the main pilot. Pilot stability also tends to be crucially dependent on the physical condition of the pilot tube tip, and significant maintenance can be necessary. Finally, if used in a heat release apparatus, a gas pilot can add noise to the baseline of the heat release measurement. Initial experimental results at NBS, using a more tractable alternative, i.e., electric spark ignition, were obtained with the NBS-II calorimeter, where a spark plug arrangement was provided for ignition. This development was successful, and so a similar electric pilot was designed for the cone calorimeter. The location of the ignitor should be at the place where the lower flammable limit begins its pyrolysis. It should, however, not be so close to the specimen surface that minor swelling of the specimen would interfere with the ignition function. In the cone calorimeter, the ignitor locations were chosen so that, when testing in the horizontal orientation, the spark plug gap is located 13 mm above the *center* of the specimen; in the

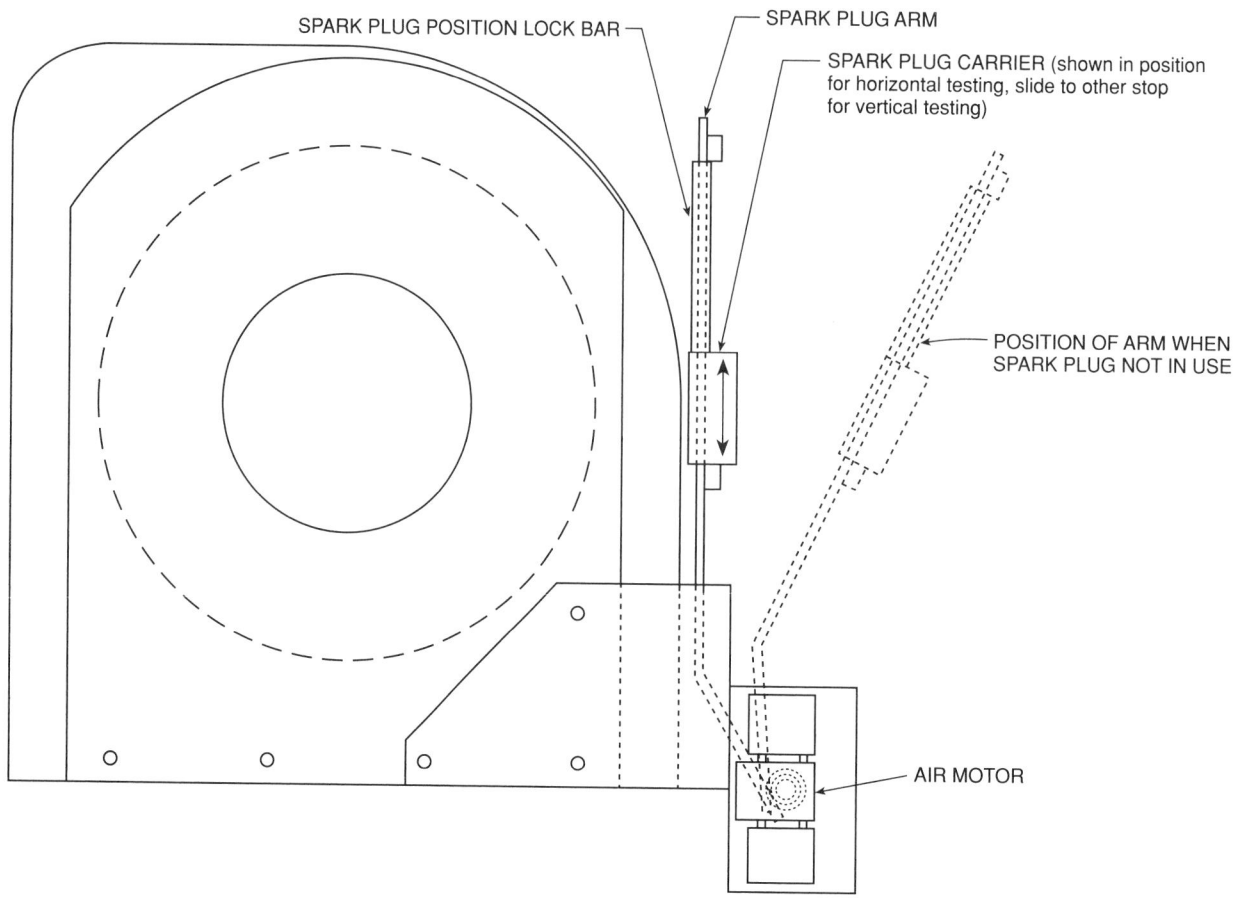

SPARK PLUG POSITION LOCK BAR

SPARK PLUG ARM

SPARK PLUG CARRIER (shown in position for horizontal testing, slide to other stop for vertical testing)

POSITION OF ARM WHEN SPARK PLUG NOT IN USE

AIR MOTOR

Fig. 3-3.9. Spark plug, carrier, and air motor.

vertical orientation, the spark plug gap is located at the specimen plane and 5 mm above the *top* of the specimen holder.

The actual spark plug arrangement is shown in Figure 3-3.9. The spark plug is provided by a special-purpose 10-kV ignition transformer. The spark plug is moved in and out by remote control, operated by an air motor that rotates the shaft on which the spark plug rests. A reversible "lock bar" is used to adjust the spark-plug-to-heater distance when changing from the horizontal to the vertical orientation (the spark gap is 13 mm away from the heater baseplate in the horizontal orientation, but 25 mm away in the vertical).

Specimen Area and Thickness

Both specimen area and thickness may be expected to have some effect on both the ignitability and the heat release rate. The main practical size and thickness limitations come from the fact that the specimens to be tested should exhibit primarily one-dimensional heat transfer; thus, the configuration should be such that excessive edge effects are not seen. If the specimen thickness is such that it is thermally thick (the heat wave penetration depth being less than the physical depth), then further increases in thickness are not expected to change ignitability results. For thinner specimens, however, there can be expected to be a thickness effect, and the backing or substrate material's thermophysical properties can be of importance.

Specimen area: Janssens[12] studied in some detail the general problem of area effect on ignition. The effect is seen to be smaller when irradiances are high rather than low. The exact magnitude of the effect is also dependent on the specimen's thermophysical properties. For specimens of area 0.01 m^2 or larger, however, his results show an increase in ignition time of only about 10 percent over what would be seen with a specimen of infinite area. Later, Nussbaum and Östman[17] studied specimens in an experimental apparatus somewhat similar to the cone calorimeter, but accommodating 200 × 200 mm specimens. Their comparison of the ignition times of these larger specimens against the standard 100 × 100 mm ones shows that quadrupling the specimen area decreases the ignition time by about 20 percent.

For heat release rate, the specimen size affects the measurement, since flame volume is larger over larger specimens; consequently the flame radiation tends to approach a value of higher emissivity. Nussbaum and Östman also examined the heat release rates from the larger size specimens; the differences were generally of the same order of magnitude as the repeatability of the results. Babrauskas, in commenting on these data,[18] discussed tests on larger size, horizontal PMMA samples, where each doubling of the specimen's area increased the heat release rate, per unit area, by about 10 percent. The more general treatment of the horizontal specimen, of course, is as a liquid pool. Equation 1 in Chapter 3-3 gives details on the size effect for burning

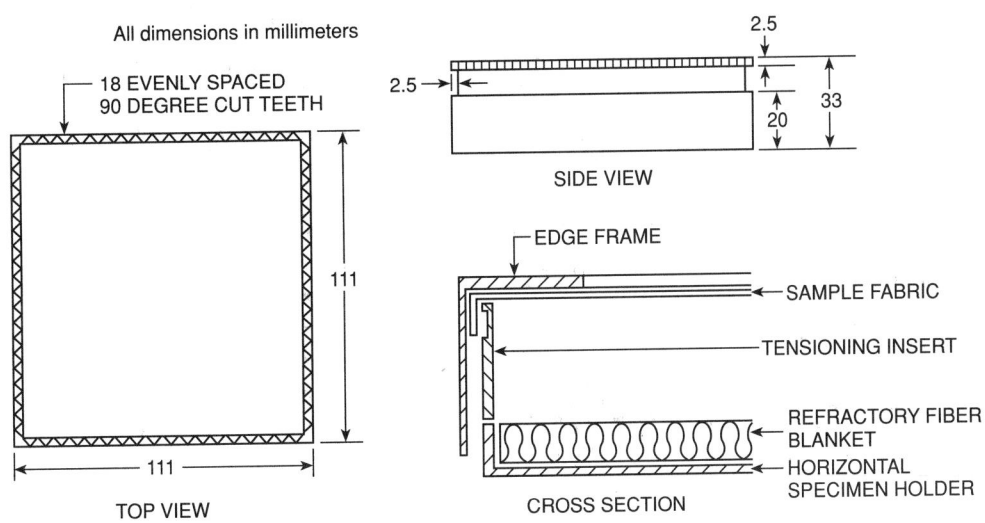

Fig. 3-3.10. Special holder for testing fabrics and similar thin materials.

pools. It can be seen that the diameter has to be greater than about 1 m before the specimen area effect becomes negligible.

The effect of specimen size for vertical samples was examined at Factory Mutual Research Corporation (FMRC) in a series of experiments on PMMA walls.[19,20] The FMRC studies showed little size effect for specimen heights up to 200 mm; beyond 200 mm there was approximately a linear dependence of $\dot{q}''$ on the height. This was true up to the maximum height tested, i.e., 3.56 m. Unlike horizontal pools, the rate of heat release was not leveling off at even these sizes, and estimates suggested that the specimen size would have to be increased by another order of magnitude before a leveling off would be seen.

The conclusion from the above studies was that 100 mm × 100 mm was a suitable size for bench-scale testing, but that the bench-scale $\dot{q}''$ rates will, perforce, always be somewhat lower than for full-scale fires.

Specimen thickness: The cone calorimeter is intended for testing actual commercial products. Thus the specimen thickness should be, as much as possible, the thickness of the finished product. There are limitations at both ends of the scale, however. The instrument is restricted to testing specimens not thicker than 50 mm. For products that in their finished state are greater than 50 mm thick, it can readily be seen that, for almost any realizable combination of thermophysical properties and incident radiant fluxes, a 50-mm specimen is thermally thick, and increasing thickness would not change the ignition times.[21,22] By making calculations for various densities and heat fluxes, it was found that for particleboard the minimum thickness required to ensure that the specimen is thermally thick can be represented by

$$\ell = 0.6 \, \rho \, / \, \dot{q}'' \qquad (1)$$

where ℓ is the thickness (mm), ρ is the density (kg · m^{-3}), and $\dot{q}''$ is the heat flux (kW · m^{-2}). This is probably a reasonable rule of thumb for other materials as well. The proportionality of the required thickness to $\rho/\dot{q}''$ is derived from classical heat conduction theory by equating the time for the front surface to reach the ignition temperature to the time for the rear surface temperature to begin to rise, assuming that the thermal conductivity is proportional to the den-

sity. Numerical calculations were necessary to determine a suitable constant because of the impact of the front surface heat losses.

For materials that are not thermally thick at the time of ignition, the nature of the backing material or substrate can influence the measured value of the ignition time. In the cone calorimeter the substrate is a blanket of refractory ceramic fiber material, having a nominal density of 65 kg · m^{-3}. In use, the material assumes a more compacted density of roughly 100 kg · m^{-3}. Whenever possible, materials, whose thicknesses are less than the minimum suggested in the above formula, should be mounted on that substrate material over which they will actually be used. As a practical guide for testing unknown commercial samples, it is desirable to specify that any specimens less than 6 mm thick should always be considered as needing to be tested over their in-use substrate.

Fabrics are a special case. Thin fabrics are sometimes used for constructing air-supported structures; these should be tested with an air space in back, simulating the usage conditions. A special holder has been constructed that allows the fabrics to be pulled taut and held above a dead-air space. (See Figure 3-3.10.)

SAMPLE TESTING SPECIFICATIONS
Specimen Orientation and Specimen Holders

As discussed, both a horizontal and vertical specimen orientation are provided; however, it is considered that the horizontal orientation is standard, while the vertical orientation is reserved for special-purpose testing only. The specimen holders in Figure 3-3.11 and 3-3.12 show the two specimen holders, respectively. With proper precautions, the horizontal orientation can be used for testing liquids and melting materials. The vertical orientation has a small melt trough that can only catch a very small amount of molten material. Also, some specimens, when tested in the vertical orientation, show a tendency to lose physical strength and fall out of the holder; this, by contrast, does not happen in the horizontal orientation.

To present a standardized heat flow boundary condition to the rear face of the specimen, all specimens are backed by

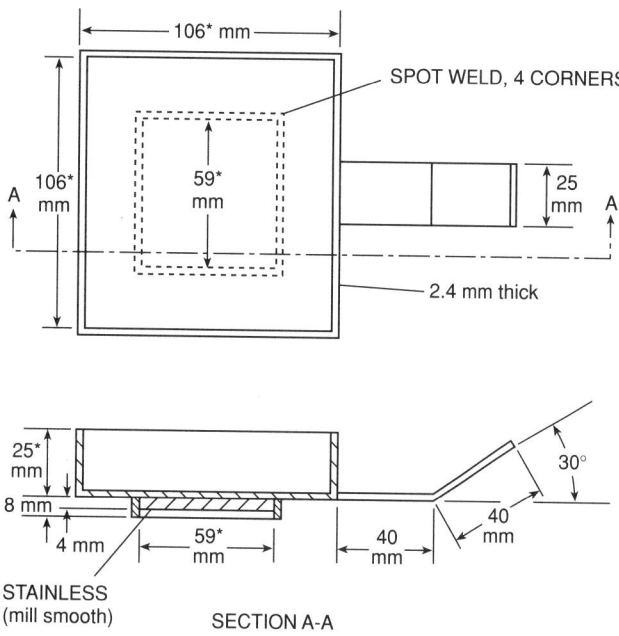

SECTION A-A

*Indicates a critical dimension

Fig. 3-3.11. Horizontal orientation specimen holder.

a 13-mm layer of low-density (nominal $65 \, kg \cdot m^{-3}$) ceramic fiber blanket. Such a blanket is the most insulating product readily available for use. In the vertical orientation, behind the blanket are several layers of rigid millboard, sufficient in thickness to fill out the depth of the specimen holder. The specimen is wrapped in a single sheet of aluminum foil, covering the sides and bottom. The aluminum foil serves to limit flow of molten material and prevent it from seeping into the refractory blanket.

Load Cell

Many ancillary measurements made in the cone calorimeter (such as yields of various gas species) require the use of a load cell. Transducers had been tried in various earlier apparatuses, but most suffered from the drawback that they were not designed for purely single-axis, linear motion. That is, if the weight of the specimen was not well balanced, or differential heating stresses occurred, it was likely that a mechanical moment (or torque) would be applied to the device, with the transducer then being prone to jamming. For the cone calorimeter, a commercial-design load cell was found that permits only up-and-down axial motion, while being insensitive to torques or forces from other directions.

The load cell has to accommodate two different orientation specimen holders, and may need to hold additional fixtures. All of these can have substantial—and different—weights, yet must allow accurate mass determination for low-density specimens. The solution adopted was a weighing system that has a large (3.5 kg) mechanical tare adjustment range, along with a sensitive weighing range (500 g). A resolution of 0.005 g is readily achievable.

Figures 3-3.6 and 3-3.7 show, respectively, how the horizontal and vertical orientation specimen holders are accommodated on the load cell. The horizontal holder has a square recess on the bottom and simply is placed straight down. The vertical holder is more conveniently inserted in a

direction of moving toward the heater; thus the specimen is correctly located by four mounting pins on the bottom. In both cases there is a positive specimen location, and the operator does not have to be concerned with how far to insert the holder.

Edge Conditions

Edge effects: In an apparatus such as the cone calorimeter, it is desired that the small-scale test specimen would behave, as much as is possible, like a correspondingly sized element of the full-scale object. If one is dealing with relatively large, flat full-scale objects, then heat and mass transfer will occur only in the direction perpendicular to the exposed face. There will be no heat or mass flow along either of the face directions. The guidance to be derived from this conceptual model in designing the bench-scale test environment is clear: there should be a minimum of heat or mass transfer at the specimen edges. The aluminum foil used to wrap the specimen usually serves to minimize any mass transfer that may occur. The heat transfer situation, however, is more complicated.

In the vertical specimen orientation, the specimen has to be restrained against falling out; thus, the vertical specimen holder incorporates a small lip extending 3 mm along the edges. In the horizontal orientation, no special measures need to be taken against falling out. Thus, for many types of specimens it is satisfactory to simply cover the edges and bottom with aluminum foil, leaving the top exposed in its entirety.

Some categories of specimens, however, present special problems. These are specimens that either have a propensity to ignite first along the outside edge, or ones that, when ignited, burn disproportionately vigorously near the edges. Such behavior is often found with wood specimens and with certain composites. This problem is alleviated by using a stainless steel edge frame for the horizontal orientation, which, in the same way as the vertical holder, provides a 3-mm lip around the edge of the specimen face. (See Figure 3-3.13.)

For specimens showing unrepresentative edge burning, the situation can be viewed as a spurious heat gain along the edges, when compared against a hypothetical ideal situation of exactly zero heat loss/gain at the edges. When an edge frame is applied, the opposite situation can tend to result, i.e., a net heat loss from the specimen is observed.[23] The ideal situation, where a specimen is prevented from showing unrepresentative increased edge burning, but is equally not sustaining any losses to an edge frame, may be difficult to approach in practice. This is still a topic of active study at several institutions.

In some cases, an edge frame is needed for thermostructural reasons. Some specimens, especially certain composites, can show pronounced edge warping and curling when subjected to heat. The burning of such a specimen would be highly non-uniform if its edges were not held down with an edge frame. In many cases, an edge frame is all that is required. In some cases, however, additional measures, such as a wire grid (see below), are required.

For the testing of electric cables, there can be a pronounced tendency for pyrolysis gases to flow along the length of the cable interior, and to burn only at the edges, not uniformly over the surface. For such specimens, it has often been found useful to coat the cable ends with a sodium silicate cement, such as Insa-Lute Adhesive Cement Paste

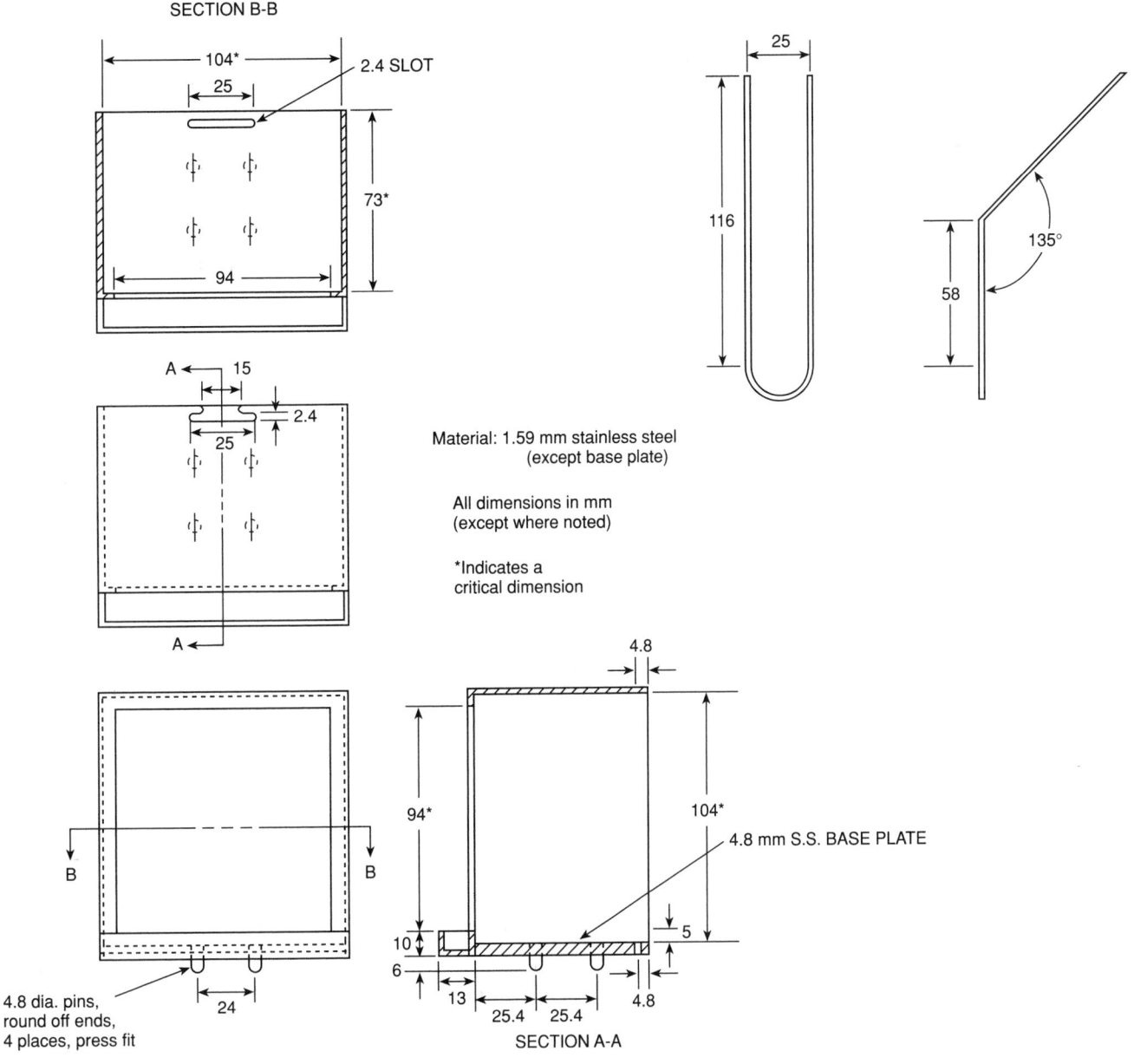

Fig. 3-3.12.   Vertical orientation specimen holder.

No. 1, produced by the Sauereisen Cements Co. When the ends are sealed in such a manner, a knife puncture must be made in the face of each piece of cable to avoid pressure buildup and rupture.

Intumescing samples: A common difficulty with fire test specimens is when they intumesce, either before ignition or during the burning. The simplest solution used in the cone calorimeter, which suffices in many, but not all, cases, is the use of a *wire grid* placed on top of the specimen. Figure 3-3.14 shows a medium-weight grid. For minimizing the effect on the measurements, the grid weight should be the smallest possible consistent with providing adequate mechanical restraint to the particular specimen being tested. Mikkola[24] demonstrated that the effects on the measurements will be negligible if the average grid mass is $< 0.6 \text{ kg} \cdot \text{m}^{-2}$ of

specimen face area. This corresponds to quite a thin, small grid and will practically be usable only in occasional cases. Additional guidance is given in the NBS "User's Guide for the Cone Calorimeter,"[25] but testing laboratories will, on occasion, be required to devise their own special schemes for mounting and restraint.

SMOKE MEASUREMENT

One of the most essential ancillary measurements performed with the cone calorimeter is smoke obscuration. This system was devised due to widespread dissatisfaction with older, closed-box types of smoke tests.[26,27] A large number of both practical and theoretical difficulties were found with closed-box systems, and these were successfully resolved by developing a flow-through smoke measuring system, using a

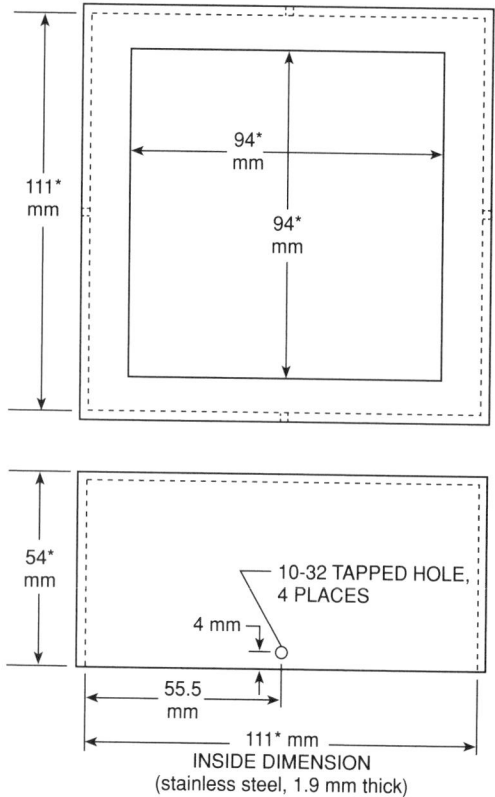

*Indicates a critical dimension

Fig. 3-3.13. Edge frame for the horizontal specimen holder.

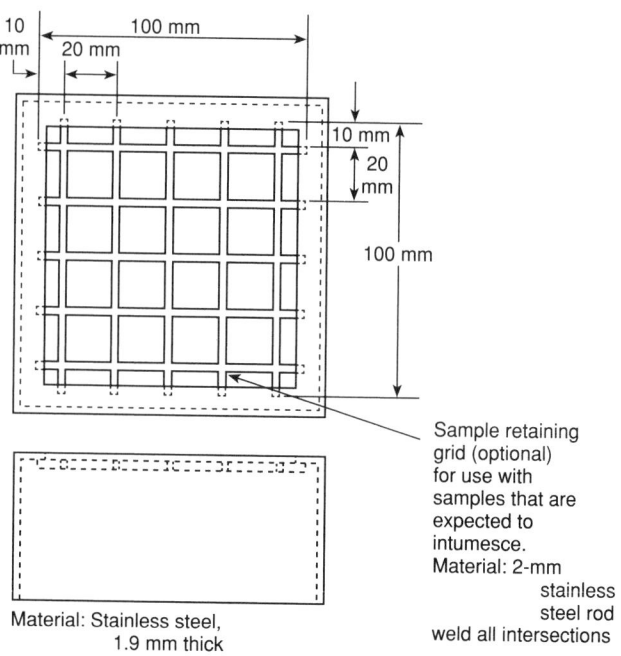

Sample retaining grid (optional) for use with samples that are expected to intumesce.
Material: 2-mm stainless steel rod weld all intersections

Material: Stainless steel, 1.9 mm thick

Fig. 3-3.14. Wire grid.

helium-neon laser as the light source, and a sophisticated quasi-dual-beam measuring arrangement. Figure 3-3.15 shows the overall arrangement of the laser photometer. It is mounted on the exhaust duct at the location shown in Figure 3-3.8. A thermocouple is also mounted nearby, since the calculations require a determination of the actual volume flow rate in the duct at the photometer location. The user should consult Reference 27 for details explaining the operation of the laser photometer. Briefly, the light from the laser goes, via two beam splitters, into two detectors. The light reaching the compensation detector is not attenuated by smoke; its signal serves as the reference to cancel out fluctuations in laser output power. The main beam detector measures a signal that is attenuated by the smoke. The optical path is purged by a minute flow of room air through a purge system. The flow is maintained by the pressure differential in the exhaust duct.

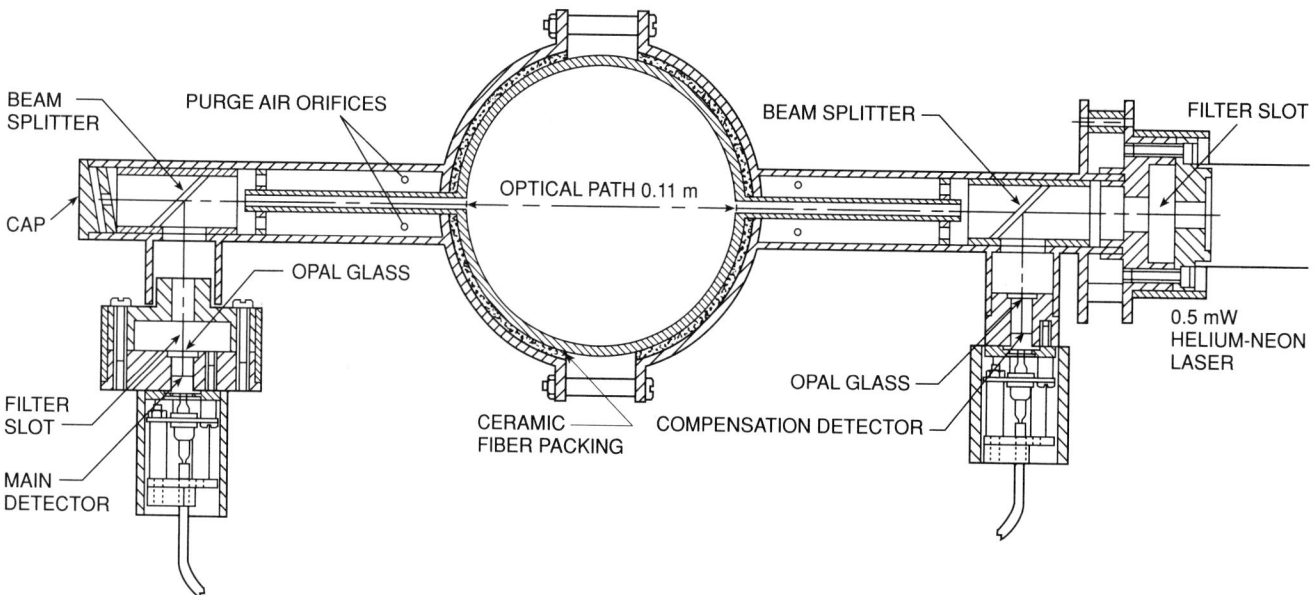

Fig. 3-3.15. Laser photometer.

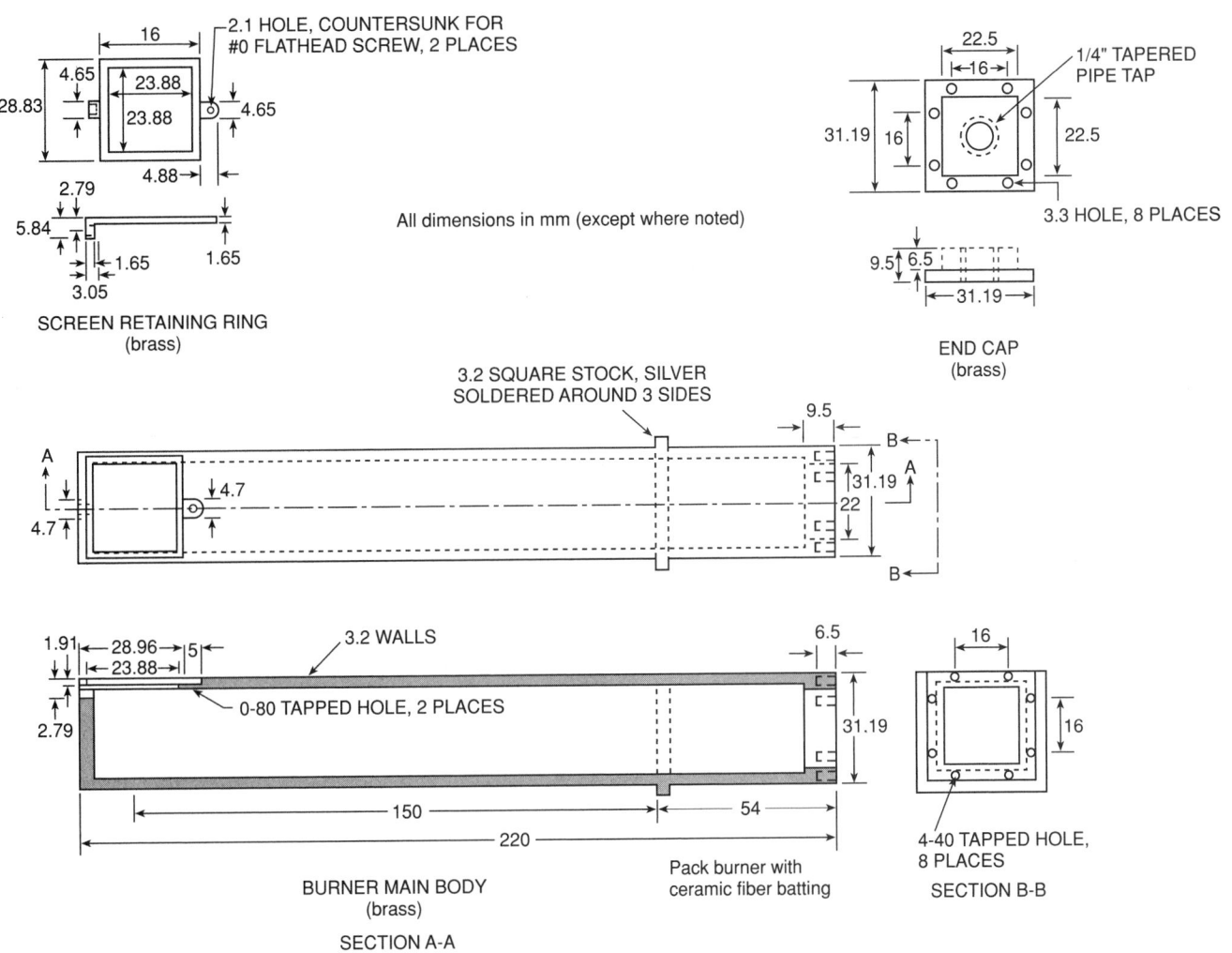

Fig. 3-3.16. Calibration burner.

For certain research purposes, it is advantageous, in addition to obtaining optical smoke obscuration measurement, also to record the gravimetric soot yield. That is, the grams of soot evolved, per gram of specimen burned, can be measured. To do this, a soot mass sampler is connected to the port indicated in Section C-C, Figure 3-3.8. A known mass fraction of the exhaust duct flow is passed through a measuring filter, and is weighed before and after the test.

CALIBRATION EQUIPMENT

Two basic calibrations are needed: (1) the calibration of the temperature controller for the conical heater, and (2) the actual heat release rate calibration. The temperature controller is calibrated using a Schmidt-Boelter-type heat flux meter. The heat flux meter is equipped with a locating collar and is inserted in place of the specimen, with its face at the same place that the specimen face would be located. No specimen holder is used for this operation. Figures 3-3.6 and 3-3.7 show the insertion of the heat flux meter.

The heat release rate is calibrated by the use of a calibration burner, again inserted into the same bracket as is used for the heat flux meter. (See Figure 3-3.16.) The cali-

bration burner, however, instead of always being inserted facing the heater, is inserted so that the discharge opening faces upward. Calibration is accomplished by controlling the flow of high-purity methane going to the burner and comparing it to a known value and using the net heat of combustion for pure CH_4 as 50 MJ $\cdot$ kg^{-1}.

The laser photometer is calibrated by neutral-density glass filters. These are inserted into a filter slot in front of the main beam detector. An auxiliary filter slot is provided in front of the laser. This serves to check the correct balancing of the dual-beam system's common mode rejection ratio.

The NBS "User's Guide to the Cone Calorimeter"[25] details how calibrations are performed.

MISCELLANEOUS DETAILS

Ring Sampler

The combustion products flowing through the exhaust system can be heavily laden in soot. This would cause rapid clogging of the oxygen measurement system if precautions were not taken. The most important precaution is the specially designed ring sampler. (See Figure 3-3.17.) The sampler is installed in the exhaust duct with the intake holes

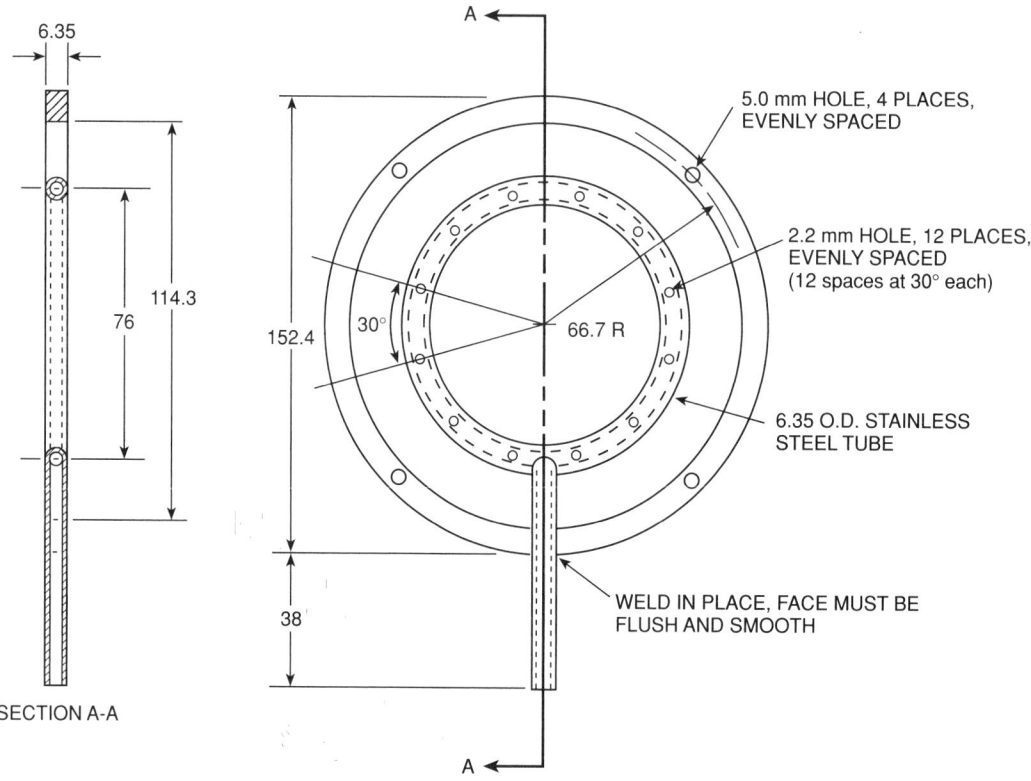

All material is stainless steel
All dimensions in millimeters
(except where noted)

6.35

114.3

76

152.4

30°

66.7 R

38

5.0 mm HOLE, 4 PLACES, EVENLY SPACED

2.2 mm HOLE, 12 PLACES, EVENLY SPACED (12 spaces at 30° each)

6.35 O.D. STAINLESS STEEL TUBE

WELD IN PLACE, FACE MUST BE FLUSH AND SMOOTH

SECTION A-A

A

A

Fig. 3-3.17. Ring sampler.

facing *away* from the direction of airflow. A number of small holes are used so as to provide a certain degree of smoothing with respect to duct flow turbulence.

Additional Gas Analyzers

Many users of cone calorimeters provide not just an oxygen analyzer, but also additional gas analyzers to help in determining the combustion chemistry and toxicity. CO and CO_2 analyzers are simply fitted into the same sampling line serving the oxygen analyzer. Other analyzers, *e.g.*, H_2O, HCl, and total unburned hydrocarbons, require a completely separate, heated sampling line system. Such a system also needs to have a heated soot filter at the front.

MEASUREMENTS TAKEN WITH THE CONE CALORIMETER

The relevant ISO or ASTM standard mandates certain minimum variables to be recorded. In practice, it is normally desired to make the data from the test be as complete as possible. Cone calorimeter data are normally handled as data tables and files standardized according to the FDMS prescription.[28] A complete set of data from the cone calorimeter are illustrated there. Here, the more important ones of these are given, somewhat augmenting the ISO and ASTM set. Note that most of these items must be reported for each test run, but a complete test consists of 3 runs.

Identification	Various data items must be included here
Preparation	Any non-standard specimen preparation details must be reported
Test no.	Serial number of test; also information on testing laboratory, operator, etc.
Irradiance	The heating flux set for the test ($kW \cdot m^{-2}$)
Exhaust flow rate	Recorded for completeness, usually the standard value of $24 \ \ell \cdot s^{-1}$
Orientation	Horizontal or vertical
Spark ignition	Yes or no
Edge frame	Yes or no
Wire grid	Yes or no
Area of specimen	(m^2), since may be non-standard in special cases
Specimen initial mass	(g)
Specimen final mass	(g)
Time to ignition	This, according to the ISO and ASTM standards, is for "sustained flaming" (s)
Time to flameout	(s)
Peak $\dot{q}''$	($kW \cdot m^{-2}$)
Peak $\dot{m}''$	($g \cdot s^{-1} \cdot m^{-2}$)
Total $\dot{q}$	($MJ \cdot m^{-2}$)

O$_2$ consumption const.	(kJ $\cdot$ kg^{-1}); this is set to a specific value if known, otherwise to 13100
Eff. heat of combustion	(MJ $\cdot$ kg^{-1}), reported for period of entire test run
Specific extinction rate	(m^2 $\cdot$ kg^{-1}), reported for period of entire test run
Avg. mass loss rate	Computed over period starting when 10 percent of the ultimate specimen mass loss rate has occurred and ending at the time when 90 percent of the ultimate specimen mass loss has occurred (g $\cdot$ s^{-1} $\cdot$ m^{-2})
Avg. $\dot{q}''$ (60 s)	Computed for the first 60 s after ignition (kW $\cdot$ m^{-2})
Avg. $\dot{q}''$ (180 s)	Computed for the first 180 s after ignition (kW $\cdot$ m^{-2})
Avg. $\dot{q}''$ (300 s)	Computed for the first 300 s after ignition (kW $\cdot$ m^{-2})

Note in the above 60, 180, and 300 s averages that, if the test is ended before having burned, say, 300 s, a proper average can still be correctly computed; i.e., at the end of the averaging period a number of zeroes are used for data points past the end of the test. Since users are often confused by this point, it must be emphasized: *It is not sensible to report an "average heat release rate" without specifying the time interval.* The reason has to do with the question of what is the *end* of the test. The ISO and ASTM standards specify that the end of the test is considered to be:

1. After all flaming and other signs of combustion cease; or
2. While there may still be vestigial combustion evidence, but the mass loss rate has become very small (less than 150 g $\cdot$ m^{-2} being lost during any 1 min); or
3. 60 min have elapsed.

These rules are needed for establishing some uniformity among testing laboratories. They do not, however, mean that it is technically sound to compare the average $\dot{q}''$ of one material that may have burned for 10 min., with another that may have burned for 5 min. It is technically sound, however, to compare their burning over the first 1, 3, etc., minutes of test.

Further information on the form, units, and usage of fire properties measured in the cone calorimeter can be found in reference 29; specific information on the smoke and soot properties measured in the cone calorimeter is given in reference 26.

Repeatability and Reproducibility

The repeatability, r, and reproducibility, R, of the cone calorimeter were studied in two sets of interlaboratory trials, one sponsored by ISO and one by ASTM. According to the ISO instructions,[30] the definitions of repeatability and reproducibility were taken as

$$r = 2.8\,\sigma_r$$

$$R = 2.8\,\sigma_R$$

where σ_r is the repeatability standard deviation, σ_R is the reproducibility standard deviation, and the 2.8 factor comes

from the fact the probability level of 95 percent is being specified.

From the results of the interlaboratory trials, values for r and R were calculated for six variables. These variables, chosen as being representative for the test results were: t_{ign}, $\dot{q}''_{max}$, $\dot{q}''_{180}$, q''_{tot}, $\Delta h_{c,\,eff}$, and σ_f. A linear regression model was used to describe r and R as a function of the mean overall replicates and overall laboratories for each of the six variables. The regression equations are given below. The range of mean values over which the fit was obtained is also indicated.

The results for time to sustained flaming, t_{ign}, in the range of 5 to 150 s were

$$r = 4.1 + 0.125\,t_{ign}$$

$$R = 7.4 + 0.220\,t_{ign}$$

The results for peak heat release rate, $\dot{q}''_{max}$, in the range of 70 to 1120 kW $\cdot$ m^{-2} were

$$r = 13.3 + 0.131\,\dot{q}''_{max}$$

$$R = 60.4 + 0.141\,\dot{q}''_{max}$$

The results for 180-s average heat release rate, $\dot{q}''_{180}$, in the range of 70 to 870 kW $\cdot$ m^{-2} were

$$r = 23.3 + 0.037\,\dot{q}''_{180}$$

$$R = 25.5 + 0.151\,\dot{q}''_{180}$$

The results for total heat released, $\dot{q}''_{tot}$, in the range of 5 to 720 MJ $\cdot$ m^{-2} were

$$r = 7.4 + 0.068\,\dot{q}''_{tot}$$

$$R = 11.8 + 0.088\,\dot{q}''_{tot}$$

The results for effective heat of combustion, $\Delta h_{c,\,eff}$, in the range of 7 to 40 kJ $\cdot$ g^{-1} were

$$r = 1.23 + 0.050\,\Delta h_{c,\,eff}$$

$$R = 2.42 + 0.055\,\Delta h_{c,\,eff}$$

The results for average specific extinction area, σ_f, in the range of 30 to 2200 m^2 $\cdot$ kg^{-1} were

$$r = 59 + 0.076\,\sigma_f$$

$$R = 63 + 0.215\,\sigma_f$$

A comparison of the cone calorimeter repeatability and reproducibility to the values obtained for the ISO 5657 radiant ignition test showed the cone calorimeter results to be about a factor of 2 better.

SPECIAL CONE CALORIMETERS

The standard cone calorimeter has been designed for using room air for its combustion. All standard testing is done under such conditions. For special combustion studies, however, it can be of interest to explore the burning of materials at oxygen levels other than 21 percent. Such a unit

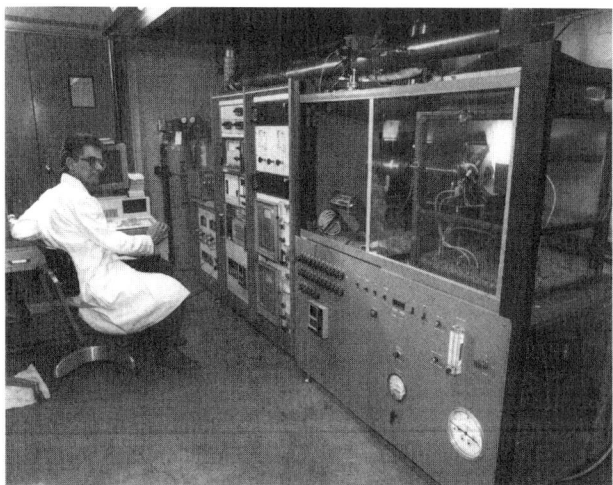

Fig. 3-3.18.   The NIST controlled-atmosphere cone calorimeter.

has been constructed at NIST; reference 14 gives details. This controlled-atmosphere NIST unit has already been used for studies of the burning of materials under conditions where the oxygen in the air supply is < 21 percent, with N_2 or CO_2 being mixed into the air stream. (See Figure 3-3.18.) It has also been used for pyrolysis studies under pure nitrogen flow conditions. In principle, it could also be used for studies of enriched-oxygen atmospheres; however, the necessary safety procedures for handling high-concentration oxygen streams are different. A unit for handling $O_2 > 21$ percent mixtures has been constructed for NASA, but data are not yet available from it. A controlled-atmosphere unit is also appropriate for use when airflow rates of less than $12 \, \ell \cdot s^{-1}$ are required.

All of the present cone calorimeter designs, both standard and not, have been designed for use only under *ambient* pressures. There is interest at this time from at least one research group to design and construct a unit for aerospace studies that would function under *non-ambient* pressures.

REFERENCES

1. ISO 5660, International Standard, "Fire Tests—Reaction to Fire—Rate of Heat Release from Building Products," International Organization for Standardization, Geneva (1993).
2. ASTM E 1354, "Standard Test Method for Heat and Visible Smoke Release Rates for Materials and Products Using an Oxygen Consumption Calorimeter," American Society for Testing and Materials, Philadelphia.
3. W.H. Twilley and V. Babrauskas, "User's Guide for the Cone Calorimeter," NBS Special Publication SP 745, U.S. Natl. Bur. Stand. (1988).
4. V. Babrauskas and S.J. Grayson, eds., *Heat Release in Fires*, Elsevier, London (1992). Distributed in the U.S. by NFPA.
5. B.J. McCaffrey and G. Cox, "Entrainment and Heat Flux of Buoyant Diffusion Flames," NBSIR 82-2473, U.S. Natl. Bur. Stand. (1982).
6. J.R. Hallman, "Ignition Characteristics of Plastics and Rubber" (Ph.D. dissertation), University of Oklahoma, Norman (1971).
7. B. Hägglund and L-E. Persson, FOA Rapport C 20126-D6(A3), "The Heat Radiation from Petroleum Fires," Försvarets Forskningsanstalt, Stockholm (1976).
8. V. Babrauskas, "Estimating Large Pool Fire Burning Rates," *Fire Technology, 19,* 251–261 (Nov. 1983).
9. J.J. Comeford, "The Spectral Distribution of Radiant Energy of a Gas-Fired Radiant Panel and Some Diffusion Flames," *Comb. and Flame, 18,* 125–132 (1972).
10. A. Tewarson, NBS-GCR-80-295, "Physico-Chemical and Combustion/Pyrolysis Properties of Polymeric Materials," U.S. Natl. Bur. Stand. (1980).
11. ISO 5657, "Fire Tests—Reaction to Fire—Ignitability of Building Products," International Organization for Standardization, Geneva (1986).
12. M.L. Janssens, "Fundamental Thermophysical Characteristics of Wood and Their Role in Enclosure Fire Growth" (Ph.D. dissertation), University of Gent, Belgium (1991).
13. V. Babrauskas, "Development of the Cone Calorimeter—A Bench-Scale Heat Release Rate Apparatus Based on Oxygen Consumption," *Fire and Materials, 8,* 81–95 (1984).
14. V. Babrauskas, W.H. Twilley, M. Janssens, and S. Yusa, "A Cone Calorimeter for Controlled-Atmospheres Studies, *Fire and Materials, 16,* 37–43 (1992).
15. ASTM E 906, "Standard Test Method for Heat and Visible Smoke Release Rates for Materials and Products," American Society for Testing and Materials, Philadelphia.
16. V. Babrauskas, NBSIR 80-2186, "Combustion of Mattresses Exposed to Flaming Ignition Sources, Part II. Bench-Scale Tests and Recommended Standard Test," U.S. Natl. Bur. Stand. (1981).
17. R.M. Nussbaum and B. A.-L. Östman, "Larger Specimens for Determining Rate of Heat Release in the Cone Calorimeter," *Fire and Materials, 10,* 151–160 (1986).
18. V. Babrauskas, Letter to the editor, *Fire and Materials, 11,* 205 (1987).
19. L. Orloff, J. deRis, and G.H. Markstein, "Upward Turbulent Fire Spread and Burning of Fuel Surfaces," pp. 183–192, in *Fifteenth Symp. (Intl.) on Combustion,* The Combustion Institute, Pittsburgh (1974).
20. L. Orloff, A.T. Modak, and R.L. Alpert, "Burning of Large-Scale Vertical Wall Surfaces," pp. 1345–54 in *Sixteenth Symp. (Intl.) on Combustion,* The Combustion Institute, Pittsburgh (1976).
21. W.D. Weatherford, Jr., and D.M. Sheppard, "Basic Studies of the Mechanism of Ignition of Cellulosic Materials," *Tenth Symp. (Intl.) on Combustion,* pp. 897–910, The Combustion Institute, Pittsburgh (1965).
22. H.R. Wesson, J.R. Welker, and C.M. Sliepcevich, "The Piloted Ignition of Wood by Thermal Radiation," *Comb. and Flame, 16,* 303–310 (1971).
23. J. Urbas and H. Sand, "Some Investigations on Ignition and Heat Release of Building Materials Using the Cone Calorimeter," pp. 183–192 in *INTERFLAM '90: Fifth Intl. Fire Conf. Proc.,* Interscience Communications, Ltd., London (1990).
24. E. Mikkola, unpublished study, VTT, Espoo, Finland (1989).
25. W.H. Twilley and V. Babrauskas, "User's Guide for the Cone Calorimeter," NBS Special Publication SP 745, U.S. Natl. Bur. Stand. (1988).
26. V. Babrauskas and G. Mulholland, "Smoke and Soot Data Determinations in the Cone Calorimeter," pp. 83–104, in ASTM STP 983, *Mathematical Modeling of Fires,* American Society for Testing and Materials, Philadelphia (1987).
27. P.J. Geake, "Smoke Characterisation by Laser Diffraction," Ph.D. dissertation, Polytechnic of the South Bank, London (1988).
28. V. Babrauskas, R.D. Peacock, M. Janssens, and N.E. Batho, "Standardizing the Exchange of Fire Data—The FDMS," *Fire and Materials, 15,* 85–92 (1991).
29. V. Babrauskas, "Effective Measurement Techniques for Heat, Smoke, and Toxic Fire Gases," pp. 4.1–4.10 in *Fire: Control the Heat . . . Reduce the Hazard,* QMC Fire & Materials Centre, London (1988).
30. ISO 5725, "Precision of Test Methods—Determination of Repeatability and Reproducibility for a Standard Test Method by Inter-Laboratory Tests," International Organization for Standardization, Geneva (1986).

GENERATION OF HEAT AND CHEMICAL COMPOUNDS IN FIRES

Archibald Tewarson

INTRODUCTION

Fire hazard is characterized by the generation of calorific energy and products, per unit of time, as a result of the chemical reactions between surfaces and vapors of materials and oxygen from air. If heat is the major contributor to hazard, it is defined as thermal hazard.[1] If fire products (smoke, toxic, corrosive, and odorous compounds) are the major contributors to hazard, it is defined as nonthermal hazard.[1] Various tests are used to determine the generation per unit of time of: (1) the calorific energy, defined as the heat release rate; and (2) fire products. The heat release rate and generation rates of fire products normalized by the generation rate of material vapors, airflow, etc., defined as fire properties, are used in models to predict: (1) heat release rate to assess the thermal hazard and fire protection needs; and (2) generation rates of fire products to assess the nonthermal hazard due to reduced visibility, and smoke damage, toxicity, corrosivity, and protection needs.

The region where vapors are generated is defined as the pyrolysis region and its leading edge as the pyrolysis front. The initiation of flaming fire is defined as ignition. Ignition is a process where vapors generated by heating the surface of a material mix with air, form a combustible mixture, ignite, and a fire is initiated. The region where the ignition process occurs is defined as the *ignition zone*. Minimum heat flux at or below which a material cannot generate the combustible mixture is defined as the *Critical Heat Flux* (CHF).[1-4] The resistance of a material to generate a combustible mixture is defined as the *Thermal Response Parameter* (TRP).[1-4] The higher the CHF and TRP values, the longer it takes for the material to heat up, ignite, and initiate a fire, and thus lower the fire propagation rate.

Depending on the magnitude of the heat flux provided by external sources and the flame of the material burning in the ignition zone, the pyrolysis front and flame can move beyond the ignition zone. The movement of the pyrolysis front is defined as fire propagation. The rate of movement of the pyrolysis front on the surface is defined as the fire propagation rate.

Heat and chemical compounds are generated as a result of the chemical reactions between: (1) pyrolyzing material vapors and oxygen in the gas phase, and (2) pyrolyzing material surface and oxygen in the solid phase. Heat generated in chemical reactions is defined as the *chemical heat*.[2-4] The rate of generation of chemical heat is defined as the *chemical heat release rate*. The chemical heat release rate distributes itself into a convective component, defined as the *convective heat release rate*, and into a radiative component, defined as the *radiative heat release rate*.[2-4] Convective heat release is associated with the flow of a hot products-air mixture, and radiative heat release is associated with the electromagnetic emission from the flame.

In a majority of cases, hazards to life and property are due to fires in enclosed spaces, such as in buildings. In general, fires in enclosed spaces are characterized by an upper and a lower layer. The main constituents of the upper layer are the hot fire products, and the main constituent of the lower layer is fresh air. In early stages, a building fire is well-ventilated, and is easy to control and extinguish. However, if the fire is allowed to grow, especially with limited enclosure ventilation and large material surface area, the chemical reactions between oxygen from air and products of incomplete combustion (smoke, CO, hydrocarbons, and other intermediate products) remain incomplete, resulting in an increase in nonthermal hazard. Rapid increase in the generation rates of products of incomplete combustion and growth rate of the fire, due to sudden and dramatic involvement of most of the exposed material surfaces, is termed flashover. Flashover is the most dangerous condition in a fire.

Heat release rate and generation rates of fire products as well as their nature are governed by: (1) fire initiation within the ignition zone; (2) fire propagation rate beyond the ignition zone; (3) fire ventilation; (4) external heat sources; (5) presence or absence of the fire suppression/extinguishing agents; and (6) materials: (a) their shapes, sizes, and arrangements; (b) their chemical natures; (c) types of additives mixed in; and (d) presence of other materials. In this handbook most of these areas have been discussed from a fundamental as well as applied views. For example, the mechanisms of thermal decomposition of polymers, which govern

Dr. Archibald Tewarson is Senior Research Specialist and Manager, Flammability Section, Factory Mutual Research Corporation. His research has focused on chemical kinetics, chemiluminescence, and chemical aspects of fires.

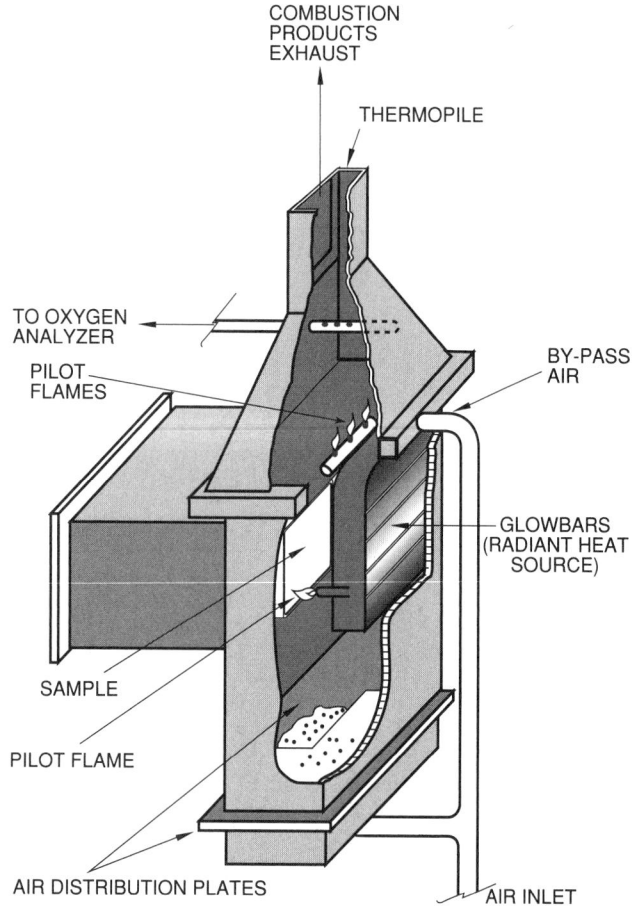

Fig. 3-4.1. *Ohio State University's (OSU) heat release rate apparatus.*[5-8]

the generation rates of material vapors, is discussed by Beyler in Section 1, Chapter 7; generation rate of heat (or heat release rate) from the viewpoint of thermochemistry is discussed by Drysdale in Section 1, Chapter 5; and its relationship with flame height by McCaffrey in Section 2, Chapter 1. Flaming ignition of the mixture of material vapors and air is discussed by Kanury in Section 2, Chapter 13; and surface flame spread by Quintiere in Section 2, Chapter 14.

This chapter presents the applications of the principles discussed in several chapters in this handbook to determine the fire properties of materials. Simple calculations have been included in the chapter to show how the properties can be used for various applications.

CONCEPTS GOVERNING GENERATION OF HEAT AND CHEMICAL COMPOUNDS IN FIRES

Fire Initiation (Ignition)

The fundamental ignition principles are described in detail by Kanury in Section 2, Chapter 13. The principles suggest that, for fire initiation, a material has to be heated above its CHF value (CHF value is related to the fire point). The CHF value can be determined in one of the several heat release rate apparatuses, e.g., the Ohio State University's (OSU)

heat release rate apparatus,[5-8] shown in Figure 3-4.1; the Flammability Apparatus,[1-4,9-16] shown in Figures 3-4.2, parts (a) and (b); and in the Cone Calorimeter,[17-19] shown in Figure 3-4.3. The design features, test conditions, and types of measurements for the three apparatuses are listed in Table 3-4.1.

Typically the CHF values are determined by exposing the horizontal sample (e.g., about 100-mm diameter or about 100×100-mm square and up to about 100-mm in thickness with blackened surface in the Flammability Apparatus) to various external heat flux values until a value is found at which there is no ignition for about 15 min.

As the surface is exposed to heat flux, initially most of the heat is transferred to the interior of the material. The ignition principles suggest that the rate with which heat is transferred depends on the ignition temperature (T_{ig}), ambient temperature (T_a), material thermal conductivity (k), material specific heat (c_P), and the material density (ρ). (See Section 2, Chapter 13.) The combined effects are expressed by a parameter defined as the Thermal Response Parameter (TRP) of the material[1-4,9-15]

$$\text{TRP} = \Delta T_{ig}\sqrt{k\rho c_P} \qquad (1)$$

where ΔT_{ig} ($= T_{ig} - T_a$) is the ignition temperature above ambient (K), k is in kW/m-K, ρ is in g/m^3, c_P is in kJ/g-K, and

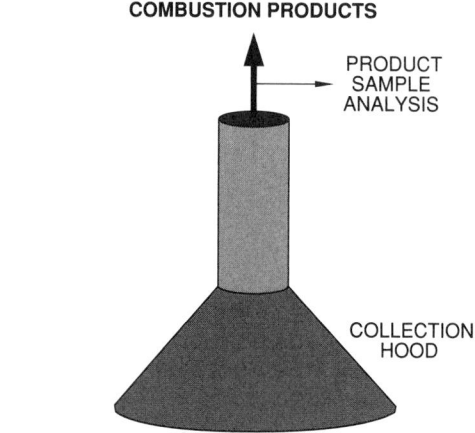

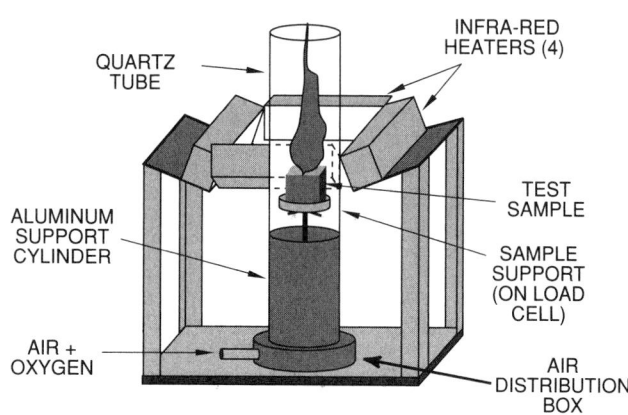

Fig. 3-4.2(a). *Flammability Apparatus designed by the Factory Mutual Research Corporation (FMRC). Sample configuration for ignition, pyrolysis, and combustion tests.*[1-4, 9-16]

TRP is in kW-s$^{1/2}$/m^2. TRP is a very useful parameter for the engineering calculations to assess resistance to ignition and fire propagation.

The ignition principles (see Section 2, Chapter 13) suggest that, for thermally thick materials, the inverse of the square root of time to ignition is expected to be a linear function of the external heat flux away from the CHF value

$$\sqrt{\frac{1}{t_{ig}}} = \frac{\sqrt{4/\pi}\,(\dot{q}''_e - \text{CHF})}{\text{TRP}} \qquad (2)$$

where t_{ig} is time to ignition (sec), $\dot{q}''_e$ is the external heat flux (kW/m^2), and CHF is in kW/m^2. Most commonly used materials behave as thermally thick materials and satisfy Equation 2, such as shown by the data in Figures 3-4.4 for polymethylmethacrylate (PMMA); in Figure 3-4.5 for heavy corrugated paper sheets, measured in the Flammability Apparatus; and in Figure 3-4.6 for non-blackened samples, measured in the Cone Calorimeter. The Cone Calorimeter data are taken from reference 20.

The value of the Thermal Response Parameter is determined, for example, in the Flammability Apparatus, by: (1) measuring the time to ignition for 100 × 100-mm square or 100-mm diameter and up to 25-mm-thick samples at differ-

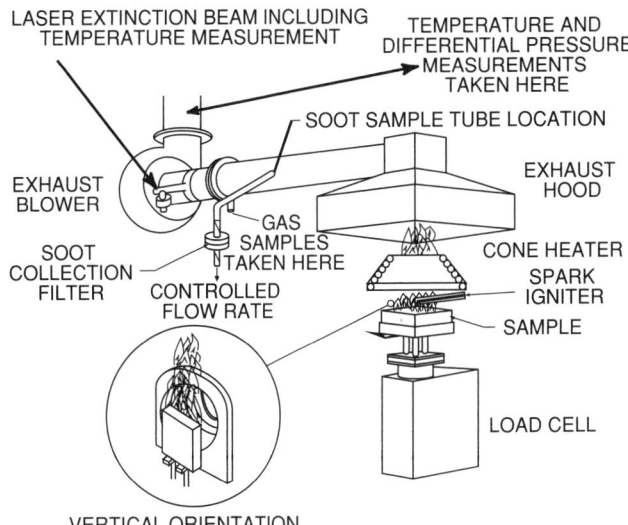

Fig. 3-4.3. *The Cone Calorimeter designed at the National Institute of Standards and Technology (NIST).*[17-19]

ent external heat flux values for samples with surfaces blackened with a very thin layer of black paint or fine graphite powder to avoid errors due to differences in the radiation absorption characteristics of the materials, and (2) performing a linear regression analysis of the data away from critical heat flux, following Equation 2, and recording the inverse of the slope of the line.

The value of the Thermal Response Parameter for a surface that is not blackened is higher than the value for the blackened surface. For example, for non-blackened and blackened surfaces of polymethylmethacrylate (PMMA), TRP = 383 and 274 kW-s$^{1/2}$/m^2, respectively, from the Flammability Apparatus.[2] The value for the Thermal Response Parameter for a blackened surface of PMMA is close

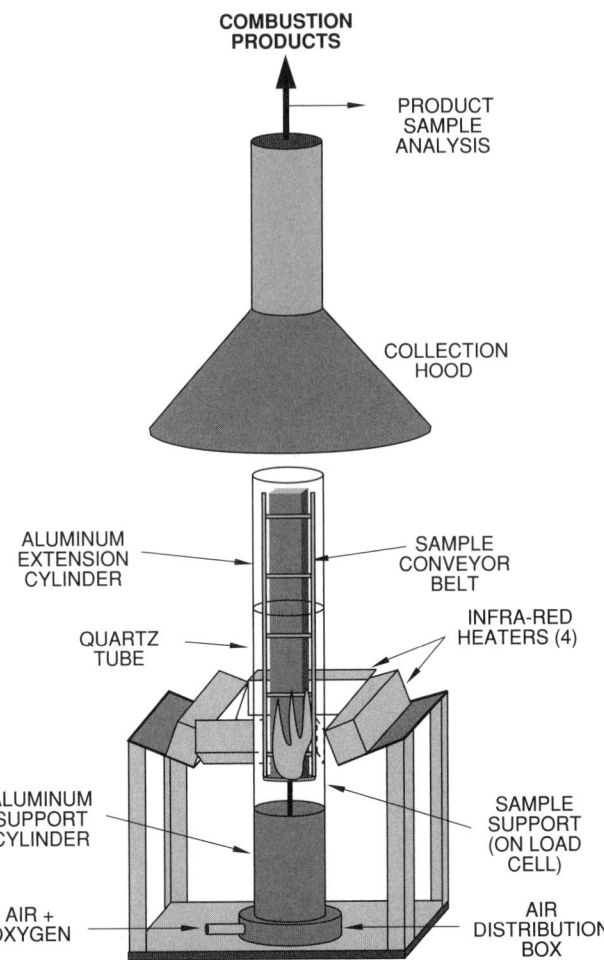

Fig. 3-4.2(b). *Flammability Apparatus designed by the Factory Mutual Research Corporation (FMRC). Sample configuration for fire propagation tests.*[1-4,9-16] *A conveyor belt sample is shown.*

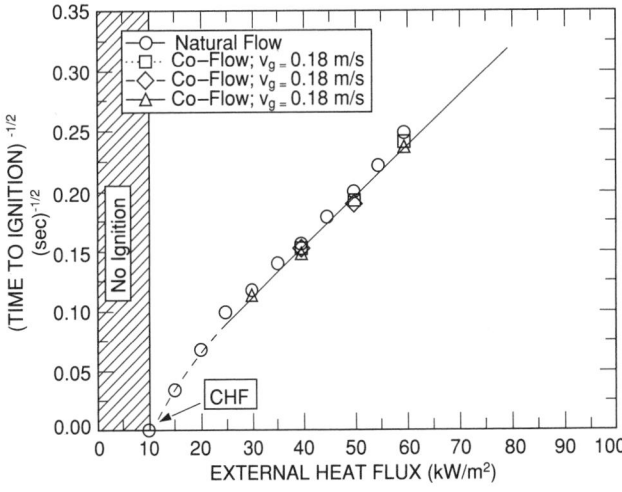

Fig. 3-4.4. *Square root of the inverse of time to ignition vs external heat flux for 100 × 100 × 25-mm-thick polymethylmethacrylate (PMMA) slab with blackened surface. Data measured in Flammability Apparatus and reported in reference 2 are shown.*

TABLE 3-4.1 *Design Features, Test Conditions, and Types of Measurements for OSU and Flammability Apparatuses, and NIST Cone Calorimeter*

Design and Test Conditions	OSU*	Flammability Apparatus	Cone[†]
Inlet gas flow	Co-flow	Co-flow/natural	Natural
Oxygen concentration (%)	21	0 to 60	21
Co-flow gas velocity (m/s)	0.49	0 to 0.146	NA
External heaters	Silicon Carbide	Tungsten-Quartz	Electrical Coils
External heat flux (kW/m²)	0 to 100	0 to 65	0 to 100
Exhaust product flow (m³/s)	0.04	0.035 to 0.364	0.012 to 0.035
Horizontal sample dimensions (mm)	110 × 150	100 × 100	100 × 100
Vertical sample dimensions (mm)	150 × 150	100 × 600	100 × 100
Ignition source	Pilot flame	Pilot flame	Spark plug
Heat release rate capacity (kW)	8	50	8
Measurements			
Time to ignition	yes	yes	yes
Material gasification rate	no	yes	yes
Fire propagation rate	no	yes	no
Generation rates of fire products	yes	yes	yes
Light obscuration by smoke	yes	yes	yes
Optical properties of smoke	no	yes	no
Electrical properties of smoke	no	yes	no
Gas-phase corrosion	no	yes	no
Chemical heat release rate	yes	yes	yes
Convective heat release rate	yes	yes	no
Radiative heat release rate	no	yes	no
Flame extinction			
By water	no	yes	no
By halon	no	yes	no
By halon alternates	no	yes	no

*As specified in ASTM E 906-83[7] and by DOT/FAA[8]
[†]As specified in ASTM E 1354-90[19]

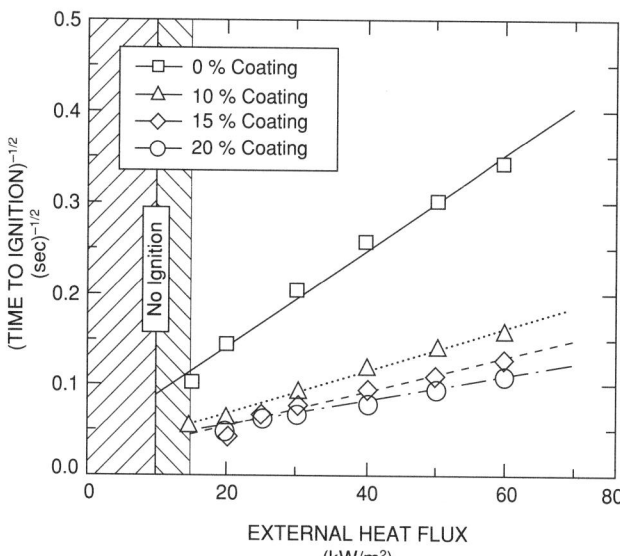

Fig. 3-4.5. *Square root of the inverse of time to ignition vs external heat flux for two 100 × 100 × 11-mm-thick sheets of heavy corrugated paper with blackened surface. Data measured in Flammability Apparatus.*

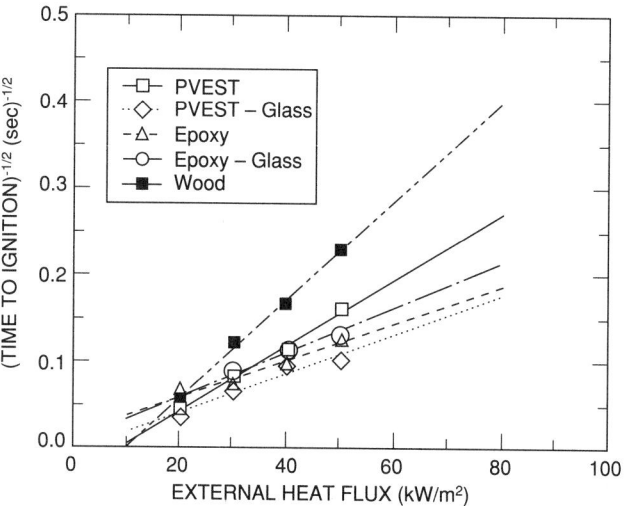

Fig. 3-4.6. *Square root of the inverse of time to ignition vs external heat flux for 100 × 100-mm non-blackened surfaces of: 10- × 11-mm-thick polyvinyl ester (PVEST), 11-mm-thick epoxy, and 6-mm-thick wood (hemlock). Data measured in the Cone Calorimeter as reported in reference 20 are shown.*

to the value calculated from the known T_{ig}, k, ρ, and c_P values for PMMA.[2]

The Thermal Response Parameter depends on the chemical as well as the physical properties of materials, such as the chemical structure, fire retardants, thickness, etc. For example, Figure 3-4.7 shows that the Thermal Response Parameter increases with sample thickness and increases in the amount of passive fire protection agent used, such as provided by a surface coating to a heavy corrugated paper sheet.

The Critical Heat Flux and the Thermal Response Parameter values for materials derived from the ignition data measured in the Flammability Apparatus and the Cone Calorimeter (as reported in reference 20) are listed in Table 3-4.2. In the Cone Calorimeter, the surface was not blackened, and thus the values of the Thermal Response Parameter may be somewhat higher than expected from the T_{ig}, k, ρ, and c_P values.

EXAMPLE 1:

In a fire, newspaper and polypropylene are exposed to a heat flux value of 50 kW/m². Estimate which material will ignite first, assuming physical conditions to be very similar for both the materials.

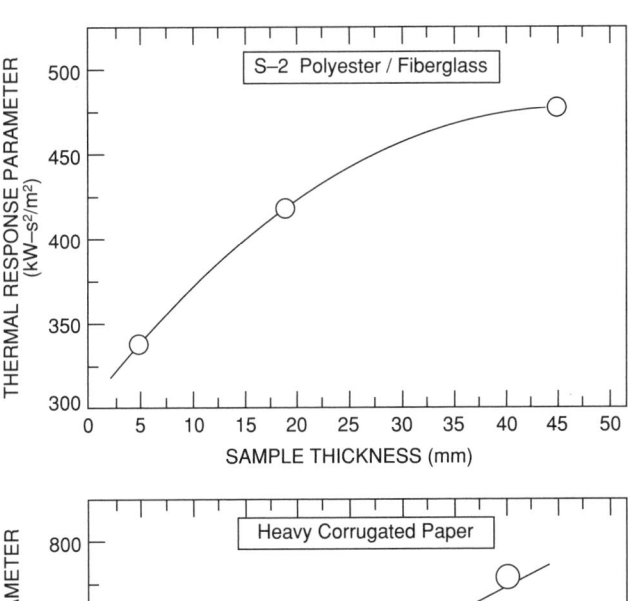

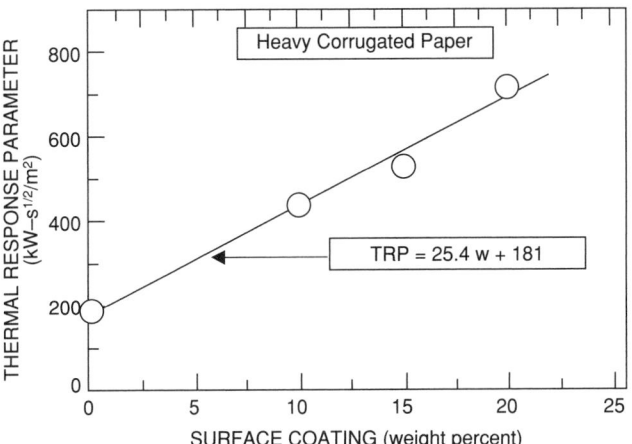

Fig. 3-4.7. *Thermal Response Parameter vs thickness for S-2 polyester/fiberglass sample and weight percent of surface coating for the heavy corrugated paper. Data measured in the Flammability Apparatus. w is weight %.*

SOLUTION:

From Table 3-4.2, for newspaper and polypropylene, CHF = 10 and 15 kW/m², respectively, and TRP = 108 and 193 kW-s$^{1/2}$/m², respectively. Substituting these values in Equation 2 with $\dot{q}''_e = 50$ kW/m², the times to ignition are calculated to be 6 and 24 sec for newspaper and polypropylene, respectively. Thus, newspaper will ignite first.

EXAMPLE 2:

Halogenated materials are obtained by replacing hydrogen atoms by halogen atoms in the chemical structures of the materials. For example, a unit in polyethylene (PE) consists of C_2H_4. If a hydrogen atom (H) is replaced by a chlorine atom (Cl) in a PE unit, it becomes a unit of rigid polyvinyl-chloride (PVC), i.e., C_2H_3Cl. If two H atoms are replaced by two fluorine atoms (F) in a PE unit, it becomes a unit of Tefzel™ (ethylene tetrafluorethylene), i.e., $C_2H_2F_2$. If all the hydrogen atoms are replaced by four F atoms in a PE unit, it becomes a unit of Teflon™ (polytetrafluoroethylene), i.e., C_2F_4. Show how the replacement of hydrogen atoms by the halogen atoms affects the ignitability of the materials.

SOLUTION:

From Table 3-4.2, for PE (high density), PVC (rigid), Tefzel™, and Teflon™, the CHF values are 15, 15, 27, and 38 kW/m², respectively, and the TRP values are 321, 406, 356, and 682 kW-s$^{1/2}$/m², respectively. In the calculations, it is assumed that these materials are exposed to a uniform heat flux of 60 kW/m² in a fire under very similar physical conditions. From Equation 2, using $\dot{q}''_e = 60$ kW/m², the times to ignition for PE (high density), PVC (rigid), Tefzel™, and Teflon™ are calculated to be 40, 64, 91, and 755 sec, respectively. Thus, resistance to ignition increases as the hydrogen atom is replaced by the halogen atom in the chemical structure of PE. The higher the number of hydrogen atoms replaced by the halogen atoms in the structure, the higher the resistance to ignition. When all the hydrogen atoms are replaced by the fluorine atoms, the material becomes highly resistant to ignition.

Fire Propagation

The fundamental surface flame spread principles are described by Quintiere in Section 2, Chapter 14. According to these principles, the fire propagation process, as indicated by surface flame spread, can be explained as follows.

As a material is exposed to heat flux from internal and/or external heat sources, a combustible mixture is formed that ignites, and a flame anchors itself on the surface in the ignition zone. As the vapors of the material burn in the flame, they release heat with a certain rate, defined as the chemical heat release rate.* Part of the chemical heat release rate is transferred beyond the ignition zone as conductive heat flux through the solid and as convective and radiative heat fluxes from the flame. If the heat flux transferred beyond the ignition zone satisfies the Critical Heat Flux, Thermal Response Parameter, and gasification requirements of the material, the pyrolysis and flame fronts move beyond the ignition zone and the flame anchors itself over additional surface. Due to increase in the burning surface area, flame

---

*In earlier papers, it was defined as the actual heat release rate, $\dot{Q}_A$.

TABLE 3-4.2 *Critical Heat Flux and Thermal Response Parameter of Materials*

Materials	CHF (kW/m²) Flammability Apparatus	TRP (kW-s$^{1/2}$/m²)	
		Flammability Apparatus	Cone Calorimeter*
Natural Materials			
Flour	10	218	—
Sugar	10	255	—
Tissue paper	10	95	—
Newspaper	10	108	—
Wood (red oak)	10	134	—
Wood (Douglas fir)	10	138	—
Corrugated paper (light)	10	152	—
Corrugated paper (heavy)			
No coating	10	189	—
Coating (10% by weight)	15	435	—
Coating (15% by weight)	15	526	—
Coating (20% by weight)	15	714	—
Wood (hemlock)	—	—	175
Wool 100%	—	—	252
Wood (Douglas fir/fire retardant, FR)	10	251	—
Synthetic Materials			
Epoxy resin	—	—	457
Polystyrene (PS)	13	162	—
Acrylic fiber 100%	—	—	180
Polypropylene (PP)	15	193	291
PP/FR panel	15	315	—
Styrene-butadiene (SB)	10	198	—
Crosslinked polyethylenes (XLPE)	15	224–301	—
Polyvinyl ester	—	—	263
Polyoxymethylene	13	269	—
Nylon	15	270	—
Polyamide-6	—	—	379
Polymethylmethacrylate (PMMA)	11	274	—
Isophthalic polyester	—	—	296
Acrylonitrile-butadiene-styrene (ABS)	—	—	317
Polyethylene (high density) (PE)	15	321	364
PE/nonhalogenated fire retardants	15	652–705	—
Polyvinyl ester panels	13–15	440–700	—
Modified acrylic (FR)	—	—	526
Polycarbonate	15	331	—
Polycarbonate panel	16	420	—
Halogenated Materials			
Isoprene	10	174	—
Polyvinylchloride (PVC)	10	194	—
Plasticized PVC, LOI = 0.20	—	—	285
Plasticized PVC, LOI = 0.25	—	—	401
Plasticized PVC, LOI = 0.30	—	—	397
Plasticized PVC, LOI = 0.35	—	—	345
Rigid PVC, LOI = 0.50	—	—	388
Rigid PVC1	15	406	—
Rigid PVC2	15	418	—
PVC panel	17	321	—
PVC fabric	26	217	—
PVC sheets	15	446–590	—
Ethylene tetrafluoroethylene (ETFE), Tefzel™	27	356	—
Fluorinated ethylene-propylene (FEP), Teflon™	38	682	—
Teflon fabric	50	299	—
Teflon coated on metal	20	488	—
Composite and Fiberglass-Reinforced Materials			
Polyether ether ketone—30% fiberglass	—	—	301
Isophthalic polyester—77% fiberglass	—	—	426
Polyethersulfone—30% fiberglass	—	—	256
Polyester 1—fiberglass	—	—	430
Polyester 2—fiberglass	10	275	—
Polyester 3—fiberglass	10	382	—
Polyester 4—fiberglass	15	406	—
Polyester 5—fiberglass	10	338	—
Epoxy Kevlar™ (thin sheet)	—	—	120
Epoxy fiberglass (thin sheet)	10	156	198
Epoxy graphite	15	395	—
Epoxy 1—fiberglass	10	420	—
Epoxy 2—fiberglass	15	540	—
Epoxy 3—fiberglass	15	500	—

*Calculated from the ignition data reported in reference 20.

TABLE 3-4.2 *Critical Heat Flux and Thermal Response Parameter of Materials (Continued)*

Materials	CHF (kW/m²) Flammability Apparatus	TRP (kW-s$^{1/2}$/m²) Flammability Apparatus	Cone Calorimeter*
Composite and Fiberglass-Reinforced Materials (Continued)			
Epoxy 4—fiberglass	10	388	
Epoxy resin—69% fiberglass	—	—	688
Epoxy-graphite 1	—	481	—
Epoxy-graphite 1/ceramic coating (CC)	—	2273	—
Epoxy-graphite 1/intumescent coating (IC)	—	962	—
Epoxy-graphite 1/IC-CC	—	1786	—
Polyvinyl ester 1—69% fiberglass	—	—	444
Polyvinyl ester 2—fiberglass	—	281	—
Polyvinyl ester 2—fiberglass/CC	—	676	—
Polyvinyl ester 2—fiberglass/IC	—	1471	—
Polyvinyl ester 2—fiberglass/IC-CC	—	1923	—
Graphite composite	40	400	—
Phenolic fiberglass (thin sheet)	33	105	172
Phenolic fiberglass (thick sheet)	20	610	—
Phenolic-graphite 1	20	333	—
Phenolic-graphite 2	—	—	400
Phenolic kevlar (thin sheet)	20	185	258
Phenolic kevlar (thick sheet)	15	403	—
Phenolic-graphite 1/CC	—	807	—
Phenolic-graphite 1/IC	—	1563	—
Foams (Wall-Ceiling Insulation Materials, etc)			
Polyurethane foams	13–40	55–221	—
Polystyrene foams	10–15	111–317	—
Phenolic	20	610	—
Phenolic laminate—45% glass	—	—	683
Latex foams	16	113–172	—
Materials with Fiberweb, Net-Like and Multiplex Structures			
Polypropylenes	8–15	108–417	—
Polyester-polypropylene	10	139	—
Wood pulp-polypropylene	8	90	—
Polyester	8–18	94–383	—
Rayon	14–17	161–227	—
Polyester-rayon	13–17	119–286	—
Wool-nylon	15	293	—
Nylon	15	264	—
Cellulose	13	159	—
Cellulose-polyester	13–16	149–217	—
Electrical Cables—Power			
PVC/PVC	13–25	156–341	—
PE/PVC	15	221–244	—
PVC/PE	15	263	—
Silicone/PVC	19	212	—
Silicone/cross linked polyolefin (XLPO)	25–30	435–457	—
EPR (ethylene-propylene rubber/EPR)	20–23	467–567	—
XLPE/XLPE	20–25	273–386	—
XLPE/EVA (ethyl-vinyl acetate)	12–22	442–503	—
XLPE/Neoprene	15	291	—
XLPO/XLPO	16–25	461–535	—
XLPO, PVF(polyvinylidine fluoride)/XLPO	14–17	413–639	—
EPR/Chlorosulfonated PE	14–19	283–416	—
EPR, FR	14–28	289–448	—
Electrical Cables—Communications			
PVC/PVC	15	131	—
PE/PVC	20	183	—
XLPE/XLPO	20	461–535	—
Si/XLPO	20	457	—
EPR-FR	19	295	—
Chlorinated PE	12	217	—
ETFE/EVA	22	454	—
PVC/PVF	30	264	—
FEP/FEP	36	638–652	—
Conveyor Belts			
Styrene-butadiene rubber (SBR)	10–15	336–429	—
Chloroprene rubber (CR)	20	760	—
CR/SBR	15	400	—
PVC	15–20	343–640	—

*Calculated from the ignition data reported in reference 20.

height, chemical heat release rate, and heat flux transferred ahead of the pyrolysis front all increase. The pyrolysis and flame fronts move again, and the process keeps repeating itself and burning area keeps increasing. Fire propagation on the surface continues as long as the heat flux transferred ahead of the pyrolysis front (from the flame or external heat sources) satisfies the Critical Heat Flux, Thermal Response Parameter, and gasification requirements of the material.

The rate of movement of the pyrolysis front is generally used to define the fire propagation rate

$$u = \frac{dX_p}{dt} \qquad (3)$$

where u is the fire propagation rate (mm or /s), and X_p is the pyrolysis front (mm or m).

The fire propagation rate can be determined in one of the several apparatuses: (1) the LIFT described by Quintiere in Section 2, Chapter 14; (2) the Flammability Apparatus (50- and 500-kW scale); the 50-kW scale apparatus is shown in Figure 3-4.2(b); and (3) the Fire Products Collector (10,000-kW scale Flammability Apparatus) shown in Figure 3-4.8. Examples of the type of data obtained from the Flammability Apparatus are shown in Figures 3-4.9 through 3-4.12. In Figure 3-4.12, heat release rates increase linearly with time during downward fire propagation, very similar to

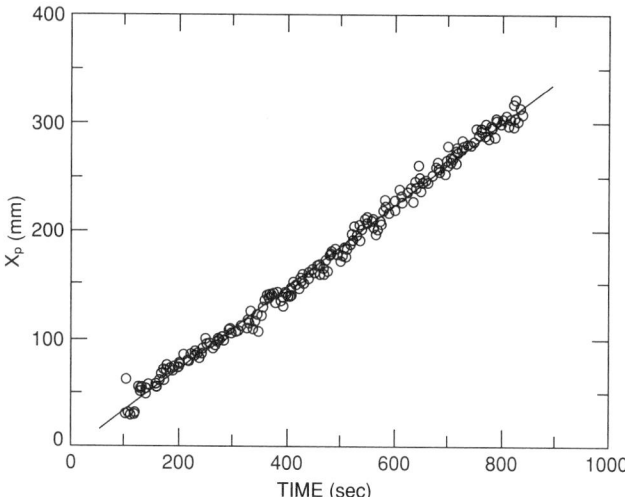

Fig. 3-4.9. Pyrolysis front versus time for the downward fire propagation for 300-mm long, 100-mm wide, and 25-mm thick PMMA vertical slab under opposed airflow condition in the Flammability Apparatus. Airflow velocity = 0.09 m/s. Oxygen mass fraction = 0.334. (Figure is taken from reference 2.)

the pyrolysis front values for the downward fire propagation in Figure 3-4.9.

The slopes of the lines in Figures 3-4.9 through 3-4.12 represent fire propagation rates. The upward fire propagation rate is much faster than the downward fire propagation rate. For downward fire propagation, linear increases in the pyrolysis front and heat release rates indicate decelerating fire propagation behavior. For upward fire propagation, nonlinear increases in the pyrolysis front indicate accelerating fire propagation behavior.

Relationship between fire propagation rate, flame height, pyrolysis front, and heat release rate: Numerous researchers have found the following relationship between the

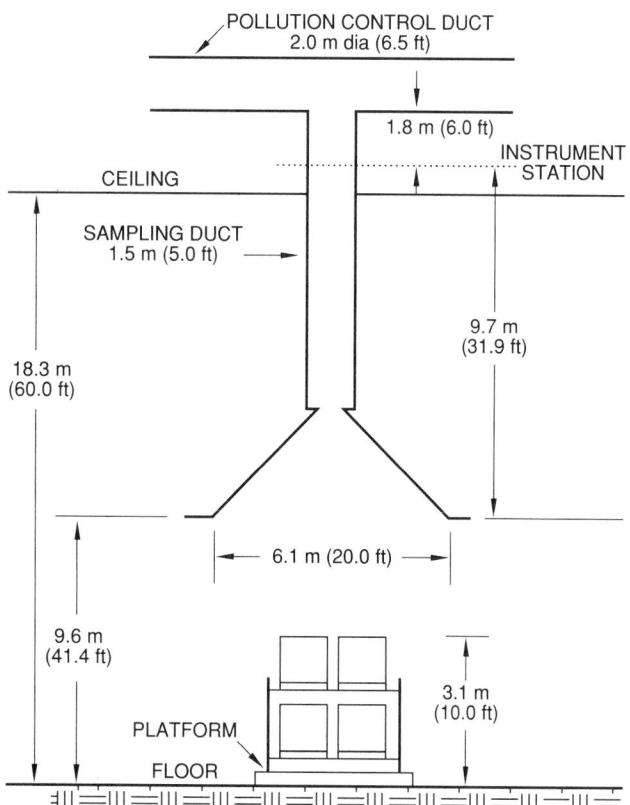

Fig. 3-4.8.   The Fire Products Collector (10,000-kW-scale Flammability Apparatus) for large-scale combustion and fire propagation tests. Corrugated boxes with various products, arranged in two-pallet loads × two-pallet loads × two-pallet loads high are shown. The Fire Products Collector is designed by the Factory Mutual Research Corporation.

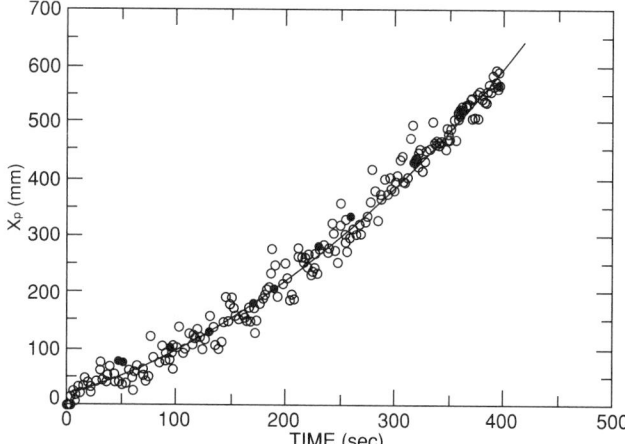

Fig. 3-4.10.   Pyrolysis front versus time for the upward fire propagation for 600-mm long, 100-mm wide, and 25-mm thick PMMA vertical slab under co-airflow condition in the Flammability Apparatus. Airflow velocity = 0.09 m/s. Oxygen mass fraction = 0.233. (Figure is taken from reference 2.)

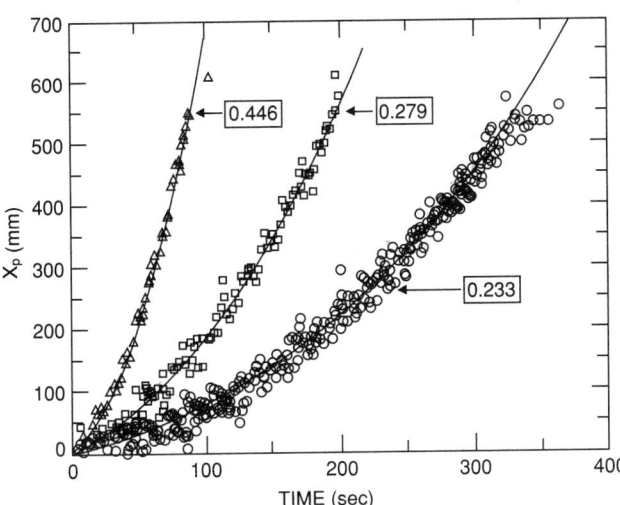

Fig. 3-4.11. *Pyrolysis front versus time for the upward fire propagation for 600-mm long and 25-mm thick diameter PMMA cylinder under co-airflow condition in the Flammability Apparatus. Airflow velocity = 0.09 m/s. Numbers inside the frames are the mass fractions of oxygen in air. (Figure is taken from reference 2.)*

flame height and pyrolysis front (as discussed by Quintiere in Section 2, Chapter 14 and reviewed in references 2 and 21)

$$X_f = aX_p^n \qquad (4)$$

where X_f is the flame height (m), X_p is in m, $a = 5.35$, and $n = 0.67$ to 0.80 for steady wall fires.[2]

Fire propagation data for PMMA from the Flammability Apparatus[2] and for electrical cables from several standard tests for cables (ICEA, CSA FT-4, and UL-1581)[9] satisfy Equation 4, as shown in Figure 3-4.13, with $a = 5.32$ and $n = 0.78$. The visual measurement of the pyrolysis front as damage length is used for the acceptance criterion in many of the standard tests for electrical cables. For example, for upward fire propagation in the CSA FT-4, damage length less than 60 percent of the total length of the cable tray for 20-min exposure time is used as the acceptance criterion.[9] For horizontal fire propagation in the UL-1581 test, flame length of less than 40 percent of the total length of the cable tray is used as the acceptance criterion.[9]

The relationship between the flame height and the chemical heat release rate, expressed as the normalized chemical heat release rate (NCHRR), has been enumerated by McCaffrey in Section 2, Chapter 1. NCHRR is defined as (see Section 2, Chapter 1)

$$\text{NCHRR} = \frac{\dot{Q}'_{ch}}{\rho c_p T_a g^{1/2} X_p^{3/2}} \qquad (5)$$

where $\dot{Q}'_{ch}$ is the chemical heat release rate per unit width (kW/m), ρ is the density of air (g/m³), c_p is the specific heat of air (kJ/g-K), T_a is the ambient temperature (K), g is acceleration due to gravity (m²/s), and X_p is in m.

Many researchers have shown that the ratio of the flame height to pyrolysis front is a function of the heat release rate, such as the following relationship (as discussed by Quintiere in Section 2, Chapter 14 and reviewed in references 2 and 21)

$$\frac{X_f}{X_p} = a(\text{NCHRR})^n \qquad (6)$$

where a and n are constants. This relationship reported in the literature (as reviewed in reference 2) for methane, ethane, and propylene is shown in Figure 3-4.14. The data for the upward fire propagation for PMMA from the Flammability Apparatus[2] and for the electrical cables from the several standard tests for cables (UL-1581, ICEA, and CSA FT-4) also satisfy this relationship as indicated in Figure 3-4.14.[9]

In Figure 3-4.14, data in the lower left-hand corner are for the low-intensity polyvinylchloride (PVC) electrical cable fire propagation in the standard tests for cables. These data show that for NCHRR < 0.2, $X_f/X_p < 1.5$, and $n = \frac{1}{10}$. This is a characteristic property of materials for which there is either no fire propagation or a limited fire propagation beyond the ignition zone. These materials are defined as

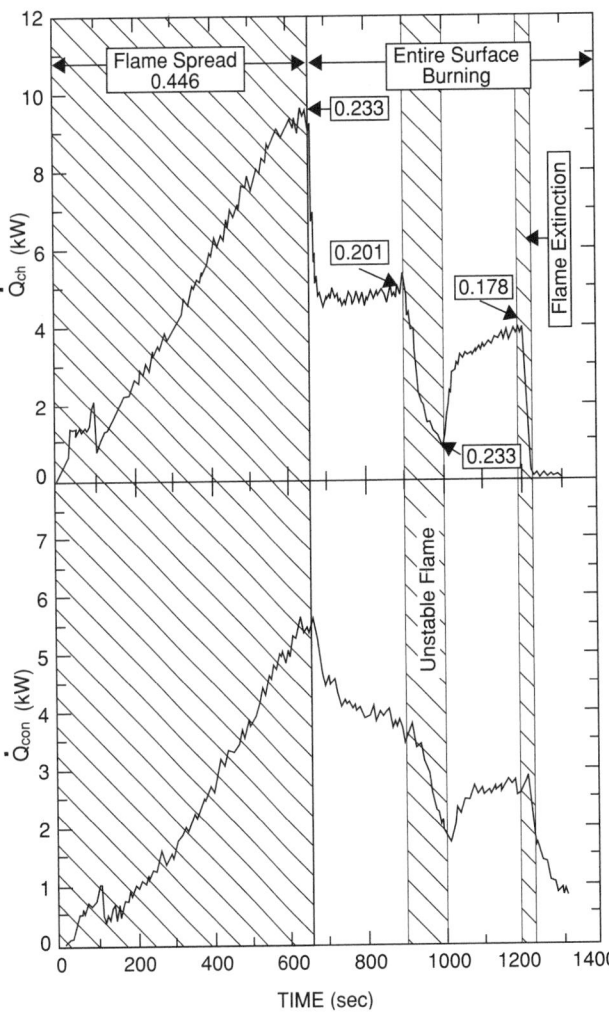

Fig. 3-4.12. *Chemical and convective heat release rate versus time for the downward fire propagation, combustion, and flame extinction for 300-mm long, 100-mm wide, and 25-mm thick PMMA vertical slab under opposed airflow condition in the Flammability Apparatus. Airflow velocity = 0.09 m/s. Numbers inside the frames are the mass fractions of oxygen in air. (Figure is taken from reference 2.)*

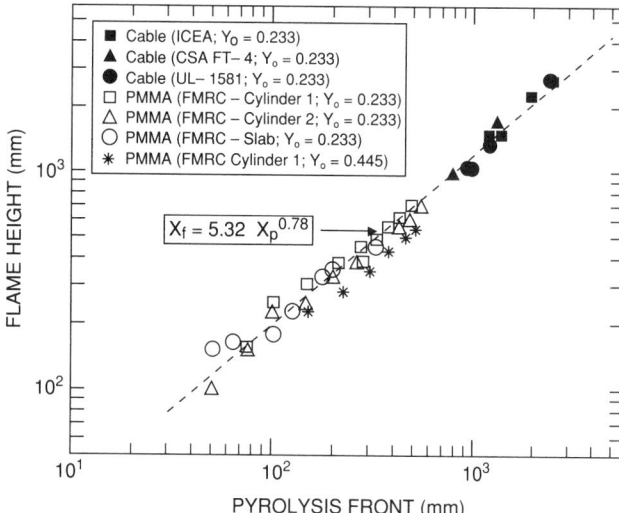

Fig. 3-4.13. Flame height versus pyrolysis front for the upward fire propagation in normal air. Data are for the vertical fire propagation for electrical cables contained in 2.44-m long, 310-mm wide, and 76-mm deep trays in standard tests for electrical cables (ICEA, CSA FT-4, and UL-1581) and for 600-mm long PMMA slabs (100-mm wide and 25-mm thick) and cylinder (25-mm diameter) in the Flammability Apparatus. Data for fire propagation in an oxygen mass fraction of 0.445 are also included. (Figure is taken from references 2 and 9.)

Group 1 materials.[4,9-15] Cables with Group 1 material characteristics pass the standard tests for cables (UL-910, CSA FT-4, UL-1581, and ICEA). The data for higher intensity fire propagation in Figure 3-4.14 show that: (1) for $0.2 > NCHRR < 5$, $n = 2/3$, and $1.5 < X_f/X_p < 20$ (PMMA fire propagation and methane combustion); and (2) for $NCHRR > 5$, $n = 1/2$, and $X_f/X_p > 20$ (ethane and propylene combustion). Thus, the ratio of the flame height to pyrolysis front is a good indicator of the fire propagation characteristics of the materials. Materials for which flame height is close to the pyrolysis front during fire propagation can be useful indicators of decelerating fire propagation behavior.

Researchers have also developed many correlations between the flame heat flux transferred ahead of the pyrolysis front and heat release rate for downward, upward, and horizontal fire propagation (as discussed by Quintiere in Section 2, Chapter 14 and reviewed in references 2 and 21). For example, small- and large-scale fire propagation test data from the Flammability Apparatus [Figure 3-4.2(b)] and Fire Products Collector (Figure 3-4.8) suggest that, for thermally thick materials with highly radiating flames, the following semi-empirical relationship is satisfied[3]

$$\dot{q}_f'' \propto \left[\frac{\chi_{rad}}{\chi_{ch}}\dot{Q}_{ch}'\right]^{1/3} \qquad (7)$$

where $\dot{q}_f''$ is the flame heat flux transferred ahead of the pyrolysis front (kW/m²) and χ_{rad} is the radiative fraction of the combustion efficiency, χ_{ch}. The fire propagation rate is expressed as[3]

$$\sqrt{u} \propto \left[\frac{\chi_{rad}}{\chi_{ch}}\dot{Q}_{ch}'\right]^{1/3}/TRP \qquad (8)$$

The right-hand side of Equation 8, multiplied by 1000 with $\chi_{rad}/\chi_{ch} = 0.42$ is defined as the *Fire Propagation Index* (FPI)[4,9-15]

$$FPI = 1000\frac{(0.42\dot{Q}_{ch}')^{1/3}}{TRP} \qquad (9)$$

FPI describes the fire propagation behavior of materials under highly flame-radiating conditions prevalent in large-scale fires. The small- and large-scale fire propagation test data and understanding of the fire propagation suggest that the FPI values can be used to classify the materials into four groups:[2-4,9-15]

1. *FPI < 7 (Non-Propagating) Group N-1 Materials*—Materials for which there is no fire propagation beyond the ignition zone. Flame is at critical extinction condition.
2. *7 < FPI < 10 (Decelerating Propagation): Group D-1 Materials*—Materials for which fire propagates beyond the ignition zone although in a decelerating fashion. Fire propagation beyond the ignition zone is limited.
3. *10 < FPI < 20 (Non-Accelerating Propagation): Group 2 Materials*—Materials for which fire propagates slowly beyond the ignition zone.
4. *FPI > 20 (Accelerating Propagation): Group 3 Materials*—Materials for which fire propagates rapidly beyond the ignition zone.

The FPI values for the upward fire propagation, under highly flame-radiating conditions, have been determined for numerous materials in the Flammability Apparatus. The highly radiating conditions are created by using a value of 0.40 for the mass fraction of oxygen. Two sets of tests are performed:

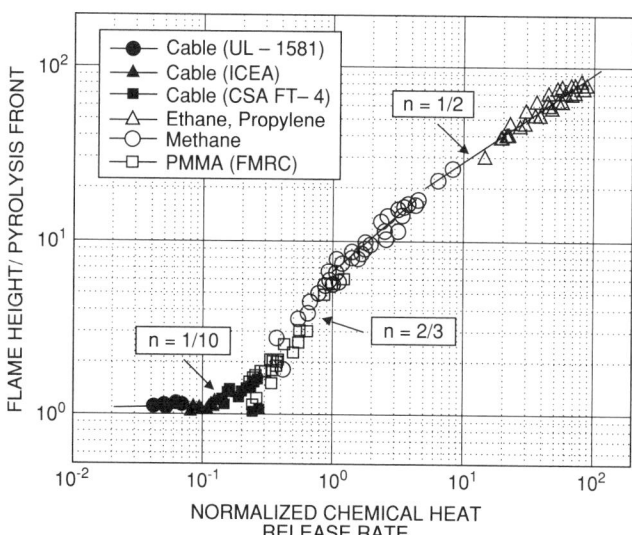

Fig. 3-4.14. Ratio of flame height to pyrolysis front versus the normalized chemical heat release rate for the upward fire propagation in normal air. Data for the diffusion flames of methane, ethane, and propylene are from the literature. Data for the cables are from the standard tests for electrical cables (ICEA, CSA FT-4, and UL-1581).[9] Data for PMMA are from the Flammability Apparatus for 600-mm-long vertical PMMA slabs (100-mm wide, 25-mm thick) and cylinders (25-mm diameter).[2] (Figure is taken from reference 9.)

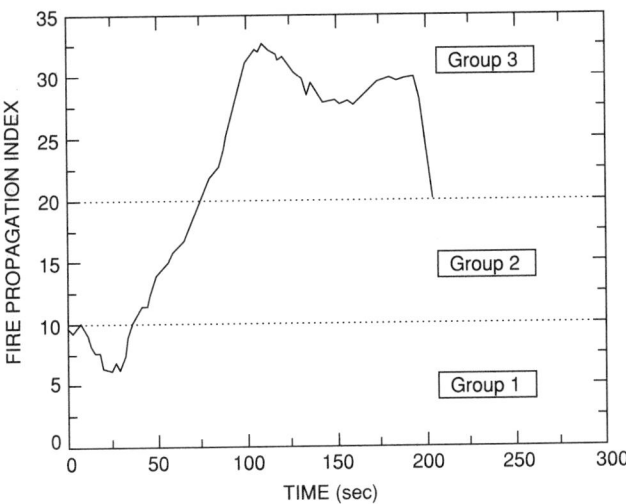

Fig. 3-4.15. Fire Propagation Index versus time for a polyethylene (PE)/polyvinylchloride (PVC) Group 3 cable determined in the Flammability Apparatus. This cable does not pass any standard tests for electrical cables.

1. **Thermal Response Parameter Test:** Ignition tests are performed in the Flammability Apparatus [Figure 3-4.2(a)], and the Thermal Response Parameter value is determined from the time to ignition *versus* external heat flux as described in the subsection on fire initiation (ignition).
2. **Upward Fire Propagation Test:** Fire propagation tests for vertical slabs, sheets, or cables are performed in the Flammability Apparatus [50- and 500-kW scale, Figure 3-4.2(b)]. About 300- to 600-mm-long, up to about 100-mm-wide, and up to about 100-mm-thick samples are used. The bottom 120 to 200 mm of the sample is in the ignition zone, where it is exposed to 50 kW/m² of external heat flux in the presence of a pilot flame. Beyond the ignition zone, fire propagates by itself, under co-airflow condition with an oxygen mass fraction of 0.40. During upward fire propagation, measurement is made for the chemical heat release rate as a function of time in each test.

The Thermal Response Parameter value and the chemical heat release rate are used in Equation 9 to calculate the Fire Propagation Index as a function of time. The Fire Propagation Index profile is used to classify materials into Group 1, 2, or 3.

Application of the Fire Propagation Index (FPI) to classify materials

Electrical Cables: The FMRC standard for cable fire propagation Class No. 3972[14] is used to classify electrical cables, based on their upward fire propagation behavior, under highly flame-radiating conditions (oxygen mass fraction = 0.40), for protection needs in noncombustible occupancies. A noncombustible occupancy is defined as an occupancy where only specific types of combustibles are present, ignition sources are relatively small, and their contributions toward thermal and non-thermal hazards are negligible compared to the contributions of the combustibles. The Thermal Response Parameter and upward fire propagation tests are performed, and Equation 9 is used to calculate the Fire Propagation Index (FPI), as described above. Figure 3-4.15

shows an example of a typical profile for the Fire Propagation Index *versus* time for a polyethylene (PE)/polyvinylchloride (PVC) cable. This cable does not pass any of the standard electrical cable tray fire tests, and the FPI profile in Figure 3-4.15 shows that it is a Group 3 cable.

The following fire protection guidelines are recommended by FMRC for grouped cables:[11,13,14]

1. Group 1 cables do not need additional fire protection in noncombustible occupancies with noncombustible construction,
2. Group 2 cables can be used without additional fire protection in noncombustible occupancies with noncombustible construction under certain conditions, and
3. Group 3 cables need fire protection.

Table 3-4.3 lists Fire Propagation Index values for selected electrical cables, composites, and conveyor belts.

EXAMPLE 3:

What type of fire behavior is represented by a 300-mm-wide, 8-m-high, and 25-mm-thick vertical sheet of a material with a Thermal Response Parameter value of 95 kW-s$^{1/2}$/m² if the peak chemical heat release during the upward fire propagation is 50 kW?

SOLUTION:

Fire propagation behavior is assessed by the FPI value. For the material, the chemical heat release rate per unit width, $\dot{Q}'_{ch} = 50/0.3 = 167$ kW/m. Substituting this value in Equation 9, with TRP = 95 kW-s$^{1/2}$/m², FPI = 43. The TRP value is greater than 20, and thus the material is a Group 3 material and represents an accelerating fire propagation behavior.

EXAMPLE 4:

A noncombustible cable spreading room has an old and a new area with a 3-hr-rated solid fire wall between the two. The old area is filled with several trays of polyethylene (PE)/polyvinylchloride (PVC) communications cables, and the new area is filled with several trays of crosslinked polyolefin (XLPO/XLPO) communications cables. In order to determine the fixed fire protection needs for these two areas, cable samples were submitted to a testing laboratory. The laboratory reported the following test data:

1. Ignition Data:

Heat Flux (kW/m²)	30	40	50	60	100
Time to Ignition (s)					
PE/PVC	76	27	14	8	2
XLPO/XLPO	—	716	318	179	45

2. Peak Chemical Heat Release Rate: During vertical fire propagation for 0.60-m-long cable sample in a highly radiating environment (oxygen mass fraction = 0.40), the following data were measured:

Cable	Peak Chemical Heat Release Rate Per Unit Cable Circumference (kW/m²)
PE/PVC	100
XLPO/XLPO	20

The data were used to calculate the FPI values, which suggested that the area with PE/PVC cable trays needed fixed fire protection, whereas the area with XLPO/XLPO cable trays did not need fixed fire protection. Do you agree?

SOLUTION:

The TRP values from the linear regression analysis of the ignition data are: 131 and 535 kW-s$^{1/2}$/m^2 for the PE/PVC and the XLPO/XLPO cable samples, respectively. The data for $\dot{Q}'_{ch}$ are given. Thus, from Equation 9, the FPI values for the PE/PVC and the XLPO/XLPO cable samples are 29 and 4, respectively. The FPI values suggest that the PE/PVC cable is a Group 3 cable and is expected to have an accelerating fire propagation behavior, and the XLPO/XLPO is a Group 1 cable and fire propagation is expected to be either limited to the ignition zone or decelerating. These calculations support that the cable spreading room area filled with the PE/PVC cable trays would need fixed fire protection, whereas it would not be needed for the area filled with the XLPO/XLPO cable trays.

Conveyor Belts: A conveyor belt standard is being developed at FMRC following the FMRC standard for cable fire propagation Class No. 3972.[14] The Thermal Response Parameter and upward fire propagation tests are performed, and Equation 9 is used to calculate the Fire Propagation Index (FPI) as described above.

Conveyor belts are classified as propagating or non-propagating. For an approximately 600-mm-long and 100-mm-wide vertical conveyor belt, the data measured in the Flammability Apparatus under highly flame-radiating conditions show that the non-propagating fire condition is satisfied for FPI ≤ 7.0 for the belts that show limited fire propagation in the large-scale fire propagation test gallery of the U.S. Bureau of Mines.[12,22]

Table 3-4.3 lists Fire Propagation Index values for selected conveyor belts taken from references 12 and 22.

EXAMPLE 5:

Conveyor belts are made of solid woven or piles of elastomers, such as styrene-butadiene rubber (SBR), poly-chloroprene rubber (CR), polyvinylchloride (PVC), reinforced with fibers made of polymers, such as nylon. In large-scale fire propagation tests in a tunnel, fire on the surface of a CR-based conveyor belt was found to be non-propagating, whereas for a CR/SBR-based conveyor belt fire was found to be propagating. Small-scale tests showed that the CR- and CR/SBR-based conveyor belts had the following fire properties, respectively: (1) CHF = 20 and 15 kW/m^2; (2) TRP = 760 and 400 kW-s$^{1/2}$/m^2; and (3) peak $\dot{Q}'_{ch}$ = 114 and 73 kW/m under highly flame-radiating conditions (oxygen mass fraction = 0.40). Show that small-scale test results are consistent with the large-scale fire propagation behaviors of the two conveyor belts, using the criterion that, for non-propagating fire behavior, the Fire Propagation Index is equal to or less than 7.

SOLUTION:

Substituting the TRP and $\dot{Q}'_{ch}$ values in Equation 9, the FPI values for the CR- and CR/SBR-based conveyor belts are 5 and 8, respectively. Thus, the CR-based conveyor belt is expected to have a non-propagating fire behavior, whereas the CR/SBR-based conveyor belt is expected to have a propagating fire behavior. The small-scale test results, therefore, are consistent with the large-scale fire propagation behaviors of the two conveyor belts.

Composites and Fiberglass-Reinforced Materials: The use of composites and fiberglass-reinforced materials is increas-

TABLE 3-4.3 *Fire Propagation Index for Cables, Composites, and Conveyor Belts, Determined in the Flammability Apparatus*

	Diameter/ Thickness (mm)	FPI	Group	Fire Propagation*
Cables				
Polymethyl-methacrylate	25	30	3	P
Power Cables				
PVC/PVC	4–13	11–28	2–3	P
PE/PVC	11	16–23	3	P
PVC/PE	34	13	2	P
Silicone/PVC	16	17	2	P
Silicone/XLPO	55	6–8	1	N-D
EP/EP	10–25	6–8	1	N-D
XLPE/XLPE	10–12	9–17	1–2	D-P
XLPE/EVA	12–22	8–9	1	D
XLPE/Neoprene	15	9	1	D
XLPO/XLPO	16–25	8–9	1	D
XLPO, PVF/XLPO	14–17	6–8	1	N-D
EP/CLP	4–19	8–13	1–2	D-P
EP, FR/None	4–28	9	1	D
Communications Cables				
PVC/PVC	4	36	3	P
PE/PVC	4	28	3	P
PXLPE/XLPO	22–23	6–9	1	N-D
Si/XLPO	28	8	1	D
EP-FR/none	28	12	2	P
PECl/none	15	18	2	P
ETFE/EVA	10	8	1	D
PVC/PVF	5	7	1	N
FEP/FEP	8	4	1	N
FEP/FEP	10	5	1	N
Composites				
Polyester-1/glass (30/70)	4.8	13	2	P
Polyester-3/glass (30/70)	4.8	10	2	P
	19	8	1	D
	45	7	1	N
Phenolic-PVB/ Kevlar™ (16/84)	4.8	8	1	D
Phenolic/Glass (20/80)	3.2	3	1	N
Epoxy-1/Glass (35/65)	4.4	9	1	D
Epoxy-2/Glass (35/65)	4.8	11	2	P
Epoxy-3/Glass (35/65)	4.4	10	2	P
Conveyor Belts†				
Styrene-butadiene rubber (SBR)	8–11	1–2		D-P
Chloroprene rubber (CR)	5	1		P
CR/SBR	8	1		D
PVC	4–10	1–2		N-P

*P: propagation; D: decelerating propagation; N: no propagation.
†3 to 25 mm thick.

ing very rapidly because of low weight and high strength in applications such as aircraft, submarines, naval ships, military tanks, public transportation vehicles including automobiles, space vehicles, tote boxes, pallets, chutes, etc. Fire propagation, however, is one of the major concerns for the composites and fiberglass-reinforced materials; the Fire Propagation Index concept thus is used.[4,10] For the determination of the Fire Propagation Index for the composites and fiberglass-reinforced materials, the Thermal Response Parameter and upward fire propagation tests are performed and Equation 9 is used, as discussed previously for electrical cables and conveyor belts.

The Fire Propagation Index concept used for the composites and fiberglass-reinforced materials is based on the knowledge gained during the development of the FMRC standard for cable fire propagation Class No. 3972[14] and FMRC studies on conveyor belts.[12,22] The non-propagating fire condition is satisfied in the Flammability Apparatus for FPI ≤ 7.0, for about 600-mm-long and 100-mm-wide vertical composites and fiberglass-reinforced materials, under highly flame-radiating conditions (oxygen mass fraction = 0.40), very similar to the conveyor belts.

Table 3-4.3 lists Fire Propagation Index values for selected composites taken from references 4 and 10.

Interior Finish Wall/Ceiling Materials: Since 1971, Factory Mutual Research Corporation has used the 25-ft corner test as a standard test. The 25-ft corner test is performed in a 7.6-m (25-ft)-high, 15.2-m (50-ft)-long and 11.6-m (38-ft)-wide building corner configuration to evaluate the burning characteristics of interior finish wall and ceiling materials.[23–26] The materials tested are typically panels with a metal skin over an insulation core material. The materials installed in the corner configuration are subjected to a growing exposure fire (peak heat release rate of about 3 MW) comprised of about 340 kg (750 lb) of 1.2-m (4-ft) × 1.2-m (4-ft) wood (oak) pallets stacked 1.5 m (5 ft) high at the base of the corner. The material is considered to have failed the test if within 15 min either: (1) fire propagation on the wall or ceiling extends to the limits of the structure, or (2) flame extends outside the limits of the structure through the ceiling smoke layer.

The fire environment within the 25-ft corner test structure has been characterized through heat flux and temperature measurements.[23,25] It has been shown that the fire propagation boundary (pyrolysis front) measured by visual damage is very close to the Critical Heat Flux (CHF) boundary for the material, as shown in Figure 3-4.16, taken from reference 24. This is in agreement with the general understanding of the fire propagation process. Through small- and large-scale fire propagation tests for low-density, highly char-forming wall and ceiling insulation materials, using the Flammability Apparatus [Figure 3-4.2(a)], Fire Products Collector (Figure 3-4.8), and 25-ft corner test (Figure 3-4.16), a semi-empirical relationship has been developed for fire propagation rate for a 15-min test in the 25-ft corner test[23–25]

$$\frac{X_p}{X_t} = \frac{\dot{Q}''_{con}}{\text{TRP}} \qquad (10)$$

where X_p is the average fire propagation length along the eaves (Figure 3-4.16) of the 25-ft corner test (pyrolysis front) measured visually (m), X_t is the total available length [11.6-m (38 ft)] in the 25-ft corner test, and $\dot{Q}''_{con}$ is the convective heat release rate (kW/m²).

The right-hand side of Equation 10 with the convective heat release rate measured at 50 kW/m² of external heat flux is defined as the convective flame spread parameter (FSP_c).[24,25] Figure 3-4.17 shows a correlation between the convective flame spread parameter obtained from the Flammability Apparatus and the normalized fire propagation

length in the FMRC 25-ft corner test. Pass/fail regions, as determined from the 25-ft corner test, are indicated in the figure. Materials for which $FSP_c ≤ 0.39$ pass the 25-ft corner test, and materials for which $FSP_c ≥ 0.47$ are judged to be unacceptable (i.e., fail).[24–26] The region where the FSP_c values are greater than 0.39 but less than 0.47 is uncertain.[24–26]

The correlation and pass/fail criterion shown in Figure 3-4.17 have been adopted in the FMRC Class No. 4880 for insulated wall or wall and ceiling panels.[26] In this standard, the 25-ft corner test has been replaced by the Flammability Apparatus [Figure 3-4.2(a)] tests. Two sets of tests are performed in the apparatus[24–26]

1. **Thermal Response Parameter Test:** Ignition tests are performed using approximately 100-mm × 100-mm and up to 100-mm-thick samples. Times to ignition at various external heat flux values are measured to determine the Thermal Response Parameter as described earlier.
2. **Convective Heat Release Rate Test:** Combustion tests are performed using about 100-mm × 100-mm and up to 100-mm-thick samples. Samples are burned in normal air under an external heat flux exposure of 50 kW/m². During the test, measurement is made for the convective heat release as a function of time.

The data for the Thermal Response Parameter and convective heat release rate at 50 kW/m² of external heat flux are used to calculate the flame spread parameter (FSP_c) that accepts or rejects the sample.

Flaming and Nonflaming Fires

During fire propagation, the surface of the material regresses in a transient fashion with a rate slower than the fire propagation rate.[2] The surface regression becomes steady after fire propagates throughout the available surfaces. The surface regression continues until all the combustible components of the material are exhausted. During fire propagation and surface regression, the material generates vapors at a transient or steady rate. The generation rate of the material vapors is measured by the mass loss rate. In the presence of a flame and/or external heat flux, the mass loss rate, under steady state, is expressed as[2,4,16]

$$\dot{m}'' = \frac{(\dot{q}''_e + \dot{q}''_{fr} + \dot{q}''_{fc} - \dot{q}''_{rr})}{\Delta H_g} \qquad (11)$$

where $\dot{m}''$ is the mass loss rate (g/m²-s), $\dot{q}''_{fr}$ is the flame radiative heat flux transferred to the surface (kW/m²), $\dot{q}''_{fc}$ is the flame convective heat flux transferred to the surface (kW/m²), $\dot{q}''_{rr}$ is the surface re-radiation loss (kW/m²), ΔH_g is the heat of gasification (kJ/g), and the total flame heat flux to the surface $\dot{q}''_f = \dot{q}''_{fr} + \dot{q}''_{fc}$.

According to Equation 11, the generation rate of material vapors is governed by the external and flame heat flux, surface re-radiation loss, and the heat of gasification.

Heat of gasification: The heat of gasification for a melting material is expressed as[27]

$$\Delta H_g = \int_{T_a}^{T_m} c_{p,s} \, dT + \Delta H_m + \int_{T_m}^{T_v} c_{p,l} \, dT + \Delta H_v \qquad (12)$$

where ΔH_g is the heat of gasification (kJ/g); $c_{p,s}$ and $c_{p,l}$ are the specific heats of the solid and molten solid in kJ/g-K, respectively; ΔH_m and ΔH_v are the heats of melting and

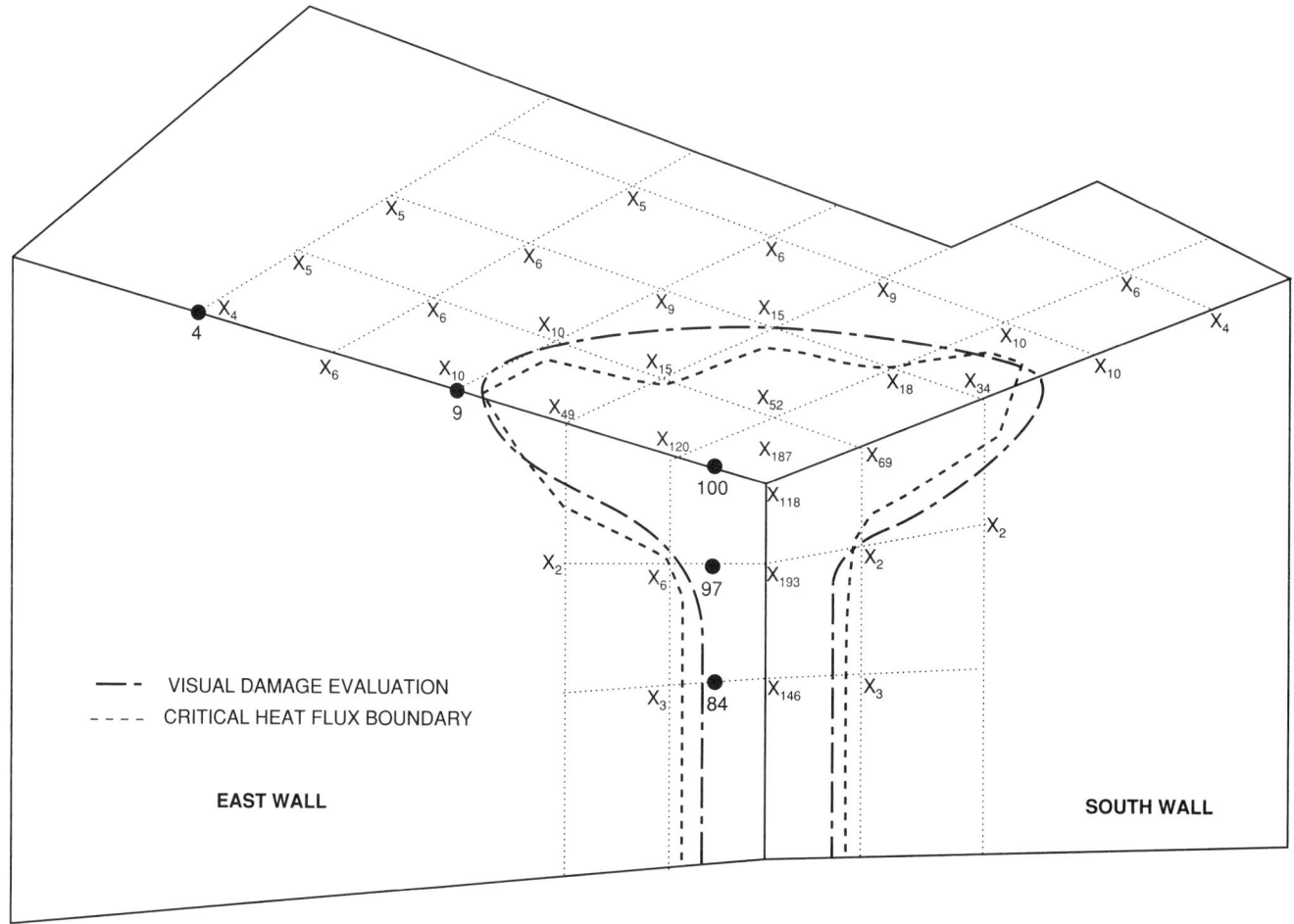

Fig. 3-4.16. *Critical Heat Flux boundary and visual observations for the extent of fire propagation in the FMRC 25-ft corner test for a product that passes the tests.*[24]

vaporization at the respective melting and vaporization temperatures in kJ/g; and T_a, T_m, and T_v are the ambient temperature, melting temperature, and vaporization temperature in K, respectively. For materials that do not melt, but sublime, decompose, or char, Equation 12 is modified accordingly. The heat of gasification can be determined from: (1) the parameters on the right-hand side of Equation 12, which can be quantified by the thermal analysis techniques or calculated from the properties listed in the literature; and (2) nonflaming tests using apparatuses, such as the OSU, the Flammability Apparatus, or the Cone Calorimeter. The following are some examples of the techniques:

1. **Heats of Gasification of Polymers from the Differential Scanning Calorimetry:** The $c_{p,s}$, $c_{p,l}$, ΔH_m, and ΔH_v values for polymers have been quantified in the FMRC laboratory.[27] The techniques involve measurement of the specific heat as a function of temperature, such as shown in Figure 3-4.18 for polymethylmethacrylate, measured in our Flammability Laboratory. The specific heat increases with temperature; a value close to the vaporization temperature of PMMA is used in Equation 12. Further measurements are made of the heats of melting and

vaporization. Some examples of the data measured in our laboratory are listed in Table 3-4.4.

2. **Heat of Gasification from the Literature Data for the Heats of Gasification for Various Molecular Weight Hydrocarbons (Alkanes):** The *CRC Handbook of Chemistry and Physics*'[28] listing for the heats of gasification for liquid and solid hydrocarbons (alkanes) satisfies the following relationship in the molecular weight range of 30 to 250 g/mole

$$\Delta H_g = -3.72 \times 10^{-6}M^2 + 0.0042 M + 0.164 \quad (13)$$

where M is the molecular weight of the hydrocarbon (g/mole).

The heats of gasification calculated from Equation 13 for various alkanes are listed in Table 3-4.4.

3. **Heat of Gasification from the Literature Data for the Specific Heats and Heats of Vaporization:** Water will be used as an example. The specific heat of liquid water, $c_{p,l} = 0.0042$ kJ/g-K,[29] and the heat of vaporization of water at 373 K is 2.26 kJ/g.[29] Assuming the ambient temperature to be 298 K and the vaporization temperature to be

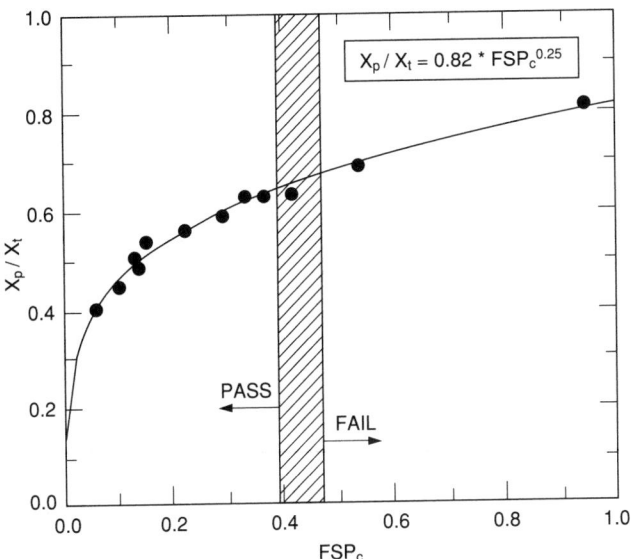

Fig. 3-4.17. Normalized fire propagation length measured in the 25-ft corner test versus the convective flame spread parameter obtained from the Flammability Apparatus. (Figure is taken from references 24 and 25.)

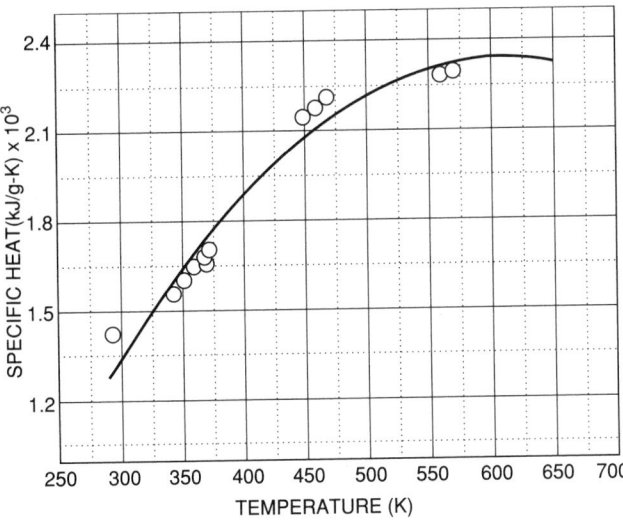

Fig. 3-4.18. Specific heat of polymethylmethacrylate versus temperature measured by a differential scanning calorimeter at the flammability laboratory of the Factory Mutual Research Corporation.

373 K, the heat of gasification of water from Equation 12 is calculated as follows

$$\int_{298}^{373} c_{p,l} \, dT = 0.0042(373 - 298) = 0.32 \text{ kJ/g};$$

$$\Delta H_{v,373} = 2.26 \text{ kJ/g};$$

$$\Delta H_g = 0.32 + 2.26 = 2.58 \text{ kJ/g}.$$

From the differential scanning calorimetry, the heat of gasification of water determined in the FMRC Flammability Laboratory is 2.59 kJ/g, which is in excellent agree-

ment with the calculated value. These two values for the heat of gasification of water are listed in Table 3-4.4.

4. **Heat of Gasification from the Nonflaming Tests Using The Flammability Apparatus:** The measurement for the heat of gasification from the nonflaming fire tests in the Flammability Apparatus was introduced in 1976.[27] In nonflaming fires, $\dot{q}_f'' = 0$, and Equation 11 becomes

$$\dot{m}'' = \frac{\dot{q}_e'' - \dot{q}_{rr}''}{\Delta H_g} \qquad (14)$$

where mass loss rate is a linear function of the external heat flux, and the heat of gasification is the inverse of the slope of the straight line. This provides a convenient method to determine the heat of gasification in the nonflaming tests, where mass loss rate of the sample is measured at various external heat flux values. The heat of gasification is determined from the linear regression analysis of the average steady-state mass loss rate as a function of the external heat flux. In the Flammability Apparatus tests, approximately 100- × 100-mm square and up to 100-mm-thick samples are used with co-flowing nitrogen or air with an oxygen mass fraction of about 0.10.

Figure 3-4.19 shows a plot of the vaporization rate of water in a 0.0072 m^2 Pyrex™ glass dish against time at 50 kW/m^2 of external heat flux, measured in the Flammability Apparatus. The figure also includes the predicted mass loss rate using Equation 14, where

$$\dot{q}_{rr}'' = \varepsilon\sigma(T_v^4 - T_a^4) \qquad (15)$$

where ε is the emissivity of water (0.95 to 0.963 in the temperature range 298 to 373 K),[30] and σ is the Stefan-Boltzmann constant (56.7 × 10^{-12} kW/m^2-deg^4). For water, $T_v = 373$ K and $T_a = 298$ K, and thus $\dot{q}_{rr}'' = 1$ kW/m^2. From Equation 14, using $\dot{q}_e'' = 50$ kW/m^2, $\dot{q}_{rr}'' = 1$ kW/m^2, and $\Delta H_g = 2.57$ kJ/g, $\dot{m}'' = 19.0$ g/m^2-s. There is excellent agreement between the measured and predicted values at the steady state in Figure 3-4.19. Water vaporization tests

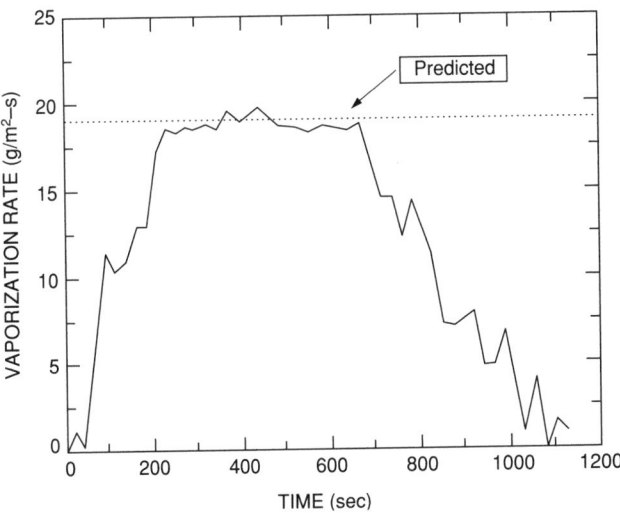

Fig. 3-4.19. Vaporization rate of water versus time measured in the Flammability Apparatus using 99.69 g of water in a Pyrex™ dish with an area of 0.0072 m^2. Water was exposed to an external heat flux of 50 kW/m^2.

TABLE 3-4.4 *Surface Re-radiation and Heats of Gasification of Various Materials*

Materials	Surface Re-radiation (kW/m²)	Heat of Gasification (kJ/g)			
		Flam. App.*	Cone†	DSC‡	Cal§
Distilled water	0.63	2.58	—	2.59	2.58
Hydrocarbons (Alkanes)					
Hexane	0.50	—	—	—	0.50
Heptane	0.63	—	—	—	0.55
Octane	0.98	—	—	—	0.60
Nonane	1.4	—	—	—	0.64
Decane	1.8	—	—	—	0.69
Undecane	2.3	—	—	—	0.73
Dodecane	2.8	—	—	—	0.77
Tridecane	3.0	—	—	—	0.81
Tetradecane	3.0	—	—	—	0.85
Hexadexane	3.0	—	—	—	0.92
Natural Materials					
Filter paper	10	3.6	—	—	—
Corrugated paper	10	2.2	—	—	—
Wood (Douglas fir)	10	1.8	—	—	—
Plywood/FR	10	1.0	—	—	—
Particleboard	—	—	3.9	—	—
Synthetic Materials					
Epoxy resin	—	—	2.4	—	—
Polypropylene	15	2.0	1.4	2.0	—
Polyethylene (PE) (low density)	15	1.8	—	1.9	—
PE (high density)	15	2.3	1.9	2.2	—
PE foams	12	1.4–1.7	—	—	—
PE/25% chlorine (Cl)	12	2.1	—	—	—
PE/36% Cl	12	3.0	—	—	—
PE/48% Cl	10	3.1	—	—	—
Rigid polyvinylchloride (PVC)	15	2.5	2.3	—	—
PVC/plasticizer	10	1.7	—	—	—
Plasticized PVC, LOI = 0.20	10	2.5	2.4	—	—
Plasticized PVC, LOI = 0.25	—	—	—	—	—
Plasticized PVC, LOI = 0.30	—	—	2.1	—	—
Plasticized PVC, LOI = 0.35	—	—	2.4	—	—
Rigid PVC, LOI = 0.50	—	—	2.3	—	—
Polyisoprene	10	2.0	—	—	—
PVC panel	17	3.1	—	—	—
Nylon 6/6	15	2.4	—	—	—
Polyoxymethylene (Delrin™)	13	2.4	—	2.4	—
Polymethylmethacrylate (Plexiglas™)	11	1.6	1.4	1.6	—
Polycarbonate	11	2.1	—	—	—
Polycarbonate panel	16	2.3	—	—	—
Isophthalic polyester	—	—	3.4	—	—
Polyvinyl ester	—	—	1.7	—	—
Acrylonitrile-butadiene-styrene (ABS)	10	3.2	2.6	—	—
Styrene-butadiene	10	2.7	—	—	—
Polystyrene (PS) foams	10–13	1.3–1.9	—	—	—
PS (granular)	13	1.7	2.2	1.8	—
Polyurethane (PU) foams					
Flexible polyurethane (PU) foams	16–19	1.2–2.7	2.4	1.4	—
Rigid polyurethane (PU) foams	14–22	1.2–5.3	5.6	—	—
Polyisocyanurate foams	14–37	1.2–6.4	—	—	—
Phenolic foam	20	1.6	—	—	—
Phenolic foam/FR	20	3.7	—	—	—
Ethylenetetrafluoroethylene (Tefzel™)	27	0.9	—	—	—
Fluorinated ethylene propylene, FEP (Teflon™)	38	2.4	—	—	—
Tetrafluoroethylene, TFE (Teflon™)	48	0.8–1.8	—	—	—
Perfluoroalkoxy, PFA (Teflon™)	37	1.0	—	—	—
Composite and Fiberglass-Reinforced Materials					
Polyether ether ketone—30% fiberglass	—	—	7.9	—	—
Polyethersulfone—30% fiberglass	—	1.8	—	—	—
Polyester 1—fiberglass	—	—	2.5	—	—
Polyester 2—fiberglass	10	1.4	—	—	—
Polyester 3—fiberglass	10	6.4	—	—	—
Polyester 4—fiberglass	15	5.1	—	—	—
Polyester 5—fiberglass	10	2.9	—	—	—
Phenolic fiberglass (thick sheet)	20	7.3	—	—	—
Phenolic Kevlar™ (thick sheet)	15	7.8	—	—	—

*From the Flammability Apparatus under nonflaming fire conditions.
†Calculated from the Cone Calorimeter data reported in references 20 and 31 for the mass loss rate at various external heat flux values in flaming fires.
‡From the flammability laboratory using the differential scanning calorimetry.
§Calculated from the data reported in the *CRC Handbook*.[28]

and calculations are routinely used for the calibration of the Flammability Apparatus.

Heats of gasification determined from the mass loss rate as a function of external heat flux in nonflaming fire conditions in the Flammability Apparatus are listed in Table 3-4.4 for selected materials. Excellent agreement can be noted between the heats of gasification determined from the Flammability Apparatus and those obtained from the differential scanning calorimetry.

Heat of gasification can also be determined from the flaming fires if high external heat flux values are used such that $\dot{q}_e'' \gg \dot{q}_{fr}'' + \dot{q}_{fc}'' - \dot{q}_{rr}''$ in Equation 11. This method has been used to calculate the heat of gasification from the Cone Calorimeter data for the mass loss rate in flaming fires reported in the literature.[20,31] The values calculated from the Cone Calorimeter data are also listed in Table 3-4.4 and show a general agreement with the values from the Flammability Apparatus.

EXAMPLE 6:

Estimate the ignition temperature of a material with a Critical Heat Flux of 11 kW/m^2. Assume its surface emissivity to be unity, ambient temperature to be 20°C, and vaporization temperature to be approximately equal to the ignition temperature.

SOLUTION:

From Equation 15,

$$11(\text{kW/m}^2) = 56.7 \times 10^{-12}(\text{kW/m}^2\text{-deg}^4)(T_v^4)(\text{deg})^4$$
$$- 56.7 \times 10^{-12}(\text{kW/m}^2\text{-deg}^4) \times (298)^4(\text{deg})^4$$

$$T_v \cong \left[\frac{11 \times 10^{12}}{56.7} + (298)^4 \right]^{1/4}$$

$$\cong (1940 \times 10^8 + 78.9 \times 10^8)^{1/4} \cong 670 \text{ K}$$

By assumption, vaporization temperature is equal to the ignition temperature, which is 670 K (397°C).

EXAMPLE 7:

A material with a surface re-radiation loss of 10 kW/m^2 and heat of gasification of 1.8 kJ/g was found to be involved in a fire with an exposed area of 2 m^2. The combined flame and external heat flux exposure to the material was estimated to be 70 kW/m^2. Estimate the peak mass loss rate at which the material may have been burning in the fire in terms of g/m^2-s and g/s.

SOLUTION:

From Equation 11,

$$\dot{m}'' = \frac{70 - 10}{1.8} = 33 \text{ g/m}^2\text{-s}$$

The estimated peak mass loss rate that the material may have been burning in the fire is 33 g/m^2-s, or 33 × 2 = 67 g/s.

Flame heat flux: For flaming fires, in the absence of external heat flux, from Equation 11

$$\dot{m}'' = \frac{(\dot{q}_{fr}'' + \dot{q}_{fc}'' - \dot{q}_{rr}'')}{\Delta H_g} \qquad (16)$$

The results from numerous small- and large-scale fire tests show that, as the surface area of the material increases, the

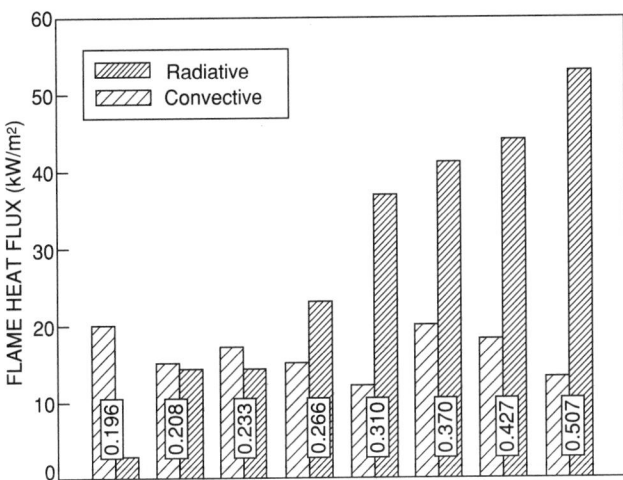

Fig. 3-4.20. *Flame radiative and convective heat fluxes at various oxygen mass fractions for the steady-state combustion of 100- × 100-mm square × 25-mm-thick slabs of polypropylene in the Flammability Apparatus under co-airflow velocity of 0.09 m/s. Data taken from reference 33. Mass fractions of oxygen are indicated by the numbers inside the frames.*

flame radiative heat flux increases and reaches an asymptotic limit, whereas the flame convective heat flux decreases and becomes much smaller than the flame radiative heat flux at the asymptotic limit in large-scale fires.[32] It is also known that, in small-scale fires of fixed size, with buoyant turbulent diffusion flames, as the oxygen mass fraction is increased, the flame radiative heat flux increases and reaches an asymptotic limit, comparable to the asymptotic limit in large-scale fires, whereas the flame convective heat flux decreases and becomes much smaller than the flame radiative heat flux.[33]

The effect of the mass fraction of oxygen on the flame radiative and convective heat fluxes in small-scale fires is shown in Figure 3-4.20 for 100- × 100-mm square × 25-mm-thick slabs of polypropylene. The data were measured in the Flammability Apparatus.[33] The increase in the flame radiative heat flux with increase in the mass fraction of oxygen is due to the increase in the flame temperature and soot formation and decrease in the residence time in the flame.[33] The oxygen mass fraction variation technique to simulate large-scale flame-radiative heat flux conditions in small-scale fires is defined as the *Flame Radiation Scaling Technique.*[4]

In the flame radiation scaling technique, the flame radiative and convective heat fluxes are determined from:[33] (1) the measurements for the mass loss rate at various oxygen mass fractions in the range of 0.12 (close to flame extinction) to about 0.60, under co-airflow conditions; (2) the convective heat transfer coefficient for the Flammability Apparatus, derived from the combustion of methanol; (3) the mass transfer number; and (4) Equation 16. In the Flammability Apparatus, the asymptotic limit is reached for the oxygen mass fraction ≥ 0.30. At the asymptotic limit, Equation 16 can be expressed as

$$\dot{m}_{asy}'' = \frac{\dot{q}_{f,asy}'' - \dot{q}_{rr}''}{\Delta H_g} \qquad (17)$$

where subscript *asy* represents the asymptotic limit. The asymptotic values for the mass loss rate and flame heat flux determined from the Flame Radiation Scaling Technique in

TABLE 3-4.5 *Asymptotic Values of Mass Loss Rate and Flame Heat Flux**

Material	$\dot{m}''$ (g/m²-s)		$\dot{q}_f''$ (kW/m²)	
	S*	L†	S*	L†
Aliphatic Carbon-Hydrogen Atoms‡				
Polyethylene	26	—	61	—
Polypropylene	24	—	67	—
Heavy fuel oil (2.6–23 m)	—	36	—	29
Kerosene (30–80 m)	—	65	—	29
Crude oil (6.5–31 m)	—	56	—	44
n-Dodecane (0.94 m)	—	36	—	30
Gasoline (1.5–223 m)	—	62	—	30
JP-4 (1.0–5.3 m)	—	67	—	40
JP-5 (0.60–17 m)	—	55	—	39
n-Heptane (1.2–10 m)	~66	75	32	37
n-Hexane (0.75–10 m)	—	77	—	37
Transformer fluids (2.37 m)	27–30	25–29	23–25	22–25
Aromatic Carbon-Hydrogen Atoms‡				
Polystyrene (0.93 m)	36	34	75	71
Xylene (1.22 m)	—	67	—	37
Benzene (0.75–6.0 m)	—	81	—	44
Aliphatic Carbon-Hydrogen-Oxygen Atoms‡				
Polyoxymethylene	16	—	50	—
Polymethylmethacrylate (2.37 m)	28	30	57	60
Methanol (1.2–2.4 m)	20	25	22	27
Acetone (1.52 m)	—	38	—	24
Aliphatic Carbon-Hydrogen-Oxygen-Nitrogen Atoms				
Flexible polyurethane foams	21–27	—	64–76	—
Rigid polyurethane foams	22–25	—	49–53	—
Aliphatic Carbon-Hydrogen-Halogen Atoms				
Polyvinylchloride	16	—	50	—
Tefzel™ (ETFE)	14	—	50	—
Teflon™ (FEP)	7	—	52	—

*S = Small-scale fires, pool diameter fixed at 0.10 m, flame radiation scaling technique was used in the Flammability Apparatus, $Y_0 \geq 0.30$.
†L = Large-scale fires in normal air.
‡ Numbers in m in parentheses are the pool diameters used in large-scale fires.
Note: Mass loss rates are from the data reported in the literature.

the Flammability Apparatus are listed in Table 3-4.5. The measured asymptotic values of the mass loss rate reported in the literature and flame heat flux in large-scale fires are also listed in Table 3-4.5. Flame heat flux values for the large-scale fires are derived from the asymptotic values of the mass loss rate and known values of surface re-radiation losses and heats of gasification.

The data in Table 3-4.5 show that the asymptotic flame heat flux values, determined in the Flammability Apparatus, using the Flame Radiation Scaling Technique, are in good agreement with the values measured in the large-scale fires. The asymptotic flame heat flux values vary from 22 to 77 kW/m², dependent primarily on the mode of decomposition and gasification rather than on the chemical structures of the materials. For example, for the liquids, which vaporize primarily as monomers or as very low molecular weight oligomer, the asymptotic flame heat flux values are in the range of 22 to 44 kW/m², irrespective of their chemical structures. For polymers, which vaporize as high molecular weight oligomer, the asymptotic flame heat flux values increase substantially to the range of 49 to 71 kW/m², irrespective of their chemical structures. The independence of the

asymptotic flame heat values from the chemical structures of materials is consistent with the dependence of flame radiation on optical thickness, soot concentration, and flame temperature in large-scale fires.

EXAMPLE 8:

Calculate the peak mass loss rate for polypropylene in large-scale fires, burning in the open, with no external heat sources in the surroundings.

SOLUTION:

In the calculation Equation 16 will be used. From Table 3-4.4, $\dot{q}_{rr}'' = 15$ kW/m² and $\Delta H_g = 2.0$ kJ/g and from Table 3-4.5, $\dot{q}_{f,asy}'' = 67$ kW/m². Using these values in Equation 16

$$\dot{m}'' = \frac{67 - 15}{2.0} = 26.0 \text{ g/m}^2\text{-s}$$

EXAMPLE 9:

Calculate the peak mass loss rate for polypropylene in large-scale fires, burning in the open, in the presence of a burning object, which provides 20 kW/m^2 of heat flux to the polypropylene surface, in addition to its own flame heat flux of 67 kW/m^2.

SOLUTION:

In the calculation Equation 11 will be used with $\dot{q}_e'' = 20$ kW/m^2. From Table 3-4.4, $\dot{q}_{rr}'' = 15$ kW/m^2 and $\Delta H_g = 2.0$ kJ/g and from Table 3-4.5, $\dot{q}_{f,asy}'' = 67$ kW/m^2. Using these values in Equation 11

$$\dot{m}'' = \frac{67 + 20 - 15}{2.0} = 36.0 \text{ g/m}^2\text{-s}$$

Heat Release Rate

The determination of heat release rate in fires has been influenced by the principles and techniques used for the controlled combustion in the heating and power industries. Heat in the flowing combustion products (convective heat) and thermal radiation are used to generate steam, heat a furnace or space, produce mechanical power in internal combustion engines or gas turbines, etc. Heat is generated by injecting fuel (gas, liquid, or solid) into a hot environment, where it undergoes evaporation, gasification, and thermal decomposition or pyrolysis. Fuel vapors react chemically with oxygen and produce heat and products, such as carbon monoxide (CO), carbon dioxide (CO$_2$), hydrocarbons, water (H$_2$O), and soot. Theoretical air requirement for complete combustion is estimated from an empirical guide, which suggests that, for every 10.6 kJ of heat in the fuel burned, 3.4 g of air are required for complete combustion.[34] Equivalently, the heat of combustion per unit mass of oxygen consumed (ΔH_0^*) is 13.4 kJ/g. Using $\Delta H_0^* \approx 13.4$ kJ/g to determine the heat release rate in fires from the mass consumption rate of oxygen is discussed in references 16 and 35. This technique is defined as the *Oxygen Consumption* (OC) *Calorimetry*.

A combustion process is characterized by its *combustion efficiency*, defined as the fraction of heat of complete combustion released in the chemical reactions, which is the ratio of the chemical heat release rate to the heat release rate for complete combustion or the ratio of the chemical heat of combustion to net heat of complete combustion. The calorific energy generated in chemical reactions leading to complete combustion per unit mass of fuel, water produced being in the vapor state, is defined as the *net heat of complete combustion*. The calorific energy generated in chemical reactions leading to varying degrees of incomplete combustion per unit mass of the fuel consumed is defined as the *chemical heat of combustion*. In the heating and power industries, combustion efficiency is determined routinely from the waste products (flue gas) analysis, especially for CO, CO$_2$, and O$_2$, and from the measurements of temperature in the combustion products-air mixture and thermal radiation. For higher combustion efficiency, *mass fuel-to-air ratio* relative to the *stoichiometric mass fuel-to-air ratio* or the *equivalence ratio* is controlled by maintaining desired primary and secondary airflow.

The net heat of complete combustion is measured in the oxygen bomb calorimeter and is calculated from the standard heats of formation of the material, CO$_2$ and H$_2$O, the standard heat of formation of O$_2$ in its standard state being zero. For example, for polymethylmethacrylate (PMMA) and polystyrene (PS), the net heats of complete combustion measured in the oxygen bomb calorimeter by the FMRC Flammability Laboratory are 25.3 and 39.2 kJ/g, respectively; and from the standard heats of formation, they are 24.9 and 39.8 kJ/g, respectively. For soot generated from the combustion of PMMA and PS, the net heats of complete combustion measured in the oxygen bomb calorimeter by the FMRC Flammability Laboratory are 33.9 and 32.1 kJ/g, respectively, and 32.8 kJ/g from the standard heats of formation of graphite and CO$_2$.

In fires, complete combustion is rarely achieved and products of incomplete combustion, such as CO and smoke, are quite common. An example of incomplete combustion is given in Table 3-4.6, where chemical heat of combustion and combustion efficiency decrease as CO, carbon, and ethylene are formed at the expense of CO$_2$ and H$_2$O with reduced O$_2$ consumption, a typical condition found in the ventilation-controlled fires.[36] The chemical heat of combustion is the ratio of the chemical heat release rate to the mass loss rate. The upper limit of the combustion efficiency is 1.00, corresponding to complete combustion, and the lower limit is 0.46, corresponding to unstable combustion leading to flame extinction for combustion efficiency ≤ 0.40.[36,37]

Chemical heat release rate: The chemical heat release rate is determined from the *Carbon Dioxide Generation* (CDG) and *Oxygen Consumption* (OC) *Calorimetries*.

The CDG Calorimetry:[2-4,16,33,36] The chemical heat release rate is determined from the following relationships

$$\dot{Q}_{ch}'' = \Delta H_{CO_2}^* \dot{G}_{CO_2}'' + \Delta H_{CO}^* \dot{G}_{CO}'' \tag{18}$$

$$\Delta H_{CO_2}^* = \frac{\Delta H_T}{\Psi_{CO_2}} \tag{19}$$

$$\Delta H_{CO}^* = \frac{\Delta H_T - \Delta H_{CO} \Psi_{CO}}{\Psi_{CO}} \tag{20}$$

where $\dot{Q}_{ch}''$ is the chemical heat release rate (kW/m^2), $\Delta H_{CO_2}^*$ is the net heat of complete combustion per unit mass of CO$_2$ generated (kJ/g), ΔH_{CO}^* is the net heat of complete combustion per unit mass of CO generated (kJ/g), ΔH_T is the net heat of complete combustion per unit mass of fuel consumed (kJ/g), Ψ_{CO_2} is the stoichiometric yield for the maximum conversion of fuel to CO$_2$(g/g), Ψ_{CO} is the stoichiometric yield for the maximum conversion of fuel to CO (g/g), $\dot{G}_{CO_2}''$

TABLE 3-4.6 *Chemical Heat of Combustion and Combustion Efficiency of Polymethylmethacrylate*

Reaction Stoichiometry	ΔH_{ch} (kJ/g)*	χ_{ch}
C$_5$H$_8$O$_2$ (g) + 6.0 O$_2$ (g) = 5CO$_2$ (g) + 4H$_2$O (g)	24.9	1.00
C$_5$H$_8$O$_2$ (g) + 5.5 O$_2$ (g) = 4CO$_2$ (g) + 4H$_2$O (g) + CO (g)	22.1	0.89
C$_5$H$_8$O$_2$ (g) + 4.5 O$_2$ (g) = 3CO$_2$ (g) + 4H$_2$O (g) + CO (g) + C (s)	18.2	0.73
C$_5$H$_8$O$_2$ (g) + 3.0 O$_2$ (g) = 2CO$_2$ (g) + 3H$_2$O (g) + CO (g) + C (s) + 0.50 C$_2$H$_4$ (g)	11.5	0.46

*Standard heat of formation in kJ/mole: PMMA (C$_5$H$_8$O$_2$) (g) = -442.7; O$_2$ (g) = 0; CO$_2$ (g) = -393.5; H$_2$O (g) = -241.8; CO (g) = -110.5; C (s) = 0; and C$_2$H$_4$ (g) = $+26.2$, where g is the gas and s is the solid.

TABLE 3-4.7 *Net Heats of Complete Combustion per Unit Mass of Fuel and Oxygen Consumed and Carbon Dioxide and Carbon Monoxide Generated for Carbon- and Hydrogen-Containing Fuels**

Fuel	Formula	ΔH_T (kJ/g)	ΔH_O^* (kJ/g)	$\Delta H_{CO_2}^*$ (kJ/g)	ΔH_{CO}^* (kJ/g)	Fuel	Formula	ΔH_T (kJ/g)	ΔH_O^* (kJ/g)	$\Delta H_{CO_2}^*$ (kJ/g)	ΔH_{CO}^* (kJ/g)
Normal Alkanes						**Normal Alkenes (Continued)**					
Methane	CH_4	50.1	12.5	(18.2)	(18.6)	Heptene	C_7H_{14}	44.6	12.9	14.1	12.2
Ethane	C_2H_6	47.1	12.7	16.2	15.4	Octene	C_8H_{16}	44.5	12.9	14.1	12.1
Propane	C_3H_8	46.0	12.9	15.3	14.0	Nonene	C_9H_{18}	44.3	12.9	14.1	12.1
Butane	C_4H_{10}	45.4	12.7	15.1	13.7	Decene	$C_{10}H_{20}$	44.2	12.9	14.1	12.2
Pentane	C_5H_{12}	45.0	12.6	14.7	13.2	Dodecene	$C_{12}H_{24}$	44.1	12.9	14.1	12.2
Hexane	C_6H_{14}	44.8	12.7	14.6	12.9	Tridecene	$C_{13}H_{26}$	44.0	12.9	14.1	12.2
Heptane	C_7H_{16}	44.6	12.7	14.5	12.8	Tetradecene	$C_{14}H_{28}$	44.0	12.9	14.1	12.2
Octane	C_8H_{18}	44.5	12.6	14.4	12.7	Hexadecene	$C_{16}H_{32}$	43.9	12.9	14.1	12.1
Nonane	C_9H_{20}	44.3	12.7	14.3	12.5	Octadecene	$C_{18}H_{36}$	43.8	12.9	14.1	12.1
Decane	$C_{10}H_{22}$	44.4	12.7	14.3	12.4						
Undecane	$C_{11}H_{24}$	44.3	12.7	14.3	12.4		**Average**		**13.2**	**14.2**	**12.4**
Dodecane	$C_{12}H_{26}$	44.2	12.7	14.2	12.3	**Cyclic Alkenes**					
Tridecane	$C_{13}H_{28}$	44.2	12.7	14.2	12.3	Cyclohexene	C_6H_{10}	43.0	13.0	13.4	11.0
Kerosene	$C_{14}H_{30}$	44.1	12.7	14.1	12.2	Methylcyclohexene	C_7H_{12}	43.1	12.9	13.4	11.1
Hexadecane	$C_{16}H_{34}$	44.1	12.7	14.2	12.3						
	Average		**12.7**	**14.6**	**12.9**		**Average**		**13.0**	**13.4**	**11.1**
Substituted Alkanes						**Dienes**					
Methylbutane	C_5H_{12}	45.0	12.6	14.7	13.1	1-3 Butadiene	C_4H_6	44.6	13.7	13.7	11.5
Dimethylbutane	C_6H_{14}	44.8	12.7	14.6	13.0	Cyclooctadiene	C_8H_{12}	43.2	13.3	13.3	10.9
Methylpentane	C_6H_{14}	44.8	12.7	14.6	12.9						
Dimethylpentane	C_7H_{16}	44.6	12.7	14.5	12.9		**Average**		**13.5**	**13.5**	**11.2**
Methylhexane	C_7H_{16}	44.6	12.6	14.4	12.7	**Normal Alkynes**					
Isooctane	C_8H_{18}	44.5	12.6	14.4	12.7	Acetylene	C_2H_2	47.8	(15.6)	14.3	12.4
Methylethylpentane	C_8H_{18}	44.5	12.6	14.4	12.7	Heptyne	C_7H_{12}	44.8	13.4	13.9	11.8
Ethylhexane	C_8H_{18}	44.5	12.6	14.4	12.7	Octyne	C_8H_{14}	44.7	13.3	14.0	11.9
Dimethylhexane	C_8H_{18}	44.5	12.7	14.5	12.8	Decyne	$C_{10}H_{18}$	44.5	13.2	13.9	11.9
Methylheptane	C_8H_{18}	44.5	12.6	14.4	12.7	Dodecyne	$C_{12}H_{22}$	44.3	13.2	14.0	12.0
	Average		**12.6**	**14.6**	**12.8**		**Average**		**13.3**	**14.0**	**12.0**
Cyclic Alkanes						**Arenes**					
Cyclopentane	C_5H_{10}	44.3	12.8	13.9	11.9	Benzene	C_6H_6	40.1	13.0	11.9	8.7
Methylcyclopentane	C_6H_{12}	43.8	12.7	13.9	11.9	Toluene	C_7H_8	39.7	12.9	12.1	9.0
Cyclohexane	C_6H_{12}	43.8	12.7	13.8	11.7	Styrene	C_8H_8	39.4	13.1	12.0	8.8
Methylcyclohexane	C_7H_{14}	43.4	12.7	13.8	11.7	Ethylbenzene	C_8H_{10}	39.4	12.9	12.3	9.4
Ethylcyclohexane	C_8H_{16}	43.2	12.7	13.8	11.7	Xylene	C_8H_{10}	39.4	13.0	12.4	9.5
Dimethylcyclohexane	C_8H_{16}	43.2	12.7	13.8	11.7	Propylbenzene	C_9H_{12}	39.4	12.9	12.5	9.6
Cyclooctane	C_8H_{16}	43.2	12.7	13.9	11.9	Trimethylbenzene	C_9H_{12}	39.2	12.9	12.5	9.7
Decalin	$C_{10}H_{18}$	42.8	12.7	13.4	11.0	Cumene	C_9H_{12}	39.2	12.9	12.9	9.6
Bicyclohexyl	$C_{12}H_{22}$	42.6	12.6	13.3	11.0	Naphthalene	$C_{10}H_8$	39.0	12.9	11.3	7.7
	Average		**12.7**	**13.8**	**11.6**	Tetralin	$C_{10}H_{12}$	39.0	12.9	12.2	9.2
Normal Alkenes						Butylbenzene	$C_{10}H_{14}$	39.0	12.9	12.7	9.9
						Diethylbenzene	$C_{10}H_{14}$	39.0	13.7	13.5	11.1
Ethylene	C_2H_4	48.0	13.8	15.0	13.6	p-Cymene	$C_{10}H_{14}$	39.0	13.0	12.5	9.6
Propylene	C_3H_6	46.4	13.4	14.6	12.9	Methylnaphthalene	$C_{11}H_{10}$	38.9	12.9	11.5	8.1
Butylene	C_4H_8	45.6	14.3	14.3	12.5	Pentylbenzene	$C_{11}H_{16}$	38.8	13.0	12.8	10.2
Pentene	C_5H_{10}	45.2	14.3	14.3	12.5	Triethylbenzene	$C_{12}H_{18}$	38.7	12.7	12.7	10.0
Hexene	C_6H_{12}	44.9	12.9	14.1	12.2		**Average**		**13.0**	**12.4**	**9.4**

*Data from references 38 and 39. Numbers in parentheses not used for averaging.

is the generation rate of CO_2 (g/m^2-s), and $\dot{G}_{CO}''$ is the generation rate of CO (g/m^2-s).

The values for the net heats of complete combustion per unit mass of fuel consumed and CO_2 and CO generated are listed in Tables 3-4.7 through 3-4.10. The values depend on the chemical structures of the materials. With some exceptions, the values remain approximately constant within each generic group of fuels. The average values are also listed in the tables. From the average values, $\Delta H_{CO_2}^* = 13.3$ kJ/g $\pm$ 11 percent, and $\Delta H_{CO}^* = 11.1$ kJ/g $\pm$ 18 percent. In the CDG calorimetry, the CO correction for well-ventilated fires is very small, because of the small amounts of CO generated. The variations of 11 and 18 percent in the $\Delta H_{CO_2}^*$ and ΔH_{CO}^* values, respectively, would reduce significantly if values for low molecular weight hydrocarbons with small amounts of O, N, and halogen were used in averaging.

TABLE 3-4.8 *Net Heats of Complete Combustion per Unit Mass of Fuel and Oxygen Consumed and Carbon Dioxide and Carbon Monoxide Generated for Carbon-, Hydrogen-, and Oxygen-Containing Fuels**

Fuel	Formula	ΔH_T (kJ/g)	ΔH_O^* (kJ/g)	$\Delta H_{CO_2}^*$ (kJ/g)	ΔH_{CO}^* (kJ/g)	Fuel	Formula	ΔH_T (kJ/g)	ΔH_O^* (kJ/g)	$\Delta H_{CO_2}^*$ (kJ/g)	ΔH_{CO}^* (kJ/g)
Alcohols						**Acids**					
Methyl alcohol	CH_4O	20.0	13.4	14.5	12.9	Formic acid	CH_2O_2	5.7	16.4	5.96	0
Ethyl alcohol	C_2H_6O	27.7	13.2	14.5	12.7	Acetic acid	$C_2H_4O_2$	14.6	13.7	9.95	5.65
n-Propyl alcohol	C_3H_8O	31.8	13.3	14.5	12.7	Benzoic acid	$C_7H_6O_2$	24.4	12.4	9.66	5.18
Isopropyl alcohol	C_3H_8O	31.8	13.3	14.5	12.7	Cresylic acid	$C_8H_8O_2$	34.0	(16.0)	13.1	10.6
Allyl alcohol	C_3H_6O	31.4	14.2	13.8	11.7	**Esters**					
n-Butyl alcohol	$C_4H_{10}O$	34.4	13.3	14.5	12.8	Ethyl formate	$C_3H_6O_2$	20.2	13.3	11.3	7.8
Isobutyl alcohol	$C_4H_{10}O$	34.4	13.3	14.5	12.8	n-Propyl formate	$C_4H_8O_2$	23.9	13.2	12.0	8.8
Sec-butyl alcohol	$C_4H_{10}O$	34.4	13.3	14.5	12.8	n-Butyl formate	$C_5H_{10}O_2$	26.6	13.0	12.3	9.4
Ter-butyl alcohol	$C_4H_{10}O$	34.4	13.3	14.5	12.8	Methyl acetate	$C_3H_6O_2$	20.2	13.3	11.3	7.8
n-Amyl alcohol	$C_5H_{12}O$	36.2	13.3	14.5	12.8	Ethyl acetate	$C_4H_8O_2$	23.9	13.2	12.0	8.8
Isobutyl carbinol	$C_5H_{12}O$	36.2	13.3	14.5	12.8	n-Propyl acetate	$C_5H_{10}O_2$	26.6	13.0	12.3	9.4
Sec-butyl carbinol	$C_5H_{12}O$	36.2	13.3	14.5	12.8	n-Butyl acetate	$C_6H_{12}O_2$	28.7	13.0	12.6	9.8
Methylpropylcarbinol	$C_5H_{12}O$	36.2	13.3	14.5	12.8	Isobutyl acetate	$C_6H_{12}O_2$	28.7	13.0	12.6	9.8
Dimethylethylcarbinol	$C_5H_{12}O$	36.2	13.3	14.5	12.8	Amyl acetate	$C_7H_{14}O_2$	30.3	13.0	12.8	10.1
n-Hexyl Alcohol	$C_6H_{14}O$	37.4	13.3	14.5	12.7	Cyclohexyl acetate	$C_8H_{14}O_2$	31.5	13.3	12.7	10.0
Dimethylbutylalcohol	$C_6H_{14}O$	37.4	13.3	14.5	12.7	Octyl acetate	$C_{10}H_{20}O_2$	33.6	12.9	13.1	10.6
Ethylbutyl alcohol	$C_6H_{14}O$	37.4	13.3	14.5	12.7	Ethylacetoacetate	$C_6H_{10}O_3$	30.3	(17.6)	(14.9)	(13.5)
Cyclohexanol	$C_6H_{12}O$	37.3	13.7	14.1	12.2	Methyl propionate	$C_4H_8O_2$	23.9	13.2	12.0	7.4
Benzyl alcohol	C_7H_8O	32.4	13.0	11.4	8.0	Ethyl propionate	$C_5H_{10}O_2$	26.6	13.0	12.3	9.4
n-Heptyl alcohol	$C_7H_{16}O$	39.8	13.7	15.0	13.6	n-Butyl propionate	$C_7H_{14}O_2$	30.3	13.0	12.8	10.1
n-Octyl alcohol	$C_8H_{18}O$	40.6	13.7	15.0	13.6	Isobutyl propionate	$C_7H_{14}O_2$	30.3	13.0	12.8	10.1
n-Nonyl alcohol	$C_9H_{20}O$	40.3	13.4	14.7	13.0	Amyl propionate	$C_8H_{18}O_2$	31.6	12.9	12.9	10.3
	Average		**13.3**	**14.5**	**12.8**	Methyl butyrate	$C_5H_{10}O_2$	26.6	13.0	12.3	9.4
Aldehydes						Ethyl butyrate	$C_6H_{12}O_2$	28.7	13.0	12.6	9.8
						Propyl butyrate	$C_7H_{14}O_2$	30.3	13.0	12.8	10.1
Formaldehyde	CH_2O	18.7	(17.5)	12.7	10.1	n-Butyl butyrate	$C_8H_{16}O_2$	31.6	12.9	12.9	10.3
Acetaldehyde	C_2H_4O	25.1	13.8	12.6	9.7	Isobutyl butyrate	$C_8H_{16}O_2$	31.6	12.9	12.9	10.3
Butyraldehyde	C_4H_8O	33.8	13.9	13.9	11.7	Ethyl laurate	$C_{14}H_{28}O_2$	37.2	13.3	13.8	11.6
Crotonaldehyde	C_4H_6O	34.8	15.2	13.8	11.8	Ethyl lactate	$C_5H_{10}O_3$	30.8	(18.9)	(16.5)	(16.0)
Benzaldehyde	C_7H_6O	32.4	13.4	11.2	7.5	Butyl lactate	$C_7H_{14}O_3$	33.3	(16.8)	(15.8)	(14.8)
Ethyl hexaldehyde	$C_8H_{16}O$	39.4	13.7	12.7	9.9	Amyl lactate	$C_8H_{16}O_3$	34.3	(16.4)	(15.6)	(14.5)
	Average		**14.2**	**13.3**	**10.6**	Ethyl benzoate	$C_9H_{10}O_2$	34.5	(15.4)	13.1	10.5
Ketones						Ethyl carbonate	$C_5H_{10}O_3$	30.8	(18.9)	(16.5)	(16.0)
						Ethyl oxalate	$C_4H_6O_4$	28.7	(20.2)	(16.6)	(20.2)
Acetone	C_3H_6O	29.7	13.4	13.1	10.5	Ethyl malonate	$C_5H_8O_4$	32.2	(17.9)	(19.3)	(20.4)
Methylethyl ketone	C_4H_8O	32.7	13.4	13.4	11.0		**Average**		**13.0**	**12.5**	**9.7**
Diethyl ketone	$C_5H_{10}O$	33.7	12.9	13.2	10.7	**Others**					
Cyclohexanone	$C_6H_{10}O$	35.9	13.8	13.3	11.0	Camphor	$C_{10}H_{16}O$	38.8	13.7	13.4	11.1
Methyl butyl ketone	$C_6H_{12}O$	35.2	12.9	13.3	11.0	Cresol	C_7H_8O	34.6	13.7	12.1	9.1
Di-acetone alcohol	$C_6H_{12}O_2$	37.3	(16.9)	(16.4)	(15.7)	Resorcinol	$C_6H_6O_2$	26.0	13.7	10.8	5.9
Dipropyl ketone	$C_7H_{14}O$	38.6	13.8	14.3	12.5	Acrolein	C_3H_4O	29.1	14.6	12.3	9.4
Phenylbutyl ketone	$C_{11}H_{14}O$	34.8	12.6	11.6	(8.4)						
	Average		**13.2**	**13.2**	**11.1**						

*Data from references 38 and 39. Numbers in parentheses not used for averaging.

For the determination of the chemical heat release rate, generation rates of CO_2 and CO are measured and either the actual values or the average values of the net heat of complete combustion per unit mass of CO_2 and CO generated are used. The measurements for the generation rates of CO_2 and CO are described in the subsection entitled "Generation Rates of Chemical Compounds and Fire Ventilation."

The OC Calorimetry:[2–4,17–19,33,35,36] The chemical heat release rate is determined from the following relationship

$$\dot{Q}_{ch}'' = \Delta H_O^* \dot{C}_O''$$ (21)

$$\Delta H_O^* = \frac{\Delta H_T}{\Psi_O}$$ (22)

where ΔH_O^* is the net heat of complete combustion per unit mass of oxygen consumed (kJ/g), $\dot{C}_O''$ is the mass consumption rate of oxygen (g/m²-s), and Ψ_O is the stoichiometric mass-oxygen-to-fuel ratio (g/g).

The values for the net heats of complete combustion per unit mass of oxygen consumed are listed in Tables 3-4.7 through 3-4.10 along with the values for the net heats of complete combustion per unit mass of fuel consumed and CO_2 and CO generated. The average values of the net heat of

TABLE 3-4.9 *Net Heats of Complete Combustion per Unit Mass of Fuel and Oxygen Consumed and Carbon Dioxide and Carbon Monoxide Generated for Carbon-, Hydrogen-, Nitrogen-, and Sulfur-Containing Fuels**

Fuel	Formula	ΔH_T (kJ/g)	ΔH_O^* (kJ/g)	$\Delta H_{CO_2}^*$ (kJ/g)	ΔH_{CO}^* (kJ/g)
C–H–N Fuels					
Acrylonitrile	C_3H_3N	24.5	8.5	9.8	5.4
Diethylamine	$C_4H_{11}N$	38.0	11.2	15.8	14.8
n-Butylamine	$C_4H_{11}N$	38.0	11.2	15.8	14.8
sec-Butylamine	$C_4H_{11}N$	38.0	11.2	15.8	14.8
Pyridine	C_5H_9N	32.2	11.0	11.6	8.2
Aniline	C_6H_7N	33.8	11.2	11.9	8.7
Picoline	C_6H_7N	33.8	11.2	11.9	8.7
Triethylamine	$C_6H_{15}N$	39.6	11.6	15.2	13.8
Toluidine	C_7H_9N	34.9	11.3	12.1	9.1
Dimethylaniline	$C_8H_{11}N$	35.7	11.5	12.3	9.3
Di-n-butylamine	$C_8H_{19}N$	40.6	11.9	14.9	13.4
Quinoline	C_9H_7N	36.1	12.4	11.8	8.5
Quinaldine	$C_{10}H_9N$	36.7	12.4	11.9	8.7
Butylaniline	$C_{10}H_{15}N$	37.0	11.7	12.5	9.7
Tri-n-butylamine	$C_{12}H_{27}N$	41.6	12.1	14.6	12.9
	Average		**11.5**	**15.4**	**14.1**
C–H–S Fuels					
Carbon disulfide	CS_2	13.6	10.8	(23.5)	(27.0)
Thiophene	C_4H_4S	31.9	14.0	15.2	14.0
Methylthiophene	C_5H_6S	33.2	13.6	14.8	13.2
Thiophenol	C_6H_6S	34.1	13.8	14.2	12.3
Hexyl mercaptan	$C_6H_{14}S$	33.0	11.6	14.8	13.2
Thiocresol	C_7H_8S	34.9	13.5	14.1	12.1
Heptyl mercaptan	$C_7H_{16}S$	33.7	11.6	14.4	12.7
Cresolmethylsulfide	$C_8H_{11}S$	36.2	13.4	15.9	15.0
Decylmercaptan	$C_{10}H_{22}S$	34.9	11.5	13.8	11.7
Dodecyl mercaptan	$C_{12}H_{26}S$	35.5	11.5	13.6	11.4
Hexyl sulfide	$C_{12}H_{26}S$	35.5	11.5	13.6	11.4
Heptyl sulfide	$C_{14}H_{30}S$	35.9	11.5	13.4	11.1
Octyl sulfide	$C_{16}H_{34}S$	36.3	11.5	13.3	10.9
Decyl sulfide	$C_{20}H_{42}S$	36.8	11.4	13.1	10.7
	Average		**11.3**	**13.1**	**11.5**

*Data from references 38 and 39. Numbers in parentheses not used for averaging.

complete combustion per unit mass of oxygen consumed are also listed in the tables. The values depend on the chemical structures of the materials. With some exceptions, the values remain approximately constant within each generic group of fuels. From the average values, $\Delta H_O^* = 12.8$ kJ/g $\pm$ 7 percent. The ΔH_O^* value of 12.8 kJ/g is close to 13.4 kJ/g used in the heating and power industries[34] and 13.1 kJ/g $\pm$ 5 percent reported in reference 35. The variation of 7 percent would reduce significantly if values for low molecular weight hydrocarbons with small amounts of O, N, and halogen were used in averaging.

For the determination of the chemical heat release rate, mass consumption rate of oxygen is measured, and either the actual values or the average values of the net heats of complete combustion per unit mass of oxygen consumed are used. The measurement for the consumption rate of oxygen is described in the subsection entitled "Generation Rates of Chemical Compounds and Fire Ventilation."

Convective heat release rate: The convective heat release rate is determined from the *Gas Temperature Rise* (GTR) *Calorimetry*, where the following relationship is used[2-7,16,33,36]

$$\dot{Q}''_{con} = \frac{\dot{W}c_P(T_g - T_a)}{A} \qquad (23)$$

where $\dot{Q}''_{con}$ is the convective heat release rate (kW/m²), c_P is the specific heat of the combustion product-air mixture at the gas temperature (kJ/g-K), T_g is the gas temperature (K), T_a is ambient temperature (K), $\dot{W}$ is the total mass flow rate of the fire product-air mixture (g/s), and A is the total exposed surface area of the material (m²).

Radiative heat release rate: Chemical heat release rate consists of a convective and a radiative component. Some fraction of the chemical heat release rate may be lost as conductive heat. In systems where heat losses are negligibly small, the radiative heat release rate can be obtained from the difference between the chemical and convective heat release rates[2-4,16,33,36]

$$\dot{Q}''_{rad} = \dot{Q}''_{ch} - \dot{Q}''_{con} \qquad (24)$$

where $\dot{Q}''_{rad}$ is the radiative heat release rate (kW/m²).

Use of GTR, CDG, and OC calorimetries: In 1972 the GTR calorimetry was used for the first time by the Ohio State University (OSU) to determine the heat release rate.[5,6] The apparatus is now known as the OSU Heat Release Rate Apparatus; it is shown in Figure 3-4.1. The OSU Apparatus is an ASTM[7] and an FAA standard test apparatus.[8] In the GTR calorimetry, it is assumed that almost all the thermal radiation from the flame is transferred to the flowing fire products-air mixture, as the flames are inside an enclosed space and heat loss by conductive heat transfer is negligibly small. The OC calorimetry has now been adapted to the OSU apparatus.[41]

The CDG, OC, and GTR Calorimetries were used for the first time during the mid-1970s by the Factory Mutual Research Corporation (FMRC) to determine the chemical, convective, and radiative heat release rates.[27,42-44] The apparatus used is now known as the Flammability Apparatus (50-kW scale); it is shown in Figure 3-4.2(a). Heat release rate from the CDG and OC calorimetries in the Flammability Apparatus was defined as the actual heat release rate until 1986,[16,33,38,42-45] but after 1986 it was changed to the chemical heat release rate to account for the effects of: (1) the chemical structures of the materials and additives; (2) fire ventilation; (3) the two dominant modes of heat release, i.e., convective and radiative; and (4) the effects of the flame extinguishing and suppressing agents.

The Flammability Apparatus is a standard test apparatus for electrical cables;[14] for wall and ceiling insulation materials, replacing the 25-ft corner test;[26] and is expected to be adopted as a standard test apparatus for conveyor belts, composites, sample storage commodities, and other applications related to the commercial and industrial fire protection needs in the future.

In 1982 the National Institute of Standards and Technology (NIST) used the OC calorimetry,[17,18] following the methodology described in reference 35. The apparatus developed to use this methodology, known as the Cone Calorimeter, is shown in Figure 3-4.3. The Cone Calorimeter became an ASTM standard test apparatus in 1990.[19]

Sampling ducts have been designed for the Flammability Apparatus and the Cone Calorimeter to measure the mass generation rates of CO_2 and CO and mass consumption rate

TABLE 3-4.10 *Net Heats of Complete Combustion per Unit Mass of Fuel and Oxygen Consumed and Carbon Dioxide and Carbon Monoxide Generated for Polymeric Materials**

Fuel	Formula[†]	$\Delta H_T{}^{\ddagger}$ (kJ/g)	$\Delta H_O{}^{*}$ (kJ/g)	$\Delta H_{CO_2}{}^{*}$ (kJ/g)	$\Delta H_{CO}{}^{*}$ (kJ/g)	Fuel	Formula[†]	$\Delta H_T{}^{\ddagger}$ (kJ/g)	$\Delta H_O{}^{*}$ (kJ/g)	$\Delta H_{CO_2}{}^{*}$ (kJ/g)	$\Delta H_{CO}{}^{*}$ (kJ/g)
Carbon-Hydrogen Atoms in the Structure						**Carbon-Hydrogen-Oxygen-Nitrogen Atoms in the Structure (Continued)**					
Polyethylene	CH_2	43.6	12.8	13.9	11.8	*Rigid polyurethane foams*					
Polypropylene	CH	43.4	12.7	13.8	11.7	GM29	$CH_{1.1}O_{0.23}N_{0.10}$	26.0	12.6	10.7	6.8
Polyisobutylene	CH_2	43.7	12.7	13.9	11.9	GM31	$CH_{1.2}O_{0.22}N_{0.10}$	25.0	11.9	10.2	6.1
Polybutadiene	$CH_{1.5}$	42.8	13.1	13.1	10.7	GM37	$CH_{1.2}O_{0.20}N_{0.08}$	28.0	12.7	11.2	7.5
Polystyrene	CH	39.2	12.7	12.2	9.2	*Rigid polyisocyanurate foams*					
Expanded polystyrene						GM41	$CH_{1.0}O_{0.19}N_{0.11}$	26.2	12.5	10.4	6.4
GM47	$CH_{1.1}$	38.1	12.4	11.3	7.7	GM43	$CH_{0.93}O_{0.20}N_{0.11}$	22.2	10.8	8.9	(4.0)
GM49	$CH_{1.1}$	38.1	12.4	11.3	7.7	**Average**			12.5	10.9	7.2
GM51	CH	35.6	11.6	10.8	7.0	**Carbon-Hydrogen-Chlorine Atoms in the Structure**					
GM53	$CH_{1.1}$	37.6	12.4	11.3	7.7	*Polyethylene with*					
Average			12.5	12.4	9.5	25% chlorine	$CH_{1.9}Cl_{0.13}$	31.6	12.7	13.4	10.8
Carbon-Hydrogen-Oxygen-Nitrogen Atoms in the Structure						36% chlorine	$CH_{1.8}Cl_{0.22}$	26.3	12.8	12.9	10.2
Polyoxymethylene	CH_2O	15.4	14.4	10.5	6.6	48% chlorine	$CH_{1.7}Cl_{0.36}$	20.6	12.8	12.3	9.4
Polymethyl- methacrylate	$CH_{1.6}O_{0.4}$	25.2	13.1	11.5	8.0	Polychloroprene	$CH_{1.3}Cl_{0.30}$	25.3	13.3	12.7	9.5
Polyester	$CH_{1.4}O_{0.22}$	32.5	13.9	12.5	9.6	Polyvinylchloride	$CH_{1.5}Cl_{0.50}$	16.4	11.7	11.7	8.2
Epoxy	$CH_{1.3}O_{0.20}$	28.8	12.1	10.8	6.9	Polyvinyl- idenechloride	$CHCl$	9.0	13.5	9.8	(5.5)
Polycarbonate	$CH_{0.88}O_{0.19}$	29.7	13.1	10.7	6.9	**Average**			12.8	12.1	9.6
Cellulose triacetate	$CH_{1.3}O_{0.67}$	17.6	13.3	9.6	5.1	**Carbon-Hydrogen-Fluorine Atoms in the Structure**§					
Polyethylene- terephthalate	$CH_{0.80}O_{0.40}$	22.0	13.2	9.6	5.1	Teflon TFE	CF_2	6.2	9.7	(7.1)	(1.1)
Rigid phenolic foam	$CH_{1.1}O_{0.24}$	36.4	(16.8)	(14.0)	(12.0)	Teflon FEP	$CF_{1.8}$	4.8	(6.9)	(5.0)	(0)
Polyacrylonitrile (PAN)	$CHN_{0.33}$	30.8	10.7	12.3	9.4	Tefzel ETFE	CHF	12.6	12.6	9.2	(4.4)
Red oak	$CH_{1.7}O_{0.72}N_{0.001}$	17.1	13.2	10.2	6.0	Teflon PFA	$CF_{1.7}O_{0.01}$	5.0	(8.0)	(5.3)	(0)
Douglas fir	$CH_{1.7}O_{0.74}N_{0.002}$	16.4	12.4	9.5	5.0	Kel-F (CTFE)	$CF_{1.5}Cl_{0.50}$	6.5	11.8	8.6	(3.5)
Nylon	$CH_{1.8}O_{0.17}N_{0.17}$	30.8	11.9	13.3	10.8	Halar (E-CTFE)	$CHF_{0.75}Cl_{0.25}$	12.0	9.8	9.8	(5.4)
Flexible polyurethane foams						Kynar (PVF$_2$)	CHF	13.3	12.4	9.1	(4.2)
GM21	$CH_{1.8}O_{0.30}N_{0.05}$	26.2	12.1	11.5	8.0	Tedlar (PVF)	$CH_{1.5}F_{0.50}$	13.5	(6.5)	(7.1)	(1.1)
GM23	$CH_{1.8}O_{0.35}N_{0.06}$	27.2	13.7	12.5	9.7	**Carbon-Hydrogen-Oxygen-Silicone Atoms in the Structure**					
GM25	$CH_{1.7}O_{0.32}N_{0.07}$	24.6	12.0	11.1	7.5	Silicone-1	$CH_{1.3}O_{0.25}Si_{0.18}$	21.7	12.6	11.0	7.4
GM27	$CH_{1.7}O_{0.30}N_{0.08}$	23.2	11.2	10.4	6.2	Silicone-2	$CH_{1.5}O_{0.30}Si_{0.26}$	21.3	13.9	12.4	9.4
						Silicone-3	$CH_3O_{0.50}Si_{0.50}$	25.1	14.5	21.0	23.0

*From the data measured in our Flammability Laboratory.
†From the data for the elemental composition of the polymeric materials measured in the FMRC Flammability Laboratory.
‡From the data measured by our Flammability Laboratory in the oxygen bomb calorimeter and corrected for water as a gas and for the residue.
§Trade names from reference 40.

of oxygen for use in the CDG and OC calorimetries. (See Equations 18 and 21.) The CDG and OC calorimetries are used in the Flammability Apparatus (50-, 500-, and 10,000-kW scale). In the OSU Apparatus and the Cone Calorimeter, only the OC calorimetry is used.

The CDG and OC calorimetries are also used in numerous large-scale fire tests, such as the CDG calorimetry in the wind-aided turbulent horizontal flame spread in large-scale fire test galleries at the Londonderry Occupational Safety Centre in Australia and Pittsburgh Research Center, U.S. Bureau of Mines.[46,47]

In the GTR calorimetry, a thermopile located in the flue gas chimney is used in the OSU Apparatus, and a thermocouple located in the sampling duct is used in the Flammability Apparatus, where heat losses by conduction are negligibly small. The Cone Calorimeter has not been designed for the GTR calorimetry.

The radiative heat release rate is determined from the difference between the chemical and convective heat release rates only in the Flammability Apparatus.

Figure 3-4.21 shows a typical example of the heat release rate profile. The profile is for the chemical heat release rate of polypropylene, determined from the CDG and OC calorimetries in the Flammability Apparatus (500-kW scale). The polypropylene sample was 100 mm in diameter and 25 mm in thickness. It was exposed to an external heat flux of 50 kW/m^2 under co-flowing normal air. In the figure, solid, molten, and boiling-liquid zones are indicated.

In the solid zone in Figure 3-4.21, combustion is at the steady state between about 400 and 900 sec. During the steady-state combustion, a very thin liquid film is present at the surface. In the molten zone, the thickness of the liquid film and chemical heat release rate increase rapidly during combustion. At the end of the zone, the entire sample is

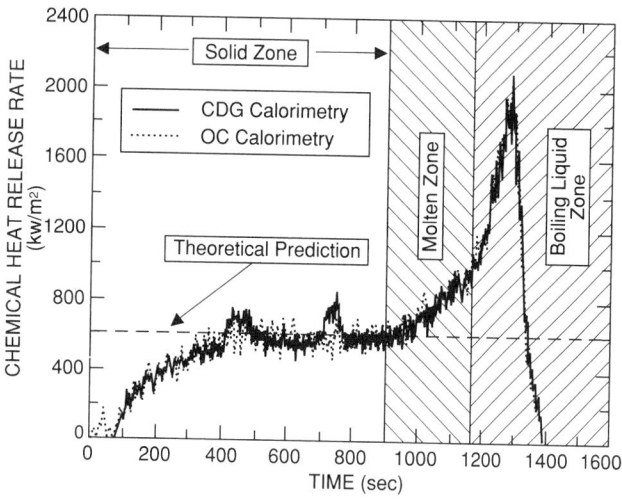

Fig. 3-4.21. *Chemical heat release rate for 100-mm diameter and 25-mm-thick slab of polypropylene exposed to an external heat flux of 50 kW/m² and 0.09 m/s co-flowing normal air in the Flammability Apparatus. The theoretical prediction is based on the Heat Release Parameter for polypropylene listed in Table 3-4.12.*

present as a liquid. In the boiling-liquid zone, the liquid boils vigorously, chemical heat release rate increases exponentially until the sample is consumed, the base diameter of the flame is considerably larger than the diameter of the sample dish (100 mm), and the flames are as high as 1.5 m (5 ft). This zone is the most dangerous zone.

The chemical heat release rate profiles from the CDG and OC calorimetries are very similar, as expected.

Energy released in a fire: The total amount of heat generated as a result of chemical reactions in the combustion of a material is defined as the chemical energy. The chemical energy has a convective and a radiative component

$$E_{ch} = E_{con} + E_{rad} \qquad (25)$$

where E_{ch} is the chemical energy (kJ), E_{con} is the convective energy (kJ), and E_{rad} is the radiative energy (kJ). The chemical energy and its convective and radiative components are calculated by the summation of the respective heat release rates

$$E_i = A \sum_{n=t_{ig}}^{n=t_{ex}} \dot{Q}_i''(t_n)\Delta t_n \qquad (26)$$

where E_i is the chemical, convective, or radiative energy (kJ), A is the total surface area of the material burning (m²), t_{ig} is the ignition time (s), and t_{ex} is the flame extinction time (s). The total mass of the material lost during combustion is measured directly from the initial and final mass and is calculated by the summation of the mass loss rate

$$W_f = A \sum_{n=t_{ig}}^{n=t_{ex}} \dot{m}''(t_n)\Delta t_n \qquad (27)$$

where W_f is the total mass of the material lost in the combustion (g).

Heat release rate can also be expressed as the product of the mass loss rate and the heat of combustion

$$\dot{Q}_i'' = \Delta H_i \dot{m}'' \qquad (28)$$

where ΔH_i is the chemical, convective, or radiative heat of combustion (kJ/g). The average chemical, convective, or radiative heats of combustion are calculated from the relationship based on Equations 26 and 27

$$\Delta\overline{H}_i = \frac{E_i}{W_f} \qquad (29)$$

where $\Delta\overline{H}_i$ is the average chemical, convective, or radiative heat of combustion (kJ/g). The average chemical heat of combustion determined in the Cone Calorimeter is defined as the effective heat of combustion.[17-19]

Heat release parameter (HRP): Heat release parameter (HRP) is defined as the amount of energy generated per unit amount of energy absorbed. From Equations 11 and 28

$$\dot{Q}_i'' = \left(\frac{\Delta H_i}{\Delta H_g}\right)(\dot{q}_e'' + \dot{q}_f'' - \dot{q}_{rr}'') \qquad (30)$$

where $\Delta H_i/\Delta H_g$ is defined as the chemical, convective, or radiative Heat Release Parameter, $(HRP)_{ch}$, $(HRP)_{con}$, or $(HRP)_{rad}$, respectively.[4] The HRP values are characteristic fire properties of materials, but depend on fire ventilation. The chemical Heat Release Parameter is independent of fire size.

In Figure 3-4.21, the theoretical prediction is from Equation 30, with chemical Heat Release Parameter = 19, external heat flux = 50 kW/m², and surface re-radiation = 18 kW/m² with negligibly small flame heat flux. The theoretical prediction is very close to the measured value in the solid zone.

Experimental data support Equation 30, as shown in Figures 3-4.22 through 3-4.24, where the average peak or steady-state chemical heat release rates are plotted against the net heat flux. Linear relationship between the chemical heat release rate and net heat flux is satisfied. For the condition $\dot{q}_e'' \gg \dot{q}_f'' - \dot{q}_{rr}''$, the average value of the Heat Release Parameter is calculated from the summation of the heat release rate and the external heat flux

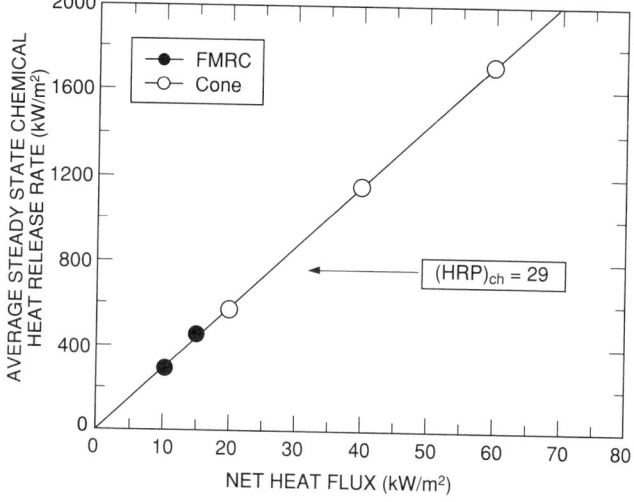

Fig. 3-4.22. *Average steady-state chemical heat release rate versus net heat flux for polystyrene slab. Net heat flux is the sum of the external and flame heat flux minus the surface re-radiation.*

$$\overline{(HRP)}_i = \frac{E_i}{A\sum_{n=t_{ig}}^{n=t_{ex}} \dot{q}_e'' \Delta t_n} \tag{31}$$

Complete and incomplete combustion: In fires, combustion is never complete. Thus, the chemical heat release rate or the chemical heat of combustion is less than the heat release rate for complete combustion or the net heat of complete combustion. The ratio of the chemical heat release rate to the heat release rate for complete combustion or the ratio of the chemical heat of combustion to net heat of complete combustion is defined as combustion efficiency[2-4,16,33,36]

$$\chi_{ch} = \frac{\dot{Q}_{ch}''}{\dot{Q}_T''} = \frac{\dot{m}''\Delta H_{ch}}{\dot{m}''\Delta H_T} = \frac{\Delta H_{ch}}{\Delta H_T} \tag{32}$$

where χ_{ch} is the combustion efficiency, and $\dot{Q}_T''$ is the heat release rate for complete combustion (kW/m^2). The convective and radiative components of the combustion efficiency are defined in a similar fashion[2-4,16,33,36]

$$\chi_{con} = \frac{\dot{Q}_{con}''}{\dot{Q}_T''} = \frac{\dot{m}''\Delta H_{con}}{\dot{m}''\Delta H_T} = \frac{\Delta H_{con}}{\Delta H_T} \tag{33}$$

$$\chi_{rad} = \frac{\dot{Q}_{rad}''}{\dot{Q}_T''} = \frac{\dot{m}''\Delta H_{rad}}{\dot{m}''\Delta H_T} = \frac{\Delta H_{rad}}{\Delta H_T} \tag{34}$$

where χ_{con} is the convective component of the combustion efficiency, and χ_{rad} is the radiative component of the combustion efficiency. From the definitions

$$\Delta H_{ch} = \Delta H_{con} + \Delta H_{rad} \tag{35}$$

$$\chi_{ch} = \chi_{con} + \chi_{rad} \tag{36}$$

The chemical, convective, and radiative heat release rates, heats of combustion or combustion efficiencies depend on the chemical structures of the materials and fire ventilation. The distribution of the chemical heat into convective and radiative components changes with fire size.

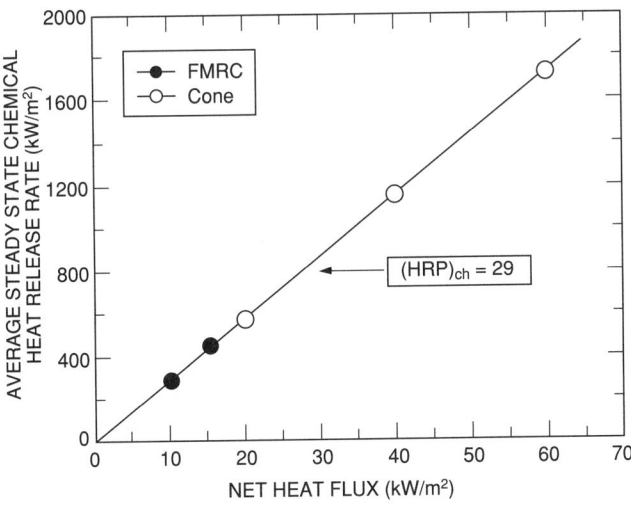

Fig. 3-4.23. *Average steady-state chemical heat release rate versus the net heat flux for high molecular weight hydrocarbon liquid burning in a 100-mm-diameter dish. The Cone Calorimeter data were measured at the research laboratory of the Dow-Corning Corporation, Midland, MI. Net heat flux is the sum of the external and flame heat flux minus the surface re-radiation.*

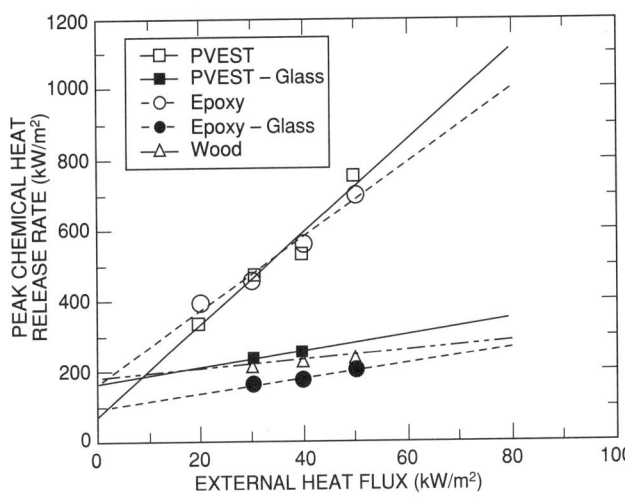

Fig. 3-4.24. *Peak chemical heat release rate versus the external heat flux for 100- × 100-mm × 3- to 11-mm-thick slab of polyvinyl ester (PVEST), PVEST/fiberglass, epoxy, epoxy/fiberglass, and wood (hemlock). Data measured in the Cone Calorimeter as reported in reference 20 are shown.*

The larger the fire size, the larger the fraction of the chemical heat distributed into the radiative component.

The chemical, convective, and radiative heats of combustion and the Heat Release Parameter values for the well-ventilated fires are listed in Tables 3-4.11 and 3-4.12, respectively. Comparisons between the limited data from the OSU Apparatus and the Flammability Apparatus and the Cone Calorimeter are satisfactory.

EXAMPLE 10:

Heptane was burned in a 2-m-diameter pan, and measurements were made for the mass loss rate, mass generation rates of CO and CO_2, and mass consumption rate of O_2. The average values in g/m^2-s for the mass loss rate, mass generation rates of CO and CO_2, and mass consumption rate of O_2 were 66, 9, 181, and 216, respectively. For large-scale fires of heptane, the literature values are: $\chi_{ch} = 0.93$, $\chi_{con} = 0.59$, and $\chi_{rad} = 0.34$. The net heat of complete combustion for heptane reported in the literature is 44.6 kJ/g. Calculate the chemical heat release rate and show that it is consistent with the rate based on the literature value of the combustion efficiency. Also calculate the convective and radiative heat release rates.

SOLUTION:

From Table 3-4.7, the net heat of complete combustion per unit mass of oxygen consumed is 12.7 kJ/g; the net heat of complete combustion per unit mass of CO_2 generated is 14.5 kJ/g; and the net heat of complete combustion per unit mass of CO generated is 12.8 kJ/g. From the CDG Calorimetry (Equation 18)

$$\begin{aligned} \dot{Q}_{ch}'' &= 14.5 \times 181 + 12.8 \times 9 \\ &= 2625 + 115 = 2740 \text{ kW/m}^2 \end{aligned}$$

(Text continued on page 3-84)

TABLE 3-4.11 *Yields of Fire Products and Chemical, Convective, and Radiative Heats of Combustion for Well-Ventilated Fires**

Material	ΔH_T (kJ/g)	y_{CO_2}	y_{CO}	y_{ch}	y_s	ΔH_{ch}	ΔH_{con}	ΔH_{rad}
			(g/g)				(kJ/g)	
Common Gases								
Methane	50.1	2.72	—	—	—	49.6	42.6	7.0
Ethane	47.1	2.85	0.001	0.001	0.013	45.7	34.1	11.6
Propane	46.0	2.85	0.005	0.001	0.024	43.7	31.2	12.5
Butane	45.4	2.85	0.007	0.003	0.029	42.6	29.6	13.0
Ethylene	48.0	2.72	0.013	0.005	0.043	41.5	27.3	14.2
Propylene	46.4	2.74	0.017	0.006	0.095	40.5	25.6	14.9
1,3-Butadiene	44.6	2.46	0.048	0.014	0.125	33.6	15.4	18.2
Acetylene	47.8	2.60	0.042	0.013	0.096	36.7	18.7	18.0
Common Liquids								
Methyl alcohol	20.0	1.31	0.001	—	—	19.1	16.1	3.0
Ethyl alcohol	27.7	1.77	0.001	0.001	0.008	25.6	19.0	6.5
Isopropyl alcohol	31.8	2.01	0.003	0.001	0.015	29.0	20.6	8.5
Acetone	29.7	2.14	0.003	0.001	0.014	27.9	20.3	7.6
Methylethyl ketone	32.7	2.29	0.004	0.001	0.018	30.6	22.1	8.6
Heptane	44.6	2.85	0.010	0.004	0.037	41.2	27.6	13.6
Octane	44.5	2.84	0.011	0.004	0.038	41.0	27.3	13.7
Kerosene	44.1	2.83	0.012	0.004	0.042	40.3	26.2	14.1
Benzene	40.1	2.33	0.067	0.018	0.181	27.6	11.0	16.5
Toluene	39.7	2.34	0.066	0.018	0.178	27.7	11.2	16.5
Styrene	39.4	2.35	0.065	0.019	0.177	27.8	11.2	16.6
Hydrocarbon	43.9	2.64	0.019	0.007	0.059	36.9	24.5	12.4
Mineral oil	41.5	2.37	0.041	0.012	0.097	31.7	—	—
Polydimethyl siloxane	25.1	0.93	0.004	0.032	0.232	19.6	—	—
Silicone	25.1	0.72	0.006	0.008	—	15.2	12.7	2.5
Natural Materials								
Tissue paper	—	—	—	—	—	11.4	6.7	4.7
Newspaper	—	—	—	—	—	14.4	—	—
Wood (red oak)	17.1	1.27	0.004	0.001	0.015	12.4	7.8	4.6
Wood (Douglas fir)	16.4	1.31	0.004	0.001	—	13.0	8.1	4.9
Wood (pine)	17.9	1.33	0.005	0.001	—	12.4	8.7	3.7
Corrugated paper	—	—	—	—	—	13.2	—	—
Wood (hemlock)†	—	—	—	—	0.015	13.3	—	—
Wool 100%†	—	—	—	—	0.008	19.5	—	—
Synthetic Materials—Solids (abbreviations/names in the nomenclature)								
ABS†	—	—	—	—	0.105	30.0	—	—
POM	15.4	1.40	0.001	0.001	—	14.4	11.2	3.2
PMMA	25.2	2.12	0.010	0.001	0.022	24.2	16.6	7.6
PE	43.6	2.76	0.024	0.007	0.060	38.4	21.8	16.6
PP	43.4	2.79	0.024	0.006	0.059	38.6	22.6	16.0
PS	39.2	2.33	0.060	0.014	0.164	27.0	11.0	16.0
Silicone	21.7	0.96	0.021	0.006	0.065	10.6	7.3	3.3
Polyester-1	32.5	1.65	0.070	0.020	0.091	20.6	10.8	9.8
Polyester-2	32.5	1.56	0.080	0.029	0.089	19.5	—	—
Epoxy-1	28.8	1.59	0.080	0.030	—	17.1	8.5	8.6
Epoxy-2	28.8	1.16	0.086	0.026	0.098	12.3	—	—
Nylon	30.8	2.06	0.038	0.016	0.075	27.1	16.3	10.8
Polyamide-6†	—	—	—	—	0.011	28.8	—	—
IPST†	—	—	—	—	0.080	23.3	—	—
PVEST†	—	—	—	—	0.076	22.0	—	—
Silicone rubber	21.7	0.96	0.021	0.005	0.078	10.9	—	—
Polyurethane (Flexible) Foams								
GM21	26.2	1.55	0.010	0.002	0.131	17.8	8.6	9.2
GM23	27.2	1.51	0.031	0.005	0.227	19.0	10.3	8.7
GM25	24.6	1.50	0.028	0.005	0.194	17.0	7.2	9.8
GM27	23.2	1.57	0.042	0.004	0.198	16.4	7.6	8.8

TABLE 3-4.11 *Yields of Fire Products and Chemical, Convective, and Radiative Heats of Combustion for Well-Ventilated Fires* (Continued)*

Material	ΔH_T (kJ/g)	y_{CO_2}	y_{CO}	y_{ch}	y_s	ΔH_{ch}	ΔH_{con}	ΔH_{rad}
		(g/g)				(kJ/g)		
Polyurethane (Rigid) Foams								
GM29	26.0	1.52	0.031	0.003	0.130	16.4	6.8	9.6
GM31	25.0	1.53	0.038	0.002	0.125	15.8	7.1	8.8
GM35	28.0	1.58	0.025	0.001	0.104	17.6	7.8	9.8
GM37	28.0	1.63	0.024	0.001	0.113	17.9	8.7	9.2
GM41	26.2	1.18	0.046	0.004	—	15.7	5.7	10.0
GM43	22.2	1.11	0.051	0.004	—	14.8	6.4	8.4
Polystyrene Foams								
GM47	38.1	2.30	0.060	0.014	0.180	25.9	11.4	14.5
GM49	38.2	2.30	0.065	0.016	0.210	25.6	9.9	15.7
GM51	35.6	2.34	0.058	0.013	0.185	24.6	10.4	14.2
GM53	37.6	2.34	0.060	0.015	0.200	25.9	11.2	14.7
Polyethylene Foams								
1	41.2	2.62	0.020	0.004	0.056	34.4	20.2	14.2
2	40.8	2.78	0.026	0.008	0.102	36.1	20.6	15.5
3	40.8	2.60	0.020	0.004	0.076	33.8	18.2	15.6
4	40.8	2.51	0.015	0.005	0.071	32.6	19.1	13.5
Phenolic Foams								
1[†]	—	—	—	—	0.002	10.0	—	—
2[†]	—	—	—	—	—	10.0	—	—
Halogenated Materials (abbreviations/names in the nomenclature)								
Polyethylene with								
25% chlorine	31.6	1.71	0.042	0.016	0.115	22.6	10.0	12.6
36% chlorine	26.3	0.83	0.051	0.017	0.139	10.6	6.4	4.2
48% chlorine	20.6	0.59	0.049	0.015	0.134	7.2	3.9	3.3
PVC	16.4	0.46	0.063	0.023	0.172	5.7	3.1	2.6
PVC-1[†] (LOI = 0.50)	—	—	—	—	0.098	7.7	—	—
PVC-2[†] (LOI = 0.50)	—	—	—	—	0.076	8.3	—	—
PVC[†] (LOI = 0.20)	—	—	—	—	0.099	11.3	—	—
PVC[†] (LOI = 0.25)	—	—	—	—	0.078	9.8	—	—
PVC[†] (LOI = 0.30)	—	—	—	—	0.098	10.3	—	—
PVC[†] (LOI = 0.35)	—	—	—	—	0.088	10.8	—	—
PVC panel	—	—	—	—	—	7.3	—	—
ETFE (Tefzel™)	12.6	0.54	0.060	0.020	0.042	5.4	—	—
PFA (Teflon™)	5.0	0.37	0.097	—	0.002	4.7	—	—
FEP (Teflon™)	4.8	0.25	0.116	—	0.003	4.1	—	—
TFE (Teflon™)	6.2	0.38	0.092	—	0.003	4.2	—	—
Building Products[‡]								
Particleboard (PB)	—	1.2	0.004	—	—	14.0	—	—
Fiberboard (FB)	—	1.4	0.015	—	—	14.0	—	—
Medium-density FB	—	1.2	0.002	—	—	14.0	—	—
Wood panel	—	1.2	0.002	—	—	15.0	—	—
Melamine-faced PB	—	0.8	0.025	—	—	10.7	—	—
Gypsumboard (GB)	—	0.3	0.027	—	—	4.3	—	—
Paper on GB	—	0.4	0.028	—	—	5.6	—	—
Plastic on GB	—	0.4	0.028	—	—	14.3	—	—
Textile on GB	—	0.4	0.025	—	—	13.0	—	—
Textile on rock wool	—	1.8	0.091	—	—	25.0	—	—
Paper on PB	—	1.2	0.003	—	—	12.5	—	—
Rigid PU	—	1.1	0.200	—	—	13.0	—	—
EPS	—	1.9	0.054	—	—	28.0	—	—

TABLE 3-4.11 *Yields of Fire Products and Chemical, Convective, and Radiative Heats of Combustion for Well-Ventilated Fires* (Continued)*

Material	ΔH_T (kJ/g)	y_{CO_2}	y_{CO}	y_{ch}	y_s	ΔH_{ch}	ΔH_{con}	ΔH_{rad}
			(g/g)				(kJ/g)	
Composite and Fiberglass-Reinforced Materials (FGR) (abbreviations/names in the nomenclature)								
PEEK/FGR[†]	—	—	—	—	0.042	20.5	—	—
IPST/FGR[†]	—	—	—	—	0.032	27.0	—	—
PES/FGR[†]	—	—	—	—	0.049	27.5	—	—
PEST1/FGR[†]	—	—	—	—	—	16.0	—	—
PEST2/FGR[†]	—	—	—	—	—	12.9	—	—
PEST1/FGR	—	—	—	—	—	19.0	—	—
PEST2/FGR	—	—	—	—	—	13.9	—	—
PEST3/FGR	—	1.47	0.055	0.007	0.070	17.9	—	—
PEST4/FGR	—	1.24	0.039	0.004	0.054	16.0	10.7	7.2
PEST5/FGR	—	0.71	0.102	0.019	0.068	9.3	9.9	6.1
Epoxy/FG[†]	—	—	—	—	0.056	27.5	6.5	2.8
PVEST/FGR	—	—	—	—	0.079	26.0	—	—
Kevlar™/Phenolic	—	1.27	0.025	0.002	0.041	14.8	11.1	3.7
Phenolic-1/FGR	—	0.98	0.066	0.003	0.023	11.9	8.9	3.0
Phenolic-2/FGR[†]	—	—	—	—	0.016	22.0	—	—
Aircraft Panel Materials								
Epoxy/FGR/paint	—	0.828	0.114	0.016	0.166	11.3	6.2	5.1
Epoxy/Kevlar™/paint	—	0.873	0.091	0.016	0.126	11.4	6.3	5.1
Phenolic/FGR/paint	—	1.49	0.027	0.002	0.059	22.9	11.5	11.4
Phenolic/Kevlar™/paint	—	1.23	0.088	0.011	0.094	18.6	8.9	9.7
Phenolic/graphite/paint	—	1.67	0.026	0.003	0.062	24.6	14.0	10.6
Polycarbonate	—	—	—	—	—	20.5	—	—
Electrical Cables (abbreviations/names in the nomenclature)								
Polyethylene/Polyvinylchloride								
1	—	2.08	0.100	0.021	0.076	31.3	11.6	19.7
2	—	1.75	0.050	0.013	0.115	25.1	11.1	14.0
3	—	1.67	0.048	0.012	—	24.0	13.0	11.0
4	—	1.39	0.166	0.038	—	22.0	14.0	8.1
5	—	1.29	0.147	0.042	0.136	20.9	10.7	10.2
EPR/Hypalon								
1	—	1.95	0.072	0.014	—	29.6	15.8	13.9
2	—	1.74	0.076	0.022	—	26.8	17.0	9.8
3	—	1.21	0.072	0.014	—	19.0	12.3	6.7
4	—	0.99	0.090	0.085	0.082	17.4	6.6	10.8
5	—	0.95	0.122	0.024	—	17.3	7.5	9.8
6	—	0.89	0.121	0.022	0.164	13.9	9.2	4.7
Silicone								
1	—	1.65	0.011	0.001	—	25.0	17.5	7.3
2	—	1.47	0.029	0.001	—	24.0	20.0	4.0
XLPE/XLPE								
1	—	1.78	0.114	0.029	0.120	28.3	12.3	16.0
2	—	0.83	0.110	0.024	0.120	12.5	7.5	5.0
XLPE/Neoprene								
1	—	0.68	0.122	0.031	—	12.6	5.9	6.7
2	—	0.63	0.082	0.014	0.175	10.3	4.9	5.5
Silicone/PVC								
1	16.4	0.76	0.110	0.015	0.111	10.0	—	—
2	16.4	1.19	0.065	0.005	0.119	15.6	—	—
PVC/Nylon/PVC-Nylon								
1	—	0.63	0.084	0.024	—	10.2	5.0	5.2
2	—	0.49	0.082	0.032	0.115	9.2	4.8	4.4

TABLE 3-4.11 *Yields of Fire Products and Chemical, Convective, and Radiative Heats of Combustion for Well-Ventilated Fires* (Continued)*

Material	ΔH_T (kJ/g)	y_{CO_2}	y_{CO}	y_{ch}	y_s	ΔH_{ch}	ΔH_{con}	ΔH_{rad}
				(g/g)			(kJ/g)	
PTFE								
1	—	0.180	0.091	0.012	0.011	3.2	2.7	0.4
2	6.2	0.383	0.103	—	0.005	5.7	—	—
Materials with Fiberweb, Netlike, and Multiplex Structure (abbreviations/names in the nomenclature)								
Olefin	—	1.49	0.006	—	—	16.5	13.3	3.2
PP-1	—	1.25	0.0029	—	—	14.0	10.8	3.2
PP-2	—	1.56	0.0048	—	—	17.2	10.5	6.7
Polyester-1	—	2.21	0.015	—	—	24.6	8.9	15.7
Polyester-2	—	1.51	0.0079	—	—	16.8	9.1	7.7
Polyester-3	—	2.55	0.020	—	—	28.5	22.6	5.9
Polyester-4	—	1.92	0.014	—	—	21.4	12.4	9.0
Rayon-1	—	1.80	0.043	—	—	20.3	14.1	6.2
Rayon-2	—	1.91	0.043	0.002	—	21.5	13.3	8.2
Rayon-3	—	1.18	0.047	—	—	13.5	8.3	5.2
Polyester-Rayon	—	1.52	0.005	—	—	16.8	9.1	7.7
Polyester-polyamide	—	1.82	0.008	—	—	20.2	10.4	9.8
Rayon-PE	—	1.50	0.027	—	—	16.9	8.72	8.2
Two to Eight 100- × 100- × 100-mm Corrugated Paper Boxes with and without the Polymers with Three-Dimensional Arrangement (abbreviations/names in the nomenclature)§								
Empty	—	1.53	0.023	0.001	—	14.2	10.7	3.5
With PVC (62%-thick)	—	1.01	0.073	0.007	0.119	10.7	9.5	1.2
With PC (59%-thick)	—	1.73	0.047	0.002	0.061	18.4	13.5	4.9
With PS (58%-thick)	—	1.40	0.138	0.026	0.285	16.2	12.5	3.7
With PS (60%-thin)	—	1.88	0.068	0.020	0.140	19.4	10.1	9.3
With PS (40%-thin)	—	1.74	0.042	0.005	0.167	18.0	11.7	6.7
With ABS (59%-thick)	—	1.53	0.089	0.006	0.143	16.1	12.7	3.4
With PET (41%-thin)	—	1.87	0.050	0.006	0.053	19.9	11.8	8.1
With PU (40%-foam)	—	1.56	0.024	—	—	14.4	8.6	5.8
High-Pressure Liquid Spray Combustion‖								
Hydraulic Fluids								
Organic polyol esters								
1	36.6	—	—	—	—	35.5	—	—
2	35.7	—	—	—	—	35.1	—	—
3	40.3	—	—	—	—	37.2	—	—
4	37.0	—	—	—	—	35.7	—	—
Phosphate esters								
1	31.8	—	—	—	—	29.3	—	—
2	32.0	—	—	—	—	29.6	—	—
Water-in-Oil Emulsions								
1	27.6	—	—	—	—	2.5	—	—
Polyglycol-in-Water								
1	11.0	—	—	—	—	10.4	—	—
2	11.9	—	—	—	—	11.1	—	—
3	14.7	—	—	—	—	12.2	—	—
4	12.1	—	—	—	—	10.6	—	—
Liquid Fuels								
Mineral oil	46.0	—	—	—	—	44.3	—	—
Methanol	20.0	—	—	—	—	19.8	—	—
Ethanol	27.7	—	—	—	—	26.2	—	—
Heptane	44.4	—	—	—	—	40.3	—	—

*Data measured in the Flammability Apparatus. Data measured in the Cone Calorimeter are identified by superscripts † and ‡. Some of the data are corrected to reflect well-ventilated fire conditions. All the data are reported for turbulent fires, i.e., materials exposed to higher external heat flux values.

Dashes: either not measured or are less than 0.001.

†Calculated from the data measured in the Cone Calorimeter as reported in references 20 and 31.

‡Calculated from the data measured in the Cone Calorimeter as reported in reference 48.

§100- × 100- × 100-mm corrugated paper boxes with and without the 99- × 99- × 99-mm polymer boxes or pieces on corrugated paper compartments. The boxes are arranged in one and two layers, about 12 mm apart, with one to four boxes in each layer, separated by about 12 mm. All the boxes are placed on a very light metal frame made of rods with screen base. Measurements made in the Flammability Apparatus; numbers in parenthesis are the weight percents.

‖Data from reference 49 measured in high-pressure liquid spray combustion in the Fire Products Collector (10,000-kW scale apparatus in Figure 3-4.8).

TABLE 3-4.12 *Chemical and Convective Heat Release Parameters*

Materials	(HRP)$_{ch}$			(HRP)$_{con}$		
	Flammability Apparatus	Cone*	Cal†	Flammability Apparatus	OSU‡	Cal†
Liquids and Gases (Hydrocarbons, Alkanes)						
Hexane	—	—	83	—	—	56
Heptane	—	—	75	—	—	50
Octane	—	—	68	—	—	46
Nonane	—	—	64	—	—	42
Decane	—	—	59	—	—	39
Undecane	—	—	55	—	—	36
Dodecane	—	—	52	—	—	34
Tridecane	—	—	50	—	—	32
Kerosene	—	—	47	—	—	17
Hexadexane	—	—	44	—	—	28
Solids (abbreviations/names in the nomenclature)						
ABS	—	14	—	—	—	—
Acrylic sheet	—	6	—	—	—	—
Epoxy	—	11	—	—	—	—
IPST	—	6	—	—	—	—
Polyamide	21	—	—	—	—	—
Polypropylene	19	—	—	—	—	—
Polyethylene	17	21	—	11	—	—
Polystyrene	16	19	—	12	—	—
Polymethylmethacrylate	15	14	—	6	—	—
Nylon	12	—	—	10	—	—
Polyamide-6	—	21	—	7	—	—
Filled phenolic foam—50% inert	—	1	—	—	—	—
Polycarbonate	9	—	—	—	—	—
Polyoxymethylene	6	—	—	5	—	—
Polyethylene/25% Cl	11	—	—	5	—	—
Plasticized-PVC-3, LOI 0.25	—	5	—	—	—	—
Plasticized-PVC-4, LOI 0.30	—	5	—	—	—	—
Plasticized-PVC-5, LOI 0.35	—	5	—	—	—	—
Polyethylene/36% Cl	4	—	—	2	—	—
Rigid PVC-1, LOI 0.50	—	3	—	—	—	—
Rigid PVC-2	2	3	—	1	—	—
PVC panel	2	—	—	—	—	—
Polyethylene/48% Cl	2	—	—	—	—	—
PVEST	—	13	—	—	—	—
ETFE (Tefzel™)	6	—	—	—	—	—
PFA (Teflon™)	5	—	—	—	—	—
FEP (Teflon™)	2	—	—	—	—	—
TFE (Teflon™)	2	—	—	—	—	—
Wood (hemlock)	—	1	—	—	—	—
Wood (Douglas fir)	7	—	—	5	—	—
Wool	—	5	—	—	—	—
Composites and Fiberglass-Reinforced Materials (FGR) (abbreviations/names in the nomenclature)						
Bismaleimide/graphite/ceramic (CC)	—	1	—	—	—	—
Epoxy/FGR	—	2	—	—	—	—
Epoxy/graphite	2	—	—	—	—	—
Epoxy/graphite/CC	2	—	—	—	—	—
Epoxy/graphite/intumescent (IC)	2	—	—	—	—	—
IPST/FGR	—	1	—	—	—	—
PEEK/FGR	—	3	—	—	—	—
PES/FGR	—	1	—	—	—	—
PEST-1/FGR	3	—	—	—	—	—
PEST-2/FGR	8	—	—	—	—	—
PEST-3/FGR	10	—	—	—	—	—
PEST-4/FGR	3	—	—	—	—	—
PEST-5/FGR	3	—	—	—	—	—

TABLE 3-4.12 *Chemical and Convective Heat Release Parameters (Continued)*

| Materials | (HRP)$_{ch}$ | | | (HRP)$_{con}$ | | |
	Flammability Apparatus	Cone*	Cal[†]	Flammability Apparatus	OSU[‡]	Cal[†]
Composites and Fiberglass-Reinforced Materials (FGR) (Continued)			—	—	—	—
PEST-6/FGR	3	—	—	—	—	—
Phenol/FGR	—	1	—	—	—	—
Phenolic/Kevlar™	2	—	—	—	—	—
Phenol/graphite	1	—	—	—	—	—
PVEST-1/FGR	3	—	—	—	—	—
PVEST-1/FGR/CC	3	—	—	—	—	—
PVEST-1/FGR/IC	1	—	—	—	—	—
PVEST-2/FGR	7	—	—	—	—	—
PVEST-3/FGR	2	—	—	—	—	—
Aircraft Panel Materials						
Epoxy fiberglass	4	5	—	2	1	—
Epoxy Kevlar™	4	4	—	2	2	—
Phenolic Kevlar™	5	4	—	2	—	—
Phenolic graphite	4	3	—	1	—	—
Phenolic fiberglass	4	3	—	2	1	—
Polycarbonate panel	9	—	—	—	—	—
Foams						
Polystyrene						
GM53	20	—	—	6	—	—
GM49	19	—	—	8	—	—
GM51	18	—	—	9	—	—
Flexible Polyurethane						
GM 21	7	—	—	3	3	—
GM 23	9	—	—	5	6	—
GM 25	14	—	—	6	4	—
GM 27	9	—	—	4	2	—
Phenolic	—	1	—	—	—	—
Electrical Cables (abbreviations/names in the nomenclature)			—	—	—	—
PVC/PVC-1 (Group 3)	15	—	—	—	—	—
PE/PVC (Group 3)	19	—	—	—	—	—
PP,PEST/PVC (Group 3)	11	—	—	—	—	—
PVC/PVC-2 (Group 3)	14	—	—	—	—	—
Chlorinated PE (Group 2)	5	—	—	—	—	—
PVC/PVC-3 (Group 2)	4	—	—	—	—	—
EPR/PVC (Group 2)	6	—	—	—	—	—
PVC/EPR (Group 2)	4	—	—	—	—	—
XLPE/XLPE (Group 2)	6	—	—	—	—	—
EPR/hypalon-1 (Group 2)	6	—	—	—	—	—
EPR/hypalon-2 (Group 2)	4	—	—	—	—	—
EPR/hypalon-3 (Group 1)	3	—	—	—	—	—
EPR/hypalon-4 (Group 1)	3	—	—	—	—	—
EPR/EPR-1 (Group 1)	3	—	—	—	—	—
EPR/EPR-2 (Group 1)	3	—	—	—	—	—
EPR/EPR-3 (Group 1)	2	—	—	—	—	—
XLPE-EVA-1 (Group 1)	3	—	—	—	—	—
XLPE-EVA-2 (Group 1)	3	—	—	—	—	—
ETFA (Group 1)	3	—	—	—	—	—
PVC/PVF$_2$ (Group 1)	1	—	—	—	—	—
FEP/FEP-1 (Group 1)	2	—	—	—	—	—
FEP/FEP-2 (Group 2)	2	—	—	—	—	—

*Calculated from the data reported in references 20 and 31.
[†]Calculated from the data in references 38 and 39.
[‡]From reference 50.

From OC Calorimetry (Equation 21)

$$\dot{Q}''_{ch} = 12.7 \times 216 = 2743 \text{ kW/m}^2$$

The chemical heat release rate from the CDG and OC calorimetries are in excellent agreement, the average being 2742 kW/m^2.

The chemical heat of combustion is the products of net heat of complete combustion (44.6 kJ/g) and the combustion efficiency (0.93), which is 41.5 kJ/g.

The chemical heat release is the product of the mass loss rate (66 g/m^2-s) and chemical heat of combustion (41.5 kJ/g), which is 2739 kW/m^2, compared to the averaged value 2742 kW/m^2 from the CDG and OC calorimetries. Thus, the chemical heat release determined from the measurements is consistent with the rate from the literature value of the combustion efficiency.

The convective heat release rate is equal to the convective heat of combustion and the mass loss rate. The convective heat of combustion is equal to the convective component of the combustion efficiency ($\chi_{con} = 0.59$) times the net of complete combustion (44.6 kJ/g). Thus, the convective heat release rate for heptane = 66 × 0.59 × 44.6 = 1737 kW/m^2. In a similar fashion, the radiative heat release rate = 66 × 0.34 × 44.6 = 1001 kW/m^2.

EXAMPLE 11:

From Radiation Scaling Technique, the asymptotic mass loss rate values in g/m^2-s, expected in large-scale fires, as listed in Table 3-4.5, for polyethylene, polystyrene, polyvinylchloride, and Teflon™ are 26, 36, 16, and 7, respectively. The chemical heats of combustion in kJ/g listed in Table 3-4.11 for these materials are 38.4, 27.0, 5.7, and 4.1, respectively. Estimate the chemical heat release rates expected in large-scale fires of polyethylene, polystyrene, polyvinylchloride, and Teflon™. (Teflon™ in this chapter refers mainly to FEP, except in cases where it is identified otherwise.)

SOLUTION:

The chemical heat release rate is calculated from Equation 28. The chemical heat release rates estimated in the large-scale fires are: (1) polyethylene: 26 × 38.4 = 998 kW/m^2; (2) polystyrene: 36 × 27.0 = 972 kW/m^2; (3) polyvinylchloride: 16 × 5.7 = 91 kW/m^2; and (4) Teflon™: 7 × 4.1 = 28 kW/m^2.

EXAMPLE 12:

Heat release rate is the product of the Heat Release Parameter and the net heat flux absorbed by the material, as indicated in Equation 30. This concept is used in various models to predict fire propagation and heat release rates, whereas values for the Heat Release Parameter are taken from a handbook, such as this handbook, and net heat flux is estimated through correlations. The lower the value of the Heat Release Parameter for a fixed value of the net heat flux, the lower the heat release rate.

The values for the surface re-radiation, flame heat flux for large-scale fires, and chemical Heat Release Parameter are listed in Tables 3-4.4, 3-4.5, and 3-4.12, respectively. Calculate the chemical heat release rates expected in large-scale fires of heptane, kerosene, polyethylene, polypropylene, polystyrene, polymethylmethacrylate, polyvinylchloride, and Teflon™.

SOLUTION:

The chemical heat release rates are calculated from the relationship $[(HRP)_{ch} \times (\dot{q}''_f - \dot{q}''_{rr})]$, which is Equation 30: (1) heptane: (75)(37 − 1) = 2700 kW/m^2; (2) kerosene: (47)(29 − 1) = 1316 kW/m^2; (3) polyethylene: (17)(61 − 15) = 782 kW/m^2; (4) polypropylene: (19)(67 − 15) = 988 kW/m^2; (5) polystyrene: (16)(75 − 13) = 992 kW/m^2; (6) polymethylmethacrylate: (15)(57 − 11) = 690 kW/m^2; (7) polyvinylchloride: (2)(50 − 15) = 70 kW/m^2; and (8) Teflon™: (2)(52 − 38) = 28 kW/m^2.

The example shows the importance of the Chemical Heat Release Parameter, flame heat flux, and surface re-radiation.

Heat release rate and fire ventilation: In the majority of fires, hazards are due to fires occurring in enclosed spaces. In early stages, a building fire is well-ventilated, and is easy to control and extinguish. However, if the fire is allowed to grow, especially with limited enclosure ventilation and large material surface area, it becomes a ventilation-controlled fire and can lead to flashover, a very dangerous condition. In ventilation-controlled fires, the chemical reactions between oxygen from air and products of incomplete combustion from the decomposed and gasified material (e.g., smoke, CO, hydrocarbons, and other intermediate products) remain incomplete and heat release rate decreases.[36]

In ventilation-controlled fires, heat release rate depends on the air supply rate and the mass loss rate, in addition to other factors. For ventilation-controlled fires, the effects of the mass flow rate of air and fuel mass loss rate are characterized, most commonly, by the local equivalence ratio

$$\Phi = \frac{S\dot{m}''A}{\dot{m}_{air}} \tag{37}$$

where Φ is the equivalence ratio, S is the stoichiometric mass air-to-fuel ratio (g/g), $\dot{m}''$ is the mass loss rate (g/m^2-s), A is the exposed area of the material burning (m^2), and $\dot{m}_{air}$ is the mass flow rate of air (g/s).

Generalized state-relationships between mass fractions of major species (O$_2$, fuel, CO$_2$, H$_2$O, CO, and H$_2$) and temperature as functions of local equivalence ratios for hydrocarbon-air diffusion flames are available.[51] The relationships suggest that the generation efficiencies of CO, fuel vapors, water, CO$_2$, and hydrogen and consumption efficiency of O$_2$ are in approximate thermodynamic equilibrium for well-ventilated combustion, but deviate from equilibrium for ventilation-controlled combustion. This concept has been used for fires of polymeric materials.[36] In the tests, chemical and convective heat release rates, mass loss rate, and generation rates of fire products have been measured for various equivalence ratios in the Flammability Apparatus [Figure 3-4.2(a)] and in the Fire Research Institute's (FRI) 0.022-m^3 enclosure in Tokyo, Japan, described in reference 36. The combustion efficiency and its convective component are found to decrease as fires become fuel rich, due to increase in the equivalence ratio. The ratio of the combustion efficiency and its convective component or chemical and convective heats of combustion for ventilation-controlled to well-ventilated combustion is expressed as[36]

$$\zeta_{ch} = \frac{(\chi_{ch})_{vc}}{(\chi_{ch})_{wv}} = \frac{(\Delta H_{ch}/\Delta H_T)_{vc}}{(\Delta H_{ch}/\Delta H_T)_{wv}} = \frac{(\Delta H_{ch})_{vc}}{(\Delta H_{ch})_{wv}} \tag{38}$$

$$\zeta_{con} = \frac{(\chi_{con})_{vc}}{(\chi_{con})_{wv}} = \frac{(\Delta H_{con}/\Delta H_T)_{vc}}{(\Delta H_{con}/\Delta H_T)_{wv}} = \frac{(\Delta H_{con})_{vc}}{(\Delta H_{con})_{wv}} \tag{39}$$

where ζ_{ch} and ζ_{con} are the ratio of the combustion efficiency

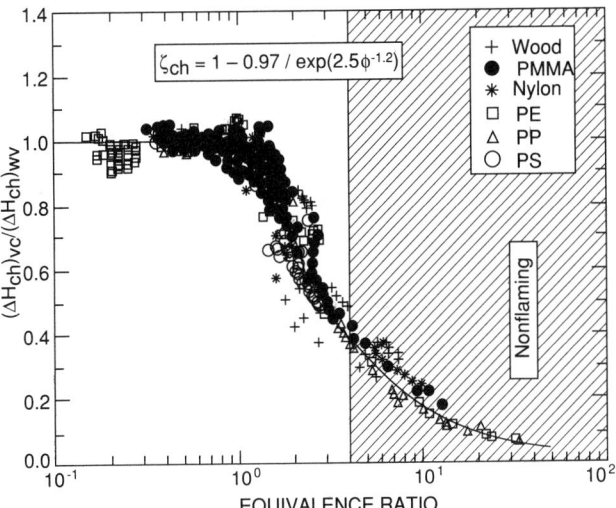

Fig. 3-4.25. Ratio of the ventilation-controlled to well-ventilated chemical heat of combustion versus the equivalence ratio. Data are measured in the Flammability Apparatus and in the Fire Research Institute's enclosure.[36] Subscript vc represents ventilation-controlled fires, and subscript wv represents well-ventilated fires.

and its convective component or chemical and convective heats of combustion for ventilation-controlled to well-ventilated combustion, subscript *vc* represents ventilation-controlled fire, and *wv* represents well-ventilated fire.

The experimental data for the ratios of the chemical and convective heats of combustion for ventilation-controlled to well-ventilated fires at various equivalence ratios are shown in Figures 3-4.25 and 3-4.26. The data are measured in the Flammability Apparatus and the FRI 0.022-m^3 enclosure, details of which are described in reference 36. The data for the polymers indicated in the figures satisfy the following general correlations, irrespective of their chemical structures[36]

$$\frac{(\Delta H_{ch})_{vc}}{(\Delta H_{ch})_{wv}} = 1 - \frac{0.97}{\exp(\Phi/2.15)^{-1.2}} \quad (40)$$

$$\frac{(\Delta H_{con})_{vc}}{(\Delta H_{con})_{wv}} = 1 - \frac{1.0}{\exp(\Phi/1.38)^{-2.8}} \quad (41)$$

The effects of ventilation on the chemical and convective heats of combustion are reflected by the magnitudes of the expressions within the parentheses on the right-hand sides of Equations 40 and 41. For a well-ventilated fire, $\Phi \ll 1.0$ and $(\Delta H_{ch})_{vc} = (\Delta H_{ch})_{wv}$ and $(\Delta H_{con})_{vc} = (\Delta H_{con})_{wv}$.

As a fire changes from well-ventilated to ventilation-controlled, equivalence ratio increases and the magnitudes of the expressions within the parentheses on the right-hand sides of Equations 40 and 41 increase. Thus with increase in the equivalence ratio, the chemical and convective heats of combustion decrease. The decrease in the convective heat of combustion is higher than it is for the chemical heat of combustion, because the coefficients for the equivalence ratios are different. The correlation thus suggests that higher fraction of the chemical heat of combustion is expected to be converted to the radiative heat of combustion as fires change from well-ventilated to ventilation-controlled. This is in general agreement with the observations for the ventilation-controlled fires in buildings.

Equations 40 and 41 can be used in models for the assessment of the ventilation-controlled fire behavior of materials, using chemical and convective heats of combustion for well-ventilated fires such as from Table 3-4.11.

EXAMPLE 13:

Calculate the chemical heats of combustion at equivalence ratios of 1, 2, and 3 for red oak, polyethylene, polystyrene, and nylon using Equation 40 and data from Table 3-4.11 for well-ventilated fires.

SOLUTION:

	Chemical Heats of Combustion (kJ/g)			
Material	$\Phi \ll 1.0$	$\Phi = 1.0$	$\Phi = 2.0$	$\Phi = 3.0$
Red oak	12.4	11.4	8.3	6.2
Polyethylene	38.4	35.3	25.9	19.3
Polystyrene	27.0	24.9	18.2	13.6
Nylon	27.1	24.9	18.2	13.6

Generation of Chemical Compounds and Consumption of Oxygen

Chemical compounds (smoke, toxic, corrosive, and odorous compounds) are the main contributors to nonthermal hazard and thus the assessments of their chemical natures and generation rates, relative to the airflow rate, are of critical importance for the protection of life and property.[1]

In fires, compounds are generated as a result of gasification and decomposition of the material and burning of the species in the gas phase with air in the form of a diffusion flame. In general, generation of the fire products and consumption of oxygen in diffusion flames occur in two zones.[36]

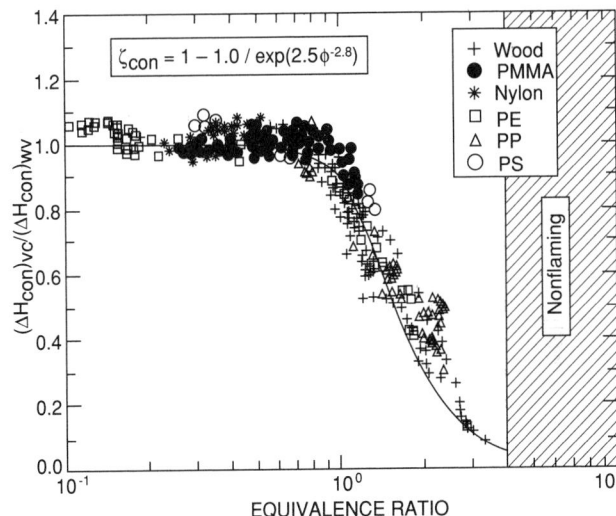

Fig. 3-4.26. Ratio of the ventilation-controlled to well-ventilated convective heat of combustion versus the equivalence ratio. Data are measured in the Flammability Apparatus and in the Fire Research Institute's enclosure.[36] Subscript vc represents ventilation-controlled fires, and subscript wv represents well-ventilated fires.

1. **Reduction Zone:** In this zone, the material melts, decomposes, gasifies, and/or generates species that react to form smoke, CO, hydrocarbons, and other intermediate products. Very little oxygen is consumed in this region. The extent of conversion of the material to smoke, CO, hydrocarbons, and other products depends on the chemical nature of the material.

2. **Oxidation Zone:** In this zone, the reduction zone products (smoke, CO, hydrocarbons, and other intermediates) react with varying degrees of efficiency with the oxygen from air and generate chemical heat and varying amounts of products of complete combustion, such as CO_2 and H_2O. The lower the reaction efficiency, the higher the amounts of reduction zone products emitted from a fire. The reaction efficiency of the reduction zone products with oxygen depends on the concentrations of the products relative to the oxygen concentration, temperature, and mixing of the products and air. For example, in laminar diffusion flames, smoke is emitted when the temperature of the oxidation zone falls below about 1300 K.

The hot ceiling layer in a building fire may be considered in terms of oxidation and reduction zone products. In building fires with plenty of ventilation, the concentrations of the reduction zone products are higher in the central region of the ceiling layer, whereas the concentrations of the oxidation zone products are higher closer to the room opening. As the air supply rate or oxygen concentration, available to the fire, decreases due to restrictions in the ventilation, the ceiling layer expands and starts occupying greater room volume with increase in the concentrations of the reduction zone products. Under these conditions, large amounts of the reduction zone products are released within the building increasing the nonthermal hazard.

The generation rate of a fire product is directly proportional to the mass loss rate, the proportionality constant being defined as the yield of the product[1-4,9-16,33,36-39,42-45]

$$\dot{G}''_j = y_j \dot{m}'' \qquad (42)$$

where $\dot{G}''_j$ is the mass generation rate of product j (g/m²-s), and y_j is the yield of product j (g/g). The total mass of the product generated is obtained by the summation of the generation rate

$$W_j = A \sum_{n=t_0}^{n=t_f} \dot{G}''_j(t_n) \Delta t_n \qquad (43)$$

where W_j is the total mass of product j generated from the flaming and/or nonflaming fire of the material (g), t_0 is the time when the sample is exposed to heat (s), and t_f is the time when there is no more vapor formation (s). From Equations 27, 42, and 43, the average value of the yield of product j is

$$\overline{y_j} = \frac{W_j}{W_f} \qquad (44)$$

The mass consumption rate of oxygen is also directly proportional to the mass loss rate[1-4,9-16,33,36-39,42-45]

$$\dot{C}''_O = c_O \dot{m}'' \qquad (45)$$

where $\dot{C}''_O$ is the mass consumption rate of oxygen (g/m²-s), and c_O is the mass of oxygen consumed per unit mass of fuel (g/g).

The mass generation rates of fire products and mass consumption rate of oxygen are determined by measuring

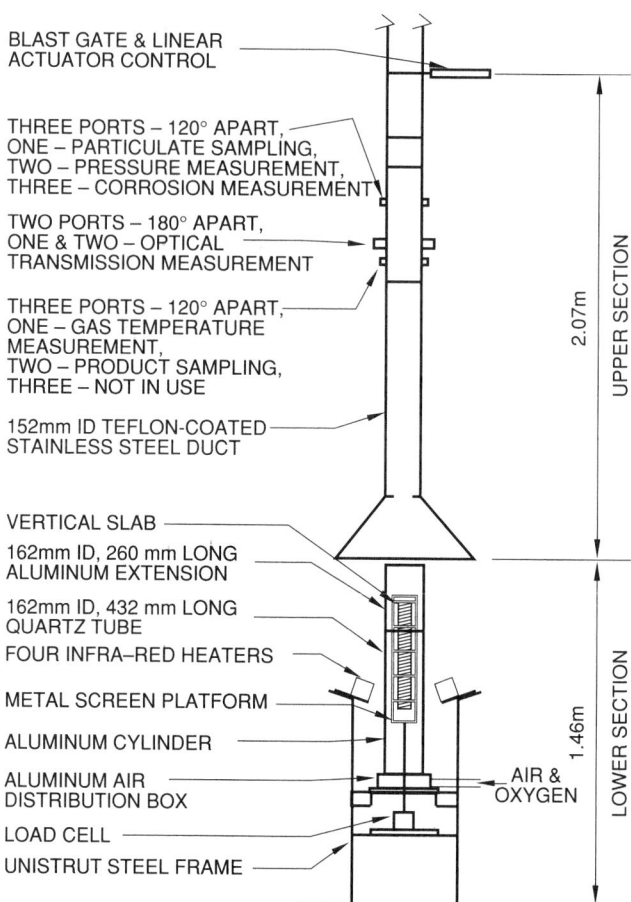

BLAST GATE & LINEAR ACTUATOR CONTROL

THREE PORTS – 120° APART, ONE – PARTICULATE SAMPLING, TWO – PRESSURE MEASUREMENT, THREE – CORROSION MEASUREMENT

TWO PORTS – 180° APART, ONE & TWO – OPTICAL TRANSMISSION MEASUREMENT

THREE PORTS – 120° APART, ONE – GAS TEMPERATURE MEASUREMENT, TWO – PRODUCT SAMPLING, THREE – NOT IN USE

152mm ID TEFLON-COATED STAINLESS STEEL DUCT

2.07m

UPPER SECTION

VERTICAL SLAB

162mm ID, 260 mm LONG ALUMINUM EXTENSION

162mm ID, 432 mm LONG QUARTZ TUBE

FOUR INFRA–RED HEATERS

METAL SCREEN PLATFORM

ALUMINUM CYLINDER

ALUMINUM AIR DISTRIBUTION BOX

LOAD CELL

UNISTRUT STEEL FRAME

AIR & OXYGEN

1.46m

LOWER SECTION

Fig. 3-4.27. *Sketch of the Flammability Apparatus showing locations where measurements are made for the product concentration, optical transmission, particulate concentration, and corrosion.*

the volume fractions of the products and oxygen and the total volumetric or mass flow rate of the fire products-air mixture[2,3,36]

$$\dot{G}''_j = \frac{f_j \dot{V} \rho_j}{A} = f_j \dot{W} \left(\frac{\rho_j}{\rho_g A} \right) \qquad (46)$$

$$\dot{C}''_O = \frac{f_O \dot{V} \rho_O}{A} = f_O \dot{W} \left(\frac{\rho_O}{\rho_g A} \right) \qquad (47)$$

where f_j is the volume fraction of product j, f_O is the volume fraction of oxygen, $\dot{V}$ is the total volumetric flow rate of the fire product-air mixture (m³/s), $\dot{W}$ is the total mass flow rate of the fire product-air mixture (g/s), ρ_j is the density of product j at the temperature of the fire product-air mixture (g/m³), ρ_g is the density of the hot fire product-air mixture (g/m³), ρ_O is the density of oxygen at the temperature of the fire product-air mixture (g/m³), and A is the total area of the material burning (m²).

For volume fraction measurements, sampling ducts are used where fire products and air are well mixed, such as in the Flammability Apparatuses [Figure 3-4.2, parts (a) and (b) and 3-4.8] and in the Cone Calorimeter (Figure 3-4.3). Figure 3-4.27 shows the measurement locations in the sampling duct of the Flammability Apparatus. The volume fractions are measured by various types of instruments; e.g., in the Flammability Apparatuses, they are measured continuously: (1) by commercial infrared analyzers for CO and CO_2; (2) by a high-sensitivity commercial paramagnetic analyzer for oxygen; (3) by a commercial flame ionization analyzer for the mixture of low molecular weight gaseous hydrocarbons; and (4) by a turbidimeter, designed by the Flammability Laboratory,[52] for smoke. The turbidimeter measures the optical density defined as

$$D = \frac{\ln\left(\frac{I_0}{I}\right)}{\ell} \quad (48)$$

where D is the optical density (1/m), I/I_0 is the fraction of light transmitted through smoke, and ℓ is the optical path length (m). The volume fraction of smoke is obtained from the following relationship[52]

$$f_s = \frac{D\lambda \times 10^{-6}}{\Omega} \quad (49)$$

where f_s is the volume fraction of smoke, λ is the wavelength of the light source (μm), and Ω is the coefficient of particulate extinction taken as 7.0.[52] In the Flammability Apparatuses, optical density is measured at wavelengths of 0.4579 μm (blue), 0.6328 μm (red), and 1.06 μm (IR). In the Cone Calorimeter, optical density is measured by using a helium-neon laser with a wavelength of 0.6328 μm (red).

From Equations 46 and 49

$$\dot{G}''_s = \frac{f_s \dot{V} \rho_s \times 10^{-6}}{A} = \left(\frac{D\lambda}{7}\right)\left(\frac{\rho_s \dot{V} \times 10^{-6}}{A}\right)$$
$$= \left(\frac{D\lambda}{7}\right)\left(\frac{\rho_s}{\rho_a}\right)\left(\frac{\dot{W} \times 10^{-6}}{A}\right) \quad (50)$$

In the Flammability Apparatuses and the Cone Calorimeter, the fire products in the sampling duct are diluted about 20 times and thus the density of air, $\rho_a = 1.2 \times 10^3$ g/m^3, and the density of smoke, $\rho_s = 1.1 \times 10^6$ g/m^3, as suggested in reference 52, are used.

$$\dot{G}''_s = \left(\frac{1.1 \times 10^6 \times 10^{-6}}{7}\right)\left(\frac{\dot{V}}{A}\right)D\lambda$$
$$= \left(\frac{1.1 \times 10^6 \times 10^{-6}}{7 \times 1.2 \times 10^3}\right)\left(\frac{\dot{W}}{A}\right)D\lambda \quad (51)$$

For blue wavelength of light ($\lambda = 0.4579$ μm)

$$\dot{G}''_s = 0.0720\left(\frac{D_{blue}\dot{V}}{A}\right) = 0.0600 \times 10^{-3}\left(\frac{D_{blue}\dot{W}}{A}\right) \quad (52)$$

For red wavelength of light ($\lambda = 0.6328$ μm)

$$\dot{G}''_s = 0.0994\left(\frac{D_{red}\dot{V}}{A}\right) = 0.0829 \times 10^{-3}\left(\frac{D_{red}\dot{W}}{A}\right) \quad (53)$$

For infrared wavelength of light ($\lambda = 1.06$ μm)

$$\dot{G}''_s = 0.1666\left(\frac{D_{IR}\dot{V}}{A}\right) = 0.1388 \times 10^{-3}\left(\frac{D_{IR}\dot{W}}{A}\right) \quad (54)$$

where D_{blue}, D_{red}, and D_{IR} are the optical densities measured at wavelengths of 0.4579, 0.6328, and 1.06 μm, respectively. These optical densities and total mass flow rate of the fire products-air mixture, $\dot{W}$, are measured continuously in the Flammability Apparatuses and the Cone Calorimeter, and A is known. The generation rates of smoke obtained from the optical densities at three wavelengths in

the Flammability Apparatus are averaged. The smoke mass generated in the test is also measured continuously in the Flammability Apparatus by a commercial smoke mass monitoring instrument. The data are used to calculate the mass generation rate of smoke. The smoke generation rates obtained from the optical density and smoke mass monitor show very good agreement.

In the Cone Calorimeter, the smoke data are reported in terms of the average specific extinction area (m^2/kg)[19]

$$\bar{\tau} = \frac{\sum_i \dot{V}_i D_i \Delta t_i}{W_f} \quad (55)$$

where $\bar{\tau}$ is the average specific extinction area determined in the Cone Calorimeter (m^2/g). Multiplying both sides of Equation 55 by $\rho_s \lambda \times 10^{-6}/7$ and rearranging

$$\bar{\tau}(\rho_s\lambda/7) \times 10^{-6} = \frac{\sum_i [(D_i\lambda \times 10^{-6}/7)\rho_s \dot{V}_i]\Delta t_i}{W_f}$$
$$= \frac{W_s}{W_f} = \overline{y_s} \quad (56)$$

In the Cone Calorimeter, $\lambda = 0.6328$ μm for red wavelength and using $\rho_s = 1.1 \times 10^6$ g/m^3, as suggested in reference 52, the average yield of smoke from the average specific extinction area determined in the Cone Calorimeter can be calculated from the following expression

$$\overline{y_s} = 0.0994 \times 10^{-3}\bar{\tau} \quad (57)$$

where $\overline{y_s}$ is the average yield of smoke (g/g).

The smoking characteristics of a material are also reported in terms of *mass optical density* (MOD)[1,4,16,50]

$$\text{MOD} = \left[\frac{\log_{10}\left(\frac{I_0}{I}\right)}{\ell}\right]\left[\frac{\dot{V}}{A\dot{m}''}\right] = \left[\frac{D}{2.303}\right]\left[\frac{\dot{V}}{A\dot{m}''}\right] \quad (58)$$

From Equations 42 and 50, with $\rho_s = 1.1 \times 10^6$ g/m^3 and $\lambda = 0.6328$ μm

$$y_s = \left(\frac{\lambda\rho_s}{7.0}\right)\left(\frac{D\dot{V} \times 10^{-6}}{A\dot{m}''}\right) = 0.0994(\text{MOD}/2.303) \quad (59)$$

MOD is generally reported with $\log_{10}$, however if it is changed to $\log_e$ and m^2/kg by multiplying it by 2.303 and dividing it by 1000, it becomes the specific extinction area, a terminology used in reporting the cone calorimeter data.

The average data for the yields of CO, CO_2, mixture of gaseous hydrocarbons, and smoke for well-ventilated fires are listed in Table 3-4.11.

EXAMPLE 14:

For a fiberglass-reinforced material, the following data were measured for combustion in normal air at an external heat flux value of 50 kW/m^2:

Total mass of the sample lost (g)	229
Total mass generated (g)	
CO	0.478
CO_2	290
Hydrocarbons	0.378
Smoke	6.31
Total energy generated (kJ)	3221

Calculate the average yields of CO, CO_2, hydrocarbons, and smoke and the average chemical heat of combustion.

SOLUTION:

The average yields are calculated from Equation 44, and the average chemical heats of combustion are calculated from Equation 29.

Average yields (g/g)
  CO   0.0021
  CO_2 1.27
  Hydrocarbons   0.002
  Smoke   0.028
Average chemical heats of combustion (kJ/g)   14.1

EXAMPLE 15:

A circular sample of polystyrene, about 0.007 m^2 in area and 25 mm in thickness, was burned in normal air in the presence of external heat flux. In the test, measurements were made for the mass loss rate and light obscuration by smoke in the sampling duct with an optical path length of 0.149 m. The total volumetric flow rate of the mixture of fire products and air through the sampling duct was 0.311 m^3/s, and the wavelength of light source used was 0.6328 μm. At the steady-state combustion of polystyrene, the measured mass loss rate was 33 g/m^2-s with smoke obscuring 83.5 percent of the light. Calculate the yield of smoke from the data using a value of 1.1×10^6 g/m^3 for the density of smoke.

SOLUTION:

Optical density from Equation 48:

$$D = \frac{\ln(I_0/I)}{\ell} = \frac{\ln(100/83.5)}{0.149} = 121 \ (1/m)$$

Smoke generation rate from Equation 51

$$\dot{G}''_s = \frac{1.1\dot{V}D\lambda}{7 \times A} = \frac{1.1 \times 0.311 \times 1.21 \times 0.6328}{7 \times 0.007} = 5.35 \ \text{g/m}^2\text{-s}$$

Smoke yield from Equation 42

$$y_s = \frac{5.35 \text{g/m}^2\text{-s}}{33 \text{g/m}^2\text{-s}} = 0.162 \ \text{g/g}$$

Efficiencies of oxygen mass consumption and mass generation of products: A chemical reaction between oxygen and a fuel monomer of a material can be expressed as

$$F + \nu_O O_2 + \nu_N N_2 + \nu_{j_1} J_1 + \nu_{j_2} J_2 + \nu_N N_2 \qquad (60)$$

where F is the fuel monomer of a material; ν_O and ν_N are the stoichiometric coefficients for oxygen and nitrogen, respectively; and ν_{j_1} and ν_{j_2} are the stoichiometric coefficients for the maximum possible conversion of the fuel monomer to products J_1 and J_2, respectively.

The stoichiometric mass oxygen-to-fuel ratio for the maximum possible conversion of the fuel monomer is expressed as

$$\Psi_O = \frac{\nu_O M_O}{M_f} \qquad (61)$$

where Ψ_O is the stoichiometric mass oxygen-to-fuel ratio for the maximum possible conversion of the fuel monomer to products; M_O is the molecular weight of oxygen (32 g/mole); and M_f is the molecular weight of the fuel monomer of the

material (g/mole), which is calculated from its elemental composition. For the elemental composition measurements, microanalytical techniques are used.

The stoichiometric yield for the maximum possible conversion of the fuel monomer of the material to a product is expressed as

$$\Psi_j = \frac{\nu_j M_j}{M_f} \qquad (62)$$

where Ψ_j is the stoichiometric yield for the maximum possible conversion of the fuel monomer of the material to product j, and M_j is the molecular weight of product (g/mole).

The stoichiometric yields for some selected materials, calculated from the elemental composition data from the flammability laboratory, are listed in Table 3-4.13 for fuel monomer conversion to CO, CO_2, hydrocarbons, smoke, HCl, and HF. The stoichiometric yields depend on the number of atoms relative to the carbon atom. The yields provide an insight into the nature of products and their amounts expected to be generated in flaming and nonflaming fires, when expressed as the stoichiometric oxygen mass consumption rate and stoichiometric mass generation rates of products

$$\dot{C}''_{\text{stoich},O} = \Psi_O \dot{m}'' \qquad (63)$$

$$\dot{G}''_{\text{stoich},j} = \Psi_j \dot{m}'' \qquad (64)$$

where $\dot{C}''_{\text{stoich},O}$ and $\dot{G}''_{\text{stoich},j}$ are the stoichiometric oxygen mass consumption rate and stoichiometric mass generation rate of product j for the maximum possible conversion of the fuel monomer to the product, respectively (g/m^2-s).

In fires, the actual oxygen mass consumption rate and the mass generation rates of products are significantly less than the stoichiometric rates. The ratio of the actual oxygen mass consumption rate to stoichiometric rates is thus defined as the *efficiency of oxygen mass consumption* or *product mass generation*[2–4,16,36]

$$\eta_O = \frac{\dot{C}''_{\text{actual},O}}{\dot{C}''_{\text{stoich},O}} = \frac{c_O \dot{m}''}{\Psi_O \dot{m}''} = \frac{c_O}{\Psi_O} \qquad (65)$$

$$\eta_j = \frac{\dot{G}''_{\text{actual},j}}{\dot{G}''_{\text{stoich},j}} = \frac{y_j \dot{m}''}{\Psi_j \dot{m}''} = \frac{y_j}{\Psi_j} \qquad (66)$$

where η_O is efficiency of oxygen mass consumption, and η_j is the generation efficiency of product j; subscript represents the actual oxygen mass consumption rate or the actual mass generation rate of a product.

EXAMPLE 16:

A material is made up of carbon, hydrogen, and oxygen. The weight of the material is distributed as follows: 54 percent as carbon, 6 percent as hydrogen, and 40 percent as oxygen. Calculate the chemical formula of the fuel monomer of the material.

SOLUTION:

From the atomic weights and the weight percent of the atoms, the number of atoms are: carbon (C): 54/12 = 4.5; hydrogen (H): 6/1 = 6.0; and oxygen (O): 40/16 = 2.5. Thus the chemical formula of the fuel monomer of the material is $C_{4.5}H_{6.0}O_{2.5}$ or dividing by 4.5, $CH_{1.33}O_{0.56}$.

TABLE 3-4.13 *Stoichiometric Yields of Major Products**

Material	Formula	Ψ_O	Ψ_{CO_2}	Ψ_{CO}	Ψ_s	Ψ_{hc}	Ψ_{HCl}	Ψ_{HF}
Carbon-Hydrogen Atoms in the Structure								
PE	CH_2	3.43	3.14	2.00	0.857	1.00	0	0
PP	CH_2	3.43	3.14	2.00	0.857	1.00	0	0
PS	CH	3.08	3.38	2.15	0.923	1.00	0	0
Expanded Polystyrene								
GM47	$CH_{1.1}$	3.10	3.36	2.14	0.916	1.00	0	0
GM49	$CH_{1.1}$	3.10	3.36	2.14	0.916	1.00	0	0
GM51	CH	3.08	3.38	2.15	0.923	1.00	0	0
GM53	$CH_{1.1}$	3.10	3.36	2.14	0.916	1.00	0	0
Carbon-Hydrogen-Oxygen-Nitrogen Atoms in the Structure								
POM	CH_2O	1.07	1.47	0.933	0.400	0.467	0	0
PMMA	$CH_{1.6}O_{0.40}$	1.92	2.20	1.40	0.600	0.680	0	0
Nylon	$CH_{1.8}O_{0.17}N_{0.17}$	2.61	2.32	1.48	0.634	0.731	0	0
Wood (pine)	$CH_{1.7}O_{0.83}$	1.21	1.67	1.06	0.444	0.506	0	0
Wood (oak)	$CH_{1.7}O_{0.72}N_{0.001}$	1.35	1.74	1.11	0.476	0.543	0	0
Wood (Douglas fir)	$CH_{1.7}O_{0.74}N_{0.002}$	1.32	1.72	1.10	0.469	0.536	0	0
Polyester	$CH_{1.4}O_{0.22}$	2.35	2.60	1.65	0.709	0.792	0	0
Epoxy	$CH_{1.3}O_{0.20}$	2.38	2.67	1.70	0.727	0.806	0	0
Polycarbonate	$CH_{0.88}O_{0.19}$	2.26	2.76	1.76	0.754	0.872	0	0
PET	$CH_{0.80}O_{0.40}$	1.67	2.29	1.46	0.625	0.667	0	0
Phenolic foam	$CH_{1.1}O_{0.24}$	2.18	2.60	1.65	0.708	0.773	0	0
PAN	$CHN_{0.33}$	2.87	2.50	1.59	0.681	0.681	0	0
Flexible Polyurethane Foams								
GM21	$CH_{1.8}O_{0.30}N_{0.05}$	2.24	2.28	1.45	0.622	0.715	0	0
GM23	$CH_{1.8}O_{0.35}N_{0.06}$	2.11	2.17	1.38	0.593	0.682	0	0
GM25	$CH_{1.7}O_{0.32}N_{0.07}$	2.16	2.22	1.41	0.606	0.692	0	0
GM27	$CH_{1.7}O_{0.30}N_{0.08}$	2.21	2.24	1.43	0.612	0.698	0	0
Rigid Polyurethane Foams								
GM29	$CH_{1.1}O_{0.23}N_{0.10}$	2.22	2.42	1.54	0.660	0.721	0	0
GM31	$CH_{1.2}O_{0.22}N_{0.10}$	2.28	2.43	1.55	0.662	0.729	0	0
GM37	$CH_{1.2}O_{0.20}N_{0.08}$	2.34	2.51	1.60	0.685	0.753	0	0
Rigid Polyisocyanurate Foams								
GM41	$CH_{1.0}O_{0.19}N_{0.11}$	2.30	2.50	1.59	0.683	0.740	0	0
GM43	$CH_{0.93}O_{0.20}N_{0.11}$	2.25	2.49	1.58	0.679	0.732	0	0
Carbon-Hydrogen-Oxygen-Silicone Atoms in the Structure								
Silicone-1[†]	$CH_{1.3}O_{0.25}Si_{0.18}$	1.98	1.97	1.25	0.537	0.595	0	0
Silicone-2[‡]	$CH_{1.5}O_{0.30}Si_{0.26}$	1.86	1.72	1.09	0.469	0.528	0	0
Silicone-3[§]	$CH_3O_{0.50}Si_{0.50}$	1.73	1.19	0.757	0.324	0.405	0	0
Carbon-Hydrogen-Oxygen-Chlorine-Fluorine Atoms in the Structure								
Fluoropolymers								
PVF (Tedlar™)	$CH_{1.5}F_{0.50}$	1.74	1.91	1.22	0.522	0.587	0	0.435
PVF$_2$ (Kynar™)	CHF	1.00	1.38	0.875	0.375	0.406	0	0.594
ETFE (Tefzel™)	$CH_{1.0}F_{0.99}$	1.01	1.38	0.880	0.377	0.409	0	0.622
E-CTFE (Halar™)	$CHF_{0.75}Cl_{0.25}$	0.889	1.22	0.778	0.333	0.361	0.257	0.417
PFA (Teflon™)	$CF_{1.7}O_{0.01}$	0.716	1.00	0.630	0.270	0	0	0.765
FEP (Teflon™)	$CF_{1.8}$	0.693	0.952	0.606	0.260	0	0	0.779
TFE (Teflon™)	CF_2	0.640	0.880	0.560	0.240	0	0	0.800
CTFE (Kel-F™)	$CF_{1.5}Cl_{0.50}$	0.552	0.759	0.483	0.207	0	0.310	0.517
Chloropolymers								
PE-25% Cl	$CH_{1.9}Cl_{0.13}$	2.56	2.38	1.52	0.650	0.753	0.254	0
PE-36% Cl	$CH_{1.8}Cl_{0.22}$	2.16	2.05	1.30	0.558	0.642	0.368	0
Neoprene	$CH_{1.25}Cl_{0.25}$	1.91	2.00	1.27	0.546	0.602	0.409	0
PE-42% Cl	$CH_{1.8}Cl_{0.29}$	1.94	1.84	1.17	0.501	0.576	0.424	0
PE-48% Cl	$CH_{1.7}Cl_{0.36}$	1.73	1.67	1.06	0.456	0.521	0.493	0
PVC	$CH_{1.5}Cl_{0.50}$	1.42	1.42	0.903	0.387	0.436	0.581	0
PVCl$_2$	$CHCl$	0.833	0.917	0.583	0.250	0.271	0.750	0

*Calculated from the data for the elemental compositions of the materials in the FMRC flammability laboratory; subscript hc is total gaseous hydrocarbons; s is soot.

[†] $\eta_{SiO_2} = 0.483$.
[‡] $\eta_{SiO_2} = 0.610$.
[§] $\eta_{SiO_2} = 0.811$.

EXAMPLE 17:

For the material in example 16, calculate the stoichiometric mass oxygen-to-fuel ratio, stoichiometric mass air-to-fuel ratio, and stoichiometric yields for maximum possible conversion of the fuel monomer of the material to CO, CO_2, hydrocarbons, water, and smoke. Assume smoke to be pure carbon, and hydrocarbons as having the same carbon atom to hydrogen atom ratio as the original fuel monomer.

SOLUTION:

1. For stoichiometric yields of CO_2 and water and the stoichiometric mass oxygen and air-to-fuel ratio for the maximum possible conversion of the fuel monomer of the material to CO_2 and H_2O, the following expression represents the maximum possible conversion of the fuel monomer of the material to CO_2 and H_2O

$$CH_{1.33}O_{0.56} + 1.06\, O_2 = CO_2 + 0.67\, H_2O$$

The molecular weight of the fuel monomer of the material is $1 \times 12 + 1.33 \times 1 + 0.56 \times 16 = 22.3$, the molecular weight of oxygen is 32, the molecular weight of CO_2 is 44, and the molecular weight of H_2O is 18. Thus,

$$\Psi_{CO_2} = \frac{44}{22.3} = 1.97,$$

$$\Psi_{H_2O} = \frac{0.67 \times 18}{22.3} = 0.54, \text{ and}$$

$$\Psi_O = \frac{1.06 \times 32}{22.3} = 1.52;$$

The stoichiometric mass air-to-fuel ratio can be obtained by dividing Ψ_O by 0.233; i.e., $1.52/0.233 = 6.52$.

2. For stoichiometric yields of CO, hydrocarbons, and smoke for the maximum possible conversion of the fuel monomer of the material to these products, the following expressions represent the maximum possible conversion of the fuel monomer of the material to these products

For CO

$$CH_{1.33}O_{0.56} + zO_2 = CO + x(HO),$$
$$\Psi_{CO} = \frac{28}{22.3} = 1.26;$$

For hydrocarbons

$$CH_{1.33}O_{0.56} + zO_2 = CH_{1.33} + x(HO),$$
$$\Psi_{hc} = \frac{13.3}{22.3} = 0.60; \text{ and}$$

For smoke

$$CH_{1.33}O_{0.56} + zO_2 = C + x(HO), \quad \Psi_s = \frac{12}{22.3} = 0.54.$$

EXAMPLE 18:

For the material in examples 16 and 17, the generation efficiencies of CO_2, CO, hydrocarbons, and smoke are 0.90, 0.004, 0.002, and 0.036, respectively; the heat of gasification is 1.63 kJ/g; the surface re-radiation loss is 11 kW/m^2; and the predicted asymptotic flame heat flux value for large-scale fires is 60 kW/m^2. Calculate the yields and asymptotic values for the generation rates of CO_2, CO, hydrocarbons, and smoke expected in large-scale fires.

SOLUTION:

1. Yields from Equation 66 and data from example 17

$$y_{CO_2} = 0.90 \times 1.97 = 1.77 \text{ g/g};$$
$$y_{CO} = 0.004 \times 1.26 = 0.005 \text{ g/g};$$
$$y_{hc} = 0.002 \times 0.60 = 0.001 \text{ g/g}; \text{ and}$$
$$y_s = 0.036 \times 0.54 = 0.019 \text{ g/g}.$$

2. Asymptotic values for the mass loss rate from Equation 11

$$\dot{m}'' = \frac{60 - 11}{1.63} = 30 \text{ g/m}^2\text{-s}$$

3. Asymptotic values for the mass generation rates of products from Equation 42 and the above data

$$\dot{G}''_{CO_2} = 1.77 \times 30 = 53 \text{ g/m}^2\text{-s};$$
$$\dot{G}''_{CO} = 0.005 \times 30 = 0.159 \text{ g/m}^2\text{-s};$$
$$\dot{G}''_{hc} = 0.001 \times 30 = 0.036 \text{ g/m}^2\text{-s}; \text{ and}$$
$$\dot{G}''_s = 0.019 \times 30 = 0.584 \text{ g/m}^2\text{-s}.$$

Generation rates of fire products and fire ventilation: As discussed previously, the effects of decrease in fire ventilation, as characterized by the increase in the local equivalence ratio, are reflected in the increase in the generation rates of the reduction zone products (smoke, CO, hydrocarbons, and others). For example, for flaming wood crib enclosure fires, as the equivalence ratio increases, the combustion efficiency decreases, flame becomes unstable, and the generation efficiency of CO reaches its peak for the equivalence ratio between about 2.5 and 4.0.[36]

The ventilation-controlled building fires are generally characterized by two layers: (1) a ceiling vitiated layer, identified as "upper layer," and (2) an uncontaminated layer below, identified as "lower layer." Incorporation of these two layers is the classical two-zone modeling of fires in enclosed spaces. Under many conditions, the depth of the "upper layer" occupies a significant fraction of the volume of the enclosed space. Eventually, the interface between the "upper layer" and the "lower layer" positions itself so that it is very close to the floor, very little oxygen is available for combustion, and most of the fuel is converted to the reduction zone products, i.e., smoke, CO, hydrocarbons, and others.

The ventilation-controlled large- and small-enclosure and laboratory-scale fires and fires in the vitiated "upper layer" under the experimental hoods have been studied in detail, and are discussed or reviewed in references 36 and 53 through 56. The results from these types of fires are very similar. Detailed studies[36] performed for the generation rates of fire products for various fire ventilation conditions in the Flammability Apparatus [Figure 3-4.2(a)], and in the Fire Research Institute's (FRI) enclosure, show that with increase in the equivalence ratio: (1) generation efficiencies of oxidation zone products, such as CO_2, and reactant consumption efficiency (i.e., oxygen) decrease, and (2) generation efficiencies of the reduction zone products, such as smoke, CO, and hydrocarbons increase.

Generalized correlations have been established between the generation efficiencies and the equivalence ratio for the oxidation and reduction zone products. The changes in the consumption or generation efficiencies of the products are expressed as ratios of the efficiencies for the ventilation-controlled (*vc*) to well-ventilated (*wv*) fires:

Reactants (Oxygen)

$$\zeta_O = \frac{(\eta_O)_{vc}}{(\eta_O)_{wv}} = \frac{(c_O/\Psi_O)_{vc}}{(c_O/\Psi_O)_{wv}} = \frac{(c_O)_{vc}}{(c_O)_{wv}} \tag{67}$$

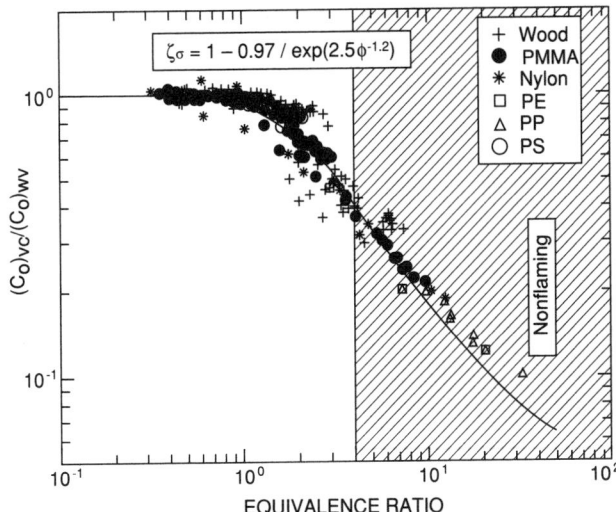

Fig. 3-4.28. *Ratio of the mass of oxygen consumed per unit mass of the fuel for ventilation-controlled to well-ventilated fires. Data are measured in the Flammability Apparatus and in the Fire Research Institute's enclosure.[36] Subscript vc represents ventilation-controlled fires, and subscript wv represents well-ventilated fires.*

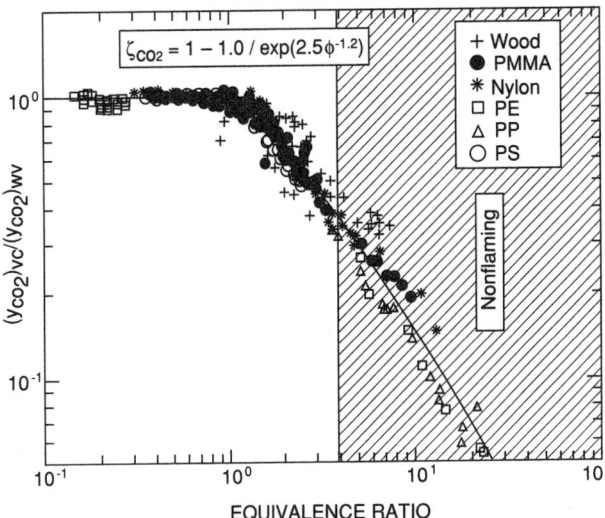

Fig. 3-4.29. *Ratio of the mass of carbon dioxide generated per unit mass of the fuel for ventilation-controlled to well-ventilated fires. Data are measured in the Flammability Apparatus and in the Fire Research Institute's enclosure.[36] Subscript vc represents ventilation-controlled fires, and subscript wv represents well-ventilated fires.*

Oxidation Zone Products (Carbon Dioxide, Water, etc.)

$$\zeta_{oxid} = \frac{(\eta_j)_{vc}}{(\eta_j)_{wv}} = \frac{(y_j/\Psi_j)_{vc}}{(y_j/\Psi_j)_{wv}} = \frac{(y_j)_{vc}}{(y_j)_{wv}} \qquad (68)$$

where ζ_{oxid} is the oxidation zone product generation efficiency ratio.

Reduction Zone Products (Smoke, Carbon Monoxide, Hydrocarbons, etc.)

$$\zeta_{red} = \frac{(\eta_j)_{vc}}{(\eta_j)_{wv}} = \frac{(y_j/\Psi_j)_{vc}}{(y_j/\Psi_j)_{wv}} = \frac{(y_j)_{vc}}{(y_j)_{wv}} \qquad (69)$$

where ζ_{red} is the reduction zone product generation efficiency ratio.

The relationships between the ratios of the mass of oxygen consumed per unit mass of fuel, the yields of the products for the ventilation-controlled to well-ventilated fires, and the equivalence ratio are shown in Figures 3-4.28 through 3-4.32. The ratios for oxygen and CO_2 (an oxidation zone product) do not depend on the chemical structures of the materials, whereas the ratios for the reduction zone products do depend on the chemical structures of the materials.

Oxygen and CO_2: The relationships for oxygen consumed and carbon dioxide generated are shown in Figures 3-4.28 and 3-4.29, respectively. The relationships are very similar to the relationships for the chemical and convective heats of combustion ratios (Equations 40 and 41), as expected

$$\frac{(c_O)_{vc}}{(c_O)_{wv}} = 1 - \frac{0.97}{\exp(\Phi/2.14)^{-1.2}} \qquad (70)$$

$$\frac{(y_{CO_2})_{vc}}{(y_{CO_2})_{wv}} = 1 - \frac{1.00}{\exp(\Phi/2.15)^{-1.2}} \qquad (71)$$

Carbon Monoxide: The relationship between the ratio of the CO yields for ventilation-controlled to well-ventilated

fires and the equivalence ratio is shown in Figure 3-4.30. The data suggest the following relationship[36]

$$\frac{(y_{CO})_{vc}}{(y_{CO})_{wv}} = 1 + \frac{\alpha}{\exp(2.5\Phi^{-\xi})} \qquad (72)$$

where α and ξ are the correlation coefficients, which depend on the chemical structures of the materials. The values for the correlation coefficients for CO are listed in Table 3-4.14.

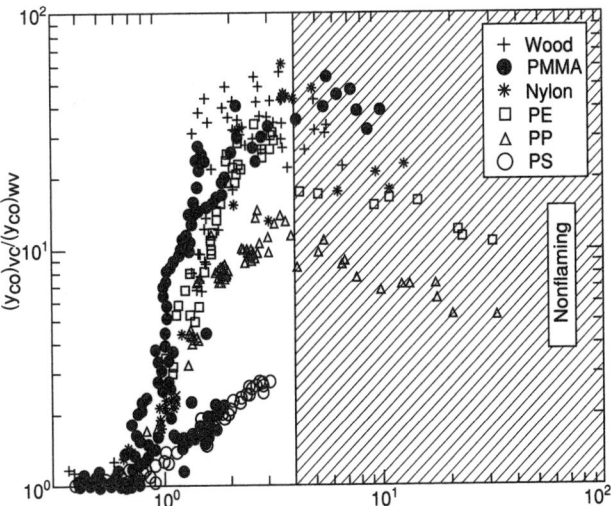

Fig. 3-4.30. *Ratio of the mass of carbon monoxide generated per unit mass of the fuel for ventilation-controlled to well-ventilated fires. Data are measured in the Flammability Apparatus and in the Fire Research Institute's enclosure.[36] Subscript vc represents ventilation-controlled fires, and subscript wv represents well-ventilated fires.*

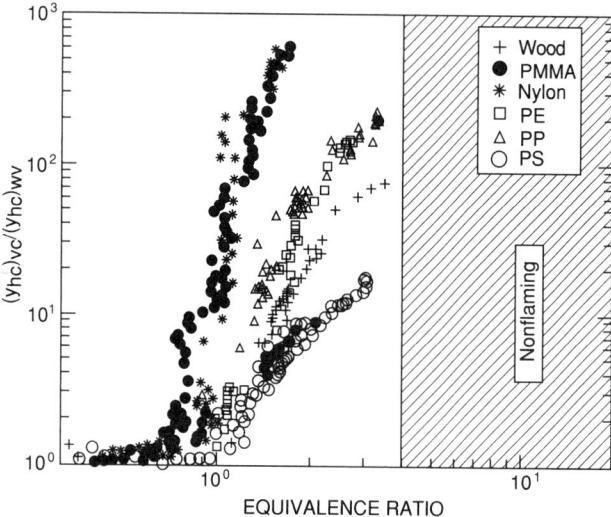

Fig. 3-4.31. Ratio of the mass of hydrocarbons generated per unit mass of the fuel for ventilation-controlled to well-ventilated fires. Data are measured in the Flammability Apparatus and in the Fire Research Institute's enclosure.[36] Subscript vc represents ventilation-controlled fires, and subscript wv represents well-ventilated fires.

The increase in the ratio of the carbon monoxide yields for the ventilation-controlled to well-ventilated fires with the equivalence ratio is due to the preferential conversion of the fuel carbon atoms to CO. The experimental data show the following order for the preferential conversion: wood (C–H–O aliphatic structure) > PMMA (C–H–O aliphatic structure) > nylon (C–H–O–N aliphatic structure) > PE (C–H aliphatic linear unsaturated structure) > PP (C–H aliphatic branched unsaturated structure) > PS (C–H aro-

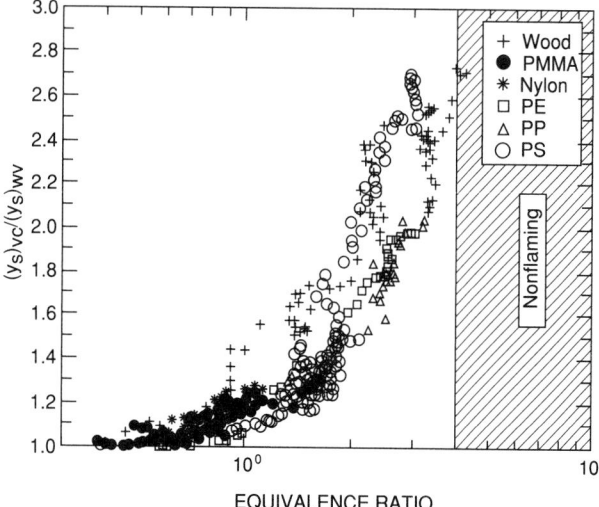

Fig. 3-4.32. Ratio of the mass of smoke generated per unit mass of the fuel for ventilation-controlled to well-ventilated fires. Data are measured in the Flammability Apparatus and in the Fire Research Institute's enclosure.[36] Subscript vc represents ventilation-controlled fires, and subscript wv represents well-ventilated fires.

TABLE 3-4.14 *Correlation Coefficients to Account for the Effects of Ventilation on the Generation Rates of CO, Hydrocarbons, and Smoke*

	CO		Hydrocarbons		Smoke	
Material	α	ξ	α	ξ	α	ξ
PS	2	2.5	25	1.8	2.8	1.3
PP	10	2.8	220	2.5	2.2	1.0
PE	26	2.8	220	2.5	2.2	1.0
Nylon	36	3.0	1200	3.2	1.7	0.8
PMMA	43	3.2	1800	3.5	1.6	0.6
Wood	44	3.5	200	1.9	2.5	1.2

matic structure). A similar trend is found for the liquid and gaseous fuels, such as shown in Table 3-4.15.[36] The presence of O and N atoms in the fuels with aliphatic C-H structure appears to enhance preferential fuel carbon atom conversion to CO.

Hydrocarbons: The relationship between the ratio of the hydrocarbon yields for ventilation-controlled to well-ventilated fires and the equivalence ratio is shown in Figure 3-4.31. The data suggest the following relationship[36]

$$\frac{(y_{hc})_{vc}}{(y_{hc})_{wv}} = 1 + \frac{\alpha}{\exp(5.0\Phi^{-\xi})} \qquad (73)$$

The correlation coefficient values for hydrocarbons are listed in Table 3-4.14. The numerator in the second term on the right-hand side of Equation 73 is 10 to 40 times that of CO, whereas the denominator is twice that for CO. This suggests that there is a significantly higher preferential fuel conversion to hydrocarbons than to CO, with increase in the equivalence ratio. The order for the preferential fuel conversion to hydrocarbons is very similar to CO, except for wood; i.e., PMMA > nylon > PE = PP > wood > PS. The exception for wood may be due to char-forming tendency of the fuel, which lowers the C to H ratio in the gas phase.

Smoke: The relationship between the ratio of the smoke yields for ventilation-controlled to well-ventilated fires and the equivalence ratio is shown in Figure 3-4.32. The data suggest the following relationship[36]

TABLE 3-4.15 *Carbon Monoxide Generation Efficiency for Ventilation-Controlled and Well-Ventilated Combustion**

Fuel	Well-Ventilated (wv)[†] $\Phi < 0.05$	Ventilation-Controlled (vc) $\Phi \approx 4.0$		$\dfrac{(y_{CO})_{vc}}{(y_{CO})_{wv}}$
		Ref. 54	Ref. 57	
Methane	0.001	0.10	—	100
Propane	0.001	—	0.12	120
Propylene	0.004	0.10	—	25
Hexane	0.002	0.10	0.52[‡]	50 (260[‡])
Methanol	0.001	0.27	1.00[‡]	270 (1000[‡])
Ethanol	0.001	0.18	0.66[‡]	180 (660[‡])
Isopropanol	0.002	0.21	—	105
Acetone	0.002	0.21	0.63[‡]	105 (315[‡])

*Table taken from reference 36.
[†]From Flammability Apparatus.
[‡]Nonflaming.

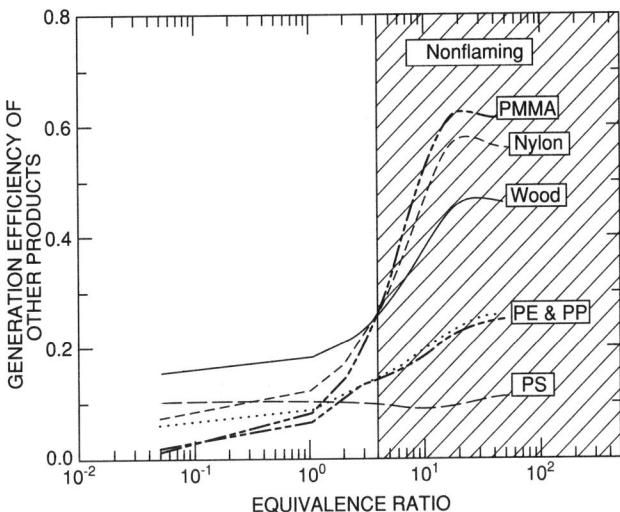

Fig. 3-4.33. *Generation efficiency of products other than CO, CO_2, hydrocarbons, and smoke versus the equivalence ratio.*

$$\frac{(y_s)_{vc}}{(y_s)_{wv}} = 1 + \frac{\alpha}{\exp(2.5\Phi^{-\xi})} \qquad (74)$$

The correlation coefficient values for smoke are listed in Table 3-4.14. The values of the correlation coefficients in the second term on the right-hand side of Equation 74 suggest that, with the increase in the equivalence ratio, the preferential fuel conversion to smoke is lower than it is to hydrocarbons and CO. Also, the order for the preferential conversion of the fuel carbon atom to smoke is opposite to the order for the conversion to CO and hydrocarbons, except for wood. The order is: PS > wood > PE = PP > nylon > PMMA, suggesting that the order is probably due to decrease in the preference for the reactions between OH and CO compared to the reactions between OH and soot.

Other Reduction Zone Products: Since the sum of the generation efficiencies of all the products for a material cannot

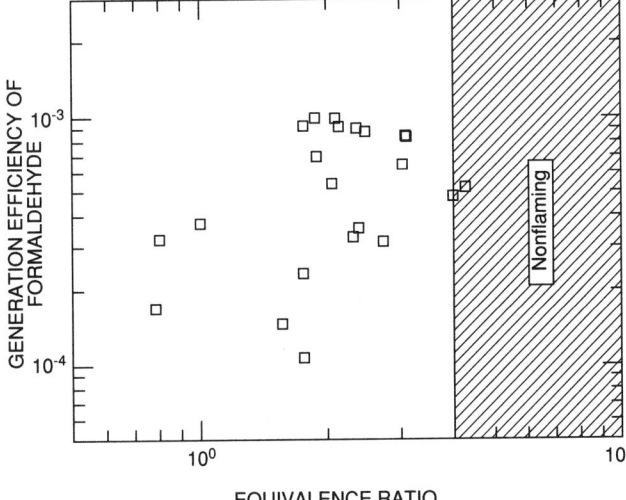

Fig. 3-4.34. *Generation efficiency of formaldehyde generated from wood versus the equivalence ratio.*

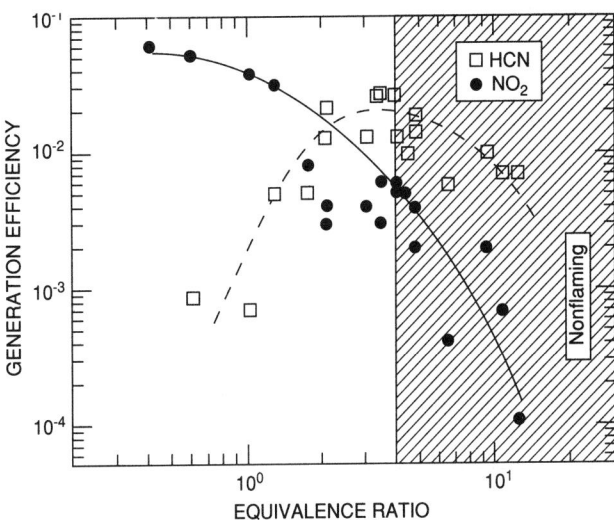

Fig. 3-4.35. *Generation efficiencies of hydrogen cyanide and nitrogen dioxide generated from nylon versus the equivalence ratio.*

exceed unity, the generation efficiency of products other than CO, CO_2, hydrocarbons, and smoke is

$$\eta_{other} = 1 - (\eta_{CO} + \eta_{CO_2} + \eta_{hc} + \eta_s) \qquad (75)$$

where η_{other} is the generation efficiency of products other than CO, CO_2, hydrocarbons, and smoke. The generation efficiency of other products can be calculated from Equations 71 through 75 using correlation coefficients from Table 3-4.14. The generation efficiency values for other products calculated in this fashion for various equivalence ratios are shown in Figure 3-4.33. The figure shows that, for equivalence ratios greater than 4, where fires are nonflaming, about 10 to 60 percent of fuel carbon is converted to products other than CO, CO_2, soot, and hydrocarbons.

The order for the preferential conversion of fuel carbon to other products in the nonflaming zone is: PS (C–H aromatic structure) < PE & PP (C–H aliphatic structure) < wood (C–H–O aliphatic structure) < nylon (C–H–O–N aliphatic structure) < PMMA (C–H–O aliphatic structure). It thus appears that, in nonflaming fire, fuels with C–H structures are converted mainly to CO, smoke, and hydrocarbons, rather than to other products, whereas fuels with C–H–O and C–H–O–N structures are converted mainly to products other than CO, CO_2, smoke, and hydrocarbons. Some of the products include formaldehyde (HCHO) and hydrogen cyanide (HCN).[36]

Generation Efficiencies of Formaldehyde, Hydrogen Cyanide, and Nitrogen Dioxide: The experimental data for the generation efficiencies of formaldehyde, hydrogen cyanide, and nitrogen dioxide *versus* the equivalence ratio are shown in Figures 3-4.34 and 3-4.35.

Formaldehyde is generated in the pyrolysis of wood (C–H–O structure). It is attacked rapidly by oxygen (O) and hydroxyl (OH) radicals in the flame, if unlimited supply of oxygen is available. Thus, only traces of formaldehyde are found in well-ventilated fires. The generation efficiency of formaldehyde, however, increases with the equivalence ratio, indicating reduced concentrations of O and OH radicals

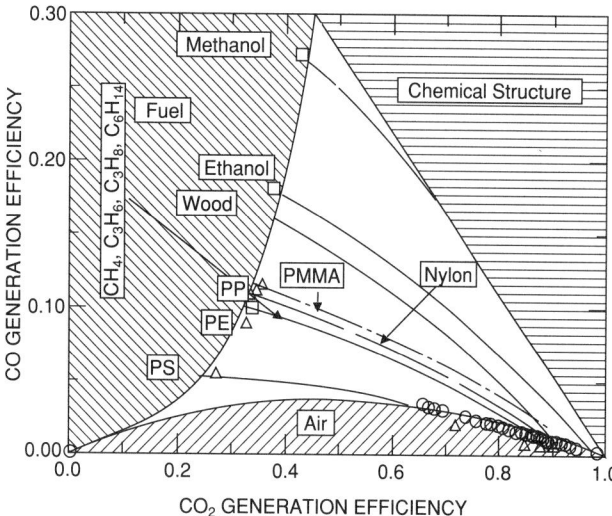

Fig. 3-4.36. Relationship between the generation efficiencies of CO_2 and CO. Data taken from reference 36.

and gas temperature due to lack of oxygen available for combustion.

In fires, hydrogen cyanide is formed in the reduction zone from materials with hydrogen and nitrogen atoms in the structure, such as nylon (C–H–O–N structure). Nitrogen dioxide (NO_2), on the other hand, is formed in the oxidation zone, as a result of the oxidation of hydrogen cyanide. The data in Figure 3-4.35 show that the generation efficiency of hydrogen cyanide increases and the generation efficiency of NO_2 decreases with the equivalence ratio. This observation supports that O and OH radical concentrations decrease with increase in the equivalence ratio. The decrease in the generation efficiency of hydrogen cyanide in the nonflaming fire suggests decrease in the fuel mass transfer rate.

Relationship Between the Generation Efficiencies of CO_2 and CO: The relationship between the generation efficiencies of CO_2 and CO is shown in Figure 3-4.36, where the data are taken from reference 36. CO is generated in the reduction zone of the flame as a result of the oxidative pyrolysis of the fuel, and is oxidized to CO_2 in the oxidation zone of the flame. The generation efficiency of CO_2 is independent of the chemical structure of the fuel (Figure 3-4.29), whereas the generation efficiency of CO depends on the chemical structure of the fuel (Figure 3-4.30). In Figure 3-4.36, the curves represent approximate predictions based on the correlation coefficients from Table 3-4.14 and Equations 71 and 72.

The relationship between the generation efficiencies of CO_2 and CO is quite complex. The boundary of the shaded region marked "air" in Figure 3-4.36 is drawn using the data for the well-ventilated combustion for equivalence ratios less than 0.05. The boundary of the "air" region may be considered as equivalent to the lower flammability limit. No flaming combustion is expected to occur in this region, as the fuel-air mixture is below the lower flammability limit; however, nonflaming combustion, generally identified as smoldering, may continue. The boundary of the shaded region marked "fuel" is drawn using the data for the ventilation-controlled combustion for equivalence ratio of 4.0, and may be considered as equivalent to the upper flammability limit. In the "fuel" region, no flaming combustion is expected to occur, as the fuel-air mixture is above the upper

flammability limit; however, nonflaming processes may continue. The shaded region marked "chemical structure," and drawn to the right of the methanol curve, is an imaginary region as it is not expected to exist, because there are no stable carbon-containing fuel structures below the formaldehyde with a structure of HCHO. For the stable fuels with C–H–O structures, formaldehyde (HCHO) and methanol (CH_3OH) have the lowest molecular weights (30 and 32, respectively); thus, data for HCHO and CH_3OH probably would be comparable.

The curves in Figure 3-4.36 show that, in flaming combustion, with increase in the equivalence ratio, the preference for fuel carbon atom conversion to CO, relative to the conversion to CO_2, follows the order: methanol (C–H–O structure) > ethanol (C–H–O structure) > wood (C–H–O structure) > PMMA (C–H–O structure) > nylon (C–H–O–N structure) > PP (C–H aliphatic unsaturated branched structure) $\geq$ (CH_4, C_3H_6, C_3H_8, C_6H_{14}) $\geq$ PE (C–H aliphatic unsaturated linear structure) > PS (C–H aromatic unsaturated structure). Thus for fires in enclosed spaces, generation of higher amounts of CO relative to CO_2 at high local equivalence ratios is expected for fuels with C–H–O structures compared to the fuels with C–H structures. The reason for higher amounts of CO relative to CO_2 for fuels with C–H–O structures is that CO is easily generated in fuel pyrolysis, but is oxidized only partially to CO_2 due to limited amounts of oxidant available.

Relationship Between the Generation Efficiencies of CO and Smoke: The relationship between the generation efficiencies of CO and smoke is shown in Figure 3-4.37, where data are taken from reference 36. CO and smoke are both generated in the reduction zone of the flame as a result of the oxidative pyrolysis of the fuel, and their generation efficiencies depend on the chemical structure of the fuel (Figures 3-4.30 and 3-4.32). In Figure 3-4.37, the curves represent approximate predictions based on the correlation coefficients from Table 3-4.14 and Equations 72 and 74.

The relationship in Figure 3-4.37 is quite complicated. The boundary of the shaded region marked "air" is drawn using the data for the well-ventilated combustion for equivalence ratios less than 0.05. The boundary of the shaded

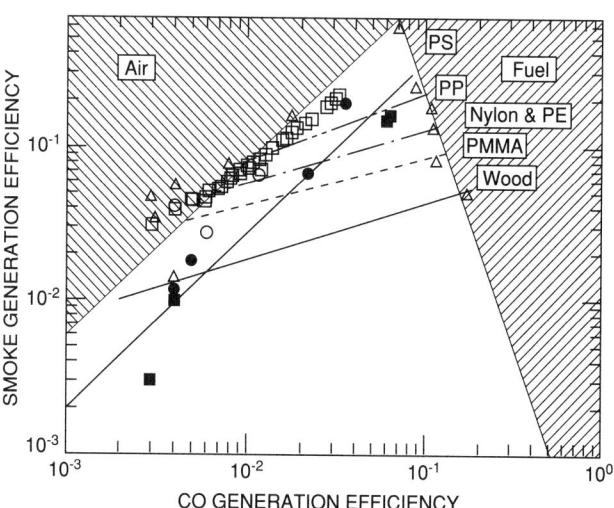

Fig. 3-4.37. Relationship between the generation efficiencies of CO and smoke. Data taken from reference 36.

region marked "fuel" is drawn using the data for the ventilation-controlled combustion for equivalence ratio of 4.0. The boundary for the region marked "air" may be considered as equivalent to the lower flammability limit, and the boundary for the region marked "fuel" may be considered as equivalent to the upper flammability limit.

In Figure 3-4.37, the order for the preference for fuel carbon atom conversion to smoke relative to conversion to CO is: wood (C–H–O structure) < PMMA (C–H–O structure) < nylon (C–H–O–N structure) < PP (C–H aliphatic unsaturated branched structure) ≈ PE (C–H aliphatic unsaturated linear structure) < PS (C–H aromatic structure). The generation efficiency of smoke for PS, which is a polymer with aromatic C–H structure, is the highest; and the generation efficiency of smoke for wood, which is a polymer with aliphatic C–H–O structure, is the lowest.

Generalized Relationships to Calculate Chemical, Convective, and Radiative Heats of Combustion and Yields of Products at Various Equivalence Ratios

The following relationship is the generalized form of Equations 40, 41, and 70 through 74

$$fp = fp_\infty \left[1 + \frac{\alpha}{\exp(\Phi/\beta)^{-\xi}} \right] \qquad (76)$$

where fp is the fire property; α, β, and ξ are the correlation coefficients characteristic of the chemical structures of the polymers, and subscript ∞ represents infinite amount of air; fp determined under turbulent flame conditions is a constant for each polymer. The fire properties are heat of combustion (or combustion efficiency) and yields (or generation efficiencies) of products. Three conditions can be identified: (1) for $\Phi \gg \beta$, $fp = fp_\infty (1 + \alpha)$; (2) for $\Phi \ll \beta$, $fp = fp_\infty$; and (3) $\Phi \approx \beta$, $fp \approx fp_\infty (1 + \alpha/2.7)$. Thus, the parameter α is associated primarily with the magnitude of the fire properties in nonflaming fires (high Φ values). The parameter β is associated with the fire properties in the transition region between the fires with an infinite amount of air and the fires with a very restricted amount of air. The parameter ξ is associated with the range of Φ values for the transition region. A high value of α is indicative of a strong effect of ventilation on the fire and its properties and *vice versa*. High values of β and ξ are indicative of rapid change of fire from flaming to nonflaming by a small change in the equivalence ratio, such as for the highly fire-retarded or halogenated materials for which flaming combustion in normal air itself is unstable.

Chemical heat of combustion *versus* equivalence ratio for the nonhalogenated polymers: From Equation 76

$$\Delta H_{ch} = \Delta H_{ch,\infty} \left[1 - \frac{0.97}{\exp(\Phi/2.15)^{-1.2}} \right] \qquad (77)$$

The values of $\Delta H_{ch,\infty}$ for several polymers are listed in Table 3-4.11.

Chemical heat of combustion *versus* equivalence ratio for the halogenated polymers (polyvinylchloride):

$$\Delta H_{ch} = \Delta H_{ch,\infty} \left[1 - \frac{0.30}{\exp(\Phi/0.53)^{-11}} \right] \qquad (78)$$

As can be noted from the terms inside the brackets in Equations 77 and 78, the effect of ventilation on the chemical heat of combustion is much stronger for PVC than it is for the nonhalogenated polymers. The effect for PVC occurs at $\Phi \geq 0.4$, which is significantly lower than $\Phi \geq 2.0$ found for the nonhalogenated polymers.[36,58] For PVC homopolymer, the flaming combustion changes to nonflaming combustion for $\Phi \geq 0.70$, which is also significantly lower than $\Phi \geq 4.0$ found for the nonhalogenated polymers. This is consistent with the highly halogenated nature of PVC and its mode of decomposition. The decomposition of PVC is characterized by the release of HCl, which is initiated at temperatures as low as about 100°C. At temperatures of up to about 200 to 220°C, HCl is the major effluent. Presence of oxygen in the air enhances HCl release. The generation of HCl from PVC leads to the formation of double bonds and release of various aromatic/unsaturated hydrocarbons (benzene, ethylene, propylene, butylene, etc.).

Convective heats of combustion *versus* equivalence ratio for the nonhalogenated polymers: From Equation 76

$$\Delta H_{con} = \Delta H_{con,\infty} \left[1 - \frac{1.0}{\exp(\Phi/1.38)^{-2.8}} \right] \qquad (79)$$

The values of $\Delta H_{con,\infty}$ for several polymers are listed in Table 3-4.11.

Radiative heats of combustion *versus* equivalence ratio for the nonhalogenated polymers: Radiative heats of combustion are obtained from the difference between the chemical and the convective heats of combustion

$$\Delta H_{rad} = \Delta H_{ch} - \Delta H_{con} \qquad (80)$$

Consumption of oxygen for the nonhalogenated polymers: From Equation 76

$$c_O = c_{O,\infty} \left[1 - \frac{0.97}{\exp(\Phi/2.14)^{-1.2}} \right] \qquad (81)$$

Yield of carbon dioxide for the nonhalogenated polymers: From Equation 76

$$y_{CO_2} = y_{CO_2,\infty} \left[1 - \frac{1.0}{\exp(\Phi/2.15)^{-1.2}} \right] \qquad (82)$$

$y_{CO_2,\infty}$ values are listed in Table 3-4.11.

Yield of carbon dioxide for the halogenated polymers (PVC): From Equation 76

$$y_{CO_2} = y_{CO_2,\infty} \left[1 - \frac{0.30}{\exp(\Phi/0.53)^{-11}} \right] \qquad (83)$$

From the terms inside the brackets in Equations 82 and 83, a stronger effect of ventilation on the yield of CO_2 for PVC than for the nonhalogenated polymers can be noted. $y_{CO_2,\infty}$ values are listed in Table 3-4.11.

Yields of carbon monoxide, hydrocarbons, and smoke for the nonhalogenated polymers: From Equation 76

Polystyrene

$$y_{CO} = y_{CO,\infty} \left[1 + \frac{2.0}{\exp(\Phi/1.44)^{-2.5}} \right] \qquad (84)$$

$$y_{hc} = y_{hc,\infty}\left[1 + \frac{25}{\exp(\Phi/2.45)^{-1.8}}\right] \tag{85}$$

$$y_s = y_{s,\infty}\left[1 + \frac{2.8}{\exp(\Phi/2.02)^{-1.3}}\right] \tag{86}$$

The $y_{CO,\infty}$, $y_{hc,\infty}$, and $y_{s,\infty}$ values are listed in Table 3-4.11.

Polyethylene and Polypropylene

$$y_{CO} = y_{CO,\infty}\left[1 + \frac{10}{\exp(\Phi/1.39)^{-2.8}}\right] \tag{87}$$

$$y_{hc} = y_{hc,\infty}\left[1 + \frac{220}{\exp(\Phi/1.90)^{-2.5}}\right] \tag{88}$$

$$y_s = y_{s,\infty}\left[1 + \frac{2.2}{\exp(\Phi/2.50)^{-1.0}}\right] \tag{89}$$

The $y_{CO,\infty}$, $y_{hc,\infty}$, and $y_{s,\infty}$ values are listed in Table 3-4.11.

Polymethylmethacrylate

$$y_{CO} = y_{CO,\infty}\left[1 + \frac{43}{\exp(\Phi/1.33)^{-3.2}}\right] \tag{90}$$

$$y_{hc} = y_{hc,\infty}\left[1 + \frac{1800}{\exp(\Phi/1.58)^{-3.5}}\right] \tag{91}$$

$$y_s = y_{s,\infty}\left[1 + \frac{1.6}{\exp(\Phi/4.61)^{-0.60}}\right] \tag{92}$$

The $y_{CO,\infty}$, $y_{hc,\infty}$, and $y_{s,\infty}$ values are listed in Table 3-4.11.

Wood

$$y_{CO} = y_{CO,\infty}\left[1 + \frac{44}{\exp(\Phi/1.30)^{-3.5}}\right] \tag{93}$$

$$y_{hc} = y_{hc,\infty}\left[1 + \frac{200}{\exp(\Phi/2.33)^{-1.9}}\right] \tag{94}$$

$$y_s = y_{s,\infty}\left[1 + \frac{2.5}{\exp(\Phi/2.15)^{-1.2}}\right] \tag{95}$$

The $y_{CO,\infty}$, $y_{hc,\infty}$, and $y_{s,\infty}$ values are listed in Table 3-4.11.

Nylon

$$y_{CO} = y_{CO,\infty}\left[1 + \frac{36}{\exp(\Phi/1.36)^{-3.0}}\right] \tag{96}$$

$$y_{hc} = y_{hc,\infty}\left[1 + \frac{1200}{\exp(\Phi/1.65)^{-3.2}}\right] \tag{97}$$

$$y_s = y_{s,\infty}\left[1 + \frac{1.7}{\exp(\Phi/3.14)^{-0.8}}\right] \tag{98}$$

The $y_{CO,\infty}$, $y_{hc,\infty}$, and $y_{s,\infty}$ values are listed in Table 3-4.11.

Yields of Carbon Monoxide, Hydrocarbons, and Smoke for the Halogenated Polymers (Polyvinylchloride)

From Equation 76

$$y_{CO} = y_{CO,\infty}\left[1 + \frac{6.5}{\exp(\Phi/0.42)^{-8.0}}\right] \tag{99}$$

$$y_s = y_{s,\infty}\left[1 + \frac{0.38}{\exp(\Phi/0.42)^{-8.0}}\right] \tag{100}$$

$$y_s = y_{s,\infty}\left[1 + \frac{2.8}{\exp(\Phi/2.02)^{-1.3}}\right] \tag{101}$$

From the above relationships for PVC, for $0.40 \geq \Phi \geq 1.0$, the maximum CO and smoke yields reach about 60 percent of the stoichiometric yields, listed in Table 3-4.13. For nonhalogenated polymers, the maximum CO and smoke yields reach ≤ 30 percent of the stoichiometric yields for $\Phi \geq 2.0$. Polystyrene is the only polymer, within the above group of polymers, for which the smoke yield exceeds that of PVC. These trends suggest that CO and smoke are generated much easier from PVC than from the nonhalogenated polymers, possibly due to the formation of double bonds, as HCl is eliminated at temperatures as low as 100°C from the PVC structure, and formation of various compounds occurs with aromatic/unsaturated bonds.

For the non-halogenated polymers considered with $\Phi \geq 4.0$, the CO yield is lowest and the smoke yield is highest for polystyrene, an aromatic ring-containing polymer; whereas, for polymethylmethacrylate, an aliphatic carbon-hydrogen-oxygen-atom-containing polymer, the CO yield is highest and smoke yield is lowest. This suggests that aromatic ring structure promotes smoke formation, whereas the strong C–O bond in the structure remains intact as ventilation is reduced.

EXAMPLE 19:

Following example 13, calculate the yields of CO and smoke at equivalence ratios of 1, 2, and 3 for polystyrene, polyethylene, wood, and nylon using Equations 84 and 86, 87 and 89, 93 and 95, and 96 and 98, respectively.

SOLUTION:

	Yield (g/g)							
	$\Phi \ll 1.0$		$\Phi = 1.0$		$\Phi = 2.0$		$\Phi = 3.0$	
Material	CO	Smoke	CO	Smoke	CO	Smoke	CO	Smoke
Polystyrene	0.060	0.164	0.070	0.202	0.137	0.331	0.162	0.417
Poly-ethylene	0.024	0.060	0.043	0.071	0.191	0.098	0.238	0.117
Wood	0.004	0.015	0.018	0.018	0.145	0.028	0.171	0.034
Nylon	0.038	0.075	0.149	0.086	1.04	0.105	1.28	0.120

PREDICTION OF FIRE PROPERTIES USING SMOKE POINT

Smoke emission characteristics of fuels have been expressed for decades by smoke point, defined as a minimum laminar axisymmetric diffusion flame height (or fuel volumetric or mass flow rate) at which smoke just escapes from the flame tip.[38,39,59-74] Smoke point values have been measured for numerous gases, liquids, and solids.[38,39,59-64]

Almost all the knowledge on smoke formation, oxidation, and emission from diffusion flames is based on the combustion of fuels containing carbon and hydrogen atoms (hydrocarbons).[61,66-69] On the basis of the chemical structure, hydrocarbons are divided into two main classes: (1) aliphatic and (2) aromatic; fuels containing both aliphatic and aromatic units are known as arenes. Aliphatic fuels have open-chain structure, and aromatic fuel structures consist of

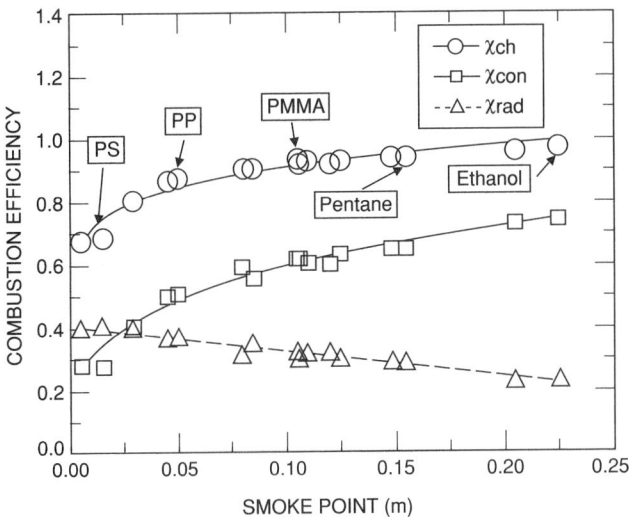

Fig. 3-4.38. Relationships between the combustion efficiency and its convective and radiative components, and the smoke point. Data were measured in the Flammability Apparatus, and reported in references 38 and 39.

benzene rings. Aliphatic hydrocarbons are divided into three families: (1) alkanes (C_nH_{2n+2}), where n is an integer; the suffix "ane" indicates a single bond; (2) alkenes (C_nH_{2n}); the suffix "ene" indicates a double bond, and "diene" two double bonds between carbon-carbon atoms; and (3) alkynes ($C_{2n}H_{2n-2}$); the suffix "yne" indicates a triple bond. The integer n can vary from one in a gas, such as methane, to several thousands in solid polymers, such as polyethylene. In cyclic aliphatic fuels, carbon atoms are also arranged as rings. Dienes are classified as: (1) conjugated—double bonds alternate with single bonds, (2) isolated—double bonds separated by more than one single bond, and (3) allens—double bonds with no separation. Conjugated dienes are more stable than other dienes.

Solid carbon particles present in smoke are defined as soot.[61,66] Soot is generally formed in the fuel-rich regions of the flame and grows in size through gas-solid reactions, followed by oxidation (burnout) to produce gaseous products, such as CO and CO_2. Time that is available for soot formation in the flame is a few milliseconds. Soot particle inception occurs from the fuel molecule *via* oxidation and/or pyrolysis products, which typically includes unsaturated hydrocarbons, especially acetylene, polyacetylenes, and polyaromatic hydrocarbons (PAH). Acetylene, polyacetylenes, and PAH are relatively stable with respect to decomposition. Acetylene and PAH are often considered the most likely precursors for soot formation in flames. PAH have the same role in diffusion flames for both aliphatic and aromatic fuels. In all flames, irrespective of the fuel, initial detection of soot particles takes place on the centerline when a temperature of 1350 K is encountered. Thus, even though the extent of conversion of a fuel into soot may significantly change from fuel to fuel, a common mechanism of soot formation is suggested.

Soot production in the flame depends on the chemical structure, concentration, and temperature of the fuel, flame temperature, pressure, and oxygen concentration.[61,66–69] The diffusion-controlled flame ends when fuel and oxidant are in stoichiometric ratio on the flame axis. The flame is followed by a soot after-burning zone, which is partially chemically controlled. The soot oxidation zone increases from about 10

to 50 percent of the visible flame length as the soot concentration increases. Flame luminosity and smoke emission in the plume depend on overall soot production and oxidation. Flames emit soot when soot temperature in the oxidation zone falls below 1300 K. The soot temperature decreases downstream because of radiation losses and diffusion of fresh cold air, both of which quench soot oxidation. At high soot concentrations, flame emissivity approaches unity, and flame luminosity becomes independent of the amount of soot.

Smoke point, carbon-to-hydrogen ratio, aromaticity, and flame temperature have been suggested as useful parameters to assess relative smoke emission characteristics of fuels in laminar diffusion flames.[38,39,59–64] The soot-forming tendency of fuels is inversely proportional to smoke point. General trends observed for smoke points for hydrocarbon fuels in laminar diffusion flames are: aromatics < alkynes < alkenes < alkanes. Smoke point values have been correlated with flame radiation, combustion efficiency and its convective and radiative components, and generation efficiencies of products.[38,39,59–64] Figures 3-4.38 through 3-4.40 show the relationships between the smoke point and the combustion efficiency and its convective and radiative components, and generation efficiencies of CO and smoke. The data were measured in the Flammability Apparatus [Figure 3-4.2(a)], and reported in references 38 and 39. The following relationships have been found from the data[38,39]

$$\chi_{ch} = 1.15 L_{sp}^{0.10} \tag{102}$$

where χ_{ch} is the combustion efficiency ($-$), and L_{sp} is the smoke point (m) as measured in the Flammability Apparatus.

$$\chi_{rad} = 0.41 - 0.85 L_{sp} \tag{103}$$

where χ_{rad} is the radiative component of the combustion efficiency ($-$). This correlation is very similar to the one reported in the literature.[62]

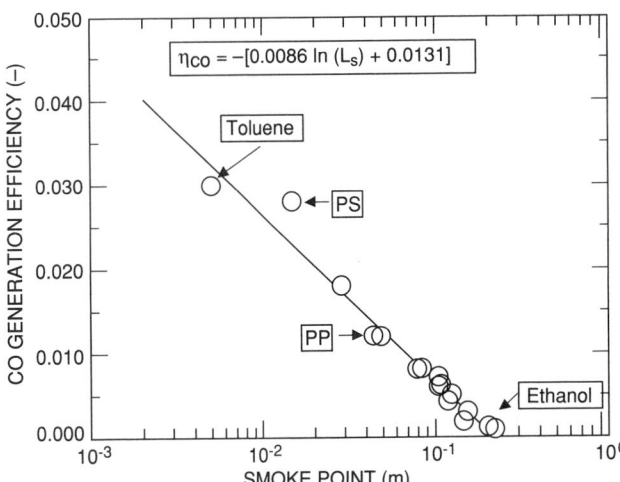

Fig. 3-4.39. Relationships between the CO generation efficiency and the smoke point. Data were measured in the Flammability Apparatus, and reported in references 38 and 39.

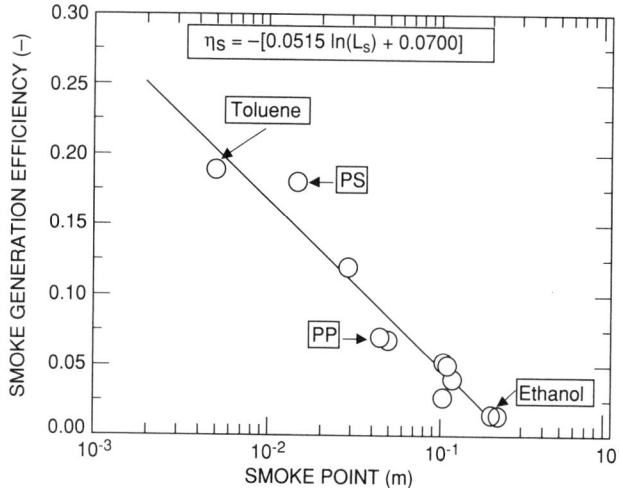

Fig. 3-4.40. Relationships between the smoke generation efficiency and the smoke point. Data were measured in the Flammability Apparatus, and reported in references 38 and 39.

$$\chi_{con} = \chi_{ch} - \chi_{rad} \qquad (104)$$

where χ_{con} is the convective component of the combustion efficiency $(-)$.

$$\eta_{CO} = -[0.0086 \ln(L_{sp}) + 0.0131] \qquad (105)$$

where η_{CO} is the generation efficiency of CO $(-)$.

$$\eta_s = -[0.0515 \ln(L_{sp}) + 0.0700] \qquad (106)$$

where η_s is the generation efficiency of smoke $(-)$.

The highest value of L_{sp} that has been measured is 0.240 m for ethane. Although methane and methanol would be expected to have smoke points higher than 0.240 m, they have not been measured experimentally. Since the combustion efficiency cannot exceed unity, and the generation efficiencies of CO and smoke cannot be negative, the relationships in Equations 102 through 106 are valid for $0 > L_{sp} \le 0.240$ m.

Smoke point decreases with increase in the molecular weight. The smoke point values for monomers and polymers, however, show different types of dependencies: (1) the smoke point values for ethylene and polyethylene are 0.097 and 0.045 m, respectively; (2) the smoke point values for propylene and polypropylene are 0.030 and 0.050 m, respectively; and (3) the smoke point values for styrene and polystyrene are 0.006 and 0.015 m, respectively. The smoke point data for polymers support the accepted vaporization mechanisms of polymers;[75] i.e., polyethylene, polypropylene, and polystyrene vaporize as higher molecular weight oligomers rather than as monomers, and thus their smoke point values are different than the values for the monomers. The smoke point values suggest that polyethylene is expected to have higher smoke emission than ethylene, whereas polypropylene and polystyrene are expected to have lower smoke emissions than propylene and styrene.

The correlations show that emissions of CO and smoke are very sensitive to changes in the smoke point values compared to combustion efficiency and its convective and radiative components. This is expected from the understanding of the relationship between the smoke point and chemical structures of fuels. For example, a decrease of 33 percent in the smoke point value of 0.15 m to 0.10 m produces a decrease of 4 and 12 percent in the combustion efficiency and its convective component, respectively, and an increase of 14 percent in the radiative component of the combustion efficiency; however, the generation efficiencies of CO and smoke increase by 89 and 67 percent, respectively.

Equations 102 through 106 can be used to estimate the fire properties of gases, liquids, and solids from their smoke point values. The smoke point values, however, depend strongly on the apparatus and cannot be used as reported. One of the approaches is to establish correlations between the smoke point values measured in different apparatuses and a single apparatus for which relationships such as given in Equations 102 through 106 are available. This type of approach has been described in references 38 and 39 for the Flammability Apparatus, where smoke point values for 165 fuels, reported in the literature, were translated to the values for the Flammability Apparatus. The fire properties (chemical, convective, and radiative heats of combustion and yields of CO and smoke) estimated in this fashion, from Equations 102 through 106, are listed in Tables 3-4.16 through 3-4.18. In the tables, molecular formula and weight, stoichiometric mass air-to-fuel ratio, and net heat of complete combustion have also been tabulated. The estimated data in the tables have been validated by direct measurements in the small- and large-scale fires using several fuels.[38,39]

The data in Tables 3-4.16 through 3-4.18 show linear dependencies on the molecular weight of the fuel monomer within each group[38,39]

$$\Delta H_i = h_i \pm \frac{m_i}{M} \qquad (107)$$

$$y_j = a_j \pm \frac{b_j}{M} \qquad (108)$$

where ΔH_i is the net heat of complete combustion or chemical, convective, or radiative heat of combustion (kJ/g); y_j is the yield of product j (g/g); M is the molecular weight of fuel monomer (g/mole); h_i is the mass coefficient for the heat of combustion (kJ/g); m_i is the molar coefficient for the heat of combustion (kJ/mole); a_j is the mass coefficient for the product yield (g/g); and b_j is the molar coefficient for the product yield (g/mole). The coefficients depend on the chemical structures of the fuel; m_i and b_j become negative with the introduction of oxygen, nitrogen, and sulfur atoms into the chemical structure. Relationships in Equations 107 and 108 support the suggestion[69] that generally smaller molecules offer greater resistance to smoke formation and emission. The relationships suggest that for gases, liquids, and solids gasifying as high molecular weight fuels, $\Delta H_i \approx h_i$ and $y_j \approx a_j$.

The variations of chemical, convective, and radiative heats of combustion and yields of CO and smoke with the chemical structures of the fuels are similar to the smoke point variations.

EXAMPLE 20:

The following smoke point values have been reported in the literature:

Polymer	PE	PP	PMMA	PS
Smoke point (m)	0.045	0.050	0.105	0.015

For well-ventilated conditions, estimate: (1) the chemical, convective, and radiative heats of combustion using Equations

TABLE 3-4.16   *Combustion Properties of Fuels with Carbon and Hydrogen Atoms in the Chemical Structure*

Hydrocarbon	Formula	M (g/mole)	S	Heat of Combustion (kJ/g)				Yield (g/g)	
				ΔH_T	ΔH_{ch}	ΔH_{con}	ΔH_{rad}	CO	Smoke
Normal Alkanes									
Ethane	C_2H_6	30	16.0	47.1	45.7	34.1	11.6	0.001	0.013
n-Propane	C_3H_8	44	15.6	46.0	43.7	31.2	12.5	0.005	0.024
n-Butane	C_4H_{10}	58	15.4	45.4	42.6	29.6	13.0	0.007	0.029
n-Pentane	C_5H_{12}	72	15.3	45.0	42.0	28.7	13.3	0.008	0.033
n-Hexane	C_6H_{14}	86	15.2	44.8	41.5	28.1	13.5	0.009	0.035
n-Heptane	C_7H_{16}	100	15.1	44.6	41.2	27.6	13.6	0.010	0.037
n-Octane	C_8H_{18}	114	15.1	44.5	41.0	27.3	13.7	0.010	0.038
n-Nonane	C_9H_{20}	128	15.0	44.4	40.8	27.0	13.8	0.011	0.039
n-Decane	$C_{10}H_{22}$	142	15.0	44.3	40.7	26.8	13.9	0.011	0.040
n-Undecane	$C_{11}H_{24}$	156	15.0	44.3	40.5	26.6	13.9	0.011	0.040
n-Dodecane	$C_{12}H_{26}$	170	14.9	44.2	40.4	26.4	14.0	0.011	0.041
n-Tridecane	$C_{13}H_{28}$	184	14.9	44.2	40.3	26.3	14.0	0.012	0.041
n-Tetradecane	$C_{14}H_{30}$	198	14.9	44.1	40.3	26.2	14.1	0.012	0.042
Hexadecane	$C_{16}H_{34}$	226	14.9	44.1	40.1	26.0	14.1	0.012	0.042
Branched Alkanes									
Methylbutane	C_5H_{12}	72	15.3	45.0	40.9	27.2	13.8	0.012	0.042
Dimethylbutane	C_6H_{14}	86	15.2	44.8	40.3	26.3	14.0	0.014	0.046
Methylpentane	C_6H_{14}	86	15.2	44.8	40.3	26.3	14.0	0.014	0.046
Dimethylpentane	C_7H_{16}	100	15.1	44.6	39.9	25.7	14.1	0.015	0.049
Methylhexane	C_7H_{16}	100	15.1	44.6	39.9	25.7	14.1	0.015	0.049
Trimethylpentane	C_8H_{18}	114	15.1	44.5	39.6	25.3	14.3	0.016	0.052
Methylethylpentane	C_8H_{18}	114	15.1	44.5	39.6	25.3	14.3	0.016	0.052
Ethylhexane	C_8H_{18}	114	15.1	44.5	39.6	25.3	14.3	0.016	0.052
Dimethylhexane	C_8H_{18}	114	15.1	44.5	39.6	25.3	14.3	0.016	0.052
Methylheptane	C_8H_{18}	114	15.1	44.5	39.6	25.3	14.3	0.016	0.052
Cyclic Alkanes									
cyclo-Pentane	C_5H_{10}	70	14.7	44.3	39.2	24.1	15.1	0.018	0.055
Methylcyclopentane	C_6H_{12}	84	14.7	43.8	38.2	23.0	15.2	0.019	0.061
Cyclohexane	C_6H_{12}	84	14.7	43.8	38.2	23.0	15.2	0.019	0.061
Methylcyclohexane	C_7H_{14}	98	14.7	43.4	37.5	22.3	15.2	0.021	0.066
Ethylcyclohexane	C_8H_{16}	112	14.7	43.2	36.9	21.7	15.3	0.021	0.069
Dimethylcyclohexane	C_8H_{16}	112	14.7	43.2	36.9	21.7	15.3	0.021	0.069
Cyclooctane	C_8H_{16}	112	14.7	43.2	36.9	21.7	15.3	0.021	0.069
Decalin	$C_{10}H_{18}$	138	14.4	42.8	36.2	20.9	15.3	0.023	0.073
Bicyclohexyl	$C_{12}H_{22}$	166	14.5	42.6	35.7	20.4	15.3	0.023	0.076
Alkenes									
Ethylene	C_2H_4	28	14.7	48.0	41.5	27.3	14.2	0.013	0.043
Propylene	C_3H_6	42	14.7	46.4	40.5	25.6	14.9	0.017	0.095
Butylene	C_4H_8	56	14.7	45.6	40.0	24.8	15.2	0.019	0.067
Pentene	C_5H_{10}	70	14.7	45.2	39.7	24.2	15.4	0.020	0.065
Hexene	C_6H_{12}	84	14.7	44.9	39.4	23.9	15.5	0.021	0.064
Heptene	C_7H_{14}	98	14.7	44.6	39.3	23.7	15.6	0.021	0.063
Octene	C_8H_{16}	112	14.7	44.5	39.2	23.5	15.7	0.022	0.062
Nonene	C_9H_{18}	126	14.7	44.3	39.1	23.3	15.8	0.022	0.062
Decene	$C_{10}H_{20}$	140	14.7	44.2	39.0	23.2	15.8	0.022	0.061
Dodecene	$C_{12}H_{24}$	168	14.7	44.1	38.9	23.1	15.9	0.023	0.061
Tridecene	$C_{13}H_{26}$	182	14.7	44.0	38.9	23.0	15.9	0.023	0.061
Tetradecene	$C_{14}H_{28}$	196	14.7	44.0	38.8	22.9	15.9	0.023	0.060
Hexadecene	$C_{16}H_{32}$	224	14.7	43.9	38.8	22.8	16.0	0.023	0.060
Octadecene	$C_{18}H_{36}$	252	14.7	43.8	38.7	22.8	16.0	0.023	0.060
Polyethylene	$(C_2H_4)_n$	601	14.7	43.6	36.8	20.6	16.2	0.027	0.077
Polypropylene	$(C_3H_6)_n$	720	14.7	43.4	37.0	21.1	15.9	0.025	0.072
Cyclic Alkenes									
Cyclohexene	C_6H_{10}	82	14.2	43.0	35.7	20.2	15.5	0.029	0.085
Methylcyclohexene	C_7H_{12}	96	14.3	43.1	35.8	19.8	16.0	0.029	0.085
Pinene	$C_{10}H_{16}$	136	14.1	36.0	33.5	18.9	14.6	0.039	0.114

TABLE 3-4.16 *Combustion Properties of Fuels with Carbon and Hydrogen Atoms in the Chemical Structure (Continued)*

Hydrocarbon	Formula	M (g/mole)	S	Heat of Combustion (kJ/g)				Yield (g/g)	
				ΔH_T	ΔH_{ch}	ΔH_{con}	ΔH_{rad}	CO	Smoke
Alkynes and Butadiene									
Acetylene	C_2H_2	26	13.2	47.8	36.7	18.7	18.0	0.042	0.096
Heptyne	C_7H_{12}	96	14.3	44.8	36.0	18.8	17.1	0.036	0.094
Octyne	C_8H_{14}	110	14.4	44.7	35.9	18.9	17.1	0.036	0.094
Decyne	$C_{10}H_{18}$	138	14.4	44.5	35.9	18.9	17.0	0.035	0.094
Dodecyne	$C_{12}H_{22}$	166	14.5	44.3	35.9	18.9	17.0	0.035	0.094
1, 3-Butadiene	C_4H_6	54	14.0	44.6	33.6	15.4	18.2	0.048	0.125
Arenes									
Benzene	C_6H_6	78	13.2	40.1	27.6	11.0	16.5	0.067	0.181
Toluene	C_7H_8	92	13.4	39.7	27.7	11.2	16.5	0.066	0.178
Styrene	C_8H_8	104	13.2	39.4	27.8	11.2	16.6	0.065	0.177
Ethylbenzene	C_8H_{10}	106	13.6	39.4	27.8	11.2	16.6	0.065	0.177
Xylene	C_8H_{10}	106	13.6	39.4	27.8	11.2	16.6	0.065	0.177
Indene	C_9H_8	116	13.0	39.2	27.9	11.3	16.6	0.065	0.176
Propylbenzene	C_9H_{12}	120	13.7	39.2	27.9	11.3	16.6	0.065	0.175
Trimethylbenzene	C_9H_{12}	120	13.7	39.2	27.9	11.3	16.6	0.065	0.175
Cumene	C_9H_{12}	120	13.7	39.2	27.9	11.3	16.6	0.065	0.175
Naphthalene	$C_{10}H_8$	128	12.9	39.0	27.9	11.3	16.6	0.065	0.175
Tetralin	$C_{10}H_{12}$	132	13.5	39.0	27.9	11.4	16.6	0.064	0.174
Butylbenzene	$C_{10}H_{14}$	134	13.8	39.0	27.9	11.4	16.6	0.064	0.174
Diethylbenzene	$C_{10}H_{14}$	134	13.8	39.0	27.9	11.4	16.6	0.064	0.174
p-Cymene	$C_{10}H_{14}$	134	13.8	39.0	27.9	11.4	16.6	0.064	0.174
Methylnaphthalene	$C_{11}H_{10}$	142	13.0	38.9	28.0	11.4	16.6	0.064	0.174
Pentylbenzene	$C_{11}H_{16}$	148	13.9	38.8	28.0	11.4	16.6	0.064	0.173
Dimethylnaphthalene	$C_{12}H_{12}$	156	13.2	38.8	28.0	11.4	16.6	0.064	0.173
Cyclohexylbenzene	$C_{12}H_{16}$	160	13.7	38.7	28.0	11.4	16.6	0.064	0.173
Diisopropylbenzene	$C_{12}H_{18}$	162	14.0	38.7	28.0	11.4	16.6	0.064	0.173
Triethylbenzene	$C_{12}H_{18}$	162	14.0	38.7	28.0	11.4	16.6	0.064	0.173
Triamylbenzene	$C_{21}H_{36}$	288	14.3	38.1	28.2	11.6	16.6	0.063	0.169
Polystyrene	$(C_8H_8)_n$	200	13.2	39.2	29.6	14.0	15.6	0.050	0.135

102 through 104 and data for the net heat of complete combustion from Table 3-4.11; and (2) yields of CO and smoke using Equations 105 and 106 and stoichiometric yields from Table 3-4.13.

SOLUTION:

(1) From Equations 102 through 104 and Table 3-4.11

Polymer	PE	PP	PMMA	PS
ΔH_T (kJ/g)	43.6	43.4	25.2	39.2
ΔH_{ch} (kJ/g)	36.8	37.0	23.1	29.6
ΔH_{con} (kJ/g)	20.6	21.1	15.0	14.0
ΔH_{rad} (kJ/g)	16.2	15.9	8.1	15.6

(2) From Equations 105 and 107 and Table 3-4.13

Polymer	PE	PP	PMMA	PS
Ψ_{CO}	2.00	2.00	1.40	2.15
Ψ_s	0.857	0.857	0.600	0.923
y_{CO} (g/g)	0.027	0.025	0.009	0.050
y_s (g/g)	0.077	0.072	0.028	0.135

NONTHERMAL DAMAGE DUE TO FIRE PRODUCTS

Damage due to heat is defined as thermal damage, and damage due to smoke, toxic, and corrosive products is defined as nonthermal damage.[1] Nonthermal damage depends on the chemical nature and deposition of products on the walls, ceilings, building furnishings, equipment, components, etc., and the environmental conditions. The severity of the nonthermal damage increases with time. Examples of nonthermal damage to property are: corrosion, electrical malfunctions, discoloration, odors, etc.

Most commercial and industrial occupancies are susceptible to nonthermal fire damage. Examples of typical commercial and industrial occupancies are telephone central offices, computer rooms, power plant control rooms, space satellites in operaton, under construction or in storage, department and grocery stores, hotels, restaurants, various manufacturing facilities, and transportation vehicles such as aircraft, ships, trains, and buses.

For this chapter, the subject of corrosion for commercial and industrial occupancies has been reviewed based on the knowledge derived from the telephone central office (TCO) experience for the deposition of atmospheric pollutants and fire products on equipment, severity of corrosion damage, and ease of cleaning the equipment.[76-79] Galvanized zinc or zinc-chromated finishes represent a major portion of the structural components of the TCO equipment as well as the HVAC ductwork.[77-79] Unfortunately all zinc surfaces are sensitive to corrosion attack by corrosive products. For example, on exposure to HCl gas, zinc forms zinc chloride, which is very hygroscopic and picks up moisture from air with relative humidity as low as 10 percent to form electrically conductive

TABLE 3-4.17 *Combustion Properties of Fuels with Carbon, Hydrogen, and Oxygen Atoms in the Chemical Structure*

Hydrocarbon	Formula	M (g/mole)	S	Heat of Combustion (kJ/g) ΔH_T	ΔH_{ch}	ΔH_{con}	ΔH_{rad}	Yield (g/g) CO	Smoke
Aliphatic Esters									
Ethyl formate	$C_3H_6O_2$	74	6.5	20.2	19.9	13.5	6.3	0.003	0.011
n-Propyl formate	$C_4H_8O_2$	88	7.8	23.9	23.4	15.4	8.0	0.005	0.019
n-Butyl formate	$C_5H_{10}O_2$	102	8.8	26.6	26.0	16.7	9.3	0.007	0.025
Methyl acetate	$C_3H_6O_2$	74	6.5	20.2	19.9	13.5	6.3	0.003	0.011
Ethyl acetate	$C_4H_8O_2$	88	7.8	23.9	23.4	15.4	8.0	0.005	0.019
n-Propyl acetate	$C_5H_{10}O_2$	102	8.8	26.6	26.0	16.7	9.3	0.007	0.025
n-Butyl acetate	$C_6H_{12}O_2$	116	9.5	28.7	28.0	17.8	10.2	0.008	0.029
Isobutyl acetate	$C_6H_{12}O_2$	116	9.5	28.7	28.0	17.8	10.2	0.008	0.029
Amyl acetate	$C_7H_{14}O_2$	130	10.0	30.3	29.5	18.6	11.0	0.009	0.033
Cyclohexyl acetate	$C_8H_{14}O_2$	142	10.2	31.5	30.6	19.1	11.5	0.010	0.035
Octyl acetate	$C_{10}H_{20}O$	172	11.2	33.6	32.6	20.2	12.5	0.012	0.039
Ethyl acetoacetate	$C_6H_{10}O_3$	130	7.4	30.3	29.5	18.6	11.0	0.009	0.033
Methyl propionate	$C_4H_8O_2$	88	7.8	23.9	23.4	15.4	8.0	0.005	0.019
Ethyl propionate	$C_5H_{10}O_2$	102	8.8	26.6	26.0	16.7	9.3	0.007	0.025
n-Butyl propionate	$C_7H_{14}O_2$	130	10.0	30.3	29.5	18.6	11.0	0.009	0.033
Isobutyl propionate	$C_7H_{14}O_2$	130	10.0	30.3	29.5	18.6	11.0	0.009	0.033
Amyl propionate	$C_8H_{16}O_2$	144	10.5	31.6	30.8	19.2	11.6	0.010	0.035
Methyl butyrate	$C_5H_{10}O_2$	102	8.8	26.6	26.0	16.7	9.3	0.007	0.025
Ethyl butyrate	$C_6H_{12}O_2$	116	9.5	28.7	28.0	17.8	10.2	0.008	0.029
Propyl butyrate	$C_7H_{14}O_2$	130	10.0	30.3	29.5	18.6	11.0	0.009	0.033
n-Butyl butyrate	$C_8H_{16}O_2$	144	10.5	31.6	30.8	19.2	11.6	0.010	0.035
Isobutyl butyrate	$C_8H_{16}O_2$	144	10.5	31.6	30.8	19.2	11.6	0.010	0.035
Ethyl laurate	$C_{14}H_{28}O$	228	12.0	37.2	35.6	26.5	9.1	0.008	0.031
Ethyl oxalate	$C_4H_6O_4$	102	6.1	28.7	27.7	21.3	6.4	0.001	0.003
Ethyl malonate	$C_5H_8O_4$	132	7.7	32.2	31.0	23.4	7.5	0.003	0.015
Ethyl lactate	$C_5H_{10}O_3$	118	7.0	30.8	29.6	22.5	7.1	0.001	0.010
Butyl lactate	$C_7H_{14}O_3$	146	8.5	33.3	32.0	24.1	7.9	0.004	0.018
Amyl lactate	$C_8H_{16}O_3$	160	9.0	34.3	32.9	24.7	8.2	0.005	0.021
Ethyl carbonate	$C_5H_{10}O_3$	118	7.0	30.8	29.6	22.5	7.1	0.001	0.010
Aliphatic Alcohols									
Methyl alcohol	CH_4O	32	6.4	20.0	19.1	16.1	3.0	0.001	0.001
Ethyl alcohol	C_2H_6O	46	9.0	27.7	25.6	19.0	6.5	0.001	0.008
n-Propyl alcohol	C_3H_8O	60	10.3	31.8	29.0	20.6	8.5	0.003	0.015
Isopropyl alcohol	C_3H_8O	60	10.3	31.8	29.0	20.6	8.5	0.003	0.015
n-Butyl alcohol	$C_4H_{10}O$	74	11.1	34.4	31.2	21.6	9.6	0.004	0.019
Isobutyl alcohol	$C_4H_{10}O$	74	11.1	34.4	31.2	21.6	9.6	0.004	0.019
Sec butyl alcohol	$C_4H_{10}O$	74	11.1	34.4	31.2	21.6	9.6	0.004	0.019
Ter butyl alcohol	$C_4H_{10}O$	74	11.1	34.4	31.2	21.6	9.6	0.004	0.019
n-Amyl alcohol	$C_5H_{12}O$	88	11.7	36.2	32.7	22.2	10.4	0.005	0.022
Isobutyl carbinol	$C_5H_{12}O$	88	11.7	36.2	32.7	22.2	10.4	0.005	0.022
Sec butyl carbinol	$C_5H_{12}O$	88	11.7	36.2	32.7	22.2	10.4	0.005	0.022
Methylpropyl carbinol	$C_5H_{12}O$	88	11.7	36.2	32.7	22.2	10.4	0.005	0.022
Dimethylethyl carbinol	$C_5H_{12}O$	88	11.7	36.2	32.7	22.2	10.4	0.005	0.022
n-Hexyl alcohol	$C_6H_{14}O$	102	12.1	37.4	33.7	22.7	11.0	0.006	0.024
Dimethylbutyl alcohol	$C_6H_{14}O$	102	12.1	37.4	33.7	22.7	11.0	0.006	0.024
Ethylbutyl alcohol	$C_6H_{14}O$	102	12.1	37.4	33.7	22.7	11.0	0.006	0.024
Allyl alcohol	C_3H_6O	58	9.5	31.4	28.6	20.4	8.2	0.003	0.014
Cyclohexanol	$C_6H_{12}O$	100	11.7	37.3	33.6	22.6	11.0	0.005	0.024
Aliphatic Ketones									
Acetone	C_3H_6O	58	9.5	29.7	27.9	20.3	7.6	0.003	0.014
Methyl ethyl ketone	C_4H_8O	72	10.5	32.7	30.6	22.1	8.6	0.004	0.018
Cyclohexanone	$C_6H_{10}O$	98	11.2	35.9	33.7	24.1	9.6	0.005	0.023
Di-acetone alcohol	$C_6H_{12}O_2$	116	9.5	37.3	35.0	24.9	10.1	0.006	0.026
Other Aliphatic Fuels									
Monoethyl ether	$C_4H_{10}O_2$	90	8.4	26.7	25.8	20.0	5.8	0.001	0.007
Monoethylether acetate	$C_6H_{12}O_3$	132	7.8	32.2	31.0	23.2	7.7	0.001	0.011
Monoethylether diacetate	$C_6H_{10}O_4$	146	6.1	33.3	32.0	24.2	7.9	0.001	0.009
Glycerol triacetate	$C_9H_{14}O_6$	218	6.0	36.9	35.4	26.3	9.1	0.002	0.011
Other Aromatic Fuels									
Benzaldehyde	C_7H_6O	106	10.4	32.4	21.2	8.1	13.2	0.062	0.166
Benzyl alcohol	C_7H_8O	108	10.8	32.6	22.9	9.8	13.1	0.050	0.137
Cresylic acid	C_8H_8O	136	9.1	34.0	25.1	11.6	13.5	0.039	0.107
Ethyl benzoate	$C_9H_{10}O_2$	150	9.6	34.5	27.4	14.1	13.3	0.030	0.084
Phenylbutyl ketone	$C_{11}H_{14}O$	162	11.9	34.8	26.3	12.6	13.7	0.041	0.115

TABLE 3-4.18 *Combustion Properties of Fuels with Carbon, Hydrogen, Nitrogen, and Sulfur Atoms in the Chemical Structure*

Hydrocarbon	Formula	M (g/mole)	S	Heat of Combustion (kJ/g)				Yield (g/g)	
				ΔH_T	ΔH_{ch}	ΔH_{con}	ΔH_{rad}	CO	Smoke
Aliphatic Fuels with Carbon, Hydrogen, and Nitrogen									
Diethylamine	$C_4H_{11}N$	73	14.6	38.0	34.0	21.3	12.6	0.012	0.039
n-Butylamine	$C_4H_{11}N$	73	14.6	38.0	34.0	21.3	12.6	0.012	0.039
sec-Butylamine	$C_4H_{11}N$	73	14.6	38.0	34.0	21.3	12.6	0.012	0.039
Triethylamine	$C_6H_{15}N$	101	14.6	39.6	35.3	22.0	13.3	0.014	0.044
Di-n-butylamine	$C_8H_{19}N$	129	14.6	40.6	36.1	22.4	13.7	0.014	0.047
Tri-n-butylamine	$C_{12}H_{27}N$	185	14.7	41.6	37.0	22.9	14.1	0.015	0.049
Aromatic Fuels with Carbon, Hydrogen, and Nitrogen									
Pyridine	C_5H_5N	79	12.6	32.2	24.0	11.5	12.5	0.037	0.104
Aniline	C_6H_7N	93	12.9	33.8	25.0	11.7	13.3	0.043	0.119
Picoline	C_6H_7N	93	12.9	33.8	25.0	11.7	13.3	0.043	0.119
Toluidine	C_7H_9N	107	13.2	34.9	25.8	11.9	13.9	0.048	0.130
Dimethylaniline	$C_8H_{11}N$	121	13.3	35.7	26.4	12.1	14.3	0.051	0.139
Quinoline	C_9H_7N	129	12.5	36.1	26.7	12.1	14.5	0.052	0.143
Quinaldine	$C_{10}H_9N$	143	12.7	36.7	27.1	12.2	14.8	0.055	0.149
Butylaniline	$C_{10}H_{15}N$	149	13.6	37.0	27.2	12.2	15.0	0.056	0.151
Aliphatic Fuels with Carbon, Hydrogen, and Sulfur									
Hexyl mercaptan	$C_6H_{14}S$	118	12.2	33.0	30.1	17.9	12.2	0.012	0.040
Heptyl mercaptan	$C_7H_{16}S$	132	12.5	33.7	30.4	18.1	12.3	0.013	0.044
Decyl mercaptan	$C_{10}H_{22}S$	174	13.0	34.9	31.1	18.4	12.7	0.016	0.051
Dodecyl mercaptan	$C_{12}H_{26}S$	202	13.3	35.5	31.4	18.6	12.8	0.017	0.054
Hexyl sulfide	$C_{12}H_{26}S$	202	13.3	35.5	31.4	18.6	12.8	0.017	0.054
Heptyl sulfide	$C_{14}H_{30}S$	230	13.4	35.9	31.6	18.7	13.0	0.018	0.057
Octyl sulfide	$C_{16}H_{34}S$	258	13.6	36.3	31.8	18.8	13.1	0.019	0.059
Decyl sulfide	$C_{20}H_{42}S$	314	13.8	36.8	32.1	18.9	13.2	0.020	0.061
Aromatic Fuels with Carbon, Hydrogen, and Sulfur									
Thiophene	C_4H_4S	84	9.8	31.9	23.4	10.8	12.6	0.031	0.086
Methylthiophene	C_5H_6S	98	10.5	33.2	24.1	10.9	13.2	0.039	0.107
Thiophenol	C_6H_6S	110	10.6	34.1	24.6	11.0	13.6	0.045	0.122
Thiocresol	C_7H_8S	124	11.1	34.9	25.0	11.0	14.0	0.050	0.135
Cresolmethyl sulfide	$C_8H_{11}S$	155	11.6	36.2	25.7	11.1	14.5	0.058	0.155

liquid zinc chloride solution. The solution flows on the surfaces, drips down or runs onto equipment, resulting in very serious electrical shorting problems. In two major TCO losses, zinc chloride played a key role in both the rate of restoration as well as the ability to salvage equipment.

In TCO fires involving PVC-based cables, contamination levels in the range of about 5 to 900 microgram/cm^2 have been observed.[77–79] In general, an electronic switch would be expected to accumulate zinc chloride levels in the range of about 5 to 9 microgram/cm^2 from the interaction with the environment over its expected lifetime of 20+

years. Clean equipment is expected to have less than about 2 microgram/cm^2 of chloride contamination, whereas contaminated equipment can have as high as 900 microgram/cm^2. Thus, equipment contamination levels and ease of restoration have been classified into four levels,[77–79] as listed in Table 3-4.19.

Corrosion

Corrosion is defined as an unwanted chemical reaction and/or destruction or deterioration of a material because of

TABLE 3-4.19 *Contamination Levels for the Surface Deposition of Chloride Ions for Electronic Equipment**

Chloride Ion (microgram/cm^2)	Level	Damage/Cleaning/Restoration
2	One	No damage expected. No cleaning and restoration required.
<30	Two	Equipment can be easily restored to service by cleaning, with little impact on long-term reliability.
30 to 90	Three	Equipment can also be restored to service by cleaning, as long as no unusual corrosion problems arise, and the environment is strictly controlled soon after the fire.
>90	Four	The effectiveness of cleaning the equipment dwindles, and the cost of cleaning quickly approaches the replacement cost. Equipment contaminated with high chloride levels may require severe environmental controls even after cleaning, in order to provide potentially long-term reliable operation.

*Data taken from reference 77.

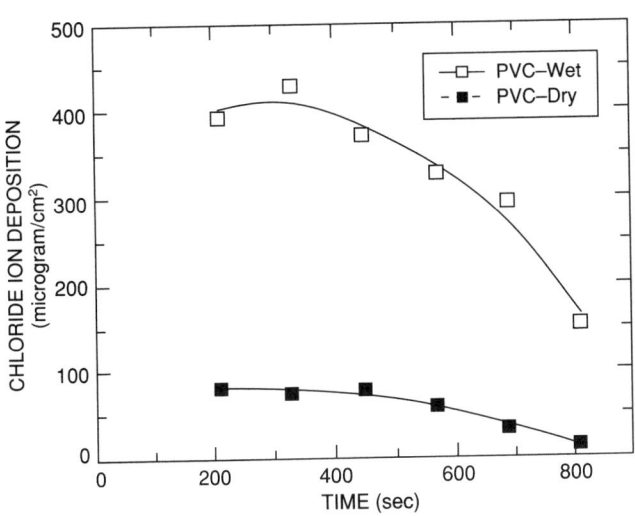

Fig. 3-4.41. *Deposition of HCl on wet and dry cellulosic filter paper during the pyrolysis of PVC at an external heat flux of 20 kW/m^2 under co-airflow with 10 percent of oxygen concentration in the Flammability Apparatus. Flow velocity 0.09 m/s with filter paper at right angle to the flow. Data used in figure are taken from reference 81.*

reaction with its environment. Factors that are considered to be important for the extent of corrosion damage are: (1) oxygen, (2) nature and concentrations of the fire products, (3) relative humidity, (4) temperature, (5) nature of the target and its orientation relative to the flow of the fire products-air mixture, (6) flow velocity of the fire products-air mixture, (7) presence of extinguishing agents, (8) techniques used for cleaning the exposed surface and their implementation time after the fire, and others.

Most of the knowledge on corrosion damage has been based on air pollution, e.g., due to acid rain. Acid deposition is generally described as "acid rain."[80] Rain usually includes all forms of precipitation (rain, snow, sleet, hail, etc.). Acid deposition is a broader term and includes the uptake of gases by surfaces, impact of fog, and settling of dust and small particles.[80] Precipitation is one of the principal removal mechanisms by which the atmosphere cleanses itself. Acids in rain precipitation result mainly from sulfuric, nitric, and hydrochloric acids, either absorbed directly into precipitation or formed in the aqueous phase from precursor compounds.

In general, all forms of pollution deposition not involving precipitation are referred to as dry, including dew and fog processes.[80] With the exception of nitric acid vapors, most gases do not readily deposit on dry, inert surfaces. However, if the gas is soluble in water, the presence of a liquid film (resulting from condensation, for example) will generally accelerate dry deposition. In these cases, the amount deposited on the surfaces will depend not only on the concentration of the pollutant, but also on the relative frequency of encountering a wet surface.

Data in Figure 3-4.41, taken from reference 81, show that the deposition of HCl on wet filter paper is almost four times as high as the deposition on dry paper, in agreement with reference 80. HCl was generated by exposing PVC to an external heat flux of 20 kW/m^2 in an inert environment in the Flammability Apparatus. The chloride ion deposition is

high in the initial stages and decreases with time, which is consistent with the decomposition mechanism of PVC. The decomposition of PVC is characterized by the release of HCl, which is initiated at a temperature as low as about 100°C. At a temperature of up to about 200 to 220°C, HCl is the major effluent. Presence of oxygen in the air enhances HCl release. The generation of HCl from PVC leads to the formation of double bonds and release of CO and various aromatic/unsaturated hydrocarbons (benzene, toluene, ethylene, propylene, butylene, etc). The yields of some of these products from the combustion and pyrolysis of PVC are listed in Table 3-4.20, taken from reference 82.

Deposition of HCl on walls of enclosures has also been quantified in larger-scale fire tests. For example, in the fire tests with PVC floor covering performed in a 2.8- × 2.8- × 2.4-m-high unventilated room, about 50 percent of the original chloride ions in PVC were deposited on the walls.[83] With the exception of vinyl film (wallpaper) and super-gloss enamel paint on polyethylene, the chloride ion deposition on all other surfaces was in the range of 30 to 90 microgram/cm^2. The differences in the chloride ion deposition on various materials on the wall appear to be related to hydrophilic (water attracting) and hydrophobic (water repelling) nature of the surfaces, i.e., filter paper is hydrophilic and vinyl film is hydrophobic, in agreement with reference 80. This deposition corresponds to the third level of contamination for TCOs. (See Table 3-4.19.)

The corrosion damage in fires follows the basic corrosion relationship

$$D_{corr} = \mu c^m t^n \qquad (109)$$

where D_{corr} is metal corrosion (penetration depth or metal loss in microns, angstroms, mils); t is the exposure time (minutes, days); c is the concentration of the corrosive product (g/m^3); and μ, m, and n are empirical constants. The constant μ may be defined as a corrosion parameter characteristic of the corrosive nature of the product. The constant n is a function of the corrosion resistance characteristics of the film at the surface. When the film on the surface protects the surface and inhibits further corrosion by diffusion, $n = \frac{1}{2}$.[80] When the film is permeable to corrosive gases and offers no protection, $n = 1$.[80]

For short-term exposure of metal surfaces to aqueous solutions of corrosive fire products, $n = 1$, and from Equation 109

$$\dot{R}_{corr} = \mu c^m \qquad (110)$$

where $\dot{R}_{corr}$ is corrosion rate (Å/min).

For long-term exposure of metal surfaces to aqueous solutions of corrosive fire products, as a protective layer of corrosion byproducts is formed at the surface, $n = \frac{1}{2}$, and from Equation 109

$$\dot{R}_{corr} = \frac{\mu c^m}{t^{1/2}} \qquad (111)$$

showing that corrosion rate decreases with time.

Figure 3-4.42 shows a plot of the corrosion rate of a mild steel probe exposed to aqueous solutions of hydrochloric and nitric acid of varying concentrations for 24 hrs. The data used in the figure are taken from reference 81. No protective layer is formed for 24 hrs, and thus Equation 110 is followed. From linear regression analysis, $\mu = 2.08$ (Å/min)(g/m^3)$^{-1/2}$ and $m = 1/2$. This relationship suggests that the corrosion

TABLE 3-4.20 *Yield of CO, HCl, Benzene, and Toluene from the Combustion/Pyrolysis of Polyvinylchloride**

Combustible[†]	Air/Inert	Yield (g/g)			
		CO	HCl	Benzene	Toluene
PVC Homopolymer					
Rigid PVC sheet (49.3% Cl)	inert	—	0.480	0.022	0.002
	air	—	0.479	0.022	0.001
Rigid PVC-1	inert	—	0.555	0.058	0.008
	air	—	0.472	0.044	0.004
Rigid PVC-2	air	0.356	0.513	—	—
PVC resin	air	—	0.486	0.048	0.001
PVC homopolymer-1	air	0.422	0.583	0.031	0.001
PVC homopolymer-2	air	0.413	0.584	0.036	0.001
PVC homopolymer-3	air	0.299	0.500	0.029	0.001
PVC homopolymer-4	air	0.429	0.580	0.043	0.004
PVC + Plasticizer					
PVC (33% Cl) + dioctylphthalate (67%)	air	0.275	0.269	—	—
PVC (31% Cl) + tricresylphosphate	air	0.248	0.269	—	—
PVC + Plasticizer + Acid Neutralizer					
PVC (%) + dioctylphthalate (%) + K_2CO_3 (%)					
42.4 + 42.4 + 15.2	N_2	—	0.171	—	—
38.2 + 38.2 + 23.6	N_2	—	0.111	—	—
32.5 + 32.5 + 35.0	N_2	—	0.029	—	—
PVC (%) + dioctylphthalate (%) + $CaCO_3$ (%)					
45.5 + 45.5 + 9.0	N_2	—	0.221	—	—
41.7 + 41.7 + 16.0	N_2	—	0.171	—	—
35.7 + 35.7 + 28.6	N_2	—	0.117	—	—
Electrical Cables					
PVC jacket	air	—	0.277–0.408	—	—
FR PVC insulation	air	—	0.204–0.285	—	—
Insulation (51% PVC + 49% Plasticizer + additives)	air	0.067	0.273	0.010	0.001
Insulation (57% PVC + 43% Plasticizer + additives)	air	0.090	0.333	0.011	0.001
PVC cable	air	—	0.263	0.033	0.001
General Products					
Floor tile (33% PVC + 70% $CaCO_3$ + inert)	air	0.031	0.073	0.001	—
PVC-nylon brattice cloth	air	—	0.174	0.048	0.001
PVC-nylon fabric	air	—	0.254	0.051	0.001
FR PVC-nylon product	air	—	0.206	0.025	0.001
FR PVC	air	—	0.300	0.020	0.001

*From reference 82.
[†]FR—fire retarded, K_2CO_3—potassium carbonate; $CaCO_3$—calcium carbonate.

rate does not increase rapidly with the concentration of the corrosive products. For example, if the concentration of the corrosive product is increased ten times, the corrosion rate would increase only by a factor of three.

For corrosion in the gas phase, the presence of water is essential or the volume fraction of water ≠ 0. The experimental data for corrosion in the gas phase suggest that $m = 1$ in Equation 110, which can be expressed in the following modified form

$$\dot{R}_{corr} = \frac{\mu y_{corr} \dot{m}'' A}{f_{water} \dot{V}} \qquad (112)$$

where y_{corr} is the yield of the corrosive product (g/g), $\dot{m}''$ is the mass loss rate of the material (g/m²-s), A is the total exposed surface area of the material (m²), f_{water} is the volume fraction of water generated in the combustion of the material and present in the humid air, and $\dot{V}$ is the total volumetric flow rate of fire product-air mixture (m³/s). All

the terms in Equation 112 can be measured, and thus the corrosion parameter, μ, can be calculated for the generalized application of the corrosion data.

Corrosion measurements: For corrosion measurements, fire products are generated in small-scale tests and the corrosion is measured by exposing metal probes to the products in the gas phase at various relative humidities or in the aqueous solutions of the products. The common tests methods are:

1. The Flammability Apparatus test method [Figures 3-4.2(a), 3-4.2(b), and 3-4.27];[1,58,81,84–86]
2. The Cone Calorimeter test method (Figure 3-4.3);[87,88]
3. The Radiant Combustion/Exposure test method;[87,89]
4. The CNET (Centre National d'Etudes des Telecommunications) corrosion test method;[87,90,91]
5. The DIN 57472 test method;[87,92] and
6. The DIN 53436 with metal sheets and CNET corrosion probe test method.[93]

Corrosion Measurements in the Gas Phase: The measurements are made in the Flammability Apparatus, the Cone Calorimeter, the CNET, and the Radiant Combustion/Exposure test methods. For the measurements, either high-sensitivity Rohrback Cosasco (RC) atmospheric metal corrosion probes or CNET metal corrosion probes are used.

The RC corrosion probes are manufactured by a vacuum deposition technique to obtain an open matrix with little resistance to in-depth diffusion of products, resulting in rapid corrosion. It is designed to monitor short-term corrosion (16 to 24 hrs) for environments with small concentrations of corrosive products. The RC probe consists of two metal strips (5,000 to 90,000 Å), embedded in an epoxy-fiberglass plate. One metal strip is coated and acts as a reference, and the other noncoated metal strip acts as a sensor. As the sensor strip corrodes and loses its thickness, its resistance changes. The change in resistance, which represents the extent of corrosion of the metal, is measured as a function of time, by the difference in the resistance between the two strips. The probe readings remain reliable up to about half the thickness of the metal strip (probes are identified as 2500 to 45,000 Å probes).

The CNET probe consists of an epoxy-fiberglass plate embedded with about 170,000 Å thick copper conductors. The change in the resistance of the probe is recorded at the beginning and at the end of the test to determine the extent of corrosion.

The corrosion in the gas phase is measured during the tests every minute and every hour after the test for 16 to 24 hrs. The corrosion rate is calculated as a function of time, using the following type of relationship

$$\dot{R}_c = \frac{D_{c_1} - D_{c_2}}{t_2 - t_1} \tag{113}$$

where $\dot{R}_c$ is the corrosion rate in Å/min, D_{c_1} is the metal thickness in angstroms at time t_1 (s), and D_{c_2} is the metal thickness in angstroms at time t_2 (s).

Data have been reported in the literature for the gas-phase corrosion, mass loss rate, and total volumetric rate of

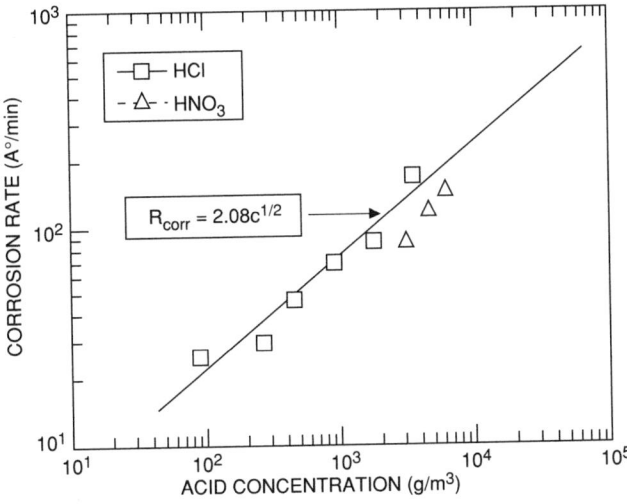

Fig. 3-4.42. Corrosion rate of a mild steel probe versus hydrochloric and nitric acid concentrations. Data used in figure are taken from reference 81.

TABLE 3-4.21 *Corrosion Rate per Unit Fuel Vapor Concentration in the Gas Phase for Flaming and Nonflaming Fires with Variable Oxygen Concentration in the Flammability Apparatus and the Radiant Combustion/Exposure Chamber*

Polymer*	O$_2$ (%)	F/NF†	Water Present‡	Corrosion Rate§ FLAM‖	RC/E#
EVA	21	F	no	nd	0.001
EVA-FR1	21	F	no	nd	0.021
PE	21	F	no	nd	0.002
PE-FR1	21	F	no	nd	0.024
PE-FR1	21	F	yes	nd	0.036
PE-FR2	21	F	no	nd	0.022
PE-FR2	21	F	yes	nd	0.024
PE-FR2	21	F	no	nd	0.014
PE-FR2	21	F	yes	nd	0.016
PE/25% Cl	10	N	yes	0.14	nd
PE/36% Cl	10	NF	yes	0.15	nd
PE/48% Cl	10	NF	yes	0.19	nd
PVC	10	NF	yes	0.15	nd
	21	NF	no	nd	0.027
	21	NF	yes	0.12	0.087
	21	F	yes	1.0	nd
TFE	0	NF	yes	0.0036	nd
	10	NF	yes	0.011	nd
	40	NF	yes	0.035	nd
	21	F	yes	0.42	nd

*See nomenclature.
†F: flaming, NF: nonflaming.
‡Increased humidity in the gas phase with water.
§Per unit fuel vapor concentration (Å/min)/(g/m^3).
‖The Flammability Apparatus test method.[1,58,81,84–87]
#The Radiant Combustion/Exposure test method.[88,90] 1500-min average.
Note: FR-1: red phosphorus fire retardant, FR-2: bromine fire retardant, EVA: ethylene-vinyl acetate copolymer, PE: polyethylene, Cl: chlorine, TFE: tetrafluoroethylene (Teflon™), and nd: not determined.

fire products-air mixture with relative humidity maintained approximately constant. From Equation 112

$$\frac{\dot{R}_{corr}}{(\dot{m}''A/\dot{V})} = \frac{\mu y_{corr}}{f_{water}} \tag{114}$$

where f_{water} is approximately constant, and thus the values of $\dot{R}_{corr}/(\dot{m}''A/\dot{V})$ can be used to assess the relative corrosion nature of the fire products generated from various materials. Tables 3-4.21 and 3-4.22 list the values of the corrosion rate per unit fuel vapor concentration, $\dot{R}_{corr}/(\dot{m}''A/\dot{V})$. The data show that:

1. For significant gas-phase corrosion, it is necessary to have hydrogen atoms in the structure of the halogenated materials as suggested by the stoichiometric yields listed in Table 3-4.13. For example, the corrosion rates per unit fuel vapor concentration for PVC (hydrogen atoms in the structure) and Teflon™ (no hydrogen atoms in the structure) differ by a factor of seven. The difference is probably due to: (a) the inefficiency of the hydrolysis process during the conversion of fluorocarbon products generated from Teflon™ to HF, and (b) the high water solubility of HCl generated from PVC.

2. The corrosion rates per unit fuel vapor concentration for halogenated materials with hydrogen atoms in the structure are high [greater than 0.14 (Å/min)/(g/m^3)], whereas,

TABLE 3-4.22 *Corrosion Rate per Unit Fuel Vapor Concentration in the Gas Phase for Flaming Fires in Air in the Radiant Combustion/Exposure Chamber**

Sample Description	Corrosion Rate†
Crosslinked polyolefin (XLPO) + metal hydrate	0.007
HD polyethylene (PE) + chlorinated PE blend	>0.098
Chlorinated PE + fillers	>0.098
Ethylvinylacetate (EVA) PO + ATH filler	0.012
Polyphenylene oxide/polystyrene (PS) blend	0.005
Polyetherimide	0.002
Polyetherimide/siloxane copolymer	0.005
Intumescent polypropylene (PP)	0.025
Polyolefin copolymer + mineral filler	0.046
XLPO + mineral filler	0.011
XLPO + ATH	0.003
XLPO + ATH	0.007
EVA-PO + mineral filler	0.013
PO + mineral filler	0.016
CLPE + chlorinated additive	>0.098
Polyvinylidene fluoride	>0.098
Polytetrafluoroethylene	>0.098
Polyvinylchloride (PVC)	>0.098
PVC wire	>0.098
PE homopolymer	0.006
Douglas fir	0.006
EVA-PO copolymer	0.003
Nylon 6,6	0.008
XLPE copolymer + brominated additives	0.091

*From reference 87.
†Per unit fuel vapor concentration $(\text{Å/min})/(g/m^3)$; average gas-phase concentration $\approx 17.0\ g/m^3$.

they are negligibly small for fires of nonhalogenated materials [less than $0.007\ (\text{Å/min})/(g/m^3)$], as expected.

3. Fire retardation of nonhalogenated materials by halogenated materials increases the corrosion rate per unit fuel vapor concentration for the nonhalogenated materials from less than 0.007 to 0.011 to $0.046\ (\text{Å/min})/(g/m^3)$. These values, however, are still about ¹⁄₁₀ the values for the halogenated materials.

4. Increase in the corrosion rate per unit fuel vapor concentration due to the presence of water is not significant for halogenated materials with hydrogen atoms in the structure, as expected, as water is generated in the combustion process.

5. Increase in the oxygen concentration in the environment increases the corrosion rate per unit fuel vapor concentration.

Corrosion Measurements in the Aqueous Solution: The measurements are made in the Flammability Apparatus and the DIN 57472 test methods. For the measurements, Rohrback Cosasco (RC) loop-type metal corrosion probes are used. The probes are exposed to the aqueous solutions of the fire products. The probe consists of a metal loop attached to an epoxy-fiberglass rod, with a built-in reference. The metal loop acts as a sensor. As the sensor loop corrodes and loses its thickness, its resistance changes. The extent of corrosion is measured by the difference in the resistance between the loop and the reference. The corrosion rate is determined from Equation 113.

The fire products are either bubbled directly into known volumes of water or are collected in the gas phase on cellulose-based filter papers of known area. After the test, the color, odor, and mass of the products deposited on the filter papers are determined. The fire products are extracted with a known volume of deionized water.

The corrosion in the aqueous solution is measured every hour for 16 to 24 hrs. In some cases, concentrations of corrosive ions, such as chloride, bromide, and fluoride, are also determined using selective ion electrodes in the Flammability Apparatus test method. In the DIN 57472 test standard, pH and conductivity of the solution are measured.

The solution-phase corrosion parameters measured by the Flammability Apparatus test method show that they are comparable for all the halogenated materials and are significantly higher than the values for the gas phase.

Smoke Damage

Smoke is a mixture of black carbon and aerosol.[94,95] Smoke is generated by many sources and is released to the environment, causing pollution, reduction in visibility, and nonthermal damage (discoloration, odor, electrical shorting and conduction, corrosion, etc.). The estimated influx of black carbon to the environment from burning is 0.5 to 2 × 10^{15} g/yr.[94] Black carbon is often called charcoal, soot, elemental carbon, etc.[94] The particulate organic matter (POM) in aerosols consists of:[94] (1) hydrocarbons—these are the alkanes, alkenes, and some aromatics, with aliphatics constituting the greatest fraction. They range from C_{17} to C_{37}; (2) polycyclic aromatic hydrocarbons; (3) oxidized hydrocarbons—these classes include acids, aldehydes, ketones, quinones, phenols, and esters, as well as the less stable epoxides and peroxides. They may be produced directly in combustion processes or through oxidations in the atmosphere; (4) organo-nitrogen compounds—the aza-arenes are the only types of this class that have been so far analyzed, and they are one or two orders of magnitude less than the polycyclic aromatic hydrocarbons; and (5) organo-sulfur compounds—heterocyclic sulfur compounds, such as benzothiazole, have been reported in urban aerosols.

The environmental behavior of black carbon introduced by combustion processes depends on the characteristics of the source, aerosol properties, chemical composition of black carbon, and meteorology.[94] The yield of black carbon depends on the material and combustion conditions, as discussed in previous sections. Table 3-4.23 lists data, taken from reference 94, for the yield of black carbon from some industrial combustion processes.

TABLE 3-4.23 *Yield of Black Carbon from Some Industrial Combustion Processes**

Fuel	Source	Yield (g/kg)
Natural gas	Steam generator	3×10^{-4}
	Domestic water heater	0.1
	Heating boiler	0.01–0.07
Gasoline	Automobile engine	0.1
Diesel	Automobile engine	2–4
	Truck/bus engine	0.6–1
Jet A	Aircraft turbine	0.5–3
Fuel oil (#2)	Utility turbine	0.08

*Data taken from reference 94.

TABLE 3-4.24 *Most Frequently Occurring Smoke Particle Radii in Fires of Some Materials**

Material	Smoke Particle Radius (microns)
Coal	0.078
Polystyrene	0.078
Kerosene	0.079
Polypropylene	0.079
Polyethylene	0.077
Propylene	0.076
Ethylene	0.072
Heptane	0.077
Propane	0.068
Nylon	0.075
PMMA	0.068
Douglas fir	0.062
Polyethylene with chlorine	0.090
Polychloroprene	0.090
PVC	0.083
Styrene-butadiene rubber with chlorine	0.073

*Data taken from reference 97.

Multi-modal distributions of black carbon issuing from flames, diesel engines, and freeway traffic show that the "nuclei mode" has a geometric mean radius between 0.0025 and 0.020 microns and probably results from the condensation of gaseous carbon moieties.[94] The "accumulation mode" encompasses particles in the size range 0.075 to 0.25 microns and apparently results from the coagulation and condensation of the "nuclei mode" particles.[94] Finally in the case of vehicular emissions there is a "coarse mode" at several microns that is attributed to the precipitation of fine particles on the walls of exhaust systems and a subsequent entrainment in the issuing gases.[94] The coal-fired utility boilers produce soot with peaks at particle radius of about 0.05 microns.[94] Long-range transport of particles shows that about 60 percent of the soot is less than 0.05 microns radius size class.[94] The larger particles are probably removed preferentially from the air during its travel.

In fires, large variations in smoke particle size, due to coagulation and condensation, have been found. As the smoke moves away from the fire origin large particles settle down to the floor, leaving small particles in the gas phase,[96] similar to the long-range transport in the atmosphere discussed in reference 94. The data from various fires show that initially the smoke particles are in the "coarse mode." The particle size decreases slowly with time, suggesting that large particles settle down from the hot layer at the ceiling.

Relationships between transport of heat and smoke generated in large enclosure fires and for smoke characterization have been developed and data have been reported for the most frequently occurring smoke particle radius.[97] These data are listed in Table 3-4.24, which shows that radii of the smoke particles vary between 0.062 to 0.09 microns, belonging to the lower end of the "accumulation mode."

It thus appears that, in fires, smoke damage in the room of fire origin is expected to be due to particles of several microns in radius in the "coarse mode," whereas smoke damage downstream of the fire is expected to be due to particles with radius less than 0.1 micron in the lower end of the "accumulation mode."

Although concentration, size, physical, and optical properties, and chemical composition of smoke particles have been studied in detail, very little is known about the charges on the particles.[98] It has been suggested that soot nucleation and growth occur near the highly ionized regions of the flames in combustion processes, possibly suggesting that some of the charges are transferred to smoke particles. In hydrocarbon-oxygen flames, the following reaction is considered to be the dominant reaction for the charge separation[98]

$$CH + O = CHO^+ + e^- \tag{115}$$

Charges on smoke particles generated in flaming and nonflaming fires of wood, cotton wick, polyurethane, heptane with 3 percent toluene, and an alcohol have been examined.[98] The results show that, in nonflaming fires, initially a very small fraction of particles is charged. During aging, the charge increases slowly. For flaming polyurethane fires, where large amounts of black carbon are generated, smoke carries a high initial charge; 70 percent of the particles in the size interval from 0.018 to 0.032 microns are charged. Similar results are found for heptane. Flaming wood fires, however, show particle charges between that of nonflaming fires of wood and cotton wick and that of flaming fires of polyurethane. In the flaming fire of alcohol, there is no smoke.

Char and black carbon are efficient absorbers of HCl. In the combustion of plasticized PVC wire, about 25 percent of the original chloride ions are retained in the char, and the ions are predominantly inorganic in nature.[99] In the combustion of PE-PVC cables in rooms, smoke particles that settle down in the room contain about 33 percent by weight of inorganic chloride ions, and less than 2 percent of the theoretically expected mass of the chloride ions leaves the enclosure.[96] In the combustion of 79.5 percent PVC-20.5 percent PE, 19 mg of HCl/g of smoke is loosely bound and 27 mg of HCl/g of smoke is tightly bound to carbon.[100]

It thus appears that, for nonthermal fire damage, the important factors are: (1) concentrations of fire products and their deposition on surfaces, (2) chemical and physical nature of the products, (3) nature of the surfaces, (4) presence of moisture, and other factors. These factors depend on: (1) fire initiation and spread, (2) generation rates of fire products and their chemical and physical natures, (3) relative humidity and temperature, (4) in-flow rate of air and its mixing with the products and the flow velocity of the mixture, (5) nature and orientation of the target relative to the flow of the products, (6) exposure duration, (7) presence or absence of fire extinguishing agents, etc.

FIRE CONTROL/SUPPRESSION/ EXTINGUISHMENT

For the prevention of loss of life and property in fires, both active and passive fire protection techniques are used.[101] Passive fire protection is provided by: (1) modifying the chemical structures of the materials for high resistance to ignition and fire propagation, (2) incorporating fire retardants within the materials, (3) coating and wrapping the surfaces, (4) separating materials by inert fire barriers, (5) modifying configuration and arrangement of materials, etc. Active fire protection is provided by the application of agents to control, suppress, and/or extinguish fires. The

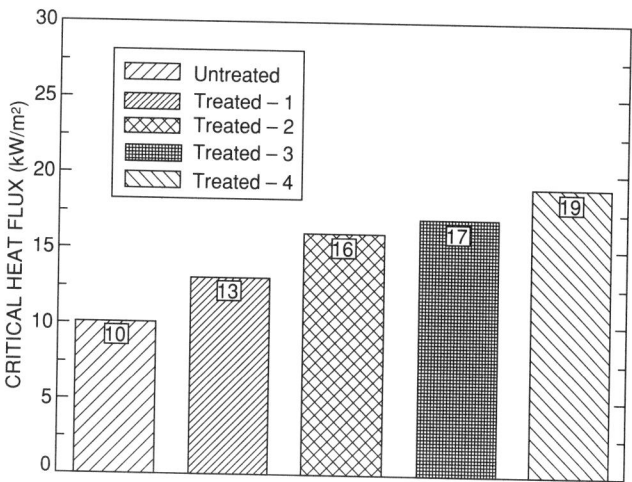

Fig. 3-4.43. Critical Heat Flux for untreated and treated tri-wall corrugated paper sheet. The amount of passive fire protection agent is increasing from Treated 1 to 4. Data obtained from the ignition experiments in the Flammability Apparatus. Numbers indicated on top of each bar are the Critical Heat Flux values.

most commonly used liquid and gaseous agents at the present time are: water, CO_2, N_2, and halons* 1211 ($CBrClF_2$), 1301 ($CBrF_3$), and 2402 ($CBrF_2CBrF_2$). Because of the contribution of halons to depletion of the stratospheric ozone layer, they will not be used in the future. There is thus an intense effort underway to develop alternative fire suppressants to replace ozone-layer-depleting halons.

The mechanisms of passive and active fire protection are generally known.[37,101-108] Flame extinction by liquid and gaseous agents is mainly due to physical processes (such as removal of heat from the flame and burning surface and creation of nonflammable mixtures) and/or chemical processes (such as termination of chemical reactions). The effectiveness of water is mainly due to removal of heat from the burning surface as a result of vaporization. The effectiveness of halons is mainly due to termination of chemical reactions. N_2 and CO_2 are effective mainly due to creation of nonflammable mixtures by reducing mass fraction of oxygen.

Passive Fire Protection

Passive fire protection is provided by various chemical and physical means, as follows.

Increasing the resistance to ignition and fire propagation by increasing the Critical Heat Flux (CHF) and Thermal Response Parameter (TRP) values: The Critical Heat Flux is expressed as

$$CHF \approx \sigma(T_{ig}^4 - T_a^4) \qquad (116)$$

where σ is the Stefan-Boltzmann constant (56.7×10^{-12} kW/m^2-K^4), T_{ig} is the ignition temperature (K), and T_a is the ambient temperature (K). TRP is defined in Equations 1 and 2, and its relationship to fire propagation in Equations 8 and 9.

The relationships between time to ignition, fire propagation rate, Fire Propagation Index, and TRP (Equations 2, 8,

*The numbers represent: First: number of carbon atoms; second: number of fluorine atoms; third: number of chlorine atoms; fourth: number of bromine atoms.

and 9) show that the time to ignition is directly proportional to the TRP value to the power two; and the fire propagation rate and the Fire Propagation Index are inversely proportional to the TRP value to the power two and one, respectively. Thus the higher the TRP value, the longer the time to ignition, the slower the fire propagation rate, and the lower the FPI value. For high TRP values with FPI < 7, there is no fire propagation beyond the ignition zone, defined as the nonfire-propagating behavior. Also, for materials with high CHF values, higher heat flux exposure is required to initiate a fire.

The CHF and TRP values can be increased by modifying the pertinent parameters, such as increase in the chemical bond dissociation energy and decrease in thermal diffusion (combination of the density, specific heat, and thermal conductivity). Figures 3-4.43 and 3-4.44 show the CHF and TRP values for a tri-wall corrugated paper sheet containing various amounts of a passive fire protection agent (identified as agent A here); the data were obtained from the ignition experiments in the Flammability Apparatus. Figure 3-4.45 shows the TRP value for a single-wall corrugated paper sheet containing various amounts of the passive fire protection agent A; the data were obtained from the ignition experiments in the Flammability Apparatus. The CHF and TRP values increase with increase in the amount of agent; thus, the passive fire protection agent would complement the active fire protection agents. Corrugated paper boxes treated with higher amounts of the passive fire protection agent are expected to require reduced amounts of the active fire protection agents for fire control, suppression, or extinguishment compared to the amounts of the active fire protection agents required for the untreated boxes.

The passive fire protection requirements for various materials can be assessed from the data for CHF and TRP listed in Table 3-4.2.

Decreasing the values of the Heat Release Parameter (HRP) and the flame heat flux: Heat release rate is equal to the Heat Release Parameter (HRP) times the net heat flux (Equation 30). HRP is the ratio of the heat of combustion to heat of gasification, and thus the HRP value can be decreased

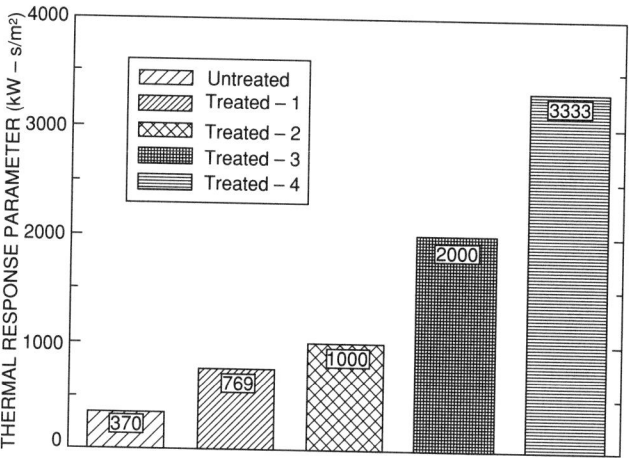

Fig. 3-4.44. Thermal Response Parameter for untreated and treated tri-wall corrugated paper sheet. The amount of passive fire protection agent is increasing from Treated 1 to 4. Data obtained from the ignition experiments in the Flammability Apparatus. Numbers indicated on top of each bar are the Thermal Response Parameter values.

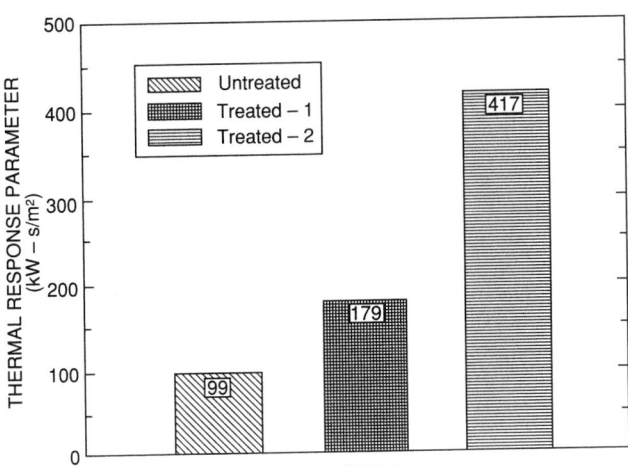

Fig. 3-4.45. Thermal Response Parameter for untreated and treated single-wall corrugated paper sheet. The amount of passive fire protection agent is increasing from Treated 1 to 2. Data obtained from the ignition experiments in the Flammability Apparatus. Numbers indicated on top of each bar are the Thermal Response Parameter values.

by decreasing the heat of combustion and/or increasing the heat of gasification by various chemical and physical means. An examination of data in Table 3-4.11 for heats of combustion shows that introduction of oxygen, nitrogen, sulfur, halogen, and other atoms into the chemical structures of the materials reduces the heat of combustion. For example, the heat of combustion decreases when the hydrogen atoms attached to carbon atoms in polyethylene are replaced by the halogen atoms, such as by fluorine in Teflon™. The chemical heat of combustion decreases from 38.4 kJ/g to 4.2 kJ/g (Table 3-4.11), and the chemical HRP value decreases from 17 to 2 (Table 3-4.12).

The HRP values can also be reduced by increasing the heat of gasification and decreasing the heat of combustion by retaining the major fraction of the carbon atoms in the solid phase, a process defined as charring. Several passive fire protection agents are available commercially to enhance the charring characteristics of materials.

Figure 3-4.46 shows the reduction in the chemical heat release rate as a result of increase in charring of a tri-wall corrugated paper sheet by the passive fire protection agent A; the data were obtained from the combustion experiments in the Flammability Apparatus. The amount of the agent A is increasing from Treated 1 to 3. There is a very significant decrease in the chemical heat release rate of the tri-wall corrugated paper sheet by the passive fire protection agent A, which will complement the active fire protection agents. Corrugated paper boxes treated with higher amounts of the passive fire protection agent are expected to require reduced amounts of the active fire protection agents for fire control, suppression, or extinguishment compared to the one required for the untreated boxes.

The effect on flame heat flux by passive fire protection is determined by using the radiation scaling technique, where combustion experiments are performed in oxygen concentration higher than the ambient values. Very little is known about this subject. Table 3-4.5 lists some of the flame heat flux values derived from the radiation scaling technique, but no systematic study has been performed for the effectiveness

of passive fire protection. For liquids that vaporize primarily as monomers or as very low molecular weight oligomers, the flame heat flux values are in the range of 22 to 44 kW/m², irrespective of their chemical structures. For solid materials, which vaporize as high molecular weight oligomers, the flame heat flux values increase substantially to the range of 49 to 71 kW/m², irrespective of their chemical structures. The independence of the asymptotic flame heat values from the chemical structures of materials is consistent with the dependence of flame radiation on optical thickness, soot concentration, and flame temperature in large-scale fires. Passive fire protection agents, which can reduce the molecular weight of the vaporized materials, would be effective in reducing the flame heat flux and complement the active fire protection agents.

Changing the molten behavior of materials: Figure 3-4.47 shows the chemical heat release rate *versus* time for the well-ventilated combustion of a 90-mm-diameter and 25-mm-thick slab of polypropylene exposed to an external heat flux of 50 kW/m². The data were measured in the Flammability Apparatus. For about 900 sec, the polypropylene slab burns as a solid with a thin liquid layer at the surface. The measured and calculated values of the heat release rate under this condition agree very well. The heat release rate was calculated from Equation 30 with $\dot{q}_e'' \gg \dot{q}_f'' - \dot{q}_{rr}''$.

Between about 900 and 1150 sec, the polypropylene slab melts rapidly. At about 1150 sec, the entire sample changes to a liquid and burns as a boiling liquid pool fire. The chemical heat release rate triples at this stage. This is the most dangerous stage in a fire and presents a serious challenge to the active fire protection agents, such as water applied as a spray from sprinklers. Inert passive fire protection agents that eliminate the boiling liquid pool fire stage will be effective in complementing the active fire protection agents, such as water.

Changing the nature of the fire products: Nonhalogenated passive fire protection agents or agents that reduce or

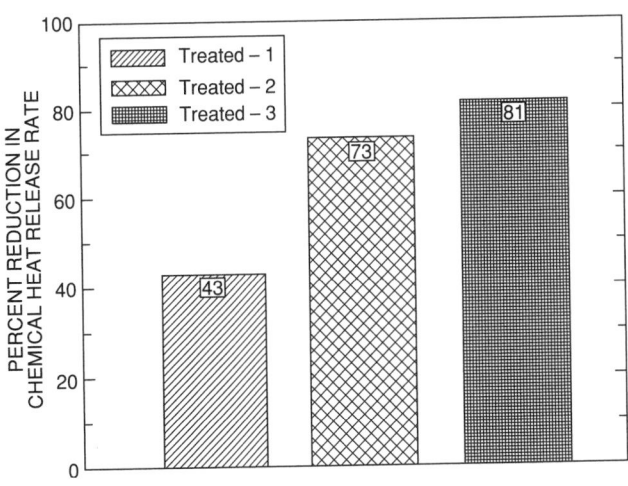

Fig. 3-4.46. Percent reduction in the chemical heat release rate of untreated tri-wall corrugated paper sheet by a passive fire protection agent. The amount of the passive fire protection agent is increasing from Treated 1 to 3. Data from the combustion experiments in the Flammability Apparatus. Numbers indicated on top of each bar are the percent reductions in the chemical heat release rate.

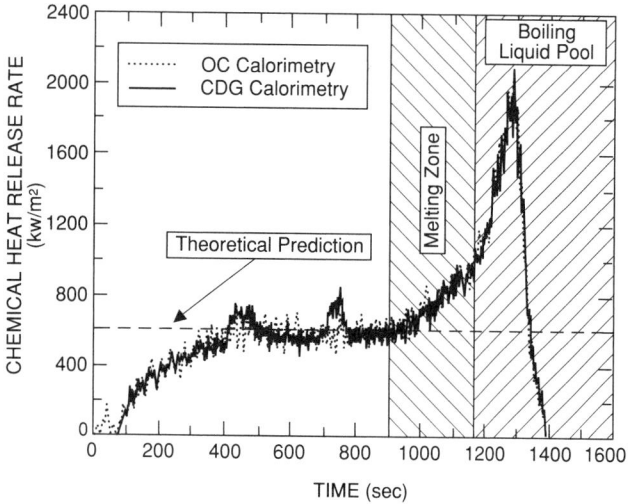

Fig. 3-4.47. *Chemical heat release rate in the well-ventilated combustion in normal air of a 90-mm-diameter and 25-mm-thick slab of polypropylene exposed to 50 kW/m² of external heat flux in the Flammability Apparatus. (See Fig. 3-4.21.)*

eliminate the release of halogenated and highly aromatic products and enhance release of aliphatic products, rich in hydrogen and oxygen atoms but poor in carbon atoms, are effective in reducing the nonthermal damage due to smoke and corrosion. Some of the passive fire protection agents, available commercially, interact with the materials in the solid as well as in the gas phase during pyrolysis and combustion.

The critical parameter that needs to be examined in the presence and absence of the passive fire protection agents is the ratio of the generation rate of products [such as for smoke, CO, corrosive products (HCl), and others] to heat release rate. The effectiveness of the passive fire protection agent is reflected in the small values of the ratios at fire control, suppression, and/or extinguishment stage.

Active Fire Protection

Active fire protection is provided by applying agents to the flame and/or to the surface of the burning material. The fire control, suppression, and extinguishment have been described by the fire point equation.[104,106] According to the fire point theory, the convective heat flux from the flame to surface as flame extinction condition is reached is expressed as[104,106]

$$\dot{q}''_{fc} = \varphi \Delta H_T \dot{m}''_{cr} \qquad (117)$$

where $\dot{q}''_{fc}$ is convective flame heat flux from the flame to the surface as the extinction condition is reached (kW/m²); φ is the maximum fraction of combustion energy that the flame reactions may lose to the sample surface by convection without flame extinction and is defined as the kinetic parameter for flame extinction; ΔH_T is the net heat of complete combustion (kJ/g); and $\dot{m}''_{cr}$ is the critical mass loss rate for flame extinction (g/m²-s). The kinetic parameter is defined as[104,106]

$$\varphi = \frac{\Delta H_{g,con}}{\Delta H_T} \qquad (118)$$

where $\Delta H_{g,con}$ is the flame convective energy transfer to the fuel per unit mass of fuel gasified (kJ/g). The kinetic parameter is expected to be higher for fast-burning material vapors

and lower for slower burning material vapors, such as materials containing halogens, sulfur, nitrogen, etc. It is suggested that, at flame extinction, combustion is controlled primarily by the convective heat transfer, and thus the critical mass loss rate would follow Spalding's mass transfer number theory[103]

$$\dot{m}''_{cr} = \frac{h}{c_P} \ln(B_{cr} + 1) \qquad (119)$$

where h is the convective heat transfer coefficient (kW/m²-K), c_P is the specific heat of air (kJ/g-K), and B_{cr} is the critical mass transfer number, defined as

$$B_{cr} = \frac{Y_O \Delta H_O^* - c_P(T_s - T_a)}{\Delta H_{g,con}} \qquad (120)$$

where Y_O is the oxygen mass fraction (−); ΔH_O^* is the net heat of complete combustion per unit mass of oxygen consumed (kJ/g), which is approximately constant (Tables 3-4.7 through 3-4.10); T_s is the surface temperature (K); and T_a is the ambient temperature (K). For ambient conditions, $Y_O \Delta H_T \gg c_P(T_s - T_a)$. From equations 118 through 120

$$\varphi = \frac{Y_O \Delta H_O^*}{\Delta H_T \cdot \exp(\dot{m}''_{cr} c_P/h) - 1} \qquad (121)$$

The fire point theory[104,106] and experimental data show that the critical mass loss rate for flame extinction is similar to the critical mass loss rate for ignition;[16,53,105,107,109] the critical mass loss rate for ignition, however, has to be measured at the time period where the sustained flame is just being established. The data for the critical mass loss rate for ignition and flame extinction and the kinetic parameter for flame extinction are listed in Table 3-4.25. The values for the critical mass loss rate for ignition from the Flammability Apparatus (reference 16) are measured at the time period where the sustained flame is just being established, and thus are higher than the values from the University of Edinburgh (reference 109). The University of Edinburgh data are probably measured just before the sustained flame is established. For polymethylmethacrylate, the critical mass loss rate for ignition from the Flammability Apparatus (reference 16) agrees with the critical mass loss rate for flame extinction from reference 107.

The data in Table 3-4.25 show that the values of the kinetic parameter are higher for the aliphatic materials than the values for the aromatic and chlorinated materials, which is opposite to the trend for the heat of combustion. The data suggest that the materials can be arranged in the following decreasing order of the kinetic parameter values (using FMRC values): polyoxymethylene ($\varphi = 0.43$) > polymethylmethacrylate ($\varphi = 0.28$) > polyethylene, polypropylene, and polyethylene foams ($\varphi = 0.27$ to 0.25) > polystyrene ($\varphi = 0.21$) > polyurethane, polystyrene, and polyisocyanurate foams and chlorinated polyethylenes ($\varphi = 0.09$ to 0.19). As expected from the fire point theory,[104,106] the reactivity of the vapors in the gas phase follows the kinetic parameter.

The combustion efficiency and product generation efficiencies follow the reactivity of the vapors in the gas phase, such as shown in Figure 3-4.48 for the combustion efficiency. The lower the value of the kinetic parameter (Equation 121), the lower the reactivity of the material vapors, which is reflected in the: (1) reduced values of the combustion efficiency (Equations 32 through 34), (2) reduced values of the generation efficiencies (Equation 66) of the oxidation zone products (such as CO_2), and (3) increased values of the

TABLE 3-4.25   *Critical Mass Loss Rate for Ignition and Kinetic Parameter for Flame Extinction*

Material	Critical Mass Loss Rate (g/m²-s)		Kinetic Parameter	
	Ref. 16*	Ref. 109†	Ref. 16*	Ref. 109†
Polyoxymethylene	4.5	1.7	0.43	1.05
Polymethylmethacrylate	3.2	1.9	0.28	0.53
Polyethylene	2.5	1.3	0.27	—
Polypropylene	2.7	1.1	0.24	0.50
Polyethylene foams				
1	2.6	—	0.24	—
2	2.6	—	0.25	—
3	2.5	—	0.25	—
4	2.6	—	0.25	—
Chlorinated polyethylenes				
25% chlorine	6.6	—	0.15	—
36% chlorine	7.5	—	0.09	—
48% chlorine	7.6	—	0.08	—
Polystyrene	4.0	0.80	0.21	0.78
Polystyrene foams				
GM47	6.3	—	0.11	—
GM49	4.9	—	0.14	—
GM51	6.3	—	0.10	—
GM53	5.7	—	0.11	—
Polyurethane foams (flexible)				
GM21	5.6	—	0.16	—
GM23	5.3	—	0.17	—
GM25	5.7	—	0.15	—
GM27	6.5	—	0.12	—
1/CaCO₃	7.2	—	0.19	—
Polyurethane foams (rigid)				
GM29	7.9	—	0.10	—
GM31	8.4	—	0.09	—
GM35	6.9	—	0.11	—
Polyisocyanurate foams (rigid)				
GM41	6.8	—	0.12	—
GM43	5.5	—	0.15	—
Phenolic foam	5.5	—	0.17	—

*Ignition data measured in the Flammability Apparatus.
†Ignition data measured at the University of Edinburgh, U.K.

generation efficiencies of the reduction zone products (such as smoke, CO, and hydrocarbons).

The flame extinction can also be expressed in terms of the critical heat release rate

$$\dot{Q}''_{cr,i} = \Delta H_i \dot{m}''_{cr} \qquad (122)$$

where $\dot{Q}''_{cr,i}$ is the critical heat release rate (chemical, convective, or radiative in kW/m²), and ΔH_i is the heat of combustion (chemical, convective, and radiative in kJ/g). Table 3-4.26 lists the critical chemical, convective, and radiative heat release rates for flame extinction, where critical mass loss rate values are taken from Table 3-4.25 and heats of combustion from Table 3-4.11.

The data in Table 3-4.26 suggest that the critical heat release rate for flame extinction is weakly dependent on the chemical nature of the material, contrary to the critical mass loss rate. The critical heat release rates thus can be averaged, which are 100 ± 7, 53 ± 9, and 47 ± 10 kW/m² for the chemical, convective, and radiative heat release rates, respectively. For materials with highly reactive vapors, such as polyethylene, large amounts of extinguishing agent are needed to reduce the heat release rate to the critical value.

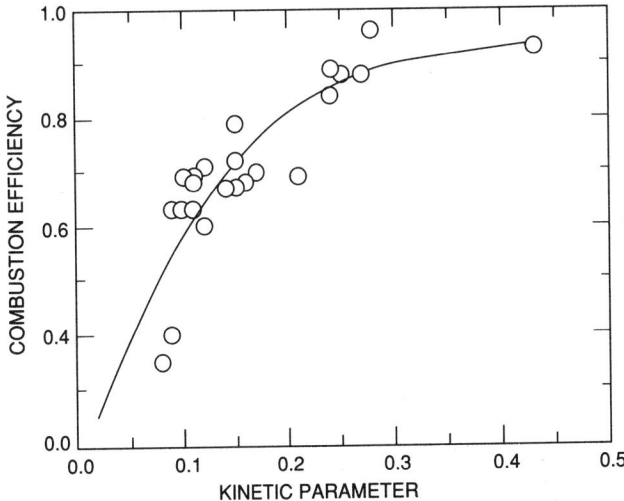

Fig. 3-4.48.   Kinetic parameter for flame extinction versus the combustion efficiency and production generation efficiencies. Data are measured in the Flammability Apparatus.

TABLE 3-4.26 *Critical Chemical, Convective, and Radiative Heat Release Rates for Flame Extinction**

Material	Critical Heat Release Rate (kW/m²)		
	Chemical	Convective	Radiative
Polyoxymethylene	(65)	50	(14)
Polymethylmethacrylate	77	53	24
Polyethylene	96	55	42
Polypropylene	104	61	43
Polyethylene foams	88	51	38
Chlorinated polyethylenes	95	48	47
Polystyrenes	108	44	64
Polyurethane foams (flexible)	101	48	53
Polyurethane foams (rigid)	102	44	58
Average	96 ± 10	51 ± 6	46 ± 12

*Critical mass loss rates from the Flammability Apparatus, and heats of combustion from Table 3-4.11.

For materials with highly non-reactive vapors, such as Teflon™, it is difficult to reach the critical heat release rate values unless high external heat flux is applied.

The energy balance at the surface as the flame extinction condition is reached is[103]

$$\dot{m}'' = \frac{\varphi \Delta H_T \dot{m}''_{cr} + \dot{q}''_e - \dot{q}''_{rr} - \dot{q}''_{agent}}{\Delta H_g} \quad (123)$$

$$\dot{Q}''_i = \frac{\Delta H_i}{\Delta H_g}(\varphi \Delta H_T \dot{m}''_{cr} + \dot{q}''_e - \dot{q}''_{rr} - \dot{q}''_{agent}) \quad (124)$$

where $\dot{q}''_e$ is the external heat flux (kW/m²); $\dot{q}''_{rr}$ is the surface re-radiation loss (kW/m²); $\dot{q}''_{agent}$ is the heat flux removed from the surface or from the flame by the agent as the flame extinction condition is reached (kW/m²); ΔH_i is the chemical, convective, or radiative heat of combustion (kJ/g); and ΔH_g is the heat of gasification (kJ/g). $\Delta H_i/\Delta H_g$ is defined as the Heat Release Parameter (HRP).

Flame suppression/extinguishment by water: The heat flux removed from the surface of a burning material by water, as a result of vaporization, is expressed as[103]

$$\dot{q}''_w = \varepsilon_w \dot{m}_w \Delta H_w \quad (125)$$

where ε_w is the water application efficiency, $\dot{m}''_w$ is the water application rate per unit surface area of the material (g/m²-s), and ΔH_w is the heat of gasification of water (2.58 kJ/g). If only part of the water applied to a hot surface evaporates and the other part forms a puddle, such as on a horizontal surface, blockage of flame heat flux to the surface and escape of the fuel from the material surface are expected. Equation 125 thus is modified as

$$\dot{q}''_w = \dot{m}''_w(\varepsilon_w \Delta H_w + \delta_w) \quad (126)$$

where δ_w is the energy associated with the blockage of flame heat flux to the surface and escape of the fuel vapors per unit mass of the fuel gasified (kJ/g).

From Equations 123 and 126

$$\dot{m}''_w = \frac{\dot{q}''_e}{\varepsilon_w \Delta H_w + \delta_w} + \frac{\varphi \Delta H_T \dot{m}''_{cr} - \dot{q}''_{rr} - \dot{m}'' \Delta H_g}{\varepsilon_w \Delta H_w + \delta_w} \quad (127)$$

At flame extinction, $\dot{m}'' = \dot{m}''_{cr}$ and from Equation 127

$$\dot{m}''_{w,ex} = \frac{\dot{q}''_e}{\varepsilon_w \Delta H_w + \delta_w} + \frac{\dot{m}''_{cr}(\varphi \Delta H_T - \Delta H_g) - \dot{q}''_{rr}}{\varepsilon_w \Delta H_w + \delta_w} \quad (128)$$

where $\dot{m}''_{w,ex}$ is the water application rate per unit surface area of the material for flame extinction (g/m²-s). As discussed in reference 103, in the absence of the external heat flux with no water puddle formation at the surface, the critical water application rate for flame extinction is

$$\dot{m}''_{w,cr} = \frac{\dot{m}''_{cr}(\varphi \Delta H_T - \Delta H_g) - \dot{q}''_{rr}}{\varepsilon_w \Delta H_w} \quad (129)$$

where $\dot{m}''_{w,cr}$ is the critical water application rate (g/m²-s), which is related to the fundamental fire property of the material. The calculated values of the critical water application rate for materials are listed in Table 3-4.27, where efficiency of water application was taken as unity. The values were calculated from Equation 128, using data from Table 3-4.11 for the net heats of complete combustion, from Table 3-4.25 for the critical mass loss rate and the kinetic parameter, from Table 3-4.4 for the heats of gasification and surface

TABLE 3-4.27 *Critical Water Application Rates for Flame Extinction*

Material	$\dot{q}''_{rr}$ (kW/m²)	ΔH_T (kJ/g)	ΔH_g (kJ/g)	$\dot{m}''_{cr}$ (g/m²-s)	φ (g/m²-s)	Critical Water Appl. Rate (g/m²-s)
Polyoxymethylene	13	15.4	2.4	4.5	0.43	2.3
Polymethylmethacrylate	11	25.2	1.6	3.2	0.28	2.5
Polyethylene	15	43.6	1.8	2.5	0.27	3.8
Polypropylene	15	43.4	2.0	2.7	0.24	3.0
Polyethylene foams						
1	12	41.2	1.7	2.6	0.24	3.6
2	13	40.8	1.4	2.6	0.25	3.8
3	12	40.8	1.8	2.5	0.25	3.5
4	12	40.8	1.5	2.6	0.25	4.1
Chlorinated polyethylenes						
25% chlorine	12	31.6	2.1	6.6	0.15	2.1
36% chlorine	12	26.3	3.0	7.5	0.12	0
48% chlorine	10	20.6	3.1	7.6	0.13	0
Polystyrene	13	39.2	1.7	4.0	0.21	5.1

re-radiation loss, and using a value of 2.59 kJ/g for the heat of gasification of water.

All the materials listed in Table 3-4.27 burn in normal air without the external heat flux, except polyethylene with 36 and 48 percent chlorine by weight. The critical water application rate for flame extinction for materials that do not burn in normal air without the external heat flux is zero. The materials in Table 3-4.27 that burn without the external heat flux can be arranged in the following order of increased water application rate required for flame extinction: polyoxymethylene, polymethylmethacrylate, and polyethylene with 25 percent chlorine (2.1 to 2.5 g/m^2-s) < polyethylene and polypropylene (3.5 to 4.1 g/m^2-s) < polystyrene (5.1 g/m^2-s).

The data in Table 3-4.27 suggest that the critical water application rate required for flame extinction, with no water puddle at the surface, can be calculated to support the experimental data. The input data for the calculation can be obtained from the measurements for the fire properties in the small-scale apparatuses, such as the oxygen bomb calorimetry, the Flammability Apparatus, the OSU Apparatus, and the Cone Calorimeter. The properties and respective tests are: (1) surface re-radiation loss [from the Critical Heat Flux (CHF) and critical mass loss rate, using ignition tests], (2) heat of gasification using the nonflaming tests, (3) net heat of complete combustion from the oxygen bomb calorimeter, and (4) kinetic parameter (Equation 121) where the ratio of the convective heat transfer coefficient to specific heat is needed. The ratio can be obtained from the methanol combustion at variable oxygen mass fractions and external heat flux for known inlet airflow rates, a procedure that has been used in the Flammability Apparatus for such applications.[33]

The first term on the right-hand side of Equation 128 can be considered as the term to account for the effects of fire size as well as the shapes and arrangements of the materials. As the fire intensity increases due to changes in the shape, size, and arrangements of the material, heat flux to the surface of the material increases, and water application rates above and beyond the critical water application rate for flame extinction thus would be required. For example, water application rate for extinguishment of fires burning at the asymptotic limits can be calculated from: (1) the values of the flame heat flux to the surface listed in Table 3-4.5, in place of the external heat flux in Equation 128; and (2) the data for the critical water application rate for flame extinction listed in Table 3-4.27. The calculated water application rates for extinguishment of fires burning at the asymptotic limits are listed in Table 3-4.28. The data show that the first term of Equation 128 becomes very dominant at the asymptotic limit compared to the second term, which is the critical water application rate. In Table 3-4.28, the water application rates at the asymptotic limits are thus calculated on the basis of flame heat flux alone.

Numerous small- and large-scale tests have been performed to assess the extinguishment of fires by water sprays.[103–108,110–113] For example, small-scale fire suppression/extinguishment tests are performed in the Flammability Apparatus [Figures 3-4.2(a) and (b)], and large-scale fire suppression/extinguishment tests are performed in the Fire Products Collector (Figure 3-4.8) and at the Test Center, mostly at the 30-ft site (Figure 3-4.49).[27,37,101,102,105,110–114]

Small-Scale Fire Suppression/Extinguishment Tests Using Water and Materials with Two- and Three-Dimensional Configurations Burning in Co- and Natural-Airflow Conditions:

TABLE 3-4.28 *Water Application Rate for the Extinguishment of Fires at the Asymptotic Limits**

Material	$\dot{m}''_{w,cr}$ (g/m^2-s)	$\dot{q}''_f$ (kW/m^2)	Water Appl. Rate (g/m^2-s)
Aliphatic Carbon-Hydrogen Atoms			
Polyethylene	3.8	61	27
Polypropylene	3.0	67	29
Heavy fuel oil (2.6–23 m)	?	29	11[†]
Kerosene (30–80 m)	?	29	11[†]
Crude oil (6.5–31 m)	?	44	17[†]
n-Dodecane (0.94 m)	?	30	12[†]
Gasoline (1.5–223 m)	?	30	12[†]
JP-4 (1.0–5.3 m)	?	40	16[†]
JP-5 (0.60–17 m)	?	39	15[†]
n-Heptane (1.2–10 m)	?	37	14[†]
n-Hexane (0.75–10 m)		37	14[†]
Transformer fluids (2.37 m)			
Aromatic Carbon-Hydrogen Atoms			
Polystyrene (0.93 m)	5.1	75	34
Xylene (1.22 m)	?	37	14[†]
Benzene (0.75–6.0 m)	?	44	17[†]
Aliphatic Carbon-Hydrogen-Oxygen Atoms			
Polyoxymethylene	2.3	50	22
Polymethylmethacrylate (2.37 m)	2.5	60	26
Methanol (1.2–2.4 m)	?	27	10[†]
Acetone (1.52 m)	?	24	9[†]
Aliphatic Carbon-Hydrogen-Halogen Atoms			
Polyvinylchloride	0	50	19
Tefzel™ (ETFE)	0	50	19
Teflon™ (FEP)	0	52	20

*For water application efficiency of unity with no water puddle at the surface.
[†]Calculated from the flame heat flux alone. Because water does not stay at the surface, the flame extinction of liquid pool fires with water is not an efficient process. The efficiency of unity used in the calculations thus may not be correct and actual water application rates would probably be higher than calculated.

Several studies have been performed for these types of configurations and airflow conditions.[101–108] For example, small-scale fire suppression/extinguishment tests using water are performed in the Flammability Apparatus, under co- and natural airflow conditions. In the tests, measurements are made, in the presence and absence of water, for the Critical Heat Flux (CHF); Thermal Response Parameter (TRP); mass loss rate; chemical, convective, and radiative heat release rates; generation rates of CO and CO_2; hydrocarbons; smoke; optical transmission through smoke; corrosion in the gas phase; and other products (depending on the need).

The test samples used, with and without the external heat flux, consist of: (1) two-dimensional samples: 100 × 100 mm square and 100-mm-diameter circular samples up to 50 mm in thickness; and (2) three-dimensional samples identified as "sample commodities": (a) cross piles of sticks, defined as the "crib"; single crib is used in the test; and (b) 50-, 75-, and 100-mm cubic boxes; one to eight boxes are arranged in one to four layers with a separation of about 12 mm between the boxes and the layers. The designation used for the arrangement of the boxes is: number of boxes along the length × number of boxes along the width × number of layers, i.e.,

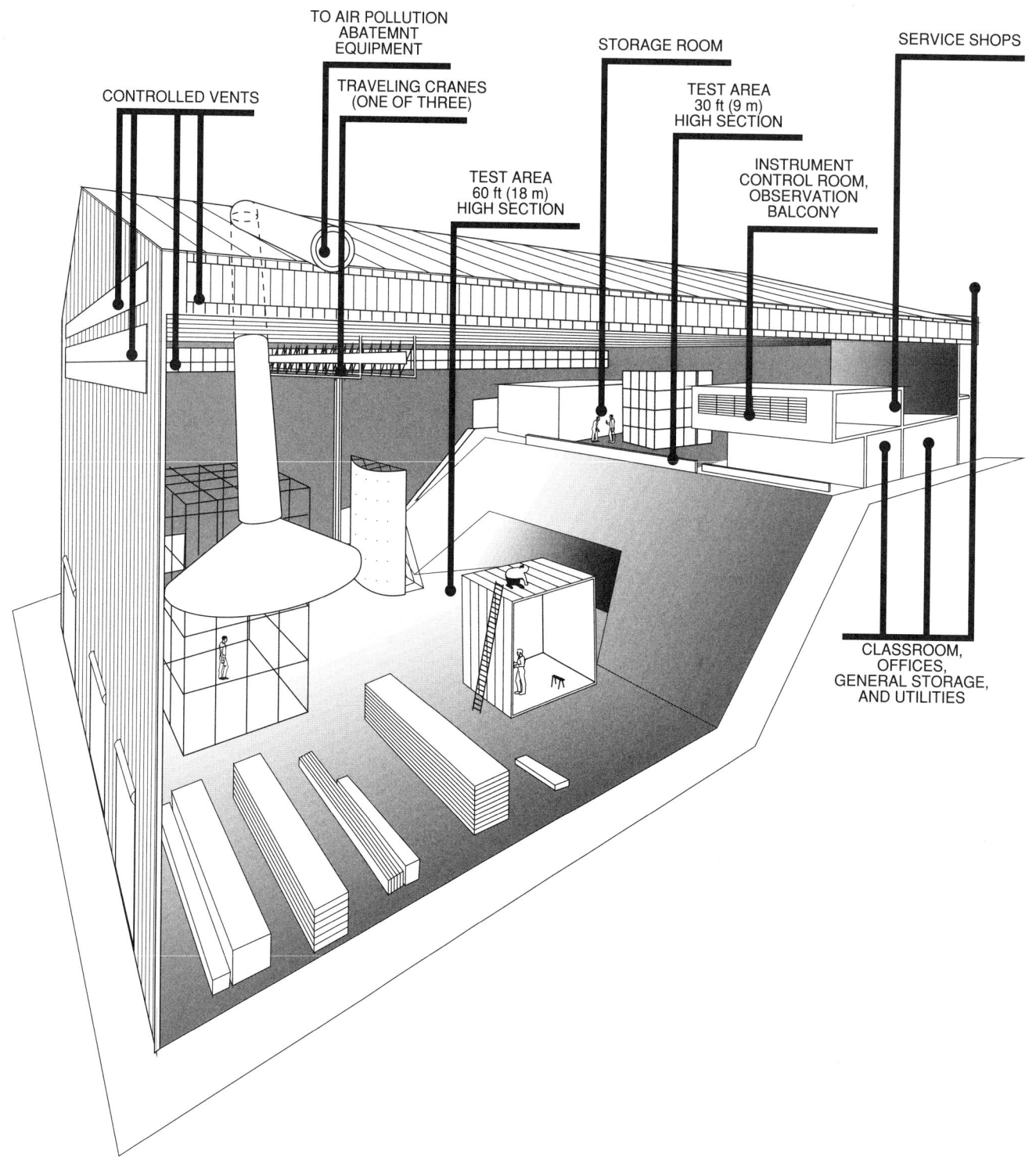

Fig. 3-4.49. *The Factory Mutual Research Corporation's Test Center at West Gloucester, RI, where large-scale fire tests are performed.*

1. One box with a single layer:
   $1 \times 1 \times 1$ sample commodity,
2. Two boxes with a single layer:
   $2 \times 1 \times 1$ sample commodity,
3. Two boxes with two layers:
   $1 \times 1 \times 2$ sample commodity,
4. Three boxes with three layers:
   $1 \times 1 \times 3$ sample commodity,

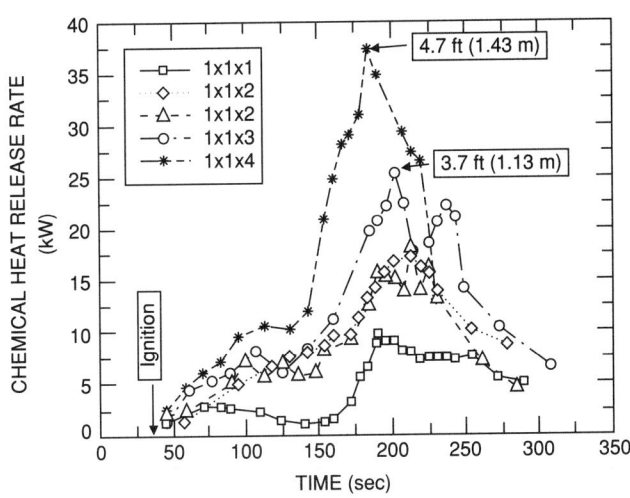

Fig. 3-4.50. *Free-burning chemical heat release rate* versus *time for 100-mm cubic empty corrugated paper boxes arranged as one box per layer for a total of four layers. Each layer is about 12 mm apart. Visual flame heights are indicated for two arrangements. Data were measured in the Flammability Apparatus.*

5. Four boxes with four layers:
   $1 \times 1 \times 4$ sample commodity,
6. Four boxes with a single layer:
   $2 \times 2 \times 1$ sample commodity,
7. Eight boxes with two layers:
   $2 \times 2 \times 2$ sample commodity.

There are provisions to use more than eight boxes and four layers. The arrangements have strong effect on the fire intensity as shown in Figure 3-4.50, where chemical heat release rate is plotted against time for 100-mm cubic box with one box to a layer for a total of four layers. Visual flame heights from the bottom of the first box are indicated for two arrangements. The data were measured in the Flammability Apparatus. The data show that the increase in the chemical heat release rate is more than expected from the increase in the surface area. For example, the surface area increases by a factor of 4 from one to four boxes, whereas the peak chemical heat release rate increases by a factor of 5, even though all the surface areas are not burning. This is indicative of the enhancement of the flame heat flux in a three-dimensional arrangement.

In the three-dimensional arrangement of the sample commodities, the water application rate for fire suppression/extinguishment is expected to be governed by the first term rather than by the second term in Equation 128 (see Table 3-4.28), due to the enhancement of the flame heat flux. With water application efficiency of unity and no water puddle at the surface, the water application rate required for flame suppression/extinguishment for the three-dimensional arrangement of sample commodities, from Table 3-4.28 for solids, is expected to be in the range of 19 to 34 g/m^2-s. These rates are about ten times the critical water application rates for flame extinction (Table 3-4.27).

Figures 3-4.51 through 3-4.53 show examples of the fire extinguishment test data from the Flammability Apparatus for 100-mm-diameter and 13-mm-thick circular Whatman No. 3 cellulosic filter paper slabs.[114] Figure 3-4.51 is a plot of the average heat flux removed from the surface of wet filter paper by the gasification of water *versus* the average heat

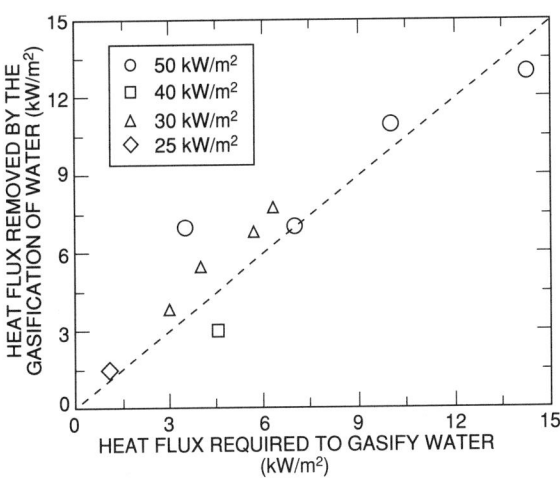

Fig. 3-4.51. *Average heat flux removed from the surface by the gasification of water versus the average heat flux required for the gasification of water. The data are for 100-mm-diameter and 13-mm-thick horizontal wetted slabs of the Whatman No. 3 cellulosic filter paper. The slabs were exposed to external heat fluxes in the range of 25 to 50 kW/m^2 in the Flammability Apparatus under co-flow conditions in normal air. The slabs were wetted with different amounts of water until saturation. Data are taken from reference 114.*

flux required to gasify the water. The average heat flux removed from the surface, during the test time period, is calculated from Equation 123 using the measured values of the mass loss rate with and without the water on the surface, and the values from Table 3-4.4 for the heats of gasification and surface re-radiation loss of filter paper. The average heat

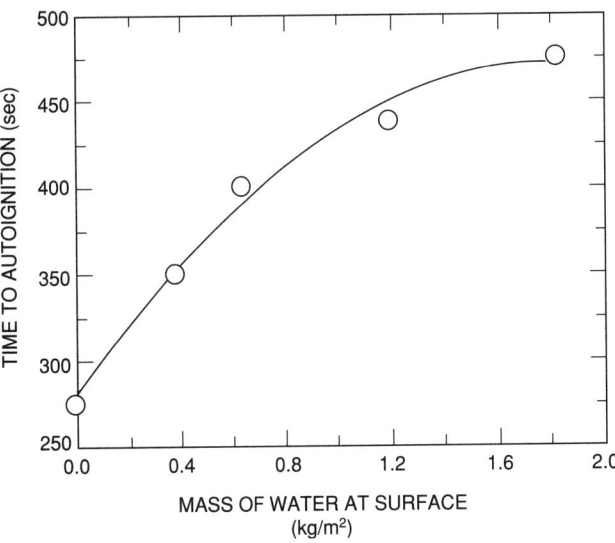

Fig. 3-4.52. *Time to autoignition versus the total amount of water used to wet the 100-mm-diameter and 13-mm-thick horizontal wetted slabs of the Whatman No. 3 cellulosic filter paper. The slabs were exposed to external heat fluxes in the range of 25 to 50 kW/m^2 in the Flammability Apparatus under co-flow conditions in normal air. The slabs were wetted with different amounts of water until saturation. Data are taken from reference 114.*

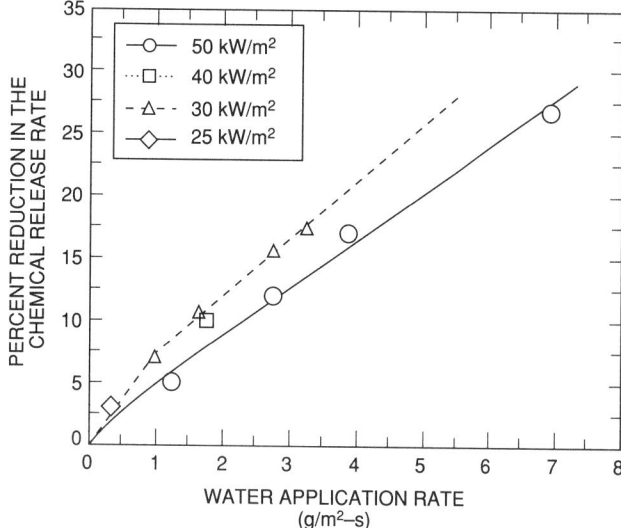

Fig. 3-4.53. *Percent reduction in the chemical heat release rate versus the water application rate for the combustion of 100-mm-diameter and 13-mm-thick horizontal wetted slabs of the Whatman No. 3 cellulosic filter paper. The slabs were exposed to external heat fluxes in the range of 25 to 50 kW/m² in the Flammability Apparatus under co-flow conditions in normal air. Data are taken from reference 114.*

flux required to gasify the water, during the test time period, is calculated from Equation 125 using the measured values of the mass of water applied to the surface, where efficiency of application is unity. As expected from the published literature on this subject,[101-108,114] there is excellent agreement between the heat flux removed from the surface by water and heat flux required to gasify it.

The data in Figure 3-4.52 show that the time to sustained autoignition for the filter paper increases with increase in the amount of water at the surface, as expected due to removal of energy by: (1) the gasification of water and (2) blockage of flame heat flux to the surface and escape of the fuel vapors. It is well known that the wetting action of water delivered from sprinklers is effective in resisting the fire jump across the aisles of stored commodities in warehouses. The wetting action of water is considered to be one of the major advantages of the sprinkler fire protection.

Figure 3-4.53 shows the percent reduction in the chemical heat release rate *versus* the water application rate for the cellulosic filter paper sample exposed to various heat fluxes. From Equation 124, the reduction in the chemical heat release rate for a fixed external heat flux value can be expressed as follows

$$\dot{Q}''_{ch} - \dot{Q}''_{ch,w} = \frac{\Delta H_{ch}}{\Delta H_g}\dot{q}''_w \qquad (130)$$

where $\dot{Q}''_{ch,w}$ is the chemical heat release rate in the presence of water (kW/m²). $\Delta H_{ch}/\Delta H_g$ is the Heat Release Parameter (HRP). From Equations 126 and 130

$$\dot{Q}''_{ch} - \dot{Q}''_{ch,w} = \text{HRP}(\varepsilon_w \Delta H_w + \delta_w)\dot{m}''_w \qquad (131)$$

In the tests, the water was applied directly to the surface and there was no puddle formation on the surface, thus $\varepsilon_w = 1$

and $\delta_w = 0$. For cellulosic filter paper, HRP = 3.6 and the heat of gasification of water is 2.6 kJ/g. Using these values in Equation 131

$$\frac{\dot{Q}''_{ch} - \dot{Q}''_{ch,w}}{\dot{Q}''_{ch}} \times 100 = \left[\frac{100 \times 3.6 \times 2.6}{\dot{Q}''_{ch}}\right]\dot{m}''_w \qquad (132)$$

or, the percent reduction in the chemical heat release rate is

$$\frac{\dot{Q}''_{ch} - \dot{Q}''_{ch,w}}{\dot{Q}''_{ch}} \times 100 = \frac{936}{\dot{Q}''_{ch}}\dot{m}''_w \qquad (133)$$

Equation 133 suggests that a plot of the percentage reduction in the chemical heat release rate *versus* the water application rate should be a straight line with a slope of $936/\dot{Q}''_{ch}$. For the external heat flux values of 25, 30, 40, and 50 kW/m², the free-burning chemical heat release rates are 120, 190, 210, and 235 kW/m², respectively. Thus the slopes at these fluxes are 7.8, 4.9, 4.5, and 4.0 (g/m²-s)$^{-1}$, respectively. The slopes of the lines from the experimental data for 30 and 50 kW/m² in Figure 3-4.53 are 4.7 and 3.9 (g/m²-s)$^{-1}$, respectively, in excellent agreement with the expected slopes from Equation 133. Thus the experimental data support the heat balance mechanism for flame extinction by the gasification of water, as long as there is no water puddle at the surface.

Small-Scale Fire Suppression/Extinguishment Tests Using Water with Horizontal and Vertical Slabs Burning under Natural Airflow Condition: Several studies have been performed in this type of configuration.[103-108] For example, fire extinguishment tests have been performed with water applied to the burning vertical and horizontal slabs of polymethylmethacrylate (PMMA), polyoxymethylene (POM), polyethylene (PE), and polystyrene (PS).[107] The horizontal slabs were 0.18-m squares and the vertical slabs were 0.18 m wide and 0.37 m high.[107] The slabs were exposed to external heat flux values in the range 0 to 17 kW/m² in normal air in the presence of water applied at a rate of 0 to 7.8 g/m²-s.[107] The water application efficiency was close to unity.

Figure 3-4.54 shows the time to flame extinction and mass loss rate for various external heat fluxes applied to the

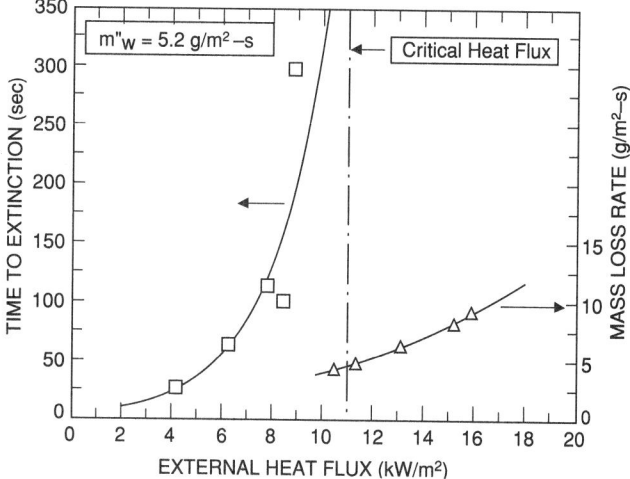

Fig. 3-4.54. *Time to flame extinction and mass loss rate at various external heat fluxes for the combustion of a 0.18-m-wide, 0.37-m-high, and 50-mm-thick vertical slab of polymethylmethacrylate in the presence of water with an application rate of 5.2 g/m²-s. Data are taken from the study reported in reference 107.*

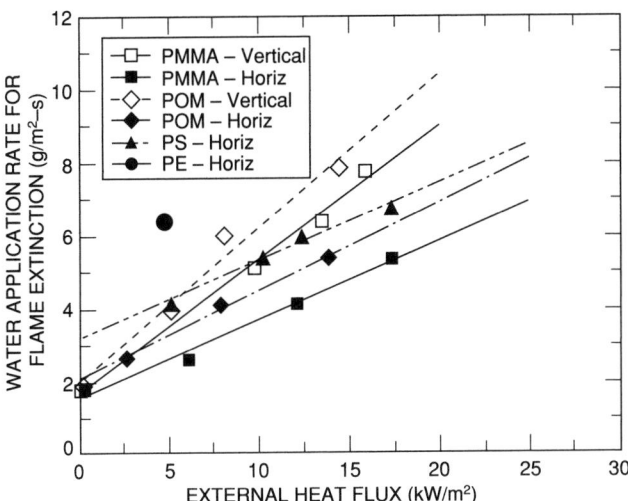

Fig. 3-4.55. *Water application rate required for flame extinction of horizontal and vertical slabs of polymethylmethacrylate (PMMA), polyoxymethylene (POM), polystyrene (PS), and polyethylene (PE) burning in normal air at various external heat fluxes. Data are taken from the study reported in reference 107.*

surface of the vertical PMMA slab burning in normal air with a water application rate of 5.2 g/m²-s; the data in the figure are taken from the study reported in reference 107. With increase in the external heat flux, the time to flame extinction increases until, close to the Critical Heat Flux (CHF) value of 11 kW/m² (Table 3-4.2), it goes to infinity (no flame extinction). The mass loss rate data in Figure 3-4.54 show that, close to the CHF value, the mass loss rate approaches the critical rate of 3.2 g/m²-s, determined from the ignition experiments (Table 3-4.26). These data support the fire point theory.[104,106]

Figure 3-4.55 shows water application rates required for flame extinction for vertical slabs of polymethylmethacrylate (PMMA), polyoxymethylene (POM), polystyrene (PS), and polyethylene (PE) burning in normal air with various external heat flux exposure. The data satisfy Equation 127:

Polymethylmethacrylate
Vertical

$$\dot{m}_w'' = 0.37\dot{q}_e'' + 1.67 \ (R^2 = 0.99) \tag{134}$$

Horizontal

$$\dot{m}_w'' = 0.22\dot{q}_e'' + 1.56 \ (R^2 = 0.99) \tag{135}$$

Polyoxymethylene
Vertical

$$\dot{m}_w'' = 0.42\dot{q}_e'' + 1.97 \ (R^2 = 0.98) \tag{136}$$

Horizontal

$$\dot{m}_w'' = 0.24\dot{q}_e'' + 2.08 \ (R^2 = 0.99) \tag{137}$$

Polystyrene
Horizontal

$$\dot{m}_w'' = 0.22\dot{q}_e'' + 3.1 \ (R^2 = 0.98) \tag{138}$$

Equations 134 and 136 show that, for vertical slabs, the inverse of the slope is equal to 2.7 and 2.3 kJ/g for PMMA and POM, respectively, which are close to the heat of gasification of water (2.6 kJ/g). Thus, the effect of water puddle at the surface is negligible as expected for the vertical surfaces. Equations 135, 137, and 138 show that, for horizontal surfaces, the inverse of the slopes for PMMA, POM, and PS are 4.6, 4.1, and 4.6 kJ/g, respectively, which are almost twice the value for the heat of gasification of water. The data for the horizontal slabs thus suggest that the blockage of flame heat flux and escape of the fuel from the surface is as important as the gasification of water. The energy associated with the blockage is about the same magnitude as the energy associated with the gasification of water.

Large-Scale Fire Suppression/Extinguishment Tests Using Water: Numerous large-scale fire suppression/extinguishment tests have been performed.[101–108,111–114] In almost all cases the materials are heterogeneous and the configurations are three dimensional, identified as "commodities." Tests are performed under natural airflow conditions with water applied from a series of sprinklers. The sprinklers are either at the ceiling or close to the top surface of the commodities. At FMRC, large-scale fire suppression/extinguishment tests are performed in the Fire Products Collector (Figure 3-4.8) and at the FMRC Test Center, mostly at the 30-ft site (Figure 3-4.49).[27,37,101,102,105,110–114]

FMRC classifies a stored commodity by its potential fire protection challenge, which is essentially dependent on the commodity's ability to release heat in a fire in the presence of water.[110] Most stored commodities are classified into one of the six classes, such as the following examples.[110]

Noncombustible: Do not burn and do not, by themselves, require sprinkler protection.

Combustibles: Class I: Example—noncombustible products on wood pallets or noncombustible products packaged in ordinary corrugated paper boxes or wrapped in ordinary paper on wood pallets. Class I commodity is simulated by glass jars in compartmented corrugated paper boxes.

Class II: Example—Class I products in more combustible packaging, such as wood crates or multiple-thickness corrugated boxes. Class II commodity is simulated by metal-lined double tri-wall corrugated paper boxes.

Class III: Example—packaged or unpackaged wood, paper, or natural-fiber cloth, or products made from them, on wood pallets. Class III commodity is simulated by using paper cups in compartmented corrugated paper boxes.

Class IV: Class I, II, and III commodities containing no more than 25 percent (by volume) or 15 percent by weight of high-heat-release-rate synthetic materials. Class IV commodity is simulated by polystyrene (15 percent by weight) and paper cups in compartmented corrugated paper boxes.

Group A Plastics: Simulated by polystyrene cups in compartmented corrugated paper boxes.

For the tests in the Fire Products Collector, the commodities are used in a 2 × 2 × 2 arrangement (two pallet loads along the length × two pallet loads along the width × two layers).[101,102,110,113] Each pallet load consists of a wood pallet with eight 0.53-cubic corrugated paper boxes, containing products under test, in a 2 × 2 × 2 arrangement (two boxes along the length × two boxes along the width in two

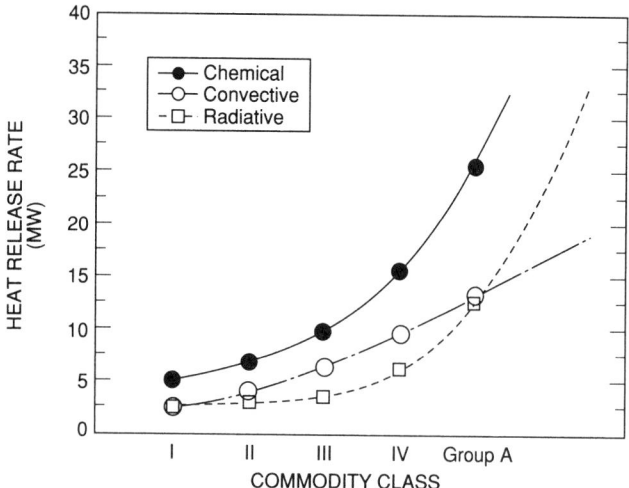

Fig. 3-4.56. *Calculated average peak heat release rates for free-burning fires of simulated commodities from the data measured in the Fire Products Collector (10,000-kW-scale Flammability Apparatus). Data are taken from references 101, 102, 110, 113, and 114.*

layers, with boxes touching each other). Each pallet load is a 1-m (42-in.) cube of product and separated by about 150 mm. This arrangement leads to the test commodity consisting of eight pallet loads with 64 corrugated paper boxes containing products with overall dimensions of 2.3 m (7.5 ft) × 2.3 m (7.5 ft) × 2.9 m (9.7 ft) high.

In the Fire Products Collector (10,000-kW-scale Flammability Apparatus) fire suppression/extinguishment tests, water is applied at the top of the commodity, in a uniform fashion, with application rates in the range of 0 to 407 g/m²-s (0 to 0.6 gpm per sq ft). The range of the water application rates is about ten times the predicted range for the three-dimensional arrangements (Table 3-4.28 with water application efficiency of unity and no water puddles at the surfaces). It thus appears that blockage of flame heat flux to the surface and escape of fuel vapors are as important as gasification of water for the fire suppression/extinguishment of the commodities, similar to the flame extinction for horizontal slabs (Equations 135, 137, and 138) in small-scale tests, discussed previously.

Figure 3-4.56 shows the calculated values of the free-burning average peak heat release rate for the simulated Class I through Group A plastic commodities. In the calculations data measured in the Fire Products Collector (10,000-kW-scale Flammability Apparatus) were used.[114] The Class I through Class III commodities were made of cellulosic materials and had lower heat release rates. This behavior is expected on the basis of the values of: (1) surface re-radiation loss and heat of gasification (Table 3-4.4), (2) flame heat flux (close to polyoxymethylene, Table 3-4.5), (3) heat of combustion (Table 3-4.11), and (4) Heat Release Parameter (Table 3-4.12) for wood and paper. The heat release rates for Class I through Class III commodities increase gradually from Class I through Class III.

Introduction of polystyrene from about 15 percent (Class IV) to 100 percent (plastics Group A) inside the corrugated paper boxes results in an exponential increase in the chemical and radiative heat release rates as indi-

cated in Figure 3-4.56. This behavior is expected on the basis of the higher values of: (1) heat of gasification (Table 3-4.4), (2) flame heat flux (Table 3-4.5), (3) heat of combustion (Table 3-4.11), and (4) Heat Release Parameter (Table 3-4.12) for polystyrene compared to the values for the cellulosic materials in Class I through III commodities. The higher intensity fire due to the introduction of polystyrene is also indicated by the higher water applications rates required for fire suppression/extinguishment in Figure 3-4.57. The higher water requirement for fire suppression/extinguishment for Class IV and plastics Group A commodities is expected from Equation 128, due to higher value of the flame heat flux which dominates the water application rate requirements.

Flame Extinction by the Processes in the Gas Phase

The process of flame extinction by gaseous, powdered, and foaming agents and by increase in the local equivalence ratio is predominantly a gas-phase process and thus is different from the process of flame extinction by water, which occurs predominantly in the solid phase at the surface of the material. The kinetic parameter for flame extinction defined in Equation 118, however, is still applicable[103]

$$\varphi = \frac{\varphi_0 - \kappa Y_{j,ex}\left[\dfrac{1 + \Delta c_P(T_{ad} - T_a) + \Delta H_D}{\Delta H_O^* Y_O}\right]}{1 - Y_{j,ex}} \quad (139)$$

where φ is the kinetic parameter in the presence of the extinguishing agent, φ_0 is the kinetic parameter in the absence of the extinguishing agent, κ is the ratio between the kinetic parameters at the flame temperature and at the adiabatic flame temperature, $Y_{j,ex}$ is the mass fraction of the

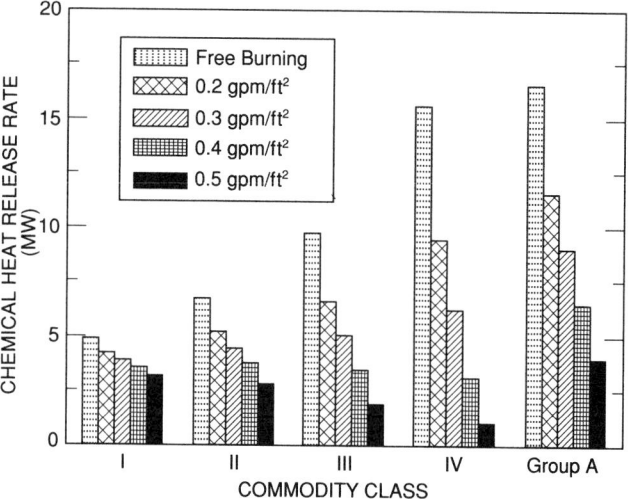

Fig. 3-4.57. *Calculated average peak chemical heat release rates at various water application rates for fires of simulated commodities from the data measured in the Fire Products Collector (10,000-kW-scale Flammability Apparatus). Data are taken from references 101, 102, 110, 113, and 114. One gpm/ft² = 769 g/m²-s.*

extinguishing agent, Δc_P is the difference between the heat capacities of the extinguishing agent and the fire products (kJ/g-K), T_{ad} is the adiabatic flame temperature at the stoichiometric limit (K), T_0 is the initial temperature of the reactants (K), and ΔH_D is the heat of dissociation (kJ/g).

Equation 139 shows that the addition of an extinguishing agent reduces the kinetic parameter from its normal value and includes the effects of four flame extinction mechanisms:[103] (1) dilution, effects are included in the $\kappa Y_{j,ex}$ term; (2) added thermal capacity, effects are included in Δc_P; (3) chemical inhibition, effects are included through increases in T_{ad} value; for most fuels the adiabatic flame temperature at the stoichiometric limit is about 1700 K;[103] more reactive fuels, such as hydrogen, have lower adiabatic flame temperature at the stoichiometric limit, and less reactive or retarded materials have higher values of the adiabatic flame temperature at the stoichiometric limit; and (4) kinetic chain breaking and endothermic dissociation through Δc_P and ΔH_D terms.

From Equation 123 in the presence of an extinguishing agent that works in the gas phase

$$\dot{m}'' = \frac{\varphi \Delta H_T \dot{m}''_{cr} + \dot{q}''_e - \dot{q}''_{rr}}{\Delta H_g} \qquad (140)$$

For a fixed value of the external heat flux, the addition of an extinguishing agent reduces the normal value of the kinetic parameter by one or more of the four mechanisms expressed by Equation 139; the mass loss rate decreases and approaches the critical value at which the flame is extinguished. Increase in the external heat flux would increase the mass loss rate, and further addition of the extinguishing agent would be needed to reduce the mass loss to its critical value and to reestablish the flame extinction condition. Continued increase in the extinguishing agent with external heat flux will result in the first term in the denominator on the right-hand side of Equation 140 to become zero, and the equation will represent a nonflaming fire.

For a fixed airflow rate, as is generally the case in enclosure fires where the extinguishing agent working in the gas phase is used, increase in the mass loss rate due to external heat flux results in an increase in the equivalence ratio, defined in Equation 36. As the equivalence ratio increases and approaches values of 4.0 and higher, the combustion efficiency approaches values less than or equal to 0.40, flames are extinguished, and nonflaming conditions become important.[36,37] Thus the upper limit for the application of the extinguishing agent working in the gas phase is dictated by the equivalence ratio ≥ 4.0 and/or the combustion efficiency ≤ 0.40. Under nonflaming conditions, increase in the external heat flux increases the generation rate of the fuel vapors and the reduction-zone products.

Flame Extinction by Reduced Mass Fraction of Oxygen

Flame extinction by reduced mass fraction of oxygen can be the result of: (1) dilution and heat capacity effects due to the addition of inert gases, such as N_2, CO_2, etc.; and (2) chemical effects due to the retardation of chemical reactions and reduction in the flame heat flux to the surface, especially the radiative component.

Theoretical and experimental analyses have been performed for flame extinction by reduced oxygen mass frac-

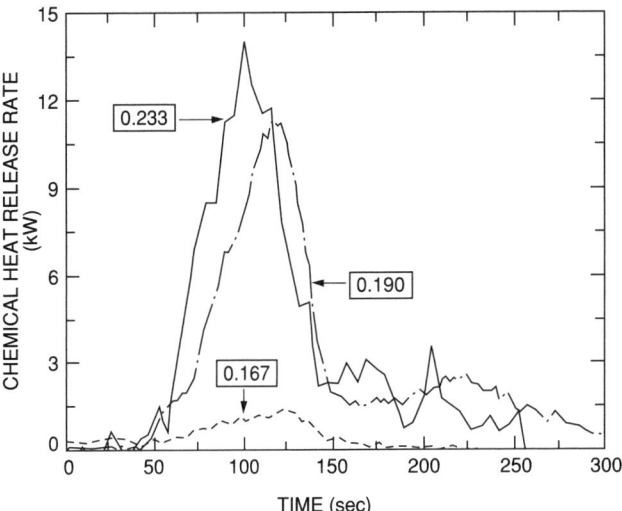

Fig. 3-4.58. Chemical heat release rate versus time for 50-mm empty corrugated paper boxes in a 2 × 2 × 2 arrangement (two boxes along the length × two boxes along the width × two layers, for a total of eight boxes separated by about 12 mm). Measurements were made in the Flammability Apparatus with no external heat flux under the co-flow condition at various oxygen mass fractions, which are indicated in the figure.

tions. For example, for polymethylmethacrylate (PMMA), an oxygen mass fraction value of 0.180 is predicted for flame extinction[115] compared to the experimental values of 0.181 for a 70-mm-wide, 190-mm-high, and 19-mm-thick vertical PMMA slab[116] and 0.178 for a 100-mm-wide, 25-mm-thick, and 300- and 610-mm-high vertical slabs of PMMA, and 25-mm-diameter and 610-mm-high vertical cylinder of PMMA.[2] The critical values of the chemical, convective, and radiative heat release for PMMA are 106, 73, and 33 kW/m^2, respectively,[2] showing a trend similar to one reported in Table 3-4.26. At oxygen mass fractions equal to or less than 0.201, flames are unstable and faint blue in color.[2]

The effect of increased external heat flux on flame extinction due to reduced oxygen mass fraction has been examined for the buoyant turbulent diffusion flames. For example, for rectangular and circular horizontal PMMA slabs, 0.06 to 0.10 m^2 in area and 0.03 to 0.05 m in thickness, exposed to external heat flux values of 0, 40, 60, and 65 kW/m^2, flame extinction is found at oxygen mass fractions of 0.178, 0.145, 0.134, and 0.128, respectively.[27] The data support Equation 140 and show that, for buoyant turbulent diffusion flames, flaming can occur up to relatively low oxygen mass fraction values; the only condition is that, in the gas phase, the reactant-oxidizer mixture is within the flammability limit.

The effect of reduced oxygen mass fraction on flame extinction of materials in the three-dimensional arrangement, where flame heat flux is enhanced, has been examined. Figure 3-4.58 shows an example where chemical heat release rates at oxygen mass fractions of 0.233, 0.190, and 0.167 *versus* time are shown for the combustion of 50-mm cubes of empty corrugated paper boxes in a 2 × 2 × 2 arrangement. The weight of each box is about 13 g (839 g/m^2). The measurements are from the Flammability Apparatus.

In Figure 3-4.58, at oxygen mass fraction of 0.167, the flame is close to the extinction condition, only 10.5 percent of the initial weight of the boxes is consumed, which is equivalent to consumption of a single box with a surface area of about 0.0155 m^2. The peak chemical heat release rate close to flame extinction, in Figure 3-4.58, is about 1.5 kW or 97 kW/m^2, using a surface area of 0.0155 m^2. This value is in excellent agreement with the average value in Table 3-4.26, derived from the critical mass loss rates for ignition. The data in Figure 3-4.58 for the three-dimensional arrangement of the corrugated boxes thus support the fire point theory,[104,106] independence of the critical heat release rate for flame extinction from the geometrical arrangement and surface areas of the materials, and Equations 139 and 140 as originally formulated in reference 103.

NOMENCLATURE

A	total exposed surface area of the material (m^2)
a_j	mass coefficient for the product yield (g/g)
b_j	molar coefficient for the product yield (g/mole)
B_{cr}	critical mass transfer number $\left(\dfrac{Y_O \Delta H_O^*}{\Delta H_{g,con}}\right)$
CHF	Critical Heat Flux (kW/m^2)
$\dot{C}_O''$	mass consumption rate of oxygen (g/m^2-s)
$\dot{C}_{stoich,O}''$	stoichiometric mass consumption rate of oxygen (g/m^2-s)
c_O	mass of oxygen consumed per unit mass of fuel (g/g)
c_P	specific heat (kJ/g-K)
Δc_P	difference between the heat capacities of the extinguishing agent and the fire products (kJ/g-K)
CDG	Carbon Dioxide Generation calorimetry
D	optical density $\{[\ln(I_0/I)]/\ell\}$ (1/m)
D_{corr}	metal corrosion (penetration depth or metal loss in microns, angstroms, mils)
E_i	total amount of heat generated in the combustion of a material (kJ)
f_j	volume fraction of a product ($-$)
fp	fire property
FPI	Fire Propagation Index 1000 $(0.42\dot{Q}_{ch}'')^{1/3}$/TRP
$\dot{G}_j''$	mass generation rate of product j (g/m^2-s)
$\dot{G}_{stoich,j}''$	stoichiometric mass generation rate of product j (g/m^2-s)
GTR	Gas Temperature Rise calorimetry
ΔH_i	heat of combustion per unit mass of fuel vaporized (kJ/g)
ΔH_{CO}	heat of complete combustion of CO (10 kJ/g)
ΔH_D	heat of dissociation (kJ/g)
ΔH_g	heat of gasification at ambient temperature (kJ/g)
$\Delta H_{g,con}$	flame convective energy transfer to the fuel per unit mass of fuel gasified (kJ/g)
ΔH_m	heat of melting at the melting temperature (kJ/g)
ΔH_T	net heat of complete combustion per unit of fuel vaporized (kJ/g)
ΔH_v	heat of vaporization at the vaporization temperature (kJ/g)
ΔH_w	heat of gasification of water (2.58 kJ/g)
ΔH_{CO}^*	net heat of complete combustion per unit mass of CO generated (kJ/g)
$\Delta H_{CO_2}^*$	net heat of complete combustion per unit mass of CO$_2$ generated (kJ/g)

ΔH_O^*	net heat of complete combustion per unit mass of oxygen consumed (kJ/g)
HRP	Heat Release Parameter ($\Delta H_i/\Delta H_g$)
h_i	mass coefficient for the heat of combustion (kJ/g)
I/I_0	fraction of light transmitted through smoke ($-$)
j	fire product
k	thermal conductivity (kW/m-K)
L_{sp}	smoke point (m)
ℓ	optical path length (m)
$\dot{m}''$	mass loss rate (g/m^2-s)
$\dot{m}_w''$	water application rate per unit surface area of the material (g/m^2-s)
M	molecular weight (g/mole)
m_i	molar coefficient for the heat of combustion (kJ/mole)
MOD	mass optical density ($D\dot{V}/A\dot{m}''$) (m^2/g)
$\dot{m}_{air}$	mass flow rate of air (g/s)
OC	Oxygen Consumption calorimetry
$\dot{q}_e''$	external heat flux (kW/m^2)
$\dot{q}_f''$	flame heat flux (kW/m^2)
$\dot{Q}_i''$	heat release rate per unit sample surface area ($\dot{m}''\Delta H_{ch}$) (kW/m^2)
$\dot{Q}_i'$	heat release rate per unit sample width (kW/m)
R_{corr}	corrosion rate (Å/min)
S	stoichiometric mass air-to-fuel ratio (g/g)
t	time (s)
t_f	time at which there is no more vapor formation (s)
t_0	time at which the sample is exposed to heat (s)
T	temperature (K)
ΔT_{ig}	ignition temperature above ambient (K)
TRP	Thermal Response Parameter $[\Delta T_{ig}(k\rho c_p)^{1/2}]$ (kW-s$^{1/2}$/m^2)
u	fire propagation rate $[dX_p/dt]$ (mm/s or m/s)
v_g	co-flow air velocity (m/s)
$\dot{V}$	total volumetric flow rate of fire product-air mixture (m^3/s)
$\dot{W}$	total mass flow rate of the fire product-air mixture (g/s)
W_f	total mass of the material lost in the flaming and nonflaming fire (g)
W_j	total mass of product j generated in the flaming and nonflaming fire (g)
X_f	flame height (m or mm)
X_p	pyrolysis front (mm or m)
X_t	total length available for fire propagation (m or mm)
y_j	yield of product j ($\dot{G}_j''/\dot{m}''$)
$Y_{j,ex}$	mass fraction of the extinguishing agent ($-$)
Y_O	mass fraction of oxygen ($-$)

Greek

α	correlation coefficient (nonflaming fire) ($-$)
β	correlation coefficient (transition region) ($-$)
δ_w	energy associated with the blockage of flame heat flux to the surface and escape of fuel vapors per unit mass of the fuel gasified (kJ/g)
ϵ_w	water application efficiency
φ	kinetic parameter for flame extinction
ξ	correlation coefficient (transition region) ($-$)
Φ	equivalence ratio ($S\dot{m}''/\dot{m}_{air}$)
χ_{ch}	combustion efficiency ($\dot{Q}_{ch}''/\dot{m}''\Delta H_T$) ($-$)

χ_{con} convective component of the combustion efficiency $(\dot{Q}''_{con}/\dot{m}''\Delta H_T)$ $(-)$

χ_{rad} radiative component of the combustion efficiency $(\dot{Q}''_{rad}/\dot{m}''\Delta H_T)$ $(-)$

η_j generation efficiency $(\dot{G}''_j/\dot{m}''\Psi_j)$ $(-)$

κ ratio between the kinetic parameters for the flame temperature and adiabatic flame temperature $(-)$

λ wavelength of light (μm)

σ Stefan-Boltzmann constant $(56.7 \times 10^{-12}$ $kW/m^2\text{-}K^4)$

τ average specific extinction area (m^2/g)

μ corrosion parameter (angstrom-minutes-ml units)

ρ density (g/m^3)

ν_j stoichiometric coefficient of product j $(-)$

ν_O stoichiometric coefficient of oxygen $(-)$

Ψ_j stoichiometric yield for the maximum conversion of fuel to product j $(-)$

Ψ_O stoichiometric mass oxygen-to-fuel ratio (g/g)

ζ ratio of fire properties for ventilation-controlled to well-ventilated combustion

ζ_{oxid} oxidation zone product generation efficiency ratio $(-)$

ζ_{red} reduction zone product generation efficiency ratio $(-)$

Subscripts

a	air or ambient
ad	adiabatic
asy	asymptotic
ch	chemical
con	convective
$corr$	corrosion
cr	critical
e	external
ex	extinguishment
f	flame or fuel
fc	flame convective
fr	flame radiative
g	gas
g,con	flame convective energy for fuel gasification
i	chemical, convective, radiative
ig	ignition
j	fire product
n	net
0	initial
$oxid$	oxidation zone of a flame
rad	radiation
red	reduction zone of a flame
$stoich$	stoichiometric for the maximum possible conversion of fuel monomer to a product
rr	surface re-radiation
s	surface
vc	ventilation-controlled fire
w	water
wv	well-ventilated fire
∞	infinite amount of air

Superscripts

$\cdot$	per unit time (s^{-1})
$'$	per unit width (m^{-1})
$''$	per unit area (m^{-2})

Definitions

Chemical heat of combustion	calorific energy generated in chemical reactions leading to varying degrees of incomplete combustion per unit fuel mass consumed
Convective heat of combustion	calorific energy carried away from the flame by the fire products-air mixture per unit fuel mass consumed
Heat of gasification	energy absorbed to vaporize a unit mass of fuel originally at ambient temperature
Heat release parameter	calorific energy generated per unit amount of calorific energy by the fuel
Kinetic parameter for flame extinction	maximum fraction of combustion energy that the flame reactions may lose to the sample surface by convection without flame extinction
Net heat of complete combustion	calorific energy generated in chemical reactions leading to complete combustion, with water as a gas, per unit fuel mass consumed
Radiative heat of combustion	calorific energy emitted as thermal radiation from the flame per unit fuel mass consumed

Abbreviations

ABS	acrylonitrile-butadiene-styrene
CPVC	chlorinated polyvinylchloride
CR	neoprene or chloroprene rubber
CSP (or CSM)	chlorosulfonated polyethylene rubber (Hypalon™)
CTFE	chlorotrifluoroethylene (Kel-F™)
E-CTFE	ethylene-chlorotrifluoroethylene (Halar™)
EPR	ethylene propylene rubber
ETFE	ethylenetetrafluoroethylene (Tefzel™)
EVA	ethylvinyl acetate
FEP	fluorinated polyethylene-polypropylene (Teflon™)
IPST	isophthalic polyester
PAN	polyacrylonitrile
PC	polycarbonate
PE	polyethylene
PEEK	polyether ether ketone
PES	polyethersulphone
PEST	polyester
PET	polyethyleneterephthalate (Melinex™, Mylar™)
PFA	perfluoroalkoxy (Teflon™)
PMMA	polymethylmethacrylate
PO	polyolefin
POM	polyoxymethylene
PP	polypropylene
PS	polystyrene
PTFE	polytetrafluoroethylene (Teflon™)
PU	polyurethane
PVEST	polyvinylester
$PVCl_2$	polyvinylidene chloride (Saran™)
PVF	polyvinyl fluoride (Tedlar™)
PVF_2	polyvinylidene fluoride (Kynar™, Dyflor™)
PVC	polyvinylchloride
SBR	styrene-butadiene rubber
TFE	tetrafluoroethylene (Teflon™)
XLPE	crosslinked polyethylene

REFERENCES CITED

1. A. Tewarson, "Non-thermal Damage," *J. Fire Science*, 10, 188–241 (1992).
2. A. Tewarson, and S.D. Ogden, "Fire Behavior of Polymethylmethacrylate," *Combustion and Flame*, 89, 237–259 (1992).
3. A. Tewarson, and M.M. Khan, "Flame Propagation for Polymers in Cylindrical Configuration and Vertical Orientation," *Twenty-Second Symposium (International) on Combustion*, pp. 1231–1240. The Combustion Institute, Pittsburgh, PA (1988).
4. A. Tewarson, "Flammability Parameters of Materials: Ignition, Combustion, and Fire Propagation," *J. Fire Science*, 10, 188–241 (1994).
5. E.E. Smith, "Measuring Rate of Heat, Smoke, and Toxic Gas Release," *Fire Technology*, 8, 237–245 (1972).
6. E.E. Smith, "Heat Release Rate of Building Materials," *Ignition, Heat Release, and Non-combustibility of Materials*, ASTM STP 502, The American Society for Testing and Materials, Philadelphia, PA, pp. 119–134 (1972).
7. ASTM E 906-83, "Standard Test Method for Heat and Visible Smoke Release Rates for Materials and Products," The American Society for Testing and Materials, Philadelphia, PA (1984).
8. C.P. Sarkos, R.A. Filipczak, and A. Abramowitz, "Preliminary Evaluation of an Improved Flammability Test Method for Aircraft Materials," Federal Aviation Administration, Atlantic City, NJ, Technical Report DOT/FAA/CT-84/22.
9. A. Tewarson, "Flame Spread in Standard Tests for Electrical Cables," Technical Report J.I.OMO2E1.RC-2, Factory Mutual Research Corporation, Norwood, MA (1993).
10. A. Tewarson, and D. Macaione, "Polymers and Composites—An Examination of Fire Spread and Generation of Heat and Fire Products," *J. Fire Sciences*, 11, 421–441 (1993).
11. A. Tewarson, and M.M. Khan, "A New Standard Test Method for the Quantification of Fire Propagation Behavior of Electrical Cables Using Factory Mutual Research Corporation's Small-Scale Flammability Apparatus," *Fire Technology*, 28, 215–227 (1992).
12. M.M. Khan, "Classification of Conveyor Belts Using Fire Propagation Index," Technical Report J.I. OT1E2.RC, Factory Mutual Research Corporation, Norwood, MA (1991).
13. A. Tewarson, and M.M. Khan, "Electrical Cables—Evaluation of Fire Propagation Behavior and Development of Small-Scale Test Protocol," Technical Report J.I. OM2E1.RC, Factory Mutual Research Corporation, Norwood, MA (1989).
14. *Specification Standard for Cable Fire Propagation, Class No. 3972*, Factory Mutual Research Corporation, Norwood, MA (1989).
15. A. Tewarson, "A Study of Fire Propagation and Generation of Fire Products for Selected Cables Used by the United States Navy," Technical Report J.I.OP3N3.RC/OP1N3.RC, Factory Mutual Research Corporation, Norwood, MA (Oct. 1988).
16. A. Tewarson, "Experimental Evaluation of Flammability Parameters of Polymeric Materials," *Flame Retardant Polymeric Materials*, M. Lewin., S.M. Atlas, and E.M. Pearce, eds., Chap. 3, pp. 97–153, Plenum Press, New York (1982).
17. V. Babrauskas, "Development of the Cone Calorimeter—A Bench-Scale Heat Release Rate Apparatus Based on Oxygen Consumption," Technical Report NBSIR 82-2611, The National Institute of Standards and Technology, Gaithersburg, MD (1982).
18. *Heat Release and Fires*, V. Babrauskas and S.J. Grayson, eds., Elsevier Publishing Company, London (1992).
19. ASTM E 1354-90, "Standard Test Method for Heat and Visible Smoke Release Rates for Materials and Products Using Oxygen Consumption Calorimeter," The American Society for Testing and Materials, Philadelphia, PA (1990).
20. M.J. Scudamore, P.J. Briggs, and F.H. Prager, "Cone Calorimetry—A Review of Tests Carried Out on Plastics for the Association of Plastics Manufacturers in Europe," *Fire and Materials*, 15, 65–84 (1991).
21. A.C. Fernandez-Pello, and T. Hirano, "Controlling Mechanisms of Flame Spread," *Combustion Science and Technology*, 32, 1–31 (1983).
22. M.M. Khan, and A. Tewarson, "Fire Propagation Behavior of Conveyor Belts," *J. Fire Sciences* (submitted).
23. J.S. Newman, and A. Tewarson, "Flame Spread Behavior of Char-Forming Wall/Ceiling Insulations," *Fire Safety Science—Proceedings of the Third International Symposium*, Elsevier Applied Science, New York, pp. 679–688 (1991).
24. J.S. Newman, "Integrated Approach to Flammability Evaluation of Polyurethane Wall/Ceiling Materials," Polyurethanes World Congress Oct. 10–13, The Society of the Plastics Industry, Washington, DC (1993).
25. J.S. Newman, "Cost-Effective Method for Flammability Characterization of Alternate Polyols and Blowing Agents," *Proceedings of the SPI 32nd Annual Technical/Marketing Conference*, San Francisco, CA, Oct. 1–4, The Society of the Plastics Industry, Washington, DC (1989).
26. *Approval Standard for Class I A) Insulated Wall or Wall and Roof/Ceiling Panels, B) Plastic Interior Finish Materials, C) Plastic Exterior Building Panels, D) Wall/Ceiling Coating Systems, E) Interior or Exterior Finish Systems, Class No. 4880*, Factory Mutual Research Corporation, Norwood, MA (March 1993).
27. A. Tewarson, and R.F. Pion, "Flammability of Plastics. I. Burning Intensity," *Combustion and Flame*, 26: 85–103 (1976).
28. *CRC Handbook of Chemistry and Physics*, 61st ed., 1980–81, (R.C. Weast and M.J. Astle, eds.), CRC Press, Inc., Boca Raton, FL (1980).
29. M.A. Paul, *Physical Chemistry*, p. 46, D.C. Heath and Company, Boston, MA (1962).
30. D.Q. Kern, *Process Heat Transfer*, p. 72, McGraw-Hill Book Company, New York (1950).
31. M.M. Hirschler, "Fire Hazard and Toxic Potency of the Smoke from Burning Materials," *J. Fire Sciences*, 5, 289–307 (1987).
32. H.C. Hottel, "Review: Certain Laws Governing the Diffusive Burning of Liquids by Blinov and Khudiakov (1957) (Dokl Akad), Nauk SSSR, Vol. 113, 1094, 1957," *Fire Research Abstract and Reviews*, 1: 41–45 (1959).
33. A. Tewarson, J.L. Lee, and R.F. Pion, "The Influence of Oxygen Concentration on Fuel Parameters for Fire Modeling," *Eighteenth Symposium (International) on Combustion*, pp. 563–570, The Combustion Institute, Pittsburgh, PA (1981).
34. J.C. Macrae, "An Introduction to the Study of Fuel," Elsevier Publishing Company, London (1966).
35. C. Hugget, "Estimation of Rate of Heat Release by Means of Oxygen Consumption Measurements," *Fire & Materials*, 4: 61–65 (1980).
36. A. Tewarson, F.H. Jiang, and T. Morikawa, "Ventilation-Controlled Combustion of Polymers," *Combustion and Flame*, 95: 151–169 (1993).
37. A. Tewarson, and M.M. Khan, "Extinguishment of Diffusion Flames of Polymeric Materials by Halon 1301," *J. Fire Sciences*, 11: 407–420 (1993).
38. A. Tewarson, "Prediction of Fire Properties of Materials Part 1: Aliphatic and Aromatic Hydrocarbons and Related Polymers," Technical Report NBS-GCR-86-521, prepared by the Factory Mutual Research Corporation, Norwood, MA, under Grant No. 60NANBA4D-0043 for the National Institute of Standards and Technology, Gaithersburg, MD, Dec. 1986.
39. A. Tewarson, "Smoke Point Height and Fire Properties of Materials," Technical Report NBS-GCR-88-555, prepared by the Factory Mutual Research Corporation, Norwood, MA, under Grant No. 60NANBA4D-0043 for the National Institute of Standards and Technology, Gaithersburg, MD, Dec. 1988.
40. *Handbook of Plastics and Elastomers*, C.A. Harper, editor-in-chief, McGraw-Hill Book Company, New York (1975).
41. Y. Tsuchiya, and J.F. Mathieu, "Measuring Degrees of Combustibility Using an OSU Apparatus and Oxygen Depletion Principle," *Fire Safety Journal*, 17, 291–299 (1991).
42. A. Tewarson, "Heat Release Rates from Samples of Polymethylmethacrylate and Polystyrene Burning in Normal Air," *Fire & Materials*, 1, 90–96 (1976).
43. A. Tewarson, and F. Tamanini, "Research and Development for a Laboratory-Scale Flammability Test Method for Cellular Plastics," Technical Report No. 22524, RC76-T-64, prepared by the

Factory Mutual Research Corporation for the Products Research Committee, Grant No. RP-75-1-33A, National Institute of Standards and Technology, Gaithersburg, MD (1976).

44. A. Tewarson, "Heat Release Rate in Fires," *J. Fire & Materials*, 8, 115–121 (1977).

45. A. Tewarson, "Physico-Chemical and Combustion/Pyrolysis Properties of Polymeric Materials," Technical Report NBS-GCR-80-295, prepared by the Factory Mutual Research Corporation for the National Institute of Standards and Technology, Gaithersburg, MD, Dec. 1980.

46. A.R. Apte, R.W. Bilger, A.R. Green, and J.G. Quintiere, "Wind-Aided Turbulent Flame Spread and Burning over Large-Scale Horizontal PMMA Surfaces," *Combustion and Flame*, 85, 169–184 (1991).

47. F.J. Perzak, and C.P. Lazzara, "Flame Spread over Horizontal Surfaces of Polymethylmethacrylate," *Twenty-Fourth Symposium (International) Combustion*, pp. 1661–1667, The Combustion Institute, Pittsburgh, PA (1992).

48. L. Tsantarides, and B. Ostman, "Smoke, Gas, and Heat Release Data for Building Products in the Cone Calorimeter," Technical Report I 8903013, Swedish Institute for Wood Technology Research, Stockholm, Sweden, March 1989.

49. M.M. Khan, "Characterization of Liquid Fuel Spray Fires," HTD-Vol. 223, *Heat and Mass Transfer in Fire and Combustion Systems*, ASME 1992, The American Society of Mechanical Engineers, New York.

50. A. Tewarson, and R.G. Zalosh, "Flammability Testing of Aircraft Cabin Materials," *73rd Symposium AGARD Conference Proceedings, No. 467, Aircraft Fire Safety*, pp. 33-1 to 33-12, National Technical Information Service, Springfield, VA (1989).

51. Y.R. Sivathanu, and G.M. Faeth, "Generalized State Relationships for Scalar Properties in Nonpremixed Hydrocarbon/Air Flames," *Combustion and Flame*, 82, 211–230 (1990).

52. J.S. Newman, and J. Steciak, "Characterization of Particulates from Diffusion Flames," *Combustion and Flame*, 67, 55–64 (1987).

53. D. Drysdale, *An Introduction to Fire Dynamics*, pp. 278–400, Wiley, New York (1985).

54. C.L. Beyler, "Major Species Production by Diffusion Flames in a Two-Layer Compartment Fire Environment," *Fire Safety J.*, 10, 47–56 (1986).

55. C.L. Beyler, *Fire Safety Science—Proceedings of the Third International Symposium*, 431–440, Elsevier Applied Science, New York (1986).

56. E.E. Zukowski, *Fire Safety Science—Proceedings of the Third International Symposium*, 1–30, Elsevier Applied Science, New York (1986).

57. T. Morikawa, "Effects of Supply Rate and Concentration of Oxygen and Fuel Location on CO Evolution in Combustion," *J. Fire Science*, 1, 364–378 (1983).

58. A. Tewarson, F. Chu, and F.H. Jiang, "Combustion of Halogenated Polymers," *Fire Safety Science Fourth International Symposium*, Elsevier Applied Science, New York, 563–574 (1994).

59. ASTM D 1322-80, *Standard Test Method for Smoke Points of Aviation Turbine Fuels*, The American Society for Testing and Materials, Philadelphia, PA (1980).

60. J. deRis, and X. Cheng, "The Role of Smoke-Point in Material Flammability Testing," *Fire Safety Science—Proceedings of the Fourth International Symposium*, Elsevier Applied Science, New York, 301–312 (1994).

61. I. Glassman, "Soot Formation in Combustion Processes," *Twenty-Second Symposium (International) on Combustion*, pp. 295–311, The Combustion Institute, Pittsburgh (1986).

62. G.H. Markstein, "Correlations for Smoke Points and Radiant Emission of Laminar Hydrocarbon Difusion Flames," *Twenty-Second Symposium (International) on Combustion*, pp. 363–370, The Combustion Institute, Pittsburgh (1986).

63. L. Orloff, J. deRis, and M.A. Delichatsios, "Radiation from Buoyant Turbulent Diffusion Flames," *Combustion and Flame*, 69, 177–186 (1992).

64. J.H. Kent, "Turbulent Diffusion Flame Sooting—Relationship to Smoke-Point Tests," *Combustion and Flame*, 67, 223–233 (1987).

65. J.H. Kent, "A Quantitative Relationship Between Soot Yield and Smoke Point Measurements," *Combustion and Flame*, 63, 349–358 (1986).

66. J.H. Kent, and Gg. Wagner, "Why Do Diffusion Flames Emit Soot," *Combustion Science and Technology*, 41, 245–269 (1984).

67. O.L. Gulder, "Influence of Hydrocarbon Fuel Structure Constitution and Flame Temperature on Soot Formation in Laminar Diffusion Flames," *Combustion and Flame*, 78, 179–194 (1989).

68. O.L. Gulder, "Soot Formation in Laminar Diffusion Flames at Elevated Temperatures," *Combustion and Flame*, 88, 74–82 (1992).

69. B.S. Haynes, and H.Gg. Wagner, "Soot Formation," *Progress in Energy and Combustion Sciences*, 7, 229–273 (1981).

70. U.O. Koylu, and G.M. Faeth, "Structure of Overfire Soot in Buoyant Turbulent Diffusion Flames at Long Residence Times," *Combustion and Flame*, 89, 140–156 (1992).

71. U.O. Koylu, Y.R. Sivathanu, and G.M. Faeth, "Carbon Monoxide and Soot Emissions from Buoyant Turbulent Diffusion Flames," *Fire Safety Science—Proceedings of the Third International Symposium*, pp. 625–634, Hemisphere Publishing Co., New York (1991).

72. U.O. Koylu, and G.M. Faeth, "Carbon Monoxide and Soot Emissions from Liquid-Fueled Buoyant Turbulent Diffusion Flames," *Combustion and Flame*, 87, 61–76 (1991).

73. Y.R. Shivathanu, and G.M. Faeth, "Soot Volume Fractions in the Overfire Region of Turbulent Diffusion Flames," *Combustion and Flame*, 81, 133–149 (1990).

74. D.B. Olson, J.C. Pickens, and Gill, "The Effects of Molecular Structure on Soot Formation, II. Diffusion Flames," *Combust. and Flame*, 62, 43–60 (1985).

75. S.L. Madorsky, *Thermal Degradation of Organic Polymers*, p. 192, Interscience Publishers, John Wiley & Sons, Inc., New York (1964).

76. "Network Reliability: A Report to the Nation, Compendium of Technical Papers," Section G, Presented by the Federal Communications Commission's Network Reliability Council, National Engineering Consortium, Chicago, IL, June 1993.

77. B.T. Reagor, "Smoke Corrosivity: Generation, Impact, Detection, and Protection," *J. Fire Sciences*, 10, 169–179 (1992).

78. B.T. Reagor, and C.A. Russell, "A Survey of Problems in Telecommunications Equipment Resulting from Chemical Contamination," *IEEE Transactions*, Vol. CHMT-9, No. 2, p. 209, June 1986.

79. B.T. Reagor, and C.A. Russell, "A Survey of Manufacturing Problems in Telecommunications Equipment," *Proceedings of the International Conference on Electrical Contacts, Electromechanical Components, and Their Applications*, Nagoya, Japan, July 1986.

80. F.W. Lipfert, "Effects of Acidic Deposition on the Atmospheric Deterioration of Materials," Paper presented during *Corrosion/86*, Paper No. 105, National Association of Corrosion Engineers, Houston, TX, 1986, *Material Performance*, pp. 12–19 (1987).

81. F.L. Chu, "Development and Application of Nonthermal Damage Assessment Techniques," Technical Report J.I. OV1J1.RC, Factory Mutual Research Corporation, Norwood, MA, Oct. 1992.

82. A. Tewarson, "The Effects of Fire-Exposed Electrical Wiring Systems on Escape Potential from Buildings, Part I: A Literature Review of Pyrolysis/Combustion Products and Toxicities—Poly(Vinyl Chloride)," Technical Report No. 22491, RC75-T-47, Factory Mutual Research Corporation, Norwood, MA, Dec. 1975.

83. K.G. Martin, and D.A. Powell, "Toxic Gas and Smoke Assessment Studies on Vinyl Floor Coverings with the Fire Propagation Tests," *Fire and Materials*, 3, 132–139 (1979).

84. A. Tewarson, "Nonthermal Damage Associated with Wire and Cable Fires," *42nd International Wire and Cable Symposium*, pp. 783–791, International Wire and Cable Symposium (IWCS), Eatontown, NJ (1993).

85. A. Tewarson, and M.M. Khan, "Generation of Smoke from Electrical Cables," *Proceedings of the ASTM Symposium on Characterization and Toxicity of Smoke*, H.K. Hasegawa, ed., ASTM STP 1082, pp.100–117, The American Society for Testing and Materials, Philadelphia, PA (1988).

86. A. Tewarson, M.M. Khan, and J.S. Steciak, "Combustibility of Electrical Wire and Cable for Rail Rapid Transit Systems, Vol. 1. Flammability," U.S. Department of Transporation Technical Report DOT-TSC-UMTA-83-4.1, National Technical Information Service, Springfield, VA, Apr. 1982.

87. S.L. Kessel, C.E. Rogers, and J.G. Bennett, "Corrosive Test Methods for Polymeric Materials, Part 5—A Comparison of Four Test Methods," *J. Fire Sciences*, 12, 196–233 (1994).

88. P.A. Dickinson, "Evolving Fire Retardant Materials Issues: A Cable Manufacturer's Perspective," *Fire Technology*, 4, 345–368 (1992).

89. A.F. Grand, "Evaluation of the Corrosivity of Smoke Using a Laboratory Radiant Combustion Exposure Apparatus," *J. Fire Sciences*, 10, 72–93 (1992).

90. P. Rio, "Presentation de l'essai Corrosivite mis au point au CNET-Lab-SER/ENV," *Centre National d'Etudes des Telecommunications* (1983).

91. M.F. Bottin, "The ISO Static Test Method for Measuring Smoke Corrosivity," *J. Fire Science*, 10, 160–168 (1992).

92. Testing of Cables, Wires, and Flexible Cords, Corrosivity of Combustion Gases, DIN 57472, Part 813 Standard, Verband Deutscher Elektrotechniker (VDE) Specification 0472, Part 813 (1983).

93. E. Barth, B. Muller, F.H. Prager, and F. Wittbecker, "Corrosive Effects of Smoke: Decomposition with the DIN Tube According to DIN 53436," *J. Fire Sciences*, 10, 432–454 (1992).

94. E.D. Goldberg, "Black Carbon in the Environment—Properties and Distribution," John Wiley & Sons, New York (1985).

95. *Particulate Carbon Formation During Combustion*, D.C. Siegla and G.W. Smith, eds., Plenum Press, New York (1981).

96. S.P. Nolan, "A Review of Research at Sandia National Laboratories Associated with the Problem of Smoke Corrosivity," *Fire Safety Journal*, 15, 403–413 (1989).

97. J.S. Newman, "Smoke Characterization in Enclosure Environments," *Proceedings of the ASTM Symposium on Characterization and Toxicity of Smoke*, H.K. Hasegawa, ed., ASTM STP 1082, pp. 123–134, The American Society for Testing and Materials, Philadelphia, PA (1988).

98. H. Burtscher, A. Reiss, and A. Schmidt-Ott, "Particle Charge in Combustion Aerosols," *J. Aerosol Science*, 17, 47– (1986).

99. J.J. Beitel, C.A. Bertelo, W.F. Carroll, R.O. Gardner, A.F. Grand, M.M. Hirschler, and G.F. Smith, "HCl Transport and Decay in a Large Apparatus, II. Variables Affecting Hydrogen Chloride Decay," *J. Fire Sciences*, 5, 105–145 (1987).

100. J.P. Stone, R.N. Hazlett, J.E. Johnson, and H.W. Carhart, "The Transmission of HCl by Soot from Burning PVC," *J. Fire and Flammability*, 4, 42–57 (1973).

101. A. Tewarson, and M.M. Khan, "The Role of Active and Passive Fire Protection Techniques in Fire Control, Suppression, and Extinguishment," *Fire Safety Science—Proceedings of the Third International Symposium*, pp. 1007–1017, Hemisphere Publishing Co., New York (1991).

102. "Small-Scale Testing: The Role of Passive Fire Protection in Commodity Classification," *FMRC Update*, Vol. 4, No. 3, Factory Mutual Research Corporation, Norwood, MA (1990).

103. C. Beyler, "A Unified Model of Fire Suppression," *Journal of Fire Protection Engineering*, 4, 5–16 (1992).

104. D.J. Rashbash, "The Extinction of Fire with Plain Water: A Review," *Fire Safety Science—Proceedings of the First International Symposium*, pp. 1145–1163, Hemisphere Publishing Co., New York (1986).

105. G. Heskestad, "The Role of Water in Suppression of Fire: A Review," *J. Fire and Flammability*, 11, 254–262 (1980).

106. D.J. Rashbash, "A Flame Extinction Criterion for Fire Spread," *Combustion and Flame*, 26, 411–412 (1976).

107. R.S. Magee, and R.D. Reitz, "Extinguishment of Radiation-Augmented Plastics Fires by Water Sprays," *Fifteenth Symposium (International) on Combustion*, pp. 337–347, The Combustion Institute, Pittsburgh, PA (1975).

108. D.J. Rashbash, "The Extinction of Fires by Water Sprays," *Fire Research Abstracts and Reviews*, 4, 28–52 (1962).

109. H.E. Thomson, and D.D. Drysdale, "Critical Mass Flow Rate at the Firepoint of Plastics," *Fire Safety Science—Proceedings of the Second International Symposium*, pp. 67–76, Hemisphere Publishing Co., New York (1989).

110. "Advances in Commodity Classification, A Progress Report," *FMRC Update*, Vol. 4, No. 1, Factory Mutual Research Corporation, Norwood, MA (1990).

111. C. Yao, "The Development of the ESFR Sprinkler System," *Fire Safety Journal*, 14, 65–73 (1988).

112. H.C. Kung, H. You, W.R. Brown, and B.G. Vincent, "Four-Tier Array Rack Storage Fire Tests with Fast-Response Prototype Sprinklers," *Fire Safety Science—Proceedings of the Second International Symposium*, pp. 633–642, Hemisphere Publishing Co., New York (1989).

113. J.L. Lee, "Extinguishment of Rack Storage Fires of Corrugated Cartons Using Water," *Fire Safety Science—Proceedings of the First International Symposium*, pp. 1177–1186, Hemisphere Publishing Co., New York (1986).

114. M.M. Khan, and A. Tewarson, "Passive Fire Protection for Materials and Storage Commodities," *Flame Retardancy, Educational Symposium No. 28, Rubber Division, American Chemical Society, Fall 1992*, Paper J, pp. 1–30, Rubber Division ACS, The John H. Gifford Library, The University of Akron, Akron, OH.

115. H. Kodama, K. Miyasaka, and A.C. Fernandez-Pello, "Extinction and Stabilization of a Diffusion Flame on a Flat Combustible Surface with Emphasis on Thermal Controlling Mechanisms," *Combustion Science and Technology*, 54, 37–50 (1987).

116. A.K. Kulkarni, and M. Sibulkin, "Burning Rate Measurements on Vertical Fuel Surfaces," *Combustion and Flame*, 44, 185–186 (1982).

COMPARTMENT FIRE MODELING

J.G. Quintiere

INTRODUCTION

An approach for predicting various aspects of fire phenomena in compartments has been called "zone" modeling. It is based on a conceptual representation for the compartment fire process, and is an approximation to reality. Any radical departure by the fire system from the basic concept of the zone model can seriously affect the accuracy and validity of the approach. The zone model simply represents the system as two distinct compartment gas zones: (1) an upper volume and (2) a lower volume resulting from thermal stratification due to buoyancy. Conservation equations are applied to each zone and serve to embrace the various transport and combustion processes that apply. The fire is represented as a source of energy and mass, and manifests itself as a plume, which acts as a "pump" for the mass from the lower zone to the upper zone through a process called "entrainment."

The zone modeling approach emerged in the mid-1970s when the effort to study the developing fire in a compartment intensified. Careful measurements and observations revealed characteristics of the compartment fire system. The upper and lower layers (zones) were deemed relatively uniform in temperature and composition. Distinct phenomena were discerned that could be studied in isolation, enabling better predictions of their roles in the compartment fire system.

Fowkes,[1] in his work with Emmons on the Home Fire Project, was the first to publish a basis for the zone model approach in his description of the "Bedroom Fire" series conducted at Factory Mutual Research Corporation. Almost simultaneously, computer models based on the zone model approach were produced by Quintiere,[2] Pape and Waterman,[3] and Mitler[4] working with Emmons. Since then the development of such computer models has been prolific. They have extended the early efforts from a single room to computer codes that can address a number of interconnected rooms, using a number of new fire phenomena and computer features. These advances in fire science, together with the development of the personal computer, have given the engineer a convenient tool for investigating the hazard of

fire in buildings. A notable illustration of this tool is the software "Hazard I," developed by the National Institute of Standards and Technology (NIST).[5] At this time numerous computer codes and software packages exist based on the zone model approach. In a recent survey Friedman[6] cited 21 zone models in use around the world.

This chapter outlines the basic conservation equations for the gas zones, and describes the various transport and combustion processes that make up the system. These processes are referred to as the submodels of the system; as such they can contribute subroutines to computer codes, which implement the mathematical solution. Discussion of submodels will be limited, but the reader will be referred to appropriate references. In most cases, other chapters of *The SFPE Handbook of Fire Protection Engineering* will be cited. No discussion of a computer code or its numerical solution algorithm will be addressed, since these are issues more of style and mathematics. The presentation will elucidate the mathematical basis of the zone model, its assumptions, its features, and its scope of application. Each user of this approach must sufficiently understand its basis to assess its accuracy and validity. When used correctly, zone models predict the average macroscopic features of compartment fires. There are many examples of comparisons to data that illustrate their level of accuracy, and these will not be repeated here. The user must be skilled in assessing the quality of the data and submodels that directly influence the variables of the problem of interest. It is hoped that the discussion that follows will make the user more knowledgeable or sensitive in making these quality assessments.

CONSERVATION EQUATIONS

The building block of the zone model is the conservation equations for the upper and lower gas zones. These equations are developed either: (1) by using fundamental equations of energy, mass, and momentum transport in control volume form as applied to the zones, or (2) by using differential equations that represent the conservation laws and integrating them over the zones. However, the momentum equation will not be explicitly applied, since information needed to compute velocities and pressures is based on assumptions and specific applications of momentum principles at vent boundaries of the compartment. An extensive review of control volume equations for mass, species, and

---

Dr. James G. Quintiere is Professor of Fire Protection Engineering at the University of Maryland. His research has focused on fire growth and flame spread.

energy conservation in a combustion system has been presented by Quintiere[7] and serves as reference for the equations that follow.

Figure 3-5.1 illustrates a typical zone model for a compartment fire process. It shows a fire plume and a door vent. The hot combustion gases that collect in the upper space of the room and spill out of the vent constitute the "upper-layer zone." A control volume, CV_1, is defined to enclose the gas in this upper layer and the fire plume. The lower interface of the upper layer moves with the control volume such that no mass is transferred across this thermally stratified region. The velocity of the control volume along this interface, $\overline{w}$, is equal to the fluid velocity, $\overline{v}$. The temperature of the upper layer is greater than that in the "lower layer" (zone) which includes all the remaining gas in the room, and is delineated by a second control volume, CV_2. It has been assumed in zone modeling that the volume of the fire plume is small relative to the gas layer or zone volumes, and therefore its effect has been ignored. In general, multiple fire plumes can occur at any height in the room, and multiple vents or mass transport can take place between the zones (CV_1 and CV_2) and the surroundings. In each case mass transport must be appropriately described in terms of the system variables; however, this may not always be easy or known. The properties of the upper and lower zones are assumed to be spatially uniform, but can vary with time. Thus, temperature, T, and species mass concentration, Y_i, are properties associated with ideal upper and lower homogeneous gas layers. Other assumptions in the application of the conservation laws to the zones are listed below.

1. The gas is treated as an ideal gas with a constant molecular weight and constant specific heats: c_p and c_v.
2. Exchange of mass at free boundaries is due to pressure differences or shear mixing effects. Generally these are caused by natural or forced convection, or by entrainment processes.
3. Combustion is treated as a source of mass and energy. No mechanism from first principles is included to resolve the extent of the combustion zone.
4. The plume instantly arrives at the ceiling. No attempt is made to account for the time required to transport mass vertically or horizontally in the compartment. Hence, transport times are not explicitly accounted for in zone modeling.
5. The mass or heat capacity of room contents is ignored compared to the enclosure wall, ceiling, and floor elements; i.e., heat is considered lost to the structure, but not to the contents. Where room contents shield boundary structural surfaces, some compensations can occur in the analysis, but for cluttered rooms this assumption may be poor.
6. The horizontal cross section of the enclosure is a constant area, A. In most cases of zone modeling rectilinear compartments have been considered. However, this is not a necessary assumption, and enclosures in which A varies with height can easily be handled.
7. The pressure in the enclosure is considered uniform in the energy equation, but hydrostatic variations account for pressure differences at free boundaries of the enclosure; i.e., $p \gg \rho gH$. In general, the enclosure pressure, p, is much greater than the variations due to hydrostatics. For example, for $p = 1$ atm $= 14.7$ psi $= 10^2$ kPa (kN/m^2) $= 10^5$ Pa, the hydrostatic variation for a height, $H = 1$ m, gives a pressure difference of $\rho gH = 1.2$ kg/m$^3 \times 9.8$ m/s^2 $\times 1$ m $= 10$ kg/m s$^2 = 10$ Pa (N/m^2).
8. Mass flow into the fire plume is due to turbulent entrainment. Entrainment is the process by which the surrounding gas flows into the fire plume as a result of buoyancy. Empirically, the inflow velocity linearly depends on the vertical velocity in the plume.
9. Fluid frictional effects at solid boundaries are ignored in the current models.

Conservation of Mass

The conservation of mass for a control volume states that the rate of change of mass in the volume plus the sum of the net mass flow rates out is zero for J flow streams; i.e.,

$$A\frac{d}{dt}(\rho z_l) + \sum_{\substack{j=1 \\ \text{(net out)}}}^{J} \dot{m}_j = 0 \qquad (1)$$

where

ρ = density of the gas in the control volume (or zone), and
z_l = the height of the zone.

For the illustration in Figure 3-5.1, applying Equation 1 to the upper layer (CV_1) would give

$$\sum_{j=1}^{3} \dot{m}_j = \dot{m} - \dot{m}_e - \dot{m}_s \qquad (2)$$

where

$\dot{m}$ = mass flow rate out of the door,
$\dot{m}_e$ = mass rate of entrainment into the fire plume, and
$\dot{m}_s$ = mass rate of gaseous fuel supplied.

Mass flows at the boundaries can occur due to many phenomena; therefore, the user or designer of a zone model must include the appropriate mass flow phenomena. For example, in addition to the mass rates in Equation 2, mass flows can occur due to forced convection from wind or ventilation effects, from shear entrainment as flows affect layer interfaces, or from cold plumes that could plunge through hot layers.

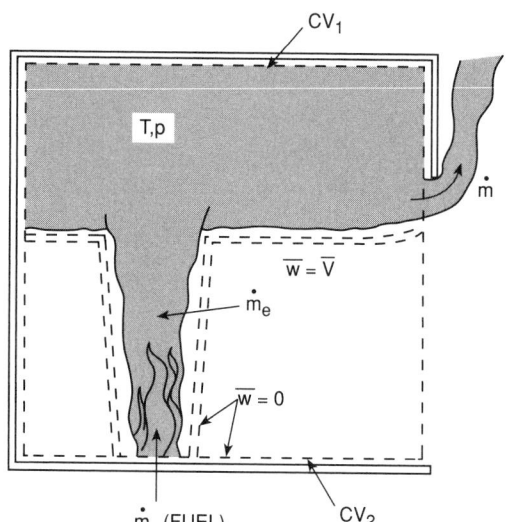

Fig. 3-5.1. Control volumes selected in zone modeling.

Conservation of Species

The mass conservation of species i is given by Y_i. By using Equation 1 and applying the conservation of mass for species i to a control volume, it follows that

$$\rho z_l A \frac{dY_i}{dt} + \sum_{j=1}^{J} \dot{m}_j (Y_{ij} - Y_i) = \dot{\omega}_i \qquad (3)$$

where

Y_{ij} = mass concentration of species i leaving the control volume through the j flow stream, and
$\dot{\omega}_i$ = mass production rate of species due to combustion.

The production term, $\dot{\omega}_i$, in principle, can be described through a knowledge of the chemical equation of the reaction or its particular stoichiometry. Thus, stoichiometric coefficients can be used to represent the production of species and the consumption of oxygen in terms of the mass rate of fuel *reacted*. Stoichiometry is not easily determined, and the fuel gases as they emerge from the pyrolysis of solids can take many chemical forms that differ from the solid fuel's original molecular composition. A partial way to overcome these complications has been to represent the mass production of species for fire in terms of the rate of mass *loss* for the pyrolyzing fuel. Hence, one must be careful to distinguish between the mass of fuel lost and that reacted, and to relate available species yield data to the particular fire conditions of the application. Yield is defined as the mass ratio of species to fuel lost. The yields or production rates may change with fire conditions, and therefore, in general, will not be consistent with data from small-scale tests. For example, the production rate of CO changes markedly with air-to-fuel ratio.

Conservation of Energy

The conservation of energy for the control volume is applied along with Equation 1 and the equation of state, $p = \rho RT$, to give

$$\rho c_p z_l A \frac{dT}{dt} - z_l A \frac{dp}{dt} + c_p \sum_{j=1}^{J} \dot{m}_j (T_j - T)$$
$$\underset{(net\ out)}{}$$
$$= \dot{\omega}_F \Delta H - \dot{Q}_{net\ loss} \qquad (4)$$

where

T = temperature of the gases within the control volume,
T_j = temperature of the gases in the j flow stream crossing the control volume boundary,
$\dot{Q}_{net\ loss}$ = net rate of heat transfer lost at the boundary,
ΔH = the heat of combustion (taken as a positive quantity), and
$\dot{\omega}_F$ = the rate at which the fuel supplied is *reacted*.

Usually in zone models it is assumed that all of the fuel supplied can react, provided there is sufficient oxygen available. One assumption on the sufficiency of oxygen is to consider that all the fuel supplied is reacted as long as the oxygen concentration in that control volume is greater or equal to zero, i.e.,

$$\dot{\omega}_F = \dot{m}_s \qquad \text{if } Y_o \geq 0 \qquad (5)$$

Thereafter, an excess rate of fuel can exist that can be transported into adjoining zones or control volumes where a de-

cision must be made about whether it can continue to react. At this condition, all of the net oxygen supplied to the control volume is reacted, so that, as long as $Y_o = 0$,

$$\dot{\omega}_F = r \times \text{(net mass rate of oxygen supplied)} \qquad (6)$$

where r is the stoichiometric fuel-to-oxygen mass ratio. This condition when $Y_o = 0$ in compartment fires is termed the ventilation-limited condition. At this moment, significant changes take place in the nature of the chemical reaction. Notably, incomplete combustion is more likely, and for hydrocarbon fuels this leads to a significant increase in the yield of carbon monoxide and soot. Thus, care must be used in interpreting the results of zone models once ventilation-limited conditions arise, particularly with respect to the prediction of species concentrations and the extent of burning. Material data used for well-ventilated conditions will no longer apply. The issue of what constitutes a flammable mixture in a compartment gas layer and combustion in a vitiated layer has not yet been resolved satisfactorily. Thus, combustion under ventilation-limited conditions has not been adequately addressed for a zone model, and needs more study.

The first term on the left-hand side of Equation 4 arises due to the change of internal energy with the control volume. If the temperature is not changing rapidly with time, this term can be small and its elimination gives rise to a quasi-steady approximation for growing fires that allows a more simple analysis. The second term arises from the rate of work done by pressure as the gas layer expands or contracts due to the motion of the thermal stratification interface. Having been rearranged, this term now is expressed as rate of pressure, p, increase for the compartment and is essentially caused by net heat or mass additions to the compartment gases. Except for the rapid accumulation of mass or energy, for compartments with small openings to the surroundings, this pressure rise is small, and the pressure nominally remains at nearly the ambient pressure. For example, an addition of 100 kW to a 40 m³ gas volume in a room with a 0.1 m² vent area gives rise to roughly an increase of 10 Pa in less than 10^{-2} s over normal ambient pressure of 10^5 Pa.[7] Any increase in pressure within the compartment could give rise to a flow of mass through a vent, and this term in Equation 4 may be associated with a "volumetric expansion" effect, as referred to by some. Conversely, a reduction in energy release rate will cause the pressure to drop relative to the ambient. This phenomenon, when cycling between heating and cooling, explains the "breathing effect" for fires in closed buildings.

The third term of Equation 4 accounts for the enthalpy flow rates and only applies to j flow streams that enter the control volume, since $T_j = T$ for all flow streams leaving, as long as the uniform temperature assumption still applies.

Summary

The zone model for the compartment fire system consists of two zones: (1) the upper and (2) lower gas layers. The solution process for the layer properties can be visualized by considering the conservation Equations 1, 3, and 4 applied to each zone. The species equation can yield the Y_i for each layer. The mass and energy equations comprise four equations that permit the determination of the two layer temperatures, one layer height (since the height of the other layer is directly found by difference from the total height of the compartment), and the compartment pressure (which is assumed uniform by Equation 7). The densities are found from the ideal gas equation of state in which approximately ρT is a constant. To complete this solution process, each source or transport term in the equations must be given in terms of the above layer properties, or auxiliary relationships must be included for each new variable introduced. The source terms are associated with the $\dot{\omega}_i$ terms, and the transport terms include the j mass flow rates and the boundary heat transfer rates. The extent to which source and transport relationships are included reflects the sophistication and scope of the zone model. Some source and transport terms are essential to a basic zone model, others can be specified as approximations to reality, and others can be ignored when physically irrelevant. These source and transport relationships can be termed "submodels" and can comprise subroutines of a zone model computer code. The nature of these submodels is discussed below.

SOURCE TERM SUBMODELS

The principal source term is the rate of fuel supplied. In an experimental fire this can be known if the fire source is simulated by a gas burner. In the other extreme, the mass of fuel supply can be a result of a spreading fire over an array of different solid fuels. In general,

$$\dot{m}_s = f \text{ (fuel properties, heat transfer)} \qquad (7)$$

in which the heat transfer to the fuel results from the flame configuration and the heated compartment. The fuel properties are still not completely defined or conventionally accepted for fire applications, since no general theory exists for pyrolysis, and theories of flame spread and ignition are couched in terms of effective fire properties, which are modeling parameters. Nevertheless, data exist for fuel fire properties and can enable approximate models for $\dot{m}_s$ of reasonable accuracy. For example, Tewarson describes how the mass supply and energy release can be determined from fuel properties, and tabulates properties for a number of solid fuels. (See Section 3, Chapter 4.) For realistic items under well-ventilated conditions, Babrauskas has compiled results that could serve as initial estimates for $\dot{m}_s$ in compartment fires. (See Section 3, Chapter 1.)

The rate of energy release, $\dot{\omega}_F \Delta H$, required by Equation 4 has already been discussed through Equations 5 and 6. The point should be made that the heat of combustion, ΔH, employed must be with respect to the mass of fuel gases pyrolyzed, given by such data as Tewarson's, and is not the theoretical oxygen bomb value for the solid fuel. (See Section 3, Chapter 4.) Due to incomplete combustion, ΔH will be less than the theoretical value, in general.

The production of species can be described in terms of species yield, γ_i, such that

$$\dot{\omega}_i = \gamma_i \dot{m}_s \qquad (8)$$

For well-ventilated fires, γ_i may be reasonably constant for a given fuel, as tabulated by Tewarson. (See Section 3, Chapter 4.) In general, it can vary with time and can significantly vary as ventilation-limited conditions are approached and achieved. For example, Tewarson shows that γ_i for CO can vary with equivalence ratio, Φ, where

$$\Phi = \left(\frac{mass\ of\ fuel\ available}{mass\ of\ oxygen\ available} \right) / r \qquad (9)$$

where r is the stoichiometric value for complete combustion. Zukoski et al[8] have shown how this relationship may be applied to compartment fires. The equivalence ratio, Φ, may be computed in a zone (or upper layer) where combustion has occurred by computing the mass concentrations of the "available" fuel and oxygen in the zone. This is done by Equation 3 in which $\dot{\omega}_i$ is set equal to zero for both the fuel and oxygen, since this yields the available Y_F and Y_o values, not their actual concentrations in the layer following combustion. The generality of considering $\gamma_i = \gamma_i(\Phi)$ for zone models is still under study, and its use must be considered as exploratory. Nevertheless, it currently offers the only practical approach for estimating species, such as CO, under ventilation-limited conditions in compartment fires.

MASS AND HEAT TRANSPORT SUBMODELS

Entrainment

An essential feature of a zone model is the mass rate of entrainment, $\dot{m}_e$, relationship for the fire plume. This allows the principal mechanism for flow between the lower and upper stratified gas layers. Considerable work has been performed to develop entrainment relationships for pool fires or axisymmetric gas burner fires. Unfortunately both the ideal theoretical plume models and correlations based on data vary widely, and no consensus exists among zone models in practice for the optimum pool fire entrainment model. Rockett[9] illustrates the variations in results he found using different fire entrainment models in the Harvard/NIST Mark VI compartment fire zone model. He found that the layer height, entrainment rate, and layer gas temperature varied by roughly a factor of 2 among the various models. Delichatsios presents a correlation for the entrainment rate of a pool fire based on dimensional analysis that agrees with the available experimental data. (See Section 2, Chapter 3.) More useful data rather than ideal mathematical models are clearly needed to resolve this issue of accuracy for a simple pool fire. Yet even a perfect entrainment relationship for an axisymmetric pool fire would not necessarily be perfect in a zone model, because a plume in a compartment can be subject to nonsymmetric airflows that can bend the plume and thus affect its entrainment rate. Usually wind effects will increase the entrainment rate.

Rockett[9] has shown that the effect of the entrainment model is crucial to predictions for the developing fire. This suggests that the entrainment model must be representative of the actual object burning. However, no entrainment models exist for a wall, corner, or item of furniture; this dramatizes the lack of much-needed research in this area. Yet, this does not mean that the zone model has a fatal flaw; it simply means much more systematic data are necessary to expand the versatility of the zone model and its accuracy. Moreover, if a zone model with its selected entrainment relationship tracks well with data from an experimental fire scenario, it can be assumed accurate for simulating the process and can be used with some assurance for that scenario. A catalogue of empirical entrainment relationships for various object fires developed from specialized entrainment apparatus would help resolve the entrainment issue. This apparatus could be developed from the large calorimeter intended to measure energy

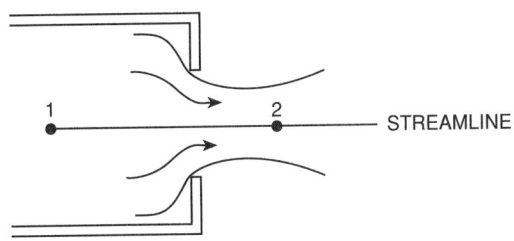

Fig. 3-5.2. Orifice flow.

release rate in which the fire plume is collected in a hood-duct system and the total flow rate is recorded.

Vent Flows Through Openings in Vertical Partitions

Classic models of fire in a room or building represent the structure with an opening, such as a door or window, to the ambient surroundings. Fire-induced flows through such openings have been well studied, and a widely accepted model exists to compute these flows based on the temperature distribution of the gases on either side of the opening. The theoretical basis of the computation is orifice flow utilizing Bernoulli's equation along a streamline, as illustrated in Figure 3-5.2. The velocity at station 2 is given by

$$v_2 = \sqrt{\frac{2(\rho_1 - \rho_2)}{\rho_1}} \qquad (10)$$

where v_1 is assumed to be zero. The mass flow rate is computed by integration over the flow area, A, adjusted by a flow coefficient, C;

$$\dot{m} = C \int \rho_1 v_2 dA \qquad (11)$$

Emmons suggests that a value of 0.68 for C has an accuracy of ±10 percent, except at very low flow rates at the beginning of a fire. (See Section 2, Chapter 5.) In general, C will depend on the Reynolds number. Figure 3-5.3 depicts examples of typical vent flows through an opening in a vertical partition. In both cases Equations 10 and 11 apply, but the pressure distribution must be described appropriately. For example, in the pure natural convection case shown in Figure 3-5.3, part a, the pressure is determined by the static pressure with respect to the floor pressure, $p(0)$. Actually it is the floor pressure that applies in Equation 4 and in the perfect gas equation of state.

The assumption is that the flow velocities are small compared to the vent flow velocities, justifying the static pressure computation. Thus, the vertical pressure distribution on either side of the opening is computed as

$$p(z) = p(0) - \int_0^z \rho g \, dz \qquad (12)$$

McCaffrey and Rockett[10] illustrate the accuracy of the hydrostatic assumption in Figure 3-5.4. The sign of the pressure difference across the opening determines the flow direction. Emmons presents the general equations that enable this computation to be included in a zone model. (See Section 2, Chapter 5.) It is by far the most accurate of the submodels, and provides the basis for linking rooms together in a zone model, which allows smoke and fire growth computations for a large building.

The flow through an opening in a horizontal partition can be compared to that for the vertical partition, provided the pressure difference is large enough. If there is only a single vent from the fire compartment through a horizontal partition, such as a ceiling, the flow must be oscillatory or bi-directional. The latter case implies a zero pressure difference, with gravity solely determining the flow. A theory for this case has been developed by Epstein[11] and has been implemented by Cooper.[12] For orifice-like vents with zero pressure difference, the volumetric exchange flow rate, $\dot{V}$, given by Epstein,[11] is approximately

$$\dot{V} = 0.055 \left[D^5 g(\rho_1 - \rho_2)\left(\frac{\rho_1 + \rho_2}{2}\right) \right]^{1/2} \qquad (13)$$

where D is the diameter of the vent and ρ_1 and ρ_2 are the corresponding fluid densities on either side of the vent. For vents of significant depth, L, the coefficient in Equation 13 depends on L/D.

Convective Heat Transfer to Surfaces

The $\dot{Q}_{net\,loss}$ term in Equation 4 is composed of the convective and radiative heat loss to the boundary surfaces of the layer control volumes. This involves both heat transfer from the gas layers at their bulk temperatures and the heat transfer from the flame. Consistent treatment of the flame and layer gas heat transfer must be carried out for the zone model. If the flame becomes large and fills the upper layer, one cannot count the flame and gas heat transfer without being redundant.

Convective heat transfer to a ceiling by a fire plume has been widely studied at modest scales, such that flame radiation may have been insignificant. Alpert[13] specifically examined only convective heating in contrast to studies by You and Faeth[14] and Kokkala[15] who included flame effects.

In general, convective effects will vary along the ceiling, walls, and floor, and depend on the nature and position of the fire. In some cases an "adiabatic wall temperature" has been appropriately introduced since the driving force for convective heat transfer locally is not the bulk gas layer temperature, but the local boundary layer temperature, which is not explicitly computed. Convective heat transfer data for the walls and floor of a fire compartment or for rooms beyond the fire compartment have not been developed. Hence, most zone models use estimates from natural convection correlations.

Radiative Heat Transfer

The theory of radiative heat transfer is adequate to develop the needed components for the zone model. However, the theory is not sufficiently developed to predict flame radiation from first principles without very sophisticated

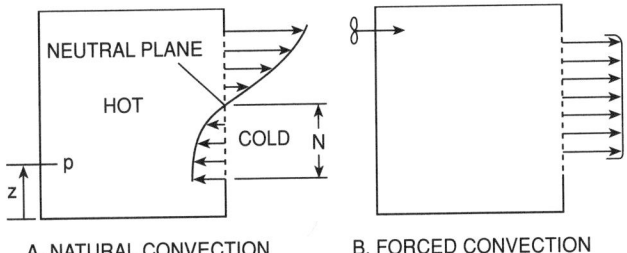

Fig. 3-5.3. Typical vent flows.

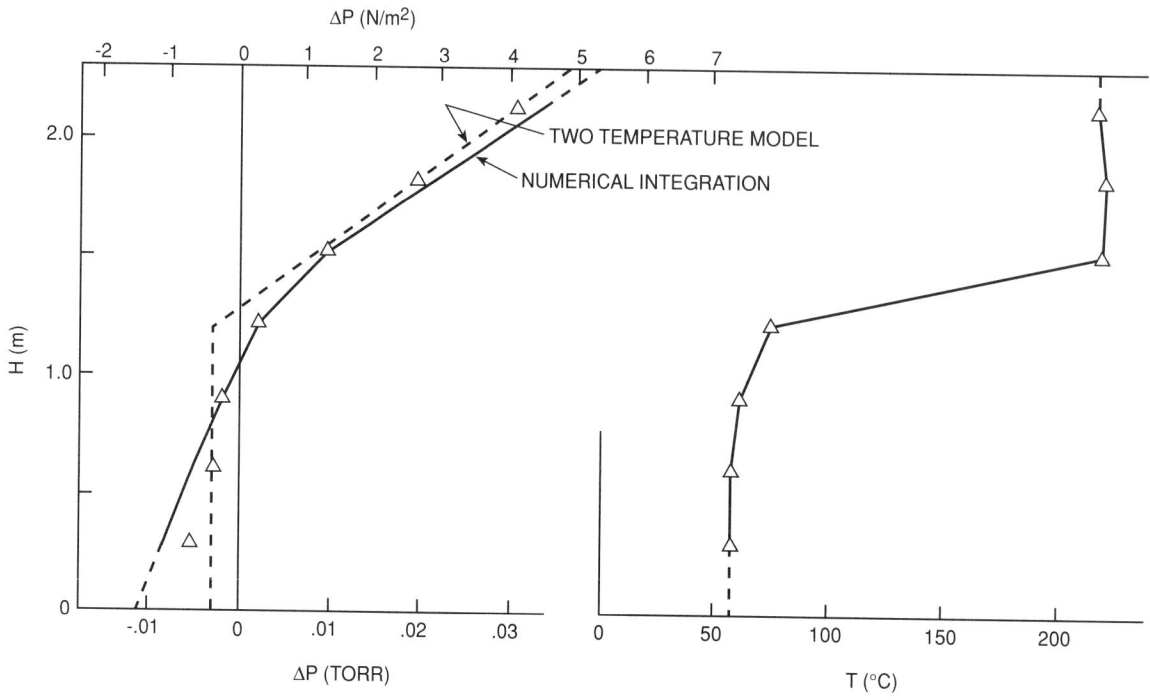

Fig. 3-5.4. *Vertical pressure difference across a room vertical partition compared to a computation based on room fire temperature distribution and a two-temperature zone model approximation using the hydrostatic pressure assumption.* [10]

modeling of the soot and temperature distributions. Hence, flame radiation is relegated to empirical practices. Radiation from a smoke layer is easier to deal with within the context of a uniform property gas layer for the zone model. One difficulty still is the availability of property data to determine the contribution of smoke particulates to the layer radiation properties. The discussion presented by Tien *et al* can be used to begin a development of the radiative equations needed by the zone model. (See Section 1, Chapter 4.) Also,

the presentation by Mudan and Croce gives empirical approaches to dealing with flame radiation. (See Section 3, Chapter 11.) The report by Forney[16] lays out the theory and equations describing radiation exchange between the gas layers and boundary surfaces.

Conduction Heat Transfer

The radiative and convective heat transfer from the gas must be balanced by conduction heat transfer through the boundary surfaces. This requires a numerical solution to a partial differential equation in conjunction with the ordinary differential equations in time describing the conservation of energy and mass for the gas layers. Usually zone models have considered only one-dimensional conduction, which should be adequate for most applications. Most multiple-compartment models do not consider communication by conduction into the next compartment, treating the structural elements as thermally thick instead. In principle, there is no difficulty with developing an accurate algorithm for conduction through the boundary elements for any conditions. For more information, the reader is referred to the discussion by Rockett and Milke. (See Section 1, Chapter 2.)

Mixing Between the Layers

The primary exchange of fluid between the lower and upper gas layers is due to the buoyant effect of the fire plume. Secondary, but significant, mixing processes can occur due to the other effects. These are shown in Figure 3-5.5 and include three phenomena:

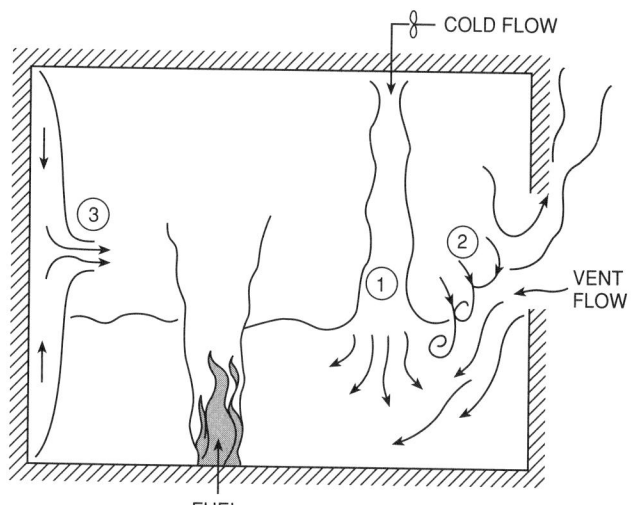

Fig. 3-5.5. *Secondary flows—mixing phenomena. 1. A cold plume descending from the upper layer into the lower layer; 2. Shear mixing of an entering vent flow stream; 3. Wall flows due to local buoyancy effects.*

1. exchange due to a cold flow injected into the hot layer,
2. exchange due to shear mixing associated with vent flows, and
3. exchange due to wall flows.

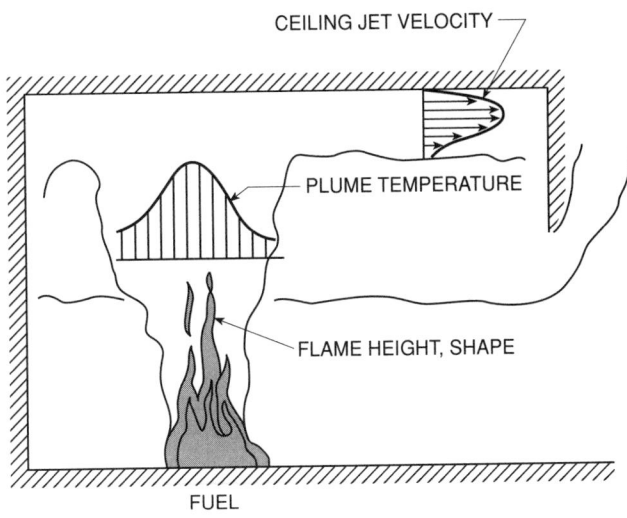

Fig. 3-5.6. *Examples of embedded phenomena.*

Phenomenon 1 is the inverse of the hot fire plume penetrating the upper layer. In both cases the fluid at the edge of the plume may not be buoyant enough to penetrate the respective layer. A comparable situation is a cold forced jet introduced vertically into the lower layer. Depending on the relative temperatures, it may not escape the lower layer and, therefore, may not penetrate into the upper layer. These are issues that can be resolved to some extent by research available in the literature on buoyant plumes and jets.

Phenomenon 2 has not been sufficiently studied. Data suggest that the flow rate of the mixed stream can be significant relative to the vent flow rate, especially for small vents.[17] A correlation for the mixing rate has been developed from saltwater simulation experiments.[18]

Phenomenon 3 has been discussed by Jaluria. (See Section 2, Chapter 6.) He presents relationships that allow the estimation of the rate of transfer of cold fluid adjacent to the wall in the hot upper gas layer into the cold lower gas layer or *vice versa*.

All of these flows tend to blur the sharp distinction between the upper and lower gas layers, reducing their degree of stratification. Obviously, if sufficient mixing occurs, the layer may appear to become well-mixed or destratified. This should occur naturally in the context of the zone model, and one should not have to switch to a well-mixed compartment model under these conditions.

Relationships for all of these secondary flows have not been developed with confidence nor with full acceptance. Although they are important for improving the accuracy of a zone fire model, little work has gone forward to establish their validity.

Forced Flow Effects

The effect of forced airflow on the fire conditions and smoke spread due to mechanical or natural wind forces has always been an issue in large building fires. Wind effects and the resultant pressure distribution around a tall building has become a standard element of design data for structural design, but this has not been utilized for fire safety design. The movement of smoke through a building due to the mechanical ventilation system has been simulated by network models that treat the compartment volume as uniform in properties, and include the pressure losses due to vents and duct friction. A two-layer zone model has not been linked to the mechanical ventilation system in a building. To create a link, one must include the full pressure-flow characteristics of fans in both directions to allow for the possibility of the backflow of smoke against the direction of airflow in the ducts. An attempt at this linkage has been presented by Klote and Cooper,[19] who hypothesize a fan characteristic relationship. Ultimately an experimental study will be needed to lay a foundation for this analysis.

Fire Growth Rate

In most all zone models, the fire source is considered an input quantity, based on some experimental or empirical data. This limits the simulation capability of a zone model, since fire growth and spread is not modeled. Also the effects of compartment feedback due to thermal and vitiation (oxygen depletion) effects are not taken into account. The versatility and utility of a zone model can only be improved by developing techniques for accommodating the fire growth of realistic building contents and architectural elements. This process will have an impact on the use and development of flammability tests for hazard analysis and product acceptability.

EMBEDDED SUBMODELS

The detailed physics that one can include in a zone model are only limited by current research and imagination. The zone model can be versatile in accommodating new phenomena, even if they appear inconsistent with the uniform property layer assumption. By analogy to the relationship between inviscid flow and boundary layer flow in the analysis of aerodynamic bodies, the layer properties can be regarded as first-order approximation for higher order analysis. Flame and boundary layer phenomena within the compartment can be computed by regarding the layer properties as infinite reservoirs. These phenomena can be computed after the primary layer properties are computed. Examples of embedded phenomena are shown in Figure 3-5.6. Although the combustion region is assumed to be of negligible volume at the zone model for mutation, the flame height can be computed along with the velocity and temperature distributions in an axisymmetric fire plume. (See Section 2, Chapters 1 and 2.) Other embedded phenomenon are: (1) the ceiling jet, (2) the computation of temperature distributions over the ceiling, (3) the deposition of soot and other products of combustion on surfaces, and (4) the heating and degradation of structural elements.

UNRESOLVED PHENOMENA

Some significant phenomena are not addressed by the zone-modeling approach for fire, such as vent flames, transient flow in corridors, and shaft flows. (See Figure 3-5.7.) These phenomena require more research and new strategies to enable them to be included into a zone model. Vent flames are significant for fire growth into the next compartment and usually follow flashover. Information about their rate of heat transfer and extent needs to be computed. Transient corridor flows are important in the analysis of smoke transport through long corridors. The current zone model methodology yields an instantaneous layer, which would descend, but the actual process produces a transient ceiling jet. Flows

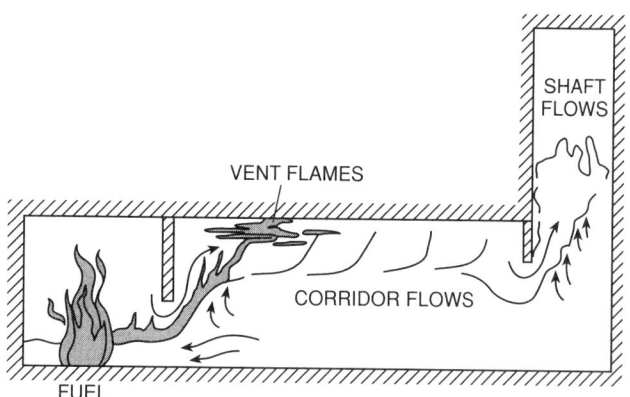

Fig. 3-5.7. *Examples of significant phenomena absent from zone models.*

in vertical shafts involve the interaction of plumes with walls, pressure-driven effects, and turbulent mixing.

SELECTED READING AND COMMENTS

Zone models provide the integrating framework for the phenomena of fire and its fire protection engineering components. Many zone models have been constructed for fire predictions in compartments. They involve the basic conservation equations, submodels describing the particular phenomena included, and the mathematical algorithm for solution. Some have developed user-friendly interfaces. Most provide documentation on the model and its use. The interested reader is referred to some published models for more detailed information.[20-22] Many zone-model computer models exist, similar both in substance and the ability to analyze fire effects in buildings. They can stimulate needed research. However, more effort appears to have gone into the computer code developments rather than the experimental research needed for improvement in the model.

NOMENCLATURE

A	area compartment floor
C	flow coefficient
c_p	specific heat at constant pressure
c_v	specific heat at constant volume
g	acceleration due to gravity
H	compartment height
J	number of flow streams in control volume
m	mass
p	pressure
Q	heat transfer
r	stoichiometric fuel-to-oxygen mass ratio
R	ideal gas constant
t	time
T	temperature
v	fluid velocity
V	volume
w	control volume velocity
Y	mass fraction
z_l	height of control volume or zone
ΔH	heat of combustion
γ_i	yield of species i
ρ	density

Φ	equivalence ratio
$\dot{\omega}_F$	consumption rate of fuel
$\dot{\omega}_i$	production rate of species

Subscripts

e	entrained
F	fuel
i	species
j	flow stream
s	supplied

Superscripts

$\dot{(\ )}$	per unit time

REFERENCES CITED

1. N.D. Fowkes, "A Mechanistic Model of the 1973 and 1974 Bedroom Test Fires," in P.A. Croce, ed., "A Study of Room Fire Development: The Second Full-Scale Bedroom Fire Test of the Home Fire Project (July 24, 1974)," Vol. II, FMRC Tech. Rept. No. 21011.4, pp. 8–50 (1975).
2. J. Quintiere "Growth of Fire in Building Compartments," in Robertson, A.F., ed., *Fire Standards and Safety*, ASTM STP614, American Soc. for Testing and Materials, pp. 131–167 (1977).
3. R. Pape and T. Waterman, "Modification to the FIRES Pre-Flashover Room Fire Computer Model," IITRI Project J6400, IIT Res. Inst., Chicago, IL (1977).
4. H.E. Mitler, "The Physical Basis for the Harvard Computer Fire Code," Home Fire Proj. Tech. Rept. No. 34, Harvard Univ., Cambridge, MA (1978).
5. R.W. Bukowski, R.D. Peacock, W.W. Jones, and C.L. Forney, "HAZARD I: Technical Reference Guide," *NIST Handbook 146*, Vol. II (1989).
6. R. Friedman, "Survey of Computer Models for Fire and Smoke," 2nd ed., Factory Mutual Research Corp., Norwood, MA, Dec. 1991.
7. J.G. Quintiere, "Fundamentals of Enclosure Fire Zone Models," *J. of Fire Protection Engr.*, Vol. 1 (3) (1989).
8. E.E. Zukoski, S.J. Toner, J.H. Morehant, and T. Kubota, "Combustion Processes in Two-Layered Configurations," *Fire Safety Science—Proc. 2nd Int. Symp.*, pp. 255–304, Hemisphere Publishing Corp., New York (1989).
9. J.A. Rockett, "Using the Harvard/NIST Mark VI Fire Simulation," NISTIR 4464, Nat. Inst. of Standards and Technology, Nov. 1990.
10. B.J. McCaffrey and J.A. Rockett, "Static Pressure Measurements of Enclosure Fires," *J. of Fire Research*, Nat. Bur. Stand., June 1977.
11. M. Epstein, "Buoyant-Driven Exchange Flow Through Small Openings in Horizontal Partitions," *Jour. of Heat Transfer*, Vol. 110, American Soc. for Testing and Materials (1988).
12. L.Y. Cooper, "An Algorithm and Associated Computer Subordinate for Calculating Flow Through a Horizontal Ceiling Flow Vent in a Zone-Type Compartment Fire Model," NISTIR 4402, Nat. Inst. of Stand. and Technology, Oct. 1990.
13. R.L. Alpert, "Convective Heat Transfer in the Impingement Region of a Buoyant Plume," *Jour. of Heat Transfer*, Vol. 109, American Soc. for Testing and Materials, Feb. 1987.
14. H.Z. You and G.M. Faeth, "Ceiling Heat Transfer During Fire Plume and Fire Impingement," *Fire and Materials*, Vol. 3, No. 3 (1979).
15. M.A. Kokkala, "Experimental Study of Heat Transfer to Ceiling from an Impinging Diffusion Flame," *Fire Safety Science—Proc. of 3rd Inter. Symp.*, G. Cox and B. Langford, ed., Elsevier Applied Science, London (1991).
16. G.P. Forney, "Computing Radiative Heat Transfer Occurring in a Zone Model," NISTIR 4709, Nat. Inst. of Stand. and Technology, Nov. 1991.

17. B.J. McCaffrey and J.G. Quintiere, "Buoyancy-Driven Counter-current Flows Generated by a Fire Source," *Heat Transfer and Turbulent Buoyant Convection*, Vol. II, D.B. Spalding and N. Afgan, ed., Hemisphere Pub. Co., pp. 457–472 (1977).

18. C.S. Lim, E.E. Zukoski, and T. Kubota, "Mixing in Doorway Flows and Entrainment in Fire Flames," Cal. Inst. of Technology, NBS Grant No. NB82NADA3033, June 1984.

19. J.H. Klote and L.Y. Cooper, "Model of a Simple Fan-Resistance Ventilation System and Its Application to Fire Modeling," NISTIR 89-4141, Nat. Inst. of Stand. and Technology, Sept. 1989.

20. H.W. Emmons, H.E. Mitler, and L.N. Trefethen, "Computer Fire Code III," Home Fire Proj. Tech Rept. No. 25, Harvard Univ. Cambridge, MA (1978).

21. T. Tanaka, "A Model of Multicompartment Fire Spread," NBSIR 83-2718, Nat. Bur. Stand., Aug. 1983.

22. R.D. Peacock, W.W. Jones, R.W. Bukowski, and C.L. Forney, "Technical Reference Guide for the HAZARD I Fire Hazard Assessment Method," Ver. 1.1, NIST Hdbk 146, Nat. Inst. of Stand. and Tech., June 1991.

ESTIMATING TEMPERATURES IN COMPARTMENT FIRES

William D. Walton and Philip H. Thomas

INTRODUCTION

The ability to predict temperatures developed in compartment fires is of great significance to the fire protection professional. There are many uses for a knowledge of compartment fire temperatures, including the prediction of (1) the onset of hazardous conditions, (2) property and structural damage, (3) changes in burning rate, (4) ignition of objects, and (5) the onset of flashover.

The fundamental principles underlying compartment fires are presented in Section 3, Chapter 5. This chapter gives a number of simplified solution techniques.

FIRE STAGES

In this chapter, compartment fires are defined as fires in enclosed spaces, which are commonly thought of as rooms in buildings, but may include other spaces such as those found in transportation vehicles such as ships, planes, trains, and the like.

Compartment fires are often discussed in terms of growth stages.[1] Figure 3-6.1 shows an idealized variation of temperature with time along with the growth stages. The growth stages are:

1. Ignition,
2. Growth,
3. Flashover,
4. Fully developed fire, and
5. Decay.

While many fires will not follow this idealization, it provides a useful framework for the discussion of compartment fires. All fires include an ignition stage but, beyond that, may fail to grow, or they may be affected by manual or automatic suppression activities before going through all of the stages listed above.

Growth Stage Definitions

Ignition stage: The period during which the fire begins.

Growth stage: Following ignition, the fire initially grows primarily as a function of the fuel itself, with little or no influence from the compartment. The fire can be described in terms of its rate of energy and combustion product generation. A discussion of energy generation or burning rate can be found in Section 3, Chapter 1. If sufficient fuel and oxygen are available, the fire will continue to grow, causing the temperature in the compartment to rise. Fires with sufficient oxygen for combustion are said to be fuel controlled.

Flashover: Flashover is generally defined as the transition from a growing fire to a fully developed fire in which all combustible items in the compartment are involved in fire. During this transition there are rapid changes in the compartment environment. Flashover is not a precise term, and several variations in definition can be found in the literature. Most have criteria based on the temperature at which the radiation from the hot gases in the compartment will ignite all of the combustible contents. Gas temperatures of 300 to 650°C have been associated with the onset of flashover, although temperatures of 500 to 600°C are more widely used.[2] The ignition of unburnt fuel in the hot fire gases, the appearance of flames from openings in a compartment, or the ignition of all of the combustible contents may actually be different phenomena.

Fully developed fire: During this stage, the heat release rate of the fire is the greatest. Frequently during this stage more fuel is pyrolized than can be burned with the oxygen available in the compartment. In this case, the fire is said to be ventilation controlled. If there are openings in the compartment, the unburned fuel will leave the compartment in the gas flow and may burn outside of the compartment. During the fully developed stage, the environment within the compartment has a significant effect on the pyrolysis rate of the burning objects.

Decay stage: Decay occurs as the fuel becomes consumed, and the heat release rate declines. The fire may change from ventilation to fuel controlled during this period.

Mr. Walton is a research fire protection engineer with the Building and Fire Research Laboratory, National Institute of Standards and Technology. Dr. Thomas is a visiting professor at Lund University in Sweden and is retired from the Fire Research Station, Borehamwood, England.

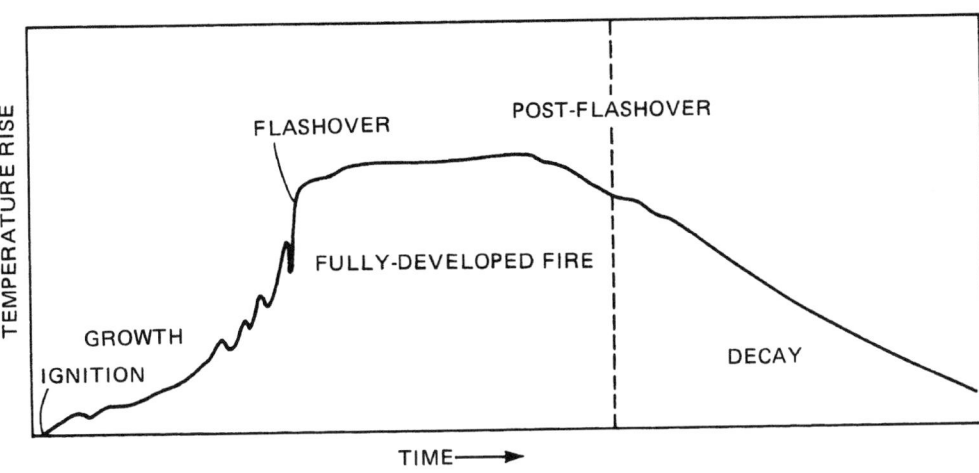

Fig. 3-6.1. General description of room fire in absence of fire control.

COMPARTMENT FIRE PHENOMENA
Compartment Fire Model

In order to calculate or predict the temperatures generated in a compartment fire, a description or model of the fire phenomena must be created. This model will be described in terms of physical equations which can be solved to predict the temperature in the compartment. Such a model is, therefore, an idealization of the compartment fire phenomena. Consider a fire which starts at some point below the ceiling and releases energy and products of combustion. The rate at which energy and products of combustion are released may change with time. The hot products of combustion form a plume, which, due to buoyancy, rises toward the ceiling. As the plume rises, it draws in cool air from within the compartment, decreasing the plume's temperature and increasing its volume flow rate. When the plume reaches the ceiling, it spreads out and forms a hot gas layer which descends with time as the plume's gases continue to flow into it. There is a relatively sharp interface between the hot upper layer and the air in the lower part of the compartment. The only interchange between the air in the lower part of the room and the hot upper layer assumed is through the plume. As the hot layer descends and reaches openings in the compartment walls (e.g., doors and windows), hot gas will flow out the openings and outside air will flow into the openings. This description of compartment fire phenomena is referred to as a two-layer or zone model. The basic compartment fire phenomena are shown schematically in Figure 3-6.2.

The two-layer model concept assumes that the compositions of the layers are uniform. That is, the temperature and other properties are the same throughout each layer. Although the temperature of the lower layer will rise during the course of the fire, the temperature of the upper layer will remain greater and is of the most importance in compartment fires. The assumptions may be less valid for very large spaces or for long, narrow spaces such as corridors and shafts.

Calculation of Compartment Fire Temperatures

The basic principle used to calculate the temperature in a compartment fire is the conservation of energy. As applied to the hot upper layer, the conservation of energy can be simply stated as: the energy added to the hot upper layer by the fire equals the energy lost from the hot layer plus the time rate of change of energy within the hot upper layer. From the time rate of change of energy within the hot layer, the temperature of the layer can be computed. Conservation of energy can also be applied to the lower layer, as well. Since the volume of the upper layer changes with time, and mass flows in and out of the upper layer, conservation of mass must be used along with the conservation of energy. Because the energy generated by the fire and the temperatures in the

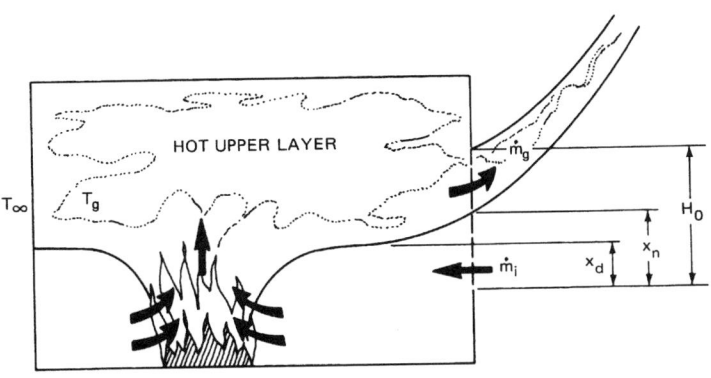

Fig. 3-6.2. Two-layer model with no exchange between layers except the plume.

compartment vary as a function of time, the application of conservation of energy will result in a series of differential equations. For the purposes of examining the components of the conservation of energy, the steady-state expressions for the conservation of energy for the hot upper layer will be used.

The transport of energy in a compartment fire is a very complex process. In order to formulate expressions for the conservation of energy in a practical way, a number of assumptions must be made. It is possible to formulate the equations for the conservation of energy in a number of ways, based on the level of detail desired. The expressions and assumptions used here are based on those commonly found in the fire research literature and represent a somewhat simplified description of the phenomena. Additional details may be found in the references cited.

The steady-state conservation of energy for the hot upper gas layer in a compartment can be simply stated as: the energy generated by the fire and added to the hot layer equals the energy lost from the hot layer through radiation and convection plus the energy convected out the compartment openings.

Energy Generated by the Fire

The energy generated by the fire is the primary influence on the temperature in a compartment fire, and much research has been conducted in predicting the energy release rate of many fuels under a variety of conditions. This discussion will focus on flaming combustion, as it is most important in generating a significant temperature rise in a compartment. A discussion of smoldering combustion is found in Section 2, Chapter 11. As a fuel is heated and releases pyrolysis products, these products react with oxygen, generating heat and producing flames. The rate of energy release is equal to the mass loss rate of the fuel times the heat of combustion of the fuel.

$$\dot{Q} = \dot{m}_f \, \Delta h_c \qquad (1)$$

where

$\dot{Q}$ = energy release rate of the fire (kW)
$\dot{m}_f$ = mass burning rate of the fuel (kg/s)
Δh_c = effective heat of combustion of the fuel (kJ/kg)

The effective heat of combustion is the heat of combustion which would be expected in a fire where incomplete combustion takes place. This is less than the theoretical heat of combustion as measured in the oxygen bomb calorimeter. The effective heat of combustion is often described as a fraction of the theoretical heat of combustion. The effect of fluctuations is largely neglected.

In fuel-controlled fires, there is sufficient air to react with all the fuel within the compartment. In ventilation-controlled fires, there is insufficient air within the compartment, and some of the pyrolysis products will leave the compartment, possibly to react outside the compartment. For calculating the temperatures produced in compartment fires, the primary interest is in the energy released within the compartment.

The pyrolysis rate of the fuel depends on the fuel type, its geometry, and on the fire-induced environment. The energy generated in the compartment by the burning pyrolysis products then depends on the conditions (temperature, oxygen concentration, etc.) within the compartment. While the processes involved are complex, and some are not well understood, there are two cases where some simplifying assumptions can lead to useful methods for approximation of the energy released by the fire.

Free-burning fires are defined as those in which the pyrolysis rate and the energy release rate are affected only by the burning of the fuel itself and not by the room environment. This is analogous to a fire burning out of doors on a calm day. Babrauskas has provided a collection of data on free-burning fires in Section 3, Chapter 1. This data is most useful for estimating burning rates of primarily horizontal fuels in pre-flashover fires, where the primary heating of the fuel is from the flames of the burning item itself. Vertical fuels, such as wall linings and fuels located in the upper hot gas layer, will likely be influenced by the pre-flashover room environment.

Ventilation-controlled fires are defined as those in which the energy release rate in the room is limited by the amount of available oxygen. The mass flow rate of air or oxygen into the room through a door or window can be calculated from the expressions described below and in Section 2, Chapter 5. For most fuels, the heat released per mass of air consumed is a constant approximately equal to 3000 KJ/kg.[3] Therefore, the rate of energy release of the fire can be approximated from the air inflow rate.

The amount of energy released by the fire which enters the hot upper layer is a function of the fire, layer conditions, and geometry. For most fires, approximately 35 percent of the energy released by the fire leaves the fire plume as radiation.[4] (A discussion of flame radiation can be found in Section 2, Chapter 14.) In a compartment fire, a fraction of the radiated energy reaches the upper layer. The majority of the remaining energy released by the fire is convected into the upper layer by the plume. As the plume rises, it entrains air from the lower layer, thus reducing its temperature and increasing the mass flow rate. For a first approximation, it can be assumed that all of the energy generated by the fire is transported to the upper layer. For a complete discussion of fire plumes see Section 2, Chapter 2.

Conservation of Mass

The mass flow into the compartment and the flow out are related by

$$\dot{m}_g = \dot{m}_a + \dot{m}_f \qquad (2)$$

where

$\dot{m}_f$ = mass burning rate of the fuel (kg/s)

The mass flow rate of hot gas out of a window or door is given by Rockett[5]

$$\dot{m}_g = \frac{2}{3} C_d W_0 \rho_\infty \left[2g \frac{T_\infty}{T_g} \left(1 - \frac{T_\infty}{T_g} \right) \right]^{1/2} (H_0 - X_N)^{3/2} \qquad (3)$$

where

$\dot{m}_g$ = mass flow rate of hot gas out an opening (kg/s)
C_d = orifice constriction coefficient (typically ≈ 0.7)
W_0 = opening width (m)
H_0 = opening height (m)
ρ_∞ = ambient air density (kg/m^2)
g = acceleration due to gravity, 9.8 m/s^2
X_N = height of neutral plane (m)
T_g = temperature of the hot upper gas layer (K)
T_∞ = ambient temperature (K)

The mass flow rate of air into a door or window is given by

$$\dot{m}_a = \frac{2}{3} C_d W_0 \rho_\infty \left[2g\left(1 - \frac{T_\infty}{T_g}\right) \right]^{1/2} (X_N - X_d)^{1/2} (X_N + X_d/2) \quad (4)$$

where

X_d = height of the interface (m)

The expressions for mass flow in and mass flow out cannot be solved directly for T_g, since the height to the neutral plane and interface are unknown. The complete solution of these equations requires expressions for plume entrainment and additional energy equations and is normally carried out only in computer fire models. If the mass burning rate of the fuel is small compared with the mass flow rate of air into the compartment, the mass flow out of the opening may be approximated as equal to the mass inflow rate. Flows out of vents in the ceiling are discussed in Section 3, Chapter 11.

For pre-flashover fires in compartments with typical doors or windows, the neutral plane and interface can be approximated at the midlevel of the opening. This approximation can only be made after the initial smoke filling of the compartment is complete, and flow in and out of the opening is established.

For fires nearing flashover and post-flashover fires the interface between the upper and lower layers is located near the floor, and the flow reaches a maximum for a given upper gas temperature. Rockett has shown the temperature dependence on the flow becomes small above 150°C, and the flow into the compartment can be approximated as a constant times[5]

$$A_0\sqrt{H_0}$$

Rockett calculated values for this constant of 0.40 to 0.61 kg/s·m$^{5/2}$, depending on the discharge coefficient of the opening. Thomas and Heselden estimate the value of this constant at 0.5 kg/s · m$^{5/2}$, which is the value most commonly found in the literature.[6] The resulting approximation is then

$$\dot{m}_a = 0.5 A_0\sqrt{H_0} \quad (5)$$

where

A_0 = area of opening (m^2)
H_0 = height of opening (m)

The term

$$A_0\sqrt{H_0}$$

is commonly known as the ventilation factor. The first use of this type of opening flow analysis for evaluating post flashover fire test data is attributed to Kawagoe.[7] From early work analyzing such data, the empirical observation was made that wood fires in rooms with small windows appeared to burn at a rate approximately stoichiometric. Although flames emerging from the windows implied that some fuel was burning outside, calculations often suggested that enough air was entering the fire for stoichiometric burning. Empirical observations on wood fires[7] led to

$$\dot{m}_f = 0.09 A_0\sqrt{H_0} \quad (6)$$

There is now a body of data[8] that modifies this simple proportionality between $\dot{m}_f$ and

$$A_0\sqrt{H_0}$$

The Conseil International du Bâtiment (CIB) experiments upon which Law[9] has based her method shows a dependance on A_T. It seems possible that the wide use of Equation 6 is a result of a concentration of experimental fires in rooms of a limited range of

$$A_T/A_0\sqrt{H_0}$$

where

A_T = total area of the compartment enclosing surfaces (m^2)

Traditionally, energy balances were often stated in terms of the energy produced by the burning fuel and, thereby, led to an effective heat of combustion of the fuel. However, this in principle leads to the same result, the energy produced is related to the air flow for ventilation-controlled fires. Kawagoe[7] and Magnusson and Thelandersson[10] used 10.75 MJ/kg for the effective heat of combustion of wood in the flaming phase for fully developed compartment fires. With 16.4 MJ/kg for the heat of combustion of wood volatiles, this corresponds to a combustion efficiency of 10.75/16.4, which is virtually identical to the 0.65 used in several computer models.

By far the majority of data are based on experiments in which the fuel was cellulosic, and much of the experimental data are based on wood in the form of cribs. For the post flashover burning of a different fuel with a different chemistry, the burning rate expressions may still be used, as long as the fuel is a hydrocarbon producing approximately 3000 kJ for each kg of air consumed in the combustion process. Because different fuels react differently to the thermal environment and will pyrolyze at different rates according to the energy requirements to produce volatiles, one can only estimate temperatures by evaluating the differences, or obtain maximum temperatures by using stoichiometry. Fuels more volatile than wood will probably produce lower temperatures inside a compartment, even if the excess fuel produces a greater hazard outside the compartment. The assumption that the energy is related to the air flow and that the fuel is in stoichiometric proportion will give an upper estimate of temperatures for ventilation-controlled fires. Since Equation 6 is close to stoichiometric, it could, coupled with the effective heat of combustion of wood, give results close to an upper temperature limit for other fuels.

Conservation of Energy

The heat generated by burning materials within a compartment is absorbed by the enclosing surfaces of the compartment and any other structural surfaces, and by the surfaces of the fuel and by the incoming air and any excess fuel. Heat is lost to the exterior in the flames and hot gases that exit from the openings in the compartment enclosing surfaces and by radiation through the openings. An example of an experimental heat balance measured in a small compartment is given in Table 3-6.1. For this compartment, unglazed windows provided ventilation from the start of the fire.

Table 3-6.1 illustrates the significant amount of heat loss in the effluent gases and shows that, with decreasing

TABLE 3-6.1 *Heat Balance Measured in Experimental Fires in a Compartment of 29 m² Floor Area with a Fire Load of Wood Cribs*

Fire load (kg)	Window area (m²)	Heat release (kcal/s)	Heat loss from hot gases (%)			
			Effluent gas	Structural surfaces	Feedback to fuel	Window radiation
877	11.2	1900	65	15	11	9
	5.6	1900	52	26	11	
1744	11.2	3200	61	15	11	11
	5.6	2300	53	26	11	13
	2.6	1600	47	30	12	9
					16	7

window area, a larger proportion of the heat released will be absorbed by the enclosing surfaces. The total heat released, assuming a complete burnout, is directly proportional to the amount of the fire load, but the rate of heat release may also be controlled by the ventilation. In this example, with the lower fire load, both window areas give sufficient ventilation for the fuel to burn at its maximum (free burning) rate but, with the doubled fire load, the burning rate is not doubled, because the window area restricts the ventilation needed.

METHODS FOR PREDICTING PRE-FLASHOVER COMPARTMENT FIRE TEMPERATURES

The solution of a relatively complete set of equations for the conservation of energy requires the solution of a large number of equations which vary with time. Although individual energy transport equations may be solved, in general there is not an explicit solution for a set of these equations. As a result, one of two approaches can be taken. The first is an approximate solution which can be accomplished by "hand" using a limiting set of assumptions. The second is a more complete solution utilizing a computer program. In either case, a number of methods have been developed. The methods presented are those which appear most widely accepted in the fire protection community. Each method employs assumptions and limitations which should be understood before employing the method. The methods presented in this chapter predict average temperatures and are not applicable to cases where prediction of local temperatures are desired. For example, these methods should not be used to predict detector or sprinkler actuation or the temperatures of materials as a result of direct flame impingement.

Method of McCaffrey, Quintiere, and Harkleroad

McCaffrey, Quintiere, and Harkleroad have used a simple conservation of energy expression and a correlation with data to develop an approximation of the upper layer temperature in a compartment.[11] Applying the conservation of energy to the upper layer yields

$$\dot{Q} = \dot{m}_g c_p (T_g - T_\infty) + q_{loss} \quad (7)$$

where

$\dot{Q}$ = energy (heat) release rate of the fire (kW)
$\dot{m}_g$ = gas flow rate out the opening (kg/s)
c_p = specific heat of gas (kJ/kg · K)
T_g = temperature of the upper gas layer (K)

T_∞ = ambient temperature (K)
q_{loss} = net radiative and convective heat transfer from the upper gas layer (kW)

The left-hand side of Equation 7 is the energy generated by the fire. On the right-hand side, the first term is the heat transported from the upper layer in the gas flow out an opening. The second term is the net rate of radiative and convective heat transfer from the upper layer, which is approximately equal to rate of heat conduction into the compartment surfaces. The rate of heat transfer to the surfaces is approximated by

$$q_{loss} = h_k A_T (T_g - T_\infty) \quad (8)$$

where

h_k = effective heat transfer coefficient (kW/m · K)
A_T = total area of the compartment enclosing surfaces (m²)

Substituting Equation 8 into Equation 7 yields the nondimensional temperature rise in terms of two dimensionless groups

$$\frac{\Delta T_g}{T_\infty} = \frac{\dot{Q}/(c_p T_\infty \dot{m}_g)}{1 + h_k A_T/(c_p \dot{m}_g)} \quad (9)$$

where

ΔT_g = upper gas temperature rise above ambient ($T_g - T_\infty$) (K).

The mass flow rate of hot gas out of a window or door can be rewritten from Equation 3.

$$\dot{m}_g = \frac{2}{3} C_d W_0 H_0^{3/2} \rho_\infty \left[2g \frac{T_\infty}{T_g} \left(1 - \frac{T_\infty}{T_g}\right) \right]^{1/2} \left(1 - \frac{X_N}{H_0}\right)^{3/2} \quad (10)$$

where

C_d = orifice constriction coefficient
W_0 = opening width (m)
H_0 = opening height (m)
ρ_∞ = ambient air density (kg/m²)
g = acceleration due to gravity, 9.8 m/s²
X_N = height of neutral plane (m)

Since X_N primarily depends on T_g, $\dot{Q}$, and geometric factors (H_0 and W_0), $\dot{m}_g$ may be replaced by

$$\sqrt{g} \, \rho_\infty A_0 \sqrt{H_0}$$

in the two dimensionless variables in Equation 10, without any loss in generality. The effects of T_g and $\dot{Q}$ are incorporated into the correlation via other terms. Based on an analysis of test data, Equation 9 was written as a power-law relationship

$$\Delta T_g = 480 \left(\frac{\dot{Q}}{\sqrt{g} c_p \rho_\infty T_\infty A_0 \sqrt{H_0}} \right)^{2/3} \left(\frac{h_k A_T}{\sqrt{g} c_p \rho_\infty A_0 \sqrt{H_0}} \right)^{-1/3} \quad (11)$$

where

A_0 = area of opening (m^2)
H_0 = height of opening (m)

The numbers 480, $\frac{2}{3}$, and $-\frac{1}{3}$ were determined by correlating the expression with the data from over 100 experimental fires. These data included both steady-state and transient fires in cellulosic and synthetic polymeric materials and gaseous hydrocarbon fuels. Compartment height ranged from 0.3 m to 2.7 m and floor areas from 0.14 m^2 to 12.0 m^2. The compartments contained a variety of window and door sizes. The term raised to the $\frac{2}{3}$ power in Equation 11 represents the ratio of the energy released to the energy convected, and the term raised to the $-\frac{1}{3}$ power represents the energy lost divided by the energy convected.

Substituting the values for ambient conditions of

g = 9.8 m/s^2
c_p = 1.05 kJ/kg·K
ρ_∞ = 1.2 kg/m^3
T_∞ = 295 K

into Equation 11 yields[12,13]

$$\Delta T_g = 6.85 \left(\frac{\dot{Q}^2}{A_0 \sqrt{H_0} h_k A_T} \right)^{1/3} \quad (12)$$

The heat transfer coefficient can be determined using a steady-state approximation when the time of exposure, t, is greater than the thermal penetration time, t_p, by

$$h_k = k/\delta \quad \text{for } t > t_p \quad (13)$$

The thermal penetration time is defined as

$$t_p = (\rho c/k)(\delta/2)^2 \quad (14)$$

where

ρ = density of the compartment surface (kg/m^3)
c = specific heat of the compartment surface material (kJ/kg·K)
k = thermal conductivity of compartment surface (kW/m·K)
δ = thickness of compartment surface (m)
t = exposure time (s)
t_p = thermal penetration time (s)

When the time of exposure is less than the penetration time, an approximation based on conduction in a semi-infinite solid is

$$h_k = (k\rho c/t)^{1/2} \quad \text{for } t \leq t_p \quad (15)$$

If there are several wall and/or ceiling materials in the compartment, an area-weighted average for h_k should be used.

The limitations as stated by McCaffrey et al on the use of this method for estimating temperatures are:

1. The correlation holds for compartment upper layer gas temperatures up to approximately 600°C,
2. It applies to steady-state as well as time-dependent fires, provided the primary transient response is the wall conduction phenomenon,
3. It is not applicable to rapidly developing fires in large enclosures in which significant fire growth has occurred before the combustion products have exited the compartment,
4. The energy release rate of the fire must be determined from data or other correlations,
5. The characteristic fire growth time and thermal penetration time of the room-lining materials must be determined in order to evaluate the effective heat transfer coefficient, and
6. The correlation is based on data from a limited number of experiments and does not contain extensive data on ventilation-controlled fires nor data on combustible walls or ceilings. Most of the fuel in the test fires was near the center of the room.

Example of McCaffrey et al method: Calculate the upper layer temperature of a room 3 × 3 m in floor area and 2.4 m high with a door opening 1.8 m high and 0.6 m wide. The fire source is a steady 750 kW fire. The wall lining material is 0.016 m ($\frac{5}{8}$ in.) gypsum plaster on metal lath. Perform the calculation at times of 10, 60, and 600 seconds after ignition. Using Equation 11

$$\Delta T_g = 480 \left(\frac{\dot{Q}}{\sqrt{g} c_p \rho_\infty T_\infty A_0 \sqrt{H_0}} \right)^{2/3} \left(\frac{h_k A_T}{\sqrt{g} c_p \rho_\infty A_0 \sqrt{H_0}} \right)^{-1/3}$$

where

c_p = 1 kJ/kg·K
T_∞ = 27°C (300 K)
ρ_∞ = 1.18 kg/m^3
A_0 = 1.8 m × 0.6 m = 1.08 m^2
g = 9.8 m/s^2
H_0 = 1.8 m
$\dot{Q}$ = 750 kW
$A_T = A_{walls} + A_{floor} + A_{ceiling} - A_{openings}$
$\quad = 4 \times (3 \times 2.4) + (3 \times 3) + (3 \times 3) - 1.08$
$\quad = 28.8 \text{ m}^2 + 9 \text{ m}^2 + 9 \text{ m}^2 - 1.08$
$\quad = 45.72 \text{ m}^2$

The wall heat loss coefficient, h_k, is a function of time.

a. Calculate the thermal penetration time, t_p.

$$t_p = (\rho c/k)(\delta/2)^2$$

where

ρ = wall material density (1440 kg/m^3)
k = 0.48 × 10^{-3} kW/m·c
c = 0.84 kJ/kg°C
δ = 0.016 m
t_p = 161.3 s

b. Calculate h_k at 10, 60, and 600 s.

For $t < t_p$ (10, 60 s)

$$h_k = (k\rho c/t)^{1/2} \quad k\rho c = 0.581$$

1. At t = 10 s.

$$h_k = (0.581/10)^{1/2} = 0.24 \text{ kW/m·K}$$

2. At $t = 60$ s.

$$h_k = (0.581/60)^{1/2} = 0.098 \text{ kW/m} \cdot \text{K}$$

3. For $t > t_p$ (600 s) at $t = 600$ s.

$$h_k = k/\delta = 0.48 \times 10^{-3}/0.016 = 0.03 \text{ kW/m} \cdot \text{K}$$

c. Calculate the compartment temperature at the three times using Equation 11.

1. At $t = 10$ s.

$$\Delta T_g = 480 \left[\frac{750}{(\sqrt{9.8})(1)(300)(1.08)(\sqrt{1.8})} \right]^{2/3}$$
$$\cdot \left[\frac{(0.24)(45.72)}{(\sqrt{9.8})(1)(1.18)(1.08)(\sqrt{1.8})} \right]^{-1/3}$$
$$= 480(0.55)^{2/3}(2.05)^{-1/3}$$
$$\Delta T_g = 254 \text{ K}$$

2. At $t = 60$ s.

$$\Delta T_g = (480)(0.55)^{2/3}(0.837)^{-1/3}$$
$$\Delta T_g = 342 \text{ K}$$

3. At $t = 600$ s.

$$\Delta T_g = (480)(0.55)^{2/3}(0.26)^{-1/3}$$
$$\Delta T_g = 506 \text{ K}$$

Method of Foote, Pagni and Alvares

This method follows the basic correlations of McCaffrey, Quintiere, and Harkleroad and adds data for forced-ventilation fires. Using Equation 9 and not introducing an expression for doorway flow results in the expression[14]

$$\frac{\Delta T_g}{T_\infty} = 0.63 \left(\frac{\dot{Q}}{\dot{m}c_p T_\infty} \right)^{0.72} \left(\frac{h_k A_T}{\dot{m}c_p} \right)^{-0.36} \quad (16)$$

where

ΔT_g = upper gas temperature rise above ambient (K)
T_∞ = ambient air temperature (K)
$\dot{Q}$ = energy (heat) release rate of the fire (kW)
$\dot{m}$ = compartment mass ventilation rate (kg/s)
c_p = specific heat of gas (kJ/kg · K)
h_k = effective heat transfer coefficient (kW/m · K)
A_T = total area of the compartment-enclosing surfaces (m²)

The coefficient and exponents are based on data from well-ventilated tests in a compartment with a 6 × 4 m floor area and a height of 4.5 m with ventilation rates of 110 to 325 g/s. The compartment exhaust was through a 0.65 × 0.65 m duct located 3.6 m above the floor. Four air inlet openings were 0.5 × 0.12 m high, with centerlines 0.1 m above the floor. A methane gas burner fire in the center of the floor with heat release rates of 150 to 490 kW resulted in upper gas temperatures of approximately 100 to 300°C.

Foote et al have shown that the correlation for forced-ventilation fires agrees well with the data presented by McCaffrey et al for free ventilation fires with

$$\dot{m} \approx 0.1(\rho_\infty \sqrt{g} A_0 \sqrt{H_0})$$

Example of Foote et al method: Estimate the temperature in a 5 × 5 m in floor area × 4 m high compartment having 0.025-m-(1-in.)-thick concrete walls. The forced-ventilation rate is 2.4 m³/s of air (5000 cfm). Perform the calculation for $t > t_p$. The fire size is given as 1000 kW; ambient air conditions at 300 K. Using Equation 16

$$\frac{\Delta T_g}{T_\infty} = 0.63(\dot{Q}/\dot{m}c_p T_\infty)^{0.72}(h_k A_T/\dot{m}c_p)^{-0.36}$$

where

$\dot{Q}$ = 1000 kW
T_∞ = 300 K
c_p = 1.0 kJ/kg · K
A_T = 4 × (5 × 4) + 2 (5 × 5) = 105 m²
$\dot{m}$ = (2.4 m³/s) (1.18 kg/m³) = 2.8 kg/s

Calculate h_k for $t > t_p$. For 0.025-m-thick concrete

δ = 0.025 m
ρ = 2000 kg/m³
k = 1.4 × 10⁻³ kW/m · K
c_p = 0.88 kJ/kg · K

$$t_p = (\rho c/k)(\delta/2)^2$$
$$= \left[\frac{(2000) \cdot (0.88)}{1.4 \times 10^{-3}} \right] \left(\frac{0.025}{2} \right)^2$$
$$= 196 \text{ s} \quad \text{for } t > t_p$$

$$h_k = k/\delta = 1.4 \times 10^{-3}/0.025 = 0.056 \text{ kW/m}^2 \cdot \text{K}$$

$$\frac{\Delta T_g}{T_\infty} = (0.63) \left[\frac{1000}{(2.8)(1)(300)} \right]^{0.72} \left[\frac{(0.056)(105)}{(2.8)(1)} \right]^{-0.36}$$

$$\Delta T_g = (0.14)(T_\infty)$$
$$\Delta T_g = 164 \text{ K}$$
$$T_g = 164 + 300 \text{ K} = 464 \text{ K}$$

METHODS FOR PREDICTING POST FLASHOVER COMPARTMENT FIRE TEMPERATURES

Method of Babrauskas

The following method is based on the work of Babrauskas.[15,16] The upper gas temperature, T_g, is expressed according to a series of factors, each one accounting for a different physical phenomenon

$$T_g = T_\infty + (T^* - T_\infty) \cdot \theta_1 \cdot \theta_2 \cdot \theta_3 \cdot \theta_4 \cdot \theta_5 \quad (17)$$

where T^* is an empirical constant = 1725 K, and the factors θ are in Equations 23, 28, 30, 31, 33, and 34.

Burning rate stoichiometry, θ_1: The dimensionless stoichiometric coefficient ϕ is defined as

$$\phi = \frac{\dot{m}_f}{\dot{m}_{f,st}} \quad (18)$$

where $\dot{m}$ is the fuel mass pyrolysis rate (kg/s) and $\dot{m}_{f,st}$ is the stoichiometric mass burning rate (i.e., no excess fuel and no excess oxygen).

$$\dot{m}_{f,st} = \frac{0.5 A_0 \sqrt{H_0}}{r} \quad (19)$$

where the ratio r is such that 1 kg fuel $+$ r kg air $\rightarrow$ $(1 + r)$ kg products. The value of r is readily computable for fuels containing carbon, hydrogen, and/or oxygen from the chemical formula of the fuel, taking the products to be CO_2, H_2O, and N_2.

$$C_xH_yO_z + wO_2 + w\left(\frac{79}{21}\right)N_2 \rightarrow$$
$$xCO_2 + \frac{y}{2}H_2O + w\left(\frac{79}{21}\right)N_2 \qquad (20)$$

where

$$w = \frac{2x + \frac{y}{2} - z}{2} \qquad (21)$$

and

$$r = \frac{[w + w(3.76)]28.97}{12.01x + 1.00y + 16.00z} \qquad (22)$$

At stoichiometry $\phi = 1$, and it is greater than 1 for fuel-rich burning and less than 1 for fuel-lean conditions.

The effect of ϕ on gas temperatures was evaluated by numerical computations using the COMPF2 computer program.[17] The efficiency factor, θ_1, accounts for deviation from stoichiometry and is shown in Figure 3-6.3. It is seen that the fuel-lean and the fuel-rich regimes exhibit a very different dependence. For the fuel-lean regime, the results can be approximated by

$$\theta_1 = 1.0 + 0.51 \ln \phi \qquad \text{for } \phi < 1 \qquad (23)$$

Similarly, in the fuel-rich regime a suitable approximation is

$$\theta_1 = 1.0 - 0.05(\ln \phi)^{5/3} \qquad \text{for } \phi > 1 \qquad (24)$$

If heat release rate, $\dot{Q}$, rather than mass loss rate, $\dot{m}$, is used, then

$$\phi = \frac{\dot{Q}}{\dot{Q}_{st}} \qquad (25)$$

And, since the stoichiometric heat release rate is

$$\dot{Q} = 1500 \, A_0\sqrt{H_0} \qquad (26)$$

then

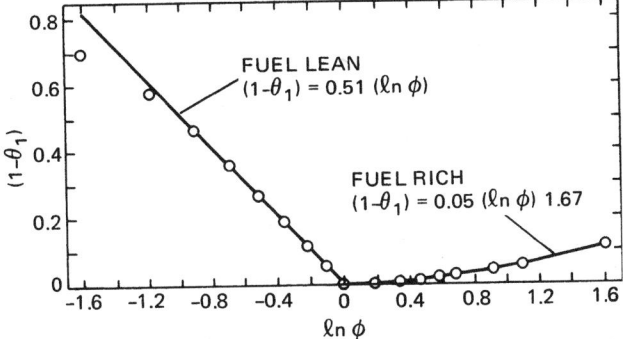

Fig. 3-6.3. Effect of equivalence ratio.

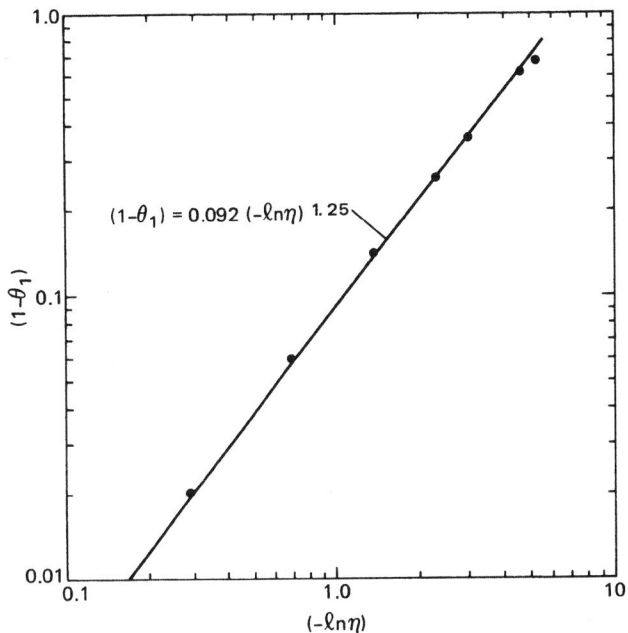

Fig. 3-6.4. Effect of pool diameter.

$$\phi = \frac{\dot{Q}}{1500A_0\sqrt{H_0}} \qquad (27)$$

The value of $\dot{Q}$ can be determined from Section 3, Chapter 1.

A separate procedure is necessary for pool fires, due to the strong radiative coupling. Here

$$\theta_1 = 1.0 - 0.092(-\ln \eta)^{1.25} \qquad (28)$$

where

$$\eta = \left(\frac{A_0\sqrt{H_0}}{A_f}\right)\frac{0.5\Delta h_p}{r\sigma(T_g^4 - T_b^4)} \qquad (29)$$

where

Δh_p = heat of vaporization of liquid (kJ/kg)

A_f = pool area (m^2)

σ = Stefan-Boltzmann constant $(5.67 \times 10^{-11} \text{ kW/m}^2 \cdot \text{K}^4)$

T_b = liquid boiling point (K)

This expression unfortunately requires an estimate for T_g to be made, so for the pool fire case, a certain amount of iteration is necessary. The relationship above is plotted in Figure 3-6.4.

Wall steady-state losses, θ_2: The next efficiency factor, θ_2, accounts for variable groups of importance involving the wall surface (which is defined to include the ceiling) properties: area A_T (m^2), thickness L (m), density ρ (kg/m^3), thermal conductivity k (kW/m $\cdot$ K), and heat capacity c_p (kJ/kg $\cdot$ K). This factor is given as

$$\theta_2 = 1.0 - 0.94 \exp\left[-54\left(\frac{A_0\sqrt{H_0}}{A_T}\right)^{2/3}\left(\frac{L}{k}\right)^{1/3}\right] \qquad (30)$$

and is shown in Figure 3-6.5.

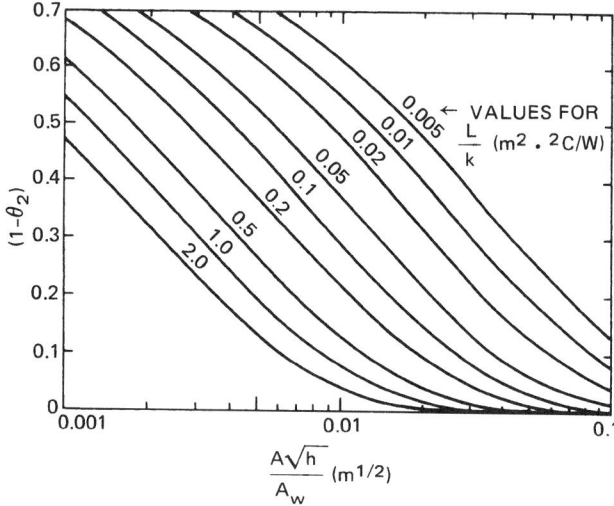

Fig. 3-6.5. Effect of wall steady-state losses.

Wall transient losses, θ_3: For the transient case, the above relationship predicts the asymptotic temperature value. An additional time-dependent factor, however, is needed. See Figure 3-6.6.

$$\theta_3 = 1.0 - 0.92 \exp\left[-150\left(\frac{A_0\sqrt{H_0}}{A_T}\right)^{0.6}\left(\frac{t}{k\rho c_p}\right)^{0.4}\right] \quad (31)$$

If only steady-state temperatures need to be evaluated, then $\theta_3 = 1.0$.

Wall effects for t just slightly greater than zero are not well modeled with the above relationships for $\theta_2 \times \theta_3$; however, this is not a serious limitation, since the method is only designed for post-flashover fires.

For transient fires, the possibility of two separate effects must be considered. First, the wall loss effect, represented by Equation 31, in all fires, exhibits a non-steady character. Second, the fuel release rate may not be constant. Since in the calculational procedure the previous results are not stored, it is appropriate to restrict consideration to fires where $\dot{m}_f$ does not change drastically over the time scale

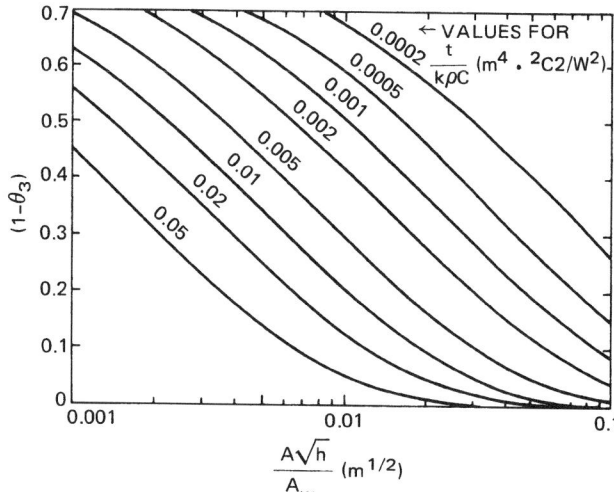

Fig. 3-6.6. Effect of wall transient losses.

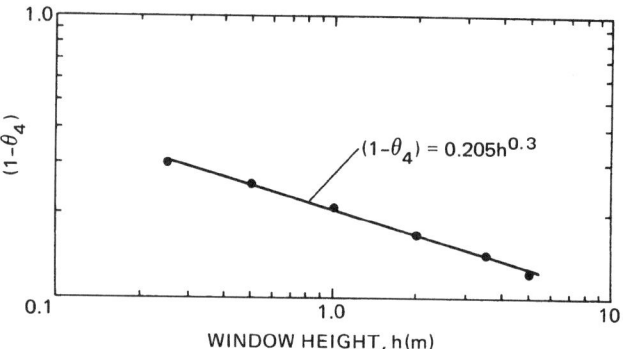

Fig. 3-6.7. Effect of window height.

established by θ_3. This "natural" time scale can be determined as the time when the response has risen to 63 percent of its ultimate value, i.e., at $\theta_3 = 0.63$, and is

$$t = 2.92 \times 10^{-6}(k\rho c_p)\left(\frac{A_T}{A_0\sqrt{H_0}}\right)^{1.5} \quad (32)$$

Opening height effect, θ_4: The normalization of burning rate and wall loss quantities with the ventilation factor

$$A_0\sqrt{H_0}$$

does not completely determine the total heat balance. An opening of a given

$$A_0\sqrt{H_0}$$

can be tall and narrow or short and squat. For the shorter opening, the area will have to be larger. Radiation losses are proportional to the opening area and will, therefore, be higher for the shorter opening. By slight simplification, a representation for θ_4 can be made as

$$\theta_4 = 1.0 - 0.205 H_0^{-0.3} \quad (33)$$

as shown in Figure 3-6.7.

Combustion efficiency, θ_5: The fire compartment is viewed as a well, but not perfectly, stirred reactor. Thus a certain "unmixedness" is present. A maximum combustion efficiency, b_p, can be used to characterize this. Since the model assumes infinitely fast kinetics, any limitations can also be included here. Data have not been available to characterize b_p in real fires, but agreement with measured fires can generally be obtained with b_p values in the range 0.5 to 0.9. The effect of b_p variation can be described by

$$\theta_5 = 1.0 + 0.5 \ln b_p \quad (34)$$

as shown in Figure 3-6.8.

Method of Law

The area of structural surface to which heat is lost is expressed by $(A_T - A_0)$. For a given fire load, compartments with different values of A_T, A_0, and height H_0 will have a different heat balance, and thus the temperatures in the compartments will differ. This is illustrated in Figure 3-6.9 which shows how temperature varies with

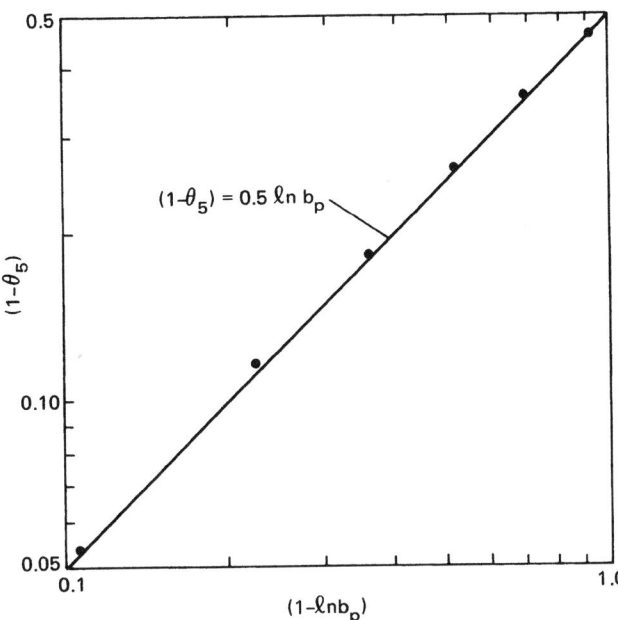

Fig. 3-6.8. Effect of b_p, the maximum combustion efficiency.

$$\Omega = (A_T - A_0)/A_0\sqrt{H_0}$$

For low values of Ω (i.e., high ventilation), the rate of heat release is at a maximum, but the heat loss from the window is also large and the resultant temperature is low. For high values of Ω (i.e., low-ventilation areas), there is little heat loss to the outside, but the rate of heat release is also small and the resultant temperature is, again, low.

The curve in Figure 3-6.10 has been derived from many experimental fires conducted internationally by CIB.[8] For design purposes, Law has defined it as follows

$$T_{g(max)} = 6000\frac{(1 - e^{-0.1\Omega})}{\sqrt{\Omega}} \tag{35}$$

where

$$\Omega = \frac{(A_T - A_0)}{A_0\sqrt{H_0}}$$

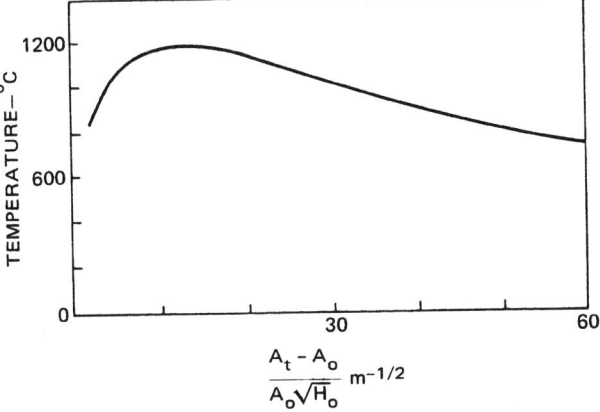

Fig. 3-6.9. Average temperature during fully developed period measured in experimental fires in compartments.

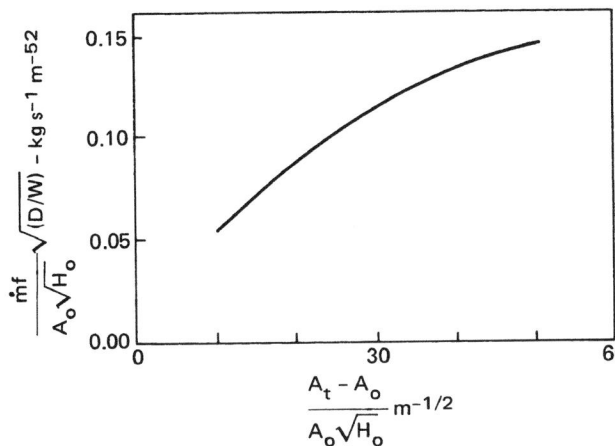

Fig. 3-6.10. Variation of a rate of burning during fully developed period measured in experimental fires in compartments.

and

A_T = total area of the compartment enclosing surfaces (m²)

A_0 = area of opening (m²)

H_0 = height of opening (m)

This represents an upper limit of fire temperature rise for a given Ω. However, if the fire load is low, this value may not be obtained. The importance of the effect of fire load also depends on A_0 and A_T, and can be expressed as

$$T_g = T_{g(max)}(1 - e^{-0.05\psi}) \tag{36}$$

where

$$\psi = \frac{L}{[A_0(A_T - A_0)]^{0.5}}$$

where

L = fire load (wood) (kg)

The effect of the fire on the structure depends not only on the value of T_g but also on the duration of heating. The effective fire duration, τ, in seconds, is given by

$$\tau = \frac{L}{\dot{m}_f} \tag{37}$$

where

$\dot{m}_f$ = rate of burning kg/s

Equation 6 implies that the smaller the value of

$$A_0\sqrt{H_0}$$

the lower the rate of burning and the longer the duration. Assuming a complete burnout, therefore, the effect on the structure tends to be more severe for large values of Ω

$$\text{small } A_0\sqrt{H_0}$$

For design purposes the following equation has been developed to express the correlation of experimental results[9]

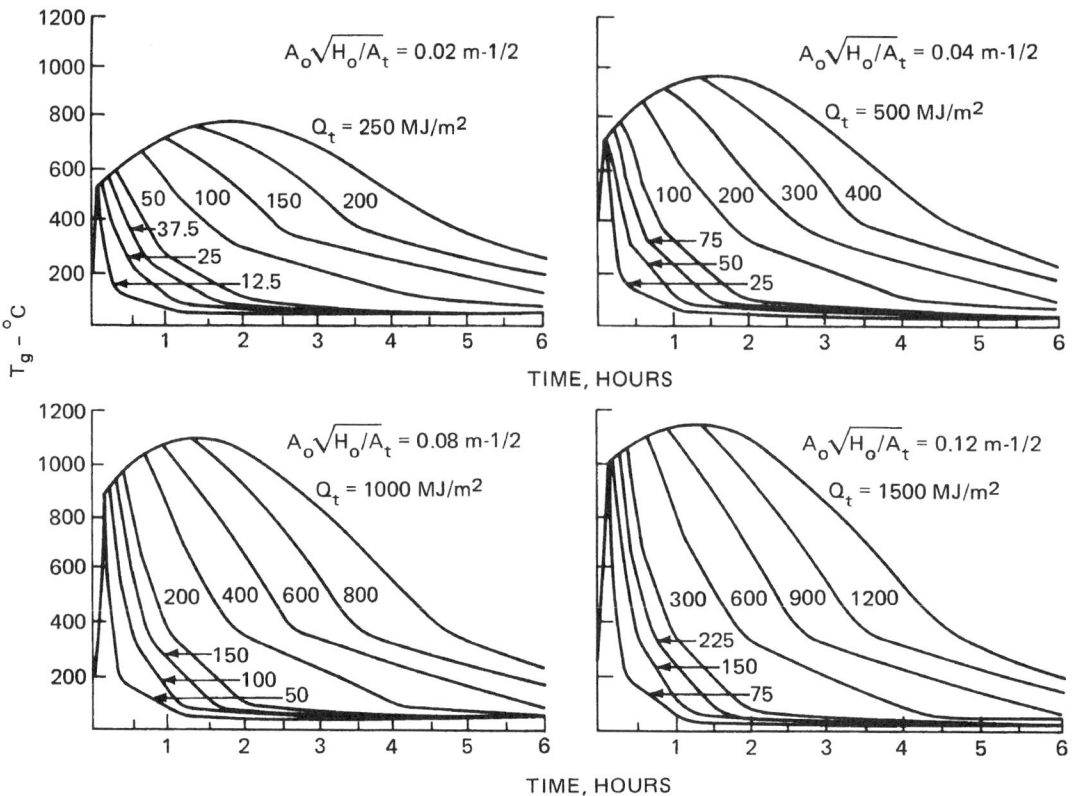

Fig. 3-6.11. *Examples of gas temperature-time curves of post-flashover compartment fires for different values of the fire load density $\dot{Q}_t$ MJ per unit of total internal surface area A_t and the opening factor $A_0 \sqrt{H_0}/A_t$. Fire compartment, type A—from authorized Swedish Standard Specifications.*[10]

$$\dot{m}_f = 0.18 A_0 \sqrt{H_0}(W/D)(1 - e^{-0.036\Omega}) \quad \xi < 60 \qquad (38)$$

where

W = compartment width (m)
D = compartment depth (m)

$$\xi = \frac{\dot{m}_f}{A_0 \sqrt{H_0}}\left(\frac{D}{W}\right)^{1/2}$$

Equation 38 is shown in Figure 3-6.10 over the range where the data lie. Both equations are for ventilation-controlled fires. When there is ample ventilation, so that the fuel is free burning, the value of $\dot{m}_f$ depends on L and the type of fuel. For example, domestic furniture has a free-burning fire duration of about 20 min, giving $\tau = 1200$ s and $\dot{m}_f = L/1200$.

The temperatures discussed above are averages measured during the fully developed period of the fire. It is assumed that all fires are ventilation controlled, with the simple relationship for rate of burning given by Equation 38, which is near stoichiometric burning, and it is assumed that combustion of 1 kg of wood releases 18.8 MJ in total.

Swedish Method

This method, developed by Magnusson and Thelandersson,[10] is based on the conventional mass and energy balance equations. The fire itself is not modeled; heat release rate curves are provided as input and, in all instances, the energy release must be less than stoichiometric. The method

does not take into account that the actual mass loss rate may be greater than stoichiometric, with the excess fuel burning outside the compartment. A computer program SFIRE (versions 1 through 3) is available to perform this method. The results from the computer program have been compared with a large number of full-scale fire experiments, both in the fuel- and ventilation-controlled regimes, with good agreement between theory and experiment. It should be added, however, that most of the experiments involved wood crib fires, which inherently burn slower and produce less excess fuel load than furnishings and other combustibles found in practical fire loads. In the Swedish method, the fire load is expressed in relation to A_T as $\dot{Q}$ 18.8 L/A_T MJ/m².

The design curves approved by the Swedish authorities were computed on the basis of systemized ventilation-controlled heat-release curves taken from reference.[10] Figure 3-6.11 shows some typical curves. The curves are calculated for wall, floor, and ceiling materials with "normal" thermal properties from an energy balance which assumes a uniform temperature in the compartment.

PREDICTING FLASHOVER

One of the uses of predicted compartment fire temperatures is the estimation of the likelihood of flashover. The methods used are similar to those used in the prediction of temperature. In one case, that of McCaffrey *et al*, the method is simply an extension of the temperature calculation.

Method of Babrauskas

Babrauskas uses the energy balance for the upper layer given in Equation 7, where the gas flow rate out of the opening is approximated by[18]

$$\dot{m}_g \approx 0.5 A_0 \sqrt{H_0} \tag{39}$$

The primary energy loss is assumed to be radiation to 40 percent of the wall area which is at approximately ambient temperature

$$q_{loss} = \varepsilon\sigma(T_g^4 - T_\infty^4)(0.40 A_T) \tag{40}$$

where

ε = emissivity of the hot gas
σ = Stefan-Boltzmann constant 5.67×10^{11} kW/m$^2 \cdot$ K^4

Combining Equations 7, 39, and 40, using a gas temperature for flashover of 873 K, a specific heat of air of 1.0 kJ/kg·K, an emissivity of 0.5, and assuming the correlation between compartment wall and opening area of

$$A_T/A_0\sqrt{H_0} \approx 50$$

yields a minimum $\dot{Q}$ required for flashover

$$\dot{Q} = 600 A_0 \sqrt{H_0} \tag{41}$$

The air flow into the compartment has been approximated as

$$0.5 A_0 \sqrt{H_0}$$

The maximum amount of fuel which can be burned completely with this air is known as the stoichiometric amount. For most fuels, the heat released per mass of air consumed is a constant approximately equal to 3000 KJ/kg. Therefore, the stoichiometric heat release rate $\dot{Q}_{stoich}$ can be calculated

$$\dot{Q}_{stoich} = 3000\, \dot{m}_g = 3000(0.5 A_0\sqrt{H_0})$$
$$= 1500 A_0\sqrt{H_0} \tag{42}$$

From this derivation, it is shown that the minimum $\dot{Q}$ required for flashover equals $0.4\,\dot{Q}_{stoich}$. Comparing these results with fire tests, Babrauskas found that the data falls within a range of $\dot{Q} = 0.3\,\dot{Q}_{stoich}$ to $\dot{Q} = 0.7\,\dot{Q}_{stoich}$. A best fit of the data suggests

$$\dot{Q} = 0.5 \dot{Q}_{stoich}$$

which, substituting into Equation 42 yields

$$\dot{Q} = 750 A_0 \sqrt{H_0} \tag{43}$$

The 33 test fires used had energy release rates from 11 to 3840 kW, with fuels primarily of wood and polyurethane. Ventilation factors

$$A_0\sqrt{H_0}$$

ranged from 0.03 to 7.51 m$^{5/2}$, and surface area to ventilation factor ratios

$$A_T/A_0\sqrt{H_0}$$

ranged from 9 to 65 m$^{-1/2}$.

Example of Babrauskas' method: Calculate the heat release rate necessary to cause flashover, using the method of Babrauskas. Assume the same room as in the McCaffrey *et al* method example for predicting compartment fire temperatures. From Equation 43

$$\dot{Q} = 750 A_0 \sqrt{H_0}$$

where

$A_0 = 1.08$ m^2
$H_0 = 1.8$ m
$\dot{Q} = (750)\,(1.08)\,(1.8)^{1/2} = 1087$ kW

Method of McCaffrey, Quintiere, and Harkleroad

The method of McCaffrey, Quintiere, and Harkleroad for predicting compartment fire temperatures may be extended to predict the energy release rate of the fire required to result in flashover in the compartment.

Equation 11 can be rewritten as

$$\dot{Q} = \left[\sqrt{g}\, c_p \rho_\infty T_\infty^2 \left(\frac{\Delta T_g}{480}\right)^3\right]^{1/2} (h_k A_T A_0 \sqrt{H_0}) \tag{44}$$

Selecting an upper gas temperature of 522°C and ambient temperature of 295 K or $\Delta T_g = 500$°C for flashover, and substituting values for the gravitational constant ($g = 9.8$ m/s^2), the specific heat of air ($c_p = 1.0$ kJ/kg · K), and the density of air ($\rho_\infty = 1.18$ kg/m^3), and rounding 607.8 to 610 yields

$$\dot{Q} = 610(h_k A_T A_0 \sqrt{H_0})^{1/2} \tag{45}$$

where

h_k = effective heat transfer coefficient [(kW/m)/K]
A_T = total area of the compartment surfaces (m^2)
A_0 = area of opening (m^2)
H_0 = height of opening (m)

Using Equation 12 yields a slightly different value, 623.6 rounded to 620, of the leading coefficient because of the difference in the value used for the specific heat of air

$$\dot{Q} = 620(h_k A_T A_0 \sqrt{H_0})^{1/2} \tag{46}$$

The use of either 610 or 620 is acceptable within the accuracy of the expression.

Example of McCaffrey *et al* method: Estimate the energy release rate required for flashover of a compartment. Assume the same room as in the McCaffrey *et al* method example for predicting compartment fire temperatures. Assuming $\Delta T_g = 500$°C as a condition for flashover, and air properties at 295 K, use Equation 45

$$\dot{Q} = 610(h_k A_T A_0 \sqrt{H_0})^{1/2}$$

where

$h_k = k/\delta = 0.48 \times 10^{-3}/0.016 = 0.03$ kW/m · K
$A_T = 45.72$ m^2

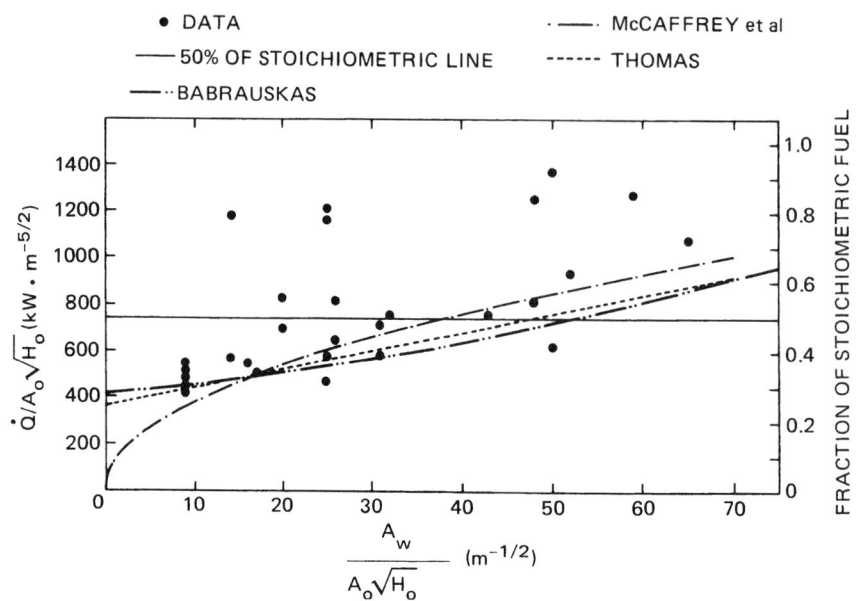

Fig. 3-6.12. The effect of room wall area (gypsum walls) on the heat required for flashover.

$A_0 = 1.08 \text{ m}^2$

$H_0 = 1.8 \text{ m}$

Therefore

$$\dot{Q} = 610[(0.03)(45.72)(1.08)(\sqrt{1.8})]^{1/2}$$

$$= 860 \text{ kW}$$

Method of Thomas

Thomas uses the energy balance for the upper layer shown in Equation 7, where the gas flow rate out of the opening is approximated by[2]

$$\dot{m}_g \approx 0.5 A_0 \sqrt{H_0} \qquad (47)$$

Thomas develops an expression for $\dot{q}_{loss}$ which assumes the area for the source of radiation for roughly cubical compartments is $A_T/6$

$$\dot{q}_{loss} \approx h_c(T_g - T_w)\frac{A_T}{2} + \varepsilon\sigma(2T_g^4 - T_w^4 - T_{flr}^4)\frac{A_T}{6} \qquad (48)$$

where

A_T = total area of the compartment-enclosing surfaces (m^2)

h_c = convective heat transfer coefficient (kW/m$^2 \cdot$ K)

T_w = temperature of the upper walls (K)

T_{flr} = temperature of the floor (K)

From experimental data, Thomas developed an average for $\dot{q}_{loss}$ of $7.8 A_T$. Using an upper layer temperature of 577°C or a ΔT_g of 600°C for flashover criterion and $c_p = 1.26$ kJ/kg $\cdot$ K, yields an expression for the minimum rate of energy release for flashover

$$\dot{Q} = 7.8 A_T + 378 A_0 \sqrt{H_0} \qquad (49)$$

Comparison of Methods for Predicting Flashover

Babrauskas has compared the effect of room wall area on the energy release required for flashover, using the above methods.[19] The results of his comparisons, along with some experimental data for rooms with gypsum board walls, are shown in Figure 3-6.12. The graph shows the energy required for flashover as a function of compartment wall area, both normalized by the ventilation factor

$$A_0 \sqrt{H_0}$$

Babrauskas observes that over the range of compartment sizes of most interest, all of the methods produce similar results. The method of McCaffrey *et al* diverts from the others for small room sizes. Babrauskas notes that all of the methods are a conservative representation of the data.

REFERENCES CITED

1. D. Drysdale, "The Pre-Flashover Compartment Fire," *An Introduction to Fire Dynamics*, John Wiley and Sons, Chichester, pp. 278–303 (1985).
2. P.H. Thomas, "Testing Products and Materials for Their Contribution to Flashover in Rooms," *F. and Matls.*, 5, 3, pp. 103–111 (1981).
3. C. Huggett, "Estimation of Rate of Heat Release by Means of Oxygen Consumption Measurements," *F. and Matls.*, 4, 2, pp. 61–65 (1980).
4. J. de Ris, *Fire Radiation—A Review*, Tech. Report FMRC, RC78-BT-27, Factory Mutual Research Corporation, Norwood, pp. 1–41 (1978).
5. J.A. Rockett, "Fire Induced Gas Flow in an Enclosure," *Comb. Sci. and Tech.*, 12, pp. 165–175 (1976).
6. P.H. Thomas and A.J.M. Heselden, "Fully Developed Fires in Single Compartments," *Fire Research Note No. 923*, Fire Research Station, Borehamwood (1972).
7. K. Kawagoe, "Fire Behaviour in Rooms," Report of the Building Research Institute, Japan, No. 27 (1958).

8. P.H. Thomas and A.J.M. Heselden, "Fully Developed Fires in Single Compartments," A Co-operating Research Programme of the Conseil International du Batiment, *CIB Report No. 20*, Joint Fire Research Organization Fire Research Note 923/197.

9. M. Law, *Structural Engr.*, 61A, 1, p. 25 (1983).

10. S.E. Magnusson and S. Thelandersson, "Temperature-Time Curves of Complete Process of Fire Development. Theoretical Study of Wood Fuel Fires in Enclosed Spaces," Acta Polytechnica Scandinavia, *Civil Engineering and Building Construction Series No. 65*, Stockholm (1970).

11. B.J. McCaffrey, J.G. Quintiere, and M.F. Harkleroad, "Estimating Room Fire Temperatures and the Likelihood of Flashover Using Fire Test Data Correlations," *Fire Tech.*, 17, 2, pp. 98–119 (1981).

12. J.G. Quintiere, "A Simple Correlation for Predicting Temperature in a Room Fire," National Bureau of Standards, *NBSIR 83-2712*, Washington (June 1983).

13. J.R. Lawson and J.G. Quintiere, "Slide-Rule Estimates of Fire Growth," National Bureau of Standards, *NBSIR 85-3196*, Washington (June 1985).

14. K.L. Foote, P.J. Pagni, and N.J. Alvares, "Temperature Correlations for Forced-Ventilated Compartment Fires," *Proceedings of the First International Symposium*, International Association for Fire Safety Science, Hemisphere Publishing, pp. 139–148 (1986).

15. V. Babrauskas, "A Closed-form Approximation for Post-Flashover Compartment Fire Temperatures," *F. Safety J.*, 4, pp. 63–73 (1981).

16. V. Babrauskas and R.B. Williamson, "Post-Flashover Compartment Fires: Basis of a Theoretical Model," *Fire and Matls.*, 2, 2, pp. 39–53 (1978).

17. V. Babrauskas, "COMPF2- A Program for Calculating Post-Flashover Fire Temperatures," National Bureau of Standards, *NBS TN 991*, Washington (1979).

18. V. Babrauskas, "Estimating Room Flashover Potential," *Fire Tech.*, 16, 2, pp. 94–104 (1980).

19. V. Babrauskas, "Upholstered Furniture Room Fires—Measurements, Comparison with Furniture Calorimeter Data, and Flashover Predictions," *J. of F. Sci.*, 2, pp. 5–19 (1984).

ZONE COMPUTER FIRE MODELS FOR ENCLOSURES

William D. Walton

INTRODUCTION

Computer programs are used in many areas of fire protection design, including suppression system design, smoke control system design, and egress analysis. The emphasis in this chapter is on zone computer fire models for enclosures. Zone fire models are computer programs designed to predict the conditions resulting from a fire in an enclosure. These models solve the equations based on the zone assumptions describing the fire-induced conditions within an enclosure.

Computer fire models can provide a faster and more accurate estimate of the impact of a fire, and the measures used to prevent or control the fire, than many of the methods previously used. While manual calculation methods provide good estimates of specific fire effects (e.g., prediction of time to flashover), they are not well suited for comprehensive analyses involving the time-dependent interactions of multiple physical and chemical processes present in developing fires.

The state of the art in computer fire modeling is changing rapidly. Understanding of the processes involved in fire growth is improving, and, thus, the technical basis for the models is improving. The capabilities, documentation, and support for a given model can change dramatically over a short period of time. In addition, computer technology itself (both hardware and software) is advancing rapidly. A few years ago, a large mainframe computer was required to use most of the computer fire models. Today, all of the zone fire models can be run on personal computers. Therefore, rather than provide an exhaustive review of rapidly changing state-of-the-art available computer models, the following discussion will focus on a representative selection. The reader is guided to references 1 and 2 for a comprehensive review of computer fire models.

ENCLOSURE FIRE MODELS

There are two major classes of computer models for analyzing enclosure fire development. *Stochastic* or *probabilistic models* generally treat fire growth as a series of sequential events or states. These models are sometimes referred to as "states transition" models. Mathematical rules

are established to govern the transition from one event to another (e.g., from ignition to established burning). Probabilities are assigned to each transfer point based on analysis of relevant experimental data, historical fire incident data, and computer model results. For a complete discussion of stochastic models, see Section 3, Chapter 16.

In contrast, *deterministic models* represent the processes encountered in a compartment fire by interrelated mathematical expressions based on physics and chemistry. These models may also be referred to as "room fire" models, "computer fire" models, or "mathematical fire" models. Ideally, such models represent the ultimate capability: discrete changes in any physical parameter could be evaluated in terms of the effect on fire hazard. While the state of the art in understanding fire processes will not yet support the "ultimate" model, a number of computer models are available that provide reasonable estimates of selected fire effects.

The newest type of deterministic fire model is the "field" model. This type of model solves the fundamental equations of mass, momentum, and energy for each element in an enclosure space that has been divided into a grid of smaller units. A complete discussion of field modeling can be found in Section 3, Chapter 8. Although field model technology and use have advanced rapidly in recent years, it should be noted that field models require substantial computer resources and are relatively complex to use.

The most common type of physically based fire model is the "zone" or "control volume" model, which solves the conservation equations for distinct and relatively large regions (control volumes). A complete discussion of the fundamental principles behind the zone fire model formulation can be found in Section 3, Chapter 5. A number of zone models exist, varying to some degree in the detailed treatment of the fire phenomena. The dominant characteristic of this type of model is that it divides the room(s) into a hot upper layer and a lower, cooler layer. (See Figure 3-7.1.) The model calculations provide estimates of key conditions for each of the layers as a function of time. Zone modeling has proved to be a practical method for providing estimates of fire processes in enclosures.

The beginnings of pre-flashover zone fire modeling can be traced to the mid 1970s with the publication of a description of the fundamental equations by Quintiere.[3] Based on these equations, the first zone fire model published was RFIRES by Pape and Waterman,[4] followed shortly by the HARVARD model by Emmons and Mitler.[5] The development of zone

Mr. Walton is a research fire protection engineer with the Building and Fire Research Laboratory, National Institute of Standards and Technology.

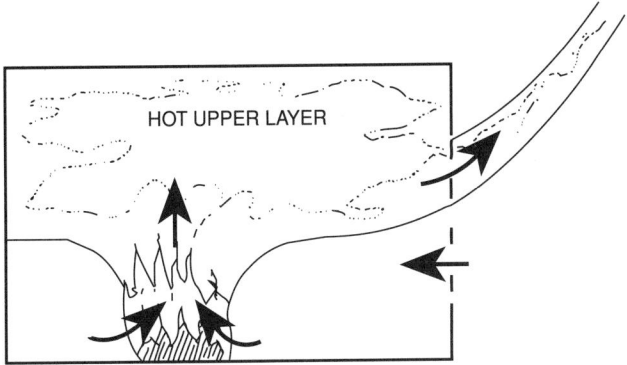

Fig. 3-7.1. Two-layer model with no exchange between layers except the plume.

fire models was facilitated both by the advancement in the understanding of the basic physics of fire growth in a compartment and advances in, and the availability of, mainframe computers. Following the publication of these two models a number of zone fire models for mainframe computers were introduced. In 1985 the first zone model, ASET-B, written specifically for the newly available IBM-compatible personal computers, was introduced by Walton.[6] Since that time additional models have been introduced, and most of the models written for mainframe computers have been converted for use in personal computers.

No zone fire model is "best" for all applications. The selection of a zone fire model for a particular application depends on a number of factors. While most of the zone fire models are based on the same fundamental principles, there is significant variation in features among the different models. The decision to use a model should be based on an understanding of the assumptions and limitations for the particular model. In general the more detailed the model outputs, the more extensive the model inputs, and the greater the computer execution time required. When using any computer fire model, it is always a good idea to test the sensitivity of the model outputs to changes in model inputs. If small changes in model inputs result in large changes in model outputs, the user must exercise great care in selecting the input values.

A key issue in selecting a model is model validation. Comparison of model results with experimental data is valuable for determining the applicability of a model to a particular situation. Comparisons of model results with experimental data are limited, and the number of comparisons varies widely among the models. The model user should carefully examine model validation comparisons before selecting a model. Frequently, experienced model users will use more than one model to evaluate a particular situation. If several models provide similar results this can increase the confidence in the results. This does not, however, guarantee that the results accurately represent the physical conditions being modeled, since most of the zone fire models are based on the same basic assumptions.

OVERVIEW OF REPRESENTATIVE ZONE FIRE MODELS

ASET Computer Program

ASET (available safe egress time) is a program for calculating the temperature and position of the hot upper smoke layer in a single room with closed doors and windows. ASET can be used to determine the time to the onset of hazardous conditions for both people and property. The required program inputs are the heat-loss fractions, the height of the fuel above the floor, criteria for hazard and detection, the room ceiling height, the room floor area, a heat release rate, and (optional) species generation rate of the fire. The program outputs are the temperature, thickness, and (optional) species concentration of the hot upper smoke layer as a function of time, and the time to hazard and detection. ASET can examine multiple cases in a single run. ASET was written in FORTRAN by Cooper and Stroup.[6]

ASET-B Computer Program

ASET-B is a program for calculating the temperature and position of the hot upper smoke layer in a single room with closed doors and windows. ASET-B is a compact version of ASET, designed to run on personal computers. The required program inputs are a heat-loss fraction, the height of the fire, the room ceiling height, the room floor area, the maximum time for the simulation, and the rate of heat release of the fire. The program outputs are the temperature and thickness of the hot upper smoke layer as a function of time. Species concentrations and time to hazard and detection, calculated by ASET, are not calculated in the compact ASET-B version. ASET-B was written in BASIC by Walton.[6]

COMPBRN III Computer Program

COMPBRN III is primarily used in conjunction with probabilistic analysis for the assessment of risk in the nuclear power industry. The model is based on the assumption of a relatively small fire in a large space, or a fire involving large fuel loads early during the pre-flashover fire growth period. The model's strengths are (1) emphasis on the thermal response of elements within the enclosure to a fire within the enclosure, and (2) model simplicity. The temperature profile within each element is computed, and an element is considered ignited or damaged when its surface temperature exceeds the user-specified ignition or damage temperature. The model outputs include the total heat release rate of the fire, the temperature and depth of the hot gas layer, the mass burning rate for individual fuel elements, the surface temperatures, and the heat flux at user-specified locations. COMPBRN III was written by Siu *et al.*[7]

COMPF2 Computer Program

COMPF2 is a computer program for calculating the characteristics of a post-flashover fire in a single building compartment, based on fire-induced ventilation through a single door or window. It is intended both for performing design calculations and for the analysis of experimental burn data. Wood, thermoplastics, and liquid fuels can be evaluated. A comprehensive output format is provided that gives gas temperatures, heat-flow terms, and flow variables. The documentation includes input instructions, sample problems, and a listing of the program. The program was written in FORTRAN by Babrauskas.[8]

CSTBZ1 Computer Program

CSTBZ1 is a computer program for post-flashover fires, based on similar basic assumptions as those of COMPF2. The equations of mass and energy conservation are written without neglecting the fuel source term, and several vertical openings in a room can be considered. The equation of heat diffusion into the walls is solved either by a classical explicit finite difference method, or by a new modal approach that offers the capability of storing in a file a few numbers characterizing a given wall, leading to a rapid calculation of the superficial wall temperatures. A sophisticated numerical algorithm was used to solve the equations through uncoupling. The program was written by Curtat and Bodart.[9]

CFAST Computer Program

CFAST is the present version of the original FAST computer program. A consolidation of the FAST[10] and CCFM.VENTS[11] models, CFAST can be configured for a maximum of either 5 or 10 rooms, depending on the computer resources available. The required program inputs are the geometrical data describing the rooms and connections; the thermophysical properties of the ceiling, walls, and floors; the fire as a rate of mass loss; and the generation rates of the products of combustion. The program outputs are the temperature and thickness of, and species concentrations in, the hot upper layer and the cooler, lower layer in each compartment. Also given are surface temperatures and heat transfer and mass flow rates. CFAST was written in FORTRAN by Jones and Forney.[12]

FIRST Computer Program

FIRST is the direct descendant of the HARVARD V[5] program developed by Emmons and Mitler. The program predicts the development of a fire and the resulting conditions within a room given a user-specified fire or user-specified ignition. It predicts the heating and possible ignition of up to three targets. The required program inputs are the geometrical data describing the rooms and openings, and the thermophysical properties of the ceiling, walls, burning fuel, and targets. The generation rate of soot must be specified, and the generation rates of other species may be specified. The fire may be entered either as a mass loss rate or in terms of fundamental properties of the fuel. Among the program outputs are the temperature and thickness of, and species concentrations in, the hot upper layer and the cooler, lower layer in each compartment. Also given are surface temperatures and heat transfer and mass flow rates. The FIRST program was written in FORTRAN by Mitler and Rockett.[13]

FPETOOL Computer Program

FPETOOL is the descendent of the FIREFORM program.[14] It contains a computerized selection of relatively simple engineering equations and models useful in estimating the potential fire hazard in buildings. The calculations in FPETOOL are based on established engineering relationships. The FPETOOL package addresses problems related to fire development in buildings and the resulting conditions and response of fire protection systems. The subjects covered include smoke filling in a room, sprinkler/detector activation, smoke flow through (small) openings, temperatures and pressures developed by fires, flashover and fire severity predictions, fire propagation (in special cases), and simple egress estimation. The largest element in FPETOOL is a zone fire model called FIRE SIMULATOR. FIRE SIMULATOR is designed to estimate conditions in both pre- and post-flashover enclosure fires. The inputs include the geometry and material of the enclosure, a description of the initiating fire, and the parameters for sprinklers and detectors being tracked. The outputs include the temperature and volume of the hot smoke layer; the flow of smoke from openings; the response of heat-actuated detection devices, sprinklers, and smoke detectors; oxygen, carbon monoxide, and carbon dioxide concentrations in the smoke; and the effects of available oxygen on combustion. FPETOOL was written in BASIC by Nelson.[15]

LAVENT Computer Program

LAVENT is a program developed to simulate the environment and the response of sprinkler elements in compartment fires with draft curtains and fusible-link-actuated ceiling vents. The zone model used to calculate the heating of the fusible links includes the effects of the ceiling jet and the upper layer of hot gases beneath the ceiling. The required program inputs are the geometrical data describing the compartment, the thermophysical properties of the ceiling, the fire elevation, the time-dependent heat release rate of the fire, the fire diameter or the heat release rate per unit area of the fire, the ceiling vent area, the fusible-link response time index (RTI) and activation temperature, the fusible-link positions along the ceiling, the link assignment to each vent, and the ambient temperature. A maximum of 5 ceiling vents and 10 fusible links are permitted in the compartment. The program outputs are the temperature and height of the hot layer, the temperature of each link, the ceiling jet temperature and velocity at each link, the radial temperature distribution along the interior surface of the ceiling, the activation tie of each link, and the area opened. LAVENT was written in FORTRAN.[16]

WPI/FIRE Computer Program

WPI/FIRE is a direct descendant of the HARVARD V[5] and FIRST[13] programs. It includes all of the features of the HARVARD program version 5.3 and many of the features of the FIRST program. WPI/FIRE also includes the following additional features: improved input routine, momentum-driven flows through ceiling vents, two different ceiling jet models for use in detector activation, forced ventilation for ceiling and floor vents, and an interface to a finite difference computer model for the calculation of boundary surface isotherms and hot spots. The WPI/FIRE program was written in FORTRAN by Satterfield and Barnett,[17] and additions to the program continue to be developed by graduate students at the Center for Fire Safety Studies, Worcester Polytechnic Institute.

REFERENCES CITED

1. R. Friedman, *An International Survey of Computer Models for Fire and Smoke*, 2nd ed. (Dec. 1991). Available from Library, Factory Mutual Research Corporation, 1151 Boston-Providence Turnpike, Norwood, MA 02062.
2. R. Friedman, "An International Survey of Computer Models for Fire and Smoke," *Jour. of Fire Prot. Engr.*, 4, 3, pp. 83–92 (1992).
3. J. Quintiere, "Growth of Fires in Building Compartments," ASTM STP 614, American Society for Testing and Materials, Philadelphia, PA (1977).
4. R. Pape, T.E. Waterman, and T.V. Eichler, "Development of a Fire in a Room from Ignition to Full Room Involvement—RFIRES," National Bureau of Standards, *NBS-GCR-81-301*, Washington (1981).
5. H.E. Mitler, "The Harvard Fire Model," *F. Safety J.*, 9, pp. 7–16 (1985).

6. W.D. Walton, "ASET-B A Room Fire Program for Personal Computers," National Bureau of Standards, *NBSIR 85-3144*, Washington, DC (1985).

7. V. Ho, N. Siu, and Apostolakis, "COMPBRN III—A Fire Hazard Model for Risk Analysis," *Fire Safety Jour.*, 13, 2&3, pp. 137–154 (1988).

8. V. Babrauskas, "COMPF2—A Program for Calculating Post-Flashover Fire Temperatures," National Bureau of Standards, NBS TN 991, Washington, DC (1979).

9. M.R. Curtat and X.E. Bodart, *1st Symposium International Association for Fire Safety Science*, Hemisphere Publications p. 637 (1986).

10. W.W. Jones, "A Multicompartment Model for the Spread of Fire, Smoke and Toxic Gases," *F. Safety J.*, 9, pp. 55–79 (1985).

11. G.P. Forney and L.Y. Cooper, "The Consolidated Compartment Fire Model (CCFM) Computer Application. VENTS, Parts I, II, III, IV," NISTIR, National Institute of Standards and Technology (1990).

12. W.W. Jones and G.P. Forney, "A Programmer's Reference Manual for CFAST, the Unified Model of Fire Growth and Smoke Transport," TN-1283, National Institute of Standards and Technology (1990).

13. H.E. Mitler and J.A. Rockett, "User's Guide to FIRST, A Comprehensive Single-Room Fire Model," CIB W14/88/22, National Bureau of Standards, Gaithersburg, MD (1987).

14. H.E. Nelson, "FIREFORM—A Computerized Collection of Convenient Fire Safety Computations," NBSIR 86-3308, National Bureau of Standards, Gaithersburg, MD (1986).

15. H.E. Nelson, "FPETOOL: Fire Protection Engineering Tools for Hazard Estimation," NISTIR 4380, National Institute of Standards and Technology, Gaithersburg, MD (1990).

16. Y.L. Cooper, "Estimating the Environment and the Response of Sprinkler Links in Compartment Fires with Draft Curtains and Fusible-Link-Actuated Ceiling Vents Theory," *Fire Safety J.*, Vol. 16, pp. 137–163 (1990).

17. D.B. Satterfield and J.R. Barnett, "User's Guide to WPI-FIRE Version 2 (WPI-2)—A Compartment Fire Model," Worcester Polytechnic Institute, Center for Fire Safety Studies, Worcester, MA (Aug. 1990).

ADDITIONAL READING

L.Y. Cooper and D.W. Stroup, "ASET—A Computer Program for Calculating Available Safe Egress Time," *F. Safety J.*, 9, pp. 29–45 (1985).

B. Hagglund, "A Room Fire Simulation Model," National Defense Research Institute, Sweden, *FOA C 20501-D6* (June 1983).

B. Hagglund, "Simulating the Smoke Filling in Single Enclosures," National Defense Research Institute, Sweden, *FOA C 20513-D6* (October 1983).

J.A. Swartz *et al*, *Final Technical Report on Building Fire Simulation Model*, Vols. I and II, National Fire Protection Association, Quincy (1983).

T. Tanaka, "A Model of Multiroom Fire Spread," National Bureau of Standards, *NBSIR 83-2718*, Washington (1983).

E.E. Zukoski and T. Kubota, "Two-Layer Modeling of Smoke Movement in Building Fires," *F. and Matls.*, 4, 1, p. 17 (1980).

USING FIELD MODELING TO SIMULATE ENCLOSURE FIRES

David W. Stroup

INTRODUCTION

Most building and fire safety codes contain equivalency clauses. These clauses allow for the use of alternative methods and materials when their equivalency can be proved to the *authority having jurisdiction*. In the past, the determination of equivalency has been based primarily on subjective judgment. Continuing research into fire phenomena has made it possible to perform an engineering analysis of the fire safety performance of a building. This building can differ widely from current perceptions of a code-conforming building. Using mathematical fire modeling, the development and impact of fire in a building can be assessed and the building's conformance to code-specified fire safety objectives evaluated.

There are two fundamentally different approaches to fire modeling: (1) probabilistic and (2) deterministic. The probabilistic or stochastic approach involves the assessment of the probable fire risk in an enclosure by associating finite probabilities with all fire-influencing parameters, such as distribution of fuel, number and extent of enclosure openings, and human behavior. Little or no physics is included in probabilistic-based models. This approach while useful in suggesting likelihood of a fire in a given enclosure provides little information about the production and distribution of fire products, such as toxic product concentrations, temperature, and smoke movement.

In deterministic models, the laws of physics and chemistry are used to calculate the impact of fire in a well-defined physical scenario. Deterministic fire models can range from simple one-line correlations of data to highly complex models requiring hours of computing time on mainframe computers. The more complex models are typically divided into two classes, i.e., (1) zone models and (2) field models, based on the strategy used to solve the equations representing the physical processes associated with the fire.

The zone modeling concept divides the hypothetical burning enclosure into several distinct zones with uniform fire characteristics. The number of zones can be as few as one but typically ranges between three and five. Examples of possible zones in a fire compartment include the combustion zone, the rising thermal plume and ceiling jet, the hot gas layer accumulated under the ceiling, and the lower layer of cooler air. Ordinary differential equations expressing conservation of mass and energy are applied to each zone and solved numerically. Typically, zone models provide results in one dimension (e.g., upper layer thickness and temperature, lower layer thickness and temperature). The calculations associated with a zone model do not necessarily have to be performed using a computer. A computer makes using the zone models more efficient and practical.

COMPUTATIONAL FLUID DYNAMICS THEORY

Field models represent the second class of deterministic models. This type of model solves the fundamental equations of mass, momentum, and energy. In order to facilitate the solution of the equations, the compartment is divided into a three-dimensional grid of tiny cubes. The field model calculates the physical conditions in each cube, as a function of time. The calculations account for physical changes generated within each cube and changes in the cube resulting from changes in surrounding cubes.

The field model is a complex fluid mechanical model of turbulent flow derived from classical fluid dynamics theory. Classical fluid dynamics is concerned with the mathematical description of the physical behavior of fluids (gases or liquids). The equations governing fluid behavior have been known for over 150 years. They consist, in general, of a set of three-dimensional, time-dependent, nonlinear partial differential equations, referred to as the Navier-Stokes equations.[1] These equations express conservation of mass, momentum, and energy.

Conservation of mass is described by the continuity equation

$$\frac{\partial \rho}{\partial t} + \nabla \cdot (\rho \mathbf{U}) = S \qquad (1)$$

Where t is time, ρ is the local density of the fluid, and $\mathbf{U}$ is its velocity vector with components in the x, y, and z directions. The S on the right-hand side corresponds to sources or sinks of mass; for almost all cases of interest, S is zero. The Navier-Stokes equation describing conservation of momentum is

$$\frac{\partial \rho \mathbf{U}}{\partial t} + \nabla(\rho \mathbf{U} \otimes \mathbf{U}) = \mathbf{B} + \nabla\sigma \qquad (2)$$

David W. Stroup, P.E., is a fire protection engineer with the U.S. General Services Administration in Washington, DC.

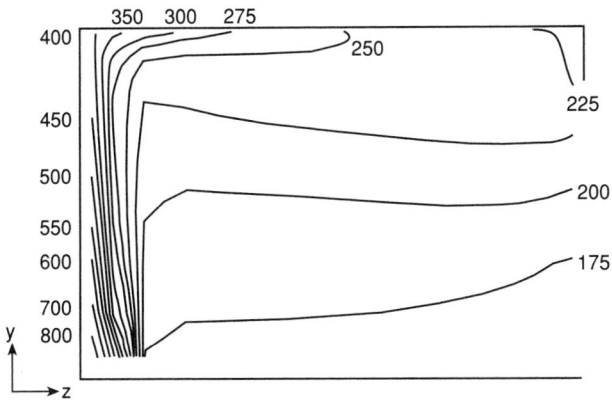

Fig. 3-8.1. *Temperature contours in a room, 3.66 m × 2.44 m × 2.44 m, with a fire against the wall. Results from ref. 18 calculated using* CLYTIE.

where σ is represented as

$$\sigma = -p\delta + \mu[\nabla U + (\nabla U)^T] \qquad (3)$$

when p is the pressure, μ is the molecular viscosity, T is temperature, and $\mathbf{B}$ is a body force. In fire problems, the body force of most importance is that due to gravity (buoyancy). Finally, the energy equation is represented as

$$\frac{\partial \rho H}{\partial t} + \nabla(\rho U H) - \nabla(\lambda \nabla T) = \frac{\partial p}{\partial t} \qquad (4)$$

where H is the total enthalpy, in terms of the static (thermodynamic) enthalpy h, given by

$$H = h + \tfrac{1}{2}\mathbf{U}^2$$

and λ is the thermal conductivity.

These equations represent five transport equations (i.e., continuity equation, three momentum equations, and an energy equation) with seven unknowns [U(U,V,W), p, T, ρ, and h]. In order to form a closed set, two additional equations are required. They are obtained from thermodynamic principles. The equation of state, $\rho = \rho(p,T)$, relates density to pressure and temperature, and the constitutive equation, $h = h(p,T)$, relates static enthalpy to pressure and temperature.

Analytical solutions of these equations exist for a limited number of special cases that have simple boundary conditions. In their most general form, the Navier-Stokes equations cannot be solved by analytical methods. Therefore, solving the equations usually requires the use of numerical techniques. Computational fluid dynamics (CFD) involves the numerical solution of the Navier-Stokes equations using computers.[2] Field models solve the fundamental partial differential equations of motion and conservation numerically at a discrete moment in time and point in space. Using a set of grids, the compartment under study is divided into as many small volume elements or cells as is practicable. The resulting difference equations are solved simultaneously in each cell to obtain various parameters of interest.

The fire simulation problem represents one of the most difficult areas in computational fluid dynamics: the numerical solution of recirculating, three-dimensional turbulent buoyant fluid flow with heat and mass transfer. Field models can display quantitative differences in physical parameters

throughout the computational grid. Typical field models calculate the velocity and temperature fields in an enclosure, given a prescribed heat source. Figures 3-8.1 and 3-8.2 illustrate the temperature and velocity fields that are obtainable with field models, respectively. Some models also calculate gas species and particle concentrations as a function of position and time.

Turbulence

The flows occurring in room fires are turbulent, generating eddies or vortices of many sizes. The energy contained in large vortices cascades down to smaller and smaller vortices, until it diffuses into heat. Eddies exist down to the size where the viscous forces dominate over inertial forces and energy is dissipated into heat. For typical fires, this scale is on the order of a millimeter or so. The control volume size used to make the Navier-Stokes equations discrete should be consistent with this scale. Applying this criteria would result in problems with many more control volumes than could possibly be solved with today's computers (or computers in the foreseeable future). As a result, turbulence models have been developed to account for the effect of small-scale fluid motion on motion in the larger-scale control volumes. A turbulence model estimates the effect of small-scale or subgrid phenomena on motion in the larger scale.

The first step in constructing a turbulence model is to modify the fluid flow equations by averaging them over both time and space. The unknown pressures, momentums, and energies are then interpreted as average rather than instantaneous values. Averaging results in the addition of turbulent flux terms (called Reynolds stresses and Reynolds fluxes) and more unknowns than equations. Turbulence models are used to achieve the necessary closure.

Several turbulence models have been developed. These models can be separated into two broad classes: (1) eddy viscosity models and (2) second-order closure models. Eddy viscosity models specify the Reynolds stresses and fluxes algebraically in terms of known mean quantities. Second-order closure models solve differential transport models for the turbulent fluxes.

The k-ε model, probably the best known turbulence model, is an example of an eddy viscosity model.[3] This

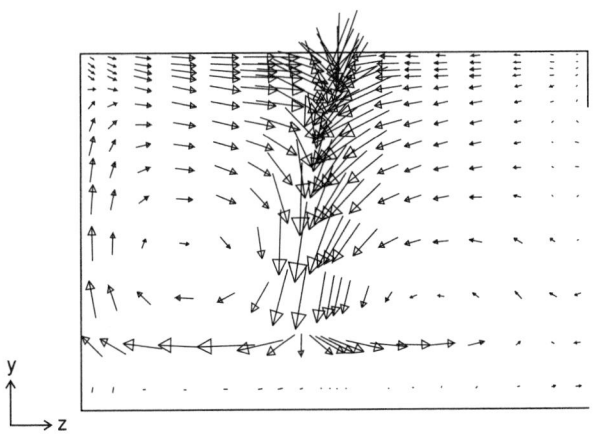

Fig. 3-8.2. *Velocity vectors in a room, 3.66 m × 2.44 m × 2.44 m, with a fire in the center of the room and an activated sprinkler. Results from ref. 18 calculated using* CLYTIE.

model results in two additional differential equations per control volume. The first equation governs the distribution of turbulent kinetic energy, k, while the second describes the dissipation of the local energy, ε. The equations used in the k-ε model contain several empirical constants. These values are chosen to cause a good fit between a set of experiments and a corresponding set of modeling results. When a new application area is being examined, the model needs to be run against a set of experiments to verify that reasonable results are being obtained. The standard k-ε model assumes turbulence has no preferred direction. Since gravitational forces occur only in the vertical direction, this assumption results in the effects of gravity being ignored. In a fire plume, gravity gives rise to buoyancy, which in turn accelerates flows and can cause turbulence. Frictional forces occurring near solid wall boundaries can cause a similar violation of the turbulence model assumptions. In one fire problem analysis, Forney et al[4] examined this problem and found its impact to be negligible.

Boundary and Initial Conditions

In order to fully specify a problem, a set of boundary and initial conditions must be provided. Boundary conditions place limits on the physical environment. These limits usually take the form of a specified parameter value (e.g., zero velocity), a flux value (e.g., kW per sq m), or a time rate of change (e.g., 2°C change in temperature per sec) at a specified position. Initial conditions are important for transient problems and specify the status of the physical environment at the start of the simulation.

Since most of the simulation volume is surrounded by walls, the specification of wall boundary conditions is critical in analysis of fire problems. Wall boundary conditions are used to specify the fluid velocities adjacent to the wall surfaces, the wall shear stresses (related to fluid viscosity), the velocity of the wall (if it is moving), and the heat transfer characteristics. Since large viscosity gradients can occur next to the wall, many of the fluid properties will vary rapidly in the vicinity of walls. In order to accurately predict the fluid properties near the wall, very fine meshes are required. An alternative is to use wall functions that relate the wall shear stress to the turbulent kinetic energy.[4]

Doors, windows, and other types of vents are usually specified as either inlet or pressure boundaries. When using an inlet boundary condition, the user must specify the fluid properties (velocity, temperature, etc.) at the inlet. This can be difficult or impossible when using a field model to analyze a fire problem. For a vent being considered as part of a fire analysis, a pressure boundary condition is the more viable alternative. With a pressure boundary condition, the pressure is set equal to ambient, and the derivatives of the velocity components normal to the vent surface are set equal to zero. This permits the flow to enter or leave the computational grid as required.

A final type of boundary condition is a plane or axis of symmetry. All variables are mathematically symmetric, with thus no diffusion across the boundary. Using this boundary condition, the number of cells used to make a problem discrete can be increased. For a fire located in the center of a room, symmetry planes can be used to model a quarter of the room. This would allow a four-fold increase in the number of cells used to model the problem.

Software and Hardware Requirements

A number of general-purpose computer programs have been developed that permit the solution of the equations describing fluid flow. Upon developing the k-ε model of turbulence, Spalding[5] embedded it in a general-purpose fluid dynamics code which is commercially available as PHOENICS.[6] Comparable codes such as FLOW3D[7] and FLUENT[8] exist, and there are others available. Several field models have been developed specifically to address fire problems. The best known of these are BF3D,[9–13] UNDSAFE,[14,15] and JASMINE.[16]

These computer codes tend to be very complex and require a great deal of user sophistication. A field model divides a space of interest into potentially thousands of cells. At a minimum, the x, y, and z coordinates for one point in each cell must be specified. The user could have to type a thousand or more entries. These models provide output results for a number of fluid properties in each cell. The user could be faced with interpreting several thousand printed numbers.

Fortunately, the usability of field models and computational fluid dynamics software has advanced significantly over the last few years. Most commercially available software packages have interfaces that allow the user to specify the problem and interpret the results graphically. Post-processor interfaces provide capabilities for viewing velocity vectors, streamlines, line and filled contour plots, and particle/droplet trajectories using two- and three-dimensional color graphics.

The implementation of multiblock grids and body-fitted coordinates systems has also improved the usability of field models. Field models require the geometry of interest to be divided into a number of cells using a fixed number of divisions along each of the three principle axes. It can be very difficult to map a real-problem geometry into a number of fixed-size cells. Multiblock capability enables the user to build a problem using several blocks. For example, a T-junction could be represented by joining three blocks (one for the upright and one for each side bar). Within certain limitations, each block's grid structure could be specified independent of the others.

Body-fitted coordinate systems simplify the specification of arbitrary two- and three-dimensional geometries. The basic idea is to use a curvilinear coordinate transformation to map the complex flow domain in physical space to a simple (i.e., rectangular) flow domain in computational space. Body-fitted coordinate grids can be viewed as regular grids that have been squeezed, stretched, bent, and twisted into the desired shape.

The significant costs involved will continue to limit the widespread use of field models in fire protection engineering applications. The computational demands and memory requirements of field models are beyond the capabilities of most currently available personal computers. In a survey of computer models, Friedman identified 10 field models.[17] Of these, only two could be used with existing personal computer hardware, and these two have very limited capabilities. Most of these computer codes require mainframe or mini computers to operate efficiently. Some will operate on powerful desktop computers such as those manufactured by Sun, Silicon Graphics, IBM, and Hewlett-Packard. Prices for these machines start at around $15,000 with annual maintenance fees of approximately $1,000 or more. Table 3-8.1 summarizes the capabilities and hardware requirements of the field models identified by Friedman.

TABLE 3-8.1 *Summary of Selected Field Models*

Model Name	Description	Hardware Required	Availability
BF3D	Field model of three-dimensional time-dependent, buoyant convection induced by a heat source (fire) in an enclosure	Large memory requirements for adequate resolution: Mainframe, minisuper computer, or super computer generally required	Research code—not generally available
FISCO-3L	A personal computer-type field model for fire development in single compartment environment	80386 chip-based computer, running MS-DOS operating system with math co-processor, EGA or VGA graphical display and minimum 640 kB of memory	RESTRICTED. Copyright by INTELLEX and Germany, The Ministry of Research and Technology
FLOW3D	A general-purpose fluid-dynamics code	Super computer, mainframe, mini, or workstation	May be purchased or leased
JASMINE	Field model for analysis of smoke movement in enclosure	Mini-supermini-VAX preferable	Fire Research Station Technical Consultancy undertakes commissions based on its use
KAMELEON FIRE E-3D	A three-dimensional field model for transient calculation of pool fires in an enclosure	Standard FORTRAN. Post-processor needs MS-DOS and VGA display or UNIX X-Windows and at least 640 kB	RESTRICTED. SINTEF is performing calculation at a price of about $4,000 for each calculation. Color diagrams are included
KAMELEON II	Three-dimensional field model for calculation of smoke and toxic gas movement in complex geometries (e.g., multi-rooms, process areas and in open landscapes, etc.)	Post- and pre-processor are MS-DOS and VGA display or UNIX X-Windows	SINTEF is operating the code commercially for a price of about $4,000 to $8,000 for each calculation
KOBRA-3D	A three-dimensional field model for the determination of the hydrodynamical flow in a single fire compartment	IBM-compatible PC, MS-DOS 3.0 or higher, EGA graphics (co-processor Intel 80 × 87 suggested but not required)	Purchasable from INTELLEX GmbH, Beffinastr
PHOENICS	A general-purpose, three-dimensional transient fluid dynamics code	Super computer, mainframe, mini, or workstation	May be purchased or leased
RMFIRE	A two-dimensional field model for the transient calculation of smoke movement in room fires	Silicon Graphics Personal IRIS	Model will become available in the future. Calculations can be made by NRCC, at present
SPLASH	A quasi-field model describing the interaction of sprinkler sprays with fire gases	VAX	Available within Brandforsk, FRS & South Bank Polytechnic
UNDSAFE	Predicts fire environment in open space and in enclosures, using three-dimensional finite difference scheme (field model)	Mainframe (more than 1 MB of memory size is needed)	Program listings only are available

In order to utilize a field model, it will be necessary to purchase a commercially available computer software package. The cost of the field modeling software is equal to or greater than the hardware cost. The price for an annual license can range from $10,000 to $100,000 per year, depending on central processing unit type and speed. Perpetual licenses are available at a cost approximately three times that of an annual license, plus an annual maintenance fee equal to about 15 percent of the perpetual license fee. Both types of licenses include technical support and software updates. Many commercial software vendors offer training courses and support user groups to facilitate use of their product.

COMPARISON OF ZONE MODELS

Using computer-based fire models to analyze fire protection engineering problems is becoming imperative. Zone models have been successfully applied to a wide range of these problems. As the problems grow more complex, zone models will be inadequate to fully address them. Zone models provide very limited detail, with bulk average values being predicted at a few select locations within an enclosure. Zone models utilize equations employing empirical rela-

tionships and constants obtained from experiments. The empirical expressions used to describe physical behavior in zone models will break down as the geometries become more complex. Therefore, the use of the zone models for problems that lie outside the range of experiments is limited.[18]

Field models avoid the simplifications inherent in zone models. In solving the basic equations for conservation of mass, momentum, and energy, the compartment is divided into thousands of computational cells. Mass, enthalpy, flow speeds, turbulent energy, etc., are calculated for each cell in the grid. Field models can be used to determine detailed velocity and temperature distributions; zone models calculate average or maximum temperatures and velocities for a few points within a room. The geometry of the room and its furnishings can have significant effects on the nature of the recirculation patterns. For these situations, the higher spatial resolution of field models can be important. In many cases, detailed knowledge of the temperature and/or flow fields near smoke detectors or sprinklers is required in order to accurately predict activation. Figures 3-8.3 and 3-8.4 illustrate the detailed temperature and velocity fields that can be obtained near objects in rooms, respectively.

Fluid dynamic considerations are automatically built into field models, rather than being forced into oversimplified

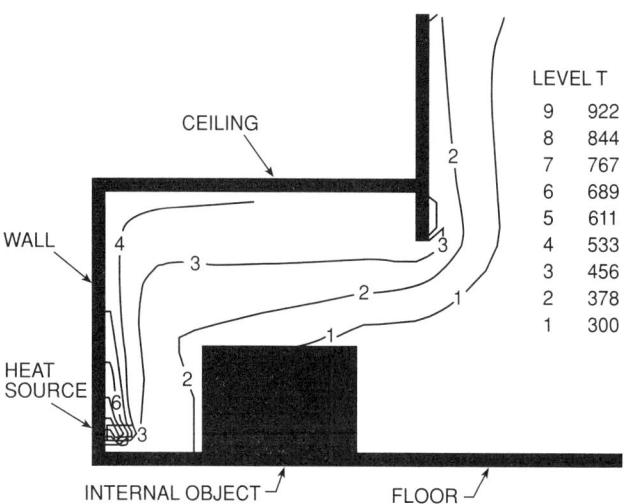

LEVEL	T
9	922
8	844
7	767
6	689
5	611
4	533
3	456
2	378
1	300

Fig. 3-8.3.   Temperature contours for a room containing an object. (From ref. 19.)

approximations. Thus, field models follow the movement of the thermal plume, rather than assuming that deposition of mass and energy from the plume/combustion zone into the upper layer is instantaneous. Similarly, they describe the spread of the ceiling jet to the entire upper layer, rather than assuming instantaneous mixing.

The coarseness of the grid used to define the room determines the physical scale lengths that can be modeled. Techniques for modeling physical phenomena whose natural scale sizes are smaller than the grid size are used in both field and zone modeling. Plume and vent flow models are

examples of this in zone models, and turbulence models are examples of this in field models. While empirical modeling is required even in these very complex field models, the empiricisms are made at a more fundamental level. Field models can be applied to a greater variety of problems with minor modifications.[20]

APPLICATIONS

Recent significant advances in computer capabilities have made the application of field models to fire protection problems practical. However, field models have been available and applied to fire problems for almost twenty years. UNDSAFE,[14,15,21,22] developed by the University of Notre Dame in Indiana; BF3D,[9-13] developed by the National Institute of Standards and Technology; JASMINE,[16,23-29] developed by the Fire Research Station (FRS) and CHAM Ltd; and SAFEAIR[30-33] and CLYTIE[34] developed at Thames Polytechnic in London, represent several field models that have been developed specifically for fire scenarios.

The original version of UNDSAFE was limited to fires in two-dimensional single compartment enclosures. The computer program was modified to simulate the in-flight fire scenario; however, these investigations continued to be limited to two-dimensional studies. Satoh and Kurioshi performed a three-dimensional simulation of an aircraft cabin fire using a further modified version of the UNDSAFE code.[15]

In the late 1970s, Baum and Rehm[9] launched a project to model the combustion and convective flow processes that occur in enclosure fires, from first principles. They developed a computational model of three-dimensional buoyant convection in an enclosure induced by a weak volumetric source of heat and mass. They used hydrodynamics based on the time-dependent inviscid Euler equations. No turbulence

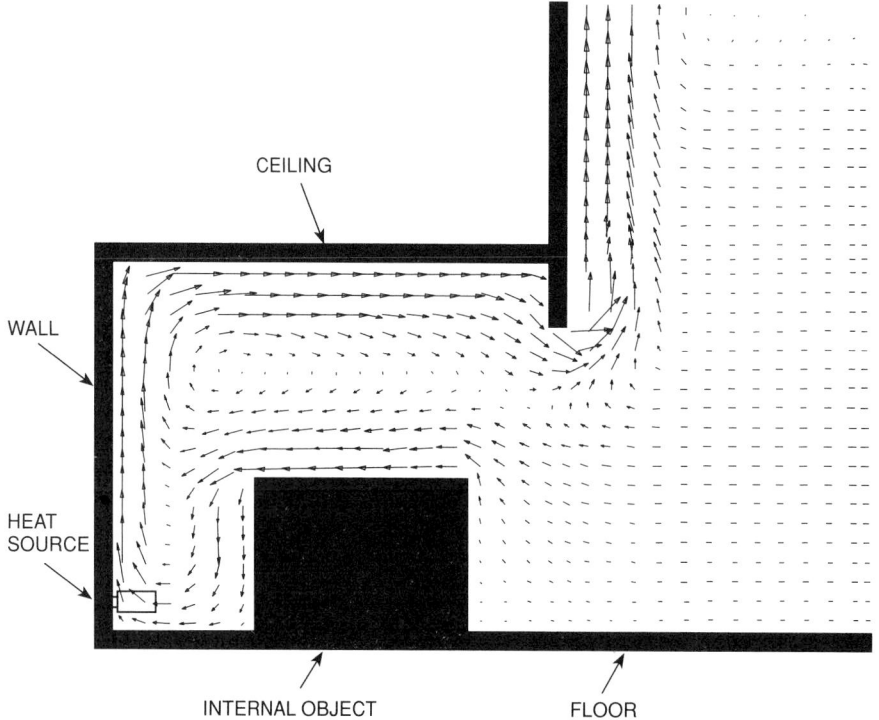

Fig. 3-8.4.   Velocity vectors for a room containing an object. (From ref. 19.)

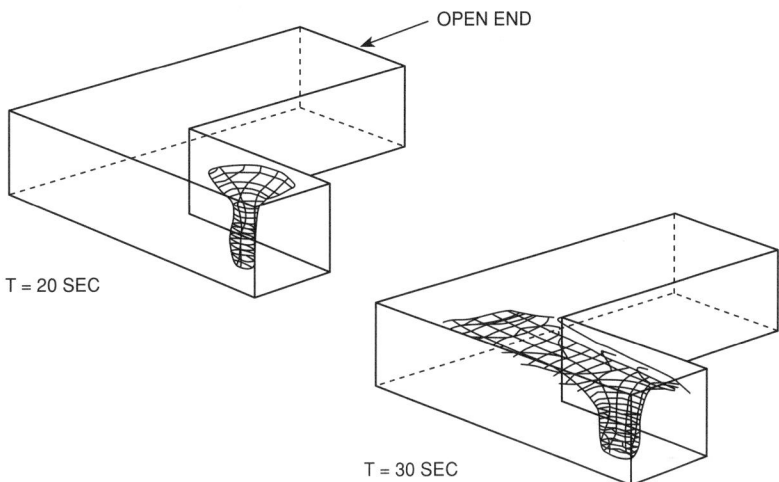

Fig. 3-8.5. Temperature contours at 20 and 30 sec for a fire in a shopping mall. (Results from ref. 18, calculated using JASMINE.)

model or other empirical parameters were used. The algorithms were verified by comparisons with exact solutions to the equations in simple, special cases. In addition, some experimental verification of the results for the formation and buoyant rise of a thermal plume due to a heat source in an enclosure was conducted. The use of Lagrangian particle tracking allowed the visualization of the three-dimensional flow patterns.

Development of JASMINE, which uses PHOENICS to solve the fluid flow equations, has been underway for over ten years. During that time, it has been used in a wide range of applications, including the study of fires in rooms, hospital wards, road tunnels, airport terminal buildings, and large sports stadia and shopping malls. In addition, it has been validated against a number of experimental fire tests. JASMINE was used to examine the smoke layer produced by a fire in a shopping mall.[27] Figures 3-8.5 and 3-8.6 show how the smoke layer flows through the mall area. The mall area does not fill instantaneously as a zone model would predict.

Another application of JASMINE by FRS and CHAM Ltd. was the simulation of fire in a large air-supported

structure.[29] The building, a 1/6th scale model of a 60,000-seat sports stadium being planned for construction in Japan, was covered by an air-supported dome, measuring 34 by 28 by 11.6 m high (111 by 92 by 38 ft high). A series of fire tests were performed within the structure. Results from JASMINE simulations reproduced the qualitative nature of the experimental data. In the far field, numerical and experimental results agreed to within 15 percent. However, in the flaming region results differ greatly. This was due, in part, to the coarse nature of the solution mesh in this region (only 12 or so cells). Due in part to a strong radial ceiling jet, the familiar well-mixed hot upper layer did not form.

Ove Arup and Partners, using JASMINE and PHOENICS, developed a code for use in the analysis of smoke movement in large buildings. The field modeling approach is particularly useful in predicting the movement of smoke in nonstandard buildings for which empirical data is nonexistent. Field modeling has been used to study the movement of smoke in large atria structures, such as the 74-m (243-ft) high atrium in the new Lloyds of London building.[35]

Using PHOENICS, SAFEAIR simulates the spread of smoke and heat in three dimensions within a burning aircraft

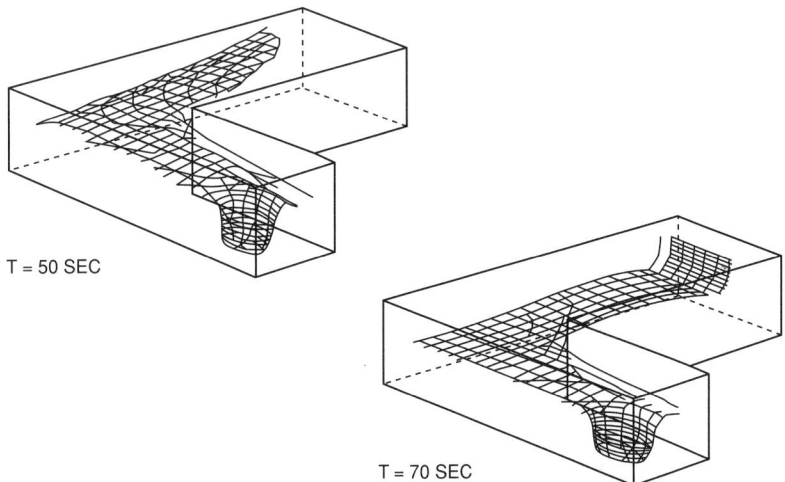

Fig. 3-8.6.   Temperature contours at 50 and 70 sec for a fire in a shopping mall. (Results from ref. 18, calculated using JASMINE.)

fuselage. The code uses body-fitted coordinates to construct relatively realistic aircraft shapes in addition to accounting for the contents of the cabin, such as passenger seats, hat racks, overhead bins, and cabin dividers. SAFEAIR has been used to reproduce the full-scale experimental results from fire tests conducted by NASA, in 1982, in an empty BOEING 737 fuselage.[31,32] Temperatures generated by SAFEAIR agreed reasonably well with the experimental values throughout the cabin. Maximum deviations of about 15 percent in the far field were observed when comparing numerical and experimental data. These calculations involved 20,328 computational cells, and required in excess of 40 hrs of CPU time on a computer comparable to a VAX 11/785.

SAFEAIR has been used to study the effects of openings in the fuselage and internal partitions on the temperature distributions within burning aircraft. Studies have also been performed to assess the interaction of the aircraft's air-conditioning system with cabin obstacles in the event of an onboard fire.[30,33]

Field modeling is being applied to the sprinkler-fire interaction problem. The presence of sprinklers complicates the problem by adding an additional phase that must be considered. The general fluid circulation of the hot combustion products represents the gas phase, while water droplets injected into the fire compartment by the sprinkler must be handled in the liquid phase. CLYTIE, using PHOENICS, is one example of a field model being developed to solve this problem.[34] Many of the commercially available computational fluid dynamics software packages can handle two-phase flow, to some extent.

A general-purpose field model, FLOW3D, has been developed and is being applied to a number of fire simulations. The model was used by Harwell, the developer, to simulate the fire at the King's Cross underground station.[36] On the evening of November 18, 1987, a fire erupted from the Piccadilly line escalator tunnel into the booking hall of King's Cross underground station in London, killing 31 people and injuring many others. The field model, surprisingly, predicted that the fire would lay down (instead of rising straight up) in the trench formed by the walls and steps of the wooden escalator. This phenomenon became known as the "trench effect." Although this prediction was met at first with much skepticism, it was shown by both the model and by experiments that the fire did behave as predicted. The "trench effect" was the result of two other effects. The first was the high-velocity, fire-induced wind speeds (14.5 m/s predicted by the model) along the base of the escalator. This high-velocity wind forced the plume to bend toward the escalator. The second effect, known as the "Coanda" effect, was the preferential entrainment of air into the plume from the side opposite the escalator floor. Traditional zone models were used in an attempt to analyze the fire. Since zone models did not account for fluid dynamic effects in sufficient detail, they could not accurately predict the observed fire conditions.

Under the sponsorship of the National Fire Protection Research Foundation, the U.S. General Services Administration, and the U.S. Fire Administration, the Building and Fire Research Laboratory at the National Institute of Standards and Technology is conducting a four-year study using FLOW3D.[4] The objective of this research study is to examine the effect of ceiling geometry on the distribution of heat and smoke in order to evaluate existing requirements for automatic fire detector location found in NFPA 72, *National Fire Alarm Code*, and for the activation of automatic sprinklers as covered by NFPA 13, *Standard for the Installation of Sprinkler Systems*; NFPA 13D, *Standard for the Installation of Sprinkler Systems in One- and Two-Family Dwellings and Manufactured Homes*; and NFPA 13R, *Standard for the Installation of Sprinkler Systems in Residential Occupancies up to and Including Four Stories in Height*. The study is examining the effects of fire size, fire location, ceiling height, ceiling slope, typical ceiling beam or joist configurations, and the effect of stratification on the distribution of heat and smoke from the fire. Expanding the study to include the effects of heating, ventilating, and air-conditioning systems on fire detectors and sprinkler activation is being considered. Initial results of the study indicate favorable comparisons with experimental data.

In another study, jointly funded by GSA and NASA, the Building and Fire Research Laboratory was given the opportunity to make measurements during fire calibration tests of the heat detection system in an aircraft hangar with a nominal 30.4-m (100-ft) ceiling height, near Dallas, TX. Fire gas temperatures resulting from an approximately 8250-kW isopropyl alcohol pool fire were measured above the fire and along the ceiling. For large spaces such as this, hot gas transport time can be very important. FLOW3D was used to model the hot gas movement in the space. Reasonable agreement was found between the temperatures predicted by the model calculations and the temperatures measured in the aircraft hangar. Also, as part of this study, an existing NASA high bay space was modeled using the FLOW3D model. The NASA space was a clean room, 27.4 m (90 ft) high with forced horizontal laminar flow. Using assumed fire sizes of 32 MW, 400 kW, and 40 kW, an analysis was conducted to determine how the existing fire detection devices would respond to various size fires in the space.[37]

In addition to ongoing work at NIST, field models are being developed, enhanced, and utilized by a number of research organizations and universities throughout the world. Researchers at Factory Mutual Research Corporation have used field models to simulate thermal plumes generated by pool and heptane spray fires,[38] and the interaction of sprinkler spray with a buoyant plume.[39] Work is underway at the National Fire Laboratory of the Institute for Research in Construction, National Research Council of Canada, to develop a field model to simulate fire in a room.[19] The model is based on a computer model developed by Lai for the simulation of industrial furnaces and burners. To make it suitable for room fire simulations, Hadjisophocleous has extensively modified Lai's original model. In an experimental study of ceiling jet and ceiling beam interaction, Motevalli at Worcester Polytechnic Institute is planning to use a field model to assist in placing measurement instrumentation before conducting tests to better analyze the data obtained from the tests.

CONCLUSION

The use of field modeling to analyze fire protection engineering problems is growing dramatically. Its use will become imperative as the complexity of problems increases. Field models use computational fluid dynamics software to solve the Navier-Stokes equations for fluid flow. The details about fluid flow and heat transfer provided by field models can prove vital in analyzing problems involving far-field smoke flow, complex geometries (e.g., sprinkler links), and impact of fixed ventilation flows. Field models have been applied to a wide range of scenarios, including aircraft terminals and atria, air-supported structures, electronic generating stations, aircraft cabins, tunnels, hospital wards, and shopping malls.

While field models provide more detail than zone models, they do have limitations. Field models do not have a direct simulation of turbulent diffusion flames. Except for limited cases, the fire source must be prescribed by the user. Other major phenomena that can only be approximated include turbulence, particularly large eddies associated with strong plumes and flames, and thermal radiation interchange between soot, gases, and solid surfaces. As the application of these models continues, these limitations should gradually be eliminated.

Cost and complexity will continue to limit the widespread use of field models in general engineering practice. Currently, field modeling set-up, computer, and software costs are approximately $100,000, with annual maintenance fees of about $10,000. This is a significant outlay of funds for an engineering firm with one or two potential applications per year. However, computer software is becoming more efficient as hardware costs continue to drop. The application of field models requires a great deal of user sophistication to specify the problem and interpret results. Most field modeling packages are commercial products, with fire being one of numerous potential applications. Work on enhanced user interfaces and simpler problem specifications can be expected to make field models easier to use in the future.

REFERENCES CITED

1. G.K. Batchelor, *An Introduction to Fluid Dynamics*, Cambridge University Press, London, England (1967).
2. S.V. Patankar, *Numerical Heat Transfer and Fluid Flow*, McGraw Hill, New York (1980).
3. B.E. Launder and D.B. Spalding, "The Numerical Computation of Turbulent Flows," *Computer Methods in Applied Mechanics and Engineering*, 3, pp. 269–289 (1974).
4. G.P. Forney, R.W. Bukowski, and W.D. Davis, *Field Modeling: Effects of Flat Beamed Ceilings on Detector and Sprinkler Response*, International Fire Detection Research Project, National Fire Protection Research Foundation, Boston, MA (1993).
5. D.B. Spalding, "A General-Purpose Computer Program for Multi-Dimensional One- and Two-Phase Flow," *Mathematics and Computers in Simulations*, North Holland (IMACS), XXIII, pg. 267 (1981).
6. N.C. Markatos, M.R. Malin, and G. Cox, "Mathematical Modeling of Buoyancy-Induced Smoke Flow in Enclosures," *International Journal of Heat and Mass Transfer*, 25, No. 1, pp. 63–75 (1982).
7. A.D. Burns, D. Ingrams, I.P. Jones, J.R. Knightly, S. Lo, and N.S. Wilkes, "FLOW3D: The Development and Applications of Rebase," *Harwell Report AERE/R/12693* (1987).
8. B. Hutchings, "Solution of Natural Convection Problems Using Fluent," *Fluent Users Newsletter*, 1, No. 1, pg. 6 (1986).
9. R.G. Rehm and H.R. Baum, "The Equations of Motion for Thermally Driven, Buoyant Flows," *Journal of Research of the NBS*, 83, pp. 297–308 (1978).
10. H.R. Baum, R.G. Rehm, H.R. Barnett, and D.M. Corley, "Finite Difference Calculations of Buoyant Convection in an Enclosure, 1. The Basic Algorithm," *SIAM J. Sci. Stat. Computing*, 4, pp. 117–135 (1983).
11. R.G. Rehm, H.R. Barnett, H.R. Baum, and D.M. Corley, "Finite Difference Calculations of Buoyant Convection in an Enclosure: Verification of the Non-Linear Algorithm," *Applied Numerical Mathematics*, 1, pp. 515–529 (1985).
12. H.R. Baum and R.G. Rehm, "Calculations of Three-Dimensional Buoyant Plumes in Enclosures," *Combustion Science and Technology*, 40, pp. 55–77 (1984).
13. R.G. Rehm, H.R. Baum, D.W. Lozier, and D.M. Corley, "A Model of Three-Dimensional Buoyant Convection Induced by a Room Fire," 1st National Fluid Dynamics Congress, AIAA-88-3723-CP (1988).
14. K.T. Yang, J.R. Lloyd, A.M. Kanury, and K. Satoh, "Modeling of Turbulent Buoyant Flows in Aircraft Cabins," *Combustion Science and Technology*, 39, p. 107 (1984).
15. K. Satoh and T. Kurioshi, "Three-Dimensional Numerical Simulations of Fires in Aircraft Passenger Compartments," *24th Jap. Aviation Symp.*, 2C-8, 1.
16. G. Cox and S. Kumar, *Combustion Science and Technology*, 52, p. 7 (1987).
17. R. Friedman, "An International Survey of Computer Models for Fire and Smoke," *Journal of Fire Protection Engineering*, 4, No. 3, pp. 81–92 (1992).
18. E. Galea, "On the Field Modeling Approach to the Simulation of Enclosure Fires," *Journal of Fire Protection Engineering*, 1, No. 1, pp. 11–22 (1989).
19. G.V. Hadjisophocleous and M. Cacambouras, "Computer Modeling of Compartment Fires," *Journal of Fire Protection Engineering*, 5, No. 2, pp. 39–52 (1993).
20. W.D. Davis, G.P. Forney, and J.H. Klote, "Field Modeling of Room Fires," NISTIR 4673, National Institute of Standards and Technology (1991).
21. Y.T. Yang and L.C. Chang, "UNDSAFE-I: A Computer Code for Buoyant Flow in an Enclosure," NBS-GCR-77-84 (1977).
22. V.K. Liu and K.T. Yang, "UNDSAFE-I: A Computer Code for Buoyant Turbulent Flow in an Enclosure with Thermal Radiation," NBS-GCR-76-150 (1978).
23. S. Kumar, N. Hoffman, and G. Cox, "Some Validation of Jasmine for Fires in Hospital Wards," *Numerical Simulation of Fluid Flow and Heat/Mass Transfer Processes*, Springer-Verlag, Berlin, p. 159 (1986).
24. N.C. Markatos and K.A. Pericleous, "An Investigation of Three-Dimensional Fires in Enclosures," *Revue Generale de Thermique*, 266, pg. 67 (1984).
25. S. Kumar and G. Cox, "Mathematical Modeling of Fires in Road Tunnels," *5th Int. Conf. on Aerodynamics and Ventilation of Vehicle Tunnels*, Lille, p. 61 (1985).
26. N.C. Markatos and G. Cox, "Hydrodynamics and Heat Transfer in Enclosures," *Physico-Chem. Hydrody.*, 5, No. 1, p. 53 (1984).
27. N.C. Markatos, M.R. Malin, and G. Cox, "Mathematical Modeling of Buoyancy Induced Smoke Flow in Enclosures," *Int. J. of Heat Mass Transfer*, 25, No. 1, p. 63 (1982).
28. G. Cox and S. Kumar, "The Mathematical Modeling of Fire in Forced Ventilated Enclosures," *18th DOE Nuclear Airborne Waste Manage. and Air Cleaning Conf.*, US Dept. of Energy Conf. 840806, p. 629 (1985).
29. K.A. Pericleous, D.R.E. Worthington, and G. Cox, "The Field Modeling of Fire in an Air-Supported Structure," *2nd Int. Symp. on Fire Safety Science*, Tokyo, pp. 13–17 (1988).
30. E.R. Galea and N.C. Markatos, "Aircraft Cabin Fires: A Numerical Simulation," *12th IMACS World Congress*, Paris (1988).
31. E.R. Galea and N.C. Markatos, "Prediction of Fire Development in Aircraft," *2nd Int. Phoenics Users Conference*, London (1987).
32. E.R. Galea and N.C. Markatos, "Modeling of Aircraft Cabin Fires," *2nd Int. Symp. on Fire Safety Science*, Tokyo, pp. 13–17 (1988).
33. E.R. Galea and N.C. Markatos, "Progress in Mathematical Modeling of Aircraft Cabin Fires," *Disaster Management*, 1, No. 1 (1988).
34. N. Hoffman, E.R. Galea, and N.C. Markatos, "A Computer Simulation of Fire-Sprinkler Interaction: A Two-Phase Phenomena," *12th IMACS World Congress*, Paris (1988).
35. R. Waters, "Air and Smoke Movement within a Large Enclosure," *Numerical Simulation of Fluid Flow and Heat/Mass Transfer Process*, Springer-Verlag, Berlin (1986).
36. S. Simcox, N.S. Wilkes, and I.P. Jones, "Computer Simulations of the Flows of Hot Gases from the Fire at King's Cross Underground Station," *Institution of Mechanical Engineers (ImechE) King's Cross Underground Fire: Fire Dynamics and the Organization of Safety*, pp. 19–25 (1989).
37. K.A. Notarianni and W.D. Davis, "The Use of Computer Models to Predict Temperature and Smoke Movement in High Bay Spaces," NISTIR 5304, National Institute of Standards and Technology (1993).
38. S. Nam and R.G. Bill, "Numerical Simulation of Thermal Plumes," *Fire Safety Journal*, 21, No. 3, pp. 231–256 (1993).
39. R.G. Bill, "Numerical Simulation of Actual Delivered Density (ADD) Measurements," *Fire Safety Journal*, 20, No. 3, pp. 227–240 (1993).

SMOKE AND HEAT VENTING

P. L. Hinkley

INTRODUCTION

Venting is the removal of hot smoky gases from the upper parts of a compartment partially or completely involved in fire and the introduction of air from outside the compartment into its lower parts. The process may involve natural convection through openings that occur fortuitously or are provided purposely or it may involve mechanical (powered) extract or inlet, or both. This chapter is concerned with the basic engineering concepts underlying the design of complete venting systems, including the provision of openings (vents and inlets) or fans and allied features, such as the provision of screens (curtains) to limit the spread of smoke beneath the ceiling.

Integrated Design

It is important that a venting system (vents, inlets, screens, and means of control) be designed as a whole, taking into account other fire protection features, such as fire resisting structural elements, escape routes, and sprinklers.

Objectives of Venting

The main objectives of venting are:

1. To facilitate escape of people by restricting spread of smoke and hot gases in escape routes.
2. To facilitate fire fighting by enabling fire fighters to enter the building and to see the seat of the fire.
3. To reduce damage due to smoke and hot gases.

Venting design achieves the above objectives through one of two general methods:

1. Usually venting is designed to control the spread of hot smoky gases within the compartment containing the fire or to an adjacent compartment connected by large openings, e.g., a shopping mall with a fire in a shop. The hot smoky gases are confined to a layer beneath the ceiling from which they flow out through vents (or are mechanically extracted), and at the same time replacement air

flows beneath the hot gases. The extent of the hot gas layer may be limited by screens. In effect the flow in the compartment is "open-circuited," and recirculation of smoke to the lower levels within the compartment is minimized.

2. Alternatively, venting may be designed to control the pressure distribution in a fire compartment so as to restrict the flow of smoke, hot gases, and flames through small openings to adjacent areas. Examples include venting the stage in a theatre to prevent smoke flow to the auditorium from a fire on stage, or venting an atrium to prevent the flow of smoke to adjacent accommodations or escape routes. This method is considered in the last section of this chapter.

VENTING ZONE MODEL

The flow of smoke in a compartment is a very complex phenomenon, but for the purpose of the design of roof venting systems it is often adequate to employ a grossly simplified "zone" model.[1] However, it is necessary for the engineer to understand its limitations.

The zone model is illustrated in Figure 3-9.1. The plume of hot gases and flames produced by a fire entrains air as it rises, and by the time it reaches the ceiling the mass rate of flow of entrained air is much greater than the mass rate of flow of gases produced from the burning material. After reaching the ceiling, the hot gases flow radially outward as a ceiling jet,[2] having a thickness that increases with radius but typically is of the order of $1/10$ of the ceiling height. (See Section 2, Chapter 4.) The ceiling jet entrains the underlying air, and its velocity decreases with increasing radius from the plume. The jet soon reaches the confines of the compartment (or the roof screens), and the hot gases then flow back toward the fire to form a layer above the underlying cool air, which deepens with time.

The zone model infers the ceiling jet to be completely immersed in the layer of hot gases and generally disregards the circulation within the layer, which is assumed to be essentially stagnant and at a uniform temperature. Mixing between the hot gas layer and the cool air beneath is inhibited by the density difference and is neglected in many situations.

Venting is employed to remove gases from the hot layer at a rate comparable with that at which they are "produced"

Peter Hinkley was a member of the UK Fire Research Station from 1951 to 1983 where he worked on smoke control among many subjects. He is now employed by Colt International on a joint project with the Fire Research Station on roof venting.

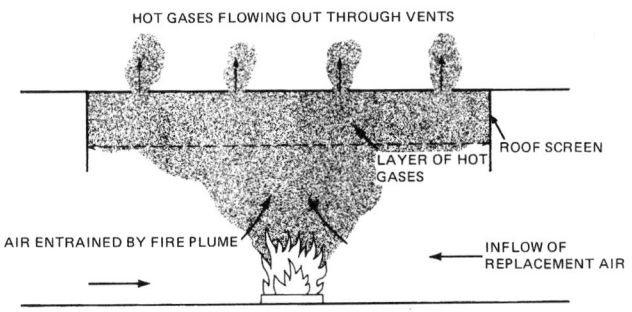

HOT GASES FLOWING OUT THROUGH VENTS

LAYER OF HOT GASES

ROOF SCREEN

AIR ENTRAINED BY FIRE PLUME

INFLOW OF REPLACEMENT AIR

Fig. 3-9.1.   Zone model of roof venting.

by decomposition or evaporation of the fuel and by subsequent entrainment of air. Often the fuel gases are neglected and it is only the air entrained into the rising plume that is considered in the calculations. Some situations where it is necessary to consider entrainment of air beyond the fire plume are considered later in this chapter.

PRESSURES CAUSING FLOW THROUGH VENTS

The most important pressures acting to control the flow of gases out of or into a compartment containing a fire are those due to buoyancy, wind, and constraints to the expansion of gases that are rising in temperature.[3] Occasionally the effects of the compartment's HVAC system may also have to be taken into account.

Buoyancy

Standard pressure of the atmosphere at sea level is 101 300 Pa, and it decreases by about 12 Pa/m rise in height. The pressure within a layer of hot gases beneath a ceiling decreases more slowly with height, as illustrated in Figure 3-9.2. If it is assumed that there are openings to the outside air below the level of the hot gases, then the pressure at the base of the layer is effectively the same as that of the atmosphere at the same level. The difference in pressure between inside and outside at ceiling level is

$$P_B = g\rho_0 \theta d / (T_0 + \theta) \tag{1}$$

where

P_B = pressure due to buoyancy (Pa)
d = depth of layer of hot gases (m)
T_0 = absolute ambient temperature (K)
θ = temperature above ambient of the layer of hot gases (K)
ρ_0 = density of ambient air (kg/m^3).

EXAMPLE:

Taking d = 2 m, T_0 = 290 K, ρ_0 = 1.2 kg/m^3, θ = 100 K, and g = 9.81 m s^{-2}, then P_B = 6.0 Pa.

Wind Pressure

The effects of wind are a function of the shape and size of the building,[4,5] and of surrounding buildings and other objects. The exterior pressures on a building due to wind are related to the wind velocity by the expression

$$P_W = C_p \rho_0 v_w^2 / 2 \tag{2}$$

where

P_W = pressure increase due to wind above that in the undisturbed free wind stream (Pa)
C_p = pressure coefficient
v_w = velocity of undisturbed free wind stream (m/s).

C_p, which may be positive or negative (suction), varies over the surface of the building and depends on the wind direction, the arrangement of surrounding buildings, and the local topography. It may be measured in wind tunnel experiments on a scale model of the building and its environment.

Typical values for a wind blowing perpendicular to one side of an isolated low square building with a flat roof would be +0.7 to windward, −0.5 to leeward, −0.5 on the sides, and a mean of −0.8 (locally up to −2) on the roof. (See Figure 3-9.3.) With pitched roofs, C_p is likely to be positive with a 30-degree windward-facing pitch and may be so for even less steep pitches. It will be negative for leeward-facing pitches.

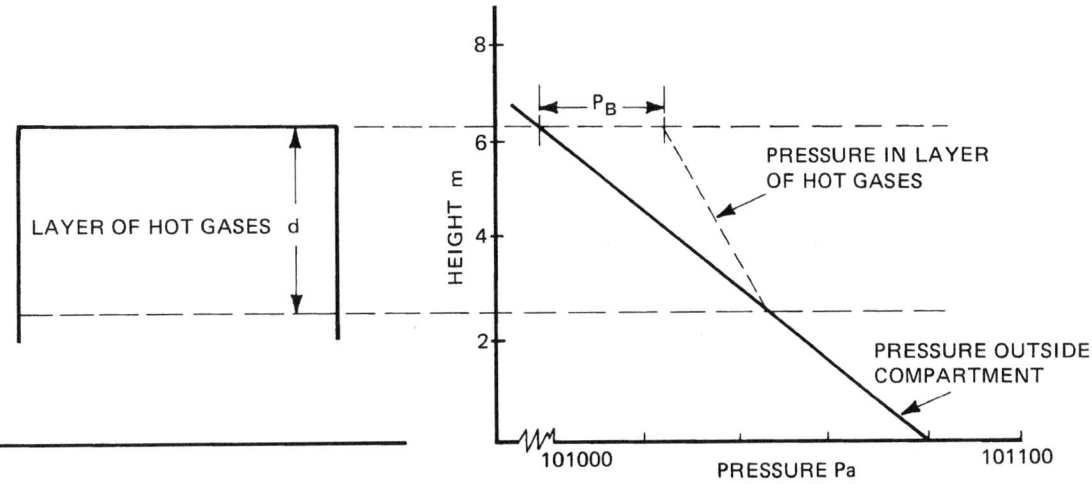

LAYER OF HOT GASES   d

P_B

PRESSURE IN LAYER OF HOT GASES

PRESSURE OUTSIDE COMPARTMENT

101000   PRESSURE Pa   101100

Fig. 3-9.2.   Pressure on a ceiling due to buoyancy, P_B.

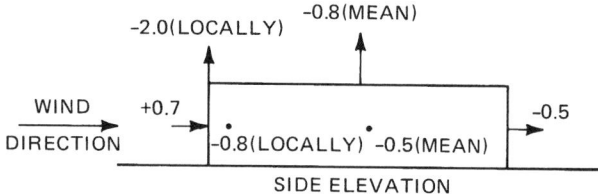

Fig. 3-9.3. Pressure coefficients for an isolated, low, square building.

EXAMPLE:

Taking a wind of 10 m/s and $C_p = 1$, then from Equation 2, $P_w = 60$ Pa. For the building illustrated in Figure 3-9.3 the suction pressure on the roof would be $60 \times 0.6 = 36$ Pa. From Equation 1 this pressure is equivalent to that produced by a 12-m-deep layer of gases at 200 K above ambient. It is therefore evident that wind pressures can be greater than buoyancy pressures in many situations.

The internal pressure in an undivided building depends on the disposition and size of openings, as well as the external pressures. If the building in the previous example has small openings uniformly distributed around its periphery, then the internal pressure will be below ambient, with C_p about -0.2, thus lessening the suction effect on the roof. A preponderance of openings on the leeward side will decrease the pressure within the building, while a preponderance of openings to windward will increase it.

In divided buildings, the internal pressures due to wind depend also on the arrangement of the internal walls and floors and on the openings (including leakage) between them. These pressure calculations, in effect, involve balancing the mass flow into every compartment with the corresponding outflow. The calculation is simple in principle but complex in practice.

Pressure Rise Due to Constraints on Expansion

The pressure will rise in a compartment with a limited area of openings to the outside when the mean temperature of the gases in it is rising. Following an abrupt change in the rate of rise in temperature, a quasi-steady pressure will be attained in a few seconds.

For pressure increases below about 10^4 Pa the pressure rise is related to both the mean temperature within the compartment and the rate of rise in temperature by the engineering equation

$$P_E = 180(BhA_c)^2/[C_d^2 A_L^2 (T_0 + \theta)^3] \qquad (3)$$

where

P_E = pressure rise due to constraints on expansion (Pa)
B = rate of rise of temperature (K/s)
h = compartment height (m)
A_c = area of compartment (m²)
A_L = total area of all openings into compartment (including leakage) (m²)
C_d = coefficient of discharge (0.6)
T_0 = absolute ambient temperature (K)
θ = temperature above ambient of the gases in compartment (K).

If the temperature rises at a constant rate, the maximum pressure occurs when the gases in the compartment are cold.

EXAMPLE:

A compartment has an area of 1000 m², a height of 10 m, and a total area of leakage of 5 m². There is a fire in the compartment producing a mean temperature of 200 K above ambient, rising at 4 K s⁻¹. From Equation 3 the pressure is 272 Pa.

The pressure is high compared with those previously calculated for either buoyancy or wind, but the assumed rate of rise in temperature was high and the "leakage" area was fairly small. Although in theory high pressures could be attained in sealed buildings, small leakage areas ensure that pressures are low unless temperatures are rising rapidly, as in the example.

Pressures Used in Calculations of Vent Areas

Calculations of vent areas usually involve buoyancy pressures only; wind pressures are, if possible, arranged to augment buoyancy flow by design and siting of vents. (See the section "Miscellaneous Design Provisions.") If adverse wind effects are likely to be severe, mechanical extract is necessary.

If is often implicitly assumed that, if natural leakage is insufficient, venting will provide sufficient area of openings to relieve pressures due to a rising temperature, except with "flash" fires.

MASS RATES OF FLOW THROUGH VENTS

Flow Through a Vent or Inlet with Pressure

The mass rate of flow of hot gases through a vent or inlet with pressure is given[1] by

$$M = C_d A [2T_0 \rho_0 P_d/(T_0 + \theta)]^{1/2} \qquad (4)$$

where

M = mass flow rate (kg/s)
P_d = pressure difference across vent or inlet (Pa)
A = area of the opening (m²)
C_d = coefficient of discharge
T_0 = absolute ambient temperature (K)
θ = temperature of the gases above ambient (K)
ρ_0 = density of the ambient air (kg/m³).

In most practical situations the Reynolds number of the flow is sufficiently high for the mass flow to be sensibly independent of it. In the absence of other data C_d may be taken to be 0.6 which is a reasonable value for a square-edged opening. (See section "Factors Affecting Discharge Through a Vent.")

EXAMPLE:

If T_0 is taken as 290 K, then ρ_0 is 1.2 kg/m³. Taking $C_d = 0.6, A = 1$ m², $P_d = 10$ Pa, $\theta = 100$ K above ambient, then $M = 2.53$ kg/s. If $\theta = 500$ K above ambient, then $M = 1.78$ kg/s.

Flow out of a Vent Due to Buoyancy

Assuming the layer of hot gases to be deep compared with the linear dimension of the vent and to be effectively

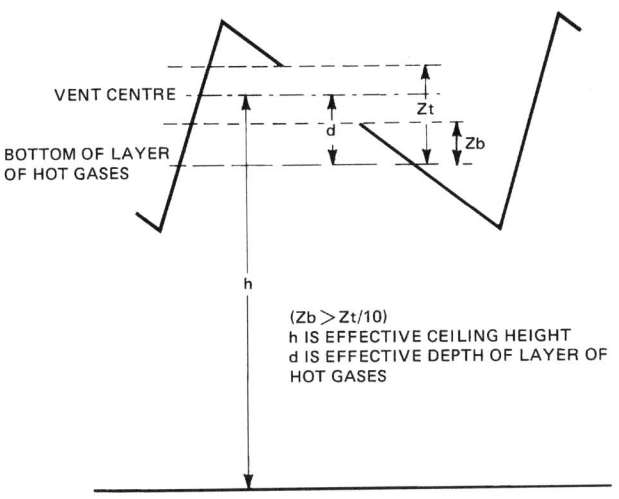

Fig. 3-9.4. *Design position of gas layer vs vent.*

stagnant, the area of inlets to be large compared to the area of the vents, and the external air to be effectively stagnant

$$M_v = C_d A_v \rho_0 (2gd\theta T_0)^{1/2}/(T_0 + \theta) \qquad (5)$$

where

M_v = mass rate of flow through vent (kg/s)
A_v = free vent area (m^2)
d = depth of the layer of hot gases (m).

EXAMPLE:

Taking $d = 1$ m and $\theta = 100$ K above ambient, the other quantities being the same as in the example for Equation 4, then $M_v = 1.39$ kg/s. If $\theta = 500$ K above ambient, then $M_v = 1.53$ kg/s.

Vent in a Sloping Roof

Equation 5 assumes that the vent is in a horizontal flat roof. If the vent is in a sloping roof or a wall, then Equation 5 can still be used with the following provisions:

1. The design position of the bottom of the layer of hot gases should be at a vertical distance below the lowest part of the vent greater than 10 percent of the vertical distance from the top of the vent. (See Figure 3-9.4.)
2. h and d are measured to the center of the vent.

Effect of Temperature on Mass Flow Due to Buoyancy

The effect of temperature on mass flow due to buoyancy is shown in Figure 3-9.5. The mass flow is a maximum when $\theta = T_0$ and it is within 80 percent of its maximum value over the approximate range of temperature $75 < 0 < 1150$ K above ambient. If the temperature of the layer of hot gases falls below about 75 K above ambient, there is a serious decrease in the mass rate of flow through a vent.

Effect of Limited Area of Inlets

The pressure drop across the inlets due to the inflow of replacement air must be subtracted from the buoyancy pressure causing the gases to flow through the vents. The effect of inlet pressure drop may be included[1] in Equation 5 by replacing A_v by an effective vent area (A_v^*) where

$$1/A_v^{*2} = (1/A_v^2) + (1/A_I^2)[T_0/(T_0 + \theta)] \qquad (6)$$

where

A_v^* = effective area of vents (m^2)
A_I = area of inlets (m^2).

(A_v^*/A_v) is plotted against (A_I/A_v) in Figure 3-9.6 for $T_0 = 290$ K and $T = 290$ K, 690 K, and 1090 K. If the capacity of the venting system is to be within 90 percent of its computed value, then the ratio between the inlet area and the vent area must be unity if the layer of hot gases has a temperature of 800 K above ambient and over 2 if the layer is anticipated to be at a low temperature.

EXAMPLE:

Taking values from the previous example, if the area of inlet is 1 m^2 from Equation 6 and if $\theta = 500$ K above ambient, then $A_v^* = 0.86$ m^2, and $M_v = 1.31$ kg/s; and if $\theta = 100$ K above ambient, then $A_v^* = 0.76$ m^2 and $M_v = 1.05$ kg/s.

If the inlet area is substantially less than the vent area, some of the vents far from the fire may act as inlets for fresh air, and in extreme cases there may be two-way flow through individual vents. (See section "Factors Affecting Discharge Through a Vent.)

RATE OF PRODUCTION OF HOT GASES

The rate of production of hot gases is primarily dependent on the rate of entrainment of air into the column of flames and hot gases rising above the fire over the distance between the floor and the base of the layer of hot gases ($h - d$). Although the entrainment process continues above that level, it effectively results only in mixing within the layer of hot gases.

A full discussion of entrainment into plumes is given in Section 2, Chapter 2. However, for the purpose of the design of roof venting systems some approximate engineering formulae have been developed on the basis of experimental

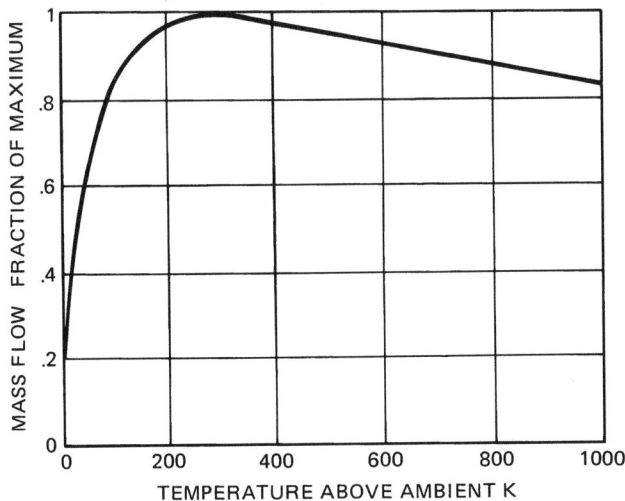

Fig. 3-9.5. *Effect of temperature on mass flow through a vent.*

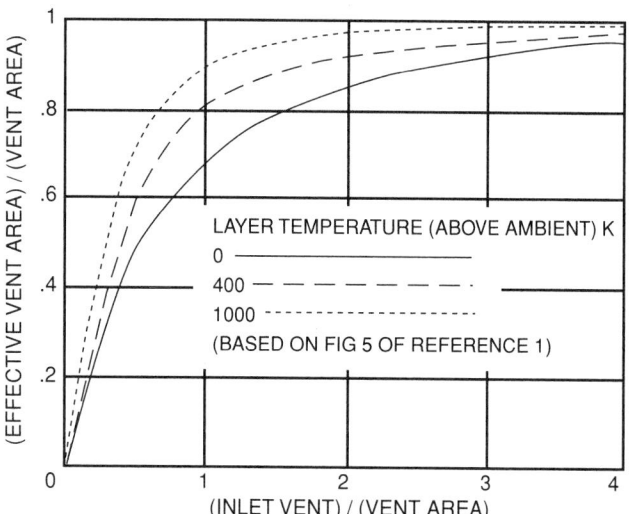

Fig. 3-9.6.　*Plot of (A_v^*/A_v) against (A_I/A_v).*

results. Different formulae have been adopted for National Fire Protection Association (NFPA) codes[6,7] and for design purposes in Europe;[8,9] the latter are based on the formulae in the UK Fire Research Station Technical Paper No. 7.[1] In general, they result in broadly similar estimations of vent area.

For fires of small dimensions and low heat output, the rising column of flames and hot gases is approximated to a thermal plume rising from a point source of heat, although modifications are required to allow for deviations from ideal theoretical conditions. Empirical formulae based on experimental results are adopted for fires that are so large that this approximation is no longer valid.

UK Fire Research Station (FRS) Technical Paper No. 7[1]

Small fires:　Small fires are defined as those having a mean diameter less than $(h - d)/2$. It is assumed that the plume from small fires behaves as though it originates from an "effective point source" at a distance (y_0) below the floor, i.e.

$$y_0 = 1.5 A_f^{1/2} \qquad (7)$$

where A_f = the area of the fire (m²).

The mass rate of production of hot gases by entrainment is given by

$$M_p = 0.15 \rho_1 [Q_c g/(\rho_1 S T_0)]^{1/3} (h + y_0 - d)^{5/3} \qquad (8)$$

where

Q_c = convective heat output of fire (W)

S = specific heat of air at constant pressure (J kg⁻¹ k⁻¹)

ρ_1 = density of the hot gases (kg/m³) at height h (m)

ρ_1 may be approximated to the density of air at the same temperature, i.e.

$$\rho_1 = \rho_0 M_p S T_0/(Q_c + M_p S T_0) \qquad (9)$$

M_p may be obtained by simultaneous solution of Equations 8 and 9 numerically.

EXAMPLE:

If $T_0 = 290$ K, then $\rho_0 = 1.2$ kg/m³ and $S = 1000$ J kg⁻¹ K⁻¹. Taking $h = 5$ m, $A_f = 1$ m², $Q_c = 500000$ W, and $d = 1$ m, then from Equation 7 $y_0 = 1.5$ m, and by trial and error solution of Equations 8 and 9, $M_p = 6.3$ kg s⁻¹.

Large fires:　Large fires are defined as those for which $D > (h - d)/2$, and have a sufficiently large heat output for flames to extend into the layer of hot gases. The engineering equation that applies to a fire roughly square or circular in plan is

$$M_p = 0.188\, P\, (h - d)^{3/2} \quad \text{(kg/s)} \qquad (10)$$

where P = perimeter of the base of the fire (m), and h and d are in m. Note that Equation 10 does not explicitly include the heat output of the fire, Q_c.

Analysis[10] shows that Equation 10 is as good a fit to the available results of experiments on roof venting as any engineering formula and is a reasonable approximation with fires of heat output in the range 0.2 to 0.75 MW m⁻² and having $P > h/2$.

A modified form of Equation 10 may be applied[9] to small area spaces, such as unit offices or small shops with ventilation openings predominantly on one side of the fire (e.g., a shop window).

$$M_p = 0.34 P (h - d)^{3/2} \quad \text{(kg/s)} \qquad (11)$$

EXAMPLE:

Taking the same conditions as for the previous example ($P = 4$ m, and $h - d = 4$ m), then $M_p = 6$ kg/s. Taking a larger fire 3 m × 3 m with a heat output of 5 MW in the same compartment, $M_p = 18$ kg/s. If the smaller fire occurred in an open-front shop, then the corresponding rate of flow would be 11 kg s⁻¹.

NFPA 204M[6]

The formulae are based on the heat output of the fire; its physical dimensions do not appear explicitly.

Small fires:　The criterion for small fires is that the *continuous* flaming region does not extend into the layer of hot gases, and this means that the convective heat output (assuming complete combustion) does not exceed a critical value

$$Q_{crit} = 2.3 \times 10^5 (h - d)^{5/2} \quad \text{(W)} \qquad (12)$$

where

Q_{crit} = critical convective heat output of the fire (W).

For small fires ($Q_f < Q_{crit}$)

where

Q_f = convective heat output of the fire assuming complete combustion (W)

$$M_p = 7.1 \times 10^{-3} Q_f^{1/3}[(h - d)^{5/3} + 2.6 \times 10^{-4} Q_f^{2/3}] \quad (13)$$

EXAMPLE:

Taking the same figures as for the example to Equation 9 ($h = 5$ m, $d = 1$ m) and $Q_f = 500000$ W, then $M_p = 6.6$ kg/s.

Large fires: The criterion for large fires is $Q_f > Q_{crit}$

$$M_p = 5.2 \times 10^{-4}(h - d)Q_f^{3/5} \quad \text{(kg/s)} \quad (14)$$

EXAMPLE:

Taking the same figures as for the example to Equation 10 ($Q_f = 5$ MW, $h - d = 4$ m), then $M_p = 22$ kg/s.

Although in the above examples there is little difference between the FRS and the NFPA figures, this is not always true. Differences can be considerable with fires of high heat output per unit area. Where there is doubt, reference should be made to experimental evidence, if it is available, or to more exact theory.

Entrainment Beyond the Fire Plume

Generally, mixing between a relatively stagnant layer of hot gases and the air beneath is small enough to be neglected in the design of venting systems. However, this may not be true for compartments having complicated geometry. In some situations, the hot gases spread laterally under a lower level ceiling before rising vertically again as a complex thermal plume (a "spill plume"). Examples of such situations occur with a fire beneath an internal balcony or mezzanine floor or in a shop opening onto a mall with a higher ceiling. If it is necessary to maintain the balcony or mezzanine floors smoke free, the design of the venting system may be complicated and involve the strategic positioning of inlets to provide a flow of fresh air to the upper floors.[8]

Generally the mass of additional air entrained is reduced by restricting lateral spread beneath the balcony or lower level ceiling with screens. (See Figure 3-9.7.)

Entrainment into a Spill Plume

Several formulae have been proposed for calculating the entrainment into a spill plume. Generally they regard the plume as a line plume but include a correction for entrainment into the ends.

One (due to Thomas[11]) applies to free plumes when there is free access of air behind the plume, e.g., if the layer of hot gases emerges from beneath a deep balcony. (See Figure 3-9.7.)

$$M_p = 0.58 \, \rho_1 (gQ_w w^2/\rho_1 ST_0)^{1/3} \\ \cdot (x + D)[1 + 0.22(x + 2D)/w]^{2/3} \quad (15)$$

where

M_p = total rate of production of hot gases (kg/s)
w = width of opening (m)
x = height of clear layer above the compartment/balcony edge ($h_B - d$ in Figure 3-9.7) (m)

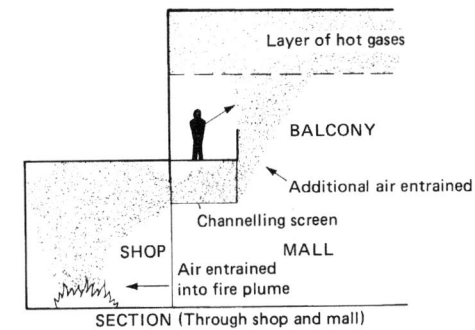

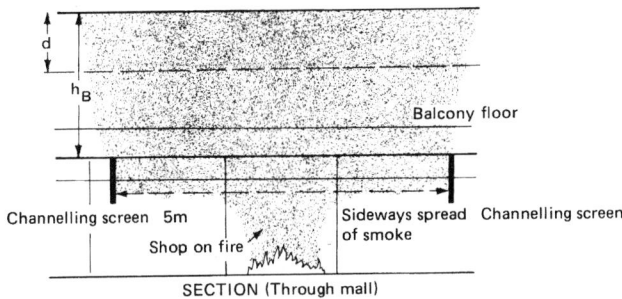

Fig. 3-9.7.   *Entrainment reduced by screens.*

Q_w = horizontal heat flow (kW)
T_0 = absolute ambient temperature (K)
ρ_1 = density of gases in the upper layer (kg/m³)
D = distance of the effective line source below compartment/balcony edge (m)
S = specific heat of air at constant pressure (kJ kg⁻¹ K⁻¹).

$$\rho_1 = \rho_0 T_0/[T_0 + Q_w/(M_p S)] \quad (16)$$

where ρ_0 is the density of air at T_0 K (kg/m³).

D apparently depends on the geometry of the compartment, but there is some evidence[9] for taking $D = 0.3 H_w$, where H_w is the height of the opening (m).

NFPA 92B[7] suggests a modification of the above formula

$$M_p = 0.41(Q_w w^2)^{1/3}(x + 0.3H_w) \\ \cdot [1 + 0.063(x + 0.6H_w)/w]^{2/3} \quad (17)$$

The Building Research Establishment (BRE) Spill Plume

The UK Building Research Establishment[9] developed a method of calculation of entrainment into spill plumes above a certain height (about 3 m, or 4.2 m when the ends of the plume are confined by walls) when the temperature of the approach flow is less than 350°C above ambient. It is applicable to both free plumes and (with modifications) adhered plumes. A simplified version is given here.

Define

$$A_d = M_E(1 + 0.847/C_d^{3/2}) + Q_E(1 + 1.159/C_d^{3/2})/290 \quad (18)$$

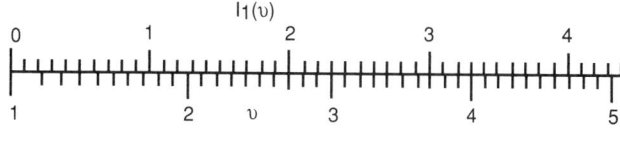

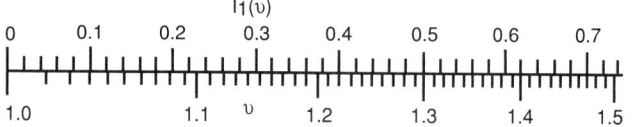

Fig. 3-9.8. $I_1(v)$ as a function of v.

where

M_E = horizontal mass flow per unit width of opening (kg m^{-1} s^{-1})

Q_E = horizontal heat flow per unit width of opening (kW/m)

C_d = coefficient of discharge at opening.

For a deep downstand $C_d = 0.6$ and $A_d = 2.822\,M_E + 0.01205\,Q_E$. With no downstand $C_d = 1$ and $A_d = 1.847\,M_E + 0.00744\,Q_E$.

Calculate

$$I_1(v) = (0.228Q_E^{1/3}x/A_d) + 0.21 \qquad (19)$$

where $I_1(v)$ is the integral defined by Lee and Emmons.[12]

$$I_1(v) = \int_1^v \frac{v\,dv}{(v^3 - 1)^{1/3}} \qquad (20)$$

In Figure 3-9.8, $I_1(v)$ is presented as a function of v in two alignment diagrams, one covering values of $I_1(v)$ up to 4, and an expanded one for values less than 0.7,

then

$$m_x = 0.930A_d v - Q_E/290 \qquad (21)$$

where m_x is the mass flow into the layer per m width of the line plume (kg m^{-1} s^{-1}). Entrainment into the ends of the plume is calculated separately.

Replace x in Equation 19 by $x/2$, and thus determine a new value of v, say v_{end}, then

$$M_{end} = 0.336v_{end}A_d x \qquad (22)$$

where

M_{end} = total rate of entrainment into both ends (kg/s).

Therefore

$$M_p = m_x w + M_{end} \text{ (kg s}^{-1}) \qquad (23)$$

Single-Sided (Adhered) Plumes

When a plume emerges from a wide opening, such as a shop front with a wall above, the plume will adhere to the wall and there will be entrainment into one side only. In this case the following procedure applies.

Calculate A_d from Equation 18 then,

$$A_{dA} = 2A_d. \qquad (24)$$

The value of $I_1(v)$ applicable to adhered plumes is

$$I_1(v) = (0.166Q_E^{1/3}x/A_{dA}) + 0.043 \qquad (25)$$

Determine v from the alignment diagram (Figure 3-9.8)

$$m_x = 0.492A_{dA}v - Q_E/290 \qquad (26)$$

To calculate the entrainment into the ends, replace x in Equation 25 by $x/2$, and thus determine a new value of v, say v_{end}, then

$$M_{end} = (0.336v_{end}A_{dA}x)/2 \qquad (27)$$

Finally, calculate M_p from Equation 23.

EXAMPLE:

Taking $w = 10$ m, $H_w = 5$ m, $M_E = 2$ kg/s, $x = 5$ m, and $Q_E = 300$ kW.

Using Thomas' Equations 15 and 16, putting $D = 0.3H_w$, then $M_p = 95$ kg/s. Taking $D = 0.67H_w$, then $M_p = 131$ kg/s.

Using NFPA 92B[7] Equation 17, $M_p = 180$ kg/s.

Using BRE spill plume Equations 18 through 23, if the gases flow beneath a downstand ($C_d = 0.6$), then $M_p = 162$ kg/s. If there is no downstand ($C_d = 1$), then $M_p = 126$ kg/s.

There are considerable differences among the results of the different calculation methods. For the example given here, with no downstand, the BRE method gives approximately the same flow rate as putting $D = 0.67\ H_w$ in Thomas' equation. With a deep downstand, the BRE method gives a flow rate approaching that given by the NFPA 92B equation.

REQUIRED VENT AREAS

Vent Areas as a Function of Size of Fire

The required vent area is calculated by equating the mass rate of flow through vents with the mass rate of production.

The following formulae apply when the rate of production of hot gases is determined only by the rate of entrainment into the fire plume. They all assume a coefficient of discharge through the vents of 0.6; areas can be reduced *pro rata* if vents have a higher discharge coefficient. They also assume no pressure drop across fresh air inlets. Areas must be increased according to Equation 6 if inlets are not much larger than vents.

If the vents are in a sloping roof or are in a wall, then h and d are measured to a horizontal plane through the center of the vents. It is implicit that the method of control ensures that the required area of vents opens.

UK Fire Research Station (FRS) Technical Paper No. 7[1]

These equations from the work of Thomas *et al*[1] are based on the dimensions of the fire; the heat output is not explicitly involved.

Small fires: Where mean diameter is less than $(h - d)/2$

$$A_v = 0.072(h + y_0 - d)^{5/2}/d^{1/2} \text{ (m}^2) \qquad (28)$$

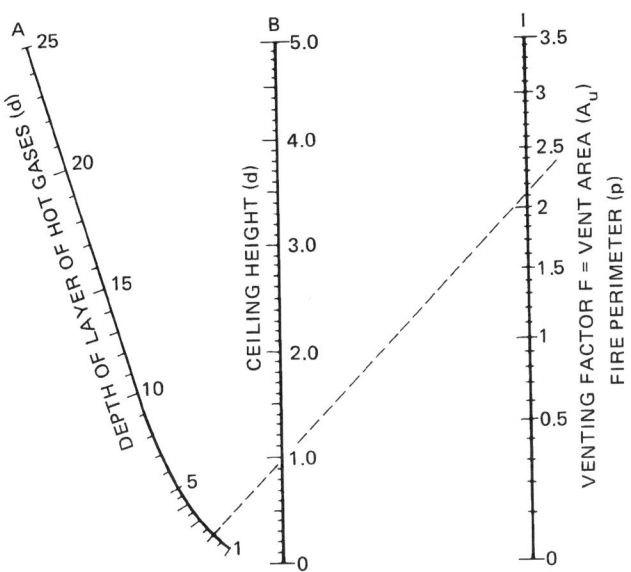

Fig. 3-9.9. *Nomogram for determining venting factor.*

where

A_v = area of vents (m^2)
h = height of compartment (m)
d = design depth of layer of hot gases (m)
y_0 = distance of effective point source beneath floor level (m).

$$y_0 = 1.5 A_f^{1/2} \quad \text{(m)} \tag{29}$$

where A_f = floor area of the fire (m^2).

Large fires: Where mean diameter is greater than $(h - d)/2$, and flames extend into the layer of hot gases

$$A_v = 0.13 \, P(h - d)^{3/2}/d^{1/2} \quad \text{(m}^2\text{)} \tag{30}$$

where P = perimeter of the fire (m).

Equation 30 has also been found[10] to apply to the results of large-scale venting experiments where the mean diameter was only $(h - d)/5$.

Figure 3-9.9 is a nomogram for determining a venting factor $F_v = (A_v/P)$ given the ceiling height, h, and the design depth of the layer of hot gases, d.

NFPA 204M[6]

These equations are based on the heat output of a fire; dimensions are not explicitly involved.

Small fires: These are defined as those where the heat output of the fire $(Q_f) < Q_{crit}$. (Q_{crit} is given by Equation 12.)

For small fires

$$A_v = 2.7 \times 10^{-3} Q_f^{1/3}[(h - d)^{5/3} + \\ 2.6 \times 10^{-4} Q_f^{2/3}]/d^{1/2} \quad \text{(m}^2\text{)} \tag{31}$$

Large fires may be defined (there is some overlap with small ones) if $Q_f > 0.2 \, Q_{crit}$

$$A_v = 1.94 \times 10^{-4} Q_f^{3/5}(h - d)/d^{1/2} \quad \text{(m}^2\text{)} \tag{32}$$

THE "DESIGN FIRE" FOR VENT AREAS

The above equations assume a foreknowledge of the fire size, either its dimensions or its heat output. In some (generally rare) circumstances such as isolated fire hazards, e.g., a tank of flammable liquids with little chance of fire spread to adjacent hazards, it may be possible to predict the fire size with a high probability.

In general, however, there are few restrictions to fire spread, and it is necessary to estimate the likely size of the fire at the time effective fire fighting is taking place. This is the "design fire" on which the calculations of vent area are based.

In some circumstances it may be possible for an experienced engineer to make a reasonable estimate for the size of the design fire. Such an estimate should take into account such features as the design and construction of the building, the nature and distribution of the contents, the effect of sprinklers, the means of detection of the fire, and the means of calling and the proximity of the fire brigade. (See NFPA 92B[7] for data on the choice of a design fire.)

Where adequate fire statistics are available, they may be used to provide information upon which to base the decision of the size of the design fire. NFPA 92B[7] states that there is a lack of data in North America to indicate that the design fire is exceeded, except in only a limited number of cases. However, based on UK statistics, a fire in 10 m^2 in area with a heat output of 5 MW is generally accepted as the design fire within sprinklered retail premises in the UK.[8] Other design fires commonly used for the design of smoke control systems in UK atrium buildings[9] are: (1) sprinklered offices,[13] 16 m^2 area, 1 MW heat output; (2) unsprinklered offices, 47 m^2 area, 6 MW heat output; and (3) unsprinklered hotel bedrooms,[14] the area of the design fire is taken to be the area of the largest bedroom with 1 MW heat output.

UK Fire Research Station (FRS) Technical Paper No. 7[1]

Given the size of the design fire, vent areas are computed from Equations 28 through 30 as appropriate. Equation 30 or Figure 3-9.9 are most often used.

EXAMPLE:

Taking $h = 10$ m, $d = 2$ m, and a design fire with a floor area of 3.16 m × 3.16 m, then $P = 12.6$ m; and from Equation 30, $A_v = 26$ m^2. From Figure 3-9.9, $F_v = 2.1$ m which, again, gives $A_v = 26$ m^2. Alternatively from Equation 29 $y_0 = 4.7$ m; and from Equation 28 $A_v = 29$ m^2.

NFPA 204M[6]

Equations 31 and 32 require a knowledge of the heat output per unit area of the design fire as well as its area. These are generally determined experimentally through measurements of fires in single stacks of materials. NFPA 204M[6] gives rates of heat release per unit floor area for a number of common types of stored goods in Btu s^{-1} ft^{-2} (1 Btu s^{-1} ft^{-2} = 11357 W m^{-2}).

NFPA 204M also enables an estimate to be made of the heat output of a fire (Q_{max}) [W], which, if exceeded, may result in flashover within the area enclosed by screens (curtains) or failure of unprotected steelwork (based on a gas temperature of 540°K above ambient).

$$Q_{max} = 1.3 \times 10^6 (h - d)^{5/2} \qquad (33)$$

EXAMPLE:

Calculate the vent area required for a fire 10 m^2 area involving wood pallets stacked 1.5 m high in a 10-m-high compartment with 2-m-deep screens (curtains). From Table 4-1 of NFPA 204M the heat output is 135 Btu s^{-1} ft^{-2} = 1.5 MW/m^2, and Q_f = 15 MW. From Equation 12, taking $(h - d)$ = 8 m, Q_{crit} = 41 MW; i.e., Equation 32 can be used. Then, from Equation 32, A_v = 22 m^2. Alternatively Equation 31 gives A_v = 23 m^2. From Equation 33, Q_{max} = 235 MW so that flashover is unlikely.

NFPA 204M method of design with growing fires:[6] The NFPA 204M method of design with growing fires does not explicitly include a reference to a design fire. It assumes a square law increase in heat output with time, following an induction period.

$$Q_t = 10^6 (t/t_s) \qquad (34)$$

where

Q_t = heat output at time t (W)
t_s = time (measured from an effective ignition time t_1) at which the heat output is 1 MW (s).

t_1 is determined by extrapolating the parabolic portion of the heat output/time curve backward to zero heat output.

t_s has been determined experimentally for various types of combustibles, and is given in Table 4-2 of NFPA 204M.

Then from Equations 32 and 34

$$A_v = 0.81[(t_d + t_r)/t_s]^{6/5}(h - d)/d^{1/2} \qquad (35)$$

where

t_r = "intervention time" following detection (i.e., the time for which clear visibility is required) (s)
t_d = "detection" time (s).

NFPA 204M calculates the detection time as the time of operation of the "first vent in a square matrix (vent farthest possible from the fire location)."[6] The vents are assumed to be operated by heat-sensitive devices having a time constant of 233 s at 1.53 m/s gas velocity and various operating temperatures. (See NFPA 204M[6] for vent area calculations.)

SMOKE RESERVOIRS

Purpose of Smoke Reservoirs

Generally the space beneath the ceiling of a large building should be subdivided into smoke reservoirs by screens (draft curtains) extending from the ceiling part way toward the floor. This restricts the unwanted excessive cooling of the smoke layer that leads to reduced efficiency of vents and mixing of smoke into the clear air beneath.

Maximum and Minimum Area of Smoke Reservoirs

Little quantitative information is available; recommendations are usually based on experience. NFPA 204M[6] recommendations are for a maximum distance between screens of 8 times the ceiling height and minimum distance of twice the ceiling height, except for special hazards surrounded by deep screens. A maximum area of 1000 to 1300 m^2 is commonly allowed in shopping malls in the UK.[8] Storage buildings commonly use 2000 m^2.

Depth of Screens

In a narrow channel (e.g., a covered shopping mall or a clerestory type of roof construction), the channel must be deep enough to contain the total flow of hot gases from the fire toward the vents.

For a rectangular cross-section channel, such as a corridor or shopping mall,

$$d_c = K[M(\theta + T_0)/W]^{2/3}/\theta^{1/3} \qquad (36)$$

where

d_c = channel depth measured from lowest point of structural beams, etc. (m)
W = channel width (m)
K = constant (m$^{1/3}$ K$^{-1/3}$)
θ = "mixed" temperature of layer of hot gases above ambient (K)
M = mass flow rate of hot gases entering smoke reservoir (kg/s).

If the flow is unconstrained (i.e., all the gases flow out through vents along the corridor or out of an open end),[9] K = 0.055. If the flow is constrained (e.g., by a curtain at the edge of a balcony), K = 0.09. If the flow is symmetrical in two directions, M is half the total flow rate.

EXAMPLE:

Taking a vented shopping mall of rectangular cross section (W = 10 m) with a fire having a convective heat output of 5 MW and M = 30 kg/s, then θ = 167°C above ambient, and from Equation 36, d_c is 1.2 m.

In a compartment with a flat ceiling, screens should be substantially deeper than the ceiling jet, which has a depth of the order of $h/10$.[2] NFPA 204M[6] recommends a minimum depth of $h/5$. Isolated high hazards may be surrounded by much deeper screens.

MISCELLANEOUS DESIGN PROVISIONS

Inlets for Fresh Air

Inlets are essential to ensure the layer of clear air near the floor. Sometimes it can be more difficult to provide adequate inlets than vents. The efficiency of vents decreases with a limited area of inlets.

Inlets must be below the smoke layer, e.g., in basements, and shafts must be provided to duct fresh air below the smoke layer. In some circumstances, vents in smoke reservoirs remote from the fire may function as inlets.

Siting of Vents and Inlets to Reduce Adverse Wind Effects

Vents and inlets should be situated in such positions that the rate of discharge through the vents is not reduced by the effects of wind, whatever its direction. Vents should be situated where the wind will create a suction effect. Inlets should be situated in such positions that a wind results in an increase in pressure inside the building.

The literature should be consulted for the distribution of pressure coefficients over simple shapes of a building. The presence of other buildings may modify the pressures due to winds.

Vents may be sited in flat roofs provided there are no taller buildings nearby, but adverse wind pressures may be significant with pitched roofs. Vents should not be situated in outward-facing roof slopes having an angle to the horizontal greater than about 30 degrees.

To reduce the possibility of certain wind directions creating an overall reduction in pressure within an undivided isolated building, inlets should be evenly distributed around the perimeter of the building.

With complex buildings such as shopping centers, particularly where venting is necessary to keep escape routes free from smoke, wind tunnel tests may be necessary to decide if natural venting is feasible, and if so, where vents and inlets may be sited.

Where adverse wind effects cannot be avoided, mechanical extract systems are necessary.

Vent Control Methods

Manual control is generally recommended to supplement automatic operation, and not as a replacement for it.

Vent opening may be initiated by the operation of fusible links or by heat-sensitive or other detectors. Section 4, Chapter 1 in this handbook provides information on the conditions necessary to operate detectors.

Since it is difficult to clear a building once it has become smoke-logged, vents should open at an early stage in the fire, and this is essential when vents are used to ensure escape routes are smoke-free. Generally vents are best operated in zones (e.g., all vents within a single smoke reservoir).

FACTORS AFFECTING DISCHARGE THROUGH A VENT

In the absence of other information the coefficient of discharge of a vent may be taken to be 0.6, which strictly applies to the flow under pressure through a sharp-edged orifice at high Reynolds numbers. It is inferred that there are infinite reservoirs of stagnant gases at the same temperature on either side of the vent.

The effects of departures from the theoretical conditions, the ways in which adverse effects can be minimized by design, and siting of ventilators are still under investigation but some information is available.

Local Wind Effects

The "global" effects of wind on pressure differences between the inside and outside of a building are significant. In addition, there are local effects that depend on the design of a ventilator and its orientation in relation to the wind direction.[15] Wind blowing across a shallow orifice in the roof will reduce its discharge coefficient. The degree of reduction depends on the buoyancy of the layer of hot gases and is greatest with layers of low buoyancy. If a vent is mounted on an upstand (typically of the order of ½ the smaller dimension of the vent in height), the resulting updraft may cause a local decrease in pressure, enhancing the discharge.

Open vent flaps may enhance or reduce discharge, according to the orientation with respect to the wind. For example, a top-hinged flap in a steeply sloping roof may be inefficient as a vent, whereas a flap hinged at its bottom edge may be better; however, individual designs need to be tested in a wind tunnel.

In general, a ventilator design should be tested in a crosswind at a range of angles. Ideally there should be a layer of hot gases beneath, although a test with cold air should give results erring on the side of safety.[15]

Effects of the Plume and Ceiling Jet

A local increase in pressure in the stagnation zone will occur where the plume impinges on the ceiling, resulting in an increased flow through a vent compared with that calculated on the assumption of a stagnant layer of hot gases.

Near the axis of the fire the velocity of the ceiling jet may be sufficiently high to reduce the flow through a vent. There is, at present, little information on this effect, but to some extent it may be counteracted by the increased flow through vents within the stagnation zone.

The Onset of Two-way Flow Through a Vent

The air above a roof vent is denser than the gases beneath, and this is an inherently unstable situation. If there are no low-level inlets, there is no net pressure drop across the vent and there will be a two-way flow of air into the vent and hot gases out of it; this is analogous to the flow from an inverted open bottle containing a liquid. If inlets are opened and the resulting pressure drop across the vent, P_v, exceeds a critical value, P_{fl}, the flow will be uni-directional of hot gases out of the vent; so-called "flooding flow." For a shallow circular vent, $C_d = 0.18$ when $P_v = P_{fl}$, and with any vent $C_d = 0.6$ only if $P_v \gg P_{fl}$.

Sufficient information on shallow circular vents was available to enable Cooper to develop a model and a consequent algorithm VENTCL2 for those vents.[16]

"Plugholing"

The flow into a vent will result in a small local reduction in pressure at the base of the layer of hot gases, and there is a critical rate of extract through a vent, above which air from beneath the layer of hot gases is drawn into the vent. The onset of this phenomenon, termed "plugholing," depends[17] on a Froude number, F_c.

$$F_c = V_v / [(g\theta/T_0)^{1/2} d^{5/2}] \tag{37}$$

where

V_v = volume rate of flow through a single isolated orifice (m^3/s)

d = depth of layer of hot gases (m)

θ = temperature above ambient (K)

T_0 = absolute ambient temperature (K).

For a circular orifice $F_c = 1.6$.[14] This seems to be only weakly dependent on the orifice size. Work on vents[18] has been reappraised[8] and suggests that 1.5 is applicable for vents near the center of a smoke reservoir, and 1.1 for vents near the sides.

EXAMPLE:

Taking a layer 2 m deep at a temperature of 300 K above ambient, the maximum rate of extract through a circular orifice from Equation 37, taking F_c as 1.6, is 28.8 m^3/s (12 kg/s).

For buoyancy flow through a shallow circular vent, Equation 37 reduces to

$$d_p/D_v = 1.05(C_d/F_c)^{1/2} \qquad (38)$$

where

d_p = critical layer depth beneath which cold air may be drawn into the vent (m)
D_v = diameter of vent (m).

Plugholing is an important consideration with venting systems employing mechanical extract.

MECHANICAL EXTRACT

Mechanical extract is generally used where adverse pressures due to wind could seriously reduce the efficiency of a natural convection venting system. It may also be required where the temperature of the smoke is likely to be low or where extract is via ducts required to have minimum cross section.

The general design of a mechanical extract system is similar to that of a natural venting system. Having decided on a design fire, calculations of mass rates of production and temperature of hot gases with the required layer depth are made, as for natural venting systems, and provide the basis for the selection of suitable fans.

The fan characteristics at the calculated mixed layer temperature must be known. As an approximation, the volume flow through a fan is independent of temperature and, therefore, the mass rate of flow will decrease with increasing temperature.

Neglecting heat losses in the layer

$$\theta = Q_f/M_pS \qquad (39)$$

where

θ = mixed layer temperature above ambient (K)
M_p = rate of production of hot gases (kg/s)
Q_f = convective heat output of fire (w)
S = specific heat of air at constant pressure (J kg^{-1} K^{-1}).

The design back pressure against the fan is

$$P_F = (P_W + P_I - P_B) \qquad (40)$$

where

P_W = anticipated maximum adverse wind pressure (Equation 2) (Pa)
P_I = pressure drop across inlet which may be obtained from Equation 4 (Pa)
P_B = buoyancy pressure (Equation 1) (Pa).

The fan must be "hardened" to withstand the calculated temperature for the required period.

EXAMPLE:

What is the specification for fans required to maintain a 3-m clear layer in a 5-m-high shopping mall given a fire in a shop? Wind tunnel experiments have shown that at the fan position the pressure coefficient is 0.5 and the design wind speed is 20 m/s. The inlet is through doors with an area of 4 m^2. The design fire (following UK practice) is 3 m × 3 m, with a heat output of 5 MW.

SOLUTION:

The rate of flow of hot gases out of the shop is (from Equation 10) 12 kg/s. The flow in the mall will be approximately twice this rate, i.e., 24 kg/s, which is the required extract rate.

The gas temperature is given by Equation 39 as approximately 210 K above ambient; therefore, the volume flow rate is 35 m^3/s. From Equation 2, P_W = 120 Pa. From Equation 4, taking C_d = 0.6, P_I = 8 Pa; and from Equation 1, P_B = 10 Pa. Thus the design back pressure (from Equation 40) is to the required accuracy: 120 Pa.

From Equation 37 (taking V_v = 35 m^3/s, θ = 210 K above ambient, and d = 2 m) the Froude number for the extract flow $F_c \approx$ 32.4. Thus if cold air is not to be sucked into the extract, the minimum number of fans is 2 so that F_c for each is 1.2. Each fan should be capable of extracting 12 kg/s of gases at a temperature of 210 K above ambient (i.e., 17 m^3/s) against a back pressure of 120 Pa.

It must be noted that these specifications include no factor of safety other than that implicit in the choice of the design fire. Should the heat output of the fire be greater than that anticipated in the design of the system, then the mass rate of "production" of hot gases may be greater and, because their temperature is higher, the mass rate of extract by the fans would be lower; this results in a deeper layer and an even higher temperature. It can be concluded that a mechanical venting system is more vulnerable to design errors than is a natural venting system, because the mass rate of flow through vents is not greatly dependent on temperature.

VENTING TO CONTROL SMOKE FLOW THROUGH SMALL OPENINGS TO ADJACENT AREAS

The Concept of the Neutral Plane[3]

In the limiting case a compartment with a fully developed fire approximates to one full of hot gases. Where openings occupy only a small proportion of the wall area the gases may be treated as a first approximation as stagnant within the compartment.

If buoyancy forces are the only ones acting on the gases, then the pressure will be above atmospheric near the top of the compartment and below atmospheric near the bottom. There will be a neutral pressure plane where the pressure is the same inside the compartment as outside. (See Figure 3-9.10.)

The Calculation of the Position of the Neutral Plane

The position of the neutral plane is determined by the condition that

$$\Sigma m_0 = \Sigma m_i + m_f \qquad (41)$$

where

Σm_0 = sum of mass rates of flow of hot gases out of the compartment through all the openings below the neutral plane (kg/s)
Σm_1 = sum of mass rates of flow of air into the compartment through all openings above the neutral plane (kg/s)
m_f = rate of generation of hot gases by the fire (kg/s).

If buoyancy is the only force acting, the flow out of a vertical opening above the neutral plane is

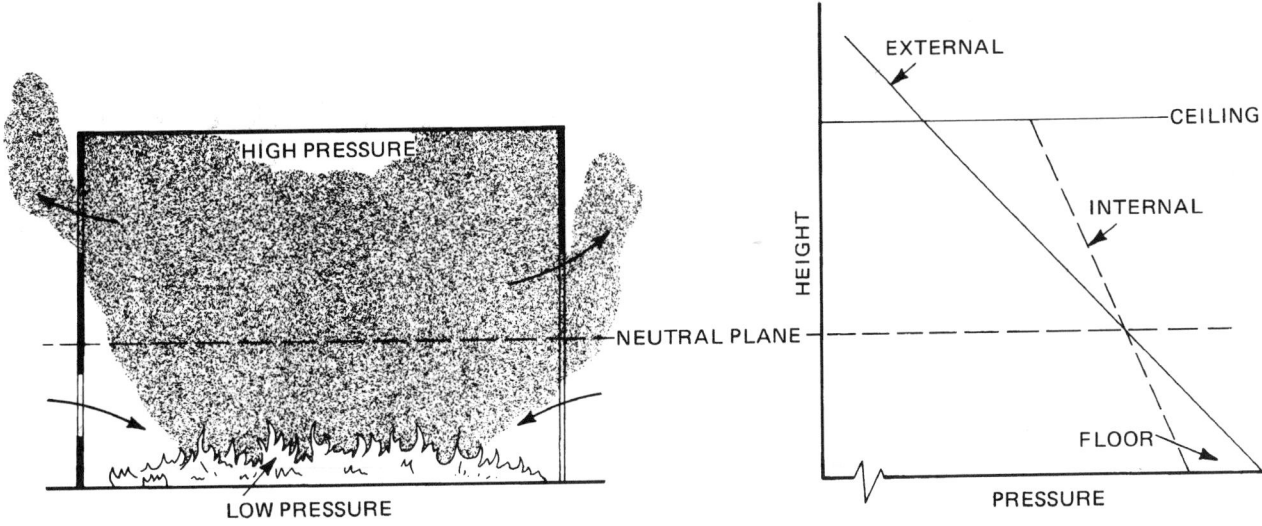

Fig. 3-9.10. Neutral pressure plane.

$$m_0 = \frac{2}{3}C_d\rho_0(h_t^{3/2} - h_b^{3/2})W(2g\theta T_0)^{1/2}/T \qquad (42)$$

where

h_t = height of top of the opening above the neutral plane (m)

h_b = height of bottom of opening above the neutral plane (m)

W = width of opening (m)

T = absolute temperature of gases (K).

The flow into a vertical opening below the neutral plane is

$$m_i = \frac{2}{3}C_d\rho_0(d_b^{3/2} - d_t^{3/2})W(2g\theta/T)^{1/2} \qquad (43)$$

where

d_b = height of neutral plane above the bottom of the opening (m)

d_t = height of neutral plane above top of opening (m).

If the fuel/air ratio is ⅕, then

$$\Sigma m_0 = 1.2\Sigma m_i \qquad (44)$$

The position of the neutral plane can be determined from the above equations by a simple computer program and, hence, the rates of flow into and out of the various openings can be calculated.

If there are two adjoining compartments with openings between them, the flow of smoke from one on fire to the other can be prevented by ensuring that the neutral plane is above the top of the highest common opening. The area of vent necessary to achieve this can readily be calculated from the equations in this chapter.

EXAMPLE:

A theater has a proscenium opening 6 m high × 10 m wide, and the roof is 10 m above the top of the opening. (See Figure 3-9.11.) The area of each door from the auditorium to the open air is 4 m², and the stage door is 1 m wide × 2 m high. Assuming a 0.05-m gap around the safety curtain, what area of vent would be required to ensure that no hot gases would pass from the stage to the auditorium, assuming a fully developed fire at a temperature of 500 K above ambient?

SOLUTION:

The pressure distribution in the theater is shown diagrammatically in Figure 3-9.11. The neutral pressure plane between the auditorium and the stage is lower than that between the stage and outside because of the pressure drop caused by the restriction to flow through the auditorium doors.

Assuming the neutral pressure plane between the stage and auditorium to be situated at the level of the top of the proscenium opening, the rates of flow of air from the auditorium to the stage through the gaps at the sides, top, and bottom of the safety curtain are calculated separately from Equation 43. The flow through the top gap is 0.19 kg/s, through the bottom gap is 3.10 kg/s, and through the gaps at the sides (equivalent to a single 0.1-m-wide gap) is 2.48 kg/s, so the total flow is 5.77 kg/s.

The pressure drop across the auditorium doors resulting from the airflow of 5.77 kg/s through them is calculated from Equation 4 to be 2.4 Pa, which is equivalent to a head of gases at 500°C above ambient of 0.32 m from Equation 1. Thus, the neutral pressure plane between the stage and the outside air is 0.44 m above the proscenium opening.

The flow through the stage door with the neutral plane in that position is calculated from Equation 43 to be 11.68 kg/s, and the total inflow to the stage is 17.45 kg/s. With an air/fuel ratio of 5:1, the mass of fuel gases is 3.49 kg/s, giving a total mass of gases to be vented of 20.94 kg/s.

The area of vent required to maintain that flow rate with a head of gases at 500°C of (10 − 0.32) = 9.68 m is calculated from Equation 5 to be 4.4 m².

If the stage door were closed, the required vent area would be only 1.45 m². This illustrates the necessity of taking all possible flow paths into consideration in the calculations. If the doors to the auditorium were closed, any area of vent would be ineffective. If the layer of hot gases did not completely fill the stage, then the area of vent required would be reduced.

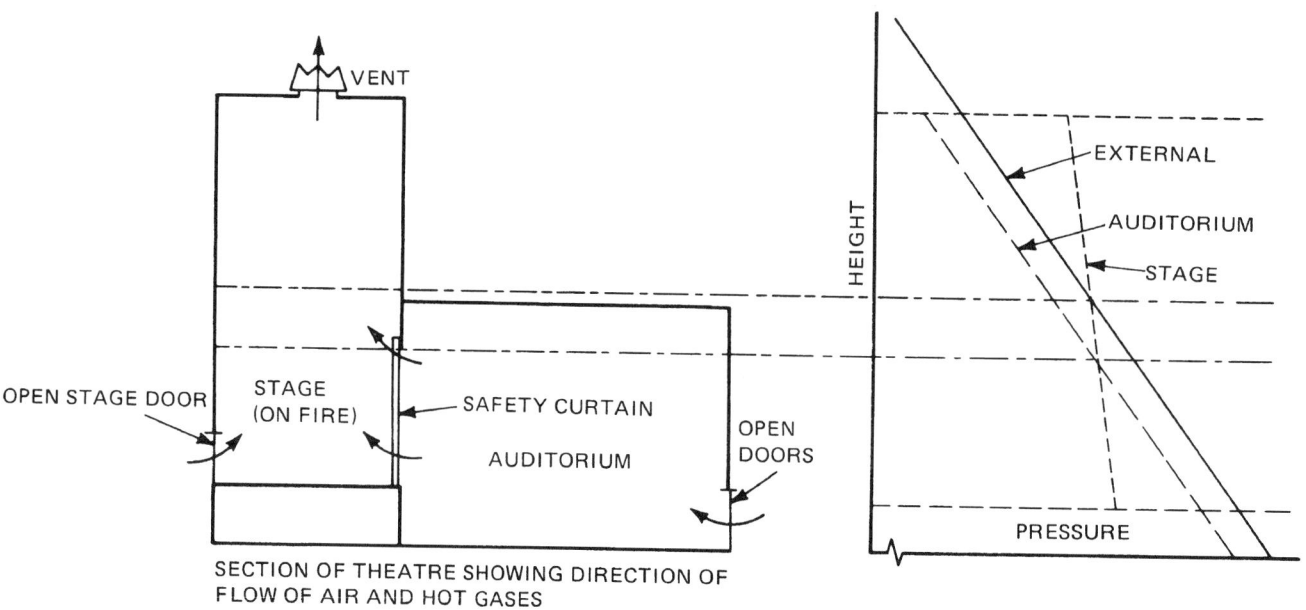

Fig. 3-9.11.   Example theater pressure distribution.

The above example does not consider the effects of wind. For the vent to be effective under any wind conditions, the negative wind pressure coefficient at the position of the vent would have to be greater than the negative coefficient at the auditorium doors, whatever the direction of the wind.

Effects of Wind

The simple concept of a neutral plane does not apply when wind pressures are taken into account, since they vary from point to point over the face of the building. Hence, for example, a situation could in theory arise where the flow is into all openings on the windward side of the building and out of all openings on the leeward side.

This greatly complicates calculations, but the flows can be calculated by a computer program that derives a mass balance by taking into account flows resulting from the sum of the pressures due to buoyancy and wind, according to the equations given in this chapter.

NOMENCLATURE

A area of vent or inlet (m^2)
A_c floor area of the compartment (m^2)
A_d defined by Equation 18 (kg m^{-1} s^{-1})
A_{dA} $2 A_d$ (kg m^{-1} s^{-1})
A_f fire area (m^2)
A_I area of inlets (m^2)
A_L total area of openings in a compartment (m^2)
A_v area of vents (m^2)
A_v^* effective vent area (m^2)
B rate of rise of temperature (K/s)
C_p wind pressure coefficient
C_d coefficient of discharge of a vent
d depth of layer of hot gases (m). (This assumes the layer to be undisturbed and uniform in temperature. Owing to the disturbed nature of the layer, some gases may extend below this depth.)

d_b height of neutral plane above bottom of opening (m)
d_c channel depth (m)
d_p critical layer depth for "plugholing" (m)
d_t height of neutral plane above top of opening (m)
D distance of line source below balcony edge (m)
D_v diameter of a circular vent (m)
F_c critical Froude number for "plugholing"
F_v venting factor $= A_v/P$ (m)
g   acceleration due to gravity (9.81 m/s^2)
h compartment height (m)
h_b height of bottom of opening above neutral plane (m)
h_t height of top of opening above neutral plane (m)
h_B height of ceiling above balcony (m)
H_w height of open shop front or similar opening (m)
$I_1(v)$ integral defined by Lee and Emmons (Equation 19)
K constant
m_f total mass rate of generation of fuel gases by the fire (kg/s)
m_i mass rate of flow into compartment (kg/s)
m_0 mass rate of flow out of the compartment (kg/s)
m_x mass flow into the upper layer per unit width of the line plume (kg m^{-1} s^{-1})
M mass rate of flow through a vent or inlet (kg/s)
M_E horizontal mass flow per unit width of opening (kg m^{-1} s^{-1})
M_{end} total rate of entrainment into ends of the line plume (kg/s)
M_p mass rate of production of hot gases (kg/s)
M_v mass rate of flow through vents (kg/s)
P perimeter of fire (m)
P_B pressure due to buoyancy (Pa)
P_d pressure drop across a vent or inlet (Pa)
P_E pressure rise due to constraints on expansion (Pa)
P_F design back pressure on fan (Pa)
P_{fl} critical pressure drop across a vent for flooding flow (Pa)

P_I	pressure drop across an inlet (Pa)
P_w	pressure due to wind (Pa)
P_v	pressure drop across a vent (Pa)
Q_c	actual convective heat output of fire (W)
Q_{crit}	critical convective heat output of the fire for continuous flaming in the layer of hot gases (W)
Q_E	horizontal heat flow per unit width of opening (kW/m)
Q_f	convective heat output of fire assuming complete combustion (W)
Q_{max}	maximum heat output for hot gas layer temperature < 540°C (W)
Q_t	heat output at time t (W)
Q_w	horizontal heat flow (kW)
S	specific heat of air at constant pressure (1000 J kg^{-1} K^{-1})
t	time (s)
t_d	time of detection (s)
t_i	effective ignition time (s)
t_r	"intervention time," measured from time of detection (s)
t_s	time at which heat output of fire = 10^6 W (s)
T_0	absolute ambient temperature (290 K)
T	absolute temperature of gases (K)
v_w	wind velocity (m/s)
V_v	volume rate of flow of hot gases (m³/s)
w	channel width (m)
W	width of opening (m)
x	height of clear layer above the balcony edge (m)
y_0	distance of effective point source beneath floor level (m)

Greek Symbols

θ	temperature above ambient (K)
ν	transformed reciprocal of buoyancy
ν_{end}	value of ν used in calculating entrainment into ends of plume
ρ_0	density of ambient air (1.2 kg/m³ at 290 K)
ρ_1	density of layer of hot gases (kg/m³)

REFERENCES CITED

1. P.H. Thomas *et al*, "Investigations into the Flow of Hot Gases in Roof Venting," Fire Research Technical Paper No. 7, HMSO, London (1963).
2. R.L. Alpert, "Turbulent Ceiling Jet Induced by Large-Scale Fires," *Combustion Science and Technology*, 11, 197–213 (1975).
3. J.H. McGuire, "Smoke Movement in Buildings," *Fire Technology*, 3 (3), 163–74, Aug. 1967.
4. "Assessment of Wind Loads," UK Building Research Establishment, Digest No. 119 (1984).
5. "Wind Environment Around Tall Buildings," UK Building Research Establishment, Digest No. 141 (1972).
6. NFPA 204M, *Guide for Smoke and Heat Venting*, National Fire Protection Association, Quincy, MA (1991).
7. NFPA 92B, *Guide for Smoke Management Systems in Malls, Atria, and Large Areas*, National Fire Protection Association, Quincy, MA (1991).
8. H.P. Morgan and J.P. Gardiner, "Design Principles for Smoke Ventilation in Enclosed Shopping Centres," Building Research Establishment Report, BRE Bookshop ref. BR186, Garston, BRE (1990).
9. G.O. Hansell and H.P. Morgan, "Design Approaches for Smoke Control in Atrium Buildings," Building Research Establishment Report, BRE Bookshop ref. BR258, Garston, BRE (1994).
10. P.L. Hinkley, "Rates of 'Production' of Hot Gases in Roof Venting Experiments," *Fire Safety Journal*, 10 (1), 57–65 (1986).
11. P.H. Thomas, "On the Upward Movement of Smoke and Related Shopping Mall Problems," *Fire Safety Journal*, 12 (1) 83–84 (1987).
12. Shao-Lin Lee and H.W. Emmons, "A Study of National Convection above a Line Fire," *Journal of Fluid Mechanics*, 11 (3) 353–368 (1961).
13. H.P. Morgan and G.O. Hansell, "Fire Sizes and Sprinkler Effectiveness in Offices: Implications for Smoke Control Design," *Fire Safety Journal*, 8, 187–198 (1984–85).
14. G.O. Hansell and H.P. Morgan, "Fire Sizes in Hotel Bedrooms: Implications for Smoke Control Design," *Fire Safety Journal*, 8, 177–186 (1984–85).
15. B.K. Ghosh, "Some Effects of Crosswind on Ventilators," *Journal of Wind Engineering and Industrial Aerodynamics*, 45, 247–270 (1993).
16. L.Y. Cooper, "Combined Buoyancy- and Pressure-Driven Flow Through a Horizontal Vent," NISTIR 5384, National Institute of Science and Technology, Gaithersburg, MD (1994).
17. J.S. Turner, *Buoyancy Effects in Fluids*, Cambridge University Press, Cambridge (1973).
18. D. Spratt and A.J.M. Heselden, "Efficient Extraction of Smoke from a Thin Layer Under a Ceiling," UK Joint Fire Research Organisation Fire Research Note No. 1001/1974.
19. S. Yokoi, "A Study on Dimensions of Smoke Vents in Fire-Resistive Construction," Japanese Ministry of Construction Building Research, Institute Report No. 29, Tokyo, March 1959.

COMPARTMENT FIRE-GENERATED ENVIRONMENT AND SMOKE FILLING

Leonard Y. Cooper

FIRE SAFETY OF BUILDING DESIGNS

Introduction

The following generic problem must be solved if one is to be able to establish the fire safety of building designs:

Given: Initiation of a fire in a compartment or enclosed space.

Predict: The environment that develops at likely locations of occupancy, at likely locations of fire/smoke sensor hardware (e.g., detectors and sprinkler links), and in locations of safe refuge and along likely egress paths.

Compute: The time of fire/smoke sensor hardware response and the time of onset of conditions untenable to life and/or property. This computation would be carried out from the above predictions, using known response characteristics of people, hardware, and materials.

The above is only a simple sketch of the overall problem that is likely to be associated with the interesting details of many real fire scenarios. A long-term challenge of fire science and technology is to solve the above type of problem, even when it is formulated in elaborate detail. Compartment fire modeling is the branch of fire science and technology which develops the necessary tools to address this generic problem.

This chapter will describe some of the key phenomena that occur in compartment fires, and it will focus on smoke filling which is one of the simplest quantitative global descriptions of these phenomena. A specific smoke-filling model will be presented, and solutions to its model equations will be discussed along with example applications.

Dr. Leonard Y. Cooper is a Research Engineer in the Building and Fire Research Laboratory of the National Institute of Standards and Technology. Since 1978 his research has focused on the development of mathematical models of fire phenomena and on the assembly of these into compartment fire models and associated computer programs for practical and research-oriented compartment fire simulations.

Compartment Fire-Generated Environment

Figures 3-10.1 through 3-10.8 depict the various phenomena that make up the compartment fire-generated environment to be predicted when compartment fire modeling is adopted. These figures are intended to illustrate the representative conditions at different instants of time in two generic compartment spaces: (1) an almost-fully enclosed single-room compartment of fire origin and (2) an almost-fully enclosed, freely connected, two-room compartment made up of the room of fire origin and an adjacent space. A description of the phenomena depicted in these figures follows. The physical bases of assumption that can be used to simplify descriptions of some of these phenomena are included in the discussion. Some of these will be important in placing the simple smoke-filling model into the perspective of the overall complex dynamic fire environment.

Room of Fire Involvement

Fire growth in the combustible of fire origin: An unwanted ignition leading rapidly to flaming is assumed to occur within an enclosed space. This ignition is depicted in Figure 3-10.1 as occurring on the cushion of a couch in, say, a residential type of occupancy. It is, however, important to realize that all of the discussion to follow, and Figure 3-10.1 itself, is also relevant to fire scenarios which may develop in other kinds of occupancies, e.g., as a result of ignitions in stacked commodity warehouse enclosures, places of assembly, etc.

Within a few seconds of ignition, early flame spread quickly leads to a flaming fire with a power output of the order of a few tens of kW (a power level characteristic of a small wastepaper basket fire). The fire continues to grow. Besides releasing energy, the combustion process also yields a variety of other products, including toxic and nontoxic gases and solids. Together, all of these products are referred to as the "smoke" produced by the fire.

With an adequate description of the ignition source and the involved combustible (e.g., ignited paper match on the corner of a couch whose frame, cushioning and finishing materials, and construction were well defined), one would hope that fire science and technology would provide methods to predict the fire spread and growth process from onset of ignition. Toward this end, ongoing research on flame spread and combustion is under way in a variety of fire research institutions throughout the world. (Examples of such research include boundary layer analyses and experiments with flame spread on idealized materials and geometries; and flame spread tests and rate of heat release tests on small samples of real material composites.) However, for the present and the foreseeable future, it is beyond the state-of-the-art of fire technology to make the required fire growth prediction with any generality. This situation leads to a dilemma for the modeler of compartment fire environments, because the physical and chemical mechanisms which govern the dynamics of the combustion zone actually drive the basic intra-compartment smoke migration phenomena whose simulations are being sought.

A practical engineering solution to the above dilemma, proposed and supported by Cooper,[1,2] lies in the following compromise in simulation accuracy:

> Prior to the time of potential flashover, it is reasonable to neglect the effect of the enclosure on flame spread and to assume that, from the time of ignition to the time shortly before potential flashover, the combustion zone in a particular grouping of combustibles develops as it would in a free-burn situation.

[Free burn here is defined as a burn of the combustibles in a large (compared to the combustion zone), ventilated space with relatively quiescent atmosphere.] To implement these ideas, one simply uses empirical, free-burn test data (which may or may not be presently available) to describe the combustion physics of a fire whose hazard is being evaluated. This compromise would, in the course of time, be supplanted by analytic models of flame spread and fire growth to the extent the future results of research lead to satisfactory methods for predicting such phenomena.

The implementation of the above compromise is, in principle, relatively simple. But for general use, it must be supported by an extensive data base acquired from a series of actual full-scale free-burn tests. This kind of data is being acquired with some regularity at fire test laboratories such as those of the National Institute of Standards and Technology and the Factory Mutual Research Corporation.[3,4,5]

Development of the plume: As depicted in Figure 3-10.2, a large fraction, λ_r, of the rate of energy released in the high-temperature combustion zone is transferred away by radiation. The transferred energy, $\lambda_r \dot{Q}(t)$, irradiates nearby surfaces of the combustible and faraway wall, ceiling, etc., surfaces which are in the line-of-sight of the combustion zone. The actual value of λ_r associated with the free-burn of a specific array of combustibles is often deduced from data acquired during the aforementioned type of free-burn tests. For typical hazardous flaming fires, λ_r is usually of the order of 0.35.

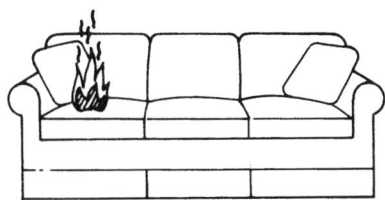

Fig. 3-10.1. *Events immediately after ignition.*

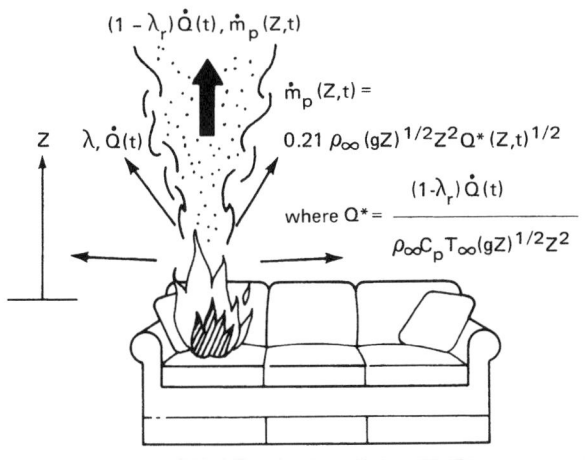

Fig. 3-10.2. *Development of the plume.*

Because of the elevated temperature of the products of combustion, buoyancy forces drive them out of the growing combustion zone and up toward the ceiling. In this way, a plume of upward-moving elevated-temperature gases and particulates is formed above the fire. For the full height of the plume and at its periphery, relatively quiescent and cool gases are entrained laterally and mixed with the plume as it continues its ascent to the ceiling. As a result of this entrainment, the total mass flow in the plume continuously increases, and the average temperature and average concentration of products of combustion in the plume continuously decrease with increasing height. With reasonable accuracy, the plume dynamics at any instant of time can be quantitatively described as a function of the rate of energy, $(1 - \lambda_r)\dot{Q}(t)$, convected up from the combustion zone. A description of the concentration of combustion products in the plume would require, in addition, the combustion zone's rate of product generation. With regard to predictions of the dynamics of the plume, results can be provided at a variety of different levels of detail.[6,7,8] For example, Zukoski *et al* provide the formula in Figure 3-10.2 as an estimate of the mass flux in the plume, $\dot{m}_p$, at a distance Z above the combustion zone.[7]

Plume-ceiling interaction: As depicted in Figure 3-10.3, when the hot plume gases impinge on the ceiling, they spread across it forming a relatively thin radial jet. This jet of hot gases contains all of the smoke generated from the combustion zone, and all the ambient air which was entrained along the length of the plume.

As the hot jet moves outward under the ceiling surface, it entrains ambient air from below. It transfers energy by conduction to the relatively cool adjacent ceiling surface, and by convection to the entrained air. It is retarded by frictional forces from the ceiling surface above, and by turbulent momentum transfer to the entrained air from below. As a result of all this flow and heat transfer activity, the ceiling jet continuously decreases in temperature, smoke concentration, and velocity, and increases in thickness with increasing radius.

Research reported in the literature has led to results for predicting the quantitative aspects of ceiling-jet dynamics that can be used for selecting and locating smoke detectors and

Fig. 3-10.3. The plume-ceiling interaction.

fusible link sprinkler head actuators, and for the mathematical modeling of overall compartment fire environments.[9-18]

With regard to detectors and fusible links, knowledge of the properties of the ceiling jet is the key to the prediction of the response of properly deployed devices in real fire scenarios. With regard to the overall modeling of compartment fire environments, the basic information that must be extracted from the ceiling-jet properties is the rate of heat transfer to the ceiling surface. Experiments have shown that this heat transfer can be significant, of the order of several tens of percent of the total energy released by the combustion zone, and, as a result, is key to predicting the temperature of the smoke which ultimately spreads throughout the enclosure. Also, a reasonable estimate of this rate of heat transfer is required for estimating temperatures of the ceiling surface material itself.

Ceiling jet-wall interaction: The ceiling jet continues to move radially outward under the ceiling surface, and it eventually reaches the bounding walls of the enclosure. As depicted in Figure 3-10.4, the ceiling jet (now somewhat reduced in temperature from its highest levels near the plume) impinges and turns downward at the ceiling-wall juncture, thereby initiating a downward-directed wall jet.

The downward wall jet is of higher temperature and lower density than the ambient air into which it is being driven. The jet is, therefore, retarded by buoyancy in its downward descent, and at some distance below the ceiling the downward motion of the smoky jet is eventually halted. The wall jet is also retarded (probably to a lesser degree) by frictional forces at the wall surface, and it is cooled by conductive/convective heat transfer to relatively cool wall surfaces. Momentum and heat transfer from the jet occur away from the wall as the jet's outer flow is sheared off and driven back upward on account of buoyancy. In its turn, the now upward-moving flow entrains ambient air in a manner which is reminiscent of entrainment into the original fire plume. Eventually a relatively quiescent upper gas layer is formed below the continuing ceiling-jet flow activity.

The strength of the wall-jet flow activity will be determined by the characteristics of the ceiling jet at the position of its impingement with the wall. For fire scenarios where the proximity of the walls to the fire is no greater than a room

height or so, it is reasonable to speculate that, based on test results,[19,20] rates of conductive/convective heat transfer to wall surfaces can be significant, of the order of tens of percent of the fire's energy release, and that entrainment to the upward moving, reverse portion of the wall flow can lead to significant variations of the early rate of thickening of the upper gas layer. On the other hand, if walls are several room heights from the fire, then it is possible that the ceiling jet will be relatively weak by the time it reaches the walls, in the sense that ceiling jet-wall interactions may not play an important role in the dynamics of the overall fire environment.

Besides being important in the prediction of the overall fire environment, knowledge of the flow and temperature environments local to vigorous ceiling jet-wall interaction zones would be the key to predicting the response of wall-mounted smoke detectors, fusible links, etc.

Development and growth of the upper layer—"smoke filling": The gases in the ceiling and wall jets redistribute themselves across the upper volume of the room. Eventually, a relatively quiescent, elevated temperature upper smoke layer of uniform thickness is formed below the continuing ceiling-jet flow activity. As the thickness of this layer grows, it eventually submerges the flows generated by the ceiling jet-wall interactions. The bottom of the layer is defined by a distinctive material interface which separates the lower ambient air from the upper, heated, smoke-laden gases. With increasing time the level of the smoke layer interface continues to drop, and the temperature and smoke concentration of the upper layer continue to rise.

In general, one would hope that fire detection, successful occupant alarm, and, if appropriate, successful intervention hardware response would occur during the state of fire growth described above. As suggested earlier, rationally engineered design in this regard would be possible with predictions of the dynamic fire environments local to deployed devices, and with predictions of the resulting response of such devices.

For reasons, already mentioned, some detail in the description of plume, ceiling-jet, and ceiling jet-wall flow dynamics is required. However, for the purpose of understanding the impact of the overall fire environment on life and property safety, simplified descriptions compatible with the

Fig. 3-10.4. Ceiling jet-wall interaction.

aforementioned detail would suffice and, for a variety of practical reasons, would actually be preferable. For this reason, available predictive models of compartment fire environments commonly describe the bulk of upper layer environment in terms of its spatially averaged properties. Thus, at any instant of time, it is typically assumed that the portion of the room which contains the fire-generated products of combustion, i.e., the smoke, is confined to an upper layer of the room. This upper layer is described as having changing thickness and changing, but spatially uniform, temperature and concentration of combustion products. Actual full-scale testing of compartment fire environments has indicated that such a simple means of describing the distribution of products of combustion represents a reasonable compromise between accuracy in simulation and practicability in implementation.

Figure 3-10.5 is a generic depiction of the compartment fire environment at the stage of fire development under discussion. At this stage, whether or not the space of fire involvement is fully enclosed (i.e., all doors and windows are closed and only limited leakage occurs at bounding partitions) or is freely communicating with adjacent space(s) (e.g., open or broken windows, open doors to the outside environment or to an adjacent enclosed space of limited or virtually unlimited extent) becomes very important in the subsequent development of the fire environment. In the sense that the upper layer thickness and temperature would grow most rapidly, the fully enclosed space with most leakage near the floor would lead to the most rapid development of potentially life and property threatening conditions.

Referring again to the depiction in Figure 3-10.5, the fire plume below the smoke layer interface continues to entrain air as it rises to the ceiling. However, as the hot plume gases penetrate the layer interface and continue their ascent, additional entrainment is from an elevated temperature, smoke-laden environment. Also, once the plume gases enter the smoke layer, they are less buoyant relative to this layer than they were relative to the cool lower layer of ambient air. Thus, the continued ascent of plume gases is less vigorous than it would otherwise be in the absence of the upper layer.

The new and more complex two-layer state of the enclosure environment requires that some modification of the

Fig. 3-10.6. Further "smoke filling."

earlier referenced quantitative descriptions of the plume, ceiling-jet, and ceiling jet-wall flow dynamics be introduced. Modifications along these lines are proposed by Cooper.[21,22]

As depicted in Figure 3-10.6, potentially significant wall flow can develop during the descent of the layer interface. This flow is expected to occur away from regions of vigorous ceiling jet-wall interactions. It is distinct from the previously described upper wall jet and develops because of relatively cool upper wall surfaces which bound the elevated-temperature upper smoke layer. The smoke which is adjacent to these wall surfaces is relatively cool and, therefore, more dense than its surroundings. As a result of this density difference, a continuous, downward-directed wall flow develops which is injected at the smoke interface into the lower, relatively smoke-free layer. Once in the lower layer, the smoke-laden wall flow, now of higher temperature than its surroundings, will be buoyed back upward to either mix with and contaminate the lower layer or to entrain additional (i.e., in addition to the fire plume) lower layer air into the upper layer.

It is noteworthy that the wall effect just described has been observed in full- and reduced-scale fire tests, and that it appears to be particularly significant in enclosures with relatively large ratios of perimeter-to-ceiling height (e.g., in corridors).[23]

As a result of its elevated temperature, the smoky upper layer transfers energy by radiation to the ceiling and upper wall surfaces which contain it. As depicted in Figures 3-10.4 and 3-10.5 by the downward-directed arrows, the layer also radiates to the lower surfaces of the enclosure and its contents. Initially, the only significant role of this downward radiation is its effect on human tissue. Indeed, only for downward-directed radiation fluxes significantly in excess of life-threatening levels (characterized by smoke layer temperature levels of the order of 200°C, or by flux levels of the order of 2.5 kW/m^2) would the radiant energy feedback to enclosure surfaces and combustibles have a significant impact on fire growth and spread, and on the overall fire environment.[1,24]

Once radiation feedback becomes of general significance, e.g., when the average upper layer temperature reaches 300 to 400°C, it is likely that the potential for flashover will develop within a relatively short time interval (compared to the time interval between ignition and the

Fig. 3-10.5. Fully enclosed space with developed growing upper layer.

Fig. 3-10.7.   Smoke and fresh air exchange between a room of fire involvement and an adjacent space.

onset of a life-threatening environment). The events which develop during this time interval will be referred to here as the "transition stage" of fire development.

The onset of life-threatening conditions, which could be caused by any one of a number of reasons, would occur prior to the transition stage of fire development. Before this event can be predicted, quantitative criteria defining a "life-threatening environment" must be established. These criteria must be defined in terms of those physical parameters for which predictive models of compartment fire environments can provide reasonable estimates. Consistent with the earlier discussion, such parameters include the smoke layer thickness (or vertical position of its interface) and temperature. The concentration of potentially hazardous components of the smoke in the upper layer would also be of basic importance. Criteria for the onset of a life-threatening environment could, for example, be based on the following consideration (which would neglect the effect of lower layer contamination due to wall flows):

When the smoke layer interface is above some specified, characteristic face-elevation, an untenable environment would occur if and when a hazardous radiation exposure from the upper layer is attained. Such an exposure could be defined by a specified upper layer critical temperature. If the interface is below face-elevation, then untenability would occur if and when a second critical smoke layer temperature is attained. However, the latter temperature would be lower than the former one, and untenable conditions would result from burns or the inhalation of hot gases. Once the interface dropped below face elevation, untenability would also occur if and when a specified critical concentration (or a specified exposure dosage) of some hazardous product of combustion was attained.[2]

Transition stage of fire development: Any detailed analysis and prediction of the fire environment during the transition stage of fire development must, of necessity, take account of the effects of upper layer and upper surface reradiation, in general, and of the complex effects of radiation-enhanced fire growth, in particular. Such an analysis would require a mathematical model of compartment fire phenom-ena which would be significantly more sophisticated than that which would be required to predict the fire environment prior to, and possibly even following, the transition stage.

Regarding the potential difficulty, uncertainty, inconvenience, and/or cost of carrying out transition stage analysis, it is noteworthy that conservative designs for life and property safety may be possible by implementing a strategy of fire environment analysis which avoided the details of the transition stage entirely. This is done by conservatively assuming that the relatively brief time interval associated with the transition stage shrinks to a flashover jump condition at a relatively early time in the fire scenario.[1]

Smoke Spread from the Room of Fire Involvement to Adjacent Spaces

Smoke and fresh air exchange between a room of fire involvement and an adjacent space: Under this and the following subheading, smoke spread phenomena associated with a fire-involved room and a communicating adjacent space will be discussed. Reference here will be made to a fully enclosed, two-room space with relatively large common penetrations (e.g., open doors or windows) through which smoke and ambient air exchange will be so significant as to render inadequate an analysis which treats the room of fire involvement as an isolated enclosure. Regarding the two-room spacial configuration, Figures 3-10.1 through 3-10.5 are still relevant to the early development of conditions within the fire room. As the smoke layer interface drops to the level of the soffit of the communicating doorway(s) or window(s), significant amounts of smoke start to move into the adjacent space from above, while significant amounts of ambient air are driven out of the adjacent space (and into the fire room) from below. From that time on, as depicted in Figure 3-10.7, an interdependent smoke-filling process in each of the two spaces is initiated, and the adjacent space starts to develop a two-layer type of environment.

Throughout the course of typical real-fire scenarios, changes in the absolute pressures of facility spaces are, at the most, of the order of one percent. Yet, dynamic elevation-dependent pressure differences that exist between the rooms

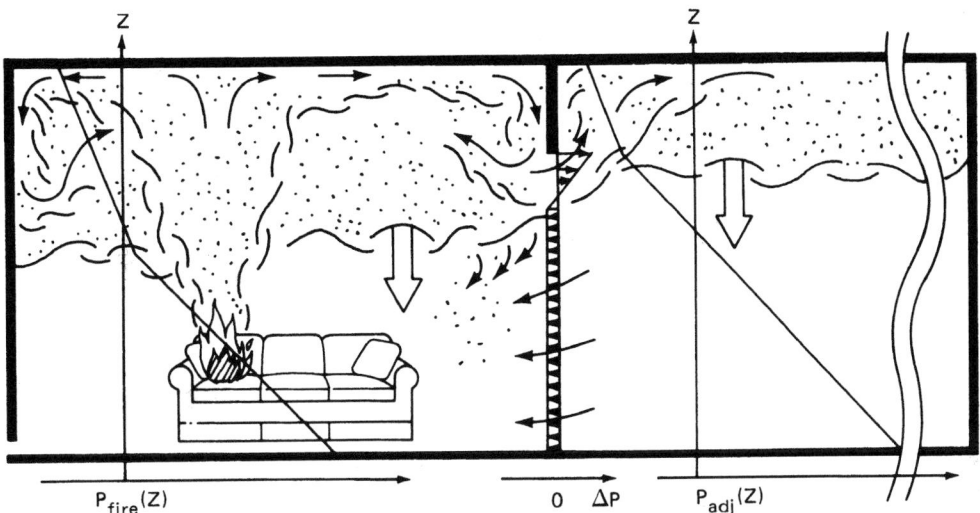

Fig. 3-10.8. *Figure 3-10.7 with sketches of pressure distributions.*

of fire involvement and adjacent spaces are large enough to drive a significant cross-door exchange of smoke and ambient air.

Toward the left side of Figure 3-10.8 is a sketch of the vertical static pressure distribution, $P_{fire}(Z)$, of the room of fire involvement. This is the pressure distribution that is measured in the bulk of the relatively quiescent-environment room, away from vigorous door and plume flows. Notice that the rate of change of pressure with elevation is uniform and relatively large between the floor and the smoke interface, and is uniform and relatively small within the smoke layer. The reason for this is that the temperature-dependent density throughout each of the two layers is assumed to be uniform, and the lower layer is more dense (i.e., of lower temperature) than the upper one. The pressure at the floor is designed as $P_{fire}(Z = 0) = P_{fire,0}$.

Toward the right side of Figure 3-10.8 is a sketch of the vertical static pressure, $P_{adj}(Z)$, in the adjacent space. There, the change of slope occurs at the elevation of the adjacent space's smoke interface, which is above the smoke interface in the fire room. Also, the slope of the pressure distribution above the interface is consistent with a smoke layer somewhat more dense or cooler than the smoke layer in the fire room. Finally, the pressure at the floor is designated as $P_{adj}(Z) = P_{adj,0}$. The two pressure distributions can be compared by the plot of the pressure difference, $\Delta P(Z) = P_{fire}(Z) - P_{adj}(Z)$, which is sketched in the doorway of Figure 3-10.8. At Z elevations below the soffit where ΔP is positive, gases are driven from the fire room into the adjacent space. At elevations where ΔP is negative, gases are driven from the adjacent space into the fire room. At the unique elevation, called the neutral plane, where ΔP is zero, the gases tend to remain stagnant in both spaces. This is the elevation in the doorway which divides outgoing fire room smoke above from inflowing adjacent space air below.

At any given elevation, it is typical in the modeling of fire-generated doorway/window flows to use Bernoulli's equation to estimate the velocity, $V(Z)$, of the flow coming out of or into the fire room. The flow is assumed to be accelerated from rest to a dynamic pressure, $\rho V^2(Z)/2 = \Delta P(Z)$, where ρ is the density of the gas from which the streamlines originate. Also, at any elevation, the flow is as-

sumed to be constricted at the *vena contracta* of the inlet/outlet jet, as with an orifice, to a fraction, C, of the width $W(Z)$ of the doorway. Then the total rate of mass flow across the doorway per unit height at any elevation would be

$$\rho C V(Z) W(Z) = C W(Z) \sqrt{2\rho \Delta P(Z)}$$

Imposing conservation principles at any instant of time when the layer thicknesses and densities (temperatures) in the two rooms are known, leads to the instantaneous values of $\Delta P(Z = 0)$ and neutral plane elevation.

An in-depth presentation of results for calculating total inlet and outlet mass flows and for general application of the above considerations is presented by Zukoski and Kubota.[25] Results on most appropriate values to use for C have been obtained from full-scale fire experiments.[26]

Door plume, ceiling jet, and smoke filling of the adjacent space: Having been driven into the adjacent space by the cross-door pressure differential, the doorway smoke jet is buoyed upward toward the ceiling due to its relatively low density (high temperature). The upward buoyant flow, depicted in Figure 3-10.7, is analogous to the previously discussed fire plume and, with minor modifications, can be quantitatively described by the same kinds of equations. In using these equations, the enthalpy flow rate of the inflowing smoke jet replaces the strength, $\dot{Q}(t)$, of the fire plume, and the smoke jet buoyancy source elevation, taken to be at or near the neutral plane elevation, replaces the elevation of the fire's combustion zone. Further quantitative details on one possible set of door-plume flow calculations are available.[25]

Just as the doorway smoke jet rises up in the adjacent space, is diluted by entrained fresh ambient air, and is mixed with the upper layer in the manner of a fire plume, so the relatively cool and dense ambient doorway jet enters the fire room, drops down past the upper layer, is contaminated by entrained smoke, and is mixed with the lower layer. This mechanism of lower layer smoke contamination in the room of fire involvement is in addition to the previously described wall flow mechanism which was depicted in Figure 3-10.6.

Figure 3-10.7 depicts the fire environment after the adjacent space upper layer is already well established. At earlier times, adjacent space smoke movement phenomena are closely related to those effects described above (i.e., Figures 3-10.3 and 3-10.4 and associated text) for the room of fire involvement. Thus, the doorway smoke jet plume impinges on the adjacent space ceiling, leads to the development of a ceiling jet which interacts with wall surfaces, and eventually redistributes itself to form a growing, upper layer of uniform thickness.

As was the case for the room of fire involvement, knowledge of adjacent space ceiling and upper wall properties is of fundamental importance in predicting the response of adjacent-space-deployed fire detection/intervention hardware, and the temperature of the adjacent space environment. Also, contamination of the lower layer by smoke injection from downward-directed wall flows can play a relatively more important role in adjacent spaces than in the fire room itself.[23]

All the above adjacent room effects must be predicted quantitatively with reasonable accuracy, since the fire-generated environments in the fire room and in adjacent spaces are strongly coupled by cross-door mass and energy exchanges. Also, of key importance is the ability to predict the onset of adjacent space environmental conditions which are untenable for life or property.

Multiroom and multilevel fire/smoke compartments: The discussion in the last two subparagraphs was related to the two-room illustration of Figure 3-10.7. However, the general principles of smoke migration are no different in fire/smoke compartments of more than two connected spaces.

In multiroom or even multilevel compartments, smoke migration occurs as smoke in successive rooms fills to the door/window soffits, and then starts to "spill out" into the adjacent spaces. At the same time, in each room where filling has been initiated, the phenomena related to plumes, ceiling jets, different wall flows, and upper layer/lower layer mixing are also taking place. In each of the spaces, these various phenomena are generally coupled together through the connecting door/window flows. For this reason, all effects must be analyzed simultaneously. For example, in a multiroom fire/smoke compartment one needs to satisfy the principle of conservation of mass when it is applied not just to a single doorway but to all envelopes which completely bound each compartment. To do so, one needs to solve for the pressure difference distributions and the resulting inflows and outflows across all intercompartment penetrations.

Some Special Classes of Multiroom Fire Scenarios

Single room vented to the outside: One practical, special class of the multiroom fire scenario is the single room of fire involvement which is vented to the outside ambient environment. One can carry out an analysis of the fire environment in such vented spaces by bringing to bear all considerations relevant to the Figure 3-10.7 discussion and by assuming the adjacent space to be arbitrarily large, i.e., large enough so that it would never be filled with smoke to the point where such smoke would interact with the fire room itself. The pressure distribution of the adjacent space from the floor to the top of the door/window would be specified to be the same as that of an outside ambient environment.

Dynamics of the plume, which is driven by the smoke flow entering the adjacent space from the fire room, would not be affected by the adjacent space ceiling or far wall surfaces. All inflow to the fire room would be uncontaminated ambient air.

Treating the adjacent space in the above manner leads to considerable simplification in modeling mathematically the room fire environment. It is noteworthy in this regard that the only mathematical models developed specifically to predict post-flashover fire environments are related to this configuration of a single room of fire involvement vented to the outside ambient environment.

Single room vented to large space: Another important class of fire scenario, which is directly related to the last one, is the single room of fire involvement which is actually vented to a very large space. Such is the configuration, for example, when a room of fire involvement is vented to a large atrium.

Under these circumstances one could analyze the fire environment which develops in the large containing space (the atrium) as one would analyze the environment in a space with a single isolated fire (e.g., see Figures 3-10.4 through 3-10.6). Here, the energy and products of combustion release rates of the fire would be taken to be the enthalpy and combustion products' flow rates of the effluent from the doorway/window jet of the fire room. As before (i.e., independent of changes in the large, but finite, adjacent space), and at least for some significant time into the fire, the development of the environment in the fire room itself and the resulting door/window smoke flow could hopefully be predicted analytically. Short of analytic predictions, however, actual measurements of the door/window effluent acquired in full-scale free-burn tests of the fire room, up to and even beyond flashover, could be used as data input in the analysis of the large adjacent space problem.

The combined experimental/analytic approach has been used to predict the environment which develops in large prison cell blocks during fires in single cells of different design.[27]

Single room and freely connected multiroom fire compartments: For those times of fire development when the compartment of fire involvement consists of a single enclosed space, analysis of the fire environment is considerably simplified. This is because an accounting of inflow and outflow at windows and doors (which are presumably closed) is not required.

When the fire compartment is partitioned into separate but freely connected spaces, the relatively simple, single-enclosed-space analysis, where the area of the single space is taken to be the total area of the fire compartment, can continue to be relevant. Here, "freely connected" refers to fire scenarios and spacial configurations where common openings between rooms are large enough, and/or the energy release rate of the fire is small enough, so that smoke layers remain reasonably uniform in thickness, temperature, and product concentration through the bulk of the compartment area.

Quantitative criteria for establishing whether a specific fire compartment is freely connected relative to a specified fire threat are not yet available. However, the concept of the freely connected, multiroom fire/smoke compartment has been shown to be valid during full-scale multiroom fire experiments.[2,28]

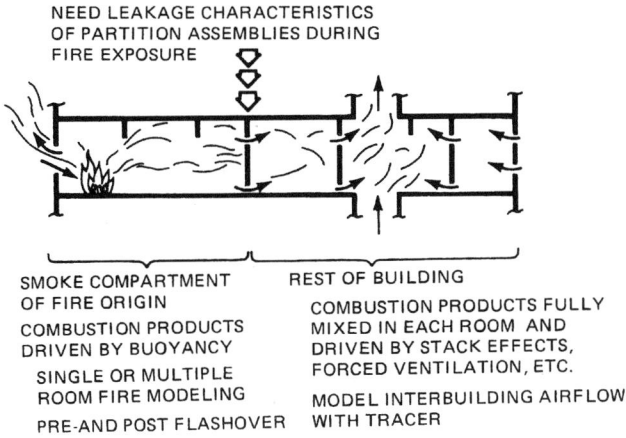

NEED LEAKAGE CHARACTERISTICS
OF PARTITION ASSEMBLIES DURING
FIRE EXPOSURE

SMOKE COMPARTMENT
OF FIRE ORIGIN

COMBUSTION PRODUCTS
DRIVEN BY BUOYANCY

SINGLE OR MULTIPLE
ROOM FIRE MODELING

PRE-AND POST FLASHOVER

REST OF BUILDING

COMBUSTION PRODUCTS FULLY
MIXED IN EACH ROOM AND
DRIVEN BY STACK EFFECTS,
FORCED VENTILATION, ETC.

MODEL INTERBUILDING AIRFLOW
WITH TRACER

Fig. 3-10.9. A concept for modeling smoke spread throughout complex facilities.

Smoke Spread Outside the Smoke Compartment of Fire Involvement

The above paragraphs addressed the development of the fire-generated environment by describing fire/smoke compartments of fire involvement. Yet, the original outline of the generic firesafety problem is also relevant to the general problem of predicting smoke environments throughout an entire facility.

Figure 3-10.9 illustrates a practical concept for modeling the development of smoke environments both inside and outside the smoke compartment of fire involvement. Facility spaces that would be included in the smoke compartment (on the left of the figure) are distinguished from those included in the rest of the building or facility (on the right) by the detail which is required to describe or model mathematically the fire-generated environments within them. In the smoke compartment of fire involvement, smoke would spread within a room, and would be driven from room to room by strong buoyancy forces which lead to layered smoke environments. These environments must be analyzed in the context of (at least) a two-layer model with associated phenomena of plume flow, surface flows, etc. In the rest of the building, it is reasonable to describe the smoke in each space as being uniformly dispersed. Here, dynamic changes in the smoke distribution in the environment come about from room-to-room pressure differences which are generated by stack effects, wind effects, and forced ventilation, leading to smoke movement, mixing, and dilution.

The fire compartment is the source of smoke to the rest of the building. The rate of introduction of this smoke depends on the pressure differences across common partition assemblies, and on their leakage characteristics.[29] Once the rate of smoke leakage across common portions can be expressed quantitatively, the rest of the building problem can be analyzed with a model of smoke movement similar to those presented by Wakamtsu[30] and Evers and Waterhouse.[31]

Mathematical Models and Computer Codes for Predicting the Compartment Fire Environment

In recent years, many mathematical models and associated computer codes for predicting dynamic compartment

fire environments have been developed. These can be divided into two types, field models and zone models.

Incorporating global partial differential equations which describe the relevant combustion, flow, and heat transfer processes, field models formulate and solve initial/boundary value problems for the unknown variables in compartment fire scenarios. Zone models, however, describe the compartment fire phenomena in terms of coupled submodel algorithms or sets of equations. Each equation set describes a single fire-generated process associated with an actual physical zone of the compartment space. The processes and corresponding zones typically correspond to the ones identified in Figures 3-10.1 through 3-10.8, and as discussed above.

There is a good deal of variation between all types of compartment fire models. Significant differences tend to be in (1) the number and detail of the individual physical phenomena that are taken into account; (2) the number and complexity of interconnected fire compartment spaces that can be analyzed; and (3) in the most common situation, when a computer is required to solve the model equations, the capability of the computer hardware that is required to carry out the calculations, the user-friendliness of the computer program, and its available documentation.

The intended use for which a given model was developed is probably the most important feature leading to its uniqueness. Such uses can differ widely; for example, at one extreme: to understand and predict coupled, compartment fire-generated processes with the greatest possible accuracy and generality; and at the other: to provide a common-use, firesafety-practitioner's tool for analysis and design. As will be seen, a set of equations which describes the dynamic smoke filling phenomenon in and of itself would constitute a compartment fire model of the simplest variety whose use could fall squarely at the latter extreme of the spectrum.

ASET—A MODEL FOR PREDICTING THE SMOKE FILLING PROCESS IN A ROOM OF FIRE ORIGIN

The smoke filling process is an essential feature of any zone-type compartment fire model. It basically involves three zones: the fire's combustion zone, the plume, and the upper smoke layer. The last section presented a relatively detailed qualitative description of many of the processes which make up the overall dynamic compartment fire environment. This section will formulate a mathematical model of the smoke filling process.

The model to be presented was originally developed within the context of life safety in fires.[1,2,24] In particular it was developed to provide estimates of the Available Safe Egress Time (ASET) in compartments of fire origin, where the available safe egress time is defined as the length of the time interval between fire detection/successful alarm and the onset of life safety hazard. Accordingly, the model has been given the name ASET.

Since life safety considerations are primary, the model focuses attention on phenomena which develop between the times of fire ignition and the onset of hazardous conditions. This allows significant simplifications in the modeling which would not be otherwise justified, *viz.*, the use of the simplest possible smoke filling process to describe the fire-generated environments of interest.

The basic phenomena of the smoke filling process are outlined as follows:

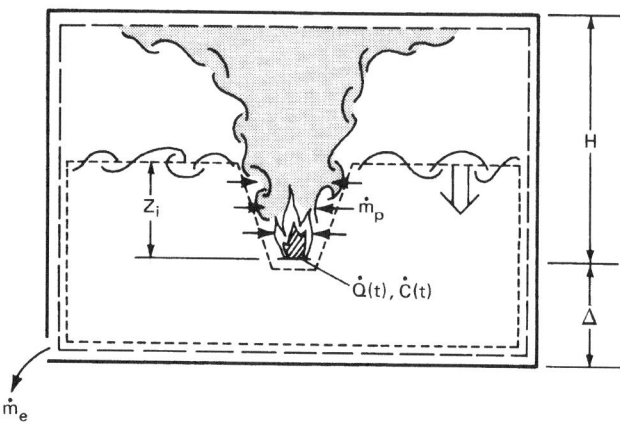

Fig. 3-10.10. *Simple illustration of fire-in-enclosure flow dynamics.*

The fire starts at some position below the ceiling of the enclosure and releases energy and products of combustion in some time-dependent manner. As the fire develops from ignition, buoyancy forces drive the high-temperature products of combustion upward toward the ceiling. In this way, a plume of upward-moving elevated temperature gases is formed above the fire. All along the axis of the plume relatively quiescent and cool ambient air is laterally entrained and mixed with the plume gases as they continue their ascent to the ceiling. As a result of this entrainment, the total mass flow rate in the plume continuously increases, and the average temperature and average concentration of products of combustion in the plume continuously decrease with increasing height. When the plume gases impinge on the ceiling they spread across it, forming a relatively thin, stably stratified upper layer. As the plume gas upward-filling process continues, the upper gas layer grows in depth, and the relatively sharp interface between it and the cool ambient air layer below continuously drops.

In this section, a simple mathematical model of these phenomena, which captures the essential features of the dynamic fire environment, is constructed. The major elements of the model include the turbulent buoyant plume theory[32] together with experimental plume results,[33] the theory of the dynamics of such plumes in confined spaces,[34] and the application of the plume dynamics theory to the fire problem as presented.[35] Figure 3-10.10 presents a simple illustration of the model's smoke filling flow dynamics. The variables introduced there will be defined in this section.

Initial Value Problem for the Temperature of the Upper Layer and the Position of the Interface

To take a conservative approach, the partitions of the room of fire origin are assumed to have all major penetrations (e.g., doors, windows, and vents) closed. Any leakage from the room resulting from fire-driven gas expansion is assumed to occur near the floor level. The sketch of Figure 3-10.7 is compatible with these assumptions, both of which lead to some conservatism in the eventual prediction of the time for onset of untenability.

The fire's combustion zone is modeled as a point source of energy release which is effectively located at or above the floor level. The mass flow rate of fuel introduced from this

zone into the plume is neglected compared with the mass flow rate of entrained air. Except for the buoyancy forces that they produce, density variations in the flow field are neglected (i.e., the Boussinesq approximation is invoked). Using the fact that the absolute pressure throughout the space varies only insignificantly from a constant uniform value, the density, ρ, can be related to the absolute temperature, T, at any time and spatial position through the perfect gas law according to

$$\rho T = \text{Constant} = \rho_a T_a \qquad (1)$$

where ρ_a and T_a are the density and absolute temperature, respectively, of the ambient air.

The time-varying total energy release rate of the combustion zone is defined by $\dot{Q}(t)$. It is assumed that $\dot{Q}(t)$ can be approximated by the free-burn energy release rate of the characteristic fuel assembly whose hazard-producing characteristics are under investigation and for which $\dot{Q}(t)$ is known. This assumption is consistent with the fact that onset of hazardous conditions within the enclosure will occur at temperature and depleted-oxygen levels which are low compared with those levels at which variations from free-burn will begin to be significant.

The fraction of $\dot{Q}$ which effectively acts to heat the plume gases and to ultimately drive the plume's upward momentum is $(1 - \lambda_r)$, where λ_r is approximately the fraction of $\dot{Q}$ lost by radiation from the combustion zone and plume.

The total mass flow in the plume, $\dot{m}_p$, and the mass mixing cup temperature of the plume, $\bar{T}_p$, at a distance Z above the fire (but below the layer interface) can be estimated by[7,35]

$$\bar{T}_p/T_a - 1 = (\dot{Q}^*)^{2/3}/0.210, \quad 0 < Z \le Z_i(t) \qquad (2)$$

$$\dot{m}_p = 0.210\rho_a(gZ)^{1/2}Z^2(\dot{Q}^*)^{1/3}, \quad 0 < Z \le Z_i(t) \qquad (3)$$

where $\dot{Q}^*$ is defined as

$$\dot{Q}^* = (1 - \lambda_r)\dot{Q}/[\rho_a C_p T_a (gZ)^{1/2}Z^2]$$

and where g is the acceleration of gravity, C_p is the specific heat at constant pressure, assumed to be constant and uniform throughout the space, and $Z_i(t)$ is the time-varying distance above the fire of the interface which separates a growing upper layer of elevated-temperature (product of combustion-laden gas) and a lower shrinking layer of ambient air. The mass flow rate of gas, $\dot{m}_e$, leaking out of the room's floor-level leakage paths can be estimated from[35]

$$\dot{m}_e = \begin{cases} (1 - \lambda_c)\dot{Q}/(C_p T_a), & -\Delta < Z_i(t) \\ (1 - \lambda_c)\dot{Q}/(C_p \bar{T}_h), & -\Delta = Z_i(t) \end{cases} \qquad (4)$$

where Δ is the height of the fire above the floor, and λ_c is the instantaneous fraction of $\dot{Q}$ lost to the bounding surfaces of the room and its contents (i.e., $\lambda_c = \dot{Q}_{loss}/\dot{Q}$). Also, assuming that the upper layer is well mixed, $\bar{T}_h$ is taken to be its absolute temperature. By using Equation 2, $\bar{T}_h$ can be related to the average upper layer density, $\bar{\rho}_h$, which is defined by

$$\bar{\rho}_h = \frac{1}{(H-Z_i)}\int_{Z_i}^{H} \rho \, dZ \qquad (5)$$

The total rate of energy loss characterized by λ_c occurs as a result of a variety of convective and radiative heat transfer exchanges between the room's gases and the above-mentioned surfaces. Equation 6 brings attention to the fact that, by the time the layer interface drops to the floor, i.e., when $Z_i = -\Delta$, all ambient air has been pushed out of the room. At all subsequent times, the entire room is filled with, and defines the bounds of, the upper layer, and the room's leakage gases are at upper layer rather than ambient conditions.

A mass balance for the lower, shrinking volume of ambient air results in

$$\rho_a A \frac{dZ_i}{dt} = \begin{cases} -\dot{m}_e - \dot{m}_p(Z = Z), & 0 < Z_i(t) \le H \\ -\dot{m}_e, & -\Delta < Z_i(y) \le 0 \\ 0, & -\Delta = Z_i(t) \end{cases} \quad (6)$$

where A is the area and H is the height of room of fire origin, and where estimates for $\dot{m}_e$ and $\dot{m}_p$ are provided in Equations 3 and 4.

Using Equations 2 and 7 in an energy balance for the upper layer results in

$$1 - \bar{\rho}_h/\rho_a = 1 - T_a/\bar{T}_h$$
$$= \left[\int_0^t (1 - \lambda_c)\dot{Q}\, d\xi\right]/[\rho_a C_p T_a A(H - Z_i)], \quad (7)$$
$$-\Delta < Z_i < H$$

$$(1 - \lambda_c)\dot{Q} = (\Delta + H)\rho_a C_p T_a A(1/\bar{T}_h)d\bar{T}_h/dt, \quad Z_i = -\Delta \quad (8)$$

Equations 4 and 6 are now used in Equation 6, and Equation 7 is recast into differential form. After some manipulation, the following pair of governing equations for Z_i and $\bar{T}_h$ result

$$\frac{dZ_i}{dt} = \begin{cases} -C_1\dot{Q} - C_2\dot{Q}^{1/3}Z_i^{5/3}, & 0 < Z_i \le H \\ -C_1\dot{Q}, & -\Delta < Z_i \le 0 \\ 0 & Z_i = -\Delta \end{cases} \quad (9)$$

$$\frac{d\bar{T}_h}{dt} = \begin{cases} \bar{T}_h[C_1\dot{Q}-(\bar{T}_h/T_a-1)C_2\dot{Q}^{1/3}Z_i^{5/3}]/(H-Z_i), & 0 < Z_i \le H \\ \bar{T}_h C_1\dot{Q}/(H-Z_i), & -\Delta \le Z_i \le 0 \end{cases}$$
$$(10)$$

$$\left.\begin{array}{l} C_1 = (1 - \lambda_c)/(\rho_a C_p T_a A) \\ C_2 = (0.21/A)[(1 - \lambda_r)g/(\rho_a C_p T_a)]^{1/3} \end{array}\right\} \quad (11)$$

The problem now becomes one of simultaneously solving Equations 9 and 10 subject to the appropriate initial conditions. For the present purpose, these initial conditions can be taken as those relating to one of two different cases.

Case 1: $\dot{Q}(t = 0) \equiv \dot{Q}_0 \ne 0$.
Here assume

$$\lim_{t \to 0} \dot{Q} \approx \dot{Q}_0 + \dot{Q}_0' t \quad (12)$$

where

$$\dot{Q}_0' = \frac{d\dot{Q}}{dt} \quad \text{at } t = 0$$

Then, solve Equations 9 and 10 subject to the initial conditions

$$Z_i(t = 0) = H$$
$$\bar{T}_h(t = 0) = T_a[1 + (C_1\dot{Q}_0^{2/3})/(C_2H^{5/3})] \quad (13)$$
$$= T_a + [(1 - \lambda_c)/(1 - \lambda_r)][\bar{T}_p(t = 0) - T_a]$$

where the value for $\bar{T}_h(t = 0)$ was obtained with the use of Equation 7. Using Equation 7 further, an analysis of the apparent singularity of Equation 10 at $t = 0$ leads to the result

$$\lim_{t \to 0} \frac{d\bar{T}_h}{dt} = T_a(C_1/C_2)[\dot{Q}_0^{2/3}/(6H^{8/3})]$$
$$\cdot [2\dot{Q}_0'H/\dot{Q}_0 + 5(C_1\dot{Q}_0 + C_2H^{2/3}\dot{Q}_0^{1/3})] \quad (14)$$

Case 2: $\dot{Q}(t = 0) = 0$
Here assume

$$\lim_{t \to 0} \dot{Q} \approx \dot{Q}_0' t \quad (15)$$

Then solve Equations 9 and 10 subject to the initial conditions

$$Z_i(t = 0) = H, \bar{T}_h(t = 0) = T_a \quad (16)$$

In this case, analysis of the problem leads to the following small time estimates

$$\lim_{t \to 0} Z_i \approx H - (3/4)C_2\dot{Q}_0'^{1/3}H^{5/3}t^{4/3} \quad (17)$$

$$\lim_{t \to 0} \bar{T}_h \approx T_a + (2/3)T_a(\dot{Q}_0'^{2/3}/H^{5/3})(C_1/C_2)t^{2/3} \quad (18)$$

Safe Available Egress Time from the Solution to the Initial Value Problem for Upper Layer Thickness and Temperature

The above initial value problem for Z_i and $\bar{T}_h$ would be solved by a numerical integration procedure. For the purpose of using the equations to determine onset of hazardous conditions, the solution would be terminated in a given problem at the time, t_{HAZ}, when

$$\bar{T}_h \ge \bar{T}_{h(HAZ)} \quad (19)$$

(layer temperature reaches a hazardous value associated with an untenable flux of thermal radiation)

or

$$Z_i \le Z_{i(HAZ)} \quad (20)$$

(interface reaches a characteristic face elevation, $Z_{i(HAZ)}$, and the upper layer gases are assumed to be hazardous for human ingestion or significantly impairing to human vision).

From the computed history of Z_i and $\bar{T}_h$, and compatible with the detection criterion which is invoked, the time of detection could also be obtained. This would be defined as that time, t_{DET}, when, for example,

$$\bar{T}_h \ge \bar{T}_{h(DET)} \quad (21)$$

(layer temperature detection criterion)

and/or

$$d\bar{T}_h/dt \ge (d\bar{T}_h/dt)_{DET} \quad (22)$$

(layer rate of temperature rise detection criterion).

The time of detection corresponding to other detection criteria which were similarly related to Z_i and $\bar{T}_h$, etc., could also be obtained. Finally, the time of detection could be explicitly specified, e.g., "immediate" detection, $t_{DET} = 0$, as a result of the guaranteed presence of alert occupants.

From all the above, the desired value for ASET is computed from

$$\text{ASET} = t_{HAZ} - t_{DET}$$

The computer program ASET has been developed to carry out the solution to the above problem for arbitrarily specified $\dot{Q}(t)$. The program is written in American National Standards Institute (ANSI) FORTRAN and it is supported by a user's manual.[24]

A simplified version of the program, ASET-B, written in BASIC and containing all necessary equation-solving software, has also been developed, and it is supported by its own user's manual.[36]

Initial Value Problem for the Concentration of Products of Combustion

In this subsection, equations for estimating the concentration of products of combustion in the upper layer are developed.

The time-varying rate at which a combustion product of interest is generated within the combustion zone is designated by $\dot{C}(t)$. The dimensions of $\dot{C}(t)$ are u_c per unit time, where u_c is a dimensional unit appropriate for the particular product. For example, u_c could have the dimensions of mass, number of particles, number of particles with mass between m and $(m + dm)$, etc.

Just as $\dot{Q}(t)$ is approximated by free-burn energy release rate data, so it is assumed that $\dot{C}(t)$ can be approximated by the free-burn product generation rate of the fuel assembly under investigation. As is the case with $\dot{Q}$, $\dot{C}$ is assumed to be known, say, from experimental free-burn measurements.

The average concentration of product in the upper layer is defined as the average amount of product (dimension u_c) per unit mass of upper layer mixture. The concentration is designated by $M(t)$. It is assumed that the mass fraction of the product in the upper layer is always small compared to 1.

Conservation of the product results in

$$\frac{d}{dt}[\bar{\rho}_h MA(H - Z_i)] = \dot{C}, \qquad -\Delta < Z_i \leq H \quad (23)$$

$$\frac{d}{dt}[\bar{\rho}_h MA(H + \Delta)] = \dot{C} - \dot{m}_e M, \qquad Z_i = -\Delta \quad (24)$$

Manipulation of Equations 23 and 24 with the use of Equations 9 and 10 leads to the following equation for M

$$\frac{dM}{dt} = \begin{cases} [\bar{T}_h/(T_a \rho_a A)] & \\ [\dot{C} - \rho_a A C_2 \dot{Q}^{1/3} Z^{5/3} M]/(H - Z_i), & 0 < Z_i < H \\ [\bar{T}_h/(T_a \rho_a A)]\dot{C}/(H - Z_i), & -\Delta \leq Z_i \leq 0 \end{cases} \quad (25)$$

With solutions for Z_i and $\bar{T}_h$ from earlier considerations, Equation 25 can be solved for M once appropriate initial conditions are established. For this purpose, the two cases must be considered again.

Case 1(a): $\dot{Q}(t = 0) \equiv \dot{Q}_0 \neq 0$; $\dot{C}(t = 0) \equiv \dot{C}_0 \neq 0$. Here assume

$$\lim_{t \to 0} \dot{C} \approx \dot{C}_0 + \dot{C}_0' t \quad (26)$$

where

$$\dot{C}_0' = \frac{d\dot{C}}{dt} \quad \text{at } t = 0$$

Then, solve Equation 25 subject to

$$M(t = 0) = \dot{C}_0/(C_2 A H^{5/3} \rho_a \dot{Q}_0^{1/3}) = \dot{C}_0/\dot{m}_p(t = 0) \quad (27)$$

Here, an analysis of the apparent zero time singularity of Equation 25 leads to the result that

$$\lim_{t \to 0} \frac{dM}{dt} = \left(\frac{5}{6}\right)(C_1/C_2)[\dot{Q}_0^{2/3}\dot{C}_0/(\rho_a A H^{8/3})]$$
$$\cdot [1 + H(3\dot{C}_0'/\dot{C} - \dot{Q}_0'/\dot{Q}_0)/(5\dot{Q}_0\dot{C}_1) + (C_2/C_1)(H^{5/3}/\dot{Q}_0^{2/3})] \quad (28)$$

Case 1(b): $\dot{Q}(t = 0) \equiv \dot{Q}_0 \neq 0$; $\dot{C}(t = 0) \equiv \dot{C}_0 \neq 0$. Here assume

$$\lim_{t \to 0} \dot{C} \approx \dot{C}_0' t \quad (29)$$

Then, solve Equation 25 subject to

$$M(t = 0) = 0 \quad (30)$$

In this case, analysis of Equation 25 leads to the following small time estimate

$$\lim_{t \to 0} M = \dot{C}_0' t/(2\rho_a A^{5/3} Q^{1/3} C_2) = \dot{C}_0' t/[2\dot{m}_p(t = 0)] \quad (31)$$

Case 2(b): $\dot{Q}(t = 0) = 0$; $\dot{C}(t = 0) = 0$. (Note that the condition $\dot{Q}(t = 0) = 0$, $\dot{C}(t = 0) \neq 0$, i.e., nonzero product generation rate with a zero heat release rate, is not allowed.) Here assume Equation 29. Then solve Equation 25 subject to Equation 30. In this case, analysis of Equation 25 leads to

$$\lim_{t \to 0} M = 2\dot{C}_0' t^{2/3}/(3\rho_a A C_2 H^{5/2} \dot{Q}_0^{'1/3}) \quad (32)$$

Using Combustion Product Concentrations to Establish the Time of Detection and the Onset of Untenability

When a fire's rate of generation of products of combustion is known, the upper layer concentrations can be estimated from the considerations of the previous section. Under such circumstances, it would be possible to apply detection and hazard criteria which are more detailed than those discussed earlier.

In the case of detection, the response of a detection device which is sensitive to the presence of the predictable combustion product can be simulated. For example, the time of detection, t_{DET}, would be predicted to be the time when the upper layer concentration of the product attained a detectable level, M_{DET}.

In the case of hazard, a criterion for the onset of untenability could depend on a variety of possible conditions involving all of the environmental parameters, Z_i, $\bar{T}_h$, and M. For example, assume that estimates of the fire's generation rate of water and CO are available. Then, the time-varying values for Z_i, $\bar{T}_h$, M_{water}, and M_{CO} could be computed, and the time of onset of untenability could be estimated to be the earliest time when (1) the interface was still above a characteristic elevation, Z_F, and $\overline{T}_h$ exceeded a specified hazardous overhead value (associated with an untenable flux of thermal radiation), or (2) the interface was below Z_F, and the upper layer CO concentration or temperature and humidity conditions were such as to be hazardous for human ingestion.

All of the above considerations are taken into account in the ASET computer program. Predictions of product of combustion concentration are not yet included in ASET-B.

AVAILABLE SAFE EGRESS TIME FROM ROOMS OF FIRE ORIGIN— SOME EXAMPLE CALCULATIONS

Assumptions on the Disposition of Energy Release and Their Implications

In order to use the proposed fire model for a specified free-burn fire, values of λ_r and λ_c are required. While appropriately chosen constant-λ values should prove to be adequate for most engineering applications, the model described can, through specified dynamic variations in these λs, readily accept more detailed characterizations of the gas-to-room surface heat transfer phenomena.

Depending on the fuel and its configuration, the total radiant power output in fire combustion zones is in the range of 15 to 40 percent of the total rate of heat release.[38,39] Based on this and, for example, on data presented by Cooper,[40] it appears that $\lambda_r = 0.35$ is a reasonable choice for the type of growing hazardous fires under consideration. Except where noted otherwise, this value will be used in all calculations described in this chapter.

Using the 0.35 value for λ_r, and taking account of convective heat transfer considerations, an appropriate value for λ_c was developed.[41] It was found to lie in the approximate range 0.6 to 0.9. The lower value, 0.6, would relate to high aspect ratio spaces (ratio of ceiling span to room height) with smooth ceilings and with fires positioned far away from walls. The intermediate values and the high, 0.9, value for λ_c would relate to low aspect ratio spaces, fire scenarios where the fire position is within a room height or so from walls and/or spaces with highly irregular ceiling surfaces. In the latter types of situations, which are representative of most realistic fire scenarios, it is not presently possible to provide general rules to accurately estimate λ_c within this 0.6 to 0.9 range. This fact has strong implications on the capability for establishing accurate estimates for the average upper layer temperature. This can be seen from Equation 10, where, early in the fire and at times of relatively cool upper layer temperatures ($\overline{T}_h/T_a$ close to 1), $d\overline{T}_h/dt$ and, ultimately, $\overline{T}_h - T_a = \Delta\overline{T}_h$ are seen to be proportional (through the factor C_1) to $(1 - \lambda_c)$. In contrast to the upper layer temperature estimate and at times of relatively small values of $\Delta\overline{T}_h/T_a = \overline{T}_h/(T_a - 1)$, the upper layer-lower layer interface position history is not nearly as sensitive to inaccuracies in λ_c. At such times, the second term on the right-hand side of the first line of Equation 9, which is independent of λ_c, will dominate the first term, the two terms being in the ratio of $\dot{m}_p$ to $\dot{m}_e$ (compare to the first line of Equation 6).

The above discussion leads to the following guidelines for selection and use of a value for λ_c, when a reliable estimate of its actual value is not otherwise available:

1. For the purpose of computing a conservative estimate of the time when a hazardous temperature or a hazardous interface elevation will be attained (i.e., the predicted t_{HAZ} will be less than the observed t_{HAZ}), one should select $\lambda_c = 0.6$.

2. For the purpose of a conservative estimate of detection time when detection is by temperature or rate of temperature rise of the upper layer (i.e., the predicted t_{DET} will be greater than the actual t_{DET}), one should select $\lambda_c = 0.9$.

3. When fire detection is by temperature or rate of temperature rise, a reasonably accurate (as compared with a conservative) estimate of detection time is achievable only (1) in large aspect ratio, smooth ceiling spaces where detection is based on a $\lambda_c = 0.6$ computation of average upper layer temperature, and (2) in other configurations where detectors are deployed near the ceiling in some regular grid array, and where the time of detection is based on estimates of actual maximum ceiling-jet temperature (i.e., predictions of average upper layer temperature are not the basis for determining likely time of detection). For such estimates, the reader is referred to Section 4, Chapter 1 of this handbook.

Available Safe Egress Time in a Semi-Universal Fire

For the smoke filling model to have utility to practitioners of fire safety, it is necessary that the significant elements of potentially threatening fire scenarios be identified. It is also necessary for the results of fire hazard analyses to be presented in a concise and practical manner. This subsection provides an example of how the whole concept might proceed in practice.

First, one must identify quantitative characteristics of a particular, potentially threatening, free-burn fire of concern. Cooper deals with some practical considerations that would be useful in deducing such characteristics.[1] For the present, a composite, semi-universal-type fire has been constructed from the data of Friedman.[41] The fire's energy release history is plotted in Figure 3-10.11. The fire is assumed to be initiated from a 10 kW ignition source. Initially, it grows exponentially at a rate which is characteristic of a fire initiated in a polyurethane mattress with bedding. This early growth rate would be characteristic of the early growth of fires in a variety of occupancies which typically contain upholstered polyurethane cushioning, e.g., hospital patient and lobby rooms, residential spaces, and auditoriums. It is also consistent with the (unreported) early growth state of fires in large assemblies of commodities stacked on pallets. Beyond 400 kW, the fire of Figure 3-10.11 is assumed to grow at a rate which is similar to and/or which bounds the anticipated growth of fires initiated in a variety of different types of commodities stacked on pallets. The portion of the semi-universal fire beyond 400 kW is no doubt also representative of other threatening fires in large mercantile and/or business occupancies.

The fire of Figure 3-10.11 was assumed to be initiated in a variety of different-size spaces. The geometries of these spaces are characterized by areas ranging from 28 to 929 m^2 and by heights ranging from 2.4 to 6.1 m.

Two possible criteria for fire detection are considered in the analysis of available safe egress time. These include using ASET to calculate instantaneous detection (by whatever means) and detection when the upper gas layer reaches an average temperature of 57°C. The utility of the latter detection criterion is at present strictly speculative. It is included here only to illustrate the type of results which one

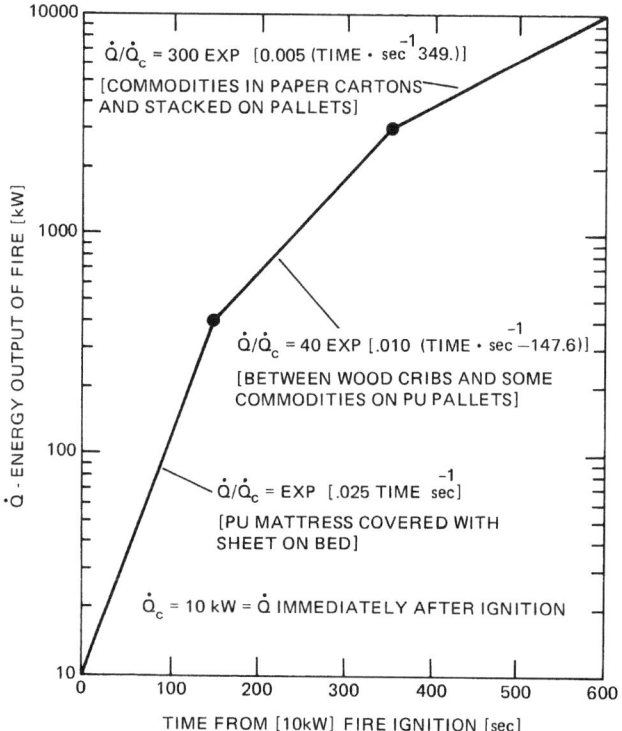

$\dot{Q}/\dot{Q}_c = 300 \text{ EXP } [0.005 \text{ (TIME} \cdot \text{sec}^{-1} 349.)]$
[COMMODITIES IN PAPER CARTONS AND STACKED ON PALLETS]

$\dot{Q}/\dot{Q}_c = 40 \text{ EXP } [.010 \text{ (TIME} \cdot \text{sec}^{-1} -147.6)]$
[BETWEEN WOOD CRIBS AND SOME COMMODITIES ON PU PALLETS]

$\dot{Q}/\dot{Q}_c = \text{EXP } [.025 \text{ TIME sec}^{-1}]$
[PU MATTRESS COVERED WITH SHEET ON BED]

$\dot{Q}_c = 10 \text{ kW} = \dot{Q}$ IMMEDIATELY AFTER IGNITION

Fig. 3-10.11.   Free-burn energy release rate from a semi-universal fire. A fictitious construction from the data of O'Neill and Hayes.[44]

might hope to generate by solution of the model equations and use of the ASET computer program. (The results could also be obtained with ASET-B.)

The criterion adopted for the onset of hazardous conditions is an upper layer interface position 0.91 m above the floor, or an average upper layer temperature of 183°C (corresponding to a heat flux of 0.25 W cm^{-2} at the floor), whichever comes first.

It is assumed that 35 percent of the fire's instantaneous energy release rate is radiated from the combustion zone (λ_r = 0.35) and that a total of 60 percent of this energy release rate is transferred to the interior surfaces of the room and its contents, i.e., 40 percent of this energy is retained in the upper layer products of combustion (λ_c = 0.60). Recall that the latter choice of λ_c would be appropriate for large aspect ratio spaces with smooth ceiling, but, in any event, the choice of λ_c would have a minor impact on estimated egress times in cases where criteria of detection or hazard are not dependent on upper layer gas temperature.

With the above range of parameters, the quantitative details of the last section were used to estimate available safe egress times with the ASET program. The results of these computations are presented in Figure 3-10.12. In this figure, ASET = $t_{HAZ} - t_{DET}$ is plotted as a function of room area for different parametric values of room height and for different detection criteria.

As an example of the utility of Figure 3-10.12, consider a scenario where a fire is initiated in an occupied, 500 m^2, nominal, 6.1 m-high ceiling auditorium outfitted with poly-urethane cushion seats (which are assumed to be the most significant fuel load). Then, from Figure 3-10.12 one would estimate an available safe egress time of approximately 450 s.

This assumes immediate detection as a result of occupant recognition and verbal alarm to fellow occupants at the time of fire initiation. If the auditorium is to be considered safe relative to successful egress, then a further study would have to reveal that the time required for a capacity crowd to evacuate the auditorium is less than 450 s.

The following general features of the results of Figure 3-10.12 are worth noting:

1. As is well known, for life safety as it relates to safe egress, temperature detectors are not particularly effective.
2. For a given curve, increasing room area eventually leads to an abrupt reduction in the curve's slope. This is the result of a shift in the triggering mechanism for onset of hazardous conditions. On the left side of the change in slope (smaller areas), untenability occurs as a result of the layer interface dropping to the 0.91 m level. On the right side (larger areas), untenability occurs as a result of thermal radiation from a hot upper layer.

Based on the previously developed model equations, the calculation procedure described in this section has been generalized and incorporated with other example calculations into the ASET computer program user's manual.[24] For a fire scenario of interest, this computer program carries out ASET calculations corresponding to user-supplied inputs which describe the fire threat, room size, and appropriate user-specified detection and hazard criteria.

A POSSIBLE EXTENSION IN THE MODEL'S UTILITY

An Experimental, Full-Scale, Multi-Room Fire Scenario

This section compares the results of a full-scale, multi-room fire experiment with calculations based on ASET. The experiment was one of a series of tests in a mockup hospital patient room/corridor building space.[44]

A plan view of the building space is presented in Figure 3-10.13. The space is made up of a room of area 14.4 m^2 connected by an open doorway to a corridor-lobby configuration of area 74.3 m^2.

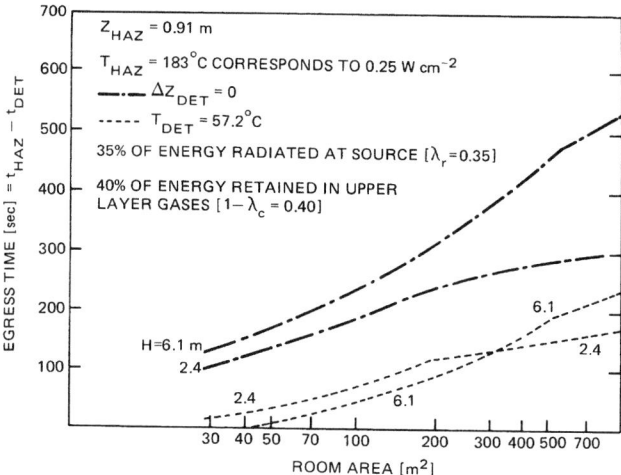

$Z_{HAZ} = 0.91$ m

$T_{HAZ} = 183°$C CORRESPONDS TO 0.25 W cm^{-2}

—·— $\Delta z_{DET} = 0$

---- $T_{DET} = 57.2°$C

35% OF ENERGY RADIATED AT SOURCE [λ_r=0.35]

40% OF ENERGY RETAINED IN UPPER LAYER GASES [$1 - \lambda_c = 0.40$]

Fig. 3-10.12.   Estimates of available egress times from the semi-universal fire of Figure 3-10.11.

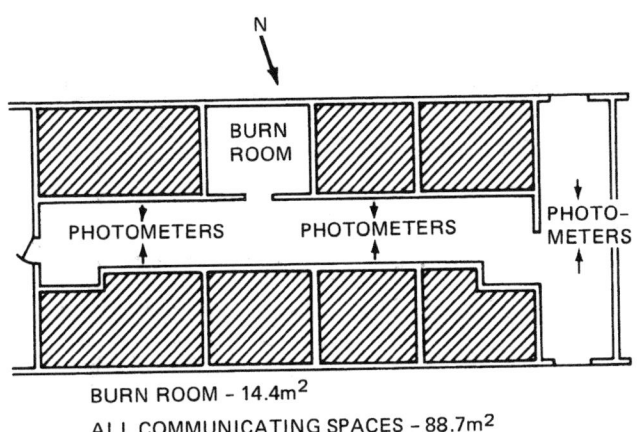

Fig. 3-10.13. *Plan view of hospital room/corridor mockup space.*

A fire is initiated in a wastepaper basket next to the corner of a polyurethane mattress covered with bedding. The burn characteristics of this assembly were studied prior to the room burns of this series.[5] The wastepaper basket/mattress fuel energy release rate, as derived from weight loss measurements, is plotted in Figure 3-10.14. For the purpose of the present analysis, it is assumed that this energy release rate was reproduced in the actual test run under review.

The model, which has been quantitatively described so far, is a single-room or room-of-fire-origin model. Thus, it may not be immediately obvious at what point, if any, it will have relevance to the present fire scenario. It would appear that a two-room or multi-room flow dynamics model would generally be required to study the room-corridor-lobby scenario under consideration. (For example, a two-room example flow calculation for a set of fire and room size parameters which somewhat corresponds to the present scenario has been considered.)[25] Nevertheless, it is possible that a simple, single-room modeling approach to fire scenarios involving relatively free-flowing multi-space configurations can be adequate for the purpose of obtaining engineering estimates of available egress times in the range of conditions that occurred in the referenced hospital patient room/corridor test.

Model Predictions Compared with Experimental Results

Room of fire origin: For early times into the fire, prior to the time when the upper layer interface drops to the level of the connecting doorway soffit, the single-room model is completely relevant. Up to that moment, the open doorway acts as the lower leakage path referred to earlier. Using the energy release data of Figure 3-10.14, and taking the fire source to be effectively at the floor, the model was used to compute the product of combustion-filling history of a 14.4 m^2 room up to the time that the upper layer thickness exceeded the existing ceiling-to-soffit dimension of 0.41 m. λ_c was estimated at 0.72.[40] The time for the interface to reach the soffit was computed to be 21 s, following ignition.

Adjacent space: Once the smoke flows under the soffit and starts to fill the large corridor-lobby space, a two-room model is required to describe the gas migration and exchange between the two spaces. This would continue to be true for at least some intermediate time interval. Following this, the single-room model can again be relevant.

If the fire is small enough or the doorway is large enough so that flows through the doorway remain relatively weak, the adjacent space will eventually attain and maintain a smoke layer thickness essentially identical to that of the room of fire origin. After some time interval, the histories of the elevations of the layer interfaces in both the corridor/lobby and room of fire origin spaces will be similar and can be computed from a single-room model, where the single room has an area equal to the combined area of both spaces. Time intervals when the upper layer thickness of the two rooms is not similar would encompass (1) the initial time when the room of fire origin fills up with smoke to the level of the doorway soffit, and (2) the subsequent time interval when the upper layer thickness of the adjacent space grows from zero to a value close to that of the room of fire origin.

For fire scenarios where the above is applicable, significant simplifications occur in that the relatively simple single room of fire origin model can be used to study the effects of fire growth when far more complicated multi-room models would, at first hand, appear to be required.

To test the above ideas, the single-room model was used to predict the history of a single interface elevation and the average upper layer temperature within the combined patient room/corridor space. Using the energy release rate of Figure 3-10.14 and a total room area of 88.7 m^2, the history of the interface elevation and of the average upper layer temperature was computed. An effective λ_c for this combined space scenario is expected to be greater than the above $\lambda_c = 0.72$ value used for the single room of fire origin because of the additional heat transfer to the corridor surfaces. A value of $\lambda_c = 0.85$ was selected for the calculation. This was done with the anticipation that comparisons between computed and experimental average upper layer temperatures would reveal an appropriate correction to this λ_c value. (Recall that the upper layer temperature difference, $\Delta \overline{T}_h$, is

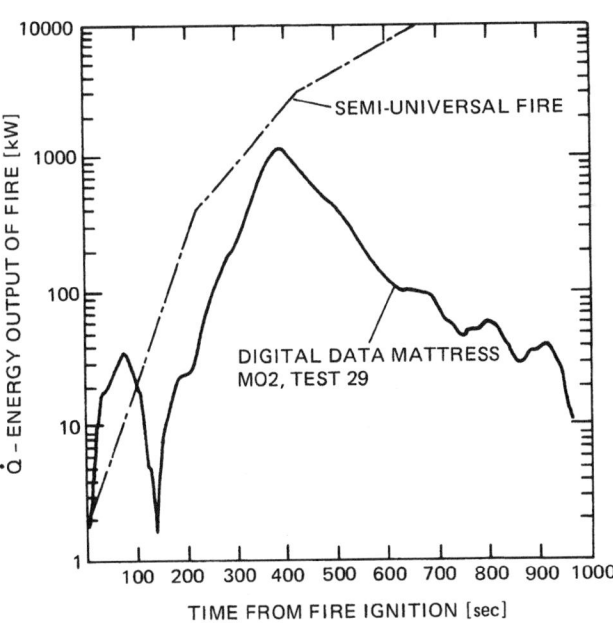

Fig. 3-10.14. *Energy release rate of wastepaper basket/mattress fuel assembly.*

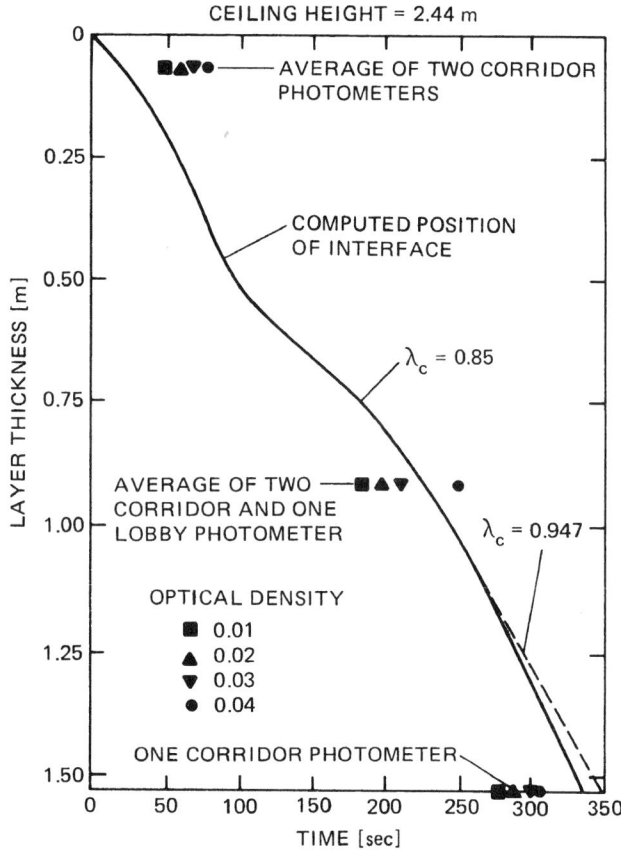

Fig. 3-10.15. *History of interface position.*

approximately proportional to $1 - \lambda_c$.) The results of the computation for interface position and upper layer temperature are presented in Figures 3-10.15 and 3-10.16, respectively.

For the purpose of comparing results for the analytic and experimental interface position, an operational definition of the experimental interface position was required. This definition was based on the outputs of a total of six photometers placed at three different elevations and at one to three different positions (see Figure 3-10.13) in the corridor and lobby. The measured optical density (OD) outputs of these photometers are indicated in Figure 3-10.17. From these outputs, and for the purpose of defining a time when the smoke layer interface position passes the elevations of these photometers, there is still ambiguity as to what value of OD should constitute the presence of a smoke layer. Four different OD values, 0.01, 0.02, 0.03, and 0.04, were used as possible definitions for a minimum upper layer OD. Using the photometer outputs, the result of these four possible interface definitions leads to four possible sets of experimental data points for the interface elevation versus time. These are plotted in Figure 3-10.15 together with the theoretical results of the interface motion.

The favorable agreement between the results of theoretical and experimental interface position at the lower two of the three photometer locations illustrates the capability of the single-room model, in the present multi-room fire scenario, to predict the growth of the potentially hazardous upper smoke layer thickness. A favorable comparison at the uppermost photometer elevation located 0.06 m from the

ceiling was not to be expected. This is because of the fact that, in the present multi-space configuration, the single-room model implemented in the manner described is clearly not adequate to predict the early growth in the corridor-lobby portion of the test space. Subsequent testing in the Figure 3-10.13 test space has corroborated the potential utility of the model in providing practical simulations of multi-room fire environments.[28]

Gas temperatures were measured by two thermocouple trees located in the center of the corridor 4.6 m on either side of the room of fire origin doorway. No temperature data were acquired in the lobby space. At any given time, the equi-elevation thermocouples of these two trees measured temperature differences, $(T - T_a)$, which agreed to within 20 percent of one another. For the purpose of comparing analytic and experimental average upper layer temperature histories, an appropriate instantaneous weighting of the measured temperatures of the limited number of these corridor thermocouples was required. At a given instant of time, this weighing has to be consistent with the estimate/measurement of the interface position as well as with the relative position of the thermocouples in question. A plot of the measured average upper layer temperature history deduced from such a data reduction scheme is presented along with the plot of the computed temperature history in Figure 3-10.16. As noted earlier, if a different λ_c has been used in the computation then, to a first approximation (i.e., using the principle of proportionality between $\Delta \overline{T}_h$ and $1 - \lambda_c$), one would anticipate a shift from the originally computed $\Delta \overline{T}_h(t)$ (with $\lambda_c = 0.85$) to a new temperature history $\Delta \overline{T}_h^{(new)}(t)$ [with $\lambda_c = \lambda_c^{(new)}$], where

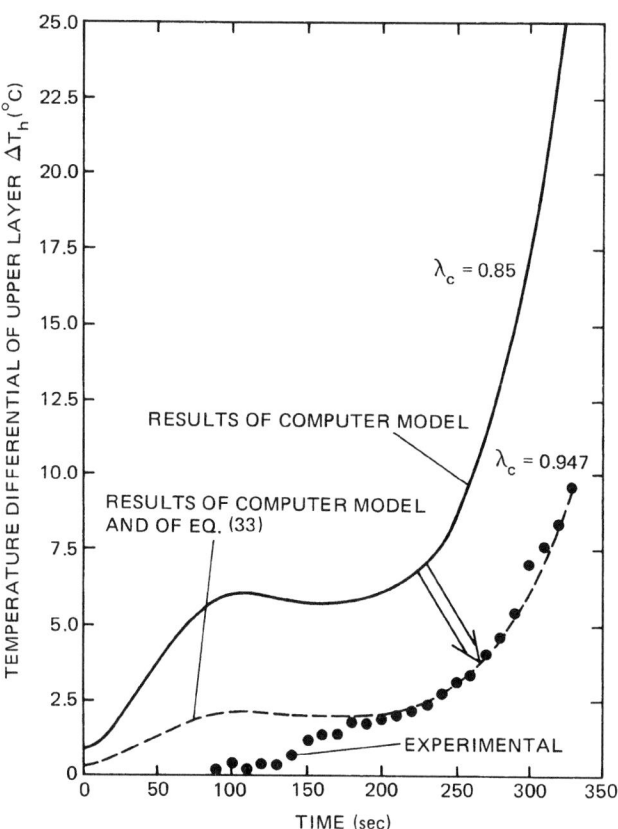

Fig. 3-10.16. *History of average upper layer temperature.*

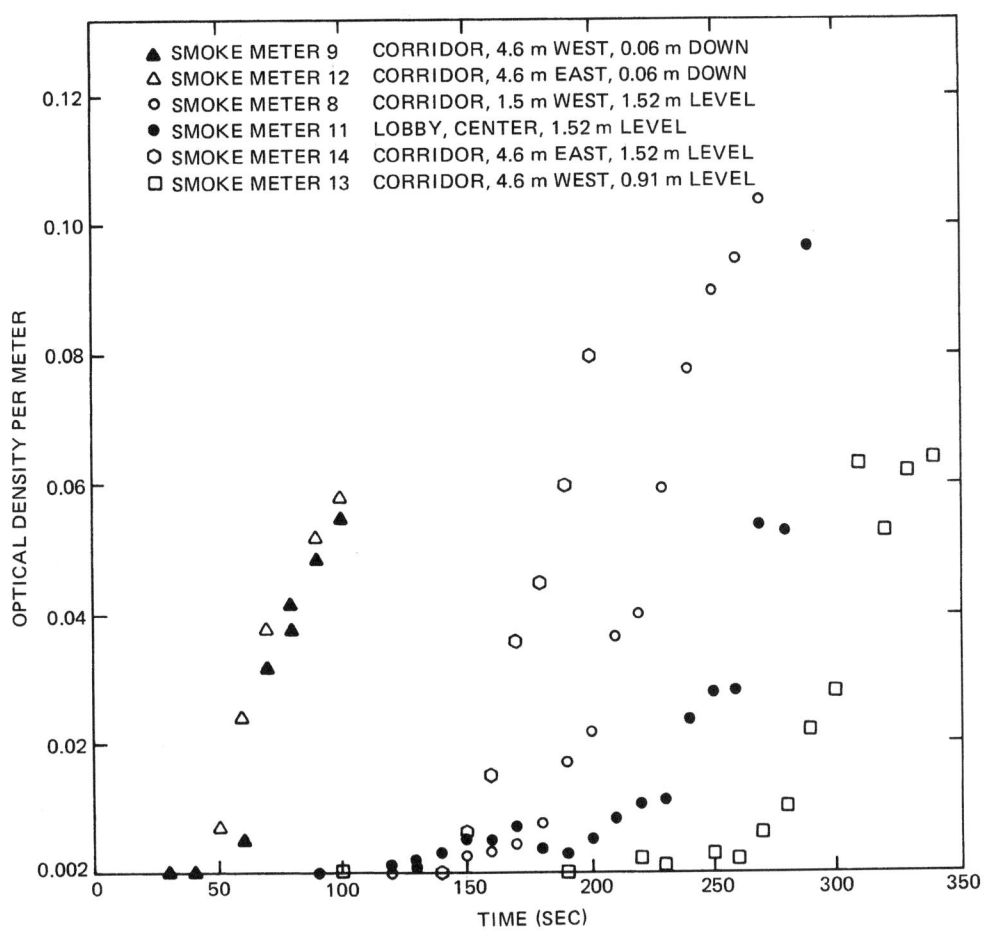

Fig. 3-10.17.   Optical densities measured by the photometers.

$$\Delta \overline{T}_h^{(new)}(t) = [(1 - \lambda_c^{(new)})/(1 - 0.85)] \, \Delta \overline{T}_h(t) \quad (33)$$

In view of the above, it is possible to bring the predicted analytic $\overline{T}_h$ plot into coincidence with the experimental $\overline{T}_h$ plot for at least one instant of time by a new choice, $\lambda_c^{(new)}$, for λ_c. Such coincidence is attained at $t = 330$ s (when the computed layer interface is at the potentially hazardous position 0.91 m from the corridor floor) by the specific choice of $\lambda_c^{(new)} = 0.947$. Using this latter value of $\lambda_c^{(new)}$ in Equation 33, an adjusted average upper layer temperature history was computed and plotted. (See Figure 3-10.16.)

The single-room model was also used to recompute the interface position and temperature histories corresponding to $\lambda_c = 0.947$. These are plotted in Figures 3-10.15 and 3-10.16, respectively. For the parameters of the present scenario, the proximity of the $\lambda_c = 0.85$ and $\lambda_c = 0.947$ plots of Figure 3-10.15 illustrates the relative insensitivity of the interface position history to changes in λ_c. The variations between the two $\lambda_c = 0.947$ temperature history estimates are so small that they cannot be discerned in most of the Figure 3-10.15 plot. For this scenario, this illustrates the insensitivity of interface position on λ_c.

As can be noted in Figure 3-10.16, the experimental and newly calculated estimates for the upper layer temperature history are in good agreement in the time interval 175–330 s but in poor agreement at earlier times. Besides the fact that

earlier times are likely to require analysis by a multi-room model, it is worth noting that the relatively complicated nature of the energy transfers which are being simulated may preclude sharper estimates of $\overline{T}_h$ with a single, constant value of λ_c. In this regard, results indicate that for fires in the present test space, λ_c can vary in time over a wide range of values.[28]

SOLUTIONS TO THE MODEL EQUATIONS FOR A SPECIAL CLASS OF GROWING FIRES

As was indicated in the last two sections, the ASET smoke filling model equations are easily solved with the use of the ASET or ASET-B computer programs for any particular fire specified by $\dot{Q}(t)$, $\dot{C}(t)$, and parameters H, A, Δ, λ_r, and λ_c. However, the approach of solving the equations for one set of conditions at a time does not lead readily to insight into solutions of generic problems of interest. For example, to obtain the results of Figure 3-10.12, solutions to the equations were required over a range of input values of A for each of the two input values of H. If more H values were of interest, and if, for example, one wished to study the effect of varying λ_c and/or Δ, then the volume of computer output would quickly become massive and unwieldy. In general, insight into the environment generated by a class of fire scenario (e.g., the semi-universal fire of Figure 3-10.11 in a room of arbitrary H, A, Δ, λ_r, and λ_c) is most clear when solutions can be obtained and displayed by means of limited numbers of graphs, charts, or tables. In this section, features of such a solution for an important, practical class of fire scenario will be displayed graphically, and explanation on how to extract practical results from this will be presented by way of examples. Some very useful and suprising time-of-smoke-filling estimates are obtained from this solution, and these will also be presented.

$Q \propto t^n$ Fire and Its Governing Equations

This subsection will present and solve Equations 9 through 11 and 25 for the broad class of fires whose $\dot{Q}(t)$ can be reasonably approximated by growth rates proportional to t^n for arbitrary $n \geq 0$ and whose product of combustion generation rates, $C(t)$, are approximately proportional to $\dot{Q}(t)$. This class of fire includes the constant fire, $n = 0$, and the t^2 growing fires, $n = 2$, both of which have been used in a variety of different references to describe the burning of many practical assemblies of combustibles. To be definite, it is assumed that $\dot{Q}$ and C can be approximated by

$$\dot{Q}(t) = \dot{Q}_0(tH^{3/2}g^{1/2}/A)^n; \quad C(t) = \beta\dot{Q}(t) \qquad (34)$$

where $\dot{Q}_0$ represents a characteristic energy release rate, n is any non-negative integer, and β is a constant of proportionality of appropriate dimension.

Notice that for the constant fire problem, $n = 0$ in Equation 34 and $\dot{Q}_0$ is simply the specified constant energy release rate of the fire. Also, the energy release rate in many practical fires is simulated by $n = 2$-type fires and is approximated by[4]

$$\dot{Q}(t) = (1000/t_g^2)t^2 \text{ kW} \qquad (35)$$

where t_g, the growth time of the fire, is defined as the time for the fire to grow in a t^2-type manner from a small flaming fire to a fire of approximately 1000 kW. Equation 34 for $n = 2$ and Equation 35 lead to the result that for these "t-squared" fires, $\dot{Q}_0$ should be chosen as[4]

$$\dot{Q}_0 = [1000A^2/(t_g^2gH^3)] \quad \text{kW for } \dot{Q} \sim t^2 \text{ fires} \qquad (36)$$

It is convenient to introduce the following dimensionless variables and parameters:

$$\begin{rcases} \zeta = Z_i/H \quad \text{(interface elevation)} \\[6pt] \phi = T/T_a \quad \text{(upper layer temperature)} \\[6pt] \mu = (1 - \lambda_c)M/(\beta C_pT_a) \\ \qquad \text{(upper layer product concentration)} \\[6pt] \tau = 3[(1 - \lambda_r)Q_0^*]^{1/3}(tH^{3/2}g^{1/2}/A)^{(n+3)/3}/(n + 3) \\ \qquad \text{(time)} \\[6pt] \varepsilon = (1-\lambda_c)[(n+3)/3]^{2n(n+3)}Q_0^{*2/(n+3)}/(1 - \lambda_r)^{(n+1)/(n+3)} \\ \qquad \text{(fire strength)} \\[6pt] \dot{Q}_0^* = \dot{Q}_0/(\rho_aC_pT_ag^{1/2}H^{5/2}) \\ \qquad \text{(characteristic energy release rate)} \\[6pt] \delta = \Delta/H \quad \text{(fire elevation)} \end{rcases} \qquad (37)$$

Using the above definitions in the model Equations 7, 9, 10, 13, and 16 through 18, eventually leads to the following equations for τ, ϕ, and μ

$$\frac{d\zeta}{d\tau} = \begin{cases} -\varepsilon\tau^{2n(n+3)} - 0.210\zeta^{5/3}; & 0 < \zeta \leq 1 \\[6pt] -\varepsilon\tau^{2n(n+3)}; & -\delta < \zeta \leq 0 \\[6pt] 0; & \zeta = -\delta \end{cases} \qquad (38)$$

$$\phi = \left\{1 - \frac{(n+3)\varepsilon\tau^{3(n+1)/(n+3)}}{3(n+1)(1-\zeta)}\right\}^{-1}; \quad -\delta < \zeta \leq 1$$

$$\frac{d\phi}{d\tau} = \frac{\varepsilon\phi\tau^{2n/(n+3)}}{(1 + \delta)}; \qquad \zeta = -\delta \qquad (39)$$

$$\mu = \phi - 1; \quad -\delta \leq \zeta \leq 1 \qquad (40)$$

where Equation 38 must be solved subject to

$$\zeta(\tau = 0) = 1 \qquad (41)$$

and where early time estimates for ζ, ϕ, and μ are

for $n = 0$:

$$\begin{rcases} \lim_{\tau \to 0} (\zeta - 1)/(1 + \varepsilon/0.210) \\ \qquad = -0.210\tau + \text{higher order terms in } \tau \\[6pt] \lim_{\tau \to 0} \phi/(1 + \varepsilon/0.210) \\ \qquad = \lim_{\tau \to 0}(\mu + 1)/(1 + \varepsilon/0.210) \\ \qquad = 1 + 5\varepsilon\tau/6 + \text{higher order terms in } \tau \end{rcases} \qquad (42)$$

for $n > 0$:

$$\begin{rcases} \lim_{\tau \to 0} (\zeta - 1) = -0.210\tau + \text{higher order terms in } \tau \\[6pt] \lim_{\tau \to 0} (\phi - 1) = \lim_{\tau \to 0} \mu \\[6pt] \qquad = \frac{(n+3)\varepsilon\tau^{2n/(n+3)}}{3(n+1)(0.210)} + \text{higher order terms in } \tau \end{rcases} \qquad (43)$$

Equation 38 describes the rate of descent of the interface as it passes through the regions above the fire ($0 < \zeta < 1$), below the fire ($-\delta < \zeta < 0$), and at the floor ($\zeta = -\delta$). Equations 39 and 40 describe the corresponding upper layer temperature and product concentration. Equations 42 and 43 are useful in starting a numerical solution to Equations 38 and 39.

Discussion of the Equations

The last subsection presented the equations which govern the dynamics of the interface, ζ, the upper layer temperature, ϕ, and the upper layer product concentration, μ. From Equation 40, the solution for μ would follow directly from the solution for ϕ. From the time of ignition to the time that the interface drops to the floor of the enclosure, a solution for ϕ could be obtained from Equations 39, 42, and 43, provided a solution for ζ was available. Beyond that time, the solution for ϕ could be determined by a direct integration of the second line of Equation 39.

With the above observations, attention is drawn to the solution for ζ. From ignition at $\tau = 0$ until $\tau = \tau_0 \equiv \tau(\zeta = 0)$, corresponding to the time when the interface drops to $\zeta = 0$, ζ is governed by Equation 41 and the first line of Equation 38. No general closed form solution is possible, and a numerical solution for $\zeta(\tau; \varepsilon, n)$ is in order. Once this has been obtained, the solution can be extended beyond τ_0 by direct integration of the second and third lines of Equation 38.

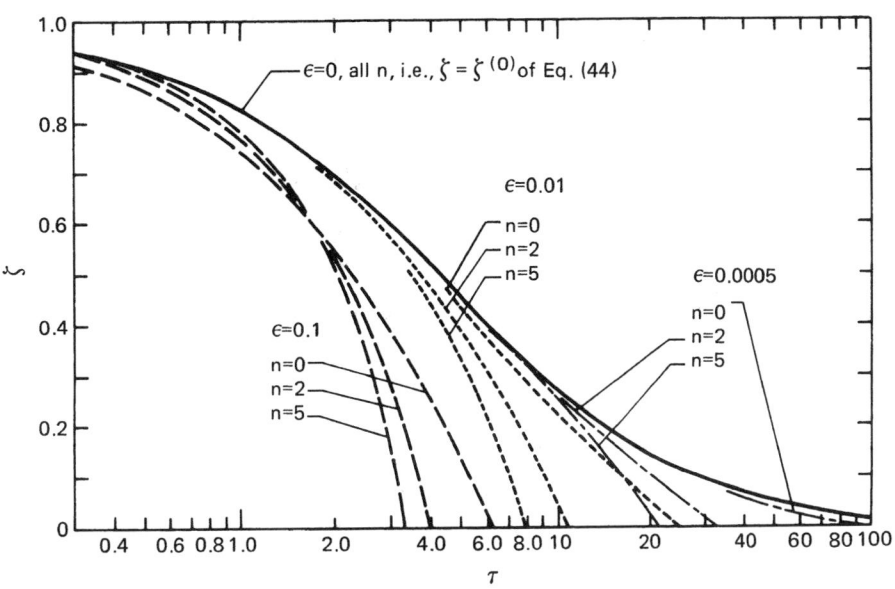

Fig. 3-10.18. *Plots of $\zeta(\tau)$ for different values of n and ε, $0 < \tau \leq \tau_0$.*

Solutions from Ignition to τ_0

In general, there is no particular problem in using a computer to integrate Equation 38 numerically and obtain ζ. However, in terms of generating a display of working graphical solutions which include times when ζ is small and positive, a problem does arise in the limit as ε approaches 0 (e.g., for small, dimensionless fire strength, $\dot{Q}^*$). Applying such a limit to the first line of Equation 38 leads, in a first approximation, to the total neglect of the earlier referenced (left-hand) expansion term in comparison to the (right-hand) entrainment term. This corresponds, physically, to the situation of an interface that approaches the elevation of the fire, $\zeta = 0$, asymptotically in time. In the present nomenclature, and for a source whose strength grows as t^n, the solution for $\varepsilon = 0$ is found to be

$$\zeta(\tau; \varepsilon = 0, n) = \zeta^{(0)}(\tau) = [1 + 0.210(2/3)\tau]^{-3/2} \quad (44)$$

This result is plotted in Figure 3-10.18 along with numerically obtained, non-zero ε solutions for ζ.

From Equation 44 it is clear that for $\varepsilon = 0$, $\zeta \to 0$ as $\tau \to \infty$. But, for a fixed n and an arbitrarily small but non-zero ε, a $\zeta = 0$ position of the interface will, in fact, be attained at some finite, large $\tau = \tau_0$.

This small ε behavior of ζ and its proximity to the $\varepsilon = 0$ solution can be observed in Figure 3-10.18. As can be seen, the smaller the value of ε and the closer the value of n to 0, the longer in time the actual solution is accurately approximated by the $\varepsilon = 0$ solution.

The small ε limit is very important in problems of physical interest. As an example, consider a constant ($n = 0$) smolder source of 0.5 kW, located a distance of 2 m below a ceiling with $\lambda_r = 0.1$ and $\lambda_c = 0.75$. This leads to $\varepsilon = 5.3$ (10^{-4}). As an example of a relatively strong fire, consider a constant flaming fire of 5,000 kW (e.g., a burning gasoline spill approximately 1 m in radius) located 5 m below a ceiling, with $\lambda_r = 0.35$ and $\lambda_c = 0.75$. This leads to $\varepsilon = 6.0$ (10^{-2}). In terms of a "small ε" criterion, the latter fire is still relatively weak.

Time, Temperature, and Concentration when the Smoke Drops to the Fire Elevation

Numerically computed τ_0, ε pairs were obtained and plotted by Cooper[43] for a variety of different n values. The corresponding values for $\phi_0 = \phi(\tau_0; \varepsilon, n)$ were also obtained and plotted. All these results are reproduced here in Figure 3-10.19. From these plots and for arbitrary ε and n, it is possible to find the time, t_0, which corresponds to τ_0, for the smoke layer to drop to the fire elevation at $Z = 0$. The plots also provide an estimate for $\phi_0 = \phi(t_0)$, from which it is possible to obtain the $t = t_0$ upper layer temperature, T_0, and (if applicable) the product of combustion concentration, M_0. The most interesting general feature of Figure 3-10.19 is that, for a given n, the value of the ordinate

$$\varepsilon^{2(n+3)/[3(n+5)]}\tau_0(\varepsilon; n) = f(\varepsilon; n) \quad (45)$$

is relatively uniform over a broad ε range of interest. For example, for $n = 0$, 1, and 2

$$\varepsilon^{2/5}\tau_0(\varepsilon; n = 0) = f(\varepsilon; n = 0) = 4.3(1 \pm 0.15)$$

$$\varepsilon^{1/3}\tau_0(\varepsilon; n = 1) = f(\varepsilon; n = 1) = 3.4(1 \pm 0.16) \quad (46)$$

$$\varepsilon^{10/33}\tau_0(\varepsilon; n = 2) = f(\varepsilon; n = 2) = 3.0(1 \pm 0.17)$$

for ε in the range

$$0.2(10^{-4}) < \varepsilon < 0.2(10^{-1}) \quad (47)$$

[With somewhat larger errors, Figure 3-10.19 indicates that estimates of τ_0 coming from Equation 46 would remain valid even for ε significantly smaller than $0.2(10^{-4})$.] This result can be expressed in practical terms as a general solution for t_0

$$t_0 = [f(\varepsilon; n)]^{3/(n+3)}\left\{\frac{[A(n+3)/3]^5[\rho_a C_p T_a t_0^n/Q(t_0)]^3}{(1-\lambda_c)^2(1-\lambda_r)g}\right\}^{1/(3n+5)}$$

$$(48)$$

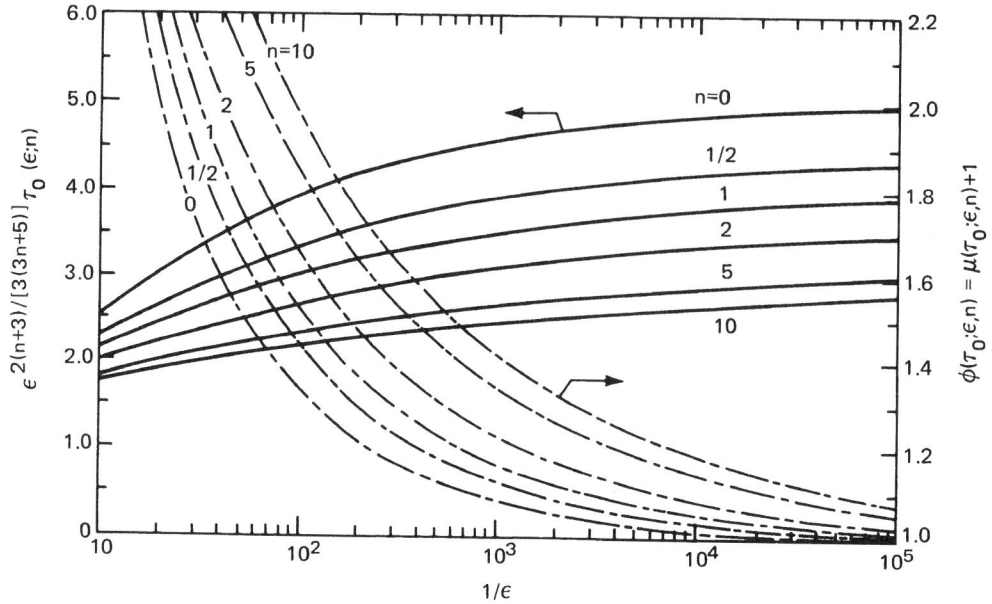

Fig. 3-10.19. *Plots of $\varepsilon^{2(n+3)/[3(n+5)]}\tau_0$ and ϕ_0 as functions of $1/\varepsilon$ for different values of n.*

where the $f(\varepsilon;n)$ are provided in Equation 46, or can be found from Figure 3-10.19. Also,

$$\varepsilon = (1 - \lambda_c)\left\{\frac{[A(n+3)/3]^{2n}[Q(t_0)/t_0^n]^2}{(1-\lambda_r)^{(n+1)}(\rho_a C_p T_a)^2 g^{(n+1)}H^{(3n+5)}}\right\}^{1/(n+3)} \tag{49}$$

Thus, the analysis has led to a remarkable practical result, namely, for fires which grow at rates which are approximately proportional to t^n and for a wide range of fire elevations and room heights of practical interest, the time for a smoke layer to drop from the ceiling to the elevation of the fire is relatively independent of H.

Taking ρ_a, T_a, C_p, and g to be

$$\rho_a = 1.18 \text{ kg/m}^3, \quad T_a = 294 \text{ K}, \quad g = 9.8 \text{ m/s}^2.$$
$$C_p = 240 \text{ cal/(kgK)} = 1.005 \text{ Ws/m}^3 \tag{50}$$

Equations 48 and 49 for $n = 0$, 1, and 2 become

$n = 0$:

$$t_0 = 91.4(1 \pm 0.15)(A/Q^{3/5})/[(1 - \lambda_c)^2(1 - \lambda_r)]^{1/5}$$
$$\varepsilon = 9.43(10^{-3})(Q^2/H^5)^{1/3}(1 - \lambda_c)/(1 - \lambda_r)^{1/3} \tag{51}$$

(A in m^2, Q in kW, t_0 in sec, H in m)

$n = 1$:

$$t_0 = 20.2(1 \pm 0.12)\{A^5[t_0/Q(t_0)]^3/[(1 - \lambda_c)^2(1 - \lambda_r)]\}^{1/8}$$
$$\varepsilon = 1.98(10^{-2})\{A[Q(t_0)/t_0]/(1 - \lambda_r)\}^{1/2}(1 - \lambda_c)/H^2 \tag{52}$$

(A in m^2, Q in kW, t_0 in sec, H in m)

$n = 2$:

$$t_0 = 10.5(1 \pm 0.10)\{A^5[t_0^2/Q(t_0)]^3/[(1-\lambda_c)^2(1-\lambda_r)]\}^{1/11}$$
$$\varepsilon = 3.68(10^{-2})(1-\lambda_c)\{A^4[Q(t_0)/t_0^2]^2/[H^{11}(1-\lambda_v)^3]\}^{1/5} \tag{53}$$

(A in m^2, Q in kW, t_0 in sec, H in m)

or, in terms of t_g of Equation 35,[4]

$n = 2$:

$$t_0 = 1.60(1 \pm 0.10)\{A^5 t_g^6/[(1 - \lambda_c)^2(1 - \lambda_r)]\}^{1/11}$$
$$\varepsilon = 0.583(1 - \lambda_c)\{(A/t_g)^4/[H^{11}(1 - \lambda_r)^3]\}^{1/5} \tag{54}$$

(A in m^2, Q in kW, t_0 and t_g in sec, H in m)

where all the above t_0 estimates are subject to the ε range of Equation 47.

Some Solution Results for $Z_i(t)$, $T(t)$, and $\mu(t)$

Plots of general solutions for $Z_i(t)$, $T(t)$, and $\mu(t)$ are presented in Figure 3-10.20 for $n = 0$, 1, and 2. These plots are useful up to the times when the interface either drops to the floor of the compartment or to an elevation $0.2\,H$ below the fire, whichever event occurs first.

As can be noted, the abscissa of the Figure 3-10.20 plots, which have been taken from Cooper,[43] are in the form

$$\sigma = \{\text{constant}\} \, t^{n+1}$$

The σ of Figure 3-10.20 corresponds to

$$\sigma = \{(n + 3)/[3(n + 1)]\}\varepsilon\tau^{3(n+1)/(n+3)} \tag{55}$$

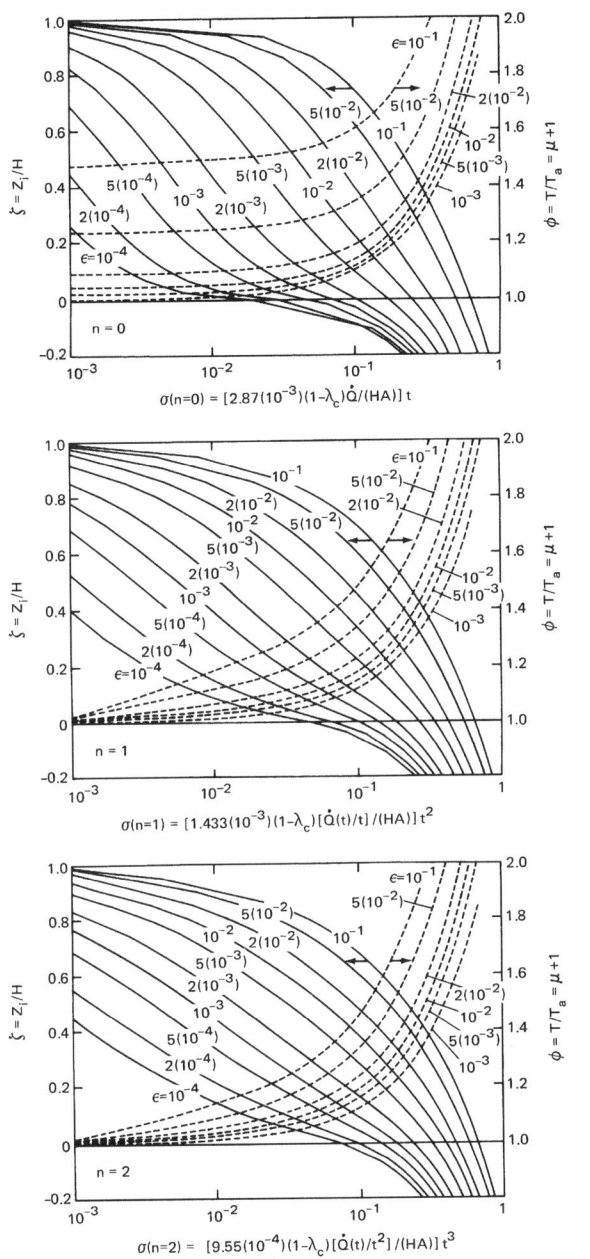

Fig. 3-10.20. *Plots of $Z_i(t)$, $T(t)$, and $\mu(t)$ up to the time when $Z_i = -0.2\,H$ or $Z_i = Z_{floor}$, whichever comes first, for different values of ϵ and for $n = 0$, 1, and 2. ($\dot{Q}$ in kW; t and t_g in s; H in m; and A in m^2.)*

under the ambient property assumptions of Equation 50. Cooper found σ to be a convenient variable for describing the upper layer environment after the interface had dropped below the fire elevation, $Z_i = 0$, and even subsequent to the time that the interface drops to the compartment floor, i.e., when $Z_i = -\Delta$.[43] A description of solutions to our problem at these latter times is beyond the scope of this chapter, and the reader is referred to Cooper's work for a full discussion of the relevant results.[43]

USING THE $\dot{Q} \sim t^n$ SOLUTION PLOTS OF FIGURES 3-10.19 and 3-10.20 TO PREDICT CHARACTERISTICS OF COMPARTMENT FIRE-GENERATED ENVIRONMENTS

To illustrate the use of the solution plots of Figures 3-10.19 and 3-10.20, they will now be applied to two example problems. The first example will involve a problem of smoldering combustion. The second example illustrates the use of the theory in predicting the environment produced in an enclosure which contains a specific large-scale flaming fire hazard.

EXAMPLE 1:

Smoldering combustion: Smoldering experiments reported by Quintiere *et al* were carried out in a single-room compartment of height 2.44 m and floor area 8.83 m^2.[45] The opening to the enclosure was formed by a closed undercut door, where the undercut formed a 0.76 m $\times$ 0.025 m open horizontal slit at floor level. A smoldering ignition source was placed in the enclosure with the top surface of the source at an elevation of 0.33 m. Gas analysis was carried out at four equidistant elevations, the inlets of sampling tubes extending horizontally approximately 0.5 m from the walls.

The tests evaluated two different smolder sources: a loosely packed bed of cotton, and blocks of flexible polyurethane foam. Mass loss rates, $\dot{m}$, were found to be approximately linear in time throughout the first hour of the two tests, i.e.,

$$\dot{m} = \alpha t \qquad 0 < t < 60 \text{ min} \tag{56}$$

where

$$\alpha = \begin{cases} 0.21 \text{ g/min}^2 & \text{for polyurethane} \\ 0.33 \text{ g/min}^2 & \text{for cotton} \end{cases}$$

The heats of combustion, H_c, of the materials as well as the ratios, γ, of mass-of-CO produced to mass of material lost were obtained in a separate small-scale apparatus. These were found to be

$$H_c = \begin{cases} (11 \pm 1) \text{ kJ/g}_{cotton} \\ (15 \pm 8) \text{ kJ/g}_{polyurethane} \end{cases} \tag{57}$$

$$\gamma = \begin{cases} 0.11 \text{ g}_{CO}/\text{g}_{cotton} \\ \left(0.10 \begin{smallmatrix} +0.01 \\ -0.04 \end{smallmatrix}\right) \text{g}_{CO}/\text{g}_{polyurethane} \end{cases} \tag{58}$$

The results of the previous section will be used to predict the environment which developed in the enclosure during the course of the two different material evaluations.

Comparing Equations 56 through 58 to Equation 34 leads to

$$\dot{Q} = \alpha H_c t = \dot{Q}_0 [tH^{3/2}g^{1/2}/A]^n \tag{59}$$
$$\dot{C}_{CO} = \alpha\gamma t = \beta\dot{Q}$$

where $\dot{C}_{CO}$ is measured in g_{CO} per unit time. From the above, it is concluded that

$$n = 1; \quad Q(t)/t = \alpha H_c; \quad \beta = \gamma/H_c \tag{60}$$

Also

$$H = 2.11 \text{ m}, \quad \Delta = 0.33 \text{ m}, \quad A = 8.83 \text{ m}^2 \quad (61)$$

Radiant losses from the combustion zone are neglected, i.e., $\lambda_r = 0$. Considerations by Cooper,[40] together with the experimental results of Mulholland et al and Veldman et al,[46,13] indicate that for a $\lambda_r = 0$ combustion zone in an enclosure with proportions similar to the present one, $\lambda_c \approx 0.6$. This λ_c will be used here.

Using the λ_c value 0.6 and Equations 50, 57, 58, and 60 in the μ definition of Equation 37 and the result of Equation 40 leads to

$$\phi = 1 + 135 M_{cotton} = 1 + 204 M_{polyurethane} \quad (62)$$

where, e.g., M_{cotton} is the upper layer concentration of CO ($g_{CO}/g_{upper\ layer}$) during smoldering of the cotton source.

From Equations 56 and 57

$$\frac{Q(t)}{t} = \alpha H_c$$

$$\approx \begin{cases} 0.21(15) \text{ kJ/min}^2 = 9(10^{-4}) \text{ kW/s for polyurethane} \\ 0.33(11) \text{ kJ/min}^2 = 9(10^{-4}) \text{ kW/s for cotton} \end{cases} \quad (63)$$

All parameters of the problem required to estimate t_0 and ε from Equation 52 are now available. These are found to be

$$t_0 = 1400 \text{ sec}; \quad \varepsilon = 1.6(10^{-4}) \quad (64)$$

The value of ε satisfies the ε range of Equation 47 thereby establishing the validity of the t_0 estimate.

From the value of ε, it is now possible to use Figure 3-10.19 to obtain ϕ_0. Thus, for $1/\varepsilon = 0.61(10^4)$ and for $n = 1$ it is found from the Equation 37 definition of ϕ that

$$\phi_0 = 1.07 \rightarrow T(t_0) = 1.07 T_a = 315 \text{ K} = 42°C \quad (65)$$

Also, from Equation 62, this result for ϕ_0 yields

$$\left.\begin{array}{l} M_{cotton}(t_0) = (1.07 - 1)/135 \\ \qquad = 5(10^{-4}) g_{CO}/g_{upper\ layer} \\ \qquad = 5(10^2) \text{ ppm CO} \\ M_{polyurethane}(t_0) = (1.07 - 1)/204 \\ \qquad = 3(10^{-4}) g_{CO}/g_{upper\ layer} \\ \qquad = 3(10^2) \text{ ppm CO} \end{array}\right\} \quad (66)$$

The smoke interface reaches the floor at the time, t_f, corresponding to

$$\zeta_f = Z_i(t_f)/H = -\Delta/H = \delta = -0.16 \quad (67)$$

From the $n = 1$ plots of Figure 3-10.20, it is found that this occurs at t_f corresponding to

$$\sigma = 0.3 = \sigma_f(\zeta = \zeta_f; n = 1)$$

$$= \{1.4(10^{-3})(1 - \lambda_c)[\dot{Q}(t)/t]/(HA)\} t_f^2$$

$$(\dot{Q} \text{ in kW}; t \text{ in s}; H \text{ in m}; A \text{ in m}^2) \quad (68)$$

$$= \{1.4(10^{-3})(1 - 0.6)[9(10^{-4})]/[(2.1)(8.8)]\} t_f^2 = 0.3(10^{-7}) t_f^2$$

(for both the cotton and polyurethane) at which time

$$\phi(t_f) = \phi_f \approx 1.25 \quad (69)$$

The following results are obtained from Equations 68 and 69 and with the use of Equation 62.

$$t_f = 3(10^3) \text{ sec}$$

$$M_{cotton}(t_f) = (1.25 - 1)/135 = 1.9(10^3) \text{ ppm CO} \quad (70)$$

$$M_{polyurethane}(t_f) = (1.25 - 1)/204 = 1.2(10^3) \text{ ppm CO}$$

Thus, the above estimate indicates that the interface reached the floor elevation somewhat prior to the 60-min duration of the tests. The reader is referred to Cooper's studies for further discussion of comparisons between calculated and experimental results.[43]

EXAMPLE 2:

Hazard development in enclosures containing some larger scale fires: NFPA 204M, *Guide for Smoke and Heat Venting*, provides a catalogue of experimentally determined energy release rates for the growth stages of flaming fires in practical fuel assemblies.[4] The $\dot{Q}$ of all items in this listing is proportional to t^2. For example, the $\dot{Q}$ of many items can be estimated by

$$t_g = 100 \text{ sec} \quad (71)$$

Using Equation 35, this corresponds to

$$\dot{Q} = 0.10 t^2 \text{ kW/s}^2 \quad (72)$$

The latter items include wood pallets stacked 3.0 to 4.6 m high, many different types of polyethylene, polypropylene, polystyrene and PVC commodities in cartons stacked 4.6 m high, and a horizontal polyurethane mattress.

The results of Figures 3-10.19 and 3-10.20 will be used to characterize the hazard development in enclosures which contain Equation 71-type fires.

From Equations 54 and 71 and the abscissa for $n = 2$ of Figure 3-10.20

$$t_0 \approx 20.\{A^5/[(1 - \lambda_c)^2(1 - \lambda_r)]\}^{1/11}$$

$$\varepsilon = 1.5(10^{-2})(1 - \lambda_c)A^{4/5}/[(1 - \lambda_r)^{3/5}H^{11/5}]$$

$$\sigma(n = 2) = 0.96(10^{-4})[(1 - \lambda_c)/(HA)]t^3 \quad (73)$$

$$(t \text{ in sec}; H \text{ in m}; A \text{ in m}^2)$$

With Equation 73, Figures 3-10.19 and 3-10.20 can now be used to answer a wide variety of hazard-related questions. For illustrative purposes, two such questions will be addressed here.

QUESTION 1:

Flaming ignition is initiated in stacked commodities of the "$tg = 100$ s variety" which are contained in a warehouse of height 6 m and floor area 1500 m^2. At what time does the upper layer attain the potentially untenable temperature (due to downward radiation) of 183°C, and what is the elevation of the layer interface at this time?[1] At what time does the upper layer completely fill the warehouse?

ANSWER:

Consistent with recommendations by Cooper,[2] assume $\lambda_r = 0.35$, and, for the purpose of a hazard analysis of this type, conservatively assume that $\lambda_c = 0.6$. Take H to be the floor-to-ceiling dimension, 6 m, and Δ to be zero. Then, for $A = 1500 \ m^2$, Equation 73 leads to

$$t_0 = 680 \ sec$$
$$\varepsilon = 5.2(10^{-2}) \qquad (74)$$
$$\sigma(n = 2) = 4.7(10^{-9})t^3$$

Notice that the above value for $\varepsilon = 5.2 \ (10^{-2}) > 0.2 \ (10^{-1})$ is somewhat outside the Equation 47 range. As a result, the above $t_0 = 680$ s estimate is not reliable. A better value for t_0, estimated from Equation 48 and Figure 3-10.19, is found to be $t_0 \approx 600$ s.

The $\varepsilon = 5.2 \ (10^{-2})$ value corresponds to $1/\varepsilon = 19$ which, for $n = 2$ in Figure 3-10.19, is found to correspond (somewhat off-scale) to

$$\phi_0 \approx 2.4 \rightarrow T(t_0) = 2.4 T_a = 710 \ K = 433°C \qquad (75)$$

At the time, t_u, of potential untenability, $T_u = T(t_u)$ is assumed to be 183°C (456 K). Thus

$$\phi_u = \phi(t_u) = T_u/T_a = 456/294 = 1.55 \qquad (76)$$

For $\varepsilon = 5.2 \ (10^{-2})$, the $n = 2$ plots of Figure 3-10.20 can be interpolated at $\phi = \phi_u = 1.55$ to yield

$$\sigma_u = \sigma(t_u) \approx 0.2 = 4.7(10^{-9})t_u^3 \rightarrow t_u = 350 \ sec \qquad (77)$$

which, in turn, is seen to correspond to

$$\zeta_u = \zeta(t_u) = Z_i(t_u)/H = 0.45 \qquad (78)$$

Using this last value for ζ_u along with $H = 6$ m leads to

$$Z_i(t_u) = 0.45(6) \ m = 2.7 \ m \qquad (79)$$

The above results are summarized as follows: the upper smoke layer will fill the compartment at $t_0 = 600$ s, at which time its average temperature will be approximately 430°C. The potentially untenable condition of $T = 183°C$ will occur at $t_u = 350$ s, at which time the layer interface is 2.7 m above the floor.

QUESTION 2:

Flaming ignition is initiated in a polyurethane mattress 0.6 m above the floor of a hospital ward with floor-to-ceiling dimension of 3 m and floor area 100 m^2. At what time, t_u, does the upper layer interface reach the potentially untenable elevation, $Z_u = 1.5$ m, and what is the upper layer temperature, T_u, at this time?

ANSWER:

Take $\lambda_c = 0.8$, and $\lambda_r = 0.35$. Also, $H = 2.4$ m, $\Delta = 0.6$ m, $Z_u = 0.9$ m, and $A = 100 \ m^2$. Then, Equation 73 leads to

$$t_0 = 230 \ sec$$
$$\varepsilon = 2.3(10^{-2}) \qquad (80)$$
$$\sigma(n = 2) = 8.0(10^{-8})t^3$$

Also, at the time of untenability

$$Z_i(t_u)H = Z_u/H = 0.9/2.4 = 0.38 \qquad (81)$$

For $\varepsilon = 2.3(10^{-2})$, the $n = 2$ plots of Figure 3-10.20 can be interpolated to obtain the desired values of $\sigma(t_u)$, and then $\phi(t_u)$ corresponding to $Z_i/H = 0.38$. Thus

$$\sigma(t_u) = 0.18 = 80(10^{-9})t_u^3 \rightarrow t_u = 130 \ sec \qquad (82)$$
$$\phi(t_u) = 1.33 \rightarrow T(t_u) = 1.33 T_a = 391 \ K = 118°C$$

In Equation 80, t_0 is the time for the smoke interface to drop to the level of the mattress which is 0.6 m above the floor. As an additional point of information, for $\varepsilon = 2.3(10^{-2})$, corresponding to $1/\varepsilon = 44$, and for $n = 2$, Figure 3-10.19 provides the result

$$\phi_0 = 1.94 \rightarrow T(t_0) = 1.94 T_a = 570 \ K = 297°C \quad (83)$$

Notice that this result can also be obtained approximately from Figure 3-10.20. To do so, select the value of $\sigma = \sigma(t_0)$ when $\zeta = Z_i/H = 0$, and find the corresponding value for $\phi = \phi(t_0)$, all on the $\varepsilon = 2(10^{-2})$ curves. This leads to $\phi(t_0) = 1.88 \approx 1.94$.

The above results are summarized as follows: the smoke layer interface will drop to the 1.5-m elevation at $t = 130$ s, at which time its average temperature will be approximately 118°C. Also, the interface will reach the mattress elevation at $t = 230$ s and have an average temperature of 297°C.

REFERENCES CITED

1. L.Y. Cooper, "A Concept for Estimating Available Safe Egress Time in Fires," *F. Safety J.*, 5, pp. 135–144 (1983).
2. L.Y. Cooper, "A Mathematical Model for Estimating Available Safe Egress Time in Fires," *F. and Matls.*, 6, pp. 135–144 (1982).
3. E.G. Butcher and A.C. Parnell, *Smoke Control and Fire Safety Design*, Spon, London (1979).
4. G. Heskestad, "Appendix A," NFPA 204M, *Guide for Smoke and Heat Venting*, National Fire Protection Association, Quincy (1991).
5. V. Babrauskas, "Combustion of Mattresses Exposed to Flaming Ignition Sources, Part I, Full-Scale Tests and Hazard Analysis," *NBSIR 77-1290*, National Bureau of Standards, Gaithersburg (1977).
6. G. Heskestad, "Engineering Relations for Fire Plumes," *F. Safety J.*, 7, pp. 25–32 (1984).
7. E.E. Zukoski, T. Kubota, and B. Cetegen, "Entrainment in Fire Plumes," *F. Safety J.*, 3, pp. 107–121 (1980/81).
8. B.J. McCaffrey, "Purely Buoyant Diffusion Flames: Some Experimental Results," *NBSIR 79-1910*, National Bureau of Standards, Gaithersburg (1979).
9. R.L. Alpert, "Turbulent Ceiling-Jet Induced by Large-Scale Fires," *Comb. Science Tech.*, 11, pp. 197–213 (1975).
10. G. Heskestad, "Similarity Relations for the Initial Convective Flow Generated by Fire," *Paper 72-WA/HT-17*, Winter Annual Meeting, ASME (1972).
11. G. Heskestad and M.A. Delichatsios, "Environments of Fire Detectors—I: Effect of Fire Size, Ceiling Heights, and Material," Vol 1: Measurements, *FMRC Report NBS-GCR-77-86*, Vol. 2: "Analysis," *FMRC Report NBS-GCR-77-95*, National Bureau of Standards, Gaithersburg (1977).
12. G. Heskestad and M.A. Delichatsios, "Environments of Fire Detectors—II: Effect of Ceiling Configuration," Vol 1: Measurements, *FMRC Report NBS-GCR-78-128*, Vol. 2: "Analysis," *FMRC Report NBS-GCR-78-129*, National Bureau of Standards, Gaithersburg (1978).
13. C.C. Veldman, T. Kubota, and E.E. Zukoski, "An Experimental Investigation of Heat Transfer from a Buoyant Plume to a Horizontal Ceiling—Part I: Unobstructed Ceiling," *C.I.T. Report NBS-GCR-77-97*, National Bureau of Standards, Gaithersburg (1975).

14. L.Y. Cooper, "Heat Transfer from a Buoyant Plume to an Unconfined Ceiling," *J. of Heat Transfer*, 104, pp. 446–451 (1982).

15. L.Y. Cooper and A. Woodhouse, "The Buoyant Plume-Driven Adiabatic Ceiling Temperature Revisited," *J. of Heat Transfer*, 108, pp. 822–826 (1986).

16. L.Y. Cooper and D.W. Stroup, "The Thermal Response of Unconfined Ceilings Above Growing Fires and the Importance of Convective Heat Transfer," *J. of Heat Transfer*, 109, pp. 172–178 (1987).

17. R.L. Alpert, "Calculation of Response Time of Ceiling Mounted Fire Detectors," *Fire Tech.*, 3, pp. 181–195 (1972).

18. I. Benjamin, G. Heskestad, R. Bright, and T. Mayes, *An Analysis of the Report on Environments of Fire Detectors*, Fire Detection Institute (1979).

19. D. Goldman and Y. Jaluria, "Effect of Opposing Buoyancy on the Flow in Free and Wall Jets," *J. of Fluid Mech.*, 166, pp. 41–56 (1986).

20. L.Y. Cooper, "Ceiling Jet Properties and Wall Heat Transfer Near Regions of Ceiling Jet-Wall Impingement," *NBSIR 86-3307*, National Bureau of Standards, Gaithersburg (1986).

21. L.Y. Cooper, "Convective Heat Transfer to Ceilings Above Enclosure Fires," *Proceedings of 19th Symposium (International) on Combustion*, Combustion Institute, Haifa (1982).

22. L.Y. Cooper, "A Buoyant Source in the Lower of Two, Homogeneous, Stably Stratified Layers—A Problem of Fire in an Enclosure," *Proceedings of 20th Symposium (International) on Combustion*, Combustion Institute, Pittsburgh, pp. 1567–1573 (1984).

23. L.Y. Cooper, "On the Significance of a Wall Effect in Enclosures with Growing Fires," *Comb. Science Tech.*, 40, pp. 19–39 (1984).

24. L.Y. Cooper and D.W. Stroup, "Calculating Available Safe Egress Time (ASET)—A Computer Program and User's Guide," *NBSIR 82-2578*, National Bureau of Standards, (1982). Also, a condensed version: "ASET—A Computer Program for Calculating Available Safe Egress Time," *F. Safety Jour.*, 9, pp. 29–45 (1985).

25. E.E. Zukoski and T. Kubota, "Two-Layer Modeling of Smoke Movement in Building Fires," *F. and Matls.*, 4, 1, pp. 17–27 (1980).

26. K. Steckler, J. Quintiere, and W. Rinkinen, "Flow Induced by Fire in a Compartment," *Proceedings of 19th Symposium (International) on Combustion*, Combustion Institute, Pittsburgh (1982).

27. NFPA *101, Life Safety Code*, (A-15-3.1.3), National Fire Protection Association, Quincy (1994).

28. L.Y. Cooper, M. Harkelroad, J. Quintiere, and W. Rinkinen, "An Experimental Study of Upper Hot Layer Stratification in Full-Scale Multiroom Fire Scenarios," *J. of Heat Transfer*, 104, pp. 741–749 (1982).

29. L.Y. Cooper, "The Need and Availability of Test Methods for Measuring the Smoke Leakage Characteristics of Door Assemblies," *Fire Safety: Science and Engineering*, ASTM STP 882, ASTM, Philadelphia, pp. 310–329 (1985).

30. T. Wakamatsu, "Calculation of Smoke Movement in Buildings," *Res. Paper 34*, Building Research Institute, Tokyo (1968).

31. E. Evers and A. Waterhouse, "A Computer Model for Analyzing Smoke Movement in Buildings," *SCS Ltd. Report CP 68/78* for Fire Research Station, Borehamwood (1978).

32. B.R. Morton, G.I. Taylor, and J.S. Turner, "Turbulent Gravitational Convection from Maintained and Instantaneous Sources," *Proceedings of Royal Society (London)*, Ser. A, 234, pp. 1–23 (1956).

33. S. Yokoi, "On the Heights of Flames from Burning Cribs," *BRI Report 12*, Ministry of Construction, Japanese Government (1963).

34. W.D. Baines and J.S. Turner, "Turbulent Buoyant Convection from a Source in a Confined Region," *J. of Fluid Mech.*, 37, Part 1, pp. 51–80 (1969).

35. E.E. Zukoski, "Development of a Stratified Ceiling Layer in the Early Stages of a Closed-Room Fire," *F. and Matls.*, 2, pp. 54–62 (1978).

36. W.D. Walton, "ASET-B: A Room Fire Program for Personal Computers," *NBSIR 85-3144-1*, National Bureau of Standards, Gaithersburg (1985).

37. D. Burgess and M. Hertzberg, "Radiation from Pool Fires" in *Heat Transfer in Flames*, ed. by N.H. Afgan and J.M. Beer, Chap. 27, Wiley and Sons, New York (1974).

38. J. DeRis, "Fire Radiation—A Review," *Tech. Report FMRC RC 78-BT-27*, Factory Mutual Research Corp., Norwood (1978).

39. A.T. Modok and P.A. Croce, "Plastic Pool Fires," *Comb. Flame*, 30 (1977).

40. L.Y. Cooper, "Estimating Safe Available Egress Time from Fires," *NBSIR 80-2172*, National Bureau of Standards, Gaithersburg (1981).

41. R. Friedman, "Quantification of Threat from a Rapidly Growing Fire in Terms of Relative Material Properties," *F. and Matls.*, 2, pp. 27–33 (1978).

42. T. Tanaka, "A Model on Fire Spread in Small-Scale Building," *Third Joint Meeting US-Japan Panel on Fire Research and Safety*, UJNR, Washington (1978).

43. L.Y. Cooper, "The Development of Hazardous Conditions in Enclosures with Growing Fires," *Comb. Science and Tech.*, 22, pp. 279–297 (1983).

44. J.G. O'Neill and W.D. Hayes, "Full-Scale Fire Tests with Automatic Sprinklers in a Patient Room," *NBSIR 79-1749*, National Bureau of Standards, Gaithersburg (1979).

45. J. Quintiere, M. Birky, and G. Smith, "An Analysis of Smoldering Fires in Closed Compartments and Their Hazard Due to Carbon Monoxide," *F. and Matls.*, 6, p. 99 (1982).

46. G. Mulholland, T. Handa, O. Sugawa, and H. Yamamoto, "Smoke Filling in an Enclosure," *20th National Heat Transfer Conference*, Milwaukee (1981).

FIRE HAZARD CALCULATIONS FOR LARGE OPEN HYDROCARBON FIRES

Krishna S. Mudan and Paul A. Croce

INTRODUCTION

A major area for fire protection in the industrial sector deals with controlling the impact from large open hydrocarbon fires. The primary mechanism for injury or damage from such fires is thermal radiation. Depending upon the circumstances and conditions leading to such an event, a different type of open fire may result. For example, ignited releases can produce pool fires, jet flames, vapor cloud fires, or fireballs, all of which behave differently and exhibit markedly different radiation characteristics.

This chapter presents detailed techniques for calculating impacts from large open hydrocarbon fires. Some of the fire types addressed in this chapter have received considerable attention in the experimental and theoretical research area while others are still evolving. In all cases, the best available models—theoretical or empirical, and supported by full-scale or reduced-scale experimentation or by accident data—are used to develop expressions for calculating hazard impacts. Examples are included throughout this chapter to illustrate the application of these expressions.

The first section of this chapter discusses hydrocarbon pool fires, an area in which considerable work has been done, including various important geometric parameters (e.g., flame height), thermal radiation models, and atmospheric absorption of radiation. The second section deals with turbulent jet flames and flares, first presenting significant geometric effects and thermal radiation models, then discussing aerodynamic effects on radiant energy and flame stability.

While the above cases involve primarily steady-state thermal radiation, the third section considers two very important cases that involve unsteady radiant effects—burning vapor clouds and fireballs. Although developmental work in both of these areas is still ongoing, limited data are available to lend some confidence to the use of these models. The fourth section, on thermal radiation hazards, discusses and develops various impact criteria for injury to people and damage to structures for both steady and transient heating.

Finally, the second part of this introduction presents an event tree for the release of a flammable material to guide the user's selection of appropriate potential impacts and model application. It cannot be overstressed that prudent judgment should be exercised in the application of any of these calculation schemes to yield a safe and fair evaluation of scenarios of interest.

Event Tree for Flammable Material Release

Figure 3-11.1 depicts a typical event tree for the release of a flammable material, showing the pathways that lead to the various types of open fires. For this purpose, assume the release occurs from a pressurized container since all types of open fires may be realized from a pressurized release. In this example, the pressurized container may be either a large vessel (for storage, reaction, batching, etc.) or a pipeline (for transfer, fittings, instruments, etc.). The pressure may be the result of either normal operations or abnormal external events. For example, a tank may be pressurized because it contains a compressed liquid or because it has been exposed to an external fire; a pipeline may be pressurized because of a pumping operation or because of steam tracing on a blocked-in segment. Finally, the release may be due to a major failure (e.g., spontaneous tank failure) or a minor accident (e.g., breakage of a fitting).

While tracing the pathways in the event tree, note that a release may or may not be accompanied by immediate ignition. With immediate ignition, i.e., following the left branch of the tree, a jet flame will result if the release is from a relatively small opening. Such a release could be either vapor or liquid, and, if liquid, could also involve flashing of liquid into vapor and/or accumulation of liquid. If the release is the result of a major spill and there is immediate ignition, the result is usually a fireball, the size of which is strongly affected by the amount of flash vaporization and liquid entrainment that occurs upon release.

Dr. Krishna Mudan is a principal at Technica Inc., an international consulting firm specializing in safety assessment and risk analysis. Prior to joining Technica Inc., Dr. Mudan was a senior member of the safety assessment group at Arthur D. Little Inc. For over a decade he has participated in several studies related to theoretical and experimental aspects of thermal radiation from pool fires, flame propagation, and explosion. Dr. Paul A. Croce is Manager of Protection and Risk Analysis Research at Factory Mutual Research Corp. He has been actively involved in applied fire research, accident investigation, and professional society activities.

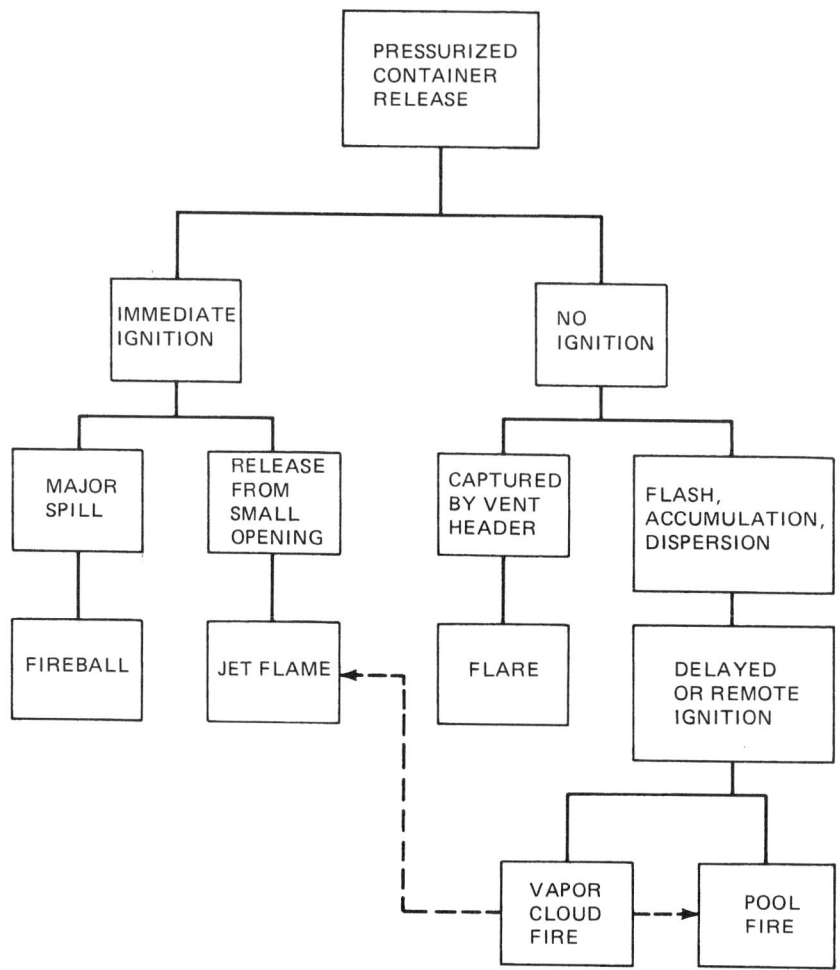

Fig. 3-11.1.   Example event tree for release of flammable material.

If ignition does not occur immediately upon release, the right branch of the event tree is followed. Releases through relief valves, either accidental or intentional, may be captured by a vent header system and directed to a flare. Flares must accordingly be designed for expected accident or operational flow capacities, and the design should include consideration of resulting hazard zones.

If the unignited release is to the atmosphere, the spill will be accompanied by flash vaporization, liquid entrainment, accumulation, and/or vapor dispersion. A delayed local ignition following accumulation results in a pool fire whose characteristics are strongly influenced by the geometry of containment (or lack thereof).

Absence of a local ignition source, either immediate or delayed, allows a vapor cloud to form as the vapors disperse downwind. A portion of this vapor cloud will be flammable, and depending upon the size of the release, the flammable region could extend significantly downwind. A remote ignition source can ignite the cloud, resulting in a vapor cloud fire that burns from the point of ignition back toward the source of the cloud, i.e., the release point. Note also that the event tree shows dashed pathways from the vapor cloud fire to a pool fire or a jet flame. This is intended to show that the transient burning of the vapor cloud fire back to the release point can initiate a subsequent steady burning jet flame or

pool fire, and consideration must be given to potential impacts from both types of fire.

Finally, this event tree represents an example of potential pathways leading to hazardous open fires. Scenarios will be encountered that will be similar to the event tree depicted in Figure 3-11.1, or to parts thereof, but it is important to recognize that in all cases, the event tree must be structured to reflect the actual scenario under consideration.

HYDROCARBON POOL FIRES

The thermal radiation hazards from hydrocarbon spill fires depend on a number of parameters, including the composition of the hydrocarbon, the size and shape of the fire, the duration of the fire, its proximity to the object at risk, and the thermal characteristics of the object exposed to the fire. The objectives of this section of the chapter are to review the literature on this subject and to identify the best available techniques for determining the thermal radiation hazards from liquid hydrocarbon pool fires under various credible spill conditions.

The quantification of the thermal properties of fires from basic principles (which consider the mixing dynamics and the chemical processes of burning a fuel with the oxygen in air) is perhaps the best and most accurate approach.

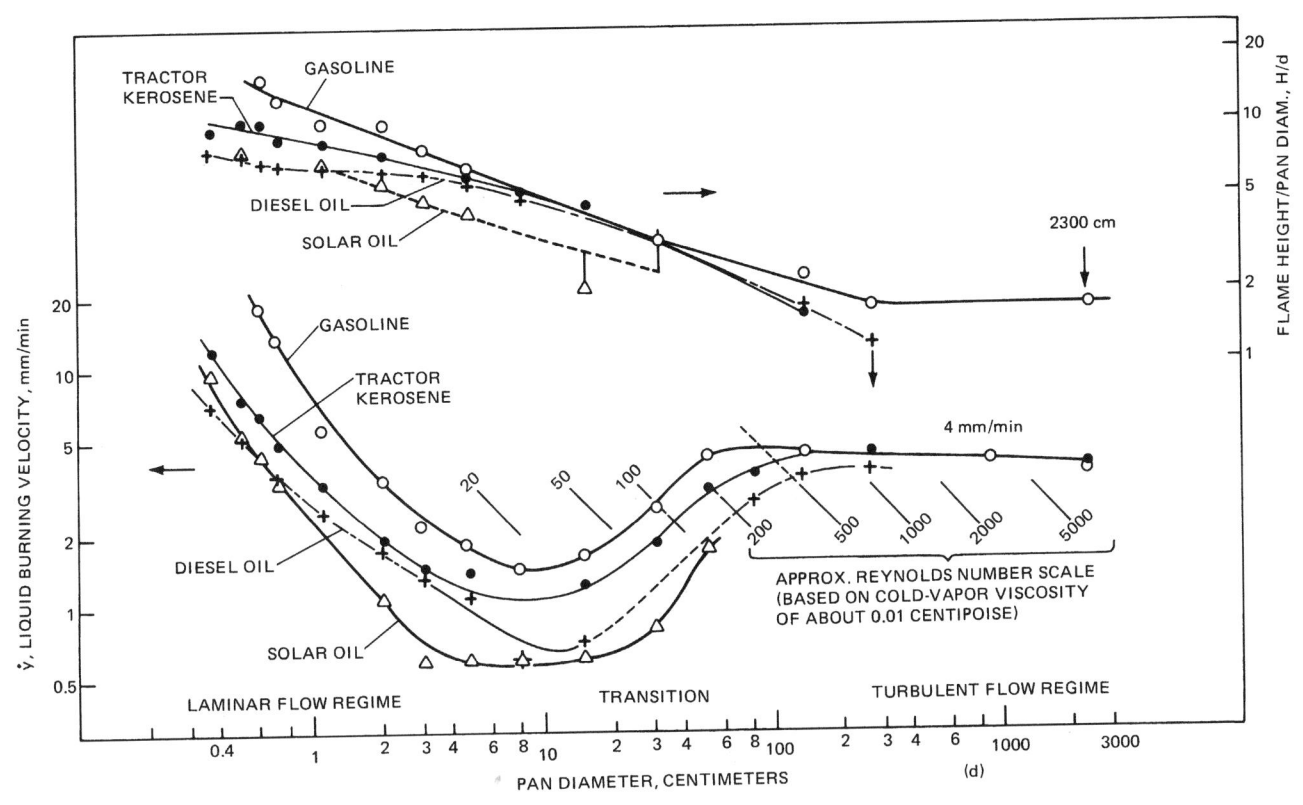

Fig. 3-11.2.   Burning rates and flame heights for various hydrocarbon pool fires.[3]

However, such an approach is formidable and beset by serious obstacles, such as an inadequate understanding of the structure of turbulent diffusion flames and of the nonequilibrium kinetics of soot formation. The state of the art of predicting the thermal environment of hydrocarbon spill fires consists essentially of semiempirical methods, some of which are based on experimental data from small- and medium-scale tests. Needless to say, such semiempirical methods are always subject to uncertainties when experimental data from small-scale fires are extrapolated to predict the thermal properties of very large-scale fires.

Estimating the thermal radiation field surrounding a fire involves the following three major steps:

1. Geometric characterization of the pool fire; that is, the determination of burning rate and the physical dimensions of the fire. In calculating thermal radiation, the size of the fire implies the time-averaged size of the visible envelope. It has been shown that the nonvisible parts of a fire radiate less than 10 percent of the radiation of the visible parts.[1]
2. Characterization of the radiative properties of the fire; that is, the determination of the average irradiance of the flames. The intensity of thermal radiation emitted by pool fires depends upon a number of parameters, including fuel type, fire size, flame temperature, and composition. The major sources of emission in large hydrocarbon fires are the water vapor, carbon dioxide, and soot.
3. Calculation of radiant intensity at a given location. This can be accomplished once the geometry of the fire, its radiation characteristics, and the location, geometry, and orientation of the receiver are known. For large distances (hundreds of meters), the absorption of thermal radiation

in the intervening atmosphere becomes appreciable. This is dependent on the path length, flame temperature, and the relative humidity in the atmosphere.

Pool Fire Geometry

The flame geometry for the solid flame model is generally determined by assuming that the flame is a solid, gray emitter having a regular well-defined shape such as a circular or a tilted cylinder. The dimensions of the flame are characterized by the flame base diameter, visible flame height, and the flame tilt. The flame diameter is dependent on the spill size (spill volume and/or spill rate) and the rate of burning. The flame height appears to depend on the flame diameter and the type of fuel (characterized by the burning rate). These factors, which influence the flame geometry, are discussed in this chapter.

Burning rate for pool fires: A systematic study of liquid hydrocarbon pool fires over the widest range of pool diameters was conducted by Blinov and Khudiakov.[2] Gasoline, tractor kerosene, diesel oil, and solar oil (and, to a limited extent, household kerosene and transformer oil) were burned in cylindrical pans (depth not indicated) of diameters 0.37 cm to 22.9 meters. Liquid burning rates and flame heights were measured, and visual and photographic observations of the flames were recorded. Hottel[3] plotted these data and the results are shown in Figure 3-11.2. The lower curve of this figure shows the variation of burning velocity, $\dot{y}$, as a function of the pan diameter. The upper curves give the ratio of flame height to flame diameter as a function of the pan diameter. The diagonal lines on the lower curves represent lines of constant Reynolds numbers, based on pan diameter.

It is interesting to note that the burning velocity-pan diameter relationship has the same general structure for all types of oils. It first decreases with increasing pan diameter, with an almost constant product of the two. This decrease is in the laminar flow regime, with a Reynolds number of less than about 20 (based on properties of nonburning fuel vapor). With further increases of pan diameter, the burning velocity begins to increase. This sharp rise in the burning rate occurs in the range of Reynolds numbers of 20 to 200 and burning velocity levels off at a Reynolds number of about 500. Above that value, the burning of the fire is turbulent and its burning velocity is substantially uninfluenced by the pan diameter or the fuel type.

Hottel[3] demonstrated that the above behavior can be related to the heat transfer rate, which determines the rate of fuel vaporization. The heat transfer from the fire to the liquid pool can be represented by

$$\dot{q}/\frac{\pi d^2}{4} = \frac{4K}{d}(T_f - T_p) + H(T_F - T_p)$$
$$+ \sigma_F(T_F^4 - T_p^4)(1 - e^{-kd}) \qquad (1)$$

where the left-hand side of the equation represents the mean heat flux to the liquid pool from the fire. The first term on the right represents the conductive heat transfer rate through the pan rim; the second term, the convective heat flux; and the last term, the radiative transfer rate. The mean heat flux to the pool divided by the heat of vaporization of the liquid gives the burning rate, $\dot{y}$.

The conduction term in Equation 1 is an edge effect that is important only for fires with extremely small dimensions. As can be seen from Figure 3-11.2, this term is not significant for pool diameters greater than about a few centimeters. The conduction term decreases linearly with increasing pool diameter, and therefore conduction heat transfer contribution in larger fires is virtually nonexistent. The second term in Equation 1, convection, pertains most clearly to the burning rate and has a minimum value at pool diameters of about 10 cm. Many flames of this dimension sweep back and forth across the liquid surface and assume wide variations in their shape. These oddities, however, disappear for larger diameter fires.

For most liquid fuels, the radiative heat transfer and the burning rate increase as the diameter increases.[4] For pool diameters greater than about 1 m, the radiative term in Equation 1 dominates the heat flux to the pool, primarily because the flame becomes a large, optically thick, radiating blackbody. Since we are dealing with thermal radiation from large-scale turbulent-diffusion flames, this is the region of interest. If we assume that the flame's geometric view factor is constant, then Equation 1 may be simplified to yield the following equation for determining the burning rate.

$$\dot{y} = \dot{y}_{max}(1 - e^{-kd}) \qquad (2)$$

where

$\dot{y}$ = burning velocity of a finite diameter pool, m/s
$\dot{y}_{max}$ = burning velocity of an infinite dimension pool, m/s

A similar study of burning of pool fires was also conducted by the U.S. Bureau of Mines, although over a smaller scale. For pool diameters in the radiation dominant region (pan diameters of about 1 m), Burgess et al[5] have shown that the burning rate of liquid fuel (specifically: methanol, liquid

hydrogen, liquid natural gas, butane, hexane, benzene, xylene, etc.) can be accurately correlated by the following simple relationship:

$$\dot{y}_{max} = 1.27 \times 10^{-6}\frac{\Delta H_c}{\Delta H_v} \qquad (3)$$

where $\dot{y}_{max}$ is the rate at which the liquid pool level decreases with time (in the absence of external supply), ΔH_c and ΔH_v are respectively the net heat of combustion and the heat of vaporization at the boiling point of the liquid fuel.

Fires in blended fuels, especially those with components that differ widely in their volatility, do not burn at a uniform rate. In the beginning, the burning rate is characteristic of the highly volatile component. During the middle portion of burning, the less volatile component still must be brought to the boiling point of the blend. Finally, as the fractionation proceeds, the burning rate becomes characteristic of the higher boiling fraction. Grumer et al[6] conducted a series of tests on uncontrolled diffusive burning of mixtures of unsymmetrical dimethyl hydrazone (UDMH) and diethylene triamine (DETA). The boiling point of UDMH is 63°C, and that of DETA is 207°C. They observed that the initial burning rate was close to that of UDMH and the final burning rate corresponded to that of DETA. Based on extensive experimental data, they suggest the following relationship to determine the burning rates of blended fuels.

$$\dot{y}_{max} = 1.27 \times 10^{-6}\frac{\sum_{i=1}^{N}n_i\Delta H_{ci}}{\sum_{i=1}^{N}n_i\Delta H_{vi} + \sum_{i=1}^{N}m_i\int_{T_a}^{T_b}C_p(T)\,dT} \qquad (4)$$

where

n_i, m_i = mole-fraction composition in the vapor and liquid phases, respectively
T_a, T_b = initial and boiling temperatures, respectively
C_p = specific heat capacity

The integrated heat capacity in the denominator determines the temperature dependence of the burning rate, normally about 0.5 percent per degree Celsius variation of the initial liquid temperature, T_a. For fuel blends such as gasoline, whose specific heats of combustion and of vaporization of the components are comparable and $n_i > m_i$, Equation 4 can be approximated by

$$\dot{y}_{max} = n_1y_1 + n_2y_2 + \cdots \qquad (5)$$

Equations 4 and 5 yield good estimates of burning rates even for blends with components of widely separated boiling points, except during the initial and final stages of burning. For example, Grumer et al[6] measured an average burning rate of 0.54 cm/min for unleaded gasoline in a 1.22 m diameter tray. The distillation data furnished by the supplier was used in Equation 5 to estimate a burning rate of 0.57 to 0.60 cm/min.

Based on extensive burning rate measurements, Burgess and Zabetakis[7] proposed use of Equation 4 to determine the burning rate of radiation dominated liquid hydrocarbon pool fires. Figure 3-11.3 shows the relationship between the burning rate and the thermochemistry of the fuel. The modified heat of vaporization, ΔH_v, is given by the denominator indicated in Equation 4. Included in Figure 3-11.3 are some of the recent measurements on burning rates of liquefied natural gas and liquefied propane. The correlation given by Equation 4 is in excellent agreement with all liquid hydrocarbon burning rates, except those for liquefied gases. For

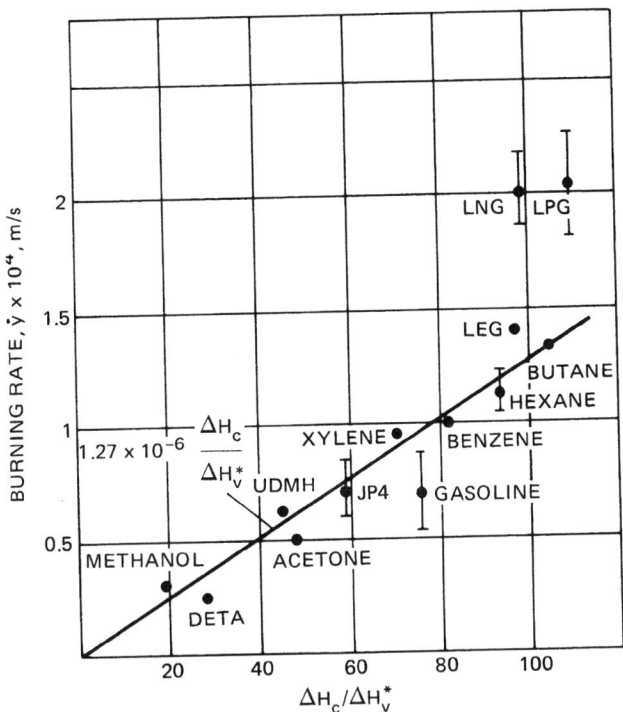

Fig. 3-11.3. *Relationship between burning rate and thermochemical properties of the fuel.*

liquefied gases, Equation 4 underestimates the burning rate by almost a factor of 2.

The mass burning rate (in kg/m² · s) can be determined by multiplying the burning rate by the liquid fuel density. In Figure 3-11.4 the relationship between the mass burning rate and the thermochemistry property of fuel is shown. The correlation for the mass burning rate is as follows:

$$\dot{m}'' = 1 \times 10^{-3} \frac{\Delta H_c}{\Delta H_v^*} \tag{6}$$

where $\dot{m}''$ = mass burning rate, kg/m² · s.

Although the correlation given by Equation 6 does not fit the data as well as that of Equation 4, it covers a wider range of fuels, including liquefied gases.

It should be noted that the burning rate data indicated in Figures 3-11.3 and 3-11.4 are for liquid hydrocarbon pool fires on land. The burning rate for pool fires on water will be similar to the burning rate for pool fires on land for all hydrocarbons with normal boiling points above ambient temperature. For liquefied gases, however, the boiling point is below ambient temperature. Therefore, considerable heat transfer takes place between the large body of water and the pool of liquefied gas. Large-scale experiments have shown that the burning rate, $\dot{y}$, for liquefied natural gas (LNG) spills on water is nearly three times the burning rate on land. For liquefied petroleum gas (LPG), the burning rate on water is typically twice that on land. The typical physical properties and burning rates on land for various hydrocarbons are shown in Table 3-11.1.

Data on large-scale, turbulent diffusion, radiation-dominated pool fires are summarized in Table 3-11.2 for various hydrocarbons. In each case, the experimental setup is described briefly; the pool diameter, flame height/diame-

ter ratio, burning rate, and a number of radiation parameters that will be discussed later in this chapter are also tabulated. A more detailed review of pool fires is given in Mudan[8] and Gollahalli and Sullivan.[9]

Pool fire diameter: The spectrum of hydrocarbon liquid spill scenarios is very wide. Spills can be classified on the basis of the activity in which the spill occurs (from storage tanks, from processing, or during transfer or transportation operations); the environment into which the liquid is spilled (onto land or onto the surface of a large body of water); or on the basis of rate, quantity, and duration of spill. The classification based on the rate of release and duration will fall into one of the following three categories:

1. Instantaneous spills—in which all of the spill occurs in a very "short time."
2. Continuous spills—in which the spill continues at a specified finite rate for a "long time."
3. Quasicontinuous spills—where a given volume of liquid is spilled over a given duration of time. This rate of release is finite, but can vary with time.

The distinction between "short time" and "long time" depends on a number of factors, including the spill size, liquid properties, and environment. For example, the rupture of a storage tank or the bursting of a tank truck by a road accident may be construed as an "instantaneous" release. The spill from a leaky pipe joint or a broken pipeline can be thought of as a "continuous" spill. Although most spills may fall into the third category of quasicontinuous spills, analytical expressions describing such a scenario are not available. When the spilled liquid is ignited, a pool fire develops. The diameter of the pool fire will depend upon the release mode, release quantity (or rate), and the burning rate.

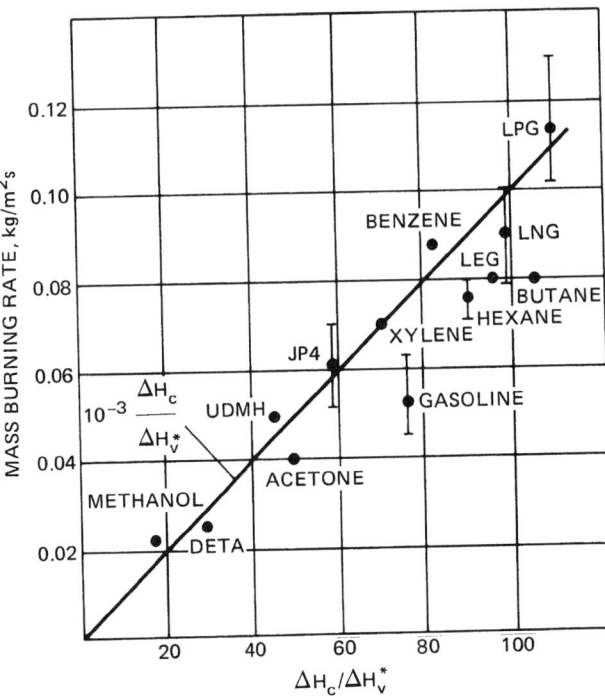

Fig. 3-11.4. *Relationship between mass burning rate on land and thermochemical property of the fuel.*

TABLE 3-11.1 *Physical Properties and Burning Rates for Hydrocarbon Pool Fires on Land*

Normal Paraffins	Net Heat Combustion (J/kg)	Heat of Vaporization (J/kg)	Boiling Point (°K)	Liq. Specific Heat (J/kg °K)	Burning Rate* (m/s)
Methane	500.2×10^5	5.1×10^5	111.7	—	2.08×10^{-4}
Ethane	472.0×10^5	4.9×10^5	264.6	—	1.22×10^{-4}
Propane	460.1×10^5	4.3×10^5	231.1	2.4×10^3	1.37×10^{-4}
Butane	453.9×10^5	3.9×10^5	272.7	2.3×10^3	1.32×10^{-4}
Pentane	450.1×10^5	3.6×10^5	309.7	—	1.43×10^{-4}
Hexane	447.7×10^5	3.4×10^5	341.9	2.5×10^3	1.22×10^{-4}
Heptane	445.9×10^5	3.2×10^5	371.9	2.1×10^3	1.13×10^{-4}
Octane	444.3×10^5	3.0×10^5	398.9	—	1.05×10^{-4}
Nonane	443.2×10^5	3.0×10^5	424.0	—	9.67×10^{-5}
Other Paraffins					
Isobutane	452.6×10^5	3.7×10^5	261.4	2.3×10^3	1.55×10^{-4}
Isohexane	445.4×10^5	3.2×10^5	333.5	—	1.37×10^{-4}
Isopentane	449.2×10^5	3.4×10^5	301.1	—	1.23×10^{-4}
Alkenes					
Ethylene	471.9×10^5	4.8×10^5	169.5	—	1.23×10^{-4}
Propylene	458.0×10^5	3.4×10^5	225.5	—	1.33×10^{-4}
Butylene	453.3×10^5	3.9×10^5	266.9	—	1.47×20^{-4}
Napthenes					
Cyclohexane	434.6×10^5	3.6×10^5	353.9	—	1.15×10^{-4}
Cyclopentane	465×10^5	3.9×10^5	322.5	—	1.32×10^{-4}
Methylcyclopentane	440×10^5	3.8×10^5	345.0	—	1.18×10^{-4}
Aromatics					
Benzene	406.0×10^5	3.9×10^5	353.3	1.4×10^3	1.00×10^{-4}
Toluene	405.5×10^5	3.6×10^5	383.8	1.5×10^3	9.5×10^{-5}
Xylene (o)	408.4×10^5	3.5×10^5	417.6	—	9.67×10^{-5}
Ethylbenzene	413.5×10^5	3.4×10^5	409.4	1.6×10^3	9.67×10^{-5}

*Note: Burning rates listed for C_1 to C_4 hydrocarbons are appropriate only for spills onto land, since their boiling points are typically below ambient temperatures and heat transfer from a body of water may be appreciable. All burning rates are estimated from property data and are not actual measurements.

It was established in the previous section that the burning rates for large, radiation-dominated pool fires are essentially dependent on the thermochemical properties of the fuel. In the case of continuous spills, the liquid will spread and increase the burning area until the total burning rate is equal to the spill rate. This condition is given by the following equation:

$$D_{eq} = 2[\dot{V}_L/\pi\dot{y}]^{1/2} \qquad (7)$$

where

D_{eq} = steady-state (equilibrium) diameter of the pool, m
$\dot{V}_L$ = liquid spill rate, m³/s
$\dot{y}$ = liquid burning rate, m/s

When Equation 7 is satisfied, the lateral pool spreading will stop and a steady pool fire will result as long as the release is maintained. Equation 7 assumes that the dominant mode of heat transfer to the liquid pool comes from the flame and that the burning rate is constant. This is a valid assumption for all liquid hydrocarbons with boiling temperatures above ambient. This is also a valid assumption for liquefied hydrocarbon spills on water (like LNG or LPG) where heat transfer from water to the pool is relatively constant and leads to a higher burning rate. For liquefied hydrocarbon spills on land, where heat transfer decreases with time, the equation ignores the time-dependent heat transfer from the substrate. It is also assumed in deriving Equation 7 that the mass balance is maintained within the burning pool; hence, the loss of liquid due to percolation through the soil or dissolution in the water column is not included. For most hydrocarbons, these terms tend to be small compared to the burning rate.

The equilibrium diameter given by Equation 7 is reached over a time given by the following equation.[10]

$$t_{eq} = 0.564\frac{D_{eq}}{[g'\dot{y}D_{eq}]^{1/3}} \qquad (8)$$

where g', the effective acceleration due to gravity (m/s²) equals

$$g' = g(1 - \rho_L/\rho_W)$$

If the duration of spill is less than the time to reach equilibrium diameter, the spill size is also less than the equilibrium diameter given by Equation 7. The nondimensional time-diameter relationship is given by the following equation:

$$\tau = \left(\frac{2}{3}\right)^{1/3}\beta\left(\frac{2}{3},\frac{2}{3}\right)I_{\xi^{2/2}}\left(\frac{2}{3},\frac{2}{3}\right)$$

where β is the complete beta function and I is the incomplete beta function, and the nondimensional time and diameter are as follows:

$$\left.\begin{array}{l} \xi = D/D_{eq} \\ \tau = t/t_{ch} \\ t_{ch} = 1.587\, D_{eq}/[g'\dot{y}D_{eq}]^{1/3} \end{array}\right\} \qquad (9)$$

When the value of $\xi = 1$, the spill has reached its equilibrium diameter and the corresponding value of time is

TABLE 3-11.2 *Summary of Radiation Data on Hydrocarbon Pool Fires*

Reference	Pool size (m)	Fuels	H/d	$\dot{y}$ (m/s $\times 10^4$)	T (K)	Radiation
1. Blinov and Khudiakov[2]	Circular, 0.004–22.9	gasoline, kerosene, diesel oil, solar oil	~1.7	~0.67	~1100	not measured
2. Burgess and Zabetakis[7] and other Bureau of Mines tests	0.07–2.4	UDMH, benzene, methanol, butane, LNG, hydrogen, and hexane	3–7	correlated to fuel property	~1300	fraction radiated is estimated and varies from 9.51 to 40.1
3. Fu[26]	0.09–2.4	aviation gasoline, JP-4, JP-5	—	—	—	radiation as a function of distance emissive power ~40 kW/m²
4. Hägglund[19]	0.5–10.1	JP-4	Thomas' correlation		1200	measured and correlated 20 kW/m² to 120 kW/m²
5. NASA[27] field tests	7.6–15.2	JP-4	—	—	1400 max at center	soot concentration was measured
6. NWC, China Lake field tests, not published, Lind	~36	gasoline	—	~0.82	—	maximum emissive power of 110 kW/m²
7. Raj. Mudan and Moussa[18]	10–15	LNG on water	Thomas' correlation	4–10	1400	average emissive power of 220 kW/m²
8. Alger et al[25]	3.05	JP-5, methanol	—	—	800–1100	average emissive power of 110 kW/m² for JP-5 and 70 kW/m² for methanol
9. Moorhouse[13]	rectangular 6.1–13.7	LNG on land	—	2.2	—	
10. May and McQueen	13	LNG on land	—	2.1	—	radiation measured
11. Minzer and Eyre[56]	20 (dia)	LNG, LPG	—	2.4	—	emissive powers estimated from radiometer measurements
12. Japan Institute for Safety Eng.[28]	30–80	gasoline-kerosene	estimated	0.8	1650 max	measured using wide-angle radiometers
13. American Gas Association[15]	1.8–6.1	LNG on land	1.5–3.0	1.5–2.2	—	emissive power estimated using radiometer measurements

given by Equation 8. The nondimensional relationship given by Equation 9 is shown in Figure 3-11.5.

An expression for spreading and burning of an instantaneous release of liquid can be developed by equating the gravitational spreading force to the inertial resistance and accounting for the loss of mass continuously during the process. The radius of the pool increases with time until all of the fuel is consumed by the fire. The radius of the pool fire, maximum diameter, and the time to reach the maximum diameter are given by the following expressions.[11]

$$\left(\frac{D}{D_m}\right)^2 = \frac{\sqrt{3}}{2}\left(\frac{t}{t_m}\right)\left[1 + \left(\frac{2}{\sqrt{3}} - 1\right)\left(\frac{t}{t_m}\right)^2\right] \quad (10)$$

$$D_{m^s} = 2[V_L^3 g'/\dot{y}^2]^{1/8} \quad (11)$$

$$t_m = 0.6743[V_L/g'\dot{y}^2]^{1/4} \quad (12)$$

where V_L = total volume of spilled liquid, m³.

It should be noted that an instantaneous unconfined pool fire grows in size until a barrier is reached or until all the fuel is consumed. Therefore, the maximum diameter predicted by Equation 11 will exist only for a very short duration. Use of the maximum pool diameter will, therefore, lead to very conservative (i.e., overestimate of hazards) results. A time-averaged pool diameter can be obtained by integrating Equation 10 up to time, t_m, given by Equation 12. Use of an average pool size provides more realistic results for

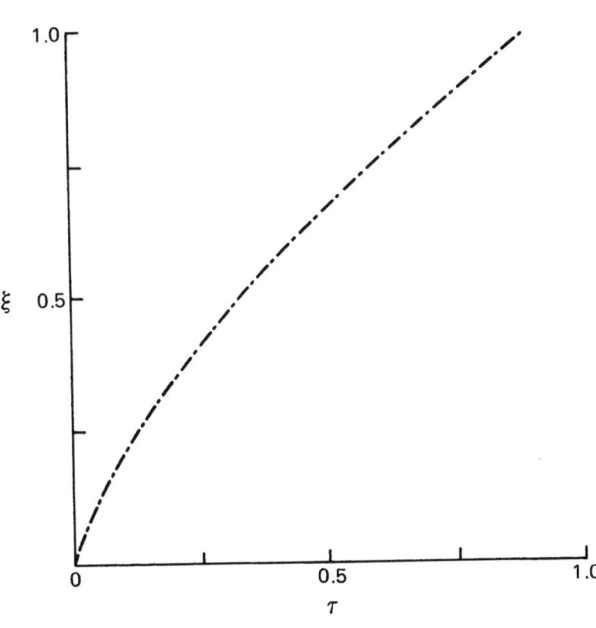

Fig. 3-11.5. *Nondimensional time-diameter relationship.*

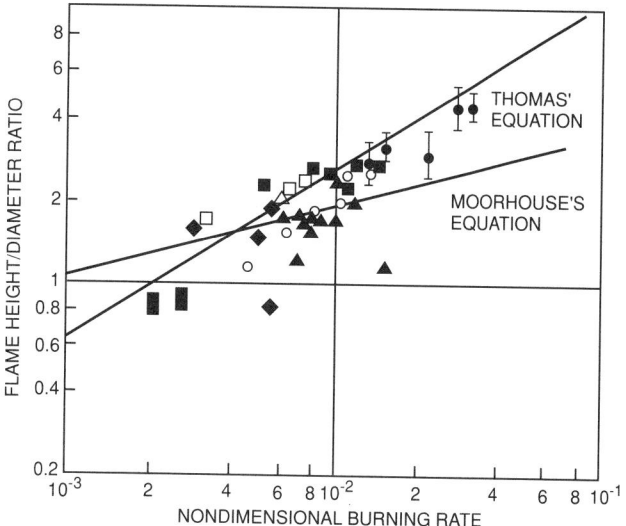

Fig. 3-11.6. *Flame heights for various hydrocarbon pool fires on land and water.[8]*

thermal radiation hazards to objects outside the fire (given by D_{max}). This, however, overlooks the additional short-term hazards to objects within the fire due to higher thermal radiation levels in the later stages of the fire.

Flame height for pool fires: Thomas[12] has developed a correlation for the mean visible height of turbulent diffusion flames (in the absence of wind), based on experimental data of laboratory-scale wooden crib fires and dimensional analysis considerations. The correlation for a circular fire is

$$H/D = 42[\dot{m}''/\rho_a\sqrt{gD}]^{0.61} \tag{13}$$

where $\dot{m}''$ = mass burning rate per unit pool area (in kg/m^2 · s); ρ_a = ambient air density, kg/m^3.

For rectangular fires of small aspect ratios (i.e., length-to-width ratios less than about 5), an area equivalent circular fire diameter may be used.

The presence of wind may also alter the visible length of flames. The correlation developed by Thomas[12] based on wood crib fires is as follows:

$$H/D = 55(\dot{m}''/\rho_a\sqrt{gD})^{0.67}u^{*-0.21} \tag{14}$$

where u^* is the nondimensional wind velocity given by

$$u^* = u_w/\left(\frac{g\dot{m}''D}{\rho_v}\right)^{1/3} \tag{15}$$

Moorhouse[13] conducted several large-scale tests of LNG pool fires. The crosswind and downwind motion picture data were analyzed to determine the exact flame length. The correlation given by Moorhouse is as follows:

$$H/D = 6.2[\dot{m}''/\rho_a\sqrt{gD}]^{0.254}u_{10}^{*-0.044} \tag{16}$$

where u_{10}^* is the nondimensional wind speed determined using Equation 15 with measured wind speed at a height of 10 m. In both Equations 14 and 16, u_{10}^* is assigned a value of unity if it is less than 1.

In Figure 3-11.6 the measured flame height (or length) to diameter ratios are shown as a function of nondimensional

burning rate.[8] The data include pool fires of LNG, LPG, kerosene, gasoline, JP-4, and acetone. Both spills on land and water are included in the data. Also shown in Figure 3-11.6 are the correlations given by Thomas (Equation 13) and Moorhouse (Equation 16). As can be seen from Figure 3-11.6, Thomas' correlation (Equation 13) predicts the visible flame heights better than the other correlation.

Flame tilt: Three correlations relating the angle of tilt of the flame from the vertical are given in the literature. The correlation given by Welker and Sliepcevich[14] was derived from small-scale experiments (0.3 to 0.6 m pans) containing liquid fuels and is as follows:

$$\frac{\tan\theta}{\cos\theta} = 3.3\left[\frac{du_w}{\nu}\right]^{0.07}\left[\frac{u_w^2}{gD}\right]^{0.8}\left[\frac{\rho_v}{\rho_a}\right]^{-0.6} \tag{17}$$

where u_w = wind speed, m/s, and ν = kinematic viscosity of air, m^2/s.

This correlation does not compare well with the measured values of flame tilt for large-scale fires.[15] Thomas[12] gave the following correlation for flame tilt based on data from two-dimensional wood cribs:

$$\cos\theta = 0.7\left[\frac{u_w}{(g\dot{m}''D/\rho_a)^{1/3}}\right]^{-0.49} \tag{18}$$

Based on measured values, the American Gas Association (AGA)[15] proposed the following correlation to determine the flame tilt:

$$\cos\theta = \begin{cases} 1 & \text{for } u^* \leq 1 \\ 1/\sqrt{u^*} & \text{for } u^* \geq 1 \end{cases} \tag{19}$$

where u^* is the nondimensional wind velocity given by Equation 15 with a wind velocity measured at a height of 1.6 m. In Figure 3-11.7 a comparison of observed flame tilt is shown for various hydrocarbon pool fires with correlations given by Equations 18 and 19. Although there is considerable scatter in the measured flame tilt, the correlation given

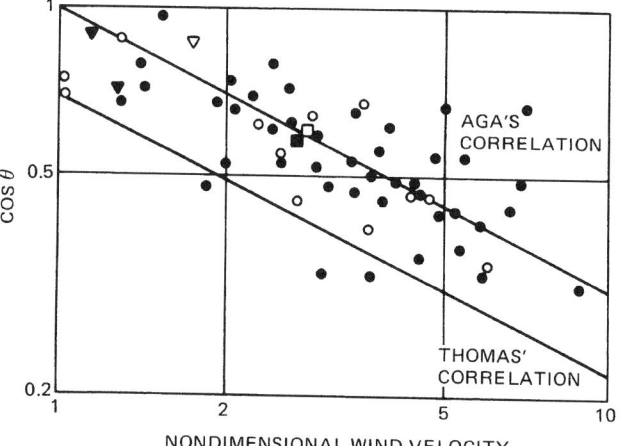

Fig. 3-11.7. *Relationship between nondimensional wind velocity and flame tilt angle.[8]*

by Equation 19 represents the flame tilt more accurately than Thomas' correlation.

Flame drag: The flame drag represents essentially an extension of the base of the flame downwind of the pool, while the upwind edge of the flame and the flame width remain unchanged. Flame drag is strongly dependent upon the wind speed. Both Thomas' experiments with wood crib fires and AGA experiments with LNG did not report the presence of flame drag. Welker and Sliepcevich[14] measured flame drag in a wind tunnel for several hydrocarbons and correlated their results with the following equation.

$$D'/D = 2.1(\mathrm{Fr})^{0.21}(\rho_v/\rho_a)^{0.48} \qquad (20)$$

where Fr = Froude number given by u_w^2/gd.

Moorhouse[13] measured flame drag for various sizes of LNG pool fires for various wind speeds. The correlation given by Moorhouse is

$$D'/D = 1.5(\mathrm{Fr})^{0.069} \qquad (21)$$

where the Froude number is calculated based upon the wind velocity at a height of 10 m. Equation 21 agrees very well with the measured results and can be generalized to other hydrocarbon fuel fires with the following equation:

$$D'/D = 1.25(\mathrm{Fr}_{10})^{0.069}(\rho_v/\rho_a)^{0.48} \qquad (22)$$

The effect of flame drag on square and rectangular fires is to increase the flame base dimension in the wind direction. The flame shape remains rectangular and the thermal radiation in the downwind direction increases because of the proximity of the flame to the surrounding objects. In case of circular flames, the flame shape changes from circular to elliptical, thus changing the view factor from the flame to the receiver.

Geometry of trench fires: Trench fires have received little attention to date. Small laboratory-scale *n*-hexane and heptane fires have been conducted under wind-free conditions. Moorhouse[13] conducted limited large-scale LNG trench fires with aspect ratios ranging from 1.5 to 2.5. More recently, Mudan and Croce[16] reported on large-scale tests with LNG trenches having aspect ratios of up to 30.0. All the data seem to indicate that flame geometry of trench fires is more sensitive to wind conditions than the conventional pool fires. Flame height was found to be a strong function of trench width rather than length. Indeed, for large-aspect ratios, the trench fire seemed to break up into small flamelets having a typical base dimension of the trench width, W.

Based on an extensive analysis of the motion picture data of LNG trench fires, Mudan and Croce[16] suggested that the trench fire geometry can be represented by a modified Froude number. The definition of the modified Froude number is

$$\mathrm{Fr}' = u_w/2\sqrt{gW} \qquad (23)$$

where

u_w = wind speed, m/s
W = trench width, m
g = acceleration due to gravity, m/s^2.

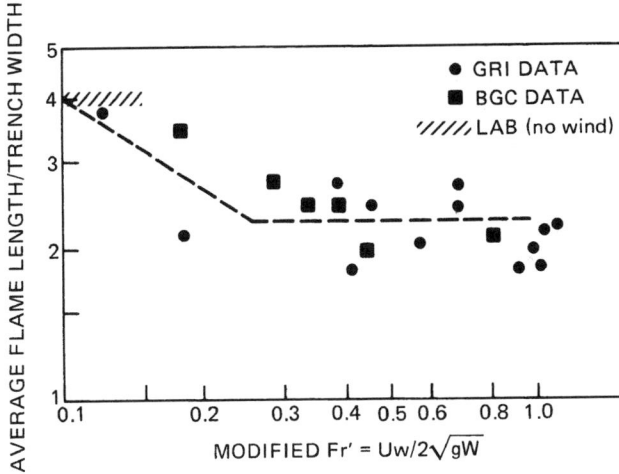

Fig. 3-11.8. *Nondimensional flame height as a function of Froude number.*[16]

The measured flame length as a function of the modified Froude number is shown in Figure 3-11.8. The flame length correlation is

$$\frac{H}{W} = \begin{cases} 2.2 & \mathrm{Fr}' \geq 0.25 \\ 0.88(\mathrm{Fr}')^{-0.65} & 0.25 < \mathrm{Fr}' \leq 0.1 \\ 4.0 & \mathrm{Fr}' \leq 0.1 \end{cases} \qquad (24)$$

The flame drag and flame tilt are given by the following expressions:

Flame drag:

$$\frac{W'}{W} = \begin{cases} 3.5 & \mathrm{Fr}' \geq 0.25 \\ 23.3(\mathrm{Fr}')^{1.37} & 0.25 < \mathrm{Fr}' \leq 0.1 \\ 1 & \mathrm{Fr}' < 0.1 \end{cases} \qquad (25)$$

Flame tilt:

$$\cos\theta = \begin{cases} 0.56 & \mathrm{Fr}' \geq 0.25 \\ 0.36(\mathrm{Fr}')^{-0.32} & 0.25 \leq \mathrm{Fr}' \leq 0.042 \\ 1 & \mathrm{Fr}' \leq 0.042 \end{cases} \qquad (26)$$

The essential features of the suggested correlations defined by Equations 24, 25, and 26 are

1. The flame geometry (length, drag, and tilt) are independent of the ambient conditions for Froude numbers greater than about 0.25. For example, the critical wind speed for a 4 m wide trench will be about 3 m/s. For wind speeds greater than 3 m/s, the trench fire geometry does not change significantly.
2. For very low Froude numbers (extremely low wind to no wind cases), the flame geometry parameters are dependent on the trench width.
3. For in-between Froude numbers, the flame length decreases with increasing wind speed and the flame drag and tilt increase.

Caution must be exercised in using the correlations given by Equations 24, 25, and 26 since they are based only on LNG fire data.

Thermal Radiation from Pool Fires

There are two basic types of thermal radiation models, namely, the point source model and the plume fire model. These two models are introduced in this section. Key thermal radiation parameters such as emissive power and view factors are also described, with methods to calculate these variables.

Point source radiation model: The point source thermal radiation model is based on the assumption that:

1. the flame can be represented by a small source of thermal energy;
2. the energy radiated from the flame is a specified fraction of energy released during combustion; and
3. the thermal radiation intensity varies proportionately with the inverse square of the distance from the source.

Expressed mathematically, the radiant intensity at any distance from the source is given by

$$\dot{q}_r'' = \dot{Q}/4\pi x^2 \qquad (27)$$

and

$$\dot{Q} = \eta \dot{m} \Delta H_c \qquad (28)$$

where

$\dot{Q}$ = total energy radiated per unit time, W
$\dot{m}$ = fuel mass burning rate, kg/s
η = fraction of combustion energy radiated
x = radial distance from source to the observer, m

While the above model is elegant in its simplicity, two important limitations should be recognized. The first limit involves the modeling of radiative output and the second is the description of the variation of the intensity as a function of distance from the source.

The most important parameter in the point source model is the fraction of combustion energy radiated to its surroundings. This fraction cannot easily be estimated theoretically. It is normally estimated using measured radiometer data. There is extensive laboratory data that suggest that the radiation from buoyant diffusion flames remains proportional to the overall heat release rate provided the flame is fully turbulent and the flame geometry is properly scaled.

The suggestion that the radiative fraction is constant for geometrically similar, buoyant turbulent diffusion flames is due to an essentially invariant Kolmogorov microscale.[17] This scale assumes that the flames are both optically thin and chemically similar and hence radiative fraction is determined by the thermochemical properties of the fuel. Both these assumptions break down as one goes from a "moderate-scale" laboratory experiment to a "large-scale" field experiment. At present, however, there is insufficient composition data on larger fires to clearly delineate the dependence between moderate- and large-scale fires.

Table 3-11.3 shows the estimated fraction of combustion energy radiated for various hydrocarbon pool fires. As can be seen from the table, there is considerable variation in the radiative fraction. This appears to be particularly true for

LNG pool fires on water[18] and gasoline fires.[5,19] In the case of LNG fires, the reduction in the radiative fraction was attributed to incomplete combustion in large fires. For gasoline pool fires, smoke obscuration resulted in a significant reduction in the incident radiation recorded by the radiometers.

Moorhouse and Pritchard[20] equated the point source model and uniformly radiated cylindrical source model and concluded that the fraction of combustion energy radiated may be expressed as follows:

$$\eta = \frac{E}{\dot{m}'' \Delta H_c} \left[1 + 4\frac{H}{d} \right] \qquad (29)$$

where E is the constant emissive power of the flame in W/m^2.

Since the flame height is influenced by the flame diameter (with larger diameter flames resulting in shorter flame heights), they concluded that the fraction of energy should decrease for larger fires. However, in large-scale hydrocarbon pool fires, smoke obscuration contributes to a reduction in measured flame emissive powers. Therefore, the effect of flame geometry on the radiative fraction may be of the second order.

The second limitation to be observed in a point source model is that the model overestimates the intensity of thermal radiation at observer locations close to the fire. This is primarily because the near-field radiation is greatly influenced by the flame size, shape, tilt, and relative orientation of the observer. The model, however, predicts far-field thermal radiation intensities with reasonable accuracy.

In summary, a point source model provides a simple and elegant means of estimating thermal radiation intensity at far-field, where the effects of flame geometry are not significant. It can be used to determine thermal radiation hazards to personnel, where normally a conservative estimate of hazard is acceptable. Caution must be exercised in using the model to determine siting criteria such as spacing between two storage tanks.

Solid flame radiation model: The solid flame model is based on the proposition that the entire visible volume of the flame emits thermal radiation and that the nonvisible gases

TABLE 3-11.3 *Radiation Fraction of Combustion Energy for Hydrocarbon Pool Fires*

Hydrocarbon	Pool Size (m)	% Radiative Output/Combustion Output
Methanol	1.2	17.0
LNG on land	18.0	16.4
	0.4–3.05	15.0–34.0
	1.8–6.1	20.0–25.0
	20.0	36.0
LNG on water	8.5–15.0	12.0–31.0*
LPG on land	20.0	7.0
Butane	0.3–0.76	19.9–26.9
Gasoline	1.22–3.05	40.0–13.0*
	1.0–10.0	60.1–10.0*
Benzene	1.22	36.0–38.0
Hexane	—	40
Ethylene	—	38

*In these cases, the smaller diameter fires were associated with higher radiative outputs.

do not emit much radiation. While it is true that all of the heat energy liberated in combustion is ultimately radiated to the surroundings, the rate of energy emission is significant only from the visible flame envelope. Measurements by Markstein[1] have confirmed that the irradiance of the burnt gas (nonvisible) plume above a fire accounts for less than 10 percent of the mean irradiance of the visible fire.

In the plume flame model, the thermal radiation intensity, $\dot{q}_r''$, to an element outside of the flame envelope is calculated by the equation

$$\dot{q}'' = \tau E_f F \tag{30}$$

Procedures for estimating the necessary values for atmospheric transmissivity, τ, the view factor, F, the average emissive power of the flame, E_f, and the experimental bases for these procedures are discussed below.

Emissive power: Once the shape and size of the pool fire are calculated, the radiative characteristics of the fire must be determined to compute the thermal radiation hazards. Emissive power of a flame is the single most important input to the solid flame model. In this section, the available data and models to compute the surface emissive power from large hydrocarbon pool fires will be discussed.

The thermal radiation from a fire emanates from both gaseous species such as water vapor, carbon dioxide, and carbon monoxide as well as from luminous soot particles. The gaseous species emit radiation in certain spectral bands, whereas the soot radiation is continuous over the entire spectral range of importance. The theories of gas radiation and the models developed to describe the band emission from various gases are described in most texts on radiative heat transfer.[21] From the point of view of fire radiation predictions, deRis[17] has reviewed these models. Modak[22] has developed simplified calculation procedures for obtaining the gas emissivities. Markstein[1] has proposed a model that considers a fire as a two species emitter whose total radiance is equal to the weighted sum of the radiance due to gas emissions and that due to luminous soot.

The emissive power of a large turbulent fire may often be approximated by the following expression:

$$E_f = E_b \varepsilon \tag{31}$$

where E_b = blackbody emissive power, kW/m^2, and ε = emissivity.

If the mean radiation temperature of the fire is known (which is significantly less than the adiabatic flame temperature), it can be converted to irradiance using Planck's law of radiation. Thus, the blackbody emissive power is given by

$$E_b = \sigma(T_f^4 - T_a^4) \tag{32}$$

where

T_f = radiation temperature of the flame, K
T_a = ambient temperature, K
σ = Stefan-Boltzmann constant, kW/m^{2}K^4.

The emissivity accounts for the fact that the flame is a gray emitter, i.e., not an ideal blackbody radiator. Calculation of combined emissivity of the products of combustion (soot, water vapor, and carbon dioxide) is extremely difficult and involved, even when the concentrations are uniform and the temperature is constant. The emissivity also de-

pends on the path length through the fire. The combined emissivity is given by:

$$\varepsilon = \varepsilon_s + \varepsilon_w + \varepsilon_c - \Delta\varepsilon_{c,w} \tag{33}$$

where

ε_s = spectrally averaged emissivity of soot
$\varepsilon_c, \varepsilon_w$ = molecular band integrated emissivity of CO_2 and H_2O (carbon monoxide, water)
$\Delta\varepsilon_{c,w}$ = correction factor for the CO_2-H_2O band overlap.

Modak[22] has shown, using the results from the analysis of Felske and Tien[23] and other experimental results, that the soot emissivity can be determined by the following expression:

$$\varepsilon_s = 1 - \frac{15}{\pi^4}\psi^{(3)}\left(1 + \frac{7}{C_2}T_f CL\right) \tag{34}$$

where

C = effective soot concentration parameter
L = path length, m
C_2 = Planck's second constant
$\psi^{(3)}$ = Penta gamma function.

Yuen and Tien[24] reviewed many experimental data on soot emission from luminous flames of gaseous, polymer, and wood fuels and demonstrated that Equation 34 can be approximated by the following expression for a gray emitter:

$$\varepsilon_s = 1 - e^{-k_s L} \tag{35}$$

where

$$k_s = 3.6\frac{CT_f}{C_2} \tag{36}$$

and is the effective soot emission parameter. Hottel and Sarofim[21] have provided charts for evaluating the gas emissivities ε_c, ε_w, and $\Delta\varepsilon_{c,w}$.

In recent years there has been considerable discussion on whether luminous flames can be regarded as being spectrally gray. In general, the emissivity of a homogeneous mixture of path length x is given by

$$\varepsilon_m(x) = \frac{1}{\sigma T_f^4}\int_0^\infty E_b(\lambda)(1 - e^{-k_{m\lambda}x})\, d\lambda \tag{37}$$

If the mixture is spectrally gray, $k_m = k_{m\lambda}$ and the integral becomes

$$\varepsilon_m(x) = \varepsilon = 1 - e^{-k_m x} \tag{38}$$

where k_m = extinction coefficient, m^{-1}.

From a theoretical point of view, the assumption of a homogeneous gray mixture permits the calculation of radiative heat transfer in terms of the characteristic flame shape, and the single dimensionless optical depth parameter, $k_m L$, where L is some characteristic length for a particular geometry. This assumption dramatically simplifies such calculations. The extinction coeffecient, however, needs to be determined based on experimental data. In Figures 3-11.9 and 3-11.10, measured emissive powers for LNG and LPG pool fires, respectively, are shown. Based on these values, data extinction coefficients of 0.4 for LNG and 0.335 for LPG were determined.

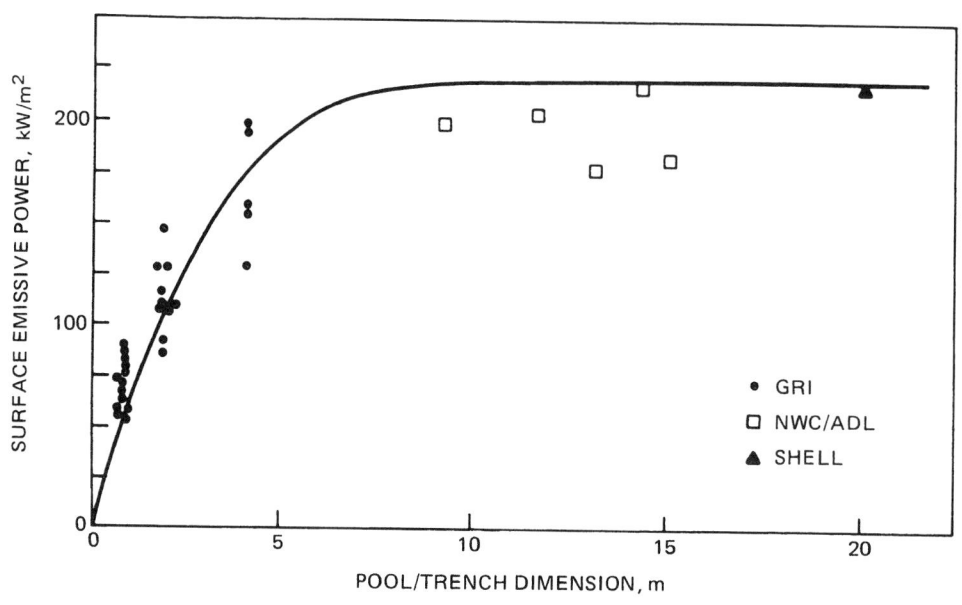

Fig. 3-11.9.   Measured emissive power of LNG pool fires as a function of pool and trench dimensions. [16]

For large fires, the numerical value of emissivity approaches unity. Therefore, the emissive power can be determined using the mean radiation temperature. However, the radiation temperatures for many liquid fuels are not available. In fact, measurements of radiative flux using narrow angle radiometers are often used to predict the emissive power at the flame surface. Table 3-11.4 shows the measured emissive powers and radiation temperatures for some liquid hydrocarbon fuels.

Most hydrocarbon fuel fires become optically thick when the diameter is about 3 m or larger. Under these conditions, the maximum emissive power that has been mea-

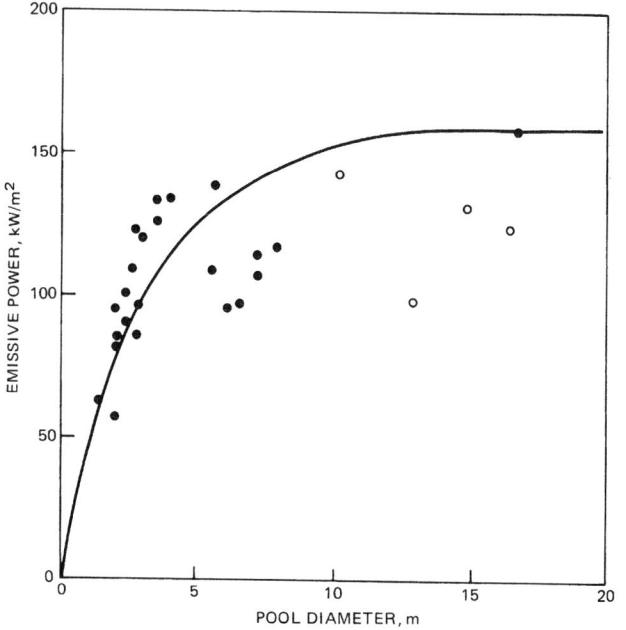

Fig. 3-11.10.   Emissive power data for LPG pool fires. [55]

sured for gasoline fires is in the range of 110 to 130 kW/m^2. Hagglund and Persson[19] have reported emissive powers of 130 kW/m^2 for 1.5 m diameter fires and 20 kW/m^2 for 10 m diameter fires. Alger *et al*[25] measured average emissive powers of 110 kW/m^2 and 70 kW/m^2 for JP-5 and methanol fires of 3 m diameter. The measured heat flux appears to decrease for larger fires, indicating that the emissive power is decreasing. This is an unintuitive result that has been reported numerous times in the literature. [19,26]

It has been observed that in large liquid hydrocarbon fuel fires with a carbon-to-hydrogen ratio greater than about 0.3, a substantial part of the fire is obscured by a thick black smoke on the outer periphery. This smoke layer absorbs a significant part of the radiation and results in very little emission to the surroundings. However, the smoke layer occasionally opens up, exposing the hot flame, and a pulse of radiation is emitted to the surroundings.

Although the thermal radiation from black soot is low, the hot spots appearing on the flame surface due to turbulent mixing have a higher emissive power. Large eddies within the flame bring fuel to the outer edges of the fire plume and a more efficient combustion takes place on the flame surface. Based on a qualitative observation of the movie records of kerosene fires on land[27,28] and gasoline fires on water,[55] it appears that the luminous zones cover approximately 20 percent of the flame surface area on a time averaged basis. These luminous spots have an emissive power of about 110 to 130 kW/m^2. However, it is not possible to calculate the radiation field surrounding a fire with intermittent luminous spots. Hagglund and Persson[19] observed that the emissive power of the black smoke is about 20 kW/m^2 and the temperature is about 800 K.

From a hazard prediction point of view, the thermal radiation from the black soot can be combined with the radiation from luminous spots on an equivalent area basis to arrive at an average emissive power for the fire. For example, if it is assumed that 80 percent of the surface area is covered with black smoke and 20 percent with luminous spots, the time average emissive power is given by the following expression:

TABLE 3-11.4 *Measured Emissive Powers and Radiation Temperatures for Various Liquid Hydrocarbon Pool Fires*

Hydrocarbon	Pool Fire Dimensions (m)	Emissive Power kW/m^2	Radiation Temperature K	Comments
LNG on water	8.5 to 15.0	210 to 280	1500	Estimated using narrow-angle radiometer data and spectral data.
LNG on land	20.0	150 to 220	—	Estimated using wide-angle and narrow-angle radiometer data.
LPG on land	20.0	48	—	LPG fires appear to be very smoky.
Gasoline	1.0 to 10.0	60 to 130 (max)	1240	The larger fires radiate less than smaller fires.
JP-4	5.8	—	1200	The flame surface is covered with black smoke.
Kerosene	30.0 to 80.0	10 to 25 (average)	1600	The larger fires radiate less than smaller ones. Emissive powers are estimated using wide-angle radiometer data.
JP-5	1.0 to 30.0	30 to 50	—	The emissive power of the 30 m diameter fire was about 30 kW/m^2.
UDMH	0.5	—	1300	
Pentane	1.0	61	—	
Ethylene	2.5	130	—	

$$E_{average} = 0.2(130) + 0.8(20) = 42 \text{ kW/m}^2 \quad (39)$$

Such an estimate is consistent with the wide-angle radiometer measurements of JP-4, JP-5, and gasoline fires.

In Figure 3-11.11 the measured (or estimated) average emissive power as a function of pool fire diameter is shown. For the sake of convenience (and lack of information), the data on gasoline, kerosene, and JP-5 are combined. The combined data indicate that the average emissive power decreases with increasing pool diameter. For very large pools of kerosene (80 m diameter), the average emissive power estimated using wide-angle radiometer data is approximately 10 kW/m^2. Based on this estimation the following correlation is suggested to determine the average emissive power of large, sooty hydrocarbon fires.

$$E_{av} = E_m e^{-SD} + E_S(1 - e^{-SD}) \quad (40)$$

where

E_m = maximum emissive power of luminous spots (approximately 140 kW/m^2)

E_s = emissive power of smoke (approximately 20 kW/m^2)

S = a parameter determined using experimental data ($= 0.12 \text{ m}^{-1}$).

The correlation given by Equation 40 leads to the following results:

Pool diameter of 1 m—average emissive power 126 kW/m^2

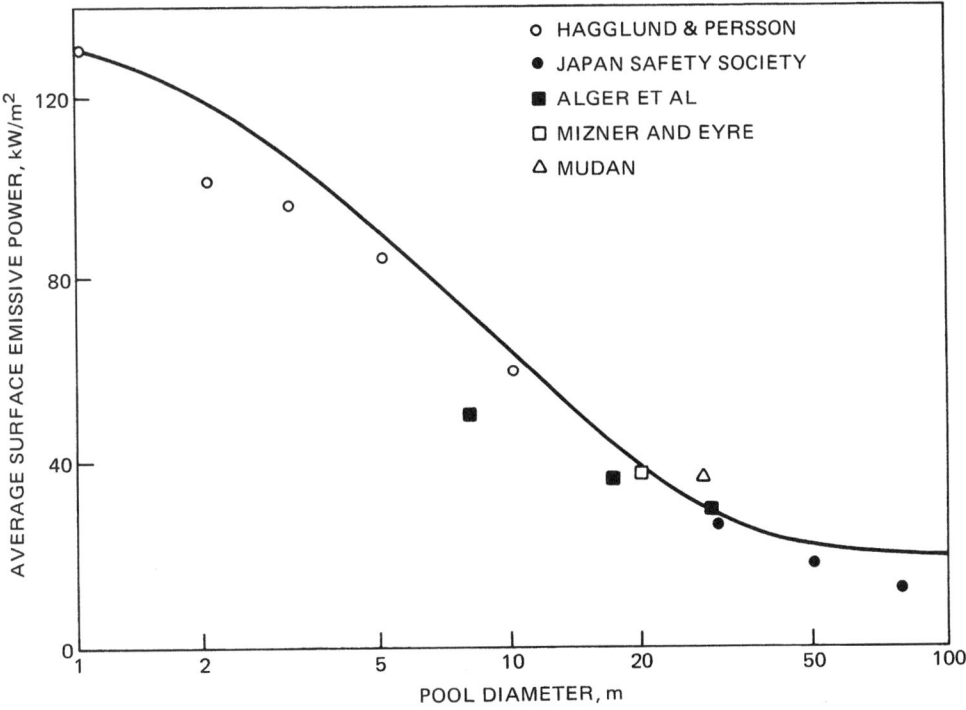

Fig. 3-11.11. Average surface emissive power for gasoline, kerosene, and JP-4 pool fires.[8]

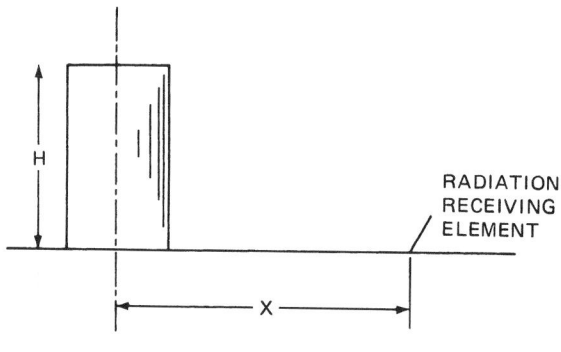

(a) RIGHT CIRCULAR SOURCE

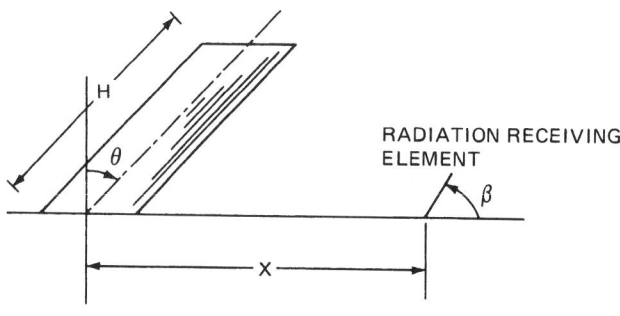

(b) INCLINED CYLINDRICAL SOURCE

Fig. 3-11.12. View factor calculation geometrics for right and inclined cylinders.

$$\pi F_v = \frac{a}{b} \frac{a^2 + b^2 + 1}{[a^2 + (b+1)^2]^{1/2}[a^2 + (b-1)^2]^{1/2}}$$

$$\times \tan^{-1} \left[\frac{a^2 + (b+1)^2}{a^2 + (b-1)^2} \right]^{1/2} \left(\frac{b-1}{b+1} \right)^{1/2}$$

$$+ \frac{1}{b} \tan^{-1} \frac{a}{b^2 - 1} - \frac{a}{b} \tan^{-1} \left(\frac{b-1}{b+1} \right)^{1/2} \quad (42)$$

$$\pi F_h = \tan^{-1} \left(\frac{b+1}{b-1} \right)^{1/2} - \frac{a^2 + b^2 - 1}{[a^2 + (b+1)^2]^{1/2}[a^2 + (b-1)^2]^{1/2}}$$

$$\times \tan^{-1} \left[\frac{a^2 + (b+1)^2}{a^2 + (b-1)^2} \right]^{1/2} \left(\frac{b-1}{b+1} \right)^{1/2} \quad (43)$$

where a is flame height (or length)/flame radius and b is observer distance/flame radius.

The maximum view factor, which is the vectorial sum of horizontal and vertical view factors, is as follows:

$$F_m = \sqrt{F_v^2 + F_h^2} \quad (44)$$

In Figure 3-11.13 the maximum view factor for various cylinder height-to-radius ratios and nondimensional observer locations is plotted. At locations very close to the fire, the view factor is not very sensitive to the flame height. This is because the observer already "sees" the maximum amount of flame surface and the increasing flame height does not increase the view factor significantly. However, at greater distances from the flame surface, the view factor is sensitive to the flame height. For example, at a distance of

Pool diameter of 10 m—average emissive power
56 kW/m^2
Pool diameter of 100 m—average emissive power
20 kW/m^2.

The correlation given by Equation 40 is also shown in Figure 3-11.11.

Geometric view factors: The radiation exchange factor between a fire and an element outside of the fire depends on the flame's shape, the relative distance between the fire and the receiving element, and the relative orientation of the element. In general, the view factor is represented by the following equation:

$$F_{A_1 \rightarrow dA_2} = \oint_{A_1} \frac{\cos \theta_1 \cos \theta_2}{\pi r^2} dA_1 \quad (41)$$

where 1 and 2 are, respectively, the angles made by the normals and dA on the fire and dA_2 on the receiving element; and where r is the distance between the fire element and the receiving element. The integration is carried out over the entire surface of the flame.

In the solid flame model the turbulent flame is approximated by a cylinder. Under wind-free conditions, the cylinder is vertical. Under the influence of wind, the cylinder is assumed to be tilted. These two configurations of the solid flame model are shown in Figure 3-11.12. The horizontal and vertical view factors for a vertical cylinder are as follows:

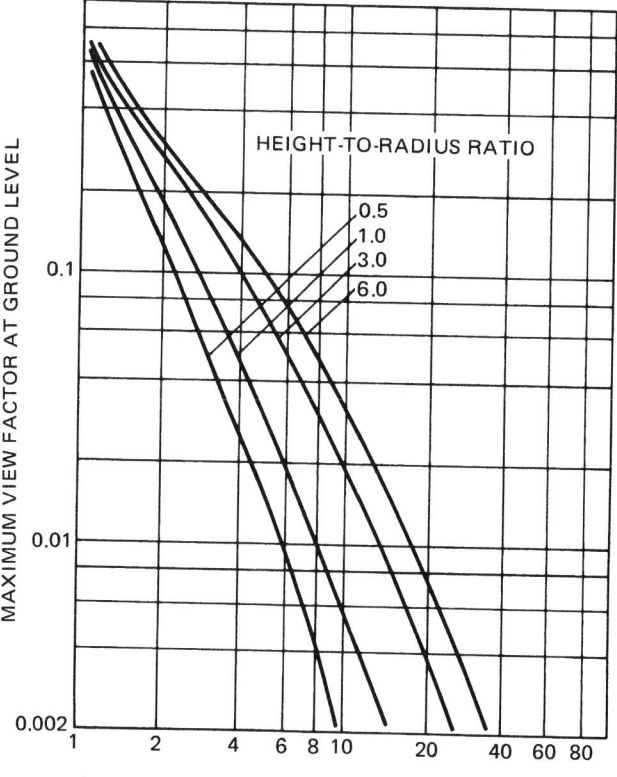

Fig. 3-11.13. Maximum view factors for a ground-level object from a right circular cylinder.

ten times the flame radius, the view factor varies from 0.003 to 0.034 when the flame height-to-radius ratio is increased from 0.5 to 6.

Rein et al[29] calculated the radiation view factors for tilted cylinders. In their calculations, they replaced the view factor integral by a summation and numerically evaluated the view factor by dividing the cylinder into a number of incremental rectangular areas. Raj and Kalelkar[11] employed a numerical integration approach to determine the view factors for tilted cylinders. More recently, Mudan[30] employed a contour integral approach developed by Sparrow[31] to determine closed form equations for view factors from a tilted cylinder. The expressions for horizontal and vertical factors are

$$\pi F_v = \frac{a \cos \theta}{b - a \sin \theta} \frac{a^2 + (b+1)^2 - 2b(1 + \sin \theta)}{\sqrt{AB}}$$

$$\times \tan^{-1} \sqrt{\frac{A}{B}} \left(\frac{b-1}{b+1}\right)^{1/2} + \frac{\cos \theta}{\sqrt{C}}$$

$$\times \left[\tan^{-1} \frac{ab - (b^2-1)\sin \theta}{\sqrt{b^2-1}\sqrt{C}} + \tan^{-1} \frac{(b^2-1) \sin \theta}{\sqrt{b^2-1}\sqrt{C}}\right]$$

$$- \frac{a \cos \theta}{(b - a \sin \theta)} \tan^{-1} \sqrt{\frac{b-1}{b+1}} \qquad (45)$$

and

$$\pi F_h = \tan^{-1} \sqrt{\frac{b+1}{b-1}} - \frac{a^2 + (b+1)^2 - 2(b+1 + ab \sin \theta)}{\sqrt{AB}}$$

$$\times \tan^{-1} \sqrt{\frac{A}{B}} \left(\frac{b-1}{b+1}\right)^{1/2} + \frac{\sin \theta}{\sqrt{C}} \qquad (46)$$

$$\times \left[\tan^{-1} \frac{ab - (b^2-1) \sin \theta}{\sqrt{b^2-1}\sqrt{C}} + \tan^{-1} \frac{(b^2-1)^{1/2} \sin \theta}{\sqrt{C}}\right]$$

where

$$A = a^2 + (b+1)^2 - 2_a (b+1) \sin \theta$$
$$B = a^2 + (b-1)^2 - 2_a (b-1) \sin \theta$$
$$C = 1 + (b^2 - 1)\cos^2 \theta.$$

When the angle of tilt is zero, Equations 45 and 46 reduce to Equations 42 and 43, respectively. The view factors given by Equations 45 and 46 were also compared with experimentally derived view factors from Rein et al.[29] In Figure 3-11.14 a comparison of predicted and measured view factors is shown and the agreement is indeed good.

The maximum view factors may, once again, be determined by using Equation 44. In Figure 3-11.15 the effect of tilt on maximum view factors is shown. For example, the incident flux at a nondimensional distance of 5 for a flame titled at 45° will be almost 2.5 times the incident flux computed using a zero tilt. In the upwind direction, the incident flux will be about 40 percent of the flux computed using a zero tilt.

The view factors for rectangular geometry may also be computed using the contour integral procedure. The horizontal view factor[32] for a tilted rectangular flame is

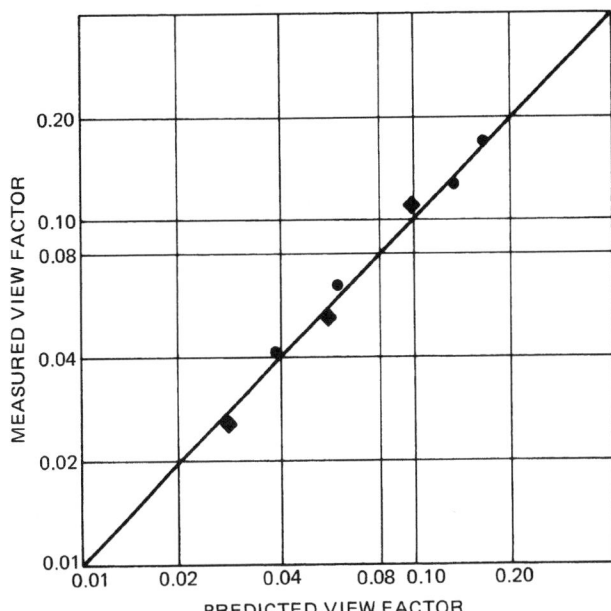

Fig. 3-11.14. *Comparison of predicted and experimentally determined view factors for tilted cylinders[29] (circles, tilt = 30°; diamonds, tilt = −30°).*

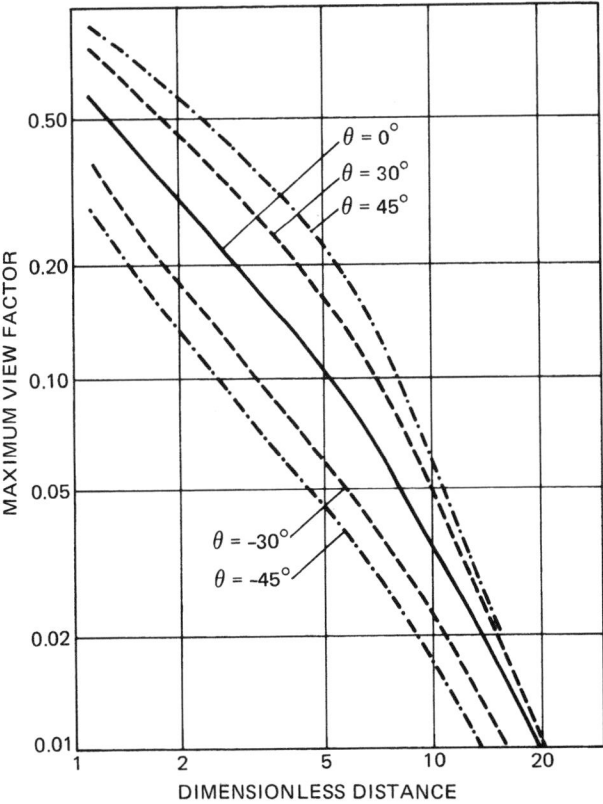

Fig. 3-11.15. *Maximum view factors for tilted circular cylinders.[30]*

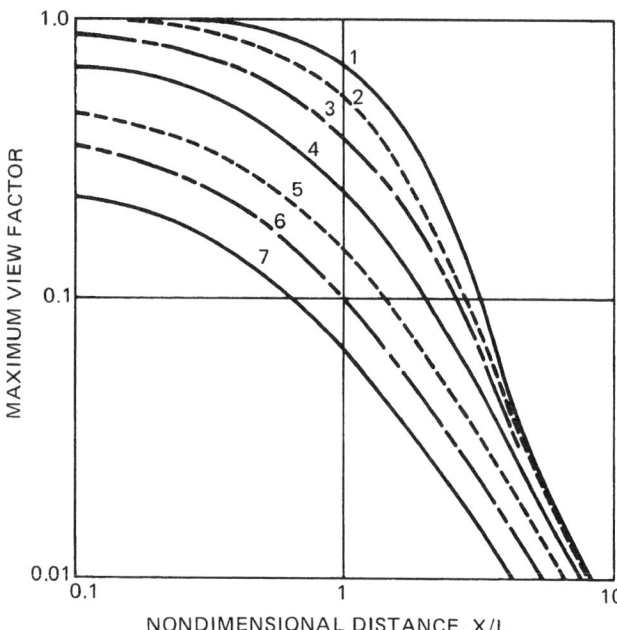

*Fig. 3-11.16. **Maximum view factors for a tilted rectangular surface.***

$$\pi F_h = \tan^{-1}\frac{1}{Q} + \frac{(P\cos\beta - Q)}{(P^2 + Q^2 - 2PQ\cos\beta)^{1/2}}$$

$$\times \tan^{-1}\frac{1}{(P^2 + Q^2 - 2PQ\cos\beta)^{1/2}} + \frac{\cos\beta}{(1 + Q^2\sin^{-2}\beta)^{1/2}}$$

$$\times \left\{ \tan^{-1}\left[\frac{P - Q\cos\beta}{(1 + Q^2\sin^{-2}\beta)^{1/2}}\right] \right.$$

$$\left. + \tan^{-1}\left[\frac{Q\cos\beta}{(1 + Q^2\sin^{-2}\beta)^{1/2}}\right] \right\} \tag{47}$$

where

$P = 2H/L$

$Q = 2X/L$

β = angle with respect to horizontal;

where the distance, X, is measured from the surface of the flame and the tilt is with respect to horizontal. The above equation may also be used with a slight geometric modification to determine the vertical view factor. The influence of tilt on maximum view factors is illustrated in Figure 3-11.16.

Atmospheric Absorption

The radiation from the fire to surrounding objects will be partially attenuated by absorption and scattering along the intervening path. The principal constituents of the atmosphere that absorb thermal radiation are water vapor (H_2O) and carbon dioxide (CO_2). Table 3-11.5 indicates the composition of various gases in the atmosphere. The CO_2 content in the atmosphere is generally constant at about 330 ppm by volume. The water vapor content varies strongly with temperature and humidity. Figure 3-11.17 indicates the relationship between atmospheric temperature, relative humidity, and the amount of precipitable water vapor in a given path length.

The principal absorption bands for water vapor are at 1.8, 2.7, and 6.27 μm. Minor absorption bands also exist at

TABLE 3-11.5 *Composition of Constituent Gases in the Atmosphere and Their Concentrations*

Constituent Gas	Concentration in Atmosphere (% by volume)
Nitrogen	78.088
Oxygen	20.949
Argon	0.93
Carbon Dioxide	0.033
Neon	1.8×10^{-3}
Helium	5.24×10^{-4}
Methane	1.4×10^{-4}
Krypton	1.14×10^{-4}
Nitrous Oxide	5.0×10^{-5}
Carbon Monoxide	2.0×10^{-5}
Xenon	8.6×10^{-6}
Hydrogen	5.0×10^{-6}
Ozone	Variable
Water Vapor	Variable (depends on temperature and relative humidity)

0.94, 1.1, 1.38, and 3.2 μm. Strong absorption by CO_2 exists in the 2.7 μm region, the 4.3 μm region, and the region between 11.4 and 20 μm. Weaker absorption bands are present at 1.4, 1.6, 2.0, 4.8, 5.2, 9.4, and 10.4 μm. As the temperature of the emitting or absorbing species increases, the bands tend to broaden.

A useful concept for the quick estimation of atmospheric absorption of continuum radiation is the "equivalent bandwidth of complete absorption." One calculates the integral of absorption over an absorption band and interprets the result as the width of a "rectangular" *complete* absorption band equivalent to the real bank profile. For a continuum source, the effect of such opaque bands is then easy to estimate. Three absorption bands in the range of interest (1.5 through 5.5 μm) can be described in this way. These are the water vapor bands at 1.87 and 2.7 μm and the 4.3 μm CO_2 band. The water absorption beyond about 4.7 μm is not as readily dealt with, since the band structure is not narrow compared to the range of interest. However, the

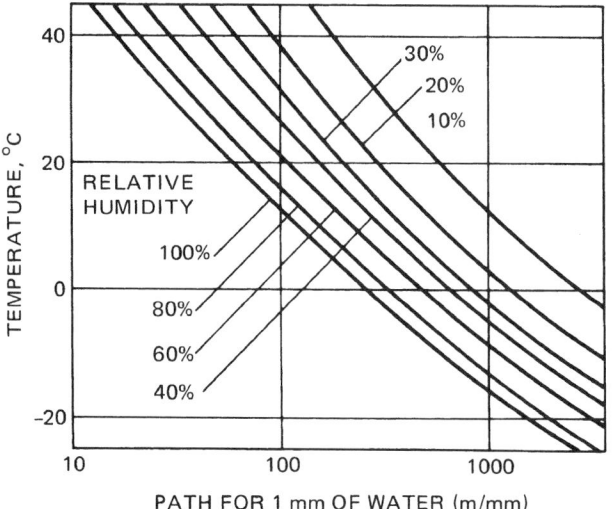

*Fig. 3-11.17. **Variation of precipitable water content of the atmosphere with temperature, humidity, and path length.***

TABLE 3-11.6 *Total Absorption Bandwidth for 1.87 μm and 2.7 μm Bands at 300 K*

| Precipitable Water (mm) | 1.87 μm Band | | 2.7 μm Band | |
	Total Absorption Bandwidth, Δλ (μm)	Fraction of 1300 K Blackbody Energy Absorbed	Total Absorption Bandwidth, Δλ (μm)	Fraction of 1300 K Blackbody Energy Absorbed
5	0.16	0.04	0.58	0.16
2	0.12	0.03	0.51	0.14
1	0.1	0.03	0.45	0.12
0.1	0.033	0.01	0.22	0.06
0.01	0.01	0.003	0.07	0.02

fraction of total energy from a 1300 K blackbody that lies beyond 4.7 μm is about 25 percent and that beyond 5.5 μm is only 19 percent. The results of total absorption bandwidth calculations for the above three bands of interest are given in Tables 3-11.6 and 3-11.7. These calculations are based on the data available in the *Infrared Handbook.*[33] Also given in the tables are the fractions of a 1300 K blackbody energy that will be absorbed in each of these bands.

The absorption by the water vapor and carbon dioxide in a certain length of the atmosphere of blackbody radiation from a source can also be calculated using the emissivity charts published by Hottel and Sarofim.[21] The procedure to calculate the absorption in the water vapor band is as follows:

1. Determine the partial pressure of water vapor, in atmosphere based on

$$p'_W = \frac{RH}{100} \exp\left(14.4114 - \frac{5328}{T_a}\right) \qquad (48)$$

where RH is the relative humidity.

2. Define a path length, L, (in m) from the flame surface to observer. Determine the partial pressure-path length parameter

$$p_W L = p'_W L(T_s/T_a) \qquad (49)$$

where

T_s = source surface temperature (K)
T_a = ambient temperature (K)

3. For the source temperature, and $p_W L$, determine the water vapor emissivity ε_W, using emissivity plots given in Figure 3-11.18.

4. Calculate the water vapor absorption coefficient.

$$\alpha_W = \varepsilon_W (T_a/T_s)^{0.45} \qquad (50)$$

TABLE 3-11.7 *Total Absorption Bandwidth for 4.3 μm CO_2 Band at 300 K*

Path Length Through the Atmosphere (m)	Total Absorption Bandwidth Δλ (μm)	Fraction of 1300 K Blackbody Energy Absorbed
1000	0.28	0.04
100	0.22	0.03
10	0.17	0.02
1	0.065	0.01
0.3	0.033	0.004

The procedure to determine the absorption by carbon dioxide is very similar. The partial pressure of CO_2 remains relatively constant at about 3×10^{-4} atm. The absorption coefficient is given by

$$\alpha_c = \varepsilon_c (T_a/T_s)^{0.65} \qquad (51)$$

The emissivity of the carbon dioxide band is shown in Figure 3-11.19. There is also a correction factor due to spectral overlap for the calculation of emissivity of a CO_2-H_2O mixture. This effect, however, accounts for a change in emissivity of about 5 percent at 1200 K and even less at higher temperatures.[21]

The transmissivity is given by

$$\tau = 1 - \alpha_W - \alpha_c \qquad (52)$$

and is used in determining the thermal radiation hazard. The procedure outlined in this section may be simplified further if it is assumed that the flame temperature and the ambient temperature remain constant. For most hydrogen fuels, the flame temperature is approximately 1400 K. If it is assumed that the typical ambient temperature is 293 K (20°C), the transmissivity may be plotted as a function of path length. In Figure 3-11.20 the transmissivity is shown as a function of path length for various relative humidities. Figure 3-11.20 provides a quick estimate of atmospheric absorption of thermal radiation.

Calculation Procedure for Pool Fires

In order to compute thermal radiation hazard distances, the flame geometry of the fire must be first characterized (i.e., pool diameter, height, tilt, drag). Then the geometric view factors, transmissivity, and average effective emissive power must be estimated to obtain thermal radiation levels at a given location. In this section a sample calculation is presented to illustrate the calculation procedure for thermal radiation from pool fires.

EXAMPLE:

Instantaneous rupture of a gasoline storage tank; tank volume is 5000 m³ (1.3 million gallons), and dike size is 67 m diameter. Determine maximum incident flux onto the closest tank, located 18 m from the dike wall.

SOLUTION:

Step 1: Define fuel properties and ambient conditions:

Gasoline Molecular Weight   =   100

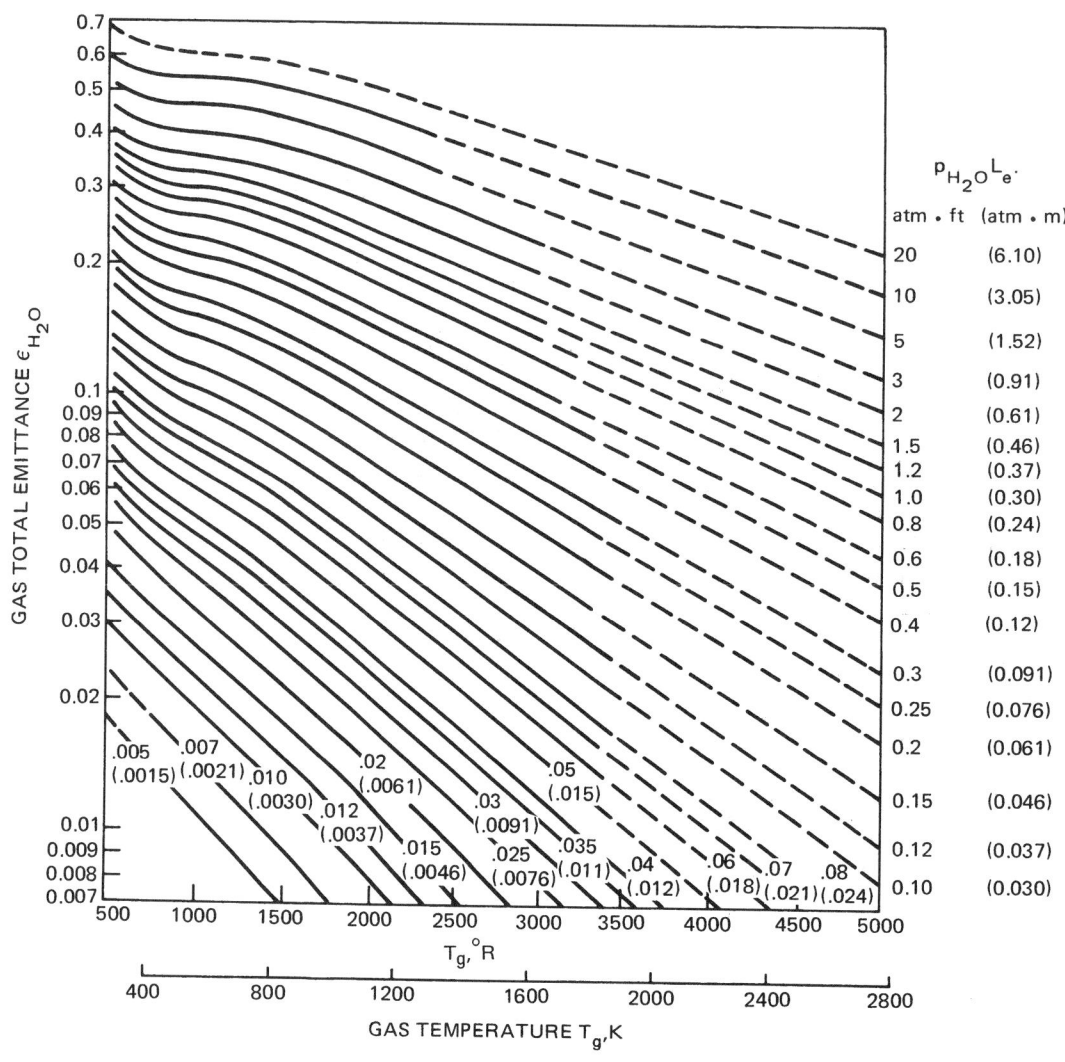

Fig. 3-11.18. *Total emissivity of water vapor in a mixture of total pressure of 1 atm.* [21]

Liquid Density	=	870 kg/m^3
Maximum Emissive Power (from Figure 3-11.11)	=	140 kW/m^2
Flame Temperature	=	1300 K
Burning Rate on Land	=	0.8 × 10^{-3} kg/m$^2 \cdot$ s
S factor in Emissive Power Calculation	=	2.0 m^{-1}
Vapor Density at Boiling Point	=	3.49 kg/m^3
Fuel Mass Burning Rate	=	0.058 kg/m$^2 \cdot$ s
Ambient Temperature	=	293 K
Wind Speed	=	2 m/s
Relative Humidity	=	50 percent
Air Density at Ambient Temperature	=	1.2 kg/m^3

Step 2: Determine flame geometry parameters:

1. Maximum Pool Diameter; use Equation 11 for instantaneous spills.

D_{max} = 680 m for confined spills. Since this is greater than the diameter of the dike, maximum diameter is equal to dike diameter of 67 m.

2. Flame Height; use Thomas' correlation (Equation 14)
H_{max} = 68.5 m.
3. Flame Tilt; compute nondimensional wind velocity, u^*, using Equation 18.
u^* = 0.85
Therefore, cos θ = 1.0; θ = 0.0 (no tilt)
4. Flame Drag; use Equation 22
Fr_{10} = 6.09 × 10^{-3}
D/D_{max} = 1.47

Step 3: Compute thermal radiation parameters.

1. Average Emissive Power, E_p; use Equation 40 for large sooty hydrocarbon fires.
E_p = 20 kW/m^2
2. Maximum View Factor, F, for Upright Cylinder:
(a) Nondimensional distance of neighboring tank, S_{ND}, from flame center
S_{ND} = 2χ/D_{max} = 1.54
(b) Height-to-radius ratio:
H/R = 2 H/D_{max} = 2.0

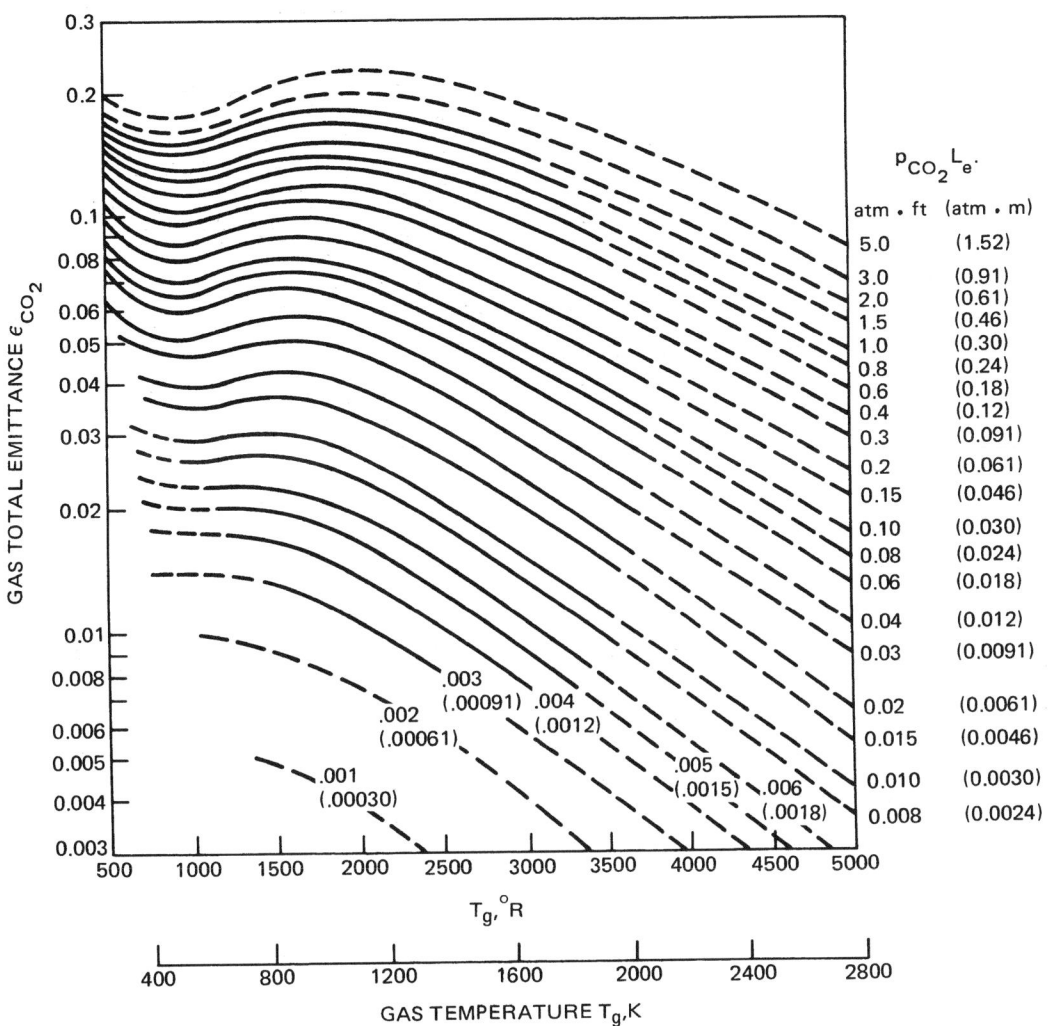

Fig. 3-11.19. *Total emissivity of carbon dioxide in a mixture of total pressure of 1 atm.*[21]

(c) From Figure 3-11.13 or Equations 43 and 44, compute F_{max}.
$F_{max} = 0.38$

Step 4: Estimate transmissivity using Figure 3-11.20, with a path length = 18 m.

$\tau = 0.77$

Step 5: Compute maximum radiant flux, $\dot{q}''_{max}$ incident upon neighboring tank:

$$\dot{q}''_{max} = E_p \, \tau \, F_{max}$$
$$= 5.85 \text{ kW/m}^2$$

TURBULENT JET FLAMES AND FLARES

Large turbulent diffusion flames are encountered in a processing environment as a result of an accidental release of hydrocarbon vapors or the intentional disposal of unwanted gases in a flare. Flaring is the combustion process that has been the traditional method for safe disposal of large quantities of unwanted flammable gases and vapors in the petroleum industry. With the advent of air quality stan-

dards, flaring has also taken on an added importance as a method of industrial environmental control, since most gases that could previously be vented to the atmosphere must now be burned in a flare. The flaring of gases in the petroleum industry occurs in three ways:

1. Production Flaring: In a producing oil field where no provision exists for collecting and processing gas, there is a requirement for safe disposal of flammable gases. There was a time when almost all gas released was flared, but the great value now placed on natural gas has made gas recovery economical for some fields. Nevertheless, if gas occurs in small quantities that are uneconomical to process or if the gas is so sour that processing is expensive, it can still be flared.

2. Process Flaring: Flaring also takes place in petrochemical plants, oil refineries, and gas processing plants where the flare system is one of the off-site facilities. In process flaring, the gas that leaks past safety valves protecting various process units is brought to the flare and burned. This gas feeds the small flames that burn almost continuously on refinery flare stacks. Process flaring can occur at much greater rates when process units are evacuated

during a shutdown or when off specification products are produced during startup.

3. Emergency Flaring: This occurs when large volumes of volatile liquids or flammable gases have to be disposed of safely in an emergency such as a fire, power failure, or overpressure in a process vessel.

The flaring process involves the release of a tremendous amount of energy. Since a portion of this energy release is in the form of thermal radiation, it represents a substantial hazard to personnel, equipment, and the environment. The sizing of flares, both in diameter and height, is of major importance to ensure personnel safety during flaring operations. The ability to predict the thermal radiation field from flares is essential in the design of a reasonably sized, safe operating flare. Experimental data on thermal radiation from full-scale flares are rare and when available at all, the flow rate and composition of the flared gases are usually unknown. However, several scale-model studies have been conducted to examine the geometric and radiative characteristics of flares.

The geometric characteristics of turbulent jet flames are similar to those of industrial flares. In fact, many of the geometric descriptions of flares are based on small-scale turbulent jet flame experiments. The base diameter of a flare stack, height of the stack, and composition of burning substance are often known to the user. In modeling accidental releases of hydrocarbon gases, these relevant data may have to be estimated.

The analytical models describing the geometric characteristics of turbulent diffusion flames are described in the following section. The models describe parameters such as the flame length, flame width, and flame tilt. A description of the thermal radiation models, the aerodynamic effects on radiation and blowout stability of jet flames, and the calculation procedure are described in succeeding parts of the chapter.

Geometric Characteristics of Turbulent Jet Flames

It is now recognized that combustion in a jet fire occurs in the form of a strong turbulent diffusion flame in a cross-wind. Such a flame presents a number of challenging phenomena for study, including the effect of crosswind on flame shape and size; radiation and formation, and dispersion of smoke and other gaseous pollutants. While applying these models to industrial flares, it is also important to recognize the effects of steam in suppressing the smoke formation and thermal radiation. A review of flaring in the energy industry has been published by Brzustowski.[34]

Turbulent flame length: A reasonable measure of progress of burning of a diffusion flame is its height. A jet fuel is assumed to issue from a nozzle of known diameter in a vertical direction upward into an ambient medium containing an oxidant. For a given fuel and burner, beginning at a low flow rate, the flame height increases with increasing flow rate, reaches a maximum and then shortens. Shortly before the flame height reaches its maximum, the flame begins to flicker (becomes turbulent) at the top. This occurrence separates "laminar diffusion flames" from flames of the "transitional type." With further increase of flow rate, flickering spreads in a downward direction and stops at a distance only a few diameters from the burner. At this point, the flame height becomes independent of flow rate and the "transition flame" passes over into the "turbulent flame."

Predicting the length of the diffusion flame of gas jets burning in still air has long been considered one of the classical solved problems in combustion. The landmark papers on this subject were published more than three decades ago by Hottel and Hawthorne[35] and Hawthorne et al.[36] Hottel and Hawthorne[35] considered the case of a primary fuel jet of higher velocity issuing into an infinite atmosphere of air with allowance for primary air in the fuel jet. They observed the progressive change in the flame shape and size as the nozzle velocity is increased. Figure 3-11.21 shows a sketch of flame height as a function of nozzle velocity. In the initial laminar region, the flame height increases with nozzle velocity. The flame length in the laminar region for hydrocarbon fuels is given by the following equation:

$$L/d_j = Q_j C_0/(4\pi D_\nu d_j C_T) \tag{53}$$

where

Q_j = volumetric flow rate, m^3/s

C_0 = original mole fraction of fuel in the jet

C_T = mole fraction for stoichiometric combustion

d_j = jet diameter, m

L = flame length, m

D_ν = diffusivity, m^2/s

Equation 53 indicates that the flame length increases directly with flow rate of fuel. The flame length continues to increase with flow velocity until a critical Reynolds number is reached. Based on extensive data, Hottel and Hawthorne[35] suggested the following Reynolds number criteria to determine transition from laminar flame to turbulent flames:

Hydrogen	~ 2000
City gas	~ 3600
Carbon monoxide	~ 4900
Propane	~ 8500
Acetylene	~ 9500

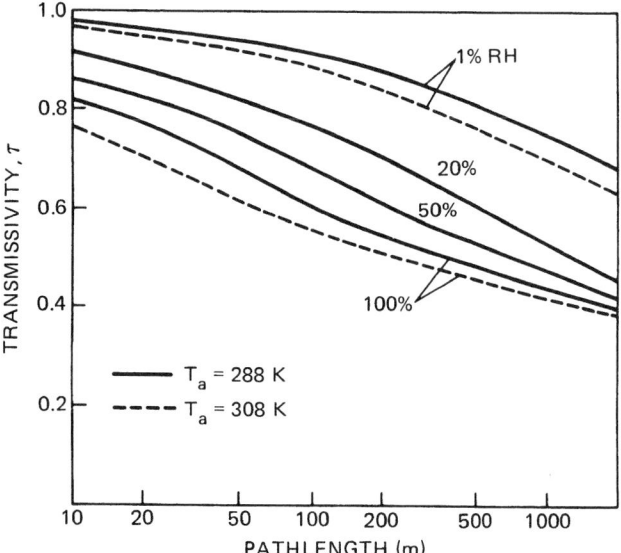

Fig. 3-11.20. Transmissivity as a function of path lengths; source temperature 1400 K.

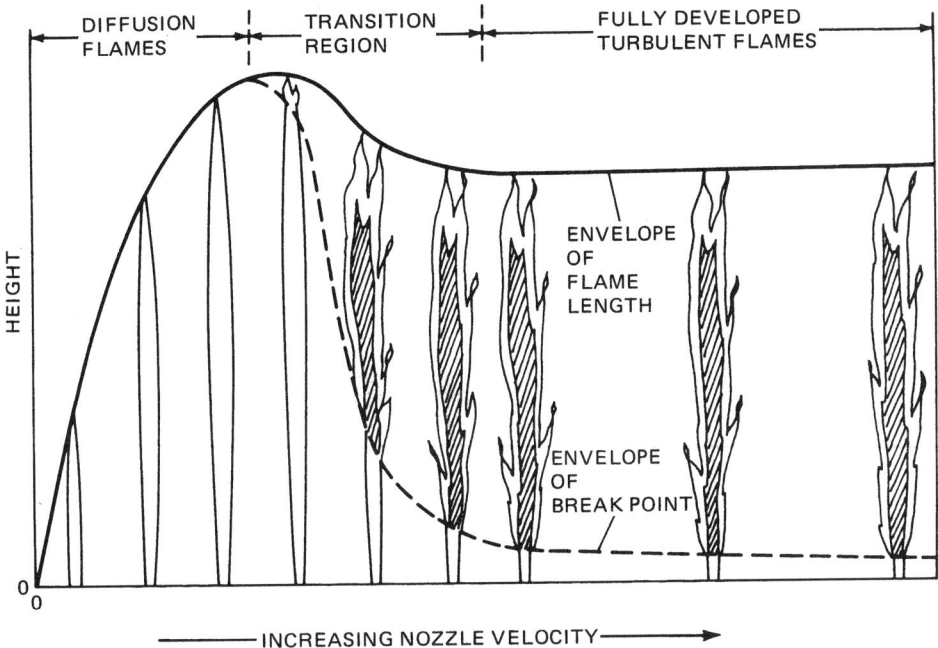

Fig. 3-11.21. Progressive change in flame type with increasing jet velocity.

For example, the critical velocity for pure propane issuing out of a 10 mm nozzle is approximately 5 m/s. For velocities above 5 m/s, the flame jet will be turbulent. Since accidental releases of flammable gases invariably occur at relatively high velocities, the discussion is limited to the theory of turbulent flames.

The first set of experimental data and a theoretical model for flame length for turbulent flame jets were developed by Hawthorne *et al.*[36] Since that time, many tests and reviews have cited as definitive the following equation for the flame length of a turbulent gas jet burning in still air.

$$\frac{L}{d_j} = \frac{5.3}{C_T}\left\{\frac{T_F}{\alpha_T T_j}\left[C_T + (1 - C_T)\frac{M_a}{M_f}\right]\right\}^{1/2} \quad (54)$$

where

L = length of visible turbulent flame measured from the break point

α_T = moles of reactant per mole of product for stoichiometric fuel-air mixture

C_T = fuel concentration in stoichiometric fuel-air mixture

M_a, M_f = molecular weight of air and fuel

T_F, T_j = adiabatic flame temperature and temperature of jet fluid (absolute).

The 5.3 factor appearing in Equation 54 is the ratio of visible length to the width of the flame at the point where stoichiometric air has been entrained (determined from experimental data).

Equation 54 for determining the flame length reduces to a much simpler expression for most hydrocarbon gases. The value of the parameters C_T, α_T, and T_F/T_j for various hydrocarbons are:

For methane:

$C_T = 0.091$
$\alpha_T = 1.00$
$T_F/T_j = 7.4$

For ethane:

$C_T = 0.074$
$\alpha_T = 1.04$
$T_F/T_j = 9.0$

For propane:

$C_T = 0.038$
$\alpha_T = 0.96$
$T_F/T_j = 7.6$

Since C_T is typically much less than unity, α_T is approximately unity and T_F/T_j varies between 7 and 9. Equation 54 may be approximated by the following equation:

$$\frac{L}{d_j} = \frac{15}{C_T}\left[\frac{M_a}{M_f}\right]^{1/2} \quad (55)$$

The concept of flame length is more elusive than it first seems. Visible flame length, for example, depends not only on all the factors that affect flame luminosity but also on the observer's ability to discern it. Background lighting in the laboratory, the type of photographic film used, and the control of photographic processing are some of the factors that must be considered in any analysis of data on visible flame lengths. For these reasons, it would be far preferable to be able to compare the proposed criterion with flame lengths defined in terms of the concentration at the flame tip. Hawthorne et al[36] also included measurements of fuel concentration along the flame axis in their paper. These measurements show that L extended some 10 to 20 percent beyond the point on the flame axis where a long-time sample indicated 99 percent complete combustion. At the 99 percent point, the concentration of fuel on the axis was lower than the stoichiometric concentration by a factor between 2 and 3. The authors related this difference to the so-called "un-mixedness factor," i.e., the rms concentration fluctuation at a point.

Ricou and Spalding[37] measured entrainment of ambient air by buoyant jets and flames. The correlation they proposed for the increase in mass of a buoyant jet or a flame is

$$\frac{\dot{m}}{\dot{m}_0} = 0.32\left(\frac{x}{d_j}\right)\left(\frac{\rho_j}{\rho_a}\right)^{1/2} \qquad (56)$$

where

ρ_j, ρ_a = density of jet and ambient air
d_j = jet diameter

Putnam and Speich[38] modified the above relationship to

$$\frac{\dot{m}}{\dot{m}_0} = 0.165\left(\frac{x}{d_j}\right)^{5/3}\left(\frac{\rho_j}{\rho_a}\right)^{5/6}\text{Fr}^{-1/3} \qquad (57)$$

where Fr is a modified form of the Froude number and is defined as

$$\text{Fr} = \frac{u_j^2}{gd_j}\left(\frac{\rho_a}{\rho_j}\right)^{1/2}\frac{C_aT_a}{Y_{fj}\Delta h + C_j(T_j - T_a)} \qquad (58)$$

where

u_j = jet velocity
g = acceleration due to gravity
Y_{fj} = mass fraction of fuel in jet fluid
T_a = ambient temperature
C_a = specific heat of air
Δh = heat of combustion per unit mass of fuel

For buoyancy controlled flames of gaseous jets, Putnam and Speich[38] correlated their data by the following equation:

$$\frac{L}{d_j} = 29\left(\frac{Q_j^2}{gd_j^5}\right)^{1/5} \qquad (59)$$

where Q_j = volumetric flow of fuel gas at ambient pressure.

Since

$$Q_j = \frac{\pi}{4}d_j^2u_j$$

Equation 59 becomes

$$\frac{L}{d_j} = 26\left(\frac{u_j^2}{gd_j}\right)^{1/5} \qquad (60)$$

Steward[39] correlated the heights of buoyant turbulent diffusion flames, which had been measured by many investigators in a great variety of experimental arrangements. He defined the flame height as the sum of two components—the height required for the entrainment of stoichiometric air and the additional height required for the entrainment of some amount of excess air. A good correlation of the data showed that about 400 percent excess air is entrained up to the flame tip of the very large number of different flames studied.

Brzustowski[40,41] proposed that the end of a turbulent diffusion flame at very high Reynolds number occurs at that point on the axis of maximum fuel concentrations where the fuel concentration equals the lean limit. In theory, this criterion can be applied to any flow configuration for which cold-flow concentration data at sufficiently high Reynolds number are available. Brzustowski[40] has also used some full-scale flare test data (conducted by Battelle Memorial Institute, Columbus, OH, 1970) to support the lean limit

criterion. With this criterion, the flame length for momentum dominated jets is given by the following equation:

$$\frac{L}{d_j} = \frac{Y_{fj}}{0.32}\left(\frac{\rho_j}{\rho_a}\right)^{1/2}\left[1 + \frac{\dot{M}_a}{M_f}\left(\frac{1}{0.297C_L} - 1\right)\right] \qquad (61)$$

For a flame where buoyancy is the dominating force, the flame length is given by

$$\frac{L}{d_j} = 2.96\left(\frac{\rho_j}{\rho_a}\right)^{1/2}\text{Fr}^{1/5}Y_{fj}^{3/5}$$
$$\times \left[1 + \frac{M_a}{M_f}\left(\frac{1}{0.297C_L} - 1\right)\right]^{3/5} \qquad (62)$$

where C_L = fuel concentration at the lean flammability limit, by volume.

Fr is the Froude number defined by Equation 58. Since C_L is about 5 percent or less for most hydrocarbons and M_a/M_f is approximately unity, Equations 61 and 62 may be simplified to the following two equations for momentum and buoyancy dominated flames, respectively.

$$\frac{L}{d_j} = \frac{10.5}{C_L}\left(\frac{M_a}{M_f}\right)^{1/2} \qquad (63)$$

$$\frac{L}{d_j} = 6.1\left(\frac{u_j^2}{d_jg}\right)^{1/5}\left(\frac{1}{C_L}\right)^{3/5}\left(\frac{\bar{C}_aT_a}{\Delta\bar{H}}\right)^{1/5} \qquad (64)$$

where $\bar{C}_a$ = molar heat capacity of air and $\Delta\bar{H}$ = molar heat of combustion.

Equations 63 and 64 are functionally very similar to the expressions obtained by Hawthorne et al[36] and Putnam and Speich.[38] These expressions indicate the similarity in three independent experiments.

Effect of crosswind on flame length: Brzustowski et al[42] and Gollahalli et al[43] conducted a series of wind tunnel tests involving hydrogen and propane diffusion flames in a crosswind. It was observed that the initial effect of crosswind was to shorten the flame, after which increases in a crossflow velocity caused increases in the flame length. Shortly before blowoff conditions were reached, flame length was observed to decrease with increases in crosswind.

This behavior can be explained in terms of entrainment and flame stability. Initially, the effect of crosswind is to create shearing forces along most of the length of the jet, since the jet bends over only slightly. This results in improved entrainment and mixing, formation of smaller amounts of condensed species, and hence a shorter flame. However, if the crosswind velocity is increased to a value such that the flame bends over quickly and is convected with the crosswind, the lower entrainment of air in the bent-over portion would appear to cancel the effect of increased entrainment near the stack. Hence, once again the flame length increases with crosswind. As the crossflow velocity is increased to very high values approaching blowoff conditions, the very high rate of mixing in the initial regions may dilute much of the fuel-air mixture to values outside the lean flammability limit. Hence, this fuel is just convected with the crosswind without burning, resulting in shorter flame lengths. A further increase in wind speed causes the flame to blow off. The results obtained with zero wind conditions are consistent with the model equations given in the previous section.

Based on wind tunnel data and limited comparison with full-scale data, Brzustowski[34] has proposed the following procedure to determine the flame shape in the presence of crosswind.

1. Calculate the dimensionless lean limit concentration.

$$\overline{C}_L = C_L \left(\frac{u_j}{u_w}\right)\left(\frac{M_f}{M_a}\right) \tag{65}$$

2. If $\overline{C}_L \le 0.5$, $\overline{S}_L = 2.04\,(\overline{C}_L)^{-1.03}$

   If $\overline{C}_L > 0.5$, $\overline{S}_L = 2.71\,(\overline{C}_L)^{-0.625}$

3. If $\overline{S}_L > 2.35$, $\overline{X}_L = \overline{S}_L - 1.65$

   If $\overline{S}_L \le 2.35$, then determine $\overline{X}_L$ by following the equation

$$\overline{S}_L = 1.04\overline{X}_L^2 + 2.05\overline{X}_L^{0.28} \tag{66}$$

4. Determine the dimensionless rise, $\overline{Z}_L$, of the flame tip above flame tip.

$$\overline{Z}_L = 2.04\overline{X}_L^{0.28} \tag{67}$$

5. Calculate dimensional coordinates of the flame tip using the following equation:

$$X = \overline{X}_L d_j \left(\frac{\rho_j}{\rho_a}\right)^{1/2}\left(\frac{u_j}{u_w}\right)$$
$$Z = \overline{Z}_L d_j \left(\frac{\rho_j}{\rho_a}\right)^{1/2}\left(\frac{u_j}{u_w}\right) \tag{68}$$

Kalghatki[44] of the Shell Research Center conducted a series of 103 wind tunnel tests to determine the size and shape of turbulent hydrocarbon jet diffusion flames in a crosswind. The tests were conducted with methane, propane, ethylene, and commercial butane. The burner diameters ranged from 6 to 22 mm and the range of velocities were between 13 and 200 m s. The crosswind velocities were varied from 2.6 to 8.1 m s. Based on these tests, Kalghatki[44] concluded that the turbulent jet flame can be described by a frustum of a cone. The five parameters describing the shape of the flame are:

1. Angle α_B, subtended by the burner tip and the tip of the flame with respect to vertical;
2. Angle α, subtended by the flame with respect to vertical;
3. Vertical length, L_{BV}, of the flame tip from the plane of the burner;
4. The flame base width, W_1; and
5. The flame tip width, W_2.

These parameters are expressed in terms of a dimensional variable called "effective source diameter" and a nondimensional velocity, R. The definitions of these two parameters are

$$D_s = D\left(\frac{\rho_j}{\rho_a}\right) \tag{69}$$

where

D = source diameter, m
ρ_j = density of jet fuel, kg/m^3
ρ_a = density of ambient air, kg/m^3

$$R = U/U_j \tag{70}$$

where

U = crosswind speed, m/s
U_j = jet velocity, m/s

In Figure 3-11.22 the variation of the vertical flame length parameter is shown as a function of the nondimensional velocity, R. Also shown in Figure 3-11.22 are the correlations suggested by Brzustowski for propane, methane, and ethylene. In order to predict the actual flame length, the flame tilt with respect to the burner axis must be known. Figure 3-11.23 shows the data on flame tilt and the comparison with Brzustowski's calculation procedure. It is seen that the calculation procedure underestimates the flame tilt. Therefore, the measured flame lengths are slightly larger than the ones predicted using Brzustowski's model. Based on these results, Kalghatki[44] suggests the following correlations to determine the flame length and flame tilt parameters.

$$\alpha_B = 94 - 1.6/R - 35R \tag{71}$$

$$\alpha = 94 - 1.1/R - 30R \tag{72}$$

$$L_{BV}/D_s = 6 + 2.35/R + 20R \tag{73}$$

Here the angles α are in degrees. The range of validity of these correlations are for the values of R greater than 0.02 and less than 0.25. The upper limit for R is not a serious limitation to the applicability of the model. For values of R less than 0.02, the wind-free data may be used to determine the flame lengths and the tilt may be assumed to be zero. It should be noted that the flame length given by Equation 73 determines only the vertical component. Actual inclined flame height is given by dividing Equation 73 by the cosine of the angle of tilt given by Equation 71. It is worth noting that the actual inclined flame length ratio is independent of the velocity ratio and is relatively constant at a value of about 120. This indicates that the majority of the tests conducted in this program were momentum dominated turbulent jets. Recently, Sonju and Hustad[45] conducted an experimental study on turbulent diffusion flames. The flare diameters ranged from 2.3 to 80 mm and the velocities ranged from 5 to 250 m/s. Their data indicate that the flame length is proportional to the one-fifth power of the Froude number. For Froude numbers greater than 100,000, the flame lengths appear to be independent of the Froude number. These results are consistent with the data of Putnam and Speich.[38]

Flame diameter: Hawthorne et al[36] also observed that the jet diameter increases as a function of distance. The measured spreading angles were in the range of 3 to 8 degrees (one-half angle). The equivalent diameter for thermal radiation calculations can be calculated from the following equation:

$$D_c/d_j = \sec\theta + L/d_j \sin\theta \sec^2\theta \tag{74}$$

The data of Kalghatki[44] for the base and tip widths of the flames indicate the spreading angle is a function of the nondimensional velocity ratio. The suggested correlations are

$$W_2/D_s = 80 - 0.57/R - 570R + 1470R^2 \tag{75}$$

$$W_1/D_s = 49 - 0.22/R - 380R + 950R^2 \tag{76}$$

From these calculations, it can also be deduced that the cone half angle for the frustum decreases from a value of about 5 degrees at $R = 0.025$ to a value of about 2.8 degrees at $R = 0.2$. Therefore, at large relative wind speeds, the diffusion flame takes an almost cylindrical shape. The data of

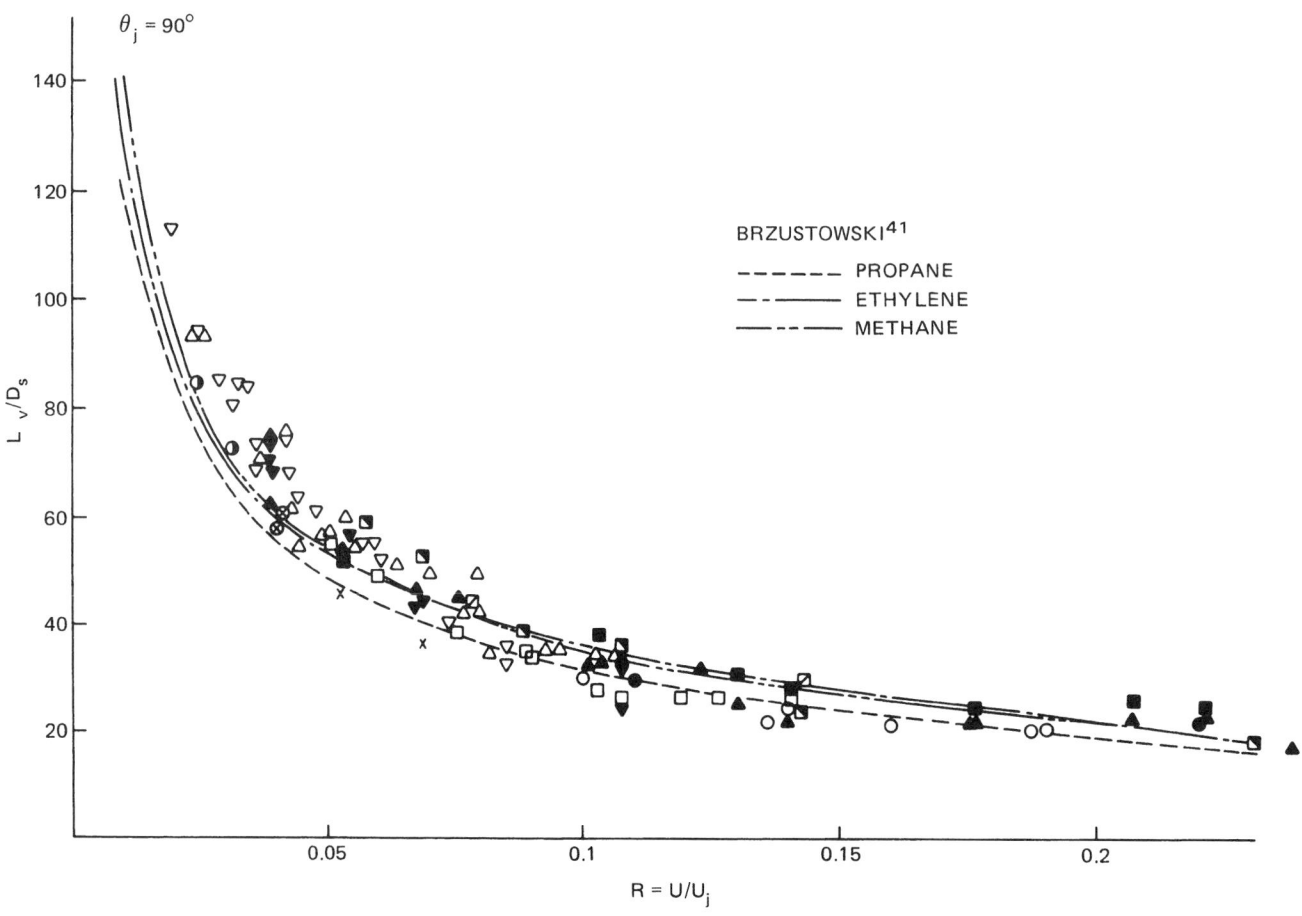

Fig. 3-11.22. *Variation of the flame length above burner tip with velocity ratio, R.*[44]

Sonju and Hustad[45] indicates that the flame diameter increases as one-fifth power of the flame Froude number. The suggested constants of proportionality for methane and propane flames are 2.5 and 4, respectively.

Inclined turbulent jets: Kalghatki[44] also conducted some limited tests with nonorthogonal jet flames and concluded that the flame lengths are also dependent on wind direction. The tests were conducted with relative wind angles varying from 45 to 135 degrees (with 90 degrees representing orthogonal cross flow). The value of the nondimensional velocity ratio was greater than 0.025. The data indicate that for a given angle of tilt, the flame length remains relatively constant; however, the flame length decreases with increasing angle of tilt. The data of flame length and wind direction over the entire velocity ratios are shown in Figure 3-11.24. As can be seen from the figure, there appears to be a linear relationship between flame length and wind direction. The correlation suggested by Kalghatki is

$$L_B/d_s = 163 - 0.64\theta_j \qquad (77)$$

where θ_j = angle between wind and jet, in degrees.

When normalized widths were plotted against the velocity ratio, R, all the data points fell within the scatter of the points for corresponding plots for $\theta_j = 90°$. Therefore, the flame width correlations given by Equations 75 and 76 can be used to compute the lower and upper flame dimensions.

Thermal Radiation from Jet Flames

Radiant energy from a hydrocarbon/air flame basically originates from three sources, viz, carbon dioxide, water vapor, and solid carbon particles or soot. The gaseous sources of radiation are extremely spectral dependent. Carbon dioxide has strong emission bands at wavelengths of 2.7, 4.3, and 15 μm, and water vapor has strong emission bands at wavelengths of 2.7 and 6.3 μm. Since the visible spectrum extends only to 0.7 μm, these sources of radiation are termed nonluminous. The solid carbon particles radiate effectively as blackbodies with a portion of their energy within the visible spectrum, and are quite luminous. The blue, chemically luminescent radiation characteristic of a well-mixed hydrocarbon-air reaction zone contributes almost nothing to the radiant heat transfer from the flame.

This section will initially examine the analytical models available to determine thermal radiation from hydrocarbon flames. First, modeling of the thermal radiation field will be discussed and then the emissive power (or the fraction of combustion energy radiated). A brief discussion of atmospheric transmissivity is presented later.

Radiation field surrounding a turbulent flame: To predict the thermal radiation field of flames, it is customary to model the flame by a point source located at the center of the real flame (that is, halfway along the flame centerline from the nozzle to the tip of the flame). The radiant flux is given by

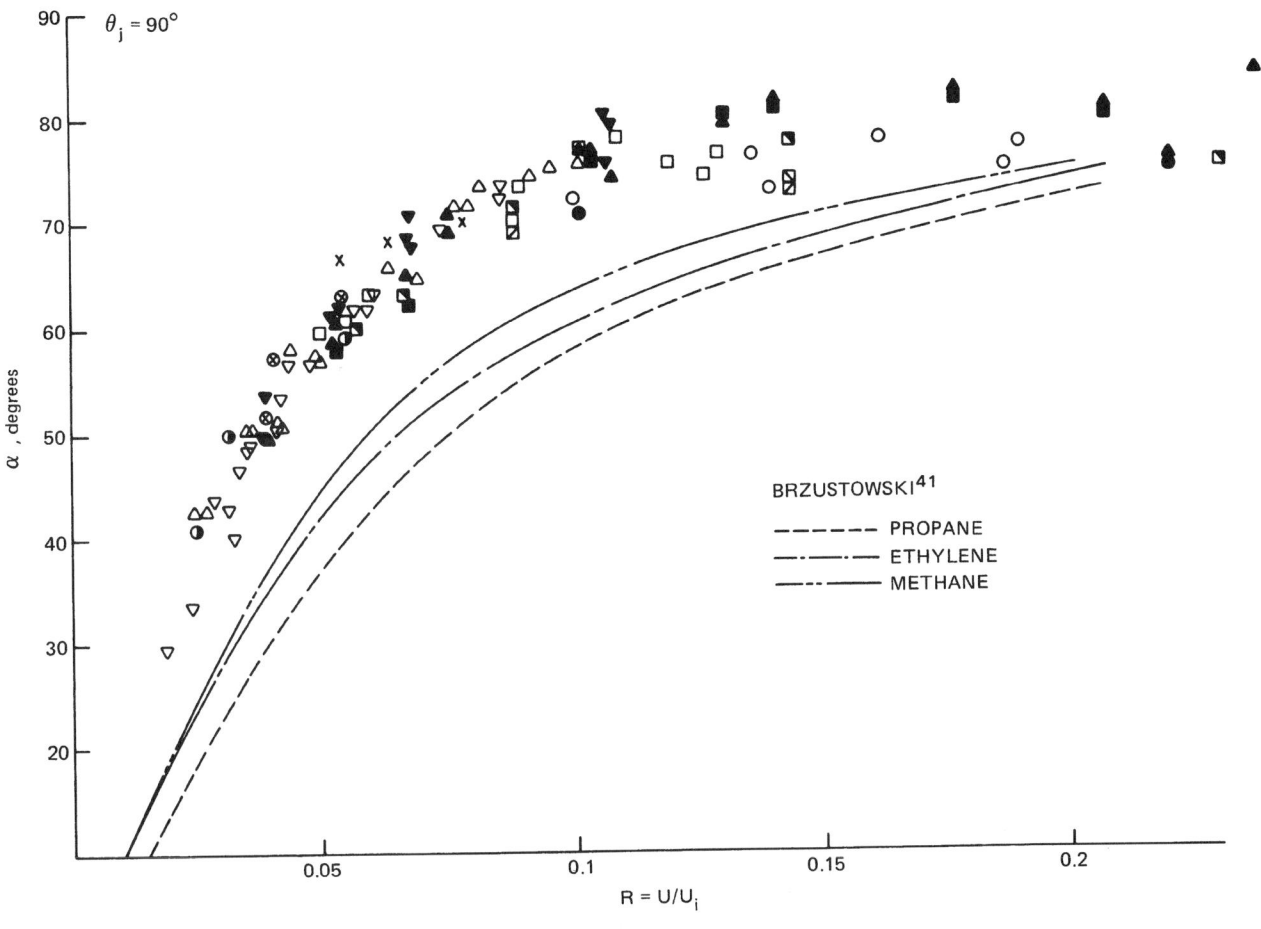

Fig. 3-11.23. Variation of angle of tilt with velocity ratio, R.[44]

$$\dot{q}'' = F\dot{Q}_{rel}/(4\pi D^2) \qquad (78)$$

where

$\dot{q}''$ = incident radiant flux, kW/m^2

D = distance from flame center to observer, m

$\dot{Q}_{rel}$ = energy release rate, kW

F = fraction of combustion energy resulting in radiation

A detailed discussion of the fraction of the combustion energy resulting in radiation is given in the following section. The geometrical aspects of the representation of thermal radiation field given by Equation 78 are surprisingly accurate for flare stacks, even in the case of long wind-blown flames.

The value of $\dot{q}''$ predicted by the point source model applies only when the illuminated surface is normal to the beam and is very small compared to $4\pi D^2$. The direction of the beam is the line of sight from the flame center to the illuminated surface. For any other orientation, the radiant heat flux incident on the surface is reduced by the cosine of the angle between the beam and the normal to the surface. The rate at which a surface receives radiation from a point source, for a general orientation is given by

$$\dot{q}'' = F\dot{Q}_{rel}\cos\theta/(4\pi D^2) \qquad (79)$$

where θ = angle between normal to the surface and line of sight from flame center.

It is customary to set $\theta = 0$ when calculating thermal radiation hazards to personnel because of the difficulty in defining the orientation. To determine the heat flux incident on buildings and structures, it is possible to define a proper value of θ.

Brzustowski[40] considered a flare (in the presence of high crosswind) as a horizontal cylinder, radiating uniformly. He observed that the thermal radiation flux given by Equation 79 is similar to those predicted using a cylindrical source model except at distances very close to the source. Brzustowski[40] has also computed thermal radiation flux for vertical elements parallel and perpendicular to the wind direction and has concluded that the corrected point source model (Equation 78) and the uniform cylinder radiation model essentially yield very similar results.

Brzustowski *et al*[42] conducted a series of laboratory and pilot scale experiments to determine the radiative characteristics of turbulent flares. Laboratory data were taken at jet velocities from 6.8 to 70 m/s, jet Reynolds numbers from 7500 to 94,000, and ratios of crosswind velocity to jet velocity from 0 to 0.113. Commercial grade methane and propane

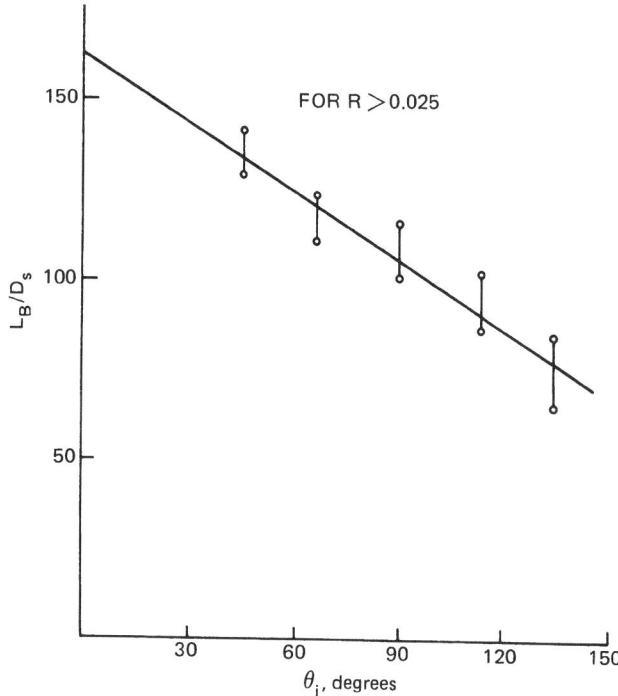

Fig. 3-11.24. *Variation of flame length with jet axis orientation.*[44]

flames were used. Pilot scale studies were performed on an outdoor site with the flare modeled by a vertical 25 mm inner diameter pipe 1.5 m high. In these tests, provision was made for injecting steam into the gas below the tip of the flare. The incident flux at various distances was measured with and without steam.

More recently, Oenbring and Sifferman[46] compared the point source model predictions with full-scale measurements. The full-scale data consisted of radiation measurements at Conoco's Ponca City, OK, refinery facility, and the Gillis, LA gas plant facility. The Gillis flare stack was 16 in. (40 cm) in diameter and 75 ft (23 m) high. The gas velocity ranged from a Mach number of 0.2 to 0.49. All these data indicate that the inverse square law predicts the thermal radiation accurately.

Radiant energy: In the previous section, we discussed the adequacy of using a point source approximation in describing the thermal radiation field surrounding a turbulent flame. In order to predict the incident thermal flux accurately, it is necessary to determine the fraction of total combustion energy resulting in thermal radiation.

Radiant energy from a hydrocarbon-air flame originates principally from carbon dioxide, water vapor, and soot. The carbon dioxide and water vapor radiation are dependent on the spectrum as well as on the partial pressures of these species in the flame. The maximum partial pressures of CO_2 and H_2O in an open air hydrocarbon flame are on the order of 0.1 and 0.2 atmospheres, respectively. The emissivities in these bands can be accurately determined if the partial pressures and the mean beam lengths are known, using Hottel's charts. The carbon particles radiate as blackbodies and, depending on their number density (soot concentration), may be the major contributors to the flame radiation. The overall average carbon particle concentration depends directly on

the quantity of hydrocarbon-air mixture within the flame that is above the fuel mass fraction limit where there are more carbon atoms than oxygen atoms. This depends on the following parameters:

1. The fuel type, which determines that fuel mass fraction limit at which carbon particles begin to form;
2. The entrainment rate of ambient air, which governs the mean fuel mass fraction at any flow distance above the flare exit; and
3. The mixing rate, which governs the distribution about the mean.

Quantitative information on these parameters is not available. Hence, most thermal radiation prediction models tend to ignore the details of the combustion process and concentrate on the overall combustion efficiency, or the fraction of the combustion energy that is radiated to the environment.

Markstein[1] conducted a series of radiation measurements on propane turbulent diffusion flames. The total radiative powers of the flames were determined using wide-angle radiometers. The flow rates varied from 44 to 412 cm^3/s. A collimated beam radiometer was used to measure the radiation characteristics of different parts of the flame. Based on these measurements, Markstein concluded that the thermal radiation from diffusion flames is at a maximum at approximately the center of the flame and tapers off on either side, forming a Gaussian distribution. The total radiative power of the flame was observed to be directly proportional to the total heat release rate. Figure 3-11.25[8] shows that the fraction of combustion energy released in the form of radiation is approximately 20 percent for propane diffusion flames. Burgess and Hertzberg[47] measured the fraction of combustion energy radiated to the surrounding for several

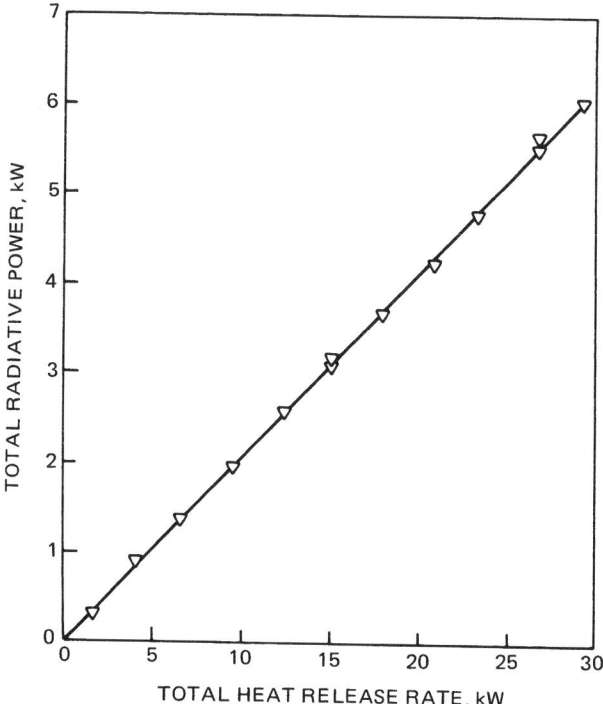

Fig. 3-11.25. *Radiative power for propane diffusion flames.*[8]

TABLE 3-11.8 *Comparison of F Values*

Gas	Chemical Formula	f_s	Tan[48]	Kent[49]	Burgess and Hertzberg[47]	Brzustowski[40]
Hydrogen	H_2	1	—	—	0.17	0.2
Methane (C_1)	CH_4	0.189	0.20	0.19	0.23	0.2
Ethylene (C_2)	C_2H_4	0.170	0.26	0.25	0.36	0.25
Propane (C_3)	C_3H_8	0.176	0.32	0.32	—	0.3
Butane (C_4)	C_4H_{10}	0.175	0.37	0.37	0.30	0.30
C_5 and Higher	—	—	—	—	—	0.40

gaseous fuels. Tan[48] and Kent[49] have also suggested values for the radiated energy for a variety of fuels.

Table 3-11.8 compares the values of F suggested by various investigators. The third column gives the fraction f_s, which represents the fuel mass fraction at which carbon particles begin to form. For any hydrocarbon, C_nH_m, burning in air, the fraction f_s is given by

$$f_s = \frac{12n+m}{12n+m+n/2(137.3)} \qquad (80)$$

The higher the value of f_s, the more fuel rich the hydrogen-oxygen-nitrogen-carbon system must be for the carbon particles to exist in the equilibrium state of the system. Thus, for the same entrainment/mixing history, a gas with a higher value of f_s has less tendency to form solid carbon particles than a gas with a lower value of f_s. Higher values of f_s, therefore, correspond to lower radiation levels.

Examination of Table 3-11.8 shows good qualitative agreement. Butane and propane have similar values of f_s and their F values are comparable. Methane and hydrogen have lower values of F and higher f_s values. But, ethylene has a lower value of f_s and except in one study[47] the F values are also lower. This may be partially due to straight molecular weight corrections applied by Tan[48] and Kent.[49]

Tilted cylinder model: It was pointed out in the previous section that a point source model can be used to determine thermal radiation from large turbulent diffusion flames. The point source model is often recommended because of its simplicity in usage. However, the point source model breaks down at locations close to the flame. This is particularly important when one is evaluating safe separation distances for storage of other hydrocarbon fuels.

An alternate model that can be used to determine the thermal radiation from large diffusion flames is the tilted cylinder model. Details of cylinder view factor calculations are given in Mudan[8] and in the second section of this chapter. The procedure involves approximating the diffusion flame by a cylindrical radiating surface. The dimensions of the tilted cylinder can be determined by the flame geometry correlations discussed in the previous section. The incident thermal radiation at any location is given by the following equation:

$$\dot{q}'' = E\tau F \qquad (81)$$

where

E = surface emissive power of the flame, W/m^2
F = geometric view factor of each segment
τ = atmospheric transmissivity.

The surface emissive power may be approximated by the following equation:

$$E = E_{bb}(1 - e^{-\kappa L}) \qquad (82)$$

where

E_{bb} = equivalent blackbody emissive power
κ = extinction coefficient, m^{-1}
L = effective path length, m.

For most hydrocarbons, the extinction coefficient is between 6 and 10 m^{-1}. Therefore, even modest diameter turbulent diffusion flames (of about 0.5 to 1.0 m) may be assumed to be optically thick. The emissive power will then be equal to the blackbody emissive power.

The blackbody emissive power is given by the following equation:

$$E_{bb} = \sigma T_R^4 \qquad (83)$$

where

σ = Stefan-Boltzmann constant, $W/m^2 \, K^4$
T_R = flame radiation temperature, K.

The mean temperature of diffusion flames and the corresponding emissive powers are shown in Table 3-11.9.[24]

Line source model: Fumarola *et al*[50] suggested a line source model to compute radiation from jet flames. Here, an elemental length of the flame is assumed to radiate similar to a point source. The total incident flux at any observer location is computed by integrating the flux due to an elemental source over the flame length. They compared their results with Brzustowski's[40] and American Petroleum Institute[51] models and observed that their model predicted lower incident fluxes at ground level.

Recently, Galant *et al*[52] proposed a three-dimensional numerical model to estimate the flame geometry and thermal radiation from large diffusion flames. The model considers

TABLE 3-11.9 *Flame Radiation Temperature and Emissive Powers*

Hydrocarbon	Temperature (K)	Emissive Power (kW/m²)
Methane*	1289	157
Ethane	1590	362
Ethylene	1722	498
Propane	1561	336
Isobutane	1554	330
Normal Butane	1612	383
Propylene	1490	279
Isobutylene	1409	223

*Most recent data on LNG fires suggest a flame temperature of 1500 K and a corresponding emissive power of 287 kW/m².

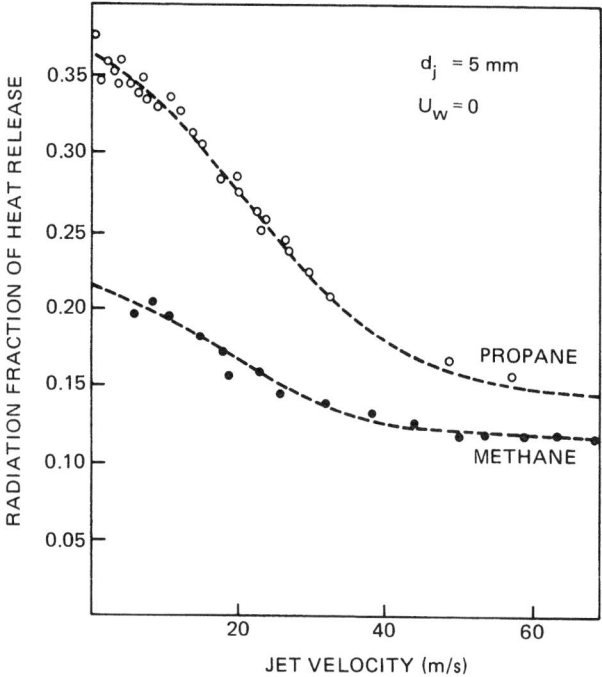

Fig. 3-11.26.   Effect of jet velocity on radiated fraction of combustion energy.

the variation in flow conditions based on a pseudo-stream-function formulation and includes effects of turbulence, combustion, and soot concentration. The model has been validated with field experiments of up to 254 mm (10-in.) diameter methane jets and the agreement between predicted and measured flux is within 15 percent.

Aerodynamic Effects

Aerodynamic effects on radiant energy: The values of F given in the previous section assume that these are properties of the fuel only. They do not take into account the variation of the operating parameters such as stack exit velocity, crosswind velocity, and the presence of steam. However, these parameters have a profound influence on the temperature profiles and affect the fraction of combustion energy radiated.

Brzustowski *et al*[76] conducted a series of laboratory scale tests on the effects of jet velocity and free stream velocity on the fraction of energy radiated from turbulent flames. Figure 3-11.26 shows the effect of jet velocity on radiation in the absence of crosswind for methane and propane flames. Also drawn in the figure are the suggested F factors of Tan[48] and Kent.[49] As can be seen from the figure, the fraction of energy radiated is strongly dependent on jet velocity and decreases with increasing jet velocity. Figure 3-11.27 shows the effect of crosswind velocity on the radiant energy. In general, increasing crosswind velocity appears to increase the fraction of energy radiated.

The significant departures of measured values of F from the values previously published (which do not take into account the aerodynamic effects) can be understood in relation to variation of the detailed temperature profiles in the flames. The underlying explanation deals with the competing processes by which the products of hydrocarbon pyrolysis near the flare stack oxidize directly or form soot which burns in the downstream portion of the flame. Quite obviously, predictions based on the traditional values of F would have overestimated the thermal radiation in all these laboratory scale experiments. Brzustowski *et al*[76] also measured thermal radiation from a full-scale flare. The 0.406 m diameter flare was operating at about 25 percent of the design flaring rate. The best estimates of jet velocity and wind velocity were 28 and 4 m/s, respectively. The flame length was measured to be 25 m and the flame tip was about 10 m above

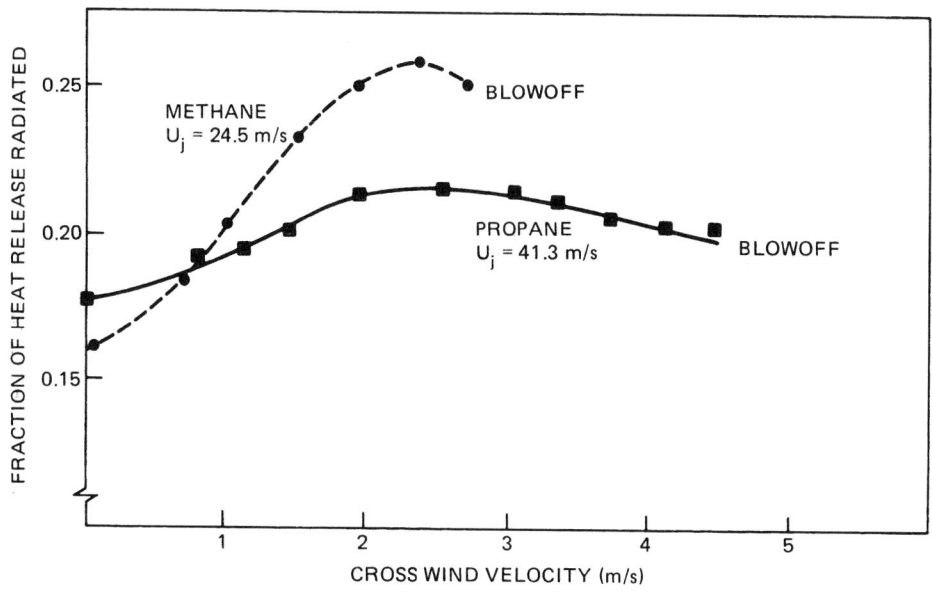

Fig. 3-11.27.   Effect of crosswind velocity on radiated fraction of combustion energy.

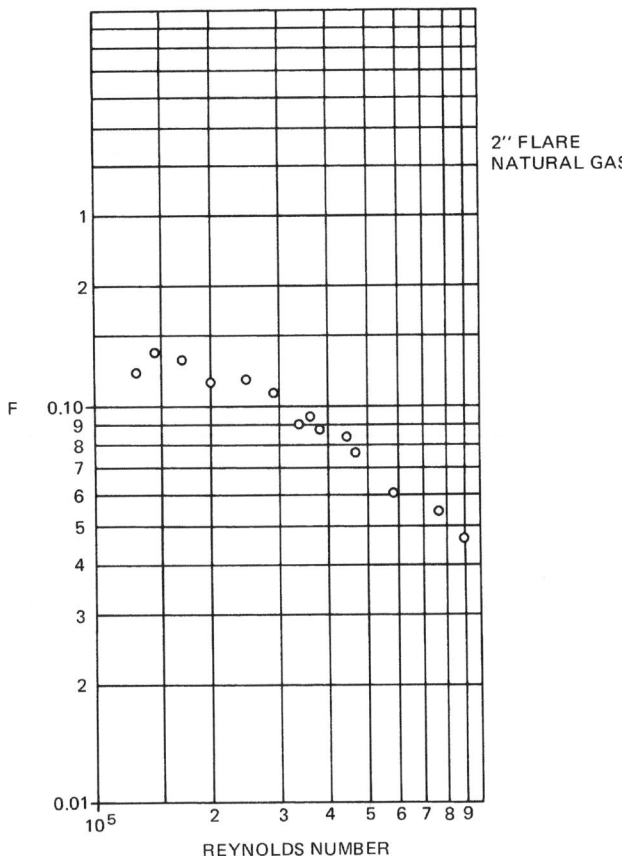

Fig. 3-11.28. *Radiative fraction for natural gas flames full-scale tests.*[53]

the flare tip level. The value of F calculated from the radiation measurements at two ground sections was 0.223, about 30 percent lower than the values predicted using Kent's[49] or Tan's[48] approach.

Figure 3-11.28 shows the fraction of net heat release radiated as a function of the flare Reynolds number for a 5 cm (2 in.) natural gas flare.[53] The Reynolds numbers in these tests are comparable to full-scale Reynolds numbers (of the order of 105 to 106). The F factor shows a significant departure at higher Reynolds numbers from its traditionally assumed value of 0.2.

It is quite evident that the aerodynamics of the flow have a significant effect on the radiation from a large turbulent flame. However, the radiation data described above were obtained on the laboratory scale, and their validity for large flames encountered in an offshore environment cannot be taken for granted.

Aerodynamic effects on flame stability: A jet diffusion flame in still air will lift off the tip of the burner and form a stable lifted flame when the flow velocity through the burner is increased beyond a limiting value known as the lift-off stability limit. If the flow velocity is increased further, the flame is extinguished at some limiting rate known as the "blowout stability" limit.

At the base of a lifted diffusion flame, the local turbulent burning velocity will be equal to the local flow velocity. If the flow rate through the burner is increased, the flow velocity will also increase and the base of the flame will be blown

downstream to a new position where the turbulent burning velocity equals local flow velocity. The flame will blow out when the change in the burning velocity cannot keep up with the flow velocity anywhere in the jet as one moves downstream from the base of the jet flame.

The distance along the burner axis where the mean concentration equals the stoichiometric level is independent of the flow velocity and is given by the following equation:

$$\frac{H}{d_e} = 4\frac{\bar{\theta}_e}{\bar{\theta}_s}\left(\frac{\rho_e}{\rho_a}\right)^{1/2} + 5.8 \tag{84}$$

where

H = height along the jet axis, m
d_e = effective jet diameter, m
$\bar{\theta}_e$ = fuel mass fraction at jet exit
$\bar{\theta}_s$ = stoichiometric fuel mass fraction
ρ_e = jet mixture density, kg/m^3
ρ_a = ambient air density, kg/m^3.

The effective jet diameter is defined as follows:
For subsonic jets:

$$d_e = d_j \text{ for } M < 1$$

For choked flow:

$$d_e = d_j\left[\frac{2 + (\gamma - 1)M^2}{\gamma + 1}\right]^{(\gamma+1)/(\gamma-1)}\frac{1}{\sqrt{M}} \tag{85}$$

where

d_j = jet diameter, m
M = Mach number after expansion to ambient pressure
γ = ratio of specific heats.

All things being equal, the larger the value of H, the more scope there will be for the base of the flame to seek a new stable position as the flow velocity is increased, and therefore it will be more difficult to blow out the flame. Similarly, larger values of burning velocity will lead to larger flow velocities to blow out the flame. The critical velocity at the burner exit for blowout will depend upon the burning velocity, density ratio, and the Reynolds number based on H.

$$\frac{U_e}{S_u} = f\left(R_H, \frac{\rho_e}{\rho_a}\right) \tag{86}$$

where

U_e = critical velocity at jet exit, m/s
S_u = maximum burning velocity, m/s
R_H = Reynolds number given by

$$R_H = HS_u\frac{\mu_e}{\mu_e} \tag{87}$$

where μ_e = dynamic viscosity.

The typical values of the relevant parameters for typical fuels are given in Table 3-11.10.

Kalghatki[54] conducted a systematic study of the blowout stability of jet diffusion flames in still air. The fuel gases used were methane, propane, ethylene, acetylene, and commercial butane. The burner diameters ranged from 0.2 to 12 mm. The "universal" stability limit is given by the following equation:

$$\bar{U}_e = 0.017R_H(1 - 3.5 \times 10^{-6}R_H) \tag{88}$$

TABLE 3-11.10   *Relevant Properties of Hydrocarbon Gases to Determine Blowout Stability*

Gas	Molecular Weight	Dynamic Viscosity at 0°C (micropoises)	Maximum Burning Rate S_u (m/s)	Ratio of Specific Heats	Stoichiometric Air-Fuel Ratio
Methane	16	102.7	0.39	1.31	17.2
Propane	44	74	0.45	1.13	15.7
Ethylene	28	91	0.75	1.255	14.9
Acetylene	26	93.5	1.63	1.25	13.3
Butane	54	80	0.44	1.1	15.7
Hydrogen	2	84	3.06	1.33	34.7

where

$$\overline{U}_e = \frac{U_e}{S_u}\left(\frac{\rho_e}{\rho_a}\right)^{1.5} \qquad (89)$$

The validity of Equation 88 is shown in Figure 3-11.29. It should be noted that Equation 88 is valid only up to a Reynolds number of 100,000.

Calculation Procedure

The calculation of thermal radiation hazards from a turbulent jet flame involves two main steps: geometric characterization of the flame, and estimation of thermal radiation parameters. In this section, a sequential computation procedure is outlined for a vertical jet flame along with a sample calculation.

EXAMPLE:

One inch diameter circular nozzle on butane tank pressurized at 75 psig. Assume a vertical jet in a crosswind; wind speed about 2 m/s; jet exit velocity = 228 m/s.

SOLUTION:

I.   Initialization

1.   Fuel type and properties:

$MW = 54.0$
$\Delta H_C = 45\,000$ kJ/kg
$F = 0.3$
$C_{LFL} = 0.02$

2.   Jet source specifications:

$D_j = 0.0254$ m (1 in.)
$U_j = 228$ m/s
$H = 10$ m
$C_D = 0.61$

3.   Ambient conditions:

$T_a = 293$ K
$U_W = 2$ m/s
$R_H = 50$ percent

Observer: at ground level 15 m away

4.   Miscellaneous:

air density = 1.20 kg/m³
saturated butane density = 2.25 kg/m³

II.   Flame Geometry

1.   Lean limit concentration (dimensionless) $\overline{C}_L$; use Equation 65:

$\overline{C}_L = 4.27$

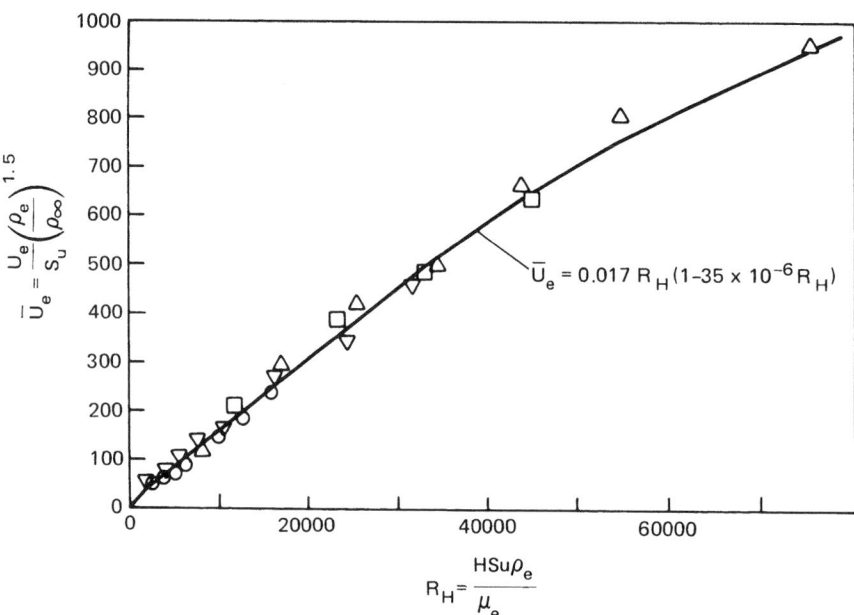

$$\overline{U}_e = 0.017\,R_H(1-35 \times 10^{-6}\,R_H)$$

$$R_H = \frac{HS_u\rho_e}{\mu_e}$$

Fig. 3-11.29.   Universal blowout stability curve for diffusion flames.[54]

2. Modified flame length, $\bar{S}_L$:

   Since $\bar{C}_L > 0.5$, $\bar{S}_L = 1.09$
3. Dimensionless distance along flame axis, $\bar{X}_L$; use Equation 66:

   Since $\bar{S}_L < 2.35$, $\bar{X}_L = 0.1$
4. Dimensionless flame rise, $\bar{Z}_L$; use Equation 67:

   $\bar{Z}_V = 1.08$
5. Flame length, S:

   $S = 4.34$ m
6. Flame tilt, α_B:

   $\alpha_B = 5.3°$ from vertical

III. Thermal Radiation

1. Calculate total energy released, $\dot{Q}_{rel}$:

   $\dot{m} = 0.563$ kg/s (based on choked flow)

   $\dot{Q}_{rel} = 25\ 335$ kW
2. Determine coordinates of the center of flame jet.

   $X_c = 0.2$ m

   $Z_c = 2.15$ m
3. For observer at (15, 0) determine radial distance:

   $r = 19.2$ m
4. Determine incident flux:

   $\dot{q}'' = 1.64$ kW/m^2

UNSTEADY THERMAL RADIATION ANALYSIS

In recent years, liquefied fuel gases having boiling points below normal ambient temperatures have come to be stored and transported in large quantities. Liquefied natural gas (LNG) is stored for peak demand use in many U.S. facilities. It is also transported by sea in bulk carriers suitably designed for cryogenic cargos. Liquefied petroleum gas (LPG) is stored under pressure and is transported by trucks, railroad tank cars, and by sea in bulk carriers. While liquefied hydrogen has been used in limited quantities as a rocket fuel, serious consideration is being given to its use as a fuel for aircraft and possibly highway vehicles. Because volatile fuels are being transported in rapidly increasing volumes, speculation is being devoted to the kinds of accidents that could result from the release of these fuels.

The failure of a container carrying a pressurized cargo will result in the flash evaporation of a portion of the released liquid and the sudden formation of a vapor cloud from the evolved vapors. Upon contact with an ignition source, one of two situations may occur: the generation of a propagating plume flame, or the formation of a fireball. If the vapor puff is ignited immediately after its formation, it may burn as a rising sphere, usually referred to as a "fireball." The rapid combustion of vapor clouds in the form of fireballs has been observed in several accidents involving vehicles carrying liquid propane. Here some of the accidents where a fireball has been reported to be observed are reviewed.

In the General Accounting Office report to the U.S. Congress regarding liquefied energy gases safety, an accident is cited involving a tractor-semitrailer carrying 34 m^3 (9000 gallons) of LPG. About two minutes after the accident, a fireball of about 123 m (135 yards) in diameter was observed. The radiant heat from the fireball burned several people, a house, several other buildings, and some 12 acres of woods.

The National Fire Protection Association has maintained descriptions of several accidents involving LPG where a rising fireball was observed. One such accident happened in Oneonta, NY, where a freight train derailment occurred involving 27 cars, seven of which contained 120 m^3 (33,000 gallons) of liquid propane. Seconds after the derailment, a huge fireball erupted from the area where the tank cars were piled up. It is believed that this fireball was the result of ignition of LP-Gas when one of the tank cars split open. The fireball heated other tanks carrying LPG, which resulted in several BLEVEs (boiling liquid expanding vapor explosions).

One of the most cited fireballs occurred in Crescent City, IL, when a freight train carrying 15 cars derailed, 10 of them containing 130 m^3 (34,000 gallons) of LPG each. One of the derailed tank cars rode up and over the pile and tore a hole in another tank car containing propane, causing the release of gas that produced the first fireball. There were several subsequent explosions which lasted for hours and destroyed 24 individual living quarters and 18 businesses.

The fireballs resulting from such accidents are large—usually of the order of about 100 m in diameter. The duration of the fireball is on the order of a few seconds because of rapid mixing with the surrounding air. During this brief period, a fraction of the combustion energy present in the initial mass of vapor is radiated as thermal energy to the surroundings. The adverse effect of this thermal radiation to population and property depends on the intensity and the duration of the radiation.

If, however, the vapor cloud is allowed to travel with the wind and is ignited at a location away from the source, the resulting vapor fire assumes the form of a propagating plume flame. In both cases, an unsteady diffusion flame is produced. However, the flame geometry is defined by the particular mode of burning. Accordingly, the levels of resulting thermal radiation differ significantly for each mechanism. The unique behavior of these vapor cloud fires is discussed in this chapter—the following sections present an analysis of burning vapor clouds that define a plume fire; a discussion of the formation and burning of a hydrocarbon fireball; and a sample calculation procedure for burning of a vapor cloud in the form of a fireball.

Thermal Radiation from Burning Vapor Clouds

Estimating the thermal radiation field surrounding a burning vapor cloud involves geometric characterization of the cloud, i.e., the time-averaged size of the visible envelope. It also requires estimation of the radiative properties of the fire, i.e., the average emissive power, etc. Finally, the radiant intensity at a given location must be determined. Since the burning behavior of a moving vapor cloud can be best described as unsteady, the standard equations for pool fires do not apply. In the discussion that follows, the flame geometry and effective thermal radiation parameters that characterize a burning vapor cloud are identified.

Given a spill of a volatile, flammable chemical, initially a pool is formed. As the pool vaporizes due to heat transfer from the medium surface (land or water), a vapor cloud is formed above the pool. These vapors are entrained by the ambient wind, and are dispersed in the downwind direction. Two conditions must be met for a burning cloud to be produced; first, there must be an ignition source located away from the spill point; second, the concentration within the vapor cloud must be within the flammability limit range for that material. Assuming these conditions exist, the fire that results is in the form of a propagating plume flame.

Based on experiments with spills of LNG on water, Mudan[55] and Raj *et al*[18] identified three stages of vapor fire development. First, a transient turbulent flame spreads through the cloud. The flame propagates in both upwind and downwind directions. The second stage in the development of a vapor fire is the steady-state propagation toward the liquid pool. At this location, there appears to be a stationary diffusion flame. The third and final stage of burning results in a small pool fire at the source location. Based on limited experimental data on vapor cloud fires, the burning behavior and resultant flame geometry can be analyzed.

Flame propagation velocity: Within a few seconds after ignition, flames tend to spread quickly both upwind and downwind of the ignition source. Flame travel in both directions is consistent if the ignition occurs after the flammable vapor cloud travels over it. The flames are initially contained within the cloud, but subsequently extend in the form of a flame plume above the cloud. This is consistent with premixed burning of the regions in the cloud that are within flammable limits prior to flame arrival, followed by diffusive burning of the richer regions in the cloud. After consuming the flammable vapors downwind of the ignition source, the downwind edge of the flame starts moving toward the spill point. Generally, the flame zone is normal to the wind direction.

During this transient flame growth, an *average* flame propagation velocity with respect to ground can be determined by noting the location of the upwind edge of the flame at various time intervals. The flame speed with respect to gases may be obtained by adding the wind speed to the flame speed with respect to ground. The initial, rapid propagation of the flame in the premixed vapor cloud can also be measured by the same technique.

Wind speed plays a significant role in the vapor cloud propagation. The flame velocity tends to increase with wind speed. Also, an increase in wind velocity increases the dispersion process. The ignition delay is also affected by the wind speed. Clearly, for a fuel-rich vapor cloud, an increase in mixedness will increase the flame propagation velocity. However, if the fuel concentration is well below stoichiometry, a further increase in ignition delay may, in fact, cause a decrease in flame propagation speed.

Mizner and Eyre[56] conducted vapor fire tests with propane spilled on water. The spill rates of propane varied from 2.1 to 5.6 m^3/min and the ignition source was located approximately 130 m from the spill point. The wind speeds varied between 6 and 7 m/s. The flame propagation velocities were measured by locating the upwind edge of the flame as a function of time. Their analysis indicates that the flame propagation velocity (with respect to ground) varies between 3.75 and 4.8 m/s.

In Figure 3-11.30 the measured flame propagation velocities (with respect to unburnt gases) are shown for various wind speeds. The data indicate that there is no significant variation in the flame propagation velocities for methane (LNG) and propane (LPG). The maximum laminar burning velocity for methane is 0.45 m/s, laminar flame speed is 3.5 m/s, and the typical expansion ratio is 7.4. The corresponding properties for propane are 0.52 m/s, 4.0 m/s, and 7.6, respectively. Since these properties are somewhat

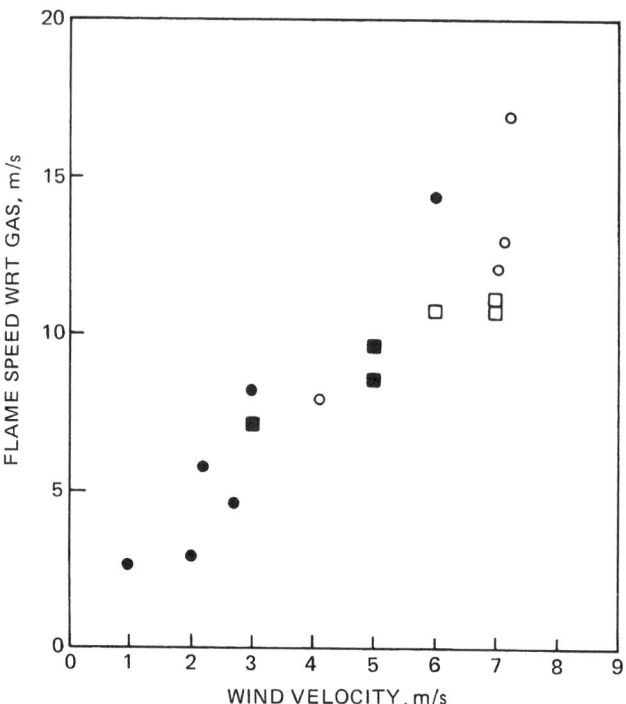

Fig. 3-11.30. Flame propagation velocities for LPG and LNG vapor fires.[55]

similar for methane and propane, it is reasonable to expect the turbulent flame propagation velocities to be similar.

Flame geometry model: Fay and Lewis[57] proposed a model for unsteady burning of unconfined fuel vapor clouds. Based on small-scale experiments with methane, ethane, and propane, and a simple entrainment model, they gave expressions to compute the maximum diameter, height, and duration for complete combustion. The model suggested by Fay and Lewis[57] assumes that the unsteady, turbulent diffusion flame is in the form of a fireball. The correlations given by the authors are validated over a range of small-scale experimental data (up to 200 cm^3) with methane, ethane, and propane gases at room temperature. However, experiments conducted with cold propane vapors ignited in an open environment do not show evidence of a fireball. In fact, the experiments performed by Shell[56] with LNG and LPG, and earlier tests involving LNG vapor fires, fail to confirm Fay and Lewis'[57] proposition that diffusive burning in unconfined vapor clouds takes place only in the form of a fireball.

Raj and Emmons[58] presented a theoretical analysis to estimate the ground level width of a large combustible vapor cloud. The model is based on the principle that the plume above a heat source is characterized by the strength of the heat source. In the case of a burning vapor plume, the rate of burning controls the plume characteristics, and the rate of burning itself is a function of the gas velocity within the plume.

The essential features of the Raj and Emmons[58] model are illustrated in Figure 3-11.31. The assumptions made in the model development are as follows:

1. The geometry of the burning vapor cloud is two dimensional.

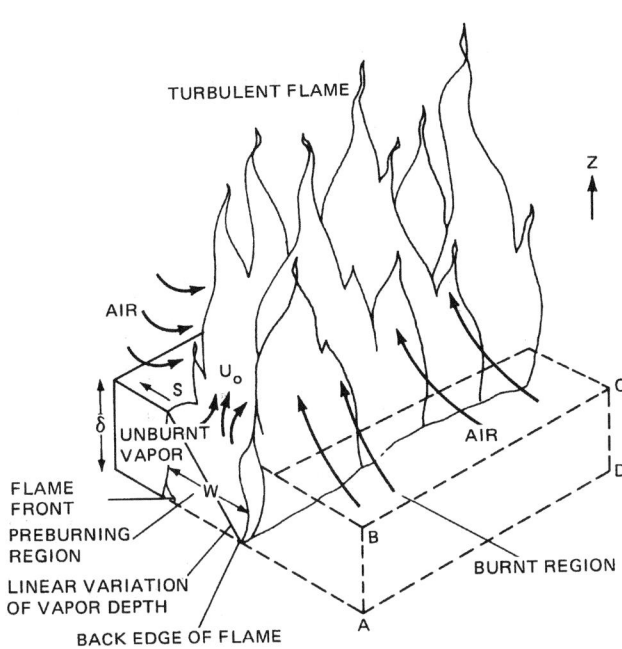

Fig. 3-11.31. *Schematic diagram showing the unconfined burning of a pure flammable vapor cloud.*

2. The burning is controlled by natural convection (buoyancy).
3. The flame propagation velocity with respect to unburnt gases is relatively constant.
4. The depth of the vapor cloud is uniform and is not affected by the flame.
5. The variation of the depth of vapor in the preburning zone is linear.
6. The steady-state turbulent flame correlation for the ratio of visible flame height to base width is valid.

Using experimentally derived values for flame height-to-width ratio and flame propagation velocity, Raj and Emmons[58] gave the following equation to determine the flame width as a function of time:

$$\tau = \left(\frac{F_f}{F}\right)^{1/3} \left\{ \frac{\pi}{3\sqrt{3}} + \frac{2}{3} \ln \left[\frac{\sqrt{\chi^2 + \chi + 1}}{(1 - \chi)} \right] \right.$$

$$\left. - \frac{2}{\sqrt{3}} \tan^{-1} \left[\frac{2}{\sqrt{3}} \left(\chi + \frac{1}{2}\right) \right] \right\} \tag{90}$$

and

$$\chi = \left[\left(\frac{F}{F_f}\right)^{1/3} \xi \right]^{1/2}$$

where

$\xi = W/\delta$
$\tau = 2 St/\delta$
$F_f = S^2/g\delta$ = flame Froude number
F = Froude number = U_0^2/gW
S = flame propagation velocity, m/s
g = acceleration due to gravity, m/s^2
δ = unburnt vapor cloud thickness, m
U_0 = upward velocity at flame base, m/s
W = flame width, m

Raj and Emmons[58] estimated the Froude number based on Steward's[39] data on flame heights for hydrocarbon diffusion flames. The analysis indicates the following relationship for the flame height in a linear heat source.

$$\left(\frac{H}{W}\right)_{stoichiometric} = 4.98 N_{co}^{1/3} \tag{91}$$

where N_{co} is the combustion number defined as follows:

$$N_{co} = F \frac{\rho_0'^2 \omega [r + \omega/\rho_0']^2}{(1 - \omega)^3} \tag{92}$$

where ρ_0' = density ratio = density of vapor at flame base/density of air

r = stoichiometric air/fuel mass ratio
ω = inverse volumetric expansion ratio and is defined as follows:

$$\omega = \frac{1}{(1 + Q_c/rC_p T_a)} \tag{93}$$

where

Q_c = heat of combustion, J/kg
C_p = specific heat, J/kg K
T_a = ambient temperature, K.

The maximum width of the vapor fire is given by the following equation:

$$\xi_\infty = \frac{W_\infty}{\delta} = (F/F_f)^{-1/3} \tag{94}$$

Steward's[39] data indicates that nearly 400 percent excess air is entrained in the fire plume. The typical height-to-width ratios measured in Steward's[39] data range between 5 and 50. Raj and Emmons assumed a height-to-width ratio of 2 for LNG vapor fires and demonstrated that Equation 87 predicts the observed behavior of flame width.

Experimental data on methane and propane vapor fires indicate that the flame width varies as the cloud propagates back to the spill point. Typically, it has been observed that the flame width increases as a function of time until all the flammable vapor is consumed. The width of the fire reduces to the dimension of the pool. The rate of increase in flame width appears to be slightly less than the flame propagation velocity (with respect to ground).

The flame length variation can also be estimated as a function of flame width. It is interesting to note that flame length also increases slightly with time, but the ratio of the flame length to flame width is relatively constant. A plot of flame length-to-width ratios for propane vapor flames is shown in Figure 3-11.32.[55] In general, flame length is about 40 percent of the flame width. The data of Mizner and Eyre[56] show that the typical flame length-to-width ratio varies between 20 percent and 40 percent. It is worth noting that vapor fire flame length-to-width ratios are significantly less than flame height-to-diameter ratios for pool fires.

The time-dependent flame width may be calculated using Equation 87. Figure 3-11.33 compares the computed flame width to measured flame widths as a function of time. Although there is considerable scatter in the data, the overall agreement between predicted and observed growth rates is good.

Thermal radiation: The incident flux received by a stationary observer from a propagating vapor fire is a complex function of several factors. First, the emissive power, which defines the radiative properties of the fire, should be determined.

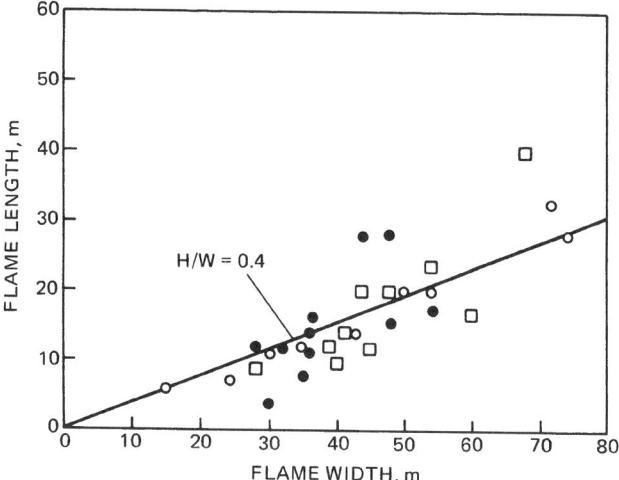

Fig. 3-11.32. *Flame length as a function of flame width.* [55]

Since the duration of a vapor fire is short, and the steady burning period is even shorter, it is difficult to assign an averaging time for determining emissive powers and average incident fluxes. There is, however, a short period over which the thermal radiation appears to have less fluctuation. This duration can be used in determining average incident fluxes and corresponding emissive powers.

Another important geometrical parameter influencing the thermal radiation from vapor fires is the area of the visible flame. If the flame is optically thick, the thermal radiation increases with an increase in the flame surface area. The area increases rapidly immediately following ignition because both the flame width and flame height increase with time. Therefore, the flame area increases approximately like the square of time. Once the flammable vapors are consumed, the flame area decreases rapidly. The incident flux also increases rapidly due to increasing flame area and drops off as the burnout process begins. The distance to the flame surface is also a key parameter. Since the flame is in motion, the distance varies continuously until the cloud approaches the spill points where a pool fire is formed. Coupled with the variation in distance is the changing effect of absorption by the water vapor and carbon dioxide in the atmosphere. And finally, the geometry of the flame relative to the observer influences the view factor, i.e., that portion of flame "seen" by the observer. Therefore, it is evident that the transient nature of the burning process, effected by the changing geometry, severely limits a detailed characterization of thermal radiation from a vapor cloud fire.

For a simple rectangular flame geometry, the centerline horizontal and vertical view factors can be determined using the following equations: [59]

$$F_h = \frac{1}{2\pi}\left(\tan^{-1}\gamma + \frac{X}{\sqrt{1+X^2}}\tan^{-1}\frac{Y}{\sqrt{1+X^2}}\right) \quad (95)$$

$$F_v = \frac{1}{2\pi}\left(\frac{X}{\sqrt{1+X^2}}\tan^{-1}\frac{Y}{\sqrt{1+X^2}} + \frac{Y}{\sqrt{1+Y^2}}\tan^{-1}\frac{X}{\sqrt{1+Y^2}}\right) \quad (96)$$

where X = flame length divided by observer distance, and Y = flame width divided by observer distance.

For asymmetric configurations, trigonometric variations of Equations 95 and 96 can be used to determine the appropriate view factors. The leading edge of the flame (with respect to the observer) may be calculated using the ignition location and the flame propagation velocity. The time-dependent flame width may be calculated using Equation 87. Since flame height is related to flame width, the cross-wind radiation may be calculated using appropriate view factors. The incident thermal flux is given by the following equation:

$$\dot{q}''_{crosswind} = EF_{V,H}\tau \quad (97)$$

where τ represents the atmospheric transmissivity.

A similar procedure may be adopted to calculate incident thermal radiation in the downwind direction. Here the flame is moving away from the observer at flame propagation speed. Therefore, the downwind incident flux will be at its maximum at the time of ignition (assuming ignition occurs at the downwind edge of the cloud) and will decrease rapidly.

Because of the complex phenomena of a vapor fire, a simple calculation procedure cannot be developed to determine the incident thermal flux. A numerical program based on equations described in this section may be used to determine the time-dependent thermal flux.

Thermal Radiation from Hydrocarbon Fireballs

The thermal flux model for fireballs includes empirical correlations from experiments involving pure methane, ethane, and propane vapor. An unconfined volume of an initially pressurized vapor (such as propane), if ignited, can burn as an unsteady turbulent diffusion flame in which buoyancy is induced by the burning process. The ambient air is entrained at the edges of the burning cloud and thus increases the size of the cloud and the volume of the combustion process. The cloud also rises en masse because of

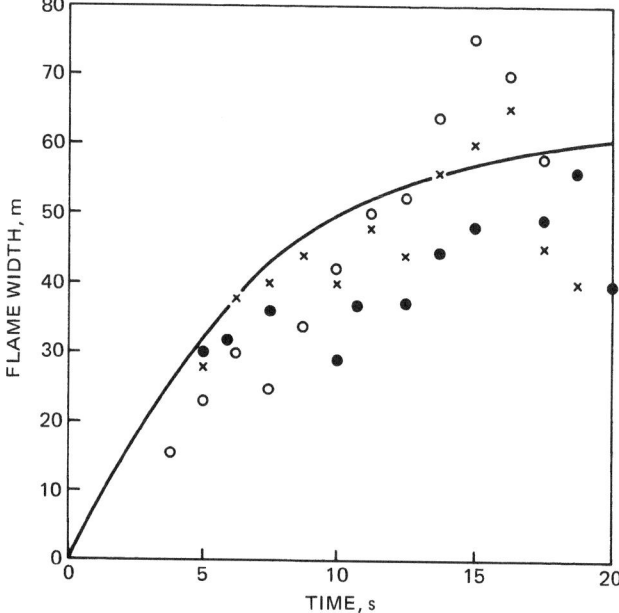

Fig. 3-11.33. *Comparison of predicted and measured flame widths for LPG vapor fires.* [55]

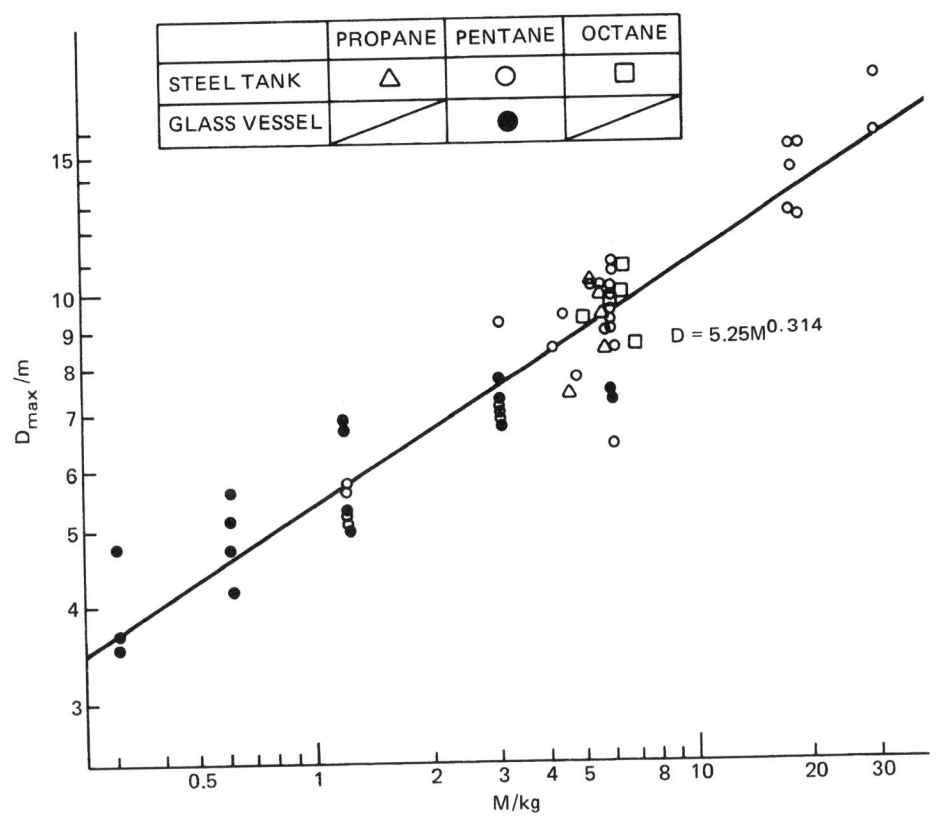

	PROPANE	PENTANE	OCTANE
STEEL TANK	△	○	□
GLASS VESSEL		●	

$D = 5.25M^{0.314}$

Fig. 3-11.34. Maximum fireball diameter.[62]

buoyancy-induced forces. Quantitative research on the behavior of propane fireballs exists for a limited range of laboratory scale experiments[57,60] and care must be taken in the application of these results for larger sizes. However, several of the relationships for fireball behavior which were developed from small-scale experiments have been compared to actual large-scale accidents with favorable results. These will be discussed in the following section.

Fireball geometry: A series of experiments which were conducted to measure the motion, growth, and thermal radiation from fireballs involving pure vapor samples of methane, ethane, and propane over a range of initial fuel volumes (20 to 200 cm^3) suggest geometric and dynamic scaling relationships that are appropriate for larger sizes.[57] The first of these involve the maximum diameter, height of rise, and total duration of burning of the fireball according to the following:

$$\left.\begin{array}{l} D = 7.71V_f^{1/3} \\ Z_p = 12.7V_f^{1/3} \\ t_p = 2.8V_f^{1/6} \end{array}\right\} \qquad (98)$$

where

D = maximum diameter (m)
Z_p = height of maximum visible flame (m)
t_p = duration of the fireball (s)
V_f = initial vapor volume of the fuel (m^3).

Gayle and Bransford[61] conducted several large-scale tests with propellants to determine the size and duration of

fireballs. Hasegawa and Sato[62] performed a series of fireball experiments involving up to 30 kg of propane, *n*-pentane, and *n*-octane. Figure 3-11.34 shows the maximum fireball diameter as a function of released mass.

The development of the propane fireball was investigated as a function of time by Fay and Lewis,[57] and the fireball diameter, D, and height of the fireball center, Z, are given as

$$D = 0.806\eta^{1.12}V_f^{1/3} \qquad (99)$$

and

$$Z = 1.97\eta^{0.86}V_f^{1/3} \qquad (100)$$

where η is a dimensionless time since ignition, defined by the following:

$$\eta = \sqrt{g}t/V_f^{1/6} \qquad (101)$$

where

g = gravitational acceleration (9.8 m/s^2)
t = real time since ignition (s).

The dimensionless time is related to the buoyancy of the system and is a measure of the rate of vertical acceleration of the combustion products. These relationships will be used in a later part of this analysis.

The thermal radiation measurements were nondimensionalized with respect to initial fuel volume and expressed empirically as a function of dimensionless time on a single

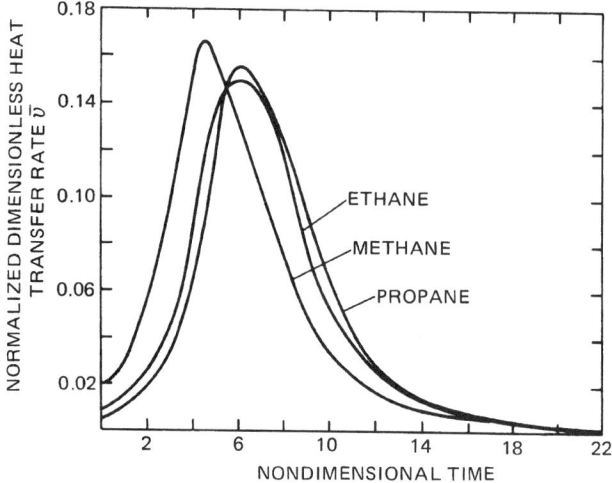

Fig. 3-11.35. *Normalized heat transfer rate.*[64]

curve.[62] This is shown in Figure 3-11.35 for the three fuel types. This suggests a dimensional scaling relationship that is appropriate for larger sizes. The normalized dimensionless heat transfer rate of Figure 3-11.35 can be used to estimate the actual heat transfer rate, q_d, as a function of time at any radial location in the vicinity of the fireball according to the following:

$$\dot{q}''(\eta) = \bar{v} f \tau g^{1/2} \rho_f h_f V_f^{5/6} / \{4\pi[d^2 + (Z - h)^2]\} \quad (102)$$

where

f = fraction of combustion energy which is radiated to the environment
τ = atmospheric transmissivity
ρ_f = density of vapor at ambient conditions (kg/m^3)
h_f = heat of combustion of vapor (kJ/kg)
d = horizontal distance to location of interest (m)
h = height above ground level of location of interest (m).

The empirically determined dimensionless heat transfer rate can be explained in terms of the physics of the fireball growth model. This is done by assuming a buoyant thermal fireball model in which the rate of entrainment of air at the burning edges is proportional to the rise in velocity of the fireball products[64] according to the following:

$$m(t) = \rho_a \pi D^2 \beta \frac{dZ}{dt} \quad (103)$$

where

m = air entrainment rate into the fireball at time t (kg/s)
D = fireball diameter at time t (m)
Z = fireball height as a function of time (m)
ρ = dimensionless entrainment constant
ρ_a = density of air (kg/m^3).

If the assumption is made that the air mixes with the fuel and burns in stoichiometric proportions as it is entrained (which appears reasonable) and that some fraction of the combustion energy is radiated to the surroundings, then the radiant heat transfer rate must be proportional to the entrainment rate ($\bar{v} \propto m$), until the time when all the fuel is consumed. This can be shown to be approximately true by first examining the dependence of m as a function of t. From

Equation 99, D goes by the 1.12 power of time; and from Equation 100, the derivative of Z with respect to time goes by the -0.14 power of time. Combining these in Equation 103, the rate of entrainment of air increases by the 2.10 power with time ($m \sim vt^{2.1}$) during the period of combustion. This can be compared with the heat transfer rate in Figure 3-11.35. From experimental observation, the combustion time lasts through dimensionless time $\eta = 6$.

The denominator of Equation 102 represents the surface area of the sphere, which is located at the center of the fireball and passes through the location of interest as shown in Figure 3-11.36. This is simply an inverse square relationship, assuming isotropic radiation and no reflectivity of the ground. The surface element perpendicular to the line of sight of the fireball receives the maximum radiation. In order to obtain a conservative estimate of thermal radiation hazards, it is assumed that the surface element is always normal to the fireball line of sight. This element is denoted by the view angle with respect to the horizontal and changes as the height of the fireball increases.

The fraction of the enthalpy of combustion, which is radiated as thermal energy to the surroundings, f, has been measured for steady turbulent flames of propane[1] at about 20 percent. It is reasonable to assume that the radiant heat loss in the fireball mode will not be greatly different, and 20 percent is used as a constant in the present model but for one exception. For extremely large propane fireballs, the use of $f = 20$ percent may overpredict the surface heat flux of the fireball at the beginning of the burn cycle when compared with a blackbody radiator at the mean temperature of a propane flame of 1561 K.[24] For these cases, the blackbody radiator value will be used as a maximum, and the value of f will be scaled down as required.

The atmospheric transmissivity accounts for the attenuation by atmospheric water vapor of radiation from a fire of any arbitrary temperature and composition.

Finally, the time-dependent heat flux at any horizontal and vertical location from the point at which the fireball is formed can be estimated from Equation 98. In Figure 3-11.37, the incident thermal flux is plotted as a function of time since ignition for a 100 te propane fireball at two horizontal ground level distances. The peak of the curve takes place at the time when the denominator, or equivalent radiation area, of Equation 102 begins increasing faster than the dimensionless heat transfer rate in the numerator. After this

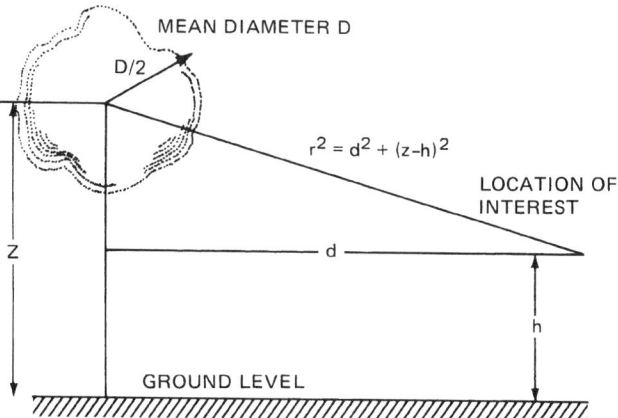

Fig. 3-11.36. *Schematic diagram of fireball radiation model.*

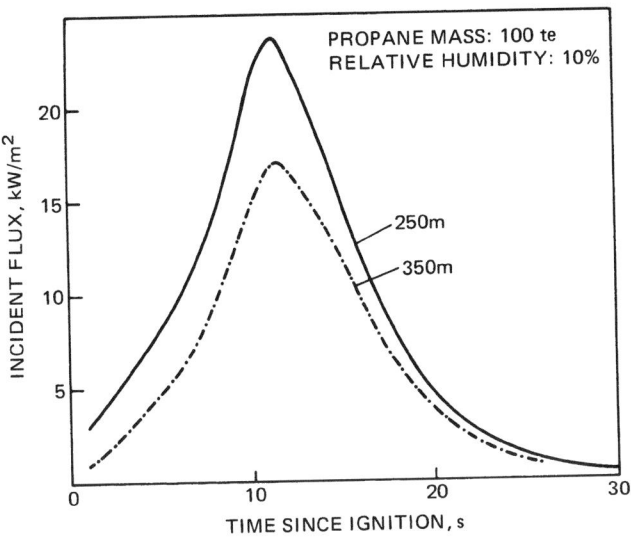

Fig. 3-11.37. *Time-dependent incident flux for a 100 te propane fireball.*

time, an inflection in the curve occurs when the peak of the dimensionless heat transfer rate is reached, and declines steadily beyond that time.

The fuel mass or amount of propane involved in the fireball refers to the vapor fraction of propane accidentally released. For example, if a pressurized bulk container with 100 000 kg of propane at an ambient temperature of 21°C (70°F) were ruptured, the liquid contents of the tank would be released instantaneously and approximately 36 percent would flash immediately to vapor; the rest would remain as aerosols and liquid droplets. A similar release with butane would result in approximately 17 percent flash, and therefore a lesser fraction of entrained liquid aerosols. The general rule of thumb suggested for mass of fuel for fireball computation is as follows:

1. If the adiabatic flash fraction exceeds 30 percent, it should be assumed that the entire mass of fuel is contained in the vapor cloud.
2. If the adiabatic flash fraction is less than 15 percent, it may be assumed that the remaining liquid will burn in the form of a pool fire.
3. If the flash fraction ranges from 15 to 30 percent, a linear interpolation is assumed for the liquid fraction.

It is recognized that this procedure is somewhat arbitrary. Since the mass of fuel alone determines the fireball size, duration, and hence the hazard, a more refined computation scheme would be desirable.

Skin response to thermal flux: The previous section described a method for estimating the time response of the thermal flux from a pressurized vapor release. A consistent criterion will now be developed for approximating the response of human skin to an energy pulse.

The effect of radiant heat on human skin has been adequately documented for the case of exposure to a steady-state flux.[65,66] In that situation, the only important parameter is the duration of exposure to a specified constant flux. The present fireball model, however, predicts a rapidly changing thermal flux where the total duration of the fire is only a few seconds, and the definition of a hazardous expo-

sure zone for this case requires an examination of the temperature response of human skin.

The effect of a thermal dose on human skin is to bring about a thermal denaturation of skin proteins in the epidermis, or outer layer, and to destroy the cell structure and collagen protein in the underlying dermis layer. The severity of the burn depends on the extent of destruction of the tissues. A "critical energy model" was developed to correlate the results of experimental data on pig skin, and it has since been verified on human tissue.[67]

The critical energy model states that the severity of the burn depends upon the amount of energy that is absorbed by the skin after the surface temperature reaches 55°C. If the amount of excess energy is 1 cal/cm^2, pain or mild second-degree burns will be experienced. For an additional exposure of more than 2 cal/cm^2, a blister or severe second-degree burn will become evident. Finally, for an exposure of greater than 3.88 cal/cm^2, severe third-degree burns will result in permanent injury.

Human skin, and the underlying tissue and blood flow, presents a complex thermal system if modeled completely; thus, for the present purposes, the situation will be simplified. The human tissue will be modeled as a semi-infinite slab with the normal to the surface directed at the center of the fireball. No attempt has been made to directly model the internal blood flow or the external perspiration; the result of these simplifications will be a conservative estimation for risk assessment purposes.

A semi-infinite slab initially at uniform temperature, T_0, is subjected to a thermal flux, F_0. The temperature profile in the semi-infinite slab will be assumed linear with a slope equal to the slope of the actual temperature profile at the surface. The depth of penetration, L, is defined as the depth at which the temperature equals the initial uniform temperature. The surface temperature of a semi-infinite slab, T_s, subjected to a uniform flux is given by[68]

$$T_s = T_0 + \frac{2F_0}{K}\left(\frac{\alpha t}{\pi}\right)^{1/2} \quad (104)$$

where

K = thermal conductivity of the medium (W/mK)
α = thermal diffusivity (m^2/s)
t = time (s).

The slope of the temperature profile at the surface is

$$\frac{dT}{dx}\bigg|_{x=0} = -F_0/K \quad (105)$$

Equations 104 and 105 completely define the location of the linear temperature profile at any time and the depth of penetration, L, is

$$L = 2\sqrt{\alpha t}/\pi \quad (106)$$

The application of the energy conservation equation to the control volume surrounding the heated skin results in the following:

$$q(t) = \frac{d}{dt}\left[\frac{T(t)-T_0}{2}L(t)\rho C_p\right] \quad (107)$$

where

$q(t)$ = energy flux at surface (kW/m^2)
ρ = density of skin (kg/m^3)
C_p = specific heat of skin (kJ/kgK)

TABLE 3-11.11 *Properties of Skin*

	Epidermis	Dermis	Average*
Thickness (m)	$8 \cdot 10^{-5}$	$2 \cdot 10^{-3}$	$2.08 \cdot 10^{-3}$
Conductivity (kJ/m s °C)	$2.092 \cdot 10^{-4}$	$3.69 \cdot 10^{-4}$	$3.63 \cdot 10^{-4}$
Specific Heat (kJ/kg °C)	3.60	3.22	3.23
Density (kg/m³)	1,200	1,200	1,200
Temperature (°C)	37	37	37

*Volumetric average as determined by the thickness of the skin layer.

If the time steps are sufficiently small with respect to the changing energy flux, the integral of Equation 107 can be approximated by the trapezoidal rule and the temperature at any time is determined, using Equation 106, by

$$T(t_2) = T_0 + [T(t_1) - T_0]\sqrt{\frac{t_1}{t_2}} + \frac{[\bar{q}(t_2) + \bar{q}(t_1)](t_2 - t_1)}{2\sqrt{K\rho C_p t_2/\pi}} \quad (108)$$

The solution of Equation 108 requires data on properties of the human skin. Table 3-11.11 shows the physical and thermal properties of the epidermis and the dermis layer. A weighted average determined by the skin layer thickness should be used in the analysis.

The time-dependent incident thermal energy flux values determined in a previous section may be applied to Equation 108 to determine the response of the human skin. The initial response of the human skin is indicated by an increase in the temperature. Once the skin temperature exceeds 55°C, the additional energy flux is integrated to determine the extent of skin damage. An example of this analysis is shown in Figure 3-11.38. The increase in skin temperature as a function of time is shown for two different observer locations for a 100 te propane fireball. The amount of energy absorbed is shown by the two curves on the right side of Figure 3-11.38. For an observer 250 m from the initial location of the fireball, the skin temperature increases to 55°C in about 4 s. An additional energy of about 5.5 cal/cm² is absorbed in the next 16 s, resulting in severe third-degree burns and permanent tissue damage. An observer at a distance of 350 m is likely to suffer minor third-degree burns due to thermal radiation from the fireball.

Calculation Procedure

It was pointed out in the section on thermal radiation that a simple calculation procedure for a vapor fire is not available because of rapid variation of view factors. The skin response to an unsteady-state thermal radiation also involves numerical integration of the incident flux. Therefore, a simple step-by-step calculation procedure may be developed only to determine incident flux due to a hydrocarbon fireball.

Step-by-step calculation procedure: The input parameters are as follows:

M_f = fuel mass (kg)
ρ_f = fuel vapor density (kg/m³)
h_f = heat of combustion (kJ/kg)
f = fraction of radiated energy
RH = relative humidity (percent)
T_a = ambient temperature (K)
T_f = flame temperature (K)
x = observer location—distance from center (m)
h = observer location—height above ground (m)

Step 1: Determine the volume of fuel

$$V_f = M_f/\rho_f$$

Step 2: Determine the maximum height, diameter, and the time for complete combustion using the following equations:

$$D_{max} = 7.71 V_f^{1/3}$$
$$Z_p = 12.7 V_f^{1/3}$$
$$t_p = 2.8 V_f^{1/3}$$

Step 3: For times less than t_p, the radiation from the fireball may be calculated. First, determine a nondimensional time using the following equation:

$$\eta = \sqrt{g}t/V_f^{1/6}$$

Step 4: Determine the diameter and the height of fireball for the chosen time

$$D = 0.806\eta^{1.12}V_f^{1/3} + d_0$$
$$Z = 1.97\eta^{0.86}V_f^{1/3} + d_0/2$$

where d_0 is the initial diameter based on fuel volume.

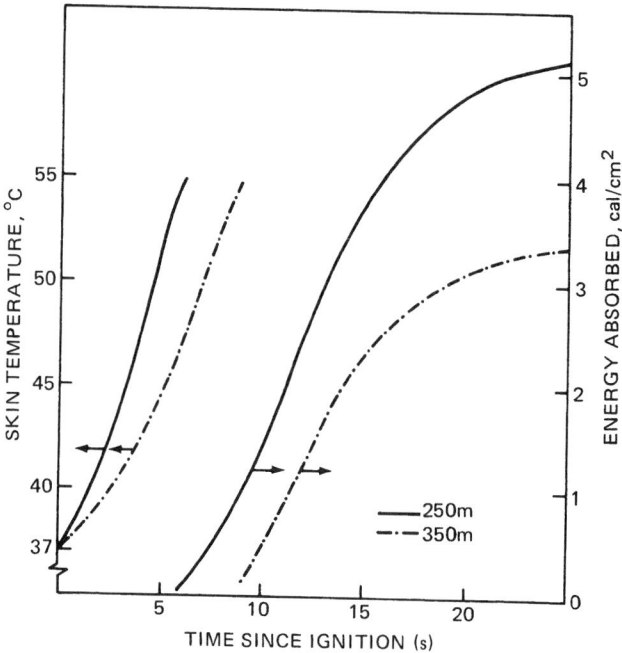

Fig. 3-11.38. Thermal radiation damage due to a 100 te propane fireball.

Step 5: If the observer is within or directly below the fireball, further calculations are not applicable. If the observer is outside the fireball, the incident flux may be computed.

Step 6: From Figure 3-11.35, determine the normalized heat transfer rate. Also determine the distance between the fireball and the observer.

$$r = [x^2 + (Z - h)^2]^{1/2}$$

Step 7: For the given relative humidity and pathlength determined in Step 6, calculate the atmospheric transmissivity. (This procedure is similar to the transmissivity calculations discussed earlier and will not be repeated here.)

Step 8: Determine the incident flux using the following equation:

$$\dot{q}'' = \frac{\bar{\nu} f \tau g^{1/2} \rho_f h_f \cdot V_f^{5/6}}{4\pi r^2}$$

Illustration of the Calculation Procedure

Sample Input

    Fuel mass = 10 000 kg
    Vapor density (butane) = 2.41 kg/m^3
    Heat of combustion = 45 720 kJ/kg
    Fraction of energy radiated = 0.2
    Relative humidity = 1 percent
    Ambient temperature = 288 K
    Flame temperature = 1612 K
    Observer location —distance from center = 100 m
                      —height above ground = 2 m

Calculations

    Fuel volume = 4149.4 m^3
    Maximum dimensions —diameter = 123.9 m
                      —height = 204.1 m
                      —duration = 11.2 s.

1. Assume that the radiation is to be calculated at 5 s. The nondimensional time is equal to 3.91.
2. The initial diameter of the fireball is equal to 19.9 m. The diameter and height at 5 s are 79.5 and 112.2 m, respectively.
3. Since the observer is outside the fireball, the thermal radiation calculations may be performed.
4. The normalized heat transfer rate (assuming butane properties are similar to propane) is equal to 0.06. For sake of simplicity, the atmospheric transmissivity is assumed to be unity. The flux is 15.4 kW/m^2.

THERMAL RADIATION HAZARDS

Thermal radiation from hydrocarbon fires may pose significant hazards to both personnel and property. Hazards to personnel result from exposure to intense thermal radiation, causing severe burn injury. Thermal radiation can also endanger various types of structural components and process equipment that may be exposed to the flame. Temperature response of metal structures becomes important in determining and providing adequate fire protection systems. In the following two subsections the criteria for thermal radiation hazard assessment for determining safe separation distances for personnel and the models to determine temperature response of structural components are discussed.

Criteria for Thermal Radiation Hazard Assessment

The thermal radiation from a fire may cause burns on bare skin if the intensity of radiation is sufficiently large and if the exposure is of sufficient duration. Skin burns occur over a continuous range of severity, starting from a burn so minor that the skin is barely damaged and extending through complete destruction of all skin layers to the underlying tissues or bone. Several classifications of skin burn severity have been proposed, each depending on the degree of skin damage. The most familiar classification is to divide skin burns into three degrees. Even with these three degrees, there are several recognized sublevels. For present purposes, the following levels of burn severity can be used, with the attached simple descriptions:

1. First Degree: The mildest level of skin burn, characterized by erythema (reddening), but no formation of blisters. The mildest of first-degree burns are not particularly painful and commonly present no medical problem. They may, in fact, not even cause symptoms other than a mild impression of warmth. More severe first-degree burns will produce some pain, but no permanent damage. Flaking or scaling of the skin will occur several days after exposure because of damage to the outer skin layer.
2. Second Degree: An intermediate level of skin burn, characterized by formation of blisters. Blister depth may be shallow, with only the surface layers of the skin damaged, resulting in a moderate second-degree burn, or with nearly the full depth of the skin destroyed, i.e., a severe second-degree burn.
3. Third Degree: Deep burns, characterized by destruction of all skin layers. The underlying tissue may also be destroyed.

The medical problems of burns covering large areas of the body include the severe loss of fluid and the extreme potential for infection following the loss of a large portion of the protective layers of the skin. Survival of healthy adults and teenagers can normally be expected if less than 20 percent of the body surface has second- and third-degree burns (percent body areas: head 7 percent, arms 14 percent, and hands 5 percent). Survivability decreases rapidly for persons who have more than 50 percent of body surface covered by full-thickness burns, and even with intensive medical care, it is unlikely to find survivors.

Pain and tissue damage are both related to heating of the skin. The skin consists of two main layers: the epidermis, which is a thin (0.05 to 0.1 mm) outer layer, and the derma, which is an inner layer (1 to 2 mm thick). Since the skin is a complex system, there is no perfect mathematical model available for describing its response to heating under all conditions. However, simplified models can be used as an aid in predicting skin response to heating. The simplest analysis begins with the assumption that the skin and underlying tissue behave as a one-dimensional medium with constant thermal properties. Heat transfer is due to conduction only, and the temperature field anywhere in the medium is given by the following equation:[66]

$$\frac{\delta T}{\delta t} = a^2 \frac{\delta^2 T}{\delta X^2} \tag{109}$$

where

T = temperature at time, t, and distance, x, below the skin surface
a^2 = $k/\rho c$ = thermal diffusivity, m²/s
k = thermal conductivity, w/m·K
ρ = density, kg/m³
c = specific heat, J/kg·K

The solution to Equation 109 depends on the initial and boundary conditions. If it is assumed that the skin and the underlying tissue are at a uniform temperature, T_0, and that at time zero a constant heat flux of $\dot{q}''$ is applied, the initial and boundary conditions are as follows:

$$\left. \begin{array}{l} t \leq 0; \quad x \geq 0 \quad T = T_0 \\ t > 0; \quad x = 0 \quad \dfrac{\delta T}{\delta x} = -\dfrac{\dot{q}''}{k} \end{array} \right\} \quad (110)$$

Equation 110 indicates that the heat flux, $\dot{q}''$, produces a temperature gradient in the epidermis which is inversely proportional to the value of thermal conductivity. The solution to Equation 109 with the initial and boundary conditions indicated in Equation 110 is as follows:

$$T = \frac{\dot{q}''}{k}\left[\frac{2a\sqrt{t}}{\sqrt{\pi}}\exp\left(-\frac{x^2}{4a^2t}\right) - x \, erfc\left(\frac{x}{2a\sqrt{t}}\right)\right] \quad (111)$$

The *erfc* is the complementary error function. The increase in skin surface temperature ($x = 0$) is given by the following equation:

$$T_s - T_0 = \frac{2\dot{q}''\sqrt{t}}{\sqrt{\pi k \rho c}} \quad (112)$$

The data of Buettner[66] obtained by having volunteers expose their forearms to varying degrees of thermal radiation indicate that the threshold pain is felt by human beings when the average temperature of 0.1 mm depth of skin is increased to about 45°C. The data indicate that the time required for pain can be correlated with the intensity of radiation by the following equation:

$$t_p = [35/\dot{q}'']^{1.33} \quad (113)$$

where

t_p = time required for pain, s
$\dot{q}''$ = incident thermal radiation, kW/m².

The deviation of Equation 113 from Equation 112 is attributed to slightly higher surface temperature for higher thermal flux for pain threshold.

In Figure 3-11.39 the time required to cause pain is shown as a function of the incident thermal flux. The sources of data used in Figure 3-11.39 include "pricking" and "threshold" pain.[66,69–71] All the sources of data show general agreement on the time required for pain at low fluxes. No pain was shown, regardless of the exposure duration, for thermal fluxes below 1.7 kW/m² (the solar constant is about 1 kW/m² on a clear summer day). At higher fluxes, the time required for pain diverges for the several studies. However, Equation 113 appears to predict the time required for pain with reasonable accuracy.

When the skin surface temperature reaches about 55°C, blistering of the skin occurs. Mehta et al[67] determined that the severity of the burn depends on the energy absorbed after the skin temperature has reached a temperature of 55°C. If the amount of energy absorbed is 41.8 kJ/m², pain or mild

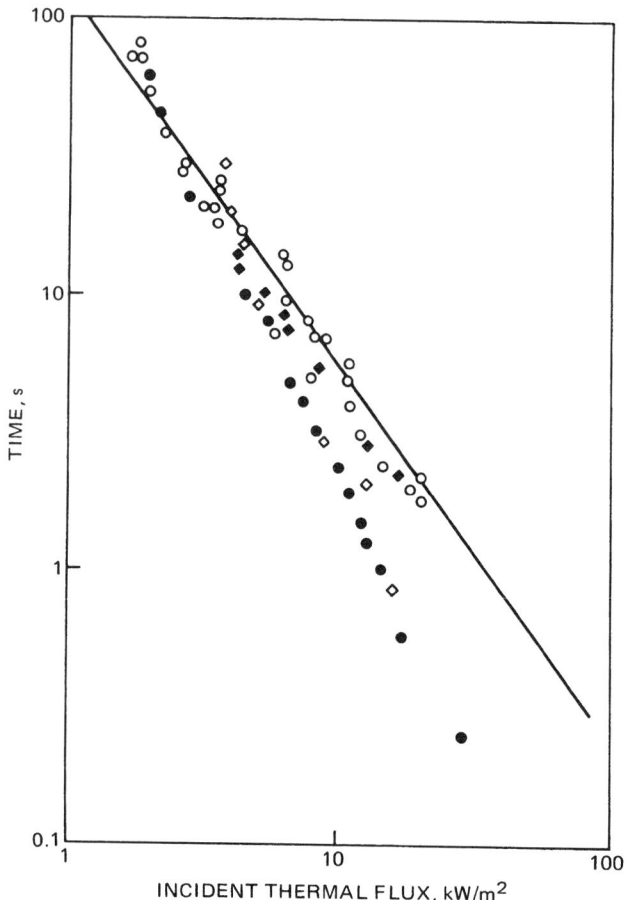

Fig. 3-11.39. *Time required for pain due to exposure to thermal radiation.*[8]

second-degree burns will be experienced. For an additional exposure of 83.6 kJ/m², a blister or severe second-degree burn will become evident. Finally, for an exposure of greater than 162.2 kJ/m², severe third-degree burns will result and the skin tissue will be permanently damaged. Figure 3-11.40 shows the time required for blister formation for human skin as measured by Stoll and Greene[71] and Mehta et al.[67] The times required are quite similar for threshold blisters and full blisters, showing that there is little practical difference between the two. Mixter[73,74] has compared the degree of burns to human skin with that of pigs and found a good correlation. Mixter's[73] data for second-degree burns to white pigs are also shown in Figure 3-11.40. The difference in Mixter's data and other sources of data is attributed primarily to the source of thermal radiation. Stoll and Greene used a 1000 W projection lamp and Mixter used a carbon arc source.

The data shown in Figures 3-11.39 and 3-11.40 are useful for estimating the time for threshold pain and how rapidly the human skin burns will result at various levels of radiant exposure. These data, however, do not aid in determining the radiant flux levels at which fatalities may be expected. The only source of data on large-scale deaths from thermal radiation are analyses of fatalities from nuclear weapons. Since the exposure times are typically very short, interpretation of these data is somewhat subjective.

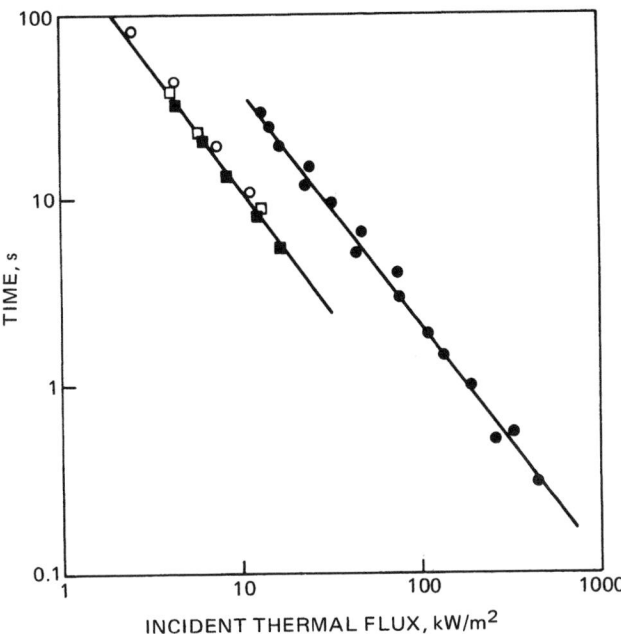

Fig. 3-11.40. *Time required for skin burns by thermal radiation; data sources.* [8]

Eisenberg *et al*[72] analyzed the data on the relation between thermal radiation intensity and burn injury for nuclear explosions at different yields. The results of their analysis are shown in Figure 3-11.41 for significant injury threshold, 1 percent lethality, near 50 percent lethality, and near 100 percent lethality for various incident radiation intensities. Also shown in Figure 3-11.41 are the second-degree burn data collected by Mixter.[73] These data correspond very closely to the significant injury threshold.

The *United States Federal Safety Standards for Liquefied Natural Gas Facilities* (49 *CFR*, Part 193, 1980) suggest an acceptable level of 5 kW/m^2 for direct exposure of human beings. At this incident flux, exposure time on bare skin before unbearable pain is about 13 seconds and second-degree burns may occur in about 40 seconds. This level can, therefore, be used as a criterion for "injury." The level at which "fatality" is likely to occur is more difficult to define. If we assume the duration of exposure to be the same, at about 40 seconds, Figure 3-11.41 can be used to determine a fatality threshold of about 10 kW/m^2. This level may, therefore, be used in determining the hazard zone for fatality.

Thermal Response of Structures

All structural materials are adversely affected by the elevated temperatures caused by a fire. The degree and significance of this adverse behavior depend largely on the function of the elements and on the degree of protection afforded. The mechanical properties of strength and stiffness decrease as the temperature increases. Other adverse behavior, such as excessive expansion and accelerated creep, also develops with increasing temperatures.

This section addresses the analytical heat transfer models and experimental data that may be used to predict temperature-time dependence of these components when exposed to a fire. These models can be used in conjunction with material property data to determine the extent of damage.

Heating of walls and structural beams: Several textbooks on heat transfer address the problem of transient temperature distribution within bodies of various geometrical configurations and under different conditions of heating. The classic book by Carslaw and Jaeger[68] is a major source of analytical solutions to a number of cases of heating of solids having infinite, semi-infinite, and finite thicknesses. Heating is achieved by various means such as by changing the temperature of one or both surfaces of a body suddenly or by applying a constant or cyclical radiant or convective heat flux, etc. All of these analytical solutions start with the basic heat conduction equation

$$\frac{\delta T}{\delta t} = \alpha \nabla^2 T \tag{114}$$

which assumes that the body is thermally isotropic (i.e., the thermal conductivity of the body is the same in all directions). The equation is integrated by applying the appropriate boundary conditions.

In a number of cases, the equation is amenable to an explicit series solution, which gives the temperature distribution within the body as a function of time. The solution is generally expressed in terms of the following dimensionless groups:

$$\frac{T_{final} - T}{T_{final} - T_{initial}} = \phi\left(\frac{\alpha t}{r_0^2}, \frac{hr_0}{k}, \frac{x}{r_0}\right) \tag{115}$$

In many cases, especially where combined convective and radiative heating may be involved, Equation 114 cannot be solved explicitly and numerical procedures must be used.

By examining the explicit solutions of various cases involving slab heating, it can be seen that when the Biot number, hr_0/k, is less than 0.1, the temperature distribution within the slab may be neglected. The slab may then be considered to have a uniform temperature at any moment in time. This simplifies the analytical procedures considerably. Most structures and walls of concern are made of metal

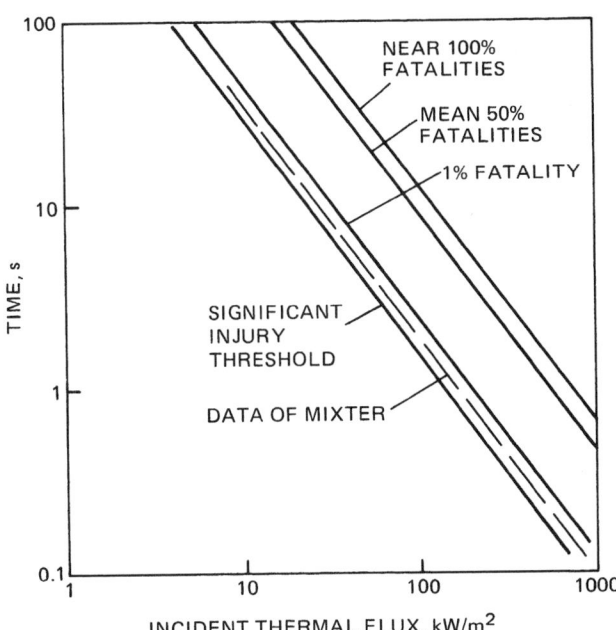

Fig. 3-11.41. *Fatality levels for thermal radiation.* [72]

having a thickness of the order of 0.02 m (i.e., $r_0 = 0.01$ m) and a thermal conductivity, k, of about 45 W/m K. The convective heat transfer coefficient between the metal surface and fire gases or cold air is typically not more than 30 W/m^{2}K. The Biot number calculated from these data for typical steel structures on offshore platforms is approximately 0.0067—considerably lower than 0.1. This justifies the use of the simple slab model to examine the heating of metal structures and walls exposed to a fire on the platform.

Metal plate heated from one side: For a relatively thin metal plate or beam of thickness, δ, exposed to thermal radiation from one side, the following heat balance can be written:

$$\dot{q}_r'' \alpha_s = \sigma(2 - F)(\varepsilon_s T_s^4 - \alpha_s T_\alpha^4)$$
$$+ 2h_a(T_s - T_a) + \delta C_s \rho_s \frac{dT_s}{dt} \quad (116)$$

The convective heat transfer coefficient, h_a, between the steel plate and the surrounding air is given by the following dimensional equations[75]

Vertical Plate:

$$h_a = 1.31(\Delta T)^{1/3} \quad (117)$$

Horizontal Plate:

$$h_a = 1.52(\Delta T)^{1/3} \quad (118)$$

where h_a is in W/m^{2}K and ΔT is in K.

Equation 116 can be rewritten as follows:

$$t = C_s \rho_s \delta \int_{T_a}^{T_s} \frac{dT_s}{[\dot{q}_r'' \alpha_s - \sigma(2 - F)(\varepsilon_s T_s^4 - \alpha_s T_\alpha^4) - 2k_a(T_s - T_a)]} \quad (119)$$

To determine the time to failure, this equation may be integrated numerically between T_a and T_s, the temperature at which metal failure is expected to take place.

Should the flames impinge on one side of the metal plate, Equation 116 is modified into the following:

$$\dot{q}'' = \delta C_s P_s \frac{dT_s}{dt} + h_a(T_s - T_a) + \sigma(\varepsilon_s T_s^4 - \alpha_s T_a^4) \quad (120)$$

which may be rearranged as follows:

$$t = C_s \rho_s \delta \int_{T_a}^{T_s} \frac{dT_s}{[\dot{q}_{total}'' - h_a(T_s - T_a) - \sigma(\varepsilon_s T_s^4 - \alpha_s T_a^4)]} \quad (121)$$

In this equation, $\dot{q}_{total}''$ is the total heat flux (radiative plus convective) measured within the specific hydrocarbon fuel flame, whereas $\dot{q}_r''$ in Equations 116 and 119 is the radiative flux reaching the metal surface from the fire.

For very large fires with a flame emissivity close to unity, the total heat flux from an impinging flame, $\dot{q}_{total}''$, is given by

$$\dot{q}_{total}'' = \sigma(\alpha_s T_f^4 - \varepsilon_s T_s^4) + h_a(T_f - T_a) \quad (122)$$

In this equation, T_f is the actual temperature measured within the fire and is not to be confused with the effective blackbody radiative temperature of the flame. The convective heat transfer coefficient, h_a, is that measured between

the flame and the steel surface and may be predicted approximately from Equations 117 or 118.

Flames impinging on both sides of the plate: When a fire impinges on both sides of the plate, and if the fire is large enough so that its emissivity (= absorptivity) is equal to unity, the heat balance on the plate takes the form

$$2\dot{q}_{total}'' = C_s \rho_s \delta \frac{dT_s}{dt} \quad (123)$$

which can be rewritten in the form

$$t = C_s P_s \delta \int_{T_a}^{T_s} \frac{dT_s}{2\sigma(\alpha_s T_f^4 - \varepsilon_s T_s^4) + 2h_a(T_f - T_s)} \quad (124)$$

SUMMARY

In the previous sections, we have given detailed techniques for computing impacts of large, open hydrocarbon fires. In particular, we have addressed steady state thermal radiation from pool fires and flame jets and unsteady state thermal radiation from vapor fires and fireballs. Particular emphasis has been placed in supporting the assessment methodology with available experimental data. These models can be used with the appropriate impact criteria to evaluate the fire and flammability hazards associated with hydrocarbon releases.

REFERENCES CITED

1. G.H. Markstein, "Radiative Energy Transfer from Turbulent Diffusion Flames," *Comb. and Flame*, 27, pp. 51–63 (1976).
2. V.I. Blinov and G.N. Khudiakov, "Certain Laws Governing Diffusive Burning of Liquids," *Academiia Nank*, SSSR Doklady, pp. 1094–1098 (1957).
3. H.C. Hottel, "Certain Laws Governing Diffusive Burning of Liquids," *F. Res. Abs. and Rev.*, 1, p. 41 (1959).
4. J. DeRis and L. Orloff, "A Dimensionless Correlation of Pool Burning Data," *Comb. and Flame*, 18, pp. 381–388 (1972).
5. D.S. Burgess, A. Strasser, and J. Grumer, "Diffusive Burning of Liquid Fuels in Open Trays," *Fire Res. Abs. and Rev.*, 3, p. 177 (1961).
6. J. Grumer, A. Strasser, T.A. Kubala, and D.S. Burgess, "Uncontrolled Diffusive Burning of Some New Liquid Propellants," *Fire Res. Abs. and Rev.*, 3, p. 159 (1961).
7. M.G. Zabetakis and D.S. Burgess, "Research in Hazards Associated with the Production and Handling of Liquid Hydrogen," U.S. Bureau of Mines Report, *RI 5707* (1961).
8. K.S. Mudan, "Thermal Radiation Hazards from Hydrocarbon Pool Fires," *Prog. Energy Comb. Sci.*, 10, pp. 59–80 (1984).
9. S.R. Gollahalli and H.F. Sullivan, "Liquid Pool Fires—A Review," *Research Report #23*, University of Waterloo, Canada (1973).
10. P.K. Raj, "Models for Cryogenic Liquid Spill Behavior on Land and Water," *J. Haz. Mat.* (1981).
11. P.K. Raj and A.S. Kalelkar, "Assessment Models in Support of the Hazard Assessment Handbook (CG-446-3)," Chap. 9, Technical Report prepared for the U.S. Coast Guard, *NTIS publication #AD776617*, January (1974).
12. P.H. Thomas, "The Size of Flames from Natural Fires," *9th Int. Combustion Symposium*, Comb. Inst., Pittsburgh, PA, pp. 844–859 (1963).
13. J. Moorhouse, "Scaling Criteria for Pool Fires Derived from Large-Scale Experiments," *I. Chem. Sym*, 71, pp. 165–179 (1982).
14. J.R. Welker and C.M. Sliepcevich, "Wind Interaction Effects on Free Burning Fires," *Tech. Report #1441-3* to Office of Civil Defense of U.S. Bureau of Standards (1967).
15. American Gas Association, "LNG Safety Research Program," *Report IS 3-1* (1974).

16. K.S. Mudan and P.A. Croce, "Thermal Radiation Model for LNG Trench Fires," ASME Winter Annual Meeting, New Orleans, LA (1984).

17. J. DeRis, "Fire Radiation—A Review," *17th Int. Symp. on Combustion*, Comb. Inst., Pittsburgh, PA, p. 1003 (1978).

18. P.K. Raj, K.S. Mudan, and A.N. Moussa, "Experiments Involving Pool and Vapour Fires from Spills of LNG on Water," *Report #CG-D-55-79, NTIS AD77073*, U.S. Coast Guard (1979).

19. B. Hagglund and L. Persson, "The Heat Radiation from Petroleum Fires," *FOA Rapport*, Forsvarets Forskningsanstalt, Stockholm (1976).

20. J. Moorhouse and M. Pritchard, "Thermal Radiation Hazards From Large Pool Fires and Fireballs—A Literature Review," *I. Chem. Sym*, 71, pp. 123–125, (1982).

21. H.C. Hottel and A.F. Sarofim, *Radiative Transfer*, McGraw-Hill, New York (1967).

22. A. Modak, "Radiation from Products of Combustion," *Tech. Rep #040E6, BU-1*, Factory Mutual Research Corp., Norwood, MA (1978).

23. J.D. Felske and C.L. Tien, "Calculation of Emissivity of Luminous Flames," *Comb. Sci. Tech.*, 7, pp. 25–31 (1973).

24. W.W. Yuen and C.L. Tien, "Simple Calculation Scheme for the Luminous Flame Emissivity," *17th Int. Symp. Comb.*, Comb. Inst., Pittsburgh, PA, p. 1481, University of Leeds, England (1976).

25. R.S. Alger, R.C. Corlett, A.S. Gordon, and F.A. Williams, "Some Aspects of Turbulent Pool Fires," *J. Fire Tech.*, 15, 2, pp. 142–156 (1979).

26. T.T. Fu, "Heat Radiation from Fires of Aviation Fuels," *Combustion Institute Eastern Section Fall Meeting*, Princeton University, NJ (1972).

27. NASA, On the Experiments with 7.5m and 15m JP-4 Fuel Pool Fire Measurements, NASA Ames Research Center, Moffet Field, California (1979).

28. Japan Safety Society, *Report on Burning Petroleum Fires*, (in Japanese) (1982).

29. R.G. Rein, Jr., C.M. Sliepcevich, and J.R. Welker, "Radiation View Factors for Tilted Cylinders," *J. Fire and Flam.*, 1., p. 140 (1970).

30. K.S. Mudan, "Geometric View Factors for Thermal Radiation Hazard Assessment," *F. Safety J.*, 12, 89–96 (1987).

31. E.M. Sparrow, "A New Simpler Formulation for Radiative Angler Factors," *J. Heat Transfer*, Trans., ASME, 85, pp. 81–88 (1963).

32. J.R. Howell, *A Catalog of Radiation Configuration Factors*, McGraw-Hill, New York (1982).

33. W.L. Wolfe, editor, *Handbook of Military Infrared Technology*, Office of National Research, Washington, DC (1965).

34. T.A. Brzustowski, "Flaring in the Energy Industry," *Progress in Energy and Comb. Sci.*, 2, pp. 129–141 (1976).

35. H.C. Hottel and W.R. Hawthorne, "Diffusion in Laminar Flame Jets," *3rd Symp. on Comb., Flame and Explosions*, pp. 254–266 (1949).

36. W.R. Hawthorne, D.S. Weddell, and H.C. Hottell, "Mixing and Combustion on Turbulent Gas Jets," *3rd Int. Comb. Symp.*, Comb. Inst., Pittsburgh, PA, pp. 266–288 (1949).

37. F.P. Ricou and D.B. Spalding, "Measurements of Entrainment by Axisymmetrical Turbulent Jets," *J. of Fluid Mech.*, II, p. 21 (1961).

38. A.A. Putnam and C.F. Speich, "A Model Study of the Interaction of Multiple Turbulent Diffusion Flames," *9th Int. Symp. on Comb.*, Comb. Inst., Pittsburgh, PA, p. 867 (1963).

39. F.R. Steward, "Prediction of the Height of Turbulent Diffusion Buoyant Flames," *Comb. Sci. Tech.*, 2, pp. 203–212 (1970).

40. T.A. Brzustowski, "Predicting Radiant Heating from Flares" Esso Engineering Research and Development Report, *EE 15ER.71* (1971).

41. T.A. Brzustowski, "A New Criterion for the Length of a Gaseous Turbulent Diffusion Flame," *Comb. Sci. and Tech.*, 6, pp. 313–319 (1973).

42. T.A. Brzustowski, S.R. Gollahalli, and H.F. Sullivan, "The Turbulent Hydrogen Diffusion Flame in a Cross-Wind," *Comb. Sci. and Tech.*, II, pp. 29–33 (1975).

43. S.R. Gollahalli, T.A. Brzustowski, and H.F. Sullivan, "Characteristics of a Turbulent Propane Diffusion Flame in a Cross-Wind," *Transactions of CSME*, 3, pp. 205–214 (1975).

44. G.T. Kalghatki, "The Visible Shape and Size of a Turbulent Hydrocarbon Jet Diffusion Flame in a Crosswind," *Comb. and Flame*, 52, pp. 91–106 (1983).

45. O.K. Sonju and J. Hustad, *An Experimental Study of Turbulent Jet Diffusion Flames*, Norwegian Maritime Research, pp. 2–11 (1984).

46. P.R. Oenbring and T.R. Sifferman, "Flare Design—Are Current Methods Too Conservative?" *Hydrocarbon Processing*, May, pp. 124–129 (1980).

47. D.S. Burgess and M. Hertzberg, "Radiation from Pool Fires," in *Heat Transfer in Flame*, edited by Afgan, N.H., and Beer, J.R., Scripta Book Co., Washington, DC (1974).

48. S.H. Tan, "Flare System Design Simplified," *Hydrocarbon Processing*, 46, pp. 172–176 (1967).

49. G.R. Kent, "Practical Design of Flare Stacks," *Hydrocarbon Processing*, 43, pp. 121–125 (1964).

50. Fumarola, *et al*, "Determining Safety Zones for Exposure to Flare Radiation," *I. Chem. E Symposium Series No. 82*, pp. G23–G30 (1983).

51. P.R. Oenbring and T.R. Sifferman, "Flare Design Based on Full-Scale Plant Data," *45th Midyear Meeting, API Refining Department Proceedings*, 59, May 12-15, Houston, TX (1980).

52. S. Galant, *et al.*, "Three-Dimensional Parabolic Calculations of Large-Scale Methane Turbulent Diffusion Flames to Predict Radiation under Crosswind Conditions," *20th Comb. Symp.*, Comb. Inst., Pittsburgh, PA, pp. 531–540 (1984).

53. J.F. Straitz III, J.A. O'Leary, J.E. Brennan, and C.J. Kardan, "Flare Testing and Safety," *Loss Prevention Volume II*, American Institute of Chemical Engineers, pp. 23–30 (1977).

54. G.T. Kalghatki, "Blowout Stability of Gaseous Diffusion Flames in Still Air," *Comb. Sci. and Tech.*, 26, pp. 233–239 (1981).

55. K.S. Mudan, "Hydrocarbon Pool and Vapor Fire Data Analysis," *USDOE Report DE-AC01-83EP16008*, October (1984).

56. G.A. Mizner and J.A. Eyre, "Radiation from Liquefied Gas Fires on Water," *Comb. Sci. and Tech.*, 35, pp. 33–57 (1983).

57. J.A. Fay and D.H. Lewis, "Unsteady Burning of Unconfined Fuel Vapor Clouds," *16th Intl. Symp. on Comb.*, Comb. Inst., Pittsburgh, PA, p. 1387 ff (1976).

58. P.K. Raj and H. Emmons, "On the Burning of a Large Flammable Vapor Cloud," *Western and Central States Meeting*, The Comb. Inst., San Antonio, TX (1975).

59. E.M. Sparrow and R.D. Cess, *Radiation Heat Transfer*, Augmented Edition, McGraw-Hill Book Company, New York (1978).

60. K.S. Mudan and G.J. Desgroseilliers, "Thermal Radiation from Hydrocarbon Fireballs," *HSE Symposium on Risk Assessment*, London, September (1981).

61. J.B. Gayle and J.W. Bransford, "Size and Duration of Fireballs from Propellant Explosions," *NASA TMX-53314* (1965).

62. K. Hasegawa and K. Sata, "Experimental Investigation of the Unconfined Vapor Cloud Explosions of Hydrocarbons," *Technical Memorandum of the Fire Research Institute*, 12, Japan (1978).

63. J.A. Fay, G.J. Desgroseilliers, and D.H. Lewis, "Radiation from Burning Hydrocarbon Clouds," *Comb. Sci. and Tech.*, 20, pp. 141–151 (1979).

64. B. Morton, G. Taylor, and J. Turner, "Turbulent Gravitational Convection from Maintained and Instantaneous Sources," *Proc. Roy. Soc.* (London), 234, A.1 (1956).

65. K. Buettner, "Effects of Extreme Heat and Cold on Human Skin, I. Analysis of Temperature Changes Caused by Different Kinds of Heat Application," *J. Applied Physiology*, 3, pp. 691–702 (1951).

66. K. Buettner, "Effects of Extreme Heat and Cold on Human Skin, II. Surface Temperature, Pain and Heat Conductivity in Experiments with Radiant Heat," *J. Ap. Phys.*, 3, 703 (1951).

67. A.K. Mehta, F. Wong, and G.C. Williams, "Measurement of

Flammability and Burn Potential of Fabrics," *Summary Report to NSF—Grant #GI-31881*, Fuels Research Laboratory, Massachusetts Institute of Technology, Cambridge, MA (1973).

68. H.S. Carslaw and J.C. Jaeger, *Conduction of Heat in Solids*, 2nd ed. Oxford University Press, New York (1959).

69. J.D. Hardy, I. Jacobs, and M.D. Meizner, "Thresholds of Pain and Reflex Contraction as Related to Noxious Stimulation," *J. Ap. Phys.*, 5, pp. 725–739.

70. N. Bigelow, I. Harrison, H. Goodell, and H.G. Wolf, "Studies on Pain: Quantitative Measurements of Two Pain Sensations of the Skin, with Reference to the Nature of the Hyperalgesia of Peripheral Neuritis," *J. of Clinical Invest.*, 24, 503–512 (1945).

71. A.M. Stoll and L.C. Greene, "Relationship Between Pain and Tissue Damage Due to Thermal Radiation," *J. Ap. Phys.*, 14, pp. 373–382.

72. N.A. Eisenberg *et al*, "Vulnerability Model. A Simulation System for Assessing Damage Resulting from Marine Spills," *NTIS AD-A015-245*, Springfield, VA (1975).

73. G. Mixter, "The Empirical Relation Between Time and Intensity of Applied Thermal Energy in Production of 2 + Burns in Pigs," *University of Rochester Report No. UR-316*, Contract W-7041-eng-49 (1954).

74. G. Mixter, "Thermal Radiation Burns Beneath Fabric Systems," *Annals of the New York Academy of Sciences*, 82, pp. 701–713 (1959).

75. W.H. McAdams, *Heat Transmission*, 3rd ed., McGraw-Hill Book Company, New York (1954).

76. T.A. Brzustowski, S.R. Gollahalli, M.P. Gupta, M. Kaptein, and H.F. Sullivan, "Radiant Heating from Flares," *ASME Paper 75-HT-4* (1975).

BEHAVIORAL RESPONSE TO FIRE AND SMOKE

John L. Bryan

INTRODUCTION

The determination of the behavioral response of individuals in fire incidents has been examined for approximately 40 years by research studies in which individuals were administered a questionnaire by fire department personnel at the time of the fire incident,[1,2] or by mailed questionnaires or personal interviews following the incident. The individual's behavioral response in a fire incident appears to be affected by

1. the variables of the building in which the fire incident occurs; and
2. the perceived physical cues of the fire severity at the time the individual becomes aware of the fire.

The behavioral response of the building occupants may vary if they perceive physical cues (an odor of smoke) in contrast to observed cues (smoke obscuring the means of egress). There is some evidence that the recognition of fire protection systems provided within the building may be a factor in an individual's perception of the severity of the fire incident threat. It would appear that the most important individual decisions and behavioral responses usually involve perceived life-threatening situations that occur in the initial stages of the fire incident prior to fire department arrival. In their studies of health care facilities, Lerup *et al*[3] have indicated the importance of the participants initial behavior in the following manner:

> In the process of investigating these case studies we have come to believe that the period between detection of the fire and the arrival of the fire department is the most crucial life saving period in terms of the first compartment. (The area in direct contact with the room of origin and the fire.)[3]

The behavioral response of the individuals intimately involved with the initiation of the fire incident or awareness of the initial fire incident cue often appeared to be a determinant outcome of the fire incident. It should be realized that altruistic behavior observed in most fire incidents, with the behavioral response of the occupants in a deliberate, purposeful manner, appears to be the most frequent mode of behavioral response. The nonadaptive flight or panic type behavioral response appears to be an infrequent, unusual, or unique participant behavioral response in most fire incidents.

Awareness of Cues

The manner in which an individual is alerted to the occurrence of a fire may predispose the perception of the threat involved. The alerting means, communication mode, and message content were discussed by Keating and Loftus,[4] in their study concerned with vocal alerting systems in buildings. It would appear that variations in voice quality, pitch, or volume, as well as the content of the message, may provide reinforcing threat cues to occupants.

Proulx and Sime,[5] in their study involving evacuation drills in an underground rapid transit station, found the use of directive public announcements with an alerting alarm bell was the most conducive to creating an immediate effective evacuation. Ramachandran,[6] in his review of the research on human behavior in fires in the United Kingdom since 1969, has summarized the effectiveness of alarm bells as awareness cues in the following manner:

> The response to fire alarm bells and sounders tends to be less than optimum. There is usually skepticism as to whether the noise indicated a fire alarm and, if so, is the alarm merely a system test or drill?[6]

Ramachandran[7] indicated that the development of "informative fire warning systems," which utilize a graphic display with a computer-generated message and a high-pitched alerting tone, has reduced the observed delay times in the initiation of practice evacuations. Cable,[8] in his study of the response times of staff personnel to the fire alarm signal in Veterans Administration hospitals, found the greatest delay in response time with the coded alarm-bell-type systems. Kimura and Sime[9] found, in the evacuation of two lecture halls with college students, the verbal instructions of the lecturer were the determining factor in the choice of the use of the fire exit over the normal entrance and exit. The research literature developed from practice evacuations tends to indicate that the use of verbal directive

Dr. John L. Bryan is Professor Emeritus, Department of Fire Protection Engineering at the University of Maryland. Currently a consultant located in Frederick, MD, his major interest is in the field of human behavior in fire situations.

informative messages may be the most effective in reducing the delay in evacuation initiation.

However, it should be noted that, if the verbal directive messages are in conflict with other awareness cues, e.g., the odor or sight of smoke, the credibility of the message may be questioned and the information disregarded by the occupants. One of the few documented cases of this type of situation occurred in the South Tower of the World Trade Center on April 17, 1975. As reported by Lathrop,[10] the fire occurred at approximately 9:04 a.m. in a trash cart in a storage area on the 5th floor, adjacent to an open stairway door; this allowed the smoke to infiltrate the 9th through 22nd floors. The occupants of these floors moved into the core area of the building. At 9:10 a.m., a verbal message from the building communications center monitoring these core lobby areas advised the occupants to remain calm and return to their office areas. In spite of this announcement, the occupants remained in the core lobby areas and became more concerned about the smoke conditions. Thus, with the occupants on the affected floors becoming more anxious, an evacuation message was announced at 9:16 a.m.

Burns[11] reported on the explosion and fire of February 26, 1993, which severely affected both towers and the Vista Hotel of the World Trade Center, wherein simultaneous occupant evacuations occurred. The explosion disrupted the structure's communications center, and the occupants, having experienced the explosion, the loss of power, and floor areas infiltrated with smoke within minutes, evacuated without the established verbal directional announcements utilized in practice evacuations.

Bryan[1] found most of the participants in the study of residential fire incidents to have become aware of the fire incident by the odor of the smoke. However, if the categories of "notified by others" and "notified by family" are combined in this residential fire incident study, the process of personal notification becomes the most frequent means of the initial awareness of the fire incident. (See Table 3-12.1.) The category of "noise" included sounds generated from persons moving downstairs, persons moving through corridors, and other related fire incident sounds, including the breaking of glass and the movement of fire department apparatus.

Table 3-12.2 compares the means of awareness of the British population from Wood's study[2] and the U.S. population from Bryan's study.[1] The number of categories for the

TABLE 3-12.1 *Means of Awareness of the Fire Incident by the Study Population*[1]

Means of Awareness	Participants	Percent
Smelled Smoke	148	26.0
Notified by Others	121	21.3
Noise	106	18.6
Notified by Family	76	13.4
Saw Smoke	52	9.1
Saw Fire	46	8.1
Explosion	6	1.1
Felt Heat	4	0.7
Saw/Heard Fire Department	4	0.7
Electricity Went Off	4	0.7
Pet	2	0.3
N = 11	569	100.0

TABLE 3-12.2 *Means of Awareness of the Fire Incident for the British and the U.S. Study Population*[1,2]

Means of Awareness	British Percent	U.S. Percent	$P_1 - P_2$	$SE_{P_1 - P_2}$	CR
Flame	15.0	8.1	6.9	1.64	4.21*
Smoke	34.0	35.1	1.1	2.27	0.48
Noises	9.0	11.2	2.2	1.41	1.56
Shouts and Told	33.0	34.7	2.7	2.25	1.20
Alarm	7.0	7.4	0.4	1.23	0.33
Other	2.0	2.8	0.8	0.70	1.14
N = 6	2193	569			

*Critical ratio significant at or above the 1 percent level of confidence.

means of awareness were reduced from the 11 categories indicated in Table 3-12.1 because the British study had fewer categories. U.S. population responses were adapted to the British categories. There was only one significant difference in the means of awareness between the two populations, with 15 percent of the British population having become aware of the fire incident upon observing flame as contrasted with 8.1 percent of the U.S. population.

Concerning the awareness of the occupants to smoke detectors, Berry[12] indicated in his study of the National Fire Protection Association-recommended smoke detector noise level of 85 dBA[13] that individuals with hearing impairments, or those taking sleeping pills or medication, may require a detector noise level exceeding 100 dBA. With hearing impaired occupants, Cohen[14] has indicated that flashing or activated visual light signals are effective indicators of fire alarm system activation in occupancies populated by hearing impaired persons. NFPA *101*®, *Life Safety Code*®, in 1981, initially permitted the flashing of the exit signs with the activation of an audible fire alarm system. This provision has been continued in the 1994 edition.[15]

Kahn[16] conducted a study with 24 male subjects relative to their being awakened by an audible smoke detector alarm signal and their identification of supplemental fire cues. Kahn found these subjects slept through the alarm signals with a signal-to-noise ratio of 10 dBA at their ears, and consistently failed to identify the awakening smoke detector cue as well as radiant heat and smoke odor cues as fire warnings. Noble *et al*[17] have indicated the alarm signal-to-noise ratio is attenuated by physical surroundings; thus a smoke detector or alarm audible signal may be reduced by 40 dBA in passing through a ceiling or wall and by 15 dBA in passing through a door. In addition, it was found that the signal could be masked to an ineffective level by a typical residential air conditioner noise level of 55 dBA.

The recognition of ambiguous fire incident cues as indicating a possible emergency condition appears to be inhibited by the presence of other persons in some occupancies. Latane and Darley,[18] in their experimental studies of the inhibition of adaptive behavioral responses to emergencies, created an experimental situation involving college students. While the students were completing a written questionnaire, the experimenters would introduce smoke into the room through a small vent in the wall. If the subject left the room and reported the smoke, the experiment was terminated. If the subject had not reported the presence of the smoke within a six-minute interval from the time the smoke was first noticed, the experiment was considered completed.

Subjects alone in the room reported the smoke in 75 percent of the cases. When two passive confederates were provided in the room with each subject, only 10 percent of the groups reported the smoke. When the total experimental group consisted of three naive subjects, one of the individuals reported the smoke in only 38 percent of the groups. Of the twenty-four persons involved in the eight naive subject groups, only one person reported the smoke within the first four minutes of the experiment. In the situations involving subjects alone in a room, 55 percent of the subjects had reported the smoke within two minutes and 75 percent reported smoke in four minutes.

Latane and Darley reported that noticing the smoke was apparently delayed by the presence of other persons, with the median delay of 5 seconds for single subjects and 20 seconds for both of the group conditions. These results would appear to indicate the inhibiting influences that individuals may accept as being imposed upon their behavior in public places. Latane and Darley reported the behavioral response of nine of the naive subjects in the ten passive confederate research situations as follows:

> The other nine stayed in the waiting room as it filled up with smoke, doggedly working on their questionnaire, and waving the fumes away from their faces. They coughed, rubbed their eyes, and opened the window but did not report the smoke.[18]

Latane and Darley suggest that, while trying to interpret ambiguous threat cues as to whether a situation requires a unique response, the individual is influenced by the behavioral response of others who are exposed to the identical cues. If these other individuals remain passive and appear to interpret the situation as a nonemergency, this inhibiting social influence may reinforce this nonemergency interpretation for an individual. This behavioral experiment may help explain the reported tendency of persons (1) to disregard initial ambiguous fire incident cues, or (2) to interpret the cues as indicating a nonemergency condition when the fire incident occurs in an occupancy with a social audience of other persons, as in a restaurant, motion picture theater, or department store. This experimental study may be helpful in understanding the reported incidents of alarms to fire departments that have been delayed by occupants for periods of minutes or even hours. In the report of the Arundel Park fire,[19] several of the participants indicated that, when they reentered the hall after observing the fire from outside the building, they warned other participants and suggested they should leave, but were laughed at and the warning disregarded.

Latane and Darley indicated the processes of social inhibition, diffusion of responsibility, and mimicking appear to be primarily responsible for the inhibition of adaptive and assistance behavior responses by participants in emergency situations. It would appear that the inhibition of behavioral responses in the early stages of a fire incident (when the fire incident cues are relatively ambiguous) may predispose participants to a nonadaptive type of flight behavior since the available evacuation time has been expended. In some fire incidents it appears to be difficult to get occupants of a building to evacuate because of the variables of social inhibition and diffused responsibility. The tendency to mimic the interpretation of cues and behavior response from others (as established by Latane and Darley) appears to be a frequent occurrence in fire incidents in restaurants, hotels, and other places of public assembly.

Perception of the Fire Incident

Withey[20] has examined seven psychological and physical processes that an individual may utilize in an attempt to perceive, identify, structure, and evaluate the situational fire incident cues. It would appear that six of these individual perceptual processes as presented in Figure 3-12.1 may be critical factors in the perception of a fire incident.[21]

Recognition: The process of recognition occurs when the individual identifies the ambiguous fire cues as indicating the occurrence of a fire incident and thus becomes aware of the fire. The initial physical perceptual cues may be very ambiguous, and not positively indicate a fire incident. The cues are produced by a continuous dynamic physical process with an increasing intensity due to the fluid mechanics properties of flame, heat, and smoke production. Withey also indicated the usual mental state and predisposition of an individual is to recognize threat cues in terms of the most probable occurrence, typically as related to a prior personal experience and in the form of an optimistic favorable outcome. The optimistic outcome aspect of an individual's perception of fire incident cues may be a result of the influence of the individual's concept of a personal invulnerability to risk.

The concept of threat recognition appears to be a very important problem in fire protection, since the decision that initiates the action involved in the activation of the fire alarm, the evacuation of the building occupants, or related to the suppression of the fire may be delayed or postponed if the individuals involved do not perceive the cues as indicative of an emergency fire situation. Apparently, the ambiguous nature of the fire cues, and the unstructured nature of many public and social groups requires the appearance of significant amounts of smoke, or sudden and threatening

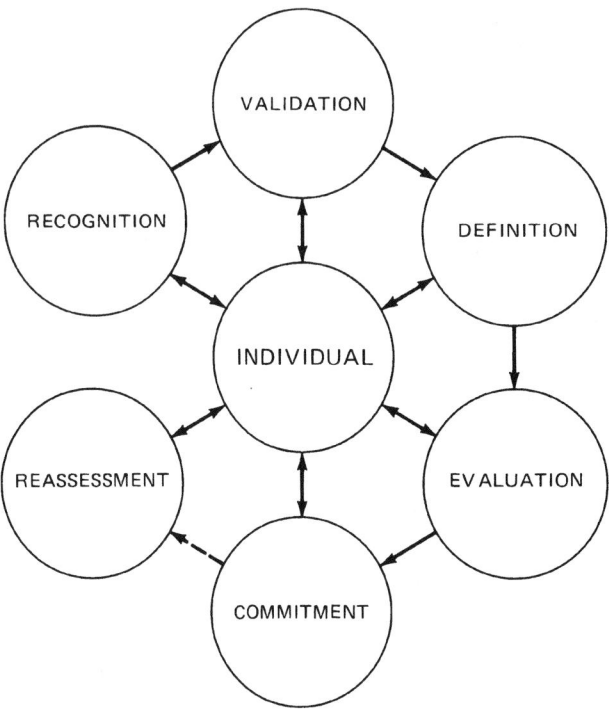

Fig. 3-12.1. The decision processes of the individual in a fire incident.

flames before most individuals without specialized fire prevention instruction or prior fire incident experiences perceive a threatening fire situation to be present.

Validation: The process of validation apparently consists of the individual attempting to validate an initial perception of the fire cues, primarily by seeking verbal reassurance of the minor and insignificant character of the fire incident. When the perceived cues are significantly ambiguous, however, the individual often attempts to obtain additional information. Thus, the person is aware that something is happening in the immediate environment but is not sure exactly how the cues define the event. The process of validation is often conducted by persons exposed to the identical cues. In a study concerned with the individual's perception of cues from the explosion of a fireworks plant, Killian[22] found from the study population of 139 persons that 85 individuals or 61 percent of the population obtained definitive information as to the source and nature of the explosion or smoke from someone in person or someone who telephoned.

Latane and Darley[18] established that the physical presence of others during the recognition and validation processes inhibit and structure the behavioral responses of an individual.

Definition: The definition process is usually considered to be the procedure whereby the individual attempts to relate the information concerning the fire to the perceived and contextual variables including: the qualitative nature of the fire relative to the location of the individual; and the magnitude of deprivation implied by the fire occurrence, with the time context and sequence of the implied deprivation. The generation of stress and anxiety in the individual appears to be most rapid and severe before the individual defines the initial ambiguous cues with structure for the situation. It is often apparent to the individual that the situation requires structure and interpretation before the cues can be defined and assimilated. The role concept of the individual appears to be one of the critical factors in the situation relative to the personalization of the fire threat, in addition to the physical environment which the fire incident has created. The most important physical aspects in relation to the definition process appear to be the generation, intensity, and propagation of the smoke, flames, and thermal exposure.

Evaluation: The individual's process of evaluating the fire incident may be described as the cognitive and psychological activities necessary for the individual to respond to the threat. The individual's ability to develop alternate strategies to cope with the fire incident—through psychological and physiological mechanisms that are designed to reduce the stress and anxiety levels of the individual to react to the fire by fight or flight—provide a basis for an initial decision to be formulated for an overt behavioral response. Because of the brief time span involved in the generation and propagation of the fire, it should be remembered that these cognitive processes (including the process of evaluation) may have to be accomplished within a time period of several seconds.

Sime[23] has emphasized the importance of the individual's perception of the time available for evacuation or to obtain a refuge area as being an estimation by the individual of the fire threat. Thus, he indicates the "perceived time available" is dependent upon the information and communication provided to the occupants concerning the location and development of the fire.

The individual's evaluation process may include the formulation of adaptation, escape, or defense procedures. Therefore, the variables of the physical environment in which the fire incident occurs may be a critical variable in the evaluation process. These determinants and critical variables may be (1) the physical location of the individual in relation to the means of egress; (2) the location and proximity of other members of the population at risk; (3) the physically perceived untenable effects of the fire incident; and (4) the overt behavioral response of other individuals in the population. During the cognitive process of evaluation, the individual may decide to evacuate the building (flight) or to obtain a fire extinguisher or activate a manual fire alarm station (fight). During the evaluation process, the individual is very perceptive to the overt actions and communications of the other members of the at-risk population. Thus, the behavioral responses of observed individuals may be mimicked, resulting in mass adaptive or nonadaptive behavior instead of selective individualized behavior.

In his classic studies of nonadaptive group behavior, Mintz[24] developed the concept of this mode of behavior being directly dependent upon the individual's perception of the reward structure of the situation. A heterogeneous population within a building, having perceived and defined a fire threat situation, would probably initially perceive a reward structure conducive to group and individual cooperative and adaptive behavior responses. Theoretically, all occupants should be able to proceed to and reach the provided means of egress. However, because of the diverse location of the exits and the relative position of the individuals within the area, the perceived reward structure for the individuals more remotely located from the means of egress could provide for the initiation of competitive behavior between occupants. With cooperative behavior, some individuals could perceive that it would be impossible for them to reach an exit in time to escape the deprivation effects of the fire incident. Once the pattern of competitive behavior becomes observable by one or more individuals, this competitive behavior pattern may become the norm for the group and result in selective, individual competition to reach the means of egress. Such competitive behavior may be normalized in the group by each individual's perception of the reward structure or their probability of obtaining the reward, which in this fire incident is to obtain the means of egress.

In the individual's evaluation process, cultural, sociological or economic influences, or the assumption of a particular individual psychological role may be critical to the formulation of behavioral response strategies. The individual in a familiar role, suitable for the fire incident situation and in familiar surroundings, may experience less anxiety and will probably select more adaptive behavior responses than individuals in an unfamiliar role who are confronted with the occurrence of an unfamiliar threat, in unfamiliar surroundings.

Jones and Hewitt[25] conducted detailed interviews with 40 occupants of a 27-story office building who had evacuated the building during a fire incident. It should be noted that the fire occurred at 9:00 p.m. when the fire management plan was not in effect due to the reduced occupancy of the building. In this situation it appeared that the leadership and the evacuation group formation were related to the fire training and the normal roles of the occupants. These investigators found the relationship between the occupancy roles and the normal or emergent leadership of the occupants to be the critical factor in a successful evacuation, with the following variables:

the social and organizational characteristics of the occupancy, including what a person knows (or believes) of the situation, whether the person is alone or part of a group, the normal roles that people hold within the occupancy, and the organizational structure or framework. One factor that appears to be related to the chosen evacuation strategy of an occupant is the presence of leadership and the form which that leadership takes.[25]

Horiuchi, Murozaki, and Hokugo[26] reported on a questionnaire study of 458 occupants of an eight-story office building involved in a fire incident. These researchers found significant differences between the normal occupants of the building (familiar with the building) and those occupants attending training sessions in the building (not familiar with the building), relative to their actions, selection of evacuation routes, and effectiveness in achieving an exit. The regular occupants of the building engaged in fire-fighting actions and the alerting or assisting of other occupants, while the occupants not familiar with the building primarily engaged in evacuation behavior.

Commitment: The process of commitment consists of the mechanisms utilized by the individual to initiate the behavioral responses necessary to achieve the behavioral response strategies that were formulated in the evaluation process. This overt behavioral response to the perceived threat of the fire incident results in completion, partial completion, or noncompletion of the response strategy. Thus, if the response strategy is not completed, the individual immediately becomes involved in the cognitive process of reassessment and commitment. If the behavioral response results in success, however, the anxiety and stress created by the situation are relieved for the individual, although the fire incident may have increased in severity.

Reassessment: The process of reassessment and overcommitment may be the most stressful of the individual's cognitive processes because of the failure of previous attempts to achieve the formulated response strategies to the fire incident situation. More intense psychological and physiological energy is allocated to the behavioral responses and the individual tends to become less selective in the risks involved in the behavioral response. If successive failures are encountered, the individual becomes more frustrated, anxiety levels increase, and the probability of success decreases. At the Arundel Park fire incident, the number of persons who selected windows as a means of escape from the building increased when the individuals were involved in a secondary evacuation response.[19]

This analysis has been an attempt to understand the cognitive processes of the individual through an examination of the variables related to the processes of recognition, validation, definition, evaluation, commitment, and reassessment. It should be noted that these cognitive processes are dynamic and are constantly being modified in relation to the magnitude, velocity, and intensity of their covert and overt responses. The behavioral responses of the individuals (relative to their psychological and physiological dynamic activity) will probably be below normal during the recognitive process, when the individual is concentrating on the ambiguous perceptual cues. During the process of validation

and the definition of the threat, there is usually overt communication and verbalization with adjacent members of the risk population. The period of above normal activity appears to occur initially during the process of commitment, and often transfers to a hyperactive level during the process of reassessment and recommitment. It should be remembered that the stress and anxiety generated will tend to increase for the individual as these cognitive and physical behavioral responses result in failure to achieve risk reduction through evacuation or fire control.

The physical variables of the fire incident consist of the flame appearance, the flame, smoke or heat proximity, and the velocity of flame or smoke propagation. These fire incident variables will tend to predispose the individual to a higher level of behavioral response activity, considering the individual's perception of these variables as increasing the personal risk. However, it should be noted that during the process of reassessment and commitment, the individual's physical activity may reach the hyperactive level, or at the other extreme be expressed as a state of complete physical immobility with a subnormal activity level and a complete loss of the ability to communicate in a coherent manner. Apparently, these individuals may appear to perceive the fire incident as too severe for their capabilities of adaptability. They appear to be overwhelmed by the stress generation and abandon their attempts to formulate a response strategy. Thus, in ceasing to attempt any cognitive or physical adaptive behavioral response, they instead adopt a complete cognitive withdrawal from the fire incident environment through psychological withdrawal from reality. A schematic presentation of the dynamics of the behavioral response activity levels of the individual are illustrated in Figure 3-12.2.

Breaux et al[27] have developed a conceptual model of the cognitive decision processes of the individual in the fire incident that is similar to some of these concepts. Instead of the six processes adopted from Withey,[20] Breaux et al have utilized only three processes: recognition/interpretation; behavior, with either action or inaction; and the outcome of the action that involves the evaluation and long term effects of the behavior. The evaluation of the behavior is similar to the process of reassessment in Withey's conceptual model. Both the recognition/interpretation process and the behavior process have cognitive inputs that are critical to the decision-making processes. These cognitive inputs involve past experience, the factors immediately arising, and the current state factors, which all impact on the recognition/interpretation process. Breaux et al have emphasized that the individuals in the fire incident may not know precisely at an early stage that they are involved in a fire, and may not know where the fire has developed in relation to their location in the building or their specific location relative to the means of egress in the building. The conceptual model developed by Breaux et al is shown in Figure 3-12.3.

Bickman et al[28] have modified the conceptual model of Breaux et al into one that involves fire as a physical event, with the individual processes of the detection of cues, the definition of the situation, and the coping behavior. Bickman et al have developed assumptions relative to the behavior responses that increase the probability of detection and the probability of fire suppression, and thus affect the various activities in the coping behavior aspects of their conceptual model.

Proulx[29] has developed a stress model to demonstrate the various levels and the generation of stress within the individual involved in the decision process during a fire

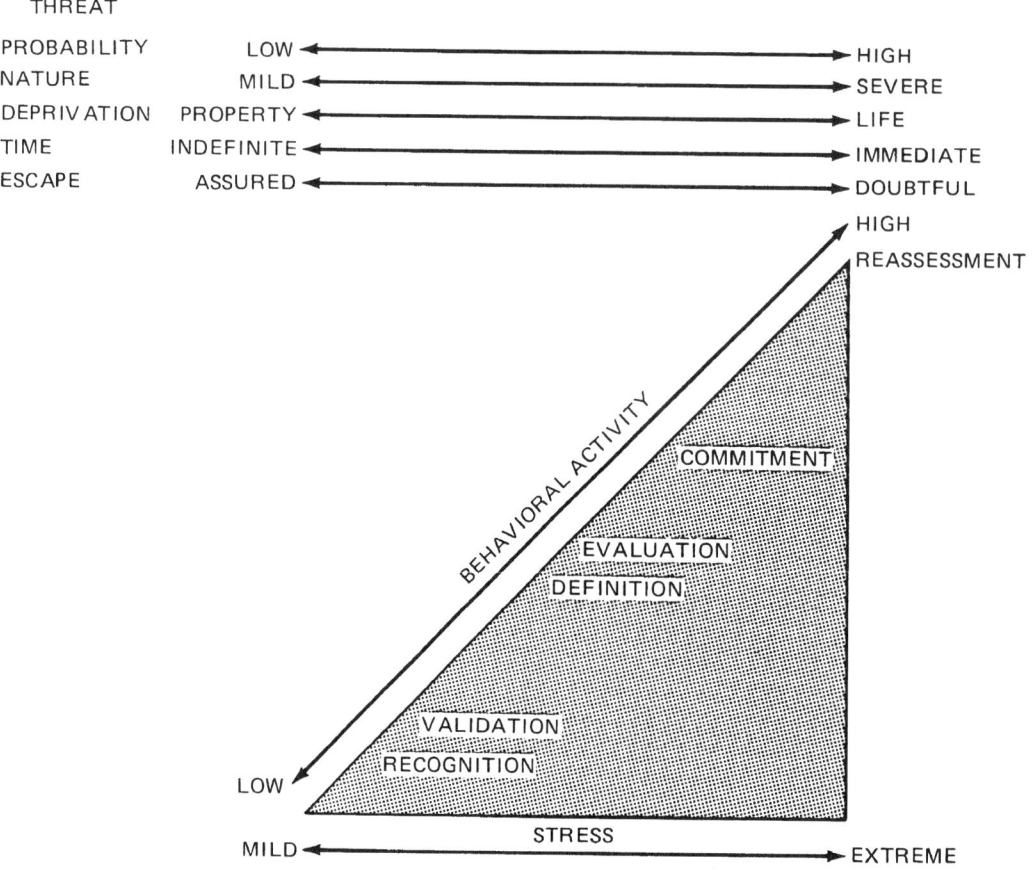

Fig. 3-12.2. The behavioral activity dynamics of the individual in a fire incident.

incident. This stress model is illustrated in Figure 3-12.4, and should be compared with the behavioral activity dynamics of the individual in a fire incident presented in Figure 3-12.2. The left side of Figure 3-12.4 indicates the information to be processed by the individual, and the right side indicates the resulting emotional state. The five loops in the model are described by Proulx in the following manner:

The first loop of the stress model starts with the perception of ambiguous information. This information is decoded in the processing system (PS in the figure) for interpretation. Given that the available information may not allow for a straightforward assessment of the situation, people will at first minimize or deny the situation. These defensive strategies of avoidance lead to an absence of reaction.

Although individuals may vary considerably in their appraisal of the same event, the repeated perception of ambiguous information will eventually generate a state of uncertainty which will then induce a feeling of stress. Some time can be spent going repeatedly through this second loop of the stress model.

The third loop of the stress model is related to the interpretation of the situation as an emergency. The thicker line around the processing system expresses the pressure of the overload of information with which the person tries to deal at once. The fear felt by the person is a manifestation of a specific appraisal of the environment.

The fourth loop of the model relates to the person's processing of irrelevant information and is represented by the very thick line around the processing system. This irrelevant information creates worry and more stress. The irrelevant information, created by the person, is caused by concern for his or her own performance in coping with the situation. Perceived feelings of arousal and fear, uncertainties regarding how to proceed with the problem, difficulties in interpreting what exactly is going on, and self-estimation of the efficiency of already-applied actions will become additional information to process.

The fifth and last loop of the model supposes an investment of more mental effort to master the problem, momentarily reducing the pressure on the processing system but resulting in fatigue and inefficiency manifested in a state of confusion.[29]

Proulx indicates that definitive, valid, and directive information provided to occupants of a building in a fire incident is the most effective stress reducer and thus tends to minimize the response delays created in the first and second loops of the stress model.

Chubb[30] identified a model of incident command procedure decision processes used by fire department officers, with the possibility that it be used for the improvement and training of the decision process for building occupants in a fire situation. The decision model was developed from the theory of naturalistic decision-making, which has evolved

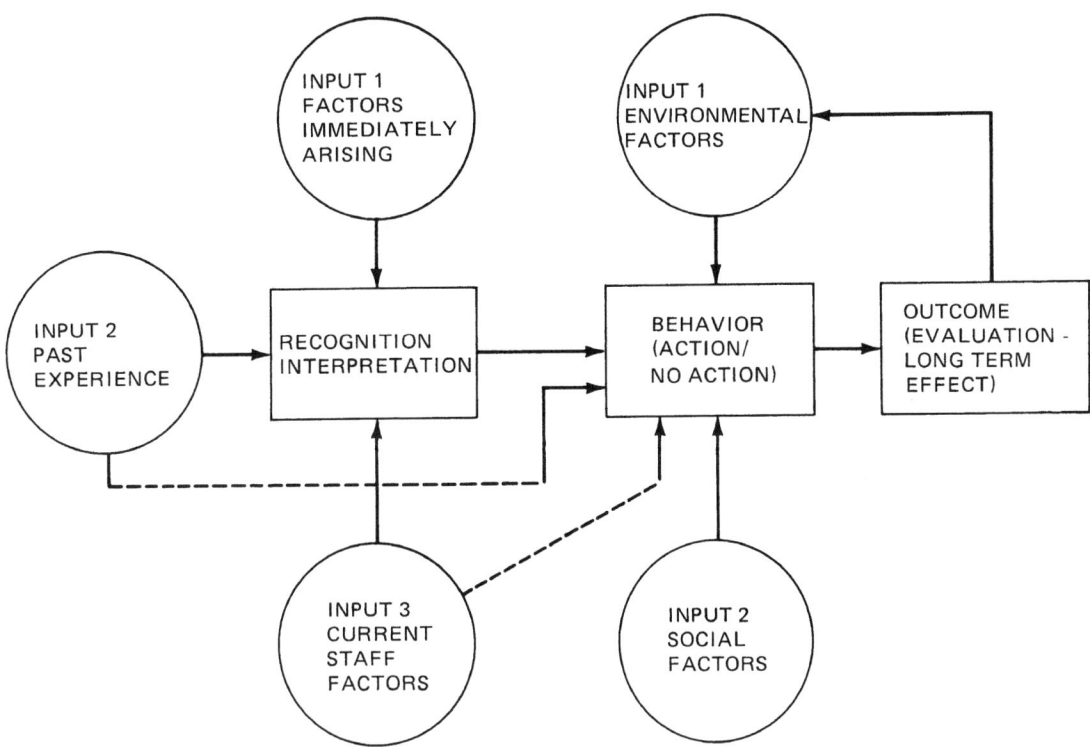

Fig. 3-12.3.   Suggested heuristic systems model of human behavior in fire incidents.

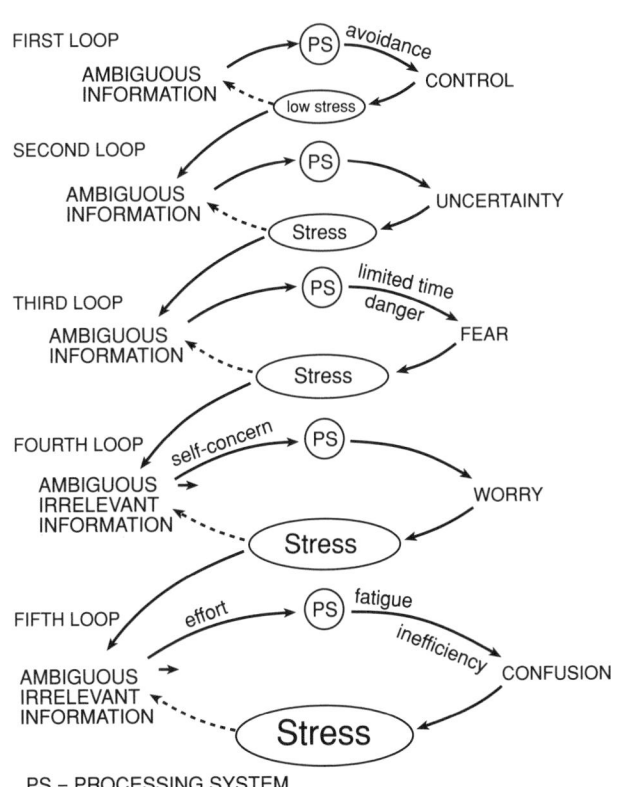

Fig. 3-12.4.   Stress model of people in a fire situation. (Source: reference 29.)

from studies of decision-makers in complex, time-critical situations. The critical variables of the naturalistic decision-making theory appear to have many of the environmental and psychological features of the fire situation involving building occupants. Chubb has identified these critical variables in the following manner:

- Ill-defined goals and ill-structured tasks
- Uncertainty, ambiguity, and missing data
- Shifting and competing goals
- Dynamic and continually changing conditions
- Real-time reactions to changed conditions
- Time stress
- High stakes
- Organizational goals and norms
- Experienced decision-makers[30]

Figure 3-12.5 is an illustration of the recognition-primed decision (RPD) model developed by Klein[31] from the studies of fire department officers. Chubb has correctly indicated that the limitation of this model, when applied to building occupants, is the lack of the fire officers' training and experience in building fire situations. That is, the static abilities relative to the mental and physical capabilities of building occupants are expected to be more varied and limited than those of fire officers. Chubb indicates that successful recognition-primed decision-making is dependent on the occupant training and practice of the firesafety plans, with the decision support system in the building consisting of egress signs, emergency lighting, and verbal communication systems.

TABLE 3-12.3 *First Actions of the British and U.S. Study Populations[1,2]*

Actions	British Percent	U.S. Percent	$P_1 - P_2$	$SE_{P_1-P_2}$	CR
Notified Others	8.1	15.0	6.9	1.38	5.00**
Searched for Fire	12.2	10.1	2.1	1.51	1.39
Called Fire Department	10.1	9.0	1.1	1.40	0.79
Got Dressed	2.2	8.1	5.9	0.85	6.94**
Left Building	8.0	7.6	0.4	1.27	0.31
Got Family	5.4	7.6	2.2	1.11	1.98*
Fought Fire	14.9	10.4	4.5	1.63	2.76**
Left Area	1.8	4.3	2.5	0.70	3.57**
Nothing	2.1	2.7	0.6	0.69	0.87
Had Others Call Fire Department	2.8	2.2	0.6	0.76	0.79
Got Personal Property	1.2	2.1	0.9	0.55	1.64
Went to Fire Area	5.6	2.1	3.5	1.01	3.47**
Removed Fuel	1.2	1.7	0.5	0.53	0.94
Enter Building	0.1	1.6	1.5	0.30	5.00**
Tried to Exit	1.6	1.6	0	0	0
Closed Door to Fire Area	3.1	1.0	2.1	0.76	2.76**
Pulled Fire Alarm	2.7	0.9	1.8	0.70	2.57*
Turned Off Appliances	4.1	0.9	3.2	0.85	3.20**
N = 18	2193	580			

*Critical ratio significant at or above the 5 percent level of confidence.
**Critical ratios significant at or above the 1 percent level of confidence.

BEHAVIORAL RESPONSES OF OCCUPANTS

A study by Wood[2] involved 952 fire incidents and 2,193 individuals interviewed by fire department personnel at the fire incident scene in Great Britain. Wood found the most frequent behavioral responses to fire could be categorized as involving the evacuation of the building, fighting or containing the fire, and the notification of other individuals or the fire brigade. Bryan[1] found similar types of broad categorization of behavioral response in a United States study primarily concerned with residential occupancies. This residential fire incident study involved interviews of 584 participants in 335 fire incidents by fire department personnel who used a structured questionnaire at the scene of the fire incident.

Table 3-12.3 presents the initial first actions in the studies of both the British and United States populations. The behavior of the individuals in both studies varied relative to

their gender, with the female and male behavior being primarily divided along culturally determined primary group roles. Thus, the males were predominantly more active in fighting the fire, while the females were predominantly concerned with alerting others and assisting others in evacuating the building.

There were ten statistically significant differences between the British and United States populations. The United States population was predominant at the five or one percent level of confidence for the first actions of "notified others," "got dressed," "got family," "left area," and "entered the building." A greater percentage of the British population engaged in the first actions of "fought fire," "went to fire area," "closed door to fire area," "pulled fire alarm," and "turned off appliances."

The general classification of the three actions for both the British and United States populations were categorized as "evacuation," "reentry," "fire fighting," "moved through smoke," and "turned back" behavior. The behavioral comparison of the two populations is presented in Table 3-12.4. There was a statistically significant difference between the

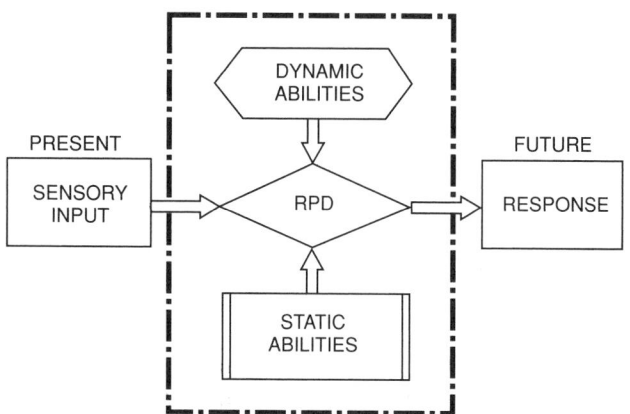

Fig. 3-12.5. Recognition-primed decision model. (Source: reference 30.)

TABLE 3-12.4 *Human Behavior of the British and U.S. Study Population[1,2]*

Behavior	British Percent	U.S. Percent	$P_1 - P_2$	$SE_{P_1-P_2}$	CR
Evacuation	54.5	80.0	25.5	2.30	11.09**
Reentry	43.0	27.9	15.1	2.30	6.57**
Fire Fighting	14.7	22.9	8.2	1.74	4.71**
Moved Through Smoke	60.0	62.7	2.7	2.29	1.18
Turned Back	26.0	18.3	7.7	2.01	3.83**
N = 5	2193	584			

*Critical ratio significant at or above the 5 percent level of confidence.
**Critical ratios significant at or above the 1 percent level of confidence.

populations in every behavioral response category except the "moved through smoke" behavior.

An interesting aspect of the actions of the participants involved in the United States study are the modifications in the first, second, and third behavior response actions of the participants. Table 3-12.5 presents these three actions for the total population of 584 individuals in the U.S. study. It should be noted how the action of "notifying others" accounted for 15 percent of the first actions and by the time of the third actions, accounted for only 5.8 percent of the behavioral actions. A similar reduction in the frequency of the behavioral response actions can be observed with the action of "searching for the fire" which is involved with a reduction in the activity from 10.1 percent as the first action to 0.8 percent for the third action. The behavioral actions of "getting dressed" and "got family" presented this same tendency toward a reduction in the frequency of the action with the progression during the fire incident. In contrast, an increase in the frequency of the action from the first to the third actions may be noted with the actions of "called fire department," "left the building," and "fought the fire."

Canter, Breaux, and Sime[32] developed a decomposition diagram of the acts of 41 persons in 14 domestic fires. This study was conducted in the United Kingdom, and the domestic occupancies were similar to those studied by Wood[2] and

TABLE 3-12.5 *Compilation of the First, Second, and Third Action of the U.S. Study Populations[1]*

Actions	1st Action Percent	2nd Action Percent	3rd Action Percent
Notified Others	15.0	9.6	5.8
Searched for Fire	10.1	2.4	0.8
Called Fire Department	9.0	14.6	12.7
Got Dressed	8.1	1.8	0.3
Left Building	7.6	20.9	35.9
Got Family	7.6	5.9	1.4
Fought Fire	4.6	5.7	11.5
Got Extinguisher	4.6	5.3	1.6
Left Area	4.3	2.8	1.1
Woke Up	3.1	0	0
Nothing	2.7	0	0
Had Others Call Fire Department	2.2	4.0	4.1
Got Personal Property	2.1	3.8	0.8
Went to Fire Area	2.1	1.0	0
Removed Fuel	1.7	1.0	1.1
Enter Building	1.6	0.8	1.1
Tried to Exit	1.6	2.4	0.5
Went to Fire Alarm	1.6	1.8	1.1
Telephoned Others—Relatives	1.2	0.6	1.1
Tried to Extinguish	1.2	1.8	1.9
Closed Door to Fire Area	1.0	0.2	0.3
Pulled Fire Alarm	0.9	0.6	0.5
Turned Off Appliances	0.9	0.6	0.3
Check on Pets	0.9	1.4	0.5
Await F.D. Arrival	0	1.0	3.6
Went to Balcony	0.2	0.8	2.7
Removed by Fire Department	0	0	1.6
Open Doors—Windows	0.2	0.4	1.1
Other	3.9	8.8	6.6
N = 29	100.0	100.0	100.0

Bryan.[1] The decomposition diagram is shown in Figure 3-12.6, and should be compared with Tables 3-12.3 through 3-12.5. The sequence of the first, second, and third actions of the U.S. study population is generally similar to the action sequence in Figure 3-12.6.

Behavior According to Gender

The differences between the first behavioral response actions of the occupants according to the gender of the occupants has been examined by Bryan.[1] Table 3-12.6 presents the initial actions of the United States study population relative to the gender of the participants.

There were significant statistical differences between males and females in the categories of "searched for fire," "called fire department," "got the family," and "got extinguishers." Male participants were predominant in fire fighting activities. Thus, 14.9 percent of the males participated in the behavioral response of "search for fire" as opposed to 6.3 percent of the females, and 6.9 percent of the males were involved in the action of "got extinguishers" as opposed to 2.8 percent of the females. In the United States population, females differed significantly from the males in the warning and evacuation activities—11.4 percent of the females "called the fire department" as their initial behavioral response action as opposed to 6.1 percent of the males. In relation to the evacuation behavior, 10.4 percent of the females left the building as the first behavioral response action, contrasted with 4.2 percent of the males. The cultural role influence on female participants is probably explicitly indicated in the concern for other family members, with the indication that 11 percent of the females "got the family" as the first behavioral response, while only 3.4 percent of the males engaged in this behavioral response. It should be noted that the male actions of "searched for fire" or "fought the fire" were matched by the female actions of "alarm initiating" and "evacuation behavior." This identical pattern of behavioral responses has also been observed in fire incidents in health care and educational occupancies.

BEHAVIOR IN HOTEL FIRE INCIDENTS

The fire protection engineering concepts related to the protection of the occupants of high-rise buildings have been examined and analyzed following the fire incident in the MGM Grand Hotel fire in Clark County, NV, on November 21, 1980.[33] This hotel fire involved both injuries and fatalities among the guests.

The management of the MGM Grand Hotel, and the Clark County Fire Department under Chief Roy L. Parrish, in cooperation with the National Fire Protection Association (NFPA),[34] conducted an intensive study of the guests registered in the hotel for the evening of November 20 to 21, 1980, to determine how the occupants became aware of the fire incident and their behavioral responses.

The MGM Grand Hotel fire was discovered by an employee of the hotel who entered the deli-restaurant located on the casino level of the hotel at approximately 7:10 a.m. on November 21, 1980. This restaurant area was unoccupied at the time, and the hotel operator was immediately notified to call the fire department. The Clark County Fire Department received the notification by the direct telephone line call from the hotel at 7:17 a.m., and the first fire company arrived on the scene from a fire station directly across the street on

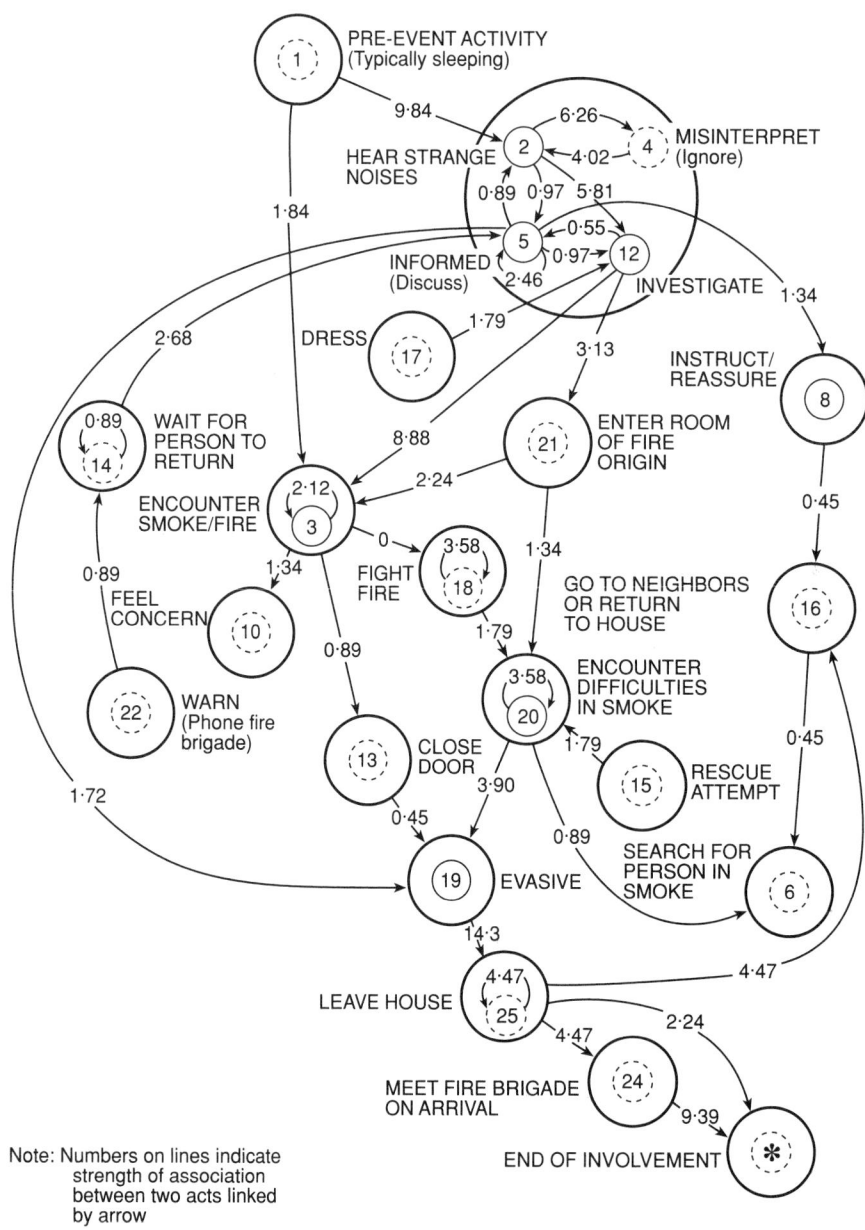

Fig. 3-12.6. *Decomposition diagram—domestic fires. (Source: reference 32.)*

Flamingo Road at approximately 7:18 a.m. The hotel telephone operators were forced from their switchboard positions by the smoke immediately after they had initiated an announcement on the public address system at approximately 7:20 a.m. for the evacuation of the casino area. The fire reached a flashover condition in the deli area, immediately spread from east to west through the main casino area, and extended out the west portico doors on the casino level immediately following the arrival of the initial fire department personnel.

An addition to the hotel was being constructed adjacent to the west end of the building, and construction workers at this addition participated in warning guests, assisted in the fire fighting, and evacuation of guests. The heat and smoke extended from the casino area through seismic joints, elevator shafts, and stairways throughout the 21 residence floors

of the hotel. The heat was intense enough on the 26th and top floor to activate automatic sprinkler heads located in the lobby area adjacent to the elevator shafts.

Due to the rapid early evacuation of the telephone staff, guests in their rooms were not alerted by the hotel public address system or the local fire alarm system. Guests who became alerted early in the fire incident, or guests already awake and dressed, were able to escape prior to the smoke conditions becoming untenable on the residential floors. Guests alerted later in the progression of the fire incident remained in their rooms or moved to other rooms, often with other occupants. The flame propagation did not extend above the casino level, with the exception of very minor extension into two guests rooms on the 5th floor. The fire resulted in 85 fatalities to guests and hotel employees, with the fatalities apparently occurring in the following areas of the hotel:[33]

TABLE 3-12.6 *First Actions of the U.S. Study Population Classified as to the Gender of the Participant[1]*

First Action	Male Percent	Female Percent	$P_1 - P_2$	$SE_{P_1 - P_2}$	CR
Notified Others	16.3	13.8	2.5	2.98	0.83
Searched for Fire	14.9	6.3	8.6	2.51	3.43**
Called Fire Department	6.1	11.4	5.3	2.41	2.19**
Got Dressed	5.8	10.1	4.3	2.30	1.87
Left Building	4.2	10.4	6.2	2.22	2.79**
Got Family	3.4	11.0	7.6	2.22	3.42**
Fought Fire	5.8	3.8	2.0	1.77	1.13
Got Extinguishers	6.9	2.8	4.1	1.77	2.31**
Left Area	4.6	4.1	0.5	1.70	0.29
Woke Up	3.8	2.5	1.3	1.45	0.90
Nothing	2.7	2.8	0.1	1.38	0.72
Had Others Call Fire Department	3.4	1.3	2.1	1.23	1.71
Got Personal Property	1.5	2.5	1.0	1.17	0.85
Went to Fire Area	1.9	2.2	0.3	1.20	0.25
Removed Fuel	1.1	2.2	1.1	1.08	1.02
Enter Building	2.3	0.9	1.4	1.02	1.37
Tried to Exit	1.5	1.6	0.1	1.05	0.09
Went to Fire Alarm	1.1	1.9	0.8	1.02	0.78
Telephoned Others—Relatives	0.8	1.6	0.8	0.91	0.87
Tried to Extinguish	1.9	0.6	1.3	0.91	1.43
Closed Door to Fire Area	0.8	1.3	0.5	0.87	0.57
Pulled Fire Alarm	1.1	0.6	0.5	0.75	0.66
Turned Off Appliances	0.8	0.9	0.1	0.79	0.12
Check on Pets	0.8	0.9	0.1	0.79	0.12
Other	6.5	2.5	4.0	1.70	2.35*
N = 25	262	318			

*Critical ratios significant at or above the 5 percent level of confidence.
**Critical ratios significant at or above the 1 percent level of confidence.

14 persons were found on the casino level, 29 persons were found in guest rooms, 21 individuals were found in corridors and lobbies, 9 persons were found in the stairways, and 5 persons were found in elevators. The victims were located on the casino level, and the 16th through 25th floors, with the majority of fatalities found between the 20th and the 25th floors.

Figure 3-12.7 is a diagram of the guest floor arrangement and layout of the MGM Grand Hotel which was used in the occupant questionnaire study conducted by the National Fire Protection Association.[34] Of the nine individuals found in the stairways it should be noted that two persons were found in Stairway number 1 at the extreme south end of the south wing, at the 17th floor; six persons were found between the 20th and 23rd floors in Stairway number 2, at the central end of the south wing; and one individual was found at the ground-floor level of Stairway number 4 at the extreme west end of the west wing.

Various estimates have been provided of the number of guests and fire department personnel that suffered injuries at the MGM Grand Hotel fire. Morris[35] indicated that 619 persons were transported to hospitals from the fire scene, and another 150 guests were treated at the Las Vegas Convention Center, where the survivors had been transported from the hotel. It should be realized that the MGM Grand Hotel fire was a unique fire incident in several aspects. First, it was the second most serious hotel fire in the United States, being surpassed in terms of the loss of life only by the Winecoff Hotel fire in Atlanta, GA on December 7, 1946. Second, it was the first high-rise fire in the United States in which helicopter evacuation was involved for about 300 guests, while the fire department rescued approximately 900 guests.

The Clark County Fire Department obtained from the management of the MGM Grand Hotel a list of the guests registered in the hotel for the evening of November 20 to 21, 1980. This list was transmitted to the NFPA which prepared a three-page, 28-item questionnaire, with the floor plan of the guest rooms attached as the fourth and last page of the questionnaire. A total of 1,960 questionnaires were mailed on December 19, 1980, and 554 questionnaires were returned. Included with the questionnaire was an interview request form by which the respondents indicated their willingness to be interviewed in person about their experience. Of the 554 questionnaires returned, a response rate of approximately 28 percent, 455 individuals (or 82 percent of the response study population) indicated they would be willing to be interviewed.

The age of the questionnaire population ranged from 20 to 84 years, with an average age of 45 years. The population consisted of 331 males and 222 females, with one respondent not indicating a gender classification. The guest population included 103 individuals who indicated they were alone at the time they became aware of the occurrence of the fire within the hotel. The presence of other persons, especially if members of the individual's primary group, has previously appeared to determine the response of some individuals in residential fire incidents.

The initial five behavioral responses of the 554 guests as elicited from the NFPA questionnaire study are presented in Table 3-12.7. The five most frequent first behavioral responses were "dressed," "opened door," "notified roommates," "partially dressed," and "looked out window." The guests involved in the first responses were predominantly

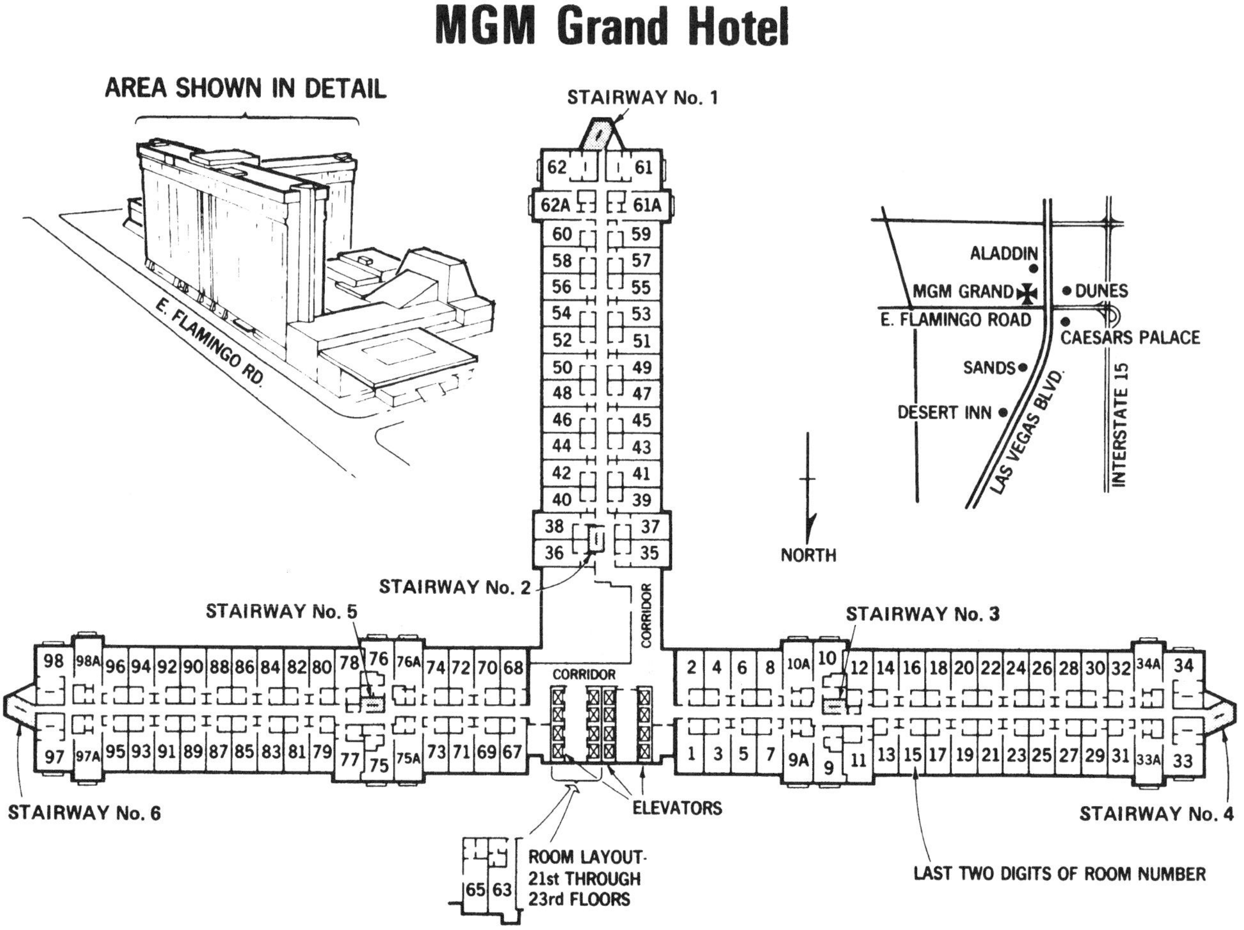

MGM Grand Hotel

AREA SHOWN IN DETAIL

Fig. 3-12.7. Residential floor diagram of the MGM Grand Hotel.

engaged in attempting to define and structure the fire incident cues relative to the severity of the threat to themselves. Approximately 8 percent of the study population initiated or attempted to initiate their evacuation behavior with the first response, as indicated by the actions of "attempted exit," "went to exit," and "left room." Sixteen individuals, 2.9 percent of the population, initiated actions to improve the room as an area of refuge with the actions "wet towels for face" and "put towels around door." The behavioral responses of the guests in this questionnaire study could be classified as evacuation responses or refuge preparation responses. The responses relating to the evacuation behavior appeared to be initiated early if the means of egress were clear of smoke, or the smoke was determined to be nonthreatening by the guests. However, if the smoke was heavy, the guests appeared to initiate the behavioral response of staying in their rooms or moving to more suitable rooms with responses designed to prevent smoke migration into the rooms and to protect themselves from the smoke.

Examination of Table 3-12.7 indicates the five most frequent behavioral responses reported by guests as second actions were "opened door," "dressed," "went to exit," "partially dressed," and "secured valuables." Approximately 19 percent of the study population reported they were still involved in the dressing actions prior to initiating evacuation or refuge procedures.

Examination of the third behavioral responses of the 537 guests in the study population indicated the responses of the guests generally progressed to evacuation, attempted evacuation, and notification responses. Thus, approximately 25 percent of the MGM Grand Hotel fire incident study population was involved in evacuation-related behavioral responses, and approximately 10 percent of the guests were involved in attempted evacuations as identified by their third responses of "attempted to exit" and "returned to room." The alerting and notification actions of the guests were involved with the third behavioral responses of "notified occupants" and "notified other room."

The fourth behavioral responses of the guests in the study population indicated a progression of the guests to evacuation, attempted evacuation, and self-protection or room refuge procedural responses. The most frequent action of the guests in their fourth responses was the behavior of "went to exit" (approximately 16 percent of this population). However, combining the guests involved with this action with the guests utilizing the actions of "went down stairs," "went to another exit," "left hotel," and "left room," there were 151 guests (approximately 30 percent of the fourth action guest population) involved in evacuation actions. The process of the guests forming convergence clusters was noted in this hotel fire. This action involved individuals clustering together in rooms as areas of refuge, with the

TABLE 3-12.7 *Compilation of the Initial Fire Actions of Guests in the MGM Grand Hotel Fire Incident[34]*

Actions	Percent of Population				
	First	Second	Third	Fourth	Fifth
Dressed	16.8	11.6	6.5	—	—
Opened Door	15.9	11.7	6.7	3.4	—
Notified Roommates	11.6	3.0	—	—	—
Dressed Partially	10.1	7.5	4.5	—	—
Looked Out Window	9.7	5.7	—	—	—
Got Out of Bed	4.5	—	—	—	—
Left Room	4.3	5.4	8.1	2.1	2.0
Attempted to Phone	3.4	3.6	—	2.8	—
Went to Exit	2.5	10.3	9.5	16.1	6.7
Put Towels Around Door	1.6	2.5	3.0	6.8	7.7
Felt Door for Heat	1.3	2.3	—	—	—
Wet Towels for Face	1.3	3.7	6.3	4.6	7.9
Got Out of Bath	1.1	—	—	—	—
Attempted to Exit	1.1	3.0	5.8	—	—
Secured Valuables	—	6.8	4.3	—	—
Notified Other Room	—	3.4	2.2	—	—
Attempted to Exit	—	—	—	4.3	—
Returned to Room	—	—	3.9	8.1	4.1
Went Down Stairs	—	—	3.9	5.4	21.3
Left Hotel	—	—	3.4	2.6	2.0
Notified Occupants	—	—	3.0	—	—
Went to Another Exit	—	—	—	3.6	4.8
Went to Other Room	—	—	—	3.6	4.6
Went to Other Room/Others	—	—	—	3.4	8.7
Looked for Exit	—	—	—	2.4	—
Broke Window	—	—	—	—	4.3
Offered Refuge in Room	—	—	—	—	1.8
Went Up Stairs to Roof	—	—	—	—	2.9
Went to Balcony	—	—	—	—	1.8
Other	14.8	19.5	28.9	30.2	20.1
Total (percent):	100.0	99.1	96.9	90.6	79.6
No. of guests:	554	549	537	501	441

clustered individuals characterized as individuals usually not known to each other prior to the occurrence of the fire incident. The fourth responses of "went to other room" and "went to other room/others," are explicit indicators of the initiation and formation of convergence clusters.

The fifth behavioral responses of the guests were primarily for self-protection, including the improvement of the room as an area of refuge and evacuation behavior. The evacuation behavioral responses would consist of the fifth responses of "went up stairs to roof," "left hotel," and "left room." Thus, the guests involved with evacuation actions consisted of 175 individuals (approximately 40 percent of this study population). The guests who decided not to evacuate and were thus concerned with refuge procedures utilized the fifth actions of "went to other room/others," "wet towels for face," "put towels around door," "broke window," "returned to room," "went to other room," "offered refuge in room," and "went to balcony." Approximately 40 percent of the fifth response study population was involved in the refuge procedures and self-protection actions.

Convergence Clusters

The phenomenon of occupant convergence cluster formation in a fire incident was initially noticed in a study of the occupant behavior in a 1979 high-rise apartment building fire.[36] Convergence clusters appear to involve the convergence of the occupants of the building involved in the fire into specific rooms selected as being more ideally suited as areas of refuge. In the MGM Grand Hotel fire, the guests tended to select rooms on the north side of the east and west wings, and rooms on the east side of the south wing due to the prevailing atmospheric conditions and the external smoke migration. In addition, guests reported that people had converged in rooms that had balconies and doors leading to the balconies because of the ease of ventilation, the reduced-smoke exposure, improved visibility, and the communication advantages the balconies offered. The guests who reported their participation in the convergence behavior in rooms with other persons provided either numerical estimates of the persons occupying the room or suite, or indicated only that "others" or "other persons" were present.

Table 3-12.8 lists the rooms that were identified by guests as being areas of refuge for a total of three or more persons with individuals other than the original occupants of the room. This table also presents the estimates of the length of time the convergence cluster was maintained in the rooms. The duration of the cluster was usually maintained until assistance was obtained for evacuation, or until the

TABLE 3-12.8 *Compilation of the Time Duration, Room Numbers, and Number of Guests Involved in Convergence Clusters in the MGM Grand Hotel Fire Incident[34]*

Floor	Room Number	Time (Hours)	Persons	
			No.	Percent
7	731	0.6	3	0.7
8	827, 840	1.5–1.75	14*	3.3
9	927	2.5	5	1.2
10	1009A, 1025, 1034, 1060	1.2	53	12.7
11	1129, 1115	1.5–2	30*	7.2
12	1261, 1225, 1233A	2–3	53	12.7
14	1433A, 1461A, 1451, 1416A	1.5–2	8*	1.9
15	1501, 1533A, 1510	2–3	38*	9.1
16	1643, 1625, 1633, 1629, 1627, 1615	2–3.5	35*	8.4
17	1725, 1775, 1731, 1719, 1762, 1756, 1733A	2–2.5	84	20.1
18	1819, 1802, 1850	2–3	20	4.8
19	1929, 1919, 1962A, 1961, 1925	2–3.5	13*	3.1
20	2027, 2013, 2030	2–3	13	3.1
22	2213, 2221, 2229	2.5–3.5	25	6.0
23	2329, 2314, 2342, 2331, 2308, 2340	2.5–3.25	20*	4.8
24	2446	3.5	4	0.9
25	2512, 2509A	3.5	*	0
Total: 17	57		418	100.0
Range 7–25	1–7	0.6–3.5	3–84	0–20.1

*Persons indicated only as "Others."

occupants were notified by fire or rescue personnel to initiate evacuation. The number of persons shown in the table indicates the total number of persons in the clusters for the total number of rooms identified on the floor. The smallest number of people identified as a single cluster involved three persons, and the largest was 35 persons.

The greatest number of rooms used for convergence clusters, and of course the largest population participating in convergence clusters, was located on the 17th floor of the hotel. No convergence clusters were identified by guests as occurring on the 6th, 21st, or 26th floors. It would appear that the convergence clusters may serve as an anxiety and tension-reducing mechanism for individuals confronted with a fire incident they perceived as life threatening. The

action of "offered refuge in room," which was previously identified in the discussion of the fifth behavioral responses, is a definitive indication of the formation of a convergence cluster.

In addition to the detailed human behavior study of the MGM Grand Hotel fire,[37] the NFPA conducted a similar questionnaire study of the guest's behavior in the Westchase Hilton Hotel fire.[38]

Figure 3-12.8 presents the decomposition diagram for eight multiple-occupancy fires with the acts of 96 persons.[32] These multiple-occupancy fires in the United Kingdom involved hotel occupancies. Figure 3-12.8 and Table 3-12.7 should be compared to illustrate the similarity of the occupants' behavior in both of these studies.

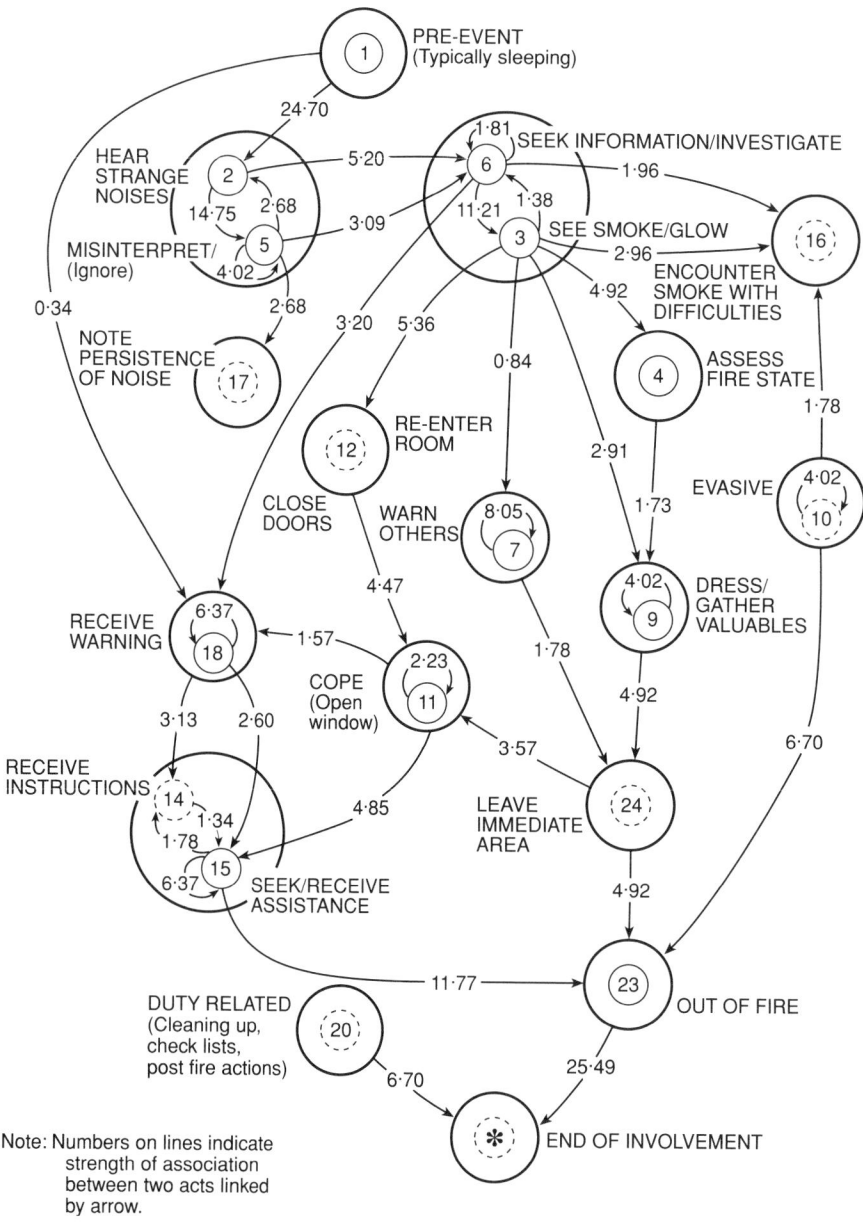

Note: Numbers on lines indicate strength of association between two acts linked by arrow.

Fig. 3-12.8. *Decomposition diagram—multiple-occupancy fires. (Source: reference 32.)*

NONADAPTIVE BEHAVIOR

The classic types of nonadaptive behavior in a fire incident involve the disregard of adaptive actions or behavior that might facilitate the evacuation of others or inhibit the propagation of the smoke, heat, or flame from a fire incident. Nonadaptive behavior may consist of a single behavioral response of leaving the room of fire origin without closing the door to the room, thus allowing the fire to propagate throughout the structure and endanger the lives of other individuals. However, the more generalized concept of nonadaptive behavior consists of an audience population fleeing from the fire incident without regard for others and inflicting physical injuries on themselves and others.

Nonadaptive behavior responses may be an omission, such as forgetting to close a door, or may involve a response that, although well meaning and positive in intent, results in negative consequences. When the action of a behavioral response results in the extinguishment of the fire and the reduction of the threat, the behavior may be said to be adaptive. However, when such a behavioral response is ineffective because the fire was more developed than perceived, the response might have been more adaptive to warn others and notify the fire department. Thus, it would appear that some behavior which appears to be nonadaptive, in reality, is unsuccessful behavior that would have appeared to be the most adaptive only if it had been successful. Injuries suffered by persons in relation to fire incidents may be cues to nonadaptive or risky behavior by the individual.

Panic Behavior

The concept of panic behavior is the nonadaptive behavioral response that is always discussed following fire incidents such as the Beverly Hills Supper Club fire[39] where multiple fatalities occurred.

According to most definitions, panic is a flight or fleeing type of behavioral response that also involves extravagant and injudicious effort. Panic is not likely to be limited to a single individual but may be mimicked and adopted by a body of persons. Schultz[40] has defined a panic type of behavior reaction from his simulation experiments in the following manner:

> A fear-induced flight behavior which is nonrational, nonadaptive, and nonsocial, which serves to reduce the escape possibilities of the group as a whole.[40]

The concept of panic is often used to explain the occurrence of multiple fatalities in fires even when there is no physical, social, or psychological evidence indicating that competitive, injudicious flight behavior actually occurred. Representatives of the media and public officials often label various types of fire-incident behavioral responses as panic. The evidence accumulated from interviews with participants and the questionnaires completed by occupants provided no evidence of the classical group type of panic behavior with competitive flight for the exits in the Beverly Hills Supper Club fire.[41]

Sime[42] has indicated that panic as a concept is primarily a description of the behavior, not an explanation of the behavior. He pointed out that the concept is used to support the introduction of requirements in building laws or ordinances to provide for the firesafety of the building occupants. Sime has also very aptly shown the difference between the use of the concept to describe other persons' behavior in a fire incident, and the use of the concept to describe an individual's behavior which is accompanied by a high state of concern and anxiety. As Sime has indicated, simply because an individual identifies behavior as being associated with a panic reaction, this does not necessarily identify the behavior as being the classic panic type behavioral response. Sime also indicates the outcome of the behavior, as previously discussed, affects its labeling and that the actual behavior of people in a fire is most likely to be misinterpreted when the outcome of the fire incident has been unfortunate.

As Sime indicated, the use of the concept of panic must be separated from the use of the terms "anxiety" or "fear." The concept of self-destructive or animalistic panic type behavioral responses to fire incident stimuli such as the presence of flames or smoke has not been supported by the research on human behavior in fire incidents. As indicated by Sime,[42] Quarantelli,[43] and others,[1,2,4,44] panic behavior in which the flight response is characterized by actual physical competition between the participants and personal injuries is rare.

In his interview study of 100 participants of single-dwelling residential fires, Keating[44] reported no instances of panic behavior and instead found primarily altruistic, helpful behavioral responses.

Ramachandran,[6] in his review of studies on human behavior in fire in the United Kingdom, has developed the following conclusion relative to nonadaptive behavior:

> In the stress of a fire, people often act inappropriately but rarely panic or behave irrationally. Such behavior, to a large extent, is due to the fact that information initially available to people regarding the possible existence of a fire and its size and location is often ambiguous or inadequate.[6]

Reentry Behavior

The study of the 1956 Arundel Park fire documented the initial examination of the phenomenon of reentry behavior.[19] Some codes and regulations affecting the design of the means of egress appeared to be predicated on the assumption that pedestrian traffic will only move away from the fire area, and away from the fire area or floor of the building involved. The Arundel Park study[19] indicated that approximately one-third of the individuals interviewed had reentered the building.

TABLE 3-12.9 *Reasons for Reentry Behavior by Occupants in the Arundel Park Fire, Classified by Gender of the Occupants*[19]

Gender	Reentered and left same exit	Reentered and left different exit	Stated reason for reentrance
M	1		Turn Off Kitchen Stoves
M	1	1	Tell People to Leave
M	3	1	To Help
M	1		Assist People
M	2	3	Find Wife
M	2	2	Assist Fire Fighting
M & 1 F		5	No Stated Reason
21 M & 1 F	10	12	

It is apparent that the means of egress components—primarily doors, stairways, and corridors—may be subjected to two-way movement from building occupants or other personnel. The behavioral response of the occupant who after safely leaving the building turns around and reenters has been observed most frequently in the residential fire incident. The occupant is often completely aware of the occurrence of the fire in the building and of the specific portions of the building involved in the area of fire origin and smoke propagation. Table 3-12.9 presents the number of individuals who reentered in the Arundel Park fire from the interviewed population of 61 persons. The reasons for the reentry behavior and the fact that the reentry participants were predominantly male in this fire incident should be noted.

The Arundel Park fire incident occurred in an assembly occupancy being utilized for a church-sponsored oyster roast, which is a family type of event. Thus, the primary group cultural role of father or husband may have been a critical variable in the reentry behavior in the population interviewed and may have resulted in the fact that the reentry participants were mostly male. Reentry behavior should not be considered as nonadaptive behavior since this type of behavior response is often engaged in to assist or rescue persons remaining or believed to be remaining in the fire incident building. The reentry type of behavior is often engaged in by parents when children are missing during a fire incident. This reentry behavior is often undertaken in a rational, deliberate, and purposeful manner, without the emotional anxiety or self-anxiety characteristics often associated with nonadaptive behavior. However, reentry behavior has usually been considered nonadaptive since it negatively affects the efficient and effective egress of other persons in the building (i.e., those initially leaving the building are impeded at any egress selected for reentry).

The reasons elicited from participants in reentry behavior in residential occupancies[1] in the United States are presented in Table 3-12.10. It would appear from the total study population of 584 persons that 162 people or approximately 27.9 percent engaged in reentry behavior. The most frequent reason for reentry behavior in this study was "to fight the fire," followed by "to obtain personal property," "to check on the fire," "to notify others," "to assist the fire department," and "to re-

TABLE 3-12.10 *Reasons for Reentry of the Population in the Project People Study*[1]

Reasons	Participants	Percent
Fight Fire	36	22.2
Obtain Personal Property	28	17.2
Check on Fire	18	11.0
Notify Others	13	8.0
Assist Fire Department	12	7.4
Retrieve Pets	12	7.4
Call Fire Department	9	5.5
Assist Evacuation	4	2.5
Taken to Hospital	3	1.8
Turn Power Back On	2	1.2
Rescue from Balcony	1	0.6
Help Injured Family Member	1	0.6
Turn Off Gas	1	0.6
Open Windows	1	0.6
Close Door	1	0.6
No Apparent Danger	1	0.6
Entered Non-Danger Area	1	0.6
Responsibility	1	0.6
Due to Fire	1	0.6
Told to by Others	1	0.6
Not Reported	16	9.8
N = 21	163	100.0

Range = 1 – 36   Percent of Participant Population = 27.9

trieve pets." These six reentry behavioral response reasons accounted for approximately 73 percent of reentry behavior.

In Table 3-12.11, which compares the reentry behavior of the British[2] and the United States[1] study populations, it should be noted that all of the behavioral response reasons were significantly different statistically, with the exception of the reason "save personal effects." The United States population was predominant with the reentry reasons of "save personal effects," "call the fire department," "rescue pets," "notify others," "assist fire department," and "assist the evacuation." The British population was predominant with the reentry reasons of "fight the fire," "observe the fire," "shut doors," "await fire department," and "fire not severe."

TABLE 3-12.11 *Comparison of Reasons for Reentry Behavior of British and U.S. Study Populations*[1,2]

Reasons	British Percent	U.S. Percent	$P_1 - P_2$	$SE_{P_1 - P_2}$	CR
Fight Fire	36.0	22.2	13.8	4.02	3.34**
Observe Fire	19.0	11.0	8.0	3.25	2.46*
Save Personal Effects	13.0	17.2	4.2	2.91	1.44
Shut Doors	10.0	0.6	9.4	2.38	3.95**
Await Fire Department	9.0	0	9.0	2.26	3.98**
Call Fire Department	2.0	5.5	3.5	1.32	2.65**
Rescue Pets	2.0	7.4	5.4	1.40	3.86**
Fire Not Severe	5.0	1.2	3.8	1.74	2.18*
Notify Others	0	8.0	8.0	0.92	8.69**
Assist Fire Department	0	7.4	7.4	0.88	8.41**
Assist Evacuation	0	2.5	2.5	0.54	4.63**
N = 11	943	163			

*Critical ratios significant at or above the 5 percent level of confidence.
**Critical ratios significant at or above the 1 percent level.

TABLE 3-12.12 *Gender and Age of U.S. Participants Engaging in Fire Fighting Behavior[1]*

Gender	Participants	Percent
Male	84	62.7
Female	50	37.3
Total	134	100.0
Age		
7–17	8	5.9
18–27	31	23.1
28–37	41	30.6
38–47	27	20.1
48–57	16	11.9
58–67	2	1.5
68–80	3	2.2
Unknown	6	4.7
Total	134	100.0

Percent of Participant Population = 22.9

OCCUPANT FIRE FIGHTING BEHAVIOR

Occupants who engaged in fire fighting behavior during fire incidents have been predominantly male. This behavior is now believed to be primarily a culturally determined and assumed aspect of the male role in certain social and occupational situations.

However, in the study of 335 primarily residential fire incidents,[1] it was found that 37.3 percent of those who chose to fight the fire were females, with the youngest participant being a girl seven years old. The fire fighting population in this study of 134 persons consisted of 50 females and 84 males. The age range of those who engaged in the fire fighting behavior varied from the 7-year-old girl to an 80-year-old man. The distribution of the participants relative to sex and age is presented in Table 3-12.12. Approximately 23 percent of the total study population were involved in occupant fire fighting behavior.

The majority of those involved in the fire fighting behavior were between 28 and 37 years old, consisting of approximately 30 percent of the fire fighting behavior population. Statistically significant differences in the behavioral responses of the males and females were shown in the responses of "got extinguisher" and "fought fire." Approximately 15 percent of the female population reacted by obtaining extinguishers. Approximately 26 percent of the male population reported they fought the fire when they first became aware of the fire incident, as opposed to approximately 10 percent of the female occupants. The female members of the population predominantly notified the fire department before initiating the fire fighting type of behavioral response. Approximately 33 percent of the females versus 25 percent of the males reacted to the fire incident by initially notifying the fire department as indicated in Table 3-12.13.

The occupant fire fighting behavioral responses appear most prevalent in occupancies in which the individuals are emotionally or economically involved (primarily their homes) or where such behavior is a result of training or an assigned occupational role. A total of 285 individuals at some time during a fire incident engaged in one of the six actions defined as a fire fighting behavioral response, and a total of 252 individuals participated in one of the four actions relative to notification of the fire department.

In the study by Crossman *et al*[45] of residential fire incidents in Berkeley, CA, a total of 180 persons were involved in fire fighting behavioral responses. The total of 208 fire incidents for this study included approximately 167 fire incidents, or approximately 80 percent that had not been reported to the fire department. The majority of the 167 unreported fire incidents had been extinguished by the occupants of the building involved in the fire incident or the

TABLE 3-12.13 *Comparison of the Gender of the Participants Engaging in Fire Fighting and Notification of the Fire Department[1]*

Action	Male Percent	Female Percent	$P_1 - P_2$	$SE_{P_1-P_2}$	CR
Searched for Fire	17.2	9.1	8.1	4.23	1.91
Got Extinguisher	15.6	6.0	9.6	3.95	2.43*
Fought Fire	25.6	9.7	15.9	4.83	3.29**
Removed Fuel	3.4	3.1	0.3	2.17	0.14
Tried to Extinguish	5.3	2.8	2.5	2.49	1.00
Went to Fire Area	3.1	2.8	0.3	2.07	0.14
Total	70.2	33.5	36.7	6.01	6.11**
N =	184	101			
Called Fire Department	25.6	33.0	7.4	5.83	1.27
Had Others Call Fire Department	9.2	7.5	1.7	3.27	0.52
Went to Fire Alarm	3.8	3.8	0	0	0
Pulled Fire Alarm	1.9	1.6	0.3	1.65	0.18
Total	40.5	45.9	5.4	6.31	0.85
N =	106	146			

*Critical ratio significant at or above the 5 percent level of confidence.
**Critical ratio significant at or above the 1 percent level of confidence.

TABLE 3-12.14 *Percentage of Occupants Extinguishing Residential Fires in Berkeley, CA[45]*

Fire Suppressed by	Percent
Person Engaged in Heat-Using Activity	52.8
Other Member(s) of Household	28.3
Friends and Neighbors	8.9
Fire Department	18.9
Burnout	6.1
Total	115.0
Single Individual	80.7
Group Effort	19.3
Total	100.0

occupants assisted by neighbors. Table 3-12.14 presents the study's percentage distribution of the individuals responsible for extinguishing the fire. Six percent of these fire incidents self-extinguished, and 52 percent of the fires were extinguished by the individual engaged in the heat-using activity that created the fire incident. As a means of comparison, the fire incidents in the Project People Study[1] may seem to have consisted of incidents that were judged uncontrollable by the occupants and resulted in the notification of the fire department. It should be remembered that approximately 85 percent of the fire occurrences in the National Fire Prevention and Control Administration National Household Fire Incident Survey[46] had also not been reported to the fire department.

The types of occupancies in which equipment provided within the occupancy was used to fight the fire are shown in Table 3-12.15. The apparently high frequency of residential occupancies, with 64 percent of the occupancies being either single-family dwellings or apartments, may be a variable created by the fire incident population of this study. This residential occupancy distribution may also be representative of many urban areas where the fire problems are essentially concentrated in residential occupancies.

TABLE 3-12.15 *Occupancies in Which Fire Fighting Equipment Was Utilized by Participants in Fire Fighting Behavior[1]*

Occupancy	Incidents	Percent
Dwelling (1 Family)	23	35.9
Apartment (< 20 Units)	18	28.1
Restaurant	3	4.8
Apartment (> 20 Units)	3	4.8
Manufacturing	2	4.8
Hotel and Motel	2	3.2
School	3	3.2
Billiard Center	1	1.5
City Club	1	1.5
Hospital	1	1.5
Dwelling (2 Family)	1	1.5
College Dormitory	1	1.5
Service Station	1	1.5
Office	1	1.5
Photographic Laboratory	1	1.5
Other	2	3.2
N = 16	64	100.0

TABLE 3-12.16 *Elicited Reasons of Participants for Not Voluntarily Leaving the Fire Incident Building[1]*

Reason	Participants	Percent
Fight Fire	52	48.7
Notify Others	7	6.5
Blocked by Smoke	7	6.5
Blocked by Fire	5	4.7
Overcome by Smoke	5	4.7
Search for Fire	3	2.8
Needed Help	2	1.9
Secure Property	2	1.9
Afraid of Fire Spread	2	0.9
No Fire in Area	1	0.9
Help Others	1	0.9
Does Not Know	1	0.9
No Response to Fire Department	1	0.9
Home	1	0.9
Return to Area	1	0.9
Not Reported	16	15.0
N = 15	107	100.0

Range = 1 − 52   Percent of Participant Population = 15.6

In the Project People Study,[1] 107 of the 584 participants did not voluntarily leave the building after becoming aware of the fire incident. The reasons given for their remaining in the building are presented in Table 3-12.16. Fifty-two occupants, or approximately 49 percent of the population staying in the building, reported that they remained because they intended to engage in fire control or fire fighting activities. The other most frequent reasons for remaining in the fire incident building were to notify others of the fire occurrence or because the means of egress were obscured by smoke. Approximately 15 percent of the study population voluntarily remained within the fire incident building.

OCCUPANT MOVEMENT THROUGH SMOKE

The movement of the occupants through smoke is sometimes related to the fire fighting behavior and the alerting of others, and is often a component of the evacuation behavior in many fire incidents.[1,2] The principal variables influencing an occupant's decision to move through smoke appear to be recollection of the location of the exit, and ability to estimate the travel distance required; secondary variables are the perception of the severity of the smoke (determined by observation of the appearance of the smoke); the smoke density; and the presence or absence of heat with smoke.[37,38] It should be recognized that to achieve evacuation, occupants have moved through smoke for extended distances (over 20 m) under conditions of extremely limited visibility (less than 4 m) at personal risk. Occupants sometimes have also been forced to turn back and not complete the evacuation.

Jin and Yamada[47] reported on a study involving 31 subjects (14 males and 17 females), traveling a maximum distance of 10.5 m in a corridor exposed to smoke from

TABLE 3-12.17 *Compilation of the Distance Moved Through Smoke for Participants in Both the British and U.S. Study Populations[1,2]*

Distance Moved (Feet)	British Percent	U.S. Percent	$P_1 - P_2$	$SE_{P_1 - P_2}$	CR
0–2	3.0	2.3	0.7	1.02	0.69
3–6	18.0	8.4	9.6	2.23	4.30**
7–12	30.0	17.1	12.9	2.71	4.76**
13–30	19.0	45.5	26.5	2.62	10.11**
31–36	5.0	2.0	3.0	1.25	2.40**
37–45	4.0	4.1	0.1	1.19	0.08
46–60	5.0	11.0	6.0	1.47	4.08**
>60	15.0	9.6	5.4	2.10	2.57**
N =	1316	322			

**Critical ratios significant at or above the 1 percent level of confidence.

TABLE 3-12.19 *Comparison of the Movement Through Smoke with the Visibility Distance Significance of These Differences in the Participant Population[1]*

Distance Moved	Participants	Percent
Greater than Visibility	170	46.4
Equal to Visibility	128	35.0
Less than Visibility	68	18.6
N = 3	366	100.0

Greater than Visibility Percent	Equal to Visibility Percent	Less than Visibility Percent	$P_1 - P_2$	$SE_{P_1 - P_2}$	CR
46.4	35.0		11.4	5.77	1.97*
46.4		18.6	27.8	6.98	3.98**
	35.0	18.6	16.4	6.83	2.40*
170	128	68			

*Critical ratios significant at or above the 5 percent level of confidence.
**Critical ratio significant at or above the 1 percent level of confidence.

smoldering cedar crib chips. The smoke extinction coefficient varied from 0.1 to 1.2 l/min. The subjects were also exposed to an increasing heat exposure from radiant heaters at the end of the corridor, with a mean temperature at the end of the corridor of 82°C. At five points in the corridor the subjects were stopped and asked arithmetic questions to be solved mentally. Both walking speed in the corridor and the mental arithmetic capability decreased with the increase in smoke density and the increased radiant heat exposure.

Table 3-12.17 compares the distance moved through smoke for the 1,316 persons in the British study[2] and the 322 persons in the United States study[1] who reported that they moved through smoke. It may be of interest to note that 60 percent of the population in the British study and 62.7 percent of the population in the United States study reported that they moved through smoke; apparently building occupants will move through smoke in an evacuation process. An important variable may be both the smoke density and

the visibility distance available to the occupants during the evacuation process, as well as their familiarity with the means of egress.

Table 3-12.18 presents the visibility distance of the British and the United States occupants as they moved through the smoke in evacuating the fire incident buildings. Occupants reported their movement through smoke in relatively high smoke-density conditions, with visibility below 4 m for 64 percent of the British population and for 47.6 percent of the United States population.

Table 3-12.19 relates the distance moved through smoke for the United States population to the visibility distance.[1]

The visibility distance for both the British and the United States populations at the time the participants were forced to turn back is presented in Table 3-12.20. It is interesting to compare Table 3-12.20 with Table 3-12.18 because

TABLE 3-12.18 *Compilation of the Visibility Distance for the British and the U.S. Populations When They Moved Through Smoke[1,2]*

Visibility Distance (Feet)	British Percent	U.S. Percent	$P_1 - P_2$	$SE_{P_1 - P_2}$	CR
0–2	12.0	10.2	1.8	1.99	0.90
3–6	25.0	17.2	7.8	2.65	2.94**
7–12	27.0	20.2	6.8	2.73	2.49*
13–30	11.0	31.7	21.7	2.24	9.69**
31–36	3.0	2.2	0.8	1.03	0.78
37–45	3.0	3.7	0.7	1.08	0.65
46–60	3.0	7.4	4.4	1.21	3.64**
>60	17.0	7.4	9.6	2.24	4.29**
N =	1316	322			

*Critical ratio significant at or above the 5 percent level of confidence.
**Critical ratios significant at or above the 1 percent level of confidence.

TABLE 3-12.20 *Compilation of the Visibility Distance for the British and the U.S. Populations at the Time They Initiated the Turned Back Behavior[1,2]*

Visibility Distance (Feet)	British Percent	U.S. Percent	$P_1 - P_2$	$SE_{P_1 - P_2}$	CR
0–2	29.0	31.8	2.8	5.31	0.53
3–6	37.0	22.3	14.7	5.57	2.64**
7–12	25.0	22.3	2.7	5.02	0.54
13–30	6.0	17.6	11.6	3.07	3.78**
31–36	0.5	1.3	0.7	0.90	0.77
37–45	1.0	0	1.0	1.10	0.91
46–60	0.5	4.7	4.2	1.16	3.62**
>60	1.0	0	1.0	1.10	0.91
N =	570	85			

**Critical ratios significant at or above the 1 percent level of confidence.

very few of the participants turned back when the visibility distance exceeded 10 m, with the greater percentage of occupants having turned back at the reduced visibility levels. Comparing the visibility distance below 4 m in Table 3-12.19, it is obvious that 91 percent of the British population who turned back and 76.4 percent of the United States population initiated their behavior at visibility distances of less than 4 m.

HANDICAPPED OR IMPAIRED OCCUPANTS

The problems involving fires in occupancies designed for permanently or temporarily disabled persons, such as nursing homes and hospitals, appear to have been properly alleviated in recent years due to building design, adequate staff training, and preparation to protect the occupants in place until evacuation is possible.[48] An extensive study of human behavior in health care facilities[49,50] indicated the nursing staff performed their professional roles even in situations with a high degree of personal risk.

The few fire incidents that have been studied in which handicapped persons have been involved in other occupancies have primarily been in residential occupancies. In both of these cases the handicapped individuals were assisted in a successful evacuation by other occupants. The one instance involved a wheelchair occupant[37] and the other situation involved a blind occupant.[43]

Pauls[51] has indicated from a number of practice evacuations in high-rise office buildings in Canada that approximately 3 percent of the occupants will be unable to use the stairs due to conditions that permanently or temporarily limit mobility. Paul's population included occupants with heart conditions and individuals recovering from surgery, illnesses, and accidents.

Isner and Klem[52,53] in their reports of the explosion and fire in the World Trade Center on February 26, 1993, indicated that normal power was lost with the occurrence of the explosion at approximately 12:18 p.m., and the emergency generators failed about 20 min. later, with all the remaining power to the World Trade Center complex being disconnected at approximately 1:32 p.m. Thus, the simultaneous evacuation of both able and disabled occupants from Towers 1 and 2 were conducted in darkness with varying smoke conditions in the stairways. These simultaneous evacuations may have involved the largest number of occupants and the longest evacuation times of any fire-induced evacuations of buildings in the United States.

Juillet,[54] in one of the first documented studies of this type, reported on the interview study of 27 occupants with disabilities who were evacuated from one of the two towers in the World Trade Center during the explosion and fire of February 26, 1993. Of those interviewed: fourteen had mobility impairments, three had sight or hearing impairments, three were pregnant, two had cardiac conditions, and seven had a respiratory condition. Juillet[55] indicated it was believed the total disabled population in both Towers 1 and 2 at the time of the incident was between 100 and 200 persons, with approximately 100 occupants having been identified previously. The average evacuation time of the 27 study participants was 3.34 hrs, with a reported range of evacuation times from 40 min. to over 9 hrs. The predominate means of evacuation was through the stairs,

with assistance from other evacuees or emergency personnel. The altruistic behavior, characteristic in many fire incidents with large populations,[36,37] appeared to have been exhibited in this fire incident with the disabled occupants, as reported by Juillet in the following manner:

> However, in the absence of communications by authorities, they gladly accepted assistance—from colleagues and even from complete strangers—in evacuating. These caring groups of people who assisted the disabled protected their "charges" until they were safety evacuated and moved away from the building.[54]

Klote, Alvord, Levin, and Groner[56] examined the design considerations needed to enable the elevators in tall buildings to be utilized for the evacuation of disabled occupants. In the explosion and fire in the World Trade Center with the loss of power in both Towers 1 and 2, including the emergency power, occupants were trapped in elevators in both buildings. Burns[11] indicated Tower 1 had 99 elevator cars, with many of them occupied. One 6- by 8-ft car, when opened, revealed 9 unconscious occupants, after an estimated exposure to the smoke in the shaft for approximately 2 hrs at the 9th floor level. Sherwood[57] reported that one 9- by 12-ft elevator car was stuck for 6 hrs at the 41st floor level of Tower 2 with 72 occupants: 62 elementary-school children and 10 adults.

SUMMARY

Canter et al[58] have stated the crux of the behavioral response in fire incidents in the following manner: "Behavior in fires can be understood as a logical attempt to deal with a complex, rapidly changing situation in which minimal information for action is available." It is suggested by Swartz[59] that the goals of codes should be "reoriented to increase the likelihood of informed decisions being made by people in fires." The examinations of the behavior in the Beverly Hills Supper Club fire led to the recommendation by Pauls and Jones[60] that "firesafety education should consider and be based on people's erroneous conceptions about distance being related to safety, and the time needed to escape from a fire emergency." Thus, more than a decade of detailed systematic research on human behavior in fires has resulted in the following consensus of the behavior of most persons by Sime:[42] "Despite the highly stressful environment, people generally respond to emergencies in a 'rational' often altruistic manner, insofar as is possible within the constraints imposed on their knowledge, perceptions and actions by the effects of the fire. In short, 'instictive,' 'panic' type reactions are not the norm."

There is a complex relationship between the physical and social environment in which the behavior occurs, which is complicated by the individual's perception of the ambiguous fire cues and primarily influenced by the person's relevant training and previous fire experience. It must be recognized that the fire cues are a product of a rapidly changing dynamic process which is constantly altering the decision choices of the occupant within the building. Pauls and Jones[60] have summarized this decision dilemma as follows: "What is an appropriate action at one stage may be quite inappropriate a minute later."

Paulsen[48] has emphasized the limited time constraints imposed on the occupant in a fire incident building as follows: "With very limited time available in which to decide on a course of action, people involved in fires often face difficult decisions. Decisions may be intellectually difficult in the context of limited knowledge of the engineered safety or the basic configuration of the occupied structure or limited knowledge of the development of the fire itself. Decisions may be difficult because of the sometimes counterinstinctive nature of the correct response, because some additional risk to one's self is incurred by a decision to alert or assist others."

REFERENCES CITED

1. J.L. Bryan, *Smoke at a Determinant of Human Behavior in Fire Situations*, University of Maryland, College Park (1977).
2. P.G. Wood, *Fire Research Note 953*, Building Research Establishment, Borehamwood (1972).
3. L. Lerup, D. Conrath, and J. Koh Liu, *NBS-GCR-77-93*, National Bureau of Standards, Gaithersburg (1978).
4. J.P. Keating and E.F. Loftus, *Psych. Today*, 14, (1981).
5. Guylene Proulx and Jonathan D. Sime, "To Prevent Panic in an Underground Emergency: Why Not Tell People the Truth?" in *Fire Safety Science—Proceedings of the Third International Symposium*, Elsevier Applied Science, New York, 843–852 (1991).
6. G. Ramachandran, "Human Behavior in Fires—A Review of Research in the United Kingdom," *Fire Technology*, 46, 2, 149–155 (May 1990).
7. G. Ramachandran, "Informative Fire Warning Systems," *Fire Technology*, 47, 1, 66–81 (Feb. 1991).
8. Eugene A. Cable, "Cry Wolf Syndrome: Radical Changes Solve the False Alarm Problem," Department of Veterans Affairs, Albany, NY (Jan. 1994).
9. Michiharu Kimura and Jonathan D. Sime, "Exit Choice Behavior During the Evacuation of Two Lecture Theatres," in *Fire Safety Science—Proceedings of the Second International Symposium*, Hemisphere Publishing Corp., Washington, 541–550 (1989).
10. James K. Lathrop, "Two Fires Demonstrate Evacuation Problems in High-Rise Buildings," *Fire Journal*, 70, 1, 65–70 (Jan. 1976).
11. Donald J. Burns, "The Reality of Reflex Time," *WNYF*, 54, 3, 26–29 (1993).
12. C.H. Berry, *Fire J.*, 72, 105 (1978).
13. NFPA *74, Household Fire Warning Equipment*, National Fire Protection Association, Quincy (1975).
14. H.C. Cohen, *Fire J.*, 76, 70 (1982).
15. NFPA *101, Life Safety Code*, National Fire Protection Association, Quincy (1994).
16. M.J. Kahn, *Fire Tech.*, 20, 20 (1984).
17. E.H. Nober, H. Pierce, A. Well, C.C. Johnson, and C. Clifton, *Fire J.*, 75, 86 (1981).
18. B. Latane and J.M. Darley, *J. of Person. and Soc. Psych.*, 10, 215 (1968).
19. J.L. Bryan, *A Study of the Survivors' Report on the Panic in the Fire at Arundel Park Hall, Brooklyn, Maryland, on January 29, 1956*, University of Maryland, College Park (1957).
20. S.B. Withey, in *Man and Society in Disaster*, Basic, New York (1962).
21. J.L. Bryan, *Human Behavior Factors and the Fire Occurrence*, University of Maryland, College Park (1971).
22. R.M. Killian, R. Quick, and F. Stockwell, *A Study of Response to the Houston, Texas, Fireworks Explosion*, National Academy of Science, Washington (1956).
23. Jonathan D. Sime, "Perceived Time Available: The Margin of Safety in Fires," in *Fire Safety Science—Proceedings of the First International Symposium*, Hemisphere Publishing Corp., Washington, 561–570 (1986).
24. A. Mintz, *J. of Abn. and Soc. Psych.*, 66, 150 (1950).
25. B.K. Jones and J. Ann Hewitt, "Leadership and Group Formation in High-Rise Building Evacuations," in *Fire Safety Science—Proceedings of the First International Symposium*, Hemisphere Publishing Corp., Washington, 513–522 (1986).
26. S. Horiuchi, Y. Murozaki, and A. Hokugo, "A Case Study of Fire and Evacuation in a Multi-Purpose Office Building, Osaka, Japan," in *Fire Safety Science—Proceedings of the First International Symposium*, Hemisphere Publishing Corp., Washington, 523–532 (1986).
27. J. Breaux, D. Canter, and J. Sime, *Psychological Aspects of Behavior of People in Fire Situations*, University of Surrey, Guilford (1976).
28. L. Bickman, P. Edelman, and M. McDaniels, *NBS-GCR-78-120*, National Bureau of Standards, Gaithersburg (1977).
29. Guylene Proulx, "A Stress Model for People Facing a Fire," *Journal of Environmental Psychology*, 13, 137–147 (1993).
30. Mark Chubb, "Human Factors Lessons for Public Fire Educators: Lessons from Major Fires," Phoenix: Education Section, National Fire Protection Association (Nov. 1993).
31. G.A. Klein and D. Klinger, "Naturalistic Decision Making," CSERIAC Gateway, Wright-Patterson AFB, Crew System Ergonomics Information Analysis Center, 1–4 (1991).
32. David Canter, John Breaux, and Jonathan Sime, "Domestic, Multiple-Occupancy, and Hospital Fires," in *Fires and Human Behaviour*, David Canter (ed.), John Wiley & Sons, New York, 117–136 (1980).
33. R.L. Best and D.P. Demers, *Fire J.*, 76, 19 (1982).
34. J.L. Bryan, *Fire J.*, 76, 37 (1982).
35. G.P. Morris, *F. Command*, 68, 20 (1981).
36. J.L. Bryan and P.J. DiNenno, *NBS-GCR-79-187*, National Bureau of Standards, Gaithersburg (1979).
37. J.L. Bryan, *An Examination and Analysis of the Dynamics of the Human Behavior in the MGM Grand Hotel Fire*, National Fire Protection Association, Quincy (1983).
38. J.L. Bryan, *An Examination and Analysis of the Dynamics of the Human Behavior in the Westchase Hilton Hotel Fire*, National Fire Protection Association, Quincy (1983).
39. R.L. Best, *Fire J.*, 72, 18 (1978).
40. D.P. Schultz, *Contract Report NR 170-274*, University of North Carolina, Charlotte (1968).
41. Kentucky State Police, *Investigative Report to the Governor, Beverly Hills Supper Club Fire*, Kentucky State Police, Frankfort (1977).
42. J.D. Sime, in *Fire and Human Behavior*, John Wiley and Sons, New York (1980).
43. E.L. Quanrantelli, *Panic Behavior in Fire Situations: Findings and a Model from the English Language Research Literature*, Ohio State University, Columbus (1979).
44. J.P. Keating, *Fire J.*, 76, 57 (1982).
45. E.R.F. Crossman, W.B. Zachary, and W. Pigman, *UCBFRG/WP 75-5*, University of California, Berkeley (1975).
46. National Fire Prevention and Control Administration, *Highlights of the National Household Fire Survey*, U.S. Fire Administration, Washington (1976).
47. Tadahisa Jin and Tokiyoshi Yamada, "Experimental Study of Human Behavior in Smoke-Filled Corridors," in *Fire Safety Science—Proceedings of the Second International Symposium*, Hemisphere Publishing Corp., Washington, 511–519 (1989).
48. R.L. Paulsen, *Fire Tech.*, 20, 15 (1984).
49. J.L. Bryan, P.J. DiNenno, and J.A. Milke, *NBS-GCR-80-297*, National Bureau of Standards, Gaithersburg (1979).
50. J.L. Bryan, J.A. Milke, and P.J. DiNenno, *NBS-GCR-80200*, National Bureau of Standards, Gaithersburg (1979).
51. J.L. Pauls, in *Human Response to Tall Buildings*, Dowden, Hutchinson and Ross, Stroudsburg (1977).
52. Michael S. Isner and Thomas J. Klem, "Fire Investigation Report World Trade Center Explosion and Fire, New York, New York, February 26, 1993," National Fire Protection Association, Quincy (1993).

53. Michael S. Isner and Thomas J. Klem, "Explosion and Fire Disrupt World Trade Center," *NFPA Journal*, 87, 6, 91–104 (Nov./Dec. 1993).
54. Edwina Juillet, "Evacuating People with Disabilities," *Fire Engineering*, 126, 12, 100–103 (Nov. 1993).
55. Edwina Juillet, personal communication, Jan. 18, 1994.
56. J.H. Klote, D.M. Alvord, B.M. Levin, and N.E. Groner, "Feasibility and Design Considerations of Emergency Evacuation by Elevators," NISTIR 4870, NIST, Building and Fire Research Laboratory, Gaithersburg, MD (1992).
57. James Sherwood, "Darkness and Smoke," *WNYF*, 54, 3, 56–60.
58. D. Canter, J. Breaux, and J. Sime, *Human Behavior in Fires*, University of Surrey, Guilford (1978).
59. J.A. Swartz, *Fire J.*, 73, 73 (1979).
60. J.L. Pauls and B.K. Jones, *Fire J.*, 74, 35 (1980).

MOVEMENT OF PEOPLE

Jake Pauls

INTRODUCTION

This chapter opens with some basic precautions about the topic and then provides a survey of literature on people's movement, with some emphasis on building evacuation. Basic crowd movement characteristics (density, speed, and flow) are explained, and key relationships are shown in graphic form and equation form (based largely on work by Fruin and Pauls). Extensive descriptions are given of building evacuation models developed by Pauls, based on empirical studies in Canada. The effective-width model for evacuation flow is highlighted especially in relation to prediction formulas for total evacuation time of large buildings. Complementing these movement efficiency topics is the topic of movement safety, especially on stairs. The chapter concludes with a discussion of time-based egress analyses. Included in this discussion is an introduction to the matter of egress-time criteria.

Qualitative Understanding Needed for Proper Use of Quantitative Methods

A precaution must be given regarding limitations in the quantitative methods currently available for people's movement in buildings. Some traditional assumptions about people's behavior in fires have been shown to be erroneous by research conducted especially over the last two decades. (See Section 3, Chapter 12.) Some models of evacuation behavior, such as the so-called hydraulic model, although applicable to certain situations, should not be applied indiscriminately to any situation. Valuable here are the views of John Archea, abstracted from his remarks (summarized by Pauls),[67] at the International Life Safety and Egress Seminar in 1981.

> Most egress rules were developed in their present form some twenty years before research was done.

Recent research falls into two schools: the carrying capacity school which examines exit flow capacity, and the human response school which says that exit capacity may be a necessary condition for safe egress, but it is not a sufficient condition. In the former, the "safe end" of the egress route is emphasized as the key point where egress is to be evaluated. The human response school looks at what happens at the other end of the route—the threatened end of the egress route. The former assumes that people will, upon hearing an alarm, drop what they are doing and immediately evacuate in an orderly fashion, without interacting with each other. Actually, people investigate conditions, compare with their experience, and then decide on actions that may have little to do with what is assumed in code rules for egress. Such activities take time. Another finding is that familiar entry routes are most often used for egress. Two different systems are commonly found in buildings: the familiar normal route in and out, and the protected, dedicated egress route. People may not be familiar with the routes that are being counted on to "drain" the building of its occupants.

Traditional exit technology also relies on what is called the "hydraulic model." There are three assumptions in the hydraulic model: occupants are alert, able-bodied, and ambulatory; firesafety depends on the safe end of the evacuation system; and there is a high-density building population that will tax the capacity of the exit system. There are two phases to evacuation: the startup phase and the evacuation phase. The hydraulic model deals only with the latter. Liquid flow or ball-bearing models do not account for the fact that people help one another or have degrees of familiarity with particular routes.

Archea's remarks raise basic questions about long-standing approaches to design of means of egress. A related chapter, "Behavioral Response to Fire and Smoke" (Section 3, Chapter 12), should be consulted for some of the background research that is helping to describe people's actual behavior when encountering fire. Additional research on the topic of fire-related human behavior, especially the behavior associated with egress, is found in the following literature review.

Jake Pauls is a consultant based in Silver Spring, MD. He worked on building use and safety with the National Research Council of Canada from 1967 to 1987. In addition to directing field studies of evacuation in tall buildings and of people's behavior at large-scale public events, his activities included research, networking, and technology transfer—especially for codes and standards—in egress design, stair safety, emergency behavior of people, safety needs of people with disabilities, etc.

LITERATURE REVIEW

The following should be regarded as an indicative survey rather than a complete review of technical contributions to the subject of evacuation. Some additional background is provided in reviews by Stahl and Archea,[90] Stahl, Crosson, and Margulis,[91] Paulsen,[75] and Pauls.[69]

Early Committee Documents

The 1935 report "Design and Construction of Building Exits"[52] described committee deliberations; however it is sometimes incorrectly spoken of as a research document. Because it contains a mix of traditional practice with some empirical studies, the report is often misinterpreted regarding qualitative and quantitative aspects of exit use by crowds; e.g., the unit exit width and unrealistically high flow figures, such as 45 persons per min per unit of exit width down stairs—concepts that predated the 1935 work. These errors influence time-based egress design calculations even today.

Sharing many of the characteristics of the 1935 report, and building upon some of its contents, was its British counterpart, "Post-War Building Studies, No. 29."[79] This report helped to establish the nominal 2.5-min clearing time for a space (based on a reported successful evacuation time of 2.5 min in the Empire Palace Theatre fire in Edinburgh in 1911), and it suggested the use of very high flows to perform the population capacity calculation (i.e., 40 persons per minute per 530 mm or 21 inches of exit width).

General Research on Crowd Movement

The post-war era also marked the beginning of modern studies of crowd movement, notably in Japan by Kikuji Togawa, whose many technical insights and empirical data were reported in 1955.[93] Among his mathematical presentations we find an equation for "time required for escape." It takes into account the flow time for an egress element plus the time needed to traverse some distance in the egress system. This general form of equation appears often in the egress literature.

Post-war publications by Russian experts included highly abstract treatment of evacuation, but did not provide the kind of empirical detail presented by Togawa. Most often referenced is the book titled (in its 1978 English translation) *Planning for Foot Traffic Flow in Buildings*, by Predtechenskii and Milinskii.[80] Their work is little known (and used even less) in North America; however, some of their abstract treatment and graphical techniques have been applied by Kendik,[38,39] using data from evacuation observations in Germany in 1977.

Often mentioned in some egress literature is a small-scale study by an operations research team that worked for the London Transport Board. Their observations and tests were described in an unpublished research report,[44] and highlights were published by Hankin and Wright.[28] The former, unpublished report has been widely circulated, referenced, and misapplied by people who compounded some original defects in the report; e.g., the failure to distinguish between maximum and mean flows. Again, as with earlier widely referenced but not critically read documents, this has led to overly optimistic predictions of egress time in some calculations. There are other problems inherent as well in applying data from a special-population transit context to the context of evacuation via (unfamiliar) routes, such as

exit stairs. Caution on this should also be heeded when applying some of the work by Togawa[93] and Fruin.[19]

John Fruin is a prominent researcher in North America whose well-known book, *Pedestrian Planning and Design*,[19] is now available in a revised edition. It is a comprehensive reference book on crowd movement. Beyond the book there are at least two articles that summarize many of the central concepts and design data from the book.[22,33] Fruin's work is often referred to in time-based analyses; however, these sometimes tend either to misuse the data for levels of service E and F (which Fruin recommends be used rarely, if at all, in analysis or design situations) or to combine high flow density assumptions with relatively high speed assumptions (an unrealistic combination). Used conservatively (including width deductions for the edge effect also reported by Fruin), there is much similarity between movement characteristics recommended by Fruin and those coming from Pauls' studies.

Between 1972 and 1982, many field observations—mostly unanalyzed and unpublished—were conducted by the National Research Council of Canada. These concentrated on people's movement in large assembly-occupancy buildings and large-scale events, such as the 1976 Olympic Games in Montreal and the 1978 Commonwealth Games in Edmonton. This work provided further empirical underpinning to the effective-width model for crowd flow on egress routes.

Research on Tall Buildings

Near the end of the 1960s the matter of high-rise fire-safety became a rapidly growing concern to safety officials and committees working on standards and codes. Among the papers prepared at this time on the matter of evacuation time of tall buildings and the design of exit stairs were ones by Galbreath,[23] Melinek and Booth,[51] and Melinek and Baldwin.[50] These papers contained a reworking of some of the "classical" reports from the 1930s and 1950s, noted previously in this chapter, to provide formulas relating exit stair width and minimum total evacuation time. Few new data were presented, and the formulas in general seriously underestimated total evacuation times of such buildings.

Beginning in 1969, Pauls began comparatively systematic and detailed observations of many evacuation drills in tall buildings, especially in Ottawa, Canada. Ignoring relatively inaccessible early references, the work is represented in general conference publications and periodical articles[57,58,59] and in published analyses and applications, especially on the "effective-width model."[4,56,61,62,65,68,74] Although dealing largely with evacuation situations where a "hydraulic model" is valid (e.g., there is queuing by people waiting to use egress routes, and much of the activity is relatively simple crowd movement dedicated to egress), this work bridged between movement studies and concurrent studies examining people's behavior in fires (especially in Great Britain and the U.S.). The behavioral studies (referenced in the following subsection) identified the role played by nonevacuation behavior before, during, and even in place of simple evacuation movement—all of which tend to increase significantly the actual evacuation times of buildings.

Mention should be made of more recent studies of evacuation of tall buildings, mostly in drills but some in connection with serious fires. Included are: German evacuation tests;[83] Japanese studies of behavior in fires and evacuation tests;[31,32,37,43] a Canadian study after a fire in a 27-story office tower;[36] a Swedish study[17] of movement in an evacuation drill (specifically testing some of Pauls' findings);

plus a current study—mainly in Australia—of evacuation time for high-rise buildings.[45,46] This study, largely funded by the Australian Uniform Building Regulations Coordinating Council, includes observations of evacuation exercises plus detailed questionnaires for evacuees.

Studies of Evacuation in Fires

Although addressed comprehensively in Section 3, Chapter 12, it is useful to mention here important evacuation technology contributions from studies of people's behavior in fires. Especially noteworthy studies are:

1. Wood's pioneer survey of behavior by over 2000 people in nearly 1000 fires in Great Britain;[99,100]
2. Bryan's replication and extension of such surveys in the U.S.,[7,11] plus a recent article highlighting an important social aspect of fire-related human behavior in high-rise fires;[10]
3. Subsequent case studies of major U.S. fires, such as the Beverly Hills Supper Club fire[92] and the MGM Grand Hotel fire;[8,9]
4. A wide range of research collected together in the first book on the topic of fires and human behavior, edited by Canter;[12]
5. Canter's paper summarizing findings and recommendations;[13] and
6. Sime's analyses of the Summerland fire plus others in Great Britain.[84,85]

Sime is also engaged in a study, funded by the U.K. Home Office, focused on movement and other behavior during the early stages of evacuation. This study specifically addresses travel distance and time—a topic that has traditionally figured heavily in requirements for means of egress but a topic that is poorly developed technically. For this reason, little coverage is given to the topic in this chapter.

Related to requirements for travel distance and illumination are recent studies of the relative ease and speed with which people can move, under various conditions of low illuminance, through familiar and unfamiliar egress routes.[5,6,96,97] These studies show that relatively normal speeds of movement can be maintained, even if illuminance levels are as low as 1.1 lux (0.1 footcandle), i.e., one-tenth of the common minimum requirement in North American codes for emergency illumination of egress routes: 10.7 lux (1.0 footcandle).

By way of very general and incomplete summary, some conclusions can be drawn about fire-related human behavior:

1. Panic is very rare even in very serious fires. Normal patterns of behavior, movement route choices, and relationships with others tend to persist during emergency situations. Behavior tends to be altruistic and reasonable (especially in light of the limited and often ambiguous information available to people at the time).
2. A central motivation and activity in fires is to seek information about the nature and seriousness of the situation, about other people, etc.
3. Evacuation, and response to fire generally, is often a social response; people tend to act in group fashion, not individually.
4. Problems that are encountered during normal building use will tend to persist and exacerbate situations in emer-

gencies. Included are: faulty communication, circulation hazards, etc.

A key assumption, based on such findings, is that the movement of people observed in normal building use and in many simulated emergency evacuations (drills) is a good basis for predicting their movement in a fire emergency. Specifically, people should not be expected to react faster or move more efficiently in a fire emergency than they do normally. Therefore, much of the evacuation technology, derived from careful documentation of realistic evacuation drills (e.g., without prior warning), is a good basis for developing guidelines for the design and use of emergency egress systems. This is a key to the validity of much of the technology presented in this chapter as well as to the validity of evaluation procedures for egress systems developed from the principles and details presented in this chapter.

Individual Capability to Evacuate

Studies dealing with situations where persons with disabilities would have to evacuate or be assisted to evacuate during a fire include those by Groner;[25] Groner, Levin, and Nelson;[26] Johnson;[34] Pearson and Joost;[76] Pauls and Juillet;[101] and Juillet.[102] The last two of these involved laboratory studies.

Canadian evacuation studies provide a rough guide to the number of people who might require special attention in evacuations of tall office buildings. Observations in office buildings suggest that about 3 percent of people in such buildings cannot or should not evacuate down multiple flights of exit stairs in a crowd situation with other evacuees.[58] This group has invisible as well as visible disabilities, including circulatory system disorders and some phobias. Most people within this 3 percent can use the stairs but should preferably do so behind, not among, other evacuees. Less than one percent of the generally active population, outside institutions, use movement aids such as wheelchairs, walkers, canes, and crutches, and some of this group are able to negotiate stairs unassisted, but with limited speed. As well as being affected by the trend to a greater proportion of active elderly persons, these figures should also be considered in light of seasonal factors, such as skiing accidents, leading to temporary disabilities.

Research on Evacuation Models

One attempt to summarize various evacuation models should also be noted. Nelson[54] has summarized key features of four models: EVACNET+, the product of a networking optimization approach, developed by Francis; the BFIRES-II computer model of fire-related behavior, developed by Stahl; the Escape and Rescue Model, a computer model, developed by Alvord that addresses evacuation of board-and-care homes; and the Effective-Width Model, developed by Pauls. These are represented in the literature by many papers, including the following: Stahl;[88,89] Francis,[18] which incorporates some of the Effective-Width Model; Chalmet, Francis, and Saunders,[14] which includes results of an evacuation exercise in a tall office building; Kisko and Francis;[40,41] and Alvord.[2] An especially useful review, dealing strictly with computer models for evacuation analysis, was done by Watts[95] for the SFPE Symposium on Quantitative Methods for Life Safety Analysis.

CROWD BEHAVIOR
AND MANAGEMENT

Crowd incidents, in which people are seriously injured or killed due to crushing or trampling, are not restricted to emergencies, such as fire, to conditions of crowd violence, or even simply to exuberance of some members of a crowd. Such events can occur, and have occurred, at sports events, religious gatherings, and rock music concerts. Serious injury and even death can occur during entry, occupancy, and evacuation of a building. It can happen under conditions that might, in every other respect, appear to be nearly normal even to people in close proximity to those hurt in the incident.

An introduction to some problems in and solutions for crowd safety has been provided by Pauls[68] and Fruin.[20,21] The former notes crowd incidents in Britain, Canada, and the U.S. and refers to reports such as the one by SCICON,[82] done after 66 football fans died in a crowd crush on a stairway in 1971 in Ibrox Park, Glasgow (which influenced a U.K. standard for sports grounds, Home Office, 1976);[30] the report of a special committee set up after 11 people were crushed in a crowd waiting to get into a rock concert (City of Cincinnati, 1980);[15] and a record of a meeting of experts called together by the U.S. National Bureau of Standards (Ventre, Stahl, and Turner)[94] in response to the Cincinnati committee. Recent contributions to the literature include the reports of British Inquiries on the Bradford stadium fire plus crowd incidents in Brussels and Sheffield.[42,77,78,103] Crowd safety engineering is also the subject of a book.[104]

Among the abovementioned NBS report's design recommendations, mostly in relation to ingress, there were some having wide applicability:

1. Strive for simplicity in all access and movement routes; this lessens the need for directional graphics and ushers.
2. Capacity-handling channels should be continuous walking surfaces, such as ramps. Stairs are satisfactory to shorten channels not subject to heavy pedestrian loads.
3. To the greatest extent possible, ingress systems should be "reversible" and usable whenever emergency egress is necessary.

Among technical papers appended to the NBS report[94] was one by Fruin, titled "Crowd Disasters—A Systems Evaluation of Causes and Countermeasures" (which was subsequently republished).[2,20] This paper discussed four fundamental elements—time, space, information, and energy—in relation to the following aspects of serious crowd incidents:

1. Rapid accumulation of queuing persons as demand for a facility outstrips its capacity.
2. Pedestrian densities which approach the critical density of about 8 persons/m² (less than 1.5 ft²/person) leave no space between people. Shock waves, causing individuals to move involuntarily as much as 3 m (10 ft) laterally, can be seen moving through crowds in this situation.
3. Competitive rushing by a crowd away from something is termed "panic" by Fruin, and competitive rushing toward some objective (such as in the Cincinnati incident) is termed a "craze."

In relation to the second item of this list and to the unusual physical forces in crowd incidents, Fruin notes:

The combined pressures of massed pedestrians and shock-wave effects through crowds at the critical density level produce forces which are impossible for individuals or even small groups of individuals to resist. Reports of persons being literally lifted out of their shoes and of clothes being torn off are a common result of the forces involved in crowd incidents. Survivors of crowd disasters report difficulty in breathing due to crowd pressures, and asphyxia is a more typical cause of death than trampling by the crowd. In the Glasgow, Scotland soccer stadium incident in which 66 persons died, the failure of a steel railing under crowd pressures contributed to the piling up of pedestrians. The bending of a steel-pipe railing under crowd pressures was reported at the Cincinnati Coliseum incident. The force required to bend a 50 mm (2 in.) diameter steel railing, applied 0.75 m (30 in.) above the base, is estimated at 500 kg (1100 lb).

Fruin lists countermeasures for critical density levels, such as provision of adequate pedestrian processing capacity and control of demand (e.g., arrival process). Also recommended are dispersion of routes as well as separation of waiting pedestrians into smaller groups. The U.K. "Guide to Safety at Sports Grounds"[30] calls for systems of rails placed across the tops of wide stairs to break up large mass flows onto stairs into smaller flows that will not tax the stair's capacity so severely, thus reducing crowd forces to safe levels, a method sometimes termed "metering." Under this U.K. guide—made mandatory retroactively through reference in British regulations—such crowd control rails, as well as all guardrails, must be designed to resist loads that are much greater than called for in North American code requirements for guardrails.

Crowd incidents often exhibit what can be termed a failure of front-to-back communication. People at the back of a crowd or bulk queue may contribute unknowingly to the forces that can build up in the crowd, forces that can reach crushing levels in the middle of the crowd or at the front, especially where forward movement is stopped by a barrier. The people being crushed are unable to communicate their plight to those at the back. For example, this was exhibited at a well-attended rock concert in Toronto's Exhibition Stadium in July 1980.[64] Thousands of people immediately in front of the stage were at critical densities for some hours; some 30,000 people farther from the stage were insufficiently aware of the importance of their behavior on the potentially catastrophic buildup of forces for those people near the stage. In addition to modifying the geometry of the stage to reduce the chance of crushing, a solution to the problem would benefit from the provision and effective use of communication systems.

There is a distinction between "crowd management" and "crowd control." Designers and managers of places where crowds assemble should be aware of this distinction, which is carefully drawn, at least by leading North American crowd behavior experts. To manage a crowd is to make use of design and operating features based on the subtle and beneficial exploitation of people's natural behavior. This requires a good understanding of a crowd's reason for being, the collective motivation of its members. Crowd control, on the other hand, is a more extreme, disruptive line of defense when crowd management is not successful; it might include dramatic police actions to subdue mob violence with force against force. Unfortunately, little literature exists that can be referenced here for guidance; designers or consultants, working on a project subject to use by crowds, should seek the advice of experienced facility and event managers as

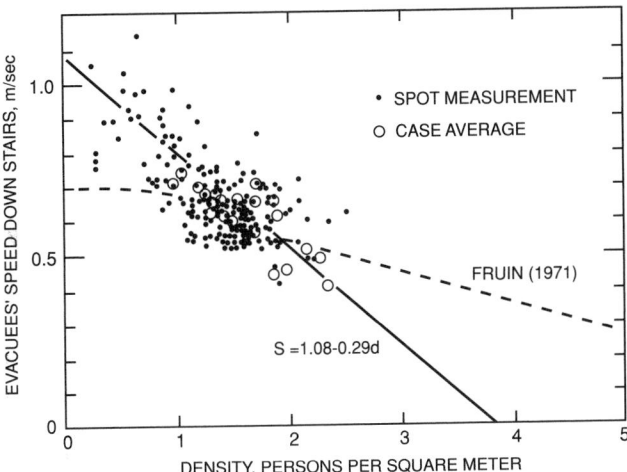

Fig. 3-13.1. *Relation between speed and density on stairs in uncontrolled total evacuations.*

well as using their own powers of observation. Examples of crowd control—which have little to offer designers and managers—may be viewed on television news accounts of riots. Examples of generally very good crowd management can be seen at the Disney complexes.

Basic Movement Characteristics and Relationships

Crowd movement is quantitatively specified using three fundamental characteristics, all of which are expressed as rates. These are density, speed, and flow. Density is the number of persons in a unit area of walkway, e.g., 2.0 persons/m². Often this characteristic is referred to by using the inverse of density, i.e., the area per person or pedestrian module, e.g., 0.5 m² (5.4 square ft²) per person. Speed is simply the distance covered by a moving person in a unit of time, e.g., 1.0 m/s (3.3 ft/s). The term "flow" is often used in a casual, nontechnical way when the general term "movement" is implied or when speed is actually being specified. Flow is specifically the number of people that pass some reference point in a unit of time, e.g., 2.0 persons/second. These three characteristics are related, along with path width, in the fundamental traffic equation (which, incidently, applies to motor vehicles as well as pedestrians)

$$\text{flow} = \text{speed} \times \text{density} \times \text{width} \qquad (1)$$

For example, the density, speed, and flow values shown above are consistent with what would be obtained with this equation, assuming the walkway's width is 1.0 m (3.3 ft). Note that, to be correct, a consistent set of units must be used, and speed must be measured along the slope of the walkway.

Also important is that speed is dependent on density. People can move quickly with a normal gait if there is a great deal of space between them. The closer they are to each other the more constrained is their movement until, when very close together, they can only shuffle along slowly. Aside from these speed implications, high density situations are also uncomfortable to varying degrees, depending on culture, social setting, and the relationship to those nearby. Expressed quantitatively, when the pedes-

trian density is less than about 0.5 persons/m² (21 ft²/person), people are able to move along walkways at about 1.25 m/s (4.1 ft/s), an average unrestricted walking speed. With greater density, speed decreases, and it decreases very markedly with very high densities, reaching a standstill when density reaches 4 or 5 persons/m² (2.1 to 2.6 ft²/person), equivalent to a fairly crowded elevator situation. This is also similar to the situation in a closely packed bulk queue of people anxiously waiting or competing to get through an entrance.

On stairs the speeds of movement are slightly lower and, at low densities, relatively fit people can average about 1.1 m/s (3.6 ft/s) along the stair slope [a horizontal speed component of about 0.8 m/s (2.6 ft/s)]. The speed-density relation data from a study by Pauls[61] in uncontrolled total evacuations of tall office buildings are shown in Figure 3-13.1. A curve, representing similar speed-density findings reported by Fruin,[19] is included in Figure 3-13.1 along with a regression equation for Pauls' data. (Note that Fruin's data were not derived from observations of evacuation in buildings.)

Given these dynamics as illustrated in Figures 3-13.1 and 3-13.2 there is a relatively complex relationship between flow and density. As shown in Figure 3-13.2, flow is small at both low and very high densities, but it attains a peak or optimum value at some intermediate density ranging around 2.0 persons/m², depending on whether they are on a level walkway or on a stairway. Equation 2 describes the flow-density relation obtained empirically for stairs used in total evacuations of tall office buildings.[61]

$$\text{flow} = 1.26(\text{density}) - 0.33(\text{density})^2 \qquad (2)$$

These basic characteristics and relationships are often described in publications on pedestrian movement.[19] Little should be made of the differences in the curves at the high-density end in Figure 3-13.2. As seen from the data points in Figure 3-13.1, these conditions are rarely or never observed and the differences are mainly of academic interest.

Putting into simple terms the optimum flow conditions observed in uncontrolled total evacuation drills[61] on a typical 1120 mm (44 in.) wide exit stair in a well-populated office building evacuation:

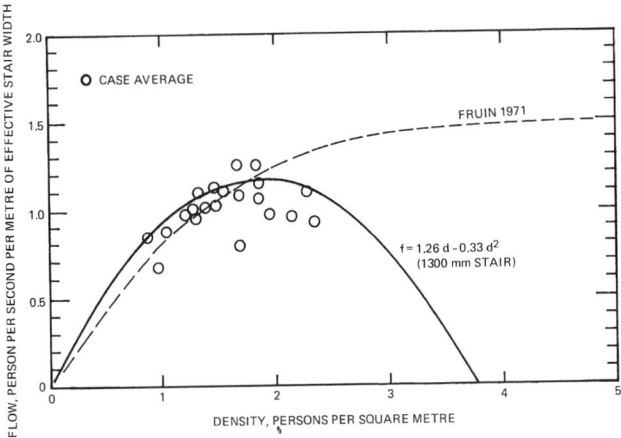

Fig. 3-13.2. *Relation between flow and density on stairs in uncontrolled total evacuations.*[61]

1. Each person would occupy slightly less than two treads;
2. There would be a descent of one story every 15 seconds; and
3. One person per second would pass a fixed point.

Going further, for evacuees on such an optimally loaded stairway, there would be one evacuee on every other stair, staggered left and right. These conditions are similar to those expected with level of service E, the busiest level of service recommended for stairs by Fruin.[19]

Expressed abstractly, the optimum flow conditions, for evacuation down stairs, include:

1. A density of 2.0 persons/m^2;
2. A speed of 0.5 m/s along the stair slope; and
3. A flow of 1.18 persons/(m·s) of effective stair width.

Note that a stair's effective width is 300 mm (12 in.) less than its nominal width (credited in building codes); e.g., a nominal width of 1120 mm (44 in.) equates to an effective width of 812 mm (32 in.).[61,62,68] This is described more completely in the following subsection.

Fruin's Levels of Service

Chapter 4 of Fruin's *Pedestrian Planning and Design*[19] is much used as a basis for deciding on appropriate pedestrian values to be used in dynamic exit calculations. The book's central concepts and data are also available in several other publications.[22,33]

Fruin describes six levels of service (A through F) for walkways, stairways, and queuing. Level A provides the highest standard with the least chance of congestion; level F provides the lowest—Fruin notes an unacceptably low— standard with much congestion. Chapter 4 describes the levels of service in terms of flow as a function of area per person, or "pedestrian module"—the inverse of crowd density, while Chapter 3 describes how speed varies with the module. For emergency movement and limited space situations, usually of concern to a fire protection engineer, levels of service C, D, and E should be used. Expressed in Fruin's original imperial-units form, pedestrian modules range, respectively, from about 25 to 5 ft^2/person on walkways and 10 to 4 ft^2/person on stairways. Corresponding flows range from 10 to 25 PFM [persons per foot (of effective width) per minute] on walkways and 7 to 17 PFM on stairways (plus or minus 1 PFM, respectively, depending on whether descent or ascent is required). Speeds of movement, which are more variable at low densities, range in average value from 250 to 100 ft/min. on walkways and from 115 to 70 ft/min. (along the slope) on stairs. It is estimated that ramps with slopes up to about 5 percent do not decrease movement speed; a 10 percent upward slope decreases speed by about 10 percent; an unusually steep 20 percent upward slope leads to a 25 percent reduction in speed.

Fruin also reports reductions in walkway effective width due to edge effects; however, he did not carry his work on this as far as Pauls in relation to stairs, nor did he carry out detailed documentation of the effective width of corridors as did Habicht and Braaksma,[27] who reported edge effects of about 150 mm (6 in.) at each corridor wall, edge effects similar to what Pauls found at stair walls.

The Time Variable and Its Subcomponents

The level-of-service characteristics, described previously, are stated in terms of rates. Most important, however, is the total time needed for crowd movement to occur. The total time taken for people to move past or through one part of a circulation system must be distinguished from the total movement time taken to go from a point of origin to some destination, such as a remote place of safety (e.g., the exterior of the burning structure at grade). The former is referred to here as "flow time" and the latter as "evacuation time." The term "total evacuation time" is used in cases where all occupants are evacuated. Flow time is simply a function of the crowd flow capacity of the usable width of a particular circulation element and the population or number of people to be moved through it. Population, capacity, and flow time are related as follows

$$\text{population} = \text{flow capacity} \times \text{flow time} \qquad (3)$$

Evacuation time is relatively complex, and it is more difficult to control and predict than is flow time. Two major components of evacuation time are: (1) the time needed for the movement directed to egress or escape and (2) the time taken up by relatively complex behavior that preceeds or accompanies egress. At least four subcomponents should be considered.

The first two subcomponents of evacuation time are relatively simple, and they are usually the only components directly affected by the means-of-egress requirements found in codes and related documents. These first two subcomponents are:

1. The flow times through the various flow elements of the egress system, especially the least efficient element with the longest flow time; and
2. The travel time for some individual in the evacuating crowd to move along the most direct egress route.

The complex way in which these two subcomponents are added together is influenced by the layout of the total space or system of spaces being evacuated and by the distribution of people in the space(s).

Even less simple, and requiring extensive information and judgment to predict well, are the third and fourth subcomponents, which may be quite substantial and generally should not be ignored:

3. The premovement time; i.e., the time between the onset of the cue or condition that is supposed to initiate an evacuation response and the decision by each evacuee to begin moving, not necessarily directly to an exit; and
4. The time component due to any behavior that diverts an individual from the most direct egress route once that person's egress movement is initiated.

(For example, in fires, people's first actions are often not simply to evacuate but to seek information, inform others, assist others, fight the fire, etc.)[84,85,86]

Subcomponents (3) and (4) are more influenced by social aspects of the occupancy than they are by building construction aspects. Hence, they are often ignored in simple engineering analyses of evacuation, sometimes with serious consequences for accurate evacuation time prediction, because the latter subcomponents can be as long or longer than the first two subcomponents. If these latter subcomponents have to be ignored in a prediction, that prediction should be stated only as the minimum possible (but improbable) evacuation time.

SOME EQUATIONS FOR TOTAL EVACUATION TIME

Several equations for total evacuation time are reproduced in the following subsections in the exact form that they were originally published in the evacuation research literature. Some of the equations appear formidably complex; the calculation benefits they promise are hidden behind scientific notation systems.

In using such formidable equations, there is a danger of getting sidetracked by their apparent sophistication. Like computer models, they may tend to keep us from understanding the world as it actually exists; the equations and models can take on a reality of their own. Therefore, the emphasis in this section is on minimizing the complexity of the mathematics and maximizing the awareness, using simple language and graphs, of real processes that many existing models depict incompletely.

Equations by Togawa

The first two of these total evacuation time equations were derived by Kikuji Togawa.[93] The first of his equations, Equation 4, is a general equation describing time required for escape. The second, Equation 5, is a simplified calculation formula. Equation 5 basically has two time components: flow time and travel time.

Time required for escape

$$T_e = \frac{1}{N'B'}\left[N_a - \sum_{i=1}^{n}\int_0^{T_0} N_i(t)B_i\phi_i(t)\,dt\right] + T_0 \qquad (4)$$

In the approximate calculation formula, if the shortest distance from the last doorway to the multitude is expressed as k_s, and the walking velocity of the multitude as v, T_e can be obtained by the following formula, considering the outflow of the multitude as continuous after the arrival of the head of the multitude to the last doorway.

$$T_e = \frac{N_a}{B'N'} + \frac{k_s}{v} \qquad (5)$$

where:

y_1 = Formula of gathering people formed to a mass

$$y_1 = \sum_{i=1}^{n}\int_0^T N_i(t)B_i\phi_i(t)\,dt$$

y_2 = Formula of outflowing crowd
ψ = Formula of delaying crowd
N = Number of outflowing people from the first doorways (persons/m · s)
N' = Number of outflowing people from the second doorways (persons/m · s)
N_a = Total number of the escaping people
n = Total number of the first doorways
T = Time
T_e = Time required for escape
T_0 = The time when the crowd formed the normal current at the second doorways or the staircases from the panic place
T_{max} = The time when the delaying crowd formed the maximum mass
t = Optional time
t' = The time required for opening the first doors from the panic place

t_e = The time when the crowd currents from the first doorways ended
k = Distance from the first doorways to the gathering point of the crowd (m)
k_s = Distance from the last doorways to the head of the escaping crowd (m)
v = Walking velocity of the crowd
B = Breadth of the first doorway (m)
B' = Breadth of the second doorway (m)
ϕ = Percentage of the number of the people gathered to one point to the number of those who went other ways (percent)
ρ = Density of the crowd (number of people/m²)

The first component of Equation 5 is simply derived from Equation 3, where flow time is obtained by dividing the population (N_a) by the flow capacity, which is the width of the most limiting passageway (B') multiplied by the mean flow capacity of that passageway (N'). The second component is simply the travel time for the first person in the evacuating crowd who moves from a point of origin to the destination where his or her evacuation is considered complete, i.e., the travel distance (k_s) divided by speed of movement (v).

Togawa's simplified equation is useful in calculating minimum evacuation times (i.e., only subcomponents 1 and 2 as described in the subsection "The Time Variable and Its Subcomponents") for buildings where stairs are extensively used. For example, consider a tall office building with two exit staircases (each having an effective width of 0.8 m) and 1000 actual occupants or 500 people per staircase. Assume a travel distance of 40 m for the first person to move from his/her workplace to an exit staircase (assumed to be close to the workplace) on through to the discharge point of the exit at grade. A mean flow of 1.1 persons/m · s of effective stair width and an unimpeded speed of 1.0 m/s (along the stair slope) would be reasonable first approximations for the case of an uncontrolled total evacuation of the building. Then the flow time is simply 455 seconds and the travel time is 40 seconds, giving a minimum total evacuation time of 495 seconds or 8.25 minutes (not including the time needed for subcomponents 3 and 4). The empirically based input assumptions for this calculation come from Pauls,[61] and they are discussed more completely in the subsection "Empirical Method by Pauls," along with further discussion of what is meant by "uncontrolled total evacuation" and the "effective-width" model.

Equations by Melinek and Booth

Melinek and Booth[51] suggested a method similar to Togawa's. However, theirs is specifically for calculation of the minimum total evacuation time of a tall building. Two different conditions are taken into account: (1) a low-population building where the travel time between two floors exceeds the flow time for entry of all the people on the floor into the exit; and (2) a higher-population building where the flow time for entry of people from a single floor exceeds the travel time between floors. Like most others who came up with prediction equations, Melinek and Booth assumed traditional flow figures that were too high (1.1 persons/m · s of nominal stair width), and they underestimated evacuation times in eight out of twelve cases, with the most inaccurate estimated at only about half of the observed times. Most of the observed times apparently were ones noted by Galbreath,[23] leaving Melinek and Booth's accuracy to be questioned. Notably, Melinek and Booth underestimated the times of the two U.K. buildings with which they were familiar, predicting their evacuation times to be 7.3 and 9.9 minutes, respectively, when observed times were about 12 minutes each.

Melinek and Booth's equations and notations are reproduced as follows:

Consider a simple model of a multistory building. It is assumed that all the people are waiting at the exit stairs initially and that people leaving the ground floor do not reduce the rate of flow from the upper floors.

The following symbols will be used:

Q_r = Population of floor r
b_r = Staircase width between floor $r - 1$ and r
N' = Rate of flow of people per unit width down the stairs
t_s = Time for member of unimpeded crowd to descend one story, typically about 16 sec.

Population of floor r and above

evacuation time = flow time + travel time

$$= \sum_{i=r}^{n} Q_i \qquad (6)$$

Minimum time for the population to enter the staircase leading down from floor r

$$= \left(\sum_{i=r}^{n} Q_i\right)/(N'b_{r-1}) \qquad (7)$$

Time for tail of crowd to reach ground floor

$$= rt_s \qquad (8)$$

Therefore, from Equations 7 and 8

$$T_r = \left(\sum_{i=r}^{n} Q_i\right)/(N'b_{r-1}) + rt_s \qquad (9)$$

where T_r is the minimum evacuation time for the population of floor r and above.

Equation 9 gives n values of T_r ($r = 1$ to n). The minimum evacuation time, T_e, for the whole building is equal to the highest of these values of T_r.

If the populations and the staircase widths are the same for all floors, then $Q_r = Q$ and $b_r = b$ for all r, and Equation 9 becomes

$$T_r = (n - r + 1)Q/(N'b) + rt_s \qquad (10)$$

If $Q/(N'b) \geq t_s$, then T_r is a maximum when $r = 1$. In this case

$$T_e = nQ/(N'b) + t_s \qquad (11)$$

If $Q/(N'b) < t_s$, then T_r is a maximum when $r = n$. In this case

$$T_e = Q/(N'b) + nt_s \qquad (12)$$

Russian Methods

A characteristic of the Russian approach to crowd movement calculation (as represented by Predtechenski and Milinski[80] or Kendik[38,39]) is the use of a density measure based on the ratio of the projected horizontal area of people

in a crowd divided by the area of walkway surface. (The units for their D, or density, variable are therefore square meter per square meter rather than persons per square meter.) Their work emphasizes modeling and prediction of times at which various merging activities, congestion, and queuing occur in a variety of crowd movement situations. Kendik has provided an equation, based on this more detailed examination of crowd dynamics, reproduced here as Equation 13, with her original notation (influenced by translation from a German paper).

$$t(Ges) = t(TR;STAU) + (n - 1) \\ \times 1(TR)/v(TR;n - 1) + (n - 2) \times dt \qquad (13)$$

where:

$t(TR; STAU)$ = Length of time required for the flow to leave the floor level ($n - 1$)
$l (TR)$ = Travel distance on the stairs between adjoining stories
$v(TR; n - 1)$ = Velocity of the flow emanating from the congested area at the floor level ($n - 1$)
dt = Delay time due to congestion
n = Number of the upper floors in the building

Obviously, use of this equation requires that Kendik's complete approach be applied and, to do this, one of Kendik's papers should be consulted.[38,39]

Empirical Method by Pauls

While early Canadian research on evacuation did not get into the mathematical abstraction found in the Russian-based work, it did utilize many simple graphic representations of people's movement, merging, and queuing in evacuations of tall buildings.

In contrast to the previous contributors to evacuation technology, Pauls began by observing as many evacuation drills as possible in tall office buildings, being careful to record many aspects of each exercise in great detail, and then developing relations that best described what was actually observed.[61,73] This method revealed defects in earlier literature on matters of crowd configuration on stairs, maximum and mean flows, and actual evacuation populations—all of which were sources of error in other predictions.

Altogether, 29 drills were observed in buildings ranging from 8 to 21 stories high, in which traditional, uncontrolled total evacuation procedures were used. Generally all stairways in each building were monitored at ground-floor discharge points and at selected heights, often through the use of instrumented moving observers. An average of two stairways were documented for each drill, thus giving 58 cases for analysis. (A smaller number of drills were documented where the procedure involved a selective or sequenced controlled evacuation procedure, with either partial or total evacuation. These are not covered by any of the equations provided here and are separately discussed under "Evacuation Procedures in Tall Office Buildings.")

By way of background to the prediction equations developed from Pauls' study, the effective-width model for crowd movement must be defined. This empirically based model describes flow as a linear function of a stair's effective width—the width remaining once edge effects are deducted [150 mm (6 in.) in from each wall boundary and 90 mm (3.5 in.) in from each handrail centerline]. (See Figure 3-13.3.) The effective-width model takes into account the propensity

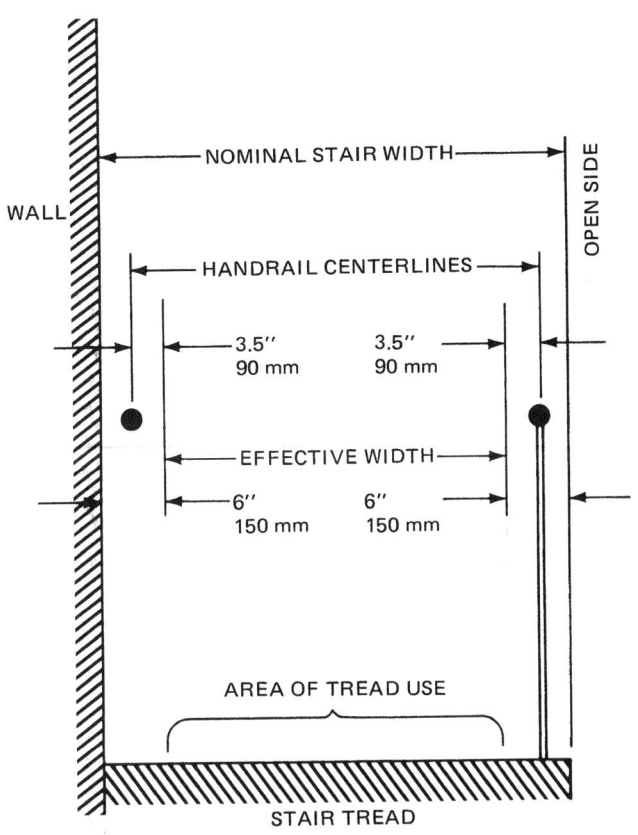

Fig. 3-13.3. Measurement of effective stair width in relation to walls and handrails.

of people to sway laterally—especially when walking slowly in a crowd—and, therefore, to arrange themselves in a staggered configuration, not in regular lanes as assumed in the traditional unit-width model based on presumed static dimensions of people's shoulders.[61] For example, a stair designed to provide two of the traditional units of exit width has a nominal width of 1120 mm (44 in.) and an effective width of 820 mm (32 in.), causing crowds to take up a staggered formation.

Another finding underlying the effective-width approach is that mean evacuation flow (per meter of effective stair width) varies in a nonlinear fashion with evacuation population (per meter of effective stair width) as shown in Figure 3-13.4. The regression equation is designated Equation 14.

$$f = 0.206p^{0.27} \qquad (14)$$

where f and p have the metric units shown in Figure 3-13.4.

Other factors, besides effective stair width, influencing the mean flow in evacuations are people's use of extra clothing (for protection against precipitation or cold conditions outside a building), plus various building design and use factors (including normal stair use).[61] Regarding building design, the assumed influence of stair-step geometry on crowd flow is considered below.

Equation 15 (with dimensions in mm) and Equation 16 (with dimensions in inches) relate effective stair width, w, actual total population, P, and the expected flow time, t, in seconds.

$$w = \frac{8040}{t^{1.37}} \ P \qquad (15)$$

$$w = \frac{317}{t^{1.37}} \ P \qquad (16)$$

These equations can be translated into a somewhat more useful form and graphed as shown in Figure 3-13.5, which relates effective stair width per person and the resulting flow time for a crowd moving down stairs. Three curves are provided to show the assumed effect of various step geometries on crowd movement efficiency. It should be noted that some of Pauls' early publications of equations similar to Equations 15 and 16 differ slightly, because they were based on stairs with step geometries approximately described by the highest of the three curves. Equations 15 and 16 relate to the middle curve.

In relation to evacuation time prediction, the preceding equations and graphs have dealt only with flow time. For uncontrolled total evacuations of tall office buildings, there is a simple way of predicting the startup time needed for flow to build up to half its mean level. This empirically derived startup time is shown in Figure 3-13.6. When added to flow time, based on empirically determined mean flows, the 41 seconds (0.68 min.) accounts for travel time plus all or part of the other subcomponents of minimum evacuation time discussed previously. The extent to which the 41 seconds is adequate depends on the experience of the building occupants and the manner in which the evacuation is initiated and run. Evacuations in cases of actual fire or other emergencies are assumed to take longer. More will be said about this in the following subsection.

From Equation 14, and from the 41 seconds startup time shown in Figure 3-13.6, the first of two prediction equations shown in Figure 3-13.7 is derived. The prediction equations apply to cases where there are no more than 800 persons/m of effective stair width. (The upper equation in Figure 3-13.7 is Equation 17 and the lower one, Equation 18.) Note that there is a good match between the prediction curve (Equation 17) and the observed total evacuation times for the uncontrolled evacuations, especially in those cases where

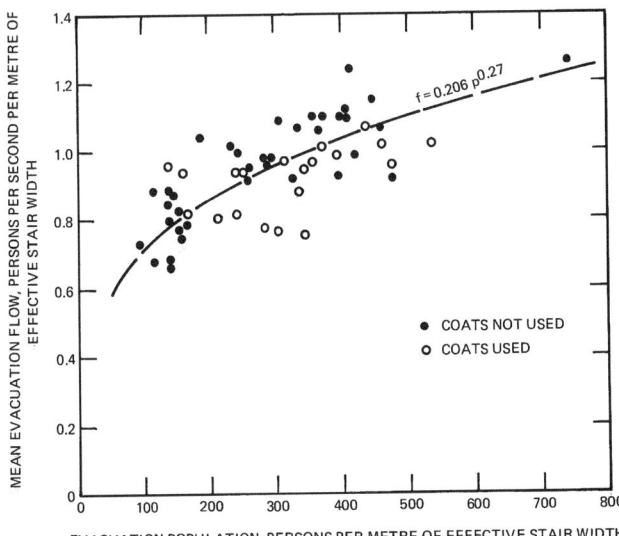

Fig. 3-13.4. Effect of population on flow down stairs in uncontrolled total evacuations.[61]

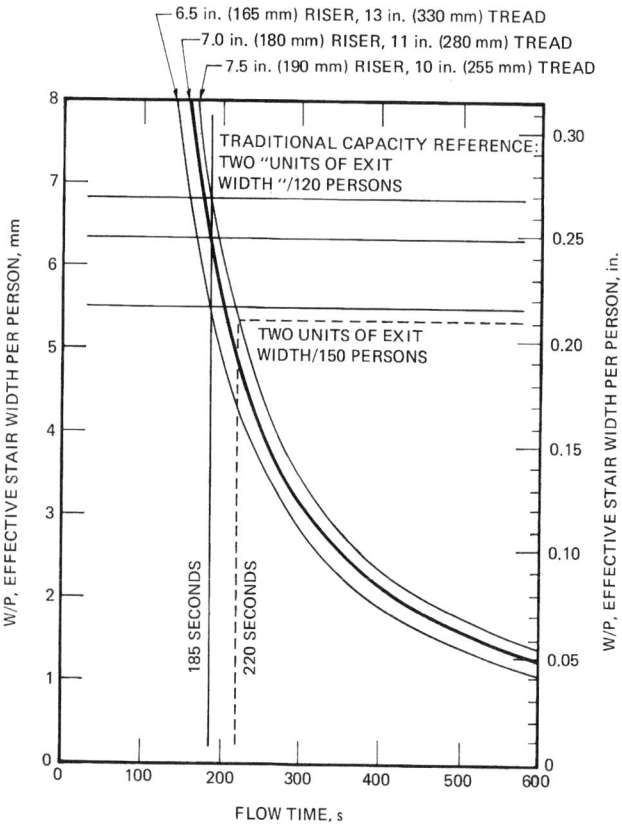

Fig. 3-13.5. *Relationship between effective stair width per person and flow time for several step geometries.*

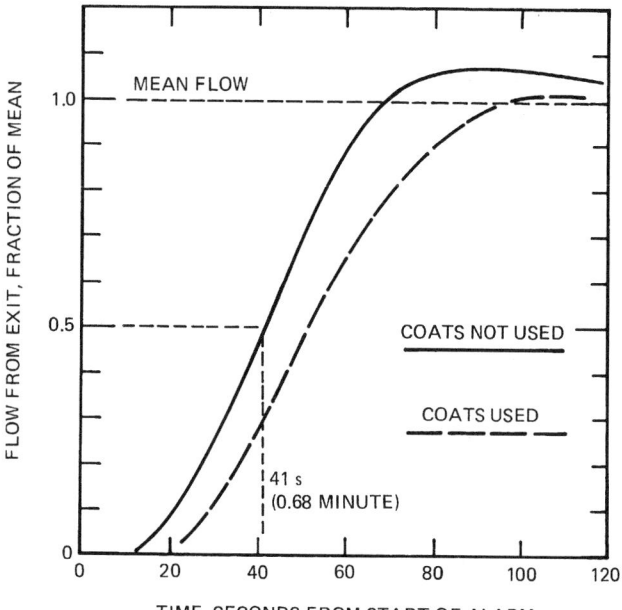

Fig. 3-13.6. *Buildup of flow from exits in uncontrolled total evacuations of tall office buildings.*

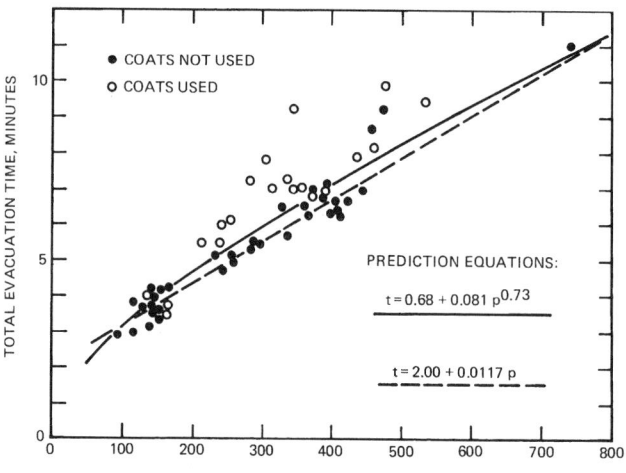

Fig. 3-13.7. *Predicted and observed total evacuation times for tall office buildings.*

extra outdoor clothing was not used. Using Equation 17, the net error in predicting total evacuation times for 50 cases in buildings 8 to 15 stories high was 0.2 percent. The simplified linear equation, Equation 18, also fits these data very well.

$$T = 0.68 + 0.081p^{0.73} \qquad (17)$$

$$T = 2.00 + 0.0117p \qquad (18)$$

where:

T = minimum time, in minutes, to complete an uncontrolled total evacuation by stairs, and

p = the actual evacuation population per meter of effective stair width, measured just above the discharge level of the exit.

(Note the upper limit of 800 persons per meter of effective stair width.) This applies to Equation 14.

Buildings that will be less accurately predicted with Equations 17 and 18 are the taller ones with very low populations on each floor. With such buildings, the total evacuation time is influenced by the travel distance and people's ability to descend stairs quickly, i.e., at about 10 seconds per story rather than the 15 to 20 seconds observed in evacuations with higher populations. The observed times departing most from the prediction lines in Figure 3-13.7 were for buildings with 18 to 20 stories.

For buildings with more than 800 persons per meter of effective stair width, Equation 19 provides a good basis for predicting times for uncontrolled total evacuations in tall office buildings.

$$T = 0.70 + 0.0133p \qquad (19)$$

Figure 3-13.8 shows Equations 18 and 19, along with prediction lines based on equations proposed by Melinek and Booth[51] and Galbreath,[23] plus a cross-hatched area showing where observed times lie.

Another indication of the accuracy of these prediction equations for uncontrolled total evacuations of office buildings is provided in Figure 3-13.9. This compares predicted and observed evacuation times for approximately 1700 people evacuating an 8-story, 6-exit Canadian office building in

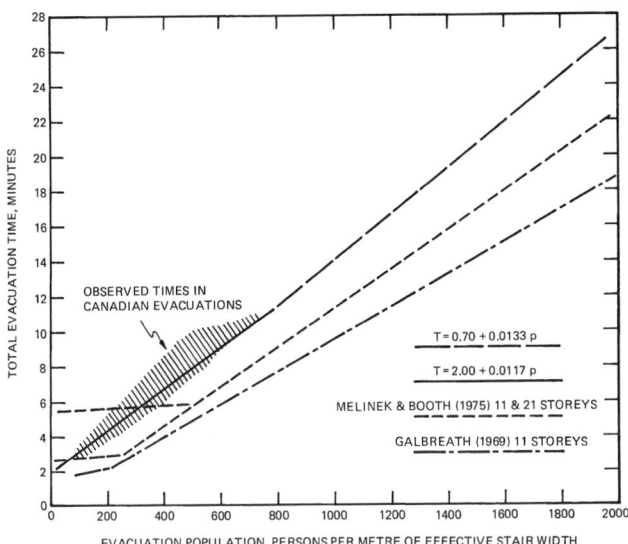

Fig. 3-13.8. *Predicted and observed total evacuation times for tall office buildings incorporating results from other investigators.*

March 1983 (in cold-weather conditions), as documented by Public Works Canada.

Actual Populations in Office Buildings

A cautionary note must be given here regarding the prediction of actual populations in buildings. In many buildings with assembly occupancies the prediction of maximum population will be relatively straightforward, based on seat counts and floor area of waiting spaces [where an area of 0.3 to 0.6 m^2/person (3 to 6 ft^2/person) is a reasonable assumption]. However, studies in Canadian office buildings strongly support a case for assuming that each actual office building occupant has an average of about 25 m^2 (268 ft^2) of gross rentable area.[35,61] This contrasts sharply with the traditional occupant-load assumption in codes and standards

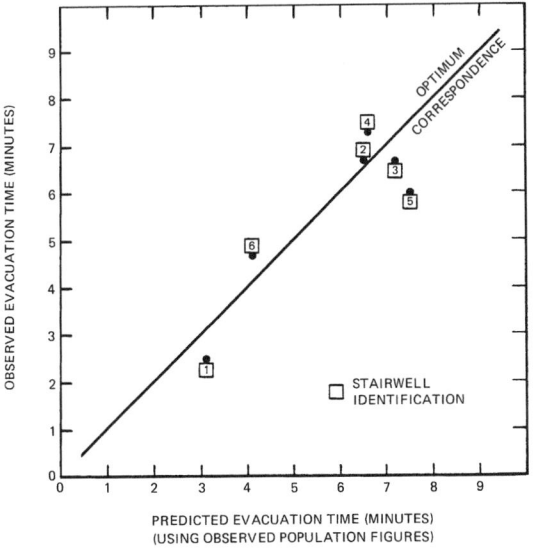

Fig. 3-13.9. *Comparison of predicted and observed evacuation times for an office building.*

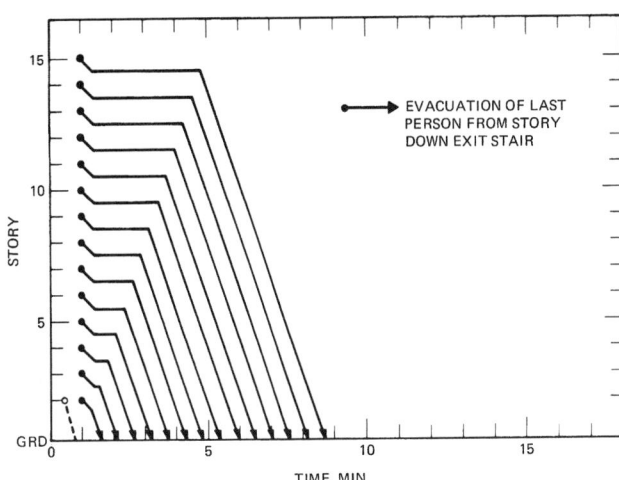

Fig. 3-13.10. *Hypothetical uncontrolled total evacuation of a 15-story office building.*

which assume 9.3 m^2 (100 ft^2) of gross rentable area per occupant. Codes and standards give estimates of actual occupants that may be incorrect by a factor of more than 2. Typical workspace counts also overestimate office populations (sometimes by 15 percent).

In the foregoing discussion of evacuation movement, only one procedure has been considered. This is traditional, total, uncontrolled evacuation. The following sections examine two basically different evacuation procedures in greater detail, and cover hypothetical and actual evacuations of office buildings. The evacuation of the MGM Grand Hotel, Las Vegas, during the course of the disastrous fire in 1980, is included to underline further the extent to which real-life evacuations can take much more time than is predicted by formulas presented here and elsewhere.

EVACUATION PROCEDURES IN TALL OFFICE BUILDINGS

Evacuations of multistory office buildings can be thought of as being one of two types: "uncontrolled total evacuation" and "controlled selective evacuation." The former is dependent largely on the nature of evacuation sequencing or deference behavior required, and the latter on the type of control imposed on the evacuation by management.

Figures 3-13.10 and 3-13.11 illustrate patterns of evacuee movement, over time and space, for a hypothetical traditional ("uncontrolled") total evacuation and a controlled selective evacuation. Traces represent the movement of the last persons to leave each floor. The slopes of the traces represent the speed of movement down an exit stairway; horizontal lines represent queuing. For a more complete discussion of modeling evacuation using such diagrams, including combined use of elevators and stairs, see references 57 and 58.

Based on Canadian observations of 29 evacuation drills in office buildings, ranging from 8 to 21 stories in height, Figure 3-13.10 shows what can be reasonably expected in an uncontrolled total evacuation of a 15-story building occupied by 70 able-bodied persons per floor—a fairly high-population condition that would be expected with about 1400 m^2 (15,000 ft^2) on each floor. It is assumed that there is

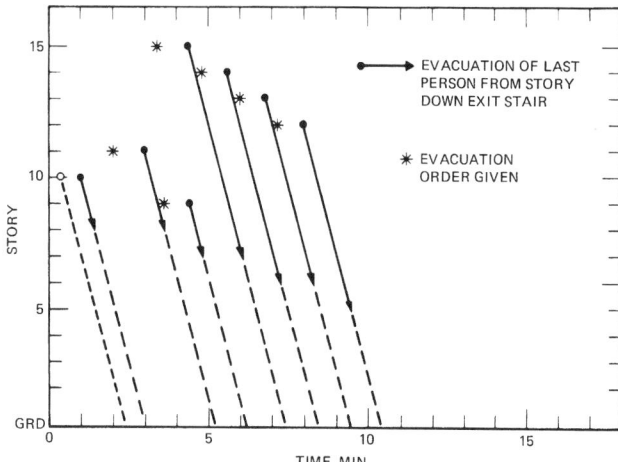

Fig. 3-13.11. Hypothetical controlled selective evacuation of 15-story office building.

an equal division of evacuation demand for two standard 1120 mm (44 in.) exit stairs and a mean egress flow of approximately one person per second discharging from each exit. Descent speed is 3.6 floors per minute (about 0.5 m/s along the stair slope), and each evacuee has slightly less than two treads of stair area (0.5 m^2/person).

Based on Canadian observations of ten evacuation drills, Figure 3-13.11 depicts a controlled selective evacuation, in this case, a partial evacuation of only the tenth-floor fire area, the floors above, and the ninth floor. Compared to the evacuation depicted in Figure 3-13.10, the movement traces are steeper, indicating faster descent speeds; however, there is also greater space between traces, indicating that densities are lower than depicted in Figure 3-13.10. Resulting flows in each exit are considerably lower.

In both cases, about nine minutes are required to move all occupants to areas below a presumed fire area—the tenth floor in the case of Figure 3-13.11. Because of the time required to initiate and control the selective sequential evacuation, it takes approximately the same time to move 490 persons to safety (below the ninth floor) as it does to move all 980 occupants to the ground in the uncontrolled total evacuation depicted in Figure 3-13.10. Rather than jump to the conclusion that uncontrolled total evacuation is better or more efficient, it must be considered that, if there really was a fire on the tenth floor, and if all occupants attempted simultaneous egress, and if the usual rules of deference shown in Figure 3-13.10 applied, occupants of the most immediately threatened floors would have to queue in the vicinity of those floors for several minutes.

An increasingly common conclusion is that, although uncontrolled total evacuation is not demanding of building and management systems, it is too upsetting for occupants who are indiscriminately evacuated (if indeed they can be encouraged, with ambiguous alarm systems, to evacuate at all). On the other hand, the uncontrolled total evacuation process is simple, and there is no need for relatively sophisticated systems and training, which are unreliable and expensive (if the observed experiences in these Canadian office building examples apply generally).

Finally, it must be stressed, whether they are planned to happen or not, uncontrolled total evacuations do occur for a variety of reasons. Some idea of the extent to which evac-

ations succeed can be gained by examining two well-studied evacuation drills held in government occupied buildings in Ottawa at a time when, significantly, there was a great deal of concern by officials and laypersons about firesafety in high-rise buildings. Case studies of the two drills, described in the next two subsections, were published by Pauls and Jones.[74]

Uncontrolled Total Evacuation

One drill case study entailed the 1972 evacuation of 1453 persons, using uncontrolled total evacuation procedures, from a 15-story office building via four 1140 mm exit stairs. The total evacuation time ranged between 6.6 and 9.3 minutes. The time variation was largely due to unbalanced demand on the four exits—the result of unequally sized catchment areas and an overcompensating attempt to redistribute some of the demand to predesignated exits on several floors. The average total evacuation time was 7.8 minutes, a figure close to the predicted 7.0 minute time. (See calculations below.) The difference is largely explained by the fact that evacuees experienced slight delays and less efficient movement down stairs because outdoor clothing was used. The attempt to have people on several floors move to more remote, predesignated, but less familiar exits, also added minor delays.

Figure 3-13.12 shows the traces of movement of instrumented observers, using the exit stairs with evacuees, during this uncontrolled total evacuation.

In this evacuation, some people were on the fifteenth floor up to 4 minutes, and their queuing on exit stairs, near this level, lasted up to 4 minutes from the initiation of the general fire alarm system. It was usually less than 2 minutes at lower levels. Conditions were much like those noted in relation to Figure 3-13.8; each evacuee occupied an average area equivalent to two stair treads, and descent speed was typically 3.5 stories per minute below the seventh floor—conditions that would be unduly arduous or impossible for only a few percent of typical office workers. In the case of this drill, such persons and some assistants (73 persons altogether) descended a central, fifth stairway reserved, in this evacuation, for their use.

The building was recently rebuilt with the same floor areas, but with only the central exit stairway and another one in the largest wing left intact. However, its occupants still are supposed to utilize the uncontrolled total evacuation precedure. Given the difficulties of obtaining a balanced use of the remaining exits, it is not unreasonable to predict that the total evacuation time will now be approximately twice what was achieved in the 1972 drill.

Sample computations of uncontrolled total evacuation time: Here it is useful to consider the evacuation time prediction for this building in somewhat greater detail.[62] This building has a cruciform plan with a total of 33,000 m^2 (350,000 ft^2) of gross rentable area on 14 office floors. One of the four wings is larger in area than the other three; thus, there is unlikely to be an equal population using each exit stairway during an evacuation. For the first computations, the unmodified stairway arrangement will be assumed, with one exit stair at the center of the building and one exit stair at the end of each of the four wings. Subsequent computations will consider the modified case with only two stairways.

For the original, unmodified situation we will assume that the exit stairs have extensive normal use, partly because the building is occupied largely by a single government department. The center stairway has a relatively high level of

TIME, MINUTES FROM BEGINNING OF ALARM

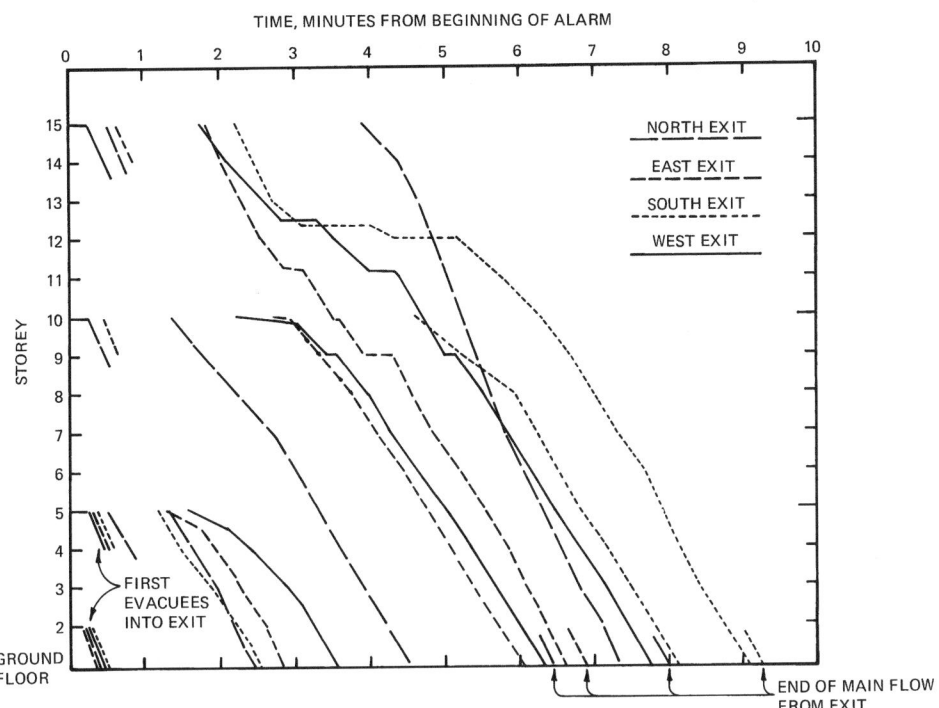

Fig. 3-13.12. **Movement of observers with evacuees during an uncontrolled total evacuation.**

normal use because of its location near the elevators and its entry door in the groundfloor lobby. The fire emergency procedures call for this stairway to be used only by handicapped persons and others assisting them in emergencies. The stairs have a nominal width of 1140 mm (45 in.) and a dogleg configuration, with two 180-degree turns per story. The step geometry provides a riser height of 180 mm (7 in.) and a tread depth of 250 mm (10 in.). Stair wall finishes have a semirough texture. Assume that the building is slightly more densely occupied than the average noted earlier (specifically, assume one actual occupant for every 22 m^2 instead of the 25 m^2 noted in the subsection "Actual Populations in Office Buildings"). Given this kind of information about the office building, as well as the knowledge that the occupants are accustomed to periodic evacuation drills, a prediction of the flow and evacuation time performance when the traditional uncontrolled total evacuation procedure is used can be attempted.

First, an estimate is made of the total number of occupants to be evacuated. At 22 m^2 per actual occupant and a total gross-rentable office area of 33,000 m^2, the anticipated evacuation population is 1500 people. It can be assumed that 2 percent of these (i.e., 30 people) cannot or should not evacuate down the stairs in a high-flow situation with relatively able-bodied evacuees and should elect to use the central stairway. Typically, only a few of these people would have such severe mobility impairments that they would need to be assisted by more than one other person to get down the stairs. If an average of one person to assist each of these 30 handicapped persons is included, the central stairway will then be used by approximately 60 persons; this leaves 1440 persons using the four remaining exit stairs at the building perimeter. As a first approximation, it can be assumed that these stairs will be used equally, even though

one of the four wings is larger than the others. Each of the four 1140 mm (45 in.) stairs would thus carry 360 evacuees.

Next, by deducting 300 mm (12 in.) from the nominal stairway width, it can be calculated that each of the stairs has an effective width of 840 mm (33 in.). The anticipated mean evacuation flow can be indirectly calculated using any of a number of methods, such as set out in Equations 14, 15, or 16 and in Figure 3-13.5. The easiest method is to use Figure 3-13.5, interpolating between the two upper curves to account for the step geometry. Each stair provides 2.33 mm (0.09 in.) of effective width per person; thus, we can read off an expected flow time of 390 seconds, estimated to the closest ten seconds (and having two significant figures). The mean flow is obtained simply by dividing the population per stairway (360 persons) by the time (390 seconds), giving a mean flow of 0.92 person per second.

This uncontrolled evacuation prediction can be adjusted to take into account some modifying factors in addition to the step geometry adjustment[68] already built into Figure 3-13.5 (and further discussed in Pauls).[63] For example, the stair walls are slightly rough; therefore, it is reasonable to expect some reduction in flow.[61] A 4 percent reduction in flow would be a reasonable assumption. The fact that this evacuation was held during somewhat cool weather, and therefore some of the evacuees wear or carry outdoor clothing, should be taken into account. Assume this leads to a 6 percent reduction in the flow. On the positive side, an adjustment for the fact that the building occupants are familiar with evacuation drills and that there is fairly extensive normal use of the stairs (although this is likely to be more the case for the center stairway than for the perimeter stairs) should be considered. There might, in fact, be some confusion as relatively able-bodied people go first to the normally used center stairway only to be redirected to one of the perimeter stairs. There may also be some confusion and delay as the exit stairway in the largest wing develops more extensive queuing at exit doors and, as a result, people either decide on their own to try another stairway or are directed by fire wardens to use another stairway. Thus, on the positive side, it is prudent to assume an adjustment of only 2 percent for normal stair use and familiarity with evacuation drills.

It is sufficiently accurate (given the underdeveloped state of the art) simply to add up the negative and positive factors to give a net adjustment of minus 8 percent for the expected flow. This results in an adjusted flow prediction of 0.83 person per second (50 persons per minute) on each of the four perimeter stairs. A well-documented, cool-weather evacuation drill in the same setting had the following mean flows, in person per second: 0.85, 0.85, 0.83, and 0.80—remarkably similar to the prediction. The differences are largely explainable by the fact that the four stairs were not equally used in the evacuation drill, despite efforts by the building's fire emergency staff to balance usage. Too many evacuees were, in fact, diverted to one of the stairways, which thereby served 448 people. They had been diverted from the largest wing where the stair was used by 385 people. The other two stairs served 329 and 291 people, respectively, for a total of 1453, compared with the initial assumption of 1440.

Repeating the above computations, using Figure 3-13.5 and the actual stair populations (448, 385, 329, 291), with the adjustments described above, mean flows of 0.88, 0.84, 0.81, and 0.78 person per second, respectively—all within 4 percent of the observed flows—can be predicted. Thus, it is possible to make some very accurate predictions for the relatively straightforward, traditional, uncontrolled total evacuations of tall office buildings, if we understand some basic factors of building design and use.

As to the matter of minimum total uncontrolled evacuation time, a simple prediction can be made with the 390 second (6.5 min) flow time (read from Figure 3-13.5), which assumes equal usage by 360 people of each of the four perimeter stairs. Adjusting this time by the 8 percent figure gives an adjusted flow time of 7.0 minutes. Adding the suggested 0.68 minute startup time (the time for flow to build up to half its mean value, read from Figure 3-13.6), a predicted total evacuation time is 7.7 minutes. Alternatively, without including any adjustments, Figure 3-13.7 or Equations 17 or 18 could be used to calculate an approximate, slightly underestimated total evacuation time in the range of 7.4 to 7.0 min.

The observed total evacuation times were 9.3, 8.2, 7.0, and 6.6 minutes, respectively, for the stairs, listed in descending order of population. This range of observed times is mainly a result of the nonuniform distribution of evacuees among the four perimeter exits. The average observed time was 7.8 minutes, compared with the (adjusted) prediction of 7.7 minutes. In fact, the adjusted predicted times, using the actual populations, are found (using Figure 3-13.5) to be 9.2, 7.9, 7.2, and 6.6 minutes, respectively, in order of decreasing population, all within 4 percent of observed times.

Finally, the predicted minimum total uncontrolled evacuation time should be computed for the building in its modified form, with only two exit stairs and with a similar total population. Having 1400 people (assuming 100 per floor) equally distributed between the two stairs, each with an effective width of 840 mm (33 in.), would lead to a flow time in excess of the 10-minute (600-second) limit of Figure 3-13.5. Because the population per meter of effective stair width, 833, slightly exceeds the 800 limit for most of the

formulas suggested, Equation 19 should be used here. Doing so gives an unadjusted minimum total evacuation time of 11.8 minutes. However, there will be great difficulty achieving balanced usage between the two remaining exit stairs. The central one could easily have a usage 40 percent higher than the other stairway.[61] In this case, (assuming a 817- and 583-person split between the exit stairs) the minimum, unadjusted total evacuation time would be approximately 13.7 minutes or about twice what was the case when the building had 5 exits.

These examples demonstrate that traditional total uncontrolled evacuations can be relatively easily understood and predicted in the case of tall office buildings. The same is likely to be true of large buildings used for public assembly. (Figure 3-13.5 is especially useful for these.)

Controlled Selective Evacuation

The other actual evacuation to be described here occurred in a 21-story office building in 1971 and was one of the first large-scale attempts at a controlled, selective evacuation in Ottawa. (Figure 3-13.13 shows heavier traces depicting actual movement of observers; lighter traces depict the movement pattern predicted several days before the drill, based on estimated populations on individual floors. Given that this was done very early on in the Ottawa study, it shows remarkably accurate prediction capability.) Although the evacuation drill consisted of the total clearing of the building—some 2100 persons evacuating over a 30-minute period—it did utilize the selective, sequential procedure which could have been terminated after 6 minutes, with the clearing of only the presumed fire floor and two adjacent floors. The drill simulated a worst-case scenario with a fire on the third floor. For this case, a concern for smoke movement to upper floors suggested a procedure that first clears the fire-floor area, then all floors above this, starting with the twenty-first floor and progressing downward.

This evacuation drill, also described in detail by Pauls and Jones,[74] included a dramatic example of what can go wrong when firesafety management personnel misuse the communication systems needed to manage such sequential evacuations. There had been no earlier large-scale attempt to test all aspects of such an evacuation procedure. Due to an incorrectly set switch on a control console, there was a delay in getting the first announcement over the public-address system, following approximately 1.5 minutes of a standard fire alarm. The first announcement successfully carried over the public address, nearly 3 minutes into the drill, was made by an obviously flustered, inadequately prepared person who said, "Ladies and gentlemen. We have to evacuate this building. The alarm has been set on the third floor. Please evacuate. Other floors stand by."

This ambiguous announcement was followed by a slightly different one in French—bilingualism being *de rigueur* in government-occupied buildings in Ottawa. Confusion followed. As many as 350 persons left their floors before they were supposed to, most going down one of the two 1040 mm (41 in.) wide exit stairs. This premature evacuation confused and delayed the intended earlier evacuation of floors 3, 4, and 2. According to a questionnaire returned shortly after the drill by a sample of 176 evacuees (an 88 percent mailback return rate), some people thought they heard the announcer say that a fire has been reported on the third floor. Of the 176 respondents, 43 percent reported interpreting the situation as an actual fire after hearing the

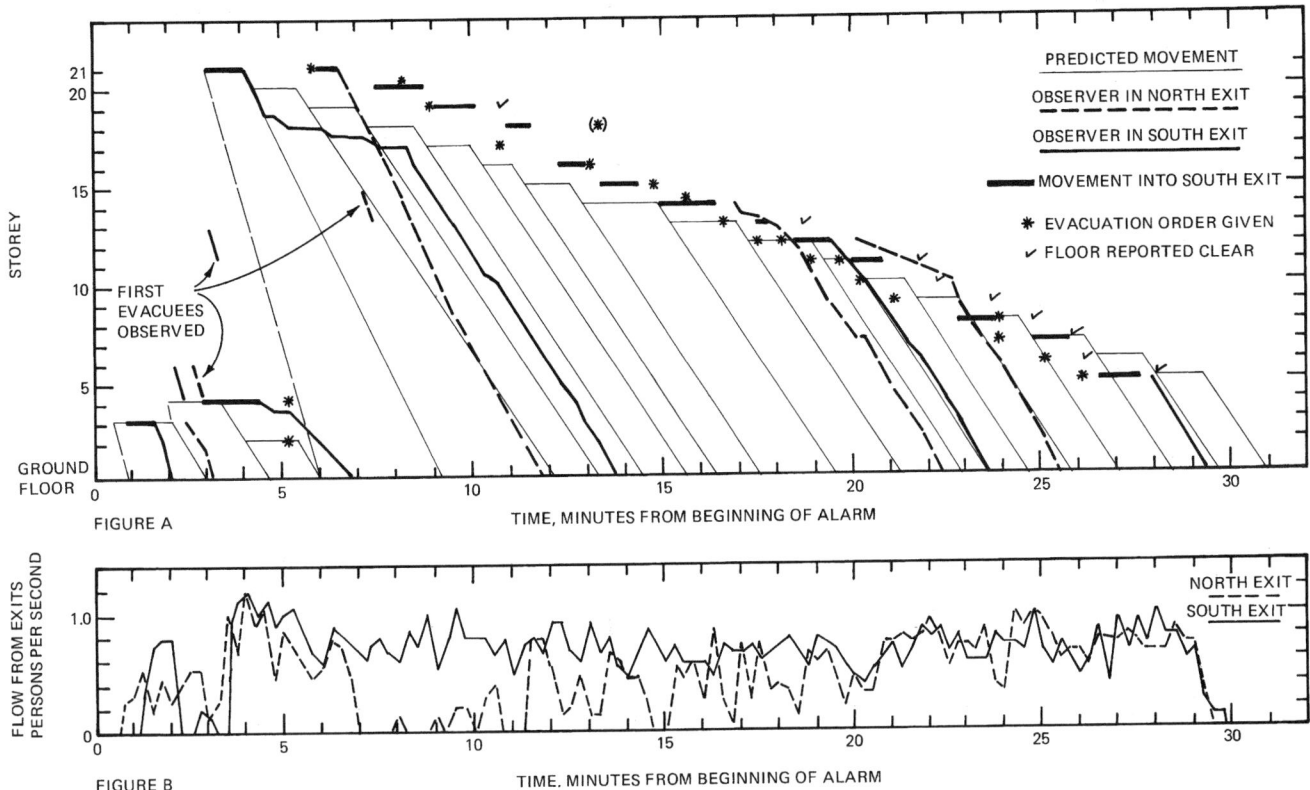

Fig. 3-13.13. *Predicted and observed movement, plus flow, in a sequential evacuation exercise.*

first public-address announcement. Regarding their interpretation of the situation before this announcement, with only the ringing of the building's fire alarm system, only 17 percent of the questionnaire respondents thought it was a fire, nearly 60 percent thought it was due to a circuit malfunction, and 16 percent thought it was a drill.

Turning to the conditions faced by evacuees in the two exit stairs, with the confused evacuation procedure, there was extensive queuing and slow progress by some evacuees who prematurely left the twenty-first floor via the south exit. In this exit, some 4 minutes were required for them to move between the nineteenth and seventeenth floors, and they took 10 minutes to descend the full distance to the ground. In the other exit, where there was minimal premature evacuation, descent speed was fairly consistent at about 3.8 stories per minute, resulting in a 5.3 minute descent time from the twenty-first floor.

It is useful to contrast these conditions with what would likely occur in a traditional, uncontrolled total evacuation of the building's 2100 able-bodied occupants via the two 1040 mm stairs. According to Equation 19, an uncontrolled evacuation would take a minimum of nearly 20 min. If all 2100 people were simultaneously standing at their respective floor's exits, there would be only 0.17 m² of stair area or about one-half tread per person. For evacuees from the highest floors, such queuing might last for 14 of the 20 min.

Whether queuing and evacuation delays create an upsetting situation for the people involved is a good question, and it raises the broader question about how aware typical occupants of high-rise buildings are of the kinds of queuing and movement conditions they might face in a major evacuation. Should they be made aware of such aspects of build-

ing safety? By what means? Who has the responsibility for doing it? An important attempt to address this issue was included in the Ontario Public Inquiry into Fire Safety in Highrise Buildings.[98] The report is highly recommended to anyone concerned with firesafety in tall buildings.

Recent Evacuation Observed in a Tall Office Building in Australia

Although only a few preliminary results are available from the first of a series of evacuation studies in tall office buildings in Australia, it is useful to share them here. They help underline the great delays that can occur in evacuations of office buildings.

A Sydney building with eight floors of offices, above three floors of above-grade parking, had a well-documented evacuation drill in August 1985. With its two standard exit stairs and a relatively limited population, the building would be predicted (by Equation 18) to have a minimum time—for uncontrolled total evacuation—of approximately 5 min (assuming 200 evacuees per exit stair). Given its size and occupancy, the building had not yet been equipped with a general alarm system. Notification about this particular evacuation was first communicated by regular telephone from the chief floor fire warden of floor 5 (the presumed fire floor) to a building manager on the second floor. As well as initiating evacuation on their floors, these individuals also informed other floors, in accordance with a prearranged sequence. The time line for these communications and for the ensuing evacuation provided a striking lesson in how long this process can take. Notification of individual floors took nearly five minutes to complete from time zero, the discovery of the hypothetical fire.

Based on a sample of eight observed exit stair doors, the time of first entry by evacuees into the exit stairs ranged between 1.45 and 9.35 minutes; completion of stair entry ranged between 4.75 and 10.65 min; and on the "fire floor" the stair-entry phase lasted between 1.27 and 4.77 min. Times for the last evacuee to reach the grade-level exit discharge were 15 min for one stair (used by only 159 persons) and 16 min for the other (used by 221 persons).

For this exercise, a prediction of total evacuation time that considered only the usually expected flow time and travel time would be grossly incorrect. The observed total evacuation time was three times longer than estimated because of the often-neglected components of evacuation time, listed in the subsection "The Time Variable and Its Subcomponents." Such long evacuation times could also occur in a real fire situation. As demonstrated in this example, a serious life-threatening fire in a tall building does not necessarily lead to a rapid evacuation.

EVACUATION OF A HIGH-RISE HOTEL DURING A FIRE

The 1980 fire in the MGM Grand Hotel was relatively well documented in terms of the related occupant behavior.[8] Approximately 3400 guests were registered at the hotel. The fire was reported to the fire department at 07:17 a.m.—a time when occupancy in the 21 floors of guest rooms (floors 5 through 25) could easily be close to 3400 people. Partly because of the assembly spaces on the 26th floor, there was, before the fire, a great deal of exit-stair capacity from the guest floors. Due to extensive smoke, only one of the six exit stairs remained tenable for the entire time during which evacuation occurred, from 06:45 to approximately 13:00, and this stair was the one used by 45 percent of questionnaire respondents who described their egress as being by stairs. Two other stairs were usable for all but the first 15 min and 90 min, respectively.

Although it is hypothetical, it would be (mathematically) possible, given a simultaneous mass use of the three generally usable exit stairs [supposedly "smokeproof towers," each reportedly 1830 mm (72 in.) wide], to clear the guest floors of 3400 people in 10 to 11 min. (Equation 18 can be used to do this calculation.) However, based on the responses of 381 guests (of 554 returned questionnaires), it is estimated that the progress of the actual evacuation was as follows (where the time, 07:19, signifies the arrival of the first fire fighters—the first indication for many that there was a possible fire):

06:45–07:18	15 percent used stairs
07:19–07:30	14 percent used stairs
07:30–08:00	10 percent used stairs
08:00–08:30	5 percent used stairs
08:30–09:00	8 percent used stairs
09:00–09:30	15 percent used stairs
09:30–10:00	15 percent used stairs
10:00–10:30	10 percent used stairs
10:30–13:00	remaining 8 percent used stairs.

These estimates, as well as other findings from the questionnaire, suggest that the exit stairs were rarely crowded during the evacuation. For example, of 299 respondents indicating obstructions to evacuation, only 13 persons (4.3 percent) suggested this was due to crowded conditions in the exits—a level of reported difficulty similar to that reported for darkness on stairs and objects (such as abandoned luggage) on stairs. Smoke was the most common obstruction—reported by 60 percent of 299 respondents reporting obstructions. Locked exit-stair doors (preventing reentry to floors) were noted by 12.7 percent of the 299 as obstruction.

In addition to the large loss of life due to smoke (responsible for most of the 85 fatalities), there was much additional evidence that smoke was serious. Nearly 60 percent of the study population moved through smoke, and half of these reported moving through more than 21 m (70 ft) of smoke-filled space. Half of those moving through smoke estimated visibility was 1.2 m (4 ft) or less. Two-thirds reported turning back when visibility was 1.5 m (5 ft) or less. Refuge action, often in rooms where other guests had gathered in "convergence clusters"—a feature covered again by Bryan[10]—was a major activity during the fire. Approximately one-quarter of the guests responding to the questionnaire sought refuge for two or three hours before leaving. The most frequently mentioned evacuation assistance was that provided by fire department personnel; hotel staff provided very little assistance.

It is interesting to note that only 42.7 percent of the 554 questionnaire respondents reported an initial interpretation of the situation as a serious fire; 18.1 percent indicated they did not know, and 35.6 percent indicated believing the situation was not serious. (There is a parallel here with what was learned from the questionnaires returned after the sequential evacuation of the 21-story office building described in the subsection "Controlled Selective Evacuation," reported by Pauls and Jones.)[74]

MOVEMENT SAFETY AND THE DESIGN OF CIRCULATION ROUTES

Statistics clearly identify movement safety, specifically falls, as a major problem in many countries. In this section, we examine some recent findings related to the most frequent type of falling accidents—those on stairs. Serious nonevacuation stair accidents, leading to emergency room treatment in hospitals, occur to about one out of four persons (over their lifetimes), or one million times each year in the U.S. (many times the number due to fires).

Stair Safety and Design

Reviews are available for a considerable body of research on stair safety and design (conducted over the last 15 years in North America, Europe, and Japan) and some of the design implications.[70,71,72] Among important references are the last two (of four) publications resulting from extensive studies funded by the U.S. Consumer Product Safety Commission, conducted at or through the U.S. National Bureau of Standards.[1,3]

Stair accidents result from a combination of factors. One succinct summation, applying to many stair accidents, is that they are due to architecturally triggered human error. For example, the largest class of stair accidents involves misjudging where a tread nosing is and misplacing the foot too far forward on the tread. This occurs when a person cannot see the tread edge clearly (due to lighting, surface features, or personal factors) and/or where the tread is too small to get adequate footing.

More generally, if somewhat simplistically, we can summarize three of the key objectives of stair design for safety as follows: (1) stairs must be readily seen; (2) tread size must provide adequate footing; and (3) reachable and graspable handrails are needed.

Expanded discussions of these three central criteria are found in a widely published two-part article[71] and an earlier guidance note.[66] The matter of handrail design is explored in detail, in relation to recent Canadian studies, by Pauls.[72] Some highlights of this, related to stairs used by crowds, are presented in the following subsections.

In the U.S., the most controversial of the design changes arising from stair design research, is the recommendation for less-steep stairs; i.e., lower risers and deeper treads. Specifically, U.S. codes and NFPA standards are increasingly requiring risers to be no higher than 178 mm (7 in.) and tread depths ("run" or "going" dimensions) to be at least 279 mm (11 in.). In the U.S., as in England and Japan where similar geometries are required for certain public stairs, this has not yet been required for dwellings that traditionally have been permitted a lower standard of design despite the high accident rates there. Debates on instituting the higher standard for dwellings have been especially lively in one of the three U.S. model building codes.[73]

Handrail Height, Lateral Spacing, and Graspability

Handrail use and design has been a special focus of Canadian research, especially beginning with the study of spectator behavior in the new Edmonton Commonwealth Stadium during the Commonwealth Games in 1978. In addition to unpublished studies contracted by the National Research Council of Canada (NRCC), products of this 1978 field study include the 18-min documentary film, "The Stair Event,"[60] and a complementary paper, "The Stair Event: Some Lessons for Design."[63] The film provides the best visual documentation available on people's use of long-aisle stairs (some with and some without central handrails); indeed the film appears to be the only one available on general stair use safety.

Partly as a result of such field observations, including the earlier series of studies on evacuation drills in tall office buildings[74] and a large field study of spectator behavior at the 1976 Olympic Games in Montreal, anthropometric aspects of handrail height and lateral spacing began to be examined. One such analysis is illustrated in Figure 3-13.14. It strongly suggests that handrails, located in accordance with traditional North American code requirements, are too low and sometimes too far apart to be effectively reached by stair users.

Observations of interpersonal spacing in crowds (at an optimum density of 2.0 persons/m²) and the anthropometric analysis suggested that handrails should be laterally spaced so that no part of the stair width is more than 825 mm (32.5 in.) from the centerline of a handrail and that handrails should be positioned about 950 mm (37.5 in.) above stair tread nosings.[72] Because handrail reachability depends on the combination of the handrails' lateral spacing and height, the prevailing U.S. standards, allowing handrails as low as 762 mm (30 in.) and 2235 mm (88 in.) apart, permit a situation where many of the people on a crowded stair are beyond the reach of a handrail.

To complement the field observations and anthropometric analysis, West Park Research, a research center affiliated with the University of Toronto, was contracted by NRCC to examine how people's functional capability and

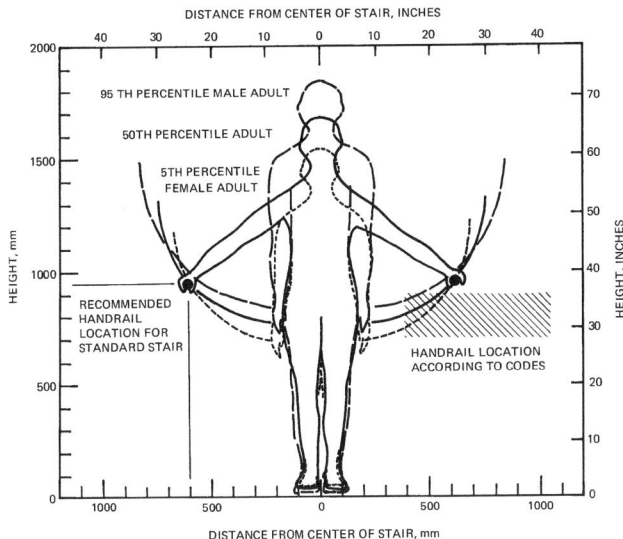

Fig. 3-13.14. Ability of people in the middle of a stair to reach a handrail.

preference was related to handrail height and pitch.[48,49] Using an experimental method with younger and older adults, this study confirmed that handrails should be higher than set out in North American codes and standards. Forces and moments, useful in regaining balance in a misstep, generally increased with increasing handrail height for both younger and older people.

Appropriate handrail centerline heights for younger adults were found to be 910 mm (36 in.) and higher; the appropriate range for older adults was found to be 910 to 1020 mm (36 to 40 in.). The mean preferred height for both age groups was 910 mm (36 in.). A slightly greater height was appropriate with steeper stair pitches; 33, 41, and 49 degrees were the pitches tested. From these and other considerations, a smaller height range, 910 to 970 mm (36 to 38 in.), was recommended as being most appropriate for these steeper conditions. Requirements for handrail heights were subsequently revised in U.S. model building codes and the NFPA *Life Safety Code®*;[55,56] specifically a lower limit of 864 mm (34 in.) was adopted with an upper limit of 965 mm (38 in.), 1067 mm (42 in.) if the handrail serves as a guardrail component.

It should be borne in mind that the foregoing research findings are stated in terms of the handrail centerlines, and that a 45 mm (1.75 in.) diameter handrail was used for the height experiments. West Park Research was recently contracted to extend their functional capability experiments to consider handrail size, shape, and surface texture.[47] Altogether 16 handrail shapes and sizes were tested along with different surface treatments. Not surprisingly, large handrails came out poorly in terms of both function and user preference. The best shape and size was a round, 38 mm (1.5 in.) diameter section; other round sections between 32 and 44 mm (1.25 and 1.75 in.) in diameter, along with other shapes having similar circumferences (100 to 140 mm), also fared well in the testing. These recommendations influenced recent revisions to requirements in U.S. model building codes and the NFPA *Life Safety Code*[55,56] to tighten up on traditionally lax or nonexistent requirements for handrail graspability.

Finally, it should be noted that a decorative handrail section, 56 mm (2.2 in.) wide, often installed on top of more elegant oak balustrade systems in dwellings, tested poorly in terms of both function and preference. This illustrates that an expensive stair could be worse for safety than a more economical one; the decorative section cost nearly ten times more than the simple round section that fared best in the tests.

Perceptual Aspects of Stairs

The importance of stair-tread-nosing visibility has already been mentioned, and it is worth quoting a related general finding: "Stairway accidents are caused by human perceptual errors, which are frequently triggered by some correctable flaw in the design or construction of stairways themselves."[3] Based on the most comprehensive research to date, the investigators from this 1979 report go on to stress "the importance of accurate perceptual cues in successful stair use. These cues include visual perception of the approach to, and use of, the first step, and tactile-kinesthetic perception during the remainder of the flight. . . . There should not be any deceptive visual cues, inadequate lighting levels, glare, or any sort of visual misinformation present in the stairway" or its surroundings.[3] A more recent study, addressing both perceptual aspects and motor coordination, has been conducted by Small.[87] This study linked environmental health problems with the hazard of falls. Even a small deterioration in perceptual or motor coordination, due to environmental factors, could mean the difference between successful stair use and a fall.

Putting these study results into practice requires additional detailed information and good judgment. The fact that these principles cannot be easily translated into the usual code specification form should not lull anyone into thinking that they can be ignored. For example, distractions deliberately designed into the stairway environment, in the mistaken notion by some designers that these are needed to heighten the drama of human movement, may lead to the wrong kind of drama—first in a fall, secondly in a hospital emergency department, and finally perhaps even in court where the designer and owner learn what is meant by "architecturally triggered human error."

Tread materials and carpet patterns must be selected with a major concern for visibility of each tread as a separate walking surface. Another concern in surface selection is attainment of adequate slip resistance. In the case of carpeting, the use of thick underpadding should be avoided, and the covering should be secured so that there is not a great curvature at the nosing, with a resulting reduction of effective tread depth and an inclined slip plane at the very point of first contact of people's feet in descent. As a guide to slip resistance, keep in mind that, if kept dry, the materials and finishes that have proven safe for level flooring are usually adequate in terms of slip resistance for stair treads. Beware of badly installed (and misnamed) nonslip or nonskid strips that may trip feet or visually confuse tread nosing positions. Keep friction characteristics generally uniform over the tread surface and throughout the stair system.

Uniformity of step geometry is also important; stairway users consciously "test" the first steps of a flight and use this information to traverse the remaining steps in such a way as to minimize energy expenditure. Subsequent nonuniformities in the flight—as little as 5 mm in the height of risers or in the depth of treads—can lead to a misstep and fall.

Finally, lighting should contribute to, not detract from, the visibility achieved through careful surface selection. (Here it is worth noting that there has been important research recently on the matter of emergency illumination levels; work by Boyce[5,6] at the U.K. Electricity Council is complemented by work by Webber[96,97] at the U.K. Building Research Station.)

Other Movement Safety Concerns

The foregoing treatment of stair safety is not intended to suggest that most of the problems of movement safety will be solved simply by following the few recommendations provided here. The focus on such matters in this section is justified because of, first, the leading role of stair accidents, and the special importance of avoiding them, during emergency evacuations; second, the recent nature of much of the research; and third, the current deliberations on these topics in U.S. organizations responsible for building safety standards. One additional reason should be mentioned: with the increasing average age of populations in most countries, movement safety problems take on even greater importance. Keeping down the number of accidents and controlling the escalating costs of liability insurance and health care (for people who would be capable of self care were it not for the disabilities resulting from accidents) will require improved standards of building design and construction.

EVACUATION-TIME CRITERIA

A review of standards, research papers, and other literature related to evacuation of buildings reveals that criteria for the acceptable time to evacuate all or part of a building are not well founded in relation to either fire hazards or human behavior factors. Until the last decade, there were generally only a few specific time criteria in common use, and these were often only implicit in code requirements presented in terms of width, distance, and population. When such criteria were explicitly mentioned, they were often unjustifiably presented as products of sound technical reasoning rather than merely as traditional artifacts. During the last decade, with the early development of systems approaches to firesafety, some slightly more sophisticated criteria were introduced. Currently, a more-fundamental basis for evacuation-time criteria is evolving as part of a rapid growth in the ability to model accurately and to predict the growth and spread of fire and products of combustion. The matter of human factors is receiving much less attention despite its importance, along with fire physics, in the choice of appropriate egress-time criteria.

Time Criteria Implied by Codes and Standards

The most influential egress-time criteria in North America were, in effect, given a stamp of approval in the 1935 report "Design and Construction of Building Exits."[52] Perhaps mainly a product of earlier traditional capacity rules and badly documented beliefs about crowd flow in relation to the traditional "unit exit width" concept, the usual time criterion of one minute and forty seconds—to clear a story for example—was accepted with little criticism (until recently) in the *Life Safety Code*.[55] Implicit in this criterion was the expectation that typical building occupants could move on stairs with average flows of 45 persons per minute per 558-mm (22-in.) unit of exit width and on level passageways with average flows of 60 persons per minute per 558-mm (22-in.) unit of exit width. The explicit manifestations of these assumptions were the capacity rules of 60 or 75 persons per unit of exit width for stairs and 100 persons per unit of exit width for level or ramped passageways. These capacity rules generally applied to egress from individual floors, hence the resulting flow times relate to individual floors, not an entire building.

Some four decades after the 1935 NBS report, its assumptions and criteria were challenged directly through the work of Pauls. As well as providing evidence for a linear (not unit-width) relationship between flow and width, errors and confusions were noted regarding crowd movement information in earlier literature. Pauls suggested that egress times, actually achieved with traditional capacity rules, were about twice as long as previously assumed. Rather than the traditional 1.67 minutes, the derived flow times were about 3.5 minutes.[56,68]

The *Uniform Building Code* did not use the 558-mm (22-in.) unit of exit width for specifying egress capacity; before its revision in 1991 it required 0.30 m (12 in.) of exit width for every 50 persons, along with a population calculation based on 100 percent of occupants of one story plus 50 percent of an adjacent one and 25 percent of the next one. These rules led to designs providing flow times ranging between 2.5 and 4 minutes, depending on the adjacent stories.

In Great Britain there was a parallel development of an egress-time criterion of, in theory, 2.5 minutes flow time. In part this was reportedly based on a successful evacuation, in 2.5 minutes, of the Empire Palace Theatre in 1911 in Edinburgh, Scotland (noted as one of three examples of relatively successful evacuation in serious fires discussed in a 1934 report reviewed by Read and Morris).[81] This criterion, coupled with a flow assumption of 40 persons per minute per 533-mm (21-in.) unit of exit width—the slightly smaller unit used in Britain—led to rules permitting a capacity of approximately 100 persons per unit. If applied to stairs, such rules would, according to Pauls' findings, lead to flow times of 4 minutes or more.

Another British examination of egress-time criteria came after a series of serious crowd incidents, the worst of which led to the deaths of 66 spectators in a crowd crush at Ibrox Park, Glasgow, in 1971. As part of an extensive series of studies and rule-making, a technical inquiry suggested that, on behavioral grounds, a maximum flow time of 8 minutes should be the basis of egress design for stands where there was no risk of fire and 2.5 minutes, the traditional figure, should be retained for those stands of a lower standard where fire could develop.[82] These criteria were combined with traditional flow assumptions of 40 and 60 persons per minute per 533-mm (21-in.) unit of exit width. With more realistic conservative flow assumptions (used by Pauls) the egress flow times actually to be expected would be closer to 11 minutes and 4 minutes, respectively, for the two qualities of stands.

Early in the 1970s, partly because of the concern for firesafety in high-rise buildings, the U.S. General Services Administration set new criteria for egress time as part of a systems approach to firesafety.[24] For ordinary office buildings, all occupants exposed to the fire environment should be able to evacuate to a safe place within 90 seconds of alarm. A portion of this time, not to exceed approximately 15 seconds, could involve movement toward the base of a fire. Upon leaving the immediate area of fire origin, occupants should be able to reach an ultimate area of refuge within 5 minutes of downward travel and 1 minute of upward travel. These criteria were coupled with flow assumptions: horizontally 60 persons per minute, single file and unimpeded; down stairs 45 persons per minute, single file and unimpeded; and up stairs 40 persons per minute, single file and unimpeded. The file width was 609 mm (24 in.). Again, questions can be raised about how realistic these flow assumptions were and whether longer egress times, actually achievable in practice, are acceptable when buildings are actually occupied to the capacities permitted by the original rules.

Egress-time criteria were spelled out explicitly in the recently developed *Standard for Fixed Guideway Transit Systems*, NFPA 130-1986. Platforms must be designed to permit clearing in 4 minutes or less, and stations must be designed to be evacuated to a point of safety in 6 minutes or less. Again, these criteria are combined with stipulated flow assumptions that are too optimistic: for example, 40 persons per minute per 558-mm (22-in.) unit down stairs. Stations designed to these criteria and accompanying flow figures may, in reality, require approximately 6 and 8 min, respectively, to clear platforms and stations—even in the case of everyday commuter populations.

Related to transportation systems, brief mention should be made of one of the criteria governing the preparedness of commercial airliners to evacuate passengers quickly. U.S. Federal Air Regulations (FAR Part 25.803) require that a "demonstration" be done showing that a normal mix of passengers and crew can evacuate an airliner in 90 seconds or less, under nighttime conditions, using only half the available exits. Real-life evacuations (as contrasted with demonstrations) could take several times longer, but evidence on this is sketchy.

Recent Developments in Codes and Standards

A recent development within the *Life Safety Code*[56] also included time criteria and a requirement that evacuation capability be demonstrated to meet the criteria. Chapter 21 of the *Code*, titled "Residential Board and Care Occupancies," refers to three classes of evacuation capability: prompt, slow, and impractical, related, respectively, to total evacuation times of 3 minutes or less, 3 to 13 minutes, and more than 13 minutes.

The latest major development in capacity requirements in egress standards is the replacement of the 558-mm (22-in.) "unit of exit width" measurement with a more-linear method in which, for each person assumed to use the egress route, 5 mm (0.2 in.) width of ramped or level egress facility and 8 mm (0.3 in.) width of stairs must be provided. The width of the facility must also satisfy requirements for minimum width based on movement criteria other than egress capacity. These widths are still measured as nominal widths, e.g., handrails are permitted to project into the required width, and no deduction is made for edge effect (the phenomenon addressed in the effective-width model described earlier in this chapter and covered by Chapter 2 of the *Manual on Alternative Approaches to Life Safety*.[105]

The change from the traditional unit of exit width to the small, per-person incremental widths was spearheaded by the Board for the Coordination of the Model Codes.[4] The board did not wish to tackle the problem of egress capacity in more fundamental terms; i.e., it did not first address the basic question of how long the egress time should be and then derive suitable egress capacity rules. The new measurement units were simply derived from existing requirements for 558 mm (22 in.) of nominal egress for every 75 and 100 persons, respectively, for stepped and level egress routes. Dividing 558 mm (22 in.) by 75 and by 100, then rounding off to one figure, gave the 7.4 mm (0.3-in.) and 5.6 mm (0.2-in.) results. The approximate flow times resulting from these capacity rules are similar to those resulting from earlier methods, i.e., 3.5 minutes.

Another major development occurring in the 1988 NFPA *Life Safety Code* provides a range of flow times and egress capacity rules for buildings with assembly occupancies that fall in the range between theaters and large stadia. For these intermediate-size assembly buildings, capacity rules result in egress flow times between 3.5 minutes for smaller occupancies and 11 minutes for the largest ones (those providing "smoke-protected assembly seating" according to NFPA 102-1986, *Standard for Assembly Seating, Tents, and Air-Supported Structures*). Use of the longer egress flow times requires that a "life safety evaluation" be done to demonstrate that suitable protection is provided for those facilities having the longer flow times.

It is important to explain that a "life safety evaluation" is a written review dealing with the adequacy of life safety features relative to fire, storm, collapse, crowd behavior, and other related safety considerations. This review should be done by a competent technical agency and be acceptable to the authority having jurisdiction. Such an evaluation includes, for example, documented cases in which products of combustion in all conceivable fire scenarios will not significantly endanger occupants using means of egress in the facility (because of fire detection, automatic suppression, smoke control, large-volume space, management procedures, etc.). Moreover, means-of-egress facilities plus facility management capabilities should be adequate to cope with scenarios where certain egress routes are blocked for some reason.

In addition to making realistic assumptions about the capabilities of persons in the facility (e.g., an assembled crowd including many disabled persons or persons unfamiliar with the facility), the life safety evaluation should include a factor of safety of at least 2 in all calculations relating hazard development time and required egress time—a combination of flow time and other time needed to detect and assess an emergency condition, initiate egress, move along the egress routes, etc. This takes into account the possibility that half of the egress routes may not be used (or usable) in certain situations.

TIME-BASED EGRESS ANALYSIS

Basic Assumptions and Calculations

The need to have realistic, verifiable estimates of egress-time criteria and accompanying movement assumptions should be recognized by fire protection engineers and other consultants conducting time-based egress analyses (sometimes termed "timed exit analyses" or "dynamic exit analyses"). Such analyses are done to help get official approval for a building design that otherwise might not meet specific egress requirements in a code. In some cases, the assumptions and calculation methods used in such analyses should be seriously questioned. Currently, authorities, having such analyses thrust on them, may have difficulty judging their value; they can only fall back on the reputation of the con-

sulting firm and/or the bulk of a report and the apparent (perhaps illusory) sophistication of its calculations.

One should be especially critical of discussions in which egress-time criteria are equated in simple fashion to hazard development times. As noted above, in relation to the "life safety evaluation," there should be a factor of safety, especially in view of the incomplete technical grasp of both egress and fire issues at the present. For example, in a conservative approach, the "time available" should be at least twice as long as the "time required."

This chapter has emphasized egress capacity or flow issues more than travel distance or speed issues. The former are generally more important in building spaces occupied by more than a few persons. Time-based egress analyses usually address both sets of issues and state the total evacuation time for a space as the sum of a flow time and a travel time. In some situations, the travel time component is simply the time taken by the person closest to the exit to move from his or her point of origin to the point considered a place of safety or refuge. Careful judgment is required to predict this speed and the actual flow of those following the first person. A simple, conservative approach would assume a modest speed for the first person, and a speed similar to the congested speed for those following behind. In this case, it is reasonable to predict the minimum total evacuation time (not including communication and decision-making times) as the sum of this first person's travel time and the flow time of those following, based on an assumed mean-flow calculation. (Note that the term "minimum total evacuation time" is used here; some consultants have erroneously reported that their time-based exit analyses predict maximum total evacuation times.)

For example, if the first person is 18.3 m (60 ft) from a 914-mm (36-in.) wide (nominal width) exit door and is followed by 100 people, the two time components are travel time, 0.3 minute [18 m (60 ft) divided by 61 m (200 ft) per min], and flow time through the doorway, 2.0 minutes (100 persons divided by 50 persons per min), for a total of 2.3 minutes for the minimum total evacuation time. Note that, in reality, the first person might walk at a free walking speed somewhat higher than the assumed 61 m (200 ft) per minute; however, then it is very unlikely that other people would be close behind. Therefore, it is unreasonable to assume that those immediately behind (the fast-walking first person) would achieve a mean flow of 50 persons per minute, a conservative figure for sustained mean flow through a nominal 914-mm (36-in.) doorway. Further information on reasonable movement rates is provided in Table 3-13.1.

One of the errors in some time-based exit analyses occurs as inconsistent, unrealistic assumptions are made about simultaneous high speed and high density of crowd movement, a combination that appears, according to Equation 1, to give a high flow. (Equation 1, the fundamental equation for traffic movement, describes flow as the product of speed, density, and path width.)

Due to interference among closely spaced people, high densities do not permit high speeds of movement, a fact illustrated by Figure 3-13.1 in relation to crowd movement down stairs. Moreover, optimum flows only occur at speeds that are about 60 percent of the speeds at which individuals can move freely. (See Figure 3-13.2 in relation to stairs and, generally, the subsection "General Research on Crowd Movement.")

The moderate conditions, shown in Table 3-13.1, are reasonable approximations for predicting speeds, densities, and flows in calculations of minimum egress time for many

TABLE 3-13.1 *Crowd Movement Parameters for Various Facilities and Conditions*

Facility	Crowd Condition	Density Per Sq Ft	Speed Ft/Min	Flow Per Min/Ft
Stair	Minimum	<0.05	150	<5
Stair	Moderate	0.10	120	12
Stair	Optimum	0.19	95	18
Stair	Crush	0.30	<40	<12
Corridor	Minimum	<0.05	250	<12
Corridor	Moderate	0.10	200	20
Corridor	Optimum	0.20	120	24
Corridor	Crush	0.30	<60	<18
Doorway	Moderate	0.10	170	17
Doorway	Optimum	0.22	120	26
Doorway	Crush	0.30	<50	<15

For SI units: 1 ft = 0.304 m; 1 ft^2 = 0.092 m^2.

situations (especially in view of the fact that other behavior, not involving simple movement directed to the exit, will often be a larger factor in determining evacuation time). The figures given for corridors apply to all walkways with level or moderate slope (less than 1:12). The figures for stairs assume relatively good step geometry and handrail provision. The figures are based on work by Fruin[19] and by Pauls (as discussed earlier in this chapter); however, they are simplified and optimistic because there are no reductions for edge effects. Nominal per-foot measurements are used here. The resultant errors will be acceptable so long as egress times calculated using these assumptions are considered minimum times for egress movement only. (Figure 3-13.5 is an example of a more sophisticated approach to stair flow capacity.)

Sample Calculation Using Table 3-13.1

Given a crowd of 170 people using a corridor 1520 mm (5.0 ft) wide leading to a doorway 914 mm (3.0 ft) in nominal width and then a stairway 1220 mm (4.0 ft) in nominal width: what mean flow should be assumed for evacuation purposes; which of the egress facilities governs this flow; what is the expected minimum flow time; and what crowd conditions can be expected in these three facilities?

Using the moderate conditions, it is predicted that the corridor will serve 100 persons per minute, the doorway will serve 51 persons per minute, and the stair will serve 48 persons per minute. Therefore, the stair capacity governs the flow. The flow time is expected to be 170 divided by 48, or 3.5 minutes, a time similar to the implied standard in current egress standards. At a flow of 48 persons per minute through each facility and without any queuing, the corridor will be minimally crowded with an average crowd density of about 0.5 person/m^2 (0.05 person/ft^2), and the doorway and stair will be used at comfortable, moderate levels.

From Figures 3-13.4 and 3-13.5 it can be understood that, with greater population per width of egress facility (and with queuing for the facility), there may be a higher, more efficient flow. In these cases, it may be appropriate to use the optimum values presented in Table 3-13.1; however, caution must be exercised with the localized crowding conditions that may result, especially on stairs and at doors. For example, if there were 800 people using the facilities described in this subsection, the stair would continue to govern the flow time but, with a higher mean flow, the flow time

could be 11.1 minutes. Movement would be restricted to a shuffling pace, with extensive queuing, and a few percent of the people might have difficulty dealing with the sustained high-density conditions.

Movement Assumptions for Simple, First-Approximation Calculations

Although very incomplete, the following will be useful in doing rough, preliminary calculations. Differing crowd compositions and abilities may alter these values up or down by about one-third. Adverse design conditions will reduce effectiveness by as much as one-third. (These values are comparable with those described as "Levels of Service D and E.")[19]

Stairs: A high-quality stairway that allows convenient counterflow and two-abreast movement, with a width of 1220 mm (4 ft) between handrail centerlines—giving an effective width of just over 1 m—will carry a flow of about one person per second under moderate flow conditions. Speed along the slope will be approximately 0.5 m (2 ft) per second or one typical office building story every 15 seconds. Each person will occupy an average of two stair treads.

Level passageways and moderate-slope ramps: A clear width of 1.22 m (4 ft) will permit a flow of 1.33 persons per second under moderate flow conditions. Speed will be approximately 1.0 m (3.33 ft) per second. Density will be approximately one person per 1 m^2 (10 ft^2).

Doorways: A common 910 mm (3 ft) nominal width doorway will permit a flow of one person per second under moderate to optimum flow conditions.

The ratio of clear widths for similar flow, comparing stairs and the other facilities, is 4:3; i.e., a 1.22 m (4 ft) clear stair width is well matched to a 1 m (3 ft) clear doorway width. Hence the ratios of 4:3 and 3:2 used in common code rules for egress widths are approximately correct and, in the case of the latter 3:2 ratio, err on the side of safety because the code rules are based on stairs' nominal width (not counting handrail incursions or other edge effects).

Other circulation facilities: For completeness, a few other circulation facilities can be noted even though they are not necessarily given egress capacity credit by codes and standards because of a variety of use and maintenance difficulties. A 1.22 m (4 ft) nominal width escalator will carry 1.5 persons per second. A typical revolving door will permit a flow of 0.5 person per second. A turnstile will permit a flow of 0.5 to 1.0 person per second depending on ticket or coin collection procedures.

SUMMARY

A quantitative approach to the movement of people must be balanced by qualitative understanding of the context within which the movement takes place. In cases of fire or other emergencies, egress movement is part of a complex behavior pattern. Calculations addressing only movement directed to egress from a space or building must therefore be considered as providing only minimum evacuation times. Nonetheless, such calculations are useful for making comparisons among design options and for using equivalency approaches to satisfy legal requirements for means of egress.

Calculation methods developed by a number of investigators have been presented. Emphasis has been given to the methods developed by Pauls as a result of his field studies of evacuation and other movement of people in buildings. Much additional work is required to develop such methods and to revise requirements in codes and standards so that an integrated systematic approach to fire protection, egress provision, and everyday movement safety is the norm.

REFERENCES CITED

1. D. Alessi, et al, *NBS-GCR 78-156*, National Bureau of Standards, Gaithersburg (1978).
2. D.M. Alvord, *NBS-GCR 85-496*, National Bureau of Standards, Gaithersburg (1985).
3. J.C. Archea, et al, *NBS-BBS 120*, National Bureau of Standards, Gaithersburg (1979).
4. Board for the Coordination of the Model Codes, *Report on means of egress*, Council of American Building Officials, Falls Church (1985).
5. P.R. Boyce, *Light. Res. and Tech.*, 17, 51 (1985).
6. P.R. Boyce, *ECRC 973*, Electricity Council Research Centre, Carpenhurst, U.K. (1985).
7. J.L. Bryan, *NBS-GCR 77-94*, National Bureau of Standards, Gaithersburg (1977).
8. J.L. Bryan, *An examination and analysis of the dynamics of the human behavior in the MGM Grand Hotel fire*, National Fire Protection Association, Quincy (1982).
9. J.L. Bryan, *Fire J.*, 76, 39 (1982).
10. J.L. Bryan, *Fire J.*, 79, 27 (1985).
11. J.L. Bryan, P.J. DiNenno, and J.A. Milke, *The Determination of Behavior Response Patterns in Fire Situations*, National Bureau of Standards, Gaithersburg (1980).
12. D. Canter, ed., *Fires and Human Behavior*, John Wiley and Sons, New York (1980).
13. D. Canter, *Studies of human behavior in fire: Empirical results and their implications for education and design*, University of Surrey, Guilford (1983).
14. L.G. Chalmet, R.L. Francis, and P.B. Saunders, *Mgt. Sci.*, 28, 86 (1982).
15. *Report of the Task Force on Crowd Control and Safety*, City of Cincinnati, Cincinnati (1980).
16. Consumer Product Safety Commission, *NEISS Data Highlights*, Consumer Product Safety Commission, Washington (1982).
17. L. Ericson, *Utrymning*, Arkitektur I, Tekniska (1984).
18. R.L. Francis, *Technology Report 79-5*, Society of Fire Protection Engineers, Boston (1979).
19. J.J. Fruin, *Pedestrian Planning and Design*, Revised ed., Elevator World Educational Services Division, Mobile, AL (1987).
20. J.J. Fruin, *Audit. News*, 22, 4 (1984).
21. J.J. Fruin, *Crowd Dynamics and the Design and Management of Public Places*, Conference, Los Angeles (1985).
22. J.J. Fruin, *Elev. World*, 33, 52 (1985).
23. M. Galbreath, *Fire Research Note 8*, National Research Council of Canada, Ottawa (1969).
24. General Services Administration, *Building Fire Safety Criteria, Appendix D*, General Services Administration, Washington (1972).
25. N.E. Groner, *NBS-GCR-82-408*, National Bureau of Standards, Gaithersburg (1982).
26. N.E. Groner, B.M. Levin, and H.E. Nelson, *Fire J.*, 75, 44 (1981).
27. T.A. Habicht and J.P. Braaksma, *J. Trans. Eng.*, 110, 80 (1984).
28. B.D. Hankin and R.A. Wright, *Oper. Res. Quart.*, 9, 81 (1959).
29. G.A. Harrison, *CRC Crit. Rev. in Environ. Cont.*, 4, 483 (1974).
30. Home Office/Scottish Home and Health Dept., *Guide to Safety at Sports Grounds (Football)*, Her Majesty's Stationery Office, London (1976).
31. S. Horiuchi, *NBSIR 802070*, National Bureau of Standards, Gaithersburg (1978).
32. S. Horiuchi, Y. Murozaki, and A. Hokugo, in *Fire Safety Science, Proceedings of the First International Symposium of Fire Safety Science*, Hemisphere, New York (1984).
33. ITE Technical Council Committee 5-R, *Traffic Eng.*, May, 34, (1976).
34. B.M. Johnson, *NRCC 23932*, National Research Council of Canada, Ottawa (1983).
35. B.M. Johnson, and J.L. Pauls, *Study of Personnel Movement in Office Buildings; Health Impacts of the Use, Evaluation and Design of Stairways in Office Buildings*, Health and Welfare, Ottawa (1977).
36. B.K. Jones and J.A. Hewitt, in *Fire Safety Science, Proceedings of the First International Symposium on Fire Safety Science*, Hemisphere, New York (1984).
37. M. Kagawa, S. Kose, and Y. Morishita in *Fire Safety Science, Proceedings of the First International Symposium on Fire Safety Science*, Hemisphere, New York (1984).
38. E. Kendik, *Technology Report 85-4*, Society of Fire Protection Engineers, Boston, (1985).
39. E. Kendik, in *Fire Safety Science, Proceedings of the First International Symposium on Fire Safety Science*, Hemisphere, New York (1984).
40. T.M. Kisko, and R.L. Francis, *EVACNET+: A Computer Program to Determine Optimal Building Evacuation Plans*, University of Florida, Gainesville (1985).
41. T.M. Kisko, and R.L. Francis, *F. Safety J.*, 9, 211 (1985).
42. T.J. Klem, *Fire J.*, 80, 128 (1986).
43. M. Kobayashi and S. Horiuchi, *NBSIR 80-2070*, National Bureau of Standards, Gaithersburg (1978).
44. London Transport Board, *Second report of the Operational Research Team on the Capacity of Footways*, London Transport Board, London (1958).
45. H.A. MacLennan, *ASHRAE Trans.*, 91, Pt. 2 (1985).
46. H.A. MacLennan, in *Fire Safety Science, Proceedings of the First International Symposium on Fire Safety Science*, Hemisphere, New York, (1984).
47. B.E. Maki, *Contract No. OSR84-00197*, National Research Council of Canada, Ottawa (1985).
48. B.E. Maki, S.A. Bartlett, and G.R. Fernie, *Human Factors*, 26, 705 (1984).
49. B.E. Maki, S.A. Bartlett, and G.R. Fernie, *Human Factors*, 17, 355 (1985).
50. S.J. Melinek, and R. Baldwin, *Current Paper CP 95/75*, Building Research Establishment, Borehamwood (1975).
51. S.J. Melinek, and S. Booth, *Current Paper CP 96/75*, Building Research Establishment, Borehamwood (1975).
52. *Design and Construction of Building Exits*, National Bureau of Standards, Washington (1935).
53. National Safety Council, *Accident Facts*, National Safety Council, Chicago (1985).
54. H.E. Nelson, *Emergency Evacuation Flow Models*, Panel, Tokyo (1980).
55. NFPA *101, Code for Safety to Life from Fire in Buildings and Structures*, National Fire Protection Association, Quincy (1981).
56. NFPA *101, Code for Safety to Life from Fire in Buildings and Structures*, National Fire Protection Association, Quincy (1994).
57. J.L. Pauls, in *Human Response to Tall Buildings*, Dowden, Hutchinson and Ross, Stroudsburg (1977).
58. J.L. Pauls, in *Proceedings of Second Conference on Designing to Survive Severe Hazards*, IIT Research Institute, Chicago (1977).
59. J.L. Pauls, *Buildings*, May, 84 (1978).
60. J.L. Pauls, *The Stair Event (film)*, National Research Council of Canada, Ottawa (1979).
61. J.L. Pauls, in *Fires and Human Behavior*, John Wiley and Sons, New York (1980).
62. J.L. Pauls, *Effective-Width Model for Evacuation Flow*, Workshop, Gaithersburg (1980).
63. J.L. Pauls, *The Stair Event: Some Lessons for Design*, Conference, Sydney, (1980).
64. J.L. Pauls, *Building Research Note No. 185*, National Research Council of Canada, Ottawa (1982).
65. J.L. Pauls, *Effective-Width Model for Crowd Evacuation Flow on Stairs*, Seminar, Karlsruhe (1982).
66. J.L. Pauls, *Building Practice Note 35*, National Research Council of Canada, Ottawa (1982).
67. J.L. Pauls, *F. Safety J.*, 5, 213 (1983).
68. J.L. Pauls, *Fire Tech.*, 20, 27 (1984).

69. J.L. Pauls, *Fire Tech.*, 20, 28 (1984).
70. J.L. Pauls, *Stair Safety: A Review of Research and Design Recommendations*, Conference, Toronto (1984).
71. J.L. Pauls, *Southern Bldg.*, April/May, 14 (1984).
72. J.L. Pauls, *Ergonomics*, 28, 999 (1985).
73. J.L. Pauls, *The Bldg. Official and Code Admin.*, May-June 26 (1985).
74. J.L. Pauls, and B.K. Jones, in *Fires and Human Behavior*, John Wiley and Sons, New York (1980).
75. R.L. Paulsen, *NBSSIR 81-2438*, National Bureau of Standards, Gaithersburg (1981).
76. R.G. Pearson, and M.G. Joost, *NBS-GCR-83-429*, National Bureau of Standards, Gaithersburg (1983).
77. Popplewell, *Committee Report 9585*, Her Majesty's Stationery Office, London (1985).
78. Popplewell, *Committee Report 9710*, Her Majesty's Stationery Office, London (1986).
79. *Fire Grading of Buildings, Part III, Personal Safety, Post-War Building Studies*, Her Majesty's Stationery Office, London (1952).
80. V.M. Predtechenskii, and A.I. Milinskii, *Planning for Foot Traffic Flow in Buildings*, Amerind Publishing, New Delhi (1978).
81. R.E.H. Read and W.A. Morris, *Aspects of Fire Precautions in Buildings*, Her Majesty's Stationery Office, London (1983).
82. SCICON, *Safety in Football Stadia: A Method of Assessment, Scientific Control Systems*, London (1972).
83. P.G. Seeger, and R. John, *NBSIR 80-2070*, National Bureau of Standards, Gaithersburg (1978).
84. J.D. Sime, *Escape from Building Fires: "Panic" or Affiliation?*, thesis, Surrey (1984).
85. J.D. Sime, *The Outcome of Escape Behavior in the Summerland Fire: Panic or Affiliation?*, Conference, Los Angeles (1985).
86. J.D. Sime, in *Fire Safety Science, Proceedings of the First International Symposium on Fire Safety Science*, Hemisphere, New York (1984).
87. B.M. Small, *Environmental Health Factors in Falling Accidents*, Conference, Los Angeles (1985).
88. F.I. Stahl, *NBSIR 80-1982*, National Bureau of Standards, Gaithersburg (1980).
89. F.I. Stahl, *Arch. Sci. Rev.*, 23, 85 (1980).
90. F.I. Stahl, and J.C. Archea, *NBSIR 77-1313*, National Bureau of Standards, Gaithersburg (1977).
91. F.I. Stahl, J.J. Crosson, and S.T. Margulis, *NBSIR 82-2480*, National Bureau of Standards, Gaithersburg (1982).
92. J.A. Swartz, *Fire J.*, 73, 73 (1979)
93. K. Togawa, *Report No. 14*, Building Research Institute, Tokyo (1955).
94. T.F. Ventre, F.I. Stahl, and G.E. Turner, *NBSIR 81-2361*, National Bureau of Standards, Gaithersburg (1981).
95. J.M. Watts, *Computer Models for Evacuation Analysis*, Symposium, College Park (1986).
96. G.M.B. Webber, *Emergency Lighting Recommendations*, Conference, Los Angeles (1985).
97. G.M.B. Webber, *Current Research at BRE, England, on Emergency Lighting on Stairs*, Seminar, Ottawa (1985).
98. G.M.B. Webber, *Report of the Ontario Public Inquiry into Fire Safety in High-Rise Buildings*, Queen's Printer of Ontario, Ontario (1983).
99. P.G. Wood, *Fire Research Note No. 953*, Fire Research Station, Borehamwood (1972).
100. P.G. Wood, in *Fires and Human Behavior*, John Wiley and Sons, New York (1980).
101. J. Pauls and E. Juillet, "Life Safety of People with Disabilities: How far have we progressed?" *Proceedings* of CIB W14 Symposium and Workshops: Fire Safety Engineering in the Process of Design, Part 1: Symposium: Engineering Fire Safety for People with Mixed Abilities, University of Ulster, Northern Ireland, pp. 17–40, (September 1993).
102. E. Juillet, "Evacuating People with Disabilities," *Fire Engineering*, 146, 12, pp. 100–103, (December 1993).
103. The Rt Hon Lord Justice Taylor, *The Hillsborough Stadium Disaster*, interim and final reports, Her Majesty's Stationery Office, London (1989).
104. R.A. Smith and J.F. Dickie (editors), *Engineering for Crowd Safety*, Elsevier, Amsterdam (1993).
105. NFPA 101M, *Manual on Alternative Approaches to Life Safety*, NFPA, Quincy, MA.

EMERGENCY MOVEMENT

Harold E. "Bud" Nelson and Hamish A. MacLennan

INTRODUCTION

This chapter contains data and methods used to estimate emergency evacuation rates and times. Evacuation time is defined as the elapsed time between the instant that occupants receive an alarm (or otherwise become aware of a possible fire) and their arrival at a destination. The destination is normally a safe location inside or outside of a facility.

ELEMENTS OF EMERGENCY MOVEMENT

Research-based methods for predicting the flow of groups of persons in emergencies have emerged in recent years. The major contributors include Predtechenskii and Milinskii,[1] Fruin,[2] and Pauls.[3,4] The methods developed are in most cases compatible and supportive of each other. All are based on the relationship between speed of movement and the population density of the evacuating stream of persons. In general, these methods assume that:

1. All persons will start to evacuate at the same instant.
2. Occupant flow will not involve any interruptions caused by decisions of the individuals involved.
3. All or most of the persons involved are free of disabilities that would significantly impede their ability to keep up with the movement of a group.

The approach is often referred to as a hydraulic model of emergency egress.

Separate works by investigators, such as Wood,[5] Bryan,[6] and Keating and Loftus,[7] have concentrated on the decisions and resulting actions taken by individuals in actual fire situations. Sime[8] and MacLennan[9] have examined the impact of occupant decisions and choices of actions on evacuation time.

In this chapter, the product of hydraulic model calculations is termed modeled evacuation time. Actual egress time is the time required for the occupants to actually leave a building. Generally, the actual egress time will exceed the modeled time. Since the modeled evacuation time is an approximation based on data from evacuation drills and fire experience, it is therefore possible that the modeled evacuation time can exceed the actual evacuation time. The difference between modeled evacuation time and actual evacuation time can be expressed in terms of an apparent evacuation efficiency using the relationship

$$T_{ae} = T_{me}e + T_d \qquad (1)$$

where:

T_{ae} = actual evacuation time (s),
T_{me} = modeled evacuation time (s),
e = apparent evacuation efficiency, and
T_d = delays in initiating evacuation.

Apparent evacuation efficiency, e, is a function of elements that interfere with the assumed hydraulic evacuation flow. Typical examples of efficiency elements are:

1. Delays caused by egress management activities of wardens or others directing the evacuation.
2. Time delays involved in stopping and restarting of flows at merging points.
3. Delays, self-instituted by individuals, that retard their start or slow their progress.
4. Inefficient balance in the use of exit facilities, where some emergency routes are overtaxed while others are underutilized.

Typical examples of delays in initiating evacuation include:

1. Time involved in making a decision.
2. Time involved in investigation. This includes actions to determine the source, reality, or importance of a fire alarm or other warning.
3. Time involved in other activities not fully contributing to effective evacuation.
4. Time involved in determining the appropriate exit route (i.e., "way finding").

Harold E. "Bud" Nelson, a Senior Research Engineer at Hughes Associates, Inc. in Columbia, MD, is Past-President of the SFPE. He has focused on hazard analysis and application of scientific principles to the solution of real-world fire safety problems. Hamish MacLennan is a member of the faculty of the University of New South Wales. He has been a visiting scientist at the National Bureau of Standards. His studies have been concentrated in the area of human response to fire emergencies.

All of these factors can reduce evacuation efficiency. However, all of the elements will seldom come to bear on a single evacuation.

The first step in appraising emergency movement is to calculate the modeled egress time. The use of model calculations provides a reproducible base of reference in appraising the impact of overall systems, individual components, or changes in systems. If, however, the results of the modeled evacuation time are to be compared to a realistically expected evacuation time or to expected fire growth, it is important that the user understand that the modeled evacuation time is seldom achieved in reality. Accurate estimation of expected evacuation time requires both the calculation of the modeled evacuation time and an appraisal of evacuation efficiency. (See Equation 1.) The organization of the rest of this chapter separately addresses the use of hydraulic flow calculations to estimate modeled evacuation time and delays in initiation and evacuation efficiency factors.

HYDRAULIC FLOW CALCULATIONS

The estimation of modeled evacuation time utilizes a series of expressions that relate data acquired from tests and observations to a hydraulic approximation of human flow. While the expressions indicate absolute relationships, there is considerable variability in the data. Figure 3-14.1, abstracted from Section 3, Chapter 13, shows a typical relationship between the source data and the derived equation. The equations and relationships presented in the following paragraphs can be used independently or collected to solve a complex egress problem. Such a coordinated collection of equations is demonstrated in the sample problem.

Effective Width, W_e

Persons moving through the exit routes of a building maintain a boundary layer clearance from walls and other stationary obstacles they pass. This clearance is needed to accommodate lateral body sway and assure balance.

Discussion of this crowd movement phenomena is found in the works of Pauls,[3] Fruin,[2] and Habicht and Braaksma.[10] The useful (effective) width of an exit path is the clear width of the path less the width of the boundary layers. Figures 3-14.2 and 3-14.3 depict effective width and boundary layer. Table 3-14.1 is a listing of boundary layer

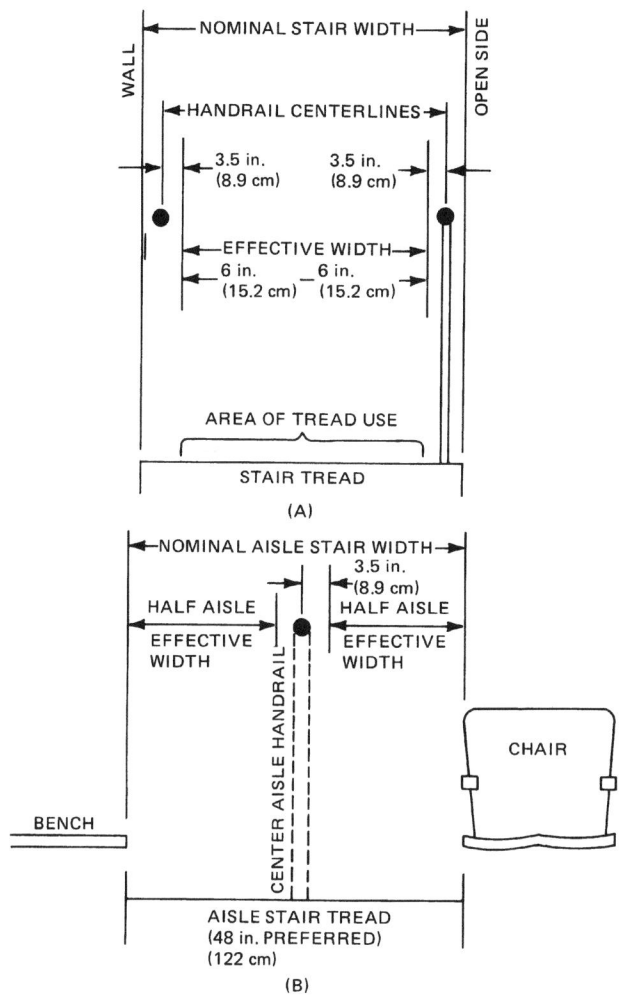

Fig. 3-14.2. Measurements of effective width of stairs in relation to walls, handrails, and seating.

widths. The effective width of any portion of an exit route is the clear width of that portion of an exit route less the sum of the boundary layers.

Clear width is measured:

1. From wall to wall in corridors or hallways.
2. As the width of the treads in stairways.
3. As the actual passage width of a door in its open position.
4. As the space between the seats along the aisles of assembly arrangement.
5. As the space between the most intruding portions of the seats (when unoccupied) in a row of seats in an assembly arrangement.

The intrusion of handrails is considered by comparing the effective width without the handrails, and the effective width using a clear width from the edge of the handrail. The smaller of the two effective widths then applies. Using the values in Table 3-14.1, only handrails that protrude more than 2.5 inches need be considered. Minor midbody height or lower intrusions such as panic hardware are treated in the same manner as handrails. Where an exit route becomes either wider or narrower, only that portion of the route has the appropriate greater or lesser clear width.

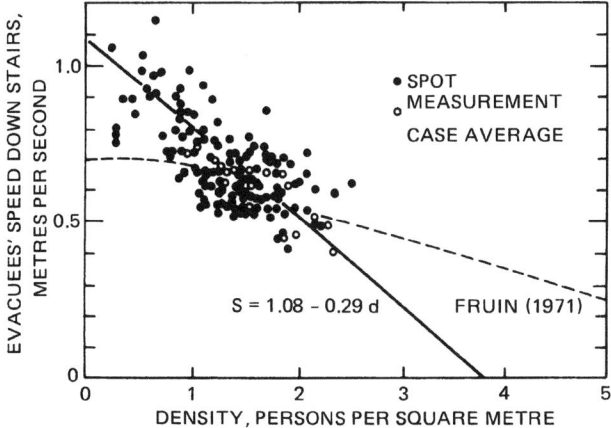

Fig. 3-14.1. Relation between speed and density on stairs in uncontrolled total evacuations. Dashed line from Fruin.[2]

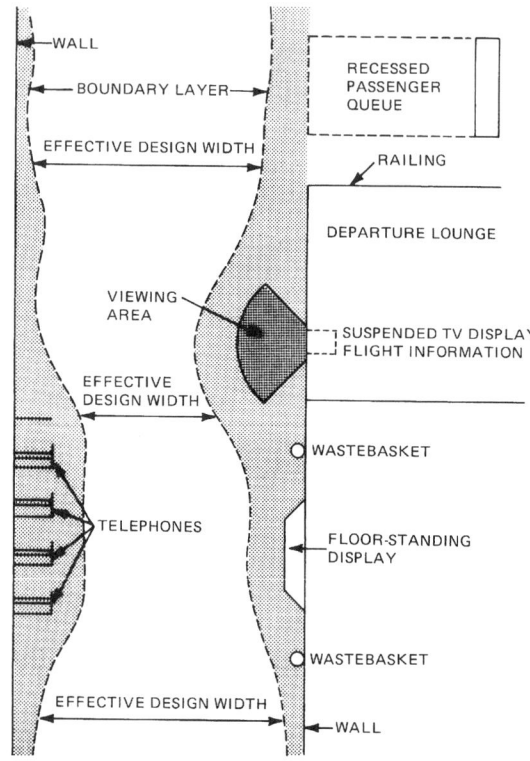

Fig. 3-14.3. Public corridor effective width.

Density, *D*

Density is the measurement of the degree of crowdedness in an evacuation route and is expressed in persons per unit area. The calculations in this chapter are based on density expressed in persons per square foot (or persons per square meter).

Unless information on the dispersion of occupants indicates otherwise, the density of the first exit element (aisle, corridor, ramp, etc.) is based on all of the served occupants. This will demonstrate the capacity limits of the route element and produce a value representing the maximum capacity of the element.

Conversely, if the egressing population is widely dispersed, in terms of reaching the exit route element, the calculation is based on an appropriate time step. At each time increment, the density of the exit route is based on those that have entered the route minus those that have passed from it.

TABLE 3-14.1 *Boundary Layer Widths*

	Boundary layer	
Exit route element	(in.)	(cm)
Stairways—wall or side of tread	6	15
Railings, handrails*	3.5	9
Theater chairs, stadium benches	0	0
Corridor, ramp walls	8	20
Obstacles	4	10
Wide concourses, passageways	up to 18	46
Door, archways	6	15

* Where handrails are present, use the value if it results in a lesser effective width.

TABLE 3-14.2 *Constants for Equation 2, Evacuation Speed*

Exit route element		k_1	k_2
Corridor, Aisle, Ramp, Doorway		275	1.40
Stairs			
Riser (inches)	Tread (inches)		
7.5	10	196	1.00
7.0	11	212	1.08
6.5	12	229	1.16
6.5	13	242	1.23

1 in. = 25.4 mm.

The density factors in subsequent portions of the egress system are determined by calculation. The calculation methods involved are contained in the section of this chapter titled "Transitions."

Speed—Movement Velocity of Exiting Individuals, *S*

Observations and experiments have shown that evacuation flow speed of a group is a function of the population density. The relationships presented in this section have been derived from Fruin,[2] Pauls,[3] and Predtechenskii and Milinskii.[1]

If the population density is less than about 0.05 persons per square foot (0.54 persons per square meter) of exit route (20 square feet per person; 1.85 square meters per person), individuals will move at their own pace, independent of the speed of others. If the population density exceeds about 0.35 persons per square foot (3.8 persons per square meter), no movement will take place until enough of the crowd has passed from the crowded area to reduce the density.

Between the density limits of 0.05 and 0.35 persons per square foot (0.54 and 3.8 persons per square meter) the relationship between speed and density can be considered as a linear function. The equation of this function is

$$S = k - akD \qquad (2)$$

where:

S = speed along the line of travel;
D = density in persons per unit area; and
k = constant, as shown in Table 3-14.2
$\quad k = k_1$; and $a = 2.86$ for speed in ft/min and density in persons/sq ft.
$\quad k = k_2$; and $a = 0.266$ for speed in m/s and density in persons/sq m.

Table 3-14.2 shows evacuation speed constant.

Figure 3-14.4 is a graphic representation of the relationship between speed and density. The speeds determined from Equation 2 are along the line of movement; for stairs this is along the line of the treads. Table 3-14.3 provides convenient multipliers for converting vertical rise of a stairway to a distance along the line of movement. The travel on landings must be added to the values derived from Table 3-14.3.

The maximum speed is that occurring when the density is less than 0.05 persons per square foot (0.54 persons per square meter). These maximum speeds are listed in Table 3-14.4.

Within the range listed in Tables 3-14.2, 3-14.3, and 3-14.4, the evacuation speed on stairs varies approximately

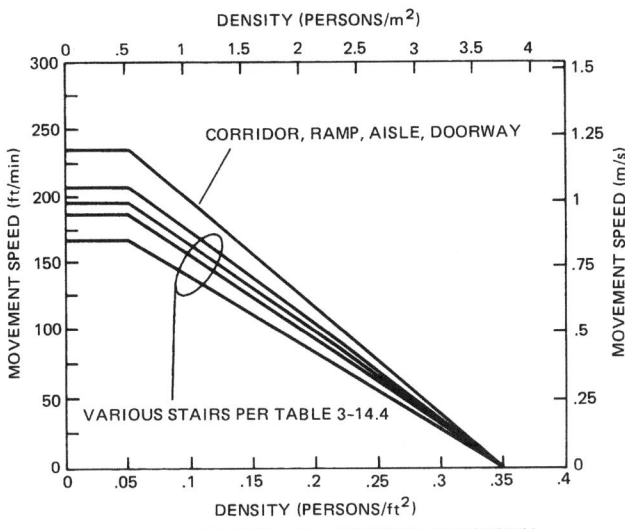

Fig. 3-14.4. *Evacuation speed as a function of density. S =*
k − a kD, where D = density in persons per square foot and k
is given in Table 3-14.2. Note that speed is along line of travel.

as the square root of the ratio of tread width to tread height.
There is not sufficient data to appraise the likelihood that
this relationship holds outside this range.

Specific Flow, F_s

Specific flow, F_s, is the flow of evacuating persons past
a point in the exit route per unit of time per unit of effective
width, W_e, of the route involved. Specific flow is expressed
in persons per minute per foot of effective width (if the value
of $k = k_1$ from Table 3-14.2), or persons per second per
meter of effective width (if the value of $k = k_2$ from Table
3-14.2). The equation for specific flow is

TABLE 3-14.3 *Conversion Factors for Relating Line of*
Travel Distance to Vertical Travel for
Various Stair Configurations

Stairs Riser (inches)	Tread (inches)	Conversion factor
7.5	10.0	1.66
7.0	11.0	1.85
6.5	12.0	2.08
6.5	13.0	2.22

TABLE 3-14.4 *Maximum (Unimpeded) Exit Flow Speeds*

		Speed—along line of travel	
Exit route element		(ft/min)	(m/s)
Corridor, Aisle, Ramp, Doorway		235	1.19
Stairs Riser (inches)	Tread (inches)		
7.5	10	167	0.85
7.0	11	187	0.95
6.5	12	196	1.00
6.5	13	207	1.05

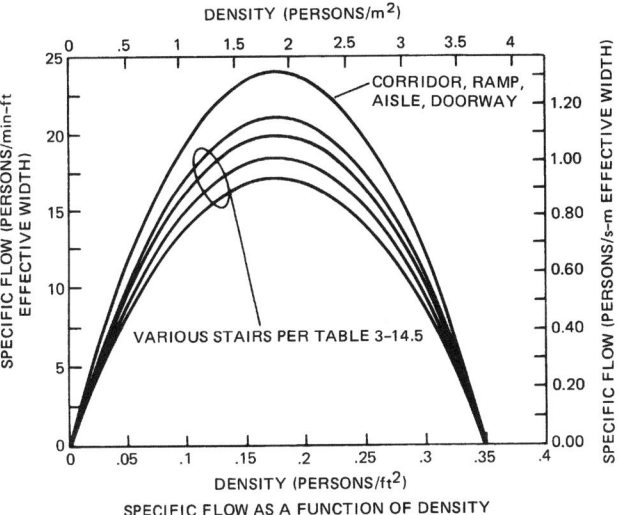

Fig. 3-14.5. *Specific flow as a function of density.*

$$F_s = SD \qquad (3)$$

where:

F_s = specific flow,
D = density, and
S = speed of movement.

F_s is in persons/min/sq ft when density is in persons/sq ft
and speed in ft/min; F_s is in persons/s/sq m when density is
in persons/sq m and speed in m/s.

Combining Equations 2 and 3 produces

$$F_s = (1 - aD)kD \qquad (4)$$

where k is as listed in Table 3-14.2.

The relationship of specific flow to density is shown in
Figure 3-14.5. In each case the maximum specific flow occurs when the density is 0.175 persons per square foot (1.9
persons per square meter) of exit route space. There is a
maximum specific flow associated with each type of exit
route element; these are listed in Table 3-14.5.

Calculated Flow, F_c

The calculated flow, F_c, is the predicted flow rate of
persons passing a particular point in an exit route.

TABLE 3-14.5 *Maximum Specific Flow, F_{sm}*

	Maximum specific flow	
Exit route element	(persons/min/ft of effective width)	(persons/s/m of effective width)
Corridor, Aisle, Ramp, Doorway	24.0	1.3
Stairs Riser (inches)	Tread (inches)		
7.5	10	17.1	0.94
7.0	11	18.5	1.01
6.5	12	20.0	1.09
6.5	13	21.2	1.16

The equation for actual flow is

$$F_c = F_s W_e \qquad (5)$$

where:

F_c = calculated flow,
F_s = specific flow, and
W_e = effective width.

Combining Equations 4 and 5 produces

$$F_c = (1 - aD)kDW_e \qquad (6)$$

F_c is in persons per minute when: $k = k_1$ (from Table 3-14.2), D is persons per square foot, and W_e is feet.
F_c is in persons per second when: $k = k_2$ (from Table 3-14.2), D is persons per square meter, and W_e is meters.

Time for Passage, T_p

Time for passage, T_p, i.e., time for a group of persons to pass a point in an exit route, can be expressed as

$$T_p = P/F_c \qquad (7)$$

where T_p is time for passage (T_p is in minutes where F_c is persons per minute; T_p is in seconds where F_c is persons per second). P is population in persons.

Combining Equations 6 and 7 yields

$$T_p = P/(1 - aD)kDW_e \qquad (8)$$

Transitions

Transitions are any point in the exit system where the character or dimension of a route changes or where routes merge. Typical examples of points of transition include:

1. Any point where an exit route becomes wider or narrower. For example, a corridor may be narrowed for a short distance by an intruding service counter or similar element. The calculated density, D, and specific flow, F_s, differ before reaching, while passing, and after passing the intrusion.
2. The point where a corridor enters a stairway. There are actually two transitions: one occurs as the egress flow passes through the doorway, the other as the flow leaves the doorway and proceeds onto the stairs.
3. The point where two or more exit flows merge, for example, the meeting of the flow from a cross aisle into a main aisle that serves other sources of exiting population. It is also the point of entrance into a stairway serving other floors.

The following rules apply to determining the densities and flow rates following the passage of a transition point:

1. The flow after a transition point is a function, within limits, of the flow(s) entering the transition point.
2. The calculated flow, F_c, following a transition point cannot exceed the maximum specific flow, F_{sm}, for the route element involved multiplied by the effective width, W_e, of that element.
3. Within the limits of rule 2, the specific flow, F_s, of the route departing from a transition point is determined by the following equations:

(a) For cases involving one flow into and one flow out of a transition point:

$$F_{s(out)} = F_{s(in)} W_{e(in)}/W_{e(out)} \qquad (9a)$$

where:

$F_{s(out)}$ = specific flow departing from transition point,
$F_{s(in)}$ = specific flow arriving at transition point,
$W_{e(in)}$ = effective width prior to transition point, and
$W_{e(out)}$ = effective width after passing transition point.

(b) For cases involving two incoming flows and one outflow from a transition point, such as that which occurs with the merger of a flow down a stair and the entering flow at a floor:

$$F_{c(out)} = \{[F_{c(in-1)}W_{e(in-1)}] + [F_{c(in-2)}W_{e(in-2)}]\}/W_{e(out)} \qquad (9b)$$

where the subscripts $(in-1)$ and $(in-2)$ indicate the values for the two incoming flows.

(c) For cases involving other merger geometries the following general relationship applies:

$$[F_{c(in-1)}W_{e(in-1)}] + \cdots + [F_{c(in-n)}W_{e(in-n)}]$$
$$= [F_{c(out-1)}W_{e(out-1)}] + \cdots + [F_{c(out-n)}W_{e(out-n)}] \qquad (9c)$$

where the letter n in the subscripts $(in-n)$ and $(out-n)$ is a number equal to the total number of routes entering $(in-n)$ or leaving $(out-n)$ the transition point.

4. Where the calculated specific flow, F_s, for the route(s) leaving a transition point, as derived from the equations in rule 3, exceeds the maximum specific flow, F_{sm}, a queue will form at the incoming side of the transition point. The number of persons in the queue will grow at a rate equal to the calculated flow, F_c, in the arriving route minus the calculated flow leaving the route through the transition point.
5. Where the calculated outgoing specific flow, $F_{s(out)}$, is less than the maximum specific flow, F_{sm}, for that route(s), there is no way to predetermine how the incoming routes will merge. The routes may share access through the transition point equally, or there may be total dominance of one route over the other. For conservative calculations, assume that the route of interest is dominated by the other route(s). If all routes are of concern, it is necessary to conduct a series of calculations to establish the bounds on each route under each condition of dominance.

EXAMPLE :

Consider an office building (Figure 3-14.6) with the following features:

1. Nine floors, 300 by 80 ft.
2. Floor to floor height is 12 ft.
3. Two stairways, located at ends of building (no dead ends).
4. Each stair is 44 inches wide (tread width) with handrails protruding 2.5 inches.
5. Stair risers are 7 inches wide, treads are 11 inches high.
6. There are two 4-foot by 8-foot landings per floor of stairway travel.

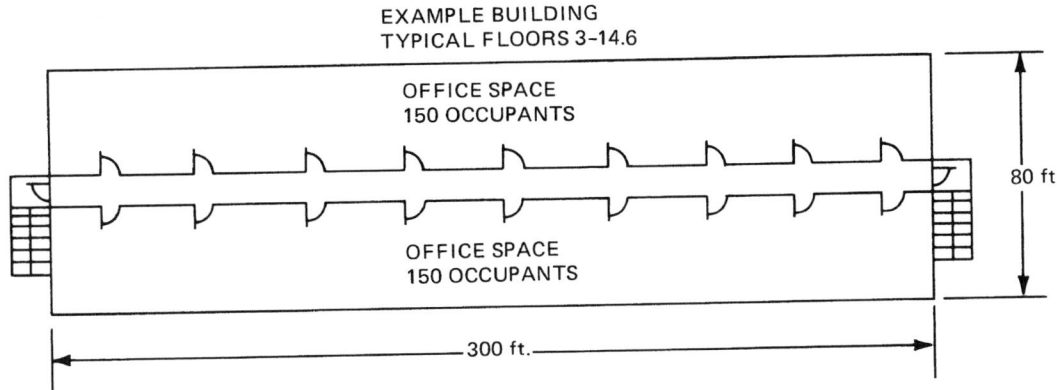

Fig. 3-14.6. *Floor plan for example.*

7. There is one, 36 inch clear width, door at each stairway entrance and exit.
8. The first floor does not exit through stairways.
9. Each floor has a single 8 ft wide corridor extending the full length of each floor. Corridors terminate at stairway entrance doors.
10. There is a population of 300 persons per floor.

SOLUTION A—First Order Approximation:

1. Assumptions.

The prime controlling factor will be either the stairways or the door discharging from them. Queuing will occur; therefore the specific flow, F_s, will be the maximum specific flow, F_{sm}. All occupants start egress at the same time. The population will use all facilities in the optimum balance.

2. Estimate flow capability of a stairway.

From Table 3-14.1, the effective width, W_e, of each stairway is $44 - 12 = 32$ in. (2.66 ft). Also, the effective width, W_e, of each door is $36 - 12 = 24$ in. (2 ft). The maximum specific flow, F_{sm}, for the stairway (from Table 3-14.5) is 18.5 persons/min/foot effective width. Specific flow, F_s, equals maximum specific flow, F_{sm}. Therefore, using Equation 5, the flow from each stairway is limited to $18.5 \times 2.66 = 49.2$ persons/min.

3. Estimate flow capacity through a door.

Again from Table 3-14.5, the maximum specific flow through any 36 in. door is 24 persons/min/foot effective width. Therefore, using Equation 5, the flow through any door is limited to $24 \times 2 = 48$ persons/min. Since the flow capacity of the doors is less than the flow capacity of the stairway served, the flow is controlled by the stairway exit doors (48 persons/stairway exit door/min).

4. Estimate the speed of movement for estimated stairway flow.

From Equation 2 the speed of movement down the stairs is $212 - (2.86 \times 212 \times 0.175) = 105$ ft/min. The travel distance between floors (using the conversion factor from Table 3-14.3) is $12 \times 1.85 = 22.2$ ft on the stair slope plus 8 feet travel on each of the two landings, for a total floor-to-floor travel distance of $22.2 + (2 \times 8) = 38.2$ ft. The travel time for a person moving with the flow is $38.2/105 = 0.36$ min/floor.

5. Estimate building evacuation time.

If all of the occupants in the building start evacuation at the same time, each stairway can discharge 48 persons per min. The population of 2400 persons above the first floor will require approximately 25 min to pass through the exit. An additional 0.36 min travel time is required for the movement from the second floor to the exit. The total minimum evacuation time for the 2400 persons located on floors 2 through 9 is estimated at 25.4 min.

SOLUTION B—More Detailed Analysis:

1. Assumptions.

The population will use all exit facilities in the optimum balance; all occupants start egress at the same time.

2. Estimate flow density, D, speed, S, specific flow, F_s, effective width, W_e, and initial calculated flow, F_c, typical for each floor.

Divide each floor in half, to produce two exit calculation zones, each 150 feet long. Determine the density, D, and speed, S, if all occupants try to move through the corridor at the same time. That is, 150 persons moving through 150 feet of an 8 ft wide corridor.

Density, $D = 150$ persons/1200 sq ft corridor area $= 0.125$ persons/sq ft.
From Equation 2, speed, $S = k - akD$.
From Table 3-14.2, $k = 275$.
$S = 275 - (2.86 \times 275 \times 0.125) = 177$ ft/min.
From Equation 4, specific flow, $F_s = (1 - aD)kD$.
$F_s = [1 - (2.86 \times 0.125)] \times 275 \times 0.125 = 22$ persons/ft effective width/min.
From Table 3-14.5, the specific flow, F_s, is less than the maximum specific flow, F_{sm}; therefore, F_s is used for the calculation of calculated flow.
From Table 3-14.1, the effective width of the corridor is $8 - (2 \times 0.5) = 7$ ft.
From Equation 6, calculated flow, $F_c = (1 - aD)kDW_e$.
$F_c = [1 - (2.86 \times 0.125)] \times 275 \times 0.125 \times 7 = 154$ persons/min.

Note: At this stage in the calculation, calculated flow, F_c, is termed initial calculated flow for the exit route element (i.e., corridors) being evaluated. This is because the calculated flow rate can be sustained only if the discharge (transition point) from the route can also accommodate the indicated flow rate.

3. Estimate impact of stairway entry doors on exit flow.

Each door has a 36 inch clear width. From Table 3-14.1, effective width, W_e, is 30 − 12 = 24 inches (2 ft).

From Table 3-14.5, the maximum specific flow, F_{sm}, is 24 persons/min/ft effective width.

From Equation 9, $F_{s(door)} = [F_{s(corridor)}W_{e(corridor)}]/W_{e(door)}$ $F_{s(door)} = (22 \times 7)/2 = 77$ persons/min/ft effective width.

Since F_{sm} is less than the calculated F_s, the value of F_{sm} is used. Therefore, the effective value for specific flow is 24.

From Equation 5 the initial calculated flow, $F_c = F_s W_e$ = 24 × 2 = 48 persons/min through a 36 inch door.

Since F_c for the corridor is 154 while F_c for the single exit door is 48, queuing is expected. The calculated rate of queue buildup will be 154 − 48 = 106 persons/min.

4. Estimate impact of stairway on exit flow.

From Table 3-14.1, effective width, W_e, of the stairway is 44 − 12 = 32 inches (2.66 ft).

From Table 3-14.5, the maximum specific flow, F_{sm}, is 18.5 persons/ft effective width.

From Equation 9, the specific flow for the stairway, $F_{s(stairway)}$, is 24 × 2/2.66 = 18.0 persons/ft effective width. In this case, F_s is less than F_{sm} and F_s is used.

The value of 18.0 for F_s applies until the flow down the stairway merges with the flow entering from another floor.

Using Figure 3-14.4 or Equation 4 and Table 3-14.2, the density of the initial stairway flow is approximately 0.146 persons per sq ft of stairway exit route.

From Equation 2 the speed of movement during the initial stairway travel is 212 × (2.86 × 212 × 0.146) = 123 ft/min.

From Solution A, the floor-to-floor travel distance is 38.2 feet. The time required for the flow to travel one floor level is 38.2/123 = 0.31 min (19 s).

Using Equation 5, the calculated flow, F_c, is 18.0 × 2.66 = 48 persons/min.

After 0.31 min, 48 × 0.31 = 15 persons will be in the stairway from each floor feeding to it. If floors 2 through 9 exit all at once, there will be 15 × 8 = 120 persons in the stairway. After this time the merging of flows between the flow in the stairway and the incoming flows at stairway entrances will control the rate of movement.

5. Estimate impact of merger of stairway flow and stairway entry flow on exit flow.

From Equation 9, $F_{s(out-stairway)} = \{[F_{s(door)} \times W_{e(door)}]$ $+ [F_{s(in-stairway)} \times W_{e(in-stairway)}]\}/W_{e(out-stairway)} = [(24 \times 2) + (18 \times 2.66)]/2.66 = 36$ persons/ft effective width.

From Table 3-14.5, F_{sm} for the stairway is 18.5 persons/min/ft effective width. Since F_{sm} is less than the calculated F_s, the value of F_{sm} is used.

6. Track egress flow.

Assume all persons start to evacuate at time zero. Initial flow speed is 177 ft/min. Assume that congested flow will reach the stairway in 30 s. At 30 s, flow starts through stairway doors. F_c through doors is 48 persons/min for the

next 19 s. At 49 s, 120 persons are in each stairway and 135 are waiting in a queue at each stairway entrance door.

Note: Progress from this point on depends on which floors take dominance in entering the stairways. Any sequence of entry may occur. To set a boundary, this example estimates the result of a situation where dominance proceeds from the highest to the lowest floor.

The remaining 135 persons waiting at each stairway entrance on the ninth floor enter through the door at the rate of 48 persons/min. The rate of flow through the stair is regulated by the 48 persons/min rate of flow of the discharge exit doors. The descent rate of the flow is 19 s per floor.

Thus:

at 218 s (3.6 min)	all persons have evacuated the 9th floor.
at 237 s (4.0 min)	the end of the flow reaches the 8th floor.
at 401 s (6.7 min)	all persons have evacuated the 8th floor.
at 420 s (7.0 min)	the end of the flow reaches the 7th floor.
at 584 s (9.7 min)	all persons have evacuated the 7th floor.
at 603 s (10.1 min)	the end of the flow reaches the 6th floor.
at 767 s (12.8 min)	all persons have evacuated the 6th floor.
at 786 s (13.1 min)	the end of the flow reaches the 5th floor.
at 950 s (15.8 min)	all persons have evacuated the 5th floor.
at 969 s (16.2 min)	the end of the flow reaches the 4th floor.
at 1133 s (18.9 min)	all persons have evacuated the 4th floor.
at 1152 s (19.2 min)	the end of the flow reaches the 3rd floor.
at 1316 s (21.9 min)	all persons have evacuated the 3rd floor.
at 1335 s (22.3 min)	the end of the flow reaches the 2nd floor.
at 1499 s (25.0 min)	all persons have evacuated the 2nd floor.
at 1518 s (25.3 min)	all persons have evacuated the building.

EVACUATION EFFICIENCY FACTORS

Decisions

Humans require time to make decisions. In general, people are hesitant to undertake overt actions unless they clearly accept the need for such action.

In group situations, group interaction is extremely important to decision-making. Latane and Darley[11] pointed out the tendency of many individuals to defer emergency decision-making until action is clearly required. More recently MacLennan[9] in his experiments in Australia has classified and is now quantifying this as a factor he and Sime call "associative." MacLennan has noted that persons in groups often delay response to a warning alarm until it is clear that the group accepts the need to take emergency action. Prior training, organization, and real-time fire information can reduce the delay in the time to take emergency action.

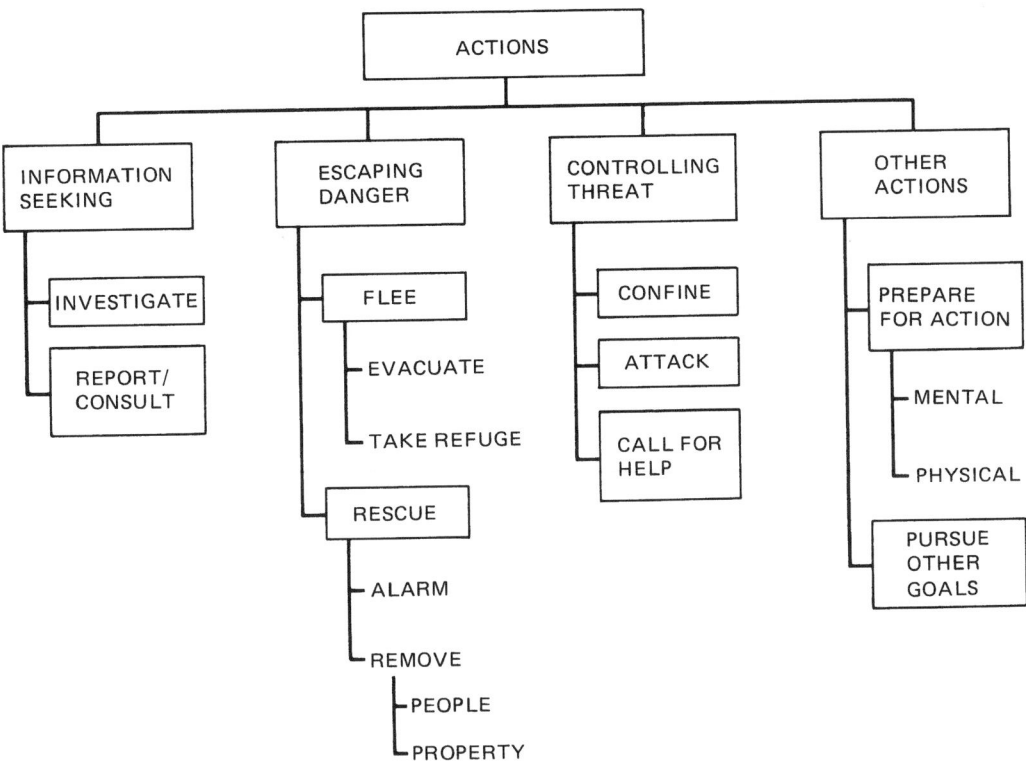

Fig. 3-14.7. Types of actions.

Levin,[12] in analyzing data from many sources related to emergency action in residences, has proposed the division of speed of response into four categories. These categories, which Levin terms ambiguity classes, are based on the initial interpretation of the cues by the individual involved. Levin's classes are:

A. Respondent believes that there may be a fire (but is not certain).
B. Respondent believes that it is likely there is a fire.
C. Respondent is sure there is a fire and has seen sufficient smoke to believe that it is a dangerous fire.
D. Respondent has seen flames.

With one major exception, the universal response to ambiguity Class A is to seek information. In the sample studied by Levin, the exception to the general rule occurred with those responsible for persons incapable of taking care of themselves (small children, invalids). These individuals immediately took action to evacuate or otherwise safeguard their charges.

Conversely, in an ambiguity Class D situation emergency action was always taken.

In ambiguity Classes B and C some persons continued to seek information while others undertook other types of emergency action. These actions included attacking the fire, calling the fire department, giving the alarm, and evacuating the building. For the limited number of cases that Levin classified in ambiguity Classes B and C (20 in Class B, and 19 in Class C), the portion of the persons who sought information rather than take emergency action was approximately 50 percent in ambiguity Class B and 35 percent in Class C.

Investigation Time

As noted in the preceding paragraphs, individuals often seek information to clarify the ambiguities inherent in fire situations. This is particularly the case for persons alone or in small groups. Sime[8] has tracked individuals involved in hotel fires, and his data demonstrate significant amounts of apparently nonproductive movement. This movement is assumed to be prompted by individuals seeking information they feel is necessary to make a proper decision. From an engineering standpoint, this emphasizes the potential of increasing evacuation efficiency by providing clear information to the occupants of a building that a fire has occurred, the location of the fire, and the condition of the exit routes.

Other Actions

When an individual has decided to take action in response to perceived fire danger, that action may or may not contribute to a speedy evacuation. Figure 3-14.7 diagrammatically shows a variety of the types of actions that are likely to occur. In actual fire situations many individuals undertake actions they believe will mitigate the fire, help others, or aid in making decisions. Often these actions do not contribute to their evacuation; these alternative actions may be contributing to safety, detrimental to safety, laudable, or mistaken. All actions, however, impact on evacuation efficiency.

Providing real-time information to building occupants and ensuing rapid response of emergency forces can mitigate delays with a minimum of impact on the desirable effects of nonevacuation actions.

Way Finding

Way finding is most important in situations involving a relatively small number of individuals evacuating a location where they are not familiar with the emergency exit system. This typically occurs in hotels, multitenant office buildings, and similar structures where persons seldom use the stairways.

The classic solution to the problem of way finding is the provision of exit signs and exit directional signs. Ozel[13] has summarized the current view of researchers in way finding problems and related these findings to fire evacuation conditions. A person's ability to find his or her way in an emergency relates to how well that person perceives his or her position and surroundings. The term frequently used is *cognitive mapping*. Most people maintain their cognitive mapping images in simple, regularized forms such as straight lines and rectangles. In this concept exit signs are only part of the overall ability of a person to perceive a suitable cognitive map.

Other factors that are rated of high importance to way finding include: (1) The complexity of the space as related to common layouts or similar types of buildings (e.g., double-loaded corridors terminating at exits are simple while arrangements involving complex curves or unusual angles are complex); and (2) The presence of distinguishing marks or other indications of points of special attention. Of particular importance are both exit routes and dead ends or other spaces that should not be entered in exiting the building.

Way finding affects evacuation efficiency in facilities where the population density is low and the occupants are unfamiliar with the evacuation routes. In such facilities the efficiency is greatest when: (1) The routes are simple, and (2) the exit points are evident in both their location and assurance (to the evacuee) that they truly lead to safety.

Way finding efficiency decreases as the layout complexity increases. Some factors increasing complexity are: unusual arrangements of corridors, obtuse angles, concentric corridors, undifferentiated enclosures (where one cannot orient oneself with respect to the exterior), and exit access doors that appear the same as every other door.

Conversely, all aspects that tend to simplify and identify the exit route decrease the time involved in way finding.

Merging Conflicts

Modeled evacuation time calculations assume that the flow of people is similar to that of a hydraulic fluid. As such, the merger of two flows (persons entering at a floor into a stairwell that is already flowing with evacuees from higher floors) is assumed to regulate itself according to the capabilities of the stairwell and the amount of flow from each of the sources.

In emergency evacuation drills MacLennan[9] has observed significant interruption of the continuity of flow when one flow is stopped and another is started at a merger point. The greatest merging flow efficiency occurs when once a flow uses an exit route to its full capacity, it blocks all other entry until it has cleared past the point of merger. In actual evacuations, however, there is normally a sharing of access at merger points. This results in breaks in the egress flow that can have a significant affect on the capacity. MacLennan has seen effects as high as 30 to 50 percent reduction in the stairwell flow from this cause alone.

Wardens

MacLennan's work indicates that the most efficient flow occurs in a situation where all occupants are fully trained and promptly evacuated upon signal without the assistance of wardens.

Where occupants are not so trained it is often necessary to have a trained warden system. While this system is essential, MacLennan's observations indicate that wardens impose a reduction in egress efficiency. The amount of reduction will depend upon the type of organization and the evacuation procedures. The impact on evacuation efficiency is the greatest when wardens hold occupants in a ready position until the wardens are directed from another point to initiate evacuation.

Self-Regulation

In high-density situations (e.g., heavily populated offices, auditoriums, etc.) individuals often withhold themselves from the evacuation procedure until the crowd lessens (i.e., density is reduced). When these persons arrive at critical exit points so that the main crowded path, normally the stairway, is continuously fed by a short queue, their action will have no impact. However, when the self-induced delay reduces the feed of the critical point, a reduction in egress efficiency will occur.

Uneven Use of Exit Facilities

If some exit facilities are used proportionally more than others, the efficiency of the egress system will decrease. Some exit paths will be utilized while others will be overtaxed.

The impact of uneven exit use can be estimated as a function of: (1) the distribution of exits relative to the distribution of population, (2) the degree that the exits will be used for either building entry or for common use within the structure, and (3) all of the cognitive mapping factors discussed under "Way Finding."

Evacuation Efficiency Factors—Summary

In his analysis, MacLennan found that large multifloor office buildings (some high-rise) demonstrated evacuation times in the range of twice the modeled time where a highly organized evacuation system was present; and up to three times the modeled evacuation time when there had been no training and no organization.

It is expected that MacLennan's findings represent reasonable norms, but are subject to many variations (e.g., education level, cultural background, age, and gender) that need to be considered in any individual evaluation.

NOMENCLATURE

D density in persons per unit area
e apparent evacuation efficiency
F flow
k constant from Table 3-14.2
P population
S speed along the line of travel
T time
W width

Subscripts

1	speed in ft/min and density in persons/sq ft
2	speed in m/s and density in persons/m^2
ae	actual evacuation
c	calculated
(*corridor*)	corridor
(*door*)	door
d	delay
e	effective
(*in* − *n*)	prior to point *n*
me	modeled evacuation
(*out* − *n*)	after point *n*
p	passage
s	specific
sm	maximum specific
(*stairway*)	stairway

REFERENCES CITED

1. V.M. Predtechenskii and A.I. Milinskii, *Planning for Foot Traffic in Buildings* (translated from the Russian), Russian publication, Stroizdat Publishers, Moscow, 1969; English translation published for the National Bureau of Standards and the National Science Foundation, Washington, by Amerind Publishing Co. Pvt., Ltd., New Delhi, India (1978).
2. J.J. Fruin, *Pedestrian Planning Design*, Metropolitan Association of Urban Designers and Environmental Planners, Inc., New York (1971).
3. J.L. Pauls, "Effective-Width Model for Evacuation Flow in Buildings," *Proceedings, Engineering Applications Workshop*, Society of Fire Protection Engineers, Boston (1980).
4. J.L. Pauls, "Calculating Evacuation Time for Tall Buildings," presented at *SFPE Symposium: Quantitative Methods for Life Safety Analysis*, Society of Fire Protection Engineers, Boston (March 1986).
5. P.G. Wood, "The Behavior of People in Fires," *Fire Research Note 953*, Building Research Establishment, Fire Research Station, Borehamwood (1972).
6. J.L. Bryan (see reference citations 6 through 14 of Section 3, Chapter 12, Behavioral Response to Fire and Smoke).
7. J.P. Keating and E.F. Loftus, "Post Fire Interviews: Development and Field Validation of the Behavioral Sequence Interview Technique," *Report GCR-84-477*, National Bureau of Standards, Gaithersburg (1984).
8. J.D. Sime, "Escape from Building Fires: Panic or Affiliation?," doctoral thesis, Psychology Department, University of Surrey, (1984).
9. H.A. MacLennan, "Towards an Integrated Egress/Evacuation Model Using an Open System Approach, Fire Safety Science," *Proceedings of the First International Symposium*, Hemisphere Publishing Corporation, Washington (1986).
10. A.T. Habicht and J.P. Braaksma, "Effective Width of Pedestrian Corridors," *Journal of Transportation Engineering*, 110, 1 (Jan. 1984).
11. B. Latane and J.M. Darley, "Group Inhibition of Bystander Intervention in Emergencies," *Journal of Personality Psychology*, 10, 3, pp. 215-221 (Nov. 1968).
12. B.M. Levin, "Design as a Function of Response to Fire Cues," *Proceedings of American Institute of Architects "Research and Design 85: Architectural Applications of Design and Technology Research,"* Los Angeles (Mar. 14-15, 1985).
13. F. Ozel, *Way Finding and Route Selection in Fires*, School of Architecture, New Jersey Institute of Technology, Newark (1986).

STOCHASTIC MODELS OF FIRE GROWTH

G. Ramachandran

INTRODUCTION

Apart from changes in environmental conditions, such as wind velocity and direction, humidity, and temperature, the spread of fire in a building is governed by physical and chemical processes evolved from a variety of burning materials arranged in different ways. Multiple interactions among these processes at different times during fire growth cause uncertainties in the pattern of fire development. Although different patterns of fire development can be simulated by varying the input values to the parameters of a deterministic model, there is a need to determine the uncertainty (probability) with which each pattern is likely to occur in a real fire in any type of building considered. The likely pattern of fire spread can only be predicted within limits of confidence expressed in probabilistic terms.

Non-deterministic models[1] (or indeterministic models as defined by Kanury[2]) rather than deterministic techniques offer rational methods of evaluating the uncertainties in the pattern of fire growth and are of two types: (1) probabilistic and (2) stochastic. The first type generally deals with a final outcome, such as area damaged, financial loss, or fire severity, and is considered as a continuous random variable reaching various levels in a fire according to a probability distribution.[3] Large values of the variable follow an extreme value distribution. (See Section 5, Chapter 3.) A semi-probabilistic approach is provided by a fault tree[4] in which the probability of occurrence of a top event, e.g., fire spreading beyond the room of origin, is estimated by assigning discrete (not continuous) probabilities to sub-events leading to the top event. Models of the first type, i.e., probability distributions and fault trees, do not consider in detail the underlying physical processes and their variation over the duration (time) of fire growth. Such "static" models can provide sufficient tools for fire protection and insurance problems concerned with "collective risk" in a group of buildings.

Stochastic models constitute the subject matter of this chapter, and may be regarded as "dynamic," since they are capable of predicting the course of fire development in a particular building. In these models, the various states, realms, or phases occurring sequentially in space and time during fire growth are specified together with the associated probability distributions. Depending on the nature of these distributions, a fire stays in each state for a random length of time and moves randomly from state to state. The "sojourn" and "transition" probabilities may be regarded as "noise" terms superimposed over a deterministic pattern of fire growth.

After describing the basic features of stochastic modeling of fire spread, two types of stochastic models are discussed in detail: (1) Markov chains and (2) networks. Attention is also drawn to the application of other stochastic models, such as random walk, diffusion processes, percolation theory, epidemic models, and branching processes.

The models discussed in this chapter mainly relate to the growth of fire and not to the spread of smoke or other combustion products.

BASIC FEATURES

Probability Distributions

Consider the burning of a particular object in a room as a random (stochastic) process, with $Q(t)$ denoting the probability that the object is still burning at time t. In a simple model the process may be assumed to be Poisson,[5] so that $Q(t)$ has the exponential form

$$Q(t) = \exp(-\mu t) \tag{1}$$

In this model the probability of extinction of fire during the short period $(t, t + \partial t)$ denoted by $\mu(t)$ has been assumed to have the constant value μ independent of t. The function $Q(t)$ can also be interpreted as the probability that the duration of burning of the object considered is greater than t. Then, the (cumulative) probability distribution function

$$F(t) = 1 - Q(t) \tag{2}$$

is the probability that the duration of burning is less than or equal to t. The parameter μ is the "instantaneous probability" of extinction of fire, whereas $\lambda (= 1 - \mu)$ is the "instantaneous probability" of fire surviving.

Dr. G. Ramachandran retired in November 1988 as Head of the Operations Research Section at the Fire Research Station, U.K. Since then he has been practicing as a consultant in risk evaluation and insurance. He is a visiting professor at the University of Hertfordshire and Glasgow Caledonian University. His research has focused on statistical and economic problems in fire protection and actuarial techniques in fire insurance.

The value of μ will vary depending on whether it is a free-burning fire or a fire extinguished by fire fighting, e.g., by fire brigade or sprinklers. The value of μ for any object burning under specified conditions can be estimated by carrying out replicated extinction experiments with the object. If $\bar{t}$ is the mean (average) duration of burning of the object

$$\mu = 1/\bar{t} \tag{3}$$

according to the properties of an exponential probability distribution. This distribution was implied in Kida's probabilistic analysis of extinction experiments.[6]

Spread to Another Object

In a simple model, as discussed earlier, it can be assumed that a fire involving one object spreads to another object if it survives (i.e., not extinguished) with the probability λ.

$$\lambda + \mu = 1 \tag{4}$$

The model can be expanded to include an (instantaneous) probability w to denote the event of fire not spreading, even though it is not extinguished; i.e., the fire continues to burn without spreading. In this case[5]

$$\lambda + \mu + w = 1 \tag{5}$$

and, following the derivation of Equation 1,

$$\begin{aligned} Q_1(t) &= \exp[-(\lambda + \mu)t] \\ &= \exp(-t + wt) \end{aligned} \tag{6}$$

where $Q_1(t)$ is the probability that the fire is burning at time t without spreading. The duration of burning in this case follows an exponential probability distribution with mean duration given by the reciprocal of $(\lambda + \mu)$ or of $(1 - w)$.

The length of time a fire involving an object burns affects future fire spread to another object: heat output (fire severity) increases with time. Equation 5 can, therefore, be modified such that, during a short time interval immediately after time t,

$$\lambda(t) + \mu(t) + w(t) = 1 \tag{7}$$

The instantaneous probabilities $\lambda(t)$, $\mu(t)$, and $w(t)$ are functions of the continuous random variable t. However, in practical problems, one can consider t in minutes and one minute as a short time interval such that the probabilities are denoted by $\lambda(i)$, $\mu(i)$, and $w(i)$ with $i = 1, 2, 3, \ldots$. The probability $w(i)$ applies to a single minute and, hence, it is likely to be small. One can, therefore, write

$$Q_1^1(t + 1) = w(1) \cdot w(2) \cdot \ldots w(t) \tag{8}$$

to denote the probability of fire burning during the time period $(t + 1)$ without spreading. Equation 8 follows from Equation 6. If $w(i)$ is a constant w as in Equation 6, it may be seen that

$$Q_1^1(t) = w^t \tag{9}$$

General Model

The probability distribution for duration of burning can have other forms, such as uniform and log-normal, although an exponential distribution has been postulated for the sake of simplicity. Following Ramachandran[5] and Aoki,[7] the following probabilities can be defined in the general case for the fire involving the object ignited first.

$q_1(t + 1)$ = probability of burning at the beginning of the $(t + 1)^{\text{th}}$ period or end of the tth period without spreading before that period.

$P_1(t + 1)$ = cumulative probability of becoming extinguished before the end of the $(t + 1)^{\text{th}}$ period.

$S_1(t + 1)$ = cumulative probability of spreading before the end of the $(t + 1)^{\text{th}}$ period.

With subscript 1 denoting the object ignited first, the following equations are easily derived:

$$q_1(t + 1) = q_1(t) \cdot w_1(t) = \prod_{r=1}^{t} w_1(r)$$

as in Equation 8, since $q_1(1) = 1$.

$$P_1(t + 1) = \sum_{r=1}^{t+1} q_1(r) \cdot \mu_1(r)$$

$$S_1(t + 1) = \sum_{r=1}^{t+1} q_1(r) \cdot \lambda_1(r)$$

The parameters $\lambda_1(r)$, $\mu_1(r)$, and $w_1(r)$ are probabilities of spreading, becoming extinguished, and burning without spreading during the rth period.

The following equations with similar definitions can be derived for the second object to which fire can spread from the object first ignited. (The subscript 2 has been used for the second object.)

$$q_2(t + 1) = q_2(t) \cdot w_2(t) + q_1(t) \cdot \lambda_1(t), q_2(1) = 0$$

$$P_2(t + 1) = \sum_{r=2}^{t+1} q_2(r) \cdot \mu_2(r)$$

$$S_2(t + 1) = \sum_{r=2}^{t+1} q_2(r) \cdot \lambda_2(r)$$

The probabilities $q_2(t + 1)$, $P_2(t + 1)$, $S_2(t + 1)$ and their associated parameters $w_2(t)$, $\lambda_2(t)$, and $\mu_2(t)$ pertain to the second object and are "conditional," given that the fire has spread to the second object.

In the first of the equations mentioned above it has been assumed that the second object starts burning in the second minute after the ignition of the first object in the first minute. The probability of this event is likely to be low such that $q_2(2)$ has a small value. The time of occurrence of sustained or established burning of the second object depends upon the "incubation" or "latent" period beyond which the fire involving the first object becomes capable of spreading to the second object.[8] This time and the spread probability $\lambda_1(r)$ also depend upon the distance between the two objects.[8] Radiation of heat to an object generally decreases in inverse proportion to the square of the distance of the object from the burning object. Fire spread may not occur if the unignited object is located beyond a "critical distance" from the burning object.[9]

As r varies, $q_1(r)$ gives rise to a discrete or continuous probability distribution for the object first ignited. This distribution can be ascertained by carrying out experiments with the two objects considered by varying the distance between them. In the continuous case, if the instantaneous probabilities are assumed as constants, the average duration of burning of the first object before spreading to the second object is an estimate of $(1/1 - w_1)$, assuming an exponential distribution. Applying this result to the discrete case with constant probabilities, for any specified distance between the two objects,

$$q_1(t + 1) = w_1^t$$

as in Equation 9. An exponential distribution with constant probabilities need not be assumed if the mathematical form of $w_1(t)$ and its variation with the distance between the two objects can be established from an analysis of the experimental data. These data would also provide an estimate of $\lambda_1(t)$ for any distance between the object ignited first and the second object. The value of $\mu_1(t)$ can then be obtained from the equation:

$$\mu_1(t) = 1 - \lambda_1(t) - w_1(t)$$

It can be verified that, since $q_1(1) = 1$,

$$
\begin{aligned}
P_1(t + 1) + S_1(t + 1) &= \sum_{r=1}^{t+1} q_1(r)[\mu_1(r) + \lambda_1(r)] \\
&= \sum_{r=1}^{t+1} q_1(r)[1 - w_1(r)] \\
&= \sum_{r=1}^{t+1} [q_1(r) - q_1(r + 1)] \\
&= 1 - q_1(t + 2)
\end{aligned}
$$

As t increases, $q_1(t)$ will decrease toward zero such that the sum of the two cumulative probabilities will tend to unity. Hence, as one would have expected, the probability of burning without spreading will tend to zero with the passage of time, and the fire involving the first object would have either been extinguished or spread to the second object.

In the language of stochastic modeling, the spread probability, $\lambda_1(r)$, is the "transition probability" at time r, and may be redefined as $\lambda_{12}(r)$ to denote the spread from the object first ignited to the second object. Other objects in a room may be considered as the first or second object such that, in the general case, $\lambda_{ij}(r)$ is the probability of spreading from the ith object to the jth object at time r. Based upon the distances between them, the objects can be arranged in order in a diagram to analyze the sequential spread of fire from object to object at any time r. (See Figure 3-15.1.) This simple analysis may be sufficient for all practical purposes, although a fire from one object can spread to another object directly or indirectly through the ignition of another object.

As t increases, $q_i(t)$ will tend to zero and the cumulative probability of extinguishment, $P_i(t + 1)$, to a limiting value E_i. This value, E_i, denotes the probability of fire being extinguished ultimately, with the spread limited to the ith object. Correspondingly, the cumulative probability of spreading, $S_i(t + 1)$, will tend to $(1 - E_i)$. The following equations may be specified in the limiting case:

$$
\begin{aligned}
E_1 &= \mu_1; \lambda_1 = 1 - \mu_1 \\
E_2 &= \lambda_1 \cdot \mu_2; \lambda_2 = 1 - \mu_2 \\
E_3 &= \lambda_1 \cdot \lambda_2 \cdot \mu_3; \lambda_3 = 1 - \mu_3
\end{aligned}
$$

and so on, such that, in general terms

$$E_i = \lambda_1 \cdot \lambda_2 \ldots \lambda_{i-1} \cdot \mu_i; \lambda_i = 1 - \mu_i \tag{10}$$

$$P_i = \sum_{j=1}^{i} E_j \tag{11}$$

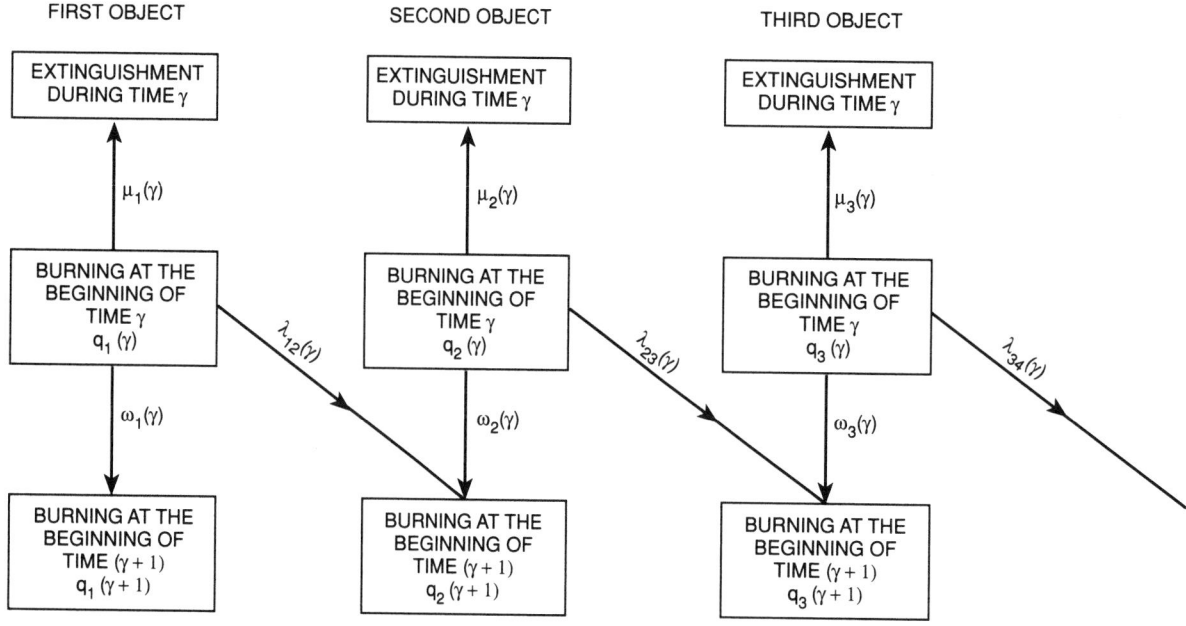

Fig. 3-15.1. Probability diagram for spread of fire from object to object at time γ.

The equations aforementioned provide values for μ_i for $i = 1, 2, 3 \ldots$ which may be regarded as the limiting probability of extinguishment for the ith object, given (conditional) that the fire has spread to this object. The parameter λ_i is the limiting and conditional probability of spread beyond the ith object.

The model mentioned above for objects in the room of fire origin can be extended to include structural barriers and objects in corridors and other rooms on the same floor or different floors of a building. This procedure involves complex calculations, since a fire in a room can spread to an adjoining room in the same or upper floor; a fire can also spread through several paths. The problem of fire spread throughout a building can be simplified to some extent by applying network models, which are discussed later.

MARKOV MODEL

Mathematical Representation

A considerable amount of statistical and experimental data are needed for applying the model for fire spread from object to object. For practical purposes it may be sufficient to consider fire spreading through a number of spatial modules,[10] phases,[11] or realms.[12] These stages of fire growth can generally be defined as "states" such that fire spreads, moves, or makes a transition from state to state.

The movement of fire from state to state is governed by a "transition probability," which is a function of time since the start of the fire. The fire also spends a certain length of time in each state before making the transition; this duration follows a "temporal" probability distribution. The state occupied by fire at any moment in time is governed by transition probabilities and temporal probability distributions. Represented mathematically, if the fire is in state a_i at the nth minute, it can be in state a_j at the $(n + 1)^{th}$ minute, according to the transition probability $\lambda_{ij}(n)$. The transition probabilities are most conveniently handled in matrix form. One may write, dropping (n) for convenience, with m states,

$$P = \begin{bmatrix} \lambda_{11} & \lambda_{12} & . & . & \lambda_{1m} \\ \lambda_{21} & \lambda_{22} & . & . & \lambda_{2m} \\ . & . & . & . & . \\ . & . & . & . & . \\ \lambda_{m1} & \lambda_{m2} & . & . & \lambda_{mm} \end{bmatrix}$$

where

$$\sum_{j=1}^{m} \lambda_{ij} = 1 \qquad i = 1, 2, \ldots m$$

The probability distribution of the system at time n can be expressed as the vector

$$\mathbf{p} = (q_1 q_2 q_3 \ldots q_m)$$

where q_i is the probability of the fire burning in the ith state at time n. Since a fire can be in one of the m states at a given time

$$\sum_{i=1}^{m} q_i = 1$$

The mth state may denote the state of fire having been extinguished if such a state is included in the model considered. The vector given by the product $p \cdot \mathbf{P}$ expresses the probabilities of burning in different states one transition (minute) later.

As an example, consider a model of fire growth in a room in which the ith state represents i objects burning. Suppose, with $m = 4$ and no extinguishment, the process stops with the occurrence of flashover when all four objects are ignited. There is no recession in growth and, hence, there is no transition to a lower state from a higher state. With these assumptions, let the transition matrix be

$$P = \begin{bmatrix} 0.4 & 0.3 & 0.2 & 0.1 \\ 0 & 0.5 & 0.3 & 0.2 \\ 0 & 0 & 0.6 & 0.4 \\ 0 & 0 & 0 & 1 \end{bmatrix}$$

If, at time n, the probabilities of fire burning in different states is given by

$$p_n = (0.1 \quad 0.2 \quad 0.3 \quad 0.4)$$

it can be seen by performing the matrix multiplication, that the probability of fire burning in different states at time $(n + 1)$ is given by

$$p_{n+1} = (0.04 \quad 0.13 \quad 0.26 \quad 0.57)$$

Hence, at time $(n + 1)$, the probability of the fire being in the third state, for example, is 0.26, and the probability of flashover (4th state) is 0.57.

Markov Chains

Markov chains are used for repetitive situations in which there is a set of probabilities that define the likelihood of transition from one state to another. A chain comprises a sequence of such transitions. In a Markov chain, the transition probabilities satisfy the following properties.[4]

1. Each state belongs to a finite set of all possible states.
2. The characteristics of any state do not depend upon any other previous state.
3. For each pair of states (i, j) there is a probability λ_{ij} that state j occurs immediately after state i occurs.

The transition probabilities can be specified in a matrix form P as discussed previously, with the aid of a hypothetical example. Berlin[4] and Watts[10] have illustrated the use of this matrix by modeling a "random walk" among five adjacent spaces.

Markov chains may possess a number of special characteristics, one of which is called an "absorbing state." The system remains in an absorbing state once it enters this state. A fire burning out (self-termination) and a fire getting extinguished by an extinguishing agent are examples of absorbing states. State i is an absorbing state if row i of the transition matrix has a value of $\lambda_{ij} = 1$ and all other values in the row are zero.

Markov Process

The next step is to consider a slightly more complex model called the "Markov process," a stochastic or random process where the probability of occurrence of some future state of the system, given its present state, is not altered by information concerning past states. That is, the history of the process has no influence on its future. This lack of a historical influence is often referred to as a "memoryless" or "Markovian" property of a process.

TABLE 3-15.1 *Transition Descriptors for a Typical Room in a Residential Occupancy*

Fire Type: Smoldering fire in a couch with cotton cushions.

Realm Transition		Transition Probability	Temporal Distribution		
From	To		Type	Mean	Standard Deviation
II	I	0.33	Uniform	2.0	5.0
II	III	0.67	Log-normal	8.45	0.78
III	II	0.25	Uniform	1.0	2.0
III	IV	0.75	Normal	5.55	3.22
IV	III	0.25	Uniform	1.5	9.0
IV	V	0.75	Uniform	0.5	3.5
V	IV	0.08	Uniform	0.6	6.0
V	VI	0.92	Log-normal	5.18	4.18

TABLE 3-15.2 *Maximum Extent of Flame Development*

Maximum Flame Extent	Probability of Flame Extent
Realm II	0.33
Realm III	0.07
Realm IV	0.02
Realm V	0.58

In a Markov process with stationary transition probabilities, the value of $\lambda_{ij}(n)$ is a constant independent of the time variable n. Following this process Berlin[12] estimated stationary transition probabilities for six realms (states) for residential occupancies: (i) nonfire state, (ii) sustained burning, (iii) vigorous burning, (iv) interactive burning, (v) remote burning, and (vi) full room involvement. These realms were defined by critical events, such as heat release rate, flame height, and upper room gas temperature. Development of fire over time was considered as a "random walk" through these realms.

Based on data from over a hundred full-scale fire tests, Berlin[12] calculated transition probabilities as in Table 3-15.1. The information in this table indicates that, when a fire is in Realm III, there is a 75 percent chance of growth to Realm IV, and a 25 percent chance of recession to Realm II. Figure 3-15.2 is the transition diagram defined by the transition probabilities in Table 3-15.1. Realm I, no fire, is an absorbing state, since all fires eventually terminate in this state. The process also ends when Realm VI (full room involvement) is reached; for this reason, this state also is an absorbing state. Berlin[12] used uniform, normal, and log-normal distributions to describe temporal probability distributions for the different states.

Among many questions asked about fire development using the Markov model is what maximum extent of fire growth represents the most extreme condition. The portion of fires that do not grow beyond Realm II is the probability (0.33) of transition from Realm II to I. If M_3 is the long-run (limiting) probability of fire reaching Realm III, but not growing beyond, then[12]

$$M_3 = \frac{\lambda_{21} + \lambda_{23} \cdot \lambda_{31}}{1 - \lambda_{23} \cdot \lambda_{32}} - \lambda_{21} \qquad (12)$$

Using the figures in Table 3-15.1 and noting that $\lambda_{31} = 0$, it may be seen that $M_3 = 0.07$. Beyond Realm III is more difficult, as described by Berlin.[12] Probabilities of maximum extent of flame development as estimated by Berlin are given in Table 3-15.2. Berlin has also discussed other fire effects, such as probability of self-termination and distribution of fire intensity.

One of the major weaknesses of the Markov model regards the "stationary" nature of the transition probabilities. It is assumed that these probabilities remain unchanged regardless of the number of transitions representing the passage of time. The length of time a fire burns in a given state affects future fire spread. For example, the probability of a wall burn-through increases with fire severity which is a function of time. The time spent by fire in a particular state may also depend on how that state was reached, i.e., whether the fire was growing or receding. Some fires may grow quickly and some grow slowly, depending on high or low heat release. In a Markov model with stationary transition probabilities no distinction is made between a growing fire and a dying fire.

Berlin[12] has estimated that 99 percent of all fires will terminate within twelve transitions. This result is based on the assumption of stationary transition probabilities that may be nearly true for a few fluctuations between the same realms where different materials would be contributing to the burning process. However, the fire will eventually consume all fuels, in which case the probabilities of termination from all realms will be equal to one. Therefore, Berlin's approach represents a worst-case analysis.

State Transition Model

According to Berlin's Markov model, a fire in a particular realm can either grow to a higher realm or recede to a lower realm. There is no transition to the nonfire (absorbing) state (Realm I) from any realm higher than Realm II, except Realm VI (full room involvement) which is also an absorbing state. Receding to a lower state may be true to some extent when describing fire growth in terms of flame spread, but such an assumption is not possible in the case of spatial spread of fire in which, as discussed previously, fire spreads sequentially from one object to another. According to this model, if fire spreads to an object it cannot spread backwards to the object from which it spread. The fire involving an

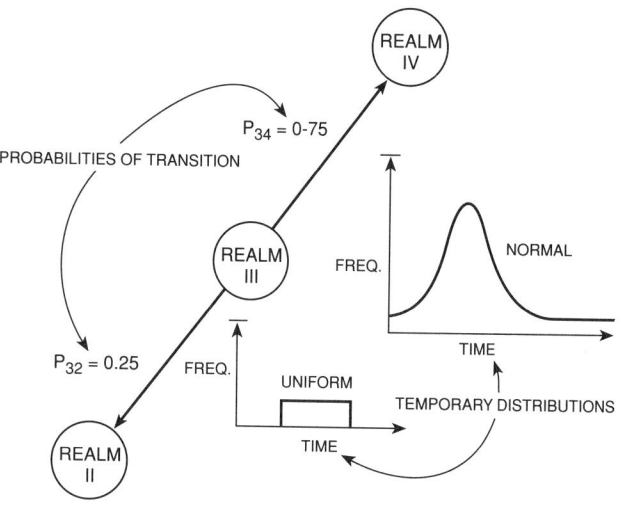

Fig. 3-15.2. Realm transition descriptors.

object either spreads forward to other objects or gets extinguished or stays with the object without spreading.

Complex computational procedures would be involved in a stochastic model for fire spread from object to object in a room or to different rooms. Hence, consistent with the fire statistics available, particularly in U.K. and U.S., a simplified model based on the following three main states can be considered for fire development in a room.

S_1 Fire confined to the object first ignited.
S_2 Fire spreading beyond the object first ignited but confined to the contents of the room.
S_3 Fire spreading beyond room of origin but confined to the building.

A fourth state may be added to denote extinguishment or burning out (self-termination) of fire; this is an "absorbing state," since a fire process cannot leave this state after entering it. The third state, S_3, is also an absorbing state, since a spreading fire will eventually terminate within the building of origin; spreading beyond the building is not considered.

The three states ($i = 1, 2, 3$) mentioned above generate a state transition model, distinct from Berlin's Markov model. This model was used by Ramachandran[5] for evaluating the transition (spread) probabilities $\lambda_i(t)$ and the probabilities $\mu_i(t)$ for extinguishment or transition to the fourth state. The value of $\lambda_3(t)$ was taken as zero, since fire spread beyond the building was not considered. The probability of burning in a state without spreading was also considered with the aid of the parameter $w_i(t)[= 1 - \lambda_i(t) - \mu_i(t)]$. The duration of burning was divided into subperiods, each of a fixed length of five minutes.

Statistics furnished by fire brigades in U.K. related to fires that were extinguished during each time period since ignition. Hence, Ramachandran[5] used the extreme value technique, with some assumptions, to estimate the number of fires that were burning in a particular stage at the beginning of each subperiod. With the aid of these estimates and

TABLE 3-15.4 *Bedroom Bedding (average times)*

	First State	Second State	Third State
Duration of burning in the state (min)	24.5	5.9	8.5
Extinguishment time in the state (min)	26.2	32.2	44.1
Time for spreading beyond the state (min)	27.9	38.9	*

*Spread beyond the third state (building) is not considered.

the actual numbers of fires that were extinguished, approximate values were obtained for the extinguishment and spread probabilities (as functions of time) and probability distributions of duration of burning in each state. The equations given previously for the general model were utilized for this purpose. Four materials ignited first in the bedroom of a dwelling were considered for illustrating the application. Tables 3-15.3 and 3-15.4 and Figures 3-15.3 and 3-15.4 are examples extracted from this study. Aoki[7] described fire growth with similar states based on the spatial extent of spread; his analysis was similar to that of Ramachandran[5] and Morishita[11] and considered eight phases of spatial spread of fire, including spread to the ceiling.

In a later study, Ramachandran[3] added another state between S_2 and S_3 to denote the event of fire involving the structural barriers of a room, assumed to occur after fire has spread beyond S_1 and S_2 but still confined to the room. This intermediate state was considered as generally consecutive to S_2, although a fire can spread directly from S_1 and involve the structural boundaries. Fire statistics available in U.K. permit the incorporation of this additional state into a state

TABLE 3-15.3 *Bedroom Bedding (probabilities)*

Period (t) (min)	First State $\mu(t)$	First State $\lambda(t)$	Second State $\mu(t)$	Second State $\lambda(t)$	Third State $\mu(t)$
1–5	0.0055	0.0745	—	—	—
6–10	0.0385	0.1170	0.7026	0.1747	—
11–15	0.0631	0.0602	0.8507	0	0.4894
16–20	0.0407	0.0780	0.8319	0	0.3750
21–25	0.0406	0.1887	0.7576	0.1688	0.8667
26–30	0.1071	0.1730	0.8474	0.0329	0.4878
31–35	0.0723	0.1131	0.8206	0.0676	0.7143
36–40	0.0449	0.1071	0.8218	0.0172	0.5455
41–45	0.0205	0.0758	0.7293	0.2632	0.8333
46–50	0.0160	0.1451	0.875	0	0.4737
51–55	0.0476	0.2508	0.7350	0.1282	0.5500
56–60	0.0860	0.2308	0.7816	0.0862	0.5417
61–65	0.0861	0.2417	0.7840	0.0800	0.5769
66–70	0.0739	0.1478	0.7444	0.1556	0.6190
71–75	0.0316	0.1519	0.8205	0.1795	0.5455
76–80	0.0233	0.1860	0.6250	0.0833	0.6316
81–85	0.0294	0	0.6774	0	0.4286
86–90	0.0101	0.4950	0.2500	0.3182	0.7500

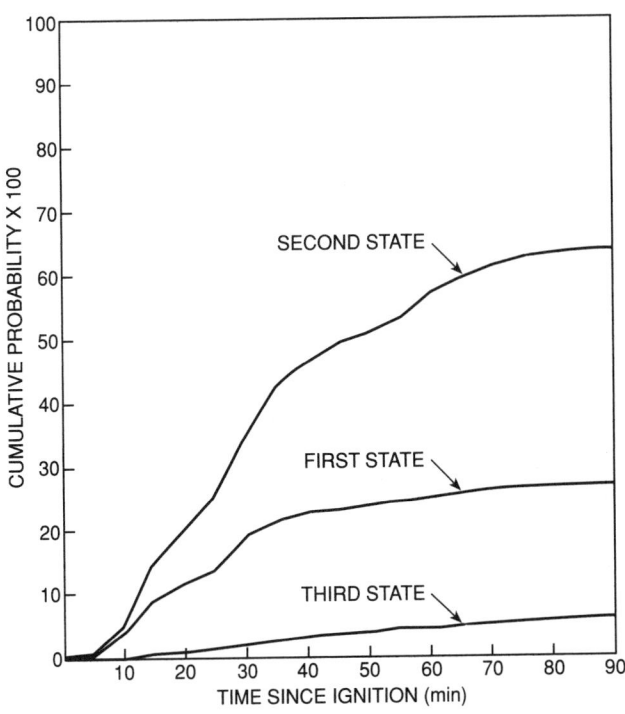

Fig. 3-15.3. Bedding—cumulative probability of extinguishment.

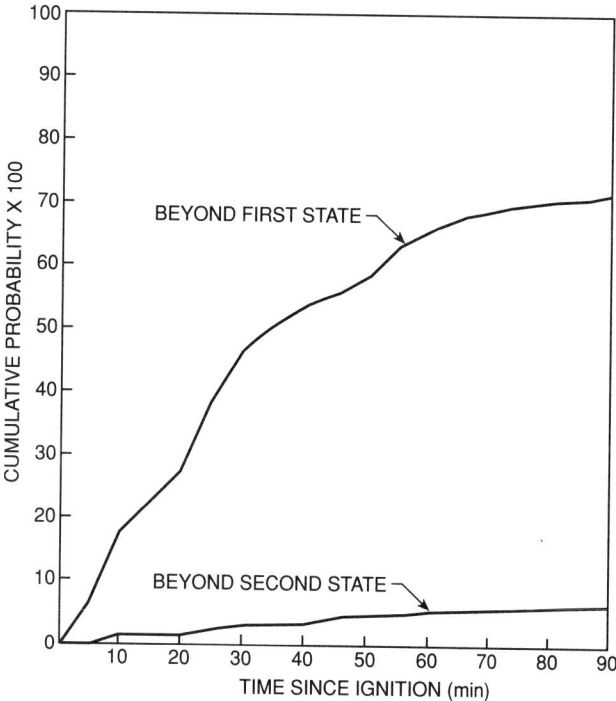

Fig. 3-15.4. Bedding—cumulative probability of spreading.

transition model. As shown in Figure 3-15.5, only the limiting probabilities λ_i and μ_i specified in Equation 10 were estimated.

Fire statistics provided estimates for $E_i (i = 1$ to 4), the proportion of fires extinguished in the ith state. The condition that

$$\sum_{j=1}^{4} E_j = 1$$

follows from the assumption that $\mu_4 = 1$ and, hence, $\lambda_4 = 0$; fire spread beyond the building of origin was not considered.

The parameter μ_1 is the same as λ_{21} in Berlin's model in which the first state (Realm I) is the nonfire state and the second state and above are fire states (Realms II through VI). Also λ_1 is the same as Berlin's λ_{23}, and μ_2 corresponds to λ_{31}. With these changes and the assumption that there is no recession of fire growth ($\lambda_{32} = 0$), the value of M_3 in Equation 12 is equal to

$$E_2 = \lambda_1 \cdot \mu_2$$

as derived from Equation 10.

In Figure 3-15.5, the product $\lambda_1 \cdot \lambda_2 (= E_3 + E_4)$ may be regarded as the probability of "flashover," and λ_3 as the probability of failure of the structural boundaries of the room.[13] For the following reasons fire statistics do not provide a valid estimate of λ_3. Figures for the number of fires that spread beyond the room of origin include: (1) fires that spread by destruction of barrier elements (wall, floor, ceiling) as well as (2) those that spread by convection through a door or window left open or through some other opening. In the latter case, the barrier elements would still be structurally sound. A "room," as recorded in fire brigade reports, is not necessarily a "fire compartment." Using a probabilistic model[13] or other methods, the value of λ_3 for any compart-

ment of given fire resistance can be estimated and multiplied by the probability of flashover to provide an estimate of probability of spread beyond the compartment of origin.

The probabilities provided by a stochastic model can be regarded as "noise" terms superimposed over a deterministic trend in fire growth over space and time. The trend can be predicted by a deterministic model, such as an exponential model.[3] Table 3-15.5 is an example based on such a model and fire incidence statistics.[13] The estimates of time in this table have been measured from the time of occurrence of "established burning" at the end of the first state, denoting confinement of fire to object first ignited. The percentage figures in Table 3-15.5 have provided the probabilities for constructing the probability tree in Figure 3-15.5.

To represent the interaction between human behavior and fire dynamics, Beck[14] developed a series of stochastic

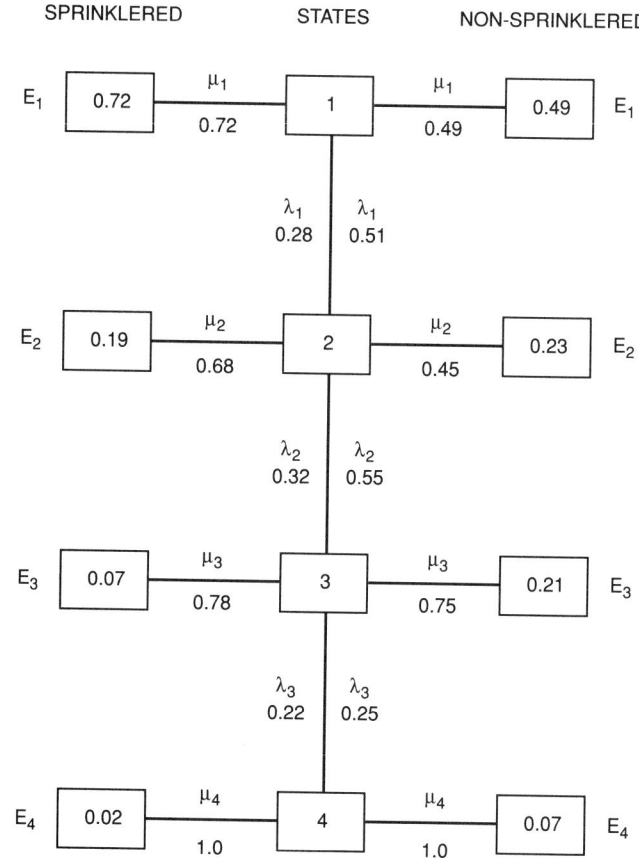

E_1 = Probability of confinement to items first ignited = μ_1

E_2 = Probability of spreading beyond item first ignited but confinement to contents of room of fire origin = $\lambda_1 \cdot \mu_2$

E_3 = Probability of spreading beyond item first ignited and other contents but confinement to room of fire origin and involvement of structure = $\lambda_1 \cdot \lambda_2 \cdot \mu_3$

E4 = Probability of spreading beyond room of fire origin but confinement in the building = $\lambda_1 \cdot \lambda_2 \cdot \lambda_3 \cdot \mu_4$, ($\mu_4 = 1$)

Fig. 3-15.5. Probability tree—textile industry.

TABLE 3-15.5 *Textile Industry, U.K., Extent of Fire Spread and Average Area Damaged*

	Sprinklered*			Non-Sprinklered		
Extent of Spread	Average Area Damaged (m²)	Percentage of Fires	Time (min)	Average Area Damaged (m²)	Percentage of Fires	Time (min)
Confined to item first ignited	4.43	72	0	4.43	49	0
Spread beyond item but confined to room of fire origin						
Contents only	11.82	19	8.4	15.04	23	6.2
Structure involved	75.07	7	24.2	197.41	21	19.4
Spread beyond room	1000.00	2		2000.00	7	
Average	30.69	100		187.08	100	

*System operated.

state transition models and interrelated deterministic models. His sequential fire growth model was based on the six realms defined by Berlin,[12] with the remote burning state denoting flashover. His results, reproduced in Table 3-15.6, are applicable to office buildings. P_i in this table is the same as E_i in Equation 10. Adopting a different notation and starting with $P_1 = \mu_1$, the conditional probabilities of extinguishment, μ_i, and conditional probabilities of spread, $\lambda_i (= 1 - \mu_i)$, were calculated according to Equation 10. The probability of a fully developed fire, given a fire defined by P_{FDF}/F, is given by the product $\lambda_1 \cdot \lambda_2 \cdot \lambda_3 \cdot \lambda_4$. The probability of spread beyond the compartment of fire origin, P_{VI}, is given by the product of P_{FDF}/F and λ_5 or by $\lambda_1 \cdot \lambda_2 \cdot \lambda_3 \cdot \lambda_4 \cdot \lambda_5$.

Beard[15] proposed a state transition model by considering a number of "critical events" with directional characteristics that a fire may pass through and the times between critical events. For example, critical heat event CHE2U referred to "fire passing through 2 kW on the way up," whereas CHE2D referred to "fire passing through 2 kW on the way down." The time between two critical events was assumed to have a "temporal" probability distribution independent of the time between earlier critical events. A particular succession of critical events formed a "chain"; specific times between critical events were referred to as a "sequence" within the chain. Based on assumed forms for "transition probabilities" and "temporal" probability distributions, "Monte Carlo" simulation was employed to generate randomly particular chains and sequences. A generated sequence to smoke and toxic gases was related to the corresponding sequence for the burning rate. Based on the concentration of carbon monoxide, Beard employed the concept of "fraction of fatality," with fatality resulting at a unit value for this fraction. He applied the model to a particular case involving flaming ignition on a bed in a hospital ward. He concluded that there would be a very large (greater than 80 percent) likelihood of having multiple fatalities if a fire exceeds 50 kW. One of the several assumptions used by him was that the fire did not spread beyond the ward.

Williamson[16] introduced a state transition model for analyzing and reporting the results of fire growth experiments performed under conditions resembling actual fire conditions. Three pre-flashover states were defined, as follows.

J The period of time from the beginning of the experiment to ignition of the specimen.

K The period of time from ignition of the specimen until flames touch the ceiling.

L The period of time from when the flames first touch the ceiling until full involvement (flashover) occurs.

Histograms and cumulative distribution functions of the state durations provided a graphical representation of fire performance. Examples were chosen to illustrate the method. Traditional cellulosic and cementitious walls and ceilings were compared to plastic materials in the same configuration.

NETWORKS

State Transition Model

There is a probability p_f for flashover occurring in a room or compartment which depends on the objects in the room and their spatial arrangement apart from ventilation and other factors. Given flashover, the fire can breach the structural boundaries of the room with a probability p_b and spread beyond the room with a probability $p_s (= p_f \cdot p_b)$. The value of p_b depends on the level of fire severity attained after flashover and the fire resistance of the structural elements, such as walls, ceilings, and floors. The probability of failure of a room or compartment of given fire resistance, p_b, can be estimated from the joint probability distribution of fire severity and fire resistance expressed in units of time.[13] Fire resistance of a compartment will be reduced and the failure probability p_b increased by weakness caused by penetrations, such as piping or cables through walls, doors, windows, or other openings in the structural barriers.

Each room or corridor in a building has, therefore, an independent probability p_s of fire spreading beyond its boundaries. Using these probabilities for different rooms and corridors, fire spread in a building can be considered as a discrete propagation process of burning among points that abstractly express the rooms, spaces, or elements of a building. In a simple analysis, states classified by the burning situation of individual points can be incorporated in a state transition model.[17]

TABLE 3-15.6 *Probabilities of Extinguishment: Fire-Growth and Suppression Model*

System Configuration	P_I	P_{II}	P_{III}	P_{IV}	P_V	P_{VI}	P_{FDF}/F
No Sprinkler	0.5673	0.0038	0.0017	0.3282	0.0666	0.0324	0.0990
Sprinkler	0.5673	0.3827	0.0201	0.0232	0.0045	0.0022	0.0067

Consider, for example, three adjoining rooms, R_1, R_2, and R_3, that provide the following four states with the fire commencing with the ignition of objects in R_1.

1. Only R_1 is burning.
2. R_1 and R_2 are burning (and not R_3).
3. R_1 and R_3 are burning (and not R_2).
4. All three rooms are burning.

There is no transition from the

1. First to the fourth state
2. Second to the third state
3. Third to the second state
4. Second or third or fourth to the first state (recession of fire growth).

A transition from the second to the fourth state involves the spread of fire to R_3 from R_1 or R_2. The probability for this transition is, therefore, the sum of probabilities for spread from R_1 to R_3 and R_2 to R_3. Likewise, the probability of transition from the third to the fourth state is the sum of probabilities for spread from R_1 to R_2 and R_3 to R_2. A fire can burn in the same state without transition to another state. The process terminates when the fourth state is reached.

With the assumptions mentioned above, the following transition matrix can be formed with p_{ij} denoting the probability, p_s, of fire spread from room i to room j per unit time.[17] The unit may be longer than one minute, say, five minutes since one is considering spread from room to room after the occurrence of flashover. The values of p_{ij} may be considered as constants in a state transition model with stationary transition probabilities.

$$\mathbf{P} = \begin{bmatrix} 1-p_{12}-p_{13} & p_{12} & p_{13} & 0 \\ 0 & 1-p_{13}-p_{23} & 0 & p_{13}+p_{23} \\ 0 & 0 & 1-p_{12}-p_{32} & p_{12}+p_{32} \\ 0 & 0 & 0 & 1 \end{bmatrix}$$

During the initial period, since only R_1 is burning, the probability of burning in different states is given by the probability distribution of the system

$$\mathbf{p_0} = (1 \quad 0 \quad 0 \quad 0)$$

The probability distribution of the system after one unit of time is given by

$$\mathbf{p_1} = \mathbf{p_0} \cdot \mathbf{P} = (1 - p_{12} - p_{13} \quad p_{12} \quad p_{13} \quad 0)$$

and after two units of time by the following row vector which, for convenience, has been written as a column vector

$$\mathbf{p_2} = p_1 \cdot P = \begin{bmatrix} (1-p_{12}-p_{13})^2 \\ 2p_{12}(1-p_{13})-p_{12}^2-p_{12}{\cdot}p_{23} \\ 2p_{13}(1-p_{12})-p_{13}^2-p_{13}{\cdot}p_{32} \\ p_{12}{\cdot}p_{23}+2p_{12}{\cdot}p_{13}+p_{13}{\cdot}p_{32} \end{bmatrix}$$

The probability distributions of the system for later periods can be obtained by repeating the matrix multiplication as described above. This process will generate for each state a probability distribution for burning in that state as a function of time. For the first state (only R_1 burning), for example, the probability after n units of time is $(1 - p_{12} - p_{13})^n$. The probability distributions for the other three states can be

obtained by performing the calculations on a computer. The distribution for any state will provide an estimate of average transition time to that state. For estimating the average transition time to the fourth state, denoting the burning of all three rooms, Morishita[17] has proposed a method based on partitioning the matrix **P**. He has also discussed the stochastic process for a system in which extinguishment is attempted. For purposes of illustration he has applied the model to a hypothetical small house.

By carrying out further calculations and adding the corresponding probabilities, cumulative probabilities over time can be estimated for burning in the four states. The cumulative probabilities would generally tend to some limits as the value of n denoting time increases. The limiting value of the cumulative probability for the fourth state (all three rooms burning) and the corresponding time would be of special interest. This probability can be reduced and the associated time increased by: (1) increasing the structural fire resistance of the rooms to reduce the probability of barrier failure, p_b, and (2) installing sprinklers to reduce the probability of flashover, p_f. With these safety measures probability of fire spread from room to room, p_s, will be reduced. Consequently the cumulative probability of fire being confined to the room of origin, R_1, will increase; this probability for a duration of t minutes is given by

$$\sum_{n=1}^{t}(1 - p_{12} - p_{13})^n$$

where p_{12} and p_{13} are the probabilities of spreading from R_1 to R_2 and R_3 per unit time.

Network Models

The model described above can be extended to provide cumulative probabilities, at time n or limiting, for the burning of more than three rooms, but this will involve tedious and complex calculations. It would be simpler to consider fire spread between two given rooms through intermediate rooms and corridors in terms of discrete values attached to the probability p_s of spread beyond a room. This probability may be the limiting value for the cumulative probability given by, say, $E_4(= \lambda_1 \cdot \lambda_2 \cdot \lambda_3)$ in Figure 3-15.5. Alternatively, the time taken by fire to breach the boundaries of a room may be ascertained from deterministic (scientific) models and a probability assigned to this time, t_s, used for p_s in the stochastic model. The duration t_s is the sum of t_f representing the time to the occurrence of flashover after the start of established burning and t_b representing the time for which the barriers of the room can withstand fire severity after flashover. The latter time may be the endurance of barrier elements as measured by a standard fire resistance test, such as ASTM E119, *Standard Test Methods for Fire Tests of Building Construction and Materials*. As mentioned earlier $p_s = p_f \cdot p_b$.

Consider, as an example, the simple layout of Figure 3-15.6, part (a), relating to four rooms and the corresponding graph shown in part (b), which also shows the probability (p_{ij}) of fire spread between each pair of rooms (i, j). The probability p_{ij} refers to p_s as defined herein, whereas Dusing et al[18] and Elms and Buchanan[19] have considered only the barrier failure probabilities denoted by p_b, ignoring the probability of flashover denoted by p_f. The specific problem considered by these authors was to compute the probability of fire spread from room 1 to 4, which might follow any of the four paths, i.e., $(1) \rightarrow (2) \rightarrow (4)$; $(1) \rightarrow (3) \rightarrow (4)$; $(1) \rightarrow (2) \rightarrow (3) \rightarrow (4)$; and $(1) \rightarrow (3) \rightarrow (2) \rightarrow (4)$.

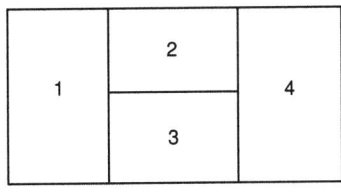

(a) ROOM LAYOUT

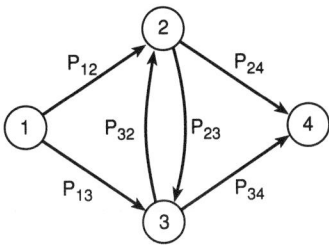

(b) CORRESPONDING GRAPH

Fig. 3-15.6. *Simple diagram for fire spread.*

Using the event space method, Elms and Buchanan[19] considered first all possible "events" or combinations of fire spreading or not spreading along various links. If a_{ij} represents spread of fire along link ij, and $\bar{a}_{ij}$ represents fire not spreading along the link, then one event might be

$$[a_{12}, \bar{a}_{13}, a_{23}, \bar{a}_{32}, \bar{a}_{24}, a_{34}]$$

There will be $2^6 = 64$ events that will be all-exclusive, as any pair of events will contain at least one link for which fire spreads in one event and does not spread in the other. The probability of each event occurring is the product of the probabilities of its elements, assuming that the elements are independent. Thus, for the example given above, the event probability will be

$$p_{12}(1 - p_{13})p_{23}(1 - p_{32})(1 - p_{24})p_{34}$$

and the overall probability is the sum of all 64 event probabilities.

The complete event space can be represented as a tree with 64 branches. The probability of fire spread for each event (branch) is obtained by multiplying all the link probabilities in a branch. However, not all branches have to be computed in full. The computation can be curtailed, while still allowing for all cases. For this purpose, Elms and Buchanan[19] have described a method of constructing the tree and its ordering to identify or search possible paths of links leading to node (room) 4 from node 1. This procedure is known as a "depth-first search" of a graph. In this algorithm, a path is a series of nodes or rooms in a building; the construction of a branch, which is part of a particular event, is based on an underlying path. Each branch allowing fire spread must contain at least one path. Figure 3-15.7 shows the actual tree as it would be computed by the algorithm. The total probability of spread from node 1 to node 4 is given by the sum of all the branch probabilities. The calculation is carried out for each pair of rooms, and the results assembled in a "fire spread matrix." The diagonal elements of the matrix are unity.

Various means have been employed to curtail the algorithm to prevent the computer developing excessively lengthy branches that would, as the branch probability decreases with branch length, have little effect on the result. The first means is to restrict the length of a fire spread path to a maximum number of compartments. The second approach is to terminate development of a branch, if the cumulative branch probability drops below a certain fraction of the running total of the branch spread probabilities calculated up to that point. The third means is to terminate development of a branch if the underlying path length becomes greater than a specified amount more than the length of the shortest possible path between the two rooms being considered.

In the computer-based technique of Elms and Buchanan[19] as described above, a building is represented as a network by defining compartments as nodes and the links between these nodes as possible paths for fire spread from compartment to compartment in a multi-compartment building. The core of this model is a probabilistic network analysis to compute the probability of fire spreading to any compartment within the building. A series of further refinements were added when the model was applied to analyze the effects of fire resistance ratings on the likely fire damage to buildings.[20]

Time Dimension

Elms and Buchanan[19] did not consider the dimension of time explicitly although it was implicit in many of their functions. The probability of fire spreading from one compartment to another was considered irrespective of how long it might take. As a result, the analysis did not take into account any intervention, e.g., the fire service. In this respect the model represented a worst-case scenario and assumed that the fire would eventually burn itself out.

The model of Elms and Buchanan[19] was not concerned with the process of fire growth and assumed that the spread was solely a function of the probable effects of a fully developed fire. The probability of flashover was not considered in this model. Platt[21] has proposed a new network model in which fire resistance and severity are related to real time instead of equating these two parameters to the time based on the ISO standard fire which is not necessarily representative of the real time. The model computes the probability that fire will have spread to any part of a building after an elapsed time t. The essential features of Platt's model[21] are described in the following paragraphs.

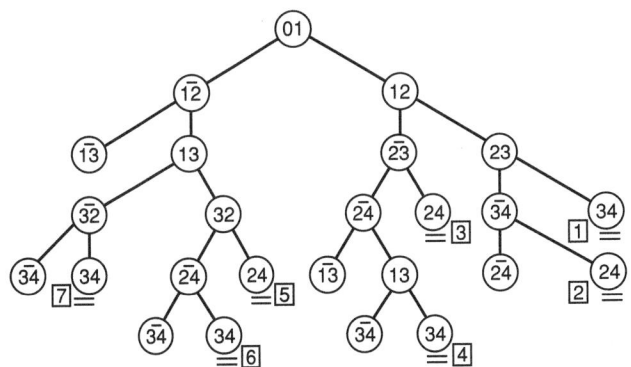

Fig. 3-15.7. *Modified event space tree.*

The spread of fire to an adjacent compartment may be via the following paths:

1. Through an open doorway;
2. Vertical spread via windows; or
3. Through a barrier, such as a wall, closed door, and ceiling.

Two models are considered for estimating fire growth as a function of time. The first model is based on the exponential relationship between fire area and growth time as suggested by Ramachandran.[22] The second model uses the parabolic relationship between rate of heat release and growth time as proposed by Heskestad.[23] This model is used in conjunction with the relationship between compartment temperature and rate of heat flow to provide an estimate of flashover time, which is taken as the time when ceiling temperature reaches 600°C. The temperature is a function of the ventilation characteristics of the compartment, the type of fuel material, its configuration, and the thermal properties of the structural boundaries.

A figure of 49 percent is used to represent the probability that the initial fire will not result in flashover. Subsequent ignitions caused by fire spread are given a 100 percent probability of reaching flashover. This assumption may overestimate the spread of fire, since a barrier may "fail" in the very latter stages of a fully developed fire which may not then have the "momentum" to initiate further ignition.

The real time of the fire duration, t, representing fire severity, s, is estimated by the ratio of fuel load to rate of combustion which is a function of ventilation and dimensional characteristics of the compartment. A formula suggested by CIB[24] is used to estimate the "equivalent time," t_e involving the "real" parameters of the compartment. The approved fire resistance rating (FRR) of a barrier element, modified for "weakness" and another factor, is multiplied by the ratio (t/t_e) to yield an "equivalent FRR" denoted by R. R and fire severity, s, are not independent but quite rightly they have been assumed to be independent random variables with log-normal probability distributions. Under this assumption, the probability of fire spread through a barrier is estimated through the safety factor (R/s) which is also a log-normal variate. The probability of fire spread via an open door is assumed to be 100 percent. The probability of fire spreading vertically up the facade of a building via windows is equated to the probability that the height of the external flames is greater than or equal to the height of the spandrel.

A comparison is then made between these values and the design values of the barrier and door fire resistance and the spandrel heights. The output from these comparisons is a series of probabilities that fire will spread via each of the three possible paths described earlier. Combining these individual probabilities gives an overall probability of fire spreading to an adjacent compartment. Repeated for each compartment within the building, these values collectively form the adjacency fire spread matrix whose values represent the probability that fire will spread from compartment i to an adjacent compartment j. The expected time for fire to spread to an adjacent compartment, given that fire does spread, provides values for the adjacency fire spread time matrix.

By combining the two matrices providing probabilities and expected times for fire spread between adjacent compartments, the analysis computes the probability of fire spreading from an initial compartment i to any compartment j. The fire may spread along any path, but is conditional on having arrived at compartment j in a given time. The result-

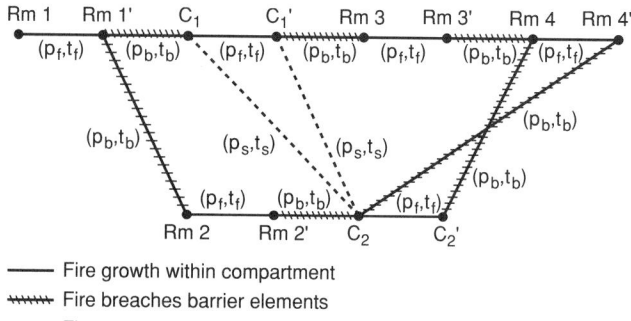

— Fire growth within compartment
〰〰 Fire breaches barrier elements
----- Fire spread along corridor

Fig. 3-15.8. Probabilistic network of fire spread of Rm 1 to C_2.

ing matrix, i.e., global fire spread matrix, may be considered as a three-dimensional matrix with each layer being evaluated at a different time. Once the fire spread matrix has been formed, Platt's model[21] is very similar to that of Elms and Buchanan[20] except that, in the former model, the probability of spread is dependent on time, whereas, in the latter model the probability of fire spreading is irrespective of the time taken.

Ling and Williamson[25] have proposed a model in which a floor plan is first transformed into a network similar to the process described by Dusing *et al*[18] and Elms and Buchanan.[19] Each link in their network represents a possible route of fire spread, and those links between nodes corresponding to spaces separated by walls with doors are possible exit paths similar to those developed by Berlin *et al*.[26] The space network is then transformed into a probabilistic fire spread network as in the example in Figure 3-15.8 with four rooms, Rm 1 to Rm 4, and two corridor segments C_1 and C_2. In this figure, Rm 1 has been assumed as the room of fire origin, but it would be a simple modification to reformulate the problem for another room of origin. With Rm 1 and Rm 1′, with a "prime" denoting the pre-flashover and post-flashover stages, the first link is represented by

$$\text{Rm 1} \underset{(p_f,\ t_f)}{\rightarrow} \text{Rm 1}'$$

where p_f represents the probability of flashover and t_f represents the time to flashover. The nodes denoted by a prime represent a fully developed (i.e., post-flashover) fire in the compartment.

In Figure 3-15.8, three different types of links are identified. The first corresponds to the fire growth in a compartment, the second to the fire breaching a barrier element, and the third to fire spread along the corridors. To each link i, a pair of numbers (p_i, t_i) is assigned, with p_i representing the distributed probability that a fire will go through link i, and t_i representing the time distribution that it will take for such a fire to go through link i. The section of the corridor, C_1, opposite Rm 1, is treated as a separate fire compartment and is assigned a (p_f, t_f) for the link from C_1 to C_1'. The number pair (p_s, t_s) represents the probability and time for the pre-flashover spread of fire along the corridor from C_1 to C_2. As a first approximation, p_s may be considered to be governed by the flame spread classification of the corridor's finish materials on the walls and ceiling, as measured by a test method such as the tunnel test in ASTM E84, *Standard Test Method for Surface Burning Characteristics of Building Materials*.

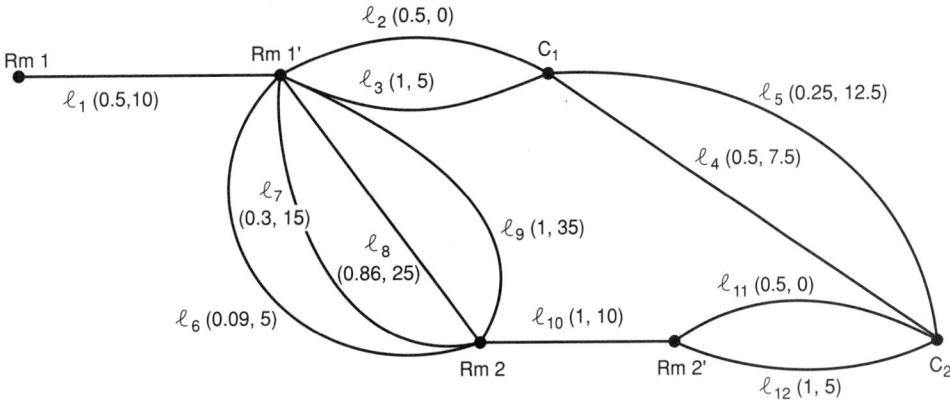

Fig. 3-15.9. Equivalent fire spread network with 5-min. unrated doors.

Once full involvement occurs in the section C_1 of the corridor outside Rm 1 (i.e., node C_1' is reached) the fire spread in the corridor is influenced more by the ventilation in the corridor and by the contribution of Rm 1 than by the material properties of the corridor itself. Thus there is a separate link, C_1' to C_2, that has its own (p_s, t_s). The number pair (p_b, t_b) represents the probability of failure of the barrier element, with t_b representing the endurance of the barrier element.

Once one has constructed the probabilistic network, the next step is to "solve" it by obtaining a listing of possible paths of fire spread with quantitative probabilities and times associated with each path. For this purpose, Ling and Williamson[25] have adopted a method based on "emergency equivalent network," developed by Mirchandani[27] to compute the expected shortest distance through a network. (The word "shortest" has been used instead of "fastest" to be consistent with the literature.) This new "equivalent" network would yield the same probability of connectivity and the same expected shortest time as the original probabilistic network. In this method, each link has a Bernoulli probability of success and the link delay time is deterministic.

It must be noted that there are multiple links between nodes in the equivalent fire spread network. For example, the door between Rm 1 and the corridor could be either open or closed at the time the fire flashed over in Rm 1. Ling and Williamson[25] assumed, as an example, that there is a 50 percent chance of the door being open and that an open door has zero fire resistance. Furthermore, they assumed that the door if closed would have a 5-min. fire rating. With further

assumptions they constructed the equivalent fire spread network (Figure 3-15.9) with twelve possible paths for the example in Figure 3-15.8 to find the expected shortest time for the fire in Rm 1 to spread to the portion of corridor C_2. This network changes to Figure 3-15.10 with ten possible paths if self-closing 20-min. fire-rated doors had been installed in the corridor, assuming that the reliability of the self-closures is perfect and that doorstops had not been allowed. Note that the links have been renumbered for Figure 3-15.10.

For the two equivalent networks shown in Figures 3-15.9 and 3-15.10, all of the possible paths are listed in Tables 3-15.7 and 3-15.8 with increasing time and with all the component links identified. Each of these paths can be described by a fire scenario; for instance path 1 in Table 3-15.7 consisting of links ℓ_1, ℓ_2, and ℓ_4, would be:

"The fire flashes over, escapes from Rm 1 through an open door into the corridor C_1 and spreads along the corridor to C_2."

The probability of that scenario (0.13) is strongly dependent on the probability (0.5) for the occurrence of flashover in Rm 1 and of the probability (0.5) that the door will be open. The time of 17.5 min. is composed of the times of 10 min. for flashover and 7.5 minutes for fire to spread in the corridor from C_1 to C_2.

Ling and Williamson[25] have derived a formula for calculating from the figures in Tables 3-15.7 and 3-15.8, the probability of connectivity R, which is 0.5 for both the networks (Figures 3-15.9 and 3-15.10). This probability is a direct result of the assumed probability of 0.5 for flashover in

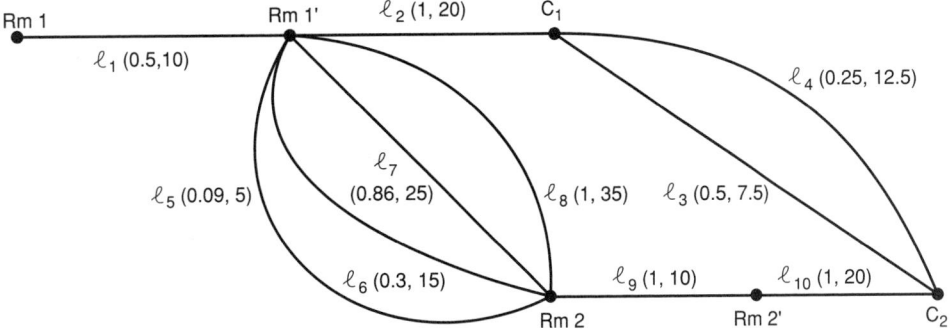

Fig. 3-15.10. Equivalent fire spread network with self-closing 20-min. fire-rated doors.

TABLE 3-15.7 *Pathways Through the Example Fire Spread Equivalent Network Assuming 5-Min. Unrated Corridor Doors, as Shown in Figure 3-15.9*

Paths	Component Links	Probability p_i	Time t_i (min.)
1	1-2-4	1/8 = 0.13	17.5
2	1-2-5	1/16 = 0.06	22.5
3	1-3-4	1/4 = 0.25	22.5
4	1-6-10-11	1/44 = 0.02	25.0
5	1-3-5	1/8 = 0.13	27.5
7	1-6-10-12	1/22 = 0.05	30.0
8	1-7-10-12	3/40 = 0.08	35.0
9	1-8-10-11	3/14 = 0.21	40.0
10	1-8-10-12	3/7 = 0.43	50.0
11	1-9-10-11	1/4 = 0.25	55.0
12	1-9-10-12	1/2 = 0.50	60.0

TABLE 3-15.8 *Pathways Through the Example Fire Spread Equivalent Network Assuming Self-Closing 20-Min. Fire-Rated Corridor Doors, as Shown in Figure 3-15.10*

Paths	Component Links	Probability p_i	Time t_i (min.)
1	1-2-3	1/4 = 0.25	37.5
2	1-2-4	1/8 = 0.13	42.5
3	1-5-9-10	1/22 = 0.5	45
4	1-6-9-10	3/20 = 0.15	55
5	1-7-9-10	3/7 = 0.43	65
6	1-8-9-10	1/2 = 0.50	75

the room of fire origin and the occurrence of unity probabilities in the remaining links that make up certain paths through the network. According to another formula, the expected shortest time is 29.6 min. for Figure 3-15.9, and increases to 47.1 min. for Figure 3-15.10 due to the presence of the 20-min. fire-rated door. The equivalent fire spread network thus facilitates an evaluation of design changes and affords ready comparison of different strategies to effect such changes.

RANDOM WALK

In a simple stochastic representation, the fire process involving any single material or number of materials can be regarded as a random walk. The fire takes a random step every short period either to spread with a probability λ or to get extinguished (or burnout) with a probability $\mu (= 1 - \lambda)$. The parameter λ denotes the success probability of the fire, whereas μ denotes the success probability of an extinguishing agent. The problem is similar to two gamblers, A (fire) and B (extinguishing agent), playing a sequence of games, the probability of A winning any particular game being λ. If A wins a game he/she acquires a unit stake by destroying, say, a unit of the floor area; if he/she loses the game, no stake is gained. In the latter case, A does not lose his/her own stake to B; an already burned out area is a loss that cannot be regained. Extinguishment can also be considered as an "absorbing boundary" to the random walk, just as an "absorbing state" in a state transition model.

A random walk will lead to an exponential model described in Equations 1 through 3. In these equations, if one writes $c = \mu - \lambda$ such that $\mu = (1 + c)/2$, since $\mu + \lambda = 1$

$$Q(t) = \exp[-(1 + c)t/2] \tag{13}$$

The fire-fighting effort is adequate if c is positive with μ greater than λ and, hence, greater than $\frac{1}{2}$; it is inadequate if c is negative with μ less than λ and, hence, less than $\frac{1}{2}$. If $c = 0$, such that $\mu = \lambda = \frac{1}{2}$, there is an equal balance between fire-fighting efforts and the propensity of fire to spread.

Associated with the random variable t denoting time, there is another random variable x denoting damage which may be expressed in terms of, say, area destroyed. Damage in fire has an exponential relationship with duration of burning,[22] such that the logarithm of x is directly proportional to t as a first approximation. This assumption would lead to Pareto distribution

$$\phi(x) = x^{-w}, \quad x > 1 \tag{14}$$

denoting the probability of damage exceeding the value x. This distribution is used in economic problems concerned with, for example, income distribution to describe the fact that there are a large number of people with low incomes and a small number of people with high incomes. The damage is small in most of the fires, with high levels of damage occurring only in a small number of fires.

The use of Pareto distribution for fire damage originally proposed by Benckert and Sternberg[28] was later supported by Mandelbrot[29] who derived this distribution following a random walk process. For all classes of Swedish houses outside Stockholm, the value of the exponent w was found to vary between 0.45 and 0.55. A value of $w = 0.5$ in Equation 14 would imply, as discussed with reference to Equation 13, an equal balance between fire-fighting efforts and the propensity of fire to spread and cause damage.

The parameter μ in Equations 1 and 2 is known as the "hazard" or "failure rate" given by the ratio

$$f(t)/Q(t)$$

where $f(t)$ is the probability density function obtained as the derivative of $F(t)$ in Equation 2. A constant value for μ would denote a "random failure." For a Pareto distribution in Equation 14, the failure rate is w/x, such that, with a constant value for w, the failure rate would decrease as x increases indicating that, in terms of damage, a fire can burn forever without getting extinguished. A constant value for μ or w is somewhat unrealistic, particularly for a fire that is fought and extinguished at some stage. A fire will also burn out when all the available fuel is consumed or when it stops spreading due to the arrangement of objects in a room or building. For the reasons mentioned above, although the failure rate can decrease in the early stages of fire development denoting a success for fire spread, it would eventually increase ("wear out failure"), since fire extinguishing efforts would succeed ultimately.[30] There may be a long intermediate stage of steady growth of fire with μ remaining as a constant. The failure rate as a function of time would, hence, resemble a "bathtub" whose cross-section is composed of two curves sloping downwards representing the early and final stages of fire growth, connected at the bottom by a long straight line representing the intermediate stage.

Erving et al[31] presented an approach to the theory of burning velocity in which a flame front moves forward at a rate determined by the random walk of chemical energy. The flame velocity is estimated by the value of a parameter which is a "collision rate" divided by a "reaction rate," both determined at the point of maximum reaction velocity. Empirical activation energies were given by the authors for certain hydrocarbon flames.

In the context of fire spread, random walk is a one-dimensional process describing the damage by random functions of time rather than by a random function of time and space. The random walk indicates the position of the fire, i.e., damage at any time. At every unit of time, there is a change in position indicated by an increment to the damage or no change due to absorption (extinguishment or burning out). Generally the walk is considered in discrete time. If the walk is continuous in time such that the increments are Gaussian, this leads to a diffusion process.[32] (A diffusion process is an approximation to Brownian motion, a phenomenon well known in many branches of science and technology.) The normal, or diffusion, term is one of two possible components of a general additive stochastic process, the other component being a discontinuous or transition term arising from occurrences of events at random times. The Markov chains discussed earlier belong to the second type of component. A linear superposition of the two components provides a solution to an equation governing a general additive process.

PERCOLATION PROCESS

In random walk and diffusion models, randomness is a property of the moving object, whereas in a percolation process, randomness is a property of the space in which the object moves.[33] Thus the transition the object suffers when at a particular point is random but, if the object ever returns to this point, it would suffer the same transition as before. The process is described by a stochastic field on the space, i.e., a vector field of transition numbers. Percolation process deals with deterministic flow in a random medium, in contrast with random walk and diffusion models which are concerned with random flow in a deterministic medium.

Broadbent and Hammersley[34] considered the walk as taking place on a graph consisting of a number of sites, connected by directed "bonds," passage being possible only along such a "bond." If such a graph obeys certain connectivity requirements, it is termed a "crystal." In a "randomized" version each bond of the crystal has an independent probability of being blocked, and it is desired to know what effect this has on the probability of communication from one site A to another site B; this is not the same as from B to A, since communication has a direction.

If fire is considered as the moving object, the movement takes place in a space or medium that has a certain random property although the object (fire) itself has some randomness associated with it. Buildings in an area, for example, are somewhat randomly distributed. Buildings are also connected by directed bonds, with spread (flow) of fire being possible only along the bonds. Each bond has an independent probability of blocking or preventing fire spread; this depends on the nature of a building and its contents, wind conditions, and the distances between buildings. A percolation problem also arises when one considers a network some of whose links, chosen at random, may be blocked, and one wishes to know the effect of this random blockage on flow through the network. Such a problem would be encountered in predicting fire spread in a forest or from building to building in an urban area.

Apparently for the reasons mentioned above, Hori[35] considered percolation process for the modeling of fire spread from building to building. Sasaki and Jin[36] were concerned with the actual application of this model and estimation of probabilities of fire spread. By using the data contained in the fire incidence reports for Tokyo, urban fires were simulated and the average number of burnt buildings per fire estimated. Apart from distances between buildings and wind velocity, the following factors were also regarded as having some effect on the probability of fire spread: building construction, building size and shape, window area, number of windows, indoor construction materials, furniture, walls, fences, gardens, and trees.

For the model the first factor was classified into three groups: (1) wood construction, (2) mortar (slow burning) construction, and (3) concrete (fireproof) construction. The wind velocity was classified into two groups: (1) 0 to 2.5 m/s and (2) 2.5 to 5.0 m/s. In the former case, fire spread was assumed to be undirectional (isotropic) whereas, in the latter case, the data were subdivided into smaller groups according to the directions of fire spread: the windward direction, the leeward direction, or direction perpendicular to that of the wind (the sideward direction). Fire incidents in which wind velocity was larger than 5.1 m/s were excluded due to their small number. If the number of burnt buildings was i and that of unburnt ones was j, the probability of fire spread was expressed as $i/(i + j)$.

The data were divided by every meter of the distance between buildings, or every 2 m or more in case the data were few. The exponential function $\exp(-cd)$ was used for estimating the probability of spread between two buildings as a function of distance d between them. Using the least square method, the values of c were obtained for different building construction and wind velocity categories. The analysis revealed that building construction was the main factor responsible for fire spread. The simulations did not evaluate the changes in the pattern of spread according to time.

In a recent study, Nahmias et al[37] have examined the feasibility of applying percolation theory to the spread of fires in forests. For studying the effect of randomness on the propagation of fire the authors have built a square network model containing combustible and noncombustible blocks randomly distributed, with a variable concentration, the parameter q denoting the fraction of noncombustible elements. In the absence of wind, the propagation was found to be consistent with a model of invasion percolation on a square site lattice with nearest neighbor interaction leading to a threshold not far from the theoretical value $q = 0.39$. The threshold was larger with wind blowing on the model. The largest threshold value obtained was $q = 0.65$. The final state of the model after combustion was represented for different values of wind velocity and fraction values q. The observation of this state can bring out the directed, nonlocal, and correlated characters of the contagion.

EPIDEMIC THEORY

For predicting fire spread in a large urban area, Albini and Rand[38] have proposed a model that has some similarity with chain-binomial models of Reed and Frost[39] for the spread of an epidemic. The authors envisaged fires in "locales," which may be single buildings or blocks of buildings. A number of these are presumed to be alight initially and randomly distributed and to stay alight for a time T in the absence of fire fighting. At time T, this "generation" of fires can spread fire and then die out, leaving a second "generation" to burn for a second period T, and so on.

Fire spread is assumed to take place only at the end of each fire interval. For the $(n + 1)^{th}$ interval, the *a priori* probability that any locale is burning is P_n and that it has not yet been burnt is A_n. It follows that

$$A_n = (1 - P_0)(1 - P_1) \ldots (1 - P_n)$$
$$P_{(n+1)} = A_n \cdot B_n$$

where B_n is the probability that during the $(n + 1)^{th}$ interval, fire spreads into a "locale" previously unburnt. To obtain B_n, Albini and Rand[38] introduced parameters defining the following three probabilities:

1. Probability that during the $(n + 1)^{th}$ interval there are just m locales burning out of N possible locales adjacent to a given locale.
2. Probability that at least one of the m burning locales spreads fire.
3. *A priori* probability that fire will spread during any interval of duration T from a burning locale to an unburnt neighbor.

Based on the aforementioned parameters the authors obtained an upper and lower approximation for B_n and narrow limits for $(1 - A_n)$, the probability of a locale being burnt.

The Albini and Rand[38] model allowing for fire fighting was based on a number of idealizations. First, fire-fighting effort was assumed to be constant. The authors introduced a parameter M for the fraction of burning locales wherein all fire fighters in a city could extinguish that fraction during the given time interval out of all possible burning locales. Fire fighting was assumed to be continuous throughout the time interval. A fire not extinguished may or may not spread; if extinguished it cannot spread. Under the assumptions mentioned above, the authors have derived an expression for the probability $P_{(n+1)}$ defined earlier. Albini and Rand considered directional spread of fire assuming that from an isolated locale the probability of spread forward and backward was the same and the directional element in the spread arose only from the initial condition. Spatial variation was included in the model by connecting the probability of spread to the probability that any building was itself burning and separated from any of its neighbors not yet burning by less than the appropriate "safe" distance for radiation or brand transfer.

Thomas[40] drew attention to the possible relevance of epidemic theory to fire spread in a building, and compared the model of Albini and Rand[38] with a deterministic epidemic model based on a continuous propensity of fire to spread. He found the results of both models to be in reasonable agreement as to their basic features, but concluded that neither would be appropriate for addressing spread in a single building where the number of "locales" is not large. For such a situation a stochastic treatment is necessary to allow for the finite chance that the initiating fires can burn out before spreading, a chance that is negligible when the number of initial fires is large.

OTHER MODELS

Branching processes[41] can be relevant for fire spread in a building in which a material first ignited (first generation) ignites one or more other materials (second generation), which ignite other materials (third generation), and so on, leading to the spread of fire throughout the building. The number of offsprings (burning materials) vary randomly from one generation to another, depending on the distances between ignited and unignited materials, ventilation, and other factors affecting fire spread. Hence, there is a need to develop a branching process model applicable to random environment.[42]

For predicting the damage to buildings and other properties resulting from incendiary or nuclear attacks, Phung and Willoughby[43] considered two types of stochastic models. In the first model, the entire fire front was regarded as a random walker moving along a linear row of cells or small square areas. In each short time interval the front may be in one of three states: (1) die or stop permanently, (2) spread or move one cell forward, or (3) pause or stay where it is. Simple probability considerations provide an estimate of the probability P_n that, at time t, the fire will be n cell units long after an initial condition of being lit at time zero

$$P_n = \exp[-(\lambda + \mu)t]\lambda^n t^n$$

The parameters λ and μ are, respectively, the probabilities for forward spread and burnout during a short time interval. The fire will stay where it is, with probability $[1 - (\mu + \lambda)]$.

The second stochastic model of Phung and Willoughby was called "fuel-state" model, because it dealt explicitly with the state of the fuel in each cell. In the burning process, the fuel changes from the unignited to the burnout state passing through the flaming state. A cell will be in one of these three states at any time with probabilities U, F, and B for unignited, flaming, and burnout states, respectively. In a two-dimensional array of cells, the cell dimension can be so chosen that a burning cell can ignite the immediate neighbor cells but not those that are farther away. Under this assumption, an unignited cell can be ignited by one or more of its eight (8) immediate burning neighbors with probability

$$P = 1 - (1 - P_1)(1 - P_2) \ldots (1 - P_8)$$

where $P_1, P_2 \ldots P_8$ denotes the chances of ignition by the neighbors. These eight spread probabilities are not necessarily symmetrical, due to factors such as wind and topography. Using the formulation described, differential equations are derived for U, F, and B for each cell with the condition $(U + F + B) = 1$, solutions of which can be obtained by numerical calculations using computers, if necessary.

REFERENCES CITED

1. G. Ramachandran, "Non-Deterministic Modeling of Fire Spread," *Journal of Fire Protection Engineering*, 3(2), 37–48 (1991).
2. A.M. Kanury, "On the Craft of Modeling in Engineering and Science," *Fire Safety Journal*, 12, 65–74 (1987).
3. G. Ramachandran, "Probabilistic Approach to Fire Risk Evaluation," *Fire Technology*, 24, 3, 204–226 (1988).
4. G.N. Berlin, "Probability Models in Fire Protection Engineering," *SFPE Handbook of Fire Protection Engineering*, DiNenno, P.J. (ed.) *et al*, National Fire Protection Association, Quincy, MA (1988).

5. G. Ramachandran, "Stochastic Modeling of Fire Growth," *Fire Safety: Science and Engineering*, ASTM STP 882, T.A. Harmathy (ed.), American Society for Testing and Materials, Philadelphia, PA, 122–144 (1985).
6. H. Kida, "On the Fluctuations of the Time Required to Extinguish Small Liquid Diffusion Flames with Sprays of Several Salt Solutions," Report of Fire Research Institute of Japan, No. 29, Mitaka, Japan, 25–33 (1969).
7. Y. Aoki, "Studies on Probabilistic Spread of Fire," Research Paper No. 80, Building Research Institute, Tokyo, Japan (1978).
8. G. Ramachandran, "Stochastic Modeling of Fire Growth," CIB Workshop on Mathematical Modeling of Fire Growth, Paris, France, July 1981.
9. C.R. Theobald, "The Critical Distance for Ignition from Some Items of Furniture," Fire Research Note No. 736, Fire Research Station, Boreham Wood, Herts, U.K. (1968).
10. J.M. Watts, Jr., "Dealing with Uncertainty: Some Applications in Fire Protection Engineering," *Fire Safety Journal*, II, 127–134 (1986).
11. Y. Morishita, "Establishment of Evaluating Method for Fire Safety Performance," Report, Research Project on Total Evaluating System on Housing Performances, Building Research Institute, Tokyo, Japan (1977).
12. G.N. Berlin, "Managing the Variability of Fire Behavior," *Fire Technology*, 16, 287–302 (1980).
13. G. Ramachandran, "Probability-Based Fire Safety Code," *Journal of Fire Protection Engineering*, 2(3), 75–91 (1990).
14. V.R. Beck, "A Cost-Effective Decision-Making Model for Building Fire Safety and Protection," *Fire Safety Journal*, 12, 121–138 (1987).
15. A.N. Beard, "A Stochastic Model for the Number of Deaths in a Fire," *Fire Safety Journal*, 4, 169–184 (1981/82).
16. R.B. Williamson, "Fire Performance Under Full-Scale Test Conditions—A State Transition Model," Sixteenth *Symposium* (International) on Combustion, The Combustion Institute, Pittsburgh, PA, 1357–1371 (1976).
17. Y. Morishita, "A Stochastic Model of Fire Spread," *Fire Science and Technology*, 5, 1, 1–10 (1985).
18. J.W.A. Dusing, A.H. Buchanan, and D.G. Elms, "Fire Spread Analysis of Multi-Compartment Buildings," Department of Civil Engineering Research Report 79/12, University of Canterbury, New Zealand (1979).
19. D.G. Elms and A.H. Buchanan, "Fire Spread Analysis of Buildings," Research Report R35, Building Research Association of New Zealand (1981).
20. D.G. Elms and A.H. Buchanan, "The Effects of Fire Resistance Ratings on Likely Fire Damage in Buildings," Department of Civil Engineering, Research Report 88/4, University of Canterbury, New Zealand (1988).
21. D.G. Platt, "Modeling Fire Spread: A Time-Based Probability Approach," Department of Civil Engineering Research Report 89/7, University of Canterbury, New Zealand (1989).
22. G. Ramachandran, "Exponential Model of Fire Growth," *Fire Safety Science: Proceedings of the First International Symposium*, C.E. Grant and P.J. Pagni, eds., Hemisphere Publishing Corporation, New York, 657–666 (1986).
23. G. Heskestad, "Engineering Relations for Fire Plumes," Technology Report 82-8, Society of Fire Protection Engineers (1982).
24. "Design Guide: Structural Fire Safety, CIB W14 Workshop," *Fire Safety Journal*, 10, 2, 81–138 (1986).
25. W.T.C. Ling and R.B. Williamson, "The Modeling of Fire Spread Through Probabilistic Networks," *Fire Safety Journal*, 9 (1986).
26. G.N. Berlin, A. Dutt, and S.M. Gupta, "Modeling Emergency Evacuation from Group Homes," Annual Conference on Fire Research, National Bureau of Standards (1980).
27. P.B. Mirchandani, *Computations and Operations Research*, 3, Pergamon Press, Elmsford, NY, 347–355 (1976).
28. L.G. Benckert and I. Sternberg, "An Attempt to Find an Expression for the Distribution of Fire Damage Amount," *Transactions of the Fifteenth International Congress of Actuaries*, 11, 288–294 (1957).
29. B. Mandelbrot, "Random Walks, Fire Damage Amount and Other Paretian Risk Phenomena," *Operations Research*, 12, 582–585 (1964).
30. G. Ramachandran, "The Poisson Process and Fire Loss Distribution," *Bulletin of the International Statistical Institute*, 43, 2, 234–236 (1969).
31. H. Evring, J.C. Giddings, and L.G. Tensmeyer, "Flame Propagation: The Random Walk of Chemical Energy," *The Journal of Chemical Physics*, 24(4), 857–861 (1956).
32. S. Karlin, *A First Course in Stochastic Processes*, Academic Press, New York (1966).
33. J.M. Hammersley and D.C. Handscomb, "Percolation Processes," Chapter II, *Monte Carlo Methods*, Methuen & Co. Ltd., London (1964).
34. S.R. Broadbent and J.M. Hammersley, "Percolation Processes, 1, Crystals and Mazes," *Proceedings* of the Cambridge Philosophical Society, 53, 629–41 (1957).
35. M. Hori, "Theory of Percolation and Its Applications," Nippon Tokeigakkai-shi, 3, 19 (1972).
36. H. Sasaki and T. Jin, "Probability of Fire Spread in Urban Fires and Their Simulations," Report No. 47, Fire Research Institute, Tokyo, Japan (1979).
37. J. Nahmias, H. Tephany, and E. Guyon, "Propagation de la Combustion sur un Reseau Heterogene Bidimensionnel," *Revue Phys. Appl.*, 24, 773–777 (1989).
38. F.A. Albini and S. Rand, "Statistical Considerations on the Spread of Fire," IDA Research and Engineering Support Division, Washington, DC (1964).
39. N.J.T. Bailey, "Reed and Frost Model," *The Elements of Stochastic Processes*, Chapter 12, Section 5, John Wiley, New York (1964).
40. P.H. Thomas, "Some Possible Applications of the Theory of Stochastic Processes to the Spread of Fire," Internal Note No. 223, Fire Research Station, Boreham Wood, Herts, U.K. (1965).
41. T.E. Harris, *The Theory of Branching Processes*, Springer-Verlag, Berlin (1963).
42. W.E. Wilkinson, "Branching Processes in Stochastic Environments," PhD thesis, University of North Carolina, Chapel Hill (1968).
43. P.D. Phung and A.B. Willoughby, "Prediction Models for Fire Spread Following Nuclear Attacks," Report No. URS641-6, URS Corporation, Burlingame, CA (1965).

EXPLOSION PROTECTION

Robert G. Zalosh

INTRODUCTION

Industrial applications of fire protection engineering often include explosion hazard evaluations and selection of appropriate explosion protection measures. Explosion hazard evaluations usually entail: (1) recognizing the potential formation of a flammable fuel-air mixture; (2) identifying potential ignition sources present after fuel-air mixture formation; and (3) determining the resulting explosion pressures and their implications with regard to structural damage potential. Explosion protection measures interrupt this sequence of explosion events either by (1) preventing fuel-air mixture formation (e.g., by ventilation or inerting); (2) eliminating potential ignition sources (e.g., by grounding equipment to prevent electrostatic discharges); or (3) limiting pressure buildup (e.g., by explosion venting or explosion suppression).

This chapter provides an engineering framework for performing this type of explosion hazard and protection measure evaluation. The emphasis is on gas explosions and dust explosions since they represent the explosion hazards encountered most often by fire protection engineers.

FLAMMABILITY, EXPLOSIBILITY, AND IGNITABILITY

Gases and Vapors

Flammable gases and vapors represent explosion hazards when and if they are mixed with air (or some other oxidant) in proportions between the lower and upper flammable limits. Flammable limit concentrations are traditionally determined in a laboratory apparatus consisting of a 5 to 10 cm diameter, 1.5 m long transparent tube. The gas-air mixture in the tube is usually ignited by a spark at the open lower end. If ignition results in flame propagation to the top of the tube, the mixture is deemed flammable. The lower

Dr. Robert Zalosh is a Professor of Fire Protection Engineering at Worcester Polytechnic Institute. He was formerly an Assistant Vice President and Manager of the Applied Research Department at Factory Mutual Research Corporation. His research has included work on gas explosion venting, vapor cloud explosions, and various special industrial fire protection projects.

flammable limit is the lowest gas/vapor concentration capable of supporting flame propagation in this apparatus. Similarly, the upper flammable limit is the highest gas/vapor concentration for which flame propagation is observed.

Flammability limit testing is also sometimes performed in pressure vessels that allow pressure increases to indicate flame propagation. Since this type of test is a more direct measure of explosion hazards, the data are usually called explosion limits. Comparisons of flammability limit data and explosion limit data for methane-air and hydrogen-air mixtures are shown in Table 3-16.1.

The two columns of explosibility data in Table 3-16.1 correspond to two different overpressure $(P - P_0)$ threshold criteria. In some cases (methane lower limit and hydrogen upper limit) the flammability limits and explosibility limits are virtually the same, but in other cases (methane upper limit and hydrogen lower limit) there are substantial differences. One problem with the use of the explosibility limit data is that the data are sensitive to the experimental configuration. For example, the data shown in Table 3-16.1 were obtained with static mixtures in an 8 L vessel. When the hydrogen tests were repeated with a fan in the vessel operating at ignition, the lower limits decreased. Furthermore, tests with a static mixture in a much larger vessel yielded a lower limit of 8.0 vol percent hydrogen, i.e., substantially higher than the corresponding values in Table 3-16.1.

In view of the occasional differences between flammability limits and explosibility limits, the prudent approach would be to use the most conservative value available. For most gases and vapors, only flammability data are available. The most comprehensive compilation of flammability data is the Bureau of Mines Bulletin 627 prepared by Zabetakis in 1965.[1] Flammability limit data for the more common gases and vapors are shown here in Table 3-16.2 based on the Zabetakis values.

Flammability/explosibility limits widen with increasing temperature and pressure. For example, Bartknecht[4] reports that the lower explosive limit of carbon monoxide decreased from about 16.2 vol percent to about 11.5 vol percent as the temperature increased from 20°F to 400°F. Flammable vapor-air mixtures do not exist at temperatures below the flash point for a particular liquid, but fine droplet mists (particularly when droplet diameters are on the order of 10 micrometers or less) and foams or froths can be flammable and/or explosible at temperatures below the flash

TABLE 3-16.1 *Comparison of Flammability Limits and Explosion Limits*

Gas Limit	Flammability Limit* (vol %)	Explosibility Limit** (vol %) @$(P - P_0)/P_0$	
		= 1.0	= 0.1
Methane Lower Limit	5.0	5.0	4.6
Methane Upper Limit	15.0	17.5	20.0
Hydrogen Lower Limit	4.0	7.5	5.0
Hydrogen Upper Limit	75.0	75.0	77.0

*From Zabetakis.[1]
**From Hertzberg.[2]

point. Upper flammable limits are more sensitive to pressure variations than lower limits, except at very low pressures (typically less than one-tenth of an atmosphere) where the two limits suddenly narrow and the gas or vapor becomes nonflammable.[3,4]

The maximum safe oxygen concentration for explosion prevention via inerting is the value corresponding to the nose of the flammability diagram.[3] This concentration, which is usually slightly lower than the oxygen concentration at the upper flammable limit, pertains to the inerting of a worst-case vapor-air or vapor-oxygen mixture, as opposed to a fuel-rich mixture. For example, the oxygen concentration corresponding to the 15 vol percent methane upper flammable limit is 18 percent O_2, but the maximum safe

oxygen concentration for any methane-oxygen-nitrogen mixture is 12 to 13 percent O_2. Tabulated values for the maximum safe oxygen concentration for nitrogen inerting and carbon dioxide inerting of various flammable gases and vapors are provided in NFPA 69, *Standard on Explosion Prevention Systems* (hereinafter referred to as NFPA 69).[5] General guidelines for inerting systems are also given in NFPA 69.

It is important to recognize that practical applications of flammability/explosibility data for explosion hazard evaluations should account for nonuniform or stratified vapor-air mixtures. Even though the volume-average vapor concentration in a large enclosure is well below the flammable limit, ignition of a locally flammable mixture could lead to destruction of the enclosure. Quantification of explosion pressures for uniform and locally uniform gas-air mixtures is provided in the section of this chapter on closed vessel explosions.

EXAMPLE 1:

What is the maximum quantity of hydrogen allowable in a poorly ventilated 100 m³ laboratory in order to prevent room-average hydrogen concentrations from reaching the lower flammable limit in the event of a worst-case release?

SOLUTION:

The volume of hydrogen, V_{H_2}, (at room temperature and pressure) corresponding to the lower flammable limit concentration, χ_{lfl}, is given by

TABLE 3-16.2 *Flammability Data for Common Gases and Vapors*

Gas/Vapor	Lower Flammable Limit* (vol %)	Upper Flammable Limit* (vol %)	Minimum Ignition Energy** (mJ)	Autoignition Temperature* (°C)
Acetone	2.6	13.0	1.15	465
Acetylene	2.5	100.0	0.02	305
Ammonia	15.0	28.0	—	—
Benzene	1.3	7.9	0.2	560
1,3-Butadiene	2.0	12.0	0.13	420
n-Butane	1.8	8.4	0.25	405
Carbon Monoxide	12.5	74.0	—	—
Diethyl Ether	1.9	36.0	0.19	160
Dimethyl Ether	3.4	27.0	0.29	350
Ethane	3.0	12.4	0.24	515
Ethyl Alcohol	3.3	19.0	—	365
Ethylene	2.7	36.0	0.07	490
Ethylene Oxide	3.6	100.0	0.06	—
Gasoline (typical)	1.3	7.1	—	440
n-Heptane	1.05	6.7	0.24	215
n-Hexane	1.2	7.4	0.24	225
Hydrogen	4.0	76.0	0.02	400
Methane	5.0	15.0	0.26	540
Methyl Alcohol	6.7	36.0	0.14	385
Methyl Ethyl Ketone	1.9	10.0	0.53	—
n-Pentane	1.4	7.8	0.22	260
Propane	2.1	9.5	0.25	450
n-Propyl Alcohol	2.2	14.0	0.65	440
Propylene	2.4	11.0	0.28	460
Toluene	1.2	7.1	—	480

*From Zabetakis.[1]
**From NFPA 68.[8]

$$V_{H_2} = \chi_{lfl} V_{lab}$$

where V_{lab} is the laboratory volume. Substituting 0.04 and 100 m^3 for χ_{lfl} and V_{lab}, respectively, V_{H_2} must not exceed 4 m^3. The mass of hydrogen, m_{H_2}, corresponding to this volume is

$$m_{H_2} = \rho_{H_2} V_{H_2} = \rho_a (M_{H_2}/M_{air}) V_{H_2}$$

where ρ_{H_2} and ρ_a are the densities of hydrogen and air, respectively, and M_{H_2} and M_{air} are the molecular weights of hydrogen (2) and air (28.6). Using $\rho_a = 1.2$ kg/m^3 at normal temperature and pressure, the preceding equation yields $m_{H_2} = 0.33$ kg.

Combustible Dusts and Powders

Combustible dusts and powders represent an explosion hazard when their characteristic particle size is smaller than about 400 micrometers and they are suspended in air at a concentration between the lower and upper explosive limits. Typical lower explosive limits for dusts with characteristic particle sizes less than about 100 micrometers are in the 30 to 60 g/m^3 range, which is roughly equal to the range for flammable gases and vapors when expressed in these units.[3] The upper explosive limit for dusts is typically[4] between 2000 and 6000 g/m^3, but it is a difficult measurement to make and an even more difficult measurement to apply as a practical explosion prevention measure.

A brief description of the experimental facilities used to measure dust explosibility properties can help illustrate the difficulties inherent in the practical application of this data. The traditional dust explosibility apparatus is the Hartmann tube, named after the Bureau of Mines researcher who developed it. The Hartmann tube, which has an internal volume of about 1.3 L, is shown schematically in Figure 3-16.1. A weighed dust sample is placed into a small dispersion cup at the bottom of the tube. The dust sample is suspended by directing a sudden blast of compressed air into the dispersion cup. The momentarily suspended dust cloud is ignited

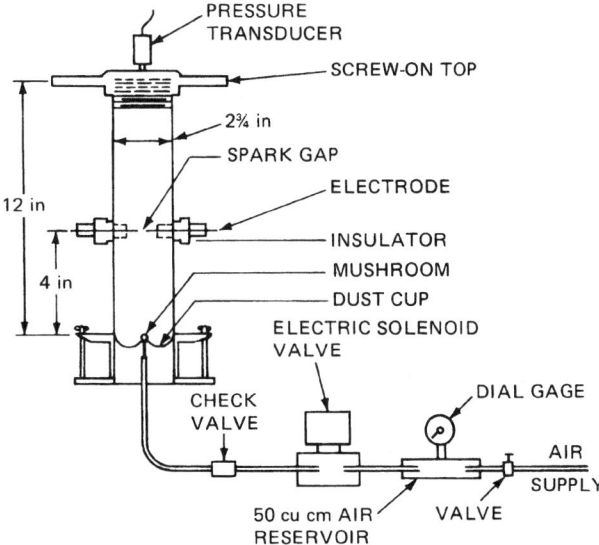

Fig. 3-16.1. *Hartmann apparatus for determining explosibility of dust.[14]*

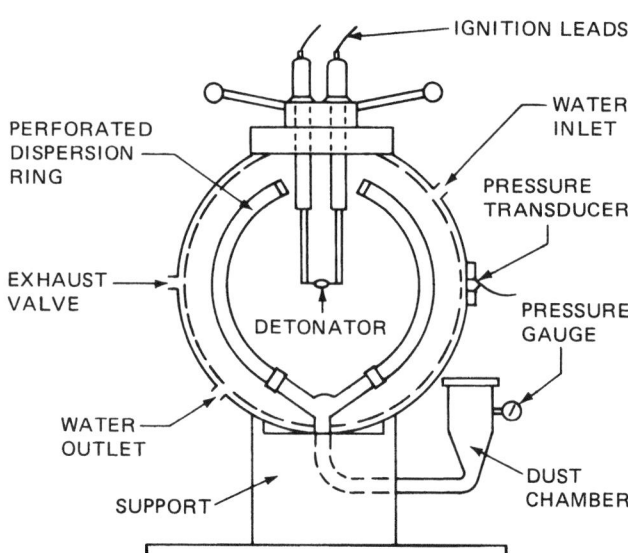

Fig. 3-16.2. *Twenty-liter spherical explosion apparatus.[7]*

by a spark ignition source located 5 to 10 cm above the bottom of the tube. The usual criterion for ignition is a minimum pressure increase (typically about 0.2 atm), sometimes observed from the bursting of a filter paper diaphragm at the top of the tube.

Optical measurements of the dust dispersion process in Hartmann-type tubes[6] reveal that the dust cloud is never uniform throughout the tube and is only uniform in the vicinity of the spark electrodes for a period roughly from 0.25 s to 0.5 s after air injection. Fluctuations in dust concentration near the electrodes are of the order of 50 percent of the mean concentration during this interval. These nonuniform, fluctuating dust concentrations are an inherent feature of dust explosions that is not usually relevant to gas/vapor explosibility.

The current trend in laboratory testing to determine dust explosibility is to use a spherical vessel with a volume of 20 to 30 liters. Figure 3-16.2 is a schematic drawing of a typical 20 L spherical apparatus. The weighed dust sample is placed into an auxiliary dust chamber and is air-injected into the sphere via a perforated dispersion ring. An electrical spark or pyrotechnic igniter located in the center of the sphere ignites the dust cloud after a suitable time delay for dispersion. Vessel wall effects on dust dispersibility and on explosion pressure increases are apparently less of a problem with the 20 to 30 L sphere than with the Hartmann tube. Field[7] has provided a comprehensive description and interpretation of these and other dust explosibility testing apparatuses and procedures as used in various countries.

Lower explosive limit dust concentrations, as determined in a 1 m^3 vessel, are shown in Table 3-16.3 for representative industrial and agricultural dusts. More comprehensive tabulations are available;[7,8] some effects of particle size and vessel apparatus on the reported limit concentrations are included in these references.

Two important difficulties in using lower explosive limit dust concentrations are (1) unknown and/or deceivingly minimal dust concentrations in many industrial facilities; and (2) highly nonuniform, layered dust concentrations during explosion propagation. An example of the first difficulty might be a situation in which there is concern

TABLE 3-16.3 *Explosibility Data for Representative Powders and Dusts**

Material	Median Particle Size (μm)	Minimum Explosive Concentration (g/m^3)	P_{max} (bar)	K_{ST}** (bar m/s)	Dust Hazard Class
Acrylonitrile	25	—	8.5	121	1
Aluminum	29	30	12.4	415	3
Calcium Stearate	12	30	9.1	132	1
Coal, Bituminous	24	60	9.2	129	1
Corn Starch	7	—	10.3	202	2
Magnesium	28	30	17.5	508	3
Milo, Powdered	83	60	5.8	28	1
Phenolic Resin	<10	15	9.3	129	1
Polyethylene, l.d.	<10	30	8.0	156	1
Polymethylmethacrylate	21	30	9.4	269	2
Polypropylene	25	30	8.4	101	1
Polyvinylchloride	107	200	7.6	46	1
Soy Flour	20	200	9.2	110	1
Sugar	30	200	8.5	138	1
Sulfur	20	30	6.8	151	1
Wheat Starch	22	20	9.9	115	1
Zinc	10	250	6.7	125	1
Zinc	<10	125	7.3	176	1

*Field.[7]

**$K_{ST} = \left(\dfrac{dP}{dt}\right)_{max} V^{1/3}$.

about possible explosive dust concentrations in a plastic powder silo. Even if suspended dust concentrations were measured during normal loading and unloading operations, considerably higher concentrations can be generated during upset conditions such as the use of vibrations or air blasts to dislodge powder flow obstructions. Secondary dust explosions represent a more ominous problem in that dust deposits on floors and walls would never be in suspension unless and/or until a primary explosion initiated elsewhere in the plant generated blast waves sufficiently strong to blow the dust into suspension, such that it could be ignited by the flame vented from the primary explosion. Secondary explosion propagation tests conducted by Tamanini[9] demonstrate that explosion propagation can occur even when the equivalent dust deposit concentration (defined as the dust loading divided by the enclosure volume) is below the reported lower explosive limit. The explanation for this apparent anomaly is that the dust clouds created by the primary explosion only extend over a relatively short distance above the floor.

EXAMPLE 2:

Coal feed rates and airflow rates at two coal pulverizers are as follows. At Pulverizer A the coal feed rate is 1.9 kg/s and the airflow rate is 7.1 m^3/s. At Pulverizer B the coal feed rate is 13.9 kg/s and the airflow rate is 7.1 m^3/s.

Are the pulverizer outlet coal dust concentrations within the explosive range?

SOLUTION:

The pulverizer outlet coal dust concentration during steady-state operation is equal to the coal feed rate divided by the airflow rate. The concentrations for these two pulverizers are: Pulverizer A: 0.27 kg/m^3; Pulverizer B: 1.96 kg/m^3. The Pulverizer A concentration is well within the explosive limits for typical bituminous coals. The Pulverizer B concentration is close to the reported upper explosive limit for

coal (2 to 4 kg/m^3). Coal dust explosions have actually occurred in these two pulverizers, but as in most pulverizer incidents, they occurred during shutdown and restart for reasons discussed in Zalosh.[10]

Hybrid Mixtures

Hybrid mixtures are ternary mixtures of combustible dust, flammable gas, and air with the dust suspended in the gas/air. They are of particular concern when the gas and dust concentrations are below their respective limits for binary fuel-air mixtures, but their combined effect renders the hybrid mixture combustible. Examples include methane added to coal dust suspensions in coal pulverizers and silos, and flammable solvent vapor additions to plastic or pharmaceutical powder suspensions in fluidized bed driers.

Data correlations to relate the lower explosive limit of a combustible dust in a hybrid mixture to the limit concentrations of the dust and gas in binary fuel-air mixtures have taken the form

$$L_{DH} = L_D(1 - \chi_G/L_G)^n \qquad (1)$$

where:

L_{DH} = lower explosive limit of dust in hybrid mixture

L_D = lower explosive limit of dust in air

L_G = lower explosive limit of flammable gas in air

χ_G = concentration of flammable gas mixed with air/dust, ($\chi_G < L_G$)

n = an empirically determined exponent

Experimental data presented by Bartknecht[4] for polyvinyl chloride (PVC) dust in methane-air mixtures and in propane-air mixtures suggest that the exponent, n, in Equation 1 should be approximately equal to 2. Data for other hybrid mixtures tested by Gaug et al[11] indicate that n falls in the range 0.37 to 1.0, depending on the size of the test vessel as well as the particular hybrid mixture. The value $n = 1$ renders Equation 1 analogous to Le Chatelier's law, used to calculate the lower explosive limit of a mixture of two or more flammable gases.[4] In view of the uncertainty regarding the best value of n for a specific combination of mixture, enclosure volume, and ignition source, the most prudent approach at present would be to use $n = 2$, which is the most conservative value in the sense that it yields the lowest value of L_{DH}.

EXAMPLE 3:

A coal storage silo is equipped with a flammable gas detector to monitor methane concentration accumulation at the top of the silo, and an optical dust concentration probe to monitor suspended coal dust concentrations during loading operations. The dust probe is intended to shut down operations when the lower explosive limit (LEL) of coal dust is reached. What is the coal dust LEL when the methane concentration is equal to 2.5 percent, i.e., half the methane lower limit in dust-free air?

SOLUTION:

Substituting $\chi_G/L_G = \frac{1}{2}$ and $n = 2$ into Equation 1, $L_{DH} = 0.25\ L_D$. Since $L_D = 60\ g/m^3$ for bituminous coal dust, $L_{DH} = 15\ g/m^3$ at this methane concentration. One complication to be considered in the actual application is that carbon monoxide (generated via spontaneous heating of the coal) may be present as well as methane and coal dust in this particular hybrid mixture. Lower limit concentrations for methane-carbon monoxide-air mixtures can be estimated from Le Chatelier's law.

Ignitability

Electrical spark energies required to ignite flammable gas-air mixtures are usually below 1 mJ at near-stoichiometric concentrations and increase rapidly toward the lower and upper flammable limits. Minimum ignition energies for the most readily ignitable concentrations are listed in Table 3-16.2 for common gases and vapors. Data for other concentrations can be found in Lewis and von Elbe.[12]

Minimum spark ignition energies for combustible dusts are typically two to three orders of magnitude higher than those for gases. However, new spark discharge techniques and energy measurements[7] have shown that several dusts (e.g., zirconium) have minimum ignition energies below 10 mJ. Therefore, avoidance of electrostatic ignition sources through equipment grounding is particularly difficult for these dusts, and may be deceivingly unreliable for many dusts.

The most readily spark-ignitable dust concentrations are often in the range of 200 to 500 gm/m^3. Small particle dusts are significantly more ignitable than large particles. Minimum ignition energy data illustrating this effect, as well as dust composition and hybrid mixture effects, are shown in Figure 3-16.3. A sixfold decrease in PVC particle size (from 125 μm to 25 μm) results in a three order-of-magnitude decrease in minimum ignition energy.

Flammable gas autoignition temperature data shown in Table 3-16.2 represent temperatures corresponding to the ignition of uniformly heated near-stoichiometric gas-air mixtures. Hot-spot temperatures required for ignition may be significantly higher than those shown in Table 3-16.2 if

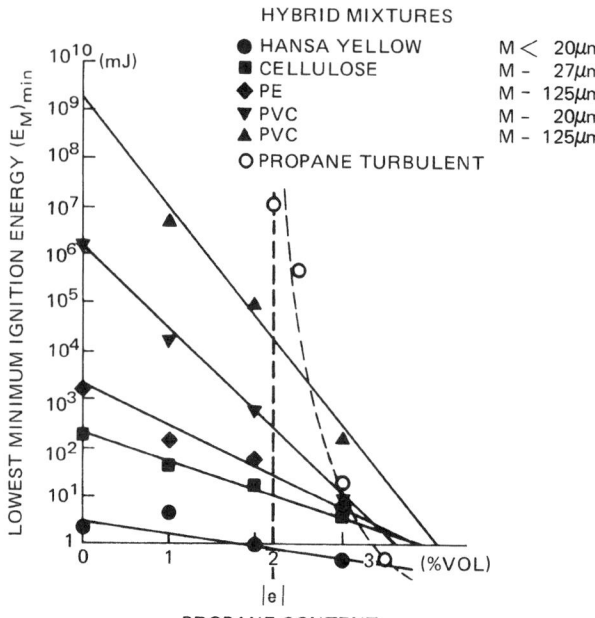

Fig. 3-16.3. Lowest minimum ignition energy of hybrid mixtures vs propane content.[4]

the heated area is relatively small or if the gas concentration approaches the extremes of the flammable range. Dust minimum ignition temperatures[7] generally lie within the range indicated in Table 3-16.2; i.e., 200 to 600°C.

Electrical equipment designated and approved as intrinsically safe is tested to verify that sparks and internal hot spots are not capable of igniting a gas-air or dust-air mixture. It is obvious from the wide range of minimum ignition energies and autoignition temperatures shown in Table 3-16.2 that approval for one gas or specific group of flammable gases does not assure approval for other gases.

CLOSED VESSEL DEFLAGRATIONS

Ignition of a flammable gas-air mixture in a closed vessel will usually result in a deflagration, i.e., flame propagation at subsonic speeds away from the ignition site. Conservative estimates of the associated peak pressure rise can be made from flame temperature calculations/data as follows.

The ratio of peak pressure to initial pressure can be calculated from the ideal gas equations at the end of combustion and prior to ignition. Thus,

$$P_m/P_0 = (M_0 T_b)/(M_b T_0) = \frac{n_b T_b}{n_0 T_0} \qquad (2)$$

where:

P_m = maximum pressure at completion of combustion
P_0 = initial pressure prior to ignition
M_0 = molecular weight of gas-air mixture
M_b = molecular weight of combustion products
T_b = burned gas temperature at the end of combustion
T_0 = initial temperature of gas-air mixture
n_b = number of moles of gaseous combustion products
n_0 = number of moles in gas-air mixture

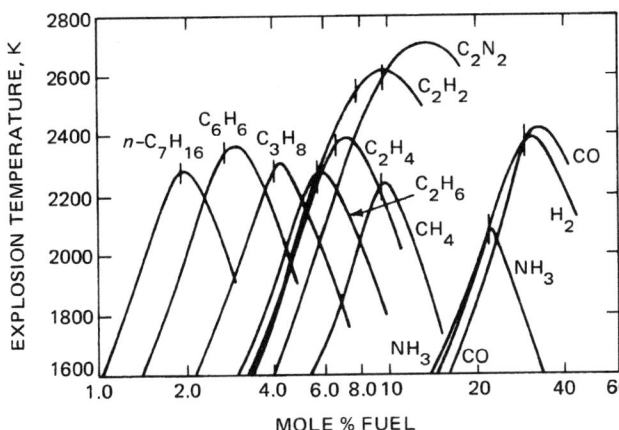

Fig. 3-16.4. The adiabatic constant-pressure explosion temperatures for a number of fuel-air mixtures. Initial conditions: T = 298.15 K, P = 101.325 kPa. The vertical bars represent stoichiometric mixtures. The species H_2 and CO share the same bar.[13]

If the explosion occurs sufficiently fast to neglect heat dissipation, the value of T_b should correspond to the adiabatic constant volume flame temperature of the gas-air mixture. Thermochemical equilibrium calculations of adiabatic flame temperatures are described in most combustion textbooks[12,13] and in Section 1, Chapters 5 and 6. Results obtained by Strehlow[13] for adiabatic constant-pressure flame temperatures are shown in Figure 3-16.4. Adiabatic constant-volume temperatures are generally slightly higher than the corresponding constant-pressure temperatures.

The constant-pressure adiabatic flame temperatures shown in Figure 3-16.4 are in the 1600 to 2400 K range for most of the hydrocarbon gases. Molecular weights and/or the number of moles needed in Equation 2 are approximately the same prior to and following combustion for many flammable gas-air mixtures (hydrogen being one exception), provided there is not much dissociation of combustion products. Therefore, use of this adiabatic temperature data in Equation 2 would indicate peak/initial pressure ratios in the 5.3 to 8.0 range for gas-air mixtures initially at 300 K. Use of constant-volume adiabatic flame temperatures and accurate calculations of mixture molecular weights would probably produce pressure ratios about 10 percent higher for many mixtures.

Few structures can withstand pressures of five to eight times atmospheric pressure. The most common forms of explosion protection therefore are explosion suppression and explosion venting, both requiring intervention in the closed vessel deflagration process before pressures have increased to the structural damage threshold. An understanding of closed vessel explosion rates of pressure rise is therefore critical to the successful design and application of these protection measures.

Theoretical estimates of rates of pressure rise are usually based on the following assumptions. First, it is assumed that the flame front propagates into the unburned gas mixture at a prescribed velocity which is small compared to the speed of sound in the gas mixture. This assumption, which is valid for most deflagrations, is tantamount to assuming that the pressure in the enclosure is uniform at any given time during the explosion. Designating the flame propagation velocity relative to unburned gas as the burning velocity, S_u, the mass burning rate is

$$\frac{dm_b}{dt} = -\frac{dm_u}{dt} = \rho_u A_f S_u \tag{3}$$

where:

m_b = mass of burned gas in enclosure at time t
m_u = mass of unburned gas at time t
ρ_u = unburned gas mass density
A_f = flame front surface area

Flame propagation into a near-stoichiometric gas-air mixture at rest will start as a laminar spherically symmetric process. Laminar burning velocities have been measured for most gases, and representative values (at worst-case gas concentrations) are shown in Table 3-16.4. Turbulence during the explosion can increase the burning velocity to values several times the laminar value, and also increases the burning rate by generating wrinkled, convoluted flame fronts with relatively large surface areas.

A second key assumption invoked in most theoretical models of closed vessel deflagrations is that the fractional pressure rise is proportional to the mass burned, i.e.,

$$\frac{P-P_0}{P_m-P_0} = \frac{m_b}{m_0} \tag{4}$$

where:

P = pressure at time t
P_0 = initial pressure
m_0 = total mass in vessel

An important implication of Equation 4 is that enclosures that are partially filled with flammable gas-air mixtures will be subjected to explosion pressures proportional to the mass fraction of the enclosure occupied by the flammable mixture.

The last assumption needed to model transient pressure development concerns the thermodynamic process governing the compression of the unburned gas ahead of the flame front. The instantaneous rate of pressure rise corresponding to an assumption of isentropic compression is

$$\frac{dP}{dt} = \frac{3S_u}{a}\left(\frac{P}{P_0}\right)^{1/\gamma_u}(P_m - P_0)\left[1 - \left(\frac{P_0}{P}\right)^{1/\gamma_u}\left(\frac{P_m - P}{P_m - P_0}\right)\right]^{2/3}$$

$$\tag{5}$$

where a is the radius of a sphere with a volume equal to the enclosure volume, and γ_u is the ratio of specific heats of the unburned gas mixture.

TABLE 3-16.4 *Closed Vessel Gas Explosion Pressure Data*

Gas or Vapor	P_{max} (bar)	S_u (cm/s)	K_G (bar × m/s)
Butane*	8.0	45	92
Ethane*	7.8	47	106
Ethylbenzene*	6.6	—	94
Ethylene**	8.0	80	150
Hydrogen*	6.9	312	660
Methane*	7.05	40	64
Pentane*	7.65	46	104
Propane*	7.9	46	96
Propane**	8.3	—	70–165

*From Table C-1 of NFPA 68.[8]
**From Pineau.[46]

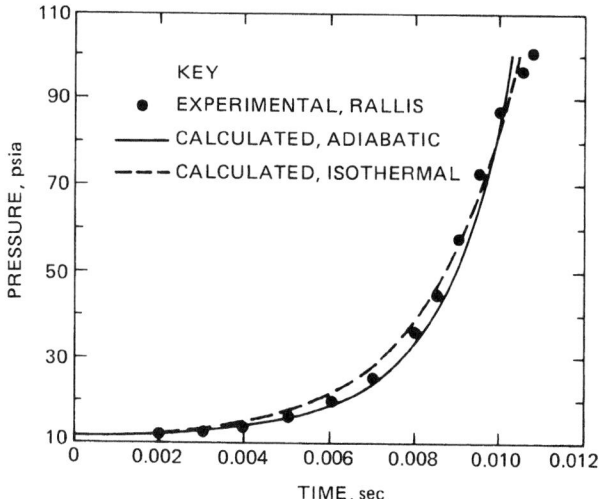

Fig. 3-16.5. Calculated and experimental data for pressure and time during an acetylene-air explosion in a 6.3 in. diameter spherical vessel.[14]

Numerical integration of Equation 5 provides the pressure development as a function of time following ignition. Results obtained by Nagy and Verakis[14] using the isentropic compression assumption and an alternative isothermal compression assumption are shown in Figure 3-16.5 for an acetylene-air explosion and in Figure 3-16.6 for a carbon monoxide-oxygen explosion. Agreement with experimental data is quite good, particularly for the isentropic (adiabatic) model.

Experimental determinations of the violence of a closed vessel deflagration are usually characterized in terms of the maximum rate of pressure rise. Theoretically, this occurs at the end of the explosion when the pressure is at its maximum value. From Equation 5, then

$$\left(\frac{dP}{dt}\right)_{max} = \frac{3S_u}{a}\left(\frac{P_m}{P_0}\right)^{1/\gamma_u}(P_m - P_0) \qquad (6)$$

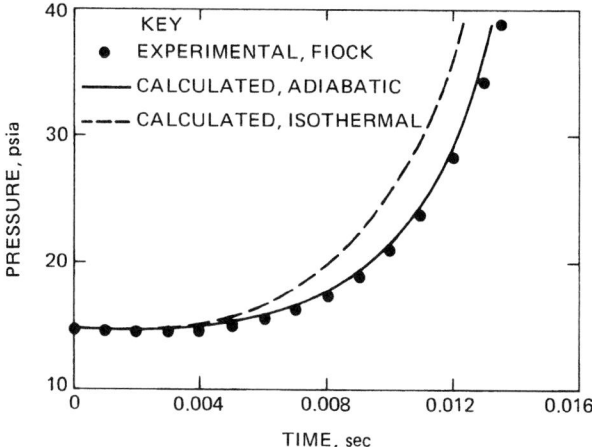

Fig. 3-16.6. Calculated and experimental data for pressure and time during a carbon monoxide-oxygen explosion in a 9.6 in. diameter spherical vessel.[14]

Actual maximum rates of pressure rise differ from theoretical values predicted with Equation 6 for two important reasons. First, heat losses as the flame approaches the vessel wall cause the experimental maximum rate of rise to occur before the end of the explosion. Second, turbulence in the vessel (often induced experimentally by fan-stirring the gas mixture) causes effective burning velocities and rates of pressure rise to be substantially higher than the corresponding laminar values given by Equation 6. The effect of turbulence is evident in the rate-of-rise data shown in Figure 3-16.7.

Since rates of pressure rise vary inversely with vessel radius according to theory (Equation 6), compilations of data for various gases and dusts are now commonly reported in terms of the product of the maximum rate of rise and the vessel volume to the ⅓rd power. The parameter for gases, denoted by K_G, is defined as

$$K_G = \left(\frac{dP}{dt}\right)_{max} V^{1/3} \qquad (7)$$

and the corresponding definition for combustible dusts is denoted by K_{ST}. Values of K_{ST} and K_G are given in Table 3-16.3 and Table 3-16.4, respectively.

Experimental determinations of K_{ST} are sensitive to the method of dispersing the dust and to the strength of the ignition source. This sensitivity has led to the development of ASTM E 1226[49] as a standard test procedure for determining K_{ST} in an apparatus of the type shown in Figure 3-16.2. A test vessel volume of 20 liters is being used because the value of K_{ST} is relatively independent of vessel volume beyond a threshold of about 10 to 15 liters, as indicated by the results of Figure 3-16.8. Experimental determinations of K_G also depend on vessel size/shape and ignition source. The recommended test procedure in NFPA 68 involves using a spherical vessel with a volume of at least 5L, and a 100 mJ spark ignition source.

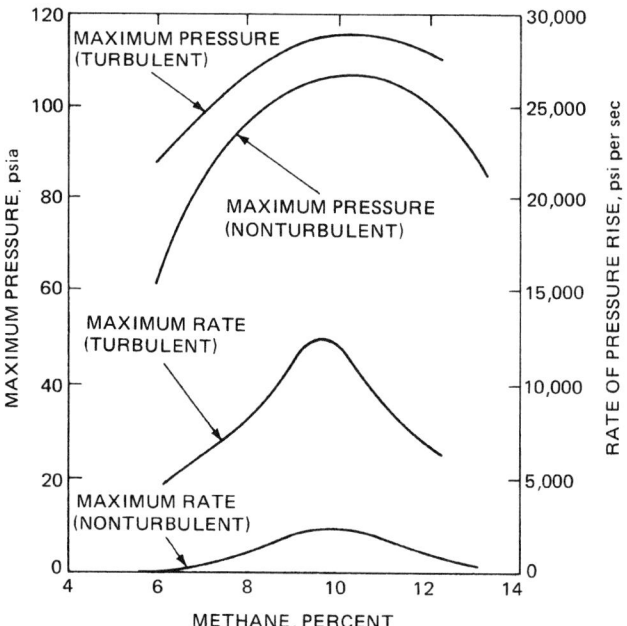

Fig. 3-16.7. Maximum pressure and rate of pressure rise for turbulent and nonturbulent methane/air mixtures in a 1 cu ft closed vessel.[4]

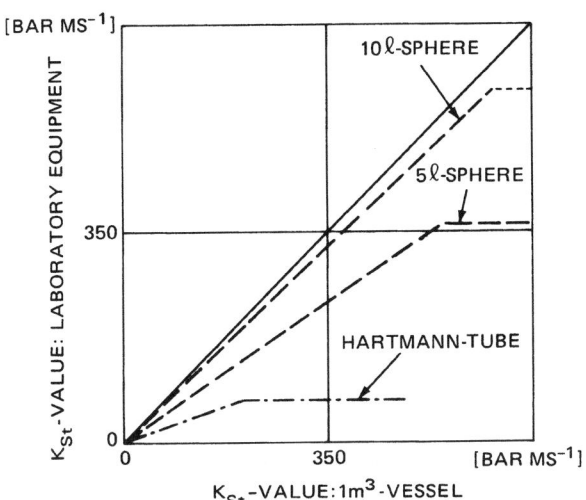

Fig. 3-16.8. *Comparison of K_{ST}-values measured in laboratory equipment with those obtained in a large vessel.[4]*

Explosion venting guidelines for gases and dusts are currently given for groups of gases and dusts with similar values of K_G and K_{ST}. In the case of dusts, the guidelines are written for three classes of dust defined according to the range of K_{ST} shown in Table 3-16.5. The majority of dusts are in the ST-1 class.

EXAMPLE 4:

Approximately 200 g of isobutane are released into an aerosol can-filling room due to a valve misalignment on one large deodorant can. Suppose the isobutane mixes in stoichiometric proportions with a portion of the air in the 4 × 4 × 3 m room and is subsequently ignited. What will the peak pressure be in the absence of any explosion venting?

SOLUTION:

The solution can be obtained from Equation 3 after first calculating the appropriate values for P_m, m_b, and m_0. The value for P_m is calculated from Equation 2 using $T_0 = 298$ K, $T_m = 2300$ K (from Figure 3-16.4), and $M_0/M_b = 1.01$ (from isobutane-air stoichiometry). Substitution into Equation 2 yields $P_m/P_0 = 7.8$.

The ratio m_b/m_0 is equal to the mass of butane actually burned divided by the mass of butane in the entire enclosure at the stoichiometric concentration (3.1 vol percent for butane). Thus,

$$m_b/m_0 = 0.2/[2(1.2)(48)(0.031)] = 0.056$$

Using this result in Equation 3,

$$P - P_0 = (6.8)(0.056) = 0.38 \text{ atmg} = 5.6 \text{ psig}$$

TABLE 3-16.5 *Dust Classes*

Dust Class	K_{ST} (bar m/s)
ST-1	<200
ST-2	201 to 300
ST-3	<300

This pressure is sufficiently high to damage most filling rooms and other explosion-resistant structures.

EXAMPLE 5:

Using Equation 6 and the data in Table 3-16.4, calculate the theoretical value of K_G for butane and compare it to the experimental value.

SOLUTION:

From Equation 6,

$$K_G = \left(\frac{dP}{dt}\right)_{max} V^{1/3} = (4\pi/3)^{1/3} 3 S_u (P_m/P_0)^{1/\gamma_u}(P_m - P_0)$$

Using $S_u = 0.45$ m/s, $\gamma_u = 1.4$, and $P_m = 8.6$ bar,

$$K_G = 76.9 \text{ bar m/s}$$

The experimental value for K_G in Table 3-16.4 is 92 bar · m/s. The higher experimental value is presumably due to turbulence or combustion instabilities.

DETONATIONS

A detonation is an explosion in which the combustion wave, i.e., flame, propagates at supersonic speeds through the unburned fuel. Detonations are fundamentally different than the closed vessel deflagrations described in the previous section of this chapter. Since flames in a deflagration propagate at speeds well below the speed of sound (which is about 340 m/s in room temperature air), the pressure increase during a deflagration occurs virtually uniformly throughout the enclosure as the explosion evolves. In contrast, the pressure rise during a detonation is highly nonuniform and occurs virtually instantaneously as the shock wave propagates through the gas-air mixture. If the flame speed is slightly less than the speed of sound, such that the pressure rise is nonuniform but shock waves do not occur, the explosion is called a quasidetonation.

The practical significance of this fundamental difference between detonations and deflagrations is that they require different approaches to explosion protection. The sudden, spatially nonuniform pressure rise during a detonation or quasidetonation precludes the use of explosion venting or explosion suppression systems. Furthermore, the high peak, short duration detonative pressure loads warrant special considerations in the evaluation of structural resistance.

The peak pressure during a detonation can be calculated from the classical Chapman-Jouguet theory, which is a combination of thermochemical equilibrium and gas dynamic conservation equations across the detonation front.[12,13] Figure 3-16.9 shows calculated detonation pressures as a function of fuel concentration for seven different flammable gases.

Burgess *et al*[16] and others have suggested that a very good approximation to the Chapman-Jouguet detonation pressure, P_D, is

$$P_D = 2P_m \tag{8}$$

where P_m is the closed vessel deflagration pressure calculated from Equation 2. Use of Equations 2 and 8 represents a much simpler alternative to Chapman-Jouguet theory of calculating detonation pressures. However, estimates of peak pressures for detonations in equipment and structures

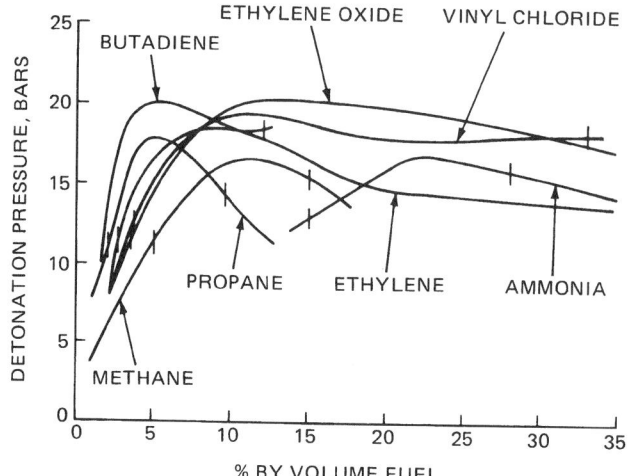

Fig. 3-16.9. Calculated detonation pressure.[15]

should also account for the presence of reflected shock waves with pressures at least twice as high as the incident detonation wave pressure given by Equation 3. Thus, the peak pressure in a detonation is often more than four times as high as the closed vessel deflagration pressure for the same fuel-air mixture.

Structural response evaluations for detonations entail considerations of the impulse (area under the pressure-versus-time curve) as well as the peak pressure developed during the detonation. The impulse depends on the size and geometry of the enclosure (or the vapor cloud). In the case of pipeline detonations, the impulse is proportional to the distance to the open end of the pipe. Burgess et al[16] note that the periodic breaks observed in pipeline detonations occur because each break serves as a pressure relief expansion, requiring the pressure duration/impulse to rebuild to the structural failure threshold by propagating over another length of pipe. Another aspect of this phenomenon is that the flame requires a certain propagation distance to accelerate to the Chapman-Jouguet detonation velocity (1800 to 1900 m/s for most stoichiometric gas-air mixtures).

Gas dynamic calculation methods to determine the pressure histories and impulses associated with detonations in various structures are described by Strehlow[13] and by Fickett and Davis.[17] Methods to assess the structural damage potential from these impulsive loads generated during detonations are described by Baker et al.[18] One after-the-fact indication of structural failure, due to detonative loads is the occurrence of fragmented structural debris associated with brittle failure, as opposed to the bulging and more ductile failure of structural steel subjected to deflagration pressures beyond the yield point.

In view of the drastically different explosion protection considerations for detonations, it is important to assess the potential for a detonation to occur as opposed to a deflagration. Some guidance, as described below, can be offered for this assessment, but there are no exact criteria to provide an unequivocal answer.

Two different characteristic length scales are currently being used to assess the relative detonability of different flammable gases. One length scale is the chemical induction length, defined loosely as the chemical reaction time multiplied by the velocity of the detonation wave. Chemical in-

duction lengths, which have orders of magnitude ranging from 0.01 to 1 cm for stoichiometric hydrocarbon-air mixtures, are determined either by shock tube measurement or by chemical kinetics calculations.

The other detonability length scale is the detonation cell width, λ, which is the transverse dimension of diamond-shaped cells generated by the transverse wave structure at a detonation front.[13] Detonation cell widths are usually measured by the traces deposited on smoke foils inserted in test vessels. Gas-air mixtures with small detonation cell widths are more easily detonated than mixtures with relatively large cell widths. The same is true for chemical induction length as a qualitative measure of detonability.

Detonation cell widths measured for nine different stoichiometric gas-air mixtures are listed in Table 3-16.6 in descending order of detonability. Values range from about 1 cm for acetylene to about 28 cm for methane. These values were obtained at atmospheric pressure and room temperature. Higher initial pressures and temperatures would result in smaller cell widths. Furthermore, some of the gases listed in Table 3-16.6 have smaller cell widths at fuel-rich concentrations. Minimum values for a given fuel can be as much as 40 percent smaller than the values shown in the table.

Figure 3-16.10 provides a pictorial representation of how detonation cell size data can be used to assess whether or not a specific gas-air mixture can sustain a detonation in a particular geometry. The general concept is that for a given gas-air mixture and geometry, a sustained detonation can only occur if the characteristic length scale of the structure is greater than some multiple of the detonation cell width. The approximate value of the multiplication factor is shown in the various sketches. For example, in the case of detonation propagation from an open or ruptured pipe into a surrounding gas mixture, the detonation will not propagate into the surrounding mixture if the pipe diameter is less than 13 λ. The detonation will not even propagate down the pipe if the pipe diameter is less than about $\lambda/3$.

Figure 3-16.11 shows the results of this type of analysis in the case of a hydrogen-air mixture with a uniform concentration in the range 10 to 70 vol percent. Curves for minimum cloud/enclosure size are shown for three different geometries: a pipe (fully confined), one degree of confinement (a vapor cloud on the ground), and an unconfined spherically symmetric geometry. It is clear from Figure 3-16.11 that it is much more difficult to sustain a detonation in hydrogen-rich or hydrogen-lean mixtures than it is near

TABLE 3-16.6 *Detonation Cell Widths and Critical Initiation Energies for Stoichiometric Gas-Air Mixtures**

Gas	Cell Width (cm)	Initiation Energy (kJ)
Acetylene (C_2H_2)	0.98	5.3
Hydrogen (H_2)	1.5	4.7
Ethylene (C_2H_4)	2.8	43–63
n-Butane (C_4H_{10})	5.0–6.2	210–340
Ethane (C_2H_6)	5.4–6.2	130–170
Propylene (C_3H_6)	5.4	53
Propane (C_3H_8)	6.9	210–340
Hydrogen Sulfide (H_2S)	10.0	>80
Methane (CH_4)	28.0	93 000

*From Sulmistras et al.[19]

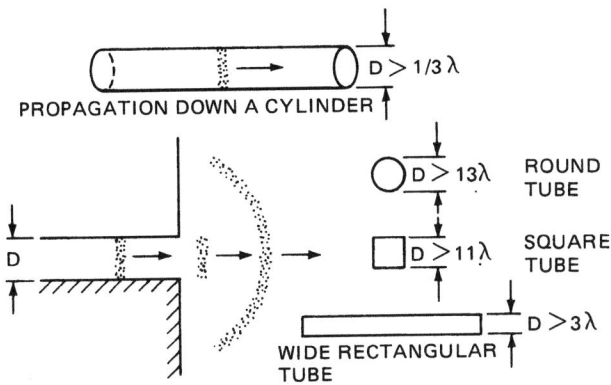

PROPAGATION DOWN A CYLINDER

PROPAGATION FROM A TUBE INTO A LARGE OPEN SPACE
"CRITICAL TUBE DIAMETER"

MINIMUM CLOUD THICKNESS FOR PROPAGATION CONFINED
ON ANY ONE SIDE

Fig. 3-16.10. *Schematic illustration of the developing empirical understanding of the effects of geometry and scale on detonation propagation.* [20]

the stoichiometric concentration. For tube or cloud diameters on the order of 1 m or less, detonations will not occur at hydrogen concentrations less than about 10 percent or greater than about 70 percent, even though these concentrations are well within the flammable limits (4 percent to 75 percent).

What is the likelihood of a detonation occurring in an enclosure or cloud larger than the critical size indicated in Figure 3-16.10? The answer depends on the strength of the ignition source and the presence of either a highly elongated geometry or an exceptionally high level of turbulence for promoting flame acceleration. The minimum ignition source energy required for the direct initiation of a detonation is shown in the last column of Table 3-16.6. It ranges from a low of about 5 kJ for acetylene and hydrogen in air to a high estimated to be 93 000 kJ for methane.

It is interesting to compare the initiation energies required for detonation to the minimum ignition energies shown in Table 3-16.2. Typically, the detonation initiation energies are nine orders of magnitude larger than the minimum ignition energies. Thus, unless ignition is due to an unusually energetic source such as several grams of high explosive, the most likely mode of combustion will be a deflagration. Whether or not the deflagration escalates into a detonation depends on the extent of flame acceleration.

In the case of ignition in a pipe or some other elongated enclosure, deflagration-to-detonation transition (DDT) distances usually require length to diameter ratios of 30 to 100 unless there is some unusual source of turbulence.[4] The presence of regularly spaced obstructions in the pipe or enclosures represents one such source of turbulence that can cause much shorter deflagration-to-detonation runup distances. Berman[20] has recently provided several other examples of rapid transitions. He suggests that the appropriate length scale to extrapolate DDT distances from experimental data is the detonation cell width. If this concept is substan-

tiated and further developed with additional data, the prediction of DDT distances (and DDT criteria in general) may become more definitive than the current judicious combination of experience, luck, and expert judgment.

EXAMPLE 6:

A large process oven is heated with a burner utilizing a 1 cm diameter, 5 m long fuel line containing a stoichiometric propane-air mixture. What is the likelihood of a detonation occurring in the oven and the fuel line as a result of a delayed ignition after the oven has been inadvertently filled with fuel-air mixture? How would the situation change if a stoichiometric hydrogen-air mixture replaced the propane-air mixture in the fuel line?

SOLUTION:

From Table 3-16.6, $\lambda = 6.9$ cm for a stoichiometric propane-air mixture. According to Figure 3-16.10, the minimum pipe diameter for detonation propagation in this case is $\lambda/3 = 2.3$ cm. Therefore, a sustained propane-air detonation will not propagate through the 1 cm fuel line. If the propane is replaced by hydrogen, $\lambda/3 = 0.50$ cm, and a fuel line detonation would be possible, particularly in view of the 500:1 length to diameter ratio.

In order for the detonation to be transmitted into the oven, the fuel line would have to be larger than $13\lambda = 20$ cm for hydrogen. Thus, the oven should not experience a detonation. Explosion venting could be a viable form of explosion protection for the oven.

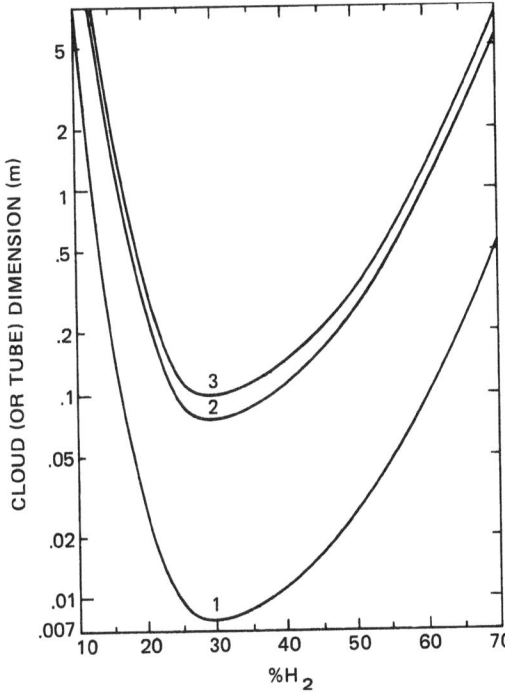

Fig. 3-16.11. *Minimum cloud dimensions for self-sustained detonation propagation under various degrees of confinement in atmospheric, hydrogen-air mixtures: (1) fully confined detonation, (2) one degree of confinement, and (3) zero degree of confinement.* [21]

EXPLOSION VENTING

General Guidelines

Explosion venting is the provision of devices such as blowoff panels and rupture discs to discharge combustion gases during a deflagration and thereby maintain pressures below the enclosure damage threshold. The enclosure can be a room, a building, or a piece of equipment, and the fuel may be a combustible gas, dust, mist, or hybrid mixture.

The most effective explosion venting systems are those that deploy and/or open early in the deflagration, have as large a vent area as possible, and allow unrestricted venting of combustion gases. Early vent deployment requires that the vent release at the lowest possible pressure without interfering with normal operations in the enclosure. In the case of vents on exterior walls and roofs of buildings, the minimum feasible vent release pressure is usually the highest expected differential pressure associated with wind loads (typically 0.14 to 0.21 psig; i.e., 0.96 to 1.44 kPa).

Unrestricted venting requires that the combustion gases be vented directly into an unconfined area. Sometimes vent ducting is needed to discharge flame and combustion gases to a safe location. Pressure increases associated with the use of vent ducting can be estimated from data and guidelines in the literature.[4,8] One other aspect of vent restrictions is the effect of vent inertia in the form of heavy vents impeding rapid vent deployment. NFPA 68 guidelines[8] recommend that the vent weight per unit area not exceed 2.5 lb/ft^2 (12 kg/m^2) for most applications. Factory Mutual Research Corporation recommends limiting vent panel weight to 3 to 4 lb/ft^2 (15 to 20 kg/m^2), depending on the fuel.[47]

The amount of vent area needed for effective explosion venting depends on the size of the enclosure and the rate of pressure rise within it. The pressure will continue to increase until the volumetric rate of vent flow (vented gas velocity multiplied by vent area) exceeds the volumetric generation rate of combustion products. The volume generation rate of combustion products is influenced by complicated flame wrinkling/folding, flame instability, and turbulent flame propagation processes. Vent flow rates are influenced by the occurrence of both unburned and burned gas venting at different stages of the explosion, and by combustion occurring immediately outside the vent as well as inside the enclosure. These phenomena are responsible for the multiple peaks observed in many vented explosion pressure-time traces, such as shown in Figure 3-16.12. Depending on conditions (as described by Cooper et al[22]), the maximum pressure during a vented explosion can corre-

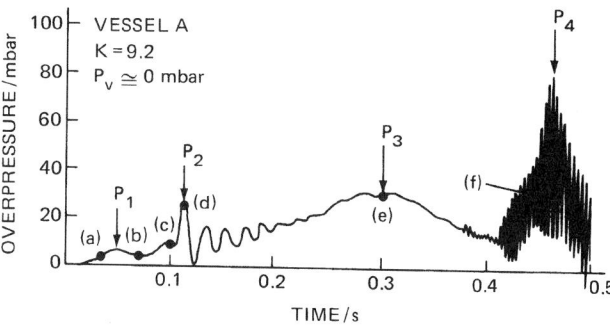

Fig. 3-16.12. Pressure-time profile typical of an explosion in a near-cubic vessel with a low failure pressure explosion relief.[22]

TABLE 3-16.7 *Fuel Characteristic Constant for Use in Equation 9*

Fuel	C (psig$^{1/2}$)	C (kPa$^{1/2}$)
Methane	0.14	0.37
Propane and other gases (except methane) with S_u < 60 cm/s	0.17	0.45
Ammonia	0.05	0.13
ST-1 Dusts	0.10	0.26
ST-2 Dusts	0.12	0.30
ST-3 Dusts	0.20	0.51

spond to any one of the four types of pressure peaks shown in Figure 3-16.12.

The present state of explosion venting technology does not allow peak pressures to be reliably predicted from either first principle mathematical models or small-scale testing. Hence, vent area guidelines should be based largely on test data obtained in test configurations that are similar to the actual application.

The following vent area guidelines are based on data obtained from numerous tests in relatively large-scale enclosures with near worst-case gas-air or dust-air mixture compositions. The guidelines were adopted by the NFPA Explosion Protection Committee for inclusion in NFPA 68.[8] Separate guidelines are offered for low-strength and high-strength enclosures, with the nominal demarcation between the two corresponding to a damage threshold pressure of 1.5 psig (0.1 bar gage).

Low-Strength Enclosures

The premise in these guidelines for low-strength enclosures is that the maximum rate of combustion occurs when the flame surface area is a maximum, i.e., when the flame fills the entire enclosure. Thus the vent area required to maintain pressures below the damage threshold is proportional to the internal surface area of the enclosure. This premise is consistent with test observations for enclosures with low vent opening pressures (for example, the third and fourth pressure peaks in Figure 3-16.12 occur when the flame has virtually filled the enclosure) and with several theoretical models, such as the Swift-Epstein model.[23]

The general form of this venting equation is

$$A_V = CA_s/\sqrt{P_{red}} \qquad (9)$$

where:

A_V = vent area (ft^2 or m^2)
C = fuel characteristic constant (psig$^{1/2}$ or kPa$^{1/2}$) specified in Table 3-16.7
A_s = internal surface area of enclosure (ft^2 or m^2)
P_{red} = overpressure damage threshold (psig or kPa)

The intended range of applicability of Equation 9 is:

- 0.35 psig < P_{red} < 12.5 psig (P_{red} < 1.5 psig in NFPA 68[8])
- $P_{red} - P_v$ > 0.35 psi, where P_v = vent opening pressure
- enclosure length/diameter (hydraulic) ≤ 3.

The recommended values of C in Table 3-16.7 represent the highest values inferred on the basis of test data presented in 18 references cited in NFPA 68.[8] Test conditions encompassed both quiescent and initially turbulent fuel-air mixtures in enclosures with volumes ranging from 0.18 to 82.3 m^3.

For a given fuel, most of the data correspond to C-values well below those given in Table 3-16.7. For example, the most likely value of C for propane is about one-third of the recommended value of 0.17. On the other hand, it is entirely possible for the recommended C value as used in Equation 9 to underestimate the pressure developed in a vented explosion with an extremely high level of turbulence.

Values of C for other gases should be based on test data. In lieu of such data, a reasonable guess that is consistent with the Swift-Epstein model could be made by multiplying the value for propane (for which the most data are available) by the ratio of the burning velocity of the other gas to that for propane.

The enclosure surface area, A_s, in Equation 9 includes the floor and roof as well as wall surface areas. In complex structures with corrugated surfaces or sawtooth roofs, simple geometric projections (i.e., ignoring the corrugations, sawtooths, etc.) can be used to estimate A_s.

Many flammable gases are vulnerable to acoustically driven flame instabilities, which are responsible for the amplified pressure oscillations in the later stages of the pressure-time trace in Figure 3-16.12. The beneficial effects of acoustically absorbing wall cladding, such as ceramic fiber blanketing and mineral wool, in preventing these flame instabilities and associated amplified pressure oscillations have been demonstrated in several test programs.[24,25] Although the use of these materials dramatically reduces or eliminates these amplified pressure oscillations, the other pressure peaks in Figure 3-16.12 are not necessarily affected. Hence, there should not be any overall reduction in vent area requirements as calculated from Equation 9 because the values of C in Table 3-16.7 account for other flame acceleration mechanisms and associated pressure peaks that are not necessarily mitigated by using these wall materials.[22]

High-Strength Enclosures

NFPA 68 vent area guidelines for high-strength enclosures are based on tests by Bartknecht[4] and his colleagues in vessels with volumes in the range of 1 to 250 m^3. These guidelines are usually presented in the form of nomographs, relating vent area to damage threshold pressure, enclosure volume, vent opening pressure, and the gas/dust rate of pressure rise as characterized by K_G or K_{ST}. Simpson[26] has recently developed empirical equations for the Bartknecht nomographs. The equations have the advantages of being easier to read and to interpolate than the nomographs.

The Simpson equation for the dust nomographs (with K_{ST} as a parameter rather than ST class) is

$$A_v = aV^{2/3}K_{ST}^b P_{red}^c \qquad (10)$$

where:

$a = 0.000\,571\,\exp\,(2\,P_{stat})$
$b = 0.978\,\exp\,(-0.105\,P_{stat})$
$c = -0.678\,\exp\,(0.226\,P_{stat})$

The analogous equation for gas explosions is

$$A_v = dV^f \exp(g P_{stat}) P_{red}^h \qquad (11)$$

with values for d, f, g, and h listed in Table 3-16.8 for four gases.

The nomenclature for Equations 10 and 11 is

A_v = vent area (m^2)
K_G, K_{ST} = gas/dust cubic law parameter (bar · m/s)

TABLE 3-16.8 Coefficients/Exponents for Equation 11

Gas	K_G	d	f	g	h
Methane	55.0	0.105	0.770	1.23	−0.823
Propane	75.0	0.148	0.703	0.942	−0.671
Coke Gas	140.0	0.150	0.695	1.38	−0.707
Hydrogen	550.0	0.279	0.680	0.755	−0.393

P_{red} = damage threshold pressure for vessel (bar g)
P_{stat} = vent static opening pressure (bar g)
V = vessel volume (m^3)

The limits of validity of Equations 10 and 11, which are the same as the limits on the original nomographs, are:

$0.1 < P_{stat} < 0.5$,
$P_{stat} + 0.1 < P_{red} < 2$,
$1 < V < 1000$,
$50 < K_{ST} < 600$ (for dusts), and
Vessel length/diameter < 5.

Equation 11 and the associated coefficient and exponent values were developed from tests with the gases listed in Table 3-16.8 in an initially quiescent state in the vessel. Situations involving other gases and/or gases in motion at ignition can often be treated as the equivalent gas in Table 3-16.8 with the closest value of K_G. Bartknecht[4] and NFPA 68 recommend that the values for hydrogen be used for initially turbulent mixtures of propane-air and gases with comparable values of K_G (or S_u).

EXAMPLE 7:

Suppose the $4 \times 4 \times 3$ m room of Example 4 has a damage threshold pressure of 1.0 psig. What would be the minimum vent area to maintain pressures under this value for a worst-case butane-air mixture explosion? Assume the gas mixture is initially at rest and the vent release pressure is 0.2 psig.

SOLUTION:

For this room,

$$A_s = 2[4(4) + 4(3) + 4(3)] = 80 \text{ m}^2$$

Since butane has a burning velocity similar to that for propane, from Table 3-16.7, $C = 0.17$. Therefore, using Equation 9,

$$A_v = 80(0.17)/1.0 = 13.6 \text{ m}^2 = 146 \text{ ft}^2$$

EXAMPLE 8:

A 400 m^3 electrostatic precipitator used in a power plant firing bituminous coal is to be equipped with explosion vents. If the damage threshold pressure for the precipitator is 1.2 bar (17.4 psig) and the vent static release pressure is 0.4 bar (5.8 psig), what is the minimum vent area needed?

SOLUTION:

For a value of $P_{stat} = 0.4$ bar, the auxiliary equations for Equation 10 have the following values:

$a = 0.00127$
$b = 0.938$
$c = -0.752$

Substitution into Equation 10, along with $K_{ST} = 129$ bar·m/s for bituminous coal dust (from Table 3-16.3), $V = 400 \text{ m}^3$, and $P_{red} = 1.2$ bar, yields

$$A_v = 0.00127(400)^{2/3}(129)^{0.938}(1.2)^{-0.752} = 5.7 \text{ m}^2$$

EXPLOSION SUPPRESSION SYSTEMS

An explosion suppression system is designed to detect and suppress an incipient explosion before the pressure rises to the enclosure damage threshold. Suppression is achieved by the rapid discharge of an extinguishing agent from pressurized containers mounted on the protected enclosure. The sequence of events is pictured in Figure 3-16.13. A pressure or flame radiation detector senses the incipient explosion while only a small fraction of the flammable gas or dust has burned. The sensor signal triggers the discharge of a liquefied gas or powdered agent into the enclosure. When the agent reaches the expanding flame front, it quenches the flame and thereby suppresses the explosion.

One important advantage of explosion suppression systems versus explosion venting is that there is no discharge of flame or fuel. Thus suppression systems can often be used more readily on indoor equipment and on equipment containing toxic materials. One major disadvantage of the system is the high cost associated with both the installation of a complex system and the refilling and resetting of the system after a discharge.

The basic components of an explosion suppression system include detectors, an electrical power supply and control system, and a set of rapid action extinguishing units. Pressure sensing detectors are usually used in applications involving dust explosion hazards, while gas explosion applications can employ either pressure sensors or ultraviolet radiation detectors, depending on the required response time (UV detectors provide the fastest response time) and concerns for spurious activation and/or optical shielding of the detectors. Electrical power supplies and control systems include appropriate interlocks for self-monitoring and for shutting down processes and triggering alarms at system activation. Extinguishing agents commonly used in explosion suppression systems have been either halogenated hydrocarbons (halons) or chemical powders. Dry chemicals, such as non-ammonium phosphate and sodium bicarbonate, are still being used to protect some new facilities. However, the recent phaseout of halon production, because of ozone depletion concern, has triggered an ongoing search for new liquid and gaseous agents (primarily perfluorocarbons and hydrofluorocarbons), and has generated renewed interest in water and aqueous salt solutions for explosion suppression. (See Section 4, Chapter 7.)

Several test programs have provided data on the quantity of agent, activation times and pressures, and method of agent dispersal needed for effective suppression. These programs have encompassed a variety of flammable gases and dusts, and have ranged in scale from 8-L laboratory vessels[27] to a 500 m³ tanker pump room.[28] Bartknecht[4] has systematically investigated the effect of enclosure size on the quantity of dry chemical agent needed for effective suppression. His data indicate that the required quantity of agent is proportional to the enclosure volume (i.e., there is a certain

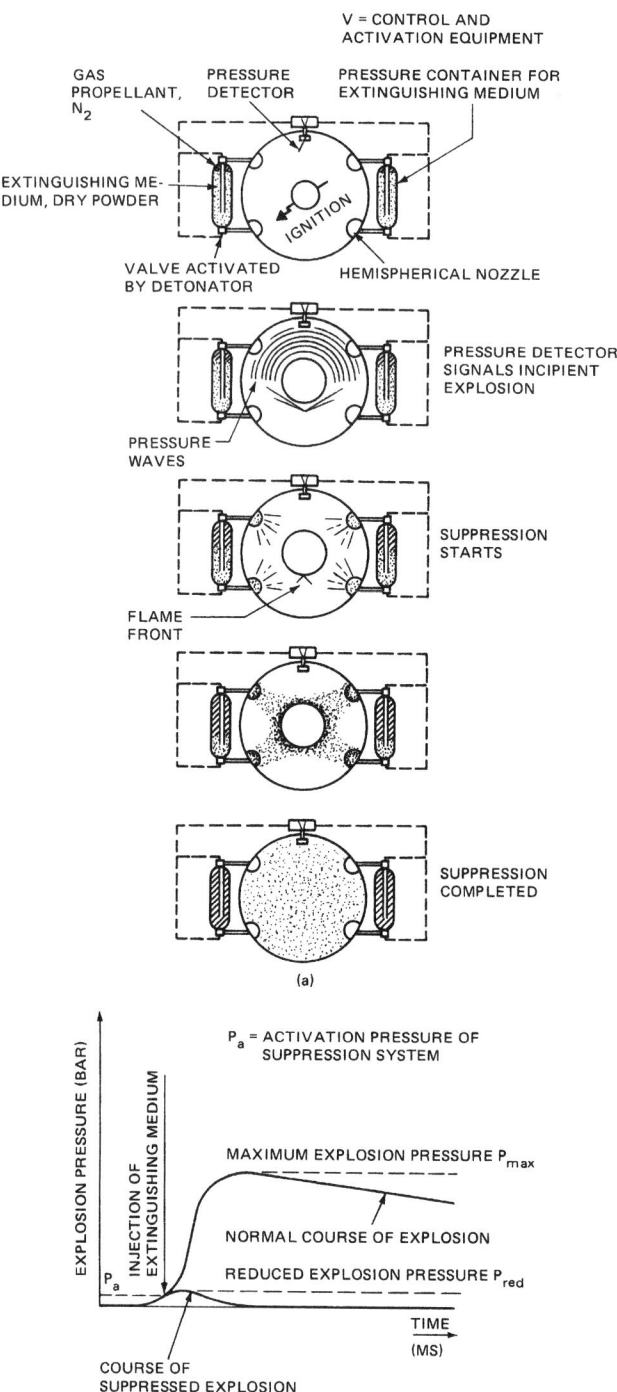

Fig. 3-16.13. (a) Explosion suppression system. (b) Pressure vs time diagram of a normal and a suppressed explosion.[4]

critical agent concentration) for volumes under about 5 m³, but that the required amount of agent is proportional to the ⅔rd power of the enclosure volume beyond 5 m³. This change in scaling suggests that the dry chemical suppression mechanism changes from advance inerting of unburned fuel ahead of the flame front in small vessels, to actual quenching of the advancing flame front in large vessels. Some agents provide chemical inhibition effects (most likely via free radical scavenging) in addition to diluent and thermal benefits, but this chemical inhibition effectiveness is both fuel dependent,[27] and dependent on the advancing flame front speed.[29]

Most of the suppression test data suggest that the various agents have comparable effectiveness for slow to moderate deflagrations, but that ammonium phosphate (and to a lesser extent potassium bicarbonate) becomes decidedly more effective for rapid deflagrations. However, Bartknecht concludes that none of these agents, as presently used in suppression systems, can suppress explosions in gases with K_G values exceeding 200 bar·m/s, or in dusts with K_{ST} values greater than 300 bar·m/s.

Recent tests at NIST[50] in a shock tube generating highly turbulent flames and quasidetonations demonstrate that these high-challenge explosions *can* be suppressed, provided (1) agent can be dispersed uniformly ahead of the shock wave, and (2) gaseous agent concentrations are around 10 vol percent, i.e., about twice as high as the Halon 1301 volumetric concentration used for more conventional, less challenging, explosion suppression applications.

The choice of agent must involve other considerations besides suppression effectiveness as determined by test data. Other relevant considerations include agent retention time to cope with repeated ignitions, agent compatibility with process materials, environmental impact regulations, and potential toxicity effects at the agent design concentration. U.S. regulations that define acceptable and unacceptable suppression agents, from environmental and toxicity considerations, are described in a significant new alternative policy for ozone-depleting chemicals.[51]

General guidelines for the design, installation, and maintenance of a reliable and effective explosion suppression system can be found in the literature[5,30,48] and in the manuals provided by system manufacturers. In addition, system manufacturers and approval organizations have a wealth of unpublished test and incident data that are often essential in developing system specifications and designs for specific applications.

VAPOR CLOUD EXPLOSIONS

Release of a large quantity of flammable gas or vapor into the atmosphere will result, at least temporarily, in the formation of a flammable vapor cloud. Ignition of the vapor cloud may, under certain vaguely defined conditions, result in sufficiently rapid flame propagation to generate destructive overpressures and blast waves. Qualitatively, the conditions required for a vapor cloud explosion are (1) a large quantity of detonation-prone gas/vapor; and (2) either a highly energetic ignition source or a highly obstructed environment supportive of turbulence-induced flame accelerations.

Historically,[31,32] all reported vapor cloud explosions have involved the release of at least 100 kg of flammable gas, with a quantity of 1000 to 10,000 kg being most common. The gases most often involved have been ethylene, propane, and butane. According to Wiekema's compilation of incident data,[32] all of the reported vapor cloud explosions have occurred in "semiconfined" environments such that buildings or other large structures were within the vapor cloud at the time of ignition. Wiekema's data suggest that the presence of a large building or structure within the cloud is a necessary, but not sufficient, condition for an explosion to

occur, since at least 15 of 68 (22 percent) reported ignitions in semiconfined environments resulted in flash fires as opposed to explosions (37 other ignitions did result in explosions). Damage surveys indicate that many of the vapor cloud explosions were deflagrations rather than detonations. On the other hand, analyses of pressure waves generated from flame propagation through vapor clouds (e.g., Lee *et al*[33]) indicate that flame speeds of at least 100 m/s are necessary to generate potentially destructive overpressures greater than about 0.1 atm. Thus, the most likely scenario is that flame speeds on the order of a few hundred m/s (corresponding to so-called quasidetonations) were generated in the actual incidents as a result of flame acceleration around buildings and structures.

The most commonly used method* to assess blast wave effects from vapor cloud explosions is to employ ideal (point source) blast wave correlations based on the blast wave energy, i.e., the TNT equivalent energy. This energy is given by

$$E = \alpha \Delta H_c m_F \qquad (12)$$

where:

E = blast wave energy (kJ)
α = yield, i.e., the fraction of available combustion energy participating in blast wave generation
ΔH_c = theoretical net heat of combustion (kJ/kg)
m_F = mass of flammable vapor released (kg)

The corresponding TNT equivalent mass, kg, W_{TNT} is

$$W_{TNT} = E/4500 \text{ kg} \qquad (13)$$

Figure 3-16.14 is the ideal blast wave overpressure versus distance correlation used in conjunction with Equations 12 and 13. Distances in Figure 3-16.14 are scaled by the cube root of W_{TNT} in accordance with ideal blast wave theory.[34] The overpressures in Figure 3-16.14 are reflected shock wave overpressures associated with reflections of the incident shock wave off a solid surface perpendicular to the wave propagation direction. Nominal building damage and personnel injury thresholds are also indicated in Figure 3-16.14 and in Table 3-16.9. More accurate and comprehensive damage assessments should be based on actual structural dynamic loading calculations leading to impulse-overpressure damage thresholds as described, for example, by Fickett and Davis.[17]

Before Equations 12 and 13 can be used effectively, some guidance is needed on the selection of appropriate values of the yield, α. Data compiled by Gugan[35] and Davenport[31] on the effective yields from approximately 20 vapor cloud explosions showed a spread of four orders of magnitude, with the highest value in one particularly devastating incident being 25 to 50 percent. Wiekema's compilation[32] shows the effective yield to be about one percent for releases of 1,000 to 10,000 kg vapor, and to be in the range of 1 to 10 percent when more than 10,000 kg is released. The yield in the Flixborough explosion (one of the most destructive and the most thoroughly investigated and reported vapor cloud explosion to date) is 4 to 5 percent based on the 30 to 40 metric tons of cyclohexane released prior to ignition.[36] Thus, the specification of yields for blast damage predictions is an exercise in risk assessment, with

---

*Although the TNT equivalency method is most common in the United States, Europeans often use other methods.[32,39]

OVERPRESSURE, psig

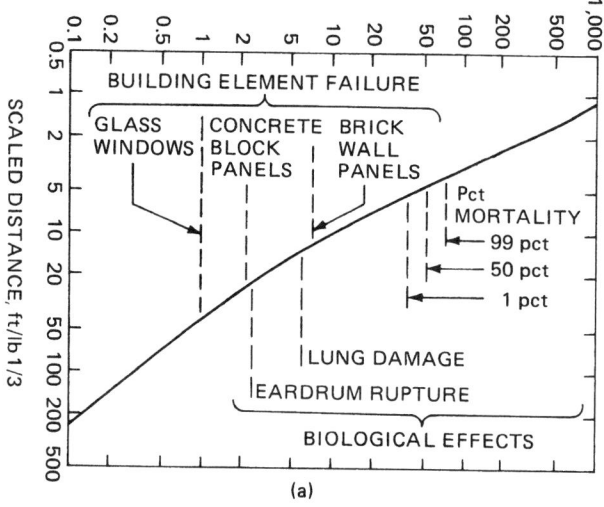

(a)

OVERPRESSURE, (gauge)

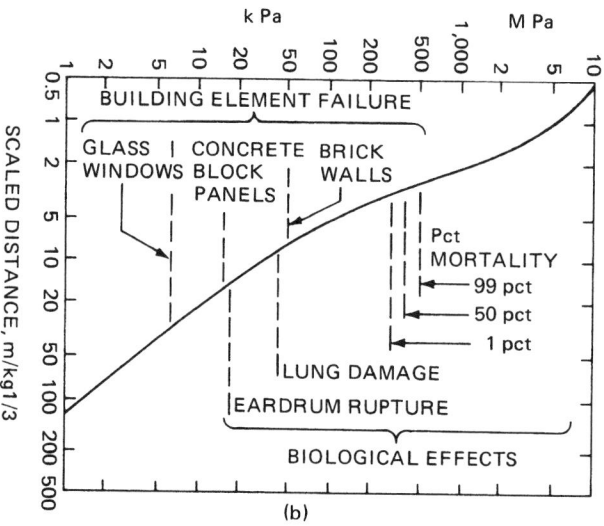

(b)

Fig. 3-16.14. *(a) Ideal blast wave overpressure vs scaled distance (English units). (b) Ideal blast wave overpressure vs scaled distance (metric units).*

conservative (but not necessarily worst-case) values falling in the range of 1 to 5 percent. Since blast wave distances scale as yield to the one-third power, a factor of 5 difference in yield corresponds to a factor of 1.7, i.e., 70 percent difference, in the actual distance to a specified overpressure.

Besides yield estimates, three other questions must be addressed in the use of Equations 12 and 13 and Figure 3-16.14. These are: (1) how much flammable vapor will be released prior to ignition; (2) what are the limits of applicability of the ideal blast wave correlations in Figure 3-16.14; and (3) where is the effective origin of the energy release? Davenport's approach[31] to the first question is that the maximum credible release is equal to the contents of the largest *process* vessel or train of vessels not readily isolated, storage

vessels not included. Another possible approach is to hypothesize that the gas/vapor is released through a valve, rupture disc, pipe break, etc., of specified orifice size, and to calculate the mass of material released in a specified ignition delay time. (According to Wiekema's data compilation,[32] the most likely ignition delay time is in the range of 1 to 5 minutes.) A third approach is to calculate the mass of gas/vapor in the flammable region of the resulting turbulent jet. If the third approach is used, allowances should be made for two-phase flow in the orifice and jet, and for the fact that the effective yield should be higher than the previous quoted values based on the total amount of material released. For example, Sadee *et al*[36] estimated that the effective yield based on turbulent free jet formation in the Flixborough explosion is either 20 or 40 percent, depending on whether one or two jets formed.

As for the range of applicability of Figure 3-16.14, it is clear that ideal (point source) blast wave correlations cannot be valid within or near the flammable vapor cloud. The length of the flammable region of the compressible jet formed from a high pressure release is in the range 170 to 220 orifice diameters of many hydrocarbons.[37] Whether the ideal blast wave pressures are higher or lower than the actual pressures within this near-field region depends on the flame speed and acceleration in the cloud. Although it is virtually impossible to predict flame speeds and accelerations, parametric calculations have been performed using a variety of nonideal, blast wave models[33,38,39] developed for vapor cloud explosions.

As for the effective origin of the energy release, the most logical guess (based on vapor cloud or jet dispersal calculations) would be that it is at the midpoint of the flammable cloud or jet, which is often also roughly the location of the stoichiometric concentration. Two other possible choices for the origin are the most likely ignition site (the ignition source closest to the release site) or the release site itself.

EXAMPLE 9:

Suppose that 1000 kg of ethylene are released through a 50 cm diameter orifice in a process vessel. How far away from the release site would broken windows be expected if ignition did indeed result in a vapor cloud explosion?

SOLUTION:

Using Equation 12 with a yield of 5 percent and a heat of combustion of 47.2 kJ/g for ethylene, $E = 2,360$ MJ, and from Equation 13, $W_{TNT} = 524$ kg. From Figure 3-16.14, broken windows can be expected at a nominal blast wave pressure of 7 kPa (1 psig), which occurs at a scaled distance of about 30 m/kg$^{1/3}$. The distance corresponding to a TNT equivalent mass of 524 kg is $30(524)^{1/3} = 240$ m. For a 1 percent yield, the corresponding distance is 140 m.

Compressible free jet calculations[37] show that the flammable portion of an ethylene jet extends to a distance of about 184 orifice diameters from the release site. For a 0.50 m orifice, this distance is 92 m. Since this distance is only about 38 percent of the 240 m distance calculated (for a 5 percent yield) for the extent of window breakage, the ideal blast wave calculation can be utilized. However, the window damage threshold distance should not be taken too literally because it does not account for (1) specific window strengths (large, thin pane plate glass windows can break at pressures as low as 2 to 3 kPa); (2) the remote possibility of a significantly higher yield; or (3), the fact that the cloud may drift a

TABLE 3-16.9 *Peak Blast Overpressure and Building Damage*

Damage	Peak Blast (psi)	Overpressure (kPa)	(External) bars
Shattering glass windows	0.5–1	3–7	0.03–0.07
Shattering corrugated asbestos siding	1–2	7–14	0.07–0.14
Buckling and connection failure: steel or aluminum paneling	1–2	7–14	0.07–0.14
Shattering concrete or cinder block walls, 400 to 600 mm thick, not reinforced	2–3	14–20	0.14–0.2
Shearing and flexure failure: brick walls, 400 to 600 mm thick	7–8	48–55	0.48–0.55
For Wood Frame Dwellings:			
Minor damage similar to that resulting from a high wind	1	7	0.07
Slight damage: doors, sashes, or frames removed, plaster or wallboard broken, shingles or siding off	2	14	0.14
Moderate damage: walls bulged, roof cracked or bulged, studs or rafters broken	3	20	0.2
Severe damage: standing, but substantially destroyed, some walls gone	6	40	0.4
Demolished, not standing	15	100	1.0
Collapse: self-framing steel panel building	3–4	20–27	0.20–0.27
Rupture: oil storage tanks	3–4	20–27	0.20–0.27
Snapping: wooden utility poles	5	33	0.33
Overturning rail cars	7	47	0.48

considerable distance downwind before being ignited. At the very least, an incremental distance of 92/2 = 46 m should be added to the 240 m calculated distance, to account for an effective energy release at the midjet location.

BLEVEs

A boiling liquid expanding vapor explosion (BLEVE) is the violent rupture of a pressure vessel containing saturated liquid/vapor at a temperature well above its normal (atmospheric pressure) boiling point. The resulting flash evaporation of a large fraction of the liquid produces a large vapor cloud. If the vapor is flammable, and if an ignition source is present at the time of vessel rupture, the vapor cloud burns in the form of a large rising fireball. If ignition is delayed a few seconds, such that the vapor has time to form a large flammable vapor-air mixture, a vapor cloud explosion ensues. If ignition is delayed still longer (actual times can be estimated from the data and analysis in Baurer et al[40]), the cloud is substantially diluted below the lower flammable limit, and only the residual liquid burns in the form of a pool fire.

The concern here is with a scenario in which vessel rupture is caused by fire exposure such that the exposure fire also ignites the vapor generated by flash evaporation. Thus, this is the fireball combustion regime, and of primary concern are the flame radiation hazards associated with the rising fireball. Of course, the blast wave and shrapnel (vessel fragments) generated at vessel rupture are also major concerns.

The vessel rupture blast wave generated during a BLEVE is characterized by the energy released in the fluid expansion from the vessel rupture pressure to atmospheric pressure. This energy, E_e, is given by

$$E_e = m(u_r - u_a) \qquad (14)$$

where m is the mass of fluid in the vessel, u_r is the fluid internal energy at the vessel rupture conditions, and u_a is the fluid internal energy after expansion to ambient pressure. The expansion process is usually assumed to occur isentropically, and thermodynamic data are used to deter-

mine the expanded vapor mass fraction, which is needed to evaluate u_a. Calculated examples for refrigerant tank bursts are given by Baker et al,[41] and blast pressures calculated from ideal blast wave correlations using the calculated value of E_e are compared to measured blast pressures. The comparisons reveal that the measured pressures are significantly less than equivalent ideal blast wave pressures, while the measured blast wave durations are significantly longer than the energy equivalent ideal blast wave durations. This would suggest that either the evaporation/expansion process may take place over too long a period to generate a strong blast wave, or that a large part of the expansion energy contributes to the propulsion of vessel fragments. Statistical methods to evaluate fragment hazards have also been described by Baker et al.[41]

Fireballs produced from the immediate ignition of the vapor cloud generated at vessel rupture usually consume both the flammable vapor and liquid droplets dispersed into the vapor cloud. Tests conducted by Hasegawa and Sato[42] indicate that the entire mass of fluid in the vessel will be burned in the BLEVE fireball if the percentage of fluid vaporized is greater than 36 percent. This will be the case in a liquefied petroleum gas (propane-butane) BLEVE. Hasegawa and Sato[43] developed the following correlations between the mass of hydrocarbon in the vessel and the fireball size, duration, and radiant intensity.

$$D_{max} = 5.25m^{0.314} \qquad (15)$$

$$\tau^{1/2} = 1.07m^{0.181} \qquad (16)$$

$$q_{rmax} = 828m^{0.771}/R^2 \qquad (17)$$

where m (kg) is the mass of fuel in the vessel, D_{max} (m) is the maximum diameter of the fireball, $\tau^{1/2}$ (s) is the half-width of the thermal radiation pulse from the fireball, and q_{rmax} (kW/m^2) is the peak thermal radiation received at a distance R (m) from the fireball.

The Hasegawa and Sato correlations for fireball diameter and duration are similar to correlations developed by Fay and Lewis[44] based on fireballs produced from the burning of small vapor-filled soap bubbles. Fay and Lewis also obtained the following empirical relationship for fireball rise height, z_p (m):

$$z_p = 12.73 V_{va}^{1/3} \tag{18}$$

where V_{va} is the fuel vapor volume (m^3) at atmospheric temperature and pressure. Since virtually all the fuel in a LP-Gas vessel BLEVE contributes to the fireball, the vapor volume used in Equation 18 should presumably be based on the entire mass of fuel in the vessel.

The Hasegawa and Sato radiant heat flux and fireball diameter correlations (Equations 15 and 17) are equivalent to fireball blackbody flame temperatures of about 850 K for values of m on the order of 1 kg. This temperature is approximately equal to the fireball average temperature obtained by Fay and Lewis[44] by matching their fireball rise and duration data to their semiempirical model of fireball buoyant rise. This consistency of the two data correlations provides some measure of confidence to estimates of BLEVE fireball radiation calculations such as those in the following sample problem.

EXAMPLE 10:

Reports of BLEVEs in the San Juan Izhuatepec (Mexico City) 1984 disaster[45] included at least one involving a 420,000 gallon capacity LPG sphere approximately one-half full at the time of the incident. How large would the fireball in this BLEVE be, and what would be the peak radiant heat flux at a distance of 200 m from the ruptured vessel?

SOLUTION:

The mass of LPG (propane) in a half-full 420,000 gallon sphere is 0.5(420,000 gal)(3.785 × 10^{-3} m^3/gal)(510 kg/m^3) = 4 × 10^5 kg. From Equation 15, $D_{max} = 5.25(4 × 10^5)^{0.314}$ = 300 m. This diameter (1000 ft) is only slightly less than the actual observed fireball diameter (1200 ft) reported by Johansson.[45]

Before Equation 17 can be used to estimate the peak radiant heat flux, the actual distance from the fireball must be calculated using the fireball rise height. Equation 18 requires that the volume of vapor equivalent to m be calculated. This, in turn, requires the propane vapor density at 1 atmosphere and 0°C, which is equal to (44/29)(1.2 kg/m^3) = 1.8 kg/m^3. Thus V_{va} = (4 × 10^5 kg)(1.8 kg/m^3) = 2.22 × 10^5 m^3. From Equation 18, z_p = 12.73 (2.22 × 10^5)$^{1/3}$ = 770 m. The distance, R, in Equation 17 is therefore equal to 800 m, and q_{rmax} = 828(4 × 10^5)$^{0.771}$/(800)^2 = 27 kW/m^2. This radiant heat flux is larger than the threshold value for piloted ignition of wood for exposure times of a few minutes or longer. The actual exposure time for this fireball, as calculated from twice the value of $\tau^{1/2}$ from Equation 18, is about 11 s. Specific flammability data (critical absorbed energy for ignition) for the wood in question would be needed to ascertain whether ignition would be expected for this exposure time.

NOMENCLATURE

A	area (m^2)
C	vent area coefficient in Equation 9 (kPa$^{1/2}$)
D	fireball diameter (m)
E	energy released in blast wave or fireball (kJ)
ΔHc	theoretical net heat of combustion (kJ/kg)
K	cubic low constant in Equation 7 (bar · m/s)
L	lower explosive limit concentration (vol percent or kg/m^3)
m	mass of gas or vapor (kg)
M	molecular weight (kg/kg mol)
P	pressure generated during explosion (kPa)
P_0	initial pressure at ignition (kPa)
q	radiant heat flux (kW/m^2)
R	distance (m)
T	temperature (K)
t	fireball duration (s)
u	thermodynamic internal energy (kJ/kg)
V	volume of gas or vapor (m^3)
W_{TNT}	weight of TNT (kg)
z_p	fireball peak rise height (m)
α	blast energy yield in Equation 12 ($-$)
ρ	density (kg/m^3)
χ	volume fraction concentration ($-$)
λ	detonation cell size (m)

Subscripts

a	ambient conditions (air)
b	burned gas
d	detonation
D	dust
DH	dust in hybrid mixture
e	effective (energy released)
f	flame
F	fuel (in vapor cloud explosion)
G	gas
H_2	hydrogen
lab	laboratory
lfl	lower flammable limit
M	maximum (pressure at end of explosion)
0	initial (pressure/temperature prior to explosion)
r	vessel rupture
red	reduced (pressure) due to venting or suppression
$rmax$	maximum radiant (heat flux)
S	enclosure surface
$stat$	static (vent opening pressure)
ST	dust (Staub)
u	unburned gas
v	vent
va	vapor

REFERENCES CITED

1. M.G. Zabetakis, "Flammability Characteristics of Combustible Gases and Vapors." *U.S. Bureau of Mines Bulletin 627*, Washington (1965).
2. M. Hertzberg, "The Flammability Limits of Gases, Vapors, and Dusts: Theory and Experiment," *Fuel-Air Explosions*, p. 3, Univ. of Waterloo Press (1982).
3. D. Drysdale, *An Introduction to Fire Dynamics*, John Wiley and Sons, New York (1985).
4. W. Bartknecht, *Explosions: Course, Prevention, Protection*, Springer-Verlag, New York (1981).
5. NFPA 69, *Standard on Explosion Prevention Systems*, National Fire Protection Association, Quincy (1992).
6. P. Wolanski, "Fundamental Problems of Dust Explosions," *Fuel-Air Explosions*, p. 603, Univ. of Waterloo Press (1982).
7. P. Field, *Dust Explosions*, Elsevier, New York (1982).
8. NFPA 68, *Guide for Venting of Deflagrations*, National Fire Protection Association, Quincy (1994).

9. F. Tamanini, "Dust Explosion Propagation in Simulated Grain Conveyor Galleries," National Grain and Feed Association Research Report ESV-83-067, *FMRC J.I. OF1R2.RK*, Factory Mutual Research Corp, Norwood (1983).

10. R.G. Zalosh, "Review of Coal Pulverizer Fire and Explosion Incidents," *Symposium on Industrial Dust Explosions*, American Society of Testing and Materials, STP 958, (1987).

11. M. Gaug, R. Knystautas, J.H.S. Lee, W.B. Benedick, L. Nelson, and J. Shepherd, "The Lean Flammability Limits of Hybrid Mixtures," *Proceedings of the 10th Colloquium of Dynamics of Explosions and Reactive Systems*, Berkeley (Aug. 1985).

12. B. Lewis and G. von Elbe, *Combustion, Flames, and Explosions of Gases*, Academic Press, New York (1961).

13. R. Strehlow, *Combustion Fundamentals*, McGraw-Hill, New York (1984).

14. J. Nagy and H.C. Verakis, *Development and Control of Dust Explosions*, Marcel Dekker, Inc., New York (1983).

15. "Vapor Explosions," Encyclopedia of Explosives, 10, p. V6 (1983).

16. D.S. Burgess, J.N. Murphy, N.E. Hanna, and R.W. Van Dolah, "Large-Scale Studies of Gas Detonations," *U.S. Bureau of Mines RI 7196*, Washington (1971).

17. W. Fickett and W.C. Davis, *Detonation*, Univ. of California Press, Berkeley (1979).

18. W.E. Baker, P.A. Cox, P.S. Westine, J.J. Kulesz, and R.A. Strehlow, *Explosion Hazards and Evaluation*, Elsevier, New York (1983).

19. A. Sulmistras, I.O. Moen, and A.J. Saber, "Detonations in Hydrogen Sulphide-Air Clouds," Defense Research Establishment Suffield, Ralston, Alberta *Report No. 1140* (May 1985).

20. M. Berman, "A Critical Review of Recent Large-Scale Experiments on Hydrogen-Air Detonations," *Proceedings of the 23rd ASME/AIChE/ANS Heat Transfer Conference* (Aug. 1985).

21. C.M. Guirao, R. Knystautas, J.H. Leew, W. Benedick, and M. Berman, "Hydrogen-Air Detonations," *Nineteenth International Symposium on Combustion*, pp. 583–590, The Combustion Institute (1982).

22. M.G. Cooper, M. Fairweather, and J.P. Tite, "On the Mechanisms of Pressure Generation in Vented Explosions," *Combustion and Flame*, 65, pp. 1–14 (1986).

23. I. Swift and M. Epstein, "The Performance of Low Pressure Explosion Vents," *20th AIChE Loss Prevention Symposium*, (1986); also in, Plant/Operations Progress, 6, pp 98–105 (1987).

24. C.J.M. van Wingerden and J.P. Zeeuwen, "Venting of Gas Explosions in Large Rooms," *4th International Symposium on Loss Prevention and Safety Promotion in the Process Industries*, EFChE Publication Series No. 33, 3, F38–F47 (1983).

25. P.F. Thorne, Z.W. Rogowski, and P. Field, "Performance of Low Inertia Explosion Reliefs Fitted to a 22 m³ Cubical Chamber," EFChE Publication Series No. 33, 3, F1–F10 (1983).

26. L.L. Simpson, "Equations for the VDI and Bartknecht Nomograms," *Plant/Operations Progress*, 5, 1 (Jan. 1986).

27. M. Hertzberg, K. Cashdollar, I. Zlochower, and D. Ng, "Inhibition and Extinction of Explosions in Heterogeneous Mixtures," *Twentieth International Symposium on Combustion*, pp. 1691–1700 (1984).

28. R. Richards, "Development of Explosion Suppression System Requirements for Shipboard Pump Rooms," *U.S. Coast Guard Report CG-D-79-76, NTIS Document #ADA031308*, (Jan. 1976).

29. I. Moen, S. Ward, P. Thibault, J. Lee, R. Knystautas, T. Dean, and C. Westbrook, "The Influence of Diluents and Inhibitors on Detonations," *Twentieth International Symposium on Combustion*, pp. 1717–1725 (1984).

30. "Explosion Suppression Systems," *Factory Mutual Data Sheet 7-17*, Factory Mutual Research Corp., Norwood (Sept. 1981).

31. J. Davenport, "A Survey of Vapor Cloud Incidents," *Chemical Engineering Progress*, pp. 54–63 (Sept. 1977).

32. B.J. Wiekema, "Vapour Cloud Explosions—An Analysis Based on Accidents," *J. of H. Matls.*, 8, pp. 295–329 (1984).

33. J.H. Lee, C.M. Guirao, K.W. Chiu, and G.G. Bach, "Blast Effects from Vapor Cloud Explosions," *AIChE Loss Prevention Symposium*, 11, pp. 59–70 (1977).

34. W.E. Baker, *Explosions in Air*, Univ. of Texas Press (1973).

35. K. Gugan, *Unconfined Vapour Cloud Explosions*, The Institution of Chemical Engineers, George Godwin Ltd., Reading, U.K. (1979).

36. C. Sadee, D.E. Samuels, and T.P. O'Brien, "The Characteristics of the Explosion of Cyclohexane at the Nypro (UK) Flixborough Plant on 1st of June 1974," *Journal of Occupational Accidents*, 1, pp. 203–235 (1976/77).

37. R.G. Zalosh, "A Dispersal and Blast Wave Analysis for Vapor Cloud Explosions," *Proceedings of the 3rd International Symposium on Loss Prevention and Safety Promotion in the Process Industries* (1980).

38. R.A. Strehlow, "The Blast Wave from Deflagrative Explosions–An Acoustic Approach," *AIChE Loss Prevention Symposium*, 14, pp. 145–152 (1981).

39. A.C. van den Berg, "The Multi-Energy Method: A Framework for Vapour Cloud Explosion Blast Prediction," *J. of Haz. Matls.*, 12, pp. 1–10 (1985).

40. B. Baurer, K. Hess, H. Giesbrecht, and W. Leuckel, "Modelling of Vapour Cloud Dispersion and Deflagration after Bursting of Tanks Filled with Liquefied Gas," *Proceedings of 2nd International Symposium on Loss Prevention and Safety Promotion in the Process Industries*, pp. 305–321 (1977).

41. W.E. Baker *et al*, "Workbook for Estimating Blast Effects of Accidental Explosions in Propellant Ground Handling and Transport Systems," *NASA CR 3023* (1978).

42. K. Hasegawa and K. Sato, "Study on the Fireball Following Steam Explosion of n-Pentane," *Proceedings of the 2nd International Symposium on Loss Prevention and Safety Promotion in the Process Industries*, pp. 297–304 (1977).

43. K. Hasegawa and K. Sato, "Experimental Investigation of the Unconfined Vapour-Cloud Explosions of Hydrocarbons," *Technical Memorandum No. 12*, Japanese Fire Research Institute (1978).

44. J.A. Fay and D.H. Lewis, Jr. "Unsteady Burning of Unconfined Vapor Clouds," *16th International Symposium on Combustion*, pp. 1397–1405 (1977).

45. O. Johansson, "The Disaster of San Juanico," *F. Jour.*, p. 32 (Jan. 1986).

46. J. Pineau, "Gas and Dust Explosions in Closed and Vented Vessels," *A Practical Introduction to Gas and Dust Explosion Venting*, EuropEx (1985).

47. FMRC, "Damage-Limiting Construction," Factory Mutual Loss Prevention Data Sheet 1-44, Factory Mutual Research Corp., Norwood, MA (1991).

48. I. Swift, "Design of Deflagration Protection Systems," *J. Loss Prevention in the Process Industries*, 1, pp. 5–15 (1988).

49. ASTM E1226-88, "Standard Test Method for Pressure and Rate of Pressure Rise for Combustible Dusts," American Society for Testing and Materials, Philadelphia, PA (1988).

50. G.W. Gmurczyk and W.L. Grosshandler, "Suppression Effectiveness Studies of Halon-Alternative Agents in a Detonation/Deflagration Tube," *Proceedings* 1994 Halon Options Technical Working Conference, New Mexico Engineering Research Institute, May 1994.

51. Environmental Protection Agency Notice of Final Rulemaking, "Protection of Stratospheric Ozone," 40 CFR Part 82, Appendix B, Summary of Listing Decisions, March 1994.

Section Four
Design Calculations

Section 4 Design Calculations

Chapter 4-13 Smoke Management in Covered Malls and Atria

DESIGN OF DETECTION SYSTEMS

Robert P. Schifiliti,
Brian J. Meacham, and Richard L.P. Custer

INTRODUCTION

Fire detection and alarm systems are recognized as key features of a building's fire prevention and protection strategy. This chapter presents a systematic technique to be used by fire protection engineers in the design and analysis of detection and alarm systems. The majority of discussion is directed toward systems used in buildings. However, many of the techniques and procedures also apply to systems used to protect planes, ships, outside storage yards, and other nonbuilding environments.

Scientific research on fire growth and the movement of smoke and heat within buildings provides fire protection engineers with information and tools that are useful in the design of fire detection systems. Also, studies of sound production and transmission allow communication systems to be engineered, thus eliminating the uncertainty in locating fire alarm sounders. All of this allows engineers and designers to design systems that meet specific, identifiable goals.

Sections 1, 2, and 3 of this handbook introduced and discussed a series of concepts and tools for use by fire protection engineers. This chapter shows how some of these tools can be used collectively to design and evaluate detection and alarm systems.

OVERVIEW OF DESIGN AND ANALYSIS

To design a fire detection and alarm system, it is first necessary to establish the system's goals. These goals are established by model codes, the property owner, risk manager, insurance carriers, and/or the authority having jurisdiction. Ultimately, the goals of the system can be put in four basic categories:

Robert Schifiliti, P.E. is a fire protection consultant based in Reading, MA. He specializes in fire detection and fire alarm system design. Brian Meacham, P.E. is a fire protection consultant in Meilen, Switzerland, specializing in performance-based design of detection and general fire protection systems. Richard Custer, M.Sc. is a fire protection consultant in Wrentham, MA and an Adjunct Associate Professor of Fire Protection Engineering at Worcester Polytechnic Institute. Mr. Custer is a Fellow of the Society of Fire Protection Engineers specializing in fire detection, failure analysis, and performance-based fire safety design.

1. Life safety,
2. Property protection,
3. Business protection, and
4. Environmental concerns.

When designing for life safety, it is necessary to provide early warning of a fire condition. The fire detection and alarm system must provide a warning early enough to allow complete evacuation of the danger zone before conditions become untenable. The fire detectors or fire alarm system may be used to activate other fire protection systems, such as special extinguishing systems and smoke control systems, that are used to help maintain a safe environment during a fire.

Property protection goals are principally economic. The objective is to limit damage to the building structure and contents. Maximum acceptable losses are established by the property owner or risk manager. The goal of the system is to detect a fire soon enough to allow manual or automatic extinguishment before the fire exceeds acceptable damage levels.

Goals for the protection of a mission or business are determined in a manner similar to that used in property protection. Here, fire damages are limited to prevent undesirable effects on the business or mission. Some items that need to be considered are the effects of loss of raw or finished goods, loss of key operations and processes, and the loss of business to competitors during downtime. Other concerns include the availability and lead time for obtaining replacement parts. If the equipment to be protected is no longer available, or requires several months for replacement, the ability to stay in business during and after an extended period of downtime may be jeopardized.

Protection of the environment is also a fire protection concern. Two examples are: (1) toxicity of products of combustion and (2) contamination by fire protection runoff water. Should large quantities of contaminants be expected from a large fire, the goal of the system may well be to detect a fire and initiate appropriate response prior to reaching a predetermined mass loss from burning materials or quantity of fire suppression agent discharged.

Once the overall goals have been set, specific performance and design objectives for a performance-based design can be established.[1,2] Performance-based fire protection design requires that specific performance objectives, rather than generic prescriptive requirements, be met. A typical prescriptive requirement would be to provide a smoke detector for every 84 m² (900 sq ft) or 9-m (30-ft) spacing. In prescriptive design, speed of detection and the fire size at detection for such an installation are not known or considered explicitly. In addition, if some action must be taken in response to the alarm in order to control the fire, the expected damage is also unknown.

Implementation of a firesafety performance objective requires that the objective first be stated by the client in terms of acceptable loss. The client loss objectives must then be: (1) expressed in engineering terms that can be quantified using fire dynamics and (2) related to design fires, design fire environments, and the performance characteristics of fire suppression equipment. For example, the client loss objective may be to prevent damage to essential electronic equipment in the compartment of origin. To meet this objective, one must first define what "damage" is. This could be expressed in terms of the thickness of the smoke layer. Other criteria, such as temperature or concentration of corrosive combustion products, or a combination of criteria, could also be used.

Based on a study of the likelihood of ignition and fire growth scenarios, a design fire needs to be established. The design fire is characterized by its heat release rate, $\dot{Q}$, at any moment in time; its growth rate, dq/dt; and a combustion product rate dcp/dt such as smoke particulate, toxic or corrosive species, etc. production rate, dp/dt. The design fire may be determined by: (1) a combination of small- and large-scale testing specific to the application or (2) analysis of data taken from studies reported in the literature.

For a given firesafety design objective, there will be a point, $\dot{Q}_{do}$, on the design fire curve where the energy and product release rates will produce conditions representative of the design objective. Given that there will be delays in detecting the fire, notifying the occupants, accomplishing evacuation, or initiating suppression actions, the fire will need to be detected at some time in advance of $\dot{Q}_{do}$. In order to account for these delays, a critical fire size, $\dot{Q}_{cr}$, can be defined as the point on the design fire curve at which the fire must be detected in order to meet the design objectives for a given spacing or radial distance from the fire.

There are two types of delays that influence the size of the fire at detection: (1) those that are variable and (2) those that are fixed. Variable delays represent transport lag and are related to radial distance of the detector from the fire, ceiling height, and the convective heat release rate of the fire. Fixed delays are associated with system characteristics, such as alarm verification time. Adding the fixed delays to $\dot{Q}_{cr}$ defines another point on the design fire curve: $\dot{Q}_i$; or the "ideal" fire; i.e., the fire that would be detected with no transport delay.

The design fire, $\dot{Q}_{do}$, has been defined as the fire size (in terms of peak heat release and given growth rate history) that corresponds to the maximum acceptable loss fire, and the critical fire, $\dot{Q}_{cr}$, as the maximum fire size at time of detection that allows actions to be taken to limit the continually growing fire to the design fire limit. The time needed to take the limiting actions is the response lag. The total system response time, then, is the amount of time required between the critical fire and the design fire for all the actions to take place before $\dot{Q}_{do}$ is reached, and is the sum of the fixed and variable delays and the response lag. The various design and evaluation points on a design fire curve are shown in Figure 4-1.1.

For example, if the design fire is determined to be 1500 kW, and manual suppression will be employed, the critical fire can be selected at a moment in time that permits detection, notification, and response before the 1500 kW fire

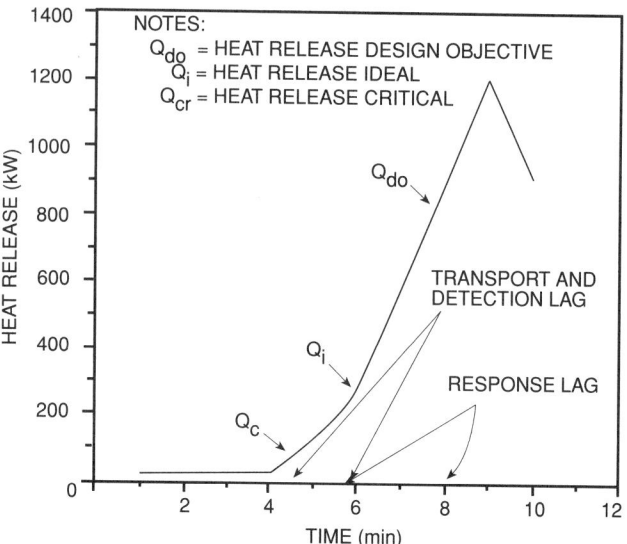

Fig. 4-1.1. Design fire curve.

size is reached. If the total system response time is estimated to be 3 min, the critical fire would be at the size determined at 3 min prior to reaching 1500 kW using the estimated fire growth rate.

Expressing fire size or fire load as an energy release and growth rate may be thought of in the same way that structural engineers use earthquake zone maps to design for potential earthquakes. Electrical engineers might compare fire loads to fault currents used in designing overcurrent protection devices. At the present time, design fire, critical fire, and total system response time requirements are not established by any building codes. It is the job of the design engineer to work with the building owner and local code officials to establish the performance requirements for a given system application.

Once the goals of a system have been established, several probable fire scenarios should be outlined. The occupancy of the building and the expected fuels should be analyzed to establish an expected fire growth rate and an expected maximum heat release rate. Fire loss reports and fire test data can be used to help estimate heat release rates and the production of smoke and fire gases. It is important that different fire scenarios be evaluated to establish how the system design or response might change as a result of varying fire conditions. Several possible fire scenarios should be outlined using the techniques presented elsewhere in this handbook.

When designing a system, select the most likely fire scenario as the basis of the design. Once the design requirements for spacing and detector type are established, the system's response can be analyzed using the other possible fire scenarios. If the alternate fire scenarios cause the design to not meet the established goals, design changes can be made and retested, if warranted.

The several fire scenarios used when analyzing a system will produce upper and lower bounds or a range of system performance characteristics. The fire scenarios selected should include best and worst case fires as well as several likely scenarios for the particular building characteristics and occupancy.

TABLE 4-1.1 *Fire Signatures and Commercially Available Detectors*

Fire Signature/Detector Type	Electromagnetic Radiation Wave Length 1700 to 2900 Angstroms	Electromagnetic Radiation (Thermal) 6500 to 8500	Invisible Products of Combustion Less than 0.1 Micron	Visible Smoke and Products of Combustion More than 0.1 Micron	Rapid Change in Temperature	High Temperature
Ultraviolet Detector	X					
Infrared Detector		X				
Sub-micron Particle Detector			X			
Wilson Cloud Chamber						
Infrared Particle Detector						
Smoke Detector						
Photoelectric				X		
Ionization			X			
Photo Beam				X		
Rate-of-Rise Heat Detector					X	
Rate Anticipation Heat Detector						X
Fixed Temperature Heat Detector						X

For the purposes of design or analysis, detection and alarm systems have three basic elements: detection, processing, and signaling. The first, detection, is that part of the system that senses fire. The second element involves the processing of signals from the detection portion of the system. Finally, the processing section of the system activates the signaling portion in order to alert occupants and perform other auxiliary signaling operations. Auxiliary functions may include smoke control, elevator capture, fire department signaling, and door closing.

This chapter focuses on the detection and signaling elements of a fire alarm system. Engineering methods for the design and analysis of heat detector response are presented along with several examples. A method to calculate the audibility of fire alarm sounders is also presented. The selection of a system's control panel and the design of auxiliary functions is beyond the scope of this chapter.

DETECTION

To design the detection portion of a fire alarm system it is necessary to determine where fire detectors should be placed in order to respond within the goals established for the system. Several different detector types might respond to the expected fire, so it may be necessary to develop several candidate system designs, using various combinations of detector types in order to optimize the system's performance and cost.

A fire signature[3] is some measurable or sensible phenomenon present during combustion. Table 4-1.1 is a cross-reference of fire signatures and commercially available detector types. The table shows the predominant fire signature to which the detector responds.

HEAT DETECTION

This section discusses an engineering method for determining the placement of heat detectors on a large flat ceiling.

The present practice in designing fire detection systems using heat detectors is to space the detectors at intervals equal to spacings established by tests at Underwriters Laboratories Inc. Listed spacings are determined in full-scale fire tests.[4]

In the Underwriters Laboratories Inc. (UL) test, a burning pan of 190 proof denatured alcohol is located in the center of a test room. Sprinkler heads having a 160°F (71°C) rated operating temperature are located on the ceiling in a square array having 10-ft sides. The fire is in the center of the square. The distance between the fire and the ceiling is varied so that the 160°F (71°C) sprinkler head being used operates in approximately 2 min. Detectors of the type being tested are located at the corners of squares having 20, 30, 40, and 50 ft sides (6.1, 9.1, 12.2, and 15.2 m). (See Figure 4-1.2.) The spacing of the last detector to operate prior to a sprinkler

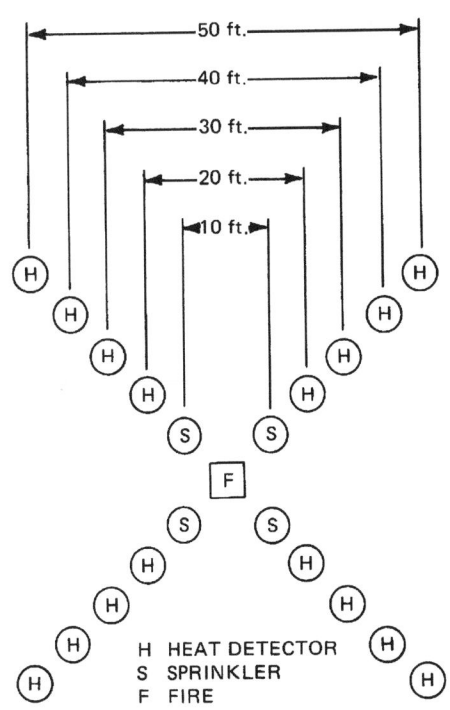

Fig. 4-1.2. Detector test layout.

head operating becomes the detector's listed spacing. A similar test procedure is employed by Factory Mutual Research Corporation (FMRC) to arrive at an approved detector spacing.

Most codes require that detectors be spaced at intervals equal to the UL or FMRC spacing. NFPA 72, *National Fire Alarm Code*,[5] Chapter 5, Initiating Devices, requires that the installed spacing be less than the listed spacing to compensate for high ceilings, beams, and air movement. High ceilings mean that the fire plume will entrain more ambient air as it rises. This has the effect of cooling the gases and reducing the concentration of fire products. Beams, joists, walls, or sloped ceilings alter the flow of combustion products. This can serve to restrict or enhance the operation of a fire detector. For instance, consider the case where a fire detector is located on a ceiling between two parallel beams and a fire occurs at floor level between the beams. If the distance between the beams is small compared to the horizontal distance from the fire to the detector, the beams will act as a channel directing the flow of hot gas to the detector, thus speeding operation. NFPA 72 allows detector spacing to be increased beyond the listed spacing in areas such as corridors with narrow walls to confine the smoke and heat produced by the fire. Systems can be designed using this type of code approach; however, this will not permit quantitative assessment of detector response or measure the ability of a given system design to meet specific design goals relating to fire size, allowable damage, or hazard.

The best possible location for a heat detector is directly over the fire. If there are specific hazards to be protected, the design should include detectors directly overhead or inside of the hazard. In areas without specific hazards, detectors should be spaced evenly across the ceiling. When detectors are evenly spaced, the point that is farthest from any detector will be in the middle of four detectors. (See Figure 4-1.3.) The spacing between detectors is

$$S = 2^{1/2}r \qquad (1)$$

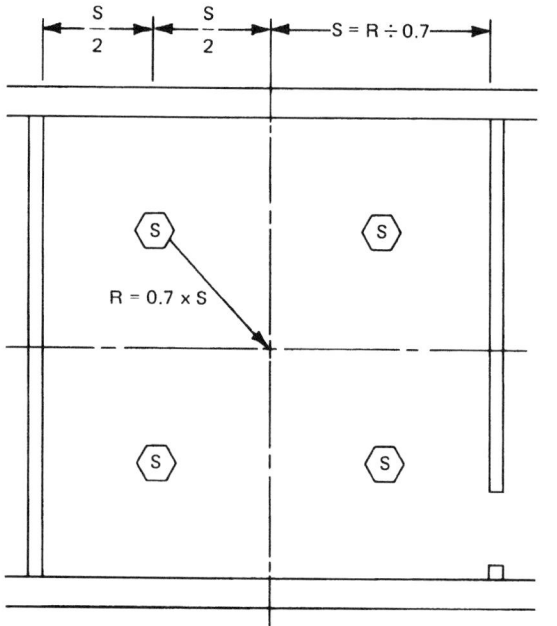

Fig. 4-1.3. Detector spacing.

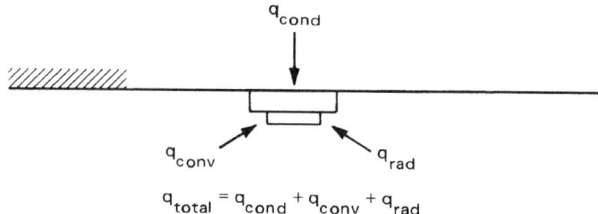

Fig. 4-1.4. Heat transfer to a ceiling-mounted detector.

For a given detector, the problem is to determine the maximum distance the detector can be located from the fire and still respond within the design goals of the system. This requires a method for predicting detector response, based on fire size and growth rate, ceiling height, and detector characteristics.

Fire plume and ceiling-jet models can be used to estimate the temperature and velocity of fire gases flowing past a detector. The heat transfer can be calculated, and the response of the detector can be modeled.

Figure 4-1.4 describes the heat transfer taking place between a heat detector and its environment. The total heat transfer rate to the unit, $\dot{q}_{total}$, can be expressed by the relationship

$$\dot{q}_{total} = \dot{q}_{cond} + \dot{q}_{conv} + \dot{q}_{rad} \quad \text{(kW or Btu/s)} \qquad (2)$$

where $\dot{q}_{cond}$, $\dot{q}_{conv}$, and $\dot{q}_{rad}$ represent conduction, convection, and radiation heat transfer rates, respectively. During the initial stage of fire growth, radiation heat transfer can be neglected. Also, the elements of most commercially available heat detectors are thermally isolated from the remainder of the unit. In these cases, it can be assumed that the heat lost from the heat-sensitive element by conduction to other parts of the detector, and to the ceiling by conduction, is negligible in comparison to the convection heat transfer taking place. This leaves a net rate of heat transfer to the detector equal to $\dot{q}_{conv}$. The convective heat transfer rate to the detector is described by

$$\dot{q} = \dot{q}_{conv} = hA(T_g - T_d) \quad \text{(kW or Btu/s)} \qquad (3)$$

The convective heat transfer coefficient is h and has units of kW/(m² °C) or Btu/(s ft² °F). A is the area being heated. T_d and T_g are the detector temperature and the temperature of the gas heating the detector, respectively. Treating the detector element as a lumped mass, m (kg or lbm), the change in its temperature is found by

$$dT_d/dt = \dot{q}/mc \quad \text{deg/sec} \qquad (4)$$

where c is the specific heat of the element being heated and has units of kJ/(kg °C) or Btu/(lbm °F) and $\dot{q}$ is heat release rate. This leads to the following relationship for the change in temperature of the detector with respect to time:

$$dT_d/dt = hA(T_g - T_d)/mc \qquad (5)$$

Heskestad and Smith[6] have proposed use of a time constant, τ, to describe the convective heat transfer to a particular detector element.

$$\tau = mc/hA \quad \text{seconds} \qquad (6)$$

$$dT_d/dt = (T_g - T_d)/\tau \qquad (7)$$

Note that τ is a function of the mass, area, and specific heat of the particular detector element being studied. For a given fire gas temperature and velocity and a particular detector design, an increase in mass increases τ. A larger τ results in slower heating of the element.

The convective heat transfer coefficient, h, is a function of the velocity of the gases flowing past the detector element and the shape of the detector element. For a given detector, if the gas velocity is constant, h is constant. It has been shown[7] that the convective heat transfer coefficient for spheres, cylinders, and other objects similar to a sprinkler or heat detector element is approximately proportional to the square root of the Reynolds number, Re.

$$\text{Re} = ud/v \qquad (8)$$

where u is the gas velocity, d is the diameter of a cylinder or sphere exposed to convective heating, and v is the kinematic viscosity of the gas. For a given detector, this means that h and, hence, τ, is proportional to the square root of the velocity of the gases passing the detector. This relationship can be expressed as a characteristic response time index, RTI, for a given detector.

$$\tau u^{1/2} \simeq \tau_0 u_0^{1/2} = \text{RTI} \qquad (9)$$

Thus, if τ_0 is measured in the laboratory at some reference velocity u_0, this expression is used to determine the τ at any other gas velocity, u, for that detector. The product, $\tau u^{1/2}$ is the response time index, RTI.

Heskestad and Smith[6] developed a test apparatus at Factory Mutual Research Corporation to determine the RTI of sprinkler heads. In the test, called a "plunge test," the sprinkler head is suddenly lowered into the flow of a hot gas. The temperature and velocity of the gas are known and are constant during the test. The equation for the change in the detector temperature is then

$$dT_d/dt = (1/\tau)(T_g - T_d) \qquad (10)$$

Since the gas temperature is constant during the test, the solution to this equation is

$$T_d - T_a = (T_g - T_a)[1 - \exp(-t/\tau)] \qquad (11)$$

where T_a is the ambient temperature or initial temperature of the sprinkler or detector at time $t = 0$. T_d is the temperature of the detector at time t. Rearranging the equation gives

$$\tau = t/\ln[(T_g - T_a)/(T_g - T_d)] \qquad (12)$$

By measuring the response time, t_r, of the unit in the plunge test this equation can be used to calculate τ_0 at the test velocity u_0. This is done by substituting the response temperature and time for T_d and t. The sensitivity of the detector or sprinkler can then be expressed as

$$\tau_0(\text{at } u_0) = t_r/\ln[(T_g - T_a)/(T_g - T_r)] \quad (\text{s}) \qquad (13)$$

In terms of the response time index this equation becomes

$$\text{RTI} = t_r u_0^{1/2}/\ln[(T_g - T_a)/(T_g - T_r)] \qquad (14)$$

The RTI has units of $m^{1/2}s^{1/2}$ or $ft^{1/2}s^{1/2}$.

A plunge test can be used to determine the RTI for a heat detector or a sprinkler. Knowing the RTI, the change in temperature of similar units can be calculated for any history of fire gases flowing past it. The form of the heat transfer equation is

$$dT_d/dt = u^{1/2}(T_g - T_d)/\text{RTI} \qquad (15)$$

This equation is used to calculate the temperature of a fixed-temperature heat detector or sprinkler exposed to fire gases. The equation can be used to determine the time at which the unit reaches its operating temperature.

The use of a lumped mass model may not hold for rate-of-rise heat detectors and rate-compensated heat detectors. The heat transferred to a fixed-temperature heat detector either heats a sensing element until it melts, or it heats two dissimilar metals of a snap disk. In each case, the element itself is exposed to the hot gases. This is not true for rate-of-rise heat detectors or rate-compensated heat detectors.

Most commercial rate-of-rise heat detectors operate when the expansion of air in a chamber exceeds the rate at which the air can escape through a small vent hole. For this type of detector, it is also necessary to model heat transfer from the detector body to the air in its chamber. Then the expansion of the air and its escape through a vent hole must be accounted for. The response time index determined in a plunge test may not be constant as fire gas velocities or temperatures vary. Hence, RTI is only an approximation of how the detector responds.

A rate-compensated detector consists of a metallic shell surrounding two bowed metal struts. There are electrical contacts on the struts. The struts and shell expand at different rates as the detector is heated. When heated fast, the outer shell expands and causes the bowed struts to straighten and close the contacts, signaling an alarm. This usually occurs at temperatures below the rated operating temperature. However, if the unit is heated more slowly, the difference between the expansion rates of the inner and outer parts is such that the contacts close at or near the unit's rated temperature.

Obviously, the rate-compensated type of heat detector cannot be treated as a lumped mass when calculating its response to a fire. As with rate-of-rise heat detectors, there are more heat transfer components to the response formula than a simple lumped mass. More research must be done to determine good working response models for rate-of-rise and rate-compensated heat detectors.

From the discussion above, it is evident that the response of fixed-temperature heat detectors can be modeled. The response of rate-anticipation heat detectors and rate-of-rise heat detectors can only be approximated using a lumped mass heat transfer model. It is necessary to know the temperature at which the detector is rated to operate. For rate-of-rise heat detectors, it is necessary to know the rate of change in the detector's temperature at which it will alarm. The response time index or τ_0 and u_0 for the detector are also needed.

In order to calculate the response of a heat detector, it is necessary to know the temperature and velocity of the gases flowing past it. Some fire plume models or ceiling-jet models may give functional relationships for temperature and velocity that can be substituted into the heat transfer equation and integrated. Other models may not be suitable for an analytical solution. In this case, the fire model should be used to produce data on time-*versus*-temperature and time-*versus*-gas velocity. This data can then be used to numerically solve the detector heat transfer equation.

Heskestad and Delichatsios[8] presented functional relationships for modeling the temperature and velocity of fires whose heat release rate grows according to the power-law relationship

$$\dot{Q} \ \text{(kW)} = \alpha \ \text{(kW/s}^2) \ t^p \ \text{(s)} \qquad (16)$$

where α is a constant for a particular fuel describing the growth of the fire, t is time, p is a positive exponent, and $\dot{Q}$ is the heat release rate. NFPA 72, Appendix B, uses a constant called "the critical time," t_c, in lieu of α to describe the fire intensity. The critical time is defined as the time at which a power-law fire would reach a heat release rate of 1055 kW (1000 Btu/s). In terms of t_c, the power-law equation becomes

$$\dot{Q} \ \text{(kW)} = (1055/t_c^p)t^p \qquad (17)$$

The relationships given by Heskestad and Delichatsios[8] for temperature and velocity of fire gases in a ceiling jet are

$$u_p^* = u/[A^{1/(3+p)}\alpha^{1/(3+p)}H^{(p-1)/(3+p)}]$$
$$= f(t_p^*, r/H) \qquad (18)$$

$$\Delta T_p^* = g(t_p^*, r/H)$$
$$= \Delta T/[A^{2/(3+p)}(T_a/g)\alpha^{2/(3+p)}H^{-(5-p)/(3+p)}] \qquad (19)$$

where

$$t_p^* = t/[A^{-1/(3+p)}\alpha^{-1/(3+p)}H^{4/(3+p)}] \qquad (20)$$

$$A = g/(C_p T_a \rho_0) \qquad (21)$$

All variables are described in this chapter's "Nomenclature."

Using these functional relationships, Heskestad and Delichatsios[8] presented the following correlations for fires whose release rates vary according to the power-law equation, with $p = 2$. These fires are referred to as t^2 fires. It has been shown[9,10] that the $p = 2$ power-law fire growth model can be used to model the heat release rate of a wide range of fuels.

$$t_{2f}^* = 0.95(1 + r/H)$$
$$\Delta T_2^* = 0 \text{ for } t_2^* < t_{2f}^* \qquad (22)$$

$$\Delta T_2^* = [(t_2^* - t_{2f}^*)/(0.188 + 0.313r/H)]4/3 \text{ for } t_2^* \geq t_{2f}^* \quad (23)$$

$$u_2^*/(\Delta T_2^*)^{1/2} = 0.59(r/H)^{-0.63} \qquad (24)$$

Beyler found that these correlations for temperature and velocity could be substituted into the heat transfer equation and integrated.[11] Beyler's analytical solution was published in *Fire Technology*[12] and is repeated here.

$$T_d(t) = T_d(0) = (\Delta T/\Delta T_2^*)\Delta T_2^*[1 - (1 - e^{-Y})/Y] \quad (25)$$

$$dT_d(t)/dt = [(4/3)(\Delta T/\Delta T_2^*)(\Delta T_2^*)^{1/4}(1 - e^{-Y})]/[(t/t_2^*)D] \quad (26)$$

where

$$Y = (3/4)(u/u_2^*)^{1/2}[u_2^*/(\Delta T_2^*)^{1/2}]^{1/2}(\Delta T_2^*/\text{RTI})(t/t_2^*)D \quad (27)$$

and

$$D = 0.188 + 0.313r/H \qquad (28)$$

In a design situation, the objective is to determine the spacing of detectors required to respond to a specific fire scenario. The detector must respond when the fire reaches a certain threshold heat release rate or in a specified amount of time. Time and heat release rate are interchanged using the fire growth model. The steps in solving this type of problem are outlined below and are discussed in more detail in the examples following this section.

1. Determine the environmental conditions of the area being considered.
   a. ambient temperature, T_a
   b. ceiling height or height above fuel, H
2. Estimate the fire growth characteristic α or t_c for the fuel expected to be burning.
3. Establish the goals of the system: t_r or $\dot{Q}_T$.
4. Select the detector type to be used. For fixed-temperature units this establishes the detector response temperature and its RTI or τ_0 and u_0.
5. Make a first estimate of the distance, r, from the fire to the detector.
6. Assume that the fire starts obeying the power-law model at time $t = 0$.
7. Set the initial temperature of the detector and its surroundings at ambient temperature.
8. Using Equation 22, calculate the nondimensional time, t_{2f}^*, at which the initial heat front reaches the detector.
9. Calculate the factor A defined in Equation 21.
10. Use the required response time along with Equation 20 and $p = 2$ to calculate the corresponding value of t_2^*.
11. If t_2^* is greater than t_{2f}^*, continue with step 12. If not, try a new detector position, r, and return to step 8.
12. Calculate the ratio u/u_2^*, using Equation 18.
13. Calculate the ratio $\Delta T/\Delta T_2^*$, using Equation 19.
14. Use Equation 23 to calculate ΔT_2^*.
15. Equation 24 is used to calculate the ratio $u_2^*/(\Delta T_2^*)^{1/2}$.
16. Use Equations 27 and 28 to calculate Y.
17. Equation 25 can now be used to calculate the resulting temperature of the detector.
18. If the temperature of the detector is below its operating temperature, this procedure must be repeated using a smaller r. If the temperature of the detector exceeds its operating temperature, a larger r can be used.
19. Repeat this procedure until the detector temperature is about equal to its operating temperature. The required spacing of detectors is then $S = 1.41r$.

This same procedure is used to estimate the response of rate-of-rise heat detectors. The difference is that in step 17, Equation 26 is used to calculate rate of change of the detector temperature. This is then compared to the rate at which the detector is designed to respond.

Discussion so far has centered around the solution of a design problem. The question asked was: How far apart must detectors of a specific design be spaced to respond within specific goals to a certain set of environmental conditions and a specific fire scenario?

The second type of problem that must be addressed is the analysis of an existing system or the analysis of a proposed design. Here the spacing of detectors or sprinklers is known.

The engineer must still estimate the burning characteristics of the fuel and the environmental conditions of the space being analyzed. The equations can then be solved in a reverse fashion to determine the rate of heat release or the time to detector response. The technique is as follows:

1. Determine the environmental conditions of the area being considered.
   a. ambient temperature, T_a
   b. ceiling height or height above the fuel, H
2. Estimate the fire growth characteristic α or t_c for the fuel expected to be burning.
3. Determine the spacing of the existing detectors or sprinklers. The protection radius is then: $r = S/(2^{1/2})$.
4. Determine the detector's rated response temperature and its RTI or τ_0 and u_0.
5. Make a first estimate of the response time of the detector or the fire size at detector response. They are related through the power-law fire growth equation: $\dot{Q} = \alpha t^2$.
6. Assume that the fire starts obeying the power-law model at time $t = 0$.
7. Set the initial temperature of the detector and its surroundings at ambient temperature.
8. Using Equation 22, calculate the nondimensional time, t_{2f}^*, at which the initial heat front reaches the detector.
9. Calculate the factor A defined in Equation 21.
10. Use the estimated response time along with Equation 20 and $p = 2$ to calculate the corresponding reduced time t_2^*.
11. If t_2^* is greater than t_{2f}^*, continue with step 12. If not, try a longer estimated response time, and return to step 8.
12. Calculate the ratio u/u_2^* using Equation 18.
13. Calculate the ratio $\Delta T/\Delta T_2^*$ using Equation 19.
14. Use Equation 23 to calculate ΔT_2^*.
15. Equation 24 is used to calculate the ratio $u_2^*/(\Delta T_2^*)^{1/2}$.
16. Use Equations 27 and 28 to calculate Y.
17. Equation 25 can now be used to calculate the resulting temperature of the detector.
18. If the temperature of the detector is below its operating temperature, this procedure is repeated using a larger estimated response time. If the temperature of the detector exceeds its operating temperature, a smaller response time is used.
19. Repeat this procedure until the detector temperature is about equal to its operating temperature.

As in the design problem, this technique can be used to estimate the response of existing systems of rate-of-rise heat detectors. The difference is that in step 4 the set point or rate of temperature rise at which the detector will respond must be determined from the manufacturer's data. In step 17, Equation 26 is used to determine the rate at which the temperature of the detector is changing.

The data in the 1993 edition of NFPA 72, Appendix B,[5] were generated using slightly different correlations. Appendix C has since been revised to include a new set of tables produced using the equations presented here. The 1993 edition contains tables that list the required detector spacing for a given set of conditions and goals. The latest revisions include the addition of tables that list the fire size at response for existing systems. A complete set of design and analysis tables, in English units, is currently available in "Use of Fire Plume Theory in the Design and Analysis of Fire Detector and Sprinkler Response."[10] That study also lists computer programs for solving design or analysis problems and for generating tables to be used in solving problems.

"Evaluating Thermal Fire Detection Systems" by Stroup, Evans, and Martin contains a complete set of analysis tables and a computer program and is available in English or metric versions.[13]

When the exact conditions of velocity and temperature of fire gases flowing past a detector is not known, errors are introduced in the design and analysis of fire detector response. Graphs in Heskestad and Delichatsios' report show the errors in calculated fire gas temperatures and velocities.[9] An exact treatment of these errors is beyond the scope of this chapter, though some discussion is warranted.

Plots of actual data and calculated data show that errors in ΔT_2^* can be as much as 50 percent, though generally there appears to be much better agreement.[9] The maximum errors occur at r/H values of about 0.37. All other plots of actual and calculated data, for various r/H, show much smaller errors. In terms of the actual change in temperature over ambient, the maximum errors are on the order of 5 to 10°C. The larger errors occur with faster fires and lower ceilings.

At $r/H = 0.37$, the errors are conservative when the equations are used in a design problem. That is, the equations predicted lower temperatures. Plots of data for other values of r/H indicate that the equations predict slightly higher temperatures.

Errors in fire gas velocities are related to the errors in temperatures. The equations show that the velocity of the fire gases is proportional to the square root of the change in temperature of the fire gases.[9] In terms of heat transfer to a detector, the detector's change in temperature is proportional to the change in gas temperature and the square root of the fire gas velocity. Hence, the expected errors bear the same relationships.

Based on the discussion above, errors in predicted temperatures and velocities of fire gases will be greatest for fast fires and low ceilings. Sample calculations simulating these conditions show errors in calculated detector spacings on the order of plus or minus one meter, or less.[10]

The $p = 2$ power-law fire model is not always the best model for a fire's heat release rate. This is especially true for steady-state fires, such as flammable liquid fires. A model giving velocity and temperature of a ceiling-jet for different heat release rate histories is needed for fires that do not follow the power-law model.

In 1972, R.L. Alpert of Factory Mutual Research Corporation presented a paper entitled "Calculation of Response Time of Ceiling-Mounted Fire Detectors" at the May meeting of the National Fire Protection Association (NFPA). That paper was later published in *Fire Technology*.[14] In the paper, Alpert presented a series of equations that can be used to calculate the temperature and velocity of fire gases in a ceiling jet for fires with a constant heat release rate.

Those equations can be used to model a growing fire by assuming the fire to be composed of a series of increasing steady heat release rates. The problem with this type of quasisteady-state modeling is that the temperature and velocity of the fire gases at a point away from the source are assumed to be related to the instantaneous heat release rate of the fire. This neglects the time required for transport of the fire gases from the source to the detector. Despite this shortcoming, the quasisteady-state model for fire gas temperatures and velocities can be used to estimate the magnitude of the difference in temperatures and velocities resulting from different heat release rate histories. More importantly, the effects on the design and analysis of detector response can be estimated.

Evans and Stroup[15] published a computer program called DETACT-QS, which uses Alpert's equations to calculate the response of heat detectors. That program requires the following input: ceiling height, H, ambient temperature, T_a, distance from fire axis to detector, r, detector activation temperature, T_S, and detector response time index (RTI). The user must also input a time-*versus*-heat-release-rate history for the fire. Their publication also includes a program called DETACT-T2, which analyzes the response of heat detectors using the equations presented in this chapter.

SELECTION OF DATA FOR DESIGN AND ANALYSIS

In order to calculate the required spacing of heat detectors or sprinklers to respond to a given fire, the following information is required:

1. System Goals: desired fire size (heat release rate) at response or time to detector response from the start of open flaming;
2. Fire growth constant α or t_c;
3. Ambient temperature; and
4. Height above the fuel or ceiling height.

In addition to the above, the heat capacity of air at constant pressure, C_p, the density of air, ρ_0, and the gravitational constant, g, are used in the calculations. It is also necessary to know the characteristics of the detector for which the spacing calculations are being made. Specifically, the response temperature and the RTI of the detector must be known.

Establishing system goals is not within the scope of this chapter. However, it should be pointed out that no matter what the goals are they must be expressed in terms of heat release rate or time to detector response. The system's goals may actually be to limit damages to some dollar value, provide adequate escape time, or limit the production of toxic gases. In order to calculate required detector spacing using this system, these goals would have to be translated. For instance, as the fire grows, at what time or heat release rate must the detector respond so that the fire department can be summoned and extinguish the fire before damage levels are exceeded or conditions become untenable due to toxic gases?

Table 4-1.2 is a list of furniture calorimeter tests done at the National Bureau of Standards.[13,14] The tests provide a database of heat release rate, particulate production, and radiation from a variety of common furnishings. Table 4-1.3 provides the corresponding α and t_c for the calorimeter tests.[10] The virtual time data in the table is the approximate time at which the heat release rate in the test began to follow the $p = 2$ power-law model ($\dot{Q} = \alpha t^2$). Prior to this time, the behavior of the fire cannot be predicted with this model. Figure 4-1.5 shows some test data along with a power-law curve superimposed.

The data in Table 4-1.3 can be used to select α or t_c for use in spacing calculations. However, in many cases the data in this table will not match the scenario being studied. If the heat release rate *versus* time history can be obtained or approximated for the expected fuel, the α or t_c can be calculated using curve fitting techniques.[10]

In most cases, since the exact fuel that will be involved in a fire cannot be known, the rigorous calculation of α is not warranted. Engineering judgement can be used to select an α or t_c that approximates the severity of the fire. The data in

TABLE 4-1.2 *Summary of NBS Calorimeter Tests*

Test No.	Description
Test 15	Metal wardrobe 41.4 kg (total)
Test 18	Chair F33 (trial loveseat) 39.2 kg
Test 19	Chair F21 28.15 kg (initial stage of fire growth)
Test 19	Chair F21 28.15 kg (later stage of fire growth)
Test 21	Metal wardrobe 40.8 kg (total) (average growth)
Test 21	Metal wardrobe 40.8 kg (total) (later growth)
Test 21	Metal wardrobe 40.8 kg (total) (initial growth)
Test 22	Chair F24 28.3 kg
Test 23	Chair F23 31.2 kg
Test 24	Chair F22 31.9 kg
Test 25	Chair F26 19.2 kg
Test 26	Chair F27 29.0 kg
Test 27	Chair F29 14.0 kg
Test 28	Chair F28 29.2 kg
Test 29	Chair F25 27.8 kg (later stage of fire growth)
Test 29	Chair F25 27.8 kg (initial stage of fire growth)
Test 30	Chair F30 25.2 kg
Test 31	Chair F31 (loveseat) 39.6 kg
Test 37	Chair F31 (loveseat) 40.40 kg
Test 38	Chair F32 (sofa) 51.5 kg
Test 39	½-in. Plywood wardrobe w/fabrics 68.8 kg
Test 40	½-in. Plywood wardrobe w/fabrics 68.32 kg
Test 41	⅛-in. Plywood wardrobe w/fabrics 36.0 kg
Test 42	⅛-in. Ply. wardrobe w/fire-ret. (int. fin. initial)
Test 42	⅛-in. Ply. wardrobe w/fire-ret. (int. fin. later)
Test 43	Repeat of ½-in. plywood wardrobe 67.62 kg
Test 44	⅛-in. Ply. wardrobe w/F-R., latex paint 37.26 kg
Test 45	Chair F21 28.34 kg (large hood)
Test 46	Chair F21 28.34 kg
Test 47	Chair adj. back metal frame, foam cush. 20.8 kg
Test 48	Easychair CO7 11.52 kg
Test 49	Easychair 15.68 kg (F-34)
Test 50	Chair metal frame minimum cushion 16.52 kg
Test 51	Chair molded fiberglass no cushion 5.28 kg
Test 52	Molded plastic patient chair 11.26 kg
Test 53	Chair metal frame w/padded seat and back 15.5 kg
Test 54	Loveseat metal frame w/foam cushions 27.26 kg
Test 55	Group chair metal frame w/foam cushion 6.08 kg
Test 56	Chair wood frame w/latex foam cushions 11.2 kg
Test 57	Loveseat wood frame w/foam cushions 54.60 kg
Test 61	Wardrobe ¾-in. particleboard 120.33 kg
Test 62	Bookcase plywood w/aluminum frame 30.39 kg
Test 64	Easychair molded flexible urethane frame 15.98 kg
Test 66	Easychair 23.02 kg
Test 67	Mattress and boxspring 62.36 kg (later fire growth)
Test 67	Mattress and boxspring 62.36 kg (initial fire growth)

Table 4-1.3 suggest a range of 50 to 500 s for t_c. Only a few rapidly developing fires had a t_c below 50 s. Three slow fires had values above 500 s for t_c.

When doing a design or analysis, try several different fire growth rates to determine the effect of their variance on the calculations. In some cases, the effect will be minimal. In other cases, this type of sensitivity analysis will show that a more thorough analysis of the possible fuels and fire scenarios is warranted.

The selection of an ambient temperature can have a measurable effect on the calculations. The calculations assume that the detector or sprinkler starts out at the same temperature as the ambient air when the fire starts. Hence, if

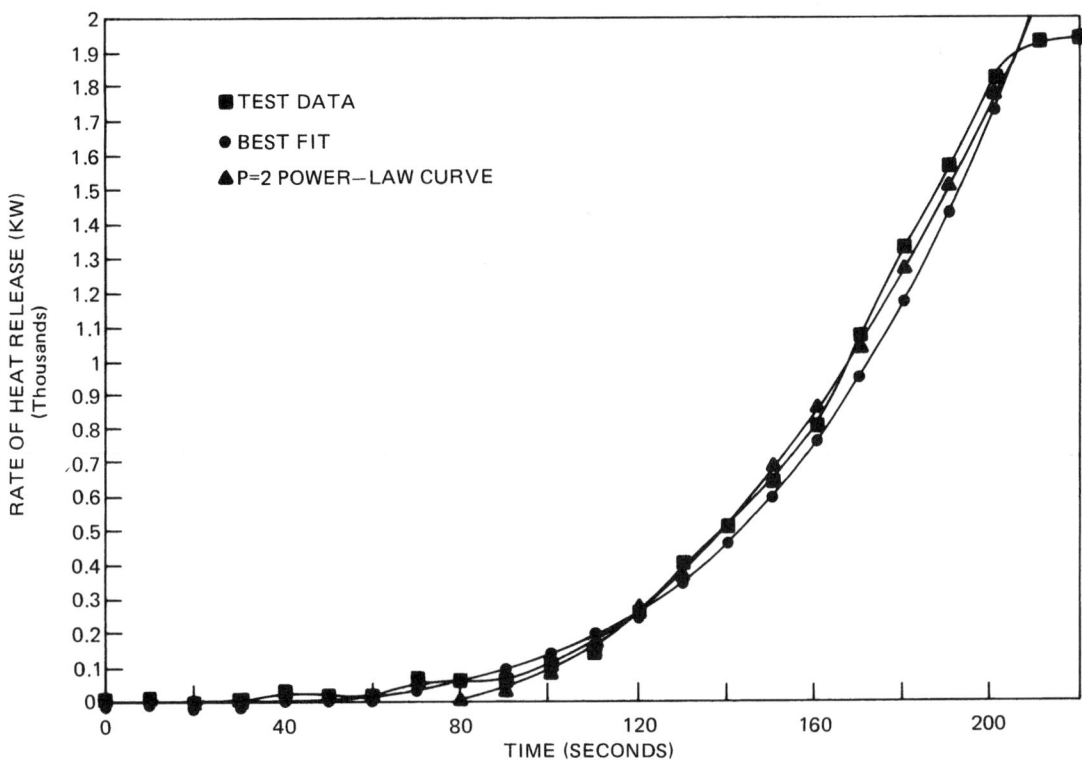

Fig. 4-1.5. Test 27 chair.

a temperature of 20°C is assumed for the spacing calculations and the actual temperature at the time of the fire is 10°C, the system's goals will not be met. For design calculations to be conservative, the lowest expected ambient temperature should be used.

The relationships presented by Heskestad and Delichatsios[8] are correlated to fire test data using the ceiling height above the fuel surface for H. If this height varies, the larger value of H will produce more conservative results in the calculations for detector spacing or response. The most conservative results are obtained when the floor-to-ceiling height is used, since this is the maximum vertical distance from fuel to detector.

The values for C_p, ρ_0, and g should be 1.040 kJ/(kg K), 1.1 kg/m^3, and 9.81 m/s^2, respectively. Slight variations in these constants have negligible effects on the calculations.

As previously mentioned, the design or analysis calculations are done for a particular detector or sprinkler. Therefore, it is necessary to know the unit's operating temperature. The response time index or τ_0 and u_0 are also needed. Operating temperature is obtained from manufacturer's data. The detector's sensitivity is best determined by conducting a plunge test.[6]

In the absence of plunge test data, a detector's UL-listed spacing can be used as a measure of detector sensitivity. Heskestad and Delichatsios analyzed UL test data and calculated time constants, τ_0, for various combinations of UL-listed spacing and detector operating temperature.[9] The Appendix Subcommittee of NFPA 72 expanded that table to include a larger selection of detectors.[5] That table is reproduced here as Table 4-1.4.

HEAT DETECTION DESIGN AND ANALYSIS EXAMPLES

Analysis and design problems will be used to show how fire protection engineers can use the techniques presented in this chapter. The examples show the sensitivity of these techniques to changes in variables and input parameters. A design problem and an analysis problem are first worked by hand to solve the equations presented earlier in the section on heat detection. The remaining examples were worked using a computer program written to solve the equations.[10]

EXAMPLE 1:

A fire detection system is being designed for an unsprinklered manufacturing building. The area being considered has a large, flat ceiling 3 m (9.8 ft) high. Ambient temperature is normally 20°C (68°F), but on weekends it is cut back to 10°C (50°F). It will be assumed that the fire scenario involves the ignition of a stack of wood pallets. The pallets are stacked 1.5 m (5 ft) high. Fire tests[5] show that this type of fire follows the $p = 2$ power-law equation with a t_c of approximately 150 s. Using Equation 16 the equivalent α can be calculated.

$$\dot{Q} \quad (\text{kW}) = \alpha \quad (\text{kW/s}^2) \quad t^p \quad (\text{s})$$
$$1055 \quad (\text{kW}) = \alpha \quad (\text{kW/s}^2) \quad 150^2 \quad (\text{s})$$
$$\alpha = 1055/150^2$$
$$\alpha = 0.047 \quad (\text{kW/s}^2)$$

TABLE 4-1.3 *Summary of Data Used to Produce Power–Law, p = 2 Curves to Fit NBS Calorimeter Tests*

Test No.	Critical Time Seconds (t_c)	α kW/s^2	Virtual Time Seconds
Test 15	50	0.4220	10
Test 18	400	0.0066	140
Test 19	175	0.0344	110
Test 19	50	0.4220	190
Test 21	250	0.0169	10
Test 21	120	0.0733	60
Test 21	100	0.1055	30
Test 22	350	0.0086	400
Test 23	400	0.0066	100
Test 24	2000	0.0003	150
Test 25	200	0.0264	90
Test 26	200	0.0264	360
Test 27	100	0.1055	70
Test 28	425	0.0058	90
Test 29	60	0.2931	175
Test 29	100	0.1055	100
Test 30	60	0.2931	70
Test 31	60	0.2931	145
Test 37	80	0.1648	100
Test 38	100	0.1055	50
Test 39	35	0.8612	20
Test 40	35	0.8612	40
Test 41	40	0.6594	40
Test 42	70	0.2153	50
Test 42	30	1.1722	100
Test 43	30	1.1722	50
Test 44	90	0.1302	30
Test 45	100	0.1055	120
Test 46	45	0.5210	130
Test 47	170	0.0365	30
Test 48	175	0.0344	90
Test 49	200	0.0264	50
Test 50	200	0.0264	120
Test 51	120	0.0733	20
Test 52	275	0.0140	2090
Test 53	350	0.0086	50
Test 54	500	0.0042	210
Test 55	Never exceeded 50 kW heat release rate		
Test 56	500	0.0042	50
Test 57	350	0.0086	500
Test 61	150	0.0469	0
Test 62	65	0.2497	40
Test 64	1000	0.0011	750
Test 66	75	0.1876	3700
Test 67	350	0.0086	400
Test 67	1100	0.0009	90

The plant fire brigade feels that they can control and extinguish the fire if the fire can be detected in under 3 min. Using Equation 17 with $p = 2$, the heat release rate of the fire 3 min after open flaming can be calculated.

$$\dot{Q} \quad \text{(kW)} = (1055/t_c^p)t^p$$
$$\dot{Q} \quad \text{(kW)} = (1055/150^2)180^2$$
$$\dot{Q} \quad \text{(kW)} = 1519 \text{ kW}$$

TABLE 4-1.4 *Time Constants for Any Listed Detector (DET TC) (s)**

Listed Spacing (ft)	UL (°F)						FMRC All Temp.
	128°	135°	145°	160°	170°	196°	
10	400	330	262	195	160	97	195
15	250	190	156	110	89	45	110
20	165	135	105	70	52	17	70
25	124	100	78	48	32		48
30	95	80	61	36	22		36
40	71	57	41	18			
50	59	44	30				
70	36	24	9				

NOTE: These time constants are based on an analysis of the Underwriters Laboratories Inc. and Factory Mutual Research Corporation listing test procedures. Plunge test results performed on the detector to be used will give a more accurate time constant.
*At a reference velocity of 5 ft/s.
(Reproduced from NFPA 72-1993, Appendix B.[5])

The goal of the system is then to respond in 180 s or when the fire reaches a heat release rate of approximately 1500 kW.

Fixed-temperature heat detectors will be used. The detectors have a 57°C (135°F) operating temperature and a UL-listed spacing of 30 ft (9.1 m). From Table 4-1.4 the time constant is found to be 80 s. This time constant is referenced to a gas velocity of 1.5 m/s (5 ft/s) and can be used with Equation 9 to calculate the detector's RTI.

$$\text{RTI} = \tau_0 u_0^{1/2}$$
$$\text{RTI} = 80(1.5)^{1/2}$$
$$\text{RTI} = 98 \text{ m}^{1/2}\text{s}^{1/2}$$

As described previously in step 5 for design of a proposed system it is necessary to make a first guess at the required detector spacing. In this case, try using $r = 5$ m (16.5 ft). Use Equation 22 to calculate the nondimensional time, t_{2f}^*, at which the initial heat front reaches the detector. Use the distance from the top of the fuel package to the ceiling for H.

$$t_{2f}^* = 0.954(1 + r/H)$$
$$t_{2f}^* = 0.954[1 + 5/(3 - 1.5)]$$
$$t_{2f}^* = 4.134$$

Next, Equation 21 is used to calculate A. Note that in this equation the ambient temperature, T_a, must be expressed as an absolute temperature. In this case add 273 to °C to get K (Kelvin). In English units add 460 to °F to get R (Rankine).

$$A = g/(C_p T_a \rho_0)$$
$$A = 9.8/[1.040(10 + 273)1.1]$$
$$A = 0.030$$

t_p^* is now calculated using $p = 2$ and a response time of 180 s.

$$t_p^* = t/[A^{-1/(3+p)}\alpha^{-1/(3+p)}H^{4/(3+p)}]$$
$$t_2^* = 180/(0.030^{-1/5}0.047^{-1/5}1.5^{4/5})$$
$$t_2^* = 35.058$$

t_p^* is greater than the value of t_{2f}^*, calculated above. This indicates that the heat front from the fire has reached the detector before the fire size has exceeded the established goals. Had the opposite been true, a smaller trial value for r would be necessary.

The ratio u/u_p^* is now calculated using Equation 18.

$$u_p^* = u/[A^{1/(3+p)}\alpha^{1/(3+p)}H^{(p-1)/(3+p)}]$$
$$u/u_p^* = [A^{1/(3+p)}\alpha^{1/(3+p)}H^{(p-1)/(3+p)}]$$
$$u/u_2^* = 0.030^{1/5}0.047^{1/5}1.5^{1/5}$$
$$u/u_2^* = 0.292$$

In the next step, Equation 19 is used to calculate the ratio $\Delta T/\Delta T_p^*$. As mentioned previously, T_a must be expressed as an absolute temperature.

$$\Delta T_p^* = \Delta T/[A^{2/(3+p)}(T_a/g)\alpha^{2/(3+p)}H^{-(5-p)/(3+p)}]$$
$$\Delta T/\Delta T_2^* = 0.030^{2/5}(283/9.8)0.047^{2/5}1.5^{-3/5}$$
$$\Delta T/\Delta T_2^* = 1.643$$

Next, calculate ΔT_2^* using Equation 23.

$$\Delta T_2^* = [(t_2^* - t_{2f}^*)/(0.188 + 0.313r/H)]^{4/3}$$
$$\Delta T_2^* = [(35.058 - 4.13)/(0.188 + 0.313(5/1.5))]^{4/3}$$
$$\Delta T_2^* = 73.546$$

The ratio $u_2^*/(\Delta T_2^*)^{1/2}$ and D can now be calculated using Equations 24 and 28.

$$u_2^*/(\Delta T_2^*)^{1/2} = 0.59(r/H)^{-0.63}$$
$$u_2^*/(\Delta T_2^*)^{1/2} = 0.59(5/1.5)^{-0.63}$$
$$u_2^*/(\Delta T_2^*)^{1/2} = 0.276$$
$$D = 0.188 + 0.313r/H$$
$$D = 0.188 + 0.313(5/1.5)$$
$$D = 1.231$$

Next, calculate Y, using Equation 27.

$$Y = (3/4)(u/u_2^*)^{1/2}[u_2^*/(\Delta T_2^*)^{1/2}]^{1/2}(\Delta T_2^*/\text{RTI})(t/t_2^*)D$$
$$Y = (3/4)(0.292)^{1/2}(0.276)^{1/2}(73.456/98)(180/35.058)1.231$$
$$Y = 1.011$$

The resulting temperature of the detector can now be calculated using Equation 25. Assume that the temperature of the detector at the start of the fire, $T_d(0)$, is the same as ambient temperature, T_a.

$$T_d(t) - T_d(0) = (\Delta T/\Delta T_2^*)\Delta T_2^*[1 - (1 - e^{-Y})/Y]$$
$$T_d(t) = (\Delta T/\Delta T_2^*)\Delta T_2^*[1 - (1 - e^{-Y})/Y] + T_d(0)$$
$$T_d(t) = 1.643(73.546)[1 - (1 - e^{-1.011})/1.011] + 10$$
$$T_d(t) = 55°C$$

After 3 min, at a distance of 5 m (16 ft) from the axis of this particular fire, the detector temperature is 55°C. The detector actuation temperature is 57°C. This indicates that for the detector to respond within the established goal of 3 min it must be located slightly closer to the fire. If the calculated temperature were higher than the actuation temperature, a larger r could be tried. The calculation should be repeated until the calculated detector temperature is approximately equal to the actuation temperature.

In this example, the difference of 2°C can be considered negligible, since the estimate of ambient temperature and other input parameters could produce errors of this magnitude. Use the final value of r along with Equation 1 to calculate the maximum installed detector spacing that will result in response of the detector within the established goals.

$$S = 2^{1/2}r$$
$$S = 2^{1/2}(5)$$
$$S = 7 \text{ m}$$

EXAMPLE 2:

This example will show how an existing heat detection system or a proposed design can be analyzed to determine its response time or fire size at response. The scenario used in Example 1 will be repeated, except that the manufacturing building has an existing system of heat detectors, which are spaced evenly on the ceiling at 9.2 m (30 ft) intervals. The detector characteristics are the same as above. The actuation temperature is 57°C (135°F) and the RTI is 98 $m^{1/2}s^{1/2}$. The ceiling height is 3 m (10 ft) and the height of the pallets is 1.5 m. Ambient temperature is 10°C (50°F). α is 0.047 kW/s^2 (t_c = 150 s).

The maximum radial distance from the fire axis to a detector is calculated using Equation 1.

$$S = 2^{1/2}r \quad \text{or} \quad r = S/2^{1/2}$$
$$r = 9.2/1.414$$
$$r = 6.5 \text{ m}$$

The next step in the analysis is to estimate the response time of the detector or the fire size at response. In the design above, the fire grew to 1519 kW in 180 s when the detector at a distance of 5 m (16.4 ft) responded. The radial distance in this example is larger and should result in a slower response and larger fire size at response. A first guess at response time might be 4 min. The remaining calculations for the resulting detector temperature are similar to those in Example 1. The corresponding fire size is found using the p = 2 power-law fire growth equation

$$\dot{Q} = \alpha t^2 \quad \text{or} \quad \dot{Q} = (1055/t_c^2)t^2 \quad \text{kW}$$
$$\dot{Q} = (1055/150^2)(4 \times 60)^2$$
$$\dot{Q} = 2700 \text{ kW}$$

Use Equation 22 to calculate the nondimensional time, t_{2f}^*, at which the initial heat front reaches the detector. As in Example 1, the distance from the top of the fuel package to the ceiling will be used for H.

$$t_{2f}^* = 0.954(1 + r/H)$$
$$t_{2f}^* = 0.954[1 + 6.5/(3 - 1.5)]$$
$$t_{2f}^* = 5.088$$

Next, Equation 21 is used to calculate A. Note that in this equation the ambient temperature must be expressed as an absolute temperature.

$$A = g/(C_p T_a \rho_0)$$
$$A = 9.8/[1.040(10 + 273)1.1]$$
$$A = 0.030$$

t_p^* is now calculated using $p = 2$ and a response time of 240 s.

$$t_p^* = t/[A^{-1/(3+p)}\alpha^{-1/(3+p)}H^{4/(3+p)}]$$
$$t_2^* = 240/(0.030^{-1/5}0.047^{-1/5}1.5^{4/5})$$
$$t_2^* = 46.740$$

t_p^* is greater than the value of t_{2f}^* calculated above. This indicates that the heat front from the fire has reached the detector before the fire size has exceeded the established goals. Had the opposite been true, a smaller trial value for r would be necessary.

The ratio u/u_p^* is now calculated using Equation 18.

$$u_p^* = u/[A^{1/(3+p)}\alpha^{1/(3+p)}H^{(p-1)/(3+p)}]$$
$$u/u_p^* = [A^{1/(3+p)}\alpha^{1/(3+p)}H^{(p-1)/(3+p)}]$$
$$u/u_2^* = 0.030^{1/5}0.047^{1/5}1.5^{1/5}$$
$$u/u_2^* = 0.292$$

In the next step, Equation 19 is used to calculate the ratio $\Delta T/\Delta T_p^*$. As mentioned above, T_a must be expressed as an absolute temperature.

$$\Delta T_p^* = \Delta T/[A^{2/(3+p)}(T_a/g)\alpha^{2/(3+p)}H^{-(5-p)/(3+p)}]$$
$$\Delta T/\Delta T_2^* = 0.030^{2/5}(283/9.8)0.047^{2/5}1.5^{-3.5}$$
$$\Delta T/\Delta T_2^* = 1.643$$

Next, calculate ΔT_2^*, using Equation 23.

$$\Delta T_2^* = [(t_2^* - t_{2f}^*)/(0.188 + 0.313r/H)]^{4/3}$$
$$\Delta T_2^* = \{(46.740 - 5.088)/[0.188 + 0.313(6.5/1.5)]\}^{4/3}$$
$$\Delta T_2^* = 80.884$$

The ratio $u_2^*/(\Delta T_2^*)^{1/2}$ and D can now be calculated using Equations 24 and 28.

$$u_2^*/(\Delta T_2^*)^{1/2} = 0.59(r/H)^{-0.63}$$
$$u_2^*/(\Delta T_2^*)^{1/2} = 0.59(6.5/1.5)^{-0.63}$$
$$u_2^*/(\Delta T_2^*)^{1/2} = 0.234$$

$$D = 0.188 + 0.313r/H$$
$$D = 0.188 + 0.313(6.5/1.5)$$
$$D = 1.544$$

Next, calculate Y using Equation 27.

$$Y = (3/4)(u/u_2^*)^{1/2}[u_2^*/(\Delta T_2^*)^{1/2}]^{1/2}(\Delta T_2^*/RTI)(t/t_2^*)D$$
$$Y = (3/4)(0.292)^{1/2}(0.234)^{1/2}(80.884/98)(240/46.740)1.544$$
$$Y = 1.284$$

The resulting temperature of the detector can now be calculated using Equation 25. Assume that the temperature of the detector at the start of the fire, $T_d(0)$, is the same as ambient temperature T_a.

$$T_d(t) - T_d(0) = (\Delta T/\Delta T_2^*)\Delta T_2^*[1 - (1 - e^{-Y})/Y]$$
$$T_d(t) = (\Delta T/\Delta T_2^*)\Delta T_2^*[1 - (1 - e^{-Y})/Y] + T_d(0)$$
$$T_d(t) = 1.643(80.884)[1 - (1 - e^{-1.284})/1.284] + 10$$
$$T_d(t) = 68°C$$

At a distance of 6.5 m (21 ft) from the axis of this fire, the detector temperature is 68°C (154°F) after 4 min exposure.

The detector actuation temperature is 57°C (135°F). This indicates that the detector response time is less than the estimated 4 min. If the calculated temperature were lower than the actuation temperature, a larger t would be tried. The calculations should be repeated until the calculated detector temperature is approximately equal to the actuation temperature. In this case, the response time converges at 220 s. This corresponds to a fire size at response of 2300 kW. It is at this time and heat release rate that the detector temperature reaches its actuation temperature.

This assumes that the fire continues to follow the power-law relationship through the burning period. If there is not enough fuel available, it is possible for the heat release rate curve to flatten out before reaching 2300 kW. These calculations do not predict when this will happen. These calculations also do not predict how the detector temperature changes after the fire stops following the power-law relationship. It may be that sufficient heat continues to be released and the detector eventually responds. It is also possible for the fire gases to cool sufficiently to preclude detector actuation unless additional fuel becomes involved in the fire.

Comparing Example 1 with Example 2 shows how detector spacing affects response time. A difference in spacing of 2.2 m (7 ft) results in a difference of 40 s in the detector response time. Because the fire is accelerating according to the $p = 2$ power-law relationship, the resulting difference in fire size at response is 800 kW.

EXAMPLE 3:

A warehouse is used to store sofas and other furniture. The sofas are similar to one tested by the National Bureau of Standards in their furniture calorimeter.[16] Burning characteristics are assumed to be similar to the sofa used in Test 38:[10,16] $\alpha = 0.1055$ kW/s^2 (0.1000 Btu/s^2), $t_c = 100$ s; peak heat release rate = 3000 kW. The sofas are stored one or two high.

The building has a flat roof and ceiling. The distance from the floor to the ceiling is 4.6 m (15 ft). When the sofas are stacked two high, the distance from the top of the fuel package to the ceiling is 2.4 m (8 ft). Ambient temperature in the warehouse is kept above 10°C (50°F). (See Figure 4-1.6.)

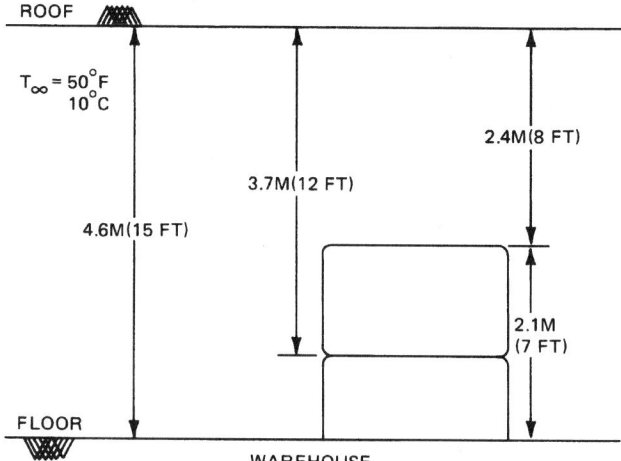

Fig. 4-1.6.   Example 3—Warehouse.

Based on maximum allowable property loss goals established by the owner, it is desirable to detect a fire and notify the fire department prior to a second fuel package becoming involved. The original NBS report[16] contains data on radiation measured during Test 38. This information can be used along with other techniques presented in this handbook to determine when a second item might ignite. For instance, it might be determined that furniture across a 6-ft aisle might ignite when the fire reaches a heat release rate of 3000 kW. The objective would then be to detect the fire soon enough so that the fire can be extinguished or controlled before the fire reaches 3000 kW. In this example it is assumed that the fire must be detected when it reaches a heat release rate of about 2000 kW.

The fire detection system will consist of fixed-temperature heat detectors connected to a control panel that is, in turn, connected to the local fire department. The detector to be used has a fixed temperature rating of 57°C (135 °F) and an RTI of 42 m$^{1/2}$s$^{1/2}$ (77 ft$^{1/2}$s$^{1/2}$).

The problem is to determine the spacing of detectors required to detect this fire. When the computer program runs, the user is prompted for all of the above information. In this example, the data is fixed except for the distance from the ceiling to the flame origin. If the distance between the top of the fuel and the ceiling (2.4 m) is used, the program calculates that the detectors must be spaced 8.6 m apart to respond when the fire reaches a heat output of 2000 kW.

For a worst-case analysis, the distance from the floor to the ceiling (4.6 m) is used. This results in a required detector spacing of 6.6 m (22 ft).

A more realistic worst-case scenario would be when the sofas are not stacked two high. With one sofa on the floor, the distance from the fuel to the ceiling would be about 3.7 m (12 ft). The required detector spacing would then be 7.5 m (25 ft).

These results are summarized in Table 4-1.5. This table clearly shows the relationship between ceiling height and detector response. The greater the distance from the fire to the ceiling, the closer the detectors must be spaced to respond within the goals of the system. Designs based on the floor-to-ceiling distance are conservative and representative of a worst-case condition. More realistic designs are based on the most probable or the greatest expected vertical clearance between fuel and detector.

EXAMPLE 4:

This example will show how to select a detector type to economically meet the system's goals. The fire scenario and goals used in Example 3 will be used with H = 2.4 m (8 ft).

In Example 3, it was found that heat detectors with a fixed temperature rating of 57°C (135°F) and an RTI of 42 m$^{1/2}$s$^{1/2}$ must be spaced 8.6 m (28 ft) apart to meet the system's goals. Here, the spacing of rate-of-rise heat detectors will be estimated.

TABLE 4-1.5 *Example 3. Ceiling Height or Height Above Fuel Versus Detector Spacing*

Height		Required Spacing	
(m)	(ft)	(m)	(ft)
2.4	(8)	8.6	(28)
3.7	(12)	7.5	(25)
4.6	(15)	6.6	(22)

The detector to be used is rated to respond when its temperature increases at a rate of 11°C/min (20°F/min) or more. The detector's RTI will be assumed to be the same as the detector in Example 3. Solving the equations, it is found that the rate-of-rise heat detectors must be spaced no more than 18.3 m apart.

If the total area of the warehouse is 5000 m^2, approximately 68 fixed-temperature heat detectors would be required to meet the established goals. The same goals can be met with only 16 rate-of-rise heat detectors. Additional detectors might be required because of obstructing beams or walls. By trying different detector types or detectors with higher sensitivities, project goals might be met with a fewer number of detectors.

The scenario in this example shows that, to detect the same fire, a much greater number of fixed-temperature heat detectors than rate-of-rise heat detectors is required. This is not always the case. Many fires will develop slowly and cause high ceiling temperatures without ever exceeding the rate of temperature rise necessary to actuate a rate-of-rise heat detector. As a backup, most commercially available rate-of-rise heat detectors have a fixed-temperature element also. The rate-of-rise element and the fixed-temperature element should be considered separately when designing or analyzing a system.

EXAMPLE 5:

In this example, the effects of varying fire growth rate will be examined. The scenario used in the previous two examples will be used again.

In Examples 3 and 4, the rate of fire growth followed the power-law equation with an α of 0.1055 kW/s^2 (0.1000 Btu/s^3) or t_c = 100 s. If the fire were to grow at a faster rate, a smaller spacing would be required to meet the system's goals. If the fire were to grow at a slower rate, a larger detector spacing is allowed. For instance, if t_c = 50 s (α = 0.4220 kW/s^2), the required spacing would be 6.0 m. If t_c = 300 s, the spacing is increased to 12.7 m. Calculations were done for several values of t_c. The results for this scenario are summarized in Figure 4-1.7.

EXAMPLE 6:

This example shows how existing systems or proposed designs are analyzed. Again, the scenario used in Examples 3, 4, and 5 will be assumed. The height of the ceiling above the fire is 2.4 m (8 ft). The detectors are 57°C (135°F) fixed-temperature heat detectors. The system to be analyzed has detectors spaced 2.8 m (9.2 ft) on center. The detectors have an RTI of 42 m$^{1/2}$s$^{1/2}$ (77 ft$^{1/2}$s$^{1/2}$). Ambient temperature is 10°C (50°F).

The detection system being analyzed is designed to respond to a 2000 kW fire that is growing according to $\dot{Q} = \alpha t^2$, with α = 0.1055 kW/s^2. What would happen if there were an occupancy change and the new fuel loading had different burning characteristics than the fuel for which the system was designed? If the fuel burns faster or slower, what will be the fire size when the detector responds?

Once again, a computer program[10] was used to solve the equations. In this case, the equations were solved for the fire size at detector response. The results are summarized in Table 4-1.6.

Table 4-1.6 shows that at faster fire growth rates the detector responds sooner, but the fire size at response is larger. At slower growth rates, the detector responds when the fire is much smaller. With the faster rates, ceiling temperatures

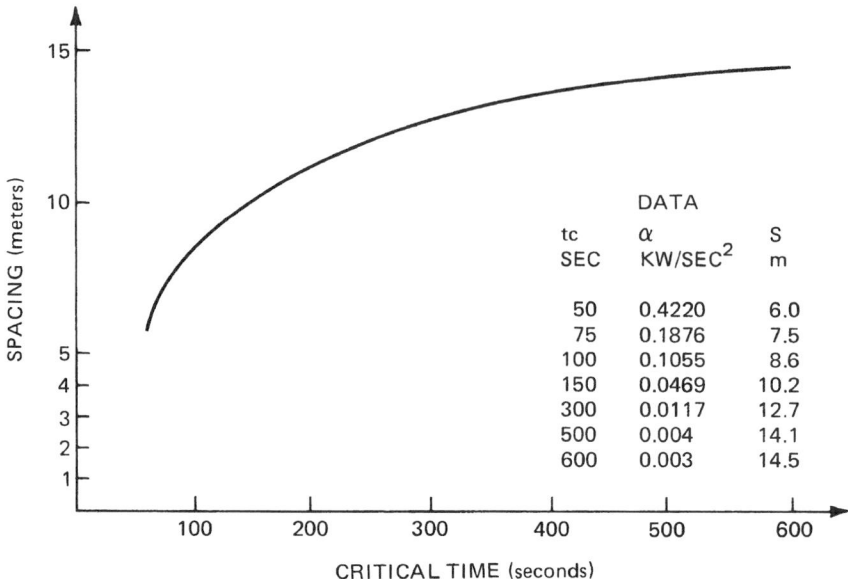

Fig. 4-1.7.   Example 5—Effects of varying fire growth.

quickly exceed the response temperature of the detectors. However, the inherent thermal lag of the detector delays response until the detector absorbs enough heat to reach its operating temperature.

When the fire grows at a slow rate, gas temperature at time of operation is closer to the detector's rated operating temperature. The thermal lag of the detector is not as significant as the fire's ability to increase the ceiling-jet gas temperatures.

Examples 5 and 6 show that changes in fuels or the burning characteristics of a fuel will alter the response of the system. This type of analysis illustrates the importance of designing a system for its expected occupancy.

As the use of a building changes, so will the characteristics of the fuels in the building. Analyses such as this can be conducted to determine whether the system requires any modifications to continue meeting its goals.

EXAMPLE 7:

A sprinkler system is being installed in a large exhibition hall. The building has a flat roof deck supported by open space frame trusses. The distance from the underside of the roof deck to the floor is 12 m (39.3 ft). Ambient temperatures do not usually fall below 5°C (41°F).

Three different designs for the sprinkler system have been proposed. All three are designed to provide the same water density over a specified area. Each proposal uses a sprinkler with a temperature rating of 74°C (165°F) and an

RTI of 110 m$^{1/2}$s$^{1/2}$ (200 ft$^{1/2}$s$^{1/2}$). The only difference among the three systems is the spacing of the sprinklers and the branch lines that feed them. The first proposal uses a square array with a spacing of 3 m (10 ft). The second and third proposals are based on square array spacings of 3.7 m (12 ft) and 4.6 m (15 ft), respectively.

What effect will the three different spacings have on the size of the fire when the system responds? Assume two different fire scenarios. In the first, the fire grows at a moderate rate with t_c = 200 s. The second fire scenario has a slower fire growth rate with t_c = 500 s. Results of the calculations are shown in Table 4-1.7.

The calculations show an increase of about 25 percent in the fire size at response when the spacing is increased 50 percent from 3 to 4.6 m (10 to 15 ft). The increased spacing may result in a lower system cost. However, closer spacings mean that the sprinkler system will probably respond sooner. The fire protection engineer can use this type of analysis to assist in choosing a system that best meets the project's overall goals.

EXAMPLE 8:

Example 8 illustrates the effect of temperature difference on the response time of fixed-temperature detectors and sprinkler actuation. The difference between a unit's operating temperature and the ambient temperature affects response time.

TABLE 4-1.6   *Example 6. Fire Growth Rate Versus Fire Size at Response*

t_c (s)	α (kW/s^2)	$\dot{Q}_T$ (kW)	t_r (s)
50	0.422	3193	84
100	0.1055	2000	138
150	0.0469	1564	180
300	0.0117	1106	306
600	0.0029	867	546

TABLE 4-1.7   *Example 7. Effect of Spacing on Fire Size at Response*

S		t_c = 200 seconds		t_c = 500 seconds	
(m)	(ft)	$\dot{Q}_T$ (kW)	t_R (min)	$\dot{Q}_T$ (kW)	t_R (min)
3	10	5131	7.4	4340	16.9
3.7	12	5690	7.7	4810	17.8
4.6	15	6439	8.2	5445	18.9

TABLE 4-1.8 *Example 8. Effect of Sprinkler Temperature Rising on Fire Size at Response*

T_s (°C)	T_a (°C)	ΔT_d (°C)	t_c = 200 seconds		t_c = 500 seconds	
			$\dot{Q}_T$ (kW)	t_r (min)	$\dot{Q}_T$ (kW)	t_r (min)
57	5	52	3655	6.2	2970	14.0
74	5	69	5131	7.4	4340	16.9
93	5	88	6961	8.6	6058	20.0
100	5	95	7680	9.0	6737	21.1

When selecting fixed-temperature heat detectors and automatic sprinklers, it is desirable to select a temperature rating that is as close as possible to the expected maximum ambient temperature. This reduces the response time of the detector in a fire condition. The closer the response temperature is to ambient temperature, the less heat the detector must absorb to respond.

If the operating temperature of the detector is too close to ambient temperatures, false detector actuations can occur. NFPA 72[5] recommends a detector rating of 25°F (14°C) above the expected maximum ambient temperature.

The fire scenario used in Example 7 will be used to quantify the effects of temperature difference on response time and fire size at response. The question asked is: What effect would the use of sprinkler heads with different temperature ratings have on the response time and the size of the fire at response?

Calculations are done for a sprinkler head spacing of 3 m (10 ft). Sprinkler heads having temperature ratings of 57, 74, 93, and 100°C (135, 165, 200, and 212°F) are analyzed. The results of the computer calculations are shown in Table 4-1.8.

Table 4-1.8 shows that there is a large difference in fire size at response when high-temperature heads are used in lieu of the lower temperature heads. If this were a detection system, the lower temperature units would be the obvious choice.

With a sprinkler system, other factors, such as the number of heads opening, must be considered. While the lower temperature rating means quicker response, it also means that more heads may open. However, quicker response might mean that the sprinkler system can control or extinguish the fire before additional heads open. These factors should also be considered by the design engineer.

Examples 7 and 8 show how the design and analysis techniques presented in this chapter should be incorporated in all phases of a building's fire protection design. These techniques can be used to show that designs that might appear to be equal, really are not. This provides the fire protection engineer with a way to measure the effectiveness of detection systems and provides a quantitative scale that can be used to compare various system designs or proposed modifications.

SMOKE DETECTION

In order to determine whether or not a smoke detector will respond to a given $\dot{Q}_{cr}$, a large number of factors must be evaluated. These include: smoke aerosol characteristics, aerosol transport, detector aerodynamics, and sensor response.

Smoke aerosol characteristics at the point of generation are a function of the fuel composition, the combustion state (smoldering or flaming), and the degree of vitiation of the combustion air. The characteristics considered include particle size and distribution, composition, color, and refractive

index. Given the dynamic nature of fire growth and spread and fuels involved, ventilation conditions will change over time, thus affecting the smoke produced.

Transport considerations include: (1) changes to the aerosol characteristics that occur with time and distance from the source and (2) transport time. Changes in the aerosol largely relate to the particle size distribution, and result from the processes of sedimentation and agglomeration. Transport time is a function of the characteristics of the travel path from the source to the detector, and include ceiling height and configuration (sloped, beamed, etc.), intervening barriers such as doors, and buoyancy effects such as layering and thermal inversions.

Once smoke reaches the detector, other factors become important, namely the aerodynamic characteristics of the detector and the type of sensor. The aerodynamics of the detector relate to the ease with which smoke can pass through the detector housing and enter the sensor. In addition, the location of the entry portion of the housing relative to the velocity profile of the detector normal to the plane of the ceiling is also a factor. Finally, different sensing modes (e.g., ionization or photoelectric) will respond differently, depending on the characteristics of the transported aerosol. Within the family of photoelectric devices, there will be variations depending upon the wavelengths of light and the scattering angles employed.

Smoke Detection System Design Calculation Examples

Standard practice for the design of smoke detection systems is much the same as that for heat detection systems. Recommended spacing criteria are established based on detector response to a specific parameter, such as the optical density within an enclosure. A variety of smoke tests are used to verify that the detector responds between defined upper and lower activation thresholds and within required response times. This information translates into an installed or listed spacing criteria that is intended to ensure that the detector responds within defined parameters. In some cases, the listed spacing can be increased, or must be decreased, depending on factors such as compartment configuration and air flow velocity.[5]

In applications where estimating the response of a detector is not critical, the installed spacing criteria provides sufficient information for design of a basic smoke detection system. If the design requires detector response within a certain time frame, optical density, specified heat release rate, or temperature rise, additional analysis may be required. In this case, information concerning the expected fuel, fire growth, sensor, and compartment characteristics is required. The following examples show various performance-based approaches to evaluating smoke detector response.

Smoke production and characteristics: The fuel characteristics of primary concern for smoke detection are: (1) material and (2) mode of combustion. These two parameters are important for determining pertinent features of expected products of combustion, such as particle size, distribution, concentration, and refractive index. The importance of these features with regard to smoke detection are well documented[3,17,18] and are discussed by Mulholland in Section 2, Chapter 15. Assuming a well-mixed smoke-filled volume, data on smoke characteristics for given fuels can provide an evaluation of detector response.

EXAMPLE 9:

The design objective is to detect the smoke from a flaming 200-g (0.5-lb) polyurethane pillow in less than 2 min. The pillow is located in a 36 m^2 room with a ceiling height of 2.5 m (8 ft). Assume that the pillow is burning at a steady rate of 50 g/min. Can the design objective be met? What assumptions are required?

SOLUTION:

The total mass loss at 2 min is 100 g. Given this, the optical density in the room can be calculated from the relationship (see Section 2, Chapter 15)

$$D = D_m M/V_c \tag{29}$$

where D_m [mass optical density (m^2/g)] can be taken from Table 2-15.5 in Section 2, Chapter 15.

$$D = (0.22 \text{ m}^2/\text{g})(100 \text{ g})/(36 \text{ m}^2)(2.5 \text{ m}) = 0.244 \text{ m}^{-1}$$

Assuming the detector will respond at the UL upper sensitivity limit of 0.14 m^{-1} (black smoke),[19] it can be assumed that the detector will respond within 2 min. This is a simplified approach, however, and assumes that the smoke is confined to the room, is well mixed, can reach the ceiling level, and can enter the detector.

Stratification: In the context of this chapter, smoke dilution refers to a reduction in the quantity of smoke available for detection at the location of the detector. This dilution can occur either through natural convection (entrainment in the plume or the ceiling jet) or by effects of a heating or ventilation system. In many cases, forced ventilation systems with high exchange rates cause the most concern. In the early stages of fire development, when smoke production rate is small and the plume is weak, smoke can easily be drawn out of the room and away from area smoke detectors. In addition, high velocity air flows out of supply and into return vents creating defined patterns of air movement within a room. Such flows can either keep smoke away from detectors that are located outside of these paths, or, in some cases, inhibit smoke from entering a detector located directly in the air flow path.

Although there currently are no quantitative methods for estimating either smoke dilution or air flow effects on smoke detector siting, these factors must be considered qualitatively. It should be clear, however, that the air flow effects become larger as the required fire size at detection, $\dot{Q}_{cr}$, gets smaller. If the application warrants, it may be useful to obtain velocity profiles of the air movement within a room or to perform small-scale smoke tests under various conditions to aid in the smoke detector placement analysis.

The potential for smoke stratification is another concern in the detection of low-energy fires and fires in rooms or volumes with very high ceilings. Stratification occurs when the temperature within the plume equals that of the surrounding air, and there is insufficient thermal energy from the fire to force the smoke higher. Once this point of equilibrium is reached, the smoke layer will maintain its height above the fire, regardless of the ceiling height, until additional energy is provided.

Unlike the effects of air flow on smoke dilution, stratification effects can be calculated using the relationship[20]

$$\dot{q}_{conv} > 0.352 \, H^{5/2} \, T_s^{3/2} \tag{30}$$

where $\dot{q}_{conv}$ is the convective heat release rate in W, H is the distance from the top of the fuel package (base of the fire) to the ceiling level in m, and T_s is the difference in ambient gas temperature in °C between the fuel location and ceiling level. This same relationship can also be found in NFPA 92B, *Guide for a Smoke Management System in Malls, Atria, and Large Areas.*[21]

EXAMPLE 10:

The design objective is to detect the pyrolysis of overheated PVC cable insulation in a 4-m (13-ft) high, 100-m^2 (1076-ft^2) room. The room is air conditioned, with a temperature differential of 10°C (18°F) between the base of the switch equipment and the ceiling. The proposed design has smoke detectors mounted at the ceiling level. Assuming the critical fire size is 1000 W, will there be sufficient thermal energy to force the smoke to the ceiling level?

SOLUTION:

In this case, one can rearrange Equation 30 and solve for H

$$H < (\dot{q}_{conv}/0.352 T_s^{3/2})^{2/5}$$

where $\dot{Q}_{cr} = 1000$ W, and $T_s = 10$°C (18°F). This indicates that the highest level of smoke rise is estimated to be 3.8 m (12 ft). As a result, the design objective may not be achieved by the proposed design. This approach is also valid for evaluating the effects of stratification in a high-ceiling room where a larger fire might be expected. However, the effects of heating and air conditioning systems and warm or cold walls are not considered.

EXAMPLE 11:

The design objective is to detect the flaming combustion of a chair located in the lobby of an office building in order to initiate smoke management functions. The lobby is located at the lowest level of a 20-m (64-ft) high atrium. The atrium has offices on three sides and a glass facade to the outside on the other. The atrium is air conditioned, with a temperature differential of 20°C (36°F) between the lobby and the ceiling level. The proposed design is for smoke detectors to be mounted at the ceiling level. Is there sufficient thermal energy to force the smoke to the ceiling level?

SOLUTION:

First, a value for $\dot{Q}_{cr}$ must be selected for the burning chair. From an analysis of the chair and a review of published heat release data, it is determined that the chair most closely resembles the metal frame chair with padded seat used in Test 53 of the NIST furniture heat release rate tests.[5] This chair had a maximum heat release rate of 280 kW, which can be used as $\dot{q}_{conv}$ (or in this case $\dot{Q}_{cr}$, the critical fire) in Equation 30. Equation 30 can then be rearranged to solve for H

$$H < \dot{Q}_{cr}/0.352 T_s^{3/2})^{2/5}$$

where $\dot{Q}_{cr} = 280{,}000$ W, and $T_s = 20$°C (36°F). In this case, the highest point of smoke rise is estimated to be 24 m (77 ft). Thus, the smoke would be expected to reach the ceiling-mounted detector.

It should be noted that air flow concerns were not considered in Examples 9, 10, and 11. In some cases, a system supplying air at a low level and exhausting at an upper level may actually help transport the smoke to the upper levels of a room, where in other cases it may serve to inhibit smoke movement. It should also be noted that, simply because the smoke reaches the level of the detector, there is no guarantee that it can enter the sensor chamber.

Spot-type smoke detectors, whether commercial or residential, ionization, or light-scattering type, all require smoke to enter the detection chamber in order to be sensed. This is another factor that must be considered when attempting to estimate smoke detector response. Smoke entry into the detector can be affected in several ways, e.g., due to insect screens, chamber configuration, and proximity of the detector to the ceiling.

In an attempt to quantify this problem, Heskestad[22] introduced the concept of smoke detector lag to explain the difference between the optical density outside (D_{ur}) and inside (D_{uo}) of a detector at the time of activation. It was demonstrated that the difference between D_{ur} and D_{uo} could be explained by the use of a correction factor, D_{uc}, where

$$D_{uc} = Ld(D_u)/dt/V \qquad (31)$$

in which L, a characteristic length for a given detector design, relates to the ease of smoke travel into the sensing chamber, $d(D_u)/dt$ is the rate of increase of optical density outside of the detector, and V is the velocity of the smoke at the detector. Although studies of this relationship have provided useful information concerning smoke detector lag,[23,24] the difficulty in quantifying L for different detectors and relating it to siting requirements has limited usefulness. In its stead, the concept of critical velocity (u_c) shows new promise.[2,25]

Velocity analog: Critical velocity, in this context, refers to the lowest gas velocity required for smoke entry into the sensor chamber at a level to sound an alarm at a given threshold. Recent experimental work has shown this to be in the range of 0.15 m/s for the detectors tested in one study.[25] When velocities fell below this value, the smoke level outside the detector at the time a specified analog output level was reached rose dramatically compared to levels when the velocity was above the critical value. This figure can be useful for design and evaluation purposes, as it is close to the low-velocity value (0.16 m/s) at which a detector must respond in the UL smoke detector sensitivity chamber in order to be listed.[19] Thus, the location of a velocity of 0.16 m/s in the ceiling jet for a given fire and ceiling height can be considered as a first approximation design radius for detector siting purposes. It should be noted that the ceiling jet velocity correlations assume a horizontal smooth ceiling. A detailed discussion of ceiling jet flows by Evans is presented in Section 2, Chapter 4. The critical velocity approach can be illustrated with a simplified example.

EXAMPLE 12:

The new owners of a hotel have established a fire detection design objective that the smoke detection system in the grand ballroom must be able to detect a 50 kW fire. The ballroom is 50 m (160 ft) long by 30 m (96 ft) wide with a 7.1-m (23-ft) high smooth ceiling. The existing smoke detectors are installed at a listed spacing of 10 m on center

and have a critical velocity of 0.15 m/s. Assuming the fire starts at a point equally spaced between the existing smoke detectors, will the velocity of the ceiling jet from a 50 kW fire be sufficient to force smoke into the detection chamber? Assume there will be no ventilation system effects.

SOLUTION:

The stated design objective is to detect a 50 kW fire. Because it is not stated whether the fire is steady-state or growing, this solution will assume a steady-state fire of 50 kW. This allows the use of Alpert's[14] velocity correlations for a steady-state fire. Alpert provides two equations that can be used: one for $r/H \leq 0.15$, and the other for $r/H > 0.15$. This correlation is generally considered to be valid when r/H is between 0.15 and 2.1. Therefore, the ratio r/H must be determined first. In addition, the fire source should be at a distance of at least 1.8 times the ceiling height from the nearest enclosure wall.

The installed spacing is 10 m (32 ft) on center. Using the relationship $S = 2^{1/2}r$, the radial distance is found to be approximately 7.1 m (23 ft). Given that H is also 7.1 m (23 ft), the ratio r/H is found to be 1.0. This value is greater than 0.15, thus the following equation can be used

$$U = 0.195 \, \dot{Q}^{1/3} \, H^{1/2}/r^{5/6} \qquad (32)$$

By entering the values of $\dot{Q} = 50$ kW, $H = 7.1$ m (23 ft), and $r = 7.1$ m (23 ft), a velocity of 0.37 m/s is calculated. This indicates that, for a steady-state 50 kW fire, there will be sufficient velocity to force smoke into the detectors at their existing locations.

However, if the 50 kW fire as stated is the design fire, $\dot{Q}_{do}$, and it was determined that the critical fire, $\dot{Q}_{cr}$, was only 5 kW, the resulting velocity using the steady-state correlation at 5 kW would be 0.17 m/s—very close to the critical velocity of 0.16 m/s. Furthermore, with a relatively small fire and a relatively high ceiling, stratification is likely to be a factor and should be considered. Assuming the room is air conditioned, with a temperature differential of 10°C from the top of the fuel package to the ceiling level, Equation 31 indicates that the smoke from a 5 kW fire would stratify at a level of about 7.3 m (23.4 ft)—very close to the ceiling height of 7.1 m (23 ft). Given probable dilution of smoke and errors in approximations, it could be considered unlikely that a 5 kW fire would be detected under the defined conditions.

In addition to illustrating how the concept of critical velocity can be used for the design of smoke detection systems, it clearly points out the need to adequately define performance and design objectives, and to select correlations that fit those objectives. First, the objectives should be stated in terms of both the design fire and the critical fire. A 50 kW design fire is significantly different from a 50 kW critical fire, and the design for one may not meet the requirements for the other. Second, care should be taken in selecting a ceiling jet velocity correlation that most closely fits the design objectives. Unless the hazard analysis indicates that the maximum fire size of $\dot{Q}_{do}$ will be 50 kW, it may be better to apply a ceiling jet velocity correlation, based on a growing fire. In this case, the fire growth rate must also be estimated as part of the evaluation. The following example shows the importance of these factors by using the same ballroom as described in Example 12, and provides more specific performance and design parameters.

EXAMPLE 13:

After additional consultation, the owners of the hotel described in Example 12 have modified their objectives as follows: assuming that a fire will begin in a chair, the smoke detection system for the grand ballroom must be able to detect the fire and initiate an internal response before it spreads beyond the chair of origin. The typical fuel load within the room consists of metal-framed chairs with padded seats and backs, and plywood tables with cotton tablecloths.

The response time from when the alarm signal is indicated at the annunciator until the first staff member arrives is estimated to be 60 sec. The delay time from detector activation until alarm initiation, as measured at the sensor, is 10 sec. Because of the potential for nuisance alarms, the detection system employs an alarm verification feature that has a minimum delay time of 15 sec and a maximum delay time of 60 sec.

The existing smoke detectors are installed at a UL-listed spacing of 10 m on center and have a critical velocity of 0.15 m/s. Assuming the fire starts at a point equally spaced between the existing smoke detectors, and there are no ventilation system effects, can the existing smoke detection system be expected to meet the design objectives?

SOLUTION:

The complete solution to a problem like this may require several steps, i.e., determination of the design fire, determination of the critical fire, estimation of ceiling jet velocity at $\dot{Q}_{cr}$, estimation of smoke production or optical density, and analysis of possible stratification effects. In all cases, however, determination of the design fire and critical fire is essential.

Given that the goal is to detect the fire while in the chair of origin, a first step might be to estimate the fire size within the chair that could ignite the cotton tablecloth.

From analysis of the chair and a review of published heat release data, it is determined that the chair most closely resembles the metal frame chair with padded seat and back used in Test 53 of the NIST furniture heat release rate tests.[5] This chair had a maximum heat release rate of 280 kW; a fire growth rate of ≈ 0.0086 kW/s^2; a growth time, t_g of 350 s; and a virtual start time, t_v, of 50 s.

Assuming that the fire would likely grow up the seatback of the chair, and that the seatback is located approximately 0.5 m from the tablecloth, an estimate of the energy output required for ignition of the tablecloth can be made. In this case, using the radiant ignition routine in FIREFORM,[26] and assuming the fuel being easy to ignite (ignition flux of 10 kW/m^2) with a separation distance of 0.5 m, it is estimated that the tablecloth will ignite when the total energy output from the burning chair reaches 139 kW. These parameters define the design fire.

The next step is to calculate the time for the design fire to reach the threshold limit of 139 kW. Using the relationship $\dot{Q} = \alpha t^2$, a time of 118 sec (about 2 min) is calculated. This is growth time of the fire after it begins to follow an exponential growth rate until the design fire size is reached. Given that the fire would probably start as smoldering combustion, the actual growth time could be considerably larger (1 to 2 hrs possible).

The critical fire size can then be estimated by subtracting the various response times and estimating the heat release rate at that moment in time. In this regard, reasonable time delays should be used based on the information pro-

vided. The focus should be on obtaining the "most reasonable" worst-case delay for the situation. From the problem statement, this delay is estimated based on the response times given, using the following equation

$$t_{response} = t_{transport} + t_{verify} + t_{system} + t_{staff} \quad (33)$$

where the following assumptions are made:

$$
\begin{aligned}
t_{transport} &= \text{smoke transport time (unknown)} \\
t_{verify} &= \text{verification time (60 sec maximum)} \\
t_{system} &= \text{system response time (10 sec)} \\
t_{staff} &= \text{staff response time (60 sec).}
\end{aligned}
$$

Momentarily ignoring the smoke transport time and assuming prompt staff response, the result is a maximum detection system response time of 130 sec. However, in an actual fire situation, the smoke detector verification time should be at its minimum of 15 sec, and not at its maximum of 60 sec. Making this assumption, the total response time (still ignoring smoke transport time) is 85 sec. This is less than the 127-sec time to ignition of the tablecloth, and is used to help define the critical fire size ($\dot{Q}_{cr}$).

Here, the 85 sec is subtracted from the 127 sec (that defines the design fire), and the relationship $\dot{Q} = \alpha t^2$ is used to calculate the heat release rate at that moment in time. The result is a heat release rate of 15 kW. Assuming no smoke transport time, this would be the critical fire size at which detection must occur in order to detect the fire and cause the required response before the design fire size is reached.

The next step is to factor in a lag due to the smoke transport time. In order to account for smoke transport lag, Brozovsky[25] suggests a "safety factor" that is equivalent to a heat release rate that is 80 percent of the maximum fire size at the time of detection. This would result in a critical fire size of 12 kW and a corresponding response time of 37 sec. These values can then be used to determine if the ceiling jet velocity will exceed 0.16 m/s.

In this case, it may be more appropriate to use the Heskestad and Delichatsios[8] ceiling jet velocity correlation for a growing fire instead of Alpert's steady-state correlation. This correlation holds for $r/H \geq 0.3$.

$$
\begin{aligned}
U = \alpha^{1/5}H^{1/5}(r/H)^{-0.63}[\alpha^{1/5}H^{1/5}t \\
- 1.95(H + r)/2.48H + 4.13r]^{2/3}
\end{aligned}
\quad (34)
$$

Entering the parameters as previously defined, a velocity of 0.11 m/s is calculated. This is well below the critical velocity of 0.16 m/s, and detection could not be expected early enough to meet the objectives. Given these results, it could be concluded that the smoke detector may not be expected to activate in time to meet the defined goals.

Although several simplifications have been made, this example outlines a methodology for estimating the potential for detector response, given the concepts of design fire and critical fire. In addition, the cross-checking utilized points out the importance of understanding the limitations and boundary conditions of correlations and empirical relationships; i.e., simply because one condition can be met, it does not automatically mean that all others will be met as well, and the complete scenario should be considered. Engineering of smoke detection, especially for low-energy fires, can be a difficult task, and the application of any method for this purpose should include clear statements of all assumptions made.

Another approach to transport lag has been presented by Mowrer[27] based on t-squared fires and correlations for plume and ceiling jet velocities of Newman.[28] It is suggested that Newman's correlations can be applied to power-law fires, using the following relationships

$$lag_{plume} = 0.66H^{4/3}/\dot{Q}^{1/3}$$
$$lag_{ceiling\ jet} = 4.61r/(\dot{Q}/H)^{1/3}$$

where:

H = height of the ceiling above the base of the fire (m)
$\dot{Q}$ = total heat release rate of the fire (kW)
r = radial distance from the plume centerline (m).

Thus the total lag time is

$$lag_{total} = lag_{plume} + lag_{ceiling\ jet}$$

Mowrer reduced Newman's correlations for the transport time lag for power-law fires to the following relationship

$$t_l = (1.4r + 0.2H)/(0.028a_tH)^{1/5}$$

where:

t_l = time lag (sec)
a_t = power-law fire growth coefficient.

The time lag, t_l, represents a time shift between actual heat release rate at a given time and the heat release rate sensed at the detector.

Temperature rise surrogate: One other correlation that can be useful for estimating smoke detector response for higher energy fires indicates that smoke detectors often respond when the temperature increase at the detector location is between 10°C (18°F) and 15°C (27°F) above ambient.[29] Although not valid for all fuels, this is typically assumed valid for flaming fires where such a temperature rise can be expected. Recent modeling of smoke detector activation using the HARWELL FLOW 3-D computational fluid dynamics fire code should also be noted.

RADIANT ENERGY DETECTION

During the combustion process, electromagnetic radiation is emitted over a broad range of the spectrum. Currently, however, fire detection devices operate only in one of three bands: ultraviolet (UV), visible, or infrared (IR), where the wavelengths are defined within the following ranges[5]

Ultraviolet	0.1	to	0.35 microns
Visible	0.35	to	0.75 microns
Infrared	0.75	to	220 microns

Selection of a specific sensor type for fire detection is based on a number of factors, including fuel characteristics, fire growth rate, ambient conditions, resulting control or extinguishing functions, and environmental conditions in the detection area. More specifically, it includes evaluation of the radiant energy absorption of the atmosphere, presence of nonfire-related radiation sources, the electromagnetic energy of the spark, ember or fire to be detected, the distance from the fire source to the sensor, and characteristics of the sensor.

These factors are important for several reasons. First, a radiation sensor is primarily a "line-of-sight" device, and must "see" the fire source. If there are other radiation sources in the area, or if atmospheric conditions are such that a large fraction of the radiation may be absorbed in the atmosphere, the type, location, and spacing of the sensors may be affected. In addition, the sensors react to specific wavelengths, and the fuel must emit radiation in the sensors' bandwidth. For example, an infrared detection device with a single sensor tuned to 4.3 microns (the CO_2 emission peak) cannot be expected to detect a noncarbon-based fire. Furthermore, the sensor must be able to respond reliably within the required time, especially when activating an explosion suppression system or similar fast-response extinguishing or control system.

Once the background information has been determined, the detection system can be designed. Standard practice for the design of radiant energy detection devices is based on application of generalized fire size *vs.* distance curves that are derived using the inverse square law[5]

$$S = kP\exp^{\zeta d}/d^2 \tag{35}$$

where:

S = radiant power reaching the detector (W)
k = proportionality constant for the detector
P = radiant power emitted by the fire
ζ = the extinction coefficient of air
d = the distance between the fire and the detector.

This relationship is used to produce sensor response information for specific fuels. By then plotting the normalized fire size *vs.* the normalized distance, the resulting curve defines the maximum distance at which the tested sensor can be expected to consistently detect a fire of a defined size (usually provided in m^2). By testing a sensor using various fuels, a family of curves can be developed to assist in system design. These curves (sometimes given in tabular form) are usually provided by the sensor manufacturer.

Before applying the distance obtained from such a curve, one must also consider the sensors' field of view. Because the radiation sensor is a line-of-sight device, the sensitivity of the device to a defined fire size decreases as the fire location is moved off the optical axis of the device. This means that a fire of $X\,m^2$, which is detectable at a distance Y m on axis from the sensor, may not be detectable at the same distance Y m if it is located 30 degrees off axis. Limitations of viewing angles are also provided by manufacturers.

Ambient conditions should also be considered as part of the evaluation and design process. Factors, such as humidity and dust, can affect the absorption of radiation in the atmosphere, thus limiting the amount of radiation reaching the sensor for a given fire size. Similarly, temperature can affect the relative sensitivity of a sensor. As the ambient temperature increases, the relative sensitivity can decrease. Even if the decrease is small, it can affect the response of the sensor to the expected fire.

Radiation Detection Example

EXAMPLE 14:

The design objective is to detect a 1.0 m^2 (11 ft^2) pool fire of JP4 aircraft fuel in a large hangar in order to activate a fixed suppression system. The hangar dimensions are 50 m (160 ft) by 80 m (257 ft) with a 20-m (64-ft) ceiling height. The ambient temperature at the ceiling level varies between

15°C (59°F) and 60°C (140°F), depending on time of day and season. The humidity also varies by season, with relative humidity of 90 percent possible. What steps should be taken during system design?

SOLUTION:

The first step should be selection of a detection device. Because the hazard is carbon based, IR detection at 4.3 microns is suitable. Also, because IR detectors generally provide a larger surveillance area per device than UV detectors, they could be more cost effective than UV detection in this case.

One should then determine possible sources of interfering radiation and select a device that is resistant to these extraneous sources. Such resistance to false response can be obtained by filtering, use of multiple sensors (e.g., two- or three-channel detector), or a combination.

The next step is to review the manufacturers data to determine mounting criteria based on the size of the critical fire [$1.0 \, m^2$ ($11 \, ft^2$)]. Generally, this begins with the fire size *vs.* distance curve or table. If only a curve is provided, one must then determine the mounting height and lateral distance limits of the detector. Lateral distances are important as related to the sensors' field of view.

Given this information, a device layout design can be made. This should consider all possible obstructions, and result in all parts of the hangar being monitored. One such design is illustrated in Figure 4-1.8.

As part of the layout, one should consider the possible effects of reduced device sensitivity due to angular displacement, temperature, and humidity. Because manufacturers' criteria vary on these parameters, typical values are used in this solution to illustrate their effects.

For example, the proposed layout has devices utilizing a field of view of 45 degrees. Assuming the nominal sensitivity is such that a $1.0 \, m^2$ ($11 \, ft^2$) fire can be detected at 40 m (128 ft), and there is a reduction in sensitivity of 30 percent due to angular displacement, the distance at which a $1.0 \, m^2$ ($11 \, ft^2$) fire can be detected at 45 degrees is reduced to 28 m (90 ft). If the manufacturers' data indicate a further reduction in sensitivity for temperature, for example 3 percent at 50°C (122°F), the distance is reduced to about 26.8 m (86 ft). If there are further reductions due to humidity, for example a 3 percent reduction at 90 percent relative humidity, the resulting detection distance at 45 degrees is about 25.6 m (82 ft).

In this example, the viewing distance at 45 degrees is a maximum of 20 m (64 ft), and the design can be considered

valid. Had the sensitivity decreased such that the distance dropped below 20 m (64 ft), an alternative layout or different devices must be used. In all cases, the manufacturers' literature should be consulted to determine all pertinant increases or reductions in detector sensitivity due to fuel, distance, angular displacement, and environmental conditions.

DESIGNING FIRE ALARM AUDIBILITY

In most cases, the purpose of a fire detection and alarm system is to alert the occupants of a building that an emergency exists and to initiate evacuation. In situations such as high-rise or industrial buildings, it may be desirable to provide the occupants with more information, such as the nature and location of the fire. In either case the purpose of the system is defeated if the signal is not heard and understood by the occupants.

This section demonstrates a method for fire protection engineers to estimate the relative effectiveness and cost of various fire alarm alerting systems during the design process. In the past, the selection and location of fire alarm devices has been based on experience and engineering judgement. The use of this simplified methodology can save thousands of dollars in retrofit costs required to correct deficiencies in an alarm system.

The transmission of sound from a source to a target is a function of many factors, such as: humidity, air viscosity and temperature, the frequency of the signal, the location of the source relative to the target, the construction of walls, floors, and ceilings, and the furnishings in the area. *Architectural Acoustics*[30] contains a good discussion of these and many other factors affecting sound transmission and loss.

Sound power and sound pressure levels are expressed in decibels (dB) relative to a reference. It is assumed that the reader is familiar with this system of measurement. Throughout this chapter sound power level (SWL or L_W) in decibels is referenced to 10^{-12} W. Sound pressure level (SPL or L_P) in decibels is referenced to 2×10^{-5} Pa. This discussion also assumes that the reader is familiar with the concept of A-weighting. The purpose of A-weighting is to adjust sound pressure level measurements to correspond as closely as possible to the way humans perceive the loudness of the many different frequencies we hear. For instance a 1000 Hz signal at an SPL of 20 dB would be clearly audible. A 100 Hz signal at the same SPL would not be heard. A-weighting allows a single number to describe the SPL

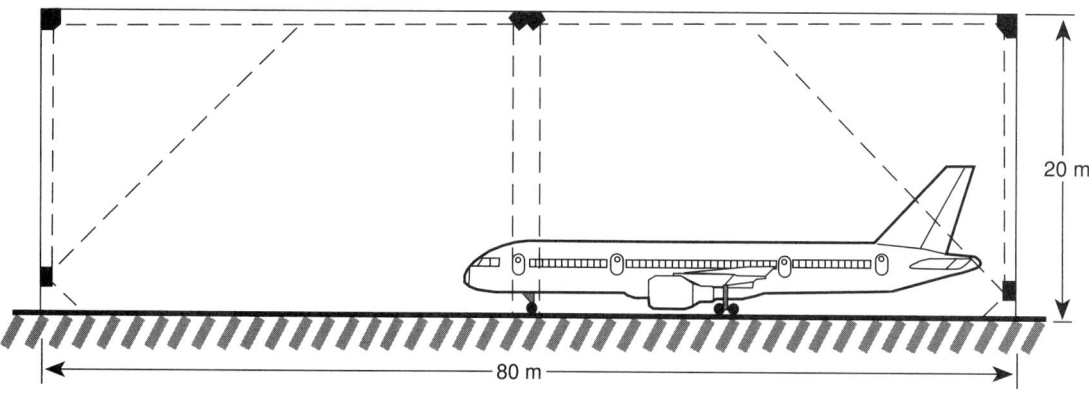

Fig. 4-1.8. IR detector layout for an aircraft hangar.

20 m

80 m

TABLE 4-1.9 *Adjustment for Mounting Position of Sounder (C_1)*

Sounder Position	C_1
Wall/ceiling mounted (more than 1 m from any other major surface)	+5
Wall/ceiling mounted (closer than 1 m to one other major surface)	+7

produced by a signal containing frequencies between 20 and 20,000 Hz. The weighting of the various frequencies is established by an internationally accepted A-weighting curve.[31]

Typical fire alerting systems consist of a combination of audible and visual signals activated by fire detection systems. The audible devices are usually horns, bells, chimes, or speakers. The visual indicators are usually strobe lights, incandescent lamps, or, occasionally, revolving beacons.

In residential occupancies, fire alerting systems should be capable of awakening a sleeping person and informing him or her that a fire emergency exists. Several studies have been done to establish the sound pressure level required to achieve this.[32,33] These studies suggest an SPL between 55 and 70 dBA will awaken a college-age person with normal hearing. The minimum required SPL is also a function of the background noise or signal-to-noise ratio. These levels establish the SPL required to alert or be audible. They do not address the problem of how the person will perceive the sound or react to it.

Until recently, fire codes did not set forth the SPL that a fire alarm system must produce within a building. NFPA 72[5] requires signals to be 15 dBA above ambient in areas where people may be sleeping. British standards require fire alarm signals to produce a sound pressure level of 65 dBA or 5 dBA above ambient noise in areas where occupants are not sleeping.[34] A sound pressure level of 75 dBA at the head of the bed is required in occupancies where people may be sleeping.

Visual signals are located to assist people in deciphering potentially confusing alarm signals. The visual signals also help alert occupants in high background noise environments. They are not usually intended to alert occupants who are sleeping or who have hearing impairments.

Butler, Bowyer, and Kew[31] have described a method to estimate sound pressure levels at some location remote from the sound source. Formulas presented in their study are analogous to standard sound attenuation formulas found in other references.[30,35] They have been simplified by replacing complex terms with constants for which they have provided tables of data. The equations and data presented in their study provide a straightforward method for analyzing proposed designs. The same equations and data can be used to determine the power requirement and maximum allowable spacing of signaling devices required to achieve a specified sound pressure level. The technique presented in their study is suitable for acoustically simple buildings only. Complex building arrangements and materials may require a more rigorous analysis using other methodologies which are beyond the scope of this chapter.

To demonstrate how signaling systems can be designed and analyzed, two scenarios will be considered. Both scenarios are based on a typical dormitory or office layout. The building has long corridors with rooms of equal size on each side. Each room is approximately 5 m wide by 6 m deep. The walls consist of two layers of sheetrock (total 25.4 mm thick) separated by wood studs. The wall cavities contain 75-mm-thick mineral fiber insulation. The floors are concrete with carpeting. The ceiling is 3 m high and consists of acoustical tiles. The room doors will be solid core with good edge seals. The alerting systems will be designed to achieve a 75 dBA sound pressure level at the farthest point in the rooms.

In the first scenario, wall-mounted fire alarm speaker/light combinations are spaced equally in the corridor. Calculations determine the maximum allowable spacing of the speakers in order to achieve the design goal of 75 dBA in the rooms.

In the second scenario, speakers are placed in each room as well as in the corridor. Calculations determine the size speaker and the power needed to drive that speaker to achieve the design goal of 75 dB. Calculations are also presented to determine the required spacing of speakers in the corridor to achieve a sound level of 65 dB.

Unless otherwise noted, the following formulas and data are from Butler, Bowyer, and Kew.[31]

Scenario A

In this scenario, the fire alerting system or sounder will consist of wall-mounted speaker/light combinations in the corridors only.

L_W is the sound power level of a horn, bell, speaker, or any sounder (dBA referenced to 10^{-12} W).

$$L_W = L + 20 \log_{10} r + 11 \text{ dB}$$

where L is the manufacturer's stated output in dBA at a distance r meters. A typical compression driver-type fire alarm speaker powered at 2 W has an L equal to 94 dBA at 3.05 m.[36]

Therefore

$$L_W = 94 + 20 \log_{10}(3.05) + 11$$
$$L_W = 115 \text{ dB}$$

L_{P1} is the sound pressure level (dBA referenced to 2×10^{-5} Pa) produced outside of a room wall from one speaker.

$$L_{P1} = L_W + C_3 + C_4 + C_5$$

TABLE 4-1.10 *Adjustment for Distance (C_2)*

Distance from Source (m)	C_2
1	−11
2	−17
3	−21
6	−27
12	−33
15	−35
20	−37
25	−39
30	−41
40	−43
50	−45
60	−47
80	−49
100	−51

TABLE 4-1.11   *Adjustment for Number of Directions of Sound Propagation (C_3)*

Number of Directions	C_3
Single direction, e.g., positioned at one end of a corridor	0
Two-directional, e.g., positioned in the length of a corridor	−3
Three-directional, e.g., positioned at a 'T' junction of corridors	−5

TABLE 4-1.12   *Adjustment Based on the Finishes in the Corridor (C_4)*

	Surface Finishes	C_4
Hard	(e.g.) walls and ceiling with solid surfaces and tarazzo floor	0
Medium	(e.g.) acoustic ceiling, plastered solid walls with 5% coverage of soft surfaces and floor of composite tiles	−8
Soft	(e.g.) acoustic ceiling, plastered solid walls with 5% coverage of soft surfaces and carpets on felt on concrete floor	−9

TABLE 4-1.13   *Adjustment for Distance from Source to Mid-Point of the Partition (C_5)*

Distance	C_5
1	0
3	−4
6	−8
10	−10
12	−11
15	−12
20	−14
30	−15
50	−17

where C_3 is a correction for the number of directions that the sounder propagates, C_4 is a correction for the characteristics of the corridor walls, ceiling, and floor, and C_5 is a function of the distance from the sounder to the center of the bedroom wall. From Table 4-1.11:[31] C_3 is −3 dB, because the speaker propagates in two directions along the corridor; from Table 4-1.12 C_4 is −9 dB, because the floor and ceiling are acoustically soft; and C_5 is unknown since the required spacing of the corridor speakers has not yet been determined.

A worst-case condition exists for a room located farthest from a speaker. In this situation the room is located equally between two speakers. Since each unit propagates sound to the room, the sound pressure level outside of the room is higher than if there were only one speaker. The sound pressure level is not double that for a single speaker. For equally spaced sounders, add 3 dB to the level expected from a single unit. Therefore

$$L_{P1} = 115 - 3 - 9 + C_5 + 3$$
$$L_{P1} = 106 + C_5$$

L_{P2} is the sound pressure level at the farthest point in a room. To achieve the established goals, L_{P2} must be 75 dBA. In this situation, with the speaker located outside of the occupied space

$$L_{P2} = L_{P1} - R + C_2 + C_6 + C_7 + 11 \text{ dBA}$$

where R is the average sound reduction index for the wall, C_2 is a function of the distance from the wall to the point of interest, C_6 is a function of the area of the room wall, and C_7 is a function of the frequency of the sound reaching the wall.

In this case, from data presented by Butler, Bowyer, and Kew,[31] the sound reduction index R for the wall is about 40 dB. (See Table 4-1.17.) This is based on incident sound in the range of 100 to 3150 Hz. Sound attenuation through the door is about 26 dB. (See Table 4-1.17.) The average sound reduction index, R, for the combined door and wall is 34 dB, if the door is 10 percent of the area. (See Table 4-1.21.) C_2 is found to be −27 dB, because there are 6.5 m from the center of the wall to the corner of the room. (See Table 4-1.10.) Since the wall is 15 m², C_6 is +11.5 dB. (See Table 4-1.14.) If it is assumed that the sound reaching the wall is a maximum at a frequency of 2000 Hz, $C_7 = -5$ dB. (See Table 4-1.15.) Therefore

$$L_{P2} = (106 + C_5) - 34 - 27 + 11.5 - 5 + 11 \text{ dBA}$$
$$L_{P2} = 62.5 + C_5 \text{ dBA}$$

TABLE 4-1.14   *Factor for Area of Partition Between Sounder and Receiver (C_6)*

Partition area (m²)	C_6
2	+3
4	+6
8	+9
10	+10
15	+11.5
20	+13
30	+15
50	+17
80	+19
100	+20
200	+23

TABLE 4-1.15   *Adjustments for Frequency of Maximum Output of Sounders (C_7)*

Frequency of sounder	C_7
500 Hz	0
1000 Hz	−3
2000 Hz	−5
4000 Hz	−9

TABLE 4-1.16   *Addition of Two Sound Pressure Levels*

Difference between the two levels— dB to be added	Add to the higher level—dB
0	3
1	2
2	2
3	2
4	2
5	1
6	1
7	1
8 or more	0

TABLE 4-1.17 *Second Reduction Indices (dB) for a Selection of Typical Structures (100–3150 Hz Frequency Range)*

Building element	Weight of partition (kg/m²)	Average attenuation (dB)
WALLS AND PARTITIONS		
1. 100 mm dense concrete with or without plaster	250	45
2. 150 mm 'no fines' concrete with 12 mm plaster on both faces	250	45
3. 115 mm brickwork with 12 mm plaster on both faces	250	45
4. 115 mm brickwork unplastered	195	42
5. 300 mm lightweight concrete precast blocks with well-grouted joints	190	42
6. 75 mm clinker blockwork with 12 mm plaster on both faces	115	40
7. 50 mm dense concrete	120	40
8. 25.4 mm plasterboard (2 layers) separated by timber studding < 75 mm and mineral fiber blanket	—	40
9. 200 mm lightweight concrete precast blocks with well-grouted joints	122	40
10. 150 mm lightweight concrete precast blocks with well-grouted joints	93	37
11. 50 mm clinker blocks with 12 mm plaster on both faces	—	35
12. 63 mm hollow clay blocks with 12 mm plaster on both faces	—	35
13. 9.5 mm plasterboard (2 layers) separated by timber studding < 75 mm with 12 mm of plaster on both faces	—	35
14. 6 mm plywood/hardboard (2 layers) separated by timber studding < 50 and 50 mm mineral fiber blanket	—	30
15. 19 mm chipboard on a supporting frame	—	25
16. 0.8 mm sheet steel	—	25
17. 21 mm tongued and grooved softwood boards tightly clamped on a support frame	—	20
18. 3.2 mm hardboard (2 layers) separated by 44 mm polystyrene core	—	20
DOORS		
19. Flush panel, hollow core, hung with one large air gap	9	14
20. Flush panel, hollow core hung with edge sealing	9	20
21. Solid hardwood, hung with edge sealing	28	26
WINDOWS		
22. Single glass in heavy frame	15	24
23. Double glazed 9 mm panes in separate frames 50 mm cavity	62	34
24. Double glazed 6 mm panes in separate frames 100 mm cavity	112	38
25. Double glazed 6 mm and 9 mm panes in separate frames 200 mm cavity, absorbent blanket in reveals	215	58

If there were no loss of sound pressure level between the speaker and the room wall due to distance, C_5 would be zero and L_{P2} would be 62.5 dBA. This shows that even if the two speakers were right outside the room, the goal of 75 dBA in the room would not be met. In fact, the resultant noise level in the room would be slightly less than the 65 dBA required by British standards[34] to alert nonsleeping persons. The sound level of 62.5 dBA would exceed the 55 dBA reported by Nober *et al*[32] to alert sleeping college-age persons in a quiet ambient setting.

To meet the goal of 75 dBA in the room, either the sound system or the environment would have to be changed. Fire alarm speakers are normally available with multiple power taps such as 4, 2, 1, ½, and ¼ W. A single unit may allow choice of two or three different power levels, which allows balancing of the system after installation.

TABLE 4-1.18 *Average Sound Reduction Indices (dB) of Partitions Incorporating Single Glazing (100–3150 Hz Frequency Range)*

Percentage of glazing (24 dB)	Sound reduction index of partition without glazing					
	25 dB	30 dB	35 dB	40 dB	45 dB	50 dB
100%	24	24	24	24	24	24
75%	24	25	25	25	25	25
50%	24	26	27	27	27	27
33%	25	27	28	29	29	29
25%	25	27	29	30	30	30
10%	25	29	31	33	34	35
5%	25	29	33	35	36	37
2½%	25	30	34	37	39	40
nil	25	30	35	40	45	50

TABLE 4-1.19 *Average Sound Reduction Indices (dB) of Partitions Incorporating a Door of 14 dB Attenuation (i.e., One with Large Air Gaps) (100–3150 Hz Frequency Range)*

Door representing percentage of total area of partition	Sound reduction index of partition without door					
	25 dB	30 dB	35 dB	40 dB	45 dB	50 dB
100%	14	14	14	14	14	14
50%	16	16	16	16	17	17
25%	19	19	19	19	20	20
10%	21	23	23	23	23	23
5%	23	25	26	26	26	26

If a 4-W power input were used, this would be a doubling of the 2 W originally tried in the previous calculation. Because decibels are logarithmic, a doubling of power results in a change of 3 dB in L_W ($10 \times \log_{10} 2 = 3$). This alone would not be sufficient to meet the 75 dBA goal. In addition, the higher sound pressure level in the immediate vicinity of the speaker might be discomforting. If the fire alarm system were also used for voice communication, a speaker tapped at 4 W in a small corridor might sound very distorted and be unintelligible.

It is also possible to change the sound pressure level in dBA by changing the frequency of the source. In general, the higher the frequency, the higher the attenuation as the sound waves pass through a wall. Hence, a lower frequency would increase the sound pressure level in the room. In the calculations above, it was assumed that the predominant frequency of the source was 2000 Hz. This resulted in a C_7 of -5 dBA. According to Table 4-1.15 if this frequency were 500 Hz, C_7 would be 0 dBA. This would increase the SPL in the room by 5 dBA.

Changes could be made to the building design that would make it possible to meet the design goal. For instance, the use of a lighter-weight door or one without good edge sealing could increase sound transmission to the room by as much as 12 dBA. However, changes such as this would tend to defeat other goals such as fire resistance and resistance to smoke spread. If the floor and ceiling were hard surfaces without carpeting or tiles, C_4 could be increased from -9 to 0 dBA. Changes such as this would probably be resisted for reasons other than firesafety.

The only remaining alternative is to provide speakers in each of the rooms.

Scenario B

In this case, a speaker in each room powered at only ¼ W will be tried in addition to the speaker in the corridor. The problem, then, is to select a speaker with a sound power output that can meet the goal of 75 dB at the pillow.

$L = ?$ at $r = 3.05$ m (3.05 m is a commonly used reference point.)

$$L_W = L + 20 \log_{10} r + 11 \text{ dB}$$
$$L_W = L + 20 \log_{10}(3.05) + 11 \text{ dB}$$
$$L_W = L + 21 \text{ dB}$$

L_{P2} is the sound level at the bed. In this case, with the speaker in the occupied space

$$L_{P2} = L_W + C_1 + C_2 \text{ dBA}$$

TABLE 4-1.20 *Average Sound Reduction Indices (dB) of Partitions Incorporating a Door of 20 dB Attenuation (i.e., Light Door with Edge Sealing) (100–3150 Hz Frequency Range)*

Door representing percentage of total area of partition	Sound reduction index of partition without door					
	25 dB	30 dB	35 dB	40 dB	45 dB	50 dB
100%	20	20	20	20	20	20
50%	21	22	22	22	22	23
25%	23	24	25	25	25	26
10%	24	27	28	29	29	29
5%	24	28	30	32	32	32

TABLE 4-1.21 *Average Sound Reduction Indices (dB) of Partitions Incorporating a Door of 26 dB Attenuation (i.e., Heavy Door with Edge Sealing) (100–3150 Hz Frequency Range)*

Door representing percentage of total area of partition	Sound reduction index of partition without door					
	25 dB	30 dB	35 dB	40 dB	45 dB	50 dB
100%	26	26	26	26	26	26
50%	25	27	28	28	28	28
25%	25	28	30	31	31	31
10%	25	28	32	34	35	35
5%	25	28	33	36	38	38

TABLE 4-1.22 *Combined Sound Reduction Indices for Combination of Standard Doors and Glazing (100–3150 Hz Frequency Range)*

Area of (24 dB) glazing (m²)	Sound reduction index for standard size door (1.54 m²)		
	14 dB	20 dB	26 dB
	Insulation values for combined door and glazing		
1	16	21	25
2	17	22	25
4	18	22	24
6	19	23	24
8	20	23	24
10	20	23	24
12	21	23	24
16	21	23	24
20	22	23	24

where C_1 is a correction for how close the sounder is to an adjacent surface, and C_2 is a correction for the distance from the speaker to the bed. In this case, the speaker is on the wall and close to the ceiling. Therefore,[31] C_1 is +7 dB, and C_2 is −27 dB (approximately 6.5 m from the speaker to the bed). (See Tables 4-1.9 and 4-1.10.) Therefore

$$L_{P2} = (L + 21) + 7 - 27 \text{ dBA}$$
$$L_{P2} = L + 1 \text{ dBA}$$

To get L_{P2} = 75 dBA, L must be at least 74 dBA. The smallest and least expensive fire alarm speaker available is a 4 in., paper cone speaker. A typical speaker of this size and type, powered at ¼ W, has an L equal to 75 dB at 3.05 m.[36] This speaker would meet the design goal in the room, without even considering any sound contribution from corridor-mounted speakers.

For the corridor speakers in Scenario B, L_{P1} is the sound pressure level at a point farthest from a speaker.

$$L_{P1} = L_W + C_3 + C_4 + C_5 \text{ dBA}$$

where C_3 and C_4 are the same as in Scenario A (−3 and −9 dB, respectively). C_5 is a function of the spacing, which is to be determined. If a single corridor speaker tapped at only ¼ W is used, with an L of 85 dB at 3.05 m[36]

$$L_W = L + 20 \log_{10} r + 11 \text{ dB}$$
$$L_W = 85 + 20 \log_{10}(3.05) + 11 \text{ dB}$$
$$L_W = 106 \text{ dB}$$
$$L_{P1} = 106 - 3 - 9 + C_5 \text{ dBA}$$
$$L_{P1} = 94 + C_5$$

The goal is to maintain a 65 dBA sound pressure level in the corridors (L_{P1}).

$$65 = 94 - C_5$$

Therefore, C_5 must be −29 dBA or more for L_{P1} to be 65 dBA or higher. From Table 4-1.13,[31] it is found that distance of 50 m between source and target in the corridor could be exceeded and still meet the 65 dBA goal.

COST ANALYSIS

Scenario A

For comparison purposes, assume that sufficient changes could be made to the building and alarm system to allow speakers to be mounted in the corridor only at a spacing of 3 m. A typical dormitory with about 30 bedrooms per floor requires approximately 24 speakers per floor in the corridors. In a building with seven floors, this amounts to 168 speakers. At 2 W per speaker, this would be 336 W. This requires three 125 W power amplifiers at an installed cost of about $1,400.00 each. This does not include other fixed costs, such as control equipment and detectors, that are the same for each of the scenarios.

Assume each corridor unit to be a speaker/light combination. The average installed cost, including backbox, wiring back to a control panel on the first level, and conduit, would total to about $135.00 per unit. The total cost is then

TOTAL = (3 × $1,400.00) + (168 × $140.00)
TOTAL = $27,720.

Scenario B

In this case, there are thirty 4-in. paper cone speakers per floor at an average cost of $100.00, installed. Assume a total of four speaker/light units per floor in the corridors. The

TABLE 4-1.23 *Average Sound Reduction Indices for a Partition the Surface of which Is a Combination of Glass, Door and Wall Partition (100–3150 Hz Frequency Range)*

Door + glazing as percentage of total partition area	Sound reduction value of partition without glazing or door											
	30 dB			35 dB			40 dB			45 dB		
	Insulation value of combined door and glazing dB (from Table 4-1.22)											
	15	20	25	15	20	25	15	20	25	15	20	25
5	26	28	30	28	31	33	28	32	36	28	33	37
10	24	27	29	24	29	32	25	30	34	25	30	35
20	22	25	28	21	26	31	22	27	32	22	27	32
30	20	24	28	20	25	29	20	25	30	20	25	30
50	18	23	27	18	23	28	18	23	28	18	23	28
75	16	21	26	16	21	26	16	21	26	16	21	26
100	15	20	25	15	20	15	15	20	25	15	20	25

calculations show that the system goals are met with only one or two units in the corridors. However, the halls may be split by smoke doors or they may be irregular in shape. Also, system reliability is increased by using more than one unit.

Each bedroom speaker and corridor speaker is powered at ¼ W. For seven floors, this gives a total power requirement of 59.5 W. Therefore, one 60 W amplifier, at a unit cost of $1,125.00, is needed. The total cost is then

TOTAL = $1,125.00 + (7 × 30 × $100.00) + (7 × 4 × $135.00)
TOTAL = $25,905.00.

SUMMARY

The estimates show the relative costs of the different scenarios, not the actual costs. The real costs of the systems are affected by factors such as whether the building is new or existing. If existing, the price is affected by the extent of other renovations. Also, the estimates do not reflect the cost of other parts of the system. The balance of the system includes such items as smoke and heat detectors, equipment for elevator capture, and air handler controls.

The relative costs of the two systems in Scenarios A and B under "Cost Analysis" differ by only 7 percent. In a building of this size and type, such a small margin cannot be considered significant enough to conclude that one system is more economical than the other.

The small difference in the costs of the two systems is due to the additional cost of amplifiers needed to power the system that has only corridor units. The total number of units (corridor + room) in Scenario B is 70 more than in Scenario A. The reduced power requirement offsets the added cost of their installation.

Scenario A has a higher equipment cost but a lower installation cost than Scenario B. This means that the relative costs of the two systems will be slightly sensitive to the type of equipment used and the cost of installation labor. By changing the figures used in the cost estimates, it can be shown that the variance is only a few percent and probably not significant.

If the building were four stories or less in height, the difference in relative cost reduces to about two percent. Again, this is not considered to be a significant difference.

By increasing the size of the building to twelve stories, Scenario B becomes significantly less expensive than Scenario A. Above this height, the combined use of room and corridor units becomes increasingly economically attractive.

Changing the size of each floor has about the same effect as changing the height of the building. Therefore, increasing the floor area makes Scenario B more viable. A reduction in floor area and building height does not make the corridor-only system attractive, unless the building is only a few stories in height. Then, a voice system is probably not needed. From an economics standpoint, a corridor-only horn/light system is probably best, since the cost of these units is generally less than that of speakers. Again, this assumes that sufficient changes could be made to the building design to increase the level of sound penetrating the corridor walls.

Obviously, if the sound loss from the corridor to the individual rooms is less, Scenario A starts to look better. This has the effect of raising the height above which Scenario B becomes significantly less expensive. However, changing construction features to reduce sound loss may

reduce the passive fire resistance of the structure below an acceptable level as well as decrease the privacy level.

There are other factors to consider when choosing between different systems. In Scenario A, the quantity of speakers in the corridors and the high power levels driving each speaker (2 W each) can cause sound distortion. Voice messages may not be intelligible in the bedrooms even though there is enough sound to wake a sleeping occupant. Also, the high sound levels (106 dBA plus) in the corridors approach uncomfortable levels.

It is clear from the discussions above that a system with room speakers in conjunction with corridor units is the most desirable case. That system has the added advantage of eliminating most of the uncertainties in the design of the system. It is easier and more accurate to calculate sound levels at a point in the same room as the sound source than it is to estimate sound losses through composite walls.

This cost-benefit analysis shows that a fire alarm alerting system with units in each office or bedroom can be installed at about the same cost or less than a corridor-only system. In addition, there is a higher confidence level that the system with the sounders in each room will perform its intended function: to awaken and alert sleeping occupants.

DESIGNING FIRE ALARM VISIBILITY

Visual alarm notification is an important part of a fire alarm system. This is especially true in cases where the ambient noise level is high, building occupants may be sleeping, or building occupants or their visitors may have hearing impairments. In these cases, it should be expected that the visual alarm will be required to alert occupants and initiate evacuation or relocation. As such, one first needs to determine a suitable intensity required to obtain this function.

In many cases, a suitable intensity can be obtained from regulatory documents, such as building codes, fire codes, or the Americans with Disabilities Act. If additional guidance is needed, reference can be made to appropriate documentation on alerting of persons by visual means.[37] It should be noted that such documents often provide the "effective intensity" required to alert someone. This should not be confused with the intensity of the lamp providing the signal. The two are related by the inverse square law

$$E = I/d^2 \qquad (36)$$

where E is the effective intensity (lumens per unit area), I is the intensity of the incident light source (candela), and d is the on-axis distance between the incident light source and the point where the effective intensity is measured.

As part of a test program to determine signaling applications for the hearing impaired, UL determined that an effective intensity of 0.398 lumens/m² (0.037 lumens/ft²) as viewed on axis from a single light source located in the center of one wall of a 6.1-m by 6.1-m (20-ft by 20-ft) room was the minimum required by their objective. It was also determined that, by increasing the "square" dimensions in increments of 3 m (10 ft) in both directions (length and width), the minimum effective intensity value of 0.398 lumens/m² could be used to extrapolate the required signal intensity as the room size increased.

For example, if the room size were increased to 12.2 m by 12.2 m (40 ft by 40 ft), the intensity of the signal could be determined using the inverse square law and solving for I

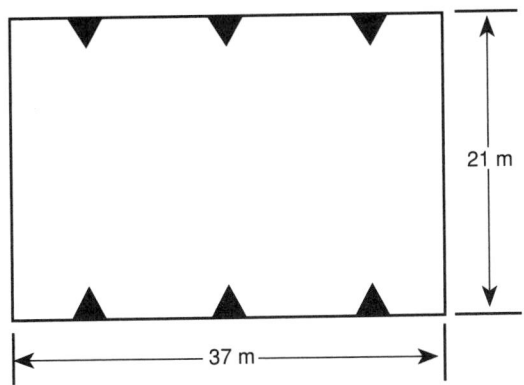

Fig. 4-1.9. Notification appliance (▲) locations.

$E = I/d^2$; therefore,
$I = Ed^2 = (0.398 \text{ lumens/m}^2)(12.2 \text{ m}^2) = 59.2 \text{ candela}$

Thus, one signal rated at 60 cd would be sufficient for the space. Using the same approach, but smaller squares, one would also find that two signals rated at 30 cd, or four signals rated at 15 cd each, would also be applicable.

These relationships can be used: (1) during design, when the intensity of incident light is known, the effective intensity is given, and the spacing of the signals needs to be determined, or (2) as part of an analysis, where signals of known intensity are installed throughout a space, and the effective intensity needs to be determined. Designers should check with the authority having jurisdiction or the current edition of NFPA 72, *National Fire Alarm Code* regarding the use of multiple flashing lights.

EXAMPLE 15:

The design objective is to evaluate the visual alarm notification system installed in a large open space for suitability in providing signals for the hearing impaired. The space is 21 m (70 ft) by 37 m (120 ft), with a 6.5-m (20-ft) ceiling height. The notification appliances are located 2 m (6.5 ft) above floor level and are spaced as shown in Figure 4-1.9. The signals are rated at 45 cd each. Is the required effective intensity of 0.398 lumens/m² currently provided?

SOLUTION:

The first step is to section off the space into blocks that are anticipated to be covered for each signal. In this case, the result is six blocks, each 12.2 m (40 ft) long by 10.5 m (34 ft) wide. This is illustrated in Figure 4-1.10.

Given these dimensions, one could calculate the effective intensity at point A, where

$$E = 45 \text{ cd}/(10.5 \text{ m})^2 = 0.41 \text{ lumens/m}^2$$

This is greater than the minimum required effective intensity of 0.398 lumens/m². However, application of this method requires the blocks of coverage by a signal to be square, with the lateral distance (90 degrees) being equal to one-half the coverage distance on-axis. In this case, the lateral distance is 12.2 m (40 ft), and this is the figure that should be used to calculate the effective intensity throughout the entire block. In doing this, one finds that the effective intensity provided is

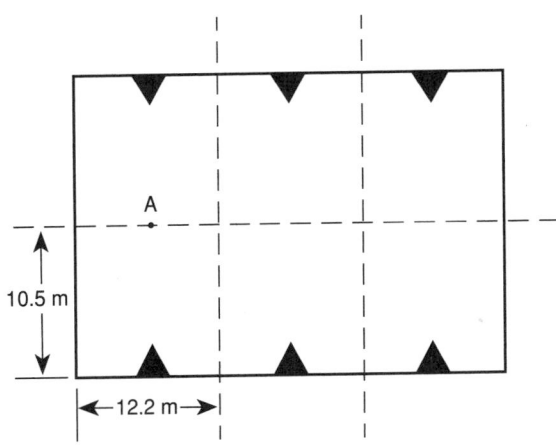

Fig. 4-1.10. Sections for anticipated signal (▲) coverage.

$$E = 45 \text{ cd}/(12.2 \text{ m})^2 = 0.29 \text{ lumens/m}^2$$

which is below the minimum required 0.398 lumens/m². This results in areas of the space not having the required intensity. This is illustrated in Figure 4-1.11.

To determine what intensity is required for the signals in order to provide the required 0.398 lumens/m², the inverse square law can be applied using the value $d = 12.3$ m. This results in a required incident intensity of 60 cd for each existing signal location.

By applying this method of dividing spaces into squares and applying the inverse square law, the incident intensity of signals and their required spacing can be calculated for spaces of any shape and size. Tradeoffs can be made between the number of signals and the intensity of signals to best fit the application (for example, one signal of 60 cd *vs.* four properly spaced signals of 15 cd each).

In cases where a minimum required effective intensity at all points in a space is specified (as opposed to the minimum effective intensity on-axis within a square), the intensity can be calculated using the inverse square law, the cosine law, and the cosine cubed law. In this case, the inverse square law provides the effective intensity on-axis, application of the cosine law provides the effective intensity at a perpendicular surface within the same plane as the signal, and application of the cosine cubed law provides the effective intensity at parallel surfaces within the same plane as the signal.

In cases where flashing signals are required, the effective intensity (as related to a steady signal of the same size, color, and shape) can be calculated using the relationship[38]

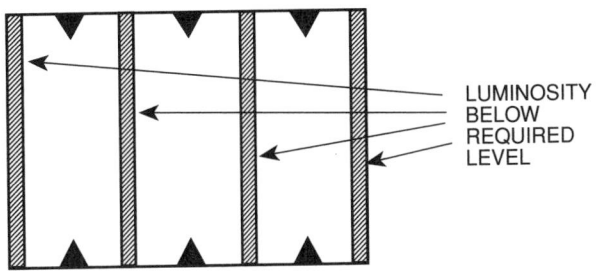

Fig. 4-1.11. Diagram of subadequate luminosity intensity.

$$I_e = \left(\int_{t_1}^{t_2} I \, dt \right) / (a + t_2 - t_1) \qquad (37)$$

where I_e is the effective intensity, I is the instantaneous intensity, and t_1 and t_2 are the times (in sec) of the beginning and ending of that part of the flash where I exceeds I_e. In the United States, the value of 0.2 is usually used for the constant a.

If the duration of the flash is less than one millisecond, Equation 37 can be further simplified to[38]

$$I_e = 5 \int I \, dt \qquad (38)$$

where the integration is performed over the complete flash cycle.

With this information, it should be possible to calculate visual fire alarm signals for most situations. In all cases, a value for the required effective intensity at some point within the room is required. If not provided at the beginning of the design process, one should determine an effective intensity based on the specific application and the condition of the occupants being alerted.

NOMENCLATURE

α	fire intensity coefficient, Btu/s^3 or kW/s^2
A	area, m^2 or ft^2
A	$g/(C_p T_a \rho_0)$, m^4/(s^{2}kJ) or ft^4/(s^{2}Btu)
c	specific heat of detector element, Btu/(lbm R) or kJ/(kg K)
C_p	specific heat of air, Btu/(lbm R) or kJ/(kg K)
d	diameter of sphere or cylinder, m or ft
D	$0.188 + 0.313 r/H$
Δt	change in time, s
ΔT	increase above ambient in temperature of gas surrounding a detector, °C or °F
ΔT_d	increase above ambient in temperature of a detector, °C or °F
ΔT_p^*	change in reduced gas temperature
f	functional relationship
g	functional relationship
g	gravitational constant, m/s^2 or ft/s^2
h_c	convective heat transfer coefficient, kW/(m^2°C) or Btu/(ft^{2}s°F)
H	ceiling height or height above fire, m or ft
ΔH_c	heat of combustion, kJ/mol
H_f	heat of formation, kJ/mol
L_P	sound pressure level
L_W	sound power level
m	mass, lbm or kg
p	positive exponent
$\dot{q}$	heat release rate, Btu/sec or kW
$\dot{q}_{cond}$	heat transferred by conduction, Btu/s or kW
$\dot{q}_{conv}$	heat transferred by convection, Btu/s or kW
$\dot{q}_{rad}$	heat transferred by radiation, Btu/s or kW
$\dot{q}_{total}$	total heat transfer, Btu/s or kW
Q	heat release rate, Btu/s or kW
Q_{cr}	critical heat release rate
Q_{do}	design heat release rate
Q_i	ideal heat release rate
Q_p	predicted heat release rate, Btu/s or kW
Q_T	threshold heat release rate at response, Btu/s or kW
r	radial distance from fire plume axis, m or ft
ρ_0	density of ambient air, kg/m^3 or lb/ft^3
Re	Reynolds number
RTI	response time index, m$^{1/2}$s$^{1/2}$ or ft$^{1/2}$s$^{1/2}$
S	spacing of detectors or sprinkler heads, m or ft
t	time, s
t_c	critical time—time at which fire would reach a heat release rate of 1000 Btu/s (1055 kW), s
t_r	response time, s
t_v	virtual time of origin, seconds
t_{2f}	arrival time of heat front (for $p = 2$ power-law fire) at a point r/H, s
t_{2f}^*	reduced arrival time of heat front (for $p = 2$ power-law fire) at a point r/H, s
t_p^*	reduced time
T	temperature, °C or °F
T_a	ambient temperature, °C or °F
T_d	detector temperature, °C or °F
T_g	temperature of fire gases, °C or °F
T_s	rated operating temperature of a detector or sprinkler, °C or °F
U	velocity, m/s
u	instantaneous velocity of fire gases, m/s or ft/s
u_0	velocity at which τ_0 was measured, m/s or ft/s
u_p^*	reduced gas velocity
v	kinematic viscosity, m^2/s or ft^2/s
x	vectorial observation point, m or ft
Y	defined in Equation 27
τ	detector time constant—$mc/(hA)$, s
τ_0	measured at reference velocity u_0, s

REFERENCES CITED

1. "Selection and Specification of the 'Design Fire' for Performance-Based Fire Protection Design," *Proceedings*, SFPE Engineering Seminar, Phoenix, AZ, Nov. 15, 1993.
2. R. Custer, B. Meacham, and C. Wood, "Performance-Based Design Techniques for Detection and Special Suppression Applications," *Proceedings* of the SFPE Engineering Seminars on Advances in Detection and Suppression Technology, San Francisco, May 18, 1994.
3. R. Custer and R. Bright, "Fire Detection: The State-of-the-Art," *NBS Tech. Note 839*, National Bureau of Standards, Washington (1974).
4. UL 521, *Standard for Safety Heat Detectors for Fire Protective Signaling Systems*, Underwriters Laboratories Inc., Northbrook (1993).
5. NFPA 72, *National Fire Alarm Code*, National Fire Protection Association, Quincy, MA (1993).
6. G. Heskestad and H. Smith, *FMRC Serial Number 22485*, Factory Mutual Research Corp., Norwood (1976).
7. J.P. Hollman, *Heat Transfer*, McGraw-Hill, New York (1976).
8. G. Heskestad and M.A. Delichatsios, *The Initial Convective Flow in Fire*, 17th Symposium on Combustion, Pittsburgh (1978).
9. G. Heskestad and M.A. Delichatsios, *NBS-GCR-77-86 and NBS-GCR-77-95*, Technical Information Service, Springfield (1977).
10. R.P. Schifiliti, "Use of Fire Plume Theory in the Design and Analysis of Fire Detector and Sprinkler Response," MS Thesis, Worcester (1986).
11. C. Beyler, Private Communication.
12. C. Beyler, *Fire Tech.*, 20, 4 (1984).
13. D.W. Stroup, D.D. Evans, and P. Martin, *NBS Special Publication 712*, National Bureau of Standards, Gaithersburg (1986).
14. R. Alpert, *Fire Tech.*, 8, 3 (1972).
15. D.D. Evans and D.W. Stroup, *NBSIR 85-3167*, National Bureau of Standards, Gaithersburg (1985).
16. J.R. Lawson, W.D. Walton, and W.H. Twilley, *NBSIR 83-2787*, National Bureau of Standards, Washington (1983).
17. Brian J. Meacham, "Characterization of Smoke from Burning Materials for the Evaluation of Light Scattering-Type Smoke

Detector Response," MS Thesis, WPI Center for Firesafety Studies, Worcester, MA (1991).

18. B.J. Meacham and V. Motevalli, "Characterization of Smoke from Smoldering Combustion for the Evaluation of Light Scattering-Type Smoke Detector Response," *J. of Fire Protection Engineering*, SFPE, Vol. 4, No. 1 (1992).

19. UL 268, *Standard for Safety Smoke Detectors for Fire Protective Signaling Systems*, Underwriters Laboratories, Inc., Northbrook, IL (1989).

20. M.A. Delichatsios, "Categorization of Cable Flammability, Detection of Smoldering, and Flaming Cable Fires," Interim Report, Factory Mutual Research Corporation, Norwood, MA NP-1630, Nov. 1980.

21. NFPA 92B, *Guide for Smoke Management Systems in Malls, Atria, and Large Areas*, National Fire Protection Association, Quincy, MA (1991).

22. G. Heskestad, FMRC Serial Number 21017, Factory Mutual Research Corp., Norwood, MA (1974).

23. C.E. Marrion, "Lag Time Modeling and Effects of Ceiling Jet Velocity on the Placement of Optical Smoke Detectors," MS Thesis, WPI Center for Firesafety Studies, Worcester, MA (1989).

24. M. Kokkala *et al*, "Measurements of the Characteristic Lengths of Smoke Detectors," *Fire Technology*, Vol. 28, No. 2, National Fire Protection Association, Quincy, MA (1992).

25. E.L. Brozovsky, "A Preliminary Approach to Siting Smoke Detectors Based on Design Fire Size and Detector Aerosol Entry Lag Time," MS Thesis, WPI Center for Firesafety Studies, Worcester, MA (1991).

26. Scott Deal, "Technical Reference Guide for FPEtool Version 3.2," NISTIR 5486, National Institute for Standards and Technology, U.S. Department of Commerce, Gaithersburg, MD, Aug. (1994).

27. F.W. Mowrer, "Lag Times Associated with Detection and Suppression," *Fire Technology*, Vol. 26, No. 3, pp. 244–265 (1990).

28. J.S. Newman, "Principles for Fire Detection," *Fire Technology*, Vol. 24, No. 2, pp. 116–127 (1988).

29. G. Heskestad and M.A. Delichatsios, *Environments of Fire Detectors, Phase I: Effects of Fire Size, Ceiling Heights, and Material*, Volume II—Analysis Technical Report Serial Number 11427, RC-T-11, Factory Mutual Research Corp., Norwood, MA (1977).

30. K.B. Ginn, *Architectural Acoustics*, Bruel and Kjaer (1978).

31. H. Butler, A. Bowyer, and J. Kew, *Locating Fire Alarm Sounders for Audibility*, Building Services Research and Information Association, Bracknell (1981).

32. E.H. Nober, H. Pierce, A. Well, and C.C. Johnson, *NBS-GCR-83-284*, National Bureau of Standards, Washington (1980).

33. M.J. Kahn, *NBS-GCR-83-435*, National Bureau of Standards, Washington (1983).

34. *British Standard Code of Practice CP3*, British Standards Institution, London (1972).

35. C. Davis and D. Davis, *Sound System Engineering*, Howard H. Sams and Co., Inc., Indianapolis (1975).

36. *Product Catalog*, Fire Control Instruments, Newton (1986).

37. UL 1971, *Standard for Safety Signaling Devices for the Hearing Impaired*, Underwriters Laboratories, Inc., Northbrook, IL (1992).

38. "Nomenclature and Definitions for Illuminating Engineering," IES RP-16-1987, Illuminating Society of North America, 345 East 47th Street, New York, NY 10017.

ADDITIONAL READING

V. Babrauskas, J.R. Lawson, W.D. Walton, and W.H. Twilley, *NBSIR 82-2604*, National Bureau of Standards, Washington (1982).

HYDRAULICS

John J. Titus

INTRODUCTION

Hydraulics may be regarded as the application of knowledge about how fluids behave to the solution of practical problems in fluid flow. It is generally held to describe the behavior and effects of water in motion in both closed conduits and open channels. In the field of fire protection we are concerned primarily with the closed conduit flow regime. In this chapter we will restrict our discussion to the behavior and properties of water flowing in pipes as the phenomenon of paramount interest.

FLUID STATICS

Physical Properties of Fluids

Solution of any flow problem requires a basic knowledge of the physical properties of the fluid being considered. A brief description of the most basic properties follows.

1. Density:

$$\rho = \frac{m}{V}$$

The density of a fluid is the mass of the fluid per unit volume, expressed in S.I. units as kg/m^3 and in English, or U. S. customary, units as $slugs/ft^3$ (or $lbf \cdot s^2/ft^4$). The density of water at 4°C ($\sim$40°F) is 1000 kg/m^3 (1.94 $lbf \cdot s/ft^4$).

2. Specific Weight:

$$\gamma = \rho g$$

As the representation of the force exerted by gravity on a unit volume of fluid, specific weight takes on units of weight per unit volume. At 4°C, the specific weight of water is 9.81 kN/m^3 (62.4 lb/ft^3).[3]

3. Specific Gravity (relative density): The ratio of a liquid's density or specific weight to that of water at 4°C.

John J. Titus is a Research Associate at the Center for Firesafety Studies at Worcester Polytechnic Institute, Worcester, MA. He has been active in engineering design and contracting of fire suppression and detection systems since 1971. Mr. Titus is also a partner in MBS Fire Technology, Inc. a fire protection engineering consulting firm based in central Massachusetts.

4. Viscosity: The term viscosity refers to a proportionality constant in the equation relating cross-sectional velocity variations (or rate of fluid deformation) to shear stresses developed in the fluid flow. (See the section of this chapter titled "Fluid Energy Losses in Pipe Flows.") Viscosity can be considered a measure of a fluid's resistance to deformation or shear or, alternatively, its readiness to flow when acted upon by an external force. In engineering analyses it is useful to think of viscosity as a momentum diffusivity term.

Viscosity is commonly expressed in one of two forms: absolute (or dynamic) viscosity, μ, which is the proportionality constant referred to above, or kinematic viscosity, ν, which is related to the absolute viscosity by the equality:

$$\nu = \frac{\mu}{\rho}$$

A wide variety of units is used to express viscosity, depending not only on U. S. customary or S. I. formulations but also on older English and metric conventions as well as on the type of instrument used to measure this fluid property. A unit based on the c.g.s. (centimeter, gram, second) convention of the old metric system has gained wide favor in the representation of absolute viscosity. This unit, called the poise, has dimensions of dyne $\cdot$ seconds per square centimeter or grams per centimeter $\cdot$ second. The centipoise, which equals 0.01 poise, is the form of preference since the viscosity of water at 20°C (68°F) equals one centipoise to a very close approximation. In the English system the unit of viscosity is pound-seconds per square foot. One lb $\cdot$ s/ft^2 equals 478.8 poise.

5. Fluid Pressure: Pressure is a force per unit area that arises when a fluid is subjected to a compressive stress. Units may be Newtons/m^2, lb/ft^2, lb/in^2, or the equivalent. Pascal's law states that the pressure in a fluid at rest is the same in all directions, a condition different from that for a stressed solid where the stress on a plane depends upon the orientation of that plane. For an infinitesimal fluid element in a larger static body of fluid, a free body diagram of the vertical forces may be drawn as in Figure 4-2.1. The pressure difference $[(p + dp) - p]$ is due only to the weight of the fluid element. Since the weight of the element is given by $mg = \rho g\,dz\,dA$, a summing of forces in the vertical direction gives

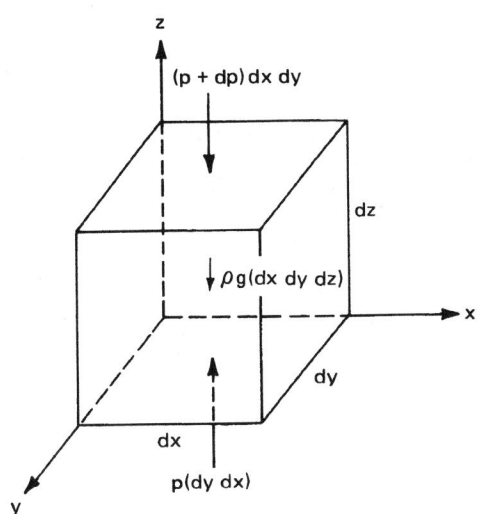

Fig. 4-2.1. Notation for basic equation of fluid statics.

$$dp\, dA = -\rho g\, dz\, dA \qquad (1a)$$

$$dp = -\rho g\, dz \qquad (1b)$$

In integral form, Equation 1b becomes:

$$\int_1^2 \frac{dp}{\rho g} = -\int_1^2 dz = -(z_2 - z_1) \qquad (2)$$

where the path endpoints 1 and 2 refer to different elevation levels.

To integrate Equation 2, it is necessary to establish a functional relation between pressure p and the term ρg. Where density varies with pressure the fluid is considered compressible, and the functional relation may be complex. For fluids that may be considered incompressible, such as water, ρ is a constant at any specified temperature. Equation 2 then becomes

$$p_2 - p_1 = -\rho g(z_2 - z_1) \qquad (3)$$

The term $(z_2 - z_1)$ may be called a static "pressure head," which can be expressed in feet, inches, or meters of water, or some height unit of any liquid. A simplified form of Equation 3 is often written

$$p = \gamma h \qquad (4)$$

where h = elevation of the column of liquid above a reference surface (i.e., $z_2 - z_1$). For water at 60°F, γ is taken equal to 62.4 lb/ft^3. The pressure corresponding to a head of h feet is, then, 0.433h lb/in^2 (psi), or approximately 3 kPa per meter elevation. The head corresponding to a pressure of one psi is, inversely, 2.3 feet. Note that Equation 4 is valid only for a homogeneous, noncompressible fluid at rest, and that regardless of the shape of the container, points in the same horizontal plane experience the same pressure.

The vertical distance h is termed the "head" of a fluid. A pressure due only to the weight of a column of fluid is called a static pressure and can be measured by a standard Bourdon-type gage (see Figure 4-2.4). Such a measure is generally referred to as gage pressure. The term absolute pressure takes into account the pressure exerted by the atmosphere as well, which at sea level is approximately 14.7 psi, equivalent to a 33.9 foot high column of water. A pressure less than atmospheric is called a vacuum pressure, a perfect vacuum being zero absolute pressure. Since most fluid properties of interest are not significantly affected by small changes in atmospheric pressure, most fluids calculations are in terms of gage pressure, although this fact is not often indicated in standard calculation nomenclature. When they are explicitly identified, gage pressure is denoted by the term psig and absolute pressure by psia. If not stated otherwise, psi may be taken to designate gage pressure.

Pressure Measuring Devices

1. Manometer tube: Pressure measurement in a manometer tube is obtained by measuring the vertical displacement of a relatively heavy fluid (usually mercury), which will rise a smaller vertical distance than water in proportion to the ratio of its specific weight to that of water. Depending on the actual arrangement of the manometer tubing, a gage equation can be written to solve for the pressure head. For the manometer shown in Figure 4-2.2, the gage equation is written by proceeding from the open end through the tube to point A', adding terms when descending a column and subtracting when ascending. Using mercury as the manometer fluid, we can write

$$(y + z)\gamma_{Hg} - z\gamma_{Hg} - xy + (x + z)\gamma = p_A \qquad (5)$$

Combining terms, generalizing the result, and expressing in terms of feet of water (head),

$$\frac{p_A}{\gamma} = ys + z \qquad (6)$$

where s is the specific gravity of the manometer fluid.

2. Piezometer tube (Figure 4-2.3): Literally a "pressure measuring tube," a piezometer consists essentially of a narrow tube rising from a container enclosing a fluid under pressure. Through the relation among pressure, height, and specific weight, the height to which the fluid rises in the tube represents the pressure of the contained fluid. While useful

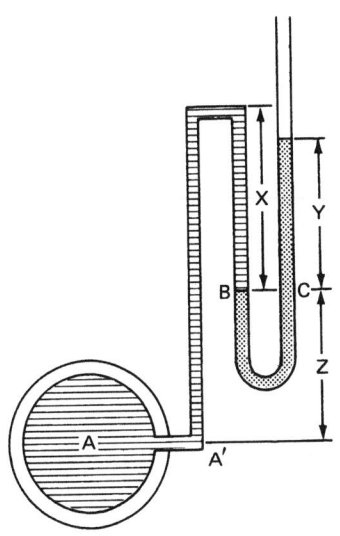

Fig. 4-2.2. Manometer.

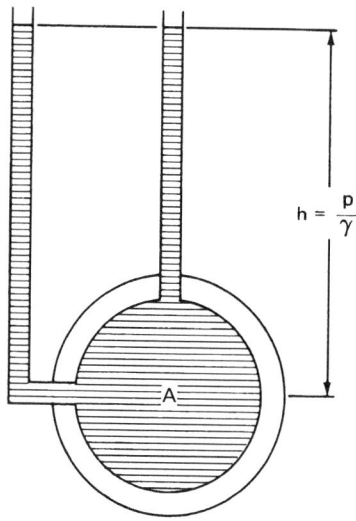

Fig. 4-2.3. Piezometer.

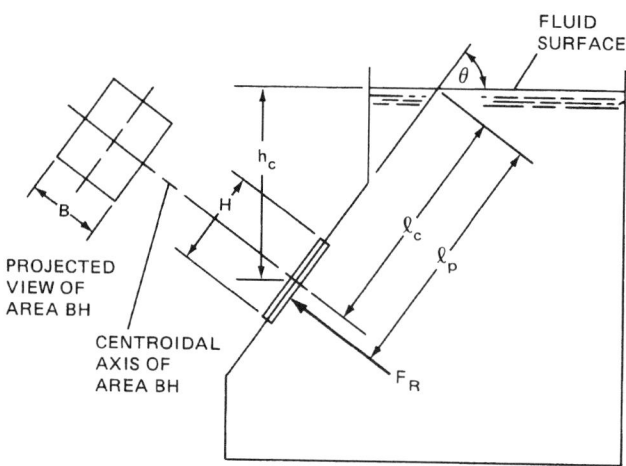

Fig. 4-2.5. Tank with a rectangular window in an inclined wall.

for some laboratory work, piezometer tubes are not generally feasible in practical applications.

3. Bourdon gage (Figure 4-2.4): The standard pressure measuring device used in a wide variety of fluid pressure measurement applications is the Bourdon gage. The gage contains a curved tube of elliptical cross-section that undergoes a change in curvature with change in pressure. A dial hand, connected to the inner tube through a linkage system, indicates gage pressure on a numerical dial face. Bourdon gages are factory calibrated and reasonably accurate instruments if not damaged by pressure surge or impact force. A field reading, unless known to be correct, cannot be assumed to be accurate and should be checked by independent means.

Forces on Submerged Plane Areas Due to Fluid Pressure

It is sometimes of interest to determine the magnitude of the resultant force on a submerged area and the location of the center of pressure where the resultant force can be as-

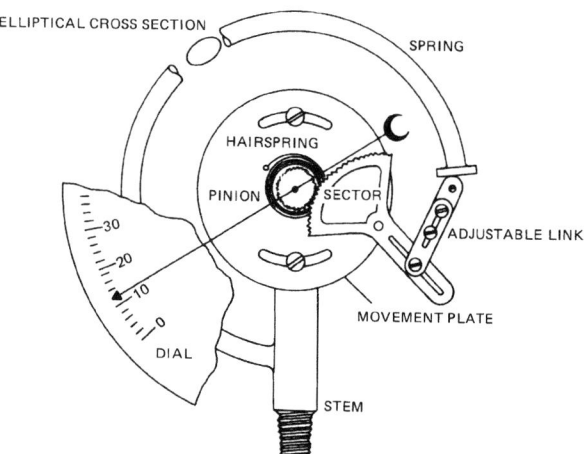

Fig. 4-2.4. Standard Bourdon gage.

sumed to act. Consider the following example of a tank that has a rectangular window in an inclined wall (Figure 4-2.5). The magnitude of the resultant force can be determined from

$$F_R = \gamma h_c A \qquad (7)$$

The center of pressure of the area is the point at which the resultant force can be considered to act. Its location is determined from the relation

$$l_p = l_c + \frac{I}{l_c A} \qquad (8)$$

where I = the moment of inertia of the area about its centroidal axis.

For a vertical wall, the center of pressure is located at a distance $d/3$ from the bottom (where d is the total depth below the free surface), and at a distance $l/3$ from the bottom of an inclined wall or area (where l is the length of the face of the wall).

CONSERVATION LAWS IN FLUID FLOWS

Fluid flow may be characterized as uniform or nonuniform, steady or unsteady, compressible or incompressible, laminar or turbulent, rotational or irrotational, and one-, two-, or three-dimensional or some combination thereof. Real flows may be modeled as approximations of ideal flows when real properties do not depart significantly from the ideal characteristics defined by these terms.

For example, uniform flow occurs when the average velocity of a fluid does not change in either magnitude or direction anywhere along the flow path. Thus, liquid flow in a constant head pipeline of unchanging diameter is considered uniform flow. Steady flow, on the other hand, is determined with reference to a stationary point in the flow path; for steady flow to occur, the velocity of flow at that point must remain constant with time. This condition implies that the fluid density, the pressure head, and the volume rate of flow also are invariant with time. Thus, liquid flow in a constant head pipeline of varying diameter may be considered steady, nonuniform flow. It is important to note that a flow may be considered uniform (no change in magnitude or direction of the velocity) in a curved pipeline as long as the reference direction of the velocity vector is taken in the direction of the flow. We can then say that the velocity of the fluid does not change direction with respect to its enclosing boundaries.

We can also consider this flow one-dimensional whenever it is permissible to say that velocities or accelerations normal to the general direction of the flow are negligible. Clearly, real flow in a real-world structure has three dimensions, but a one-dimensional analysis is highly desirable as it represents a considerable mathematical simplification. Fortunately, a very large number of practical engineering flow problems involving water can be modeled as one-dimensional, steady flow problems, particularly many pipeline flows. In such cases it is possible to apply basic physical principles of conservation of mass and conservation of energy in the direction of flow to obtain the energy balance at any point in the flow. In fire flow hydraulics, it is common practice to introduce additional simplifying assumptions, such as the requirements that the fluid be incompressible and that flow properties be invariant with temperature and pressure. It then follows directly that with no flow additions or subtractions, the volumetric flow rate at any point in a fluid stream must be a constant. This statement of mass conservation, known as the equation of continuity, can be expressed mathematically as

$$\rho_1 A_1 v_1 = \rho_2 A_2 v_2 = \text{constant} \tag{9}$$

If the fluid is considered incompressible, as is the case with water, Equation 9 becomes

$$A_1 v_1 = A_2 v_2 = \text{constant} = Q \tag{9a}$$

By applying the principal of conservation of energy to a flowing fluid, an expression can be derived that gives the theoretical net energy balance of the fluid at any point along its flow path. This is known as the Bernoulli equation, which can be written as

$$\frac{p_1}{\gamma} + \frac{v_1^2}{2g} + z_1 = \frac{p_2}{\gamma} + \frac{v_2^2}{2g} + z_2 \tag{10}$$

In this form, units are ft lb/lb of fluid or, simply, ft of fluid. Each term thus represents a fluid "head." Multiplying each term by the specific weight, γ, converts the equation to units of pressure. Changes in internal energy of the fluid are ignored and are assumed to be negligible. The form of Equation 10 suggests that the flow of liquid (or transport of fluid energy) results from three principal causes: pressure difference, gravity, and inertia. Equation 10 expresses an ideal condition fulfilled by the three components of "head" corresponding to these three causes. The assumption of incompressibility (i.e., constant density) requires that the product of the velocity of flow and the cross-sectional area of the flow of any conserved portion of the stream be constant; the ideal flow streamlines, therefore, converge as the velocity increases and diverge as the velocity decreases. If it could be assumed that the total Bernoulli head were, indeed, constant or, equivalently, if it were possible to obtain total head simply as a function of the coordinates of the moving fluid element, then many hydrokinetic problems could be solved theoretically by mathematically manipulating and extrapo-

lating the Bernoulli equation. Unfortunately, this is not the case. Other energy transfers are possible, and these require use of a more general form of the equation. In addition to the pressure, velocity, and position (elevation) energies possessed by the fluid at Sections 1 and 2, energy may be added to the fluid (work done on the fluid by a pump), lost by the fluid (through friction), or extracted from the fluid (work done by the fluid). Therefore, we write the Bernoulli energy conservation expression in the more general form

$$\underbrace{\left[\frac{p_1}{\gamma} + \frac{v_1^2}{2g} + z_1\right]}_{\substack{\text{Energy at}\\\text{Section 1}}} + \underbrace{h_A}_{\substack{\text{Energy}\\\text{Added}}} - \underbrace{h_L}_{\substack{\text{Energy}\\\text{Lost}}} - \underbrace{h_E}_{\substack{\text{Energy}\\\text{Extracted}}} = \underbrace{\left[\frac{p_2}{\gamma} + \frac{v_2^2}{2g} + z_2\right]}_{\substack{\text{Energy at}\\\text{Section 2}}} \tag{10a}$$

EXAMPLE 1:

Water flows from a reservoir through a pipeline as shown in the following diagram. The flow is considered frictionless and discharges freely at point C.

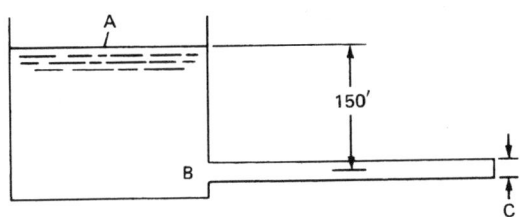

(a) What is the total head (total specific energy) at point B?

(b) What is the discharge velocity at point C?

SOLUTION:

(a) At A, both the velocity and gage pressures are considered to be zero. By Bernoulli, then, the total head would be written as:

$$h_A = 0 + 0 + 150 \text{ ft} = 150 \text{ ft}$$

At B, the fluid has a nonzero velocity head and is under hydrostatic pressure. As long as we consider the flow frictionless, the total head is constant. Therefore

$$h_B = h_A = 150 \text{ ft}$$

(b) At C, the pressure head is again zero, since the discharge is at atmospheric pressure. Once more by Bernoulli

$$h_C = 0 + \frac{v^2}{2g} + 0$$

$$v^2 = 2(32.2)(150)$$

$$v = 98.3 \text{ ft/s}$$

Note that we could calculate the actual values of the pressure and velocity heads at point B if we had more information about the system. For example, we could determine $\dot{Q}$, the discharge at point C, knowing the area and type of discharge opening (see the section of this chapter titled "Free Discharge at an Opening"). This is simply an application of the continuity equation. Knowing the pipeline diameter at point B allows us to apply continuity constraints once again

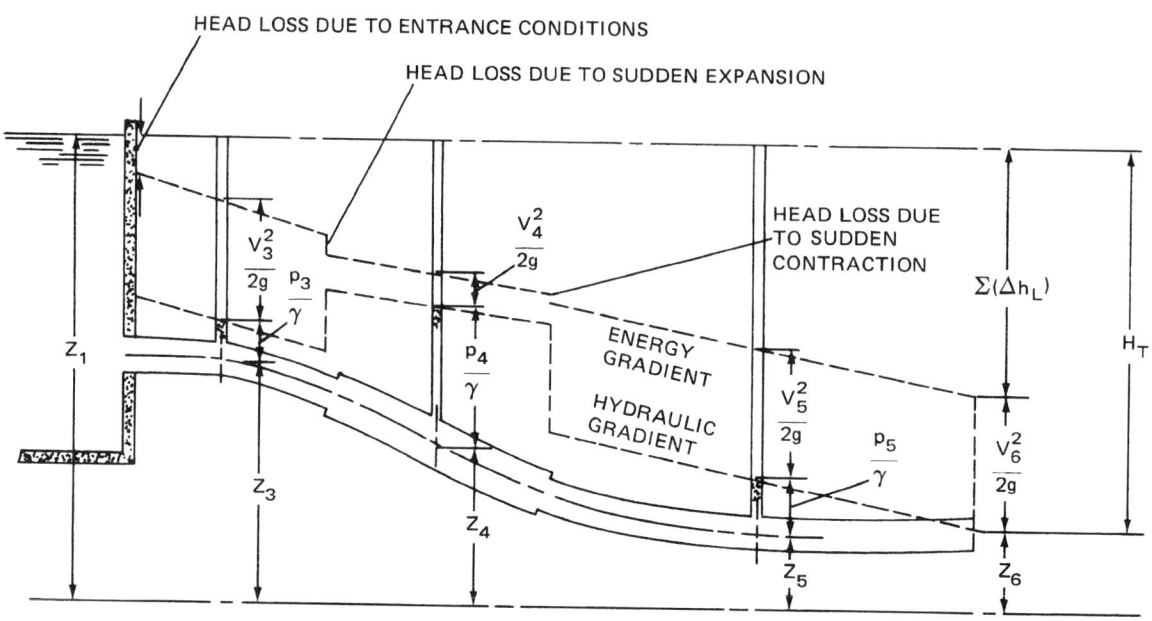

Fig. 4-2.6. Realistic flow characteristics.

to calculate v_B from which the velocity head may be determined. The pressure head at B is simply a function of the weight of the vertical column of water.

The components of the Bernoulli equation may be expressed graphically in terms of energy levels existing at any points in the flow regime. In Figure 4-2.6 a simple system representing a realistic flow is shown. Water flows from a reservoir (with presumed constant surface elevation) to atmosphere. The flow is accompanied by losses of energy represented by h_L. The losses may occur in many places such as at valves, bends, and sudden changes in pipe diameter. Generally, the most important loss is that due to friction between the moving fluid and the pipe wall. Since there are always energy losses in real flows, the total energy of the system decreases in the direction of flow. The line connecting all points representing the total energy is called the "energy gradient." It must always decrease in the direction of flow unless energy is added to the system, such as by a pump. The "hydraulic gradient" connects the points representing the sum of static pressure and elevation energies (i.e., the heights to which water in piezometer tubes would rise). Note that the hydraulic gradient may increase in the direction of flow if velocity head is converted to pressure head at any point (such as at an increase in pipe diameter). Thus, the relationship between the energy and hydraulic gradients can be written as

$$\text{E.G.} = \text{H.G.} + \frac{v^2}{2g} \qquad (11)$$

FLUID ENERGY LOSSES IN PIPE FLOWS

General Considerations

That part of hydraulics which treats fluid energy losses due to friction in piping is well known and well utilized in

fire protection engineering. Losses to friction are due to shear stresses set up within a moving fluid in a conduit by an imposed pressure gradient. Flow driven by the pressure force is restrained by drag forces acting at the conduit wall. To better visualize this phenomenon, it is useful to introduce the concept of the boundary layer. For many fluids, such as air or water, motion through a stationary conduit or pipe is characterized in most practical situations by a nearly constant velocity cross-section everywhere except in a very thin layer near the wall of the pipe. This layer thickness may be as little as 0.1 mm but may vary significantly with the nature of the fluid, the velocity of flow, and the surface roughness of the conduit. We may visualize boundary layer flow (see Figure 4-2.7) in terms of a velocity profile. Theory* holds that a very thin (molecular) layer of fluid sticks to the conduit wall. The tendency to motion of the next fluid layer due to an imposed force creates a shearing stress, τ, between the layers. If the boundary is thought of as many fluid lamina sliding on each other, then we can expect the velocities of these lamina to increase with distance y from the wall until, at the edge of the boundary layer, the local velocity reaches the free stream velocity of the fluid. The factor relating the velocity profile to the developed stress in the fluid is termed the fluid viscosity. The relationship was expressed mathematically by Newton as

$$\tau = \mu \frac{du}{dy} \qquad (12)$$

The smaller the value of fluid viscosity, the thinner the boundary layer will be. The first layer of fluid sticks or adheres to the surface of the conduit while lamina above it successively slide on each other, exerting drag forces which, for most fluids, are proportional to the viscosity (so-called Newtonian fluids). The rate of change of the velocity between successive lamina is a measure of the unit shearing

---

* Early theoretical development of the boundary layer concept is due primarily to Prandtl.[1,9]

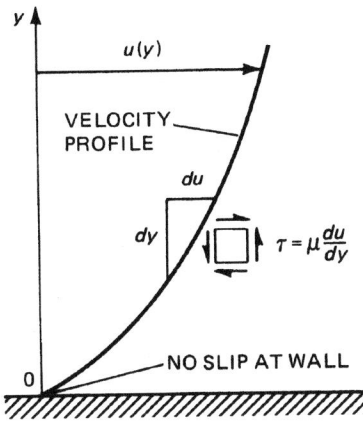

Fig. 4-2.7. *Velocity profile.*

force between them. A curve joining the tips of velocity vectors plotted for the different lamina in the boundary layer is called a velocity profile. Laminar (smooth, streamline) flow (see Figure 4-2.8a) is characterized by a parabolic velocity profile with maximum velocity attained at the theoretical centerline of the flow. Turbulent flow, by contrast, is rough (nonstreamline) flow (see Figure 4-2.8b), characterized by an essentially uniform average velocity across the flow section, with only a very thin boundary layer close to the wall where viscous forces predominate. The velocities associated with laminar flows are generally so low as to be impracticable from a design standpoint. Most flows of interest, therefore, are turbulent and the use of an assumed uniform or average velocity to calculate kinetic energy and velocity pressures does not introduce significant error. In those atypical situations where relatively large velocity heads are involved (such as where a pump adds a large amount of energy), a correction factor may be used to relate the actual average kinetic energy to the kinetic energy calculated using average velocity. From continuity considerations,

$$\text{KE} = \int_A \rho u^3 \, dA = \alpha \rho \int_A v^3 \, dA \qquad (13)$$

where

KE = true kinetic energy of the flow,
v = average velocity of flow, and
α = kinetic energy correction factor.

For incompressible fluids

$$\alpha = \frac{1}{A} \int_A \left(\frac{u}{v}\right)^3 \, dA \qquad (14)$$

and has the value of approximately 1.1 for most turbulent flow problems. Since the velocity head in water distribution piping is small, this correction factor is usually ignored.

While the development of boundary layer theory and the theory of viscous forces has led to an improved theoretical understanding of the mechanics of pipe flows, most flows of interest in fire protection cannot be fully analyzed from theoretical considerations alone. Fire protection flows are almost always turbulent flows. Despite a great expenditure of effort to develop a general predictive theory of turbulent flow phenomena, a fully descriptive theory does not yet exist. While it is postulated that head losses arise because of friction between the fluid and the pipe wall, there is an additional head loss contribution due to turbulence within the fluid. In turbulent flows the rate of head loss, unfortunately, is not simply a function of velocity but depends also on pipe wall roughness. It is further complicated by the changing interaction between these variables at different velocities and roughness element sizes. Within the last century, however, a large body of empirical flow data has been collected, analyzed, and reduced by several investigators. The major features and limits of applicability of the more important results are presented in the following paragraphs.

Fluid Flow Energy Loss Equations

Theoretical development of the physical relationships describing pipe flows dates from about the middle of the nineteenth century, when Chezy postulated a fundamental proportionality between volumetric flow and pipe size based on the continuity equation. His formula is commonly given as

$$Q = \frac{\pi D^2}{4} v = \frac{\pi D^2}{4} \frac{C}{2} \sqrt{DS} \qquad (15)$$

and may also be written as

$$S = \left(\frac{8Q}{\pi C}\right)^2 D^{-5} \qquad (16)$$

where D and S are pipe diameter and slope of the energy grade line, respectively. The factor C is a proportionality factor incorporating a significant degree of physical uncertainty. Since, by definition

$$S = \frac{h_L}{L}$$

we may write

$$h_L = \left(\frac{8}{\pi C}\right)^2 \frac{L}{D^5} Q^2 \qquad (17)$$

as an expression for pipe flow head loss as a function of pipe diameter and discharge. Use of this equation was limited by uncertainties relating to evaluations of the C-factor, which is not, in fact, a constant for a given size conduit or wall condition as was originally thought.

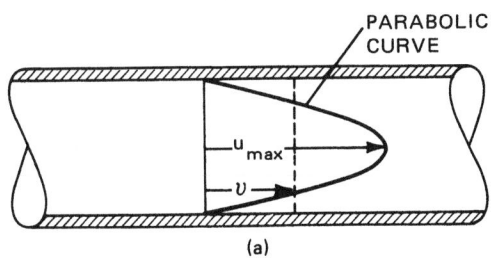

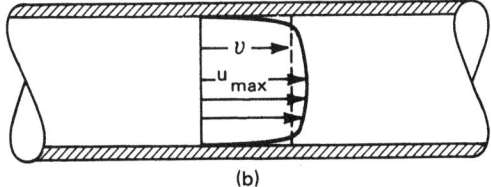

Fig. 4-2.8. *Laminar (a) and turbulent (b) pipe flow velocity profiles for the same volume.*

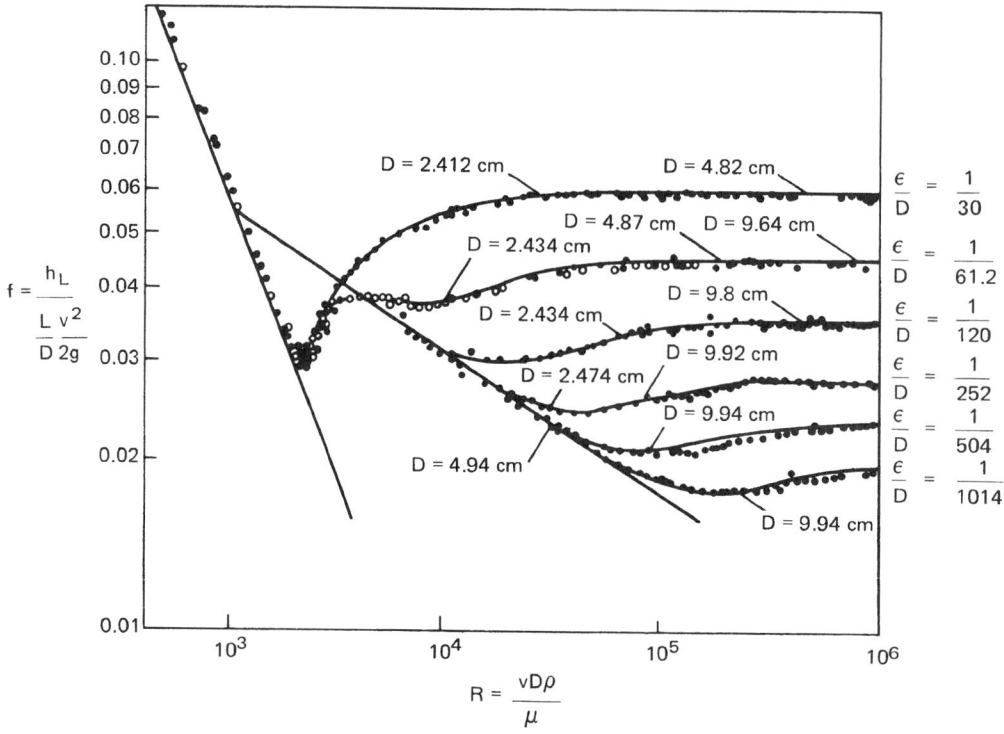

Fig. 4-2.9. Nikuradse's sand-roughened-pipe tests.

A theoretically more satisfying approach was taken by Darcy, Weisbach, and others. Their formula, which bears the names of the two primary investigators, is generally written as

$$h_L = f\frac{L}{D}\frac{v^2}{2g} \qquad (18)$$

It postulates a basic proportionality between head loss and the kinetic energy of the flow, as well as to pipe length and diameter. The proportionality factor f, known as the friction factor, became the subject of extensive theoretical and experimental investigation. The value of f for laminar flow can be shown theoretically to be a simple linear function of the Reynolds number, Re, where Re = $D_e v\rho/\mu$. The term D_e is the equivalent flow diameter, which is the actual inside diameter of a circular pipe flowing full. The equivalent diameter, D_e can be found from the hydraulic radius, r_h, which is defined as the area in flow divided by the wetted perimeter. The wetted perimeter does not include the free fluid surface.

$$D_e = 4r_h$$

For Re less than about 2000 (corresponding to low velocity flows or fluids of high viscosity) the relation is

$$f = \frac{64}{\text{Re}} \qquad (19)$$

In turbulent flows the roughness of the pipe walls becomes a much more significant factor, and a simple expression to determine f is unavailable.

A systematic investigation of the actual characterisitics of piping inner wall surfaces was first performed by Nikuradse in 1933. To simulate varying degrees of roughness in commercial pipes due to corrosion or surface finish, Nikuradse glued sand grains of known sizes to the inside walls

of test pipes. The resulting logarithmic plot of friction factor versus Re is shown in Figure 4-2.9. Although the tests are from Nikuradse, the plot is called Stanton's diagram in recognition of his earlier (1914) elucidation of the relation between friction factor and Reynolds number. Note that at sufficiently high Re, the friction factor depends almost entirely on pipe roughness and is essentially independent of Re. In these plots the roughness parameter is expressed as the ratio of the root mean square grain diameter to the pipe diameter. The resulting ratio is termed the relative roughness and is represented mathematically as ε/D. Typical roughnesses of new commercial pipe is shown in Table 4-2.1. Several later investigators developed mathematical formulas for expressing Nikuradse's results. Chief among them were Colebrook, Moody, and VonKarman.

Moody plotted various equations on a graph similar to the earlier work of Stanton. The resulting Moody diagram (Figure 4-2.10) is widely used today in conjunction with the Darcy-Weisbach equation to compute friction losses for water flowing in pipe. Figure 4-2.11 presents relative roughness values for use with the Moody diagram over a wide range of conditions. Other diagrams have been developed for use with the Darcy-Weisbach equation[7,11] when parameters other than h_L are sought. Essentially, the alternative graphical formulations employ a rearrangement of variables to facilitate solving for some other unknown such as Q or D.

Both experimental and theoretical investigations have yielded uncertain results in the region known as the "critical zone," wherein the flow changes from laminar to turbulent. This may be expected since the transition point is difficult to define precisely and, in fact, varies over a considerable range of Re depending upon the direction of the transition (i.e., flow going from laminar to turbulent or from turbulent to laminar) and the local conditions affecting flow stability. As

TABLE 4-2.1 *Values of Absolute Roughness of New Clean Commercial Pipes*

Type of Pipe or Tubing	ε ft (0.3048 m) × 10⁶ Range	Design	Probable maximum variation of f from design (%)
Asphalted cast iron	400	400	−5 to −5
Brass and copper	5	5	−5 to +5
Concrete	1,000 to 10,000	4,000	−35 to 50
Cast iron	850	850	−10 to +15
Galvanized iron	500	500	0 to +10
Wrought iron	150	150	−5 to 10
Steel	150	150	−5 to 10
Riveted steel	3,000 to 30,000	6,000	−25 to 75
Wood stave	600 to 3,000	2,000	−35 to 20

Source: *Pipe Friction Manual, 3d ed.* Hydraulic Institute, 1961.

a practical consideration, however, this uncertainty is of little importance, since most real flows of interest fall well into the turbulent range.

Colebrook developed an empirical transition function for the region between smooth flow and complete turbulence. Flow in this region is sometimes referred to as hydraulically smooth or turbulent smooth. The equation has been presented in various forms, the following expression being commonly used

$$\frac{1}{\sqrt{f}} = -0.86 \ln\left(\frac{\varepsilon/D}{3.7} + \frac{2.51}{\mathrm{Re}\,\sqrt{f}}\right) \tag{20}$$

An alternate and equivalent expression is

$$f = \left[1.14 - 2\log\left(\frac{\varepsilon}{D} + \frac{9.35}{\mathrm{Re}\,\sqrt{f}}\right)\right]^{-2} \tag{20a}$$

This relation forms the primary basis for the Moody diagram.

VonKarman used boundary layer theory to derive an expression characterizing the friction factor for fully turbulent flow within rough-walled pipes. The final numerical form of the equation,

$$\frac{1}{\sqrt{f}} = 1.4 + 2\log\frac{D}{\varepsilon} \tag{21}$$

was adjusted to agree more closely with Nikuradse's experimental results. As pipe roughness decreases, this expression approaches Colebrook's equation.

Perhaps the most widely used flow-energy loss relation is the empirically based Hazen-Williams formula, developed near the turn of the century from observations of a very large number of pipeline flows. The Hazen-Williams equation was originally written in the form

$$V = 0.113 C D^{0.63} S^{0.54} \tag{22}$$

where V is the average velocity in feet per second, S is the slope of the energy grade line—i.e., the loss of energy per

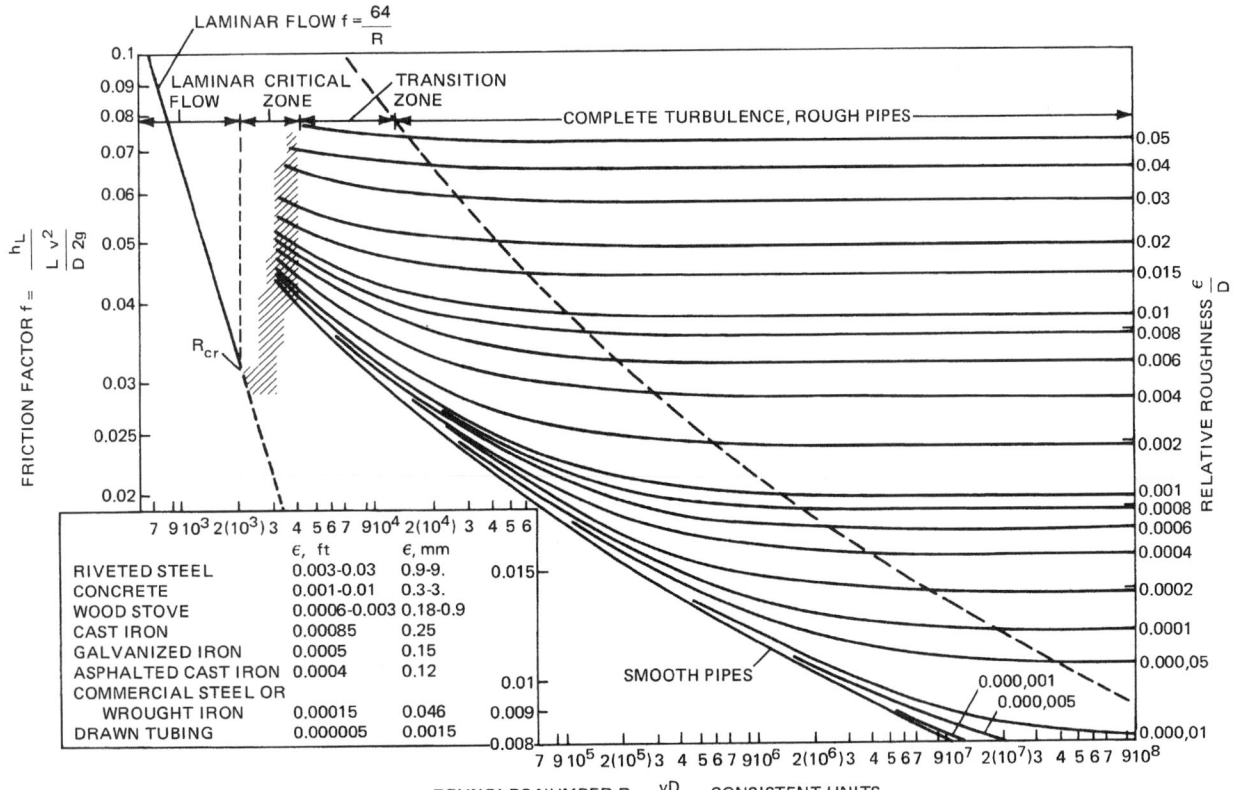

Fig. 4-2.10. Moody diagram.

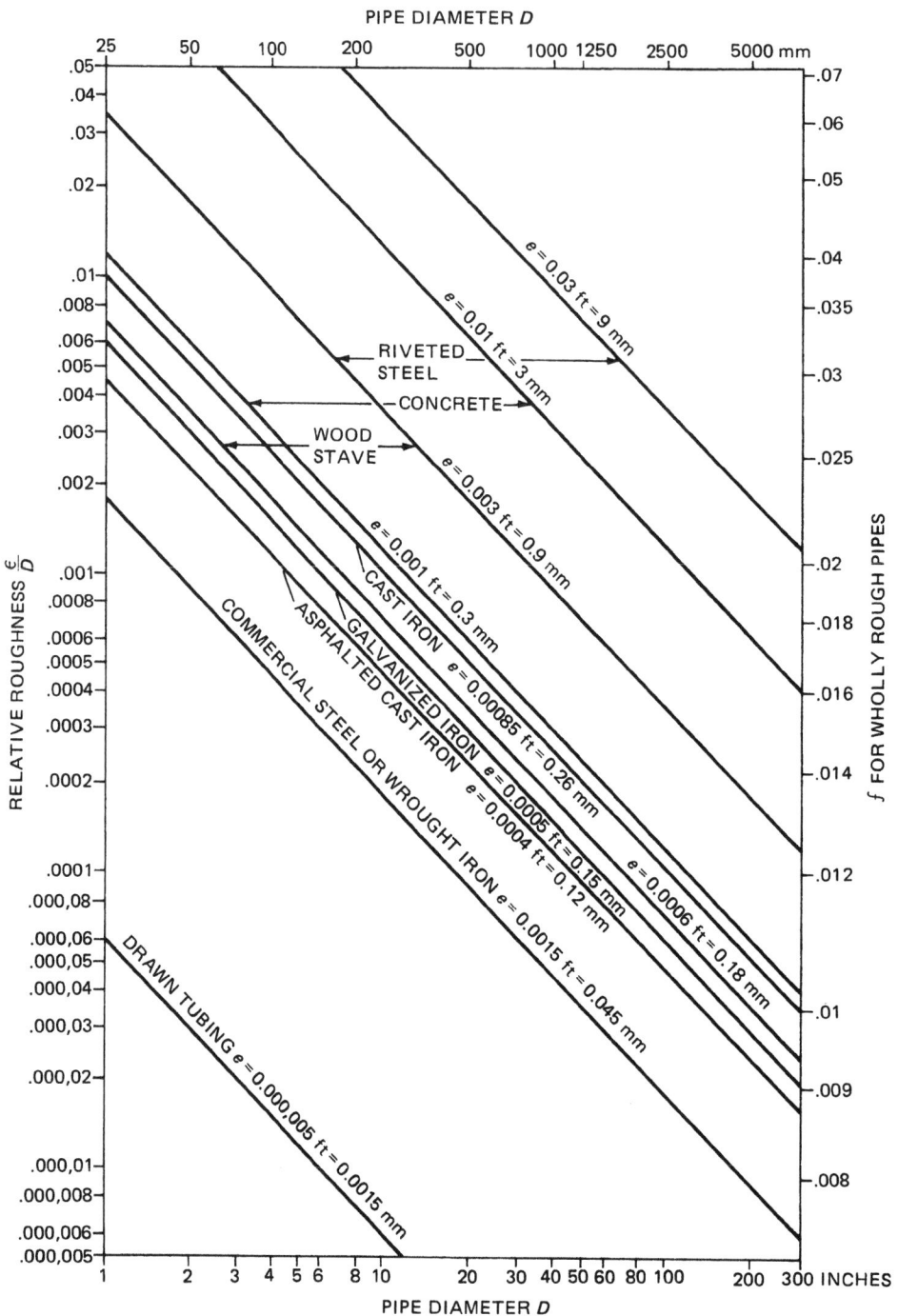

Fig. 4-2.11. Relative roughness chart.

unit length of the pipe—and D is the actual internal pipe diameter in inches. The coefficient C is a "friction" factor introduced as a constant to represent the roughness of the pipe walls. Table 4-2.2 presents a representative list of C coefficients for various piping materials. Note that the value of C can vary significantly with the piping material, the age of the pipe, and the corrosive qualities of the water.

The Hazen-Williams formula is also encountered in the form

$$Q = 0.285CD^{2.63}S^{0.54} \qquad (22a)$$

where Q is volumetric flow rate in gpm and D is in inches. Yet another form, also in the same units for Q and D, is

TABLE 4-2.2 *Values of C in Hazen-Williams Formula**

Type of Pipe	C Values for Certain Pipe Diameters					
	2.5 cm (1 in.)	7.6 cm (3 in.)	15.2 cm (6 in.)	30.5 cm (12 in.)	61 cm (24 in.)	122 cm (48 in.)
Uncoated cast iron—smooth and new		121	125	130	132	134
Coated cast iron—smooth and new		129	133	138	140	141
30 years old						
Trend 1—slight attack		100	106	112	117	120
Trend 2—moderate attack		83	90	97	102	107
Trend 3—appreciable attack		59	70	78	83	89
Trend 4—severe attack		41	50	58	66	73
60 years old						
Trend 1—slight attack		90	97	102	107	112
Trend 2—moderate attack		69	79	85	92	96
Trend 3—appreciable attack		49	58	66	72	78
Trend 4—severe attack		30	39	48	56	62
100 years old						
Trend 1—slight attack		81	89	95	100	104
Trend 2—moderate attack		61	70	78	83	89
Trend 3—appreciable attack		40	49	57	64	71
Trend 4—severe attack		21	30	39	46	51
Miscellaneous						
Newly scraped mains		109	116	121	125	127
Newly brushed mains		97	104	108	112	115
Coated spun iron—smooth and new		137	142	145	148	148
Old—take as coated cast iron of same age						
Galvanized iron—smooth and new	120	129	133			
Wrought iron—smooth and new	129	137	142			
Coated steel—smooth and new	129	137	142	145	148	148
Uncoated steel—smooth and new	134	142	145	147	150	150
Coated asbestos-cement—clean		142	149	150	152	
Uncoated asbestos-cement—clean		142	145	147	150	
Spun cement-lined and spun bitumen-lined—clean		147	149	150	152	153
Smooth pipe (including lead, brass, copper, polythene, and smooth PVC)—clean	140	147	149	150	152	153
PVC wavy—clean	134	142	145	147	150	150
Concrete—Scobey						
Class 1—$C_s = 0.27$; clean		69	79	84	90	95
Class 2—$C_s = 0.31$; clean		95	102	106	110	113
Class 3—$C_s = 0.345$; clean		109	116	121	125	127
Class 4—$C_s = 0.37$; clean		121	125	130	132	134
Best—$C_s = 040$; clean		129	133	138	140	141
Tate relined pipes—clean		109	116	121	125	127
Prestressed concrete pipes—clean				147	150	150

*The above table has been compiled from an examination of 372 records. It is emphasized that the Hazen-Williams formula is not suitable for the coefficient C values appreciably below 100, but the values in the above table are approximately correct at a velocity of 0.9 m/s (3 ft/s).

For other velocities the following approximate corrections should be applied to the values of C in the table above.

Values of C at 0.9 m/s	Velocities Below 0.9 m/s for Each Halving, Rehalving of Velocity Relative to 0.9 m/s	Velocities Above 0.9 m/s for Each Doubling, Redoubling of Velocity Relative to 0.9 m/s
C below 100	add 5 percent to C	subtract 5 percent from C
C from 100 to 130	add 3 percent to C	subtract 3 percent from C
C from 130 to 140	add 1 percent to C	subtract 1 percent from C
C above 140	subtract 1 percent from C	add 1 percent to C

widely used in automatic sprinkler system design. It is arranged to solve for the pressure drop in psi per linear foot of pipe

$$p = \frac{4.52 Q^{1.85}}{C^{1.85} D^{4.87}} \qquad (22b)$$

In SI units

$$p = \frac{6.05 Q^{1.85}}{C^{1.85} D^{4.87}} \times 10^5 \qquad (22c)$$

where the units of Q are L/min, D is in mm, and p is in bars per meter of pipe.

Many manufacturers of fire protection equipment, many fire underwriters, and others have published Hazen-Williams-based pipe friction loss data (usually in tabular

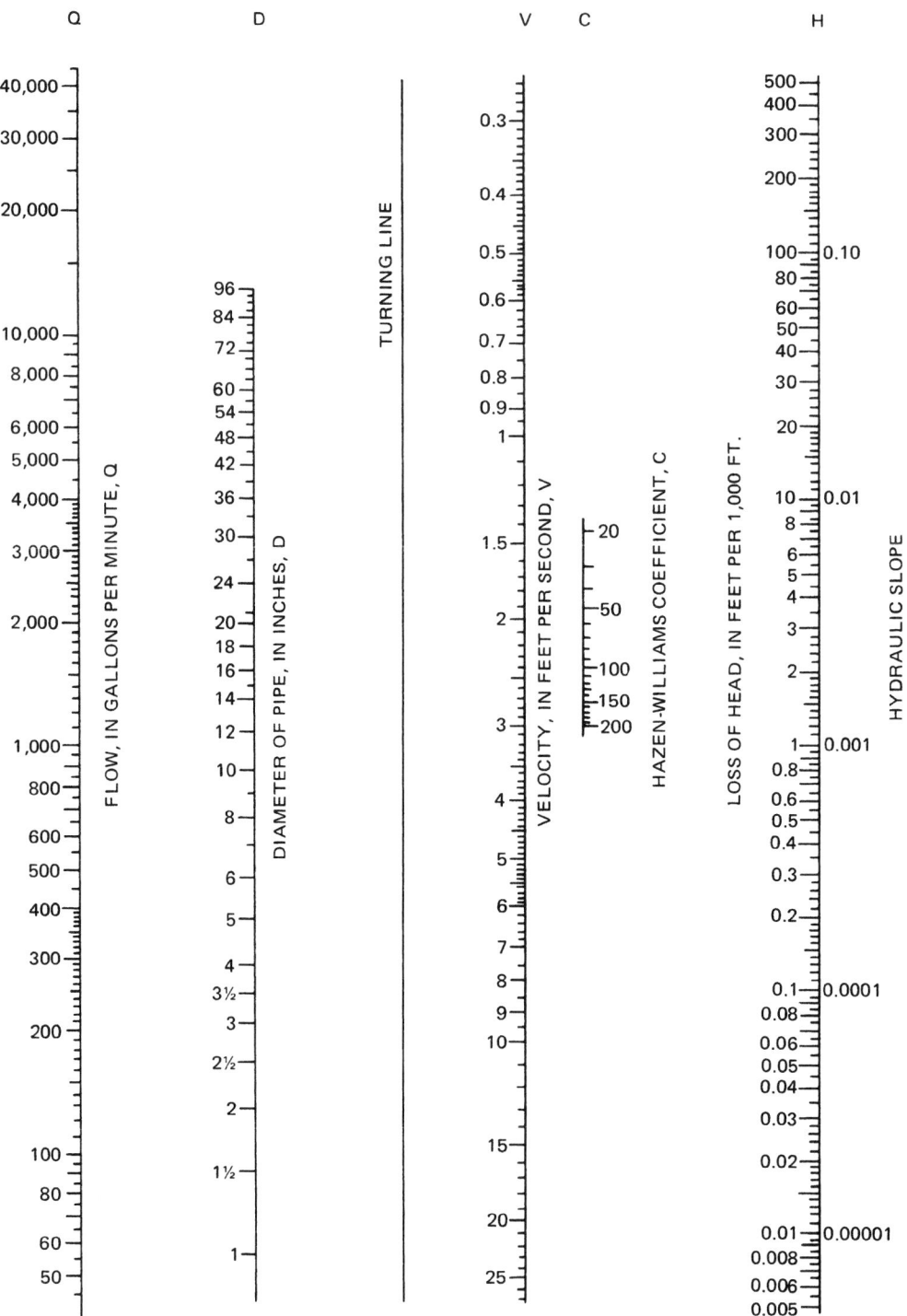

Fig. 4-2.12.   Nomograph for solution of the Hazen-Williams formula.

format) over applicable ranges of pipe sizes, flow rates, and C-factors. A useful calculation aid in a more compact format is the Hazen-Williams nomograph (Figure 4-2.12), which is reproduced here in its generalized form.

The Hazen-Williams formula may be used only for water at or around 60°F, as it does not contain any terms relating to the physical properties of the fluid. The formula does not actually have a sound theoretical basis, but still gives

acceptable results in practice with a judicious choice of the C coefficient. Fundamentally, the C-factor is a proportionality constant and, as such, its "true" value depends as much upon the values chosen for the exponents as it does upon actual pipe roughness. The suggested values are the result of curve fitting exercises and cannot be expected to accurately and evenly represent flow parameter relationships across the full range of observed flow velocities. Allowing the desirability of retaining constant exponents for D and S (i.e., a presumed theoretically stable correlation among all flow parameters in the equation), the value of C for any given flow scenario becomes a narrowly bounded variable that reflects the pipe roughness. Although as in the Chezy formula C is not actually a constant, for practical use it is assigned a constant value for a given presumed roughness. Unfortunately, as Table 4-2.2 shows, the Hazen-Williams equation is a much better model of smooth pipe flow than of rough pipe flow. As long as the flow velocity is close to that at which C was measured and as long as the pipe roughness is not excessive, the Hazen-Williams relation can be expected to give reliable results. It has been noted, however, that in rough pipes head loss varies with flow (and velocity) to the power of 2 rather than the 1.85 power characteristic of smooth pipes.[10] This observation introduces a significant element of uncertainty into the hydraulic analysis of rough pipe with higher velocity flows.

EXAMPLE 2:

Water at 50°F flows through 4 in. Schedule 40 welded steel pipe at a rate of 500 gpm. Compare the friction head losses calculated by the Darcy-Weisbach and Hazen-Williams equations for flow through 100 ft of pipe.

SOLUTION:

Basic Data:
For 50°F water, $\nu = 1.41 \times 10^{-5}$ ft²/s
pipe flow area = 0.0884 ft²
$\varepsilon = 0.0002$
pipe inside diameter = 0.3355 ft = 4.026 in.

Using the Darcy approach, we first determine Re, ε/D, and then enter the Moody diagram (Figure 4-2.10).

$$\text{Flow quantity} = Q = 500 \text{ gpm} \equiv 1.1140 \text{ cfs}$$

$$\text{Velocity} = v = \frac{Q}{A} = \frac{1.1140}{0.0884} = 12.60 \text{ fps}$$

$$\text{Re} = \frac{Dv}{\nu} = \frac{0.3355(12.60)}{1.41 \times 10^{-5}} = 3.0 \times 10^5$$

$$\frac{\varepsilon}{D} = \frac{0.0002}{0.3355} = 0.0006$$

From the Moody friction chart, $f = 0.0188$

From Equation 18

$$h_L = \frac{0.0188(100)(12.60)^2}{2(0.3355)(32.2)} = 13.8 \text{ ft} = 5.98 \text{ psi}$$

The Hazen-Williams approach—Equation 22b—does not take into account any variability in the physical properties of water. It is more straightforward to apply but will likely be less accurate. If we assume a C-factor of 100 and solve directly for pressure drop in psi per 100 ft we obtain

$$\Delta p = \frac{4.52(100)(500)^{1.85}}{(100)^{1.85}(4.026)^{4.87}} = 10.06 \text{ psi}$$

If we take $C = 140$,

$$\Delta p = 5.40 \text{ psi}$$

a drop of nearly 50 percent and much closer to the Darcy result.

Accuracy in using Hazen-Williams clearly depends on a careful choice of C-factor. The Darcy result is not nearly so sensitive to choice of specific roughness ε.

Minor Losses

Flows through pipe fittings, valves, or other pipeline fixtures generate additional turbulence and, therefore, additional energy losses. These losses, although termed minor, can be very significant fractions of the total energy loss due to friction in a piping system. In particular, losses due to pipeline obstructions such as swing-type check valves and certain types of flow meters are equivalent to many feet (or meters) of straight pipe losses. Thus, in some instances "minor" losses may have to be considered major, particularly in systems where there are many fittings, valves, or other appurtances. Empirical methods are used to determine these losses for a range of flow or obstruction geometries. One common method is to define a "minor loss coefficient" to express head loss as a function of velocity head. Thus,

$$h_L = k\frac{v^2}{2g} \qquad (23)$$

where k is a dimensionless loss coefficient. It is sometimes convenient to express such losses in terms of "equivalent length of straight pipe," or as pipe diameters that produce the same head loss. Thus, by Darcy-Weisbach,

$$\frac{L}{D} = \frac{k}{f} \qquad (24)$$

Table 4-2.3 shows local loss coefficients for a number of fittings and flow patterns. Wherever possible manufacturers' data should be used, particularly for valves because of the wide variety of designs even for valves of the same generic type. Such data is often published in the form of flow coefficient or C_v values, which may be used in the equation

$$Q = C_v\sqrt{h_L} \qquad (25)$$

C_v is determined from the relation

$$C_v = \pi D^2 \sqrt{\frac{g}{8k}} \qquad (26)$$

which results directly from a combination of the continuity equation with the equations above.

EXAMPLE 3:

Table 4-2.4 lists a number of "equivalent lengths" of standard Schedule 40 pipe for screwed steel fittings and valves. Using the table determine the equivalent length of the 2 in. diameter pipe network shown below.

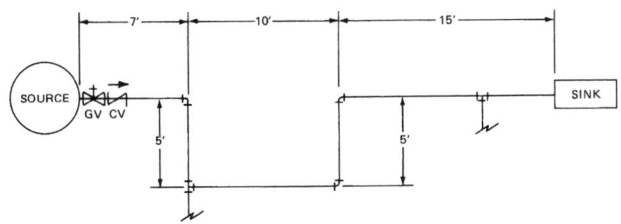

TABLE 4-2.3 *Local Loss Coefficients*

Use the equation $h_L = kv^2/2g$ unless otherwise indicated. Energy loss E_L equals h_v head loss in feet.

1	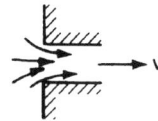	Perpendicular square entrance: $k = 0.50$ if edge is sharp

2

Perpendicular rounded entrance:

$R/d =$	0.05	0.1	0.2	0.3	0.4
$k =$	0.25	0.17	0.08	0.05	0.04

3
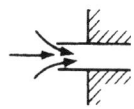
Perpendicular re-entrant entrance:
$k = 0.8$

4

Additional loss due to skewed entrance:
$k = 0.505 + 0.303 \sin \alpha + 0.226 \sin^2 \alpha$

5
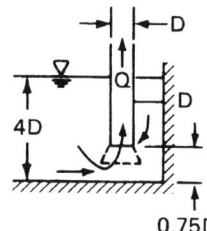
Suction pipe in sump with conical mouthpiece:

$$E_L = D + \frac{5.6Q}{\sqrt{2g}\, D^{1.5}} - \frac{v^2}{2g}$$

Without mouthpiece:

$$E_L = 0.53\,D + \frac{4Q}{\sqrt{2g}\, D^{1.5}} - \frac{v^2}{2g}$$

Width of sump shown: $3.5D$

(After I. Vágás)

6
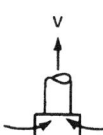
Strainer bucket:
$k = 10$ with foot valve
$k = 5.5$ without foot valve

(By Agroskin)

7
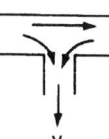
Standard Tee, entrance to minor line:
$k = 1.8$

8
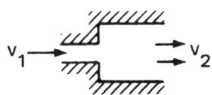
Sudden expansion:

$$E_L = \left(1 - \frac{v_2}{v_1}\right)^2 \frac{v_1^2}{2g}$$

or

$$E_L = \left(\frac{v_1}{v_2} - 1\right)^2 \frac{v_2^2}{2g}$$

9
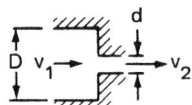
Sudden contraction:

$(d/D)^2 =$	0.01	0.1	0.2	0.4	0.6	0.8
$k =$	0.5	0.5	0.42	0.33	0.25	0.15

use v_2 in equation (4.15)

TABLE 4-2.3 *Local Loss Coefficients (Continued)*

10

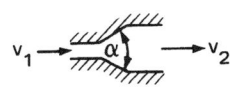

Diffusor:

$E_L = k(v_1^2 - v_2^2)/2g$

$\alpha° =$	20	40	60	80
$k =$	0.20	0.28	0.32	0.35

11

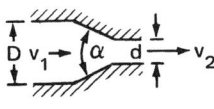

Confusor:

$E_L = k(v_1^2 - v_2^2)/2g$

$\alpha° =$	6	10	20	40	60	80	100	120	140
k for $D = 3d$	0.12	0.16	0.39	0.80	1.0	1.06	1.04	1.04	1.04
$D = 1.5d$	0.12	0.16	0.39	0.96	1.22	1.16	1.10	1.06	1.04

12

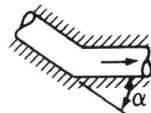

Sharp elbow:
$k = 67.6 \times 10^{-6}(\alpha°)^{2.17}$

(By Gibson)

13

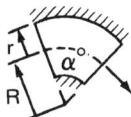

Bends:
$k = [0.13 + 1.85(r/R)^{3.5}] \sqrt{\alpha°/180°}$

(By Hinds)

14

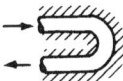

Close return bend:
$k = 2.2$

15

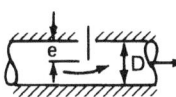

Gate valve:

$e/D =$	0	1/4	3/8	1/2	5/8	3/4	7/8
$k =$	0.15	0.26	0.81	2.06	5.52	17.0	97.8

16

Globe value:
$k = 10$ when fully open

17

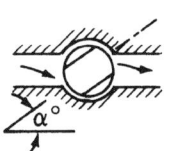

Rotary valve:

$\alpha° =$	5	10	20	30	40	50	60	70	80
$k =$	0.05	0.29	1.56	5.47	17.3	52.6	206	485	∞

(By Argroskin)

18

Check valves:
Swing type $k = 2.5$ when fully open
Ball type $k = 70.0$
Lift type $k = 12.0$

19

Angle valve:
$k = 5.0$ if fully open

20

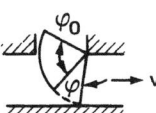

Segment gate in rectangular conduit:

$k = 0.3 + 1.3\left[\left(\dfrac{1}{n}\right)\right]^2$

where $n = \varphi/\varphi_0 =$ the rate of opening with respect to the central angle

(By Abelyev)

21

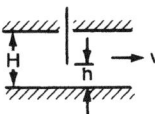

Sluice gate in rectangular conduit:

$k = 0.3 + 1.9\left[\left(\dfrac{1}{n}\right) - n\right]^2$

where $n = h/H.$

(By Burkov)

TABLE 4-2.4 *Typical Equivalent Lengths of Schedule 40 Straight Pipe for Screwed Steel Fittings and Valves (For any fluid in turbulent flow)*

Fitting Type	Equivalent Length, ft Pipe Size		
	1″	2″	4″
Regular 90° Elbow	5.2	8.5	13.0
Long Radius 90° Elbow	2.7	3.6	4.6
Regular 45° Elbow	1.3	2.7	5.5
Tee, flow through line (run)	3.2	7.7	17.0
Tee, flow through stem	6.6	12.0	21.0
180° Return Bend	5.2	8.5	13.0
Globe Valve	29.0	54.0	110.0
Gate Valve	0.84	1.5	2.5
Angle Valve	17.0	18.0	18.0
Swing Check Valve	11.0	19.0	38.0
Coupling or Union	0.29	0.45	0.65

SOLUTION:

The line comprises:

1 check valve		19.0 ft
3 90° standard elbows	3 × 8.5 =	25.5 ft
1 tee (flow through run)		7.7 ft
1 tee (flow through branch or stem)		12.0 ft
1 gate valve		1.5 ft
straight pipe		42.0 ft
		L_e = 107.7 ft

The Darcy equation for determining friction losses through the network would then have the form

$$h_L = \frac{f L_e v^2}{2Dg}$$

Alternately, the loss coefficient approach may be used, where

$$h_L = k \frac{v^2}{2g}$$

This method must be used to find entrance and exit losses. For this example, however, we either refer to manufacturer's data for valve and fitting C_v values or calculate k from the relation

$$k = \frac{f L_e}{D}$$

Energy Losses in Pipe Networks

Flow networks may consist of pipes arranged in series, parallel, or some more complicated configuration. In any case, an evaluation of friction losses for the flows is based on energy conservation principles applied to the flow junction points. Methods of solution depend on the particular piping configuration. In general, however, they involve establishing a sufficient number of simultaneous equations or employing a friction loss formula where the friction coefficient depends only on the roughness of the pipe (e.g., Darcy or Hazen-Williams).

Pipes in series: When two pipes of different sizes or roughnesses are connected in series [Figure 4-2.13(a)], head loss for a given discharge, or discharge for a given head loss, may be calculated by applying the energy equation between

the bounding points, taking into account all losses in the interval. Thus, head losses are cumulative.

Series pipes may be treated as a single pipe of constant diameter to simplify the calculation of friction losses. The approach involves determining an "equivalent length" of a constant diameter pipe which has the same friction loss and discharge characteristics as the actual series pipe system. Minor losses due to valves and fittings are also included. Using the previous example once again, we note that application of the continuity equation to the solution allows the head loss to be expressed in terms of only one pipe size.

The lost head in equivalent feet of 6 in. pipe is then given in Darcy-Weisbach form by

$$h_L = f\left(\frac{L_e}{D}\right)\left(\frac{v^2}{2g}\right)$$

L_e can be obtained if f is known. Exact hydraulic equivalence in the velocity head terms depends upon f being a constant over the range of velocities applicable to the problem. In fact, f is not a constant over wide ranges of velocity. Since it varies only slightly with Reynolds number, however, solutions are sufficiently accurate.

Pipes in parallel: Two or more pipes connected as in Figure 4-2.13(b), so that flow is first divided among the pipes and is then rejoined, comprise a parallel pipe system. Flows in pipes arranged in parallel are also determined by application of energy conservation principles—specifically, energy losses through all pipes connecting common junction points must be equal. Each leg of the parallel network is treated as a series piping system and converted to a single equivalent length pipe. The friction losses through the equivalent length parallel pipes are then considered equal and the respective flows determined by proportional distribution. For a given Q, an outline of the procedure is as follows:

1. Express each branch of the parallel system as an equivalent length of a single pipe size, including all minor losses between the bounding junction points.

2. Assume a discharge Q_1' through pipe branch 1.

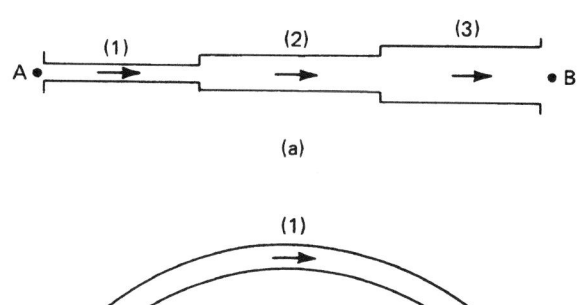

(a)

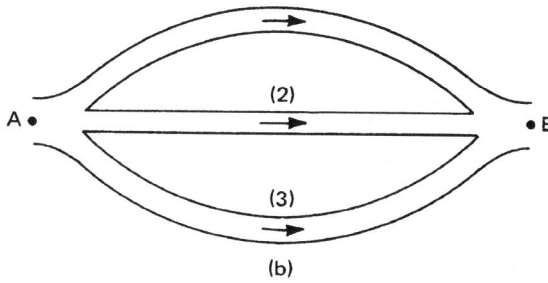

(b)

Fig. 4-2.13. Energy losses in pipe network: (a) pipes in series, (b) pipes in parallel.

3. Solve for h_L, using Q_1'.

4. Using h_L, find Q_2' and Q_3' for the remaining branches.

5. Knowing the proportional distribution of flow among the legs, Q_1', Q_2', and Q_3' are adjusted so that their sum equals the known Q; thus

$$Q_1 = \frac{Q_1'}{\Sigma Q'}Q, \quad Q_2 = \frac{Q_2'}{\Sigma Q'}Q, \quad Q_3 = \frac{Q_3'}{\Sigma Q'}Q \quad (27)$$

6. h_{L_1}, h_{L_2}, and h_{L_3} are computed for the values of Q_1, Q_2, and Q_3 as a check for correctness.

For judicious choice of assumed discharges, solutions are obtained rapidly that agree within a few percent, well within the range of accuracy of the assumed friction factors.

In the case where the head loss is known between points A and B, Q for each branch is found simply by solution of the equation for pipe discharge. The discharges are added to obtain the total flow through the system.

Compound piping networks: Energy loss calculations in compound piping configurations or networks employ the same basic physical principles as for single pipes. That is, conservation of energy and conservation of mass (continuity) must be satisfied throughout the network. In particular, at each pipe junction

$$\sum Q = Q_1 + Q_2 + \cdots + Q_n = 0 \quad (28)$$

and around each closed loop or circuit

$$\sum h_L = h_{L_1} + h_{L_2} + \cdots + h_{L_n} = 0 \quad (29)$$

The general solution procedure involves setting up a sufficient number of independent equations of these two types and solving simultaneously for the unknowns. For complicated networks straightforward algebraic solution is clearly impractical. A very widely used relaxation method for systematic solution of large networks was developed by Hardy Cross in 1928. The method is well-suited for solution by hand and is readily adaptable for machine computation.

We have seen that loss of head in a pipe may be represented generally by a equation of the form, $h_L = KQ^n$ (where, for the Hazen-Williams formula, $n = 1.85$). For any single pipe in a network, we may write

$$Q = Q_0 + \Delta \quad (30)$$

where Q is the corrected flow, Q_0 is an assumed flow, and Δ is the flow correction. The problem, so stated, reduces to finding Q to a desired degree of accuracy by successive evaluations of Δ based on updated estimates of Q_0. We solve for Δ as follows

$$\begin{aligned} h_L = KQ^n &= K(Q_0 + \Delta)^n \\ &= K(Q_0^n + nQ_0^{n-1}\Delta + \cdots) \end{aligned} \quad (31)$$

If Δ is small relative to Q_0, the higher order terms in the expansion may be neglected. Since, for any circuit, $\Sigma h_L = 0$, we may write

$$\sum KQ^n = 0 = \sum [KQ_0^n + KnQ_0^{n-1}\Delta] \quad (32)$$

to a good approximation. Solving for Δ we have

$$\Delta = \frac{-\Sigma KQ_0^n}{n\Sigma KQ_0^{n-1}} = \frac{\Sigma h_{L_0}}{n\Sigma(h_{L_0}/Q_0)} \quad (33)$$

The overall formulation is made algebraically consistent by designating clockwise flows positive and counter-clockwise flows negative. The calculation procedure is controlled by the requirement that the algebraic sum of all assumed flows must equal zero at each pipe junction. The originally assumed flows are repeatedly and cyclically corrected until the Δ values are negligible, indicating that a hydraulic balance has been reached. Note that pipes common to two circuits are corrected twice in each cycle, once for each circuit. For a system where total head loss is known, flows can be balanced by correcting assumed head losses instead of flows.

Several other methods exist for determining flows and head losses in compound pipe networks. Many can be performed manually, although computer analysis is desirable and necessary for the more complex methods, particularly those involving unsteady flow. For a review of alternative methods, the reader is referred to Stephenson[8] and Walski.[10]

FLOW MEASUREMENT AND DISCHARGE

Flow Measuring Devices

This section deals primarily with the basic principles of operation of some flow measuring devices in common use and, in particular, with the pitot tube and the pipeline differential flow meters that have been standardized by the ASME (American Society of Mechanical Engineers): namely, the Venturi, the flow nozzle, and the square-edge thin-plate concentric orifice.

In general, an incompressible fluid of density ρ, viscosity μ, flows with average velocity v through a metering element of diameter d. The metering element is located in a horizontal metering tube of roughness ϵ and diameter D. The flow through the element produces a pressure differential Δp sensed by pressure taps located a distance L apart. It can be shown by dimensional analysis that the fundamental parameters involved in fluid metering, namely L, ϵ, v, ρ, μ, d, D and Δp, yield relational solutions conventionally formulated as follows

$$\frac{d\rho v}{\mu} = \text{Re}_d \quad \text{metering element Reynolds Number}$$

$$\frac{L}{D} \quad \text{tap location ratio}$$

$$\frac{d}{D} = \beta \quad \text{beta ratio}$$

$$\frac{\epsilon}{D} \quad \text{relative roughness}$$

$$\frac{v}{\sqrt{2g\Delta p/\rho}} = \overline{K} \quad \text{flow coefficient (pressure coefficient)}$$

Since $v = \overline{K}\sqrt{2g\Delta p/\rho}$, the continuity equation allows the volumetric flow rate measured by the meter to be expressed as

$$Q = \overline{K}A_d\sqrt{2g\Delta p/\gamma} \quad (34)$$

where A_d is the flow area of the metering element.

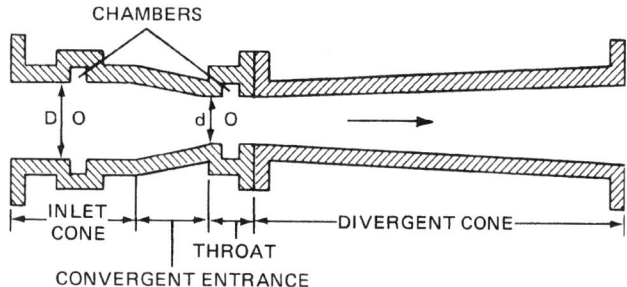

CHAMBERS

INLET
CONE

THROAT

CONVERGENT ENTRANCE

DIVERGENT CONE

Fig. 4-2.14. Venturi tube.

Typically, flow meter calculations are based on the idealized flow of a one-dimensional, frictionless, incompressible fluid in a horizontal metering tube. Real conditions require corrections to the ideal formulation. Conventional corrections for the effects of variations from ideal geometry and flow velocity profile are achieved through the use of modification factors. Thus, in Equation 34 above $\overline{K}$ includes pressure and flow modifications which are conventionally defined as

$$\overline{K} = CE \tag{35}$$

where C = coefficient of discharge defined as the ratio of actual flow rate to ideal rate and

$$E = \frac{1}{\sqrt{1 - \beta^4}}$$

and is known as the velocity of approach factor, since it accounts for the one-dimensional kinetic energy at the inlet tap.

The general volumetric flow metering equation is, then,

$$Q = \overline{K}A_d \sqrt{2g\Delta p/\gamma} = CEA_d \sqrt{2g\Delta p/\gamma} \tag{36}$$

Venturi flow meter: Figure 4-2.14 shows a schematically typical Venturi-type flow tube. The divergent cone section reduces the overall pressure loss of the meter. Pressure is sensed through a series of holes in the inlet cone and throat. These holes lead to an annular chamber, and the two chambers are connected to a pressure differential sensor such as a U-tube manometer. ASME standardized discharge coefficients are given in Table 4-2.5. Venturi tubes must be individually calibrated to obtain these coefficients outside the tabulated limits.

Determination of volumetric flow rate is a simple calculation employing the general flow metering formula—Equation 36—where C is the applicable tabulated value based on Re_d, and E is calculated directly from the beta ratio.

TABLE 4-2.5 *ASME Coefficients for Venturi Tubes*

| | Re₂ | | | |
	Minimum	Maximum	Value of C	Tolerance %
Type of Inlet Cone				
Machined		1,000,000	0.995	± 1.00
Rough welded sheet metal	500,000	2,000,000	0.985	± 1.50
Rough cast			0.984	± 0.70

ASME flow nozzle: This nozzle is depicted in Figure 4-2.15. The pressure differential is sensed by either throat taps or pipe wall taps appropriately located. Coefficients of discharge for ASME flow nozzles may be accurately computed from the following equation:

$$C = 0.9975 - 0.00653 \left(\frac{10^6}{Re_d}\right)^a \tag{37}$$

where
$$\begin{cases} a = \frac{1}{2} \text{ for } Re_d < 10^6 \\ a = \frac{1}{5} \text{ for } Re_d > 10^6 \end{cases}$$

Volumetric flow rates are calculated in the same manner as for the Venturi tube.

ASME orifice meters: Fluid flowing through a thin, square-edged orifice plate experiences a contraction of the flow stream some distance downstream from the orifice. The minimum area of flow is called the *vena contracta* and its location is a function of the beta ratio. Figure 4-2.16 shows the relative pressure difference due to the presence of the orifice plate and the location of the *vena contracta* with respect to beta. By inspection of Figure 4-2.16 it is clear that the actual location of the pressure taps is critical. Three distinct arrangements for tap locations are specified by the ASME for accurately measuring the pressure differential. These types of tap arrangements are called the flange, *vena contracta*, and the 1D and ½D. Each has certain advantages and disadvantages and affects the value of the discharge coefficient.

Discharge coefficients for orifice metering plates may be calculated from the equation

$$C = C_o + \frac{\Delta C}{Re_d^\alpha} \tag{38}$$

where C_o and α are obtained from Table 4-2.6. Since the jet contraction downstream of the orifice can amount to nearly half of the orifice area, orifice discharge coefficients are in the order of 0.6 compared to the near-unity coefficients obtained with Venturi tubes and flow nozzles.

Pitot tube: A pitot tube is a device designed to sense stagnation or total pressure for the determination of velocity and volumetric flow rate. A number of commercial devices are available, some of which include a static pressure tap, that are designed for insertion into a water main under pressure

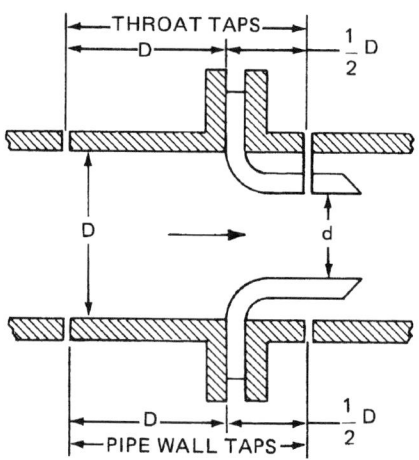

THROAT TAPS

PIPE WALL TAPS

Fig. 4-2.15. ASME flow nozzle.

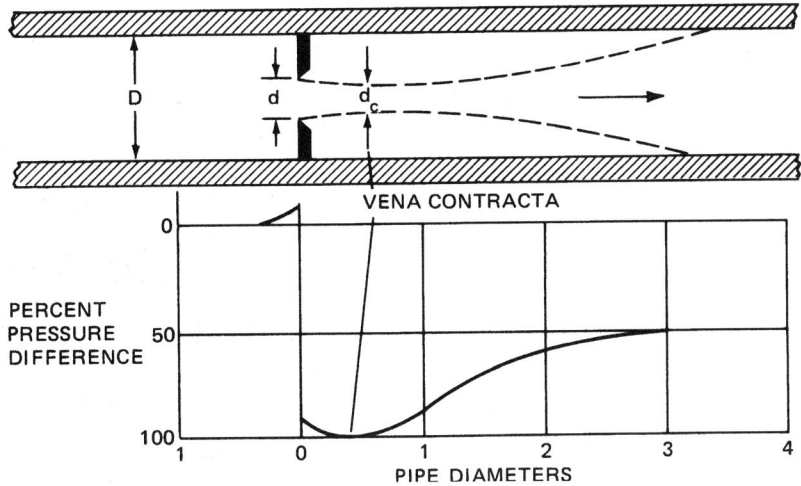

Fig. 4-2.16. *Relative pressure changes due to flow through an orifice.*

through a standard pipe tap or corporation cock. The installed pitot tube measures velocity at a point in the fluid. Conventional practice assumes that the conversion of kinetic energy to flow work in the tube is frictionless. Thus, applying the energy equation to the generalized pitot tube diagram (Figure 4-2.17) we obtain

$$\frac{u_s^2 - u_i^2}{2g} + \frac{p_s - p_o}{\rho_o g} = 0 \qquad (39)$$

where u_s = stagnation point velocity
u_i = ideal streamtube velocity
p_s = stagnation pressure
p_o = static pressure

Since, by definition $u_s = 0$, solving for u_i we obtain

$$u_i = \sqrt{2g(p_s - p_0)/\gamma_0} = \sqrt{2g\Delta p/\gamma_0} \qquad (40)$$

Typically, a pipe coefficient, C_p, which is independent of the geometry of the velocity profile, is defined as

$$C_p = \frac{\text{average velocity}}{\text{centerline velocity}}$$

For typical velocity profiles, C_p varies from about 0.75 to 0.97 but usually lies within a narrower range of about 0.80 to 0.90. Knowing the centerline velocity, the flow can be obtained simply by

TABLE 4-2.6 Values of C_o, ΔC and a for use in Equation 38

β	D = 2 in. = 50 mm		D = 4 in. = 100 mm		D = 8 in. = 200 mm		D = 16 in. = 400 mm	
	C_o	ΔC	C_o	ΔC	C_o	ΔC	C_o	ΔC
				Flange Taps α = 1				
0.20	0.5972	127	0.5946	200	0.5951	327	0.5955	551
0.30	0.5978	144	0.5977	209	0.5978	307	0.5980	457
0.40	0.6014	181	0.6005	256	0.6002	362	0.6001	514
0.50	0.6050	260	0.6034	386	0.6026	584	0.6022	903
0.60	0.6078	392	0.6055	622	0.6040	1015	0.6032	1710
0.70	0.6068	573	0.6030	953	0.6006	1637	0.5991	2898
				Vena Contracta Taps α = ½				
0.20	0.5938	1.61	0.5928	1.61	0.5925	1.61	0.5924	1.61
0.30	0.5939	1.78	0.5934	1.78	0.5933	1.78	0.5932	1.78
0.40	0.5970	2.01	0.5954	2.01	0.5953	2.01	0.5953	2.01
0.50	0.5994	2.29	0.5992	2.29	0.5992	2.29	0.5991	2.29
0.60	0.6042	2.68	0.6041	2.68	0.6041	2.69	0.6041	2.70
0.70	0.6069	3.34	0.6068	3.37	0.6067	3.44	0.6068	3.57
				1D and ½D Taps α = ½				
0.20	0.5909	2.03	0.5922	1.41	0.5936	1.10	0.5948	0.94
0.30	0.5915	2.02	0.5930	1.50	0.5944	1.24	0.5956	1.12
0.40	0.5936	2.17	0.5951	1.72	0.5963	1.49	0.5974	1.38
0.50	0.5979	2.40	0.5978	1.99	0.5999	1.79	0.6007	1.69
0.60	0.6036	2.67	0.6040	2.31	0.6044	2.12	0.6048	2.11
0.70	0.6078	3.19	0.6072	2.98	0.6068	3.07	0.6064	3.51

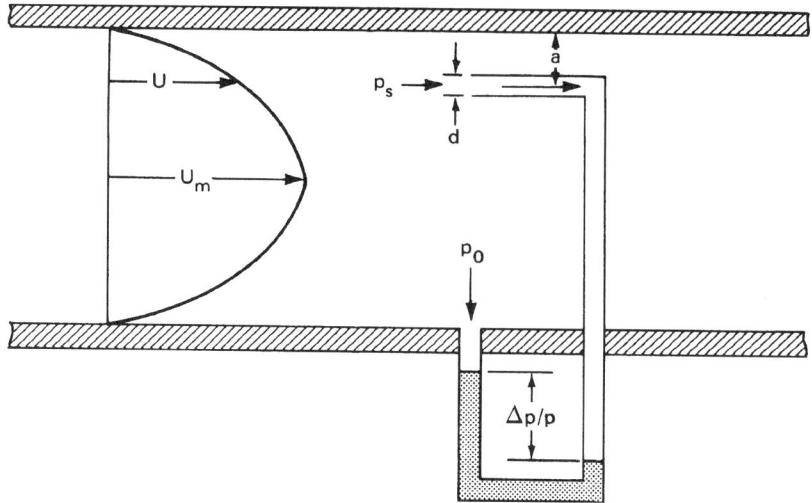

Fig. 4-2.17. Pitot tube study.

$$Q = C_p A v_{C.L.} \qquad (41)$$

In situations where pipe velocity profiles are unknown, and therefore average velocities are not available, it may be necessary to obtain velocity measurements at many individual points. Given n velocities, the flow is then

$$Q = \sum_{i=1}^{n} v_i A_i \qquad (42)$$

where v_i = velocity at the ith point
A_i = area of annular ring of flow cross-section for which velocity v_i is accurate.

Detailed procedures for obtaining accurate pitot traverses are available in the literature along with suggestions for assessing the reliability of water audits, C-factor tests, etc., based on pitot gage measurements.[7,10] See the section "Free Discharge at an Opening" for a discussion of discharge measurements using pitot tubes.

Free Discharge at an Opening

Flow discharging to the atmosphere from a tank, hydrant, nozzle, or open conduit is affected by the area and shape of the opening. The total energy of the fluid is converted into kinetic energy at the orifice according to an appropriate form of the Bernoulli equation. In the most general case of a closed pressurized tank,

$$\frac{v_0^2}{2g} = z_1 + \frac{p_1}{\rho} \qquad (43)$$

$$v_o = \left[2g\left(z_1 + \frac{p_1}{\rho}\right)\right]^{1/2} \qquad (44)$$

Accounting for losses at the point of discharge

$$v_o = C_v \sqrt{2gh} \qquad (45)$$

where C_v, the coefficient of velocity, is determined from the coefficients of discharge and contraction

$$C_v = \frac{C_d}{C_c}$$

Commonly used values of orifice coefficients for water are given in Table 4-2.7. The orifice discharge can then be expressed as

$$Q_o = C_d A_o \sqrt{2gh} \qquad (46)$$

and the head loss due to turbulence at the orifice as

$$h_L = \left(\frac{1}{C_v^2} - 1\right)\frac{v_0^2}{2g} \qquad (47)$$

where

$\left(\dfrac{1}{C_v^2} - 1\right)$ is the "minor loss" k-factor.

For the general case of a tank of varying cross-sectional area being replenished with inflow, $\dot{Q}_{IN}$, the time to empty from height, z_1 to z_2 is given by:

$$t = \int_{z_1}^{z_2} \frac{A_t \, dz}{C_d A_o \sqrt{2gh} - \dot{Q}_{IN}} \qquad (48)$$

where A_t is expressed as a function of z.

For a tank of constant cross-section this simplifies to

$$t = \frac{2A_t(\sqrt{z_1} - \sqrt{z_2})}{C_d A_o \sqrt{2g}} \qquad (49)$$

EXAMPLE 4:

A 15 ft diameter tank discharges water at 50°F through a 2 in. diameter sharp-edged orifice. If the initial water depth in the tank is 10 ft and the tank is continuously pressurized to 50 psig, how long will it take to empty the tank?

TABLE 4-2.7 Orifice Coefficients for Water

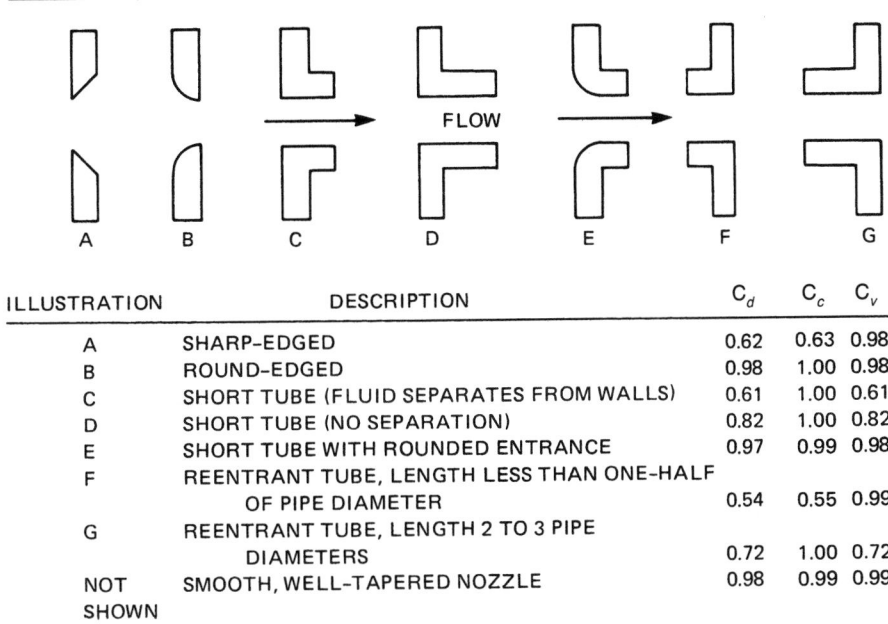

ILLUSTRATION	DESCRIPTION	C_d	C_c	C_v
A	SHARP-EDGED	0.62	0.63	0.98
B	ROUND-EDGED	0.98	1.00	0.98
C	SHORT TUBE (FLUID SEPARATES FROM WALLS)	0.61	1.00	0.61
D	SHORT TUBE (NO SEPARATION)	0.82	1.00	0.82
E	SHORT TUBE WITH ROUNDED ENTRANCE	0.97	0.99	0.98
F	REENTRANT TUBE, LENGTH LESS THAN ONE-HALF OF PIPE DIAMETER	0.54	0.55	0.99
G	REENTRANT TUBE, LENGTH 2 TO 3 PIPE DIAMETERS	0.72	1.00	0.72
NOT SHOWN	SMOOTH, WELL-TAPERED NOZZLE	0.98	0.99	0.99

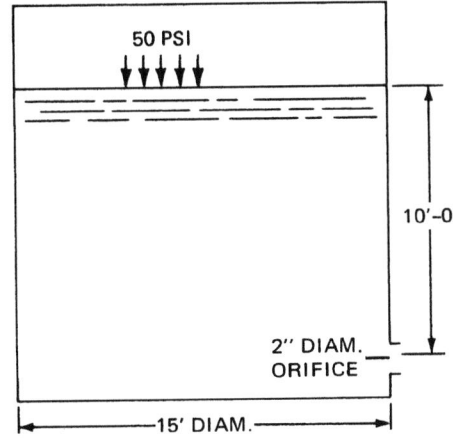

SOLUTION:

At 50°F, $\gamma = 62.4$ lbm/ft^3

For the orifice:

$$A_0 = \frac{\pi D^2}{4} = 3.14 \text{ ft}^2$$

$$C_d = 0.62 \text{ (sharp-edged orifice)}$$

For the tank:

$$A_t = \frac{\pi D^2}{4} = 176.7 \text{ ft}^2$$

$$h_0 = 10 + \frac{50(144)}{62.4} = 125.38 \text{ ft}$$

$$h_1 = 0 + \frac{50(144)}{62.4} = 115.38 \text{ ft}$$

The total pressure head on the discharging fluid results from both an elevation and a static pressure head. Therefore

$$t = \frac{2A_t\left[\left(z_0 + \frac{p_o}{\gamma}\right)^{1/2} + \left(z_1 + \frac{p_1}{\gamma}\right)^{1/2}\right]}{C_d A_o \sqrt{2g}}$$

$$t = 10.4 \ s$$

Discharge stream coordinates are given by:

$$x = v_o t = v_o\sqrt{\frac{2y}{g}} = 2C_v\sqrt{zy} \tag{50a}$$

$$y = \frac{gt^2}{2} \tag{50b}$$

For the simpler case of a hydrant discharging to atmosphere, the flow can be determined by an appropriate form of Equation 36

$$Q = 29.8D^2 C_d\sqrt{p} \tag{51}$$

where Q = discharge, gpm
 D = outlet diameter, in.
 p = pressure detected by pitot gage, psi
 C_d = coefficient based on hydrant outlet geometry (usually taken to be 0.90 for full flow across a standard 2½ in. outlet)

In the absence of a pitot gage, hydrant flows may be estimated by observing the trajectory of the discharge stream. The horizontal component of the velocity does not change appreciably over time, thus allowing calculation of the velocity based on the height of the outlet and the distance traveled by the stream. Figure 4-2.18 presents the basic

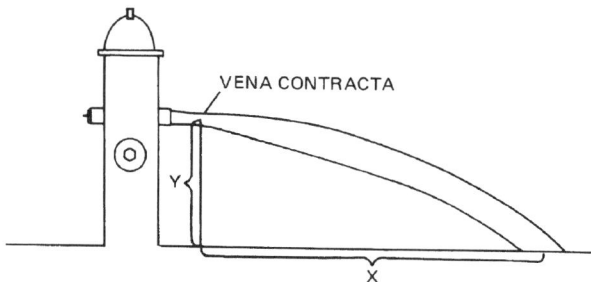

Fig. 4-2.18. *Determining discharge by the trajectory method.*

parameters. The velocity determined in this manner is at the *vena contracta* and is given by

$$v = \frac{x}{\sqrt{2y/g}} \qquad (52)$$

The discharge is simply the product of this velocity and the area of the *vena contracta*. The method is relatively inaccurate due to the obvious difficulty of measuring the required area and the distance x. It is a useful bounding guide, however, in the absence of precision measuring devices.

WATER HAMMER†

Water hammer in a pipeline is caused by a sudden stoppage of flow and is characterized by loud noise and vibration. The kinetic energy from the interrupted flow is transferred to the walls of the enclosing pipe or equipment, which expand under the increased pressure. Such pressures, or shock waves, can be severe enough to destroy the equipment and the pipeline itself.

Density changes due to pressure are assumed zero for nearly all hydraulic calculations, as water is considered incompressible for practical purposes even though it is about 100 times more compressible than steel. When shock waves arise in confined water, however, the compressibility of water becomes very significant, and water's elastic properties must be taken into account. The primary property of interest is the bulk modulus of elasticity, E, which is defined as the ratio of pressure change to the corresponding change of volume as determined by compression tests on volumes. (The bulk modulus is analogous to Young's modulus in solid mechanics, which is the ratio of linear stresses to linear strains as determined by tension tests.) The formula expressing the relationship between pressure and volume is

$$\Delta p = -E \frac{\Delta V}{V_0} \qquad (53)$$

where the minus sign indicates a positive change in pressure produces a decrease in volume. A modulus of compressibility, K, is also defined as the inverse of E.

Under normal conditions, water confined and flowing under pressure in a pipeline exerts pressure on the pipe walls according to the pressure-energy term of the energy equation. Any change in discharge within the system (due to valve closure, pump stoppage, etc.) results in a change of flow momentum. By virtue of the impulse-momentum relation, the momentum change will cause an impulse force to

---

† This discussion is patterned after the theory of water hammer as developed by N. J. Zhukovsky and as presented in Andrew L. Simon's *Practical Hydraulics*, 2nd ed.[7]

be created. This force in a pipeline is commonly referred to as water hammer.

The theory of water hammer, as developed by Zhukovsky, can be briefly illustrated as follows: a valve in a pipeline is closed instantaneously, the fluid impacts the closed gate, and is decelerated to zero velocity, thereby creating a pressure shock. By Newton, pressure shocks in fluids of infinite extent travel at a velocity given by

$$c^* = \sqrt{\frac{E}{\rho}} \qquad (54)$$

where c^* is called the celerity (velocity) of the shock wave, KE is the kinetic energy of the fluid, and ρ is the fluid density. The pipe, however, is also elastic. Therefore, if the fluid in the pipe is compressed, the pipe will expand. The modulus of elasticity, E_c, of a system composed of fluid and pipe may be determined from the equation

$$\frac{1}{E_c} = \frac{1}{E} + \frac{D}{E_p w} \qquad (55)$$

where D is the pipe diameter, w is the thickness of the pipe wall, and E_p is the modulus of elasticity of the pipe material. Table 4-2.8 gives the modulus of elasticity for common pipe materials. The celerity of a shock wave in a pipe system of finite extent can then be computed from

$$\frac{c}{c^*} = \frac{1}{\sqrt{1 + ED/(E_p w)}} \qquad (56)$$

which is plotted in Figure 4-2.19. The graph indicates the considerable influence of pipe rigidity on the velocity of the shock.

The shock waves that travel upstream and downstream from the valve closure eventually reach points in the system that correspond to large stationary energy stores (e.g., reservoirs) or other sudden closure points, which may vary in their ability to absorb or reflect the shock wave. If the shock is absorbed into a larger energy field it will disappear, and it will do so in time

$$t = \frac{L}{c} \qquad (57)$$

where L is the distance from the energy reservoir to the shock wave point of origin. At the instant of shock absorption the compressed fluid, no longer balanced, begins to flow backward, creating a relief pressure shock that travels back to the valve. The time period T that the initial shock or impulse

TABLE 4-2.8 *Modulus of Elasticity E_p of Various Pipe Materials*

Pipe Material	E_p	
	(in million psi)	(in lb/ft²)
Lead	0.045	6.48×10^6
Lucite (at 73°F)	0.4	57.6×10^6
Rubber (vulcanized)	2	288×10^6
Aluminum	10	1440×10^6
Glass (silica)	10	1440×10^6
Brass, bronze	13	1872×10^6
Copper	14	2016×10^6
Cast iron, grey	16	2304×10^6
Cast iron, malleable	23	3312×10^6
Steel	28	4032×10^6

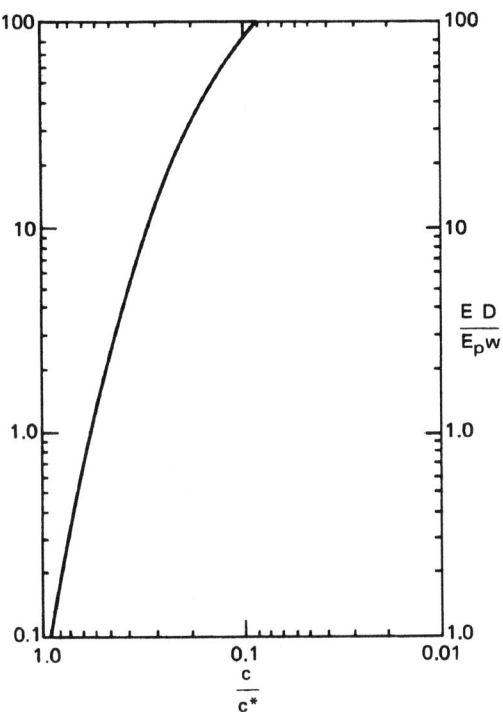

$$\frac{E\,D}{E_p w}$$

$$\frac{c}{c^*}$$

Fig. 4-2.19. *Celerity of pressure waves in pipes, c equals celerity in elastic pipe; c* equals celerity in fluid of infinite extent.*

pressure acts on the valve is, therefore, the time required for the pressure wave to travel away from and back to the valve

$$T = 2t = \frac{2L}{c} \tag{58}$$

At time T, all the fluid is moving backward at some velocity v. Since the valve is closed, there is no supply for this flow. A negative pressure shock is created at the valve. The shock travels to and back from the reservoir, as the flow is reversed. Such oscillations of pressure and periodic flow reversals persist until the kinetic energy is dissipated by friction. The process described will occur both upstream and downstream from the point of origin, though the initial shock will be positive upstream and negative downstream and the periodicities would not likely be equal.

The theoretical magnitude of the pressure shock at instantaneous valve closure can be determined directly from

$$p^* = \rho c \Delta v \tag{59}$$

and the pressure will oscillate in the pipe within the range

$$p = p_0 \pm p^* \tag{60}$$

In actuality, the time of closure of a valve is not zero but some finite time period which we may call T_c. The water hammer pressure increases gradually with the rate of closure of the valve. Depending on whether T_c is smaller or larger than T, we distinguish between quick and slow closure. For $T_c < T$, the shock pressure will attain its maximum value p^*. (In this sense, quick closure is equivalent to instantaneous closure.) For $T_c > T$, maximum pressure will be somewhat less than p^* and may be calculated by the Allievi formula

$$p = p_0 \left[\frac{N}{2} + \sqrt{\frac{N^2}{4} + N} \right] \tag{61}$$

in which

$$N = \left(\frac{Lv\rho}{p_0 T_c} \right)^2 \tag{61a}$$

In general the calculation of water hammer pressure rises, regardless of method, will tend to underestimate the actual values. Real systems will tend to experience superimposition of positive or negative pressure waves due to complex piping configurations. Discontinuities introduced by a variety of auxiliary valving and metering equipment complicate the analysis considerably. Other methods are available for analyzing water hammer effects on systems that may not be reasonably handled by the above idealized method.[8] Since water hammer can be extremely detrimental, often resulting in complete loss of the system, it is desirable to perform an analysis wherever such effects are of concern. Control over the development of damaging shock waves is achieved through use of slow-closing valves, pressure relief valves, or shock absorbing devices.

PUMPS

Pump Operating Characterisitics

Pumps are mechanical devices that convert electrical or mechanical energy into hydraulic energy. There are many classes of pumps—e.g., reciprocating, rotary, jet, ram, centrifugal—each class referring to different ways pumps move liquids. A common class of pump is the centrifugal and it is usually the only type we are concerned with in fire protection applications. Based on the way in which the impeller (the rotating component) imparts energy to the water, centrifugal pumps may be divided into several categories. Turbine or radial flow centrifugal pumps force water outward at right angles to the rotating axis. Mixed flow pumps force water in both radial and axial directions. Propeller pumps move water in the axial direction only. Any of these types may be single or multistage, where stage refers to the number of impellers on the pump's rotating shaft. The orientation of the shaft may be vertical or horizontal. The following discussion, while broadly applicable, is directed mainly to centrifugal pumps.

Figure 4-2.20 illustrates several of the terms commonly used to describe pump performance conditions. In general, pumping of liquids requires that the pressure at any point in the intake line be greater than the vapor pressure of the liquid to avoid loss of prime and the highly destructive phenomenon known as cavitation. The pressure gradient that causes a liquid to move through the intake line to the pump impeller is termed the net positive suction head (NPSH). In pump selection, it is essential to determine that the *available* NPSH of the system exceeds the *required* NPSH for the pump under consideration. Required NPSH depends upon many factors relating to pump geometry and construction and intake system operating conditions, but it is defined simply as the difference between net suction head and vapor pressure at a given flow, or the energy needed to fill the pump on the intake side and overcome intake system head losses. If the net suction head is less than the vapor pressure of the water, the water will vaporize in the pump, producing cavitation. Small vapor bubbles formed in the low pressure

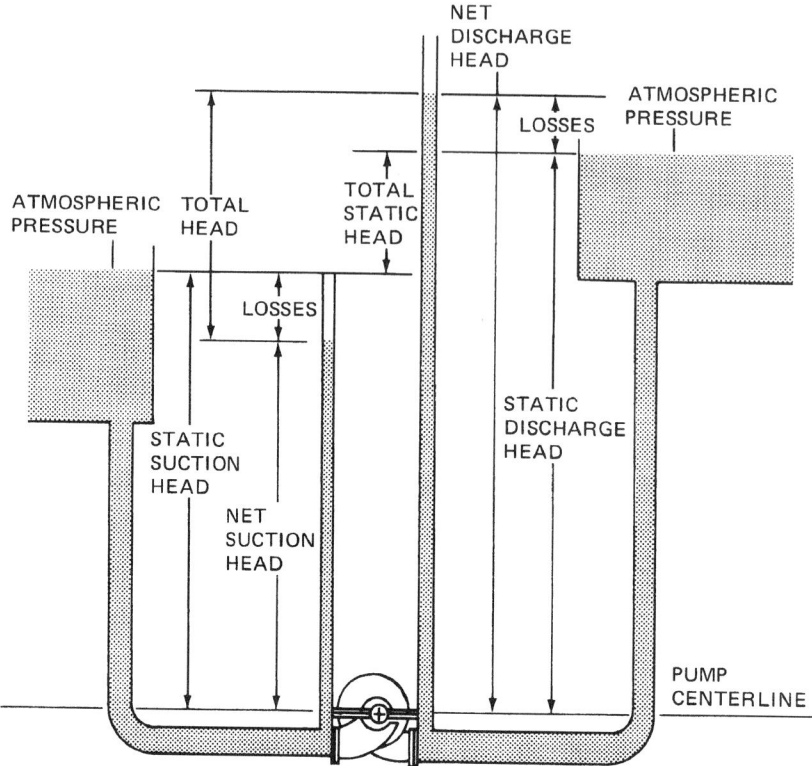

Fig. 4-2.20. Pump head definitions.

region will collapse violently upon entering regions of high pressure, causing localized stress concentrations and vibrations, ultimately leading to mechanical failure.

The required net positive suction head (NPSH_{req}) for any pump can be obtained from the manufacturer. The available net positive suction head (NPSH_{av}) must be calculated for each system. Because the total energy of a system is constant, the available NPSH may be determined at any point in the system. The general expression at the pump centerline follows from Bernoulli as

$$\text{NPSH}_{av} = \frac{(p_{gage} + p_{atm})}{\rho g} + z - h_L - \frac{p_{vp}}{\rho g} \quad (62)$$

where h_L = friction head loss in intake system piping (in feet of water)

p_{vp} = vapor pressure (0.256 psia for water at 20°C)

Knowing the pressure and pipe friction loss terms, the pump can be set at a height, z, which will ensure that $\text{NPSH}_{av} > \text{NPSH}_{req}$.

Where a free surface exists on the intake side (such as at the surface of an intake reservoir) and the velocity at a point on the surface is negligible, the above expression simplifies to

$$\text{NPSH}_{av} = z - h_L + \frac{(P_{atm} - p_{vp})}{\rho g} \quad (63)$$

For pumps of relatively low heads and large discharge capacities (common in fire protection applications) the available NPSH may be less than zero (h_L is large). These pumps should be installed well below the reservoir water level to eliminate the possibility of cavitation. For this reason and also to avoid accidental loss of prime, authorities

having jurisdiction generally require "positive suction" installation. In such instances the pump should be of the vertical shaft type so that the motor can be installed at an elevation above any possible flood level.

The useful work done by a pump is the product of the weight of the liquid pumped and the head developed by the pump. The work per unit time in this context is the hydraulic horsepower, commonly called the water horsepower (WHP). For discharge, Q, in gpm, total dynamic head, h, in feet, and specific weight, γ, for water at 20°C (68°F)

$$\text{WHP} = \frac{Qh}{3960} \quad (64)$$

The power required to actually drive the pump is the brake horsepower (BHP). The difference between water horsepower and brake horsepower is the power lost within the pump due to mechanical and hydraulic friction. The ratio of WHP to BHP is the pump efficiency, η_p. Similarly, the ratio of BHP to electrical or engine horsepower (EHP) is the motor efficiency, η_m. The overall efficiency is, then, the pump efficiency multiplied by the motor efficiency

$$\eta = (\eta_p)(\eta_m) = \frac{\text{WHP}}{\text{BHP}} \cdot \frac{\text{BHP}}{\text{EHP}} \quad (65)$$

Although WHP should be calculated using the specific weight of the fluid at known conditions of temperature and pressure, the variation for water is very small; it should be noted that pump motor sizes are chosen from standard available sizes in any case.

The interrelations of head, capacity, power, and efficiency for a given pump are known as the characteristics of

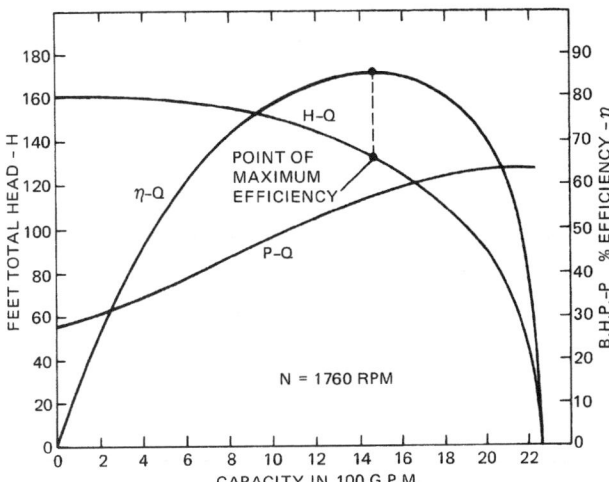

Fig. 4-2.21. *Centrifugal pump characteristics.*

the pump. They can be expressed graphically in the form of pump characteristic curves. Figure 4-2.21 shows a standard plot of the several variables at constant impeller speed (N). Note that the point of maximum operating efficiency on the head-capacity curve corresponds to the maximum value of the efficiency curve. The actual operating point of the pump, however, depends on the system demand (or system head) curve. The system head loss for any flow rate is the sum of the system friction head loss at that rate plus the total static head to be overcome in the system. Figure 4-2.22 illustrates the relationship. Recall from Figure 4-2.20 that the total static head is the difference in elevation between the discharge level and the suction level. System friction losses may be determined by calculations methods given in previous sections.

Pump Selection

Economical pump selection for fire protection applications requires consideration of the following factors:

1. the maximum discharge rate required under the most demanding design conditions;

2. the total head-capacity relation (characteristic curve);

3. the suction head—in particular, the net positive suction head available;

4. pump speed and power source requirements;

5. pump spatial and environmental requirements; and

6. the maximum allowable system head downstream of the pump discharge.

The usual design condition is that a system will be given or will be chosen from a very limited range of possibilities, and the proper pump must be selected. As shown in Figure 4-2.22, when the system demand curve and the pump head-capacity curve are superimposed, their intersection will determine the operating point of the pump. This point also locates the efficiency and, therefore, the power requirements. It is often economically desirable to select a pump such that its operating point is at or near its peak efficiency. In many fire protection applications, however, a pump may be called upon to operate very infrequently. Power con-

sumption may, therefore, not be a significant factor relative to initial cost. Common practice in fire protection applications is to select a pump to operate at 150 percent of rated capacity at 65 percent of rated head (see NFPA 20[12])—i.e., an operating point farther out along the characteristic curve. A pump is chosen such that its operating point so defined meets or exceeds the system demand curve at that point.

If the pump is to be used as a "booster" to increase supply mains pressure, care must be exercised to select a pump having a maximum discharge head at zero flow (also known as "churn" head) which, when added to the maximum mains supply head, does not exceed the maximum allowable working pressure on the system. The maximum allowable working pressure prescribed by NFPA 13, for example, is 175 psig.[13]

Centrifugal Pump Affinity Relations

The abstract concept of "pump specific speed" has been developed to simplify the description of pump performance characteristics. It consolidates the discharge, head, and speed (rpm) at optimum performance into a single number. For a single stage, single suction pump, specific speed may be calculated from

$$N_s = \frac{NQ^{1/2}}{H^{3/4}} \qquad (66)$$

where Q (in gpm) is taken at pump rpm, N, and total dynamic head, H. The specific speed of a pump is not actually a speed for that pump in any physical sense; it is defined as the speed in revolutions per minute at which a homologous (geometrically similar) pump would run if constructed to deliver 1 gpm against 1 ft total head at its peak efficiency. For pump impeller designs of identical proportions but different sizes, the specific speed is a constant performance index. That is, the performance of any impeller can be predicted from knowledge of the performance of any other geometrically similar impeller.

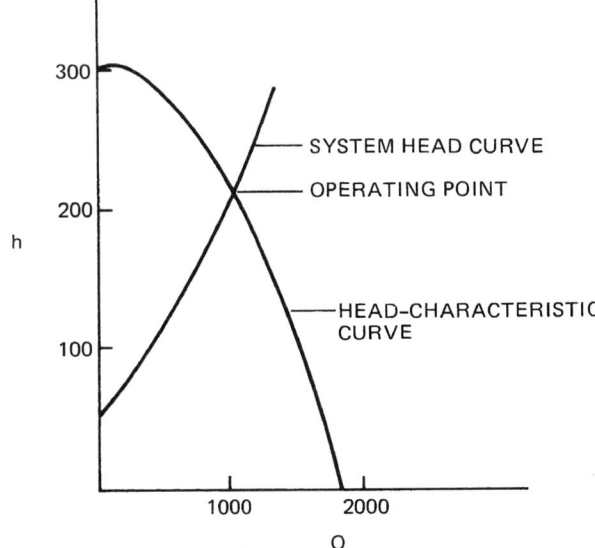

Fig. 4-2.22. *Graphical determination of operating point.*

Changing the impeller diameter results in changes in discharge, total head, and delivered power. These changes occur according to the follow relations

$$\frac{Q_1}{Q_2} = \left(\frac{D_1}{D_2}\right)^3 \frac{n_1}{n_2} \tag{67a}$$

$$\frac{H_1}{H_2} = \left(\frac{D_1}{D_2}\right)^2 \left(\frac{n_1}{n_2}\right)^2 \tag{67b}$$

$$\frac{Q_1}{Q_2} = \left(\frac{H_1}{H_2}\right)^{1/2} \left(\frac{D_1}{D_2}\right)^2 \tag{67c}$$

$$\frac{\text{BHP}_1}{\text{BHP}_2} = \left(\frac{D_1}{D_2}\right)^5 \frac{\rho_1}{\rho_2} \frac{n_1^3}{n_2^3} \tag{67d}$$

Since

$$\frac{N_1}{N_2} = \frac{D_1}{D_2} \tag{68}$$

a change in motor speed only will yield similar results. That is, a change in impeller size has the same effect on pump performance as a change in speed provided, of course, that there is no marked change in operating efficiency.

EXAMPLE 5:

A 6 in. pump operating at 1770 rpm discharges 1500 gpm of water at 40°F against a 120 foot head.

(a) What discharge capacity and total head can be expected from a homologous 8 in. pump operating at 1170 rpm?

(b) If the pumps operate at an overall 80 percent efficiency, what is the 8 in. pump power requirement?

SOLUTION:

(a) From Equation 67b

$$H_2 = \left[\frac{8^2(1170)^2}{(6)^2(1770)^2}\right] (120) = 93.2 \text{ ft}$$

From Equation 67a

$$Q_2 = \left[\frac{(8)^3(1170)}{(6)^3(1770)}\right] (1500) = 2350 \text{ gpm}$$

(b) From Equation 64

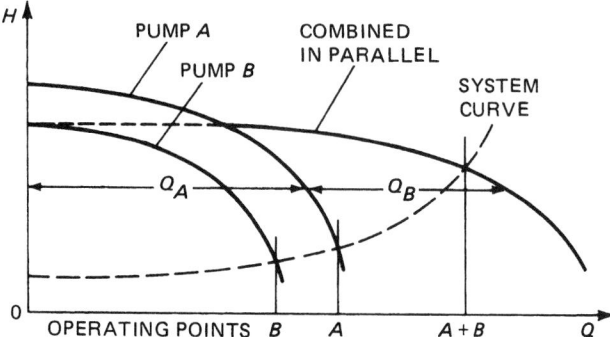

Fig. 4-2.23. *Two pumps combined in parallel.*

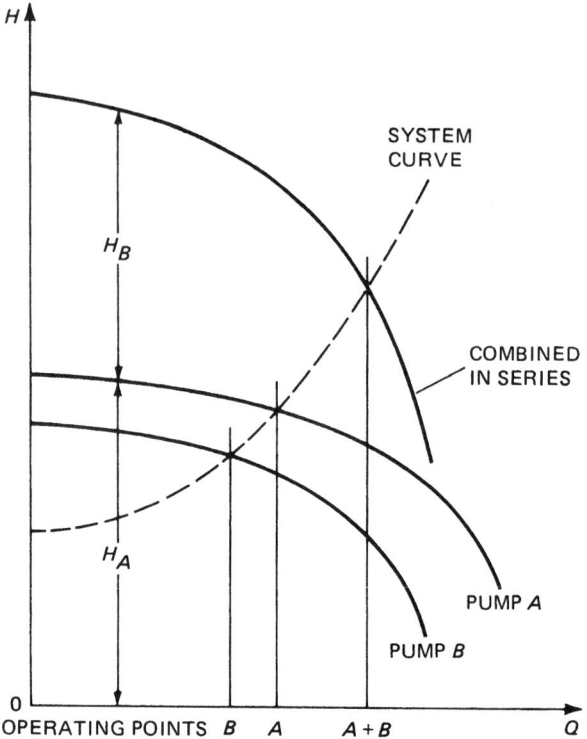

Fig. 4-2.24. *Two pumps combined in series.*

$$\text{WHP} = \frac{2350(93.2)}{3960} = 55.3 \text{ HP}$$

Therefore

$$\text{BHP} = \frac{55.3}{0.8} = 69.1 \text{ HP}$$

The motor chosen would be the next highest standard horsepower rating.

If more discharge or more head is required than a single pump can provide, two or more pumps may be combined to provide the necessary output. For example, when discharge is too little, pumps may be installed in parallel, sharing the same suction and inlet conditions. Figure 4-2.23 illustrates the principle. If a pump provides sufficient discharge but too little head, a second pump may be installed in series, the output of the first pump being fed directly into the suction of the second pump. Figure 4-2.24 depicts the series arrangement. A variety of compound arrangements are possible, depending on details of actual supply and demand, with economics being the prime arbiter.

NOMENCLATURE

A area
C proportionality constant or flow coefficient, Hazen-Williams C-factor
c celerity of a shock wave
D pipe diameter
d element diameter
E velocity of approach factor, bulk modulus of elasticity
f Darcy-Weisbach friction loss factor
g gravitational acceleration constant, 9.8 m/s²
H head of water
h head

h_c height of centroid
h_L head loss
I moment of inertia
K proportionality constant or flow coefficient
k proportionality constant or flow coefficient
L length of conduit (in friction loss equations)
l length or distance
m mass
N pump rpm
p pressure
Q volumetric discharge rate
Re   Reynolds number
S slope of energy gradient
s specific gravity
u stream velocity at a given point in flow cross-section
V volume
v average stream velocity
z height above a reference datum (potential head)
α kinetic energy correction factor
β beta ratio
γ specific weight
Δ increment
ε pipe wall absolute roughness
η efficiency
μ absolute (dynamic) viscosity
ν kinematic viscosity
ρ density
τ fluid shear stress

REFERENCES CITED

1. R.P. Benedict, *Fundamentals of Pipe Flow*, Wiley-Interscience, New York (1980).
2. R.A. Granger, *Fluid Mechanics*, Holt, Rinehart and Winston (CBS College Pub.), New York (1985).
3. H.E. Hickey, *Hydraulics for Fire Protection*, National Fire Protection Association, Quincy (1980).
4. N.H.C. Hwang, *Fundamentals of Hydraulic Engineering Systems*, Prentice-Hall, Englewood Cliffs (1981).
5. I.J. Karassik, ed., *Pump Handbook*, McGraw-Hill, New York (1986).
6. J.W. Murdock, *Fluid Mechanics and Its Applications*, Houghton Mifflin, Boston (1976).
7. A.L. Simon, *Practical Hydraulics*, John Wiley & Sons, New York (1981).
8. D. Stephenson, *Pipeflow Analysis*, Elsevier, Amsterdam (1984).
9. V.L. Streeter and E.G. Wylie, *Fluid Mechanics*, McGraw-Hill, New York (1979).
10. T.M. Walski, *Analysis of Water Distribution Systems*, Van Nostrand Reinhold, New York (1984).
11. F.M. White, *Fluid Mechanics*, McGraw-Hill, New York (1986).
12. NFPA 20, *Installation of Centrifugal Fire Pumps*, National Fire Protection Association, Quincy (1993).
13. NFPA 13, *Installation of Sprinkler Systems*, National Fire Protection Association, Quincy (1994).

AUTOMATIC SPRINKLER SYSTEM CALCULATIONS

Russell P. Fleming

INTRODUCTION

Applications Where Water Is Appropriate

Water is the most commonly used fire extinguishing agent, mainly due to the fact that it is widely available and inexpensive. This is not to discount the very desirable fire extinguishing characteristics of water, however, such as a high specific heat and high latent heat of vaporization. A single gallon of water can absorb 9,280 Btus of heat as it increases from a 70°F room temperature to become steam at 212°F.

Water is not the perfect extinguishing agent, however, and is considered inappropriate for the protection of certain water reactive materials. In some cases, the use of water can produce heat, flammable or toxic gases, or explosions. The quantities of such products must be considered, however, because application of sufficient water can overcome the reaction of minor amounts of such materials.

Another drawback of water is that it is more dense than most hydrocarbon fuels, and immiscible as well. This means that water will not provide an effective cover for burning hydrocarbons, or mix with them and dilute them to the point of not sustaining combustion. Instead, the hydrocarbons will float on top of the water, continuing to burn and possibly spread. To combat such fires, foam solutions can be introduced into the water so as to provide an effective cover and smother the fire. Success has also been achieved by applying water in a fine mist.

However, even in cases where water from sprinklers will not suppress the fire, the cooling ability of water spray can protect structural elements of a building, containing the fire until it can be extinguished by other means.

Types of Sprinkler Systems

Automatic sprinkler systems are considered to be the most effective and economical way to apply water to suppress a fire. There are four basic types of sprinkler systems:

1. A *wet pipe system* is by far the most common type of sprinkler system. It consists of a network of piping containing water under pressure. Automatic sprinklers are connected to the piping such that each sprinkler protects an assigned building area. The application of heat to any sprinkler will cause that single sprinkler to operate, permitting water to discharge over its area of protection.

2. A *dry pipe system* is similar to a wet system, except that water is held back from the piping network by a special dry pipe valve. The valve is kept closed by air or nitrogen pressure maintained in the piping. The operation of one or more sprinklers will allow the air pressure to escape, causing operation of the dry valve, which then permits water to flow into the piping to suppress the fire. Dry systems are used where the water in the piping would be subject to freezing.

3. A *deluge system* is one that does not use automatic sprinklers, but rather open sprinklers. A special deluge valve holds back the water from the piping, and is activated by a separate fire detection system. When activated, the deluge valve admits water to the piping network, and water flows simultaneously from all of the open sprinklers. Deluge systems are used for protection against rapidly spreading, high hazard fires.

4. A *preaction system* is similar to a deluge system except that automatic sprinklers are used, and a small air pressure is usually maintained in the piping network to ensure that the system is air tight. As with a deluge system, a separate detection system is used to activate a deluge valve, admitting water to the piping. Because automatic sprinklers are used, however, the water is usually stopped from flowing unless heat from the fire has also activated one or more sprinklers. Some special arrangements of preaction systems permit variations on detection system interaction with sprinkler operation. Preaction systems are generally used where there is special concern for accidental discharge of water, as in valuable computer areas.

These four basic types of systems differ in terms of the most fundamental aspect of how the water is put into the area of the fire. There are many other "types" of sprinkler systems, classified according to the hazard they protect (such as residential, in-rack, or exposure protection); additives to the system (such as antifreeze or foam); or special

Russell P. Fleming, P.E., is Vice President of Engineering, National Fire Sprinkler Association, Patterson, NY. Mr. Fleming has served as a member of more than a dozen NFPA technical committees, including the Committee on Automatic Sprinklers. He currently serves as Chairman of the NFPA Standards Council.

connections to the system (such as multipurpose piping); but all sprinkler systems can still be categorized as one of the basic four types.

Applicable Standards

NFPA 13, *Standard for the Installation of Sprinkler Systems* (hereinafter referred to as NFPA 13), is a design and installation standard for automatic sprinkler systems, referenced by most building codes in the U.S. and Canada.[6] This standard, in turn, references other NFPA standards for details as to water supply components, including NFPA 14, *Standard for the Installation of Standpipe and Hose Systems*; NFPA 20, *Standard for the Installation of Centrifugal Fire Pumps* (hereinafter referred to as NFPA 20); NFPA 22, *Standard for Water Tanks for Private Fire Protection* (hereinafter referred to as NFPA 22); and NFPA 24, *Standard for the Installation of Private Fire Service Mains and Their Appurtenances*.

For protection of warehouse storage, NFPA 13 references design criteria contained in special storage standards, NFPA 231, *Standard for General Storage* (hereinafter referred to as NFPA 231); and NFPA 231C, *Standard for Rack Storage of Materials* (hereinafter referred to as NFPA 231C).

Other NFPA standards contain design criteria for special types of occupancies or systems, including NFPA 13D, *Standard for the Installation of Sprinkler Systems in One- and Two-Family Dwellings and Manufactured Homes* (hereinafter referred to as NFPA 13D); NFPA 15, *Standard for Water Spray Fixed Systems for Fire Protection*; NFPA 16, *Standard for the Installation of Deluge Foam-Water Sprinkler and Foam-Water Spray Systems*; NFPA 16A, *Standard for the Installation of Closed-Head Foam-Water Sprinkler Systems*; NFPA 30B, *Code for the Manufacture and Storage of Aerosol Products*; NFPA 214, *Standard on Water-Cooling Towers*; and NFPA 409, *Standard on Aircraft Hangars*.

Limits of Calculation in an Empirical Design Process

Engineering calculations are best performed in areas where an understanding exists as to relationships between parameters. This is not the case with the technology of automatic sprinkler systems. Calculation methods are widely used with regard to only one aspect of sprinkler systems: water flow through piping. There are only very rudimentary calculation methods available with regard to the most fundamental aspect of sprinkler systems, i.e., the ability of water spray to suppress fires.

The reason that calculation methods are not used is simply the complexity of the mechanisms by which water suppresses fires. Water-based fire suppression has to this point not been thoroughly characterized to permit application of mathematical modeling techniques. As a result, the fire suppression aspects of sprinkler system design are empirical at best.

Some, but not all, of the current sprinkler system design criteria are based on full-scale testing, including the criteria within NFPA 231C and 13D, and parts of NFPA 13 such as the material on the use of large drop and ESFR (early suppression fast response) sprinklers. Most of the NFPA 13 protection criteria, however, are the result of evolution and application of experienced judgment. In the 1970s, the capabilities of pipe schedule systems, which had demonstrated a hundred years' satisfactory performance, were codified into a system of area/density curves. This permitted the

introduction of hydraulic calculations to what had become a "cookbook" method of designing sprinkler systems. It allowed system designers to take advantage of strong water supplies to produce more economical systems. It also permitted the determination of specific flows and pressures available at various points of the system, opening the door to the use of "special sprinklers." Special sprinklers are approved for use on the basis of their ability to accomplish specific protection goals, but are not interchangeable since there is no standardization of minimum flows and pressures.

Because of this history, the calculation methods available to the fire protection engineer in standard sprinkler system design are only ancillary to the true function of a sprinkler system. The sections that follow in this chapter address hydraulic calculations of flow through piping, simple calculations commonly performed in determining water supply requirements, and optional calculations that may be performed with regard to hanging and bracing of system piping. The final section of this chapter deals with the performance of a system relative to a fire, and the material contained therein is totally outside the realm of standard practice. Due to the state of the art, this material is not sufficiently complete to permit a full design approach, but only isolated bits of total system performance.

HYDRAULIC CALCULATIONS

Density-Based Sprinkler Demand

Occupancy hazard classification is the most critical aspect of the sprinkler system design process. If the hazard is underestimated, it is possible for fire to overpower the sprinklers, conceivably resulting in a large loss of property or life. Hazard classification is not an area in which calculation methods are presently in use, however. The proper classification of hazard requires experienced judgment and familiarity with relevant NFPA standards.

Once the hazard or commodity classification is determined and a sprinkler spacing and piping layout has been proposed in conformance with the requirements of the standard, the system designer can begin a series of calculations to demonstrate that the delivery of a prescribed rate of water application will be accomplished for the maximum number of sprinklers that might be reasonably expected to operate. This number of sprinklers, which must be supplied regardless of the location of the fire within the building, is the basis of the concept of the remote design area. The designer needs to demonstrate that the shape and location of the sprinkler arrangement in the design area will be adequately supplied with water in the event of a fire.

Prior to locating the design area, there is the question of how many sprinklers are to be included. This question is primarily addressed by the occupancy hazard classification, but the designer also has some freedom to decide this matter.

Figure 2-2.1(b) of the 1987 edition of NFPA 13, and corresponding figures in NFPA 231 and 231C contain area/density curves from which the designer can select a design area and density appropriate for the occupancy hazard classification. Any point on or to the right of the curve in the figure(s) is acceptable. The designer may select a high density over a small area, or a low density over a large area. In either event, the fire is expected to be controlled by the sprinklers within that design area, without opening any additional sprinklers.

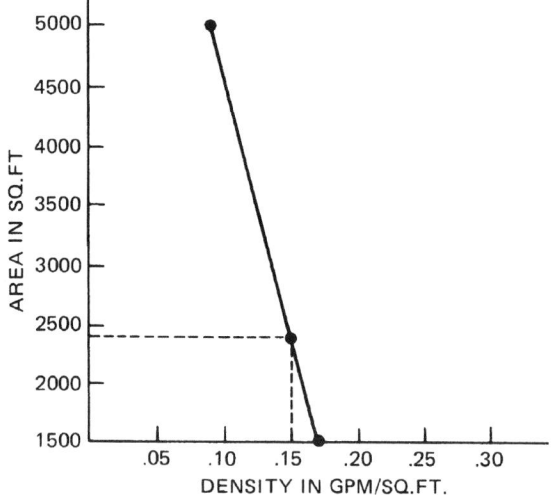

Fig. 4-3.1. Sample area/density curve.

EXAMPLE 1:

Using the sample area/density curve shown in Figure 4-3.1, many different sets of design criteria could be selected, ranging from a density of 0.1 gpm/sq ft over 5000 square feet to 0.17 gpm/sq ft over 1500 square feet. Either of these two points, or any point to the right of the curve (such as 0.16 gpm/sq ft over 3000 sq ft), would be considered acceptable. A selection of 0.15 gpm/sq ft over 2400 square feet is indicated.

Water is provided only for the number of sprinklers in the design area, since no water is needed for the sprinklers that are not expected to open. The actual number of sprinklers in the design area depends, of course, on the spacing of the sprinklers. NFPA 13 requires that the design area be divided by the maximum sprinkler spacing used, and that any fractional result be rounded up to the next whole sprinkler.

EXAMPLE 2:

Based on the point selected from the sample area/density curve above and the proposed maximum spacing of sprinklers, the number of sprinklers to be included in the design area can be determined. If sprinklers are spaced at 12 × 15 feet so as to each protect an area of 180 square feet, the design area of 2400 square feet would include

$$2400/180 = 13.33 = 14 \text{ sprinklers}$$

The remote design area is required to have a rectangular shape, with the long side along the run of the branch lines. The length of the design area (needed to determine how many sprinklers along a branch line are contained within it) is found by multiplying the square root of the design area by a factor of 1.2. Again, any fractional result is rounded to the next whole sprinkler.

EXAMPLE 3:

If the 14 sprinklers from Example 2 were spaced 12 feet along branch lines 15 feet apart, the length of the rectangular area along the branch lines would be

$$\frac{1.2(2500)^{1/2}}{12} = \frac{1.2(50)}{12} = 5$$

If the sprinklers were spaced 15 feet along branch lines 12 feet apart, the same length of the design area would include only 4 sprinklers.

Figure 5-2.3 of the 1994 edition of NFPA 13 contains some exceptions to this method of locating a remote design area and determining the number of sprinklers to be supplied. Chapter 5 of the standard has special modifications to the design area based on factors such as the use of a dry system and the existence of unsprinklered combustible concealed spaces within the building. The chapter also contains a "room design method," which can reduce the number of sprinklers expected to operate in a highly compartmented occupancy. Also, beginning in 1985, the standard adopted a 4 sprinkler design area for dwelling units and their adjacent corridors when residential sprinklers are installed in accordance with their listing requirements.

Figures A-6-2.2(a) through A-6-2.2(c) in the appendix of NFPA 13 (1994 edition) show the step-by-step calculation procedure for a sample sprinkler system. The starting point is the most remote sprinkler in the design area. For tree systems, in which each sprinkler is supplied from only one direction, the most remote sprinkler is generally the end sprinkler on the farthest branch line from the system riser. This sprinkler, and all others as a result, must be provided with a sufficient flow of water to meet the density appropriate for the point selected on the area/density curve.

Where a sprinkler protects an irregular area, NFPA 13 prescribes that the area of coverage for the sprinkler must be based on the largest sides of its coverage. In other words, the area which a sprinkler protects for calculation purposes is equal to

$$\text{area of coverage} = S \times L$$

where S is twice the larger of the distances to the next sprinkler (or wall for an end sprinkler) in both the upstream and downstream directions, and L is twice the larger of the distances to adjacent branch lines (or wall in the case of the last branch line) on either side. This reflects the need to flow more water with increasing distance from the sprinkler, since increased flow tends to expand the effective spray umbrella of the sprinkler.

The minimum flow from a sprinkler must be the product of the area of coverage multiplied by the minimum required density

$$Q = \text{area of coverage} \times \text{density}$$

Most of the special listed sprinklers and residential sprinklers have a minimum flow requirement associated with their listings, which is often based on the spacing at which they are used. These minimum flow considerations override the minimum flow based on the area/density method.

EXAMPLE 4:

If a standard ½-inch orifice sprinkler protects an area extending to 7 feet on the north side (half the distance to the next branch line), 5 feet on the south side (to a wall), 6 feet on the west side (half the distance to the next sprinkler on the branch line), and 4 feet on the east side (to a wall), the minimum flow required for the sprinkler to achieve the density requirement selected in Example 1 can be found by completing two steps. The first step involves determining the area of coverage. In this case

$$S \times L = 2(6 \text{ ft}) \times 2(7 \text{ ft}) = 12 \text{ ft} \times 14 \text{ ft} = 168 \text{ sq ft}$$

The second step involves multiplying this coverage area by the required density

$$Q = A \times \rho = 168 \text{ sq ft} \times 0.15 \text{ gpm/sq ft} = 25.2 \text{ gpm}$$

Pressure Requirements of the Most Remote Sprinkler

When flow through a sprinkler orifice takes place, the energy of the water changes from the potential energy of pressure to the kinetic energy of flow. A formula can be derived from the basic energy equations to determine how much water will flow through an orifice based on the water pressure inside the piping at the orifice

$$Q = 29.83 c_d d^2 P^{1/2}$$

However, this formula contains a factor, c_d, which is a discharge coefficient characteristic of the orifice and which must be determined experimentally. For sprinklers, the product testing laboratories determine the orifice discharge coefficient at the time of listing of a particular model of sprinkler. To simplify things, all factors other than pressure are lumped into what is experimentally determined as the K-factor of a sprinkler, such that

$$Q = K \times P^{1/2}$$

where K has units of gpm/(psi)$^{1/2}$.

If the required minimum flow at the most remote sprinkler is known, determined by either the area/density method or the special sprinkler listing, the minimum pressure needed at the most remote sprinkler can easily be found

Since $\qquad Q = K(P)^{1/2}, \qquad$ then $P = (Q/K)^2$

NFPA 13 sets a minimum pressure of 7 psi at the end sprinkler in any event, so that a proper spray umbrella is ensured.

EXAMPLE 5:

The pressure required at the sprinkler in Example 4 can be determined using the above formula if the K-factor is known. A midrange K-factor for a standard ½ in. orifice sprinkler is 5.6.

$$P = (Q/K)^2 = (25.2/5.6)^2 = 20.2 \text{ psi}$$

Once the minimum pressure at the most remote sprinkler is determined, the hydraulic calculation method proceeds backward toward the source of supply. If the sprinkler

TABLE 4-3.1 *C Values for Pipes*

Type of Pipe	Assigned C Factor
Steel pipe—dry and preaction systems	100
Steel pipe—wet and deluge systems	120
Galvanized steel pipe—all systems	120
Cement lined cast or ductile iron	140
Copper tube	150
Plastic (listed)	150

spacing is regular, it can be assumed that all other sprinklers within the design area will be flowing at least as much water, and the minimum density is assured. If spacing is irregular or sprinklers with different K-factors are used, care must be taken that each sprinkler is provided with sufficient flow.

As the calculations proceed toward the system riser, the minimum pressure requirements increase, because additional pressures are needed at these points if elevation and friction losses are to be overcome while still maintaining the minimum needed pressure at the most remote sprinkler. These losses are determined as discussed below, and their values added to the total pressure requirements. Total flow requirements also increase backward toward the source of supply, until calculations get beyond the design area. Then there is no flow added other than hose stream allowances.

It should be noted that each sprinkler closer to the source of supply will show a successively greater flow rate, since a higher total pressure is available at that point in the system piping. This effect on the total water demand is termed hydraulic increase, and is the reason why the total water demand of a system is not simply equal to the product of the minimum density and the design area. When calculations are complete, the system demand will be known, stated in the form of a specific flow at a specific pressure.

Pressure Losses Through Piping, Fittings, and Valves

Friction losses resulting from water flow through piping can be estimated by several engineering approaches, but NFPA 13 specifies the use of the Hazen-Williams method. This approach is based on the formula developed empirically by Hazen and Williams

$$p = \frac{4.52 Q^{185}}{C^{1.85} d^{4.87}}$$

where

p = friction loss per foot of pipe in psi
Q = flow rate in gpm
d = internal pipe diameter
C = Hazen-Williams coefficient.

The choice of C is critical to the accuracy of the friction loss determination, and is therefore stipulated by NFPA 13. The values assigned for use are intended to simulate the expected interior roughness of aged pipe. (See Table 4-3.1.)

Rather than make the Hazen-Williams calculation for each section of piping, it has become standard practice, when doing hand calculations, to use a friction loss table, which contains all values of p for various values of Q and various pipe sizes. In many cases the tables are based on the use of Schedule 40 steel pipe for wet systems, and the use of other pipe schedules, pipe materials, or system types may require the use of multiplying factors.

Once the value of friction loss per foot is determined using either the previous equation or friction loss tables, the total friction loss through a section of pipe is found by multiplying p by the length of pipe, L. Since NFPA 13 uses p to designate loss per foot, total friction loss in a length of pipe can be designated by p_f, where

$$p_f = p \times L$$

In the analysis of complex piping arrangements, it is sometimes convenient to lump the values of all factors in the

TABLE 4-3.2 *Equivalent Pipe Length Chart (For C = 120)*

Fittings and Valves	Fittings and Valves Expressed in Equivalent Feet of Pipe													
	¾ in.	1 in.	1¼ in.	1½ in.	2 in.	2½ in.	3 in.	3½ in.	4 in.	5 in.	6 in.	8 in.	10 in.	12 in.
45° Elbow	1	1	1	2	2	3	3	3	4	5	7	9	11	13
90° Standard Elbow	2	2	3	4	5	6	7	8	10	12	14	18	22	27
90° Long Turn Elbow	1	2	2	2	3	4	5	5	6	8	9	13	16	18
Tee or Cross (Flow Turned 90°)	3	5	6	8	10	12	15	17	20	25	30	35	50	60
Butterfly Valve	—	—	—	—	6	7	10	—	12	9	10	12	19	21
Gate Valve	—	—	—	—	1	1	1	1	2	2	3	4	5	6
Swing Check*	—	5	7	9	11	14	16	19	22	27	32	45	55	65

For SI Units: 1 ft = 0.3048 m.
*Due to the variations in design of swing check valves, the pipe equivalents indicated in the above chart are to be considered average.

Hazen-Williams equation (except flow) for a given piece of pipe into a constant, K, identified as a friction loss coefficient. To avoid confusion with the nozzle coefficient K, this coefficient can be identified as FLC, friction loss coefficient.

$$FLC = (L \times 4.52)/(C^{1.85}d^{4.87})$$

The value of p_f is therefore equal to

$$p_f = FLC \times Q^{1.85}$$

EXAMPLE 6:

If the most remote sprinkler on a branch line requires a minimum flow of 25.2 gpm (for a minimum pressure of 20.2 psi) as shown in Examples 4 and 5, and the second sprinkler on the line is connected by a 12 foot length of 1 in. Schedule 40 steel pipe, with both sprinklers mounted directly in fittings on the pipe (no drops or sprigs), the minimum pressure required at the second sprinkler can be found by determining the friction loss caused by a flow of 25.2 gpm through the piping to the end sprinkler. No fitting losses need to be considered if it is a straight run of pipe, since NFPA 13 permits the fitting directly attached to each sprinkler to be ignored.

Using the Hazen-Williams equation with values of 25.2 for Q, 120 for C, and 1.049 for d (the inside diameter of Schedule 40 steel 1 in. pipe) results in a value of $p = 0.20$ psi per foot of pipe. Multiplying by the 12-foot length results in a total friction loss of $p_f = 2.4$ psi. The total pressure required at the second sprinkler on the line is therefore 20.2 psi + 2.4 psi = 22.6 psi. This will result in a flow from the second sprinkler of $Q = K(P)^{1/2} = 26.6$ gpm.

Minor losses through fittings and valves are not friction losses but energy losses, caused by turbulence in the water flow which increase as the velocity of flow increases. Nevertheless, it has become standard practice to simplify calculation of such losses through the use of "equivalent lengths," which are added to the actual pipe length in determining the pipe friction loss. NFPA 13 contains a table of equivalent pipe lengths for this purpose. (See Table 4-3.2.) As an example, if a 2 in. 90-degree long turn elbow is assigned an equivalent length of 3 feet, this means that the energy loss associated with turbulence through the elbow is expected to approximate the energy loss to friction through 3 feet of 2 in. pipe. As with the friction loss tables, the equivalent pipe length chart is based on the use of steel pipe with a C-factor of 120, and the use of other piping materials requires mul-

tiplying factors. The equivalent pipe length for pipes having C values other then 120 should be adjusted using the following multiplication factors: 0.713 for a C value of 100; 1.16 for a C value of 130; 1.33 for a C value of 140; 1.51 for a C value of 150.

EXAMPLE 7:

If the 12 foot length of 1 in. pipe in Example 6 had contained 4 elbows so as to avoid a building column, the pressure loss from those elbows could be approximated by adding an equivalent length of pipe to the friction loss calculation. Table 4-3.2 gives a value of 2 feet as the appropriate equivalent length for standard elbows in 1 in. Schedule 40 steel pipe. For 4 elbows, the equivalent fitting length would be 8 feet. Added to the actual pipe length of 12 feet, the total equivalent length would be 20 feet. This results in a new value of $p_f = 20\text{-ft} \cdot 0.20$ psi/ft $= 4.0$ psi. The total pressure at the second sprinkler would then be equal to 20.2 psi + 4.0 psi = 24.2 psi. The total flow from the second sprinkler in this case would be $Q = K(P)^{1/2} = 27.5$ gpm.

Some types of standard valves, such as swing check valves, are included in the equivalent pipe length chart, Table 4-3.2. Equivalent lengths for pressure losses through system alarm, dry, and deluge valves are determined by the approval laboratories at the time of product listing.

Use of Velocity Pressures

The value of pressure, P, in the sprinkler orifice flow formula can be considered either the total pressure, P_t, or the normal pressure, P_n, since NFPA 13 permits the use of velocity pressures at the discretion of the designer. Total pressure, normal pressure, and velocity pressure, P_v, have the following relationship:

$$P_n = P_t - P_v$$

Total pressure is the counterpart of total energy or total head, and can be considered the pressure that would act against an orifice if all of the energy of the water in the pipe at that point were focused toward flow out of the orifice. This is the case where there is no flow past the orifice in the piping. Where flow does take place in the piping past an orifice, however, normal pressure is that portion of the total pressure which is actually acting normal to the direction of flow in the piping, and therefore acting in the direction of

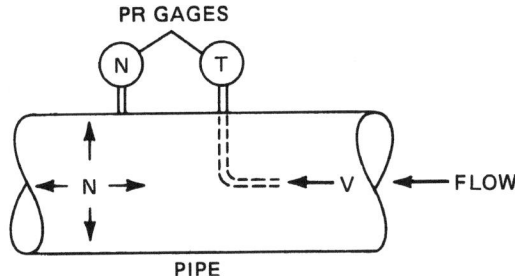

Fig. 4-3.2. Velocity and normal pressures in piping.

flow through the orifice. The amount by which normal pressure is less than total pressure is velocity pressure, which is acting in the direction of flow in the piping. Velocity pressure corresponds to velocity energy, which is the energy of motion. There is no factor in the above expression for elevation head, because the flow from an orifice can be considered to take place in a datum plane.

When velocity pressures are used in calculations, it is recognized that some of the energy of the water is in the form of velocity head, which is not acting normal to the pipe walls (where it would help push water out the orifice), but rather in the downstream direction. Thus, for every sprinkler (except the end sprinkler on a line), slightly less flow takes place than what would be calculated from the use of the formula $Q = K(P_t)^{1/2}$. (See Figure 4-3.2.)

NFPA 13 permits the velocity pressure effects to be ignored, however, since they are usually rather minor, and since ignoring the effects of velocity pressure tends to produce a more conservative design.

If velocity pressures are considered, normal pressure rather than total pressure is used when determining flow through any sprinkler except the end sprinkler on a branch line, and through any branch line except the end branch line on a cross main. The velocity pressure, P_v, which is subtracted from the total pressure in order to determine the normal pressure, is determined as

$$P_v = v^2/2g \times 0.433 \text{ psi/ft}$$

or

$$P_v = 0.001123Q^2/d^4$$

where Q is the upstream flow through the piping to an orifice (or branch line) in gpm and d is the actual internal diameter of the upstream pipe in inches.

Because NFPA 13 mandates the use of the upstream flow, an iterative approach to determining the velocity pressure is necessary. The upstream flow cannot be known unless the flow from the sprinkler (or branch line) in question is known, and the flow from the sprinkler (or branch line) is affected by the velocity pressure resulting from the upstream flow.

EXAMPLE 8:

If the pipe on the upsteam side of the second sprinkler in Example 6 were 1¼ in. Schedule 40 steel pipe with an inside diameter of 1.38 in., the flow from the second sprinkler would be considered to be 26.6 gpm as determined at the end of Example 6, if velocity pressures were not included.

If velocity pressures were to be considered, an upstream flow would first be assumed. Since the end sprinkler had a minimum flow of 25.2 gpm and the upstream flow would consist of the combined flow rates of the two sprinklers, an estimate of 52 gpm might appear reasonable. Substituting this flow and the pipe diameter into the equation for velocity pressure gives

$$P_v = 0.001123Q^2/d^4$$
$$= 0.001123(52)^2/(1.38)^4$$
$$= 0.8 \text{ psi}$$

This means that the actual pressure acting on the orifice of the second sprinkler is equal to

$$P_n = P_t - P_v$$
$$= 22.6 \text{ psi} - 0.8 \text{ psi}$$
$$= 21.8 \text{ psi}$$

This would result in a flow from the second sprinkler of

$$Q = K(P)^{1/2}$$
$$= 26.1 \text{ gpm}$$

Combining this flow with the known flow from the end sprinkler results in a total upstream flow of 51.3 gpm. To determine if the initial guess was close enough, determine the velocity pressure that would result from an upstream flow of 51.3 gpm. This calculation also results in a velocity pressure of 0.8 psi, and the process is therefore complete. It can be seen that the second sprinkler apparently flows 0.5 gpm less through the consideration of velocity pressures.

Elevation Losses

Variation of pressure within a fluid at rest is related to the density or unit (specific) weight of the fluid. The unit weight of a fluid is equal to its density multiplied by the acceleration of gravity. The unit weight of water is 62.4 lbs/ft³.

This means that one cubic foot of water at rest weighs 62.4 pounds. The cubic foot of water, or any other water column one foot high, thus results in a static pressure at its base of 62.4 pounds per square foot. Divided by 144 square inches per square foot, this is a pressure of 0.433 pounds per square inch per foot of water column.

A column of water 10 feet high similarly exerts a pressure of 10 ft × 62.4 lbs/ft² × 1 ft/144 in.² = 4.33 psi. The static pressure at the top of both columns of water is equal to zero (gauge pressure), or atmospheric pressure.

On this basis, additional pressure must be available within a sprinkler system water supply to overcome the pressure loss associated with elevation. This pressure is equal to 0.433 psi per foot of elevation of the sprinklers above the level where the water supply information is known.

Sometimes the additional pressure needed to overcome elevation is added at the point where the elevation change takes place within the system. If significant elevation changes take place within a portion of the system that is likely to be considered as a representative flowing orifice (such as a single branch line along a cross main that is equivalent to other lines in the remote design area), then it is considered more accurate to wait until calculations have been completed, and simply add an elevation pressure increase to account for the total height of the highest sprinklers above the supply point.

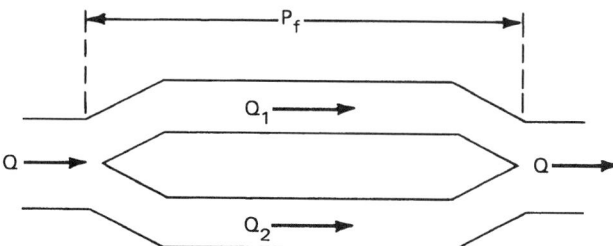

Fig. 4-3.3. Example of a simple loop configuration.

EXAMPLE 9:

The pressure that must be added to a system supply to compensate for the fact that the sprinklers are located 120 feet above the supply can be found by multiplying the total elevation difference by 0.433 psi/ft.

$$120 \text{ ft} \times 0.433 \text{ psi/ft} = 52 \text{ psi}$$

Loops and Grids

Hydraulic calculations become more complicated when piping is configured in loops or grids, such that water feeding any given sprinkler or branch line can be supplied through more than one route. A number of computer programs that handle the repetitive calculations have therefore been developed specifically for fire protection systems, and are being marketed commercially.

Determining the flow split that takes place in the various parts of any loop or grid is accomplished by applying the basic principles of conservation of mass and conservation of energy. For a single loop, it should be recognized that the energy loss across each of the two legs from one end of the system to the other must be equal. Otherwise, a circulation would take place within the loop itself. Also, mass is conserved by the fact that the sum of the two individual flows through the paths is equal to the total flow into (and out of) the loop. (See Figure 4-3.3.)

Applying the Hazen-Williams formula to each leg of the loop

$$p_f = L_1 \frac{4.52 Q_1^{1.85}}{C_1^{1.85} d_1^{4.87}} = L_2 \frac{4.52 Q_2^{1.85}}{C_2^{1.85} d_2^{1.85}}$$

Substituting the term FLC for all terms except Q,

$$p_f = \text{FLC}_1 Q_1^{1.85} = \text{FLC}_2 Q_2^{1.85}$$

This simplifies to become

$$(Q_1/Q_2)^{1.85} = \text{FLC}_2/\text{FLC}_1$$

Since Q_1 and Q_2 combine to create a total flow of Q, the flow through one leg can be determined as

$$Q_1 = Q/[(\text{FLC}_1/\text{FLC}_2)^{0.54} + 1]$$

For the simplest of looped systems, a single loop, hand calculations are not complex. Sometimes, seemingly complex piping systems can be simplified by substituting an "equivalent pipe" for two or more pipes in series or in parallel.

For pipes in series

$$\text{FLC}_e = \text{FLC}_1 + \text{FLC}_2 + \text{FLC}_3 + \cdots$$

For pipes in parallel

$$(1/\text{FLC}_e)^{0.54} = (1/\text{FLC}_1)^{0.54} + (1/\text{FLC}_2)^{0.54} + \cdots$$

For gridded systems, which involve flow through multiple loops, computers are generally used since it becomes necessary to solve a system of nonlinear equations. When hand calculations are performed, the Hardy Cross method of balancing heads is generally employed. This method involves assuming a flow distribution within the piping network, then applying successive corrective flows until differences in pressure losses through the various routes are nearly equal.

The Hardy Cross[1] solution procedure applied to sprinkler system piping is as follows:

1. Identify all loop circuits and the significant parameters associated with each line of the loop, such as pipe length, diameter, and Hazen-Williams coefficient. Reduce the number of individual pipes where possible by finding the equivalent pipe for pipes in series or parallel.
2. Evaluate each parameter in the proper units. Minor losses through fittings should be converted to equivalent pipe lengths. A value of all parameters except flow for each pipe section should be calculated (FLC).
3. Assume a reasonable distribution of flows that satisfies continuity, proceeding loop by loop.
4. Compute the pressure (or head) loss due to friction, p_f, in each pipe using the FLC in the Hazen-Williams formula.
5. Sum the friction losses around each loop with due regard to sign. (Assume clockwise positive, for example.) Flows are correct when the sum of the losses, dp_f, is equal to zero.
6. If the sum of the losses is not zero for each loop, divide each pipe's friction loss by the presumed flow for the pipe, p_f/Q.
7. Calculate a correction flow for each loop as

$$dQ = -dp_f/[1.85 \Sigma (p_f/Q)]$$

8. Add the correction flow values to each pipe in the loop as required, thereby increasing or decreasing the earlier assumed flows. For cases where a single pipe is in two loops, the algebraic difference between the two values of dQ must be applied as the correction to the assumed flow.
9. With a new set of assumed flows, repeat Steps 4 through 7 until the values of dQ are sufficiently small.
10. As a final check, calculate the pressure loss by any route from the initial to the final junction. A second calculation along another route should give the same value within the range of accuracy expected.

NFPA 13 requires that pressures be shown to balance within 0.5 psi at hydraulic junction points. In other words, the designer (or computer program) must continue to make successive guesses as to how much flow takes place in each piece of pipe until the pressure loss from the design area back to the source of supply is approximately the same (within 0.5 psi) regardless of the path chosen.

EXAMPLE 10:

For the small two-loop grid shown in Figure 4-3.4, the total flow in and out is 100 gpm. It is necessary to determine the flow taking place through each pipe section. The system

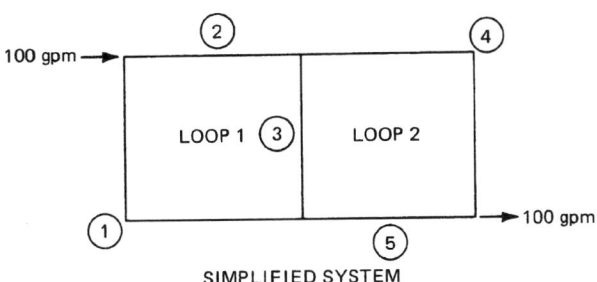

Fig. 4-3.4. Simplified system, pipe in series.

has already been simplified by finding the equivalent pipe for all pipes in series and in parallel, and the following values of FLC have been calculated.

Pipe 1 FLC = 0.001
Pipe 2 FLC = 0.002
Pipe 3 FLC = 0.003
Pipe 4 FLC = 0.001
Pipe 5 FLC = 0.004

Under Step 3 of the Hardy Cross procedure, flows that would satisfy conservation of mass are assumed. (See Figure 4-3.5.)

Steps 4 through 9 are then carried out in a tabular approach. (See Table 4-3.3.)

Making these adjustments to again balance flows, a second set of iterations can be made. (See Table 4-3.4.) For pipe segment 3, the new flow is the algebraic sum of the original flow plus the flow corrections from both loops. (See Figures 4-3.6 and 4-3.7.)

In starting the third iteration, it can be seen that the pressure losses around both loops are balanced within 0.5 psi. (See Table 4-3.5.) Therefore, the flow split assumed after two iterations can be accepted. As a final check, Step 10 of the above procedure calls for a calculation of the total pressure loss through two different routes, requiring that they balance within 0.5 psi:

Route through Pipes 1 and 5:

$$FLC_1, (Q_1)^{1.85} + FLC_2(Q_2)^{1.85}$$

$$= 0.001(54.0)^{1.85} + 0.004(35.9)^{1.85} = 1.6 + 3.0 = 4.6 \text{ psi}$$

Route through Pipes 2 and 4:

$$0.002(46.0)^{1.85} + 0.001(64.1)^{1.85} = 2.4 + 2.2 = 4.6 \text{ psi}$$

This is acceptable. Note that it required only two iterations to achieve a successful solution despite the fact that the initial flow assumption called for reverse flow in pipe 3. The initial assumption was for a clockwise flow of 5 gpm in pipe 3, but the final solution shows a counterclockwise flow of 18.1 gpm.

WATER SUPPLY CALCULATIONS
Determination of Available Supply Curve

Testing a public or private water supply permits an evaluation of the strength of the supply in terms of both quantity of flow and available pressures. The strength of a water supply is the key to whether it will adequately serve a sprinkler system.

Each test of a water supply must provide at least two pieces of information: a static pressure and a residual pressure at a known flow. The static pressure is the "no flow" condition, although it must be recognized that rarely is any public water supply in a true no flow condition. But this condition does represent a situation where the fire protection system is not creating an additional flow demand beyond that which is ordinarily placed on the system. The residual pressure reading is taken with an additional flow being taken from the system, preferably a flow that approximates the likely maximum system demand.

Between the two (or more) points, a representation of the water supply, termed a water supply curve, can be made. For the most part, this water supply curve is a fingerprint of the system supply and piping arrangements, since the static pressure tends to represent the effect of elevated tanks and operating pumps in the system, and the drop to the residual pressure represents the friction and minor losses through the piping network that result from the increased flow during the test.

The static pressure is read directly from a gauge attached to a hydrant. The residual pressure is read from the same gauge while a flow is taken from another hydrant, preferably downstream. A pitot tube is usually used in combination with observed characteristics of the nozzle though which flow is taken in order to determine the amount of flow. Appendix B to NFPA 13 provides more thorough information on this type of testing.

Figure A-6-2.2(d) of NFPA 13 (1994 edition) is an example of a plot of water supply information. The static pressure is plotted along the *y*-axis, reflecting a given pressure under zero flow conditions. The residual pressure at the known flow is also plotted, and a straight line is drawn between these two points. Note that the *x*-axis is not linear, but rather shows flow as a function of the 1.85 power. This corresponds to the exponent for flow in the Hazen-Williams equation. Using this semi-exponential graph paper demonstrates that the residual pressure effect is the result of friction loss through the system, and permits the water supply curve to be plotted as a straight line. Since the drop in residual pressure is proportional to flow to the 1.85 power, the available pressure at any flow can be read directly from the water supply curve.

For adequate design, the system demand point, including hose stream allowance, should lie below the water supply curve.

EXAMPLE 11:

If a water supply is determined by test to have a static pressure of 100 psi and a residual pressure of 60 psi at a flow

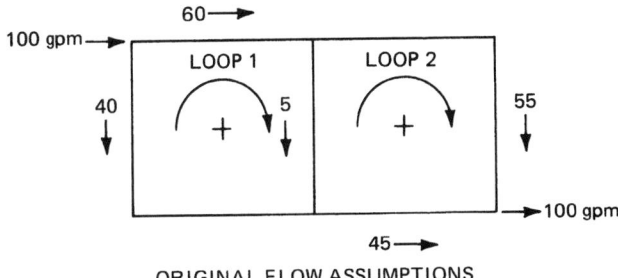

Fig. 4-3.5. Original flow assumptions.

TABLE 4-3.3 1st Iteration

Loop	Pipe	Q	FLC	p_f	dp_f	(p_f/Q)	$dQ = -dp_f/1.85[\Sigma(p_f/Q)]$	$Q + dQ$
1	1	-40	0.001	-0.92		0.023	$dQ = -16.4$	-56.4
	2	60	0.002	3.90		0.065		$+43.6$
	3	5	0.003	0.06		0.012		-11.4
					$= 3.04$	0.100		
2	3	-5	0.003	-0.06		0.012	$dQ = 11.2$	$+6.2$
	4	55	0.001	1.66		0.030		$+66.2$
	5	-45	0.004	-4.58		0.102		$+33.8$
					$= -2.98$	0.144		

TABLE 4-3.4 2nd Iteration

Loop	Pipe	Q	FLC	p_f	dp_f	(p_f/Q)	$dQ = -dp_f/1.85[\Sigma(p_f/Q)]$	$Q + dQ$
1	1	-56.4	0.001	-1.74		0.031	$dQ = 2.4$	-54.0
	2	43.6	0.002	2.16		0.050		$+46.0$
	3	-22.6	0.003	-0.96		0.042		-20.2
					$= -0.54$	0.123		
2	3	22.6	0.003	0.96		0.042	$dQ = -2.1$	$+20.5$
	4	66.2	0.001	2.34		0.035		$+64.1$
	5	-33.8	0.004	-2.69		0.080		$+35.9$
					$= 0.61$	0.157		

TABLE 4-3.5 3rd Iteration

Loop	Pipe	Q	FLC	p_f	dp_f	(p_f/Q)	$dQ = -dp_f/1.85[\Sigma(p_f/Q)]$	$Q + dQ$
1	1	-54.0	0.001	-1.60				
	2	46.0	0.002	2.38				
	3	-18.1	0.003	-0.64				
					$= 0.14$			
2	3	18.1	0.003	0.64				
	4	64.1	0.001	2.20				
	5	-35.9	0.004	-3.01				
					$= -0.17$			

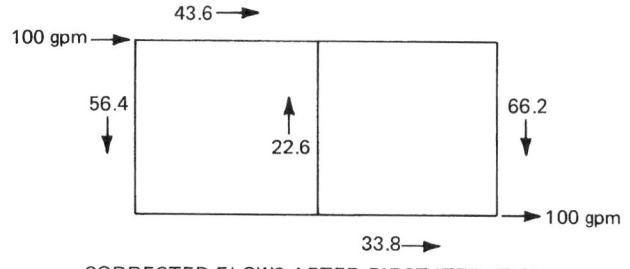

CORRECTED FLOWS AFTER FIRST ITERATION

Fig. 4-3.6. Corrected flows after first iteration.

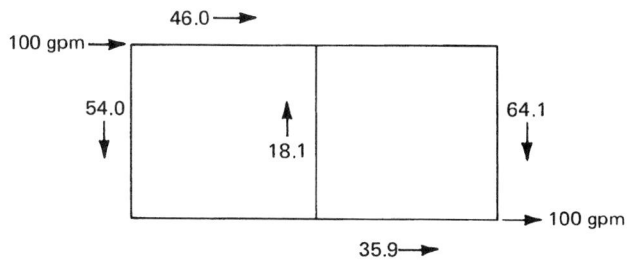

CORRECTED FLOWS AFTER 2nd ITERATION

Fig. 4-3.7. Corrected flows after second iteration.

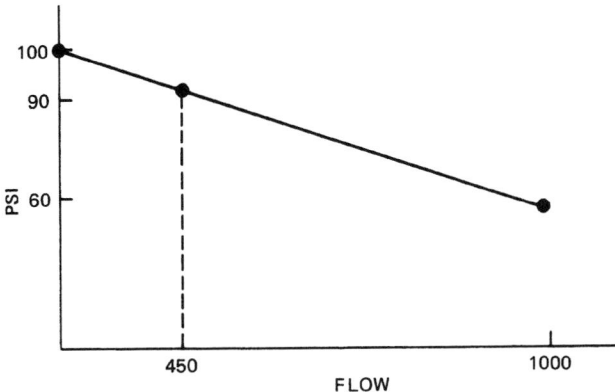

Fig. 4-3.8. Pressure available from 450 gpm flow water supply.

of 1000 gpm, the pressure available at a flow of 450 gpm can be approximated by plotting the two known data points on the hydraulic graph paper as shown in Figure 4-3.8. At a flow of 450 gpm, a pressure of 90 psi is indicated.

Pump Selection and Testing

Specific requirements for pumps used in sprinkler systems are contained in NFPA 20, which is cross-referenced by NFPA 13.

Fire pumps provide a means of making up for pressure deficiencies where an adequate volume of water is available at a suitable net positive suction pressure. Plumbing codes sometimes set a minimum allowable net positive suction pressure of 10 to 20 psi. If insufficient water is available at such pressures, then it becomes necessary to use a stored water supply.

Listed fire pumps are available with either diesel or electric drivers, and with capacities ranging from 25 to 5000 gpm, although fire pumps are most commonly found with capacities ranging from 250 to 2500 gpm in increments of 250 up to 1500 gpm and 500 gpm beyond that point. Each pump is specified with a rated flow and rated pressure. Rated pressures vary extensively, since manufacturers can control this feature with small changes to impeller design.

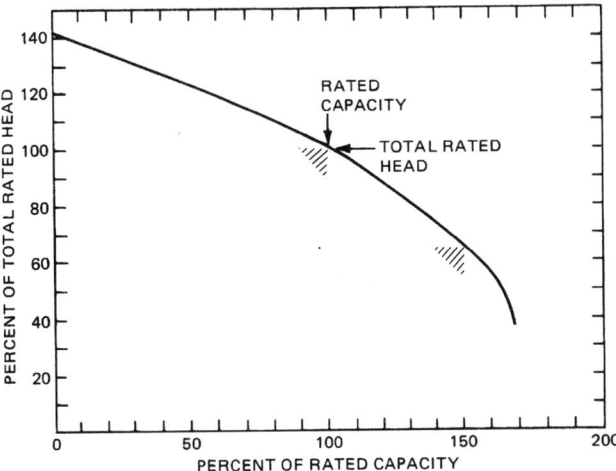

Fig. 4-3.9. Pump performance curve.

Pump affinity laws govern the relationship between impeller diameter, D, pump speed, N, flow, Q, pressure head, H, and brake horsepower, bhp. The first set of affinity laws assumes a constant impeller diameter.

$$\frac{Q_1}{Q_2} = \frac{N_1}{N_2} \qquad \frac{H_1}{H_2} = \frac{N_1}{N_2} \qquad \frac{\text{bhp}_1}{\text{bhp}_2} = \frac{N_1}{N_2}$$

These affinity laws are commonly used when correcting the output of a pump to its rated speed.

The second set of the affinity laws assumes constant speed with change in impeller diameter, D.

$$\frac{Q_1}{Q_2} = \frac{D_1}{D_2} \qquad \frac{H_1}{H_2} = \frac{D_1}{D_2} \qquad \frac{\text{bhp}_1}{\text{bhp}_2} = \frac{D_1}{D_2}$$

Pumps are selected to fit the system demands on the basis of three key points relative to their rated flow and rated pressure. (See Figure 4-3.9.) NFPA 20 specifies that each horizontal fire pump must meet these characteristics, and the approval laboratories ensure these points are met:

1. A minimum of 100 percent of rated pressure at 100 percent of rated flow.
2. A minimum of 65 percent of rated pressure at 150 percent of rated flow.
3. A maximum of 140 percent of rated pressure at 0 percent of rated flow (churn).

Even before a specific fire pump has been tested, therefore, the pump specifier knows that a given pump can be expected to provide certain performance levels. It is usually possible to have more than one option when choosing pumps, since the designer is not limited to using a specific point on the pump performance curve.

There are limits to flexibility in pump selection, however. For example, it is not permitted to install a pump in a situation where it would be expected to operate with a flow exceeding 150 percent of rated capacity, since the performance is not a known factor, and indeed available pressure is usually quick to drop off beyond this point.

NFPA 20 gives the following guidance on what part of the pump curve to use:[6]

"A centrifugal fire pump should be selected in the range of operation from 90 percent to 150 percent of its rated capacity. The performance of the pump when applied at capacities over 140 percent of rated capacity may be adversely affected by the suction conditions. Application of the pump at capacities less than 90 percent of the rated capacity is not recommended.

The selection and application of the fire pump should not be confused with pump operating conditions. With proper suction conditions, the pump can operate at any point on its characteristic curve from shutoff to 150 percent of its rated capacity."

For design capacities below the rated capacity, the rated pressure should be used. For design capacities between 100 and 150 percent of rated capacity, the pressure used should be found by the relationship made apparent by similar triangles.

$$\frac{0.35P}{0.5Q} = \frac{P' - 0.65P}{1.5Q - Q'}$$

where P and Q are the rated pressure and capacity, and P' is the minimum available pressure at capacity, Q', where $Q < Q' < 1.5 Q$.

EXAMPLE 12:

A pump is to be selected to meet a demand of 600 gpm at 85 psi. To determine whether a pump rated for 500 gpm at 100 psi would be able to meet this point without having an actual pump performance curve to work from, the above formula can be applied, with $P = 100$, $Q = 500$, and $Q' = 600$.

Inserting these values gives

$$(0.35)(100)/(0.5)(500) = [P' - (0.65)(100)]/[(1.5)(500) - 600]$$
$$35/250 = (P' - 65)/(750 - 600)$$
$$P' = 65 + 21 = 86 \text{ psi}$$

Since the value of P' so calculated is greater than the 85 psi required, the pump will be able to meet the demand point.

Tank Sizing

Tank selection and sizing are relatively easy compared to pump selection. The most basic question is whether to use an elevated storage (gravity) tank, a pressure tank, or a suction tank in combination with a pump. Following that is a choice of materials. NFPA 22 is the governing standard for water tanks for fire protection, and includes a description of the types of tanks as well as detailed design and connection requirements.

From a calculation standpoint, tanks must be sized to provide the minimum durations specified by NFPA 13 or other applicable standards for the system design. Required pressures must still be available when the tanks are at the worst possible water level condition, i.e., nearly empty.

If the tank is intended to provide the needed supply without the use of a pump, the energy for the system must be available from the height of a gravity tank or the air pressure of a pressure tank.

An important factor in gravity tank calculations is the requirement that the pressure available from elevation (calculated using 0.433 psi per foot) must be determined using the lowest expected level of water in the tank. This is normally the point at which the tank would be considered empty.

In sizing pressure tanks, the percentage of air in the tanks must be controlled so as to ensure that the last water leaving the tank will be at an adequate pressure. While a common rule of thumb has been that the tank should be one-third air at a minimum pressure of 75 psi, this rule does not hold true for systems with high pressure demands or where the tank is located a considerable distance below the level of the highest sprinkler.

For pipe schedule systems, two formulas have traditionally been used, based on whether the tank is located above the level of the highest sprinkler or some distance below.

For the tank above the highest sprinkler:

$$P = \frac{30}{A} - 15$$

For the tank below the highest sprinkler:

$$P = \left(\frac{30}{A} - 15\right) + \left(\frac{0.434H}{A}\right)$$

where

A = proportion of air in the tank
P = air pressure carried in the tank in psi

H = height of the highest sprinkler above the tank bottom in feet.

It can be seen that these formulas are based simply on the need to provided a minimum pressure of 15 psi to the system at the level of the highest sprinkler, and an assumption of 15 psi atmospheric pressure.

Using the same approximation for atmospheric pressure, a more generalized formula has come into use for hydraulically designed systems

$$P_i = \frac{P_f + 15}{A} - 15$$

where

P_i = tank air pressure to be used
P_f = system pressure required per hydraulic calculations
A = proportion of air in the tank.

EXAMPLE 13:

A pressure tank is to be used to provide a 30 min water supply to a system with a hydraulically calculated demand of 140 gpm at a pressure of 118 psi. Since there are sprinklers located adjacent to the tank, it is important that air pressure in the tank not exceed 175 psi. To determine the minimum size tank that can be used, it is important not only to consider the total amount of water needed, but also the amount of air necessary to keep the pressures within the stated limits.

The above equation for hydraulically designed systems can be used to solve for A.

If $\qquad\qquad P_i = [(P_f + 15)/A] - 15$

then $\qquad\qquad A = (P_f + 15)/(P_i + 15)$

$$A = (118 + 15)/(175 + 15) = 133/190 = 0.70$$

This means that the tank will need to be 70 percent air if the air pressure in the tank is to be kept to 175 psi.

The minimum water supply required is 30 min × 140 gpm = 4200 gallons.

Thus, the minimum tank volume will be such that 4200 gallons can be held in the remaining 30 percent of volume.

$$0.3V = 42000 \text{ gal} \qquad V = 4200/0.3 = 14{,}000 \text{ gal tank}$$

HANGING AND BRACING METHODS

Hangers and Hanger Supports

NFPA 13 contains a great deal of specific guidance relative to hanger spacing and sizing based on pipe sizes. It should also be recognized that the standard allows a performance-based approach. Different criteria exist for the hanger itself and the support from the building structure.

Any hanger and installation method is acceptable if certified by a registered professional engineer for the following:

1. Hangers are capable of supporting five times the weight of the water-filled pipe plus 250 pounds at each point of piping support.
2. Points of support are sufficient to support the sprinkler system.
3. Ferrous materials are used for hanger components.

The building structure itself must be capable of supporting the weight of the water-filled pipe plus 250 pounds applied at the point of hanging.

The 250 pound weight is intended to represent the extra loading that would occur if a relatively heavy individual were to hang on the piping.

Trapeze Hangers

Trapeze hangers are used where structural members are not located, so as to provide direct support of sprinkler lines or mains. This can occur when sprinkler lines or mains are run parallel to structural members such as joists or trusses.

A special section of NFPA 13 addresses the sizing of trapeze hangers. Because they are considered part of the support structure, the criteria within NFPA 13 are based on the ability of the hangers to support the weight of 15 feet of water-filled pipe plus 250 pounds applied at the point of hanging. An allowable stress of 15,000 psi is used for steel members. Two tables are provided in the standard, one of which presents required section moduli based on the span of the trapeze and the size and type of pipe to be supported, and the other of which presents the available section moduli of standard pipes and angles typically used as trapeze hangers.

In using the tables, the standard allows the effective span of the trapeze hanger to be reduced if the load is not at the midpoint of the span. The equivalent length of trapeze is determined from the formula

$$L = 4ab/(a + b)$$

where L is the equivalent length, a is the distance from one support to the load, and b is the distance from the other support to the load.

EXAMPLE 14:

A trapeze hanger is required for a main running parallel to two beams spaced 10 feet apart. If the main is located 1 foot 6 inches from one of the beams, the equivalent span of trapeze hanger required can be determined by using the above formula

$$L = 4(1.5 \text{ ft})(8.5 \text{ ft})/(1.5 \text{ ft} + 8.5 \text{ ft}) = 5.1 \text{ ft}$$

Earthquake Braces

Protection for sprinkler systems in earthquake areas is provided in several ways. Flexibility and clearances are added to the system where necessary to avoid the development of stresses that could rupture the piping. Too much flexibility could also be dangerous, however, since the momentum of the unrestrained piping during shaking could result in breakage of the piping under its own weight or upon collision with other building components. Therefore, bracing is required for large piping (including all mains) and for the ends of branch lines.

Calculating loads for earthquake braces is based on the assumption that the normal hangers provided to the system are capable of handling vertical forces, and that horizontal forces can be conservatively approximated by a constant acceleration equal to one-half that of gravity.

$$a_h = 0.5g$$

NFPA 13 contains a table of factors that can be applied if building codes require the use of other horizontal acceleration values.

Since the braces can be called upon to act in both tension and compression, it is necessary not only to size the brace member to handle the expected force applied by the weight of the pipe in its zone of influence, but also to avoid a member that could fail as a long column under buckling.

The ability of the brace to resist buckling is determined through an application of Euler's formula with a maximum slenderness ratio of 300. This corresponds to the maximum slenderness ratio generally used under steel construction codes for secondary framing members. This is expressed as

$$\ell/r \le 300$$

where ℓ = length of the brace and r = least radius of gyration for the brace.

The least radius of gyration for some common shapes is as follows:

pipe	$r = (r_0^2 + r_i^2)^{1/2}/2$
rod	$r = r/2$
flat	$r = 0.29h$

Special care must be taken in the design of earthquake braces so that the load applied to any brace does not exceed the capability of the fasteners of that brace to the piping system or the building structure, and that the braces are attached only to structural members capable of supporting the expected loads.

PERFORMANCE CALCULATIONS

Sprinkler Response as a Detector

Automatic sprinklers serve a dual function as both heat detectors and water distribution nozzles. As such, the response of sprinklers can be estimated using the same methods as for response of heat detectors. (See Section 4, Chapter 1.) Care should be taken, however, since the use of such calculations for estimating sprinkler actuation times has not been fully established. Factors, such as sprinkler orientation, air flow deflection, radiation effects, heat of fusion of solder links, and convection within glass bulbs, are all considered to introduce minor errors into the calculation process. Heat conduction to the sprinkler frame and distance of the sensing mechanism below the ceiling have been demonstrated to be significant factors affecting response, but are ignored in some computer models.

Nevertheless, modeling of sprinkler response can be useful, particularly when used on a comparative basis. Beginning with the 1991 edition, an exception within Section 4-1.1 of NFPA 13 permitted variations from the rules on clearance between sprinklers and ceilings ". . . provided the use of tests or calculations demonstrate comparable sensitivity and performance."

EXAMPLE 15:

Nonmetallic piping extending 15 in. below the concrete ceiling of a 10-ft-high basement 100 ft by 100 ft in size makes it difficult to place standard upright sprinklers within the 12 in. required by NFPA 13 for unobstructed construction. Using the LAVENT computer model, and assuming RTI values of 400 ft$^{1/2}$s$^{1/2}$ for standard sprinklers and 100 ft$^{1/2}$s$^{1/2}$ for quick-response sprinklers, it can be demonstrated that the comparable level of sensitivity can be maintained at a distance of 18 in. below the ceiling. Temperature rating is assumed to be 165°F, and maximum lateral distance to a sprinkler is 8.2 ft (10-ft × 13-ft spacing). Assuming the default fire (empty wood pallets stored 5 ft high), for example, the time of actuation for the standard sprinkler is calculated to be 200 s, as compared to 172 s for the quick-response sprinkler. Since the noncombustible construction minimizes concern relative to the fire control performance for the structure, the sprinklers can be located below the piping obstructions.

Dry System Water Delivery Time

Total water delivery time consists of two parts. The first part is the trip time taken for the system air pressure to bleed down to the point where the system dry valve opens to admit water to the piping. The second part is the transit time for the water to flow through the piping from the dry valve to the open sprinkler. In other words,

$$\text{water delivery time} = \text{trip time} + \text{transit time}$$

where water delivery time commences with the opening of the first sprinkler.

NFPA 13 does not contain a maximum water delivery time requirement if system volume is held to no more than 750 gal. Larger systems are permitted only if water delivery time is within 60 s. As such, the rule of thumb for dry system operation is that no more than a 60 s water delivery time should be tolerated, and that systems should be divided into smaller systems if necessary to achieve this 1-min response. Dry system response is simulated in field testing by the opening of an inspector's test connection. The test connection is required to be at the most remote point of the system from the dry valve, and is required to have an orifice opening of a size simulating a single sprinkler.

The water delivery time of the system is recorded as part of the dry pipe valve trip test that is conducted using the inspector's test connection. However, it is not a realistic indication of actual water delivery time for two reasons:

1. The first sprinkler to open on the system is likely to be closer to the system dry valve, reducing water transit time.
2. If additional sprinklers open, the trip time will be reduced since additional orifices are able to expel air. Water transit time may also be reduced since it is easier to expel the air ahead of the incoming water.

Factory Mutual Research Corporation (FMRC) researchers have shown[2] that it is possible to calculate system trip time using the relation

$$t = 0.0352 \frac{V_T}{A_n T_0^{1/2}} \ln(p_{a0}/p_a)$$

where

t = time in seconds
V_T = dry volume of sprinkler system in cubic feet
T_0 = air temperature in Rankine degrees
A_n = flow area of open sprinklers in square feet
p_{a_0} = initial air pressure (absolute)
p_a = trip pressure (absolute).

Calculating water transit time is more difficult, but may be accomplished using a mathematical model developed by FMRC researchers. The model requires the system to be divided into sections, and may therefore produce slightly different results, depending on user input.

Droplet Size and Motion

For geometrically similar sprinklers, the median droplet diameter in the sprinkler spray has been found to be inversely proportional to the ⅓ power of water pressure and directly proportional to the ⅔ power of sprinkler orifice diameter such that

$$d_m \propto D^{2/3}/P^{1/3} \propto D^2/Q^{2/3}$$

where

d_m = mean droplet diameter
D = orifice diameter
P = pressure
Q = rate of water flow.

Total droplet surface area has been found to be proportional to the total water discharge rate divided by the median droplet diameter

$$A_S \propto Q/d_m$$

where A_S is the total droplet surface area.

Combining these relationships, it can be seen that

$$A_S \propto (Q^3 p D^{-2})^{1/3}$$

When a droplet with an initial velocity vector of **U** is driven into a rising fire plume, the one-dimensional representation of its motion has been represented as[3]

$$m_1 d\mathbf{U}/dt = m_1 g - C_D \rho_g (\mathbf{U} + V)^2/2S$$

where

U = velocity of the water droplet
V = velocity of the fire plume
m = mass of the droplet
ρ_g = density of the gas
g = acceleration of gravity
C_D = coefficient of drag
S_f = frontal surface area of the droplet.

The first term on the right side of the equation represents the force of gravity, while the second term represents the force of drag caused by gas resistance. The drag coefficient for particle motion has been found empirically to be a function of the Reynolds number as[4]

$$C_D = 18.5 \, \text{Re}^{-0.6} \quad \text{for Re} < 600$$
$$C_D = 0.44 \quad \text{for Re} > 600$$

Spray Density and Cooling

The heat absorption rate of a sprinkler spray is expected to depend on the total surface area of the water droplets, A_s, and the temperature of the ceiling gas layer in excess of the droplet temperature, T. With water temperature close to ambient temperature, T can be considered excess gas temperature above ambient.

Chow[7] has developed a model for estimating the evaporation heat loss due to a sprinkler water spray in a smoke layer. Sample calculations indicate that evaporation heat loss is only significant for droplet diameters less than 0.5 mm. For the droplet velocities and smoke layer depths analyzed, it was found that the heat loss to evaporation would be small (10 to 25 percent) as compared to the heat loss from convective cooling of the droplets.

Factory Mutual Research Corporation researchers[5] have developed empirical correlations for the heat absorption rate of sprinkler spray in room fires, as well as convective heat loss through the room opening, such that

$$\dot{Q} = \dot{Q}_{cool} + \dot{Q}_c + \dot{Q}_l$$

where

$\dot{Q}$ = total heat release rate of the fire

$\dot{Q}_{cool}$ = heat absorption rate of the sprinkler spray

$\dot{Q}_c$ = convective heat loss rate through the room opening

$\dot{Q}_l$ = sum of the heat loss rate to the walls and ceiling, $\dot{Q}_s$, the heat loss rate to the floor, $\dot{Q}_f$, and the radiative heat loss rate through the opening, $\dot{Q}_r$.

Test data indicated that

$$\dot{Q}_{cool}/\dot{Q} = 0.000039\Lambda^3 - 0.003\dot{\iota}^2 + 0.082\dot{\iota}$$

for $\quad 0 < \Lambda \le 33 \; l/(\min \times kW^{1/2} \times m^{5/4})$

where Λ is a correlation factor incorporating heat losses to the room boundaries and through openings as well as to account for water droplet surface area.

$$\Lambda = (AH^{1/2}\dot{Q}_l)^{-1/2}(W^3PD^{-2})^{1/3}$$

for $\quad P = p/(17.2 \; kPa) \quad$ and $\quad D = d/0.0111 \; m$

where A and H are the area and height of the room opening in meters, P is the water pressure at the sprinkler in kPa, d is the sprinkler nozzle diameter in meters, and W is the water discharge in liters per minute.

The above correlations apply to room geometry with length-to-width ratio of 1.2 to 2 and opening size of 1.70 to 2.97 m².

SUPPRESSION BY SPRINKLER SPRAYS

Researchers at the National Institute of Standards and Technology (NIST) have developed a "zeroth order" model of the effectiveness of sprinklers in reducing the heat release rate of furnishing fires.[8] Based on measurements of wood crib fire suppression with pendant spray sprinklers, the model is claimed to be conservative. The model assumes that all fuels have the same degree of resistance to suppression as a wood crib, despite the fact that tests have shown furnishings with large burning surface areas can be extinguished easily compared to the deep-seated fires encountered with wood cribs.

The recommended equation, which relates to fire suppression for a 610-mm-high crib, has also been checked for validity with 305-mm crib results. The equation is

$$\dot{Q}(t - t_{act}) = \dot{Q}(t_{act})\exp[-(t - t_{act})/3.0(\dot{w}'')^{-1.85}]$$

where

$\dot{Q}$ = the heat release rate (kW)

t = any time following t_{act} of the sprinklers (s)

$\dot{w}''$ = the spray density (mm/s)

The NIST researchers claim the equation is appropriate for use where the fuel is not shielded from the water spray, and the application density is at least 0.07 mm/s [4.2 mm/min. (0.1 gpm/ft²)]. The method does not account for variations in spray densities or suppression capabilities of individual sprinklers.

The model must be used with caution, since it was developed on the basis of fully involved cribs. It does not consider the possibility that the fire could continue to grow in intensity following initial sprinkler discharge, and, for that reason, should be restricted to use in light hazard applications.

Sprinklers are assumed to operate within a room of a light hazard occupancy when the total heat release rate of the fire is 500 kW. The significance of an initial application rate of 0.3 gpm/ft² as compared to the minimum design density of 0.1 gpm/ft² can be evaluated by the expected fire size after 30 s. With the minimum density of 0.07 mm/s (0.1 gpm/ft²), the fire size is conservatively estimated as 465 kW after 30 s. With the higher density of 0.205 mm/s (0.3 gpm/ft²), the fire size is expected to be reduced to 293 kW after 30 s. Corresponding values after 60 s are 432 kW and 172 kW, respectively.

NOMENCLATURE

C coefficient of friction

FLC   friction loss coefficient

Q flow (gpm)

$\dot{w}''$ spray density (mm/s)

REFERENCES CITED

1. H. Cross, *Analysis of Flow in Networks of Conduits or Conductors*, Univ. of Illinois Engineering Experiment Station, Illinois (1936).
2. G. Heskestad and H. Kung, *FMRC Serial No. 15918*, Factory Mutual Research Corp., Norwood (1973).
3. C. Yao and A.S. Kalelkar, *Fire Tech.*, 6, 4 (1970).
4. C.L. Beyler, *The Interaction of Fire and Sprinklers*, National Bureau of Standards, Washington (1977).
5. H.Z. You, H.C. Kung, and Z. Han, *Spray Cooling in Room Fires*, National Bureau of Standards, Washington (1986).
6. *NFPA Codes and Standards*, National Fire Protection Association, Quincy, MA.
7. W.K. Chow, "On the Evaporation Effect of a Sprinkler Water Spray," *Fire Tech.*, pp. 364–373 (1989).
8. D.D. Evans, "Sprinkler Fire Suppression Algorithm for HAZARD," NISTIR 5254, National Institute of Standards and Technology, Gaithersburg, MD (1993).

FOAM AGENTS AND AFFF SYSTEM DESIGN CONSIDERATIONS

Joseph L. Scheffey

INTRODUCTION

Foams have been developed almost entirely from experimental work. While the technologies are rather mature, no fundamental explanations of foam extinguishment performance have been developed based on first principles. As a result, foams are characterized by (1) fire tests for which there is no general international agreement and (2) physical and chemical properties which may or may not correlate with empirical results. This chapter reviews the important parameters associated with foam agents, test methods used to evaluate foams, and the relevant data in the literature that can be used to evaluate designs for special hazards. Because of their superior performance in extinguishing hydrocarbon fuel fires, the emphasis is on the film-forming foams and thin pool fires (e.g., from spills). Situations involving fuels "in depth" are not included here.

Fire-fighting foam consists of air-filled bubbles formed from aqueous solutions. The solutions are created by mixing a foam concentrate with water in the appropriate proportions (typically 1, 3, or 6 percent concentrate to water). The solution is then aerated to form the bubble structure. Some foams, notably those that are protein-based, form thick, viscous foam blankets on hydrocarbon fuel surfaces. Other foams, such as film-formers, are much less viscous and spread rapidly on the fuel surface. The film-formers are capable of producing a vapor-sealing film of surface-active water solution on most of the hydrocarbon fuels of interest.

Since the foam is lighter than the aqueous solution that drains from the bubble structure, and lighter than flammable or combustible liquids, it floats on the fuel surface. This produces an air-excluding layer of aqueous agent, which suppresses and prevents combustion by halting fuel vaporization at the fuel surface. If the entire surface is covered with foam, the fuel vapor will be completely suppressed, and the fire will be extinguished. Low-expansion foams (i.e., foam volume : solution volume of ≤ 10:1) are quite effective on two-dimensional (pool) flammable and combustible liquid fires, but not particularly effective on three-dimensional fuel fires. This is particularly true of three-dimensional fires involving a low-flash-point fuel. Typically, an auxiliary

agent, such as dry chemical, is used with foam where a three-dimensional fire (running fuel or pressurized spray) is anticipated.

DESCRIPTION OF FOAM AGENTS

There are no universally agreed-on definitions of foam agents or terms associated with fire-fighting foam. For example, where foam is referenced in NFPA standards, definitions vary from document to document. Because foams vary in performance, in terms of application rates and quantities required for extinguishment, agent definitions can be cast to accentuate positive attributes, such as "rapid knockdown" or "superior burnback resistance." Geyer *et al* have described the composition of various foam agents, paraphrased as follows.[1]

1. Protein Foam. Protein foam is a "mechanical" foam produced by combining (proportioning) foam concentrate and water and discharging the resulting solution through a mixing chamber. The mixing chamber introduces (aspirates) air, which expands the solution to create foam bubbles. The liquid concentrate consists primarily of hydrolyzed proteins in combination with iron salts. Hoof and horn meal, and hydrolyzed feather meal are examples of proteinaceous materials used in protein-foam concentrates. No aqueous film is formed on the fuel surface with this type of agent.

2. Fluoroprotein. These agents are basically protein foams with fluorocarbon surface-active agents added. The varying degrees of performance are achieved by using different proportions of the base protein hydrolyzates and the fluorinated surfactants. While fluoroprotein foams generally have good fuel shedding capabilities and dry chemical compatibility, the solution that drains out from the expanded foam does not form a film on hydrocarbon fuels. However, the addition of the fluorinated surfactants may act to reduce the surface tension of the solution. This reduction may, in turn, decrease the viscosity of the expanded solution, thus promoting more rapid fire control when compared to protein foams.

3. Aqueous Film-forming Foam (AFFF). These agents are synthetically formed by combining fluorine-free hydrocarbon foaming compounds with highly fluorinated surfactants. When mixed with water, the resulting solution

Joseph L. Scheffey, P.E., is Director of Fire Protection RDT&E for Hughes Associates, Inc., Columbia, MD.

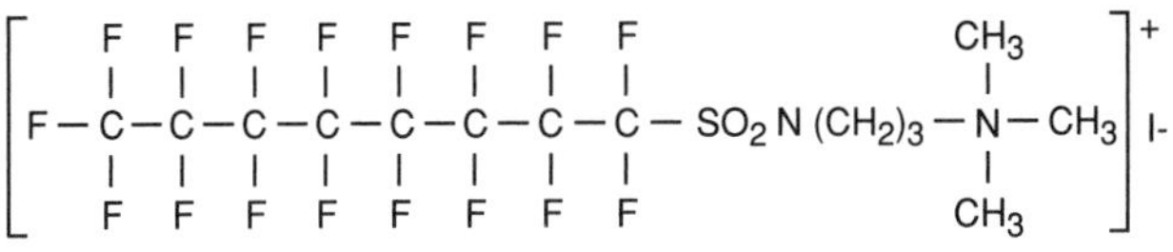

$$\begin{bmatrix} & F & F & F & F & F & F & F & F & & & CH_3 \\ & | & | & | & | & | & | & | & | & & & | \\ F- & C-C-C-C-C-C-C-C & - & SO_2\,N\,(CH_2)_3 & - & N-CH_3 \\ & | & | & | & | & | & | & | & | & & & | \\ & F & F & F & F & F & F & F & F & & & CH_3 \end{bmatrix}^{+} I^{-}$$

Perfluoroctylsulfommide - N - Propyltrimethylammonium Iodide

Fig. 4-4.1. Typical AFFF fluorosurfactant molecule.[3]

achieves the optimum surface and interfacial tension characteristics needed to produce a film that will spread across a hydrocarbon fuel. The foam produced from this agent will extinguish in the same vapor-excluding fashion as other foams. Further, the solution that results from normal drainage or foam breakdown produces an aqueous "film" that spreads rapidly and is highly stable on the liquid hydrocarbon fuel surface. It is this film formation characteristic that is the significant distinguishing feature of AFFF.

These definitions are by no means all-inclusive. For example, film-forming fluoroprotein (FFFP) foam is an agent that is produced by increasing the quantity and quality of the surfactants added to a protein hydrolyzate. By doing this, the surface tension of the resulting solution, which drains from the expanded foam, is reduced to the point where it may spread across the surface of a liquid hydrocarbon fuel. An alcohol-resistant concentrate is formulated to produce a floating polymeric skin for foam buildup on water-miscible fuels. This polymeric skin protects the foam from breakdown by polar solvents, e.g., acetone, methanol, and ethanol.

The descriptions show that there are distinct chemical differences between protein-based foams and AFFF. In general, the surfactants used in aqueous foams are long-chained compounds that have a hydrophobic or hydrophilic (i.e., water repelling or water attracting, respectively) group at one end.[2] The molecular structure of a typical AFFF fluorinated surfactant is shown in Figure 4-4.1.[3] In this molecule, the perfluoroctyl group on the left is the hydrophobic group, while the propyltrimethylammonium group is the hydrophilic group. When these compounds are dissolved into solution with water, they will tend to group near the surface of the solution, aligned so that their hydrophobic ends are facing toward the air/solution interface. The advantage of this is that the perfluoroctyl group found in these compounds is also oliophobic (i.e., oil repelling) as well as hydrophobic.[4]

AFFF concentrates also contain hydrocarbon surfactants. These compounds are less hydrophobic than those containing the perfluoroctyl group. However, they do provide greater stability once the solution is expanded into a foam. As a result, the surface tension of the solution is reduced below that of water; the expanded foam produced from the solution is resistive to breakdown from heat, fuels, or dry chemical extinguishing agents; and the solution that drains out from the expanded foam is able to form a film on hydrocarbon fuels.

The importance of both the film formation and foam bubble characteristics of AFFF, resulting from the combination of fluorocarbon and hydrocarbon surfactants, was evaluated in early work by Tuve *et al*.[5] When a highly expanded, stiff formulation of AFFF was used, these researchers found it difficult to obtain good fire extinguishment and vapor

sealing characteristics. The foam resisted flow, and drainage of the aqueous solution (film) was slow. This was corrected by expanding the foam to a lesser degree. This pioneering AFFF formulation, with an expansion ratio of 8:1 and 25 percent drainage time of 6 min, appeared to offer the best compromise in characteristics. It provided a readily flowable foam that sealed up against obstructions, promoted the rapid formation of a surface-active film barrier on the fuel, and provided a sufficiently stable foam to resist burnback.

FIRE EXTINGUISHMENT AND SPREADING THEORY

As noted by Friedman in his review of suppression theory, the mechanisms of foam fire extinguishment on two-dimensional pool fires have not been developed.[6] Usually, the fire extinguishment is described simply as a factor of the cessation of fuel vaporization at the fuel surface. As the fuel vapor decreases, the size of the combustion zone decreases. When the area is totally covered, extinguishment occurs. Cooling must occur to bring the vapor pressure of the fuel below that of its boiling point. Once the fuel is cooled, a layer of foam must then be applied either manually, or by spreading, to prevent combustion. Hanauska *et al* have proposed fundamental extinguishment parameters, summarized in the following text.[7]

Foam Loss Mechanisms

Fire extinguishment by foams can be summarized as shown in Figure 4-4.2. Foam having a temperature, T_i, and depth, h, spreads at a rate of V_s along a fuel of temperature, T_s, and vapor pressure, P_v. Fuel is volatized by the fire at a rate of $\dot{m}_{fuel}$, which is a function of the radiative feedback, $\dot{q}_{rad}$. The foam is added by the discharge application, $\dot{m}_{add}$, and lost through evaporation, $\dot{m}_{vap}$, and drop-through, $\dot{m}_{drop}$.

The total mass loss of the foam is a function of the loss due to drop-through and the mass loss due to vaporization.

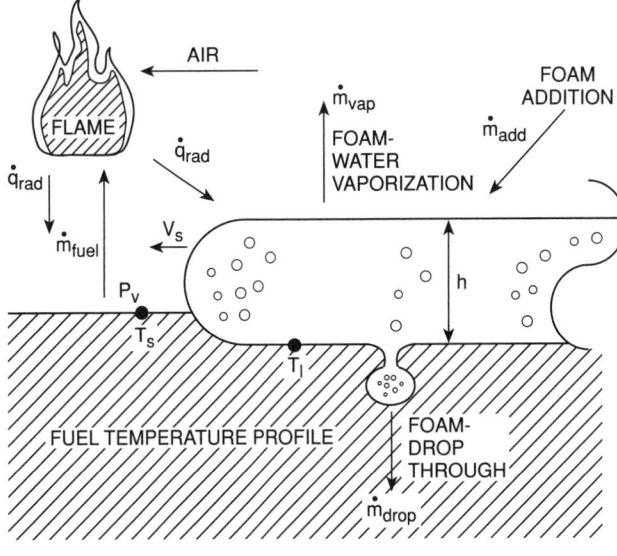

Fig. 4-4.2. Illustration of the significant parameters affecting a foam's hydrocarbon fuel fire extinguishment capability.

The mass loss due to drop-through is at least partially dependent on the drainage of liquid from the foam. Evaporation of the liquid results primarily from radiant energy from the fire. Assuming that most of the radiation results in direct evaporation of the foam, the evaporation of foam can be characterized by

$$\dot{m}''_{vap} = \frac{\dot{q}''_{rad}}{\Delta H_v} \qquad (1)$$

where ΔH_v is the combined latent and sensible heats of vaporization. Using a rough estimate of $\dot{q}''$ from large pool fires of 45–185 kW/m^2 yields an evaporation rate of 18 and 72 g · m^2/s, assuming a heat of vaporization of 2563 kJ/kg. (See Section 2, Chapter 4.) To account for reflective and absorbed losses, Persson[8] has proposed a calculation method

$$\dot{m}''_{vap} = \dot{q}''_{rad} k_e \qquad (2)$$

where k_e is an experimentally derived constant using different fluxes from a radiant exposure. For $\dot{q}''_{rad}$ values of 45 and 185 kW/m^2, Equation 2 yields values for $\dot{m}''_{vap}$ of 11 and 46 g · m^2/s, respectively. Since the estimated $\dot{m}''_{vap}$ values based on Equation 1 at the same heat fluxes were 18 and 72 g · m^2/s, the experimental mass loss rate results are about 62 percent lower than the theoretical loss. The difference between values is attributable to neglecting the reflected and absorbed losses in Equation 1. This indicates that about 48 percent of the radiant flux to the foam surface is either reflected from or absorbed into the foam blanket. The division between these two heat-transfer mechanisms is not clear and is an area for further study.

Foam loss can likewise be described theoretically, based on the downward force of gravity and the opposing forces due to surface tension and buoyancy. Alternately, a model mass loss due to drainage can be expressed as a time-averaged constant

$$\dot{m}''_{drain} = k_d \qquad (3)$$

where k_d is an experimentally determined drainage coefficient. From the data of Persson, the drainage coefficient can be estimated to be 17 to 25 g · m^2/s.[8] The drainage rate was found to be relatively independent of the radiant heat flux to the foam, but highly dependent on the expansion ratio. Foams with lower expansion ratios will drain faster. For example, decreasing the expansion ratio by about half (11.3 to 5.3) increased the drainage rate by a factor of about 2 (55 to 105 g/min). Decreasing the expansion ratio changes fundamental parameters of the foam, which allows it to drain faster.

Foam drainage is a complicated phenomenon that is highly time dependent. Besides the forces associated with the bubble structure, drainage is dependent on the continual changing geometries of the cells and other variable conditions, such as collapsing cells. Even though all aspects of this problem cannot be fully detailed, simplified models have been created that predict the drainage rate for foams. Kraynik has developed one such model that considers the drainage from a column of persistent foam.[9] The model contains no empirical parameters and assumes the foam is dry with very thin walls such that the liquid contained in the cell walls is negligible. In relaxing the assumptions, this basic model might ultimately be used to assess the effect of various fundamental parameters on foam drainage.

Foam Spread over Liquid Fuels

In order to predict the extinguishment of a liquid pool fire by fire-fighting foam, it is necessary to describe the process of spreading the foam over the liquid fuel surface. This process of foam spread on a liquid fuel is similar to the spread of a less dense liquid (such as oil) on a more dense liquid (such as water). This phenomenological approach to the spread of foam on a liquid pool fire is appropriate to the extent that foam can be treated as a liquid. Kraynik characterizes foams macroscopically as being Bingham fluids with a finite shear stress and non-newtonian viscosity.[10] That is, foam displays an infinite viscosity up to some initial shear rate above which it displays a shear-rate dependent viscosity.

Since fuels typically have low viscosities (especially compared to foam viscosities at relatively low shear rates), it may be appropriate to model foam spread across a fuel surface using models developed for oil spread on water. These models assume that the oil spreads as a fluid with a viscosity much higher than the water on which it is spreading. The process of oil spread on water has been described in detail by Fay,[11] and Fay and Hoult.[12] Their phenomenologically based model describes three regimes of spread as characterized by combinations of spreading forces and retarding forces. The first regime is the gravity-inertia regime, where the outward spread of the oil is driven by a gravity force and retarded by the inertia required to accelerate the oil. The second regime is the gravity-viscous regime, where the gravity-induced spreading is retarded by viscous dissipation in the water. Since the oil is much more viscous than the water, they assume that there is slug flow in the oil and that the viscous drag force is dominated by the velocity gradient in the water. The third regime is characterized by a surface-tension spreading force opposed by the viscous retarding force. By setting the spreading and retarding forces equal in each of the regimes, they developed equations to estimate the length of the spread as a function of time.

By treating the spread of foam on fuel as similar to the spread of oil on water, the equations developed by Fay and Hoult might be used to describe the spread of a foam blanket over a fuel pool as a function of time.[12] Since foam generally has a much higher viscosity than the fuel on which it is spreading, the assumption of slug flow made for the oil by Fay and Hoult should be reasonably valid for foam spread on fuel as well.[12] The equations are

gravity-inertia regime: $\quad l = (\Delta g V t^2)^{1/4}$

gravity-viscous regime: $\quad l = \left(\dfrac{\Delta g V^2 t^{3/2}}{\nu^{1/2}}\right)^{1/6}$ $\qquad$ (4)

surface tension–viscous regime: $\quad l = \left(\dfrac{\sigma^2 t^3}{\rho^2 \nu}\right)^{1/4}$

where

l = length of spread (cm),
Δ = $(\rho_{fuel} - \rho_{foam})/\rho_{fuel}$,
g = acceleration of gravity (981 cm/s^2),
V = foam volume (cm^3),
t = time (s),
ν = kinematic viscosity of fuel (cm^2/s),
σ = spreading coefficient (dynes/cm), and
ρ = density of fuel or foam (g/cm).

Equation 4 represents an untested theoretical model of foam spread. The equation includes the parameters that are known or suspected to affect foam spread. They are presented here as an initial effort to understand foam flow based on first principles. They are not yet developed for engineering use. The following discussion expands on this theory.

The transition from gravity-dominated spread to surface-tension-dominated spread can be shown to occur at a critical thickness of the foam layer, h_c, given by

$$h_c = \left(\frac{\sigma}{g\Delta\rho_{foam}}\right)^{1/2} \tag{5}$$

The transition from inertia to viscous-dominated retarding force occurs when the foam thickness, h, is equal to the viscous boundary layer thickness, δ, of the fuel, with

$$h = \frac{V}{l^2}$$
$$\delta = (\nu t)^{1/2} \tag{6}$$

The equations for length of spread can be used to generate preliminary estimates of the spread distance and area coverage for the placement of a volume of foam on a fuel surface. The equations are only estimates because they consider a force balance between just the dominant forces for each regime. All forces are actually present in each regime. Also, the densities of both fluids are considered to be very nearly equal for the development of the equation for the gravity-viscous regime. This is the case for oil spread on water, but may not be the case for foam on fuel.

Using approximations for fuel and foam characteristics, it can be shown that a positive spreading coefficient does not begin to affect the spread of foam until the foam layer has become very thin. For the placement of a volume of foam on a fuel, this may not occur until after significant time has passed, relative to the time scale for knockdown desired in many fire protection situations.

The equations for foam spread on fuel include many of the parameters known to be important to foam spread. However, the equations are independent of the foam viscosity. Observations indicate that low-viscosity non-rigid foams, such as AFFF, spread faster than high-viscosity rigid protein foams. The inclusion in the model of a term to account for this is desirable.

TABLE 4-4.1 *Surface Tension of Hydrocarbon Liquids and Fuels[14]*

Hydrocarbon Liquid	Grade	Surface Tension at 25°C (dynes/cm)
Cyclohexane	Certified A.C.S.	24.2
n-Heptane	Certified Spectroanalyzed	19.8
n-Heptane	Commercial	20.9
Isooctane	Certified A.C.S.	18.3
Avgas	115/145	19.4[†] 19.5[‡]
JP-4	Navy Specification	22.4[†] 22.8[‡]
JP-5	Navy Specification	25.6[†] 25.8[‡]
Motor Fuel	Regular	20.5[†] 21.5[‡]
Naphtha	Stove and Lighting	20.6

[†]Sample 1.
[‡]Sample 2.

The equations for spread length so far have assumed that the foam spreads over the fuel as plug flow, with no relative movement within the foam itself. It is easy to conceive that the foam has the capability to flow over itself. The relative movement within the foam is equivalent to the foam flowing over a solid surface. The total foam flow might ultimately be modeled as the combination of the foam plug over the fuel and the flow within the foam layer itself.

According to Cann *et al*, several regimes exist for spread of a liquid on a solid that are similar to those described for spread of a liquid on a liquid.[13] Most of this spread occurs in a gravity-viscous force regime, where the spread is given by

$$l = \frac{kt}{\mu} \tag{7}$$

where k is an empirically determined constant, and μ is the foam viscosity.

Thus, the spread of foam over fuel can be characterized by two scenarios: (1) high-viscosity liquid spreading over a low-viscosity liquid and (2) a liquid spreading over a "solid." The spread of foam can be described by modifying Equation 4, as follows

gravity-inertia regime: $\quad l = (\Delta g V t^2)^{1/4} + \dfrac{kt}{\mu}$

gravity-viscous regime: $\quad l = \left(\dfrac{\Delta g V^2 t^{3/2}}{\nu^{1/2}}\right)^{1/6} + \dfrac{kt}{\mu} \qquad (8)$

surface tension regime: $\quad l = \left(\dfrac{\sigma^2 t^3}{\rho^2 \nu}\right)^{1/4} + \dfrac{kt}{\mu}$

Kraynik describes foams as being characterized by a yield stress and shear thinning viscosity.[10] Thus, the foam viscosity in the equations above is not a constant but is a function of the shear rate. The stress in the foam is a result of the gravity-induced pressure gradient. As the foam flows out and becomes thin, the stress will be reduced. When the stress falls below the yield stress, the viscosity will become infinite and the second term, kt/μ, in the spread length equations will go to zero. The foam will flow simply as plug flow. Above the yield stress, the foam will have a finite viscosity, but this viscosity will be dependent on the yield stress.

An AFFF agent that is very free flowing will have a relatively small yield stress and would retain the second term in the spread length equations until it had flowed out to a very thin layer. A protein foam that is relatively stiff will have a large yield stress, and the second term will go to zero before the foam has spread very far. Above the yield stress, the viscosity of the AFFF will be lower than that of a protein foam, and the second term will provide a greater contribution to foam spread. The rheological properties described appear to have a significant impact on foam spread; however, the properties are not a part of any current specification and are rarely measured.

Foam Extinguishment Modeling

At present, modeling of foam extinguishment cannot be performed because of the large number of remaining uncertainties. A model would have to take into account the addition of foam to the fuel surface, the spread of foam on the fuel surface, and the foam loss mechanisms of evaporation and drop-through. The foam spread length equations can be used to estimate the area of foam coverage at a specific time and for a specific quantity of foam. Modeling at this time is limited because of the lack of established values for k_e

TABLE 4-4.2 *Interfacial Tensions, Spreading Coefficients, and Film Formation Observations for Various Surfactant Solution-Hydrocarbon Liquid Combinations[14]*

Surfactant Solution	Hydrocarbon Liquid	Interfacial Tension (dynes/cm)	Spreading Coefficient (dynes/cm)	Film Formed
FC-194 (lot 107) (solution surface tension of 15.5 dynes/cm at 25°C)	Cyclohexane	4.3	4.4	Yes
	n-Heptane, certified	5.5	−1.2	No
	n-Heptane, commercial	4.3	1.1	Yes (very slow spread)
	Avgas*	4.6	−0.7	No
	JP-4*	3.6	3.3	Yes
	JP-5*	4.9	5.2	Yes
	Motor Fuel*	3.7	1.3	Yes
FC-195 (lot 9) (solution surface tension of 15.6 dynes/cm at 25°C)	Cyclohexane	3.2	5.4	Yes
	n-Heptane, certified	4.2	0.0	Yes (slow spread)
	Isooctane	2.5	0.2	Yes (slow spread)
	Avgas*	0.5	3.3	Yes
	JP-4†	3.6	3.6	Yes
	JP-5†	4.9	5.3	Yes
	Motor Fuel*	2.6	2.3	Yes
	Naphtha	2.8	2.2	Yes
FC-195 (lot 10) (solution surface tension of 16.4 dynes/cm at 25°C)	Cyclohexane	1.5	6.3	Yes
	n-Heptane, certified	3.2	0.6	Yes
	Isooctane	2.8	−1.3	No
	Avgas*	2.1	1.0	Yes
	JP-4*	2.7	3.3	Yes
	JP-5*	4.2	5.0	Yes
	Motor Fuel*	1.2	2.9	Yes
	Naphtha	0.8	3.4	Yes (slow spread)

*Sample 1.
†Sample 2.

(Equation 2) and k_d (Equation 3). Also, the yield stress and viscosity relationships for fire-fighting foams have not been quantified. Experimental work is needed to complete this modeling effort. Also, the actual method of application (e.g., from a handline nozzle or fixed device such as a sprinkler) must ultimately be taken into account. Even so, preliminary calculations using this methodology are encouraging and support continued development.[7]

Surface Tension and Spreading Coefficient

Film-forming foams are defined by the ability of the aqueous solution draining from the foam to spread spontaneously across the surface of a hydrocarbon fuel. The fundamental relationship used to describe the spreading coefficient is

$$S_{a/b} = \gamma_b - \gamma_a - \gamma_l \qquad (9)$$

where

$S_{a/b}$ = spreading coefficient (dynes/cm),
γ_b = surface tension of the lower liquid phase of a hydrocarbon fuel (dynes/cm),
γ_a = surface tension of the upper layer of liquid using AFFF solution (dynes/cm), and
γ_l = interfacial tension between liquids a and b (dynes/cm).

Surface tension and interfacial tension can be measured using methods such as those described in ASTM D-1331, *Standard Test Methods for Surface and Interfacial Tension of Solutions of Surface-active Agents*. Reagent-grade cyclohexane is typically used as a reference fuel. A du Nouy tensiometer, having a torsion balance with a 4- or 6-cm circumference platinum-iridium ring, is lowered into the liquid and slowly

pulled out until the liquid detaches from the ring's surface. The force recorded at the point where this separation occurs is the surface tension (dynes/cm) of the pure liquid. Similarly, the interfacial tension is the measurement of tension when the ring is pulled through the boundary layer between two liquids.

The Naval Research Laboratory developed some of the earliest quantitative data on the spreading coefficient of AFFF on hydrocarbons, as shown in Tables 4-4.1 and 4-4.2.[14] As fuel temperature increases, the surface tensions of both the fuel and solution decrease. The spreading coefficient may go to zero or go negative.[14,15]

While it has been shown that film-forming foams are superior fire extinguishing agents compared to other foams, there are no one-to-one correlations between bench-scale surface tension/spreading coefficient data and fire control, extinguishment, and burnback resistance times. Both Scheffey *et al*[16] and Geyer[17] have demonstrated that there is no direct correlation between fire extinguishment and spreading coefficient. As such, spreading coefficient data alone cannot be used as a relative predictor of fire performance.

Since the surface tensions of most AFFFs are approximately equal, there must be a balance between the surface tension of the fuel and the interfacial tension of the two liquids to create a positive spreading coefficient. It can be seen then that, while both the surface tension of the solution and the interfacial tension between the liquids have an impact on the spreading coefficient, the interfacial tension is usually the determining factor. For fuels, such as avgas or n-heptane, which have surface tensions in the range of 19 to 20 dynes/cm, either the foam surface tension or the interfacial tension, or both, must be reduced. Normally, the changes resulting from a modification of the formulation will be more significant for the interfacial tension value than they will be for the foam surface tension value. Still, a relationship between the two values does exist.[4] Therefore, in reducing the sum of the values to obtain a positive spreading coefficient, a delicate balance must be maintained.

Maintaining this balance and achieving a positive spreading coefficient is accomplished by controlling the amount and type of fluorinated surfactants used to formulate the agent. This at first seems beneficial, since a positive number on a low surface tension fuel will ensure an even larger value with higher surface tension fuels (e.g., JP-5 or motor gasoline). But, in reducing the interfacial tension, the agent may lose some of its fuel shedding capabilities. The effects of adding too much fluorosurfactant to an aqueous solution and the result on foam bubble stability are described by Rosen[4] and Aubert et al.[2] This could be a problem that manifests itself only during actual fire testing. The type and amount of fluorosurfactants also affect spreading coefficient.[16]

Despite the lack of one-to-one correlations between surface tension spreading coefficient data and fire control, extinguishment, and burnback results, this criteria is useful in categorizing film-forming agents. The spreading coefficient test is used throughout the world as a standard indicator of aqueous film-forming foams. Although undocumented, it is believed that film formation results in improved viscosity (or associated mechanisms that improve spreading), ultimately resulting in superior extinguishing performance.

ASSESSMENT OF FIRE EXTINGUISHING AND BURNBACK PERFORMANCE

Standard Test Methods

Since a theory of foam spreading has not been developed, performance of foams is measured using fire tests. The use of bench-scale burning fuel trays (e.g., less than 1 m diameter) results in varying fuel burning rates. This was observed by Chiesa and Alger when they attempted to use a 15-cm by 45-cm pan for foam performance evaluation.[18] Data from their experiments are shown in Figure 4-4.3, which correlates control times observed when foam samples were tested using the bench-scale apparatus (laboratory) and 4.6 m² (50 ft²) fire tests (field method). Equal control times correspond to a 45-degree line. Since the majority of the points fall below this line, the laboratory test is more severe (about 35 percent) than the field test.

Fire test methods used by regulatory authorities for certification are usually on the order of 2.6 to 9.3 m² (28 to 100 ft²). Foams must also meet additional test parameters related to storage, proportioning, and equipment factors.

Underwriters Laboratories Standard 162: Underwriters Laboratories (UL) 162, *Standard for Foam Equipment and Liquid Concentrates*, is the principle test standard for the listing of foam concentrates and equipment in the United States. Test procedures outlined in this standard have been developed to evaluate specific agent/proportioner/discharge device combinations. When a foam concentrate is submitted for testing, it must be accompanied by the discharge device and proportioning equipment with which it is to be listed. Listed products, including the agent, discharge device, and proportioner, are then described in the UL *Fire Protection Equipment Directory*.

Listed with a system, foam liquid concentrates are associated with discharge devices classified as Type I, II, or III. Type I devices deliver foam gently onto the flammable liquid

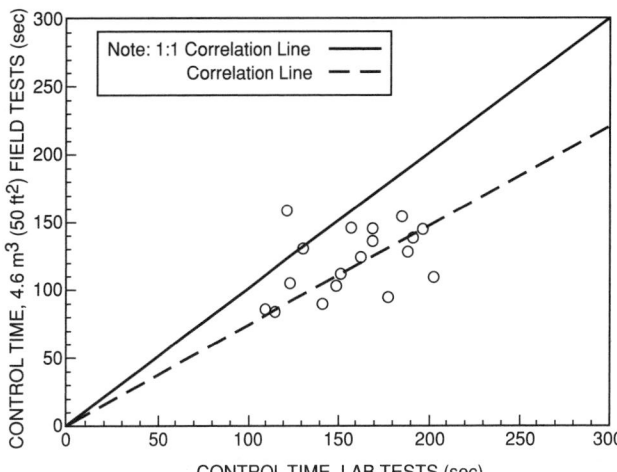

Fig. 4-4.3. Correlation of control times observed in laboratory and field tests of foam.[18]

fuel surface, e.g., a foam trough along the inside of a tank wall. A proposed revision to UL 162, *Standard for Foam Equipment and Liquid Concentrates*, would delete the Type I discharge device category. Type II discharge devices deliver foam onto the liquid surface in a manner that results in submergence of the foam below the fuel surface, and restricted agitation at the fuel surface. Examples include subsurface injection systems, tank-wall-mounted foam chambers, and applications where foam is bounced off the wall of a tank. Type III discharge devices deliver foam directly onto the liquid surface and cause general agitation at the fuel surface, e.g., by using hand-held nozzles. The flammable liquid fire tests in UL 162 include methods for sprinklers, subsurface injection, and topside discharge devices, including nozzles.

Class B fire test requirements for Types II and III discharge devices and sprinklers are shown in Table 4-4.3. Commercial-grade n-heptane is placed in a square test pan. The area of the pan is a minimum of 4.6 m² (50 ft²). The application rates ("densities" in UL 162, *Standard for Foam Equipment and Liquid Concentrates*) for various concentrates are outlined in Table 4-4.3.

In the test fire, the fuel is ignited and allowed to burn for 60 sec. Foam is then discharged for the duration specified in Table 4-4.3. The foam blanket resulting from the foam discharge must spread over and completely cover the fuel surface, and the fire must be completely extinguished before the end of the foam discharge period.

After all of the foam is discharged, the foam blanket formed on top of the fuel is left undisturbed for the period specified in Table 4-4.3. During the time the foam blanket is left undisturbed, a lighted torch is passed approximately 25.4 mm (1 in.) above the entire foam blanket in an attempt to reignite the fuel. The fuel must not reignite, candle, flame, or flash over while the torch is being passed over the fuel. However, candling, flaming, or flashover that self-extinguishes is acceptable, provided that the phenomenon does not remain in one area for more than 30 sec.

After the attempts to reignite the fuel with the lighted torch are completed, a 305-mm (12-in.) diameter section of stovepipe is lowered into the foam blanket. The portion of the foam blanket that is enclosed by the stovepipe is removed with as little disturbance as possible to the blanket

TABLE 4-4.3   *Foam Application Rates and Duration to Burnback Ignition in UL 162 for Hydrocarbon Fuels*

Application	Foam Concentrate	Fuel Group	Test Application Density [Lpm/m^2 (gpm/ft^2)]	Time of Foam Application (min)	Maximum Extinguishment Density [L/m^2 (gal/ft^2)]	Duration until Burnback Ignition (min)	Minimum Application Rate [Lpm/m^2 (gpm/ft^2)]
Type III Discharge	P, FP, S, FFFP*	Hydrocarbon	2.5 (0.06)	5	12.2 (0.03)	15	6.6 (0.16)
Outlets	AFFF, FFFP*	Hydrocarbon	1.6 (0.04)	3	4.9 (0.12)	9	4.1 (0.10)
Type II Discharge	P, FP, S, FFFP*	Hydrocarbon	2.5 (0.06)	5	12.2 (0.3)	15	4.1 (0.10)
Outlets	AFFF, FFFP*	Hydrocarbon	1.6 (0.04)	3	4.9 (0.12)	9	4.1 (0.10)
	All	Polar	†	5	—	15	‡
Foam-Water Sprinklers	P, FP, S	Hydrocarbon	6.6 (0.16)	5	30 (0.8)	15	6.6 (0.16)
Standard Orifice Sprinkler and Spray Systems	AFFF, FFFP	Hydrocarbon	4.1 (0.10)	5	20.4 (0.5)	15	6.6 (0.16)
		Polar	†	5	—	15	§

Note: P = Protein   FFFP = Film-forming fluoroprotein   FP = Fluoroprotein   AFFF = Aqueous film-forming fluoroprotein   S = Synthetic

*Film-forming fluoroprotein is to be tested at application densities of 2.5 and 1.6 Lpm/m^2 (0.06 and 0.04 gpm/ft^2).

†Application rate may vary among polar groups, as specified by the manufacturer.

‡0.10 or 1.67 times test application rate, whichever is greater.

§0.16 or 1.6 times test application rate, whichever is greater.

Source: UL 162, *Standard for Foam Equipment and Liquid Concentrates*, July 1993.

outside the stovepipe. The cleared fuel area inside the stovepipe is ignited and allowed to burn for 1 min. The stovepipe then is slowly removed from the pan while the fuel continues to burn. After the stovepipe is removed, the foam blanket must either restrict the spread of fire for 5 min to an area not larger than 0.9 m^2 (10 ft^2), or flow over and reclose the burning area.

When the UL 162 test is passed, the agent, proportioning device, and discharge device become listed. The fact that foam concentrate has a UL label does not mean it has been tested under all potential end-use conditions. The UL *Fire Protection Equipment Directory* must be referenced to determine with what equipment the concentrate has been tested and approved.

UL 162, *Standard for Foam Equipment and Liquid Concentrates*, is not an agent specification; therefore, there are no requirements for physical properties, such as film formation and sealability, corrosion resistance, and spreading coefficient. Neither are there any provisions to test, on a large scale, the degree of dry chemical compatibility of an agent, or the effects of aging or mixing with agents of another manufacturer. UL is considering such requirements. In particular, requirements for a positive spreading coefficient (greater than zero using cyclohexane) for film-forming foams have been proposed and are being implemented.[19]

U.S. Military Specification: The U.S. Military Specification, MIL-F-24385, is the AFFF procurement specification for the U.S. military and federal government. The U.S. military, in all likelihood, is the largest user of foam in the world. It is important to recognize that MIL-F-24385 is a procurement specification as well as a performance specification. Hence, there are requirements for packaging, initial qualification inspection, and quality conformance inspection, in addition to fire performance criteria. Equipment designs unique to the military, in particular U.S. Navy ships, also impact on the specification requirements (e.g., use of seawater solutions and misproportioning-related fire tests). These requirements have been developed based on research and testing at the Naval Research Laboratory and actual operational experience with protein and film-forming foams.

Table 4-4.4 summarizes the important fire extinguishment, burnback resistance, film formation, and foam quality requirements established by MIL-F-24385. The fire tests are conducted using 2.6 m^2 (28 ft^2) and 4.6 m^2 (50 ft^2) circular fire test pans. There are specific requirements to conduct a fire test of the agent after it has been subjected to an accelerated aging process (simulating prolonged storage) and after intentionally misproportioning the concentrate with water. In particular, the requirement to conduct a fire test of the agent at one-half of its design concentration is one of the most difficult tests. The 2.6 m^2 (28 ft^2) half-strength fire test must be extinguished in 45 sec, only 15 sec greater than allowed when the full-strength solution is used.

The physical and chemical properties evaluated for MIL-F-24385 agents are outlined in Table 4-4.5, along with the rationale for each test. These procedures have been developed based on experience and specific military requirements. For example, MIL-F-24385 requires that the agent be compatible with dry chemical agents. Dry chemical agents may be used as "secondary" agents in aviation and shipboard machinery space fires, e.g., to combat three-dimensional fuel fires, where AFFF alone may have limited effectiveness. MIL-F-24385 requires that an agent's compatibility with potassium bicarbonate dry chemical agent (PKP) be demonstrated. The burnback time of the foam in the presence of the dry chemical is measured. Also, the concentrate of one manufacturer must be compatible with concentrates of the same type furnished by other manufacturers, as determined by fire tests and accelerated aging tests.

Standards outside the United States: The number of standards developed for foams outside the United States is quite substantial. A brief review of the literature yielded over 17 different standards and test methods.[20] Developments in the European community are reviewed here to provide examples of differences in test standards.

The International Civil Aviation Organization (ICAO) develops crash fire-fighting and rescue documents for its member bodies. The ICAO *Airport Services Guide*, Part 1—Rescue Firefighting, describes airport levels of protection

TABLE 4-4.4 *Summary of the U.S. Military AFFF Specification (MIL-F-24385, Revision F) Key Performance Requirements*

Test Parameter	Revision F
Fire Extinguishment	
2.6 m² (28 ft²) fire test	
Application rate	2.9 Lpm/m² (0.071 gpm/ft²)
Maximum extinguishment time	30 s
Maximum extinguishment density	1.45 L/m² (0.036 gal/ft²)
4.6 m² (50 ft²) fire test*	
Application rate	1.6 Lpm/m² (0.04 gpm/ft²)
Minimum 40-s summation	320 s
Maximum extinguishment time	50 s
Maximum extinguishment density	1.34 L/m² (0.033 gal/ft²)
Fire Extinguishment—Over- and Under-Proportioning [2.6 m² (28 ft²) Test]	
One-half strength	
Maximum extinguishment time	45 s
Maximum extinguishment density	2.2 L/m² (0.054 gal/ft²)
Quintuple (5×) Strength	
Maximum extinguishment time	55 s
Maximum extinguishment density	2.7 L/m² (0.066 gal/ft²)
Burnback Resistance	
2.6 m² (28 ft²) fire test	25% maximum at 360 s[†]
4.6 m² (50 ft²) fire test	25% maximum at 360 s
Foam Quality	
Expansion ratio	6.0 : 1 minimum
25% drainage time	150 s minimum
Film Formation	
Spreading coefficient	
Fuel	Cyclohexane
Minimum value	3 dynes/cm
Ignition resistance test	
Fuel	Cyclohexane
Pass/fail criteria	No ignition

*Saltwater only.

[†]300 s for one-half-strength solutions; 200 s for quintuple-strength solutions.

to be provided and extinguishing agent characteristics. Minimum usable amounts of extinguishing agents are based on two levels of performance: Level A and Level B. The amounts of water specified for foam production are predicted on an application rate of 8.2 Lpm/m² (0.20 gpm/ft²) for Level A, and 5.5 Lpm/m² (0.13 gpm/ft²) for Level B. Agents that meet performance Level B require less agent for fire extinguishment. ICAO foam test criteria are described in Table 4-4.6. Foams meeting performance Level B have an extinguishment application density of 2.5 L/m² (0.061 gal/ft²). There are no surface tension, interfacial tension, and spreading coefficient requirements.

The International Organization for Standardization (ISO) has issued a draft specification for low-expansion foams, ISO/DIS 7203-1.[21] The specification includes definitions for protein, fluoroprotein, synthetic, alcohol resistance, AFFF, and FFFP concentrates. A positive spreading coefficient is required for film-forming foams when cyclohexane is used as the test fuel. There are toxicity, corrosion,

sedimentation, viscosity, expansion, and drainage criteria. The fire test uses a 2.4-m (8-ft) diameter circular pan with heptane as the fuel. The UNI 86 foam nozzle is used for either a "forceful" or "gentle" application method at a flow rate of 11.4 Lpm (3 gpm). The application rate is 2.4 Lpm/m² (0.06 gpm/ft²). For the greatest performance level, a 3-min extinguishment time is required. This results in an extinguishing application density of 7.6 L/m² (0.19 gal/ft²).

The proposed ISO requirements for extinguishing and burnback are summarized in Table 4-4.7. There are three levels of extinguishment performance and four levels of burnback performance. For extinguishing performance, Class I is the highest class and Class III the lowest class. For burnback resistance, Level A is the highest level and Level D the lowest level. Foam concentrates can be compared for each factor

TABLE 4-4.5 *Physical/Chemical Properties and Procurement Requirements of the AFFF Mil Spec*

Requirement	Rationale
Refractive Index	enable use of refractometer to measure solution concentrations in field; this is most common method recommended in NFPA 412*
Viscosity	ensures accurate proportioning when proportioning pumps are used; e.g., balanced pressure proportioner or positive displacement injection pumps
pH	ensures concentrate will be neither excessively basic or acidic; intention is to prevent corrosion in plumbing systems
Corrosivity	limits corrosion of, and deposit buildup on, metallic components (various metals for 28 days)
Total Halides/ Chlorides	limits corrosion of, and deposit buildup on, metallic components
Environmental Impact	biodegradability, fish kill, BOD/COD[†]
Accelerated Aging	film formation capabilities, fire performance, foam quality; ensures a long shelf life
Seawater Compatibility	ensures satisfactory fire performance when mixed with brackish or saltwater
Interagent Compatibility	allows premixed or storage tanks to be topped off with different manufacturers' agents, without affecting fire performance
Reduced- and Over-Concentration Fire Test	ensures satisfactory fire performance when agents are proportioned inaccurately
Compatibility with Dry Chemical (PKP) Agents	ensures satisfactory fire performance when used in conjunction with supplementary agents
Torque to Remove Cap	able to remove without wrench
Packaging Requirements	strength, color, size, stackable, minimum pour, and vent-opening tamperproof seal; ensures uniformity of containers and ease of handling
Initial Qualification Inspection	establish initial conformance with requirements
Quality Conformance Inspection (each lot)	ensures continued conformance with requirements

*NFPA 412, *Standard for Evaluating Aircraft Rescue and Fire Fighting Foam Equipment*, 1993 edition.

[†]BOD/COD (Biological Oxygen Demand/Chemical Oxygen Demand)

TABLE 4-4.6 *ICAO Foam Test Requirements*

Fire Tests	Performance Level A	Performance Level B
1. Nozzle (air aspirated)		
(a) Branch pipe	UNI 86 foam nozzle	UNI 86 foam nozzle
(b) Nozzle pressure	700 kPa (100 psi)	700 kPa (100 psi)
(c) Application rate	4.1 Lpm/m^2 (0.10 gpm/ft^2)	2.5 Lpm/m^2 (0.06 gpm/ft^2)
(d) Discharge rate	11.4 Lpm (3.0 gpm)	11.4 Lpm (3.0 gpm)
2. Fire Size	$\cong$ 2.8 m^2 ($\cong$ 30 ft^2) (circular)	$\cong$ 4.5 m^2 ($\cong$ 48 ft^2) (circular)
3. Fuel (on water surface)	Kerosene	Kerosene
4. Preburn time	60 s	60 s
5. Fire Performance		
(a) Extinguishing time	$\leq$ 60 s	$\leq$ 60 s
(b) Total application time	120 s	120 s
(c) 25% reignition time	$\geq$ 5 min	$\geq$ 5 min

separately but not necessarily in combination. For example, a IC concentrate is superior to a ID or a IIC concentrate, but it is not possible to say that it is superior to a IIB concentrate, since it is superior in extinguishing performance but inferior in burnback resistance.

Typical performance classes and levels for different concentrates are provided. Typical anticipated performance for AFFF is noted as ID and for FFFP as IA/B. For alcohol-resistant foams, both AFFF and FFFP are typically IA.

Comparison of small-scale tests: Table 4-4.8 outlines the large number of variables associated with foam performance and testing. These include factors such as foam bubble stability and fluidity, actual fire test parameters (e.g., fuel, foam application method and rate), and environmental effects. Even the fundamental methods of measuring foam performance (i.e., knockdown, control, and extinguishment) vary. For example, Johnson reported that FFFP fails the proposed ISO gentle application tests because small flames persist along a small area of the tray rim.[22] As a result, the foam committees have proposed redefining extinction to include flames.

Given the variations and lack of fundamental foam spreading theory, it follows that tests and specifications for various foams and international standards have different requirements. This is reflected in Table 4-4.9, which compares four key parameters of MIL-F-24385, UL 162, ICAO,

and ISO standards for manual application (e.g., handline or turret nozzles). There is no uniform agreement among test fuel, application rate, the allowance to move the nozzle, and the extinguishment application density for AFFF. There is a factor of six difference between the lowest permitted extinguishment application density (MIL-F-24385) and the highest (ISO). This significant difference is attributed, at least in part, to the fixed nozzle requirement in the ISO specification.

No study has been performed to correlate test methods; given the significant differences in performance characteristics and requirements, it is unlikely that correlation between these test methods could be established, even when considering AFFF only. An AFFF that meets the ICAO standard could not be said to meet MIL-F-24385 without actual test data. The problem of correlating differences in small-scale tests was demonstrated by UL in a comparison of UL, MIL-F-24385, O-F-555B (U.S. government protein foam specification), and United Kingdom test methods.[23] In those tests, differences between different classes of agents (protein *vs* AFFF) and between agents within a class (e.g., AFFF) were demonstrated. No correlations between test standards could be established.

The problem of correlation is compounded when a single test method is used in an attempt to assess different classes of foam, e.g., protein and AFFF. Attempts to use a single test method are problematic because of the inherent difference between these foams. That is, protein foams require air aspiration

TABLE 4-4.7 *Maximum Extinction Times and Minimum Burnback Times from Proposed ISO Specification*

Extinguishing Performance Class	Burnback Resistance Level	Gentle Application Test		Forceful Application Test	
		Extinction Time (min) Not More than	Burnback Time (min) Not Less than	Extinction Time (min) Not More than	Burnback Time (min) Not Less than
I	A	not applicable		3	10
	B	5	15	3	not tested
	C	5	10	3	not tested
	D	5	5	3	not tested
II	A	not applicable		4	10
	B	5	15	4	not tested
	C	5	10	4	not tested
	D	5	5	4	not tested
III	B	5	15	not tested	not tested
	C	5	10	not tested	not tested
	D	5	5	not tested	not tested

TABLE 4-4.8 *Variables Associated with Foam Performance and Testing*

I. Physical/chemical properties of foam solution
  A. Bubble stability
    1. Measures
      a. Expansion ratio
      b. Drainage rate
    2. Variables
      a. Water temperature
      b. Water hardness/salinity
      c. Water contamination
  B. Fluidity of foam
    1. Measures
      a. Viscosity
      b. Spreading rate
      c. Film formation
    2. Variables
      a. Fuel type and temperature
      b. Foam bubble stability
  C. Compatibility with auxiliary agents
    1. Measures—fire and burnback test
    2. Variables
      a. Other foam agents
      b. Dry chemical agents
  D. Effects of aging
    1. Measures—fire and burnback test
    2. Variable—shelf life of agent
II. Test methods to characterize foam performance
  A. Fuel
    1. Measures
      a. Vapor pressure
      b. Flash point
      c. Surface tension
      d. Temperature
    2. Variables
      a. Volatility
      b. Depth and size
      c. Initial temperature of air and fuel temperature
      d. Time fuel has been burning (e.g., short *vs* long, and depth of hot layer)
  B. Foam application method
    1. Measures
      a. Stream reach
      b. Aspiration of foam
      c. Foam stability, e.g., contamination by fuel
      d. Water content of foam
      e. Proportioning rate
    2. Variables
      a. Aspiration
        (1) Effect on stream reach
        (2) Degree to which foam is aspirated and the need to aspirate based on foam type

II. B. 2. Variables (continued)
      b. Fixed *vs* mobile device
      c. Application technique
        (1) Indirect, e.g., against backboard or sidewall
        (2) Direct
          (a) Gentle
          (b) Forceful
          (c) Subsurface injection
      d. Application location
        (1) High—need to penetrate plume
        (2) Low
      e. Application rate of foam
      f. Wind (as it affects stream reach)
        (1) Crosswind
        (2) With and against
      g. Effect of reduced or increased concentration due to improper proportioning
  C. Fire configuration
    1. Measures
      a. Fuel burning rate, radiation feedback to fire
      b. Propensity for reignition
      c. Surface tension
    2. Variables
      a. Pan/containment geometry
      b. Two-dimensional (pool) *vs* three-dimensional (running fuel/atomized spray)
      c. Presence and temperature of freeboard
      d. Wind (as it affects flame tilt and reradiation)
      e. Surface on which there is fuel
        (1) Rough
        (2) Smooth
        (3) Water substrate—"peeling" effect of fuel
  D. Measurement of results
    1. Measures
      a. Time to knockdown, control, extinguish, and burnback
        (1) Actual or estimated time by visual observations
        (2) Summation values, i.e., summation of control at 10, 20, 30, and 40 sec
      b. Heat flux during extinguishment and burnback
    2. Variables—qualitative and quantitative methods to determine fire knockdown, extinguishment, and burnback
      a. 90 percent control—measure of ability of foam to quickly control the fire
      b. 99 percent (virtual extinguishment)—all but the last flame or edge extinguished
      c. Extinguishment—100 percent
      d. Burnback—25 percent, 50 percent

so that the foam floats on the fuel surface. This stiff, "drier" foam is viscous and does not inherently spread well without outside forces (e.g., nozzle stream force). AFFF, because of its film-formation characteristics, does not require the degree of aspiration that protein foams require. This heavier, "wetter" foam is inherently less viscous, which contributes to improved spreading and fluidity on fuel surfaces. This is related, at least in part, to the degree of aspiration of the foam. A more exact description of foam aspiration is appropriate. Thomas has described two levels of foam aspiration: (1) primary aspirated and (2) secondary aspirated.[24] Primary aspirated foam occurs when a foam solution is applied by means of a special nozzle designed to mix air with the solution within the nozzle. The consequence is foam bubbles of general uniformity. "Air-aspirated" foam refers to this primary aspirated foam. Secondary aspirated foam results when a foam solution is applied using a nozzle that does not mix air with the solution within the nozzle. Air is, however, drawn into the solution in-flight or at impact at the fire. Secondary aspirated foam is more commonly referred to as "non-air aspirated" foam.

The correlation between foam solution viscosity and extinguishment time has been shown by Fiala, but the entire foam spreading and extinguishment theory has yet to be demonstrated based on first principles.[25] Thus, the test standards reference bench-scale methods that measure a factor of foam fluidity (e.g., spreading coefficient), but fail to recognize the total foam spreading system, including viscous effects. Fundamental understanding of foam mechanisms would promote the development of bench- and small-scale test apparatus that potentially have greater direct correlation for predicting large-scale results.

There has been some criticism of the human element involved in many of the test methods. The human factor occurs when an operator is allowed to apply foam from a hand-held nozzle onto the burning test fire. Personnel are also called upon in some tests to qualitatively assess the percentage of fire involvement in the test pan during the burnback procedure. Using a fixed nozzle during a specification test eliminates the human element during extinguishment. For sprinkler applications, this is entirely appropriate and should yield results comparable to actual installations. For applications where movement is actually involved (e.g., fire-fighting handlines, crash-rescue truck turrets, and movable monitors on ships and at petrochemical facilities), the extinguishment densities in the fixed test application will generally exceed the densities found in actual applications in the field. (See Table 4-4.9 for differences in extinguishing densities for manual *vs* fixed applications.) Removal of the human element is certainly advisable from a test repeatability standpoint. The Canadian government has revised its military foam procurement test method to eliminate the human operator, based on criticisms of the repeatability where manual human intervention is involved.

Quantitative methods for evaluating burnback performance have been described by Scheffey et al[16] and been adopted in ISO and Scandinavian (NORDTEST) test methods. These methods involve the use of radiometers to establish a heat flux during full test pan involvement. After extinguishment, the radiometers measure the increasing flux as the burnback fire grows. This increasing flux due to burnback is compared against the original flux. A cutoff is established so that the maximum burnback time is the time for the burnback flux to reach some percentage (e.g., 25 percent) of the original full-burning flux.

Critical Application Rates and Correlations Between Small- and Large-scale Tests

The previous text described the application rate differences in standard test methods between AFFF, fluoroprotein, and protein foams. These application rate differences were established based on full-scale testing. For sprinklers, much of the fundamental application rate differences were established during testing conducted by Factory Mutual Research Corporation (FMRC). (See section on foam-water sprinkler systems.) For manual applications, tests in the aviation fire protection field provide the basis for the fundamental application rates. The application rates specified in test standards are usually rates lower than those used in actual practice. (See Table 4-4.3.) There are two reasons for this: (1) a factor of safety is used when specifying rates in actual practice and (2) differences between individual foam agents are more readily apparent at critical application rates. To demonstrate how application rates are developed and how specification tests correlate with large-scale results, an example from aviation fire tests will be used. This is based

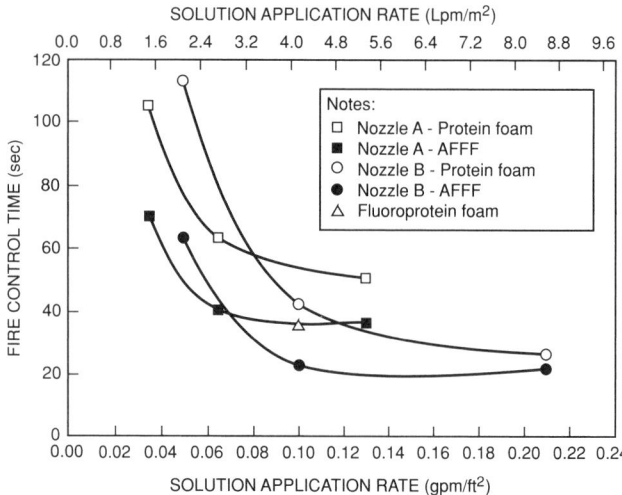

Fig. 4-4.4.   Fire control time as a function of solution application rate using protein foam and AFFF on JP-4 pool fires.[26]

on a review of foam fire test standards performed by Scheffey et al for the Federal Aviation Administration (FAA).[20]

Tests were conducted by the FAA to determine application rates for a single-agent attack to achieve fire control (e.g., 90 percent extinguishment of a fire area) within 1 min under a wide variety of simulated accident conditions. Two factors are important in addition to the application rate required for 1-min fire control: (1) the critical application rate, below which fires will not be extinguished independent of the amount of time an agent is applied; and (2) application density, which is the amount of foam per unit area to control or extinguish a fire.

Minimum application rates were originally developed by Geyer in tests of protein and AFFF agents.[26] These tests involved "modeling" tests with JP-4 pool fires of 21-, 30-, and 43-m (70-, 100-, and 140-ft) diameter. Large-scale verification tests with a B-47 aircraft and simulated shielded fires (requiring the use of secondary agents) were conducted with 34- and 43-m (110- and 140-ft) JP-4 pool fires. All tests were conducted with air-aspirating nozzles. The protein foam conformed to the U.S. government specification, O-F-555b, while the AFFFs used were in nominal conformance with MIL-F-24385 for AFFF. These tests were being performed at the time when the seawater-compatible version of MIL-F-24385 had just been adopted based on large-scale tests.

Figure 4-4.4 illustrates the results of the "modeling" experiments. This shows that, for a fire control time of 60 sec, the application rate for AFFF was on the order of 1.6 to 2.4 Lpm/m² (0.04 to 0.06 gpm/ft²), while the application rate for protein foam was 3.3 to 4.1 Lpm/m² (0.08 to 0.10 gpm/ft²). The data indicated that the application rate curves become asymptotic at rates of 4.1 Lpm/m² (0.1 gpm/ft²) and 8.2 Lpm/m² (0.2 gpm/ft²) for AFFF and protein foam, respectively. Above these rates, fire control times are not appreciably improved. Likewise, critical application rates for fire control are indicated when control times increase dramatically. The single test with a fluoroprotein agent indicated that this agent, as expected, fell between AFFF and protein foam.

Large-scale auxiliary agent tests were conducted to identify increases in foam required when obstructed fires

TABLE 4-4.9 *Examples of Extinguishment Application Densities of Various Test Standards*

Test Standard	Fuel	Application Rate [Lpm/m² (gpm/ft²)]	Nozzle Movement Permitted	Extinguishment Application Density [L/m² (gal/ft²)]
MIL SPEC	motor gasoline	1.6 (0.04)	yes	1.34 (0.033)
UL 162	heptane	1.6 (0.04)	yes	4.9 (0.12)
ICAO	kerosene	2.5 (0.06)	yes (horizontal plane)	2.6 (0.061)
ISO—Forceful	heptane	2.5 (0.06)	no	7.6 (0.19)

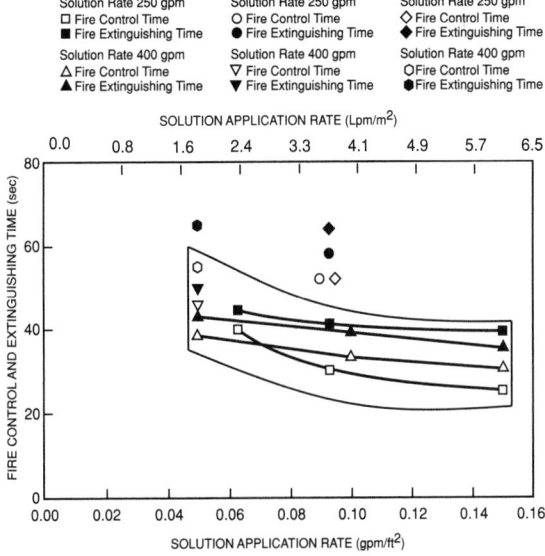

Fig. 4-4.5. *Fire control and extinguishing times as functions of the foam solution application rate using AFFF at 250 gpm (946 Lpm), 400 gpm (1514 Lpm), and 800 gpm (3028 Lpm) on JP-4, JP-5, and avgas fires.*[27]

with an actual fuselage were added to the scenario. The results indicated that fire control times increased by a factor of 1 to 1.9 for AFFF and 1.5 to 2.9 for protein foams. It was estimated that the most effective foam solution application rates were 4.9 to 5.7 Lpm/m² (0.12 to 0.14 gpm/ft²) for AFFF and 7.5 to 9 Lpm/m² (0.18 to 0.22 gpm/ft²) for protein foam. This is the original basis of the recommendations adopted by ICAO of 5.5 Lpm/m² (0.13 gpm/ft²) for AFFF and 8.2 Lpm/m² (0.20 gpm/ft²) for protein foam. A rate of 7.5 Lpm/m² (0.18 gpm/ft²) was subsequently established for fluoroprotein foam. These application rate values are still used by FAA, NFPA, and ICAO to establish minimum agent supplies at airports.

Tests of AFFF alone were conducted by Geyer.[27] These agents, selected from the U.S. Qualified Products List (MIL-F-24385 requirements), were tested on JP-4, JP-5, and aviation gasoline (avgas) fires. Air-aspirating nozzles were used with different AFFF agents. Example results are shown in Figure 4-4.5. Similar data were collected by holding the JP-4 fuel fire size constant at 743 m² (8000 ft²) and varying the flow rates to develop application rate comparisons. These data are shown in Figure 4-4.6.

Additional tests were conducted by Geyer *et al* to verify the continuation of the reduction of water when AFFF agents were substituted for protein foam in aviation situations.[1] In 25-, 31-, and 44-m (82.4-, 101-, and 143-ft) diameter Jet A pool

fires, AFFF, fluoroprotein, and protein foams were discharged with air-aspirating and non-air-aspirating nozzles. The data, summarized in Figure 4-4.7, validated the continued allowance of a 30 percent reduction in water requirement at certified U.S. airports when AFFF is substituted for protein foam.

Although some test criteria in standardized methods do not necessarily correlate directly with actual fire and burnback performance, small-scale test data for AFFF formulated to the U.S. military specification (MIL-F-24385) has been shown to correlate with large-scale fire test results. This is based on a comprehensive review of small- and large-scale test data.[20] In these data, a key variable was controlled; i.e., all AFFF agents were formulated to meet MIL-F-24385. Ninety percent fire control times were used as the most accurate measure of fire knockdown performance, which were reported in all tests. The use of 90 percent control times eliminates the variability of total extinguishment, which might be dependent on test bed edge effects or running fuel fire scenarios. Data for tests using air-aspirated or non-air-aspirated nozzles were combined. Low-flash-point [less than 0°C (32°F)] fuels were evaluated. The evaluation included only tests where manual application was used, eliminating the variable of fixed *vs* manual application.

The effects of application rate on control and extinguishment times, as demonstrated in Figures 4-4.4 through

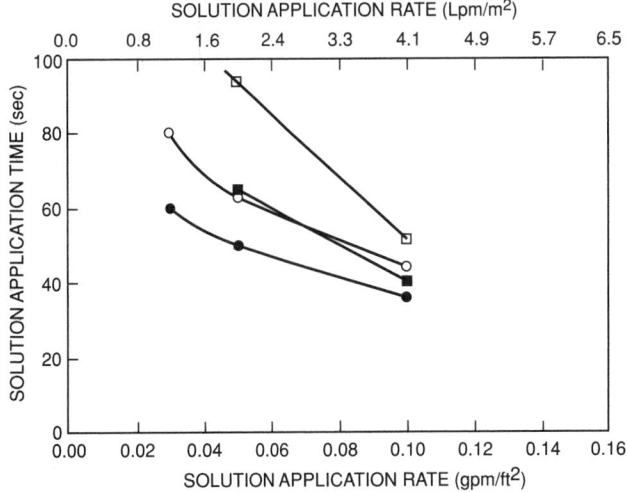

Note: No fire control or extinguishment at 0.03 gpm/ft²

● CONTROL TIME - AGENT A
○ FIRE EXTINGUISHING TIMES - AGENT B

■ CONTROL TIME - AGENT A
□ FIRE EXTINGUISHING TIMES - AGENT B

Fig. 4-4.6. *Fire control and extinguishing times as a function of solution application rate using AFFF at 250, 400, and 800 gpm on 743 m² (8000 ft²) JP-4 fuel fires.*[27]

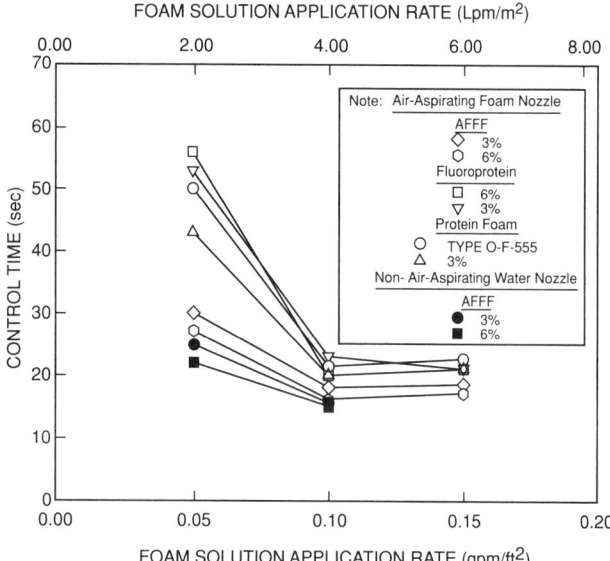

Fig. 4-4.7. *Fire control time as a function of solution application rate for AFFF, fluoroprotein, and protein foams for Jet A pool fires.*[1]

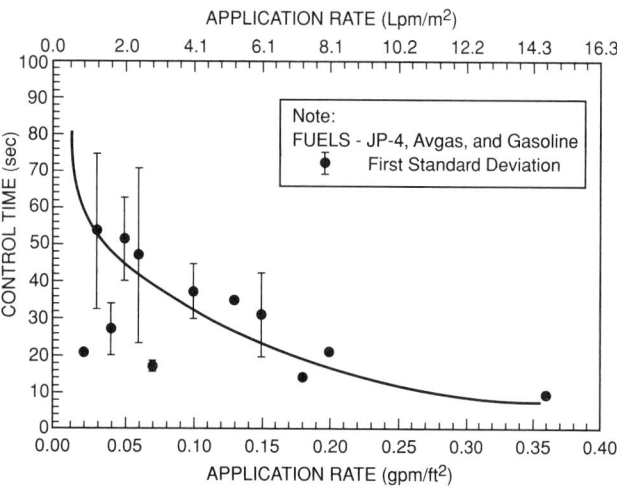

Fig. 4-4.8. *AFFF control time as a function of application rate.*[20]

4-4.7, were reconfirmed as shown in Figure 4-4.8. Control time increases exponentially as application rate decreases, particularly below 4.1 Lpm/m² (0.10 gpm/ft²). Variability of the data is shown by the first standard deviation.

The scaling of small fires with large fires is shown in Figures 4-4.9 and 4-4.10, which relate the time needed to control the burning fuel surface as a function of fire size. The time needed to control a unit of burning area [sec/ft² (s/m²)], designated as the specific control time, is plotted as a function of fire size. For low [1.2 to 2.5 Lpm/m² (0.03 to 0.06 gpm/ft²)] and intermediate [2.8 to 4.1 Lpm/m² (0.07 to 0.10 gpm/ft²)] application rates, the specific control times decrease linearly as a function of fire area. These data are in agreement with data from Fiala, which also indicate decreasing specific extinguishment control times as a function of burning area for increasing application rates of AFFF.[25]

Also, Fiala showed that, for a constant application rate, AFFFs have lower specific extinguishment times as a function of burning area than those of protein and fluoroprotein foams. Obviously, this linear relationship must change at very large areas; otherwise, the specific control/extinguishment time would go to zero. This is evidenced in Figure 4-4.9, where the curve flattens at the high-area end of the plot.

Figures 4-4.9 and 4-4.10 show that higher specific control times are required for MIL-F-24385 test fires [2.6 and 4.7 m² (50 and 20 ft²) compared to large fires. This is readily apparent as actual/control extinguishment times for the small fires are on the same order as results from large fires. FAA and NFPA criteria for minimum quantities of agent are also shown in Figures 4-4.9 and 4-4.10. These criteria are expressed in terms of specific control time as a function of area by using the required control time of 60 sec and the practical critical fire areas for airports serving different sizes of aircraft. The data indicate that specific control times with MIL-F-24385 agents are roughly equivalent or less than the specific control times established by NFPA and FAA requirements for large fire areas. This is true even with the AFFF discharged at rates 25 to 75 percent below the minimum NFPA/FAA discharge rate of 5.5 Lpm/m² (0.13 gpm/ft²).

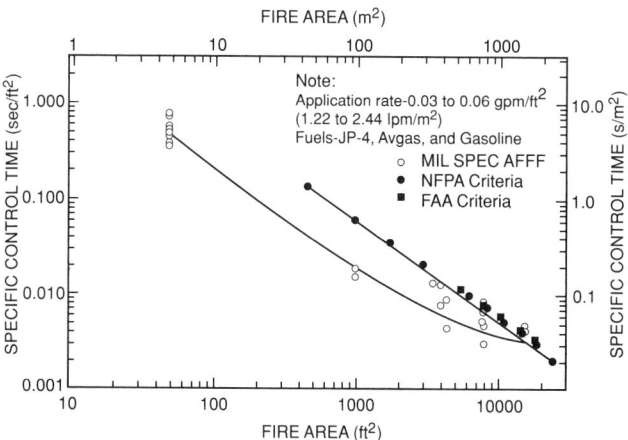

Fig. 4-4.9. *Specific control times for AFFF at low application rates.*[20]

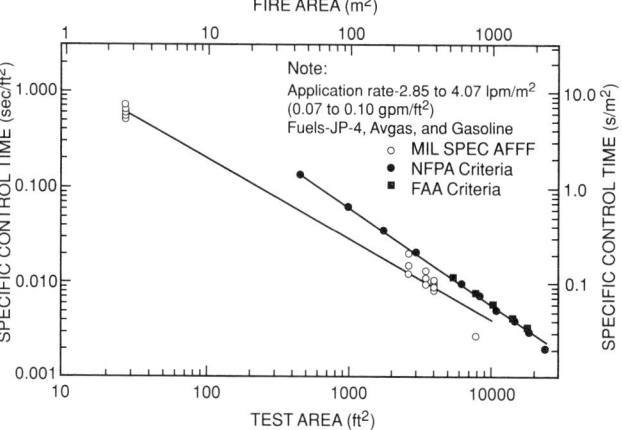

Fig. 4-4.10. *Specific control times for AFFF at intermediate application rates.*[20]

From these data, it can be concluded that a scaling relationship exists between MIL-F-24385 small-scale tests and actual large-scale crash rescue and fire fighting applications. The MIL-F-24385 tests are more challenging than the larger tests in terms of specific control time, but this challenging test produces an agent that can meet NFPA and FAA requirements at less than the design application rate. This factor of safety accounts for variables in actual aviation crash situations, e.g., running fuel fires, debris that may shield fires, and cross winds that may limit foam stream reach.

AVIATION FIRE PROTECTION CONSIDERATIONS

Historical Basis for Foam Requirements

The underlying principle in aviation fire protection is to temporarily maintain the integrity of an aircraft fuselage after a mishap to allow passenger escape or rescue. When an aircraft is involved in a fuel spill fire, the aluminum skin will burn through in about 1 min. If the fuselage is intact, the sidewall insulation will maintain a survivable temperature inside the cabin until the windows melt out in approximately 3 min. At that time, the cabin temperature rapidly increases beyond survivable levels.

Aircraft rescue and fire-fighting (ARFF) vehicles are designed to reach an incident scene on the airport property in 2 to 3 min, depending on the standard enforced by the authority having jurisdiction (AHJ). Having reached the scene in this time frame, agent must be applied to control a fire in 1 min or less. The 1-min critical time for fire control is recognized by FAA, NFPA, and ICAO.

Minimum agent requirements on ARFF vehicles are established using the 1-min critical control time plus the anticipated spill area for the largest aircraft using the airport. A "theoretical critical fire area" has been developed, based on tests, and is defined as the area adjacent to the fuselage, extending in all directions to the point beyond which a large fuel fire would not melt an aluminum fuselage regardless of the duration of the exposure. A function of the size of an aircraft, the theoretical critical fire area was amended to a "practical critical fire area" after evaluation of actual aircraft fire incidents. The practical critical area, two-thirds the size of the theoretical critical area, is widely recognized by the aviation fire safety community, including FAA, NFPA, and ICAO. Vehicles must be equipped with sufficient agent and discharge devices to control a fire in the practical critical area within 1 min. Vehicles must also be equipped with secondary agent (dry chemical or Halon 1211) for use in combating three-dimensional fuel fires.

Agent Quantities and Standards

The previous text on critical application rates described the rationale used to develop design application rates used in aviation fire protection. These rates are 5.5 Lpm/m^2 (0.13 gpm/ft^2) for AFFF, 7.5 Lpm/m^2 (0.18 gpm/ft^2) for fluoroprotein foam, and 8.2 Lpm/m^2 (0.28 gpm/ft^2) for protein foam. Using these rates, the practical critical fire area and the 60-sec control time criteria, minimum agent quantities are established for airports serving different size aircraft. These criteria are contained in NFPA 403, *Standard for Aircraft Rescue and Fire Fighting Services at Airports*, and the FAA Advisory Circular 150/5210-6C, "Aircraft Fire and Rescue Facilities and Extinguishing Agents." ICAO uses similar criteria.

TABLE 4-4.10 *Foam Quality Requirements from NFPA 412*

Agent	Minimum Expansion Ratio	Minimum Solution 25% Drainage Time (min)
AFFF or FFFP		
Air Aspirated	5:1	2.25
Non-air Aspirated	3:1	0.75
Protein	8:1	10
Fluoroprotein	6:1	10

NFPA 403 recently adopted the 4.6 m^2 (50 ft^2) fire extinguishment and burnback criteria from MIL-F-24385 for AFFF agents. UL test criteria are acceptable for protein and fluoroprotein foams. Most airports in the United States use AFFF as the primary fire-fighting agent. Recognizing the limitations of its test methods for aviation applications, UL has deleted references to crash rescue fire fighting from the scope of UL 162, *Standard for Foam Equipment and Liquid Concentrates*. NFPA 403 recognizes that the standards for foam that it references are widely recognized throughout North America, but may not be recognized in other areas of the world. In particular, the ICAO test method has significantly different test parameters, including test fuel, application rate, and extinguishment density. The NFPA notes that it is incumbent on the national authority having jurisdiction to determine that alternative test methods meet the level of performance established by NFPA 403 test criteria.

NFPA 412, *Standard for Evaluating Aircraft Rescue and Fire Fighting Foam Equipment*, provides field test methods to determine the adequacy of foam equipment on crash rescue vehicles. It includes criteria for foam expansion and drainage, and methods to determine foam solution concentration.

Expansion and drainage: Foam expansion and drainage requirements of the current version of NFPA 412, *Standard for Evaluating Aircraft Rescue and Fire Fighting Foam Equipment*, are shown in Table 4-4.10.

NFPA 412 references a 1600-mL foam sample collector, which was originally adopted by ICAO and ISO. This single method is used to obtain expansion and drainage measurements for all types of foams in hope that similar success could be obtained in using a single fire test method for all foams. The multiple categories of foam test classification in Table 4-4.7 for the ISO method show how difficult this has been to achieve. Given the different methods of foam flow over a fuel surface, it may not be practical to use a common fire test method predicated on the current means of testing. Further development of fundamental foam extinguishing principles is recommended.

The 1600-mL expansion and drainage test method replaced two other methods where a 1000-mL cylinder or 1400-mL pan was used as the collection device. MIL-F-24385 still uses the 1000-mL collection method. This situation, plus other different test methods, makes direct comparison of expansion and drainage data difficult. Tests performed by Underwriters Laboratories (UL) identified differences among the three test methods based on expansion and drainage results.[28] UL found that expansion ratios remained the same but that drainage was quicker using the 1600-mL method compared to the 1000-mL method for film-forming foams. Drainage time increased (i.e., doubled) for the protein foams when the 1600-mL method was used compared to the 1400-mL pan method.

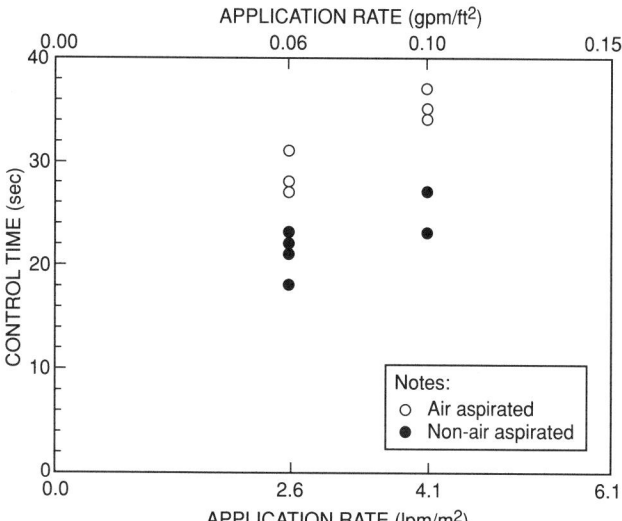

Fig. 4-4.11. Effects of AFFF aspiration on JP-4 pool fire control times.[29]

No direct correlations have been established between expansion, drainage, and fire-extinguishing performance. There is a relationship between foam drainage and burnback. Longer drainage times generally result in longer burnback times.

The expansion and drainage data in Table 4-4.10 indicate the inherent differences between air-aspirated and non-air-aspirated film-forming foams. The data in Figure 4-4.7 showed that non-air-aspirated AFFF was more effective at critical application rates than air-aspirated AFFF. This was verified by Jablonski in tests with U.S. Air Force crash trucks as shown in Figure 4-4.11.[29] Even so, there continues to be debate over aspirated and non-air-aspirated foam for manual applications involving aviation fuel spills.

Under certain conditions, non-air-aspirated AFFF is not as effective as air-aspirated AFFF. The results of the foam tests in the UK[30,31] and the results from DiMaio et al[32] described situations where air-aspirated AFFF resulted in better fire extinguishment performance than non-air-aspirated foam.

Given that one-to-one correlation between expansion, drainage, and fire extinguishing performance is difficult to identify, there appears to be a lower limit where non-air-aspirated AFFF becomes ineffective. This has not been quantified, but it is speculated that poor performance occurs when AFFF expansion ratio is less than 2.5 to 3.0, and drainage is difficult to measure, i.e., nearly instantaneous. This is based in part on unpublished data from the Naval Research Laboratory on shipboard bilge AFFF sprinklers[33] and the results of the UK tests.[30,31] The importance of this lower limit of foam aspiration is recognized in NFPA 412 criteria.

Foam concentration determination: The most common method of determining foam concentration in the field is by use of a hand-held refractometer. The refractive index, n, is defined as

$$n = \frac{\sin i}{\sin r} \qquad (10)$$

where

$\sin i$ = angle of incidence, and
$\sin r$ = angle of refraction.

This is depicted graphically in Figure 4-4.12.

Manufacturers report the glycols in AFFF formulations create the necessary refractive characteristics to determine concentration. However, they also report that glycol has a potential detrimental impact on overall agent performance. Elimination of this compound might improve (slightly) the performance of AFFF, but the glycol is also needed as a fundamental component of agent mixing.

The refractive index of water at 20°C (68°F) is 1.333 (air has a refractive index of 1.0002926). Since the refractive index of a solution is proportional to the inverse of the solution density, and density is proportional to temperature, then

$$n \propto \frac{1}{T} \qquad (11)$$

where T is the temperature. This is illustrated in Figure 4-4.13. Any refractive index measurements must be made considering temperature. Some hand-held measurement devices are temperature compensated. It is good procedure to conduct concentration measurements at a constant temperature.

Other scales may be used. For example, the Brix scale is used as a measure of sucrose weight percent concentration.

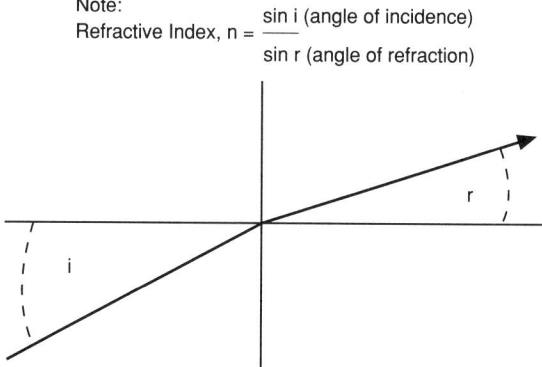

Note:
Refractive Index, $n = \dfrac{\sin i \text{ (angle of incidence)}}{\sin r \text{ (angle of refraction)}}$

Fig. 4-4.12. Refractive index of solutions.

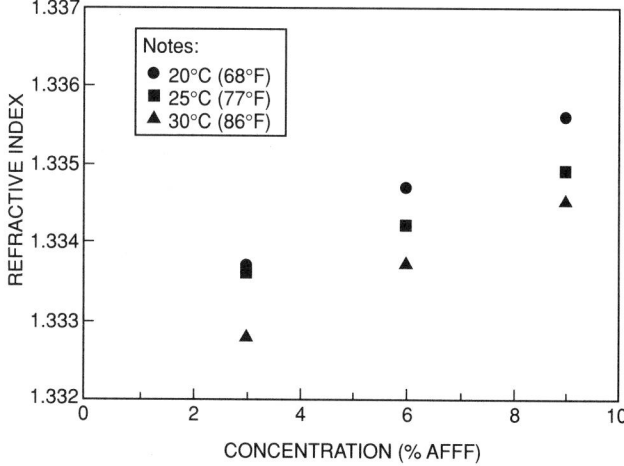

Fig. 4-4.13. Effects of temperature on refractive index.[34]

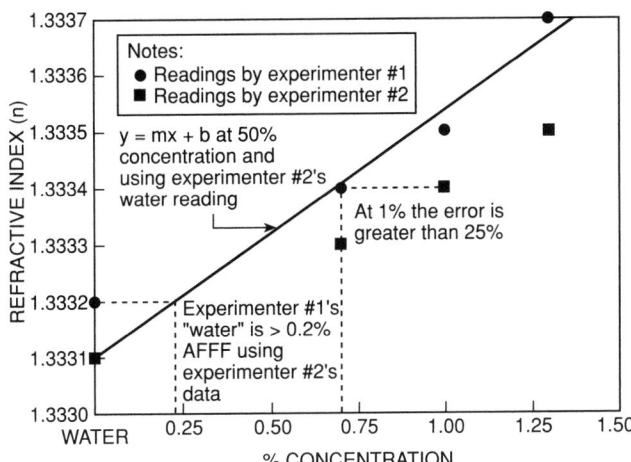

Fig. 4-4.14. *Refractive indices of 1 percent AFFF solutions in tap water.*

Units with this scale, commonly found in the food product industry, can be used to measure foam concentration. A typical range of a bench or hand-held refractometer is 1.3000 to 1.7000.

NFPA 412, *Standard for Evaluating Aircraft Rescue and Fire Fighting Foam Equipment*, describes a method to determine foam concentration using the refractive method. In NFPA 412, the preparation of three standard solutions is recommended: one at the nominal concentration, one at one-third more than the nominal concentration, and one at one-third less than the nominal concentration. A plot of the refractive scale reading against the known foam concentration is made on graph paper. This establishes a "calibration" curve against which foam samples from a vehicle or system can be judged. Since refractive index is linear, a calibration curve can be created by

$$\text{AFFF \%}_{sample} = \frac{n_{foam} - n_{water}}{n_{concentrate} - n_{water}} \times 100 \qquad (12)$$

This method is used by the U.S. Navy for checking proportioning system accuracy onboard ships.

The limitations of the refractive index technique are described by Timms and Haggar.[34] The accuracy of the refractometer can become poor due to the focusing and setting of the refracted light junction on the cross hairs of the viewing window, and the reading of the graduated scale to four decimal places (where the scale is graduated only to three places). This is illustrated in Figure 4-4.14, where a calibration curve for a 1 percent AFFF concentrate was established using a straight line through the 50 percent concentration point and the "water" reading by one of the experimenters. Note that the error between readings by the two experimenters at 1 percent concentration exceeds 25 percent. In this example, differences in the baseline water reading will create substantial error in the calibration curve. These differences are exaggerated with 1 percent concentrates. At 3 percent or 6 percent, the experimental error in reading the refractometer, for field testing, is generally accepted as adequate.

Alternative methods for measuring AFFF concentration include total fluorine content, optical absorption methods, and electrical conductivity. Since neither the total fluorine

content method nor optical absorption method is suited to field use, the conductivity method has been proposed. Since foams contain electrolytes, their conductance, G, can be measured and described as

$$G = \frac{1}{R} \text{ (mhos)} \qquad (13)$$

where

R = resistance (ohms).

Conductivity, σ, is conductance per unit length:

$\sigma = G/$unit length
  = mhos/cm
  = siemens/cm.

Since conductivity is directly proportional to temperature, conductivity increases with temperature. (See Figure 4-4.15.) Temperature compensation is appropriate when using this method.

Timms and Haggar showed the influence of the substrate water on both refractive index and conductivity.[34] (See Figures 4-4.16 and 4-14.17.) It is important to note the difference of the characteristic curve for a salt solution. AFFF actually reduces the conductivity of this highly conductive water. Note also that, while conductance may exhibit straight-line characteristics in the area of interest (0 to 10 percent), the overall curves from 0 to 100 percent are nonlinear.

The "sensitivity" of the two methods (i.e., refractive index and conductivity) was shown by these researchers by comparing the difference between readings for solutions of 3 percent and 6 percent divided by the reading at 6 percent. The sensitivities for tap water show that the conductivity method is more sensitive than the refractive index measure. (See Table 4-4.11.) In repeated readings of refractive index and conductivity, the foam concentration accuracy using conductivity was ± 0.1 percent, where the accuracy of the refractive index method was ± 0.8 percent. (See Table 4-4.12.)

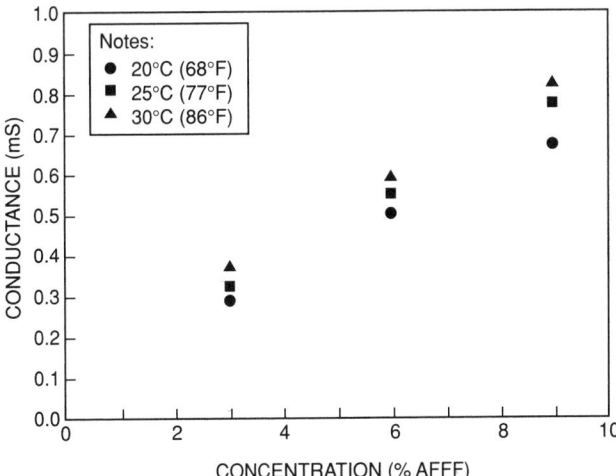

Fig. 4-4.15. *Effects of temperature on the conductance of AFFF solutions.*[34]

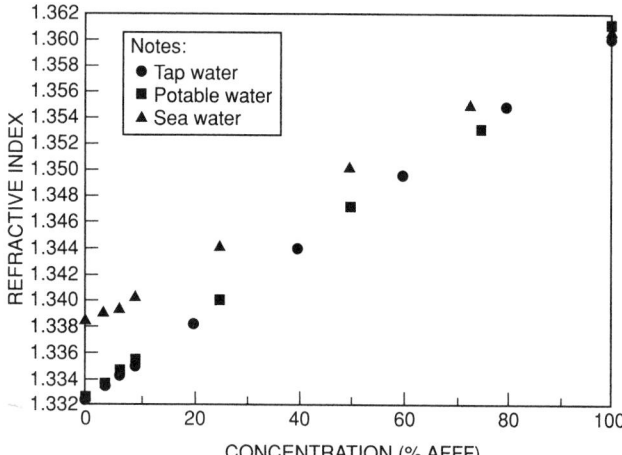

Fig. 4-4.16. Effect of substrate water on refractive index of AFFF solutions. [34]

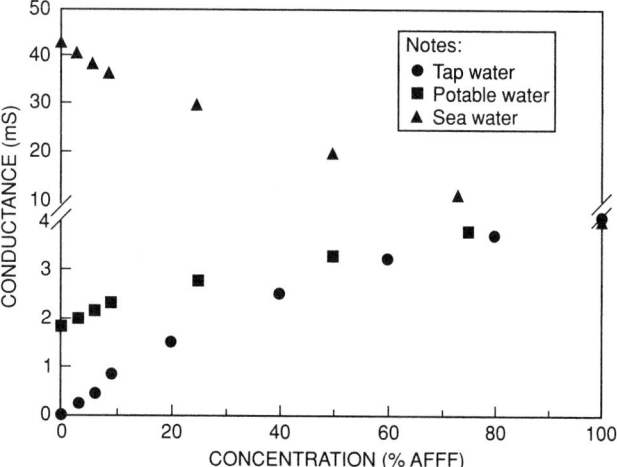

Fig. 4-4.17. Effect of substrate water on conductivity of AFFF solutions. [34]

The electrical conductivity method is now recognized in NFPA standards. NFPA 412 cautions against the use of this method for seawater applications. The electrical conductivity method, used for process control in the chemical industry, has recently been adapted for use as a proportioning controller for AFFF systems.

Aircraft Hangar Protection

The two objectives of aircraft hangar protection are: (1) protect aircraft and (2) prevent collapse of the hangar roof structure, which is usually unprotected steel. The protection of the aircraft is the principle concern, since its value is generally many times that of the structure. This is particularly true for advanced military aircraft. These protection systems are deluge-type sprinkler systems with open-head nozzles. They are activated by rapid-response detection systems. Before the development of foam, water-deluge systems were used. The original foam-water sprinkler systems used protein foam. With the development of AFFF, research was performed to determine appropriate application rates and

types of discharge devices. The research work, performed primarily by Factory Mutual Research Corporation (FMRC), provides the basis not only for current aircraft hangar protection criteria, but also for other sprinkler suppression system criteria.

Overhead sprinkler protection: Before the advent of foam, hangars were protected by conventional spray sprinklers using water. Water-deluge systems having discharge rates on the order of 10.4 Lpm/m^2 (0.25 gpm/ft^2) were used in conjunction with sloped floors and drains to protect aircraft. Even with these systems, activated by detection systems, burnthrough protection of aircraft fuselages (e.g., 1 min) could not be assured. Ceiling temperatures in an 18.3-m (60-ft) high space on the order of 427 to 816 °C (800 to 1500 °F) have been recorded for fuel spill fires where this protection was provided. For a 121 m^2 (1300 ft^2) JP-4 fuel fire, 927 °C (1700 °F) ceiling temperatures have been recorded within 30 sec of ignition prior to deluge system discharge.

Protein foam systems, discharging at a rate of 8.2 Lpm/m^2 (0.20 gpm/ft^2), were an improvement on the water systems. Air-aspirating sprinklers were required to make effective protein foam. Because of the high centerline velocities of a pool fire plume, the foam flow from the perimeter toward the center of the fire was thought to be the dominant suppression mechanism. [35]

With the development of AFFF, FMRC conducted a series of tests for the U.S. military to establish appropriate design parameters. In a series of baseline comparison tests, FMRC compared AFFF with protein foam. The tests consisted of 83.6-m^2 (900-ft^2) JP-4 pool fires in an 18.3-m (60-ft) high space. Air-aspirating, standard upright, and old-style upright sprinklers were evaluated at application rates of 4.1 to 8.2 Lpm/m^2 (0.10 to 0.20 gpm/ft^2). In one test, a low-level turret nozzle discharging AFFF was used in conjunction with sprinklers discharging water. Table 4-4.13 summarizes the results of the AFFF tests. A comparison of Tests 4 and 5 with Test 3 indicates improved results from the use of standard sprinklers compared to foam-water sprinklers. At application rates of 6.6 Lpm/m^2 (0.16 gpm/ft^2), the standard sprinklers were 1.3 to 1.6 times as effective in achieving extinguishment compared to air-aspirating foam-water sprinklers. At an application rate of 8.2 Lpm/m^2 (0.20 gpm/ft^2),

TABLE 4-4.11 Sensitivity of Refractive Index and Conductivity Methods for Determining Foam Concentration[34]

	Refractive Index	Conductance (mS)
3%	1.3337	0.318
6%	1.3343	0.558
Difference	0.0006	0.240
"Sensitivity"	0.0005	0.43
	(0.5 in 1000)	(430 in 1000)

TABLE 4-4.12 Accuracy of Foam Test Measurements[34]

Solution	Refractive Index	Electrical Conductance	Actual
A	4.3% ± 0.8%	3.5% ± 0.1%	3.50 ± 0.01%
B	5.1% ± 0.8%	5.5% ± 0.1%	5.50 ± 0.01%
C	8.7% ± 0.8%	8.5% ± 0.1%	8.50 ± 0.01%

TABLE 4-4.13 *Hangar Deluge System Tests by Factory Mutual Research Corporation*[35]

Test Conditions	Test No. 2	Test No. 3	Test No. 4	Test No. 5	Test No. 6	Test No. 7 (turret nozzle)
Type of head	Foam-water	Foam-water	Standard	Standard	Standard	Old-style Sprinkler
Spacing [m^2 head^{-1} (ft^2 head^{-1})]	7.4 (80)	9.3 (100)	12.1 (130)	12.1 130	9.3 (100)	9.3 (100)
Application rate [(Lpm/m^2 (gpm/ft^2)]	8.2 (0.20)	6.6 (0.16)	6.6 (0.16)	6.6 (0.16)	5.2 to 4.4 (0.125 to 0.105)	6.6 (0.16) (water system)
End head pressure [kPa (psi)]	193 (28)	193 (28)	97 (14)	97 (14)	35 (5)	55 (8) (water system)
25% Drainage time (min)	2.5	2.1	0.5–0.8	1.0–1.3	0.5–0.7	No data recorded
50% Drainage time (min)	5.0	4.4	1.3–1.8	1.8–2.3	1.2–1.6	No data recorded
Expansion ratio	4.3 : 1	3.4 : 1	2.2 : 1	2.3 : 1	2.2 : 1	12 : 1
Extinguishment time (min : sec)	2 : 22	2 : 15	1 : 45	1 : 25	3 : 05	≈0 : 33

the extinguishment times with AFFF from foam-water sprinklers were comparable to results from protein foam tests. Rapid suppression with the turret nozzle [at 8.3 Lpm/m^2 (0.22 gpm/ft^2)] combined with an overhead water system was demonstrated in Test 7. No adverse effects were evident from the water discharged from the overhead sprinklers after the foam ran out.

The superior performance of the standard sprinklers was attributed to more effective plume penetration by higher density foam particles. The maximum centerline velocities measured were 23.2 m/s (76 ft/s), with 15.2 m/s (50 ft/s) at the centerline of the fire. The fire plumes tended to bend due to air currents within the test building. Since the terminal velocity of the foam agents was estimated to be on the order of 9.1 m/s (30 ft/s) maximum, the droplets near the centerline never reached the fire. This supports the theory that extinguishment occurs from the outside perimeter inward. Since foam droplets from standard sprinklers are about twice as dense as air-aspirated particles, the terminal velocities are greater. This allows greater penetration of the fire plume. The same mechanisms explain why air-aspirated AFFF provides similar performance to protein foam. When the AFFF is air-aspirated, there is no longer any advantage of increased droplet terminal velocity.

Additional work by FMRC established estimates for the terminal velocity of foam, as shown in Table 4-4.14.[36,37] Plume theory was used to roughly estimate that velocity on the order of 18.3 m/s (60 ft/s) could be expected in an 18.3-m (60-ft) high space with an 83.6-m^2 (900-ft^2) JP-4 fire. This estimate was in good agreement with the experimental results. Based on an average foam particle diameter of 6.3 mm (0.25 in.), a maximum terminal velocity of 7.3 m/s (24 ft/s) could be expected. For a JP-4 pool fire, this translates into a 0.7-m^2 (8-ft^2) maximum fire size before plume penetration would not be possible.

The practical significance of AFFF discharged through non-air-aspirating sprinklers was demonstrated by Breen *et al.*[37] Air-aspirating sprinklers require 207 kPa (30 psi) nozzle pressure to be effective. Standard sprinklers can discharge effective AFFF solution at pressures as low as 69 kPa (10 psi). This had important retrofit considerations where foam proportioning system losses could be made up through reduced sprinkler pressures.

Additional tests were conducted with closed-head sprinklers in an 18.3-m (60-ft) high hangar.[38] Potential cost benefits would have resulted from reduced hardware costs and unwanted discharges from deluge systems. These tests demonstrated that this concept was not feasible for the hangar scenario because of the large number of sprinklers that opened during the 83.6-m^2 (900-ft^2) fire tests.

The superior performance of standard sprinklers compared to air-aspirating sprinklers is reflected in the criteria of NFPA 409, *Standard on Aircraft Hangars*. If standard sprinklers are used with AFFF, the design application rate for overhead deluge systems may be reduced to 6.6 Lpm/m^2 (0.16 gpm/ft^2) from 8.2 Lpm/m^2 (0.20 gpm/ft^2) required for air-aspirated sprinklers. This represents a 20 percent reduction in foam required when standard sprinklers are used.

TABLE 4-4.14 *Estimated Particle Diameter vs Terminal Velocity*[37]

Particle Diameter [mm (in.)]	Terminal Velocity [m/s (ft/s)]			
		Foam		
	Water	Expansion Ratio 2 : 1	Expansion Ratio 5 : 1	Expansion Ratio 10 : 1
12.7 (0.5)	*	10.1 (33)	6.7 (22)	4.6 (15)
6.3 (0.25)	10.4 (34)	7.3 (24)	4.6 (15)	3.4 (11)
2.5 (0.1)	6.7 (22)	4.6 (15)	2.7 (9)	—

*The breakup of water drops greater than about 6.3-mm (0.25-in.) diameter is highly probable due to instability.

TABLE 4-4.15 Fire Test Data for Low-level Application of AFFF

Reference	Test No.	Test Area [m² (ft²)]	Fuel	Nozzle	Nozzle k Factor (gal/psi[0.5])	Maximum Spray Height* [m (ft)]	Spray Diameter* [m (ft)]	Nominal Application Rate [Lpm/m² (gpm/ft²)]	Control and Extinguishment Times
FMRC 1975[36]	3	83.6 (900)	JP-4	Turret Nozzle (monitor)	50.3	50-degree arc, 8-s cycle time, 15-degree angle of elevation, 25.9 m (85 ft) from the center of the test pool		4.1 (0.10)	90% in 10 to 15 s / 100% in 35 to 40 s
	4	83.6 (900)	JP-4	Turret Nozzle (monitor)	50.3	50-degree arc, 8-s cycle time, 15-degree angle of elevation, 25.9 m (85 ft) from the center of the test pool		4.1 (0.10)	90% in 1 min 30 s[†] / 100% in ≈ 2 min
	6	83.6 (900)	JP-4	Turret Nozzle (monitor)	50.3	50-degree arc, 8-s cycle time, 15-degree angle of elevation, 25.9 m (85 ft) from the center of the test pool		4.1 (0.10)	90% in 20 s / 100% in 25 s
FMRC 1973[35]	7	83.6 (900)	JP-4	Overhead OSS[‡] + Turret Nozzle	5.6 / 5.0	NA	NA	6.6 (0.16)[§] + 9.0 (0.22) / 9.5 (0.38)	Control in 17 s[§] / 100% in 33 s
Australia[39]	1	78.5 (846)	Aviation Kerosene	P10 Pop-up	4.1	0.8 (2.6)	4.3 (14.1)	5.5 (0.13)	95% in 30 s
	2	78.5 (846)	Aviation Kerosene	W-1 Pop-up	3.6	1.5 (4.9)	3.3 (10.8)	4.9 (0.12)	≈90% in 25 s[‖]
	3	78.5 (846)	Aviation Kerosene	P10	4.1	0.8 (2.6)	3.3 (10.8)	5.5 (0.13)	98% in 30 s
Naval Weapons Center Phase III 1972[40]	5	697 (7500)	JP-5	Type S Flush Deck	5.5	1.8 (6)	12.2 (40)	1.6 (0.04)	50% in 30 s / 90% in 60 s
	11	697 (7500)	JP-5	Type S Flush Deck	5.5	1.8 (6)	12.2 (40)	2.4 (0.06)	70% in 30 s / 95% in 60 s
	9	697 (7500)	Avgas	Type S Flush Deck + Deck Edge	5.5 [114 Lpm (30 gpm)]	1.8 (6)	12.2 (40)	2.4 (0.06) + 1.6 (0.04) / 4.1 (0.10)	15% in 30 s / 50% in 60 s
	15	697 (7500)	Avgas	Type S Flush Deck + Deck Edge	5.5 [114 Lpm (30 gpm)]	1.8 (6)	12.2 (40)	2.4 (0.06) + 1.6 (0.04) / 4.1 (0.10)	40% in 30 s / 70% in 60 s

TABLE 4-4.15 Fire Test Data for Low-level Application of AFFF (continued)

Reference	Test No.	Test Area [m² (ft²)]	Fuel	Nozzle	Nozzle k Factor (gal/psi^0.5)	Maximum Spray Height* [m (ft)]	Spray Diameter* [m (ft)]	Nominal Application Rate [Lpm/m² (gpm/ft²)]	Control and Extinguishment Times
Naval Weapons Center Pop-up 1984[41]	10 and 10R	372 (4000)	JP-5	Type SB Flush Deck	5.1	1.8 (6)	9.1 to 12.2 (30 to 40)	2.4 (0.06)	60 to 90 s for 90% control; 99% in 2 min
	5, 5R, and 5R1	372 (4000)	JP-5	Bete Pop-up	5.5	1.8 (6)	9.8 (32)	2.4 (0.06)	60 to 90 s for 90% control; 99% in ≈2 min
Naval Weapons Center Weapons Staging Area 1986[42]	I8	48.3 (520)	JP-5	Overhead Side-mounted Spray Nozzles	1.9	NA	NA	8.6 (0.21)	90% in 15 s 99% in 52 s 100% in 57 s
	I11	48.3 (520)	JP-5	Overhead Side-mounted Spray Nozzles	1.9	NA	NA	21.6 (0.53)	90% in 8 s 99% in 15 s 100% in 27 s
	I16#	66.9 (720)	JP-5	Low-Level Fan	4.7	NA	NA	11.8 (0.29)	90% in 24 s 99% in 52 s 100% in 79 s
	II12#	66.9 (720)	JP-5	Low-Level Fan	4.7	NA	NA	20.4 (0.50)	90% in 9 s 99% in 16 s

*Spray height and diameter at the pressure/flow used in the test.

†An unplanned 69 kPa (10 psi) pressure drop in FMRC Test 4 caused a 4.6-m (15-ft) reduction in nozzle range, resulting in 90 percent control and extinguishment times 3 to 4 times those observed in Tests 3 and 6.

‡No wing obstruction over fire test area.

§The overhead deluge system discharging ordinary water was accidentally activated 12 sec later than the turret nozzle (5 sec before control was attained). The contribution, if any, of the overhead deluge system toward complete extinguishment was judged to be quite small compared to the turret nozzle.

‖Wind-affected results.

#Deck pool fire area was obstructed with simulated weapons carts.

Low-level application of AFFF: With the increase in wingspan areas of large aircraft, it was recognized that significant damage could occur before extinguishment of the pool fire underneath the wing. Using overhead sprinklers only, FMRC demonstrated the times required for the foam to spread and extinguish fires. (See Table 4-4.13.) The concept of low-level application of foam, using monitors or turret nozzles, was developed to reduce extinguishment time where shielded fires may occur. This concept was later extended to include side-mounted nozzles and discharge outlets, and flush-mounted nozzles installed in a floor or deck.

These systems are effective because AFFF solution droplets do not have to penetrate the fire plume. They also typically deliver, at spot locations, high densities of foam. This allows the foam to gain a "bite" or toehold on the fire. Low-level AFFF systems have been used successfully for over two decades on U.S. Navy air-capable ships, protecting flight decks and special hazard areas.

Table 4-4.15 summarizes fire test data for low-level application of AFFF. As seen, control and extinguishment times are quite rapid. NFPA 409, *Standard on Aircraft Hangars*, criterion of 4.1 Lpm/m^2 (0.10 gpm/ft^2) for low-level applications is based on a fire control time of 30 sec and extinguishment in 60 sec. Data indicate that a JP-5 pool fire can be 90 percent controlled in 60 to 90 sec and 99 percent extinguished in 2 min when an application rate of 2.4 Lpm/m^2 (0.06 gpm/ft^2) is used. The system can be effective at rates as low as 1.6 Lpm/m^2 (0.04 gpm/ft^2). For low-flash-point fuels (e.g., avgas), control time increases. Control and extinguishment times can be reduced by increasing the application rates on JP-5 fuel fires. Based on these results, the U.S. Navy adopted an AFFF application rate of 2.4 Lpm/m^2 (0.06 gpm/ft^2) for protecting aircraft carrier flight decks.[43]

While they may help control a three-dimensional (spill) fire, low-level application systems cannot be assumed to totally suppress a running fuel fire. Running fuel fires at a spill rate of 189 Lpm (50 gpm) are typically used in U.S. Navy flight-deck suppression tests using the flush-deck system. The running fuel fire, shielded by simulated aircraft debris, requires aggressive handline attack for extinguishment.[44]

Obstructions, such as parked vehicles, may block low-level nozzles. Testing for a flight-deck weapons staging area showed that a side-mounted low-level system could be effective even when nozzles are obstructed.[42] In these tests, 5 of the 12 deck-edge nozzles were obstructed to simulate vehicle tires blocking edge-mounted nozzles. Even with 40 percent reduction, the fire was controlled and extinguished in less than 1 min (compared to 15 to 30 sec when unobstructed).

Cost of installation, maintainability, and reliability are factors when considering a low-level application system. Reliability issues with turrets/monitors have been identified by both FMRC and the U.S. Navy. The flush-deck system adopted by the U.S. Navy took considerable effort before a high degree of reliability and maintainability could be achieved. This open deluge nozzle, originally installed as a water washdown nozzle, incorporates a ball-check feature in the nozzle orifice to prevent debris from clogging the nozzle. Clean-out traps are installed in system piping for maintenance. Pop-up nozzles have been proposed as an alternative to flush-deck nozzles. These nozzles have their own reliability and maintainability issues. Unless there are very high costs associated with the loss of an aircraft, in-floor or flush-deck nozzles are generally cost-prohibitive for commercial aviation facilities. For high risk/cost applications, in-floor nozzles may be

justified. This is the case for advanced military aircraft; research is underway on an inverted deluge system that can not only suppress a pool fire, but also cool exterior combustible components of the airframe.

Side-mounted nozzles are the most reliable systems, consisting of an open pipe or spray nozzles. The spreading rate of foam from an aspirated open pipe system increases control and suppression time. Open spray nozzles can be very effective, but their reach is limited.

FOAM-WATER SPRINKLER SYSTEMS

This chapter has dealt with foam characteristics, foam concentrate, test standards, and manual application techniques. In particular, applications in the aviation industry were described. The text on aircraft hangar protection introduced the concept of fixed foam protection systems. In particular, much of the foam-water sprinkler system test data was originally developed for aircraft hangars. Herein, additional foam-water sprinkler system design criteria are described. Again, emphasis is placed on AFFF systems since they are more effective for extinguishment than protein or fluoroprotein systems.

Codes, Standards, and Regulations

Overhead foam-water sprinkler systems, as specified in the NFPA standards, are generally designed to serve dual purposes: (1) control and/or suppression of a fuel spill fire; and (2) when the foam runs out, cooling of materials with water. Since the systems are designed to provide protection for flammable/combustible liquid hazards and ordinary combustibles, the specified application rates reflect this dual-protection approach. Table 4-4.3 shows the fundamental application rates used by Underwriters Laboratories on hydrocarbon fuel fires to evaluate sprinklers. The fire must be extinguished within 5 min for AFFF discharged at 4.1 Lpm/m^2 (0.10 gpm/ft^2) for standard sprinklers and 6.6 Lpm/m^2 (0.16 gpm/ft^2) for agents discharged from foam-water sprinklers (air aspirating). However, since most deluge and closed-head sprinkler systems are installed in industrial occupancies, they must meet "highly protected risk (HPR)" insurance requirements. As a result, the NFPA standards for deluge (NFPA 16, *Standard on the Installation of Deluge Foam-Water Sprinkler and Foam-Water Spray Systems*) and closed-head (NFPA 16A, *Standard for the Installation of Closed-Head Foam-Water Sprinkler Systems*) AFFF systems require 6.6 Lpm/m^2 (0.16 gpm/ft^2) minimum water application. This also provides a safety factor over the 4.1 Lpm/m^2 (0.10 gpm/ft^2) rate at which AFFF discharged from sprinklers is effective on pool fires. This is reflected in Table 4-4.3 under the "minimum design application rate."

Table 4-4.16 summarizes current requirements from NFPA standards and guidelines. NFPA 11, *Standard for Low-Expansion Foam*, is geared toward petroleum and chemical industry protection. Previous requirements from NFPA 11 allowed 4.1 Lpm/m^2 (0.10 gpm/ft^2) for loading racks, e.g., tank truck loading facilities. The latest requirements for NFPA 11 eliminate this design criteria and reference NFPA 16 and NFPA 16A requirements, which require 6.6 Lpm/m^2 (0.16 gpm/ft^2). In special situations, 4.1 Lpm/m^2 (0.10 gpm/ft^2) is permitted by NFPA 11, but only where there is low-level or manual application for a hydrocarbon fuel spill scenario. NFPA 16 and NFPA 16A are consistent in requiring 6.6 Lpm/m^2 (0.16 gpm/ft^2); they reference

TABLE 4-4.16 *NFPA Standards Related to AFFF Sprinkler Systems*

Standard*	Minimum AFFF Application Rate [Lpm/m² (gpm/ft²)]	Duration (min)
NFPA 16, *Deluge Foam-Water Sprinklers*	6.6 (0.16)	10 min; 7 min if above minimum design
NFPA 16A, *Closed-Head Foam-Water Sprinklers*	6.6 over 139 m² (0.16 over 1500 ft²)	10 min; 7 min if above minimum design
NFPA 11, *Low-Expansion Foam*	Indoor Storage Tank Greater than 37 m² (400 ft²) 6.6 (0.16)	30
	Loading Rack Monitors 4.1 (0.10)	15
	Diked Areas Fixed Low Level (Class II hydrocarbon) 4.1 (0.10)	20
	Monitor 6.6 (0.16)	20
	Undiked Areas for AFFF Handlines 4.1 (0.10)	15
NFPA 409, *Aircraft Hangars*	Overhead Deluge 8.2 (0.20) for aspirated AFFF 6.6 (0.16) for non-air aspirated AFFF	10 min; 7 min if above minimum design
	Supplemental Low Level (for shielded wing areas) 4.1 (0.10)	10 min

*See Additional Reading for complete titles and dates.

other NFPA standards for special exceptions, e.g., NFPA 409, *Standard on Aircraft Hangars*, and NFPA 30, *Flammable and Combustible Liquids Code*. NFPA 409 requirements were previously discussed. Section 4/Chapter 5 provides an example for calculating foam quantities based on design application rates and areas to be protected.

Model building and fire codes in the U.S. are in the process of adopting AFFF protection criteria for the storage of flammable and combustible liquids. Criteria of insuring authorities [e.g., Industrial Risk Insurers (IRI) and Factory Mutual (FM)] are similar to the NFPA requirements. Insurance authority guidelines should be referenced for specific projects, since there are differences in protection criteria.

Protection of Stored Flammable and Combustible Liquids

Flammable and combustible liquids are stored in containers ranging in size from less than a quart to several hundred gallons. These liquids may be stored for display in a retail outlet or "super store," stored for distribution in a general-purpose warehouse housing many different combustibles, or stored in "liquid" warehouses containing large quantities of the liquid. NFPA 30, *Flammable and Combustible Liquids Code*, is the applicable NFPA protection document. This code includes requirements for tank storage, piping systems, containers, and operations. Criteria for suppression system protection is addressed in the sections dealing with container storage.

The protection of flammable and combustible liquids is a function of many factors, including the liquid properties, the ignition scenario (which can be a factor of the storage occupancy), the packaging system (e.g., stored in cardboard cartons), the container design and material (e.g., steel, plastic, glass, fiberboard), and the arrangement of storage (e.g., rack *vs* pallet, storage height, aisle width, mixture of other combustibles in the array). Based on these factors, a suppression system is provided to control or suppress the anticipated fire and protect the structure. The system may be designed to (1) control a fire so that the fire department can ultimately extinguish or suppress the burning material or (2) suppress the fire. Variables in suppression system design include sprinkler application rate, agent, orifice size, spacing, response time index (RTI), temperature rating, and provision of in-rack protection.

Fire test data to support the protection criteria in NFPA 30, *Flammable and Combustible Liquids Code*, requirements are not well documented. The appendices of NFPA 30 give example test data and anecdotal accounts of significant fires. Some of the data in NFPA 30 has been derived from protection standards for ordinary combustibles. For example, protection of Class I-III liquids in general-purpose warehouses is permitted when the automatic sprinkler protection is designed to protect Class IV ordinary commodities as described in NFPA 231, *Standard for General Storage*. A suppression system design of 12.2 Lpm over 186 m² (0.30 gpm over 2000 ft²) may be used to protect these liquids in a general-purpose warehouse where high-temperature-rated sprinklers are installed.

Until recently, protection of flammable and combustible liquids with water sprinklers was considered more than adequate, based on the low loss history and infrequent ignitions in these occupancies. Some facilities had no protection and were considered "sacrificial" if a fire occurred. Given adequate segregation from other hazards and acceptance of the loss, this approach was considered acceptable.

Tests conducted for storage occupancies used a "point" ignition source. This was predicated on the assumption that the most likely ignition sources in storage occupancies were small (e.g., spark from a forklift truck) and ignited ordinary combustibles (e.g., storage cartons) or small spills. For flammable and combustible liquids stored in metal containers, this was a reasonable scenario, particularly where cans were cartoned. The water suppression systems were designed to

TABLE 4-4.17 *Closed-Head Sprinkler Tests[46]*

Sprinkler Temperature Rating [°C (°F)]	Nominal Application Rate [Lpm/m² (gpm/ft²)]	Total Heads Opened	Sprinkler Operation and Control Times (min:sec)
71 (160)	4.5 (0.11)	34	First Sprinkler—0:27 Final Sprinkler—1:01 3:50 Control Time
71 (160)	7.4 (0.18)	32	First Sprinkler—0:22 Final Sprinkler—1:08 1:00 to 1:20 for Knockdown 2:20 Control Time
138 (280)	7.4 (0.18)	7	First Sprinkler—0:33 Final Sprinkler—0:53 1:50 Control Time
138 (280)	7.4 (0.18)	15	First Sprinkler—0:28 Final Sprinkler—1:44 2:20 Control Time
138 (280)	7.4 (0.18)	17 to 19	First Sprinkler—0:22 to 0:24 Final Sprinkler—1:03 to 1:13 2:00 Control Time
141 (286)	12.3 (0.30)	10	First Sprinkler—0:24 Final Sprinkler—1:19 2:25 Control Time

control the fire before the involvement of the flammable or combustible liquid in the metal container. The metal container afforded a degree of protection before sufficient heat resulted in rupture of the can and discharge of its combustible contents. The fire was an ordinary combustible fire, not a large liquid spill fire.

In the early 1970s, there was a recognition that larger and more catastrophic fires were occurring in warehouses storing combustible and flammable liquids. Concerns were raised among industry, insurers, codes and standards groups, and regulatory agencies about the capacity of water sprinkler systems to control these liquid warehouse fires. The reason for the increase in losses, and in particular total loss of structures, has not been specifically quantified but can be attributed to several factors: introduction of plastic containers for liquid storage; increase in storage height and quantities of liquids; and mixture of flammable liquids with other combustible products, such as aerosol products.

Related to these issues was the apparent increase in the size and growth rate of the igniting fires. Instead of single-point, moderate-growth ignitions, large-loss fires were occurring as a result of larger spill scenarios and rapidly growing fires.

Protection of drum storage: Some of the earliest work using AFFF sprinklers involved the protection of 208 L (55-gal) drums. In work conducted at Factory Mutual Research Corporation, sponsored by Allendale Insurance, Factory Insurance Association (FIA), and the 3M Company, the effectiveness of standard sprinklers supplied with AFFF for controlling drum fires was determined.[45] Five fire tests were conducted in simulated flammable liquid drum storage using two types of storage arrangements. Three tests were conducted with two-, three-, and four-high palletized drum storage, respectively. Two tests were conducted with five-tier high-rack storage of palletized drums.

In all tests, a heptane fuel supply simulated leakage from the upper level of storage. Except for one rack-storage test that used a 57-Lpm (15-gpm) spill rate, fuel spillage was 7.6 Lpm (2 gpm). Ceiling protection employed high-

temperature sprinklers at discharge rates of either 12.3 or 24.6 Lpm/m² (0.30 or 0.60 gpm/ft²). In-rack supplemental protection for the rack storage tests was provided at three levels with ordinary temperature sprinklers each discharging 113 Lpm (30 gpm). The success of each test was based on storage stability, i.e., no pile collapse, and limitation of drum pressure to 104 kPa (15 psig).

AFFF was effective in controlling spill fires on the floor. The exception was in areas not reached by the discharge from operating sprinklers, where the flow of foam was blocked by pallets. Protection was not effective on the three-dimensional spill fires. Fire exposure and resultant pressure development within drums was more severe with increased clearances between storage and sprinklers due to greater delays in sprinkler operation.

Generally, results were considered good in the rack-storage tests, where in-rack sprinklers were provided in each tier. For palletized storage, the AFFF protection controlled the floor fire although pallets hindered the spread of foam. Ceiling sprinklers alone did not adequately protect palletized storage where an elevated spill resulted in a three-dimensional fire within the pile.

The results of these tests are reflected in NFPA 30, *Flammable and Combustible Liquids Code*, criteria that recommend AFFF sprinkler protection of 12.3 Lpm/m² (0.30 gpm/ft²) at the ceiling plus in-rack protection at each tier of drum storage in racks up to 7.6 m (25 ft) high.

Liquid spill and container storage: Table 4-4.17 summarizes early closed-head AFFF sprinkler testing on a flammable liquid spill scenario.[46] In a 9.1-m (30-ft) high ceiling room, n-heptane was discharged in a simulated spill to create a three-dimensional spill and two-dimensional pool fire. Fuel spill rate was varied up to 113 Lpm (30 gpm). AFFF application rates were 4.5 to 12.3 Lpm/m² (0.11 to 0.30 gpm/ft²). The primary variables were the temperature rating of the sprinkler and the application rate. Non-air-aspirating sprinklers were used. The data show that high-temperature-rated sprinklers activated at about the same time as ordinary temperature sprinklers, controlled the fire in comparable times (roughly 2 min control time), and resulted in significantly fewer sprinklers operating (7 *vs* 32). An increase in application rate when the high-temperature sprinklers were used resulted in fewer heads operating, but did not decrease overall control and extinguishment time. Fires were controlled, but not totally extinguished as a result of the three-dimensional spill fire. These tests showed the advantage of using high-temperature-rated sprinklers in AFFF closed-head suppression systems.

In response to the concerns related to flammable liquid warehouse protection, the National Fire Protection Research Foundation (NFPRF) initiated the "International Foam-Water Sprinkler Research Project." The objectives were to document the performance of foam-water sprinkler systems designed for real-world storage and ignition scenarios and provide a design basis and minimum design parameters for foam-water sprinkler systems. Five tasks were performed, including a literature search, range-finding tests, and large-scale tests involving palletized and rack-storage of liquids.

The literature search identified over 1100 sources of information related to flammable liquid fires and foam protection, but a dearth of data related to water and foam-water sprinkler suppression of liquid storage fires.[47] The range-finding tests indicated that the Class IB flammable liquids (heptane) provided a greater challenge than water-miscible fuels (e.g., isopropanol).[48] Breach of steel containers exposed to a flammable liquid pool fire without sprinkler protection occurred over a range of times between 2 and 7.5 min, depending on the particular type of container. Plastic containers were quickly breached and discharged their contents to the exposing pool fire.

Large-scale tests were conducted under an 8.2-m (27-ft) high ceiling at the Underwriters Laboratories fire test facility in Northbrook, IL.[49] A series of 14 fire tests involving the protection of 3.8- and 18.9-L (1- and 5-gal) metal and 18.9-L (5-gal) plastic containers filled with heptane (Class IB flammable liquid) were conducted. The use of closed-head foam-water sprinkler systems for the protection of these fuel packages was investigated. Quantities of fuel used in the fire tests varied from 605 to 7260 L (160 to 1920 gal); fuel storage densities ranged from 160 to 1907 L/m² (3.9 to 46.5 gal/ft²); and storage heights ranged from 4.3 to 42.7 m (1.3 to 13 ft).

Each fire test was initiated using a 37.8-L (10-gal) flammable liquid (heptane) spill, recognizing the larger spill ignition scenarios observed in large-loss fires.

Fire tests involving palletized storage of 3.8-L (1-gal) metal F-style containers of heptane, packaged four containers in a corrugated cardboard carton, were conducted. The results indicated that the 37.8-L (10-gal) flammable liquid spill fire could be suppressed by a closed-head foam-water sprinkler system at a 16.4 Lpm/m² (0.40 gpm/ft²) design application rate for storage heights up to 3.3 m (10.7 ft) under the 8.2-m (27-ft) ceiling prior to any container breach or fuel loss. Fires involving 18.9-L (5-gal) metal containers of heptane could be suppressed by a closed-head foam-water sprinkler system application rate of 12.3 Lpm/m² (0.30 gpm/ft²) for a palletized storage height of up to 3.6 m (12 ft). Plastic pour spouts in the 18.9-L (5-gal) tight-head metal containers safety vented and prevented container breaching.

Fires involving 18.9-L (5-gal) plastic containers of heptane could not be suppressed by a preprimed, closed-head foam-water sprinkler system with an application rate of 12.3 Lpm/m² (0.30 gpm/ft²), where containers were stacked one-high [483 mm (19 in.)], due to container breaching and flammable liquid spillage prior to foam-water discharge.

Rack-storage tests also conducted in the NFPRF "International Foam-Water Sprinkler Research Project" did not lead to conclusive results.[50]

Based on the results of the NFPRF foam-water sprinkler testing, NFPA 30, *Flammable and Combustible Liquids Code*, adopted the protection criteria as appendix material, as shown in Table 4-4.18, when AFFF sprinklers are used. Where the hazard involves water-miscible fuels, an alcohol-type AFFF should be used.

AFFF protection of flammable and combustible liquids should be used where large spills of low flash point fuels is a realistic scenario. Other protection options are available and have recently been adopted or are currently being considered by NFPA 30 and the model building/fire prevention codes. Designers of warehouse protection should have a thorough knowledge of these criteria and the available test data (including water-only protection) when considering design options for the protection of stored combustible and flammable liquids.

TABLE 4-4.18 *AFFF Sprinkler Protection Requirements in NFPA 30 for Solid-Pile and Palletized Storage of Flammable and Combustible Liquids in Metal Containers of 18.9-L (5-gal) Capacity or Less*

Package Type	Cartoned	Uncartoned
Class Liquid	IB, IC, II	IB, IC, II
Application Rate [Lpm/m² (gpm/ft²)]	16.4 (0.40)	12.3 (0.30)
Area [m² (ft²)]	186 (2000)	186 (2000)
Temperature Rating [°C (°F)]	141 (286)	141 (286)
Maximum Spacing [m²/head (ft²/head)]	9.3 (100)	9.3 (100)
Orifice Size [mm (in.)]	13.3 (0.53)	12.5 or 13.3 (0.5 or 0.53)
Maximum Height [m (ft)]	3.4 (11)	3.7 (12)
Hose [Lpm (gpm)]	1891 (500)	1891 (500)
Water Supply Duration (min)	120	120
Foam Supply Duration (min)	15	15

FOAM ENVIRONMENTAL CONSIDERATIONS

There has been increasing concern on the consequences of the discharge of foam to the environment. This affects the users of foam, the manufacturers of foam agents, the fire safety authority having jurisdiction, and environmental authorities. Quantitative data and methods to evaluate environmental impact are not widely published or well developed. The issue is not a new or unique development, but has received increased notice as a result of increased attention to environmental impact of fire-fighting agents.

Factors related to the impact of fire-fighting foam on the environment include:

1. Discharge of foam solutions and fuel-contaminated foam solutions to waterways and the potential toxicity to aquatic life,
2. Effects on water treatment facilities,
3. Persistence and biodegradability of chemicals in foam concentrates and solutions, and
4. Combustion products of fuel/foam solutions.

Perspective on the Use of Foam Agents

In order to assess the impact of foam on the environment, the likely scenarios under which AFFF may be discharged should be considered. Based on these scenarios, the overall impact can be assessed and, where appropriate, potential mitigation strategies can then be developed. Likely scenarios include uncontrolled fire situations, potential hazardous situations, fire-fighting training evolutions, and fixed or mobile vehicle suppression system discharge testing.

Uncontrolled fire situations: There are many fire scenarios where foam may be used, including flammable liquid storage, process industry protection, aviation protection, and marine applications. In most situations, the elimination of foam as a suppression agent results in the potential for dramatically increased environmental impact. This results from the potential increase in hydrocarbon fuel effluent to the environment (due to smoke from uncontrolled burning and fuel/fire-fighting water effluent). Consider the example shown in Figure 4-4.18. A 929-m^2 (10,000-ft^2) section of a warehouse containing combustible and flammable liquids may be protected using traditional water sprinklers discharging at a rate of 12.3 Lpm/m^2 (0.30 gpm/ft^2). If these sprinklers fail to control a large spill fire, the fire may develop and spread past the design area of the sprinklers. The example assumes the fire is contained within the fire wall; this may not always be the case for high-challenge fires. If the fire department aggressively combats the fire, a rough estimate of fire-fighting water that may be used is 15 to 50 times the minimum anticipated agent required for suppression.[6,51,52] A rough estimate of the potential fuel-contaminated effluent (neglecting the actual quantities of hydrocarbon liquid) is shown in Figure 4-4.18. In the alternative situation, a properly specified foam-water sprinkler system designed for a high degree of reliability can control or suppress the fire. Using application rates and discharge times based on recent tests and building code requirements, the anticipated fuel/foam/water effluent for this scenario can be estimated. (See Figure 4-4.18.) The use of the foam-water system reduces the potential effluent by a factor of nearly 500 compared to the "unsuccessful" water sprinkler scenario where handlines are used. This neglects the impact of smoke discharged to the atmosphere during the uncontrolled burning in the water-only scenario.

In some cases, it may be possible to collect the effluent from an uncontrolled fire. In other situations, it may not be possible. Any foam solution that has been used in fire suppression is likely to be contaminated with fuel and diluted with water.

Potential hazardous situations: Potential hazardous situations may result from a fuel spill where there is a likely ignition source. In this situation, foam may be applied for ignition prevention. The potential impact of ignition and resulting uncontrolled fire must be assessed against the potential additional environmental impact by discharging foam for ignition prevention. The potential environmental effects from an uncontrolled fire scenario should be considered as described in the previous text. Another consideration is the assessment of any additional impact of foam when applied to a fuel spill. For example, would the resulting fuel with foam have any greater impact on the environment than the fuel alone? If so, how is this impact quantitatively determined?

Training evolutions: Fire-fighting training is usually conducted under conditions conducive to collection of fuel, water, and foam. A separation process might be used to recover fuel. Water/foam solution may then be treated or reused. Alternately, simulated hydrocarbon fuel spill scenarios might be used, with a simulated foam agent. Propane-fired burners are typically used. The disadvantage of these systems is the potential loss of realism of the simulated fire/agent interaction. This may potentially reduce training effectiveness. Quantitative comparisons have not been performed to assess these differences.

System discharge testing: Facilities protected by foam systems may have containment systems that can hold effluent. Requirements for these containment systems are becoming more widespread in model building and fire codes. An alternative to discharge testing with foam is the use of a simulant that can be measured using concentration determination methods. For example, salt solutions can be used as the "concentrate" to test AFFF systems, with the simulant concentration measured using the conductivity method. Simulators may be more difficult to use for protein-based systems, where viscosity factors influence proportioning system accuracy.

Methods of Assessment

Biodegradability: The primary component of AFFF solution is water. Examples of other components are non-fluorinated surfactants, glycol ethers, and fluorinated surfactants. Freeze-resistant concentrate may contain ethylene or propylene glycol. Alcohol-type foams may contain xanthan or similar gums. The fluorinated surfactants are particularly resistant to biodegradation. Further, the less-effective protein-based foams were largely assumed to be non-polluting because of their "natural" organic base. An early review of the available literature by Factory Mutual Research Corporation indicated that both types of agents, i.e., AFFF and protein-based, present inherent environmental issues and that effluents containing either should be processed in some form of sewage treatment facility or diluted prior to discharge into a stream.[36]

A conventional method used to determine the biodegradability of a material is comparison of the chemical oxygen demand (COD) of the material with its biological oxygen demand (BOD). This is particularly important for waste treatment facilities where the stability of the treatment process may be upset. The method typically used is the "Standard Methods for the Examination of Water and Wastewater."[53] BOD measures the amount of oxygen consumed by microorganisms in breaking down a hydrocarbon. COD measures the maximum amount of oxygen that could theoretically be consumed by microorganisms. Therefore, a BOD/COD ratio is representative of the ability of microorganisms to biodegrade the components in a foam. The higher the BOD/COD ratio, the more biodegradable the foam. Results reported for BOD/COD of AFFF range from 0.60 to 0.99. MIL-F-24385 requires a maximum COD of 500,000 mg/L and a minimum 20-day BOD/COD ratio of 0.65 for 6 percent concentrate. AFFF agents have been reported to have higher BOD and COD values than protein foams.[36] AFFF solutions are high-BOD materials compared to the normal influent to treatment plants. Large quantities can "shock load" wastewater treatment facilities.

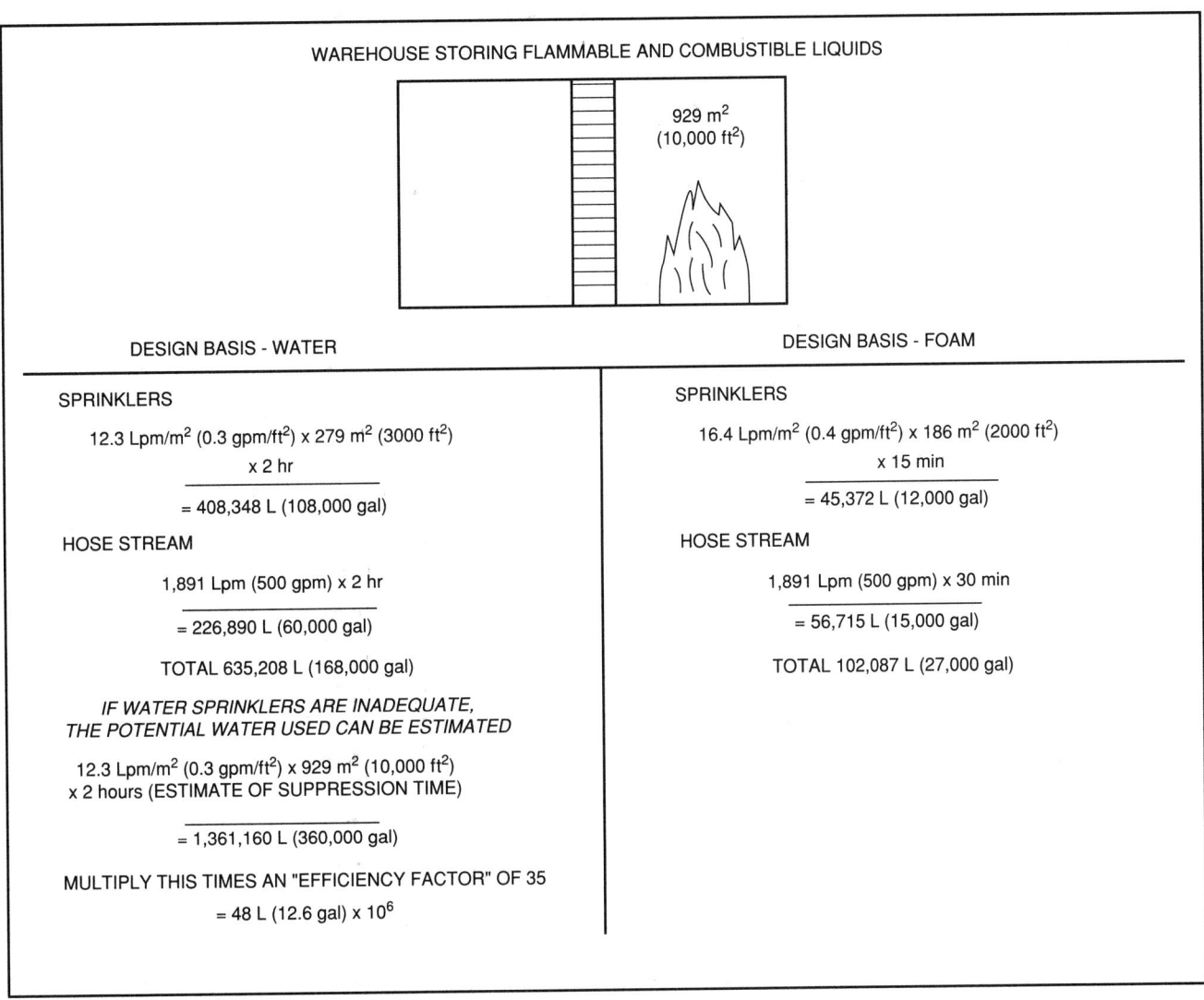

Fig. 4-4.18.   Example of Potential Effluent from Flammable Liquid Warehouse Fire.

The fluorochemical-based surfactants in AFFF have a carbon-fluorine chain that apparently does not break down in either the BOD or the COD test. The AFFF might then appear to be completely "biodegradable," even though the carbon-fluorine chain remains.

If non-biodegradability concerns are based on the persistence of the fluorochemical surfactants, then the environmental impact tests currently used to assess foams do not address this concern. There is speculation that the undegradable material is biologically inert, but no published data confirms this. Since the fluorinated surfactants are required to create surface tension reduction of the solution, replacement with less-persistent chemicals is problematic. There is a need for a more thorough understanding and testing related to the environmental impact of fluorosurfactants and possible alternatives.

Toxicity: In sufficient concentrations, foams may affect aquatic life. A number of fish toxicity studies have been performed. In tests using fathead minnows, the U.S. Air Force found that these fish could live in a simulated effluent stream containing 250 ppm (V/V) AFFF without fatality for up to eight days. LC_{50} values (i.e., the concentration causing deaths of 50 percent of the fish exposed) at 96 and 24 hours were 398 and 650 ppm, respectively.[54] MIL-F-24385 requires AFFF toxicity testing in accordance with ASTM E-729, using dynamic procedures with killifish. LC_{50} of 1000 mg/L for 6 percent concentrate is permitted.

Alone, these values may be considered as having a low degree of fish toxicity using environmental regulation rating scales. Localized concentrations in ponds or streams may exceed the values cited, if there is limited water movement.

Published data do not exist for the phytotoxicity of foam solutions; however, there have been no published reports of plant kills resulting from foam solution discharges.

Manufacturers report that thermal decomposition products from AFFF do not present a health hazard during firefighting. Again, there are no data published in the literature. Manufacturers' product environmental data for AFFF include references to a test where a layer of AFFF was burned in a pan of gasoline inside an enclosure. Two measurements of hydrogen fluoride recorded above the sample were 0.23 and 0.16 ppm.[55]

Foaming and emulsification of fuels: The surfactants in AFFF solutions can cause foaming in treatment aeration ponds. This foaming process may suspend high BOD solids in the foam. If these are carried over to the outfall of the treatment facility, nutrient loading in the outfall waterway may result. Foam aeration may also cause foam bubble backup in sewer lines.

In uncontrolled fires, spills, and live fire training scenarios, foams may contain suspended fuels. The fuel may become emulsified in the foam-water solution.

Mitigation Strategies

Foam discharges are more easily handled where there is an in-place collection capability. This may be the situation at warehouses, tank farms, and fire-fighting training facilities. Where these facilities are not available, temporary diking is an alternative where time and resources permit.

Discharge to water treatment facilities is recommended by many foam vendors when the solution is uncontaminated by fuel. Metering or dilution may be required to prevent levels of foam that will upset treatment facility reactions or cause excessive foaming. The use of defoamers to reduce aeration has been suggested.

Where fuels contaminate foam solutions, fuel/water separators might be used to skim-off the hydrocarbon fuel. AFFF solutions have a tendency to form emulsions with fuels, potentially reducing the effectiveness of fuel/water separators. An alternative is to hold the solution in a pond or tank until the emulsion breaks and the separation process can be used. Agitation should be avoided to prevent the emulsion from reforming. In some situations (e.g., training), the fuel and treated water have been reused. Many fire training facilities collect foam solution for ultimate discharge to water treatment facilities.

To ensure that unbalanced conditions do not occur in water treatment facilities, foam discharge should be carefully monitored. Different ranges of discharge rates have been suggested. This is an area requiring further investigation. Manufacturers of the foam solution should be consulted in conjunction with the wastewater treatment operator.

The entire area of environmental aspects of foam discharge requires additional evaluation and development of generally recognized guidance. Until generally recognized guidance is promulgated, users must rely on manufacturers' data and guidance. In all situations, discussions with the operator of the wastewater treatment facility and the environmental regulatory authorities are appropriate.

REFERENCES

1. Geyer, G.B., Neri, L.M., and Urban, C.H., "Comparative Evaluation of Fire Fighting Foam Agents," Report FAA-RD-79-61, Aug. 1979, Federal Aviation Administration, Washington, DC.
2. Aubert, J.H., Kraynik, A.M., and Rand, P.B., "Aqueous Foams," *Scientific American*, Vol. 19, No. 1, 1988, pp. 74–82.
3. Francen, V.L., "Fire Extinguishing Composition Comprising a Fluoroaliphatic and a Fluorine-Free Surfactant," U.S. Patent 3,562,156, 1971.
4. Rosen, M.J., *Surfactants and Interfacial Phenomena*, Chapters 1, 5, 7, John Wiley & Sons, New York, 1989.
5. Tuve, R.L., Peterson, H.B., Jablonski, E.J., and Neill, R.R., "A New Vapor-Securing Agent for Flammable Liquid Fire Extinguishment," NRL Report 6057, Mar. 1964, Washington, DC.
6. Friedman, R., "Theory of Fire Extinguishment," *Fire Protection Handbook*, 17th ed. National Fire Protection Association, Quincy, MA, 1991, pp. 1–74.
7. Hanauska, C.P., Scheffey, J.L., Roby, R.J., and Gottuk, D.T., "Improved Formulations of Firefighting Agents for Hydrocarbon Fuel Fires," SBIR Phase I Final Report for the U.S. Air Force, January 1994, Hughes Associates, Inc., Columbia, MD.
8. Persson, H., "Fire Extinguishing Foams—Resistance Against Heat Radiation," Brandforsk Project 609–903, SR Report 1992:54, 1992, Swedish National Testing and Research Institute, Sweden.
9. Kraynik, A.M., "Foam Drainage," Sandia Report SAND-83-0844, Nov. 1983, Sandia National Laboratories, Albuquerque, NM.
10. Kraynik, A.M., "Foam Flows," *Annual Review of Fluid Mechanisms*, 20, 1988, pp. 325–357.
11. Fay, J.A., "The Spread of Oil Slicks on a Calm Sea," *Oil on the Sea* (D.P. Hoult, ed.), Plenum, NY, 1964, pp. 53–64.
12. Fay, J.A., and Hoult, D.P., "Physical Processes in the Spread of Oil on a Water Surface," Coast Guard Final Report, Contract DOT-CG-01, 381-A, Project No. 714107/A/001, 1971, Department of Mechanical Engineering, Massachusetts Institute of Technology, Cambridge, MA.
13. Cann, P., Spikes, H.A., and Caporico, G., "Spreading of Perfluorinated Fluids on Metal Surfaces," *4th International Colloquium* on Synthetic Lubricants and Operation Fluids, Oatfildern, Germany, 1984, pp. 631–638.
14. Moran, H.E., Burnett, J.C., and Leonard, J.T., "Suppression of Fuel Evaporation by Aqueous Films of Fluorochemical Surfactant Solutions," Naval Research Laboratory Report 7247, Apr. 1971, Washington, DC.
15. Briggs, T., and Abdo, B., "Emphasis on Spreading Quality of AFFF Could Be Misleading," *FIRE*, Vol. 80, No. 94, June 1988.
16. Scheffey, J.L., Darwin, R.L., Leonard, J.T., Fulper, C.R., Ouellette, R.J., and Siegmann, C.W., "A Comparative Analysis of Film-Forming Fluoroprotein Foam (FFFP) and Aqueous Film-Forming Foam (AFFF) for Aircraft Rescue and Fire Fighting Services," Hughes Associates, Inc. Report 2108-A01-90 for the NFPA Aviation Committee, June 1990, Hughes Associates Inc., Columbia, MD.
17. Geyer, G.B., "Status Report on Current Foam Fire Fighting Agents," presented at the International Conference on Aviation Fire Protection, Interlaken, Switzerland, 22–24 September 1987.
18. Chiesa, P.J., and Alger, R.S., "Severe Laboratory Fire Test for Fire Fighting Foams," *Fire Technology*, Vol. 16, No. 1, Feb. 1980, pp. 12–21.
19. Underwriters Laboratories Inc., "Proposed Requirements for the Sixth Edition of the Standard for Foam Equipment and Liquid Concentrates," *Report* of Industry Advisory Conference, Northbrook, IL, Sept. 1992.
20. Scheffey, J.L., Wright, J., and Sarkos, C., "Analysis of Test Criteria for Specifying Foam Firefighting Agents for Aircraft Rescue and Firefighting," FAA Technical Report, DOT/FAA/CT-94/04, FAA Technical Center, Atlantic City, NJ, August 1994.
21. International Organization for Standardization, "Fire Extinguishing Media—Foam Concentrates—Part 1: Specification for Low-Expansion Foam Characteristics for Top Application to Water-Immiscible Liquids," ISO *Draft Standard*, ISO/DIS 7203-1, 1992.
22. Johnson, B.P., "A Comparison of Various Foams when Used Against Large-scale Petroleum Fires," Home Office Fire Research and Development Office, *Pub. 2/93*, 1993.
23. Carey, W.M., and Suchomel, M.R., "Testing of Fire Fighting Foam," Underwriters Laboratories Report No. CG-M-1-81, Nov. 1980, Underwriters Laboratories Inc., Northbrook, IL.
24. Thomas, M.D., "UK Home Office Research into Domestic Fire Fighting," First International Conference on Fire Suppression Research *Proceedings*, Stockholm and Boras, Sweden; Swedish Fire Research Board, Stockholm; and National Institute of Standards and Technology, Gaithersburg, MD, May 19, 1993, pp. 283–289.

25. Fiala, R., "Aircraft Post-crash Firefighting/Rescue," from AGARD Aircraft Fire Safety Lecture Series No. 123, June 1982.

26. Geyer, G.B., "Evaluation of Aircraft Ground Firefighting Agents and Techniques," Technical Report AGFSRS 71-1, Tri-Service System Program Office for Aircraft Ground Fire Suppression and Rescue, Wright-Patterson AFB, OH.

27. Geyer, G.B., "Firefighting Effectiveness of Aqueous Film-forming Foam (AFFF) Agents," FAA Technical Report FAA-NA-72-48, prepared for the DOD Aircraft Ground Fire Suppression and Rescue Unit (ASD-TR-73-13), Apr. 1973, Washington, DC.

28. Carey, W.M., "Improved Apparatus for Measuring Foam Quality," Dec. 1983, Underwriters Laboratories Inc., Northbrook, IL.

29. Jablonski, E.J., "Comparative Nozzle Study for Applying Aqueous Film-forming Foam on Large-scale Fires," U.S. Air Force Report, CEEDO-TR-78-22, Apr. 1978, U.S. Air Force, Tyndall AFB, Florida.

30. Foster, J.A., "Additions for Hosereel Systems: Trials of Foam on 40 m² Petrol Fires," Home Office Scientific Research and Development Branch Report 40/87, 1987, London.

31. Parsons, P.L., "Trials of Foam on Petrol Fires at the Fire Service Technical College," Home Office Scientific Advisory Branch Report No. 14/75, 1976, London.

32. DiMaio, L.R., Lange, R.F., and Cone, F.J., "Aspirating vs. Non-Aspirating Nozzles for Making Fire Fighting Foams—Evaluation of a Non-Aspirating Nozzle," *Fire Technology*, Vol. 20, No. 1, Feb. 1984, pp. 5–10.

33. Scheffey, J.L., "Submarine Bilge AFFF Sprinklers," Naval Research Laboratory Project, 64561N-S1946-2024000, 1988, unpublished data.

34. Timms, G., and Haggar, P., "Foam Concentration Measurement Techniques," *Fire Technology*, Vol. 26, No. 1, Feb. 1990, pp. 41–50.

35. Breen, D.E., "Hangar Fire Protection with Automatic AFFF Systems," *Fire Technology*, Vol. 9, No. 2, May 1973, pp. 119–131.

36. Krasner, L.M., Breen, D.E., and Fitzgerald, P.M., "Fire Protection of Large Air Force Hangars," [AFWL-TR-75-119]. Factory Mutual Research Corporation, Norwood, MA.

37. Breen, D.E., Krasner, L.M., Vincent, B.G., and Chicarello, P.J., "Evaluation of Aqueous Film Forming Foam for Fire Protection in Aircraft Hangars," FMRC Technical Report Ser. No. 21032, Sept. 1974, Factory Mutual Research Corporation, Norwood, MA.

38. Krasner, L.M., "Closed-Head AFFF Sprinkler Systems for Aircraft Hangars," FMRC S.I. 0C6N3.RG, RC 79-T-58, Dec. 1979, Factory Mutual Research Corporation, Norwood, MA.

39. "Pop-up-Type Floor-Mounted Foam Sprinklers for Aircraft Hangars," Technical Record TR44/153/14(L), Department of Housing and Construction, Commonwealth of Australia.

40. Unpublished data, China Lake CVA Fire Fighting Tests, Phase III, February 1972.

41. Scheffey, J.L., "Flow, Pattern, and Fire Performance Characteristics of a Prototype Pop-up Nozzle for Use on Aircraft Carrier Flight Decks," Report 2429-17 of 20, May 1985, Hughes Associates, Inc., Columbia, MD.

42. Scheffey, J.L., and Leonard, J.T., "AFFF Protection for Weapons Staging Areas," *Fire Safety Journal*, Vol. 14, No. 14 (1 & 2), July 1988, pp. 47–63.

43. Department of the Navy, Section 555, "General Specifications for Ships of the United States Navy, 1991 Edition," NAVSEA S9AA0-AA-SPN-010/GEN-SPEC, Jan. 2, 1991, Department of the Navy, Washington, DC.

44. Carhart, H.W., Leonard, J.T., Darwin, R.L., Burns, R.E., Hughes, J.T., and Jablonski, E.J., "Aircraft Carrier Flight Deck Fire Fighting Tactics and Equipment Evaluation Tests," NRL Memorandum Report 5952, Feb. 26, 1987, Washington, DC.

45. Newman, R.M., Fitzgerald, P.M., and Young, J.R., "Fire Protection of Drum Storage Using 'Light Water' Brand AFFF in a Closed-Head Sprinkler System," FMRC Technical Report Ser. No. 22464, RC 75-T-16, Mar. 1975, Factory Mutual Research Corporation, Norwood, MA.

46. Young, J.R., and Fitzgerald, P.M., "The Feasibility of Using 'Light Water' Brand AFFF in a Closed-Head Sprinkler System for Protection Against Flammable Liquid Spill Fires," FMRC Technical Report RC 75-T-4, Ser. No. 22352, Jan. 1975, Factory Mutual Research Corporation, Norwood, MA.

47. Schirmer Engineering Corporation, "International Foam-Water Sprinkler Research Project: Task 1—Literature Search," Technical Report No. 10-90001-04-00, Feb. 1992, prepared for the National Fire Protection Research Foundation, Deerfield, IL.

48. Hill, J.P., "International Foam-Water Sprinkler Research Project: Task 3—Range Finding Tests," Technical Report OTOR6.RR, July 1991, prepared for the National Fire Protection Research Foundation, Quincy, MA.

49. Underwriters Laboratories Inc., "International Foam-Water Sprinkler Research Project: Task 4— Palletized Storage Fire Tests 1 Through 13," Technical Report 91NK14873/NC987, Feb. 1992, prepared for the National Fire Protection Association, Quincy, MA.

50. Carey, W.M., "International Foam-Water Sprinkler Research Project: Task 5—Rack Storage Fire Tests," National Fire Protection Research Foundation, Technical Report, Oct. 1992, National Fire Protection Research Foundation, Quincy, MA.

51. Rasbash, D.J., "The Extinction of Fire with Plain Water: A Review," *Proceedings* from the First International Symposium on Fire Safety Science, Grant, C.E., and Pagni, P.J., eds., National Institute of Standards and Technology, Gaithersburg, MD, 1986, pp. 1145–1163.

52. Scheffey, J.L., and Williams, F.W., "The Extinguishment of Fires Using Low-Flow Water Hose Streams—Part II," *Fire Technology*, Vol. 27, No. 4, Nov. 1991, pp. 291–320.

53. American Public Health Association, "Standard Methods for the Examination of Water and Wastewater," 18th ed., Washington, DC, 1992.

54. National Environmental Health Laboratory, "Biological Treatment of Fire Fighting Foam Waste," Report No. REHL(K) 67–14, Sept. 1967, U.S.A.F., Kelly AFB, Texas.

55. "Light Water Brand AFFF Waste Disposal Recommendations and Hazard Evaluation," 3M Product Environmental Data Sheet, Feb. 19, 1991, 3M Company, St. Paul, MN.

ADDITIONAL READING

ASTM D-1331, *Standard Test Methods for Surface and Interfacial Tension of Solutions of Surface-active Agents*, American Society for Testing and Materials, Philadelphia, PA, 1989.

ASTM E-729, "Standard Practice for Conducting Acute Toxicity Tests with Fish, Macroinvertebrates, and Amphibians," American Society for Testing and Materials, Philadelphia, PA.

FAA, "Aircraft Fire and Rescue Facilities and Extinguishing Agents," FAA Advisory Circular 150/5210-6C, Jan. 28, 1985, Federal Aviation Administration, Washington, DC.

ICAO, *Airport Services Guide*, Part 1—Rescue and Firefighting, 3rd ed., International Civil Aviation Organization, Montréal, Québec, 1990.

NFPA 11, *Standard for Low-Expansion Foam*, National Fire Protection Association, Quincy, MA, 1994.

NFPA 16, *Standard on the Installation of Deluge Foam-Water Sprinkler and Foam-Water Spray Systems*, National Fire Protection Association, Quincy, MA, 1991.

NFPA 16A, *Standard for the Installation of Closed-Head Foam-Water Sprinkler Systems*, National Fire Protection Association, Quincy, MA, 1994.

NFPA 30, *Flammable and Combustible Liquids Code*, National Fire Protection Association, Quincy, MA, 1993.

NFPA 231, *Standard for General Storage*, National Fire Protection Association, Quincy, MA, 1990.

NFPA 403, *Standard for Aircraft Rescue and Fire Fighting Services at Airports*, National Fire Protection Association, Quincy, MA, 1993.

NFPA 409, *Standard on Aircraft Hangars*, National Fire Protection Association, Quincy, MA, 1990.

NFPA 412, *Standard for Evaluating Aircraft Rescue and Fire Fighting Foam Equipment*, National Fire Protection Association, Quincy, MA, 1993.

UL 162, *Standard for Foam Equipment and Liquid Concentrates*, Underwriters Laboratories Inc., Northbrook, IL, July, 1993.

UL Fire Protection Equipment Directory, Underwriters Laboratories Inc., Northbrook, IL.

US Military Specification, MIL-F-24385, Department of the Navy, Washington, DC, January 7, 1992.

US Government, O-F-555B (U.S. Government protein foam specification), General Services Administration, "Federal Specification, Foam Liquid, Fire Extinguishing, Mechanical," O-F-555B, Washington, DC, March 11, 1964.

FOAM SYSTEM CALCULATIONS

Harry E. Hickey

INTRODUCTION

Foam agent fire protection is especially suited for the control and extinguishment of flammable and combustible liquid-type fire protection problems.

An impressive array of fire protection problems can be properly addressed using foam classification agents. A detailed examination of the 1984 NFPA *Fire Codes*® indicates that 21 separate standards discuss foam fire protection systems as suitable for the protection of numerous fire protection problems.[1] It is important to note that other classifications of fire extinguishing systems, including dry chemicals, wet chemicals, carbon dioxide, halon, and some special agents, may be suitable for similar hazards.

Table 4-5.1 identifies special hazards that may be suitable for adequate fire protection by different types of foam systems. Each special hazard is cross-referenced to one or more classifications of foam fire protection systems that are identified in NFPA standards as being suitable for the stated hazard. This table is useful for examining the scope and limitations of different types of foam application systems for special hazard fire protection. The referenced standards should be consulted concerning specification and design considerations for each specific problem condition.

Foam agent fire protection systems are suitable for Class A fires in ordinary combustible materials in addition to Class B fires (flammable and combustible liquids). Historically, portable foam fire extinguishers provided important fire protection for both Class A and Class B problems. The dual consideration of evaluating foam fire protection systems for both Class A and Class B fire protection problems is important. This consideration is especially important for mixed occupancy storage, which may be suitably protected by foam spray systems, foam water sprinkler or spray systems, or closed head sprinkler systems using aqueous film forming foam (AFFF) type foam agents.

Dr. Harry E. Hickey is retired from the University of Maryland where he served as Associate Professor, Department of Fire Protection Engineering. He now serves as Chief Fire Protection Engineer and Navy Fire Chief at the Johns Hopkins University, Applied Physics Laboratory.

Objective Classification of Fire Problems for Foam Agent Fire Protection

The following objectives identify five performance areas for evaluating foam agent fire protection. Performance objectives may be combined for specific problem situations. These objectives form important considerations for the hydraulic design of foam agent systems.

Objective 1: Secure the surface of a flammable or combustible liquid that is not burning.

Flammable and combustible liquids emit vapors that may create a hazard condition. Flammable vapors may be suppressed by providing an adequate foam blanket over the surface area. Flammable liquid spills present a fire hazard. This hazard may be mitigated and/or controlled by the use of appropriate foam agent application.

Objective 2: Control and extinguish fires in flammable and combustible liquid hazardous locations in local areas within buildings.

Flammable and combustible liquids are often stored inside buildings in 55 gallon drums and other types of containers. These liquids are also used in association with manufacturing processes, industrial machinery, heating equipment, experimental activities, etc. The use of flammable and combustible liquids inside buildings may result in liquid spill and fuel ignition. Foam agents are appropriate for protecting localized flammable and combustible liquid problems inside buildings.

Objective 3: Extinguish fires in atmospheric storage tanks.

For nearly 100 years, foam agent fire protection has successfully extinguished fires in outdoor vertical atmospheric storage tanks. This type of protection still represents one of the most successful uses of foam agents; in fact, foam agent fire protection is the primary means for the proper protection of atmospheric storage tanks. Foam agents have been successfully used to extinguish flammable and combustible liquid fires in atmospheric storage tanks with diameters up to 200 feet.

Objective 4: Extinguish fires in outdoor and indoor processing areas.

A large variety of industrial processes utilize flammable and combustible liquids. In most processing plants, these liquids pass through pipelines and are captured in

TABLE 4-5.1 *Special Hazard Identification[1]*

	Protection Reference					
	NFPA 11			NFPA 11A High Expansion Foam	NFPA 16 Foam-Water Sprinkler Spray	NFPA 16A Closed Head Sprinkler
	Monitors	Foam Spray	Other Systems			
Aircraft Protection						
1. Aircraft Hangars (See NFPA 409)	X	X	X	X	X	X
2. Rooftop Heliport Construction and Protection (See NFPA 418)		X	X		X	
3. Aircraft Engine Test Facilities (See NFPA 423)		X		X	X	
Enclosed Stockpiles						
1. Flammable Liquids—Flash Point Below 100°C (See NFPA 30)				X*		
2. Combustible Liquids—Flash Point of 100°C and Above (See NFPA 30)				X*		
3. Low-Density Combustibles (Foam Rubber, Foam Plastics, Rolled Tissue or Crepe Paper)				X*		
4. High-Density Combustibles—Rolled Paper (See NFPA 231F)				X*		
5. Combustibles in Containers (Cartons, Bags, Fiber Drums)				X*		
Nuclear Power Plants						
1. Fire Protection for Light Water Nuclear Power Plants (See NFPA 803)		X		X	X	
2. Nuclear Research Reactors (See NFPA 802)		X		X	X	
Protection of Commodity Storage						
1. Indoor General Storage (See NFPA 231)				X		
2. Rack Storage of Materials (See NFPA 231C)				X		
3. Storage of Rubber Tires (See NFPA 231D)				X		
4. Storage of Rolled Paper (See NFPA 231F)				X		
5. Archives and Record Center Storage (See NFPA 232AM)				X		
6. Rack Container Storage of Liquids (See NFPA 30)						X
Special Problems						
1. Production, Storage, and Handling of Liquefied Natural Gas (LNG) (See NFPA 59A)		X		X	X	
2. Fire and Dust Explosions in Chemical, Dye, Pharmaceutical, and Plastics Industry (See NFPA 654)		X		X	X	
3. Fires and Explosions in Wood Processing and Wood Working Facilities (See NFPA 664)		X		X		
4. Ovens and Furnaces: Design, Location, and Equipment (See NFPA 86)		X				
5. Mobile Surface Mining Equipment (See NFPA 121)		X		X		
6. Tank Vehicle and Tank Car Loading and Unloading (See NFPA 30)	X	X	X		X	X
7. Automotive Service Station Filling Areas (See NFPA 30)		X				
8. Dipping and Coating Processes Using Flammable or Combustible Liquids (See NFPA 34)		X				
9. Manufacturer of Organic Coatings (Protection of Equipment Mixers, Solvent Tanks, and Open Containers— See NFPA 35)		X		X*		
10. Laboratories Using Chemicals (See NFPA 45)		X			X	X
11. Storage and Handling of Liquefied Petroleum Gases at Utility Gas Plants (See NFPA 59)		X		X	X	
Special Problems Identified in Manufacturers' Literature and/or Identified Foam Standards and Recommended Practices						
1. Process Structures and Equipment		X				
2. Horizontal Atmospheric Tanks		X				
3. Pump Rooms		X				
4. Dip Tanks		X				
5. Engine Test Cells		X				
6. Transformer Rooms				X		
7. Dike Areas	X		X			

* Consider in conjunction with automatic sprinklers.

holding tanks. Foam agents, properly selected for the specific hazard, are suitable for controlling and extinguishing fires in process equipment. However, foam agents are not suitable for coping with vertical running fires or fires where the flammable material is flowing from an orifice under pressure.

Objective 5: Protect, prevent, control, and extinguish fire problems in selected special hazards.

In addition to the other specified objectives, foam fire extinguishing agents provide appropriate fire protection for numerous special hazard problems. Some important special hazard conditions that are suitable for foam fire protection are included in the following list. (See also Table 4-5.1.)

Dike areas
Engine test cells
Transformers
Engine rooms
Laboratories using chemicals
Aircraft hangars
Nuclear research reactors
High density storage of combustibles
Rubber tires
Rack container storage of aerosols
Loading racks
Automotive service station filling areas
Ovens and furnaces

BASIC TYPES OF FOAM SYSTEM PROTECTION

Foam fire protection systems are divided into four basic classifications by the National Fire Protection Association. Each of these classifications is briefly identified below. Conditions exist where it may be proper to use more than one classification of protection on a given fire problem. Examples of fixed foam systems are covered under supplemental topics later in this chapter on the hydraulic design of low expansion and high expansion foam systems.

Fixed Foam Systems

These systems are complete installations piped from a central foam station, discharging through fixed delivery outlets to the hazard to be protected. Any required pumping equipment is permanently installed. For example, a fixed system for a vertical atmospheric cone roof storage tank would include the following permanently installed equipment: water supply lines, foam proportioning equipment, a foam liquid storage tank, foam solution lines to the storage tank, all necessary control valves, a tank solution riser pipe, and one or more topside foam chambers. Other equipment may be added based on the complexity of the problem and associated hydraulic conditions.

Semi-fixed Foam Systems

Two separate classifications of semifixed foam systems are identified below. The first classification is more predominantly used in the United States, although one major oil company does use applications of the second classification. Common to both classifications is the concept that part of the total system is permanently installed and part of the system is provided by portable elements.

The first classification of semifixed systems indicates a type in which the fire hazard is equipped with fixed dis-

charge outlets connected to piping that terminates at a safe distance. The fixed piping installation may or may not include a foam maker. Necessary foam producing materials are transported to the scene after the fire starts and are connected to the piping.

The second classification of semifixed systems indicates a type in which foam solutions are piped through the area from a central foam station, the solution being delivered through hose lines to portable foam makers such as monitors, foam towers, hose lines, etc.

Mobile Systems

Mobile systems basically consist of a unit on wheels that transports all of the required equipment and foam liquid necessary for making finished foam. This concept includes any foam producing unit which is mounted on wheels, and which may be self-propelled, or towed by a vehicle. These units may be connected to an available water supply or may use a premixed foam solution. The original concept of a mobile foam system was called a "foam house on wheels," a mobile piece of fire apparatus with a UL rated fire pump, an integral part of the pumping network, a foam liquid tank, and fire hose. Essentially this unit can double as a structural fire suppression unit. NFPA 11C, *Standard for Mobile Foam Apparatus*, covers the specifications and performance criteria for mobile systems.[1]

Portable Systems

Portable systems represent a rather economical approach to providing basic foam fire protection for small hazards. This classification considers that the foam producing equipment and materials, including the foam liquid, the proportion device(s), the discharge nozzle, the hose, and other required appliances, are transported by hand from a storage location to the incident scene. While portable systems are simple to operate, they are limited by their foam discharge rate capability; they may also be labor intensive to maintain a continuous foam supply over the required duration of discharge. Foam equipment manufacturers can provide technical information on a range of portable equipment.

PROTECTION OF INCIPIENT SPILLS AND RELATED HAZARDS

Portable fire extinguishers provide one method of protection for small flammable liquid storage hazards, fuel transfer hazards, and incipient spill fires.

As of 1986, *Underwriters Laboratories Inc.* (UL) lists two classifications of portable foam-type fire extinguishers. One classification is concerned with chemical foam extinguishers. These extinguishers are currently available in the following sizes: 1½ gal, 2½ gal for hand portables, and wheeled units from 17 to 40 gal. Correspondingly, the UL rating ranges from 10A:12B to 10A:40B. However, these extinguishers are now generally considered obsolete since their manufacture in the United States was discontinued in 1969. The National Fire Protection Association recommends that these units be replaced with currently available models.

Aqueous film forming foam agent portable fire extinguishers are used to replace the chemical-type foam extinguisher and provide protection for hazard conditions where this type of extinguishing agent is recommended. Extinguishers of this type are usually available in hand portable models of 2½ gallon capacity and in wheeled models having a liquid capacity of 33 gallons. These extinguishers have ratings of 3A:20B and 20A:160B, respectively. The AFFF portable model closely resembles the stored pressure water extinguisher except for the special type of nozzle. NFPA 10, *Standard for Portable Fire Extinguishers*, should be consulted concerning the selection and placement of portable extinguishers.[1]

PROTECTION FOR FIXED ROOF ATMOSPHERIC STORAGE TANKS

Fixed or cone roof atmospheric storage tanks for the storage of flammable liquids can be protected by fire fighting foam. Several techniques are available for correctly applying foam to a cone roof tank fire. Each technique should be carefully considered with reference to the size of the storage tank, the flammable or combustible liquid being stored in a given tank, and the foam agent classification that is suitable for the hazard. Some fundamental concepts associated with the proper protection for fixed roof atmospheric storage tanks are discussed below. Each individual topic is further developed through design problems on cone roof atmospheric storage tanks.

Foam Monitors

One or more foam monitors may be positioned around the periphery of a cone roof tank to project foam over the tank shell and onto the surface of a burning liquid. This technique has been successfully used on numerous fires. Foam monitors for foam protection of tanks are specified to have a discharge in excess of 300 gallons per minute. However, NFPA 11, *Standard for Low-Expansion Foam* (hereinafter referred to as NFPA 11), clearly indicates that foam monitors may be considered the primary means of protection for fixed roof tanks when the tank is less than 60 feet in diameter. This indicates a severe limitation on the recommended use of foam monitors for the protection of cone roof tanks.

Foam Handlines

Similar to the concept of providing foam protection with foam monitors, foam handlines may be positioned around the periphery of a cone roof tank to project foam over the tank shell and onto the surface of a burning liquid. Foam handlines have a flow range from 50 gpm to less than 300 gpm and are only suitable for possible protection of fixed roof tanks with a diameter of less than 30 feet and a height not greater than 20 feet. The selection of foam handline nozzles must be carefully considered to provide the correct total discharge for the flammable liquid problem to be protected.

Foam handlines are very important for supplemental fire protection requirements. Handlines delivering a minimum of 50 gallons per minute are very useful for extinguishing small spill fires and dike fires in the vicinity of a storage tank. The number of such supplemental foam hose streams is dependent on the diameter of the largest storage tank in the compound to be protected. (NFPA 11 should be consulted.)

Surface Application of Foam

One common and acceptable method of applying foam to the flammable liquid surface of a single roof storage tank is through fixed discharge outlets installed on the tank shell.

Two distinct types of foam discharge outlets are available based on the hazard problem and the foam agent. Each type of device may be distinguished as follows:

Type I outlets: These approved discharge outlets will conduct and deliver foam gently onto the liquid surface without submergence of the foam below the flammable liquid surface and agitation of the surface. This type of device was originally intended to apply special alcohol resistant foams to polar solvent fuels. Today, Type I discharge outlets may be used with hydrocarbon fuels. (NFPA 11 should be consulted.)

Two classifications of Type I outlets are commercially available where this device is suitable. A porous tube is a Type I foam discharge outlet. The tube is coarsely woven and rolled up into a foam chamber so that there is an attached end at the foam maker and a free end. When foam is admitted to the tube at the foam chamber, the tube unrolls, dropping into the tank. The buoyancy of the foam causes the tube to rise to the surface, and foam flows out through the pores of the fabric directly onto the liquid surface.

A foam trough represents a second variety of a Type I discharge outlet. The trough consists of sections of steel sheet formed into a chute which is securely attached to the inside of the tank wall so that it forms a descending spiral from the top of the tank to within 4 feet of the bottom.

Based upon advances in foam fire protection, porous tubes and foam chutes are rarely installed on new installations for the proper protection of cone roof atmospheric storage tanks.

Type II outlets: These approved discharge outlets do not deliver foam gently onto the liquid surface without submergence of the foam or agitation of the surface. An air foam chamber with a Type II outlet may be attached to the tank shell at the weak seam line. The Type II discharge outlet is positioned on the inside of the tank to permit discharge of the foam down the inside of the tank wall surface onto the flammable liquid surface. The number and discharge capacity of Type II foam chambers for a given size (diameter) cone roof storage tank are presented in NFPA 11.

Portable foam towers: These devices represent specialized portable equipment that may be fitted with either a Type I or a Type II foam discharge outlet. A tower is a device that is brought to the scene of the fire, erected, and placed in operation for delivering foam to the burning surface of a tank after the fire starts. The Type II discharge outlets are shaped to apply foam inward toward the tank shell. The erection of foam towers adjacent to a burning tank and the operation of the foam towers may present a safety hazard to the personnel working with this equipment. The number of persons required to place foam towers in service is also a problem associated with these devices.

Subsurface Application of Foam

An alternative method of applying foam to a cone roof atmospheric storage tank is through subsurface injection, usually near the base of the tank but above the water bottom in the tank. This application technique involves injecting expanded foam into the flammable liquid near the bottom of the liquid level under controlled velocity conditions. The buoyancy of the foam allows the foam to slowly rise to the flammable liquid surface and spread across the surface to effect fire control and then total extinguishment.

There are three important conditions to consider in subsurface application of foam in fixed and semifixed systems.

1. Subsurface foam application is not considered suitable for the protection of Class IA hydrocarbon liquids.
2. Subsurface foam application is not currently suitable for polar solvent.
3. Subsurface and semisubsurface injection systems are not recommended for *open top* or *covered floating* roof tanks.

Semi-subsurface Injection Method

A modified form of subsurface foam injection for cone roof tanks is used in a number of European countries. The modified technique is designated the semi-subsurface injection method, based on the equipment used to insert the expanded foam into the tank shell. The semisurface injection method has not found any particular application in the United States.

PROTECTION OF FLOATING ROOF STORAGE TANKS

In contrast to single roof tanks, floating roof tanks have a cover or roof over the flammable liquid that floats on the surface of the liquid and moves vertically with the liquid level in the tank. The floating roof may be open to the atmosphere. This physical arrangement of the tank is classified as an "open top floating roof tank." A permanently installed cover may be placed over the entire tank; this second designation is classified as a "covered floating roof tank."

The floating roof has a perimeter seal between the roof cover perimeter and the tank shell. The seal is necessary to prevent flammable vapors from escaping into the atmosphere and collecting over the floating-roof. Three types of seal devices may be found on floating roof-type storage tanks: (1) a mechanical shoe seal or a pantograph-type seal; (2) a metal weather shield; and (3) a metal secondary seal.

Appendix A of NFPA 11 should be consulted for a description of each seal device and the physical arrangement of these devices. Some devices also require the use of a foam dam when protected by fixed foam fire protection systems. The requirements for foam dams are also given in the referenced standard.

The fire experience with floating roof tanks appears to be very good. Consequently, fixed foam outlets are not generally required on either open top floating roof tanks or covered floating roof tanks. When an oil company elects to protect these types of tanks or the local fire protection authority requests protection for these types of tanks, three different application techniques may be used for proper protection of open top floating roof tanks. A brief description of each technique follows.

Portable Nozzle Method

The basic fire problem associated with floating roof tanks is a fire burning in the seal area between the cover and the tank shell. Typically, the surface area of this fire is quite small. One technique to extinguish this type of fire is to advance a portable hose line to the top of the tank, supply foam to this hose line, and manually apply foam to the seal area. Personnel operating this hose line should be highly trained in this type of operation and follow established safety practices.

Catenary System Method

The catenary system consists of a series of foam makers at evenly spaced points in the roof near the seal. These foam makers are connected to a common section of piping which in turn is attached to a flexible hose that rides up and down with the access stairway to the roof cover. The stairway is fixed to the top of the tank shell, and the bottom portion of the stairway rides on a set of tracks attached to the floating cover. This arrangement allows the stairway to move both horizontally and vertically as the cover moves with the flammable liquid level.

At the time of a fire, foam solution is pumped under pressure through a vertical pipe and flexible hose to the foam makers. This system can be designed to discharge foam under the seal directly onto the flammable liquid, or foam can be discharged above the seal. Foam equipment manufacturers producing this type of equipment should be consulted for engineering data on design requirements, installation techniques, and hydraulic calculations.

Fixed Foam Maker Method

The fixed foam maker method consists of installing piping around the outside wall of the tank and connecting to it a series of foam makers installed on special mounting shields above the storage tank rim. The circumference of the tank will determine the number of points needed for foam application. This method requires a foam dam to retain the foam over the seal or weather shield. This dam is normally 12 to 24 inches in height. Complete construction details of the foam dam may be found in the appendix of NFPA 11.

Covered floating roof tanks generally do not require fixed foam fire protection systems. There may be some cases of substandard installations or locations where local codes require proper protection for this classification of storage tank. The standards for fixed roof tanks should apply where it is required to protect covered floating roof-type tanks.

PROTECTION OF STORAGE OR HIGH VOLUME HAZARDS WITH HIGH EXPANSION FOAM

High expansion foam is an agent for the control and extinguishment of both Class A and Class B fires. The classification of foam makes it particularly suitable as a flooding agent for use in confined spaces.

The development and application of high expansion foams for fire fighting purposes started with the work of the Safety in Mines Research Establishment in England concerning the difficult problem of fires in coal mines. It was found that by expanding an aqueous surface active agent solution to a semistable foam of about 1000 times the volume of the original solution, it was possible to force the foam down relatively long corridors, thus providing a means for transporting water to a fire inaccessible to ordinary hose streams. This work was expanded upon by the United States Bureau of Mines immediately after World War II.

Developmental work in the United States on high expansion foam has led to the refinement of specialized high expansion foam generating equipment for fighting fire in confined spaces, for specific applications to fire control problems in both municipal and industrial fire fighting, and for the protection of special hazard occupancies. Medium expansion foam was developed to cover the need for a more wind-resistant foam than high expansion foam for outdoor applications.

Concepts and Suitability for Medium and High Expansion Foams

Medium and high expansion foams are aggregations of bubbles that are mechanically generated by the passage of air or other bases through a net, screen, or other porous medium which is wetted by an aqueous solution of surface active foaming agents. Under proper conditions, fire fighting foams of expansions from 20:1 to 1000:1 can be generated. Such foams provide a unique agent for transporting water to inaccessible places, for total flooding of confined spaces such as basements, and for volumetric displacement of vapor, heat, and smoke. Extensive tests have demonstrated that under certain circumstances high expansion foam, when used in conjunction with water sprinklers, will produce more positive fire control and extinguishment than either extinguishment system by itself; this appears to be especially true with high rack storage of mixed commodities (e.g., high piled storage of paper stock and mixed storage of Class A and Class B materials). Optimum efficiency of high expansion foam in any one type of hazard is dependent to some extent on the rate of application and also the foam expansion and stability.

Personal Safety

Persons should not enter a space filled with high expansion foam without wearing full protective gear, self-contained breathing apparatus, an attached lifeline, and operating in a "buddy" system. A person who is immersed in high expansion foam can experience disorientation and other psychological and personal discomforts. Foam entering any of the body cavities may cause severe irritation and membrane swelling.

Special Considerations

The proper design and application of high expansion foam systems are directly related to a number of special system considerations. Special design factors include maximum submergence time and location of foam generating equipment.

A maximum time needs to be specified for filling the enclosed space to the proper depth with expanded foam. The time, expressed in minutes, is a function of the type of combustible material and the arrangement of the combustible material. An important consideration in maximum submergence time is whether stock material should be considered at a constant storage level in the space to be protected. Also of importance is whether the stock is protected by automatic sprinklers in addition to the high expansion foam. The basic objective is to control a developing fire before the fire has an opportunity to spread vertically over the face of a storage pile.

Fixed installations using high expansion foam fire protection will probably involve the use of customized foam generating equipment to produce the cubic feet per minute requirement of foam discharge. The following points should be observed in the selection and placement of high expansion foam generating equipment:

1. Two generators positioned remotely from each other are more effective and efficient than a single generator.
2. Generating equipment should be top mounted to avoid

back pressures on the foam making equipment. Generators are normally mounted on external towers or special roof supports.
3. Generating equipment should be so positioned as to avoid product-of-combustion air intake. Induced smoke into the generating equipment can significantly reduce the quality and quantity of the foam produced.
4. To effectively dampen convection currents from a developing fire in an area to be protected, the capacity of each required foam generator should be the same.

All of the above information on types and classifications of foam systems serves as the background for actually designing a specific foam fire protection system. Some important limitations concerning both low and high expansion foam systems are presented in the following section before some problem examples.

LIMITATIONS OF FOAM FIRE PROTECTION SYSTEMS

This section discusses both low and high expansion foam systems. The limitations of foam fire protection must be addressed relative to each of these two system classifications. The following points should be reviewed during the proper selection and design of foam-type fire protection.

Limiting Factors for Low Expansion Foam Systems

1. Low expansion foam application is limited to the extinguishment of horizontal or two-dimensional fire problems. This type of foam application is not suitable for three-dimensional fires.
2. Low expansion foam systems are limited by foam agent suitability for the defined flammable or combustible liquid. Basically, foam agents are suitable for either hydrocarbon fuels or polar solvents. Alcohol resistant-type foams may be approved for both hydrocarbons and polar solvents.
3. Different types and brands of foam concentrates may be incompatible and should not be mixed in storage.
4. Foam solution consists of 90 percent or more water. Foam system limitations should be evaluated with respect to the proper use of aqueous based agents on flammable materials and the electrical conductivity of the application method.
5. Foam systems are limited by the equipment appliances and devices used to proportion the foam and to deliver the finished foam onto a given hazard or fire problem. Equipment limitations pertaining to flow rate, operating pressure ranges, and proportioning ranges should be carefully considered in the selection and application of foam systems.

Limiting Factors for High Expansion Foam Systems

1. Medium and high expansion foams are finding applications for a broad range of fire protection problems. However, unlike low expansion foam systems, medium and high expansion foam fire protection systems should be specifically evaluated for each type of hazard condition. The fact that each system requires a feasibility study and individual design may be considered a form of limitation when contrasted to the design concepts for low expansion foam systems.

2. NFPA 11A, *Standard for Medium- and High-Expansion Foam Systems*,[1] states that "under certain circumstances it may be possible to utilize medium and high expansion foam systems for control of fires involving flammable liquids or gases under pressure, but no general recommendations can be made in this standard due to the infinite variety of particular situations which can be encountered in actual practice." This statement is considered to be a design limitation.

3. Medium and high expansion foam systems should not be used on fires in the following hazards unless competent evaluation, including tests, indicates acceptability:

   (a) Chemicals, such as cellulose nitrate, which release sufficient oxygen or other oxidizing agents to sustain combustion.
   (b) Energized unenclosed electrical equipment.
   (c) Water reactive metals, such as sodium and potassium (Na, K).
   (d) Hazardous water reactive materials, such as triethylaluminum and phosphorous pentoxide.
   (e) Liquefied flammable gas.

HYDRAULIC CALCULATION FOR ATMOSPHERIC STORAGE TANKS PROTECTED BY LOW EXPANSION FOAM SYSTEMS

This section of the chapter is concerned with the proper design and associated hydraulic calculations for foam fire protection systems protecting atmospheric storage tanks with low expansion foam systems. The material presented is limited to fixed protection systems using either top mounted foam chambers or subsurface injection as discussed in the first section of this chapter. A single flammable liquid storage tank problem is presented for developing the appropriate methods and techniques for computing the foam agent requirements, system hardware requirements, and the necessary hydraulic calculations to properly deliver the required rate of foam to the subject hazard. The single example will be calculated using topside application of foam and subsurface injection of foam. This approach permits comparison and contrast of the system design and hydraulic requirements between the two foam application methods.

EXAMPLE 1:

Problem Statement: The single outside storage tank depicted in Figure 4-5.1 is to be protected by a completely fixed foam system. The topside foam chamber arrangement is used in this problem. Note that the foam system is connected to a domestic water supply. The water supply curve is illustrated in Figure 4-5.2. For this problem, consider that the water available for the foam system is limited to the street main flow characteristics. A foam system job work sheet and a complete set of hydraulic calculations are to be prepared for this problem.

SOLUTION:

Procedure Statements: A systematic outline follows for the proper design and hydraulic assessment associated with the stated problem; reference is made to criteria established in NFPA 11. This standard should serve as a companion guide to the systematic evaluation of each problem scenario.

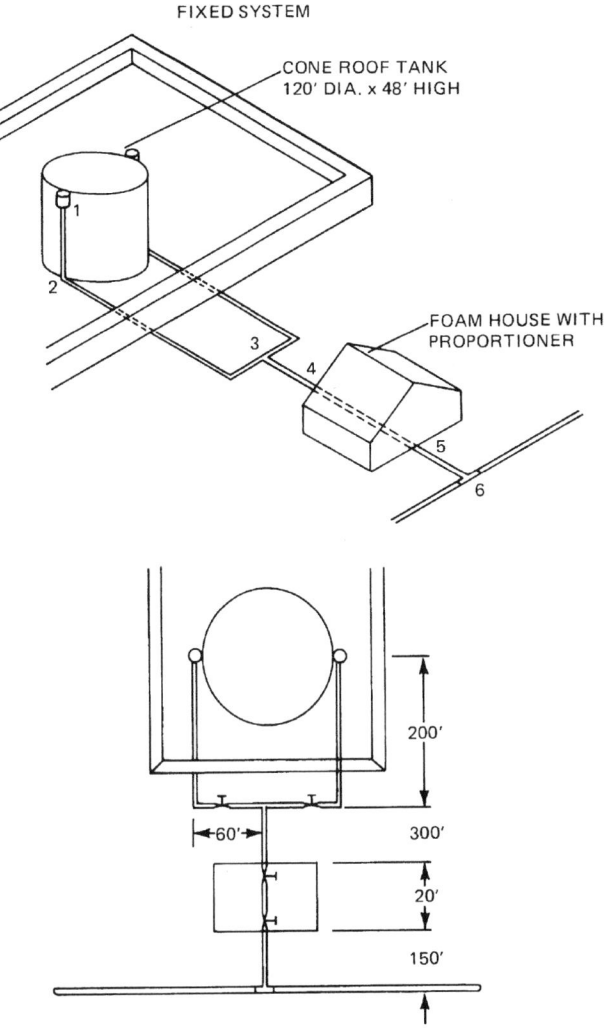

Fig. 4-5.1. *Single storage tank with fixed foam system protection. The numbers 1 through 6 in the top portion of the drawing represent the reference points for hydraulic calculations for Example 1.*

Individual item information is transferred to the referenced problem job sheet (Figure 4-5.3) and the associated hydraulic calculation sheet (Figure 4-5.4).

Problem Assessment: In addition to the physical layout of the design problem, information is required on the hazard to be protected. The nature of the hazard drives the problem design. The following steps identify the hazard and standard requirements associated with the hazard.

Step 1: Installation Identification.

Refer to Figure 4-5.1. One vertical atmospheric storage tank is positioned in a dike area. The tank is protected by a fixed foam fire protection system and connected to the domestic water supply.

Step 2: Hazard classification.

Flammable liquid atmospheric storage tank.

Step 3: Type of protection.

Fixed protection systems.

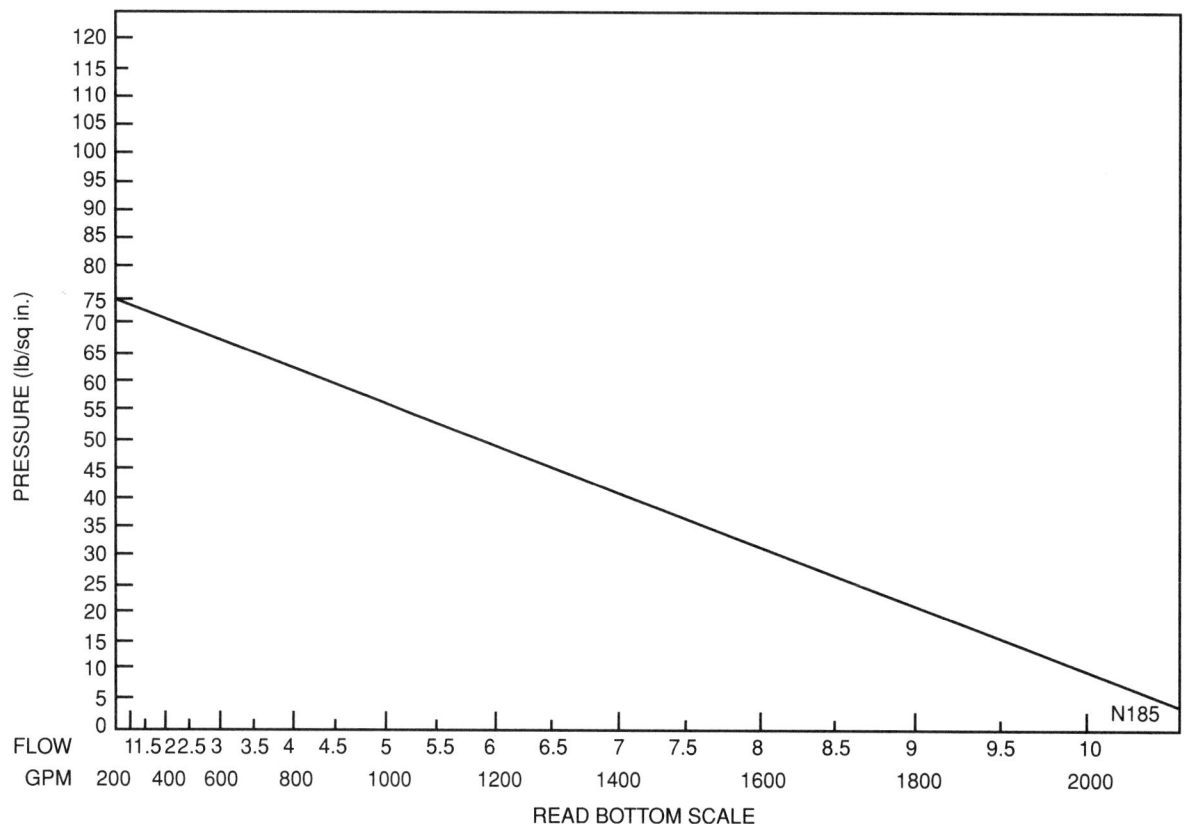

Fig. 4-5.2.   Water supply curve for fixed foam system for Examples 1 and 2.

Step 4:   Hazard description.
120 ft diameter outdoor cone roof flammable liquid storage tank.

Step 5:   Flammable or combustible liquid area to be protected.
Calculate the flammable liquid surface area: Area $= 0.7854d^2$.

$$A = 0.7854(120)^2 = 11,310 \text{ ft}^2$$

Step 6:   Flammable liquid or combustible liquid identification.
Gasoline—Sg. 0.72.

Step 7:   Foam application method.
Top mounted fixed foam chambers—Type II.

Step 8:   Description, number, and placement of foam application devices.
Several factors need to be simultaneously considered in responding to this step. The description should reference a given manufacturer's foam chamber because the flow and pressure characteristics of the foam chamber *may be* a factor or present options for the system design. Manufacturer's literature and the UL listing of foam equipment should be consulted on this matter.

For proprietary reasons, the selected foam chamber for this problem will be identified as an FMA chamber with an operating pressure range of 40 to 80 psi and a flow range from 300 to 700 gpm of foam solution.

The foam standard (NFPA 11) requires a minimum of two foam chambers for a 120 ft diameter tank. More foam makers may be used based on hydraulic considerations, or

economies of scale based on equipment costs. Individual manufacturers of foam equipment must be consulted on these options.

Placement of foam makers should consider equal spacing around the upper tank perimeter and placement of the foam solution feed lines. If possible, maintain constant flow and pressure requirements of each device.

Step 9:   Foam agent selected.
A 3 percent fluoroprotein foam is selected for the defined hazard. Note that the type of foam agent selected for a particular design problem may affect other variables or consideration in the foam system design. This caution is reflected in Step 10.

Step 10:   Foam solution application rate.
Foam solution application rates for storage tanks containing liquid hydrocarbons should be at least 0.1 gpm/ft² of liquid surface area of the tank to be protected. It should be noted that other types of foam protection (e.g., portable nozzles) may require different application rates. Also, flammable and combustible liquids not classified as hydrocarbons may require different foam solution application rates. Reference should be made to NFPA 11 concerning design application rates.

The foam solution application rate for the stated problem is calculated as follows:

$$\text{rate (gpm)} = 0.1 \text{ gpm/ft}^2 \times 11,310 \text{ ft}^2 = 1,131 \text{ gpm}$$

However, the total rate is divided equally between two foam makers so it is appropriate to specify a rate of 1,132 gpm, or 566 gpm per foam maker.

FOAM SYSTEM JOB WORK SHEET

Designer: _____ Staff _____

Sheet: __1__ of: __1__

Date: _____ 1986 _____

Installation Identification: _____ Ourville Oil Company _____

Hazard Classification: _Flammable Liquid Atmospheric Storage Tank_

Type of Protection: _Fixed Protection System_

Hazard Description: _120 ft. diameter outdoor cone roof flammable liquid storage tank_

Flammable or Combustible Liquid Area to be Protected: _11,310 ft²_

Flammable or Combustible Liquid Identification: _Gasoline_
Sq. 0.72

Foam Application Method: _Type II — Fixed Chambers_

Description, Number, and Placement of Foam Application Devices: _____
2 — Chambers Equally Spaced

Foam Agent Selected: _Fluoroprotein — 3%_

Foam Solution Application Rate: _0.1 GPM per sq. ft. or 1,131 GPM_

Foam Concentration Rate: _34 GPM_

Water Application Rate: _1,098 GPM_

Duration of Discharge: _55 Min._

Gallons of Foam Required: _1,870 gallons_

Gallons of Water Required: _60,390 gallons_

Water Supply Information: _See Figure 4-5.2_

Special Foam Design Considerations: _____

Fig. 4-5.3. Foam system job work sheet.

Step 11: Foam concentrate rate.

The foam concentrate rate is based on the foam agent proportioning rate. A 3 percent fluoroprotein foam is selected for this problem (see Step 9). In other words, 3 percent of the calculated solution rate is the foam concentrate rate. This rate may be determined as follows:

foam concentrate rate = 0.03(%) × 1,132 gpm = 34 gpm

Note that a continuous supply of foam agent (concentrate) must be available at a rate of 34 gpm for the required duration of discharge (see Step 13).

Step 12: Water application rate.

Quite simply, the water application rate is the foam solution rate minus the foam concentrate rate. The water application rate proceeds as follows:

water application rate = 1,132 gpm − 34 gpm = 1,098 gpm

The water application rate can also be determined as 97 percent of the "solution" rate when using a 3 percent foam concentrate.

Step 13: Duration of discharge.

A minimum foam solution discharge time is specified in NFPA 11 to control and extinguish a fire in a cone roof atmospheric storage tank. Duration of discharge is dependent on the classification of flammable or combustible liquid and the *type* of discharge outlet. Information for the given problem is contained in NFPA 11. The requirement for protecting gasoline with Type II foam chambers is 55 min. of continuous foam solution discharge.

Step 14: Gallons of foam required.

The required foam supply for any given problem should properly consider a primary supply and a reserve supply. The primary supply is computed by multiplying the determined rate of foam agent required by the duration of discharge as follows:

foam agent required = 34 gpm × 55 min = 1,870 gal

The authority having jurisdiction may require that equal quantity of foam be placed in reserve for a second fire.

HYDRAULIC CALCULATIONS

Subject: _____ Example Problem 1 _____   Job No: _____

_____   Sheet No: _____ Of: _____

_____   By: _____ Date: _____

_____   Chkd By: _____ Date: _____

Application Rate: ___ 0.1 ___ GPM per sq ft   Area: ___ 11,130 sq. ft. ___

Minimum Solution Rate: __ 1131 GPM __   Actual Solution Rate: __ 1132 GPM __

Foam Maker Pressure and Rate: __ 50 PSI — 566 GPM __ Foam System: __ Chamber Type 2 __

Water Data: _____   Ref. Drawing: __ Figure 4-5.1 __

Foam Maker Type and Location	Added GPM	Total GPM	Pipe Size In.*	Pipe & Equivalent Fitting Lgth. (Feet)**	Friction PSI/Ft C=100	Friction Total PSI	Static PSI	Proportioner PSI	Req'd Pres PSI
Starting Point: 1				Elevation: 48 ft.		Pressure at Foam Maker:=		50 PSI	
1	566	566							50
1–2		566	5.047	48' + (1E) 8.6' = 56.6'	0.0420	2.4	20.8		73.2
2–3		566	5.047	260' + (1GV) 2' = (1T) 25'					—
(2–3)				Σ = 287'	0.0420	12.1	–		85.3
3–4	566	1132	8.071	300' + (1GV) 2' = 302'	0.154	4.7	–		90.0
4–5		1132	8.071	18' + (1GV) 2' = 20'	0.154	0.3		4.0	94.3
5–6		#1098	8.071	150' = (1T) 35' = 185'	0.145	2.7			97.0
Σ at 6		1098							97.0

Water Only

* "X" Indicates Extra Hvy. — Std. Wt. Otherwise

** See Sheet for Tabulation of Pipe and Fittings

Fig. 4-5.4. Hydraulic calculation work sheet.

Step 15:   Gallons of water required.

The basic procedure follows the concept presented in Step 14. The water requirement is the product of the water rate times the time:

water required = 1,098 gpm × 55 min = 60,390 gal

While the above calculation is straightforward, the answer may come as a surprise. Fixed foam protection systems for atmospheric storage tanks require a large quantity of water. The total quantity of water must be available at the site to assure foam delivery over the required time.

Step 16:   Special foam system design considerations.

The first 15 steps in this problem analysis focus on fundamental considerations required to determine foam agent and water supply requirements. It is now appropriate to examine a series of special considerations that directly relate to the hydraulic design factors associated with this problem. Other design factors may be appropriate for different problems. However, the basic considerations outlined below should serve to guide similar calculations for other design problems.

1. Pipe size selection. Water supply pipe and foam solution pipe may be sized to minimize head loss between identified supply and demand points; pipe may also be designed on the basis of a mean velocity flow in a given section of pipe. A flow velocity of 10 feet per second may be used in the absence of other specific criteria for the determination of both water supply pipe and foam solution pipe. Pipe will be sized in this manner for the stated problem.

2. Valves in the pipe system.
   (a) The laterals of each foam chamber or fixed roof tank should be separately valved outside of the dike installation.
   (b) The water line to each proportioner inlet should be separately valved.
      Note: Valves are properly shown in Figure 4-5.1.

3. Foam proportioner selection.
   Several different foam proportioners are available from equipment manufacturers. It is very important to select foam proportioning equipment that meets the following requirements:

   (a) Proper proportioning over the range of desired flows.
   (b) Minimum or acceptable head loss across the proportioning device.
   (c) Suitability for the foam agent selected.
   (d) Capability of overcoming any back pressure limitations.

4. Water pumps. The water supply in a given case may require a pressure boost to meet the foam chamber discharge requirements. Where a water pump is required, consideration must be given to the pump capacity, the pressure profile, the pump horsepower requirement, and the pump intake-discharge positions with respect to the total installation of pipe.

Hydraulic analysis for Example 1: The supporting documentation above provides a foundation for conducting a hydraulic design for the problem depicted in Figure 4-5.1. The design parameters and associated calculations are presented in sequential steps below. All reference points conform to Figure 4-5.1. Computations are also charted on a hydraulic calculation sheet as shown in Figure 4-5.4.

Step 1: Starting point.
   The design objective is to provide each of the two foam chambers with the required pressure to discharge the calculated quantity of foam solution. The stated problem requires a foam maker discharge of 566 gpm at 50 psi. An orifice plate is supplied by the foam equipment manufacturer to provide the correct discharge at the design pressure (50 psi). The design pressure is a function of the range of pressures that can be used with a specific manufacturer's foam chamber.

Step 2: Design tank riser.
   The vertical pipe supplying the foam chamber (Ref. 1-2) is sized on the basis of a maximum flow velocity of 10 feet per second. The pipe size is determined as follows:

Formula:
$$\text{Velocity} = \frac{0.4085 \times \text{gpm}}{d^2}$$

   a. Solve for d

   b. $10 \text{ fps} = \dfrac{0.4085 \times 566 \text{ gpm}}{d^2}$

   c. $d^2 = 23.12$ in.

   d. $d = 4.8$ in.

   e. 5 in. pipe is selected. (Note: The authority having jurisdiction may require a 6 in. pipe.)

Step 3: Enter hydraulic calculation from Ref. 1-2.
1. Friction loss, FL, is determined by the Hazen-Williams formula
$$\text{FL} = \frac{4.52 \times Q^{1.85}}{C^{1.85} \times d^{4.87}}$$
where:

   $Q = 566$ gpm
   $C = 100$
   $d = 5.047$ (internal diameter of pipe)
   FL $= 0.0420$ psi/ft

2. Note: All friction losses and head losses are summed in the required pressure column.
3. The head loss, H_L, for 48 feet of elevation difference between reference points is computed as follows:

$$H_L = 0.433 \text{ psi/ft} \times 48 \text{ ft} = 20.8 \text{ psi}$$

4. The pipe section includes one standard elbow at Ref. 2. For hydraulic calculations, fittings are treated as equivalent feet of pipe in accordance with NFPA 13, *Standard for the Installation of Sprinkler Systems*.[1]

Step 4: Enter hydraulic calculations from Ref. 2 (bottom of tank) to Ref. 3.

1. The flow is constant (566 gpm) so the pipe size remains the same at 5 in.
2. Pipe fittings include a globe valve outside the dike area and a standard tee at Ref. 3.

Step 5: Enter hydraulic calculations from Ref. 3 to the foam house (Ref. 4).

1. The total foam solution flow (1,132 gpm) is supplied by line 2-3.
2. Determine the pipe size based on a maximum flow velocity of 10 fps.

3.
$$\text{velocity} = \frac{0.4085 \times \text{gpm}}{d^2}$$
$$10 \text{ fps} = \frac{0.4085 \times 1{,}132 \text{ gpm}}{d^2}$$
$$d = 6.8 \text{ in.}$$

4. An 8 in. pipe is recommended between the foam house and Ref. 3.
5. The friction loss in the stated line includes the linear distance plus the gate valve.
6. It should be observed that the required pressure at the discharge side of the foam house is 90.0 psi.

Step 6: Enter the hydraulic calculations in the foam house.

1. To the left of the top view, bottom half, Figure 4-5.1, is an illustration of the foam proportioning ratio controller inserted into the water line. The proportioning device selected for this problem has a friction loss of 4 psi at a solution flow rate of 1,132 gpm. Foam equipment manufacturers should be consulted on head loss characteristics for specific devices. The ratio controller takes up a

lineal distance of 2 ft leaving 18 ft of straight run pipe in the foam house.

2. The calculations provided do not include provisions for a water pump.

Step 7: Enter the hydraulic calculations from the foam house to the street main.

1. The flow rate in line 5-6 is 1,098 gpm. Note the change in friction loss.
2. An 8 in. main is used to connect the street main to the foam house.

Step 8: Summary.

1. The water demand requirement at Ref. 6 is 1,098 gpm at 97.0 psi.
2. The hydraulic demand has been calculated to provide a foam solution flow of 566 gpm at 50 psi for *each* designated foam chamber.
3. The water supply curve referenced as Figure 4-5.2 shows 1,098 gpm available at 53 psi. Therefore, a water pump is required in the pump house to boost the pressure from 50 psi (loss of pressure from Ref. 6 to 5 is approximately 3 psi) to 97 psi or approximately 50 psi. A pump can be engineered for this specific application.

EXAMPLE 2:

Overview Statement: The flammable liquid storage tank presented as Example 1 is also presented as Example 2; only the method of fire protection changes. Example 2 considers the depicted 120-ft diameter storage tank to be protected with subsurface application of foam to the described hazard. The subsurface application technique requires new design considerations with respect to foam equipment and hydraulic calculations. The following principles should be observed in the design of subsurface application foam systems.

1. Foam solution is expanded outside of the dike area by a "high back pressure" foam maker. A typical foam expansion of 4:1 is achieved at the foam maker.
2. The expanded foam flows through a carefully designed pipe-line from the foam maker to an opening in the tank shell just above the water bottom in the tank. In accordance with NFPA 11, *Standard for Low-Expansion Foam*, the foam velocity at the point of discharge into the tank contents shall not exceed 10 fps for Class IB liquids and 20 fps for other type liquids. An excessive input velocity to the tank can cause the foam to be saturated with fuel pickup.
3. Foam entering the product rises to the fuel surface by natural buoyancy.

Problem Statement: A single outside storage tank illustrated in Figure 4-5.5 is to be protected by a completely fixed foam system. The subsurface foam application method is used for this problem. The Figure 4-5.2 water supply curve is to be applied with this problem. Consider that the water available for the foam system is limited to the street main flow characteristics. A foam system job work sheet and a complete set of hydraulic calculations are to be prepared for this problem.

Procedure Statements: A systematic outline follows for the proper design and assessment associated with Example 2. Reference is made to criteria established in NFPA 11 by paragraph reference. NFPA 11 should serve as a companion guide to the systematic evaluation of each problem scenario. Individual item information is transferred to the referenced

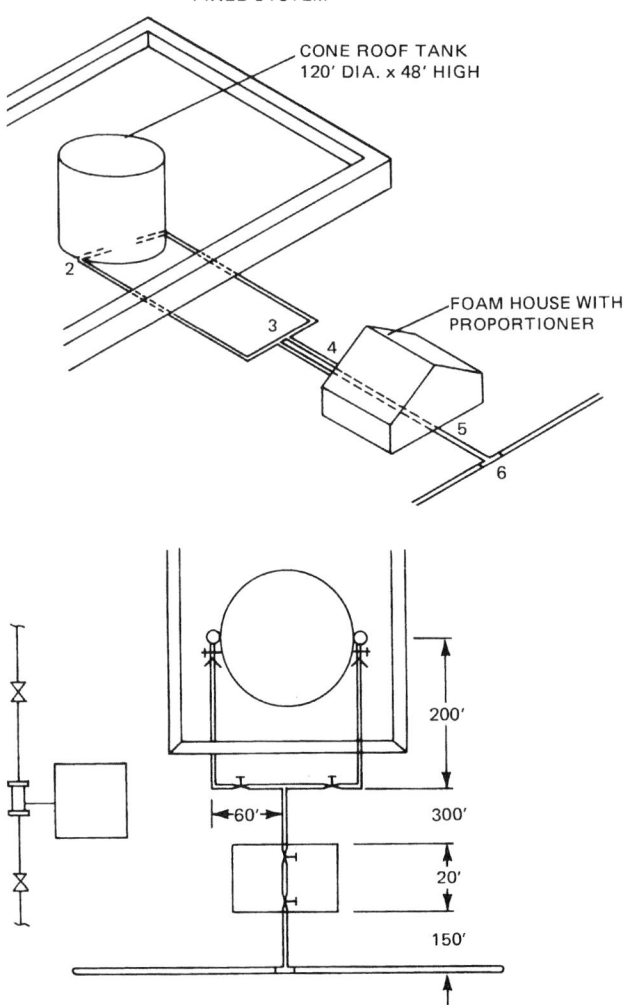

Fig. 4-5.5. *Subsurface application foam system used in Example 2.*

problem sheet (Figure 4-5.6) and the associated hydraulic calculation sheet (Figure 4-5.7).

Problem Assessment: In addition to the physical layout of the design problem, information is required on the hazard to be protected. The nature of the hazard is identical to Example 1; only the method of protection changes. Some of the following steps are repeated for maintaining a sequence to the problem solution. New material will be explained in detail. Calculations for similar material are referenced to Example 1.

Step 1: Installation identification.
Refer to Figure 4-5.5. One vertical atmospheric storage tank is positioned in the dike area. The tank is protected by a fixed subsurface injection foam fire protection system and connected to the domestic water supply.

Step 2: Hazard classification.
Flammable liquid atmospheric storage tank.

Step 3: Type of protection.
Subsurface application to fixed roof storage tanks. (Re: NFPA 11, Section 3-2.6.)

FOAM SYSTEM JOB WORK SHEET

Designer: _____Staff_____ Sheet: ___1___ of: ___1___

Date: _____1986_____

Installation Identification: _____Ourville Oil Company_____

Hazard Classification: _Flammable Liquid Storage Tank_

Type of Protection: _Subsurface application to fixed roof storage tank_

Hazard Description: _120 ft. diameter outdoor cone roof flammable liquid storage tank_

Flammable or Combustible Liquid Area to be Protected: _11,310 ft²._

Flammable or Combustible Liquid Identification: _Gasoline — Sq. 0.72_

Foam Application Method: _Subsurface application to a liquid hydrocarbon_

Description, Number, and Placement of Foam Application Devices: _Two subsurface injection_
points positioned equal and opposite on the tank shell. A PHB Foam Maker is used.

Foam Agent Selected: _____3%_____

Foam Solution Application Rate: _1,132 GPM_

Foam Concentration Rate: _34 GPM_

Water Application Rate: _1098 GPM_

Duration of Discharge: _55 min._

Gallons of Foam Required: _1,870_

Gallons of Water Required: _60,390_

Water Supply Information: _See Figure 4-5.2_

Special Foam Design Considerations: _Foam injection piping to be sized for a maximum_
fluid velocity of 10 fps.

Fig. 4-5.6. Foam system job work sheet.

Step 4: Hazard description.

120-ft-diameter outdoor cone roof flammable liquid storage tank.

Step 5: Flammable of combustible liquid area to the protected.

11,310 ft². (See Step 5—Example 1.)

Step 6: Flammable liquid or combustible liquid identification.

Gasoline—Sq. 0.72.

Step 7: Foam application method.

Subsurface application to a liquid hydrocarbon. (Re: NFPA 11, Section 3-2.6.3.)

Step 8: Description, number, and placement of foam application devices.

Several equipment design variables must be considered simultaneously in this step. Again, reference must be made to a specific foam manufacturer's equipment offerings or the conduct of a comparative analysis between two or more manufacturers of suitable equipment. The manufacturer's literature and the UL listing of foam equipment should be consulted on this matter.

The following substeps outline the key features of the design to be evaluated and computed at this overall step in the design development. Each substep may impact on other substeps and therefore all substeps must be evaluated before selecting a set a equipment and associated calculations.

Step 8(a): Velocity of approach into the storage tank.

Gasoline is a Class IB liquid, and therefore the injection velocity of the expanded foam into the product tank should not exceed 10 fps. This does not mean that the velocity of foam between the foam maker and the injection point has to be controlled to a maximum of 10 fps. Rather, the foam velocity at the physical point of entry to the product is the key consideration.

HYDRAULIC CALCULATIONS

Subject: _____ Example Problem 2 _____ Job No: _____

_____ Sheet No: _____ Of: _____

_____ By: _____ Date: _____

_____ Chkd By: _____ Date: _____

Application Rate: _____ 0.1 _____ GPM per sq ft   Area: _____ 11,130 sq. ft. _____

Minimum Solution Rate: __ 1132 GPM __   Actual Solution Rate: 1132 _____

Foam Maker Pressure and Rate: PHB – 159 psi — 566 gpm   Foam System: __Subsurface__

Water Data: _____ Figure 4-5.2 _____   Ref. Drawing: _____

Foam Maker Type and Location	Added GPM	Total GPM	Pipe Size In.*	Pipe & Equivalent Fitting Lgth. (Feet)**	Friction PSI/Ft C=100	Friction Total PSI	Static PSI	Proportioner PSI	Req'd Pres PSI
Starting Point: 1 Elevation: 48 ft. Pressure at Foam Maker:= 50 PSI									
1		2 264	10"				15		15
1–2		2 264	10"	20	0.20	0.4			15
2–3				560 + 1CV (55)1GV (5)	0.40	24.8			40
				620					
Σ 3									40

NOTE: The demand pressure at 3 is less than the allowable pressure of 64.0 psi

* "X" Indicates Extra Hvy. — Std. Wt. Otherwise

** See Sheet for Tabulation of Pipe and Fittings

Fig. 4-5.7. Hydraulic calculations.

Remember that the foam is expanded at the entry point to the product. Special flow curves must be examined to determine velocity characteristics with expanded foam. These curves are not available in the current edition of NFPA 11; one must turn to manufacturer's literature on this topic.

A set of flow curves for expanded foam in various pipe diameters are included in the Appendix of the chapter with approval of National Foam System, Inc. Figure A-4-5.3 is consulted for this problem element.[2]

Table 3-2.6.3 of NFPA 11 indicates that two discharge outlets must be provided for a 120-ft-diameter tank. The foam solution rate for each outlet is given in Step 10 of this example and is equal to the calculations for Example 1: 566 gpm per outlet. However, the foam is expanded at the high back pressure foam maker using a ratio of 4:1. The expanded foam flow rate at each outlet is 2,264 gpm. This is the value used when checking foam velocity.

A 10-in. pipe is required to maintain a foam velocity less than 10-fps when the rate of expanded foam is 2,264 gpm (see Figure A-4-5.3). The pipe length upstream from the discharge point must be at least 20 times the diameter of the pipe to establish uniform velocity. Therefore, a straight run of 10-in. pipe at least 17 ft in length is necessary.

The foam outlet is not required to be at the tank shell.

Note that the 10-in. pipe is actually inserted into the tank. This design approach permits economizing the pipe sizes between the tank and the high back pressure foam maker. The high back pressure foam maker is to be positioned outside of the dike area. A gate valve and a check valve are installed adjacent to the tank shell.

Step 9: Foam agent selected.

A 3 percent fluoroprotein foam is selected for the defined hazard. Note that NFPA 11 states that "only fluoroprotein and AFFF foams have the necessary tolerance to contamination by liquid hydrocarbon fuels required when application is by subsurface injection." (Re: NFPA 11, Section 3-2.6.5.)

Step 10: Foam solution application rate.

For tanks containing liquid hydrocarbons, the foam solution rate must be at least 0.10 gpm/sq ft of liquid surface area of the tank to be protected. The maximum rate must be 0.20 gpm/sq ft. (Re: NFPA 11, Section 3-2.6.6.)

The foam solution application rate for Example 2 is the same foam solution rate calculated for Example 1; the application rate is 1,132 gpm.

Step 11: Foam concentrate rate.

The foam concentrate rate is determined in the same manner as set forth for Example 1. Using a 3 percent fluoroprotein foam, the requirement is 34 gpm for a total solution rate of 1,132 gpm. This requirement impacts on Step 14 of this example.

Step 12: Water application rate.

The water application rate is also determined in the same manner as set forth in Example 1. The water application rate is the foam solution rate minus the foam concentrate rate: the water rate is 1,098 gpm.

Step 13: Duration of discharge.

The minimum discharge time for subsurface application of foam is identical to the requirement for Type II application. (Re: NFPA 11, Section 3-2.6.7.) Gasoline product requires a total discharge time of 55 minutes.

Step 14: Gallons of foam required.

The gallons of foam required is computed in the same manner as set forth in Example 1. The primary foam supply is computed by multiplying the determined rate of foam agent by the duration of flow, which indicates a requirement of 1,870 gal.

Step 15: Gallons of water required.

The water requirement is the product of the water rate times the discharge time, or 60,390 gal.

Step 16: Special foam system design considerations.

One special design consideration is presented in Step 8(a) and involves the pipe requirements for injecting foam into the base of a storage tank under controlled velocity conditions. Other special conditions that apply specifically to subsurface injection of foam to storage tanks are given below.

1. High Back Pressure Foam Maker. A high back pressure foam maker is designed for capability to make foam and discharge the foam against considerable back pressure. The high back pressure foam maker selected for the example problem is designed to operate satisfactorily at inlet pressures of 100 to 300 psi and produce foam of 2:4 expansion against back pressures not exceeding 40 percent of the inlet pressure. With an inlet pressure of 150

psi, for example, 60 psi is available at the discharge for forcing the foam through a hose and/or piping into the storage tank and to overcome the fuel head in the tank. Manufacturers of high back pressure foam equipment should be consulted for obtaining flow and pressure characteristics and back pressure limitations. Two high-back pressure foam makers are used with Example 2. The two foam makers are located in the foam house and are arranged for parallel operations. (See Figure 4-5.5.)

2. Pipe Size Selection. Expanded foam flowing in conduit (pipe) does not follow the head loss characteristics expressed in the Hazen-Williams formula. A set of flow curves have been developed for determining friction loss for expanded foam discharge by a high back pressure foam maker.[2] A set of these curves is provided in the Appendix to this chapter with the permission of National Foam System, Inc. (See Figures A-4-5.1 and A-4-5.2.)

A flow velocity of 10 fps is used for the determination of pipe sizes flowing foam solution and water. If necessary, water supply pipe and foam solution pipe may be sized to minimize head loss between identified supply and demand points.

3. Valves in the Pipe System. (Re: NFPA 11, Section 3-2.6.3.) For subsurface application, each foam delivery line must be provided with a valve and check valve, unless the latter is an integral part of the high back pressure foam maker or pressure generator to be connected at the time of use. When product lines are used for foam, product valving must be arranged to ensure foam enters only the tank to be protected.

4. Foam Proportioner Selections. The practices and procedures outlined in Example 1 apply to Example 2. However, to accommodate the pressure requirements associated with a high back pressure foam maker, a balanced proportioner would appear to provide the best level of constant proportioning over designated pressures.

5. Water Pumps. The required pressure at the intake to the high back pressure device is approximately 150 psi. The static pressure on the water system is only 75 psi. Therefore, a water pump is required to boost the water-solution pressure in the foam hose. The most efficient approach to designing a required water pump installation is to select or design a pump-driver combination that will boost the available residual pressure to the required residual pressure at the demand flow. In other words, with the right capacity pump, the driver horsepower is calculated to raise the pressure over the differential range. Other criteria must be used if a standard fire pump is required by the authority having jurisdiction.

Hydraulic analysis for Example 2: The information above provides a foundation for conducting a hydraulic design for the problem depicted in Figure 4-5.5. The design parameters and associated calculations are presented in sequential steps below. All reference points conform to Figure 4-5.2. Computations are also charted on a hydraulic calculation sheet, designated as Figure 4-5.7.

Step 1: Precalculation for high back pressure foam maker.

The hydraulic characteristics of the high back pressure foam maker must be considered before the computations start. A high-back pressure foam maker delivering 550 gpm at 150 psi is selected for each finished foam line to the tank.

a. Determine a K value for the specified unit:

$$Q = K \sqrt{150}$$
$$K = 44.9$$

b. Required discharge per foam maker is 566 gpm.
c. Determine the required input pressure for a flow of 566 gpm.
    566 gpm = 44.9 $\sqrt{P}$
    P = 159 psi
d. The available back pressure becomes: 40 percent of 159 or 64.0 psi.

Remember: The head loss converted to psi between the foam maker discharge and the foam discharge outlet to the tank, plus the product head, must not exceed 64 psi.

Step 2: Size foam injection pipe to tank.
    Step 8(a) under problem assessment for Example 2 establishes that a 10-in. pipe is required to maintain a flow velocity under 10 fps.

Step 3: Determine the head loss from the production storage.
    Finished foam rising through the product must overcome the product head. Gasoline is the product for this series of problems with a Sq. = 0.72.
    psi loss = 48 ft × 0.433 psi/ft × 0.72 Sq.
    psi loss = 15

Step 4: Size the foam supply line from the tank shell to the foam house.
    The stated pipe is selected on the basis of the allowable friction loss of 64 psi minus the product head loss which equals 15 psi. Therefore, 49 psi (64 psi − 15 psi) can be dissipated from the tank to the foam maker through 500 ft of pipe and be used as an initial estimator; the flow rate is 2,264 gpm. Figure 4-5.8 indicates a 6-in. pipe is required. A 6-in. check valve and a 6-in. gate valve will be installed on the foam supply line adjacent to the tank in the dike area. The required friction loss calculations are presented in Figure 4-5.7.

Calculation note: Subsurface foam system hydraulics actually divide into two separate calculation sets, as follows: (1) the hydraulics between the high back pressure foam maker and the storage tank, and (2) the hydraulics between the street main supply and the high back pressure foam maker.

Step 5: Street main to fire pump calculation.
    The lateral supply line will be designated at a velocity of 10 fps. Recall that only water is moving through this line.

$$\text{Velocity} = \frac{0.4085 \times \text{gpm}}{d^2}$$

a. Solve for d

b. 10 fps = $\dfrac{0.4085 \times 1098 \text{ gpm}}{d^2}$

c. d^2 = 44.85 in.
d. d = 6.69 in.
e. Use an 8-in. pipe

Step 6: Piping in foam hose.
    8-in.-diameter pipe will be used in the foam house to connect between the water pump, the foam proportioner, and the high back pressure foam maker.

High Expansion Foam Systems

Many of the fundamental hydraulic concepts presented with low expansion foam system design problems also apply to high expansion foam systems. Some similarities and differences between the hydraulic design for high expansion foam systems and low expansion foam systems are presented in Table 4-5.2. In this analysis, a low expansion foam system using top-mounted foam chambers is compared to an elevated high expansion foam generator installation.

Example 3 considers the use of a high expansion foam system in conjunction with automatic sprinkler protection for fire control and suppression in a specified government warehouse. Be careful to note the generator flow rates, foam concentrate rates, and water supply rates. One of the advantages for considering high expansion foam for the protection of confined space hazards is the low rate of foam application and associative water rate when compared to other aqueous types of systems.

EXAMPLE 3:

Problem Statement: A General Services Administration unit, for the U.S. government has elected to protect a number of warehouse complexes with a combination of automatic sprinklers and high expansion foam. A typical four-bay warehouse complex is illustrated in Figure 4-5.8. The storage item is crude rubber piled 12 ft 6 in. high in 2,000 sq ft pile areas. The installed sprinkler design is 0.2 gpm/ft². The location of the high expansion foam generators is illustrated on the 12-in. wide brick fire walls. Each foam generator is equipped with a set of remote-controlled baffels that permits directional flow of foam into adjacent fire areas. Custom

TABLE 4-5.2 *Comparison of Design Criteria for Low-Expansion and High-Expansion Foam Systems*

Design/Hydraulic Step Function	Low-Expansion Foam System — Top Chamber	High-Expansion Foam System — Top Generator
#1: Starting Point	Foam Chamber(s)	Foam Generator(s)
#2: Second Determination	Foam Solution Requirement per Chamber (GPM)	Expanded Foam Requirement per Chamber (CFM)
#3: Third Determination	Foam Solution Delivery Rate Between Foam Maker and Foam House	Same Determination
#4: Fourth Determination	Size Pipe from Foam Maker(s) to Foam House	Size Pipe from Foam Generator(s) to Foam House
#5: Fifth Determination	Determine Type and Size of Foam Proportioner	Same Determination
#6: Sixth Determination	Determine Hydraulic Requirements in Foam House	Same Determination
#7: Seventh Determination	Evaluate Water Supply/Demand Requirement at Foam House	Same Determination
#8: Eighth Determination	Assess Requirement for Water Pump in Foam House; Recalculate Hydraulic Requirements in Foam House	Same Requirement

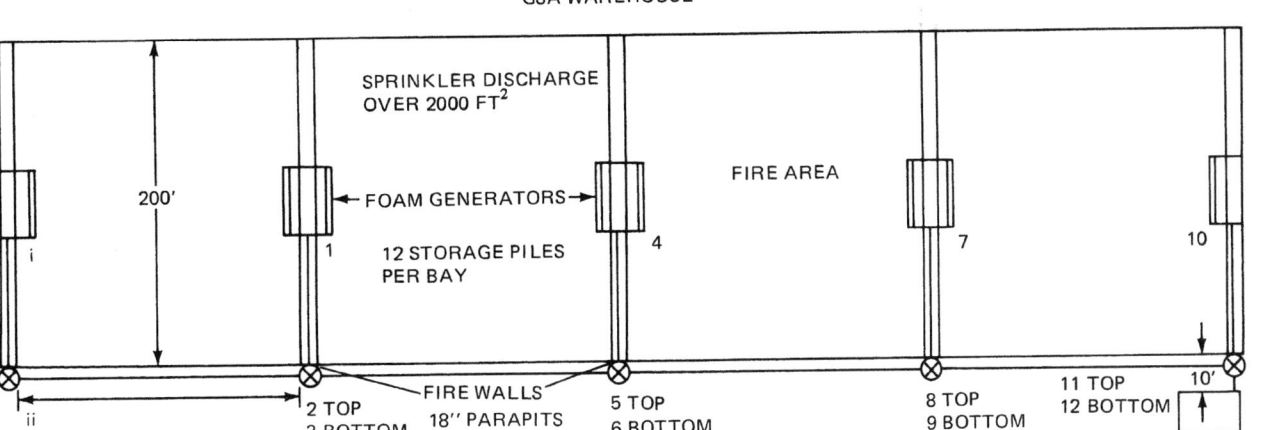

GSA WAREHOUSE

RATE OF FOAM DISCHARGE

$R = (V/T + R_s) \times C_N \times C_L$ WHERE:

R = RATE OF DISCHARGE–cfm
V = SUBMERGENCE VOLUME–CUBIC FEET
T = SUBMERGENCE TIME–MINUTES
R_s = RATE OF FOAM BREAKDOWN BY SPRINKLERS
C_N = COMPENSATION FOR NORMAL SHRINKAGE (1.15)
C_L = COMPENSATION FOR LEAKAGE (1.2)

$R_s = S \times Q$ WHERE:

S = FOAM BREAKDOWN IN cfm PER gpm OF
SPRINKLER DISCHARGE. S SHALL BE
10 cfm/gpm
Q = ESTIMATED TOTAL DISCHARGE FROM
MAXIMUM NUMBER OF SPRINKLERS
OPERATING

Fig. 4-5.8. Typical four-bay warehouse complex.

generators are used that have a foam solution rate requirement of 1.83 gpm per 1,000 cu ft of foam production. Three percent foam proportion with a UL-listed high expansion foam is used for this system.

Procedure Statements: The key consideration in high expansion foam system design is the proper sizing of the foam-generating equipment to be used for a specific application. A special job work sheet is provided to systematically calculate the foam generation requirements. (See Figure 4-5.9.) Individual item information is transferred from the referenced job sheet to the associated hydraulic calculation sheet. (See Figure 4-5.10.) Reference is made to criteria established in NFPA 11A, *Standard for Medium- and High-Expansion Foam Systems*. This standard should serve as a companion guide to the systematic evaluation of the warehouse protection problem.

Problem Assessment: The fundamental considerations of the hazard to be protected establish the elements of the problem design. In the case of high expansion foam, some subjective criteria needs to be established due to the lack of specific information in NFPA 11A. Subjective criteria will be fully noted in the problem development. The following steps identify the hazard and the standard calculations associated with the hazard.

Step 1: Installation identification.
Refer to Figure 4-5.9, a defense materials warehouse.

Step 2: Hazard classification.
High density combustibles. Note: The actual storage material is crude rubber provided in irregular flat sheets. This commodity is not specifically specified in Table 2-3.4 of NFPA 11A. Therefore, some judgment must be made when

selecting foam submergence time as required for the calculations below.

Step 3: Type of protection.
The warehouse is protected by a drypipe automatic sprinkler system with a maximum discharge capability of 0.2 gpm/ft^2 over 2,000 ft^2. This discharge density is not considered adequate protection for crude rubber in 2,000 ft^2 piles. The automatic sprinkler protection is supplemented by a fixed high expansion foam system. The foam generators are mounted on the coping section to the fire walls that divide the warehouse into fire areas. Generators positioned on the internal fire walls are arranged to discharge foam into whichever compartment, as required.

Step 4: Hazard description.
The fundamental considerations associated with the hazard are given under Step 1. It should be further noted that 12 storage piles of 2,000 ft^2 each are located in the designated fire areas. Each individual pile is approximately 12 ft 6 in. high. Due to the piling arrangement of the rubber and the burning characteristics of rubber, no deduction is made for "stock" in the rate discharge determination.

Step 5: Rate of discharge determination.
The basic design objective is to determine the rate of expanded foam discharge in cubic feet per minute to submerge the hazard in a defined period of time. This can be accomplished by applying a rate formula developed by the NFPA Foam Committee. The formula is given in Figure 4-5.9, under sub-item 6. The formula can be applied by first calculating and then assigning values to the formula variables.

HIGH EXPANSION FOAM SYSTEM JOB WORK SHEET

Sheet: _____1_____ of: _____1_____

Designer: _____Staff_____

Date: _____1986_____

Installation Identification: _____GSA Defence Materials Warehouse_____

Hazard Classification: _____High density combustibles_____

Type of Protection: _____Dry Pipe Automatic Sprinkler — Fixed Hi-X Foam_____

Hazard Description: _____Crude Rubber in Piles_____

Rate of Discharge Determination:

  1. Submergence Volume (cubic feet)

   V = Floor area _____40,000_____ sq. ft. × Foam depth _____14.5 ft._____ = _____580,000_____ cu. ft.

  2. Submergence Time (Minutes) T = _____5_____

  3. Rate of Foam Breakdown by Automatic Sprinklers: $R_s = S \times Q$ where
   S shall be 10 cfm/gpm *and* Q shall be the total discharge from operating sprinklers

   R_s = 10 cfm × _____400_____ gpm = _____4000_____ cfm

  4. Compensation for Normal Foam Shrinkage — C_N: C_N = _____1.15_____

  5. Compensation for Leakage — C_L; C_L range is from 1 to 1.2: C_L = _____1.1_____

  6. Rate of Discharge CFM = $(V/T = R_s) \times C_N \times C_L$ _____151,800_____

Description, Number and Placement of Foam Generators: _____2 — 80,000 cfm foam generators per_____
_____storage bay. Placement on fire walls as shown_____

Foam Solution Rate: _____146 gpm/generator × 2 = 292 gpm_____

Foam Concentration Rate: _____3% proportion × 292 gpm = 9 gpm_____

Duration of Discharge: _____15 minutes of full operation_____

Gallons of Foam Required: _____Main and Reserve = 270 gals_____

Gallons of Water Required: _____4,245_____

Water Supply Information: _____Adequate for Demand Curve_____

Special Foam System Design Considerations: _____System is activated by automatic sprinkler_____
_____system dry pipe trip._____

Fig. 4-5.9. High expansion foam system job work sheet.

1. Submergence volume. (Re: NFPA 11A, Section 2-3.3.)
    Floor area: 200 ft × 200 ft = 40,000 ft^2
 Foam Depth:
    (a) 1.1 × height = 1.1 × 12.5 = 13.75 ft
    (b) Height + 2 ft − 12.5 ft + 2 ft = 14.5 ft
  Use the larger of the two values or 14.5 ft for calculations.
 Volume = Area × depth
 Volume = 40,000 sq ft × 14.5 ft = 580,000 cu ft
 (See Step 4—no deduction is made for stock)
2. Submergence time. (Re: NFPA 11A, Section 2-3.4.)
    5 minutes for high density materials with sprinkler protection.
3. Rate of foam breakdown for sprinklers. (Re: NFPA 11A, Section 2-3.5.)
   (a) Discharge from sprinklers:

   Q = 0.2 gpm/ft^2 × 2,000 ft^2 = 400 gpm
   (b) Apply formula:

   R_s = 10 cfm × 400 gpm = 4,000 cfm

4. Compensation for shrinkage: Set at 1.15 as a constant. (Re: NFPA 11A, Section 2-3.5.)
5. Compensation for leakage: Use 1.1 to allow for some leakage around doors. (Re: NFPA 11A, Section 2-3.5.)
6. Apply formula:

   cfm = $(V/T + R_s) \times c_N \times C_L$

   cfm = (580,000 ft^3/5 min + 4,000 cfm) × 1.15 × 1.1 = 151,800 cfm
 Note: The foam breakdown from sprinklers is a relatively small value compared to the total cfm rate.

Step 6: Description, number, and placement of generators.

HYDRAULIC CALCULATIONS

Subject: GSH Warehouse Job No: Sample Problem 3

Sheet No: 1 Of: 1

By: Staff Date: 1986

Chkd By: ___ Date: ___

Application Rate: NH GPM per sq ft Area: 40,000 ft

Minimum Solution Rate: 1.83 gpm/1000 ft³ Actual Solution Rate: 185 gpm/1000 ft³

Foam Maker Pressure and Rate: 50 psi @ 146 gpm Foam System: Fixed

Water Data: ___ Ref. Drawing: ___

Foam Maker Type and Location	Added GPM	Total GPM	Pipe Size In.*	Pipe & Equivalent Fitting Lgth. (Feet)**	Friction PSI/Ft C=100	Total PSI	Static PSI	Proportioner PSI	Req'd Pres PSI
Starting Point: 1				Elevation: 22 ft.	Pressure at Foam Maker:= 50 psi				
1	146	146							50
1-2-3		146	3.068	212' +E(7')=1 T(15)	.0387	9.1	9.5		68.6
				234'					
2 @ 3									68.6
3-6		146	4.026	200'	.0103	2.06			70.7
6-9-12	148	294	4.026	400'	.0375	15.0			85.7
Foam House		294						4.0	89.7
System demand		294							90.0

* "X" Indicates Extra Hvy. — Std. Wt. Otherwise

** See Sheet for Tabulation of Pipe and Fittings

Fig. 4-5.10. Hydraulic calculations.

Custom-built foam generators will be required for this problem. Each generator will have a capacity of 80,000 cfm with a foam solution rate of 146 gpm. (See given information with problem statement.) Five generators will be required to protect the entire warehouse. Generators mounted on interior fire walls will be equipped with baffels arranged to discharge foam into either adjacent compartment; electrical controls will be operated from the foam house. Generators are actually mounted 22 ft above the finished floor.

Step 7: Foam solution rate.

The foam solution rate per generator is given in Step 6. The solution rate requirement is 1.83 gpm per 1000 cu ft of foam production.

Solution rate = 80,000 ft³/1000 ft³ × 1.83 gpm = 146 gpm

Two generators require 292 gpm.

Step 8: Foam concentrate rate.

The foam selected for this problem proportions at 3 percent. Therefore, the concentrate rate is 3 percent × 292 gpm = 9 gpm.

Step 9: Duration of discharge.

Duration of discharge for the foam systems should be checked with the authority having jurisdiction. A basic minimum discharge time is 15 minutes of continuous operation.

Step 10: Gallons of foam required.

It is assumed that enough foam will be placed in storage to meet both a main and a reserve requirement: 9 gpm × 15 min × 2 = 270 gal.

Step 11: Gallons of water required.

The primary water supply must provide a rate of 283 gpm for 15 minutes or 4,245 gal. A like amount must be supplied for the secondary demand.

Step 12: The problem considers that the water suply to the foam house is adequate to meet the calculated demand for the system.

Step 13: The foam system is arranged to be activated by the automatic sprinkler system when the drypipe valve trips due to a sprinkler head operating. The system can also be activated manually.

Hydraulic analysis for Example 3:

Step 1: The inlet pressure requirement for the foam generator is 50 psi.

Step 2: The foam solution line supplying each foam generator and the riser pipe to the top of the fire wall are sized on the basis of a maximum flow velocity of 10 fps.

$$\text{Velocity} = \frac{0.4085 \times \text{gpm}}{d^2}$$

a. $10 \text{ fps} = \dfrac{0.4085 \times 146 \text{ gpm}}{d^2}$

b. $10d^2 = 59.64$

c. $d^2 = 5.96$

d. $d = 2.44$

e. Use a 3-in. pipe

Since the same size pipe is used from the foam generator to ground level, the hydraulic analysis can go from Reference point 1 to Reference point 3. The elevation head to be considered is 22 ft.

Step 3: The flow and pressure demand at the base of each riser supplying a foam generator is the same, since the generator sizes are equal. It is necessary to calculate a flow constant at this location so the pressure points upstream can be correctly adjusted for higher pressure values developed by friction loss between supply points. The demand constant is calculated as follows:

(Reference Point 3)

$Q = K\sqrt{P}$

$146 \text{ gpm} = K\,68.6$

Step 4: The ground-level cross-main connecting the foam generator risers is sized on the basis of a maximum flow velocity of 10 fps. The flow from two generators is used for the flow computations.

$$\text{Velocity} = \frac{0.4085 \times \text{gpm}}{d^2}$$

a. $V = \dfrac{0.4085 \times 292 \text{ gpm}}{d^2}$

b. $10d^2 = 119.282$

c. $d^2 = 11.928$

d. $d = 3.45$

e. Use a 4-in. pipe

Step 5: Determine the actual flow characteristics for the high expansion foam generator at Reference Point 4.

Use the K value (constant) determined in Step 3 to calculate the actual supply to the second foam generator at Reference Point 4. The new pressure at the riser base is (Reference 6) 70.7 psi from the hydraulic calculation sheet.

The higher pressure is used with the K value to determine the actual flow for the second high expansion foam unit.

$Q = K\sqrt{P}$

$Q = 17.6\sqrt{70.7}$

$Q = 148$

The actual flow increases by 2 gpm for the second generator.

Step 6: Determine the flow and pressure requirements at the foam house. Determine the friction loss for the total flow back to the foam hose and add in 4 psi for the foam proportioner.

Step 7: System demand.

The final system demand is 294 gpm at 90.0 psi at the foam proportioner inlet to supply the two high expansion foam generators. The water supply to the foam house must meet this demand.

THE ADVENT OF CLASS A FOAMS

Class A foams have been used extensively in wildland fire suppression. The success of Class A foam for the confinement, control, and extinguishment of natural cover fuel fires suggests that this type of foam may be effective for structural fire protection as foam solution in fire streams. Initial research has been conducted to quantify the fire fighting efficiency of Class A foams to improve the operating efficiency of these foams when compared to plain water fire streams. The National Fire Protection Research Foundation has published research findings on Class A foam effectiveness: one in December 1993[3] and one in November 1994[4]. A synopsis of the findings are presented below.

The National Fire Protection Research Foundation (NFPRF) sponsored a research program with Underwriters Laboratories, Inc. (UL) to investigate the effectiveness of Class A foams by means of three discharge devices: 1) a standard spray nozzle, 2) an air-aspirated spray nozzle, and 3) by injecting compressed air into the Class A foam solution. This research investigation has two objectives: 1) to develop test data related to the fire fighting effectiveness of Class A foam solutions as compared to water only, and 2) to conduct laboratory analysis of the Class A foam concentrate used in the performance tests.

Briefly, the initial fire test plan included a Class 20A wood crib fire with foam solution concentrates selected at 0.1, 0.3, and 0.5 percent. Adjunct variables included nominal expansion ratios of 5 for a standard nozzle at 15 gpm, 7.5 for an air-aspirated nozzle at 15 gpm, and 7.5 for injecting compressed air into the Class A foam solution.

The wood crib fire tests were conducted at UL's test facility located in Northbrook, IL, and are reported in the December 1993 publication.

In summary, the initial set of fire tests provide support of the following conclusions by the Technical Advisory Committee (TAC):

- Handheld hoselines supplied with Class A foam solutions provide enhanced fire fighting performance when compared to handheld hose lines supplied with water only.
- The best foam quality, as measured by retention and exposure protection tests, was achieved with compressed air foam.
- Results of the wood crib fire tests indicated superior characteristics in terms of fire control time for Class A foams when compared to water application only.

- Fire tests conducted with the air-aspirated test nozzle had the longest reignition times, while tests conducted with the Compressed Air Foam had the lowest crib weight losses.
- Exposure protection test results demonstrated the ability of the Class A foam to lengthen the ignition time of a combustible surface when compared to cribs protected by the same rate and duration of water.
- Retention of weight tests demonstrated that wood cribs exposed to Class A foam retain more weight than cribs treated with water.

The testing program outline above was very controlled in a laboratory setting. Foam applications were not subjected to many real world variables that could include wind, weather conditions, fuel geometry, pre-burn times, and human factors in the foam application. Despite such conditions, the reported testing program clearly supports a number of advantages for using Class A foam on structural type fires.

The Phase II research project report of 1994 reviews the conduct of structural fire suppression tests. These tests were also conducted at UL's test facilities in Northbrook, IL. A test cell measuring 30 by 36 by 30 feet was used for the Class A foam comparative analysis tests. Two fuel package scenarios were used as follows:

- The Series I UL 1626 residential fuel package consisted of a wood crib and simulated furniture positioned in one corner of the enclosure.
- The Series II fuel package consisted of a corner upholstered sofa scenario.

Fire test monitoring of the enclosure included measurements of the Class A foam solution or water flow rate; room temperature gradients at distances of 2, 9, 18, 24, 33, 48, and 72 inches below the ceiling; rate of heat release, oxygen content, smoke density, and heat flux. In each test series, observations were made of fire knockdown and damage to the walls of the enclosure and the fuel package.

Upon ignition, the fuel package was allowed to burn until **flashover** was achieved in the enclosure. Five seconds after flashover, a water application **or** a Class A foam solution was applied to suppress the fire using either a direct or indirect application method. The direct application method consisted of discharging the agent directly onto the walls of the enclosure **and** the fuel package. In contrast, the indirect application method consisted of discharging the agent first onto the ceiling and walls **and then** onto the fuel package.

The 1994 Class A Foam Study Report divides the summary information according to the Series I and II testing programs. The Series I abbreviated findings are summarized as follows:

- Class A foam using a direct application method took less time and quantity of agent to lower heat release to 500 KW than plain water.
- Class A Compressed Air Foam (CAF) using the indirect application method was more effective in reducing heat release values down to 500 KW.

The Series II abbreviated findings are limited to the following selective observations.

- The test results using Class A foam solutions generally provided for a reduced amount of total heat release from the fire and less damage to the sofa.
- Class A Compressed Air Foam applied at 7 gpm using the direct application method demonstrated the shortest time period and the lowest quantity of agent required to reduce the rate of heat released to 500 KW.
- The direct application method provides for a reduced amount of total heat release and less damage to the sofa when compared to the same tests conducted using the indirect application method.

Both reports recommend additional research on the application of Class A Foams with special attention given to hardware devices that include handheld fixed nozzles, proportioning equipment, and foam-generating equipment.

REFERENCES CITED

1. *NFPA Codes and Standards*, National Fire Protection Association, Quincy, MA.
2. "Flammable Liquid Storage Tank Protection," *National Foam Engineering Manual*, Section 6, National Foam, Lionville, PA (no date).
3. Carey, William M., *National Class A Foam Research Project Technical Report*, National Fire Protection Research Foundation, Quincy, MA, (December 1993).
4. Carey, William M., *National Class A Foam Research Project Technical Report, Phase II*, National Fire Protection Research Foundation, Quincy, MA (December 1994).

APPENDIX

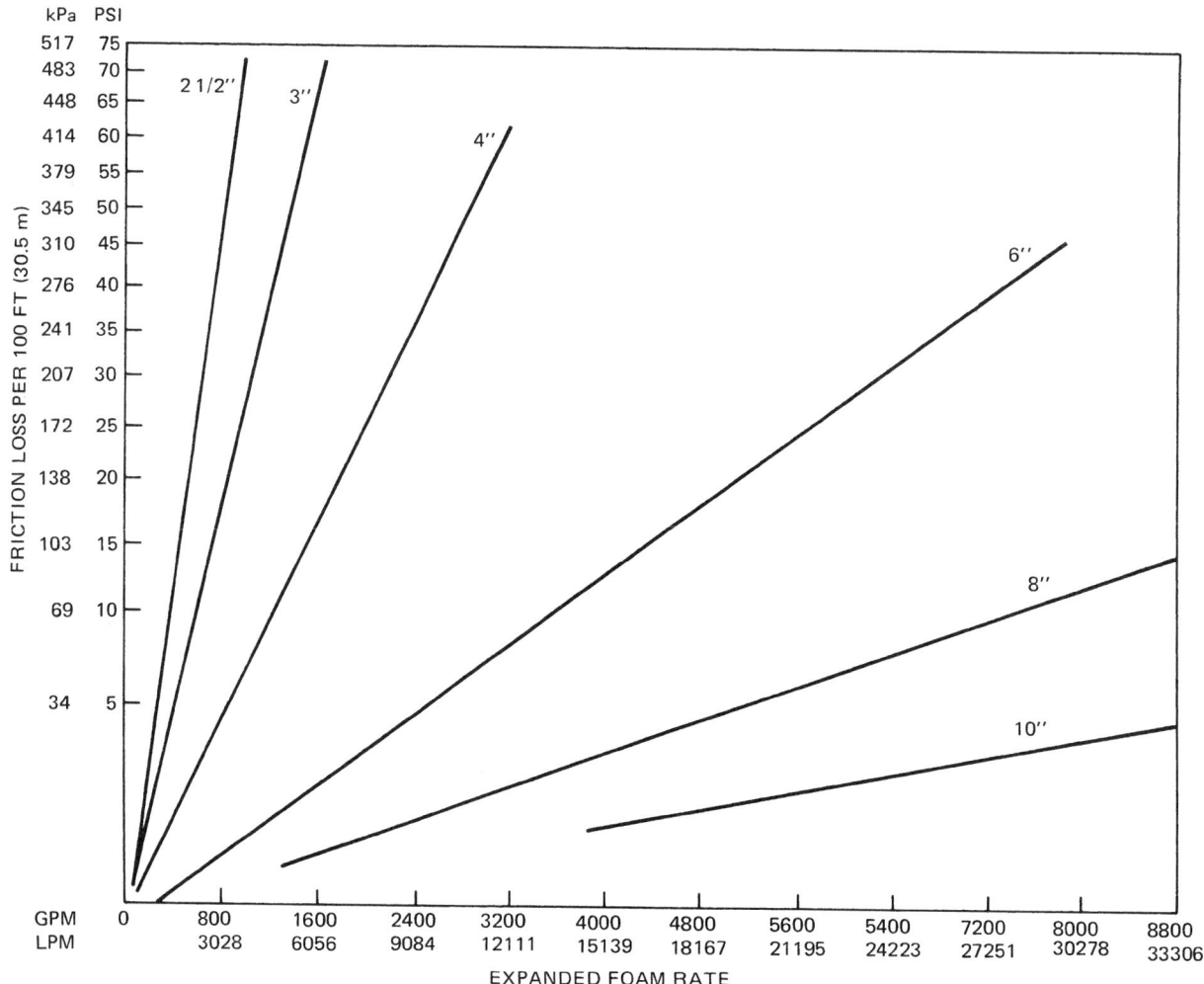

Fig. A-4-5.1.   *Foam friction losses—4:1 Expansion (2½″, 3″, 4″, 6″, 8″, and 10″ pipe).*

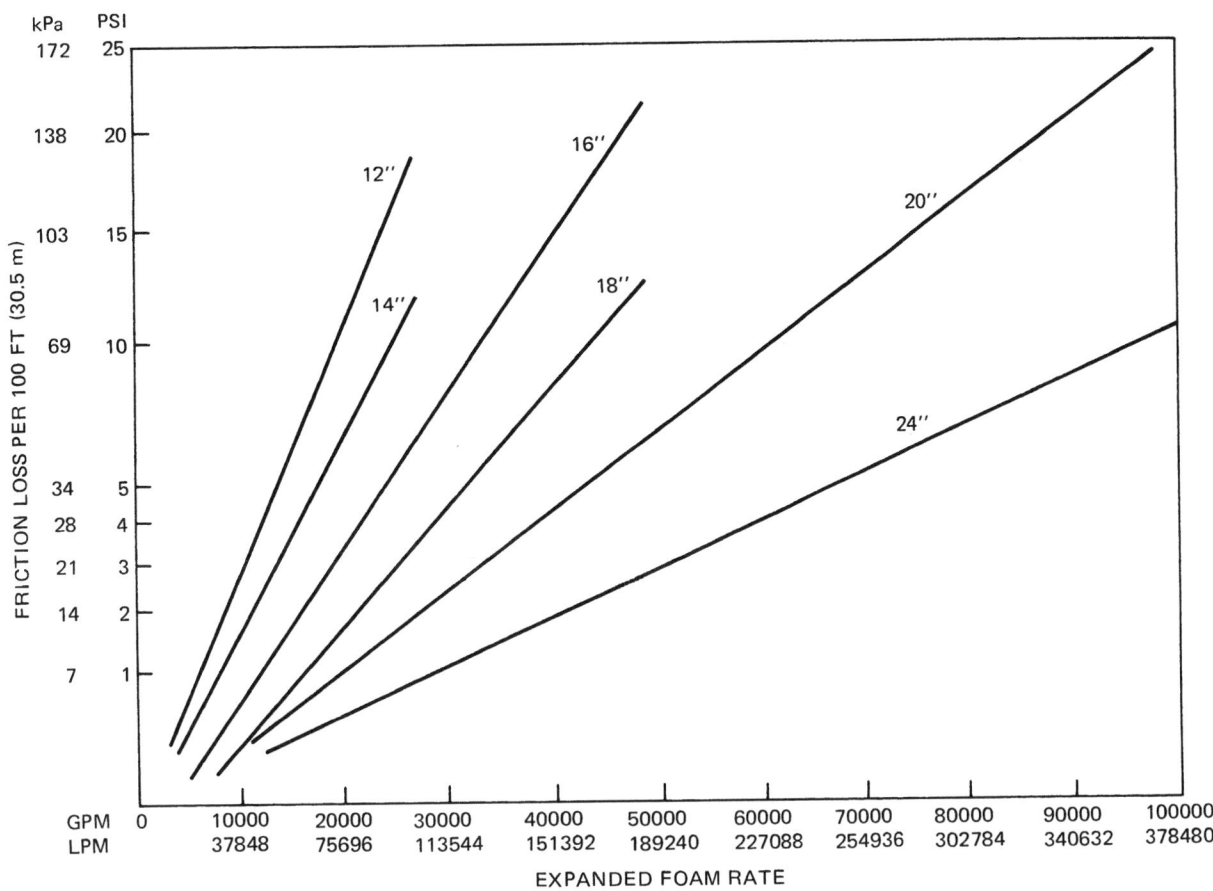

Fig. A-4-5.2. *Foam friction losses—4:1 Expansion (12″, 14″, 16″, 18″, 20″, and 24″ pipe).*

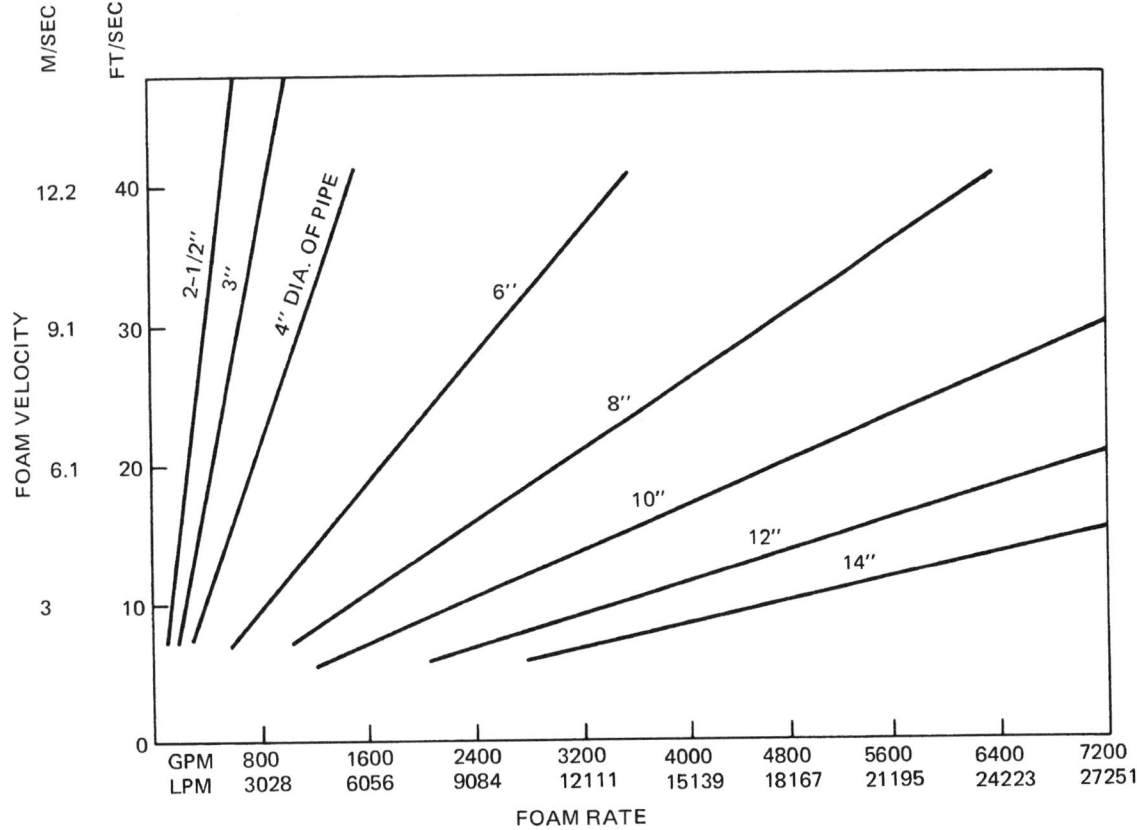

Fig. A-4-5.3. *Foam velocity vs. pipe size (2½″, 3″, 4″, 6″, 8″, 10″, 12″, and 14″ pipe).*

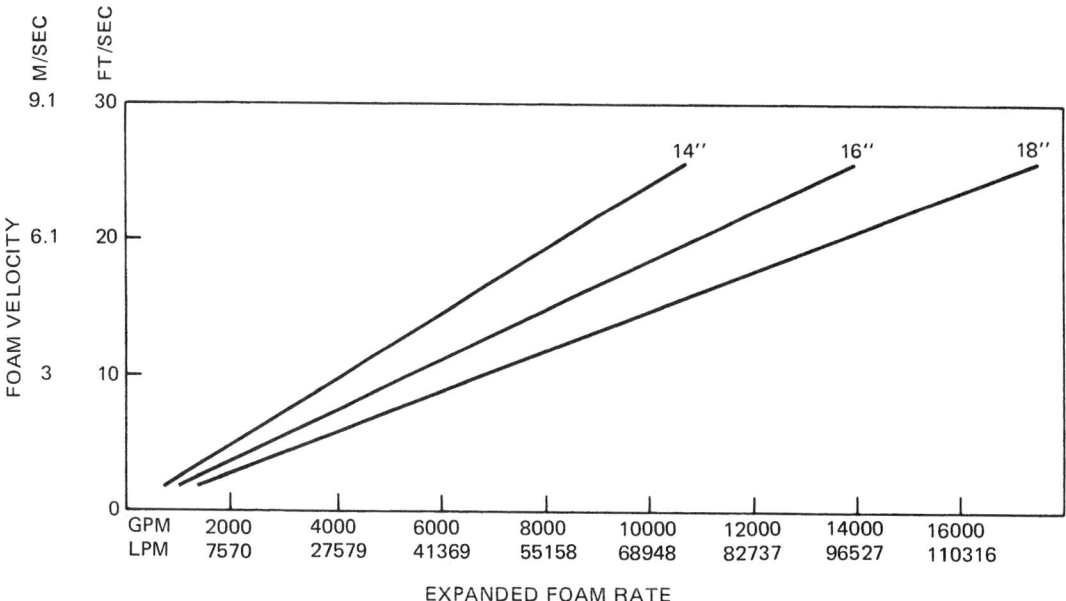

Fig. A-4-5.4.   Foam velocity vs. pipe size—Schedule 40 pipe (14″, 16″, and 18″ pipe).

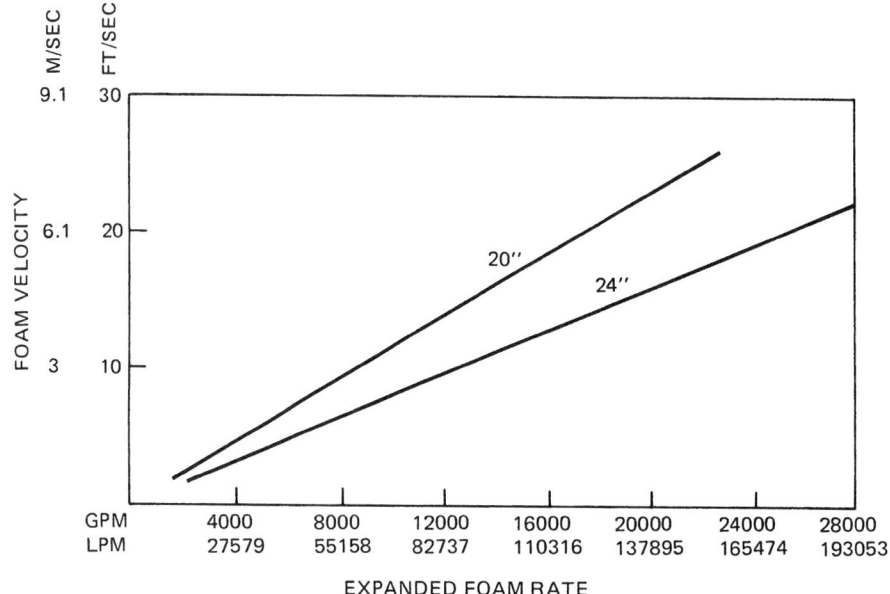

Fig. A-4-5.5.   Foam velocity vs. pipe size—Schedule 40 pipe (20″ and 24″ pipe).

HALON DESIGN CALCULATIONS

Casey C. Grant

INTRODUCTION

Halogenated agent extinguishing systems are a relatively recent innovation in fire protection, but, despite this, they already face extinction. As of January 1, 1994, the global production of fire protection halons in many countries ceased.

The obvious question is, "Why maintain this chapter on halon design calculations?" Although global production of fire protection halons essentially ceased on January 1, 1994, it is expected that this technology will continue for an extended period of time to address the modification and maintenance of existing systems, and new essential systems that will use recycled surplus stock.

The stratospheric ozone layer depletion issue is a problem confronting the global community unlike any other. Late in 1987, the United States and 24 other countries (including the European Economic Community) signed the Montréal Protocol to protect stratospheric ozone.[1] Originally, the protocol restricted the consumption of ozone-depleting chlorofluorocarbons (CFCs) to 50 percent of the 1986 use levels by 1998, and halon production was to be frozen in 1993 at 1986 production levels. But the November, 1992, Copenhagen revision to the Montréal Protocol accelerated this, such that all production of the chemicals ceased worldwide as of January 1, 1994.

The Montréal Protocol is based on unprecedented trade restrictions and is the first time nations of the world have joined forces to address an environmental threat in advance of fully established effects. The trade restrictions concern nations not participating in the agreement (the nonsignatories). Within one year of the agreement taking effect, each party shall ban the import of the bulk chemicals from the nonsignatory nations. About four years after the effective date of the agreement, imports of products containing the identified chemicals from nonsignatory nations would be banned. Within five years, products made with the chemicals (but not containing them) would be banned or restricted. This is truly significant since many products, including many electronic components, are currently manufactured using CFCs.

Today, high technology demands new and different fire protection techniques for which halon systems have proved ideal.

CHARACTERISTICS OF HALON

Background, Definition, and Classifications of Halon Compounds

Although there are a variety of methods available for applying halogenated agents, the most common is the total flooding system. The most popular halogenated agent is Halon 1301, with its superior fire extinguishing characteristics and low toxicity. Halogenated agent extinguishing systems are a promising tool for the fire protection engineer and have great potential for solving many of our fire protection problems now and in the future.

Halogenated extinguishing agents are hydrocarbons in which one or more hydrogen atoms have been replaced by atoms from the halogen series: fluorine, chlorine, bromine, or iodine. This substitution confers flame extinguishing properties to many of the resulting compounds that make them ideal for certain fire protection applications.

The halogenated extinguishing agents are currently known simply as halons, and are described by a nomenclature that indicates the chemical composition of the materials without the use of chemical names. This simplified system was proposed by James Malcolm at the U.S. Army Corps of Engineers Laboratory in 1950 and avoids the use of possibly confusing names.[2] The United Kingdom and parts of Europe still use the initial capital "alphabet" system, i.e., bromotrifluoromethane (Halon 1301) is BTM and bromochlorodifluoromethane (Halon 1211) is BCF. The number definition for the chemical composition of Halon 1301, perhaps the most widely recognized halogenated extinguishing agent, is: 1 (carbon), 3 (fluorine), 0 (chlorine), 1 (bromine), and 0 (iodine).

By definition, the first digit of the number represents the number of carbon atoms in the compound molecule; the second digit, the number of fluorine atoms; the third digit, the number of chlorine atoms; the fourth digit, the number of bromine atoms; and the fifth digit, if any, the number of iodine atoms. Trailing zeros in this system are not expressed. Figure 4-6.1 graphically demonstrates this concept by illustrating Halon 1301 in comparison to methane.

Casey Grant, P.E., is NFPA's Technical Director of Codes and Standards. He is a former member of the NFPA Technical Committee on Halogenated Fire Extinguishing Agent Systems and was previously Supervisor of Systems Design Engineering at Fenwal Incorporated.

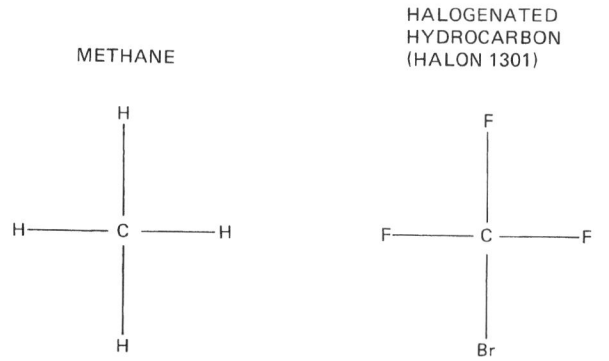

Fig. 4-6.1. Molecular composition of methane and Halon 1301.

There are three halogen elements commonly found in halon extinguishing agents used for fire protection: fluorine (F), chlorine (Cl), and bromine (Br). Compounds containing combinations of fluorine, chlorine, and bromine can possess varying degrees of extinguishing effectiveness, chemical and thermal stability, toxicity, and volatility. In general, the relevant properties of these three halogen elements are characterized as shown in Table 4-6.1.

Due to the many chemical combinations available, the characteristics of halogenated fire extinguishing agents differ widely. It is generally agreed that the agents most widely used for fire protection applications are Halon 1301, Halon 1211, Halon 1011, and Halon 2402. Also somewhat common is Halon 122, which has been used as a test gas because of its economic advantages. However, because of its widespread use as a test agent, many individuals have wrongly assumed that Halon 122 is an effective fire extinguishing agent. Table 4-6.2 illustrates the halogenated hydrocarbons most likely to be used today.

History

The earliest halogenated fire extinguishing agent known to be used for industrialized fire protection was carbon tetrachloride (Halon 104).[3] First becoming available as early as 1907, it was most widely used in handpump portable extinguishers and was popular due to its low electrical conductivity and lack of residue following application. Also referred to as "pyrene" extinguisher fluid, Halon 104 caused a number of accidental deaths and serious injuries due to its toxicity, and eventually its use was halted during the 1950s.

Methyl bromide (Halon 1001) gained popularity after it was discovered in the late 1920s to be a more effective extinguishing agent than carbon tetrachloride. Due to its high toxicity, it was never used in portable extinguishers even though it was used extensively in British and German air-

craft and ships during World War II. Interestingly, methyl bromide possesses a narrow "flammability" range between 13.5 and 14.5 percent in air, though above and below this range it is an efficient fire extinguishant. Germany developed bromochloromethane (Halon 1011) in the late 1930s to replace methyl bromide, but it failed to enjoy widespread use until after World War II.[4]

Thus, prior to World War II, three halogenated fire extinguishing agents were available: Halon 104, Halon 1001, and Halon 1011. Yet because of their inherently high toxic nature, these agents slowly disappeared from typical system applications. By the mid-1960s Halon 104 and Halon 1001 were no longer being used, and Halon 1011 was only in limited use for specialized explosion suppression applications. Figure 4-6.2 represents a chronology chart that indicates the usage of these early halons as well as the halons more commonly used today.

Joint research was undertaken in 1947 by the U.S. Army Chemical Center and the Purdue Research Foundation to evaluate the fire suppression effectiveness and toxicity of the large number of available agents.[2] After testing more than 60 new agents, 4 were selected for further study: dibromodifluoromethane (Halon 1202), bromochlorodifluoromethane (Halon 1211), bromotrifluoromethane (Halon 1301), and dibromotetrafluoromethane (Halon 2402). Further testing revealed that Halon 1202 was the most effective yet also most toxic, while Halon 1301 was the second most effective and least toxic. As a result of this testing, the use of halon to provide fire protection for modern technology took on new dimensions. Halon 1202 was used by the U.S. Air Force for military aircraft engine protection while the Federal Aviation Administration (FAA) selected Halon 1301 for a similar application in commercial aircraft engine nacelles.[5] Portable extinguishers using Halon 1301 were implemented by the U.S. Army. The use of total flooding systems originated in 1963 and in the following five years several total flooding systems were installed based on carbon dioxide system technology.

In 1966, attention began to focus on the use of Halon 1301 for the protection of electronic data processing equipment. That year, the NFPA organized a Technical Committee (NFPA 12A) to standardize the design, installation, maintenance, and use of halon systems. Their resulting work was officially adopted by the NFPA membership as a standard in 1968.[6] Subsequent recognition that there were differences among the halon agents made it apparent that separate standards would be necessary. The initial halon standard, NFPA 12A, *Standard for Halon 1301 Fire Extinguishing Systems* (hereinafter referred to as NFPA 12A), focused on the use of

TABLE 4-6.1 *Contributing Characteristics of Fluorine, Chlorine, and Bromine*

	Fluorine	Chlorine	Bromine
Stability to Compound	Enhances	—	—
Toxicity	Reduces	Enhances	Enhances
Boiling Point	Reduces	Enhances	Enhances
Thermal Stability	Enhances	Reduces	Reduces
Fire Extinguishing Effectiveness	—	Enhances	Enhances

TABLE 4-6.2 *Halons Commonly Used for Fire Protection*

Chemical Name	Formula	Halon Number
Methyl Bromide	CH_3Br	1001
Methyl Iodide	CH_3I	10001
Bromochloromethane	CH_2BrCl	1011
Dibromodifluoromethane	CF_2Br_2	1202
Bromochlorodifluoromethane	CF_2BrCl	1211
Dichlorodifluoromethane*	CF_2Cl_2	122
Bromotrifluoromethane	CF_3Br	1301
Carbon Tetrachloride	CCl_4	104
Dibromotetrafluoroethane	$C_2F_4Br_2$	2402

* A popular test gas without substantial fire extinguishing properties.

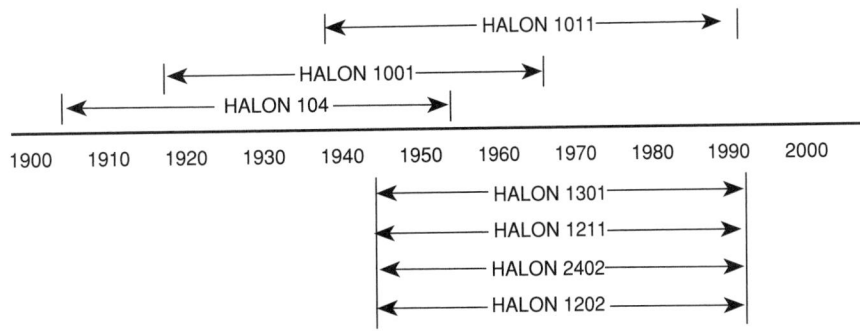

Fig. 4-6.2.   Time span usage of selected halons.

Halon 1301 due to its high desirability and growing popularity.[7] Work on an additional standard, NFPA 12B, *Standard on Halon 1211 Fire Extinguishing Systems*, concerning the use of Halon 1211, was started in 1969 and was officially adopted by the NFPA as a standard in 1972.[8] A tentative standard on the use of Halon 2402 (NFPA 12CT) was established, but has not been officially adopted.[9]

Another NFPA committee directly concerned with the use of halon is the NFPA Committee on Electronic Computer/Data Processing Equipment (NFPA 75, *Standard for the Protection of Electronic Computer/Data Processing Equipment*).[10] Even though this standard was adopted in 1961, the use of halon was not considered until after 1972, when extensive testing by several major companies demonstrated that the use of Halon 1301 was suitable for protecting electronic computer and data processing equipment.[11] Today, Halon 1301 is the most widely used extinguishing agent for this purpose in the United States and throughout much of the world. However, certain areas of Europe have preferred Halon 1211 and 2402.

In anticipation of the worldwide production phase-out of fire protection halons, which eventually settled at January 1, 1994 for developed countries, a new committee was established during 1992 within the NFPA standards-making system designated as the Technical Committee on Alternative Protection Options to Halon. The committee's first document is NFPA 2001, *Standard on Clean Agent Fire Extinguishing Systems*, which addresses the design, installation, maintenance, and operation of total-flooding fire extinguishing systems that use halon replacement agents.[12]

Halon 1301

Attributes and limitations: Of all the halogenated extinguishing agents used in fire protection today, Halon 1301 is by a wide margin the most commonly used. The primary use of this agent is for the protection of electrical and electronic equipment, flammable liquids and gases, and surface-burning flammable solids such as thermoplastics. Areas normally or frequently occupied, air and ground vehicle engines, and other areas where rapid extinguishment is important or where damage to equipment or materials or cleanup after use must be minimized are also ideally protected by this agent. However, Halon 1301 is not a panacea, and it is appropriate to recognize its limitations as well as its attributes. The benefits of Halon 1301 are: fast chemical suppression, penetrating vapor, clean (no residue), noncorrosive, compact storage volumes, nonconductive, and colorless (no obscuration). There are also limitations to using

Halon 1301: it has minimal extinguishing effectiveness on reactive metals and rapid oxidizers, it may have unfavorable side effects on deep-seated Class A fires, the agent is expensive, and it is potentially harmful to the environment. Obviously, the most significant limitation is the detrimental effect that the halons have on the earth's stratospheric ozone layer.

Because Halon 1301 inhibits the chain reaction of the combustion process, it chemically suppresses the fire very quickly, unlike other extinguishing agents that work by removing the fire's heat or oxygen. Stored as a liquid under pressure and released at normal room temperature as a vapor, Halon 1301 gets into blocked and baffled spaces readily and leaves no corrosive or abrasive residue after use. A high liquid density permits compact storage containers, which on a comparative weight basis, makes Halon 1301 approximately 2.5 times more effective as an extinguishing agent than carbon dioxide. Since it is virtually free of electrical conductivity, Halon 1301 is highly suitable for electrical fires. Halon 1301 is a colorless vapor when discharged into a hazard volume, though it sometimes temporarily clouds the volume due to the chilling of any moisture in the air. But of all its attributes, the most promising is that of people compatibility, for unlike other extinguishing agents, Halon 1301 is essentially nontoxic in the concentrations usually required for fire suppression.

There are several types of flammable materials on which Halon 1301 is ineffective and not recommended. Reactive metals such as potassium, Nak eutectic alloy, magnesium, sodium, titanium, and zirconium burn so intensely that they overpower the agent's extinguishing abilities.[5] Included with these are the metal hydrides such as lithium hydride, and petroleum solvents such as butyl-lithium. Autothermal decomposers and fuels that contain their own oxidizing agent will also burn freely in the presence of halon agents. These latter substances, such as gunpowder, rocket propellants, and cellulose nitrate, have an oxidizer physically too close to the fuel and the agent cannot penetrate the fire zone fast enough. Halon is also not effective in preventing the combustion or reaction of chemicals capable of autothermal decomposition such as hydrazine or organic peroxides. Even though Halon 1301 is effective with certain surface-burning flammable solids such as thermoplastics, deep-seated Class A fires typically require relatively high agent concentrations for long soaking periods. When exposed to deep-seated fires for long periods of time, Halon 1301 may decompose into toxic and corrosive products of decomposition. Therefore, it is important that the agent be dispersed

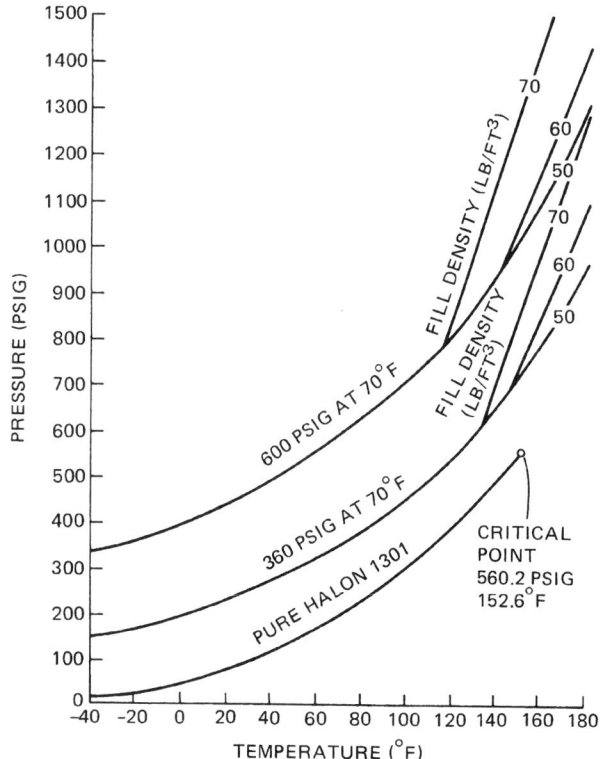

Fig. 4-6.3. *Temperature pressure relationship for pure Halon 1301.*

while the fire is small. The expense necessary to purchase, install, and maintain a properly functioning Halon 1301 system for more specific Class A hazards is often not economically justified. Halon 1301 fire suppression systems are usually not associated with everyday commodities, but instead are found in applications pertaining to highly valued risks.

Properties

Physical properties: On the average, Halon 1301 requires 10 percent less agent on a gas-volume basis than does Halon 1211 to extinguish any given fuel.[2] However, both agents are approximately 2.5 times more effective on a weight-of-agent basis than carbon dioxide. Halon 1301 is a gas at 70°F with a vapor pressure of 199 psig. Although this pressure would adequately expel the material, it decreases rapidly to 56 psig at 0°F and to 17.2 psig at −40°F. Therefore, it is necessary to increase the container pressure with dry nitrogen either to 360 or 600 psig at 70°F, ensuring adequate performance at all temperatures. Figure 4-6.3 demonstrates the temperature

TABLE 4-6.3 *Selected Physical Properties of Halon 1301*

Boiling Point	−72.0°F
Freezing Point	−270.4°F
Specific Gravity of Liquid (@70°F)	1.57
Specific Gravity of Vapor (@70°F)	5.14
Liquid Density @70°F	98.0 lb/ft³
Vapor Density @70°F	7.49 lb/ft³ (standard)
Critical Temperature	152.6°F
Critical Pressure	575 PSIA

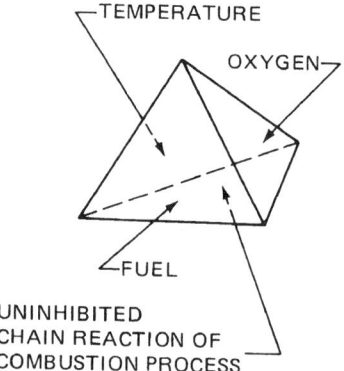

Fig. 4-6.4. *The tetrahedron of fire.*

pressure profile for Halon 1301 and Halon 1301 superpressurized with dry nitrogen.

Halon 1301 is normally stored in a pressure vessel as a liquid before it is released to occupy the hazard volume as a vapor. With a boiling point of −72°F, it is approximately 1.5 times more dense than water in its liquid phase and approximately 5 times heavier than air in its vapor phase. Thus, Halon 1301 vapor will typically escape through openings in the low portions of a totally flooded volume. Other physical properties are shown in Table 4-6.3.

Traditionally, there were three distinct elements assumed for combustion: heat, fuel, and oxygen. Known as the fire triangle, this theory had to be modified as halons became more widely used and better understood. Typical fire extinguishment involves either removing the fuel from the fire, limiting oxygen to the fire (smothering), or removing the heat (quenching). The halons do not extinguish fire in any of these ways, but instead breakup the uninhibited chain reaction of the combustion process. The tetrahedron of the fire, as it is now called, is shown in Figure 4-6.4.

The extinguishing mechanism of the halogenated agents is not completely understood, yet there is definitely a chemical reaction that interferes with the combustion process. The halogen atoms act by removing the active chemical species involved in the flame chain reaction. While all the halogens are active in this way, bromine is much more effective than chlorine or fluorine. With Halon 1301 (54 percent by weight bromine), it is the bromine radical that acts as the inhibitor in extinguishing the fire. Yet the fluorine in the molecule also serves a specific task since it is the fluorine that gives the agent thermal stability and keeps Halon 1301 from decomposing until approximately 900°F.[13]

Extinguishing effectiveness: As shown in Figure 4-6.5, the four types of fire are ordinary combustibles (Class A), flammable liquids and gases (Class B), electrical (Class C), and reactive metals (Class D).[5]

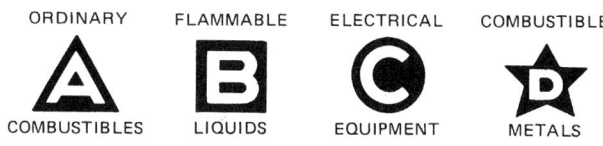

Fig. 4-6.5. *The four classes of fire.*

It was previously mentioned that Halon 1301 is ineffective on Class D fires and is not as desirable as other agents in extinguishing deep-seated Class A fires. The effectiveness of Halon 1301 on Class A fires is not as predictable as with other classes of fire. It depends to a large extent upon the burning material, its configuration, and how early in the combustion cycle the agent is applied. Most plastics behave as flammable liquids and can be extinguished rapidly and completely with 4 to 6 percent concentrations of Halon 1301.[7]

Other materials, particularly cellulosic products, can in certain forms develop deep-seated fires in addition to flaming combustion. The flaming portion of such fires can be extinguished with low 4 to 6 percent Halon 1301 concentrations, but the glowing deep-seated portion of the fire may continue under some circumstances. Even so, the deep-seated fire can be controlled since its rate of burning and consequent heat release will be reduced. Considerably higher concentrations (18 to 30 percent) of Halon 1301 are required to achieve complete extinguishment, but these levels are seldom economical to apply and their application may result in unwanted products of decomposition. However, the concept of controlling deep-seated fires with halogenated agents has been accepted in the respective NFPA standards.[7]

It is Class B and Class C fires for which halon is particularly well suited. The most common applications involve Class C electrical hazards, with the increase in popularity of Halon 1301 keeping well in stride with the development of high technology. Typically, electrical and electronic equipment are protected with a concentration of 5 percent Halon 1301 by volume, though a significantly lower concentration will suitably extinguish a potential fire.[15] The concentrations necessary to extinguish Class B fires have been the subject of much testing with results that vary widely. The effectiveness of halogenated agents on flammable liquid and vapor fires is quite dramatic, especially in total flooding systems. Rapid and complete extinguishment is obtainable with low concentrations of the agent.[7] To be effective, the fire must be contained (such as inside a building) so that the agent can react with it; Halon 1301 applied to large exterior running pool fires dissipates into the atmosphere without penetrating the flame zone.

Corrosive effects of undecomposed halons: Unlike Halon 1301 and Halon 1211, the early nonfluorinated halogenated agents had significant corrosive problems. Laboratory tests by DuPont in a 44-month exposure period with aluminum, magnesium, steel, stainless steel, titanium, and brass exposed to undecomposed Halon 1301 support the fact that this agent will not corrode these metals, which may all commonly be used in fixed fire extinguishing systems.[13] This is not surprising from a chemical standpoint because the presence of the fluorine atom in a molecule generally reduces its chemical reactivity and corrosive properties and increases its stability.

The presence of free water in systems containing Halon 1301 should be avoided. Free water is defined as the presence of a separate water phase in the liquid halon. When present in a small quantity, free water can provide a site for concentrating acid impurities into a corrosive liquid.[16] This should not be confused with dissolved water, which is not a problem in a Halon 1301 system.

Halon 1301 is inert toward most elastometers and plastics. In general, rigid plastics that are normally unaffected include polytetrafluorethylene, nylon, and acetal copoly-

TABLE 4-6.4 *Permitted Exposure Time for Halon 1301*

Concentration Percent by Volume		Permitted Time of Exposure
0 – 7%	Normally Occupied Areas	15 minutes
7 – 10%		1 minute
Above 10%		Not permitted
0 – 7%	Normally Unoccupied Areas	15 minutes
7 – 10%		1 minute
10 – 15%		30 seconds
Above 15%		Prevent exposure

mers. Most of the commonly used plastics undergo little, if any, swelling in the presence of Halon 1301, with the exception of ethyl cellulose and possibly cellulose acetate/butyrate. Elastomers are particularly suitable when exposed to Halon 1301 for extended periods of time with the notable exception of silicone rubber.[13] Halons decomposed at high temperatures during suppression produce halogen acids and free halons that can be corrosive.

Toxicity

General toxic properties: The relative safety of Halon 1301 has been established through more than 30 years of medical research involving both humans and test animals. No significant adverse health effects have been reported from the proper use of Halon 1301 as a fire extinguishant since its original introduction into the marketplace.[7]

Early studies by the U.S. Army Chemical Center on Halon 1301 determined the approximate lethal concentration for a 15-minute exposure to be 83 percent by volume.[2] Animals exposed to concentrations below lethal levels exhibit two distinct types of toxic effects. Concentrations greater than 10 percent by volume produce cardiovascular effects such as decreased heart rate, hypotension, and occasional cardiac arrythmias.[17] Concentrations of Halon 1301 greater than 30 percent by volume result in central nervous system changes including convulsions, tremors, lethargy, and unconsciousness. Effects are considered transitory and disappear after exposure.[18]

Human exposure to concentrations of Halon 1301 greater than 10 percent by volume have shown indications of pronounced dizziness and a reduction in physical and mental dexterity.[19] With concentrations between 7 and 10 percent by volume, subjects experienced tingling of the extremities and dizziness, indicating mild anesthesia. Exposure to Halon 1301 concentrations less than 7 percent by volume have little effect, with the exception of a deepening in the tone of voice caused by a higher density in the medium between the vocal chords. The effects at all levels of concentration disappear quickly after removal from the exposure. Testing of Halon 1301 for potential teratogenic (i.e., altering the normal process of fetal development) and mutagenic (a carcinogen in humans) effects have indicated that no serious problems exist.[5]

Most fire protection applications today have a design concentration of 5 percent by volume, thus the question of toxicity is not usually a serious concern. Exposure limitations for Halon 1301 (indicated by NFPA 12A) are summarized in Table 4-6.4.[7]

In addition to possible toxic effects, liquid Halon 1301 (including the spray in the immediate proximity of a discharge) may freeze the skin on contact and cause frostbite. However, direct contact is necessary for this to occur and is unlikely, since with engineered Halon 1301 fire extinguishing systems the discharge nozzles are typically distant from all occupants.

Products of decomposition: Consideration of the life safety of Halon 1301 must also include the effects of breakdown products which have a relatively higher toxicity than the agent itself. Upon exposure to flames or hot surfaces above approximately 900°F, Halon 1301 decomposes to form primarily hydrogen bromide (HBr) and hydrogen fluoride (HF).[20] Trace quantities of bromine (Br_2), carbonyl fluoride (COF_2), and carbonyl bromide ($COBr_2$) have been observed, but the quantities are generally too small to be of concern. Although small amounts of carbonyl halides (COF_2 and $COBr_2$) were reported in early tests, more recent studies have failed to confirm the presence of these compounds. Table 4-6.5 summarizes the predominant products of decomposition for Halon 1301.[21]

The primary toxic effect of the decomposition products is irritation. Even in concentrations of only a few parts per million, the decomposition products have characteristically sharp, acrid odors. This characteristic provides a built-in warning system since the irritation becomes severe well in advance of truly hazardous levels. In addition, the odor also serves as a warning that carbon monoxide and other potentially toxic products of combustion may be present. Prompt detection and rapid extinguishment of a fire will produce the safest post-extinguishment atmosphere.

Other Halons

Physical properties: The predominant halogenated agent in existence today for total flooding fire extinguishing systems is Halon 1301, though some areas of Europe have utilized Halon 1211 for this purpose. One reason for this use of Halon 1301 (besides toxicity) is the ability of the agent to vaporize and penetrate all portions of the hazard volume. Table 4-6.6 shows that Halon 1301 has the lowest boiling point and Halon 1211 has the second lowest.

With the discharge of a halon system at ambient temperature, Halon 1301 flashes to a vapor almost instantaneously, while Halon 1211 tends to pool momentarily. Agents with boiling points exceeding the temperature of the hazard volume will stay liquid until heated by the fire itself. These high boiling point halogenated agents have two distinct attributes: they can be projected in a liquid stream and they have a quenching effect in addition to breaking the unin-

TABLE 4-6.6 *Selected Physical Properties of Typical Halogenated Fire Extinguishing Agents*

Halon Number	Type of Agent	Approx. Boiling Point (°F)	Approx. Freezing Point (°F)	Specific Gravity of Liquid (@70°F)
104	liquid	170	−8	1.59
1001	liquid	40	−135	1.73
1011	liquid	151	−124	1.93
1202	liquid	76	−223	2.28
1211	liquefied gas	25	−257	1.83
1301	liquefied gas	−72	−270	1.57
2402	liquid	117	−167	2.17

hibited chain reaction. Thus, portable extinguishers generally use Halon 1301 as a propellant for other halon agents.

Toxicity

One of the primary reasons that Halon 1301 is the most preferred of the halogenated agents is its relatively low toxicity, as discussed earlier. Table 4-6.7 compares the approximate lethal concentration of both the natural and decomposed vapors for a variety of fire extinguishing agents. Included with this list of halon agents is carbon dioxide for sake of comparison. As a natural vapor, Halon 1301 is the least toxic halogenated agent. Carbon dioxide may appear to compare favorably with Halon 1301, yet high concentrations of carbon dioxide are necessary for fire extinguishment, which also makes the hazard volume lethal to human occupants.

Halon in the Fire Protection Spectrum

Halogenated agent extinguishing systems are only one segment of the total fire protection spectrum. Good engineering judgement is necessary when trying to determine the applicability of halon and whether it should be used instead of, or in addition to, other fire protection measures. It must be clearly understood that halogenated agent extinguishing systems are not the panacea for all fire hazards, yet they do offer a safe method to extinguish certain fires in their very early stages. Thus, these systems are commonly applied to situations where even the smallest fire is absolutely unthinkable.

As an example, total computer room fire protection might involve several different control measures addressing different possible fire conditions. Table 4-6.8 illustrates this concept, based on the different stages of a growing fire. This is not a rigid

TABLE 4-6.5 *Predominant Halon 1301 Decomposition Products*

Compound	Formula	ALC* for 15 Minute Exposure PPM by Volume in Air
Hydrogen Fluoride	HF	2,500
Hydrogen Bromide	HBr	4,750
Bromine	Br_2	550
Carbonyl Fluoride	COF_2	1,500
Carbonyl Bromine	$COBr_2$	—

* Approximate Lethal Concentration.

TABLE 4-6.7 *Approximate Lethal Concentrations (ppm) for 15 Minute Exposure to Vapors of Various Fire Extinguishing Agents*

Formula	Halon Number	Natural Vapor	Decomposed Vapor
CCl_4	104	28,000	300
CH_3Br	1001	5,900	9,600
CH_2ClBr	1011	65,000	4,000
CF_2Br_2	1202	54,000	1,850
CF_2ClBr	1211	324,000	7,650
CF_3Br	1301	832,000	14,000
$C_2F_4Br_2$	2402	126,000	1,600
CO_2	—	658,000	658,000

TABLE 4-6.8 *Necessary Control Measures for Computer Room Fire Stage Sequence*

Fire Stage	Control	Serious Danger Concern
1. Pre-Ignition	Good Housekeeping Practices, Control Combustible Furnishings and Interior Finish	
2. Initial Pyrolysis	Smoke Detection System	Occupants and Business Interruption
3. Incipient	Portable Fire Extinguishers, Halon 1301 Automatic Suppression System	Occupants and Contents
4. Pre-Flashover	Automatic Sprinklers	Occupants and Structure
5. Post-Flashover	Fire Walls, Compartmentalization	Surrounding Structures

description of the fire protection requirements of every computer room, but instead an example of how total fire protection is the overall objective when approaching a design situation.

An important factor of developing halogenated agent extinguishing systems is the interaction of all concerned individuals. To design, install, maintain, and operate a halon system requires a cooperative effort from a number of different groups. As shown in Figure 4-6.6, these individuals include the end users, consultants, manufacturers, installers, insurance representatives and other selected authorities. Representatives from all these groups work together to develop and enhance model codes, which provide guidance and understanding for proper halon system usage.

SYSTEM CONFIGURATIONS

Detection

The three primary parts of every halogenated agent extinguishing system are detection, control panel, and agent delivery. Since there is no single type of detector that offers the ultimate for every application, consideration must be given to the types of combustibles and combustion that are likely to occur in the protected area.

Photoelectric and ionization smoke detectors have different response characteristics to fires and can be susceptible to false or unwanted alarms. Thermal detectors, although more reliable, react more slowly to fire conditions. In certain applications, speed is critical and optical detectors would be required.

To optimize the speed and reliability of detection systems, it is important to use two different types of detectors on two separate detection loops within the hazard area. This method is referred to as cross-zone detection. Each detection loop functions independently to provide both added reliability with a comforting degree of redundancy.[22]

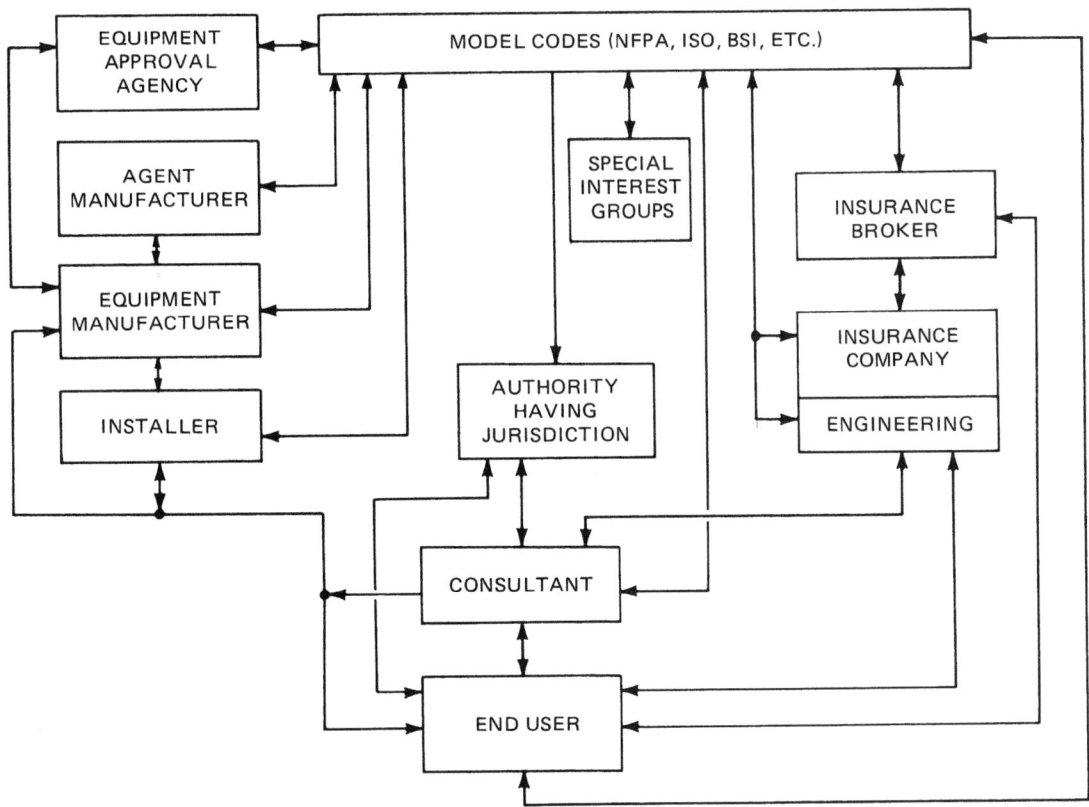

Fig. 4-6.6. Typical inter-relationship of halon fire protection interests.

TABLE 4-6.9 *Typical Control Unit Features*

	Halon Zone	Fire Alarm Zone
Initiating Circuit	Two Cross Zone Detection Circuits	One Circuit for Detection
Signaling Circuit	Multiple Signaling Sequence	Multiple Signaling Sequence
Release Circuit	One Circuit	None

Control Panels

Features: As its name implies, the control panel is the device that controls system operation and allows the system to function as designed. When a control panel protects more than one area, each individual area is referred to as a zone of protection. Each zone of every halon control panel has three different types of circuits: initiating, signaling, and release. A fire alarm zone and halon zone are compared in Table 4-6.9 to illustrate the differences between these circuit types. It is unusual for a single halon control panel to protect more than five zones at once due to the high number of circuits required. Fire alarm control panels, on the other hand, may have dozens of individual zones.

Initiating circuits provide the input into the panel and support automatic detectors, manual pull stations, and other initiating devices. Automatic detectors are normally cross-zoned, which implies two separate detection circuits. One circuit is required for prealarm and both circuits are necessary for halon release. The signaling circuits, sometimes referred to as bell or auxiliary circuits, are used for audible/visual alarms and other auxiliary functions. The release circuits allow the halon to release from the containers and are sometimes referred to as firing, solenoid, initiator, dump, or halon circuits.

Modes of operation: At any time, a halon control panel and the halon system could be in one of four modes of operation; as shown in Table 4-6.10 these include unpowered, normal, alarm, and trouble condition. The alarm condition is further definable with prealarm, prerelease, release, and postrelease condition. Typical systems utilizing cross-zoning detection activate, when required, into prealarm and/or release condition, but this often becomes more complicated with time delays, abort switches, and other auxiliary functions. Unless otherwise specified, manual pull stations activate all alarm conditions, override abort switches, if present, and immediately release the halon. These different alarm conditions provide a convenient mechanism for sequential operation of audible/visual signaling, equipment shutdown, fire service notification, and other auxiliary functions.

TABLE 4-6.10 *Modes of Control Panel Operation*

Unpowered Condition	Off.
Normal Condition	On.
Alarm Condition:	
Prealarm	One detector activates.
Prerelease	Two cross-zoned detectors activate. Time delay starts.
Release	Time delay ends or manual pull station activates. Halon is released.
Postrelease	Halon has been released.
Trouble Condition	Failure or disruption of field wiring. Insufficient power input.

TABLE 4-6.11 *Comparison of Different Methods of Agent Delivery*

	Central Storage	Modular	Shared Supply
Hardware Cost	Moderate	High	Moderate
Installation Cost	Moderate	Low	Moderate
Design Simplicity	Difficult	Simple	Difficult
Installation Simplicity	Difficult	Medium	Difficult
Operation and Maintenance Simplicity	Medium	Medium	Medium
Reliability	Moderate	High	Low
Future Flexibility	Low	High	Low

Control panel economics: As halogenated agent extinguishing systems become more numerous, the frequency of large-scale projects with multiple halon zones in a single facility is increasing. Today, entire data processing centers and telecommunications buildings are protected with Halon 1301 systems. To protect a large building with many halon zones, it may appear that the most effective way of configuring the system is by using a single large control panel with the capacity for all required halon zones. This is not true, since there is a limitation to the number of halon zones that any one halon panel can effectively manage. Figure 4-6.7 illustrates an alternative method, where the individual halon zones of a large building each have their own halon panel wired to give an alarm or trouble signal to a central fire alarm panel.

A typical halon zone requires an average of 12 wires to support all the necessary system functions. Thus, the cost of running multiple wires and large conduit instead of only two wires (for interpanel communication) often offsets the cost of smaller, more numerous panels located near the halon zones. This configuration offers flexibility for future consolidations or additions, which are common for high technology facilities. Aesthetics are enhanced at the master control location and system operation is simplified. Installation checkout and servicing is easier when the halon control panel is within the hazard area. Finally, the overall system is more reliable due to less wiring, lack of design complexity, simplified maintenance, and multi-source dependence.

Agent Delivery

In addition to the control panel and detection, the other primary part of every halogenated agent extinguishing system is agent delivery. The agent delivery includes the discharge nozzles, agent storage container(s), release mechanism, and associated piping. As shown in Table 4-6.11, three methods of agent delivery exist: (1) central storage, (2) modular, and (3) shared supply. Central storage has the container(s) centrally located, with the agent piped accordingly. This method is popular due to its similarity with carbon dioxide system technology (which helped develop early systems), along with usually having the lowest initial cost. Modular systems use smaller containers strategically located throughout the hazard area, with minimal piping. The high reliability of modular systems is based on lack of dependancy on piping integrity, negligible piping calculations, total system supervision, multi-source dependence, and the inherent ability to be heat actuated regardless of catastrophic system failure. Modular systems are simple to

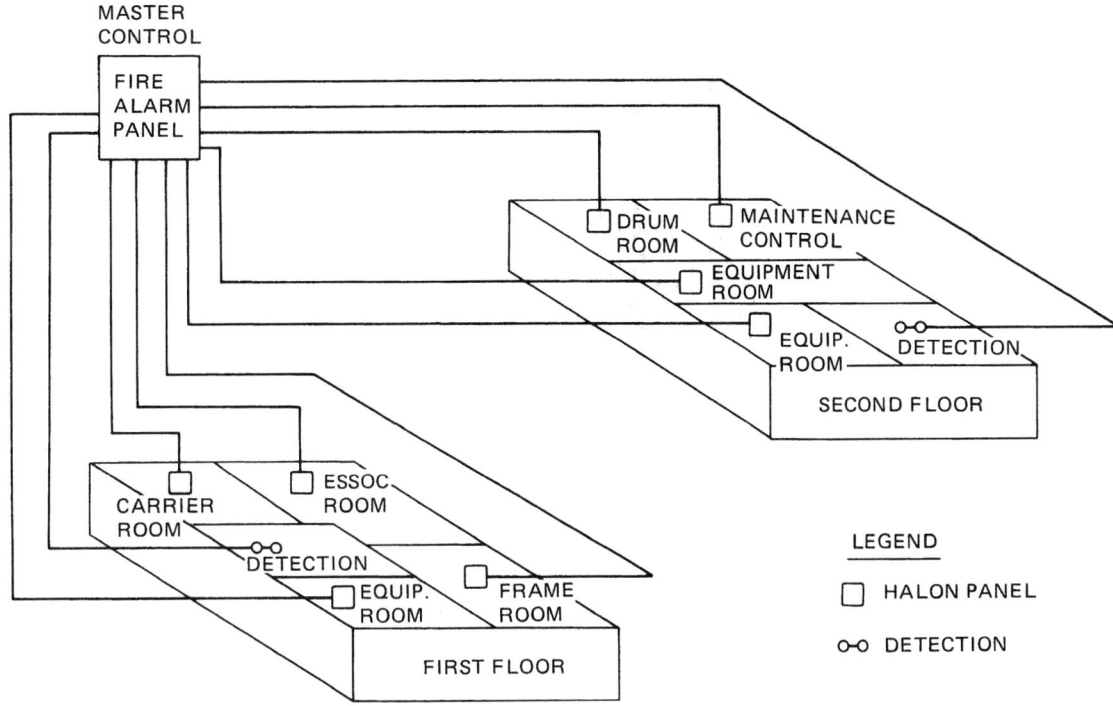

Fig. 4-6.7. The network concept of control panel interface for a typical halon application.

design, are relatively easy to install, and have a high degree of future flexibility. Systems utilizing shared supply are essentially central storage systems with container(s) shared by more than one hazard volume. Even though fewer containers are used, directional valves and extensive piping do not often allow shared supply systems to be cost effective. Adding to its unpopularity are its design and installation complexity, low reliability, and impaired future flexibility. When a shared supply halon system activates for one hazard, the remaining hazards become unprotected until the system is completely recharged.

DESIGN CONCEPTS AND METHODOLOGY

Definitions and Terminology

Halogenated agent extinguishing systems are typically classified as either total flooding or local application systems. A total flooding system is designed to develop and maintain a concentration of halon that will extinguish fires in combustible materials located in an enclosed space. Local application systems are designed to apply the agent directly to a fire that may occur in an area or space that is not immediately enclosed. In addition to these, there are specialized applications, which may include combination total flooding/local application or partial flooding. The vast majority of installed halon systems today are the total flooding type using Halon 1301.

The definitions of halon system and halon zone are often confusing. This is especially true to individuals closely associated with the fire alarm industry, since fire alarm terminology is similar. Figure 4-6.8 defines the basic features of a halon system and halon zone and offers a comparison with each respective fire alarm counterpart.

A halon zone usually equates to an area of halon coverage functioning on a single release circuit, while the zones in a fire alarm system typically are each detection circuit. As an example, one halon zone could be a single computer room, whereas a fire alarm zone could be the entire floor of a building. A halon system also has much fewer (though more comprehensive) zones than a fire alarm system.

Halon Design Guidelines

The design process necessary for total flooding systems is easily quantified. The procedure can be separated into five definable steps: (1) hazard identification, (2) determination of agent quantity, (3) specification of operating requirements, (4) determination of hardware requirements, and (5) generation of postdesign information.

The initial step is to provide a definition of the hazard. This includes determining the fuels involved, the dimensions and configuration of the enclosure, the maximum and

HALON SYSTEM	FIRE ALARM SYSTEM
• 1 CONTROL UNIT	• 1 CONTROL UNIT
• 1 - 5 ZONES	• 1 - 100 ZONES
• ~ 12 WIRES PER ZONE	• ~ 4 WIRES PER ZONE

HALON ZONE	FIRE ALARM ZONE
• VOLUME OF HALON ZONE COVERAGE	• AREA OF DETECTION ZONE COVERAGE
• RELEASE CIRCUIT EQUALS HALON ZONE	• DETECTION CIRCUIT EQUALS FIRE ALARM ZONE

Fig. 4-6.8. Halon/fire alarm differences.

minimum net volumes, the status of occupancy, the expected hazard area temperature range, and possible unclosable openings. Based on this information, the minimum design concentration can be established. Next, the agent quantity is determined based upon the design concentration, the volume, minimum expected temperature, leakage due to ventilation or unclosable openings, and altitude above sea level. Usually, the gross volume is used to calculate the agent quantity to allow for extra agent to replace that lost through normal building leakage. However, agent concentrations must conform with the applicable toxicity criteria with respect to the minimum net volume and maximum temperature. The operating specifications are then required if they have not already been established. These will indicate how the system is to operate, the modes of operation, the type of agent delivery, etc. When these are known, the necessary hardware requirements must be obtained and the design of the system completed. The final step is to generate the postdesign information necessary for others to install, test, operate, and maintain the system. Postdesign information should contain all design calculations (including hydraulic calculations), complete blueprint drawings, and detailed information describing the testing, operation, and maintenance of the system.

Local Application and Special Systems

Local application systems are often installed to extinguish fires involving flammable liquids, gases, and surface burning solids. Such systems are designed to apply the agent directly to a fire that may occur in an area or space not immediately enclosed. They must be designed to deliver halon agent to the hazard being protected in such a manner that the agent will cover all burning surfaces during discharge of the system. Because of its lower volatility, Halon 1211 may be better suited than other forms of halon for local application systems. The lower volatility, plus a high liquid density, permits the agent to be sprayed as a liquid and thus propelled into the fire zone to a greater extent than is possible with other vaporized agents. Examples of areas protected by local application are spray booths, dip and quench tanks, oil-filled electric transformers, printing presses, heavy construction equipment, vapor vents, etc. An example of a local application system is shown in Figure 4-6.9.

Currently, NFPA standards do not set a minimum limit on the discharge time for a local application design. The rate of discharge and the amount of agent required for a given application must be determined by experimentation and evaluation. The most critical components of these systems are the discharge nozzles; the discharge velocity and rate must be sufficient to penetrate the flames and produce ex-

tinguishment but not be so great as to cause splashing or spreading of fuel and thus increase the fire hazard. The minimum design discharge quantity should not be less than 1.5 times the minimum quantity required for extinguishment at any selected design rate.[20] Also of critical importance are type and location of detectors.

As with other types of gaseous suppression systems, local application systems can be designed according to the rate-by-volume method or the rate-by-area method. The rate-by-area method determines nozzle discharge rates based on the exposed surface area of the hazard being protected. This method is less popular than the rate-by-volume method, which requires discharge rates sufficient to fill (within the discharge time) a volume whose imaginary boundaries extend a limited distance from the protected hazard. This method is favored since it performs similarly to total flooding systems. Important factors to be considered in the design of a local application system are the rate of agent flow, the distance and area limitations of the nozzles, the quantity of agent required, the agent distribution system, and the placement of detectors.

Unlike total flooding systems, only the liquid portion of the discharge is effective for local application systems. The computed quantity of agent needed for local application must be increased to compensate for the residual vapor in the storage container at the end of liquid flow. An additional 25 percent storage capacity is required in the absence of an enclosure that would prevent gas dissipation. Systems should also compensate for any agent vaporized in the pipe lines due to heat absorption from the piping. The heat transfer is important when the piping is at a higher temperature than the agent. The following equation determines the amount of agent increase necessary to compensate for this effect.[7]

$$W_x = \frac{2\pi k L (T_p - T_a)(t)}{3600 h (\ln r_o / r_i)} \qquad (1)$$

where:

W_x = amount of agent increase, kg (lbs)
k = thermal conductivity of the piping, W/m · K (Btu · t/hr · ft² · °F)
L = linear length of the piping, m(ft)
T_p = pipe temperature, °C (°F)
T_a = agent temperature, °C (°F)
t = system discharge time
h = heat of vaporization of the agent at T_a, kJ/kg (Btu/lb)
r_o = outside pipe radius, mm (inches)
r_i = inside pipe radius, mm (inches)

Specialized systems using a variety of agents are in wide use throughout the world to protect hazards such as aircraft engine nacelles, military vehicles, emergency generator motors, earth moving equipment, racing cars, etc. The characteristic common to all these systems is that they can only be applied to the specific hazard for which they were designed and tested. One unusual concept used to protect aircraft flight simulator areas is known as partial flooding, where only the volume containing the simulator equipment receives the total flooding concentration, and not the expansive open areas above it. A design concentration of 7 percent is recommended to achieve a 5 percent concentration in the hazard area and should provide for a minimum agent height level relative to the agent concentration of approximately 1.5 m (5 ft) above the highest part of the hazard. The placement of the nozzle is critical and should be designed to direct agent discharge approximately 30 degrees

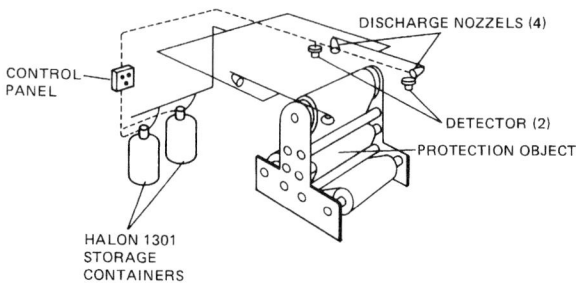

Fig. 4-6.9.   Local application system.

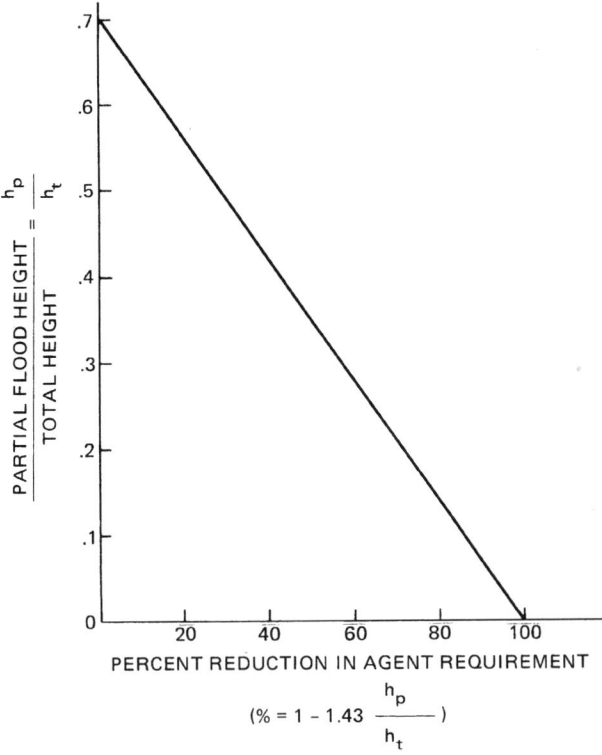

Fig. 4-6.10. *Agent reduction associated with partial flooding systems.*

below the horizontal plane. As shown in Figure 4-6.10, the savings associated with partial flooding systems can be substantial, especially in areas with very high ceilings.[20]

AGENT REQUIREMENTS: TOTAL FLOODING

Design Concentrations: Solid Fuels

Flammable solids may be classified as those that do not develop deep-seated fires and those that do. Class A combustible solids that develop deep-seated fires do so after exposure to flaming combustion for a certain length of time, which varies with the material. Some materials may begin as deep seated through internal heating such as spontaneous ignition. With respect to Halon 1301 total flooding systems, a fire is considered deep seated if a 5 percent concentration will not extinguish the fire within ten minutes after agent discharge.[7] Materials that do not become deep seated undergo surface combustion only and may be treated much the same as those in a flammable liquid fire.

The presence of Halon 1301 in the vicinity of a deep-seated fire will extinguish the flame and reduce the rate of burning, yet the quantity of agent required for complete extinguishment of all embers is difficult to assess. Often it is impractical to maintain an adequate concentration of Halon 1301 for a sufficient time to ensure the complete extinguishment of a deep-seated fire. Factors affecting this concentration include:

1. Nature of fuel,
2. Time during which it has been burning,
3. Availability of oxygen within the enclosure,
4. Ratio of burning surface area to the volume of the enclosure,
5. Geometric characteristics of the fuel, and
6. Fuel distribution within the enclosure.

Table 4-6.12 illustrates the extinguishing concentrations of selected flammable solid fires as indicated by six different halon industry groups.[23]

Even where the fire has inadvertently become deep seated, application of a low Halon 1301 concentration has

TABLE 4-6.12 *Extinguishing Concentrations of Selected Flammable Solid Fires*

	Halon 1301 Concentration, Percent Volume					
	Factory Mutual	Fenwal	Ansul	DuPont	Safety First	Underwriters Labs
SURFACE FIRES						
Polyvinyl chloride		2.0		2.6	3.8	
Polystyrene		3				
Polyethylene		3				
Stacked computer printout			5.1			
Polyester computer tape		5			3.8	
Wood crib 30 pcs. ¾″ × ⅞″	3					
Wood crib 24 pcs. 2″ × 2″ × 18″					3.8	
Wood crib 1A 50 pcs. 2″ × 2″ × 18″					3.8	3.88
Excelsior loose on floor					3.8	6.0
Shredded paper loose on floor					3.8	
Polyurethane foam				3	3.8	
Cotton lint					3.8	
Crumpled paper	3	6			3.8	
Wood pallets—stack of 10	3					
DEEP-SEATED FIRES						
Shredded paper in wire basket					20	18.0
Polyester computer tape loose in open wire basket		10				
Charcoal	13					
Parallel wood blocks	20					
Glazed fox fur					6.5	

TABLE 4-6.13 *Design Concentration for Flame Extinguishment*

Fuel	Minimum Design Concentration Percent by Volume
Acetone	5.0
Benzene	5.0
Ethanol	5.0
Ethylene	8.2
Methane	5.0
n-Heptane	5.0
Propane	5.2

two benefits. First, all flaming combustion is halted, preventing rapid spread of the fire to adjacent fuels. Second, the rate of combustion is drastically reduced. These two characteristics justify the ability of Halon 1301 to control, if not extinguish, deep-seated fires. However, Halon 1301 systems that are specifically designed to extinguish deep-seated fires are seldom economical to apply and may not be as effective in these fires as other types of extinguishing systems.

Design Concentrations: Liquid and Gas Fires

There are two general types of flammable liquid or gas fires. First, a flammable or explosive mixture of vapors exists that must be prevented from burning and second, fuel is burning that must be extinguished. Associated with each of these conditions is a minimum level of Halon 1301 extinguishing concentration, respectively known as inerting and flame extinguishment. When determining the halon design concentration, proper consideration must be given to the quantity and type of fuel involved, the conditions under which it normally exists in the hazard, and any special conditions of the hazard itself. If certain hazards have explosion potential either before or following a fire due to the presence of volatile, gaseous, or atomized fuel, then special consideration should be given to vapor detection and explosion suppression measures.

As its name implies, the flame extinguishment concentration assumes that the given fuel is burning and that Halon 1301 injected into the air surrounding the fuel at the stated concentration will extinguish the fire.[7] Design concentrations for flame extinguishment are given in Table 4-6.13. These concentrations are not considered effective with premixed flames or explosive mixtures of fuel vapor in air, but instead apply to diffusion flames, where the flames emanate from pure fuel vapor and oxygen suffuses into the flame zone from the outside. If the possibility of a subsequent reflash or explosion exists, then the flame extinguishing concentration is not sufficient. NFPA 12A[7] defines these conditions as "when both:

1. The quantity of fuel permitted in the enclosure is sufficient to develop a concentration equal to or greater than one-half of the lower flammable limit throughout the enclosure, and
2. The volatility of the fuel before the fire is sufficient to reach the lower flammable limit in air (maximum ambient temperature or fuel temperature exceeds the closed cup flash point temperature) or the system response is not rapid enough to detect and extinguish the fire before the volatility of the fuel is increased to a dangerous level as a result of the fire."

Most fuels exhibit about a 30 to 40 percent higher concentration for inerting than for flame extinguishment. The minimum inerting concentration suppresses the propagation of the flame front at the "flammability peak" or stoichiometric fuel/air composition and inerts the enclosure so that any fuel/air mixture will not burn. The higher inerting concentration is often considered safer to use even if the flame extinguishment concentration is feasible, yet the sacrifices include higher system cost and higher concentrations to which personnel may be exposed. (See Table 4-6.14.)

It is possible to calculate whether the flame extinguishing concentration is acceptable by determining if the fuel present in the hazard will permit attainment of the one-half lower flammable limit of the fuel. The equation to determine the maximum allowable fuel loading (*MFL*) for flame extinguishment concentrations is

$$MFL = \frac{(K_c)(LFL)(MW)}{T} \qquad (2)$$

where:

MFL = maximum allowable fuel loading, kg/m^3 (lb/ft^3)
K_c = conversion factor, 0.06 093 (0.00685)
LFL = lower flammable limit of fuel in air, percent volume
MW = molecular weight of fuel
T = temperature, K (R).

This can be compared with the actual fuel loading (*FL*) which is calculated by

$$FL = \frac{(VF)(W_{H_2O})(SG)}{V} \qquad (3)$$

where:

FL = fuel loading, kg/m^3 (lb/ft^3)
VF = volumetric quantity of fuel, m^3 (ft^3)
W_{H_2O} = specific weight of water, 997.9 kg/m^3 (62.3 lb/ft^3)
SG = specific gravity of fuel
V = volume of enclosure, m^3 (ft^3).

If the fuel loading, *FL*, exceeds the maximum allowable fuel loading, *MFL*, then the inerting concentration for the particular fuel should be used. Most applications involve a variety of fuels within a single enclosure. If the sum of the actual fuel loadings, *FL*, is greater than any single maximum allowable fuel loading, *MFL*, then the most stringent inerting concentration is recommended. If it is determined that a

TABLE 4-6.14 *Halon 1301 Design Concentrations for Inerting*

Fuel	Minimum Concentration Percent by Volume
Acetone	7.6
Benzene	5.0
Ethanol	11.1
Ethylene	13.2
Hydrogen	31.4
Methane	7.7
n-Heptane	6.9
Propane	6.7

NOTE: Includes a safety factor of 10 percent added to experimental values

TABLE 4-6.15 *Correction Factors for Altitudes*

Altitude		Correction Factor
Feet	Meters	
3000	914	0.90
4000	1219	0.86
5000	1524	0.83
6000	1829	0.80
7000	2134	0.77
8000	2438	0.74
9000	2743	0.71
10000	3048	0.69
11000	3353	0.66
12000	3658	0.64
13000	3962	0.61
14000	4267	0.59
15000	4572	0.56

flame extinguishment concentration is sufficient, the value for the fuel requiring the greatest concentration is most applicable.

Calculation of Agent Quantity

The calculations necessary for determing the Halon 1301 total flooding quantity are dependent on temperature, volume of the enclosure, agent concentration, altitude with respect to sea level, and losses due to ventilation and leakage. Most applications are based on a static volume enclosure with all openings sealed and all ventilation systems shut down prior to discharge. This simplifies the calculation significantly. Often the ventilation system does not shut down but instead is dampered to allow recirculating air (without makeup air) to continue cooling sensitive electronic equipment and promote the mixing of halon and air. Total flooding quantities are still based on a static volume for these applications. However, in this instance, it may be necessary to include the volume of the ventilation ductwork in addition to the volume of the enclosure. The equation to determine the Halon 1301 total flooding quantity is

$$W = \frac{(V)(C)(A_c)}{S(100-C)} \tag{4}$$

where:

W = weight of Halon 1301 required, kg (lb)
C = Halon 1301 concentration, percent by volume
A_c = altitude correction factor—(Refer to Table 4-6.15)
S = specific vapor volume based on temperature, m³/kg (ft³lb)
$S = 0.14\,781 + 0.000\,567T$; T = Temperature °C
$S = 2.2062 + 0.005046T$; T = Temperature °F.

Application Rate

Discharge time and soaking period: When designing a Halon 1301 total flooding system, it is important to determine the system discharge time and soaking period.

As indicated in NFPA 12A, "the agent shall be completed in a nominal 10 seconds or as otherwise required by the authority having jurisdiction."[7] The reasons for a rapid discharge time include keeping unwanted products of decomposition to a minimum and achieving complete dispersal of agent throughout the enclosure. Sometimes a much faster application rate is required due to the possibility of a fast spreading fire; yet, discharge times longer than 10 seconds are sometimes necessary for areas such as museums requiring that turbulence be kept to a minimum, or areas with unavoidably difficult piping configurations.

The soaking time is another important requirement for a Halon 1301 total flooding system. This is especially true for deep-seated fire or fires that may reflash. The most common application today for total flooding systems is the protection of valuable electronic equipment. Fires in these applications are almost always extinguished within a few seconds by the Halon 1301 agent, yet a 10-minute soaking period is usually required. This estimated time period allows responsible individuals to arrive at the scene to take followup action. It is important to remember that halogenated agent extinguishing systems in most cases have only a single chance to control an unwanted fire.

Effects of ventilation: When Halon 1301 is discharged into a total flooding enclosure that is ventilated, some agent will be lost with the ventilating air. Assuming that ventilation must continue during and after discharge, a greater amount of agent is required to develop a given concentration. Also, to maintain the concentration at a given level requires continuous agent discharge for the duration of the soaking period. If an enclosure initially contains pure air, the Halon 1301 discharge rate required to develop a given concentration for agent at any given time after the start of discharge is[7]

$$R = \frac{(C)(E)}{(S)(100-C)[1-e^{(-Et_1/V)}]} \tag{5}$$

where:

R = Halon 1301 discharge rate, kg/s (lb/s)
E = ventilation rate, m³/s (ft³/s)
t_1 = discharge time, s
e = natural logarithm base, 2.71 828.

The Halon 1301 discharge rate necessary to maintain a given concentration of agent is[7]

$$R = \frac{(C)(E)}{(S)(100-C)} \tag{6}$$

After the agent discharge is stopped, the decay of the agent concentration with respect to time is[7]

$$C = C_0 e^{(-Et_2/V)} \tag{7}$$

where:

C_0 = agent concentration at end of discharge, percent volume
t_2 = time after stopping discharge, s.

Compensation for leakage: Occasionally a Halon 1301 total flooding system is designed for an enclosure that has openings that cannot be closed. An example may be a conveyor belt penetrating an enclosure wall, yet even these openings can sometimes be closed using inflatable seals. Halon 1301 discharged into an enclosure for total flooding will result in an air/agent mixture that has a higher specific gravity than the air surrounding the enclosure. Therefore, any openings in the lower portions of the enclosure will allow the heavier air/agent mixture to flow out and the lighter outside air to flow in. Fresh air entering the enclosure will collect toward the top, forming an interface between the air/agent mixture and fresh air. As the leakage proceeds, the interface will descend toward the bottom of the enclosure.

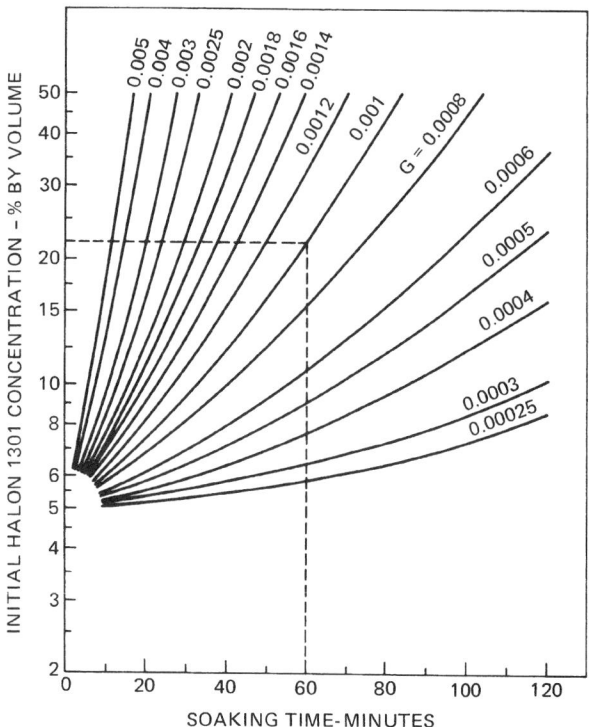

Fig. 4-6.11. *Initial amount of Halon 1301 to produce a 5 percent residual concentration in enclosures equipped for mechanical mixing.*

The space above the interface will be completely unprotected, whereas the lower space will essentially contain the original extinguishing concentration. There are two methods of compensating for unclosable openings: initial overdose and extended discharge.

The initial overdose method provides for an adequate overdose of Halon 1301 to ensure a pre-established minimum of agent at the end of the desired soaking period. Mechanical mixing is required within the enclosure to prevent stratification of agent concentration and a descending interface. Also caution must be used to prevent personnel exposure to the high initial concentrations. The necessary initial concentration depends upon the extended protection time required, the opening height, the opening width, and the volume of the enclosure. Referring to Figure 4-6.11, the equation used to determine the initial concentration for a final concentration of 5 percent is[7]

$$G = \frac{(K)(W_o)(2g_cH^3)^{1/2}}{3V} \qquad (8)$$

where:

G = geometric constant
K = orifice discharge coefficient, 0.66
W_o = opening width, m (ft)
g_c = acceleration due to gravity, 9.81 m/s^2 (32.2 ft/s^2)
H = opening height, m (ft).

The other method used to compensate for unclosable openings is extended discharge. This involves at least two separate piping systems: one to achieve the initial agent concentration, and the other to provide a continuous addi-tion of Halon 1301 at a rate which will compensate for leakage out of the enclosure during the soaking period. The agent must be discharged in such a way that uniform mixing of agent and air is obtained. This mixing is often difficult due to the extremely low flow rates being discharged over the entire soaking period, occasionally resulting in small nozzles freezing due to air moisture. Based on the design concentration and opening height, Figure 4-6.12 can be used to determine the Halon 1301 makeup rate per unit opening width.

Assuming the design concentration of Halon 1301 is established in the enclosure initially, the time required for the interface to reach half-way down the enclosure height can be calculated. Referring to Figure 4-6.13, the geometric constant previously calculated for intial overdose is used to find the soaking time based on the initial design concentration.

FLOW CALCULATIONS

Piping Theory

The overall objective of designing a Halon 1301 piping system is to properly disperse the required concentration of Halon 1301 throughout the hazard volume within the specified time period. Systems must be engineered to operate quickly and effectively. The discharge time (usually a nominal 10 s as indicated by NFPA 12A) is a critical system constraint and is measured as the interval between the first appearance of liquid at the nozzle and the time when the discharge becomes predominantly gaseous.[7] The hydraulic calculations are considered to be the most difficult part of the entire design process, and are almost always calculated with the aid of computer programs due to the tedious nature of manual calculations.

As illustrated in Figure 4-6.14, the primary components of a Halon 1301 piping system are the agent storage container, the discharge nozzle, and the pipe. Often, more than one nozzle is required, complicating the calculations significantly. An attempt should always be made to keep the piping system simple and if possible, balanced. A balanced system has the actual and equivalent pipe lengths from container to each nozzle within ± 10 percent of each other and has equal design flow rates at each nozzle.[7]

As with sprinkler systems or other systems involving fluid flow, the methodology for solving Halon 1301 piping calculations involves seeking terminal characteristics based on property changes encountered due to the movement of the fluid. The system hydraulics are controlled by the selection of the orifice area at the discharge nozzle. This orifice area is calculated from the nozzle pressure, which is based on the starting pressure in the container and pressure losses in the pipe. Because the flow of Halon 1301 is nonsteady and has a change in phase from liquid to vapor, the calculations become highly complex. To simplify calculations, the average discharge conditions are determined so that they might reasonably represent the entire discharge time span. This time-independent model is based on the moment in time when half the liquid phase of the agent has left the nozzle. All the calculations for a 10 s discharge condition shown in Figure 4-6.15 would be solved at the mid-discharge condition (5 s). Hence, the critical characteristics that vary with discharge, such as the storage container pressure and the pressure-density relationship in the pipeline, are replaced with average time-independent values.[24]

By the time half of the liquid agent is out of the nozzle, the original pressure in the storage container has dropped

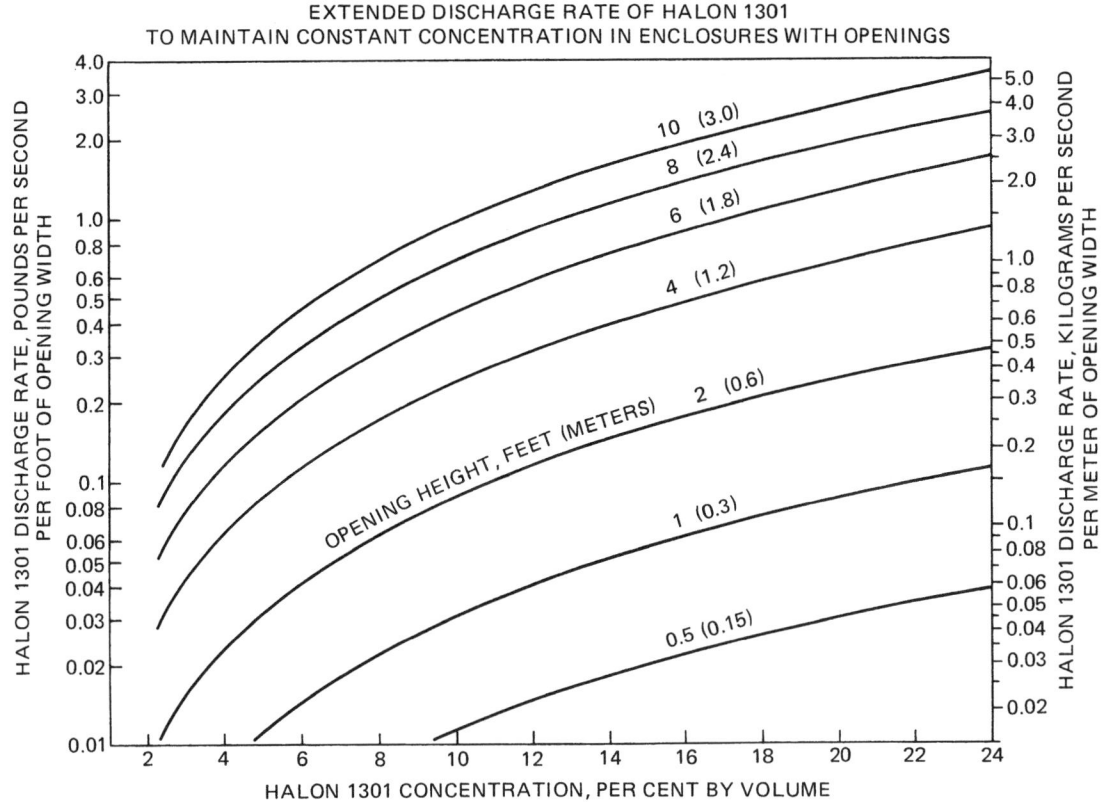

Fig. 4-6.12. *Extended discharge rate of Halon 1301 to maintain constant concentrations in enclosures with openings.*

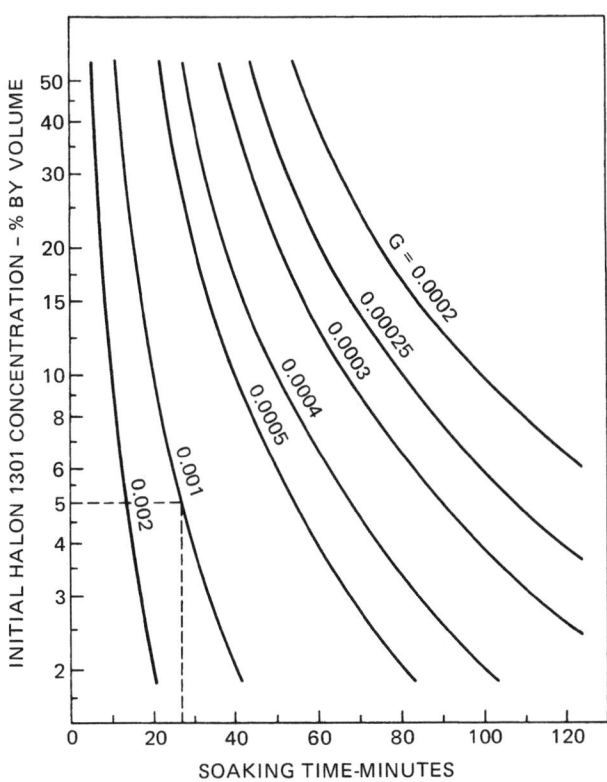

Fig. 4-6.13. *Time required for interface between effluxing Halon 1301/air mixtures and influxing air to descend to center of enclosures not equipped for mixing.*

considerably. To calculate the mid-discharge storage container pressure, the percent of agent still within the pipe must be determined. Also, the initial drop in pressure immediately after the start of discharge is nonlinear. As seen in Figure 4-6.16, the pressure recovery is due to the nitrogen vigorously boiling out of the halon/nitrogen mixture within the storage container.

Unlike water-based fluid flow, the pressure drop occurring when Halon 1301 flows through a pipe is nonlinear and is dependent on the pipeline agent density, not the distance traveled. The pipeline flow is two phase, with a mixture of liquid and vapor agent. As the agent travels in the pipe, the pressure and density decrease, which increases the velocity and the amount of halon vapor. Interestingly, the evolution

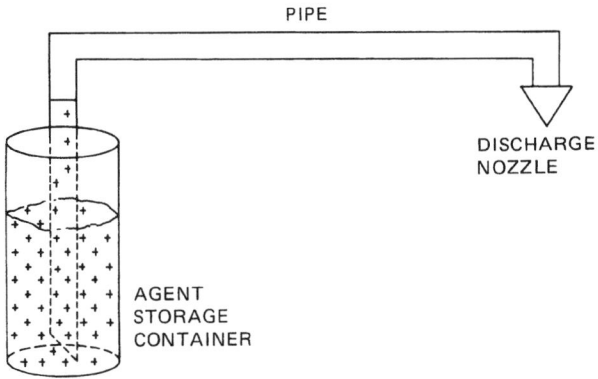

Fig. 4-6.14. *Primary components of a Halon 1301 piping system.*

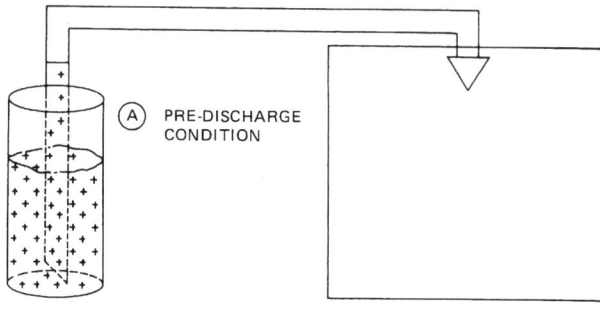

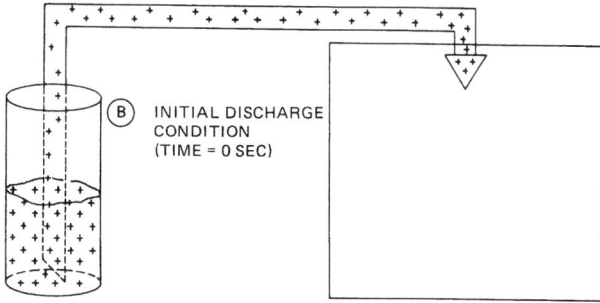

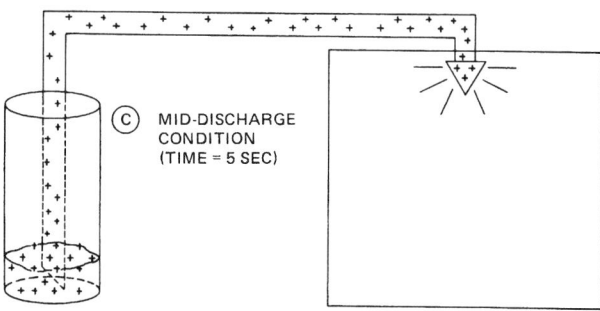

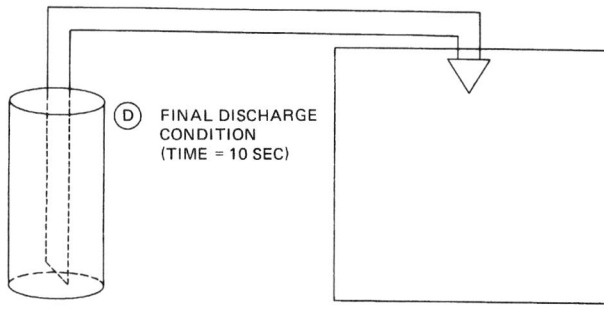

Fig. 4-6.15. Summary of Halon 1301 discharge conditions based on a 10 s discharge.

of the nitrogen from the halon/nitrogen mixture in the storage container causes the halon to drop in temperature and become more dense. This fortunately is not a factor in the calculations since a time-independent model is being used. The increase in density at any one location over the entire time span should not be confused with the decrease in density that occurs when the agent flows from one location to another.

Guidelines and Limitations

Unrealistic distribution networks often fail to perform to specifications and are difficult if not impossible to predict from a calculation standpoint. As the piping system becomes more unrealistic, the calculations become more unreliable. To aid in the development of accurate calculations, certain fundamental limitations are necessary to ensure proper system design. These limitations are especially important with respect to computer programs since these programs have a tendency to be operated abusively with high expectations. Summarized below are the design constraints for Halon 1301 hydraulic calculations.[25]

1. Good design practice.
2. Discharge time $\leq$ 10 s.
3. Favorable system temperature.
4. Initial container pressure = 2482.2 kPa (360 psig) or 4137.0 kPa (600 psig).
5. Initial container fill density $\leq$ 1121.4 kg/m^3 (70 lb/ft^3).
6. Percent in pipe $\leq$ maximum value.
7. Turbulent flow $\geq$ minimum value.
8. Nozzle pressure $\geq$ minimum value.
9. Actual nozzle area $\leq$ percentage of feed pipe area.
10. Actual nozzle area = calculated nozzle $\pm$ 5 percent.

Good design practice includes such items as favoring balanced systems, keeping the degree of flow/split imbalance below a maximum value, avoiding vertically installed tees, and avoiding nozzles on different floor levels which may separate the halon gas/vapor mixture. The values for some of the constraints are determined by the individuals developing computer programs that are verified by approval agencies through testing.

Calculation Procedure

The piping calculations comprise four steps:

1. Determining the necessary input data,
2. Calculating the average storage container pressure,
3. Calculating the nozzle pressure at each nozzle, and
4. Calculating the nozzle orifice areas.

Pipeline calculations are performed for each segment of pipe having both a constant flow rate and a uniform pipe diameter; thus the piping network is divided into sections

TABLE 4-6.16 *Internal Volume of Steel Pipe*

Nominal Pipe Diameter (in.)	Schedule 40 Inside Diameter (in.)	ft^3/ft	Schedule 80 Inside Diameter (in.)	ft^3/ft
$\frac{1}{4}$	0.364	0.0007	0.302	0.0005
$\frac{3}{8}$	0.493	0.0013	0.423	0.0010
$\frac{1}{2}$	0.622	0.0021	0.546	0.0016
$\frac{3}{4}$	0.824	0.0037	0.742	0.0030
1	1.049	0.0060	0.957	0.0050
$1\frac{1}{4}$	1.380	0.0104	1.278	0.0089
$1\frac{1}{2}$	1.610	0.0141	1.500	0.0123
2	2.067	0.0233	1.939	0.0205
$2\frac{1}{2}$	2.469	0.0332	2.323	0.0294
3	3.068	0.0513	2.900	0.0459
$3\frac{1}{2}$	3.548	0.0687	3.364	0.0617
4	4.026	0.0884	3.826	0.0798

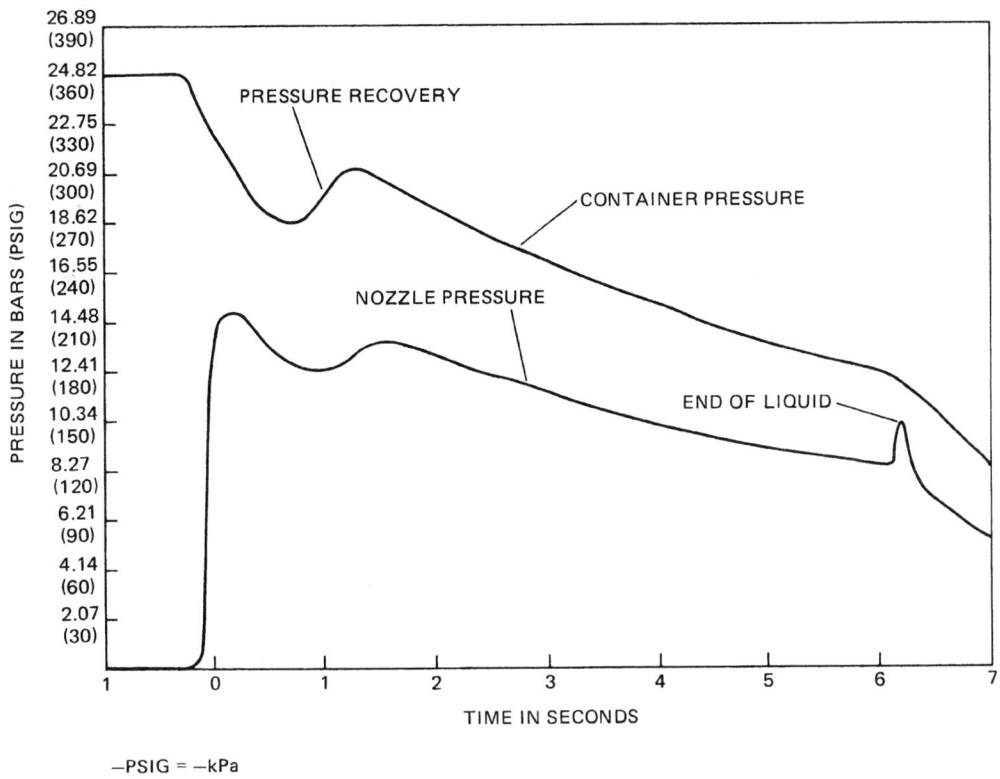

—PSIG = —kPa

Fig. 4-6.16.   Pressure profile during system discharge.

called junctions. Each discharge nozzle is also identified. The forms necessary for the input data, pressure calculations, and nozzle calculations are contained in Figures 4-6.17 and 4-6.18. Assuming the appropriate input data is known, the average storage container pressure is determined from Figure 4-6.19 based on the percent agent in pipe, which itself is determined by[7]

$$\% \text{ in pipe} = \frac{K_1}{(W_i/V_p) + K_2} \quad (9)$$

where:

W_i = initial charge weight of Halon 1301, lb
V_p = internal pipe volume, ft^3 (See Table 4-6.16)
K_1 and K_2 = constants (See Table 4-6.17)

TABLE 4-6.17   Constants to Determine Percent of Agent in Piping

Storage (psig)	Filling Density	K_1	K_2
600	70	7180	46
600	60	7250	40
600	50	7320	34
600	40	7390	28
360	70	6730	52
360	60	6770	46
360	50	6810	40
360	40	6850	34

Once the average storage container pressure is known, Figures 4-6.18 and 4-6.20, and Equations 10 through 22 can be used to determine the nozzle orifice areas for a 360 psig system. Usually the calculations are based on a 10 s discharge time, though this is sometimes changed slightly to produce flow rates in accordance with Table 4-6.18. Turbulent pipeline flow can also be achieved by using smaller pipe

TABLE 4-6.18   Minimum Design Flow Rates to Achieve Turbulent Pipeline Flow

Nominal Pipe Diameter (in.)	Schedule 40 Minimum Flow Rate (lb/s)	Schedule 80 Minimum Flow Rate (lb/s)
1/8	0.20	0.11
1/4	0.34	0.24
3/8	0.68	0.48
1/2	1.0	0.79
3/4	2.0	1.9
1	3.4	2.8
1 1/4	5.8	4.8
1 1/2	8.4	7.5
2	13	13
2 1/2	19.5	17
3	33	26
4	58	48
5	95	81
6	127	109

For SI Units: 1 lb/s = 0.454 kg/s

SYSTEM HALON WEIGHT _____ LBS
CONTAINER FILL DENSITY _____ LBS/FT3
DISCHARGE TIME _____ SECONDS

FORM I : SYSTEM SUMMARY

N1: _____ LBS   N3: _____ LBS   N5: _____ LBS
N2: _____ LBS   N4: _____ LBS   N6: _____ LBS

A	B	C	D	E	F	G	H	I	W	X	Y
				INPUTS						OUTPUTS	
JUNCTION NUMBER	NOZZLE NUMBER	FLOW RATE Q	PIPE TYPE	PIPE DIAMETER D	ACTUAL PIPE LENGTH L	FITTINGS, EQUIVALENT LENGTH L	TOTAL LENGTH L	ELEVATION CHANGE h	JUNCTION PRESSURE P (STARTING OR FROM FORM II)	DENSITY AT ORIFICE r (FIG. 4-6.20)	ORIFICE AREA F (EQ. 22)

Fig. 4-6.17. Halon 1301 piping calculation summary form.

sizes. Pipe diameters that are too small result in unacceptably high pressure losses; therefore, care must be used in pipe size selection. It is important to recognize that approximations have been made for Y and Z factors and nozzle coefficients. The calculation procedure presented here is only intended to demonstrate the current methodology and not to provide a rigorous solution. The necessary equations are[7,26]

$$P_e = \frac{rL_e}{144} \qquad (10)$$

where:

P_e = elevation pressure, psig
r = agent density, lb/ft^3
L_e = pipe elevation length, ft

$$A = 1.013D^{5.25} \qquad (11)$$

where:

A = pipe size factor
D = actual pipe diameter, inches

$$B = 7.97/D^4 \qquad (12)$$

where:

B = pipe size factor

$$Y_1 = -\left(\frac{a}{3} P_0^3 + \frac{b}{2} P_0^2 + cP_0 + d\right) \qquad (13)$$

where:

Y_1 = first Y factor
P_0 = junction starting pressure, psig
a, b, c and d = constants (See Table 4-6.19)

J	K	L	M	N	O	P	Q	R	S	T	U	V
INITIAL PRESSURE				FINAL PRESSURE								
JUNCTION PRES-SURE	ELEVATION		CORRECT-ED START-ING PRESSURE P_0	PIPE SIZE FACTORS		1ST Y FACTOR Y_1 (TABLE 4-6.19, EQ.13)	1ST Z FACTOR Z_1 (EQS. 14-17)	TEMPOR-ARY Y FACTOR Y_T (EQ. 18)	TEMPOR-ARY PRES-SURE P_T (TABLE 4-6.19 EQ. 19)	2ND Z FACTOR Z_2 (EQS. 14-17)	2ND Y FACTOR Y_2 (EQ. 20)	FINAL JUNCTION PRES-SURE P (TABLE 4-6.19, EQ. 21)
	DENSITY r (FIG. 4-6.20)	PRESS-URE P_e (EQ. 10)		A (EQ. 11)	B (EQ. 12)							

Fig. 4-6.18. Halon 1301 pressure calculation summary form.

AVERAGE STORAGE CYLINDER PRESSURE VS
PERCENT OF THE AGENT SUPPLY NEEDED TO FILL THE PIPELINE

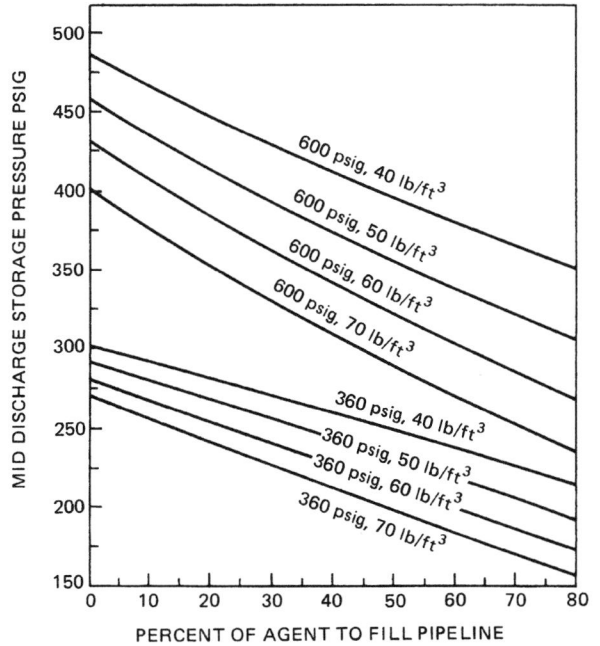

Fig. 4-6.19. Mid-discharge storage container pressure.

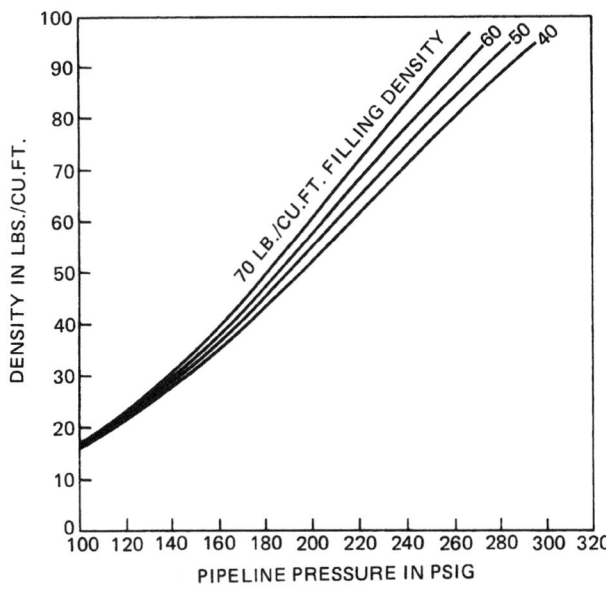

Fig. 4-6.20. Pipeline density/pressure relationship for a 360 psig system.

TABLE 4-6.19 *Constant for Y Factor/Pressure Equations*

P Storage (psig)	Fill Density lb/ft^3	a	b	c	d
360	70	3.571×10^{-4}	0.6971	−63.50	−5921
360	60	4.018×10^{-4}	0.6913	−64.01	−6333
360	50	3.125×10^{-4}	0.6238	−56.90	−7386
360	40	3.720×10^{-4}	0.6187	−55.55	−8120

$$Z = 1.01790 - 0.01179(P - 160) \quad \text{(14)}$$
for 70 lb/ft^3 fill density

$$Z = 0.96913 - 0.01098(P - 170) \quad \text{(15)}$$
for 60 lb/ft^3 fill density

$$Z = 0.96412 - 0.01051(P - 175) \quad \text{(16)}$$
for 50 lb/ft^3 fill density

$$Z = 0.95900 - 0.01008(P - 180) \quad \text{(17)}$$
for 40 lb/ft^3 fill density

where:

$Z = Z$ factor
$P = $ pressure, psig

$$Y_T = Y_1 + L(Q^2/A) \quad \text{(18)}$$

where:

$Y_T = $ temporary Y factor
$Q = $ flow rate, lb/s

$$P_T^3 + \left(\frac{3b}{2a}\right)P_T^2 + \left(\frac{3c}{a}\right)P_T = -\left(\frac{3}{a}\right)Y_T - \left(\frac{3d}{a}\right) \quad \text{(19)}$$

where:

$P_T = $ temporary pressure, psig

$$Y_2 = Y_T + B(Z_2 - Z_1)Q^2 \quad \text{(20)}$$

where:

$Y_2 = $ second Y factor

$$P^3 + \left(\frac{3b}{2a}\right)P^2 + \left(\frac{3c}{a}\right)P = -\left(\frac{3}{a}\right)Y_2 - \left(\frac{3d}{a}\right) \quad \text{(21)}$$

$$F = 1.5Q(1/f(rp)^{1/2}) \quad \text{(22)}$$

where:

$F = $ nozzle orifice area, in^2
$f = $ nozzle coefficient (approximation $= 0.7$).

POST DESIGN CONSIDERATIONS

Post design considerations are divided into two categories: system documentation and inspection/acceptance practices. Good halon system design is not complete until full documentation is provided for installation, acceptance, and eventual end user operation. Proper documentation is especially important to prevent the inadvertent discharge of a halon system for other than a fire, since replacement of the halon agent could be very difficult with future availability being dependent on recycled stock.

System Documentation

System documentation should include the items listed below. This material is necessary for others to install, test, operate, and maintain the system. Information can be recorded entirely on system drawings or in both a written manual and system drawings.

System manual

1. Design Data
   (a) Functional and operational description
   (b) Halon 1301 weight calculations
   (c) Hydraulic piping calculations
   (d) Special considerations

2. Installation, Maintenance, and Inspection Instructions

As-built system drawings

1. Floor Plan Layout
   (a) Suitable dimensions
   (b) Equipment locations
   (c) Special installation details

2. Electrical Schematic
3. Equipment Identification
4. Special Notes

Inspection and Acceptance

After installation, each system should be inspected and tested by technicians trained by the equipment manufacturer covering the items listed below:

1. Test system wiring for proper connection, continuity, and resistance to ground.
2. Check system control unit in accordance with factory recommended procedures.
3. Calibrate and test each detector in accordance with factory recommended procedures.
4. Test each releasing circuit for proper resistance by means of a current-limiting meter.
5. Test the operation of all ancillary devices such as alarms, dampers, magnetic closers, etc.
6. Obtain a certificate of inspection signed and dated by the installing contractor and the authority having jurisdiction.

An installation checklist is often used, which expands on the above items in complete detail.[27] These checklists are available from agent and equipment manufacturers, installers, insurance groups, and consultants.

When accepting a newly installed halon system, it is important to determine compliance with design specifications. In previous years, a full discharge test was required to provide unquestionable evidence of performance, yet this could be a costly and sometimes unnecessary burden carried by the end user. End users with multiple systems would often prove system acceptance based on the performance characteristics of their other systems.

The primary reason for discharge test failure, when it was performed, was because the hazard enclosure would not hold the design concentration over the entire soaking period.[28] Checking the enclosure for possible halon leakage points has always been difficult and is the only questionable part of the acceptance/inspection procedure. A method referred to as the enclosure integrity test has proved to be very effective for this problem, and validates the integrity of the protected enclosure.[7] This technique shows much promise and has potential for substantially enhancing the reliability of proper system operation.

The most immediate and effective use of fan pressurization techniques is for leakage path indication.[29] This involves pressurizing or depressurizing the enclosure with the fan pressurization apparatus and using an indicating device, such as a smoke pencil or acoustic sensor, to determine leakage paths. The installers visual inspection of the enclosure now becomes very effective since even the smallest cracks can be located. Due to low cost and simplicity, a smoke source is usually the most desirable method for locating leaks, but an excellent alternative is the use of a directional acoustic sensor that can be selectively aimed at different sound sources.[30] Highly sensitive acoustic sensors are available that can detect air as it flows through an opening and are sensitive enough to clearly hear a human eye blink.[31] Openings can also be effectively detected by placing an acoustic source on the other side of the barrier and searching for acoustic transmission. Another method is to use an infrared scanning device if temperature differences across the boundary are sufficient.[32] These techniques are not quantitative, but they are effective, inexpensive, and easily performed.

ENVIRONMENTAL CONSIDERATIONS

Scientific evidence indicates that fire protection Halon 1301 is one of several man-made substances adversely effecting the earth's ozone layer.[33] Ozone exists naturally as a thin layer of gas in the stratosphere that blocks the sun's harmful ultraviolet rays and thus is vital to life on earth. Several adverse environmental and direct health effects are linked to ozone layer depletion, and its preservation is of paramount concern to mankind. It's believed that Halon 1301 (and other chlorofluorocarbon's) chemically destroy ozone when emitted into the atmosphere.

Earlier, the phase-out of full system discharge tests that were used to verify enclosure integrity received special attention since they accounted for a proportionately large percentage of fire protection halon emissions. Fortunately, the amount of fire protection Halon 1301 released for actual fires is relatively small. Testing a system by performing a full discharge test allows the release of Halon 1301 which on a cumulative basis may be potentially harmful to the environment and depletes relatively precious stocks of halon agent that should be dedicated to suppressing fires. The release of Halon 1301 should be minimized.

With regard to ozone layer depletion, halons used for fire protection are different than halons used for other industrial applications.[34] Fire protection halons are unique because of their essential mission to prevent the loss of life, minimize the loss of irreplaceable property, assure the continuity of vital operations, and reduce the amount of fire byproducts polluting the atmosphere. Efforts have been made to minimize the release of fire protection halons for noncritical tasks like training, testing, and research. It is assumed that halon systems will remain in existence for many more years, despite the present worldwide restriction on their production.

NOMENCLATURE

a = constant (see Table 4-6.19)
A = pipe size factor
A_c = altitude correction factor—(refer to Table 4-6.15)
b = constant (see Table 4-6.19)
B = pipe size factor
c = constant (see Table 4-6.19)
C = Halon 1301 concentration, percent by volume
C_0 = agent concentration at end of discharge, percent by volume
d = constant (see Table 4-6.19)
D = actual pipe diameter, inches
e = natural logarithm base, 2.71828
E = ventilation rate, m³/s (ft/s)
f = nozzle coefficient (approximation = 0.7)
F = nozzle orifice area, in²
FL = fuel loading, kg/m³(1b/ft³)
g_c = acceleration due to gravity, 9.81 m/s² (32.2 ft/s²)
G = geometric constant
h = heat of vaporization of the agent at T_a, kJ/kg (Btu/lb)
H = opening height, m (ft)
k = thermal conductivity of the piping, W/m · K (Btu-t/hr-ft²-f)
K = orifice discharge coefficient, 0.66
K_c = conversion factor, 0.06093 (0.00685)
L = linear length of piping, m (ft)
L_e = pipe elevation length, ft
LFL = lower flammable limit of fuel in air, percent volume
MFL = maximum allowable fuel loading, kg/m³ (lb/ft³)
MW = molecular weight of fuel
P = pressure, psig
P_0 = junction starting pressure, psig
P_e = elevation pressure, psig
P_T = temporary pressure, psig
Q = flow rate, lb/s
r = agent density, lb/ft³
R = Halon 1301 discharge rate, kg/s (lb/s)
r_i = inside pipe radius, mm (inches)
r_o = outside pipe radius, mm (inches)
S = specific vapor volume of Halon 1301 based on temperature, m³/kg (ft³/lb)
SG = specific gravity of fuel
t = system discharge time
T = temperature, K (R)
t_1 = discharge time, s
t_2 = time after stopping discharge, s
T_a = agent temperature, C (F)
T_p = pipe temperature, C (F)
V = enclosure volume, m³ (ft³)

VF = volumetric quantity of fuel, m³ (ft³)
V_p = internal pipe volume, ft³ (See Table 4-6.16)
W_x = amount of agent increase, kg (lbs)
W_{h_2O} = specific weight of water, 997.9 kg/m³ (62.3 lb/ft³)
W = weight of Halon 1301 required, kg (lb)
W_o = opening width, m (ft)
W_i = initial charge weight of Halon 1301, lb
Y_1 = first Y factor
Y_2 = second Y factor
Y_T = temporary Y factor
Z = Z factor

REFERENCES CITED

1. Grant, C.C., "Fire Protection Halons and the Environment: An Update Symposium," *Fire Technology*, 24, 1, Feb. 1988.
2. "The Halogenated Extinguishing Agents," *NFPA Quarterly*, 48, 8, Part 3, Boston, 1954.
3. Wharry, David and Hirst, Ronald, *Fire Technology: Chemistry and Combustion*, Inst. of F. Engrs., Leicester, England, 1974.
4. Strasiak, Raymond, "The Development of Bromochloromethane (CB)," *WADC Technical Report 53-279*, Wright Air Development Center, Ohio, Jan. 1954.
5. *Fire Protection Handbook*, Seventeenth Edition, National Fire Protection Association, Quincy, MA (1991).
6. NFPA 12A-T, *Standard on Halogenated Fire Extinguishing Agent Systems*, National Fire Protection Association, Quincy, MA (1968).
7. NFPA 12A, *Standard on Halon 1301 Fire Extinguishing Systems*, National Fire Protection Agency, Quincy, MA (1992).
8. NFPA 12B, *Standard on Halon 1211 Fire Extinguishing Systems*, National Fire Protection Association, Quincy, MA (1990).
9. NFPA 12C-T, *Tentative Standard on Halon 2402 Fire Extinguishing Systems*, National Fire Protection Association, Quincy, MA (1983).
10. NFPA 75, *Standard for the Protection of Electronic Computer/Data Processing Equipment*, National Fire Protection Association, Quincy, MA (1992).
11. Ford, Charles, *Halon 1301 Computer Fire Test Program—Interim Report*, DuPont Co., Wilmington, DE, Jan. 10, 1972.
12. NFPA 2001, *Standard on Clean Agent Fire Extinguishing Systems*, National Fire Protection Agency, Quincy, MA (1994).
13. "DuPont Halon 1301 Fire Extinguishant," *Technical Bulletin B-29E*, DuPont, Co., Wilmington, DE.
14. Ford, Charles, "Overview of Halon 1301 Systems," *Symposium on the Mechanism of Halogenated Extinguishing Agents*, ACS Symposia Series, April 1962.
15. *Evaluation of Telephone Frame Fire Protection*, GTE/Fenwal, Holliston, MA (1970).
16. "Handling and Transferring 'Freon' FE 1301 Fire Extinguishing Agent," *Technical Bulletin FE-2*, DuPont Co., Wilmington, DE, 1969.
17. Clark, D.G., *The Toxicity of Bromotriflouromethane (FE 1301) in Animals and Man*, Ind. Hyg. Res. Lab., Imperial Chemical Industries, Alderley Park, Cheshire, England (1970).
18. Stewart, Richard D., Newton, Paul E., Wu, Anthony, Hake, Carl L. and Krivanek, Neil D., *Human Exposure to Halon 1301*, Medical College of Wisconsin, Milwaukee, unpublished (1978).
19. The Hine Laboratories, Inc., *Clinical Toxicologic Studies on Freon Fe-1301, Report No. 1*, San Francisco, CA, unpublished (1968).
20. Bryan, John L., *Fire Suppression and Detection Systems*, Macmillan, NY (1982).
21. Sax, N., *Dangerous Properties of Industrial Materials*, Section 12, 2nd ed, Reinhold, NY (1963).
22. Grabowski, George J., *Fire Detection and Actuation Devices for Halon Extinguishing System, An Appraisal of Halogenated Fire Extinguishing Agents*, National Academy of Sciences, Washington, D.C. (1972).
23. Ford, Charles "Extinguishment of Surface and Deep-Seated Fires with Halon 1301," *Symposium of an Appraisal of Halogenated Fire Extinguishing Agents*, National Academy of Sciences, Washington, D.C. (1972).
24. Williamson, H.V., *Halon 1301 Flow Calculations—An Analysis of a Series of Tests Conducted by FEMA at the Fenwal Test Site*, Chemetron Corp., Hanover (1975).
25. Grant, C.C., "Computer Aided Halon 1301 Piping Calculations," *Fire Safety Journal*, May/July 1985.
26. *Flow in Pipes—Pyroforane Halon 1301*, Produits Chimiques Ugine Kuhlmann, Corbevoie (France).
27. Brenneman, James J., and Charney, Marvin, "Testing a Total Flooding Halon 1301 System in a Computer Installation," *Fire Journal*, 68, 6, Nov. 1974.
28. Chines, S.A., "Halon System Discharge Testing—An Authority Having Jurisdiction Point of View," *Seminar Paper for Fire Protection Halons and the Environment, NFPA Annual Meeting*, Cincinnati, May 17, 1987.
29. Grant, C.C., "Controlling Fire Protection Halon Emissions," *Fire Technology*, 24, 1, Feb. 1988.
30. Keast, D.N., and Pei, H.S., "The Use of Sound to Locate Infiltration Openings in Buildings," *Proceedings of the ASHRAE-DOE Conference on the Thermal Performance of the Exterior Envelope of Buildings*, Orlando, FL, p. 85, Dec. 1979.
31. *Ultraprobe 2000 Data Sheet (acoustic sensor)*, UE Systems, Elmsford, NY.
32. Blomsterberg, A.K., and Harrje, D.T., "Approaches to Evaluation of Air Infiltration Energy Losses in Buildings," *ASHRAE Transactions, Vol. 85, Pt. 2*, p. 797 (1979).
33. Andersen, Stephen O., "Halons and the Stratospheric Ozone Issue," *Fire Journal*, May/June 1987.
34. Taylor, Gary, "Achieving the Best Use of Halons," *Fire Journal*, May/June 1987.

HALON REPLACEMENT CLEAN AGENT TOTAL FLOODING SYSTEMS

P.J. DiNenno

INTRODUCTION

The regulation of Halon 1301 under the Montréal Protocol and its amendments culminated in the phaseout of production of halons in the developed countries on December 31, 1993. This regulation engendered tremendous research and development efforts across the world in a search for replacements and alternatives. Over the past several years, several total flooding, clean agent alternatives to Halon 1301 have been commercialized, and development continues on others. In addition to clean total flooding gaseous alternatives, new technologies, such as water mist and fine solid particulate, are being introduced. This chapter focuses on total flooding clean agent halon replacements.

Table 4-7.1 is a summary of the most important halocarbon and inert gas extinguishing agents developed to date. The table gives the chemical name, trade name, American Society of Heating, Refrigerating, and Air-Conditioning Engineers, Inc. (ASHRAE) designation (for halocarbons), and chemical formula.

CHARACTERISTICS OF HALON REPLACEMENTS

Clean fire suppression agents are defined as fire extinguishants that vaporize readily and leave no residue.[1] Clean agent halon replacements fall into two broad categories: (1) halocarbon compounds, and (2) inert gases and mixtures. Halocarbon replacements include compounds containing carbon, hydrogen, bromine, chlorine, fluorine, and iodine. They are grouped into five categories: (1) hydrobromofluorocarbons (HBFC), (2) hydrofluorocarbons (HFC), (3) hydrochlorofluorocarbons (HCFC), (4) perfluorocarbons (FC or PFC), and (5) fluoroiodocarbons (FIC).

While the characteristics of halocarbon clean agents vary widely, they share several common attributes:

1. All are *electrically nonconductive*;
2. All are *clean agents*; i.e., they vaporize readily and leave no residue;

3. All are *liquefied gases* or display analogous behavior (e.g., compressible liquid);
4. All can be stored and discharged from typical Halon 1301 hardware [with the possible exception of HFC-23, which more closely resembles 600 psig (40 bar) superpressurized halon systems];
5. All (except HFC-23) use nitrogen superpressurization in most applications for discharge purposes;
6. All are less efficient fire extinguishants than Halon 1301 in terms of storage volume and agent weight. The use of most of these agents requires increased storage capacity;
7. All are total flooding gases after discharge. Many require additional care relative to nozzle design and mixing;
8. All produce more decomposition products (primary HF) than Halon 1301, given similar fire type, fire size, and discharge time; and
9. All are more expensive at present than Halon 1301 on a weight (mass) basis.

Inert gas alternatives include nitrogen and argon, and blends of these. One inert gas replacement has a small fraction of carbon dioxide. Carbon dioxide is not an inert gas, because it is physiologically active and toxic at low concentrations (approximately 9 percent). Inert gas clean agents are stored as pressurized gases. They are electrically nonconductive, form stable mixtures in air, and leave no residue.

Extinguishing Mechanisms

Halocarbon clean agents extinguish fires by a combination of chemical and physical mechanisms, depending upon the compound. Chemical suppression mechanisms of HBFC and HFIC compounds are similar to Halon 1301; i.e., the Br and I species scavenge flame radicals, thereby interrupting the chemical chain reaction. Other replacement compounds suppress fires primarily by extracting heat from the flame reaction zone, thereby reducing the flame temperature below that which is necessary to maintain sufficiently high reaction rates by a combination of heat of vaporization, heat capacity, and the energy absorbed by the decomposition of the agent.

Oxygen depletion also plays an important role in reducing flame temperature. The energy absorbed in decomposing the agent by breaking fluorine and chlorine bonds is quite

Philip J. DiNenno, P.E. is Vice President of Hughes Associates Inc., a fire protection engineering research and development firm. He has been actively involved in the testing and development of halon replacement chemicals and alternative fire suppression technologies.

TABLE 4-7.1 *Commercialized Halon Replacement Nomenclature*

Chemical Name	Trade Name	ASHRAE Designation	Chemical Formula
Perfluorobutane	CEA-410	FC-3-1-10	C_4F_{10}
Heptafluoropropane	FM-200	HFC-227ea	CF_3CHFCF_3
Trifluoromethane	FE-13	HFC-23	CHF_3
Chlorotetrafluoroethane	FE-24	HCFC-124	$CHClFCF_3$
Pentafluoroethane	FE-25	HFC-125	CHF_2CF_3
Dichlorotrifluoroethane (4.75%)	NAF-SIII	HCFC Blend A	$CHCl_2CF_3$
Chlorodifluoromethane (82%)			$CHClF_2$
Chlorotetrafluoroethane (9.5%)			$CHClFCF_3$
Isopropenyl-1-methylcyclohexene (3.75%)			
$N_2/Ar/CO_2$	Inergen	IG-541	N_2 (52%)
			Ar (40%)
			CO_2 (8%)
N_2/Ar	Argonite	IG-55	N_2 (50%) Ar (50%)
Argon	Argon	IG-01	Ar (100%)

important, particularly with respect to decomposition production formation. There is undoubtedly some degree of "chemical" suppression action in flame radical combustion with halogens, but it is considered to be of minor importance since it is not catalytic (e.g., one F radical combines with one H flame radical).

The lack of significant chemical reaction inhibition in the flame zone by HCFC, HFC, and FC compounds results in higher extinguishing concentrations relative to Halon 1301. The relative importance of the energy sink represented by breaking halogen species bonds results in higher levels of agent decomposition relative to Halon 1301.

Inert gas acts by reducing the flame temperature below thresholds necessary to maintain combustion reactions. This is done by reducing the oxygen concentration and by raising the heat capacity of the atmosphere supporting the flame. The addition of a sufficient quantity of inert gas to reduce the oxygen concentration below 12 percent (in air) will extinguish flaming fires. The agent concentration required is also a function of the heat capacity of the inert gas added. Hence, there are differences in minimum extinguishing concentration between inert gases.

Flame Suppression Effectiveness

Flame suppression effectiveness of total flooding halon replacement agents has been evaluated in a number of ways. The predominant small-scale test method for establishing flame extinguishing concentrations for liquid and gaseous fuels is the ICI cup burner or variations thereof.

Figure 4-7.1 is a schematic of the ICI cup burner. A small laminar flame is established above a "cup" of fuel surrounded by a cylindrical chimney. An air/agent mixture flows up the chimney surrounding the flame. The minimum concentration of agent (in air) at which the flame is extinguished is the minimum extinguishing concentration (MEC). There are many variations of the basic device as used by different laboratories; these variations include cup and chimney diameter, different mixing and measuring methods, chimney height, and agent/air mixture velocity past the flame.[2] Table 4-7.2 gives some indication of the variation in cup burner extinguishing concentration for a range of extinguishment concentrations for a range of agents with n-heptane as the fuel.

Given the wide variation in test methods, the minimum extinguishing concentrations measured from these different devices are reasonably close.

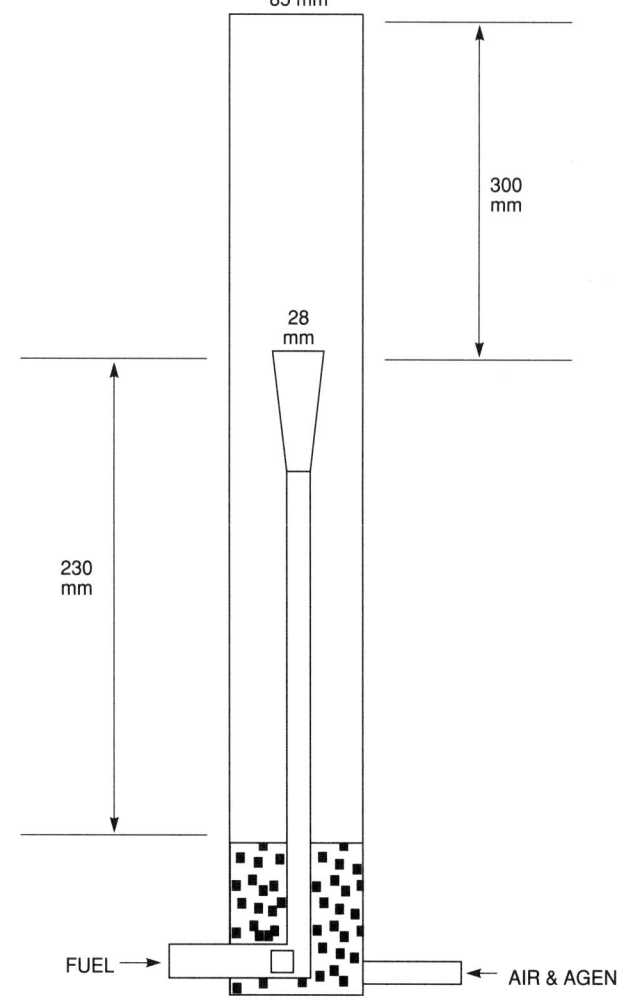

Fig. 4-7.1.   Schematic of ICI cup burner apparatus.[2]

TABLE 4-7.2 *n-Heptane Cup Burner Extinguishing Values from Various Investigators (from NFPA 2001,[1] except as noted)*

Reference	Nitrogen	FC-3-1-10	HFC-227ea	HFC-23	Halon 1301
Sheinson	30[3]	5.2	6.6	12	3.1
3M		5.9	—	—	3.9
NMERI*		5.0	6.3	12.6	2.9
Senecal, Fenwal		5.5	5.8	12 (13)	3 (3.5)
Robin			5.9	12.7	3.5
NIST[†]	32	5.3	6.2	12	3.2

*New Mexico Engineering Research Institute.
[†]National Institute of Standards and Technology.

The cup burner can also be used to establish the MEC for inert gases; Ansul obtained a value of 2.91 percent for IG-541. This is in contrast to concentrations of 32, 41, and 23 percent by volume measured by the National Institute of Standards and Technology (NIST) for nitrogen, argon, and carbon dioxide, the components of Inergen.

NIST has conducted investigations on a wide range of halon replacement chemicals for aviation use. In order to give a wider perspective on the type and range of chemicals being evaluated for fire suppression use, Table 4-7.3 from reference 3 is included. The table gives cup burner n-heptane flame extinction data for a wide range of potential halon replacements.

Table 4-7.4 presents cup burner MEC for a range of fuels and agents taken from various sources. Where multiple values of the MEC were found, they are given. The nitrogen data are presented as representative inert gas values. Argon N_2 blend MEC values would be higher. These data should not be used for design purposes without ensuring that the concentrations are consistent with system manufacturer requirements and third-party approvals.

In addition to the cup burner apparatus, researchers at NIST have utilized an opposed-flow diffusion flame (OFDF) apparatus to rank halon replacements for fire extinguishing effectiveness. The OFDF burner is commonly used for combustion research. It has many advantages as a research tool for fundamental combustion studies. Its primary advantage is in the ability to relate the results to fundamental predictions of flame structure and conditions at flame extinction. The OFDF apparatus is shown schematically in Figure 4-7.2. The oxidizer (and suppressant) stream is forced down onto the fuel surface, exhaust gases are drawn down through an annulus or jacket around the fuel cup, and a flat flame is established. Water cooling is provided for the fuel cup and exhaust gas.

The OFDF burner can vary the turbulence intensity or strain rate of the flame. For most applications of clean agent fire suppression, this is not a major concern; but in specialized applications, such as engine nacelles with high fuel and oxidizer flow rates or in high-pressure spray or jet fires, the strain rate will substantially impact the minimum condition for extinguishment. Figure 4-7.3 is a sample plot from reference 3 showing the variation of the mole fraction of extinguishing agent *versus* the strain rate at extinction for n-heptane fuels for a range of suppressants. For typical natural fires, the strain rate is approximately 25 s^{-1}. At high strain rates, the flame is extinguished at lower agent concentrations.

Figure 4-7.4 from reference 3 shows the relationship between MEC for the cup burner and OFDF apparatus. As expected, the cup burner concentration is quite similar to

the OFDF concentration at a low strain rate (25 s^{-1}), typical of natural fires. In all cases, the MEC of agent is much lower for high strain rate flames. This further reinforces the value of the cup burner and OFDF apparatus for evaluation of minimum extinguishing concentration.

Tables 4-7.5 and 4-7.6 summarize the fire suppression capability of halon replacement agents relative to Halon 1301 for halocarbon and inert gas agents, respectively. Agent mass and storage volume ratio equivalents for Halon 1301 are given as 1. All comparisons are based on 120 percent of the agent manufacturer's recommended MEC, based on the cup burner. The design values used are as provided by the manufacturer. Design values for HCFC Blend A and Argon are problematic, because these values are below the MEC as measured by the cup burner.

Note that all clean agent alternatives require at least 60 percent more agent by weight and storage volume. This is a consequence of the elimination of bromine in the compounds and subsequent level of catalytic recombination of flame radicals. These data should be taken as representative values, as there are variations between hardware manufacturers.

TABLE 4-7.3 *Agent Fraction in the Oxidizer Stream at Extinction of n-Heptane Cup Burner Flames (from reference 3)*

Agent Type	Agent	Mass Percent	Volume Percent
Inert	N_2	31	32
	CO_2	32	23
	He	6.0	31
	Ar	38	41
Nitrogen containing	NF_3	*	*
Silicon containing	SiF_4	36	13
Sodium containing	$NaHCO_3$ (10–20 μm)	3.0	[†]
Hydrofluorocarbons	CF_3H	25	12
	CF_2H_2	[‡]	[‡]
	CF_2H_2/C_2HF_5	30	15
	CH_2FCF_3	29	10
	CHF_2CF_3	29	8.7
	$CF_3CH_2CF_3$	27	6.5
	C_3HF_7	28	6.2
Fluorocarbons	CF_4	37	16
	C_2F_6	30	8.1
	C_3F_6	29	7.3
	C_3F_8	30	6.3
	c-C_4F_8	32	6.3
	C_4F_{10}	32	5.3
Chlorine containing	CHF_2Cl	28	12
	$CHCl_2F$	32	11
	CH_3CF_2Cl	[‡]	[‡]
	$CF_2 = CHCl$	[‡]	[‡]
	$CF_2 = CFCl$	31	10
	$CHFClCF_3$	26	7.0
Bromine containing	CF_3Br	14	3.1
	CF_2Br_2	16	2.6
	CH_2BrCF_3	17	3.5
	$CH_2 = CHBr$	[‡]	[‡]
	$CF_2 = CFBr$	27	6.3
	$CF_2 = CHBr$	24	6.0
Iodine containing	CF_3I	18	3.2

*Acted as an oxidizer, promoted flame stability.
[†]Solid powder not expressed in volume percent.
[‡]Agent observed to be flammable.

TABLE 4-7.4 *Cup Burner Minimum Extinguishing Concentrations*

Fuel	Cup Burner Extinguishment Concentration (% by Volume)				
	HFC-227ea (ref. 4)	FC-3-1-10	HFC-23	HCFC Blend A	N_2
Acetone	6.8	5.5 (ref. 7)			
Acetonitrile	3.7				
AV Gas	6.7				
n-Butanol	7.1				
n-Butyl Acetate	6.6				
Cyclopentanone	6.7				
Diesel No. 2	6.7				
Ethanol	8.1	6.8 (ref. 7)			
Ethyl Acetate	5.6				
Ethylene Glycol	7.8				
Gasoline (unleaded)	6.5				
n-Heptane	6.0 (ref. 3) 5.8–6.6 (ref. 5)	5.2 (ref. 5) 5.0 (ref. 6)	12.0 (ref. 5) 12.6 (ref. 6)	12.6 (ref. 6)	32 (ref. 3)
Hydraulic Fluid	5.8	4.3–4.5 (ref. 3)			22–26 (ref. 3)
i-Propanol	7.3				
JP-4	6.6				
JP-5	6.0 (ref. 3) 6.6 (ref. 5)	4.8 (ref. 3)			27 (ref. 3)
Methane	6.2				
Methanol	10.0	9.4 (ref. 7)			
Methyl Ethyl Ketone	6.7				
Methyl Isobutyl Ketone	6.6				
Morpholine	7.3				
Propane	6.3	6.0 (ref. 3)			32.5 (ref. 3)
Pyrrolidine	7.0				
Tetrahydrofuran	7.2				
Toluene	5.8				
Turbo Hydraulic Oil 2380	5.1				
Xylene	5.3				

TABLE 4-7.5 *Weight and Storage Volume Equivalent Data for New Technology Halocarbon Gaseous Alternatives*

Trade Name	Designation	Formula		BP (°C)	Cup Burner (% V/V)**	Min. Design Conc. (% V/V)	Ratio Agent Mass Req'd to H 1301[∥]	Ratio Agent Storage Vol. Req'd to H 1301[#]	Storage Pressure (PSI) 20°C
Halon 1301	Halon 1301	CF_3Br		−58	2.9–3.9	5[§]	1.0	1.0	360
CEA-410	FC-3-1-10	C_4F_{10}		−2	5.0–5.9	6[*†]	1.9	1.7	360
FM-200	HFC-227ea	C_3F_7H		−16.4	5.8–6.6	7[*†]	1.7	1.6	360
FE-13	HFC-23	CHF_3		−82.1	12–13	16[*]	1.7	2.2	609
FE-25	HFC-125	H_2HF_5		−48.5	8.1–9.4	10.9[*]	1.9	2.3	166
NAF-S-III	HCFC Blend A	HCFC-22 HCFC-123 HCFC-124 Organic	82% 4.75% 4.5% 3.75%	—	>11%	8.6[‡]	1.1	1.4	360
CF_3I	Halon 1301	CF_3I		−22.5	2.7–3.2	5.0	~1	~1	—

* Based on 120 percent of cup burner value for n-Heptane.
† Based on 120 percent of cup burner value verified by listing/approval tests.
‡ Based on listing/approval tests.
§ Minimum design concentration per NFPA 12A, *Standard on Halon 1301 Fire Extinguishing Systems*; cup burner value approx. 3 percent.
∥ Ratio of halon design concentration 20.6 lb/ft³ at 70°C to new agent.
Ratio of halon storage volume required at minimum design concentration (max. fill density 70 lb/ft³) to new agent.
**Range of independently established values.

The storage volume equivalents are based on the maximum fill density permitted in a storage cylinder with a pressure rating as recommended by the manufacturer. The approximate 10:1 storage volume requirement for inert gases is a consequence of the inability to liquefy these gases at ambient temperature.

The storage volume equivalent does not translate directly to a required area or volume for storage cylinders. The

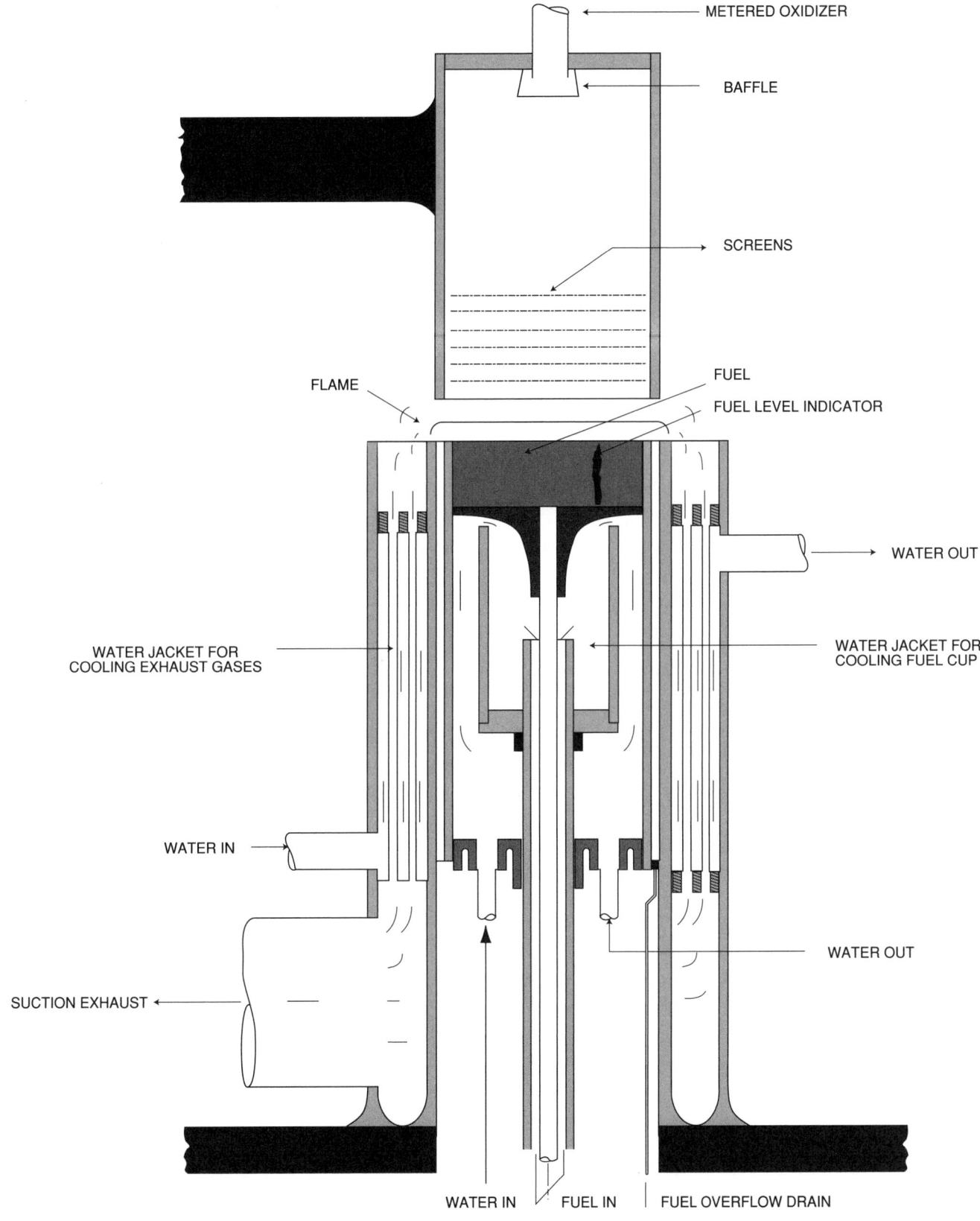

Fig. 4-7.2. Schematic illustration of the OFDF burner.[3]

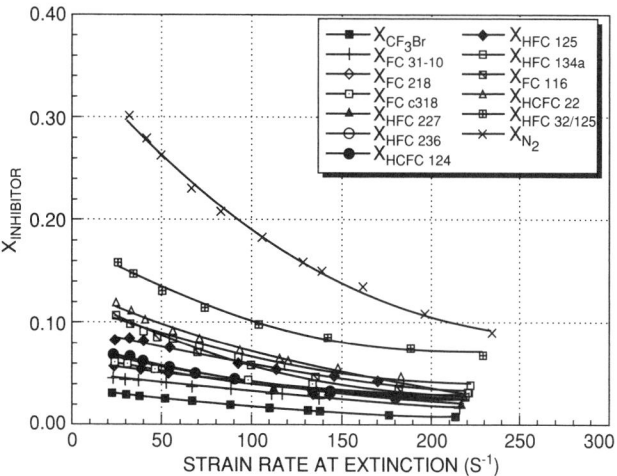

Fig. 4-7.3. Mole fraction of various suppressants as a function of strain rate at extinction for n-Heptane.[3]

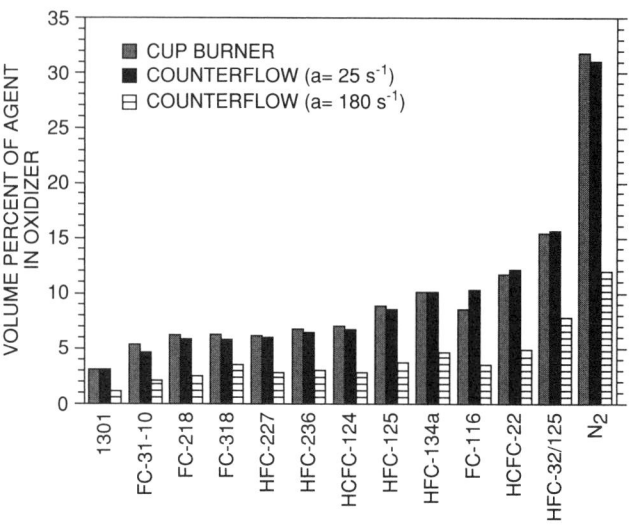

Fig. 4-7.4. Comparison of n-Heptane extinction results for the cup burner and OFDF apparatus at two strain rates.[3]

relative "footprint" of these storage volume equivalents will vary with the volume of the space protected and the maximum storage cylinder size offered by a manufacturer for a particular gas. In general, the floor area required for storage of inert gases exceeds 10:1 for large protected volumes.

Explosion Inerting

One of the most important application areas of total flooding fire suppressants is explosion inertion. The inerting concentration of an agent is the concentration required to prevent unacceptable pressure increases in a premixed fuel/air/agent mixture subjected to an ignition source. Inertion concentrations are typically measured in small laboratory-scale spheres, with an electric spark initiator.

The measured inerting concentration of an agent is dependent on the details of the test apparatus used, particularly the ignition source strength and "allowable" pressure rise. The "allowable" pressure rise is a surrogate measurement of the distance the flame front travels inside the constant volume sphere prior to suppression. Inerting concen-

tration is not appropriate for use in deflagration or detonation (explosion) suppression.

Small-scale sphere data are used to develop flammability diagrams for various fuel/oxidizer/agent concentrations. Section 2, Chapter 9, which addresses flammability limits, gives an excellent introduction to the subject. There is a wealth of data in the combustion literature on flammability limits for inert gases, such as nitrogen and argon, for a variety of fuels available.

Table 4-7.7 provides inerting concentration data for several agents and fuels, taken from small-scale inertion spheres. There are some substantial differences in results. Heinonen[11] has identified both ignition source type and strength as important variables with differences of ±40 percent for Halon 1301 inerting concentrations reported. Figure 4-7.5[13] shows flammability diagrams derived from small-scale inertion data such as that presented in Table 4-7.7, along with points taken from a large-scale (22.5 m³) explosion vessel. While the small- to large-scale agreement is reasonable, there are scale effects.

TABLE 4-7.6 *Agent Weight and Storage Volume Equivalent Data for New Technology Inert Gas Alternatives*

Trade Name	Designation	Formula		Cup Burner (% V/V) (n-Heptane)	Min. Design Conc. (% V/V)	Ratio Agent Mass Req'd to H 1301[†]	Ratio Agent Storage Vol. Req'd to H 1301[§]	Storage Pressure [PSI (bar)] 20°C
Halon 1301	Halon 1301	CF₃Br		2.9–3.9[‡]	5*	1	1	360 (25)
Inergen	IG-541	N₂	52%	29.1[‡]	35[#]	2.0	10.0[‖]	2180 (150)
		Ar	40%					
		CO₂	8%					
Argonite	IG-55	N₂	50%	30[‖]	~36[#]	2.0	10.0[‖]	2220 (153)
		Ar	50%					
Argon	IG-01	Ar	100%	30[‖]	~36[#]	2.0	8.0[‖]	2370 (163)

* Minimum design concentration per NFPA 12A, *Standard on Halon 1301 Fire Extinguishing Systems*; cup burner value approx. 3 percent.
[†] Ratio of halon design concentration 20.6 lb/ft³ at 70°C to new agent.
[‡] Range of independently established values.
[§] Ratio of halon storage volume required at minimum design concentration (max. fill density 70 lb/ft³) to new agent.
[‖] Manufacturers' values. MEC for Argon is approximately 41 percent (see ref. 3).
[#] Based on 120 percent of cup burner value for n-Heptane.

TABLE 4-7.7 *Explosion Inerting Concentrations, Small-scale Inertion Sphere*

Agent	Inerting Concentration (% by Volume) of Fuel			
	Propane	Methane	i-Butane	Pentane
FC-3-1-10	10.3 (ref. 8) 9.5 (ref. 10)	~7.8 (ref. 10)	—	
HFC-227ea	12.0 (ref. 10) 11.6 (ref. 4)	8.0 (ref. 10)	11.3 (ref. 8)	11.6 (ref. 4)
HFC-23	20.2 (ref. 8) 19.8 (ref. 10)	20.2 (ref. 8) 14.0 (ref. 10)	—	
IG-541	49.0 (ref. 9)	43.0 (ref. 9)	—	
HCFC Blend A	18.0 (ref. 10)	13.3 (ref. 10)		

Explosion Suppression

Explosion suppression systems employ rapid delivery of agent following very early detection of an ignition. Such systems employ significantly higher agent quantities (than flame suppression or inertion), delivered at higher rates. The total agent delivery time is on the order of 100 milliseconds.

Explosion suppression systems must be specifically designed for a particular application. There are no generic design requirements or standards currently available for such systems.

Senecal[12] and Senecal *et al*[14] report on explosion suppression testing in occupied armored fighting vehicles and aerosol filling rooms. Results were obtained on premixed fuel droplet sprays. In contrast to flame suppression or inerting, suppression or deflagration requires significantly more agent. The aerosol filling room tests employed 20 kg of HFC-227ea, FC-3-1-10, HFC-236fa, and 10 kg of water in a 80 m^3 test room to suppress a 90-g propane release in a simulated aerosol filling station. Suppression of the propane-air deflagration was achieved, and the maximum flame front extension was approximately 4 ft. Suppression tests of heated diesel fuel droplet cloud deflagrations were conducted in simulated armored fighting vehicle crew compartments.

Table 4-7.8 summarizes typical data for flame suppression, inertion, and deflagration suppression concentrations. Note these values are for comparison purposes only. They should not be used in any way for design purposes.

Suppression of detonations requires substantially higher agent concentrations. An excellent discussion is given in reference 3.

Toxicity

A major factor in the use of a clean agent fire suppressant in a normally occupied area is toxicity. While all halocarbon agents are tested for long-term health hazards, the primary endpoint is acute or short-term exposure. The primary acute toxicity effects of the halocarbon agents described in this chapter are anesthesia and cardiac sensitization. For inert gases, the primary physiological concern is reduced oxygen concentration.

Halocarbon agents: Cardiac sensitization is the primary short-term toxicity problem for fire suppression applications. Cardiac sensitization is a term describing the sudden onset of cardiac arrhythmia in the presence of a concentration of an agent caused by sensitization of the heart to epinephrine. The presence of epinephrine is critical to the onset of arrhythmia. This is important in fire protection applications due to the increased production of epinephrine by the body under stress.

The two toxicity endpoints used to describe cardiotoxicity and allowable exposure levels are: (1) no observed adverse effect level (NOAEL) and (2) the lowest observed adverse effect level (LOAEL). The NOAEL is the highest concentration of an agent at which no "marked" or adverse effect occurred. The LOAEL is the lowest concentration at which an adverse effect was measured.

The procedures used to evaluate cardiac sensitization vary somewhat. The procedure involves intravenous dosing of male beagle hounds with epinephrine for 5 min. Continuous inhalation exposure to the agent follows for 5 min. Following this inhalation exposure, the hound is dosed again with epinephrine and monitored for 5 min to determine the effect of the agent and epinephrine. The protocol is performed at higher doses until an effect occurs.

Effects are monitored by electrocardiograph (EKG) measurements. An adverse effect is generally considered to be the appearance of five or more arrhythmias or ventricular fibrillation. The data from these tests are evaluated by medical experts, and the appropriate NOAEL and LOAEL values are reported by the Environmental Protection Agency (EPA) under the Significant New Alternatives Policy (SNAP) program.

There is no direct correlation between the experimental results from hounds to humans. It is generally accepted, due to the combination of the high doses of epinephrine in the tests and the similarity in cardiovascular function between hounds and humans, that the results can be applied to humans.

In addition to the short-term chronic exposure limits of interest in fire suppression system design, the EPA evaluates longer-term inhalation data for these compounds. Table 4-7.9 summarizes NOAEL, LOAEL, and LC_{50} values. Note that the LC_{50} (the concentration lethal to 50 percent of a population) values greatly exceed the NOAEL at typical fire extinguishing concentrations.

The use of halocarbon agents in occupied areas is subject to the constraint that the design concentration must be less than the NOAEL. While it is recommended that all systems employ predischarge alarms and that personnel

TABLE 4-7.8 *Comparison of Concentrations for Flame Extinguishment, Inertion, and Deflagration Suppression*

Agent	Volume (%)		
	Typical Value Flame Suppression	Inerting Concentration in Propane	Diesel Fuel Droplet Deflagration Suppression
Halon 1301	3	6–7	12
FC-3-1-10	5.5	10.3	8
HFC-227ea	5.8	~12	11
HFC-23	12	20.2	—
IG-541	29	49.0	—

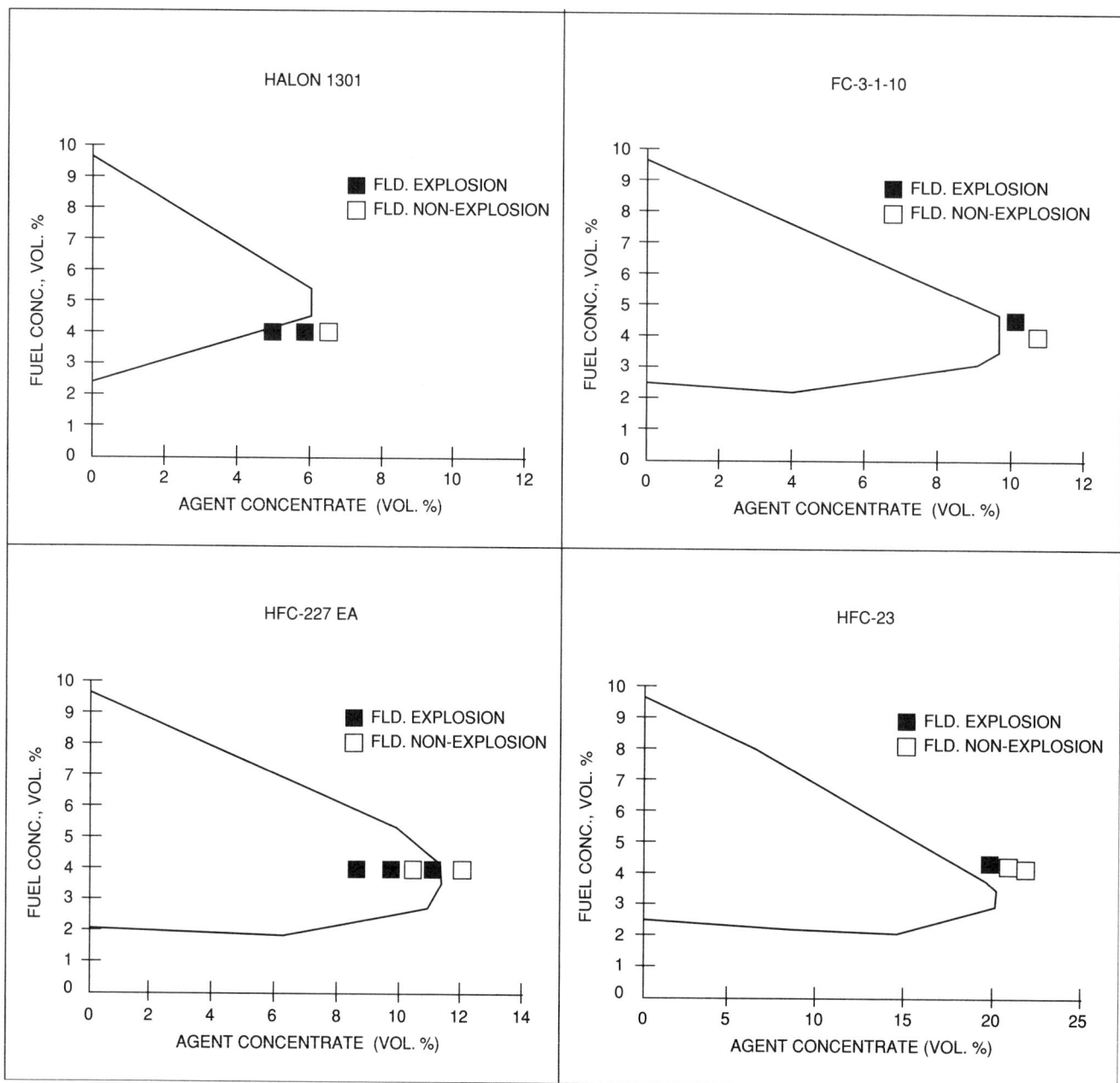

Fig. 4-7.5. Small-scale flammability diagrams for propane and several replacement agents. Squares denote explosion/no explosion points in large-scale 22.5 m³ chamber.[13]

evacuate prior to system actuation, it is understood that inadvertent discharges and short-term exposures will occur, hence the limitation. It is expected that emergency exposures for up to several minutes at or below the NOAEL are reasonably safe. In no case should systems be designed or installed where intentional exposure of any duration is anticipated.

It has been proposed by the EPA that agents be permitted for use at concentrations up to the LOAEL where evacuation will occur in less than 60 sec. This has not been integrated into design standards to date due to the uncertainty of accidental exposure conditions.

Based on the limitation that the design concentration must be below the NOAEL, it can be seen from Table 4-7.9 that three agents are acceptable for use in normally occupied areas for flame extinguishant purposes. These are: HFC-227ea, HFC-23, and FC-3-1-10.

Inert gas agents:　Inert gas agents are in effect physiologically inert. The primary physiological problem with these agents is the reduced oxygen concentration caused by the high agent design concentrations. One inert gas blend employs a low concentration of CO_2 (which is not physiologically inert) in order to counter the effects of the reduced

TABLE 4-7.9 *Toxicity Data for Halocarbon Clean Agent Fire Suppressants*

Trade Name	Designation	Formula		NOAEL % V/V*	LOAEL % V/V*	LC$_{50}$ or ALC†
CEA-410	FC-3-1-10	C_4F_{10}		40	>40‡	>80%
FM-200	HFC-227ea	C_3F_7H		9.0	>10.5	>80%
FE-13	HFC-23	CHF_3		30	>50‡	>65%
FE-24	HCFC-124	C_2HClF_4		1	2.5	23–29%
FE-25	HFC-125	H_2HF_5		7.5	10.0	>70%
NAF-S-III	HCFC Blend A	HCFC-22	82%	10	>10	64%
		HCFC-123	4.75%			
		HCFC-124	4.5%			
		Organic	3.75%			
CF_3I	Halon 1301	CF_3I		0.2	—	—

*From EPA SNAP documents.
†From NFPA 2001, *Standard on Clean Agent Fire Extinguishing Systems*.
‡Maximum concentration before oxygen depletion.

oxygen concentration. The mechanism of this effect is discussed in reference 15.

The EPA has determined that inert gases can be used in concentrations up to 43 percent in normally occupied areas subject to similar limits placed on halocarbon agents when used at the NOAEL. This results in a residual oxygen concentration of 12 percent. The analog to the LOAEL is 52 percent agent concentration, with a 10 percent residual oxygen concentration. NFPA 2001, *Standard on Clean Agent Fire Extinguishing Systems*, requirements mirror the EPA limitations at 43 percent agent concentration.

Environmental Factors

The evaluation of clean agent fire suppressants includes a consideration of environmental factors. International, national, and local government regulations control the use of any alternatives in this regard. The primary environmental consideration is ozone depletion potential (ODP). This is a measure of a chemical's ability to deplete stratospheric ozone, with CFC-12 as a basis with an ODP of 1. All chemicals with non-zero ODP are subject to phaseout under the Montréal Protocol and its amendments. Table 4-7.10 summarizes environmental impact data for halocarbon alternatives. Note that FC and HFC compounds have zero ozone depletion potential. The HCFC compounds have quite low ODP. Other HCFC compounds are widely used as CFC replacements for refrigerants.

Other environmental factors that are potentially important in a regulatory context are global warming potential (GWP) and atmospheric lifetime. GWP is a measure of the contribution of a gas to the so-called "greenhouse effect." It is a function of atmospheric lifetime and the ability of the gas to absorb infrared radiation. The evaluation of GWP is an extremely complex issue, and, currently, none of these compounds are regulated on that basis in the United States. Long atmospheric lifetime, a measure of the persistence of a chemical in the atmosphere, is of concern not only as it relates to GWP, but also due to the uncertainty of the effects of chemicals for long time periods in the atmosphere. The EPA currently has use restrictions on FC-3-1-10 based primarily on its long atmospheric lifetime. These restrictions permit the use of this chemical in applications where no other alternative is technically feasible.

Thermophysical Properties

Tables 4-7.11 and 4-7.12 give thermophysical properties of clean agent replacements from NFPA 2001,[1] in English and SI units, respectively. Table 4-7.13, extracted from reference 16, gives independent data and estimates for some thermophysical properties. Additional thermophysical and transport property data can be found in reference 4 for FM-200, and reference 16 for a range of halocarbon alternatives.

Isometric diagrams for halocarbon agents HFC-227ea, pressurized at 360 and 600 psig at 70°F with nitrogen, and

TABLE 4-7.10 *Environmental Factors for Halocarbon Clean Agents*

Trade Name	Designation	Formula		ODP	GWP (100 yr)	Atmospheric Lifetime (yr)
Halon 1301	Halon 1301	CF_3Br		16	5800	100
CEA-410	FC-3-1-10	C_4F_{10}		0	5500	2600
FM-200	HFC-227ea	C_3F_7H		0	2050	31
FE-13	HFC-23	CHF_3		0	9000	280
FE-24	HCFC-124	C_2HClF_4		0.022	440	7
FE-25	HFC-125	H_2HF_5		0	3400	41
NAF-S-III	HCFC Blend A	HCFC-22	82%	0.05	1600	16
		HCFC-123	4.75%			
		HCFC-124	4.5%			
		Organic 3	3.75%			
CF_3I	Halon 1301	CF_3I		<0.2	0	—

TABLE 4-7.11 *Thermophysical Properties of Clean Halocarbon Agents (English Units)*

	Units	FC-3-1-10	HCFC Blend A	HCFC-124	HFC-125	HFC-227ea	HFC-23	IG-541	IG-55	IG-01
Molecular Weight	N/A	238.03	92.90	136.5	120.2	170.03	70.01	34.0	33.95	39.9
Boiling Point @ 760 mm Hg	°F	28.4	−37.0	12.2	−55.3	2.6	−115.7	−320	−310.2	−302.6
Freezing Point	°F	−198.8	< −161.0	−326.0	−153	−204	−247.4	−109	−327.5	−308.9
Critical Temperature	°F	235.8	256.0	252.0	158.8	215.0	78.6	—	−210.5	−188.1
Critical Pressure	psia	337	964	524.5	521	422	701	—	602	711
Critical Volume	ft³/lbm	0.0250	0.0280	0.0283	0.0281	0.0258	0.0305	N/A	N/A	N/A
Critical Density	lbm/ft³	39.3	36.00	35.28	35.68	38.76	32.78	N/A	N/A	N/A
Specific Heat, liquid @ 77°F	Btu/lb-°F	0.25	0.30	0.270	0.301	0.2831	0.370	N/A	N/A	N/A
Specific Heat, vapor @ constant pressure (1 atm) & 77°F	Btu/lb-°F	0.192	0.16	0.177	0.191	0.1932	0.176	0.195	0.187	0.125
Heat of Vaporization at Boiling Point @ 77°F	Btu/lb	41.4	97	83.2	70.8	57.0	103.0	94.7	77.8	70.1
Thermal Conductivity of Liquid @ 77°F	Btu/h ft°F	0.0310	0.052	0.0417	0.0376	0.040	0.0450	N/A	N/A	N/A
Viscosity, liquid @ 77°F	lb/ft hr	0.783	0.508	0.723	0.351	0.443	0.201	N/A	N/A	N/A
Relative dielectric strength @ 1 atm @ 734 mm Hg 77°F (N_2 = 1.0)	N/A	5.25	1.32	1.55	0.955 @ 70°F	2.00	1.04	1.03	1.01	1.01
Solubility of water in agent @ 70°F	N/A	0.001% by weight	0.12% by weight	0.07% by weight @ 77°F	0.07% by weight @ 77°F	0.06% by weight	500 ppm @ 50°F	0.015%	0.006%	0.006%
Vapor Pressure @ 77°F	psi	42.0	1.37	56	199	66.4	686.0	2207	N/A	N/A

TABLE 4-7.12 *Thermophysical Properties of Clean Halocarbon Agents (SI Units)*

	Units	FC-3-1-10	HCFC Blend A	HCFC-124	HFC-125	HFC-227ea	HFC-23	IG-541	IG-55	IG-01
Molecular Weight	N/A	238.03	92.90	136.5	120.2	170.03	70.01	34.0	33.95	39.9
Boiling Point @ 760 mm Hg	°C	−2.0	−38.3	−11.0	−48.5	−16.4	−82.1	−196		
Freezing Point	°C	−128.2	< −107.2	−198.9	−102.8	−131	−155.2	−78.5		
Critical Temperature	°C	113.2	124.4	122.2	66.0	101.7	25.9	—		
Critical Pressure	kPa	2323	6647	3614	3595	2912	4836	—	N/A	N/A
Critical Volume	cc/mole	371	162	241.6	210	274	133	N/A	N/A	N/A
Critical Density	kg/m³	629	577	565	571	621	525	N/A	N/A	N/A
Specific Heat, liquid @ 25°C	kJ/kg°C	1.047	1.256	1.13	1.260	1.184	1.549	N/A	N/A	N/A
Specific Heat, vapor @ constant pressure (1 atm) & 25°C	kJ/kg°C	0.804	0.67	0.741	0.800	0.8082	0.737	0.574		
Heat of Vaporization at Boiling Point @ 25°C	kJ/kg	96.3	225.6	194	164.7	132.6	239.6	220		
Thermal Conductivity of Liquid @ 25°C	W/m°C	0.0537	0.0900	0.0722	0.0651	0.069	0.0779	N/A	N/A	N/A
Viscosity, liquid @ 25°C	centipoise	0.324	0.21	0.299	0.145	0.184	0.083	N/A	N/A	N/A
Relative dielectric strength @ 1 atm @ 734 mm Hg 25°C (N_2 = 1.0)	N/A	5.25	1.32	1.55	0.955 @ 21°C	2.00	1.04	1.03	1.01	1.01
Solubility of water in agent @ 21°C	N/A	0.001% by weight	0.12% by weight	0.07% by weight @ 25°C	0.07% by weight @ 25°C	0.06% by weight	500 ppm @ 10°C	0.015%	0.006%	0.006%
Vapor Pressure @ 25°C	kPa	289.6	948	386	1371	457.7	4,730	15,200	N/A	N/A

TABLE 4-7.13 *Selected Properties of Agents (extracted from reference 16)*

	FC-3-1-10	HFC-227ea	HCFC-124	HFC-125	HFC-23
Boiling Point @ 0.101 MPa (°C)	−2.0	−16.4	−13.2	−48.6	−82.1
Critical Temperature (°C)	113.2	101.7	122.5	66.3	25.6
Critical Pressure (MPa)	2.32	2.9	3.65	3.62	4.82
Vapor Pressure @ 25°C (MPa)	0.27	0.47	0.38	1.38	4.69
Liquid Density at 25°C (kg/m³)	1497	1395	1357	1190	0.685
Liquid Heat Capacity @ Boiling Point (kJ/kg K)	0.951	1.074	1.080	1.107	1.269
Liquid Heat Capacity at 25°C (kJ/kg K)	1.017	1.177	1.111	1.358	—
Latent Heat of Vaporization at Boiling Point (kJ/kg)	96	131	162	160	240

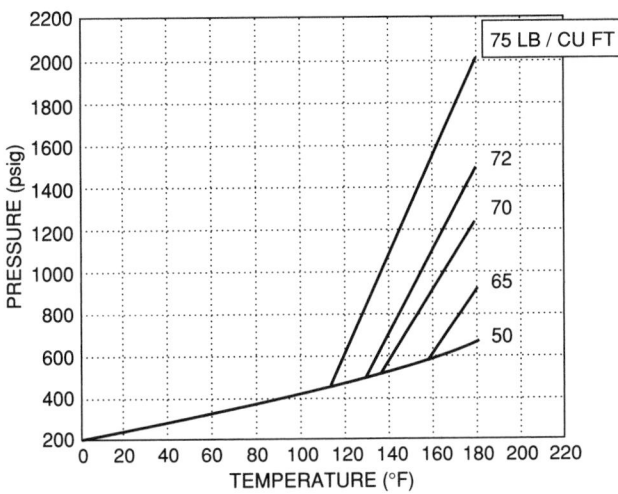

Fig. 4-7.6. Isometric diagram of HFC-227ea, pressurized to 360 psig with N_2, at 70°F.[4]

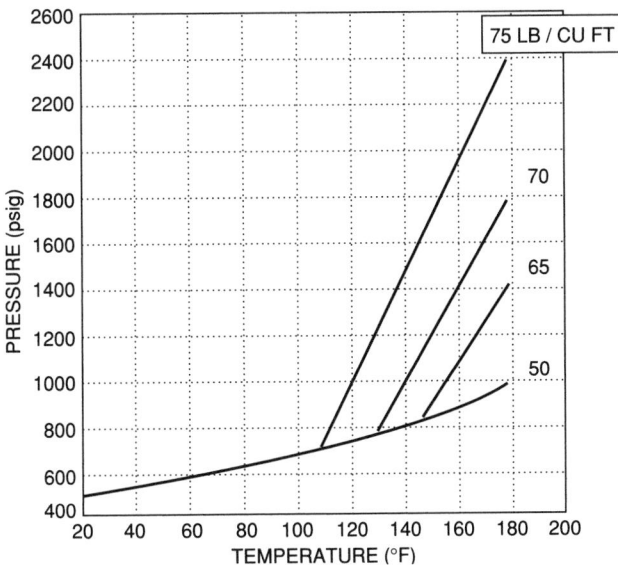

Fig. 4-7.7. Isometric diagram of HFC-227ea, pressurized to 600 psig with N_2, at 70°F.[4]

HFC-23 are given in Figures 4-7.6, 4-7.7, and 4-7.8, respectively. Note that HFC-23 is not pressurized with nitrogen. Figure 4-7.9 gives the pressure/temperature relationship for inert gases IG-541, IG-55, and IG-01, pressurized to 2175 psig, at 70°F. This is the pressure/temperature relationship for an ideal gas.

CLEAN AGENT SYSTEM DESIGN

Once the agent has been selected, the general discussion on clean agent system design presented by Grant in Section 4, Chapter 6 should be reviewed. The basic process is outlined below.

1. Determine the design concentration,
2. Determine the total agent quantity,
3. Establish the maximum discharge time,

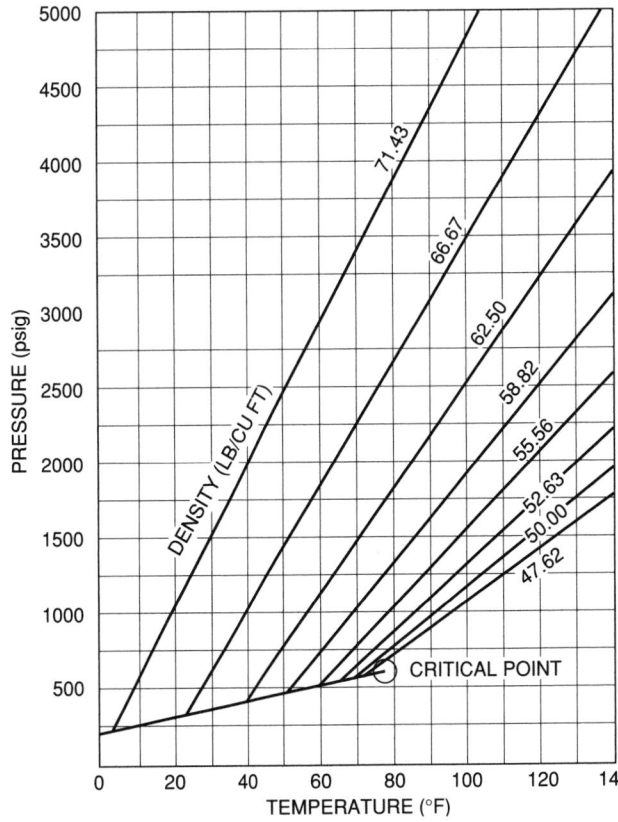

Fig. 4-7.8. Isometric diagram of HFC-23.[1]

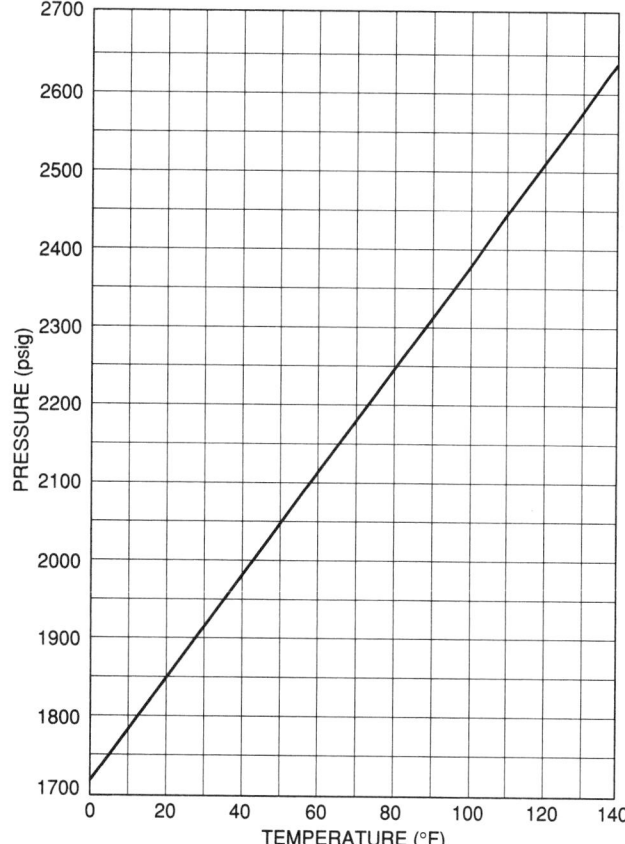

Fig. 4-7.9. *Isometric diagram of inert gases and blends, treated as ideal gases, pressurized to 2175 psig, at 70°F.* [1]

4. Select piping material and thickness consistent with pressure rating requirements,
5. Design piping network and select nozzles to deliver required concentration at required discharge time to ensure mixing,
6. Evaluate compartment over/underpressurization and provide venting if required, and
7. Establish minimum agent hold requirements and evaluate compartments for leakage.

These attributes apply only to the mechanical design of the system.

The detection and actuation systems are critical and integral parts of a clean agent system design. The detection system should be designed to actuate the system, with appropriate pre-discharge alarms, before unacceptable thermal or nonthermal damage occurs. This is particularly important where the thermal decomposition products of halocarbon clean agents are a concern. Section 4, Chapter 1 provides engineering methods and calculation procedures for this purpose.

In addition to the detection, actuation, and alarm systems, the enclosure itself is critical in the design of any total flooding suppression system. The most important considerations are that the enclosure be of sufficient integrity to: (1) prevent preferential agent loss during discharge and (2) prevent excessive agent/air mixture loss after discharge to ensure adequate hold time.

As a general rule, all openings, notably doors and ventilation fans and/or openings, must be secured prior to discharge in conjunction with the detection and alarm systems. Agent system installation in rooms with unclosable openings should not be attempted unless sufficient test data are available to ensure adequate concentrations. Some enclosures, such as very tightly sealed low EMF emission electronics spaces, require additional care to avoid compartment damage due to over/underpressurization during agent discharge.

Design Concentration

Flame extinguishment: Design concentrations for various agents and fuel combinations are generally determined by a combination of small-scale testing, large-scale testing, independent laboratory approval of hardware, and addition of design safety factors.

Historically, minimum design concentrations for Halon 1301 were set by the cup burner extinguishing concentration plus a 20 percent safety factor. A minimum Halon 1301 design concentration of 5 percent was also established for all applications. For heptane, the cup burner value was approximately 3 percent; with a 20 percent safety factor, a design concentration of 3.6 percent is obtained. At the minimum design concentration set by NFPA 12A, *Standard on Halon 1301 Fire Extinguishing Systems*, of 5 percent, a 66 percent safety factor was achieved. For fuels with cup burner extinguishing concentrations greater than 4.2 percent, the safety factor remained at 20 percent.

The basic requirement for determining the design concentration of clean agents in NFPA 2001, *Standard on Clean Agent Fire Extinguishing Systems*, is two-fold. First, the minimum extinguishing concentration as determined by the cup burner must be established. Second, after this minimum is established by the *system* manufacturer, full-scale third-party approval testing is conducted using the manufacturer's hardware on heptane, wood crib, and selected flammable liquids. These tests are performed at the cup burner minimum extinguishing concentration, not the design concentration. Further, they are conducted with flooding factors lower than utilized in design. Hence, the minimum set by the cup burner or equipment manufacturer, whichever is higher, is tested in full scale as part of the approval/listing process for the agent/system combination. Often, hardware manufacturers will establish a minimum concentration greater than the cup burner value to account for nozzle inefficiency.

Reliance on the cup burner to set a minimum value is supported by the observation that no full-scale test conducted at the cup burner concentration plus 80 percent safety factor has failed to extinguish a flammable liquid fire in an enclosure. It is critical to recognize that, in real system design, the agent and the hardware delivering it, particularly the nozzle, are not independent. They must be taken as a system and designed and installed per the listing/approval guidelines and limitations.

There has been some full-scale test work that indicates that the 20 percent safety factor may be insufficient. Sheinson *et al* noted significant improvement in extinguishing time performance with a safety factor of 40 percent.[17] Brockway noted similar results, with no performance improvement beyond a safety factor of 40 percent.[18]

It has also been noted by several investigators that higher safety factors result in lower thermal decomposition products.[17,18,19] None of the above referenced investigations utilized listed or approved hardware for the specific agents tested; as in most cases, the tests were performed before such hardware was available.

There is an exception in NFPA 2001, *Standard on Clean Agent Fire Extinguishing Systems*, to the general rule that a minimum extinguishing concentration be established by the cup burner method. It was alleged that reliable cup burner data were not available for HCFC Blend A due to the fact that (1) the agent was a blend, and (2) one of the blend components heats at a low vapor pressure. In the case of this agent, a minimum extinguishing concentration of 7.2 percent and, hence, a design concentration of 8.6 percent was established through limited full-scale testing. Since at the time insufficient data were available to evaluate the claim, the exception that requires full-scale testing at minimum extinguishing concentration consistent with UL 1058, *Halogenated Agent Extinguishing System Units*,[20] was invoked. Since that time, reliable cup burner data were obtained for the blend from several laboratories. The data are consistent with MEC values for the blend components, primarily HCFC 22. Furthermore, some full-scale testing has indicated that the design concentration of 8.6 percent may be inadequate.[21] This issue is, however, unresolved at the present time.

For Class A fires, NFPA 2001, *Standard on Clean Agent Fire Extinguishing Systems*, requires full-scale testing and third-party approval for evaluating design concentration on solid polymeric materials. In many cases, the MEC for heptane is used as a practical minimum.

There has been no systematic evaluation of these agents under so-called "deep-seated" fire scenarios. Part of the problem is the circular definition of deep-seated fires in NFPA 12A, *Standard on Halon 1301 Fire Extinguishing Systems*. However, the Underwriters Laboratories Inc. (UL) and Factory Mutual Research Corporation (FMRC) listing procedures require testing on wood cribs subsequent to long preburn times (approximately 5 min). Under these tests, surface oxidation and char reactions do occur.

Design concentrations for fire scenarios involving long preburn times in thick arrays of cellulosic fuels will require additional testing. For most applications where incidental quantities of cellulosic materials may be involved and preburn times are relatively short (<5 min) time frames (i.e., automatic actuation), the flame extinguishing concentrations for Class A fuels will be less than, or equal to, that of n-heptane and can be used. Surface oxidation or charring reactions do not occur with most polymers; hence, so-called "deep-seated" fires are not a concern where Class A fuels are involved.

IG-541 is used at 37.5 percent minimum design concentration where Class A materials are involved. Other inert gases should have similar or higher minimum design concentrations.

As previously discussed, the minimum design concentration is a function of the fuel, the agent, and the delivery system. Design concentrations for specific hazards must be determined in accordance with the system manufacturer's approval or listing.

Agent Quantity

Once the design concentration is established, the quantity of agent necessary to achieve that concentration is determined. The quantity of halocarbon agent necessary is determined by the following equation

$$w = \frac{V}{S}\left(\frac{C}{100 - C}\right) \qquad (1)$$

where:

V = net volume of protected space
C = design concentration (%),
w = specific weight of agent required, and
S = specific volume [ft³/lb (m³/kg)], and is determined by

$$S = k_1 + k_2(T) \qquad (2)$$

where:

T = minimum ambient temperature of the protected space, and k_1 and k_2 are constants.

Values for k_1 and k_2 used in Equation 2 are given in Table 4-7.14.

The flooding factor in Equation 1 [$C/(100 - C)$] implies that the agent/air mixture "lost" during discharge is well mixed and has an agent concentration of C. This formula makes no assumption regarding leakage of the enclosure. During UL/FMRC approval testing, the agent is evaluated with a flooding factor of ($C/100$), essentially assuming that losses during discharge are 100 percent air.

For inert gases, the following formula is used

$$X = 2.303\frac{V}{S} \log\left(\frac{100}{100 - C}\right)V_s \qquad (3)$$

where:

X = volume of inert gas required at 70°F,
V_s = specific volume at 70°F,
V = net protected hazard volume, and
S = specific volume at ambient temperature in protected volume (from Equation 2) = $k_1 + k_2 (T)$.

The flooding factor used here, log [$100/(100 - C)$], is derived assuming that leakage from the compartment during discharge occurs with a varying concentration of agent from zero to C from beginning to the end of discharge. It is identical to the expression used in CO_2 system design. It assumes that the displaced atmosphere is freely vented from the enclosure.

Discharge Time

The maximum discharge time permitted for halocarbon clean agent systems is 10 sec. This discharge time is taken to be the moment where all liquid agent has cleared the nozzle. The total discharge time will be longer as agent vapor and nitrogen are expelled from the system.

TABLE 4-7.14   Specific Volume Constants

Agents	°F		°C	
	k_1	k_2	k_1	k_2
FC-3-1-10	1.409	0.0031	0.0941	0.0003
HCFC Blend A	3.612	0.0079	0.2413	0.00088
HCFC-124	2.352	0.0057	0.1578	0.0006
HFC-125	2.724	0.0063	0.1701	0.0007
HFC-227ea	1.885	0.0046	0.1269	0.0005
HFC-23	4.731	0.0107	0.2954	0.0012
IG-541	9.7261	0.0211	0.649	0.00237
IG-01	8.514	0.0185	0.5685	0.00208
IG-55	10.0116	0.0217	—	—

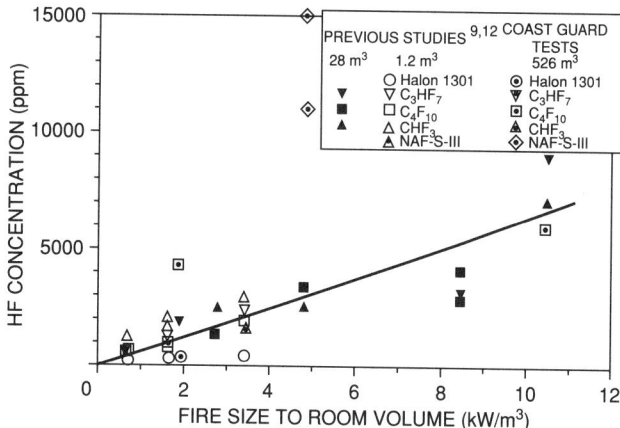

Fig. 4-7.10. *Maximum HF concentration resulting from extinguishment of heptane fires with nominal 10–15 sec total discharge times.*[21]

The 10-sec discharge time limitation for halocarbon agents is designed to aid 4 objectives:

1. Provide high flow rates through nozzles to ensure adequate mixing of agent with air inside the enclosure;
2. Provide sufficient velocity through pipes to ensure homogeneous flow of liquid and vapor;
3. Limit the formation of agent thermal decomposition products; and
4. Minimize direct and indirect fire damage, particularly in fast-developing fire scenarios.

The most important of these objectives relative to discharge time is the minimization of agent thermal decomposition product formulation. Items 1 and 2 alone are determined by the piping system design.

The discharge time requirement for inert gases is currently 60 sec.[1] Longer discharge times are typically used for these systems in Europe. The two primary reasons to constrain the discharge time of inert gas agents that form no thermal decomposition products are: (1) to limit the direct and indirect fire damage and (2) to minimize the length of time that the fire burns in a depleted oxygen atmosphere. As more information is developed on the effect of discharge time on inert gas agent performance, this 60-sec limit may be increased for certain applications.

In some applications, such as flammable liquid hazards and explosion inerting, it is necessary to discharge the agent quickly to minimize direct fire damage or to ensure that the agent concentration is achieved prior to the lower explosive limit (LEL) being reached.

Thermal Decomposition Products

All of the halocarbon replacement agents form higher levels of thermal decomposition products than Halon 1301 under similar conditions. For a given fuel, the two primary variables determining the level of decomposition products are: (1) the size of the fire at the time of discharge, and (2) the time required to reach an extinguishing concentration in the compartment.

The dependence of thermal decomposition product formulation on discharge time and fire size has been extensively evaluated.[17–19,21–24] Figure 4-7.10 from reference 21 is a plot of peak HF concentration as a function of fire size to

room volume ratio. Similar data for 10-min average HF concentrations are given in Figure 4-7.11. Data are given for Halon 1301, HFC-23, HFC-227ea, FC-3-1-10, and HCFC Blend A from three series of fire tests done at different room scales. The data are for a total discharge time of 15 sec, which is analogous to a 10-sec discharge time based on nozzle liquid runout. The data are for heptane pool or heptane pool and spray fires.

The first observation is that the quantity of HF formed is approximately five to ten times higher for all halocarbon replacements relative to Halon 1301. There may be differences between the various HFC/HCFC compounds tested, but it is not clear from these data whether (1) such differences occur; or (2) are the differences attributable to agent mixing and distribution, or (3) attributable to locally high velocities or concentrations of agent from the nozzle. In all of the data reported, the fire source, i.e., heptane pans of varying sizes, was baffled to prevent direct interaction with the agent jet.

These data were taken with an FTIR spectrometer at a location approximately 1 m from the floor, or about mid flame height, near the wall. The method used was correlated to grab sample and ion-specific electrode (ISE) methods.[22] In all cases, the agreement was good, except for the HCFC blend. In this case, the HF concentration inferred from the treated grab sample was significantly (> 50 percent) higher than that measured using the FTIR. Since the HCFC blend contains an HF "scrubber," it is postulated that treatment of the grab sample with a basic solution, as required for the ISE measurements, caused formation of additional HF by reentry with F loosely bound up by reaction with the scrubber. Hence, the FTIR data presented for HCFC Blend A represent a significantly lower quantity of HF than would actually be expected if the product was hydrolyzed. This is also consistent with the fact that the agent was tested at the manufacturer's recommended design concentration, which is approximately 40 percent lower than the basis for all other agents.

The effect of long discharge times or delayed extinguishing times is shown in Figure 4-7.12.[22] The variation between the HFC/HCFC alternatives and Halon 1301, relative to HF production, is approximately the same as that shown in Figure 4-7.11 for different fire sizes.

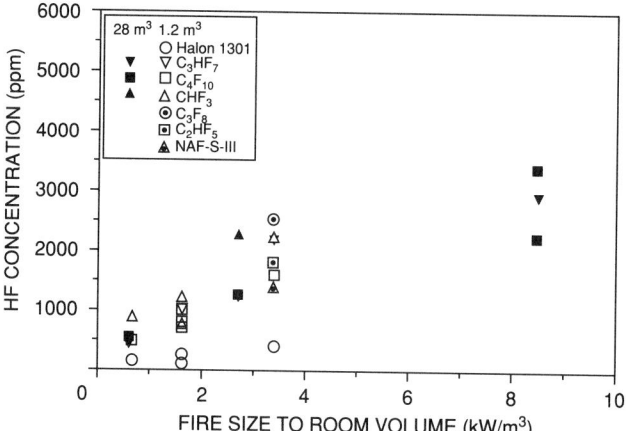

Fig. 4-7.11. *Average HF concentration resulting from extinguishment of heptane pool or heptane pool and spray fires with nominal 15-sec total discharge time.*[22]

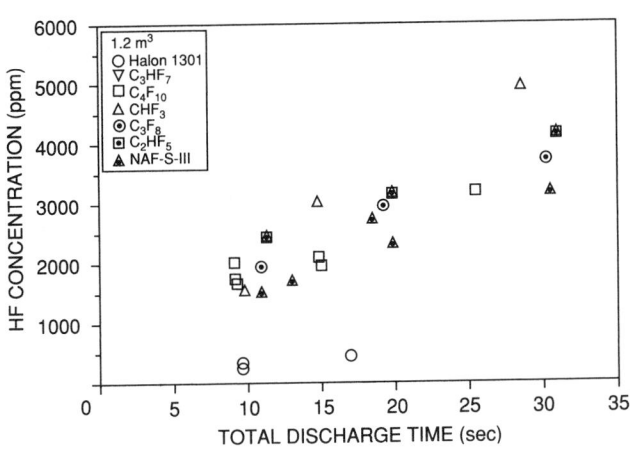

Fig. 4-7.12. *Maximum HF concentration resulting from extinguishment of 4.0-kW heptane fires.*[22]

Although other thermal decomposition products have been identified in some cases, it appears that HF is the primary thermal decomposition product of interest relative to human safety and equipment damage.

HF, like HCl, is an irritant gas, detectable at very low concentrations. For HF there are very large differences between the approximate lethal concentration (ALC) and human detection and severe sensory irritant thresholds (approximately two and three orders of magnitude, respectively).

Fire size necessary to generate short-term lethal concentrations of HF in an enclosure (on the order of > 1000 ppm) can, in some cases, pose a greater hazard to personnel in the protected space during a discharge in a fire incident, due to the fire and its effects, than the secondary impact of agent thermal decomposition products. This, however, should be verified for a particular application under a range of fire scenarios, using engineering methods discussed by Hanauska[25] and Hanauska *et al.*[26]

The impact of thermal decomposition products on electronics equipment is another area of concern. There are not sufficient data at present to predict the effects of a given HF exposure scenario on all electronics equipment. Several evaluations of the impact of HF on electronics equipment have been performed relative to the thermal decomposition of Halon 1301, where decomposition products include HF and HBr. One of the more notable was a NASA study where the shuttle orbital electronics were exposed to 700, 7000, and 70,000 ppm HF and HBr.[25] In these tests, exposures up to 700 ppm HF and HBr caused no failures. At 7000 ppm, severe corrosion was noted; there were some operating failures at this level.

Dumayas exposed IBM-PC-compatible multifunction boards to environments produced by a range of fire sizes as part of an evaluation program on halon alternatives.[27] He found no loss of function of these boards following a 15-min exposure to postfire extinguishment atmosphere up to 5000 ppm HF, with unconditioned samples stored at ambient humidity and temperature conditions for up to 30 days. Forssell *et al*[28] exposed multifunction boards for 30 min in the postfire extinguishment environment; no failures were reported up to 90 days posttest. HF concentrations up to 550 ppm were evaluated.

While no generic rule or statement can be made at present, it appears that short-term damage (< 90 days) re-sulting in electronics equipment malfunction is not likely for exposures of between 500 to 1000 ppm HF for up to 30 min. This, however, is dependent on the characteristics of the equipment exposed, postexposure treatment, exposure to other combustion products, and relative humidity. Important equipment characteristics include its location in the space, existence of equipment enclosures, and the sensitivity of the equipment to damage.

All HCFC and HFC clean agents form more thermal decomposition products than Halon 1301, given similar fire sizes and discharge times. The primary variable controlling the quantity of thermal decomposition products is the size of the fire at the time of agent discharge. Through evaluation of the fire size at the time of system actuation, using engineering methods described in Section 4, Chapter 1, and subsequent design of the detection system, the potential hazard posed can be managed adequately.

Hanauska[25] and Hanauska *et al*[26] have indicated that the degree of thermal decomposition products of agents can be managed safely. Full-scale testing with typical Class A fuel packages, in conjunction with typical detection system installation,[28] has shown that the level of thermal decomposition products is acceptable in typical computer/electronics spaces. For installation in hazard areas where very rapidly developing large fires are likely, the degree of thermal decomposition formation should be evaluated in the context of the hazard posed by the fire and the performance of alternative fire protection systems.

Hydraulic Flow Characteristics

All halocarbon replacement agents exhibit two-phase flow behavior. Since all, except HFC-23, are used in cylinders pressurized to 360 or 600 psig, they are also multiple-component flows. Inert gas mixtures are single-phase gas flows with one or more components. As in the case of engineered Halon 1301 systems, all flow calculation procedures must be listed or approved by the authority having jurisdiction, and within the limitations of the flow calculation method determined during the engineered system approval process.

The characteristic that differentiates two-phase pipe flow from incompressible fluid (e.g., water) pipe flow is the existence of gas and liquid phases simultaneously in the pipe network. This, coupled with the relatively short flow times, results in significant challenges to correctly predicting the flow. Among the important factors are the change in density of the fluid with pressure, the release of nitrogen in the cylinder and pipe as the fluid pressure and temperature change, differences in agent mass delivered caused by the flow time imbalances between nozzles, and preferential distribution of phases (and subsequently agent mass) at tee splits.

The need for accurate flow predictions is driven by three design requirements:

1. Control of agent discharge time;
2. Maintenance of adequate nozzle flow and pressure to ensure agent distribution and mixing at the listed coverage area; and
3. Delivery of adequate, but not excessive, agent quantities to different rooms within the same protected area, when such rooms are flooded simultaneously.

In addition, agent flow rate and thermodynamic state properties are necessary for estimating compartment pressurization levels during agent discharge.

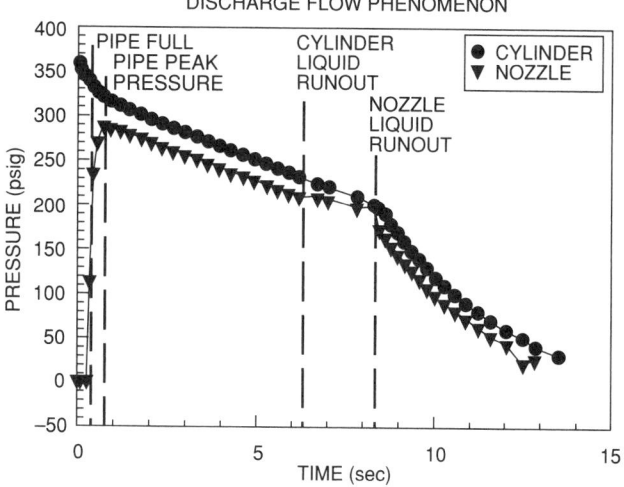

Fig. 4-7.13. Idealized cylinder and nozzle pressure time curves for halocarbon agents.

For pre-engineered systems, limits on discharge time and nozzle pressure are built into the limits on piping system geometry. Agent distribution is handled by constraining pre-engineered systems to balanced flow conditions (i.e., the same agent mass is distributed from each nozzle). For adequate design of engineered systems, accurate methods for predicting these elements are required.

Figure 4-7.13 is an idealized plot of cylinder and nozzle pressure during discharge. Throughout the discharge process the amount of agent vapor and liquid, as well as dissolved and gaseous nitrogen, varies. As the pressure decreases in the cylinder and piping system, more agent is vaporized, and nitrogen is released from the solution in the agent. The formation of additional vapor and nitrogen bubbles lowers the average density of the fluid. The rapid vaporization of agent is more pronounced in low boiling point/high vapor pressure agents. The fluid temperature also varies with time and along the length of the piping network. The fluid temperature also impacts the degree of agent vaporization and nitrogen release as well as liquid agent density. The discharge process can be divided into five sections.

The first is the process of filling the pipe with agent. The rate at which this occurs is driven by the speed of the agent interface moving through the network. This is driven by either the sonic velocity at the agent interface or the discharge of displaced air through the nozzle. This phase determines the time at which the agent discharges from each nozzle. For systems with high degrees of imbalance in terms of flow path length or large pipe volume differences between nozzles, there can be significant delay in agent reaching one nozzle before another. This has a dramatic effect on the distribution of mass from each nozzle.

Once the agent reaches the nozzle and is compressed in the pipeline, the so-called nozzle peak pressure is reached. At this moment, agent is discharging from each nozzle.

The next step in the discharge process is the so-called quasi-steady agent flow regime. This is generally the longest portion of the discharge, particularly for systems with low pipe volume to agent volume ratios. This period of the discharge process is the basis for the simplified pressure drop

calculations embodied in NFPA 12A, *Standard on Halon 1301 Fire Extinguishing Systems*, for balanced Halon 1301 systems.

The next milestone during the discharge process is cylinder liquid runout, where no liquid agent remains in the cylinder. At this moment, an interface between liquid agent and nitrogen/agent vapor forms and travels through the network.

When the trailing liquid/vapor interface reaches the first nozzle, nozzle liquid runout occurs. This is important in two ways. First, this liquid runout occurs at different times during the discharge for each nozzle and can significantly impact the quantity of agent flowing from any given nozzle. Second, it is possible in many circumstances to discharge sufficient vapor/gas mixture from the first nozzle at NLRO (nozzle liquid runout) to reduce the pressure in the piping below that necessary to flow the remaining agent in the network. This is especially important for low vapor pressure agents and for nozzle designs requiring relatively high minimum operating pressures.

Once all of the nozzles have been cleared of liquid agent, the system is discharging a combination of nitrogen and agent vapor. This regime is usually ignored since most (>95 percent) of the agent has already been delivered through the nozzles.

The importance of the pipe filling and nozzle runout with these alternatives is relatively more critical with low vapor pressure alternatives due to: (1) the inability of the agent to deliver significant pressure to the system by boiling and (2) the higher fluid densities that occur in the piping relative to Halon 1301.

Figures 4-7.14(a) through (d) illustrate the stages of the agent discharge network.

Flow regime: If the flow velocity of the agent in the piping is not high enough, the flow may separate into two distinct phases in the piping. This causes severe problems at tee splits and in evaluating pressure drops. Therefore, minimum flow rates that ensure a homogeneous mixture of agent liquid and vapor/nitrogen bubbles must be maintained. Various flow regimes are illustrated in Figures 4-7.15 and 4-7.16 for horizontal and vertical pipe, respectively.[29] One of the objectives of approval testing of flow calculation procedures is to ensure that homogeneous flow regimes are maintained in the piping throughout the discharge process. In Figures 4-7.15 and 4-7.16, these are denoted as dispersed bubble and bubble flow regimes, respectively.

Flow division at tees: For a single component, single-phase flow condition, the flow split at a tee junction is determined by the flow rate of the nozzles downstream of the tee. For two-phase fluids, flow distribution occurs at tees that is sensitive to the velocity of the flow along each branch of the tee, the orientation of the tee, the pressure at the tee, and the phase distribution of the fluid (gas or liquid) entering the tee.

The primary cause of preferential flow splits at tees is the inertia of the liquid *versus* vapor/gas phase. This is most readily envisioned for side-flow tees where one branch of the flow is required to turn 90 degrees. Gas/vapor bubbles with lower momentum relative to the liquid agent will make this change of direction more readily. This results in relatively less mass flow down the side-flow branch at approximately the same volumetric flow rate or velocity. For bullhead tees, the same problem applies, except that it is more subtle and involves velocity differences through each branch of the tee. For evenly split (50 percent/50 percent) flows, the velocity is

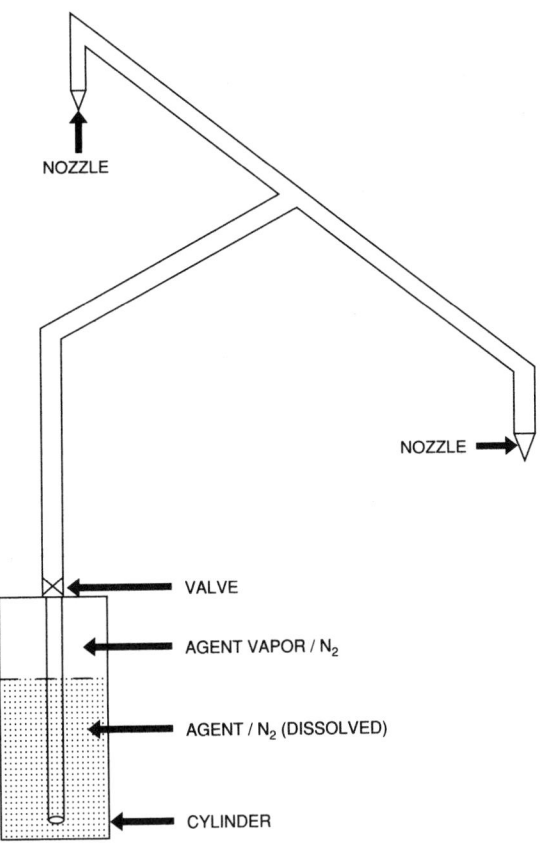

Fig. 4-7.14(a). *Initial conditions.*

NOZZLE

NOZZLE

VALVE

AGENT VAPOR / N$_2$

AGENT / N$_2$ (DISSOLVED)

CYLINDER

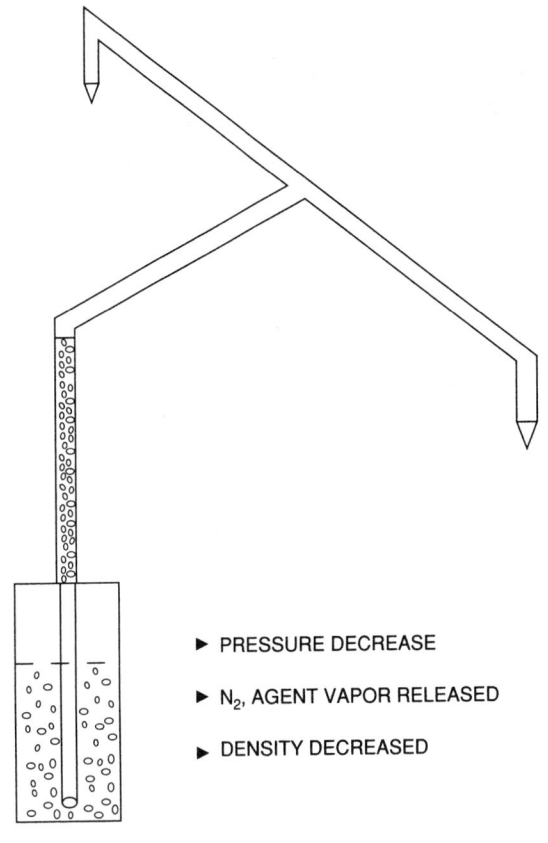

Fig. 4-7.14(b). *Valve open, pipe filling.*

► PRESSURE DECREASE

► N$_2$, AGENT VAPOR RELEASED

► DENSITY DECREASED

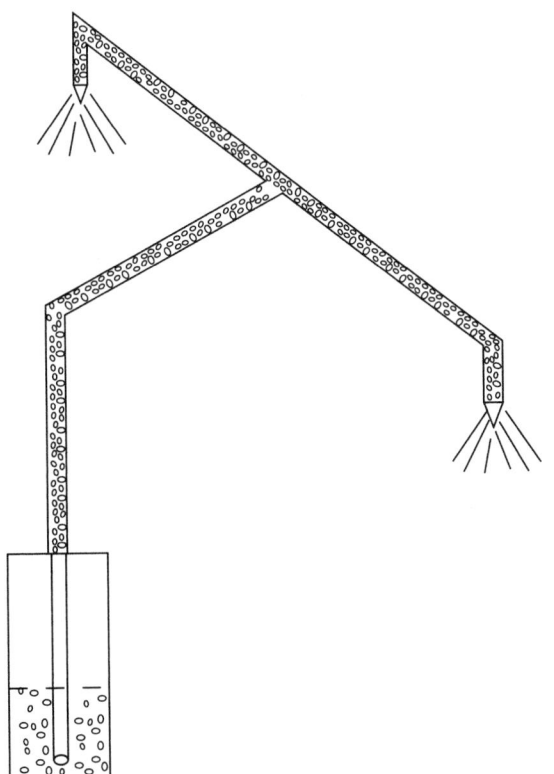

Fig. 4-7.14(c). *Quasi-steady flow, liquid throughout network.*

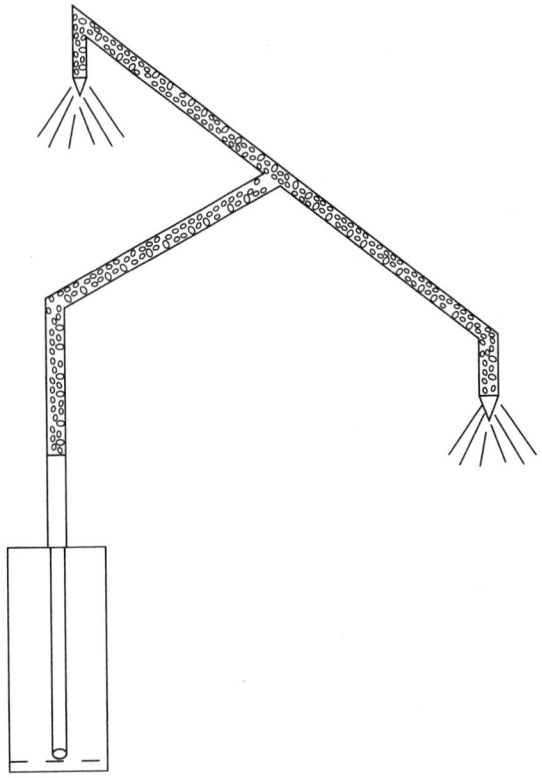

Fig. 4-7.14(d). *Cylinder liquid runout.*

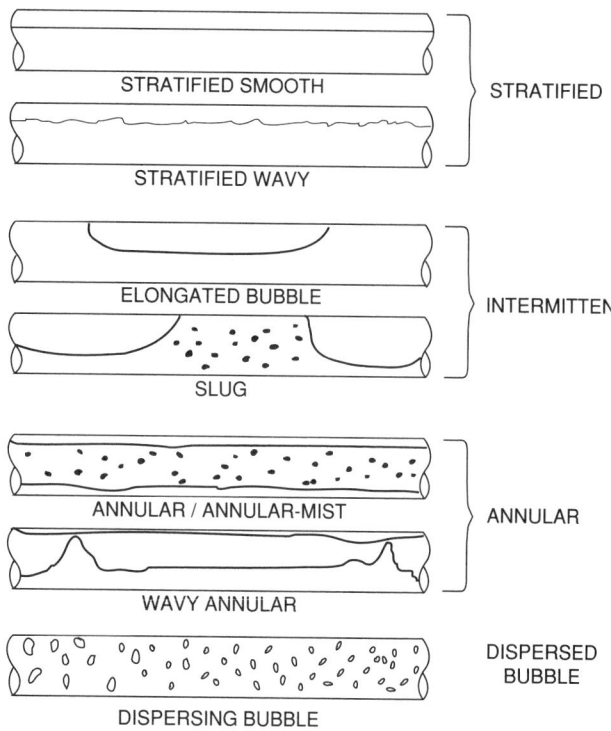

Fig. 4-7.15. Horizontal pipe flow regimes. [29]

identical in both directions, resulting in no flow split correction; as the split becomes greater the velocity differences are greater, and inertial effects of the gas/vapor relative to the liquid cause significant redistribution of mass through each branch of the tee.

The dependence was understood for Halon 1301 and described in detail by Williamson.[30] Similar processes occur in all two-phase flows including air/water, steam/water, and refrigerant flows. In the context of clean agent system design calculations, this flow distribution is dealt with using empirical factors that redistribute the flow relative to the pure pressure-driven flow distribution which would occur without preferential phase distribution at tees.

Figures 4-7.17 and 4-7.18 illustrate these correction factors for Halon 1301 flows in bullhead and side-flow tees, respectively.[30] All of the halocarbon agent flow predictions require similar treatment. Side-flow tees and bullhead tees require independent empirical correction factors. One of the most important limitations to any flow calculation procedure is the maximum flow split allowed for each type of tee. For a bullhead tee, as one moves farther away from 50 percent/50 percent splits, the correction factor becomes greater, and at some point usually in the range of 80 percent/20 percent, it becomes so large that the prediction becomes unreliable. For side-allowable flow splits, ranges between 75 percent/25 percent and 90 percent/10 percent are typical. This correction of flow splits at tees is one reason that final approval of engineered system designs should be constrained to calculation methods that have undergone testing within the range of flow splits required.

Pressure drop due to friction loss: The pressure drop caused by friction in the pipeline is calculated differently for two-phase fluids. The presence of agent vapor and gas affects

the pressure drop per unit length of pipe. There are numerous methods for dealing with two-phase fluid pressure drop.[31,32] Those typically used for fire suppression agent calculations involve either: (1) correcting the pressure drop estimated for single-phase fluid as a function of liquid to vapor/gas volume fraction or (2) empirical correlations of the

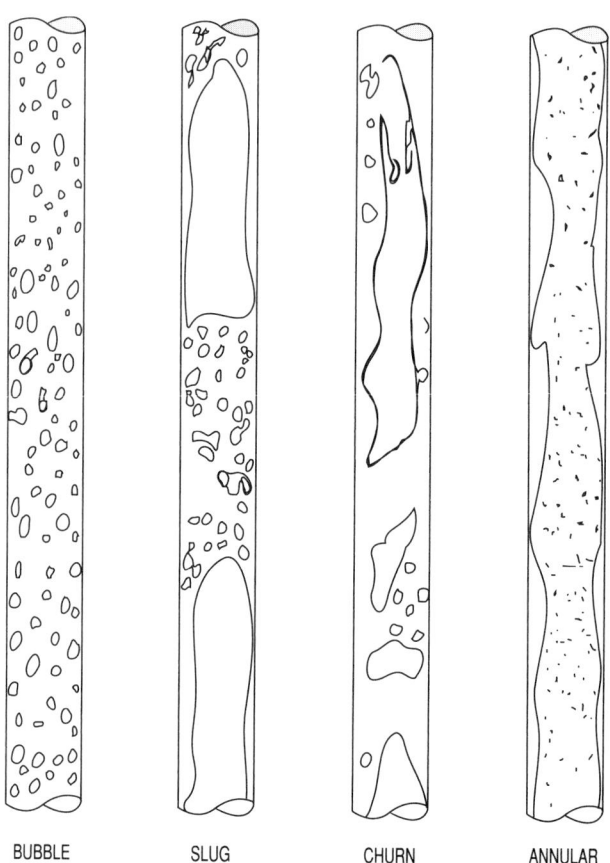

Fig. 4-7.16. Vertical pipe flow regimes. [29]

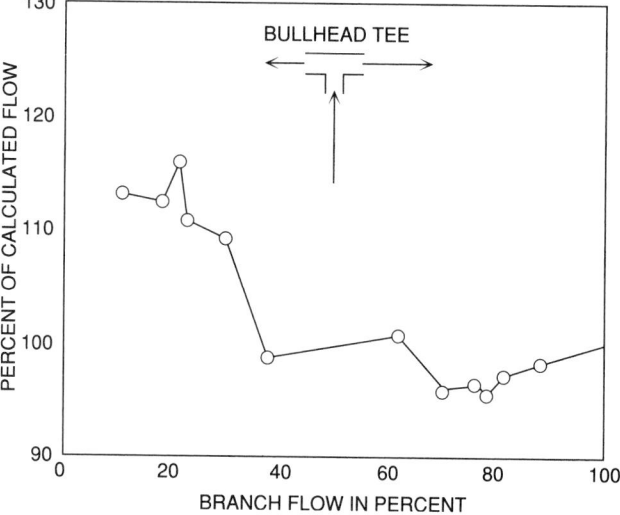

Fig. 4-7.17. Bullhead tee flow split corrections for Halon 1301. [30]

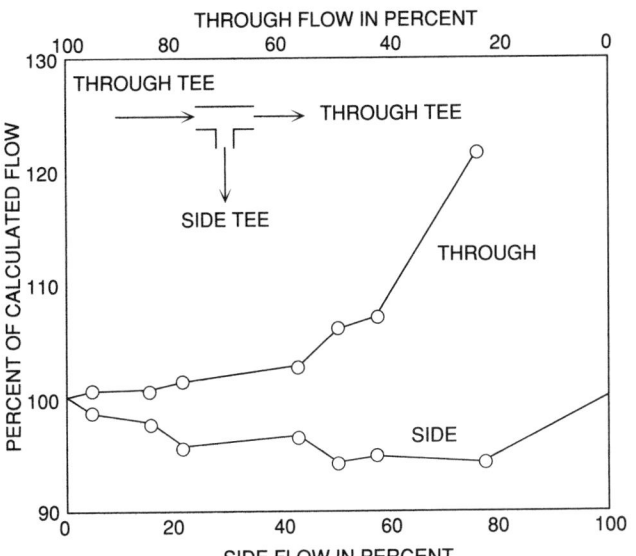

Fig. 4-7.18. Side- and through tee-flow split corrections for Halon 1301.[30]

pressure drop to average fluid density. Figure 4-7.19 illustrates the dependence of pressure drop on liquid volume fraction. In all cases for purposes of design of fire protection systems, the pressure drop is calculated on the basis of a homogeneous flow assumption where changes in the liquid fraction are seen as density changes in the homogeneous fluid.

Testing and approval of design methods: The approval or listing of a two-phase flow calculation procedure is part of the approval granted for engineered systems. Since some aspects of two-phase flow calculations are empirically based (e.g., flow regime, pressure drop, flow splits) and all calculation procedures have some bounds on their validity, testing is performed to verify the predictions and establish the limits of the calculation procedure. These limitations are crucial in helping to ensure that system designs do not exceed verified limits of calculation.

One of the most rigorous approval procedures used in verifying design methods is outlined by Underwriters Laboratories Inc.[20] UL 1058, *Halogenated Agent Extinguishing System Units*, was used for evaluating engineered Halon 1301 systems, but the same approach is taken for all clean agent alternatives. Design method limitations are described by the following ten parameters:

1. Percent of agent in piping (maximum);
2. Minimum and maximum discharge times;
3. Minimum pipeline flow rates;
4. Variance of piping volume to each nozzle;
5. Maximum variance of nozzle pressures within a piping arrangement;
6. Maximum ratio of nozzle diameter to inlet pipe diameter;
7. Arrangement most likely to exhibit vapor time-imbalance condition at nozzle;
8. All types of tee splits, including through tees, bullhead tees, etc.;
9. Minimum and maximum container fill density; and
10. Minimum and maximum flow split for each type of tee.

These parameters are related to the important attributes of the agent discharge process previously discussed. Full-scale testing is performed to evaluate the performance of the design method. The limits on flow calculation method performance are as follows:

1. Actual *versus* predicted discharge time ± 1 sec;
2. Actual *versus* predicted nozzle pressure ± 10 percent; and
3. Actual *versus* predicted mass flow through a nozzle, − 5 + 10 percent.

Testing in conjunction with a particular manufacturer's hardware is important. Ensuring that pressure drop through a particular valve assembly is calculated properly and nozzle orifice discharge coefficient evaluation are two critical hardware-dependent verifications.

Several generic flow calculation routines have been developed.[22,33–36] Of these, two are directed at single-nozzle systems with very short discharge times[35] or relatively simple balanced networks.[34] It is not recommended that any generic calculation procedure be used for final design purposes, unless it has been tested with the specific hardware to be installed and the system is within the limitations derived by tests.

In order to preserve a 10-sec discharge time, the mass flow rate of these clean agents must be higher than Halon 1301. The increased density of some of the alternative agents in the piping, caused by lower vapor pressures and nitrogen solubility differences, may result in high enough mass flow rates to retrofit existing Halon 1301 systems. While agent cylinders and nozzles will require replacement, it is often possible to preserve the existing Halon 1301 pipe network. This often requires the use of lower fill density cylinders to increase the average system pressure throughout the discharge time. Any such retrofit using existing Halon 1301 piping must be carefully evaluated with respect to hydraulic

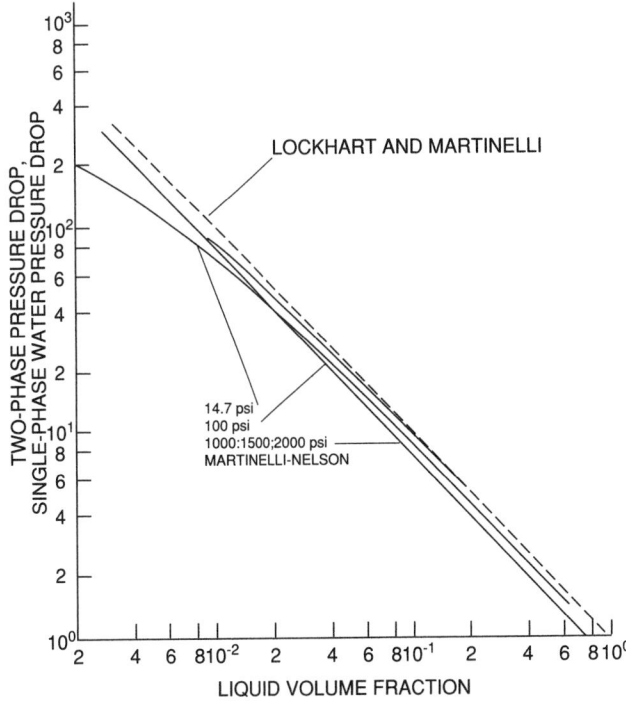

Fig. 4-7.19. Pressure drop versus liquid volume fraction.[32]

performance, with particular care given to preserving minimum required nozzle pressures and flow divisions at tees.

Nozzle Area Coverage and Height Limitations

One of the most important requirements of a gaseous total flooding fire suppression system is the ability of the system to deliver a uniform concentration of agent throughout the protected enclosure. The nozzle design and minimum nozzle pressure are critical in ensuring this distribution of agent. The performance of the nozzle is evaluated by full-scale approval testing, such as UL 1058.[20] The basic testing performed to evaluate nozzles is as follows:

1. Establish minimum nozzle pressure and maximum nozzle height by ensuring extinguishment of heptane fires located throughout a space with a height equal to the maximum allowable, at the minimum allowable nozzle pressure; and
2. Establish maximum nozzle coverage area by extinguishing tests in a plenum at the minimum height (generally less than 0.5 m) at the maximum nozzle coverage area (on the order of 100 m^2) and minimum nozzle operating pressure.

There are substantial differences between hardware manufacturers relative to minimum nozzle pressure, maximum ceiling height, and maximum average coverage. All nozzle orientations should be evaluated. In general, maximum nozzle heights are on the order of 4 to 5 m, nozzle area coverage on the order of 9 to 10 m^2, and minimum nozzle pressure between 3 and 6 bar. It is critical to ensure that the nozzle spacing, height, and minimum pressure limits are not exceeded for a particular manufacturer's hardware in a specific design.

The flow, mixing, and distribution of an agent from a nozzle into an enclosure can be predicted theoretically for relatively simple nozzle designs using sophisticated computer models.[34] Further development of such methods for complex nozzle designs and compartment geometries may eventually form the basis of a design procedure. At present, however, the primary means of ensuring adequate nozzle performance is the hardware approval process and real-scale testing.

Since many of the halocarbon replacements have lower vapor pressures than Halon 1301, there is often a much higher percentage of liquid at the nozzle. This makes the task of vaporizing and mixing the agent in the compartment more difficult. In general, nozzle designs used for Halon 1301 systems are not adequate for the halocarbon replacement agents. Due to the increased liquid fraction at the nozzle, it is critical to ensure that no unenclosed openings exist along the trajectory of the nozzle orifices. This may result in significant preferential loss of agent through these openings. This further emphasizes the need for third-party approval testing of nozzle performance. In any retrofit situation, the nozzles will need to be replaced even if the piping is adequately sized to deliver adequate agent flow rates.

Compartment Pressurization

The rapid discharge of agent into a compartment will cause rapid changes in the compartment pressure. Depending on the agent and rate of discharge, the initial pressure change may be negative. Figure 4-7.20 is a plot of compart-

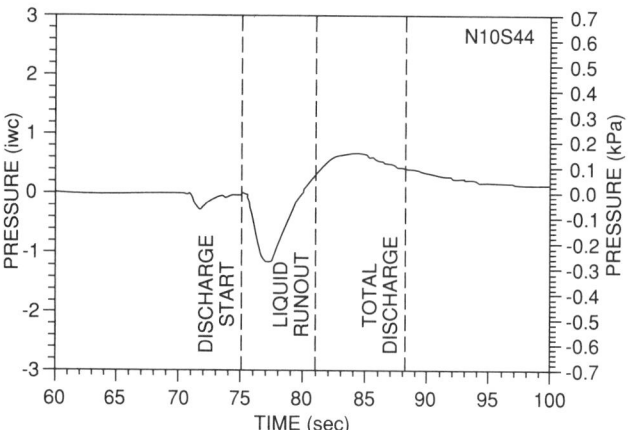

Fig. 4-7.20. Pressure measured in 28-m^3 enclosure during C$_3$F$_7$H discharge, with nominal 15-sec discharge time and 5-cm pan of n-heptane.[22]

ment pressure *versus* time for the discharge of HFC-227ea into a 28-m^3 room with a 360-cm^2 (56-in.^{2}) leakage area.[22] Immediately after discharge, the pressure in the compartment drops below ambient to a minimum of -0.3 kPa; at approximately 1.5 sec after discharge began, the pressure then begins to increase to a maximum of approximately 0.14 kPa after nozzle liquid runout. Similar results were obtained for FC-3-1-10. HFC-23 discharge exhibited much higher compartment overpressurization, without the marked initial negative pressure. The maximum overpressure for HFC-227ea and FC-3-1-10 discharge was similar to that of Halon 1301.

As the halocarbon agent is discharged into the space, it vaporizes rapidly, cooling the compartment and lowering the pressure. As the agent/air mixture gains heat from the walls or other objects in the space, the pressure recovers and, as additional agent is added, the pressure increases over ambient as mass is added to the compartment.

The expected maximum and minimum compartment pressure during discharge will be a function of:

1. Thermodynamic state of the agent at the nozzle,
2. Nozzle design,
3. Compartment volume and wall surface area,
4. Size of fire,
5. Initial conditions in space,
6. Leakage area from compartment, and
7. Agent flow rate.

For inert gases, significant compartment overpressurization can occur during discharge, unless adequate free vent area is provided. Calculation of required open area for venting is a part of the design manual for IG-541 systems.[37]

No generalized design procedure for calculating under/overpressurization has been established. Forssell and DiNenno[22] have developed a procedure for estimating the compartment pressure as a function of agent, agent flow rate, agent thermodynamic state at the nozzle, compartment volume, and surface area and leakage area. The method has not been sufficiently tested for general application.

Agent Hold Time and Leakage

Traditionally, total flooding gas systems were required to maintain a minimum concentration for a specified time period (10 to 20 min) after discharge. The minimum required hold time was based on:

1. Soak time required for deep-seated Class A fuels;
2. Response time of emergency personnel; and
3. Prevention of reflash due to presence of hot surfaces and other reignition sources, particularly in flammable and combustible liquid applications.

Currently, there is no specified minimum hold or soak time for clean agents. The variables described above will vary between installations, and there is no significant database on the performance of these agents on deep-seated fires other than wood cribs. The designer will be required to specify the minimum soak time consistent with the requirements of the hazard being protected.

The ability of a compartment to maintain adequate agent concentrations is a function of the leakage of the compartment. Historically, this was done with Halon 1301 through the use of discharge tests. Discharge testing for this purpose was rendered unnecessary by the introduction of door fan pressurization leakage tests. Appendix B of NFPA 2001, *Standard on Clean Agent Fire Extinguishing Systems*, describes a complete procedure for evaluating agent hold time as a function of compartment leakage measured by the door fan pressurization method.

The only difference between alternative agents and Halon 1301 in this regard is the density of the agent/air mixture, which is the driving force for leakage in quiescent environments. The mixture density can be estimated as follows[1]

$$\rho_m = V_d \frac{C}{100} + \left[\frac{\rho_a(100 - C)}{100} \right]$$

where:

ρ_m = clean agent/air mixture density (kg/m^3),
ρ_a = air density (1.202 kg/m^3),
C = clean agent concentration (%), and
V_d = agent vapor density at 21°C (kg/m^3).

Agent vapor densities at 21°C are given below.

FC-3-1-10	9.85 kg/m^3 (0.615 lb/ft^3)
HBFC-22B1	5.54 kg/m^3 (0.346 lb/ft^3)
HCFC Blend A	3.84 kg/m^3 (0.240 lb/ft^3)
HFC-124	5.83 kg/m^3 (0.364 lb/ft^3)
HFC-125	5.06 kg/m^3 (0.316 lb/ft^3)
HFC-227ea	7.26 kg/m^3 (0.453 lb/ft^3)
HFC-23	2.915 kg/m^3 (0.182 lb/ft^3)
IG-541	1.43 kg/m^3 (0.089 lb/ft^3)
Halon 1301	6.283 kg/m^3 (0.392 lb/ft^3)

All agents, except inert gases, have higher mixture densities than Halon 1301 at 5 percent when used at their design concentrations. This will require slightly more leaktight enclosures to maintain the same hold time.

SUMMARY

A wide range of inert gas and halocarbon total flooding clean agents has been introduced over the past several years. More will be commercialized in the near future. The use of an agent must be consistent with applicable environmental regulations. The selection of an agent is driven by its fire performance characteristics; agent and system space and weight concerns; toxicity, particularly for use in occupied areas; and the availability of approved system hardware.

The design of clean agent systems must be carefully done in accordance with third-party listing and approval limitations on both agent and hardware. Given the relative lack of experience with systems employing these new agents, particular care in design, installation, inspection, testing, and maintenance is warranted. Design and installation standards, such as NFPA 2001, *Standard on Clean Agent Fire Extinguishing Systems*, form the minimum requirements for these new technologies.

As generalized design methods and more detailed requirements evolve, the ability to design and install systems on a performance basis will increase. A critical part of the installation process is post-installation inspection and testing. NFPA 2001, *Standard on Clean Agent Fire Extinguishing Systems*, contains requirements for the approval and post-installation inspection and test of clean agent systems. Bearing in mind the relative complexity of these systems and the importance of the detection system and enclosure integrity, post-installation inspection and testing should be rigorously performed.

REFERENCES CITED

1. NFPA 2001, *Standard on Clean Agent Fire Extinguishing Systems*, National Fire Protection Association, Quincy, MA (1994).
2. T.A. Moore, "Cup Burner Analysis," *Halon Substitute Program Review*, Albuquerque, NM, May 14, 1993.
3. A. Hamins *et al*, "Flame Suppression Effectiveness," in *Evaluation of Alternative In-flight Fire Suppressants for Full-Scale Testing in Simulated Aircraft Engine Nacelles and Dry Bays*, Grosshandler *et al*, eds., NIST SP 861, National Institute of Standards and Technology, Gaithersburg, MD, Apr. 1994.
4. M.L. Robin, "Properties and Performance of FM-200™," *Proceedings of the Halon Options Technical Working Conference 1994*, Albuquerque, NM, pp. 531–542, May 3–5, 1994.
5. R. Sheinson *et al*, "Halon 1301 Total Flooding Fire Testing, Intermediate Scale," *Proceedings of the Halon Options Technical Working Conference 1994*, Albuquerque, NM, pp. 43–53, May 3–5, 1994.
6. T.A. Moore *et al*, "Intermediate Scale (645 ft^3) Fire Suppression Evaluation of NFPA 2001 Agents," *Proceedings of the Halon Options Technical Working Conference 1993*, Albuquerque, NM, pp. 115–127, May 11–13, 1993.
7. M.J. Ferreira *et al*, "Thermal Decomposition Product Results Utilizing PFC-410," *Proceedings of the Halon Options Technical Working Conference 1992*, Albuquerque, NM, May 13, 1992.
8. J.A. Senecal, "Agent Inerting Concentrations for Fuel-Air Systems," Fenwal Safety Systems, CRC Technical Note No. 361, May 27, 1992.
9. F. Tamanini, "Determination of Inerting Requirements for Methane/Air and Propane/Air Mixtures by an Ansul Inerting Mixture of Argon, Carbon Dioxide, and Nitrogen," Factory Mutual Research Corp., Norwood, MA, Aug. 24, 1992.
10. E. Heinonen, "Laboratory-Scale Inertion Results," *Halon Substitutes Program Review*, CGET/NMERI, Albuquerque, NM, May 14, 1993.
11. E.W. Heinonen, "The Effect of Ignition Source and Strength on Sphere Inertion Results," *Proceedings of the Halon Options Technical Working Conference 1993*, Albuquerque, NM, pp. 565–576, May 11–13, 1993.
12. J.A. Senecal, "Explosion Protection in Occupied Spaces: The Status of Suppression and Inertion Using Halon and Its Descendants," *Proceedings of the 1993 International CFC and Halon Alternatives Conference*, Washington, DC, pp. 767–772, Oct. 20–22, 1993.
13. T.A. Moore, "Large-Scale Inertion Evaluation of NFPA 2001

Agents," *Proceedings of the 1993 International CFC and Halon Alternatives Conference*, Washington, DC, Oct. 20–22, 1993.

14. J.A. Senecal, D.N. Ball, and A. Chattaway, "Explosion Suppression in Occupied Spaces," *Proceedings of the Halon Options Technical Working Conference 1994*, Albuquerque, NM, pp. 79–86, May 3–5, 1994.

15. "Research Basis for Improvement of Human Tolerance to Hypoxic Atmospheres in Fire Prevention and Extinguishment," EBRDC Report 10.30.92, Environmental Biomedical Research Data Center, Institute for Environmental Medicine, University of Pennsylvania, Philadelphia, PA, Oct. 1992.

16. J.C. Yang and B.D. Bruel, "Thermophysical Properties of Alternative Agents," in *Evaluation of Alternative In-flight Fire Suppressants for Full-Scale Testing in Simulated Aircraft Engine Nacelles and Dry Bays*, NIST SP 861, National Institute of Standards and Technology, Gaithersburg, MD, Apr. 1994.

17. R.S. Sheinson *et al*, "Halon 1301 Replacement Total Flooding Fire Testing, Intermediate Scale," *Proceedings of Halon Options Technical Working Conference 1994*, Albuquerque, NM, pp. 43–53, May 3–5, 1994.

18. J.C. Brockway, "Recent Findings on Thermal Decomposition Products of Clean Extinguishing Agents," 3M Report presented to NFPA 2001 Committee, Ft. Lauderdale, FL, Sept. 19–22, 1994.

19. T.A. Moore *et al*, "Intermediate Scale (645 ft^3) Fire Suppression Evaluation of NFPA 2001 Agents," *Proceedings of Halon Options Technical Working Conference 1993*, Albuquerque, NM, pp. 115–128, May 11–13, 1993.

20. UL 1058, *Halogenated Agent Extinguishing System Units*, Underwriters Laboratories, Inc., Northbrook, IL (1984).

21. G.G. Back *et al*, "Draft Report: Full-scale Machinery Space Testing of Gaseous Halon Alternatives," USCG R&D Center, Groton, CT, Sept. 1994.

22. E.W. Forssell and P.J. DiNenno, "Evaluation of Alternative Agents for Use in Total Flooding Fire Protection Systems," Contract NAS 10-1181, National Aeronautics and Space Administration, John F. Kennedy Space Center, FL, Oct. 1994.

23. P.J. DiNenno *et al*, "Thermal Decomposition Testing of Halon Alternatives," *Proceedings of the Halon Alternatives Technical Working Conference 1993*, Albuquerque, NM, May 1993.

24. D.S. Dierdorf *et al*, "Decomposition Product Analysis During Intermediate Scale (645 ft^3) Testing of NFPA 2001 Agents," *Proceedings of the Halon Alternatives Technical Working Conference 1993*, Albuquerque, NM, May 1993.

25. C.P. Hanauska, "Hazard Assessment of HFC Decomposition Products," presented at *1994 International CFC and Halon Alternatives Conference*, Washington, DC, Oct. 1994.

26. C.P. Hanauska *et al*, "Hazard Assessment of Thermal Decomposition Products of Halon Alternatives," *Proceedings of the Halon Alternatives Technical Working Conference 1993*, Albuquerque, NM, May 1993.

27. W.A. Dumayas, "Effect of HF Exposure on PC Multifunction Cards," Senior Research Project, Department of Fire Protection Engineering, University of Maryland, College Park, MD (1992).

28. E.F. Forssell *et al*, "Draft Report: Performance of FM-200 on Typical Class A Computer Room Fuel Packages," Hughes Associates, Inc., Columbia, MD, Oct. 1994.

29. D. Barnea and Y. Taitel, "Flow Pattern Transition in Two-phase Gas-liquid Flows," *Encyclopedia of Fluid Mechanics, Vol. 3*, N.P. Cheremisinoff, ed., Gulf Publishing Company, Houston, TX (1986).

30. H.V. Williamson, "Halon 1301 Flow in Pipelines," *Fire Technology*, 13 (1), pp. 18–32 (1976).

31. D. Chisholm, "Predicting Two-phase Flow Pressure Drop," *Encyclopedia of Fluid Mechanics, Vol. 3*, N.P. Cheremisinoff, ed., Gulf Publishing Company, Houston, TX (1986).

32. Y.Y. Hsu and R.W. Graham, *Transport Processes in Boiling and Two-phase Systems*, Hemisphere Publishing Corporation, Washington, DC (1976).

33. P.J. DiNenno *et al*, "Modeling the Flow Properties and Discharges of Halon Replacement Agents," *Proceedings of the Halon Options Technical Working Conference 1994*, Albuquerque, NM, May 1994.

34. E.B. Bird *et al*, "Development of Computer Model to Predict the Transient Discharge Characteristics of Halon Alternatives," *Proceedings of the Halon Options Technical Working Conference 1994*, Albuquerque, NM, May 1994.

35. T.G. Cleary *et al*, "Flow of Alternative Agents in Piping," *Proceedings of the Halon Options Technical Working Conference 1994*, Albuquerque, NM, May 1994.

36. W.M. Pitts *et al*, "Fluid Dynamics of Agent Discharge," in *Evaluation of Alternative In-flight Fire Suppressants for Full-Scale Testing in Simulated Aircraft Engine Nacelles and Dry Bays*, Grosshandler *et al*, eds., NIST SP 681, National Institute of Standards and Technology, Gaithersburg, MD, Apr. 1994.

37. Ansul Co., "Inergen System Design Installation and Maintenance Manual," Ansul Co., July 1994.

ADDITIONAL READING

NFPA 12A, *Standard on Halon 1301 Fire Extinguishing Systems*, National Fire Protection Association, Quincy, MA (1992).

FIRE TEMPERATURE-TIME RELATIONS

T. T. Lie

INTRODUCTION

The intensity and duration of fire in buildings can vary in a wide range, and several studies have been carried out to investigate the determining factors. At present it is possible to estimate the temperature course of fire in enclosures under various conditions, provided the values of the parameters that determine it are known.

Several of these parameters, however, such as amount and surface area of the combustible materials, are unpredictable as they change with time and often vary from compartment to compartment in a building. It is not possible, therefore, to know at the time a building is erected the temperature course of a fire to which objects in that building might be exposed during its service life.

It is possible, however, to indicate for any enclosure a temperature-time curve that, with reasonable likelihood, will not be exceeded during the lifetime of the building. Such curves are useful as a basis for the fire-resistive design of buildings. They can also facilitate studies of fire resistance of building components exposed to fires of various intensity and duration.

In this chapter, analytical expressions will be given that describe characteristic temperature curves as a function of the significant parameters for various fire conditions commonly met with in practice.

Expressions will also be given for the standard fire curve used in North America, and for the fire curve adopted by the International Organization for Standardization (ISO).

FIRE TEMPERATURES

The temperature course of a fire in an enclosure may be divided into three periods:

1. The growth period,
2. The fully developed period, and
3. The decay period.

Dr. T.T. Lie is Research Officer with the Institute for Research in Construction, National Research Council of Canada. He has carried out research related to fire resistance design, which includes evaluation of the fire resistance of building constructions by calculation and testing.

These periods are illustrated in Figure 4-8.1, where an idealized fire temperature course is shown. During the growth period, heat produced by the burning materials is accumulated in the enclosure. As a result, other materials may be heated so severely that they also ignite. At this stage of the fire, the gas temperatures rise very quickly to high values. The rather sudden ignition of materials in all parts of the room is called "flashover." After the flashover, the fully developed period starts. Because the temperatures in the enclosure are relatively low in the growth period, their influence on the fire resistance of structural members is negligible. In fire resistance studies, therefore, the growth period can be disregarded. Actual risk of failure of structural members or fire separations begins when the fire reaches the fully developed stage. In this stage, temperatures of about 1000°C or higher can be reached, and the heat transferred from the fire to structural members may substantially reduce their strength. This risk also exists in the decay period.

Parameters Determining the Fire Temperature Course

The most important parameters that determine the temperature course of a fire were first shown by Kawagoe and Sekine[1] and by Odeen,[2] who estimated the heat balance for fires in enclosed spaces. Usually part of the heat produced during a fire in an enclosure will be absorbed by the walls and contents, a part by the gases, and a part will be lost by radiation and convection from windows. (See Figure 4-8.2.) There is also loss of chemical energy that could have been released as heat because of outflow of unburned gases, which burn outside the enclosure. In addition, there is loss of unburned particles.

To be able to determine the temperature course, it is necessary to know at each moment during a fire, the rate at which heat is produced and the rate at which heat is lost to exposed materials and surroundings. Several of the parameters that determine heat production and heat losses, such as material properties, room dimensions, wall construction, window area, and emissivity of the flames and exposed materials, can be determined with reasonable accuracy. Others that are known approximately are the amount of gases that burn outside the room, the loss of unburned particles through windows, and the temperature differences in the room.

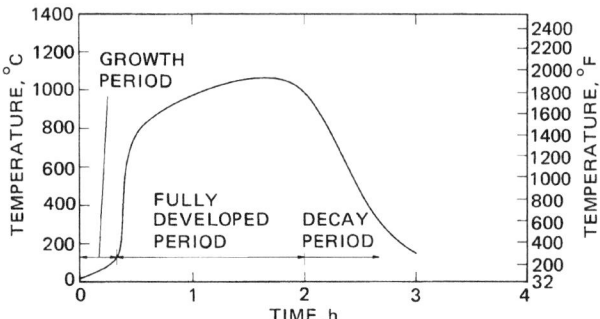

Fig. 4-8.1. Idealized temperature course of fire.

There are several parameters, however, whose magnitude cannot be predicted. Usually they change with time, and therefore, their value at the time of occurrence of a fire is determined by chance. Such parameters include the amount, surface area, and arrangement of the combustible contents, velocity and direction of wind, and the outside temperature. The influence of wind[3] and that of fire load can be substantial. Surveys show, for instance, that the variability of fire loads in various types of buildings is such that deviations in the order of 50 percent or more from the most probable fire load are common.[4] As a consequence, variability of fire load alone may easily cause deviations from the most probable temperature course of hundreds of degrees centigrade in temperature and 50 percent or more in fire duration.

POSSIBLE FIRE SEVERITIES

Owing to the substantial influence of uncertain factors, it is impossible to predict accurately the temperatures to which building components will be exposed during their service life. Even if the analysis to predict fire temperature courses in enclosures is perfect, it is very improbable that a certain predicted temperature course will occur.

The fire temperature to which building components will most likely be exposed during the use of a building is the relatively low temperature of a fire that has been extinguished before it reaches the fully developed stage. There is a small although not insignificant chance of occurrence of a fully developed fire. In this case, and assuming that the fire cannot be influenced by action of the fire brigade, the fire will be controlled either by the surface area of the materials that can participate in the burning or by the rate of air supply through the openings.[2,5]

Whether the fire will be largely controlled by surface area or ventilation depends on the amount of combustible contents. Unless its quantity, surface area, and arrangement are controlled, or the size of the windows and floor area made such that the possibility of a ventilation-controlled fire becomes remote,[6,7] the type of fire that may occur is unpredictable. According to statistical data, combustible contents of 10 to 60 kg per m² of floor area are normal, and there is a considerable probability of enclosures having a combustible content of 40 to 100 kg/m.[2,4] It is probable that in the latter range, as confirmed by experiments,[5,8] the fire will be mainly ventilation controlled, even when large window openings are present. It is likely that the greater the space behind the windows, or to a certain extent, the deeper the enclosure, the more material or surface area it will contain

and therefore the greater will be the probability of a ventilation-controlled fire. Usually a ventilation-controlled fire is the more severe fire, and because of the substantial probability of its occurrence, it is common to base fire resistance requirements for buildings on the assumption that fire severities will be controlled by ventilation.

CHARACTERISTIC TEMPERATURE CURVES

It is possible to indicate for any enclosure a characteristic temperature-time curve whose effect, with reasonable likelihood, will not be exceeded during the lifetime of the building. Such curves are useful as a basis for the fire-resistance design of buildings. They can also facilitate studies of fire resistance of building components exposed to fires of different severity.

There are several reports which present the temperature course of fires in fully developed and decay periods.[1,2,7,9,10] In all of these studies a procedure is followed in which the fire temperatures are determined by solving a heat balance for the enclosure under consideration.

For the fully developed period and ventilation-controlled fires, there is reasonable agreement in the temperatures found in the various studies, except for rather shallow rooms of limited size. In the latter case, the amount of combustible gases that burn outside may increase in such a way with increasing ventilation that the temperature decreases.[7]

There is less agreement in the results of the various studies for the decay period, partly due to the complexity of the processes that determine the temperature in that period. So far, rates of decay of temperature can only be established empirically or by making conservative or highly idealized assumptions. Because of the different approaches in deriving the rates of decay, there is a rather wide spread in the results of the various studies. Fortunately the influence of

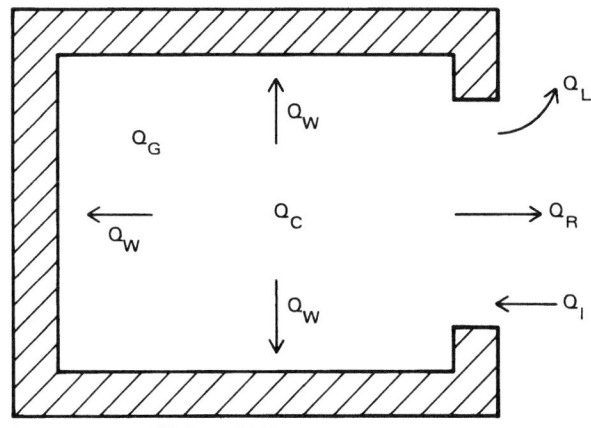

Q_R = RADIATION LOSSES

Q_I = HEAT CONTENT OF INFLOWING AIR

Q_L = HEAT CONTENT OF OUTFLOWING GASES

Q_W = HEAT LOSSES TO THE WALLS

Q_C = HEAT PRODUCED BY COMBUSTION

Q_G = RISE OF THE HEAT CONTENT OF THE GASES IN THE ENCLOSURE

Fig. 4-8.2. Heat balance for an enclosure during a fire.

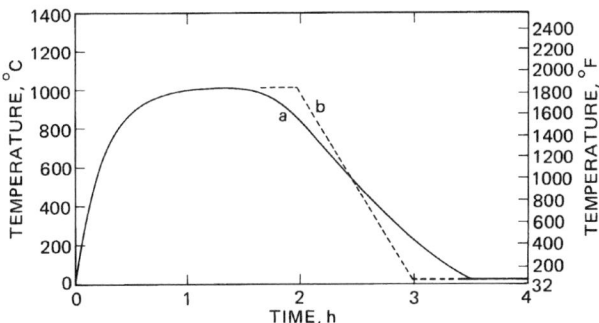

Fig. 4-8.3. *Temperature curves for fire resistance design.*

temperature variation in the decay period on the maximum temperatures reached in building components is relatively small.[11] For the purpose of deriving a temperature-time curve that, with reasonable probability, will not be exceeded during the lifetime of the building, it will be sufficient to use a curve that only approximately reflects the effect of heating in the decay period. This is further explained in Figure 4-8.3.

In Figure 4-8.3 Curve a illustrates a fire temperature curve derived theoretically for a certain building. The probability of occurrence of a fire with a more severe effect than shown by the curve is once in 50 years. Curve b illustrates a fire temperature curve for the same building, but it is assumed that the rate of burning remains constant until all combustible materials are consumed, whereupon the fire temperature drops linearly to room temperature. Although Curve b differs in shape from Curve a, their heating effect is approximately the same. If Curve b is used instead of Curve a, the probability of occurrence of a more severe fire than that represented by the relevant curve may change somewhat, for instance, from once in fifty years to somewhat more or less than fifty years. In practice this means that virtually the same fire safety will be provided whether Curve a or Curve b is used for the fire-resistance design of a building. The use of Curve b instead of Curve a has the advantage that it is easier to define.

Expressions for Characteristic Temperature Curves

In the following, analytical expressions are given that describe characteristic temperature curves as a function of the significant parameters for various fire conditions commonly met with in practice. For the fully developed period, the derivation of these curves will be based on the temperature curves for ventilation-controlled fires calculated according to the method described by Kawagoe and Sekine.[1]

The temperatures attained in ventilation-controlled fires are described (in addition to the thermal properties of the material bounding the enclosure) by a parameter, known as the opening factor F

$$F = \frac{A\sqrt{H}}{A_T} \tag{1}$$

where A is area of the openings in the enclosure, H is height of the openings, and A_T is area of the bounding surfaces (walls and floor and ceiling). The method of calculating $A\sqrt{H}$ for openings of unequal height is described in References 9 and 11.

The rate of burning, R, of the combustible materials in the enclosure is given by

$$R = 330 A\sqrt{H} \tag{2}$$

and, thus, if Q is the fire load per unit area of the surfaces bounding the enclosure, the duration of the fire, τ, is determined by

$$\tau = \frac{QA_T}{330 A \sqrt{H}} = \frac{Q}{330 F} \tag{3}$$

For given thermal properties of the material bounding the enclosure, the heat balance can be solved for the temperature as a function of the opening factor F. Besides depending on F, the temperature course is also a function of the thermal properties of the material bounding the enclosure.

In this study, two materials have been chosen as representative bounding materials: one with thermal properties resembling those of a heavy material (high heat capacity and conductivity) and one representing those of a light material (low heat capacity and conductivity). The thermal properties of these materials are given in Table 4-8.1. In practice, materials with a density of approximately 1600 kg/m² or more, e.g., normal-weight concretes, sand lime brick, and most clay bricks, can be considered as belonging to the group of heavy material. Those with a density of less than 1600 kg/m², e.g., lightweight and cellular concretes and plasterboard, can be regarded as belonging to the group of light materials.

Using the method described in Reference 11, the temperature course of fires in enclosures has been calculated for the two chosen bounding materials and for various values of the opening factor.[12] The conditions for which the calculations have been performed are shown in Table 4-8.1 and the

TABLE 4-8.1 *Thermal Properties of the Enclosure*

Factor	Description
k	Thermal conductivity of bounding material: 1.16 W/m K for a heavy material ($\rho \geq$ 1600 kg/m³), 0.58 W/m K for a light material (r < 1600 kg/m³)
ρc	Volumetric specific heat of bounding material: 2150×10^3 J/m³K for a heavy material ($\rho \geq$ 1600 kg/m³), 1075×10^3 J/m³K for a light material ($\rho <$ 1600 kg/m³)
A_T	Total inner surface area bounding the enclosure, including window area: 1000 m²
H	Window height: 1.8 m
ε	Emissivity for radiation transfer between hot gases and inner bounding surface of the enclosure: 0.7
α_c	Coefficient of heat transfer by convection between fire and inner bounding surface area: 23 W/m²K
a_u	Coefficient of heat transfer between outer bounding surface area and surroundings: 23 W/m²K
c	Specific heat of combustion gases: 1340 J/Nm³°C
G	Volume of combustion gas produced by burning 1 kg of wood: 4.9 Nm³/kg
q	Heat released in the enclosure by burning 1 kg of wood: 10.77×10^6 J/kg
T_0	Initial temperature: 20°C
V	Volume of enclosure:* 1000 m³
Δx	Thickness of elementary layers of bounding material: 0.03 m
Δt	Time increment: 0.0004167 hr
D	Thickness of bounding material: 0.15 m

*It can be shown that the influence of the volume of the enclosure on the fire temperature is negligible.

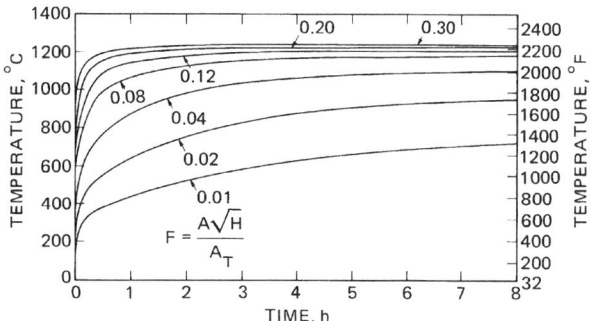

Fig. 4-8.4. Temperature-time curves for ventilation-controlled fires in enclosures bounded by dominantly heavy materials ($\rho \geq$ 1600 kg/m³), calculated for various opening factors by solving a heat balance for the enclosure.

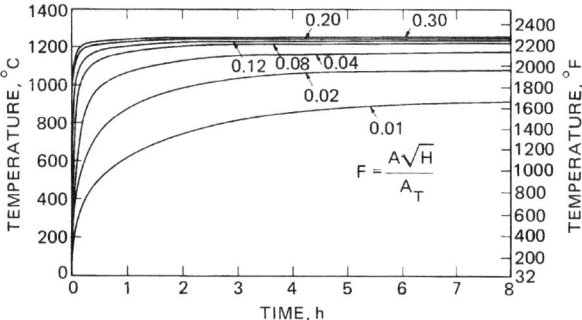

Fig. 4-8.5. Temperature-time curves for ventilation-controlled fires in enclosures bounded by dominantly light materials ($\rho <$ 1600 kg/m³), calculated for various opening factors by solving a heat balance for the enclosure.

results of the calculations in Figures 4-8.4 and 4-8.5. The curves in these figures were used as a basis for the derivation of temperature curves for fire-resistance design. It was found that these temperature curves could be reasonably described by the expression

$$T = 250(10F)^{0.1/F^{0.3}}e^{-F^2t}[3(1 - e^{-0.6t})$$
$$- (1 - e^{-3t}) + 4(1 - e^{-12t})] + C\left(\frac{600}{F}\right)^{0.5} \qquad (4)$$

where T = the fire temperature in °C, t = time in hr, F = opening factor in m$^{1/2}$, and C = a constant taking into account the influence of the properties of the boundary material on the temperature. C = 0 for heavy materials ($\rho \geq$ 1600 kg/m²), and C = 1 for light materials ($\rho <$ 1600 kg/m).

The expression is valid for

$$t \leq \frac{0.08}{F} + 1 \qquad (5)$$

and

$$0.01 \leq F \leq 0.15 \qquad (6)$$

If $t > (0.08/F) + 1$, a value of $t = (0.08/F) + 1$ should be used. If $F > 0.15$, a value of $F = 0.15$ should be used.

The temperature-time curves evaluated from Equation 4 and those obtained by solving the heat balance for the en-

closure are shown in Figures 4-8.6 and 4-8.7 for various values of the opening factor.

It is seen that with the aid of the analytical expression, temperature curves can be developed that reasonably describe the curves derived from solving the heat balance.

As discussed previously, the temperatures in the decay period are more difficult to calculate due to the complexity of the processes that determine the temperature in this period. On the other hand, if the temperature variations are not very large, the influence of such variations in the decay period on the temperature attained in exposed building components is in general relatively small. Therefore, describing the temperature course in the decay period by a temperature-time relation that approximately reflects the decrease of temperature in this period is sufficient.

According to experimental data of Kawagoe[8] the rate of temperature decrease of a fire with a fully developed period of less than one hour is roughly 10°C per minute, and that of a fire with a fully developed period of more than one hour is 7°C per minute. The Swedish code assumes a rate of decrease of 10°C per minute irrespective of the duration of the fully developed period of the fire.[9] A comparison with semi-empirical data developed by Magnusson and Thelandersson[9] shows that the assumption of a rate of decrease of 10°C per minute is too fast for fires of long duration and too slow for fires of short duration. According to Harmathy,[7] who studied several experimental fires of relatively short

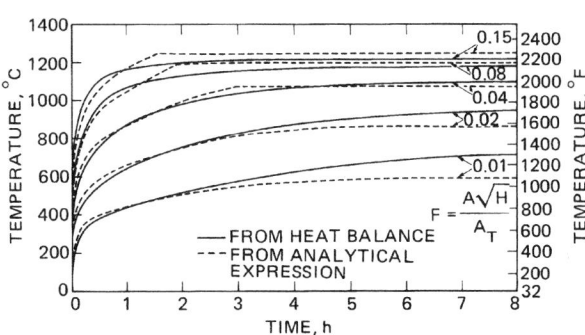

Fig. 4-8.6. Comparison between temperature-time curves obtained by solving a heat balance and those described by an analytical expression for ventilation-controlled fires in enclosures bounded by dominantly heavy materials ($\rho \geq$ 1600 kg/m³).

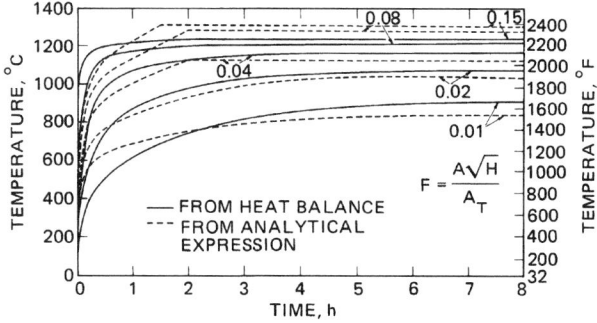

Fig. 4-8.7. Comparison between temperature-time curves obtained by solving a heat balance and those described by an analytical expression for ventilation-controlled fires in enclosures bounded by dominantly light materials ($\rho <$ 1600 kg/m³).

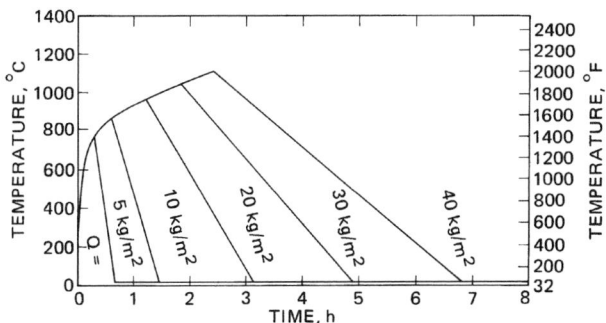

Fig. 4-8.8. *Characteristic temperature curves for various fire loads, Q, (opening factor, F = 0.05 m$^{1/2}$, heavy bounding material).*

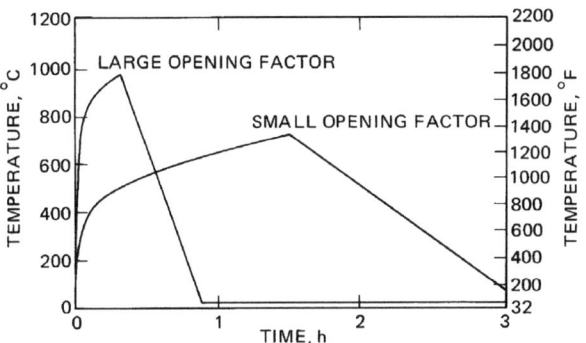

Fig. 4-8.9. *Influence of opening factor on fire temperature course.*

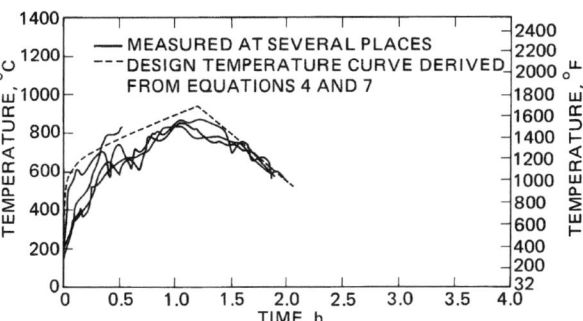

Fig. 4-8.10. *Comparison of design temperature curves derived from analytical expressions with temperatures measured during an experimental fire (fire load per unit internal area of the bounding surfaces Q = 18.75 kg/m^3, opening factor F = 0.047 m$^{1/2}$, heavy bounding material).*

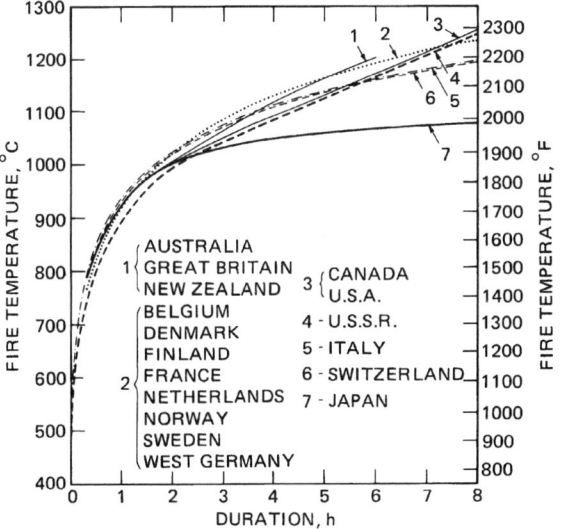

Fig. 4-8.11. *Standard fire temperature-time relations used in various countries for testing of building elements.*

duration,[13,14] the rate of decrease of temperature for such fires is in the order of 15 to 20°C per minute.

In general, the longer the duration of the fully developed period the lower the rate of decrease of temperature. Using this information the following expressions have been derived for the temperature course of fire in the decay period

$$T = -600\left(\frac{t}{\tau} - 1\right) + T_\tau \tag{7}$$

with the condition

$$T = 20 \quad \text{if } T < 20°C \tag{8}$$

In the above equations T = fire temperature, τ = time at which the decay starts as given by Equation 3, t = time under consideration ($t > \tau$), and T_τ = temperature given by Equation 4 at the time $t = \tau$.

The temperature curves obtained from Equations 4 and 7 are illustrated in Figure 4-8.8 for various fire loads and an opening factor of 0.05 m$^{1/2}$. In Figure 4-8.9 the influence is shown of the openings on the fire temperature course. It can be seen that the fire load determines the duration of the fire, whereas the openings determine both the duration and the intensity of the fire. In Figure 4-8.10 a characteristic temperature curve is compared with the temperatures measured at several places in a room during an experimental fire.[1] It is seen that the curve developed from the analytical expression reasonably characterizes the temperatures obtained during the experimental fire. It is somewhat conservative but satisfactory to use as a design curve for fire resistance.

STANDARD FIRE CURVE

In studies of fire resistance, it is common to expose building elements to heating in accordance with a standard temperature-time relation. The standard temperature-time curves used in various countries are shown in Figure 4-8.11. It can be seen that there are no significant differences between the various standard curves. The values of the curve adopted by ISO[15] are given in Table 4-8.2. Those used in North America[16] are given in Table 4-8.3.

There are also analytical expressions for several of the standard curves. The expression that describes the ISO curve is

$$T - T_0 = 345 \log_{10}(8t + 1) \tag{9}$$

where t = time in minutes, T = fire temperature in °C, and T_0 = initial temperature in °C.

For the curve used in North America, several analytical expressions exist.[17] One of the expressions is of the form of a sum of exponential functions

TABLE 4-8.2 *Standard Temperature-Time Relation According to ISO 834*[15]

Time h:min	Temperature °F	Temperature °C	Time h:min	Temperature °F	Temperature °C
0:00	68	20	3:10	1,938	1,059
0:05	1,000	538	3:20	1,950	1,066
0:10	1,300	704	3:30	1,962	1,072
0:15	1,399	760	3:40	1,975	1,079
0:20	1,462	795	3:50	1,988	1,086
0:25	1,510	821	4:00	2,000	1,093
0:30	1,550	843	4:10	2,012	1,100
0:35	1,584	862	4:20	2,025	1,107
0:40	1,613	878	4:30	2,038	1,114
0:45	1,638	892	4:40	2,050	1,121
0:50	1,661	905	4:50	2,062	1,128
0:55	1,681	916	5:00	2,075	1,135
1:00	1,700	927	5:10	2,088	1,142
1:05	1,718	937	5:20	2,100	1,149
1:10	1,735	946	5:30	2,112	1,156
1:15	1,750	955	5:40	2,125	1,163
1:20	1,765	963	5:50	2,138	1,170
1:25	1,779	971	6:00	2,150	1,177
1:30	1,792	978	6:10	2,162	1,184
1:35	1,804	985	6:20	2,175	1,191
1:40	1,815	991	6:30	2,188	1,198
1:45	1,826	996	6:40	2,200	1,204
1:50	1,835	1,001	6:50	2,212	1,211
1:55	1,843	1,006	7:00	2,225	1,218
2:00	1,850	1,010	7:10	2,238	1,225
2:10	1,862	1,017	7:20	2,250	1,232
2:20	1,875	1,024	7:30	2,262	1,239
2:30	1,888	1,031	7:40	2,275	1,246
2:40	1,900	1,038	7:50	2,288	1,253
2:50	1,912	1,045	8:00	2,300	1,260
3:00	1,925	1,052			

$$T - T_0 = a_1(1 - e^{a_4 t}) + a_2(1 - e^{a_5 t}) + a_3(1 - e^{a_6 t}) \quad (10)$$

where $a_1 = 532$ for °C, 957 for °F; $a_2 = -186$ for °C, -334 for °F; $a_3 = 820$ for °C, 1476 for °F; $a_4 = -0.6$; $a_5 = -3$; $a_6 = -12$.

The extreme deviation from the values given in Table 4-8.2 are -26°C at 45 min; $+48$°C at 3.5 hours; and -78°C at 8 hours.

TABLE 4-8.3 *Standard Fire Temperature-Time Relation Used in North America (NFPA No. 251)*[16]

Time in Minutes	Temperature rise of fire (°C)
0	0
5	556
10	659
15	718
30	821
60	925
90	986
120	1,029
180	1,090
240	1,133
360	1,193

This form is suitable for use in analytical heat flow calculations, because when it is used as a boundary condition the heat transfer equations are integrable.

A set of expressions that more accurately approximate the values given in Table 4-8.2 is

$$T - T_0 = a_1 \tanh a_4 t + a_2 \tanh a_5 t + a_3 \tanh a_6 t, \quad t < 2 \quad (11)$$

$$T - T_0 = 906.7 + 41.67t, \quad t \geq 2 \text{ for °C} \quad (12)$$

$$T - T_0 = 1632 + 75t, \quad t \geq 2 \text{ for °F} \quad (13)$$

where $a_1 = 580$ for °C, 1044 for °F; $a_2 = -276.8$ for °C, -498.2 for °F; $a_3 = 714.4$ for °C, 1286 for °F; $a_4 = 0.8429$; $a_5 = 0.9736$; $a_6 = 8.910$.

The maximum deviation of the temperature after 20 min, given by Equations 11, 12, and 13, from the values tabulated in Table 4-8.2 is -7°C at 40 min.

Another temperature-time relation, given in Reference 18, has the form

$$T - T_0 = a[1 - \exp(-3.79553\sqrt{t})] + b\sqrt{t} \quad (14)$$

where $a = 750$ for °C, 1350 for °F; $b = 170.41$ for °C, 306.74 for °F; and $t = $ time in hrs.

This expression is frequently used and is a reasonably accurate approximation of the relation between temperature and time given in Table 4-8.2.

NOMENCLATURE

A area of the openings in the enclosure, m^2

A_T area of the internal bounding surfaces, m^2

C constant

F opening factor, $m^{1/2}$

H height of openings in the enclosure, m

Q fire load per unit area of the internal bounding surfaces, kg/m^2

R rate of burning, kg/hr

T fire temperature, °C

T_τ fire temperature at the time τ, °C

t time, hr

τ time at which the temperature starts to decline, hr

REFERENCES CITED

1. K. Kawagoe and T. Sekine, "Estimation of Fire Temperature-Time Curve in Rooms," *B.R.I. Occasional Report No. 11*, Building Research Institute, Ministry of Construction, Tokyo (1963).

2. K. Odeen, "Theoretical Study of Fire Characteristics in Enclosed Spaces," *Bulletin 10*, Division of Building Construction, Royal Institute of Technology, Stockholm (1963).

3. P.H. Thomas and A.J.M. Heselden, "Fully Developed Fires in Single Compartments," *Fire Research Note No. 923*, Building Research Establishment, Fire Research Station, Borehamwood (1972).

4. T.T. Lie, *Fire and Buildings*, Applied Science Publishers Limited, London, pp. 19–22 (1972).

5. P.H. Thomas, A.J.M. Heselden, and M. Law, "Fully Developed Compartment Fires; Two Kinds of Behavior," *Fire Research Technical Paper No. 18*, Her Majesty's Stationery Office, London (1967).

6. T.T. Lie, *Fire and Buildings*, Applied Science Publishers Limited, London (1972), pp. 9–11.

7. T.Z. Harmathy, "A New Look at Compartment Fires, Part I and Part II," *Fire Tech.*, 8, 3 and 4 (1972).

8. K. Kawagoe, "Fire Behavior in Rooms," *Report No. 27*, Building Research Institute, Ministry of Construction, Tokyo (1958).

9. S.E. Magnusson and S. Thelandersson, "Temperature-Time Curves of Complete Process of Fire Development. Theoretical Study of Wood Fuel Fires in Enclosed Spaces," Acta Polytechnica Scandinavica, Civil Engineering and Building Construction Series No. 65, Stockholm (1970).

10. Y. Tsuchiya and K. Sumi, "Computation of the Behavior of Fire in an Enclosure," *Comb. and Flame*, 16 (1971).

11. K. Kawagoe, "Estimation of Fire Temperature-Time Curve in Rooms," *Research Paper No. 29*, Building Research Institute, Japan (1967).

12. T.T. Lie, "Characteristic Temperature Curves for Various Fire Severities," *Fire Tech.*, 10, 4 (1974).

13. E.G. Butcher, T.B. Chitty, and L.A. Ashton, "The Temperatures Attained by Steel in Building Fires," *Fire Research Technical Paper No. 14*, Her Majesty's Stationery Office, London (1966).

14. E.G. Butcher, G.K. Bedford, and P.J. Fardell, "Further Experiments on Temperatures Reached by Steel in Buildings," *Symposium No. 2, Behavior of Structural Steel in Fire*, Paper No. 1, Her Majesty's Stationery Office, London (1968).

15. "Fire Resistance Tests—Elements of Building Construction," *International Standard ISO 834*, (1975).

16. NFPA 251, *Standard Methods of Fire Tests of Building Construction and Materials*, National Fire Protection Association, Quincy, MA (1990).

17. G. Williams-Leir, "Analytical Equivalents of Standard Fire Temperature Curves," *Fire Tech.*, 9, 2 (1973).

18. J.P. Fackler, "Concernant la Resistance au Feu des Elements de Construction," *Cahier 299*, Centre Scientifique et Technique du Batiment (1959).

ANALYTICAL METHODS FOR DETERMINING FIRE RESISTANCE OF STEEL MEMBERS

James A. Milke

INTRODUCTION

Traditionally, fire resistance has been evaluated by subjecting a structural member to a standard test for a specified duration.[1] All members performing acceptably are rated and listed for the duration period of the test, e.g., one hour, two hours, etc. Assemblies not listed are assumed to be unable to meet the test criteria and, thus, have no rating, unless proved otherwise. Providing proof of acceptable performance can be accomplished in one of three manners:

1. Conduct the standard test,[1]
2. Conduct a special experiment, or[20]
3. Apply an analytical technique.[25]

The standard test can involve an appreciable turnaround time in order to specify, schedule, and analyze the results of the test. An experimental program can require a substantial amount of effort in order to obtain accurate data. The costs involved in sponsoring a standard test or experimental program can be appreciable. In the case of archaic structural assemblies, materials may no longer be available to reconstruct the design for possible testing.

Because of these drawbacks, calculation methods have been developed to analyze structural designs for fire conditions. The calculation methods have been formulated based on analyses of data from standard tests, experimental programs, and theoretically based investigations.

Analytical methods for fire resistance must consider three basic aspects of the problem:

1. Fire exposure,
2. Heat transfer, and
3. Structural response.

The fire exposing the structure must be characterized. This can be accomplished using methods described in Sections 1, 2, and 3 for the case of a "real" fire or by assuming the fire exposure specified in the standard test. The heating of the structural member can be addressed using principles of convection and radiation heat transfer. Heating within the member is treated by conduction heat transfer analysis (ra-

diation and convection heat transfer may also need to be considered, if voids are present within the assembly). Finally, the structural response is examined by comparing some or all of the following: deflections, strains, and stress levels to established limits.

All of the calculation methods do not address the fire exposure, heat transfer, and structural response aspects in a similar manner. The calculation methods can be grouped as follows:

1. Empirical correlations,
2. Heat transfer analyses, and
3. Structural analyses.

Empirical correlations are based on the analysis of data resulting from performing the standard test numerous times. A limitation of the empirical correlations is that they can only be applied when considering the fire exposure, loading, and span provided in the standard test. If other conditions apply, then another approach is needed.

The second group of calculation methods consists of heat transfer analyses. Because of the difficulty in obtaining an exact solution to the governing heat transfer equations, a numerical method is often used. The heat exposure conditions may be those associated with the standard test or a specified fire. The purpose of applying a heat transfer analysis is to determine the time required for the structural member to attain a predetermined critical temperature. The temperature endpoint criteria cited by ASTM E-119[1] are often accepted as the critical temperatures. Typically, inaccuracies of this method are related to the temperature dependence of the material properties.

Structural analysis-based calculations comprise the third group. The calculations are similar to those conducted for structural engineering purposes, except the material properties are evaluated at elevated temperatures and thermal expansion is considered. As in structural analysis, the loading and end conditions must be known or assumed. Limitations of this method result from uncertainties in characterizing the end conditions and, again, in determining the material properties at elevated temperatures.

This chapter provides an overview of the available calculation methods for determining the fire resistance of steel structural members. The basis of each method will be presented along with a sample application.

James A. Milke is an assistant professor in the Department of Fire Protection Engineering at the University of Maryland. His recent research activities have included the impact of fires on the structural response of steel and composite members.

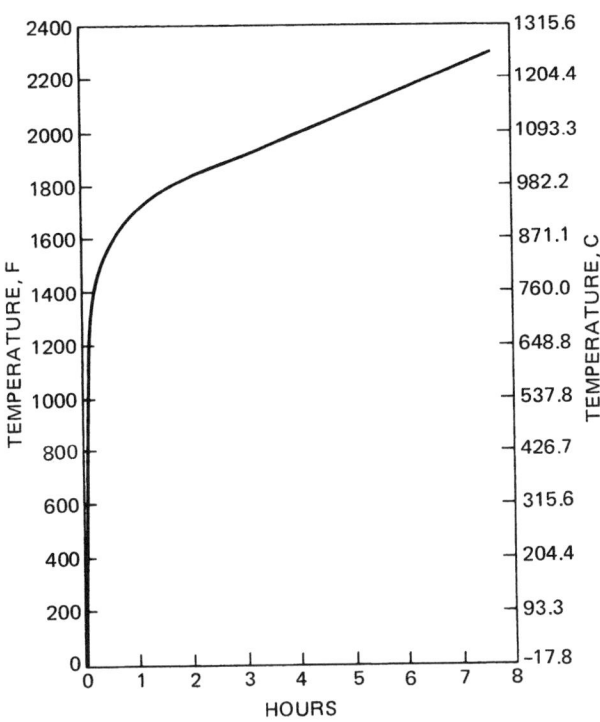

Fig. 4-9.1.   ASTM E-119 standard time-temperature curve.[1]

STANDARD TEST FOR FIRE RESISTANCE OF STRUCTURAL MEMBERS

The standard test method in the U.S. for determining the fire resistance of columns, floor and roof assemblies, and walls is ASTM E-119.*[1] Basically, the test involves subjecting the structural component to a heated furnace environment for the desired duration. If the endpoint criteria are not reached prior to the end of the test period, the assembly "passes" the test and is rated.

Currently, in testing laboratories throughout North America, gas burners are used to heat the furnace. The furnace is heated in such a manner so that the temperature inside the furnace follows the time-temperature curve illustrated in Figure 4-9.1. In principle, the time-temperature curve is intended to relate to a severe exposure from a room fire. Thus, the applicability of the test method to examine the fire resistance of exterior structural members exposed to fires outside of the building is questionable.[2]

Assemblies may be tested with or without load. If tested under load, the assembly is loaded to induce maximum design stress levels calculated based on theory. Floor and roof assemblies and bearing walls are always tested under load. Columns are tested with or without any loading. Steel beams and girders may be tested without load, if the considered design loading cannot be achieved in the laboratory.

Structural assemblies may be restrained or unrestrained against thermal expansion. The effect of restraint on the fire resistance of assemblies has been investigated by Bletzacker.[2] The degree of restraint in structural members varies

with the geometry, connection method, and framing system, among other factors. The descriptions presented in Table 4-9.1 relate actual construction conditions to the restrained and unrestrained designation noted in the ASTM E-119 test method.

The minimum dimensions of the structural components for testing are specified in ASTM E-119. A maximum set of dimensions is established by the size of available test furnaces. It is because of the maximum permissible dimensions that tested members can only be considered large-scale and not full-scale. The consequence of not testing full-scale members means that continuous beams, actual floor/roof ASTM assemblies, and long columns, among others, are not specifically tested. Consequently, this test is truly a comparative test and is not a measure of actual performance.

The ASTM E-119 endpoint criteria for the variety of assemblies can be grouped into three categories: structural integrity, temperature, and ignition of cotton waste. For the tests without loading, the structural integrity endpoint criterion is relaxed to require the component only to remain in place. The structural integrity criterion addresses the need for members to remain in place (supporting self-weight of member) and to continuously support any applied loads. The ignition of cotton waste endpoint addresses the ability of the structural assembly to prevent the transmission of flame and hot gases to the side not exposed to the furnace fire.

The temperature endpoint criteria are noted in Table 4-9.2. The endpoint temperatures are selected according to conservative estimates of the maximum allowable reduction in load-bearing capacity of the structural member, based on an average reduction in strength due to elevated temperatures.

Fire Resistance of Steel Members

Several calculation techniques are available to determine the fire resistance of steel members.[52,53] The available techniques include all three types that are generally available, i.e., empirically derived correlations, heat transfer analyses, and structural analyses. Using these techniques, the fire endurance of steel columns, beams in floor and roof assemblies, and trusses can be predicted. In addition, temperature profiles, stress levels, and deflections can be estimated.

The equations and models do not eliminate the need for all future testing. Testing is still required for numerous reasons, including the validation of calculation techniques. In addition, the mechanical behavior of protection materials used in an assembly, such as a steel structural member with insulating materials or other components, can only be examined by test. However, the calculation techniques can be used to extend the application of test results and/or reduce the number of required tests.

Steel Material Properties

The material properties of interest are yield strength, ultimate strength, modulus of elasticity, coefficient of thermal expansion, density, specific heat, and thermal conductivity. With the exception of density, all of the properties are strongly influenced by temperature. The effect of temperature on steel properties has been examined by many researchers.[3]

The thermal properties of ASTM A36 steel for temperatures between 0 and 2000°F are presented in Table 4-9.3.[4–6] The influence of temperature on the mechanical properties of A36 steel is presented in Figure 4-9.2.[4,7] At 538°C (1000°F), the yield strength is approximately 60 percent of

*Versions of the test method are also published as NFPA 251[49] and UL 263.[50]

TABLE 4-9.1 *Restrained and Unrestrained Construction Systems (from ASTM E-119 Table X3.1)*[1]

I. Wall bearing:
Single span and simply supported end spans of multiple bays:[a]

  (1) Open-web steel joists or steel beams, supporting concrete slab, precast units, or metal decking .... unrestrained
  (2) Concrete slabs, precast units, or metal decking .... unrestrained

Interior spans of multiple bays:

  (1) Open-web steel joists, steel beams or metal decking, supporting continuous concrete slab .... restrained
  (2) Open-web steel joists or steel beams, supporting precast units or metal decking .... unrestrained
  (3) Cast-in-place concrete slab systems .... restrained
  (4) Precast concrete where the potential thermal expansion is resisted by adjacent construction[b] .... restrained

II. Steel framing:

  (1) Steel beams welded, riveted or bolted to the framing members .... restrained
  (2) All types of cast-in-place floor and roof systems (such as beam-and-slabs, flat slabs, pan joists, and waffle slabs) where the floor or roof system is secured to the framing members .... restrained
  (3) All types of prefabricated floor or roof systems where the structural members are secured to the framing members and the potential thermal expansion of the floor or roof system is resisted by the framing system or the adjoining floor or roof construction[b] .... restrained

III. Concrete framing:

  (1) Beams securely fastened to the framing members .... restrained
  (2) All types of cast-in-place floor or roof systems (such as beam-and-slabs, flat slabs, pan joists, and waffle slabs) where the floor system is cast with the framing members .... restrained
  (3) Interior and exterior spans of precast systems with cast-in-place joints resulting in restraint equivalent to that which would exist in condition III(1) .... restrained
  (4) All types of prefabricated floor or roof systems where the structural members are secured to such systems and the potential thermal expansion of the floor or roof systems is resisted by the framing system or the adjoining floor or roof construction[b] .... restrained

IV. Wood construction:

  All types .... unrestrained

---

[a]Floor and roof systems can be considered restrained when they are tied into walls with or without tie beams, the walls being designed and detailed to resist thermal thrust from the floor or roof system.

[b]For example, resistance to potential thermal expansion is considered to be achieved when:
1. Continuous structural concrete topping is used,
2. The space between the ends of precast units or between the ends of units and the vertical face of supports is filled with concrete or mortar, or
3. The space between the ends of precast units, and the vertical faces of supports, or between the ends of solid or hollow core slab units, does not exceed 0.25% of the length for normal-weight concrete members or 0.1% of the length for structural lightweight concrete members.

---

the value at normal room temperature. The American Institute for Steel Construction *Specification for the Design, Fabrication, and Erection of Structural Steel for Buildings*[51] limits the maximum permissible design stress to approxi-

TABLE 4-9.2 *ASTM E-119 Temperature Endpoint Criteria*[1]

Structural Member	Location	Maximum Temperature °F (°C)
Walls/Partitions (Bearing and Non-Bearing)	1. Unexposed Side	250 (139)*
Steel Columns	1. Average	1000 (530)
	Single Point	1200 (649)
Floor/Roof Assemblies and Loaded Beams	1. Unexposed Side	250 (139)*
	2. Steel Beam (average)	1100 (593)
	(single point)	1300 (704)
	3. Pre-stressing steel	800 (426)
	4. Reinforcing steel	1100 (593)
	5. Open-web steel joists	1100 (593)
Steel Beams/Girders (Not loaded)	1. Average	1000 (530)
	Single point	1200 (649)

*Maximum temperature cited refers to the maximum temperature rise above initial conditions.

mately 60 percent of the yield strength. Thus, for structural members at 538°C (1000°F) designed to carry the maximum permissible stress, the applied stress is approximately the same as the strength of the member. It should also be noted that, at 538°C (1000°F), the modulus of elasticity has decreased appreciably from the value at normal room temperature.

Mathematical expressions describing the relationship of the yield strength, modulus of elasticity, and coefficient of thermal expansion on temperature are[8,9]

$$\sigma_y = \sigma_{y0}(1 - 0.78\theta - 1.89\theta^4) \qquad \theta < 0.63$$
$$E = E_0(1 - 2.04\theta^2) \qquad \theta < 0.63$$
$$\alpha = (6.1 + 0.0019T) \times 10^{-6} \qquad \theta < 0.68$$

where

σ_y = yield strength at elevated temperature, (psi) (MPa)

σ_{y0} = yield strength at 68°F (20°C), (psi) (MPa)

E = modulus of elasticity at elevated temperature, (psi) (MPa)

E_0 = modulus of elasticity at 68°F (20°C), (psi) (MPa)

α = coefficient of thermal expansion at temperature, T, m/m°C (in./in.°F)

T = steel temperature (°F)

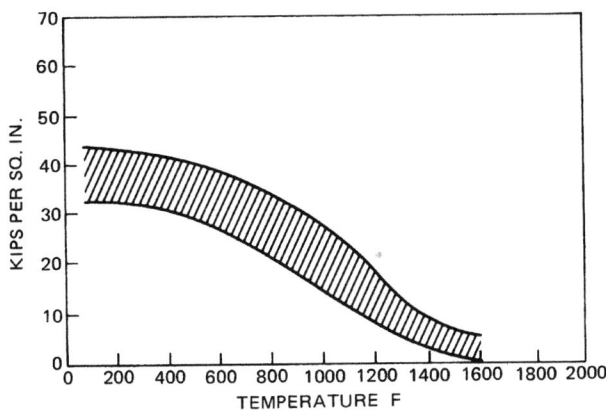

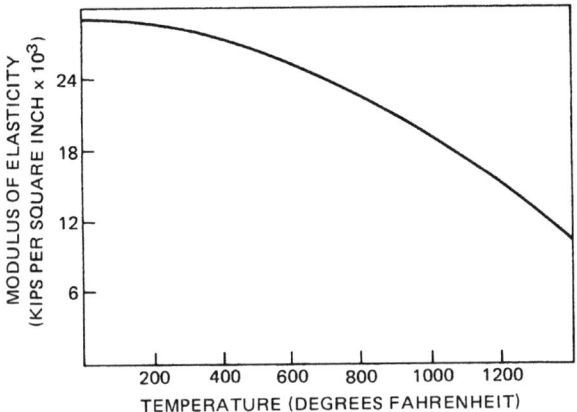

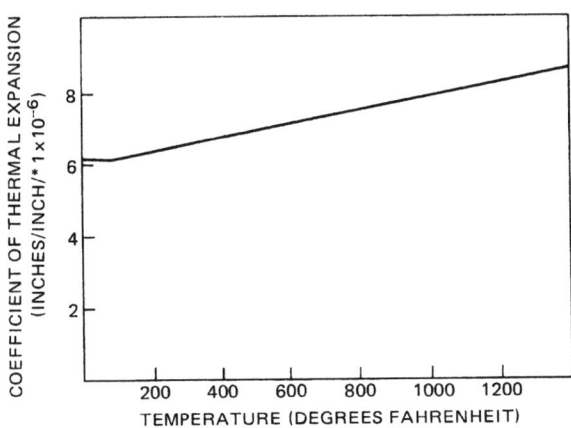

Fig. 4-9.2. Temperature effects on properties of ASTM A36 steel.[4]

TABLE 4-9.3 *Thermal Properties of A36 Steel at Elevated Temperatures[4-6]*

Thermal Conductivity (Btu/ft hr°F)	Specific Heat (Btu/lb °F)	Density (lb/ft³)
30.0 @ 0°F	0.107 @ 0°F	480.0
24.7 @ 600°F	0.144 @ 750°F	
20.1 @ 1100°F	0.172 @ 1100°F	
15.0 @ 2000°F	0.172 @ 2000°F	

$$\theta = \frac{T' - 68}{1800} \quad T' \text{ in } °F$$

$$ \frac{T' - 20}{1000} \quad T' \text{ in } °C$$

$$T' = \text{steel temperature}$$

In addition to the changes in material properties that occur at elevated temperatures, the crystalline structure of steel also changes as noted in Figure 4-9.3.[10] However, for the carbon steels typically used in building construction, significant changes in crystalline structure occur only at temperatures in excess of 600 to 650°C (1100 to 1200°F).[11] Since the yield strength of steel is reduced to 30 to 40 percent of its normal room temperature yield strength at 600 to 650°C (1100 to 1200°F), failure is likely to occur before the crystalline structure changes become significant, if the member is stressed to a level near the maximum allowable stress. Thus, changes in the crystalline structure of steel are typically not a factor.

Creep, the time-dependent deformation of a material, is significant in structural steel at temperatures in excess of 850°F.[12] The rate of creep increases approximately 300 times for ASTM A36 structural steel, when the steel temperature is increased from 850 to 950°F. Since creep is a complex phenomenon depending on the stress level and rate of heating, among other factors, often it is not considered in fire resistance calculations.[7,11] In-depth discussions of creep have been prepared by Harmathy.[13,14]

METHODS OF PROTECTION

The basic intent of the various methods of protection is to reduce the rate of heat transfer to the structural steel. This is accomplished by using insulation, membranes, flame shielding, and/or heat sinks.

Insulation

Insulation of the steel is achieved by surrounding the steel with materials that preferably have the following characteristics:[15]

1. Noncombustibility and the added attribute of not producing smoke or toxic gases when subjected to elevated temperatures;
2. Thermal protective capability when tested in accordance with the standard fire test, ASTM E-119;
3. Product reliability giving positive assurance of consistent uniform protection characteristics;
4. Availability in a form that permits efficient and uniform application;
5. Sufficient bond strength and durability to prevent either dislodgement or surface damage during normal construction operations; and
6. Resistance to weathering or erosion resulting from atmospheric conditions.

In addition to the insulating qualities of the protection materials, chemical reactions may occur in the insulation to further reduce the rate of heat transfer. The chemical reactions include calcination, ablation, intumescence, thermal hydrogeneration, and sublimation.

Insulating methods include the use of board products, spray-applied materials, and concrete encasement. A brief review of each method is presented below.

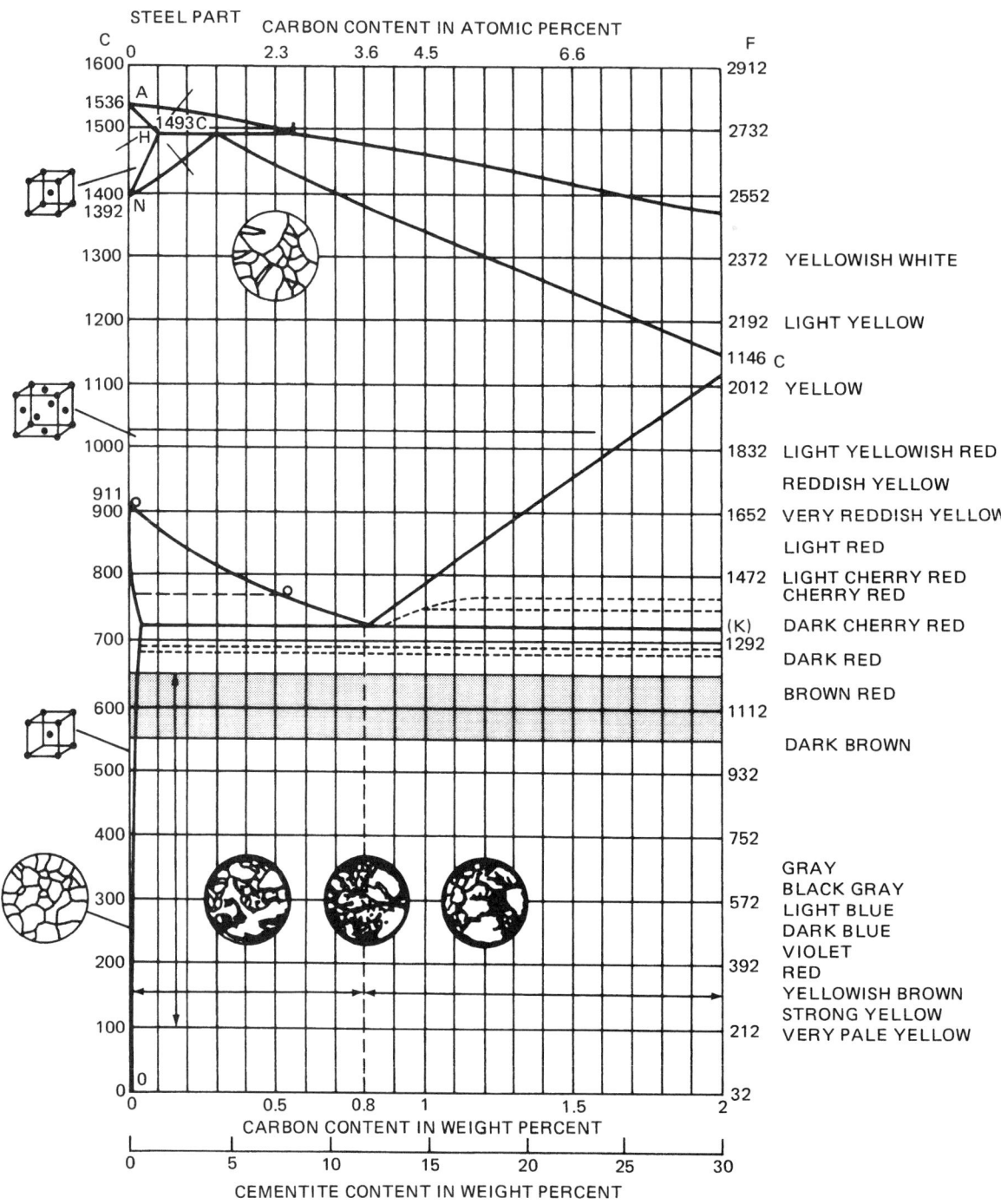

Fig. 4-9.3. *Influence of elevated temperatures vs. carbon content in steel.*[10]

Board products: Four types of board products are commonly used to protect structural steel: gypsum board, fiber-reinforced calcium silicate board, vermiculite-sodium silicate board, and mineral fiber board. In each case, the means of attachment of the boards surrounding the steel is a critical parameter affecting the performance of the assembly. Two commonly used methods of attachment of gypsum wallboard with and without steel covers are illustrated in Figure 4-9.4.[16] Detailed descriptions of the attachment mecha-nisms for the other board products are provided elsewhere.[16–18] Also, board products can be used in wall assemblies to provide an envelope around steel trusses, as is described later.

Spray-applied materials: Several types of spray-applied materials are commonly used. These include cementitious plasters, mineral fibers, magnesium oxychloride cements, and intumescents. Sufficient data has been obtained to characterize

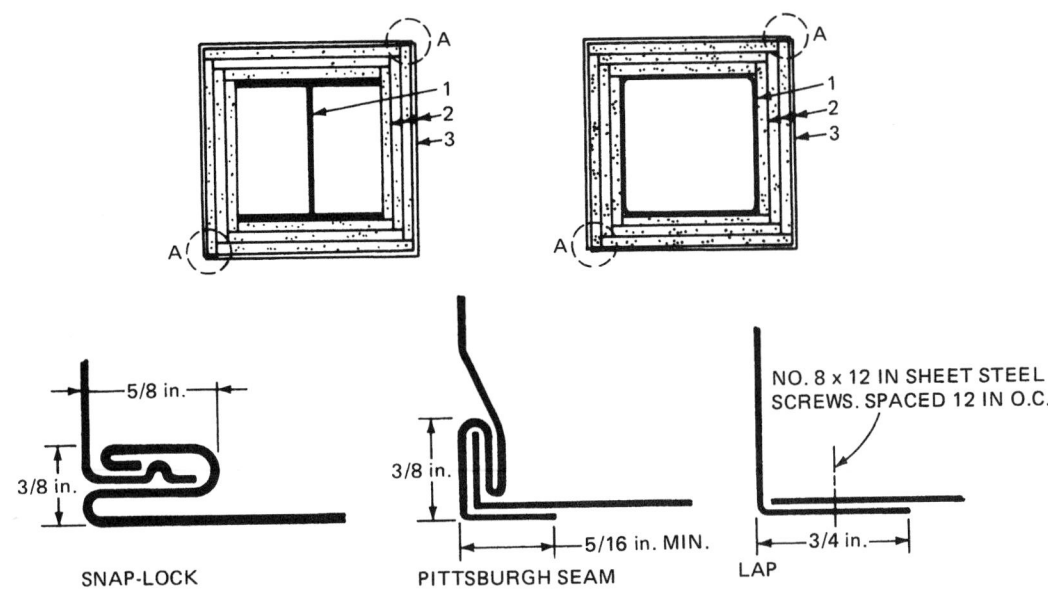

CORNER JOINT DETAILS (A)

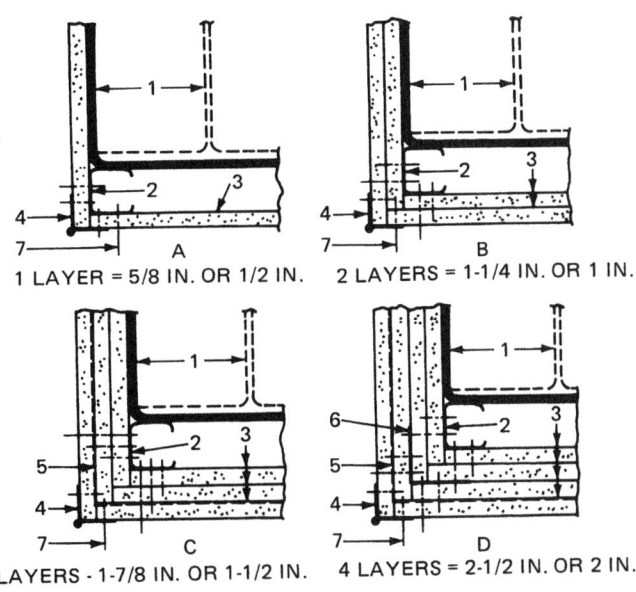

Fig. 4-9.4. Attachment mechanisms of gypsum wallboard to steel columns.[16]

spray-applied cementitious and mineral fiber materials for the purpose of estimating the fire endurance of structural steel protected with these materials. An illustration of a steel column protected by a spray-applied material is presented in Figure 4-9.5.

Concrete encasement: Concrete encasement of steel members to surround and insulate the steel is illustrated in Figure 4-9.6. As indicated in Figure 4-9.6, the concrete is cast to fill in all re-entrant spaces. Alternatively, concrete column covers may be used as illustrated in Figure 4-9.7. The concrete is assumed to act only to thermally protect the steel.

The load-bearing capacity of the concrete and possible steel-concrete composite action are neglected for calculation purposes.

Membrane: Suspended ceiling assemblies are used as membranes to protect structural steel in floor and roof assemblies. The ceiling panels and tiles comprising the ceiling assembly may consist of gypsum, perlite, vermiculite, and/or mineral fibers.

The membrane method of protection is illustrated in Figure 4-9.8. Heat transfer to the structural steel is reduced due to the air space above the membrane and the insulating

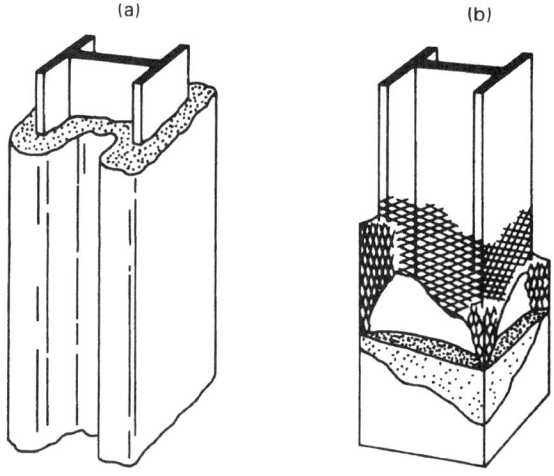

Fig. 4-9.5. (a) Sprayed insulation; (b) Metal lath and plaster encasement.[15]

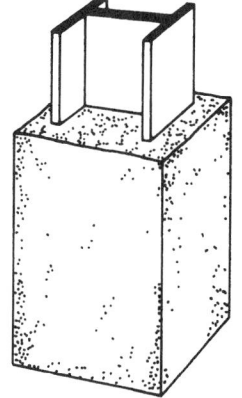

Fig. 4-9.6. Steel column with concrete encasement.[15]

characteristics of the membrane. Also, membranes help prevent the direct impingement of flame on the structural steel.

Flame shield: Flame shields are useful in preventing direct flame impingement. Without direct flame impingement, heating of the steel is by radiation only. The effectiveness of flame shields to protect exposed spandrel beams was first examined by Seigel.[19,20] In this instance, 14-gage sheet steel was used as the flame shield.

Heat sinks: The heat sink approach delays the heating of steel by absorbing heat transferred through the steel. The heat sink approach usually involves liquid- or concrete-filling of the interior of hollow steel members (tubular and pipe sections). Liquid-filling can be used to provide a sufficient level of protection for the columns, without any externally applied coating. The liquid used for protection is an aqueous solution. Additives are provided primarily for antifreeze, corrosion protection, and biological reasons.

A diagram of a typical design for a liquid-filled column fire protection system is presented in Figure 4-9.9. The components of this system include the hollow structural steel columns, piping to connect the columns, a water storage tank, and associated valves.

In many tests with liquid-filling, steel temperatures have been observed to be well below those required for failure, as long as the column remains full of the liquid. The system operates on the principle that heat incident on the column is removed by circulation of the liquid. If sufficient heat is delivered to the liquid, boiling can be expected, which enhances the efficiency of the heat-removal process.

Another heat-sink approach consists of filling the interior of hollow steel columns with concrete. Being a load-bearing material, load transfer from the steel to the concrete can be accommodated as the steel weakens with increasing temperature. Several calculation methods to determine the fire resistance of concrete-filled steel columns are available.[4]

EMPIRICALLY DERIVED CORRELATIONS

Numerous, easy-to-use, empirically derived correlations are available to calculate the fire resistance of steel columns, beams, and trusses. The correlations are based on

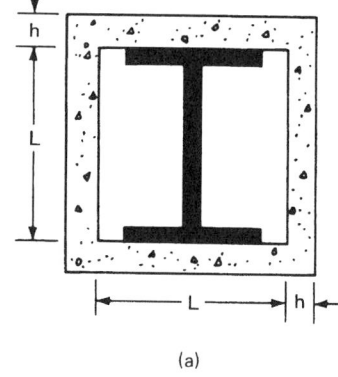

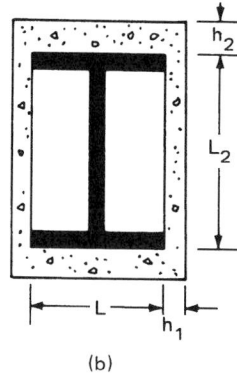

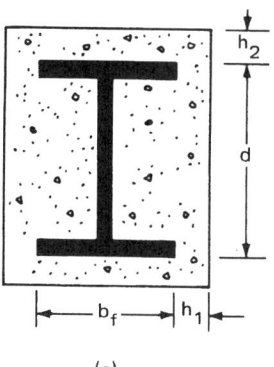

Fig. 4-9.7. Concrete-protected structural steel columns. (a) Square shape protection with a uniform thickness of concrete cover on all sides; (b) Rectangular shape with varying thickness of concrete cover; and (c) Encasement having all re-entrant spaces filled with concrete.

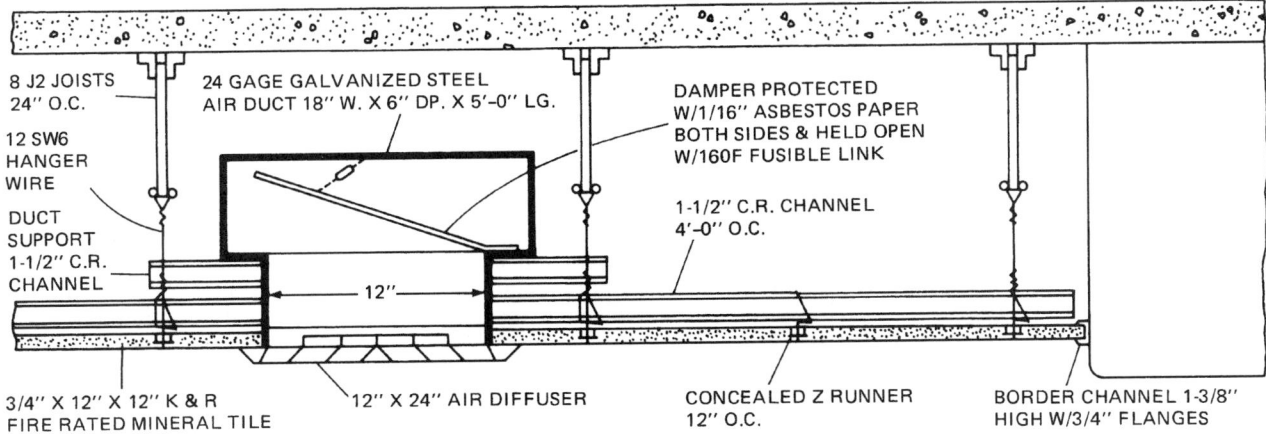

(a) CROSS-SECTION OF A FLOOR/CEILING SYSTEM WITH CONVENTIONAL SHEET STEEL FUSIBLE-LINK
DAMPER FOR PROTECTING TYPICAL CEILING OUTLETS IN GLAVANIZED SHEET DUCTS

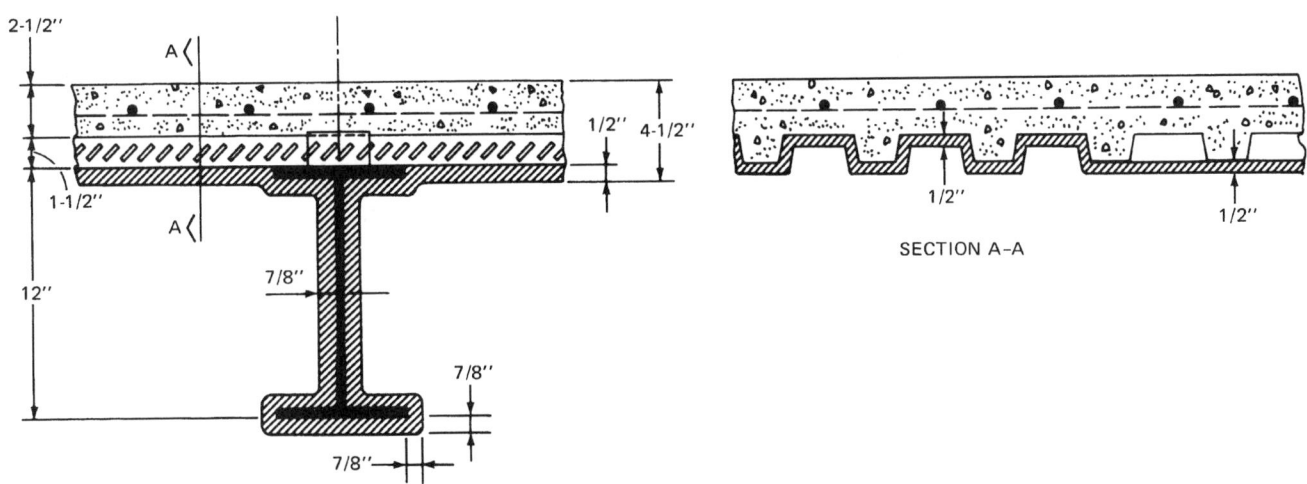

(b) SPRAYED CONTACT FIREPROOFING APPLIED DIRECTLY TO THE UNDERSIDE OF
FORMED-STEEL DECKING AND TO A SUPPORTING STEEL BEAM

Fig. 4-9.8.   Membrane method of protection.[15]

data from performing the standard test numerous times on variations of a standard assembly. Curve-fitting techniques are used to establish the various correlations. In some cases, a "best-fit" line has been drawn for the data points, whereas in other cases, lines were placed to provide conservative estimates of the fire endurance by connecting the two "lowest" points.[21]

Steel Columns

The equations to estimate the fire endurance of unprotected and protected steel columns are given in Table 4-9.4. Present in each of the equations is W/D for wide-flange sections and A/P for hollow sections. The W/D and A/P ratios are comparable. The W/D ratio is the weight per lineal foot to the heated perimeter of the steel at the protection interface (or the perimeter of the steel if unprotected). The A/P ratio is the cross-sectional area divided by the heated perimeter. Essen-

tially, the W/D ratio relates to the product of the density of the steel and the A/P ratio.

The relevance of the W/D and A/P ratios was first noted by Lie and Stanzak.[22] W/D ratios for commonly used wide-flange and tubular shapes for columns and beams are available elsewhere.[16,23,24] The two factors in the W/D ratio that affect the heat transfer rate to the steel (and consequently its rise in temperature) are: (1) shape of the fire protection system, D; and (2) steel mass per unit of length, W.

The parameter that characterizes the shape of the fire protection system is the heated perimeter, D, expressed in inches, which is defined as the inside perimeter of the steel at the fire protection material interface. Figure 4-9.10 illustrates the method for determining D in four typical cases. As can be seen from the figure, the heated perimeter depends on the size of the column and also on the profile of the protection system. Two different commonly used profiles are: (1) contour profile, where all surfaces of the steel column are in

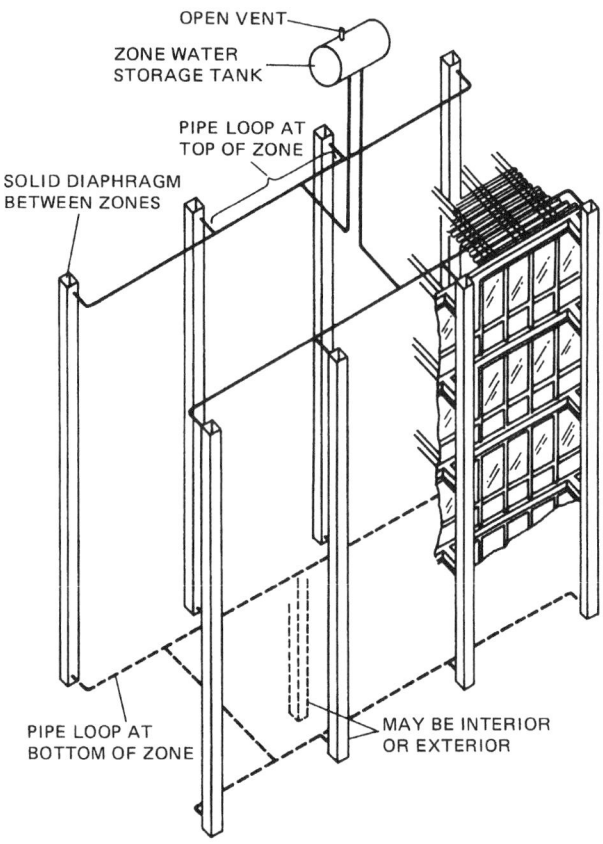

Fig. 4-9.9. Schematic layout of a typical piping arrangement used in a liquid-filled column fire-protection system.[2]

contact with the protection material; and (2) box profile, where a rectangular box of protection material is built around the column.

The greater the W/D ratio, the greater is the inherent fire resistance of the assembly. A large value of W refers to a column with a large weight per lineal foot. A given amount of energy will raise the temperature of the massive column to a lesser degree than that of a light column. A small value of the heated perimeter, D, means that less surface area is available for heat transfer, thereby inhibiting the temperature rise in the steel.

The fact that steel elements with larger W/D ratios are inherently more fire resistant is the basis for permitting the substitution of shapes with W/D ratios greater than the shapes identified in the listed designs in the UL *Fire Resistance Directory*,[25] while maintaining the same thickness of protection. However, such substitution yields inefficient designs, since shapes with large W/D ratios actually require less fire protection material than shapes with small W/D ratios for the same level of fire resistance.

The equation for gypsum wallboard protection is nonlinear. The weight of the gypsum wallboard is included, because the heat capacity of gypsum has a considerable impact on the fire resistance of the assembly.

For some board products and spray-applied materials, the equations are of the form[16,18]

$$R = (C_1 W/D + C_2)h$$

where

R = fire endurance (hr),

W = steel weight per lineal foot (lb/ft),

D = heated perimeter of the steel at the insulation interface (in.), and

h = thickness of insulation (in.).

The constants C_1 and C_2 need to be determined for each protection material. The constants take into account the thermal conductivity and heat capacity of the insulation material. Included in Table 4-9.4 is a list of values of C_1 and C_2 for those materials that have been characterized.

Considering the equation for the concrete cover column protection method (see Table 4-9.4), R_0 is the fire endurance of the assembly if the concrete has no moisture content. However, since the fire resistance of concrete-cover over steel columns is known to increase approximately by 3 percent for each 1 percent of moisture, R_0 is multiplied by the $(1 + 0.03m)$ factor where m is the equilibrium moisture content of concrete. The parameters h and L noted in the equation are shown in Figure 4-9.7. If the protection thickness and/or column dimensions are not the same in the vertical and horizontal directions, average values are used for h and L.

The heat capacity of the concrete must be accounted for in the determination of H if all re-entrant spaces are filled. (See Figure 4-9.7.) If specific data on the concrete's thermal properties are not available, values given in Table 4-9.5 may be used. Typical densities for normal-weight and lightweight concrete are 145 and 110 lb/cu ft (2320 and 1760 kg/m^3). Also, the typical equilibrium moisture content (by volume) for normal-weight concrete is 4 percent and lightweight concrete is 5 percent.

Many of the equations cited in Table 4-9.4 are limited to a range of shapes or protection thickness. Before applying

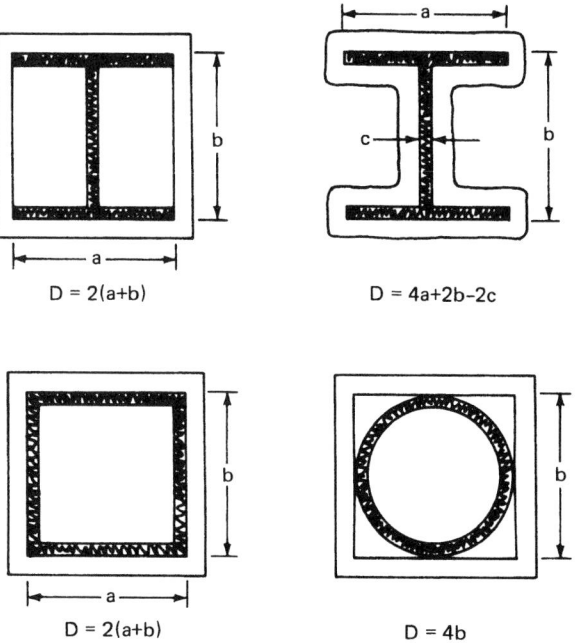

Fig. 4-9.10. Heated perimeter for steel columns.[16]

TABLE 4-9.4 *Empirical Equations for Steel Columns*[11,17,18]

Member/Protection	Solution	Symbols
Column/Unprotected	$R = 10.3\,(W/D)^{0.7}$, for $W/D < 10$ $R = 8.3\,(W/D)^{0.8}$, for $W/D \geq 10$ (for critical temperature of 1000°F)	R = fire endurance time, min. W = weight of steel section per linear foot, lb/ft. D = heated perimeter, in.
Column/Gypsum Wallboard	$R = 130\left(\dfrac{hW'/D}{2}\right)^{0.75}$ where $W' = W + \left(\dfrac{50hD}{144}\right)$	h = thickness of protection, in. W' = weight of steel section and gypsum wallboard, lb/ft.
Column/Spray-Applied Materials and Board Products—Wide-Flange Shapes	$R = [C_1(W/D) + C_2]h$	C_1 & C_2 = constants for specific protection material
Column/Spray-Applied Materials and Board Products—Hollow Sections	$R = C_1\left(\dfrac{A}{P}\right)h + C_2$	C_1 & C_2 = constants for specific protection material The A/P ratio of a circular pipe is determined by $$A/P \text{ pipe} = \frac{t(d-t)}{d}$$ where d = outer diameter of the pipe (in.) t = wall thickness of the pipe (in.) The A/P ratio of a rectangular or square tube is determined by $$A/P \text{ tube} = \frac{t(a+b-2t)}{a+b}$$ where a = outer width of the tube (in.) b = outer length of the tube (in.) t = wall thickness of the tube (in.)
Column/Concrete Cover	$R = R_0(1 + 0.03m)$ where $R_0 = 10(W/D)^{0.7} + 17\left(\dfrac{h^{1.6}}{k_c^{0.2}}\right)$ $\cdot\left\{1 + 26\left[\dfrac{H}{\rho_c c_c h(L+h)}\right]^{0.8}\right\}$ $D = 2(b_f + d)$	R_0 = fire endurance at zero moisture content of concrete, min. m = equilibrium moisture content of concrete, % by volume b_f = width of flange, in. d = depth of section, in. k_c = thermal conductivity of concrete at ambient temp., Btu/hr ft °F
Column/Concrete Encased	for concrete-encased columns use: $H = 0.11W + \dfrac{\rho_c c_c}{144}(b_f d - A_s)$ $D = 2(b_f + d)$ $L = (b_f + d)/2$	H = thermal capacity of steel section at ambient temp., $= 0.11W$ Btu/ft.°F. c_c = specific heat of concrete at ambient temp., Btu/lb °F L = inside dimension of one side of *square* concrete box protection, in. A_s = cross-sectional area of steel column, in.[2]

any equation from this table, users should consult the original reference and confirm that the equation is being applied properly.

TABLE 4-9.5 *Thermal Properties of Concrete at 70°F*

	Normal-Weight Concrete	Structural Lightweight Concrete
Thermal Conductivity (k)*	0.95	0.35
Specific Heat (c)**	0.20	0.20

*Expressed as Btu/hr ft°F.
** Expressed as Btu/lb°F.

EXAMPLE 1:

Determine the thickness of spray-applied cementitious material to obtain a 2-hr fire endurance when applied to a W 12 × 106 column.

SOLUTION:

From UL X772, the applicable equation is

$$R = (63W/D + 36)h$$

Solving for h,

$$h = \frac{R}{63W/D + 36}$$

where

R = 2 hrs = 120 min, and
W/D = 1.44 lb/ft-in. for a $W\,12 \times 106$ with contour profile protection.

Substituting,

$$h = \frac{120}{63 \times 1.44 + 36} = 0.95 \text{ in.}$$

EXAMPLE 2:

Determine the fire endurance of a $W\,8 \times 28$ column encased in lightweight concrete (density of 110 lb/ft³) with all re-entrant spaces filled. The concrete cover thickness is 1.25 in.

SOLUTION:

From Table 4-9.4, the appropriate equation is

$$R = R_0(1 + 0.03m)$$

where

$$R_0 = 10(W/D)^{0.7} + 17(h^{1.6}/k_c^{0.2})\{1 + 26[H/\rho_c c_c h(L + h)]^{0.8}\}$$

Referring to Figure 4-9.7,

$h_2 = h_1 = h = 1.25$ in.
$b_f = 6.535$ in.
$d = 8.060$ in.
$W/D = 0.67$ lb/ft-in. (contour profile)
$A = 8.25$ in.²

From Table 4-9.5,

$k_c = 0.35$ Btu/hr ft°F
$c_c = 0.20$ Btu/lb°F
$\rho_c = 110$ lb/ft³
$L = \frac{1}{2}(b_f + d) = 7.30$ in.
$H = 0.11W + \frac{\rho_c c_c}{144}(b_f D - A_s)$
$H = 0.11 \times 28 + \frac{110 \times 0.20}{144}(6.535 \times 8.060 - 8.25)$
$= 9.87$

$R_0 = 10(0.67)^{0.7} + 17\left(\frac{1.25^{1.6}}{0.35^{0.2}}\right)$

$\times \left\{1 + 26\left[\frac{9.87}{110 \times 0.2 \times 1.25(7.30 + 1.25)}\right]^{0.8}\right\}$

$R_0 = 99$ minutes.

Assuming a moisture content of 5 percent for lightweight concrete,

$$R = 99(1 + 0.03 \times 5) = 114 \text{ minutes}$$

Steel Beams

As in the case of columns, the W/D ratio is an important parameter affecting the fire resistance of a beam. Beams with larger W/D ratios may be substituted for beams with lesser W/D ratios for an equivalent rating with no change in the protection thickness. However, as with columns, designs resulting from the direct substitution of larger beams without reducing the protection thickness may not be efficient.

In 1984, an empirically derived correlation was developed to calculate the required thickness of spray-applied material protection.[23] However, unlike the equations for steel columns, the thickness of protection is determined on the basis of a scaling relationship. The beam substitution equation is

$$h_1 = \left(\frac{W_2/D_2 + 0.6}{W_1/D_1 + 0.6}\right)h_2 \qquad (1)$$

where

h = thickness of spray-applied fire protection (in.),
W = weight of steel beam (lb/ft),
D = heated perimeter of the steel beam (in.). (See Figure 4-9.11.);

and where the subscripts

1 = substitute beam and required protection thickness, and
2 = the beam and protection thickness specified in the referenced tested design or tested assembly.

Limitations of this equation are noted as follows:

1. $W/D \geq 0.37$,
2. $h \geq \frac{3}{8}$ in., and
3. The unrestrained beam rating in the referenced tested design or tested assembly is at least 1 hr.

It should be noted that the above equation only pertains to the determination of the protection thickness for a beam in a floor or roof assembly. All other features of the assembly, including the protection thickness for the deck, must remain unaltered.

EXAMPLE 3:

Calculate the thickness of spray-applied fire protection required to provide a 2 hr fire endurance for a $W\,12 \times 16$ beam to be substituted for a $W\,8 \times 17$ beam requiring 1.44 in. of protection for the same rating.

SOLUTION:

The beam substitution correlation, presented as Equation 1, is used.

$$h_1 = \left(\frac{W_2/D_2 + 0.6}{W_1/D_1 + 0.6}\right)h_2$$

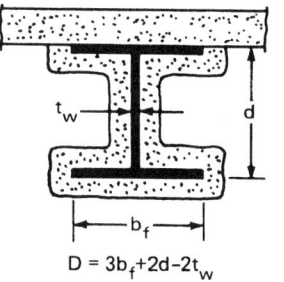
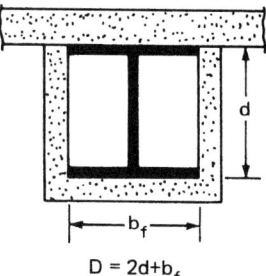

A. CONTOUR PROTECTION   B. BOX PROTECTION

Fig. 4-9.11. Heated perimeter for steel beams.[23]

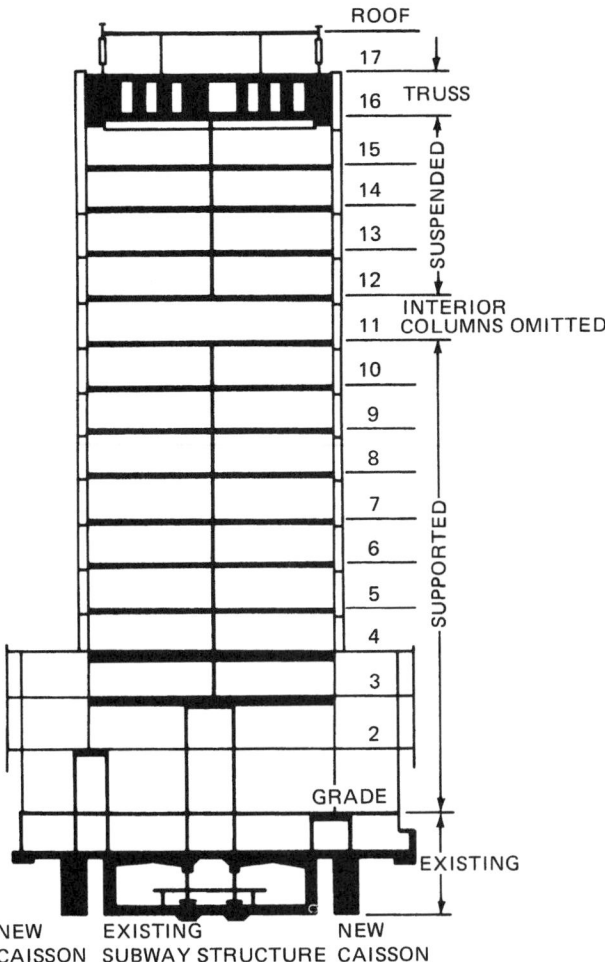

Fig. 4-9.12. Vierendell truss providing support from above and below.[26]

where

$$W_2/D_2 = 0.54 \quad \text{for } W\ 8 \times 17$$
$$W_1/D_1 = 0.45 \quad \text{for } W\ 12 \times 16$$
$$h_2 = 1.44$$
$$h_1 = \left(\frac{0.54 + 0.6}{0.45 + 0.6}\right) \times 1.44 = 1.6 \text{ in.}$$

Steel Trusses

There are three types of trusses used in buildings: transfer, staggered, and interstitial trusses. Because of the inherent features of each type of truss, some fire protection systems are more appropriate than others.[26]

A load-transfer truss supports loads from more than one floor. The loads may be suspended from a transfer truss or the transfer truss can be used to eliminate columns on lower floors. An example of a transfer truss is illustrated in Figure 4-9.12.

A staggered truss is illustrated in Figure 4-9.13. Generally, staggered trusses are used in residential occupancy buildings. Staggered trusses carry loads from two floors.

Interstitial trusses are used to create deep floor/ceiling concealed spaces containing mechanical and electrical equipment, as shown in Figure 4-9.14. Interstitial trusses

support only those loads from the equipment enclosure and the floor above. Interstitial trusses are typically used in healthcare facilities with heavy mechanical equipment needs.

Three methods of fire protection are often used for trusses: membrane, envelope, and individual element protection. Some fire protection methods are more appropriate than others for the specific truss types. The fire protection method(s) typically used for each truss type are indicated in Table 4-9.7. Membrane protection is accomplished through the use of a fire-resistant ceiling assembly. Design parameters for such an assembly can be determined from listings of fire-rated designs.[25,27] No empirical correlations are available to assess the design of membrane protection systems.

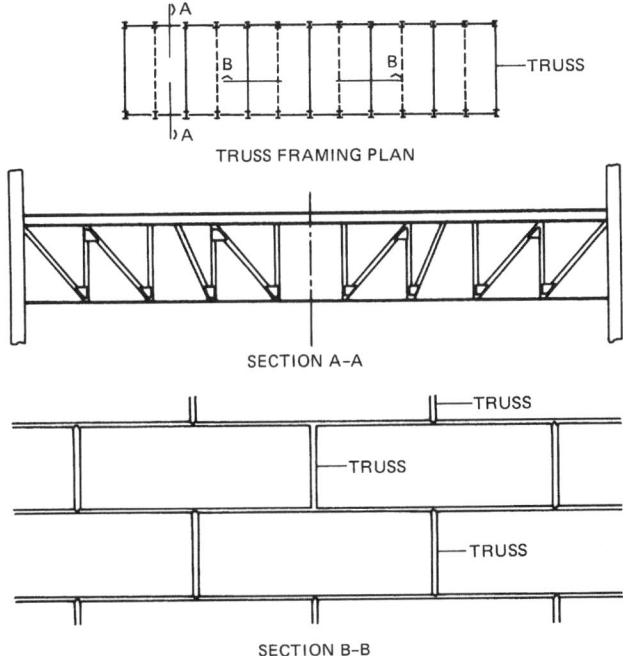

Fig. 4-9.13. A typical truss and positionings in a staggered truss system.[26]

Fig. 4-9.14. Hospital interstitial truss system.[26]

TABLE 4-9.6 *Practical Guidelines for Thickness of Gypsum Wallboard for Steel Truss Envelope Protection*

Fire Endurance	Gypsum X	Wallboard Type C
60	⅝″	⅝″
120	1¼″	—
180	—	1½″

The envelope means of protection is illustrated in Figure 4-9.15. The truss is enclosed in layers of a board product, with the number of layers determined by the required fire endurance. Some practical rules of thumb based on test results are noted in Table 4-9.6.[26]

Individual element protection is generally accomplished using a spray-applied material. Since critical truss elements perform structurally as columns, i.e., in tension or compression (as opposed to bending), the applicable equations for determining the thickness of spray-applied material for columns is used. In order to use these equations, the W/D ratio must be calculated for each element. Unlike columns and beams, the ratio may not be readily available. The diagrams in Figure 4-9.16 are provided for assistance in calculating the heated perimeter.

HEAT TRANSFER ANALYSES

Heat transfer analyses are applied to determine the time period required to heat structural members to a specified critical temperature. The required time period is then defined as the fire endurance time of the member.

TABLE 4-9.7 *Typical Fire Protection Methods for Steel Trusses*

Truss Type	Fire Protection Method		
	Membrane	Envelope	Individual Element
Transfer		X	X
Staggered		X	X
Interstitial	X	X	X

The critical temperature of a structural member can be determined by referring to the temperature endpoint criteria cited in ASTM E-119[1] or by a structural assessment, as is discussed later in this chapter.

The available types of heat transfer analyses can be grouped into the following categories:

1. Numerical methods,
2. Graphical solutions, and
3. Computer programs.

Numerical Methods

Numerous numerical methods are available to estimate the temperature rise in steel structural elements. The equations are derived from simplified heat transfer approaches.

Unprotected steel members: The temperature in an unprotected steel member can be calculated using a quasi-steady-state, lumped heat capacity analysis. The equation for temperature rise during a short time period, Δt, is[12]

$$\Delta T_s = \frac{\alpha}{c_s(W/D)}(T_f - T_s)\Delta t \qquad (2)$$

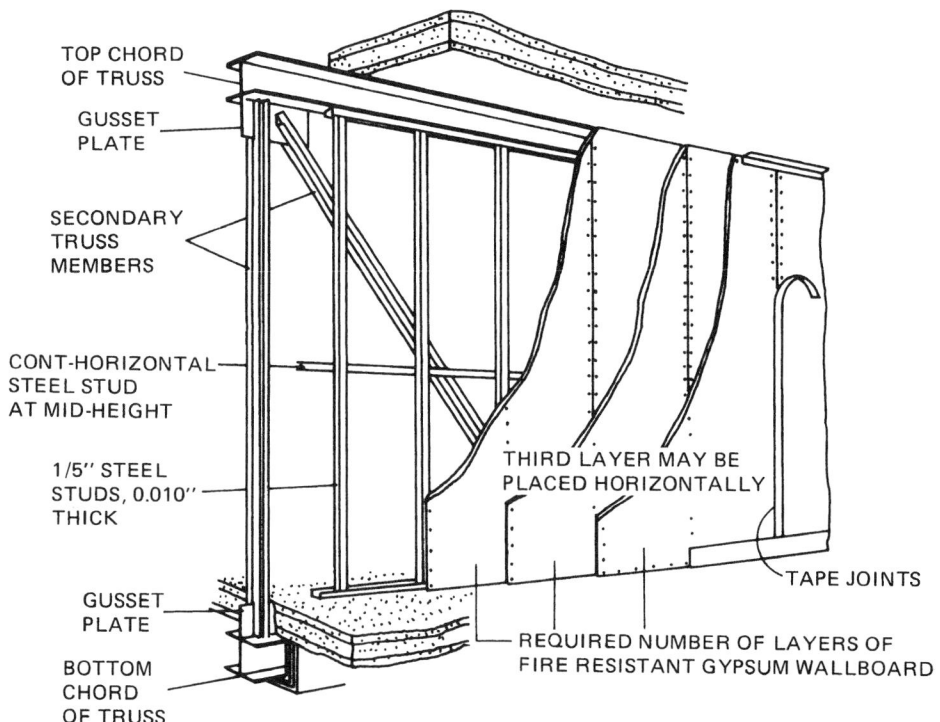

TOP CHORD OF TRUSS

GUSSET PLATE

SECONDARY TRUSS MEMBERS

CONT-HORIZONTAL STEEL STUD AT MID-HEIGHT

1/5″ STEEL STUDS, 0.010″ THICK

GUSSET PLATE

BOTTOM CHORD OF TRUSS

THIRD LAYER MAY BE PLACED HORIZONTALLY

TAPE JOINTS

REQUIRED NUMBER OF LAYERS OF FIRE RESISTANT GYPSUM WALLBOARD

Fig. 4-9.15. Staggered truss protection with envelope protection.[26]

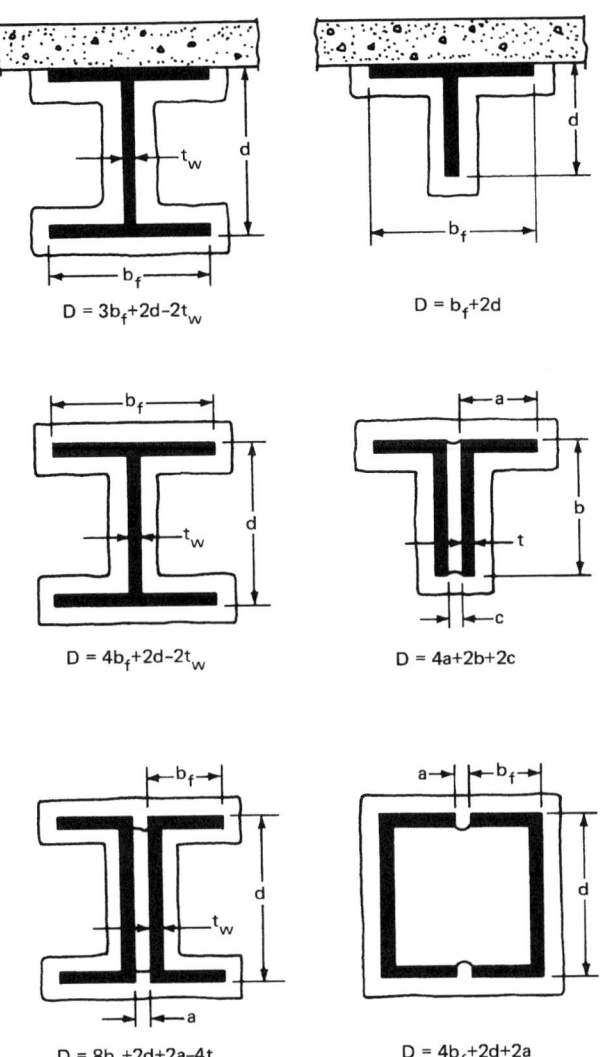

Fig. 4-9.16. *Heated perimeter for steel truss shapes.*[26]

where

ΔT_s = temperature rise in steel (°F)/(°C),
α = heat transfer coefficient from exposure to steel member (Btu/ft²-sec)/(W/m),
D = heated perimeter (ft)/(m),
c_s = steel specific heat (Btu/lb °F)/(J/kg °C),
W = steel weight per lineal foot (lb/ft)/(kg/m),
T_f = fire temperature (R)/(K),
T_s = steel temperature (R)/(K), and
Δt = time step (s).

where

$$\alpha = \alpha_r + \alpha_c,$$

α_r = radiative portion of heat transfer, and

$$\alpha_r = \frac{C_1 \varepsilon_f}{T_f - T_s}(T_f^4 - T_s^4)$$

where

$C_1 = 4.76 \times 10^{-13}$ Btu/s ft² R⁴ (5.77 × 10⁻⁸ W/m³ K⁴), and

ε_f, the flame emissivity, can be evaluated from Table 4-9.8.

α_c = convective portion of heat transfer
$$9.8 \times 10^{-4} - 1.2 \times 10^{-3} \text{ Btu/ft}^2\text{-s}$$
$$20\text{--}25 \text{ W/m}^2 \text{ K}$$

The quasi-steady assumption dictates that the time step should not be "large." Malhotra suggests a maximum time step to be determined from the following relationship:[12]

$$\Delta t < \begin{cases} 15.9 \times W/D & \text{(English)} \\ 3.25 \times W/D & \text{(metric)} \end{cases}$$

Equation 2 is successively applied up to the time duration of interest. The fire temperature, T_f, is evaluated at the midpoint of each time step.[12,28] If the exposure under consideration is that associated with the ASTM E-119 test, T_f at any time, t, is given by the following expression[12]

$$T_f = C_1 \log_{10}(0.133t + 1) + T_0 \qquad (3)$$

where

C_1 = 620 with T_f, T_0 in °F
345 with T_f, T_0 in °C

Protected steel members: For protected members, the thermal resistance provided by the insulating material must be considered. If the thermal capacity of the insulation layer is neglected,[12]

$$\Delta T_s = \frac{k_i}{c_s h W/D}(T_f - T_s)\Delta t \qquad (4)$$

where all parameters are as defined in Equation 2, and

k = thermal conductivity of insulation material (Btu/ft-s °F), (W/m°C), and
h = protection thickness (ft), (m).

TABLE 4-9.8 *Determination of the Maximum Temperature in the Event of Fire in Uninsulated Steel Structures*[30]

Calculation procedure:

Determine the resultant emissivity
Determine the F/V ratio
Determine the maximum temperature

Resultant emissivity, ϵ_f, for different constructions

Type of Construction	Resultant Emissivity
1. Column exposed to fire on all sides	0.7
2. Column outside façade	0.3
3. Floor girder with floor slab of concrete, only the underside of the bottom flange being directly exposed to fire	0.5
4. Floor girder with floor slab on the top flange	
Girder of I section for which the width-depth ratio is not less than 0.5	0.5
Girder of I section for which the width-depth ratio is less than 0.5	0.7
Box girder and lattice girder	0.7

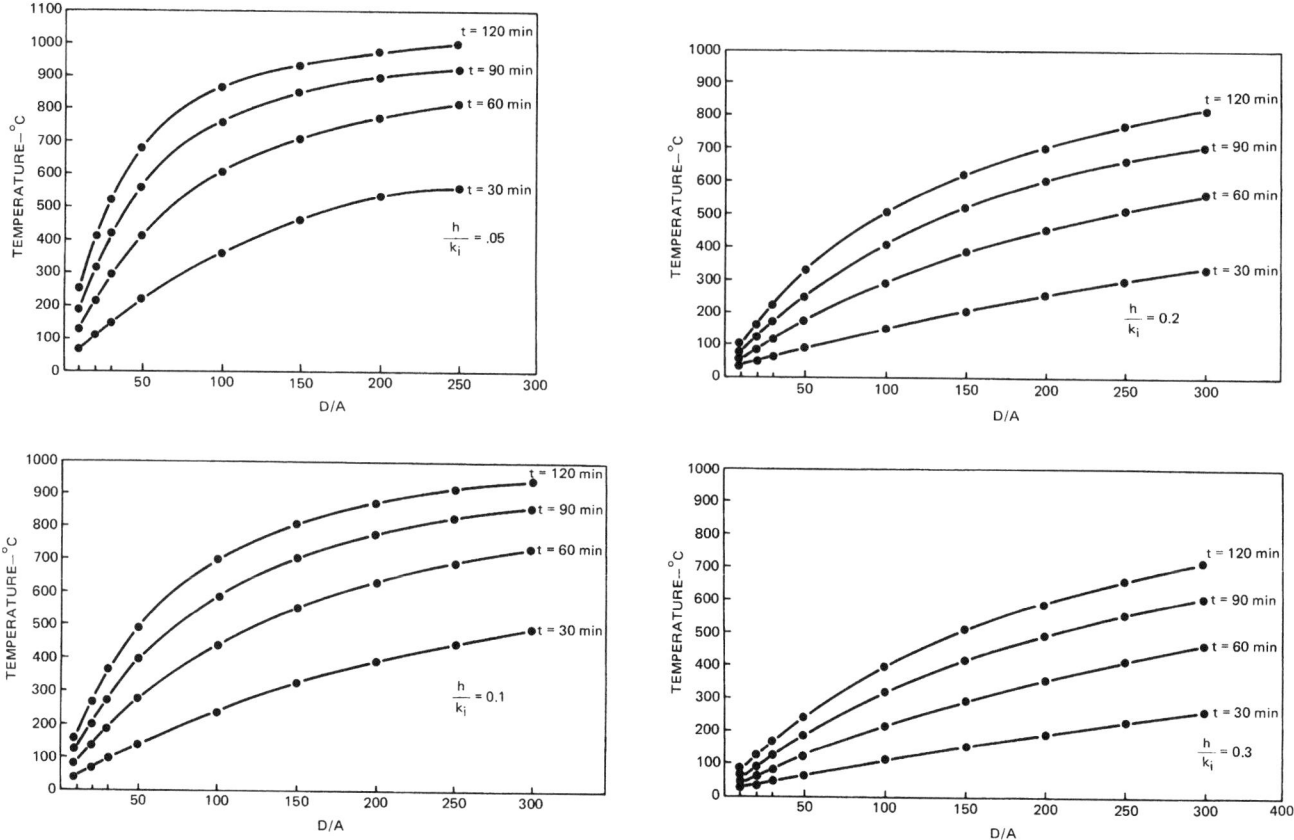

Fig. 4-9.17. *Relationship of heated area to steel weight with temperature.*[12]

The thermal capacity of the insulation material may be neglected if the following inequality is true (see parameter definitions for Equation 2):

$$c_s W/D > 2c_i\rho_i h$$

If the thermal capacity must be accounted for, as in the case of gypsum and concrete insulating materials,

$$\Delta T_s = \frac{k_i}{h}\left[\frac{T_f - T_s}{c_s(W/D) + \frac{1}{2}c_i\rho_i h}\right]\Delta t \qquad (5)$$

where all parameters are as defined for Equation 2, and

c_i = specific heat of insulating material (Btu/lb °F), (J/kg°C), and

ρ_i = density of insulating material (lb/ft^3), (kg/m^3).

Equations 2 through 5 can be used to predict the temperature rise of steel columns and steel beams. Application of the lumped heat capacity approach in deriving the equations is readily accepted for the four-sided exposure typical of columns. However, the equations would not appear to be applicable for the case of a beam, due to the heat sink characteristics of the deck or slab supported by the beam.[29] It has been suggested that the heat sink effect can be accounted for by using a reduced flame emissivity value (0.5), as noted in Table 4-9.8.[30] Also, the heated perimeter, D, of the beam is

evaluated as indicated in Figure 4-9.11, without considering the width of the top flange.

Exterior steel columns and steel spandrel beams: A design guide is available for calculating the exposure of exterior steel columns and steel spandrel beams.[31] The guide is based on research by Law and basic radiation heat transfer principles.[32]

The temperature of the steel member is calculated from a steady-state conduction analysis. The exposure boundary conditions consist of radiant heating from a fully developed room fire and flames emitting from windows near the steel member. For this method, a specific design is considered unacceptable if the steel temperature exceeds 1000°F.

Liquid-filled columns: The design calculations for liquid-filled columns are based on the thermal capacity of the liquid. The design of a liquid-filled column fire protection system consists of three major steps:

1. Heat transfer analysis,
2. Determination of volume of liquid required, and
3. Pipe network design.

The heat transfer analysis is used to assess the impact of fire exposure on the liquid-filled column. The heat transfer analysis considers radiation and convection heat transfer from the fire to the column surface, conduction through the column wall, and convection with localized boiling into the liquid. Both temperature of the steel column and total

AVERAGE SECTION TEMPERATURE OF STEEL BEAM, W 12 X 14 (W/D = 0.40), FOR VARIOUS THICKNESSES OF DIRECT-APPLIED FIRE PROTECTION.

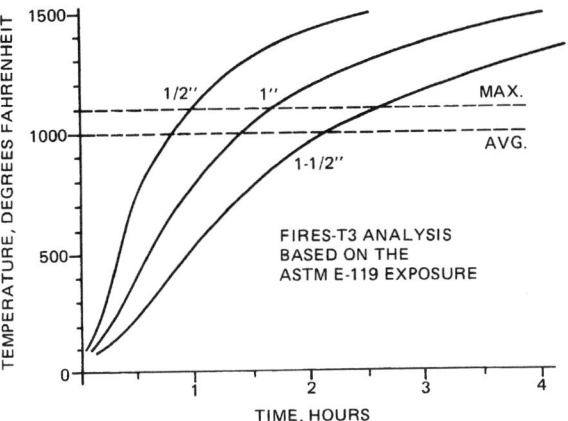

MAXIMUM STEEL BEAM TEMPERATURE, W 12 X 14 (W/D = 0.40), FOR VARIOUS THICKNESSES OF DIRECT-APPLIED FIRE PROTECTION

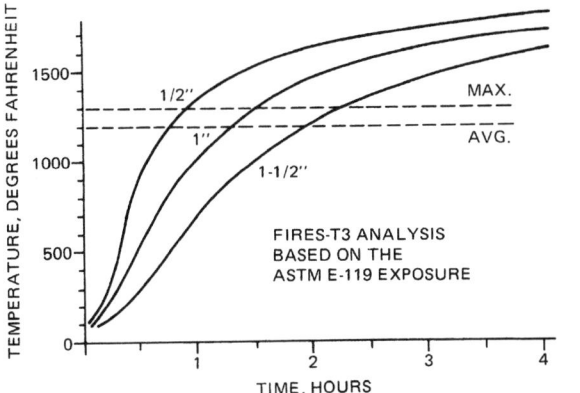

Fig. 4-9.18.   Predicted steel beam temperature by FIRES-T3.[4]

amount of heat transferred to the liquid causing evaporation are determined as a result of this analysis.

The liquid volume calculation is important to ensure the column remains full of liquid for the entire fire exposure period. Since heat transferred to the liquid will cause some evaporation, a supplemental amount of liquid must be provided in a storage tank.

The final step in the design method is a hydraulic analysis of the tubular column and pipe network. This analysis assesses the ability of the liquid to circulate based on friction losses, elevation changes, and buoyancy of the heated liquid.

A comprehensive design aid for liquid-filled columns is available.[33] Since the procedure is rather lengthy, it will not be reviewed here.

Graphical Solutions

Since heat transfer analyses can be very tedious and/or involve the use of complex computer programs, graphic solutions have been formulated to simplify the estimation of steel temperature. Graphs of the temperature of protected steel members have been developed by Malhotra,[12] Jeanes,[4] Lie,[7] and others.

The series of graphs presented in Figure 4-9.17 for estimating steel temperature, developed by Malhotra,[12] are based on the lumped heat capacity approach described in

the previous section. Steel temperatures are plotted *versus* the D/A ratio (analogous to the inverse of W/D) for selected time periods of exposure and thermal resistances of the insulating material. Time periods of 30 to 120 min are noted in the graphs. The range of thermal resistances of the insulating material covered by these graphs is 0.01 to 0.30 $(W/m^2 \circ C)^{-1}$ (0.003 to 0.10) $(Btu/ft^2 \cdot hr \cdot \circ F)^{-1}$.

Based on the application of FIRES-T3, a heat transfer computer program which will be described in the next section, Jeanes formulated a series of time-temperature graphs of protected steel beams.[4] The steel beams are protected by a proprietary specific spray-applied cementitious material with a range of thicknesses of 0.5 to 1.5 in. Graphs are available for a variety of common wide-flange beam shapes.[4] Examples of these graphs are presented in Figure 4-9.18 with graphs addressing the average and single-point steel temperatures relating to the maximum endpoint criteria from ASTM E-119.[1] Average and single-point steel temperatures are represented by the dashed lines. These graphs can be used to determine the thickness of protection material required to provide a desired level of fire resistance. Alternatively, the fire endurance can be estimated for a particular steel beam and insulation thickness design which has not been tested.[4]

Information from numerous applications of FIRES-T3 examining the time-temperature response of steel beams protected with a spray-applied cementitious material is summarized in Figure 4-9.19.[4] Using this graph, the fire endurance of protected steel beams with a W/D ratio of 0.4 to 2.5 lb/ft-in. can be determined for thicknesses of the spray-applied protection between 1.3 to 3.8 cm (0.5 to 1.5 in.).

Lie provides graphical representations of the exact solutions of the governing differential equations for the temperature of protected steel members.[7] The heat transfer is assumed to be one-dimensional through the insulation layer. A uniform temperature throughout the steel cross-section is assumed. The two graphs presented in Figure 4-9.20 are applicable to a wide range in the Fourier Number, *Fo*, for the insulation layer. In order to use the graphs, the following dimensionless parameters must be defined

$$Fo = \frac{\alpha t}{h^2}$$

$$N = \frac{\rho_i c_i h}{c_s (W/D)}$$

Fig. 4-9.19.   Fire endurance of steel beams vs. fire protection thickness for average section temperature of 1000°F. (Based on FIRES-T3 analysis of ASTM E-119 fire exposure.)[4]

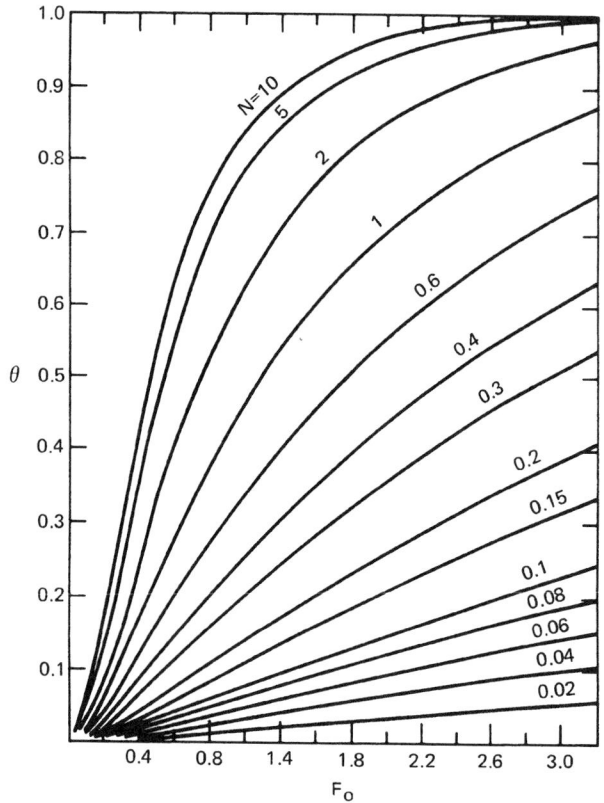

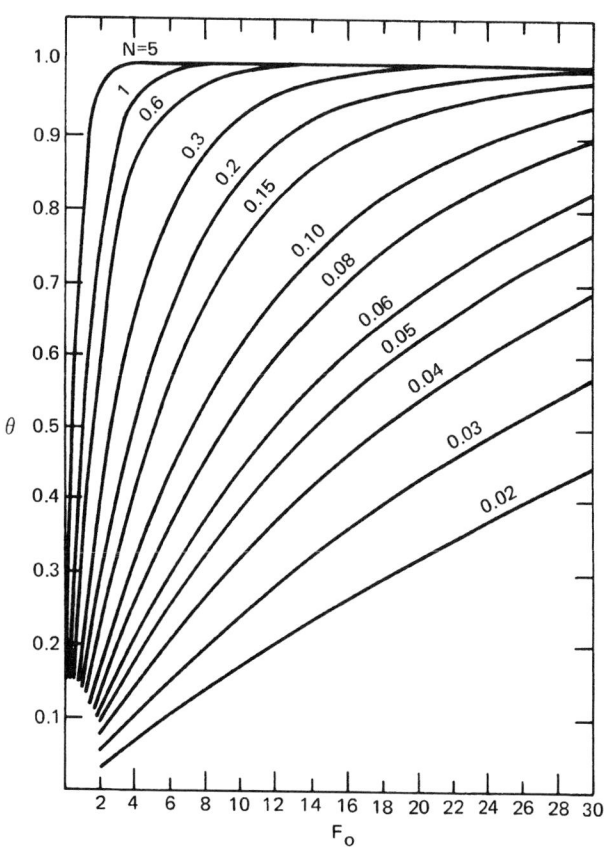

Fig. 4-9.20. **Dimensionless steel temperature vs. Fourier numbers.**[7]

$$\theta = \frac{T - T_0}{T_m - T_0}$$

where

α = thermal diffusivity of insulation (ft²/hr) (m²/hr),
t = heating time (hr),
h = thickness of insulation (ft) (m),
ρ_i = density of insulation (lb/ft³), (kg/m³),
c_i = specific heat of insulation (Btu/lb °F),
c_s = specific heat of steel (Btu/lb °F),
T = temperature of steel at time t (°F) (°C),
T_0 = initial temperature of steel (°F) (°C), and
T_m = mean fire temperature (°F) (°C).

The mean fire temperature associated with a heating time, t, for these graphs is calculated from the standard time-temperature curve, where:

$$T_m = \begin{cases} 150(\ln 480t - 1) - 30/t, & T°C \\ 270(\ln 480t) - 238 - 54/t, & T°F \end{cases}$$

EXAMPLE 4:

Determine the fire endurance of a $W\,24 \times 76$ steel beam, protected with 0.50 in. of spray-applied cementitious material, by three methods:

1. Graphical approach from Jeanes,[4]
2. Graphical approach by Lie,[7] and
3. Quasi-steady-state approach by Malhotra.[12]

SOLUTION:

A $W\,24 \times 76$ steel beam has a W/D ratio of 1.03 lb/ft-in. or 12.36 lb/ft². The material properties are evaluated at mean temperatures expected during the exposure. Mean temperatures of 500°F and 750°F are selected (arbitrarily) for the steel and insulation, respectively. The following material property values are assumed:[4]

	Steel	Insulation
Thermal conductivity (Btu/ft-hr °F)	25.6	0.067
Specific heat (Btu/lb°F)	0.132	0.304
Density (lb/ft³)	480	15

Jeanes' graph: Using Figure 4-9.19 with a W/D of 1.03 lb/ft-in. and an insulation thickness of 0.50 in., the fire endurance is estimated to be 1.33 hr or 80 min.

Lie's graph: Evaluating the dimensionless parameters,

$$Fo = \frac{\alpha t}{h^2}$$

where

$$\alpha = \frac{k_i}{\rho_i c_i} = \frac{0.067}{15 \times 0.304} = 0.0147 \text{ ft}^2/\text{h}$$

$$Fo = \frac{0.0147t}{(0.5/12)^2} = 8.47t \quad (t \text{ in h})$$

$$N = \frac{\rho_i c_i h}{c_s (W/D)} = \frac{15 \times 0.304 \times 0.5/12}{0.132 \times 12.36} = 0.116$$

Referring to Figure 4-9.20 and using a trial and error approach with a critical temperature selected as 1000°F, the fire endurance time is estimated as approximately 75 min.

Quasi-steady-state approach: First, a check is performed to determine if the thermal capacity of the insulation material must be considered.

$$c_s\ W/D > 2c_i\rho_i h$$
$$0.132 \times 12.36 > 2 \times 0.304 \times 15 \times 0.50/12$$
$$1.63 > 0.38$$

Neglecting thermal capacity of insulation, Equation 4 is used to predict the steel temperature rise for each time step.

$$\Delta T_s = \frac{0.067/3600}{0.0132 \times \dfrac{0.50}{12} \times 12.36}(T_f - T_s)\Delta t$$

$$= 2.74 \times 10^{-4}(T_f - T_s)\Delta t$$

The maximum allowable time step is

$$\Delta t = 15.9\ W/D$$
$$\Delta t = 15.9 \times 12.36 = 196\ \text{s}.$$

Selecting a time step of 3 min,

Time (min.)	$T_f - T_s$ (°F)	ΔT_s (°F)	T_s (°F)
0			70
3	691	34	104
6	938	46	150
9	1026	51	201
54	814	40	923
57	789	39	962
60	764	38	1000

Thus, the fire endurance is 60 min.

Comparing the calculated fire endurance by the three methods:

Jeanes (FIRES-T3)   80 min.
Lie   75 min.
Quasi-steady-state   60 min.

The agreement between the fire endurance times determined by Jeanes' and Lie's graphs is very good. The significantly reduced fire endurance calculated using the quasi-steady-state approach may be attributable to the approximate nature of the method as a result of the basic assumptions, i.e., one-dimensional heat transfer.

Computer-Based Analyses

Several computer-based analyses are available to estimate the temperature rise of steel members. The analyses range from a spreadsheet procedure to perform the iterative calculations for the quasi-steady-state approach to finite element models.

Spreadsheets are one example of providing a framework to perform the iterative, quasi-steady calculations.[54,55] Typically, the spreadsheet procedures mimic the quasi-steady analysis procedure described previously, including the evaluation of material properties at a mid-range temperature for the exposure of interest. Although temperature-dependent material properties can be included within the spreadsheet framework, the accuracy implied by considering temperature-dependent properties is not consistent with the first-order nature of the quasi-steady approach.

Another framework for conducting computer-based analyses includes the numerous mathematical equation-solver software packages. This software can be used to conduct the iterations associated with the quasi-steady approach or to solve the partial differential equations exactly.

A two-dimensional finite difference model was developed by Harmathy and Lie to predict the temperature rise in protected steel columns.[34] The two-dimensional network is formulated over the cross-section of the insulation layer, assuming the temperature to be independent of length. The steel is assumed to be a perfect conductor, i.e., the temperature is uniform throughout the steel. Heat transfer via radiation is considered across any air spaces enclosed by the insulation and steel.

The boundary conditions included by Harmathy are those associated with the ASTM E-119 test.[1] To simplify the model, convection is neglected, since convection comprises a minor portion of the heat transfer process. A flame emissivity of 0.9 is selected in characterizing the radiation heat transfer.

A comparison between the calculated and experimental steel temperatures is presented in Figure 4-9.21. As is evident, the agreement is very good for three insulating materials.

Pettersson et al[30] include a finite difference formulation to predict the temperature rise of steel beams protected with a suspended ceiling exposed to a specified fire. The formulation uses a one-dimensional approximation accounting for conduction through the suspended ceiling and floor slab (above the beam), and radiation and convection in the air space between the slab and beam. The temperature of the steel is assumed to be uniform. The assembly is divided into several elements as depicted in Figure 4-9.22.

A system of simultaneous equations is derived for the temperature rise in each of the assembly elements. A numerical integration technique, i.e., Runge-Kutta, is used to obtain the solution. A comparison of the calculated versus experimentally observed temperatures for a steel beam is presented in Figure 4-9.23.

General heat transfer finite element programs have been available for numerous years.[35] FIRES-T3, TASEF-2, and SUPER-TEMPCALC, among others, have been developed specifically to address the heating of assemblies exposed to fire conditions.[36-38]

TASEF-2 examines the conduction heat transfer through assemblies.[36] Assemblies may include internal voids, in which convection and radiation heat transfer modes are considered. Two time-temperature curves are available: (1) the ISO 834 standard time-temperature curve, and (2) one from a ventilation-controlled fire.

SUPER-TEMPCALC can also be used to analyze the conduction heat transfer through assemblies with air gaps. Numerous fire curves are included within the software.

FIRES-T3 was specifically developed to examine the heating of structural members exposed to fire conditions.[37] FIRES-T3 has been applied successfully to predict the temperature rise in protected steel beams and columns.[4,39] Bardell used a finite difference heat transfer model to estimate the protection thickness of spray-applied cementitious material required for tubular steel columns.[40]

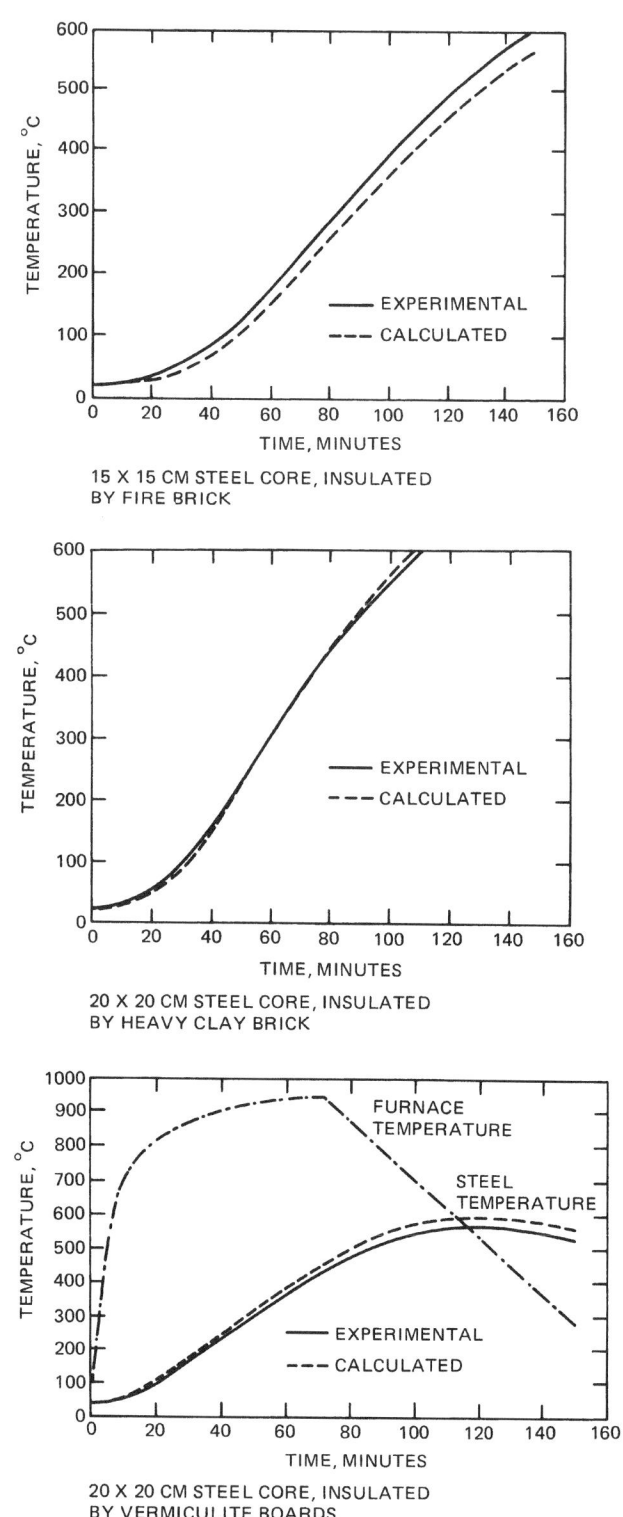

15 X 15 CM STEEL CORE, INSULATED
BY FIRE BRICK

20 X 20 CM STEEL CORE, INSULATED
BY HEAVY CLAY BRICK

20 X 20 CM STEEL CORE, INSULATED
BY VERMICULITE BOARDS

Fig. 4-9.21. Comparison of calculated and measured steel temperatures.[34]

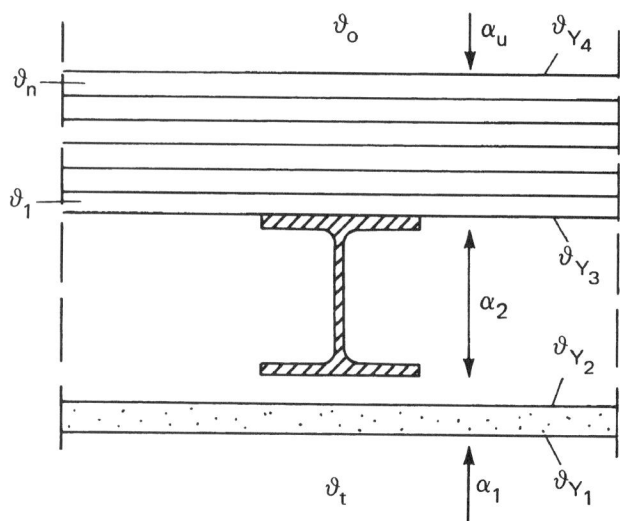

Fig. 4-9.22. Division of the floor slab into elements.[30]

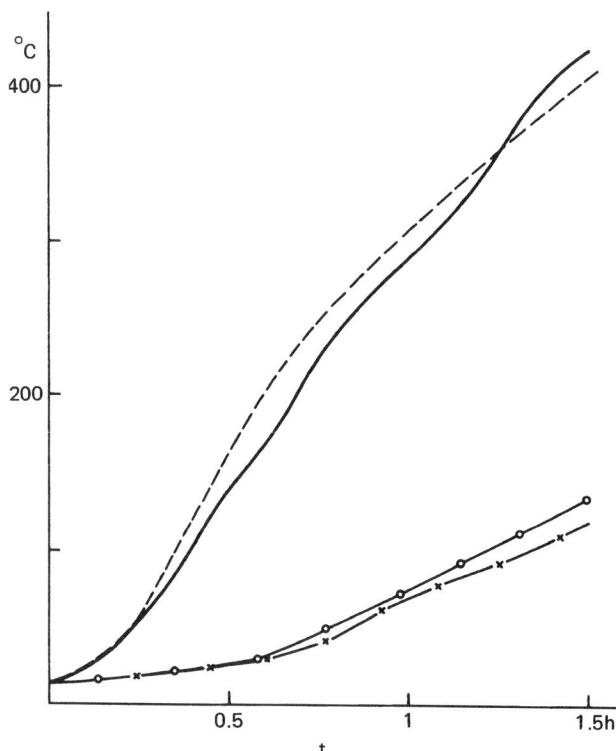

Fig. 4-9.23. Calculated (--) and measured (—) steel temperature-time $(\theta_s - t)$ curve for a floor girder IPE 140 with insulation in the form of a suspended ceiling of 40 mm thick mineral wool slabs of density $\gamma = 150\ kg/m^3$. The figure also gives the calculated (-∘-) and measured (—x—) temperature-time curve for the top of the 50-mm-thick concrete floor slab.[30]

TABLE 4-9.9 *Critical Stress Equations*[12]

Design Basis	Critical Yield Stress
Elastic Design	$\dfrac{\sigma_{YT}}{\sigma_Y} = \dfrac{1}{F_e}\,\dfrac{Z_e}{Z_p}$
Plastic Design	$\dfrac{\sigma_{YT}}{\sigma_Y} = \dfrac{1}{F_p}$

where
σ_{YT} = critical yield stress at elevated temperature, T
σ_Y = yield stress at ordinary room temperature
F_e = factor of safety, elastic design
F_p = factor of safety, plastic design
Z_e = elastic section modulus
Z_p = plastic section modulus.

The input data requirements for the heat transfer computer models can be grouped into two categories:

1. A description of the assembly, and
2. A description of the fire exposure.

The information necessary to describe the assembly includes geometric factors (dimensions, shape of member) and material property values (thermal conductivity, specific heat, and density). The fire exposure is characterized in terms of the temperature of the surrounding environment and appropriate heat transfer coefficients. The geometry of the assembly is established by formulating an element mesh for the assembly of interest. Required material property data consists of the density, specific heat, and thermal conductivity of the steel and insulation. Material property data is available for a limited number of insulation materials.[4,41] The exposure associated with the ASTM E-119 test[1] may be selected as the fire exposure to be simulated by FIRES-T3. The input data for this fire exposure used by Jeanes and Milke is presented in Table 4-9.9.[4,39] A pre-processor routine for FIRES-T3 was recently developed by Stubblefield and Edwards.[56] TASEF and SUPER-TEMPCALC both include user-interfaces to simplify assembly of the input files, including automatic mesh generation. SUPER-TEMPCALC also includes a post-processor to generate graphs of the output.

For models using an explicit transient solution technique, such as FIRES-T3, caution must be exercised in selecting the time step and mesh size to obtain correct results that are numerically stable. TASEF-2 internally determines a numerically stable time step. Most heat transfer models do not address the effects of phase changes or chemical reactions that may influence the heating process. Phase changes and chemical reactions have been accounted for by altering the value of the material properties. Milke addressed the evaporization of free water in a spray-applied cementitious material by increasing the specific heat in a narrow temperature region around 100°C (212°F).[39]

Agreement between the predicted and experimental average steel temperatures is quite good in both applications of FIRES-T3 by Jeanes and Milke. A comparison of the temperature history for a steel column protected with a spray-applied cementitious material subjected to the ASTM E-119 test is presented in Figure 4-9.24. A similar comparison is presented in Figure 4-9.25 for steel beams protected with the same material.[4]

FIRES-T3 has also been used to conduct a preliminary analysis of the heating of partially protected steel columns, i.e., where a portion of the spray-applied protection is missing.[57] The analysis indicated that even a small portion

of missing protection significantly decreased the fire resistance of the column, especially for cases involving small columns. Results of the analysis are indicated in Figure 4-9.26.

STRUCTURAL ANALYSES

The structural analysis methods calculate one of three parameters: deflection, critical temperature, or critical load. In several of the methods, all three of the parameters may be considered, since they are interrelated. Algebraic equations, graphs, and computer programs are available to perform a structural analysis for the purpose of addressing fire resistance.

General Discussion of Three Parameters Addressed in Structural Analysis

Deflection: The total deflection and rate of deflection can be calculated for loaded and heated steel beams by considering all sources of strain. The total strain comprises components of the elastic and/or plastic strains due to the applied loads, thermal strain (due to thermal expansion), and creep strain.

The calculated deflection and rate of deflection can be compared with established maximum limits of each. The Robertson-Ryan criteria have been widely accepted for this

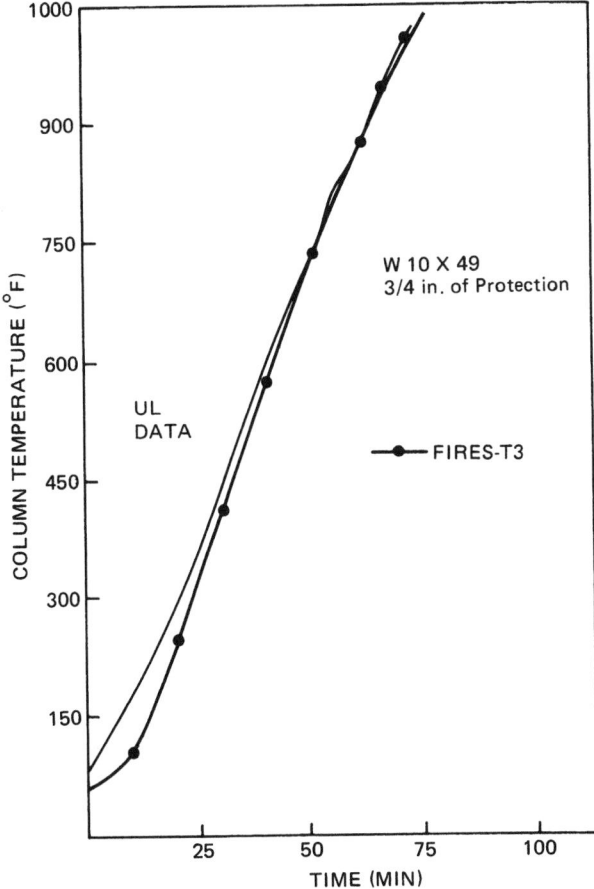

Fig. 4-9.24. Comparison of predicted and measured average steel column temperature.[37]

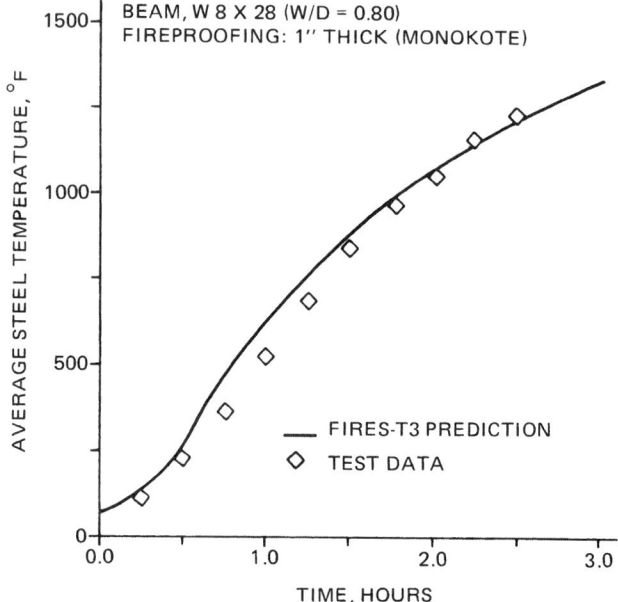

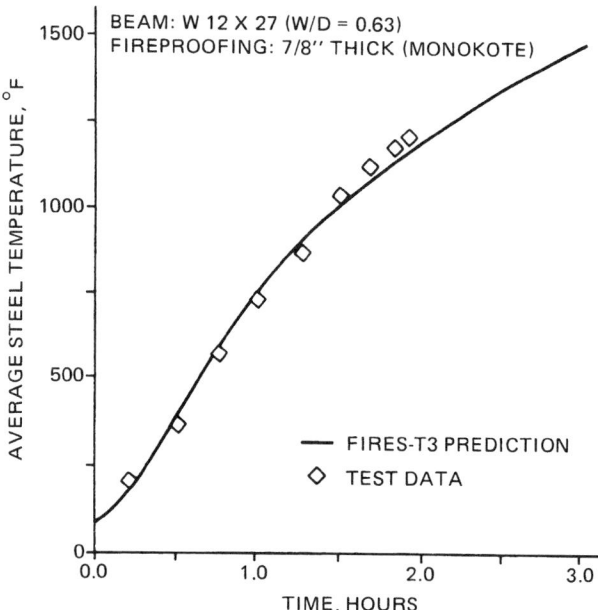

Fig. 4-9.25. *Comparison of experimental data and FIRES-T3 analysis.*[4]

purpose.[11,42,43] However, calculation of the deflection of unheated beams is difficult except for simple loadings, geometries, and end conditions. Adding the thermal expansion and creep components further complicates the calculation, virtually requiring computer solution.

Critical temperature: As mentioned earlier in the chapter, the material properties of steel change with increasing temperature. The two most important material properties for critical temperature calculations are yield strength and modulus of elasticity. The critical temperature is defined as the temperature at which the material properties have decreased

to the extent that the steel structural member is no longer capable of carrying a specified load or stress level. In this context, the factor of safety of the member is considered to be reduced if the member reaches unacceptable stress levels, buckling becomes imminent, or deflections exceed maximum limits. The critical temperature can be calculated as long as the dependence of the material properties with temperature is known. Numerous algebraic equations to calculate the critical temperature of steel structural members are described later.

Critical load: The critical load is defined as the minimum applied load that will result in failure if the structural member is heated to a temperature, T. The critical load can be expressed as a point load or distributed load. As with critical temperature, the critical load calculation requires a knowledge of the temperature dependence of the material properties of steel. Critical load calculations can be conducted with algebraic equations, by a trial and error approach, or with a computer program.

The critical load and critical temperature are closely related. In the case of critical load, temperature is the independent variable and load is the dependent variable. Conversely, with critical temperature, the loading is the independent variable and temperature is the dependent variable.

ALGEBRAIC EQUATIONS

Critical Temperature

There are two approaches for determining the critical temperature of a steel structural member, using algebraic equations. First, numerous equations are available to directly calculate the critical temperature of steel structural members. The second approach calculates a critical temperature based on an allowable reduction in material properties (yield stress or modulus of elasticity) compared to the temperature-related reduction in material properties.

The algebraic equations including the direct solution of the critical temperature have been developed for use in Switzerland.[28] These unique equations address conditions specific to Swiss conditions, including steel type, material property temperature relationships, assumed stress levels, and deflection limits.

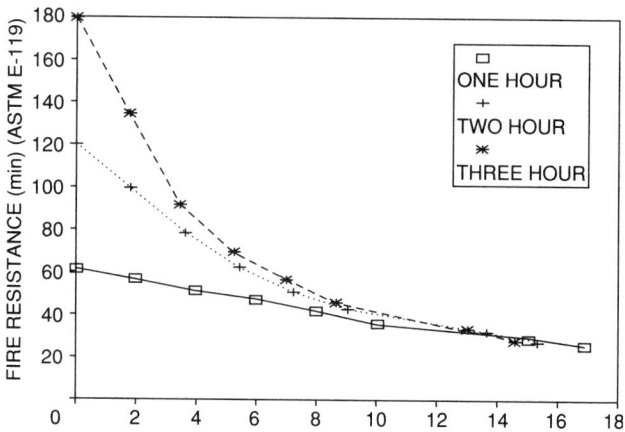

Fig. 4-9.26. *Fire resistance vs percent protection loss for W 10 × 49 column, flange exposure.*

TABLE 4-9.10 *Critical Temperature of Steel Beams*[45]

			factor $\frac{kq^*}{q_e}$ resp. $\frac{kq^*}{q_p}$				
		Static System	0.3	0.4	0.5	0.6	0.7
Base of structural design at room temperature	Theory of plasticity	Statically determinate Statically indeterminate	585	540	490	430	360
	Theory of elasticity	Statically determinate	605	565	525	475	425
		Statically indeterminate					
		$\Theta = 1.33$	640	605	575	545	510
		$\Theta = 1.0$	605	565	525	475	425
		$\Theta = 1.47$	650	615	590	560	535
		$\Theta = 1.12$	615	580	545	505	465
		$\Theta = 1.47$	650	615	590	560	535

q_p = Ultimate plastic load
q^* = Applied load
k = Load multiplier
Θ = Factor addressing plastic reserve of beam from redistribution of moments.

Beams

Representative of the second approach are equations by Lie and Stanzak, based on Grade 250 steel, with an allowable stress of 20,000 psi (138 MPa).[22] The Lie and Stanzak equations account for creep strain. Joists, trusses, and beams are all assumed to be simply supported and thermally unrestrained.

The approaches taken by Malhotra,[12] Vinnakota,[43] and Kruppa[44] are similar. Differences in the percent reduction in yield stress or modulus of elasticity are related to design method (elastic or plastic), factor of safety, and end conditions. Equations for the ratio of yield stress at elevated temperature with yield stress at ordinary room temperature are presented in Table 4-9.9. Typical values of Z/S are between 1.13 and 1.15 for I sections,[12] and 1.5 for rectangular sections.

Another example of the second approach is the analysis of the critical temperature of beams by European Convention for Constructional Steelwork (ECCS).[45] The ECCS guide addresses the maximum allowable reduction in yield strength by considering the applied loading, beam geometry, structural end conditions, and whether the applied loading results in stresses in the elastic or plastic range. Critical temperature calculations based on the ECCS analysis are presented in Table 4-9.10.

Columns

Lie and Stanzak calculated a critical temperature of 941°F (505°C) for slender, axially loaded columns.[22] The calculation was based on the temperature for the onset of elastic buckling for columns under maximum permissible applied stress conditions.

The Euler buckling stress at which elastic buckling is imminent is given by

$$\sigma_{cr} = \frac{\pi^2 E_T}{\lambda^2} \tag{6}$$

where

σ_{cr} = Euler buckling stress (psi)(MPa),
E_T = modulus of elasticity at temperature T (psi)(MPa),
λ = slenderness ratio ℓ/r,
r = radius of gyration (ft)(m), and
ℓ = effective length of column (ft)(m).

If the applied stress is equal to the maximum allowed,[44]

$$\sigma_a = \frac{\pi^2 E_0}{1.92\lambda^2}$$

where

σ_a = maximum allowable stress (psi)(MPa), and
E_0 = modulus of elasticity at 20°C (68°F) (psi)(MPa),

assuming:
$E_T = E_0 (1 - 2.04\theta^2)$.

Equating σ_{cr} and θ is determined as 0.486, which relates to a temperature of 505°C (941°F). In general, for any factor of safety, f,

$$\theta = \sqrt{\frac{f-1}{2.04f}} \tag{7}$$

Included in the ECCS guide[45] are dimensionless buckling curves for steel columns at elevated temperatures. These curves are presented in Figure 4-9.27.

Equation 6 is only valid for columns that buckle in the elastic range. Generally, slender columns having a slenderness ratio in excess of 100 can be expected to buckle elastically. Buckling stresses for stout columns (slenderness ratio

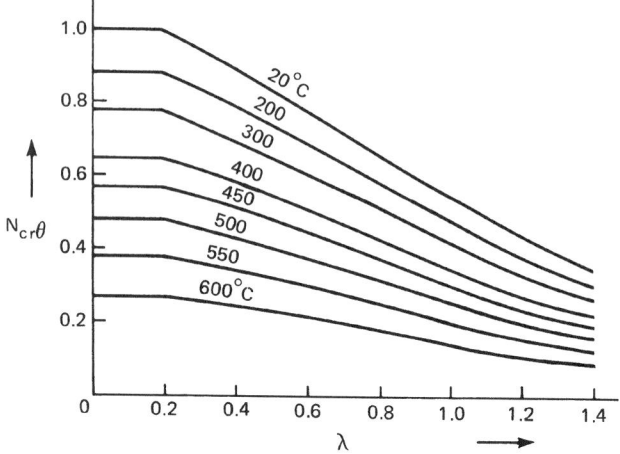

Fig. 4-9.27. *Dimensionless buckling curves for steel columns.*[45]

less than 80) are in the plastic range, requiring a more complex analysis. The failure mode for columns with a slenderness ratio between 80 and 100 cannot be reliably predicted.[47] The tangent modulus can be used instead of the modulus of elasticity in Equation 6 for stout columns. However, predictions of the critical temperature using Equation 6 may not be accurate due to characteristics of the steel fabrication process.[47] Thus, for stout columns, a conservative estimate for the critical temperature of steel columns may be obtained by determining the temperature at which the yield stress is equal to the applied stress.

If the applied stress is

$$\frac{\sigma_{y0}}{1.92}, \text{ and}$$

$$\sigma_y = \sigma_{y0}(1 - 0.78\theta - 1.89\theta^4)$$

then the allowable reduction in the yield stress is

$$\frac{\sigma_{yT}}{\sigma_{y0}} = 1/1.92$$

Solving for θ:

$$\theta = 0.48, \quad \text{and}$$

$$T = 932°F \ (500°C)$$

General

Malhotra has observed that critical temperatures determined from the structural analysis algebraic equations will be somewhat low when compared to experimental data.[12] Thus, the following correction factors, V, are suggested by Malhotra to improve the prediction capabilities of the approach:

1. Columns: $V = 0.85$
2. Statically determinate beams: $V = 0.77 + 0.15\ P_s/P_u$
3. Statically indeterminate beams: $V = 0.25 + 0.77\ P_s/P_u$

where

P_s = service (applied) load (N or N/m) (lb or lb/ft), and
P_u = load to induce ultimate stress at midspan (N or N/m) (lb or lb/ft).

Determine the critical temperature of a simply supported, 15 ft long, W 8 × 28 steel column based on the approach by Lie and Stanzak. The applied load is 12,000 psi. Assume the yield stress is 36,000 psi and the modulus of elasticity is 30,000,000 psi. The characteristics of the column are

$$A = 8.23 \text{ in.}^2$$

$$I = 21.6 \text{ in.}^4$$

SOLUTION:

Calculate the slenderness ratio to determine the failure mode.

$$\lambda = \frac{\ell}{\sqrt{I/A}} = \frac{12 \times 15}{\sqrt{21.6/8.23}} = 111$$

Since 111 > 100, the column may be considered "slender" in Lie and Stanzak's approach.

Stanzak and Lie

$$\sigma_a = \frac{E_T \pi^2}{\lambda^2}$$

$$12000 = \frac{30 \times 10^6(1 - 2.04\theta^2)\pi^2}{111^2}$$

$$\theta = 0.495$$
$$T = 960°F$$

EXAMPLE 6:

Determine the critical temperature of a simply supported W 12 × 26 steel beam supporting a 53 in. thick rectangular slab. The applied moment is 41,750 ft-lb. The rectangular slab is 8 ft wide. The section properties of the beam are

$$S = 33.4 \text{ in.}^3$$

$$I = 204 \text{ in.}^4$$

Assume $\sigma_y = 36,000$ psi.

SOLUTION:

Using Lie and Stanzak's equation for a beam,

$$T_{cr} = \frac{70,000}{45.62 - 4.23\dfrac{I_d}{I}} - 460$$

$$I_d = \frac{3^3 \times 96}{12} = 216 \text{ in.}^4$$

$$T = \frac{70,000}{45.62 - 4.23\left(\dfrac{216}{204}\right)} - 460 = 1,240°F$$

Critical Stress

Columns: Sample expressions for determining the critical stress for steel columns[22] are noted below.

$$P_{cr}^2 - P_{cr}\left[\sigma_{yT} + \pi^2 E_T\left(4.8 \times 10^{-5} + \frac{1}{\lambda^2}\right)\right] + \sigma_{yT}A\frac{\pi^2 E_T}{\lambda^2} = 0$$

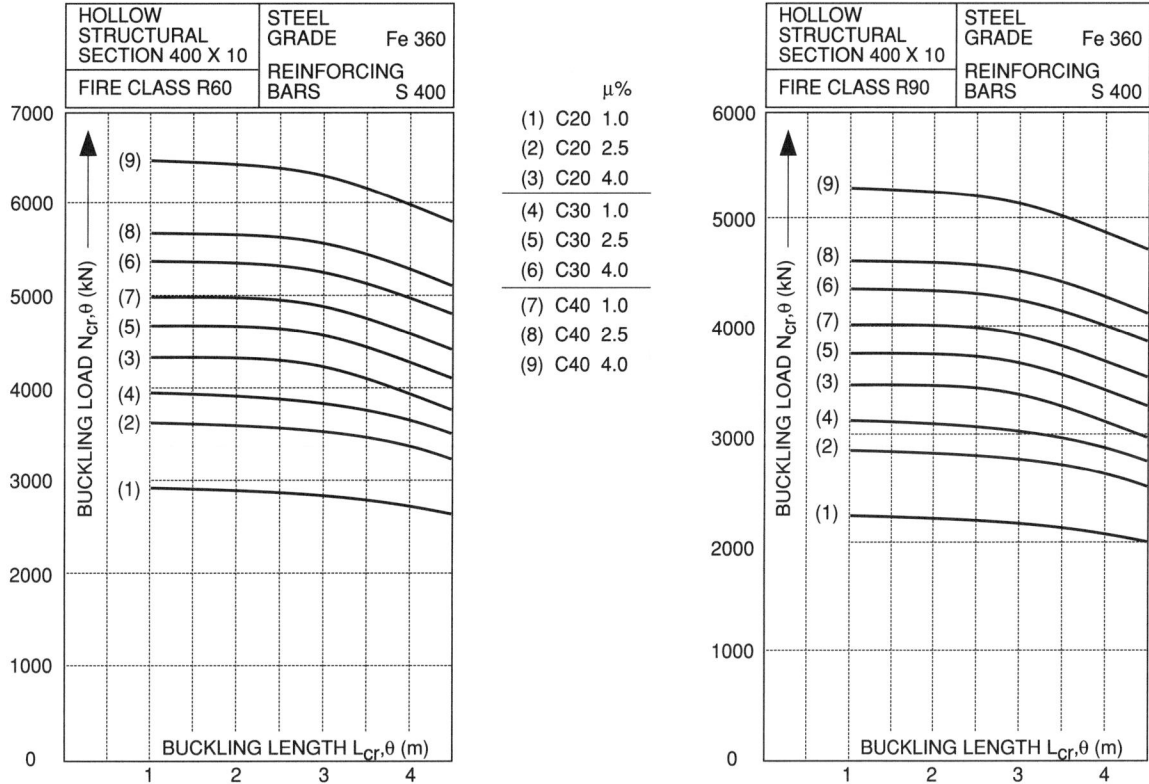

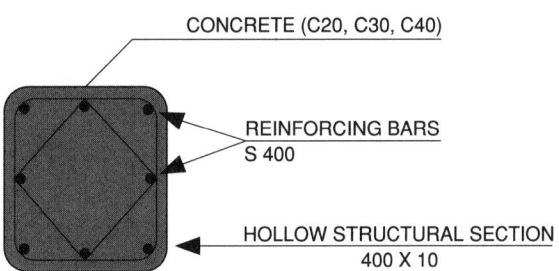

Fig. 4-9.28. *Design graphs for ISO fire resistance requirements R60 and R90. For the concrete-filled square hollow structural section 400 × 10, the axial buckling load is a function of the buckling length, of the concrete quality, and of the percentage* μ *of reinforcement; this design diagram is based on a simple calculation model.*[53]

where

P_{cr} = critical point load (N)(lb),
σ_{yT} = yield stress at temperature T (MPa) (psi),
E_T = modulus of elasticity at temperature T (MPa) (psi), and
λ = slenderness ratio.

In order to improve the prediction capabilities of the critical stress approach for slender columns, the modulus of elasticity should be replaced by the reduced modulus of elasticity.[7] The reduced modulus is defined as

$$E_r = \frac{4EE_T}{(\sqrt{E} + \sqrt{E_t})^2}$$

where

E_t = tangent modulus.

In addition, the 0.2 percent proof stress may be replaced by the 0.5 percent proof stress in the yield stress parameter.[46]

Results of a buckling analysis on concrete-filled square hollow sections are provided in Figure 4-9.28.[53]

Beams: The expressions for the critical loads for beams assume at failure that the beam is in a state of full plasticity at the location of the maximum moment.[46] Obviously, in order to calculate the critical stress, the material property-temperature relationships must be known.

The critical distributed load for a simply supported beam is[43]

$$q_{cr} = \frac{8\sigma_{yT}Z}{L^2}$$

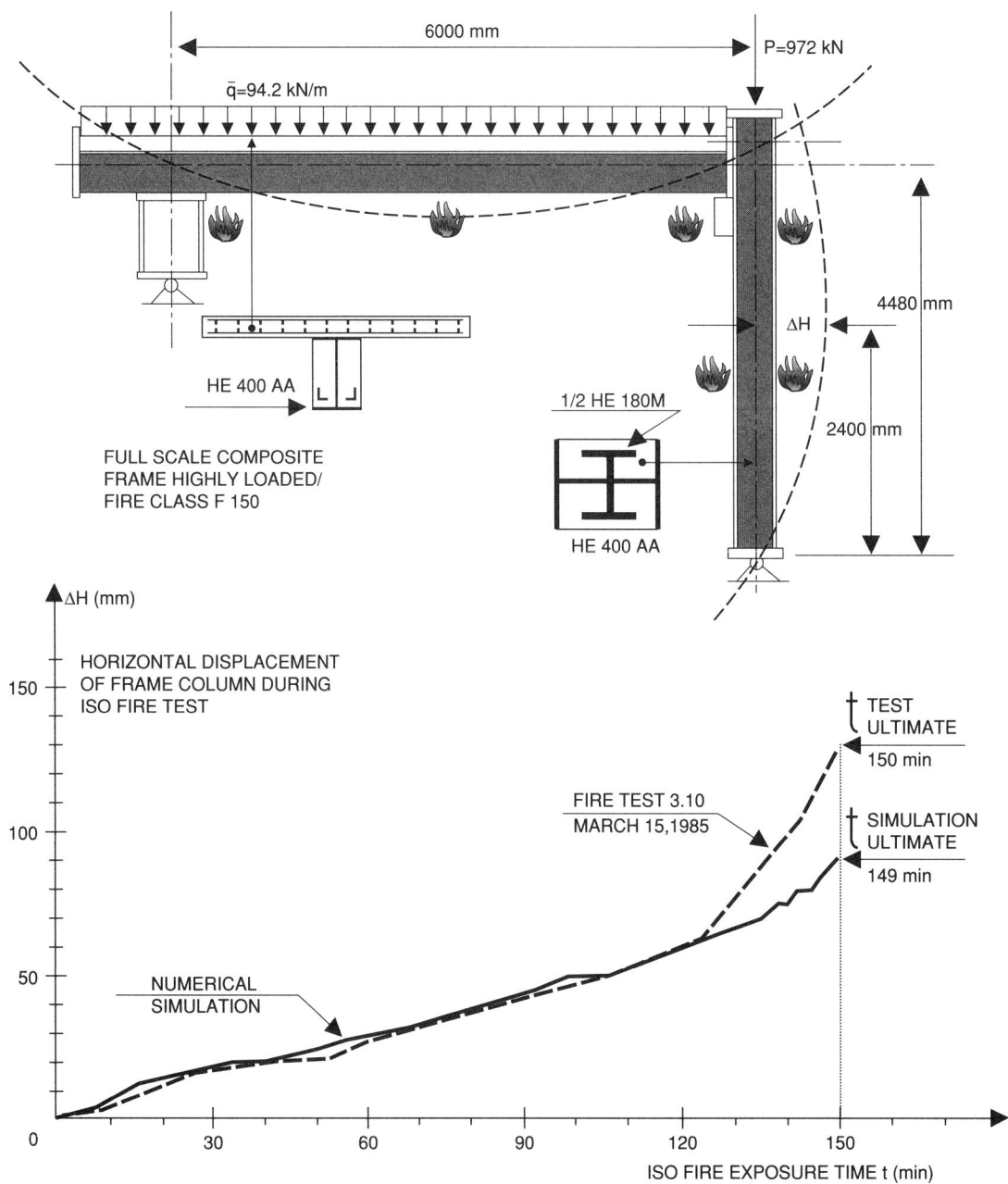

Fig. 4-9.29. *Deformations measured and calculated by a numerical model for a composite frame.*[53]

where

q_{cr} = critical distributed load (N/m) (lb/ft),
Z = plastic section modulus (m³) (in.³),
L = span of beam (m) (ft), and
σ_{yT} = yield stress at elevated temperature (MPa) (psi).

Considering a cantilever beam with a point load applied one-third of the span from the fixed end, plastic hinges can be expected at the point of load application and at the fixed end. The critical load can be determined by

$$p_{cr} = \frac{7.5\sigma_{yT}Z}{L}$$

The above equations in this section do not account for creep strain. Based on an analysis of the deflection history of heated, loaded beams, Pettersson *et al*[30] include a load ratio, β, to determine the critical distributed stress.[30]

$$q_{cr} = \beta\frac{8\sigma_{yT}Z}{L^2}$$

where the yield stress is evaluated at ordinary room temperature, relaxing the need to know the yield stress-temperature relations. β is defined as the ratio of the load causing a maximum allowable deflection under fire conditions to the load inducing stresses equal to the yield stress at ordinary

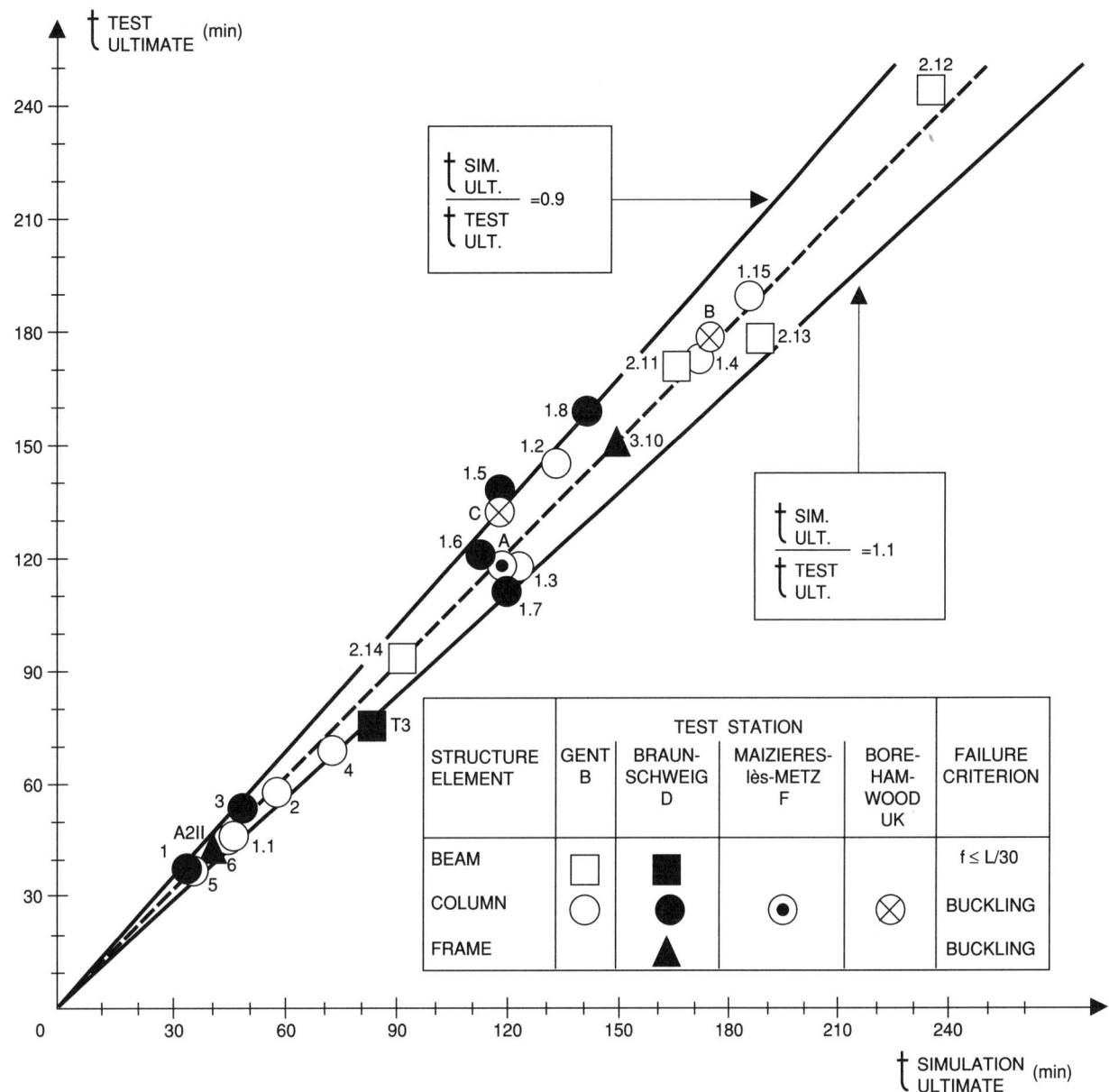

***Fig. 4-9.30.** Fire resistance times measured and calculated by a numerical model for columns, beams, or frames of any cross-section types (bare steel, protected steel, composite).[53]*

room temperature. Thus, the parameter β takes into account the dependence of both the yield stress and creep on temperature. Graphs of β are available for a variety of thermal restraint and structural end conditions.

Computer Programs

Both Pettersson *et al*[30] and Harmathy[13] describe the basis for a computer program to iteratively calculate the deflection history of a beam. Creep, thermal, and elastic strains are all considered in the models. Redistribution of the loads to the supporting deck can also be considered.

Required input information for these models include: material property-temperature relationships, applied loads, a description of the geometry, and end conditions.

FIRE DESIGN by Anderberg interfaces with TEMPCALC to evaluate the structural response of heated concrete or steel beams and columns. The analysis uses a finite element approach to determine the load-bearing capacity of the structural member.

One other program that may be used to estimate the fire resistance of steel-framed floor systems is FASBUS II.[48] FASBUS II is a nonlinear finite element program. The model uses an iterative procedure to calculate the displacement, rotations, and stress and strain distributions in the floor system. The model accounts for material property changes and thermal expansion.

Input information for the model consists of:

1. Applied loads,

2. Thermal restraint conditions,
3. Structural end conditions,
4. Temperature profile in floor system at selected time intervals,
5. Material property values over the temperature range of interest (e.g., yield stress, modulus of elasticity, coefficient of thermal expansion), and
6. Geometric description of structural members, deck, and floor slab.

FASBUS II has the following limitations according to Jeanes:[48]

1. The material models do not provide for "cool down" conditions; therefore, if the assembly is modeled into the fire decay period, original material properties will be effectively restored;
2. The beam element model does not recognize the member to have any lateral resistance to applied forces—the beam is assumed to have continuous lateral support;
3. The stresses (strains) applied to a slab element at the node are distributed over the entire element, and, therefore, areas of high stress concentrations which could result in localized cracking and crushing of the slab are effectively distributed over the model element; and
4. The vertical frame of the structure is assumed to continuously support the floor assembly throughout the exposure period.

In addition, convergence of the output is very sensitive to the time-step and element mesh. Thus, care must be exercised in selecting an appropriate time step for the element mesh selected.

The output of FASBUS II consists of the deflection, rotation, and strain in each element comprising the beam, slab, and reinforcing steel at each time step. This information can be compared to critical limits of each output parameter. The fire endurance of the assembly is defined as the first time step at which any of the output parameters exceed a critical limit.

Traditionally, analysis of the response of the structure exposed to fire has been limited to an analysis of the response of single members. However, in structural frames comprising many members, load transfer often occurs. Load transfer allows stronger members to support additional loads not capable of being carried by heated, weak members. In order to capture this phenomenon, a frame analysis is required.[54] Numerous software packages are available to conduct the frame analysis. Results of a frame analysis are presented in Figures 4-9.29 and 4-9.30.[53]

REFERENCES CITED

1. ASTM E-119-88, *Standard Test Methods for Fire Tests of Building Construction and Materials*, American Society for Testing and Materials, Philadelphia (1988).
2. R.W. Bletzacker, *Effect of Structural Restraint on the Fire Resistance of Protected Steel Beam Floor and Roof Assemblies*, Ohio State Univ., Columbus (1966).
3. D. Boring, J. Spence, and W. Wells, *Fire Protection Through Modern Building Codes*, American Iron and Steel Institute, Washington (1981).
4. D.C. Jeanes, *Technical Report 84-1*, Society of Fire Protection Engineers, Boston (1984).
5. M.S. Abrams, *ASTM STP 685*, American Society for Testing and Materials, Philadelphia (1979).
6. T.Z. Harmathy, *NRCC 20956 (DBR Paper No. 1080)*, National Research Council of Canada, Ottawa (1983).
7. T.T. Lie, *Fire and Buildings*, Applied Science, London (1972).
8. T.T. Lie and W.W. Stanzak, *Eng. J.*, 57, 5/6 (1974).
9. D.F. Boring, *An Analytical Evaluation of the Structural Response of Simply Supported, Thermally Unrestrained Structural Steel Beams Exposed to the Standard Fire Endurance Test*, Master's Thesis, Ohio State University, Columbus (1979).
10. R.A. Lindberg, *Processes and Materials of Manufacture*, Allyn and Bacon, Inc., Boston, p. 46 (1978).
11. D.C. Jeanes, *Methods of Calculating Fire Resistance of Steel Structures*, Engineering Applications of Fire Technology Workshop, SFPE, Boston (1980).
12. H.L. Malhotra, *Design of Fire-Resisting Structures*, Chapman and Hall (1982).
13. T.Z. Harmathy, *ASME J. of Basic Eng.*, 89 (1967).
14. T.Z. Harmathy, *ASTM STP 422*, American Society for Testing and Materials, Philadelphia (1967).
15. *Fire Resistant Steel Frame Construction*, American Iron and Steel Institute, Washington (1974).
16. *Designing Fire Protection for Steel Columns*, American Iron and Steel Institute, Washington (1980).
17. W.W. Stanzak and T.T. Lie, *Fire Tests on Protected Steel Columns with Different Cross-Sections*, National Research Council of Canada, Ottawa (1973).
18. PABCO, *Pabco Super Firetemp Fireproofing Board Fire Protection Guide*, PABCO, Ruston, Louisiana (1984).
19. L.G. Seigel, *Fire Tech.*, 6 (1970).
20. L.G. Seigel, *Matls. Res. and Standards*, 4, (February 1970).
21. Standard Building Code Congress, *Southern Standard Building Code*, SSBC, Birmingham (1985).
22. T.T. Lie and W.W. Stanzak, *Eng. J. Amer. Inst. Steel Const.*, 3rd Qtr., (1973).
23. *Designing Fire Protection for Steel Beams*, American Iron and Steel Institute, Washington (1985).
24. *Manual of Steel Construction*, American Institute of Steel Construction, New York (1981).
25. *Fire Resistance Directory*, Underwriters Laboratories, Northbrook (1994).
26. *Designing Fire Protection for Steel Trusses*, American Iron and Steel Institute, Washington (1980).
27. *Fire Resistance Design Manual*, Gypsum Association, Evanston (1984).
28. W.W. Stanzak, trans., *Technical Trans. 1425*, National Research Council of Canada, Ottawa (1971).
29. W.W. Stanzak and T.Z. Harmathy, *Fire Tech.*, 4, 4 (1968).
30. O. Pettersson, S. Magnusson, and J. Thor, *Bulletin 52*, Lund Institute of Technology, Lund (1976).
31. *Fire-Safe Structural Steel, A Design Guide*, American Iron and Steel Institute, Washington (1979).
32. M. Law, *AISC Eng. J.*, 2nd Qtr., (1978).
33. G.V.L. Bond, *Fire and Steel Construction—Water Cooled Hollow Columns*, Constrado (1974).
34. T.T. Lie and T.Z. Harmathy, *Fire Study No. 28*, National Research Council of Canada, Ottawa (1972).
35. O.C. Zienkewicz, *The Finite Element Method*, McGraw-Hill, New York (1983).
36. M. Paulsson, *TASEF-2*, Lund Institute of Technology, Lund (1983).
37. R.H. Iding, Z. Nizamuddin, and B. Bresler, *UCB FRB 77-15*, University of California, Berkeley (1977).
38. A. Anderberg, *PC-TEMPCALC*, Institutet for Brandtekniska Fragor, Sweden (1985).
39. J.A. Milke, *Estimating Fire Resistance of Tubular Steel Columns*, Proceedings of Symposium on Hollow Structural Sections in Building Construction, ASCE, Chicago (1985).
40. K. Bardell, *ASTM STP 826*, American Society for Testing and Materials, Philadelphia (1983).
41. D. Gross, *NBSIR 85-3223*, National Bureau of Standards, Gaithersburg (1985).
42. A.F. Robertson and J.V. Ryan, *J. of Res.*, 63C, 2 (1959).

43. S. Vinnakota, *Calculation of the Fire Resistance of Structural Steel Members*, ASCE Spring Meeting (1978).
44. J. Kruppa, *J. of Struc. Div.*, ASCE, 105 (1979).
45. European Convention for Constructional Steelwork, Technical Committee 3, *European Recommendations for the Fire Safety of Steel Structures*, Elsevier, Amsterdam (1983).
46. T.T. Lie and W.W. Stanzak, *AISC Eng. J.*, 13, 2 (1976).
47. A. Chajes, *Principles of Structural Stability Theory*, Prentice-Hall, Englewood Cliffs (1974).
48. D.C. Jeanes, *F. Safety J.*, 9, 1 (1985).
49. NFPA 251, *Standard Methods of Fire Tests of Building Construction and Materials*, National Fire Protection Association, Quincy (1990).
50. UL 263, *Fire Tests of Building Construction and Materials*, Underwriters Laboratories, Northbrook (1992).
51. *Specification for the Design, Fabrication, and Erection of Structural Steel for Buildings*, American Institute of Steel Construction, New York (1978).
52. T.T. Lie, (ed.), *Structural Fire Protection*, American Society of Civil Engineers, New York (1992).
53. *International Fire Engineering Design for Steel Structures: State of the Art*, International Iron and Steel Institute, Brussels (1993).
54. W.L. Gamble, "Predicting Protected Steel Member Fire Endurance Using Spreadsheet Programs," *Fire Technology*, 25, 3, pp. 256–273 (1989).
55. G.S. Berger, "Estimating the Temperature Response of Wide Flange Steel Columns in the ASTM E-119 Test," Department of Fire Protection Engineering, University of Maryland, College Park, 1987 (unpublished).
56. R. Stubblefield and M.L. Edwards, "NODES-T3: Making FIRES-T3 a Little Easier," Department of Fire Protection Engineering, University of Maryland, College Park, 1991 (unpublished).
57. D.V. Tomecek and J.A. Milke, "A Study of the Effect of Partial Loss of Protection on the Fire Resistance of Steel Columns," *Fire Technology*, 29, 1, pp. 3–21 (1993).

ANALYTICAL METHODS FOR DETERMINING FIRE RESISTANCE OF CONCRETE MEMBERS

Charles Fleischmann

INTRODUCTION

Analytical methods developed to predict the fire resistance of structural assemblies can be divided into two groups: (1) standard and (2) nonstandard fire exposure. For the case of the standard fire exposure, a large data base exists from referenced standard tests. The analytical methods use empirically based correlations and minimum dimensions to determine the fire resistance. For nonstandard fire exposure, the analysis is more complicated, requiring both heat transfer and structural analyses. Analytical methods are an alternative to conventional methods that require destructive testing of exemplar systems in accordance with standard testing procedures, e.g., ASTM E 119 or ISO 834.

Fire resistance calculations typically use the same acceptance criteria specified in standard test methods, i.e., heat transmission and structural integrity. The analysis can be broadly divided into two parts: (1) heat transfer and (2) structural analysis. Heat transfer calculations are used to evaluate unexposed surface temperature and temperature distribution required to evaluate material strength. The structural integrity analysis applies the strength theory[1] used to design reinforced concrete members. The reduced strength of the concrete and steel resulting from elevated temperature is taken into account by using experimental results for the compressive and yield strengths as a function of temperature. This procedure is known as the rational design method.

As the fire protection field advances into performance-based engineering, techniques like the rational design method are more likely to be used. In the rational design approach, a design time-temperature curve, based on the expected fire, is specified. The engineer then performs the heat transfer analysis to determine the temperature profile and unexposed surface temperature. Knowing the temperature distribution of the member, a structural analysis is conducted to determine the fire endurance.

This chapter presents an overview of the analytical methods for calculating the fire resistance of concrete structural members, and provides a description of the mechanical properties for concrete and steel at elevated temperatures. A brief discussion of heat transfer for a concrete assembly is given, along with temperature profiles from ASTM E 119 test results. The structural calculations for simply supported and continuous members are explained. A simple example is shown to further demonstrate the basics of the design concept. Fire resistance for columns and walls is also presented. A more comprehensive discussion of the rational design method can be found elsewhere.[2,3,4]

MATERIAL PROPERTIES OF CONCRETE AND STEEL

Most of the material properties for concrete and steel change significantly at elevated temperatures. In order to accurately predict the structural fire endurance of concrete members, these changes must be taken into account. Temperature-dependent values of strength, modulus of elasticity, and thermal expansion have been presented in a graphical format to aid in the design process. The thermophysical properties required for a heat transfer analysis, i.e., thermal conductivity and specific heat, also change significantly. The thermophysical properties have been investigated at elevated temperatures by Harmathy and others.[5-9]

Strength

The strength of the reinforcing steel changes significantly with temperature, and must be taken into account in any structural calculation. Figure 4-10.1 shows the strength-temperature relationship for hot-rolled, cold-drawn, and high-strength alloy steels. Yield strength *versus* temperature relationship is given for hot-rolled steel, used for reinforcing bars. Tensile strength *versus* temperature relationship is shown for the cold-drawn steel and high-strength alloy steel, used for prestressing bars, wire, or strands. The change from yield strength to tensile strength for the two steel types relates to the design parameters used for reinforced *versus* prestressed concrete assemblies.

Like steel, the strength of concrete is also diminished at elevated temperatures. Figure 4-10.2 shows the strength-temperature relationship for carbonate, siliceous, and sand-lightweight aggregate concretes. The compressive strength is not only a function of temperature but is also affected by the applied load.

The results shown in Figure 4-10.2 were obtained from specimens loaded to 40 percent of their compressive

Charles Fleischmann is a member of the Department of Civil Engineering, University of Canterbury, in Christchurch, New Zealand.

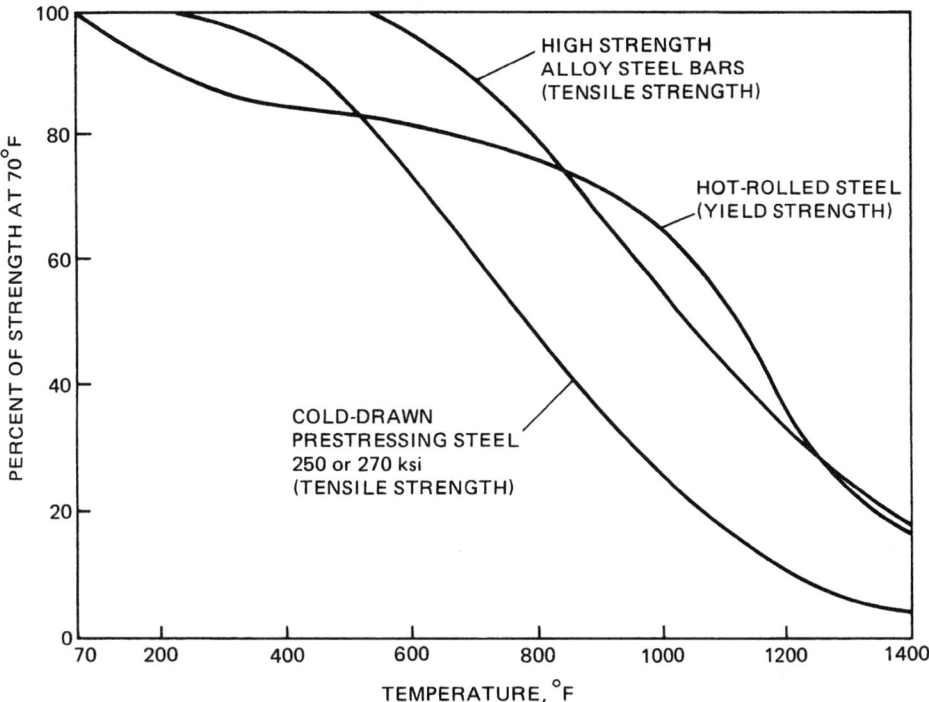

Fig. 4-10.1. *Strength-temperature relationships for hot-rolled, cold-drawn, and high-strength alloy steels.* [9-11]

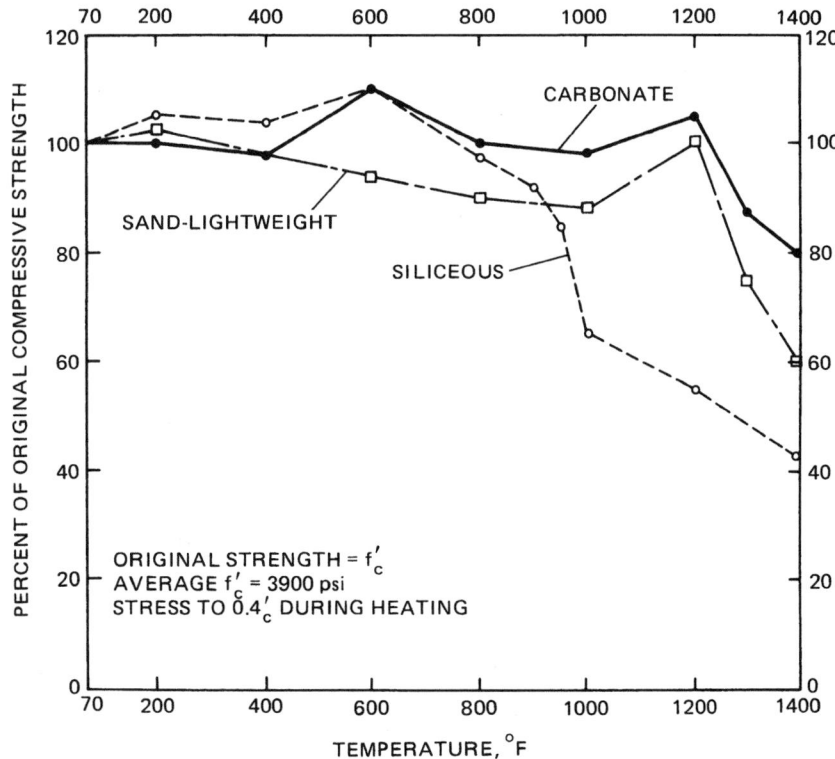

Fig. 4-10.2. *Strength-temperature relationships for carbonate, siliceous, and sand-lightweight aggregate concretes.* [12]

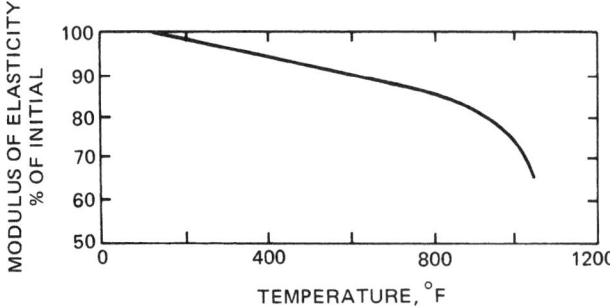

Fig. 4-10.3. *Modulus of elasticity for hot-rolled steel at elevated temperatures.*[13]

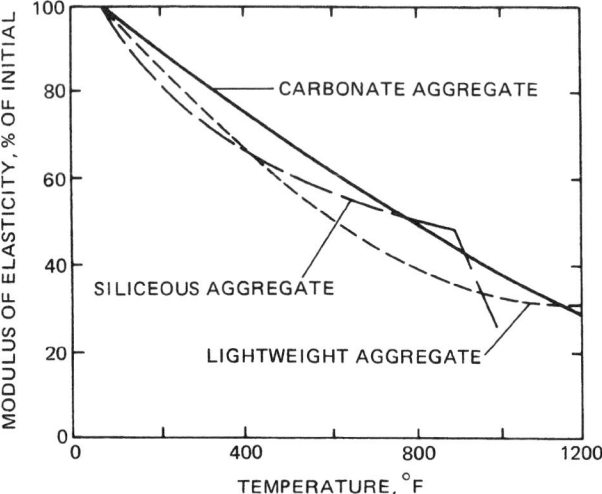

Fig. 4-10.4. *Modulus of elasticity, at elevated temperatures, for carbonate, siliceous, and lightweight concretes.*[14]

strength during the heating process. Once the test temperature of interest was reached, the specimens were loaded to failure. Figure 4-10.2 illustrates that the compressive strength of concrete remains relatively unchanged up to 900°F. Above 900°F, the compressive strength of the siliceous aggregate concrete starts to decrease rapidly and is considered ineffective at temperatures above 1200°F, where the compressive strength has been reduced by approximately 50 percent of the value at normal temperatures. However, for carbonate and lightweight aggregates, compressive strength remains relatively unchanged up to 1200°F and is not considered to be ineffective until it reaches a temperature of 1400°F. The experimental method used may influence the reported compressive strength. Specimens heated without compressive loads and then loaded to failure while hot have lower compressive strengths than those heated while loaded.[3]

Modulus of Elasticity

The modulus of elasticity for steel decreases as the temperature increases, as shown in Figure 4-10.3. Figure 4-10.4 shows the modulus of elasticity-temperature curve for three different concrete aggregates. In each case, the modulus of elasticity of concrete is greatly reduced at elevated temperatures. This large reduction of the elastic modulus is helpful in reducing induced thermal stresses in concrete members due to fire.[3]

Thermal Expansion

Both concrete and steel expand when heated. This thermal expansion can actually increase the fire resistance. The effects of thermal expansion are discussed later in this chapter. Figure 4-10.5 shows the coefficient of thermal expansion for concrete and steel. The values for the steel were calculated using the following equation taken from the *Manual of Steel Construction*.[15]

$$\alpha = (6.1 + 0.0019\theta) \times 10^6 \qquad (1)$$

where:

α = coefficient of thermal expansion (in./in.°F), and
θ = temperature (°F).

HEAT TRANSMISSION

The temperature of the unexposed side of concrete floors, roofs, and walls are usually limited to prevent ignition of combustibles in contact with the unexposed surface. In ASTM E 119 the criteria are 250°F average and 325°F single-point temperatures. These criteria often govern the fire resistance of the assembly. In addition to the unexposed surface temperature, the temperature distribution throughout the member is required in order to evaluate the material strengths in the structural calculations.

The temperature distribution within concrete slabs exposed to the ASTM E 119 time-temperature curve is illustrated in Figures 4-10.6 through 4-10.8. This data applies to any slab thickness, as long as the slab is at least one inch thicker than the point in question.

Heat is mainly transferred through a solid concrete member by conduction. The temperature of the unexposed side of the slab is mainly a function of the slab thickness and the type of aggregate used. The fire endurance *versus* slab thickness is presented in Figure 4-10.6 for three types of concrete typically used in building construction. The data is based on actual fire tests of concrete slabs.[3] For the normal-weight concretes used in the fire tests, the maximum aggregate size was ¾ in. and the air content was about 6 percent. The maximum aggregate size for the structural lightweight concretes was slightly less than ¾ in. and the air content was about 7 percent. Although the slab thickness and type of

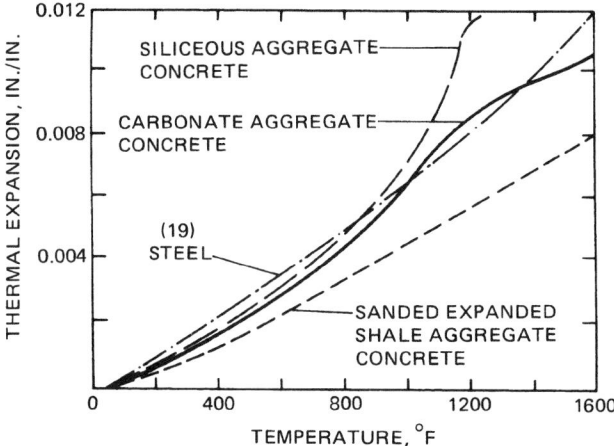

Fig. 4-10.5. *Thermal expansion of concrete and steel at elevated temperatures.*[3]

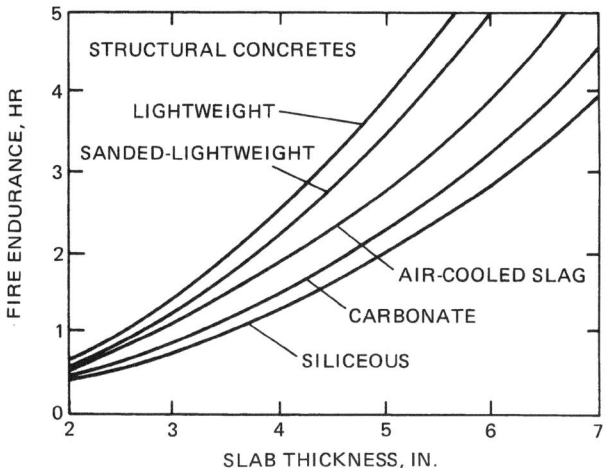

Fig. 4-10.6. *Fire endurance of concrete slabs—effect of thickness and type of aggregate, based on heat transmission.*[3]

aggregate are the main factors that affect heat transmission through the concrete, other factors do have some impact. These factors include: moisture content, unit weight, air content, and maximum aggregate size. Within the usual range of values, water-cement ratio, strengths, and age have been shown to have insignificant effects on the heat transfer process.[3]

Floor and roof slabs are often composites of materials, for example, a concrete base slab with overlays or undercoatings of either insulating materials or other types of concrete. Research has been conducted on two-course composite assemblies. An example of a composite slab of normal and lightweight concrete is shown in Figure 4-10.7. Similar plots for different composite assemblies can be found in reference 16.

The temperature on the unexposed side is not the only temperature of concern. The temperature distribution within the member is used to determine the temperature of the reinforcing or prestressing steel. The temperature of the reinforcing bars is approximately equal to the temperature of the concrete at the level of the center of the bar;[3] i.e., the

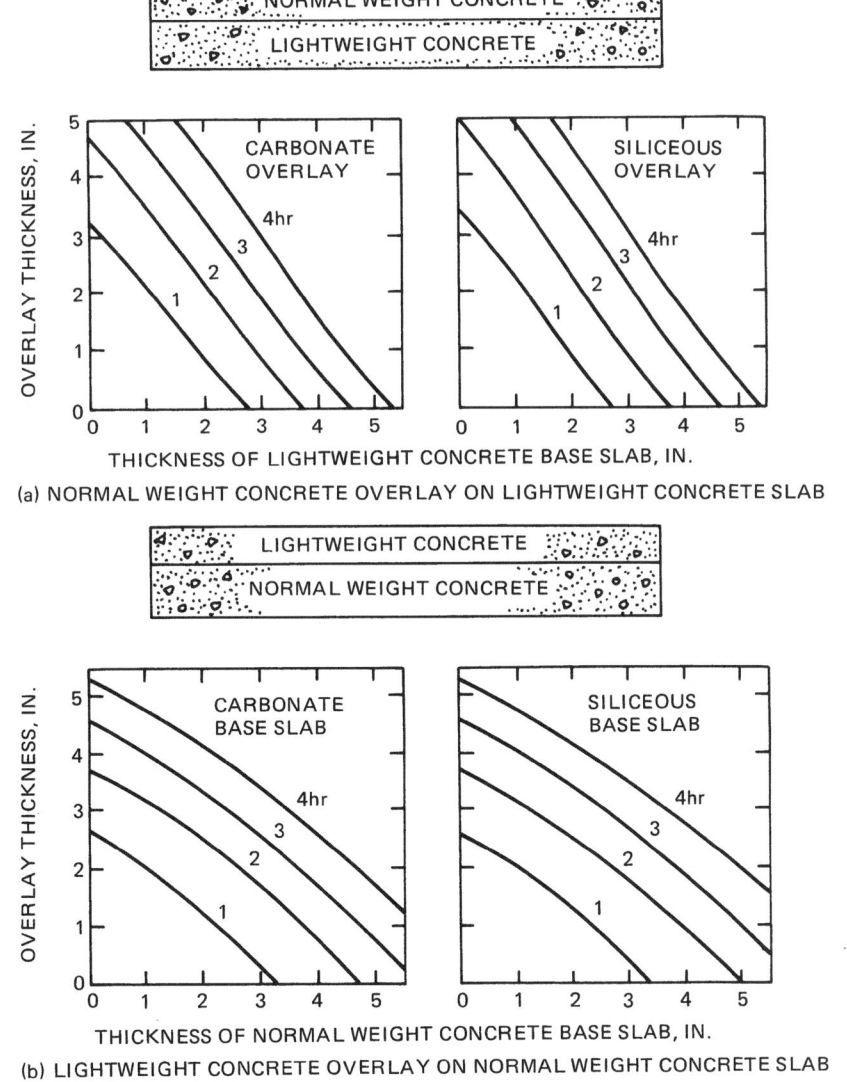

Fig. 4-10.7. *Fire endurance of base slabs and overlays of normal weight or lightweight concretes, based on heat transmission.*[16]

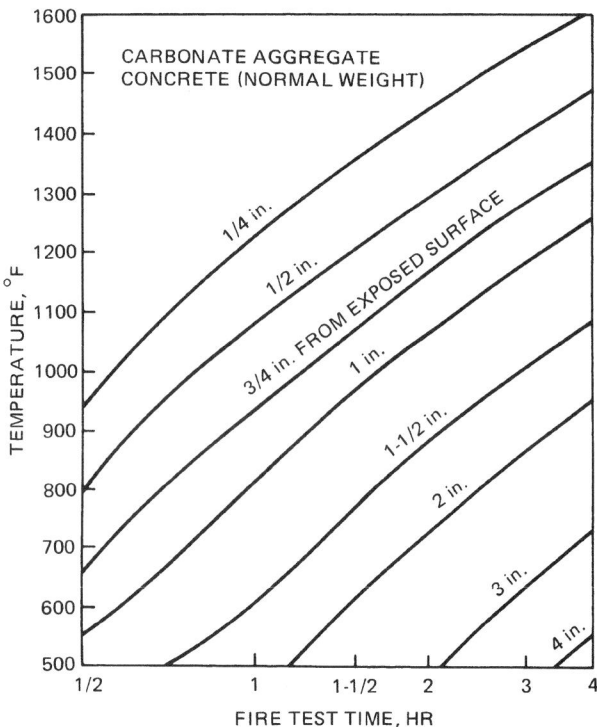

Fig. 4-10.8. *Temperatures within solid or hollow-core concrete slabs during fire tests, carbonate aggregate.*[17]

presence of the steel is neglected in the heat transfer analysis. Thus, temperature distribution is primarily affected by the type of concrete, shape of the member, and exposure conditions. During fire tests, slabs and walls are typically heated on one side only; beams are heated from one, two, or three sides; and columns are heated on all four sides. Data on the temperature distribution within concrete members is available from results of fire tests.

The temperature distribution within a 6- × 12-in. rectangular concrete beam exposed to the ASTM E 119 standard time-temperature curve is illustrated in Figure 4-10.11. Because the temperature distribution is a function of the beam size, it is not practical to present a complete set of figures. A procedure has been developed in which the temperature distribution can be constructed.[3] The procedure, while completely empirical, gives reasonable results.

Computer models exist that can accurately predict the temperature distribution in various types of concrete members. FIRES-T3[18] is one model that can handle one-, two-, or three-dimensional heat flows, time-dependent non-linear boundary conditions, and temperature-dependent thermophysical properties, but it does not accurately analyze materials with latent heat, e.g., humid concrete. TASEF-2,[19] a two-dimensional model, is able to calculate the latent heat and is more accurate for humid concrete, but is limited to two dimensions. Both programs are finite-element based. These programs are somewhat difficult to use and require well-defined thermophysical properties. Use of such models is not necessary for typical analyses assuming a standard time-temperature exposure. When a nonstandard fire is expected, a heat transfer model is required to calculate the temperature distribution within the member.

SIMPLY SUPPORTED SLABS AND BEAMS

Simply supported, unrestrained members are not typically cast in place. However, a discussion of simply supported members will make the discussion of continuous members easier to understand. A simply supported, reinforced concrete slab is illustrated in Figure 4-10.12.

The slab is supported by "frictionless" rollers, so that the slab is free to expand without resistance but should not deflect at the support. The load, w, is evenly distributed over the surface of the slab and the reinforcing steel runs the entire length of the slab. Considering these conditions without a fire, the moment diagram for the slab is illustrated in Figure 4-10.12, part (b). The moment strength of the slab will be constant along the entire length[15]

$$M_n = A_s f_y (d - a/2) \qquad (2)$$

where:

A_s = the area of the reinforcing steel,
f_y = the yield stress of the reinforcing steel,
d = the distance from the extreme compression fiber to the centroid of the reinforcing steel,
a = depth of the equivalent rectangular stress block,[1]

$$a = \frac{A_s f_y}{0.85 f'_c b} \qquad (3)$$

f'_c = compressive strength of the concrete, and
b = width of the beam or slab.

During exposure to a fire, the temperature of the reinforcing steel will increase. As the temperature of the steel increases, the yield strength decreases. (See Figure 4-10.1.) This reduction in the steel strength causes a reduction of the moment strength of the slab[3]

$$M_{n\theta} = A_s f_{y\theta} (d - a_\theta/2) \qquad (4)$$

Where θ denotes the effects of elevated temperature.

The reduced moment strength diagram is shown in Figure 4-10.12, part (c).

With a reduction in the yield stress, $f_{y\theta}$, there is a corresponding reduction in the size of the equivalent stress block, a_θ.[3]

$$a_\theta = \frac{A_s f_{y\theta}}{0.85 f'_c b} \qquad (5)$$

Typically, the temperature at the top of a slab remains relatively unchanged from normal conditions even after 2 hr of fire exposure, since the concrete is a good insulating medium. (See Figures 4-10.8 through 4-10.10.) Thus, the values for f'_c and d are not affected. However, if the temperatures in the compression zone exceed 900°F for a siliceous aggregate or 1200°F for a carbonate aggregate, the concrete compressive strength, f'_c, should be reduced. (See Figure 4-10.2.)

As previously noted, the compressive strength of concrete is reduced significantly at a critical temperature, selected here as 1200°F for a siliceous aggregate or 1400°F for a carbonate aggregate. To account for this substantial reduction in strength, regions of concrete in the compression zone

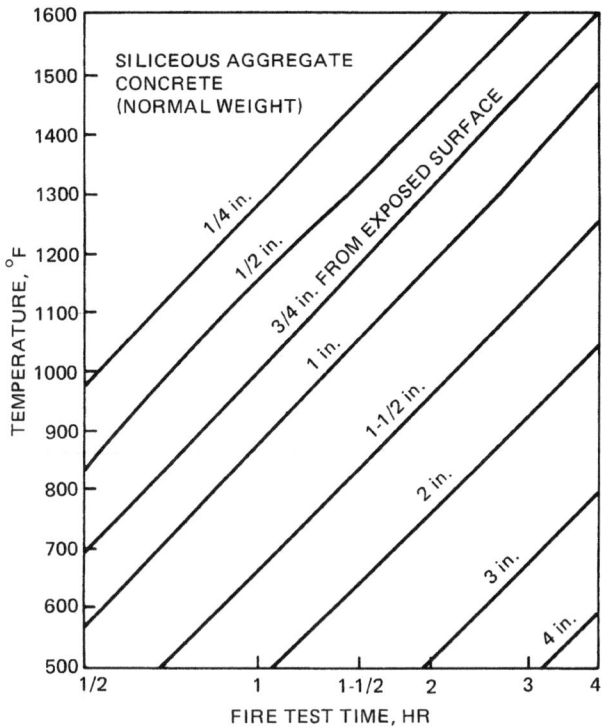

Fig. 4-10.9. *Temperatures within solid or hollow-core concrete slabs during fire tests, siliceous aggregate.*[17]

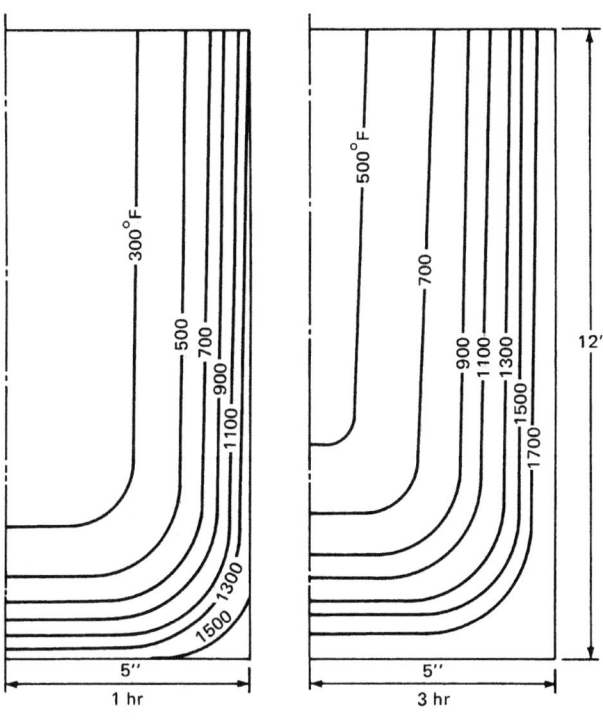

Fig. 4-10.11. *Temperature distribution within a 6- × 12-in. lightweight concrete beam, 1- and 3-hr exposure time.*[3]

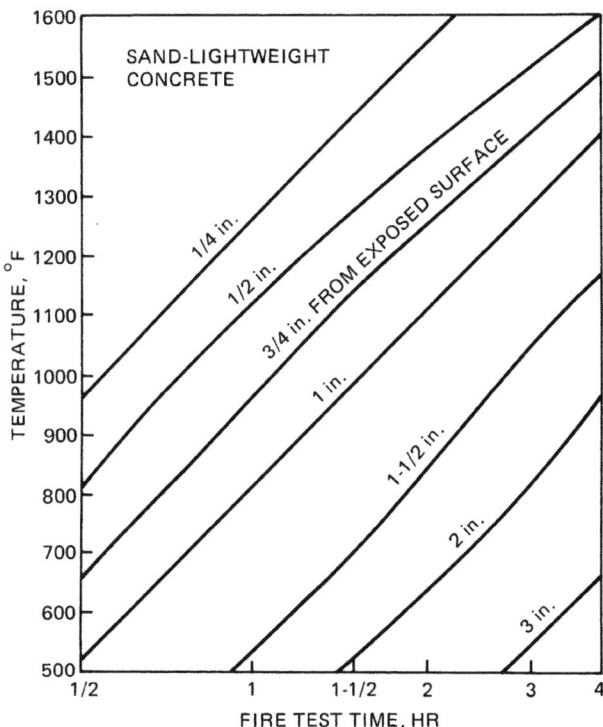

Fig. 4-10.10. *Temperatures within solid or hollow-core slabs during fire tests, sand-lightweight concrete.*[17]

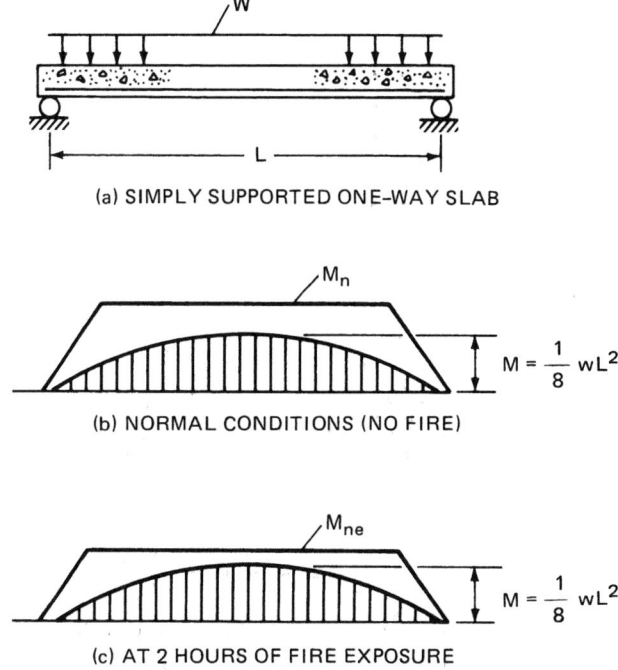

Fig. 4-10.12. *Applied moments and reduced moment strength diagrams for simply-supported one-way slab.*[3]

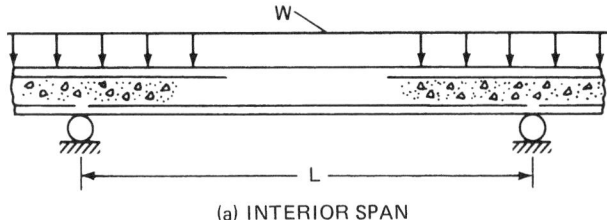

(a) INTERIOR SPAN

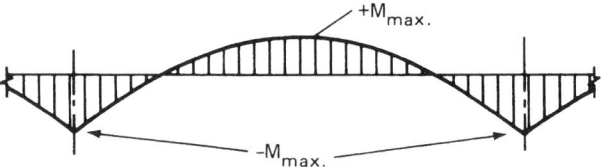

(b) NORMAL CONDITIONS (NO FIRE), APPLIED MOMENTS

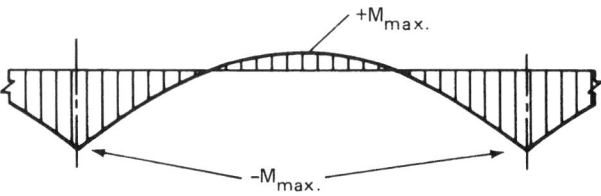

(c) AT 2 HOURS OF FIRE EXPOSURE, APPLIED MOMENTS

Fig. 4-10.13. Moment redistribution in interior span of continuous unrestrained one-way slab due to fire exposure.[3]

at temperatures above the critical points are neglected in the design process. As a result, the depth and/or width of the compression zone are reduced by subtracting the area of the concrete which is heated in excess of the critical temperature.

For a simply supported slab it is unlikely that the compression zone would be heated to above the critical temperature without the steel failing first, but it should be noted that, if the section of concrete is reduced, the value of d in Equation 4 must be adjusted accordingly.

As the slab is being heated from below, the bottom of the slab expands, causing the slab to deflect downward. Flexural failure occurs when the moment strength is reduced to the applied service load moment, M, at the center of the span[3]

$$M = wL^2/8 \qquad (6)$$

where:

M = applied service load moment,
L = length of the span, and
w = applied live load plus dead load, "factor of safety" = 1.0.

The factor of safety used in fire endurance calculations is equal to 1.0, which yields the actual applied moment. The factors of safety used in structural design (1.7 live load, 1.4 dead load, etc.) do not apply to fire endurance calculations.[2]

As indicated in Equation 4, the structural fire endurance of a simply supported one-way slab or beam is a function of the load intensity, strength-temperature characteristics of the reinforcing steel, and the depth of protection given to the reinforcement by the concrete cover. There is no benefit of continuity and/or restraint of thermal expansion with

a simply supported slab, the total slab depth, based on heat transmission, h_t, required to obtain the desired fire rating is probably as small or smaller than the total slab thickness, h_s, required for gravity loads.[3] Therefore, there is no advantage of doing a structural fire endurance analysis for unrestrained, simply supported structural members.[3]

CONTINUOUS UNRESTRAINED FLEXURAL MEMBERS

Continuous unrestrained members have a considerably longer fire endurance than simply supported members because of their ability to redistribute the applied moments. Figure 4-10.13, part (a), shows an interior span of a continuous unrestrained slab. The applied moment diagram for a normal condition, with no fire, is shown in Figure 4-10.13, part (b). The maximum positive moment occurs near the center of the span, and the maximum negative moments are located over the supports.

When the slab is exposed to fire conditions from below, the moments will be redistributed within the slab. This redistribution may be sufficient to cause the negative moment reinforcement to yield. This yielding generally occurs within the first half hour of the fire, based on observation made during standard fire tests.[3] Figure 4-10.13, part (c), shows the redistribution of moments after 2 hr of fire exposure (2 hr was selected at random). The American Concrete Institute (ACI) warns that increasing the negative reinforcement will increase the attracted negative moment, possibly leading to a compressive failure. It is important that flexural tension governs the design of concrete members. Thus, to avoid compressive failure in the negative moment region, the negative reinforcement should be small enough so that[2]

$$\frac{A_s f_{y\theta}}{b_\theta d_\theta f'_{c\theta}} < 0.30 \qquad (7)$$

and Equation 7 should be reduced due to the elevated temperature within the concrete member.

Flexural failure of continuous members occurs when three hinges are formed within a span. One of the hinges will form near the midspan and the other two at the adjacent supports. A hinge is formed at the point where the applied moment is equal to the moment strength at that point.

The moment strength at any point can be calculated using Equation 4 for simply supported members. Figure 4-10.14 shows the moment diagram for a one-way span with unequal end moments, i.e., when the spans are of unequal lengths. This represents the general case and can be used for other conditions, e.g., end spans, and slabs with equal

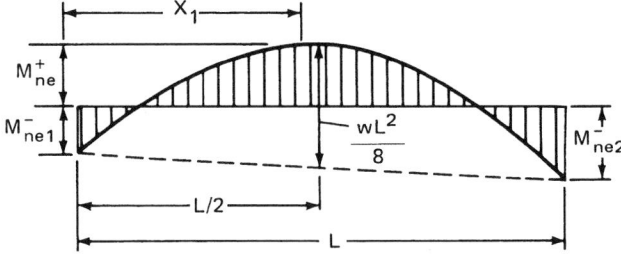

Fig. 4-10.14. Redistributed applied moment diagram at structural endpoint for span of a uniformly loaded continuous one-way slab or beam with unequal end moments.[3]

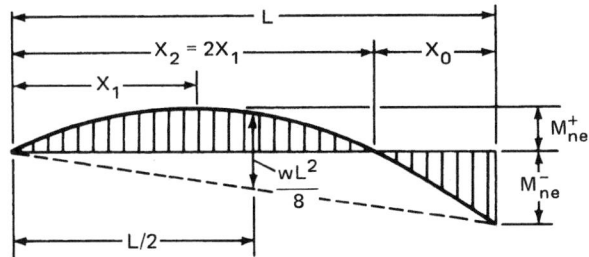

Fig. 4-10.15. Redistributed applied moment diagram at structural endpoint for end span of a continuous one-way slab or beam.[3]

spans. The member fails when the sum of the moment strengths is less than the applied moment, $w L^2/8$. The negative moments are calculated at the supports, and the positive moment strength is calculated at the center of the span. The negative moment strength is then used in the following equation for the minimum positive moment strength[3]

$$\text{Minimum required } M_{n\theta} = \frac{M_{n\theta_1}^- - M_{n\theta_2}^-}{2wL^2}$$
$$- \frac{M_{n\theta_1}^-}{2} - \frac{M_{n\theta_2}^-}{2} + \frac{WL^2}{8} \qquad (8)$$

If the minimum positive moment strength required is less than the positive moment strength, the member has the calculated fire endurance. The location of the maximum positive moment, X_1, is calculated

$$X_1 = L/2 + \frac{(M_{n\theta_1}^- - M_{n\theta_2}^-)}{wL} \qquad (9)$$

End Span

Equations 8 and 9 can be modified and used for the end span of a continuous member. (See Figure 4-10.15.) For the end span, $M_n = 0$, leaving

$$\text{Minimum required } M_{n\theta}^+ = \frac{(M_{n\theta}^-)^2}{2wL^2} - \frac{M_{n\theta}^-}{2} + \frac{wL^2}{8} \qquad (10)$$

$$X_1 = \frac{L}{2} - \frac{M_{n\theta}^-}{wL} \qquad (11)$$

Interior Span with Equal End Moments

Equations 8 and 9 can also be modified for spans with equal end moments, as indicated in Figure 4-10.16. For this case, $M_{n\theta_1}^- = M_{n\theta_2}^-$, changing Equation 8 to

$$\text{Minimum required } M_{n\theta}^+ = \frac{wL^2}{8} - M_{n\theta}^- \qquad (12)$$

Equation 9 becomes

$$X_1 = \frac{L}{2} \qquad (13)$$

The location of the points of inflection, X_0, is dependent on the magnitude of the negative moment strengths and can be calculated using

$$X_0 = \frac{L}{2} - \sqrt{\frac{2M_{n\theta}^+}{w}} \qquad (14)$$

The negative moment reinforcement must be extended a sufficient distance beyond the point of inflection to allow the bar strength to become fully developed. Design criteria for the development length are outlined in the ACI *Building Code Requirements for Reinforced Concrete.*[1] It is further recommended that at least 20 percent of the maximum negative moment reinforcement in the span extends throughout the entire length of the span.

FIRE ENDURANCE OF CONCRETE STRUCTURAL MEMBERS RESTRAINED AGAINST THERMAL EXPANSION

When a fire occurs beneath an interior portion of a floor or roof slab, the heated portion of the slab tends to expand. As this portion of the slab expands, the surrounding cooler portions resist the expansion and exert a resistive force on the heated portion of the slab. This resistive force is referred to as the "thermal thrust force."

Most U.S. fire tests of floor slabs are conducted with the specimen mounted within a restraining frame which restricts the thermal expansion.[20] The amount of restraining force provided by the restraining frame varies from one laboratory to another, based on factors such as frame design, specimen design, and specimen tightness.

Prior to 1960, no research had been conducted to measure the magnitude of the thermal thrust force. In 1960, the Portland Cement Association (PCA) began operation of its floor furnace.[21] This furnace allowed for both variable and monitored restraint during the fire test. Restraining the slab against expansion greatly affects the thermal thrust, as indicated in Figure 4-10.17. Notice that with no expansion allowed, the thermal thrust force would be very high, which would cause compression failure of the concrete. However, with only a slight increase in the allowed expansion, there is a significant decrease in the thermal thrust force. It should also be noted that the thermal thrust force developed in lightweight concrete is considerably less than is developed within normal-weight concrete. This is believed to be due to the lower modulus of elasticity and the lower coefficient of expansion of the lightweight concrete.[3]

As a result of the fire research done at PCA, the thermal thrust force was found to vary with the initial modulus of elasticity and the heated perimeter.[22] The heated perimeter, S, is defined as that portion of the perimeter of a section of the specimen, normal to the direction of the thermal thrust, that is exposed to fire. Having assembled a large data base of "reference specimens," the thermal thrust from these specimens can be used to predict the thermal thrust within a concrete member.[23]

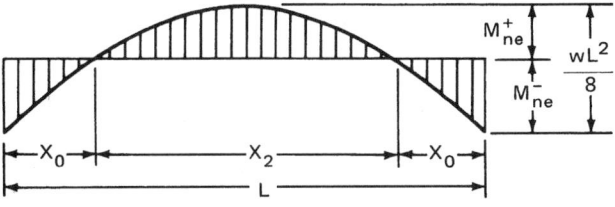

Fig. 4-10.16. Redistributed applied moment diagram at structural endpoint for symmetrical interior span of a uniformly loaded continuous one-way slab or beam.[3]

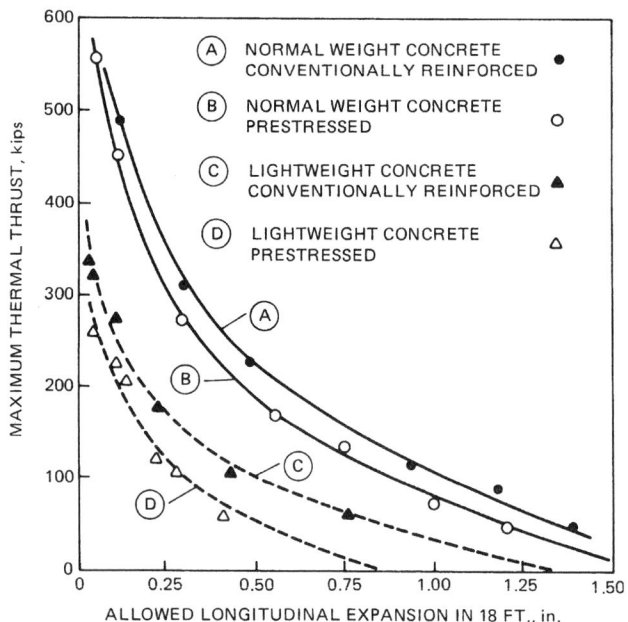

Fig. 4-10.17. *Maximum thrust for allowed expansion of reference specimens.[23]*

$$\frac{T_1}{A_1E_1} = \frac{T_0}{A_0E_0}\frac{Z_0}{Z_1} \qquad (15)$$

where:

$Z_0 = A_0/S_0$,
$Z_1 = A_1/S_1$,
S_0 = heated perimeter of the reference member,
S_1 = heated perimeter of the member in question,
T_1 = maximum thermal thrust (lbs) of the member in question,
T_0 = maximum thermal thrust (lbs) of the reference member,

A_0 = cross-sectional area normal to the direction of thermal thrust (sq in.) of the reference member,
A_1 = cross-sectional area normal to the direction of thermal thrust (sq in.) of the member in question,
E_0 = modulus of elasticity of the member in question, (psi), and
E_1 = modulus of elasticity of the reference member, (psi).

Nomographs, presented in Figure 4-10.18, are used to solve Equation 15.

For any given partially restrained expansion of a concrete member exposed to fire, there is a compatible thermal thrust developed in the fire-exposed portion. Similarly, there is a force within the unheated portion of the member which acts to resist the thermal thrust. The actual thermal thrust, caused by heating, must equal the resisting force of the unheated portion of the member.

The effect of the thermal thrust on the structural behavior of a reinforced concrete slab is the same as that of a prestressing force along the line of action of the thrust. In structural fire endurance calculations, the moment strength is the primary interest, for which case the thermal thrust can be considered a "fictitious reinforcement" along the line of an action of the thrust.[24]

The moment due to the thermal thrust, referred to as the thrust moment, is equal to the thrust force multiplied by the distance between the line of action of the thermal thrust and the centroid of the compression block[23]

$$M_T = T(d_t - \Delta - a_\theta^+/2) \qquad (16)$$

where:

$$a_\theta^+ = \frac{T+A_s^+ f_{y_\theta}^+}{0.85f_{c_\theta}'b_\theta} \qquad (17)$$

T = the magnitude of the thermal thrust,
d_t = the distance from extreme compression fiber to the line of action of the thermal thrust, T,
Δ = the deflection of the slab at the point in question, and
M_T = the thrust moment strength.

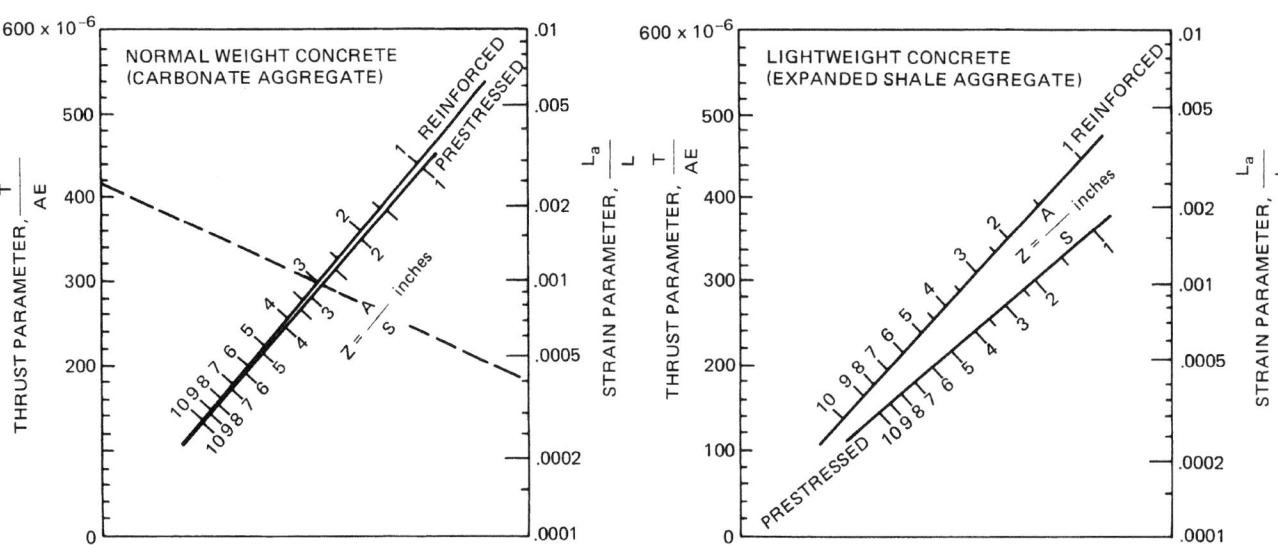

Fig. 4-10.18. *Nomographs relating thrust parameter, strain parameter, and ratio of cross-sectional area to heated perimeter.[23]*

TABLE 4-10.1 *Location of Thermal Thrust Line*[2]

Type of Construction	Fire Exposure (hr)	Location of Thrust Line at Supports*
Solid Slab	2	1 in.
	3	1¼ in.
	4	1½ in.
Slab-and-Joist	≤2	0.1h
	2–4	0.15h

*Distance above bottom of member where h = overall depth of the joist and slab.

The line of action of the thermal thrust must act below the resultant of the equivalent rectangular stress block in order to contribute to the fire endurance of the slab. Results from fire tests have shown that the line of action for the thermal thrust is near the bottom of the member throughout the fire test in most cases, particularly when the thrust is small.[23] Although the line of action acts near the bottom, the actual position changes during the fire test. The exact location of the line of action depends on the shape of the member, type of concrete, amount of reinforcement, stiffness of the restraining frame, and the amount of expansion permitted. Table 4-10.1 is used to locate the line of action of the thermal thrust for floor systems developing a minimal restraint to thermal expansion. The guidelines presented in Table 4-10.1 are based on results from standard fire tests.[23]

In order to calculate the thrust moment, the deflection must be estimated. Since the deflections at the supports are assumed to be zero, the only other deflection of interest is at the midspan. The midspan deflection can be approximated using the following equation derived from the deflection equation for simply supported members[23]

$$\Delta_1 = \frac{L_1^2 \Delta_0}{3500 y_{b1}} \qquad (18)$$

Δ_1 = deflection for the member (in.),
Δ_0 = deflection for the reference member (in.) (see Figure 4-10.19),

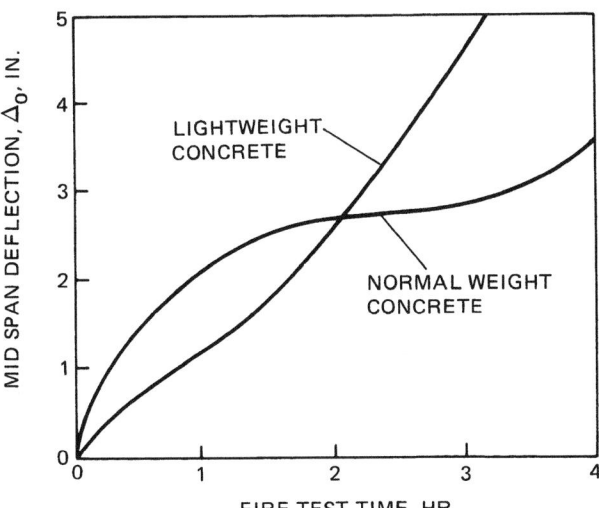

Fig. 4-10.19. Idealized midspan deflection, Δ_0, of reference specimens with minimal restraint.[3]

L_1 = length of the span of the member (in.), and
y_{b1} = distance from the centroidal axis to the extreme fiber (in.).

Equation 18 is for members with minimal restraint to thermal expansion. Another equation should be used when the thrust is greater than minimal.[3]

In order to summarize and illustrate how to apply this information to calculate the structural fire endurance for reinforced concrete members, a step-by-step procedure is presented. This procedure was taken from the Concrete Reinforcing Steel Institute (CRSI), *Reinforced Concrete Fire Resistance*. (See Table 4-10.2.)[3]

EXAMPLE OF CONTINUOUS ONE-WAY SPAN

This example has been included to illustrate the step-by-step procedure for structural analysis for fire endurance that is used. The example problem is for a one-way continuous slab with no thermal restraint assumed. The slab is found to have the desired fire endurance, but the development length of the steel bars required for the negative moment strength is significantly longer than is required for standard gravity loading. The development length is then recalculated assuming minimal thermal restraint.

Given: A one-way, multispan continuous slab supported on beams as shown in Figure 4-10.20. The slab is 4 in. thick with 12-ft beam spacing. The concrete for the slab is made from siliceous aggregate with a compressive strength of 3,000 psi. The slab is subjected to an 80 psf superimposed live load and a 5 psf dead load.

The reinforcement consists of No. 4 bars that meet the requirements of ASTM for A615 Grade 60. Reinforcing bars are placed in accordance with the 1975 *CRSI Handbook*.[25]

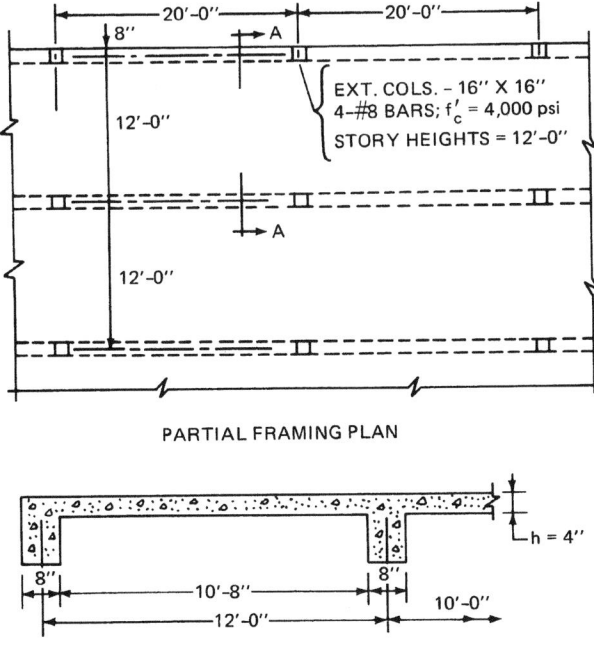

Fig. 4-10.20. One-way continuous slab, supported on beams.[3]

TABLE 4-10.2   *Step-by-Step Procedure—Structural Analysis for Fire Endurance*

STEP NO.	DESCRIPTION
1	From the building code governing the project—model, municipal, state, etc.—look up the required fire ratings.
2	Determine the total depths of slabs, h_t, based on heat transmission to provide the required fire ratings.
3	Compare h_t vs. h_s.
4	If $h_t < h_s$, no further fire endurance considerations are necessary.
4a	If the governing building code permits a reduced fire rating for heat transmission as long as the required structural fire rating is provided, then proceed to Step 5.
5	Only if $h_t > h_s$ (or as in step 4a), compute the structural fire endurance, in hours, based on continuity and/or restraint to thermal expansion.

STRUCTURAL FIRE ENDURANCE BASED ON CONTINUITY ONLY

6	*Solid slabs.* Compute the reduced nominal positive and negative moment strengths, $M_{n\theta}^+$ and $M_{n\theta}^-$, available at the required fire rating, e.g., 3 hours.
7a	*Interior spans.* If the absolute sum of available nominal moment strengths is equal to or greater than the applied moment, the fire endurance $\geq$ than the required fire rating, i.e., if $M_{n\theta}^+ + M_{n\theta}^- \geq \dfrac{wL^2}{8}$.
7b	*Exterior spans.* Using either the reduced nominal negative or positive moment strength available at the specified fire endurance, compute the minimum required nominal moment strength.
8	If the nominal moment strength available $\geq$ minimum required nominal moment strength, the structural fire endurance is adequate—go to Step 9.
8a	If the nominal moment strength available $<$ required nominal moment strength, the structural fire endurance based on continuity only is not sufficient—go to Step 10.
9	If continuity only is considered in the structural fire endurance calculations, and restraint to thermal expansion is neglected, check the lengths of the top reinforcing bars to make sure the bars are long enough to develop the required nominal negative moment strength.

Note: The procedure for analyzing continuous beams and joist systems is the same as for the solid slab above, except that isothermal diagrams would be required for determining the available nominal moment strengths.

STRUCTURAL FIRE ENDURANCE BASED ON RESTRAINT TO THERMAL EXPANSION

10	Estimate the deflection, Δ_1, of the heated bay, assuming minimal restraint occurs.
11	Locate the line of action of the thermal thrust force at the supports.
12	Compute the moment, M_T, which the thermal thrust force has to develop to provide the required additional nominal positive moment strength for the specified fire endurance: $$M_T = (\text{Min. req'd } M_{n\theta}^+ - \text{Available } M_{n\theta}^+)$$
13	Compute the thermal thrust, T_1, required to produce M_T.
14	Compute the "thrust parameter," T_1/A_1E_1.
15	Compute the value of $z = A_1/s$.
16	With T_1/A_1E_1 and z, determine the "strain parameter."
17	Compute the expansion, ΔL, by multiplying the "strain parameter" by the heated length, L_h, of the member.
18	Determine if the restraining elements, i.e., spandrel or effective edge beams, columns, walls, etc., can withstand the thermal thrust, T_1, with a displacement no greater than the expansion, ΔL.

Note: For interior bays, the above procedure can be shortened significantly by using a "thick-walled cylinder" analysis to account for the beneficial effects of restraint to thermal expansion.

Problem: Determine if the slab has a 2 hr fire endurance.

Step 1:   Determine the required fire rating: 2 hr, as stated in the problem.

Step 2:   Determine the total depth of the slab, h_t, based on heat transmission: From Figure 4-10.6, $h_t = 5$ in.

Step 3:   Compare h_t vs. h_s: $h_t = 5 > 4 = h_s$.

Step 4:   In this example, the authority having jurisdiction has waived the requirements for heat transmission, as long as the required structural fire endurance is provided.

Step 5:   Because 5 in. > 4 in., the fire endurance for the end span must be computed based on continuity only.

Structural Fire Endurance Based on Continuity Only

Step 6:   Compute the reduced positive and negative moment strengths, M_n^+ and M_n^-, respectively, available after 2 hr.

6a: $M_{n\theta}^+$ available at 2 hr.

U^+ for bottom bars, $U^+ = 0.75 + 0.25 = 1.0$ in. (See Figure 4-10.21.)

At 2 hr, $U^+ = 1.0$ in., $\theta_s^+ = 1170°F$. (See Figure 4-10.9.)
$f_y = 0.42(60) = 25.2$ ksi. (See Figure 4-10.1.)
$A_s^- = 0.27$ in.^{2}.
A_s^+ is calculated from the rebar spacing requirements.

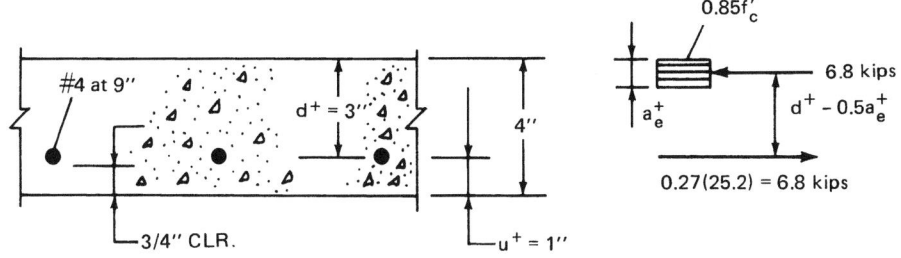

Fig. 4-10.21. M_n^+ calculation for bottom bars.

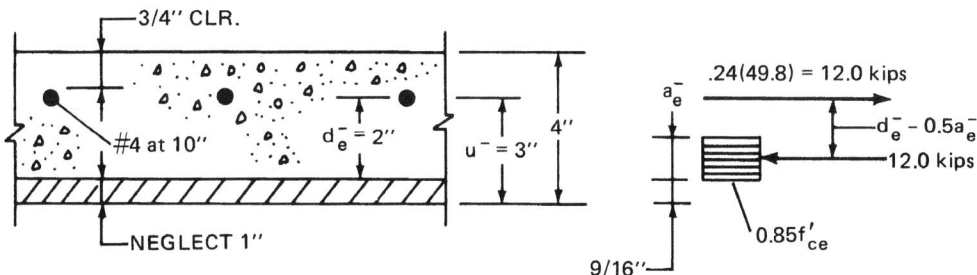

Fig. 4-10.22. M_n^- calculation for top bars.

$$a_\theta^+ = \frac{A_s f_{y\theta}^+}{0.85 f_c' b} = \frac{0.27(25.2)}{0.85(3)(12)} = 0.22 \text{ in.} \quad \text{[from Eq. 5]}$$

$$M_{n\theta}^+ = A_s^+ f_{y\theta}^+ (d - a_\theta^+/2) = 0.27(25.2)(3 - 0.22/2)/12.0$$
$$= M_{n\theta}^+ = 1.64 \text{ ft.} - \text{kips/ft.} \quad \text{[from Eq. 4]}$$

6b: $M_{n\theta}^-$ available at 2 hr.

The bottom 1 in. has been neglected, because the concrete temperature is above 1200°F with a significantly reduced f_c'.

Top bars, $U^- = 4.0 - (0.75 + 0.25) = 3.0$ in. (See Figure 4-10.22.)

At 2 hr, $U^- = 3.0$ in., $\theta = 520$°F. (See Figure 4-10.9.)

$f_{y\theta}^- = 0.83(60.0) = 49.8$ ksi. (See Figure 4-10.1.)

The $f_{y\theta}^-$ stress block, a_θ^-, is estimated to be about ⅝ in. with a temperature ranging from 1200 to 900°F.

Temperature values are estimated from Figure 4-10.9. The average temperature is approximately 1050°F.

$f_{c\theta}' = 0.65(3) = 1.95$ ksi. (See Figure 4-10.2.)

$d_\theta = 4.0 - (0.75 + 0.25 + 1.0) = 2.0$ in.

$$a_\theta^- = \frac{A_s^- f_{y\theta}^-}{0.85 f_{c\theta}' b} = \frac{0.24(49.8)}{0.85(1.95)(12)} = 0.60 \text{ in.} \quad \text{[from Eq. 5]}$$

$$\left\{ \begin{array}{l} A_s^- = 0.24 \text{ in.}^2 \\[6pt] M_n^- = A_s^- f_{y\theta}^- (d_\theta^- - a_\theta/2) = 0.24(49.8)(2.0 - 0.60/2)/12 \\[6pt] \qquad = Mn_\theta^- = 1.69 \text{ ft} - \text{kips/ft.} \quad \text{[from Eq. 4]} \end{array} \right.$$

Step 7: Calculate the minimum positive moment required at 2 hr.

Minimum required $M_{n\theta}^+ = \dfrac{(M_{n\theta}^-)^2}{2wL^2} - \dfrac{M_{n\theta}^-}{2} + \dfrac{wL^2}{8}$

$w = 4/12(150) + 5 + 80 = 135$ lbs/ft $= 0.135$ kips/ft.

Minimum required $M_n^+ = \dfrac{(1.69)^2}{2(0.135)(12)^2} - \dfrac{1.69}{2} + \dfrac{0.135(12)^2}{8}$

Minimum required $M_n^+ = 1.66$ ft $-$ kips/ft. [from Eq. 10]

Step 8: If the positive moment strength available is greater than the required positive moment strength, structural fire endurance is adequate. Because the moment strength available, 1.64 ft—kips/ft, is for practical purposes equal to the required moment strength, 1.66 ft—kips/ft, the structural fire endurance for the end span is 2 hr.

Step 9: Check the lengths of the top reinforcing bars to make sure the bars are long enough to develop the required negative moment strength. The length of the top bars under normal conditions, considering only gravity loads and no fire, is taken from the *CRSI Handbook*.[25] (See Figure 4-10.23.)

9a: Top Bar Lengths, at First Interior Support, Neglecting Restraint to Thermal Expansion.

The distance to the point of inflection at first interior support for structural fire endurance is calculated using

$$X_0 = \frac{2M_{n\theta}^-}{wL}$$

Because the negative reinforcement generally yields early in the fire, as discussed previously, within the first half

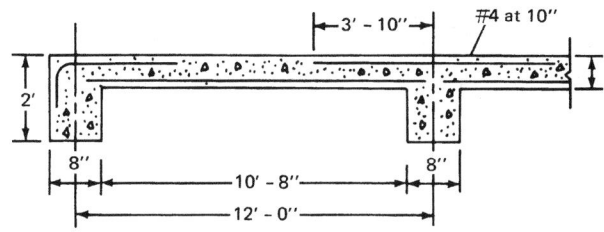

Fig. 4-10.23. Top bar lengths at 2 hr of fire exposure.

hour, the value for the negative moment used in Equation 14 should be the maximum negative moment that the beam can support.

$$a = \frac{A_s f_y}{0.85 f'_c b} = \frac{0.24(60)}{0.85(3)(12)} = 0.47 \text{ in.}$$

$$M_n^- = A_s f_y (d - a/2) - 0.24(60)(3 - 0.47/2)/12$$
$$= M_n^- = 3.32 \text{ ft} - \text{kips/ft.}$$

The value used for w is left to engineering judgment based on the expected loading during a fire. For this example, the full dead load and one-half the live load is used.

$$w = \frac{4}{12}(150) + 5 + 40 = 95 \text{ psf} = 0.095 \text{ kips/ft}$$

$$X_0 = \frac{2(3.32)}{0.095(12)} = 5.82 \text{ ft} \approx 5 \text{ ft } 10 \text{ in.}$$

The distance the top bars have to be embedded beyond the point of inflection is given in the ACI *Building Code*.[1] At least one-third of the bars should be embedded $\frac{1}{16}$ of the clear span, d, or $12d_b$, whichever is greater. In this example, the $\frac{1}{16}$ of the clear span criterion governs (10 in.). Thus, some of the top bars must extend 6 ft 6 in. into the end span. The length of the top steel, 6 ft 6 in., is nearly twice the required length for the gravity load (3 ft 10 in.). The maximum negative moment strength, M_n, used in Equation 14, represents the most severe condition for the development length. However, the assumption of frictionless roller bearing supports used in the above example neglected the restraining force in all calculations. The restraining force, or thermal thrust, T, is developed early in the fire, producing a moment opposite the support moment, which acts to reduce the magnitude of the support moment. The net support moment will then be less than the moment strength, M_n^-, used in the calculation above, thereby, overestimating the development length required for the desired fire endurance.[3]

The restraint criteria discussed will be used to determine if there is sufficient restraint developed in the longitudinal direction, to reduce the development lengths to that required for the gravity load.

9b: Top Bar Lengths, at First Interior Support Including Restraint to Thermal Expansion.

Using $X_0 = 3.83$ ft, the length required for gravity loading, we can determine the net moment at the support required.

$$X_0 = \frac{2M_n^-}{wL} \tag{19}$$

$$M_n^- = \frac{X_0 wL}{2} = \frac{3.83(0.095)(12)}{2} = 2.18 \text{ ft} - \text{kips/ft}$$

The thermal thrust must produce a moment equal to

$$M_T = 3.32 - 2.18 = 1.14 \text{ ft} - \text{kips/ft.}$$

Early in the fire, T will act at or near the bottom of the slab. (See Table 4-10.1.) T is assumed to act $\frac{1}{2}$ in. above the bottom of the slab (taking the fire exposure as approximately one-half hour)

$$d_T = 4 - 1/2 = 3.5 \text{ in.}$$
$$\Delta = 0 \quad \text{at the support.}$$

The depth of the stress block, a_θ^+, is assumed initially to be zero because the required thrust is small.

$$T = \frac{M_T}{d_T - \Delta - a_\theta^+} = \frac{1.14(12)}{3.5 - 0 - 0} = 3.91 \text{ kips}$$

Recalculating a_θ^+

$$a_\theta^+ = \frac{T}{0.85 f'_c b} = \frac{3.91}{0.85(3)(12)} = 0.13 \text{ in.}$$

$$T = \frac{1.14(12)}{3.5 - 0 - 0.13/2} = 3.98 \text{ kips/ft}$$

Compute the expansion, L, that corresponds to

$$T = 3.98 \text{ kips/ft}$$
$$E_1 = 57,000 \quad \sqrt{f_c} = 57,000 \quad \sqrt{3000} = 3,122,000 \text{ psi}$$
$$A_1 = 12(4.0) = 48 \text{ in.}^2$$
$$\frac{T_1}{A_1 E_1} = \frac{3.98}{48(3.122)} = 26.6 \times 10^{-6}$$
$$Z = \frac{A_1}{s} = \frac{48}{12} = 4 \text{ in.}$$
$$\frac{\Delta L}{L_h} = 0.006$$
$$L = 0.006(10.67 \times 12) = 0.77 \text{ in.}$$

In order to maintain equilibrium of the horizontal forces and compatibility of the displacements, the restraining elements must withstand $T = 3.98$ kips/ft and not deflect more than $\Delta L = 0.77$ in. The next step would be to check the strength and stiffness of the restraining elements, i.e., the exterior spandrel beams and columns of the exterior support, and the plane floor area of the first interior support. In this example, it is not necessary to check the strength and stiffness toward the interior of the structure, because there is considerable restraint from the large unheated floor area and many columns to provide the thrust moment at the first interior support.[3] However, the spandrel beams and columns at the exterior support should be checked to ensure that there is sufficient strength and stiffness to resist the thrust moment. Determining the strength and stiffness of the spandrel beams and columns requires a long and complex structural analysis and is not shown here. An explanation of the structural analysis of spandrel beams and columns can be found in the literature.[3]

Assuming there is sufficient restraint in the spandrel beams and columns to resist the thrust moment, the required length of the top bars over the first interior support at 2 hr of fire exposure must be determined. Neglecting restraint to thermal expansion

$$X_0 = \frac{2M_{n\theta}^-}{wL} = \frac{2(1.69)}{(0.095)(12)} = 2.96 \text{ ft}$$

As previously discussed, at least one-third of the top bars should be embedded $\frac{1}{16}$ of the clear span at the point of inflection, X_0, therefore the top steel should extend 3 ft 8 in. into the end span. This is less than the top bar length required for gravity loads, so no adjustment in the length of the reinforcement steel is required to obtain the desired fire endurance.

REINFORCED CONCRETE COLUMNS

Throughout the history of concrete construction reinforced concrete columns have performed well when exposed to fire. The reason for this is threefold:

TABLE 4-10.3 *Fire Endurance Proposed by Hull and Ingberg*[26]

Aggregate type	Minimum area of round or square cross section, cm^2 (sq in.)	Concrete cover, mm (in.)	Fire endurance classification, hr
Siliceous	710 (110)	38 (1½)	1½
Siliceous	1,290 (200)	38 (1½)	2½
Siliceous	1,290 (200)*	38 (1½)	3½
Siliceous	1,613 (250)	64 (2½)	3
Siliceous	1,613 (250)*	64 (2½)	6
Traprock & slag†	1,290 (200)	38 (1½)	4
Carbonate	1,290 (200)	38 (1½)	6

*Mesh in cover
†Air-cooled slag.

1. Columns are generally large enough to prevent the center core from losing a significant amount of strength even in prolonged fire exposure,
2. Ties or spirals contain the concrete within the core, and
3. The vertical reinforcing bars are generally protected by at least 1⅞ in. of concrete cover, thereby insulating the steel bars.[3]

Most of the building codes in the United States assign 3- and 4-hr fire resistance to reinforced concrete columns larger than 12 × 12 in. for square shapes, or a diameter of at least 12 in. for round columns.

ACI suggests that the information in Table 4-10.3, by Hull and Ingberg,[26] be used for designing reinforced concrete columns for exposure to fire. The information presented in Table 4-10.3 is based on the results of a series of fire tests on concrete columns reported in 1925.[26]

Recently, analytical procedures have been developed for estimating the temperature distributions within concrete columns exposed to fire and for designing concrete columns for a specific load and fire endurance.[27–30] These models are based on a stability and/or strength analysis.

REINFORCED CONCRETE WALLS

Typically, the fire endurance of concrete and concrete masonry walls is determined by heat transmission criteria as opposed to structural performance.[2,3] As a result, estimating the fire resistance of walls can be accomplished using a heat transfer analysis only. For this reason, the discussion of thickness requirements presented in the heat transmission section can be used. The required thickness can be determined graphically or by applying a heat transfer computer model.

The distinction between bearing and nonbearing walls is based on building code structural requirements and not fire endurance. For example, some building codes require bearing walls to be thicker than nonbearing walls. Such a requirement has not been justified by results of a fire test.[3] ASTM E 119 requires that a superimposed load be applied and maintained at a constant magnitude throughout the test of a bearing wall. When testing nonbearing walls, there is no applied load; however, the edges of the walls are restrained against thermal expansion. Therefore, a thermally induced load is applied during the fire test to the non-bearing wall. Generally, this thermally induced load is of much greater magnitude than the load applied to bearing walls.[3]

PRESTRESSED CONCRETE ASSEMBLIES

Procedures are also available to calculate the fire resistance of prestressed concrete members. The reader is directed to *Design for Fire Resistance of Precast Prestressed Concrete*.[31]

REFERENCES CITED

1. *Building Code Requirements for Reinforced Concrete*, ACI 318-83, American Concrete Institute, Detroit (1983).
2. *Guide for Determining the Fire Endurance of Concrete Elements*, ACI 216-81, American Concrete Institute, Detroit (1982).
3. *Reinforced Concrete Fire Resistance*, Concrete Reinforcing Steel Institute, Chicago (1980).
4. A.H. Gustaferro and T.D. Lin, "Rational Design of Reinforced Concrete Members for Fire Resistance," *Fire Safety Journal*, 11, 85–98 (1986).
5. M.S. Abrams, "Behavior of Inorganic Materials in Fire," *ASTM STP 685*, American Society for Testing and Materials, Philadelphia (1974).
6. T.Z. Harmathy, "Variable-State Methods of Measuring the Thermal Properties of Solids," *Journal of Applied Physics*, 35, 4 Apr. (1964).
7. T.Z. Harmathy, "Thermal Properties of Concrete at Elevated Temperatures," *Journal of Materials*, 5, 4 Mar. (1970).
8. T.Z. Harmathy and L.W. Allen, "Thermal Properties of Selected Masonry Unit Concretes," *ACI Journal*, 70, 2 (1973).
9. K. Odeen, "Fire Resistance of Prestressed Concrete Double-T Units," *Civil Engineering and Building Construction Series No. 48*, ACTA Polytechnica Scandinavica, Stockholm (1968).
10. A.H. Gustaferro et al, "Fire Resistance of Prestressed Concrete Beams; Study C: Structural Behavior During Fire Tests," *PCA Research and Development Bulletin*, RD009.01b, Portland Cement Association, Skokie (1971).
11. M.S. Abrams and C.R. Cruz, "The Behavior at High Temperatures of Steel Strands for Prestressed Concrete," *Journal of the PCA Research and Development Laboratories*, 3, 3 (1968).
12. M.S. Abrams, "Compressive Strength of Concrete at Temperatures to 1600°F," *Temperature and Concrete, Special Publication SP-25*, American Concrete Institute, Detroit (1971).
13. R.L. Brockenbrough and B.G. Johnston, *Steel Design Manual*, U.S. Steel Corporation, Pittsburgh (1968).
14. C.R. Cruz, "Elastic Properties of Concrete at High Temperatures," *PCA Research Bulletin, 191*, Portland Cement Association, Skokie (1966).
15. *Manual of Steel Construction*, American Institute of Steel Construction, Chicago (1980).
16. M.S. Abrams and A.H. Gustaferro, "Fire Endurance of Two-Course Floors and Roofs," *Journal of American Concrete Institute*, 66, 2 (1969).

17. M.S. Abrams and A.H. Gustaferro, "Fire Endurance of Concrete Slabs as Influenced by Thickness, Aggregate Type, and Moisture," *PCA Research Bulletin*, 223, Portland Cement Association, Skokie (1968).

18. R.H. Iding, Z. Nizamuddin, and B. Bresler, "FIRES T3, A Computer Program for the Fire Response of Structures—Thermal-Three-Dimensional Version," UCB FRG 77-15, University of California, Berkeley (1977).

19. U. Wickstrom, "TASEF-2—A Computer Program for Temperature Analysis of Structures Exposed to Fire," Report No. 79-2, Lund Institute of Technology, Lund, Sweden (1979).

20. *Symposium* on Fire Resistance of Concrete, *ACI Publication SP 5*, American Concrete Institute, Detroit (1962).

21. C.C. Carlson and J.B. Hubbell, "Design and Operation of the PCA Floor Furnace," *PCA Publication RR001*, Portland Cement Association, Skokie.

22. S.L. Selvaggio and C.C. Carlson, "Restraint in Fire Tests of Concrete Floors and Roofs," *ASTM STP 422*, American Society for Testing and Materials, Philadelphia; also *PCA Research Department Bulletin 220*, Portland Cement Association, Skokie (1967).

23. L.A. Issen *et al*, "Fire Tests of Concrete Members: An Improved Method for Estimating Restraint Forces," *Fire Performance*, ASTM STP 464, American Society for Testing and Materials, Philadelphia.

24. E.A.B. Salse and A.H. Gustaferro, "Structural Capacity of Concrete Beams During Fires as Affected by Restraint and Continuity," *Proceedings* 5th CIB Congress, Paris (1971).

25. *CRSI Handbook*, Concrete Reinforcing Steel Institute, Chicago (1975).

26. W.A. Hull and S.H. Ingberg, "Fire Resistance of Concrete Columns," *NBS Technological Papers No. 272* (1925).

27. T.T. Lie and D.E. Allen, "Calculation of the Fire Resistance of Reinforced Concrete Columns," *Technical Paper No. 378*, Division of Building Research, National Research Council of Canada, Ottawa.

28. T.T. Lie and D.E. Allen, "Further Studies on the Fire Resistance of Reinforced Concrete Columns," *Technical Paper No. 416*, Division of Building Research, National Research Council of Canada, Ottawa.

29. T.T. Lie and T.Z. Harmathy, "Fire Endurance of Concrete-Protected Steel Columns," *ACI Journal*, 71, 1 (1974).

30. A. Haksever and Y. Anderberg, "Comparison Between Measured and Computed Structural Response of Some Reinforced Concrete Columns in Fire," *Fire Safety Journal*, 4 (1981).

31. *Design for Fire Resistance of Precast Prestressed Concrete*, Prestressed Concrete Institute, Chicago (1977).

ANALYTICAL METHODS FOR DETERMINING FIRE RESISTANCE OF TIMBER MEMBERS

Robert H. White

INTRODUCTION

The fire resistance ratings of wood members and assemblies, as of other materials, have traditionally been obtained by testing the assembly in a furnace in accordance with American Society for Testing and Materials (ASTM) Standard E 119.[1] These ratings are also published in listings, such as the Underwriters Laboratories *Fire Resistance Directory*[2] or the Gypsum Association's *Fire Resistance Design Manual*,[3] and in publications of the model building code organizations. The ratings listed are limited to the actual assembly tested and normally do not permit modifications such as adding insulation, changing member size, changing or adding interior finish, or increasing the spacing between members. Code interpretation of the test results sometimes allows the substitution of larger members, thicker or deeper assemblies, reduction in member spacing, and thicker protection layers, without reducing the listed rating. Two fire-endurance design procedures for wood that allow greater flexibility have U.S. and Canadian building code acceptance. In addition, other procedures and models have been proposed or are being developed.

When attention is given to all details, the fire endurance of a wood member or assembly depends on three items:

1. Performance of its protective membrane (if any),
2. Extent of charring of the structural wood element, and
3. Load-carrying capacity of the remaining uncharred portions of the structural wood elements.

The following sections review the methods available for determining the contribution of each item and discuss the major properties of wood that affect the thermal and structural response of wood assemblies or components.

CONTRIBUTION OF THE PROTECTIVE MEMBRANE

Gypsum wallboard and plywood paneling are two common types of protective membrane, which is the first line of resistance to a fire in wood construction. In a protected assembly, the fire resistance rating is largely determined by the type and thickness of the protective membrane. The effects of the protective membrane on the thermal performance of an assembly are included in Harmathy's ten rules of fire endurance rating.[4] These ten rules (Figure 4-11.1) provide guidelines to evaluate the relative effects of changes in materials on the fire resistance rating of an assembly. The rules apply primarily to the thermal performance of the assembly.

The contribution of the protective membrane to the fire resistance rating of a light-frame assembly is clearly illustrated in the component additive calculation procedure discussed in the following subsection. Brief discussions of direct protection of wood members and numerical heat transfer models are also presented.

Component Additive Calculation Procedure

The component additive calculation procedure is a method to determine conservatively the fire resistance ratings of load-bearing light-frame wood floor and roof assemblies and of load-bearing and nonload-bearing wall assemblies. With this procedure, as with Harmathy's rules 1 and 2, one assumes that times can be assigned to the types and thicknesses of protective membranes and that an assembly with two or more protective membranes has a fire resistance rating at least that of the sum of the times assigned for the individual layers and the times assigned to the framing. The procedure was developed by the National Research Council of Canada (NRCC), and has gained code approval in both the U.S. and Canada.

The times assigned to the protective membranes (Table 4-11.1), the framing (Table 4-11.2), and other factors (Table 4-11.3) are added together to obtain the fire resistance rating for the assembly. The times are based on empirical correlation with actual ASTM E 119 tests of assemblies. The ratings obtained in these tests ranged from 20 to 90 min. The times given in Table 4-11.1 are based on the membrane's ability to remain in place during fire tests. The times assigned to the protective membranes are not the "finish ratings" of the material cited in test reports or listings. [A finish rating is defined as the time to reach either an average temperature rise of 250°F (139°C) or a maximum rise of 325°F (181°C), on the unexposed side of the material.] The building codes include requirements for fastening the protective membranes to the frame. The addition of insulation to a wall

Dr. Robert H. White is a Supervisory Wood Scientist at the USDA, Forest Service, Forest Products Laboratory. His research has primarily been in the areas of wood charring and fire endurance of wood assemblies.

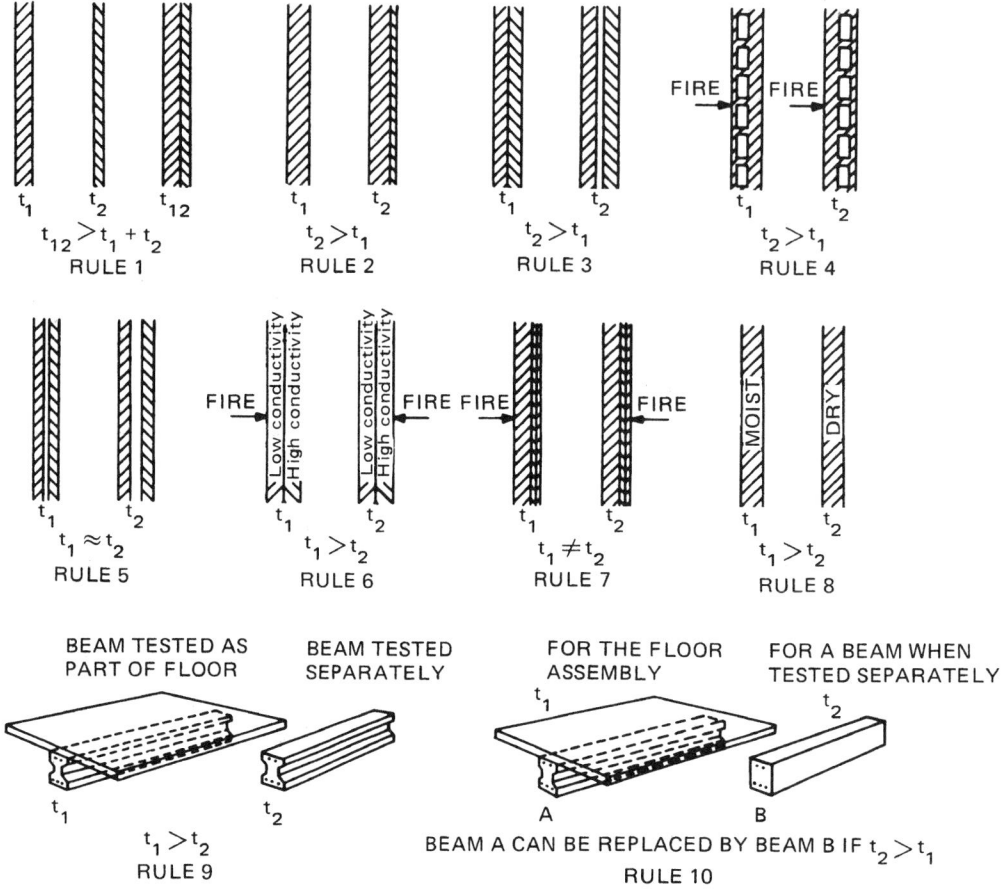

Fig. 4-11.1.   Harmathy's ten rules of fire endurance.[4]

assembly can increase its fire resistance. (See Table 4-11.3.) Adding insulation to a floor or roof assembly can decrease its fire resistance, depending on its location within the assembly and the method of attachment.

For asymmetrical wall assemblies, the rating is based on the side with the lesser fire resistance. For exterior walls rated only from the interior and floor/roof assemblies, there are minimal requirements for the membrane on the side or top of the assembly not exposed to the fire (Tables 4-11.4 and 4-11.5), in order to ensure that the wall or floor/roof assembly does not fail because of fire penetration or heat transfer

TABLE 4-11.1   *Time Assigned to Protective Membranes**†

Description of Finish	Time (min.)
⅜-in. Douglas fir plywood, phenolic bonded	5
½-in. Douglas fir plywood, phenolic bonded	10
⅝-in. Douglas fir plywood, phenolic bonded	15
⅜-in. gypsum board	10
½-in. gypsum board	15
⅝-in. gypsum board	20
½-in. type X gypsum board	25
⅝-in. type X gypsum board	40
Double ⅜-in. gypsum board	25
½-in. + ⅜-in. gypsum board	35
Double ½-in. gypsum board	40
Double ½-in. gypsum board‡	50

*On walls, gypsum board must be installed with the long dimension parallel to framing members, with all joints finished. However, ⅝-in. type X gypsum board may be installed horizontally with the horizontal joints unsupported.
†On floor/ceiling or roof/ceiling assemblies, gypsum board must be installed with the long dimension perpendicular to framing members, and must have all joints finished.
‡Wire mesh with 0.06-in.-diameter wire and 1-sq-in. openings must be fastened between the two sheets of gypsum board.

TABLE 4-11.2   *Time Assigned for Contribution of Wood Frame**

Description of Frame	Time Assigned to Frame (min.)
Wood studs, 16 in. on center	20
Wood floor and roof joists, 16 in. on center	10
Wood roof and floor truss assemblies, 24 in. on center	5

*Minimum size for studs is nominal 2 in. by 4 in.. Wood joists and members of trusses also must not be less than nominal 2 in. by 4 in. The listing for truss assemblies does not apply to trusses with metal-tube or bar webs. The spacing between studs or joists should not exceed 16 in. on center. The spacing between trusses should not exceed 24 in. on center.

TABLE 4-11.3 *Time Assigned for Additional Protection*

Description of Additional Protection	Time Assigned to Insulation (min.)
Adds to the fire endurance rating of wood stud walls if the spaces between the studs are filled with rock wool or slag mineral wool batts weighing not less than ¼ lb/sq ft of wall surface.	15
Adds to the fire endurance rating of nonload-bearing wood stud walls if the spaces between the studs are filled with glass fiber batts weighing not less than ¼ lb/sq ft of wall surface.	5

through the assembly. Instead of being one of the combinations listed in Tables 4-11.4 and 4-11.5, the membrane on the side not exposed to fire (the outside or top) may be any membrane listed in Table 4-11.1 with an assigned time of 15 min. or greater.

The component additive calculation procedure is in the *Supplement to the National Building Code of Canada* (SNBCC)[5] and some U.S. building codes. The application of the method is generally limited to 60 or 90 min. The tables presented in this chapter are based on publications of the American Forest & Paper Association[6] and the Canadian Wood Council.[7] For specific situations, the applicable building code should be checked for acceptance of, modifications to, and limitations on the procedure as presented in this chapter. There are differences between the codes in what is accepted. There are individual items in the tables that are not accepted by all the codes that otherwise accept the procedure.

TABLE 4-11.4 *Alternative Membranes on Face of Wood Stud Walls Not Exposed to Fire (Exterior)**

Sheathing	Paper	Exterior Finish
⅝-in. tongue-and-groove lumber		Lumber siding
⁵⁄₁₆-in. exterior-grade plywood	Sheathing paper	Wood shingles and shakes
½-in. gypsum board		¼-in. exterior-grade plywood ¼-in. hardboard Metal siding Stucco on metal lath Masonry veneer
None	None	⅜-in. exterior-grade plywood

*Membrane may be any combination of sheathing, paper, and exterior finish in table or any other membrane listed at 15 min. or greater in Table 4-11.1.

The component additive calculation procedure gives flexibility, for example, in calculations for plywood and gypsum board combined as an interior finish.

EXAMPLE 1:

The calculated fire resistance rating of a wood stud exterior wall (2-in. × 4-in. studs, 16 in. on center) with ⅝-in. Douglas fir phenolic-bonded plywood over ½-in. type X gypsum wallboard on the side exposed to fire with fiberglass insulation in the cavity is:

From Table 4-11.1:	
⅝-in. Douglas fir plywood, phenolic bonded	15 min.
½-in. type X gypsum board	25 min.
From Table 4-11.2:	
Wood stud framing	20 min.
Calculated rating (total)	60 min.

The fiberglass insulation provides no additional fire resistance time for a load-bearing wall. The other side of the exterior wall, if it has no fire resistance requirement, can be ⅜-in. exterior-grade plywood (Table 4-11.3) or any panel with an assigned time of 15 min. (Table 4-11.1).

Direct Protection of Wood Members

The steel industry improves the fire endurance of steel members by directly covering them with fire-resistive panels or coatings. Currently, the marketing of fire-resistive coatings for use on wood is very limited or nonexistent. The fire retardant coatings marketed for wood are only designed and recognized for use to reduce the spread of flames over a surface (flamespread).

Depending upon its thickness and durability under fire exposure, a coating may merely delay ignition of the wood for a few minutes or may provide an effective insulative layer that reduces the rate of charring. Both for fire-retardant coatings and fire-resistive coatings, the performance as a fire resistant membrane on wood has been evaluated.[8–10] Tests on coated timber members have also been reported in Finland and U.S.S.R.[11]

There is limited published data on the protection provided by directly covering a wood member with gypsum board or other nonwood panel products. Finish ratings listed for panel products used in ASTM E 119 tests of assemblies have been used to estimate the delay in the onset of char formation provided by the panel product. Gardner and Syme[12] found that gypsum board not only delayed the onset of char formation but also reduced the subsequent rate of char formation. In their two-hour tests, ½-in.-thick gypsum board on wood beams reduced the depth of char by approximately

TABLE 4-11.5 *Flooring or Roofing over Wood Framing**

Assembly	Subfloor or Roof Deck	Finish Flooring or Roofing
Floor	½-in. plywood or ¹¹⁄₁₆-in. tongue-and-groove softwood lumber	Hardwood or softwood flooring on building paper; or Resilient flooring, parquet floor, felted-synthetic-fiber floor coverings, carpeting, or ceramic tile on ⅜-in.-thick panel-type underlay; or Ceramic tile on 1¼-in. mortar bed.
Roof	½-in. plywood or ¹¹⁄₁₆-in. tongue-and-groove softwood lumber	Finish roofing material with or without insulation

*Upper membrane consists of a subfloor and finish floor, roof deck and roofing, or any other membrane listed at 15 min. or greater in Table 4-11.1.

40 percent. Of the 40 percent, only 17 percent was credited to the initial delay in char formation.

Numerical Heat Transfer Models

The protective membrane contributes to fire resistance by providing thermal protection. Numerical heat transfer methodologies are available to evaluate this thermal protection. Fung[13] developed a one-dimensional finite difference model and computer program for thermal analysis of construction walls. Gammon[14] developed a two-dimensional finite element heat transfer model for wood stud wall assemblies. Difficulties in modeling the charring of wood and the physical deterioration of the panel products complicate these numerical methodologies. New models are being developed in North America, Sweden, New Zealand,[15] and Australia.

Numerical heat transfer models are used not only to model the performance of the protective membranes but also to model the charring of the structural wood members, the second major factor in the fire endurance of a wood member or assembly.

CHARRING OF WOOD

Wood undergoes thermal degradation (pyrolysis) when exposed to fire. (See Figure 4-11.2.) The pyrolysis and combustion of wood have been studied extensively. Literature reviews include articles by Browne,[16] Schaffer,[17,19] Hall *et al*,[18] and Hadvig.[20] By converting the wood to char and gas, pyrolysis results in a reduction in the wood's density. The pyrolysis gas undergoes flaming combustion as it leaves the charred wood surface. Glowing combustion and mechanical disintegration of the char eventually erode or ablate the outer char layer.

The charring rate generally refers to the linear rate at which wood is converted to char. Under standard fire expo-

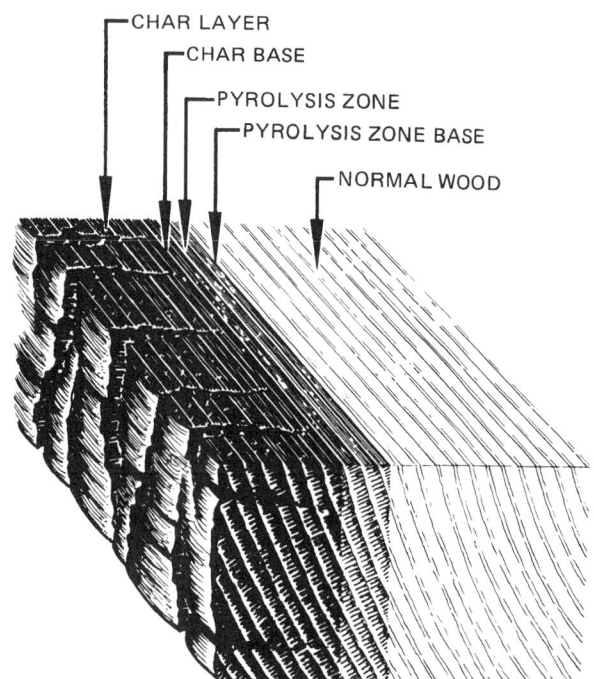

Fig. 4-11.2. Degradation zones in a wood section.

sure, the charring rates tend to be fairly constant after a higher initial charring rate.

Establishing the charring rate is critical to evaluating fire resistance, because char has virtually no load-bearing capacity. There is a fairly distinct demarcation between char and uncharred wood. The base of the char layers is wood reaching a temperature of approximately 290°C (550°F). To determine the charring rate, we use both empirical models based on experimental data and theoretical models based on chemical and physical principles.

EMPIRICAL MODELS
Standard ASTM E 119 Fire Exposure

Expressions for charring rate in the standard ASTM E 119 test are the result of many experimental studies. The empirical model that is most generally used assumes a constant transverse-to-grain char rate of 0.6 mm/min. (1½ in./hr) for all woods, when subjected to the standard fire exposure. There are differences among species associated with their density, chemical composition, and permeability. In addition, the moisture content of the wood affects the charring rate. (See also reference 21.)

Schaffer[22] reported transverse-to-grain charring rates as a function of density and moisture content for white oak, Douglas fir, and southern pine. The regression equations for B (min. per in., the reciprocal of charring rate) were

$$B = 2[(28.726 + 0.578M)\rho + 4.187] \quad \text{for Douglas fir} \quad (1)$$

$$B = 2[(5.832 + 0.120M)\rho + 12.862] \quad \text{for southern pine} \quad (2)$$

$$B = 2[(20.036 + 0.403M)\rho + 7.519] \quad \text{for white oak} \quad (3)$$

where

M = percent moisture content, and
ρ = dry specific gravity.

White[23] developed an empirical model based on eight species. The char rate equation was of the form

$$t = mx_c^{1.23} \quad (4)$$

where

t = time (min.),
m = char rate coefficient, and
x_c = char depth (mm).

The char rate coefficient could be estimated with the equation

$$m = -0.147 + 0.000564\rho + 0.0121u + 0.532f_c \quad (5)$$

where

ρ = oven-dry density (kg/m³),
u = moisture content (percent), and
f_c = char contraction factor (dimensionless).

The char contraction factor was the thickness of the char layer at the end of the fire exposure divided by the original thickness of the wood layer that was charred (char depth).

Recent char rate experiments have been reported in Australia,[12] Europe,[24] and New Zealand.[25]

Assumption of a constant charring rate is reasonable when the member or panel product is thick enough to be treated as a semi-infinite slab. For smaller dimensions, the charring rate increases once the temperature has risen above the initial temperature at the center of the member or at the unexposed surface of the panel.

Kanury and Holve[26] suggest the model

$$\frac{\ell}{t} \approx \left(\frac{2}{a}\right)\left(1 - \frac{b\ell}{a}\right) \tag{6}$$

where

ℓ = thickness of slab,
t = fire endurance time, and
a,b = constants.

They consider the $(2/a)$ factor an ideal charring rate and the ratio $(b\ell/a)$ as a correction factor accounting for thickness and thermal diffusion effects.

Noren and Ostman[27] provided the equation

$$b_m = 1.128t + 0.0088t^2 \tag{7}$$

where

b_m = contribution to fire resistance (min.), and
t = panel thickness (mm).

The equation is based on data for various wood-based panel products. Differences in the fire resistance at equal thickness depended on panel density, moisture content, type of adhesive, and the structural composition of the panel.

The charring rate parallel to the grain of wood is approximately twice that transverse to the grain.[18] As a beam or column chars, the corners become rounded. The rounding is generally considered to have a radius equivalent to the char depth on the sides.

In Europe, structural Eurocodes are being developed for the design of structures. As currently written, the draft of Eurocode 5 (Timber Structures), Part 1.2 on structural fire design is largely based on calculation methods.[28] Specific design values for char rate are included in the document.

The effect of fire-retardant treatment and adhesives on fire resistance depends on the type of adhesive or treatment. Lumber bonded with phenolic or resorcinol adhesives has a charring rate consistent with that of solid wood. Fire-retardant treatments are designed to reduce flamespread. The fire retardant's effect on the charring rate may be to only slightly increase the time until ignition of the wood. Some fire retardants reduce flammability by lowering the temperature at which charring occurs. This may increase the charring rate. However, a few fire retardants have been found to improve charring resistance.[29]

Nonstandard Fire Exposures

The above equations were stated to apply to the standard ASTM E 119 fire exposure.[1] Data on charring rates for other fire exposures have been limited. Schaffer[22] provided data for constant temperatures of 538°C (1000°F), 815°C (1500°F), and 927°C (1700°F). As a result of increased testing with heat release rate calorimeters, char rate data as a function of external heat flux are becoming available.[30-32]

Hadvig[20] has developed equations for nonstandard fire exposure. The charring rate in a real fire depends upon the severity of the fire to which the wood is exposed. The fire severity depends upon such factors as the available combustible material (fire load) and the available air supply (design opening factor).

The design fire load is

$$q = k \cdot \frac{Q}{A_t} \tag{8}$$

where

q = design fire load (MJ/m^2);
k = transfer coefficient (dimensionless);
Q = sum of the products of mass and lower calorific value of materials to be found in the compartment (MJ); and
A_t = total internal area of the compartment, including floor, walls, ceiling, windows, and doors (m^2).

The transfer coefficients are given in Table 4-11.6 for different types of compartments and geometrical opening factors. In the case of fire compartments whose bounding structures do not come under any of the types A-H, k is usually determined by a linear interpolation in the table between appropriately chosen types of compartments.

The geometrical opening factor is

$$F' = \frac{A\sqrt{h}}{A_t} \tag{9}$$

where

F' = geometrical opening factor (m$^{1/2}$),
A = total area of windows, doors, and other openings in walls (i.e., vertical openings only) (m^2), and
h = weighted mean value of the height of vertical openings, weighted against the area of the individual openings (m).

TABLE 4-11.6 *The Transfer Coefficient, k[20,33]*

Type of fire com-partment*	Geometrical opening factor, F'					
	0.02	0.04	0.06	0.08	0.10	0.12
A	1.0	1.0	1.0	1.0	1.0	1.0
B	0.85	0.85	0.85	0.85	0.85	0.85
C	3.0	3.0	3.0	3.0	3.0	2.5
D	1.35	1.35	1.35	1.50	1.55	1.65
E	1.65	1.50	1.35	1.50	1.75	2.00
F†	1.0–0.5	1.0–0.5	0.8–0.5	0.7–0.5	0.7–0.5	0.7–0.5
G	1.50	1.45	1.35	1.25	1.15	1.05
H	3.0	3.0	3.0	3.0	3.0	2.5

*A: (Standard fire compartment). The average consisting of brick, concrete, and gas concrete.
B: Concrete, including concrete on the ground.
C: Gas concrete (density 500 kg/m^3).
D: 50 pct concrete, 50 pct gas concrete (density 500 kg/m^3).
E: 50 pct gas concrete (density 500 kg/m^3), 33 pct concrete, and 17 pct laminate consisting of (taken from the inside) 13-mm plasterboard (density 500 kg/m^3), 10-cm mineral wool (density 50 kg/m^3), and brick (density 1,800 kg/m^3).
F: 80 pct steel plate, 20 pct concrete. The fire compartment is comparable to a storehouse or other building of a similar kind with an uninsulated roof, walls of steel plate, and floor of concrete.
G: 20 pct concrete and 80 pct laminate consisting of a double plasterboard (2 × 13 mm) (density 790 kg/m^3), 10-cm air space, and another double plasterboard (2 × 13 mm) (density 790 kg/m^3).
H: Steel plate on either side of 100-mm mineral wool (density 50 kg/m^3).
†The higher values apply to $q < 60$ MJ/m; the lower values apply to $q > 500$ MJ/m^2. Intervening values are found by interpolation.

The design opening factor is

$$F = F' \cdot k \cdot f \tag{10}$$

where

F = design opening factor ($m^{1/2}$),
F' = geometrical opening factor ($m^{1/2}$),
k = transfer coefficient of bounding structure (dimensionless), and
f = coefficient (dimensionless) to account for horizontal openings.

The dimensionless coefficient, f, (Figures 4-11.3 and 4-11.4) increases the opening factors when there are horizontal openings. For only vertical openings, f is equal to 1.
Hadvig's[20] equations are

$$\theta = 0.0175 \frac{q}{F} \tag{11}$$

$$\beta_0 = 1.25 - \frac{0.035}{F + 0.021} \quad \text{for } 0.02 \le F \le 0.30 \tag{12}$$

$$X = \beta_0 \cdot \tau \quad \text{for } 0 \le \tau \le \frac{\theta}{3} \tag{13}$$

$$X = \beta_0 \left(-\frac{1}{12} \theta + \frac{3}{2} \tau - \frac{3}{4} \frac{\tau^2}{\theta} \right) \quad \text{for } \frac{\theta}{3} \le \tau \le \theta \tag{14}$$

where

θ = time at which maximum charring is reached for the values used for F and q (min.),
β_0 = initial value of rate of charring (mm/min.),
X = charring depth (mm),
F = design opening factor ($m^{1/2}$) (defined in Equation 10),
q = design fire load (MJ/m^2) (defined in Equation 8), and
τ = time (min.).

These equations are valid for fire exposures less than 120 min. and a room where the combustible material is wood. Plastic burns more intensely and for a shorter time than wood. When the combustible materials in the room are plastics, Equations 11 and 12 are therefore modified for faster char rate (β_0 is 50 percent higher), shorter time is allowed for maximum charring (θ is cut in half), and Equation 13 is applicable for $\tau < \theta$.[20]

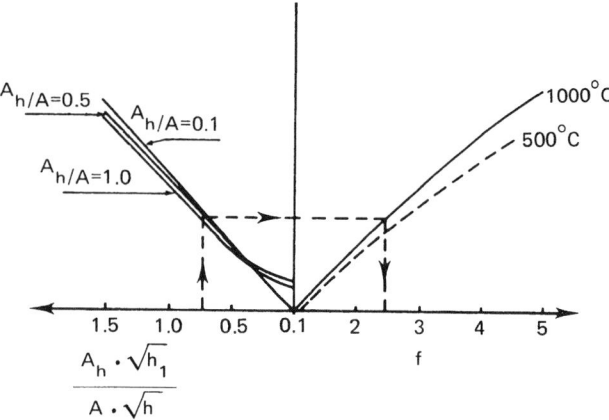

Fig. 4-11.3. Diagram for the determination of f for fire temperatures of 500°C and 1000°C.

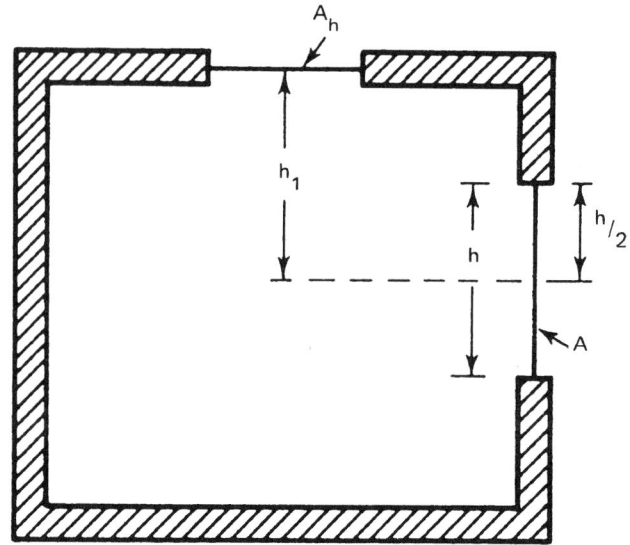

Fig. 4-11.4. Simplified sketch of vertical cross section of ventilated compartment with notation.[20]

Equations 11 through 14 are for glued timber with a density of 470 kg/m^3 including a moisture content of 10 percent and minimum width of 80 mm or greater or square members of minimum 50 × 50 mm. Equations 13 and 14 are valid only for $0 < X < b/4$, where b is the dimension of the narrow face of a rectangular member. For dimensions of nonsquare cross sections between 30 and 80 mm, the ratio of the original dimensions must be equal to or greater than 1.7, the charring depth perpendicular to the wide face is X, and the charring depth perpendicular to the narrow face is determined by multiplying Equation 13 or 14 times the dimensionless quantity

$$1.35 - 0.0044(b) \tag{15}$$

where

b = dimension of narrow face (mm).

EXAMPLE 2:

The room is a standard fire compartment consisting of brick, concrete, and gas concrete. The floor area is 5 × 10 m, and the height is 3 m. The openings are one window 1.5 m high and 2 m wide, three windows 1.5 m high and 1 m wide, and one skylight 1.5 × 3 m. The skylight is 2 m above the midheight of the windows. The fire load is 6 m^3 of wood.

Assuming a fire temperature of 1000°C, a wood density of 500 kg/m^3, and lower calorific value of 17 MJ/kg, describe the charring of a 38- × 250-mm wood beam exposed on three sides after 8 min. of the fire.

The geometrical opening factor (Equation 9) is

$$F' = \frac{A\sqrt{h}}{A_t} = \frac{[1(1.5 \times 2) + 3(1.5 \times 1)]\sqrt{1.5}}{[2(5 \times 10) + 2(3 \times 5) + 2(3 \times 10)]}$$
$$= \frac{7.5\sqrt{1.5}}{190} = 0.048 \text{ m}^{1/2}$$

The design opening factor (Equation 10) is

$$F = F' \cdot k \cdot f$$

The k is obtained from Table 4-11.6 ($k = 1.0$ for type A, $F' = 0.048$). The f is obtained from Figures 4-11.3 and 4-11.4.

$$\frac{A_h\sqrt{h_1}}{A\sqrt{h}} = \frac{(1.5 \times 3)\sqrt{2}}{7.5\sqrt{1.5}} = \frac{4.5\sqrt{2}}{7.5\sqrt{1.5}} = 0.69$$

$$\frac{A_h}{A} = \frac{4.5}{7.5} = 0.6$$

For $A_h\sqrt{h_1}/A\sqrt{h}$ of 0.69 and A_h/A of 0.6,

the f from Figure 4-11.3 is 2.4.

$$F = (0.048)(1.0)(2.4) = 0.115 \text{ m}^{1/2}$$

The design fire load (Equation 8) is

$$q = k \cdot \frac{Q}{A_t} = (1.0)\frac{(6 \times 500 \times 17)}{190} = \frac{51,000}{190} = 268 \text{ MJ/m}^2$$

Maximum charring will be reached at θ min. (Equation 11).

$$\theta = 0.0175\frac{268 \text{ MJ/m}^2}{0.115 \text{ m}^{1/2}} = 41 \text{ min.}$$

The initial charring rate (Equation 12) will be

$$\beta_0 = 1.25 - \frac{0.035}{0.115 + 0.021} = 1 \text{ mm/min.}$$

At 8 min., the char depth (Equation 13) will be

$$X = 1 \times 8 = 8 \text{ mm} \quad \text{for } 0 \leq 8 \leq \frac{41}{3}$$

The smaller dimension b of the beam is 38 mm. The charring depth criterion $0 < x < b/4$ is $0 < 8 < 9.5$ mm, so Equations 13 and 14 are valid. The ratio of the original dimensions is 25/3.8 or 6.6. Since 38 mm is less than 80 mm, the multiplying factor (Equation 15) is

$$1.35 - 0.0044(38) = 1.18$$

At 8 min., the uncharred area of the beam will be approximately

$$38 \text{ mm} - 2(8 \text{ mm}) = 22 \text{ mm wide}$$

and

$$250 \text{ mm} - (1.18 \times 8 \text{ mm}) = 240 \text{ mm high}$$

As the charring proceeds after (9.5 mm)/(1 mm/min.) or 9.5 min., the $b/4$ criterion of the equations no longer holds. This is because the charring rate increases as the temperature at the center of the beam starts to increase.

For situations for which no empirical models exist, solutions may be found by the use of theoretical models. Most theoretical models have the flexibility to be used for any desired fire exposures.

Oleson and Konig[34] noted that, compared to conditions at standard exposure, the mechanical behavior at natural fire exposure is different due to the changes of temperature in the residual cross section during the cooling period. The influence of elevated temperature is no longer concentrated to the outer layer of the residual cross section.

Theoretical Models

Considerable efforts have gone into developing theoretical models for wood charring. Theoretical models allow calculation of the charring rate for geometries other than a semi-infinite slab and for nonstandard fire exposures. Unfortunately, no completely satisfactory model has yet been developed. Roberts[35] reviewed the problems associated with the theoretical analysis of the burning of wood, including structural effects and internal heat transfer, kinetics of the pyrolysis reactions, heat of reaction of the pyrolysis reactions, and variations of thermal properties during pyrolysis. He considered the major problems to be in the formulation of a mathematical model for the complex chemical and physical processes occurring and in the acquisition of reliable data for use in the model.

Many models for wood charring are based on the standard conservation of energy equation. The basic differential equation includes a term for each contribution to the internal energy balance. An early model for wood charring was given by Bamford et al.[36] The basic differential equation used by Bamford was

$$c\rho\frac{\partial T}{\partial t} = K\frac{\partial^2 T}{\partial X^2} - q\frac{\partial w}{\partial t} \tag{16}$$

where

K = thermal conductivity,
T = temperature,
X = location,
w = weight of volatile products per cubic centimeter of wood,
t = time,
q = heat liberated at constant pressure per gram of volatile material evolved,
c = specific heat, and
ρ = density.

In Equation 16, the term on the left side of the equal sign represents the energy stored at a given location as indicated by the increase or decrease of the temperature with time at that location. The first term on the right side of the equal sign represents the thermal conduction of energy away from or into the given location. The second term on the left side represents the energy absorbed (endothermic reaction) or the energy given off (exothermic reaction) as the wood undergoes pyrolysis or thermal degradation. Numerical solutions using computers are normally used to solve these differential equations.

In Bamford's calculations using Equation 16, the rate of decomposition was given by an Arrhenius equation. The heat of decomposition, q, was the difference between the heat of combustion of the wood and that of the products of decomposition. Thermal constants for wood and char were assumed to be the same, and the total thickness of char and wood was assumed to remain constant.

Thomas[37] added a convection term to Bamford's equation to obtain

$$\rho c\frac{\partial T}{\partial t} = K\frac{\partial^2 T}{\partial X^2} + Mc_g\frac{\partial T}{\partial X} - q\frac{\partial w}{\partial t} \tag{17}$$

where

M = local mass flow of pyrolysis gases, and
c_g = specific heat of the gases.

The convection term represents the energy transferred in or out of a location as a result of the convection of the pyrolysis gases through a region with a temperature gradient.

The Factory Mutual Research Corporation model (SPYVAP) includes terms for internal convection of volatiles and thermal properties as functions of temperature and density. It was developed by Kung[38] and later revised by Tamanini.[39] Atreya[40] has further revised this model to include moisture absorption. His energy conservation equation is

$$(\rho_a C_{pa} + \rho_c C_{pc} + \rho_m C_{pm})\frac{\partial T}{\partial t} = \frac{\partial}{\partial X}\left(K\frac{\partial T}{\partial X}\right) + i\left(1 - j\frac{\rho_c}{\rho_f}\right) \cdot$$

$$M_g\frac{\partial H_g}{\partial X} - \frac{\partial \rho_s}{\partial t}\left[-Q + \left(H_a - H_c\frac{\rho_f}{\rho_w}\right)\middle/\left(1 - \frac{\rho_s}{\rho_w}\right) - H_g\right] -$$

$$\frac{\partial \rho_m}{\partial t}(-Q_m + H_m - H_g) \tag{18}$$

where

C_p = specific heat [J/(kg K)],
K = thermal conductivity [W/(m K)],
T = temperature (K),
t = time (s),
X = distance (m),
ρ = density (kg/m^3),
M_g = outward mass flux of volatile gases (kg/m^2 s),
H = thermal-sensible specific enthalpy (J/kg),
Q = endothermic heat of decomposition of wood for a unit mass of volatiles generated (J/kg at T_∞), and
i,j = parameters to simulate cracking, between 0 and 1;

subscripts:

∞ = ambient,
w = virgin wood,
c = char,
g = volatile gases,
a = unpyrolyzed active material,
m = moisture,
f = final value, and
s = solid wood.

Equation 18 is similar to the previous equations except the material has been broken up into its components (wood, water, and char). The parameter j eliminates the convection term if the pyrolysis gases are escaping through cracks or fissures in the wood. The last term represents the heat absorbed with vaporization of the water. The conservation of mass equation is

$$\frac{\partial M_g}{\partial X} = \frac{\partial \rho_s}{\partial t} + \frac{\partial \rho_m}{\partial t} \tag{19}$$

and ensures that the mass of the gases equals the mass loss due to thermal degradation of the wood and vaporization of the moisture.

As noted before, the decomposition kinetics equation for wood is the Arrhenius equation

$$\frac{\partial \rho_s}{\partial t} = -A\frac{(\rho_s - \rho_f)}{\left(1 - \frac{\rho_f}{\rho_w}\right)}\exp(-E/RT) \tag{20}$$

where

A = frequency factor (1/s),
E = activation energy (J/mole), and
R = gas constant.

Atreya[40] uses a moisture desorption kinetics equation for vaporization of the water in the wood, which is

$$\frac{\partial \rho_m}{\partial t} = -A_m\rho_m\exp(-E_m/RT) \tag{21}$$

Parker[41] has taken char shrinkage parallel and normal to the surface into account in the model. Parker also includes different Arrhenius equations for each of the three major components of wood: (1) cellulose, (2) hemicellulose, and (3) lignin.

Kanury and Holve[26] have presented dimensional, phenomenological, approximate analytical, and exact numerical solutions for wood charring. Other models include those of Havens,[42] Knudson and Schniewind,[43] Kansa et al,[44] Hadvig and Paulsen,[45] and Tinney.[46]

Moisture desorption and surface recession were not considered until recently. There may be not only moisture desorption but also an increase in moisture content behind the char front caused by moisture movement away from the surface.[47] The CMA model[48] developed for NASA provides good results for oven-dry wood, because it includes surface recession but does not take into account moisture desorption. A model of Fredlund[49] includes mass transfer as well as heat transfer and provides for surface recession due to char oxidation. A major problem in the use of the more sophisticated models is the lack of adequate data to use as input.

Most theoretical models for wood charring not only define the charring rate but provide results for the temperature gradient. This temperature gradient is important in evaluating the load-carrying capacity of the wood remaining uncharred.

LOAD-CARRYING CAPACITY OF UNCHARRED WOOD

During the charring of wood caused by fire, the temperature gradient is fairly steep in the wood section remaining uncharred. Some loss of strength undoubtedly results from elevated temperatures. Schaffer et al[50] have combined parallel-to-grain strength and stiffness relationships with temperature and moisture content and the gradients of temperature and moisture content within a fire-exposed slab to obtain graphs of relative modulus of elasticity, compressive strength, and tensile strength as a function of distance below the char layer. (See Figure 4-11.5.) The theoretical models discussed previously can be used to determine the temperature gradient within the wood remaining uncharred. In tests of sawn timber, Noren[51] found no significant difference between low-grade and high-grade material. For equal stress ratios, the time to failure in fire established for clear wood can be applied to lumber with knots.

There are basically two approaches to evaluating the load-carrying capacity: to evaluate the remaining section either as a single homogeneous material or as a composite of layers with different properties.

Empirical α Models

In the standard ASTM E 119 test, structural failure is assumed to occur when the member is no longer capable of supporting its design load, the design load being a fraction of the ultimate load of the original beam. Failure occurs when the cross-sectional area of the member has been reduced by

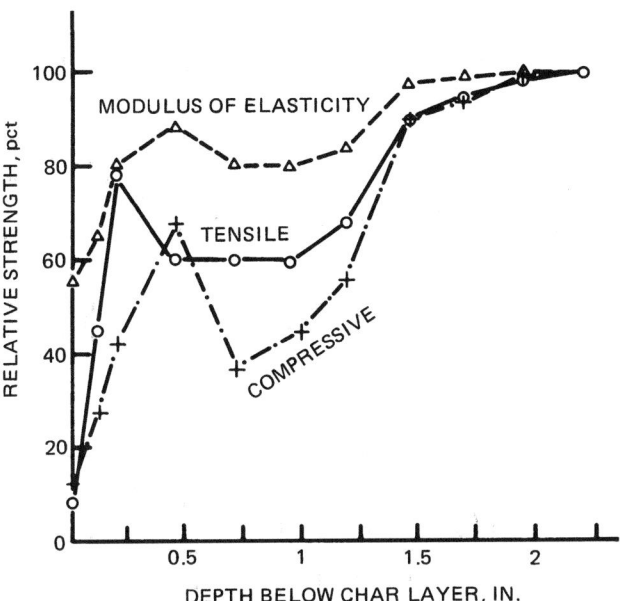

Fig. 4-11.5. *Relative modulus of elasticity and compressive and tensile strength as a function of distance below char layer in softwood section under fire exposures. (Expressed in percent of that at 25°C and initial moisture content of 12 percent.) Duration of fire exposure should be equal to or greater than 20 min. to apply results of this figure.*

the charring of the wood. One common approach in accounting for the loss in strength in the section remaining uncharred is to assume that the strength and stiffness of the entire uncharred region are fractions α of their room temperature values.

For bending rupture of a beam, an equation of this type would be

$$\frac{M}{S(t)} = \alpha\sigma_0 \qquad (22)$$

where

M = applied moment (design load),
S = section modulus of charred member,
σ_0 = modulus of rupture at room temperature, and
t = time.

Assuming the residual cross-section is rectangular in shape before and during fire exposure, the section modulus of the charred member is[52]

$$S(t) = \frac{1}{6}[(B - 2C_1t)(D - jC_2t)^2] \qquad (23)$$

where

B = original breadth of beam,
D = original depth of beam,
C_1 = charring rate in breadth direction,
C_2 = charring rate in depth direction, and
j = 1 for three-sided fire exposure or 2 for four-sided fire exposure (Figure 4-11.6).

Alternative to Equations 22 and 23 are the following, Equations 24 through 26:

$$\frac{k}{\alpha} \frac{B/D}{[d/D - (1 - B/D)]} = \left(\frac{d}{D}\right)^2 \qquad (24)$$

for exposure on all four sides,[53] and

$$\frac{k}{\alpha} \frac{B/D}{[B/D - 2(1 - d/D)]} = \left(\frac{d}{D}\right)^2 \qquad (25)$$

for exposure on three sides,[54,55]

where

k = load, as fraction of room temperature ultimate load of original member, and

d = critical depth of the uncharred beam.

The fire resistance is equal to the time to reach the critical depth, or

$$t = (D - d)/jC \qquad (26)$$

Proposed α values ranged from 0.5 in New Zealand to 0.83 in France.[52] The differences in α values are due to uncertainty, differences in design load, and desired level of safety. In the proposed Eurocode 5, this approach is called the "reduced strength and stiffness method."[28] The reduction factors are a function of the perimeter of the fire-exposed residual cross section divided by the area of the cross section.

The effect of the rounding of the charred member can be taken into account by increasing the value for char rate or including the effect in the empirical α parameter of Equation 22. In addition to bending rupture, the fire resistance of a beam may depend on lateral buckling of the beam.[53,56] Similar expressions can be developed for columns and tension members.[21,52,54,55,57]

The application of the above equations is generally limited to large wood members. Other reviews of fire resistance design methodologies for large wood members include those of Schaffer,[52] Pettersson,[58] and Barthelemy and Kruppa.[59] Kirpichenkov and Romanenkov[60] discussed the calculation procedures in the Soviet Union. The fire resistance of wood structures is also briefly discussed by Odeen.[61]

Fig. 4-11.6. *Fire exposure of beams on three or four sides.*

In developing a model for fire-exposed unprotected wood joist floor assemblies, Woeste and Schaffer[62,63] evaluated various time-dependent geometric terms that could be used to modify the strength reduction factor, α. The selected term was

$$\alpha = \frac{1}{1 + \left(\dfrac{B + 2D}{BD}\right)\gamma t_f} \qquad (27)$$

where

t_f = failure time, and
γ = empirical thermal degrade parameter.

The model has been experimentally evaluated,[64,65] extended to floor-truss assemblies,[63,67] and used as part of a first-order second-moment reliability analysis of floor assemblies.[62,63] In a model for metal-plate-connected wood trusses,[66] the strength degradation factors for the wood are calculated as a function of the duration of exposure and the temperature profile within the wood component.

Composite Models

A second approach to evaluating the fire endurance of a wood member is to assume that the uncharred region consists of layers. In one model with layers, the compressive and tensile strengths and modulus of elasticity of each layer are assumed to be fractions of the room temperature values. Using one 38-mm (1.5-in.) heated layer with reduced properties, Schaffer et al[50] analyzed a beam using transformed section analysis. In the similar elastic transformed section model of King and Glowinski,[68] the heated zone of the remaining wood section is divided into two layers at elevated temperatures.

For a second model with layers, an equivalent zero-strength layer, δ, was calculated.[50] For bending, the δ was estimated to be 8 mm (0.3 in. thick). This zero-strength layer, δ, was added to the char depth, βt, to obtain the total zero-strength layer. The rest of the member was then evaluated using room temperature property values. This zero-strength layer model was incorporated within a reliability-based model to predict the strength of glued-laminated beams with individual laminates of various grades of lumber.[69] This zero-strength layer approach is called the "effective cross-section method" in Eurocode 5.[28]

For fire-damaged members, Williamson[70] recommended δ of 6 mm (0.25 in.) for designs controlled by compression [16 mm (0.625 in.) if design is controlled by tension] and the use of 100 percent of the original basic allowable stresses in calculation of load capacity.

Do and Springer[71-73] have proposed a fire resistance model for wood beams based on mass loss *versus* strength data. The work included a program to predict the temperatures and mass loss within the wood member. The input data came from small-scale tension, compression, and shear tests done on specimens that had previously been heated in a muffle oven.

ONE-HOUR FIRE-RESISTIVE EXPOSED WOOD MEMBERS

Lie[54] developed simple formulas for calculating the fire resistance of large wood beams and columns, based on theoretical studies involving experimental data and equations similar to Equations 22 through 26. These formulas are contained within model building code documents and the *Supplement to the National Building Code of Canada*.[5] The methodology is discussed in two wood industry publications.[74,75] These formulas give the fire resistance time, t, in minutes, of a wood beam or column with minimum nominal dimension of 6 in. The net finish width for a nominal 6-in. glued-laminated member is 5⅛ inches.

For beams, the equations are

$$t = 2.54ZB[4 - 2(B/D)] \text{ for fire exposure on four sides} \qquad (28)$$

$$t = 2.54ZB[4 - (B/D)] \text{ for fire exposure on three sides} \qquad (29)$$

where

B = width (breadth) of a beam before exposure to fire (in.),
D = depth of a beam before exposure to fire (in.), and
Z = load factor. (See Figure 4-11.7.)

For columns, the equations are

$$t = 2.54ZD[3 - (D/B)] \text{ for fire exposure on four sides} \qquad (30)$$

$$t = 2.54ZD[3 - (D/2B)] \text{ for fire exposure on three sides} \qquad (31)$$

where

B = larger side of a column (in.), and
D = smaller side of a column (in.).

For columns, the load factor, Z, (see Figure 4-11.7) includes the effect of the effective length factor, K_e, (see Figure 4-11.8) and the unsupported length of the column, ℓ, (in.). Currently, the codes do not permit the wide side of the column to be the unexposed face (Equation 30). The full dimensions of the column are used even if the column is recessed into a wall.

Connectors and fasteners relating to support of the member must be protected for equivalent fire-resistive construction. Where minimal 1-hr fire endurance is required, connectors and fasteners must be protected from fire exposure by 1½ in. of wood, fire-rated gypsum board, or any coating approved for a 1-hr rating. The American Forest &

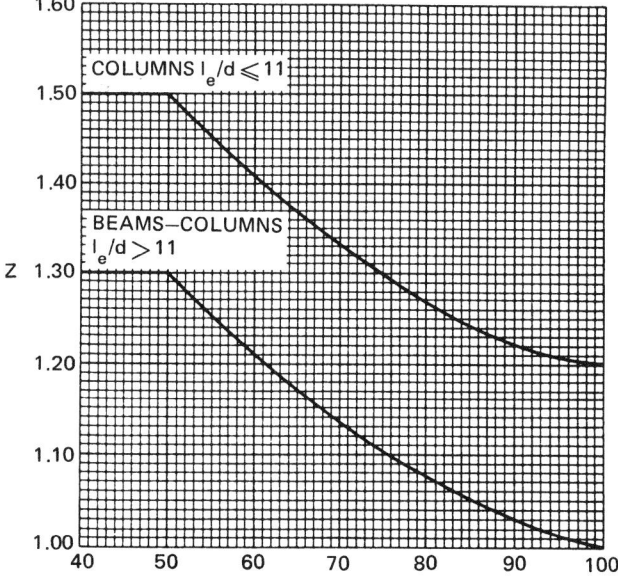

Fig. 4-11.7. Load factor versus load on member as percent of allowable. (NBCC uses 12 instead of 11 as criterion for two curves.)

EFFECTIVE COLUMN LENGTH FOR VARIOUS END CONDITIONS

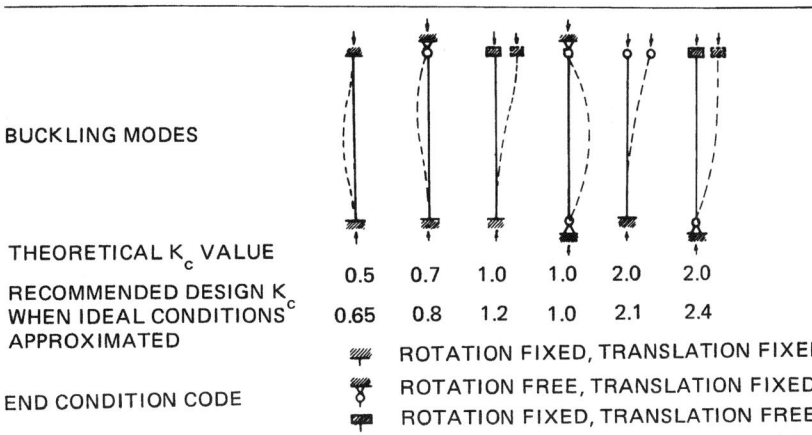

Fig. 4-11.8. Effective column length.

Paper Association publication[74] on the procedure includes diagrams giving typical details of such protection. Carling[76] summarized work done in Europe on the fire resistance of joint details in load-bearing wood construction. The new Eurocode 5[28] also includes information on calculating the fire endurance of connections and protecting connections in fire-rated timber members.

There is often a high-strength tension laminate on the bottom of glued-laminated timber beams. As a result, it is required that a core lamination be removed, the tension zone moved inward, and the equivalent of an extra nominal 2-in.-thick outer tension lamination be added to ensure that there is still a high-strength laminate left after fire exposure.

EXAMPLE 3:

Determine the fire resistance rating for a 5⅛-in. × 21-in. beam exposed to fire on three sides and loaded to 75 percent of its allowable load.

D = 21 in.
B = 5.125 in.

From Figure 4-11.7, Z for a beam loaded to 75 percent of allowable is 1.1. From Equation 29,

$$t = 2.54(1.1)(5.125)[4 - (5.125/21)]$$
$$t = 53.8 \text{ min.}$$

PROPERTY DATA

Proper input data are critical to the use of any model. For the models discussed in this section, property data include strength and stiffness properties and thermal properties. Property data for wood can be found in the *Wood Handbook: Wood as an Engineering Material*.[77] Equations and graphs of the strength and stiffness of wood as functions of temperature and moisture content are available,[78–80] but additional research is needed to better understand these relationships. Thermal properties can also be found in the various references for charring models and in other sources.[81] Thermal properties are needed for char and wood at the higher temperatures.

While it is often less complicated to assume constant property values, these properties are very often a function of other properties or factors. Most wood properties are functions of density, moisture content, grain orientation, and temperature.[77] Chemical composition may also be a factor. Since an understanding of these factors is important to the application of property data, the factors are defined in the rest of this section.

The oven-dry density of wood can range from 160 kg/m³ (10 lb/ft³) to over 1040 kg/m³ (65 lb/ft³), but most species are in the 320 to 720 kg/m³ (20 to 45 lb/ft³) range.[77] The density of wood relative to the density of water, i.e., specific gravity, is normally used to express the density. The specific gravity of wood is normally based on the oven-dry weight and the volume at some specified moisture content, but in some cases the oven-dry volume is used. As the empirical equations for charring rate show, the materials with higher density have slower char rate.

Wood is a hygroscopic material, which gains or loses moisture depending upon the temperature and relative humidity of the surrounding air. Moisture content of wood is defined as the weight of water in wood divided by the weight of oven-dry wood. Green wood can have a moisture content in excess of 100 percent. However, air-dry wood comes to equilibrium at a moisture content less than 30 percent. Under the conditions stated in ASTM E 119 (23°C, 50 percent relative humidity), wood has an equilibrium moisture content of 9 percent. At 23°C, 65 percent relative humidity, the equilibrium moisture content is 12 percent.[77] Moisture generally reduces the strength of wood but also reduces the charring rate.

Both density and moisture content affect the thermal conductivity of wood. The average thermal conductivity perpendicular to the grain for moisture contents below 40 percent[77] is

$$k = S(0.00020 + 0.000004M) + 0.024$$

where

k = thermal conductivity (W/m °C),
S = density based on volume at current moisture content and oven-dry weight (kg/m³), and
M = moisture content (percent).

The fiber (grain) orientation is important because wood is an orthotropic material. The longitudinal axis is parallel to the fiber or grain. The two transverse directions (perpendicular to the grain) are the radial and tangential axes. The radial axis is normal to the growth rings, and the tangential axis is tangent to the growth rings. For example, the longitudinal strength properties are usually about 10 times the transverse properties, and the longitudinal thermal conductivity is 2.0 to 2.8 times the transverse property.

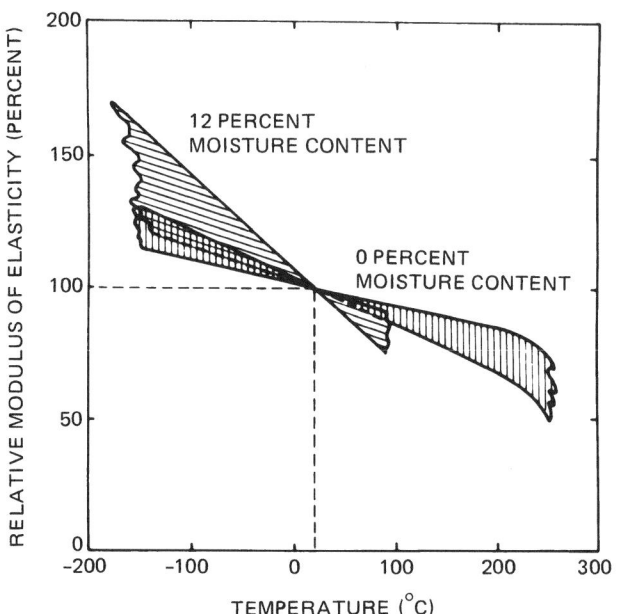

Fig. 4-11.9. *The immediate effect of temperature on modulus of elasticity parallel to the grain at two moisture contents relative to value at 20°C. The plot is a composite of results from several studies. Variability in reported trends is illustrated by the width of bands.*[77]

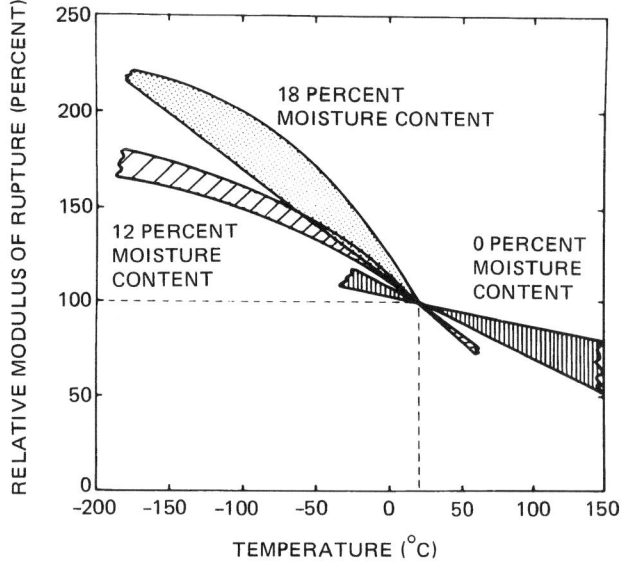

Fig. 4-11.10. *The immediate effect of temperature on modulus of rupture in bending at three moisture contents relative to value at 20°C.*[77]

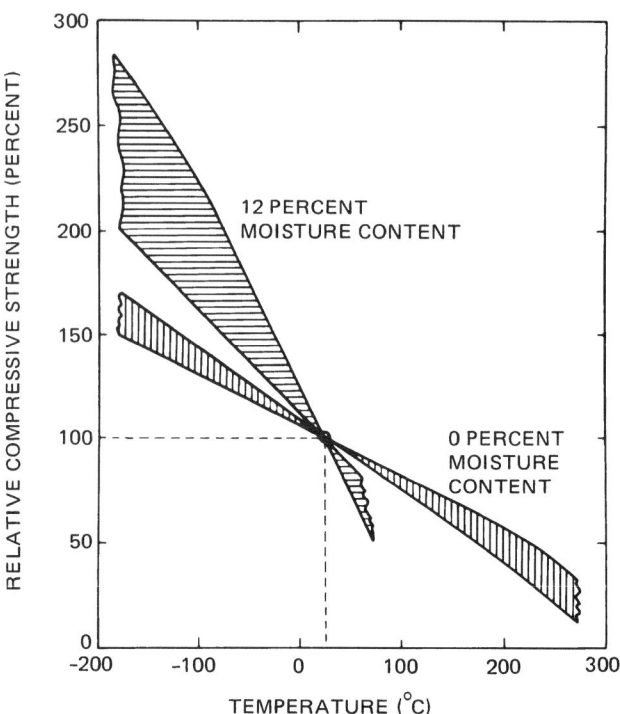

Fig. 4-11.11. *The immediate effect of temperature on compressive strength parallel to the grain at two moisture contents relative to the value at 20°C.*[77]

In fire resistance analysis, temperature can have a significant influence on the properties of wood. The preponderance of property data is often limited to temperatures below 100°C. The effect of temperatures on the strength properties of wood is shown in Figures 4-11.9 through 4-11.11. The specific heat capacity (kJ/kg °C) of dry wood is approximately related to temperature, *t*, in °C by[77]

$$\text{Specific heat capacity} = 1.125 + 0.00452\,t$$

The major components of wood are cellulose, lignin, hemicellulose, extractives, and inorganic materials (ash). Softwoods have lignin contents of 23 to 33 percent, while hardwoods have only 16 to 25 percent. The types and amounts of extractives vary. Cellulose content is generally around 50 percent by weight. The component sugars of hemicellulose are different for the hardwood and softwood species. Chemical composition can affect the kinetics of pyrolysis (Equation 20) and the percentage weight of the residual char. In the degradation of wood, higher lignin content results in greater char yield.

REFERENCES CITED

1. ASTM E 119-88, *Standard Test Methods for Fire Tests of Building Construction and Materials*, American Society for Testing and Materials, Philadelphia (1988).
2. *Fire Resistance Directory*, Underwriters Laboratories, Northbrook (annual).
3. *Fire Resistance Design Manual*, Gypsum Association, Evanston (1988).
4. T.Z. Harmathy, *Fire Tech.*, 1, 93 (1965).

5. *Supplement to the National Building Code of Canada*, National Research Council of Canada, Ottawa (1990).
6. *DCA No. 4: Component Additive Method (CAM) for Calculating and Demonstrating Assembly Fire Endurance*, American Forest & Paper Association, Washington, DC (1991).
7. *Wood and Fire Safety*, Canadian Wood Council, Ottawa (1991).
8. R.H. White, in *ASTM STP 826*, American Society for Testing and Materials, Philadelphia (1983).
9. L.R. Richardson and A.A. Cornelissen, *Fire and Materials*, 11, 191 (1987).
10. R.H. White, *J. of Test. and Eval.*, 14, 97 (1986).
11. *Fire Resistance of Wood Structures*, Technical Research Centre of Finland, Helsinki (1980).
12. W.D. Gardner and D.R. Syme, *Technical Report No. 5*, N.S.W. Timber Advisory Council Ltd., Sydney (1991).
13. F.C.W. Fung, *NBSIR 77-1260*, National Bureau of Standards, Washington, DC (1977).
14. B.W. Gammon, "Reliability Analysis of Wood-Frame Wall Assemblies Exposed to Fire," Dissertation, University of California, Berkeley (1987).
15. J.R. Mehaffey and M.A. Sultan, in *Proc. of First International Fire and Materials Conference*, Interscience Communications Ltd., London (1992).
16. F.L. Browne, *Rep. No. 2136*, USDA Forest Service, Forest Product Lab., Madison (1958).
17. E.L. Schaffer, *Res. Note FPL-145*, USDA Forest Service, Forest Product Lab., Madison (1966).
18. G.S. Hall, R.G. Saunders, R.T. Allcorn, P.E. Jackman, M.W. Hickey, and R. Fitt, *Fire Performance of Timber—A Literature Survey*, Timber Research and Development Association, High Wycombe (1971).
19. E.L. Schaffer, *Wood and Fiber*, 9, 145 (1977).
20. S. Hadvig, *Charring of Wood in Building Fires*, Technical University of Denmark, Lyngby (1981).
21. H.L. Malhotra, *Design of Fire-Resisting Structures*, Surrey University Press, London (1982).
22. E.L. Schaffer, *Res. Pap. FPL 69*, USDA Forest Service, Forest Product Lab., Madison (1967).
23. R.H. White, *Fire Technology*, 28, 5 (1992).
24. M. Lache, *Holz-Zentralblatt*, 117, 473 (1991).
25. P.C.R. Collier, *Study Report No. 42*, Building Research Association of New Zealand, Judgeford (1992).
26. A.M. Kanury and D.J. Holve, *NBS-GCR 76-50*, National Bureau of Standards, Washington, DC (1975).
27. B.J. Noren and B.A.-L. Ostman, in *Fire Safety Science—Proceedings of the First International Symposium*, Hemisphere, New York (1986).
28. M. Kersken-Bradley, in *Proc. of Oxford Fire Conference*, Timber Research and Development Association, High Wycombe (1993).
29. E.L. Schaffer, *J. Fire and Flamm.*, 1, 96 (1974).
30. H.C. Tran and R.H. White, *Fire and Materials*, 16, 197 (1992).
31. E. Mikkola, in *Fire Safety Science—Proceedings of the Third International Symposium*, Elsevier Applied Science, London (1991).
32. R.M. Nussbaum, *J. Fire Sciences*, 6, 290 (1988).
33. O. Pettersson, S.E. Magnusson, and J. Thor, *Publication 50*, Swedish Institute of Steel Construction, Sweden (1976).
34. F.B. Oleson and J. Konig, *Report No. I 9210061*, Swedish Institute for Wood Technology Research (Tratek), Stockholm (1991).
35. A.F. Roberts, in *Thirteenth Symposium (Int.) on Combustion*, The Combustion Institute, Pittsburgh (1971).
36. C.H. Bamford, J. Crank, and D.H. Malan, *Proc. of Camb. Phil. Soc.*, 46, 166 (1946).
37. P.H. Thomas, *Fire Research Note No. 446*, Fire Research Station, Borehamwood (1960).
38. H. Kung, *Combustion and Flame*, 18, 185 (1972).
39. F. Tamanini, in *Appendix A of Factory Mutual Research Corporation Report No. 21011.7*, Factory Mutual Research Corp., Norwood (1976).
40. A. Atreya, *Pyrolysis: Ignition and Fire Spread on Horizontal Surfaces of Wood*, Ph.D. Thesis, Harvard University, Cambridge (1983).
41. W.J. Parker, in *Fire Safety Science—Proceedings of the Second International Symposium*, Hemisphere, New York (1989).

42. J.A. Havens, *Thermal Decomposition of Wood*, Dissertation, University of Oklahoma (1969).
43. R.M. Knudson and A.P. Schniewind, *Forest Prod. J.*, 25, 23 (1975).
44. E.J. Kansa, H.E. Perlee, and R.F. Chaiken, *Comb. and Flame*, 29, 311 (1977).
45. S. Hadvig and O.R. Paulsen, *J. Fire and Flamm.*, 1, 433 (1976).
46. E.R. Tinney, in *Tenth Symposium (Int.) on Combustion*, The Combustion Institute, Pittsburgh (1965).
47. R.H. White and E.L. Schaffer, *Wood and Fiber*, 13, 17 (1981).
48. R.H. White and E.L. Schaffer, *Fire Tech.*, 14, 279 (1978).
49. B. Fredlund, *Fire Safety J.*, 20, 39 (1993).
50. E.L. Schaffer, C.M. Marx, D.A. Bender, and F.E. Woeste, *Res. Pap. FPL 467*, USDA Forest Service, Forest Product Lab., Madison (1986).
51. J. Noren, *Report I 8810066*, Swedish Institute for Wood Technology (Tratek), Stockholm (1988).
52. E.L. Schaffer, *Res. Pap. FPL 450*, USDA Forest Service, Forest Product Lab., Madison (1984).
53. C. Imaizumi, *Norsk Skogind*, 16, 140 (1962).
54. T.T. Lie, *Can. J. of Civil Engg.*, 4, 161 (1977).
55. K. Odeen, in *Fire and Structural Use of Timber in Buildings*, Her Majesty's Stationery Office, London (1970).
56. B. Fredlund, *Report No. 79-5*, Lund Institute of Technology, Lund (1979).
57. C. Meyer-Ottens, in *Three Decades of Structural Fire Safety*, Building Research Establishment, Fire Research Station, Borehamwood, England (1983).
58. O. Pettersson, in *Three Decades of Structural Fire Safety*, Building Research Establishment, Borehamwood (1983).
59. B. Barthelemy and J. Kruppa, *Resistance au Leu des Structures*, Editions Eyrolles, Paris (1978).
60. G.M. Kirpichenkov and I.G. Romanenkov, *NBSIR 80-2188*, National Bureau of Standards, Washington, DC (1980).
61. K. Odeen, *Fire Tech.*, 21, 34 (1985).
62. F.E. Woeste and E.L. Schaffer, *Fire and Matls.*, 3, 126 (1979).
63. F.E. Woeste and E.L. Schaffer, *Res. Pap. FPL 386*, USDA Forest Service, Forest Product Lab., Madison (1981).
64. R.H. White, E.L. Schaffer, and F.E. Woeste, *Wood and Fiber*, 16, 374 (1984).
65. E.L. Schaffer, R.H. White, and F.E. Woeste, in *Proc. 1988 International Conference on Timber Engineering*, Forest Products Research Society, Madison (1988).
66. R.H. White, S.M. Cramer, and D. Shrestha, *Res. Pap. FPL 522*, USDA, Forest Service, Forest Products Lab., Madison (1993).
67. E.L. Schaffer and F.E. Woeste, in *Proceedings, Metal Plate Wood Truss Conference*, Forest Products Research Society, Madison (1981).
68. E.G. King and R.W. Glowinski, *Forest Prod. J.*, 38(10), 31 (1988).
69. D.A. Bender, F.E. Woeste, E.L. Schaffer, and C.M. Marx, *Res. Pap. FPL 460*, USDA Forest Service, Forest Prod. Lab., Madison (1985).
70. T.G. Williamson, in *Evaluation, Maintenance, and Upgrading of Wood Structures*, American Society of Civil Engineers, New York (1982).
71. M.H. Do and G.S. Springer, *J. of Fire Sci.*, 1, 271 (1983).
72. M.H. Do and G.S. Springer, *J. of Fire Sci.*, 1, 285 (1983).
73. M.H. Do and G.S. Springer, *J. of Fire Sci.*, 1, 297 (1983).
74. *DCA No. 2, Design of Fire-Resistive Exposed Wood Members*, American Forest & Paper Association, Washington, DC (1985).
75. American Institute of Timber Construction, *Timber Construction Manual*, John Wiley and Sons, New York (1985).
76. O. Carling, *Study Report No. 18*, Building Research Association of New Zealand, Judgeford (1989).
77. *Wood Handbook: Wood as an Engineering Material (USDA Agr. Hdbk. No. 72)*, Superintendent of Documents, Washington, DC (1987).
78. C.C. Gerhards, *Wood and Fiber*, 14, 4 (1982).
79. F.C. Beall, in *Structural Use of Wood in Adverse Environments*, Van Nostrand Reinhold, New York (1982).
80. B.A.-L. Ostman, *Wood Sci. Tech.*, 19, 103 (1985).
81. K.W. Ragland, D.J. Aerts, and A.J. Baker, *Bioresource Technology*, 37, 161 (1991).

SMOKE CONTROL

John H. Klote

INTRODUCTION

In building fire situations, smoke often flows to locations remote from the fire, threatening life and damaging property. Stairwells and elevators frequently become smoke-logged, thereby blocking and/or and inhibiting evacuation. Today smoke is recognized as the major killer in fire situations.[1]

In the late 1960s, the idea of using pressurization to prevent smoke infiltration of stairwells started to attract attention. This was followed by the idea of the "pressure sandwich," i.e., venting or exhausting the fire floor and pressurizing the surrounding floors. Frequently, the building's ventilation system is used for this purpose. The term "smoke control" was coined as a name for such systems that use pressurization produced by mechanical fans to limit smoke movement in fire situations.

Research in the field of smoke control has been conducted in Australia, Canada, England, France, Japan, the United States, and West Germany. This research has consisted of field tests, full-scale fire tests, and computer simulations. Many buildings have been built with smoke control systems and numerous others have been retrofitted for smoke control.

In this chapter the term smoke is defined in accordance with the American Society for Testing and Materials (ASTM)[2] and the National Fire Protection Association (NFPA)[3] definitions which state that smoke consists of the airborne solid and liquid particulates and gases evolved when a material undergoes pyrolysis of combustion.

SMOKE MOVEMENT

A smoke control system must be designed so that it is not overpowered by the driving forces that cause smoke movement. For this reason, an understanding of the fundamental concepts of smoke movement and of smoke control is a prerequisite to intelligent smoke control design. The major driving forces causing smoke movement are stack effect, buoyancy, expansion, wind, and the heating, ventilating, and air conditioning (HVAC) system. Generally, in a fire situation, smoke movement will be caused by a combination of these driving forces. The following subsections are a discussion of each driving force as it would act independent of the presence of any other driving force.

Stack Effect

When it is cold outside, there is often an upward movement of air within building shafts such as stairwells, elevator shafts, dumbwaiter shafts, mechanical shafts, or mail chutes. This phenomenon is referred to as normal stack effect. The air in the building has a buoyant force because it is warmer and less dense than the outside air. This buoyant force causes air to rise within the shafts of buildings. The significance of normal stack effect is greater for low outside temperatures and for tall shafts. However, normal stack effect can exist in a one-story building.

When the outside air is warmer than the building air, a downward airflow frequently exists in shafts. This downward airflow is called reverse stack effect. At standard atmospheric pressure, the pressure difference due to either normal or reverse stack effect is expressed as

$$\Delta P = K_s\left(\frac{1}{T_0} - \frac{1}{T_I}\right)h \qquad (1)$$

where:

ΔP = pressure difference, in. H_2O (Pa)

T_0 = absolute temperature of outside air, R (K)*

T_I = absolute temperature of air inside shaft, R (K)*

h = distance above neutral plane, ft (m)**

K_s = coefficient, 7.64 (3460).

For a building 200 ft (60 m) tall, with a neutral plane at the midheight, an outside temperature of 0°F (−18°C) and an inside temperature of 70°F (21°C), the maximum pressure difference due to stack effect would be 0.22 in. H_2O (55 Pa).

Dr. John H. Klote is Leader of Smoke Management Research at the Building and Fire Research Laboratory of the National Institute of Standards and Technology.

* Because the Fahrenheit and Celsius temperature scales are so commonly used by design engineers, these scales are used exclusively in the discussions in the text and in figures. However, the reader is cautioned to use absolute temperatures in calculations where such temperatures are stipulated.

** The neutral plane is the horizontal plane where the hydrostatic pressure inside equals that outside.

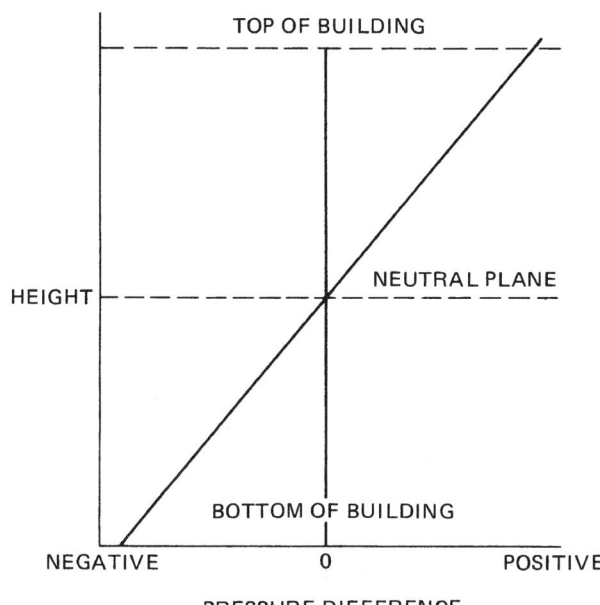

Fig. 4-12.1. *Pressure difference between an inside shaft and the outside due to normal stack effect.*

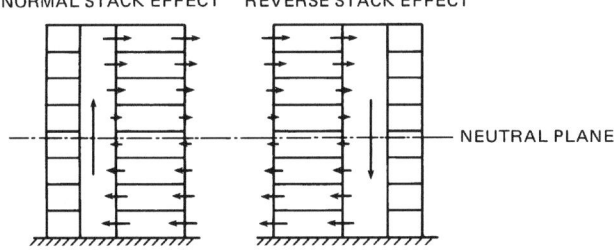

Fig. 4-12.2. *Air movement due to normal (left) and reverse stack effect (right). Note: arrows indicate direction of air movement.*

This means that at the top of the building, a shaft would have a pressure of 0.22 in. H_2O (55 Pa) greater than the outside pressure. At the bottom of the shaft, the shaft would have a pressure of 0.22 in. H_2O (55 Pa) less than the outside pressure. Figure 4-12.1 is a diagram of the pressure difference between a building shaft and the outside. In the diagram, a positive pressure difference indicates that the shaft pressure is higher than the outside pressure, and a negative pressure difference indicates the opposite.

Stack effect is usually thought of as existing between the inside of a building and the outside atmosphere. The air movement in buildings caused by both normal and reverse stack effect is illustrated in Figure 4-12.2. In this case, the pressure difference expressed in Equation 1 would actually refer to the pressure difference between the shaft and the outside of the building.

Figure 4-12.3 can be used to determine the pressure difference due to stack effect. For normal stack effect, the term $\Delta P/h$ is positive, and the pressure difference is positive above the neutral plane and negative below it. For reverse stack effect, the term $\Delta P/h$ is negative, and the pressure difference is negative above the neutral plane and positive below it.

In unusually airtight buildings with exterior stairwells, reverse stack effect has been observed even with low outside air temperatures.[4] In this situation, the exterior stairwell temperature was considerably lower than the building temperature. The stairwell was the cold column of air and other shafts within the building were the warm columns of air.

When considering stack effect, if the air leakage paths between a building and the outside are fairly uniform with height, the neutral plane will be located near the midheight of the building. However, when the leakage paths are not uniform, the location of the neutral plane can vary considerably, as in the case of vented shafts. McGuire and Tamura[5] provide methods for calculating the location of the neutral plane for some vented conditions.

Smoke movement from a building fire can be dominated by stack effect. In a building with normal stack effect, the existing air currents (as shown in Figure 4-12.2) can move smoke considerable distances from the fire origin. If the fire is below the neutral plane, smoke moves with the building air into and up the shafts. This upward smoke flow is enhanced by any buoyancy forces on the smoke existing due to its temperature. Once above the neutral plane, the smoke flows out of the shafts into the upper floors of the building. If the leakage between floors is negligible, the floors below the neutral plane, except the fire floor, will be relatively smoke free until the quantity of smoke produced is greater than can be handled by stack effect flows.

Smoke from a fire located above the neutral plane is carried by the building airflow to the outside through openings in the exterior of the building. If the leakage between floors is negligible, all floors other than the fire floor will remain relatively smoke-free, again, until the quantity of smoke produced is greater than can be handled by stack

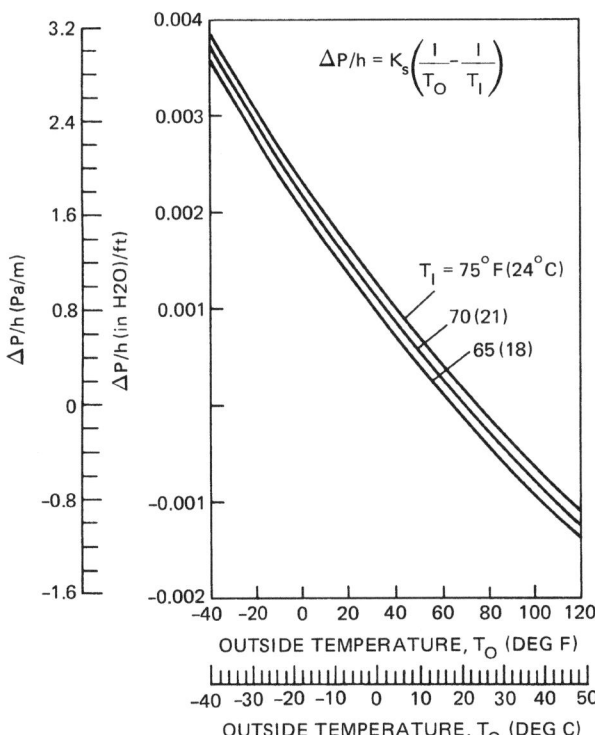

Fig. 4-12.3. *Pressure difference due to stack effect.*

effect flows. When the leakage between floors is considerable, there is an upward smoke movement to the floor above the fire floor.

The air currents caused by reverse stack effect are also shown in Figure 4-12.2. These forces tend to affect the movement of relatively cool smoke in the reverse of normal stack effect. In the case of hot smoke, buoyancy forces can be so great that smoke can flow upward even during reverse stack effect conditions.

Buoyancy

High-temperature smoke from a fire has a buoyancy force due to its reduced density. The pressure difference between a fire compartment and its surroundings can be expressed by an equation of the same form as Equation 1.

$$\Delta P = K_s\left(\frac{1}{T_0} - \frac{1}{T_F}\right)h \qquad (2)$$

where:

ΔP = pressure difference, in. H_2O (Pa)
T_0 = absolute temperature of the surroundings, R (K)
T_F = absolute temperature of the fire compartment, R (K)
h = distance above the neutral plane, ft (m)
K_s = coefficient, 7.64 (3460).

The pressure difference due to buoyancy can be obtained from Figure 4-12.4 for the surroundings at 68°F (20°C). The neutral plane is the plane of equal hydrostatic pressure between the fire compartment and its surroundings. For a fire with a fire compartment temperature of 1470°F (800°C), the pressure difference 5 ft (1.52 m) above the neutral plane is 0.052 in. H_2O (13 Pa). Fang[6] has studied pressures caused by room fires during a series of full-scale fire tests. During these tests, the maximum pressure difference reached was 0.064 in. H_2O (16 Pa) across the burn room wall at the ceiling.

Much larger pressure differences are possible for tall fire compartments where the distance, h, from the neutral plane can be larger. If the fire compartment temperature is 1290°F (700°C), the pressure difference 35 ft (10.7 m) above the neutral plane is 0.35 in. H_2O (88 Pa). This amounts to an extremely large fire, and the pressures produced by it are beyond the state-of-the-art of smoke control. However, the example is included here to illustrate the extent to which Equation 2 can be applied.

In a building with leakage paths in the ceiling of the fire room, this buoyancy-induced pressure causes smoke movement to the floor above the fire floor. In addition, this pressure causes smoke to move through any leakage paths in the walls or around the doors of the fire compartment. As smoke travels away from the fire, its temperature drops due to heat transfer and dilution. Therefore, the effect of buoyancy generally decreases with distance from the fire.

Expansion

In addition to buoyancy, the energy released by a fire can cause smoke movement due to expansion. In a fire compartment with only one opening to the building, building air will flow into the fire compartment and hot smoke will flow out of the fire compartment. Neglecting the added mass of the fuel (which is small compared to the airflow), the ratio of volumetric flows can simply be expressed as a ratio of absolute temperatures.

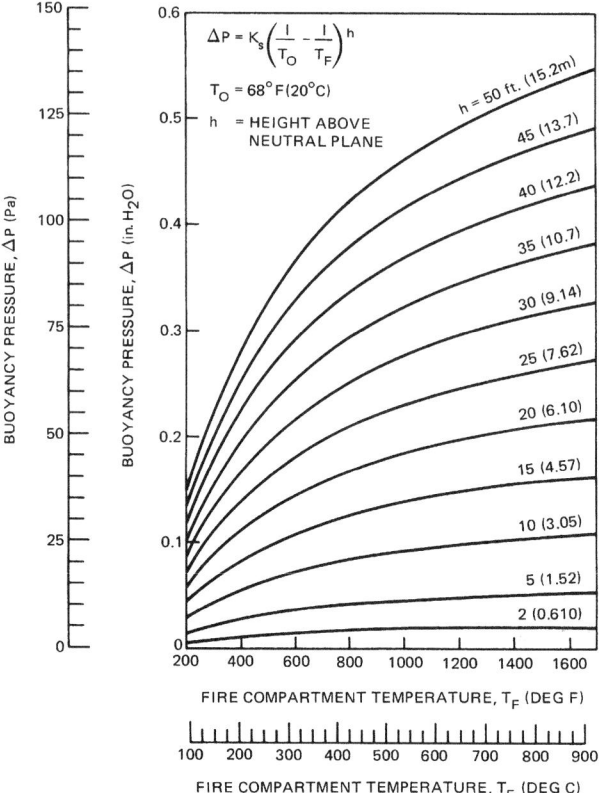

Fig. 4-12.4.   *Pressure difference due to buoyancy.*

$$\frac{Q_{out}}{Q_{in}} = \frac{T_{out}}{T_{in}}$$

where:

Q_{out} = volumetric flow rate of smoke out of the fire compartment, cfm (m³/s)
Q_{in} = volumetric flow rate of air into the fire compartment, cfm (m³/s)
T_{out} = absolute temperature of smoke leaving fire compartment, R (K)
T_{in} = absolute temperature of air into fire compartment, R (K).

For a smoke temperature of 1290°F (700°C) the ratio of volumetric flows would be 3.32. The reader is reminded to use absolute temperatures for calculation. In such a case, if the air flowing into the fire compartment is 3180 cfm (1.5 m³/s), then the smoke flowing out of the fire compartment would be 10,600 cfm (4.98 m³/s). In this case, the gas has expanded to more than three times its original volume.

For a fire compartment with open doors or windows, the pressure difference across these openings due to expansion is negligible. For a tightly sealed fire compartment, however, the pressure differences due to expansion may be important.

Wind

In many instances, wind can have a pronounced effect on smoke movement within a building. The pressure, P_w, that the wind exerts on a surface can be expressed as

$$P_w = \frac{1}{2}C_w\rho_o V^2 \qquad (3)$$

where:

C_w = dimensionless pressure coefficient
ρ_O = outside air density
V = wind velocity.

For an air density of 0.075 lb/ft³ (1.20 kg/m³) this relation becomes

$$P_w = C_w K_w V^2 \qquad (3a)$$

where:

P_w = wind pressure, in. H_2O (Pa)
V = wind velocity, mph (m/s)
K_w = coefficient, 4.82×10^{-4} (0.600).

The pressure coefficients, C_w, are in the range of -0.8 to 0.8, with positive values for windward walls and negative values for leeward walls. The pressure coefficient depends on building geometry and varies locally over the wall surface. In general, wind velocity increases with height in the boundary layer nearest the surface of the earth. Detailed information concering wind velocity variations and pressure coefficients is available from a number of sources.[7-10] Specific information about wind data, with respect to air infiltration in buildings, has been generated by Shaw and Tamura.[11]

A 35 mph (15.6 m/s) wind produces a pressure on a structure of 0.47 in. H_2O (117 Pa) with a pressure coefficient of 0.8. The effect of wind on air movement within tightly constructed buildings with all doors and windows closed is slight. However, the effects of wind can become important for loosely constructed buildings or for buildings with open doors or windows. Usually, the resulting airflows are complicated and, for practical purposes, computer analysis is required.

Frequently in fire situations, a window breaks in the fire compartment. If the window is on the leeward side of the building, the negative pressure caused by the wind vents the smoke from the fire compartment. This can greatly reduce smoke movement throughout the building. However, if the broken window is on the windward side, the wind forces the smoke throughout the fire floor and even to other floors. This both endangers the lives of building occupants and hampers fire fighting. Pressures induced by the wind in this type of situation can be relatively large and can easily dominate air movement throughout the building.

HVAC Systems

Before the development of the concept of smoke control, HVAC systems were shut down when fires were discovered.

The HVAC system frequently transports smoke during building fires. In the early stages of a fire, the HVAC system can serve as an aid to fire detection. When a fire starts in an unoccupied portion of a building, the HVAC system can transport the smoke to a space where people can smell the smoke and be alerted to the fire. However, as the fire progresses, the HVAC system will transport smoke to every area that it serves, thus endangering life in all those spaces. The HVAC system also supplies air to the fire space, which aids combustion. These are the reasons HVAC systems traditionally have been shut down when fires have been discovered. Although shutting down the HVAC system prevents it from supplying air to the fire, this does not prevent smoke movement through the supply and return air ducts, air shafts, and other building openings due to stack effect, buoyancy, or wind.

SMOKE MANAGEMENT

The term "smoke management," as used in this chapter, includes all methods that can be used independently or in combination to modify smoke movement for the benefit of occupants and fire fighters and for the reduction of property damage. The use of barriers, smoke vents, and smoke shafts are traditional methods of smoke management.

The effectiveness of a barrier in limiting smoke movement depends on the leakage paths in the barrier and on the pressure difference across the barrier. Holes where pipes penetrate walls or floors, cracks where walls meet floors, and cracks around doors are a few possible leakage paths. The pressure difference across these barriers depends on stack effect, buoyancy, wind, and the HVAC system.

The effectiveness of smoke vents and smoke shafts depends on their proximity to the fire, the buoyancy of the smoke, and the presence of other driving forces. In addition, when smoke is cooled due to sprinklers the effectiveness of smoke vents and smoke shafts is greatly reduced.

Elevator shafts in buildings have been used as smoke shafts. Unfortunately, this prevents their use for fire evacuation and these shafts frequently distribute smoke to floors far from the fire. Specially designed smoke shafts, which have essentially no leakage on floors other than the fire floor, can be used to prevent the smoke shaft from distributing smoke to nonfire floors.

PRINCIPLES OF SMOKE CONTROL

Smoke control uses the barriers (walls, floors, doors, etc.) used in traditional smoke management in conjunction with airflows and pressure differences generated by mechanical fans.

Figure 4-12.5 illustrates a pressure difference across a barrier acting to control smoke movement. Within the barrier is a door, and the high-pressure side of the door can be either a refuge area or an escape route. The low-pressure side is exposed to smoke from a fire. Airflow through the cracks around the door and through other construction cracks prevents smoke infiltration to the high-pressure side.

When the door in the barrier is opened, air flows through the open door. When the air velocity is low, smoke can flow against the airflow into the refuge area or escape

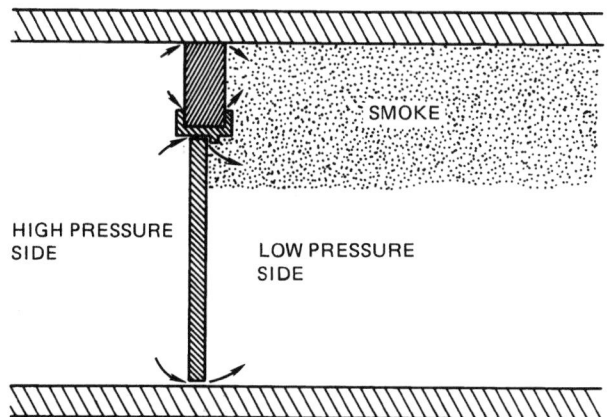

Fig. 4-12.5. *Pressure difference across a barrier of a smoke control system preventing smoke infiltration to the high-pressure side of the barrier.*

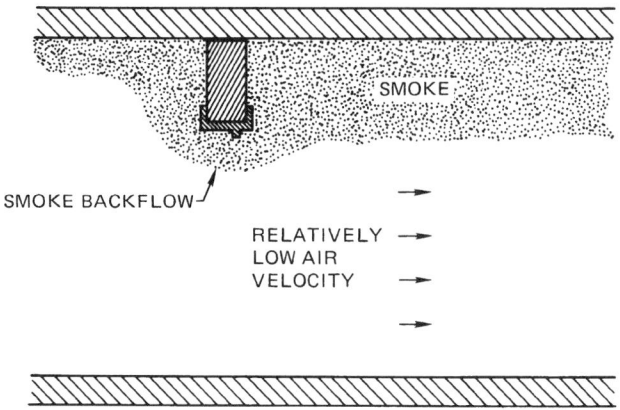

Fig. 4-12.6.   Smoke backflow against low air velocity through an open doorway.

route, as shown in Figure 4-12.6. This smoke backflow can be prevented if the air velocity is sufficiently large, as shown in Figure 4-12.7. The magnitude of the velocity necessary to prevent backflow depends on the energy release rate of the fire, as discussed in the section of this chapter regarding airflow.

The two basic principles of smoke control can be stated as follows:

1. Airflow by itself can control smoke movement if the average air velocity is of sufficient magnitude.
2. Air pressure differences across barriers can act to control smoke movement.

The use of air pressure differences across barriers to control smoke is frequently referred to as pressurization. Pressurization results in airflows in the small gaps around closed doors and in construction cracks, thereby preventing smoke backflows through these openings. Therefore, in a strict physical sense, the second principle is a special case of the first principle. However, considering the two principles separately is advantageous for smoke control design. For a barrier with one or more large openings, air velocity is the appropriate physical quantity for both design considerations and for acceptance testing. However, when there are only cracks, such as around closed doors, determination of the velocity is difficult and including it in the design is impractical. In this case, the appropriate physical quantity is pressure difference. Separate consideration of the two principles has the added advantage of emphasizing the different considerations necessary for open and closed doors.

Because smoke control relies on air velocities and pressure differences produced by fans, it has the following three advantages in comparison to the traditional methods of smoke management:

1. Smoke control is less dependent on tight barriers. Allowance can be made in the design for reasonable leakage through barriers.
2. Stack effect, buoyancy, and wind are less likely to overcome smoke control than passive smoke management. In the absence of smoke control, these driving forces cause smoke movement to the extent that leakage paths allow. However, pressure differences and airflows of a smoke control system act to oppose these driving forces.

3. Smoke control can be designed to prevent smoke flow through an open doorway in a barrier by the use of airflow. Doors in barriers are opened during evacuation and are sometimes accidentally left open or propped open throughout fires. In the absence of smoke control, smoke flow through these doors is common.

Smoke control systems should be designed so that a path exists for smoke movement to the outside; such a path acts to relieve pressure of gas expansion due to the fire heat.

The smoke control designer should be cautioned that dilution of smoke in the fire space is not a means of achieving smoke control, i.e., smoke movement cannot be controlled by simply supplying and exhausting large quantities of air from the space or zone in which the fire is located. This supplying and exhausting of air is sometimes referred to as purging the smoke. Because of the large quantities of smoke produced in a fire, purging cannot assure breathable air in the fire space. In addition, purging in itself cannot control smoke movement, because it does not provide the needed airflows at open doors and the pressure differences across barriers. However, for spaces separated from the fire space by smoke barriers, purging can significantly limit the level of smoke.

Airflow

Theoretically, airflow can be used to stop smoke movement through any space. However, the two places where air velocity is most commonly used to control smoke movement are open doorways and corridors. Thomas[12] has developed an empirical relation for the critical velocity to prevent smoke from flowing upstream in a corridor.

$$V_k = K \left(\frac{gE}{W\rho cT} \right)^{1/3} \tag{4}$$

where:

V_k = critical air velocity to prevent smoke backflow
E = energy release rate into corridor
W = corridor width
ρ = density of upstream air
c = specific heat of downstream gases
T = absolute temperature of downstream mixture of air and smoke
K = constant of the order of 1
g = gravitational constant.

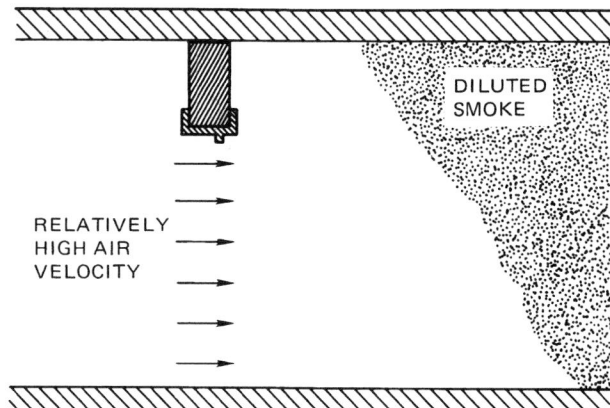

Fig. 4-12.7.   No smoke backflow with high air velocity through an open doorway.

The downstream properties are considered to be taken at a point sufficiently far downstream of the fire for the properties to be uniform across the cross-section. The critical air velocity can be evaluated at ρ = 0.081 lb/ft^3 (1.3 kg/m^3), c = 0.24 Btu/lb°F (1.005 kJ/kg°C), T = 81°F (27°C), and K = 1.

$$V_k = K_v \left(\frac{E}{W}\right)^{1/3} \tag{4a}$$

where:

V_k = critical air velocity to prevent smoke backflow, fpm (m/s)

E = energy release rate into corridor, Btu/hr (W)

W = corridor width, ft (m)

K_v = coefficient, 5.68 (0.0292).

This relation can be used when the fire is located in the corridor or when the smoke enters the corridor through an open door, air transfer grille, or other opening. The critical velocities calculated from the above relation are approximate because only an approximate value of K was used. However, critical velocities calculated from this relation are indicative of the type of air velocities required to prevent smoke backflow from fires of different sizes.

Equation 4 can be evaluated from Figure 4-12.8. For example, for an energy release rate of 0.512 × 10^6 Btu/hr (150 kW) into a corridor 4.00 ft (1.22 m) wide, the above relation yields a critical velocity of 286 fpm (1.45 m/s). However, for a larger energy release rate of 7.2 × 10^6 Btu/hr (2.1 MW), the relation yields a critical velocity of 690 fpm (3.50 m/s) for a corridor of the same width.

In general, a requirement for a high air velocity results in a smoke control system that is expensive and difficult to design. The use of airflow is most important in preventing smoke backflow through an open doorway that serves as a boundary of a smoke control system. Thomas[12] indicated that Equation 4 can be used to obtain a rough estimate of the airflow needed to prevent smoke backflow through a door. Many designers feel that it is prohibitively expensive to design systems to maintain air velocities in doorways greater than 300 fpm (1.5 m/s). A discussion of the elements of an appropriate design air velocity in a smoke control system is provided later, in this chapter.

Equation 4 is not appropriate for sprinklered fires that have small temperature differences between the upstream air and downstream gases. Shaw and Whyte[13] provide an analysis with experimental verification of a method to determine the velocity needed through an open doorway to prevent backflow of contaminated air. This analysis is specifically for small temperature differences and includes the effects of natural convection. If this method is used for a sprinklered fire where the temperature difference is only 3.6°F (2°C), then an average velocity of 50 fpm (0.25 m/s) would be the minimum velocity needed through a doorway to prevent smoke backflow. This temperature difference is small, and it is possible that larger values may be appropriate in many situations. Further research is needed in this area.

Even though airflow can be used to control smoke movement, it is not the primary control method because of the large quantities of air required for such systems to be effective. The primary means to control smoke movement is by air pressure differences across partitions, doors, and other building components.

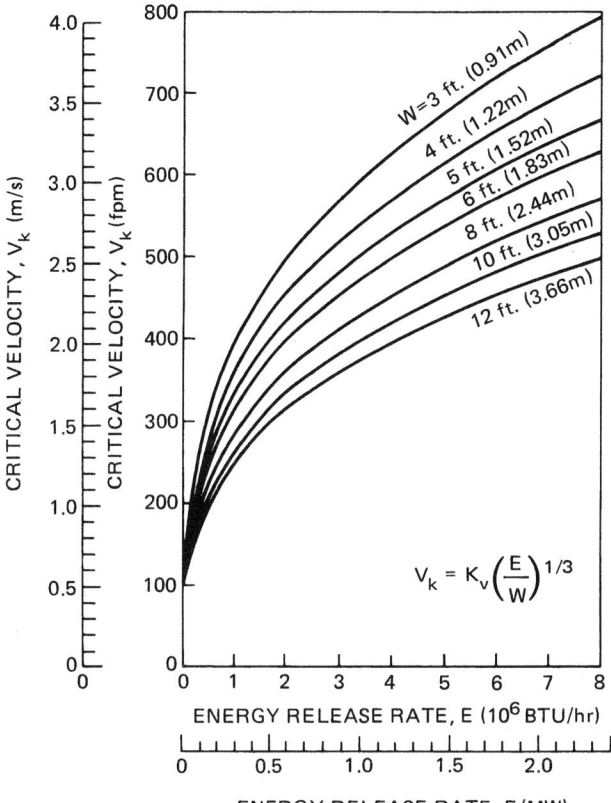

Fig. 4-12.8.   Critical velocity to prevent smoke backflow.

Pressurization

The airflow rate through a construction crack, door gap, or other flow path is proportional to the pressure difference across that path raised to the power n. For a flow path of fixed geometry, n is theoretically in the range of 0.5 to 1. However, for all flow paths except extremely narrow cracks, using n = 0.5 is reasonable and the flow can be expressed as

$$Q = CA \sqrt{\frac{2\Delta P}{\rho}} \tag{5}$$

where:

Q = volumetric airflow rate
C = flow coefficient
A = flow area (also called leakage area)
ΔP = pressure difference across the flow path
ρ = density of air entering the flow path.

The flow coefficient depends on the geometry of the flow path as well as on turbulence and friction. In the present context, the flow coefficient is generally in the range of 0.6 to 0.7. For ρ = 0.075 lb/ft^3 (1.2 kg/m^3) and C = 0.65, the flow equation above can be expressed as

$$Q = K_f A \sqrt{\Delta P} \tag{5a}$$

where:

Q = volumetric flow rate, cfm (m^3/s)
A = flow area, ft^2 (m^2)
ΔP = pressure difference across flow path, in. H$_2$O (Pa)
K_f = coefficient, 2610 (0.839).

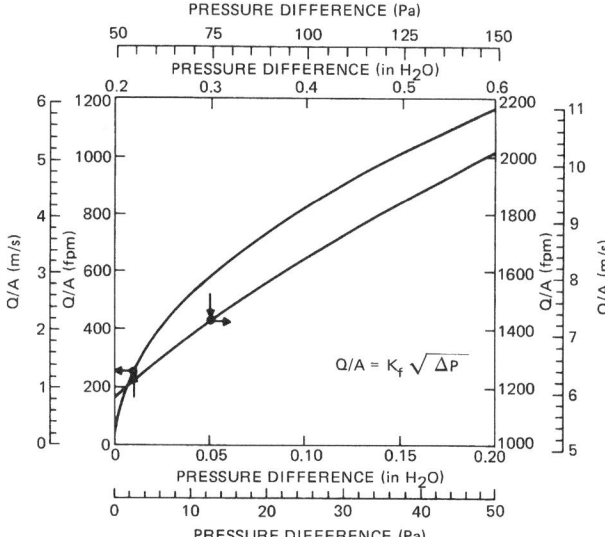

Fig. 4-12.9. *Airflow due to pressure difference.*

Airflow rate can also be determined from Figure 4-12.9. The flow area is frequently the same as the cross-sectional area of the flow path. A closed door with a crack area of 0.11 ft^2 (0.01 m^2) and a pressure difference of 0.01 in. H$_2$O (2.5 Pa) would have an air leakage rate of approximately 29 cfm (0.013 m^3/s). If the pressure difference across the door were increased to 0.30 in. H$_2$O (75 Pa), then the flow would be 157 cfm (0.073 m^3/s).

In field tests of smoke control systems, pressure differences across partitions or closed doors have frequently fluctuated by as much as 0.02 in. H$_2$O (5 Pa). These fluctuations have generally been attributed to wind, although they could have been due to the HVAC system or some other source. Pressure fluctuations and the resulting smoke movement are a current topic of research. To control smoke movement, the pressure differences produced by a smoke control system must be sufficiently large that they are not overcome by pressure fluctuations, stack effect, smoke buoyancy, and the forces of the wind. However, the pressure difference produced by a smoke control system should not be so large that door opening problems result.

PURGING

In general, the systems discussed in this chapter are based on the two basic principles of smoke control. However, it is not always possible to maintain sufficiently large airflows through open doors to prevent smoke from infiltrating a space that is intended to be protected. Ideally, such occurrences of open doors will only happen for short periods of time during evacuation. Smoke that has entered such a space can be purged, i.e., diluted by supplying outside air to the space.

Consider the case where a compartment is isolated from a fire by smoke barriers and self-closing doors, so that no smoke enters the compartment when the doors are closed. However, when one or more of the doors is open, there is insufficient airflow to prevent smoke backflow into the compartment from the fire space. To facilitate analysis, it is assumed that smoke is of uniform concentration throughout the compartment. When all the doors are closed, the concentration of contaminant in the compartment can be expressed as

$$\frac{C}{C_0} = e^{-at} \tag{6}$$

where:

C_0 = initial concentration of contaminant
C = concentration of contaminant at time, t
a = purging rate in number of air changes per minute
t = time after doors closed, in minutes
e = constant, approximately 2.718.

The concentrations C_0 and C must both be in the same units, and they can be any units appropriate for the particular contaminant being considered. McGuire, Tamura, and Wilson[14] evaluated the maximum levels of smoke obscuration from a number of tests and a number of proposed criteria for tolerable levels of smoke obscuration. Based on this evaluation, they state that the maximum levels of smoke obscuration are greater by a factor of 100 than those relating to the limit of tolerance. Thus, they indicate that an area can be "reasonably safe" with respect to smoke obscuration if its atmosphere will not be contaminated to an extent greater than 1 percent by the atmosphere prevailing in the immediate fire area. It is obvious that such dilution would also reduce the concentrations of toxic smoke components. Toxicity is a more complicated problem, and no parallel statement has been made regarding the dilution needed to obtain a safe atmosphere with respect to toxic gases.

Equation 6 can be solved for the purging rate as

$$a = \frac{1}{t} \log_e \left(\frac{C_0}{C} \right) \tag{7}$$

For example, if doors are open, the contaminant in a compartment is 20 percent of the burn room concentration, and at six minutes after the door is closed, the contaminant concentration is 1 percent of the burn room; then Equation 7 indicates that the compartment must be purged at a rate of one air change every two minutes.

In reality, it is impossible to assure that the concentration of the contaminant is uniform throughout the compartment. Because of buoyancy, it is likely that higher concentrations of contaminant would tend to be near the ceiling. Therefore, an exhaust inlet located near the ceiling and a supply outlet located near the floor would probably purge the smoke even faster than the previous calculations indicate. Caution should be exercised in the location of the supply and exhaust points to prevent the supply air from blowing into the exhaust inlet and thus short-circuiting the purging operation.

DOOR OPENING FORCES

The door opening forces resulting from the pressure differences produced by a smoke control system must be considered. Unreasonably high door opening forces can result in occupants having difficulty in, or being unable to open doors to refuge areas or escape routes.

The force required to open a door is the sum of the forces (1) to overcome the pressure difference across the door and (2) to overcome the door closer. This can be expressed as

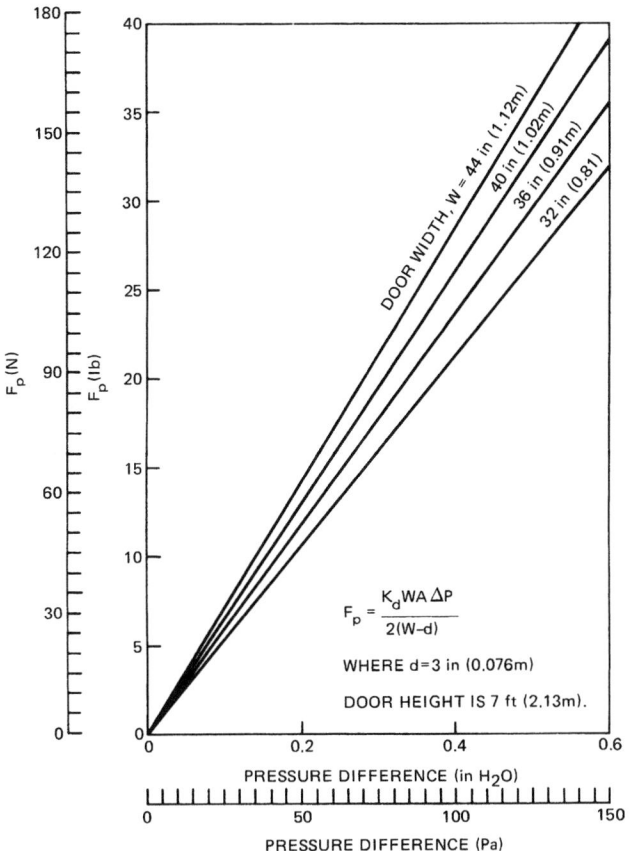

Fig. 4-12.10. *Door opening force due to a pressure difference.*

$$F = F_{dc} + \frac{K_d WA\Delta P}{2(W-d)} \qquad (8)$$

where:

    F = the total door opening force, lb (N)

    F_{dc} = the force to overcome the door closer, lb (N)

    W = door width, ft (m)

    A = door area, ft^2 (m^2)

    ΔP = pressure difference across the door, in. H$_2$O (Pa)

    d = distance from the doorknob to the edge of the knob side of the door, ft (m)

    K_d = coefficient, 5.20 (1.00).

This relation assumes that the door opening force is applied at the knob. Door opening forces due to pressure difference can be determined from Figure 4-12.10. The force to overcome the door closer is usually greater than 3 lb (13 N) and in some cases, can be as large as 20 lb (90 N). For a door that is 7 ft (2.13 m) high and 36 in. (0.91 m) wide, subject to a pressure difference of 0.30 in. H$_2$O (75 Pa), the total door opening force is 30 lb (133 N), if the force to overcome the door closer is 12 lb (53 N).

FLOW AREAS

Airflow paths must be identified and evaluated in the design of smoke control systems. Some leakage paths are obvious, such as cracks around closed doors, open doors,

elevator doors, windows, and air transfer grilles. Construction cracks in building walls are less obvious but no less important.

The flow area of most large openings, such as open windows, can be calculated easily. However, flow areas of cracks are more difficult to evaluate. The area of these leakage paths depends on workmanship, i.e., how well a door is fitted or how well weatherstripping is installed. A door that is 36 in. by 7 ft (0.9 × 2.1 m) with an average crack width of ⅛ in. (3.2 mm) has a leakage area of 0.21 ft^2 (0.020 m^2). However, if this door is installed with a ¾ in. (19 mm) undercut, the leakage area is 0.32 ft^2 (0.30 m^2). This is a significant difference. The leakage area of elevator doors has been measured in the range of 0.55 to 0.70 ft^2 (0.051 to 0.065 m^2) per door.

For open stairwell doorways, Cresci[15] found that complex flow patterns exist and that the resulting flow through open doorways was considerably below the flow calculated by using the geometric area of the doorway as the flow area in Equation 5a. Based on this research, it is recommended that the flow area of an open stairwell doorway be half that of the geometric area (door height multiplied by width) of the doorway. An alternate approach for open stairwell doorways is to use the geometric area as the flow area and use a reduced flow coefficient. Because it does not allow the direct use of Equation 5a, this alternate approach is not used here.

Typical leakage areas for walls and floors of commercial buildings are tabulated as area ratios in Table 4-12.1. These data are based on a relatively small number of tests performed by the National Research Council of Canada.[16-19] The area ratios are evaluated at typical airflows at 0.30 in. H$_2$O (75 Pa) for walls, and 0.10 in. H$_2$O (25 Pa) for floors. It is believed that actual leakage areas are primarily dependent on workmanship rather than construction materials, and in some cases, the flow areas in particular buildings may vary from the the values listed. Considerable data concerning leakage through building components is also provided in the *ASHRAE Handbook*.[20]

The determination of the flow area of a vent is not always straightforward, because the vent surface is usually covered by a louver and screen. Thus the flow area is less

TABLE 4-12.1 *Typical Leakage Areas for Walls and Floors of Commercial Buildings*

Construction Element	Wall Tightness	Area Ratio A/A_w
Exterior Building Walls (includes construction cracks, cracks around windows and doors)	Tight	0.70×10^{-4}
	Average	0.21×10^{-3}
	Loose	0.42×10^{-3}
	Very Loose	0.13×10^{-2}
Stairwell Walls (includes construction cracks but not cracks around windows or doors)	Tight	0.14×10^{-4}
	Average	0.11×10^{-3}
	Loose	0.35×10^{-3}
Elevator Shaft Walls (includes construction cracks but not cracks around doors)	Tight	0.18×10^{-3}
	Average	0.84×10^{-3}
	Loose	0.18×10^{-2}
		A/A_F
Floors (includes construction cracks and areas around penetrations)	Tight	0.66×10^{-5}
	Average	0.52×10^{-4}
	Loose	0.17×10^{-3}

A = leakage area; A_w = wall area; A_F = floor area.

than the vent area (vent height multiplied by width). Because the slats in louvers are frequently slanted, calculation of the flow area is further complicated. Manufacturers' data should be sought for specific information.

EFFECTIVE FLOW AREAS

The concept of effective flow areas is quite useful for analysis of smoke control systems. The various paths of smoke movement in the system can be parallel with one another, in series, or a combination of parallel and series paths. The effective flow area of a given system of flow paths is the area of a single opening that results in the same flow as the given system when subjected to the same pressure difference over the total system of flow paths. This concept is similar to an effective resistance of a system of electrical resistances.

The effective area, A_e, for the three parallel leakage areas of Figure 4-12.11 is

$$A_e = A_1 + A_2 + A_3 \qquad (9)$$

If A_1 is 1.08 ft^2 (0.10 m^2) and A_2 and A_3 are 0.54 ft^2 (0.05 m^2) each, then the effective flow area, A_e, is 2.16 ft^2 (0.20 m^2).

Equation 9 can be extended to any number of flow paths in parallel; i.e., it can be stated that the effective area is the sum of the individual leakage paths.

$$A_e = \sum_{i=1}^{n} A_i \qquad (10)$$

where n is the number of flow areas, A_i, in parallel.

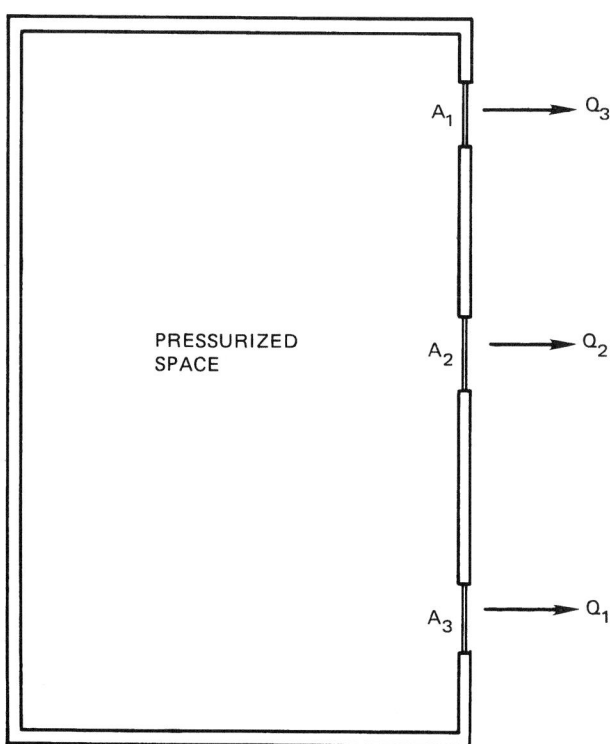

Fig. 4-12.11.   Leakage paths in parallel.

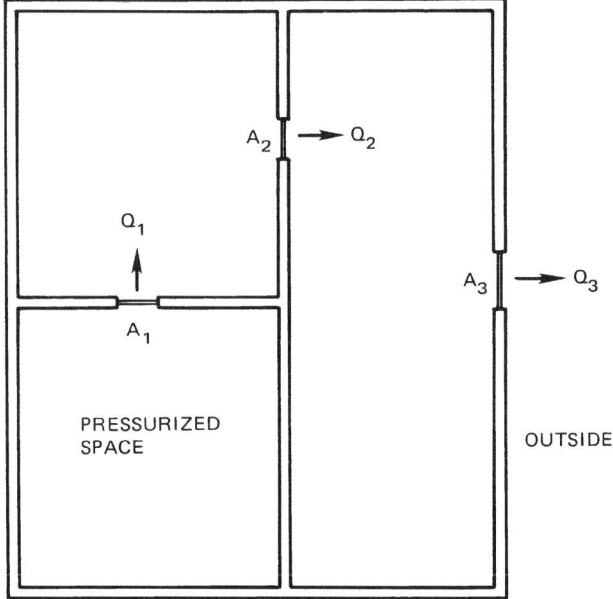

Fig. 4-12.12.   Leakage paths in series.

Three leakage areas in series from a pressurized space are illustrated in Figure 4-12.12. The effective flow area of these paths is

$$A_e = \left(\frac{1}{A_1^2} + \frac{1}{A_2^2} + \frac{1}{A_3^2} \right)^{-1/2} \qquad (11)$$

The general rule for any number of leakage areas is

$$A_e = \left(\sum_{i=1}^{n} \frac{1}{A_i^2} \right)^{-1/2} \qquad (12)$$

where n is the number of leakage areas, A_i, in series. In smoke control analysis, there are frequently only two paths in series. For this case, the effective leakage area is

$$A_e = \frac{A_1 A_2}{\sqrt{A_1^2 + A_2^2}} \qquad (13)$$

EXAMPLE 1:

Calculate the effective leakage area of two equal flow paths of 0.2 ft^2 in series. Let $A = A_1 = A_2 = 0.02$ m^2.

$$A_e = \frac{A^2}{\sqrt{2A^2}} = \frac{A}{\sqrt{2}} = 0.15 \text{ ft}^2 \quad (0.014 \text{ m}^2)$$

EXAMPLE 2:

Calculate the effective area of two flow paths in series, where $A_1 = 0.22$ ft^2 (0.02 m^2) and $A_2 = 2.2$ ft^2 (0.2 m^2).

$$A_e = \frac{A_1 A_2}{\sqrt{A_1^2 + A_2^2}} = 0.219 \text{ ft}^2 \quad (0.0199 \text{ m}^2)$$

This example illustrates that when two areas are in series and one is much larger than the other, the effective area is approximately equal to the smaller area.

The method of developing an effective area for a system of both parallel and series paths is to systemically combine groups of parallel paths and series paths. The system illustrated in Figure 4-12.13 is analyzed as an example.

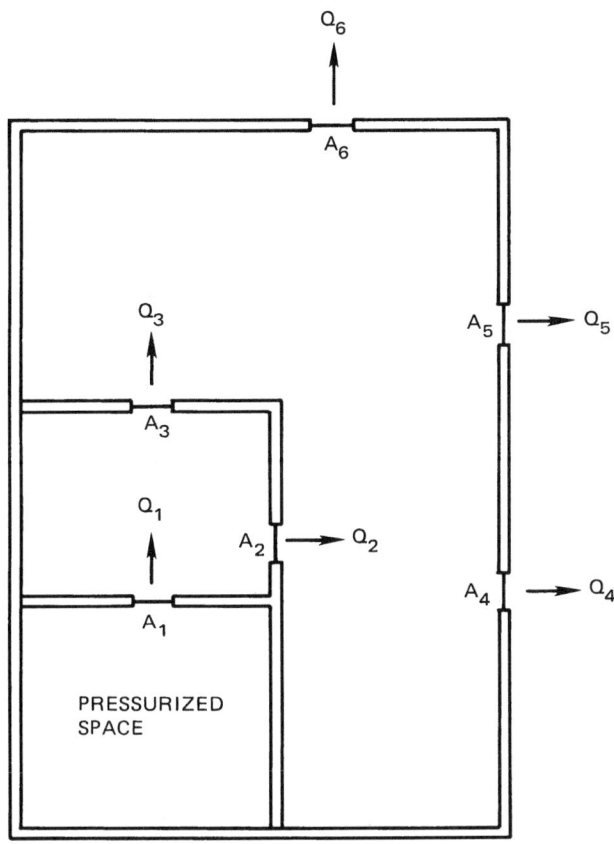

Fig. 4-12.13. Combination of leakage paths in parallel and series.

The figure shows that A_2 and A_3 are in parallel; therefore, their effective area is

$$A_{23_e} = A_2 + A_3$$

Areas A_4, A_5, and A_6 are also in parallel, so their effective area is

$$A_{456_e} = A_4 + A_5 + A_6$$

These two effective areas are in series with A_1. Therefore, the effective flow area of the system is given by

$$A_e = \left(\frac{1}{A_1^2} + \frac{1}{A_{23_e}^2} + \frac{1}{A_{456_e}^2} \right)^{-1/2}$$

EXAMPLE 3:

Calculate the effective area of the system in Figure 4-12.13, if the leakage areas are $A_1 = A_2 = A_3 = 0.22$ ft^2 (0.02 m^2) and $A_4 = A_5 = A_6 = 0.11$ ft^2 (0.01 m^2).

$$A_{23_e} = 0.44 \text{ ft}^2 \quad (0.04 \text{ m}^2)$$
$$A_{456_e} = 0.33 \text{ ft}^2 \quad (0.03 \text{ m}^2)$$
$$A_e = 0.16 \text{ ft}^2 \quad (0.015 \text{ m}^2)$$

SYMMETRY

The concept of symmetry is useful in simplifying problems and thereby easing solutions. Figure 4-12.14 illustrates the floor plan of a multistory building that can be divided in one-half by a plane of symmetry. Flow areas on one side of the plane of symmetry are equal to corresponding flow areas on the other side. For a building to be so treated, every floor of the building must be such that it can be divided in the same manner by the plane of symmetry. If wind effects are not considered in the analysis or if the wind direction is parallel to the plane of symmetry, then the airflow in only one-half of the building need be analyzed. It is not necessary that the building be geometrically symmetric, as shown in Figure 4-12.14; it must be symmetric only with respect to flow.

DESIGN PARAMETERS: A GENERAL DISCUSSION

Ideally, building and fire codes should contain design parameters leading to the design of functional and economical smoke control systems. Unfortunately, because smoke control is a new field, consensus has not yet been reached as to a definition of reasonable design parameters. Clearly, the designer has an obligation to adhere to any smoke control design criteria existing in appropriate codes or standards, but such criteria should be scrutinized to determine whether or not they will result in an effective system. If necessary, the designer should seek a waiver of the local codes, to ensure an effective smoke control system.

Five areas for which design parameters must be established are: (1) leakage areas, (2) weather data, (3) pressure differences, (4) airflow, and (5) number of open doors in the smoke control system.

Leakage areas have already been discussed in this chapter. An additional consideration affecting pressure differences and airflow is whether or not a window in the fire

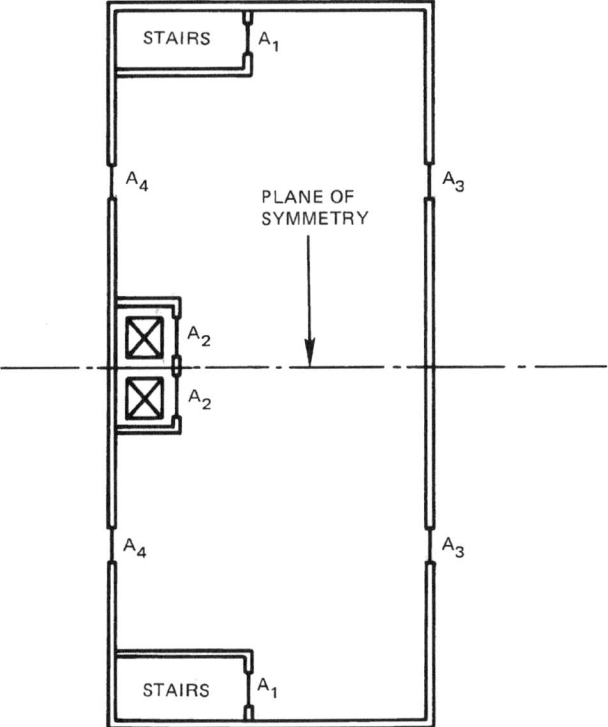

Fig. 4-12.14. Building floor plan illustrating symmetry concept.

compartment is broken. This factor is included in the following discussion of these parameters. In the absence of code requirements for specific parameters, the following discussion may be helpful to the designer.

Weather Data

The state-of-the-art of smoke control is such that little consideration has been given to the selection of weather data specifically for the design of smoke control systems. However, design temperatures for heating and cooling during winter and summer are recommended in the *ASHRAE Handbook*.[20] For example, 99 and 97.5 percent winter design temperatures have been provided. These values represent the temperatures that are equaled or exceeded in these portions of the heating season.*

A designer may wish to consider using these design temperatures for the design of smoke control systems. It should be remembered that in a normal winter, there would be approximately 22 hours at or below the 99 percent design value and approximately 54 hours at or below the 97.5 percent design value. Furthermore, extreme temperatures can be considerably lower than the winter design temperatures. For example, the ASHRAE 99 percent design temperature for Tallahassee, Florida is 27°F (−3°C), but the lowest temperature observed there by the National Climatic Center[21] was −2°F (−19°C) on February 13, 1899.

Temperatures are generally below the design values for short periods of time, and because of the thermal lag of building materials, these short intervals of low temperature usually do not result in problems with respect to heating systems. However, the same cannot necessarily be said of a smoke control system. There is no time lag for a smoke control system, i.e., a smoke control system is subjected to all the forces of stack effect that exist at the moment the system is being operated. If the outside temperature is below the winter design temperature for which a smoke control system was designed, then problems from stack effect may result. A similar situation can result with respect to summer design temperatures and reverse stack effect.

Wind data is needed for a wind analysis of a smoke control system. At present, no formal method of performing such an analysis exists, and the approach most generally taken is to design the smoke control system so as to minimize any effects of wind. The development of temperature and wind data for design of smoke control systems is an area for future effort.

Pressure Differences

It is appropriate to consider both the maximum and minimum allowable pressure differences across the boundaries of smoke control zones. The maximum allowable pressure difference should be a value that does not result in excessive door opening forces, but it is difficult to determine what constitutes excessive door opening forces. Clearly, a person's physical condition is a major factor in determining a reasonable door opening force for that person. NFPA *101®*, *Life Safety Code®*,[22] states that the force required to open any door in a means of egress shall not exceed 30 lb (133 N). In the section of this chapter on purging, a method of determining the door opening force is provided.

The criterion used in this chapter for selecting a minimum allowable pressure difference across a boundary of a smoke control system is that no smoke leakage should occur during building evaluation.** In this case, the smoke control system must produce sufficient pressure differences so that it is not overcome by the forces of wind, stack effect, or buoyancy of hot smoke. The pressure differences due to wind and stack effect can become very large in the event of a broken window in the fire compartment. Evaluation of these pressure differences depends on evacuation time, rate of fire growth, building configuration, and the presence of a fire suppression system. In the absence of a formal method of analysis, such evaluation must of necessity be based on experience and engineering judgment.

A method for determining the pressure difference across a smoke barrier resulting from the buoyancy of hot gases is provided in the section of this chapter regarding buoyancy. For a particular application, it may be considered necessary to design a smoke control system to withstand an intense fire next to a door at the boundary of a smoke control zone. Earlier in this chapter it was stated that in a series of full-scale fire tests, the maximum pressure difference reached was 0.064 in. H_2O (16 Pa) across the burn room wall at the ceiling. To prevent smoke infiltration, the smoke control system should be designed to maintain a pressure slightly higher than that generated in fire conditions. A minimum pressure difference in the range of 0.08 to 0.10 in. H_2O (20 to 25 Pa) is suggested.

If a smoke control boundary is exposed to hot smoke from a remote fire, a lower pressure difference due to buoyancy will result. For a smoke temperature of 750°F (400°C), the pressure difference caused by the smoke 5.0 ft (1.53 m) above the neutral plane would be 0.04 in. H_2O (10 Pa). In this situation, it is suggested that the smoke control system be designed to maintain a minimum pressure in the range of 0.06 to 0.08 in. H_2O (15 to 20 Pa).

Water spray from fire sprinklers cools smoke from a building fire and reduces the pressure differences due to buoyancy. In such a case it is probably wise to allow for pressure fluctuations. Accordingly, a minimum pressure difference in the range of 0.02 to 0.04 in. H_2O (5 to 10 Pa) is suggested.

Windows in the fire compartment can break due to exposure to high temperature gases. In such cases, the pressure due to the wind on the building exterior can be determined from Equation 3. If this window is the only opening to the outside on the fire floor and the window faces into the wind, the boundary of the smoke control system could be subjected to higher pressures. One possible solution is to vent the fire floor on all sides to relieve such pressures. For a building that is much longer than it is wide, it may be necessary to vent only on the two longer sides.

In addition to wind effects, stack effect can be increased in the event of a broken fire compartment window. With a fire on a lower floor during cold weather, stack effect will increase pressures of the fire floor above surrounding spaces. Even though little research has been done on the subject, the chances of a window breaking in the fire compartment are reduced by the operation of fire sprinklers.

---

* The heating season usually consists of three winter months. A more exact definition of these temperatures is available in Chapter 24 of the *ASHRAE Handbook—1985 Fundamentals*.[20]

** Other criteria might involve maintaining a number of smoke-free egress routes or preventing smoke infiltration to a refuge area. Discussion of all possible alternatives is beyond the scope of this chapter.

Airflow

When the doors in the boundaries of smoke control systems are open, smoke can flow into refuge areas or escape routes unless there is sufficient airflow through the open door to prevent smoke backflow, as discussed in the previous section. One criterion for selecting a design velocity through an open door is that no smoke backflow should occur during building evacuation.* Selection of this velocity depends on evacuation time, rate of fire growth, building configuration, and the presence of a fire suppression system. In the absence of a formal method of analysis, such an evaluation must be based on experience and engineering judgment.

At present, there is still much to be learned about the critical velocity needed to stop smoke backflow through an open door. In the absence of a specific relationship for doorways, the method of analysis presented for corridors in the earlier section regarding airflow can be used to yield approximate results. The width of the doorway may be used in place of the width of the corridor. This technique is based on the assumption that smoke properties are uniform across the cross-section. As previously illustrated, for a particular application, it may be considered necessary to design for an intensive fire, such as one with an energy release rate of 8×10^6 Btu/hr (2.4 MW). A critical velocity of approximately 800 fpm (4 m/s) would be required to stop smoke.

In another application, it may be estimated that the building would be subjected to a much less intense fire with an energy release rate of 427,000 Btu/hr (125 kW). To protect against smoke backflow during evacuation, the critical velocity would be 300 fpm (1.5 m/s).

In a sprinklered building, it might be considered that the smoke away from the immediate fire area would be cooled to near ambient temperature by the spray from the sprinklers. In such a case a design velocity in the range of 50 to 250 fpm (0.25 to 1.25 m/s) may be used. Research is needed to fully evaluate the effect of sprinklers on smoke control design parameters.

Number of Open Doors

The need for air velocity through open doors in the perimeter of a smoke control system has been discussed in this chapter. Another design consideration is the number of doors that could be opened simultaneously when the smoke control system is operational. A design that allows for all doors to be opened simultaneously may ensure that the system will always work, but it will probably add to the cost of the system.

Deciding on the number of doors that will be opened simultaneously depends largely on the building occupancy. For example, in a densely populated building, it is very likely that all the doors will be opened simultaneously during evacuation. However, if a staged evacuation plan or refuge area concept is incorporated in the building fire emergency plan, or if the building is sparsely occupied, only a few of the doors may be opened simultaneously during a fire.

PRESSURIZED STAIRWELLS

Many pressurized stairwells have been designed and built with the goal of providing a smoke-free escape route in

---

* Other criteria might include the allowance of limited smoke leakage into areas to be protected. Under such criteria, the toxicity of the smoke is a factor that must be considered.

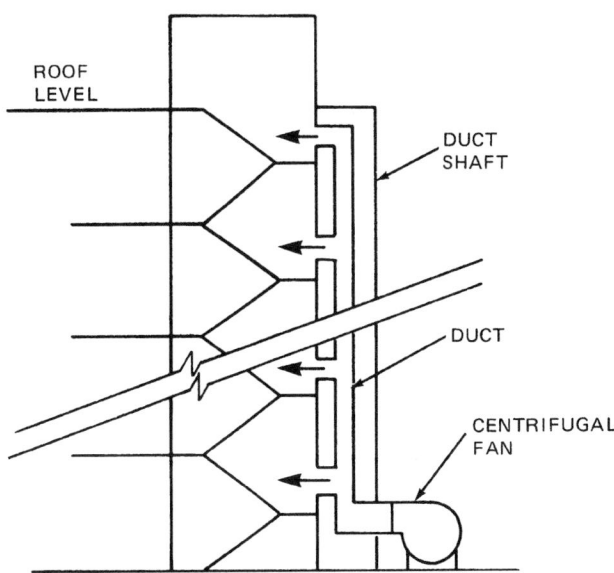

Fig. 4-12.15. Stairwell pressurization by multiple injection with the fan located at ground level.

the event of a building fire. A secondary objective is to provide a smoke-free staging area for fire fighters. On the fire floor, a pressurized stairwell must maintain a positive pressure difference across a closed stairwell door so that smoke infiltration is prevented.

During building fire situations, some stairwell doors are opened intermittently during evacuation and fire fighting, and some doors may even be blocked open. Ideally, when the stairwell door is opened on the fire floor, there should be sufficient airflow through the door to prevent smoke backflow. Designing such a system is difficult because of the large number of permutations of open stairwell doors and weather conditions that affect the airflow through open doors.

Stairwell pressurization systems are divided into two categories—single and multiple injection systems. A single injection system is one that has pressurized air supplied to the stairwell at one location; the most common injection point is at the top of the stairwell. Associated with this system is the potential for smoke feedback into the pressurized stairwell, i.e., of smoke entering the stairwell through the pressurization fan intake. Therefore, the capability of automatic shutdown in such an event should be considered.

For tall stairwells, single injection systems can fail when a few doors are open near the air supply injection point. All of the pressurized air can be lost through the few open doors, and the system can then fail to maintain positive pressures across doors farther from the injection point. Such a failure mode is especially likely with bottom injection systems when a ground level stairwell door is open.

For tall stairwells, supply air can be supplied at a number of locations over the height of the stairwell. Figures 4-12.15 and 4-12.16 are two examples of many possible multiple injection systems which can be used to overcome the limitations of single injection systems. In these figures the supply duct is shown in a separate shaft, but systems have been built that have eliminated the expense of a separate duct shaft by locating the supply duct in the stairwell

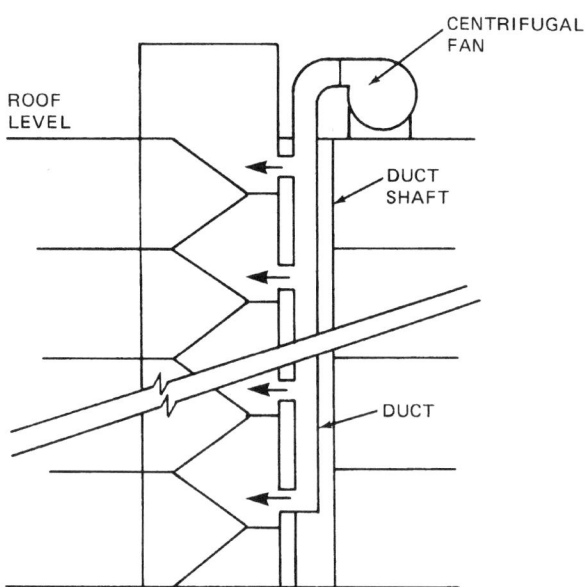

Fig. 4-12.16. Stairwell pressurization by multiple injection with roof-mounted fan.

itself. Obviously, care must be taken in such a case, that the duct does not become an obstruction to orderly building evacuation.

STAIRWELL COMPARTMENTATION

An alternative to multiple injection is compartmentation of the stairwell into a number of sections, as illustrated in Figure 4-12.17. When the doors between compartments are open, the effect of compartmentation is lost. For this reason, compartmentation is inappropriate for densely populated buildings where total building evacuation by the stairwell is planned in the event of fire. However, when a staged evacuation plan is used and when the system is designed to operate successfully when the maximum number of doors between compartments are open, compartmentation can be an effective means of providing stairwell pressurization for tall stairwells.

STAIRWELL ANALYSIS

In this section of the chapter, a method of analysis is presented for a pressurized stairwell in a building without vertical leakage. The performance of pressurized stairwells in buildings without elevators may be closely approximated by this method, which is useful for buildings with vertical leakage in that it yields conservative results. Only one stairwell is considered in the building, but the analysis can be extended to any number of stairwells by the concept of symmetry. For evaluation of vertical leakage and open stairwell doors, computer analysis is recommended.

This analysis is for buildings where the leakage areas are the same for each floor of the building and where the only significant driving forces are the stairwell pressurization system and the temperature difference between the indoors and outdoors.

The pressure difference, ΔP_{SB}, between the stairwell and the building can be expressed as

$$\Delta P_{SB} = \Delta P_{SBb} + \frac{by}{1 + \left(\frac{A_{SB}}{A_{BO}}\right)^2} \qquad (14)$$

where:

ΔP_{SBb} = the pressure difference, ΔP_{SB}, at the stairwell bottom
y = distance above the stairwell bottom
A_{SB} = flow area between the stairwell and the building (per floor)
A_{BO} = flow area between the building and the outside (per floor)

$$b = \frac{gP}{R}\left(\frac{1}{T_0} - \frac{1}{T_S}\right)$$

T_0 = absolute temperature of outside air
T_S = absolute temperature of stairwell air.

For a stairwell with no leakage directly to the outside, the flow rate of pressurization air is

$$Q = \frac{2}{3} NCA_{SB} \sqrt{\frac{2}{\rho}} \left(\frac{\Delta P_{SBt}^{3/2} - \Delta P_{SBb}^{3/2}}{\Delta P_{SBt} - \Delta P_{SBb}}\right) \qquad (15)$$

where:

N = number of floors
C = flow coefficient (See Equation 5)
ΔP_{SBt} = the pressure difference, ΔP_{SB}, at the stairwell top.

EXAMPLE 4:

Each story of a 20-story stairwell is 10.8 ft (3.3 m) in height. The stairwell has a single-leaf door at each floor leading to the occupant space and one ground level door to the outside. The exterior of the building has a wall area of 6030 ft^2 (560 m^2) per floor. The exterior building walls and stairwell walls are of average leakiness. The stairwell wall area is 560 ft^2 (52 m^2) per floor. The area of the gap around each stairwell door to the building is 0.26 ft^2 (0.024 m^2). The exterior door is well gasketed, and its leakage is negligible when it is closed.

For this example, the following design parameters are used: outside design temperatures, T_0 = 14°F (−10°C); stairwell temperature, T_S = 70°F (21°C); minimum design pressure differences when all stairwell doors are closed of 0.551 in. H$_2$O (137 Pa).

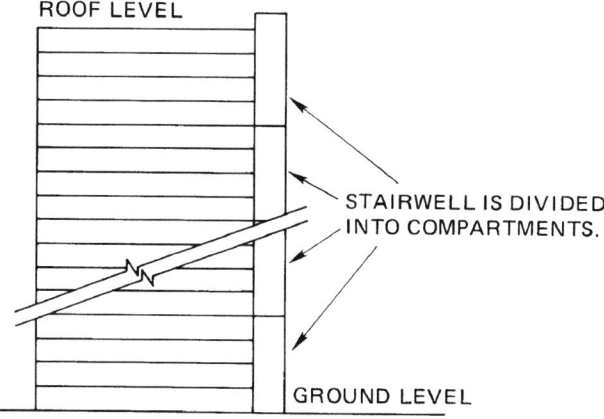

Fig. 4-12.17. Compartmentation of a pressurized stairwell. Note: each four floor compartment has at least one supply air injection point.

SOLUTION:

Using the leakage ratios for an exterior building wall of average tightness from Table 4-12.1, $A_{BO} = 6030 \times (0.21 \times 10^{-3}) = 1.27$ ft^2 (0.118 m^2). Using leakage ratios for a stairwell wall of average tightness from Table 4-12.1, the leakage area of the stairwell wall is $560 \times (0.11 \times 10^{-3}) = 0.06$ ft^2 (0.006 m^2). A_{SB} equals the leakage area of the stairwell wall plus the gaps around the closed doors. $A_{SB} = 0.06 + 0.26 = 0.32$ ft^2 (0.030 m^2). The temperature factor, 6, is calculated at 0.00170 in. H$_2$O/ft (1.39 Pa/m). The pressure difference at the stairwell bottom is selected as $\Delta P_{SBb} = 0.080$ in. H$_2$O (20 Pa) to provide an extra degree of protection above the minimum allowable value of 0.052 in. H$_2$O (13 Pa). The pressure difference, ΔP_{SBt}, is calculated from Equation 14 at 0.426 in. H$_2$O (106 Pa), using y = 217 ft (66.1 m). Thus, ΔP_{SBt} does not exceed the maximum allowable pressure. The flow rate of pressurization air is calculated from Equation 15 at 8200 cfm (3.9 m^3/s).

The flow rate is highly dependent on the leakage area around the closed doors and upon the leakage area that exists in the stairwell walls. In practice, these areas are difficult to evaluate and even more difficult to control. If the flow area, A_{SB}, were 0.54 ft^2 (0.050 m^2) rather than 0.32 ft^2 (0.030 m^2), then a flow rate of pressurization air of 13,800 cfm (6.5 m^3/s) would have been calculated from Equation 15. A fan with a sheave is one approach to allow adjustment of supply air to offset for variations in actual leakage from the values used in design calculations.

ELEVATOR SMOKE CONTROL

Elevator shaft smoke control can prevent smoke spread to floors away from the fire by way of the elevator shaft. The problems that can result from smoke migration through elevator shafts are illustrated by the fire at the MGM Grand Hotel.[23] The fire occurred on the ground floor, but smoke migrated to the upper floors where the majority of the fatalities occurred. The elevators at this hotel did not have any special smoke protection, and they were one of the major paths of smoke migration to the upper floors. This chapter does not address smoke control of elevator systems intended for fire evacuation; however, the topic is addressed by Klote and Milke.[24]

Piston Effect

The transient pressures produced when an elevator car moves in a shaft (i.e., piston effect) affect elevator smoke control. Such piston effect can pull smoke into a normally pressurized elevator lobby or elevator shaft. Klote[25] analyzed air flows and pressures produced by elevator car motion in a pressurized elevator shaft, based on the continuity equation for the contracting control volume in an elevator shaft above a moving elevator car. Piston effect experiments[26] on a hotel elevator in Mississauga, Ontario, Canada, validated the analysis.

From the analysis by Klote, an expression was developed for the critical pressure difference, ΔP_{crit}, at which piston effect cannot overcome the elevator pressurization system.

$$\Delta P_{crit} = \frac{K_{pe}\rho}{2}\left(\frac{A_s A_e V}{A_a A_{si} C_c}\right)^2 \qquad (16)$$

where:

ΔP_{crit} = critical pressure difference, in. H$_2$O(Pa)
ρ = air density in elevator shaft, lb/ft^3 (kg/m^3)
A_s = cross-sectional area of the elevator shaft, ft^2 (m^2)
A_{si} = leakage area between the lobby and the building, ft^2 (m^2)
A_a = free area around the elevator car, ft^2 (m^2)
A_e = effective area between the elevator shaft and the outside, ft^2 (m^2)
V = elevator car velocity, ft/min (m/s)
C_c = flow coefficient for flow around car, dimensionless
K_{pe} = coefficient, 1.66×10^{-6}, (1.00).

The flow coefficient, C_c, was determined experimentally to be about 0.94 for a multiple-car elevator shaft, and about 0.83 for a single-car elevator shaft.[27] Equation 16 is for elevators without enclosed elevator lobbies. The effective area from the elevator to the outside is

$$A_e = \left(\frac{1}{A_{si}^2} + \frac{1}{A_{io}^2}\right)^{-1/2} \qquad (17)$$

where A_{io} is the leakage area between the outside and the building, in ft^2 (m^2).

EXAMPLE 5:

An elevator shaft with two cars is pressurized to a minimum of 0.05 in. H$_2$O (12.4 Pa) from the elevator shaft to the building. This system is to prevent smoke movement through the elevator shaft, and there is no enclosed elevator lobby. The parameters are: $A_{si} = 1.52$ ft^2 (0.141 m^2), $A_{io} = 2.26$ ft^2 (0.210 m^2), $A_s = 121$ ft^2 (11.2 m^2), $A_a = 80$ ft^2 (7.43 m^2), $\rho = 0.075$ lb/ft^3 (1.20 kg/m^3), $V = 500$ ft/min (2.54 m/s), and $C_c = 0.94$. Is it possible for the pressure difference due to elevator piston effect to pull smoke into the elevator shaft?

SOLUTION:

From Equation 17, $A_e = 1.26$ ft^2 (0.117 m^2). From Equation 16, $\Delta P_{crit} = 0.028$ in. H$_2$O (6.9 Pa). The elevator shaft is pressurized at a level above ΔP_{crit}. Therefore, piston effect will not pull smoke into the elevator shaft.

Elevator Shaft Pressurization

These systems supply air to the elevator shaft and can produce a pressure difference sufficient to prevent smoke flow into the elevator shaft in the event of a fire. Upon fire detection, the general procedure is for elevator cars to be taken out of normal service and automatically recalled to the ground floor. A recent modification of this is the capability for recall to an alternate floor, in the event of a fire on the ground floor. Two elevator door scenarios can occur: (1) the elevator doors remain open after the car reaches the ground floor or the alternate floor, or (2) the elevator doors close after sufficient time to allow passengers to leave the car. The fire service has elevator keys that enable them to: (1) operate elevators for rescue and (2) transport personnel and equipment to fight the fire.

As with pressurized stairwells, factors that must be considered are: shaft friction, outside-to-inside temperature difference, and pressure fluctuations due to doors opening and closing. The analysis of pressurized stairwells can be applied to pressurized elevators by redefining the subscript S in the analysis from stairwell to elevator shaft. This analysis is then applicable to buildings without vertical leakage and to shafts with negligible pressure loss due to friction. This analysis can only be used where the elevator pressurization system is the only system using pressurization or operating in the building. Further, the effect of any exhaust system must be negligible. Computer analysis incorporating shaft friction and more complex building flow paths can be done by the programs cited later in this chapter.

ZONE SMOKE CONTROL

Pressurized stairwells are intended to prevent smoke infiltration into stairwells. However, in a building that has only stairwell pressurization, smoke can flow through cracks in floors and partitions and through shafts to damage property and threaten life at locations remote from the fire. The concept of zone smoke control is intended to limit such smoke movement.

With this smoke control method, a building is divided into a number of smoke control zones, and each zone is separated from the others by partitions, floors, and doors that can be closed to inhibit the smoke movement. In the event of a fire, pressure differences and airflows produced by mechanical fans are used to limit the smoke spread from the zone in which the fire was initiated. The concentration of smoke in the fire zone is unchecked and accordingly, it is intended that the building occupants evacuate this zone as soon as possible after fire detection.

Frequently, each floor of a building is chosen to be a separate smoke control zone. However, a smoke control zone can consist of more than one floor, or a floor can be divided into more than one smoke control zone. Some arrangements of smoke control zones are illustrated in Figure 4-12.18. All of the nonsmoke zones in the building may be pressurized. The term "pressure sandwich" is used to describe cases where only adjacent zones to the smoke zone are pressurized as in (b) and (d) of Figure 4-12.18.

The intent of zone smoke control is to limit smoke movement to the smoke zone by use of the two principles of smoke control. Pressure differences in the desired direction across the barriers of a smoke zone can be achieved by either supplying outside (fresh) air to nonsmoke zones, by venting the smoke zone, or by both methods.

Venting of smoke from a smoke zone is important because it prevents significant overpressures which are due to thermal expansion of gases as a result of the fire. However, venting only slightly reduces smoke concentration in the smoke zone. Venting in this zone can be accomplished by exterior wall vents, smoke shafts, and mechanical venting (exhausting).

COMPUTER ANALYSIS

Some design calculations associated with smoke control are appropriate for hand calculation. However, other calculations involve time-consuming, trial-and-error solutions that are more appropriately left to a digital computer. The National Bureau of Standards has developed a computer program[28] specifically for analysis of smoke control systems. A number of other programs applicable to smoke control have been developed. Some calculate steady-state airflow and pressures throughout a building[29,30]; other programs go beyond this to calculate the smoke concentrations that would be produced throughout a building in the event of a fire.[31-34]

Each of these programs differs from the others to some extent but all employ similar basic concepts. A building is

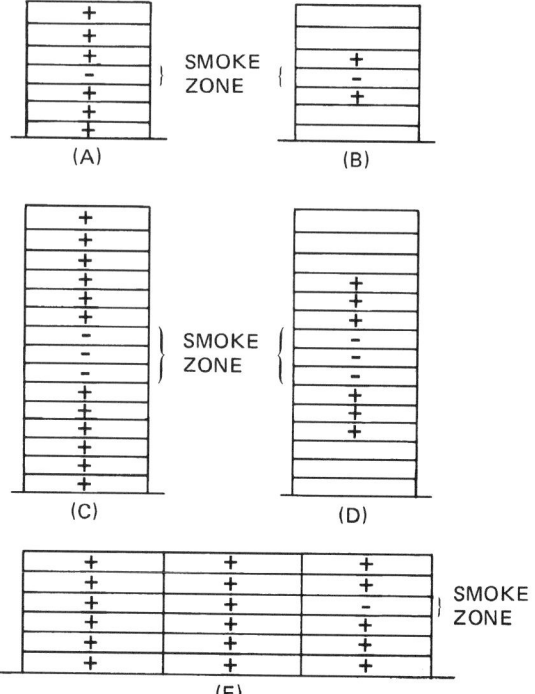

Fig. 4-12.18. Some arrangements of smoke control zones. The smoke zone is indicated by a minus (−) sign and pressurized spaces are indicated by a (+) sign. Each floor can be a smoke control zone (a, b) or a smoke zone can consist of more than one floor (c, d). All of the nonsmoke zones in a building may be pressurized as in (a) and (c), or only nonsmoke zones adjacent to the smoke zone (b, d). A smoke zone can be limited to a part of a floor (e).

represented by a network of spaces, or nodes, each at a specific pressure and temperature. The stairwells and other shafts are modeled by a vertical series of spaces, one for each floor. Air flows through leakage paths from regions of high pressure to regions of low pressure. These leakage paths are doors and windows that may be opened or closed. Leakage can also occur through partitions, floors, and exterior walls and roofs. The airflow through a flow path is a function of the pressure difference across the path as presented in Equation 5.

Air from outside the building can be introduced by a pressurization system into any level of a shaft or even into other building compartments, allowing for simulation of stairwell pressurization. In addition, any building space can be exhausted, allowing simulation of zoned smoke control systems. The pressures throughout the building and flow rates through all the flow paths are obtained by solving the airflow network, including the driving forces such as wind, the pressurization system, or a temperature difference between inside and outside air.

ACCEPTANCE TESTING

Regardless of the care, skill, and attention to detail with which a smoke control system is designed, an acceptance test is needed to ensure that the system, as built, operates as intended.

An acceptance test should be composed of two levels of testing. The first level is functional, to determine if everything in the system works as it is supposed to work, i.e., an initial check of the system components. The importance of the initial check has become apparent because of the many problems that have been encountered during tests of smoke control systems. These problems include fans operating backward, fans to which no electrical power was supplied, and controls that did not work properly.

The second level of testing is performance oriented to determine if the system, as a system, performs adequately under all required modes of operation. This testing can consist of measuring pressure differences across barriers under various modes of smoke control system operation. In cases where airflows through open doors are important, these should also be measured. Chemical smoke from smoke candles (sometimes called smoke bombs) is not recommended for performance testing because it normally lacks the buoyancy of hot smoke from a real building fire. Smoke near a flaming fire has a temperature in the range of 1000 to 2000°F (540 to 1100°C). Heating chemical smoke to such temperatures to emulate smoke from a real fire is not recommended unless precautions are taken to protect life and property. These same comments about buoyancy apply to tracer gases. Thus, it seems that pressure difference testing is the most practical performance test.

REFERENCES CITED

1. W.G. Berl and B.M. Halpin, "Human Fatalities from Unwanted Fires," *J. Hopkins APL Tech. Dig.*, 1, 129 (1980).
2. ASTM EL76-80, *Annual Book of ASTM Standards, Part 18*, American Society for Testing and Materials, Philadelphia (1980).
3. NFPA 90A, *Standard for the Installation of Air Conditioning and Ventilating Systems*, National Fire Protection Association, Quincy, MA (1993).
4. J.H. Klote, "Stairwell Pressurization," *ASHRAE Transactions*, 86, 604 (1980).
5. J.H. McGuire and G.T. Tamura, "Simple Analysis of Smoke Flow Problems in High Rise Buildings," *Fire Tech.*, 11, 15 (1975).
6. J.B. Fang, "Static Pressures Produced by Room Fires," *NBSIR 80-1984*, National Bureau of Standards, Washington (1980).
7. P. Sachs, *Wind Forces in Engineering*, Pergamon, New York (1972).
8. E.L. Houghton and N.B. Carruther, *Wind Forces on Buildings and Structures*, John Wiley and Sons, New York (1976).
9. E. Simiu and R.H. Scanlan, *Wind Effects on Structures: An Introduction to Wind Engineering*, John Wiley and Sons, New York (1978).
10. A.J. MacDonald, *Wind Loading on Buildings*, John Wiley and Sons, New York (1975).
11. C.Y. Shaw and G.T. Tamura, "The Calculation of Air Infiltration Rates Caused by Wind and Stack Action for Tall Buildings," *ASHRAE Transactions*, 83, 145 (1977).
12. P.H. Thomas, "Movement of Smoke in Horizontal Corridors Against an Air Flow," *Inst. of Fire Engg. Q.*, 30, 45 (1970).
13. B.H. Shaw and W. Whyte, "Air Movement Through Door- ways—The Influence of Temperature and its Control by Forced Air Flow," *Bldg. Serv. Engg.*, 42, 210 (1974).
14. J.H. McGuire, G.T. Tamura, and A.G. Wilson, *Factors in Controlling Smoke in High Buildings*, Symposium on Fire Hazards in Buildings, ASHRAE, San Francisco (1970).
15. R.J. Cresci, *Smoke and Fire Control in High-Rise Office Buildings—Part II, Analysis of Stair Pressurization Systems*, Symposium on Experience and Applications on Smoke and Fire Control, ASHRAE, Atlanta (1973).
16. G.T. Tamura and C.Y. Shaw, "Studies of Exterior Wall Air Tightness and Air Infiltration of Tall Buildings," *ASHRAE Transactions*, 83, 122 (1976).
17. G.T. Tamura and A.G. Wilson, "Pressure Differences for a 9-Story Building as a Result of Chimney Effect and Ventilation System Operation," *ASHRAE Transactions*, 72, 180 (1966).
18. G.T. Tamura and C.Y. Shaw, "Air Leakage Data for the Design of Elevator and Stair Shaft Pressurization Systems," *ASHRAE Transactions*, 83, 179 (1976).
19. G.T. Tamura and C.Y. Shaw, "Experimental Studies of Mechanical Venting for Smoke Control in Tall Office Buildings," *ASHRAE Transactions*, 86, 54 (1978).
20. *ASHRAE Handbook—1985 Fundamentals*, American Society of Heating, Refrigerating and Air Conditioning Engineers, Atlanta (1985).
21. *Temperature Extremes in the United States*, National Oceanic and Atmospheric Administration, Ashville (1979).
22. NFPA 101, *Life Safety Code*, National Fire Protection Association, Quincy, MA (1994).
23. R. Best and D.P. Demers, "Investigation Report on the MGM Grand Hotel Fire—Las Vegas, NV, Nov. 21, 1980," National Fire Protection Association, Quincy, MA (1982).
24. J.K. Klote and J.A. Milke, *Design of Smoke Management Systems*, American Society of Heating, Refrigerating, and Air-Conditioning Engineers, Atlanta, GA (1992).
25. J.H. Klote, "An Analysis of the Influence of Piston Effect on Elevator Smoke Control," NBSIR 88-3751, U.S. National Bureau of Standards, Gaithersburg, MD (1988).
26. J.H. Klote and G.T. Tamura, "Experiments of Piston Effect on Elevator Smoke Control," ASHRAE *Transactions*, Vol. 93, Part 2, pp. 2217–2228 (1987).
27. J.H. Klote and G.T. Tamura, "Elevator Piston Effect and the Smoke Problem," *Fire Safety Journal*, Vol. 11, No. 3, pp. 227–233, May, 1986.
28. J.H. Klote, "A Computer Program for Analysis of Smoke Control Systems," *NBSIR 82-2512*, National Bureau of Standards, Washington (1982).
29. D.M. Sander, *DBR Computer Program No. 37*, National Research Council of Canada, Ottawa (1974).
30. D.M. Sander and G.T. Tamura, *DBR Computer Program No. 35*, National Research Council of Canada, Ottawa (1973).
31. H. Yoshida, C.Y. Shaw, and G.T. Tamura, *DBR Computer Program No. 45*, National Research Council of Canada (1979).
32. E. Evers and A. Waterhouse, *A Computer Model for Analyzing Smoke Movement in Buildings*, Building Research Establishment, Borehamwood (1978).
33. T. Wakamatsu, "Calculation Methods for Predicting Smoke Movement in Building Fires and Designing Smoke Control Systems, Fire Standards and Safety," *ASTM STP 614*, American Society for Testing and Materials, Philadelphia (1977).
34. J. Rilling, *Smoke Study, 3rd Phase, Method of Calculating the Smoke Movement Between Building Spaces*, Centre Scientifique et Technique du Batiment, Champs Sur Marne (1978).

SMOKE MANAGEMENT IN COVERED MALLS AND ATRIA

James A. Milke

INTRODUCTION

The atrium is an architectural construct originating from the era of the Roman Empire.[1] Initially, the atrium was a courtyard bounded by a building, without a roof. During the latter part of this century, a roof or ceiling has been placed over the courtyard, and occasionally bounding walls around the courtyard are removed to provide openings between the courtyard and adjacent spaces.[2] Covered malls are also a relatively recent architectural development. Here, the large-volume structure comprises a pedestrian space, often with large openings into the communicating stores.

Smoke management in large-volume spaces, such as atria and covered malls, poses separate and distinct challenges from well-compartmented spaces. In particular, smoke control strategies using pressure differences and physical barriers described by Klote in Section 4, Chapter 12, and NFPA 92A, *Recommended Practice for Smoke-Control Systems*,[3] are infeasible. Without physical barriers, smoke propagation is unimpeded, spreading easily throughout the entire space. The tall ceiling heights in many large-volume spaces pose additional challenges in terms of substantial quantities of smoke production and delayed detection times. However, on the positive side, the large-volume space and tall ceiling height permit the smoke to become diluted and cooled as it spreads vertically and horizontally. Dilution acts to reduce the level of hazard posed by the smoke. The problem posed by large-volume spaces is expressed in NFPA *101®*, *Life Safety Code®* (paragraph A-6-2.4.6)[4]

Where atriums are used, there is an added degree of safety to occupants because of the large volume of space into which smoke can be dissipated. However, there is a need to ensure that dangerous concentrations of smoke are promptly removed from the atrium, and the exhaust system needs careful design.

In addition to atria and covered malls, there are many other examples of large-volume spaces, including convention centers, airport terminals, sports arenas, and warehouses. The engineering principles governing the design of smoke management systems for these various large-volume spaces are the same. However, differences in the smoke management system designs for these large-volume spaces may occur as a result of different fire scenarios and design goals reflecting changes in function, shape, and connection to other spaces, among other factors. Given the similarities in engineering principles affecting smoke management system design, the term "atrium" will be used throughout the remainder of this chapter to refer to all types of large-volume spaces.

The discussion presented in this chapter is divided into two sections. First, conditions within the atrium prior to actuation of a smoke management system are discussed. As part of this discussion, the smoke filling process is described along with the time required for actuation of a smoke management system. The second part includes a description of conditions within the atrium after actuation of the smoke management system.

As a preface to any discussion on smoke management systems, a definition of smoke must be established (NFPA 92B, *Guide for Smoke Management Systems in Malls, Atria, and Large Areas*,[5] Section 1-4)

The airborne solid and liquid particulates and gases evolved when a material undergoes pyrolysis or combustion, together with the quantity of air that is entrained or otherwise mixed into the mass.

Air is entrained along the entire length of the fire and smoke plume. The rate of air entrainment into a smoke plume is approximately proportional to the square of the clear height, i.e., the distance from the top of the burning fuel to the smoke layer. During the early stages of a fire located in an atrium involving a large clear height, a significant quantity of air is entrained into the plume. Thus, the principal component of "smoke" in atria is entrained air.

HAZARD PARAMETERS

A technically based hazard analysis of smoke conditions in an atrium expresses the level of hazard in terms of physically based parameters. In this chapter, selected hazard parameters of the smoke layer include interface position, temperature, carbon monoxide concentration, and light obscuration. The magnitude of each of these parameters can be

---

James A. Milke is an assistant professor in the Department of Fire Protection Engineering at the University of Maryland. His recent research activities have included the impact of fires on the structural response of steel and composite members.

AXISYMMETRIC PLUME

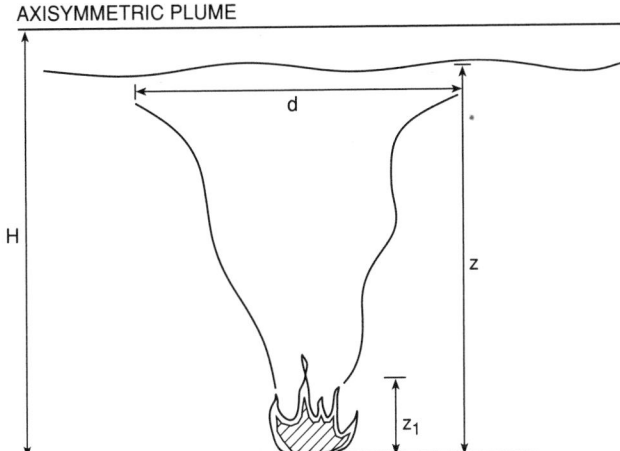

Fig. 4-13.1.   Axisymmetric plume.[5]

predicted based on engineering principles. In addition to being predictable, critical threshold values are available for the hazard parameters in order to properly assess the severity of the threat. (See Section 2, Chapter 8.) This chapter will concentrate on the life hazards posed by smoke. The hazards smoke poses to contents, property, and mission continuity are described elsewhere.[6]

Smoke Layer Interface Position

The smoke layer interface position is located a distance, z, above the top of the fuel, as indicated in Figure 4-13.1. This parameter addresses the hazard of people being immersed in a smoke layer. Sole use of this parameter to assess hazard level is conservative by considering any concentration of smoke to be unacceptable. However, even though the physiological effects due to "light" smoke levels may be minor, the psychological effects and extended evacuation time may be appreciable. Being surrounded by smoke of any nature may decrease the speed of evacuees, delaying evacuation, perhaps until the smoke is no longer relatively benign. In terms of property protection issues, any smoke may be unacceptable because of smoke staining.

Light Obscuration

As with the smoke layer depth parameter, light obscuration is not lethal by itself.[7] Associated with an increase in light obscuration is a reduction in visibility, which is likely to yield a longer evacuation time and extend the duration of the exposure to the toxins included in the smoke. In some documented fire incidents, evacuation has been terminated due to a lack of sufficient visibility.[11,13,14] A fire fighter injury in a fire incident in an atrium has been attributed to a significant reduction in visibility due to light obscuration.[12] As a result of being unable to see the edge of the balcony, the fire fighter fell from an upper balcony.

Limiting values from 0.23 to 1.2 m^{-1} have been suggested for the extinction coefficient.[11,13,14] (See Section 2, Chapter 15.) Alternatively, a critical limit may be based on a preferred minimum visibility distance; e.g., a limit of light obscuration can be suggested such that occupants can see across a room or down a corridor.

Carbon Monoxide Concentration and Temperature

The final two parameters, carbon monoxide (CO) concentration and elevated temperature of the smoke layer, can be directly related to the potential for harm. (See Section 2, Chapter 8.) Critical limits for these two parameters can be suggested based on toxicity studies. In principle, the discussion on CO concentration is also applicable for an assessment of the hazard posed by other gaseous combustion products, e.g., hydrogen cyanide (HCN) or carbon dioxide (CO_2).

SMOKE MANAGEMENT APPROACHES

The design of a smoke management system for an atrium is influenced by the following three characteristics of the atrium:

1. Geometric shape and dimensions,
2. Relative location within the building, and
3. Separation from communicating spaces.

Several approaches are available to achieve smoke management goals in an atrium. Possible approaches include limiting the fire size, providing physical barriers, and providing mechanical ventilation. Selection of the best smoke management approach for a particular atrium should consider the use, size, and arrangement of spaces.

Limiting the fire size can be accomplished by controlling the type, quantity, and arrangement of fuel. In addition, the fire size can be controlled through an automatic suppression system.

Physical barriers serve to limit smoke spread to adjacent spaces. The physical barrier should be airtight and be able to withstand both exposure of smoke and exposure to an elevated temperature environment. In atria with tall ceiling heights, the temperature of the smoke layer in the atrium is only modestly above ambient conditions.

Mechanical ventilation may be provided to exhaust smoke from the atrium. Exhausting smoke from the atrium can limit the accumulation of heat and smoke within the atrium or can arrest the descent of the smoke layer. Alternatively, ventilation can be provided in a direction opposite of the smoke movement to restrict smoke spread to communicating spaces. Gravity vents may be provided to remove smoke, though their performance may be compromised by environmental factors. Discussions of gravity vents are provided elsewhere.[15] (See Section 3, Chapter 9.)

ANALYTICAL APPROACH

Predictions of conditions within an atrium can be evaluated using analytical or scale models. Scale models provide physical representations of a space, though in a reduced scale. Scaling laws are used to formulate an experimental program. Scale models are especially useful for atria with irregular shapes or numerous projections. A review of scaling relationships was presented previously by Quintiere.[16]

Two categories of analytical models are: (1) zone and (2) field models. A description of field models is outside the scope of this chapter. Zone models divide each compartment into a limited number of control volumes, typically an upper and lower zone. Inherent in the zone approach is the assumption of uniform properties throughout each zone. In spaces with a large floor area, this assumption may be tenuous. Nonetheless, calculations associated with the zone model approach are relatively easy to perform and are often accepted for engineering purposes. Calculations following the zone model approach may be in the form of algebraic equations or a computer algorithm.

The zone approach assumes that smoke from a fire is buoyant, rises to the ceiling, and forms a smoke layer. The buoyant nature of smoke is due to the decreased density of the heated smoke. As smoke rises in a plume, air is entrained to increase the mass flow rate in the plume. Associated with an increase in the plume mass flow rate is a decrease in the velocity and temperature of the smoke plume, as dictated by conservation of momentum and energy. In addition, the entrained air dilutes the combustion products in the plume. The entire smoke layer is assumed to have uniform characteristics. As smoke is supplied to the smoke layer from the plume, the interface between the smoke layer and lower clear air zone descends. The additional smoke supplied by the plume also results in an increase in the smoke layer temperature, carbon monoxide concentration, and light obscuration.

Being a simplification, the zone model approach may not be applicable in some situations. One example includes a scenario with operating sprinklers, which may cool the layer and also entrain smoke from the upper layer into the water spray pattern descending into the lower zone. Another example consists of the case where smoke does not reach the ceiling as a result of a loss of relative buoyancy where the temperature in the upper portion of the atrium near the ceiling is greater than that at the lower portion near the floor. This situation is discussed in more detail later in this chapter. A third situation involves an atrium with a large cross-sectional area where the horizontal variation in conditions from one portion of the atrium to another is important to the analyst. Where local conditions need to be assessed, field models are more appropriate than zone models.

Two categories of fire scenarios of interest for smoke management design in atria include: (1) fires located in the atrium and (2) fires located in a space adjacent and open to the atrium. This chapter concentrates only on scenarios involving fires originating within the atrium space. In addition, methods to estimate conditions in any of the adjacent spaces resulting from fires originating in the atrium or from fires located in adjacent spaces are addressed elsewhere.[17]

SMOKE FILLING PERIOD

Both empirical correlations and theoretically based methods are available to address conditions during the smoke filling period using a zone model approach. Theoretically based methods use statements of conservation of mass and energy to determine the volume of the upper layer.[18] Conservation of mass accounts for the smoke mass supplied from the plume to the smoke layer along with any smoke leaving the zone *via* ventilation openings. Conservation of energy is applied to address the energy being supplied by the plume along with heat losses from the layer.

Generally, predictions of the smoke layer interface position by the two analytical methods differ. The principal reason for the discrepancy is the difference in definitions of the smoke layer interface used by the two methods. The empirical correlations are based on first indications of smoke, either using temperature rise or visual measurements. The theoretically based approach defines the smoke layer position as the demarcation between the upper and lower zones.

Empirical Correlations

Empirical correlations have been developed by Heskestad to determine the smoke layer interface position as a function of time for steady and t-squared fires. These correlations, included within NFPA 92B, *Guide for Smoke Management Systems in Malls, Atria, and Large Areas*,[5] are based on experimental data in large spaces. In the experimental efforts, the smoke layer interface position was established by a variety of means, including visual observations and first change in measured temperature, carbon dioxide concentration, or optical density.

The correlations are intended to be simplified expressions, with easily acquired input and minimal computations. The correlations provide conservative estimates of the smoke layer interface position, e.g., predicting the smoke layer interface to be lower than may be typically expected.[19] The correlations are applicable to simplified cases related to the fire scenario and geometry of the space. Fire scenarios either consist of steady fires or growing fires following a t-squared profile. The assumed geometrical configuration is a space of uniform cross-sectional area, e.g., rectangular or right cylindrical solids.

In addition to the noted simplified conditions, second-order parameters have been omitted, such as environmental factors (e.g., stack effect and wind). In addition, the effect of heating, ventilating, and air-conditioning (HVAC) systems is ignored.

Steady fires: The time-dependent position of the smoke layer interface for steady fires can be estimated using Equation 1.[20,21] Equation 1 is based on experimental data from fires in large-volume spaces with A/H^2 of 0.9 to 14.[22-24]

$$\frac{z}{H} = 1.11 - 0.28 \ln\left(\frac{tQ^{1/3}H^{-4/3}}{A/H^2}\right) \tag{1}$$

where $z/H \geq 0.2$.

Equation 1 is presented in nondimensional form. The quantity $tQ^{1/3}H^{-4/3}$ represents the normalized time from ignition. The significance of the normalized time parameter is to indicate that the same relative smoke layer position occurs for a long duration, low heat release rate fire in a tall ceiling height atrium, as from a short duration, large fire in an atrium with a short ceiling height. Different atrium geometries are accounted for by the nondimensional shape factor, A/H^2.[22,23]

The limits noted for A/H^2 reflect the range of shape factors for the facilities in which the experiments were performed.[22,23] Examples of atria within the noted range include atria with a cross-sectional area of 10,000 m^2 and a height of 105 m (A/H^2 = 0.9) or a height of 27 m (A/H^2 = 14). Comparisons of the predictions from Equation 1 to experimental data from fires in tall spaces are provided in Figure 4-13.2.[24-26]

The initial time period to form a smoke layer is implicitly included in Equation 1. Evidence of this characteristic is obtained for short time durations where the resulting z/H is greater than 1.0 (otherwise $z/H > 1$ would literally mean that the smoke layer interface is *above* the ceiling). The lower limit for z/H of 0.2 relates to the lowest level where data was taken in any of the referenced experiments.

T-squared fires: Equation 2 provides a correlation of the time-dependent smoke layer interface position for fires following a t-square-type profile.[20,21] Equation 2 is also based on experimental data in spaces with shape factors ranging from 0.9 to 14.[24,27]

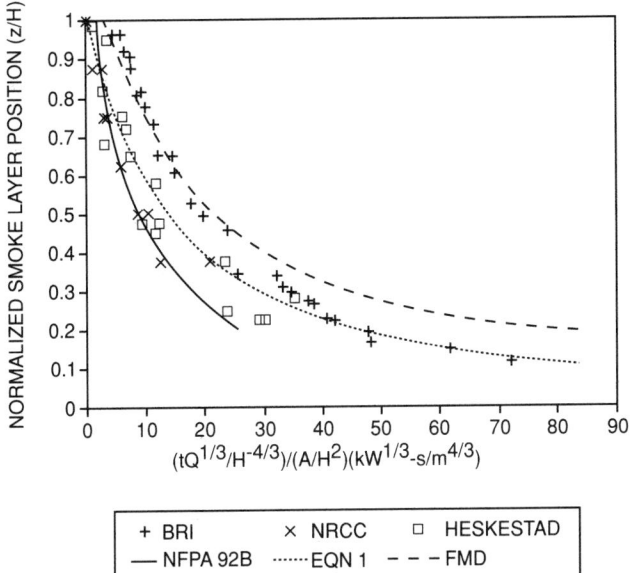

Fig. 4-13.2. Comparisons of smoke layer position—experimental data versus predictions.

$$\frac{z}{H} = 0.91[tt_g^{-2/5} H^{-4/5} (A/H^2)^{-3/5}]^{-1.45} \qquad (2)$$

Both Equations 1 and 2 assume that the fire is located near the center of the atrium floor, remote from any walls. Smoke production is greatest for the centered configuration and thereby represents the worst-case condition.

EXAMPLE 1:

For a fast, t-squared fire in an atrium with a cross-sectional area of 800 m^2 and height of 20 m, determine the position of the smoke layer interface after 120 s if the atrium cross-sectional area is 800 m^2.

SOLUTION:

Applying Equation 2 with $A/H^2 = 2.0$ and $t_g = 150$ s, z/H is 0.95 or $z = 19$ m.

EXAMPLE 2:

For a fast, t-squared fire in an atrium with a cross-sectional area of 800 m^2 and height of 20 m, determine the time for the smoke layer interface to reach 15 m above floor level.

SOLUTION:

Re-expressing Equation 2 to solve for t:

$$t = 0.94t_g^{2/5}H^{4/5}(A/H^2)^{3/5}(z/H)^{-0.69} \qquad (3)$$

Applying Equation 3 with $A/H^2 = 2.0$ and $t_g = 150$ s, t is 140 s.

Reviewing the results from Examples 1 and 2, the smoke layer barely descends below the ceiling in the first 120 s. This is indicative of the lag time required for the plume to reach the ceiling and to form a layer. Then, after only another 20 s, the smoke layer descends 4 m, demonstrating the rapid initial descent rate of the smoke layer interface. The

rapid descent is attributable to the significant quantity of smoke produced during the early stage of a fire in a tall ceiling space. The trend of rapid filling during the early stage of a fire has been reported by eyewitness accounts from four fire incidents in atria.[12,28-30]

Theoretically Based Approach

Conservation of mass and energy can be applied to provide an estimate for the position of the theoretical smoke layer interface.[18] Equation 4 expresses the conservation of mass, m_u, for the upper smoke layer, assuming no exhaust from the layer

$$\frac{dm_u}{dt} = \dot{m} \qquad (4)$$

Approximating the smoke as an ideal gas with properties of heated air and assuming that the ambient pressure and specific heat are constant, the expression for conservation of energy for the smoke layer is

$$(\rho h)_u \frac{dV_u}{dt} = \dot{Q}_c + \dot{m}h_1 \qquad (5)$$

Given the previously assumed conditions, ρh is a constant. Substituting the volumetric flow rate for the mass flow rate and simplifying

$$\frac{dV_u}{dt} = \frac{\dot{Q}_c}{\rho h} + \dot{V} \qquad (6)$$

The growth rate of the upper layer indicated in Equation 6 is dependent on two terms: (1) the volume supplied by the plume and (2) the expansion of the volume due to heating. For the case of an atrium with a constant cross-sectional area, A

$$\frac{dV_u}{dt} = A \frac{dz_u}{dt} \qquad (7)$$

As long as the smoke layer interface is well above the flaming region (see discussion later in this chapter), the plume mass entrainment rate can be estimated from the following equation[31]

$$\dot{m} = 0.071\dot{Q}_c^{1/3}(z - z_o)^{5/3} + 0.002\dot{Q}_c \qquad (8)$$

For large clear heights, i.e., clear heights in excess of 10 m, several simplifications can be made. The clear height is the distance from the top of the fuel to the bottom of the smoke layer. The magnitude of the second term is much less than the first. Generally, z is much greater than z_o. In addition, the volume increase of the upper layer supplied by the plume is appreciably greater than that due to expansion. With these simplifications and by substituting Equations 7 and 8 into Equation 6, an expression for dz_u/dt can be formulated

$$\frac{dz_u}{dt} = \frac{k_v\dot{Q}^{1/3}z^{5/3}}{A} \qquad (9)$$

In Equation 9, k_v is the volumetric entrainment constant (0.065 $m^{4/3}kW^{-1/3}s^{-1}$), defined as the ratio of the plume entrainment constant (0.071 kg $kW^{-1/3}$ $m^{-5/3}s^{-1}$) to the product of the density of ambient air (1.2 kg/m^3) and the convective heat release fraction. The convective heat release fraction is the ratio of the convective heat release rate to the total heat release rate and is typically assumed to be on the order of 0.7 to 0.8. Throughout this chapter, a value of

0.7 is selected for the convective heat release fraction.[5] One difference between Equations 8 and 9 is that the convective portion of the heat release rate is included in Equation 8, whereas the total heat release rate is included in Equation 9.

An expression for the smoke layer position resulting from a steady fire as a function of time can be obtained by integrating Equation 9

$$\frac{z}{H} = \left[1 + \frac{2k_v t Q^{1/3}}{3(A/H^2)H^{4/3}}\right]^{-3/2} \tag{10}$$

Alternatively, for a t-squared fire

$$\frac{z}{H} = \left[1 + \frac{4k_v t \, (t/t_g)^{2/3}}{(A/H^2)H^{4/3}}\right]^{-3/2} \tag{11}$$

A comparison of the predictions from Equations 1 and 10 is provided in Figure 4-13.2. One principal difference relates to the time delay for the smoke layer to form, i.e., transport lag. Transport lag is included implicitly in Equation 1. Equation 10 assumes that a smoke layer forms immediately. The transport lag can be accounted for separately.[32]

EXAMPLE 3:

For a fast, t-squared fire in an atrium with a cross-sectional area of 800 m² and height of 20 m, determine the position of the smoke layer interface after 120 s.

SOLUTION:

Applying Equation 11 with $A/H^2 = 2.0$ and $t_g = 150$ s, z/H is 0.72 or $z = 14.4$ m.

VENTILATED PERIOD

If a smoke management system includes the capability to exhaust smoke, the descent of the smoke layer can be arrested if the volumetric rate of smoke exhaust from the smoke layer equals the volumetric rate of smoke supplied to the layer. Neglecting the effect of expansion, the layer descent is stopped when the mass exhaust rate is equal to the mass entrainment rate by the plume. Algebraic equations are available to estimate the properties of the smoke layer, including:

1. Position of smoke layer interface,
2. Temperature of smoke layer,
3. Light obscuration in smoke layer, and
4. Gas concentration in smoke layer.

Equilibrium Smoke Layer Interface Position

The exhaust rate necessary to arrest the descent of the smoke layer can be estimated based on a knowledge of the mass entrainment rate into the plume. The mass entrainment rate depends on the configuration of the plume. Plume configurations reviewed in this chapter are:

1. Axisymmetric plume,
2. Wall plume,
3. Corner plume, and
4. Balcony spill plume.

Axisymmetric plume: Axisymmetric plumes are formed when fuel packages are remote from any walls, i.e., are lo-cated near the center of the atrium floor. Being remote from any walls, air is entrained around all of the plume perimeter along the entire clear height of the plume. The functional relationship of the mass entrainment rate to the heat release rate and clear height is[33]

$$\dot{m} = f\left(Q_c^{1/3} z^{5/3}\right) \tag{12}$$

The equations for the mass entrainment rate included within NFPA 92B, *Guide for Smoke Management Systems in Malls, Atria, and Large Areas*,[5] were originally derived by Heskestad.[31] One of the equations, previously presented as Equation 8 with z_o set equal to zero, is applicable when the clear height, z, is greater than the limiting height, z_f. The limiting height is defined as the height of the continuous flaming region, i.e., where flames are present 50 percent of the time. The limiting height may be estimated as[31]

$$z_f = 0.166 Q_c^{2/5} \tag{13}$$

The validity of neglecting z_o in the version of Equation 8 presented in NFPA 92B[5] was based on the observation that z_o is typically small as compared to z. The location of the virtual origin of an assumed point source can be estimated as[31]

$$z_o = 0.083 Q^{2/5} - 1.02D \tag{14}$$

For noncircular fuels, an equivalent diameter needs to be defined. The definition of an equivalent diameter is based on a circle that has an area equal to the floor area covered by the fuel. Considering a wide range of diameters and heat release rates associated with a variety of typical fuel packages, the virtual origin ranges from 0.5 to −5 m. Negative values are obtained when the second term is greater than the first, i.e., for fuel commodities with modest heat release rates spread over a large area.

For clear heights less than the limiting height, the entrainment rate is estimated using Equation 15

$$\dot{m} = 0.032 Q_c^{3/5} z \tag{15}$$

Originally, Equations 8 and 15 were developed to describe plumes from horizontal, circular flammable liquid pool fires. However, these equations have been shown to be applicable to more complex fuels, as long as the limiting height is greater than the diameter of the fuel, and the fire only involves the surface of the material, i.e., is not deep-seated.[31]

The mass rate of smoke production estimated by Equations 8 and 15 is independent of the type of materials involved in the fire, other than indirectly in terms of the heat release rate. This is due to the mass rate of entrained air being much greater than the mass rate of combustion products generated, which is true as long as sufficient air is available for combustion. As a result of the fire being approximated as a point source in the entrainment equations, even the shape or form of the fuel is not of primary importance. Thus, the material-related parameters are relegated to a level of secondary importance.

In both Equations 8 and 15, the mass entrainment rate is dependent on the clear height, where the mass entrainment rate increases with increasing values of the clear height. During the early stages of the fire, the clear height has its maximum value to provide the maximum smoke production rate, thereby

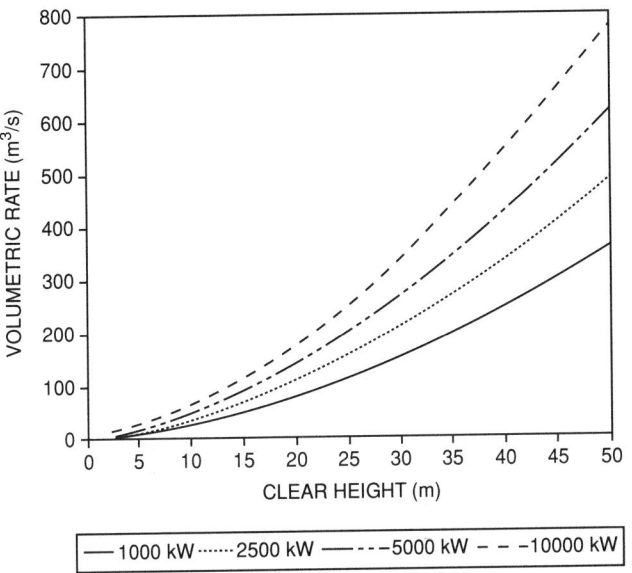

Fig. 4-13.3. *Smoke production rate for axisymmetric plumes.*

supporting the eyewitness accounts of the smoke filling process in atria. This is especially true if the flame height is well below the smoke layer, such that Equation 8 applies where the smoke production rate is proportional to $z^{5/3}$.

In most engineering applications, the smoke production (or exhaust) rate is expressed in terms of a volumetric rate rather than a mass rate. In order to accommodate this preference, the relationship between the volumetric rate and mass rate is expressed as Equation 16.

$$\dot{V} = \frac{\dot{m}}{\rho} \qquad (16)$$

Assuming smoke to have the same properties as air, the density of smoke may be evaluated as the density of air at the temperature of the smoke layer.[17] Graphs relating the volumetric smoke production rate to the clear height for selected total heat release rates ranging from 1,000 to 10,000 kW are provided in Figure 4-13.3.

EXAMPLE 4:

A fire has a total heat release rate of 5,000 kW and is located at the center of the atrium floor. The smoke layer interface is 35 m above the floor. Determine the mass and volumetric rates of smoke being supplied by the plume to the smoke layer, i.e., at the location of the smoke layer interface.

SOLUTION:

First, the limiting height is evaluated using Equation 13 to determine the applicable equation for the mass rate of entrainment. Assuming the convective heat release fraction is 0.7, $z_f = 4.3$ m. Because $z_f < z$, Equation 8 is the applicable equation for determining the mass rate of smoke production. Neglecting z_o, the mass smoke production rate is 410 kg/s. The associated volumetric rate (from Equation 16, assuming standard conditions) is 340 m^3/s.

Wall and corner plumes: Fires located near walls and corners principally entrain air only along the surface of the plume away from the walls or corner. Consequently, the amount of smoke production is reduced for these locations as compared to the axisymmetric plume remotely located from the walls. Using the concept of reflection, the smoke production rate from wall and corner plumes can be estimated.[34,35]

A plume generated by a fire located against a wall only entrains air from approximately half of its perimeter, as indicated in Figure 4-13.4. According to the concept of reflection, the smoke production rate is estimated as half of that from a fire that is twice as large (in terms of heat release rate).

Similarly, a plume generated by a fire located near a corner of a room is referred to as a "corner plume." (See Figure 4-13.4.) Using the concept of a reflection, the smoke production rate from corner plumes, where the intersecting walls form a 90-degree angle, is estimated as one-quarter of that from a fire that is four times as large.

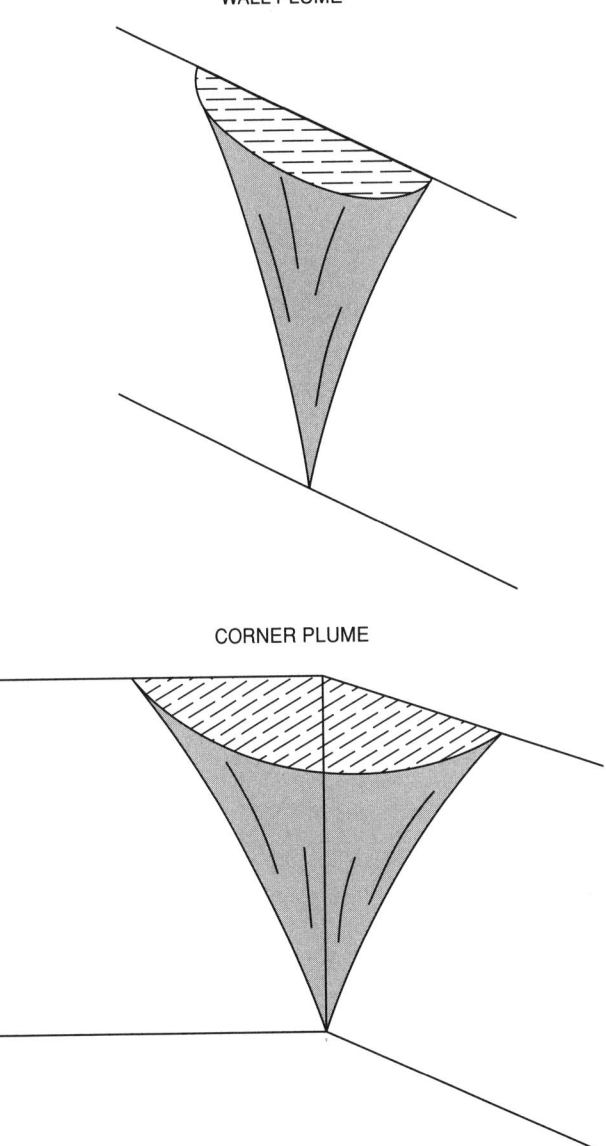

Fig. 4-13.4. *Wall and corner plume diagrams.*

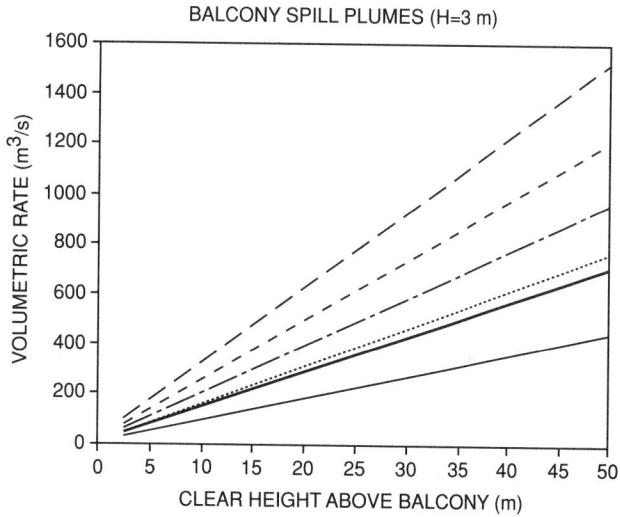

Fig. 4-13.5. Smoke production rate predictions for balcony spill plumes (H-3 m).

EXAMPLE 5:

A fire located on the floor of an atrium has a total heat release rate of 5,000 kW. The smoke layer interface is 35 m above the floor. Compare the mass rates of smoke being supplied by the plume to the smoke layer, given an axisymmetric, wall, or corner plume configuration.

SOLUTION:

In Example 4, $z_f = 4.3$ m and the smoke production rate for the axisymmetric plume using Equation 8 is 410 kg/s. Applying the same equation for the wall plume, the smoke production rate for a fire size of 10,000 kW is estimated as 520 kg/s. Dividing that rate by two provides the smoke production rate for the wall plume, i.e., 260 kg/s. Similarly, for the case of the corner plume, the smoke production rate is 170 kg/s (considering one-quarter of the smoke production rate from a 20,000 kW fire).

Comparing the smoke production rates for the three plumes (axisymmetric, wall, and corner plumes), the smoke production rate is greatest for the axisymmetric plume (410 kg/s) compared to 260 and 170 kg/s for the wall and corner plumes, respectively. Thus, conservative hazard assessments should assume an axisymmetric plume is developed from a fire that is located away from the walls, near the center of the space.

Balcony spill plume: A balcony spill plume is generated in cases where smoke reaches an intermediate obstruction, such as a balcony, travels horizontally under the obstruction and then turns and moves vertically. Scenarios with balcony spill plumes involve smoke rising above a fire, reaching a ceiling, balcony, or other significant horizontal projection, then traveling horizontally toward the edge of the "balcony." Characteristics of the resulting balcony spill plume depend on characteristics of the fire, width of the spill plume, and height of the ceiling above the fire. In addition, the path of horizontal travel from the plume centerline to the balcony edge is significant.

For situations involving a fire in a communicating space, which is immediately adjacent to the atrium, air entrainment into balcony spill plumes can be estimated using Equation 17. Equation 17 is based on Law's[36] interpretation of small-scale experimental data obtained by Morgan and Marshall.[37] Equation 17 provides an approximation of the mass flow in the plume for a complex situation.

$$\dot{m} = 0.36(QL^2)^{1/3}(z + 0.25H_b) \qquad (17)$$

The small-scale experiments simulated the situation of smoke being generated from a fire in a store, discharging out of the store *via* a single opening, traveling under a walkway serving the level above, then turning and rising once, reaching the edge of the walkway. The applicability of Equation 17 is questionable as a result of appreciable heat losses from the smoke for situations involving smoke discharging from a space connected to the atrium *via* a long corridor.

Predictions of the smoke production rate using Equation 17 for the balcony spill plume are included in Figure 4-13.5. The calculations represented in the figure consider a 3-m height to the underside of the balcony.

A comparison of the smoke production rate for axisymmetric and balcony spill plumes is provided in Figure 4-13.6. The heat release rate for both fires is 5,000 kW, $z_o = 0$ for the axisymmetric plume, and $H_b = 3$ m for the balcony spill plume. For short heights, the smoke production rate for the balcony spill plume is appreciably greater than that for the axisymmetric plume. However, with increasing height, the smoke production rates from the two plumes become comparable. Eventually, the two curves intersect, suggesting that, at some height, the balcony spill plume behaves the same (i.e., produces the same amount of smoke) as an axisymmetric plume. The point of intersection can be determined by setting Equation 8 equal to Equation 17. For large

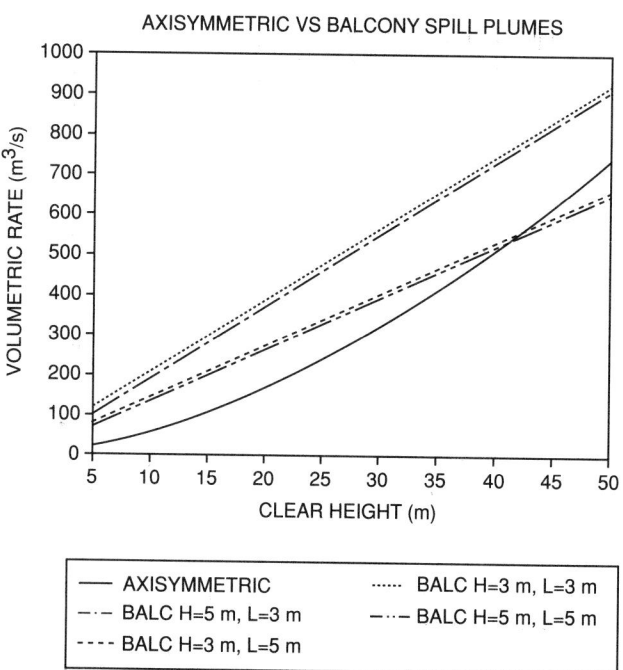

Fig. 4-13.6. Comparison of smoke production rate for axisymmetric and balcony spill plumes.

z and z much greater than H_b, the mass flow rates are equal when z is 12.5 times the width. Consequently, for greater heights the smoke production rate from a balcony spill plume should be estimated using Equation 8 with z evaluated from the balcony to the smoke layer.

The width of the plume, W, can be estimated by considering the presence of any physical vertical barriers attached to the balcony. The barriers act to restrict dispersion of the horizontal flow of smoke under the balcony. However, in the absence of any barriers, an equivalent width can be defined, based on results from visual observations of the width of the balcony spill plume at the balcony edge from the set of small-scale experiments by Morgan and Marshall.[37] The definition of an *equivalent confined plume width* is the width that entrains the same amount of air as an unconfined balcony spill plume. The equivalent width is evaluated using the following expression

$$L = w + b \qquad (18)$$

Properties of Smoke Layer

Where equilibrium conditions are established during the ventilated period as a result of an operating mechanical exhaust system, approximations of the properties of the smoke layer can be readily obtained using algebraic expressions. The properties of interest include temperature, light obscuration, and gas species concentration.

Temperature rise in smoke layer: The equilibrium smoke layer temperature is approximated by applying an energy balance to the smoke layer. The most elementary approach is to assume adiabatic conditions, i.e., no heat is lost from the layer to the ceiling, enclosing walls, or other surroundings *via* convection or radiation. Energy is supplied to the smoke layer *via* the plume from the convective portion of the heat released by the fire. The energy loss from the layer is due to smoke being exhausted from the atrium.

As a result of the adiabatic assumption, the smoke layer temperature is overestimated, provided a conservative estimate of the hazard is posed. In reality, some heat is lost from the upper smoke layer to the surrounding walls and ceiling. However, no elementary method is available to estimate the overall proportion of heat that is lost to the surroundings.[38] The only method of attempting to account for the heat losses to the boundaries is a tedious, component-by-component analysis.[39] Alternatively, zone and field computer fire models do account for heat losses to the boundary.[23,40]

Given the complexities associated with a more realistic analysis, the adiabatic temperature rise can be used as a worst-case engineering analysis. The adiabatic temperature rise is estimated as

$$\Delta T_{ad} = \frac{Q_c}{\dot{m}c_p} = \frac{Q_c}{\rho c_p V} \qquad (19)$$

The temperature rise is a function of both the heat release rate and mass flow. The mass flow can be replaced by the appropriate entrainment equation, depending on the type of plume. Considering an axisymmetric plume with the mass flow determined using Equation 8 (neglecting z_o), the temperature rise for various heat release rates and clear heights is presented in Figure 4-13.7. For tall ceiling heights, assuming adiabatic conditions is reasonable, because heat losses are minimal when the temperature rise is modest. Alternatively, for short heights, the temperature rise may be significantly

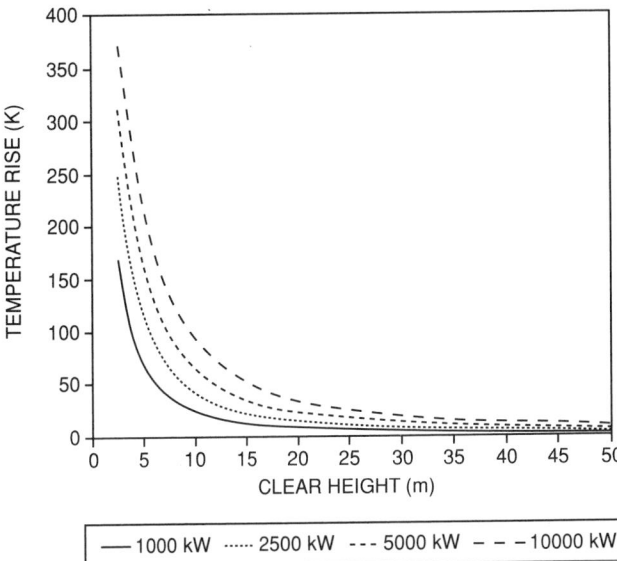

Fig. 4-13.7. *Temperature rise of smoke layer for axisymmetric plumes.*

overestimated. The degree of overestimation can be assessed by comparing the estimated smoke layer temperature with the plume centerline temperature. For thermodynamic reasons, the smoke layer temperature cannot exceed the plume centerline temperature. The plume centerline temperature, T_c, can be evaluated using Equation 20[41]

$$T_c = 0.08T_o Q_c^{2/3} z^{-5/3} + T_o \qquad (20)$$

The volumetric venting rate for other heat release rates or temperature rises may be determined considering that the specific heat is virtually constant for the expected temperature range of interest

$$\frac{Q_{c_1}}{Q_{c_2}} = \frac{V_1}{V_2} \frac{\Delta T_{ad_1}}{\Delta T_{ad_2}} \frac{T_2}{T_1} \qquad (21)$$

As can be observed from Equation 21, doubling the volumetric venting rate for the same size fire results in reducing the temperature rise by approximately 50 percent (the temperature rise is not precisely halved, since the absolute temperature of the smoke layer in both instances is not exactly the same).

Light obscuration: The visibility distance through smoke can be related to the optical density per unit pathlength *via* empirical correlations.[42,43] (See Section 2, Chapter 15.) The experimental basis for the correlations consists of tests with humans viewing objects through smoke. However, the participants were not directly exposed to the irritating effects of smoke. Consequently, the reported correlations are likely to overestimate the visibility distance.

In addition to the light obscuration quality of the smoke, the visibility of an object is dependent on the light source for the object being viewed as well as ambient lighting conditions.[43] The following correlation has been suggested by Heskestad[44] for estimating the visibility of an "illuminated sign" or "placard" and "focused lamp in black smoke" or "yellow smoke"

$$Vis = 2.0\delta^{-0.69} \qquad (22)$$

The visibility estimated by Equation 22 overestimates the visibility in diffuse light and underestimates the visibility in situations with "back-lighted signs." Alternatively, the following expression is provided by Mulholland (see Section 2, Chapter 15)

$$Vis = \frac{K}{\alpha} \qquad (23)$$

A more complete discussion of Equation 23 along with a comparison of Equations 22 and 23 is provided by Klote and Milke.[17]

The optical density per unit pathlength, δ, is a function of the fire scenario, m_f, and air flow conditions, V, as well as the fuel material, δ_{mass},

$$\delta = \delta_{mass}\left(\frac{m_f}{V}\right) \qquad (24)$$

The mass optical density is reported for a variety of common fuels. (See Section 3, Chapter 4.) For steady fires, the mass burning rate can be estimated as

$$m_f = \frac{Q_c}{H_{c,conv}} \qquad (25)$$

The portion of the heat of combustion released via convection, $H_{c,conv}$, is assumed to be 70 percent of the total heat of combustion, to be consistent with the relationship between the convective and total heat release rates. Substituting Equations 24 and 25 into Equation 22 yields

$$Vis = 2.0\left(\frac{\delta_{mass}Q_c}{H_{c,conv}V}\right)^{-0.69} \qquad (26)$$

An alternative expression for the visibility minimizing the number of variables related to the fire scenario can be derived by using Equation 19 to substitute for Q_c/V

$$Vis = 2.0\left(\frac{\delta_{mass}\rho c_p \Delta T_{ad}}{H_{c,conv}}\right)^{-0.69} \qquad (27)$$

ΔT_{ad} is the only parameter related to the fire scenario and smoke exhaust system. All other variables in Equation 27 are a function of the fuel material. Being dependent on the adiabatic temperature rise (which overestimates the actual temperature rise), the visibility through smoke will generally be underestimated.

Carbon monoxide concentration: The concentration of carbon monoxide (CO) in the smoke layer can be estimated based on the yield fraction of the fuel as long as the fuel material is burning in free air. As with light obscuration, the concentration of CO can also be related to the adiabatic temperature rise.[44]

$$\frac{C_{CO}}{\Delta T_{ad}} = c_p \frac{M_{CO}}{M_{air}} \frac{Y_{CO}}{H_{c,conv}} \qquad (28)$$

Tabulated values of Y_{CO} and $H_{c,conv}$ are available from Tewarson for a wide variety of smoldering or flaming materials. (See Section 3, Chapter 4.) For wood products, $Y_{CO}/H_{c,conv}$ is approximately 5×10^{-7} kg/kJ. With an adiabatic temperature rise of 10°C, the concentration of CO estimated from Equation 28 is 4.8 ppm. This is several orders of magnitude below that which is typically considered to be hazardous. (See Section 2, Chapter 8.) Given that an adiabatic temperature rise on the order of 10°C is reasonable for tall spaces, it is doubtful that CO will be the governing hazard parameter to assess the loss of tenability. Initially, this may be surprising, considering numerous

fire incident accounts citing CO as being the principal combustion product causing death. However, most of these accounts are related to fire incidents in spaces with short ceiling heights. In tall spaces, the combustion gases are appreciably diluted as a result of the entrainment process along the entire height of the plume. Further, being directly proportional to the adiabatic temperature rise (which overestimates the actual temperature rise of the smoke layer), the concentration of CO in the smoke layer will be overestimated.

EXAMPLE 6:

Estimate the steady-state smoke layer properties (temperature, visibility, and CO concentration), given the following situation:

1. The smoke layer interface is maintained 35 m above floor level.
2. The rate of heat release of the flaming fire is 5,000 kW.
3. The fuel comprises principally red oak.

SOLUTION:

Smoke Layer Temperature
Equation 19 can be applied to determine the adiabatic smoke layer temperature rise. In Example 4, a mass rate of smoke production of 410 kg/s was determined. Thus, assuming a convective heat release rate fraction of 0.7 and specific heat of air of 1.0 kJ/kg-K, the adiabatic temperature rise is 8.5°C.

Visibility
Visibility is estimated using Equation 27. Fuel-related parameters are obtained in Section 3, Chapter 4.

$$\delta_{mass} = 38 \text{ m}^2/\text{kg}$$

$$H_c = 12{,}400 \text{ kJ/kg}$$

Evaluating ρ and c_p at the temperature of the smoke layer, the resulting visibility is 8.5 m.

CO Concentration
CO concentration is estimated using Equation 28, with the fuel-related properties again evaluated from Section 3, Chapter 4.

$$Y_{CO} \text{ for red oak is } 0.004 \text{ kg}_{co}/\text{kg}_{fuel}$$

The resulting CO concentration in the smoke layer is 2.7 ppm.

The CO concentration of 15 ppm is slight. A very long exposure would be necessary to approach a time-integrated value of 27,000 ppm-min.

SPECIAL CONDITIONS

There are three special conditions where the previously described smoke filling and production analyses are inappropriate. The limitation is attributed to violations of the basic assumptions on which the analyses are based. The special conditions involve situations of confined flow, intermediate stratification, and limited fuel.

Confined Flow

As a plume rises, it also widens as a result of the entrainment of additional mass into the plume. For tall, narrow spaces, the plume may fill the entire cross section of the atrium prior to reaching the ceiling. Above this position, air entrainment into the plume is greatly reduced due to the limited amount of air available. In such situations, initially the bottom of the smoke layer may be assumed to be located at this point of contact. Following a delay to fill the entire volume above the point of contact, the descent of the smoke layer can be estimated by analogy with the ceiling being moved down to the point of contact. The delay to fill the upper volume can be estimated by assuming duct flow, with the velocity estimated as the plume velocity at the height of contact.

In order to determine the point of contact of the plume with the walls, the plume width must be expressed as a function of height. Typically, the width of the plume is estimated by examining the difference between the plume temperature and ambient temperature, i.e., temperature excess, at various horizontal distances from the plume centerline.[33] Generally, the plume width is defined as the position where the temperature excess is one-half of the value at the centerline.

Heskestad[45] noted that the visible plume diameter was greater than that determined from the temperature excess. Consequently, Heskestad estimated the visible plume diameter to be twice that determined by the excess temperature approach. Thus, the plume diameter is estimated as

$$d = 0.48 \left(\frac{T_c}{T_o}\right)^{1/2} z \qquad (29)$$

The plume centerline temperature, T_c, can be evaluated using Equation 20. As indicated in Equation 20, the plume centerline temperature decreases appreciably with increasing height. Thus, for tall spaces, the plume centerline temperature may be close to ambient. For example, at a height of 30 m with a fire size of 5,000 kW and T_o of 293 K, T_c is 312 K. In this case $(T_c/T_o)^{1/2}$ in Equation 29 is only 1.03. Because of the rapid decline in T_c with increasing height, for engineering purposes $(T_c/T_o)^{1/2}$ can be approximated as being 1.0. Consequently, the total plume diameter may be succinctly estimated in many cases by considering the plume diameter to be approximately one-half of the height. This rule-of-thumb method yields a reasonable estimate for the plume diameter for small- or moderate-size fires in moderate to tall atria. Specifically, if $(Q_c^{2/3} z^{-5/3})$ is less than 2.6, the rule-of-thumb provides a plume width within 5 percent of that estimated by Equation 29.

Intermediate Stratification

Smoke in the plume rises only as long as the smoke is buoyant relative to the surroundings. In a quiescent, uniform environment, the maximum rise of buoyant plume gases is given by[21]

$$z_m = 3.79 \, F^{1/4} G^{-3/8} \qquad (30)$$

where:

$F = gQ_c/(T_o \rho_o c_p)$
$G = -(g/\rho_o) \, d\rho_o/dz$

Assuming standard conditions and that air is a perfect gas, the expressions for F and G are

$F = 0.0279 \, Q_c$
$G = 0.0335 \, dT_o/dz$

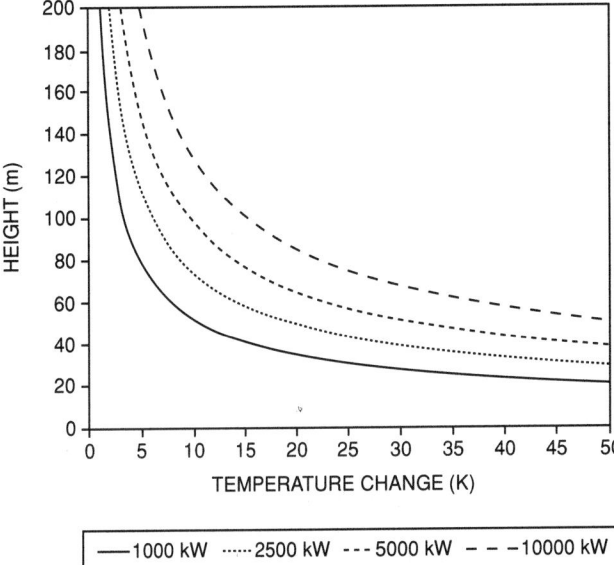

Fig. 4-13.8. Maximum rise of smoke plume, constant temperature gradient.

In the particular case where the ambient, pre-fire temperature increases uniformly along the entire height, dT_o/dz is a constant, $\Delta T_o/H$. Substituting the simplified expressions for F and G into Equation 30 yields[46]

$$z_m = 5.54 \, Q_c^{1/4} (\Delta T_o/H)^{-3/8} \qquad (31)$$

By re-formulating Equation 31 to solve for Q_c, a minimum fire size can be determined that is just large enough to force the smoke to the ceiling of an atrium without prematurely stratifying due to the increasing ambient temperature.

$$Q_c = 0.00118 H^{5/2} \Delta T_o^{3/2} \qquad (32)$$

Dillon[47] reported measurements of the difference in ambient temperature from floor to ceiling to be on the order of 50°C in some atria with glazed ceilings. The relationship of the minimum heat release rate, ambient temperature change, and height is indicated in Figure 4-13.8.

Limited Fuel

In some cases smoke management objectives may be fulfilled without a dedicated smoke management system due to the intrinsic qualities of the atrium. The intrinsic qualities of the atrium include parameters, such as the composition and quantity of fuel and geometry of the atrium. As an example, a limited amount of fuel may be present that is unable to sustain a fire for a sufficient period of time to create conditions beyond the allowable limits. The amount of fuel consumed during the time period of interest depends on whether the fire is steady or unsteady. In the case of a steady fire, the fuel mass consumed in a given period of time is determined as

$$\dot{m}_f = \frac{Qt}{H_c} \qquad (33)$$

Alternatively, for an unsteady, t-square profile fire, the fuel mass consumed during a given period of time is given as

$$\dot{m}_f = 333\frac{t^3}{H_c t_g^2} \tag{34}$$

When analyzing the inherent ability of the atrium to fulfill the smoke management design goals, the time period should relate either to the performance of a fire protection system or to the development of smoke layer conditions in excess of acceptable levels. For example, in life-safety oriented designs, the time period may be either that required for evacuation, or for untenable conditions to be generated, whichever is less. Alternatively, when considering smoke management for property protection, possible time periods of interest could be the time until fire department arrival, or the time interval necessary to produce smoke conditions in excess of allowable damage limits; again, whichever is less.

OPPOSED AIRFLOW

Opposed airflow refers to systems where airflow is provided in a direction opposite to smoke movement. Opposed airflow may be used in lieu of physical barriers to prevent smoke spread from one space to another, i.e., between the communicating space and the atrium. Opposed airflow limits smoke flow by countering the momentum of the smoke attempting to enter the communicating space. A minimum airflow velocity at all points of the opening must be provided in order to prevent smoke migration through the opening. Two empirical correlations to estimate the minimum average velocity for the entire opening are available, based on limited experimental data.[48] The calculated average velocity is greater than the actual minimum velocity required at a point to oppose smoke propagation. The excess velocity is required so that the minimum critical velocity is achieved at all points, considering the effects of turbulence caused by the edges and corners of the opening.

The minimum average velocity to oppose smoke originating in the communicating space is evaluated using Equation 35.

$$v_e = 0.64\sqrt{\frac{gH(T_s - T_o)}{T_s}} \tag{35}$$

Alternatively, if the smoke at the opening is part of a rising plume that is rising along the side of the atrium wall, then Equation 36 is applicable.

$$v_e = 0.057\left(\frac{Q}{z}\right)^{1/3} \tag{36}$$

The opposed airflow velocity should not exceed 1 m/s. Above that limit, the airflow velocity may deflect the plume away from the wall, making more plume surface area available for entrainment. The increased area for entrainment will enhance the smoke generation rate. Consequently, the problem of propagation to the communicating space may be solved by an excessive average velocity; however, other problems may be created by the increased smoke production rate and a possible increase in the depth of the smoke layer in the atrium. The volumetric capacity of the mechanical equipment required to deliver the necessary velocity for opposed airflow can be approximated as

$$V_{oa} = A_o v_e \tag{37}$$

If several openings are protected with the opposed airflow approach using the same mechanical equipment, the cross-sectional area should be the sum of the areas for all of the openings. The opposed airflow technique may be infeasible due to the substantial amount of airflow capacity required to protect numerous openings having a large total area.

Where opposed airflow is incorporated to prevent smoke migration into a communicating space from the atrium, the impact of the volume of air being introduced into the atrium must be assessed. Specifically, if smoke exhaust equipment is also provided to maintain a constant position of the smoke layer interface in the atrium, then all of the additional air used for opposed airflow must also be exhausted. The additional air can be accounted for by increasing the required mass rate of exhaust in the atrium by the amount used for the opposed airflow. The additional air being exhausted will also affect the qualities of the smoke layer within the atrium. (See Equations 19, 27, and 28.) The smoke layer temperature, T_s (K), can be determined using Equation 38, based on an analysis included elsewhere.[17]

$$T_s = 293 + \left[0.0018 + 0.072Q_c^{-2/3}z^{5/3} + \frac{712A_o\sqrt{H(T_s - 293)}}{Q_c T_s^{3/2}}\right]^{-1} \tag{38}$$

Equation 38 must be applied iteratively to determine the resulting smoke layer temperature. In cases with large clear heights, the temperature of the air used for the opposed airflow strategy will be virtually equal to the temperature of the smoke layer to permit the addition of volumetric rates of air rather than mass rates.

EXAMPLE 7:

Considering the atrium from Example 4, consider that there are five 5-m wide × 2.5-m high openings to the communicating space. The bottoms of the openings are 30 m above the floor of the atrium. Considering a 5,000 kW fire in the center of the floor of the atrium, determine the following:

1. Minimum airflow velocity required for opposed airflow,
2. Volumetric rate of air supply for opposed airflow, and
3. Capacity of the exhaust fans in the atrium to maintain the smoke layer interface at an elevation 25 m above floor level and also to accommodate the additional air from the opposed airflow approach.

SOLUTION:

The minimum opposed airflow velocity can be determined using Equation 35. However, the temperature of the smoke layer, T_s, is unknown. Thus, Equation 38 must be applied first. Solving iteratively, T_s is approximately 305 K. The minimum airflow velocity is 0.20 m/s. The volumetric supply capacity for the opposed airflow strategy for all five openings is 12.5 m³/s. The associated mass flow rate is 15.0 kg/s.

Without the opposed airflow, the mass rate of smoke exhaust required to maintain the smoke layer interface height in the atrium at a height of 25 m is determined using Equation 8 (neglecting z_o) to be 236 kg/s. Thus, the combined mass exhaust rate necessary is 251 kg/s. This mass flow rate corresponds to a volumetric rate of 209 m³/s.

As a practical issue, this exhaust rate should be compared to that required to keep the smoke layer interface above the top of the openings, i.e., 32.5 m above floor level. Based on Equations 8 and 16, the required volumetric exhaust rate is 362 kg/s. Thus, in this situation, the combined exhaust rate with the opposed airflow strategy is less than that associated with the strategy to keep the smoke layer interface above the opening.

MAKEUP AIR SUPPLY

The makeup air supplied to the atrium should be:

1. Uncontaminated,
2. Introduced below the smoke layer,
3. Introduced at a slow velocity, and
4. Supplied at a rate less than the required exhaust rate.

Air that is not contaminated by smoke can be provided by locating intakes for the makeup air remote from the smoke exhaust discharge, preventing smoke feedback. In the event that smoke is introduced into the makeup air supply, a smoke detector should be provided to shut down the makeup air supply system. Selection of a smoke detector for this application should consider the operating conditions, range of temperatures, and installation within a duct.

All makeup air should be provided below the smoke layer interface. Any makeup air provided above the smoke layer interface merely adds mass to the smoke layer, which must be added to the required capacity of the smoke exhaust to prevent an increase in the smoke layer depth. If introduced near the smoke layer interface, the makeup air may increase the amount of mixing of clean air with the smoke to further add to the smoke layer.

Makeup air should be provided at a slow velocity so that the plume, fire, and smoke layer are not adversely affected. Makeup air supplied at a rapid velocity near the plume may deflect the plume to enhance the entrainment rate, thereby increasing the rate of smoke production. In addition, the burning rate of the fire may be increased by makeup air provided at an excessive velocity. A makeup air velocity of 1 m/s in the vicinity of the plume is considered to be just capable of enhancing the burning rate of the fire, as this velocity is approximately the velocity of the air entrained into the plume. Because of the diffusion of air once past the diffuser, the makeup air velocity at the diffuser may be greater than 1 m/s. The relationship of air velocity with distance from the diffuser is outlined elsewhere.[49]

Finally, the mass rate of makeup air supplied must be less than that being exhausted. Failure to follow this guideline may lead to the atrium being pressurized relative to the communicating spaces. Being at a positive pressure, smoke movement will be forced through any unprotected openings in physical barriers into the communicating spaces.

NOMENCLATURE

A cross-sectional area of the atrium (m^2)
A_o cross-sectional area of opening (m^2)
b distance from the store opening to the balcony edge (m)
C_{CO} volumetric concentration of carbon monoxide (ppm)
c_p specific heat (kJ/kg-K)
D diameter of fire (m)
d plume diameter (based on excess temperature) (m)

g gravitational acceleration (9.8 m/s^2)
H height of ceiling above top of fuel surface (m)
H_b height of balcony above top of fuel surface (m)
H_c heat of combustion (kJ/kg)
$H_{c,conv}$ convective heat of combustion (kJ/kg)
h enthalpy
k_v volumetric entrainment constant (0.065 $m^{4/3}$ $kW^{-1/3}s^{-1}$)
L width of balcony spill plume (m)
M_{CO} molecular weight of carbon monoxide (28)
M_{air} molecular weight of air (29)
m_u mass of upper smoke layer (kg)
m mass entrainment rate in plume (kg/s)
m_f mass burning rate (kg/s)
Q heat release rate of fire (kW)
Q_c convective heat release rate of fire (kW)
T_c temperature at plume centerline (K)
T_o ambient temperature (K)
T_s heated smoke temperature (K)
ΔT_{ad} temperature difference between smoke layer and ambient air (°C)
ΔT_o temperature change from floor to ceiling of the ambient air (°C)
t time (s)
t_g growth time (s)
V volumetric smoke production rate (m^3/s)
Vis visibility distance (m)
V_{oa} volumetric capacity required for opposed airflow (m^3/s)
V_u volume of upper layer (m^3)
v_e opposed airflow velocity (m/s)
w width of the store opening from the area of origin (m)
Y_{CO} yield of CO produced per unit mass of combustible consumed (kg_{CO}/kg_{fuel})
z clear height, position of smoke layer interface above the top of fuel surface (m)
z_f limiting height above fuel (m)
z_m maximum rise of plume (m)
z_o virtual origin of plume (m)
α attenuation coefficient of the smoke layer (m^{-1})
δ optical density per unit pathlength (m^{-1})
δ_{mass} mass optical density (m^2/kg)
ρ density of smoke (kg/m^3)
ρ_o density of air (1.2 kg/m^3 at 20°C)

REFERENCES CITED

1. R. Saxon, *Atrium Buildings, Development and Design*, The Architectural Press, London, (1983).
2. P. Robinson, "Atrium Buildings—A Fire Service View," *Fire Surveyor*, 11, pp. 43–47, Aug. 1982.
3. NFPA 92A, *Recommended Practice for Smoke-Control Systems*, National Fire Protection Association, Quincy, MA (1993).
4. NFPA *101, Life Safety Code*, National Fire Protection Association, Quincy, MA (1994).
5. NFPA 92B, *Guide for Smoke Management Systems in Malls, Atria, and Large Areas*, National Fire Protection Association, Quincy, MA (1991).
6. J.L. Bryan, "Damageability of Buildings, Contents, and Personnel from Exposure to Fire," *Fire Safety J.*, 11, pp. 15–32 (1984).
7. G.E. Hartzell, "Criteria and Methods for Evaluation of Toxic Hazard," *Fire Safety J.*, 12, pp. 179–182 (1987).
8. J.L. Bryan, "Smoke as a Determinant of Human Behavior in Fire Situations (Project People I)," University of Maryland, College Park, June 1977.

9. J.L. Bryan and J.A. Milke, "The Determination of Human Response Patterns in Fire Situations—Project People II," University of Maryland, College Park, Aug. 1981.

10. H.H. Spieth, J.G. Gaume, R.E. Luoto and D.M. Klinck, "A Combined Hazard Index Fire Test Methodology for Aircraft Cabin Materials," Vol. I and II, DOT/FAA/CT-82/36-1 and DOT/FAA/CT-82-36-11, Atlantic City, Department of Transportation, Apr. 1982.

11. R.D. Peacock and E. Braun, "Fire Tests of Amtrak Passenger Rail Vehicle Interiors." NBS Technical Note 1193, National Bureau of Standards, Gaithersburg, MD, May 1984.

12. J. Morehart, "Sprinklers in the NIH Atrium: How Did They React During the Fire Last May?," *Fire Journal*, 83, pp. 56–57, Jan./Feb. 1989.

13. V. Babrauskas, "A Laboratory Flammability Test for Institutional Mattresses," *Fire Journal*, 72, pp. 35–40, 93, Nov. 1981.

14. S.W. Harpe, T.E. Waterman, and W.S. Christian, "Detector Sensitivity and Siting Requirements for Dwellings, Phase 2, Part 2 of 'Indiana Dunes Tests,'" National Bureau of Standards, Gaithersburg, MD, (1976).

15. NFPA 204M, *Guide for Smoke and Heat Venting*, National Fire Protection Association, Quincy, MA (1991).

16. J.G. Quintiere, "Scaling Applications in Fire Research," *Fire Safety J.*, 15, pp. 3–29 (1989).

17. J.H. Klote and J.A. Milke, *Design of Smoke Management Systems*, ASHRAE, Atlanta (1992).

18. J.A. Milke and F.W. Mowrer, "A Design Algorithm for Smoke Management Systems in Atria and Covered Malls," Report FP93-04, University of Maryland, College Park (1993).

19. J.A. Milke, "Smoke Management for Covered Malls and Atria," *Fire Technology*, 26, 3, pp. 223–243 (1990).

20. G. Heskestad and M.A. Delichatsios, "Environments of Fire Detectors—Phase I: Effect of Fire Size, Ceiling Height, and Materials." Vol. I—"Measurements (NBS-GCR-77-86), Vol. II—"Analysis" (NBS-GCR-77-95), National Bureau of Standards, Gaithersburg, MD (1977).

21. B.R. Morton, Sir Geoffrey Taylor, and J.S. Turner, "Turbulent Gravitational Convection from Maintained and Instantaneous Sources," *Proc.* Royal Society A, 234, pp. 1–23 (1956).

22. G. Mulholland, T. Handa, O. Sugawa, and H. Yamamoto, "Smoke Filling in an Enclosure," Paper 81-HT-8, The American Society of Mechanical Engineers, New York (1981).

23. L.Y. Cooper, M. Harkleroad, J. Quintiere, and W. Rinkinen, "An Experimental Study of Upper Hot Layer Stratification in Full-Scale Multiroom Fire Scenarios," Paper 81-HT-9, The American Society of Mechanical Engineers, New York (1981).

24. G. Heskestad, Letter to the Editor, *Fire Technology*, 27, 2, pp. 174-185, May 1991.

25. T. Yamana and T. Tanaka, "Smoke Control in Large Spaces (Part 2—Smoke Control Experiments in a Large-Scale Space)," *Fire Science and Technology*, 5, 1, pp. 41–54 (1985).

26. G.D. Lougheed, National Research Council of Canada, Mar. 20, 1991; personal communication.

27. S.P. Nowlen, "Enclosure Environment Characterization Testing for the Baseline Validation of Computer Fire Simulation Codes," NUREG/CR-4681, SAND 86-1296, Sandia National Laboratories, Mar. 1987.

28. J.A. Sharry, "An Atrium Fire," *Fire Journal*, 67, 6, pp. 39–41 (1973).

29. J. Lathrop, "Atrium Fire Proves Difficult to Ventilate," *Fire Journal*, 73, 1, pp. 30–31 (1979).

30. D.M. McGrail, "Denver's Polo Club Condo Fire: Atrium Turns High-Rise Chimney," *Fire Engineering*, pp. 67–74, Mar. 1992.

31. G. Heskestad, "Engineering Relations for Fire Plumes," SFPE TR 82-8, Society of Fire Protection Engineers, Boston (1982).

32. F.W. Mowrer, "Lag Times Associated with Fire Detection and Suppression," *Fire Technology*, 26, 3, pp. 244-265 (1990).

33. C. Beyler, "Fire Plumes and Ceiling Jets," *Fire Safety J.*, 11, pp. 53–76, Jul./Sept. 1986.

34. R.L. Alpert and E.J. Ward, "Evaluation of Unsprinklered Fire Hazards," *Fire Safety J.*, 7, pp. 127–143 (1984).

35. F.W. Mowrer and B. Williamson, "Estimating Room Temperatures from Fires along Walls and in Corners," *Fire Technology*, 23, 2, pp. 133–145 (1987).

36. M. Law, "A Note on Smoke Plumes from Fires in Multi-Level Shopping Malls," *Fire Safety J.*, 10, pp. 197–202 (1986).

37. H.P. Morgan and N.R. Marshall, "Smoke Control Measures in a Covered Two-Story Shopping Mall Having Balconies and Pedestrian Walkways," BRE CP11/79, Fire Research Station, Borehamwood (1979).

38. Lincolne Scott Australia Pty Ltd., "Jupiters Casino—Report on Atrium Smoke Tests," Toowong, Australia (1986).

39. G.O. Hansell and H.P. Morgan, "Smoke Control in Atrium Buildings Using Depressurization," PD 66/88, Fire Research Station, Borehamwood (1988).

40. R.A. Waters, "Stansted Terminal Building and Early Atrium Studies," *J. of Fire Protection Engineering*, Vol. 1., No. 2, pp. 63–76, Apr./May/June 1989.

41. G. Heskestad, "Similarity Relations for the Initial Convective Flow Generated by Fire," Paper 72-WA/HT-17, American Society of Mechanical Engineers, New York (1972).

42. T. Jin, "Visibility Through Fire Smoke (Part 2)," Report of the Fire Research Institute of Japan, Nos. 33, 31 (1971).

43. J.G. Quintiere, "An Assessment of Correlations Between Laboratory and Full-Scale Experiments for the FAA Aircraft Fire Safety Program, Part 1: Smoke," NBSIR 82-2508, National Bureau of Standards, Gaithersburg, MD, Jul. 1982.

44. G. Heskestad, "Hazard Evaluation," submitted to NFPA Task Group on Smoke Management of Atria, Covered Malls, and Large Spaces, Sept. 1988; unpublished.

45. G. Heskestad, "Fire Plume Entrainment and Related Problems in Venting of Fire and Smoke from Large Open Spaces," submitted to NFPA Task Group on Smoke Management of Atria, Covered Malls, and Large Spaces, Dec. 1987; unpublished.

46. G. Heskestad, "Note on Maximum Rise of Fire Plumes in Temperature-Stratified Ambients," *Fire Safety J.*, 15, pp. 271–276 (1989).

47. M. Dillon, "Acceptance Testing and Techniques," presented at *The Roundtable on Fire Safety in Atriums–Are the Codes Meeting the Challenge?*, Washington, DC, Dec. 16, 1988.

48. G. Heskestad, "Inflow of Air Required at Wall and Ceiling Apertures to Prevent Escape of Fire Smoke," FMRC J.I.OQ4E4.RU, Factory Mutual Research Corporation, Norwood, MA, Jul. 1989.

49. American Society of Heating, Refrigerating, and Air-conditioning Engineers, Inc., *1993 ASHRAE Handbook-Fundamentals*, ASHRAE, Atlanta, GA (1993).

Section Five
Fire Risk Analysis

Section 5 Fire Risk Analysis

COMPUTER SIMULATION FOR FIRE PROTECTION ENGINEERING

William G.B. Phillips

INTRODUCTION

Fire protection engineers are required to deal with a complex fire scenario that includes human reactions and behavior, in addition to the physical and chemical fire processes. Operations research (OR) pioneered the application of the scientific method to the management of organized systems in which human behavior is a key element. Fire protection engineering could be defined as the application of operations research to the fire system.

Systems involving human beings are difficult to study because realistic experiments may be impossible and neither past experience nor the available data provides sufficient insight. Operations research overcomes these difficulties by the use of simulation models. Simulation models are widely used in science, engineering, and mathematics in the study of problems that involve ordinary and partial differential equations (either overtly or implicitly). In fire science, for example, simulation models have been used to handle phenomena such as smoke movement and absorption of toxic substances; both computational fluid dynamics (CFD) models and simulation models are solved by similar techniques. These matters are dealt with elsewhere in this handbook. This chapter concentrates on the variety of procedural simulation models that are applied in operations research to interdisciplinary problems, specifically involving human agents and objectives.

Simulation models are dynamic and allow for continuous interaction between the processes within the system under study. The model is made to predict outcomes by actually executing procedural steps that represent events in the real world. Running the model creates the prediction. Advances in computer software and hardware have greatly reduced the cost and simplified the use of simulation models. Recent developments in virtual reality are expected to facilitate more realistic interaction between the model and its user.

Uncertainty can be handled by introducing stochastic elements into the model. Estimates of risk can be derived from simulation models by the use of Monte Carlo techniques. Regression analysis can be applied to obtain compact expressions that can be used to measure the sensitivity of the output variables to variations in the inputs.

TYPES OF MODELS

A model can represent a system as a unified and precisely definable whole, all of whose aspects are simultaneously and unambiguously accessible for assessment. Models include pictures, diagrams, and "scale models," as well as mathematical structures.

Models can be classified as descriptive, physical, and symbolic.[1] The *descriptive model* is expressed in ordinary language and is the most common tool for decision making in science, engineering, and everyday life. Descriptive models function like metaphors. For example, the flow of smoke through a vent might be compared to the flow of water in a channel, implying that buoyancy and gravity play a similar role albeit with a reversal of sign.

Physical models include scale models; examples can range from basic hydraulic models of harbors and estuaries to transparent plastic models to demonstrate the flow of smoke in buildings. They make it possible to try out alternative arrangements in the search for an optimum design or strategy. Analogue models are a special type of physical model that exploits the isomorphisms that exist between different physical processes. The behavior of voltages, currents, resistance, capacitance, and inductance in an electrical circuit has many analogies with processes in heat and fluid flow and acoustic, electromagnetic, and mechanical systems.

Science and engineering depend heavily on *symbolic models* in which algebraic symbols represent the values of variables and the relationships between them. Once a model has been cast in symbolic form, the whole mathematical apparatus can be deployed to deduce additional relationships and solve equations to find optimal solutions. Symbolic models may be static or dynamic; in the latter case they specifically include a variable representing time. A very important type of dynamic model is formulated in terms of ordinary or partial differential equations. Many problems in fire science lead to models that can be expressed in the form of one or more simultaneous differential equations. Some of

Bill Phillips, BSc (Eng), BSc (Econ), ARSM, MIMM, is a member of the Fire Risk Unit at the Fire Research Station of the Building Research Establishment at Borehamwood, Herts. His recent research has been mainly concerned with the development of new methods of fire risk assessment based on simulation models.

these have simple analytic solutions, but many interesting cases do not, e.g., the partial differential equations that arise in fluid dynamics (Stokes-Navier equations). Such equations are normally solved on computers by standard numerical methods.

A *simulation model* treats the dynamic relationships that are assumed to exist in the real situation as a series of elementary operations on the appropriate variables. A simulation model is made to predict outcomes by actually executing the procedural steps with appropriate initial data and parameters. The computer is programmed to execute the procedure that is the model. Running the model creates the prediction. The variables in a simulation may change continuously in value or take on only certain discrete values. These changes may take place at any time or only at certain times, or both. For example, a simulation of a fire incident might handle the flow of hot gases by a differential equation expressed in continuous terms, while the people would be treated as discrete individuals moving at prescribed moments in time. However, since such a model would almost certainly be implemented on a computer, the differential equations would be approximated by a difference equation for the purposes of numerical integration (a computer cannot handle "real" numbers). This tends to blur the distinction between continuous simulation models and the dynamic symbolic models discussed in the previous paragraph.

SIMULATION MODELS

A procedural simulation model is a representation of a dynamic system in which the processes or interactions bear a close resemblance or relationship to those of the specific system being simulated or studied. These models are concrete rather than abstract and may contain approximations and subjective elements. They are amenable to manipulations that would be impossible, too expensive, or impractical to perform on the entity portrayed. The operation of the model can be studied, and properties concerning the behavior of the actual system or its subsystems can be inferred. Manipulation of a procedural model requires the acceptance of inputs and the generation of outputs that are similar or analogous to those of the system represented.

Procedural simulation is essentially a technique that involves setting up a model of a real situation and then performing experiments on the model. The idea that a computer model might partly replace experimentation is both dangerous and attractive. It is, therefore, appropriate to remember a warning given by Ackoff and Beer.[2]

> Often when an operations researcher does not understand a phenomenon which he can nevertheless describe well, he can simulate it on a computer and thus conduct experiments. These are experiments on the model on which the simulation is based, not on the system involved. The fact that a simulation may reproduce history with some accuracy does not by itself establish a correspondence of structure between the model and reality. This is true for the same reason that a straight line may have been generated by a sine function, not one that is linear.

Procedural simulation modeling has been widely applied; examples are Link (aircraft pilot) trainers, military war games, business management games, space exploration, physical models of river basins and estuaries, econometric models,

electrical analog devices, and wind-tunnel tests for aircraft. It has proved particularly useful in situations involving human intervention of one kind or another. There are great opportunities for improving the interface between the model and operator, making use of virtual reality techniques. In this way, "realistic experience" of rare events can be acquired without danger and at low cost.

Types of Simulation Models

Simulation models can be classified as either discrete or continuous. In the real world there is no such distinction, yet it is possible to model some real-world systems either discretely or continuously. In both types of simulation what is of concern are the changes in the state of the model. Continuous simulations are analogous to a stream of fluid passing through a pipe. The volume may increase or decrease, but the flow is continuous. Using the pipe analogy for discrete event simulation, the pipe could either be empty or have something traveling through it. Whether anything came out of the pipe would depend on some event occurring at the other end.

Both types of simulation can be applied in the fire situation. As time progresses the state of a building changes continuously as the once small fire becomes larger, overwhelms the building, and eventually dies out as it runs out of fuel. The chemical processes in the fire and the physical processes that mediate the flow of heat and hot gases naturally lend themselves to the continuous type of model. On the other hand, discrete simulations are more appropriate to the strategies used to fight the fire and evacuate the building. For example, the fire is ignited and detected, fire fighters and equipment arrive, hoses are deployed, water is applied, more equipment may be called in, and so on. Similarly the occupants respond to the alarm, collect their belongings, locate their household members, and go toward the exits in a sequence of clear-cut stages.

In continuous models, changes in the variables are directly based on changes in time. The values of the variables reflect the state of the model at any particular time, and simulated time advances from one step to the next usually in equal increments. In discrete event models, events occur as items move through the simulation. The state of the model changes only when those events occur. Simulated time advances from one event to the next (generally in unequal increments) and the mere passing of time has no direct effect.

Discrete Event Simulation Models

Discrete event simulation models are built up from several different elements. These are given different names in different programming languages. A terminology based on *blocks*, *items*, and *events* will be used in this discussion. *Items* are characterized by *attributes*, *priorities*, and *values*.

A *block* is like a block in a block diagram. It is used to represent an action, operation, resource, or process. These blocks are connected in an activity or data flow diagram that represents the system. Information comes into the block and is processed by the program that is in the block. The block then transmits information out of the block to the next block in the simulation. Some blocks may simply represent a source of information that is passed on to other blocks. Other blocks may modify information as it passes through them. Output blocks take information from the simulation and present it to the user in graphic form.

The basic unit passed between the blocks is called an *item*. An item is a data set that carries information about the item's attributes, priorities, and values. In a manufacturing model, an item might be a part on an assembly line; in a network model, an item would be a packet of information; in an evacuation model, an item would be an occupant. Items are generated by special blocks, either according to a fixed schedule or a random distribution.

The model moves items from block to block only when an *event* occurs. Events only occur when specific blocks specify that they should. For example:

1. Blocks that depend on time cause events to happen at an appropriate time. For example, a block representing an activity, e.g., the journey from the fire station to the fire location, might introduce a delay of t minutes into the system representing the duration of the journey. If the alarm was received by the fire station at T minutes, the block would post an event in the event queue at $T + t$ minutes.
2. Blocks that have an accumulated demand for inputs cause an event immediately after they receive items. For example, a block representing a parent searching for a child, following a fire alarm, would cause an event when the child was located.
3. Blocks that do not generate events allow the blocks after them to pull items during a single event. Thus, an item can pass through many blocks after a single event if those blocks do not stop it.

Note that every event has the potential to cause every block in a model to move items. Thus, one event may cause many unrelated items to progress in the simulation.

In order to provide true discrete event simulation, the time clock that controls the simulation must move to the exact time of each event. The most common way of doing this is to have an event queue. Each block places the time of its next event in a slot in the event queue. In each cycle the event queue is checked to find the next closest time point; the current time is set to that value, and a "simulate" message is sent to every block. Most blocks ignore these messages unless the message occurs at the event time that was previously posted by the block, or there are items waiting in the block's inputs that can be pulled in. For example, a block representing a queue should pull in an available item whenever it gets a "simulate" message, no matter which block posted the event.

Attributes are an important aspect of discrete event simulation. Attributes are characteristic properties of an item that stay with it as it moves through the simulation. For example, an occupant making an escape from a building might have attributes representing mobility and the amount of carbon monoxide absorbed. *Priorities* specify the relative importance of an item. *Values* allow the model to deal with items that represent groups of identical entities.

Continuous Simulation Models

There is a close similarity between continuous simulation models and the solution of ordinary differential equations by numerical methods. Almost any phenomenon that can be represented by differential equations may be modeled by continuous simulation. This includes virtually the whole of classical physics, and the method can easily be extended to chemistry and biology and more speculatively to economics, ecology, and human behavior. The fundamental princi-ple is that the rate of change of certain variables can be expressed as a function of a set of variables (the state variables) that describe the state of the system at a given time. Therefore

$$\frac{dx_i}{dt} = f(x_1 \ldots x_i \ldots x_n), \quad i = 1 \ldots n$$

The solution to this set of equations is found by integration and yields a function that depends on time, which characterizes the dynamic behavior of the system. For example, suppose the rate at which a chemical reagent is consumed is proportional to the concentration of the reagent. This relationship could be expressed in the form of a differential equation, with the following solution

$$\frac{dx}{dt} = -kx$$

and, therefore

$$x = Ae^{-kt}$$

This result can be interpreted to mean that the concentration declines exponentially and tends toward zero as the reaction proceeds. Note that the result is a function of time and not concentration. This example leads to a first order linear differential equation that has an analytic solution. Many systems of interest, however, give rise to higher order non-linear differential equations, which, in general, do not have analytic solutions or only have solutions if drastic simplifications are imposed. Their solution is normally carried out on computers using one of the standard numerical methods (e.g., Euler, Runge-Kutta). Higher order differential equations can lead to very complex dynamic behavior, e.g., various types of oscillation. (It is useful to note that differential equations of an order above the first can always be replaced by one or more first order simultaneous differential equations.)

For a computer to handle a differential equation expressed in continuous terms, the equation has to be replaced by an equivalent difference equation. Consider the following differential equation that shows the rate of change in *volume* is equal to *flow*.

$$\frac{d(Volume)}{dt} = flow$$

This will have an analytic solution that can be found by integration; i.e.,

$$Volume = \int_{t=start}^{t=stop} flow \cdot dt$$

The original differential equation can be written as a difference equation

$$\frac{Volume_t - Volume_{t-\Delta t}}{\Delta t} = flow$$

which can be rearranged to show that

$$Volume_t = Volume_{t-\Delta t} + \Delta t \cdot flow$$

This implies that the value of *Volume* at time t is simply the sum of the value of *Volume* at time $t - \Delta t$ plus the time interval Δt multiplied by *flow*. This solution (which amounts to Euler's method) can be implemented directly on a computer. Note that the analytic integral solution and the finite difference method will only give the same result if *flow* is constant. If *flow* is changing rapidly, the finite difference method will require a large number of iterations (i.e., $\Delta t \to 0$) to converge to the analytic integral solution.

Continuous simulation models are particularly suited to the analysis of dynamic processes. Frequently the system under examination will display a particular behavior pattern that may be interesting in itself or may be causing a problem. The objective of the model designer is then to construct a model based on differential equations that will reproduce this reference pattern. Decisions need to be made as to which state variables and rates of change are significant. If the model is being used to address some well-understood physical phenomena, it may be possible to incorporate relationships that are derived from first principles and have been tested by experiment. Outside the "hard" science areas it may be necessary to introduce parameters into the model, which lack theoretical backing, but enable the model to exhibit the reference behavior either qualitatively or quantitatively.

The overall behavior of the model often depends not as much on the precise value of the parameters and the detailed algebra as on the structure of the model itself. Systems of simultaneous differential equations can usually be shown to be linked together in longer or shorter loops that can be of either positive or negative sign. Positive feedback loops tend to reinforce change and often give rise to exponential growth. Negative feedback loops tend to negate change and are usually associated with goal-seeking activity.

Model building can be thought of as passing through a sequence of iterations in which the model becomes more and more realistic. At each iteration the model designer makes observations, forms hypotheses, carries out tests, modifies the model, and tests the model again. The advantage of working within this framework is that simulation models are objective and explicit rather than subjective and implicit. This makes them particularly valuable for: (1) communicating results and (2) acting as a workbench on which alternative solutions to problems can be tested.

Monte Carlo Procedures

Monte Carlo analysis is a simulation technique applicable to problems having a stochastic or probabilistic basis. Two different types of problems give rise to the use of this technique. Firstly, there are those problems that involve some kind of stochastic process. The rate of flame spread and fire growth and the response of individuals to fire alarms are examples of variables that may be considered to be stochastic in nature. In this case, the stochastic element may be introduced at any point in the model run so that the value of a variable at any time depends in some way on its previous value and a random component. Secondly, there are problems in which the process is treated as deterministic, but the starting conditions and model parameters are randomly selected from probability distributions.

The model may be used simply to estimate the value of the endogenous variables at a given location and time. The estimate will, of course, depend on the values of the parameters supplied to the model. However, the exact value of a parameter may not be known, but it might be possible to estimate its mean and variance and the form of its distribution. This information can be encoded as a probability density. A sequence of runs of the model can then be carried out using samples drawn from this distribution. Methods have been developed for generating values from most of the well-known probability distributions as well as any empirical

distribution. (For details see Appendix A.) It is then possible to estimate probability densities representing, for example, the fire conditions or the number of casualties from the model output. If the uncertainty in the input has little effect on the uncertainty in the output, it would not be necessary to go to great expense to refine the value of the input parameter.

Everything said so far about Monte Carlo methods has been on the assumption that there is only one input variable and one output variable. This assumption can now be relaxed so that all the uncertain input parameters can be treated as random samples from probability densities. The probability density of the output variables is now generated by the probability densities of all the input variables. This is important because it allows analysis of synergistic effects in which certain combinations of symmetrically distributed input variables can give rise to strongly skewed distributions of the output variables. Distributions of this type are typical of multiplicative stochastic processes. Probability distributions of fire damage are frequently skewed and can often be described by a density of the log-normal variety with a pronounced right-hand tail.[3] This may be due to the simultaneous occurrence of several adverse factors which tend to reinforce each other due to non-linearities in the system.

The input parameters can be interpreted as exogenous variables and the model output as an endogenous variable. This suggests that it would be possible to regress the outputs on the inputs using a standard multiple linear regression routine. This is equivalent to creating a linear model of the non-linear simulation model. The relationships between the actual fire, the simulation model, and the linear approximation to the simulation model are shown in Figure 5-1.1.

It can be shown[5] that the resulting regression coefficients correspond to the partial differential coefficients (i.e., the rate of change of an endogenous variable, y, with respect to each of the exogenous variables, x).

Simulation Languages

Simulation models may be written in any of the widely used general-purpose programming languages, such as FORTRAN, Pascal, and C. These languages are widely understood and readily portable between machines. It is probable that most of the successful, practical applications of the simulation technique have been written in one or the other of these languages. This approach offers maximum flexibility in the design and formulation of the model, the format of the output, and the kinds of experiments that can be performed. However, the obstacles to the development of a successful tool using this approach should not be underestimated. A central difficulty is the problem of controlling the sequence in which the interdependent actions forming the model occur. It is easy for non-specialists to become enmeshed in the complexities of the sequencing control, which is not itself of great interest, but is fertile ground for minor errors that are difficult to correct.

Accordingly, a number of attempts have been made to develop simulation languages that simplify the task of model development. These languages provide a structure for the model and a rapid way of converting the model into a computer program. They also facilitate revisions and alterations to the model, help to identify errors, and provide useful output. The more recent special-purpose languages have a convenient user interface that allows the user to interact with the model using graphical tools. For example, the structure of the model may be communicated to the computer by connecting

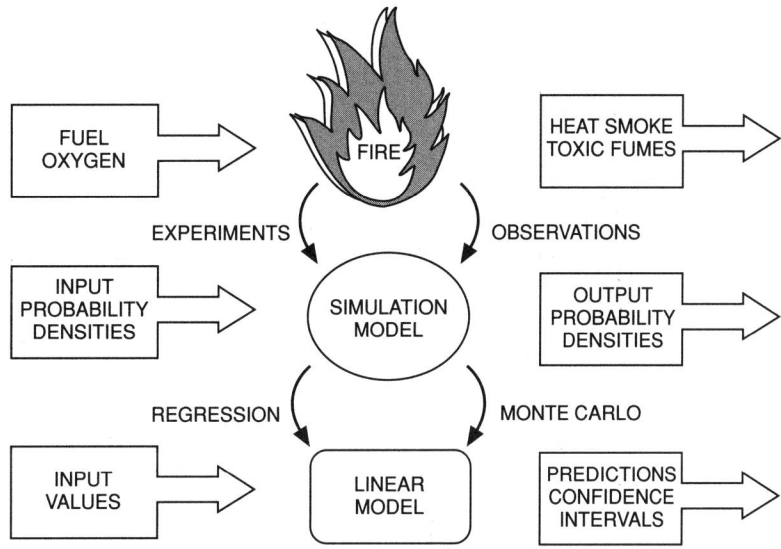

Fig. 5-1.1. Relationships between observations and models.[4]

up icons representing the components of the model. This reduces the need to write computer code and allows the user to concentrate on real problems rather than irrelevant problems thrown up by the computing methodology.

The most widely used languages for discrete event simulation are GPSS II,[6] SimScript,[7] and GASP.[8] They are all event-based languages though they use different terminology. GPSS II and SimScript have special compilers. GASP consists of a set of subroutines written in FORTRAN, which makes it more portable. Instructions for using these languages are given in the references. It should be noted that it is possible, though less convenient, to build discrete event simulation models using languages that have been developed for continuous simulation. This makes it possible to build comprehensive models of, for example, the fire situation, which includes both continuous and discrete processes.

During the last few years several application packages have appeared that greatly facilitate the development of continuous simulation models. One line of development leads from Forrester's work at MIT beginning in 1956. Forrester developed an approach to simulation modeling known as System Dynamics.[9] This technique can be implemented in any programming language, but it is more convenient to use purpose-built software. These usually contain sort routines to place the simulation equations into a computable sequence, and include functions to assist with the formulation of specific model relationships and policies.

The original System Dynamics software was DYNAMO, which was implemented on mainframe computers. This is now available as Professional DYNAMO[10] for use on the IBM PC range of computers. DYSMAP2[11] uses a similar equation format to Professional DYNAMO, and is specifically designed for use on the IBM PC range of computers. More recently, STELLA[12] was developed for the Apple Macintosh range of computers and takes advantage of their hardwired graphics interface. Its major innovation is that diagrams representing the model can be drawn directly on the monitor using a predefined tool kit. The variables of the model are represented by icons, which can be "opened" to insert parameter values and relationships between variables. STELLA is useful for demonstrating the relation between feedback structures and system behavior and for involving decision makers more closely in the model building and analysis process.

STELLA in its original form was not well-suited to the analysis of discontinuous processes, though this could be partly overcome by creating algorithms representing timers and switches. STELLA II specifically includes queues, "conveyor belts," and "ovens" designed to model the processing of discrete items. There is nothing in either version of STELLA comparable to subroutines or procedures that can be called from any point in the program. Each element of a vector or matrix must be represented explicitly by individual state variables. These features tend to limit the range of STELLA to relatively small models.

EXTEND[13] is also implemented on the Apple Macintosh family of microcomputers. With EXTEND it is possible to construct hybrid models that include both continuous and discrete processes, though it is necessary to abide by certain rules when connecting one to the other. EXTEND also has a graphic interface, which enables the model to be drawn on the monitor screen using icons representing a wide range of operations and processes. EXTEND uses an object-oriented, block diagram approach to modeling. Libraries of ready-built blocks allow the user to set up models without writing computer code. Alternatively, the user may modify blocks or create new ones using a programming language similar to C. This allows virtually any numerical procedure to be carried out including vector and matrix manipulation and integration over space. System Dynamics models can be constructed using a restricted set of EXTEND blocks, but in this environment it is less easy to relate system behavior to system structure.

APPLICATIONS

There are not many fully developed applications of procedural simulation models to fire protection engineering. CRISP would certainly qualify, but BFIRES and FIRE STATION are not complete system models because they ignore the interactions between the fire and the occupants. These examples might be better described as OR simulation models.

CRISP (Comparison of Risk Indices by Simulation Procedures)[14]

Simulation models can be used to compensate for lack of information about "real" fires and to work out the fire risk implications of new materials, building designs, and protection systems. A fire risk assessment model called CRISP is under development at the Fire Research Station, Borehamwood, which will be used to decide priorities for remedial action and to test the validity of new guidelines for building control officers. The model includes mechanisms representing the physical and chemical processes of fire development, as well as the behavior of people trying to escape from or suppress the fire. Figure 5-1.2 represents the structure of the model. The modules representing the various processes in the fire system are connected by positive and negative feedback loops. The direction of causation is indicated by the arrows. Open arrows show that the process represented by the earlier module facilitates the process in the later module; filled arrows indicate that the process in the later module is inhibited. For example, there is a positive feedback loop between COHb (the amount of hemoglobin bound to carbon monoxide) and the rate of evacuation. The relationships between heat and temperature, or combustion and oxygen, are examples of negative feedback loops. The model is capable of generating a rich variety of fully interactive behavior patterns. Monte Carlo methods on the lines described in this chapter will be used to derive probabilistic estimates of fire risk in buildings.

BFIRES-II,[15] a Behavior-based Computer Simulation of Emergency Egress During Fires

BFIRES-II is a dynamic stochastic computer simulation of emergency egress behavior by building occupants during fires. BFIRES-II is designed to simulate the movement of people within building enclosures in response to life-threatening stimuli (e.g., fire and smoke). Originally planned for use in evaluating health care facility designs, the program permits users to simulate special activities, e.g., rescuing non-ambulatory persons, in addition to simulating more frequent and general categories of emergency response (e.g., exit seeking, threat evasion, the deterioration of emergency responses due to heat, and the absorption of smoke and toxic gases).

The response-generating capability of BFIRES-II is based on an information processing explanation of human behavior, which suggests that the building occupants act in accordance with their perceptions of a constantly changing environment. When preparing a behavioral response at time t, the occupant first gathers information that describes the state of the environment at time t. Next, the occupant interprets this information and relates it to the emergency egress goals, which guide the individual's overall behavior. Finally, the occupant evaluates alternative responses and selects an action as the response at time t.

The general patterns of emergency egress behavior produced by BFIRES-II are found to agree with those found in earlier research literature, with professional opinions about such behavioral patterns and general impressions gathered from anecdotal accounts.

FIRE STATION,[16] Optimum Fire Station Location for Minimum Loss of Life and Property

This model determines the optimum location, among five alternate sites, for an urban fire station in a community of 25 wards. Optimization is the minimization of lives lost, building damage, and capital outlay. The model generates a sample population from an observed distribution by ward, structure value, and relative population density. The travel time from the station to the fire is also chosen from an observed distribution. The amount of damage and the number of lives lost is then calculated with reference to the delay in reaching the fire, which depends on the travel time.

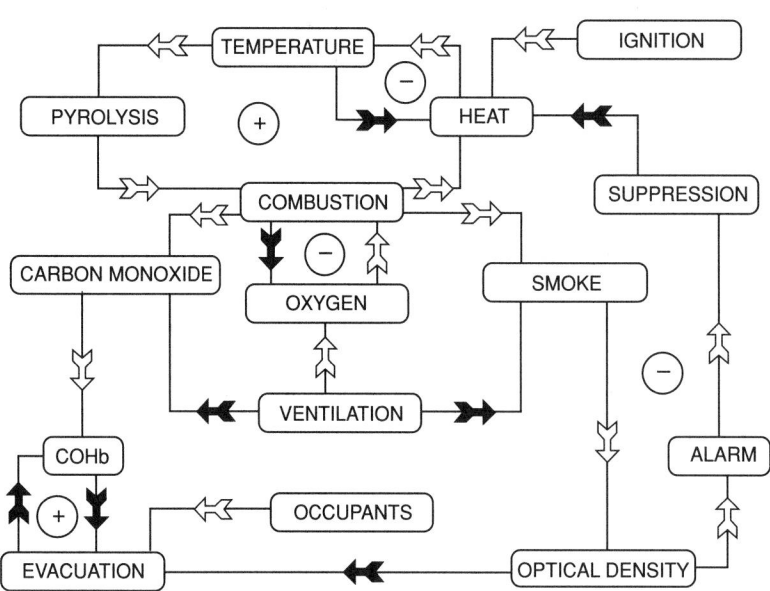

Fig. 5-1.2.   Structure of a simulation model for fire risk assessment (CRISP).

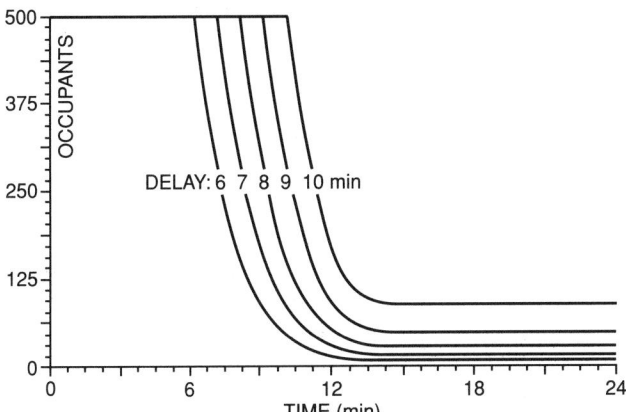

Fig. 5-1.3. *Occupants remaining in a place of assembly if the alarm is delayed for the periods shown.*

MODEL VALIDATION

Simulation models can never be validated over the whole range of their behavior. However, confidence in the reliability of a model is enhanced if the relationships built into it are based on accepted scientific theory supported by experimental evidence. The model must also stand up to tests designed to show it behaves reasonably in response to exogenous disturbance. The calibrated model should also be able to simulate time-series data from instrumented experimental fires. Finally, the model should be able to mimic the sequence of events recorded by observers at real fires.

Like other models, the validity of simulation models depends on goodness of fit and predictive power. In the fire situation, deterministic predictions are not feasible except in very simple cases. In a typical case, a simulation model would be calibrated to produce an output distribution whose mean and variance were in agreement with an observed distribution. Changes in fire protection strategy would be reflected in the model, which would then be used to generate a new output distribution with a different mean and variance. Comparisons between the output distributions before and after implementing the strategy can be used to calculate confidence intervals for statements such as "if strategy B is preferred to A, casualties will be reduced in x percent of cases." Statistical predictions of this kind are quite sufficient for the purposes of fire protection engineering.

SENSITIVITY ANALYSIS

Sensitivity analysis can be used to draw useful conclusions to guide decisions on fire safety design and priorities for research. In its simplest form, sensitivity analysis is carried out by varying one of the input parameters in steps over a prescribed range and observing the effect on chosen output parameters, either at the end of the model run or at a set time after the run commences. (Advanced methods of sensitivity analysis are available that allow more than one variable to be varied at a time.) The results of this exercise can then be presented as a table or, more usefully, the output parameter can be regressed against the input parameters to give a simple linear or non-linear expression that summarizes the results and permits interpolation and extrapolation.

The rate of change of the output parameter, with respect to the input parameter, is defined as the "sensitivity." In fire safety terms this might indicate how much the risk is reduced as fire protection measures are improved. If the cost of improved protection is also known, it is possible to estimate the cost of saving a life.

Figure 5-1.3 shows five runs of an evacuation model in which the delay in giving the fire alarm has been increased in steps of 1 min., from 6 min. to 10 min. The number of people trapped in a building at, for example, 18 min. is represented by the distances between the curves and the horizontal axis. If the number of casualties is regressed against the delay, the results shown in Figure 5-1.4 are obtained. The number of casualties that would occur if the alarm were delayed by t min. can be calculated directly from the formula. Conversely, the formula can be used to predict the number of lives that might be saved if the delay were reduced by, for example, an improved alarm system. Note that, in this case, the sensitivity is not a constant, but an exponential function of time.

The significance of calculations like these is that they enable designers to select those areas where expenditure on fire protection measures would earn the greatest return in terms of lives saved. It also makes it possible to identify those areas where additional expenditure might have little or no effect. This approach is also helpful for making decisions on research priorities. If preliminary calculations indicate that the fire risk can be substantially reduced by a small change in a parameter, it would make sense to put research effort into investigations designed to refine the measurement of that parameter and reduce the level of uncertainty associated with it.

Sensitivity analysis makes it possible to place the tools of fire protection engineering in the hands of designers and other users who might not have the time or the inclination to work directly with a simulation model. The results of the analysis could be published as tables or graphs in manuals for the guidance of draftspersons who would not need to know in detail how the figures were derived. It is vital to ensure that figures are not applied outside their range of application. This implies that the work must be verified by a fully qualified professional who would check that the guidance was appropriate to the circumstances. Similar problems of professional self-regulation have been successfully resolved in other disciplines.

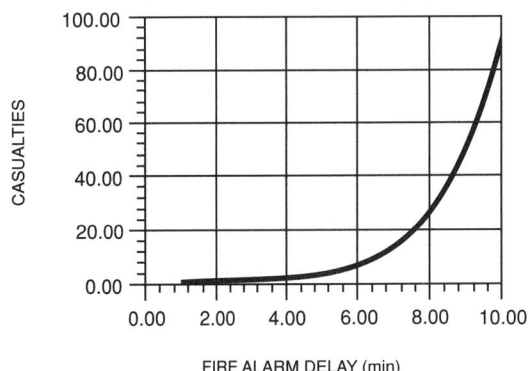

CASUALTIES = exp (−1.76 + 0.62* DELAY)

Fig. 5-1.4. *The number of casualties rises exponentially as the delay in giving the signal to evacuate is increased.*

CONCLUSION

Progress in fire protection engineering is handicapped by data limitations and the difficulty of conducting realistic experiments on complete fire systems involving human behavior. These problems may be overcome by applying a variety of procedural simulation models originally developed in operations research. These models make efficient use of the available data and can be used to test fire protection strategies. They can be complemented by Monte Carlo methods to take account of uncertainty in the data, measure the sensitivity of the casualty rate to fire protection measures, and estimate fire risk.

APPENDIX A: HANDLING UNCERTAINTY

Generating Random Numbers

In order to introduce a stochastic element into simulation modeling a source of random numbers is required. Virtually every computer is equipped with a subroutine that can generate a pseudo-random number in the interval [0, 1] on demand. These are not truly random numbers because they are generated by a deterministic algorithm that will repeat itself cyclically after an interval. The algorithm is designed to make this interval as long as possible given the capacity of the computer. A variety of methods, most of which are based on congruence relationships, are discussed by Naylor et al.[17]

Generating Probability Distributions for Monte Carlo Studies

The generation of simulated statistics is entirely of a numerical nature and is carried out by supplying pseudo-random numbers into the process or system under study and obtaining probability distributions from it as the result. As a rule, statistical simulation involves replacing an actual sample population by some assumed theoretical distribution and then sampling from this theoretical population by means of some type of random number generator. In some cases it may not be possible to find a standard theoretical distribution that describes a particular stochastic process. In these cases the process can be simulated by sampling from an empirical distribution rather than a theoretical one. Naylor et al[17] provide specific techniques for generating variates from several of the widely used probability distributions as well as general methods for generating variates from empirical distributions.

Inverse transformation method:　This method depends on the relationship between the probability density function (pdf) and the cumulative distribution function (cdf). The pdf gives the probability that the variate x lies between x and $x + dx$. The cdf gives the probability that $x \leq x$. The cdf $F(x)$ is obtained from the pdf $f(x)$, by integrating $f(x)dx$ over the interval from $-\infty$ to x.

To generate variates x_i from some particular statistical population whose pdf is given by $f(x)$, one must first obtain the cdf $F(x)$. Since $F(x)$ is defined over the range [0, 1], one can generate uniformly distributed random numbers and set $F(x) = r$. It is clear that x is uniquely determined by $r = F(x)$. It follows that for any particular value of r, say r_0, that is generated, it is possible to find the value of x, in this case x_0 corresponding to r_0 by the inverse function, if it is known. That is,

$$x_0 = F^{-1}(r_0)$$

where $F^{-1}(r)$ is the inverse transformation or mapping of r on the unit interval into the domain of x.

EXAMPLE 1:

Generate variates x with pdf $f(x) = 2x$, $0 \leq x \leq 1$.

SOLUTION:

$$
\begin{aligned}
r = F(x) &= \int_{-\infty}^{x} f(t)dt \\
&= \int_{0}^{x} 2t\, dt \quad 0 \leq x \leq 1 \\
&= x^2
\end{aligned}
$$

Then taking the inverse transformation, $F^{-1}(r)$, i.e., solving this equation for x, one obtains

$$x = F^{-1}(r) = \sqrt{r} \quad 0 \leq r \leq 1$$

Therefore, values of x with pdf $f(x) = 2x$ can be generated by taking the square root of random numbers r.

EXAMPLE 2:

Generate a variate x with density function

$$
\begin{aligned}
f(x) &= \frac{1}{4} \quad 0 \leq x \leq 1 \\
&= \frac{3}{4} \quad 1 \leq x \leq 2
\end{aligned}
$$

SOLUTION:

The pdf and cdf are illustrated graphically in Figure A-5-1.1. Using the previous results

$$
\begin{aligned}
r = F(x) &= \int_{0}^{x} \frac{1}{4}dt \qquad 0 \leq x \leq 1 \\
&= \frac{x}{4} \\
r = F(x) &= \frac{1}{4} + \int_{1}^{x} \frac{3}{4}dt \quad 1 \leq x \leq 2 \\
&= \frac{3}{4}x - \frac{1}{2}
\end{aligned}
$$

Taking the inverse transformation, i.e., solving the above equations for x, one obtains

$$
\begin{aligned}
x &= 4r & 0 \leq r \leq \frac{1}{4} \\
x &= \frac{4}{3}r + \frac{2}{3} & \frac{1}{4} \leq r \leq 1
\end{aligned}
$$

The rejection method:　In many cases it is either impossible or very difficult to express x in terms of the inverse transformation of the probability distribution, $F^{-1}(r)$. In these cases it is necessary to obtain a numerical approximation to the inverse function, F^{-1}, or make use of the rejection method. The application of the rejection method requires the following steps:

1. Normalize the range of f by a scale factor, c, such that

$$cf(x) \leq 1 \qquad a \leq x \leq b$$

2. Define x as a linear function of r

$$x = a + (b - a)r$$

3. Generate pairs of random numbers (r_1, r_2).

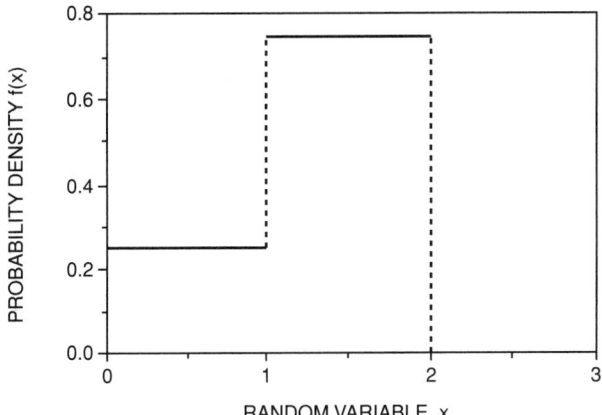

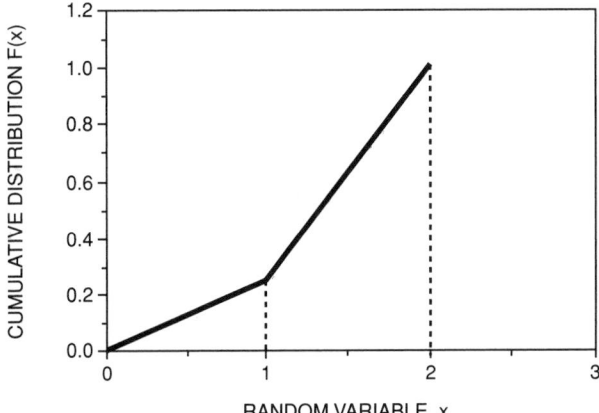

Fig. A-5-1.1. *Probability density function (top) and cumulative distribution (bottom) for empirical data.*

4. Whenever one encounters a pair of random numbers that satisfy the relationship

$$r_2 \leq cf[a + (b - a)r_1]$$

then "accept" the pair and use $x = a + (b - a)r_1$ as the variate generated.

The theory behind this method is based on the realization that the probability of r being less than or equal to $cf(x)$ is

$$P[r \leq cf(x)] = cf(x)$$

Consequently if x is chosen at random from the range $[a, b]$ according to step 2 above and then rejected if $r > cf(x)$, the probability density function of accepted x will be exactly $f(x)$. It can be shown[18] that the expected number of trials before a successful pair is found is equal to $1/c$. This implies that the method may be quite inefficient for certain probability density functions. The rejection method can also be used as a Monte Carlo technique to evaluate definite integrals. This may be particularly useful for the evaluation of multivariate functions.

EXAMPLE 3:

Use the rejection method to generate variates x with density function $f(x) = 6(x - x^2)$, where $0 \leq x \leq 1$.

SOLUTION:

Since x was defined over the unit interval, $x = r$. But $f(r) = 6(r - r^2)$ is defined over the interval $0 \leq f(r) \leq 1.5$. Scaling will transform $f(r)$ to the unit interval if $g(r) = 2/3 \, f(r)$, in which case $g(r) = 4(r - r^2)$. The rejection method then consists of the following four steps:

1. Generate r_1 and calculate $g(r_1)$.
2. Generate r_2 and compare with $g(r_1)$.
3. If $r_2 \leq g(r_1)$, accept r_1 as x from $f(x)$. If $r_2 > g(r_1)$, then reject r_1 and repeat step 1.
4. Repeat this procedure until n values of x have been generated.

A comparison between the theoretical distribution and the distribution obtained by using the rejection method is shown in Figure A-5-1.2.

EXAMPLE 4:

Generate a variate with a normal distribution.

SOLUTION:

If a random variable x has a pdf $f(x)$ given as follows

$$f(x) = \frac{1}{\sigma\sqrt{2\pi}} e^{-\frac{1}{2}\left(\frac{x-\mu}{\sigma}\right)^2}, \quad -\infty < x < \infty$$

it is said to have a normal or Gaussian distribution with parameters μ and σ. The graph of this function is the familiar bell-shaped curve. The cdf $F(x)$ does not exist in explicit form and, therefore, the inverse transformation method cannot be applied. There are several ways to avoid this difficulty, which are fully described in the literature. However, it is useful to include here a method for generating variates from a normal distribution, because data with this type of distribution is so widely encountered in practice.

Let r_1 and r_2 be two uniformly distributed random variables defined on the interval [0, 1], then

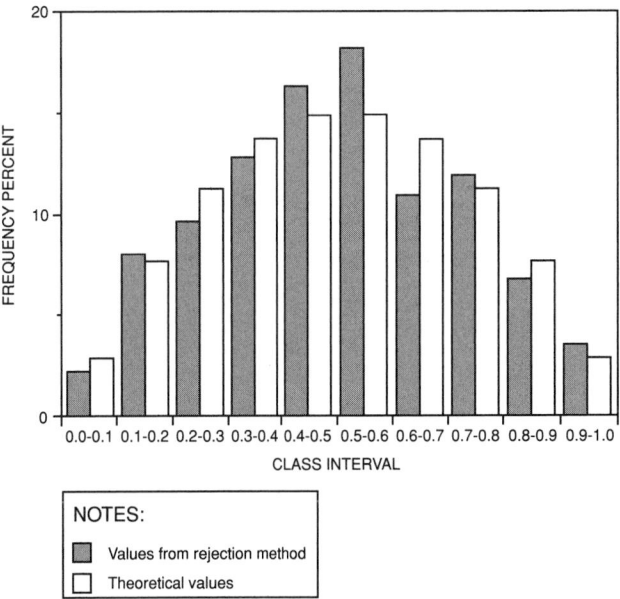

Fig. A-5-1.2. *Generating a sample from a quadratic pdf using the rejection method.*

$$x_1 = (-2 \log_e r_1)^{1/2} \cos 2\pi r_2$$
$$x_2 = (-2 \log_e r_1)^{1/2} \sin 2\pi r_2$$

are two variates from a standard normal distribution. This method produces exact results and is quite fast, subject to the efficiency of the special function subroutines.[19]

APPENDIX B: ANALYSIS OF MODEL OUTPUT

Local Linearization

Suppose that at a given time, t, the output variables are dependent on the input variables of the model, according to a set of functions

$$y_i = f_i(x_1, x_2, \ldots, x_n); \qquad i = 1, \ldots, n$$

Then, in the neighborhood of $x_1, x_2, \ldots, x_n$, one can make a local linear approximation to y_i by expanding $f_i(x_1 + \Delta x_1, x_2 + \Delta x_2, \ldots, x_n + \Delta x_n)$ as a Taylor series about $f_i(x_1, x_2, \ldots, x_n)$, and then truncating all terms after the second. The Taylor series expansion of the ith variable is

$$y_i = f_i(x_1 + \Delta x_1, x_2 + \Delta x_2, \ldots, x_n + \Delta x_n)$$

$$\therefore y_i = f_i(x_1, x_2, \ldots, x_n) + \sum_j \Delta x_j \frac{\partial y_i}{\partial x_j}$$

$$+ \frac{1}{2!} \left(\sum_j \Delta x_j^2 \frac{\partial^2 y_i}{\partial x_j^2} + 2 \sum_{j \neq k} \sum_k \Delta x_j \Delta x_k \frac{\partial^2 y_i}{\partial x_j \partial x_k} \right) + \cdots$$

$$\therefore \Delta y_i \cong \sum_j \Delta x_j \frac{\partial y_i}{\partial x_j}; \qquad i = 1, \ldots, n$$

provided that Δx_j is small. It is more convenient to express this in matrix notation

$$\Delta \mathbf{y} = \mathbf{A} \Delta \mathbf{x}$$

where $\mathbf{A}$ is the Jacobian for the ys. I.e.,

$$a_{ij} = \frac{\partial y_i}{\partial x_j}$$

Note that, if $\mathbf{S}$ is the variance-covariance matrix for the xs, then $\mathbf{V}$ is the variance-covariance matrix for the ys and is given by

$$\mathbf{V} = \mathbf{A}'\mathbf{S}\mathbf{A}$$

Additivity of Variances

It can often be shown that a model is stable, with respect to variations in any one of its input variables taken one at a time. However, it is possible that, if more than one variable were allowed to change simultaneously, the variance of the sum would be greater than the sum of the variances. This would imply that the interaction or covariance term could not be ignored. However, if the regression procedure establishes that the linear hypothesis is valid, then the variance of the output is equal to the sum of the variances of the inputs weighted by the squares of the coefficients. Note, however, that the linear hypothesis will probably hold only over a limited range, and can be estimated by examining a plot of the residuals.

If it is established that the linear hypothesis holds over a certain range, i.e.,

$$y = a_0 + a_1 x_1 + a_2 x_2 + \ldots + a_n x_n$$

and if $x_1, \ldots, x_n$ are random variables, then

$$\mathrm{var}(a_0 + a_1 x_1 + a_2 x_2 + \ldots + a_n x_n)$$

$$= \sum_i a_i^2 \mathrm{var}(x_i) + \sum_i \sum_j a_i a_j \mathrm{cov}(x_i, x_j)$$

The variables $x_1, \ldots, x_n$ have been generated so as to be independent, which implies that the covariance term vanishes, and, therefore

$$\mathrm{var}(y) = \sum_i a_i^2 \mathrm{var}(x_i)$$

The regression model can, therefore, be used to predict the variance of a dependent variable on any assumptions that are chosen about the variances of the independent variables.

If it turned out that the contribution of one variable to the total variance was large, it would be desirable to undertake further studies to establish more exact values for this parameter. A smaller variance could then be assigned to this variable and one could immediately deduce how much effect this would have on the uncertainty in the dependent variable, without having to carry out any further Monte Carlo runs. Within limits there is no need to be too particular about the variances ascribed to the independent variables in the preliminary studies, since these can be revised at a later date. It is necessary to make a distinction between (1) parameters that have a large variance due to measurement uncertainties (which can be improved) and (2) parameters drawn from populations having a large variance (where further measurement may reduce the uncertainty of the variance but not the variance itself).

REFERENCES CITED

1. M. Black, *Models and Metaphors*, Cornell University Press, Ithaca, NY.
2. R.L. Ackoff and Stafford Beer, "In Conclusion: Some Beginnings," *Progress in Operations Research*, J.S. Aronofsky, ed., John Wiley and Sons, New York (1969).
3. G. Ramachandran, "Probabilistic Approach to Fire Risk Evaluation," *Fire Technology*, Aug. 1988.
4. W.G.B. Phillips, "The Development of a Fire Risk Assessment Model," BRE Information Paper IP 8/92, Fire Research Establishment, Borehamwood (1992).
5. W.G.B. Phillips, "Monte Carlo Tests of Conclusion Robustness," *Proceedings* of the Fifth International System Dynamics Conference, Geilo, Norway, p. 19, Aug. 1976.
6. *General-Purpose Systems Simulator II, Reference Manual*, International Business Machines Corp.
7. H.M. Markowitz, B. Hausner, and H.W. Karr, *SimScript: A Simulation Programming Language*, The Rand Corporation, RM-3310 (Nov. 1962).
8. Philip J. Kiviat, *GASP—A General Activity Simulation Program*, Project No. 90, 17-019(2), Applied Research Laboratory, United States Steel, Monroeville, PA, July 8, 1963.
9. J.W. Forrester, *Principles of Systems*, MIT Press, Cambridge, MA (1968).
10. Pugh-Roberts Associates, *Professional DYNAMO Introductory Guide and Tutorial and Professional DYNAMO Reference Manual*, Pugh-Roberts Associates, Cambridge, MA (1986).
11. O. Vapenikova, "DYSMAP2/386—DYSMAP2 Implementation on 80386-based P.C.s," *Proceedings* of the European Simulation Multi-Conference, Rome, published by Society for Computer Simulation International (1989).

12. B.M. Richmond, P. Vescuso, and S. Peterson, *STELLA for Business*, High-Performance Systems Publications, Lyme, NH (1987).

13. Paul Hoffman, *EXTEND—Performance Modeling for Decision Support*, Imagine That Inc., San Jose, CA (1990).

14. W.G.B. Phillips, "Simulation Models for Fire Risk Assessment," *Fire Safety Journal*, Vol. 23, pp. 159–169 (1994).

15. Fred I. Stahl, "BFIRES-II, A Behavior-Based Computed Simulation of Emergency Egress During Fires," *Fire Technology*, 18, 1 (1982).

16. J. Oberstone, in *Management Science: Concepts and Applications*, West Publishing Company, St. Paul, MN (1990).

17. T.H. Naylor, Joseph L. Balintfy, Donald S. Burdick, and Kong Chu, *Computer Simulation Techniques*, John Wiley and Sons, New York (1966).

18. K.D. Tocher, *The Art of Simulation*, Princeton, NJ, Van Nostrand Co., New York (1963).

19. Mervin E. Muller, "A Comparison of Methods for Generating Normal Deviates on Digital Computers," *Journal of the Association for Computing Machinery*, VI, pp. 376–383 (1959).

FIRE RISK RANKING

John M. Watts, Jr.

INTRODUCTION

Fire risk ranking is a link between fire science and fire safety. As we learn more about the behavior of fire, it is important that we implement our new knowledge to meet fire safety goals and objectives. One of the barriers to implementing new technology is the lack of structured fire safety decision making. Fire risk ranking is evolving as a method of evaluating fire safety that is valuable in assimilating research results.

Fire safety decisions often have to be made under conditions where the data are sparse and uncertain. The technical parameters of fire risk are very complex and normally involve a network of interacting components; these interactions are generally non-linear and multi-dimensional. However, complexity and sparseness of data do not preclude useful and valid approaches. Such circumstances are not unusual in decision making in business or other risk venues. (The space program illustrates how success can be achieved when there is little relevant data.) However, detailed risk assessment can be an expensive and labor intensive process, and there is considerable room for improving the presentation of results. Ranking can provide a cost-effective means of risk evaluation that is both useful and valid.

Fire risk ranking systems are heuristic models of fire safety. They constitute various processes of analyzing and scoring hazard and other risk parameters to produce a rapid and simple estimate of relative fire risk. They are also known as rating schedules, point schemes, indexing, numerical grading, and scoring. Using professional judgment and past experience, fire risk ranking assigns values to selected variables representing both positive and negative fire safety features. The selected variables and assigned values are then operated on by some combination of arithmetic functions to arrive at a single value, which is then compared to other similar assessments or to a standard.

METHODS OF FIRE RISK ASSESSMENT

Formal fire risk assessment evolved with the insurance industry in the 19th century. However, in the last few decades, there has been a move to develop more wide spread analytical procedures. Methods of fire risk analysis may be classified into four categories: (1) narratives, (2) check lists, (3) ranking, and (4) probabilistic methods.[1]

Narratives

Probably the earliest fire risk assessment was simply the observation that fire was capable of destroying certain materials such as wood, fur, and flesh. This realization would have led to a communication from parent to child on the avoidance of these fire dangers. Such advice may be considered to have evolved over the years to the present thirteen volume set of the NFPA *National Fire Codes*.[2] These contain the bulk of our present day wisdom on fire safety. The information is presented in the form of descriptions of various hazardous conditions and ways to reduce or eliminate them.

Narratives do not attempt to evaluate the fire risk quantitatively; rather, a risk is judged acceptable if it complies with published recommendations. The criteria is one of pass or fail. An obvious limitation to this approach is that such narratives cannot hope to cover the myriad conditions of human activity. While there is much common ground among different fire hazard situations, there is considerable variation in detail.

In some applications, a jump is made from generalized narrative advice to an estimate of *probable maximum loss* (PML). The subjectivity of such a determination suggests that this value is more of an ordinal label than a quantitative measure of risk (which is not to say that it does not have usefulness).

Checklists

A common accessory of fire safety is a listing of hazards and recommended practices. These checklists are valuable tools for identifying fire risk factors.

It is very seldom that all criteria in a code or standard applies to a single building. The fire protection engineer must focus on only those requirements that are applicable to

Dr. Watts holds degrees in Fire Protection Engineering, Industrial Engineering, and Operations Research. He is Director of the Fire Safety Institute, a not-for-profit information, research, and educational corporation located in Middlebury, Vermont. Dr. Watts also serves as editor of NFPA's *Fire Technology*.

a specific project. A checklist can aid in this process. It also makes requirements easier to read and understand and easier to track to compliance.

While checklists attempt to identify important fire risk features, it is not always clear what is important. In general, a long checklist on the order of 50 fire safety factors contains items that are readily visible or measurable but not necessarily comparable. A short checklist, on the other hand, is usually comprised of conceptual features of fire safety, which are difficult to measure. What is lacking is a systematic approach to the generation of the checklist.

Moreover, checklists do not distinguish the importance of fire risk factors. For example, the relative value of hydrants, sprinklers, and extinguishers is not constant, but a function of other features of a structure's form and utility. This leads to the need for quantitative methods of fire risk assessment.

Ranking

Quantitative fire risk assessment originated with the insurance rating schedule. The approach has broadened to include a wide variety of applications.[3,4,5] In general, fire risk rating schedules assign values to selected variables based on professional judgment and past experience. The selected variables represent both positive and negative fire safety features and the assigned values are then operated on by some combination of arithmetic functions to arrive at a single value. This single value can be compared to other similar assessments or to a standard to rank the fire risk.

Probabilistic Methods

Growing interest in analytical fire risk assessment and an increasing data base have led to use of more sophisticated mathematical techniques. Probabilistic methods manipulate fire safety variables according to recognized theoretical principles. These methods include computer simulation, linear regression, network analysis, and stochastic modeling.

FIRE RISK RANKING

Fire risk ranking has gained widespread acceptance as a cost-effective prioritization and screening tool for fire risk assessment programs. It is a useful and powerful tool that can provide valuable information on the risks associated with fire. As defined here, fire risk ranking is the process of modeling and scoring hazard and exposure parameters to produce a rapid and simple estimate of relative risk.

Relative *vs* Absolute Risk

Within the framework of quantitative fire hazard and risk assessment principles, different levels of depth can be developed. Consider this range of depth as a continuous quantitative assessment spectrum.

At one end of this spectrum is a detailed probabilistic risk analysis. Here the assessment of hazards and the characterization of exposure are as detailed and as well-defined as possible. The analysis may rely on thousands or even millions of dollars worth of statistical studies and fire testing. It is usually complex, time-consuming, and costly; however, it is necessary when risks must be known as accurately as possible, or where legal issues demand that an exhaustive analysis must be performed. The purpose is to determine a best estimate of the absolute fire risk, which can then be compared to risks from other hazards.

Fire risk ranking can be considered to be at the other end of the quantitative assessment spectrum. For many situations, an in-depth analysis is not appropriate. It is under these circumstances that a fire risk ranking can be used. In contrast to a probabilistic risk analysis, fire risk ranking defines the relationship between hazard and exposure by using more simplistic models requiring less data and less analysis. This approach produces measures of relative risk that can be compared to one another or to a standard.

Figure 5-2.1[6] provides a graphic view of the relative power and limitations of three broad levels of risk quantification. Curves A, B, and C do not represent actual data points, but are demonstrative of a continuum of fire risk analysis possibilities.

Curve A is representative of a rigorous probabilistic risk analysis where hazard and exposure are analyzed through full quantitative analysis of the hazard and the statistics of exposure. It is clear that this is the most accurate approach to defining risks, especially where the risk is low. However, it is also clear that a large resource investment is necessary to accomplish this task.

Curve C is a simple fire risk ranking providing the ability to screen for high-risk catastrophic-type situations where the analysis can be consequence-oriented. However, for small differences in risk, the ability of the more simplistic screening system to differentiate between two more subtle risks diminishes.

A more complex and accurate assessment model will provide greater differentiation between lesser risks and an improved overall accuracy. The trade-off for this approach is increased time and resources expended for model development, implementation, and data collection.

Applications

Choosing the depth of the risk analysis is a critical decision that depends on such factors as time, resource commitment, and the intended use of the results. Each approach may have certain advantages or trade-offs when used for specific tasks. A fire risk ranking approach may be appropriate in several situations.

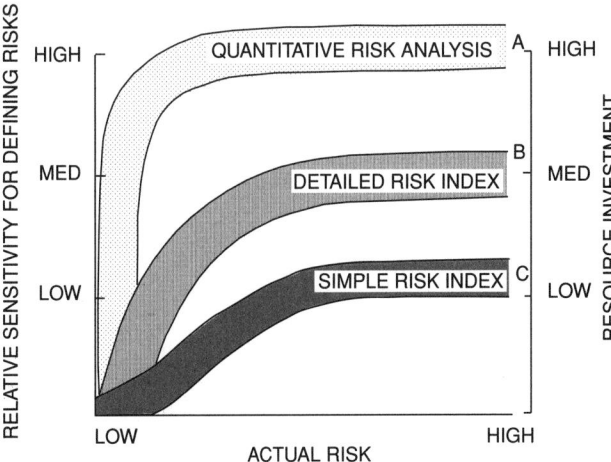

Fig. 5-2.1. Risk index systems and relative sensitivity for defining actual risk.[6]

1. Where greater sophistication is not required,
2. Where risk screening will be cost-effective, or
3. Where there is a need for risk communication.

The level of accuracy demanded for a fire risk analysis is not typically the same as for other engineering purposes. Often establishing an order of magnitude will suffice. Time and resource expenditure will increase as the depth of analysis is increased. Where resources are scarce, and efficiency is prized, maximizing the utility of the fire risk ranking is clearly desirable.

The principles of fire risk ranking have been applied to a variety of hazard and risk assessment projects to set priorities and help manage resources. Risk assessment can be an expensive and labor-intensive process; much time and money can be wasted if the products or facilities with the greatest potential for risk and associated liability are not identified and assessed first. Without a prioritization plan, it will not be known whether a risk was worth assessing until after the time and money has been spent.

Fire risk ranking also has appeal to staff charged with risk management decision-making responsibilities and those who may be unfamiliar with the details and mechanics of the risk assessment process. Because fire risk ranking simplifies basic fire risk assessment principles, it can be an effective way to acquire a global grasp of the issues.

Significance

The importance of fire risk ranking has been widely recognized. A working group of educators and researchers addressed the issue of fire risk ranking at the National Academy of Sciences 1987 Workshop on Analytical Methods for Designing Buildings for Fire Safety.[4] They concluded there is a need for a three-part system of fire safety comprised of: (1) codes, (2) the method of fire risk ranking, referred to as a numerical grading system, and (3) the means of supplying inputs to that system derived, as far as feasible, from basic principles of decision science. The working group went on to state the rationale for its conclusion:

> The advantage of keeping numerical grading systems in the trio is that they provide a coherent structure that still allows some qualitative analysis of fire safety. These systems also readily accept change associated with aspects of operations research, management science, risk analysis, and quantitative analytical solutions or models to the fire safety problem or parts of it.[7]

The importance of scientific rigor in the development of fire risk ranking methods cannot be overemphasized and will be addressed in more detail in the final section of this chapter.

Examples of Approaches to Fire Risk Ranking

It is difficult to describe a typical fire risk ranking method. The practical necessity of trying to assess dozens or hundreds of risks with limited resources has led to the creation of an array of fire risk ranking systems. Approaches to fire risk ranking are virtually limitless in their possible variations. Representative examples of fire risk ranking were selected from the literature and are summarized in the fol-

lowing sections. They provide some idea of the types of variations involved with modeling and quantifying fire risk.

INSURANCE RATING

The purpose of risk analysis is to facilitate the process of risk management. One of the most fundamental tools of risk management is transfer of risk by insurance. To be acceptable to an insurer, the risk is rated by actuarial means, applying principles of mathematics to the particular pricing problems of the insurance industry.

Fire insurance rates are promulgated as *class* rates or *specific* rates. Class rates apply to all properties that fall within a given category or classification; the most common example of class rating is for dwellings or residences. When class rates do not apply, specific rates are determined by the application of a schedule or formula designed to measure the relative quantity of fire hazard present. This process, known as *schedule rating*, is typically used for institutions, manufacturing properties, and business establishments..

The two most widely used schedules in the U.S. are known as the *mercantile schedule* and the *analytic system*.[8] At present, the analytic system is in predominant use throughout the country; it is more generally referred to as the Dean Schedule, named for A.F. Dean, author of the plan.[9] The schedules basically differ in their fundamental analysis of the factors affecting insurance rates, but they are alike in that they establish an arbitrary point from which to build up the rate, based on various physical hazards. A schedule of additions and reductions is computed and the difference applied to the arbitrary point of departure.

Schedule rating, then, is a plan by which fire hazards with respect to any particular property are measured. A schedule has been defined as ". . . an empirical standard for the measurement of relative quantity of fire hazard."[10] Schedule rating takes into consideration the various factors contributing to the peril of fire, such as construction and occupancy, with a view to determining which features either enhance or minimize the probability of loss. Credits and charges representing departures from standard conditions are incorporated in the schedules. Thus the schedule rate is typically the sum of all charges less the sum of its credits, and constitutes a standard for the measurement of the fire risk.

ISO Specific Commercial Property Evaluation Schedule

The most commonly used insurance rating schedule in the U.S. is the ISO (Insurance Services Office) Specific Commercial Property Evaluation Schedule (SCOPES). For each building, a percentage occupancy charge is determined from tabulated charges for classes of occupancy modified by factors such as the specific hazards of a particular occupancy. The basic building grade is a function of the resistance to fire of structural walls and floor and roof assemblies. The building fire insurance rate is the product of occupancy charges and building grade modified by factors such as the exposure to fire in nearby buildings, and protection provided by portable extinguishers, fire alarm systems, etc.

An important concept of insurance rating is the use of loss experience. In general, tabulated values and conversion factors are based on actuarial analysis of fire losses paid by insurers and reported to the insurance industry.

Gretener Method

In 1960, M. Gretener of the Swiss Fire Prevention Service began to study the possibility of an arithmetical evaluation of fire risk in buildings. His premise was that determining fire risk by statistical methods based on loss experience was no longer adequate for the following reasons[11]:

1. Lack of exchange of loss experience.
2. Inadequate analysis with respect to causes and factors determining the size of loss, resulting in distortion of statistical data.
3. Rapid technological change altering the credibility of previous experience.
4. Different criteria, according to country and company, for data collection and evaluation.

As a result of this work, a new approach to schedule rating, the Gretener method, has been developing in Switzerland and Austria.[12,13] The basic idea of the process consists of expressing, in relative, empirically derived numerical values, factors for a fire to start and spread and factors for fire protection. The product of the hazard factors gives a value for potential hazard, while the product of the fire protection factors expresses a value for protective measures. The ratio of these products is taken as the measure of expected fire severity.

Of immediate appeal is that the approach begins with the explicit concept of risk as the expectation of loss given by the product of hazard probability and hazard severity:

$$R = A \times B$$

where

R = fire risk,
A = probability that a fire will start, and
B = fire hazard, degree of danger, or probable severity.

Thus, the Gretener method is based on these two probabilities and combines them in accordance with probability theory.

A further departure from U.S. schedule rating is the calculation of fire hazard as a ratio rather than a sum.

Fire Hazard = Potential Hazard/Protective Measures

$$B = P/(N \times S \times F)$$

where

B = fire hazard,
P = potential hazard,
N = standard fire safety measures,
S = special measures, and
F = fire resistance of the building.

Potential hazard, P, is the product of hazard elements whose magnitudes are influenced on the one hand by the building contents, i.e., materials and merchandise present, and on the other hand, by the building itself.

As with most other schedule approaches, the values for these individual factors are not based on statistics, but are empirical figures resulting from a comparison of analyses of fire risks for which fire protection measures are either common or required by law.

DOW'S FIRE AND EXPLOSION INDEX

A need for systematic identification of areas with significant loss potential motivated Dow Chemical Company to develop the Fire and Explosion Index and risk guide.[14] The original edition issued in 1964 was a modified version of the "Chemical Occupancy Classification" rating system developed by Factory Mutual prior to 1957. It has been subsequently improved, enhanced, and simplified, and is now in its 7th edition.[15]

Today there are many risk assessment methods available that can examine a chemical plant in great detail. The Fire and Explosion Index (FEI) remains a valuable screening tool that serves to quantify the expected damage from potential fire, explosion, and reactivity incidents and to identify equipment that could likely contribute to the creation or escalation of an incident.[16] Risks associated with operations where a flammable, combustible, or reactive material is stored, handled, or processed can be evaluated with this system. The guide is intended to provide a direct and logical approach for determining the probable "risk exposure" of a process plant and to suggest approaches to fire protection and loss prevention design. An important application of the FEI is to help decide when a more detailed quantitative risk analysis is warranted, as well as the appropriate depth of such a study.[17] This section will provide an overview of the method. The source document 15 should be consulted for specific application.

The concept of the Fire and Explosion Index (FEI) is to divide a process plant into separate operations or units and consider each of these individually. The key feature of the method is to identify the dominant combustible material in the unit being studied and assess its thermodynamic properties. This basic material factor is then built up with a series of individual features concerning operation of the unit. These potential hazards are based on experience and information from incident records. They are intended to cover the majority of likely abnormal situations leading to fires, explosions, and material releases. Features considered are those upon which all hazard and reliability analyses are based, however, the FEI does not purport to be as comprehensive as a detailed hazard and reliability study. The approach identifies most of the potentially hazardous features of a unit. Quantitative measurements used in the analysis are based on historic loss data, the energy potential of the key material, and the extent to which loss prevention practices are applied.

Index Calculation

The basic procedure for calculation of the Fire and Explosion Index is shown in Figure 5-2.2. The first step is to identify those process units considered pertinent to the process and having the greatest potential impact in the event of loss by fire or explosion. A unit is considered a part of the plant that can be readily and logically characterized as a separate entity. Generally a unit consists of a segment of the process such as reactors, blenders, furnaces, storage tanks, etc. In some instances, units may be portions of a plant separated from the remainder by distance, fire walls, or other barriers. In other cases, the unit may be an area where a particular hazard exists.

The next step is the determination of the "material factor" for the dominant combustible component of the process

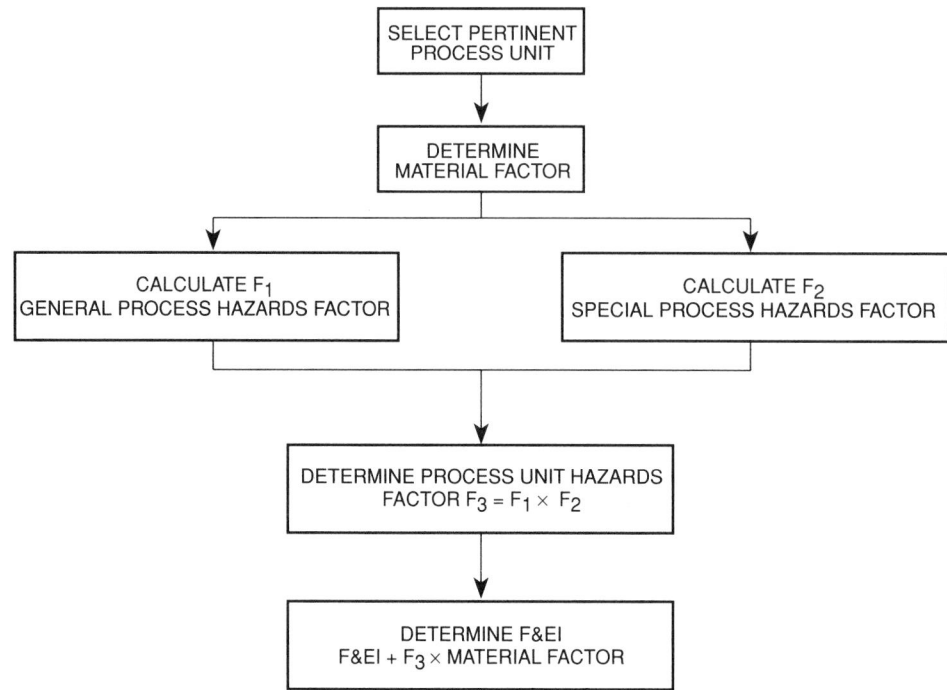

Fig. 5-2.2. *Dow procedure for calculating the Fire and Explosion Index.*

unit. The material factor is a measure of intensity of energy release from a chemical compound, mixture of compounds, or substance. It is determined by considering the flammability and reactivity of a material as described in NFPA 704, *Standard System for the Identification of the Fire Hazards of Materials*,[18] NFPA 49, *Hazardous Chemicals Data*,[19] and NFPA 325, *Guide to Fire Hazard Properties of Flammable Liquids, Gases, and Volatile Solids.*[20] Based on these values the material factor is denoted by a number from 1 to 40. This is an arbitrary ordinal ranking.

Subsequent to selecting the appropriate material factor, penalties for contributing hazards are assessed as indicated by the schedule shown in Figure 5-2.3. Items listed are considered contributing factors to an incident that may result in fire or explosion. Not every hazard is applicable to a given process unit, however, all applicable items should be evaluated and an appropriate penalty applied. The list is divided into two parts, General Process Hazards and Special Process Hazards. General Process Hazards relate to a type of process and represent conditions that may increase the magnitude or severity of an incident. Special Process Hazards are items that increase the probability of a fire or explosion.

Penalties are summed separately for each class of process hazard. These sums are then multiplied (severity × probability) to yield a Process Unit Hazards Factor. This factor has a numerical range of from 1 to 8 and is considered a measure of the probable relative damage exposure magnitude. The Fire and Explosion Index (FEI) is the product of the Material Factor and the Process Unit Hazards Factor.

Risk Analysis

The Process Unit Hazards Factor and Material Factor are also used to derive a Damage Factor. The Damage Factor represents the overall effect of fire, plus blast damage result-

ing from a fire or reactive energy release, caused by the various contributing factors associated with the process unit. It is estimated graphically.

A radius of exposure in feet is estimated as 84 percent of the FEI. The dollar value exposed is the estimated value of all equipment within the circle prescribed by the exposure radius, i.e., the defined circular area of exposure indicates those assets that may be exposed to a fire or explosion generated by the process unit being evaluated. The replacement value of equipment in this area multiplied by the Damage Factor provides the base Maximum Probable Property Damage (MPPD), i.e., the value at risk times the relative damage potential.

Three categories of loss control features have been assigned credits that can potentially reduce the base MPPD to an actual MPPD: (1) process control, (2) material isolation, and (3) fire protection. Twenty-two potential credit factors are shown in Table 5-2.1. The base MPPD reduced by the Loss Control Credit Factor gives the actual MPPD. This value represents the probable resulting loss from an incident of a reasonable magnitude given the proper functioning of protective equipment. Failure of the protective equipment could revert the probable loss to the base MPPD. Final steps include determination of the maximum probable days outage and the business interruption cost. The risk analysis procedure is summarized in Figure 5-2.4.

The most important goal of the FEI analysis is to make the engineer aware of the loss potential of each process area and to help identify ways to lessen the severity and resultant loss of potential incidents. The Dow Chemical Company requires an FEI analysis for existing plants.[16] The requirement for use of the FEI has been adopted into law in the Netherlands.[21] The FEI has been found to be a valuable screening tool that can be used in conjunction with other analyses to help determine the relative risk of process units, and provide valuable guidance to both engineering and management staffs.

FIRE & EXPLOSION INDEX

AREA/ COUNTRY	DIVISION	LOCATION		DATE
SITE	MANUFACTURING UNIT	PROCESS UNIT		
PREPARED BY:	APPROVED BY: (Superintendent)		BUILDING	
REVIEWED BY: (Management)	REVIEWED BY: (Technology Center)		REVIEWED BY: (Safety & Loss Prevention)	

MATERIALS IN PROCESS UNIT

STATE OF OPERATION ___ DESIGN ___ START UP ___ NORMAL OPERATION ___ SHUTDOWN	BASIC MATERIAL(S) FOR MATERIAL FACTOR

MATERIAL FACTOR (See Table 1 or Appendices A or B) Note requirements when unit temperature over 140 °F (60 °C)

1. General Process Hazards	**Penalty Factor Range**	**Penalty Factor Used** (1)
Base Factor	1.00	1.00
A. Exothermic Chemical Reactions	0.30 to 1.25	
B. Endothermic Processes	0.20 to 0.40	
C. Material Handling and Transfer	0.25 to 1.05	
D. Enclosed or Indoor Process Units	0.25 to 0.90	
E. Access	0.20 to 0.35	
F. Drainage and Spill Control _____ gal or cu.m.	0.25 to 0.50	
General Process Hazards Factor (F$_1$)		
2. Special Process Hazards		
Base Factor	1.00	1.00
A. Toxic Material(s)	0.20 to 0.80	
B. Sub-Atmospheric Pressure (< 500 mm Hg)	0.50	
C. Operation In or Near Flammable Range ___ Inerted ___ Not Inerted		
1. Tank Farms Storage Flammable Liquids	0.50	
2. Process Upset or Purge Failure	0.30	
3. Always in Flammable Range	0.80	
D. Dust Explosion (See Table 3)	0.25 to 2.00	
E. Pressure (See Figure 2) Operating Pressure _____ psig or kPa gauge Relief Setting _____ psig or kPa gauge		
F. Low Temperature	0.20 to 0.30	
G. Quantity of Flammable/Unstable Material: Quantity_____ lb or kg H$_c$ =_____ BTU/lb or kcal/kg		
1. Liquids or Gases in Process (See Figure 3)		
2. Liquids or Gases in Storage (See Figure 4)		
3. Combustible Solids in Storage, Dust in Process (See Figure 5)		
H. Corrosion and Erosion	0.10 to 0.75	
I. Leakage — Joints and Packing	0.10 to 1.50	
J. Use of Fired Equipment (See Figure 6)		
K. Hot Oil Heat Exchange System (See Table 5)	0.15 to 1.15	
L. Rotating Equipment	0.50	
Special Process Hazards Factor (F$_2$)		
Process Unit Hazards Factor (F$_1$ × F$_2$) = F$_3$		
Fire and Explosion Index (F$_3$ × MF = F&EI)		

Fig. 5-2.3. Dow fire and explosion index schedule.

TABLE 5-2.1 *Dow Loss Control Credit Factors*

1. Process Control Credit Factor (C₁)

Feature	Credit Factor Range	Credit Factor Used (2)	Feature	Credit Factor Range	Credit Factor Used (2)
a. Emergency Power	0.98		f. Inert Gas	0.94 to 0.96	
b. Cooling	0.97 to 0.99		g. Operating Instructions/Procedures	0.91 to 0.99	
c. Expolsion Control	0.84 to 0.98		h. Reactive Chemical Review	0.91 to 0.98	
d. Emergency Shutdown	0.96 to 0.99		i. Other Process Hazard Analysis	0.91 to 0.98	
e. Computer Control	0.93 to 0.99				

C₁ Value (3) ☐

2. Material Isolation Credit Factor (C₂)

Feature	Credit Factor Range	Credit Factor Used (2)	Feature	Credit Factor Range	Credit Factor Used (2)
a. Remote Control Valves	0.96 to 0.98		c. Drainage	0.91 to 0.97	
b. Dump/Blowdown	0.96 to 0.98		d. Interlock	0.98	

C₂ Value (3) ☐

3. Fire Protection Credit Factor (C₃)

Feature	Credit Factor Range	Credit Factor Used (2)	Feature	Credit Factor Range	Credit Factor Used (2)
a. Leak Detection	0.94 to 0.98		f. Water Curtains	0.97 to 0.98	
b. Structural Steel	0.95 to 0.98		g. Foam	0.92 to 0.97	
c. Fire Water Supply	0.94 to 0.97		h. Hand Extinguishers/Monitors	0.93 to 0.98	
d. Special Systems	0.91		i. Cable Protection	0.94 to 0.98	
e. Sprinkler Systems	0.74 to 0.97				

C₃ Value (3) ☐

Loss Control Credit Factor = C₁ × C₂ × C₃ (3) = ☐

Mond Fire, Explosion, and Toxicity Index

The Mond division of ICI, Ltd. identified that the Dow Fire and Explosion Index method had considerable scope for the evaluation of plant hazard potential at the earliest design stages. Following trials with the published Dow method, it was clear that there was a need to extend the method in a number of directions to develop its potential for new project design. The resulting Mond Fire, Explosion, and Toxicity Index[22,23] was developed and has been applied to a range of new projects within ICI.

The main contribution of the Mond method is the inclusion of *offsetting features*, allowing the effect of good design concepts, good management attitudes, and other preventative measures to reduce the overall hazard. An important outcome of using the technique is to raise questions concerning hazard potential at an early enough stage in planning to allow for adequate investigations to be made before the process is in operation. Achieving some measure of hazard early in the planning process provides information that can be used to select appropriate protection features as the planning proceeds. Early assessment of hazards has value in dealing with possible problems in obtaining approval for proposals and in predicting possible delays and problems that are likely to be encountered.

FIRE SAFETY EVALUATION SYSTEM

The Fire Safety Evaluation System (FSES),[24,25] is a schedule approach to determining equivalencies to the NFPA *101 Life Safety Code*®[26] for certain institutional occupancies. The technique was developed at the Center for Fire Research, National Bureau of Standards in cooperation with the U.S. Department of Health and Human Services (formerly Health, Education, and Welfare) in the late 1970s. It was adapted to new editions of the *Life Safety Code* and is presently published in NFPA 101A, *Alternate Approaches to Life Safety*.[27] (The tables in this section are taken from the worksheets for health care occupancies in Chapter 3 of NFPA 101A.)

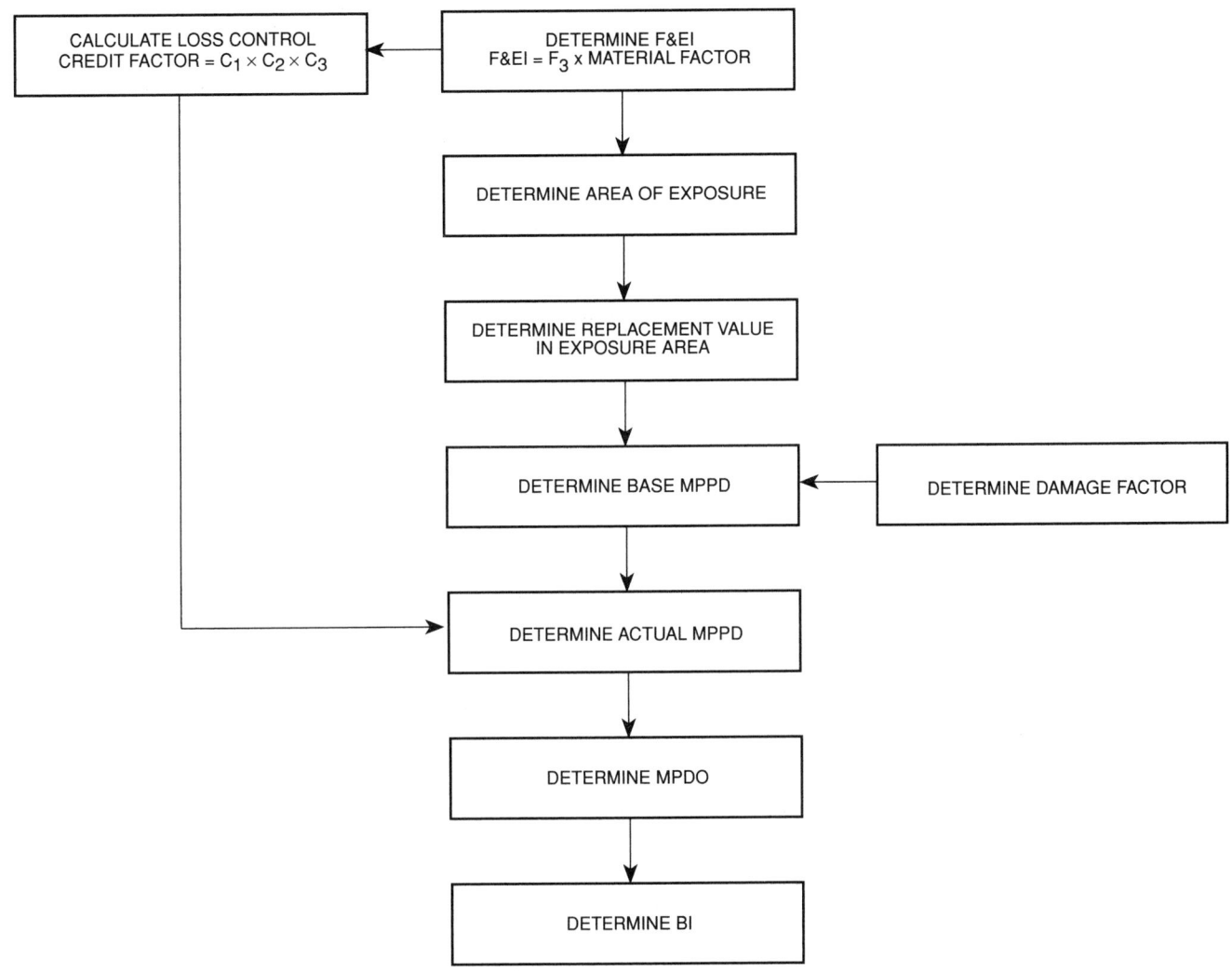

Fig. 5-2.4. Dow procedure for calculating other risk analysis information.

Equivalency Concept

In an effort to promote economical upgrading of fire safety, U.S. codes include an equivalency option. This provision allows alternative designs to satisfy regulations if they provide a level of fire safety equivalent to that called for by the regulations. The difficult decision as to what constitutes equivalency has been left to local jurisdictions. This leads to a lack of uniformity across the country in terms of what may be waived and what constitutes an adequate alternative to provide the required level of safety. The FSES was developed to provide a uniform method of evaluating health care facilities to determine what fire safety measures would provide a level of fire safety equivalent to that provided by the *Life Safety Code*. The objective was to compile an evaluation system that would be easily workable, presenting useful information for the amount of effort expended.

Fire Zone Concept

Unlike the *Life Safety Code*, the FSES subdivides a building into fire zones for evaluation. A fire zone is defined as a space separated from other parts of the building by floors, fire barriers, or smoke barriers. When a floor is not partitioned by fire or smoke barriers, the entire floor is the fire zone. In application, every zone in the facility should be evaluated. Repetitive arrangements may be evaluated by selection of a typical zone.

Risk

Also unlike the *Life Safety Code*, the FSES begins with a determination of relative risk deriving from characteristics of a health care occupancy. Five occupancy risk parameters are used: (1) patient mobility, (2) patient density, (3) fire zone location, (4) ratio of patients to attendants, and (5) average patient age. Variations of these parameters have been assigned relative weights as indicated in Table 5-2.2. These values were determined from the experienced judgment of a group of fire safety professionals and represent the opinions of that panel of experts. There is no documented process for validating or revising these values.

Occupancy risk factor for a zone is calculated as the product of the assigned values for the five risk parameters. Multiplication implicitly suggests these factors are interdependent. A hardship adjustment for existing buildings is applied to the occupancy risk factor; this modifies the risk in

TABLE 5-2.2 *FSES Occupancy Risk Parameter Factors*

RISK PARAMETERS	RISK FACTOR VALUES				
1. PATIENT MOBILITY (M)	MOBILITY STATUS	MOBILE	LIMITED MOBILITY	NOT MOBILE	NOT MOVABLE
	RISK FACTOR	1.0	1.6	3.2	4.5
2. PATIENT DENSITY (D)	NO. PATIENTS	1-5	6-10	11-30	>30
	RISK FACTOR	1.0	1.2	1.5	2.0
3. ZONE LOCATION (L)	FLOOR	1ST	2ND OR 3RD	4TH TO 6TH	7TH AND ABOVE	BASE-MENTS
	RISK FACTOR	1.1	1.2	1.4	1.6	1.6
4. RATIO OF PATIENTS TO ATTENDANTS (T)	PATIENTS / ATTENDANT	1-2 / 1	3-5 / 1	6-10 / 1	>10 / 1	ONE OR† MORE / NONE
	RISK FACTOR	1.0	1.1	1.2	1.5	4.0
5. PATIENT AVERAGE AGE (A)	AGE	UNDER 65 YEARS AND OVER 1 YEAR		65 YEARS & OVER 1 YEAR & YOUNGER	
	RISK FACTOR	1.0		1.2	

†RISK FACTOR OF 4.0 IS CHARGED TO ANY ZONE THAT HOUSES PATIENTS WITHOUT ANY STAFF IN IMMEDIATE ATTENDANCE

existing buildings to 60 percent of that for an equivalent new building.

Fire Safety Parameters

The calculated occupancy risk must be offset by safety features. Thirteen fire safety parameters were selected. These parameters and their respective ranges of values (shown in Table 5-2.3) were also developed by the same panel of experts. Table 5-2.3 is designed to be used as a survey instrument whereby appropriate values for each safety parameter can be selected by inspection of the fire zone. There is no attempt made to directly correlate these fire safety parameters to the previously defined risk parameters.

Fire Safety Redundancies

An important concept of the FSES is redundancy through simultaneous use of alternative safety strategies. The purpose is to ensure that failure of a single protection device or system will not result in a major fire loss. Three fire safety strategies are identified: (1) containment, (2) extinguishment, and (3) people movement. Table 5-2.4 indicates the expert panel's opinion of which fire safety parameters apply to each fire safety strategy. Values from Table 5-2.3 are entered in the appropriate places on Table 5-2.4 and summed for each column. The implication of addition is that

there is no interaction among the fire safety parameters. The limited value of automatic sprinklers for people movement safety is adjusted for by using one-half of the parameter value in this column. The resulting sums are considered to be the available level of each fire safety strategy.

Equivalency Evaluations

The FSES determines if the fire zone in question possesses a level of fire safety equivalent to that of the NFPA *Life Safety Code*. This is done by comparing the calculated level for each fire safety strategy to stated minimum values. These values for existing buildings range from 100 percent down to 12 percent of those for new buildings. For the column of Table 5-2.4 labeled General Safety, the sum of all available safety parameter values is compared to the occupancy risk factor calculated from the parameters in Table 5-2.2.

Supplemental Requirements

Because the thirteen selected fire safety parameters were found not to cover all requirements of the *Life Safety Code*, an addendum to the FSES was created. This consists of a list of twelve additional parameters that may be required by the Code. It should not be implied that these parameters are extraneous to the risk and safety factors of Tables 5-2.2 and 5-2.3, and to the identified fire safety strategies.

TABLE 5-2.3 *FSES Safety Parameter Values*

Parameters	Parameters Values						
1. CONSTRUCTION	COMBUSTIBLE TYPES III, IV AND V				NONCOMBUSTIBLE TYPES I AND II		
FLOOR OR ZONE	000 (U)	111	200 (U)	211 + 2HH	000 (U)	111	222, 322, 433
FIRST	-2	0	-2	0	0	2	2
SECOND	-7	-2	-4	-2	-2	2	4
THIRD	-9	-7	-9	-7	-7	2	4
4TH & ABOVE	-13	-7	-13	-7	-9	-7	4

2. INTERIOR FINISH (Corridors & Exits)	CLASS C	CLASS B	CLASS A	
	-5 (0)[f]	0 (3)[f]	3	

3. INTERIOR FINISH (Rooms)	CLASS C	CLASS B	CLASS A	
	-3 (1)[f]	1 (3)[f]	3	

4. CORRIDOR PARTITIONS/WALLS	NONE OR INCOMPLETE	<1/3 HR	≥1/3 < 1 HR	≥1 HR
	-10 (0)[a]	0	1 (0)[a]	2 (0)[a]

5. DOORS TO CORRIDOR	NO DOOR	<20 MIN FPR	≥20 MIN FPR	≥20 MIN FPR & AUTO CLOS.
	-10	0	1 (0)[d]	2 (0)[d]

6. ZONE DIMENSIONS	DEAD END			NO DEAD ENDS >30' & ZONE LENGTH IS:		
	>100'	>50' - 100'	30' - 50'	>150'	100' - 150'	<100'
	-6 (0)[b]	-4(0)[b]	-2 (0)[b]	-2	0	1

7. VERTICAL OPENINGS	OPEN 4 OR MORE FLOORS	OPEN 2 OR 3 FLOORS	ENCLOSED WITH INDICATED FIRE RESIST.		
			<1 HR.	≥1 HR. <2 HR.	≥2 HR.
	-14	-10	0	2 (0)[e]	3 (0)[e]

8. HAZARDOUS AREAS	DOUBLE DEFICIENCY		SINGLE DEFICIENCY		NO DEFICIENCIES
	IN ZONE	OUTSIDE ZONE	IN ZONE	IN ADJACENT ZONE	
	-11	-5	-6	-2	0

9. SMOKE CONTROL	NO CONTROL	SMOKE BARRIER SERVES ZONE	MECH. ASSISTED SYSTEMS BY ZONE	
	-5 (0)[c]	0	3	

10. EMERGENCY MOVEMENT ROUTES	<2 ROUTES	MULTIPLE ROUTES			
		DEFICIENT	W/O HORIZONTAL EXIT(S)	HORIZONTAL EXIT(S)	DIRECT EXIT(S)
	-8	-2	0	1	5

11. MANUAL FIRE ALARM	NO MANUAL FIRE ALARM	MANUAL FIRE ALARM	
		W/O F.D. CONN.	W/F.D. CONN.
	-4	1	2

12. SMOKE DETECTION & ALARM	NONE	CORRIDOR ONLY	ROOMS ONLY	CORRIDOR & HABIT. SPACE	TOTAL SPACE IN ZONE
	0 (3)[g]	2 (3)[g]	3 (3)[g]	4	5

13. AUTOMATIC SPRINKLERS	NONE	CORRIDOR & HABIT. SPACE	ENTIRE BUILDING	
	0	8	10	

NOTES:
[a] Use (0) when Parameter 5 is -10.
[b] Use (0) when Parameter 10 is -8.
[c] Use (0) on floor with less than 31 patients (existing buildings only).
[d] Use (0) when Parameter 4 is -10.

[e] Use (0) when Parameter 1 is based on first floor zone or on an unprotected type of construction (columns marked "U").
[f] Use ( ) if the area of Class B or C interior finish in the corridor and exit or room is protected by automatic sprinklers and Parameter 13 is 0.
[g] Use this value in addition to Parameter 13, Automatic Sprinklers value, if the entire zone is protected with quick-response automatic sprinklers.

Conversion: 1 ft x 0.3048 = m.

TABLE 5-2.4   *FSES Worksheet for Evaluating Fire Safety Strategies*

SAFETY PARAMETERS	CONTAINMENT SAFETY (S_1)	EXTINGUISHMENT SAFETY (S_2)	PEOPLE MOVEMENT SAFETY (S_3)	GENERAL SAFETY (S_4)
1. CONSTRUCTION			▨	
2. INTERIOR FINISH (Corr. & Exit)		▨		
3. INTERIOR FINISH (Rooms)		▨	▨	
4. CORRIDOR PARTITIONS/WALLS		▨	▨	
5. DOORS TO CORRIDOR		▨		
6. ZONE DIMENSIONS	▨	▨		
7. VERTICAL OPENINGS		▨		
8. HAZARDOUS AREAS			▨	
9. SMOKE CONTROL	▨	▨		
10. EMERGENCY MOVEMENT ROUTES	▨	▨		
11. MANUAL FIRE ALARM	▨		▨	
12. SMOKE DETECTION & ALARM	▨			
13. AUTOMATIC SPRINKLERS			÷ 2 =	
TOTAL VALUE	$S_1 =$	$S_2 =$	$S_3 =$	$S_4 =$

Computer Optimization

A distinct advantage of schedule approaches to fire risk assessment is that they lend themselves to computer optimization techniques. FSES has capitalized on this characteristic by creating a computerized version.[28,29] This computerized procedure is based on a linear programming algorithm. Linear programming refers to a mathematical model for allocating limited resources among competing activities subject to a set of constraints. The procedure finds the distribution of values that optimizes an objective function.

For a fire risk assessment schedule, the objective is to minimize the cost of fire protection that will meet a pre-scribed acceptable level. In the FSES, the acceptable levels are given and the variables are the safety parameters that can take on the values indicated in Table 5-2.3. By assigning a cost to each value in Table 5-2.3, an economic optimum can be calculated.

HIERARCHICAL APPROACH

Development of a hierarchical approach to fire risk ranking was initially undertaken at the University of Edinburgh, sponsored by the U.K. Department of Health and Social Services.[30,31,32] The objective of this study was to improve the evaluation of fire safety in U.K. hospitals through a systematic method of appraisal. This approach was further developed at the University of Ulster for application to dwelling occupancies.[33,34] Most recently, it has been refined and implemented for the assessment of fire risk in telecommunications facilities.[35,36]

Defining fire safety is difficult and often results in a listing of factors that together comprise the intent. These factors tend to be of different sorts. For example, fire safety may be defined in terms of goals and aims such as: fire prevention, fire control, occupant protection, etc. These broad concepts are usually found in the introductory section of building codes and other fire safety legislation. Or, fire safety may be defined in terms of more specific hardware items such as: combustibility of materials, heat sources, detectors, sprinklers, etc. These topics are more akin to items listed in the table of contents of building codes. A meaningful exercise is to construct a matrix of fire safety goals *vs* more specific fire safety features. This helps to identify the roles of these two concepts, in both theory and practice.

Decision Making Levels

As a logical extension of a single fire safety matrix, consider that there are more than two categories of fire safety factors. This suggests a hierarchy of lists of things, or decision making levels, that comprise fire safety. Such a hierarchy of fire safety decision making levels is presented in Table 5-2.5.

These represent common levels of fire safety decision making but there may be more or fewer in a particular application. For example, an even lower level dealing with individual physical items could be added, or intermediate levels could be used to better define certain relationships.

This hierarchy of levels of detail of fire safety suggests that a series of matrices is appropriate to model the relationships among various fire safety factors; i.e., a matrix of policy *vs* objectives would define a fire safety policy by identifying the specific objectives that are held most desirable. In turn, a matrix of objectives *vs* strategies would identify the relationship of these factors, and a matrix of strategies *vs* parameters would suggest where to use what. Thus, a matrix may be constructed to examine the association of any two adjacent levels in a hierarchy of fire safety factors.

An even more appealing aspect of this approach is that two or more matrices may be combined (multiplied) to produce information on the importance of specific detail of building elements to an overall fire safety policy—information not previously available. This approach is the only such grading of fire safety with an explicitly defined relationship to fire safety goals and objectives.

Generalized Procedure

A generalized procedure for ranking fire safety parameters to determine their relative importance is summarized in the four steps below.

1. Identify hierarchical levels of fire safety specification.
2. Specify items comprising each level.
3. Construct and assign values to matrices of each sequential pair of levels.
4. Combine (multiply) matrices to yield importance ranking of items.

An example of step 1 is represented by Table 5-2.5 above. Step 2 requires that lists of objectives, strategies, parameters, and survey items be developed. A list of fire safety objectives might include statements about life safety, property protection, continuity of operations, environmental protection, and heritage preservation.

No significant work has been done to identify just what it is that fire safety is trying to achieve (i.e., allocation of resources for fire safety is not generally directly associated with a specific corporate objective), so these objectives are a very subjective list. (One benefit of the hierarchical approach is facilitating the incorporation of fire safety into more global organizational objectives.)

In most applications a Delphi process is used to define fire safety policy in terms of the specified list of objectives. That is, a group of experts is asked to rank fire safety objectives with respect to their importance to policy. Each member of the Delphi group receives feedback in the form of response averages and the process repeats until an acceptable level of consensus is reached. The Delphi exercise yields a vector representing the relative importance of each objective to organizational policy. In some work the more formal Analytical Hierarchy Process (AHP) is used.[37] However, this process is unstable when there are more than six or seven factors to be ranked.

The next decision making level involves fire safety strategies. A list of strategies can be derived by taking a cut set of the NFPA Fire Safety Concepts Tree.[38] Example fire safety strategies are ignition prevention, limitation of combustibles, compartmentation, fire detection and alarm, fire suppression, and protection of exposed people or things.

Now, a matrix of objectives *vs* strategies can be constructed. Values of the cells are again supplied by Delphi or some other subjective decision making process. In this case the question to be answered is: How important is each strategy to the achievement of each objective?

In order to facilitate mathematical manipulation, the values of the matrices can be normalized. Then, multiplying

TABLE 5-2.5 *Hierarchy of Fire Safety Decision Making Levels*

Level	Name	Description
1	POLICY	Course or general plan of action adopted by an organization to achieve security against fire and its effects
2	OBJECTIVES	Specific fire safety goals to be achieved
3	STRATEGIES	Independent fire safety alternatives, each of which contributes wholly or partly to the fulfillment of fire safety objectives
4	PARAMETERS	Components of fire risk that are determinable by direct or indirect measure or estimate
5	SURVEY ITEMS	Measurable feature that serves as a constituent part of a fire safety parameter

the objectives/strategies matrix by the policy vector yields a new vector that shows the relative contribution of each strategy to overall fire safety policy. While this vector is not essential to the fire safety evaluation, it is illustrative of the matrix manipulation process that is the essence of the hierarchical approach.

Continuing this procedure, the next level of fire safety parameters is considered. The following is a typical list of these parameters.

construction	equipment	fixed suppression
height	special hazards	fire department
compartmentation	detection	egress system
building services	alarm	personnel
furnishings	smoke control	management

A matrix of strategies *vs* fire safety parameters is then constructed and evaluated. Multiplying this matrix by the previously derived vector yields a new vector that weighs each fire safety parameter according to its relative contribution to organizational fire safety policy. The significance of this vector is that it is the only such weighing of fire safety factors that has an explicit link to fire safety goals and objectives. The matrix manipulation process is summarized in Figure 5-2.5.

Evaluating Parameters

In order to use the resulting vector of parameter weights to develop a fire risk ranking of a building or facility space, the extent to which each parameter is present must be evaluated. That is, a level of functional value of each fire safety parameter must be assessed. These parameter grades may be directly observable or, more often, they are derived from various functions of a lower level of features that includes specific hardware components, e.g., fire safety survey items.

Parameters are defined as components of fire risk that are quantitatively determinable by direct or indirect measurement or estimation. They are intended to represent factors that account for an acceptably large portion of the total fire risk. In most cases they are not directly measurable. This is especially true for existing buildings where only limited information is readily available.

Each parameter has a specific relative importance that is universal for all facilities within the scope of the assessment method. Individual buildings will vary in the degree to which parameters exist or occur in a space. Parameter grades are a measure of the intensity level or degree of danger or security afforded by the parameter. Grading of parameters is facilitated by partitioning them into measurable constituent parts. Usually these parts are directly assessable survey items, the next lower level in the decision hierarchy. The determination of parameter grades is dependent on those features of a space identified as survey items.

A survey item is a measurable feature of a space that serves as a constituent part of one or more parameters. In developing means to grade parameters in a given building, each parameter has associated with it one or more survey items. These specific features evolve from analysis of the parameters. Items are chosen for contributing significantly to the effectiveness of the irrespective parameters and for being directly measurable. It is therefore necessary that survey items be defined in sufficient detail to support these traits. Detailed descriptions of the survey items are required to frame questions that provide input into the decision logic that produces the parameter grades.

In a recent application, grades were established for each fire risk parameter by associating readily measurable survey items, using logic described by decision tables.[39] Input to these tables included fire test results, fire hazard modeling, field experience from previous fire events, logic diagrams, and professional judgment.

The scalar product of the resulting parameter weights and grades yields a relative measure of fire risk. This may be used to rank facilities or it can be compared to a standard value.

CRITERIA FOR DEVELOPMENT AND EVALUATION OF FIRE RISK RANKING

The fire protection engineering community appears to be largely unconcerned with the proliferation of fire risk ranking and how it is being used. The available literature deals only with development and application of a specific method or general descriptions of several selected approaches. Like any analytical technique, risk ranking methods have their limitations and should not be used uncritically.

The purpose of fire risk ranking is to provide a useful aid to decision making. It must be easy to apply but sophisticated enough to provide a minimum of technical validity. Credibility can also be improved through consistency and transparency. The approach should be systematic, and it should be clear to all interested parties that the relevant technical issues have been appropriately covered. Based on the review of numerous fire risk ranking systems, ten criteria have been proposed to aid in the development and evaluation of other such systems.[40]

Criterion 1: *Development and implementation of the method should be thoroughly documented according to standard procedures.* One of the hallmarks of professionalism is that as a study proceeds, a record is made of assumptions, data, parameter estimates and why they were chosen, model

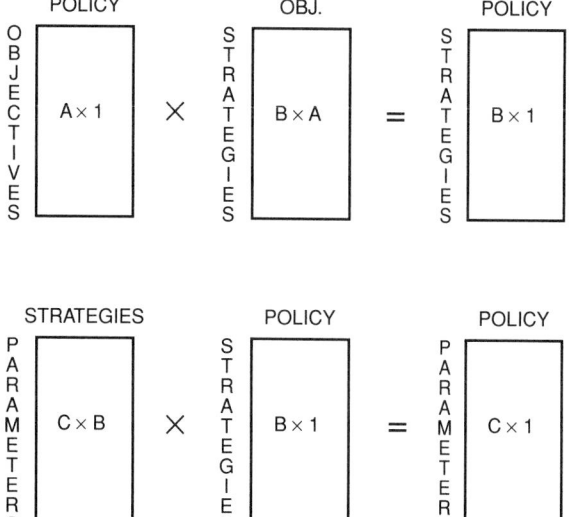

Fig. 5-2.5. Schematic summary of hierarchical approach.

structure and details, steps in the analysis, relevant constraints, results, sensitivity tests, validation, and so on. Little of this information is available for most fire risk ranking methods.

In addition to facilitating review, there are three other practical reasons not to slight the documentation: (1) if external validation is to be conducted, adequate documentation will be a prerequisite; (2) during the life cycle of a fire risk ranking system the inescapable changes and adjustments will require appropriate documentation; and (3) clear and complete documentation enhances confidence in the method; its absence inevitably carries with it the opposite effect.

The value of the documentation will be improved if it follows established guidelines. Standard formats for documentation are primarily directed at large-scale computer models[41,42] but can be readily adapted in principle to more general applications.

Criterion 2: *Partition the universe rather than select from it.* One of the least well-established procedures in fire risk ranking is the choice of parameters. In following a systemic approach it is best to be comprehensive. Using an exhaustive model of fire safety such as the NFPA Fire Safety Concepts Tree[38] is helpful in being inclusive. This logic tree branches out from the holistic concept of fire safety objectives. A cut set on the tree will then identify a group of parameters that encompasses all possible fire safety features.

Criterion 3: *Parameters should represent the most frequent fire scenarios.* In determining the level of detail of the parameters, it is necessary to look at those factors which are most significant, statistically or by experienced judgment. This criterion may also be used as an alternative to Criterion 2, providing the need for systemic comprehensiveness is satisfied.

Criterion 4: *Provide operational definitions of parameters.* If the methodology is to be used by more than a single individual, it is necessary to ensure precise communication of the intent of key terms. Many fire risk parameters are ambiguous concepts that have a wide variety of interpretation even within the fire community.

Criterion 5: *Elicit subjective values systematically.* Most fire risk ranking methods rely heavily on experienced judgment. The use of formalized, documented procedures, such as the multi-attribute utility theory, analytical hierarchy process, and Delphi, significantly increases credibility of the system. Similarly, use of recognizable scaling techniques will enhance credibility.

Criterion 6: *Parameter values should be maintainable.* One variable that is not explicitly included in fire risk ranking, but is very important, is time. It influences the fire risk both internally (e.g., deterioration) and externally (e.g., technological developments). In order for a method to have a reasonable useful lifetime, it must be amenable to updating. This implies that the procedure for generating parameter values must be repeatable. Changes over time and new information dictate that the system facilitate revisions.

Criterion 7: *Treat parameter interaction consistently.* In the majority of cases this will consist of an explicitly stated assumption of no interactive effect among parameters. Where interactions are considered, it is important that they be dealt with systematically to avoid bias.

Criterion 8: *State the linearity assumption.* While this assumption is universal in fire risk ranking, it is also well known that fire risk variables do not necessarily behave in a linear fashion. It is important to the acceptance of ranking methods and their limitations that such assumptions are understood.

Criterion 9: *Describe fire risk by a single indicator.* The objective of most fire risk ranking methods is to sacrifice details and individual features for the sake of making the assessment easier. Information should be reduced to a single score even in the most complex applications. The results should be presented in a manner that makes their significance clear in a simple and unambiguous way. Unless all those involved can understand and discuss the meaning of the ranking, there will not be general confidence in its adequacy.

Criterion 10: *Evaluate predictive capability.* Some attempt should be made to verify that the method does in fact differentiate between lesser and greater fire risks with sufficient precision. It is not feasible to validate a model *per se* but some testing should be documented. Reference 43 gives some guidance on this subject. The level of accuracy demanded here is not the same as for other engineering purposes, establishing an order of magnitude will generally suffice.

REFERENCES CITED

1. J. Watts, "Systematic Methods of Evaluating Fire Safety: A Review," *Hazard Prevention*, 18, 2, 24–27 (1981).
2. NFPA, National Fire Codes, National Fire Protection Association, Quincy (1995).
3. E.M. Marchant, "A Cost-effective Approach to Fire Safety, Paper I—Points Scheme," International Fire Security and Safety Conference (1984).
4. Harold E. Nelson, "Overview—Numerical Grading Systems," Building Research Board, Report from the 1987 Workshop on Analytical Methods for Designing Buildings for Fire Safety, National Academy Press, Washington (1988).
5. John M. Watts, Jr., "Fire Risk Rating Schedules," *Fire Hazard and Fire Risk Assessment*, ASTM STP 1150, M.M. Hirschler, ed., American Society for Testing and Materials, Philadelphia, 24–34 (1992).
6. G.R. Rosenblum and S.A. Lapp, "The Use of Risk Index Systems to Evaluate Risk," *Risk Analysis: Setting National Priorities*, *Proceedings* of the Society for Risk Analysis, Houston (1987).
7. Building Research Board, Report from the 1987 Workshop on Analytical Methods for Designing Buildings for Fire Safety, National Academy Press, Washington (1988).
8. John H. Magee and David L. Bickelhaupt, *General Insurance*, Richard D. Irwin, Homewood, IL, p. 153 (1964).
9. Albert Flandreau Dean, "Analytic System for the Measurement of Relative Fire Hazard," Western Actuarial Bureau, Chicago (1902).
10. Jay S. Glidden, "Analytic System for the Measurement of Relative Fire Hazard, an Explanation," Western Actuarial Bureau, Chicago, p. 17 (1916).
11. BVD, "Evaluation of Fire Hazard and Determining Protective Measures," Association of Cantonal Institutions for Fire Insurance (VKF) and Fire Prevention Service for Industry and Trade (BVD), Zurich (1973).
12. BVD, "Fire Risk Evaluation," Association of Cantonal Institutions for Fire Insurance (VKF), Society of Engineers and Architects (SIA), and Fire Prevention Service for Industry and Trade (BVD), Zurich (1980).
13. J. Kaiser, "Experiences of the Gretener Method," *Fire Safety Journal*, 2, 213–222 (1980).

14. "Process Safety Manual," *Chemical Engineering Progress*, 62, 6, The Dow Chemical Company (1966).

15. *Dow's Fire and Explosion Index Hazard Classification Guide*, 7th ed., AICHE Technical Manual, The Dow Chemical Company, American Institute of Chemical Engineers, New York (1994).

16. Norman E. Scheffler, "Improved Fire and Explosion Index Hazard Classification," AICHE Spring National Meeting (1994).

17. "Guidelines for Chemical Process Quantitative Risk Analysis," Center for Chemical Process Safety (CCPS), American Institute of Chemical Engineers, New York, 32 (1989).

18. NFPA 704, *Identification of the Fire Hazards of Materials*, National Fire Protection Association, Quincy (1990).

19. NFPA 49, *Hazardous Chemicals Data*, National Fire Protection Association, Quincy (1991).

20. NFPA 325M, *Fire Hazard Properties of Flammable Liquids, Gases, and Volatile Solids*, National Fire Protection Association, Quincy (1991).

21. Occupational Safety Report Guideline for Compilation, P172-2E, Directorate-General of Labour of the Ministry of Social Affairs and Employment, Voorburg, The Netherlands (1990).

22. D.J. Lewis, "The Mond Fire Explosion, and Toxicity Index—A Development of the Dow Index," paper presented at AIChE Loss Prevention Symposium, Houston, April 1-5, 1979.

23. *The Mond Index*, 2 ed., ICI (Imperial Chemical Industries), ICI PLC, Explosion Hazards Section, Technical Department, Winnington, UK (1985).

24. I.A. Benjamin, "A Firesafety Evaluation System for Health Care Facilities," *Fire Journal*, 73, 2 (1979).

25. H.E. Nelson and A.J. Shibe, "A System for Fire Safety Evaluation of Health Care Facilities," NBSIR 78-1555, Center for Fire Research, National Bureau of Standards, Washington (1980).

26. NFPA 101, *Life Safety Code*, National Fire Protection Association, Quincy (1994).

27. NFPA 101A, *Alternative Approaches to Life Safety*, National Fire Protection Association, Quincy (March 1995).

28. R.E. Chapman, "Cost-Effective Methods for Achieving Compliance to Firesafety Codes," *Fire Journal*, September, pp. 30-39, 123 (1979).

29. R.E. Chapman, W.G. Hall, and P.T. Chen, "A Computerized Approach for Identifying Cost-Effective Fire Safety Retrofits in Health Care Facilities" NBSIR 79-1929, National Bureau of Standards, Washington (1980).

30. "Fire Safety Evaluation (Points) Scheme for Patient Areas Within Hospitals," Department of Fire Safety Engineering, University of Edinburgh (1982).

31. P. Stollard, "The Development of a Points Scheme to Assess Fire Safety in Hospitals," *Fire Safety Journal*, 7, 2, 145-153 (1984).

32. E. Marchant, "Fire Safety Engineering—A Quantified Analysis," *Fire Prevention*, 210, 34-38 (1988).

33. T.J. Shields and G.W. Silcock, "An Application of the Hierarchical Approach to Fire Safety," *Fire Safety Journal*, 11, 3, 235-242 (1986).

34. H.A. Donegan, T.J. Shields, and G.W. Silcock, "A Mathematical Strategy to Relate Fire Safety Evaluation and Fire Safety Policy Formulation for Buildings," *Fire Safety Science—Proceedings* of the Second International Symposium, Takao Wakamatsu, et al., eds., Hemisphere, NY, 433-441 (1989).

35. Edward K. Budnick, Brian D. Kushler, and John M. Watts Jr., "Fire Risk Assessment: A Systematic Approach for Telecommunications Facilities," Wanda J. Duffin, ed., 1993 Annual Conference on Fire Research: Book of Abstracts, NISTIR5280, National Institute of Standards and Technology, Gaithersburg, 133-134 (1993).

36. Kathleen C. Tomaino, Lyman L. Parks, Brian D. Kushler, Dean Arrington, Lawrence A. McKenna Jr., Michael Serapiglia, Edward K. Budnick, and John M. Watts Jr., Fire Risk Assessment for Telecommunications Central Offices, in preparation.

37. T.J. Shields, G.W. Silcock, and Y. Bell, "Fire Safety Evaluation of Dwellings," *Fire Safety Journal*, 10, 1, 29-36 (1986).

38. NFPA 550, *Guide to the Fire Safety Concepts Tree*, National Fire Protection Association, Quincy (1994).

39. John M. Watts Jr., Edward K. Budnick, and Brian D. Kushler, "Fire Risk Assessment: Using Decision Tables to Quantify Parameters," to be submitted to *Fire Technology*.

40. John M. Watts Jr., "Criteria for Fire Risk Ranking," *Fire Safety Science—Proceedings* of the Third International Symposium, G. Cox and B. Langford, eds., Elsevier, London, 457-466 (1991).

41. S.I. Gass, "Documenting a Computer Based Model," *Interfaces*, 14, 3, 84-93 (1984).

42. ASTM E-1472, *Documenting Computer Software for Fire Models*, American Society for Testing and Materials, Philadelphia (1992).

43. ASTM E-1355, *Evaluating the Predictive Capability of Fire Models*, American Society for Testing and Materials, Philadelphia (1992).

EXTREME VALUE THEORY

G. Ramachandran

INTRODUCTION

The theory of extreme values, as generally known, deals with the statistical properties of the maximum or minimum value of a random variable. The theory has found practical applications in fields such as structural engineering, aeronautics, and meteorology. A comprehensive treatise on the extreme value theory and its applications was presented by Gumbel.[1] Further theoretical developments and applications are contained in *Statistical Extremes and Applications*, edited by Tiago de Oliveira.[2]

Fire protection is one such area, where figures for financial losses are generally available for large fires which, in the United Kingdom, for example, are currently defined as fires costing £50,000 or more in property damage. To make the best use of limited data provided by a small sample of large-loss fires, Ramachandran[3] has developed a general mathematical framework, the extreme order theory, which includes the largest loss as a particular case.

The *object* of this chapter is to explain the basic features of extreme order statistics and show how data on financial losses in large fires can be used for:

1. Predicting the behavior of the tail (large losses) of the probability distribution of fire loss;
2. Estimating the average loss in all fires, large and small, in a particular building or group of buildings with similar fire risks; and
3. Assessing the probable reduction in loss due to a fire protection measure.

A few other possible applications of extreme value theory in the fire protection field are also discussed briefly; these include statistical analysis of test results, fire severity, and fire resistance.

Dr. G. Ramachandran retired in November 1988 as Head of the Operations Research Section at the Fire Research Station, U.K. Since then he has been practicing as a consultant in risk evaluation and insurance. He is a visiting professor at the University of Hertsfordshire and Glasgow Caledonian University. His research has focused on statistical and economic problems in fire protection and actuarial techniques in fire insurance.

EXTREME ORDER DISTRIBUTIONS

The probability distribution of fire loss (x) is skewed, and, in general, the variable z ($= \log x$) has a distribution belonging to the "exponential type." (See Ramachandran[4–6] and Shpilberg.[7]) Among distributions of this type, normal for z (which is the same as log-normal for x) has been recommended widely for modeling fire insurance claims. Exponential for z or Pareto for x has been considered by some actuaries. The (cumulative) distribution function of z may be denoted by $F(z)$ and its density function by $f(z)$.

The logarithms of losses in n fires during a defined period of time constitute a sample of observations generated by the "parent" distribution $F(z)$. If these figures are arranged in decreasing order of magnitude, the mth value in this arrangement may be denoted by $z_{(m)n}$, which is the logarithm of the mth loss $x_{(m)n}$. For the largest value, the subscript m takes the value 1 (first rank).

In a series of samples (periods), each with a large number n of observations (fires), the values pertaining to $z_{(m)n}$ have a probability distribution with density function approximately given by

$$\frac{m^m a_{(m)n}}{(m-1)!} \exp[-my_{(m)} - m \exp(-y_{(m)})]$$
$$-\infty \leq z_{(m)n} \leq \infty \quad (1)$$

$$y_{(m)} = a_{(m)n}[z_{(m)n} - b_{(m)n}] \quad (2)$$

where $a_{(m)n}$ and $b_{(m)n}$ are solutions of

$$F[b_{(m)n}] = 1 - (m/n) \quad (3)$$

$$a_{(m)n} = (n/m)f[b_{(m)n}] \quad (4)$$

The parameter $b_{(m)n}$ is the modal value of $z_{(m)n}$. Ramachandran[3] has discussed different methods of estimating the parameters $a_{(m)n}$ and $b_{(m)n}$.

BEHAVIOR OF LARGE LOSSES

Let $\bar{z}_{(m)n}$ and $s^2_{(m)n}$ be the mean and variance of $z_{(m)n}$, respectively, in N samples. For large N, the values of these two parameters tend to

$$\mu_{(m)z} = b_{(m)n} + [1/a_{(m)n}]\bar{y}_{(m)} \tag{5}$$

$$\sigma^2_{(m)z} = \sigma^2_{(m)y}/a^2_{(m)n} \tag{6}$$

where

$$\bar{y}_{(m)} = 0.5772 + \log_e m - \sum_{\nu=1}^{m-1}(1/\nu) \tag{7}$$

$$\sigma^2_{(m)y} = 1.6449 - \sum_{\nu=1}^{m-1}(1/\nu^2) \tag{8}$$

The value of $z_{(m)n}$ for any probability level, p, is given by

$$z_{(m)np} = b_{(m)n} + [1/a_{(m)n}]y_{(m)p} \tag{9}$$

where $y_{(m)p}$ is the value of $y_{(m)}$ corresponding to the probability level, p. The values of $y_{(m)p}$ for selected probability levels have been tabulated by Gumbel.[8]

Consider, as an example, the analysis carried out by Ramachandran[4] using the top 17 fire losses ($m = 1$ to 17) in the UK textile industry during the 21-year period ($N = 21$) from 1947 to 1967. The losses (in units of £1000) were first corrected for inflation by adjusting them to 1947 values. Base e was used for calculating the logarithms of losses. The linear method (Ramachandran[3]) was then applied to the logarithms of the corrected losses for estimating the parameters $a_{(m)n}$ and $b_{(m)n}$. Since the number of fires, n, per year varied over the period, the following approximate correction was included in the estimation process

$$b_{(m)n_j} = b_{(m)n} + [1/a_{(m)n}]\log_e(n_j/n) \tag{10}$$

where $b_{(m)n_j}$ is the modal value of $z_{(m)}$ for the year with n_j fires. The results, reproduced in Table 5-3.1, pertain to a sample size of 465 fires, the frequency experienced in the base year 1947.

TABLE 5-3.1　*Textile Industry Top Fire Losses from 1947 to 1967, UK*

Extreme (m)	$a_{(m)n}$	$b_{(m)n}$
1	2.247	5.214
2	1.785	4.829
3	1.626	4.534
4	1.460	4.327
5	1.387	4.113
6	1.424	3.988
7	1.239	3.749
8	1.163	3.564
9	1.212	3.448
10	1.034	3.259
11	0.973	3.137
12	0.925	2.972
13	0.886	2.832
14	0.924	2.749
15	0.937	2.680
16	0.950	2.583
17	1.002	2.537

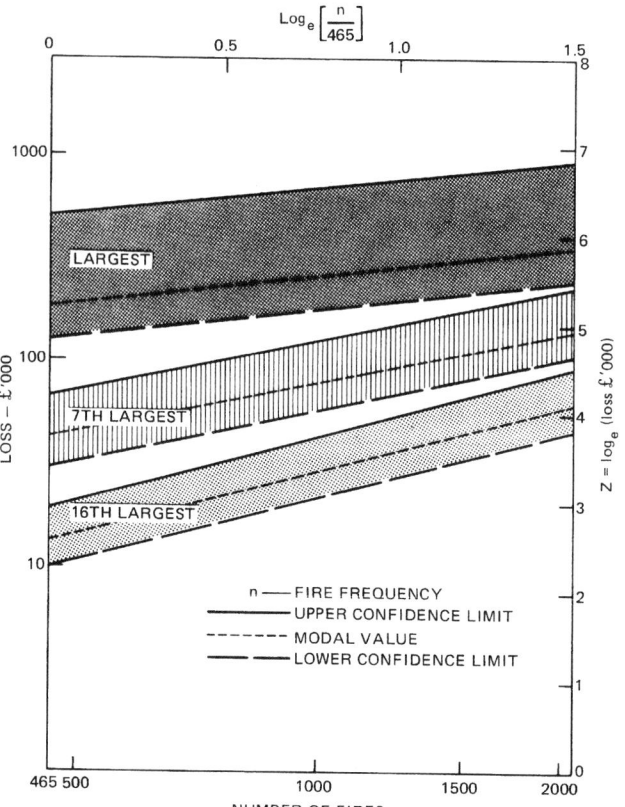

Fig. 5-3.1.　Fire frequency and large losses.

Figure 5-3.1 has been drawn on a log scale and shows the relationship between the annual frequency of fires in the textile industry and the probable loss, at 1947 prices, of the 1st ($m = 1$), 7th ($m = 7$), and 16th ($m = 16$) largest fires in a year. For each of these three ranks, the modal sizes of the losses are shown with confidence bands. For an estimated number of fires in any year, an ordinate erected at the corresponding point on the x-axis would intersect the upper and lower confidence lines at points giving the corresponding confidence limits. The probability of exceeding the upper or falling short of the lower limit is 0.1. For example, if the number of fires expected in a year in the textile industry is 1000, the most probable value of the largest loss would be £260,000, with upper and lower confidence limits of £700,000 and £180,000; all figures are at 1947 money values.

The confidence lines represent a control chart based on the current trend. The increase in the frequency, n, of fires may be partly due to the inadequacy of fire prevention measures. In addition, if some or all of the actual large losses corrected for inflation exceeded the corresponding upper limits, it may be concluded that general changes in fire fighting and fire protection methods or in the industrial processes are taking place to alter the picture for the worse. If the losses are less than the lower limits, then the changes are for the better. These arguments and the data on losses and number of fires for the period 1968 to 1978 suggested that protection measures were coping well with fire outbreaks in the UK textile industry. (See Ramachandran.[3]) This observation is true to some extent for the later period, as well.

TABLE 5-3.2 *Average Loss Per Fire at 1966 Prices (£ × 000)*

	Sprinklered single story	Sprinklered multistory	Nonsprinklered single story	Nonsprinklered multistory
Textiles	2.9	3.5	6.6	25.2
Timber and furniture	1.2	3.2	2.4	6.5
Paper, printing, and publishing	5.2	5.0	7.1	16.2
Chemical and allied	3.6	4.3	4.3	8.2
Wholesale distributive trades	—	4.7	3.8	9.4
Retail distributive trades	—	1.4	0.4	2.4

AVERAGE LOSS

If fire loss, x, is assumed to have a log-normal probability distribution, the expected or average loss is given by

$$\bar{x} = \exp\left(\mu + \frac{\sigma^2}{2}\right)$$

where μ and σ are the mean and standard deviation, respectively, of $z = \log_e x$. Estimation of the parameters μ and σ is a simple statistical problem if loss figures are available for all the n fires in a sample. If loss figures are available only for large losses above a threshold level, extreme value techniques can provide reasonably good estimates of μ and σ. The basic features of these techniques are as follows:

If z ($= \log x$) has a normal distribution, the standard variable

$$t = (z - \mu)/\sigma$$

has also a normal distribution with distribution function $G(t)$ and density function $g(t)$. As defined earlier, if $z_{(m)n}$ is the logarithm of the mth loss $x_{(m)n}$ from the top, the variable

$$t_{(m)n} = [z_{(m)n} - \mu]/\sigma \tag{11}$$

follows the extreme order distribution shown in Equation 1 with

$$y_{(m)} = A_{(m)n}[t_{(m)n} - B_{(m)n}] \tag{12}$$

instead of Equation 2. As in Equations 3 and 4, the parameters $A_{(m)n}$ and $B_{(m)n}$ are given by

$$G[B_{(m)n}] = 1 - (m/n) \tag{13}$$

$$A_{(m)n} = (n/m)g[B_{(m)n}] \tag{14}$$

The values of $B_{(m)n}$ for the standard normal distribution have been well tabulated by, for example, Fisher and Yates.[9] The corresponding values of $A_{(m)n}$ can then be obtained for Equation 14. From Equation 12 the expected value of $t_{(m)n}$ is

$$\bar{t}_{(m)n} = B_{(m)n} + [1/A_{(m)n}]\bar{y}_{(m)} \tag{15}$$

with $\bar{y}_{(m)}$ given by Equation 7. From Equation 11, the expected values of $z_{(m)n}$ is

$$\bar{z}_{(m)n} = \mu + \sigma\bar{t}_{(m)n} \tag{16}$$

The value of $z_{(m)n}$ in a single sample or its mean value over N samples (each with n fires) will provide an estimate of $\bar{z}_{(m)n}$.

If figures for losses are available for, say, r large fires ($m = 1$ to r), the straight line in Equation 16 can be used to obtain rough estimates of μ and σ by plotting the r pairs of values $[\bar{z}_{(m)n}, \bar{t}_{(m)n}]$ and drawing the best line, or by the method of least squares. However, this method is somewhat imprecise, since the residual errors in Equation 16 arise from ranked variables and, hence, vary with m (not constant) and are correlated (not independent). For dealing with this problem, Ramachandran[3,10] has developed a generalized least squares method that provides the best and unbiased estimates of μ and σ but requires the use of a complex computer program. Ramachandran also described a maximum likelihood method in these studies which involves simple calculations for estimating μ and σ for each sample.

ECONOMIC VALUE OF FIRE PROTECTION MEASURES

The economic value of a fire protection measure depends on the reduction in loss that could be expected due to the satisfactory operation of the measure in the event of a fire. This reduction could be assessed in terms of the difference in the expected (average) loss in two groups of buildings with similar fire risks—one group equipped with the fire protection measure and the other not equipped. The reduction in loss or gain is one of the main components in a cost benefit analysis of a fire protection measure.

Consider, as an example, the investigation carried out by Rogers[11] in regard to the effect of sprinkler protection of buildings. Rogers applied the generalized least square method mentioned in the previous section, assuming specifically that fire loss has a log-normal probability distribution. Table 5-3.2 contains some interesting results obtained by Rogers for average loss in all fires (estimated from large losses). The figures in Table 5-3.2 relate to fires that survived "infant mortality"; very small fires were excluded from the total sample size (n), since their inclusion would distort the shape of the loss distribution, particularly at the upper tail. It is apparent that sprinklers reduce the loss expected in multistory buildings to a considerable extent. Sprinklers also reduce the probability of loss in a fire exceeding a large amount (see Figure 5-3.2), based on the parameters μ and σ estimated from large losses in connection with the results in Table 5-3.2. (The parameters in Figure 5-3.2 pertain to $z = \log_{10}x$.)

FACTORS AFFECTING FIRE DAMAGE

A full assessment of fire risk ought to consider all the relevant factors affecting fire damage and evaluate their independent contribution to the damage. This is possible by

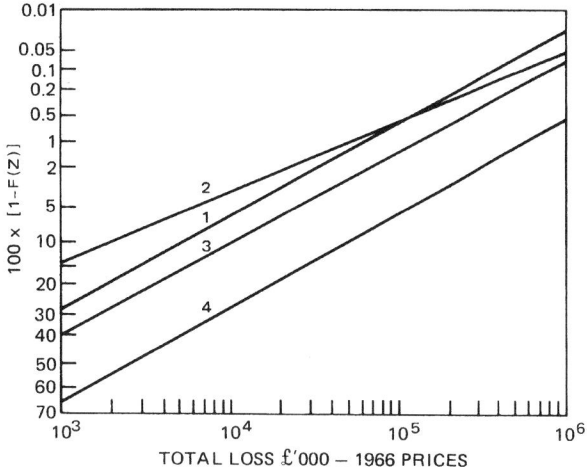

LINE	SUBPOPULATION	PARAMETERS
1	SPRINKLER/SINGLE STORY	$\mu = -0.616$ $\sigma = 1.024$
2	SPRINKLER/MULTISTORY	$\mu = -1.419$ $\sigma = 1.340$
3	NON-SPRINKLER/SINGLE STORY	$\mu = -0.334$ $\sigma = 1.062$
4	NON-SPRINKLER MULTISTORY	$\mu = 0.401$ $\sigma = 0.992$

$F(Z)$ (CUMULATIVE DISTRIBUTION FOR Z) = PROBABILITY OF LOSS LESS THAN OR EQUAL TO Z

$V(X)$ (CUMULATIVE DISTRIBUTION FOR X) = PROBABILITY OF LOSS LESS THAN OR EQUAL TO X

SURVIVOR PROBABILITY = $1 - F(Z) = 1 - V(X)$ = PROBABILITY OF LOSS EXCEEDING X OR Z

Fig. 5-3.2. The survivor probability distribution of fire loss for each class in the textile industry.

performing a multiple regression analysis. The problem is to estimate the regression parameters, using large observations of the dependent variable, logarithm of loss. (Factors affecting the damage, assigned numerical values, will be independent variables.) For tackling this problem, a multiple regression model, based on extreme order theory, has been developed by Ramachandran.[3,12] Using large losses, this model gives estimates of regression parameters approximately equivalent to estimates that would be obtained if loss figures were available for all the fires and were utilized in the calculations.

Consider, for example, the expected loss which has a "power relationship" with the size of a building such that

$$z = \alpha + \beta \log A \qquad (17)$$

where z is the logarithm of loss, and A is the total floor area of a building.

The linear relationship in Equation 17 was incorporated in an investigation concerned with the trade-offs between sprinklers and structural fire resistance (Ramachandran[3]). Based on the information about buildings given in the reports on fires furnished by the fire brigades, industrial buildings were classified into two broad groups—high and low fire resistance. Each of these groups was further divided into single story and multistory and sprinklered and nonsprin-

TABLE 5-3.3 *Textile Industry, Multistory Buildings Expected Loss (£) at 1978 Prices*

Building Category	Floor Area (sq ft)	
	100,000	1,000,000
Sprinklered		
High fire resistance	4080	5790
Low fire resistance	6460	23,070
Nonsprinklered		
High fire resistance	19,910	38,260
Low fire resistance	31,510	152,410

klered. The parameters α and β were estimated for each of the eight subgroups and for each industry using large-loss figures. As an example, expected losses for two hypothetical multistory buildings engaged in textile manufacture are given in Table 5-3.3.

ANALYSIS OF TEST RESULTS

In some fire tests, observations recorded are extreme values (maximum or minimum). In such cases, classical methods of analysis of variance, although normally carried out, are not strictly applicable. Extremes, particularly the maximum and the minimum, have highly skewed probability distributions and, hence, are not normally distributed. In these problems, the following approach may be used to test the difference between two maximum values. (For a similar test for minimum values, see Ramachandran and Rogers.[13])

If the reduced variable of the maximum is written as

$$y_{(1)} = a_{(1)}[z_{(1)} - b_{(1)}] \qquad (18)$$

and

$$u_{(1)} = \exp[-y_{(1)}] \qquad (19)$$

the variable $2u_{(1)}$ has a chi square distribution with 2 degrees of freedom. It follows, therefore, that exp $[y_{(1)2}$ and $y_{(1)1}]$ has an "F" distribution with (2, 2) degrees of freedom. The reduced maximums, $y_{(1)1}$ and $y_{(1)2}$, belong to two independent samples from exponential-type parents, with $z_{(1)1}$ and $z_{(1)2}$ as the maximum observations in the samples. From Equation 18

$$y_{(1)1} = a_{(1)1}[z_{(1)1} - b_{(1)1}] \qquad (20)$$

$$y_{(1)2} = a_{(1)2}[z_{(1)2} - b_{(1)2}] \qquad (21)$$

and from Equation 19

$$u_{(1)1} = \exp[-y_{(1)1}]$$

$$u_{(1)2} = \exp[-y_{(1)2}]$$

As in the "F" test, the values $y_{(1)1}$ and $y_{(1)2}$ should be so denoted that $u_{(1)1}$ is greater than $u_{(1)2}$. The variable

$$z = \frac{1}{2} \log[u_{(1)1}/u_{(1)2}]$$
$$= \frac{1}{2}[y_{(1)2} - y_{(1)1}] \qquad (22)$$

has the z distribution, the probability points of which have been tabulated by Fisher and Yates.[9] It may be verified that

TABLE 5-3.4 *Maximum Temperature at Vent Position*

Levels of other factors (kg)		Vent width levels				
		240 mm		700 mm		
Load	PVC	Max Temp.	y	Max Temp.	y	z
120.5	0	870	−0.6533	925	−0.8340	0.0904
120.5	95	950	0.4587	975	−0.1390	0.2989
241	0	960	0.5977	1100	1.5985	0.5004
241	95	1010	1.2927	1020	0.4865	0.4031
103.5 137.5	0	960	0.5977	1070	1.1815	0.2919

the α percent point of z in this particular case is given by the following formula derived by Ramachandran and Rogers.[13]

$$z = \frac{1}{2} \log_e\left(\frac{1-\alpha}{\alpha}\right) \qquad (23)$$

Significant difference in $y_{(1)2} - y_{(1)1}$ at α level would occur if the observed value of z given by Equation 22 is greater than the value of z given by Equation 23.

Consider, as an example, large-scale tests conducted at the Fire Research Station, U.K., to assess the products due to combustion of plastics. The three factors investigated were: (1) crib load, (2) weight of the plastic (PVC), and (3) width of ventilation opening. The observations recorded were maximum temperature, maximum concentrations of carbon dioxide and carbon monoxide, and minimum oxygen; a graphic analysis revealed that variables as such (without any log transformation) approximate an exponential-type distribution.

For purposes of illustration, the results for maximum temperature at vent position are reproduced in Table 5-3.4. (See Ramachandran and Rogers[13] for full details of the statistical analysis.) It was assumed that the scale parameter $a_{(1)}$ was the same for all the 10 maximum observations arising from 10 independent samples. The standard deviation, σ_z, of these observations was 68.06. According to Gumbel,[1] the standard deviation, σ_y, of $y_{(1)}$ is 0.9497 for $N = 10$. Hence, from Equation 18, an estimated value of $a_{(1)}$ was given by

$$a_{(1)} = \sigma_y/\sigma_z = 0.0139$$

The averages of the maximum temperatures $\bar{z}_{240}$ and $\bar{z}_{700}$ for the two vent width levels were 950 and 1018. For $N = 5$, the mean $[\bar{y}_{(1)}]$ of $y_{(1)}$ was found to be 0.4588. From Equation 18, the values of parameters $b_{(1)}$ were estimated as

$$b_{240} = \bar{z}_{240} - [\bar{y}_{(1)}/a_{(1)}] = 917$$
$$b_{700} = \bar{z}_{700} - [\bar{y}_{(1)}/a_{(1)}] = 985$$

For each pair of observations, the y values and the z test statistic were calculated using Equations 20 through 22. From Equation 23, for $\alpha = 0.10$ (10 percent point), the theoretical value of z was 1.0986. In all the 5 cases, the observed z values in Table 5-3.4 were less than 1.0986. This implied that there was no significant difference between the maximum temperature at the vent position for the two ventilation levels for any fixed level of the other two factors, load and PVC. At present, it is not possible to test the interactions between factors and the differences between overall means, such as $(\bar{z}_{240} - \bar{z}_{700})$.

FIRE SEVERITY AND FIRE RESISTANCE

For the design of a fire-resistant building, knowledge of the expected fire severity is essential. The fire severity in a room will vary with the quantity of combustible material (fire load), the area of ventilation, and the dimensions of the room. Fire severities, however, may differ from room to room in a building, because the furnishings in a building are seldom everywhere equal and may also change with time. Thus, it is difficult to predict whether a fire in a given building will be ventilation controlled. For the reasons mentioned above, fire severity must be regarded as a random variable with a probability distribution. This distribution can be assumed to be exponential such that the probability of fire severity being greater than S is given by $\exp(-\lambda_s S)$, with the variable S measured in time units, say, minutes. (See, for example, Sara-Coward.[14])

There are no data available at present to estimate the probability distribution of the fire resistance, R, of a room or a complete building. It might be possible to determine this distribution from the fire resistance of structural elements and the plan of a building showing the connections between rooms, corridors, floors, etc., using, in addition, data on factors such as doors and windows which affect the fire resistance. Purely from heuristic reasoning, exponential or log-normal distribution has been suggested for the fire resistance of a structural element. Log-normal was assumed in a statistical analysis of fire resistance of laminated timber columns (Ramachandran.[15])

In a fire, failure of the structure of, for example, a room occurs if the fire resistance, R, of the room is less than the fire severity, S, both the variables expressed in units of time. The probability of this failure event can be evaluated from the probability distributions of R and S, if this information is available.

What is ideally required is the probability of minimum fire resistance, R_p, being less than the maximum possible fire severity, S_q. If this probability is denoted by ∂ such that

$$\text{Prob}(R_p < S_q) = \partial \qquad (24)$$

then ∂ should be a small value which is allowable or acceptable, depending on the consequences of failure in terms of damage to life and property. From a political point of view, it might be possible to tolerate a certain number of deaths and injuries and a certain amount of property loss resulting from the failure of building structures due to fires. For any type of building, the acceptable level of ∂ depends on the number of buildings of this type at risk and the fire damage (life and property) experienced by these buildings.

It must be recognized that both R_p and S_q are random variables and not constants. Since fire resistance has an exponential-type distribution, the reduced variable

$$y_p = a_p(R_p - b_p) \tag{25}$$

has the probability distribution of the smallest (minimum) value with distribution function

$$1 - \exp[-\exp(y_p)] \tag{26}$$

and density function

$$\exp[y_p - \exp(y_p)] \tag{27}$$

The parameters a_p and b_p can be estimated from the parent distribution of fire resistance, if the exact nature of this distribution is known together with its location and scale parameters. In the case of the minimum, if $U(R)$ is the parent cumulative distribution function,

$$U(b_p) = 1/n \tag{28}$$

such that the probability of fire resistance being less than b_p is $(1/n)$. The value of n may be so chosen that Equation 28 gives a small value which is acceptable. The parameter a_p is given by $nu(b_p)$, where $u(R)$ is the density function corresponding to $U(R)$.

Since fire severity has an exponential-type distribution, the maximum fire severity S_q, with $m = 1$ in Equation 1, has the density function

$$\exp[-y_q - \exp(-y_q)] \tag{29}$$

and distribution function

$$\exp[-\exp(-y_q)] \tag{30}$$

where

$$y_q = a_q(S_q - b_q) \tag{31}$$

If $V(S)$ is the distribution function of fire severity, the parameter b_q, from Equation 3, is given by

$$V(b_q) = 1 - (1/n) \tag{32}$$

such that the probability of fire severity exceeding b_q is $(1/n)$.

Then, from Equation 4, $a_q = nv(b_q)$, where $v(s)$ is the density function of fire severity.

Having defined the probability distributions and parameters of minimum fire resistance and maximum fire severity, it follows to study the random behavior of the variable

$$d_{RS} = R_p - S_q \tag{33}$$

which will enable the evaluation of the probability given by Equation 24. This problem is the subject of a current research investigation by the author. In conclusion, the fire resistance of the structural elements of a building should be such that the firesafety level specified in Equation 24 will be satisfied.

REFERENCES CITED

1. E.J. Gumbel, *Statistics of Extremes*, Columbia Univ. Press, New York (1958).
2. J. Tiago de Oliveira, ed., *Statistical Extremes and Applications*, D. Reidel, Dordrecht (1984).
3. G. Ramachandran, *F. Safety J.*, 5, 59 (1982).
4. G. Ramachandran, *Fire Research Note No. 910*, Fire Research Station, Borehamwood (1972).
5. G. Ramachandran, *ASTIN Bul.*, 7, 293 (1974).
6. G. Ramachandran, "Extreme Order Statistics in Large Samples from Exponential-Type Distributions and Their Application to Fire Loss," in *Statistical Distributions in Scientific Work*, 2, 335, D. Reidel, Dordrecht (1975).
7. D.C. Shpilberg, *Tech. Rep. 22431*, Factory Mutual Research Corp., Norwood (1974).
8. E.J. Gumbel, *Probability Tables for the Analysis of Extreme Value Data—Applied Mathematics*, National Bureau of Standards, Washington (1953).
9. R.A. Fisher and F. Yates, *Statistical Tables for Biological, Agricultural, and Medical Research*, Oliver and Boyd, London (1953).
10. G. Ramachandran, *Extreme Order Statistics from Exponential-Type Distributions with Applications to Fire Protection and Insurance*, Ph.D. Thesis, London (1975).
11. F.E. Rogers, *Current Paper CP 9/77*, Building Research Establishment, Borehamwood (1977).
12. G. Ramachandran, *ASTIN Bul.*, 8, 229 (1975).
13. G. Ramachandran and F.E. Rogers, *Toxic Gases and Smoke from Polyvinyl Chloride in Fires—Statistical Analysis of Test Results*, Fire Research Note, No. 1021, Fire Research Station, Borehamwood (1974).
14. K.D. Sara-Coward, *Current Paper CP 31/75*, Building Research Establishment, Borehamwood (1975).
15. G. Ramachandran, *Fire Research Note No. 943*, Fire Research Station, Borehamwood (1972).

RELIABILITY

Mohammad Modarres and Yu-Shu Hu

INTRODUCTION

Reliability has two connotations. One is probabilistic in nature; the other is deterministic. This chapter generally deals with the probabilistic aspect. First reliability must be defined. The most widely accepted definition of reliability is the *ability of an item (product, system, etc.) to operate under designated operating conditions for a designated period of time or number of cycles*. The *ability* of an item to operate can be: (1) designated through a probability (the probabilistic connotation) or (2) designated deterministically. The deterministic approach, in essence, deals with understanding how and why an item fails, and how it can be designed and tested to prevent such failures from occurrence or recurrence. This includes analyses, such as deterministic analysis and review of field failure reports, understanding physics of failure, the role and degree of test and inspection, performing redesign, or performing reconfiguration. In practice, this is an important aspect of reliability analysis.

The probabilistic treatment of an item's reliability according to the definition above can be represented by

$$R(t) = Pr(T \geq t \mid c_1, c_2, \ldots) \qquad (1)$$

where

$$
\begin{aligned}
R(t) &= \text{reliability of the item,} \\
Pr &= \text{probability,} \\
T &= \text{time to failure or cycle to failure of the item,} \\
t &= \text{designated period of time or cycles for the item's operation (mission time), and} \\
c_1, c_2, \ldots &= \text{designated conditions, such as environmental conditions.}
\end{aligned}
$$

Often, in practice, $c_1, c_2, \ldots$ are implicitly considered in the probabilistic reliability analysis and, thus, Equation 1 reduces to

Dr. M. Modarres is a Professor of Reliability and of Nuclear Engineering and a cofounder of the Reliability Engineering Program at the University of Maryland, College Park. He is the director of the Center for Reliability Engineering at the University of Maryland. A consultant to government, industry, and international organizations, he is the author or coauthor of more than 100 professional papers in reliability and risk assessment. Y-S. Hu is a postdoctoral fellow with the Center for Reliability Engineering at the University of Maryland, College Park.

$$R(t) = Pr(T \geq t) \qquad (2)$$

This chapter addresses the basic elements of component and system reliability evaluation, including the classical frequency approach to component reliability, important aspects of component reliability analysis, system reliability models, and logic trees.

COMPONENT RELIABILITY

The probabilistic notion is used to quantitatively measure the reliability of an item and treats reliability as the conditional probability of the successful achievement of an item's intended function, given designated conditions. The probabilistic definition of reliability given in Equation 1 is its mathematical representation. The right-hand side of Equation 1 denotes the probability that a specified failure time, T, exceeds a specified mission time, t, given that conditions $c_1, c_2, \ldots$ exist (or are met). Practically, random variable T represents *time-to-failure* of the item, and conditions $c_1, c_2, \ldots$ represent conditions (e.g., design-related conditions) that are required, *a priori*, for successful performance of the item. Other representations of random variable T include *cycle-to-failure*, *stress-to-failure*, and so on. This chapter considers only time-to-failure representation. Conditions $c_1, c_2, \ldots$ are often implicitly considered; therefore, Equation 1 is written in a nonconditional form of Equation 2. Equation 2 is used in the remainder of this chapter.

RELIABILITY FUNCTION

Let's start with the formal definition given in Equation 1. Furthermore, let $f(t)$ denote a probability distribution function representing the random variable T. $f(t)$ is obviously characterized by the design (e.g., strength), operational, and environmental (e.g., stress) effects of the item. The probability of failure of the item as a function of time can be defined by

$$Pr(T \leq t) = \int_0^t f(\theta) \, d\theta = F(t), \qquad \text{for } t \geq 0 \qquad (3)$$

where $F(t)$ denotes the probability that the item will fail sometime up to time t. Based on Equation 2, Equation 3 is the

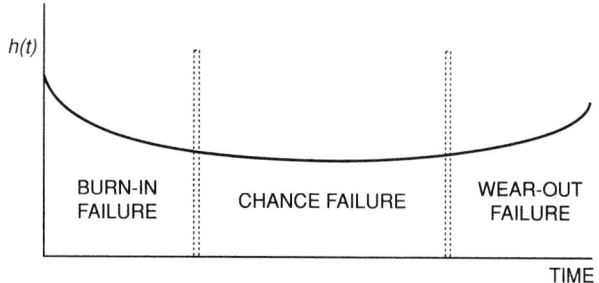

Fig. 5-4.1. Typical bathtub curve.

unreliability of the item. Formally, $F(t)$ is called the *unreliability function*. Conversely, one can find the *reliability function* by writing

$$R(t) = 1 - F(t) = \int_t^\infty f(\theta)\, d\theta \qquad (4)$$

Provided one can obtain the probability distribution function, $f(\theta)$, one can then obtain $R(t)$. Basic characteristics of the probability distribution function and $R(t)$ can be helpful to create some useful definitions. The *mean time to failure* (MTTF), for example, is defined as the expected value of $f(t)$. This illustrates the expected time during which the item will perform its function successfully (sometimes called *expected life*) as

$$MTTF = E(t) = \int_0^\infty t f(t)\, dt \qquad (5)$$

If $\lim_{t \to \infty} R(t) = 0$, then it is easy to show that a more compact form of Equation 5 is given by

$$E(t) = \int_0^\infty R(t)\, dt \qquad (6)$$

MTTF refers to the mean time from first use of a new item to first and only failure. MTTF is different from *mean time between failures* (MTBF), which is the mean time between two consecutive item failures. MTBF includes the time to discover the earlier failure, the time to repair that failure, and the time until it fails again. If discovery and repair are essentially instantaneous and repair is as good as replacement (as with changing a light bulb), then the repaired or replaced item will behave like a new item, and MTBF will equal MTTF. In general, discovery and repair are not instantaneous, repair is not perfect, and MTBF involves very complex modeling.

FAILURE RATE AND HAZARD RATE

Failures can occur due to many physical processes or mechanisms. In probabilistic reliability, all these mechanisms of failure are accounted for through a function called *instantaneous failure rate* or *hazard rate*, $h(t)$. The hazard rate can be interpreted as the probability of the first and only failure of an item in the next instant of time, given that the item is presently operating. Clearly, this only applies to non-repairable items in which only one failure can occur. For repairable items the most appropriate term is failure rate or more correctly the *rate of occurrence of failure*. Hazard rate function is obtained *via life test data*. The rate of failure for a component c is defined as $\lambda = 1/\Delta t$ (probability that c will fail between t and $t + \Delta t$, given that no failure is observed before t). The instantaneous failure rate (hazard rate) for

component c is defined as the limit of the rate of failure as the interval Δt approaches zero. $h(t)$ is obtained from

$$h(t) = \frac{f(t)}{R(t)} \qquad (7)$$

Here, $f(t)$ represents the time to failure probability distribution function of a component, and $R(t)$ is its reliability function. Hazard rate is an important function in reliability analysis, since it shows changes in the probability of failure over the lifetime of a component.

In practice, $h(t)$ often exhibits a "bathtub" shape and is referred to as a *bathtub curve*. A typical bathtub curve is shown in Figure 5-4.1.

Generally, a bathtub curve can be divided into three regions. The *burn-in* early failure region exhibits a *decreasing rate of failure*, characterized by early failures attributable to defects in design, manufacturing, or construction. The *chance failure* region exhibits a reasonably *constant rate of failure*, characterized by random failures of the component. In this period, many mechanisms of failure due to complex underlying physical, chemical, or nuclear phenomena give rise to this approximately constant rate of failure. The third region, called *wear-out failure*, exhibits an *increasing rate of failure*, characterized mainly by complex aging phenomena. Here the component deteriorates (e.g., due to accumulated fatigue) and is more vulnerable to outside shocks. One interesting observation is that these three regions are different for different types of components. Figure 5-4.2 and Figure 5-4.3 show typical bathtub curves for electrical and mechanical devices, respectively. Electrical devices exhibit a relatively higher rate of failure during the chance failure period. Figure 5-4.4 shows the effect of various levels of stress on a device. As stress level increases, the chance failure region decreases, and premature wear-out occurs. Therefore, it is important to minimize stress factors, such as harsh operating environment, to maximize reliability.

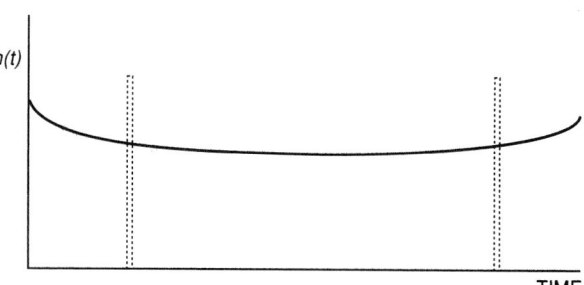

Fig. 5-4.2. Bathtub curve for typical electrical devices.

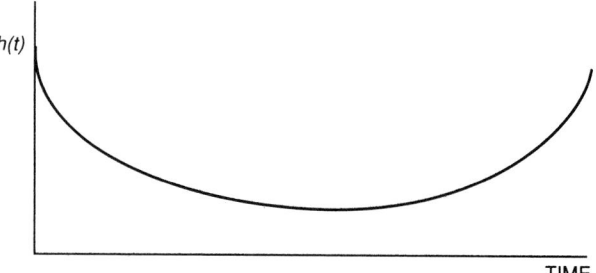

Fig. 5-4.3. Bathtub curve for typical mechanical devices.

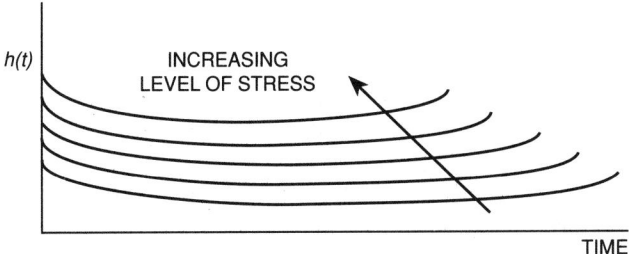

Fig. 5-4.4. **Effect of stress on a typical bathtub curve.**

Next, the mathematical relationship between $h(t)$ and $R(t)$ needs clarification. Recall that from Equation 4, $R(t) = 1 - F(t)$, or $dR(t) = -dF(t) = -f(t)$. By integrating both sides of Equation 7 from 0 to t,

$$\int_0^t h(x)\, dx = -\ln R(t) + \ln R(0)$$

By assuming that a component is totally reliable at the beginning of "life" [i.e., $R(0) = 1$)], then

$$\int_0^t h(x)\, dx = -\ln R(t) \qquad (8)$$

and

$$R(t) = \exp\left(-\int_0^t h(x)\, dx\right) \qquad (9)$$

If the beginning of "life" is not at $t = 0$, but rather at $t = t_1$, then the integrands of (3.6) will be between t_1 and t. If any one of $f(t)$, $h(t)$, or $R(t)$ is known, the other two can be computed by using Equations 7, 8, and 9.

EXAMPLE 1:

A component is known to have an instantaneous failure rate, $h(t)$, as shown below. Determine the probability distribution function and the reliability function.

SOLUTION:

For $0 \le t \le 200$, $h(t) = 5 \times 10^{-6} t^2 - 10^{-3} t + 0.1$,

$$\int_0^t h(t')\, dt' = \int_0^t (5 \times 10^{-6} t'^2 - 10^{-3} t' + 0.1)dt'$$

$$= \left[\left(\frac{5 \times 10^{-6}}{3}\right)t'^3 - \left(\frac{10^{-3}}{2}\right)t'^2 + 0.1t'\right]\Bigg|_0^t$$

$$= 1.66 \times 10^{-6} t^3 - 5 \times 10^{-4} t^2 + 0.1t$$

thus,

$$R(t) = \exp(-1.66 \times 10^{-6} t^3 + 5 \times 10^{-4} t^2 - 0.1t)$$

Using Equation 7,

$$f(t) = (5 \times 10^{-6} t^2 - 10^{-3} t + 0.1)$$
$$\times \exp(-1.66 \times 10^{-6} t^3 + 5 \times 10^{-4} t^2 - 0.1t)$$

COMMON DISTRIBUTIONS AND PARAMETER ESTIMATION

To represent component failure probabilities as a function of time, the most common distributions are exponential, lognormal, and Weibull probability distribution functions. The exponential distribution is used to model component time to failure when failures have a constant "arrival" rate. When the mean time to failure of a component is the same, no matter how much time has already passed, the exponential model is reasonable and the parameter estimation should proceed.

The lognormal distribution is useful, in part, because the product of two lognormally distributed variables also has a lognormal distribution. This is useful for a unit whose failure occurs only after the failure of several of its constituent parts. If each part has a lognormal failure probability, then so will the unit. Reflecting this underlying phenomenology, the lognormal distribution can be used for failure probability distributions with very large variances. It is not unusual in such a distribution to see probability values separated by orders of magnitude. The Weibull distribution has three parameters; this creates the greatest challenge but also the greatest flexibility in setting a specific distribution. In particular, the Weibull distribution is flexible enough to depict the distribution for a unit that shows pronounced changes in reliability during the wear-out stage or whose reliability improves over time.

This section introduces common distributions and deals with statistical methods for estimating model parameters, such as λ of the exponential probability distribution function, μ and σ of the lognormal probability distribution functions, and α and β of the Weibull probability distribution function. The objective is to find a *point estimate* and a *confidence interval* for the parameters of interest. It is important to realize why one needs to consider confidence intervals in the estimation process. In essence, this need stems from the fact that there is only a limited amount of information (e.g., times to failure), and thus one cannot state the estimation with certainty. Therefore, the confidence interval is highly influenced by the amount of data available. Of course other factors, such as diversity in the sources of data and accuracy of the selected model and the data sources, also influence the state of uncertainty regarding the estimated parameters.

Exponential Distribution

This distribution is widely used in reliability evaluation to model a random variable representing time-to-failure of a device (often a device composed of several independent units). The distribution is a one-parameter probability distribution function defined by

$$f(t) = \lambda \exp(-\lambda t) \qquad \lambda, t > 0$$
$$= 0, \qquad\qquad t \le 0 \qquad (10)$$

It is easy to show that requirements $\int_{\text{all } t} f(t)\, dt = 1$ and $f(t) > 0$ for a valid probability distribution function are met for this distribution. Figure 5-4.5 illustrates the exponential distribution for a value of λ.

EXAMPLE 2:

A component has a constant failure rate of 10^{-3}/hr. What is the probability that this component will fail before $t = 1000$ hrs? Determine the probability that it works for at least 1000 hrs.

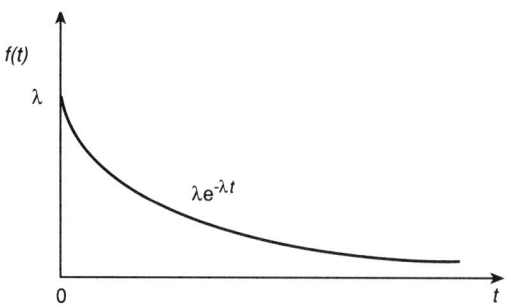

Fig. 5-4.5. *Exponential distribution.*

SOLUTION:

$$Pr(t < 1000) = \int_0^{1000} \lambda \exp(-\lambda t)\, dt = \exp(-\lambda t)\Big|_0^{1000}$$

$$= 1 - \exp(-1) = 0.632$$

$$Pr(t > 1000) = 1 - Pr(t < 1000) = 0.368$$

Epstein[1] has shown that, if the time to failure is exponentially distributed with parameter λ, the quantity $2r\lambda/\hat{\lambda} = 2\lambda T$ has a chi-square distribution with $2r$ degrees of freedom for the Type II (failure-terminated) failure data and $(2r + 2)$ degrees of freedom for the Type I (time-terminated) failure data, where r is the number of failures observed, T is the total accumulated operation time, and $\hat{\lambda}$ is the point estimator for the parameter λ.

$$\hat{\lambda} = \frac{r}{T} \qquad (11)$$

Consequently, for the two-sided confidence interval for the $1 - \alpha/2$ and $\alpha/2$ confidence levels of the chi-square distribution,

$$Pr\left[\chi^2_{\alpha/2}(2r) \le \frac{2r\lambda}{\hat{\lambda}} \le \chi^2_{1-\alpha/2}(2r)\right] = 1 - \alpha \qquad (12)$$

By rearranging and using Equation 11,

$$Pr\left[\frac{\chi^2_{\alpha/2}(2r)}{2T} \le \lambda \le \frac{\chi^2_{1-\alpha/2}(2r)}{2T}\right] = 1 - \alpha \qquad (13)$$

Equation 13 shows the *two-sided confidence interval* for the true value of λ. Another alternative to Equation 12 is the *one-sided confidence interval*,

$$Pr\left[0 \le \hat{\lambda} \le \frac{\chi^2_{1-\alpha}(2r)}{2T}\right] = 1 - \alpha \qquad (14)$$

Accordingly, confidence intervals of MTTF and $R(t)$ at $t = t_0$ can also be obtained for both one-sided and two-sided confidence intervals from Equations 13 and 14.

It is possible to show that the upper confidence limit of λ for the Type I (time-terminated) test is obtained from a chi-square distribution with $(2r + 2)$ degrees of freedom. The lower confidence limit is obtained from the chi-square distribution with $2r$ degrees of freedom. Thus for Type I tests, the two-sided confidence interval is

$$Pr\left[\frac{\chi^2_{\alpha/2}(2r)}{2T} \le \lambda \le \frac{\chi^2_{1-\alpha/2}(2r+2)}{2T}\right] = 1 - \alpha \qquad (15)$$

The one-sided confidence interval is

$$Pr\left[0 \le \lambda \le \frac{\chi^2_{1-\alpha/2}(2r+2)}{2T}\right] = 1 - \alpha \qquad (16)$$

It should be emphasized here that Equations 12 through 16 apply only when the failure rate is constant. Otherwise, a Weibull model or other appropriate distribution should be used.

If no failure is observed, $\hat{\lambda} = 0$ or MTTF $= \infty$. This cannot realistically be true, since one may have had a small or restricted test. Had the test been continued, eventually a failure would be observed. An upper level estimate for both one-sided and two-sided confidence limits can be obtained with $r = 0$. However, the lower limit for the two-sided confidence limit cannot be obtained with $r = 0$. It is possible to relax this limitation by conservatively assuming that a failure occurs in the very next instant. Then $r = 1$ can be used to evaluate the lower two-sided confidence limit. This conservative modification, although sometimes used to allow a complete statistical analysis, lacks firm statistical basis. Welker and Lipow[2] have shown methods to determine approximate point estimates in these cases.

EXAMPLE 3:

Twenty-five units are subjected to a reliability test that lasts 500 hrs. In this test, eight failures occur at 75, 115, 192, 258, 312, 389, 410, and 496 hrs. The failed units are replaced. Find $\hat{\lambda}$, one-sided and two-sided confidence limits on λ, and MTTF at the 90 percent confidence level; one-sided and two-sided 90 percent confidence limits on reliability at $t_0 = 1000$ hrs.

SOLUTION:

This is Type I data. The accumulated time T is given by:

$$T = 25 \times 500 = 12,500 \text{ component hrs.}$$

$$\hat{\lambda} = 8/12,500 = 6.4 \times 10^{-4} \text{ hr}^{-1}.$$

One-sided confidence limits on λ are

$$0 \le \lambda \le \frac{\chi^2(2 \times 8 + 2)}{2 \times 12,500}$$

$$\chi^2_{0.9}(18) = 25.99, \qquad 0 \le \lambda \le 1.04 \times 10^{-3}$$

Two-sided confidence limits on λ are

$$\frac{\chi^2_{0.05}(2 \times 8)}{2 \times 12,500} \le \lambda \le \frac{\chi^2_{0.95}(2 \times 8 + 2)}{2 \times 12,500}$$

$$\chi^2_{0.05}(16) = 7.96, \qquad \text{and} \qquad \chi^2_{0.95}(18) = 28.87$$

Thus,

$$3.18 \times 10^{-4} \le \lambda \le 1.15 \times 10^{-3}$$

One-sided 90 percent confidence limits on $R(1000)$ are

$$\exp(-1.04 \times 10^{-3} \times 1000) \le R(1000) \le 1,$$

or

$$0.35 \le R(1000) \le 1$$

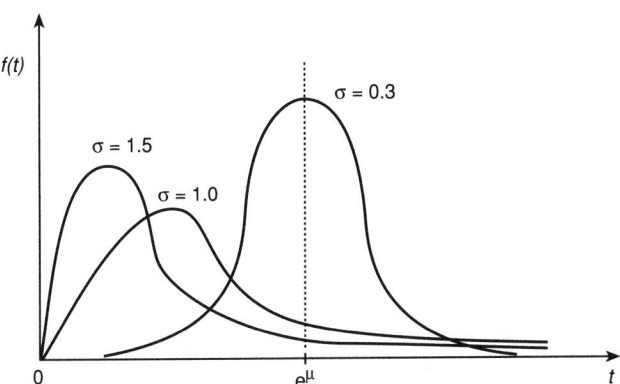

Fig. 5-4.6. Lognormal distribution.

Two-sided 90 percent confidence limits on $R(t)$ are

$$\exp(-1.15 \times 10^{-3} \times 1000) \leq R(1000)$$
$$\leq \exp(-3.18 \times 10^{-4} \times 1000),$$

or

$$0.32 \leq R(1000) \leq 0.73$$

Lognormal Distribution

A random variable is said to be lognormally distributed if its logarithm is normally distributed. The lognormal distribution has considerable application in engineering. One major application of this distribution is to present random variables that are the result of the product of many independent random variables.

The transformation $T = \ln Y$ or $Y = \exp(T)$ transfers the normal probability distribution function representing random variable T with mean μ_t and standard deviation σ_t to a lognormal probability distribution function.

$$f(y) = \frac{1}{\sigma_t y \sqrt{2\pi}} \exp\left[\frac{-1}{2\sigma_t^2}(\ln y - \mu_t)^2\right], \tag{17}$$
$$y \leq 0, \quad -\infty < \mu_t < \infty, \quad \sigma_t > 0$$

Figure 5-4.6 shows the probability distribution function of the lognormal distribution for different values of μ_t and σ_t.

A lognormal distribution is usually used to represent the occurrence of events in time. For example, a random variable representing the length of time required for repair of hardware follows a lognormal distribution. Because the lognormal distribution has two parameters, parameter estimation poses a more challenging problem than for the exponential distribution. It is easy to prove through the maximum likelihood estimation method that point estimators for the two parameters of the lognormal distribution can be obtained from

$$\hat{\mu}_t = \sum_{i=1}^{n} \frac{\ln(t_i)}{n}, \quad \text{and} \quad \hat{\sigma}_t^2 = \frac{\sum_{i=1}^{n}[\ln(t_i) - \hat{\mu}_t]^2}{n-1} \tag{18}$$

where t_i is the time that the ith failure in a set of n failures has occurred. The confidence interval for μ_t is

$$Pr\left[\chi^2_{\alpha/2}(2r) \leq \frac{2r\lambda}{\hat{\lambda}} \leq \chi^2_{1-\alpha/2}(2r)\right] = 1 - \alpha \tag{19}$$

Similarly, the confidence interval on σ_t is

$$Pr\left[\chi^2_{\alpha/2}(2r) \leq \frac{2r\lambda}{\hat{\lambda}} \leq \chi^2_{1-\alpha/2}(2r)\right] = 1 - \alpha \tag{20}$$

Weibull Distribution

This distribution is widely used to represent the *time to failure* or *life length* of the components in a system, measured from a start time to the time that a component fails. The continuous random variable T representing the time to failure follows a Weibull distribution if

$$f(t) = \frac{\beta(t-\gamma)^{\beta-1}}{\alpha^\beta} \exp\left[-\left(\frac{t-\gamma}{\alpha}\right)^\beta\right] \quad \alpha, \beta > 0; t > \gamma$$
$$= 0 \text{ otherwise} \tag{21}$$

The parameter γ serves only to reset the starting time and so is of limited value. The simplified variation of the Weibull distribution used here has $\gamma = 0$.

$$f(t) = \frac{\beta t^{\beta-1}}{\alpha^\beta} \exp\left[-\left(\frac{t}{\alpha}\right)^\beta\right], \quad t, \alpha, \beta, > 0$$
$$= 0 \text{ otherwise} \tag{22}$$

Figure 5-4.7 shows the Weibull distribution for various parameters of α and β. A careful inspection of these graphs reveals that the parameter β has a considerable effect on the shape of the distribution. Therefore, β is referred to as the *shape parameter*. The parameter α, on the other hand, controls the scales of the distribution. For this reason, α is referred to as the *scale parameter*. If $\beta = 1$, the Weibull distribution reduces to an exponential distribution with $\lambda = 1/\alpha$. For values of $\beta > 1$, the distribution becomes bell-shaped with some skew.

A Weibull distribution can be used for data believed to have an increasing, decreasing, or constant rate of failure. This distribution is a two-parameter distribution, and estimation of the parameters is rather involved. It can be shown that, under the situation where all n units under test or observation have failed, the maximum likelihood estimates of β and α parameters of the Weibull distribution can be obtained from

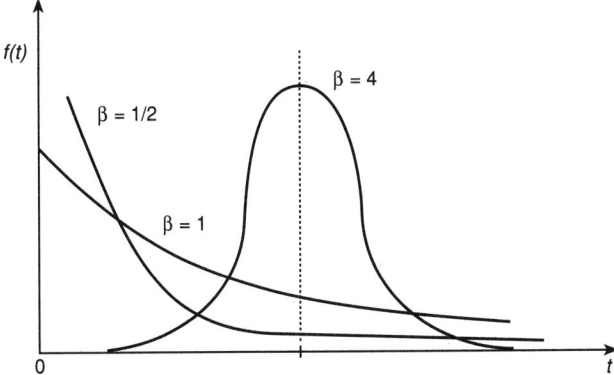

Fig. 5-4.7. Weibull distribution.

$$\frac{\sum_{i=1}^{n}(t_i)^{\hat{\beta}} \ln t_i}{\sum_{i=1}^{n}(t_i)^{\hat{\beta}}} - \frac{1}{\hat{\beta}} = \frac{1}{n}\sum_{i=1}^{n}\ln t_i,$$

$$\hat{\alpha} = \left(\frac{\sum_{i=1}^{n}(t_i)^{\hat{\beta}}}{n}\right)^{1/\hat{\beta}} \tag{23}$$

where t_i is the time that the ith failure in a set of n failures has occurred.

The solution of Equation 23 is not trivial and may require a trial and error process to obtain $\hat{\beta}$. Estimation of the confidence intervals of β and α is very involved. Readers are referred to Bain[3] and Mann et al.[4]

If a lot of failure data is available, the analyst should first fit the data into a distribution that best represents the data. The analyst should then proceed with the estimation of the distribution's parameters. However, when data are not enough, the analyst should check the adequacy of a distribution. Therefore, the exponential, lognormal, or other distribution should only be selected when the adequacy of a distribution fit can be justified. For example, when we have only five sample points for the times required for maintenance, one can theoretically fit these points to several types of distributions. However, we can assume that since the data occur in multiples of time, they can be adequately represented by a lognormal distribution. An accurate representation can only be determined if there are enough data to test the adequacy of the selected model.

SYSTEM RELIABILITY

Assessment of the reliability of a system from its basic elements is one of the most important aspects of reliability analysis. A system is a collection of items (subsystems, components, units, blocks, etc.) whose proper, coordinated function leads to the proper functioning of the system. In reliability analysis, it is therefore important to model the relationship between various items as well as the reliability of the individual items to determine the reliability of the system as a whole. Previous sections addressed the reliability analysis at a basic item level (one for which enough information is available to predict its reliability). This section addresses methods to model the relationship between system components, which allows the determination of overall system reliability.

The physical configuration of an item that belongs to a system is often used to model system reliability. In some cases, the manner in which an item fails is important for system failure, and should be considered in the system reliability analysis. For example, in a system composed of two parallel electronic units, if a unit fails, the system will fail; but for most other types of failures of the unit, the system will still be functional, since the other unit functions properly.

There are several system modeling schemes for reliability analysis. *Reliability block diagram* methods include series, parallel, standby, shared load, and complex systems. *Logic tree* methods include fault tree and success tree methods, which include the method of construction and evaluation of the tree; event tree method, which includes modeling of multisystem designs and complex systems whose individual units should work in a chronological or approxi-

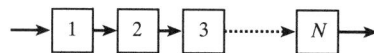

Fig. 5-4.8. Series system reliability block diagram.

mately chronological manner to achieve a mission; failure mode and effect analysis; and the master logic diagram (MLD).

Reliability Block Diagram Methods

Reliability block diagrams are frequently used to model the effect of item failures (or functioning) on system performance. They often correspond to the physical arrangement of items in the system. However, in certain cases, the arrangement may be different. For instance, when two resistors are in parallel, the system fails if one fails short. Therefore, the reliability block diagram of this system for the "fail short" mode of failure would be composed of two series blocks. However, for other modes of failure of one unit, such as "open" failure mode, the reliability block diagram is composed of two parallel blocks.

Series systems: A reliability block diagram is in a series configuration when failure of any one item (according to the failure mode of each item, on which the reliability block diagram is based) results in the failure of the system. Accordingly, for functional success of a series system, all of its items must successfully function during the intended mission time of the system. Figure 5-4.8 shows the reliability block diagram of a series system consisting of N units.

The reliability of the system in Figure 5-4.8 is the probability that all N units succeed during its intended mission time, t. Thus, probabilistically, the system reliability, $R_s(t)$ for independent units is obtained from

$$R_s(t) = R_1(t) \cdot R_2(t) \cdot \ldots R_N(t) = \prod_{i=1}^{N} R_i(t) \tag{24}$$

where $R_i(t)$ represents the reliability of the ith unit. The hazard rate (instantaneous failure rate) for a series system is also a convenient expression. The hazard rate of the system, $h_s(t)$, is

$$h_s(t) = \sum_{i=1}^{N} h_i(t) \tag{25}$$

Assume a constant hazard rate model for each unit (e.g., assume an exponential time to failure for each unit). Thus, $h_i(t) = \lambda_i$. According to Equation 24, the system constant rate of failure is

$$\lambda_s = \sum_{i=1}^{N} \lambda_i \tag{26}$$

Equation 26 can also be easily obtained from Equation 24 by using the constant failure rate reliability model for each unit, $R_i(t) = \exp(-\lambda_i t)$.

$$R_s(t) = \prod_{i=1}^{N} \exp(-\lambda_i t) = \exp\left(-\sum_{i=1}^{N} \lambda_i\right)t \tag{27}$$

Accordingly, the MTTF of the system (MTTF$_s$) can be obtained from

$$\text{MTTF}_s = \frac{1}{\lambda_s} = \frac{1}{\sum_{i=1}^{N} \lambda_i} \tag{28}$$

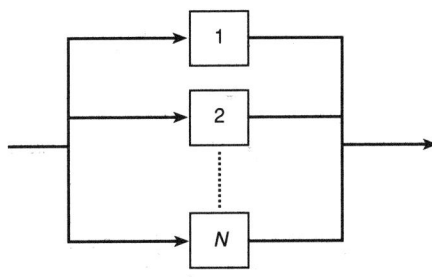

Fig. 5-4.9. *Parallel system block diagram.*

EXAMPLE 4:

A system consists of three units whose reliability block diagram is in a series. The failure rate for each unit is constant, as follows: $\lambda_1 = 4.0 \times 10^{-6}\text{hr}^{-1}$, $\lambda_2 = 3.2 \times 10^{-6}\text{hr}^{-1}$, $\lambda_3 = 9.8 \times 10^{-6}\text{hr}^{-1}$. Determine the following parameters of the system:

1. λ_s
2. R_s (1000 hrs)
3. MTTF_s.

SOLUTION:

1. According to Equation 26, $\lambda_s = 4.0 \times 10^{-6} + 3.2 \times 10^{-6} + 9.8 \times 10^{-6}$
$= 1.7 \times 10^{-5}\text{hr}^{-1}$
2. $R_s(t) = \exp(-\lambda_s t) = \exp(-1.7 \times 10^{-5} \times 1000) = 0.983$, or unreliability of $F_i(1000) = 0.017$.
3. According to Equation 28, $\text{MTTF}_s = 1/\lambda_s = 1/1.7 \times 10^{-5}$
$= 58,823.5$ hrs.

Parallel systems: A reliability block diagram is in a parallel configuration when the failure of all units in the system results in system failure. Accordingly, for a parallel system, success of only one unit would be sufficient to guarantee the success of the system. Figure 5-4.9 shows a parallel system consisting of N units.

According to the definition of a parallel system, failure of all units results in the failure of the system. Thus, for a set of N independent units,

$$F_s(t) = F_1(t) \cdot F_2(t) \cdot \ldots F_N(t) = \prod_{i=1}^{N} F_i(t) \qquad (29)$$

Since $R_i(t) = 1 - F_i(t)$ [i.e., $F_i(t)$ is the unreliability of the units], then

$$R_s(t) = 1 - F_s(t) = 1 - \prod_{i=1}^{N}[1 - R_i(t)] \qquad (30)$$

The hazard rate can be determined by using $h(t) = -d\ln R_s(t)/dt$. The resulting form of $h(t)$ is rather complex.

To address various characteristics of system reliability, consider a special case where the instantaneous failure rate is constant for each unit (exponential time to failure model), and the system is composed of only two units. Since $R_i(t) = \exp(-\lambda_i t)$, then according to Equation 30,

$$R_s(t) = 1 - [1 - \exp(-\lambda_1 t)][1 - \exp(-\lambda_2 t)]$$

$$= \exp(-\lambda_1 t) + \exp(-\lambda_2 t) - \exp[-(\lambda_1 + \lambda_2)t] \qquad (31)$$

Since $h_s(t) = f_s(t)/R_s(t)$ and $f_s(t) = -d[R_s(t)]/dt$, then using Equation 31,

$$h_s(t) = \frac{f_s(t)}{R_s(t)} = \frac{-d[R_s(t)]/dt}{R_s(t)}$$

$$= \frac{\lambda_1\exp(-\lambda_1 t) + \lambda_2\exp(-\lambda_2 t) - (\lambda_1 + \lambda_2)\exp[-(\lambda_1 + \lambda_2)t]}{\exp(-\lambda_1 t) + \exp(-\lambda_2 t) - \exp[-(\lambda_1 + \lambda_2)t]}$$

The MTTF of the system can also be obtained from

$$\text{MTTF}_s = \int_0^{\infty} R_s(t)\, dt$$

$$= \int_0^{\infty} \{\exp(-\lambda_1 t) + \exp(-\lambda_2 t)$$

$$- \exp[-(\lambda_1 + \lambda_2)t]\}\, dt$$

$$= \frac{1}{\lambda_1} + \frac{1}{\lambda_2} - \frac{1}{\lambda_1 + \lambda_2}.$$

Accordingly, one can use the binomial expansion to derive the MTTF for N parallel units

$$\text{MTTF}_s = \left(\frac{1}{\lambda_1} + \frac{1}{\lambda_2} + \cdots \frac{1}{\lambda_N}\right)$$
$$- \left(\frac{1}{\lambda_1 + \lambda_2} + \frac{1}{\lambda_1 + \lambda_3} + \cdots + \frac{1}{\lambda_{N-1} + \lambda_N}\right)$$
$$+ \left(\frac{1}{\lambda_1 + \lambda_2 + \lambda_3} + \cdots + \frac{1}{\lambda_{N-2} + \lambda_{N-1} + \lambda_N}\right) \cdots$$
$$+ (-1)^{N+1}\frac{1}{\lambda_1 + \lambda_2 + \cdots + \lambda_N} \qquad (32)$$

In the special case where all units are identical with a constant failure rate λ (e.g., in an active redundant system), Equation 30 simplifies to the following form

$$R_s(t) = 1 - [1 - \exp(-\lambda t)]^N \qquad (33)$$

and from Equation 32,

$$\text{MTTF}_s = \text{MTTF}\left(1 + \frac{1}{2} + \cdots + \frac{1}{N}\right). \qquad (34)$$

It can be seen from Equation 34 that, in the design of active redundant systems, the MTTF_s exceeds the MTTF of an individual unit. However, the contribution to the MTTF_s from the second unit, the third unit, and so on would have a diminishing return as N increases. That is, there would be an optimum number of parallel units by which a designer can balance the reliability and the cost of the component in its life cycle.

Consider a more general form of a series and parallel system—the so-called K-out-of-N system. In this type of system, if any combination of K units out of N independent units work, it guarantees the success of the system. For simplicity, assume that all units are identical (which, by the way, is often the case). The binomial distribution can easily represent the probability that the system functions

$$R_s(t) = \sum_{r=K}^{N}\binom{N}{r}[R(t)]^r[-R(t)]^{N-r}$$

$$= 1 - \sum_{r=0}^{K-1}\binom{N}{r}[R(t)]^r[-R(t)]^{N-r} \qquad (35)$$

EXAMPLE 5:

A system is composed of the same units as in Example 4. However, these units are in parallel. Find the unreliability and MTTFs of the system.

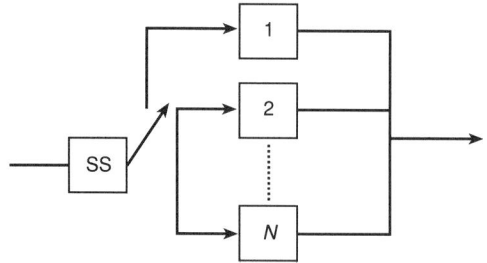

Fig. 5-4.10.   Standby redundant system.

SOLUTION:

According to Equation 30,

$$R_s(t) = 1 - (1 - e^{-\lambda_1 t})(1 - e^{-\lambda_2 t})(1 - e^{\lambda_3 t})$$
$$F_s(1000) = 1 - R_s(1000)$$
$$= 1 - (1 - e^{-4.0 \times 10^{-6} \times 1000})$$
$$\cdot (1 - e^{-3.2 \times 10^{-6} \times 1000})(1 - e^{-9.8 \times 10 - 6 \times 1000})$$
$$= 1.25 \times 10^{-7}$$

$$\mathrm{MTTF}_s = \left(\frac{1}{\lambda_1} + \frac{1}{\lambda_2} + \frac{1}{\lambda_3} \right)$$
$$- \left(\frac{1}{\lambda_1 + \lambda_2} + \frac{1}{\lambda_1 + \lambda_3} + \frac{1}{\lambda_2 + \lambda_3} \right)$$
$$+ \frac{1}{\lambda_1 + \lambda_2 + \lambda_3} = 4.35 \times 10^5 \text{ hrs}$$

EXAMPLE 6:

How many components should be used in an active redundancy design to achieve a reliability of 0.999 such that, for successful system operation, a minimum of two components is required? Assume a mission of $t = 720$ hrs for a set of components that are identical and have a failure rate of 0.00015/hr.

SOLUTION:

For each component $R(t) = \exp(-\lambda t) = \exp(-0.00015 \times 720) = 0.8976$. According to Equation 35,

$$0.999 = R_s(t) = 1 - \sum_{r=0}^{1} \binom{N}{r}(0.8976)^r(0.1024)^{N-r}$$
$$= 1 - (0.1024)^N - N(0.8976)(0.1024)^{N-1}$$

From the above equation, $N = 5$, which means that at least five components should be used to achieve the desired reliability over the specified mission time.

Standby redundant systems: A system is called a standby redundant system when some of its units remain idle until they are called for service by a sensing and switching device. For simplicity, consider a situation where only one unit operates actively and the others are in standby, as shown in Figure 5-4.10.

In Figure 5-4.10, unit 1 operates constantly until it fails. The sensing and switching device recognizes a unit failure in the system and switches to another unit. This process continues until all standby units have failed, in which case the system is considered failed. Since units 2 to N do not operate constantly (as is the case in active parallel systems), one would expect them to fail at a much slower rate. This is because the failure rate for components is usually lower when the components are operating than when they are idle or dormant.

It is clear that system reliability is totally dependent on the reliability of the sensing and switching device. The reliability of a redundant standby system is the reliability of unit 1 over the mission time, t (i.e., the probability that it succeeds the whole mission time), plus the probability that unit 1 fails at time t_1 prior to t and the probability that the sensing and switching unit does not fail by t_1 and the probability that standby unit 2 does not fail by t_1 (in the standby mode) and the probability that standby unit 2 successfully functions for the remainder of the mission in an active operation mode, and so on.

Mathematically, the reliability function for a two-unit standby device according to this definition can be obtained from

$$R_s(t) = R_1(t) + \int_0^t f_1(t_1)\,dt_1 \cdot R_{ss}(t_1) \cdot R_2'(t_1) \cdot R_2(t - t_1) \quad (36)$$

where $f_1(t)$ is the probability distribution function for the time to failure of unit 1, $R_{ss}(t_1)$ is the reliability of the sensing and switching device, $R_2'(t)$ is the reliability of unit 2 in the standby mode of operation, and $R_2(t - t_1)$ is the reliability of unit 2 after it started to operate at time t_1.

If all units are identical with perfect switching, then the reliability of the system is

$$R_s(t) = \exp(-\lambda t) + \lambda t \exp(-\lambda t) = (1 + \lambda t) \exp(-\lambda t) \quad (37)$$

Shared load systems: A shared load system refers to a parallel system whose units equally share the system function. For example, if a set of two parallel pumps delivers x gpm of water to a reservoir, each pump delivers $x/2$ gpm. If a minimum of x gpm is required at all times, and one of the pumps fails at a given time t_1, then the other pump's speed should be increased to provide x gpm alone. Other examples of load sharing are multiple load-bearing units (such as those in a bridge), and load-sharing multi-unit electric power plants. In these cases, when one of the units fails, the others should carry its load. Since these other units would then be working under a more stressful condition, they would experience a higher rate of failure.

Assume that two units share a load (i.e., each unit carries half the load), and the time-to-failure distribution for both units to work is $f_h(t)$. When one unit fails (i.e., one unit carries the full load), the time-to-failure distribution is $f_f(t)$. Further, call the corresponding reliability functions during full-load and half-load operation $R_f(t)$ and $R_h(t)$, respectively. The system will succeed if both units carry half the load, or if unit 1 fails at time t_1 and unit 2 carries a full load thereafter, or if unit 2 fails at time t_1 and unit 1 carries the full load thereafter. Accordingly, the system reliability function, $R_s(t)$, can be obtained from

$$R_s(t) = [R_h(t)]^2 + 2\int_0^t f_h(t_1)\,R_h(t_1)\,R_f(t - t_1)\,dt_1. \quad (38)$$

In Equation 38, the first term shows the contribution from both units working successfully, with each carrying a half load; the second term represents the two equal probabilities that unit 1 fails first and unit 2 takes the full load at time t_1, or *vice versa*.

If there are switching or control mechanisms involved to shift the total load to the unfailed unit when one fails,

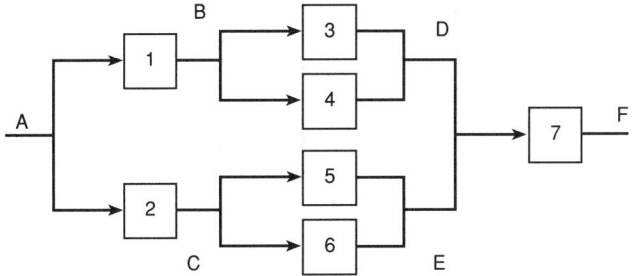

Fig. 5-4.11. Complex parallel-series system.

then similar to Equation 36, the reliability of the switching mechanism can be incorporated into Equation 38.

In the special situation where an exponential time-to-failure model with failure rates λ_f and λ_h can be used for the two units for full-and half-load cases, respectively, then Equation 38 can be simplified to

$$R_s(t) = \exp(-2\lambda_h t)$$
$$+ \frac{2\lambda_h \exp(-\lambda_f t)}{(2\lambda_h - \lambda_f)} \exp\left[-(2\lambda_h - \lambda_f)t\right] \tag{39}$$

Complex systems: Most practical systems are not parallel or series, but exhibit some hybrid combination of the two. These systems are often referred to as *parallel-series systems*. Figure 5-4.11 shows an example of such a system.

A parallel-series system can be analyzed by dividing it into its basic parallel and series modules, and then determining the reliability function for each module separately. The process can be continued until a reliability function for the whole system is determined. Another type of complex system is one that is neither series nor parallel alone, nor parallel-series. Figure 5-4.12 shows an example of such a system.

For the analysis of all types of complex systems, Shooman[5] describes several analytical methods for complex systems. These are the *inspection method*, *event space method*, *path-tracing method*, and *decomposition*. These methods are good only when there are not a lot of units in the system. For analysis of large numbers of units, fault trees would be more appropriate.

A computationally intensive method for determining the reliability of a complex system involves the use of *path set* and *cut set* methods. A *path set* (or tie set) is a set of units that include a connection between input and output when traversed in the direction of the reliability block diagram arrows. Thus, a path set provides a "path" through the graph. A minimal path set (or minimal tie set) is a path set that would not provide a connection between the input and output points if any of its units were removed. For example, in Figure 5-4.12, path set $P_1 = (1,3)$ is a minimal path set; but $P_2 = (1,3,6)$ is not, since units 1 and 3 are sufficient to guarantee a path.

A *cut set* is a set of units that interrupts all possible connections between the input and output points. A *minimal cut set* is the smallest set of units needed to guarantee an interruption of flow. In practice, minimal cut sets show a combination of unit failures that cause a system to fail. For example, in Figure 5-4.12, the minimal path sets are: $P_1 = (2), P_2 = (1,3), P_3 = (1,4,7), P_4 = (1,5,8), P_5 = (1,4,6,8)$, and $P_6 = (1,5,6,7)$. The minimal cut sets are: $C_1 = (1,2), C_2 = (4,5,3,2), C_3 = (7,8,3,2), C_4 = (4,6,8,3,2)$, and $C_5 = (5,6,$

7,3,2). If a system has m minimal path sets denoted by $P_1, P_2, \ldots, P_m$, then the system reliability is given by

$$R_s(t) = Pr(P_1 \cup P_2 \cup \cdots \cup P_m) \tag{40}$$

where each path set P_i represents the event that units in the path set survive during the mission time, t. This guarantees the success of the system. Since many path sets may exist, the union of all these sets gives all possible events for successful operation of the system. The probability of this union clearly represents the reliability of the system. It should be noted here that, in practice, the path sets, P_is, are not disjoint. This poses a problem for determining the left-hand side of Equation 40. However, a useful upper bound on the system reliability may be obtained by assuming that the P_is are highly disjoint. Thus,

$$R_s(t) \le Pr(P_1) + Pr(P_2) + \cdots Pr(P_m) \tag{41}$$

Expression 41 yields better answers when using small reliability values. Since this is not usually the case, Equation 41 is not a good bound for use in practical applications.

Similarly, system reliability can be determined through minimal cut sets. If the system has n minimal cut sets denoted by $C_1, C_2, \ldots, C_n$, then the system reliability is obtained from

$$R_s(t) = 1 - Pr(C_1 \cup C_2 \cup \cdots \cup C_n) \tag{42}$$

where C_i represents the event that units in the cut set fail sometime before the mission time, t. This guarantees system failure. The Pr($\bullet$) term on the right-hand side of Equation 42 shows the probability that at least one of all possible minimal cut sets exists before time t. Thus, it represents the probability that the system fails sometimes before t. By subtracting this probability from 1, the reliability of the system is obtained. Similar to path sets, cut sets are not usually disjoint. Again, Equation 42 can be written in the form of its lower bound, which is a much simpler expression given by

$$R_s(t) \ge 1 - [Pr(C_1) + Pr(C_2) + \cdots Pr(C_n)] \tag{43}$$

Notice that each element of a path set represents the *success* of a unit, whereas each element of a cut set represents the *failure* of a unit. Thus, for probabilistic evaluations, the reliability function of each unit should be used in connection with path set evaluations, i.e., Equation 41, while the unreliability function should be used in connection with cut set evaluations, i.e., Equation 43.

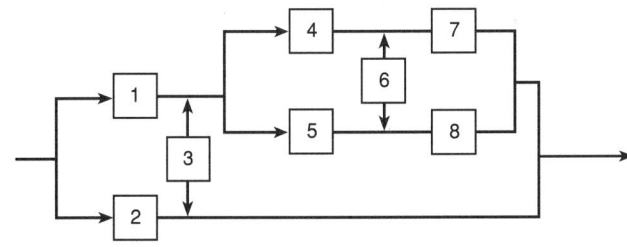

Fig. 5-4.12. Complex nonparallel-series system.

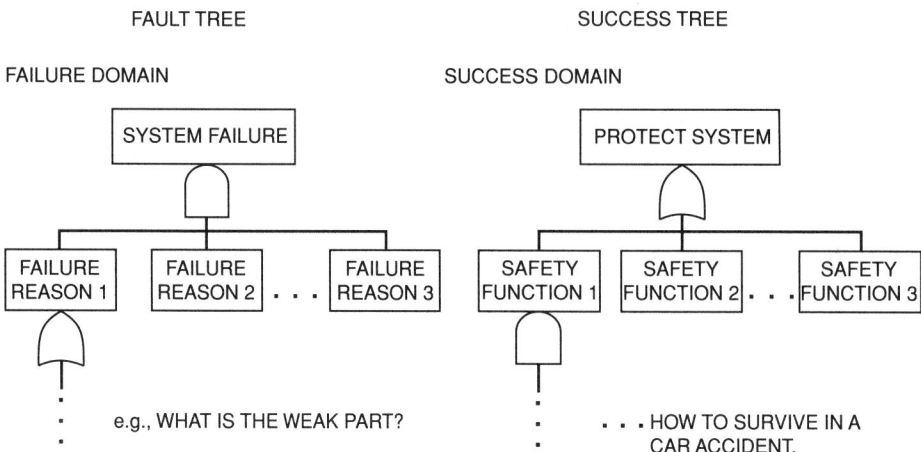

Fig. 5-4.13. *Fault tree* vs *success tree.*

The bounding technique used in Equation 43, in practice, yields a much better representation of the reliability of the system than Equation 41, because most engineering units have reliabilities greater than 0.9 over their mission time, making the use of Equation 43 appropriate.

In cases that deal with very complex systems that have multiple failure modes for each unit and complex physical and operational interactions, the use of reliability block diagrams is difficult. The method of fault tree and success tree analysis is more appropriate in this context, especially if the role of humans in the operation of the system needs to be modeled.

Logic Tree Methods

Fault tree (FT), success tree (ST), and master plant logic diagram (MLD) are three of the most popular logic tree methods. This section discusses the basic concepts of these logic tree methods by pointing out the differences between them. A simple example (a fire protection system) is discussed for illustration.

Fault tree and success tree: The fault tree (the success tree) is a deductive process that has a tree-like hierarchical structure. In such a tree-like structure, the basic underlying principles of a complex system are represented. Each node in the tree represents an event (a goal) or a function. As shown in Figure 5-4.13, the top event of "system failure" or "system success" can be decomposed to subevents, subgoals, or subfunctions that have, in turn, their own subevents, subgoals, or subfunctions, and so on, until some lowest level of elementary components is reached. A set of symbols similar to the ones shown in Figure 5-4.14 is used to develop a logic tree.

Logically, it must be recognized that logic trees can be developed equally well in success or failure space, so that the tree analysis can proceed with either success or failure orientation. However, experience has shown that success orientation is preferable, since it tends to lead to better and more succinct definition of system functions and operation.

However, when one is attempting to understand the failure characteristics of a piece of equipment or an industrial process, fault descriptions are more easily grasped than the success space descriptions of the system, because the process is typically being viewed externally, from the perspective that a failure has already occurred. This means

that the person is likely to be looking at the system in a deductive manner and tries to find in what way such a failure may occur.

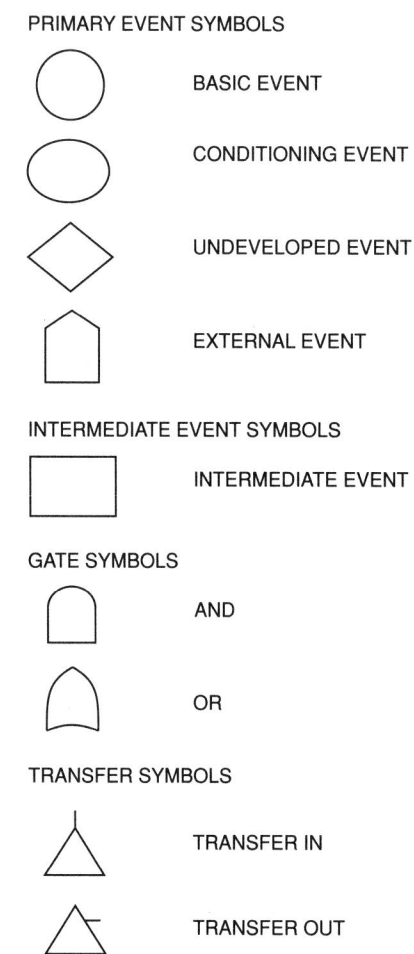

Fig. 5-4.14. *Primary event, gate, and transfer symbols.*

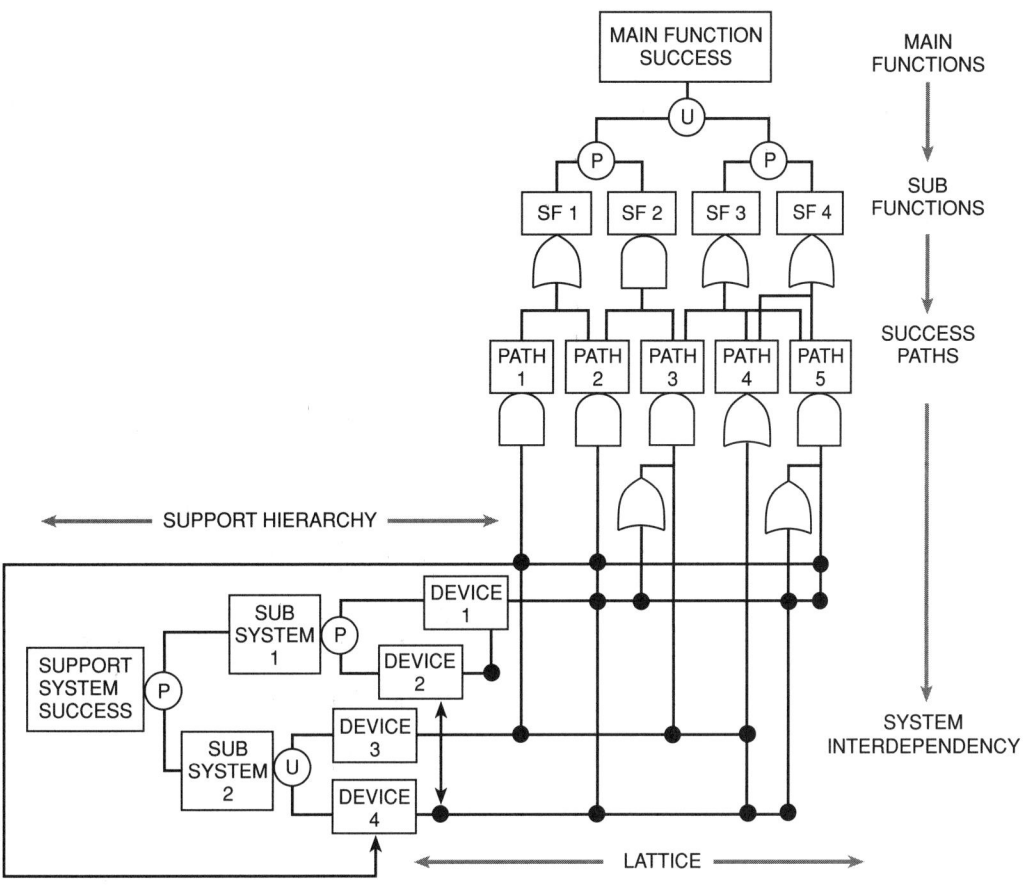

Fig. 5-4.15. Master logic diagram.

For a more detailed discussion of the construction and evaluation of fault trees, refer to the *Fault Tree Handbook*. The Goal Tree Success Tree structure was initially developed as part of a DOE sponsored research[7] to assess and improve information systems in a nuclear plant. Subsequently, it has been further developed and extended to a number of other applications.[8,9,10]

Master plant logic diagram (MLD): The hardware or human activities in the success tree often are associated with other support equipment, and since more often than not challenges to systems or processes originate from other support equipment, it is important to include these relationships in the success tree. A single visual display of the plant logical relations is represented by an MLD model.[6] This logic diagram shows all of the interrelationships among the so-called front-line systems and support systems. Front-line systems are those that perform main functions of the system or process. Support systems are those that either cool, actuate, power, lubricate, or control the front-line systems. As shown in Figure 5-4.15, the MLD diagram is also developed hierarchically in a top-down manner. Therefore, most of the characteristics of the success tree are equally applicable to this diagram. A computer software called REVEAL_W™ has been developed to perform the MLD modeling.

EXAMPLE 7:

Consider the fire protection system shown in Figure 5-4.16. This system is designed to extinguish all possible fires in a plant with toxic chemicals. Two physically independent water extinguishing nozzles are designed such that each is capable of controlling all types of fires in the plant. Extinguishing nozzle 1 is the primary method of injection. Upon receiving a signal from the detector/alarm/actuator device, pump 1 starts automatically, drawing water from the reservoir tank and injecting it into the fire area in the plant. If this pump injection path is not actuated, plant operators can start a second injection path manually. If the second path is not available, the operators will call for help from the local fire department, although the detector also sends a signal directly to the fire department. However, due to the delay in the arrival of the local fire department, the magnitude of damage would be higher than it would be if the local fire extinguishing nozzles were available to extinguish the fire. Under all conditions, if the normal off-site power is not available due to the fire or other reasons, a local generator would provide electric power to the pumps. The power to the detector-alarm-actuator system is provided through batteries, which are constantly charged by the off-site power. Even if the ac power is not available, the dc power provided through the battery is expected to be available at all times.

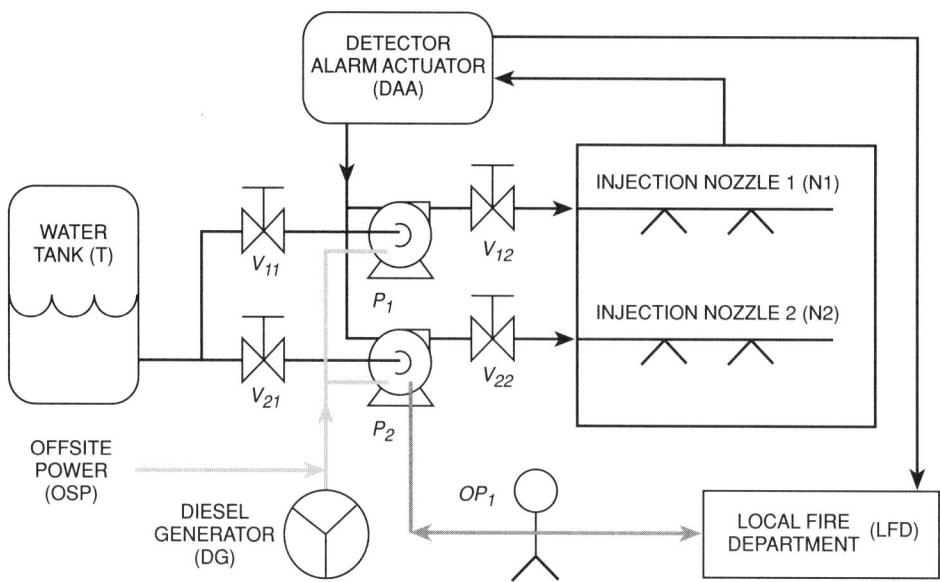

Fig. 5-4.16.   A fire protection system.

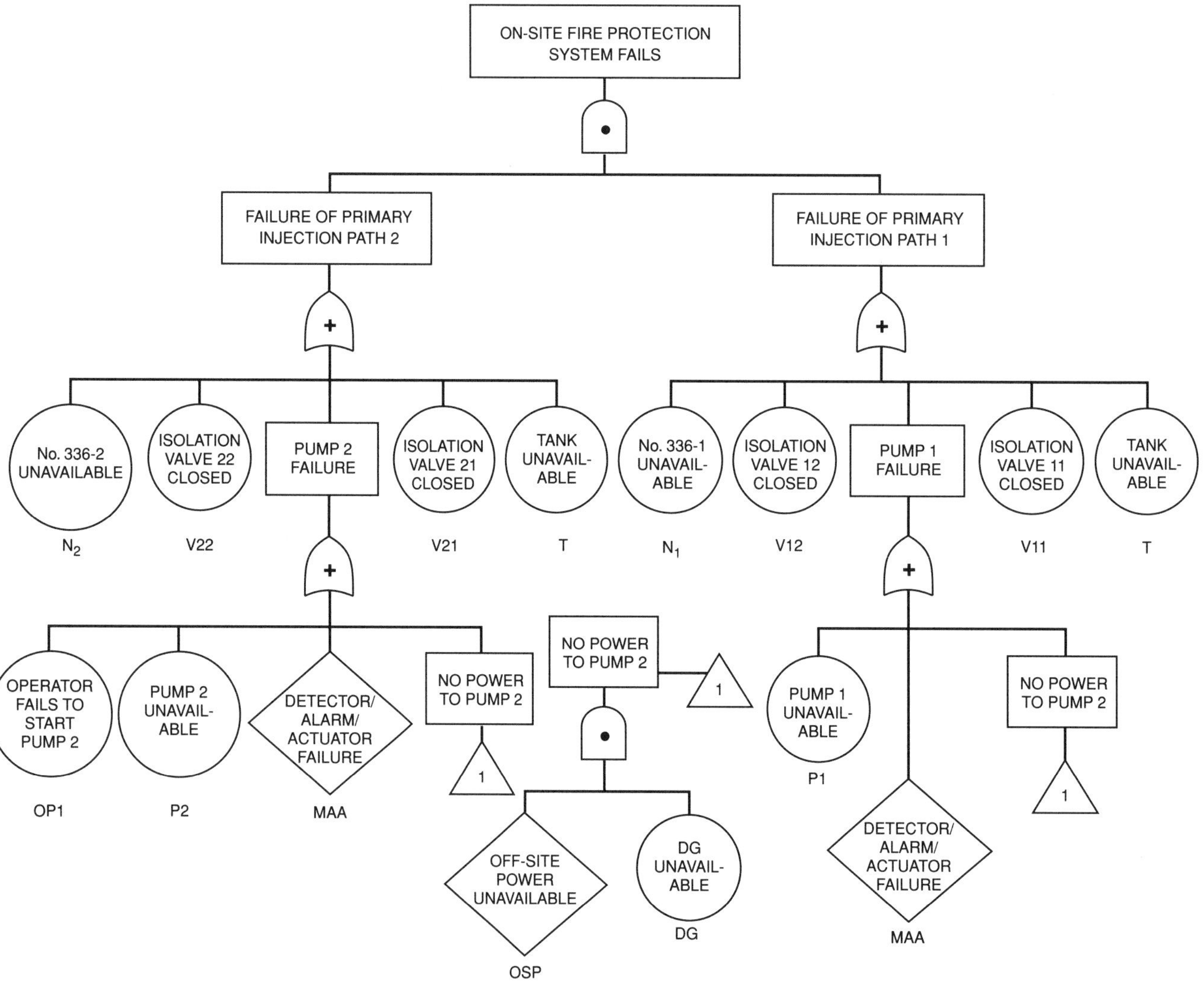

Fig. 5-4.17.   Fault tree for on-site fire protection system failure.

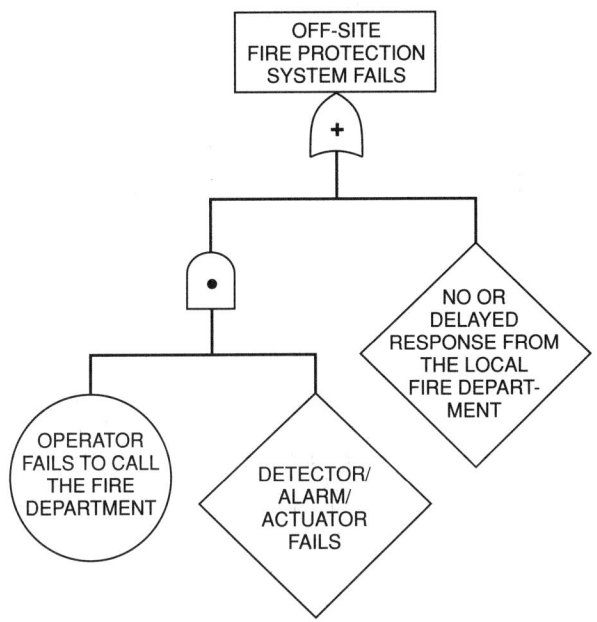

Fig. 5-4.18. *Fault tree for off-site fire protection system failure.*

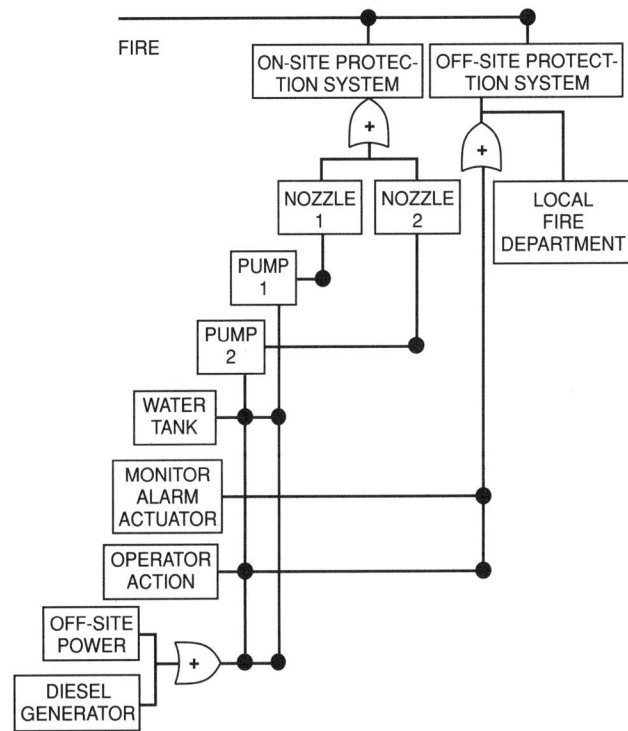

Fig. 5-4.19. *Master logic diagram for fire protection system.*

The manual valves on the two sides of pump 1 and pump 2 are normally open, and only remain closed when they are being repaired. The entire fire system and generator are located outside of the reactor compartment, and are therefore not affected by an internal fire.

SOLUTION:

For reliability analysis, one should explain the cause and effect relationship between the fire and the progression of events following the fire, and should identify all failures (equipment or human) that lead to failure of the event-tree headings (on-site or off-site protective measures).

For example, Figure 5-4.17 shows the fault tree developed for the on-site fire protection system failure. In this fault tree, all basic events that lead to the failure of the two independent paths are described. Note that MAA, electric power to the pumps, and the water tank are shared by the two paths. Clearly these are considered physical dependencies. This is taken into account in the quantification step of the risk analysis. In this tree, all external event failures and passive failures are neglected.

Figure 5-4.18 shows the fault tree for the off-site fire protection system failure. This tree is simple, since it only includes failures that do not lead to an on-time response from the local fire department. It is also possible to use the MLD for system analysis. An example of the MLD for this problem is shown in Figure 5-4.19.

REFERENCES CITED

1. B. Epstein, "Estimation from Life Test Data," *Technometrics*, 2, p. 447 (1960).
2. E.L. Welker and M. Lipow, "Estimating the Exponential Failure Rate Dormant Data with No Failure Events," *Proc. Rel. Maint. Sysmp.*, Vol. 1 (2), p. 1194 (1974).
3. L.J. Bain, *Statistical Analysis of Reliability and Life-Testing Models: Theory and Methods*, Marcel Dekker, New York (1978).
4. N. Mann, R.E. Schafer, and N.D. Singpurwalla, *Methods for Statistical Analysis of Reliability and Life Data*, John Wiley and Sons, New York (1974).
5. M.L. Shooman, *Probabilistic Reliability: An Engineering Approach*, 2nd ed., Kreiger, Melbourne, FL (1990).
6. R.N. Hunt and M. Modarres, "Performing a Plant-Specific PRA by Hand—A Practical Reality," presented at the 14th INTER-RAM Conf., Minneapolis (1987).
7. R. N. Hunt and M. Modarres. "Integrated Economic Risk Management in a Nuclear Plant," *Annual Meeting of the Society for Risk Analysis*, October 1984 .
8. I. Kim and M. Modarres, "Goal Tree–Success Tree Model as the Knowledge-Base of Operator Advisory Systems," *Topical Meeting on Artificial Intelligence and Other Innovative Computer Applications in Nuclear Industry*, Snowbird, Utah (1987).
9. L-W. Chen and M. Modarres, "A Hierarchical Decision Process for Fault Administration," *Computers and Chemical Eng. J.*, Vol. 16, No. 5: 425–448 (1992).
10. D. Sebo, D. Marksberry, and M. Modarres, "RSAS: A Reactor Safety Assessment System," *Proc. of the 7th Power Plant Dynamics, Control and Testing Symposium*, Knoxville, TN (1989).

ADDITIONAL READING

U.S. NRC, *Fault Tree Handbook*, US Naval Research Corporation, (1981).

Y-S. Hu, M. Modarres, and D. Marksberry, "A Knowledge-Based System Approach to Reactor Safety Assessment," Submitted for publication to *Reliability Eng. and Sys. Safety J.*

SUBJECTIVE MEASUREMENT IN FIRE PROTECTION ENGINEERING

F.J. Dodd and H.A. Donegan

INTRODUCTION

Rational solutions for many of today's fire safety decision-making problems are difficult to formulate. This is often due not only to the complexities of the problems themselves but also to the vagueness (also referred to as *fuzziness*) of the concepts involved. Despite the power of modern computing technology, the complexity factor can be handled only by the introduction of a structure into the elements of the problem. This chapter: (1) reviews some principal methods of dealing with the vagueness inherent in the elements of many complex problems and (2) offers definitions and approaches applicable to rational solutions.

Judgments must be made on individual fire safety attributes to enable an overall assessment. These judgments, based on local-level criteria, should be consistent if the final conclusion is to have any validity or credibility.

MEASUREMENT AND MEASURE THEORY

Measurement is defined as the assignation of a number to the entity under consideration in accordance with a rule; this number then reflects the measured property of the entity.[1] In mathematical terminology, the measurement is the image under a *monotonic*, i.e., order preserving, mapping of the dimensional property into a real line.

Measurement of the physical entities encountered in experimental work presents a large number of fundamental problems that must be recognized by researchers. For example, the *precision* of the result cannot exceed the precision of *any* of the data used in its calculation. Thus, a result to ten significant digits is pointless when even one item of the contributory data is approximated to two significant figures. This is why 10.000 is not the same as 10; the former indicates that the measuring equipment allowed for measurements of three decimal places and the "correct" result is therefore between 9.9995 and 10.0005, whereas the latter shows much less precision, i.e., 9.5 to 10.5.

Further allowance must be made for experimental error. Many external factors (e.g., traffic outside the laboratory) can introduce random elements into the measuring process. These factors, however small, must be taken into account in the result. Systematic errors, caused by a bias in the measurement process, assume an even greater significance with the introduction of opinions, which can compromise the work and should be eliminated. The opinion problem is present in most questionnaire design. These opinions, or psychological components where linearity is less measureable can jeopardize experimental results.

SCALING

The four types of scale most relevant to fire protection engineering are: (1) *nominal*, (2) *ordinal*, (3) *interval*, and (4) *ratio*.[2] Each scale, in order, represents progressively stronger properties.

Nominal Scale

The values on a *nominal scale* classify or categorize the objects represented: i.e., no information on size is implied. For example, the numbers identifying food items on a menu or the room numbers in an office block might be regarded as nominal scales. The property common to these is the *category property*. Many coding systems have this; for example, the computer system of a car dealer might store car type as a single digit: 1 for Ford, 2 for General Motors, 3 for Lincoln, 4 for Buick, etc. Such a digit does not reflect the size or power of the car; it merely indicates the manufacturer. Similarly, fire extinguishers might be classified by type: 1—water, 2—CO_2, 3—halon, etc. However, the numbers 1 to 3 have no implications with regard to quality.

Ordinal Scale

The *ordinal scale* not only characterizes the entities represented but also ranks or orders them. However, proportion is not necessarily maintained. This type of scale is exemplified by examination grades: in a descending scale, a grade 2 is better than a grade 3, but one cannot say by how much. Glenburg[2] offers the stripes on military personnel as an ordinal scale: a sergeant (3 stripes) is "superior" to a corporal (2 stripes), but it is nonsense to argue that the

Mr. Donegan is on the Faculty of Informatics at the University of Ulster, Northern Ireland. Mr. Dodd is in Bangor, County Down, Northern Ireland.

sergeant is one and a half times as authoritative. Ordinal scale examples relating to characteristics of smoke and heat are available.[3]

Interval Scale

More mathematically tractable and, hence, of greater importance in quantitative assessments are numbers on an *interval scale*. This is a continuous scale between two points, e.g., from 0 to 10, or 1 to 5. Relative difference is maintained, e.g., the difference between 1 and 2 is exactly the same as the difference between 4 and 5. An example of an interval scale is a ruler or similarly graduated device. Celsius and Fahrenheit temperature scales are interval, and the *equal intervals* or *relative difference property* is essential when one scale is converted to the other. Arithmetic-based operations involving addition, subtraction, and multiplication by a scalar (a number with no units) can be used on numbers from an interval scale.

The Likert scale is of particular interest in the type of work associated with fire protection engineering. Researchers assign a number on a scale, e.g., 0 to 5, or 0 to 10, to an issue that they consider reflects the significance of that issue. Scheibe *et al*[4] show how such scales have the equal difference property and are, therefore, subject to combinatorial calculations.

Ratio Scale

The *ratio scale* is an interval scale with the *absolute zero property*; i.e., one end is fixed so that the values on it are absolute rather than relative. For example, the Kelvin (K) scale relates temperatures to absolute zero temperature, at which all motion ceases.

Scale Application

Each type of scale has a specific application. One should not confuse scales that have no direct relationship. Further, the function of the scale must also be taken into account.

MULTI-ATTRIBUTE UTILITY ANALYSIS

Fire safety application analyses require more than one attribute to capture all relevant aspects of the decision consequences. If the attributes for a decision problem are $x_1, x_2, x_3, \ldots, x_n$, then a utility function, $u(x_1, x_2, x_3, \ldots, x_n)$, needs to be determined over these measures in order to conduct a decision analysis. Utility functions are elicited by asking questions of a person; and it is difficult to apply answers that impact on the n-dimensional function. Once this is done, an expected utility analysis can be conducted to identify the preferred alternative. Determining utility functions over multiple evaluation measures is a complex problem.

It has been shown that if tradeoffs among the attributes in any subset of $x_1, x_2, x_3, \ldots, x_n$ do not depend on the levels of the remaining attributes, a single measure of the outcome is given by

$$u(x_1, x_2, x_3, \ldots, x_n) = \sum_{i=1}^{n} \lambda_i u_i(x_i) \quad (1)$$

where λ_i is a weighting constant greater than zero. Keeny and Raiffa[5] address this model in detail, and it has been widely applied. Computer software packages are available for multiple-attribute decision problems.

HCIA Methodology

This methodology, developed at the University of Ulster[3] formalizes many aspects of the Edinburgh approach.[6] It uses a similar hierarchical cross impact analysis (HCIA) over the different levels in the system, but allows for interactions between issues at the same level by the use of perturber matrices. Again, it can be used only for hierarchical networks.

Using the terminology of Marchant,[7] the partial impact of TACTIC T_j on COMPONENT C_i relative to the rth objective, O_r, is written and defined

$$\partial I(C_i/T_j)_{O_r} = (C_i \backslash T_j)(T_j \backslash O_r) \quad (2)$$

The total impact of the collection of tactics on C_i relative to O_r is then defined as

$$\sum_{j=1}^{n} \partial I(C_i/T_j)_{O_r} = \sum_{j=1}^{n} t_{ij}\sigma_{jr} \quad (3)$$

which is clearly a matrix product. The elimination of the TACTICS collection is completed by defining the interaction

$$C_i \backslash O_r = \sqrt{\frac{1}{n}\sum_{j=1}^{n} t_{ij}\sigma_{jr}} \quad (4)$$

thus giving a COMPONENTS to OBJECTIVES interaction matrix. Similarly, the OBJECTIVES collection is eliminated leaving a COMPONENTS to POLICY vector.

The pairwise comparison perturbation matrices, utilized at all the levels in the hierarchy to eliminate the problems of interdependence of issues, are symmetric and generally sparse. These adjust for "noise" in the system resulting from vagueness in the definitions of the entities in the collections at each level.

Analytic Hierarchy Process

The analytic hierarchy process (AHP)[8,9,10] devised by T.L. Saaty in the 1970s, offers another method of prioritizing issues. It entails the comparison of all the pairs of individual issues at each level. The inherent complexity of the process prohibits a simple description.

The technique is based on the fact that for a square matrix, C, of non-negative real numbers representing, for example, pairwise comparisons of the importance of elements in the COMPONENTS collection of the hierarchy with respect to one element of the next higher TACTICS level, there is a dominant eigenvalue, $\hat{\lambda}$, and a corresponding right eigenvector x emerging from the characteristic equation

$$Cx = \hat{\lambda}x \quad (5)$$

Given that $C = (c_{ij})$

$$c_{ij} > 0; \quad i, j = 1, 2, 3, \ldots\ldots, m \quad (6)$$

and

$$c_{ij} = 1/c_{ji}; \quad i, j = 1, 2, 3, \ldots\ldots, m \quad (7)$$

where the consistency condition

$$c_{ik} = c_{ij} \times c_{jk}; \quad i, j, k = 1, 2, \ldots\ldots, m \quad (8)$$

holds, it is easily shown $\hat{\lambda}$ and, by virtue of the fact that C is of unit rank, the remaining m − 1 eigenvalues are each zero.

For such a matrix any column is essentially the dominant right eigenvector of priorities. In reality it is unreasonable to expect decision makers to be perfectly consistent, although a fair degree of consistency is expected. Variations from perfect consistency in C perturb the eigenvalues such that $\hat{\lambda}_{>m}$[8], the proximity of the maximum eigenvalue to the order of C being an indication of consistency.

The corresponding eigenvector is normalized and the procedure repeated for each tactic in the TACTICS collection. The set of dominant eigenvectors is considered as a matrix of priorities (C/T) between COMPONENTS and TACTICS. Similarly, priority matrices are established for TACTICS to OBJECTIVES (T/O) and for OBJECTIVES to POLICY (O/P). A combined matrix multiplication

$$(C/T) \times (T/O) \times (O/P) \qquad (9)$$

yields a components to policy priority vector. In the calculation of such priorities the matrix entries are selected from a statistically optimized set of Saaty weightings, i.e.

$$S = \{1/9, 1/8, 1/7, 1/6, 1/5, 1/4, 1/3, 1/2, 1, 2, 3, 4, 5, 6, 7, 8, 9\} \qquad (10)$$

Consider, for example, the situation where five components, C_1 to C_5, must be prioritized with respect to tactic T_3, say. The decision maker indicates the relative importance of C_1 and C_2, C_1 and C_3, C_1 and C_4, etc. on a scale of 1 to 9. A score of 1 on this scale implies that the two components being compared are of "equal importance," whereas a score of 9 signifies that one of the components is "of absolute importance" relative to the other. The intermediate scores indicate varying degrees of importance between the two extremes. Obviously, the importance of any component relative to itself is unity and the law of reciprocity holds; i.e., if C_1 rates a score of 3 against C_2, then C_2 rates a score of 1/3 against C_1, and so on. Thus, these pairwise criteria comparisons by the decision maker can be incorporated into a positive reciprocal decision matrix, as shown below.

$$
\begin{bmatrix}
 & C_1 & C_2 & C_3 & C_4 & C_5 \\
C_1 & 1 & 3 & 2 & 2 & 1 \\
C_2 & 1/3 & 1 & 1/4 & 1/4 & 2 \\
C_3 & 1/2 & 4 & 1 & 1/2 & 3 \\
C_4 & 1/2 & 4 & 2 & 1 & 1/5 \\
C_5 & 1 & 1/2 & 1/3 & 5 & 1
\end{bmatrix}
$$

Such a matrix is a typical response from a decision maker; if it were perfectly consistent, then the spectrum would be $\{5, 0, 0, 0\}$. However, a simple calculation shows the spectrum to have a right dominant eigenvalue, $\hat{\lambda}$ of 6.7 and a corresponding normalized dominant right eigenvector of $[0.25, 0.14, 0.19, 0.19, 0.24\}$ in this case. Saaty defines a consistency index

$$CI = (\hat{\lambda} - k)/(k - 1), \qquad (11)$$

where k is the order of the matrix, and a corresponding consistency ratio

$$CR = CI/Random\ Consistency\ Number \qquad (12)$$

Saaty provides a list of random consistency numbers.[8]

The consistency check is designed to provoke the decision maker into reconsidering the entries in the decision matrix should the consistency ratio be unacceptable, e.g., greater than 10 percent.

PANELS OF EXPERTS

The development of expert systems implies an underlying assumption as to the meaning of *expertness*. Consequently, the assessment of expert opinion using, for example, a Delphi process, may lead to conclusions that, without some formal insight, could on some occasions be considered ambiguous. A review of Delphi literature will not solve this problem, since Delphi researchers fail to define the term *expert* as applied to their study. If, in simplistic terms, an expert is regarded as one practiced or skillful within the area of consideration, then there is an intuitive extrapolation as to the notion of *expertise*, and the assessment of *expert opinion* relies to some extent on a global understanding of this extrapolation. In general, the expert responses must conform to some form of measurement criterion in relation to consensus if any conclusion is to be reached on any issue that is the subject of expert opinion.

On the assumption that "more heads are better than one" in achieving a balanced view, many different forums for assessing group opinion have been utilized over the years.[4] Some of the most common are discussed as follows.

Committee

The traditional method of forming group opinion is the committee. Under the guidance of a (impartial) chairperson the issue is isolated (if necessary) and debated, with consensus determined by a majority vote. Besides the obvious discrepancy that a majority and consensus are not synonymous, the recognized difficulties with the method are the *dominant member* and the group pressure upon the individual to conform (the *herd instinct*). Thus, the resultant group pronouncement may not be at all representative of the individuals in the group.

Nominal Group

A committee is strictly a subgroup of a larger organization from which its authority is derived. More relevant to "panels of experts," i.e., the pooling of considered opinion, are nominal groups and the variations thereon. In this the alternatives are created by the panel members outside the group context. The individual members' lists, circulated before meetings, are then pooled. The personality issues are reduced: the *dominant member* should be the member best prepared; authorship of ideas is on record and cannot be disputed; and preconsideration should reduce the herd instinct.

Delphi Panel

Of more rigorous delineation is the Delphi panel,[11,12,13] devised as a method of obviating the practical difficulties inherent in the committee concept. In this the group members never meet physically; all communication is through a group controller or coordinator who selects the members of the panel, presents the basic problem, and informs the individuals in the group of progress to date (*feedback*). When the atomic or elemental issues that constitute the main problem have been determined (the qualitative phase), consensus is sought on the quantitative value to be ascribed to each of these (*statistical response*).

A series of rounds of voting is held; after each round the group controller returns to each panel member his or her scores together with a measure of the group opinion on each issue. The panel members are then asked if they wish to revise their opinions. The process is repeated until group opinion has converged sufficiently to be described as consensus, or until it is recognized that consensus is unattainable. Unanimity is less important in the method than the absence of group pressure.

The two essential characteristics in a Delphi panel are, thus: (1) the anonymity of panelists and (2) feedback (histograms, measures of central tendency, etc.). The former eliminates the problem of the dominant member and group pressure to conform. However a new problem is introduced—a considerable time-scale is involved in the sending of questions and answers to and fro. While this (at least) theoretically guarantees a considered opinion, a substantial attrition rate is almost inevitable. Members of the group lose interest, change jobs, or become otherwise indisposed. Also, the group controller has a role of far greater significance than a committee chairperson: only a high level of awareness on his or her part can prevent the unconscious bias that might be introduced. Other doubts about, and difficulties with, the method are summarized by Marchant[7] and Sackman.[14]

Computer Conferencing

Modern technology enables remote participation in group discussion. The participants input their views to a computer system. These are then circulated to the other panelists for comment and reply. A suitable schedule allows either a quick result or a more leisurely, considered process. Anonymity is optional.

This concept can be combined with others, e.g., the Delphi method, with a reduction of time scale and consequent diminishing of attrition (i.e., drop-outs). Further, if a computer system for analyzing the results exists, these results may be obtained almost instantaneously. Hence, the group may have access to the consequences of their polling before their views are forgotten. Such a system is currently under development by the authors.

DACAM Group

A DACAM group[15] is an alternative method of overcoming the problems with committees. Here, the group meets physically but is divided (usually systematically) into subgroups for discussion sessions. In between each session, representatives from the subgroups meet to compare progress and coordinate the agenda for the next session. In this way the dominant member, being limited to one subgroup, is less influential on the group decision.

While this method guarantees a quick result (e.g., in a day), the facility to organize such a group is dependent on at least substantial goodwill, if not considerable finance.

CONSENSUS

The notion of *consensus* is very much an intuitive one. In politics there is "consensus" on a policy that is acceptable to a majority. If no consensus exists, then it can often be achieved by broadening the policy. However, the concept is very largely negative in that the emphasis is on avoidance of what is intolerable to others. A certain amount of confusion over the term "consensus" occurred until the authors made the important distinction between two cases.

The nature of consensus depends on the context or terms of reference of the problem. For example, the arrangement of a number of options in order of importance or preference (*comparative consensus*) is essentially different from assessing the importance of a single elemental issue within the problem (*definitive consensus*). As is shown later, Saaty's AHP may entail elements of both these (*compound consensus*).

Definitive Consensus

Definitive consensus involves panel members who assign to each issue a score chosen from a Likert-type scale, i.e., a range of values, e.g., (0,10) or (0,5). For example, one might say that a very definite consensus had been reached if a comfortable majority (say, 60 percent) agreed on an exact value, or if a substantial majority (say, 75 percent) agreed on a small range, or if all agreed to within a slightly larger range. Thus, there is a balance between the number conforming to the view and the range permitted.

The essential problem, however, is finding mechanisms for representing and assessing the opinions on each individual issue for the complete panel, i.e., mechanisms that provide for the set of scores for each issue:

1. A profile of the scores,
2. A focal point of agreement (if any), and
3. A measure of the agreement at that point.

Additional problems arise if the panel is divided into several groups, each of which is in internal accord. This is referred to as *split consensus* or a *multimodal* situation.

Traditionally, a histogram was generally used as a profile, and the presence, or otherwise, of consensus determined by a statistical measure of dispersion. Chatterjee and Chatterjee[16] discuss mean- and median-based methods. Trimmed and winsorized means might be regarded as a compromise between the two. Some of the assumptions and drawbacks of these measures, e.g., the non-representativeness of the mean, the instability of the mode, and the inconsistency and precision-dependence of the median, are described in reference 17.

It is worth pointing out here the difference between *consensus* and *compromise*. If, on a scale of 0 to 5, one-half of a group selects 0 and the other half selects 5, then 2.5 is a *compromise* score. It would be inaccurate, however, to describe 2.5 as a *consensus* score, since this figure is not representative of any members of the group and, therefore, not a measure of the opinion of the group itself. There is, in fact, no consensus among the group.

Alternative stability approach: A major problem is that stability may set in before consensus has been achieved, and the latter might therefore never be reached. When seeking consensus on a large number of issues, this is highly likely for some issues. This approach is more consistent with Delphi philosophy, which is more concerned with narrowing the spread of diverse opinions than enforcing an artificial "consensus."

It is possible and perhaps simpler to use a definition of stability (based on a comparison of successive rounds) as the termination criterion for the Delphi panel. This would also allow item/individual/tactic/group stability to be considered. It might be based on the ratio of total scores for two rounds, e.g., the average of the moduli of the differences between the rounds. Linstone and Turoff[11] suggest the quadratic mean as a refinement of this. Alternatively, if the data were normalized (i.e., translated and magnified to fit a uniform or rectangular distribution) it might be simpler to use Pearson's correlation coefficient.

The principal advantage of this approach is that stability is almost certain to be achieved, and fairly rapidly. It is generally recognized that this normally occurs within four polling rounds in most Delphi panels. (See reference 12.)

Comparative Consensus

Comparative consensus involves a panel of individuals who has to decide on an arrangement of n issues (or objects) into order of importance or merit. For the sake of simplicity let the issues be numbered 1 to n. Each panel member decides on an arrangement that might be expected to differ from the arrangements of others. Two problems can arise: (1) establishing a universally acceptable arrangement and (2) the amount of agreement among the panel. A trivial but typical example of this situation is the newspaper competition in which a number of elements must be ranked in order of priority, e.g., good accommodation, easy access, etc., when selecting a holiday.

Comparative consensus [18] also involves only the ranking of a number of issues (in contrast to the simpler notion of *definitive consensus*, as described above, in which the agreement of a panel on a single issue is assessed). The obvious question in the situation where several panelists judge several issues is whether: (1) it is preferable to determine the agreement on each issue on the basis of scores and rank on the basis of the consensus scores or (2) whether each panel member should rank on the issues on the basis of his or her scores and subsequently look for consensus on the set of rankings. This has never been satisfactorily resolved.

Compound Consensus

Compound consensus arises when, e.g., a panel is required to produce an agreed ranking (with weightings) for a set of issues. It can occur when AHP is used by a relatively small panel to produce a set of rankings. The obvious question is whether it is better to: (1) find (definitive) consensus among the panel on the issues and then rank the issues for the whole panel or (2) let each panel member rank the issues and then seek (comparative) consensus on the rankings. Compound consensus allows a weighting of each of the issues to be taken into account.

Current research on the problems of consensus and prioritization using the theory of "fuzzy sets" is still at the embryonic stage.[19,20] The accelerating pace of leading-edge technology, however, would appear to indicate the availability of practical developments in the near future.

EXPERT SYSTEMS

Expert systems, also called knowledge-based systems, "teach" computers to perform tasks with the knowledge and intuition of human experts. By organizing knowledge of established experts into a sophisticated "knowledge base" of patterns and rules, expert systems allow computers to act as intelligent assistants, advising the user with the insight and acumen of experts.

An example is artificial intelligence, a branch of computer science, that develops computer programs that perform tasks requiring intelligence, or systems that exhibit the characteristics associated with intelligence in human behavior, such as understanding language, learning, reasoning, solving problems, etc. The area of artificial intelligence that is developing fastest for practical application consists of computer programs that contain large masses of information and the ability to reason.

An expert system has three main components: (1) knowledge base, (2) inference engine, and (3) interfaces.

Knowledge Base

There are several common forms of knowledge representation. A rule-based system is the most appropriate for representing fire safety knowledge. Rule-based knowledge representation generally takes the form of "if . . . (condition), then . . . (consequence)." The conditions are also referred to as attributes, premises, or antecedents; the consequences may also be called conclusions or actions. In some situations, knowledge about an attribute may not be absolutely certain. In these cases, one can assign a measure of certainty or belief to the instantiated value of the attribute. One can also use a weight to represent the degree of confidence, the probability, or the strength of a rule.

Inference Engine

The inference engine is the mechanism that applies the rules to the facts or data to make decisions. It is the reasoning approach used in the program to interpret and apply the knowledge. In rule-based systems, the inference engine typically uses either forward chaining or backward chaining to find reasonable alternatives. Forward chaining is a way of reasoning toward a goal or hypothesis. Backward chaining starts with the goal and determines the approach necessary to accomplish the goal. The inference engine also uses the weights on the rules to evaluate the strength or probability of the sequence of rules by, for example, combining the probabilities of the rules.

Interfaces

The user interface is the part of the computer program that elicits information and explains its conclusions. Rapid proliferation of expert system applications has been made possible by the wide availability of low-priced general-purpose development tools called "shells." Expert system shells have shortened the typical development time, by an order of magnitude, from years to months.

A simple example of an expert system might assess the fire safety quality of buildings against known criteria. In this case, the facts used might be knowledge derived from a Delphi panel with rules and conditions to be applied. Typical questions with which the expert system might deal are:

1. Deciding whether a given building satisfies a set of minimum requirements,
2. Tradeoff between active and passive safety factors,[21]
3. Suggesting acceptable requirements of each type of measure, and
4. Allowing experimental changes in the building specification as part of a "what-if" environment.

REFERENCES CITED

1. N.C. Barford, "Experimental Measurements: Precision, Error, and Truth" 2nd ed., ISBN 0 471 90702 2, John Wiley & Sons (1987).
2. Arthur M. Glenburg, "Learning from Data: An Introduction to Statistical Reasoning," ISBN 0 15 550381 2, Harcourt Brace Jovanovich, Orlando, FL (1980).

3. H.A. Donegan, T.J. Shields, and G.W.H. Silcock, "A Mathematical Strategy to Relate Fire Safety Evaluation and Fire Safety Policy Formulation for Buildings" *Proceedings* of 2nd International Symposium on Fire Safety Science, Tokyo, pp. 433–441 (1988).
4. M. Scheibe, M. Skutsch, and J. Schofer, "The Delphi Method: Techniques and Applications," *Experiments in Delphi Methodology*, H.A. Linstone and M. Turoff, (eds.), pp. 262–282, Addison-Wesley (1975).
5. R.L. Keeny and H. Raiffa, *Decisions With Multiple Objectives*, Wiley, New York (1976).
6. E.W. Marchant, "Fire Safety Evaluation (Points) Scheme for Patient Areas within Hospitals: A Report on Its Origins and Development," Univ. of Edinburgh, Scotland (1982).
7. E.W. Marchant, "Problems Associated with the Delphi Technique," *Fire Technology*, 24, 1, pp. 59–62, Feb. 1989.
8. Thomas L. Saaty, *The Analytic Hierarchy Process*, McGraw-Hill (1980).
9. R.G. Vachnadze and N.I. Markozashvili, "Some Applications of the Analytic Hierarchy Process," *Mathl. Modeling*, 9 3–5, pp. 185–191, Pergamon Journals Ltd. (1987).
10. R.W. Saaty, "The Analytic Hierarchy Process: What It Is and How It Is Used," *Mathl. Modeling*, 9, 3–5, pp. 161–176, Pergamon Journals Ltd. (1987).
11. H.A. Linstone and M. Turoff, "The Delphi Method: Techniques and Applications," Addison-Wesley (1975).
12. Barry Render and Ralph M. Stair, "Quantitative Analysis for Management," 2nd ed., ISBN 0 205 08335 8, Bacon, Newton, MA (1985).
13. Steven C. Wheelwright and Spyros Makridakis, "Forecasting Methods for Management," 3rd ed., ISBN 0 471 05630 8, John Wiley, New York (1980).
14. H. Sackman, "Delphi Assessment: Expert Opinion, Forecasting, and Group Process," Paper R-1283-PR, Rand Corporation, Santa Monique (1974).
15. R.E. Norton, *The DACAM Handbook*, Leadership Series 67, National Center for Research in Vocational Education, Columbus, OH (1985).
16. S. Chatterjee and S. Chatterjee, "On Combining Expert Opinions," *Am. Jnl. Math and Mngt. Stud.*, 7, Nos. 3 and 4, pp. 271–295 (1987).
17. F.J. Dodd and H.A. Donegan, "The Representation and Combination of Opinions," *Proceedings* of ILIAM, 6, H.A. Donegan (ed.), ISBN 1 871206 11 1, pp. 17–29, Univ. of Ulster, (1989).
18. F.J. Dodd and H.A. Donegan, "Comparative and Compound Consensus," *App. Math. Lett*, 5, 3, pp. 31–33, Pergamon Press (1992).
19. Bonnie Spillman, Richard Spillman, and James Bezdek, "A Fuzzy Analysis of Consensus in Small Groups," *Fuzzy Sets: Theory and Applications to Policy Analysis and Information Systems*, Paul P. Wang and S.K. Chang (eds.) pp. 291–308, Plenum, New York & London (1980).
20. Didier Dubois and Henri Prade, "Possibility Theory: An Approach to Computerized Processing of Uncertainty," ISBN 0 306 42520 3, Plenum, New York (1988).
21. H.A. Donegan, I.R. Taylor, and R.T. Meehan, "An Expert System to Assess Fire Safety in Buildings," *Proceedings* of 3rd International Symposium on Fire Safety Science, Edinburgh, ISBN 1 85166 719 9, pp. 485–494, Elsevier, New York (1991).

ENGINEERING ECONOMICS

John M. Watts, Jr.

INTRODUCTION

Engineering economics is that phase of engineering which has to do with analysis of proposed engineering projects to determine the relative worth of the net economic gains to be expected from alternative proposals, in relation to their net economic costs. This chapter discusses the time value of money and other cash-flow concepts, such as compound and continuous interest. It continues with economic practices and techniques used to evaluate and optimize decisions on selection of fire protection strategies. The final section expands on the principles of benefit-cost analysis.

CASH-FLOW CONCEPTS

Cash flow is the stream of monetary (dollar) values—costs (inputs) and benefits (outputs)—resulting from a project investment.

Time Value of Money

The following are reasons why $1000 today is "worth" more than $1000 one year from today:

1. Inflation,
2. Risk, and
3. Cost of money.

Of these, the cost of money is the most predictable, and, hence, it is the essential component of economic analysis. Cost of money is represented by (1) money paid for the use of borrowed money or (2) return on investment. Cost of money is determined by an interest rate.

Time value of money is defined as the time-dependent value of money stemming both from changes in the purchasing power of money (inflation or deflation) and from the real earning potential of alternative investments over time.

Cash Flow Diagrams

It is difficult to solve a problem if you cannot see it. The easiest way to approach problems in economic analysis is to draw a picture. The picture should show three things:

1. A time interval divided into an appropriate number of equal periods;
2. All cash outflows (e.g., deposits, expenditures, etc.) in each period; and
3. All cash inflows (e.g., withdrawals, income, etc.) for each period.

Unless otherwise indicated, all such cash flows are considered to occur at the end of their respective periods.

Figure 5-6.1 is a cash-flow diagram showing an outflow or disbursement of $1000 at the beginning of year one and an inflow or return of $2000 at the end of year five.

Notation

To simplify the subject of economic analysis, symbols are introduced to represent types of cash flows and interest factors. The symbols used in this chapter conform to ANSI Z94,[1] however, not all practitioners follow this standard convention, and care must be taken to avoid confusion when reading the literature. The following symbols will be used here. (A complete list of the ANSI Z94 symbols is given in Appendix A to this chapter.)

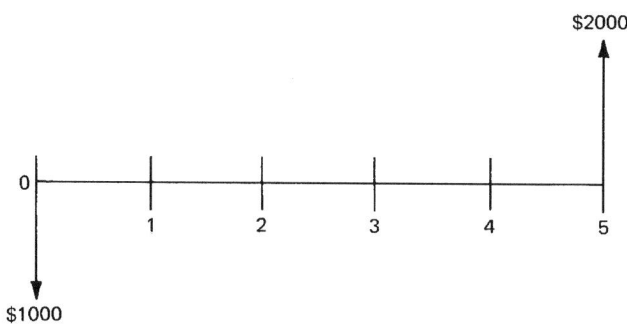

Fig. 5-6.1. *Cash-flow diagram.*

Dr. Watts holds degrees in Fire Protection Engineering, Industrial Engineering, and Operations Research. He is Director of the Fire Safety Institute, a not-for-profit information, research, and educational corporation located in Middlebury, Vermont. Dr. Watts also serves as editor of NFPA's *Fire Technology*.

P = Present sum of money ($),
F = Future sum of money ($),
N = Number of interest periods, and
i = Interest rate per period (%).

Interest Calculations

Interest is the money paid for the use of borrowed money or the return on invested capital. The economic cost of construction, installation, ownership, or operation can be estimated correctly only by including a factor for the economic cost of money.

Simple interest: To illustrate the basic concepts of interest, an additional notation

$F(N)$ = Future sum of money after N periods will be used.

Then, for simple interest

$$F(1) = P + (P)(i) = P(1 + i)$$

and

$$F(N) = P + (N)(P)(i) = P(1 + Ni)$$

For example: $100 at 10 percent per year for five years yields

$$F(5) = 100[1 + (5)(0.1)]$$
$$= 100(1.5)$$
$$= \$150$$

However, interest is almost universally compounded to include interest on the interest.

Compound interest

$$F(1) = P + (P)(i) = P(1 + i)$$

same as simple interest.

$$F(2) = F(1) + F(1)(i)$$

interest is applied to the new sum.

$$= (F)(1)(1 + i) = P(1 + i)^2$$
$$F(3) = F(2)(1 + i) = P(1 + i)^3$$

and by mathematical induction

$$F(N) = P(1 + i)^N$$

EXAMPLE:

$100 at 10 percent per year for five years yields

$$F(5) = 100(1 + 0.1)^5$$
$$= 100(1.1)^5$$
$$= 100(1.61051)$$
$$= \$161.05$$

which is over 7 percent greater than with simple interest.

EXAMPLE:

In 1626 Willem Verhulst bought Manhattan Island from the Canarsie Indians for 60 florins ($24) worth of merchan-

dise (a price of about 0.2 cents per acre). At an average interest rate of 6 percent, what is the present value (1995) of the Canarsies' $24?

$$F = P(1 + i)^N$$
$$= \$24(1 + 0.06)^{362}$$
$$= \$3.5 \times 10^{10}$$

Thirty-five billion dollars is a reasonable approximation of the present land value of the island of Manhattan.

INTEREST FACTORS

Interest factors are multiplicative numbers calculated from interest formulas for given interest rates and periods. They are used to convert cash flows occurring at different times to a common time. The functional formats used to represent these factors are taken from ANSI Z94, and they are summarized in Appendix B to this chapter.

Compound Amount Factor

In the formula for finding the future value of a sum of money with compound interest, the mathematical expression $(1 + i)^N$ is referred to as the "Compound Amount Factor," represented by the functional format $(F/P, i, N)$. Thus,

$$F = P(F/P, i, N)$$

Interest tables: Values of the compound amount, present worth, and other factors that will be discussed shortly, are tabulated for a variety of interest rates and number of periods in most texts on engineering economy. Example tables are presented in Appendix C to this chapter. Although calculators and computers have greatly reduced the need for such tables, they are often still useful for interpolations.

Present Worth

Present worth is the value found by discounting future cash flows to the present or base time.

Discounting: The inverse of compounding is determining a present amount which will yield a specified future sum. This process is referred to as discounting. The equation for discounting is found readily by using the compounding equation to solve for P in terms of F:

$$P = F(1 + i)^{-N}$$

EXAMPLE:

What present sum will yield $1000 in 5 years at 10 percent?

$$P = 1000(1.1)^{-5}$$
$$= 1000(0.62092)$$
$$= \$620.92$$

This means that $620.92 "deposited" today at 10 percent compounded annually will yield $1000 in 5 years.

Present worth factor: In the discounting equation, the expression $(1 + i)^N$ is called the "present worth factor" and is represented by the symbol $(P/F, i, N)$. Thus, for the present worth of a future sum at i percent interest for N periods

$$P = F(P/F, i, N)$$

Note that the present worth factor is the reciprocal of the compound amount factor. Also note that

$$(P/F, i, N) = 1/(F/P, i, N)$$

EXAMPLE:

What interest rate is required to triple $2000 in 10 years?

$$P = F/3 = F(P/F, i, 10)$$

therefore

$$(P/F, i, 10) = 1/3$$

from Appendix C

$$(P/F, 10\%, 10) = 0.3855$$

and

$$(P/F, 12\%, 10) = 0.3220$$

by linear interpolation

$$i = 11.6\%.$$

Interest Periods

Normally, but not always, the interest period is taken as one year. There may be subperiods of quarters, months, weeks, etc.

Nominal *versus* effective interest: It is generally assumed that interest is compounded annually. However, interest may be compounded more frequently. When this occurs, there is a "nominal interest" or annual percentage rate and an "effective interest," which is the figure used in calculations. For example, a savings bank may offer 5 percent interest compounded quarterly, which is not the same as 5 percent per year. A nominal rate of 5 percent compounded quarterly is the same as 1.25 percent every three months or an effective rate of 5.1 percent per year. If

$$r = \text{Nominal interest rate,}$$

and

$$M = \text{Number of subperiods per year,}$$

TABLE 5-6.1 *Continuous Interest (%)*

	Effective	
Nominal %	Monthly	Continuous
5	5.1	5.1
10	10.5	10.5
15	16.1	16.2
20	21.9	22.1

then the effective interest rate

$$i = (1 + r/M)^M - 1$$

EXAMPLE:

Credit cards usually charge interest at a rate of 1.5 percent per month. This is a nominal rate of 18 percent. What is the effective rate?

$$
\begin{aligned}
i &= (1 + 0.015)^{12} - 1 \\
&= 1.1956 - 1 \\
&= 19.56\%.
\end{aligned}
$$

Continuous interest: A special case of effective interest occurs when the number of periods per year is infinite. This represents a situation of "continuous interest," also referred to as "continuous compounding." Formulas for continuous interest can be derived by examining limits as M approaches infinity. Formulas for interest factors using continuous compounding are included in Appendix B. Continuous interest is compared to monthly interest in Table 5-6.1.

EXAMPLE:

Compare the future amounts obtained under various compounding periods at a nominal interest rate of 12 percent for 5 years, if $P = \$10,000$. (See Table 5-6.2.)

Series Payments

Life would be simpler if all financial transactions were in single lump-sum payments, now or at some time in the future. However, most situations involve a series of regular payments, for example, car loans and mortgages.

Series compound amount factor: Given a series of regular payments, what will they be worth at some future time?

TABLE 5-6.2 *Example of Continuous Interest N = 5 years, r = 12%*

Compounding	M	i	NM	F/P	F
Annual	1	12.000	5	1.76234	17,623.40
Semi-annual	2	12.360	10	1.79085	17,908.50
Quarterly	4	12.551	20	1.80611	18,061.10
Monthly	12	12.683	60	1.81670	18,167.00
Weekly	52	12.734	260	1.820860	18,208.60
Daily	365	12.747	1825	1.821938	18,219.38
Hourly	8760	12.749	43800	1.822061	18,220.61
Instantaneously	∞	12.750	∞	1.822119*	18,221.19

*F/P (instantaneous) $= e^{Ni} = e^{5(0.12)} = e^{0.6}$.

Let

A = The amount of a regular end-of-period payment.

Then, note that each payment, A, is compounded for a different period of time. The first payment will be compounded for $N - 1$ periods (years):

$$F = A(1 + i)^{N-1}$$

and the second payment for $N - 2$ periods:

$$F = A(1 + i)^{N-2}$$

and so forth. Thus, the total future value is

$$F = A(1 + i)^{N-1} + A(1 + i)^{N-2} + \cdots + A(1 + i) + A$$

or

$$F = A[(1 + i)^N - 1]/i$$

The interest expression in this equation is known as the "series compound amount factor," $(F/A, i, N)$, thus

$$F = A(F/A, i, N)$$

[Note that $(F/A, i, N)$ is the sum of N $(F/P, i, N)$'s.]

Sinking fund factor: The process corresponding to the inverse of series compounding is referred to as a "sinking fund;" i.e., what size regular series payments are necessary to acquire a given future amount?

Solving the series compound amount equation for A:

$$A = F\{i/[(1 + i)^N - 1]\}$$

Or, using the symbol $(A/F, i, N)$ for the "sinking fund factor"

$$A = F(A/F, i, N)$$

Here, note that the sinking fund factor is the reciprocal of the series compound amount factor, i.e., $(A/F, i, N) = 1/(F/A, i, N)$.

Capital recovery factor: It is also important to be able to relate regular periodic payments to their present worth; e.g., what monthly installments will pay for a $10,000 car in 3 years at 15 percent?

Substituting the compounding equation $F = P(F/P, i, N)$ in the sinking fund equation, $A = F(A/F, i, N)$, yields

$$A = P(F/P, i, N)(A/F, i, N)$$

And, substituting the corresponding interest factors gives

$$A = P[i(1 + i)^N]/[(1 + i)^N - 1]$$

In this equation, the interest expression is known as the "capital recovery factor," since the equation defines a regular income necessary to recover a capital investment. The symbolic equation is

$$A = P(A/P, i, N)$$

Series present worth factor: As with the other factors, there is a corresponding inverse to the capital recovery fac-

tor. The "series present worth" factor is found by solving the capital recovery equation for P:

$$P = A[(1 + i)^N - 1]/[i(1 + i)^N]$$

or, symbolically

$$P = A(P/A, i, N)$$

Other Interest Calculation Concepts

Additional concepts involved in interest calculations include: continuous cash flow, capitalized costs, beginning of period payments, and gradients.

Continuous cash flow: Perhaps the most useful function of continuous interest is its application to situations where the flow of money is of a continuous nature. "Continuous cash flow" is representative for:

1. A series of regular payments for which the interval between payments is very short, or
2. A disbursement at some unknown time (which is then considered to be spread out over the economic period).

Factors for calculating present or future worth of a series of annual amounts, representing the total of a continuous cash flow throughout the year, may be derived by integrating corresponding continuous interest factors over the number of years the flow is maintained.

Continuous cash flow is an appropriate way to handle economic evaluations of risk, e.g., the present value of an annual expected loss.

Formulas for interest factors representing continuous, uniform cash flows are included in Appendix B.

Capitalized costs: Sometimes there are considerations, such as some public works projects, which are considered to last indefinitely and thereby provide perpetual service. For example, how much should a community be willing to invest in a reservoir which will reduce fire insurance costs by some annual amount, A? Taking the limit of the series present worth factor as the number of periods goes to infinity gives the reciprocal of the interest rate. Thus, "capitalized costs" are just the annual amount divided by the interest rate. When expressed as an amount required to produce a fixed yield in perpetuity, it is sometimes referred to as an "annuity."

Beginning-of-period payments: Most returns on investment (cash inflows) occur at the end of the period during which they accrued. For example, a bank computes and pays interest at the end of the interest period. Accordingly, interest tables, such as those in Appendix C, are computed for end-of-year payments, e.g., the values of the capital recovery factor $(A/P, i, N)$ assume that the regular payments, A, occur at the end of each period.

On the other hand, most disbursements (cash outflows) occur at the beginning of the period (e.g., insurance premiums). When dealing with beginning-of-period payments, it is necessary to make adjustments. One method of adjustment for beginning-of-period payments is to calculate a separate set of factors. Another way is to logically interpret the effect of beginning-of-period payments for a particular problem, e.g., treating the first payment as a present value. The important thing is to recognize that such variations can affect economic analysis.

Gradients: It occasionally becomes necessary to treat the case of a cash flow which regularly increases or decreases at each period. Such patterned changes in cash flow are called "gradients." They may be a constant amount (linear or arithmetic progression), or they may be a constant percentage (exponential or geometric progression). Various equations for dealing with gradient series may be found in Appendix B.

COMPARISON OF ALTERNATIVES

Most decisions are based on economic criteria. Investments are unattractive, unless it seems likely they will be recovered with interest. Economic decisions can be divided into two classes:

1. Income-expansion—i.e., the objective of capitalism, and
2. Cost-reduction—the basis of profitability.

Fire protection engineering economic analysis is primarily concerned with cost-reduction decisions, finding the least expensive way to fulfill certain requirements, or minimizing the sum of expected fire losses plus investment in fire protection.

There are four common methods of comparing alternative investments: (1) present worth, (2) annual cost, (3) rate of return, and (4) benefit-cost analysis. Each of these is dependent on a selected interest rate or discount rate to adjust cash flows at different points in time.

Discount Rate

The term "discount rate" is often used for the interest rate when comparing alternative projects or strategies.

Selection of discount rate: If costs and benefits accrue equally over the life of a project or strategy, the selection of discount rate will have little impact on the estimated benefit-cost ratios. However, most benefits and costs occur at different times over the project life cycle. Thus, costs of constructing a fire-resistive building will be incurred early in contrast to benefits which will accrue over the life of the building. The discount rate then has a significant impact on measures such as benefit-cost ratios, since the higher the discount rate, the lower the present value of future benefits.

In view of the uncertainty concerning appropriate discount rate, analysts frequently use a range of discount rates. This procedure indicates the sensitivity of the analysis to variations in the discount rate. In some instances, project rankings based on present values may be affected by the discount rate as shown in Figure 5-6.2. Project A is preferred to project B for discount rates below 15 percent, while the converse is true for discount rates greater than 15 percent. In this instance, the decision to adopt project A in preference to project B will reflect the belief that the appropriate discount rate is less than or equal to 15 percent.

A comparison of benefits and costs may also be used to determine the payback period for a particular project or strategy. However, it is important to discount future costs or benefits in such analyses. For example, an analysis of the Beverly Hills Supper Club fire compared annual savings from a reduction in insurance premiums to the costs of sprinkler installation.[2] Annual savings were estimated at $11,000, while costs of sprinkler installation ranged from $42,000 to $68,000. It was concluded that the installation would have been paid back in four to seven years (depending

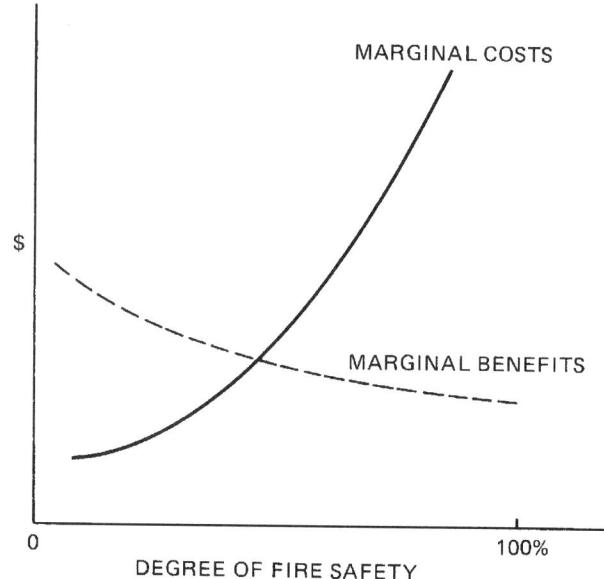

Fig. 5-6.2. Impact of discount rate on project selection.

on the cost of the sprinklers). However, this analysis did not discount future benefits, so that $11,000 received at the end of four years was deemed equivalent to $11,000 received in the first year. Once future benefits are discounted, the payback period ranges from five to eleven years with a discount rate of 10 percent.

Inflation and the discount rate: Provision for inflation may be made in two ways: (1) estimate all future costs and benefits in constant prices, and use a discount rate which represents the opportunity cost of capital in the absence of inflation or (2) estimate all future benefits and costs in current or inflated prices, and use a discount rate which includes an allowance for inflation. The discount rate in the first instance may be considered the real discount rate, while the discount rate in the second instance is the nominal discount rate. The use of current or inflated prices with the real discount rate, or constant prices with the nominal discount rate, will result in serious distortions in economic analysis.

Present Worth

In a "present worth" comparison of alternatives, the costs associated with each alternative investment are all converted to a present sum of money, and the least of these values represents the best alternative. Annual costs, future payments, and gradients must be brought to the present. Converting all cash flows to present worth is often referred to as "discounting."

EXAMPLE:

Two alternate plans are available for increasing the capacity of existing water transmission lines between an unlimited source and a reservoir. The unlimited source is at a higher elevation than the reservoir. Plan A calls for the construction of a parallel pipeline and for flow by gravity; plan B specifies construction of a booster pumping station. Estimated cost data for the two plans are as follows:

	Plan A Pipeline	Plan B Pumping Station
Construction Cost	$1,000,000	$200,000
Life	40 years	Structure 40 years Equipment 20 years
Cost of replacing equipment at the end of 20 years	0	$75,000
Operating Costs	$1000/year	$50,000/year

If money is worth 12 percent, which plan is more economical? (Assume annual compounding, zero salvage value, and all other costs equal for both plans.)

Present worth (Plan A) = $P + A(P/A, 12\%, 40)$
  = $1,000,000 + $1000(8.24378)$
  = $1,008,244$
Present worth (Plan B) = $P + A(P/A, 12\%, 40)$
  $+ F(P/F, 12\%, 20)$
  = $200,000 + $50,000(8.24378)$
  $+ $75,000(0.10367)$
  = $619,964$

Thus, plan B is the least-cost alternative.

A significant limitation of present worth analysis is that it cannot be used to compare alternatives with unequal economic lives. That is, a ten-year plan and a twenty-year plan should not be compared by discounting their costs to a present worth. A better method of comparison is annual cost.

Annual Cost

To compare alternatives by annual cost, all cash flows are changed to a series of uniform payments. Current expenditures, future costs or receipts, and gradients must be converted to annual costs. If a lump-sum cash flow occurs at some time other than the beginning or end of the economic life, it must be converted in a two-step process: first moving it to the present and then spreading it uniformly over the life of the project.

Alternatives with unequal economic lives may be compared by assuming replacement in kind at the end of the shorter life, thus maintaining the same level of uniform payment.

	System Cost	Insurance Premium	Life
Partial System	$ 8,000	$1000	15 years
Full System	$15,000	$250	20 years

EXAMPLE:

Compare the value of a partial or full sprinkler system purchased at 10 percent interest.

Annual cost (partial system) = $A + P(A/P, 10\%, 15)$
  = $1000 + $8000(0.13147)$
  = 2051.76
Annual cost (full system) = $A + P(A/P, 10\%, 20)$
  = $250 + $15,000(0.11746)$
  = 2011.90

The full system is slightly more economically desirable. When costs are this comparable, it is especially important to consider other relevant decision criteria, e.g., uninsured losses.

Rate of Return

Rate of return is, by definition, the interest rate at which the present worth of the net cash flow is zero. Computationally, this is the most complex method of comparison. If more than one interest factor is involved, the solution is by trial and error. Microcomputer programs are most useful with this method.

The calculated interest rate may be compared to a discount rate identified as the "minimum attractive rate of return" or to the interest rate yielded by alternatives. Rate-of-return analysis is particularly suitable to the selection of a number of projects to be undertaken within a fixed or limited capital budget.

EXAMPLE:

An industrial fire fighting truck costs $100,000. Savings in insurance premiums and uninsured losses from the acquisition and operation of this equipment is estimated at $60,000 per year. Salvage value of the apparatus after 5 years is expected to be $20,000. A full-time driver during operating hours will accrue an added cost of $10,000 per year. What would the rate of return be on this investment?

@ 40% Present worth
  = $P + F(P/F, 40\%, 5) + A(P/A, 40\%, 5)$
  = $- $100,000 + $20,000(0.18593)$
  $+ ($60,000 - 10,000)(2.0352)$
  = $5,478.60$
@ 50% Present worth
  = $P + F(P/F, 50\%, 5) + A(P/A, 50\%, 5)$
  = $- $100,000 + $20,000(0.13169)$
  $+ ($60,000 - $10,000)(1.7366)$
  = $- $10,536.40$

By linear interpolation, rate of return = 43 percent.

BENEFIT-COST ANALYSIS

Benefit-cost analysis, also referred to as cost-benefit analysis, is a method of comparison in which the consequences of an investment are evaluated in monetary terms and divided into the separate categories of benefits and costs. The amounts are then converted to annual equivalents or present worths for comparison.

The important steps of a benefit-cost analysis are:

1. Identification of relevant benefits and costs,
2. Measurement of these benefits and costs,
3. Selection of appropriate criteria for comparison, and
4. Treatment of uncertainty.

Identification of Relevant Benefits and Costs

The identification of benefits and costs will depend on the particular project under consideration. Thus, in the case of fire prevention or control activities, the benefits are based on fire losses prior to such activities. Fire losses may be classified as direct or indirect. Direct economic losses are property and contents losses. Indirect losses include such things as the costs of injuries and deaths, costs incurred by business or industry due to business interruption, losses to the community from interruption of services, loss of payroll or taxes, loss of market share, and loss of reputation. The direct costs of fire protection activities include the costs of constructing fire-resistive buildings, installation costs of fire protection systems, and the costs of operating fire departments. Indirect costs are more difficult to measure. They include items such as the constraints on choice due to fire protection requirements by state and local agencies.

A major factor in the identification of relevant benefits and costs pertains to the decision unit involved. Thus, if the decision maker is a property owner, the relevant benefits from fire protection are likely to be the reduction in fire insurance premiums and fire damage or business interruption losses not covered by insurance. In the case of a municipality, relevant benefits are: the protection of members of the community, avoidance of tax and payroll losses, and costs associated with assisting fire victims. Potential benefits, in these instances, are considerably greater than those faced by a property owner. However, the community may ignore some external effects of fire incidents. For example, the 1954 automobile transmission plant fire in Livonia, Michigan, affected the automobile industry in Detroit and various automobile dealers throughout the U.S. However, there was little incentive for the community to consider such potential losses in their evaluation of fire strategies, since they would pertain to persons outside the community. It might be concluded, therefore, that the more comprehensive the decision unit, the more likely the inclusion of all relevant costs and benefits, in particular, social costs and benefits.

Measurement of Benefits and Costs

Direct losses are measured or estimated statistically or by *a priori* judgment. Actuarial fire-loss data collected nationally or for a particular industry may be used, providing it is adequately specific and the collection mechanism is reliable. More often, an experienced judgment of potential losses is made, sometimes referred to as the "maximum probable loss" (MPL).

Indirect losses, if considered, are much more difficult to appraise. A percentage or multiple of direct losses is sometimes used. However, when indirect loss is an important decision parameter, a great deal of research into monetary evaluation may be necessary.

Procedures for valuing a human life are discussed elsewhere in this handbook.

In the measurement of benefits, it is appropriate to adjust for utility or disutility which may be associated with a fire loss.

Costs may be divided into two major categories: (1) costs of private fire protection services and (2) costs of public fire protection services. In either case, cost estimates will reflect the opportunity cost of providing the service. For example, the cost of building a fire-resistive structure is the production foregone due to the diversion of labor and resources to make such a structure. Similarly, the cost of a fire department is the loss of other community services which might have been provided with the resources allocated to the fire department.

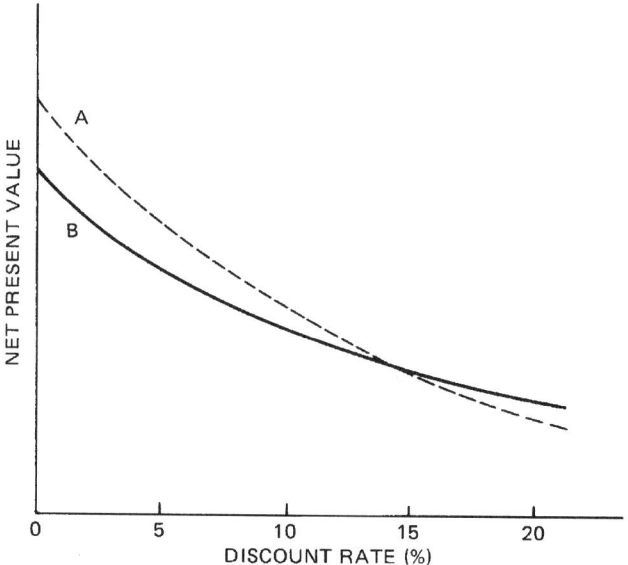

Fig. 5-6.3. Project selection.

Selection of Appropriate Criteria for Comparison

There are two considerations in determining benefit-cost criteria. The first pertains to project acceptability, while the second pertains to project selection.

Project acceptability may be based on benefit-cost difference or benefit-cost ratio. Benefit-cost ratio is a measure of project worth in which the monetary equivalent benefits are divided by the monetary equivalent costs. The first criterion requires that the value of benefits less costs be greater than zero, while the second criterion requires that the benefit-cost ratio be greater than one.

The issue is more complicated in the case of project selection, since several alternatives are involved. It is no longer a question of determining the acceptability of a single project but rather selecting from among alternative projects. Consideration should be given to changes in costs and benefits as various strategies are considered. Project selection decisions are illustrated in Figure 5-6.3. The degree of fire protection is given on the horizontal axis, while the marginal costs and benefits associated with various levels of firesafety are given on the vertical axis. As the diagram indicates, marginal costs are low initially and then increase. Less information is available concerning the marginal benefit curve, and it may, in fact, be horizontal. The economically optimum level of fire protection is given by the intersection of the marginal cost and marginal benefit curves. Beyond this point, benefits from increasing fire protection are exceeded by the costs of providing the additional safety.

A numerical example is given in Table 5-6.3. There are five possible strategies or programs possible. The first strategy, A, represents the initial situation, while the remaining four strategies represent various fire loss reduction activities, each with various costs. Strategies are arranged in ascending order of costs. Fire losses under each of the five strategies are given in the second row, while the sum of fire losses and fire reduction costs for each strategy is given in the third row.

TABLE 5-6.3 *Use of Benefit-Cost Analyses in Strategy Selection*

Category	Strategy				
	A	B	C	D	E
Fire Reduction Costs	0	10	25	45	70
Fire Losses	100	70	50	40	35
Sum of Fire Reduction Costs and Fire Losses	100	80	75	85	105
Marginal Benefits	0	30	20	10	5
Marginal Costs	0	10	15	20	25
Benefits-Costs	0	20	5	−10	−20
Benefit-Cost Ratio	—	3.0	1.33	0.5	0.2

Data in the first two rows may then be used to determine the marginal costs or marginal benefits from the replacement of one strategy by another. Thus, strategy B has a fire loss of $70 compared to $100 for strategy A, so the marginal benefit is $30. Similarly, the marginal benefit from strategy C is the reduction in fire losses from B to C or $20. The associated marginal cost of strategy C is $15. Declining marginal benefits and rising marginal costs result in the selection of strategy C as the optimum strategy. At this point, the benefit cost difference is still positive.

Benefit-cost ratios are given in the last row. It is worth noting that, while the highest benefit-cost ratio is reached at activity level B (as is the highest benefit-cost difference), project C is still optimum, since it yields an additional net benefit of $5. This finding is reinforced by examining changes in the sum of fire losses and fire reduction costs. Total cost plus loss first declines, reaching a minimum at point C, and then increases. This is not surprising, since as long as marginal benefits exceed costs, total losses should decrease. Thus, the two criteria—equating marginal costs and benefits, and minimizing the sum of fire losses and fire reduction costs—yield identical outcomes.

Treatment of Uncertainty

A final issue concerns the treatment of uncertainty. One method for explicitly introducing risk considerations is to treat benefits and costs as random variables which may be described by probability distributions. For example, an estimate of fire losses might consider the following events: no fire, minor fire, intermediate fire, and major fire. Each event has a probability of occurrence and an associated damage loss. The total expected loss (EL) is given by

$$EL = \sum_{i=0}^{3} p_i D_i$$

where

p_0 = Probability of no fire,
p_1 = Probability of a minor fire,
p_2 = Probability of an intermediate fire,
p_3 = Probability of a major fire, and
D_n = Associated damage loss, $n = 0,1,2,3$.

Expected losses may be computed for different fire protection strategies. Thus, a fire protection strategy that costs C_3 and reduces damage losses of a major fire from D_3 to D_3' will result in an expected loss

$$EL = p_0 D_0 + p_1 D_1 + p_2 D_2 + p_3 D_3' + C_3$$

Similarly, a fire control strategy that costs C_2 and reduces the probability of an intermediate fire from p_2 to p_2' has an expected loss

$$EL = p_0 D_0 + p_1 D_1 + p_2' D_2 + p_3 D_3 + C_2$$

A comparison of expected losses from alternative strategies may then be used to determine the optimal strategy.

Use of expected value has a limitation in that only the average value of the probability distribution is considered. Discussion of other procedures for evaluating uncertain outcomes is given by Anderson and Settle.[3]

APPENDIX A—SYMBOLS AND DEFINITIONS OF ECONOMIC PARAMETERS

Symbol	Definition of Parameter
A_j	Cash flow at end of period j.
A	End-of-period cash flows (or equivalent end-of-period values) in a uniform series continuing for a specified number of periods.
$\bar{A}$	Amount of money (or equivalent value) flowing continuously and uniformly during each period, continuing for a specified number of periods.
F	Future sum of money. The letter "F" implies future (or equivalent future value).
G	Uniform period-by-period increase or decrease in cash flows (or equivalent values); the arithmetic gradient.
M	Number of compounding periods per interest period.*
N	Number of compounding periods.
P	Present sum of money. The letter "P" implies present (or equivalent present value). Sometimes used to indicate initial capital investment.
$\bar{P}$ or $\bar{F}$	Amount of money (or equivalent value) flowing continuously and uniformly during given period.
S	Salvage (residual) value of capital investment.
f	Rate of price level increase or decrease per period; an "inflation" of "escalation" rate.
g	Uniform *rate* of cash flow increase or decrease from period to period; the geometric gradient.
i	Effective interest rate per interest period* (discount rate), expressed as a percent or decimal fraction.
r	Nominal interest rate per interest period,* expressed as a percent or decimal fraction.

*Normally, but not always, the interest *period* is taken as one *year*. Subperiods, then, would be quarters, months, weeks, etc.

APPENDIX B—FUNCTIONAL FORMS OF COMPOUND INTEREST FACTORS*

Name of Factor	Algebraic Formulation	Functional Format
Group A. All cash flows discrete: end-of-period compounding		
Compound Amount (single payment)	$(1 + i)^N$	$(F/P, i, N)$
Present Worth (single payment)	$(1 + i)^{-N}$	$(P/F, i, N)$
Sinking Fund	$\dfrac{i}{(1+i)^N - 1}$	$(A/F, i, N)$
Capital Recovery	$\dfrac{i(1+i)^N}{(1+i)^N - 1}$	$(A/P, i, N)$
Compound Amount (uniform series)	$\dfrac{(1+i)^N - 1}{i}$	$(F/A, i, N)$
Present Worth (uniform series)	$\dfrac{(1+i)^N - 1}{i(1+i)^N}$	$(P/A, i, N)$
Arithmetic Gradient to Uniform Series	$\dfrac{(1+i)^N - iN - 1}{i(1+i)^N - i}$	$(A/G, i, N)$
Arithmetic Gradient to Present Worth	$\dfrac{(1+i)^N - iN - 1}{i^2(1+i)^N}$	$(P/G, i, N)$
Geometric Gradient to Present Worth (for $i = g$)	$\dfrac{1 - (1+g)^N(1+i)^{-N}}{i - g}$	$(P/A, g, i, N)$
Group B. All cash flows discrete: continuous compounding at nominal rate *r* per period.		
Continuous Compounding Compound Amount (single payment)	e^{rN}	$(F/P, r, N)$
Continuous Compounding Present Worth (single payment)	E^{-rN}	$(P/F, r, N)$
Continuous Compounding Present Worth (single payment)	$\dfrac{e^{rN} - 1}{e^{rN}(e^r - 1)}$	$(P/A, r, N)$
Continuous Compounding Sinking Fund	$\dfrac{e^r - 1}{e^{rN} - 1}$	$(A/F, r, N)$
Continuous Compounding Capital Recovery	$\dfrac{e^{rN}(e^r - 1)}{e^{rN} - 1}$	$(A/P, r, N)$
Continuous Compounding Compound Amount (uniform series)	$\dfrac{e^{rN} - 1}{e^{rN} - 1}$	$(F/A, r, N)$
Group C. Continuous, uniform cash flows: continuous compounding (payments during one period only)		
Continuous, Compounding Present Worth (single, continuous payments)	$\dfrac{i(1+i)^{-N}}{\ln(1+i)}$	$(P/\bar{F}, i, N)$
Continuous Compounding Compound Amount (single, continuous payment)	$\dfrac{i(1+i)^{N-1}}{\ln(1+i)}$	$(F/\bar{P}, i, N)$
Group D. Continuous, uniform cash flows: continuous compounding (payments during a continuous series of periods)		
Continuous Compounding Sinking Fund (continuous, uniform payments)	$\dfrac{\ln(1+i)}{(1+i)^N - 1}$	$(\bar{A}/F, i, N)$
Continuous Compounding Capital Recovery (continuous, uniform payments)	$\dfrac{(1+i)^N\ln(1+i)}{(1+i)^N - 1}$	$(\bar{A}/P, i, N)$
Continuous Compounding Compound Amount (continuous, uniform payments)	$\dfrac{(1+i)^N - 1}{\ln(1+i)}$	$(F/\bar{A}, i, N)$
Continuous Compounding Present Worth (continuous, uniform payments)	$\dfrac{(1+i)^N - 1}{(1+i)^N\ln(1+i)}$	$(P/\bar{A}, i, N)$

*See Appendix A for definitions of symbols used in this table.

APPENDIX C—INTEREST TABLES

TABLE C-6.1 *Present Worth Factor   (Changes F to P)*

Yr.	2%	4%	6%	8%	10%	12%	15%	20%	25%	30%	40%	50%
1	.9804	.9615	.9434	.9259	.9091	.8929	.8696	.8333	.8000	.7692	.7143	.6667
2	.9612	.9246	.8900	.8573	.8264	.7972	.7561	.6944	.6400	.5917	.5102	.4444
3	.9423	.8890	.8396	.7938	.7513	.7118	.6575	.5787	.5120	.4552	.3644	.2963
4	.9238	.8548	.7921	.7350	.6830	.6355	.5718	.4823	.4096	.3501	.2603	.1975
5	.9057	.8219	.7473	.6806	.6209	.5674	.4972	.4019	.3277	.2693	.1859	.1317
6	.8880	.7903	.7050	.6302	.5645	.5066	.4323	.3349	.2621	.2072	.1328	.0878
7	.8706	.7599	.6651	.5835	.5132	.4523	.3759	.2791	.2092	.1594	.0949	.0585
8	.8535	.7307	.6274	.5403	.4665	.4039	.3269	.2326	.1678	.1226	.0678	.0390
9	.8368	.7026	.5919	.5002	.4241	.3606	.2843	.1938	.1342	.0943	.0484	.0260
10	.8203	.6756	.5584	.4632	.3855	.3220	.2472	.1615	.1074	.0725	.0346	.0173
11	.8043	.6496	.5268	.4289	.3505	.2875	.2149	.1346	.0859	.0558	.0247	.0116
12	.7885	.6246	.4970	.3971	.3186	.2567	.1869	.1122	.0687	.0429	.0176	.0077
13	.7730	.6006	.4688	.3677	.2897	.2292	.1625	.0935	.0550	.0330	.0126	.0051
14	.7579	.5775	.4423	.3405	.2633	.2046	.1413	.0779	.0440	.0254	.0090	.0034
15	.7430	.5553	.4173	.3152	.2394	.1827	.1229	.0649	.0352	.0195	.0064	.0023
16	.7284	.5339	.3936	.2919	.2176	.1631	.1069	.0541	.0281	.0150	.0046	.0015
17	.7142	.5134	.3714	.2703	.1978	.1456	.0929	.0451	.0225	.0116	.0033	.0010
18	.7002	.4936	.3503	.2502	.1799	.1300	.0808	.0376	.0180	.0089	.0023	.0007
19	.6864	.4746	.3305	.2317	.1635	.1161	.0703	.0313	.0144	.0068	.0017	.0005
20	.6730	.4564	.3118	.2145	.1486	.1037	.0611	.0261	.0115	.0053	.0012	.0003
21	.6698	.4388	.2942	.1987	.1351	.0926	.0531	.0217	.0092	.0040	.0009	
22	.6468	.4220	.2775	.1839	.1228	.0826	.0462	.0181	.0074	.0031	.0006	
23	.6342	.4057	.2618	.1703	.1117	.0738	.0402	.0151	.0059	.0024	.0004	
24	.6217	.3901	.2470	.1577	.1015	.0659	.0349	.0126	.0047	.0018	.0003	
25	.6095	.3751	.2330	.1460	.0923	.0588	.0304	.0105	.0038	.0014	.0002	
30	.5521	.3083	.1741	.0994	.0573	.0334	.0151	.0042	.0012			
35	.5000	.2534	.1301	.0676	.0356	.0189	.0075	.0017	.0004			
40	.4529	.2083	.0972	.0460	.0221	.0107	.0037	.0007	.0001			
45	.4102	.1712	.0727	.0313	.0137	.0061	.0019	.0003				
50	.3715	.1407	.0543	.0213	.0085	.0035	.0009	.0001				
60	.3048	.0951	.0303	.0099	.0033	.0011	.0002					
70	.2500	.0642	.0169	.0046	.0013	.0004						
80	.2051	.0434	.0095	.0021	.0005							
90	.1683	.0293	.0053	.0010	.0002							
100	.1380	.0198	.0030	.0005								

$$\frac{1}{(1+i)^y}$$

REFERENCES CITED

1. American National Standards Institute, *Industrial Engineering Terminology*, ANSI Z94.0-89, Industrial Engineering and Management Press, Atlanta (1989).
2. "Federal Fire Foci," *SFPE Bul.*, (1977).
3. L.G. Anderson, and R.E. Settle, 1977, *Benefit-Cost Analysis: A Practical Guide*, Lexington Books, Lexington (1977).

ADDITIONAL READING

L. Blank and A. Tarquin, *Engineering Economy*, McGraw-Hill, New York (1983).

P.E. DeGarmo, W.G. Sullivan, and J.R. Canada, *Engineering Economy*, Macmillian, New York (1984).

G.A. Fleischer, *Engineering Economy: Capitol Allocation Theory*, Brooks/Cole, New York (1984).

E.L. Grant, G.W. Ireson, and R.S. Leavenworth, *Principles of Engineering Economy*, 7th ed., John Wiley and Sons, New York (1982).

D.G. Newman, *Engineering Economic Analysis*, Engineering Press (1983).

J.L. Riggs and T.M. West, *Engineering Economics*, McGraw-Hill, New York (1986).

G.J. Thuesen and W.J. Fabrycky, *Engineering Economy*, Prentice-Hall, Englewood Cliffs (1984).

ASTM E833, *Standard Terminology of Building Economics*, American Society for Testing and Materials, Philadelphia (1992).

ASTM E917, *Standard Practice for Measuring Life-Cycle Costs of Buildings and Building Systems*, American Society for Testing and Materials, Philadelphia (1993).

ASTM E964, *Standard Practice for Measuring Benefit-to-Cost and Savings-to-Investment Ratios for Buildings and Building Systems*, American Society for Testing and Materials, Philadelphia (1993).

ASTM E1074, *Standard Practice for Measuring Net Benefits for Investments in Buildings and Building Systems*, American Society for Testing and Materials, Philadelphia (1993).

TABLE C-6.2 *Capital Recovery Factor* *(Changes P to A)*

Yr.	2%	4%	6%	8%	10%	12%	15%	20%	25%	30%	40%	50%
1	1.020	1.040	1.060	1.080	1.100	1.120	1.150	1.200	1.250	1.300	1.400	1.500
2	.5150	.5302	.5454	.5608	.5762	.5917	.6151	.6545	.6944	.7348	.8167	.9000
3	.3468	.3603	.3741	.3880	.4021	.4163	.4380	.4747	.5123	.5506	.6294	.7105
4	.2626	.2755	.2886	.3019	.3155	.3292	.3503	.3863	.4234	.4616	.5408	.6231
5	.2122	.2246	.2374	.2505	.2638	.2774	.2983	.3344	.3719	.4106	.4914	.5758
6	.1785	.1908	.2034	.2163	.2296	.2432	.2642	.3007	.3388	.3784	.4613	.5481
7	.1545	.1666	.1791	.1921	.2054	.2191	.2404	.2774	.3163	.3569	.4419	.5311
8	.1365	.1485	.1610	.1740	.1874	.2013	.2229	.2606	.3004	.3419	.4291	.5203
9	.1225	.1345	.1470	.1601	.1736	.1877	.2096	.2481	.2888	.3312	.4203	.5134
10	.1113	.1233	.1359	.1490	.1627	.1770	.1993	.2385	.2801	.3235	.4143	.5088
11	.1022	.1141	.1268	.1401	.1540	.1684	.1911	.2311	.2735	.3177	.4101	.5059
12	.0946	.1066	.1193	.1327	.1468	.1614	.1845	.2253	.2685	.3135	.4072	.5039
13	.0881	.1001	.1130	.1265	.1408	.1557	.1791	.2206	.2645	.3102	.4051	.5026
14	.0826	.0947	.1076	.1213	.1357	.1509	.1747	.2169	.2615	.3078	.4036	.5017
15	.0778	.0899	.1030	.1168	.1315	.1469	.1710	.2139	.2591	.3060	.4026	.5011
16	.0737	.0858	.0990	.1130	.1278	.1434	.1679	.2114	.2572	.3046	.4019	.5008
17	.0700	.0822	.0954	.1096	.1247	.1405	.1654	.2094	.2558	.3035	.4013	.5005
18	.0667	.0790	.0924	.1067	.1219	.1379	.1632	.2078	.2546	.3027	.4009	.5003
19	.0638	.0761	.0896	.1041	.1195	.1358	.1613	.2065	.2537	.3021	.4007	.5002
20	.0611	.0736	.0872	.1019	.1175	.1339	.1598	.2054	.2529	.3016	.4005	.5002
21	.0588	.0713	.0850	.0998	.1156	.1322	.1584	.2044	.2523	.3012	.4003	.5000
22	.0566	.0692	.0830	.0980	.1140	.1308	.1573	.2037	.2519	.3009	.4002	
23	.0547	.0673	.0813	.0964	.1126	.1296	.1563	.2031	.2515	.3007	.4002	
24	.0529	.0656	.0797	.0950	.1113	.1285	.1554	.2025	.2512	.3006	.4001	
25	.0512	.0640	.0782	.0937	.1102	.1275	.1547	.2021	.2510	.3004	.4001	
30	.0446	.0578	.0726	.0888	.1061	.1241	.1523	.2008	.2503	.3001	.4000	
35	.0400	.0536	.0690	.0858	.1037	.1223	.1511	.2003	.2501			
40	.0366	.0505	.0664	.0839	.1023	.1213	.1506	.2001	.2500			
45	.0339	.0483	.0647	.0826	.1014	.1207	.1503	.2001				
50	.0318	.0466	.0634	.0817	.1009	.1204	.1501	.2000				
60	.0288	.0442	.0619	.0808	.1003	.1200	.1500					
70	.0267	.0428	.0610	.0804	.1001							
80	.0252	.0418	.0606	.0802	.1000							
90	.0241	.0412	.0603	.0801								
100	.0232	.0408	.0602	.0800								

$$\frac{i(1+i)^y}{(1+i)^y - 1}$$

R.E. Chapman, "A Cost-Conscious Guide to Fire Safety in Health Care Facilities," *NBSIR 82-1600*, National Bureau of Standards, Washington (1982).

R.E. Chapman and W.G. Hall, "Code Compliance at Lower Costs: A Mathematical Programming Approach," *Fire Tech.*, 18, 77 (1982).

L.P. Clark, "A Life-Cycle Cost Analysis Methodology for Fire Protection Systems in New Health Care Facilities," *NBSIR 82-2558*, National Bureau of Standards, Washington (1982).

J.S. McConnaughey, "An Economic Analysis of Building Code Impacts: A Suggested Approach," *NBSIR 78-1528*, National Bureau of Standards, Washington (1978).

R.T. Ruegg and S.K. Fuller, "A Benefit-Cost Model of Residential Fire Sprinkler Systems," *NBS Technical Note 1203*, National Bureau of Standards, Washington (1984).

CONSEQUENTIAL/ INDIRECT LOSS

G. Ramachandran

INTRODUCTION

During the course of its development, a fire can cause damage to a building and its contents and to occupants (fatal or non-fatal casualties). These costs are known as direct losses, whereas those associated with the fire but incurred after it is extinguished are indirect or consequential losses. According to this definition, the distress and financial loss that an individual's death or injury would cause to his/her family is an indirect loss. This component of indirect loss is, however, not discussed in this chapter which is only concerned with consequential losses, such as loss of production; of trade, e.g., profits; of employment; and of exports and costs toward extra imports. These losses occur mainly in industrial and commercial sectors.

Consequential losses due to fires is an underresearched topic on which only very few investigations have been carried out. These studies and some major fires are briefly reviewed, with reference to the evaluation of consequential losses and the factors affecting them. Other topics discussed include the role of fire protection and utility theory.

LEVELS OF ECONOMIC ACTIVITY

In assessing consequential losses an important problem to be reconciled is concerned with the differences between two major levels of economic activity: (1) private sector/ community level and (2) national/societal level. The first level includes the fire-hit firm and firms supplying to or purchasing from the fire-hit firm's materials, components, or services. Costs associated with moving, temporary accommodation, and lost profits are valid costs at the private sector level but not at the national level.

At the national level, the loss of a specific unit of productive capacity may be spread among the remaining capacity in the nation such that competitors may seize the opportunity to enter the market and maintain the national rate and volume of manufacture. Consequently, it is likely that there

is only a small incremental loss to the national economy as a result of a fire in the premises of, e.g., a manufacturing firm. This is also because of the redistribution or "netting out" effect of some of the losses at the societal level.

INSURANCE STATISTICS

The effects of a fire on the earning capacity of a firm can be measured in terms of loss of profits during the period of interruption following the damage until the resumption of the activity in which the firm was engaged before the fire. Loss of profits is usually expressed as a percentage of loss of turnover. A cover against this loss can be obtained by purchasing a consequential loss insurance policy, the premium for which is a function of the period of indemnity. Loss of profits sustained by a supplier or customer of the fire-hit firm can be covered by a normal consequential loss policy based on reduction in turnover.

The form of insurance policy in more general use in the United States is known as business interruption insurance (BII) and this operates on lines similar to the United Kingdom contract of consequential loss insurance (CLI) with a turnover specification, though there are some differences. For private sector level insurance firms transacting BII or CLI there are useful sources of data for estimating consequential losses due to fires in industrial and commercial premises. Organizations, such as Insurance Institute of America, compile consequential loss data furnished by major insurance firms.

It is, however, doubtful whether insurance statistics can provide realistic estimates of consequential losses to the national economy, which is further influenced by several economic factors. These factors include level of employment or unemployment, level of capacity utilization, volume of exports and imports, exchange rates, and performance of national and international competitors. Due to the interactions of these factors, evaluation of consequential losses to the national economy is a complex problem requiring the application of econometric models, such as of the input-output type.

SPECIAL FACTORS— PRIVATE SECTOR LEVEL

Certain types of industrial or commercial activities may have special factors affecting consequential losses. These

Dr. G. Ramachandran retired in November 1988 as Head of the Operations Research Section at the Fire Research Station, U.K. Since then he has been practicing as a consultant in risk evaluation and insurance. He is a visiting professor at the University of Hertsfordshire and Glasgow Caledonian University. His research has focused on statistical and economic problems in fire protection and actuarial techniques in fire insurance.

factors were identified in a series of studies conducted by the Insurance Technical Bureau (ITB) in the United Kingdom, now defunct. The observations on consequential losses contained in these reports may or may not be applicable to conditions prevailing now in the United Kingdom or other countries. However, the following summaries extracted from three of the reports provide an indication of the various special factors one should consider in the evaluation of consequential losses due to fires and other hazards.

A few production lines of a plant manufacturing pharmaceutical products[1] may generate an abnormally high proportion of gross profits. Restrictions imposed by the licensing authority may limit possibilities for manufacture in other plants or even other lines in the same plant. Natural raw materials may be irreplaceable or out of season. Specialized plant equipment, e.g., tailor-made driers or centrifuges, may involve long delays for replacement. Loss of laboratory facilities may seriously interrupt testing and quality control programs.

Aerospace industry[2] is another example where some activities are of special importance, particularly in the development of a new aircraft-prototype assembly, untried or unproven research-and-development projects, and fatigue testing of aircraft structures. Loss of any of the above could result in a significant interruption to the program. In addition, the effect of delays in the development or supply of components or assemblies from specialist equipment manufacturers can be serious. The interactions of the many activities and firms involved in the manufacture of aerospace products makes for involved consequential loss considerations.

Resin, paint, and ink manufacture[3] would not normally be expected to give rise to unduly high consequential loss. Facilities are generally dispersed in small units throughout a given country, and there may be sufficient manufacturing capacity to absorb temporary loss at individual sites. Also few, if any, products are so special that they cannot be made elsewhere in the industry. Consequential loss, therefore, hinges primarily on the time for reinstatement of the plant and the ability of management to arrange for the supply of goods from other sources, pending a return to full production. Loss of raw materials or finished goods normally results in relatively short interruption periods. However, longer periods may be required for the replacement of tanks and pumps destroyed by fires and other hazards, such as explosion.

Due to high investment costs, specialized equipment, e.g., those which are electronically or computer controlled, are generally used at full capacity in some industrial processes. Continuous operation of these processes may reduce the chance of a fire spreading, but provides no scope for making up for lost production following a fire. Specialized equipment, if damaged by fire, cannot be replaced easily or quickly, since either they or spare parts for them may have to be imported. Industries using such equipment are liable to sustain high consequential losses.

RESEARCH STUDIES— NATIONAL LEVEL

While decades of research have provided a good deal of knowledge about the physical effects of fires, including direct damage and methods to control them, there is still very little understanding of the indirect or consequential losses due to fires particularly at the national level. The scarcity of research studies on consequential losses is partly due to the fact that such losses are considered, to some extent, as intangible costs of fires. Hence, these losses are not taken into account sufficiently in determining fire protection and insurance requirements for industrial and commercial buildings. Only some prudent business undertakings and their suppliers or customers adopt loss of profits or business interruption insurance as an essential complement of insurance against material (direct) damage. Consequential losses to the national economy are, however, rarely considered.

In the United Kingdom, the Home Office carried out during 1970–80 two research studies on consequential losses to the national economy. The first study[4] adopted an input-output-type model in which all losses were considered as output losses. They would either be: (1) losses in the type of output actually hit by fire or (2) losses in some other output, because production factors, e.g., fixed assets, entrepreneurial effort, or labor, have been less effectively employed as a result of the fire. The effects of a fire were assumed to impact most on: fire-hit firm, supplying firm, purchasing firm, parallel firm, and rest of the economy. A fire-hit firm was defined as a compartment of production covering just that type of output which had been hit by a fire and no other output. A parallel firm was defined as the compartment of a firm that produced in parallel to the fire-hit compartment (which might be in the same firm or in another firm). Any effects in a parallel firm or somewhere else in the rest of the economy were assumed to be included in the calculation of the effects in a fire-hit firm, in a supplying firm, or in a purchasing firm.

In the Home Office study,[4] consequential losses were measured by the net present values of streams of annual outputs lost by the fire-hit firm, supplying firms, and purchasing firms. In regard to the fire-hit firm, it was necessary to determine a length of time over which fixed assets destroyed by fire were assumed not to be replaced by extra investment in the economy. This time choice had to depend on a view of the future course of the economy, which depended on unknown events and influences. Hence, alternative calculations were produced that were based on the remaining lives of the assets and on a number of shorter periods. The net present values were corrected for offsetting influences within the fire-hit firm, supplying firms, and purchasing firms. These influences were due to two factors: (1) some production factors affected by fire might be used elsewhere in the economy, and (2) production factors already employed elsewhere might be used more intensively. The extent to which such off-setting influences would operate would depend largely on the level of employment and the pressure of demand in the economy. Separate calculations were made for three alternative cases: (1) slack, (2) middle, and (3) tight conditions in the economy. Results were given for each of 15 industries, including a factor by which a fixed assets valuation should be multiplied to give the sum of all the corrected output losses.

In order to verify the assumptions employed and results obtained in the study mentioned above, the Home Office commissioned a field research[5] aimed at an in-depth investigation of a small sample of fires. This study involved direct contact with fire-hit firms and concentrated on direct, consequential, and, hence, the total loss to the UK economy from industrial, distributive, and service sector fires. During the first (pilot) stage, 10 firms covering a range of industries and fire size were interviewed in order to confirm the practicality of the method adopted. Fires on multi-occupancy sites were excluded, since it was found difficult to identify the firms hit by such fires.

Using the method identified in the first stage, only 75 fires were investigated in the second stage for reasons of economy but were chosen so as to reflect certain key parameters involved in the sampling process. This was the minimum number to permit coverage of one fire from each industry in each year. Since many of these fires produced no or small consequential loss to the national economy, 10 more fires were selected to ascertain the key factors leading to large consequential losses. In all, 20 fires were identified in the second stage where the fire-hit firms had reported significant effects on one or more of the following: suppliers, customers, competitors, employment, investment, and foreign trade. Firms involved in only sixteen of these fires agreed to participate in a further investigation for the third stage. The fourth stage of the study was concerned with a postal survey to provide supplementary data and the analysis of all data, including those collected in earlier stages.

In the second stage, estimates of direct loss were based on insurance figures. Consequential losses were considered to arise from loss of exports, extra imports, the diversion of resources from other productive activities, and reduction in the efficiency of resource use following the fire. The study assumed: (1) full-capacity utilization of resources and (2) that market values of the resources reflected their true worth. Insurance estimates of losses were used as measures of the assets destroyed in fires and, by application of national capital output ratios, these asset losses were translated into losses of output from fire. Allowances were made for the secondary impact on suppliers and customers of fire-hit firms and for the impact of the level of capacity utilization. A correction factor was applied to account for the ability of the economy to "make good" the losses of the fire-hit firm by other firms. The analysis produced estimates of the ratios of consequential to direct losses to the economy for "off-peak" and "peak" years and for each industry and service sector. The main conclusion was that most fires, except those in chemical and allied industries, produced no consequential losses to the national economy. Only in one sector (chemicals) was evidence found of a statistical link between consequential losses and direct losses. The study failed to estimate this link for other sectors and a number of other possible effects on consequential losses.

RESEARCH STUDY—
PRIVATE SECTOR LEVEL

The Home Office reports aforementioned were not published, perhaps due to the unacceptability of the results and conclusions by major industrial undertakings and insurance organizations. The only published research study appears to be the study by Hicks and Liebermann,[6] which deals with costs and losses from the community/private perspective as they impact the fire victim. The property class categories addressed in this study only included commercial occupancies separated into four types: (1) mercantile, (2) non-manufacturing, (3) manufacturing, and (4) warehouses.

Indirect losses could be viewed as a type of "production process," the product being indirect fire losses. Hence, Hicks and Liebermann[6] considered first the following expression, based on a convenient formulation of the Cobb-Douglas production function[7]

$$IL = ke^{rT}E^a X^{1-a} \tag{1}$$

where:

IL = indirect loss,
k = constant,
r, a = regression coefficients,
E = expenditure for fire protection $(-)$,
X = number of fires $(+)$, and
T = time (surrogate for technological advance) $(-)$.

The signs in parentheses relate to the expected values of the coefficients for the independent variables. The term ke^{rT} is a scalar factor in which r measures increases in fire department efficiency due to technological advances in suppression equipment, training, and/or facilities as well as altered building codes, smoke alarms, and the like. Equation 1 can be converted to a multiple-regression model by taking logarithms of terms on both sides.

In principle, the parameters r and a can be estimated but, in practice, it proved to be an insurmountable task to attempt to collect statistics in the detail required for the regression analysis. Hence, due to serious data constraints and limited resources to generate original statistics, the following general form was adopted, which proved successful.

$$IL = c(DL)^b \tag{2}$$

where DL is the direct loss and c and b are constants. Equation 2 was based on the assumption that very small fires typically generate small indirect losses while large fires produce larger indirect losses. Equation 2 can be transformed into a simple regression by taking logarithms of terms on both sides.

It can be observed that the time component, T, in Equation 1 has not been included in Equation 2. This was because the test of this component did not yield satisfactory results. The values of the regression coefficients, log c and b, were estimated for six levels: local; national; and the four types of occupancies mentioned earlier, i.e., mercantile, nonmanufacturing, manufacturing, and warehouses. The estimation for the first two levels, i.e., local and national, was based on data for all four types of occupancies. Data for the national level were also provided by the Insurance Services Office (ISO), New York. For the local level, two insurance firms furnished information processed at the local insurance company level prior to transmittal to the ISO; these statistics were augmented by those provided by six case studies. The ISO data and insurance company statistics were combined for an occupancy level other than warehouses for which only ISO data were available. All the data used in the six regression models were in millions of dollars, normalized to 1976 dollar values.

Statistical tests of significance showed that the regression model fitted well with the data in all the cases except warehouses. Additional data might have, perhaps, improved the statistical significance of the warehouse model. The results obtained by Hicks and Liebermann[6] are given in Table 5-7.1. Since nationally aggregated data were utilized, it was recommended that the "occupancy-specific" models be used only at the national level and that any desired analysis of local impacts be accomplished using the "local" model.

Hicks and Liebermann have established a "power" relationship between indirect and direct fire loss. The value of this exponent, parameter b, has been estimated to be greater than unity for local and national levels and less than unity

TABLE 5-7.1 *Relationship Between Direct and Indirect Fire Loss Model* Parameters*[†]

Level	Parameters	
	c	b
Local	0.203	1.146
National	0.015	1.245
Mercantile	0.109	0.889
Non-manufacturing	0.069	0.874
Manufacturing	0.135	0.890
Warehouse	0.047	0.804

*Source: Hicks and Liebermann (reference 6 and Equation 2).
[†]Losses in millions of dollars at 1976 dollar values.

for the occupancy levels. For any increase in direct loss, the ratio of indirect to direct loss would increase if $b > 1$, and decrease if $b < 1$. The ratio would be a constant if $b = 1$. From the information given in the study, it was not possible to test whether the value of b was significantly different from unity for any of the six levels. The results of Hicks and Liebermann, however, cast doubts on the use of a constant value for the ratio between indirect and direct fire loss.

Statistical (actuarial) techniques are well developed for calculating the insurance premium for loss of profits due to fire. See, for example, Benckert.[8] The "risk premium" is a function of the period of indemnity, and is generally expressed as the product of the loss frequency and the mean amount of loss. The loss frequency is assumed to be independent of the period of indemnity. The frequency function of the period of interruption following a fire has a log-normal distribution.[8,9] An insurance company generally adds two types of "loading" to the risk premium to calculate the premium payable by a policy-holder. First, a safety loading is added toward chance fluctuations of loss beyond the expected loss. Second, another loading is imposed to cover the insurer's operating costs, which include profits, taxes, and other administrative expenses. A number of texts have been published on different types of insurance and claims concerned with consequential losses; see, for example, Riley.[10]

NATIONAL ESTIMATES

According to the Bland report,[11] productivity losses in the United States were somewhat greater than direct losses ($3.3 billion *versus* $2.7 billion per annum, respectively). In a series of papers, Wilmot[12,13,14] has produced national estimates on consequential losses for different countries for some years. According to his figures, indirect fire losses in European countries, except UK and France, were 25 percent or less of the direct losses. According to his latest figures,[14] reproduced in Table 5-7.2, the percentage was 42 for UK and 29 for France. Wilmot's estimate for the USA was very low (6 percent) compared with 122 percent as given in the Bland report.[11] This large difference was mainly due to definitions adopted for consequential losses. Wilmot's figures are the only available national totals for consequential losses due to fires in different countries. They have been produced on widely varying bases and, hence, should be regarded with some reservations.

As pointed out by Rasbash,[15] low estimates, such as those obtained by Wilmot, may be due to the assessment of indirect losses to the nation rather than to the sum of such losses for those who suffer fires. At the national level, one person's or firm's loss due to economic stress following a fire can be another person's or firm's gain, which therefore cancels it out. A question, however, arises as to whether the national picture is the one that really matters. An individual's loss and inconvenience is felt no less deeply, even if other persons can benefit as a result. Moreover, for the individual who loses, his/her loss is acute; whereas for those who gain, the gain is usually marginal and probably unnoticed. If all such consequential losses and hurt were insured as the bulk of direct losses are, it would cost the community (nation and individual) a total consequential loss that is likely to be nearer to the direct loss.[15]

The hypothesis postulated by Rasbash[15] will have some validity if the values of the parameters c and b in Equation 2 are close to unity. The value of b in Table 5-7.1 may not be significantly different from unity for the national level (and other levels), if larger samples of data had been used in the regression analysis. The value of c is, however, significantly closer to zero than unity.

TABLE 5-7.2 *Ratio of Indirect to Direct Fire Loss*

Country	Indirect Loss* (percent of GDP[†])	Direct Loss* (percent of GDP[†])	Ratio of Indirect to Direct Loss
United States	0.011 (1982/83)	0.18	0.061
Japan	0.016	0.18	0.089
Norway	0.024 (1985)	0.45	0.053
Sweden	0.025	0.28	0.089
Netherlands	0.026 (1984)	0.22	0.118
Austria	0.029 (1979/80)	0.21 (1979/80)	0.138
Germany (West)	0.035	0.18	0.194
Denmark	0.039	0.29	0.134
Finland	0.040	0.22	0.182
United Kingdom	0.076	0.18	0.422
France	0.084 (1980/81)	0.29 (1981/82)	0.290

*Source: Wilmot (reference 14).
[†]Gross Domestic Product.
NOTE: Average values relating to 1985/1986.

MAJOR FIRES (CASE STUDIES)

Major fires causing severe consequential losses have occurred in several countries during the last two or three decades. A number of catastrophic hotel fires with huge consequential losses also have occurred during the last decade.

Royal Army Ordnance Depot, UK

In the United Kingdom, a warehouse at the Royal Army Ordnance Depot in Donnington was completely destroyed by fire and collapsed in June 1983. The direct loss was estimated at £165 million, but this figure was thought not to have included stores that were destroyed and could not be replaced. In addition, there could have been consequential losses due to the destruction of irreplaceable spare parts that were not estimated. Apart from the financial loss, there was also the strategic value of the goods: i.e., the security of the United Kingdom.

Sunshine Silver Mine, U.S.

In May 1972, one of the worst mining accidents in history occurred at the Sunshine Silver Mine, located in Kellogg, ID. Ninety-one lives were lost when a fire broke out at the 3400-ft level of the deepest silver mine in North America. Toxic smoke spread rapidly throughout the working areas of the mine, rendering useless a number of self-contained breathing devices available to those trapped in the mine.

Liquefied Natural Gas Storage Tank, U.S.

A major fire incident occurred on Staten Island, NY, in February 1973 while 40 workers were repairing the mylar lining of a liquefied natural gas (LNG) storage tank facility that was down for repairs. A small fire began during the repair process and resulted in an explosive fire that trapped and killed all the workers inside the tank facility.

MGM Grand Hotel, U.S.

In November 1980, a fire broke out in the kitchen area of a restaurant located on the lobby floor of the 26-story casino and hotel, the MGM Grand Hotel in Las Vegas, NV. The fire rapidly involved the first floor public areas of the hotel, and smoke quickly coursed through elevator shafts and enclosed stairwells. Numerous victims were trapped in the high-rise tower, resulting in 84 fatal and over 600 non-fatal casualties. Property damage exceeded $150 million.

Las Vegas Hilton Hotel, U.S.

In February 1981, a fire broke out in the elevator corridor on the eighth floor of the 30-story Las Vegas Hilton Hotel. The fire soon breached the window, causing flame to impinge on the floor directly above. In a period of less than 3 minutes, the fire leaped in similar fashion to each of the elevator lobbies above the eighth floor. The fire resulted in the deaths of 8 and personal injuries to approximately 350 people. On a comparative basis the personal injuries sustained by the survivors of the Las Vegas Hilton fire were more severe than MGM Grand.

San Juan Dupont Plaza Hotel, PR

On the afternoon of New Year's Eve, 1986, a fire started in a ballroom on the ground floor of the 15-story San Juan Dupont Plaza Hotel. Within 8 minutes the fire trapped the majority of its victims in a casino located 1 story above the area of origin. The fire killed 97 people and resulted in over 2000 insurance claims.

Other Major Fires

Other major fires occurred in the Kentucky Beverley Hills Supper Club, U.S. (1977) and Isle of Man Summerland Recreation Complex, UK (1973). Further, in the UK two major fires occurred in department stores: Henderson's, Liverpool (1960) and Woolworth's, Manchester (1979). Big hotels and department stores, such as those mentioned above, and their suppliers of goods and services would have sustained heavy consequential losses.

Case studies and litigation reports on major fires provide useful data on direct and consequential losses and the factors affecting them. These data and those available from other sources can be combined and a "meta-analysis" performed for obtaining reliable estimates on the relationship between direct and consequential losses for major industrial and commercial sectors.

ROLE OF FIRE PROTECTION

Insurance, in many cases, can provide indemnity against direct and consequential losses; however, in other cases, the loss of production over a period of several months can lead to a permanent loss of markets to the competitors; and, in extreme cases, may even result in the closure of the firm and job losses. A high percentage of firms hit by large fires face bankruptcy within a short period after the fire. It is in this area, i.e., serious disruption or bankruptcy, that the justification for fire prevention and protection measures reaches its maximum level of importance, particularly for parts of an industrial or commercial property that have potentially high risk of consequential losses. It is necessary to: (1) identify such parts or areas of a property and (2) provide appropriate safety measures.

The now-defunct Insurance Technical Bureau (ITB) in the UK, previously mentioned, conducted studies on fire and explosion hazards in major manufacturing industries in the UK. Some of these investigations identified areas, e.g., plants and equipment of an industry exposed to consequential losses, and factors likely to cause such losses. (See Section entitled "Special Factors—Private Sector Level.)

The reports of the ITB also contain recommendations for necessary fire protection measures to be adopted, particularly for reducing consequential losses. In the pharmaceutical industry,[1] for example, high levels of fire detection and protection are justified for production areas of patented products responsible for the bulk of profits. Adequate fire precautions are necessary for sites where bulk chemicals and other raw materials are supplied from outside and stored. A high standard of engineering maintenance and housekeeping is important to prevent explosions due to leakage in a bulk-powder-handling plant. Drums containing highly flammable liquids should be stored in a properly designed drum park with a curb or embankment, protection from the sun, with suitable drainage and suitable arrangement for access and handling of the drums. If the drum park is unavoidably close to buildings, it should be provided with fixed fire-fighting facilities, e.g., water-drench or foam, as appropriate. Drums should never be discharged inside a storage area, but must be removed to a special area. The ITB report[1] also contains recommendations for location of solvent recovery plants; for protection of tank farms containing solvents, several types of driers, and centrifuges; and for laboratories including laboratory documents.

In buildings designed for aircraft assembly or built for aircraft assembly but used for other purposes, factors, e.g., the presence of heat sinks and ventilation arrangements, might increase the time to response of conventional sprinkler arrangements. Shorter response times are achieved by deluge systems actuated by faster detectors, such as rate-of-rise detectors.[2] Such systems are desirable for protection of buildings with large volumes of flammable solvents, and for buildings over 10 m in height used for storage, general machinery, or assembly purposes. Supplementary underwing protection is desirable for large-wing aircraft; such aircraft block substantial areas from fire extinguishment by overhead sprinklers. In tall sprinklered buildings containing high values and in areas where fuels are present, detectors enable fires to be contained more quickly, possibly before sprinklers are activated. The ITB report[2] has discussed several factors supporting the increased use of detectors in the aerospace industry, including factors that may reduce the effectiveness of the works fire brigade.

A third ITB report,[3] mentioned earlier, recommended several fire safety measures for paint and ink manufacturing industries. Measures to reduce the possibility of fire or explosion in resin plants include: the use of indirect heating systems (oil and steam) or induction heating in place of conventional gas heating, and fitting of pressure relief valves with vent pipes to safe areas or, preferably, into vessels at least 1.5 times the capacity of the kettle being vented. Runaway reactions are usually prevented by "crash cooling" procedures, whereby water is circulated through cooling coils fitted inside or around the kettle. Other measures for a resin plant include the installation of flame-proofed or intrinsically safe electrical equipment and sprinkler systems. In paint manufacturing, solvent-containing products should be made in an area separate from aqueous-based (e.g., emulsion) lines, and oil-based paints should be segregated from highly flammable materials, such as cellulose. Similarly, in the ink industry, black inks, paste inks, and liquid inks should be segregated. (Segregation of manufacture of different paint types is a common practice.) Laboratories in paint and ink factories should be segregated from manufacturing zones; where this is not possible, the laboratory should be pressurized. The ITB report contains several other safety recommendations for the paint and ink industry, including sprinkler protection in storage areas, particularly where non-aqueous products are stored.

Woolhead[16] has discussed the indirect losses to a business activity arising from electrical cable fires. For cable protection, special ready-mixed inert materials can be applied by brush, spray, or trowel to single or grouped cables and supporting trays. These coatings are also moisture and humidity resistant. Some manufacturers can also provide material for firestopping where cables pass through apertures and fire barriers for cable trenches or cable tunnels. Other recommended safety measures include fire detection systems; water spray systems, particularly where multi-layers of cable trays are involved; and gas flooding systems (halon or carbon dioxide). Apart from fire protection, a contingency plan[17] should be prepared for main exposure areas including customers, suppliers of goods and services, transport, and distribution.

UTILITY FUNCTION

The negative values for log c (Table 5-7.1) and the intercept term in the regression line based on Equation 2 indicate the fact that there will be no indirect losses for fires with direct losses less than a minimum threshold level. A direct loss in excess of a maximum might put a business out of action permanently or temporarily for a long duration. By only considering fires with direct losses between these minimum and maximum limits, Equation 2 can be modified to the form in Equation 23 of Section 5, Chapter 9 with $U(x)$ denoting disutility, i.e., negative counterpart of utility, measured in terms of consequential losses. This modified form can provide some idea of the shape of the utility function for any area of an industrial or commercial activity. But the parameter θ of the utility function also depends on other factors, such as the attitude to risk and financial strength of a particular property owner.

REFERENCES CITED

1. "Fire and Explosion Hazards in the UK Pharmaceutical Industry," The Insurance Technical Bureau, London, Apr. 1977.
2. "Fire and Explosion Hazards in the UK Aerospace Industry," The Insurance Technical Bureau, London, Nov. 1976.
3. "Fire and Explosion Hazards in the UK Paint and Ink Manufacturing Industries," The Insurance Technical Bureau, London, May 1978.
4. "The Economic Cost of Fire," Report by the Economist Intelligence Unit Ltd. to the Home Office, London, Dec. 1971 (unpublished).
5. "Investigation of Consequential Losses to the Economy from Fires," report by PA Management Consultants Ltd. to the Home Office, London (1977) (unpublished).
6. H.L. Hicks and R.R. Liebermann, "A Study of Indirect Fire Losses in Non-Residential Properties," *Fou-brand*, 1, pp. 8–15 (1979).
7. J.M. Henderson and R.E. Quandt, *Microeconomic Theory*, Chapter 3, McGraw-Hill, New York (1971).
8. Lars-G. Benckert, "The Premium for Insurance Against Loss of Profit Due to Fire as a Function of the Period of Indemnity," *Transactions* of the 15th International Congress of Actuaries, New York, pp. 297–305 (1957).
9. D. Flach, Schlunz, and J. Straub, "An Analysis of German Fire Loss of Profits Statistics," *Blatter der Deutschen Gesellschaft fur Versicherungsmathematik*, Vol. X, Part 2 (1971).
10. D. Riley, *Consequential Loss Insurance and Claims*, Sweet and Maxwell Ltd., London (1967).
11. "America Burning," Report of National Commission on Fire Prevention and Control (1973).
12. T. Wilmot, "Indirect Losses—Time for the International Apathy to Be Overcome," *Fire Protection*, 42 (501), pp. 22–23 (1979).
13. T. Wilmot, "United Nations Fire Statistics Study," *World Fire Statistics Centre Bulletin 7*, The Geneva Association, Geneva, Nov. 1989.
14. T. Wilmot, "United Nations Fire Statistics Study," *World Fire Statistics Centre Bulletin 8*, The Geneva Association, Geneva, Dec. 1990.
15. D.J. Rasbash, "Economics of Fire," Lecture given to the South Eastern Branch of the Institution of Fire Engineers, Guildford, UK, Sept. 1977.
16. F. Woolhead, "Risk Management: Preventing and Controlling Cable Fires," *Fire Surveyor*, 18, 3, pp. 24–28 (1989).
17. I. Chance, "Planning for the Event which Escapes Loss Prevention," *The Post Magazine and Insurance Monitor*, pp. 1732–1734, 19 July 1984.

VALUE OF HUMAN LIFE

G. Ramachandran

INTRODUCTION

An economic analysis of safety expenditure involves a consideration of costs of various safety measures and the benefits that can be expected by adopting these measures. For economic justification of this expenditure, the benefits should exceed the costs. In principle, the costs of fire protection devices are not too difficult to determine, but some of the benefits due to these control systems are difficult to quantify. Benefits such as tax allowances and savings in insurance premiums are realized by the property owners with 'certainty.' But a reduction in fire damage due to a fire protection measure is an 'uncertain' benefit whose 'expected value' depends on the probabilities associated with the occurrence and spread of fire. Damage to a building and its contents can be fairly assessed but it is difficult to evaluate consequential losses due to loss of profits, production, exports, and employment. Damage to life in terms of injuries and deaths is another important factor to be considered in fire protection problems. Insurance claims provide some data for the valuation of injury; an alternative method is to aggregate various components—treatment costs, the value of time lost, social costs, and the value of pain and suffering, which is the most difficult to evaluate. The ultimate cost assessment is concerned with the monetary value of human loss, which has been referred to as "value of life" in several studies on the economics of life safety. The value assigned to human life has to be a finite amount since no society can devote its entire resources to the elimination of the risk of death.

The objective of this chapter is to review briefly different methods of assessing the value of human life and their applications in some safety problems with provided numerical values. Use of human life valuation in fire protection economics is explained with the aid of a few examples. It is shown that policy makers will need to carry out a sensitivity analysis using a range of values for human life if they wish to assess economically the recommendation of any fire pro-

tection measure. The analysis presented in this chapter is concerned with the theory of decision making at the national level and the cost of a statistical fatality. A homeowner considering fire or life insurance would use a different approach.

METHODS OF VALUING HUMAN LIFE

Key aspects of the value of life and safety have been discussed by several authors who contributed to the proceedings of a conference held by the Geneva Association in 1981. This book[1] contains surveys of theoretical and empirical work and pertinent methodological and philosophical issues. Recent developments on this subject have been reviewed by Jones-Lee.[2] As discussed below, there are essentially five approaches to valuing human life.

The first method is concerned with gross *output* based on goods and services which a person can produce if not deprived, by death, of the opportunity to do so. Sometimes gross productivity is reduced by an amount representing consumption (net output). Discounted values are generally taken to allow for the lag with which the production or consumption occurs. The output approach usually gives a small value for life, especially if discounted consumption is deducted from discounted production. This must be so since the community as a whole consumes most of what it produces. It is argued that when a person dies, although the community loses that person's future output, it also saves concurrent future consumption. The person's own consumption or the utility that would be derived if the person were alive is not counted as a loss.

The *livelihood approach* to value of life, which is not fundamentally different from the output approach, assigns valuations in direct proportion to income. The present value of future earnings of an individual is estimated and reduced by an amount equal to discounted consumption;[3] this would give the net economic value of an individual to a family. This method also gives a small value for life. As in the case of output approach, deduction of consumption is to some extent unethical and not economically justifiable. The livelihood method normally favors males over females, working persons over those retired, and higher paid over lower paid persons in a way that may not reflect individual or social preferences.

Dr. G. Ramachandran retired in November 1988 as Head of the Operations Research Section at the Fire Research Station, U.K. Since then he has been practicing as a consultant in risk evaluation and insurance. He is a visiting professor at the University of Hertsfordshire and Glasgow Caledonian University. His research has focused on statistical and economic problems in fire protection and actuarial techniques in fire insurance.

The third approach assumes that if an individual has a life insurance policy for x £, then he/she implicitly values his/her life at x £. Collection of necessary data from insurance companies is not a difficult task, and this is the major advantage in adopting the *insurance method*. There are, however, two drawbacks to this method. First, a decision whether or not to purchase insurance and the amount of insurance is not necessarily made in a manner consistent with one's best judgment of the value of one's life. This decision depends largely on the premium the assured can bear from his or her income, taking into account family expenditures. Secondly, purchasing an insurance policy does not affect the mortality risk to an individual; this action is not intended to compensate fully for death or to reduce the risk of accidental death. Hence in insuring life it is not exactly a value tradeoff that is considered between mortality risks and costs.

The fourth method for assessing value of life involves *court awards* to heirs of a deceased person as restitution from a party felt to be responsible for the fatality. Here again, collection of necessary data is not a problem. Assessment of values of life could also be expected to be reasonably accurate since lawyers and judges have a massive professional expertise in the "*ex post*" analysis of accidents. The object of such an analysis is to discover whether the risk could have been reasonably foreseen and whether the risk was justified or unreasonable.

There are, however, a few problems in using court awards for valuing human life. The court should ideally be concerned with the assessment of suitable sums as compensation for an objective loss, e.g., loss of earnings of the deceased as well as for a subjective loss, e.g., damages to spouse and children for their bereavement and grief. In some countries damages can include a subjective component for pain and suffering of survivors, but certain courts are generally against such compensation for subjective losses to persons who are not themselves physically injured, believing that bereavement and grief are not losses which deserve substantial compensation. It is also difficult to value the quality of a life that has been lost. People who themselves suffer severe personal injury, of course, qualify for substantial damages for subjective losses. Resource costs such as medical and hospital expenses are significantly higher for obvious reasons in serious injury cases than in fatal cases; hence, awards for subjective losses tend to be much larger and more important in serious nonfatal cases than in fatal cases. Some courts have also limited to very low levels the damages that may be awarded for reductions of life expectancy.

Last, in court awards risks to individuals are considered relative to the plaintiff and costs to the defendant. However, value judgments are likely to vary according to whether the individuals making these judgments are associated with the plaintiff, the defendant, or the court.

The fifth approach is the one widely adopted for valuing life and this willingness to pay is based on the money people are willing to spend to increase their safety or reduce a particular mortality risk.[4,5] It is difficult to differentiate between the benefit from increasing peoples' feeling of safety and that from reducing the number of deaths. Anxiety is a disbenefit even if the risk is much smaller than believed. Likewise, if a person dies from a risk of which he/she is unaware he/she still suffers a loss. This approach to value of life rests on the principle that living is a generally enjoyable activity for which people would be willing to sacrifice other activities such as consumption.

The implied value of life revealed by a willingness-to-pay criterion would depend on a number of factors. The acceptable expenditure per life saved for involuntary risks is likely to be higher than the acceptable expenditure for voluntary risks, as people are generally less willing to accept involuntarily the same level of risk they will accept voluntarily. The sum people are prepared to pay to reduce a given risk will also depend on the total level of risk, the amount already being spent on safety, and the earnings of the individuals.

The theoretical superiority of the willingness-to-pay method consists of its connection with the principle of "consumer sovereignty," that goods should be valued according to the value individuals put on them. This consumer preference approach treats safety as a commodity like any other so that when a government carries out projects to alter safety it should estimate costs and benefits as people do. This method would provide a level of safety expenditure that people could be expected to accept or bear, thereby avoiding the disadvantages of compulsory regulations which are often complex and ineffective and can destroy an individual's sense of responsibility.[6] This approach is the most appropriate where we are considering the expenditure of government (i.e., our) money to protect our lives.

Following the willingness-to-pay criterion, people may be asked to specify the amounts they are willing to spend to avoid different risks. However, surveys carried out in this connection have shown variability and inconsistencies in the responses to questionnaires;[7–9] quite simply individuals have difficulty in answering questions involving very small changes in their mortality risks. Due to insufficient knowledge about the risk, most people find it difficult to accurately quantify the magnitude of a risk. Also, the benefits are often intangible, e.g., enjoyment, peace of mind. It is difficult to put a monetary value on these factors. As literature on compensating wage differential indicates, individual willingness to pay can be estimated by methods other than direct questioning of individuals.

A great majority of accidents that occur are the result of events or a chain of events of a simple character—a slight miscalculation in overtaking a vehicle, excessive speed while driving a car, or careless disposal of smoking materials in the home. Hence, it is doubtful whether individuals are good judges of risk and costs in the area under consideration. Since many accidents do occur, the question that must be posed is whether these accidents are generally the result of conscious and deliberate acts of risk taking, or whether they are indeed unanticipated and unforeseen accidents. It is possible that some accidents arise not from consciously taken risks but from risks that have not been (or not adequately been) perceived at all. People not involved in an accident under study cannot estimate how they would behave in that accident which to them is a hypothetical situation.

APPLICATIONS

The expected loss, L, due to the risk of death is the product of the value, V, placed on life and the probability, p, of death, which is a numerical measure of the risk. A value for life is obtained by equating expected loss with acceptable protection expenditure, E, to avoid the risk. The implied value of life V is therefore given by

$$V = \frac{L}{p} = \frac{E}{p} \qquad (1)$$

The parameters L and p depend on the type of risk considered. Equation 1 proposed by Melinek[6] has also been suggested by McGuire,[10] according to whom the value to be adopted is given by the quotient of the money a person is prepared to spend to avoid a risk divided by the level of the risk.

Consider, for example, the use of a pedestrian subway.[6] It was estimated that the risk (probability) of being killed while crossing a road was approximately 1.225×10^{-8} and that people were willing to use a subway if the additional time was less than 16 seconds. U.K. transport studies indicated that the value people put on their time was of the order of £0.24 per hour. Under these assumptions, the implied value of life for this particular risk was estimated as

$$V = £0.24 \times (16/3600)/(1.225 \times 10^{-8})$$
$$= £87,000$$

(at 1973 prices) or about £260,000 at 1981 prices. For two other examples, the estimated values of life based on mortality risks were as follows:

	1973 prices	1981 prices
Smoking	£28,000	£84,000
Employment (industrial accidents)	£200,000	£600,000

Based on these studies, Melinek[6] suggested a figure of £50,000 (at 1973 prices) for implied value of life for fire protection problems. This is equivalent to an estimate of about £150,000 at 1981 prices. There is evidence that people are willing to take higher risks in voluntary activities. This factor may account for the low implied value of life from smoking. It is also likely that people assume that if they die from smoking it will be in old age.

The above estimates of the value of life make no allowance for the risk of injury or ill health. The number of injuries (death) depends on the hazard; for example, aircraft crashes result in few nonfatal injuries. On the other hand, a large proportion of injuries are caused by events (e.g., abrasions, lifting) which cause few deaths. The number of injuries/deaths also depends on age. Fire fatalities tend to be very young or very old. Melinek[6] obtained the following figures for estimated values of life allowing for the subjective (but not financial) cost of injuries and ill health. He assumed that figures for injury are about 15 percent of the estimated values of life.

	1973 prices	1981 prices
Use of subways	£74,000	£222,000
Smoking	£24,000	£72,000
Employment	£161,000	£483,000

If the subjective cost of injuries is not subtracted, then the value of life obtained includes the risk of injury.

Blomquist[11] examined the methodology and results of various empirical studies which estimated values of life based on individual willingness to pay. The evidence on the value of life and safety comes from two different types of sources, although both yield information on individual willingness to pay. One type is implicit values, which are de-rived from observable individual behavior with respect to goods and services with well-developed markets. Much of this type of evidence comes from the labor market through the estimation of risk compensating wage differentials. Implicit values are estimated from consumption activity that also includes housing and travel choices. Another type of evidence comes from creating hypothetical markets for health and safety and asking individuals directly how much they would pay for improvements contingent on the existence of such markets. Blomquist[11] has discussed in detail the advantages and disadvantages of these methods used by several authors. Table 5-8.1 reproduced from his paper summarizes these estimates.

It should be noted that only point estimates are given in Table 5-8.1 and that each study should be consulted for a discussion of the upper and lower bounds of the values of life. It should also be remembered that value of life would vary with risk, income, age, family status, and other circumstances. When grouped by the range of estimated values of life, the contingent values show the greatest range followed by the values based on risk compensating wage differentials and values from consumption activity. When grouped by risk reduction, a clearer pattern emerges, i.e., the estimated values of life tend to increase as the risk reduction declines. This is understandable to some extent, since in Equation 1 for implied value of life, the risk factor (probability p of death) appears in the denominator. The average values of life for different risk levels are as follows:

Risk level	Average value of life
10^{-3}	$ 168,000
10^{-4}	$1,068,000
10^{-5}	$1,963,000
10^{-6}	$6,746,000

The value of life of $351,000 in Table 5-8.1 in regard to smoke detectors for residences has taken into account the purchase price of the detector, the replacement cost of batteries, and the changes in the probabilities of death and injury. Dardis[12] estimated that the average value of life changes from $274,000 to $428,000 depending on whether nonfatal injuries are weighted as equal to one-half a fatality or zero, respectively. Two shortcomings that bias this estimate in opposite directions are the omission of installation costs and the treatment of household size. The inclusion of installation costs means that the amount that residents are giving up to obtain more safety is greater than estimated and that the implied value of life is higher than estimated by Dardis. Since usually there will be more than one person per household, the implied value of life per person is perhaps ½ to ⅓ of the household value.

Graham and Vaupel[13] have compared the costs and benefits of 57 life-saving programs. Quoting surveys of expressed willingness to pay for small reductions in the probability of death, these authors have shown that values of a life ranged from $50,000 to $8 million (in 1978 dollars). Nine labor market studies of wage premiums have produced a narrower but still disparate range of values spread from $300,000 to $3.5 million. Graham and Vaupel[13] conclude that within a broad range, the monetary value assigned to the benefits of averting a death usually does not alter the policy implications of the analyses.

TABLE 5-8.1 *Values of Life from Implicit and Contingent Valuation*

Source of Evidence	Authors	Value of Life (1980 U.S. Dollars)* Thousands	Risk Reduction
Implicit Values from Labor Market Activity:			
Blue-collar workers in manufacturing and construction	Dillingham	378	10^{-4}
Workers in risky occupations	Thaler & Rosen	494	10^{-3}
Males in manufacturing industries	Smith	2,785	10^{-4}
Blue-collar workers	Viscusi	2,820	10^{-4}
Implicit Values from Consumption Activity:			
Residential housing market	Portney	180	10^{-4}
Residential smoke alarms	Dardis	351	10^{-5}
Highway speed	Ghosh, Lees & Seal**	419	10^{-4}
Auto seatbelt use	Blomquist	466	10^{-4}
Contingent Values:			
Air travel	Frankel	57	10^{-3}
		3,372	10^{-6}
	Jones-Lee	10,120	10^{-6}
Heart attack prevention	Acton	59	10^{-3}
Nuclear power	Mulligan	62	10^{-3}
		428	10^{-4}
		3,576	10^{-5}

*All values are converted to June 1980 dollars using the Consumer Price Index.
**Since the risk reduction is not specified, it is assumed to be the same order of magnitude as that in the Blomquist study.

Future risks can be discounted to the extent that people are willing to take immediate risks (e.g., a medical operation) to avoid greater risks at a future date. The rate of discount is equivalent to a rate of interest and would also reflect the fact that people tend to value the earlier years of their lives more highly than the later years. Alternatively, future risks can be discounted because the money required to reduce them can be obtained by investing a smaller sum beforehand. Dawson[3] obtained values of life by calculating discounted future earnings plus a fixed subjective loss, assuming 6 percent per annum rate of discount. Melinek[6] updated these figures using average annual earnings of £1920 for men and £1060 for women and obtained a figure of £14,740 for average discounted earnings; this estimate is considerably less than the value £50,000 provided by the willingness-to-pay approach. Average discounted consumption, assuming annual consumption of £600 and a rate of discount of 6 percent per annum, was £7800. Average discounted earnings less consumption was thus estimated as £6940 (at 1973 prices). This sum represents the average net economic value of an individual to his/her family. Melinek[6] has produced estimates for discounted earnings for different age groups and for males and females separately.

Schelling[14] estimated the subjective value of life to be 10 to 100 times one year's income, giving a value of £20,000 to £200,000, assuming an annual income of £2000. The average value of £50,000 obtained by Melinek is near the geometric mean (£60,000) of the limits estimated by Schelling.

Court awards in the U.K. for damages for fatal (or potentially fatal) injuries are based on loss of earnings plus a small sum for reduction of life expectancy. Figures for court awards and sums assured for life insurance are not readily available but can be obtained for purposes of comparison.

Maycock[15] has reproduced figures in Table 5-8.2 which give estimates (in U.S. dollars at 1979 prices) of the value of a fatality arising from the various valuation methods (based on data from developed countries). The "appropriate" valuation depends upon the economic or social objectives of the particular country applying the technique. If national output is the key objective, then "gross output" measures are the most appropriate for accident valuation. If a country's objectives are related to broader social-welfare considerations, then willingness to pay would seem more appropriate.

TABLE 5-8.2 *Estimates of the Cost of a Statistical Fatality (or Value of Avoidance of a Statistical Fatality) by Costing/ Valuation Methods*

	Costing/Valuation method	Cost/Value of one statistical fatality
(i)	Gross output approach:	
	(a) including subjective component[3]	$ 120,000
	(b) including subjective component but increased 50% and with reduced discount rate applied: (UK Dept of Transport, 1979)	$ 225,000
(ii)	Net output approach:	
	(a) excluding subjective component: (Reynolds, 1956)	$ 25,000
	(b) including subjective component: (Dawson, 1976)	$ 76,000
(iii)	Life insurance basis: (Fromm, 1965)	$ 930,000
(iv)	Court-awards basis:	
	(a) Abraham and Thedie, 1960	$ 83,000
	(b) Shepherd, 1974	$ 1,000,000
(v)	Implicit public sector valuation: (Mooney, 1977)	$3,000-60M
(vi)	Willingness-to-pay approach:	$ 2,100,000

COST-BENEFIT ANALYSIS

The main purpose of estimating a value for human life is to use it in a cost-benefit decision analysis of measures aimed at increasing safety or reducing risk. As a first example, consider the decision analysis carried out by Helzer et al[16] to evaluate alternative strategies for reducing residential upholstered furniture fire losses. Three alternatives were evaluated: no action, mandatory smoke detector installation, and an upholstered furniture standard under consideration by the U.S. Consumer Product Safety Commission. The alternatives were evaluated on the basis of minimizing the total cost plus loss to society over time. Figure 5-8.1 shows how the comparison of alternatives was affected by the value assigned to life.

According to Figure 5-8.1, if no value was placed on preventing the loss of life, no action was the most attractive alternative. At a value of approximately $60,000 per life saved, the smoke detector alternative became the most attractive strategy and at approximately $300,000 per life saved, the proposed standard became the most attractive alternative. The proposed standard was the most attractive strategy for all values greater than $300,000 per life saved. At a value of $1 million per life saved the proposed standard resulted in a reduction in present value of cost plus loss of 7 percent over the detector alternative and 16 percent over no action. These sensitivity studies showed that the smoke detector alternative and the proposed standard were the most attractive over a wide range of values assigned to life.

Chandler and Baldwin[17] carried out a statistical study of fires in the United Kingdom during 1970 that involved the ignition of furniture and furnishings in the home. The total cost of furniture fires for this year was estimated by combining property damage, deaths, and injuries. A value of £50,000 was used for each life loss and a value of £1000 for each injured case. (The latter value was based on hospitalization costs, although for more serious injuries involving permanent disability a value in excess of £1000 would have been appropriate.) Based on some unpublished data, the following values were adopted for property damage:

Fire confined to object ignited first	£50
Fire spreading beyond object ignited first but confined to the room of origin	£300
Fire spreading beyond the room of origin	£2000

Using the values mentioned previously, it was estimated that for the year 1970, the total costs were £18.8 million for the furniture fires and £54.5 million for all dwelling fires.

The cost associated with a particular type of fire represented the maximum potential benefit if all fires of that type were prevented. Using this criterion, Chandler and Baldwin[17] concluded that action to reduce fatalities would have the greatest benefit, particularly deaths resulting from smokers' materials. For furniture already in use, the remedy suggested was to either reduce or eliminate the likelihood of ignition or to reduce the toxic hazard in the immediate vicinity of the object ignited. This action would be preferable to prevention of fire spread, as most fatalities were found in the room of fire origin.

A few fires led to much controversy over the use of foamed plastics as a building material and it was proposed that such materials be replaced by some more traditional materials, e.g., plasterboard. This suggestion was subjected to a cost-benefit analysis by Appleton[18] who investigated the following three types of remedial action:

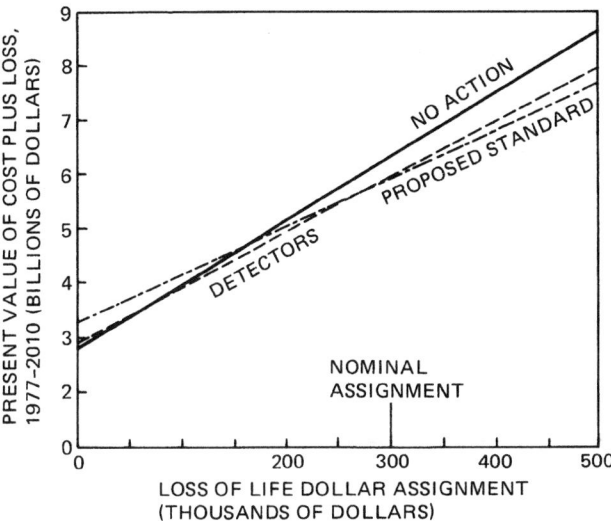

Fig. 5-8.1. Sensitivity of results to dollar assignment on loss of life.[16]

1. Replace foamed plastic by plasterboard.
2. Install plasterboard below the foamed plastic.
3. Lay fiberglass above the foamed plastic.

The upper (most dangerous) bounds for the total discounted losses (at 1976 prices) were obtained for the following two typical dwellings:

1. A top floor flat consisting of hall, kitchen, living/dining room, two bedrooms, and a bathroom.

   Total discounted loss—£52.6.

2. A large bungalow consisting of hall, kitchen, living room, dining room, 3 bedrooms, and a bathroom.

   Total discounted loss—£77.0.

For the two types of dwellings mentioned above, the average costs of remedial action per dwelling ranged from £130 to £930 at 1976 prices. Hence, the discounted losses (maximum benefits) were considerably less than the costs of remedial action, even when taking the most pessimistic view of the effect of foamed plastics ceilings. Hence, no remedial action was advocated on a cost-benefit basis.

In this analysis, a value of £100,000 was used for each life saved. In order to test the sensitivity of this parameter, the calculations were repeated with the implied value of life as a variable. (See Figures 5-8.2 and 5-8.3.) It was estimated for a bungalow that the replacement of foamed plastic by plasterboard would cost £920. Should reality lie at the upper limit of possibilities and this replacement proceed, that decision would then imply the value of life to be £5 million. Alternative propositions would yield figures of £4.2, £7.5, and £11 million. Since such large values for life were suggested to be unacceptable, the remedial actions were considered economically unjustifiable.

GENERAL DISCUSSION

The value of life discussed in the previous sections of this chapter included only items of concern to the individual and hence is a private as opposed to a social assessment.

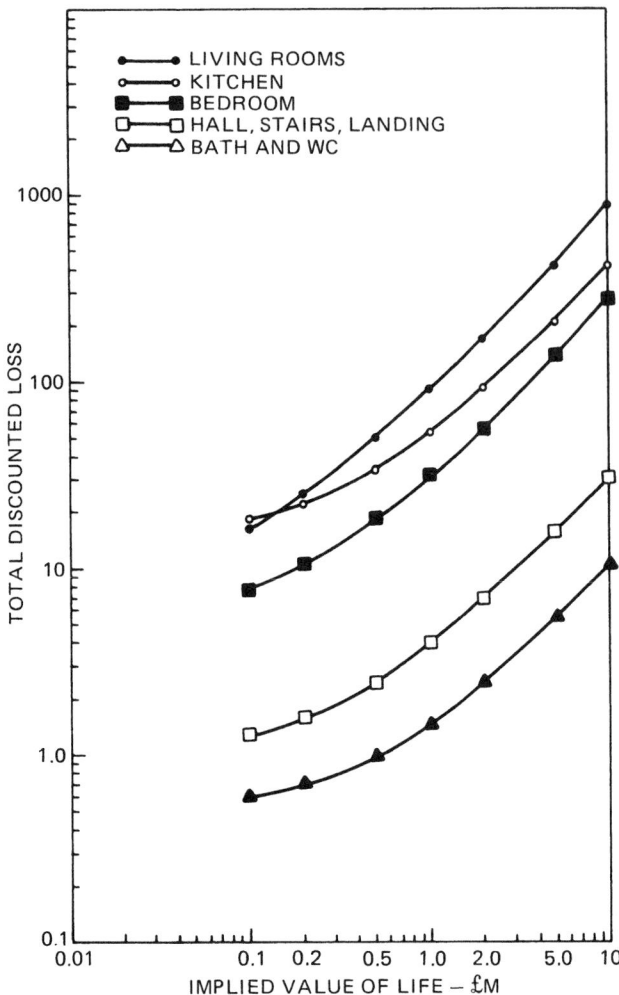

Fig. 5-8.2. Total discounted loss versus implied value of life—different types of rooms.

Distress to those not killed or injured, including those not at risk, is an additional consideration to the extent that people are willing to pay for the safety of others. People feel some concern for the safety of others. Such concern increases the value to be placed on life in calculating safety expenditures acceptable to the community as a whole. The distress and financial loss that an individual's death would cause to his/her family is included in private evaluation insofar as the individual takes account of these factors in deciding which risks are worthwhile. The financial loss to the rest of the society can be considered an additional factor in the value of an individual's life. The social value of life is the sum of the individual's value of an improvement in his/her safety and the value to others of that improvement. Bailey[19] estimates several components of this externality such as imprecise insurance classifications, including social security, indirect business taxes, and medical costs borne by others. As estimated by Bailey, the total adjustment for these external factors increases Blomquist's private (willingness-to-pay) valuation of $466,000 by 17.5 percent to $548,000.

In a safety problem such as fire protection, all such externalities should be carefully enumerated and their values included if necessary in order to derive the social values

of life which are more relevant for public policies than individual implied values. A citizen's valuation of his/her own safety will understate its value to society as a whole.

A problem arises in an organizational evaluation of mortality risks to several individuals; this pertains to the manner in which individual implicit values should be aggregated. Because the responses about willingness to pay concern funds of the individual at risk and the institutional problem concerns funds of some other entity, a direct aggregation of the individual responses would likely be inappropriate in many circumstances. For example, the implied life values of the occupants cannot be simply added in order to provide the manager of a hotel an appropriate level of firesafety expenditure for the establishment. This problem can be resolved by applying a technique suggested by McGuire.[20] According to McGuire, the desired level of protection for any activity should be such that the additional expenditure that would result in the saving of one more life would be precisely the sum considered acceptable for the saving (or protecting) of a life. The criterion is in terms of the incremental expenditure for the protection of a further life and not in terms of the average expenditure.

Many ethical issues become entwined in the life valuation by a government organization for safety expenditure. The government may have to spend less on some sections of the population in order to be able to spend more on other sections. Most of the fire victims are, for example, either young children or old people, particularly those who live alone. Again, although an individual is poor his/her life should not receive a lower valuation when pollution and fire

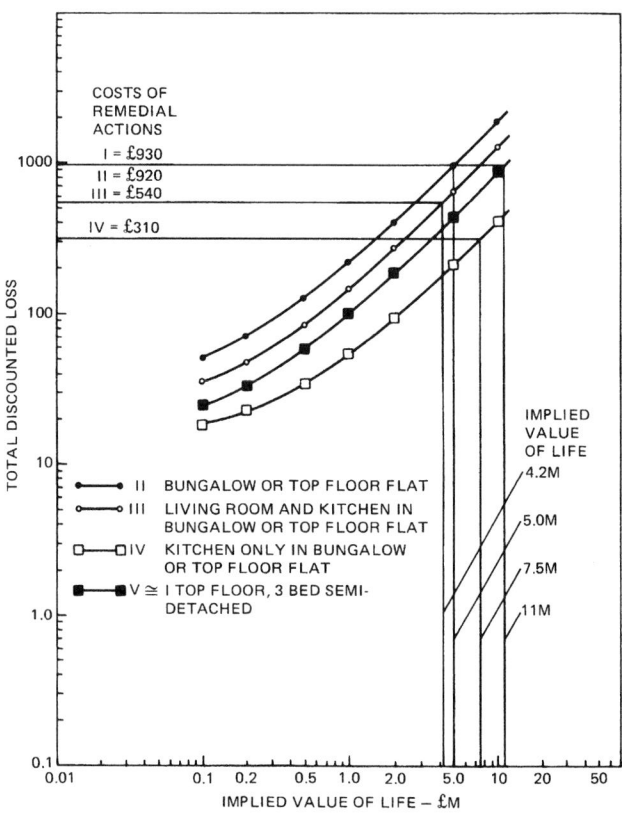

Fig. 5-8.3. Total discounted loss versus implied value of life—different types of dwellings.

control decisions are made. Governments may need to also give serious consideration to the social disutility or consequences due to deaths in fires. Disutility would be low for small fires or single-death fires in the mortality context. Disutility is high for multiple-death fires. It is important to consider not only the number of deaths but also whether they occur singly or as a result of catastrophes involving many deaths. Catastrophes normally have social and political consequences which will not be present in an equal number of deaths occurring singly (in separate events). Representing this symbolically,

$$D(10) > 10D(1)$$

where $D(x)$ is the disutility associated with x deaths. The disutility associated with a multiple-death fire would be high and hence not desirable. A small probability of a catastrophic loss of life is worse than a larger probability of a smaller loss of life, given that the expected number of fatalities are the same for each case.

REFERENCES CITED

1. M.W. Jones-Lee, ed., *The Value of Life and Safety*, North Holland, New York (1982).
2. M.W. Jones-Lee, "The Value of Life and Safety: A Survey of Recent Developments," *Geneva Papers Risk and Ins.*, 10, 36, 141 (1985).
3. R.F.F. Dawson, "Current Costs of Road Accidents in Great Britain," *Report No. RRLLR 396*, Road Research Laboratory, United Kingdom (1971).
4. J. Linnerooth, "The Evaluation of Life Saving," *Research Report RR-75-21*, International Institute of Applied Systems Analysis, Laxenbury, Austria (1975).
5. E.J. Mishan, *Cost Benefit Analysis*, Allen and Unwin, London (1971).
6. S.J. Melinek, "A Method of Evaluating Human Life for Economic Purposes," *Acc. Anal. and Preven.*, 6, 103 (1974).
7. J.P. Acton, "Measuring the Social Impact of Heart and Circulatory Disease Programs: Preliminary Framework and Estimates," *Rand R-1697/NHLI*, Rand Corporation, Santa Monica (1975).
8. G.W. Fischer and J.W. Vaupel, *A Lifespan Utility Model; Assessing Preferences for Consumption and Longevity*, working paper, Durham, (1976).
9. E. Keeler, "Models of Disease Costs and Their Use in Medical Research Resource Allocations," *P-4537*, Rand Corporation, Santa Monica (1970).
10. J.H. McGuire, "The Economics of Protecting Lives," *F. News Canada*, 32 (1985).
11. G. Blomquist, *Estimating the Value of Life and Safety: Recent Developments in the Value of Life and Safety*, North Holland, New York (1982).
12. R. Dardis, "The Value of Life: New Evidence From the Marketplace," *Amer. Econ. Rev.*, 70, 1077 (1980).
13. J.D. Graham and J.W. Vaupel, "Value of a Life: What Difference Does It Make?" *Risk Anal.*, 1, 89 (1981).
14. T.C. Schelling, "The Life You Save May Be Your Own," in *Problems in Public Expenditure Analysis*, Brookings Institution, Washington (1968).
15. G. Maycock, "Accident Modelling and Economic Evaluation," *Acc. Anal. and Preven.*, 18, 169 (1986).
16. S.G. Helzer, B. Buchbinder, and F.L. Offensend, "Decision Analysis of Strategies for Reducing Upholstered Furniture Fire Losses," *Technical Note 1101*, National Bureau of Standards, Washington (1979).
17. S.E. Chandler and R. Baldwin, "Furniture and Furnishings in the Home—Some Fire Statistics," *Fire and Matls.*, 7, 76 (1976).
18. I.C. Appleton, "A Cost-Benefit Analysis Applied to Foamed Plastics Ceilings," *Current Paper CP 50/77*, Fire Research Station, Borehamwood (1977).
19. M.J. Bailey, *Reducing Risk to Life: Measurement of the Benefits*, American Enterprise Institute for Public Policy Research, Washington, DC, (1980).
20. J.H. McGuire, "A Unified Economic Concept for the Protection of Human Life," *Can. Bldg. Off. Assoc. News and J. of Bldg. Safety*, 10, (1986).

ADDITIONAL READING

F. Warner and D.H. Slater, eds., *The Assessment and Perception of Risk*, University Press, Cambridge (1981).

J. Ferreira and L. Slesin, "Observations on the Social Impact of Large Accidents," *Technical Report No. 122*, Mass. Institute of Technology, Cambridge (1976).

M.W. Jones-Lee, *The Value of Life: An Economic Analysis*, Martin Robertson, London, (1976).

L.B. Lave, "Economic Tools for Risk Reduction," in *Societal Risk Assessment*, Plenum, New York (1980).

E.J. Mishan, "Evaluation of Life and Limb." *J. Pol. Econ.*, 79, 687 (1971).

W.D. Rowe, *An Anatomy of Risk*, John Wiley and Sons, New York (1977).

P. Slovic, B. Fisenhoff, and S. Lichtenstein, "Risk Assessment: Basic Issues," in *Managing Technological Hazard: Research Needs and Opportunities*, Univ. of Colorado, Boulder (1977).

C. Star, "Social Benefit versus Technological Risk," *Science*, 165, 1232 (1969).

UTILITY THEORY

G. Ramachandran

INTRODUCTION

The technique of benefit-cost analysis was discussed in Chapter 6 of this section to determine the acceptability of an investment project and to compare the economic value of different projects. For benefits such as reduction in fire damage (which involves chance effects), expected values are calculated by weighting them according to their probabilities.

In benefit-cost analysis as usually carried out, expected values of costs and benefits are evaluated in monetary terms. This expected monetary value approach assumes that the decision maker is risk neutral, but most decision makers would be keen to avoid risks and adopt a risk averse attitude. Some people may prefer to take risks. Any person, generally, is a "risk preferer" for ventures involving small losses and a "risk avoider" for those involving large values. Such risk preferences can be quantified by appropriate utility functions which measure the intrinsic values of positive monetary outcomes, i.e., gains. Disutility, the negative counterpart of utility, is the appropriate term in an analysis involving negative outcomes such as fire loss, cost of fire protection, and insurance. The decision analysis needs to be carried out in terms of expected disutilities associated with fire loss and costs. The object of this chapter is to explain the basic concepts concerned with the utility/disutility theory and illustrate their application to fire protection and insurance problems.

UTILITY

The concept of utility is defined as intrinsic value of money which, as shown by von Neumann and Morgenstern,[1] can be quantified in terms of rational economic behavior of people in satisfying their needs. In the context of decision analysis, utility is a number measuring the attractiveness of a consequence—the higher the utility, the more desirable the consequence (the measurement being made on a probability scale).

Dr. G. Ramachandran retired in November 1988 as Head of the Operations Research Section at the Fire Research Station, U.K. Since then he has been practicing as a consultant in risk evaluation and insurance. He is a visiting professor at the University of Hertsfordshire and Glasgow Caledonian University. His research has focused on statistical and economic problems in fire protection and actuarial techniques in fire insurance.

It will be simpler to explain the need for the utility theory approach by considering examples based on participation in a game of chance. Suppose a person is offered the following bet on toss of a coin—to win $100 if the coin comes up heads, and lose $75 if the coin comes up tails. If the coin is a fair coin the probability of heads or tails coming up is ½. The expected payoff is

$$\frac{1}{2}(\$100) + \frac{1}{2}(-\$75) = \$12.5$$

if the person playing the game takes the bet and $0 if he or she does not take the bet. According to the expected value criterion, the bet should be taken by the person since it looks intuitively advantageous (unless the individual is opposed in principle to gambling). Now suppose the amounts involved are $10,000 and $7500 rather than $100 and $75. The expected payoff is now $1250 if the bet is taken and $0 if the bet is not taken. According to the expected value criterion the bet should be taken, especially in view of the large expected payoff. Will the person playing the game do so? Probably not, unless he/she is wealthy enough so that a loss of $7500 would not seriously affect his or her financial position. The possible gain of $10,000 is tempting but there is still a 50 percent chance of a $7500 loss.

As another example, consider a choice between two bets. In the first bet, the person playing the game wins $2 million if a coin comes up heads and wins $1 million if the coin comes up tails. In the second bet $8 million can be won if the coin comes up heads but nothing will be won if the coin comes up tails. The expected payoffs of the two bets are $1.5 and $4 million. The second bet has a much larger expected payoff than the first and hence should be chosen on the basis of the expected value criterion. However, most persons would probably choose the first bet since they could certainly win at least $1 million, which is a large sum of money. By taking the second bet there is a 50 percent chance of winning nothing.

Consider a third example. Suppose there is an (annual) probability, p, of fire occurring in a building and an expected loss, $\bar{x}$, in the event of a fire. The expected annual loss due to fire in that building is $p\bar{x}$. In order to recover this loss in the event of a fire, the building owner can take out a fire insurance policy and pay an annual premium, I. In the event of a fire most of the fire damage (except for a small sum) will be compensated by the insurance company. The property owner has two options—to insure or not insure the building.

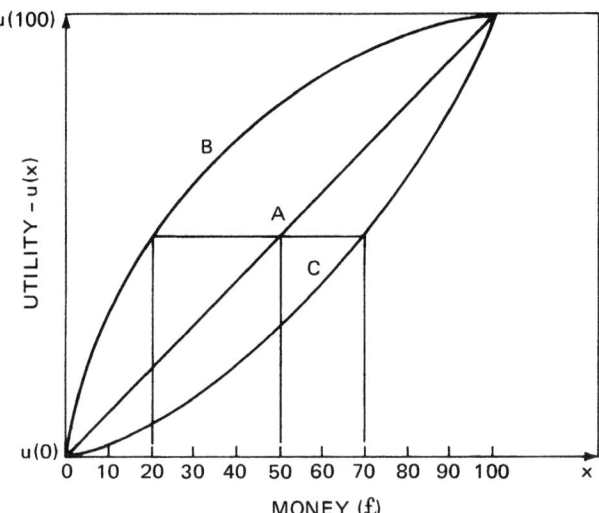

Fig. 5-9.1. Typical utility functions.

The expected loss (cost) is I if insured and $p\bar{x}$ if not insured. On the basis of the expected value principle, the owner should choose the insurance option only if I is less than $p\bar{x}$. This condition will never be satisfied, since according to basic principles of insurance, the premium has to be necessarily greater than the expected damage if it has been determined according to a sound statistical analysis of fire risk. The insurance company will be adding two types of loading to the risk premium $p\bar{x}$. A safety loading is added toward chance fluctuations of loss beyond the expected damage. Another loading is imposed to cover the insurer's operating costs, which include profits, taxes, and other administrative expenses. Insurance premium is a known (certain) cost, whereas $p\bar{x}$ is an unknown cost subject to statistical uncertainty. The property owner would prefer the certain loss of a small sum (the insurance premium) to a small chance of a loss much larger than $p\bar{x}$ in the event of a fire. This preference for a small fixed loss over a risk of large loss originates primarily from an aversion to the psychological state of uncertainty. For the reasons mentioned previously the expected monetary value is not a satisfactory criterion for decisions involving insurance.

These examples illustrate the fact that for a specific person, the "value" of gaining x dollars (or "consequence" of losing x dollars) is not necessarily x multiplied by the "value" (or "consequence") of a single dollar. The value of money may also differ from person to person depending on other economic factors such as assets. The gain or loss of a small sum may not affect the financial strength of a rich person but it could cause serious problems for a poor person. If it could somehow be possible to measure the true relative values to the decision maker of the various possible payoffs in a problem of decision making under uncertainty, expectations could be taken in terms of these "true" values instead of the monetary values. The theory of utility prescribes such a decision-making rule: the maximization of expected utility or minimization of expected disutility. The theory makes it possible to measure the relative value to a decision maker of the payoffs or consequences, including all aspects of the consequence, monetary or otherwise. Utility theory provides a means of encoding risk preferences in such a way that the risky venture with the highest expected utility or

lowest expected disutility is preferred. Symbolically, if the monetary value of the ith outcome is X_i, the utility corresponding to a gain X_i is $U(X_i)$; the disutility corresponding to a loss X_i may be denoted by $D(X_i)$.

UTILITY FUNCTIONS

The mathematical structure of the function $U(X)$ is central to the application of utility theory. Figure 5-9.1 graphically shows three typical utility functions that are usually encountered in this analysis.[2] The utility function represented by the straight line A is appropriate for a decision maker operating on an expected monetary value or, EMV, basis. This line satisfies the equation $U(X) = X$ and represents risk neutrality. The concave curve B corresponds to a risk averse (or risk avoiding) decision maker, and the convex curve C to a risk prone (or risk taking) decision maker. For a decision maker who is more risk prone than the EMV individual or who prefers a risk, the utility of a fair game exceeds the utility of not gambling and hence a fair game will always be played. On the other hand a decision maker who is more risk averse than the EMV person does not like or cannot afford risks and is a risk avoider. Some individuals could have a sigmoid form of utility function as illustrated by Figure 5-9.2. Such a person is a risk preferer for small values of X but a risk avoider for larger values.

Consider now a game with 0.5 chance of winning £100 and 0.5 chance of winning nothing, which has the expected value £50. The expected value line A in Figure 5-9.1 connects the points $[0, u(0)]$ and $[100, u(100)]$. To find the utility of the game for the risk avoider (curve B), find the utility value corresponding to the point on the straight line above the expected £50 value of the game. By reading to the left, cutting curve B, this value is equal to $U(£20)$ so that the decision maker's cash equivalent (CE) for the game is £20. He or she would be willing to pay up to £20 to be able to participate in the game. This is still below the EMV of £50 since the utility function B is that of a risk avoider.

The difference between the EMV and CE is the risk premium, which is £30 in this example. The decision maker would be willing to pay £30 to avoid the risk involved in participating in the game.

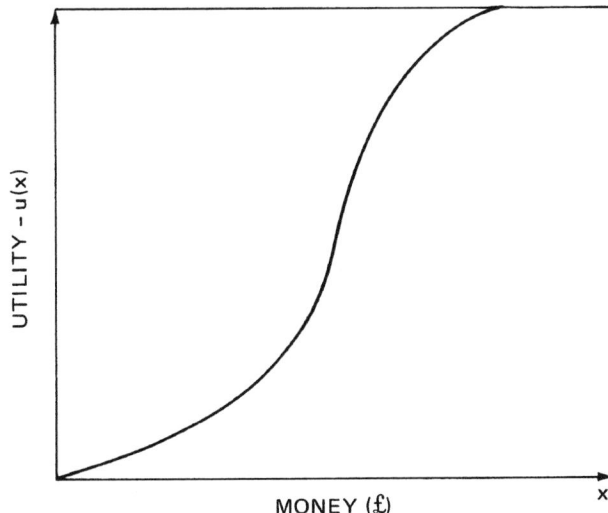

Fig. 5-9.2. Sigmoid utility function.

In the case of the risk taker denoted by curve C, the utility of the game is equal to $U(£70)$, so £70 is the cash equivalent for the game. Although the expected value is only £50, the risk taker is willing to pay up to £70 to be able to participate in the game. Hence the risk premium is $-£20$; it is negative because the decision maker, instead of being willing to pay a premium to avoid the risk in the game, is willing to pay a premium (above and beyond the expected value) to be able to participate in the game.

The risk premium, RP, discussed above is the amount which a decision maker on the basis of his/her utility function is willing to pay to avoid or participate in a risky activity. For increasing utility functions such as those shown in Figure 5-9.1, the risk premium for any risky venture is defined as

$$RP = EMV - CE \qquad (1)$$

where EMV is the expected monetary value and CE the cash equivalent. The parameter CE is also referred to as certainty monetary equivalent, CME, in the literature on utility theory.[3]

The CE is defined mathematically as

$$U(CE) = E[U(X)] = \overline{U} \qquad (2)$$

where the right-hand side is the expected value of the utility over the range of values taken by X. If $x_1, x_2, \ldots x_n$ are the values (consequences) with probabilities $p_1, p_2, \ldots p_n$

$$\overline{U} = E[U(X)] = \sum_{i=1}^{n} p_i U(x_i) \qquad (3)$$

If X is a continuous variable with probability density function $h(X)$, the expected utility is given by

$$\overline{U} = \int_x U(X)h(X) \qquad (4)$$

The CE or CME of a risky venture, V, is an amount, $\hat{x}$, such that the decision maker is indifferent between V and the certain amount $\hat{x}$. The expected value (or consequence) is given by

$$\bar{x} = E(x_i) = \sum_{i=1}^{n} p_i x_i \qquad (5)$$

or by

$$\bar{x} = \int_x xh(x) \qquad (6)$$

in the continuous case.

Consider, as an example, the utility function

$$U(X) = -e^{-cX} \qquad (7)$$

Suppose the decision maker is faced with a venture with two possible outcomes: x_1 with probability ½ and x_2 with probability ½. The expected value of the venture is

$$\bar{x} = (x_1 + x_2)/2$$

The certainty equivalent is the solution to

$$U(\hat{x}) = -e^{-c\hat{x}}$$

$$= -\frac{e^{-cx_1} + e^{-cx_2}}{2}$$

It may be verified that for $c = 1$, $x_1 = 10$ and $x_2 = 20$, the certainty equivalent is

$$\hat{x} = 10.69$$

The expected value is

$$(10 + 20)/2 = 15$$

Consider the case when x varies continuously from 0 to 20 with the exponential probability density function

$$f(x) = ke^{-x} \qquad (8)$$

Since

$$\int_0^{20} f(x) = 1$$

the value of k is unity for all practical purposes so that $f(x) = e^{-x}$. The expected value is

$$\bar{x} = \int_0^{20} xe^{-x} = 1$$

Suppose we take the utility function as

$$U(x) = -e^{-2x}$$

The certainty equivalent is given by $\hat{x}$ such that

$$-e^{-2\hat{x}} = -\int_0^{20} e^{-2x}e^{-x}$$

$$= -\int_0^{20} e^{-3x} = -\frac{1}{3} \qquad \hat{x} = 0.55$$

For a risk avoider whose increasing utility function is concave, the risk premium RP given by Equation 1 for any situation in which the outcome is uncertain is positive (EMV is greater than CE). For a risk taker whose increasing utility function is convex, RP is negative. For a risk neutral person whose utility function is linear, RP is always zero (EMV = CE).

FIRE PROTECTION AND INSURANCE

Against the background ideas described previously the possibility of applying utility theory to fire protection and insurance problems can now be investigated. It is apparent that a risk averse utility function must be chosen since the problem is one of risk avoidance. A property owner should spend more money than the expected loss toward fire protection and insurance. But how much more should he/she spend or be willing to spend? This amount depends on the attitude of the property owner toward fire risk and the extent of the owner's risk aversion based on factors such as assets and consequential losses.

Consider a property owner with an asset, W. If a loss of x in a fire is incurred the asset would reduce to

$$X = W - x \qquad (9)$$

An appropriate utility function in terms of positive X would be

$$U(X) = -e^{-\Theta X}, \quad \Theta > 0 \tag{10}$$

which is an increasing risk averse utility function.[3] Although the extent of risk aversion quantified by Θ is constant for all X, this exponential utility function is generally recommended in view of its computational simplicity. Equation 10 may be rewritten as

$$U(x) = -W' e^{\Theta x}$$

where $W' = e^{-\Theta W}$. As discussed earlier the certainty equivalent x in terms of fire loss $\hat{x}$ is given by

$$-W' e^{\Theta \hat{x}} = -W' \int_x e^{\Theta x} v(x) \tag{11}$$

where $v(x)$ is the probability density function of fire loss, x. From Equation 11 it is apparent that the certainty equivalent is independent of the parameter W', a constant related to the asset. For this reason, for the calculation of CE, it will be sufficient to use the risk averse utility function

$$U(x) = -e^{\Theta x}, \quad \Theta > 0 \tag{12}$$

in terms of the fire loss x. The value of asset W is, of course, taken into consideration in determining the extent of risk aversion Θ.

As x increases from zero, $U(x)$ decreases from a value of -1. In other words, as x increases the net asset X (Equation 9) and hence utility decrease. But as x increases the "disutility" increases. In that sense disutility is the negative of utility and the disutility function may be written as

$$D(x) = e^{\Theta x} \tag{13}$$

It may be seen that Equations 12 and 13 will both lead to the same value of the certainty equivalent. Shpilberg and Neufville[5] used the form in Equation 12 in their earlier investigation but later[4] they modified this function to

$$U(x) = e^{-\Theta x} \tag{14}$$

According to statistical studies carried out by Ramachandran,[6-8] Shpilberg[9] and other authors, loss in a fire has a skewed (nonnormal) probability distribution. In general the transformed variable z ($= \log x$), i.e., the logarithm of loss, has a probability distribution of the "exponential type." Among distributions belonging to this type, normal for z (which is same as log normal for x) has been recommended widely for modeling fire insurance claims. Exponential for z or Pareto for x have been considered by some actuaries.

Shpilberg[9] used utility functions in Equations 12 and 14 in conjunction with probabilities of fire loss falling in different ranges of magnitudes. But these utility functions are computationally difficult for performing integration if continuous forms of probability distributions of loss are used, particularly log normal. Hence, as suggested by Ramachandran,[10] z ($= \log x$) may be used in Equation 12 instead of x so that the utility function is

$$U(z) = -e^{\Theta z}$$

which is equivalent to

$$U(x) = -x^{\Theta} \tag{15}$$

The utility function in Equation 15 is a decreasing function with $\Theta = 1$ representing risk neutrality. The value of Θ should be greater than unity to express a risk averse attitude.

Consider a property worth total financial value V belonging to a risk category with fire loss x having a log-normal distribution. If μ and σ are the mean and standard deviation of z ($= \log x$), following the method described by Ramachandran,[11] the certainty equivalent for the range (o, V) is given by

$$\hat{x}^{\Theta} = \frac{1}{G(k)} \frac{1}{\sqrt{2\Pi}\sigma} \int_{-\infty}^{\log V} \exp\left[-\frac{1}{2}\left(\frac{z-\mu}{\sigma}\right)^2\right] e^{\Theta z}$$

$$= \frac{G(k-\sigma\Theta)}{G(k)} \exp\left(\mu\Theta + \frac{\sigma^2\Theta^2}{2}\right) \tag{16}$$

where

$$k = (\log_e V - \mu)/\sigma$$

$$G(k) = \frac{1}{\sqrt{2\Pi}} \int_{-\infty}^{k} \exp\left(-\frac{t^2}{2}\right)$$

$$G(k - \sigma\Theta) = \frac{1}{\sqrt{2\Pi}} \int_{-\infty}^{k-\sigma\Theta} \exp\left(-\frac{t^2}{2}\right)$$

and $G(t)$ is the standard normal distribution. The expected value of the loss is given by $\Theta = 1$ in Equation 16.

For a decreasing utility function such as Equation 15, the certainty equivalent, CE, is greater than the expected value, EMV. The risk premium, RP, is given by

$$RP = CE - EMV \tag{17}$$

while Equation 1 is true for increasing utility functions. RP is not the insurance premium; it is the loading to be added to EMV to provide an estimate of CE. As discussed earlier in connection with curve B in Figure 5-9.1, the certainty equivalent CE is the maximum insurance premium which a property owner will be prepared to pay in order to meet the uncertain consequences of a fire.[3] Fire risk cannot be avoided. The property owner has to participate in the risky game involving a loss due to a fire, but the consequences (disutility) can be mitigated by insurance coverage which converts an uncertain loss into a known cost, the insurance premium payable at a certain date. If two property owners are compared, the one with a higher value of Θ (more risk averse) will be prepared to spend more money on insurance coverage than the other property owner. The degree of risk aversion increases with Θ.

The value of the expected loss EMV will decrease with increasing levels of fire protection. By adopting efficient fire protection measures, a property owner can manage to spend on insurance an amount less than the maximum CE consistent with the extent of his or her risk aversion. Without fire protection the owner may have to spend on insurance more than the corresponding CE, depending on the risk aversion of the insurance firm reflected in the insurance premium. Of course, fire protection involves investment and maintenance costs but devices such as sprinklers may qualify for tax allowances in addition to reduced insurance premiums.

Apart from the two options, full insurance or no insurance (full self-insurance), a property owner can also consider partial insurance with partial self-insurance. This option, known as "deductible," requires the participation of the insured in a loss up to a certain limit agreed upon with the insurer in advance. The insured has to bear the entire amount of any loss up to the deductible level, D. For a loss, L, greater than D, the owner's liability is limited to D since he/she will receive the difference $(L - D)$ from the insurer. With a deductible the insured is given the advantage of reduced premiums. On the other hand, when a deductible is applied the insurer will not have to settle and pay small losses which obviously relieves the insurer of a considerable amount of work. With a deductible in the insurance contract, the insured is expected to take a good deal of interest in adopting loss prevention and reduction measures since he/she will have to bear part or whole of the loss whenever a fire occurs. At one extreme a property owner can have minimum fire protection and select a small deductible or at the other extreme, a high level of fire protection with a large deductible.

Assuming a log-normal distribution for loss x with μ and σ as the mean and standard deviation for z ($= \log x$) and using the utility function in Equation 15, the certainty equivalent, $\hat{x}_{DV}$, for a deductible level D and property worth V is given by

$$\hat{x}_{DV}^{\Theta} = \frac{1}{G(k)}\int_{-\infty}^{\log D} e^{z\Theta}f(z) + \frac{1}{G(k)}\int_{\log D}^{\log V} D^{\Theta}f(z)$$

where

$$f(z) = \frac{1}{\sqrt{2\Pi}\sigma}\exp\left[-\frac{1}{2}\left(\frac{z-\mu}{\sigma}\right)^2\right]$$

and the parameters k and $G(k)$ have already been defined. Performing the integration it may be seen that

$$\hat{x}_{DV}^{\Theta} = \frac{G(w-\sigma\Theta)}{G(k)}\exp\left(\mu\Theta + \frac{\sigma^2\Theta^2}{2}\right) + D^{\Theta}\left[1 - \frac{G(w)}{G(k)}\right] \quad (18)$$

where

$$w = (\log_e D - \mu)/\sigma$$

$$G(w) = \frac{1}{\sqrt{2\Pi}}\int_{-\infty}^{w}\exp\left(-\frac{t^2}{2}\right)$$

$$G(w - \sigma\Theta) = \frac{1}{\sqrt{2\Pi}}\int_{-\infty}^{w-\sigma\Theta}\exp\left(-\frac{t^2}{2}\right)$$

The expected value $\bar{x}_{DV}$ of the amount the property owner has to bear is given by $\Theta = 1$ in Equation 18. The certainty equivalent $\hat{x}_{DV}$ is the monetary equivalent of the disutility the property owner has to incur in making provision for self-insurance. If such a provision is not made, the owner could use the amount productively in the investment project. However, with a deductible, the insurance premium is reduced. Hence the disutility associated with self-insurance would be compensated to some extent by a decrease in the disutility associated with cost of insurance. The extent of this compensation depends on the actual costs of insurance with and without the deductible.

An approximate value of the maximum insurance premium which a property owner will be willing to pay with a deductible, D, is $\hat{P}_{DV}$, given by the following formula

$$\hat{P}_{DV}^{\Theta} = \frac{1}{G(k)}\int_{\log D}^{\log V} e^{z\Theta}f(z) - D^{\Theta}\left[1 - \frac{G(w)}{G(k)}\right]$$

$$= \frac{G(k-\sigma\Theta) - G(w-\sigma\Theta)}{G(k)}\exp\left(\mu\Theta + \frac{\sigma^2\Theta^2}{2}\right)$$

$$- D^{\Theta}\left[1 - \frac{G(w)}{G(k)}\right] \quad (19)$$

The expected value $\bar{P}_{DV}$ of the amount the insurance firm has to bear is given by $\Theta = 1$ in Equation 19. For calculating the insurance premium appropriate to the deductible D, the insurer will add to $\bar{P}_{DV}$ a safety loading and another loading toward operating costs. It may be observed that the total value given by Equations 18 and 19 is equal to the value given by Equation 16. Hence

$$\hat{x}_{DV}^{\Theta} + \hat{P}_{DV}^{\Theta} = \hat{x}^{\Theta} \quad (20)$$

The property owner incurs a total disutility $\hat{x}^{\Theta}$ part of which is toward the provision of self-insurance ($\hat{x}_{DV}^{\Theta}$) and the remaining part toward insurance ($\hat{P}_{DV}^{\Theta}$). This result, although true conceptually, is based on the approximate value given by Equation 19. The exact value of $\hat{P}_{DV}^{\Theta}$ is given by

$$\frac{1}{G(k)}\int_{\log D}^{\log V} (x - D)^{\Theta}f(z) = \frac{1}{G(k)}\int_{\log D}^{\log V} (e^z - D)^{\Theta}f(z)$$

which is difficult to compute.

Consider, as an example, a multistory textile industry building for which, according to Rogers,[12] the mean μ and standard deviation σ of the logarithm (base e) of loss (in 1966 prices) in units of thousands of pounds are as follows

With sprinklers: $\mu = -3.267$, $\sigma = 3.085$

Without sprinklers: $\mu = 0.923$, $\sigma = 2.284$

Consider a particular building with a financial value at risk of £500,000 at 1966 prices. Based on Equations 16, 18, and 19, the expected values and certainty equivalents are given in Table 5-9.1 for $\Theta = 1.2$ and 1.5 and two deductible levels, £50,000 and £100,000.

From the results in Table 5-9.1 it is apparent that whatever the insurance option, sprinklers reduce the expected loss to a considerable extent. Consequently, by installing sprinklers a property owner can expect a substantial saving in insurance premiums for any deductible level. This will reduce the disutility associated with the insurance premium, the actual reduction depending on the premium charged by the insurance firm chosen by the property owner.

Sprinklers also significantly reduce the disutility suffered by the property owner in making provisions for self-insurance in financial planning. As one might expect, the expected loss and self-insurance disutility increase with increasing deductible levels. The increase in this disutility is somewhat high for a building without sprinklers but only marginal for a sprinklered building, particularly for low values of Θ (greater than unity). Self-insurance or insurance disutility for any insurance option increases with Θ, the degree of risk aversion. It may be verified that Equation 20 is satisfied by the figures in Table 5-9.1. For example, for the sprinklered building with $\Theta = 1.5$ and £50,000 deductible

$$(3.5)^{1.5} + (5.7)^{1.5} = (7.4)^{1.5}$$

TABLE 5-9.1 *Multistory Textile Industry Building—Expected Losses and Certainty Equivalents (1966 prices) (V = £500,000)*

	Sprinklers			No sprinklers		
		Certainty Equivalent			Certainty Equivalent	
Insurance Option	Expected loss (£)	$\theta = 1.2$ (£)	$\theta = 1.5$ (£)	Expected loss (£)	$\theta = 1.2$ (£)	$\theta = 1.5$ (£)
Full insurance (no deductible)	2,200	3,800	7,400	17,700	23,500	33,200
£50,000 deductible self-insurance disutility	1,400	2,200	3,500	10,000	11,800	14,400
Insurance disutility	800	2,100	5,700	7,700	14,500	26,500
£100,000 deductible self-insurance disutility	1,700	2,700	4,700	13,100	16,000	20,700
Insurance disutility	500	1,500	4,600	4,600	10,200	21,100

In the analysis discussed thus far financial loss has been considered as the (random) variable with a log-normal probability distribution whose parameters μ and σ have been estimated. This method requires data on financial losses which may not be available in some cases. As for fires attended by fire brigades in the U.K., data may be available for extent of fire spread according to the following classifications:

1. confined to object first ignited;
2. spread beyond object but confined to room of fire origin; and
3. spread beyond room of fire origin.

The data will provide an estimate of the probability of fire falling into each of the three classes mentioned above. For any particular building such spread probabilities can also be estimated by adopting a detailed engineering method developed by Fitzgerald.[13]

Let the probabilities for the three spread categories be denoted by p_a, p_b and p_c such that

$$p_a + p_b + p_c = 1$$

Also let x_a, x_b, and x_c be the average or maximum area expected to be destroyed by fire. The expected value of area destroyed is given by

$$\bar{x} = p_a x_a + p_b x_b + p_c x_c$$

For a utility function of the form in Equation 15 the certainty equivalent is $\hat{x}$, given by

$$\hat{x}^\Theta = p_a x_a^\Theta + p_b x_b^\Theta + p_c x_c^\Theta$$

Both $\bar{x}$ and $\hat{x}$ can be converted to approximate monetary values by using financial loss per unit area.

Consider the following data for a particular building:

$$p_a = 0.5, \quad x_a = 5 \text{ m}^2$$
$$p_b = 0.4, \quad x_b = 20 \text{ m}^2$$
$$p_c = 0.1, \quad x_c = 3000 \text{ m}^2.$$

The expected value (area destroyed) is 310.5 m². With $\Theta = 1.5$ in Equation 15 the certainty equivalent is 647.4 m² such that

$$(647.4)^{1.5} = (0.5)(5)^{1.5} + (0.4)(20)^{1.5} + (0.1)(3000)^{1.5}$$

It may be verified that for $\Theta = 2$, the certainty equivalent is 948.8 m².

DECISION ANALYSIS

There are three phases to a decision analysis cycle: (1) deterministic, (2) probabilistic, and (3) informational. In the deterministic phase, the alternative choices or courses of action that are available should be enumerated. The economic parameters governing the decision problem should be identified, together with estimates of the possible variation of the values of the parameters. If that variation is large enough to create uncertainty in the selection of the "best" alternative, then a probability analysis is required. This occurs in the second phase of the analysis cycle where probabilities and risk preferences are determined and alternative choices compared. In the third phase new data are collected which might lead to a restructuring of the deterministic model either to simplify the computations or to more accurately reflect the process under study. There may be a need for updating of probablities and other parameter values. After executing the third phase the decision cycle is repeated. Only the first two phases of the decision cycle are discussed in this chapter.

Within financial constraints and restrictions imposed by fire regulations, codes, and legislation, a property owner has two basic choices in reducing the adverse effects of fire incidents: (1) invest in fire protection measures such as sprinklers, detectors, and structural fire protection or (2) a choice of insurance coverage, full insurance coverage, full self-insurance (no insurance), or partial insurance plus partial self-insurance. As discussed earlier, a number of deductible levels can be chosen for self-insurance. The possible choices to be included in the decision analysis can be conveniently represented in a decision tree as shown in Figure 5-9.3. Decision trees are normally structured as a sequence of decision forks, which represent the various choices or levels of decision, and probability forks, which represent uncertain effects of decisions. The values of the economic parameters associated with each decision level are then estimated and where a chance effect occurs, the parameters are weighed according to the probability of the chance effect.

The economic parameters are costs and losses in a fire protection and insurance problem. Symbolically, the total cost is given by

$$T = I + C + p\bar{x} \qquad (21)$$

where

I = annual insurance premium

C = annual cost of fire protection

$\bar{x}$ = loss expected to be incurred by the property owner in the event of a fire

p = annual probability of a fire occurrence.

The property owner has to meet the annual costs I and C whether a fire occurs or not, but is liable for part of the loss (x) only in the event of a fire.

As discussed earlier in this chapter, the expected value of the part of fire loss the property owner has to bear depends on the deductible level chosen and is given by Equation 18 with $\Theta = 1$. The insurance premium, I, also depends on the deductible level. For the parameter C, the net annual cost of fire protection should be used. This net cost is given by the capital cost of fire protection expressed on an annual basis, plus annual maintenance cost minus an annual value of benefits such as tax allowances.

In a conventional decision analysis the monetary values of I, C and $\bar{x}$ are used with the objective of identifying a fire protection plus insurance package that will have the least total cost. However, as discussed earlier, it would be desirable to carry out the analysis in terms of the disutility associated with the total cost and select a package that will have the minimum total disutility. The expected value of the total disutility is given by

$$\hat{T}^{\Theta} = I_{DV}^{\Theta} + C^{\Theta} + p\hat{x}_{DV}^{\Theta} \qquad (22)$$

where $\hat{T}$ is the certainty equivalent of total cost with $\hat{x}_{DV}$ given by Equation 18. (For $\Theta = 1$, Equation 22 reduces to Equation 21). The parameter I_{DV} is the insurance premium for a property with value V and deductible level D. In principle, I_{DV} should decrease with increasing D so that an increase in the disutility associated with $\hat{x}_{DV}$ would be compensated by a decrease in the disutility associated with I_{DV}. The actual reduction in the insurance premium, I_{DV}, depends on the insurance firm and its attitude toward risk. The property owner's attitude to risk and capacity to bear the required insurance premium is reflected by $\hat{P}_{DV}$ (Equation 19).

Since the value of Θ would vary from one property owner to another the insurance firm would normally take into account the value of $\bar{P}_{DV}$ in the calculation of I_{DV}. But the loadings on $\bar{P}_{DV}$ are usually not directly proportional to $\bar{P}_{DV}$, which is correlated with $\bar{x}_{DV}$. Hence, I_{DV} may be regarded as independent of $\bar{x}_{DV}$. The cost of fire protection, C, is of course independent of I_{DV} and $\bar{x}_{DV}$. For these reasons, Equation 22 appears to be a reasonable model for calculating the certainty equivalent of the disutility associated with the total cost. According to this equation the total disutility due to *independent* factors is the sum of disutilities associated with the factors.

Consider now the example discussed earlier with reference to Table 5-9.1. With a value density of £200/m² (1966 prices) this property of total value £500,000 (1966 prices) has an (estimated) total floor area of 2500 m². The probablity of fire starting in a building is given by the following formula[14,15]

$$p(A) = KA^{\alpha}$$

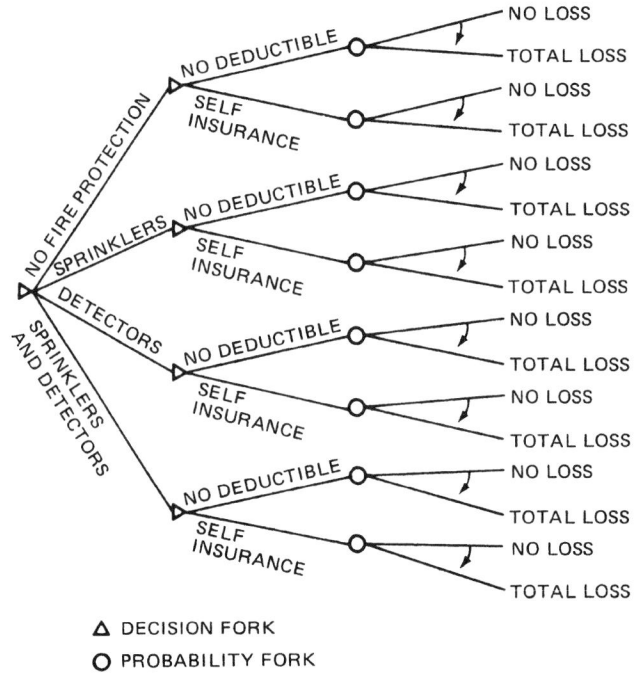

△ DECISION FORK

○ PROBABILITY FORK

↯ PROBABILITY DISTRIBUTION OF LOSS

Fig. 5-9.3. Decision tree for investment in fire protection and/or insurance.

where A is the total floor area and K and α are constants for a category of fire risk. According to Rutstein,[16] with A in m², $K = 0.0075$ and $\alpha = 0.35$ provide an estimate of the annual value of $p(A)$ for the textile industry. Hence, for $A = 2500$, $p = 0.12$ for the example considered.

Assuming an interest rate of 10 percent and 40 years for the life of a building, the annual amortized initial cost of sprinklers for the building considered is about £240 at 1966 prices. The annual maintenance cost is negligible in comparison with the initial cost. The figure of £240 may be used for C in this example although a small amount should be subtracted toward benefits such as tax allowances.

According to hypothetical calculations carried out by Ramachandran[10] for full insurance coverage, the annual premium rate would be 0.0025 for a property equipped with sprinklers and 0.0065 for a property without sprinklers. Hence, for the example considered, the annual premium at 1966 prices would be £1250 if sprinklered and £3250 if without sprinklers. Suppose the insurance firm calculates rebates for deductibles in terms of insurance disutility for $\Theta = 1.5$ in Table 5-9.1. Then, if the building is sprinklered, the annual premium for a £50,000 deductible would be

$$\frac{5700}{7400} \times 1250 = 960$$

Similarly, the annual premium for this building would be £780 for a £100,000 deductible. Also, for the nonsprinklered case the annual premium would be £2590 for a £50,000 deductible and £2070 for a £100,000 deductible.

Making use of the estimates mentioned above, the results in Table 5-9.2 have been obtained for the certainty equivalent based on the total disutility of the total annual cost. The values for $\Theta = 1.2$ and $\Theta = 1.5$ are based on Equation 22 and those for $\Theta = 1$ on Equation 21. The values

TABLE 5-9.2 *Multistory Textile Industry Building—Certainty Equivalent of the Total Annual Cost (1966 prices) (V = £500,000)*

	Sprinklers			No sprinklers		
Insurance Option	$\theta = 1$ (£)	$\theta = 1.2$ (£)	$\theta = 1.5$ (£)	$\theta = 1$ (£)	$\theta = 1.2$ (£)	$\theta = 1.5$ (£)
Full insurance	1490	1392	1319	3250	3250	3250
£50,000 deductible	1368	1357	1504	3790	4110	4863
£100,000 deductible	1224	1259	1603	3642	4288	5886

for $\hat{x}_{DV}$ (self-insurance disutility) and $\bar{x}_{DV}$ (expected value) have been taken from Table 5-9.1. For full insurance $\hat{x}_{DV}$ and $\bar{x}_{DV}$ have zero values since the property owner is compensated for the entire loss and does not suffer from any self-insurance disutility. Also $C = 0$ for a building without sprinklers. Values in Table 5-9.1 describe the situation in the event of a fire occurring and do not take into account the annual probability p (= 0.12) of fire occurrence. Hence these values are higher than those in Table 5-9.2.

The results in Table 5-9.2 clearly confirm the economic value of sprinklers. Also, for a sprinklered building, the total disutility for $\Theta = 1.2$ decreases with increasing levels of deductibles under the assumed figures used in the calculations. If actual insurance premiums are obtained and used in this analysis, the results may be somewhat different. However, it appears that by installing sprinklers in a building the property owner can take a risk and accept a large deductible provided he/she is moderately risk averse ($\Theta = 1.2$). If the owner is very risk averse ($\Theta = 1.5$), full insurance will be the best option. Full insurance will also be the safest option if the owner does not equip the building with sprinklers.

Shpilberg and Neufville[5] applied the decision analysis procedure to the problem of choosing levels of fire protection that might be best for airport facilities whose principal exposure to risk comes from fuel spills and stored cargo. In practice, the facilities are usually protected by sprinkler systems. Customarily, the insurance policy for a facility does not cover any aircraft; it is covered separately. Because of the limited statistical data that existed in this area the study was confined to two levels of fire protection:

1. No fixed fire protection equipment, some fire-resistive construction and fire walls, some portable extinguishers;
2. Average level of fixed fire protection equipment, fire-resistive construction, adequate water supplies.

The authors considered the most popular deductibles in the aircraft industry—5, 10, 20, 50, 100 and 500 thousand dollars—together with the no-deductible and self-insurance choices. To simplify the calculations, instead of the probability distribution of fire damage, nine ranges of damage and their associated probabilities were used in the analysis. Actual losses observed were all at sprinklered facilities. The probabilities of losses for similar unsprinklered facilities were developed by using industry data, which indicated that average losses in unsprinklered facilities were three to five times higher than the average losses in sprinklered facilities.

The decision tree adopted by Shpilberg and Neufville[4] for large airport facilities is reproduced in Figure 5-9.4. In the light of previous experience, the authors assumed that the property owners were constantly risk averse with the utility function

$$U(L) = e^{-4.6L}$$

The results of their evaluation of each of the sixteen possible levels of fire protection and insurance for large facilities are reproduced in Figure 5-9.5. The expected monetary costs are on the left-hand side and the certainty monetary equivalents based on the utility function on the right-hand side. Because risk averse people place a very high value on large and total losses, the CMEs are all increasingly higher than the expected monetary losses for higher deductibles. The analysis indicated that the best option for the owner of a new facility would be to build it without sprinklers and then take out an insurance policy with a $100,000 deductible. The best deductible for sprinklered aircraft facilities was found to be $20,000 for a small facility but only $5,000 for a large facility.

It must be remembered that the investigation carried out by Shpilberg and Neufville and the analysis described with reference to Tables 5-9.1 and 5-9.2 have considered only direct material damage to a property. No consideration has been given to factors such as the loss of future revenues caused by a business interruption, which are only partially insurable, or by changes in the safety image of an industry and the loss of lives. A huge loss in profits might seriously disrupt or even bankrupt a business activity. It is difficult to put a monetary value on human life although this is necessary, as discussed in the previous chapter. Multiple-death fires usually have serious social, economic, and political consequences. Fires causing large consequential losses due to loss of production, exports, employment, and extra imports could cause a good deal of stress to a national economy.

The results of the examples discussed in this chapter should not be taken as recommendations of levels of self-insurance or fire protection; every industrial or commercial property has its own economic tradeoffs and alternatives. The choice of whether to invest in fire protection devices depends on (1) the cost of fire protection and its relation to the cost of insurance, (2) the shape of the probability distribution of losses under different fire protection measures, and (3) the particular degree of risk aversion of the decision maker. Inaccurate or outdated deductible schedules adopted by insurance companies can produce unexpected results in terms of minimum insurance costs. These schedules are expected to accurately reflect the probability distribution of losses which could change over a period due to factors such as inflation and larger average exposed areas.

The main purpose of the decision analysis described in this section is to illustrate how complex decisions involving fire protection and insurance can be systematically analyzed by rationalizing the decision process and considering the implications of risk averse behavior. Although this method is specifically intended for property owners it can be used by insurance companies to analyze the implications of their rates. The decision tree approach can also be used to study the investment policies for municipal fire protection or

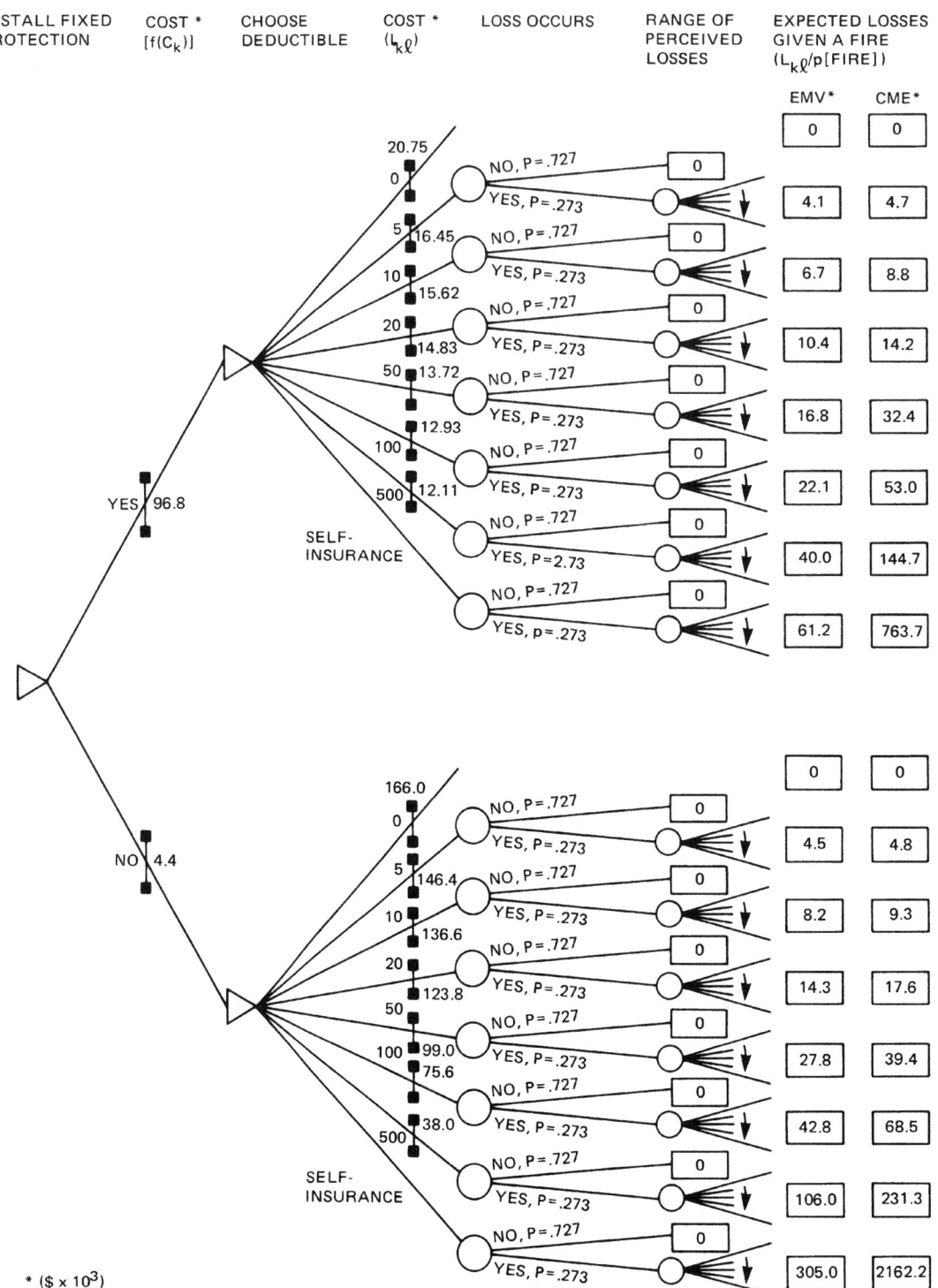

Fig. 5-9.4. *Decision tree for $35 million airport facility based on expected monetary value and certainty monetary equivalent (CME) of losses.*[4]

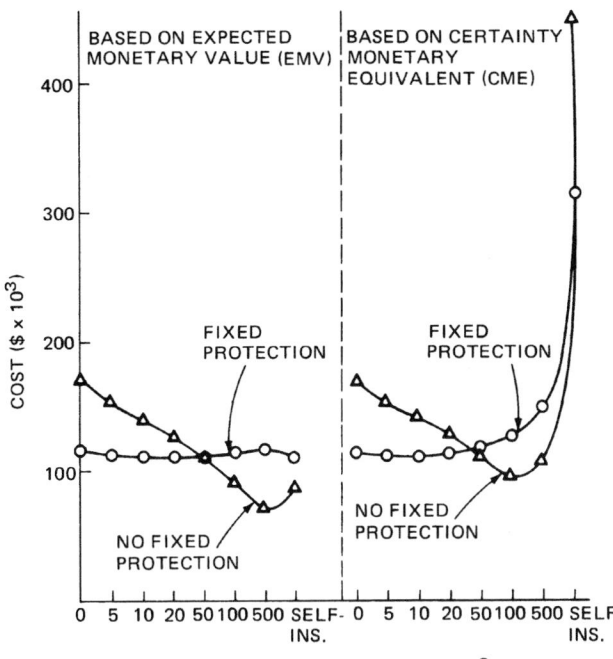

Fig. 5-9.5. *Annual cost as a function of deductible coverage and fixed protection for a $35 million airport facility.*[4]

design policies with respect to fire protection for high-rise buildings. Such problems would require the application of multiple attribute utility theory to take into consideration not only the economic risk, but also loss of life and other intangible factors that cannot be easily quantified.

CONSTRUCTION OF UTILITY FUNCTIONS

Within the last few years, attempts have been made to measure how strongly people feel about accidents and their consequences and how much they are willing to spend to protect themselves against such risks. These measures have been developed by carrying out surveys in which carefully structured questionnaires are distributed to individuals. Such data are statistically analyzed to provide quantitative estimates of the perception of and attitude to risk. The data are also used to construct utility functions which are then incorporated into decision analyses. The surveys for this purpose tend to be expensive and time consuming if sufficient statistical accuracy is desired.

The shape of the utility function could be of the mathematical form shown in Equations 12, 14, or 15, since a property owner or a firm is generally risk averse. In terms of disutility $D(x)$ [$= -U(x)$], Equation 15, for example, implies that

$$D(10x) = (10)^\theta x^\theta$$

which is $> 10\,x^\theta$ or $10\,D(x)$ for $\theta > 1$. In other words, the disutility due to a single large loss of magnitude ten times that of a small loss is greater than the total disutility caused by ten small losses. The consequences of a single large fire are worse than the consequences of ten small fires of equiv-

alent total loss. In the case of life loss, it is important to consider not only the number of deaths but also whether they occur singly or as a result of catastrophes involving many deaths.

Modifying Equation 15, it could be written

$$U(x) = -\left(\frac{x-m}{M-m}\right)^\theta \qquad (23)$$

where m is the minimum, a loss less than which will cause no disruption to the business activity. A loss in excess of the maximum, M, might put the property owner out of business temporarily, for a long duration, or permanently. By choosing intermediary values of x which cause minor disruption, major disruption, etc., an approximate value of θ can be determined.

Estimates of probable consequential losses due to fires would also provide data for constructing an appropriate utility function. The value of θ also depends on the attitude to risk and financial strength of a particular property owner. As the fire damage, x, increases from m to M, $U(x)$ decreases from 0 to -1.

Although risk averse, the utility function in Equation 15 exhibits decreasing degrees of risk aversion. The risk aversion in a range with small losses is greater than the risk aversion in a range with larger losses. This is due to the fact that the risk aversion function

$$q(x) = U''(x)/U'(x)$$
$$= \frac{\theta-1}{x}$$

decreases with increasing x. [The functions $U'(x)$ and $U''(x)$ are the first and second derivatives of $U(x)$.] For the exponential form in Equation 12 the risk aversion function ($= \theta$) is a constant. To reflect realistically the true value of avoiding or providing for a large or catastrophic loss, the decision maker may choose a risk-averse utility function with an increasing risk aversion function. The following is one such utility function

$$U(x) = \log(b - x) \qquad (24)$$

whose risk aversion function

$$q(x) = 1/(b - x)$$

increases with x in the range $x < b$. The value of b could correspond to the assets of the property owner or the value of building and contents. It is, however, mathematically complex to use Equation 24 in conjunction with a log-normal probability distribution for fire loss x.

REFERENCES CITED

1. J. von Neumann and O. Morgenstern, *Theory of Games and Economic Behavior*, Princeton Univ. Press, Princeton (1947).
2. P.G. Moore, *Risk in Business Decision*, Longman Group, London, (1972).
3. R.L. Keeney and H. Raiffa, *Decisions with Multiple Objectives: Preferences and Value Trade-Offs*, John Wiley and Sons, New York (1976).
4. D.C. Shpilberg and R.D. Neufville, "Best Choice of Fire Protection, An Airport Study," *Technical Paper RC74-TP-16*, Factory Mutual Research Corp., Norwood (1974).

5. D.C. Shpilberg and R.D. Neufville, "The Use of Decision Analysis for Optimising the Choice of Fire Protection and Insurance. An Airport Study," *Fire Tech.*, 10, 5 (1974).

6. G. Ramachandran,"Extreme Value Theory and Fire Losses—Further Results," *Fire Research Note No. 910*, Fire Research Station, Borehamwood, England (1972).

7. G. Ramachandran, "Extreme Value Theory and Large Fire Losses," *ASTIN Bul.*, 7, 293 (1974).

8. G. Ramachandran, "Extreme Order Statistics in Large Samples From Exponential Type Distributions and Their Application to Fire Loss," in *Statisitical Distributions in Scientific Work*, 355, D. Reidel, Dordrecht (1975).

9. D.C. Shpilberg,"Risk Insurance and Fire Protection; A Systems Approach. Part 1: Modelling The Probability Distribution of Fire Loss Amount," *Technical Report No. 22431*, Factory Mutual Research Corp., Norwood, (1974).

10. G. Ramachandran, *The Interaction Between Fire Protection and Insurance*, Seminar, Zurich (1984).

11. G. Ramachandran, "Properties of Extreme Order Statistics and Their Application to Fire Protection and Insurance Problems," *F. Safety J.*, 5, 59, (1982).

12. F.E. Rogers, "Fire Losses and the Effect of Sprinkler Protection of Buildings in a Variety of Industries and Trades," *BRE Current Paper CP 9/77*, Fire Research Station, Borehamwood, (1977).

13. R.W. Fitzgerald, "An Engineering Method for Building Fire Safety Analysis," *F. Safety J.*, 9, 233, (1985).

14. G. Ramachandran, "Fire Loss Indexes," *Fire Research Note No. 839*, Fire Research Station, Borehamwood, (1970).

15. G. Benktander, "Claims Frequency and Risk Premium Rate as a Function of the Size of the Risk," *ASTIN Bul.*, 7, 119 (1973).

16. R. Rutstein, "The Estimation of Fire Hazard in Different Occupancies," *Fire Surv.*, 21, April (1979).

PRODUCT FIRE RISK

John R. Hall, Jr.

INTRODUCTION

In the big picture of fire safety, there are only two ways to make changes—change things or change behavior. Every "thing" you can change could be seen as a product, in that it is a physical object that people buy, e.g., from the raw materials that are used to make furniture to the finished furniture, from the components of fire detection and suppression systems to the complete systems, from wood and steel and concrete to whole buildings. That definition of product is too broad for one chapter.

For purposes of this chapter, a product will be a finished product, not raw materials, not components in an assembly, but objects in an end-use form. And a product will be something that starts or feeds a fire as either a heat source or a fuel source. Computationally, fuel-source products involve much more elaborate calculations and effects, because the risk of heat-source products is measured solely in terms of whether or not ignition occurs. The treatment in this chapter will reflect that.

The fire risk of a product captures the range of severities of fires associated with the product and the probabilities that fires will occur having those severities. Fire risk is usually measured as expected loss, i.e., the sum over all fires of probability times severity, or as the probability of having a fire more severe than a stated threshold. Either way, fire risk analysis relies heavily on fire scenarios, which are used to set up calculations of both probabilities and severities. This is *not* the same as choosing a handful of specific fires and calculating their severities and probabilities. A valid calculation must demonstrate that each specific fire analyzed is representative of a larger class of fires, that probabilities are calculated for the larger classes of fires rather than the specific fires, and that collectively the larger classes include every type of fire there can be.

Dr. John R. Hall, Jr. is Assistant Vice President for Fire Analysis and Research at the National Fire Protection Association. He has been involved in studies of fire experience patterns and trends, models of fire risk, and studies of fire department management experiences since 1974 at NFPA, the National Bureau of Standards, the U.S. Fire Administration, and the Urban Institute.

STEPS IN A PRODUCT FIRE RISK ANALYSIS

The steps and sub-steps briefly described below parallel current thinking at U.S. standards-writing organizations, notably the American Society for Testing and Materials (ASTM), and previous global reviews of approaches to this subject, particularly as synthesized by Bukowski and Tanaka.[1] Later sections expand on techniques to be used in executing these steps.

1. **Define the Scope of Products Included, Including Context of Use.**
   a. Define the product or, more typically, the product class to be evaluated.
   b. Specify where and how the product is used. For example, a standard for floor coverings would not include all uses of carpeting, because carpeting is sometimes used as a wall covering. The specification of application will not only limit the range of product characteristics, but will also specify or limit the input parameters used to identify fire scenarios in which the product may play a role.

2. **Specify the Class of Properties in Which the Product Will Be Used.** Part of identifying the scope is categorizing and specifying the property in which the product is used. The end-use or principal activity in a property defines it as an *occupancy*, which will imply a variety of characteristics and conditions in the environment of the product. For example, a risk analysis of upholstered furniture in homes will be different from a risk analysis of upholstered furniture in offices, and both will be different from a risk analysis of upholstered furniture in hotels. The types of pieces used are different, the applicable standards are different, the mix of fires they could be exposed to are different, and the mix of people likely to be present (and their capabilities) are different.

3. **Identify Measure(s) of Loss or Harm to Be Used.** Some measures, called end-measures, are meaningful in and of themselves but are very difficult to predict in models or measure in tests (e.g., monetary damages, injuries). Some measures are easily predicted in models or measured in tests, but are not meaningful in and of themselves (e.g., temperature or toxic gas concentrations or obscuration for particular areas or volumes). Typically, models must

be used to convert readily measurable quantities to end-measures of loss.

4. **Identify and Specify the Relevant Scenarios.** A scenario is a set of details about the initiating conditions and early growth of a fire that are needed as input conditions to a test method, fire model, or probability or other calculation. This may include:
   a. Location and characteristics of the initial fuel and initial heat source. (The product will have the potential to be one or the other, depending on which type of product it is.)
   b. Proximities and characteristics of other fuel packages near the first ignited item. (The product may also belong here, if it is a fuel-source-type product.)
   c. If fire growth or effects beyond the first affected room or area are important to the estimation of the chosen measures of loss, then complete descriptions of those other areas will be needed, including spatial dimensions; fuel load; thermal properties of room linings, barriers, and openings connecting areas; occupants; and damageable property.
   d. Fire protection systems must be specified for any areas to be modeled.

5. **Identify Test Methods, Models, and Other Data Sources and Calculation Procedures.** The models needed will depend, in part, on the scenarios to be addressed, but the models listed below include the major modeling components included in most of the major modeling packages now in use. Each model has implications for data needs, including fire tests and statistical data bases. (See Reference 2 for a more detailed review of available models.)
   a. Fire growth model
      i. Model of rate of growth in terms of heat release rate, for example, as a function of fuel load and distances between items
      ii. Horizontal flame spread model
      iii. Barrier failure (e.g., door, ceiling, window)
      iv. Exterior vertical flame spread model
      v. Flame spread model in concealed spaces
      vi. Building-to-building flame spread
   b. Smoke spread model
      i. Model of room filling
      ii. Model of spread between rooms
      iii. Flashover models, including timing of flashover and post-flashover smoke spread
      iv. Model of spread via heating, ventilation, or air conditioning system
   c. Occupant behavior model
      i. Model of automatic detection equipment performance
      ii. Model of how fire is discovered in the absence of automatic detection
      iii. Model of decision-making activities leading to decisions to egress or attempt rescue
      iv. Model of egress and rescue activities
   d. Intervention models
      i. Automatic suppression models, including timing of activation and effects on fire growth
      ii. Model of other suppression or extinguishment efforts and their effects (e.g., whether fire extinguishers will be used and to what effect)
      iii. Fire fighter response models
   e. Fire effects or outcome models
      i. Predicted deaths and injuries due to fire effects in affected areas as a function of time
      ii. Structural damage or failure models
      iii. Predicted extent or monetary value of property damage
   f. Ignition probability models
      i. Fault tree, success tree, or event tree
      ii. Bayesian analysis of test results, historic fire probabilities, and other data

In practice, many of these component models are rarely used. For example, a fire risk assessment of a burnable product may not need an elaborate analysis of intervention strategies, because the dominant scenarios may be those in which no prompt, effective intervention occurs. On the other hand, the modeling components used may identify a need for data for which no standardized source exists (e.g., burning properties of products in post-flashover environments). It is not unusual, therefore, for the full calculation to require judgments by analysts, which must be checked through sensitivity analyses.

Bukowski and Tanaka[1] have proposed a conceptual scheme for standardizing the role of these expert judgments in fire hazard and risk assessment. Their scheme involves identifying groups of parameters and variables in the models and defining the acceptable sources of data for them, among which could be expert judgments.

Specifying and standardizing needed data sources is an essential part of the process of using fire hazard or risk assessment in a standard. The expectation is that, instead of stating a standard in terms of specifications, the standards-setting process would specify outcome measures, models and other calculation methods, modeling assumptions and input parameters, test methods and other data sources, and possibly the type of expertise required by those who run the models.

Define the Scope of Products Included

The scope definition should define a class of interchangeable items having a common function or application in a specified occupancy and with a range of allowable choices for composition. Specification of the product should be done in a way that facilitates use of existing data, from fire incident data to product test data.

For heat-source products, this means that initial specification of the product by function and construction should be based on the categories defined by NFPA 901, *Uniform Coding for Fire Protection*,[3] Chapter F, "Equipment Involved in Ignition," and Chapter G, "Form of Heat of Ignition."

For fuel-source products, this means that initial specification of the product by function should be based on the categories defined by NFPA 901,[3] Chapter I, "Form of Material First Ignited." Initial specification of the product by material composition should be based on the categories defined by NFPA 901,[3] Chapter H, "Type of Material First Ignited." A product, e.g., carpet, is defined as a floor covering made of certain materials, chosen to distinguish it from vinyl flooring, wood flooring, concrete slabs, etc.

Further specification of the product by function may be needed (e.g., selecting bookcases from the cabinetry group). In such cases, the nationally representative fire incident data bases will not be sufficient to estimate probabilities. Other, special fire incident data bases and expert judgment will be needed.

When calculating probabilities, be sure to include appropriate shares of fires involving products that were partially or wholly undefined (e.g., upholstered furniture fires should include shares of fires involving unknown-type furniture or unknown-type form of material first ignited, and might include shares of fires involving unclassified furniture or unclassified form of material first ignited).

The range of items defined as examples of the product—which may be referred to as members of the product class—must, for analysis purposes, be reduced to a manageable number of subgroups. Each subgroup will be *defined* by a range of characteristics (e.g., all cellulosic versions of the product) but will be *represented* by one specific set of product fire characteristics. Ordinarily, these product fire characteristics will be identified from review of results of actual fire tests on one or more representatives of the product class.

Specify the Class of Properties

For reasons similar to those already cited for product scope definitions, property classes (i.e., occupancies) should have their primary definitions stated in terms of the categories defined in NFPA 901, *Uniform Coding for Fire Protection*,[3] Chapter B, "Fixed or Specific Property Use." Whenever occupancy scenarios can be defined using nationally representative, valid fire incident data, the analyst will have the strongest possible basis for estimating probabilities. The principal weakness of this data source involves the level of detail of readily available fire incident data, which often falls well short of the detail needed to run the fire hazard analysis portion of the method.

Identify Measure(s) of Loss or Harm

The two principal measures of loss are expected loss and probability of loss exceeding a certain threshold. Both measures can be calculated directly from nationally representative fire incident data bases, without the need for modeling or testing, provided that the product class definition matches the categories used in those data bases. For example, a fire risk analysis comparison of the major types of home heating equipment is possible, because each can be identified within the fire incident data bases. However, a comparison of different designs for, say, portable electric heaters could not be easily done from statistics alone, because different designs cannot be so distinguished. And for fuel-source products, their role in fire can be identified in statistics only when they are the first item ignited.

For a variety of reasons, therefore, one is usually forced to use test methods and models to develop probability estimates and fire severity estimates more appropriate to the product class and product alternatives of interest. In such cases, much calculation effort can be saved if the problem lends itself to restatement in terms of measures of loss that can be measured in the laboratory and at the fire scene. Three examples are:

1. Probability of flashover and/or of flame spread beyond the room of origin,
2. Probability of fire ignition, and
3. Probability that time to flashover exceeds *x* minutes (where *x* is chosen to reflect the expected arrival of suppression and rescue forces).

One approach that should usually be avoided is to try to measure loss in terms of the product's share of responsibility for overall fire severity. Such measures tend to be far too subjective and require answers to inherently unanswerable questions. For example, suppose a small trash can fire leads to a large couch fire. If *either* the factors in the initial trash ignition or the burning properties of the couch are changed, no large fire would have resulted. How much loss should be assigned to the couch? There is no good answer to that question.

Instead, fire risk analysis should proceed through calculations of differences, i.e., fire risk with the product of interest *vs.* fire risk with something else substituted for the product of interest.

From this perspective, one can see how fire risk analyses can be constructed as extensions of past successful applications of fire modeling. For example, one of the earliest practical applications of the Harvard code was to the reconstruction of the 1980 MGM Grand Hotel fire. As suggested above, flashover was used as a well-defined event to focus the analysis, after it was shown that most of the fatal fire victims would have survived if flashover had been prevented. Professor Howard Emmons then used the model to rerun the fire with changes, considered individually, in the room of origin's ceiling covering; its benches and chairs; and the area's heating, ventilating, and air-conditioning (HVAC) arrangements.

If one wished to do a fire risk analysis on, say, benches and chairs for dining areas of hotels, one could define a range of possible fire scenarios, do a similar Harvard code analysis of each, weight the consequences by the scenario probabilities, and thereby calculate an overall probability of flashover with two different choices of benches and chairs. The difference between the two probabilities would be a valid product fire risk measure.

A recent fire hazard analysis of rigid non-metallic conduit in hospital emergency systems, done by Benjamin/Clarke Associates, provides a rare example of circumstances where the product's share of fire loss can be validly used for analysis. Dr. Fred Clarke devised a realistic scenario designed to maximize the likelihood of significant product involvement in fire, by placing the initiating fire directly under the product, which was assumed to be exposed due to missing ceiling tile. From a fire risk analysis perspective, this scenario was designed to put an upper bound on the product's share of fire loss in scenarios with significant loss.

If this upper bound were applied to all scenarios and if it were assumed that a substitute product could eliminate the product's role as a fuel source for the fire, then one would have all the requisite parameters for an upper bound estimate on the fire risk consequences of using that product. This is a one-sided analysis. That is, if the product's fire risk proves to be negligible under these conditions, one knows that the true fire risk is also negligible; but if the product looks bad in this analysis, one does not know whether it would still look bad in a fairer, more representative analysis.

Specify Fire Scenarios—Initiating Fire

For every scenario, each aspect of fire initiation and growth must be specified in such a way that: (1) one can model, test, or otherwise calculate the fire severity consequences of a fire with those specifications; and (2) one can calculate or estimate the probability of having a fire with those specifications. This process of specification usually requires the analyst to address three stages of fire:

1. *What are the initial heat source, the initial fuel source, and the circumstances that bring them together?* These are the basics of the initiating fire, and they need to be specified so that fire incident data bases can be used as a major source for estimating probabilities.

2. *What are the factors that will determine whether, and how quickly, fire will spread from the first item to the product, if they are not the same?*
3. *What are the characteristics of the room or area of origin and its fuel packages and surfaces that will determine how large the fire will grow and whether, and how quickly, it will reach flashover and leave the room?*

These three questions reflect the three stages at which a burnable product may become involved in a fire—as the first item ignited; as a secondary item ignited by exposure to other items ignited earlier; or as part of a room that has gone to flashover, when everything that can burn will burn. (If the product is a heat-source product, the latter two stages are not concerns and the analyst need not address them in as much detail.)

Two general approaches can be used to set up the model of these stages. One is to use surveys of typical fuel loads, room configurations, and the like. Then, one can run a fire growth model with these specifications. The drawbacks of this approach are that the magnitude of the data requirements is extremely large; that such survey data is very scarce, and, when it exists, almost never captures the variations in practice that produce different probabilities and different fire outcomes; and that the probability of ignition is probably not constant from one configuration to another nor is it susceptible to estimation from any existing fire incident data bases. If this approach is used, it will tend to force the analyst away from some of the essentials of fire risk analysis, i.e., a suitably diverse set of scenarios and an adequate attention to the role of probabilities.

The other general approach is to infer patterns of fuel loads and room configurations from fire loss experience. The logic used here is as follows: Recent fires were produced by recent fuel load and room layout practices. What would those practices have to be in order to produce the observed fires? A critical element in this approach is data on final extent of flame damage, which is captured in the major fire incident data bases, as follows:

- Confined to object of origin
- Confined to area of origin
- Confined to room of origin
- Confined to fire-rated compartment of origin
- Confined to floor of origin
- Confined to building of origin
- Extended beyond building of origin

One can assume that a fire confined to object of origin involved only the first item ignited and that a fire extending beyond the room of origin reached flashover in the room of origin.

If the product was not the first item ignited but the fire spread beyond the object of origin, then the fire could have ignited the product through radiant exposure. What is the best way to estimate the probability that this will occur? The fire risk analysis method *FRAMEworks*, developed under the auspices of the National Fire Protection Research Foundation, uses the following four elements:

1. For each type of item first ignited (e.g., trash), a set of estimated typical values for mass and burning properties, sufficient to estimate a rate of heat release curve for the product burning alone;
2. Ignitability characteristics of the product, i.e., critical radiant flux;

3. For each type of item first ignited, a probability distribution on the distance from the item to the product, as a function of the type of room, with distributions based on survey data and expert judgment; and
4. Established mathematical relationships showing the minimum distance at which ignition of a second item will occur, given the first item's burning characteristics and the second item's ignitability characteristics.

This second approach still needs the kind of property survey data required by the first approach, but far less of it because the only geometric information sought is distances between the product and other items. Even so, this is still a data-hungry approach that requires either: (1) survey data that may not exist or may be very expensive to collect or (2) expert judgment that may be especially difficult to make.

As in so many other areas, the temptation will be to reshape the analysis to bypass elements that cannot now be modeled with confidence. However, the analysis must somehow provide a valid basis for combining different product burning properties, and the phenomenon of secondary ignition is central to any evaluation of the product's relative ignitability.

Still other phenomena must be reduced to assumptions for modeling purposes. The following are examples: For a fire that does not reach flashover, what is the physical measure (e.g., temperature) of its peak size? What stops the fire, and what characteristics of fire development (e.g., burning time, detector activation, fire size) triggered fire suppression? (This is important in order to know when to stop the fire if the product is changed.) What is the fire's profile after it reaches its peak? Is there an initial smoldering phase, and, if so, how long is it and what is the fire profile during this period? Each of these questions needs to be answered through a crosswalk between the physical parameters measured in tests and used in models and the parameters recorded in fire incident data bases, because the latter is always needed to calibrate probability estimation.

Specify Fire Scenarios— The Building Beyond the First Room

The dimensions used to define the different occupancy scenarios need to be dimensions that are relevant to fire development. Most of these dimensions will be one of three types:

1. *Building Dimensions and Geometry.* Dimensions of rooms and other areas in which fire may grow or smoke may spread.
2. *Openings.* Dimensions of openings between rooms and areas relating to paths of flame or smoke spread and sources of air to feed the fire.
3. *Room Linings.* Thermal properties of rooms that may bear on burning at and after flashover.

Building dimensions and geometry: The overall building size and geometry can be structured into a series of questions on which data must be sought and decisions made. The first is the range of variation in the number of floors. After determining this point, the user must specify a number of floors for each occupancy scenario and assign a probability to each.

The second is a room layout for each floor. Room heights and the sizes of openings connecting rooms tend to be standardized by common industry practices, so there may be no need to consider variations. For other factors, e.g., the number and sizes of rooms, there usually is too much variation in practice and too little data on the relative likelihood of these variations to do much more than (1) estimate one or two values for the number of rooms or the total square feet per floor; and (2) use expert panels to develop detailed layouts for the purposes of modeling and analysis of the rooms or spaces specified in (1).

However, panels of people who are experts on buildings of a certain type are likely to think in terms of the characteristics of the particular buildings they know best. They may therefore give estimates biased toward characteristics of new construction or characteristics of the buildings they live in or frequent. Fires are more likely to occur in smaller, less prestigious units in any property class. The expert panel needs to be continually reminded to adjust their perspective to think in terms of those kind of buildings.

Openings: There usually will be some information on the sizes of doors and windows, because construction practices are highly standardized even beyond code requirements. However, in a fire, the openings will depend critically on whether and how much key doors and windows are open. There is no body of data on this point for any occupancy. It may be possible to ignore windows, because there are studies indicating that windows affect most fires only after the point in time where fire severity has been determined. (However, the few exceptions will tend to be very large fires, so the reasonableness of an assumption excluding windows will need to be rechecked for any analysis.) How does one set the assumptions for doors, short of large-scale property surveys or special fire data collection projects?

For most fire protection engineering studies, the answer would be to make conservative assumptions, i.e., those that present the greatest fire challenge. It is important to understand that this is usually not the right answer for fire risk analysis. If conservative assumptions lead to an overestimate of fire risk, then they may also lead to a gross overestimate or underestimate of the fire risk consequences of particular product choices. There is no substitute for a best estimate, without conservatism, in fire risk analysis.

However, an assumption that might be made in fire hazard analysis because it is conservative may also turn out to be a reasonable best estimate for fire risk analysis, if it reflects a pattern in actual fire experience. If a certain arrangement *could* produce more serious fires, it qualifies as a conservative assumption for fire hazard analysis. If that same arrangement *is producing* more serious fires, then it is more likely that that arrangement is present when a reported fire occurs than that it is present in buildings in general, and one could be justified in assuming that that arrangement is likely, in a fire risk analysis.

However, this line of reasoning has limits. Suppose that open doors is the conservative assumption, but that we know that doors tend to be open only 5 percent of the time. In that case, the fire risk analysis could reasonably assume that doors are open 10 to 20 percent of the time, reflecting the likelihood that open doors will be more likely in reported fires than in buildings, in general. But the typical situation would still be closed doors.

The analysis would need to have scenarios with open doors and scenarios with closed doors, because neither condition is dominant enough to justify omitting the other condition for a variable (i.e., whether doors are open) that is so influential on final fire size. Or, it might be possible to use

one condition, consisting of doors open slightly, trying to seek a single physical condition that will reproduce the appropriate average between fully open and fully closed. Either way, considerable judgment would be needed.

Remember that, if an "average" value is used, the analyst is implicitly assuming that the fire severity associated with that average value is equal to the average of the fire severities associated with all the individual values that occur. In mathematics, this is sometimes called assuming that the average of the function equals the function of the average, and it is not usually the case. The analyst has to make the case that the assumption is reasonable in the situation being analyzed.

Room linings: Linings of rooms and other areas need to be addressed in terms of the thermal properties required for calculations of time to flashover, speed of vertical flame spread, and the like.

Room and area linings for most occupancies are tightly regulated by codes. However, some of the most important occupancies, e.g., dwellings, are not so covered, and even for those that are, one must allow for a significant probability that the codes will not have been in force when fire occurs. Unfortunately, there is little or no data on the probabilities of different combinations of fuels in particular occupancies; and there is only very limited, dated information on typical or average fuel loads and only for some occupancies.

Specifying Fire Scenarios— Exposure of People or Property

In order to translate model or test outputs on the physical characteristics of fire, as a function of location and time, into end-measures of human or property loss, one must address: (1) the locations of people or property as a function of time, and (2) the damage or loss consequences to people or property of the different possible physical characteristics of fire, e.g., temperature, quantities of toxic gases by type, corrosive properties, and quantities of smoke. The methods for doing this are not extensively developed, except for deaths. Therefore, this section will focus on that outcome.

Occupant exposure depends upon: (1) initial locations of the occupants relative to the fire, and (2) their escape behavior. A complete specification of the number of occupants with their initial locations and other characteristics is called an *occupant set*. The user must first define a group of occupant sets that can validly represent all possible combinations of people and their characteristics and locations, and then must estimate probabilities for each. These must then be joined to a model of occupant behavior. (See Reference 2 for a list of evacuation models.)

Occupant behavior models consist of a set of rules for calculating the locations of occupants at a time, t, as a function of their locations, other occupant characteristics, and fire characteristics at the time stage just prior to t. Some such models track occupants individually; others give only the number of people at each location. Some, but not all, models include interactions among occupants, such as congestion or queuing effects or behavioral rules based on relationships between occupants (e.g., parents who seek to rescue babies). The more comprehensive the model may be in capturing potentially important phenomena, the more computationally demanding it will be and the more data it will demand, possibly including data that is not readily available. As in all other aspects of fire risk analysis, tradeoffs must be made in the modeling.

A brief summary of the steps required is as follows:

- Develop a probability distribution for the number of people present in the building.
- Expand the basic distribution to address relevant characteristics, including ages and relationships of occupants, time of day, and occupant conditions.
- Develop probability distributions for occupant activity as a function of time of day and of occupant characteristics, specified in the previous step.
- Develop probability distributions for occupant location given occupant activity and other occupant characteristics. (If every activity implies a unique location, this will reduce to a crosswalk.)
- Combine all probability distributions to produce a probability distribution for all occupant sets. Merge very similar occupant sets, if needed, for computational simplicity.

Specifying Fire Scenarios— Fire Protection Systems and Features

The following requirements are straightforward, in principle, but necessary models or data are often sketchy:

- For each type of fire protection system (e.g., detectors, sprinklers, smoke control systems) or feature (e.g., fire doors), identify a range of alternatives. These alternatives must address not only variations in the type and coverage of system or feature used (e.g., quick-response *vs.* conventional sprinkler), but also variations in operational status (e.g., fully operational *vs.* water turned off).
- For each alternative, probabilities will be needed. As in the other parts of the analysis, start with representative national fire incident data bases for best estimates, then add needed detail using other data bases and expert judgment.
- For each alternative, it will be necessary to specify rules for how the system or feature, under that alternative, will affect the fire development, the evacuation, or other conditions being tracked. Often, this will be fairly simple. One could assume that a fully operational sprinkler system will activate once a specified set of fire conditions are reached and, once activated, will totally and immediately stop the fire, except for certain specified fire scenarios (e.g., fire origin in concealed spaces), when its effect will be only to block fire entry into sprinklered areas. One could assume that a full-coverage automatic detection system will activate once a specified set of smoke or heat conditions is reached and, once activated, will alert everyone in the building to the fire, leading anyone not already in motion in the occupant evacuation model to begin evacuating.

Identify Test Methods and Models

Analyzing ignition probability: While historic fire data may suffice to estimate ignition probabilities for the mix of existing products, they may not suffice to estimate ignition probabilities for specific existing products, and they will not suffice for new products. The basic approach involves converting laboratory test results to ignition probabilities.

Probabilities of ignition must be estimated from frequencies of ignition in laboratory tests of existing and new products. In essence, it is assumed that, if the new product produces more or fewer ignitions in laboratory tests than the

existing products did, then the new product will have a higher or lower probability of ignition, increased or decreased by the same proportion as the ratio of the laboratory ignition frequencies. For fuel-source products, this estimation should be done separately for each of the major classes of heat sources (e.g., flaming *vs.* smoldering), because product ignitability may vary across different heat sources. For heat-source products, it should be done for suitably chosen classes of initial fuel sources.

To prepare a set of laboratory test data for use in these calculations on a fuel-source product, organize the data according to the following terminology:

- Distinguish different versions of the product ($i = 1, 2, \ldots$)
- Distinguish different heat sources ($j = 1, 2, \ldots$)
- Estimate the share of all the product now in use that is version i; call this q_i where $\sum_i q_i = 1$
- Let N_{ij} be the number of times that version i of the product has been tested in the laboratory against heat source j; let n_{ij} be the number of times that ignition occurred; and let $f_{ij} = n_{ij}/N_{ij}$.
- Let p_j be the probability of ignition of the mix of existing product by heat source j, calculated from fire experience data on the occupancy being studied.

Then assume that p_j (the product's overall probability of ignition) is proportional to $\sum_i q_i f_{ij}$ (the weighted frequency of laboratory ignitions). For a more sophisticated approach, one may wish to use Bayesian analysis, which requires estimating prior probability distributions to which one can apply the lab tests.

- Let f_{Ij} be corresponding values summarizing laboratory tests on new product I against heat source j. Note that every heat source must have its own body of test results.
- Let p_{Ij} refer to the quantities to be estimated, which are the probabilities of ignition of product I by each heat source j. Estimate as follows:

$$p_{Ij} = p_j \left[f_{Ij} / \left(\sum_i q_i f_{ij} \right) \right]$$

In general, the p values will be much smaller than the f values, which is assumed to reflect the fact that the p values incorporate all the probabilities involved in bringing the heat source and the product in contact with one another. This estimation procedure is not so reliable if the f values are equal to, or very near, 0 or 1. Bayesian analysis is definitely required in such cases.

Calibration and sensitivity analysis: Any fire risk analysis will involve complex calculations, with many unavoidable assumptions. While you should not use a more complex method than necessary, you also should not use a less complex method than is valid. A fire risk analysis model without a long list of stated assumptions is bound to be a model with many hidden assumptions, which are almost certain to be less well-founded, if examined, than a list of "shaky" but explicit assumptions.

For these reasons, running the calculations in a product fire risk analysis is easier to do well than the blend of science and art required to set up the analysis correctly (e.g., appropriate models, reasonable assumptions, best data) and to interpret the results, which includes calibration and sensitivity analysis.

The principal rule to remember is that the model is calibrated by assessing how well it reproduces recent fire experience from data on recent product use patterns and other practices. If the model captures the principal aspects of fire risk, then it will predict rates of loss that are very close to those actually experienced in the properties being analyzed. Results should be close not only from an overall perspective, but also for major groups of scenarios. Use the specifics of the scenarios that need better calibration as guides to which assumptions need to be modified. For example, if predictions are poor for fires with long smoldering periods, but good for all other fires, then one might want to adjust the assumptions on length of smoldering period or on fire profiles (e.g., rate of heat release curve) during smoldering periods.

Some options involve changes not just to the parameter values, but to the model structure. Examples include changes that would further multiply scenarios by allowing for multiple values (and associated probabilities) of walking speeds, evacuation decision rules, rules for waking people without detectors, etc.

Another approach to calibration is to use the model not to directly predict losses, but to predict percentage changes in losses due to product choices, by major scenario group. The advantage of this approach is that it allows the analyst to use the fire experience statistics to do a great deal of automatic recalibration. The disadvantage is that it does not directly address, or correct, the flawed assumptions and estimates that are preventing the model from producing accurate results without such recalibration.

Most analysts will need to both: (1) adjust the structure of the analysis, and (2) use fire statistics to recalibrate.

Identify data sources

Product and Property Survey Data: There are many good sources of national data on the characteristics of occupancies or products in general. For occupancies, this kind of data may be obtained from ongoing federal government data collection activities (typified by publications of the U.S. Census Bureau; the General Services Administration, e.g., fuel load per room, by type of room; and the U.S. Departments of Defense and Energy), from major one-time studies, or from industry association surveys (e.g., Building Owners and Managers Association, American Hotel and Motel Association, and American Restaurant Association).

For products, there may exist market surveys on patterns of composition or use. The U.S. Census Bureau and, for certain types of products, the U.S. Departments of Commerce, Energy, Housing and Urban Development, and Health and Human Services are all likely sources of information. Much of the data gathered by the U.S. government can be found in summary form in the *Statistical Abstract of the United States*, with reference to primary sources. Another source of information is the U.S. Consumer Product Safety Commission, which carries out some field surveys of product performance and usage.

For products, however, often the only source of such information is trade associations of manufacturers and sellers of the product. It is important to recall, however, that some products have a lifetime longer than that of a trade association, which means the association's knowledge of current industry practice, however accurate, may not describe use patterns as a whole. Upholstered furniture, for example, is typically cushioned today with synthetics, with previously used natural materials, such as cotton batting and horsehair, now a rarity. Nevertheless, the lifetime of a piece of furniture can be 30 years or more, during which time it may be re-covered several times and pass through several owners, often of continually diminishing economic station.

Therefore, it is reasonable to expect that a substantial fraction of the current furniture inventory retains the burning characteristics of products of a bygone era.

In general, people knowledgeable of today's product will tend to have statistics on, and think in terms of, what is currently being *sold*, not what is currently being *used*. Translating sales data into usage statistics is far from straightforward. Moreover, one might suspect that such products would be found disproportionately where fires are likely to be more common, i.e., at the lower end of the economic spectrum.

For the method to operate, all of these qualitative observations must take a quantitative form, and it is the analyst who must decide how this is to be accomplished.

Always remember that, although data on national practices is more representative of the nation than fire incident data, it is less desirable for that very reason for fire risk analysis, because it does not provide probabilities implicitly weighted by the likelihood of having a fire. For example, if you walked into a randomly selected U.S. home in 1989, the chances were better than 85 percent that you would find a smoke detector present; but if you focused on homes that had fires, the chances would drop below 50 percent. It is the latter probability that is more relevant in determining how risk will develop in home fires. Therefore, when data on national practices is used, it is necessary to review the data for any adjustments that may be needed to better estimate probabilities relevant to fires.

Code requirements: Many relevant building characteristics are covered by provisions of building and fire codes. Some product fire characteristics also are covered by regulations. In the absence of direct data from the field, one may assume that all buildings and products have the characteristics implied by compliance with these codes and regulations. This bypasses the need for probability estimation and usually provides enough detail to permit calculation of the input needs of the fire hazard analysis method.

When this approach is followed, the analyst needs to check a number of points that may undercut the central assumption of the approach, namely, that all buildings and products are as the codes and regulations would have them. In practice, many code provisions are of fairly recent vintage so that they were not in place when many or even most of the buildings and products now in use were put into place. Some jurisdictions do not follow national consensus codes, and many more lack the enforcement apparatus to ensure a high rate of compliance. Buildings and products can be altered or may deteriorate after being built, manufactured, or sold. And some building and product features may be better than code requirements because of marketplace demands.

Putting all this together, it is important that the user verify, through the expertise of people with broad familiarity with the state of old and new buildings and products around the country, that the particular characteristics of interest are ones where the code provisions are good indicators of actual practice in nearly all buildings and products in recent use.

Expert judgment: When all else fails and numbers are needed, there is no alternative but to make the best estimates possible. In some cases, it will be a judgment made by the user alone, but especially in areas where the user is least experienced, one or more true experts should be sought. For example, if the user is a maker or seller of the product, it would be wise to make use of fire scientists for assistance in assigning values to the product's fire properties.

One of the persistent potential pitfalls in the method is the danger that the typical product in use is not the typical product involved in fires. There is no foolproof way to avoid this situation, but one way to address it is to include fire service personnel in drawing up the profile of product characteristics. Fire fighters and fire marshals see the products involved in fires, whether or not they make up a substantial fraction of the statistical profile of the product.

Another pitfall is that, no matter what the expert's area of expertise, he or she may tend to underestimate the (typically) enormous variation in every characteristic of interest. Manufacturers may focus on a few best-selling versions of the product; surveys may focus on a few most widely used versions; and fire officials may focus on a few of the worst, most obsolete versions, seen in the worst (but not necessarily the most) fires. This illustrates the value of: (1) a panel of experts, where biases can be balanced, and (2) a facilitator sensitive to the variation in practice, who can steer the group away from premature or overly narrow consensus.

It also suggests that, even when data is available, expert judgment is needed to interpret the data and apply it correctly. Knowledge of common *vs.* uncommon practices, relevant codes and standards, and the length of time they have been in place, are among the kinds of information essential to spot areas of likely bias or critical uncertainty.

For all these reasons, the user can expect to make extensive use of expert judgment and should make sure that the expertise available to the project is both broad and deep.

REFERENCES CITED

1. Richard W. Bukowski and Takeyoshi Tanaka, "Toward the Goal of a Performance Fire Code," *Fire and Materials*, Vol. 15, pp. 175–180 (1991).
2. Raymond Friedman, "An International Survey of Computer Models for Fire and Smoke," *Journal of Fire Protection Engineering*, Vol. 4, No. 3, pp. 81–92 (1992).
3. NFPA 901, *Uniform Coding for Fire Protection*, National Fire Protection Association, Quincy, MA (1990).

BUILDING FIRE SAFETY RISK ANALYSIS

David Yung and Vaughan R. Beck

INTRODUCTION

The major design objectives for the effects of fire in buildings are to achieve satisfactory levels of life safety for

a) occupants of the building of fire origin,
b) occupants of adjoining buildings, and
c) fire brigade personnel.

Some have argued that the level of property protection in buildings should not be subjected to community regulation, but should be a matter for building owners and their insurers. However, the overall costs associated with providing fire safety and protection, and the expected losses from the effects of fire, should be minimized.

The level of fire safety in a building is a reflection of a complex interaction among many phenomena, including: fire initiation, fire growth and spread, the response of building components to fire, the response of occupants to the presence of fire, and the response of the fire brigade to the fire. To achieve required levels of safety from the effects of fire in buildings, it is essential that designers have at their disposal the means to predict the level of life safety for any particular building design and use. The development of the capability to predict the level of life safety requires a model to quantify the performance of the building fire safety system.

BACKGROUND

Conceptual models, such as the NFPA fire safety concepts tree,[1] have been used for qualitative fire risk assessment in buildings.

Another model called the building fire simulation model (BFSM),[2] is a comprehensive program that includes fire growth and spread, development and movement of products of combustion, and building evacuation. The simulation is driven by the fire development module, which is based on regression analysis of numerous fire tests. An earlier version used probability distributions defined by selected parame-

ters. The evacuation component of the BFSM identifies available escape routes with a deterministic, combinatorial model, and then determines which will be blocked by products of combustion.

Others who have considered approaches to the problem include Lie,[3] Kobayashi,[4] Ling and Williamson,[5] and Fitzgerald.[6]

This chapter will use the risk-cost assessment model developed in Australia and Canada[7-13] as an example of a comprehensive system model of building fire safety with regulatory application.

Cost-Effective Alternative Designs

In the design of building fire safety systems, it is appropriate that explicit consideration be given to the level of life safety afforded to occupants of buildings, and to the costs associated with such provision. Such an approach enables designers to undertake a performance-based approach to design, and to select the most appropriate cost-effective solution for the building fire safety system. For a particular building design, the effect of fire is predicted using two performance parameters[1,14]

1. Expected risk to life, and
2. fire-cost expectation.

No attempt was made to assign monetary value to either the loss of life or the value of lives saved, which is addressed in Chapter 8, Section 5.

To identify alternative designs that are considered equivalent to, and more cost-effective than, designs conforming with current regulatory provisions, the decision criterion is[8,14]

For an alternative design to be considered acceptable, the expected risk-to-life value shall be equal to or less than the risk-to-life value of a building conforming with the regulations, and the fire-cost expectation for the alternative design shall be less than or equal to the value for the conforming building.[8]

With such a comparative approach, it is not required to directly compare estimated risk-to-life values, derived from a risk assessment model, with an acceptable level of risk

Dr. D. Yung is a Senior Research Officer and Group Leader, Fire Risk Assessment, National Fire Laboratory, National Research Council Canada, Ottawa, Ontario, Canada. Dr. V. R. Beck is Professor and Director, Centre for Environmental Safety and Risk Engineering, Victoria University of Technology, Melbourne, Victoria, Australia.

derived from independent sources. This comparative approach also provides some flexibility in the required level of accuracy for the two performance parameters.

The calculated expected risk-to-life values for designs conforming with current regulatory requirements provide an estimate of current levels of risk-to-life safety. These risk levels are assumed to be acceptable to the community.

RISK-COST ASSESSMENT MODEL

A brief description of the current risk-cost assessment model and its submodels are given in this section.[15] More detailed descriptions are given for the design fire submodel, fire growth submodel, and the smoke movement submodel. As for the other submodels, more details can be found in other publications.[8,9,11–13]

The risk-cost assessment model employs an event-based modeling approach in which events are characterized by discrete times and probability of occurrence. The event-based approach is used to define the outcomes of fire growth and spread scenarios in terms of the times of occurrence of untenable conditions. The consequence of these outcomes is in terms of the number of people exposed to untenable conditions.

The risk-cost assessment model for office and apartment buildings assesses the fire safety performance of a fire protection design in terms of two decision-making parameters: (1) the expected risk-to-life (ERL) and (2) the fire-cost expectation (FCE). The ERL is the expected number of deaths over the lifetime of the building divided by the total population of the building and the design life of the building. The FCE is the total fire cost, which includes the capital cost for the passive and active fire protection systems, the maintenance cost for the active fire protection systems, and the expected losses resulting from fires in the building. The ERL is a quantitative measure of the risk-to-life from all probable fires in the building, whereas the FCE quantifies the fire cost associated with the particular fire safety system design.

To calculate the ERL and FCE values, the risk-cost assessment model considers the dynamic interaction between fire growth, fire spread, smoke movement, human behavior, and the response of fire brigades. These calculations are performed by a number of submodels interacting with each other, as shown in the flowchart in Figure 5-11.1. In Figure 5-11.1, the term "submodel" has been abbreviated as "model."

Design Fire Submodel

The risk-cost assessment model uses six design fires in the room of fire origin, and the subsequent fire and smoke spread, to evaluate life risks and protection costs in office and apartment buildings. The six design fires, representing the wide spectrum of possible fire types, are:

1. Smoldering fire with room entrance door open,
2. Smoldering fire with room entrance door closed,
3. Flaming non-flashover fire with room entrance door open,
4. Flaming non-flashover fire with room entrance door closed,
5. Flashover fire with room entrance door open, and
6. Flashover fire with room entrance door closed.

The probability of occurrence of each design fire, given that a fire has occurred, is based on statistical data. For example, in Canada, statistics show that 18 percent of all apartment fires reach flashover and become fully developed fires, 63 percent are flaming fires that do not reach flashover,

and the remaining 19 percent are smoldering fires that do not reach the flaming stage.[16] If sprinklers are installed, the model assumes that some of the flashover and non-flashover fires, depending on the reliability and effectiveness of the sprinkler system, are rendered non-lethal.

The risk-cost assessment model evaluates the effects of various fire scenarios that may occur in the building during its life. For example, in an apartment building, one fire scenario is the fire and smoke spread resulting from one design fire in any one of the apartment units in the building and during a time when the occupants are either awake or asleep. The number of fire scenarios, therefore, is the product of the number of design fires, the number of apartment units, and whether the occupants are awake or asleep.

Fire Growth Submodel

The fire growth submodel[17] predicts the development of the six design fires in the room of fire origin. The submodel calculates the burning rate, room temperature, and the production and concentration of toxic gases as a function of time. With these calculations, the model determines the time of occurrence of five important events: (1) time of fire cue, (2) time of smoke detector activation, (3) time of sprinkler activation, (4) time of flashover, and (5) time of fire burnout. The first three detection times are used by the evacuation duration submodel to estimate the time available for evacuation; the flashover time is used by the fire brigade action submodel, in combination with the arrival time of the fire brigade, to evaluate the effectiveness of fire fighting; and the burnout time is used by the smoke hazard submodel as part of the calculation for the maximum smoke hazard. The submodel also predicts the mass flow rate, the temperature, and the concentrations of CO and CO_2 in the hot gases leaving the fire room. This latter information is used by the smoke movement submodel to calculate the spread of smoke to different parts of the building as a function of time.

Smoke Movement Submodel

The smoke movement submodel[18] calculates the spread of smoke and toxic gases to different parts of the building as a function of time. The submodel also calculates the critical time when the stairs become untenable, which is considered to be the time when the occupants are trapped in the building. This critical time is used later by the evacuation duration submodel to calculate the time available for evacuation.

Fire Detection Submodel

The fire detection submodel calculates the probabilities of detection at the first three detection times mentioned under the fire growth submodel, based on the probabilities of detection by smoke detectors, sprinklers, and occupants. This information is used by the occupant warning and response submodel to calculate the probabilities of response of the occupants.

Occupant Warning and Response Submodel

The occupant warning and response submodel calculates the probabilities of warning and response at the first three detection times mentioned under the fire growth submodel. This information is used by the fire brigade action submodel to calculate the probability of response of the fire brigade, and by the egress submodel to model the evacuation of the occupants.

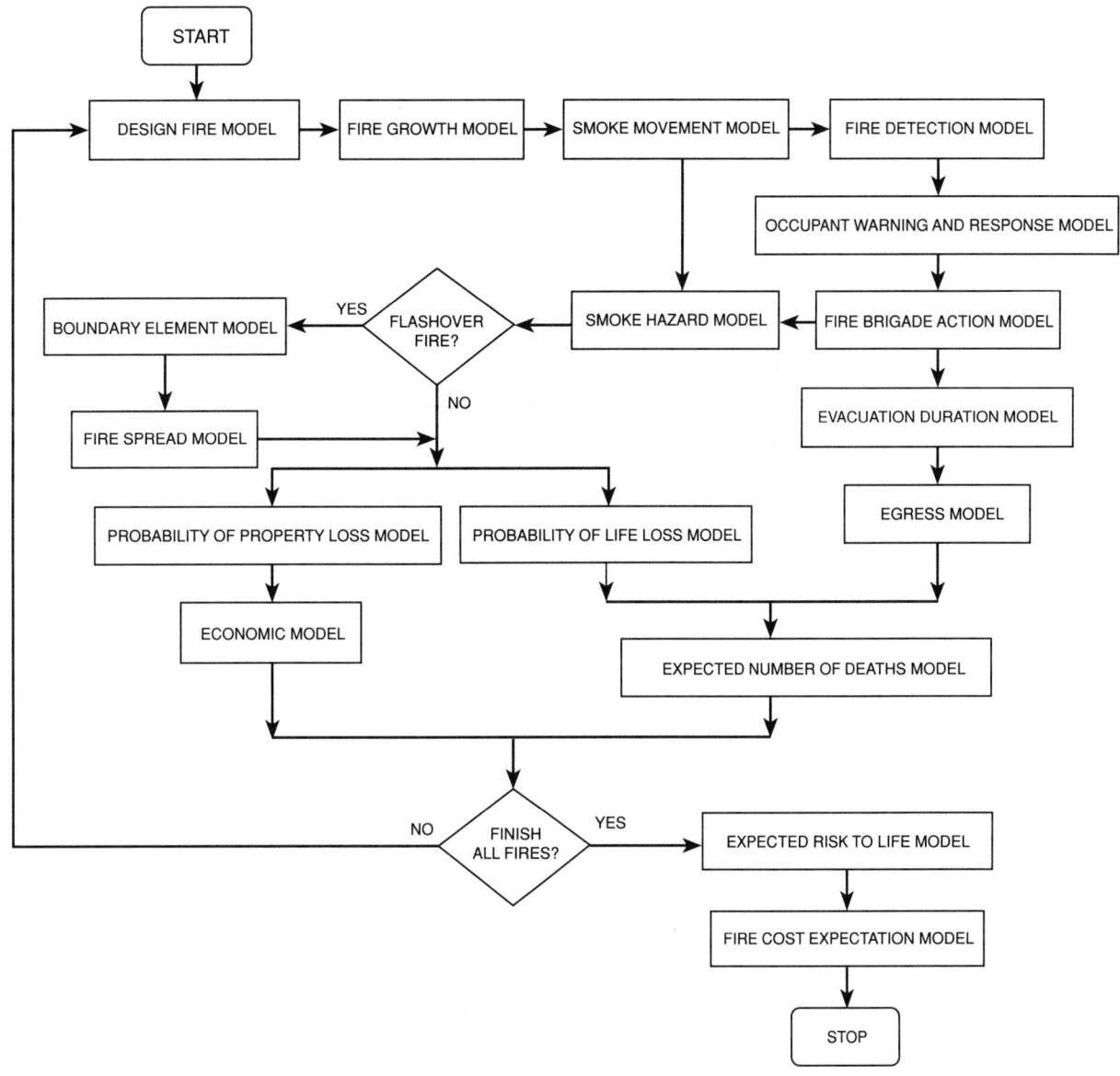

Fig. 5-11.1.   Risk-cost assessment model.

Fire Brigade Action Submodel

This submodel calculates the probability and time of arrival of the fire brigade. This submodel also evaluates the effectiveness of fire fighting, based on the flashover time from the fire growth submodel and the arrival time of the fire brigade. The information on arrival and effectiveness of the fire brigade is used by the smoke hazard submodel to calculate the maximum smoke hazard to the occupants, and by the fire spread submodel to calculate the probabilities of fire spread.

Smoke Hazard Submodel

This submodel calculates the maximum smoke hazard to the occupants based on the burnout time from the fire growth submodel and the arrival time and effectiveness of the fire brigade from the fire brigade action submodel. This information is used by the life loss submodel to calculate the probabilities of life loss.

Evacuation Duration Submodel

This submodel uses the three fire detection times from the fire growth submodel and the critical time in the stairs from the smoke movement submodel to calculate three durations available for evacuation. This information is used by the egress submodel to model the evacuation of the occupants.

Egress Submodel

Based on the evacuation time available and the probability of response of the occupants, this submodel calculates the number of occupants who have evacuated the building and the number trapped in the building. This information is used by the expected number of deaths submodel to calculate the expected number of deaths.

Boundary Element Submodel

This submodel calculates the probabilities of failure of the boundary elements (walls, floors, doors, etc.) when they are subjected to fully developed, realistic fires. The submodel comprises the following probabilistic models: fire severity, temperature distribution, thermo-mechanical material properties, failure performance for each limit state, and overall probability of failure.

Fire Spread Submodel

Based on the probabilities of failure of the boundary elements, this submodel calculates the probabilities of fire spread to each part of the building given a fully developed fire in any enclosure. A probabilistic network of the building is developed where nodes represent building volumes, links represent boundary elements between volumes, and probabilities of failure of the boundary elements are assigned to links. Allowance is made for the effectiveness of the fire brigade. The probability of fire spread information is used by both the property loss submodel and the life loss submodel to estimate fire losses and life loss.

Life Loss Submodel

Based on the probabilities of smoke hazard from the smoke hazard submodel and fire spread from the fire spread submodel, this submodel calculates the probabilities of life loss.

Expected Number of Deaths Submodel

Based on the probabilities of life loss from the life loss submodel and the number of occupants trapped in the building from the egress submodel, this submodel calculates the expected number of deaths in the building.

Property Loss Submodel

Based on the probabilities of fire spread from the fire spread submodel, this submodel calculates the expected property loss.

Economic Submodel

Based on the expected property loss and the capital and maintenance costs of the fire protection systems, this submodel calculates the expected fire costs.

Expected Risk-to-Life Submodel

This submodel calculates the overall expected risk-to-life (ERL) by summing the expected number of deaths in the building for each fire scenario and the probability of each fire scenario.

Fire-Cost Expectation Submodel

This submodel calculates the fire-cost expectation (FCE) using the capital and maintenance costs of the fire protection systems, the expected fire loss for each fire scenario, and the probability of each fire scenario.

Assumptions and Limitations

In the risk-cost assessment model, due to the complexity and the lack of sufficient understanding of fire phenom- ena and human behavior, certain conservative assumptions and approximations were made in the mathematical modeling. In addition, not all aspects of the risk-cost assessment model have been fully verified by full-scale fire experiments or actual fire experience. Only some of the submodels have been verified by experiments or statistical data.

As a result, the predictions made by the model can only be considered as approximate. The model, therefore, should not be used for absolute assessments of life risks and protection costs. For comparative assessments of life risks and protection costs, and for the selection of a cost-effective fire safety system design solution, the model is considered to be reliable.

As in many computer models, the model uses certain input parameters to describe the characteristics of various fire safety designs. These include the fire resistance rating of boundary elements, the reliability of smoke alarms and sprinklers, the probability of doors open or closed, and the response time of fire brigades. The sensitivity of these parameters on the predicted risks have been checked and found to be reasonable.[19]

APPLICATIONS

Three-Story Apartment Building

The life risks of various fire protection designs for a proposed three-story apartment building in Australia were evaluated as a case study using the risk-cost assessment model.[20] This case study was carried out to support a proposal to the Australian Uniform Building Regulations Coordinating Council (AUBRCC) for changes to the *Building Code of Australia* (*BCA*) to permit the construction of three-story timber-framed apartment buildings. In this study, the risk-cost assessment model was used to compare the fire safety performance of three-story timber-framed apartment buildings, with various fire protection designs, with that of the code-compliant concrete/masonry construction. The objective was to determine whether three-story timber-framed apartment buildings, with proper fire protection, could be as safe as concrete/masonry construction.

Figure 5-11.2 shows the floor plan of the prototype three-story apartment building being considered. Each floor

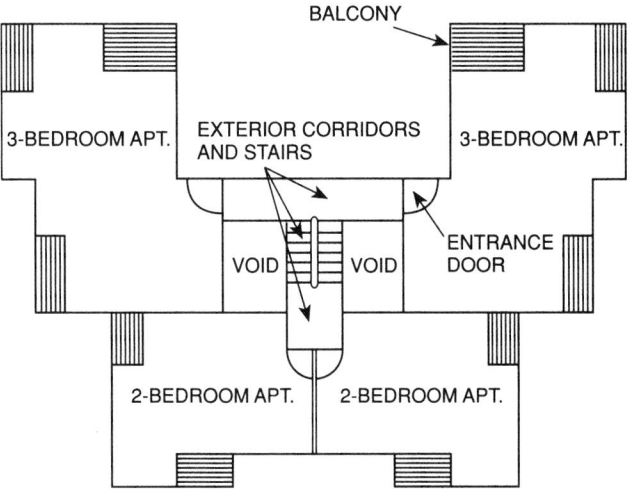

Fig. 5-11.2.   Prototype floor plan.

has four apartment units: two two-bedroom units (96.3 m² each) and two three-bedroom units (129.7 m² each). The four units on each floor are planned around an open, central staircase with direct access to it from any of the four apartment entrance doors. To ensure safe passage for evacuation, an open concrete staircase is used, and all apartment entrance doors facing it are fire-rated and with self-closing devices. In addition, all entrance doors are recessed from the staircase to prevent direct flame impingement on the staircase in the event that flames emerge from one entrance door.

Based on the prototype design, five different fire protection options were considered for the timber-framed construction and compared to the reference BCA-compliant concrete/masonry option. These options are summarized in Table 5-11.1. Option 1 is the reference BCA-compliant concrete/masonry option. Option 2 is the unprotected timber-framed option. Option 3 is the same timber-framed construction as in Option 2, but with barriers having a 1-hour fire resistance rating and a central alarm system consisting of individual smoke detectors that are connected to a central alarm. Options 4 through 6 are the same as Option 3, but with exterior brick cladding (higher fire resistance to external fire spread) and three levels of acoustic insulation (higher interior fire resistance rating).

The risk-cost assessment model was used to determine the values of the two performance parameters, i.e., the expected risk-to-life (ERL) and the fire-cost expectation (FCE), for the six design options shown in Table 5-11.1. Only the ERL values are shown herein to compare the relative fire safety performance of timber-framed and concrete/masonry construction. The FCE values are not discussed; however, they are consistent with the cost estimates that three-story timber-framed apartment buildings are more cost effective than similar concrete/masonry buildings.

The expected risk-to-life (ERL) values obtained by the risk-cost assessment model for the six design options are shown in Figure 5-11.3. In this figure, the ERL values have been normalized by that of the reference concrete/masonry option (Option 1), to provide a relative comparison of the five timber options with the concrete/masonry option. Option 2, which is the basic timber option without fire resistance cladding or an alarm system, is shown, as expected, to have the highest relative ERL value (2.32). Option 3, timber-framed construction with a 1-hour fire resistance rating and a central alarm system, is shown to reduce the relative ERL value to a level (0.93) slightly lower than that of the reference concrete/masonry option. Options 4 through 6 are timber-framed construction with exterior brick cladding and three levels of acoustic insulation. The results show that these additional features have little or no influence on the ERL.

The results in Figure 5-11.3 show that timber-framed options with proper fire resistant designs and a central alarm

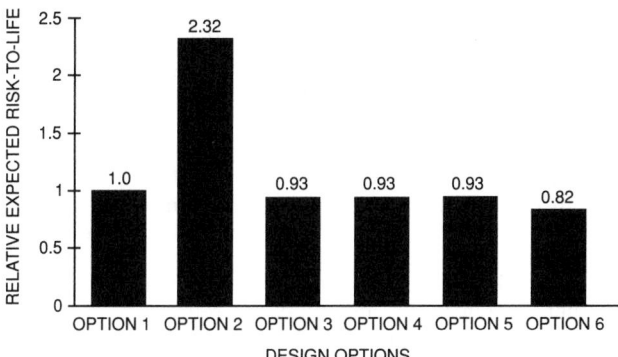

Fig. 5-11.3. *Relative expected risk-to-life for six design options shown in Table 5-11.1.*

system can be as safe as the reference concrete/masonry design without a central alarm system. With a minimum 1-hour fire resistance rating and a central alarm system, the occupants in such timber buildings would have left the building via the exterior central staircase long before the fire could spread and thus pose any significant hazard. The concrete/masonry option may have the advantage of a higher fire resistant construction, but lacks a central alarm to warn the occupants. These findings are consistent with a Japanese full-scale fire experiment.[21]

The results in Figure 5-11.3 also show that timber options, with proper fire resistant design and a central alarm system (Options 3 through 6), can reduce the risk-to-life by about 60 percent when compared with the timber option with no fire protection (Option 2). This is consistent with statistical findings that show that the installation of smoke detectors alone reduces the number of deaths from fires in buildings by about 50 percent.[22]

Regulation

During the Warren Centre Project on Fire Safety and Engineering,[23] it was estimated that the Australian economic losses and costs associated with building fires, fire protection, and insurance amounted to about $2 billion Australian dollars per year, and that savings of at least $250 million per year could be made by the development of fire safety engineering technology based on performance-based design codes.

The principal recommendations from the Warren Centre Project include:

1. The current levels of fire safety in Australia should be maintained;

TABLE 5-11.1 *Fire Protection Options*

Option	Frame	Internal Wall and Floor Acoustic Insulation	Internal Wall and Floor Fire Resistance Rating (FFR)	Exterior Cladding	Central Fire Alarm
1	Concrete/masonry	None	180 min	Brick	No
2	timber	None	20 min	Timber	No
3	timber + FRR	None	60 min	Timber	Yes
4	timber + FRR	None	60 min	Brick	Yes
5	timber + FRR	BCA-compliant	75 min	Brick	Yes
6	timber + FRR	Higher	120 min	Brick	Yes

2. Design for fire safety be treated as an engineering responsibility rather than a matter for detailed regulatory control;
3. Risk assessment models be used as a basis for identifying cost-effective combinations of fire safety subsystems for building design;
4. Designers adopt appropriate fire safety engineering design techniques for the design of fire safety systems in buildings;
5. Fire engineering design courses and training strategies be developed and implemented (up to, and including, postgraduate level); and
6. National strategy be developed for research, development, applications, and education relevant to fire safety engineering design.

The Building Regulations Review Task Force (BRRTF) was commissioned in March 1989 by the prime minister and the premiers to review Australian building regulations and standards; they recommended that:

1. *Building Code of Australia* (*BCA*) should be revised to make its fire safety objectives explicit and to use provisions based on systematic fire safety analysis, and
2. New fire safety systems code be formulated to permit those objectives to be met by sound engineering design.

The BRRTF commissioned a team, drawn from the principal participants of the Warren Centre Project, to prepare a draft performance-based *National Building Fire Safety Systems Code* (*NBFSSC*).[24] The Australian Uniform Building Regulations Coordinating Council (the organization responsible for the *BCA*), Standards Australia, and the BRRTF endorsed the draft *NBFSSC*, supporting its development as an Australian standard to be referenced by the *BCA* to provide an alternative compliance route to the design of fire safety subsystems in buildings. The feasibility of this approach has been demonstrated by the Warren Centre Project, which applied it to typical buildings, albeit without the benefit of a comprehensive scientific base.

The draft *NBFSSC*, the first performance-based engineering code for the design of fire safety systems in buildings, uses a risk assessment framework. The draft code is based on modeling fire phenomena and response to fire phenomena in buildings. It also uses a risk-assessment methodology to identify cost-effective designs that achieve acceptable levels of life safety. The table of contents of the draft code is given in Appendix A. Attached to the draft code is an outline commentary. The draft NBFSSC has been written in a format that will enable its eventual publication as an Australian standard.

The routine application of the draft code requires the availability of a user-friendly computer program of a validated fire safety system model, together with input parameters specified in the code, which are required for the model. Accordingly, the present version of the draft code provides a substantial foundation for the eventual publication of the code as an Australian standard. Furthermore, the preparation of the draft code has provided a clear focus for research and development activities.

CONCLUSIONS

The risk assessment models contain a number of deficiencies, and current research has lead to a number of improvements. However, further improvements and developments are required. The research collaboration between the National Research Council Canada and the Victoria University of Technology has identified a comprehensive research program to further improve and develop each of the submodels comprising the system model.

It must be recognized that absolute precision in the predicted levels of life safety is not required, since a comparative decision criterion has been adopted. In essence, it is necessary to arrive at an appropriate balance between precision and correct identification of the appropriate fire safety system design solution. In order to achieve this balance, and to reduce computational demands, it has been decided to adopt "simplified" models to estimate performance, which will be or have been validated against more sophisticated numerical and experimental modeling techniques. Examples include the use of full-scale experimentation and computational fluid dynamics modeling of realistic fires to validate zone models, which will be used to predict the spread of smoke and toxic products throughout a building. In addition, Monte Carlo techniques will be used to simulate the performance of the system by employing time-dependent probability distributions for the events associated with the system. (See Section 5, Chapter 1.) Results from this Monte Carlo simulation will be used to validate results obtained from a system model based on the use of events characterized by discrete values for time and probability. This approach forms the basis of the system model that will be used routinely by designers and regulatory authorities.

The development of a fire safety engineering methodology to identify cost-effective fire safety systems in buildings based on a systematic approach, which combines both physical and probabilistic modeling, is a challenging task. The essential nature of this task is to develop reliable models that: (1) predict the level of life safety and (2) are applicable to a wide range of occupancies. In addition, these models should be able to predict the expected fire costs to allow selection of cost-effective design solutions that provide a satisfactory level of safety. The risk-cost assessment model described in this chapter is capable of identifying cost-effective fire safety systems in buildings.

APPENDIX A

DRAFT
NATIONAL BUILDING FIRE SAFETY SYSTEMS CODE

Table of Contents

Appendix

REFERENCES CITED

1. NFPA 550, *Guide to the Fire Safety Concepts Tree*, National Fire Protection Association, Quincy, MA (1995).
2. R.F. Fahy, "Building Fire Simulation Model. An Overview," *Fire Safety Journal*, Vol. 9, pp. 189–203 (1985).
3. T.T. Lie, "Safety Factors for Fire Loads," *Canadian Journal of Civil Engineering*, Vol. 6, pp. 617–628 (1979).
4. M. Kobayashi, "A Methodology for Evaluating Fire/Life Safety Planning of Tall Buildings," in "Evaluation of Fire Safety in Buildings," Occasional Report of Japanese Association of Fire Science and Engineering, No. 3, Nihon Kasaigakka, pp. 204–214 (1979).
5. W-C.T. Ling and R.B. Williamson, "Using Fire Tests for Quantitative Risk Analysis," in G.T. Castino and T.Z. Harmathy (eds.), "Fire Risk Assessment," *ASTM Special Publication STP 762*, American Society for Testing and Materials, Philadelphia, PA (1982).
6. R.W. Fitzgerald, "An Engineering Method for Building Fire Safety Analysis," *Fire Safety Journal*, Vol. 9, No. 2, pp. 233–243 (1985).
7. V.R. Beck, "The Prediction of Probability of Failure of Structural Steel Elements under Fire Conditions," *Civil Engineering Transactions*, The Institution of Engineers, Australia, Vol. CE27, No. 1, pp. 111–118 (1985).
8. V.R. Beck, "A Cost-Effective Decision-Making Model for Building Fire Safety and Protection," *Fire Safety Journal*, Vol. 12, pp. 121–138 (1987).
9. V.R. Beck and S.L. Poon, "Results from a Cost-Effective Decision-Making Model for Building Fire Safety and Protection," *Fire Safety Journal*, Vol. 13, pp. 197–210 (1988).
10. V.R. Beck, "Cost-Effective Fire Safety and Protection Design Requirements for Canadian Apartment Buildings," Contract Report for the National Research Council Canada, Footscray Institute of Technology, Melbourne, Australia, 161 pp., May 1988.
11. V.R. Beck and D. Yung, "A Cost-Effective Risk-Assessment Model for Evaluating Fire Safety and Protection in Canadian Apartment Buildings," *International Fire Protection Engineering Institute, 5th Conference*, Ottawa, Ontario, Canada, Vol. 1 Papers, May 21–31, 1989.
12. D. Yung and V.R. Beck, "A Risk-Cost Assessment Model for Evaluating Fire Risks and Protection in Apartment Buildings," *Proceedings of the International Symposium on Fire Engineer-ing for Building Structures and Safety*, The Institution of Engineers, Australia, Melbourne, pp. 15–19, Nov. 14–15, 1989.
13. V.R. Beck and D. Yung, "A Cost-Effective Risk-Assessment Model for Evaluating Fire Safety and Protection in Canadian Apartment Buildings," *Journal of Fire Protection Engineering*, Vol. 2, No. 3, pp. 65–74 (1990).
14. V.R. Beck, "Cost-Effective Fire Safety and Protection Design Requirements for Buildings," Ph.D. Thesis, University of New South Wales, Australia, July 1986.
15. D. Yung, G.V. Hadjisophocleous, and H. Takeda, "Comparative Risk Assessments of 3-Storey Wood-Frame and Masonry Construction Apartment Buildings," *Proceedings of Interflam '93*, Oxford, England, pp. 499–508, Mar. 30–Apr. 1, 1993.
16. J. Gaskin and D. Yung, "Canadian and U.S.A. Fire Statistics for Use in the Risk-Cost Assessment Model," IRC Internal Report No. 637, National Research Council Canada, Ottawa, Canada, Jan. 1993.
17. H. Takeda and D. Yung, "Simplified Fire Growth Models for Risk-Cost Assessment in Apartment Buildings," *J. of Fire Protection Engineering*, Vol. 4, No. 2, pp. 53–66 (1992).
18. G.V. Hadjisophocleous and D. Yung, "A Model for Calculating the Probabilities of Smoke Hazard from Fires in Multi-Story Buildings," *J. of Fire Protection Engineering*, Vol. 4, No. 2, pp. 67–80 (1992).
19. G.V. Hadjisophocleous and D. Yung, "Parametric Study of the NRCC Fire Risk-Cost Assessment Model for Apartment and Office Buildings," IRC Internal Report, *Proceedings of the 4th International Symposium on Fire Safety Science*, Ottawa, Canada, pp. 829–840, June 13–17, 1994.
20. D. Yung and G.V. Hadjisophocleous, "The Use of the NRCC Risk-Cost Assessment Model to Apply for Code Changes for 3-Storey Apartment Buildings in Australia," *Proceedings of the Symposium on Computer Applications in Fire Protection Engineering*, Worcester, MA, pp. 57–62, June 28–29, 1993.
21. Y. Hasemi, "Wooden 3-Storey Apartment Building Shake and Burn Test Report," Building Research Institute, Tsukuba, Japan, Jan. 1992 (in Japanese).
22. "Review of Automatic Sprinkler Protection for Buildings in Canada," Professional Loss Control, Ltd., Fredericton, New Brunswick, Canada, Feb. 1992.
23. Warren Centre "Project Report" and "Technical Papers, Books 1 and 2," Fire Safety and Engineering Project, The Warren Centre for Advanced Engineering, The University of Sydney, Australia, Dec. 1989.
24. V.R. Beck, *et al*, Draft *National Building Fire Safety Systems Code*, in Building Regulation Review Task Force, "Microeconomic Reform: Fire Regulation," Department of Industry Technology and Commerce, Canberra, Australia, 165 pp., May 1991.

ADDITIONAL READING

I.R. Thomas, I.D. Bennetts, P. Dayawansa, D.J. Proe, and R.R. Lewis, "Fire Tests of the 140 William Street Office Building," BHP Research—Melbourne Laboratories, Report No. BHPR/ENG/R/92/043/SG2C, Australia, Feb. 1992.

AN INTRODUCTION TO QUANTITATIVE RISK ASSESSMENT IN CHEMICAL PROCESS INDUSTRIES

Thomas F. Barry

INTRODUCTION

Fire and explosion risk, which can involve property damage, business interruptions, life safety, environmental issues, corporate image, and future profitability, presents a major threat to corporate goals and survival.

Quantitative fire and explosion risk assessment offers the capability of being able to identify weak links in loss prevention and protection systems before an accident actually occurs. It also affords the capability of optimizing loss control investments with the greatest allocation going to the area giving rise to the highest risk.

Fire and explosion risk assessment is a function of two parameters: (1) the frequency of occurrence of an undesired loss event and (2) the consequences resulting from its occurrence.

Risk = Frequency of the Event × Expected Consequences

This means that, in order to reduce the risks from a defined fire or explosion event scenario, one can either seek to reduce the frequency of occurrence of the undesired event, the consequences of its occurrence, or both.

The application of quantitative risk assessment (QRA) techniques enforces a disciplined, analytical thinking approach to fire risk problems and provides quantification of key fire safety issues to aid management understanding and decisions. Primary reasons companies utilize risk assessment as decision support include:

1. As part of the engineering design process to reduce risk to acceptable in-house safety standards. Design risks are identified and prioritized to optimize safety investment options.
2. To evaluate code equivalency and/or the relative risk difference between loss prevention or protection design options. Relative risk reduction assessment is used as an effective cost/benefit analysis tool for establishing the optimum balance between prevention, protection, and emergency response.

Thomas F. Barry, P.E., is a fire and explosion risk analyst and project manager at HSB Professional Loss Control, Kingston, TN. He has performed and managed numerous quantitative risk assessments (QRA's) of industrial process hazards at government, chemical, and oil and gas facilities. Tom has a masters degree in Fire Protecton Engineering from Worcester Polytechnic Institute.

3. To demonstrate to the community or an insurance company that, while the hazard may be large, the risk is small. These studies quantify the probability of the occurrence of extremely severe accidents.
4. To predict gas dispersion, fire radiant heat, and explosion overpressure effect zones for use in emergency response planning and training.
5. The company is required to develop quantitative risk assessment information by federal, state, or local government agencies.

Environmental Protection Agency (EPA) regulations related to the provisions of the 1990 Clean Air Act Amendments (CAAA) may require facilities with threshold amounts of hazardous materials to develop risk assessment and risk management programs. The prevention (reduction of loss event probabilities) component of the EPA CAAA relates to the Occupational Safety and Health Administration's (OSHA) Progress Safety Management (PSM) regulation 1910.119, which focuses on process hazard analysis (PHA) and the documentation, implementation, and monitoring of management loss control programs.

In addition, quantitative risk assessment is increasingly being required by state legislation, such as California's Risk Management and Prevention Program, New Jersey's Toxic Catastrophe Prevention Act, and Delaware's Extremely Hazardous Substances Risk Management Act. The specific objectives of these regulations vary, but the emphasis is on risk reduction and emergency planning.

The hazardous materials specified by EPA, OSHA, and state regulations include many flammable and explosive materials. As a result, fire protection engineers will become involved as part of the risk assessment teams responsible for hazard analysis, management programs, and regulatory compliance.

The areas in which fire risk assessment is finding application are rapidly expanding. The credibility of these risk assessment studies in the view of the management decision makers is highly dependent on the personnel, procedures, documentation, and quality assurance controls integrated into the process. The risk assessment process can be broken down into systematic steps, as presented in Figure 5-12.1. The focus of this chapter is to familiarize the reader with risk assessment methods by providing an overview of each of the steps and a discussion of general risk assessment techniques.

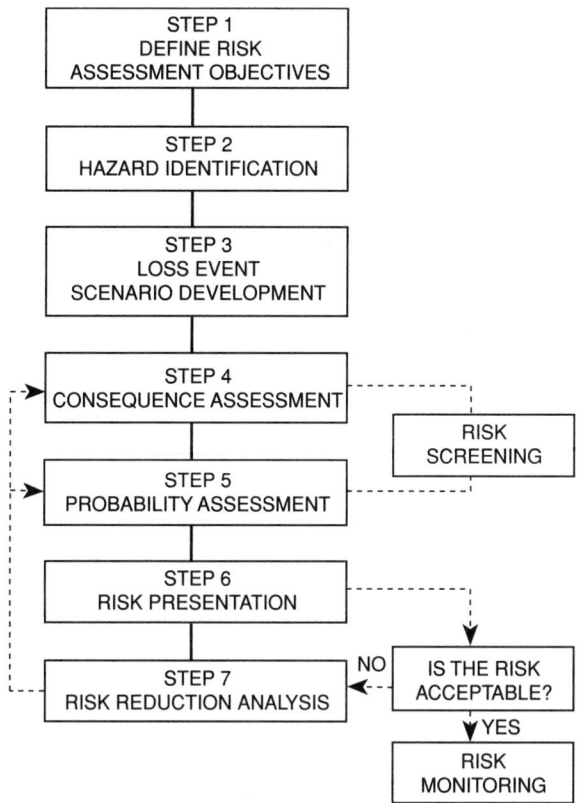

Fig. 5-12.1. *Fire and explosion risk assessment steps.*

Quantitative risk assessment is both an interesting and complex subject. The references listed at the end of this chapter will supplement and expand on the material summarized in this chapter.

The two objectives of this chapter are to:

1. Familiarize readers with the general process of fire and explosion risk assessment, and
2. Provide a basis for the readers' future development and involvement in the areas of risk assessment by providing procedural steps and associated references.

STEP 1: DEFINE RISK ASSESSMENT OBJECTIVES

Risk assessment is the process by which the results of a risk analysis (i.e., quantitative risk estimates) are used to support risk management decisions through comparison with acceptable risk levels and/or prioritization of risk reduction strategies. The key term here is *decision support*.

Prior to the start of risk assessment projects it is imperative to have a clear project scope and to explicitly state and agree upon project objectives.

1. What is the decision?
   a. Plant siting
   b. Protection options
   c. Regulatory compliance
2. What is the focus of the assessment?
   a. Casualties/injuries
   b. Property damage
   c. Loss of production/delivery

   d. Health risk to employees, public
   e. Environmental damage
   f. Legal liabilities
3. What is the level of detail?
   a. Relative application of risk assessment
   b. Absolute application of risk assessment
4. What are the acceptable risk limits?
   a. Hazard severity acceptance limits
   b. Event likelihood acceptance limits
   c. Management's risk tolerance profile
5. What are the resources required?
   a. Involvement of plant personnel
   b. Use of experienced risk assessment teams
   c. Quality control procedures and review
   d. Project scheduling and funding.

Ten major issues that should be addressed prior to conducting fire and explosion risk assessments are:

1. The type of risk profile presentation format that will be used to aid management decision making.
2. Establishing management's acceptable risk criteria for risk comparisons.
3. Methods for determining the absolute or relative risk levels.
4. Models or algorithms for determining the potential sizes of toxic or flammable vapor clouds, overpressure zones from explosions, fire intensities, etc.
5. The weather conditions that will be applied in gas dispersion consequence models.
6. Domino effects caused by the initial failure incident.
7. Appropriate sources of failure rate and reliability data and selection methods.
8. Methods for incorporating loss event time relationships into event tree scenarios.
9. Incorporating human error and management factors into failure probability ranges.
10. Procedures for conducting uncertainty (sensitivity) analysis and quality control.

STEP 2: HAZARD IDENTIFICATION

This step is the foundation for conducting credible risk assessment studies. Fire and explosion hazards that are not properly identified and defined in terms of cause and consequences cannot be properly addressed within the risk assessment framework.

The results of the hazard identification should include:

1. Identification of the physical and chemical properties of materials processed, stored, or transported on site that can harm employees, the public, property, environment, or other selected risk targets.
2. Identification of weaknesses in the design, operation, and protection of facilities that could lead to toxic exposures, fires, or explosions.
3. Evaluation of the significance of potential hazardous events associated with a process or activity to allow categorization of the consequences and ranking the risk.

The tasks associated with step 2 include:

1. Gathering technical information
   a. Plant and process data
   b. Material properties

2. Gathering historical accident data
   a. Facility tour and process review
   b. Plant personnel interviews
   c. Process safety management documentation evaluation
3. Documenting the hazard evaluation(s)
   a. Use of a selected hazard evaluation method (e.g., PHA, HAZOP, FMEA, ETA).

Plant and Process Data

The fire protection risk analyst must understand plant processes, their interdependence, and the inventories and conditions of materials. Plant and process data must describe the plant as it actually operates, which may be different from the original design. The following examples of risk assessment information required may serve as a checklist:

- Material information [material safety data sheets (MSDS)]
- General process chemistry (including side reactions under normal and abnormal conditions)
- Process flow diagrams (including process description and operating parameters such as flow rates, pressures, temperatures, and stream compositions)
- Process design bases (including external events)
- Process utilities (cooling, steam, electricity, instrument air, and utility backup systems)
- Waste treatment, pollution control systems
- Equipment specifications (including materials of construction)
- Equipment detail drawings
- Piping and instrument diagrams (P & IDs, including utilities and pressure relief systems)
- Plant layout drawings (plant and immediate surroundings, including elevations and potential ignition sources)
- Fire water and drainage system drawings
- Process, equipment modifications
- Control logic (e.g., instrument loopsheets, relay logic diagrams)
- Operating instructions
- Operating philosophy (storage inventory levels, operating schedule, manning, startup and shutdown, operator training, safety policy)
- Protection systems diagrams (fire protection, emergency relief, interlock, and alarm systems; design basis should also be included)
- Historical systems failure incident and maintenance records
- Maintenance philosophy and programs
- Emergency response procedures
- Past hazard identification information (HAZOPs, audits, surveys, etc.)
- Replacement cost estimates: structures, equipment, inventory
- Production dependency
- Employee distribution by shift in process area and on plant site
- Weather data (e.g., wind rose) and off-site population distribution
- Operational and loss incident history involving fire and explosions
- Process safety management (PSM) documentation

Material Properties

Accurate information concerning material properties, inventories, and processing and storage conditions is re-

quired to perform hazard evaluations. Detailed information is needed on both the physical and chemical properties of materials. Some of this data can be obtained from material safety data sheets (MSDS). A nonexclusive list of properties includes:

1. Thermodynamic data (including vapor pressure, boiling point, freezing point, critical temperature and pressure, specific and latent heats, heats of combustion)
2. Flammability
3. Dust explosion data (for samples reflecting process conditions)
4. Industrial hygiene and toxicity data
5. Shock sensitivity, thermal analysis data from differential scanning calorimetry, accelerating rate calorimetry, vent sizing package
6. Miscellaneous, e.g., peroxide-forming materials, susceptibility to spontaneous ignition, ability to hold static charge, effect of contaminants.

Chemical and flammability data can be extracted from sources, such as NFPA 49, *Hazardous Chemicals Data*; NFPA 491M, *Manual of Hazardous Chemical Reactions*; NFPA 325, *Guide to Fire Hazard Properties of Flammable Liquids, Gases, and Volatile Solids*; and NFPA *Fire Protection Handbook*. Much of the data available in these publications are at atmospheric temperature and pressure. Experimental data appropriate to process conditions will sometimes be needed.

Dust explosion data for explosion venting calculations are presented in NFPA 68, *Guide for Venting of Deflagrations*, which includes additional references. A considerable amount of dust explosion data can be obtained from various U.S. Bureau of Mines and National Fire Protection Association (NFPA) publications.

Historical Accident Data

An important part of fire and explosion hazard identification and risk screening is a review of the history of loss incidents similar to the hazard being analyzed. A review of the available information on loss incidents or the available loss trending data provides:

1. A relative breakdown of consequential effects, in terms of type of fire or explosion and/or in terms of resulting damage (can generally be used for estimating conditional event probabilities);
2. Identification of representative or dominant failure modes (equipment related, human error, system(s) related) that have led to fire or explosion accidents;
3. Identification of ignition sources and fire propagation contributing factors;
4. Information concerning the duration of the fire and the general effect of loss mitigation factors;
5. Information to support the generation of credible fire and explosion incident loss scenarios and the structuring of event tree analysis.

Data bases are the information foundation of hazard analysis. One of the most effective ways to determine if a system has a fire or explosion potential is to review past loss incident records. Accident data from specific plant operations (if available) is usually the best source and probably more accurate for specific equipment and operations, since the data reflects the operating and maintenance practices of the specific facility. Fire records of NFPA and American Petroleum Institute (API) provide fire loss incident data for a number of processes, plants, and equipment. Federal and state agencies also collect a wide variety of data related to safety and loss prevention issues. A few loss incident data sources are listed in Table 5-12.1.

These kinds of events are sufficiently serious to be reported fairly widely in publicly available sources (e.g., regulatory agencies, research organizations, the media). Data sources can generally be grouped into three categories:

1. Failure mechanisms and causes,
2. Consequence effects (e.g., downwind gas concentrations, fire radiation levels, explosion overpressures, etc.), and
3. Generic frequency categorization of certain types of incidents.

Data sources in the first two categories may be helpful in constructing a fault or event tree model or in understanding the consequences of a specific incident. However, they usually do not provide information on the frequency of incidents. Data sources in the third category provide generic frequency information, but should be used with caution. In most cases, frequency data derived from incident reports may not be applicable to the specific risk assessment being conducted. Historical data bases are rarely complete. Minor incidents, which could have escalated into major incidents, are sometimes not reported, and therefore may not be included in the data. Consequently, the fire and explosion risk analyst must examine sources of data very carefully to determine applicability.

TABLE 5-12.1 *Some Sources of Information for Use in Fire Risk Assessments*

Source	Nature of Information
NFPA (National Fire Protection Association)	Fire Incident Data: FIDO (Fire Incident Data Organization) NFIRS (National Fire Incident Reporting System)
DOT (Department of Transportation)	Annual reports on hazardous materials transportation accidents
NTSB (National Transportation Safety Board), U.S. DOT, Washington, DC	Accident reports: A detailed report is produced for transportation accidents involving hazardous materials
	Hazardous materials accident spill maps: These maps show the location of the spill, any airborne plume, site of fatalities and/or injured people, at one or more times after the start of the incident
API (American Petroleum Institute) Washington, DC	Annual summaries of petroleum industry loss incidents
Association of American Railroads, Federal Railroad Administration	"Railroad Facts" (annual editions) Accident/incident bulletins (annual)
EPA (Environmental Protection Agency)	Reports on various aspects of hazardous material release incidents
U.S. Department of Transportation, Research and Special Programs Administration, Office of Pipeline Safety Washington, DC	Pipeline leak reports for onshore gas transmission and gathering lines, and liquid lines

Hazard Evaluation

Hazard evaluation techniques include:[1]

- Safety review
- Checklist analysis
- Relative ranking indexes
- Preliminary hazard analysis
- "What-if" analysis
- "What-if"/checklist analysis
- HAZOP analysis
- FMEA
- Fault tree analysis
- Event tree analysis
- Cause-consequence analysis
- Human reliability analysis

Each technique has specific application benefits, limitations, resource needs (e.g., manpower, time, budget), and documentation requirements. It would be impossible in this chapter to describe each technique. An excellent reference, which addresses those techniques in detail and provides illustrative examples, is the Center for Chemical Process Safety's (CCPS) *Guidelines for Hazard Evaluation Procedures with Examples.*[2]

Documenting Hazard Evaluation

The synonym for hazard evaluation presently in wide use is process hazard analysis (PHA), as this term is used in OSHA's Process Safety Management (PSM) regulation 1910.119.[3]

The OSHA PSM regulation requires an analysis that identifies and evaluates hazards involved in a process, and must include:

1. Use of one or more of the following to perform a hazard analysis:
   a. Checklists
   b. "What-if"/checklist
   c. Hazard and operability study (HAZOP)
   d. Failure mode and effects analysis (FMEA)
   e. An appropriate equivalent methodology
2. The hazard analysis must address:
   a. Process hazards
   b. Engineering and administrative controls
   c. Consequences of failure of controls
   d. Consequence analysis of effects on employees
3. The hazard analysis should be performed by a team with engineering and operations expertise, including at least one person with knowledge specific to the process.
4. The employer is required to establish a system for documenting findings and actions taken, then communicating them to employees.
5. The hazard analysis must be reviewed and updated at least once every five years.

It should be noted that OSHA's definition of process includes the storage, handling, processing, and transportation of hazardous chemicals, and flammable and explosive materials.

A broad evaluation of hazards should include the following four general techniques:

1. Safety reviews,
2. Checklists,
3. Preliminary hazard analysis, and
4. Relative ranking indexes.

TABLE 5-12.2 *Scenario and Ranking Capabilities of Some Hazard Evaluation Techniques*

Hazard Evaluation Technique	Provides Loss Event Scenario Information?	Loss Event Ranking Possible?
Checklists/Safety Reviews	No; specific scenarios usually not identified	No
Dow and Mond Indexes	Yes, on a unit or a major system basis	Consequence ranking
Preliminary Hazard Analysis (PHA)	No; specific scenarios usually not identified	Yes
"What-If" and "What-If"/Checklist Analysis	No; specific scenarios usually not identified	Consequence ranking
HAZOP Analysis	Yes	Consequence ranking
FMEA	Yes	Consequence ranking
Fault Tree Analysis (FTA)	Yes	Frequency ranking
Event Tree analysis (ETA)	Yes	Yes
Cause-Consequence Analysis	Yes	Yes
Human Reliability Analysis (HRA)	Yes	Frequency ranking

However, a detailed analysis of the wide range of hazards during design and operational stages should include the following three fundamental methods:

1. "What-if" checklist analysis,
2. HAZOP analysis, and
3. FMEA.

Table 5-12.2 provides a general breakdown of some hazard evaluation techniques in terms of their general capability to provide scenario information and ranking of loss events.[2]

STEP 3: LOSS EVENT SCENARIO DEVELOPMENT

Structuring credible fire and explosion loss scenarios is an important aspect of the risk assessment process. The primary components that must be evaluated in the fire and explosion loss scenario development are:

1. Initiating failure event(s)
2. Intermediate event(s)
3. Incident outcome(s).

Fire exposure is very time dependent. The most widely used technique in the structure of fire scenarios is the event tree logic, which conveys the initiation, propagation, and consequences of potential fire events for probability *versus* time assessments.

An event tree provides an inductive, forward logic framework that identifies a failure process, such as fire initiation and propagation. It has a major advantage of being able to incorporate time and sequential conditionality into a scenario. In some cases, it may be weak in identifying specific details that contribute to the consequences. In these cases, fault trees are used to supplement the event tree structure and provide detail to the top events leading to the final consequence.

Risk assessment concerning the release of flammable and explosive materials and expected consequences is a major issue for a fire protection engineer when part of a risk assessment team conducting OSHA and EPA regulatory compliance projects and evaluation of risk reduction options.

Figure 5-12.2 presents a breakdown of the general event components from accidental release to consequences.

Initiating Failure Event(s)

Initiating failure events generally include one or a combination of the following:

1. Containment failures,
2. Human error,
3. Ignitions, and/or
4. External exposures (i.e., flooding, earthquake, etc.)

From a review of past fire and explosion incidents associated with the accidental release of flammable liquid and/or gas, the following causes are evident as major causes of containment failure:[4]

1. Rupture of temporary hose
2. Overfilling of tanks
3. Release from drainage and sampling valves
4. Leaks from gaskets and pump seals
5. Leaks from flanged joints and small pipe connections
6. Piping failures

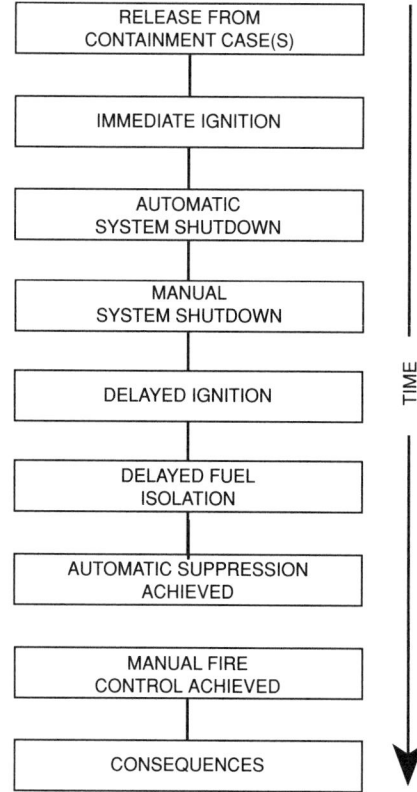

Fig. 5-12.2. General flammable liquid/gas event sequence.

7. Tank rupture
8. Internal tank overpressures.

Containment failures can be created by equipment malfunctions (mechanical/electrical breakdown, instrumentation failures, etc.), human errors, loss of utilities (electricity/cooling water, etc.), and external events (floods, earthquakes, vandalism, etc.).

Human error analysis [or human reliability analysis (HRA)] generally involves the following evaluations:

1. Describing the characteristics of the personnel, the work environment, and the tasks that are performed
2. Evaluating the human-machine interfaces
3. Performing a task analysis of intended operator functions
4. Performing a human error analysis of intended operator functions to assess human error likelihood.

Human error assessment methods and details are related to the process or equipment complexity; the operating environment; human-equipment interface factors; hardware design interfaces for emergency actions; and procedures for operations, maintenance, testing, and training.

Intermediate Event(s)

Intermediate events are events that affect the growth, propagation, and mitigation of the initiating event source. These events are a function of time, which is very important to recognize when addressing the conditional probabilities.

Propagating factors within the fire and explosion development scenario include:

1. Process parameter deviations
   a. Pressure
   b. Temperature
   c. Flow rate
   d. Concentration
   e. Phase/state change
2. Material release type
   a. Combustibles
   b. Explosive materials
   c. Toxic materials
   d. Reactive materials
3. Ignition
4. Energy release rate
5. Ventilating/weather effects
6. Operator emergency response errors
7. Domino effects of fires and/or explosions.

Mitigating factors include:

1. Safety system responses
   a. Relief valves
   b. Back-up utilities
   c. Back-up components
   d. Back-up systems
2. Barrier effectiveness with time
3. Mitigation system responses
   a. Vents
   b. Dikes
   c. Detection/alarms/shutdown
   d. Fire protection/suppression
4. Control responses/operator responses
   a. Planned
   b. Emergency
5. Contingency operations
   a. Emergency procedures

b. Personnel safety equipment
c. Evacuations
d. Security.

Incident Outcome(s)

Potential incidents of primary interest to the fire protection engineer include:

1. Radiant heat from a fire
   a. Pool fire
   b. Torch fire
   c. Flash fire
   d. Fire ball [boiling liquid expanding vapor explosion (BLEVE)]
2. Explosion overpressures
   a. Unconfined vapor cloud explosion (UVCE)
   b. Tank/equipment rupture/fragmentation
3. Corrosive smoke/fire products concentration
   a. Toxic gas concentrations.

Figure 5-12.3 presents two generic event trees associated with flammable liquid and gas releases developed by The World Bank.[4]

STEP 4: CONSEQUENCE ASSESSMENT

Consequence assessment is the estimation of the undesirable result of an accidental fire or explosion exposure to a target, usually measured in terms of health and safety effects to people, loss of property, or business interruption costs.

This assessment involves the evaluation of two off issues:

1. The susceptibility or vulnerability of people or damage to targets (structures, equipment, etc.) being evaluated in the risk assessment project, and
2. The rate of development of a hazardous environment (intensity, distance, time) within the boundaries of the predicted hazardous event (gas dispersion, fire, explosion).

When these issues are evaluated quantitatively, they provide estimates of the loss vulnerability to selected targets *versus* the predicted (distance and time) exposure from the fire or explosion scenario. The analogy within this assessment is one of source → target exposure and vulnerability, as illustrated in Figure 5-12.4. References 4, 5, 6, and 7 provide additional information on this subject.

The steps involved in the quantification of accidental flammable liquid and gas release exposures include:

1. Characterizing failure modes
2. Calculating release rates and durations
3. Evaluating ignition scenarios in terms of immediate or delayed ignition times
4. Calculating fire and explosion intensities
5. Estimating the vulnerability to the target(s) of interest
6. Plotting effect distances.

Characterizing Failure Modes

Steps 2 and 3 identify failure scenarios. Failure identification, however, must be broken down into potential failure modes to allow quantification of flammable liquid or gas release rates. For example, release from a pipe failure could

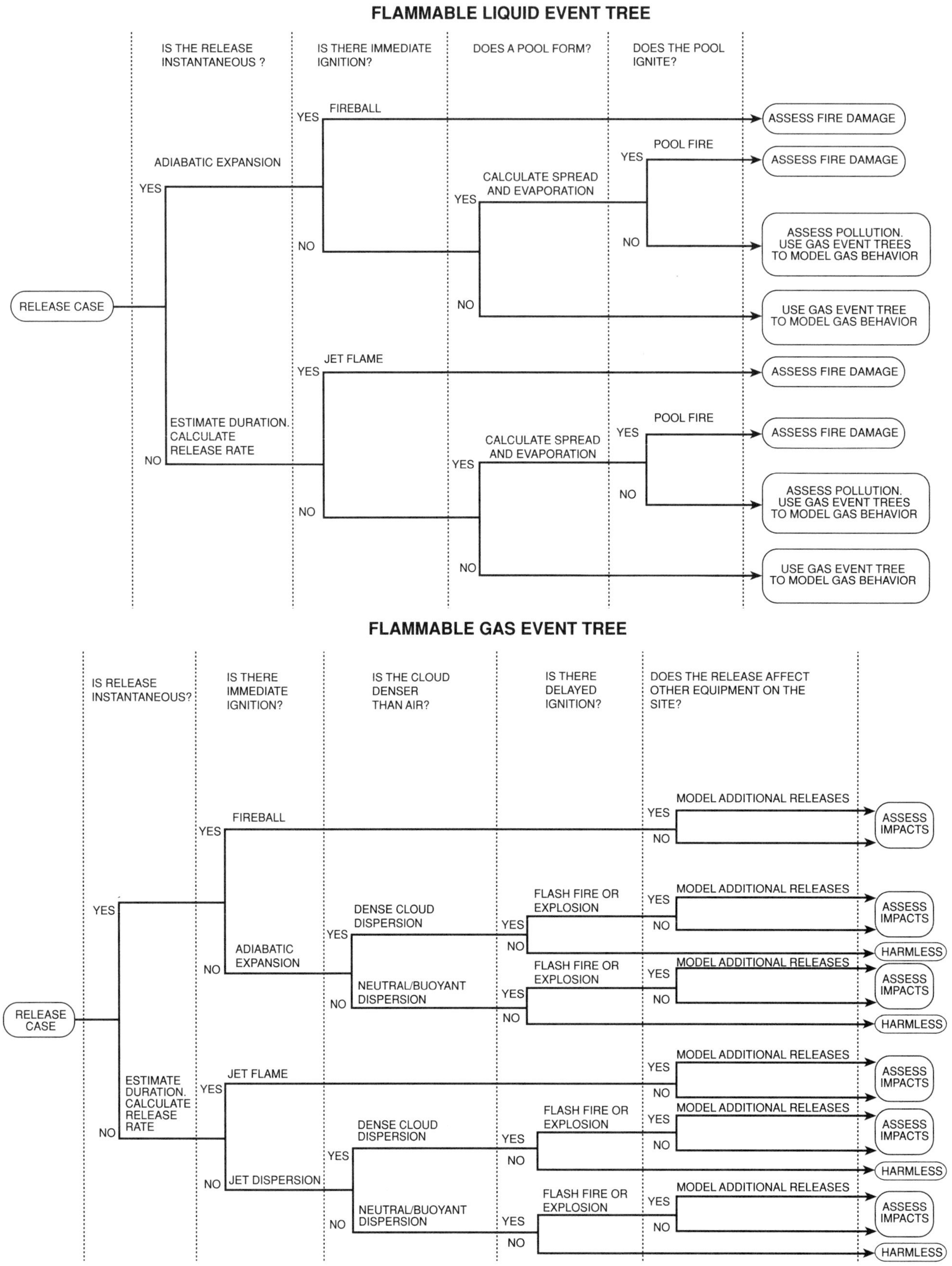

FLAMMABLE LIQUID EVENT TREE

FLAMMABLE GAS EVENT TREE

Fig. 5-12.3.   Generic event trees. (Modified from reference 4.)

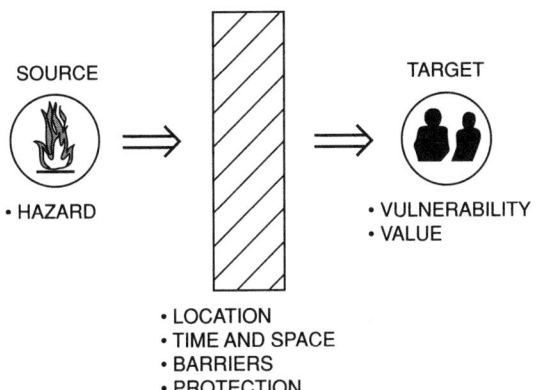

Fig. 5-12.4. *General illustration of source → target concept.*

involve a failure mode stemming from a pinhole leak, to a partial fracture of the pipe, to a full rupture of the pipe. With each failure mode there is associated relative failure probability (e.g., very small leaks are much more likely than major pipe fractures) and a release rate.

To manage failure mode evaluation within a time-constrained risk assessment project, industry risk assessment experts generally focus on three to five failure modes related to the equipment release source being studied.

Figure 5-12.5 provides some general information on some equipment-related failure cases and failure sizes, summarized from The World Bank Technical Paper No. 55, with permission from The World Bank.[4]

Calculating Release Rates and Durations

The release from the failure point may be liquid, gas, or a combination of the two (i.e., two-phase flow). It is beyond the scope of this chapter to discuss the benefits and limitations of various calculation methods, and the reader is directed to references 4, 5, and 6.

When calculating release rates and durations, it is important to document both the physical parameters and assumptions, including:

1. The type of release opening (i.e., rectangular or circular, and opening coefficient), the elevation, and size. If using a computer program, these are required input.
2. The effects of containment (diking, curbing) and drainage design reliability.
3. The estimated time to either automatic or manual shutdown to limit the duration of the release. This estimated time will be related to a probability of successful shutdown within the event-free framework.
4. For evaluations involving vapor dispersion (e.g., gas release, liquid pool vaporization), which could lead to flash fires or explosion overpressure effect zones, the topography, selected weather conditions, and any other assumptions (e.g., gas dispersion obstructions or explosion overpressure wave deflections) must be documented.

This step is critical, as the release rate and duration dictate the amount of fuel that can be available in the accident scenario. The methods to actuate the emergency shutdown system (ESS) should be described in detail; this includes gas, flame detection, and emergency isolation valves. The estimated time to shutdown and the residual fuel that can be released following shutdown are impor-

tant considerations in the justification of the volume of fuel and the potential duration of exposure if the fuel is ignited.

Isolation of a release: The behavior of a release can be very dependent on its duration. The duration depends on the amount of material available to be released, which in turn depends on the speed and effectiveness of shutdown or isolation. Therefore, isolation can affect the consequences of a release, and it is important to make a realistic estimate of the time required for isolation. This time will depend on the following:

1. **Leak Detection.** It is usual to assume that major ruptures and leaks will be detected quickly, either by process instrumentation or by operators. Smaller leaks may be detected by gas or flame detectors, if installed. The analyst should determine the position, effectiveness, and reliability of such detectors as part of this evaluation.
2. **Shutdown Activation.** The speed of shutdown will depend on whether emergency shutdown actuation is manual or automatic. The response time concerning manual actuation depends on instrumentation and alarm design reliability, operating procedures, and operator emergency response training. Response times of 3 to 15 minutes are typical, but must be judged on a plant-specific basis. The analyst should also evaluate the availability and reliability of the shut-down valves and the estimated closing time of the particular emergency shut-down valve design.

Evaluating Ignition Scenarios

Fire and explosion outcomes resulting from the accidental release of flammable liquids or gases can be categorized as follows:[4,5,6,7]

Pool fire: A fire involving a pool of confined or spreading flammable material. This can include both immediate and delayed ignition scenarios creating both near-field and remote damage effects.

Torch fire (or flame jet): A fire extended from a point of release of flammable liquid or gas, which is being discharged under pressure. This is usually associated with an immediate ignition scenario and usually includes near-field damage effects.

BLEVE fireball: BLEVE (boiling liquid expanding vapor explosion) results from the overheating of a pressurized vessel by a primary fire. This overheating raises the internal pressure and weakens the vessel shell, until it bursts open and releases its contents as a large and very intense fireball. This is generally considered a delayed event (8 to 15 minutes of exposure) and can involve both near-field damage effects and remote personnel exposure.

Flash fire: A fire involving the delayed ignition of a dispersed vapor cloud, which does not cause blast damage. That is, the flame speed is not as high as in an unconfined vapor cloud explosion, but the fire spreads quickly throughout the flammable zone of the cloud and is usually associated with near-field damage effects and remote personnel effects.

Unconfined vapor cloud explosion: A release of a large amount of flammable vapors (usually greater than 1 ton), which forms a large vapor cloud and upon delayed ignition explodes, creating an explosion overpressure exposure to

FLEXIBLE CONNECTIONS

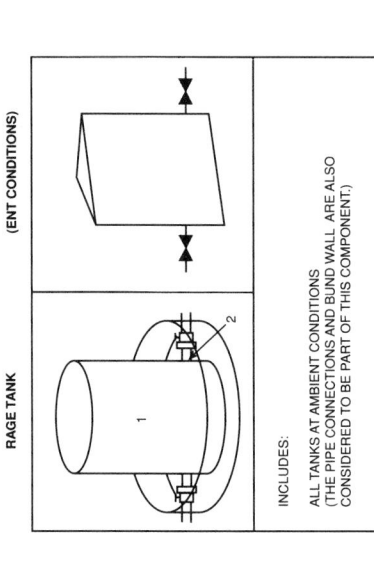

INCLUDES:

HOSES, BELLOWS, ARTICULATED ARMS.

TYPICAL FAILURES	SUGGESTED FAILURE SIZES
1. RUPTURE LEAK	100% AND 20% PIPE DIAMETER
2. CONNECTION LEAK	20% PIPE DIAMETER
3. CONNECTION MECHANISM FAILURE	100% PIPE DIAMETER

PIPE

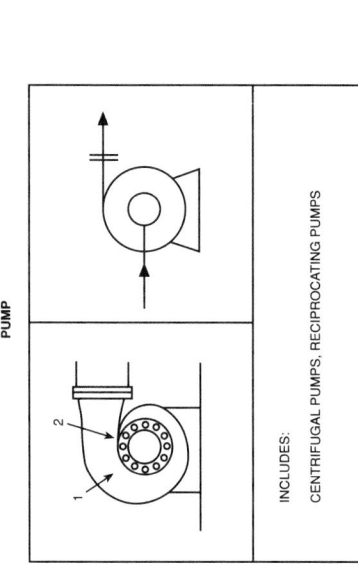

INCLUDES:

PIPES, FLANGES, WELDS, ELBOWS.

TYPICAL FAILURES	SUGGESTED FAILURE SIZES
1. FLANGE LEAK	20% PIPE DIAMETER
2. PIPE LEAK	100% AND 20% PIPE DIAMETER
3. WELD FAILURE	100% AND 20% PIPE DIAMETER

STORAGE VESSEL (PRESSURIZED OR REFRIGERATED)

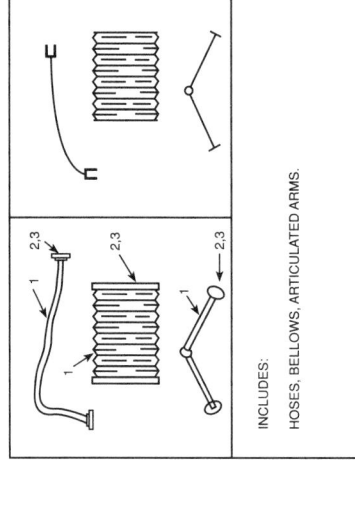

INCLUDES:

PRESSURIZED STORAGE OR TRANSPORT VESSELS, REFRIGERATED STORAGE OR TRANSPORT VESSELS, BURIED OR NON-BURIED VESSELS.

TYPICAL FAILURES	SUGGESTED FAILURE SIZES
1. BLEVE (NON-BURIED CASE ONLY)	TOTAL RUPTURE (IGNITED)
2. RUPTURE	TOTAL RUPTURE
3. WELD FAILURE	100% AND 20% PIPE DIAMETER

NOTE: THESE STORAGE VESSELS MAY HAVE BUND WALLS WHICH SHOULD BE TAKEN INTO CONSIDERATION IN THE ANALYSIS.

PRESSURE VESSEL/PROCESS VESSEL

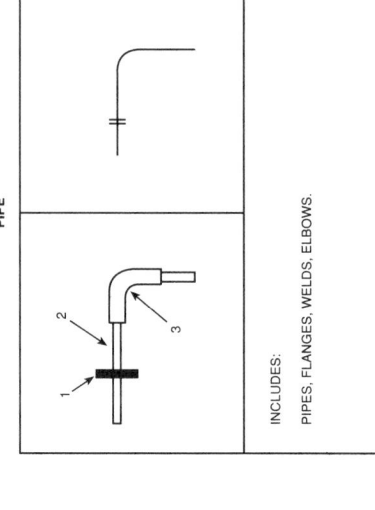

INCLUDES:

SEPARATORS, SCRUBBERS, CONTACTORS, REACTORS, HEAT EXCHANGERS, FIRED HEATERS, COLUMNS, RECEIVERS, REBOILERS.

TYPICAL FAILURES	SUGGESTED FAILURE SIZES
1. VESSEL RUPTURE; VESSEL LEAK	TOTAL RUPTURE; 100% PIPE DIAMETER OF LARGEST PIPE
2. MANHOLE COVER LEAK	20% OPENING DIAMETER
3. NOZZLE FAILURE	100% PIPE DIAMETER
4. INSTRUMENT LINE FAILURE	100% AND 20% PIPE DIAMETER
5. INTERNAL EXPLOSION	TOTAL RUPTURE

STORAGE TANK (AMBIENT CONDITIONS)

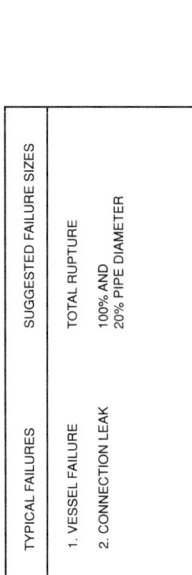

INCLUDES:

ALL TANKS AT AMBIENT CONDITIONS (THE PIPE CONNECTIONS AND BUND WALL ARE ALSO CONSIDERED TO BE PART OF THIS COMPONENT.)

TYPICAL FAILURES	SUGGESTED FAILURE SIZES
1. VESSEL FAILURE	TOTAL RUPTURE
2. CONNECTION LEAK	100% AND 20% PIPE DIAMETER

PUMP

INCLUDES:

CENTRIFUGAL PUMPS, RECIPROCATING PUMPS

TYPICAL FAILURES	SUGGESTED FAILURE SIZES
1. CASING FAILURE	100% AND 20% PIPE DIAMETER
2. GLAND LEAK	20% PIPE DIAMETER

Fig. 5-12.5. Suggested failure modes. (Source: The World Bank.[4])

the surrounding area causing both severe near-field damage effects and remote property and personnel exposure.

Calculating Fire and Explosion Intensities

Deterministic models have been developed to support fire and explosion intensities, and more sophisticated models are continuously being refined and validated. Several chapters within this handbook describe fire and explosion modeling methods and available computer programs.

Available consequence computer models generally include the capabilities to:[5,7,8]

- Estimate discharge rate of liquid or gas
- Estimate area of liquid pool
- Estimate vaporization rate of liquid pool
- Evaluate toxic vapor dispersion hazards
- Evaluate pool fire radiation hazards
- Evaluate fireball radiation hazards
- Evaluate flame jet hazards
- Evaluate vapor cloud/plume fire hazards
- Evaluate vapor cloud explosion hazards
- Evaluate tank overpressurization rupture and fragmentation hazard
- Evaluate solid/liquid explosion overpressure hazards.

In addition, various analytical assessment methods described in this handbook can be applied to consequence and risk assessment studies and include methods for:

- Estimating heat release rates and flame heights
- Estimating exposure temperatures from uncontrolled fires using fire plume modeling equations
- Estimating the time to critical damage thresholds for specified targets using heat transfer approaches
- Estimating the response of existing fire protection or improvement strategies
- Applying the results of fire modeling efforts to support loss estimates and the benefits of recommended fire protection improvements.

Target Vulnerability

In evaluating the vulnerability to selected targets, there are two levels of target vulnerability assessment that can be utilized:

1. Using various tables and references that establish generic criteria for damage to people and property from fire and explosion exposures (i.e., radiant heat, explosion overpressures). This level of vulnerability evaluation provides a good first-order review, however it is weak in terms of specific time relationships. Since time is usually a very uncertain variable, a number of risk assessment projects usually employ generic threshold damage levels.
2. Apply sophisticated heat transfer or structural impact models to support vulnerability assessments where time or specific conditions (design, layout, emergency response) warrant this level of detail, usually termed second-order vulnerability assessment.

There are various tables that give criteria for damage to people and property from fire, usually expressed in terms of radiation heat intensity. The effect on people is expressed in terms of the likelihood of death and different degrees of

TABLE 5-12.3 *Damage Caused at Different Incident Levels of Thermal Radiation*

Incident Flux (kW/m²)	Type of Damage Caused	
	Damage to Equipment	Exposure to People
37.5	Damage to process equipment	100% lethality in 1 min. 1% lethality in 10 s
25.0	Minimum energy to ignite wood at indefinitely long exposure without a flame	100% lethality in 1 min. Significant injury in 10 s
12.5	Minimum energy to ignite wood with a flame; melts plastic tubing	1% lethality in 1 min. First-degree burns in 10 s
4.0		Causes pain if duration is longer than 20 s, but blistering is unlikely
1.6		Causes no discomfort for long exposure

injury for different levels of heat radiation. In Table 5-12.3, the radiative incident flux is related to the levels of damage; this table is based on observations of large fires.[4]

Vapor cloud explosion exposures can be generically correlated with the energy of the explosion overpressure. This correlation is used to relate distances to various levels of damage and people exposure. Table 5-12.4 provides a general breakdown of damage threshold levels.[4,7]

The general methods for assessing damage and human exposure for flash fires and torch fires is to assume fatalities would occur within the flame envelope of the flash fire or torch fire zone, and to estimate secondary damage potential to equipment or structures within the flame exposure zone.

Plotting Effect Distances

Plotting the results of the fire and explosion consequences associated with defined failure scenarios provides a graphical depiction of potential exposure zones. Figure 5-12.6 provides an illustrative example for plotting effect zones from an uncontrolled liquefied gas release potentially resulting in a torch fire, flash fire, BLEVE, or vapor cloud explosion scenario.

TABLE 5-12.4 *Explosion Overpressure Damage Estimates*

Overpressure (psig)	Characteristic Damage	
	To Equipment	To People
2.5–5	Heavy damage to buildings and to process equipment	1% death from lung damage >50% eardrum rupture >50% serious wounds from flying objects
1–2.4	Repairable damage to buildings and damage to the facades of dwellings	1% eardrum rupture 1% serious wounds from flying objects
0.5–1	Glass damage	Injury from flying glass
0.15–0.30	Glass damage to about 10% of panes	Slight injury from flying glass

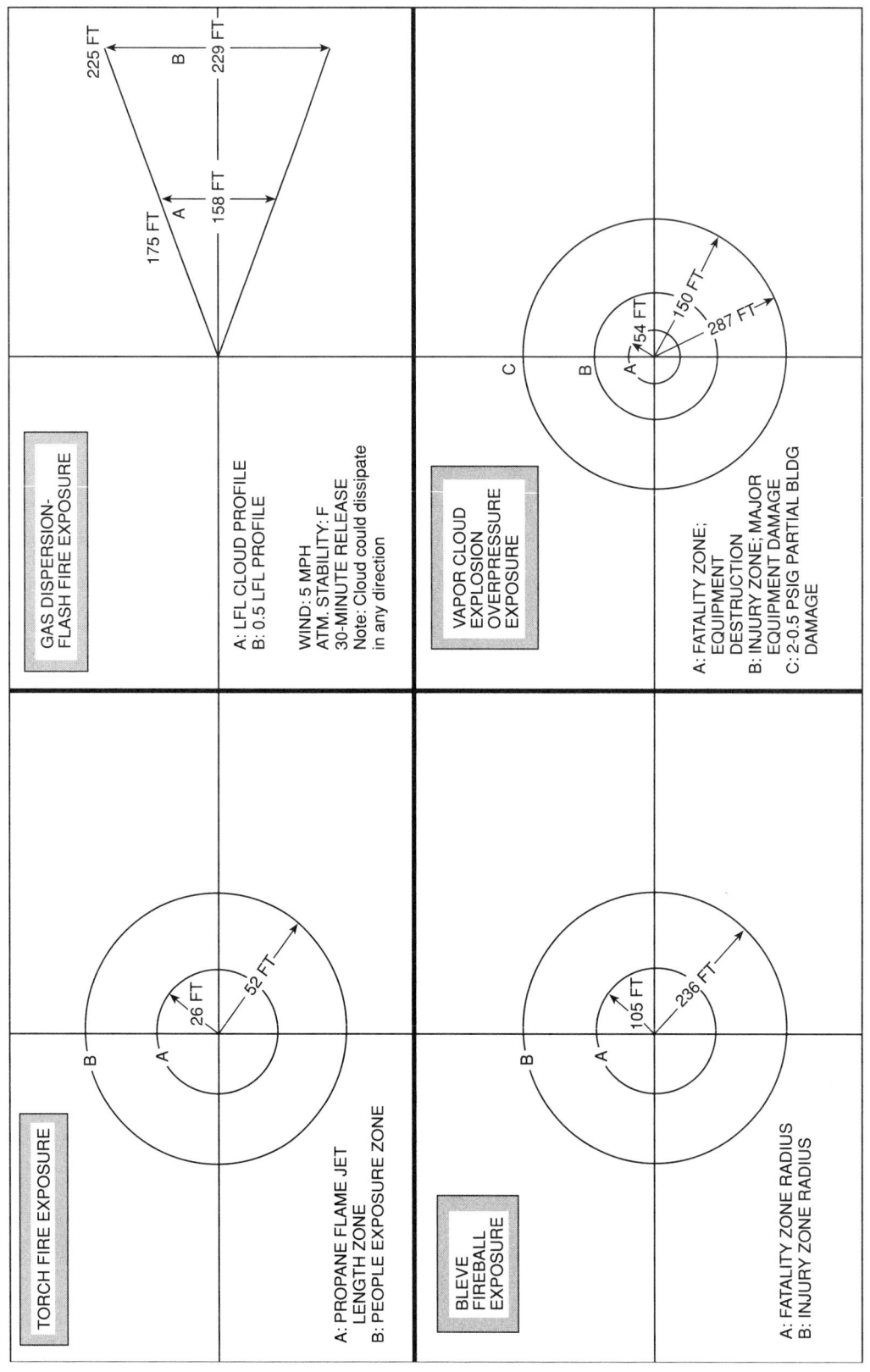

TORCH FIRE EXPOSURE

A: PROPANE FLAME JET
   LENGTH ZONE
B: PEOPLE EXPOSURE ZONE

26 FT
52 FT

**GAS DISPERSION-
FLASH FIRE EXPOSURE**

A: LFL CLOUD PROFILE
B: 0.5 LFL PROFILE

WIND: 5 MPH
ATM. STABILITY: F
30-MINUTE RELEASE
Note: Cloud could dissipate
in any direction

175 FT
158 FT
225 FT
229 FT

**BLEVE
FIREBALL
EXPOSURE**

A: FATALITY ZONE RADIUS
B: INJURY ZONE RADIUS

105 FT
236 FT

**VAPOR CLOUD
EXPLOSION
OVERPRESSURE
EXPOSURE**

A: FATALITY ZONE;
   EQUIPMENT
   DESTRUCTION
B: INJURY ZONE; MAJOR
   EQUIPMENT DAMAGE
C: 2–0.5 PSIG PARTIAL BLDG
   DAMAGE

54 FT
150 FT
287 FT

NOTE: ESTIMATED LPG FIRE - EXPLOSION ZONES FOR SELECTED BASELINE SCENARIOS. CALCULATED USING THE ARCHIE COMPUTER MODEL. DISTANCE IS FEET
FROM POINT OF LPG ACCIDENTAL RELEASE ORIGIN.

Fig. 5-12.6.   General example of plotting effect zones.

STEP 5:
PROBABILITY ASSESSMENT

This section presents an introduction to a complex subject. The mathematics and sensitivity analysis involved in the structuring of detailed event and fault trees is beyond the scope of this section and the reader is referred to references 2, 5, 9, and 10 to gain a more detailed insight into these areas.

Probability assessments must include a team approach. The team assembled to support the fire risk analyst in process risk assessments will generally include members experienced in process design and instrumentation, reliability engineering, safety, and environment issues, as well as team members from the plant knowledgeable in the specific facility hazards and operations. In some cases, the services of a statistician may be required to compile raw plant data from plant insurance reports and OSHA injury logs.

The general structure of the event tree for evaluating the consequences from the release of flammable liquids or gases is presented in Figure 5-12.7.

In mathematical terms, Figure 5-12.7 can be expressed as:

$$F_{\text{(Consequences)}} = F_{\text{(Initiating events)}} \times \Sigma P_{\text{(Intermediate event probabilities)}}$$

There are two basic approaches that are commonly employed in fire and explosion risk assessments to estimate initiating event frequencies and conditional probabilities:

1. Use of relevant historical data, and
2. Synthesis of event frequencies and probabilities using techniques, such as fault trees, human reliability analysis, and expert engineering judgment.

In many cases the two approaches are used in a complementary manner to provide an independent check on one another and to increase the validity and confidence in the final results.

The first approach examines: (1) relevant historical data in order to assess the frequency with which these events have occurred in the past and (2) the likelihood of their occurrence in the future. Where sufficient relevant past data is available, historically based frequencies may be adequate in making a reasonable assessment of fire and explosion risks. However, frequencies derived in this way represent only average values and are most applicable to simple systems where there are not a number of variables (i.e., propagation and mitigation factors) that can significantly change the consequential results.

In actual application, assessments of potential fire and explosion risks at a specific facility will usually require adjustments to historical data to reflect the particular facility management and protection deviations. When evaluating the relevance of historical data for use in a specific facility or process risk assessment, the following factors should be taken into account:

1. There are many types of hazardous industrial operations for which historical data is limited or not available to make confident predictions about consequence likelihoods.
2. The general distribution between the primary causes leading to fire explosions shows that human error is the leading cause:[11]
   - Human factors (e.g., management controls, human operator errors): 70 to 85 percent
   - Equipment failures: 10 to 20 percent
   - External factors (e.g., earthquake, floods): 10 to 20 percent.

   With the majority of causes being associated with human factors, it would be expected that the majority of risk research would be in this area; however, the quantification of the human element is very difficult and much research continues in this area.
3. Fire prevention and protection technology is continually being expanded and updated based on new codes, standards, and industry practices that are developed after loss occurrences, and continued fire and explosion research efforts.

Considering these factors, the fire risk analyst must take a systematic approach to first determine the best approach for estimating frequencies and probabilities based on available data and specific plant conditions. A general approach is illustrated in Figure 5-12.8.[5]

Risk Assessment Fundamentals

Three primary concerns that the fire risk analyst must address are:

1. Integrating failure (equipment, human, external) mechanisms,
2. Application of engineering judgment and expert opinion, and
3. Documentation methods.

Integrating failure mechanisms: The method of integrating hardware failures (e.g., equipment, emergency shutdown systems, passive and active protection systems), human errors (e.g., equipment testing and maintenance deficiencies, operational errors), and potential external upsets (e.g., windstorm, flood, earthquake) into top event frequencies must be considered by the risk assessment team. In the majority of fire and explosion risk assessments, top events will have to be supported by fault tree analysis to derive and validate the selected event tree frequencies and conditional probabilities. The general framework is illustrated in Figure 5-12.9.[2] References 5, 9, and 10 provide

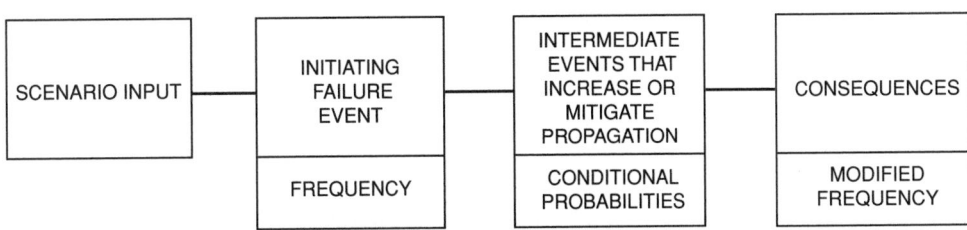

Fig. 5-12.7. General event tree sequence.

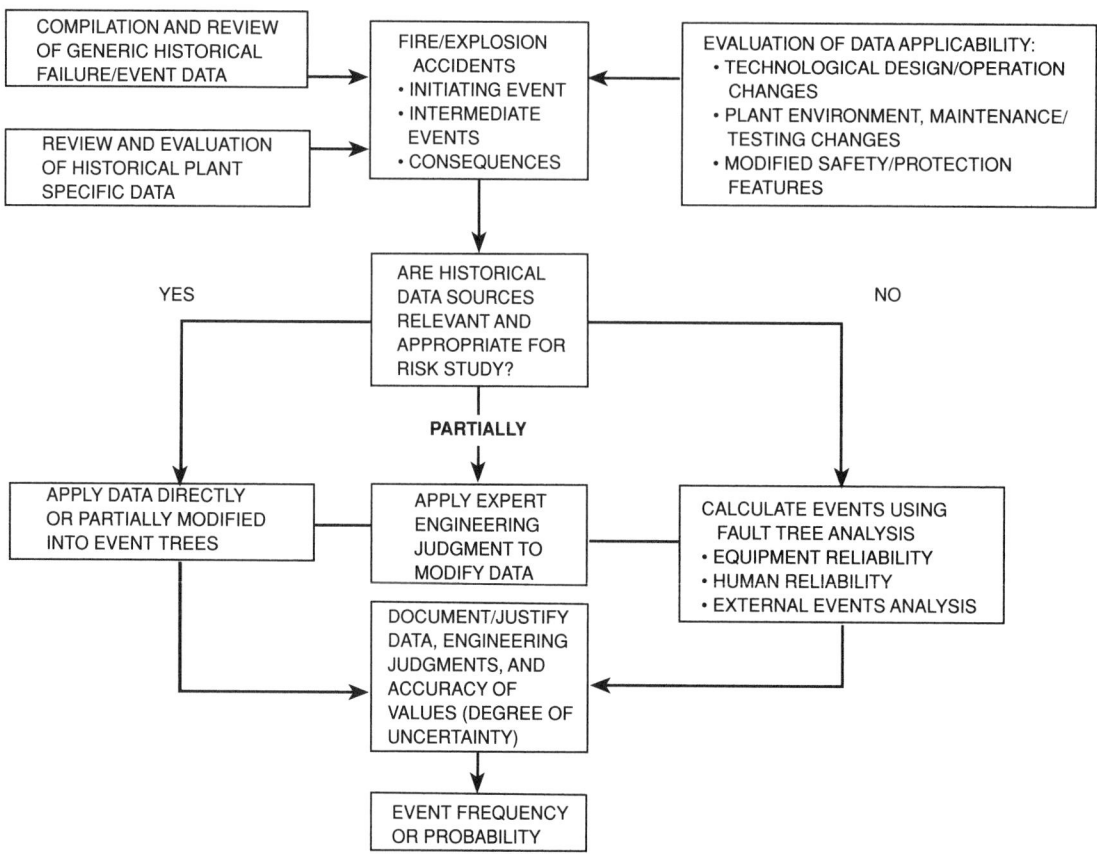

Fig. 5-12.8. General procedure for estimating event frequency or probability.

good information on fault tree structuring, Boolean algebra, cut-sets, and sensitivity analysis.

Engineering judgment and expert opinion: The method for systematic and consistent application of expert engineering judgment and expert opinions within the risk assessment process is a very important consideration.

There are many hazardous situations and potential fire events that are often encountered where data will be insufficient. When these situations occur, it is often possible, and many times necessary, to generate data using the engineering judgment and expertise of the risk assessment team and, in some cases, outside experts. Experts, through their past experience with similar hazards or operations, can provide

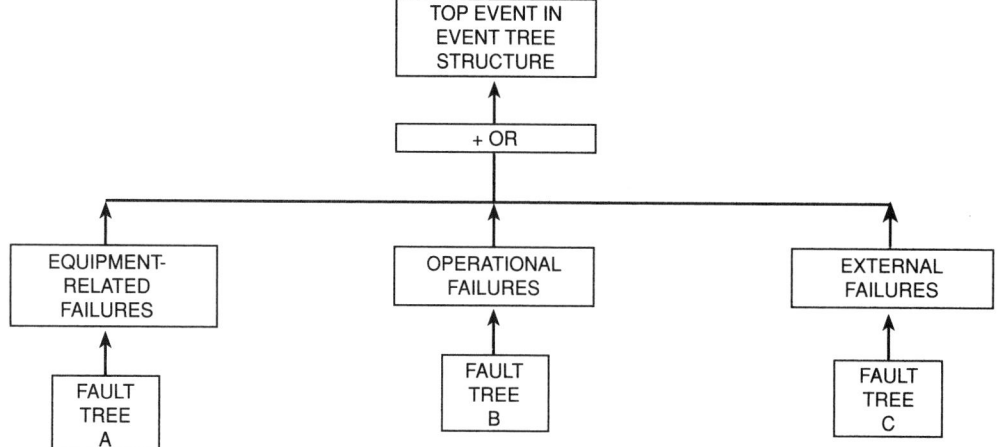

Note: Fault trees A,B, and C are generally required to be developed into separate detailed fault tree models.

Fig. 5-12.9. General fault tree support framework.

TABLE 5-12.5 *Example, Documentation Table Format*

Event Tree Designation	Event Description	Frequency (F) or Probability (P) Range	Data Source	Frequency (F) or Probability (P) Assigned to Event	Selection Basis and Comments
Event tree and event branch line identification	Description of Event Scenario • Release • Ignition • Emergency shutdown • Automatic/manual fire control	Frequency or probability range applied to event	Source of data or method used for establishing range	Assigned frequency or probability within range	Justification of Selection • Historical • Records • Fault tree • Engineering judgment • Combination

valuable input. Frequency or probability estimates obtained in this manner should combine the judgment and opinions of a group of experts, rather than rely on a single opinion.

Documentation methods: Methods for documenting the frequency and probability selection basis must be determined. To provide documentation, a table is usually constructed for each fire or explosion scenario event tree. Using the event sequences for the release from containment of a flammable liquid or gas, example information is presented in Table 5-12.5. It is essential in all risk assessment projects to document the data selection basis and the associated uncertainties.

Equipment Failure Data Sources

An excellent source for generic equipment failure data is *Guidelines for Process Equipment Reliability Data, with Data Tables*, developed by the Center for Chemical Process Safety (CCPS).[12] The failure data can also be obtained as a computerized data base on diskette. This book provides a listing and general description of all the data sources used to construct the CCPS generic failure rate data base. It is essential for the fire risk analyst to understand the sources of this data, the limitations and applicability to the particular environment, and management controls for the specific facility being assessed, and this reference provides a good summary of these concerns.

Ignition Likelihood

Ignition likelihood is a measure of the expected ignition occurrence for a defined fire scenario, and may be expressed as:

1. Frequency (i.e., number of occurrences per unit of time)
2. Annual probability (i.e., a probability of occurrence during a per-year time interval must be expressed as a number ranging from 0 to 1)
3. Conditional probability of occurrence given that a precursor event has occurred (e.g., release of flammable gas, molten metal spill, etc.).

The objective of ignition likelihood estimation is to estimate, in a systematic and credible manner, the cumulative ignition likelihood from defined fire event scenarios. The procedures involved in this method generally require the integration of plant-specific data, deterministic flammable liquid spread, and gas dispersion modeling, as well as engineering judgment. The sources listed in Table

5-12.1 and industrial insurance companies (such as Factory Mutual Research Corporation, Industrial Risk Insurers, etc.) are good sources for obtaining generic ignition data.

Ignition likelihood estimation steps include:

1. Characterize and map ignition source potentials in the area(s) being evaluated.
2. Compile available plant-specific ignition–fire data, based on plant records and interviews.
3. Review available historical incident data sources that are similar to hazards being evaluated; this should be used for reference and comparison.

Ignition potential area boundaries (i.e., effect zone radius) should include:

1. Potential pathway (area of exposure) where a flammable liquid and/or gas release could exist.
2. Potential area that may be affected by a pressured fluid released (e.g., combustible hydraulic fluid, etc.).
3. Potential radiant heat energy effect area from an exposure fire.

Figure 5-12.10 provides a general illustration of plotting ignition source distribution within a potential flammable gas dispersion area. Table 5-12.6 provides an example format for estimating cumulative ignition likelihood.

Ignition source potential(s) within the modeled effect zone area should be characterized by:

1. Type and location of ignition sources, including fixed point (FP), linear source (LS), and variable point (VP).
2. Continuous (C) or intermittent (I) source.
3. Ignition source energy potential; that is, strong (S), moderate (M), or weak (W).
4. Ignition source presence, whether: (C) continuous, or (I) intermittent. This is the estimated percentage (%) of time the ignition source is present.
5. Ignition source energy potential. This is in relation to the fuel's ignition sensitivity, and should include: chemistry, physical state, surface texture, and moisture content.

Figures 5-12.11, 5-12.12, and 5-12.13 provide some examples of potential ignition sources.

Ignition is a function of time as well as temperature; i.e., potential fuel subjected to a relatively high temperature for a short period of time may not ignite, while the same fuel can undergo ignition when exposed for a longer duration to a lower temperature. For example, wood has a normal ignition

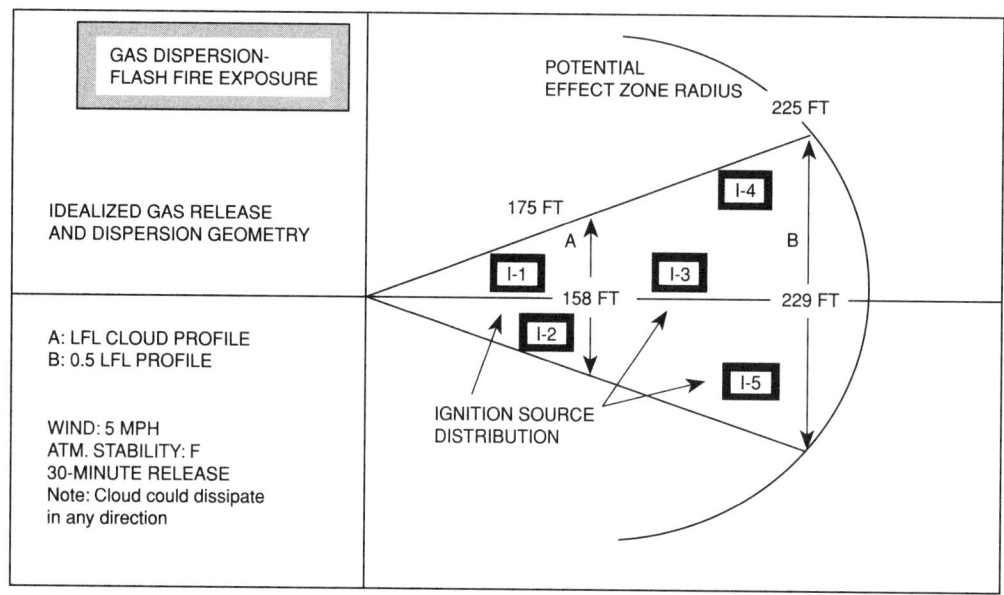

Fig. 5-12.10. *Illustration of plotting an effect zone to evaluate cumulative ignition likelihood. (See Table 5-12.6 for bold-face box definitions.)*

temperature of 400 to 500°F, but has been found to ignite when subjected to a heat source of 228°F for four days.

Examples of conditional ignition likelihood ranges for flammable liquid vapors or gases engulfing ignition sources are provided by references 5 and 13:

Qualitative Ranking	Likelihood Range
(S) Strong	0.25 to 1
(M) Moderate	0.1 to 0.24
(W) Weak	0.01 to 0.09

Ignition likelihood estimation provides for the likelihood of ignition, F_{ig}, and is primarily dependent on the following factors:

F_{te} Frequency of time the ignition source exposure is present in the immediate area of the combustible material being examined under defined scenario conditions.

P_{il} Probability the ignition source exposure exceeds the ignition threshold limits (i.e., temperature, energy) for the combustible material being examined under defined scenario conditions. This probability parameter includes both accedence of ignition threshold limits under normal operating conditions and

also under fault (e.g., equipment failure that could result in shorting and/or overheating, thus exceeding the ignition threshold limits of the material being examined).

P_{cf} Probability that the configuration (e.g., unobstructed pathway, positioning) between the ignition source and combustible is conducive to initiating ignition.

P_{ne} Probability the ignition source is *not* eliminated (e.g., automatic, manual shutdown of electrical power, etc.) prior to established ignition.

Therefore, the general sequence of ignition-related events can be represented as:

$$F_{ig} = F_{te} \times P_{il} \times P_{cf} \times P_{ne}$$

Fire Suppression Failure Probability

Fire suppression system assessment involves evaluation of three factors (see Figure 5-12.14):

1. Availability,
2. Reliability, and
3. Effectiveness.

TABLE 5-12.6 *Example of Estimating Cumulative Ignition Likelihood*

(Fig. 5-12.10) Potential Ignition Source(s)	Continuous (C) or Intermittent (I) Ignition Source	Strong (S), Moderate (M), or Weak (W) Ignition Energy Source	Likelihood of Ignition Based on Ignition Source Being Engulfed with Flammable Vapor	If Intermittent Source, Estimated Time (per year) for which Intermittent Source is Present and Active	Likelihood of Ignition, Given Release and Flammable Cloud
I-1	I	M	0.06	10%	0.006
I-2	I	S	0.9	15%	0.135
I-3	I	M	0.06	30%	0.018
I-4	C	M	0.06	100%	0.06
I-5	C	S	0.24	100%	0.24
				Cumulative likelihood of ignition:	0.46

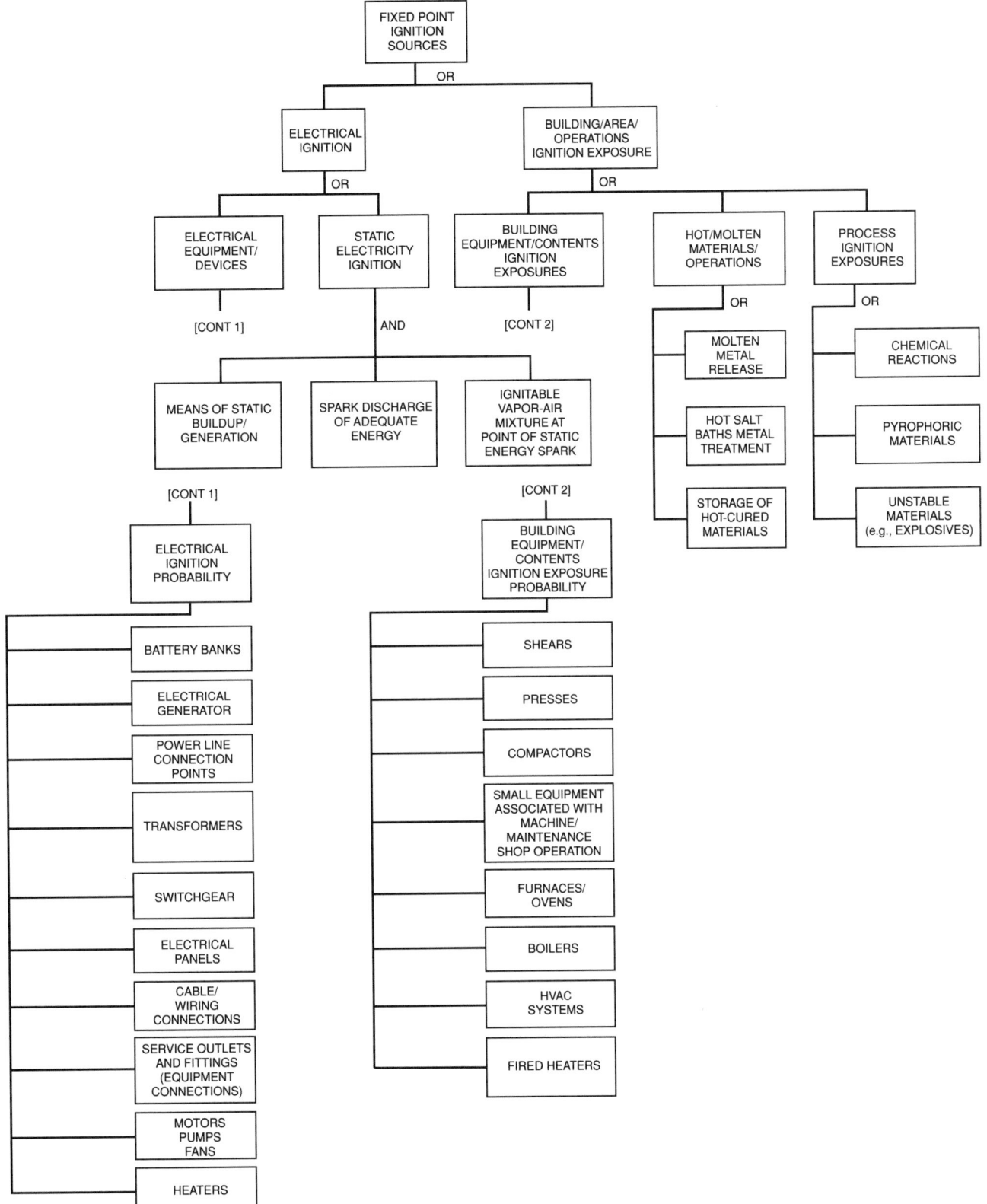

Fig. 5-12.11. Potential ignition sources—fixed point.

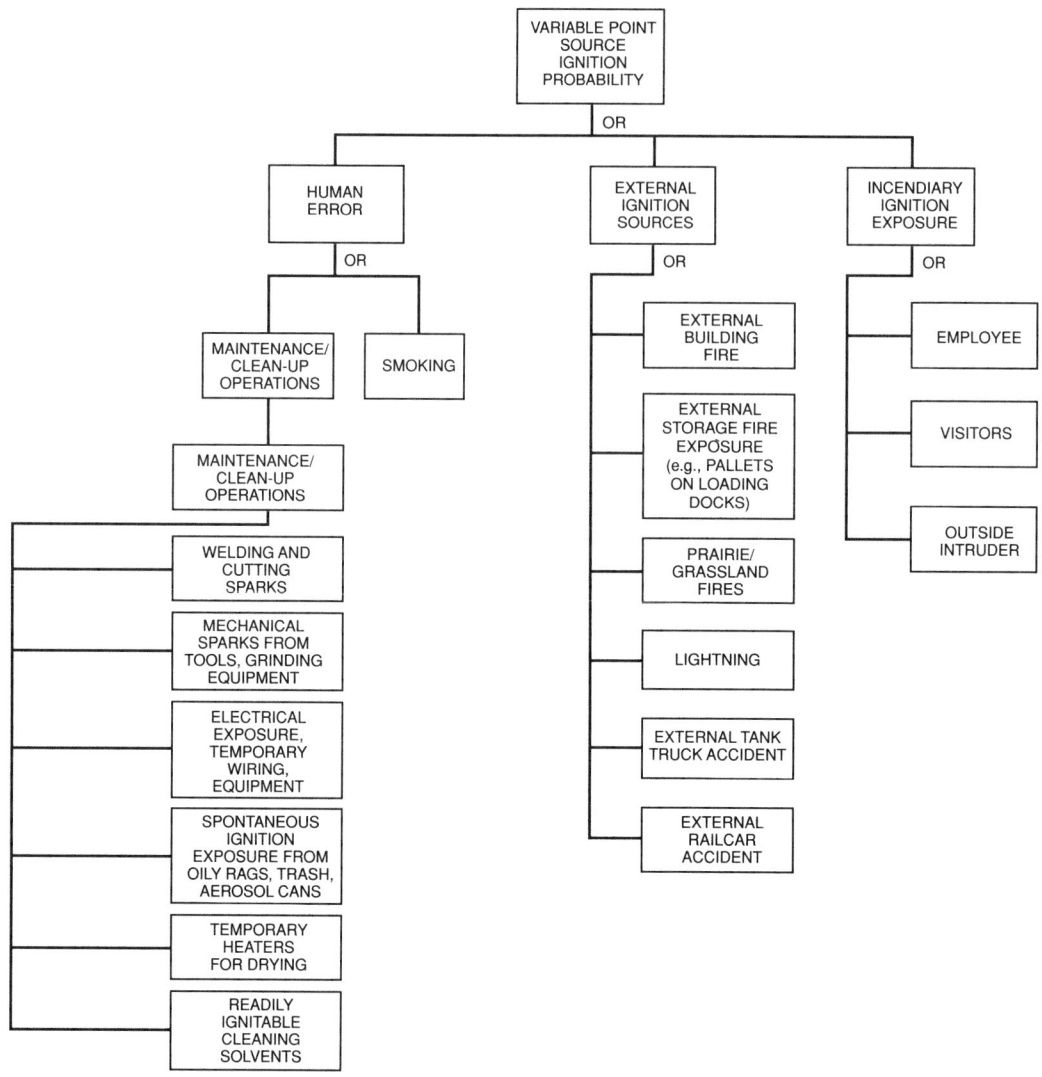

Fig. 5-12.12. Potential ignition sources—variable point.

Problems found in reviewing a number of fire risk assessments include:

1. System suppression success is only associated with the system being available. This approach assumes:
   a. Operational reliability = 100 percent
   b. Suppression effectiveness = 100 percent.
2. *Time* is not properly integrated into fire suppression success probability, including:
   a. Time at which damage starts *versus* time at which system automatically operates, and
   b. Time delays involving manually actuated suppression systems or manually applying extinguishing agent.

Figure 5-12.16 presents an example of a partial fault tree structure for a water-based fire suppression system.

Integration of Deterministic Fire Models

As identified at the bottom of Figure 5-12.16, in many cases deterministic fire models can be used to support conditional probabilities of the success or failure of existing fire systems or alternative design strategies.

The modeling of response times of detection and fire suppression systems can be used to support and validate engineering judgments concerning conditional probabilities of fire spread or control and potential property damage and human exposure.

Fire dynamics and modeling are discussed in several other chapters in this handbook and will not be elaborated on further in this chapter.

Sprinkler Performance Data

Table 5-12.7 provides a list of sprinkler performance failures based on NFPA data.[14] This type of data can be used to develop and support fault and event tree probabilities.

Manual Fire Suppression

Manual fire suppression problems can occur when:

1. Early manual suppression is not achieved by employee prior to critical damage, or

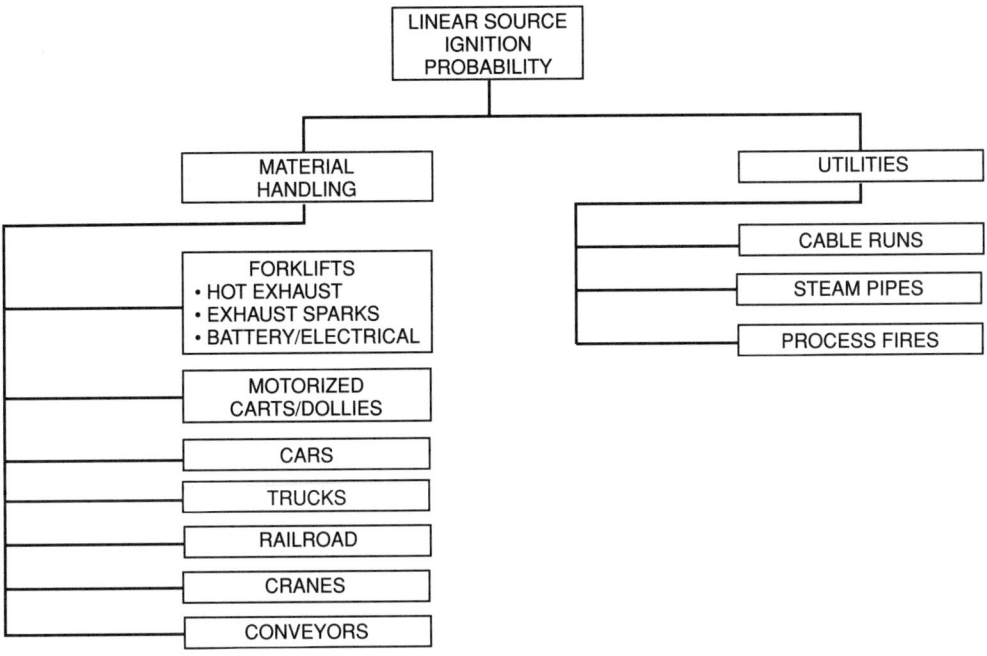

Fig. 5-12.13. Potential ignition—linear sources.

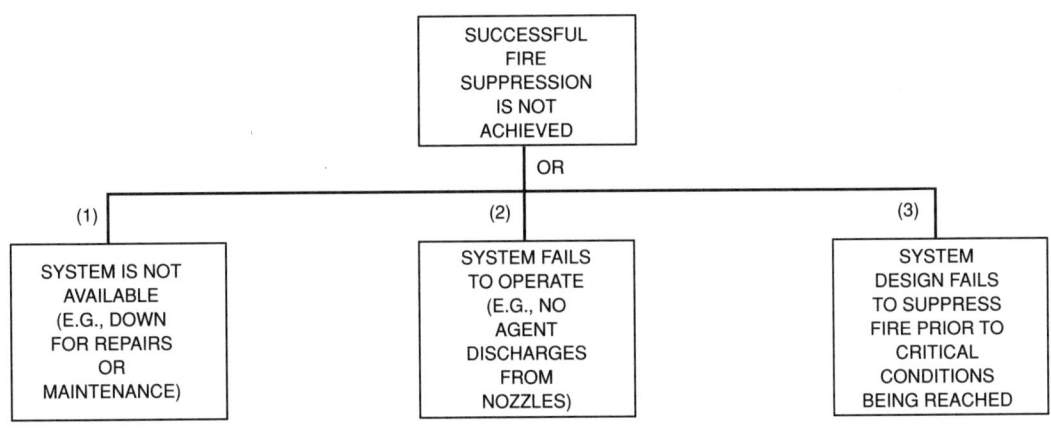

NOTE:
(1) SYSTEM IS NOT AVAILABLE
• THE FUNCTION OF TIME A SUPPRESSION SYSTEM IS NOT FULLY OPERATIONAL (DOWN FOR REPAIRS, MAINTENANCE, TESTING, ETC.)
(2) SYSTEM FAILS TO OPERATE
• NOT FUNCTIONAL AT ALL (EMERGENCY IMPAIRMENT; i.e., LOSS OF ELECTRICITY, SHUT VALVES)
• FAILING IN A DANGEROUS MODE (GENERALLY ASSOCIATED WITH MECHANICAL AND/OR ELECTRICAL COMPONENT FAILURES.)
(3) SYSTEM DESIGN FAILS TO SUPPRESS FIRE PRIOR TO CRITICAL CONDITIONS BEING REACHED
• SYSTEM FAILS TO RESPOND (AUTOMATICALLY ACTUATED, OR MANUALLY OPERATED) PRIOR TO THE TIME THAT CRITICAL HEAT EXPOSURE CONDITIONS MAY OCCUR. FIGURE 5-12.15 PROVIDES A GENERAL ILLUSTRATION OF THIS.
• DESIGN DENSITY OF EXTINGUISHING AGENT (FLOW RATE PER UNIT AREA) IS NOT SUFFICIENT.
• EXTINGUISHING AGENT COVERAGE (i.e., NOZZLE PLACEMENT) IS NOT ADEQUATE OR IS OBSTRUCTED.
• THE DURATIONAL SUPPLY OF AGENT (i.e., WATER SUPPLY) IS NOT SUFFICIENT.

Fig. 5-12.14. Fault tree analysis approach—fire suppression failure.

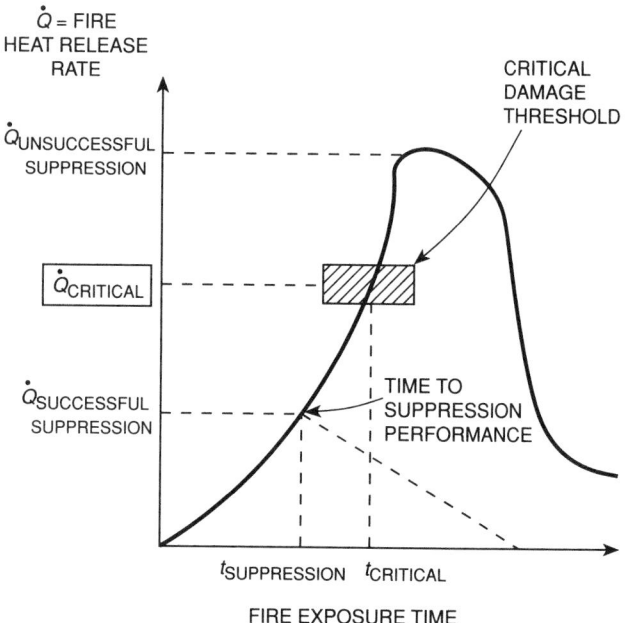

Fig. 5-12.15. System design fails to suppress fire prior to critical damage threshold being reached.

2. Successful manual suppression is not achieved by fire brigade/fire department prior to the occurrence of critical damage.

Figures 5-12.18 and 5-12.19 provide general fault tree structures for manual suppression.

Development of Probability Data

Developing probability data for protection system fault trees requires a large degree of engineering judgment.

A distinct advantage of applying an analytical fault tree approach is that it can be made highly facility or process specific. Another advantage is that it can indicate the relative likelihood of fire protection options as they affect the consequential outcomes and thus identify a framework for risk reduction by identifying the most beneficial improvement for optimal fire protection strategies. Again, it is essential to document the probability selection basis and uncertainty to support the credibility of the risk assessment process.

Consequence Probabilities

Following the assessment of the sequence of events leading to undesirable fire and explosion incidents (e.g., fire radiant heat, explosion overpressure, exposure zones, etc.) the consequential damage impact to selected target(s) must be probabilistically evaluated.

This evaluation involves the integration of the results (from the consequence assessment modeling, event tree frequency modeling, and the probabilistic assessment of the vulnerability of buildings, structures, and people) with the fire or explosive exposure in terms of distance to the fire or explosion source and the time exposed. The extent of detail can involve using damage category ranges similar to those used in risk screening and/or applying more rigorous engineering methods such as radiant heat transfer calculations where time may be a very influential factor (e.g., time to fire brigade response, time to people evacuation). In the majority of fire and explosion risk assessments, consequential impact analysis will require a combination of deterministic modeling and engineering judgment to predict probabilistic target vulnerability.

In basic terms, the general expressions for estimating primary risk concerns can be presented as shown in Figure 5-12.16. In most risk assessment, a major concern is exposure to humans from fire and explosion incidents; this element of risk presentation (i.e., individual risk, societal risk) is addressed in the next step.

TABLE 5-12.7 *Groups of Leading Reasons for Unsatisfactory Sprinkler Performance**

Problem Group	Percentage of Cases	Problem	Percentage of Cases
A. Failure to maintain operational status of system	53.4	A1. Water shutoff	35.4
		A2. Inadequate maintenance	8.4
		A3. Obstruction to water distribution	8.2
		A4. System frozen	1.4
B. Failure to ensure adequacy of system for complete coverage of current hazard	21.6	B1. System not adequate for level of hazard in occupancy	13.5
		B2. System designed for partial protection only	8.1
C. Defects affecting but not involving sprinkler system	15.9	C1. Inadequate water supply	15.9
		C2. Faulty building construction	
D. Inadequate performance by sprinkler system itself	5.6	D1. Antiquated system	2.1
		D2. Slow operation of sprinklers	1.8
		D3. Defective dry-pipe valve	1.7
E. Other	3.6	E1. Exposure fire	1.7
		E2. Other or unknown	1.9
Total	100.0		100.0

*Source: "Automatic Sprinkler Performance Tables," *Fire Journal*, July 1970, page 37. Based on 3,134 fires reported to NFPA during 1925 to 1969 for which sprinkler performance was deemed unsatisfactory. Of these, 75% were in industrial facilities, 12.0% were in storage facilities, 5.6% were in stores, and 7.4% were in other properties.

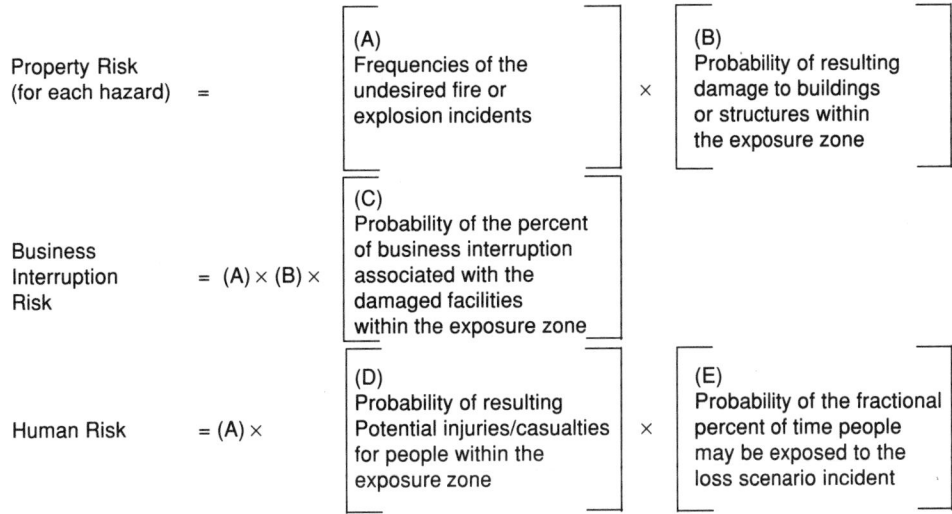

Fig. 5-12-16. General expressions for estimating primary risk concerns.

STEP 6:
RISK PRESENTATION

It is very important to integrate the large number of fire and explosion incident frequencies into a presentation format that will be easy to interpret and use by the management decision makers.

The most common risk assessment and presentation method is to multiply the frequency of fire and explosion incidents with the consequences and then sum these products for all scenarios considered in the risk assessment project. The aggregated results can be presented in terms of average exposure values; however, a better approach is to present the results as a range defined by upper and lower uncertainty bandwidths (confidence limits) that contain the best estimates that can be made.

The objectives of risk presentation are threefold:

1. Provide presentation of estimated risk results in terms of a graphical risk profile or risk contour plot to aid managements' understanding of the existing risk to the targets of interest as stated in the risk assessment objectives and acceptable risk limits.
2. Provide a graphical presentation of the differences in risk afforded by various risk reduction strategies to allow further cost/benefit assessment study, which may be a requirement of the specific risk assessment project.
3. Provide an uncertainty bandwidth (i.e., degree of confidence limits) associated with the above two items to allow management the opportunity to evaluate alternative risk management techniques (i.e., risk transfer by insurance). This is especially important for large consequence potentials which may be quantitatively classified as rare events that may also have a high degree of uncertainty in terms of the likelihood of occurrence.

The information provided in this step focuses on item 1 above, i.e., graphical presentation of risk results.

Risk Profiles

Developing a risk profile consists of structuring a graph that portrays the aggregate relationship between the ex-

pected frequencies of fire or explosion incidents and their consequences to selected targets. These graphs can be used to present property risk, business interruption risk, and human risk as illustrated in Figure 5-12.20.

Risk profiles can be used to present the potential degree of individual risk *versus* distance from the fire or explosive source. Individual risk is the probability of injury or death to a person at a specified location relative to a defined point within the fire or explosion incident effect zone. An example of an individual risk profile is shown in Figure 5-12.21.

Risk Contours

Individual risk exposure is presented in many risk assessment studies in terms of a risk contour plot. (See Figure 5-12.22.) Developing a risk contour (sometimes called a risk isopleth) consists of structuring a closed line graphical depiction connecting lines of constant potential risk. Points within the contour represent a risk greater than or equal to the risk of the contour edge. Risk contour plots provide a good way of illustrating individual risk *versus* distance from defined fire or explosion incidents.

Individual risk contours show the geographical distribution of individual risk. The risk contours show the expected frequency of fire or explosion loss events capable of causing a specified level of harm to an individual at a specified location, regardless of whether or not anyone is present at that location to suffer that harm. Reference 5 provides a good discussion on the various measures of individual risk, which includes maximum individual risk and average individual risk.

Societal Risk

Another measure of human risk is societal (sometimes called population) risk, which takes into account the likelihood of multiple casualties resulting from fire or explosion incidents. Societal risk is often presented in the form of F-N curves, which is a plot of the cumulative frequency (F) of multiple fatalities *versus* the number (N) of fatalities. Referring to Figure 5-12.23, the left axis is the frequency of exceeding a specified number of fatalities.[5,6]

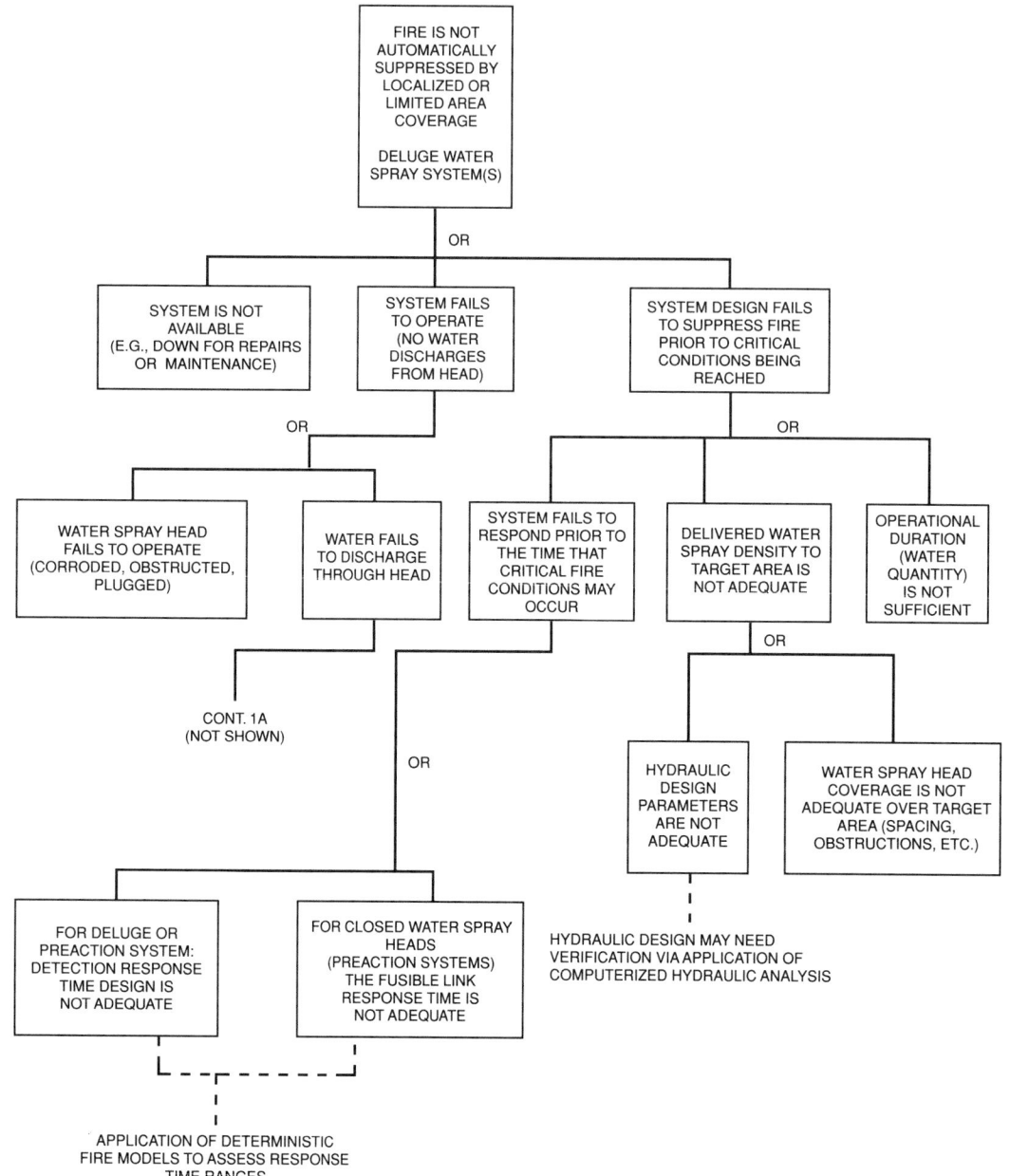

Fig. 5-12.17. Example of a partial fault tree structure.

F-N-type curves can provide useful insight into the degree of risks from a facility or hazardous process to the employees on the plant site and to the community located beyond the plant boundaries. Assessing the risk beyond the plant line boundaries requires a definition of the population at risk (industrial, residential, school, hospital, etc.), likelihood of people being present, and reliability of mitigation factors (people evacuation procedures, etc.).

In constructing F-N curves a logarithmic plot is usually used because the frequency of number of fatalities may range over several orders of magnitude. It is also sometimes useful to identify risk contributors, directly in the graph, as illustrated in Figure 5-12.23.

In addition, societal risks can be expressed in the form of various risk indices, which generally provide a summa-

tion of risks from each accident and an annual predicted fatality rate. Risk indices usually provide an easily understood, single-value number to present the acute risk and are sometimes quite useful in comparing various engineering design and protection options. A ranking of the events that contribute most to the total risk is also very useful, as this allows the analyst to focus attention on the most critical failures and facilitates efficiency in assessing prevention and mitigation risk reduction options for those events.

Reference 5 provides descriptions of various risk indices and includes:

- Fatal accident rate
- Individual hazard index
- Average rate of death

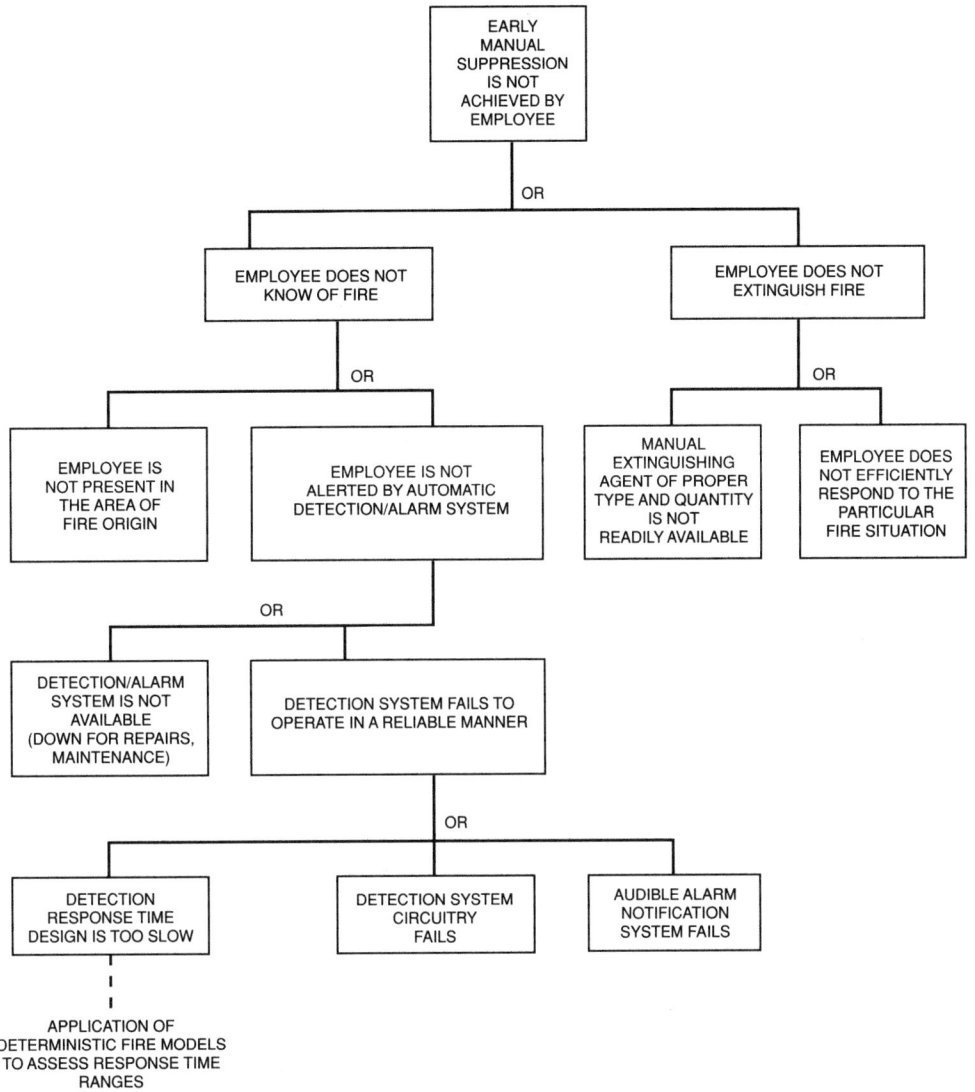

Fig. 5-12.18. Fault tree structuring: early manual suppression is not achieved by employee.

- Equivalent social cost index
- Mortality index
- Economic index.

The majority of human risk (i.e., individual risk, societal risk) quantification is conducted on the basis of fatality effects. In evaluating fatal effects, there are uncertainties involved in exactly what constitutes a fatal dose of thermal radiation from a fire, explosion blast effects, or toxic chemicals, and therefore appropriate documentation must be provided for all assumptions and judgments. The same applies when it is desired to estimate potential human injuries in addition to fatalities.

In conducting cost/risk reduction benefit studies, it may be necessary to convert the element of human risk exposure into equivalent monetary terms. Section 5, Chapter 8, provides information on this subject.

Presentation Format Selection

The specifics of the presentation format should be included in Step 1, i.e., define risk assessment objectives. (See Figure 5-12.1.) There is a major time and cost difference between generation of single-point estimates *versus* detailed generation of a series of risk contours. The following four factors should be considered in deciding which risk presentation forms are chosen:[5]

1. **User Requirements.** The user may have a specific need to see risk estimates in a certain format.
2. **User Knowledge.** Where the user is unfamiliar with presentation formats, sample formats should be presented to and approved by the user before any effort is made to secure approval for the scope of work.
3. **Effectiveness of Communicating Results.** It is vital that the presentation communicate the results in an acceptable fashion. The presentation should be as simple as necessary to ensure comprehension, but not so simple that resolution is lost or that bias is introduced.
4. **Need for Comparative Presentations.** It may be desirable to present comparisons of the results of a study with other risk assessments or risk data. This type of presentation might involve:

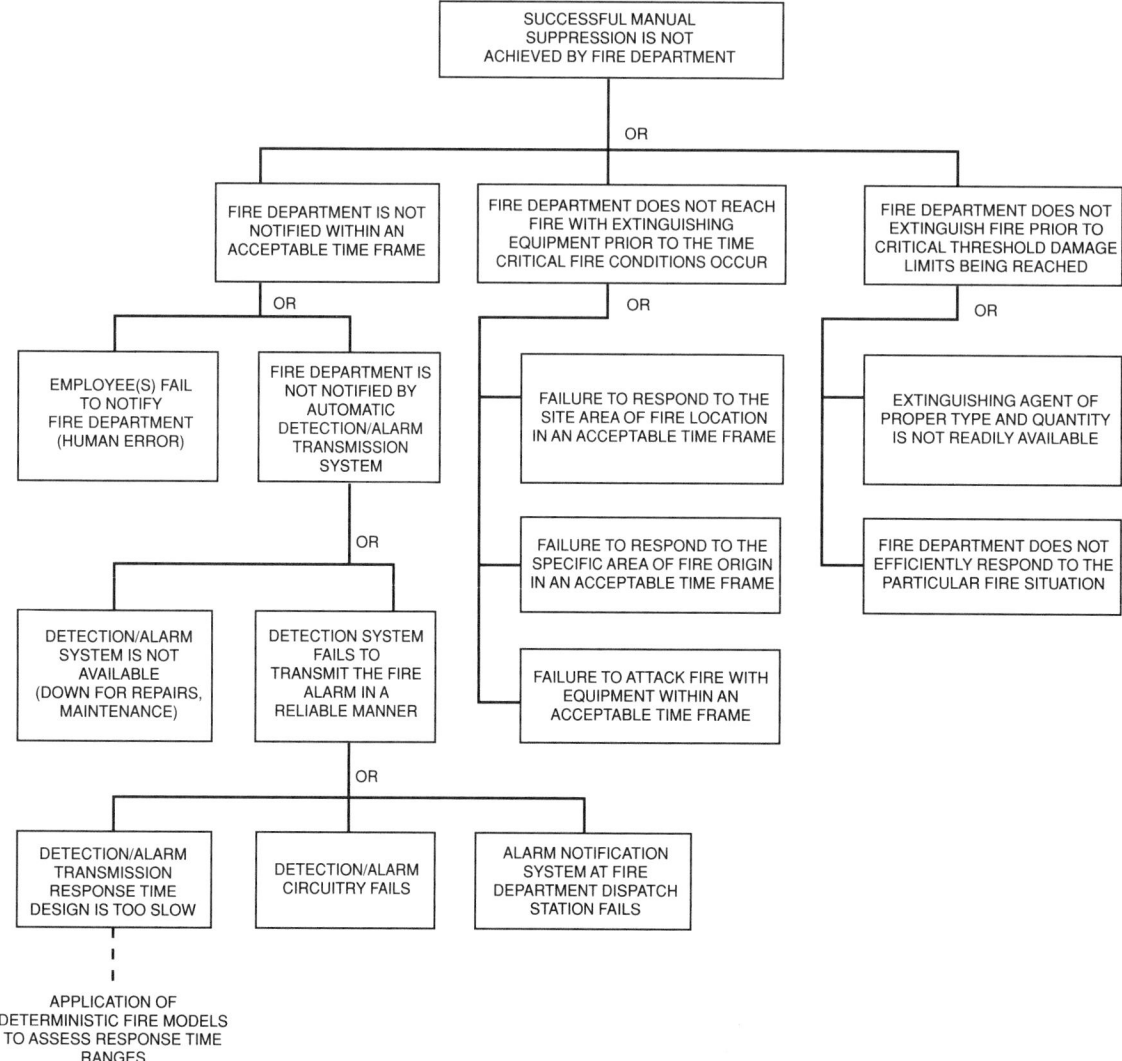

Fig. 5-12.19.   Fault tree structuring: successful manual suppression is not achieved by fire department.

a. A comparison of alternate design, operation, or protection options;
b. A comparison of the current risk estimates with risk estimates of other similar systems studied previously, to highlight areas for risk reduction or further study; and/or
c. A comparison of risk estimates with other voluntary and involuntary risks, to rank the current risk estimate among these reference values.

STEP 7: RISK REDUCTION ANALYSIS

If the risk is unacceptable or uncertain, then engineering risk reduction analysis can be conducted. *Risk Reduction can be defined as the identification and selection of cost-effective options for reducing or mitigating unacceptable risks, including technological measures such as fire protection systems and/or management safety programs such as* *mechanical integrity programs, loss prevention programs, operator training, and emergency procedures.*[5] Risk reduction analysis consists of systematic evaluation of measures to reduce the potential frequency and/or the severity of loss event occurrence.

Frequency Reduction

Incident frequency reduction methods include the following:

• Reduce equipment failure potential:
  a. design,
  b. testing, and
  c. preventive maintenance.
• Reduce ignition potential.
• Reduce human error:
  a. operator procedures and training, and
  b. process safety management programs.

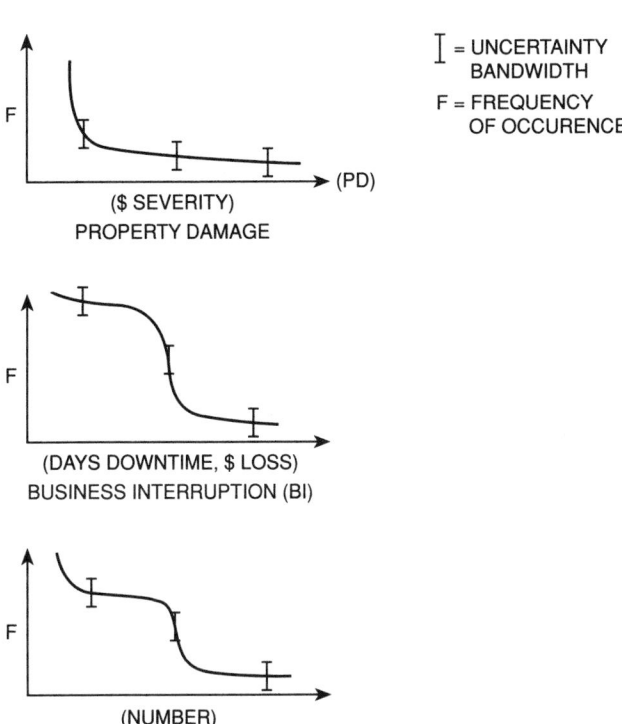

Fig. 5-12.20. *Example general risk profiles.*

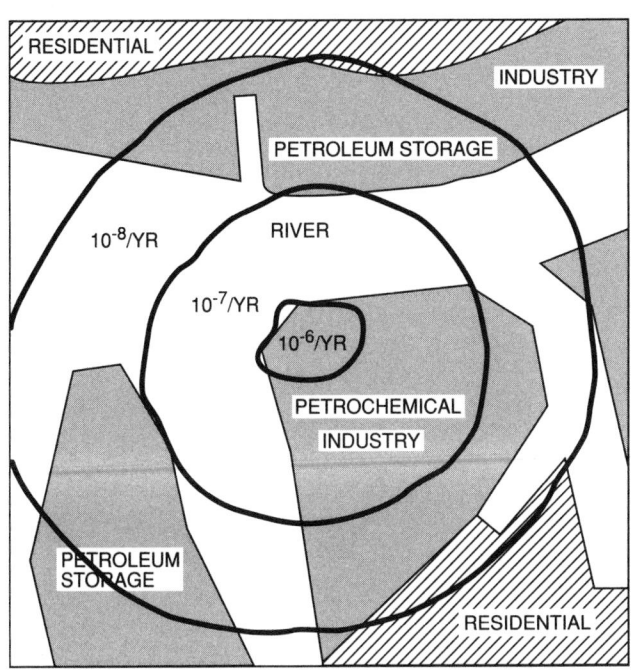

Fig. 5-12.22. *Example of an individual risk contour plot.*

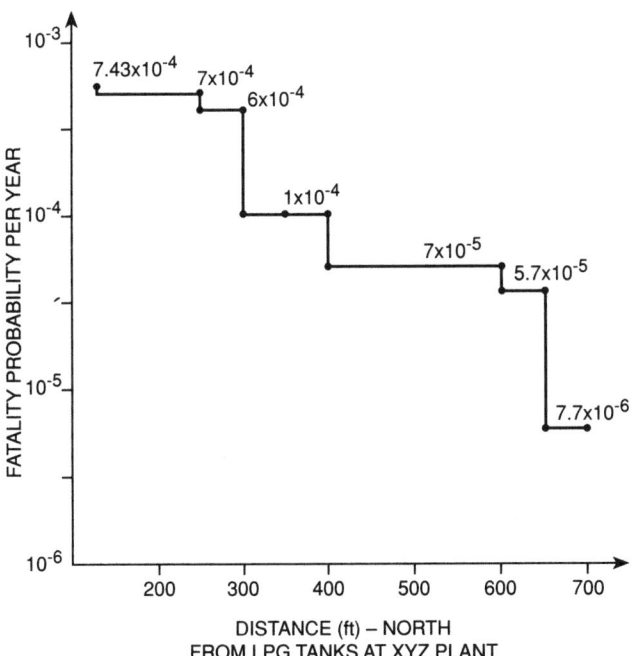

Fig. 5-12.21. *Example individual risk profile graph.*

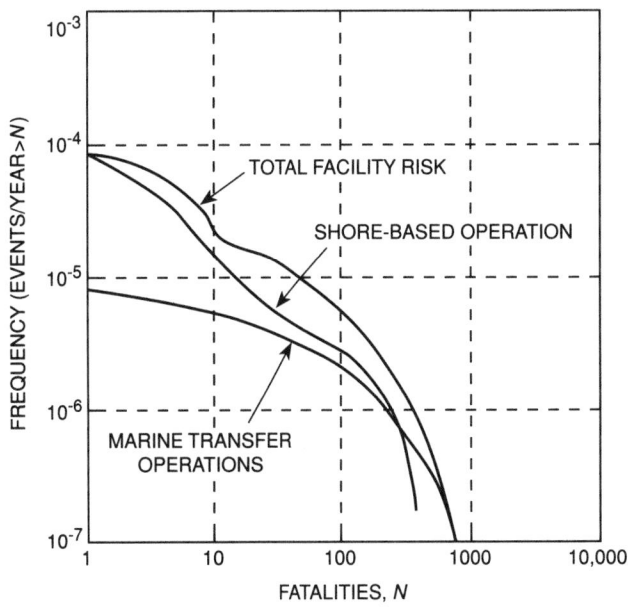

Fig. 5-12.23. *Example of a societal risk F-N curve.*

Severity Reduction

Fire detection and protection features that reduce potential severity levels can include:

- Gas and fire detection systems
- Fireproofing structural steel and vessel supports
- Providing automatic or manual water sprays
- Providing monitor nozzles or hand-hose standpipe protection
- Installing foam protection systems.

REFERENCES CITED

1. *System Safety Analysis Handbook*, System Safety Society, Albuquerque, NM (1993).
2. Center for Chemical Process Safety, *Guidelines for Hazard Evaluation Procedures with Examples*, 2nd ed., American Institute of Chemical Engineers, New York (1992).
3. 29 CFR 1910.119, *Process Safety Management of Highly Hazardous Chemicals*, Occupational Safety and Health Administration, Washington, DC (1992).
4. Technical, Ltd., *Techniques for Assessing Industrial Hazards*, World Bank Technical Paper No. 55, The World Bank, Washington, DC (1988).
5. Center for Chemical Process Safety, *Guidelines for Chemical Process Quantitative Risk Analysis*, American Institute of Chemical Engineers, New York (1989).
6. John Whithers, *Major Industrial Hazards: Their Appraisal and Control*, Halsted Press (1988).
7. *Handbook of Chemical Hazard Analysis Procedures* (ARCHIE Manual), Federal Emergency Management Agency, Washington, DC (1989).
8. *Chemical Engineering Progress, Software Directory of Hazard and Risk Computer Models*, published annually by the American Institute of Chemical Engineers, New York.
9. D.F. Haasl, *et al, Fault Tree Handbook*, USNRC, NUREG-0492, Washington, DC, Jan. 1981.
10. *PRA Procedures Guide: A Guide to the Performance of Probabilistic Risk Assessments for Nuclear Power Plants*, NUREG/OR 300, Jan. 1983.
11. D.H. Slater, R.M. Pitblado, and J.C. Williams, "Quantitative Assessment of Process Safety Programs," *Plant/Operator Progress*, Vol. 9, No. 3, AIChE, New York, July 1990.
12. Center for Chemical Process Safety, *Guidelines for Process Equipment Reliability Data, with Data Tables*, AIChE, New York (1989).
13. M. Considine, G.C. Grint, and P.L. Holden, "Bulk Storage of LPG—Factors Affecting Offsite Risk," *The Assessment of Major Hazards*, American Institute of Chemical Engineers, Pergamon Press, New York (1982).
14. John R. Hall, *U.S. Experience with Sprinklers*, Fire Analysis and Research Division Publication, National Fire Protection Association, Quincy, MA (1990).

ADDITIONAL READING

American Petroleum Institute, "Management of Process Hazards," Recommended Practice 750, 1st ed., Washington, DC (1990).

G.E. Apostolakis, *et al, Accident Sequence Modeling: Human Actions, System Response, Intelligent Decision Support*, Elsevier Science Publishing Co., Inc., New York (1988).

J. Bond, "The International Safety Rating System or the Five Star Audit System," Loss Prevention Bulletin No. 80, UK Institution of Chemical Engineers, Apr. 1988.

T.A. Braustowski, "Risk Assessment in Large Chemical Energy Projects," *Technological Risk, Proceedings of a Symposium on Risk in New Technologies, University of Waterloo, Dec. 1981*, University of Waterloo Press, Waterloo, Ontario (1982).

L. Bretherick, *Handbook of Reactive Chemical Hazards*, 4th ed., Butterworth's, London (1990).

Canvey: An Investigation, Health and Safety Executive, Her Majesty's Stationery Office, London (1978).

Canvey: A Second Report, Health and Safety Executive, Her Majesty's Stationery Office, London (1981).

M.L. Casada, *et al*, "Facility Risk Review as an Approach to Prioritizing Loss Prevention Efforts," *Chemical Engineering Progress* (1990).

Vincent T. Covello, *et al*, eds., *The Analysis of Actual Versus Perceived Risks*, Plenum Press, New York (1983).

Edmund A.C. Crouch and Richard Wilson, *Risk/Benefit Analysis*, Ballinger Publishing Company, Cambridge, MA (1982).

Dow Chemical Company, *Fire and Explosion Index Hazard Classification Guide*, 7th ed., American Institute of Chemical Engineers, New York (1989).

J.P. Drago, R.J. Borkowski, D.H. Pike, and F.F. Goldberg, "The In-Plant Reliability Data Base for Nuclear Power Plant Components: Data Collection and Methodology Report," NUREG/CR-2641, ORNL/TM-9216, Jan. 1985.

J. Gillett, "Rapid Ranking of Process Hazards," *Process Engineering*, Vol. 66, No. 219 (1985).

H.R. Greenberg and J.J. Cramer, eds., *Risk Assessment and Risk Management for the Chemical Process Industry*, ISBN 0-442-23438-4, Van Nostrand Reinhold, New York (1991).

Guide to the Collection and Representation of Electrical, Electronic, Sensing Component, and Mechanical Equipment Reliability Data for Nuclear Generating Stations, IEEE Std. 500-1984, Institute of Electrical and Electronic Engineers, New York (1984).

E.J. Henley and H. Kumamoto, *Reliability Engineering and Risk Assessment*, Prentice-Hall, Englewood Cliffs, NJ (1981).

W.G. Johnson, *MORT, The Management Oversight and Risk Tree*, U.S. Atomic Energy Commission, Washington, DC (1983).

Trevor A. Kletz, *An Engineer's View of Human Error*, The Institution of Chemical Engineers, Warwickshire, England (1983).

S.A. Lapp, "The Major Risk Index System," *Plant/Operations Progress*, Vol. 9, No. 3, July 1990.

F.P. Lees, *Loss Prevention in the Process Industries*, Butterworth's, London (1980).

Frederick A. Lercari, *RMPP, Risk Management and Prevention Program Guidance*, California Office of Emergency Services—Hazardous Materials Division.

D.J. Lewis, *Mond Fire, Explosion, and Toxicity Index: A Development of the Dow Index*, 13th Annual Loss Prevention Symposium, Houston, American Institute of Chemical Engineers, New York (1979).

D.K. Lorenzo, *A Manager's Guide to Reducing Human Errors—Improving Human Performance in the Chemical Industry*, Chemical Manufacturers Association, Washington, DC, July 1990.

D.R.T. Lowe, "Major Incident Criteria—What Risk Should Society Accept?," *I. Chem. E. Proceedings of Eurochem Symposium* (1980).

William W. Lowrance, *Of Acceptable Risk*, William Kaufmann, Inc., Los Altos, CA (1976).

A. Mazzocchi and M. Campona, "Risk Analysis of a Large Farm of Liquefied Petroleum Gases," *Reliability Data Collection and Use in Risk and Availability Assessment* (H.J. Wingender, ed.), *Proceedings* of the 5th Euredata Conference, Heidelberg, Germany, Springer-Verlag Press, New York, Apr. 1986.

K. Mudan and Richard Gustafson, "Ignition Potential Distribution for Heavy Gas Plumes," *International Conference on Vapor Cloud Modeling*, American Institute of Chemical Engineers, New York (1987).

NFPA *Fire Protection Handbook*, A.E. Cote, ed., 17th ed., National Fire Protection Association, Quincy, MA (1991).

NFPA 49, *Hazardous Chemicals Data*, National Fire Protection Association, Quincy, MA (1994).

NFPA 68, *Guide for Venting of Deflagrations*, National Fire Protection Association, Quincy, MA (1994).

NFPA 325, *Guide to Fire Hazard Properties of Flammable Liquids, Gases, and Volatile Solids*, National Fire Protection Association, Quincy, MA (1994).

NFPA 491M, *Manual of Hazardous Chemical Reactions*, National Fire Protection Association, Quincy, MA (1991).

Offshore Reliability Data Handbook, OREDA-84, P.O. Box 370, N-1322, Hovik, Norway, 1984; distributed by Pennwell Publishing Company, Tulsa, OK.

"Procedures for Performing a Failure Mode, Effects, and Criticality Analysis," MIL-STD-1629A, Department of Defense, Washington, DC, Nov. 1980.

Richard W. Prugh, "Evaluation of Unconfined Vapor Cloud Explosion Hazards," *International Conference on Vapor Cloud Modeling*, Cambridge, MA, American Institute of Chemical Engineers, New York (1987).

C.G. Ramsey, R. Evans, and M.A. English, "Siting and Layout of Major Hazardous Installation," *The Assessment of Major Hazards*, Institution of Chemical Engineers, Pergamon Press, New York (1982).

D.J. Rasbash, "Criteria for Acceptability for Use with Quantitative Approaches to Fire Safety, *Fire Safety Journal*, 8 (1984).

"Risk Analysis in the Process Industries," European Federation of Chemical Engineering Publication No. 45, The Institution of Chemical Engineers, Rugby, England (1985).

Risk Analysis of Six Potentially Hazardous Industrial Objects in the Rijnmond Area; A Pilot Study, D. Reidel Publishing Company, Dordrecht, Holland (1982).

William D. Rowe, *An Anatomy of Risk*, John Wiley and Sons, Inc., New York (1977).

John H. Shortreed and Angela Stewart, "Risk Assessment and Legislation," *Journal of Hazardous Materials*, 20, pp. 315–334 (1988).

State of California, Office of Emergency Services, *Guidance for the Preparation of a Risk Management and Prevention Program*, Sacramento, CA (1989).

J. Stephenson, *System Safety 2000—A Practical Guide for Planning, Managing, and Conducting System Safety Programs*, ISBN 0-442-23840-1, Van Nostrand Reinhold, New York (1991).

The Mond Index, 2nd ed., Imperial Chemical Industries, Winnington, Northwick, Cheshire, UK (1985).

Ray A. Waller and Vincent T. Covello, eds., *Low-Probability/High-Consequence Risk Analysis*, Plenum Press, New York (1984).

Appendices

CONTENTS

REFERENCES

Chemical Engineers Handbook, R.H. Perry and C.H. Chilton, eds., McGraw-Hill, New York (1973).

E.A. Mechtly, "The International System of Units, Physical Constants and Conversion Factors," 2nd revision, National Aeronautics and Space Administration, Washington, DC (1973). (No copyright.)

Eshbach's Handbook of Engineering Fundamentals, 4th ed., B.D. Tapley and T.R. Poston, eds., John Wiley and Sons, New York (1990).

E.R.G. Eckert and R.M. Drake, *Analysis of Heat and Mass Transfer*, McGraw-Hill, New York (1972).

Fire Protection Handbook, 17th ed., A.E. Cote and J. Linville, eds., National Fire Protection Association, Quincy, MA (1991).

Heat Transfer in Fires, P. Blackshear, ed., Scripta Book Company, Washington, DC, 1974 (Out of Print).

J.P. Holman, *Heat Transfer*, McGraw-Hill, New York (1986).

A.M. Kanury, *Introduction to Combustion Phenomena*, Gordon and Breach Science Publishers, New York (1975).

R. Siegel and J.R. Howell, *Thermal Radiation Heat Transfer*, 3rd ed., Taylor & Francis, Washington, DC (1992).

A-1 NAMES AND SYMBOLS OF SI UNITS*

Quantity	Name of Unit	Symbol	
SI BASE UNITS			
Length	meter	m	
Mass	kilogram	kg	
Time	second	s	
Electric current	ampere	A	
Thermodynamic temperature	kelvin	K	
Luminous intensity	candela	cd	
Amount of substance	mole	mol	
SI-DERIVED UNITS			
Area	square meter	m^2	
Volume	cubic meter	m^3	
Frequency	hertz	Hz	s^{-1}
Mass density (density)	kilogram per cubic meter	kg/m^3	
Speed, velocity	meter per second	m/s	
Angular velocity	radian per second	rad/s	
Acceleration	meter per second squared	m/s^2	
Angular acceleration	radian per second squared	rad/s^2	
Force	newton	N	$kg \cdot m/s^2$
Pressure (mechanical stress)	pascal	Pa	N/m^2
Kinematic viscosity	square meter per second	m^2/s	
Dynamic viscosity	newton-second per square meter	$N \cdot s/m^2$	
Work, energy, quantity of heat	joule	J	$N \cdot m$
Power	watt	W	J/s
Quantity of electricity	coulomb	C	$A \cdot s$
Potential difference, electromotive force	volt	V	W/A
Electric field strength	volt per meter	V/m	
Electric resistance	ohm	Ω	V/A
Capacitance	farad	F	$A \cdot s/V$
Magnetic flux	weber	Wb	$V \cdot s$
Inductance	henry	H	$V \cdot s/A$
Magnetic flux density	tesla	T	Wb/m^2
Magnetic field strength	ampere per meter	A/m	
Magnetomotive force	ampere	A	
Luminous flux	lumen	lm	$cd \cdot sr$
Luminance	candela per square meter	cd/m^2	
Illuminance	lux	lx	lm/m^2
Wave number	1 per meter	m^{-1}	
Entropy	joule per kelvin	J/K	
Specific heat capacity	joule per kilogram kelvin	$J/(kg \cdot K)$	
Thermal conductivity	watt per meter kelvin	$W/(m \cdot K)$	
Radiant intensity	watt per steradian	W/sr	
Activity (or a radioactive source)	1 per second	s^{-1}	
SI SUPPLEMENTARY UNITS			
Plane angle	radian	rad	
Solid angle	steradian	sr	

*Source: E.A. Mechtly, "The International System of Units, Physical Constants and Conversion Factors," 2nd revision, National Aeronautics and Space Administration, Washington, DC (1973). (No copyright.)

A-2 DEFINITIONS OF SI UNITS*

English	French

meter (m)

The *meter* is the length equal to 1 650 763.73 wavelengths in vacuum of the radiation corresponding to the transition between the levels 2 p_{10} and 5 d_s of the krypton-86 atom.

kilogram (kg)

The *kilogram* is the unit of mass; it is equal to the mass of the international prototype of the kilogram. (The international prototype of the kilogram is a particular cylinder of platinum-iridium alloy which is preserved in a vault at Sèvres, France, by the International Bureau of Weights and Measures.)

second (s)

The *second* is the duration of 9 192 631 770 periods of the radiation corresponding to the transition between the two hyperfine levels of the ground state of the cesium-133 atom.

ampere (A)

The *ampere* is that constant current which, if maintained in two straight parallel conductors of infinite length, of negligible circular cross section, and placed 1 meter apart in vacuum, would produce between these conductors a force equal to 2×10^{-7} newton per meter of length.

kelvin (K)

The *kelvin*, unit of thermodynamic temperature, is the fraction 1/273.16 of the thermodynamic temperature of the triple point of water.

candela (cd)

The *candela* is the luminous intensity, in the perpendicular direction, of a surface of 1/600 000 square meter of a blackbody at the temperature of freezing platinum under a pressure of 101 325 newtons per square meter.

mole (mol)

The *mole* is the amount of substance of a system which contains as many elementary entities as there are carbon atoms in 0.012 kg of carbon 12. The elementary entities must be specified and may be atoms, molecules, ions, electrons, other particles, or specified groups of such particles.

newton (N)

The *newton* is that force which gives to a mass of 1 kilogram an acceleration of 1 meter per second per second.

joule (J)

The *joule* is the work done when the point of application of 1 newton is displaced a distance of 1 meter in the direction of the force.

---

mètre (m)

Le *mètre* est la longueur égale à 1 650 763,73 longueurs d'onde dans le vide de la radiation correspondant à la transition entre les niveaux 2 p_{10} et 5 d_5 de l'atome krypton 86.

kilogramme (kg)

Le *kilogramme* est l'unité de masse; il est égal à la masse du prototype international du kilogramme.

seconde (s)

La *seconde* est la durée de 9 192 631 770 périodes de la radiation correspondant à la transition entre les deux niveaux hyperfins de l'état fondamental de l'atome de césium 133.

ampère (A)

L'*ampère* est l'intensité d'un courant constant qui, maintenu dans deux conducteurs parallèles, rectilignes, de longueur infinie, de section circulaire négligeable et placés à une distance de 1 mètre l'un de l'autre dans le vide, produirait entre ces conducteurs une force égale à 2×10^{-7} newton par mètre de longueur.

kelvin (K)

Le *kelvin*, unité de température thermodynamique, est la fraction 1/273,16 de la température thermodynamique du point triple de l'eau.

candela (cd)

La *candela* est l'intensité lumineuse, dans la direction perpendiculaire, d'une surface de 1/600 000 mètre carré d'un corps noir à la température de congélation du platine sous la pression de 101 325 newtons par mètre carré.

mole (mol)

La *mole* est la quantité de matière d'un système contenant autant d'entités élémentaires qu'il y a d'atomes dans 0,012 kg de carbone 12. Les entités élémentaires doivent être spécifiées et peuvent être des atomes, des molécules, des ions, des électrons, d'autres particules ou des groupements spécifiés de telles particules.

newton (N)

Le *newton* est la force qui communique à une masse de 1 kilogramme l'accélération de 1 mètre par seconde, par seconde.

joule (J)

Le *joule* est la travail effectué lorsque le point d'application de 1 newton de force se déplace d'une distance égale à 1 mètre dans la direction de la force.

English

watt (W)

The *watt* is the power which gives rise to the production of energy at the rate of 1 joule per second.

volt (V)

The *volt* is the difference of electric potential between two points of a conducting wire carrying a constant current of 1 ampere, when the power dissipated between these points is equal to 1 watt.

ohm (Ω)

The *ohm* is the electric resistance between two points of a conductor when a constant difference of potential of 1 volt, applied between these two points, produces in this conductor a current of 1 ampere, this conductor not being the source of any electromotive force.

coulomb (C)

The *coulomb* is the quantity of electricity transported in 1 second by a current of 1 ampere.

farad (F)

The *farad* is the capacitance of a capacitor between the plates of which there appears a difference of potential of 1 volt when it is charged by a quantity of electricity equal to 1 coulomb.

henry (H)

The *henry* is the inductance of a closed circuit in which an electromotive force of 1 volt is produced when the electric current in the circuit varies uniformly at a rate of 1 ampere per second.

weber (Wb)

The *weber* is the magnetic flux which, linking a circuit of one turn, produces in it an electromotive force of 1 volt as it is reduced to zero at a uniform rate in 1 second.

lumen (lm)

The *lumen* is the luminous flux emitted in a solid angle of 1 steradian by a uniform point source having an intensity of 1 candela.

radian (rad)

The *radian* is the plane angle between two radii of a circle which cut off on the circumference an arc equal in length to the radius.

steradian (sr)

The *steradian* is the solid angle which, having its vertex in the center of a sphere, cuts off an area of the surface of the sphere equal to that of a square with sides of length equal to the radius of the sphere.

French

watt (W)

Le *watt* est la puissance qui donne lieu à une production d'énergie égale à 1 joule par seconde.

volt (V)

Le *volt* est la différence de potentiel électrique qui existe entre deux points d'un fil conducteur transportant un courant constant de 1 ampère, lorsque la puissance dissipée entre ces points est égale à 1 watt.

ohm (Ω)

L'*ohm* est la résistance électrique qui existe entre deux points d'un conducteur lorsqu'une différence de potentiel constante de 1 volt, appliquée entre ces deux points, produit, dans ce conducteur, un courant de 1 ampère, ce conducteur n'étant le siège d'aucune force électromotrice.

coulomb (C)

Le *coulomb* est la quantité d'électricité transportée en 1 seconde par un courant de 1 ampère.

farad (F)

Le *farad* est la capacité d'un condensateur électrique entre les armatures duquel apparaît une différence de potentiel électrique de 1 volt, lorsqu'il est chargé d'une quantité d'électricité égale à 1 coulomb.

henry (H)

Le *henry* est l'inductance électrique d'un circuit fermé dans lequel une force électromotrice de 1 volt est produite lorsque le courant électrique qui parcourt le circuit varie uniformément à raison de 1 ampère par seconde.

weber (Wb)

Le *weber* est le flux magnétique qui, traversant un circuit d'une seule spire, y produirait une force électromotrice de 1 volt, si on l'amenait à zéro en 1 seconde par décroissance uniforme.

lumen (lm)

Le *lumen* est le flux lumineux émis dans l'angle solide unité (stéradian), par une source ponctuelle uniforme ayant une intensité lumineuse de 1 candela.

radian (rad)

Le *radian* est l'angle plan compris entre deux rayons qui, sur la circonférence d'un cercle, interceptent un arc de longueur égale à celle du rayon.

stéradian (sr)

Le *stéradian* est l'angle solide qui, ayant son sommet au centre d'une sphère, découpe sur la surface de cette sphère une aire égale à celle d'un carré ayant pour côté le rayon de la sphère.

A-3 SI PREFIXES*

The names of multiples and submultiples of SI units can be formed by application of the following prefixes:

Factor by which unit is multiplied	Prefix	Symbol
10^{12}	tera	T
10^{9}	giga	G
10^{6}	mega	M
10^{3}	kilo	k
10^{2}	hecto	h
10	deka	da
10^{-1}	deci	d
10^{-2}	centi	c
10^{-3}	milli	m
10^{-6}	micro	μ
10^{-9}	nano	n
10^{-12}	pico	p
10^{-15}	femto	f
10^{-18}	atto	a

The International Organization for Standardization (ISO) recommends the following rules for the use of SI prefixes:

1. Prefix symbols are printed in roman (upright) type without spacing between the prefix symbol and the unit symbol.
2. An exponent affixed to a symbol containing a prefix indicates that the multiple or submultiple of the unit is raised to the power expressed by the exponent.

$$\textit{Example:} \quad 1 \text{ cm}^3 = 10^{-6} \text{ m}^3$$
$$1 \text{ cm}^{-1} = 10^2 \text{ m}^{-1}$$

3. Compound prefixes, formed by the juxtaposition of two or more SI prefixes, are not to be used.

$$\textit{Example:} \quad 1 \text{ nm} \qquad \textit{but not:} \text{ 1 mμm}$$

ISO has issued additional recommendations with the aim of securing uniformity in the use of units.

According to these recommendations:

1. The product of two or more units is preferably indicated by a dot. The dot may be dispensed with when there is no risk of confusion with another unit symbol.

$$\textit{Example:} \quad \text{N} \cdot \text{m or N m} \qquad \textit{but not:} \text{ mN}$$

2. A solidus (oblique stroke, /), a horizontal line, or negative powers may be used to express a derived unit formed from two others by division.

$$\textit{Example:} \text{ m/s, } \frac{\text{m}}{\text{s}} \text{, or m} \cdot \text{s}^{-1}$$

3. The solidus must not be repeated on the same line unless ambiguity is avoided by parentheses. In complicated cases negative powers or parentheses should be used.

$$\textit{Example:} \quad \text{m/s}^2 \quad \text{or} \quad \text{m} \cdot \text{s}^{-2} \qquad \textit{but not:} \text{ m/s/s}$$
$$\text{m} \cdot \text{kg/(s}^3 \cdot \text{A) or m} \cdot \text{kg} \cdot \text{s}^{-3} \cdot \text{A}^{-1} \qquad \textit{but not:} \text{ m} \cdot \text{kg/s}^3\text{/A}$$

A-4   PHYSICAL CONSTANTS*

The following lists of physical constants are from the work of B. N. Taylor, W. H. Parker, and D. N. Langenberg (*Reviews of Modern Physics*, July 1969). Their least-squares adjustment of values of the constants depends strongly on a highly accurate (2.4 ppm) determination of e/h from the ac Josephson effect in superconductors, and is believed to be more accurate than the 1963 adjustment, which appears to suffer from the use of an incorrect value of the fine structure constant as an input datum. See also NBS Special Publication 344, issued March 1971.

Quantity	Symbol	Value	Error (ppm)	Prefix	Unit
Speed of light in vacuum	c	2. 997 925 0	0.33	$\times 10^8$	m s^{-1}
Gravitational constant	G	6. 673 2	460	10^{-11}	N m^2 kg^{-2}
Avogadro constant	N_A	6. 022 169	6.6	10^{26}	kmol^{-1}
Boltzmann constant	k	1. 380 622	43	10^{-23}	J K^{-1}
Gas constant	R	8. 314 34	42	10^3	J kmol^{-1} K^{-1}
Volume of ideal gas, standard conditions	V_0	2. 241 36	—	10^1	m^3 kmol^{-1}
Farady constant	F	9. 648 670	5.5	10^7	C kmol^{-1}
Unified atomic mass unit	u	1. 660 531	6.6	10^{-27}	kg
Planck constant	h	6. 626 196	7.6	10^{-34}	J s
	$h/2\pi$	1. 054 591 9	7.6	10^{-34}	J s
Electron charge	e	1. 602 191 7	4.4	10^{-19}	C
Electron rest mass	m_e	9. 109 558	6.0	10^{-31}	kg
		5. 485 930	6.2	10^{-4}	u
Proton rest mass	m_p	1. 672 614	6.6	10^{-27}	kg
		1. 007 276 61	0.08	—	u
Neutron rest mass	m_n	1. 674 920	6.6	10^{-27}	kg
		1. 008 665 20	0.10	—	u
Electron charge to mass ratio	e/m_e	1. 758 802 8	3.1	10^{11}	C kg^{-1}
Stefan-Boltzmann constant	σ	5. 669 61	170	10^{-8}	W m^{-2} K^{-4}
First radiation constant	$2\pi hc^2$	3. 741 844	7.6	10^{-16}	W m^2
Second radiation constant	hc/k	1. 438 833	43	10^{-2}	m K
Rydberg constant	R_∞	1. 097 373 12	0.10	10^7	m^{-1}
Fine structure constant	α	7. 297 351	1.5	10^{-3}	
	α^{-1}	1. 370 360 2	1.5	10^2	
Bohr radius	a_0	5. 291 771 5	1.5	10^{-11}	m
Classical electron radius	r_e	2. 817 939	4.6	10^{-15}	m
Compton wavelength of electron	λ_C	2. 426 309 6	3.1	10^{-12}	m
	$\lambda_C/2\pi$	3. 861 592	3.1	10^{-13}	m
Compton wavelength of proton	$\lambda_{C,p}$	1. 321 440 9	6.8	10^{-15}	m
	$\lambda_{C,p}/2\pi$	2. 103 139	6.8	10^{-16}	m
Compton wavelength of neutron	$\lambda_{C,n}$	1. 319 621 7	6.8	10^{-15}	m
	$\lambda_{C,n}/2\pi$	2. 100 243	6.8	10^{-16}	m
Electron magnetic moment	μ_e	9. 284 851	7.0	10^{-24}	J T^{-1}
Proton magnetic moment	μ_p	1. 410 620 3	7.0	10^{-26}	J T^{-1}
Bohr magneton	μ_B	9. 274 096	7.0	10^{-24}	J T^{-1}
Nuclear magneton	μ_n	5. 050 951	10	10^{-27}	J T^{-1}
Gyromagnetic ratio of protons in H$_2$O	γ'_p	2. 675 127 0	3.1	10^8	rad s^{-1}T^{-1}
	$\gamma'_p/2\pi$	4. 257 597	3.1	10^7	Hz T^{-1}
Gyromagnetic ratio of protons in H$_2$O	γ_p	2. 675 196 5	3.1	10^8	rad s^{-1} T^{-1}
corrected for diamagnetism of H$_2$O	$\gamma_p/2\pi$	4. 257 707	3.1	10^7	Hz T^{-1}
Magnetic flux quantum	Φ_0	2. 067 853 8	3.3	10^{-15}	Wb
Quantum of circulation	$h/2m_e$	3. 636 947	3.1	10^{-4}	J s kg^{-1}
	h/m_e	7. 273 894	3.1	10^{-4}	J s kg^{-1}

Unitless Numerical Ratios		Value	Error (ppm)	Prefix
(c^2)	kg/eV	5. 609 538	4.4	10^{35}
(c^2)	u/eV	9. 314 812	5.5	10^8
	u/kg	1. 660 531	6.6	10^{-27}
(c^2)	m_c/eV	5. 110 041	3.1	10^5
(c^2)	m_p/eV	9. 382 592	5.5	10^8
(c^2)	m_n/eV	9. 395 527	5.5	10^8
	eV/J	1. 602 191 7	4.4	10^{-19}
(h^{-1})	eV/Hz	2. 417 965 9	3.3	10^{14}
$(hc)^{-1}$	eVm	8. 065 465	3.3	10^5
(k^{-1})	eV/K	1. 160 485	42	10^4
(hc)	(eV m)$^{-1}$	1. 239 854 1	3.3	10^{-6}
(hc)	R_∞/J	2. 179 914	7.6	10^{-18}
(hc)	R_∞/eV	1. 360 582 6	3.3	10^1
(c)	R_∞/Hz	3. 289 842 3	0.35	10^{15}
(hc/k)	R_∞/K	1. 578 936	43	10^5
	m_p/m_e	1. 836 109	6.2	10^3
	μ_e/μ_B	1. 001 159 638 9	0.0031	
	μ'_p/μ_B	1. 520 993 12	0.066	10^{-3}
	μ_p/μ_B	1. 521 032 64	0.30	10^{-3}
	μ'_p/μ_n	2. 792 709	6.2	
	μ_p/μ_n	2. 792 782	6.2	

Other Important Constants

π = 3.141 592 653 589

e = 2.718 281 828 459

μ_0 = $4\pi \times 10^{-7}$ H/m (exact), permeability of free space

   = $1.256\ 637\ 061 \times 10^{-6}$ H/m

ε_0 = $\mu_0^{-1}c^{-2}$ F/m, permittivity of free space

   = $8.854\ 185 \times 10^{-12}$ F/m

A-5 CONVERSION FACTORS*

The following tables express the definitions of miscellaneous units of measure as exact numerical multiples of coherent SI units, and provide multiplying factors for converting numbers and miscellaneous units to corresponding new numbers and SI units.

The first two digits of each numerical entry represents a power of 10. An asterisk following a number expresses an exact definition. For example, the entry " $-02\ 2.54*$ " expresses the fact that 1 inch = 2.54×10^{-2} meter, exactly, by definition. Most of the definitions are extracted from National Bureau of Standards (NBS) documents. Numbers not followed by an asterisk are only approximate representations of definitions, or are the results of physical measurements.

The conversion factors are listed alphabetically and by physical quantity.

The listing by physical quantity (see A-5.2) includes only relationships that are frequently encountered, and deliberately omits the great multiplicity of combinations of units that are used for more specialized purposes. Conversion factors for combinations of units are easily generated from numbers given in the alphabetical listing (see A-5.1) by the technique of direct substitution or by other well-known rules for manipulating units. These rules are adequately discussed in many science and engineering textbooks and are not repeated here.

A-5.1 Alphabetical Listing

To Convert from	to	Multiply by
abampere	ampere	+01 1.00*
abcoulomb	coulomb	+01 1.00*
abfarad	farad	+09 1.00*
abhenry	henry	−09 1.00*
abmho	siemens	+09 1.00*
abohm	ohm	−09 1.00*
abvolt	volt	−08 1.00*
acre	meter^2	+03 4.046 856 422 4*
angstrom	meter	−10 1.00*
are	meter^2	+02 1.00*
astronomical unit (IAU)	meter	+11 1.496 00
astronomical unit (radio)	meter	+11 1.495 978 9
atmosphere	newton/meter^2	+05 1.013 25*
bar	newton/meter^2	+05 1.00*
barn	meter^2	−28 1.00*
barrel (petroleum, 42 gallons)	meter^3	−01 1.589 873
barye	newton/meter^2	−01 1.00*
board foot (1′ × 1′ × 1″)	meter^3	−03 2.359 737 216*
British thermal unit:		
(IST before 1956)	joule	+03 1.055 04
(IST after 1956)	joule	+03 1.055 056
British thermal unit (mean)	joule	+03 1.055 87
British thermal unit (thermochemical)	joule	+03 1.054 350
British thermal unit (39°F)	joule	+03 1.059 67
British thermal unit (60°F)	joule	+03 1.054 68
bushel (U.S.)	meter^3	−02 3.523 907 016 688*
cable	meter	+02 2.194 56*
caliber	meter	−04 2.54*
calorie (International Steam Table)	joule	+00 4.1868
calorie (mean)	joule	+00 4.190 02
calorie (thermochemical)	joule	+00 4.184*
calorie (15°C)	joule	+00 4.185 80
calorie (20°C)	joule	+00 4.181 90
calorie (kilogram, International Steam Table)	joule	+03 4.1868
calorie (kilogram, mean)	joule	+03 4.190 02
calorie (kilogram, thermochemical)	joule	+03 4.184*
carat (metric)	kilogram	−04 2.00*
Celsius (temperature)	kelvin	$t_K = t_c + 273.15$
centimeter of mercury (0°C)	newton/meter^2	+03 1.333 22
centimeter of water (4°C)	newton/meter^2	+01 9.806 38
chain (engineer or ramden)	meter	+01 3.048*
chain (surveyor or gunter)	meter	+01 2.011 68*
circular mil	meter^2	−10 5.067 074 8
cord	meter^3	+00 3.624 556 3
cubit	meter	−01 4.572*
cup	meter^3	−04 2.365 882 365*
curie	disintegration/second	+10 3.70*

*Source: E.A. Mechtly, "The International System of Units, Physical Constants and Conversion Factors," 2nd revision, National Aeronautics and Space Administration, Washington, DC (1973). (No copyright.)

To Convert from	to	Multiply by
day (mean solar)	second (mean solar)	+04 8.64*
day (sidereal)	second (mean solar)	+04 8.616 409 0
degree (angle)	radian	−02 1.745 329 251 994 3
denier (international)	kilogram/meter	−07 1.00*
dram (avoirdupois)	kilogram	−03 1.771 845 195 312 5*
dram (troy or apothecary)	kilogram	−03 3.887 934 6*
dram (U.S. fluid)	meter^3	−06 3.696 691 195 312 5*
dyne	newton	−05 1.00*
electron volt	joule	−19 1.602 191 7
erg	joule	−07 1.00*
Fahrenheit (temperature)	kelvin	$t_K = (5/9)(t_F + 459.67)$
Fahrenheit (temperature)	Celsius	$t_C = (5/9)(t_F - 32)$
faraday (based on carbon 12)	coulomb	+04 9.68 70
faraday (chemical)	coulomb	+04 9.649 57
faraday (physical)	coulomb	+04 9.652 19
fathom	meter	+00 1.828 8*
fermi (femtometer)	meter	+15 1.00*
fluid ounce (U.S.)	meter^3	−05 2.957 352 956 25*
foot	meter	−01 3.048*
foot (U.S. survey)	meter	+00 1200/3937*
foot (U.S. survey)	meter	−01 3.048 006 096
foot of water (39.2°F)	newton/meter^2	+03 2.988 98
footcandle	lumen/meter^2	+01 1.076 391 0
footlambert	candela/meter^2	+00 3.426 259
free fall, standard	meter/second^2	+00 9.806 65*
furlong	meter	+02 2.011 68*
gal (galileo)	meter/second^2	−02 1.00*
gallon (U.K. liquid)	meter^3	−03 4.546 087
gallon (U.S. dry)	meter^3	−03 4.404 883 770 86*
gallon (U.S. liquid)	meter^3	−03 3.785 411 784*
gamma	tesla	−09 1.00*
gauss	tesla	−04 1.00*
gilbert	ampere turn	−01 7.957 747 2
gill (U.K.)	meter^3	−04 1.420 652
gill (U.S.)	meter^3	−04 1.182 941 2
grad	degree (angular)	−01 9.00*
grad	radian	−02 1.570 796 3
grain	kilogram	−05 6.479 891*
gram	kilogram	−03 1.00*
hand	meter	−01 1.016*
hectare	meter^3	+04 1.00*
hogshead (U.S.)	meter^3	−01 2.384 809 423 92*
horsepower (550 foot lbf/second)	watt	+02 7.456 998 7
horsepower (boiler)	watt	+03 9.809 50
horsepower (electric)	watt	+02 7.46*
horsepower (metric)	watt	+02 7.354 99
horsepower (U.K.)	watt	+02 7.457
horsepower (water)	watt	+02 7.460 43
hour (mean solar)	second (mean solar)	+03 3.60*
hour (sidereal)	second (mean solar)	+03 3.590 170 4
hundredweight (long)	kilogram	+01 5.080 234 544*
hundredweight (short)	kilogram	+01 4.535 923 7*
inch	meter	−02 2.54*
inch of mercury (32°F)	newton/meter^2	+03 3.386 389
inch of mercury (60°F)	newton/meter^2	+03 3.376 85
inch of water (39.2°F)	newton/meter^2	+02 2.490 82
inch of water (60°F)	newton/meter^2	+02 2.4884
kayser	1/meter	+02 1.00*
kilocalorie (International Steam Table)	joule	+03 4.186 8
kilocalorie (mean)	joule	+03 4.190 02
kilocalorie (thermochemical)	joule	+03 4.184*
klilogram mass	kilogram	+00 1.00*
kilogram force (kgf)	newton	+00 9.806 65*
kilopound force	newton	+00 9.806 65*

To Convert from	to	Multiply by
kip	newton	+03 4.448 221 615 260 5*
knot (international)	meter/second	−01 5.144 444 444
lambert	candela/meter^2	+04 1/π*
lambert	candela/meter^2	+03 3.183 098 8
langley	joule/meter^2	+04 4.184*
lbf (pound force, avoirdupois)	newton	+00 4.448 221 615 260 5*
lbm (pound mass, avoirdupois)	kilogram	−01 4.535 923 7*
league (U.K. nautical)	meter	+03 5.559 552*
league (international nautical)	meter	+03 5.556*
league (statute)	meter	+03 4.828 032*
light year	meter	+15 9.460 55
link (engineer or ramden)	meter	−01 3.048*
link (surveyor or gunter)	meter	−01 2.011 68*
liter	meter^3	−03 1.00*
lux	lumen/meter^2	+00 1.00*
maxwell	weber	−08 1.00*
meter	wavelengths Kr 86	+06 1.650 763 73*
micron	meter	−06 1.00*
mil	meter	−05 2.54*
mile (U.S. statute)	meter	+03 1.609 344*
mile (U.K. nautical)	meter	+03 1.853 184*
mile (international nautical)	meter	+03 1.852*
mile (U.S. nautical)	meter	+03 1.852*
millibar	newton/meter^2	+02 1.00*
millimeter of mercury (0°C)	newton/meter^2	+02 1.333 224
minute (angle)	radian	−04 2.908 882 086 66
minute (mean solar)	second (mean solar)	+01 6.00*
minute (sidereal)	second (mean solar)	+01 5.983 617 4
month (mean calendar)	second (mean solar)	+06 2.628*
nautical mile (international)	meter	+03 1.852*
nautical mile (U.S.)	meter	+03 1.852*
nautical mile (U.K.)	meter	+03 1.853 184*
oersted	ampere/meter	+01 7.957 747 2
ounce force (avoirdupois)	newton	−01 2.780 138 5
ounce mass (avoirdupois)	kilogram	−02 2.834 952 312 5*
ounce mass (troy or apothecary)	kilogram	−02 3.110 347 68*
ounce (U.S. fluid)	meter^3	−05 2.957 352 956 25*
pace	meter	−01 7.62*
parsec (IAU)	meter	+16 3.085 7
pascal	newton/meter^2	+00 1.00*
peck (U.S.)	meter^3	−03 8.809 767 541 72*
pennyweight	kilogram	−03 1.555 173 84*
perch	meter	+00 5.0292*
phot	lumen/meter^2	+04 1.00
pica (printers)	meter	−03 4.217 517 6*
pint (U.S. dry)	meter^3	−04 5.506 104 713 575*
pint (U.S. liquid)	meter^3	−04 4.731 764 73*
point (printers)	meter	−04 3.514 598*
poise	newton second/meter^2	−01 1.00*
pole	meter	+00 5.0292*
pound force (lbf avoirdupois)	newton	+00 4.448 221 615 260 5*
pound mass (lbm avoirdupois)	kilogram	−01 4.535 923 7*
pound mass (troy or apothecary)	kilogram	−01 3.732 417 216*
poundal	newton	−01 1.382 549 543 76*
quart (U.S. dry)	meter^3	−03 1.101 220 942 715*
quart (U.S. liquid)	meter^3	−04 9.463 592 5
rad (radiation dose absorbed)	joule/kilogram	−02 1.00*
Rankine (temperature)	kelvin	$t_K = (5/9)t_R$
rayleigh (rate of photon emission)	1/second meter^2	+10 1.00*
rhe	meter^2/newton second	+01 1.00*
rod	meter	+00 5.0292*
roentgen	coulomb/kilogram	−04 2.579 76*
rutherford	disintegration/second	+06 1.00*

To Convert from	to	Multiply by
second (angle)	radian	−06 4.848 136 811
second (ephemeris)	second	+00 1.000 000 000
second (mean solar)	second (ephemeris)	Consult American Ephemeris and Nautical Almanac
second (sidereal)	second (mean solar)	−01 9.972 695 7
section	meter^2	+06 2.589 988 110 336*
scruple (apothecary)	kilogram	−03 1.295 978 2*
shake	second	−08 1.00
skein	meter	+02 1.097 28*
slug	kilogram	+01 1.459 390 29
span	meter	−01 2.286*
statampere	ampere	−10 3.335 640
statcoulomb	coulomb	−10 3.335 640
statfarad	farad	−12 1.112 650
stathenry	henry	+11 8.987 554
statohm	ohm	+11 8.987 554
statute mile (U.S.)	meter	+03 1.609 344*
statvolt	volt	+02 2.997 925
stere	meter^3	+00 1.00*
stilb	candela/meter^2	+04 1.00
stoke	meter^2/second	−04 1.00*
tablespoon	meter^3	−05 1.478 676 478 125*
teaspoon	meter^3	−06 4.928 921 593 75*
ton (assay)	kilogram	−02 2.196 666 6
ton (long)	kilogram	+03 1.016 046 908 8*
ton (metric)	kilogram	+03 1.00*
ton (nuclear equivalent of TNT)	joule	+09 4.20
ton (register)	meter^3	+00 2.831 684 659 2*
ton (short, 2000 pound)	kilogram	+02 9.071 847 4*
tonne	kilogram	+03 1.00*
torr (0°C)	newton/meter^2	+02 1.333 22
township	meter^2	+07 9.323 957 2
unit pole	weber	−07 1.256 637
yard	meter	−01 9.144*
year (calendar)	second (mean solar)	+07 3.1536*
year (sidereal)	second (mean solar)	+07 3.155 815 0
year (tropical)	second (mean solar)	+07 3.155 692 6
year 1900, tropical, Jan., day 0, hour 12	second (ephemeris)	+07 3.155 692 597 47*
year 1900, tropical, Jan., day 0, hour 12	second	+07 3.155 692 597 47

A-5.2   Listing by Physical Quantity

To Convert from	to	Multiply by
ACCELERATION		
foot/second^2	meter/second^2	−01 3.048*
free fall, standard	meter/second^2	+00 9.806 65*
gal (galileo)	meter/second^2	−02 1.00*
inch/second^2	meter/second^2	−02 2.54*
AREA		
acre	meter^2	+03 4.046 856 422 4*
are	meter^2	+02 1.00*
barn	meter^2	−28 1.00*
circular mil	meter^2	−10 5.067 074 8
foot^2	meter^2	−02 9.290 304*
hectare	meter^2	+04 1.00*
inch^2	meter^2	−04 6.4516*
mile^2 (U.S. statute)	meter^2	+06 2.589 988 110 336*

To Convert from	to	Multiply by
AREA (continued)		
section	meter^2	+ 06 2.589 988 110 336*
township	meter^2	+ 07 9.323 957 2
yard^2	meter^2	− 01 8.361 273 6*
DENSITY		
gram/centimeter^3	kilogram/meter^3	− 03 1.00*
lbm/inch^3	kilogram/meter^3	+ 04 2.767 990 5
lbm/foot^3	kilogram/meter^3	+ 01 1.601 846 3
slug/foot^3	kilogram/meter^3	+ 02 5.153 79
ENERGY		
British thermal unit:		
(IST before 1956)	joule	+ 03 1.055 04
(IST after 1956)	joule	+ 03 1.055 056
British thermal unit (mean)	joule	+ 03 1.055 87
British thermal unit (thermochemical)	joule	+ 03 1.054 350
British thermal unit (39°F)	joule	+ 03 1.059 67
British thermal unit (60°F)	joule	+ 03 1.054 68
calorie (International Steam Table)	joule	+ 00 4.1868
calorie (mean)	joule	+ 00 4.190 02
calorie (thermochemical)	joule	+ 00 4.184*
calorie (15°C)	joule	+ 00 4.185 80
calorie (20°C)	joule	+ 00 4.181 90
calorie (kilogram, International Steam Table)	joule	+ 03 4.1868
calorie (kilogram, mean)	joule	+ 03 4.190 02
calorie (kilogram, thermochemical)	joule	+ 03 4.184*
electron volt	joule	− 19 1.602 191 7
erg	joule	− 07 1.00*
foot lbf	joule	+ 03 1.355 817 9
foot poundal	joule	− 02 4.214 011 0
joule (international of 1948)	joule	+ 00 1.000 165
kilocalorie (International Steam Table)	joule	+ 03 4.1868
kilocalorie (mean)	joule	+ 03 4.190 02
kilocalorie (thermochemical)	joule	+ 03 4.184*
kilowatt hour	joule	+ 06 3.60*
kilowatt hour (international of 1948)	joule	+ 06 3.600 59
ton (nuclear equivalent of TNT)	joule	+ 09 4.20
watt hour	joule	+ 03 3.60*
ENERGY/AREA TIME		
Btu (thermochemical)/foot^2 second	watt/meter^2	+ 04 1.134 893 1
Btu (thermochemical)/foot^2 minute	watt/meter^2	+ 02 1.891 488 5
Btu (thermochemical)/foot^2 hour	watt/meter^2	+ 00 3.152 480 8
Btu (thermochemical)/inch^2 second	watt/meter^2	+ 06 1.634 246 2
calorie (thermochemical)/cm^2 minute	watt/meter^2	+ 02 6.973 333 3
erg/centimeter^2 second	watt/meter^2	− 03 1.00*
watt/centimeter^2	watt/meter^2	+ 04 1.00*
FORCE		
dyne	newton	− 05 1.00*
kilogram force (kgf)	newton	+ 00 9.806 65*
kilopound force	newton	+ 00 9.806 65*
kip	newton	+ 03 4.448 221 615 260 5*
lbf (pound force, avoirdupois)	newton	+ 00 4.448 221 615 260 5*
ounce force (avoirdupois)	newton	+ 01 2.780 138 5
pound force, lbf (avoirdupois)	newton	+ 00 4.448 221 615 260 5*
poundal	newton	− 01 1.382 549 543 76*
LENGTH		
angstrom	meter	− 10 1.00*
astronomical unit (IAU)	meter	+ 11 1.496 00
astronomical unit (radio)	meter	+ 11 1.495 978 9
cable	meter	+ 02 2.194 56*
caliber	meter	− 04 2.54*

To Convert from	to	Multiply by
LENGTH (continued)		
chain (surveyor or gunter)	meter	+01 2.011 68*
chain (engineer or ramden)	meter	+01 3.048*
cubit	meter	−01 4.572*
fathom	meter	+00 1.8288*
fermi (femtometer)	meter	−15 1.00*
foot	meter	−01 3.048*
foot (U.S. survey)	meter	+00 1200/3937*
foot (U.S. survey)	meter	−01 3.048 006 096
furlong	meter	+02 2.011 68*
hand	meter	−01 1.016*
inch	meter	−02 2.54*
league (U.K. nautical)	meter	+03 5.559 552*
league (international nautical)	meter	+03 5.556*
league (statute)	meter	+03 4.828 032*
light year	meter	+15 9.460 55
link (engineer or ramden)	meter	−01 3.048*
link (surveyor or gunter)	meter	−01 2.011 68*
meter	wavelengths Kr 86	+06 1.650 763 73*
micron	meter	−06 1.00*
mil	meter	−05 2.54*
mile (U.S. statute)	meter	+03 1.609 344*
mile (U.K. nautical)	meter	+03 1.853 184*
mile (international nautical)	meter	+03 1.852*
mile (U.S. nautical)	meter	+03 1.852*
nautical mile (U.K.)	meter	+03 1.853 184*
nautical mile (international)	meter	+03 1.852*
nautical mile (U.S.)	meter	+03 1.852*
pace	meter	−01 7.62*
parsec (IAU)	meter	+16 3.085 7
perch	meter	+00 5.0292*
pica (printers)	meter	−03 4.217 517 6*
point (printers)	meter	−04 3.514 598*
pole	meter	+00 5.0292*
rod	meter	+00 5.0292*
skein	meter	+02 1.097 28*
span	meter	−01 2.286*
statute mile (U.S.)	meter	+03 1.609 344*
yard	meter	−01 9.144*
MASS		
carat (metric)	kilogram	−04 2.00*
gram (avoirdupois)	kilogram	−03 1.771 845 195 312 5*
gram (troy or apothecary)	kilogram	−03 3.887 934 6*
grain	kilogram	−05 6.479 891*
gram	kilogram	−03 1.00*
hundredweight (long)	kilogram	+01 5.080 234 544*
hundredweight (short)	kilogram	+01 4.535 923 7*
kgf second^2 meter (mass)	kilogram	+00 9.806 65*
klilogram mass	kilogram	+00 1.00*
lbm (pound mass, avoirdupois)	kilogram	−01 4.535 923 7*
ounce mass (avoirdupois)	kilogram	−02 2.834 952 312 5*
ounce mass (troy or apothecary)	kilogram	−02 3.110 347 68*
pennyweight	kilogram	−03 1.555 173 84*
pound mass, lbm (avoirdupois)	kilogram	−01 4.535 923 7*
pound mass (troy or apothecary)	kilogram	−01 3.732 417 216*
scruple (apothecary)	kilogram	−03 1.295 978 2*
slug	kilogram	+01 1.459 390 29
ton (assay)	kilogram	−02 2.196 666 6
ton (long)	kilogram	+03 1.016 046 908 8*
ton (metric)	kilogram	+03 1.00*
ton (short, 2000 pound)	kilogram	+02 9.071 847 4*
tonne	kilogram	+03 1.00*
POWER		
Btu (thermochemical)/second	watt	+03 1.054 350 264 488
Btu (thermochemical)/minute	watt	+01 1.757 250 4

To Convert from	to	Multiply by
POWER (continued)		
calorie (thermochemical)/second	watt	+ 00 4.184*
calorie (thermochemical)/minute	watt	− 02 6.973 333 3
foot lbf/hour	watt	− 04 3.766 161 0
foot lbf/minute	watt	− 02 2.259 696 6
foot lbf/second	watt	+ 00 1.355 817 9
horsepower (550 foot lbf/second)	watt	+ 02 7.456 998 7
horsepower (boiler)	watt	+ 03 9.809 50
horsepower (electric)	watt	+ 02 7.46*
horsepower (metric)	watt	+ 02 7.354 99
horsepower (U.K.)	watt	+ 02 7.457
horsepower (water)	watt	+ 02 7.460 43
kilocalorie (thermochemical)/minute	watt	+ 01 6.973 333 3
kilocalorie (thermochemical)/second	watt	+ 03 4.184*
watt (international of 1948)	watt	+ 00 1.000 165
PRESSURE		
atmosphere	newton/meter^2	+ 05 1.013 25*
bar	newton/meter^2	+ 05 1.00*
barye	newton/meter^2	− 01 1.00*
centimeter of mercury (0°C)	newton/meter^2	+ 03 1.333 22
centimeter of water (4°C)	newton/meter^2	+ 01 9.806 38
dyne/centimeter^2	newton/meter^2	− 01 1.00*
foot of water (39.2°F)	newton/meter^2	+ 03 2.988 98
inch of mercury (32°F)	newton/meter^2	+ 03 3.386 389
inch of mercury (60°F)	newton/meter^2	+ 03 3.376 85
inch of water (39.2°F)	newton/meter^2	+ 02 2.480 82
inch of water (60°F)	newton/meter^2	+ 02 2.4884
kgf/centimeter^2	newton/meter^2	+ 04 9.806 65*
kgf/meter^2	newton/meter^2	+ 00 9.806 65*
lbf/foot^2	newton/meter^2	+ 01 4.788 025 8
lbf/inch^2 (psi)	newton/meter^2	+ 03 6.894 757 2
millibar	newton/meter^2	+ 02 1.00*
millimeter of mercury (0°C)	newton/meter^2	+ 02 1.333 224
pascal	newton/meter^2	+ 00 1.00*
psi (lbf/inch^2)	newton/meter^2	+ 03 6.894 757 2
torr (0°C)	newton/meter^2	+ 02 1.333 22
SPEED		
foot/hour	meter/second	− 05 8.466 666 6
foot/minute	meter/second	− 03 5.08*
foot/second	meter/second	− 01 3.048*
inch/second	meter/second	− 02 2.54*
kilometer/hour	meter/second	− 01 2.777 777 8
knot (international)	meter/second	− 01 5.144 444 444
mile/hour (U.S. statute)	meter/second	− 01 4.4704*
mile/minute (U.S. statute)	meter/second	+ 01 2.682 24*
mile/second (U.S. statute)	meter/second	+ 03 1.609 344*
TEMPERATURE		
Celsius	kelvin	$t_K = t_C + 273.15$
Fahrenheit	kelvin	$t_K = (5/9)(t_F + 459.67)$
Fahrenheit	Celsius	$t_C = (5/9)(t_F − 32)$
Rankine	kelvin	$t_K = (5/9)t_R$
TIME		
day (mean solar)	second (mean solar)	+ 04 8.64*
day (sidereal)	second (mean solar)	+ 04 8.616 409 0
hour (mean solar)	second (mean solar)	+ 03 3.60*
hour (sidereal)	second (mean solar)	+ 03 3.590 170 4
minute (mean solar)	second (mean solar)	+ 01 6.00*
minute (sidereal)	second (mean solar)	+ 01 5.983 617 4
month (mean calendar)	second (mean solar)	+ 06 2.628*
second (ephemeris)	second	+ 00 1.000 000 000
second (mean solar)	second (ephemeris)	Consult American Ephemeris and Nautical Almanac

To Convert from	to	Multiply by
TIME (continued)		
second (sidereal)	second (mean solar)	− 01 9.972 695 7
year (calendar)	second (mean solar)	+ 07 3.1536*
year (sidereal)	second (mean solar)	+ 07 3.155 815 0
year (tropical)	second (mean solar)	+ 07 3.155 692 6
year 1900, tropical, Jan., day 0, hour 12	second (ephemeris)	+ 07 3.155 692 597 47*
year 1900, tropical, Jan., day 0, hour 12	second	+ 07 3.155 692 597 47
VISCOSITY		
centistoke	meter^2/second	− 06 1.00*
stoke	meter^2/second	− 04 1.00*
foot^2/second	meter^2/second	− 02 9.290 304*
centipoise	newton second/meter^2	− 03 1.00*
lbm/foot second	newton second/meter^2	+ 00 1.488 163 9
lbf second/foot^2	newton second/meter^2	+ 01 4.788 025 8
poise	newton second/meter^2	− 01 1.00*
poundal second/foot^2	newton second/meter^2	+ 00 1.488 163 9
slug/foot second	newton second/meter^2	+ 01 4.788 025 8
rhe	meter^2/newton second	+ 01 1.00*
VOLUME		
acre foot	meter^3	+ 03 1.233 481 837 547 52*
barrel (petroleum, 42 gallons)	meter^3	− 01 1.589 873
board foot	meter^3	− 03 2.359 737 216*
bushel (U.S.)	meter^3	− 02 3.523 907 016 688*
cord	meter^3	+ 00 3.624 556 3
cup	meter^3	− 04 2.365 882 365*
dram (U.S. fluid)	meter^3	− 06 3.696 691 195 312 5*
fluid ounce (U.S.)	meter^3	− 05 2.957 352 956 25*
foot^3	meter^3	− 02 2.831 684 659 2*
gallon (U.K. liquid)	meter^3	− 03 4.546 087
gallon (U.S. dry)	meter^3	− 03 4.404 883 770 86*
gallon (U.S. liquid)	meter^3	− 03 3.785 411 784*
gill (U.K.)	meter^3	− 04 1.420 652
gill (U.S.)	meter^3	− 04 1.182 941 2
hogshead (U.S.)	meter^3	− 01 2.384 809 423 92*
inch^3	meter^3	− 05 1.638 706 4*
liter	meter^3	− 03 1.00*
ounce (U.S. fluid)	meter^3	− 05 2.957 352 956 25*
peck (U.S.)	meter^3	− 03 8.809 767 541 72*
pint (U.S. dry)	meter^3	− 04 5.506 104 713 575*
pint (U.S. liquid)	meter^3	− 04 4.731 764 73*
quart (U.S. dry)	meter^3	− 03 1.101 220 942 715*
quart (U.S. liquid)	meter^3	− 04 9.463 592 5
stere	meter^3	+ 00 1.00*
tablespoon	meter^3	− 05 1.478 676 478 125*
teaspoon	meter^3	− 06 4.928 921 593 75*
ton (register)	meter^3	+ 00 2.831 684 659 2*
yard^3	meter^3	− 01 7.645 548 579 84*

A-6   CONVERSION FACTOR TABLES*†

TABLE A-6.1   *Length (L)*

to Obtain ↓ \ Multiply Number of → by →	centimeters	feet	inches	kilometers	nautical miles	meters	mils	miles	millimeters	yards
centimeters	1	30.48	2.540	10^5	1.853×10^5	100	2.540×10^{-3}	1.609×10^5	0.1	91.44
feet	3.281×10^{-2}	1	8.333×10^{-2}	3281	6080.27	3.281	8.333×10^{-5}	5280	3.281×10^{-3}	3
inches	0.3937	12	1	3.937×10^4	7.296×10^4	39.37	0.001	6.336×10^4	3.937×10^{-2}	36
kilometers	10^{-5}	3.048×10^{-4}	2.540×10^{-5}	1	1.853	0.001	2.540×10^{-8}	1.609	10^{-6}	9.144×10^{-4}
nautical miles		1.645×10^{-4}		0.5396	1	5.396×10^{-4}		0.8684		4.934×10^{-4}
meters	0.01	0.3048	2.540×10^{-2}	1000	1853	1		1609	0.001	0.9144
mils	393.7	1.2×10^4	1000	3.937×10^7		3.937×10^4	1		39.37	3.6×10^4
miles	6.214×10^{-6}	1.894×10^{-4}	1.578×10^{-5}	0.6214	1.1516	6.214×10^{-4}		1	6.214×10^{-7}	5.682×10^{-4}
millimeters	10	304.8	25.40	10^6		1000	2.540×10^{-2}		1	914.4
yards	1.094×10^{-2}	0.3333	2.778×10^{-2}	1094	2027	1.094	2.778×10^{-5}	1760	1.094×10^{-3}	1

*Boldface units in the following tables are SI.
†Source: *Eshbach's Handbook of Engineering Fundamentals*, 4th ed., B.D. Tapley and T.R. Poston, eds., John Wiley and Sons, New York (1990). Reprinted by permission of John Wiley & Sons, Inc.

TABLE A-6.2 *Area (L²)*

to Obtain ↓ / Multiply Number of →	acres	circular mils	square centimeters	square feet	square inches	square kilometers	*square meters*	square miles	square millimeters	square yards
acres	1			2.296×10^{-5}		247.1	2.471×10^{-4}	640		2.066×10^{-4}
circular mils		1	1.973×10^{5}	1.833×10^{8}	1.273×10^{6}		1.973×10^{9}		1973	
square centimeters		5.067×10^{-6}	1	929.0	6.452	10^{10}	10^{4}	2.590×10^{10}	0.01	8361
square feet	4.356×10^{4}		1.076×10^{-3}	1	6.944×10^{-3}	1.076×10^{7}	10.76	2.788×10^{7}	1.076×10^{-5}	9
square inches	6,272,640	7.854×10^{-7}	0.1550	144	1	1.550×10^{9}	1550	4.015×10^{9}	1.550×10^{-3}	1296
square kilometers	4.047×10^{-3}		10^{-10}	9.290×10^{-8}	6.452×10^{-10}	1	10^{-6}	2.590	10^{-12}	8.361×10^{-7}
square meters	4047		0.0001	9.290×10^{-2}	6.452×10^{-4}	10^{6}	1	2.590×10^{6}	10^{-6}	0.8361
square miles	1.562×10^{-3}		3.861×10^{-11}	3.587×10^{-8}		0.3861	3.861×10^{-7}	1	3.861×10^{-13}	3.228×10^{-7}
square millimeters		5.067×10^{-4}	100	9.290×10^{4}	645.2	10^{12}	10^{6}		1	8.361×10^{5}
square yards	4840		1.196×10^{-4}	0.1111	7.716×10^{-4}	1.196×10^{6}	1.196	3.098×10^{6}	1.196×10^{-6}	1

TABLE A-6.3 *Volume (L³)*

to Obtain ↓ / Multiply Number of →	bushels (dry)	cubic centimeters	cubic feet	cubic inches	*cubic meters*	cubic yards	gallons (liquid)	liters	pints (liquid)	quarts (liquid)
bushels (dry)	1		0.8036	4.651×10^{-4}	28.38			2.838×10^{-2}		
cubic centimeters	3.524×10^{4}	1	2.832×10^{4}	16.39	10^{6}	7.646×10^{5}	3785	1000	473.2	946.4
cubic feet	1.2445	3.531×10^{-5}	1	5.787×10^{-4}	35.31	27	0.1337	3.531×10^{-2}	1.671×10^{-2}	3.342×10^{-2}
cubic inches	2150.4	6.102×10^{-2}	1728	1	6.102×10^{4}	46,656	231	61.02	28.87	57.75
cubic meters	3.524×10^{-2}	10^{-6}	2.832×10^{-2}	1.639×10^{-5}	1	0.7646	3.785×10^{-3}	0.001	4.732×10^{-4}	9.464×10^{-4}
cubic yards		1.308×10^{-6}	3.704×10^{-2}	2.143×10^{-5}	1.308	1	4.951×10^{-3}	1.308×10^{-3}	6.189×10^{-4}	1.238×10^{-3}
gallons (liquid)		2.642×10^{-4}	7.481	4.329×10^{-3}	264.2	202.0	1	0.2642	0.125	0.25
liters	35.24	0.001	28.32	1.639×10^{-2}	1000	764.6	3.785	1	0.4732	0.9464
pints (liquid)		2.113×10^{-3}	59.84	3.463×10^{-2}	2113	1616	8	2.113	1	2
quarts (liquid)		1.057×10^{-3}	29.92	1.732×10^{-2}	1057	807.9	4	1.057	0.5	1

TABLE A-6.4   *Plane Angle (no dimensions)*

to Obtain ↓ \ Multiply Number of → by	degrees	minutes	quadrants	*radians**	revolutions* (circumferences)	seconds
degrees	1	1.667×10^{-2}	90	57.30	360	2.778×10^{-4}
minutes	60	1	5400	3438	2.16×10^4	1.667×10^{-2}
quadrants	1.111×10^{-2}	1.852×10^{-4}	1	0.6366	4	3.087×10^{-6}
*radians**	1.745×10^{-2}	2.909×10^{-4}	1.571	1	6.283	4.848×10^{-6}
revolutions* (circumferences)	2.788×10^{-3}	4.630×10^{-5}	0.25	0.1591	1	7.716×10^{-7}
seconds	3600	60	3.24×10^5	2.063×10^5	1.296×10^6	1

*2π rad = 1 circumference = 360 degrees by definition.

TABLE A-6.5   *Linear Velocity (LT^{-1})*

to Obtain ↓ \ Multiply Number of → by	centimeters per second	feet per minute	feet per second	kilometers per hour	kilometers per minute	knots*	meters per minute	*meters per second*	miles per hour	miles per minute
centimeters per second	1	0.5080	30.48	27.78	1667	51.48	1.667	100	44.70	2682
feet per minute	1.969	1	60	54.68	3281	101.3	3.281	196.8	88	5280
feet per second	3.281×10^{-2}	1.667×10^{-2}	1	0.9113	54.68	1.689	5.468×10^{-2}	3.281	1.467	88
kilometers per hour	0.036	1.829×10^{-2}	1.097	1	60	1.853	0.06	3.6	1.609	96.54
kilometers per minute	0.0006	3.048×10^{-4}	1.829×10^{-2}	1.667×10^{-2}	1	3.088×10^{-2}	0.001	0.06	2.682×10^{-2}	1.609
knots*	1.943×10^{-2}	9.868×10^{-3}	0.5921	0.5396	32.38	1	3.238×10^{-2}	1.943	0.8684	52.10
meters per minute	0.6	0.3048	18.29	16.67	1000	30.88	1	60	26.82	1609
meters per second	0.01	5.080×10^{-3}	0.3048	0.2778	16.67	0.5148	1.667×10^{-2}	1	0.4470	26.82
miles per hour	2.237×10^{-2}	1.136×10^{-2}	0.6818	0.6214	37.28	1.152	3.728×10^{-2}	2.237	1	60
miles per minute	3.728×10^{-4}	1.892×10^{-4}	1.136×10^{-2}	1.036×10^{-2}	0.6214	1.919×10^{-2}	6.214×10^{-4}	3.728×10^{-2}	1.667×10^{-2}	1

*Nautical miles per hour.

TABLE A-6.6 Linear Acceleration* (LT^{-2})

to Obtain ↓ \ Multiply Number of → by	centimeters per second per second	feet per second per second	kilometers per hour per second	*meters per second per second*	miles per hour per second
centimeters per second per second	1	30.48	27.78	100	44.70
feet per second per second	3.281×10^{-2}	1	0.9113	3.281	1.467
kilometers per hour per second	0.036	1.097	1	3.6	1.609
meters per second per second	0.01	0.3048	0.2778	1	0.4470
miles per hour per second	2.237×10^{-2}	0.6818	0.6214	2.237	1

*The (standard) acceleration due to gravity (g_0) = 908.7 cm/sec sec, = 32.17 ft/sec sec = 35.30 km/h sec = 9.807 m/sec sec = 21.94 mph/sec.

TABLE A-6.7 Mass (M) and Weight*

to Obtain ↓ \ Multiply Number of → by	grains	grams	*kilograms*	milligrams	ounces†	pounds†	tons (long)	tons (metric)	tons (short)
grains	1	15.43	1.543×10^4	1.543×10^{-2}	437.5	7000			
grams	6.481×10^{-2}	1	1000	0.001	28.35	453.6	1.016×10^6	$\times 10^6$	9.072×10^5
kilograms	6.481×10^{-5}	0.001	1	10^{-6}	2.835×10^{-2}	0.4536	1016	1000	907.2
milligrams	64.81	1000	10^6	1	2.835×10^4	4.536×10^5	1.016×10^9	10^9	9.072×10^8
ounces	2.286×10^{-3}	3.527×10^{-2}	35.27	3.527×10^{-5}	1	16	3.584×10^4	3.527×10^4	3.2×10^4
pounds	1.429×10^{-4}	2.205×10^{-3}	2.205	2.205×10^{-6}	6.250×10^{-2}	1	2240	2205	2000
tons (long)		9.842×10^{-7}	9.842×10^{-4}	9.842×10^{-10}	2.790×10^{-5}	4.464×10^{-4}	1	0.9842	0.8929

TABLE A-6.8   *Density or Mass per Unit Volume (ML^{-3})*

Multiply Number of → to Obtain ↓ by →	grams per cubic centimeter	kilograms per cubic meter	pounds per cubic foot	pounds per cubic inch
grams per cubic centimeter	1	0.001	1.602×10^{-2}	27.68
kilograms per cubic meter	1000	1	16.02	2.768×10^4
pounds per cubic foot	62.43	6.243×10^{-2}	1	1728
pounds per cubic inch	3.613×10^{-2}	3.613×10^{-5}	5.787×10^{-4}	1
pounds per mil foot*	3.405×10^{-7}	3.405×10^{-10}	5.456×10^{-9}	9.425×10^{-6}

*Unit of volume is a volume one foot long and one circular mil in cross-section area.

TABLE A-6.9   *Force* (MLT^{-2}) or (F)*

Multiply Number of → to Obtain ↓ by →	dynes	grams	joules per centimeter	newtons or joules per meter	kilograms	pounds	poundals
dynes	1	980.7	10^7	10^5	9.807×10^5	4.448×10^5	1.383×10^4
grams	1.020×10^{-3}	1	1.020×10^4	102.0	1000	453.6	14.10
joules per centimeter	10^{-7}	9.807×10^{-5}	1	0.01	9.807×10^{-2}	4.448×10^{-2}	1.383×10^{-3}
newtons or joules per meter	10^{-5}	9.807×10^{-3}	100	1	9.807	4.448	0.1383
kilograms	1.020×10^{-6}	0.001	10.20	0.1020	1	0.4536	1.410×10^{-2}
pounds	2.248×10^{-6}	2.205×10^{-3}	22.48	0.2248	2.205	1	3.108×10^{-2}
poundals	7.233×10^{-5}	7.093×10^{-2}	723.3	7.233	70.93	32.17	1

*Conversion factors between absolute and gravitational units apply only under standard acceleration due to gravity conditions.

TABLE A-6.10 *Pressure or Force per Unit Area ($ML^{-1}T^{-2}$) or (FL^{-2})*

Multiply Number of → / by / to Obtain ↓	atmospheres*	baryes or dynes per square centimeter	centimeters of mercury at 0°C†	inches of mercury at 0°C†	inches of water at 4°C	kilograms per square meter‡	pounds per square foot	pounds per square inch	tons (short) per square foot	pascal
atmospheres*	1	9.869×10^{-7}	1.316×10^{-2}	3.342×10^{-2}	2.458×10^{-3}	9.678×10^{-5}	4.725×10^{-4}	6.804×10^{-2}	0.9450	9.689×10^{-6}
baryes or dynes per square centimeter	1.013×10^{6}	1	1.333×10^{4}	3.386×10^{4}	2.491×10^{-3}	98.07	478.8	6.895×10^{4}	9.576×10^{5}	10
centimeters of mercury at 0°C†	76.00	7.501×10^{-5}	1	2.540	0.1868	7.356×10^{-3}	3.591×10^{-2}	5.171	71.83	7.501×10^{-4}
inches of mercury at 0°C†	29.92	2.953×10^{-5}	0.3937	1	7.355×10^{-2}	2.896×10^{-3}	1.414×10^{-2}	2.036	28.28	2.953×10^{-4}
inches of water at 4°C	406.8	4.015×10^{-4}	5.354	13.60	1	3.937×10^{-2}	0.1922	27.68	384.5	4.015×10^{-3}
kilograms per square meter‡	1.033×10^{4}	1.020×10^{-2}	136.0	345.3	25.40	1	4.882	703.1	9765	0.1020
pounds per square foot	2117	2.089×10^{-3}	27.85	70.73	5.204	0.2048	1	144	2000	2.089×10^{-2}
pounds per square inch	14.70	1.450×10^{-5}	0.1934	0.4912	3.613×10^{-2}	1.422×10^{-3}	6.944×10^{-3}	1	13.89	1.450×10^{-4}
tons (short) per square foot	1.058	1.044×10^{-6}	1.392×10^{-2}	3.536×10^{-2}	2.601×10^{-3}	1.024×10^{-4}	0.0005	0.072	1	1.044×10^{-5}
pascal	1.013×10^{5}	10^{-1}	1.333×10^{3}	3.386×10^{3}	2.49×10^{2}	9.807	47.88	6.895×10^{3}	9.576×10^{4}	1

*Definition: One atmosphere (standard) = 76 cm of mercury at 0°C.
†To convert height h of a column of mercury at t degrees centigrade to the equivalent height h_0 at 0°C, use $h_0 = h[1 - (m - l)t/(1 + mt)]$ where $m = 0.0001818$ and $l = 18.4 \times 10^{-6}$ if the scale is engraved on brass; $l = 8.5 \times 10^{-6}$ if on glass. This assumes the scale is correct at 0°C; for other cases (any liquid) see *International Critical Tables*, Vol. 1 (1968).
‡1 g/cm² = 10 kg/m².

TABLE A-6.11   *Energy, Work, and Heat* (ML^2T^{-2}) or (FL)*

to Obtain ↓ / Multiply Number of → by →	British thermal units[†]	centimeter-grams	ergs or centimeter-dynes	foot-pounds	horsepower-hours	joules[‡] or watt-seconds	kilogram-calories[†]	kilowatt-hours	meter-kilograms	watt-hours
British thermal units[†]	1	9.297×10^{-8}	9.480×10^{-11}	1.285×10^{-3}	2545	9.480×10^{-4}	3.969	3413	9.297×10^{-3}	3.413
centimeter-grams	1.076×10^{7}	1	1.020×10^{-3}	1.383×10^{4}	2.737×10^{10}	1.020×10^{4}	4.269×10^{7}	3.671×10^{10}	10^{5}	3.671×10^{7}
ergs or centimeter-dynes	1.055×10^{10}	980.7	1	1.356×10^{7}	2.684×10^{12}	10^{7}	4.186×10^{10}	3.6×10^{13}	9.807×10^{7}	3.6×10^{10}
foot-pounds	778.0	7.233×10^{-5}	7.367×10^{-8}	1	1.98×10^{6}	0.7376	3087	2.655×10^{6}	7.233	2655
horsepower-hours	3.929×10^{-4}	3.654×10^{-11}	3.722×10^{-14}	5.050×10^{-7}	1	3.722×10^{-7}	1.559×10^{-3}	1.341	3.653×10^{-6}	1.341×10^{-3}
joules[‡] or watt-seconds	1054.8	9.807×10^{-5}	10^{-7}	1.356	2.684×10^{6}	1	4186	3.6×10^{6}	9.807	3600
kilogram-calories[†]	0.2520	2.343×10^{-8}	2.389×10^{-11}	3.239×10^{-4}	641.3	2.389×10^{-4}	1	860.0	2.343×10^{-3}	0.8600
kilowatt-hours	2.930×10^{-4}	2.724×10^{-11}	2.778×10^{-14}	3.766×10^{-7}	0.7457	2.788×10^{-7}	1.163×10^{-3}	1	2.724×10^{-6}	0.001
meter-kilograms	107.6	10^{-5}	1.020×10^{-8}	0.1383	2.737×10^{5}	0.1020	426.9	3.671×10^{5}	1	367.1
watt-hours	0.2930	2.724×10^{-8}	2.778×10^{-11}	3.766×10^{-4}	745.7	2.778×10^{-4}	1.163	1000	2.724×10^{-3}	1

*See note to Table A-6.12.

[†]Mean calorie and Btu used throughout. One gram-calorie = 0.001 kilogram-calorie; one Ostwald calorie = 0.1 kilogram-calorie.

The IT cal, 1000 international steam table calories, has been defined as the 1/860th part of the international kilowatt-hour (see *Mechanical Engineering*, Nov., 1935, p. 710). Its value is very nearly equal to the mean kilogram-calorie, 1 IT cal-1.00037 kilogram-calories (mean). 1 Btu = 251.996 IT cal.

[‡]Absolute joule, defined as 10^{7} ergs. The international joule, based on the international ohm and ampere, equals 1.0003 absolute joules.

TABLE A-6.12 *Power or Rate of Doing Work* (ML^2T^{-3}) *or* (FLT^{-1})

to Obtain ↓ \ Multiply Number of → by	British thermal units per minute	ergs per second	foot-pounds per minute	foot-pounds per second	horsepower*	kilogram-calories per minute	kilowatts	metric horsepower	*watts*
British thermal units per minute	1	5.689×10^{-9}	1.285×10^{-3}	7.712×10^{-2}	42.41	3.969	56.89	41.83	5.689×10^{-2}
ergs per second	1.758×10^{8}	1	2.259×10^{5}	1.356×10^{7}	7.457×10^{9}	6.977×10^{8}	10^{10}	7.355×10^{9}	10^{7}
foot-pounds per minute	778.0	4.426×10^{-6}	1	60	3.3×10^{4}	3087	4.426×10^{4}	3.255×10^{4}	44.26
foot-pounds per second	12.97	7.376×10^{-8}	1.667×10^{-2}	1	550	51.44	737.6	542.5	0.7376
horsepower*	2.357×10^{-2}	1.341×10^{-10}	3.030×10^{-5}	1.818×10^{-3}	1	9.355×10^{-2}	1.341	0.9863	1.341×10^{-3}
kilogram-calories per minute	0.2520	1.433×10^{-9}	3.239×10^{-4}	1.943×10^{-2}	10.69	1	14.33	10.54	1.433×10^{-2}
kilowatts	0.01758	10^{-10}	2.260×10^{-5}	1.356×10^{-3}	0.7457	0.06977×10^{-2}	1	0.7355	0.001
metric horsepower	2.390×10^{-2}	1.360×10^{-10}	3.072×10^{-5}	1.843×10^{-3}	1.014	9.485×10^{-2}	1.360	1	1.360×10^{-3}
watts	17.58	10^{-7}	2.260×10^{-2}	1.356	745.7	69.77	1000	735.5	1

1 Cheval-vapeur = 75 kilogram-meters per second
1 Poncelet = 100 kilogram-meters per second

*The "horsepower" used in these tables is equal to 550 foot-pounds per second by definition. Other definitions are one horsepower equals 746 watts (U.S. and Great Britain) and one horsepower equals 736 watts (continental Europe). Neither of these latter definitions is equivalent to the first; the "horsepowers" defined in these latter definitions are widely used in the rating of electrical machinery.

TABLE A-6.13 *Heat Flux (Power/Area)*

to Obtain ↓ \ Multiply by From →	Btu/(min × ft²)	Btu/(sec × ft²)	kW/m²	W/m²	W/cm²
Btu/(min × ft²)	1	1.6×10^{-2}	5.28	5.2×10^{-3}	5.2×10^{-1}
Btu/(sec × ft²)	60	1	6.81×10^{-2}	8.8×10^{-5}	8.8×10^{-3}
kW/m²	0.18923	11.3565	1	10^{-3}	10^{-1}
W/m²	189.273	1.1356×10^{4}	10^{3}	1	10^{4}
W/cm²	1.89273	1.1356×10^{2}	10	10^{-4}	1
kg-cal/sec·m²	6.135×10^{-6}	1.02×10^{-7}	8.60400×10^{5}	8.6×10^{2}	8.604×10^{4}
kg-cal/min·m²	3.681×10^{-4}	6.07×10^{-6}	1.434×10^{4}	1.4341×10^{1}	1.434×10^{3}

TABLE A-6.14   *Specific Heat ($L^2T^{-2}t^{-1}$); (t = temperature)*

To change specific heat in gram-calories per gram per degree centigrade to the units given in any line of the following table, multiply by the factor in the last column.

Unit of Heat or Energy	Unit of Mass	Temperature Scale*	Factor
gram-calories	gram	centigrade	1
kilogram-calories	kilogram	centigrade	1
British thermal units	pound	centigrade	1.800
British thermal units	pound	Fahrenheit	1.000
joules	gram	centigrade	4.186
joules	pound	Fahrenheit	1055
joules	***kilogram***	***kelvin***	4.187×10^3
kilowatt-hours	kilogram	centigrade	1.163×10^{-3}
kilowatt-hours	pound	Fahrenheit	2.930×10^{-4}

*Temperature conversion formulae:

t_c = temperature in centigrade degrees
t_f = temperature in Fahrenheit degrees
t_K = temperature in kelvin degrees

$$1°F = \frac{5}{9}°C$$

$$1K = 1°C$$

$$t_c = \frac{5}{9}(t_f - 32)$$

$$t_f = \frac{9}{5}t_c + 32$$

$$t_K = t_c + 273$$

TABLE A-6.15   *Thermal Conductivity* ($LMT^{-3}t^{-1}$)

Multiply by → / To Obtain ↓ \ From →	Btu · ft per h · ft² · °F	Btu · in per h · ft² · °F	Btu · in per sec · ft² · °F	joules per m · s · °C	kcal per m · h · °C	erg per cm · s · °C	kcal per m · s · °C	cal per cm · s · °C	W per ft · °C	*W per m · K*
Btu · ft per h · ft² · °F	1	8.333×10^{-2}	3.0×10^2	5.778×10^{-1}	6.720×10^{-1}	5.778×10^{-6}	2.419×10^3	2.419×10^2	1.895	5.778×10^{-1}
Btu · in per h · ft² · °F	12	1	3.6×10^3	6.933	8.064	6.933×10^{-5}	2.903×10^4	2.903×10^3	2.275×10^1	6.933
Btu · in per s · ft² · °F	3.333×10^{-3}	2.778×10^{-4}	1	1.926×10^{-3}	2.240×10^{-3}	1.926×10^{-8}	8.064	8.064×10^{-1}	6.319×10^{-3}	1.926×10^{-3}
joules per m · s · °C	1.731	1.442×10^{-1}	5.192×10^2	1	1.163	1.000×10^{-5}	4.187×10^3	4.187×10^2	3.281	1.0
kcal per m · h · °C	1.483	1.240×10^{-1}	4.465×10^2	8.599×10^{-1}	1	8.599×10^{-6}	3.6×10^3	3.6×10^2	2.821	8.599×10^{-1}
erg per cm · s · °C	1.731×10^5	1.442×10^4	5.192×10^7	1.0×10^5	1.163×10^5	1	4.187×10^8	4.187×10^7	3.281×10^5	1.0×10^5
kcal per m · s · °C	4.134×10^{-4}	3.445×10^{-5}	1.240×10^{-1}	2.388×10^{-4}	2.778×10^{-4}	2.388×10^{-9}	1	1.0×10^{-1}	7.835×10^{-4}	2.388×10^{-4}
cal per cm · s · °C	4.134×10^{-3}	3.445×10^{-4}	1.240	2.388×10^{-3}	2.778×10^{-3}	2.388×10^{-8}	10	1	7.835×10^{-3}	2.388×10^{-3}
W per ft · °C	5.276×10^{-1}	4.395×10^{-2}	1.582×10^2	3.048×10^{-1}	3.545×10^{-1}	3.048×10^{-6}	1.276×10^3	1.276×10^2	1	3.048×10^{-1}
W per m · K	1.731	1.442×10^{-1}	5.192×10^2	1.0	1.163	1.00×10^{-5}	4.187×10^3	4.187×10^2	3.281	1

*International Table Btu = 1.055056×10^3 joules; and International Table cal = 4.1868 joules are used throughout.

TABLE B-1 *Approximate Properties of Common Gases*

	English (FSS) Units				
	Engineering Gas Constant R (ft-lb/slug · R)	Universal Gas Constant $\mathcal{R} = mR$ (ft-lb/slug · R)	Adiabatic Exponent k —	Specific Heat at Constant Pressure c_p (ft-lb/slug · R)	Viscosity at 68°F (20°C) $\mu \times 10^5$ (lb-sec/ft²)
Carbon dioxide	1,123	49,419	1.28	5,132	0.0307
Oxygen	1,554	49,741	1.40	5,437	0.0419
Air	1,715	49,709	1.40	6,000	0.0377
Nitrogen	1,773	49,644	1.40	6,210	0.0368
Methane	3,098	49,644	1.31	13,095	0.028
Helium	12,419	49,677	1.66	31,235	0.0411
Hydrogen	24,677	49,741	1.40	86,387	0.0189

	SI Units				
	R (J/kg · K)	$\mathcal{R} = mR$ (J/kg · K)	k —	c_p (J/kg · K)	$\mu \times 10^5$ (Pa · s)
Carbon dioxide	187.8	8,264	1.28	858.2	1.47
Oxygen	259.9	8,318	1.40	909.2	2.01
Air	286.8	8,313	1.40	1,003	1.81
Nitrogen	296.5	8,302	1.40	1,038	1.76
Methane	518.1	8,302	1.31	2,190	1.34
Helium	2,076.8	8,307	1.66	5,223	1.97
Hydrogen	4,126.6	8,318	1.40	14,446	0.90

Source: *Eshbach's Handbook of Engineering Fundamentals*, 4th ed., B.D. Tapley and T.R. Poston, eds., John Wiley and Sons, New York (1990). Reprinted by permission of John Wiley & Sons, Inc.

TABLE B-2 *Thermophysical Property Values for Gases at Standard Atmospheric Pressure*

T, K	ρ, kg/m³	c_p, Ws/kg K	μ, kg/ms	ν, m²/s	k, W/m K	α, m²/s	Pr
				Air			
100	3.6010	1.0266×10^3	0.6924×10^{-5}	1.923×10^{-6}	0.009246	0.0250×10^{-4}	0.768
150	2.3675	1.0099	1.0283	4.343	0.013735	0.0574	0.756
200	1.7684	1.0061	1.3289	7.514	0.01809	0.1016	0.739
250	1.4128	1.0053	1.488	10.53	0.02227	0.1568	0.722
300	1.1774	1.0057	1.983	16.84	0.02624	0.2216	0.708
350	0.9980	1.0090	2.075	20.76	0.03003	0.2983	0.697
400	0.8826	1.0140	2.286	25.90	0.03365	0.3760	0.689
450	0.7833	1.0207	2.484	31.71	0.03707	0.4636	0.683
500	0.7048	1.0295	2.671	37.90	0.04038	0.5564	0.680
550	0.6423	1.0392	2.848	44.27	0.04360	0.6532	0.680
600	0.5879	1.0551	3.018	51.34	0.04659	0.7512	0.682
650	0.5430	1.0635	3.177	58.51	0.04953	0.8578	0.682
700	0.5030	1.0752	3.332	66.25	0.05230	0.9672	0.684
750	0.4709	1.0856	3.481	73.91	0.05509	1.0774	0.686
800	0.4405	1.0978	3.625	82.29	0.05779	1.1951	0.689
850	0.4149	1.1095	3.765	90.75	0.06028	1.3097	0.692
900	0.3925	1.1212	3.899	99.3	0.06279	1.4271	0.696
950	0.3716	1.1321	4.023	108.2	0.06525	1.5510	0.699
1000	0.3524	1.1417	4.152	117.8	0.06752	1.6779	0.702
1100	0.3204	1.160	4.44	138.6	0.0732	1.969	0.704
1200	0.2947	1.179	4.69	159.1	0.0782	2.251	0.707
1300	0.2707	1.197	4.93	182.1	0.0837	2.583	0.705
1400	0.2515	1.214	5.17	205.5	0.0891	2.920	0.705
1500	0.2355	1.230	5.40	229.1	0.0946	3.266	0.705
1600	0.2211	1.248	5.63	254.5	0.100	3.624	0.705
1700	0.2082	1.267	5.85	280.9	0.105	3.977	0.705
1800	0.1970	1.287	6.07	308.1	0.111	4.379	0.704

TABLE B-2   *(Continued)*

T, K	ρ, kg/m³	c_p, Ws/kg K	μ, kg/ms	ν, m²/s	k, W/m K	α, m²/s	Pr
Air (continued)							
1900	0.1858	1.309	6.29	338.5	0.117	4.811	0.704
2000	0.1762	1.338	6.50	369.0	0.124	5.260	0.702
2100	0.1682	1.372	6.72	399.6	0.131	5.680	0.703
2200	0.1602	1.419	6.93	432.6	0.139	6.115	0.707
2300	0.1538	1.482	7.14	464.0	0.149	6.537	0.710
2400	0.1458	1.574	7.35	504.0	0.161	7.016	0.718
2500	0.1394	1.688	7.57	543.0	0.175	7.437	0.730
Helium							
144	0.3379	5.200	125.5×10^{-7}	37.11×10^{-6}	0.0928	0.5275×10^{-4}	0.70
200	0.2435	5.200	156.6	64.38	0.1177	0.9288	0.694
255	0.1906	5.200	181.7	95.50	0.1357	1.3675	0.70
366	0.13280	5.200	230.5	173.6	0.1691	2.449	0.71
477	0.10204	5.200	275.0	269.3	0.197	3.716	0.72
589	0.08282	5.200	311.3	375.8	0.225	5.215	0.72
700	0.07032	5.200	347.5	494.2	0.251	6.661	0.72
800	0.06023	5.200	381.7	634.1	0.275	8.774	0.72
Hydrogen							
150	0.16371	12.602	5.595×10^{-6}	34.18×10^{-5}	0.0981	0.475×10^{-4}	0.718
200	0.12270	13.540	6.813	55.53	0.1282	0.772	0.719
250	0.09819	14.059	7.919	80.64	0.1561	1.130	0.713
300	0.08185	14.314	8.963	109.5	0.182	1.554	0.706
350	0.07016	14.436	9.954	141.9	0.206	2.031	0.697
400	0.06135	14.491	10.864	177.1	0.228	2.568	0.690
450	0.05462	14.499	11.779	215.6	0.251	3.164	0.682
500	0.04918	14.507	12.636	257.0	0.272	3.817	0.675
550	0.04469	14.532	13.475	301.6	0.292	4.516	0.668
600	0.04085	14.537	14.285	349.7	0.315	5.306	0.664
700	0.03492	14.574	15.89	455.1	0.351	6.903	0.659
800	0.03060	14.675	17.40	569	0.384	8.563	0.664
900	0.02723	14.821	18.78	690	0.412	10.217	0.676
Oxygen							
150	2.6190	0.9178	11.490×10^{-6}	4.387×10^{-6}	0.01367	0.05688×10^{-4}	0.773
200	1.9559	0.9131	14.850	7.593	0.01824	0.10214	0.745
250	1.5618	0.9157	17.87	11.45	0.02259	0.15794	0.725
300	1.3007	0.9203	20.63	15.86	0.02676	0.22353	0.709
350	1.1133	0.9291	23.16	20.80	0.03070	0.2968	0.702
400	0.9755	0.9420	25.54	26.18	0.03461	0.3768	0.695
450	0.8682	0.9567	27.77	31.99	0.03828	0.4609	0.694
500	0.7801	0.9722	29.91	38.34	0.04173	0.5502	0.697
550	0.7096	0.9881	31.97	45.05	0.04517	0.6441	0.700
Nitrogen							
200	1.7108	1.0429	12.947×10^{-6}	7.568×10^{-6}	0.01824	0.10224×10^{-4}	0.747
300	1.1421	1.0408	17.84	15.63	0.02620	0.22044	0.713
400	0.8538	1.0459	21.98	25.74	0.03335	0.3734	0.691
500	0.6824	1.0555	25.70	37.66	0.03984	0.5530	0.684
600	0.5687	1.0756	29.11	51.19	0.04580	0.7486	0.686
700	0.4934	1.0969	32.13	65.13	0.05123	0.9466	0.691
800	0.4277	1.1225	34.84	81.46	0.05609	1.1685	0.700
900	0.3796	1.1464	37.49	91.06	0.06070	1.3946	0.711
1000	0.3412	1.1677	40.00	117.2	0.06475	1.6250	0.724
1100	0.3108	1.1857	42.28	136.0	0.06850	1.8591	0.736
1200	0.2851	1.2037	44.50	156.1	0.07184	2.0932	0.748

TABLE B-2 *(Continued)*

T, K	ρ, kg/m^3	c_p, kJ/ kg °C	μ, kg/ms	ν, m^2/s	k, W/m °C	α, m^2/s	Pr
			Carbon dioxide				
220	2.4733	0.783	11.105×10^{-6}	4.490×10^{-6}	0.010805	0.05920×10^{-4}	0.818
250	2.1657	0.804	12.590	5.813	0.012884	0.07401	0.793
300	1.7973	0.871	14.958	8.321	0.016572	0.10588	0.770
350	1.5362	0.900	17.205	11.19	0.02047	0.14808	0.755
400	1.3424	0.942	19.32	14.39	0.02461	0.19463	0.738
450	1.1918	0.980	21.34	17.90	0.02897	0.24813	0.721
500	1.0732	1.013	23.26	21.67	0.03352	0.3084	0.702
550	0.9739	1.047	25.08	25.74	0.03821	0.3750	0.685
600	0.8938	1.076	26.83	30.02	0.04311	0.4483	0.668
			Ammonia, NH$_3$				
273	0.7929	2.177	9.353×10^{-6}	1.18×10^{-5}	0.0220	0.1308×10^{-4}	0.90
323	0.6487	2.177	11.035	1.70	0.0270	0.1920	0.88
373	0.5590	2.236	12.886	2.30	0.0327	0.2619	0.87
423	0.4934	2.315	14.672	2.97	0.0391	0.3432	0.87
473	0.4405	2.395	16.49	3.74	0.0467	0.4421	0.84
			Water vapor				
380	0.5863	2.060	12.71×10^{-6}	2.16×10^{-5}	0.0246	0.2036×10^{-4}	1.060
400	0.5542	2.014	13.44	2.42	0.0261	0.2338	1.040
450	0.4902	1.980	15.25	3.11	0.0299	0.307	1.010
500	0.4405	1.985	17.04	3.86	0.0339	0.387	0.996
550	0.4005	1.997	18.84	4.70	0.0379	0.475	0.991
600	0.3652	2.026	20.67	5.66	0.0422	0.573	0.986
650	0.3380	2.056	22.47	6.64	0.0464	0.666	0.995
700	0.3140	2.085	24.26	7.72	0.0505	0.772	1.000
750	0.2931	2.119	26.04	8.88	0.0549	0.883	1.005
800	0.2739	2.152	27.86	10.20	0.0592	1.004	1.010
850	0.2579	2.186	29.69	11.52	0.0637	1.130	1.019

Source: E.R.G. Eckert and R.M. Drake, *Analysis of Heat and Mass Transfer*, McGraw-Hill, New York (1972). Reprinted by permission of McGraw-Hill, Inc.

TABLE B-3 *Approximate Properties of Some Common Liquids at Standard Atmospheric Pressure*

	English (FSS) Units						
	Temperature T (°F)	Density, ρ (slug/ft³)	Specific Gravity s.g. —	Modulus of Elasticity K (psi)	Viscosity $\mu \times 10^5$ (lb-sec/ft²)	Surface Tension σ (lb/ft)	Vapor Pressure ρ_v (psia)
Benzene	68	1.70	0.88	150,000	1.37	0.0020	1.45
Carbon tetrachloride	68	3.08	1.59	160,000	2.035	0.0018	1.90
Crude oil	68	1.66	0.86	—	15.0	0.002	—
Ethyl alcohol	68	1.53	0.79	175,000	2.51	0.0015	0.85
Freon-12	60	2.61	1.35	—	3.10	—	—
	−30	2.91	—	—	3.82	—	—
Gasoline	68	1.32	0.68	—	0.61	—	8.0
Glycerin	68	2.44	1.26	630,000	3,120	0.0043	0.000002
Hydrogen	−431	0.143	—	—	0.0435	0.0002	3.1
Jet fuel (JP-4)	60	1.50	0.77	—	1.82	0.002	1.3
Mercury	60	26.3	13.57	3,800,000	3.26	0.035	0.000025
	600	24.9	12.8	—	1.88	—	6.85
Oxygen	−320	2.34	—	—	0.58	0.001	3.1
Sodium	600	1.70	—	—	0.690	—	—
	1000	1.60	—	—	0.472	—	—
Water	68	1.936	1.00	318,000	2.10	0.0050	0.34

	SI Units						
	T (°C)	ρ (kg/m³)	s.g. —	K (kPa)	$\mu \times 10^4$ (Pa · s)	σ (N/m)	ρ_v (kPa)
Benzene	20	876.2	0.88	1,034,250	6.56	0.029	10.0
Carbon tetrachloride	20	1,587.4	1.59	1,103,200	9.74	0.026	13.1
Crude oil	20	855.6	0.86	—	71.8	0.03	—
Ethyl alcohol	20	788.6	0.79	1,206,625	12.0	0.022	5.86
Freon-12	15.6	1,345.2	1.35	—	14.8	—	—
	−34.4	1,499.8	—	—	18.3	—	—
Gasoline	20	680.3	0.68	—	2.9	—	55.2
Glycerin	20	1,257.6	1.26	4,343,850	14,939	0.063	0.000014
Hydrogen	−257.2	73.7	—	—	0.21	0.0029	21.4
Jet fuel (JP-4)	15.6	773.1	0.77	—	8.7	0.029	8.96
Mercury	15.6	13,555	13.57	26,201,000	15.6	0.51	0.00017
	315.6	12,833	12.8	—	9.0	—	47.2
Oxygen	−195.6	1,206.0	—	—	2.78	0.015	21.4
Sodium	315.6	876.2	—	—	3.30	—	—
	537.8	824.6	—	—	2.26	—	—
Water	20	998.2	1.00	2,170,500	10.0	0.073	2.34

Source: *Eshbach's Handbook of Engineering Fundamentals*, 4th ed., B.D. Tapley and T.R. Poston, eds., John Wiley and Sons, New York (1990). Reprinted by permission of John Wiley & Sons, Inc.

TABLE B-4 *Properties of Water*

°F	°C	c_p, kJ/kg · °C	ρ, kg/m³	μ, kg/m · s	k, W/m · °C	Pr	$\dfrac{g\beta\rho^2 c_p{}^*}{\mu k}$, 1/m³ · °C
32	0	4.225	999.8	1.79×10^{-3}	0.566	13.25	
40	4.44	4.208	999.8	1.55	0.575	11.35	1.91×10^9
50	10	4.195	999.2	1.31	0.585	9.40	6.34×10^9
60	15.56	4.186	998.6	1.12	0.595	7.88	1.08×10^{10}
70	21.11	4.179	997.4	9.8×10^{-4}	0.604	6.78	1.46×10^{10}
80	26.67	4.179	995.8	8.6	0.614	5.85	1.91×10^{10}
90	32.22	4.174	994.9	7.65	0.623	5.12	2.48×10^{10}
100	37.78	4.174	993.0	6.82	0.630	4.53	3.3×10^{10}
110	43.33	4.174	990.6	6.16	0.637	4.04	4.19×10^{10}
120	48.89	4.174	988.8	5.62	0.644	3.64	4.89×10^{10}
130	54.44	4.179	985.7	5.13	0.649	3.30	5.66×10^{10}
140	60	4.179	983.3	4.71	0.654	3.01	6.48×10^{10}
150	65.55	4.183	980.3	4.3	0.659	2.73	7.62×10^{10}
160	71.11	4.186	977.3	4.01	0.665	2.53	8.84×10^{10}
170	76.67	4.191	973.7	3.72	0.668	2.33	9.85×10^{10}
180	82.22	4.195	970.2	3.47	0.673	2.16	1.09×10^{11}
190	87.78	4.199	966.7	3.27	0.675	2.03	
200	93.33	4.204	963.2	3.06	0.678	1.90	
220	104.4	4.216	955.1	2.67	0.684	1.66	
240	115.6	4.229	946.7	2.44	0.685	1.51	
260	126.7	4.250	937.2	2.19	0.685	1.36	
280	137.8	4.271	928.1	1.98	0.685	1.24	
300	148.9	4.296	918.0	1.86	0.684	1.17	
350	176.7	4.371	890.4	1.57	0.677	1.02	
400	204.4	4.467	859.4	1.36	0.665	1.00	
450	232.2	4.585	825.7	1.20	0.646	0.85	
500	260	4.731	785.2	1.07	0.616	0.83	
550	287.7	5.024	735.5	9.51×10^{-5}			
600	315.6	5.703	678.7	8.68			

Note: $^*Gr_x Pr = \left(\dfrac{g\beta\rho^2 c_p}{\mu k}\right)\Delta T$

Source: J.P. Holman, *Heat Transfer*, McGraw-Hill, New York (1986). Reprinted by permission of McGraw-Hill, Inc.

TABLE B-5 *Properties of Saturated Liquids*

t, °C	ρ, kg/m³	c_p, kJ/ kg · °C	ν, m²/s	k, W/ m · °C	α, m²/s	Pr	β, K⁻¹
			Ammonia, NH_3				
−50	703.69	4.463	0.435×10^{-6}	0.547	1.742×10^{-7}	2.60	
−40	691.68	4.467	0.406	0.547	1.775	2.28	
−30	679.34	4.476	0.387	0.549	1.801	2.15	
−20	666.69	4.509	0.381	0.547	1.819	2.09	
−10	653.55	4.564	0.378	0.543	1.825	2.07	
0	640.10	4.635	0.373	0.540	1.819	2.05	
10	626.16	4.714	0.368	0.531	1.801	2.04	
20	611.75	4.798	0.359	0.521	1.775	2.02	2.45×10^{-3}
30	596.37	4.890	0.349	0.507	1.742	2.01	
40	580.99	4.999	0.340	0.493	1.701	2.00	
50	564.33	5.116	0.330	0.476	1.654	1.99	
			Carbon dioxide, CO_2				
−50	1,156.34	1.84	0.119×10^{-6}	0.0855	0.4021×10^{-7}	2.96	
−40	1,117.77	1.88	0.118	0.1011	0.4810	2.46	
−30	1,076.76	1.97	0.117	0.1116	0.5272	2.22	
−20	1,032.39	2.05	0.115	0.1151	0.5445	2.12	
−10	983.38	2.18	0.113	0.1099	0.5133	2.20	
0	926.99	2.47	0.108	0.1045	0.4578	2.38	
10	860.03	3.14	0.101	0.0971	0.3608	2.80	
20	772.57	5.0	0.091	0.0872	0.2219	4.10	14.00×10^{-3}
30	597.81	36.4	0.080	0.0703	0.279	28.7	
			Sulfur dioxide, SO_2				
−50	1,560.84	1.3595	0.484×10^{-6}	0.242	1.141×10^{-7}	4.24	
−40	1,536.81	1.3607	0.424	0.235	1.130	3.74	
−30	1,520.64	1.3616	0.371	0.230	1.117	3.31	
−20	1,488.60	1.3624	0.324	0.225	1.107	2.93	
−10	1,463.61	1.3628	0.288	0.218	1.097	2.62	
0	1,438.46	1.3636	0.257	0.211	1.081	2.38	
10	1,412.51	1.3645	0.232	0.204	1.066	2.18	
20	1,386.40	1.3653	0.210	0.199	1.050	2.00	1.94×10^{-3}
30	1,359.33	1.3662	0.190	0.192	1.035	1.83	
40	1,329.22	1.3674	0.173	0.185	1.019	1.70	
50	1,299.10	1.3683	0.162	0.177	0.999	1.61	
			Dichlorodifluoromethane (Freon), CCl_2F_2				
−50	1,546.75	0.8750	0.310×10^{-6}	0.067	0.501×10^{-7}	6.2	2.63×10^{-3}
−40	1,518.71	0.8847	0.279	0.069	0.514	5.4	
−30	1,489.56	0.8956	0.253	0.069	0.526	4.8	
−20	1,460.57	0.9073	0.235	0.071	0.539	4.4	
−10	1,429.49	0.9203	0.221	0.073	0.550	4.0	
0	1,397.45	0.9345	0.214	0.073	0.557	3.8	
10	1,364.30	0.9496	0.203	0.073	0.560	3.6	
20	1,330.18	0.9659	0.198	0.073	0.560	3.5	
30	1,295.10	0.9835	0.194	0.071	0.560	3.5	
40	1,257.13	1.0019	0.191	0.069	0.555	3.5	
50	1,215.96	1.0216	0.190	0.067	0.545	3.5	
			Glycerin, $C_3H_5(OH)_3$				
0	1,276.03	2.261	0.00831	0.282	0.983×10^{-7}	84.7×10^3	
10	1,270.11	2.319	0.00300	0.284	0.965	31.0	
20	1,264.02	2.386	0.00118	0.286	0.947	12.5	0.50×10^{-3}
30	1,258.09	2.445	0.00050	0.286	0.929	5.38	
40	1,252.01	2.512	0.00022	0.286	0.914	2.45	
50	1,244.96	2.583	0.00015	0.287	0.893	1.63	

TABLE B-5 *(Continued)*

t, °C	ρ, kg/m^3	c_p, kJ/ kg · °C	ν, m^2/s	k, W/ m · °C	α, m^2/s	Pr	β, K^{-1}
			Ethylene glycol, $C_2H_4(OH)_2$				
0	1,130.75	2.294	57.53×10^{-6}	0.242	0.934×10^{-7}	615	
20	1,116.65	2.382	19.18	0.249	0.939	204	0.65×10^{-3}
40	1,101.43	2.474	8.69	0.256	0.939	93	
60	1,087.66	2.562	4.75	0.260	0.932	51	
80	1,077.56	2.650	2.98	0.261	0.921	32.4	
100	1,058.50	2.742	2.03	0.263	0.908	22.4	
			Engine oil (unused)				
0	899.12	1.796	0.00428	0.147	0.911×10^{-7}	47,100	
20	888.23	1.880	0.00090	0.145	0.872	10,400	0.70×10^{-3}
40	876.05	1.964	0.00024	0.144	0.834	2,870	
60	864.04	2.047	0.839×10^{-4}	0.140	0.800	1,050	
80	852.02	2.131	0.375	0.138	0.769	490	
100	840.01	2.219	0.203	0.137	0.738	276	
120	828.96	2.307	0.124	0.135	0.710	175	
140	816.94	2.395	0.080	0.133	0.686	116	
160	805.89	2.483	0.056	0.132	0.663	84	
			Mercury, Hg				
0	13,628.22	0.1403	0.124×10^{-6}	8.20	42.99×10^{-7}	0.0288	
20	13,579.04	0.1394	0.114	8.69	46.06	0.0249	1.82×10^{-4}
50	13,505.84	0.1386	0.104	9.40	50.22	0.0207	
100	13,384.58	0.1373	0.0928	10.51	57.16	0.0162	
150	13,264.28	0.1365	0.0853	11.49	63.54	0.0134	
200	13,144.94	0.1570	0.0802	12.34	69.08	0.0116	
250	13,025.60	0.1357	0.0765	13.07	74.06	0.0103	
315.5	12,847	0.134	0.0673	14.02	81.5	0.0083	
			Water, H_2O				
0	1,002.28	4.2178×10^3	1.788×10^{-6}	0.552	1.308×10^{-7}	13.6	0.18×10^{-3}
20	1,000.52	4.1818	1.006	0.597	1.430	7.02	
40	994.59	4.1784	0.658	0.628	1.512	4.34	
60	985.46	4.1843	0.478	0.651	1.554	3.02	
80	974.08	4.1964	0.364	0.668	1.636	2.22	
100	960.63	4.2161	0.294	0.680	1.680	1.74	
120	945.25	4.250	0.247	0.685	1.708	1.446	
140	928.27	4.283	0.214	0.684	1.724	1.241	
160	909.69	4.342	0.190	0.680	1.729	1.099	
180	889.03	4.417	0.173	0.675	1.724	1.004	
200	866.76	4.505	0.160	0.665	1.706	0.937	
220	842.41	4.610	0.150	0.652	1.680	0.891	
240	815.66	4.756	0.143	0.635	1.639	0.871	
260	785.87	4.949	0.137	0.611	1.577	0.874	
280.6	752.55	5.208	0.135	0.580	1.481	0.910	
300	714.26	5.728	0.135	0.540	1.324	1.019	

Source: E.R.G. Eckert and R.M. Drake, *Analysis of Heat and Mass Transfer*, McGraw-Hill, New York (1972). Reprinted by permission of McGraw-Hill, Inc.

TABLE B-6 *Properties of Metals*

Metal	Properties at 20°C				Thermal Conductivity k, W/m · °C									
	ρ, kg/m³	c_p, kJ/ kg · °C	k, W/ m · °C	α, m²/s × 10⁵	−100°C −148°F	0°C 32°F	100°C 212°F	200°C 392°F	300°C 572°F	400°C 752°F	600°C 1112°F	800°C 1472°F	1000°C 1832°F	1200°C 2192°F
Aluminum:														
Pure	2,707	0.896	204	8.418	215	202	206	215	228	249				
Al-Cu (Duralumin) 94–96% Al, 3–5% Cu, trace Mg	2,787	0.883	164	6.676	126	159	182	194						
Al-Si (Silumin, copper-bearing) 86.5% Al, 1% Cu	2,659	0.867	137	5.933	119	137	144	152	161					
Al-Si (Alusil) 78–80% Al, 20–22% Si	2,627	0.854	161	7.172	144	157	168	175	178					
Al-Mg-Si 97% Al, 1% Mg, 1% Si, 1% Mn	2,707	0.892	177	7.311		175	189	204						
Lead	11,373	0.130	35	2.343	36.9	35.1	33.4	31.5	29.8					
Iron:														
Pure	7,897	0.452	73	2.034	87	73	67	62	55	48	40	36	35	36
Wrought iron 0.5% C	7,849	0.46	59	1.626		59	57	52	48	45	36	33	33	33
Steel (C max ≈ 1.5%):														
Carbon steel C ≈ 0.5%	7,833	0.465	54	1.474		55	52	48	45	42	35	31	29	31
1.0%	7,801	0.473	43	1.172		43	43	42	40	36	33	29	28	29
1.5%	7,753	0.486	36	0.970		36	36	36	35	33	31	28	28	29
Nickel steel Ni ≈ 0%	7,897	0.452	73	2.026										
20%	7,933	0.46	19	0.526										
40%	8,169	0.46	10	0.279										
80%	8,618	0.46	35	0.872										
Invar 36% Ni	8,137	0.46	10.7	0.286										
Chrome steel Cr = 0%	7,897	0.452	73	2.026	87	73	67	62	55	48	40	36	35	36
1%	7,865	0.46	61	1.665		62	55	52	47	42	36	33	33	
5%	7,833	0.46	40	1.110		40	38	36	36	33	29	29	29	
20%	7,689	0.46	22	0.635		22	22	22	22	24	24	26	29	
Cr-Ni, chrome- nickel: 15% Cr, 10% Ni	7,865	0.46	19	0.526										
18% Cr, 8% Ni (V2A)	7,817	0.46	16.3	0.444		16.3	17	17	19	19	22	26	31	
20% Cr, 15% Ni	7,833	0.46	15.1	0.415										
25% Cr, 20% Ni	7,865	0.46	12.8	0.361										
Tungsten steel W = 0%	7,897	0.452	73	2.026										
1%	7,913	0.448	66	1.858										
5%	8,073	0.435	54	1.525										
10%	8,314	0.419	48	1.391										
Copper:														
Pure	8,954	0.3831	386	11.234	407	386	379	374	369	363	353			
Aluminum bronze 95% Cu, 5% Al	8,666	0.410	83	2.330										
Bronze 75% Cu, 25% Sn	8,666	0.343	26	0.859										
Red brass 85% Cu, 9% Sn, 6% Zn	8,714	0.385	61	1.804		59	71							
Brass 70% Cu, 30% Zn	8,522	0.385	111	3.412	88		128	144	147	147				

TABLE B-6 *(Continued)*

Metal	ρ, kg/m³	c_p, kJ/ kg · °C	k, W/ m · °C	α, m²/s × 10⁵	−100°C −148°F	0°C 32°F	100°C 212°F	200°C 392°F	300°C 572°F	400°C 752°F	600°C 1112°F	800°C 1472°F	1000°C 1832°F	1200°C 2192°F
Copper, cont'd:														
German silver 62% Cu, 15% Ni, 22% Zn	8,618	0.394	24.9	0.733	19.2		31	40	45	48				
Constantan 60% Cu, 40% Ni	8,922	0.410	22.7	0.612	21		22.2	26						
Magnesium:														
Pure	1,746	1.013	171	9.708	178	171	168	163	157					
Mg-Al (electrolytic) 6–8% Al, 1–2% Zn	1,810	1.00	66	3.605		52	62	74	83					
Molybdenum	10,220	0.251	123	4.790	138	125	118	114	111	109	106	102	99	92
Nickel:														
Pure (99.9%)	8,906	0.4459	90	2.266	104	93	83	73	64	59				
Ni-Cr 90% Ni, 10% Cr	8,666	0.444	17	0.444		17.1	18.9	20.9	22.8	24.6				
80% Ni, 20% Cr	8,314	0.444	12.6	0.343		12.3	13.8	15.6	17.1	18.0	22.5			
Silver:														
Purest	10,524	0.2340	419	17.004	419	417	415	412						
Pure (99.9%)	10,524	0.2340	407	16.563	419	410	415	374	362	360				
Tin, pure	7,304	0.2265	64	3.884	74	65.9	59	57						
Tungsten	19,350	0.1344	163	6.271		166	151	142	133	126	112	76		
Zinc, pure	7,144	0.3843	112.2	4.106	114	112	109	106	100	93				

Source: E.R.G. Eckert and R.M. Drake, *Analysis of Heat and Mass Transfer*, McGraw-Hill, New York (1972). Reprinted by permission of McGraw-Hill, Inc.

TABLE B-7 *Properties of Non-Metals*

Substance	Temperature, °C	k, W/m · °C	ρ, kg/m³	C, kJ/kg · °C	α, m²/s × 10⁷
Insulating Material					
Asbestos:					
Loosely packed	−45	0.149			
	0	0.154	470–570	0.816	3.3–4
	100	0.161			
Asbestos-cement boards	20	0.74			
Sheets	51	0.166			
Felt, 40 laminations/in	38	0.057			
	150	0.069			
	260	0.083			
20 laminations/in	38	0.078			
	150	0.095			
	260	0.112			
Corrugated, 4 plies/in	38	0.087			
	93	0.100			
	150	0.119			
Asbestos cement	. . .	2.08			
Balsam wool, 2.2 lb/ft³	32	0.04	35		
Cardboard, corrugated	. . .	0.064			
Celotex	32	0.048			
Corkboard, 10 lb/ft³	30	0.043	160		
Cork, regranulated	32	0.045	45–120	1.88	2–5.3
Ground	32	0.043	150		
Diatomaceous earth (Sil-o-cel)	0	0.061	320		
Felt, hair	30	0.036	130–200		
Wool	30	0.052	330		
Fiber, insulating board	20	0.048	240		
Glass wool, 1.5 lb/ft³	23	0.038	24	0.7	22.6
Insulex, dry	32	0.064			
		0.144			
Kapok	30	0.035			
Magnesia, 85%	38	0.067	270		
	93	0.071			
	150	0.074			
	204	0.080			
Rock wool, 10 lb/ft³	32	0.040	160		
Loosely packed	150	0.067	64		
	260	0.087			
Sawdust	23	0.059			
Silica aerogel	32	0.024	140		
Wood shavings	23	0.059			
Structural and Heat-Resistant Materials					
Asphalt	20–55	0.74–0.76			
Brick:					
Building brick, common	20	0.69	1600	0.84	5.2
Face		1.32	2000		
Carborundum brick	600	18.5			
	1400	11.1			
Chrome brick	200	2.32	3000	0.84	9.2
	550	2.47			9.8
	900	1.99			7.9
Diatomaceous earth, molded and fired	200	0.24			
	870	0.31			
Fireclay brick, burnt 2426°F	500	1.04	2000	0.96	5.4
	800	1.07			
	1100	1.09			

TABLE B-7 *(Continued)*

Substance	Temperature, °C	k, W/m·°C	ρ, kg/m³	C, kJ/kg·°C	α, m²/s × 10⁷
		Insulating Material			
Brick, cont'd:					
Fireclay brick,					
Burnt 2642°F	500	1.28	2300	0.96	5.8
	800	1.37			
	1100	1.40			
Missouri	200	1.00	2600	0.96	4.0
	600	1.47			
	1400	1.77			
Magnesite	200	3.81		1.13	
	650	2.77			
	1200	1.90			
Cement, portland		0.29	1500		
Mortar	23	1.16			
Concrete, cinder	23	0.76			
Stone 1-2-4 mix	20	1.37	1900–2300	0.88	8.2–6.8
Glass, window	20	0.78 (avg)	2700	0.84	3.4
Corosilicate	30–75	1.09	2200		
Plaster, gypsum	20	0.48	1440	0.84	4.0
Metal lath	20	0.47			
Wood lath	20	0.28			
Stone					
Granite		1.73–3.98	2640	0.82	8–18
Limestone	100–300	1.26–1.33	2500	0.90	5.6–5.9
Marble		2.07–2.94	2500–2700	0.80	10–13.6
Sandstone	40	1.83	2160–2300	0.71	11.2–11.9
Wood (across the grain):					
Balsa 8.8 lb/ft³	30	0.055	140		
Cypress	30	0.097	460		
Fir	23	0.11	420	2.72	0.96
Maple or oak	30	0.166	540	2.4	1.28
Yellow pine	23	0.147	640	2.8	0.82
White pine	30	0.112	430		

Source: J.P. Holman, *Heat Transfer*, McGraw-Hill, New York (1986). Reprinted by permission of McGraw-Hill, Inc.

TABLE C-1 Physical and Combustion Properties of Selected Fuels in Air

Fuel	Mol. wt.	Spec. grav.	T_{Boil} (°C)	Heat of vap. (kJ/kg)	Heat of comb. (mJ/kg)	Stoichiometry % Vol.	Stoichiometry f^*	Flammability Limits (% stoichio.) Lean	Flammability Limits (% stoichio.) Rich	Spont. Ign. Temp. (°C)	Fuel for Max. Flame Speed (% stoichio.)	Max. Flame Speed (cm/sec)	Flame Temp. at Max. Fl. Speed K	Ign. Energy Stoich. (10^{-5} cal.)	Ign. Energy Min. (10^{-5} cal.)	Quenching Dist. Stoich. mm	Quenching Dist. Min. mm
Acetaldehyde	44.1	0.783	−56.7	569.4	—	0.0772	0.1280	—	—	—	—	—	—	8.99	—	2.29	—
Acetone	58.1	0.792	56.7	523.0	30.8	0.0497	0.1054	59	233	561.1	131	50.18	2,121	27.48	—	3.81	—
Acetylene	26.0	0.621	−83.9	—	48.2	0.0772	0.0755	31	—	305.0	133	155.25	—	0.72	—	0.76	—
Acrolein	56.1	0.841	52.8	—	—	0.0564	0.1163	48	752	277.8	100	61.75	2,461	4.18	—	1.52	—
Acrylonitrile	53.1	0.797	78.3	—	—	0.0528	0.1028	87	—	481.1	105	46.75	2,600	8.60	3.82	2.29	1.52
Ammonia	17.0	0.817	−33.3	1,373.6	—	0.2181	0.1645	—	—	651.1	—	—	—	—	—	—	—
Aniline	93.1	1.022	184.4	432.6	—	0.0263	0.0872	—	—	593.3	—	—	—	—	—	—	—
Benzene	78.1	0.885	80.0	431.8	39.9	0.0277	0.0755	43	336	591.7	108	44.60	2,365	13.15	5.38	2.79	1.78
Benzyl alcohol	108.1	1.050	205.0	—	—	0.0240	0.0923	—	—	427.8	—	—	—	—	—	—	—
1,2-Butadiene (methylallene)	54.1	0.658	11.1	—	45.5	0.0366	0.0714	—	—	—	117	63.90	2,419	5.60	—	1.30	—
n-Butane	58.1	0.584	−0.5	385.8	45.7	0.0312	0.0649	54	330	430.6	113	41.60	2,256	18.16	6.21	3.05	1.78
Butanone (methylethyl ketone)	72.1	0.805	79.4	—	—	0.0366	0.0951	—	—	—	100	39.45	—	12.67	6.69	2.54	2.03
1-Butene	56.1	0.601	−6.1	443.9	45.3	0.0377	0.0678	53	353	443.3	116	47.60	2,319	—	—	—	—
d-Camphor	152.2	0.990	203.4	—	—	0.0153	0.0818	—	—	466.1	—	—	—	—	—	—	—
Carbon disulfide	76.1	1.263	46.1	351.0	—	0.0652	0.1841	18	1,120	120.0	102	54.46	—	0.36	—	0.51	—
Carbon monoxide	28.0	—	−190.0	211.7	—	0.2950	0.4064	34	676	608.9	170	42.88	—	—	—	—	—
Cyclobutane	56.1	0.703	12.8	—	—	0.0377	0.0678	—	—	—	115	62.18	2,308	—	—	—	—
Cyclohexane	84.2	0.783	80.6	258.1	43.8	0.0227	0.0678	48	401	270.0	117	42.46	2,250	32.98	5.33	4.06	1.78
Cyclohexene	82.1	0.810	82.8	—	—	0.0240	0.0701	—	—	—	—	44.17	—	20.55	—	3.30	—
Cyclopentane	70.1	0.751	49.4	388.3	44.2	0.0271	0.0678	—	—	385.0	117	41.17	2,264	19.84	—	3.30	—
Cyclopropane	42.1	0.720	−34.4	—	—	0.0444	0.0678	58	276	497.8	113	52.32	2,328	5.74	5.50	1.78	1.78
trans-Decalin	138.2	0.874	187.2	—	—	0.0142	0.0692	—	—	271.7	109	33.88	2,222	—	—	—	—
n-Decane	142.3	0.734	174.0	359.8	44.2	0.0133	0.0666	45	356	231.7	105	40.31	2,286	—	—	2.06	—
Diethyl ether	74.1	0.714	34.4	351.6	—	0.0337	0.0896	55	2,640	185.6	115	43.74	2,253	11.71	6.69	2.54	2.03
Ethane	30.1	—	−88.9	488.3	47.4	0.0564	0.0624	50	272	472.2	112	44.17	2,244	10.04	5.74	2.29	1.78
Ethyl acetate	88.1	0.901	77.2	—	26.8	0.0402	0.1279	61	236	486.1	100	35.59	—	33.94	11.47	4.32	2.54
Ethanol	46.1	0.789	78.5	836.8	—	0.0652	0.1115	—	—	392.2	—	—	—	—	—	—	—
Ethylamine	45.1	0.706	16.7	611.3	—	0.0528	0.0873	—	—	—	—	—	—	57.36	—	5.33	—
Ethylene oxide	44.1	1.965	10.6	581.1	—	0.0772	0.1280	53	450	428.9	125	100.35	2,411	2.51	1.48	1.27	1.02
Furan	68.1	0.936	32.2	400.0	—	0.0444	0.1098	—	—	—	—	—	—	5.40	—	1.78	—
n-Heptane	100.2	0.688	98.5	364.9	44.4	0.0187	0.0661	53	450	247.2	122	42.46	2,214	27.49	5.74	3.81	1.78
n-Hexane	86.2	0.664	68.0	364.9	44.7	0.0216	0.0659	51	400	260.6	117	42.46	2,239	22.71	5.50	3.56	1.78
Hydrogen	2.0	—	−252.7	451.0	119.9	0.2950	0.0290	—	—	571.1	170	291.19	2,380	0.36	0.36	0.51	0.51
iso-Propanol	60.1	0.785	82.2	664.8	—	0.0444	0.0969	—	—	455.6	100	38.16	—	15.54	—	2.79	—
Kerosene	154.0	0.825	250.0	290.8	43.1	—	—	—	—	—	—	—	—	—	—	—	—
Methane	16.0	—	−161.7	509.2	50.0	0.0947	0.0581	46	164	632.2	106	37.31	2,236	7.89	6.93	2.54	2.03
Methanol	32.0	0.793	64.5	1,100.9	19.8	0.1224	0.1548	48	408	470.0	101	52.32	—	5.14	3.35	1.78	1.52
Methyl formate	60.1	0.975	31.7	472.0	—	0.0947	0.2181	—	—	—	—	—	—	14.82	—	2.79	—
n-Nonane	128.3	0.772	150.6	288.3	44.6	0.0147	0.0665	47	434	238.9	—	—	—	—	—	—	—
n-Octane	114.2	0.707	125.6	300.0	44.8	0.0165	0.0633	51	425	240.0	—	—	2,251	—	—	—	—

TABLE C-1   *(Continued)*

Fuel	Mol. wt.	Spec. grav.	T_{Boil} (°C)	Heat of vap. (kJ/kg)	Heat of comb. (mJ/kg)	Stoichiometry % Vol.	Stoichiometry f*	Flammability Limits (% stoichio.) Lean	Flammability Limits (% stoichio.) Rich	Spont. Ign. Temp. (°C)	Fuel for Max. Flame Speed (% stoichio.)	Max. Flame Speed (cm/sec)	Flame Temp. at Max. Fl. Speed K	Ign. Energy (10^{-5} cal.) Stoich.	Ign. Energy (10^{-5} cal.) Min.	Quenching Dist. (mm) Stoich.	Quenching Dist. (mm) Min.
n-Pentane	72.1	0.631	36.0	364.4	45.3	0.0255	0.0654	54	359	284.4	115	42.46	2,250	19.60	5.26	3.30	1.78
1-Pentene	70.1	0.646	30.0	—	45.0	0.0271	0.0678	47	370	298.3	114	46.75	2,314	—	—	—	—
Propane	44.1	0.508	−42.2	425.5	46.3	0.0402	0.0640	51	283	504.4	114	42.89	2,250	7.29	—	2.03	1.78
Propene	42.1	0.522	−47.7	437.2	45.8	0.0444	0.0678	48	272	557.8	114	48.03	2,339	6.74	—	2.03	—
n-Propanol	60.1	0.804	97.2	685.8	—	0.0444	0.0969	—	—	433.3	—	—	—	—	—	—	—
Toulene	92.1	0.872	110.6	362.8	40.9	0.0227	0.0743	43	322	567.8	105	38.60	2,344	27.48	—	3.81	—
Triethylamine	101.2	0.723	89.4	—	—	0.0210	0.0753	—	—	252.2	—	—	—	—	—	—	—
Turpentine	—	—	—	—	—	—	—	—	—	—	—	—	—	—	—	—	—
Xylene	106.0	0.870	130.0	334.7	43.1	—	—	—	—	298.9	—	—	—	—	—	—	—
Gasoline 73 octane	120.0	0.720	155.0	338.9	44.1	—	—	—	—	468.3	106	37.74	—	—	—	—	—
Gasoline 100 octane	—	—	—	—	—	—	—	—	—	248.9	107	36.88	—	—	—	—	—
Jet fuel JP1	150.0	0.810	—	—	43.0	0.0130	0.0680	—	—	—	—	—	—	—	—	—	—
JP3	112.0	0.760	—	—	43.5	0.0170	0.0680	—	—	261.1	—	—	—	—	—	—	—
JP4	126.0	0.780	—	—	43.5	0.0150	0.0680	—	—	—	107	38.17	—	—	—	—	—
JP5	170.0	0.830	—	—	43.0	0.0110	0.0690	—	—	242.2	—	—	—	—	—	—	—

*f is the stoichiometric air/fuel ratio; i.e., $f = 1/r$.

Source: A.M. Kanury, *Introduction to Combustion Phenomena, Combustion Science and Technology, Vol. 2*, Gordon and Breach Science Publishers, New York (1975). Reprinted with modifications by permission of Gordon and Breach Science Publishers.

TABLE C-2 Heats of Combustion and Related Properties of Pure Substances

Material	Composition	W Molecular Weight	Δh_c^u Gross (MJ/kg)	Δh_c^l Net (MJ/kg)	$\Delta h_c^l/r_o$ (MJ/kg O_2)	r_o Oxygen-Fuel Mass Ratio	T_b Boiling Temp. (°C)	Δh_v Latent Heat of Vaporization (kJ/kg)	C_{pl} Liquid Heat Capacity (kJ/kg-°C)	C_{pv} Vapor Heat Capacity (kJ/kg-°C)
Acetaldehyde	C_2H_4O	44.05	27.07	25.07	13.81	1.816	20.8	—	1.94	1.24
Acetic acid	$C_2H_4O_2$	60.05	14.56	13.09	12.28	1.066	118.1	395		1.11
Acetone	C_3H_6O	58.08	30.83	28.56	12.96	2.204	56.5	501	2.12	1.29
Acetylene	C_2H_2	26.04	49.91	48.22	15.70	3.072	−84.0	—	—	1.69
Acrolein	C_3H_4O	56.06	29.08	27.51	13.77	1.998	52.5	505	—	1.17
Acrylonitrile	C_3H_3N	53.06	33.16	31.92	14.11	2.262	77.3	615	2.10	1.20
(Allene) → propadiene										
Ammonium perchlorate*	NH_4ClO_4	117.49	2.35	2.16	3.97	0.545	—	—	—	
iso-Amyl alcohol	$C_5H_{12}O$	88.15	37.48	34.49	12.67	2.723	132.0	501	2.90	1.50
Aniline	C_6H_7N	93.12	36.44	34.79	13.06	2.663	184.4	478	2.08	1.16
Benzaldehyde	C_7H_6O	106.12	33.25	32.01	13.27	2.412	179.2	385	1.61	
Benzene	C_6H_6	78.11	41.83	40.14	13.06	3.073	80.1	389	1.72	1.05
Benzoic acid*	$C_7H_6O_2$	122.12	26.43	25.35	12.90	1.965	250.8	415	—	0.85
Benzyl alcohol	C_7H_8O	108.13	34.56	32.93	13.09	2.515	205.7	467	2.00	1.19
Bicyclohexyl	$C_{12}H_{22}$	166.30	45.35	42.44	12.61	3.367	236.	263		
1,2-Butadiene	C_4H_6	54.09	47.95	45.51	13.99	3.254	10.8	—	—	1.48
1,3-Butadiene	C_4H_6	54.09	46.99	44.55	13.69	3.254	−4.4	—	—	1.47
(1,3-Butadiyne) → diacetylene										
n-Butane	C_4H_{10}	58.12	49.50	45.72	12.77	3.579	−0.5	—	2.30	1.68
iso-Butane	C_4H_{10}	58.12	48.95	45.17	12.62	3.579	−11.8	—	—	1.67
1-Butene	C_4H_8	56.10	48.44	45.31	13.24	3.422	−6.2	—	—	1.53
n-Butylamine	$C_4H_{11}N$	73.14	41.75	38.45	12.84	2.994	77.8	372	2.57	1.62
d-Camphor*	$C_{10}H_{16}O$	152.23	38.75	36.44	12.84	2.838	203.4	—	—	0.82
Carbon*	C	12.01	32.80	32.80	12.31	2.664	4200.	—	—	0.71
Carbon disulfide	CS_2	76.13	6.34	6.34	5.03	1.261	46.5	351	1.00	0.60
Carbon monoxide	CO	28.01	10.10	10.10	17.69	0.571	−191.3	—	—	1.04
Cellulose*	$C_6H_{10}O_5$	162.14	17.47	16.12	13.61	1.184	—	—	1.16	—
(Chloroethylene) → vinyl chloride										
(Chloroform) → trichloromethane										
Chlorotrifluoroethylene	C_2F_3Cl	116.47	2.00	2.00	3.64	0.549	−28.3	188	1.34	0.72
m-Cresol	C_7H_8O	108.13	34.26	32.64	12.98	2.515	202.2	399	2.00	1.13
Cumene	C_9H_{12}	120.19	43.40	41.20	12.90	3.195	152.3	312	1.77	1.26
Cyanogen	C_2N_2	52.04	21.06	21.06	17.12	1.230	−21.2	—	—	1.12
Cyclobutane	C_4H_8	56.10	48.91	45.77	13.38	3.422	12.9	—	—	1.29
Cyclohexane	C_6H_{12}	84.16	46.58	43.45	12.70	3.422	80.7	357	1.84	1.26
Cyclohexene	C_6H_{10}	82.14	45.67	42.99	12.99	3.311	82.8	371	1.80	1.28
Cyclohexylamine	$C_6H_{13}N$	99.18	41.05	38.17	12.79	2.984	134.5			
Cyclopentane	C_5H_{10}	70.13	46.93	43.80	12.80	3.422	49.3	389	2.23	1.18
Cyclopropane	C_3H_6	42.08	49.70	46.57	13.61	3.422	−32.9	—	1.92	1.33
(Decahydronaphthalene) → cis-decalin										
cis-Decalin	$C_{10}H_{18}$	138.24	45.49	42.63	12.70	3.356	195.8	309	1.67	1.21
n-Decane	$C_{10}H_{22}$	142.28	47.64	44.24	12.69	3.486	174.1	276	2.19	1.85
Diacetylene	C_4H_2	50.06	46.60	45.72	15.89	2.877	10.3	—	—	1.47
(Diamine) → hydrazine										
Diborane	H_6B_2	27.69	79.80	79.80	23.02	3.467	−92.5	—	—	1.75
Dichloromethane	CH_2Cl_2	84.94	6.54	6.02	10.65	0.565	39.7	330	1.18	0.80
Ciethyl cyclohexane	$C_{10}H_{20}$	140.26	46.30	43.17	12.58	3.422	174.		1.87	
Diethyl ether	$C_4H_{10}O$	74.12	36.75	33.79	13.04	2.590	34.6	360	2.34	1.52
(2,4 Diisocyanotoulene) → toluene diisocyanate										
(Diisopropyl ether) → iso-propyl ether										
Dimethylamine	C_2H_7N	45.08	38.66	35.25	13.24	2.662	6.9	—	—	1.80
(Dimethyl aniline) → xylidene										
Dimethyldecalin	$C_{12}H_{22}$	166.30	45.70	42.79	13.15	3.254	220.	260		
(Dimethyl ether) → methyl ether										
1,1-Dimethylhydrazine (UDMH)	$C_2H_8N_2$	60.10	32.95	30.03	14.10	2.130	25.	578	2.73	
Dimethyl sulfoxide	C_2H_6SO	78.13	29.88	28.19	15.30	1.843	189.	677	1.89	1.14

TABLE C-2 *(Continued)*

Material	Composition	W Molecular Weight	Δh_c^u Gross (MJ/kg)	Δh_c^l Net (MJ/kg)	$\Delta h_c^l/r_o$ (MJ/kg O_2)	r_o Oxygen-Fuel Mass Ratio	T_b Boiling Temp. (°C)	Δh_v Latent Heat of Vaporization (kJ/kg)	C_{pl} Liquid Heat Capacity (kJ/kg-°C)	C_{pv} Vapor Heat Capacity (kJ/kg-°C)
1,3 Dioxane	$C_4H_8O_2$	88.10	26.57	24.58	9.66	2.543	105.	404		
1,4 Dioxane	$C_4H_8O_2$	88.10	26.83	24.84	9.77	2.543	101.1	406	1.74	1.07
Ethane	C_2H_6	30.07	51.87	47.49	12.75	3.725	−88.6	—	—	1.75
Ethanol	C_2H_6O	46.07	29.67	26.81	12.87	2.084	78.5	837	2.43	1.42
(Ethene) → ethylene										
Ethyl acetate	$C_4H_8O_2$	88.10	25.41	23.41	12.89	1.816	77.2	367	1.94	1.29
Ethyl acrylate	$C_5H_8O_2$	100.12	27.44	25.69	13.39	1.918	100.	290		1.14
Ethylamine	C_2H_7N	45.08	38.63	35.22	13.23	2.662	16.5	—	2.89	1.61
Ethyl benzene	C_8H_{10}	106.16	43.00	40.93	12.93	3.165	136.1	339	1.75	1.21
Ethylene	C_2H_4	28.05	50.30	47.17	13.78	3.422	−103.9	—	2.38	1.56
Ethylene glycol	$C_2H_6O_2$	62.07	19.17	17.05	13.22	1.289	197.5	800	2.43	1.56
Ethylene oxide	C_2H_4O	44.05	29.65	27.65	15.23	1.816	10.7	—	1.97	1.10
(Ethylene trichloride) → trichloroethylene										
(Ethyl ether) → diethyl ether										
Formaldehyde	CH_2O	30.03	18.76	17.30	16.23	1.066	−19.3	—	—	1.18
Formic acid	CH_2O_2	46.03	5.53	4.58	13.15	0.348	100.5	476	2.15	0.98
Furan	C_4H_4O	68.07	30.61	29.32	13.86	2.115	31.4	398	1.69	0.96
α-D-glucose*	$C_6H_{12}O_6$	180.16	15.55	14.08	13.21	1.066	—	—	—	—
(Glycerine) → glycerol										
Glycerol	$C_3H_8O_3$	92.10	17.95	16.04	13.19	1.216	290.0	800	2.42	1.25
(Glycerol trinitrate) → nitroglycerin										
n-Heptane	C_7H_{16}	100.20	48.07	44.56	12.68	3.513	98.4	316	2.20	1.66
n-Heptene	C_7H_{14}	98.18	47.44	44.31	12.95	3.422	93.6	317	2.17	1.58
Hexadecane	$C_{16}H_{34}$	226.43	47.25	43.95	12.70	3.462	286.7	226	2.22	1.64
Hexamethyldisiloxane	$C_6H_{18}Si_2O$	162.38	38.30	35.80	15.16	2.364	100.1	192	2.01	—
(Hexamethylenetetramine) → methenamine										
n-Hexane	C_6H_{14}	86.17	48.31	44.74	12.68	3.528	68.7	335	2.24	1.66
n-Hexene	C_6H_{12}	84.16	47.57	44.44	12.99	3.422	63.5	333	2.18	1.57
Hydrazine	H_4N_2	32.05	52.08	49.34	49.40	0.998	113.5	1180	3.08	1.65
Hydrazoic acid	HN_3	43.02	15.28	14.77	79.40	0.186	35.7	690	—	1.02
Hydrogen	H_2	2.00	141.79	130.80	16.35	8.000	−252.7	—	—	14.42
(Hydrogen azide) → hydrazoic acid										
Hydrogen cyanide	HCN	27.03	13.86	13.05	8.82	1.480	25.7	933	2.61	1.33
Hydrogen sulfide	H_2S	34.08	48.54	47.25	16.77	2.817	−60.3	548	—	1.00
Maleic anhydride*	$C_4H_2O_3$	74.04	18.77	18.17	14.01	1.297	202.0	—	—	—
Melamine*	$C_3H_6N_6$	126.13	15.58	14.54	12.73	1.142	—	—	—	—
Methane	CH_4	16.04	55.50	50.03	12.51	4.000	−161.5	—	—	2.23
Methanol	CH_4O	32.04	22.68	19.94	13.29	1.500	64.8	1101	2.37	1.37
Methenamine*	$C_6H_{12}N_4$	140.19	29.97	28.08	13.67	2.054	—	—	—	—
2-Methoxyethanol	$C_3H_8O_2$	76.09	24.23	21.92	13.03	1.682	124.4	583	2.23	—
Methylamine	CH_5N	31.06	34.16	30.62	13.21	2.318	−6.3	—	—	1.61
(2-Methyl 1-butanol) → iso-amyl alcohol										
(Methyl chloride) → dichloromethane										
Methyl ether	C_2H_6O	46.07	31.70	28.84	13.84	2.084	−24.9	—	—	1.43
Methyl ethyl ketone	C_4H_8O	72.10	33.90	31.46	12.89	2.441	79.6	434	2.30	1.43
1-Methylnaphthalene	$C_{11}H_{10}$	142.19	40.88	39.33	12.95	3.038	244.7	323	1.58	1.12
Methyl methacrylate	$C_5H_8O_2$	100.11	27.37	25.61	12.33	2.078	101.0	360	1.91	—
Methyl nitrate	CH_3NO_3	77.04	8.67	7.81	75.10	0.104	64.6	409	2.04	0.99
2-Methyl propane) → iso-butane										
Naphthalene*	$C_{10}H_8$	128.16	40.21	38.84	12.96	2.996	217.9	—	1.18	1.03
Nitrobenzene	$C_6H_5NO_2$	123.11	25.11	24.22	14.90	1.625	210.7	330	1.52	—
Nitroglycerin	$C_3H_5N_3O_9$	227.09	6.82	6.34	—	—	unstable	462	1.49	—
Nitromethane	CH_3NO_2	61.04	11.62	10.54	15.08	0.699	101.1	567	1.74	0.94
n-Nonane	C_9H_{20}	128.25	47.76	44.33	12.69	3.493	150.6	295	2.10	1.65
Octamethyl-cyclotetrasiloxane	$C_8H_{24}Si_4O_4$	296.62	26.90	25.10	14.56	1.725	175.0	127	1.88	—
n-Octane	C_8H_{18}	114.22	47.90	44.44	12.69	3.502	125.6	301	2.20	1.65
iso-Octane	C_8H_{18}	114.22	47.77	44.31	12.65	3.502	117.7	272	2.15	1.65

TABLE C-2 (Continued)

Material	Composition	W Molecular Weight	Δh_c^u Gross (MJ/kg)	Δh_c^l Net (MJ/kg)	$\Delta h_c^l/r_o$ (MJ/kg O_2)	r_o Oxygen-Fuel Mass Ratio	T_b Boiling Temp. (°C)	Δh_v Latent Heat of Vaporization (kJ/kg)	C_{pl} Liquid Heat Capacity (kJ/kg-°C)	C_{pv} Vapor Heat Capacity (kJ/kg-°C)
1-Octene	C_8H_{10}	112.21	47.33	44.20	12.92	3.422	121.3	301	2.19	1.59
(1-Octylene) → 1-octene										
1,2-Pentadiene	C_5H_8	68.11	47.31	44.71	13.60	3.288	44.9	405	2.21	1.55
n-Pentane	C_5H_{12}	72.15	48.64	44.98	12.68	3.548	36.0	357	2.33	1.67
1-Pentene	C_5H_{10}	70.13	47.77	44.64	13.04	3.422	30.0	359	2.16	1.56
Phenol*	C_6H_6O	94.11	32.45	31.05	13.05	2.380	181.8	433	1.43	1.10
Phosgene	$COCl_2$	98.92	1.74	1.74	10.74	0.162	8.3	247	1.02	0.58
Propadiene	C_3H_4	40.06	48.54	46.35	14.51	3.195	−34.6	—	—	1.44
Propane	C_3H_8	44.09	50.35	46.36	12.78	3.629	−42.2	—	2.23	1.67
n-Propanol	C_3H_8O	60.09	33.61	30.68	12.81	2.396	97.2	686	2.50	1.45
iso-Propanol	C_3H_8O	60.09	33.38	30.45	12.71	2.396	80.3	663	2.42	1.48
Propene	C_3H_6	42.08	48.92	45.79	13.38	3.422	−47.7	—	—	1.52
(iso-Propylbenzene) → cumene										
(Propylene) → propene										
iso-Propyl ether	$C_6H_{14}O$	102.17	39.26	36.25	12.86	2.819	67.8	286	2.14	1.55
Propyne	C_3H_4	40.06	48.36	46.17	14.45	3.195	−23.3	—	—	1.51
Styrene	C_8H_8	104.14	42.21	40.52	13.19	3.073	145.2	356	1.76	1.17
Sucrose*	$C_{12}H_{22}O_{11}$	342.30	16.49	15.08	13.44	1.122	—	—	1.24	—
1,2,3,4-Tetrahydronaphthalene) → tetralin										
Tetralin	$C_{10}H_{12}$	132.20	42.60	40.60	12.90	3.147	207.0	425	1.64	1.19
Tetranitromethane	CN_4O_8	196.04	2.20	2.20	—	—	125.7	196	—	—
Toluene	C_7H_8	92.13	42.43	40.52	12.97	3.126	110.4	360	1.67	1.12
Toluene diisocyanate	$C_9H_6N_2O_2$	174.16	24.32	23.56	13.50	1.746	120.0	—	1.65	—
Triethanolamine	$C_6H_{15}NO_3$	149.19	29.29	27.08	15.30	1.770	360.0	—	—	—
Triethylamine	$C_6H_{15}N$	101.19	43.19	39.93	12.95	3.083	89.5	303	2.22	1.59
1,1,2-Trichloroethane	$C_2H_3Cl_3$	133.42	7.77	7.28	11.02	0.660	114.0	260	1.11	0.67
Trichloroethylene	C_2HCl_3	131.40	6.77	6.60	12.05	0.548	86.9	245	1.07	0.61
Trichloromethane	$CHCl_3$	119.39	3.39	3.21	9.60	0.335	61.7	249	0.97	0.55
Trinitromethane	CHN_3O_6	151.04	3.41	3.25	—	—	unstable	—	—	—
Trinitrotoluene*	$C_7H_5N_3O_6$	227.13	15.12	14.64	19.80	0.740	240.0	322	1.40	—
Trioxane	$C_3H_6O_3$	90.08	16.57	15.11	14.17	1.066	114.5	450	—	—
Urea*	CH_4ON_2	60.06	10.52	9.06	11.34	0.799	—	—	—	1.55
Vinyl acetate	$C_4H_6O_2$	86.09	24.18	22.65	13.54	1.673	72.5	167	2.00	1.05
Vinyl acetylene	C_4H_4	52.07	47.05	45.36	14.76	3.073	5.1	—	—	1.41
Vinyl bromide	C_2H_3Br	106.96	12.10	11.48	13.95	0.823	15.6	—	2.42	0.53
Vinyl chloride	C_2H_3Cl	62.50	20.02	16.86	11.97	1.408	−13.8	—	—	0.86
(Vinyl trichloride) → 1,1,2-trichlorothane										
Xylenes	C_8H_{10}	106.16	42.89	40.82	12.90	3.165	138–144	343	1.72	1.21
Xylidene	$C_8H_{11}N$	121.22	38.28	36.29	12.79	2.838	192.7	366	1.77	—

*Denotes substance in crystalline solid form; otherwise, liquid if $T_b > 25°C$, gaseous if $T_b < 25°C$.

Source: *Fire Protection Handbook*, 17th ed., A.E. Cote and J. Linville, eds., National Fire Protection Association, Quincy, MA (1991).

TABLE C-3 *Heats of Combustion and Related Properties of Plastics*

Material	Unit Composition	W Molecular Weight	Δh_c^u Gross (MJ/kg)	Δh_c^l Net (MJ/kg)	$\Delta h_c^l/r_o$ (MJ/kg O_2)	r_o Oxygen-Fuel Mass Ratio	C_{ps} Heat Capacity Solid (kJ/kg-°C)
Acrylonitrile-butadiene styrene copolymer	—	—	35.25	33.75			1.41–1.59
Bisphenol A epoxy	$C_{11.85}H_{20.37}O_{2.83}N_{0.3}$	212.10	33.53	31.42	13.41	2.343	
Butadiene-acrylonitrile 37% copolymer	—	—	39.94				
Butadiene/styrene 8.58% copolymer	$C_{4.18}H_{6.09}$	56.30	44.84	42.49	13.11	3.241	1.94
Butadiene/styrene 25.5% copolymer	$C_{4.60}H_{6.29}$	61.55	44.19	41.95	13.07	3.209	1.82
Cellulose acetate (triacetate)	$C_{12}H_{16}O_8$	288.14	18.88	17.66	13.25	1.333	1.34
Cellulose acetate-butyrate	$C_{12}H_{18}O_7$	274.27	23.70	22.3	14.67	1.517	1.70
Epoxy, unhardened	$C_{31}H_{36}O_{5.5}$	496.63	32.92	31.32	13.05	2.400	
Epoxy, hardened	$C_{39}H_{40}O_{8.5}$	644.74	30.27	28.90	13.01	2.221	
Melamine formaldehyde (Formica™)	$C_6H_6N_6$	162.08	19.33	18.52	12.51	1.481	1.46
Nylon 6	$C_6H_{11}NO$	113.08	30.1–31.7	28.0–29.6	12.30	2.335	1.52
Nylon 6,6	$C_{12}H_{22}N_2O_2$	226.16	31.6–31.7	29.5–29.6	12.30	2.405	1.70
Nylon 11 (Rilsan)	$C_{11}H_{21}NO$	183.14	36.99	34.47	12.33	2.796	1.70–2.30
Phenol formaldehyde -foam	$C_{15}H_{12}O_2$	224.17	27.9–31.6 21.6–27.4	26.7–30.4 20.2–26.2	11.80	2.427	1.70
Polyacenaphthalene	$C_{12}H_8$	152.14	39.23	38.14	12.95	2.945	
Polyacrylonitrile	C_3H_3N	53.04	32.22	30.98	13.70	2.262	1.50
Polyallylphthalate (Polyamides) → nylon	$C_{14}H_{14}O$	198.17	27.74	26.19	9.54	2.745	
Poly-1,4-butadiene	C_4H_6	54.05	45.19	42.75	13.13	3.256	
Poly-1-butene	C_4H_8	56.05	46.48	43.35	12.65	3.426	1.88
Polycarbonate	$C_{16}H_{14}O_3$	254.19	30.99	29.78	13.14	2.266	1.26
Polycarbon suboxide	C_3O_2	68.03	13.78	13.78	14.64	0.941	
Polychlorotrifluorethylene	C_2F_3Cl	116.47	1.12	1.12	2.04	0.549	0.92
Polydiphenylbutadiene	$C_{16}H_{10}$	202.18	39.30	38.2	13.05	2.928	
Polyester, unsaturated	$C_{5.77}H_{6.25}O_{1.63}$	101.60	21.6–29.8	20.3–28.5	11.90	2.053	1.20–2.30
polyether, chlorinated	$C_5H_8OCl_2$	154.97	17.84	16.71	12.45	1.342	
Polyethylene	C_2H_4	28.03	46.2–46.5	43.1–43.4	12.63	3.425	1.83–2.30
Polyethylene oxide	C_2H_4O	44.02	26.65	24.66	13.57	1.817	
Polyethylene terephthalate	$C_{10}H_8O_4$	192.11	22.18	21.27	12.77	1.666	1.00
Polyformaldehyde	CH_2O	30.01	16.93	15.86	14.88	1.066	1.46
Poly-1-hexene sulfone	$C_6H_{12}SO_2$	148.13	29.78	28.00	14.40	1.944	
Polyhydrocyanic acid	HCN	27.02	23.26	22.45	15.17	1.480	
(Polyisobutylene) → poly-1-butene							
Polyisocyanurate foam	—		26.3	22.2–26.2			
Polyisoprene	C_5H_8	68.06	44.90	42.30	12.90	3.291	
Poly-3-methyl-1-butene	C_5H_{10}	70.06	46.55	43.42	12.67	3.426	
Polymethyl methacrylate	$C_5H_8O_2$	100.06	26.64	24.88	12.97	1.919	1.44
Poly-4-methyl-1-pentene	C_6H_{12}	84.08	46.52	43.39	12.67	3.425	2.18
Poly-α-methylstyrene	C_9H_{10}	118.11	42.31	40.45	13.00	3.116	
Polynitroethylene	$C_2H_3O_2N$	73.03	15.96	15.06	19.64	0.767	
Polyoxymethylene	CH_2O	30.01	16.93	15.65	14.68	1.066	
Polyoxytrimethylene	C_3H_6O	58.04	31.52	29.25	13.27	2.205	
Poly-1-pentene	C_5H_{10}	70.06	45.58	42.45	12.39	3.426	
Polyphenylacetylene	C_8H_6	102.09	40.00	38.70	13.00	2.978	
Polyphenylene oxide	C_8H_8O	120.09	34.59	33.13	13.09	2.531	1.34
Polypropene sulfone	$C_3H_6SO_2$	106.10	23.82	22.58	16.64	1.357	
Poly-β-propiolactone	$C_3H_4O_2$	72.14	19.35	18.13	13.62	1.331	
Polypropylene	C_3H_6	42.04	46.37	43.23	12.62	3.824	2.10
Polypropylene oxide	C_3H_6O	58.04	31.17	28.90	13.11	2.205	

TABLE C-3 *(Continued)*

Material	Unit Composition	W Molecular Weight	Δh_c^u Gross (MJ/kg)	Δh_c^l Net (MJ/kg)	$\Delta h_c^l/r_o$ (MJ/kg O_2)	r_o Oxygen-Fuel Mass Ratio	C_{ps} Heat Capacity Solid (kJ/kg-°C)
Polystyrene	C_8H_8	104.10	41.4–42.5	39.7–39.8	12.93	3.074	1.40
Polystyrene-foam	—		39.7	35.6–40.8			
Polystyrene-foam, FR	—		41.2–42.9				
Polysulfones, butene	$C_4H_8SO_2$	120.11	24.04–26.47	22.25–25.01	14.79	1.598	1.30
Polysulfur	S	32.06	9.72	9.72	9.74	0.998	
Polytetrafluoroethylene	C_2F_4	100.02	5.00	5.00	7.81	0.640	1.02
Polytetrahydrofuran	C_4H_8O	72.05	34.39	31.85	13.04	2.443	
Polyurea	$C_{15}H_{18}O_4N_4$	318.20	24.91	23.67	13.45	1.760	
Polyurethane	$C_{6.3}H_{7.1}NO_{2.1}$	130.30	23.90	22.70	13.16	1.725	1.75–1.84
Polyurethane-foam	—		26.1–31.6	23.2–28.0			
Polyurethane-foam, FR	—		24.0–25.0				
Polyvinyl acetate	$C_4H_6O_2$	86.05	23.04	21.51	12.86	1.673	
Polyvinyl alcohol	$C_2H_4)$	44.03	25.00	23.01	12.66	1.817	1.70
Polyvinyl butyral	$C_8H_{14}O_2$	142.10	32.90	30.70	13.00	2.365	
Polyvinyl chloride	C_2H_3Cl	62.48	17.95	16.90	12.00	1.408	0.90–1.20
Polyvinyl-foam	—		22.83				1.30–2.10
Polyvinyl fluoride	C_2H_3F	46.02	21.70	20.27	10.60	1.912	
Polyvinylidene chloride	$C_2H_2Cl_2$	96.93	10.52	10.07	12.21	0.825	1.34
Polyvinylidene fluoride	$C_2H_2F_2$	64.02	14.77	14.08	11.26	1.250	1.38
Urea formaldehyde	$C_3H_6O_2N_2$	102.05	15.90	14.61	13.31	1.098	1.60–2.10
Urea formaldehyde-foam	—	—	14.80				

Source: *Fire Protection Handbook*, 17th ed., A.E. Cote and J. Linville, eds., National Fire Protection Association, Quincy, MA (1991).

TABLE C-4 *Heats of Combustion of Miscellaneous Materials*

Material	Δh_c^u Gross (MJ/kg)	Δh_c^ℓ Net (MJ/kg)
Acetate (see cellulose acetate)		
Acrylic fiber	30.6–30.8	
Blasting powder	2.1–2.4	
Butter	38.5	
Celluloid (cellulose nitrate and camphor)	17.5–20.6	16.4–19.2
Cellulose acetate fiber, $C_8H_{12}O_6$	17.8–18.4	16.4–17.0
Cellulose diacetate fiber, $C_{10}H_{14}O_7$	18.7	
Cellulose nitrate, $C_6H_9N_1O_7/C_6H_8N_2O_9/C_6H_7N_3O_{11}$	9.11–13.48	
Cellulose triacetate fiber, $C_{12}H_{16}O_8$	18.8	17.6
Charcoal	33.7–34.7	33.2–34.2
Coal—anthracite	30.9–34.6	30.5–34.2
—bituminous	24.7–36.3	23.6–35.2
Coke	28.0–31.0	28.0–31.0
Cork	26.1	
Cotton	16.5–20.4	
Dynamite	5.4	
Epoxy, $C_{11.9}H_{20.4}O_{2.8}N_{0.3}/C_{6.064}H_{7.550}O_{1.222}$	32.8–33.5	31.1–31.4
Fat, animal	39.8	
Flint powder	3.0–3.1	
Fuel oil—No. 1	46.1	
—No. 6	42.5	
Gasketing—chlorosulfonated polyethylene (Hypalon)	28.5	
—vinylidene fluoride/hexafluoropropylene (Fluorel, Viton A)	14.0–15.1	
Gasoline	46.8	43.7
Jet fuel—JP1		43.0
—JP3		43.5
—JP4	46.6	43.5
—JP5	45.9	43.0

TABLE C-4 *(Continued)*

Material	Δh_c^u Gross (MJ/kg)	Δh_c^ℓ Net (MJ/kg)
Kerosene (jet fuel A)	46.4	43.3
Lanolin (wool fat)	40.8	
Lard	40.1	
Leather	18.2–19.8	
Lignin, $C_{2.6}H_3O$	24.7–26.4	23.4–25.1
Lignite	22.4–33.3	
Modacrylic fiber	24.7	
Naphtha	43.0–47.1	40.9–43.9
Neoprene, C_5H_5Cl—gum	24.3	
—foam	9.7–26.8	
Nomex™ (polymethaphenylene isophthalamide) fiber, $C_{14}H_{10}O_2N_2$	27.0–28.7	
Oil—castor	37.1	
—linseed	39.2–39.4	
—mineral	45.8–46.0	
—olive	39.6	
—solar	41.8	
Paper—brown	16.3–17.9	
—magazine	12.7	
—newsprint	19.7	
—wax	21.5	
Paraffin wax	46.2	43.1
Peat	16.7–21.6	
Petroleum jelly ($C_{7.118}H_{12.957}O_{0.091}$)	45.9	
Rayon fiber	13.6–19.5	
Rubber—buna N	34.7–35.6	
—butyl	45.8	
—isoprene (natural) C_5H_8	44.9	42.3
—latex foam	33.9–40.6	
—GRS	44.2	
—tire, auto	32.6	
Silicone rubber (SiC_2H_6O)	15.5–16.8	
—foam	14.0–19.5	
Sisal	15.9	
Spandex fiber	31.4	
Starch	17.6	16.2
Straw	15.6	
Sulfur—rhombic		9.28
—monoclinic		9.29
Tobacco	15.8	
Wheat	15.0	
Wood—beech	20.0	18.7
—birch	20.0	18.7
—douglas fir	21.0	19.6
—maple	19.1	17.8
—red oak	20.2	18.7
—spruce	21.8	20.4
—white pine	19.2	17.8
—hardboard	19.9	
Woodflour	19.8	
Wool	20.7–26.6	

Source: *Fire Protection Handbook*, 17th ed., A.E. Cote and J. Linville, eds., National Fire Protection Association, Quincy, MA (1991).

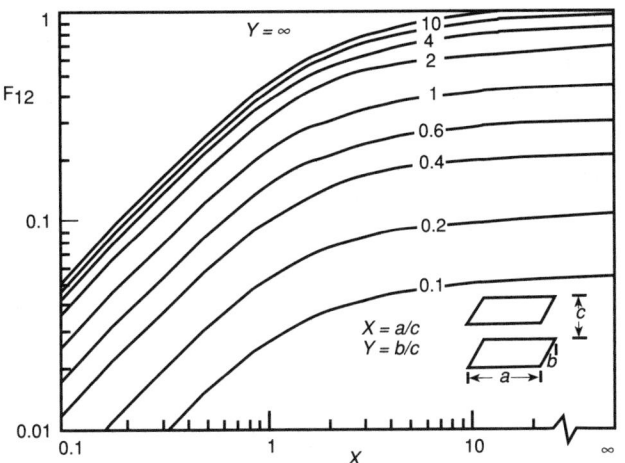

Fig. D-1. *View factor for parallel rectangular plates. [Source: Heat Transfer in Fires, P. Blackshear, ed., Scripta Book Company, Washington, DC (1974).]*

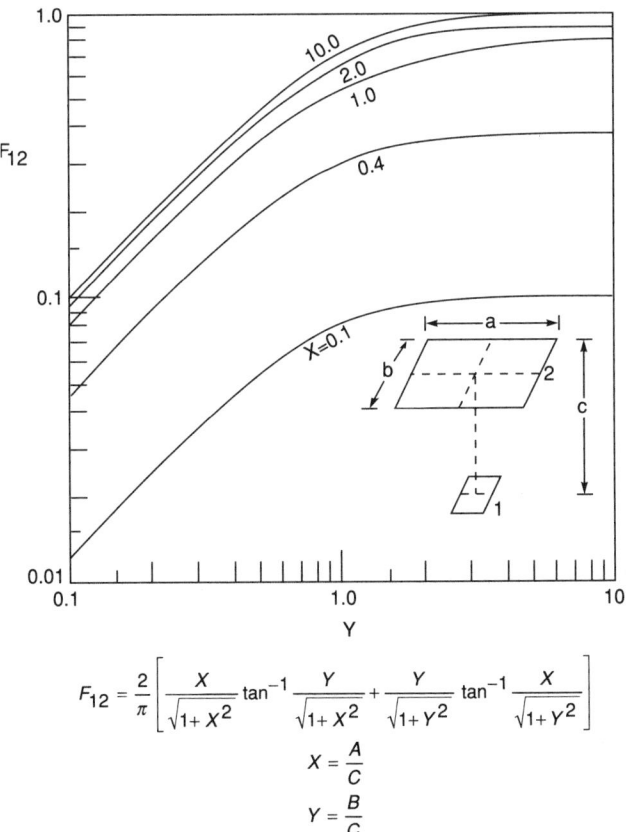

$$F_{12} = \frac{2}{\pi}\left[\frac{X}{\sqrt{1+X^2}}\tan^{-1}\frac{Y}{\sqrt{1+X^2}} + \frac{Y}{\sqrt{1+Y^2}}\tan^{-1}\frac{X}{\sqrt{1+Y^2}}\right]$$

$$X = \frac{A}{C}$$

$$Y = \frac{B}{C}$$

Fig. D-2. *View factor for parallel rectangular radiator. [Source: Heat Transfer in Fires, P. Blackshear, ed., Scripta Book Company, Washington, DC (1974).]*

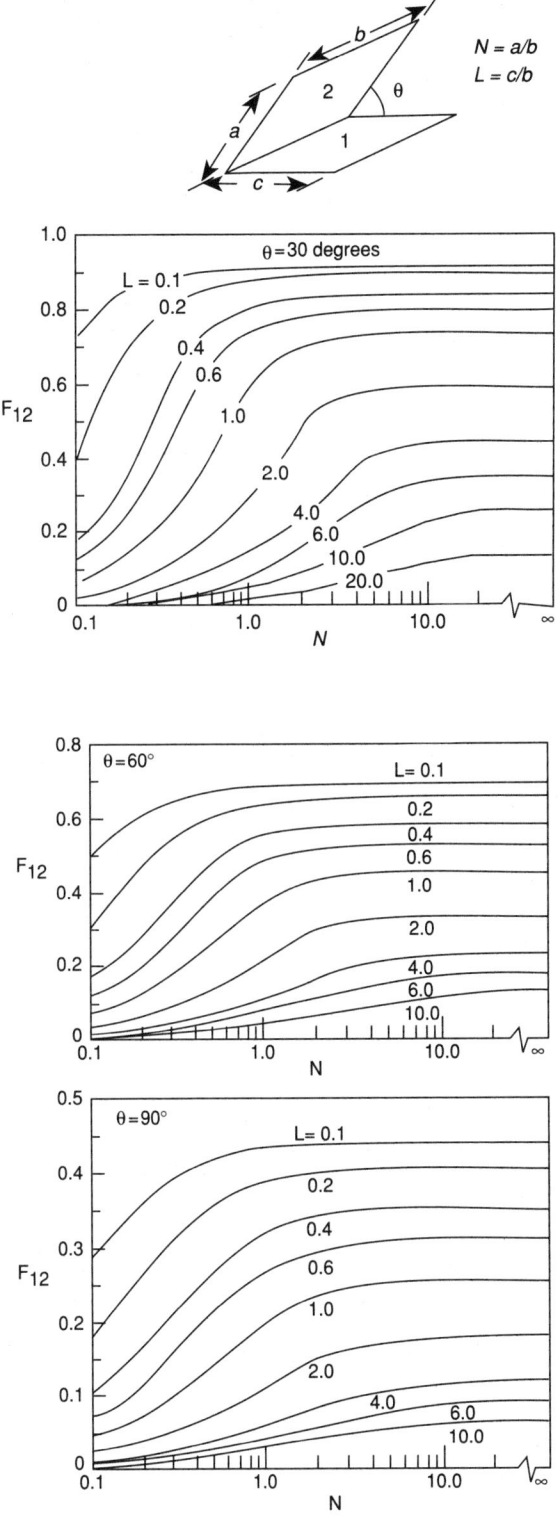

Fig. D-3. *View factor for rectangular plates at various angles. [Source: Heat Transfer in Fires, P. Blackshear, ed., Scripta Book Company, Washington, DC (1974).]*

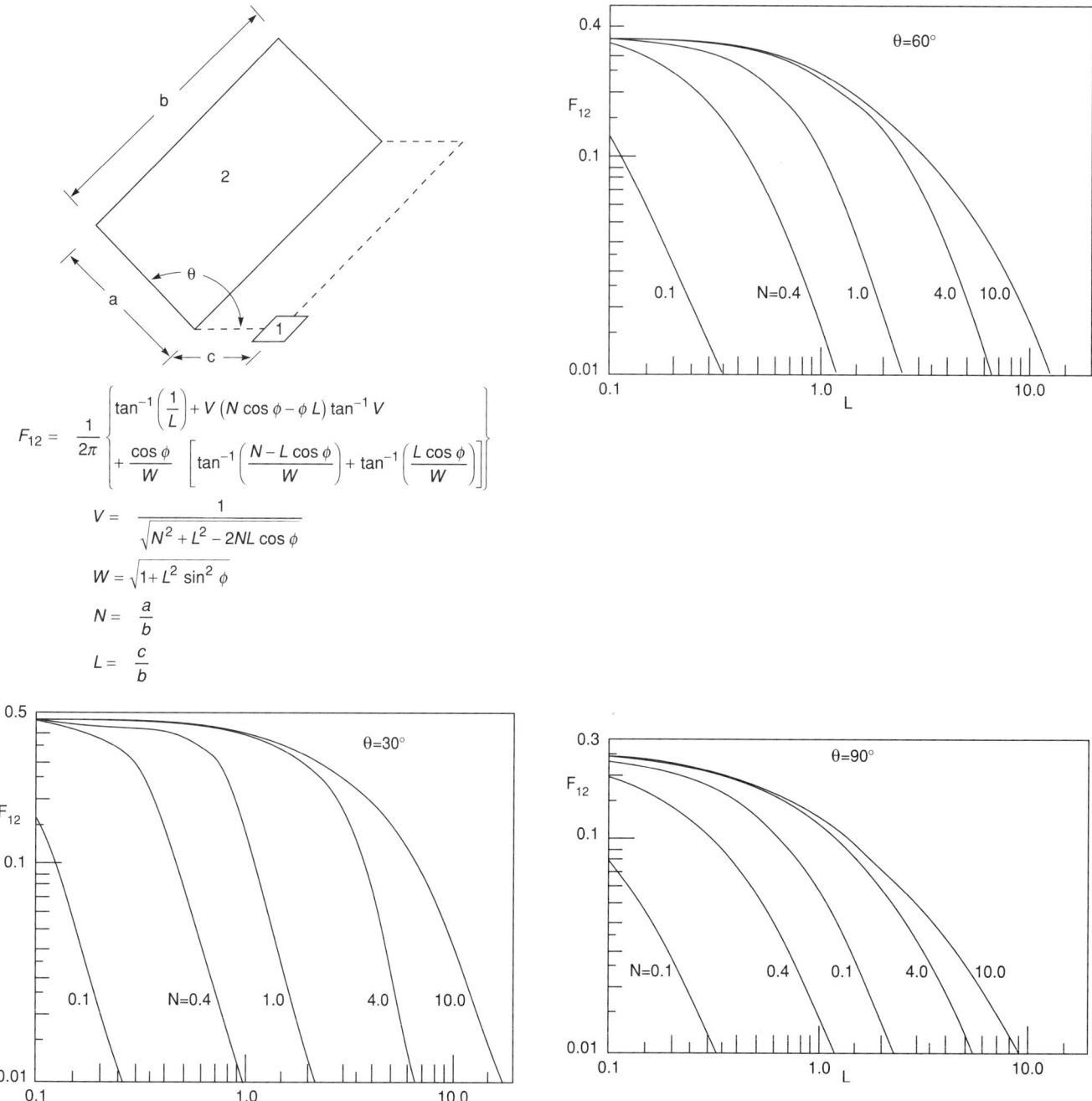

$$F_{12} = \frac{1}{2\pi}\left\{\begin{array}{l} \tan^{-1}\left(\dfrac{1}{L}\right) + V\left(N\cos\phi - \phi\,L\right)\tan^{-1}V \\[2mm] + \dfrac{\cos\phi}{W}\left[\tan^{-1}\left(\dfrac{N-L\cos\phi}{W}\right) + \tan^{-1}\left(\dfrac{L\cos\phi}{W}\right)\right]\end{array}\right\}$$

$$V = \frac{1}{\sqrt{N^2 + L^2 - 2NL\cos\phi}}$$

$$W = \sqrt{1 + L^2 \sin^2\phi}$$

$$N = \frac{a}{b}$$

$$L = \frac{c}{b}$$

Fig. D-4.   *View factor for rectangular radiator to differential area at various angles. [Source:* **Heat Transfer in Fires***, P. Blackshear, ed.,* **Scripta Book Company***, Washington, DC (1974).]*

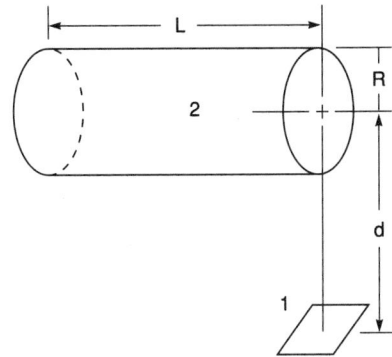

$$F_{12}= \frac{1}{\pi D} \tan^{-1}\left(\frac{L}{\sqrt{D^2-1}}\right) + \frac{L}{\pi}\left[\frac{A-2D}{D\sqrt{AB}}\tan^{-1}\sqrt{\frac{A(D-1)}{B(D+1)}} - \frac{1}{D}\tan^{-1}\sqrt{\frac{D-1}{D+1}}\right]$$

$$D= \frac{d}{r} \quad , \quad L= \frac{L}{R}$$

$$A= (D+1)^2 + L^2 \;,\; B= (D-1)^2 + L^2$$

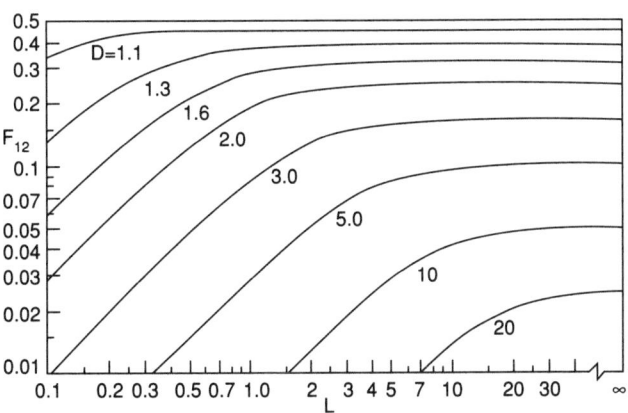

Fig. D-5. *Cylindrical radiator to parallel receiver. [Source: Heat Transfer in Fires, P. Blackshear, ed., Scripta Book Company, Washington, DC (1974).]*

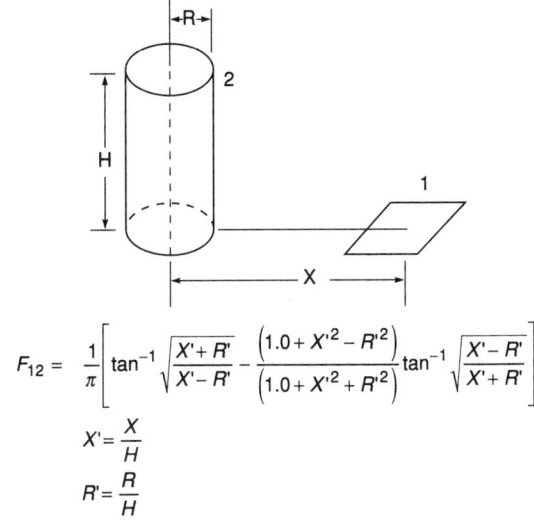

$$F_{12}= \frac{1}{\pi}\left[\tan^{-1}\sqrt{\frac{X'+R'}{X'-R'}} - \frac{(1.0+X'^2-R'^2)}{(1.0+X'^2+R'^2)}\tan^{-1}\sqrt{\frac{X'-R'}{X'+R'}}\right]$$

$$X'= \frac{X}{H}$$

$$R'= \frac{R}{H}$$

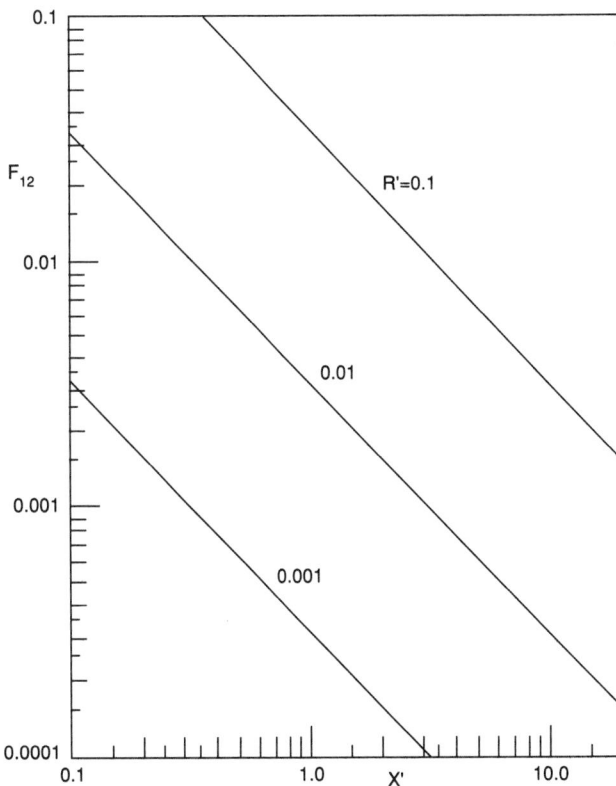

Fig. D-6. *View factor for cylindrical radiator to normal target. [Source: Heat Transfer in Fires, P. Blackshear, ed., Scripta Book Company, Washington, DC (1974).]*

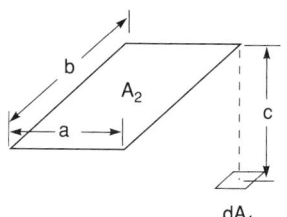

PLANE ELEMENT dA_1 TO PLANE PARALLEL RECTANGLE; NORMAL TO ELEMENT PASSES THROUGH CORNER OF RECTANGLE.

$$X = \frac{a}{c} \qquad Y = \frac{b}{c}$$

$$F_{d1-2} = \frac{1}{2\pi}\left(\frac{X}{\sqrt{1+X^2}} \tan^{-1} \frac{Y}{\sqrt{1+X^2}} + \frac{Y}{\sqrt{1+Y^2}} \tan^{-1} \frac{X}{\sqrt{1+Y^2}} \right)$$

PLANE ELEMENT dA_1 TO RECTANGLE IN PLANE 90° TO PLANE OF ELEMENT.

$$X = \frac{a}{b} \qquad Y = \frac{c}{b}$$

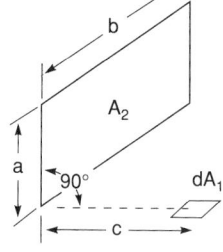

$$F_{d1-2} = \frac{1}{2\pi}\left(\tan^{-1} \frac{1}{Y} - \frac{Y}{\sqrt{X^2+Y^2}} \tan^{-1} \frac{1}{\sqrt{X^2+Y^2}} \right)$$

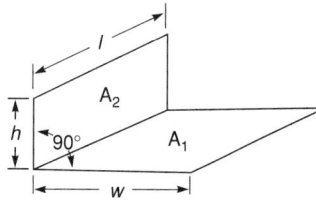

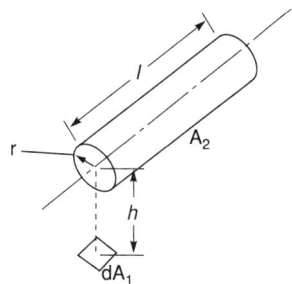

$$F_{1-2} = \frac{1}{\pi W}\left(W \tan^{-1}\frac{1}{W} + H \tan^{-1}\frac{1}{H} - \sqrt{H^2+W^2}\tan^{-1}\frac{1}{\sqrt{H^2+W^2}} \right.$$
$$\left. + \frac{1}{4}\ln\left\{ \frac{(1+W^2)(1+H^2)}{1+W^2+H^2}\left[\frac{W^2(1+W^2+H^2)}{(1+W^2)(W^2+H^2)} \right]^{W^2}\left[\frac{H^2(1+H^2+W^2)}{(1+H^2)(H^2+W^2)} \right]^{H^2} \right\} \right)$$

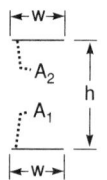

TWO INFINITELY LONG, DIRECTLY OPPOSED PARALLEL PLATES OF THE SAME FINITE WIDTH.

$$H = \frac{h}{w}$$

$$F_{1-2} = F_{2-1} = \sqrt{1+H^2} - H$$

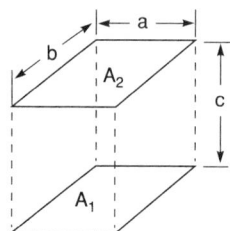

IDENTICAL, PARALLEL, DIRECTLY OPPOSED RECTANGLES.

$$X = \frac{a}{c} \qquad Y = \frac{b}{c}$$

$$F_{1-2} = \frac{2}{\pi XY}\left\{ \ln\left[\frac{(1+X^2)(1+Y^2)}{1+X^2+Y^2} \right]^{\frac{1}{2}} + X\sqrt{1+Y^2}\,\tan^{-1}\frac{X}{\sqrt{1+Y^2}} \right.$$
$$\left. + Y\sqrt{1+X^2}\,\tan^{-1}\frac{Y}{\sqrt{1+X^2}} - X\tan^{-1}X - Y\tan^{-1}Y \right\}$$

TWO INFINITELY LONG PLATES OF UNEQUAL WIDTHS h AND w, HAVING ONE COMMON EDGE AND HAVING AN ANGLE OF 90° TO EACH OTHER.

$$H = \frac{h}{w}$$
$$F_{1-2} = \frac{1}{2}\left(1+H-\sqrt{1+H^2} \right)$$

TWO FINITE RECTANGLES OF SAME LENGTH, HAVING ONE COMMON EDGE AND HAVING AN ANGLE OF 90° TO EACH OTHER.

$$H = \frac{h}{l} \qquad W = \frac{w}{l}$$

PLANE ELEMENT dA_1 TO RIGHT CIRCULAR CYLINDER OF FINITE LENGTH l AND RADIUS r; NORMAL TO ELEMENT PASSES THROUGH ONE END OF CYLINDER AND IS PERPENDICULAR TO CYLINDER AXIS.

$$L = \frac{l}{r} \qquad H = \frac{h}{r}$$

$$X = (1+H)^2 + L^2$$
$$Y = (1-H)^2 + L^2$$

$$F_{d1-2} = \frac{1}{\pi H}\tan^{-1}\frac{L}{\sqrt{H^2-1}} + \frac{L}{\pi}\left[\frac{X-2H}{H\sqrt{XY}}\tan^{-1}\sqrt{\frac{X(H-1)}{Y(H+1)}} - \frac{1}{H}\tan^{-1}\sqrt{\frac{H-1}{H+1}} \right]$$

Fig. D-7. *View factor equations for various geometries. [Source: R. Seigel and J.R. Howell,* Thermal Radiation Heat Transfer, *3rd ed., Taylor & Francis, Washington, DC (1992).] Reproduced with permission. All rights reserved.*

TABLE E-1 *Properties of Steel Pipe**

Nominal Pipe Size, in.	Outside Diam., in.	Schedule No.	Wall Thickness, in.	Inside Diam., in.	Cross-Sectional Area Metal, sq in.	Cross-Sectional Area Flow, sq ft	Circumference, ft, or Surface, sq ft/ft of length Outside	Circumference, ft, or Surface, sq ft/ft of length Inside	Capacity at 1 ft/sec Velocity U.S. gal./min.	Capacity at 1 ft/sec Velocity lb/hr water	Weight of Plain-end Pipe, lb/ft
1/8	0.405	10S	0.049	0.307	0.055	0.00051	0.106	0.0804	0.231	115.5	0.19
		40ST, 40S	.068	.269	.072	.00040	.106	.0705	.179	89.5	.24
		80XS, 80S	.095	.215	.093	.00025	.106	.0563	.113	56.5	.31
1/4	0.540	10S	0.065	.410	.097	.00092	.141	.107	.412	206.5	.33
		40ST, 40S	.088	.364	.125	.00072	.141	.095	.323	161.5	.42
		80XS, 80S	.119	.302	.157	.00050	.141	.079	.224	112.0	.54
3/8	0.675	10S	.065	.545	.125	.00162	.177	.143	.727	363.5	.42
		40ST, 40S	.091	.493	.167	.00133	.177	.129	.596	298.0	.57
		80XS, 80S	.126	.423	.217	.00098	.177	.111	.440	220.0	.74
1/2	0.840	5S	.065	.710	.158	.00275	.220	.186	1.234	617.0	.54
		10S	.083	.674	.197	.00248	.220	.176	1.112	556.0	.67
		40ST, 40S	.109	.622	.250	.00211	.220	.163	0.945	472.0	.85
		80XS, 80S	.147	.546	.320	.00163	.220	.143	0.730	365.0	1.09
		160	.188	.464	.385	.00117	.220	.122	0.527	263.5	1.31
		XX	.294	.252	.504	.00035	.220	.066	0.155	77.5	1.71
3/4	1.050	5S	.065	.920	.201	.00461	.275	.241	2.072	1036.0	0.69
		10S	.083	.884	.252	.00426	.275	.231	1.903	951.5	.86
		40ST, 40S	.113	.824	.333	.00371	.275	.216	1.665	832.5	1.13
		80XS, 80S	.154	.742	.433	.00300	.275	.194	1.345	672.5	1.47
		160	.219	.612	.572	.00204	.275	.160	0.917	458.5	1.94
		XX	.308	.434	.718	.00103	.275	.114	0.461	230.5	2.44
1	1.315	5S	.065	1.185	.255	.00768	.344	.310	3.449	1725	0.87
		10S	.109	1.097	.413	.00656	.344	.287	2.946	1473	1.40
		40ST, 40S	.133	1.049	.494	.00600	.344	.275	2.690	1345	1.68
		80XS, 80S	.179	0.957	.639	.00499	.344	.250	2.240	1120	2.17
		160	.250	0.815	.836	.00362	.344	.213	1.625	812.5	2.84
		XX	.358	0.599	1.076	.00196	.344	.157	0.878	439.0	3.66
1 1/4	1.660	5S	.065	1.530	0.326	.01277	.435	.401	5.73	2865	1.11
		10S	.109	1.442	0.531	.01134	.435	.378	5.09	2545	1.81
		40ST, 40S	.140	1.380	0.668	.01040	.435	.361	4.57	2285	2.27
		80XS, 80S	.191	1.278	0.881	.00891	.435	.335	3.99	1995	3.00
		160	.250	1.160	1.107	.00734	.435	.304	3.29	1645	3.76
		XX	.382	0.896	1.534	.00438	.435	.235	1.97	985	5.21
1 1/2	1.900	5S	.065	1.770	0.375	.01709	.497	.463	7.67	3835	1.28
		10S	.109	1.682	0.614	.01543	.497	.440	6.94	3465	2.09
		40ST, 40S	.145	1.610	0.800	.01414	.497	.421	6.34	3170	2.72
		80XS, 80S	.200	1.500	1.069	.01225	.497	.393	5.49	2745	3.63
		160	.281	1.338	1.429	.00976	.497	.350	4.38	2190	4.86
		XX	.400	1.100	1.885	.00660	.497	.288	2.96	1480	6.41
2	2.375	5S	.065	2.245	0.472	.02749	.622	.588	12.34	6170	1.61
		10S	.109	2.157	0.776	.02538	.622	.565	11.39	5695	2.64
		40ST, 40S	.154	2.067	1.075	.02330	.622	.541	10.45	5225	3.65
		80ST, 80S	.218	1.939	1.477	.02050	.622	.508	9.20	4600	5.02
		160	.344	1.687	2.195	.01552	.622	.436	6.97	3485	7.46
		XX	.436	1.503	2.656	.01232	.622	.393	5.53	2765	9.03
2 1/2	2.875	5S	.083	2.709	0.728	.04003	.753	.709	17.97	8985	2.48
		10S	.120	2.635	1.039	.03787	.753	.690	17.00	8500	3.53
		40ST, 40S	.203	2.469	1.704	.03322	.753	.647	14.92	7460	5.79
		80XS, 80S	.276	2.323	2.254	.02942	.753	.608	13.20	6600	7.66
		160	.375	2.125	2.945	.02463	.753	.556	11.07	5535	10.01
		XX	.552	1.771	4.028	.01711	.753	.464	7.68	3840	13.70

*Source: *Chemical Engineers Handbook*, R.H. Perry and C.H. Chilton, eds., McGraw-Hill, New York (1973). Reprinted by permission of McGraw-Hill, Inc.

TABLE E-1   *(Continued)*

Nominal Pipe Size, in.	Outside Diam., in.	Schedule No.	Wall Thickness, in.	Inside Diam., in.	Cross-Sectional Area		Circumference, ft, or Surface, sq. ft/ft of Length		Capacity at 1 ft/sec Velocity		Weight of Plain-end Pipe, lb/ft
					Metal, sq in.	Flow, sq ft	Outside	Inside	U.S. gal./ min.	lb/hr water	
3	3.500	5S	0.083	3.334	0.891	0.06063	0.916	0.873	27.21	13,605	3.03
		10S	.120	3.260	1.274	.05796	.916	.853	26.02	13,010	4.33
		40ST, 40S	.216	3.068	2.228	.05130	.916	.803	23.00	11,500	7.58
		80XS, 80S	.300	2.900	3.016	.04587	.916	.759	20.55	10,275	10.25
		160	.438	2.624	4.213	.03755	.916	.687	16.86	8430	14.31
		XX	.600	2.300	5.466	.02885	.916	.602	12.95	6475	18.58
3½	4.0	5S	.083	3.834	1.021	.08017	1.047	1.004	35.98	17,990	3.48
		10S	.120	3.760	1.463	.07711	1.047	0.984	34.61	17,305	4.97
		40ST, 40S	.226	3.548	2.680	.06870	1.047	0.929	30.80	15,400	9.11
		80XS, 80S	.318	3.364	3.678	.06170	1.047	0.881	27.70	13,850	12.51
4	4.5	5S	.083	4.334	1.152	.10245	1.178	1.135	46.0	23,000	3.92
		10S	.120	4.260	1.651	.09898	1.178	1.115	44.4	22,200	5.61
		40ST, 40S	.237	4.026	3.17	.08840	1.178	1.054	39.6	19,800	10.79
		80XS, 80S	.337	3.826	4.41	.07986	1.178	1.002	35.8	17,900	14.98
		120	.438	3.624	5.58	.07170	1.178	0.949	32.2	16,100	18.98
		160	.531	3.438	6.62	.06647	1.178	0.900	28.9	14,450	22.52
		XX	.674	3.152	8.10	.05419	1.178	0.825	24.3	12,150	27.54
5	5.563	5S	.109	5.345	1.87	.1558	1.456	1.399	69.9	34,950	6.36
		10S	.134	5.295	2.29	.1529	1.456	1.386	68.6	34,300	7.77
		40ST, 40S	.258	5.047	4.30	.1390	1.456	1.321	62.3	31,150	14.62
		80XS, 80S	.375	4.813	6.11	.1263	1.456	1.260	57.7	28,850	20.78
		120	.500	4.563	7.95	.1136	1.456	1.195	51.0	25,500	27.04
		160	.625	4.313	9.70	.1015	1.456	1.129	45.5	22,750	32.96
		XX	.750	4.063	11.34	.0900	1.456	1.064	40.4	20,200	38.55
6	6.625	5S	.109	6.407	2.23	.2239	1.734	1.677	100.5	50,250	7.60
		10S	.134	6.357	2.73	.2204	1.734	1.664	98.9	49,450	9.29
		40ST, 40S	.280	6.065	5.58	.2006	1.734	1.588	90.0	45,000	18.97
		80XS, 80S	.432	5.761	8.40	.1810	1.734	1.508	81.1	40,550	28.57
		120	.562	5.501	10.70	.1650	1.734	1.440	73.9	36,950	36.42
		160	.719	5.187	13.34	.1467	1.734	1.358	65.9	32,950	45.34
		XX	.864	4.897	15.64	.1308	1.734	1.282	58.7	29,350	53.16
8	8.625	5S	.109	8.407	2.915	.3855	2.258	2.201	173.0	86,500	9.93
		10S	.148	8.329	3.941	.3784	2.258	2.180	169.8	84,900	13.40
		20	.250	8.125	6.578	.3601	2.258	2.127	161.5	80,750	22.36
		30	.277	8.071	7.265	.3553	2.258	2.113	159.4	79,700	24.70
		40ST,40S	.322	7.981	8.399	.3474	2.258	2.089	155.7	77,850	28.55
		60	.406	7.813	10.48	.3329	2.258	2.045	149.4	74,700	35.66
		80XS,80S	.500	7.625	12.76	.3171	2.258	1.996	142.3	71,150	43.39
		100	.594	7.437	14.99	.3017	2.258	1.947	135.4	67,700	50.93
		120	.719	7.187	17.86	.2817	2.258	1.882	126.4	63,200	60.69
		140	.812	7.001	19.93	.2673	2.258	1.833	120.0	60,000	67.79
		XX	.875	6.875	21.30	.2578	2.258	1.800	115.7	57,850	72.42
		160	.906	6.813	21.97	.2532	2.258	1.784	113.5	56,750	74.71
10	10.75	5S	.134	10.842	4.47	.5993	2.814	2.744	269.0	134,500	15.19
		10S	.165	10.420	5.49	.5922	2.814	2.728	265.8	132,900	18.65
		20	.250	10.250	8.25	.5731	2.814	2.685	257.0	128,500	28.04
		30	.307	10.136	10.07	.5603	2.814	2.655	252.0	126,000	34.24
		40ST, 40S	.365	10.020	11.91	.5475	2.814	2.620	246.0	123,000	40.48
		80S, 60XS	.500	9.750	16.10	.5185	2.814	2.550	233.0	116,500	54.74
		80	.594	9.562	18.95	.4987	2.814	2.503	233.4	111,700	64.40
		100	.719	9.312	22.66	.4729	2.814	2.438	212.3	106,150	77.00
		120	.844	9.062	26.27	.4479	2.814	2.372	201.0	100,500	89.27
		140, XX	1.000	8.750	30.63	.4176	2.814	2.291	188.0	94,000	104.13
		160	1.125	8.500	34.02	.3941	2.814	2.225	177.0	88,500	115.65

TABLE E-1 *(Continued)*

Nominal Pipe Size, in.	Outside Diam., in.	Schedule No.	Wall Thickness, in.	Inside Diam., in.	Cross-Sectional Area		Circumference, ft, or Surface, sq. ft/ft of Length		Capacity at 1 ft/sec Velocity		Weight of Plain-end Pipe, lb/ft
					Metal, sq in.	Flow, sq ft	Outside	Inside	U.S. gal./min.	lb/hr water	
12	12.75	5S	0.156	12.438	6.17	0.8438	3.338	3.26	378.7	189,350	20.98
		10S	0.180	12.390	7.11	.8373	3.338	3.24	275.8	187,900	24.17
		20	0.250	12.250	9.82	.8185	3.338	3.21	367.0	183,500	33.38
		30	0.330	12.090	12.88	.7972	3.338	3.17	358.0	179,000	43.77
		ST, 40S	0.375	12.000	14.58	.7854	3.338	3.14	352.5	176,250	49.56
		40	0.406	11.938	15.74	.7773	3.338	3.13	349.0	174,500	54.56
		XS, 80S	0.500	11.750	19.24	.7530	3.338	3.08	338.0	169,000	65.42
		60	0.562	11.626	21.52	.7372	3.338	3.04	331.0	165,500	73.72
		80	0.688	11.374	26.07	.7056	3.338	2.98	316.7	158,350	88.57
		100	0.844	11.062	31.57	.6674	3.338	2.90	299.6	149,800	107.29
		120, XX	1.000	10.750	36.91	.6303	3.338	2.81	283.0	141,500	125.49
		140	1.125	10.500	41.09	.6013	3.338	2.75	270.0	135,000	139.68
		160	1.312	10.126	47.14	.5592	3.338	2.65	251.0	125,500	160.33
14	14	5S	0.156	13.688	6.78	1.0219	3.665	3.58	459	229,500	23.07
		10S	0.188	13.624	8.16	1.0125	3.665	3.57	454	227,000	27.73
		10	0.250	13.500	10.80	0.9940	3.665	3.53	446	223,000	36.71
		20	0.312	13.376	13.42	0.9750	3.665	3.50	438	219,000	45.68
		30, ST	0.375	13.250	16.05	0.9575	3.665	3.47	430	215,000	54.57
		40	0.438	13.124	18.66	0.9397	3.665	3.44	422	211,000	63.37
		XS	0.500	13.000	21.21	0.9218	3.665	3.40	414	207,000	72.09
		60	0.594	12.812	25.02	0.8957	3.665	3.35	402	201,000	85.01
		80	0.750	12.500	31.22	0.8522	3.665	3.27	382	191,000	106.13
		100	0.938	12.124	38.49	0.8017	3.665	3.17	360	180,000	130.79
		120	1.094	11.812	44.36	0.7610	3.665	3.09	342	171,000	150.76
		140	1.250	11.500	50.07	0.7213	3.665	3.01	324	162,000	170.22
		160	1.406	11.188	55.63	0.6827	3.665	2.93	306	153,000	189.15
16	16	5S	0.165	15.670	8.21	1.3393	4.189	4.10	601	300,500	27.90
		10S	0.188	15.624	9.34	1.3314	4.189	4.09	598	299,000	31.75
		10	0.250	15.500	12.37	1.3104	4.189	4.06	587	293,500	42.05
		20	0.312	15.376	15.38	1.2985	4.189	4.03	578	289,000	52.36
		30, ST	0.375	15.250	18.41	1.2680	4.189	3.99	568	284,000	62.58
		40, XS	0.500	15.000	24.35	1.2272	4.189	3.93	550	275,000	82.77
		60	0.656	14.688	31.62	1.1766	4.189	3.85	528	264,000	107.54
		80	0.844	14.312	40.19	1.1171	4.189	3.75	501	250,500	136.58
		100	1.031	13.938	48.48	1.0596	4.189	3.65	474	237,000	164.86
		120	1.219	13.562	56.61	1.0032	4.189	3.55	450	225,000	192.40
		140	1.438	13.124	65.79	0.9394	4.189	3.44	422	211,000	223.57
		160	1.594	12.812	72.14	0.8953	4.189	3.35	402	201,000	245.22
18	18	5S	0.165	17.670	9.25	1.7029	4.712	4.63	764	382,000	31.43
		10S	0.188	17.624	10.52	1.6941	4.712	4.61	760	379,400	35.76
		10	0.250	17.500	13.94	1.6703	4.712	4.58	750	375,000	47.39
		20	0.312	17.376	17.34	1.6468	4.712	4.55	739	369,500	59.03
		ST	0.375	17.250	20.76	1.6230	4.712	4.52	728	364,000	70.59
		30	0.438	17.124	24.16	1.5993	4.712	4.48	718	359,000	82.06
		XS	0.500	17.000	27.49	1.5763	4.712	4.45	707	353,500	93.45
		40	0.562	16.876	30.79	1.5533	4.712	4.42	697	348,500	104.76
		60	0.750	16.500	40.64	1.4849	4.712	4.32	666	333,000	138.17
		80	0.938	16.124	50.28	1.4180	4.712	4.22	636	318,000	170.84
		100	1.156	15.688	61.17	1.3423	4.712	4.11	602	301,000	208.00
		120	1.375	15.250	71.82	1.2684	4.712	3.99	569	284,500	244.14
		140	1.562	14.876	80.66	1.2070	4.712	3.89	540	270,000	274.30
		160	1.781	14.438	90.75	1.1370	4.712	3.78	510	255,000	308.55

TABLE E-1 *(Continued)*

Nominal Pipe Size, in.	Outside Diam., in.	Schedule No.	Wall Thickness, in.	Inside Diam., in.	Metal, sq. in.	Flow, sq. ft.	Outside	Inside	U.S. gal./min.	lb/hr water	Weight of Plain-end Pipe, lb./ft.
20	20	5S	0.188	19.624	11.70	2.1004	5.236	5.14	943	471,500	39.78
		10S	.218	19.564	13.55	2.0878	5.236	5.12	937	467,500	46.06
		10	.250	19.500	15.51	2.0740	5.236	5.11	930	465,500	52.73
		20, ST	.375	19.250	23.12	2.0211	5.236	5.04	902	451,000	78.60
		30, XS	.500	19.000	30.63	1.9689	5.236	4.97	883	441,500	104.13
		40	.594	18.812	36.21	1.9302	5.236	4.92	866	433,000	123.06
		60	.812	18.376	48.95	1.8417	5.236	4.81	826	413,000	166.50
		80	1.031	17.938	61.44	1.7550	5.236	4.70	787	393,500	208.92
		100	1.281	17.438	75.33	1.6585	5.236	4.57	744	372,000	256.15
		120	1.500	17.000	87.18	1.5763	5.236	4.45	707	353,500	296.37
		140	1.750	16.500	100.3	1.4849	5.236	4.32	665	332,500	341.10
		160	1.969	16.062	111.5	1.4071	5.236	4.21	632	316,000	379.14
24	24	5S	0.218	23.564	16.29	3.0285	6.283	6.17	1359	679,500	55.37
		10, 10S	0.250	23.500	18.65	3.012	6.283	6.15	1350	675,000	63.41
		20, ST	0.375	23.250	27.83	2.948	6.283	6.09	1325	662,500	94.62
		XS	0.500	23.000	36.90	2.885	6.283	6.02	1295	642,500	125.49
		30	0.562	22.876	41.39	2.854	6.283	5.99	1281	640,500	140.80
		40	0.688	22.624	50.39	2.792	6.283	5.92	1253	626,500	171.17
		60	0.969	22.062	70.11	2.655	6.283	5.78	1192	596,000	238.29
		80	1.219	21.562	87.24	2.536	6.283	5.64	1138	569,000	296.53
		100	1.531	20.938	108.1	2.391	6.283	5.48	1073	536,500	367.45
		120	1.812	20.376	126.3	2.264	6.283	5.33	1016	508,000	429.50
		140	2.062	19.876	142.1	2.155	6.283	5.20	965	482,500	483.24
		160	2.344	19.312	159.5	2.034	6.283	5.06	913	456,500	542.09
30	30	5S	0.250	29.500	23.37	4.746	7.854	7.72	2130	1,065,000	79.43
		10, 10S	0.312	29.376	29.10	4.707	7.854	7.69	2110	1,055,000	98.93
		ST	0.375	29.250	34.90	4.666	7.854	7.66	2094	1,048,000	118.65
		20, XS	0.500	29.000	46.34	4.587	7.854	7.59	2055	1,027,500	157.53
		30	0.625	28.750	57.68	4.508	7.854	7.53	2020	1,010,000	196.08

TABLE E-2 *Properties of Copper Water Tube, Types K, L, M*

Nominal Size	Actual Outside Diam., in.	Mean Outside Diam. Tolerances, in. Soft Annealed	Hard Drawn	Type K Nominal	Tolerance	Type L Nominal	Tolerance	Type M Nominal	Tolerance	Type K	Type L	Type M
¼	0.375	0.002	0.001	0.035	0.004	0.030	0.0035	. . . . .	. . . . .	0.145	0.126	
⅜	.500	.0025	.001	.049	.004	.035	.0035	0.025	0.0025	.269	.198	0.145
½	.625	.0025	.001	.049	.004	.040	.0035	.028	.0025	.344	.285	.204
⅝	.750	.0025	.001	.049	.004	.042	.0035	. . . . .	. . . . .	.418	.362	
¾	.875	.003	.001	.065	.0045	.045	.004	.032	.003	.641	.455	.328
1	1.125	.0035	.0015	.065	.0045	.050	.004	.035	.0035	.839	.655	.465
1¼	1.375	.004	.0015	.065	.0045	.055	.0045	.042	.0035	1.04	.884	.682
1½	1.625	.0045	.002	.072	.005	.060	.0045	.049	.004	1.36	1.14	.940
2	2.125	.005	.002	.083	.007	.070	.006	.058	.006	2.06	1.75	1.46
2½	2.625	.005	.002	.095	.007	.080	.006	.065	.006	2.93	2.48	2.03
3	3.125	.005	.002	.109	.007	.090	.007	.072	.006	4.00	3.33	2.68
3½	3.625	.005	.002	.120	.008	.100	.007	.083	.007	5.12	4.29	3.58
4	4.125	.005	.002	.134	.010	.110	.009	.095	.009	6.51	5.36	4.66
5	5.125	.005	.002	.160	.010	.125	.010	.109	.009	9.67	7.61	6.66
6	6.125	.005	.002	.192	.012	.140	.011	.122	.010	13.9	10.2	8.92
8	8.125	.006	+ .002 − .004	.271	.016	.200	.014	.170	.014	25.9	19.3	16.5

TABLE E-3   Properties of Copper and Red Brass Pipe

A. Dimensions and Weights of Regular Pipe

Nominal Pipe Size, in.	Nominal Dimensions, in.			Cross-Sectional Area of Bore, sq. in.	lb./ft.	
	Outside Diam.	Inside Diam.	Wall Thickness		Red Brass	Copper
1/8	0.405	0.281	0.062	0.062	0.253	0.259
1/4	.540	.376	.082	.110	.447	.457
3/8	.675	.495	.090	.192	.627	.641
1/2	.840	.626	.107	.307	.934	.955
3/4	1.050	.822	.114	.531	1.27	1.30
1	1.315	1.063	.126	.887	1.78	1.82
1 1/4	1.660	1.368	.146	1.47	2.63	2.69
1 1/2	1.900	1.600	.150	2.01	3.13	3.20
2	2.375	2.063	.156	3.34	4.12	4.22
2 1/2	2.875	2.501	0.187	4.91	5.99	6.12
3	3.500	3.062	.219	7.37	8.56	8.75
3 1/2	4.000	3.500	.250	9.62	11.2	11.4
4	4.500	4.000	.250	12.6	12.7	12.9
5	5.562	5.062	.250	20.1	15.8	16.2
6	6.625	6.125	.250	29.5	19.0	19.4
8	8.625	8.001	.312	50.3	30.9	31.6
10	10.750	10.020	.365	78.8	45.2	46.2
12	12.750	12.000	.375	113.	55.3	56.5

B. Dimensions and Weights of Extra-Strong Pipe

Nominal Pipe Size, in.	Nominal Dimensions, in.			Cross-Sectional Area of Bore, sq. in.	lb./ft.	
	Outside Diam.	Inside Diam.	Wall Thickness		Red Brass	Copper
1/8	0.405	0.205	0.100	0.033	0.363	0.371
1/4	.540	.294	.123	.068	.611	.625
3/8	.675	.421	.127	.139	.829	.847
1/2	.840	.542	.149	.231	1.23	1.25
3/4	1.050	.736	.157	.425	1.67	1.71
1	1.315	.951	.182	.710	2.46	2.51
1 1/4	1.660	1.272	.194	1.27	3.39	3.46
1 1/2	1.900	1.494	.203	1.75	4.10	4.19
2	2.375	1.933	.221	2.94	5.67	5.80
2 1/2	2.875	2.315	0.280	4.21	8.66	8.85
3	3.500	2.892	.304	6.57	11.6	11.8
3 1/2	4.000	3.358	.321	8.86	14.1	14.4
4	4.500	3.818	.341	11.5	16.9	17.3
5	5.562	4.812	.375	18.2	23.2	23.7
6	6.625	5.751	.437	26.0	32.2	32.9
8	8.625	7.625	.500	45.7	48.4	49.5
10	10.750	9.750	.500	74.7	61.1	62.4

Index